# Vector Mechanics
# For Engineers

## Statics and Dynamics

**Twelfth Edition**

# Vector Mechanics For Engineers

## Statics and Dynamics

**Ferdinand P. Beer**

Late of Lehigh University

**E. Russell Johnston, Jr.**

Late of University of Connecticut

**David F. Mazurek**

U.S. Coast Guard Academy

**Phillip J. Cornwell**

Rose-Hulman Institute of Technology

**Brian P. Self**

California Polytechnic State University—San Luis Obispo

Mc Graw Hill Education

VECTOR MECHANICS FOR ENGINEERS: STATICS AND DYNAMICS, TWELFTH EDITION

Published by McGraw-Hill Education, 2 Penn Plaza, New York, NY 10121. Copyright © 2019 by McGraw-Hill Education. All rights reserved. Printed in the United States of America. Previous editions © 2016, 2013, and 2010. No part of this publication may be reproduced or distributed in any form or by any means, or stored in a database or retrieval system, without the prior written consent of McGraw-Hill Education, including, but not limited to, in any network or other electronic storage or transmission, or broadcast for distance learning.

Some ancillaries, including electronic and print components, may not be available to customers outside the United States.

This book is printed on acid-free paper.

3 4 5 6 7 8 9 LWI 21 20 19

ISBN 978-1-259-63809-1
MHID 1-259-63809-X

Sr. Portfolio Manager: *Thomas Scaife, Ph.D.*
Product Developer: *Jolynn Kilburg*
Marketing Manager: *Shannon O'Donnell*
Sr. Content Project Managers: *Sherry Kane / Tammy Juran*
Sr. Buyer: *Laura Fuller*
Designer: *Matt Backhaus*
Content Licensing Specialist: *Shannon Mandersheid*
Cover Image: *©Zak Kendal/Getty Images; ©MichaelSvoboda/Getty Images*
Compositor: *SPi Global*

All credits appearing on page or at the end of the book are considered to be an extension of the copyright page.

### Library of Congress Cataloging-in-Publication Data

Names: Beer, Ferdinand P. (Ferdinand Pierre), 1915–2003, author. | Johnston, E. Russell (Elwood Russell), 1925–2010, author. | Mazurek, David F. (David Francis), author. | Cornwell, Phillip J., author. | Self, Brian P., 1966- author.
Title: Vector mechanics for engineers. Statics and dynamics / Ferdinand P. Beer (Late of Lehigh University), E. Russell Johnston, Jr. (Late of University of Connecticut), David F. Mazurek (U.S. Coast Guard Academy), Phillip J. Cornwell (Rose-Hulman Institute of Technology), Brian P. Self (California Polytechnic State University-San Luis Obispo).
Other titles: Statics and dynamics
Description: Twelfth edition. | New York, NY : McGraw-Hill Education, [2018]
Identifiers: LCCN 2017037987| ISBN 9781259638091 (alk. paper) | ISBN 125963809X (alk. paper)
Subjects: LCSH: Statics. | Dynamics. | Mechanics, Applied.
Classification: LCC TA350 .B3552 2018 | DDC 620.1/05—dc23 LC record available at https://lccn.loc.gov/2017037987

The Internet addresses listed in the text were accurate at the time of publication. The inclusion of a website does not indicate an endorsement by the authors or McGraw-Hill Education, and McGraw-Hill Education does not guarantee the accuracy of the information presented at these sites.

# About the Authors

**Ferdinand P. Beer.** Born in France and educated in France and Switzerland, Ferd received an M.S. degree from the Sorbonne and an Sc.D. degree in theoretical mechanics from the University of Geneva. He came to the United States after serving in the French army during the early part of World War II and taught for four years at Williams College in the Williams-MIT joint arts and engineering program. Following his service at Williams College, Ferd joined the faculty of Lehigh University where he taught for thirty-seven years. He held several positions, including University Distinguished Professor and chairman of the Department of Mechanical Engineering and Mechanics, and in 1995 Ferd was awarded an honorary Doctor of Engineering degree by Lehigh University.

**E. Russell Johnston, Jr.** Born in Philadelphia, Russ received a B.S. degree in civil engineering from the University of Delaware and an Sc.D. degree in the field of structural engineering from the Massachusetts Institute of Technology. He taught at Lehigh University and Worcester Polytechnic Institute before joining the faculty of the University of Connecticut where he held the position of chairman of the Department of Civil Engineering and taught for twenty-six years. In 1991 Russ received the Outstanding Civil Engineer Award from the Connecticut Section of the American Society of Civil Engineers.

**David F. Mazurek.** David holds a B.S. degree in ocean engineering and an M.S. degree in civil engineering from the Florida Institute of Technology and a Ph.D. degree in civil engineering from the University of Connecticut. He was employed by the Electric Boat Division of General Dynamics Corporation and taught at Lafayette College prior to joining the U.S. Coast Guard Academy, where he has been since 1990. He is a registered Professional Engineer in Connecticut and Pennsylvania, and has served on the American Railway Engineering & Maintenance-of-Way Association's Committee 15—Steel Structures since 1991. He is a Fellow of the American Society of Civil Engineers, and was elected to the Connecticut Academy of Science and Engineering in 2013. He was the 2014 recipient of both the Coast Guard Academy's Distinguished Faculty Award and its Center for Advanced Studies Excellence in Scholarship Award. Professional interests include bridge engineering, structural forensics, and blast-resistant design.

**Phillip J. Cornwell.** Phil holds a B.S. degree in mechanical engineering from Texas Tech University and M.A. and Ph.D. degrees in mechanical and aerospace engineering from Princeton University. He is currently a professor of mechanical engineering at Rose-Hulman Institute of Technology where he has taught since 1989. He served as Vice President for Academic Affairs at Rose-Hulman from July 2011 to June 2015. Phil received an SAE Ralph R. Teetor Educational Award in 1992, the Dean's Outstanding Teacher Award at Rose-Hulman in 2000, and the Board of Trustees' Outstanding Scholar Award at Rose-Hulman in 2001. Phil was one of the developers of the Dynamics Concept Inventory, and in 2012 he was one of the professors featured in The Princeton Review's book The Best 300 Professors.

**Brian P. Self.** Brian obtained his B.S. and M.S. degrees in engineering mechanics from Virginia Tech, and his Ph.D. in bioengineering from the University of Utah. He worked in the Air Force Research Laboratories before teaching at the U.S. Air Force Academy for seven years. Brian has taught in the Mechanical Engineering Department at Cal Poly, San Luis Obispo since 2006. He has been very active in the American Society of Engineering Education, serving on its Board from 2008–2010. He won the Academy Outstanding Instructor Award in 2003 and the Learn By Doing Scholar Award in 2016. With a team of five, Brian developed the Dynamics Concept Inventory to help assess student conceptual understanding. His professional interests include educational research, aviation physiology, and biomechanics.

# Brief Contents

# Contents

*Advanced or specialty topics

# Preface

## Objectives

A primary objective in a first course in mechanics is to help develop a student's ability first to analyze problems in a simple and logical manner, and then to apply basic principles to their solutions. A strong conceptual understanding of these basic mechanics principles is essential for successfully solving mechanics problems. We hope this text will help instructors achieve these goals.

## General Approach

Vector algebra is introduced at the beginning of the *Statics* volume and is used in the presentation of the basic principles of statics, as well as in the solution of many problems, particularly three-dimensional problems. Similarly, the concept of vector differentiation is introduced early in the *Dynamics* volume, and vector analysis is used throughout the presentation of dynamics. This approach leads to more concise derivations of the fundamental principles of mechanics. It also makes it possible to analyze many problems in kinematics and kinetics which could not be solved by scalar methods. The emphasis in this text, however, remains on the correct understanding of the principles of mechanics and on their application to the solution of engineering problems, and vector analysis is presented chiefly as a convenient tool.[†]

**Practical Applications Are Introduced Early.** One of the characteristics of the approach used in this book is that mechanics of *particles* is clearly separated from the mechanics of *rigid bodies*. This approach makes it possible to consider simple practical applications at an early stage and to postpone the introduction of the more difficult concepts. For example:

- In *Statics,* the statics of particles is treated first, and the principle of equilibrium of a particle is immediately applied to practical situations involving only concurrent forces. The statics of rigid bodies is considered later, at which time the vector and scalar products of two vectors are introduced and used to define the moment of a force about a point and about an axis.
- In *Dynamics,* the same division is observed. The basic concepts of force, mass, and acceleration, of work and energy, and of impulse and momentum are introduced and first applied to problems involving only particles. Thus, students can familiarize themselves with the three basic methods used in dynamics and learn their respective advantages before facing the difficulties associated with the motion of rigid bodies.

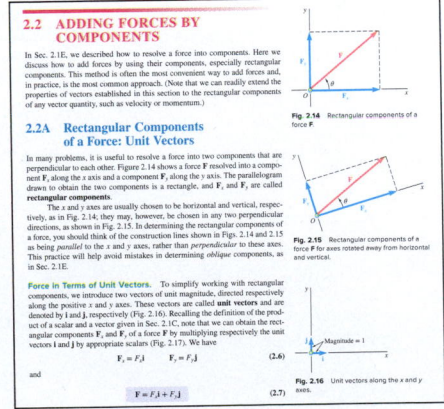

[†]In a parallel text, *Mechanics for Engineers,* fifth edition, the use of vector algebra is limited to the addition and subtraction of vectors, and vector differentiation is omitted.

**xv**

## 17.1    ENERGY METHODS FOR A RIGID BODY

We now use the principle of work and energy to analyze the plane motion of rigid bodies. As we pointed out in Chap. 13, the method of work and energy is particularly well-adapted to solving problems involving velocities and displacements. Its main advantage is that the work of forces and the kinetic energy of objects are scalar quantities.

### 17.1A    Principle of Work and Energy

To apply the principle of work and energy to the motion of a rigid body, we again assume that the rigid body is made up of a large number $n$ of particles of mass $\Delta m_i$. From Eq. (14.30) of Sec. 14.2B, we have

**Principle of work and energy, rigid body**

$$T_1 + U_{1\rightarrow2} = T_2 \tag{17.1}$$

where $T_1$, $T_2$ = the initial and final values of total kinetic energy of particles forming the rigid body

$U_{1\rightarrow2}$ = work of all forces acting on various particles of the body

Just as we did in Chap. 13, we can express the work done by nonconservative forces as $U_{1\rightarrow2}^{NC}$, and we can define potential energy terms for conservative forces. Then we can express Eq. (17.1) as

$$T_1 + V_{g_1} + V_{e_1} + U_{1\rightarrow2}^{NC} = T_2 + V_{g_2} + V_{e_2} \tag{17.1'}$$

where $V_{g_1}$ and $V_{g_2}$ are the initial and final gravitational potential energy of the center of mass of the rigid body with respect to a reference point or datum, and $V_{e_1}$ and $V_{e_2}$ are the initial and final values of the elastic energy associated with springs in the system.

We obtain the total kinetic energy

$$T = \frac{1}{2} \sum_{i=1}^{n} \Delta m_i v_i^2 \tag{17.2}$$

by adding positive scalar quantities, so it is itself a positive scalar quantity. You will see later how to determine $T$ for various types of motion of a rigid body.

The expression $U_{1\rightarrow2}$ in Eq. (17.1) represents the work of all the forces acting on the various particles of the body, whether these forces are internal or external. However, the total work of the internal forces holding together the particles of a rigid body is zero. To see this, consider two particles $A$ and $B$ of a rigid body and the two equal and opposite forces $\mathbf{F}$ and $-\mathbf{F}$ they exert on each other (Fig. 17.1). Although, in general, small displacements $d\mathbf{r}$ and $d\mathbf{r}'$ of the two particles are different, the components of these displacements along $AB$ must be equal; otherwise, the particles would not remain at the same distance from each other and the body would not be rigid. Therefore, the work of $\mathbf{F}$ is equal in magnitude and opposite in sign to the work of $-\mathbf{F}$.

**New Concepts Are Introduced in Simple Terms.**    New concepts are presented in simple terms and every step is explained in detail. On the other hand, by discussing the broader aspects of the problems considered, and by stressing methods of general applicability, a definite maturity of approach has been achieved. For example, the concept of potential energy is discussed in the general case of a conservative force. Also, the study of the plane motion of rigid bodies is designed to lead naturally to the study of their general motion in space. This is true in kinematics as well as in kinetics, where the principle of equivalence of external and effective forces is applied directly to the analysis of plane motion, thus facilitating the transition to the study of three-dimensional motion.

**Fundamental Principles Are Placed in the Context of Simple Applications.**    The fact that mechanics is essentially a *deductive* science based on a few fundamental principles is stressed. Derivations have been presented in their logical sequence and with all the rigor warranted at this level. However, the learning process is largely inductive, and simple applications are considered first. For example:

- The statics of particles precedes the statics of rigid bodies, and problems involving internal forces are postponed until Chap. 6.
- In Chap. 4, equilibrium problems involving only coplanar forces are considered first and solved by ordinary algebra, while problems involving three-dimensional forces and requiring the full use of vector algebra are discussed in the second part of the chapter.
- The kinematics of particles (Chap. 11) precedes the kinematics of rigid bodies (Chap. 15).
- The fundamental principles of the kinetics of rigid bodies are first applied to the solution of two-dimensional problems (Chaps. 16 and 17), which can be more easily visualized by the student, while three-dimensional problems are postponed until Chap. 18.

**The Presentation of the Principles of Kinetics Is Unified.**    The twelfth edition of *Vector Mechanics for Engineers* retains the unified presentation of the principles of kinetics which characterized the previous eleven editions. The concepts of linear and angular momentum are introduced in Chap. 12 so that Newton's second law of motion can be presented not only in its conventional form $\mathbf{F} = m\mathbf{a}$, but also as a law relating, respectively, the sum of the forces acting on a particle and the sum of their moments to the rates of change of the linear and angular momentum of the particle. This makes possible an earlier introduction of the principle of conservation of angular momentum and a more meaningful discussion of the motion of a particle under a central force (Sec. 12.3A). More importantly, this approach can be readily extended to the study of the motion of a system of particles (Chap. 14) and leads to a more concise and unified treatment of the kinetics of rigid bodies in two and three dimensions (Chaps. 16 through 18).

**Systematic Problem-Solving Approach.**    All the sample problems are solved using the steps of **S**trategy, **M**odeling, **A**nalysis, and **R**eflect & **T**hink, or the "SMART" approach. This methodology is intended to give students confidence when approaching new problems, and students are encouraged to apply this approach in the solution of all assigned problems.

## Free-Body Diagrams Are Used Both to Solve Equilibrium Problems and to Express the Equivalence of Force Systems.

Free-body diagrams are introduced early in *Statics,* and their importance is emphasized throughout. They are used not only to solve equilibrium problems but also to express the equivalence of two systems of forces or, more generally, of two systems of vectors. In *Dynamics* we introduce a kinetic diagram, which is a pictorial representation of inertia terms. The advantage of this approach becomes apparent in the study of the dynamics of rigid bodies, where it is used to solve three-dimensional as well as two-dimensional problems. By placing the emphasis on the free-body diagram and kinetic diagram, rather than on the standard algebraic equations of motion, a more intuitive and more complete understanding of the fundamental principles of dynamics can be achieved. This approach, which was first introduced in 1962 in the first edition of *Vector Mechanics for Engineers,* has now gained wide acceptance among mechanics teachers in this country. It is, therefore, used in preference to the method of dynamic equilibrium and to the equations of motion in the solution of all sample problems in this book.

## A Careful Balance between SI and U.S. Customary Units Is Consistently Maintained.

Because of the current trend in the American government and industry to adopt the international system of units (SI metric units), the SI units most frequently used in mechanics are introduced in Chap. 1 and are used throughout the text. Approximately half of the sample problems and 60 percent of the homework problems are stated in these units, while the remainder are in U.S. customary units. The authors believe that this approach will best serve the need of the students, who, as engineers, will have to be conversant with both systems of units.

It also should be recognized that using both SI and U.S. customary units entails more than the use of conversion factors. Since the SI system of units is an absolute system based on the units of time, length, and mass, whereas the U.S. customary system is a gravitational system based on the units of time, length, and force, different approaches are required for the solution of many problems. For example, when SI units are used, a body is generally specified by its mass expressed in kilograms; in most problems of statics it will be necessary to determine the weight of the body in newtons, and an additional calculation will be required for this purpose. On the other hand, when U.S. customary units are used, a body is specified by its weight in pounds and, in dynamics problems, an additional calculation will be required to determine its mass in slugs (or lb·s$^2$/ft). The authors, therefore, believe that problem assignments should include both systems of units.

The *Instructor's and Solutions Manual* provides six different lists of assignments so that an equal number of problems stated in SI units and in U.S. customary units can be selected. If so desired, two complete lists of assignments can also be selected with up to 75 percent of the problems stated in SI units.

## Optional Sections Offer Advanced or Specialty Topics.

A large number of optional sections have been included. These sections are indicated by asterisks and thus are easily distinguished from those which form the core of the basic course. They can be omitted without prejudice to the understanding of the rest of the text.

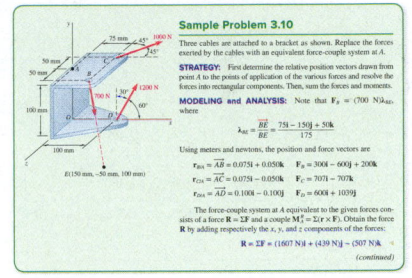

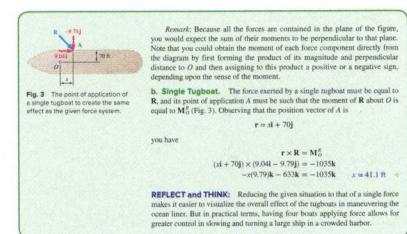

The topics covered in the optional sections in *Statics* include the reduction of a system of forces of a wrench, applications to hydrostatics, equilibrium of cables, products of inertia and Mohr's circle, the determination of the principal axes and the mass moments of inertia of a body of arbitrary shape, and the method of virtual work. The sections on the inertia properties of three-dimensional bodies are primarily intended for students who will later study in dynamics the three-dimensional motion of rigid bodies.

The topics covered in the optional sections in *Dynamics* include graphical methods for the solution of rectilinear-motion problems, the trajectory of a particle under a central force, the deflection of fluid streams, problems involving jet and rocket propulsion, the kinematics and kinetics of rigid bodies in three dimensions, damped mechanical vibrations, and electrical analogues. These topics will be of particular interest when dynamics is taught in the junior year.

The material presented in the text and most of the problems require no previous mathematical knowledge beyond algebra, trigonometry, elementary calculus, and the elements of vector algebra presented in Chaps. 2 and 3 of the volume on statics. However, special problems are included, which make use of a more advanced knowledge of calculus, and certain sections, such as Secs. 19.5A and 19.5B on damped vibrations, should be assigned only if students possess the proper mathematical background. In portions of the text using elementary calculus, a greater emphasis is placed on the correct understanding and application of the concepts of differentiation and integration, than on the nimble manipulation of mathematical formulas. In this connection, it should be mentioned that the determination of the centroids of composite areas precedes the calculation of centroids by integration, thus making it possible to establish the concept of moment of area firmly before introducing the use of integration.

# Guided Tour

**Chapter Introduction.** Each chapter begins with a list of learning objectives and an outline that previews chapter topics. An introductory section describes the material to be covered in simple terms, and how it will be applied to the solution of engineering problems.

**Chapter Lessons.** The body of the text is divided into sections, each consisting of one or more sub-sections, several sample problems, and a large number of end-of-section problems for students to solve. Each section corresponds to a well-defined topic and generally can be covered in one lesson. In a number of cases, however, the instructor will find it desirable to devote more than one lesson to a given topic. *The Instructor's and Solutions Manual* contains suggestions on the coverage of each lesson.

**Sample Problems.** The Sample Problems are set up in much the same form that students will use when solving assigned problems, and they employ the SMART problem-solving methodology that students are encouraged to use in the solution of their assigned problems. They thus serve the double purpose of reinforcing the text and demonstrating the type of neat and orderly work that students should cultivate in their own solutions. In addition, in-problem references and captions have been added to the sample problem figures for contextual linkage to the step-by-step solution. In the digital version, many Sample Problems now have simulations to help students visualize the problem. Enhanced digital content is indicated by a ⊙ within the text.

**Concept Applications.** Concept Applications are used within selected theory sections in the Statics volume to amplify certain topics, and they are designed to reinforce the specific material being presented and facilitate its understanding.

**Solving Problems on Your Own.** A section entitled *Solving Problems on Your Own* is included for each lesson, between the sample problems and the problems to be assigned. The purpose of these sections is to help students organize in their own minds the preceding theory of the text and the solution methods of the sample problems so that they can more successfully solve the homework problems. Also included in these sections are specific suggestions and strategies that will enable the students to more efficiently attack any assigned problems.

⊙ **Case Studies.** Statics and dynamics principles are used extensively in engineering applications, particularly for the designing of solutions to problems and for failure analysis when those solutions do not work as planned. Much can be learned from the historical successes and failures of past designs, and unique insight can be gained by studying how engineers developed different products and structures. To this end, real-world Case Studies have been introduced in this revision to provide relevance and application to the principles of engineering mechanics being discussed. The Case Studies are developed using the SMART problem-solving methodology to present the story. In this way, they serve as both a practical illustration of the concepts linked to some real-world situation and reinforce the consistent five-step approach to solving engineering problems.

### 1
### Introduction

The tallest skyscraper in the Western Hemisphere, One World Trade Center is a prominent feature of the New York City skyline. From its foundation to its structural components and mechanical systems, the design and operation of the tower is based on the fundamentals of engineering mechanics.

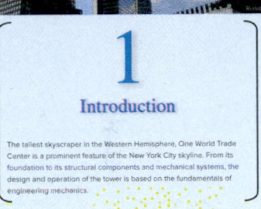

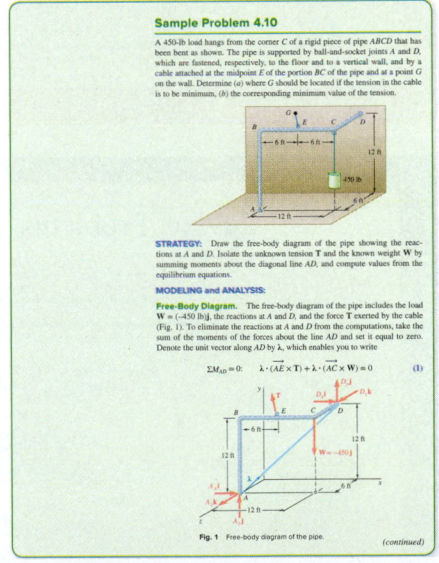

**Sample Problem 4.10**

A 450-lb load hangs from the corner $C$ of a rigid piece of pipe $ABCD$ that has been bent as shown. The pipe is supported by ball-and-socket joints $A$ and $D$, which are fastened, respectively, to the floor and to a vertical wall, and by a cable attached at the midpoint $E$ of the portion $BC$ of the pipe and at a point $G$ on the wall. Determine (a) where $G$ should be located if the tension in the cable is to be minimum, (b) the corresponding minimum value of the tension.

**STRATEGY:** Draw the free-body diagram of the pipe showing the reactions at $A$ and $D$. Isolate the unknown tension $\mathbf{T}$ and the known weight $\mathbf{W}$ by summing moments about the diagonal line $AD$, and compute values from the equilibrium equations.

**MODELING and ANALYSIS:**

**Free-Body Diagram.** The free-body diagram of the pipe includes the load $\mathbf{W} = (-450\ \text{lb})\mathbf{j}$, the reactions at $A$ and $D$, and the force $\mathbf{T}$ exerted by the cable (Fig. 1). To eliminate the reactions at $A$ and $D$ from the computations, take the sum of the moments of the forces about the line $AD$ and set it equal to zero. Denote the unit vector along $AD$ by $\boldsymbol{\lambda}$, which enables you to write

$$\Sigma M_{AD} = 0: \quad \boldsymbol{\lambda} \cdot (\overrightarrow{AE} \times \mathbf{T}) + \boldsymbol{\lambda} \cdot (\overrightarrow{AC} \times \mathbf{W}) = 0 \quad (1)$$

**Fig. 1** Free-body diagram of the pipe. *(continued)*

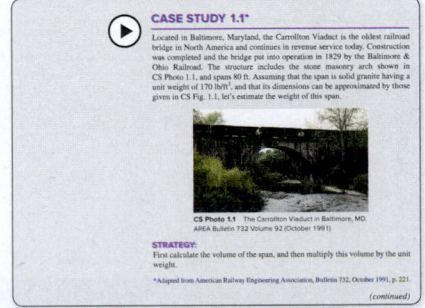

**CASE STUDY 1.1***

Located in Baltimore, Maryland, the Carrollton Viaduct is the oldest railroad bridge in North America and continues in revenue service today. Construction was completed and the bridge put into operation in 1829 by the Baltimore & Ohio Railroad. The structure includes the stone masonry arch shown in CS Photo 1.1, and spans 80 ft. Assuming that the span is solid granite having a unit weight of 170 lb/ft³, and that its dimensions can be approximated by those given in CS Fig. 1.1, let's estimate the weight of this span.

**CS Photo 1.1** The Carrollton Viaduct in Baltimore, MD.
AREA Bulletin 732 Volume 92 (October 1991)

**STRATEGY:** First calculate the volume of the span, and then multiply this volume by the unit weight.

*Adapted from American Railway Engineering Association, Bulletin 732, October 1991, p. 221.

*(continued)*

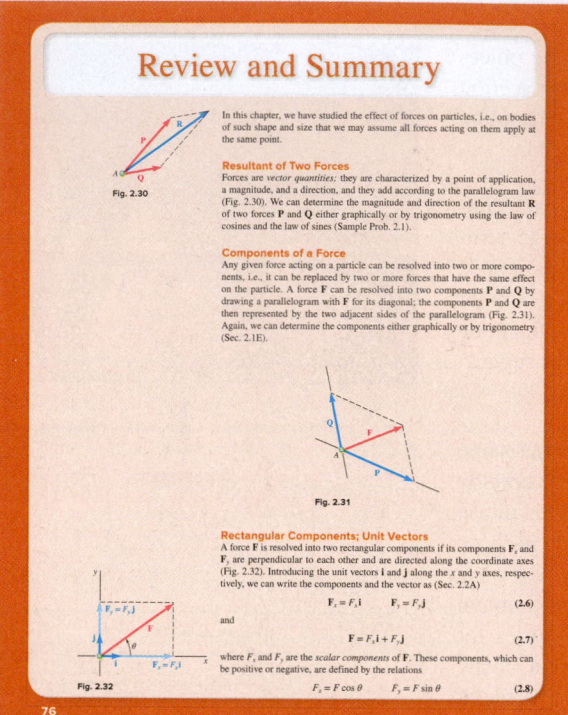

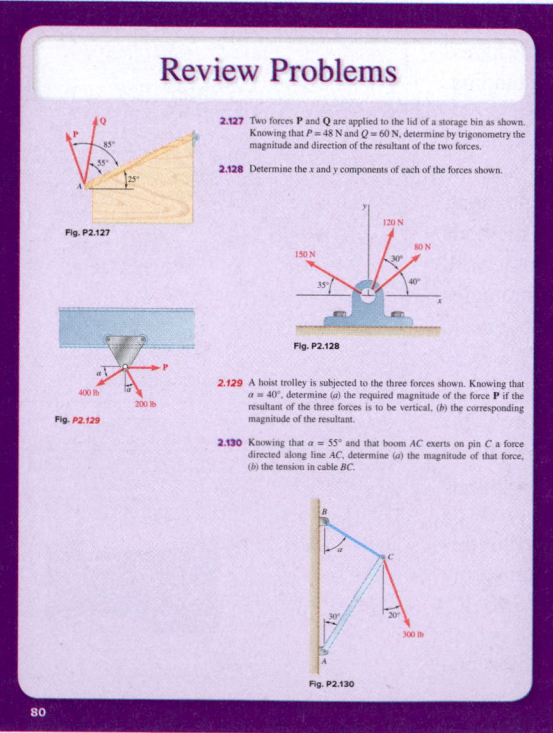

In some instances, these Case Studies are examined further in the accompanying digital content through Connect®. The digital content also provides additional cases that are developed in their entirety.

**Homework Problem Sets.** Most of the problems are of a practical nature and should appeal to engineering students. They are primarily designed, however, to illustrate the material presented in the text and to help students understand the principles of mechanics. The problems are grouped according to the portions of material they illustrate and, in general, are arranged in order of increasing difficulty. Problems requiring special attention are indicated by asterisks. Answers to 70 percent of the problems are given at the end of the book. Problems for which the answers are given are set in straight type in the text, while problems for which no answer is given are set in italic and red font color.

**Chapter Review and Summary.** Each chapter ends with a review and summary of the material covered in that chapter. Marginal notes are used to help students organize their review work, and cross-references have been included to help them find the portions of material requiring their special attention.

**Review Problems.** A set of review problems is included at the end of each chapter. These problems provide students further opportunity to apply the most important concepts introduced in the chapter.

**Computer Problems.** Accessible through Connect are problem sets for each chapter that are designed to be solved with computational software. Many of these problems are relevant to the design process; they may involve the analysis of a structure for various configurations and loadings of the structure, or the determination of the equilibrium positions of a given mechanism that may require an iterative method of solution. Developing the algorithm required to solve a given mechanics problem will benefit the students in two different ways: (1) it will help them gain a better understanding of the mechanics principles involved; (2) it will provide them with an opportunity to apply their computer skills to the solution of a meaningful engineering problem.

**Concept Questions.** Educational research has shown that students can often choose appropriate equations and solve algorithmic problems without having a strong conceptual understanding of mechanics principles.[†] To help assess and develop student conceptual understanding, we have included

[†]Hestenes, D., Wells, M., and Swakhamer, G (1992). The force concept inventory. *The Physics Teacher*, 30: 141–158.
Streveler, R. A., Litzinger, T. A., Miller, R. L., and Steif, P. S. (2008). Learning conceptual knowledge in the engineering sciences: Overview and future research directions, *JEE*, 279–294.

Concept Questions, which are multiple choice problems that require few, if any, calculations. Each possible incorrect answer typically represents a common misconception (e.g., students often think that a vehicle moving in a curved path at constant speed has zero acceleration). Students are encouraged to solve these problems using the principles and techniques discussed in the text and to use these principles to help them develop their intuition. Mastery and discussion of these Concept Questions will deepen students' conceptual understanding and help them to solve dynamics problems.

**Free Body and Impulse-Momentum Diagram Practice Problems.** Drawing diagrams correctly is a critical step in solving kinetics problems in dynamics. A new type of problem has been added to the text to emphasize the importance of drawing these diagrams. In Chaps. 12 and 16 the Free Body Practice Problems require students to draw a free-body diagram (FBD) showing the applied forces and an equivalent diagram called a "kinetic diagram" (KD) showing $m\mathbf{a}$ or its components and $\bar{I}\alpha$. These diagrams provide students with a pictorial representation of Newton's second law and are critical in helping students to correctly solve kinetic problems. In Chaps. 13 and 17 the Impulse-Momentum Diagram Practice Problems require students to draw diagrams showing the momenta of the bodies before impact, the impulses exerted on the body during impact, and the final momenta of the bodies. The answers to all of these questions can be accessed through Connect.

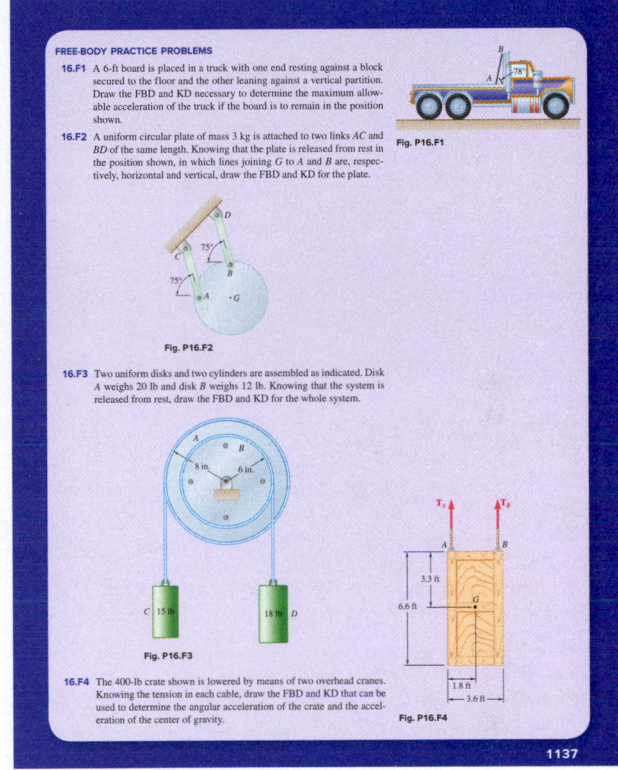

FREE-BODY PRACTICE PROBLEMS

**16.F1** A 6-ft board is placed in a truck with one end resting against a block secured to the floor and the other leaning against a vertical partition. Draw the FBD and KD necessary to determine the maximum allowable acceleration of the truck if the board is to remain in the position shown.

Fig. P16.F1

**16.F2** A uniform circular plate of mass 3 kg is attached to two links AC and BD of the same length. Knowing that the plate is released from rest in the position shown, in which lines joining G to A and B are, respectively, horizontal and vertical, draw the FBD and KD for the plate.

Fig. P16.F2

**16.F3** Two uniform disks and two cylinders are assembled as indicated. Disk A weighs 20 lb and disk B weighs 12 lb. Knowing that the system is released from rest, draw the FBD and KD for the whole system.

Fig. P16.F3

**16.F4** The 400-lb crate shown is lowered by means of two overhead cranes. Knowing the tension in each cable, draw the FBD and KD that can be used to determine the angular acceleration of the crate and the acceleration of the center of gravity.

Fig. P16.F4

1137

# Digital Resources

 **Connect**® is a highly reliable, easy-to-use homework and learning management solution that embeds learning science and award-winning adaptive tools to improve student results.

**Analytics** Connect Insight is Connect's one-of-a-kind visual analytics dashboard. Now available for both instructors and students, it provides at-a-glance information regarding student performance, which is immediately actionable. By presenting assignment, assessment, and topical performance results together with a time metric that is easily visible for aggregate or individual results, Connect InSight gives the user the ability to take a just-in-time approach to teaching and learning, which was never before available. Connect Insight presents data that empower students and help instructors improve class performance in a way that is efficient and effective.

### Autograded Free-Body Diagram Problems

**NEW!**

- Within Connect, algorithmic end-of-chapter problems include our new Free-Body Diagram Drawing tool. The Free-Body Diagram Tool allows students to draw free-body diagrams that are auto graded by the system. Student's receive immediate feedback on their diagrams to help student's solidify their understanding of the physical situation presented in the problem.

### Case Study Interactives

⊙ New digital content has been added throughout the text to enhance student learning. This includes a more in-depth discussion of the new Case Studies, as well as interactive questions embedded in these video explorations to make students *think* about the problem rather than just viewing the video. Within the text, simulations and short videos have been added to help students visualize topics, such as zero-force members and the motion of different linkages.

Find the following instructor resources available through Connect:

- **Instructor's and Solutions Manual.** *The Instructor's and Solutions Manual* that accompanies the twelfth edition features solutions to all end of chapter problems. This manual also features a number of tables designed to assist instructors in creating a schedule of assignments for their course. The various topics covered in the text have been listed in Table I and a suggested number of periods to be spent on each topic has been indicated. Table II prepares a brief description of all groups of problems and a classification of the problems in each group according to the units used. Sample lesson schedules are shown in Tables III, IV, and V, together with various alternative lists of assigned homework problems.
- **Lecture PowerPoint Slides** for each chapter that can be modified. These generally have an introductory application slide, animated worked-out problems that you can do in class with your students, concept questions, and "what-if?" questions at the end of the units.

- **Textbook images**
- **Computer Problem sets** for each chapter that are designed to be solved with computational software.
- **C.O.S.M.O.S.,** the Complete Online Solutions Manual Organization System that allows instructors to create custom homework, quizzes, and tests using end-of-chapter problems from the text.

**SMARTBOOK®**    SmartBook helps students study more efficiently by highlighting where in the chapter to focus, asking review questions and pointing them to resources until they understand.

# Acknowledgments

A special thanks to our colleagues who thoroughly checked the solutions and answers to all problems in this edition and then prepared the solutions for the accompanying *Instructor's and Solutions Manual,* Dr. Charles Birdsong and Sabrina Gough of California Polytechnic State University and Amy Mazurek.

The authors thank the many companies and individuals that provided photographs for this edition.

The authors also thank the members of the staff at McGraw-Hill Education for their support and dedication during the preparation of this new edition.

We particularly wish to acknowledge the contributions of Portfolio Manager Thomas Scaife, Ph.D., Associate Director of Digital Content, Chelsea Haupt, Ph.D., Product Developer, Jolynn Kilburg, Editorial Coordinator and SmartBook development manager, Marisa Moreno, Content Project Manager, Sherry Kane, and Program Manager, Lora Neyens.

David F. Mazurek
Phillip J. Cornwell
Brian P. Self

The authors gratefully acknowledge the many helpful comments and suggestions offered by focus group attendees and by users of the previous editions of *Vector Mechanics for Engineers:*

George Adams
Northeastern University

William Altenhof
University of Windsor

Sean B. Anderson
Boston University

Manohar Arora
Colorado School of Mines

Gilbert Baladi
Michigan State University

Francois Barthelat
McGill University

Oscar Barton, Jr.
U.S. Naval Academy

M. Asghar Bhatti
University of Iowa

Shaohong Cheng
University of Windsor

Philip Datseris
University of Rhode Island

Timothy A. Doughty
University of Portland

Howard Epstein
University of Connecticut

Asad Esmaeily
Kansas State University, Civil Engineering Department

David Fleming
Florida Institute of Technology

Jeff Hanson
Texas Tech University

David A. Jenkins
University of Florida

Shaofan Li
University of California, Berkeley

William R. Murray
Cal Poly State University

Eric Musslman
University of Minnesota, Duluth

Masoud Olia
Wentworth Institute of Technology

Renee K. B. Petersen
Washington State University

Amir G Rezaei
California State Polytechnic University, Pomona

Martin Sadd
University of Rhode Island

Stefan Seelecke
North Carolina State University

Yixin Shao
McGill University

Muhammad Sharif
The University of Alabama

Anthony Sinclair
University of Toronto

Lizhi Sun
University of California, Irvine

Jeffrey Thomas
Northwestern University

Jiashi Yang
University of Nebraska

Xiangwa Zeng
Case Western Reserve University

# List of Symbols

| | |
|---|---|
| $\mathbf{a}$, $a$ | Acceleration |
| $a$ | Constant; radius; distance; semimajor axis of ellipse |
| $\bar{\mathbf{a}}$, $\bar{a}$ | Acceleration of mass center |
| $\mathbf{a}_{B/A}$ | Acceleration of $B$ relative to frame in translation with $A$ |
| $\mathbf{a}_{P/\mathcal{F}}$ | Acceleration of $P$ relative to rotating frame $\mathcal{F}$ |
| $\mathbf{a}_c$ | Coriolis acceleration |
| $\mathbf{A}, \mathbf{B}, \mathbf{C}, \ldots$ | Reactions at supports and connections |
| $A, B, C, \ldots$ | Points |
| $A$ | Area |
| $b$ | Width; distance; semiminor axis of ellipse |
| $c$ | Constant; coefficient of viscous damping |
| $C$ | Centroid; instantaneous center of rotation; capacitance |
| $d$ | Distance |
| $\mathbf{e}_n$, $\mathbf{e}_t$ | Unit vectors along normal and tangent |
| $\mathbf{e}_r$, $\mathbf{e}_\theta$ | Unit vectors in radial and transverse directions |
| $e$ | Coefficient of restitution; base of natural logarithms |
| $E$ | Total mechanical energy; voltage |
| $f$ | Scalar function |
| $f_f$ | Frequency of forced vibration |
| $f_n$ | Natural frequency |
| $\mathbf{F}$ | Force; friction force |
| $g$ | Acceleration of gravity |
| $G$ | Center of gravity; mass center; constant of gravitation |
| $h$ | Angular momentum per unit mass |
| $\mathbf{H}_O$ | Angular momentum about point $O$ |
| $\dot{\mathbf{H}}_G$ | Rate of change of angular momentum $\mathbf{H}_G$ with respect to frame of fixed orientation |
| $(\dot{\mathbf{H}}_G)_{Gxyz}$ | Rate of change of angular momentum $\mathbf{H}_G$ with respect to rotating frame $Gxyz$ |
| $\mathbf{i}, \mathbf{j}, \mathbf{k}$ | Unit vectors along coordinate axes |
| $i$ | Current |
| $I$, $I_x$, $\ldots$ | Moments of inertia |
| $\bar{I}$ | Centroidal moment of inertia |
| $I_{xy}$, $\ldots$ | Products of inertia |
| $J$ | Polar moment of inertia |
| $k$ | Spring constant |
| $k_x, k_y, k_O$ | Radii of gyration |
| $\bar{k}$ | Centroidal radius of gyration |
| $l$ | Length |
| $\mathbf{L}$ | Linear momentum |
| $L$ | Length; inductance |
| $m$ | Mass |
| $m'$ | Mass per unit length |
| $\mathbf{M}$ | Couple; moment |
| $\mathbf{M}_O$ | Moment about point $O$ |

| | |
|---|---|
| $M_O^R$ | Moment resultant about point $O$ |
| $M$ | Magnitude of couple or moment; mass of earth |
| $M_{OL}$ | Moment about axis $OL$ |
| $n$ | Normal direction |
| $\mathbf{N}$ | Normal component of reaction |
| $O$ | Origin of coordinates |
| $\mathbf{P}$ | Force; vector |
| $\dot{\mathbf{P}}$ | Rate of change of vector $\mathbf{P}$ with respect to frame of fixed orientation |
| $q$ | Mass rate of flow; electric charge |
| $\mathbf{Q}$ | Force; vector |
| $\dot{\mathbf{Q}}$ | Rate of change of vector $\mathbf{Q}$ with respect to frame of fixed orientation |
| $(\dot{\mathbf{Q}})_{Oxyz}$ | Rate of change of vector $\mathbf{Q}$ with respect to frame $Oxyz$ |
| $\mathbf{r}$ | Position vector |
| $\mathbf{r}_{B/A}$ | Position vector of $B$ relative to $A$ |
| $r$ | Radius; distance; polar coordinate |
| $\mathbf{R}$ | Resultant force; resultant vector; reaction |
| $R$ | Radius of earth; resistance |
| $\mathbf{s}$ | Position vector |
| $s$ | Length of arc |
| $t$ | Time; thickness; tangential direction |
| $\mathbf{T}$ | Force |
| $T$ | Tension; kinetic energy |
| $\mathbf{u}$ | Velocity |
| $u$ | Variable |
| $U$ | Work |
| $U_{1-2}^{\mathrm{NC}}$ | work done by non-conservative forces |
| $\mathbf{v}, v$ | Velocity |
| $v$ | Speed |
| $\bar{\mathbf{v}}, \bar{v}$ | Velocity of mass center |
| $\mathbf{v}_{B/A}$ | Velocity of $B$ relative to frame in translation with $A$ |
| $\mathbf{v}_{P/\mathscr{F}}$ | Velocity of $P$ relative to rotating frame $\mathscr{F}$ |
| $\mathbf{V}$ | Vector product |
| $V$ | Volume; potential energy |
| $w$ | Load per unit length |
| $\mathbf{W}, W$ | Weight; load |
| $x, y, z$ | Rectangular coordinates; distances |
| $\dot{x}, \dot{y}, \dot{z}$ | Time derivatives of coordinates $x, y, z$ |
| $\bar{x}, \bar{y}, \bar{z}$ | Rectangular coordinates of centroid, center of gravity, or mass center |
| $\boldsymbol{\alpha}, \alpha$ | Angular acceleration |
| $\alpha, \beta, \gamma$ | Angles |
| $\gamma$ | Specific weight |
| $\delta$ | Elongation |
| $\varepsilon$ | Eccentricity of conic section or of orbit |
| $\lambda$ | Unit vector along a line |
| $\eta$ | Efficiency |
| $\theta$ | Angular coordinate; Eulerian angle; angle; polar coordinate |
| $\mu$ | Coefficient of friction |
| $\rho$ | Density; radius of curvature |
| $\tau$ | Periodic time |

| | |
|---|---|
| $\tau_n$ | Period of free vibration |
| $\phi$ | Angle of friction; Eulerian angle; phase angle; angle |
| $\varphi$ | Phase difference |
| $\psi$ | Eulerian angle |
| $\boldsymbol{\omega}, \omega$ | Angular velocity |
| $\omega_f$ | Circular frequency of forced vibration |
| $\omega_n$ | Natural circular frequency |
| $\Omega$ | Angular velocity of frame of reference |

# Vector Mechanics
# For Engineers

## Statics and Dynamics

©Renato Bordoni/Alamy

# 1

# Introduction

The tallest skyscraper in the Western Hemisphere, One World Trade Center is a prominent feature of the New York City skyline. From its foundation to its structural components and mechanical systems, the design and operation of the tower is based on the fundamentals of engineering mechanics.

## Objectives

- **Define** the science of mechanics and examine its fundamental principles.
- **Discuss** and compare the International System of Units and U.S. customary units.
- **Discuss** how to approach the solution of mechanics problems, and introduce the SMART problem-solving methodology.
- **Examine** factors that govern numerical accuracy in the solution of a mechanics problem.

# 1.1   WHAT IS MECHANICS?

Mechanics is defined as the science that describes and predicts the conditions of rest or motion of bodies under the action of forces. It consists of the mechanics of *rigid bodies,* mechanics of *deformable bodies,* and mechanics of *fluids.*

The mechanics of rigid bodies is subdivided into **statics** and **dynamics**. Statics deals with bodies at rest; dynamics deals with bodies in motion. In this text, we assume bodies are perfectly rigid. In fact, actual structures and machines are never absolutely rigid; they deform under the loads to which they are subjected. However, because these deformations are usually small, they do not appreciably affect the conditions of equilibrium or the motion of the structure under consideration. They are important, though, as far as the resistance of the structure to failure is concerned. Deformations are studied in a course in mechanics of materials, which is part of the mechanics of deformable bodies. The third division of mechanics, the mechanics of fluids, is subdivided into the study of *incompressible fluids* and of *compressible fluids.* An important subdivision of the study of incompressible fluids is *hydraulics,* which deals with applications involving water.

Mechanics is a physical science, because it deals with the study of physical phenomena. However, some teachers associate mechanics with mathematics, whereas many others consider it as an engineering subject. Both of these views are justified in part. Mechanics is the foundation of most engineering sciences and is an indispensable prerequisite to their study. However, it does not have the *empiricism* found in some engineering sciences, i.e., it does not rely on experience or observation alone. The rigor of mechanics and the emphasis it places on deductive reasoning makes it resemble mathematics. However, mechanics is not an *abstract* or even a *pure* science; it is an *applied* science.

The purpose of mechanics is to explain and predict physical phenomena and thus to lay the foundations for engineering applications. You need to know statics to determine how much force will be exerted on a point in a bridge design and whether the structure can withstand that force. Determining the force a dam needs to withstand from the water in a river requires statics. You need statics to calculate how much weight a crane can lift, how much force a locomotive needs to pull a freight train, or how much force a circuit board in a computer can withstand. The concepts of dynamics enable you to analyze the flight characteristics of a jet, design a building to resist earthquakes, and mitigate shock and vibration to passengers inside a vehicle. The concepts of dynamics enable you to calculate how much force you need to send a satellite into orbit, accelerate a 200,000-ton cruise ship, or design a toy truck that doesn't break. You will not learn how to do these things in this course, but the ideas and methods you learn here will be the underlying basis for the engineering applications you will learn in your work.

# 1.2    FUNDAMENTAL CONCEPTS AND PRINCIPLES

Although the study of mechanics goes back to the time of Aristotle (384–322 B.C.) and Archimedes (287–212 B.C.), not until Newton (1642–1727) did anyone develop a satisfactory formulation of its fundamental principles. These principles were later modified by d'Alembert, Lagrange, and Hamilton. Their validity remained unchallenged until Einstein formulated his **theory of relativity** (1905). Although its limitations have now been recognized, **newtonian mechanics** still remains the basis of today's engineering sciences.

The basic concepts used in mechanics are *space, time, mass,* and *force.* These concepts cannot be truly defined; they should be accepted on the basis of our intuition and experience and used as a mental frame of reference for our study of mechanics.

The concept of **space** is associated with the position of a point *P*. We can define the position of *P* by providing three lengths measured from a certain reference point, or *origin,* in three given directions. These lengths are known as the *coordinates* of *P*.

To define an event, it is insufficient to indicate its position in space. We also need to specify the **time** of the event.

We use the concept of **mass** to characterize and compare bodies on the basis of certain fundamental mechanical experiments. Two bodies of the same mass, for example, are attracted by the earth in the same manner; they also offer the same resistance to a change in translational motion.

A **force** represents the action of one body on another. A force can be exerted by actual contact, like a push or a pull, or at a distance, as in the case of gravitational or magnetic forces. A force is characterized by its *point of application,* its *magnitude,* and its *direction;* a force is represented by a *vector* (Sec. 2.1B).

In newtonian mechanics, space, time, and mass are absolute concepts that are independent of each other. (This is not true in **relativistic mechanics,** where the duration of an event depends upon its position and the mass of a body varies with its velocity.) On the other hand, the concept of force is not independent of the other three. Indeed, one of the fundamental principles of newtonian mechanics listed below is that the resultant force acting on a body is related to the mass of the body and to the manner in which its velocity varies with time.

In this text, you will study the conditions of rest or motion of particles and rigid bodies in terms of the four basic concepts we have introduced. By **particle,** we mean a very small amount of matter, which we assume occupies a single point in space. A **rigid body** consists of a large number of particles occupying fixed positions with respect to one another. The study of the mechanics of particles is therefore a prerequisite to that of rigid bodies. Besides, we can use the results obtained for a particle directly in a large number of problems dealing with the conditions of rest or motion of actual bodies.

The study of elementary mechanics rests on six fundamental principles, based on experimental evidence.

- **The Parallelogram Law for the Addition of Forces.** Two forces acting on a particle may be replaced by a single force, called their *resultant*, obtained by drawing the diagonal of the parallelogram with sides equal to the given forces (Sec. 2.1A).
- **The Principle of Transmissibility.** The conditions of equilibrium or of motion of a rigid body remain unchanged if a force acting at a given point of the rigid body is replaced by a force of the same magnitude and same direction, but acting at a different point, provided that the two forces have the same line of action (Sec. 3.1B).
- **Newton's Three Laws of Motion.** Formulated by Sir Isaac Newton in the late seventeenth century, these laws can be stated as follows:

  **FIRST LAW.** If the resultant force acting on a particle is zero, the particle remains at rest (if originally at rest) or moves with constant speed in a straight line (if originally in motion) (Sec. 2.3B).

  **SECOND LAW.** If the resultant force acting on a particle is not zero, the particle has an acceleration proportional to the magnitude of the resultant and in the direction of this resultant force.

  As you will see in Sec. 12.1, this law can be stated as

$$\mathbf{F} = m\mathbf{a} \tag{1.1}$$

  where $\mathbf{F}$, $m$, and $\mathbf{a}$ represent, respectively, the resultant force acting on the particle, the mass of the particle, and the acceleration of the particle expressed in a consistent system of units.

  **THIRD LAW.** The forces of action and reaction between bodies in contact have the same magnitude, same line of action, and opposite sense (Chap. 6, Introduction).

- **Newton's Law of Gravitation.** Two particles of mass $M$ and $m$ are mutually attracted with equal and opposite forces $\mathbf{F}$ and $-\mathbf{F}$ of magnitude $F$ (Fig. 1.1), given by the formula

$$F = G\frac{Mm}{r^2} \tag{1.2}$$

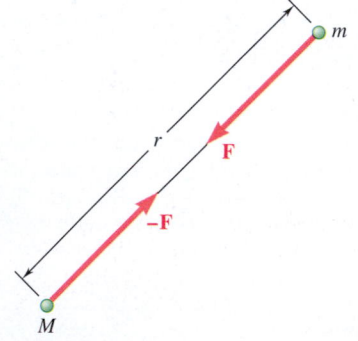

**Fig. 1.1**  From Newton's law of gravitation, two particles of masses $M$ and $m$ exert forces upon each other of equal magnitude, opposite direction, and the same line of action. This also illustrates Newton's third law of motion.

where $r$ = the distance between the two particles and $G$ = a universal constant called the *constant of gravitation*. Newton's law of gravitation introduces the idea of an action exerted at a distance and extends the range of application of Newton's third law: the action $\mathbf{F}$ and the reaction $-\mathbf{F}$ in Fig. 1.1 are equal and opposite, and they have the same line of action.

A particular case of great importance is that of the attraction of the earth on a particle located on its surface. The force $\mathbf{F}$ exerted by the earth on the particle is defined as the **weight W** of the particle. Suppose we set $M$ equal to the mass of the earth, $m$ equal to the mass of the particle, and $r$ equal to the earth's radius $R$. Then, introducing the constant

$$g = \frac{GM}{R^2} \tag{1.3}$$

we can express the magnitude $W$ of the weight of a particle of mass $m$ as[†]

$$W = mg \tag{1.4}$$

The value of $R$ in formula (1.3) depends upon the elevation of the point considered; it also depends upon its latitude, because the earth is not truly spherical. The value of $g$ therefore varies with the position of the point considered.

---

[†]A more accurate definition of the weight $\mathbf{W}$ should take into account the earth's rotation.

However, as long as the point actually remains on the earth's surface, it is sufficiently accurate in most engineering computations to assume that $g$ equals 9.81 m/s$^2$ or 32.2 ft/s$^2$.

The principles we have just listed will be introduced in the course of our study of mechanics as they are needed. The statics of particles carried out in Chap. 2 will be based on the parallelogram law of addition and on Newton's first law alone. We introduce the principle of transmissibility in Chap. 3 as we begin the study of the statics of rigid bodies, and we bring in Newton's third law in Chap. 6 as we analyze the forces exerted on each other by the various members forming a structure. We introduce Newton's second law and Newton's law of gravitation in dynamics. We will then show that Newton's first law is a particular case of Newton's second law (Sec. 12.1) and that the principle of transmissibility could be derived from the other principles and thus eliminated (Sec. 16.1D). In the meantime, however, Newton's first and third laws, the parallelogram law of addition, and the principle of transmissibility will provide us with the necessary and sufficient foundation for the entire study of the statics of particles, rigid bodies, and systems of rigid bodies.

As noted earlier, the six fundamental principles listed previously are based on experimental evidence. Except for Newton's first law and the principle of transmissibility, they are independent principles that cannot be derived mathematically from each other or from any other elementary physical principle. On these principles rests most of the intricate structure of newtonian mechanics. For more than two centuries, engineers have solved a tremendous number of problems dealing with the conditions of rest and motion of rigid bodies, deformable bodies, and fluids by applying these fundamental principles. Many of the solutions obtained could be checked experimentally, thus providing a further verification of the principles from which they were derived. Only in the twentieth century has Newton's mechanics been found to be at fault, in the study of the motion of atoms and the motion of the planets, where it must be supplemented by the theory of relativity. On the human or engineering scale, however, where velocities are small compared with the speed of light, Newton's mechanics have yet to be disproved.

**Photo 1.1** When in orbit of the earth, people and objects are said to be *weightless,* even though the gravitational force acting is approximately 90% of that experienced on the surface of the earth. This apparent contradiction will be resolved in Chapter 12 when we apply Newton's second law to the motion of particles. Source: NASA

## 1.3 SYSTEMS OF UNITS

Associated with the four fundamental concepts just discussed are the so-called *kinetic units,* i.e., the units of *length, time, mass,* and *force.* These units cannot be chosen independently if Eq. (1.1) is to be satisfied. Three of the units may be defined arbitrarily; we refer to them as **basic units**. The fourth unit, however, must be chosen in accordance with Eq. (1.1) and is referred to as a **derived unit**. Kinetic units selected in this way are said to form a **consistent system of units**.

**International System of Units (SI Units).[†]** In this system, which will be in universal use after the United States has completed its conversion to SI units, the base units are the units of length, mass, and time, and they are called, respectively, the **meter** (m), the **kilogram** (kg), and the **second** (s). All three are arbitrarily defined. The second was originally chosen to represent 1/86 400 of the mean solar day, but it is now defined as the duration of 9 192 631 770 cycles of the radiation corresponding to the transition between two levels of the fundamental state of the cesium-133 atom. The meter, originally defined as one

[†]SI stands for *Système International d'Unités* (French).

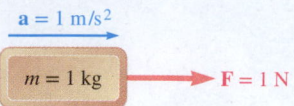

**Fig. 1.2**   A force of 1 newton applied to a body of mass 1 kg provides an acceleration of 1 m/s².

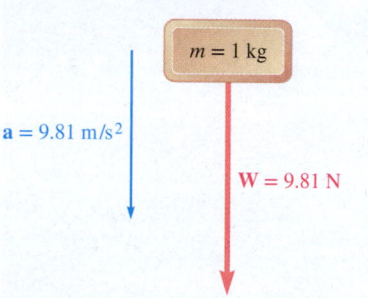

**Fig. 1.3**   A body of mass 1 kg experiencing an acceleration due to gravity of 9.81 m/s² has a weight of 9.81 N.

ten-millionth of the distance from the equator to either pole, is now defined as 1 650 763.73 wavelengths of the orange-red light corresponding to a certain transition in an atom of krypton-86. (The newer definitions are much more precise, and with today's modern instrumentation, are easier to verify as a standard.) The kilogram, which is approximately equal to the mass of 0.001 m³ of water, is defined as the mass of a platinum-iridium standard kept at the International Bureau of Weights and Measures at Sèvres, near Paris, France. The unit of force is a derived unit. It is called the **newton** (N) and is defined as the force that gives an acceleration of 1 m/s² to a body of mass 1 kg (Fig. 1.2). From Eq. (1.1), we have

$$1 \text{ N} = (1 \text{ kg})(1 \text{ m/s}^2) = 1 \text{ kg·m/s}^2 \tag{1.5}$$

The SI units are said to form an *absolute* system of units. This means that the three base units chosen are independent of the location where measurements are made. The meter, the kilogram, and the second may be used anywhere on the earth; they may even be used on another planet and still have the same significance.

The *weight* of a body, or the *force of gravity* exerted on that body, like any other force, should be expressed in newtons. From Eq. (1.4), it follows that the weight of a body of mass 1 kg (Fig. 1.3) is

$$W = mg$$
$$= (1 \text{ kg})(9.81 \text{ m/s}^2)$$
$$= 9.81 \text{ N}$$

Multiples and submultiples of the fundamental SI units are denoted through the use of the prefixes defined in Table 1.1. The multiples and submultiples of the units of length, mass, and force most frequently used in engineering are, respectively, the *kilometer* (km) and the *millimeter* (mm); the *megagram*‡ (Mg) and the *gram* (g); and the *kilonewton* (kN). According to Table 1.1, we have

$$1 \text{ km} = 1000 \text{ m} \qquad 1 \text{ mm} = 0.001 \text{ m}$$
$$1 \text{ Mg} = 1000 \text{ kg} \qquad 1 \text{ g} = 0.001 \text{ kg}$$
$$1 \text{ kN} = 1000 \text{ N}$$

The conversion of these units into meters, kilograms, and newtons, respectively, can be effected by simply moving the decimal point three places to the right or to the left. For example, to convert 3.82 km into meters, move the decimal point three places to the right:

$$3.82 \text{ km} = 3820 \text{ m}$$

Similarly, to convert 47.2 mm into meters, move the decimal point three places to the left:

$$47.2 \text{ mm} = 0.0472 \text{ m}$$

Using engineering notation, you can also write

$$3.82 \text{ km} = 3.82 \times 10^3 \text{ m}$$
$$47.2 \text{ mm} = 47.2 \times 10^{w-3} \text{ m}$$

The multiples of the unit of time are the *minute* (min) and the *hour* (h). Because 1 min = 60 s and 1 h = 60 min = 3600 s, these multiples cannot be converted as readily as the others.

---

‡Also known as a *metric ton.*

**Table 1.1   SI Prefixes**

| Multiplication Factor | Prefix[†] | Symbol |
|---|---|---|
| $1\ 000\ 000\ 000\ 000 = 10^{12}$ | Tera | T |
| $1\ 000\ 000\ 000 = 10^{9}$ | Giga | G |
| $1\ 000\ 000 = 10^{6}$ | Mega | M |
| $1\ 000 = 10^{3}$ | Kilo | k |
| $100 = 10^{2}$ | Hecto[‡] | h |
| $10 = 10^{1}$ | Deka[‡] | da |
| $0.1 = 10^{-1}$ | Deci[‡] | d |
| $0.01 = 10^{-2}$ | Centi[‡] | c |
| $0.001 = 10^{-3}$ | Milli | m |
| $0.000\ 001 = 10^{-6}$ | Micro | $\mu$ |
| $0.000\ 000\ 001 = 10^{-9}$ | Nano | n |
| $0.000\ 000\ 000\ 001 = 10^{-12}$ | Pico | p |
| $0.000\ 000\ 000\ 000\ 001 = 10^{-15}$ | Femto | f |
| $0.000\ 000\ 000\ 000\ 000\ 001 = 10^{-18}$ | Atto | a |

[†]The first syllable of every prefix is accented, so that the prefix retains its identity. Thus, the preferred pronunciation of kilometer places the accent on the first syllable, not the second.

[‡]The use of these prefixes should be avoided, except for the measurement of areas and volumes and for the nontechnical use of centimeter, as for body and clothing measurements.

By using the appropriate multiple or submultiple of a given unit, you can avoid writing very large or very small numbers. For example, it is usually simpler to write 427.2 km rather than 427 200 m and 2.16 mm rather than 0.002 16 m.[†]

**Units of Area and Volume.** The unit of area is the *square meter* (m$^2$), which represents the area of a square of side 1 m; the unit of volume is the *cubic meter* (m$^3$), which is equal to the volume of a cube of side 1 m. In order to avoid exceedingly small or large numerical values when computing areas and volumes, we use systems of subunits obtained by respectively squaring and cubing not only the millimeter, but also two intermediate submultiples of the meter: the *decimeter* (dm) and the *centimeter* (cm). By definition,

$$1\ \text{dm} = 0.1\ \text{m} = 10^{-1}\ \text{m}$$
$$1\ \text{cm} = 0.01\ \text{m} = 10^{-2}\ \text{m}$$
$$1\ \text{mm} = 0.001\ \text{m} = 10^{-3}\ \text{m}$$

Therefore, the submultiples of the unit of area are

$$1\ \text{dm}^2 = (1\ \text{dm})^2 = (10^{-1}\ \text{m})^2 = 10^{-2}\ \text{m}^2$$
$$1\ \text{cm}^2 = (1\ \text{cm})^2 = (10^{-2}\ \text{m})^2 = 10^{-4}\ \text{m}^2$$
$$1\ \text{mm}^2 = (1\ \text{mm})^2 = (10^{-3}\ \text{m})^2 = 10^{-6}\ \text{m}^2$$

Similarly, the submultiples of the unit of volume are

$$1\ \text{dm}^3 = (1\ \text{dm})^3 = (10^{-1}\ \text{m})^3 = 10^{-3}\ \text{m}^3$$
$$1\ \text{cm}^3 = (1\ \text{cm})^3 = (10^{-2}\ \text{m})^3 = 10^{-6}\ \text{m}^3$$
$$1\ \text{mm}^3 = (1\ \text{mm})^3 = (10^{-3}\ \text{m})^3 = 10^{-9}\ \text{m}^3$$

Note that when measuring the volume of a liquid, the cubic decimeter (dm$^3$) is usually referred to as a *liter* (L).

[†]Note that when more than four digits appear on either side of the decimal point to express a quantity in SI units—as in 427 000 m or 0.002 16 m—use spaces, never commas, to separate the digits into groups of three. This practice avoids confusion with the comma used in place of a decimal point, which is the convention in many countries.

Table 1.2 shows other derived SI units used to measure the moment of a force, the work of a force, etc. Although we will introduce these units in later chapters as they are needed, we should note an important rule at this time: When a derived unit is obtained by dividing a base unit by another base unit, you may use a prefix in the numerator of the derived unit, but not in its denominator. For example, the constant $k$ of a spring that stretches 20 mm under a load of 100 N is expressed as

$$k = \frac{100 \text{ N}}{20 \text{ mm}} = \frac{100 \text{ N}}{0.020 \text{ m}} = 5000 \text{ N/m or } k = 5 \text{ kN/m}$$

but never as $k = 5$ N/mm.

**U.S. Customary Units.**   Most practicing American engineers still commonly use a system in which the base units are those of length, force, and time. These units are, respectively, the *foot* (ft), the *pound* (lb), and the *second* (s). The second is the same as the corresponding SI unit. The foot is defined as 0.3048 m. The pound is defined as the *weight* of a platinum standard, called the *standard pound,* which is kept at the National Institute of Standards and Technology outside Washington, DC, the mass of which is 0.453 592 43 kg. Because the weight of a body depends upon the earth's gravitational attraction, which varies with location, the standard pound should be placed at sea level and at a latitude of 45° to properly define a force of 1 lb. Thus, the U.S. customary units do not form an absolute system of units. Because they depend upon the gravitational attraction of the earth, they form a *gravitational* system of units.

Although the standard pound also serves as the unit of mass in commercial transactions in the United States, it cannot be used that way in engineering

**Table 1.2    Principal SI Units Used in Mechanics**

| Quantity | Unit | Symbol | Formula |
|---|---|---|---|
| Acceleration | Meter per second squared | . . . | $m/s^2$ |
| Angle | Radian | rad | † |
| Angular acceleration | Radian per second squared | . . . | $rad/s^2$ |
| Angular velocity | Radian per second | . . . | $rad/s$ |
| Area | Square meter | . . . | $m^2$ |
| Density | Kilogram per cubic meter | . . . | $kg/m^3$ |
| Energy | Joule | J | N·m |
| Force | Newton | N | $kg·m/s^2$ |
| Frequency | Hertz | Hz | $s^{-1}$ |
| Impulse | Newton-second | . . . | kg·m/s |
| Length | Meter | m | ‡ |
| Mass | Kilogram | kg | ‡ |
| Moment of a force | Newton-meter | . . . | N·m |
| Power | Watt | W | J/s |
| Pressure | Pascal | Pa | $N/m^2$ |
| Stress | Pascal | Pa | $N/m^2$ |
| Time | Second | s | ‡ |
| Velocity | Meter per second | . . . | m/s |
| Volume |  |  |  |
|   Solids | Cubic meter | . . . | $m^3$ |
|   Liquids | Liter | L | $10^{-3}$ $m^3$ |
| Work | Joule | J | N·m |

†Supplementary unit (1 revolution $= 2\pi$ rad $= 360°$).
‡Base unit.

computations, because such a unit would not be consistent with the base units defined in the preceding paragraph. Indeed, when acted upon by a force of 1 lb—that is, when subjected to the force of gravity—the standard pound has the acceleration due to gravity, $g = 32.2 \text{ ft/s}^2$ (Fig. 1.4), not the unit acceleration required by Eq. (1.1). The unit of mass consistent with the foot, the pound, and the second is the mass that receives an acceleration of 1 ft/s² when a force of 1 lb is applied to it (Fig. 1.5). This unit, sometimes called a *slug,* can be derived from the equation $F = ma$ after substituting 1 lb for $F$ and 1 ft/s² for $a$. We have

$$F = ma \qquad 1 \text{ lb} = (1 \text{ slug})(1 \text{ ft/s}^2)$$

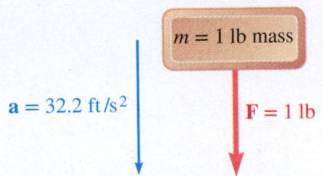

**Fig. 1.4** A body of 1 pound mass acted upon by a force of 1 pound has an acceleration of 32.2 ft/s².

This gives us

$$1 \text{ slug} = \frac{1 \text{ lb}}{1 \text{ ft/s}^2} = 1 \text{ lb·s}^2/\text{ft} \qquad \textbf{(1.6)}$$

Comparing Figs. 1.4 and 1.5, we conclude that the slug is a mass 32.2 times larger than the mass of the standard pound.

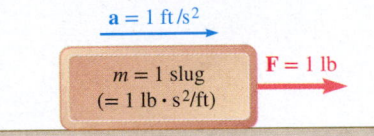

**Fig. 1.5** A force of 1 pound applied to a body of mass 1 slug produces an acceleration of 1 ft/s².

The fact that, in the U.S. customary system of units, bodies are characterized by their weight in pounds rather than by their mass in slugs is convenient in the study of statics, where we constantly deal with weights and other forces and only seldom deal directly with masses. However, in the study of dynamics, where forces, masses, and accelerations are involved, the mass $m$ of a body is expressed in slugs when its weight $W$ is given in pounds. Recalling Eq. (1.4), we write

$$m = \frac{W}{g} \qquad \textbf{(1.7)}$$

where $g$ is the acceleration due to gravity ($g = 32.2 \text{ ft/s}^2$).

Other U.S. customary units frequently encountered in engineering problems are the *mile* (mi), equal to 5280 ft; the *inch* (in.), equal to (1/12) ft; and the *kilopound* (kip), equal to 1000 lb. The *ton* is often used to represent a mass of 2000 lb but, like the pound, must be converted into slugs in engineering computations.

The conversion into feet, pounds, and seconds of quantities expressed in other U.S. customary units is generally more involved and requires greater attention than the corresponding operation in SI units. For example, suppose we are given the magnitude of a velocity $v = 30$ mi/h and want to convert it to ft/s. First we write

$$v = 30 \frac{\text{mi}}{\text{h}}$$

Because we want to get rid of the unit miles and introduce instead the unit feet, we should multiply the right-hand member of the equation by an expression containing miles in the denominator and feet in the numerator. However, because we do not want to change the value of the right-hand side of the equation, the expression used should have a value equal to unity. The quotient (5280 ft)/(1 mi) is such an expression. Operating in a similar way to transform the unit hour into seconds, we have

$$v = \left(30 \frac{\text{mi}}{\text{h}}\right)\left(\frac{5280 \text{ ft}}{1 \text{ mi}}\right)\left(\frac{1 \text{ h}}{3600 \text{ s}}\right)$$

Carrying out the numerical computations and canceling out units that appear in both the numerator and the denominator, we obtain

$$v = 44 \frac{\text{ft}}{\text{s}} = 44 \text{ ft/s}$$

## 1.4 CONVERTING BETWEEN TWO SYSTEMS OF UNITS

In many situations, an engineer might need to convert into SI units a numerical result obtained in U.S. customary units or vice versa. Because the unit of time is the same in both systems, only two kinetic base units need to be converted. Thus, because all other kinetic units can be derived from these base units, only two conversion factors need to be remembered.

**Units of Length.** By definition, the U.S. customary unit of length is

$$1 \text{ ft} = 0.3048 \text{ m} \tag{1.8}$$

It follows that

$$1 \text{ mi} = 5280 \text{ ft} = 5280(0.3048 \text{ m}) = 1609 \text{ m}$$

or

$$1 \text{ mi} = 1.609 \text{ km} \tag{1.9}$$

Also,

$$1 \text{ in.} = \frac{1}{12} \text{ ft} = \frac{1}{12}(0.3048 \text{ m}) = 0.0254 \text{ m}$$

or

$$1 \text{ in.} = 25.4 \text{ mm} \tag{1.10}$$

**Units of Force.** Recall that the U.S. customary unit of force (pound) is defined as the weight of the standard pound (of mass 0.4536 kg) at sea level and at a latitude of 45° (where $g = 9.807 \text{ m/s}^2$). Then, using Eq. (1.4), we write

$$W = mg$$
$$1 \text{ lb} = (0.4536 \text{ kg})(9.807 \text{ m/s}^2) = 4.448 \text{ kg} \cdot \text{m/s}^2$$

From Eq. (1.5), this reduces to

$$1 \text{ lb} = 4.448 \text{ N} \tag{1.11}$$

**Units of Mass.** The U.S. customary unit of mass (slug) is a derived unit. Thus, using Eqs. (1.6), (1.8), and (1.11), we have

$$1 \text{ slug} = 1 \text{ lb} \cdot \text{s}^2/\text{ft} = \frac{1 \text{ lb}}{1 \text{ ft/s}^2} = \frac{4.448 \text{ N}}{0.3048 \text{ m/s}^2} = 14.59 \text{ N} \cdot \text{s}^2/\text{m}$$

Again, from Eq. (1.5),

$$1 \text{ slug} = 1 \text{ lb} \cdot \text{s}^2/\text{ft} = 14.59 \text{ kg} \tag{1.12}$$

Although it cannot be used as a consistent unit of mass, recall that the mass of the standard pound is, by definition,

$$1 \text{ pound mass} = 0.4536 \text{ kg} \tag{1.13}$$

We can use this constant to determine the *mass* in SI units (kilograms) of a body that has been characterized by its *weight* in U.S. customary units (pounds).

To convert a derived U.S. customary unit into SI units, simply multiply or divide by the appropriate conversion factors. For example, to convert the moment of a force that is measured as $M = 47$ lb·in. into SI units, use formulas (1.10) and (1.11) and write

$$M = 47 \text{ lb·in.} = 47(4.448 \text{ N})(25.4 \text{ mm})$$
$$= 5310 \text{ N·mm} = 5.31 \text{ N·m}$$

You can also use conversion factors to convert a numerical result obtained in SI units into U.S. customary units. For example, if the moment of a force is measured as $M = 40$ N·m, follow the procedure at the end of Sec. 1.3 to write

$$M = 40 \text{ N·m} = (40 \text{ N·m})\left(\frac{1 \text{ lb}}{4.448 \text{ N}}\right)\left(\frac{1 \text{ ft}}{0.3048 \text{ m}}\right)$$

Carrying out the numerical computations and canceling out units that appear in both the numerator and the denominator, you obtain

$$M = 29.5 \text{ lb·ft}$$

The U.S. customary units most frequently used in mechanics are listed in Table 1.3 with their SI equivalents.

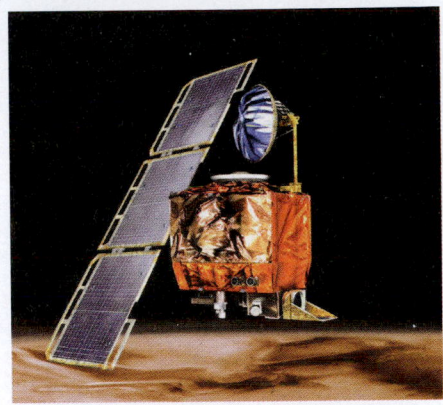

**Photo 1.2** In 1999, The *Mars Climate Orbiter* entered orbit around Mars at too low an altitude and disintegrated. Investigation showed that the software on board the probe interpreted force instructions in newtons, but the software at mission control on the earth was generating those instructions in terms of pounds.
Source: NASA/JPL-Caltech

# 1.5 METHOD OF SOLVING PROBLEMS

You should approach a problem in mechanics as you would approach an actual engineering situation. By drawing on your own experience and intuition about physical behavior, you will find it easier to understand and formulate the problem. Once you have clearly stated and understood the problem, however, there is no place in its solution for arbitrary methodologies.

**The solution must be based on the six fundamental principles stated in Sec. 1.2 or on theorems derived from them.**

Every step you take in the solution must be justified on this basis. Strict rules must be followed, which lead to the solution in an almost automatic fashion, leaving no room for your intuition or "feeling." After you have obtained an answer, you should check it. Here again, you may call upon your common sense and personal experience. If you are not completely satisfied with the result, you should carefully check your formulation of the problem, the validity of the methods used for its solution, and the accuracy of your computations.

In general, you can usually solve problems in several different ways; there is no one approach that works best for everybody. However, we have found that students often find it helpful to have a general set of guidelines to use for framing problems and planning solutions. In the Sample Problems throughout this text, we use a four-step method for approaching problems, which we refer to as the SMART methodology: **S**trategy, **M**odeling, **A**nalysis, and **R**eflect and **T**hink.

1. **Strategy.** The statement of a problem should be clear and precise, and it should contain the given data and indicate what information is required. The first step in solving the problem is to decide what concepts you have learned that apply to the given situation and to connect the data to the required information. It is often useful to work backward from the information you are trying to find: Ask yourself what quantities you need to

**Table 1.3  U.S. Customary Units and Their SI Equivalents**

| Quantity | U.S. Customary Unit | SI Equivalent |
|---|---|---|
| Acceleration | ft/s$^2$ | 0.3048 m/s$^2$ |
| | in./s$^2$ | 0.0254 m/s$^2$ |
| Area | ft$^2$ | 0.0929 m$^2$ |
| | in$^2$ | 645.2 mm$^2$ |
| Energy | ft·lb | 1.356 J |
| Force | kip | 4.448 kN |
| | lb | 4.448 N |
| | oz | 0.2780 N |
| Impulse | lb·s | 4.448 N·s |
| Length | ft | 0.3048 m |
| | in. | 25.40 mm |
| | mi | 1.609 km |
| Mass | oz mass | 28.35 g |
| | lb mass | 0.4536 kg |
| | slug | 14.59 kg |
| | ton | 907.2 kg |
| Moment of a force | lb·ft | 1.356 N·m |
| | lb·in. | 0.1130 N·m |
| Moment of inertia | | |
| Of an area | in$^4$ | $0.4162 \times 10^6$ mm$^4$ |
| Of a mass | lb·ft·s$^2$ | 1.356 kg·m$^2$ |
| Momentum | lb·s | 4.448 kg·m/s |
| Power | ft·lb/s | 1.356 W |
| | hp | 745.7 W |
| Pressure or stress | lb/ft$^2$ | 47.88 Pa |
| | lb/in$^2$ (psi) | 6.895 kPa |
| Velocity | ft/s | 0.3048 m/s |
| | in./s | 0.0254 m/s |
| | mi/h (mph) | 0.4470 m/s |
| | mi/h (mph) | 1.609 km/h |
| Volume | ft$^3$ | 0.02832 m$^3$ |
| | in$^3$ | 16.39 cm$^3$ |
| Liquids | gal | 3.785 L |
| | qt | 0.9464 L |
| Work | ft·lb | 1.356 J |

know to obtain the answer, and if some of these quantities are unknown, how can you find them from the given data.

2. **Modeling.** The first step in modeling is to define the system; that is, clearly define what you are setting aside for analysis. After you have selected a system, draw a neat sketch showing all quantities involved, with a separate diagram for each body in the problem. For equilibrium problems, indicate clearly the forces acting on each body along with any relevant geometrical data, such as lengths and angles. (These diagrams are known as **free-body diagrams** and are described in detail in Sec. 2.3C and the beginning of Chap. 4.)

3. **Analysis.** After you have drawn the appropriate diagrams, use the fundamental principles of mechanics listed in Sec. 1.2 to write equations expressing the conditions of rest or motion of the bodies considered. Each equation should be clearly related to one of the free-body diagrams and should be numbered. If you do not have enough equations to solve for the unknowns, try selecting another system, or reexamine your strategy to see if you can apply other principles to the problem. Once you

have obtained enough equations, you can find a numerical solution by following the usual rules of algebra, neatly recording each step and the intermediate results. Alternatively, you can solve the resulting equations with your calculator or a computer. (For multipart problems, it is sometimes convenient to present the Modeling and Analysis steps together, but they are both essential parts of the overall process.)

4.  **Reflect and Think.** After you have obtained the answer, check it carefully. Does it make sense in the context of the original problem? For instance, the problem may ask for the force at a given point of a structure. If your answer is negative, what does that mean for the force at the point?

You can often detect mistakes in *reasoning* by checking the units. For example, to determine the moment of a force of 50 N about a point 0.60 m from its line of action, we write (Sec. 3.3A)

$$M = Fd = (30 \text{ N})(0.60 \text{ m}) = 30 \text{ N·m}$$

The unit N·m obtained by multiplying newtons by meters is the correct unit for the moment of a force; if you had obtained another unit, you would know that some mistake had been made.

You can often detect errors in *computation* by substituting the numerical answer into an equation that was not used in the solution and verifying that the equation is satisfied. The importance of correct computations in engineering cannot be overemphasized.

## CASE STUDY 1.1*

Located in Baltimore, Maryland, the Carrollton Viaduct is the oldest railroad bridge in North America and continues in revenue service today. Construction was completed and the bridge put into operation in 1829 by the Baltimore & Ohio Railroad. The structure includes the stone masonry arch shown in CS Photo 1.1, and spans 80 ft. Assuming that the span is solid granite having a unit weight of 170 lb/ft$^3$, and that its dimensions can be approximated by those given in CS Fig. 1.1, let's estimate the weight of this span.

**CS Photo 1.1**    The Carrollton Viaduct in Baltimore, MD.
AREA Bulletin 732 Volume 92 (October 1991)

### STRATEGY:

First calculate the volume of the span, and then multiply this volume by the unit weight.

*Adapted from American Railway Engineering Association, Bulletin 732, October 1991, p. 221.

*(continued)*

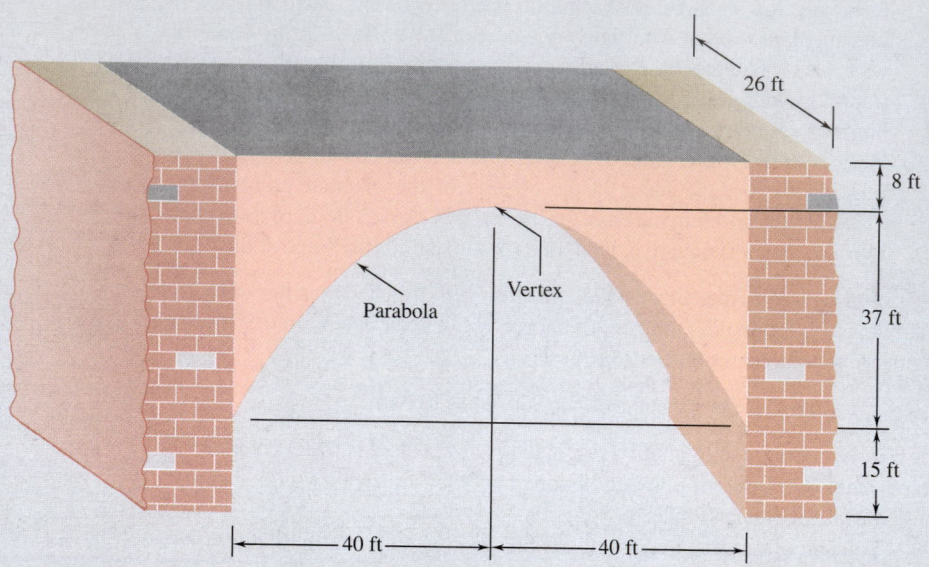

**CS Fig. 1.1** Assumed arch span geometry.

## MODELING:

The span can be represented by a body where a parabolic portion has been removed from a rectangular portion as shown in CS Fig. 1.2 (with both parts having a depth of 26 ft).

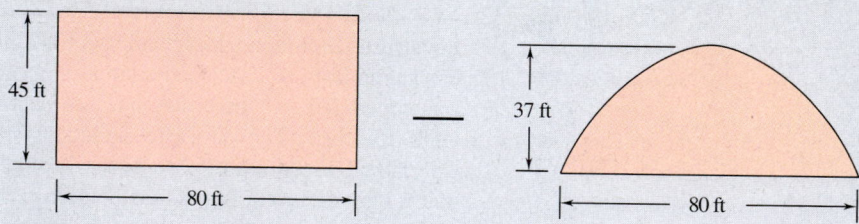

**CS Fig. 1.2** Modeling the arch span.

## ANALYSIS:

**Volume of the Span, V.** Removing the parabolic region from the rectangle,

$$V = \left[ (80 \text{ ft})(45 \text{ ft}) - \frac{2}{3} (80 \text{ ft})(37 \text{ ft}) \right] (26 \text{ ft}) = 42{,}300 \text{ ft}^3$$

**Weight of the Span, W.** Multiplying the volume by the unit weight,

$$W = (170 \text{ lb/ft}^3)(42{,}300 \text{ ft}^3) = 7.19 \times 10^6 \text{ lb}$$

## REFLECT AND THINK:

Though completed in 1829, regular locomotive usage didn't begin on this bridge until 1831 with the steam-powered *York,* which weighed approximately 7000 lb. (Up to that point, trains had been pulled by horses.) Then in 1832, there was initially concern regarding the ability of the stone arch to support a newer and heavier locomotive, the 13,000-lb *Atlantic.** As our knowledge of engineering mechanics has progressed since then, we better understand that a massive arch

*(continued)*

like this can indeed sustain such loads quite easily. This is illustrated by the modern-day coal cars shown crossing this same span in CS Photo 1.1, where each car has a rated weight of 263,000 lb. Arches derive load-carrying capacity through compression and are well-suited for stone masonry construction, because it provides high compressive strength. And while trains traversing the bridge would tend to introduce other types of effects into the span, the massiveness of the span itself (which we estimated to be $7.19 \times 10^6$ lb) far exceeds the car loads and therefore keeps the barrel (or portal) of the arch in compression.

# 1.6   NUMERICAL ACCURACY

The accuracy of the solution to a problem depends upon two items: (1) the accuracy of the given data and (2) the accuracy of the computations performed. The solution cannot be more accurate than the less accurate of these two items.

For example, suppose the loading of a bridge is known to be 75,000 lb with a possible error of 100 lb either way. The relative error that measures the degree of accuracy of the data is

$$\frac{100 \text{ lb}}{75,000 \text{ lb}} = 0.0013 = 0.13\%$$

In computing the reaction at one of the bridge supports, it would be meaningless to record it as 14,322 lb. The accuracy of the solution cannot be greater than 0.13%, no matter how precise the computations are, and the possible error in the answer may be as large as $(0.13/100)(14{,}322 \text{ lb}) \approx 20$ lb. The answer should be properly recorded as $14{,}320 \pm 20$ lb.

In engineering problems, the data are seldom known with an accuracy greater than 0.2%. It is therefore seldom justified to write answers with an accuracy greater than 0.2%. A practical rule is to use four figures to record numbers beginning with a "1" and three figures in all other cases. Unless otherwise indicated, you should assume the data given in a problem are known with a comparable degree of accuracy. A force of 40 lb, for example, should be read as 40.0 lb, and a force of 15 lb should be read as 15.00 lb.

Electronic calculators are widely used by practicing engineers and engineering students. The speed and accuracy of these calculators facilitate the numerical computations in the solution of many problems. However, you should not record more significant figures than can be justified merely because you can obtain them easily. As noted previously, an accuracy greater than 0.2% is seldom necessary or meaningful in the solution of practical engineering problems.

# 2

# Statics of Particles

Many engineering problems can be solved by considering the equilibrium of a "particle." In the case of this beam that is being hoisted into position, a relation between the tensions in the various cables involved can be obtained by considering the equilibrium of the hook to which the cables are attached.

# Objectives

- **Describe** force as a vector quantity.
- **Examine** vector operations useful for the analysis of forces.
- **Determine** the resultant of multiple forces acting on a particle.
- **Resolve** forces into components.
- **Add** forces that have been resolved into rectangular components.
- **Introduce** the concept of the free-body diagram.
- **Use** free-body diagrams to assist in the analysis of planar and spatial particle equilibrium problems.

## Introduction

In this chapter, you will study the effect of forces acting on particles. By the word "particle" we do not mean only tiny bits of matter, like an atom or an electron. Instead, we mean that the sizes and shapes of the bodies under consideration do not significantly affect the solutions of the problems. Another way of saying this is that we assume all forces acting on a given body act at the same point. This does not mean the object must be tiny—if you were modeling the mechanics of the Milky Way galaxy, for example, you could treat the sun and the entire Solar System as just a particle.

Our first step is to explain how to replace two or more forces acting on a given particle by a single force having the same effect as the original forces. This single equivalent force is called the *resultant* of the original forces. After this step, we will derive the relations among the various forces acting on a particle in a state of *equilibrium*. We will use these relations to determine some of the forces acting on the particle.

The first part of this chapter deals with forces contained in a single plane. Because two lines determine a plane, this situation arises any time we can reduce the problem to one of a particle subjected to two forces that support a third force, such as a crate suspended from two chains or a traffic light held in place by two cables. In the second part of this chapter, we examine the more general case of forces in three-dimensional space.

# 2.1 ADDITION OF PLANAR FORCES

Many important practical situations in engineering involve forces in the same plane. These include forces acting on a pulley, projectile motion, and an object in equilibrium on a flat surface. We will examine this situation first before looking at the added complications of forces acting in three-dimensional space.

## 2.1A Force on a Particle: Resultant of Two Forces

A force represents the action of one body on another. It is generally characterized by its **point of application**, its **magnitude**, and its **direction**. Forces acting on a given particle, however, have the same point of application. Thus, each force considered in this chapter is completely defined by its magnitude and direction.

The magnitude of a force is characterized by a certain number of units. As indicated in Chap. 1, the SI units used by engineers to measure the magnitude of a force are the newton (N) and its multiple the kilonewton (kN), which is equal to 1000 N. The U.S. customary units used for the same purpose are the pound (lb) and its multiple the kilopound (kip), which is equal to 1000 lb. We saw in Chap. 1 that a force of 445 N is equivalent to a force of 100 lb or that a force of 100 N equals a force of about 22.5 lb.

We define the direction of a force by its **line of action** and the **sense** of the force. The line of action is the infinite straight line along which the force acts; it is characterized by the angle it forms with some fixed axis (Fig. 2.1). The force itself is represented by a segment of that line; through the use of an appropriate scale, we can choose the length of this segment to represent the magnitude of the force. We indicate the sense of the force by an arrowhead. It is important in defining a force to indicate its sense. Two forces having the same magnitude and the same line of action but a different sense, such as the forces shown in Fig. 2.1a and b, have directly opposite effects on a particle.

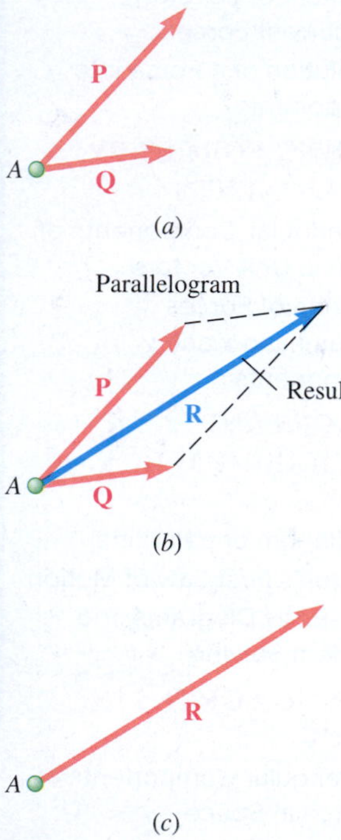

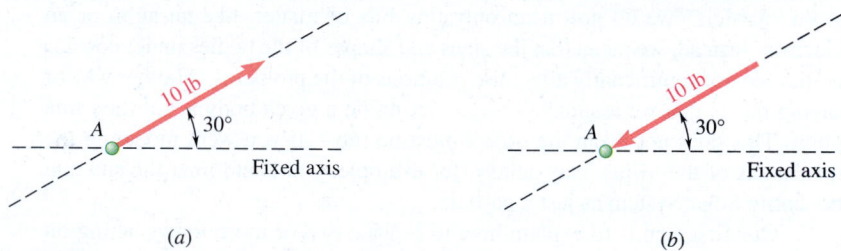

**Fig. 2.1**   The line of action of a force makes an angle with a given fixed axis. (a) The sense of the 10-lb force is away from particle A; (b) the sense of the 10-lb force is toward particle A.

Experimental evidence shows that two forces **P** and **Q** acting on a particle A (Fig. 2.2a) can be replaced by a single force **R** that has the same effect on the particle (Fig. 2.2c). This force is called the **resultant** of the forces **P** and **Q**. We can obtain **R**, as shown in Fig. 2.2b, by constructing a parallelogram, using **P** and **Q** as two adjacent sides. **The diagonal that passes through A represents the resultant.** This method for finding the resultant is known as the **parallelogram law** for the addition of two forces. This law is based on experimental evidence; it cannot be proved or derived mathematically.

## 2.1B   Vectors

We have just seen that forces do not obey the rules of addition defined in ordinary arithmetic or algebra. For example, two forces acting at a right angle to each other, one of 4 lb and the other of 3 lb, add up to a force of 5 lb acting at an angle between them, *not* to a force of 7 lb. Forces are not the only quantities that follow the parallelogram law of addition. As you will see later, *displacements, velocities, accelerations,* and *momenta* are other physical quantities possessing magnitude and direction that add according to the parallelogram law. All of these quantities can be represented mathematically by **vectors**. Those physical quantities that have magnitude but not direction, such as *volume, mass,* or *energy,* are represented by plain numbers often called **scalars** to distinguish them from vectors.

(a)

Parallelogram

(b)

(c)

**Fig. 2.2**   (a) Two forces **P** and **Q** act on particle A. (b) Draw a parallelogram with **P** and **Q** as the adjacent sides and label the diagonal that passes through A as **R**. (c) **R** is the resultant of the two forces **P** and **Q** and is equivalent to their sum.

**Photo 2.1**   In its purest form, a tug-of-war pits two opposite and almost-equal forces against each other. Whichever team can generate the larger force, wins. As you can see, a competitive tug-of-war can be quite intense. ©DGB/Alamy

Vectors are defined as **mathematical expressions possessing magnitude and direction, which add according to the parallelogram law**. Vectors are represented by arrows in diagrams and are distinguished from scalar quantities in this text through the use of boldface type (**P**). In longhand writing, a vector may be denoted by drawing a short arrow above the letter used to represent it ($\vec{P}$). The magnitude of a vector defines the length of the arrow used to represent it. In this text, we use italic type to denote the magnitude of a vector. Thus, the magnitude of the vector **P** is denoted by *P*.

A vector used to represent a force acting on a given particle has a well-defined point of application—namely, the particle itself. Such a vector is said to be a *fixed,* or *bound,* vector and cannot be moved without modifying the conditions of the problem. Other physical quantities, however, such as couples (see Chap. 3), are represented by vectors that may be freely moved in space; these vectors are called *free* vectors. Still other physical quantities, such as forces acting on a rigid body (see Chap. 3), are represented by vectors that can be moved along their lines of action; they are known as *sliding* vectors.

Two vectors that have the same magnitude and the same direction are said to be *equal,* whether or not they also have the same point of application (Fig. 2.3); equal vectors may be denoted by the same letter.

The *negative vector* of a given vector **P** is defined as a vector having the same magnitude as **P** and a direction opposite to that of **P** (Fig. 2.4); the negative of the vector **P** is denoted by −**P**. The vectors **P** and −**P** are commonly referred to as **equal and opposite** vectors. Thus, we have

$$\mathbf{P} + (-\mathbf{P}) = 0$$

## 2.1C Addition of Vectors

By definition, vectors add according to the parallelogram law. Thus, we obtain the sum of two vectors **P** and **Q** by attaching the two vectors to the same point *A* and constructing a parallelogram, using **P** and **Q** as two adjacent sides (Fig. 2.5). The diagonal that passes through *A* represents the sum of the vectors **P** and **Q**, denoted by **P** + **Q**. The fact that the sign + is used for both vector and scalar addition should not cause any confusion if vector and scalar quantities are always carefully distinguished. Note that the magnitude of the vector **P** + **Q** is *not*, in general, equal to the sum *P* + *Q* of the magnitudes of the vectors **P** and **Q**.

Because the parallelogram constructed on the vectors **P** and **Q** does not depend upon the order in which **P** and **Q** are selected, we conclude that the addition of two vectors is *commutative,* and we write

$$\mathbf{P} + \mathbf{Q} = \mathbf{Q} + \mathbf{P} \qquad (2.1)$$

From the parallelogram law, we can derive an alternative method for determining the sum of two vectors, known as the **triangle rule**. Consider Fig. 2.5, where the sum of the vectors **P** and **Q** has been determined by the parallelogram law. Because the side of the parallelogram opposite **Q** is equal to **Q** in magnitude and direction, we could draw only half of the parallelogram (Fig. 2.6*a*). The sum of the two vectors thus can be found by **arranging P and Q in tip-to-tail fashion and then connecting the tail of P with the tip of Q**. If we draw the other half of the parallelogram, as in Fig. 2.6*b*, we obtain the same result, confirming that vector addition is commutative.

We define *subtraction* of a vector as the addition of the corresponding negative vector. Thus, we determine the vector **P** − **Q**, representing the

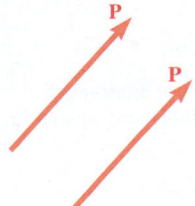

**Fig. 2.3** Equal vectors have the same magnitude and the same direction, even if they have different points of application.

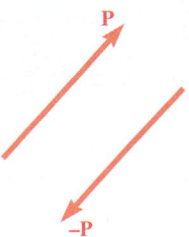

**Fig. 2.4** The negative vector of a given vector has the same magnitude but the opposite direction of the given vector.

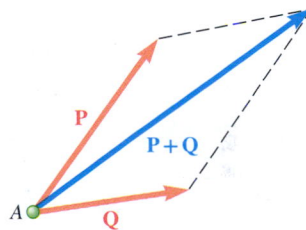

**Fig. 2.5** Using the parallelogram law to add two vectors.

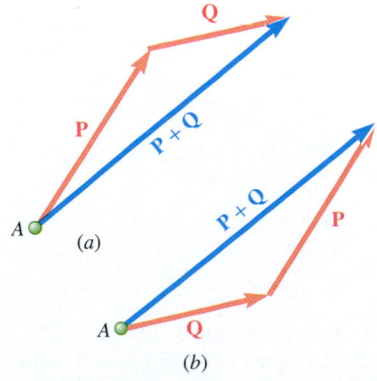

**Fig. 2.6** The triangle rule of vector addition. (*a*) Adding vector **Q** to vector **P** equals (*b*) adding vector **P** to vector **Q**.

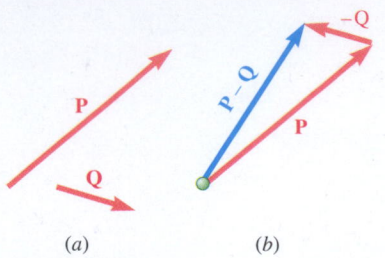

**Fig. 2.7**  Vector subtraction: Subtracting vector **Q** from vector **P** is the same as adding vector −**Q** to vector **P**.

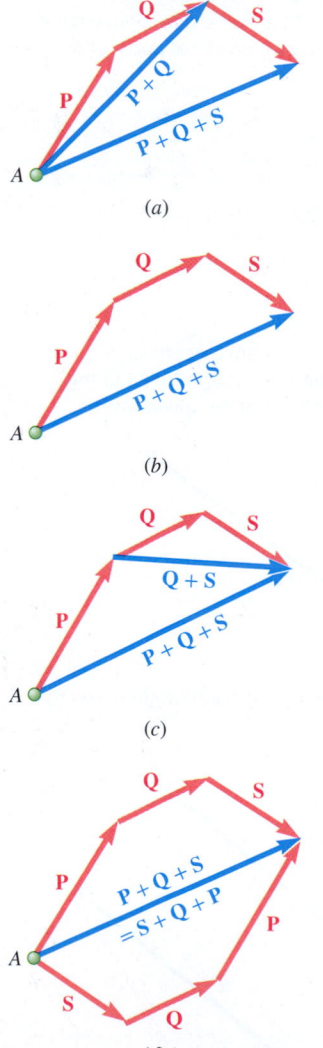

**Fig. 2.8**  Graphical addition of vectors. (*a*) Applying the triangle rule twice to add three vectors; (*b*) the vectors can be added in one step by the polygon rule; (*c*) vector addition is associative; (*d*) the order of addition is immaterial.

difference between the vectors **P** and **Q**, by adding to **P** the negative vector −**Q** (Fig. 2.7). We write

$$P - Q = P + (-Q) \tag{2.2}$$

Here again we should observe that, although we use the same sign to denote both vector and scalar subtraction, we avoid confusion by taking care to distinguish between vector and scalar quantities.

We now consider the *sum of three or more vectors*. The sum of three vectors **P**, **Q**, and **S** is, *by definition*, obtained by first adding the vectors **P** and **Q** and then adding the vector **S** to the vector **P** + **Q**. We write

$$P + Q + S = (P + Q) + S \tag{2.3}$$

Similarly, we obtain the sum of four vectors by adding the fourth vector to the sum of the first three. It follows that we can obtain the sum of any number of vectors by applying the parallelogram law repeatedly to successive pairs of vectors until all of the given vectors are replaced by a single vector.

If the given vectors are *coplanar,* i.e., if they are contained in the same plane, we can obtain their sum graphically. For this case, repeated application of the triangle rule is simpler than applying the parallelogram law. In Fig. 2.8*a*, we find the sum of three vectors **P**, **Q**, and **S** in this manner. The triangle rule is first applied to obtain the sum **P** + **Q** of the vectors **P** and **Q**; we apply it again to obtain the sum of the vectors **P** + **Q** and **S**. However, we could have omitted determining the vector **P** + **Q** and obtain the sum of the three vectors directly, as shown in Fig. 2.8*b*, by **arranging the given vectors in tip-to-tail fashion and connecting the tail of the first vector with the tip of the last one**. This is known as the **polygon rule** for the addition of vectors.

The result would be unchanged if, as shown in Fig. 2.8*c*, we had replaced the vectors **Q** and **S** by their sum **Q** + **S**. We may thus write

$$P + Q + S = (P + Q) + S = P + (Q + S) \tag{2.4}$$

which expresses the fact that vector addition is *associative*. Recalling that vector addition also has been shown to be commutative in the case of two vectors, we can write

$$\begin{aligned} P + Q + S &= (P + Q) + S = S + (P + Q) \\ &= S + (Q + P) = S + Q + P \end{aligned} \tag{2.5}$$

This expression, as well as others we can obtain in the same way, shows that the order in which several vectors are added together is immaterial (Fig. 2.8*d*).

**Product of a Scalar and a Vector.**   It is convenient to denote the sum **P** + **P** by 2**P**, the sum **P** + **P** + **P** by 3**P**, and, in general, the sum of *n* equal vectors **P** by the product *n***P**. Therefore, we define the product *n***P** of a positive integer *n* and a vector **P** as a vector having the same direction as **P** and the magnitude *nP*. Extending this definition to include all scalars and recalling the definition of a negative vector given earlier, we define the product *k***P** of a scalar *k* and a vector **P** as a vector having the same direction as **P** (if *k* is positive) or a direction opposite to that of **P** (if *k* is negative) and a magnitude equal to the product of *P* and the absolute value of *k* (Fig. 2.9).

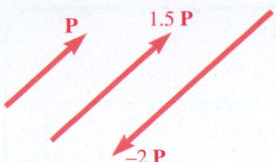

**Fig. 2.9** Multiplying a vector by a scalar changes the vector's magnitude, but not its direction (unless the scalar is negative, in which case the direction is reversed).

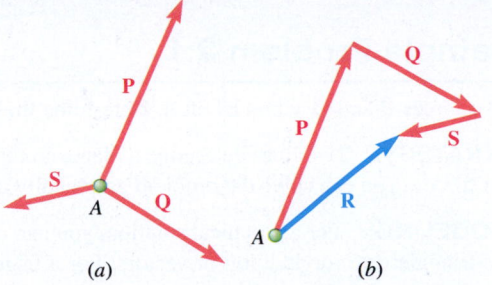

**Fig. 2.10** Concurrent forces can be added by the polygon rule.

## 2.1D Resultant of Several Concurrent Forces

Consider a particle $A$ acted upon by several coplanar forces, i.e., by several forces contained in the same plane (Fig. 2.10$a$). Because the forces all pass through $A$, they are also said to be *concurrent*. We can add the vectors representing the forces acting on $A$ by the polygon rule (Fig. 2.10$b$). Because the use of the polygon rule is equivalent to the repeated application of the parallelogram law, the vector $\mathbf{R}$ obtained in this way represents the resultant of the given concurrent forces. That is, the single force $\mathbf{R}$ has the same effect on the particle $A$ as the given forces. As before, the order in which we add the vectors $\mathbf{P}$, $\mathbf{Q}$, and $\mathbf{S}$ representing the given forces is immaterial.

## 2.1E Resolution of a Force into Components

We have seen that two or more forces acting on a particle may be replaced by a single force that has the same effect on the particle. Conversely, a single force $\mathbf{F}$ acting on a particle may be replaced by two or more forces that, together, have the same effect on the particle. These forces are called **components** of the original force $\mathbf{F}$, and the process of substituting them for $\mathbf{F}$ is called **resolving the force F into components**.

Each force $\mathbf{F}$ can be resolved into an infinite number of possible sets of components. Sets of *two components* $\mathbf{P}$ *and* $\mathbf{Q}$ are the most important as far as practical applications are concerned. However, even then, the number of ways in which a given force $\mathbf{F}$ may be resolved into two components is unlimited (Fig. 2.11).

In many practical problems, we start with a given vector $\mathbf{F}$ and want to determine a useful set of components. Two cases are of particular interest:

1. **One of the Two Components, P, Is Known.** We obtain the second component, $\mathbf{Q}$, by applying the triangle rule and joining the tip of $\mathbf{P}$ to the tip of $\mathbf{F}$ (Fig. 2.12). We can determine the magnitude and direction of $\mathbf{Q}$ graphically or by trigonometry. Once we have determined $\mathbf{Q}$, both components $\mathbf{P}$ and $\mathbf{Q}$ should be applied at $A$.
2. **The Line of Action of Each Component Is Known.** We obtain the magnitude and sense of the components by applying the parallelogram law and drawing lines through the tip of $\mathbf{F}$ that are parallel to the given lines of action (Fig. 2.13). This process leads to two well-defined components, $\mathbf{P}$ and $\mathbf{Q}$, which can be determined graphically or computed trigonometrically by applying the law of sines.

You will encounter many similar cases; for example, you might know the direction of one component while the magnitude of the other component is to be as small as possible (see Sample Prob. 2.2). In all cases, you need to draw the appropriate triangle or parallelogram that satisfies the given conditions.

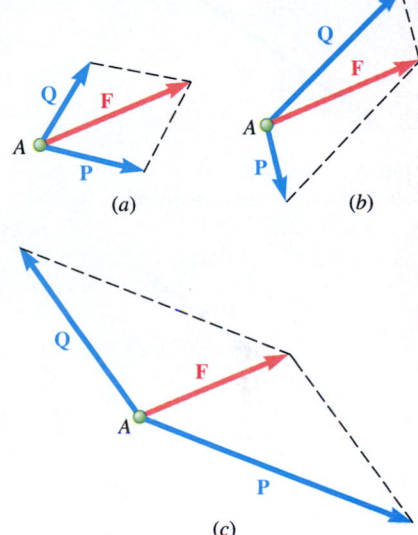

**Fig. 2.11** Three possible sets of components for a given force vector **F**.

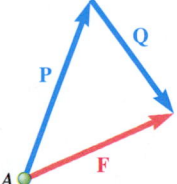

**Fig. 2.12** When component **P** is known, use the triangle rule to find component **Q**.

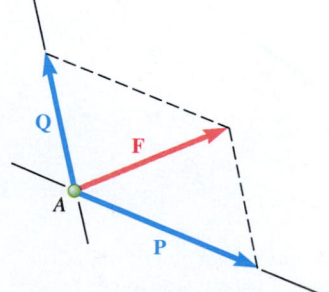

**Fig. 2.13** When the lines of action are known, use the parallelogram rule to determine components **P** and **Q**.

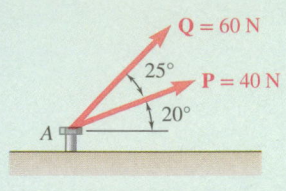

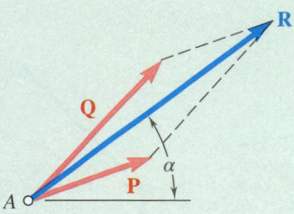

**Fig. 1**   Parallelogram law applied to add forces **P** and **Q**.

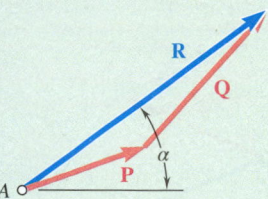

**Fig. 2**   Triangle rule applied to add forces **P** and **Q**.

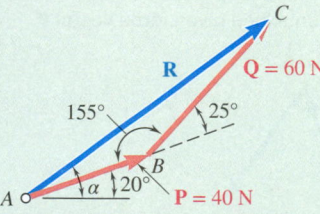

**Fig. 3**   Geometry of triangle rule applied to add forces **P** and **Q**.

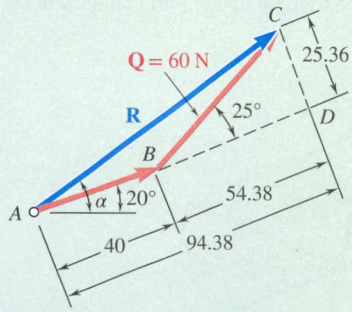

**Fig. 4**   Alternative geometry of triangle rule applied to add forces **P** and **Q**.

## Sample Problem 2.1

Two forces **P** and **Q** act on a bolt A. Determine their resultant.

**STRATEGY:**   Two lines determine a plane, so this is a problem of two coplanar forces. You can solve the problem graphically or by trigonometry.

**MODELING:**   For a graphical solution, you can use the parallelogram rule or the triangle rule for addition of vectors. For a trigonometric solution, you can use the law of cosines and law of sines or use a right-triangle approach.

**ANALYSIS:**
**Graphical Solution.**   Draw to scale a parallelogram with sides equal to **P** and **Q** (Fig. 1). Measure the magnitude and direction of the resultant. They are

$$R = 98 \text{ N} \qquad \alpha = 35° \qquad \textbf{R} = \textbf{98 N} \measuredangle \textbf{35°} \blacktriangleleft$$

You can also use the triangle rule. Draw forces **P** and **Q** in tip-to-tail fashion (Fig. 2). Again measure the magnitude and direction of the resultant. The answers should be the same.

$$R = 98 \text{ N} \qquad \alpha = 35° \qquad \textbf{R} = \textbf{98 N} \measuredangle \textbf{35°} \blacktriangleleft$$

**Trigonometric Solution.**   Using the triangle rule again, you know two sides and the included angle (Fig. 3). Apply the law of cosines.

$$R^2 = P^2 + Q^2 - 2PQ \cos B$$
$$R^2 = (40 \text{ N})^2 + (60 \text{ N})^2 - 2(40 \text{ N})(60 \text{ N}) \cos 155°$$
$$R = 97.73 \text{ N}$$

Now apply the law of sines:

$$\frac{\sin A}{Q} = \frac{\sin B}{R} \qquad \frac{\sin A}{60 \text{ N}} = \frac{\sin 155°}{97.73 \text{ N}} \qquad \text{(1)}$$

Solving Eq. (1) for sin A, you obtain

$$\sin A = \frac{(60 \text{ N}) \sin 155°}{97.73 \text{ N}}$$

Using a calculator, compute this quotient, and then obtain its arc sine:

$$A = 15.04° \qquad \alpha = 20° + A = 35.04°$$

Use three significant figures to record the answer (c.f. Sec. 1.6):

$$\textbf{R} = \textbf{97.7 N} \measuredangle \textbf{35.0°} \blacktriangleleft$$

**Alternative Trigonometric Solution.**   Construct the right triangle *BCD* (Fig. 4) and compute

$$CD = (60 \text{ N}) \sin 25° = 25.36 \text{ N}$$
$$BD = (60 \text{ N}) \cos 25° = 54.38 \text{ N}$$

*(continued)*

Then, using triangle *ACD,* you have

$$\tan A = \frac{25.36 \text{ N}}{94.38 \text{ N}} \qquad A = 15.04°$$

$$R = \frac{25.36}{\sin A} \qquad R = 97.73 \text{ N}$$

Again,

$$\alpha = 20° + A = 35.04° \qquad \mathbf{R} = 97.7 \text{ N} \, \measuredangle \, 35.0° \quad \blacktriangleleft$$

**REFLECT and THINK:** An analytical solution using trigonometry provides for greater accuracy. However, it is helpful to use a graphical solution as a check.

## Sample Problem 2.2

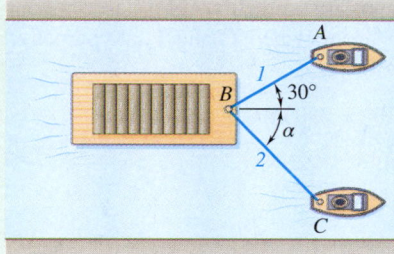

Two tugboats are pulling a barge. If the resultant of the forces exerted by the tugboats is a 5000-lb force directed along the axis of the barge, determine (*a*) the tension in each of the ropes, given that $\alpha = 45°$, (*b*) the value of $\alpha$ for which the tension in rope *2* is a minimum.

**STRATEGY:** This is a problem of two coplanar forces. You can solve the first part either graphically or analytically. In the second part, a graphical approach readily shows the necessary direction for rope *2*, and you can use an analytical approach to complete the solution.

**MODELING:** You can use the parallelogram law or the triangle rule to solve part (*a*). For part (*b*), use a variation of the triangle rule.

**ANALYSIS: a. Tension for $\alpha = 45°$.**

**Graphical Solution.** Use the parallelogram law. The resultant (the diagonal of the parallelogram) is equal to 5000 lb and is directed to the right. Draw the sides parallel to the ropes (Fig. 1). If the drawing is done to scale, you should measure

$$T_1 = 3700 \text{ lb} \qquad\qquad T_2 = 2600 \text{ lb} \quad \blacktriangleleft$$

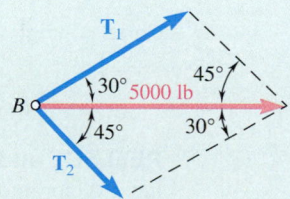

**Fig. 1** Parallelogram law applied to add forces **T**₁ and **T**₂.

*(continued)*

**Trigonometric Solution.** Use the triangle rule. Note that the triangle in Fig. 2 represents half of the parallelogram shown in Fig. 1. Using the law of sines,

$$\frac{T_1}{\sin 45°} = \frac{T_2}{\sin 30°} = \frac{5000 \text{ lb}}{\sin 105°}$$

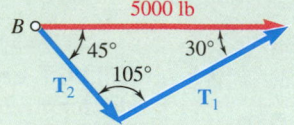

**Fig. 2** Triangle rule applied to add forces **T**$_1$ and **T**$_2$.

With a calculator, compute and store the value of the last quotient. Multiply this value successively by sin 45° and sin 30°, obtaining

$$T_1 = 3660 \text{ lb} \qquad T_2 = 2590 \text{ lb} \quad ◀$$

**b. Value of $\alpha$ for Minimum $T_2$.** To determine the value of $\alpha$ for which the tension in rope *2* is a minimum, use the triangle rule again. In Fig. 3, line *1-1'* is the known direction of **T**$_1$. Several possible directions of **T**$_2$ are shown by the lines *2-2'*. The minimum value of $T_2$ occurs when **T**$_1$ and **T**$_2$ are perpendicular (Fig. 4). Thus, the minimum value of $T_2$ is

$$T_2 = (5000 \text{ lb}) \sin 30° = 2500 \text{ lb}$$

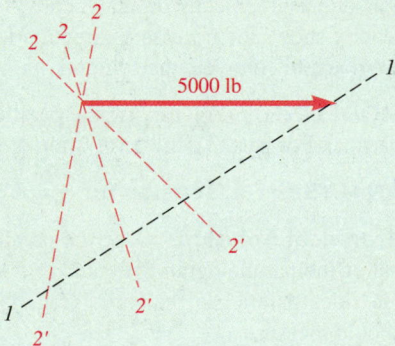

**Fig. 3** Determination of direction of minimum **T**$_2$.

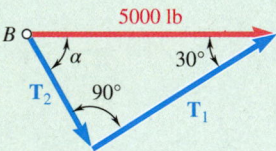

**Fig. 4** Triangle rule applied for minimum **T**$_2$.

Corresponding values of $T_1$ and $\alpha$ are

$$T_1 = (5000 \text{ lb}) \cos 30° = 4330 \text{ lb}$$
$$\alpha = 90° - 30° \qquad\qquad \alpha = 60° \quad ◀$$

**REFLECT and THINK:** Part (*a*) is a straightforward application of resolving a vector into components. The key to part (*b*) is recognizing that the minimum value of $T_2$ occurs when **T**$_1$ and **T**$_2$ are perpendicular.

# SOLVING PROBLEMS
# ON YOUR OWN

The preceding sections were devoted to adding vectors by using the parallelogram law, triangle rule, and polygon rule with application to forces.

We presented two Sample Problems. In Sample Prob. 2.1, we used the parallelogram law to determine the resultant of two forces of known magnitude and direction. In Sample Prob. 2.2, we used it to resolve a given force into two components of known direction.

You will now be asked to solve problems on your own. Some may resemble one of the Sample Problems; others may not. What all Problems and Sample Problems in this section have in common is that they can be solved by direct application of the parallelogram law.

Your solution of a given problem should consist of the following steps:

**1. Identify which forces are the applied forces and which is the resultant.** It is often helpful to write the vector equation that shows how the forces are related. For example, in Sample Prob. 2.1 you could write

$$\mathbf{R} = \mathbf{P} + \mathbf{Q}$$

You may want to keep this relation in mind as you formulate the next part of the solution.

**2. Draw a parallelogram with the applied forces as two adjacent sides and the resultant as the included diagonal (Fig. 2.2).** Alternatively, you can use the **triangle rule** with the applied forces drawn in tip-to-tail fashion and the resultant extending from the tail of the first vector to the tip of the second (Fig. 2.6).

**3. Indicate all dimensions.** Using one of the triangles of the parallelogram or the triangle constructed according to the triangle rule, indicate all dimensions—whether sides or angles—and determine the unknown dimensions either graphically or by trigonometry.

**4. Recall the laws of trigonometry.** If you use trigonometry, remember that the law of cosines should be applied first if two sides and the included angle are known (Sample Prob. 2.1), and the law of sines should be applied first if one side and all angles are known (Sample Prob. 2.2).

If you have had prior exposure to mechanics, you might be tempted to ignore the solution techniques of this lesson in favor of resolving the forces into rectangular components. The component method is important and is considered in the next section, but use of the parallelogram law simplifies the solution of many problems and should be mastered first.

# Problems

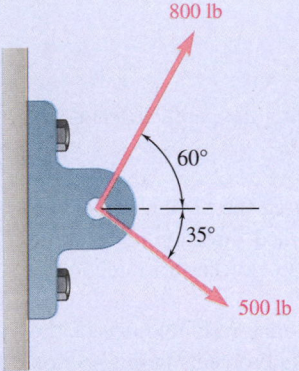

Fig. P2.2

**2.1** Two forces are applied as shown to a hook. Determine graphically the magnitude and direction of their resultant using (*a*) the parallelogram law, (*b*) the triangle rule.

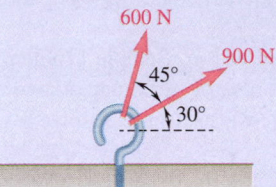

Fig. P2.1

**2.2** Two forces are applied as shown to a bracket support. Determine graphically the magnitude and direction of their resultant using (*a*) the parallelogram law, (*b*) the triangle rule.

**2.3** Two forces **P** and **Q** are applied as shown at point *A* of a hook support. Knowing that $P = 75$ N and $Q = 125$ N, determine graphically the magnitude and direction of their resultant using (*a*) the parallelogram law, (*b*) the triangle rule.

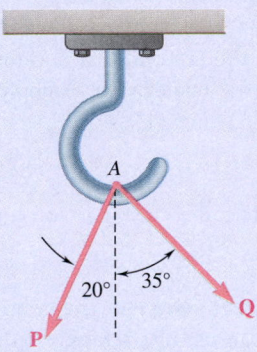

Fig. *P2.3* and P2.4

**2.4** Two forces **P** and **Q** are applied as shown at point *A* of a hook support. Knowing that $P = 60$ lb and $Q = 25$ lb, determine graphically the magnitude and direction of their resultant using (*a*) the parallelogram law, (*b*) the triangle rule.

**2.5** A stake is being pulled out of the ground by means of two ropes as shown. Knowing that $\alpha = 30°$, determine by trigonometry (*a*) the magnitude of the force **P** so that the resultant force exerted on the stake is vertical, (*b*) the corresponding magnitude of the resultant.

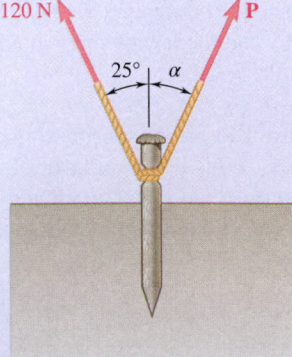

Fig. P2.5

**2.6** A telephone cable is clamped at $A$ to the pole $AB$. Knowing that the tension in the left-hand portion of the cable is $T_1 = 800$ lb, determine by trigonometry (a) the required tension $T_2$ in the right-hand portion if the resultant $\mathbf{R}$ of the forces exerted by the cable at $A$ is to be vertical, (b) the corresponding magnitude of $\mathbf{R}$.

**2.7** A telephone cable is clamped at $A$ to the pole $AB$. Knowing that the tension in the right-hand portion of the cable is $T_2 = 1000$ lb, determine by trigonometry (a) the required tension $T_1$ in the left-hand portion if the resultant $\mathbf{R}$ of the forces exerted by the cable at $A$ is to be vertical, (b) the corresponding magnitude of $\mathbf{R}$.

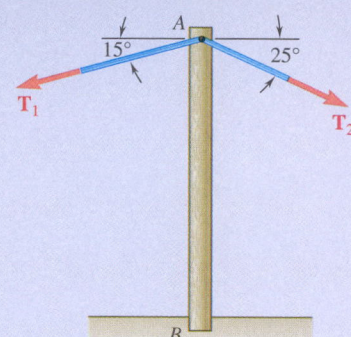

Fig. P2.6 and *P2.7*

**2.8** A disabled automobile is pulled by means of two ropes as shown. The tension in rope $AB$ is 2.2 kN, and the angle $\alpha$ is 25°. Knowing that the resultant of the two forces applied at $A$ is directed along the axis of the automobile, determine by trigonometry (a) the tension in rope $AC$, (b) the magnitude of the resultant of the two forces applied at $A$.

**2.9** A disabled automobile is pulled by means of two ropes as shown. Knowing that the tension in rope $AB$ is 3 kN, determine by trigonometry the tension in rope $AC$ and the value of $\alpha$ so that the resultant force exerted at $A$ is a 4.8-kN force directed along the axis of the automobile.

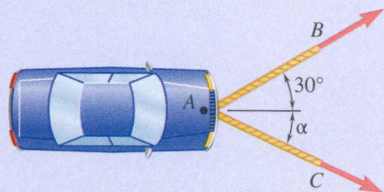

Fig. P2.8 and P2.9

**2.10** Two forces are applied as shown to a hook support. Knowing that the magnitude of $\mathbf{P}$ is 35 N, determine by trigonometry (a) the required angle $\alpha$ if the resultant $\mathbf{R}$ of the two forces applied to the support is to be horizontal, (b) the corresponding magnitude of $\mathbf{R}$.

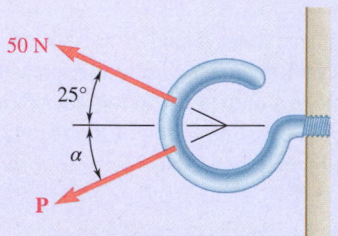

Fig. P2.10

**2.11** A steel tank is to be positioned in an excavation. Knowing that $\alpha = 20°$, determine by trigonometry (a) the required magnitude of the force $\mathbf{P}$ if the resultant $\mathbf{R}$ of the two forces applied at $A$ is to be vertical, (b) the corresponding magnitude of $\mathbf{R}$.

**2.12** A steel tank is to be positioned in an excavation. Knowing that the magnitude of $\mathbf{P}$ is 500 lb, determine by trigonometry (a) the required angle $\alpha$ if the resultant $\mathbf{R}$ of the two forces applied at $A$ is to be vertical, (b) the corresponding magnitude of $\mathbf{R}$.

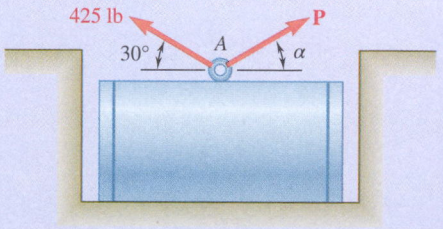

Fig. P2.11, *P2.12*, and P2.13

**2.13** A steel tank is to be positioned in an excavation. Determine by trigonometry (a) the magnitude and direction of the smallest force $\mathbf{P}$ for which the resultant $\mathbf{R}$ of the two forces applied at $A$ is vertical, (b) the corresponding magnitude of $\mathbf{R}$.

**2.14** For the hook support of Prob. 2.10, determine by trigonometry (*a*) the magnitude and direction of the smallest force **P** for which the resultant **R** of the two forces applied to the support is horizontal, (*b*) the corresponding magnitude of **R**.

**2.15** The barge *B* is pulled by two tugboats *A* and *C*. At a given instant, the tension in cable *AB* is 4500 lb and the tension in cable *BC* is 2000 lb. Determine by trigonometry the magnitude and direction of the resultant of the two forces applied at *B* at that instant.

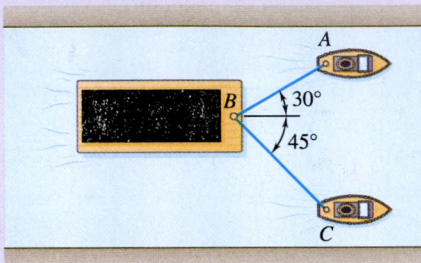

Fig. P2.15

**2.16** Solve Prob. 2.1 by trigonometry.

**2.17** Solve Prob. 2.4 by trigonometry.

**2.18** For the stake of Prob. 2.5, knowing that the tension in one rope is 120 N, determine by trigonometry the magnitude and direction of the force **P** so that the resultant is a vertical force of 160 N.

**2.19** Two structural members *A* and *B* are bolted to a bracket as shown. Knowing that both members are in compression and that the force is 10 kN in member *A* and 15 kN in member *B*, determine by trigonometry the magnitude and direction of the resultant of the forces applied to the bracket by members *A* and *B*.

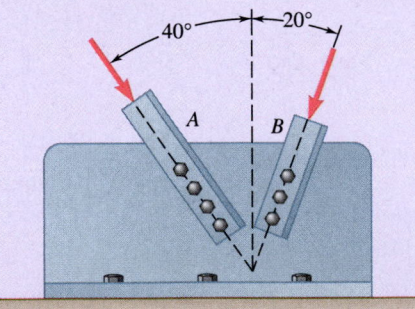

Fig. P2.19 and *P2.20*

**2.20** Two structural members *A* and *B* are bolted to a bracket as shown. Knowing that both members are in compression and that the force is 15 kN in member *A* and 10 kN in member *B*, determine by trigonometry the magnitude and direction of the resultant of the forces applied to the bracket by members *A* and *B*.

# 2.2 ADDING FORCES BY COMPONENTS

In Sec. 2.1E, we described how to resolve a force into components. Here we discuss how to add forces by using their components, especially rectangular components. This method is often the most convenient way to add forces and, in practice, is the most common approach. (Note that we can readily extend the properties of vectors established in this section to the rectangular components of any vector quantity, such as velocity or momentum.)

## 2.2A Rectangular Components of a Force: Unit Vectors

In many problems, it is useful to resolve a force into two components that are perpendicular to each other. Figure 2.14 shows a force **F** resolved into a component **F**$_x$ along the $x$ axis and a component **F**$_y$ along the $y$ axis. The parallelogram drawn to obtain the two components is a rectangle, and **F**$_x$ and **F**$_y$ are called **rectangular components**.

The $x$ and $y$ axes are usually chosen to be horizontal and vertical, respectively, as in Fig. 2.14; they may, however, be chosen in any two perpendicular directions, as shown in Fig. 2.15. In determining the rectangular components of a force, you should think of the construction lines shown in Figs. 2.14 and 2.15 as being *parallel* to the $x$ and $y$ axes, rather than *perpendicular* to these axes. This practice will help avoid mistakes in determining *oblique* components, as in Sec. 2.1E.

**Force in Terms of Unit Vectors.** To simplify working with rectangular components, we introduce two vectors of unit magnitude, directed respectively along the positive $x$ and $y$ axes. These vectors are called **unit vectors** and are denoted by **i** and **j**, respectively (Fig. 2.16). Recalling the definition of the product of a scalar and a vector given in Sec. 2.1C, note that we can obtain the rectangular components **F**$_x$ and **F**$_y$ of a force **F** by multiplying respectively the unit vectors **i** and **j** by appropriate scalars (Fig. 2.17). We have

$$\mathbf{F}_x = F_x\mathbf{i} \qquad \mathbf{F}_y = F_y\mathbf{j} \qquad (2.6)$$

and

$$\mathbf{F} = F_x\mathbf{i} + F_y\mathbf{j} \qquad (2.7)$$

The scalars $F_x$ and $F_y$ may be positive or negative, depending upon the sense of **F**$_x$ and of **F**$_y$, but their absolute values are equal to the magnitudes of the component forces **F**$_x$ and **F**$_y$, respectively. The scalars $F_x$ and $F_y$ are called the **scalar components** of the force **F**, whereas the actual component forces **F**$_x$ and **F**$_y$ should be referred to as the **vector components** of **F**. However, when there exists no possibility of confusion, we may refer to the vector as well as the scalar components of **F** as simply the **components** of **F**. Note that the scalar component $F_x$ is positive when the vector component **F**$_x$ has the same sense as the unit vector **i** (i.e., the same sense as the positive $x$ axis) and is negative when **F**$_x$ has the opposite sense. A similar conclusion holds for the sign of the scalar component $F_y$.

**Scalar Components.** Denoting by $F$ the magnitude of the force **F** and by $\theta$ the angle between **F** and the $x$ axis, which is measured counterclockwise

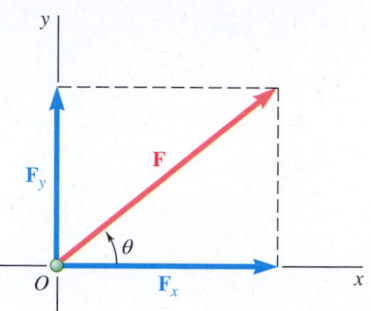

**Fig. 2.14** Rectangular components of a force **F**.

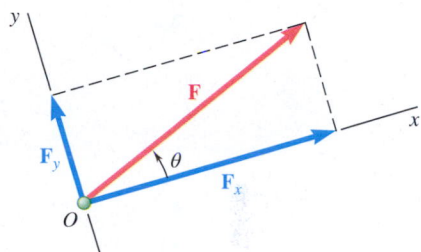

**Fig. 2.15** Rectangular components of a force **F** for axes rotated away from horizontal and vertical.

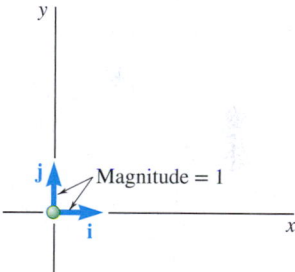

**Fig. 2.16** Unit vectors along the $x$ and $y$ axes.

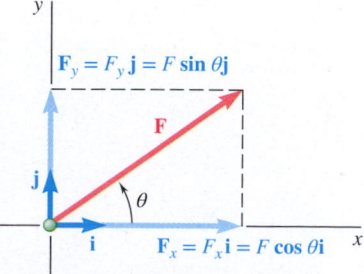

**Fig. 2.17** Expressing the components of **F** in terms of unit vectors with scalar multipliers.

from the positive $x$ axis (Fig. 2.17), we may express the scalar components of **F** as

$$F_x = F \cos \theta \qquad F_y = F \sin \theta \tag{2.8}$$

These relations hold for any value of the angle $\theta$ from 0° to 360°, and they define the signs and absolute values of the scalar components $F_x$ and $F_y$.

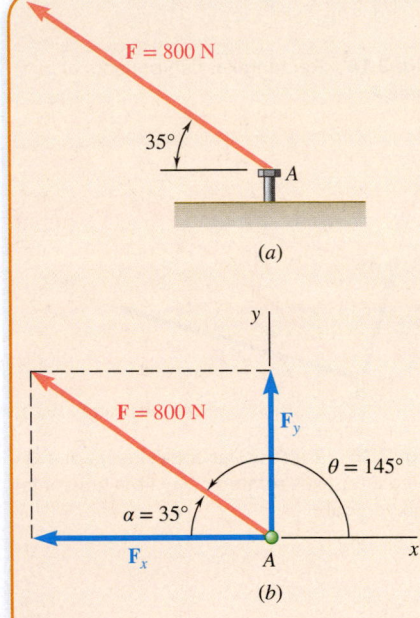

**Fig. 2.18** (a) Force **F** exerted on a bolt; (b) rectangular components of **F**.

## Concept Application 2.1

A force of 800 N is exerted on a bolt $A$, as shown in Fig. 2.18a. Determine the horizontal and vertical components of the force.

**Solution** In order to obtain the correct sign for the scalar components $F_x$ and $F_y$, we could substitute the value $180° - 35° = 145°$ for $\theta$ in Eqs. (2.8). However, it is often more practical to determine by inspection the signs of $F_x$ and $F_y$ (Fig. 2.18b) and then use the trigonometric functions of the angle $\alpha = 35°$. Therefore,

$$F_x = -F \cos \alpha = -(800 \text{ N}) \cos 35° = -655 \text{ N}$$
$$F_y = +F \sin \alpha = +(800 \text{ N}) \sin 35° = +459 \text{ N}$$

The vector components of **F** are thus

$$\mathbf{F}_x = -(655 \text{ N})\mathbf{i} \qquad \mathbf{F}_y = +(459 \text{ N})\mathbf{j}$$

and we may write **F** in the form

$$\mathbf{F} = -(655 \text{ N})\mathbf{i} + (459 \text{ N})\mathbf{j} \quad \blacktriangleleft$$

## Concept Application 2.2

A man pulls with a force of 300 N on a rope attached to the top of a building, as shown in Fig. 2.19a. What are the horizontal and vertical components of the force exerted by the rope at point $A$?

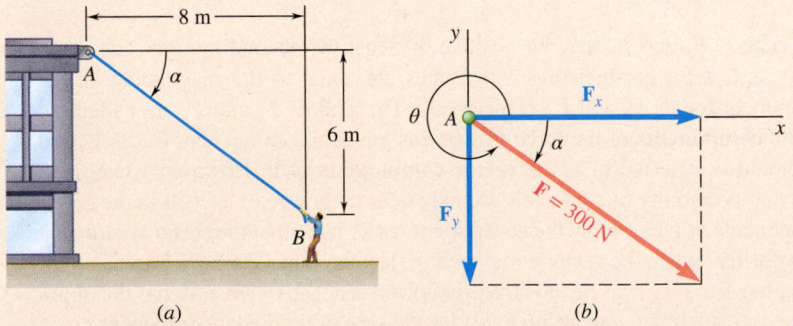

**Fig. 2.19** (a) A man pulls on a rope attached to a building; (b) components of the rope's force **F**.

*(continued)*

**Solution**  You can see from Fig. 2.19*b* that

$$F_x = +(300 \text{ N}) \cos \alpha \qquad F_y = -(300 \text{ N}) \sin \alpha$$

Observing that $AB = 10$ m, we find from Fig. 2.19*a* that

$$\cos \alpha = \frac{8 \text{ m}}{AB} = \frac{8 \text{ m}}{10 \text{ m}} = \frac{4}{5} \qquad \sin \alpha = \frac{6 \text{ m}}{AB} = \frac{6 \text{ m}}{10 \text{ m}} = \frac{3}{5}$$

We thus obtain

$$F_x = +(300 \text{ N})\frac{4}{5} = +240 \text{ N} \qquad F_y = -(300 \text{ N})\frac{3}{5} = -180 \text{ N}$$

This gives us a total force of

$$\mathbf{F} = (240 \text{ N})\mathbf{i} - (180 \text{ N})\mathbf{j} \qquad \blacktriangleleft$$

**Direction of a Force.**  When a force $\mathbf{F}$ is defined by its rectangular components $F_x$ and $F_y$ (see Fig. 2.17), we can find the angle $\theta$ defining its direction from

$$\tan \theta = \frac{F_y}{F_x} \qquad (2.9)$$

We can obtain the magnitude $F$ of the force by applying the Pythagorean theorem,

$$F = \sqrt{F_x^2 + F_y^2} \qquad (2.10)$$

or by solving for $F$ from one of the Eqs. (2.8).

## Concept Application 2.3

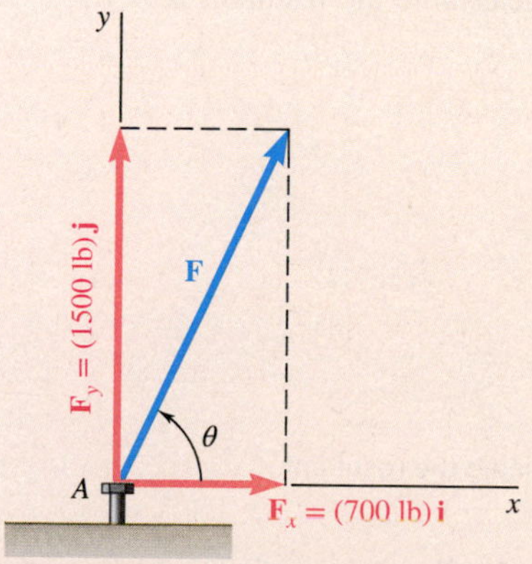

**Fig. 2.20**  Components of a force **F** exerted on a bolt.

A force $\mathbf{F} = (700 \text{ lb})\mathbf{i} + (1500 \text{ lb})\mathbf{j}$ is applied to a bolt $A$. Determine the magnitude of the force and the angle $\theta$ it forms with the horizontal.

**Solution**  First draw a diagram showing the two rectangular components of the force and the angle $\theta$ (Fig. 2.20). From Eq. (2.9), you obtain

$$\tan \theta = \frac{F_y}{F_x} = \frac{1500 \text{ lb}}{700 \text{ lb}}$$

Using a calculator, enter 1500 lb and divide by 700 lb; computing the arc tangent of the quotient gives you $\theta = 65.0°$. Solve the second of Eqs. (2.8) for $F$ to get

$$F = \frac{F_y}{\sin \theta} = \frac{1500 \text{ lb}}{\sin 65.0°} = 1655 \text{ lb}$$

The last calculation is easier if you store the value of $F_y$ when originally entered; you may then recall it and divide it by $\sin \theta$.

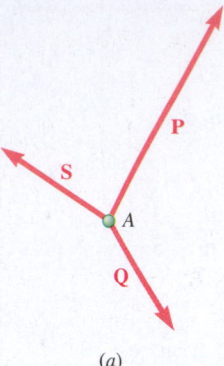

(a)

**Fig. 2.21** (a) Three forces acting on a particle.

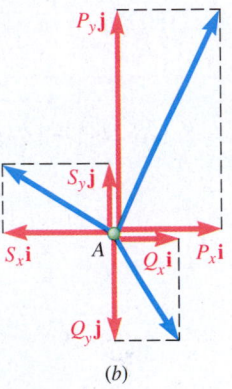

(b)

**Fig. 2.21** (b) Rectangular components of each force.

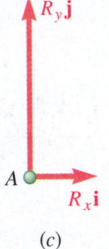

(c)

**Fig. 2.21** (c) Summation of the components.

## 2.2B Addition of Forces by Summing *x* and *y* Components

We described in Sec. 2.1A how to add forces according to the parallelogram law. From this law, we derived two other methods that are more readily applicable to the graphical solution of problems: the triangle rule for the addition of two forces and the polygon rule for the addition of three or more forces. We also explained that the force triangle used to define the resultant of two forces could be used to obtain a trigonometric solution.

However, when we need to add three or more forces, we cannot obtain any practical trigonometric solution from the force polygon that defines the resultant of the forces. In this case, the best approach is to obtain an analytic solution of the problem by resolving each force into two rectangular components.

Consider, for instance, three forces **P**, **Q**, and **S** acting on a particle *A* (Fig. 2.21*a*). Their resultant **R** is defined by the relation

$$\mathbf{R} = \mathbf{P} + \mathbf{Q} + \mathbf{S} \tag{2.11}$$

Resolving each force into its rectangular components, we have

$$
\begin{aligned}
R_x\mathbf{i} + R_y\mathbf{j} &= P_x\mathbf{i} + P_y\mathbf{j} + Q_x\mathbf{i} + Q_y\mathbf{j} + S_x\mathbf{i} + S_y\mathbf{j} \\
&= (P_x + Q_x + S_x)\mathbf{i} + (P_y + Q_y + S_y)\mathbf{j}
\end{aligned}
$$

From this equation, we can see that

$$R_x = P_x + Q_x + S_x \qquad R_y = P_y + Q_y + S_y \tag{2.12}$$

or for short,

$$R_x = \Sigma F_x \qquad R_y = \Sigma F_y \tag{2.13}$$

We thus conclude that **when several forces are acting on a particle, we obtain the scalar components $R_x$ and $R_y$ of the resultant R by adding algebraically the corresponding scalar components of the given forces.** *(This result also applies to the addition of other vector quantities, such as velocities, accelerations, or momenta.)*

In practice, determining the resultant **R** is carried out in three steps, as illustrated in Fig. 2.21.

1. Resolve the given forces (Fig. 2.21*a*) into their *x* and *y* components (Fig. 2.21*b*).
2. Add these components to obtain the *x* and *y* components of **R** (Fig. 2.21*c*).
3. Apply the parallelogram law to determine the resultant $\mathbf{R} = R_x\mathbf{i} + R_y\mathbf{j}$ (Fig. 2.21*d*).

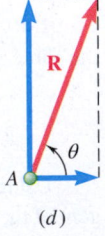

(d)

**Fig. 2.21** (d) Determining the resultant from its components.

The procedure just described is most efficiently carried out if you arrange the computations in a table (see Sample Prob. 2.3). Although this is the only practical analytic method for adding three or more forces, it is also often preferred to the trigonometric solution in the case of adding two forces.

## Sample Problem 2.3

Four forces act on bolt $A$ as shown. Determine the resultant of the forces on the bolt.

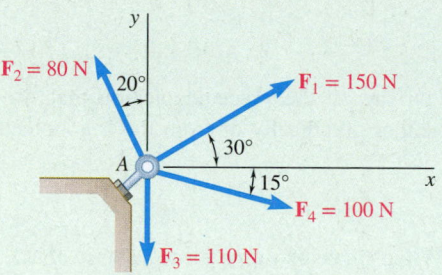

**STRATEGY:** The simplest way to approach a problem of adding four forces is to resolve the forces into components.

**MODELING:** As we mentioned, solving this kind of problem is usually easier if you arrange the components of each force in a table. In the table below, we entered the $x$ and $y$ components of each force as determined by trigonometry (Fig. 1). According to the convention adopted in this section, the scalar number representing a force component is positive if the force component has the same sense as the corresponding coordinate axis. Thus, $x$ components acting to the right and $y$ components acting upward are represented by positive numbers.

**ANALYSIS:**

| Force | Magnitude, N | $x$ Component, N | $y$ Component, N |
|-------|-------------|------------------|------------------|
| $F_1$ | 150 | +129.9 | +75.0 |
| $F_2$ | 80 | −27.4 | +75.2 |
| $F_3$ | 110 | 0 | −110.0 |
| $F_4$ | 100 | +96.6 | −25.9 |
| | | $R_x = +199.1$ | $R_y = +14.3$ |

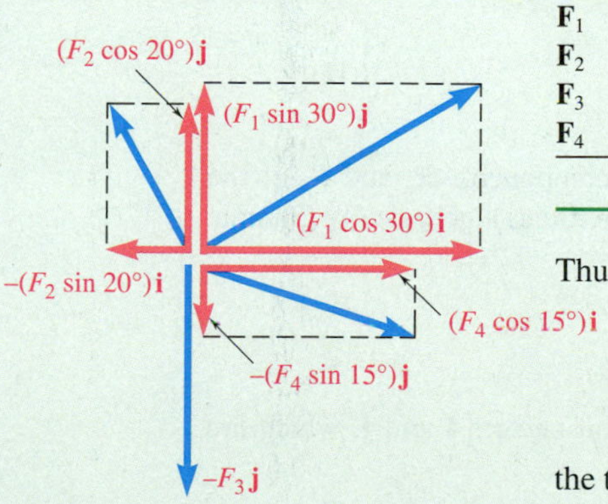

Thus, the resultant $\mathbf{R}$ of the four forces is

$$\mathbf{R} = R_x\mathbf{i} + R_y\mathbf{j} \qquad \mathbf{R} = (199.1\ \text{N})\mathbf{i} + (14.3\ \text{N})\mathbf{j}$$

You can now determine the magnitude and direction of the resultant. From the triangle shown in Fig. 2, you have

$$\tan \alpha = \frac{R_y}{R_x} = \frac{14.3\ \text{N}}{199.1\ \text{N}} \qquad \alpha = 4.1°$$

$$R = \frac{14.3\ \text{N}}{\sin \alpha} = 199.6\ \text{N} \qquad \mathbf{R} = 199.6\ \text{N} \measuredangle 4.1°$$

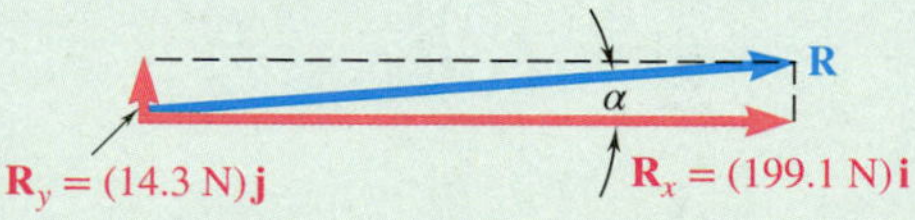

**REFLECT and THINK:** Arranging data in a table not only helps you keep track of the calculations, but also makes things simpler for using a calculator on similar computations.

$(F_2 \cos 20°)\mathbf{j}$

$(F_1 \sin 30°)\mathbf{j}$

$(F_1 \cos 30°)\mathbf{i}$

$-(F_2 \sin 20°)\mathbf{i}$

$(F_4 \cos 15°)\mathbf{i}$

$-(F_4 \sin 15°)\mathbf{j}$

$-F_3\mathbf{j}$

**Fig. 1** Rectangular components of each force.

**R**

$\alpha$

$\mathbf{R}_y = (14.3\ \text{N})\mathbf{j}$

$\mathbf{R}_x = (199.1\ \text{N})\mathbf{i}$

**Fig. 2** Resultant of the given force system.

# SOLVING PROBLEMS
# ON YOUR OWN

Y ou saw in the preceding lesson that we can determine the resultant of two forces either graphically or from the trigonometry of an oblique triangle.

**A. When three or more forces are involved, the best way to determine their resultant R** is by first resolving each force into **rectangular components.** You may encounter either of two cases, depending upon the way in which each of the given forces is defined.

**Case 1. The force F is defined by its magnitude $F$ and the angle $\alpha$ it forms with the $x$ axis.** Obtain the $x$ and $y$ components of the force by multiplying $F$ by cos $\alpha$ and sin $\alpha$, respectively (Concept Application 2.1).

**Case 2. The force F is defined by its magnitude $F$ and the coordinates of two points $A$ and $B$ on its line of action** (Fig. 2.19). Find the angle $\alpha$ that **F** forms with the $x$ axis by trigonometry, and then use the process of Case 1. However, you can also find the components of **F** directly from proportions among the various dimensions involved without actually determining $\alpha$ (Concept Application 2.2).

**B. Rectangular components of the resultant.** Obtain the components $R_x$ and $R_y$ of the resultant by adding the corresponding components of the given forces algebraically (Sample Prob. 2.3).

You can express the resultant in vectorial form using the unit vectors **i** and **j**, which are directed along the $x$ and $y$ axes, respectively:

$$\mathbf{R} = R_x\mathbf{i} + R_y\mathbf{j}$$

Alternatively, you can determine the *magnitude and direction* of the resultant by solving the right triangle of sides $R_x$ and $R_y$ for $R$ and for the angle that **R** forms with the $x$ axis.

# Problems

**2.21 and *2.22***    Determine the *x* and *y* components of each of the forces shown.

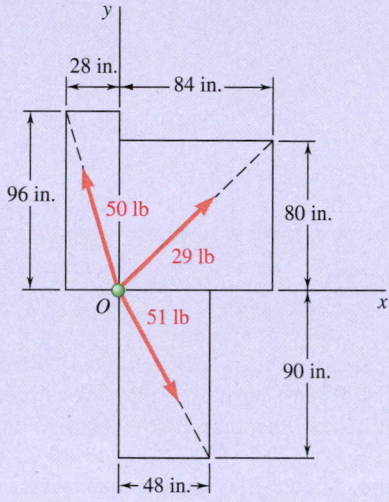

Fig. P2.21

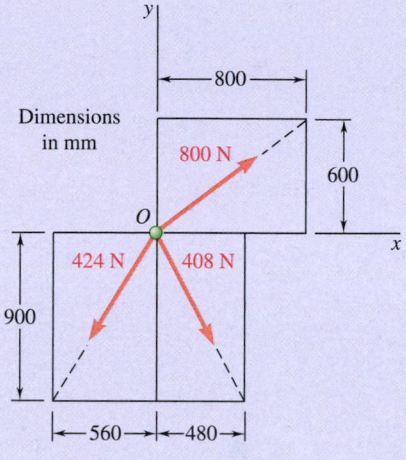

**Fig. *P2.22***

**2.23 and 2.24**    Determine the *x* and *y* components of each of the forces shown.

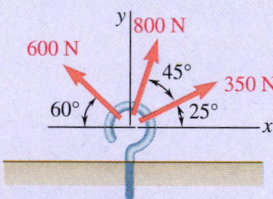

Fig. P2.23

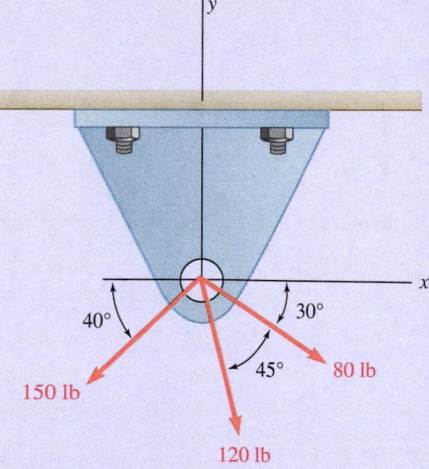

**Fig. P2.24**

***2.25***   Member *BC* exerts on member *AC* a force **P** directed along line *BC*. Knowing that **P** must have a 325-N horizontal component, determine (*a*) the magnitude of the force **P**, (*b*) its vertical component.

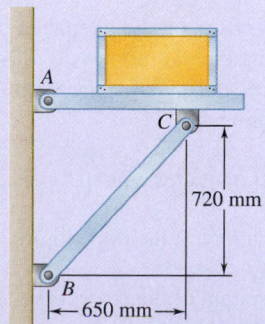

**Fig. *P2.25***

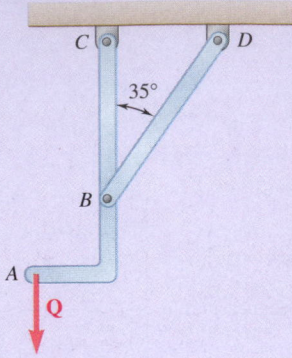

Fig. P2.26

**2.26** Member *BD* exerts on member *ABC* a force **P** directed along line *BD*. Knowing that **P** must have a 300-lb horizontal component, determine (*a*) the magnitude of the force **P**, (*b*) its vertical component.

**2.27** The hydraulic cylinder *BC* exerts on member *AB* a force **P** directed along line *BC*. Knowing that **P** must have a 600-N component perpendicular to member *AB*, determine (*a*) the magnitude of the force **P**, (*b*) its component along line *AB*.

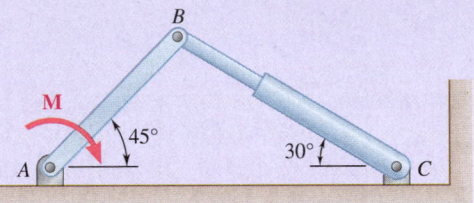

Fig. P2.27

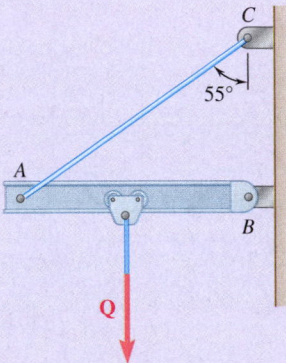

Fig. P2.28

**2.28** Cable *AC* exerts on beam *AB* a force **P** directed along line *AC*. Knowing that **P** must have a 350-lb vertical component, determine (*a*) the magnitude of the force **P**, (*b*) its horizontal component.

**2.29** The hydraulic cylinder *BD* exerts on member *ABC* a force **P** directed along line *BD*. Knowing that **P** must have a 750-N component perpendicular to member *ABC*, determine (*a*) the magnitude of the force **P**, (*b*) its component parallel to *ABC*.

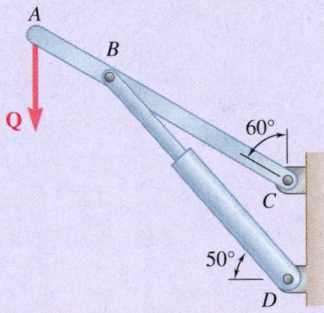

Fig. P2.29

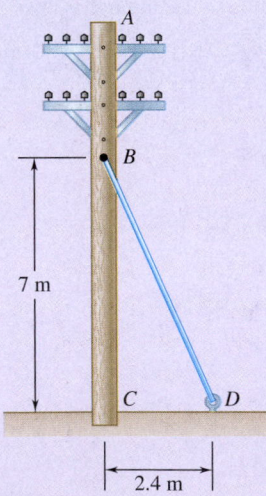

Fig. *P2.30*

**2.30** The guy wire *BD* exerts on the telephone pole *AC* a force **P** directed along *BD*. Knowing that **P** must have a 720-N component perpendicular to the pole *AC*, determine (*a*) the magnitude of the force **P**, (*b*) its component along line *AC*.

**2.31** Determine the resultant of the three forces of Prob. 2.21.

**2.32** Determine the resultant of the three forces of Prob. 2.23.

**2.33** Determine the resultant of the three forces of Prob. 2.24.

**2.34** Determine the resultant of the three forces of Prob. 2.22.

**2.35** Knowing that $\alpha = 35°$, determine the resultant of the three forces shown.

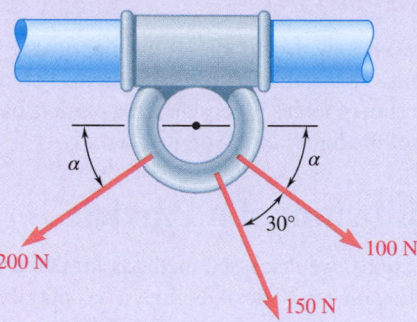

Fig. P2.35

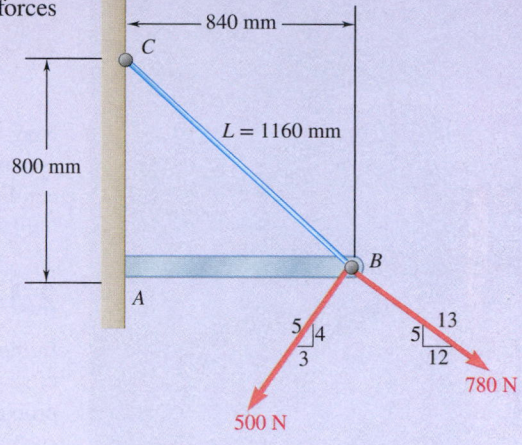

Fig. P2.36

**2.36** Knowing that the tension in cable $BC$ is 725 N, determine the resultant of the three forces exerted at point $B$ of beam $AB$.

**2.37** Knowing that $\alpha = 40°$, determine the resultant of the three forces shown.

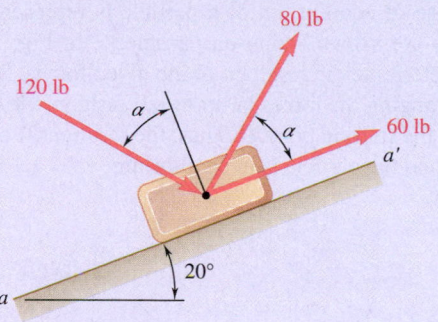

Fig. P2.37 and P2.38

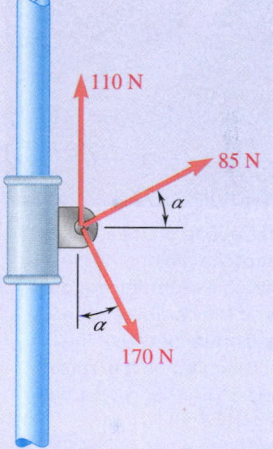

Fig. P2.39

**2.38** Knowing that $\alpha = 75°$, determine the resultant of the three forces shown.

**2.39** A collar that can slide on a vertical rod is subjected to the three forces shown. Determine (a) the required value of $\alpha$ if the resultant of the three forces is to be horizontal, (b) the corresponding magnitude of the resultant.

**2.40** For the beam of Prob. 2.36, determine (a) the required tension in cable $BC$ if the resultant of the three forces exerted at point $B$ is to be vertical, (b) the corresponding magnitude of the resultant.

**2.41** Determine (a) the required tension in cable $AC$, knowing that the resultant of the three forces exerted at point $C$ of boom $BC$ must be directed along $BC$, (b) the corresponding magnitude of the resultant.

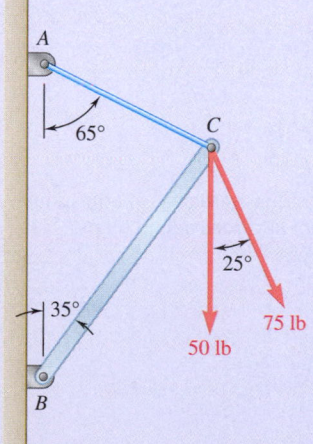

Fig. P2.41

**2.42** For the block of Probs. 2.37 and 2.38, determine (a) the required value of $\alpha$ if the resultant of the three forces shown is to be parallel to the incline, (b) the corresponding magnitude of the resultant.

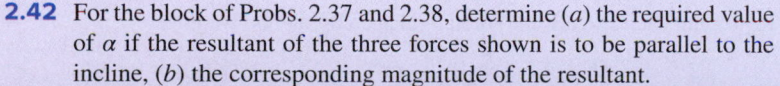

**Photo 2.2** Forces acting on the carabiner include the weight of the girl and her harness, and the force exerted by the pulley attachment. Treating the carabiner as a particle, it is in equilibrium because the resultant of all forces acting on it is zero.
©Michael Doolittle/Alamy

# 2.3 FORCES AND EQUILIBRIUM IN A PLANE

Now that we have seen how to add forces, we can proceed to one of the key concepts in this course: the equilibrium of a particle. The connection between equilibrium and the sum of forces is very direct: a particle can be in equilibrium only when the sum of the forces acting on it is zero.

## 2.3A Equilibrium of a Particle

In the preceding sections, we discussed methods for determining the resultant of several forces acting on a particle. Although it has not occurred in any of the problems considered so far, it is quite possible for the resultant to be zero. In such a case, the net effect of the given forces is zero, and the particle is said to be in **equilibrium**. We thus have the definition:

> **When the resultant of all the forces acting on a particle is zero, the particle is in equilibrium.**

A particle acted upon by two forces is in equilibrium if the two forces have the same magnitude and the same line of action but opposite sense. The resultant of the two forces is then zero, as shown in Fig. 2.22.

Another case of equilibrium of a particle is represented in Fig. 2.23a, where four forces are shown acting on particle A. In Fig. 2.23b, we use the polygon rule to determine the resultant of the given forces. Starting from point O with $\mathbf{F}_1$ and arranging the forces in tip-to-tail fashion, we find that the tip of $\mathbf{F}_4$ coincides with the starting point O. Thus, the resultant **R** of the given system of forces is zero, and the particle is in equilibrium.

**Fig. 2.22** When a particle is in equilibrium, the resultant of all forces acting on the particle is zero.

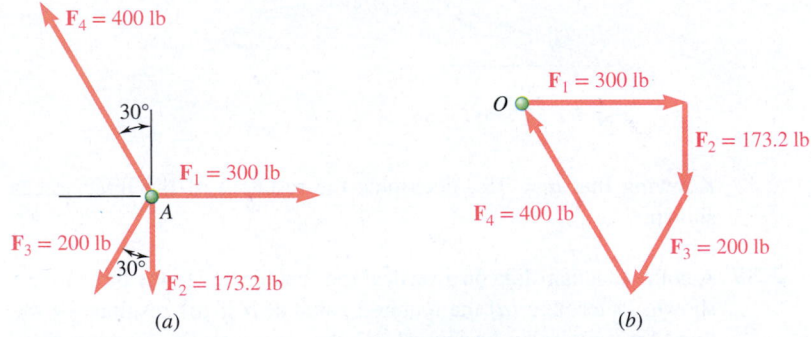

**Fig. 2.23** (a) Four forces acting on particle A; (b) using the polygon law to find the resultant of the forces in (a), which is zero because the particle is in equilibrium.

The closed polygon drawn in Fig. 2.23b provides a *graphical* expression of the equilibrium of A. To express *algebraically* the conditions for the equilibrium of a particle, we write

**Equilibrium of a particle**     $\mathbf{R} = \Sigma\mathbf{F} = 0$     **(2.14)**

Resolving each force **F** into rectangular components, we have

$$\Sigma(F_x\mathbf{i} + F_y\mathbf{j}) = 0 \qquad \text{or} \qquad (\Sigma F_x)\mathbf{i} + (\Sigma F_y)\mathbf{j} = 0$$

We conclude that the necessary and sufficient conditions for the equilibrium of a particle are

**Equilibrium of a particle
(scalar equations)**

$$\Sigma F_x = 0 \qquad \Sigma F_y = 0 \qquad \text{(2.15)}$$

Returning to the particle shown in Fig. 2.23, we can check that the equilibrium conditions are satisfied. We have

$$\Sigma F_x = 300 \text{ lb} - (200 \text{ lb}) \sin 30° - (400 \text{ lb}) \sin 30°$$
$$= 300 \text{ lb} - 100 \text{ lb} - 200 \text{ lb} = 0$$
$$\Sigma F_y = -173.2 \text{ lb} - (200 \text{ lb}) \cos 30° + (400 \text{ lb}) \cos 30°$$
$$= -173.2 \text{ lb} - 173.2 \text{ lb} + 346.4 \text{ lb} = 0$$

## 2.3B Newton's First Law of Motion

As we discussed in Sec. 1.2, Sir Isaac Newton formulated three fundamental laws upon which the science of mechanics is based. The first of these laws can be stated as:

> **If the resultant force acting on a particle is zero, the particle will remain at rest (if originally at rest) or will move with constant speed in a straight line (if originally in motion).**

From this law and from the definition of equilibrium just presented, we can see that a particle in equilibrium is either at rest or moving in a straight line with constant speed. If a particle does not behave in either of these ways, it is not in equilibrium, and the resultant force on it is not zero. In the following section, we consider various problems concerning the equilibrium of a particle.

Note that most of statics involves using Newton's first law to analyze an equilibrium situation. In practice, this means designing a bridge or a building that remains stable and does not fall over. It also means understanding the forces that might act to disturb equilibrium, such as a strong wind or a flood of water. The basic idea is pretty simple, but the applications can be quite complicated.

## 2.3C Free-Body Diagrams and Problem Solving

In practice, a problem in engineering mechanics is derived from an actual physical situation. A sketch showing the physical conditions of the problem is known as a **space diagram**.

The methods of analysis discussed in the preceding sections apply to a system of forces acting on a particle. A large number of problems involving actual structures, however, can be reduced to problems concerning the equilibrium of a particle. The method is to choose a significant particle and draw a separate diagram showing this particle and all the forces acting on it. Such a diagram is called a **free-body diagram**. (The name derives from the fact that when drawing the chosen body, or particle, it is "free" from all other bodies in the actual situation.)

As an example, consider the 75-kg crate shown in the space diagram of Fig. 2.24a. This crate was lying between two buildings, and is now being lifted onto a truck, which will remove it. The crate is supported by a vertical cable that is joined at A to two ropes, which pass over pulleys attached to the buildings at B and C. We want to determine the tension in each of the ropes AB and AC.

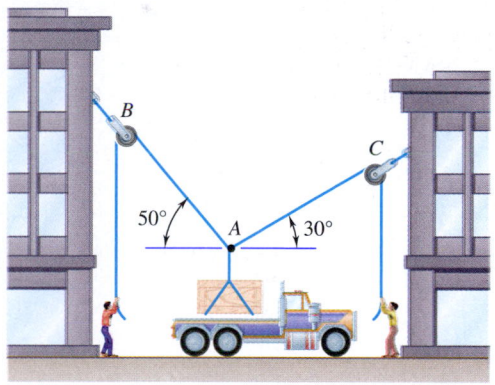

(a) Space diagram

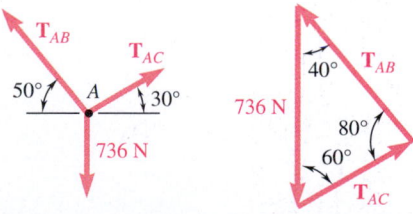

(b) Free-body diagram          (c) Force triangle

**Fig. 2.24** (a) The space diagram shows the physical situation of the problem; (b) the free-body diagram shows one central particle and the forces acting on it; (c) the force triangle can be solved with the law of sines. Note that the forces form a closed triangle because the particle is in equilibrium and the resultant force is zero.

**Photo 2.3** As illustrated in Fig. 2.24, it is possible to determine the tensions in the cables supporting the precast concrete panel shown by treating the hook as a particle and then applying the equations of equilibrium to the forces acting on the hook.
©Mack7777/Getty Images RF

In order to solve this problem, we first draw a free-body diagram showing a particle in equilibrium. Because we are interested in the rope tensions, the free-body diagram should include at least one of these tensions or, if possible, both tensions. You can see that point $A$ is a good free body for this problem. The free-body diagram of point $A$ is shown in Fig. 2.24$b$. It shows point $A$ and the forces exerted on $A$ by the vertical cable and the two ropes. The force exerted by the cable is directed downward, and its magnitude is equal to the weight $W$ of the crate. Recalling Eq. (1.4), we write

$$W = mg = (75 \text{ kg})(9.81 \text{ m/s}^2) = 736 \text{ N}$$

and indicate this value in the free-body diagram. The forces exerted by the two ropes are not known. Because they are respectively equal in magnitude to the tensions in rope $AB$ and rope $AC$, we denote them by $\mathbf{T}_{AB}$ and $\mathbf{T}_{AC}$ and draw them away from $A$ in the directions shown in the space diagram. No other detail is included in the free-body diagram.

Because point $A$ is in equilibrium, the three forces acting on it must form a closed triangle when drawn in tip-to-tail fashion. We have drawn this **force triangle** in Fig. 2.24$c$. The values $T_{AB}$ and $T_{AC}$ of the tensions in the ropes may be found graphically if the triangle is drawn to scale, or they may be found by trigonometry. If we choose trigonometry, we use the law of sines:

$$\frac{T_{AB}}{\sin 60°} = \frac{T_{AC}}{\sin 40°} = \frac{736 \text{ N}}{\sin 80°}$$

$$T_{AB} = 647 \text{ N} \qquad T_{AC} = 480 \text{ N}$$

When a particle is in equilibrium under three forces, you can solve the problem by drawing a force triangle. When a particle is in equilibrium under more than three forces, you can solve the problem graphically by drawing a force polygon. If you need an analytic solution, you should solve the **equations of equilibrium** given in Sec. 2.3A:

$$\Sigma F_x = 0 \qquad \Sigma F_y = 0 \qquad\qquad \textbf{(2.15)}$$

These equations can be solved for no more than *two unknowns*. Similarly, the force triangle used in the case of equilibrium under three forces can be solved for only two unknowns.

The most common types of problems are those in which the two unknowns represent (1) the two components (or the magnitude and direction) of a single force or (2) the magnitudes of two forces, each of known direction. Problems involving the determination of the maximum or minimum value of the magnitude of a force are also encountered (see Probs. 2.57 through 2.61).

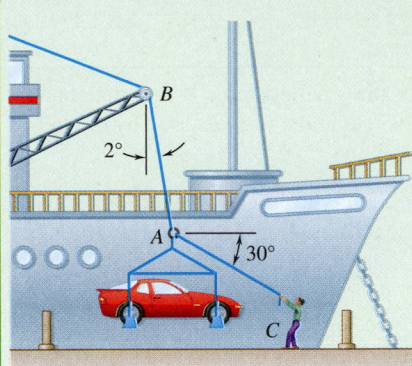

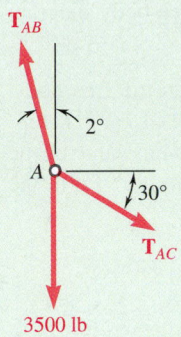

**Fig. 1**   Free-body diagram of particle *A*.

## Sample Problem 2.4

In a ship-unloading operation, a 3500-lb automobile is supported by a cable. A worker ties a rope to the cable at *A* and pulls on it in order to center the automobile over its intended position on the dock. At the moment illustrated, the automobile is stationary, the angle between the cable and the vertical is 2°, and the angle between the rope and the horizontal is 30°. What are the tensions in the rope and cable?

**STRATEGY:**   This is a problem of equilibrium under three coplanar forces. You can treat point *A* as a particle and solve the problem using a force triangle.

### MODELING and ANALYSIS:

**Free-Body Diagram.**   Choose point *A* as the particle and draw the complete free-body diagram (Fig. 1). $T_{AB}$ is the tension in the cable *AB*, and $T_{AC}$ is the tension in the rope.

**Equilibrium Condition.**   Because only three forces act on point *A*, draw a force triangle to express that it is in equilibrium (Fig. 2). Using the law of sines,

$$\frac{T_{AB}}{\sin 120°} = \frac{T_{AC}}{\sin 2°} = \frac{3500 \text{ lb}}{\sin 58°}$$

With a calculator, compute and store the value of the last quotient. Multiplying this value successively by sin 120° and sin 2°, you obtain

$$T_{AB} = 3570 \text{ lb} \qquad\qquad T_{AC} = 144 \text{ lb} \blacktriangleleft$$

**REFLECT and THINK:**   This is a common problem of knowing one force in a three-force equilibrium problem and calculating the other forces from the given geometry. This basic type of problem will occur often as part of more complicated situations in this text. As an alternative to the force triangle approach used here, a more general analytic solution could have been used as well. This second approach is especially suited for equilibrium problems involving more than three forces, and is applied in Sample Prob. 2.6.

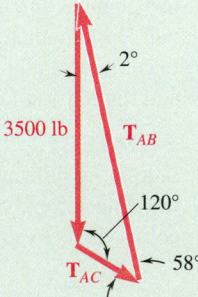

**Fig. 2**   Force triangle of the forces acting on particle *A*.

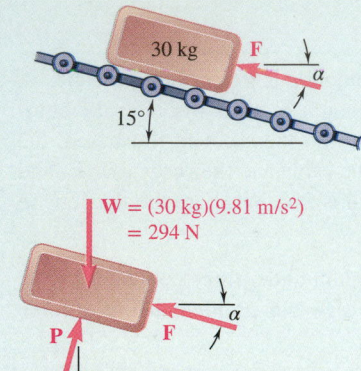

$$W = (30 \text{ kg})(9.81 \text{ m/s}^2)$$
$$= 294 \text{ N}$$

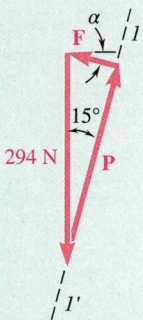

**Fig. 1** Free-body diagram of package, treated as a particle.

**Fig. 2** Force triangle of the forces acting on package.

## Sample Problem 2.5

Determine the magnitude and direction of the smallest force **F** that maintains the 30-kg package shown in equilibrium. Note that the force exerted by the rollers on the package is perpendicular to the incline.

**STRATEGY:** This is an equilibrium problem with three coplanar forces that you can solve with a force triangle. The new wrinkle is to determine a minimum force. You can approach this part of the solution in a way similar to Sample Prob. 2.2.

**MODELING and ANALYSIS:**

**Free-Body Diagram.** Choose the package as a free body, assuming that it can be treated as a particle. Then draw the corresponding free-body diagram (Fig. 1).

**Equilibrium Condition.** Because only three forces act on the free body, draw a force triangle to express that it is in equilibrium (Fig. 2). Line *1-1'* represents the known direction of **P**. In order to obtain the minimum value of the force **F**, choose the direction of **F** to be perpendicular to that of **P**. From the geometry of this triangle,

$$F = (294 \text{ N}) \sin 15° = 76.1 \text{ N} \qquad \alpha = 15°$$
$$\mathbf{F} = 76.1 \text{ N} \measuredangle 15° \blacktriangleleft$$

**REFLECT and THINK:** Determining maximum and minimum forces to maintain equilibrium is a common practical problem. Here, the force needed is about 25% of the weight of the package, which seems reasonable for an incline of 15°.

---

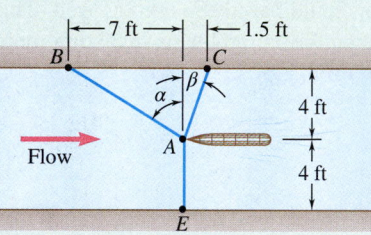

## Sample Problem 2.6

For a new sailboat, a designer wants to determine the drag force that may be expected at a given speed. To do so, she places a model of the proposed hull in a test channel and uses three cables to keep its bow on the centerline of the channel. Dynamometer readings indicate that for a given speed, the tension is 40 lb in cable *AB* and 60 lb in cable *AE*. Determine the drag force exerted on the hull and the tension in cable *AC*.

**STRATEGY:** The cables all connect at point *A*, so you can treat that as a particle in equilibrium. Because four forces act at *A* (tensions in three cables and the drag force), you should use the equilibrium conditions and sum forces by components to solve for the unknown forces.

*(continued)*

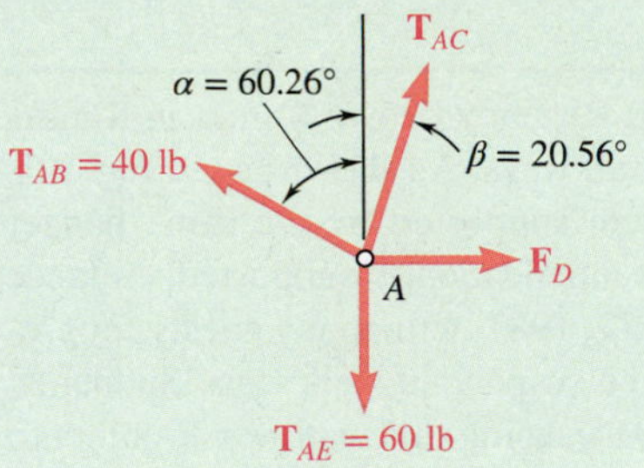

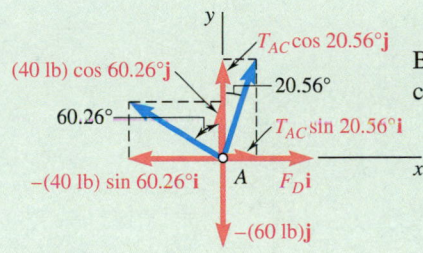

**Fig. 1** Free-body diagram of particle A.

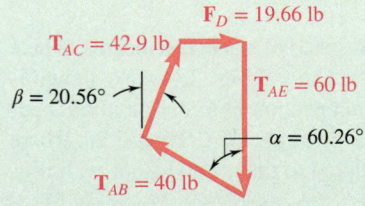

**Fig. 2** Rectangular components of forces acting on particle A.

**MODELING and ANALYSIS:**

**Determining the Angles.** First, determine the angles $\alpha$ and $\beta$ defining the direction of cables $AB$ and $AC$:

$$\tan \alpha = \frac{7 \text{ ft}}{4 \text{ ft}} = 1.75 \qquad \tan \beta = \frac{1.5 \text{ ft}}{4 \text{ ft}} = 0.375$$

$$\alpha = 60.26° \qquad \beta = 20.56°$$

**Free-Body Diagram.** Choosing point $A$ as a free body, draw the free-body diagram (Fig. 1). It includes the forces exerted by the three cables on the hull, as well as the drag force $\mathbf{F}_D$ exerted by the flow.

**Equilibrium Condition.** Because point $A$ is in equilibrium, the resultant of all forces is zero:

$$\mathbf{R} = \mathbf{T}_{AB} + \mathbf{T}_{AC} + \mathbf{T}_{AE} + \mathbf{F}_D = 0 \tag{1}$$

Because more than three forces are involved, resolve the forces into $x$ and $y$ components (Fig. 2):

$$\mathbf{T}_{AB} = -(40 \text{ lb}) \sin 60.26°\mathbf{i} + (40 \text{ lb}) \cos 60.26°\mathbf{j}$$
$$= -(34.73 \text{ lb})\mathbf{i} + (19.84 \text{ lb})\mathbf{j}$$
$$\mathbf{T}_{AC} = T_{AC} \sin 20.56°\mathbf{i} + T_{AC} \cos 20.56°\mathbf{j}$$
$$= 0.3512 T_{AC}\mathbf{i} + 0.9363 T_{AC}\mathbf{j}$$
$$\mathbf{T}_{AE} = -(60 \text{ lb})\mathbf{j}$$
$$\mathbf{F}_D = F_D\mathbf{i}$$

Substituting these expressions into Eq. (1) and factoring the unit vectors $\mathbf{i}$ and $\mathbf{j}$, you have

$$(-34.73 \text{ lb} + 0.3512 T_{AC} + F_D)\mathbf{i} + (19.84 \text{ lb} + 0.9363 T_{AC} - 60 \text{ lb})\mathbf{j} = 0$$

This equation is satisfied if, and only if, the coefficients of $\mathbf{i}$ and $\mathbf{j}$ are each equal to zero. You obtain the following two equilibrium equations, which express, respectively, that the sum of the $x$ components and the sum of the $y$ components of the given forces must be zero.

$$(\Sigma F_x = 0:) \qquad -34.73 \text{ lb} + 0.3512 T_{AC} + F_D = 0 \tag{2}$$

$$(\Sigma F_y = 0:) \qquad 19.84 \text{ lb} + 0.9363 T_{AC} - 60 \text{ lb} = 0 \tag{3}$$

From Eq. (3), you find

$$T_{AC} = +42.9 \text{ lb} \quad \blacktriangleleft$$

Substituting this value into Eq. (2) yields

$$F_D = +19.66 \text{ lb} \quad \blacktriangleleft$$

**REFLECT and THINK:** In drawing the free-body diagram, you assumed a sense for each unknown force. A positive sign in the answer indicates that the assumed sense is correct. You can draw the complete force polygon (Fig. 3) to check the results.

**Fig. 3** Force polygon of forces acting on particle A.

## Case Study 2.1

Completed in 1980, the atrium of the Hyatt Regency Crown Center in Kansas City, Missouri, featured three suspended walkways. As shown in CS Fig. 2.1, the second and fourth floor walkways were supported by the same hanger system, while the third floor walkway was independently supported. A dance competition was held in the atrium on July 17, 1981, with many guests congregating on the main floor as well as the three suspended walkways. Suddenly, the fourth floor walkway connections failed, causing this walkway to fall onto the second floor walkway, with both then crashing onto the main floor (see CS Photo 2.1). Tragically, 113 people lost their lives and another 186 were injured; in terms of human casualties, this was the worst structural failure in U.S. history up to that time.[*]

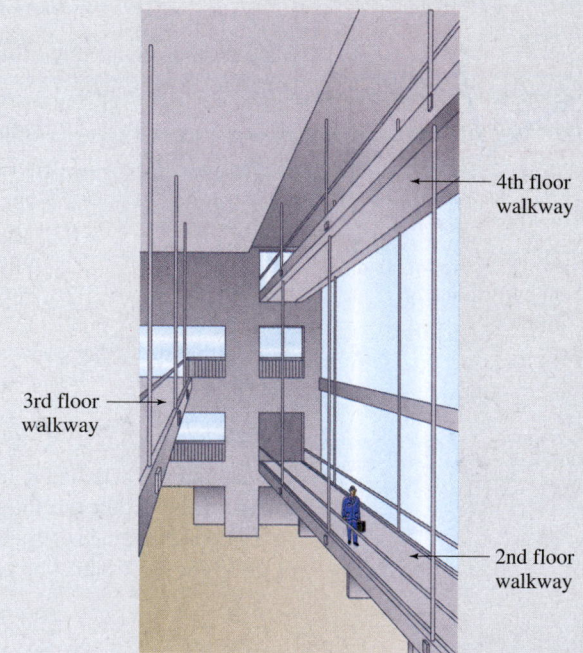

**CS Fig. 2.1**   Schematic of the atrium walkways.[*]

The support system of each walkway consisted of transverse beams, which were then attached to the hanger rods depicted in CS Fig. 2.1. Also shown is the critical fastener that was involved in the connection failure. The initial connection design for the fourth floor walkway is illustrated in CS Fig. 2.2a, where the support hanger would continue uninterrupted to the second floor walkway.

**CS Photo 2.1**   Wreckage of walkway collapse. Note that the third floor walkway remained intact.

©Pete Leabo/AP Photo

[*]Source: Marshall, R. D., E. O. Pfrang, E. V. Leyendecker, K. A. Woodward, R P. Reed, M B. Kasen, T. R. Shives *NBS Building Science Series 143: Investigation of the Kansas City Hyatt Regency Walkways Collapse,* Figure 3.4, p. 21. Washington, DC: US Department of Commerce, National Institute of Standards and Technology, May 1982.

*(continued)*

This would require turning the fastener on the threaded hanger rod all the way from the second floor end to the fourth floor level. During construction, it was realized that this would be impractical, and a new connection detail was implemented in the field as shown in CS Fig. 2.2*b*. Let's apply a static equilibrium analysis to determine the effect of this design change on the fastener.

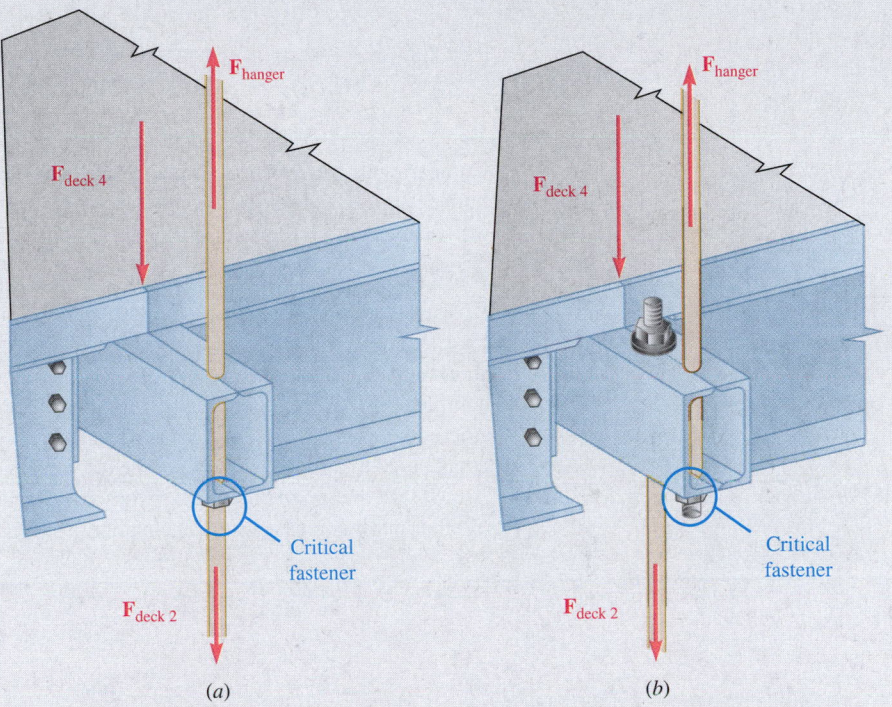

(a)                                                                  (b)

**CS Fig. 2.2**  Typical fourth-floor walkway support (*a*) original design, (*b*) as built. Source: Marshall, R. D., E. O. Pfrang, E. V. Leyendecker, K. A. Woodward, R P. Reed, M B. Kasen, T. R. Shives *NBS Building Science Series 143: Investigation of the Kansas City Hyatt Regency Walkways Collapse,* Figure 3.4, p. 21. Washington, DC: US Department of Commerce, National Institute of Standards and Technology, May 1982.

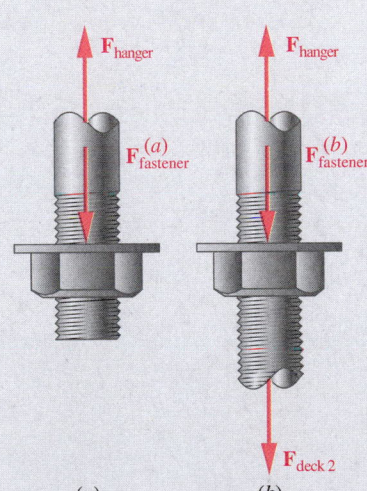

(a)                          (b)

**CS Fig. 2.3**  Free-body diagram of critical fastener (*a*) as built, (*b*) original design.

**STRATEGY:**  First, identify the loads involved. Then, treating the fastener and a small portion of the hanger as a particle, draw a free-body diagram and perform an equilibrium analysis.

**MODELING:  Free-Body Diagram.** The loads involved are shown in CS Fig. 2.2. The hanger system supports a portion of the second-floor and fourth-floor walkways, and the resulting loads are identified as $\mathbf{F}_{\text{deck 2}}$ and $\mathbf{F}_{\text{deck 4}}$. The force developed in the hanger extending from the fourth floor to the ceiling is denoted as $\mathbf{F}_{\text{hanger}}$. In both cases, this hanger must carry the loads of both walkways to the ceiling, or

$$\mathbf{F}_{\text{hanger}} = \mathbf{F}_{\text{deck 2}} + \mathbf{F}_{\text{deck 4}}$$

Treating the critical fastener and a small portion of the hanger as a particle, the free-body diagrams in each case are shown in CS Fig. 2.3.

*(continued)*

**ANALYSIS:** **As-Built Connection.** From CS Fig. 2.3*a*,

$$+\uparrow \Sigma F_y = 0: \quad F_{hanger} - F_{fastener}^{(a)} = 0$$

$$(F_{deck\ 2} + F_{deck\ 4}) - F_{fastener}^{(a)} = 0$$

$$F_{fastener}^{(a)} = +(F_{deck\ 2} + F_{deck\ 4}) \qquad F_{fastener}^{(a)} = (F_{deck\ 2} + F_{deck\ 4})\downarrow \blacktriangleleft$$

**Original Connection Design.** From CS Fig. 2.3*b*,

$$+\uparrow \Sigma F_y = 0: \quad F_{hanger} - F_{fastener}^{(b)} - F_{deck\ 2} = 0$$

$$(F_{deck\ 2} + F_{deck\ 4}) - F_{fastener}^{(b)} - F_{deck\ 2} = 0$$

$$F_{fastener}^{(b)} = +F_{deck\ 4} \qquad\qquad F_{fastener}^{(b)} = F_{deck\ 4}\downarrow \blacktriangleleft$$

Comparing the force exerted on the fastener in each case, it is apparent that the as-built design requires the fastener to support loads from both walkways, whereas the original design subjects the fastener to loads only from the fourth-floor walkway.

**REFLECT and THINK:** The change in the connection, completed "on the fly," resulted in the unintended consequence of subjecting the critical fastener to loads from two floors instead of just one. In the same manner, one should avoid shortcuts in analyzing engineering mechanics problems, and instead employ the complete SMART methodology, even for very simple situations like the one considered here. It should also be noted that there were other important factors that contributed to this tragedy besides the circumstance examined in this Case Study. Along with the report cited earlier, these factors have been discussed in a number of other publications as well. The reader is strongly encouraged to study further.

# SOLVING PROBLEMS
# ON YOUR OWN

When a particle is in **equilibrium**, the resultant of the forces acting on the particle must be zero. Expressing this fact in the case of a particle under *coplanar forces* provides you with two relations among these forces. As in the preceding sample problems, you can use these relations to determine two unknowns—such as the magnitude and direction of one force or the magnitudes of two forces.

**Drawing a clear and accurate free-body diagram is a must in the solution of any equilibrium problem.** This diagram shows the particle and all of the forces acting on it. Indicate in your free-body diagram the magnitudes of known forces, as well as any angle or dimensions that define the direction of a force. Any unknown magnitude or angle should be denoted by an appropriate symbol. Nothing else should be included in the free-body diagram. Skipping this step might save you pencil and paper, but it is very likely to lead you to a wrong solution.

**Case 1. If the free-body diagram involves only three forces,** the rest of the solution is best carried out by drawing these forces in tip-to-tail fashion to form a **force triangle**. You can solve this triangle graphically or by trigonometry for no more than two unknowns (Sample Probs. 2.4 and 2.5).

**Case 2. If the free-body diagram indicates more than three forces,** it is most practical to use an *analytic solution*. Select $x$ and $y$ axes and resolve each of the forces into $x$ and $y$ components. Setting the sum of the $x$ components and the sum of the $y$ components of all the forces to zero, you obtain two equations that you can solve for no more than two unknowns (Sample Prob. 2.6).

We strongly recommend that, when using an analytic solution, you write the equations of equilibrium in the same form as Eqs. (2) and (3) of Sample Prob. 2.6. The practice adopted by some students of initially placing the unknowns on the left side of the equation and the known quantities on the right side may lead to confusion in assigning the appropriate sign to each term.

Regardless of the method used to solve a two-dimensional equilibrium problem, you can determine at most two unknowns. If a two-dimensional problem involves more than two unknowns, you must obtain one or more additional relations from the information contained in the problem statement.

# Problems

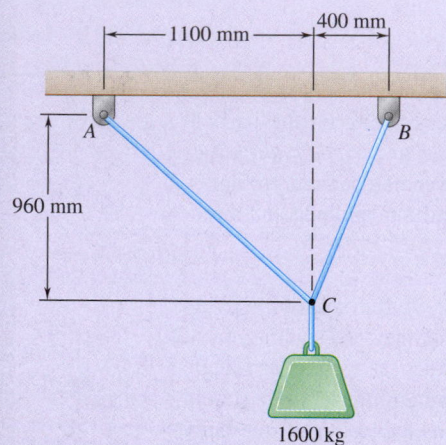

Fig. P2.F1

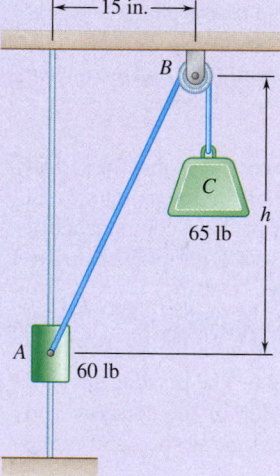

Fig. P2.F3

FREE-BODY PRACTICE PROBLEMS

**2.F1** Two cables are tied together at $C$ and loaded as shown. Draw the free-body diagram needed to determine the tension in $AC$ and $BC$.

**2.F2** Two forces of magnitude $T_A = 8$ kips and $T_B = 15$ kips are applied as shown to a welded connection. Knowing that the connection is in equilibrium, draw the free-body diagram needed to determine the magnitudes of the forces $T_C$ and $T_D$.

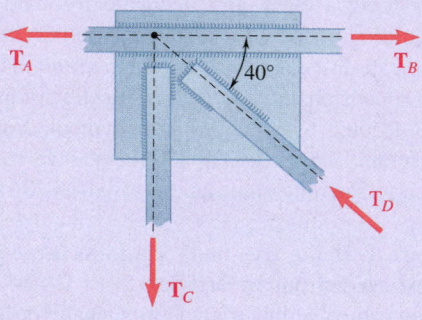

Fig. P2.F2

**2.F3** The 60-lb collar $A$ can slide on a frictionless vertical rod and is connected as shown to a 65-lb counterweight $C$. Draw the free-body diagram needed to determine the value of $h$ for which the system is in equilibrium.

**2.F4** A chairlift has been stopped in the position shown. Knowing that each chair weighs 250 N and that the skier in chair $E$ weighs 765 N, draw the free-body diagrams needed to determine the weight of the skier in chair $F$.

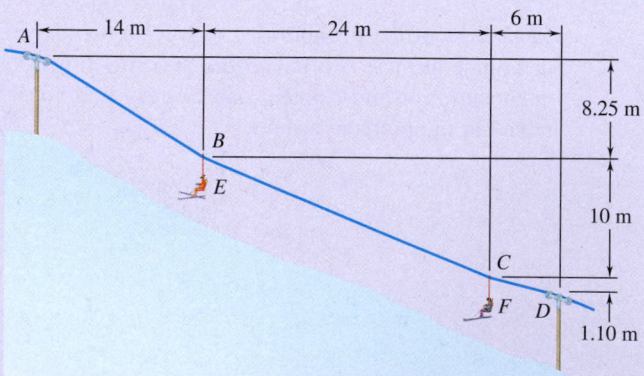

Fig. P2.F4

**2.43** Two cables are tied together at *C* and are loaded as shown. Determine the tension (*a*) in cable *AC*, (*b*) in cable *BC*.

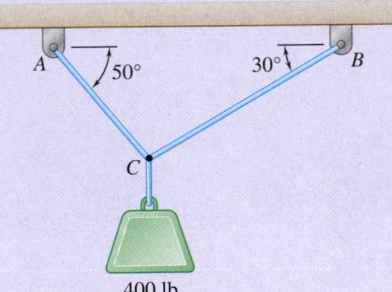

400 lb

**Fig. P2.43**

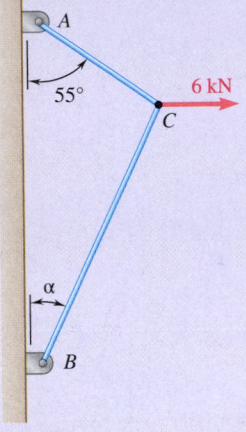

6 kN

**Fig. P2.44**

**2.44** Two cables are tied together at *C* and are loaded as shown. Knowing that $\alpha = 30°$, determine the tension (*a*) in cable *AC*, (*b*) in cable *BC*.

**2.45** Two cables are tied together at *C* and loaded as shown. Determine the tension (*a*) in cable *AC*, (*b*) in cable *BC*.

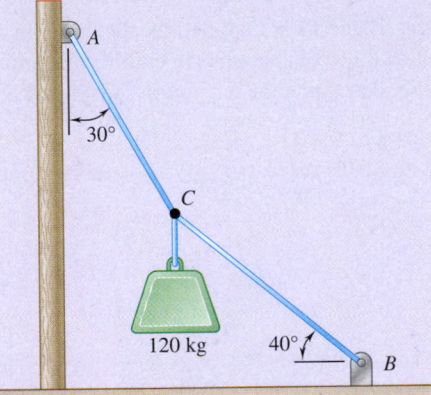

120 kg

**Fig. *P2.45***

**2.46** Two cables are tied together at *C* and are loaded as shown. Knowing that *P* = 500 N and $\alpha = 60°$, determine the tension (*a*) in cable *AC*, (*b*) in cable *BC*.

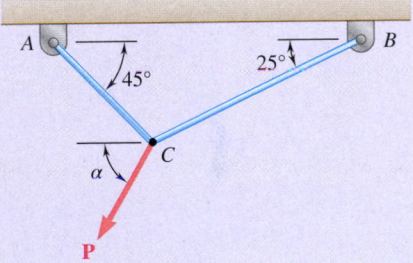

P

**Fig. P2.46**

**2.47** Two cables are tied together at *C* and are loaded as shown. Determine the tension (*a*) in cable *AC*, (*b*) in cable *BC*.

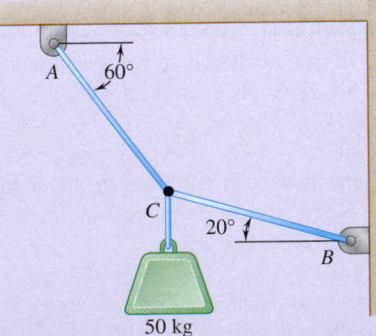

50 kg

**Fig. *P2.47***

**2.48** Knowing that $\alpha = 20°$, determine the tension (*a*) in cable *AC*, (*b*) in rope *BC*.

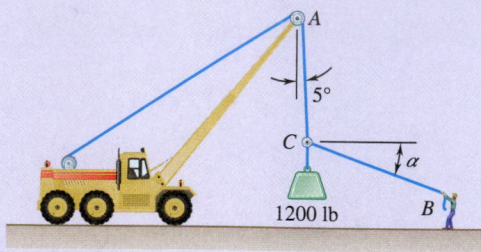

**Fig. P2.48**

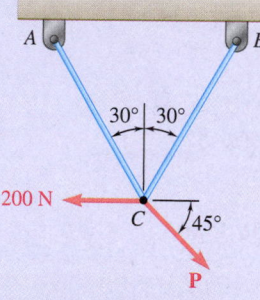

**Fig. P2.49 and P2.50**

**2.49** Two cables are tied together at *C* and are loaded as shown. Knowing that $P = 300$ N, determine the tension in cables *AC* and *BC*.

**2.50** Two cables are tied together at *C* and are loaded as shown. Determine the range of values of **P** for which both cables remain taut.

**2.51** Two forces **P** and **Q** are applied as shown to an aircraft connection. Knowing that the connection is in equilibrium and that $P = 600$ lb and $Q = 800$ lb, determine the tension in rods *A* and *B*.

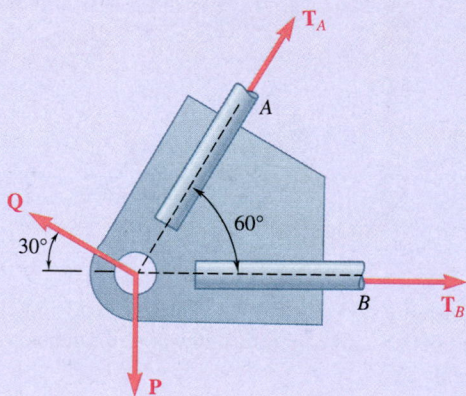

**Fig. P2.51 and *P2.52***

**2.52** Two forces **P** and **Q** are applied as shown to an aircraft connection. Knowing that the connection is in equilibrium and that the tensions in rods *A* and *B* are $T_A = 240$ lb and $T_B = 500$ lb, determine the magnitudes of **P** and **Q**.

**2.53** A welded connection is in equilibrium under the action of the four forces shown. Knowing that $F_A = 8$ kN and $F_B = 16$ kN, determine the magnitudes of the other two forces.

**2.54** A welded connection is in equilibrium under the action of the four forces shown. Knowing that $F_A = 5$ kN and $F_D = 6$ kN, determine the magnitudes of the other two forces.

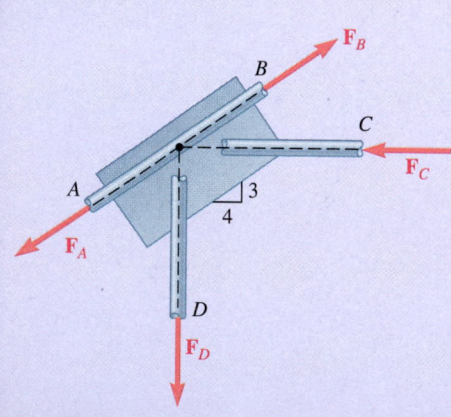

**Fig. P2.53 and P2.54**

**2.55** A sailor is being rescued using a boatswain's chair that is suspended from a pulley that can roll freely on the support cable $ACB$ and is pulled at a constant speed by cable $CD$. Knowing that $\alpha = 30°$ and $\beta = 10°$ and that the combined weight of the boatswain's chair and the sailor is 200 lb, determine the tension $(a)$ in the support cable $ACB$, $(b)$ in the traction cable $CD$.

**2.56** A sailor is being rescued using a boatswain's chair that is suspended from a pulley that can roll freely on the support cable $ACB$ and is pulled at a constant speed by cable $CD$. Knowing that $\alpha = 25°$ and $\beta = 15°$ and that the tension in cable $CD$ is 20 lb, determine $(a)$ the combined weight of the boatswain's chair and the sailor, $(b)$ the tension in the support cable $ACB$.

**2.57** For the cables of Prob. 2.44, find the value of $\alpha$ for which the tension is as small as possible $(a)$ in cable $BC$, $(b)$ in both cables simultaneously. In each case, determine the tension in each cable.

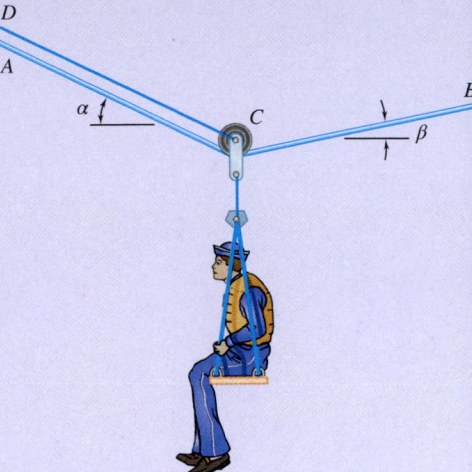

Fig. P2.55 and *P2.56*

**2.58** For the cables of Prob. 2.46, it is known that the maximum allowable tension is 600 N in cable $AC$ and 750 N in cable $BC$. Determine $(a)$ the maximum force **P** that can be applied at $C$, $(b)$ the corresponding value of $\alpha$.

**2.59** For the situation described in Fig. P2.48, determine $(a)$ the value of $\alpha$ for which the tension in rope $BC$ is as small as possible, $(b)$ the corresponding value of the tension.

**2.60** Two cables tied together at $C$ are loaded as shown. Determine the range of values of $P$ for which both cables remain taut.

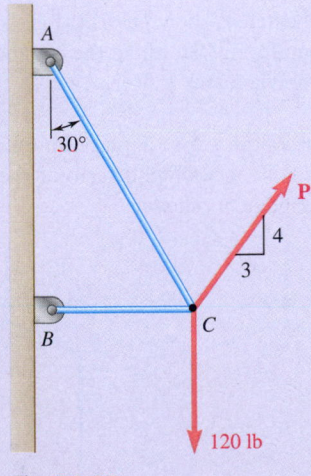

Fig. *P2.60*

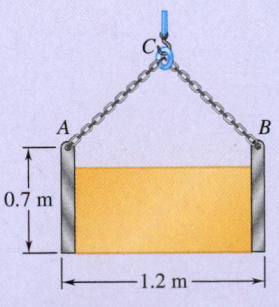

**2.61** A movable bin and its contents have a combined weight of 2.8 kN. Determine the shortest chain sling $ACB$ that can be used to lift the loaded bin if the tension in the chain is not to exceed 5 kN.

Fig. P2.61

**51**

**2.62** For $W = 800$ N, $P = 200$ N, and $d = 600$ mm, determine the value of $h$ consistent with equilibrium.

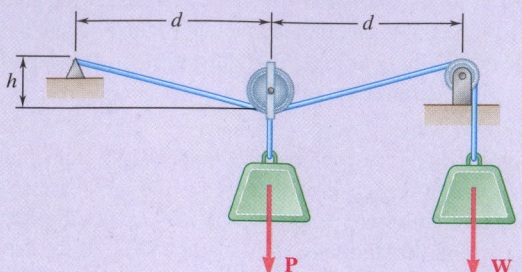

**Fig. P2.62**

**2.63** Collar $A$ is connected as shown to a 50-lb load and can slide on a frictionless horizontal rod. Determine the magnitude of the force **P** required to maintain the equilibrium of the collar when (*a*) $x = 4.5$ in., (*b*) $x = 15$ in.

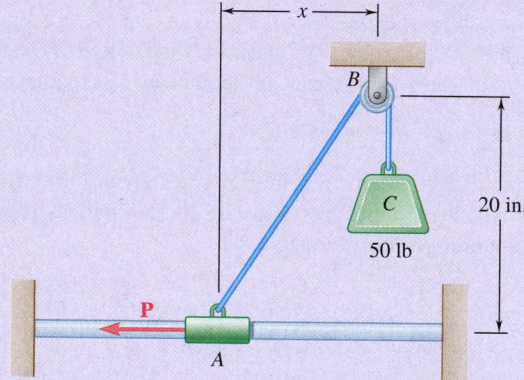

**Fig. P2.63 and *P2.64***

**2.64** Collar $A$ is connected as shown to a 50-lb load and can slide on a frictionless horizontal rod. Determine the distance $x$ for which the collar is in equilibrium when $P = 48$ lb.

**2.65** A cable loop of length 1.5 m is placed around a crate. Knowing that the mass of the crate is 300 kg, determine the tension in the cable for each of the arrangements shown.

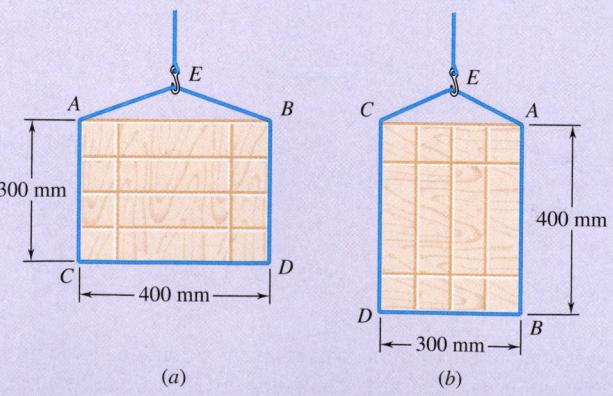

(*a*)        (*b*)

**Fig. P2.65**

**2.66** A 200-kg crate is to be supported by the rope-and-pulley arrangement shown. Determine the magnitude and direction of the force **P** that must be exerted on the free end of the rope to maintain equilibrium. (*Hint:* The tension in the rope is the same on each side of a simple pulley. This can be proved by the methods of Chap. 4.)

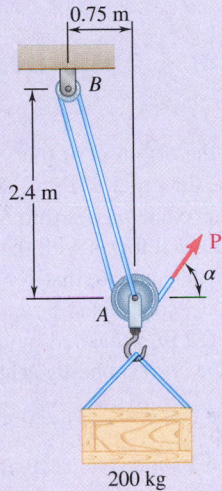

Fig. *P2.66*

**2.67** A 600-lb crate is supported by several rope-and-pulley arrangements as shown. Determine for each arrangement the tension in the rope. (See the hint for Prob. 2.66.)

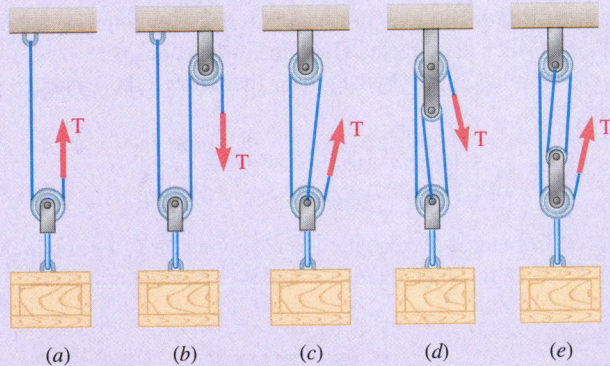

Fig. P2.67

**2.68** Solve parts *b* and *d* of Prob. 2.67, assuming that the free end of the rope is attached to the crate.

**2.69** A load **Q** is applied to pulley *C*, which can roll on the cable *ACB*. The pulley is held in the position shown by a second cable *CAD*, which passes over the pulley *A* and supports a load **P**. Knowing that *P* = 750 N, determine (*a*) the tension in cable *ACB*, (*b*) the magnitude of load **Q**.

**2.70** An 1800-N load **Q** is applied to pulley *C*, which can roll on the cable *ACB*. The pulley is held in the position shown by a second cable *CAD*, which passes over the pulley *A* and supports a load **P**. Determine (*a*) the tension in cable *ACB*, (*b*) the magnitude of load **P**.

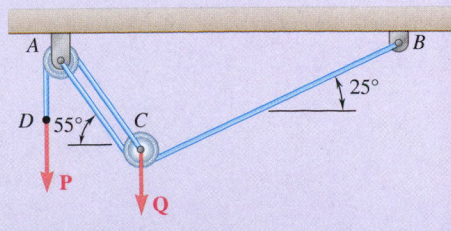

Fig. P2.69 and *P2.70*

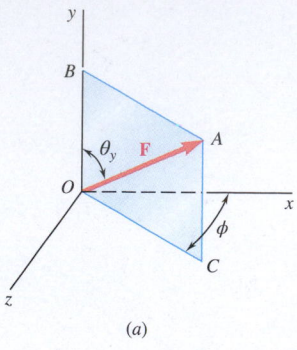

*(a)*

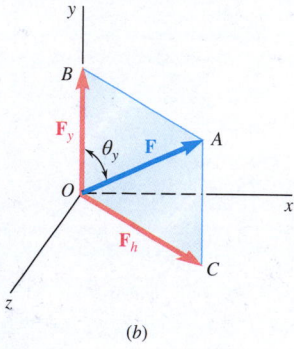

*(b)*

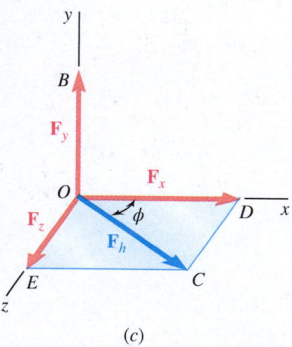

*(c)*

**Fig. 2.25** *(a)* A force **F** in an *xyz* coordinate system; *(b)* components of **F** along the *y* axis and in the *xz* plane; *(c)* components of **F** along the three rectangular axes.

# 2.4   ADDING FORCES IN SPACE

The problems considered in the first part of this chapter involved only two dimensions; they were formulated and solved in a single plane. In the last part of this chapter, we discuss problems involving the three dimensions of space.

## 2.4A   Rectangular Components of a Force in Space

Consider a force **F** acting at the origin *O* of the system of rectangular coordinates *x*, *y*, and *z*. To define the direction of **F**, we draw the vertical plane *OBAC* containing **F** (Fig. 2.25*a*). This plane passes through the vertical *y* axis; its orientation is defined by the angle $\phi$ it forms with the *xy* plane. The direction of **F** within the plane is defined by the angle $\theta_y$ that **F** forms with the *y* axis. We can resolve the force **F** into a vertical component $\mathbf{F}_y$ and a horizontal component $\mathbf{F}_h$; this operation, shown in Fig. 2.25*b*, is carried out in plane *OBAC* according to the rules developed earlier. The corresponding scalar components are

$$F_y = F \cos \theta_y \qquad F_h = F \sin \theta_y \tag{2.16}$$

However, we can also resolve $\mathbf{F}_h$ into two rectangular components $\mathbf{F}_x$ and $\mathbf{F}_z$ along the *x* and *z* axes, respectively. This operation, shown in Fig. 2.25*c*, is carried out in the *xz* plane. We obtain the following expressions for the corresponding scalar components:

$$F_x = F_h \cos \phi = F \sin \theta_y \cos \phi$$
$$F_z = F_h \sin \phi = F \sin \theta_y \sin \phi \tag{2.17}$$

The given force **F** thus has been resolved into three rectangular vector components $\mathbf{F}_x$, $\mathbf{F}_y$, $\mathbf{F}_z$, which are directed along the three coordinate axes.

We can now apply the Pythagorean theorem to the triangles *OAB* and *OCD* of Fig. 2.25:

$$F^2 = (OA)^2 = (OB)^2 + (BA)^2 = F_y^2 + F_h^2$$
$$F_h^2 = (OC)^2 = (OD)^2 + (DC)^2 = F_x^2 + F_z^2$$

Eliminating $F_h^2$ from these two equations and solving for *F*, we obtain the following relation between the magnitude of **F** and its rectangular scalar components:

**Magnitude of a force in space**

$$F = \sqrt{F_x^2 + F_y^2 + F_z^2} \tag{2.18}$$

The relationship between the force **F** and its three components $\mathbf{F}_x$, $\mathbf{F}_y$, and $\mathbf{F}_z$ is more easily visualized if we draw a "box" having $\mathbf{F}_x$, $\mathbf{F}_y$, and $\mathbf{F}_z$ for edges, as shown in Fig. 2.26. The force **F** is then represented by the main diagonal *OA* of this box. Figure 2.26*b* shows the right triangle *OAB* used to derive the first of the formulas (2.16): $F_y = F \cos \theta_y$. In Fig. 2.26*a* and *c*, two other right triangles have also been drawn: *OAD* and *OAE*. These triangles occupy positions in the box comparable with that of triangle *OAB*. Denoting by $\theta_x$ and $\theta_z$, respectively, the angles that **F** forms with the *x* and *z* axes, we can derive two formulas similar to $F_y = F \cos \theta_y$. We thus write

**Scalar components of a force F**

$$F_x = F \cos \theta_x \qquad F_y = F \cos \theta_y \qquad F_z = F \cos \theta_z \tag{2.19}$$

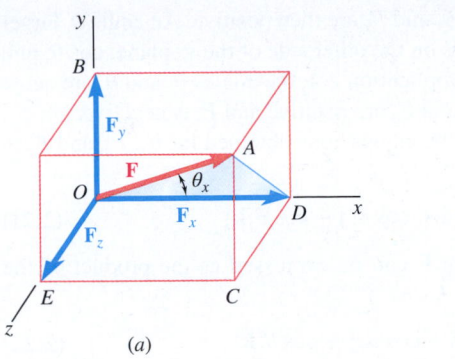

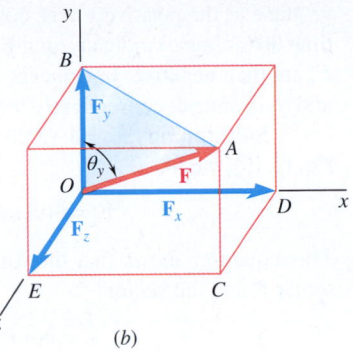

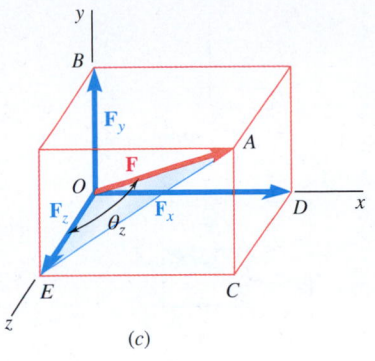

**Fig. 2.26**   (*a*) Force **F** in a three-dimensional box, showing its angle with the *x* axis; (*b*) force **F** and its angle with the *y* axis; (*c*) force **F** and its angle with the *z* axis.

The three angles $\theta_x$, $\theta_y$, and $\theta_z$ define the direction of the force **F**; they are more commonly used for this purpose than the angles $\theta_y$ and $\phi$ introduced at the beginning of this section. The cosines of $\theta_x$, $\theta_y$, and $\theta_z$ are known as the **direction cosines** of the force **F**.

Introducing the unit vectors **i**, **j**, and **k**, which are directed respectively along the *x*, *y*, and *z* axes (Fig. 2.27), we can express **F** in the form

**Vector expression of a force F**

$$\mathbf{F} = F_x\mathbf{i} + F_y\mathbf{j} + F_z\mathbf{k} \qquad \textbf{(2.20)}$$

where the scalar components $F_x$, $F_y$, and $F_z$ are defined by the relations in Eq. (2.19).

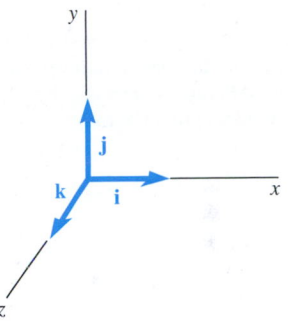

**Fig. 2.27**   The three unit vectors **i**, **j**, **k** lie along the three coordinate axes *x*, *y*, *z*, respectively.

## Concept Application 2.4

A force of 500 N forms angles of 60°, 45°, and 120°, respectively, with the *x*, *y*, and *z* axes. Find the components $F_x$, $F_y$, and $F_z$ of the force and express the force in terms of unit vectors.

**Solution**   Substitute $F = 500$ N, $\theta_x = 60°$, $\theta_y = 45°$, and $\theta_z = 120°$ into formulas (2.19). The scalar components of **F** are then

$$F_x = (500 \text{ N}) \cos 60° = +250 \text{ N}$$
$$F_y = (500 \text{ N}) \cos 45° = +354 \text{ N}$$
$$F_z = (500 \text{ N}) \cos 120° = -250 \text{ N}$$

Carrying these values into Eq. (2.20), you have

$$\mathbf{F} = (250 \text{ N})\mathbf{i} + (354 \text{ N})\mathbf{j} - (250 \text{ N})\mathbf{k}$$

As in the case of two-dimensional problems, a plus sign indicates that the component has the same sense as the corresponding axis, and a minus sign indicates that it has the opposite sense.

The angle a force **F** forms with an axis should be measured from the positive side of the axis and is always between 0 and 180°. An angle $\theta_x$ smaller than 90° (acute) indicates that **F** (assumed attached to *O*) is on the same side of the

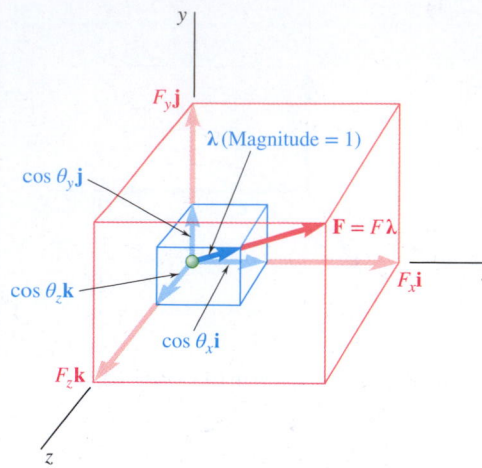

**Fig. 2.28** Force **F** can be expressed as the product of its magnitude *F* and a unit vector λ in the direction of **F**. Also shown are the components of **F** and its unit vector.

*yz* plane as the positive *x* axis; cos $\theta_x$ and $F_x$ are then positive. An angle $\theta_x$ larger than 90° (obtuse) indicates that **F** is on the other side of the *yz* plane; cos $\theta_x$ and $F_x$ are then negative. In Concept Application 2.4, the angles $\theta_x$ and $\theta_y$ are acute and $\theta_z$ is obtuse; consequently, $F_x$ and $F_y$ are positive and $F_z$ is negative.

Substituting into Eq. (2.20) the expressions obtained for $F_x$, $F_y$, and $F_z$ in Eq. (2.19), we have

$$\mathbf{F} = F(\cos \theta_x \mathbf{i} + \cos \theta_y \mathbf{j} + \cos \theta_z \mathbf{k}) \tag{2.21}$$

This equation shows that the force **F** can be expressed as the product of the scalar *F* and the vector

$$\lambda = \cos \theta_x \mathbf{i} + \cos \theta_y \mathbf{j} + \cos \theta_z \mathbf{k} \tag{2.22}$$

The vector λ is a vector whose magnitude is equal to 1 and whose direction is the same as that of **F** (Fig. 2.28). The vector λ is referred to as the **unit vector along the line of action** of **F**. It follows from Eq. (2.22) that the components of the unit vector λ are respectively equal to the direction cosines of the line of action of **F**:

$$\lambda_x = \cos \theta_x \qquad \lambda_y = \cos \theta_y \qquad \lambda_z = \cos \theta_z \tag{2.23}$$

Note that the values of the three angles $\theta_x$, $\theta_y$, and $\theta_z$ are not independent. Recalling that the sum of the squares of the components of a vector is equal to the square of its magnitude, we can write

$$\lambda_x^2 + \lambda_y^2 + \lambda_z^2 = 1$$

Substituting for $\lambda_x$, $\lambda_y$, and $\lambda_z$ from Eq. (2.23), we obtain

**Relationship among direction cosines**
$$\cos^2 \theta_x + \cos^2 \theta_y + \cos^2 \theta_z = 1 \tag{2.24}$$

In Concept Application 2.4, for instance, once the values $\theta_x = 60°$ and $\theta_y = 45°$ have been selected, the value of $\theta_z$ *must* be equal to 60° or 120° in order to satisfy the identity in Eq. (2.24).

When the components $F_x$, $F_y$, and $F_z$ of a force **F** are given, we can obtain the magnitude *F* of the force from Eq. (2.18). We can then solve relations in Eq. (2.19) for the direction cosines as

$$\cos \theta_x = \frac{F_x}{F} \qquad \cos \theta_y = \frac{F_y}{F} \qquad \cos \theta_z = \frac{F_z}{F} \tag{2.25}$$

From the direction cosines, we can find the angles $\theta_x$, $\theta_y$, and $\theta_z$ characterizing the direction of **F**.

## Concept Application 2.5

A force **F** has the components $F_x = 20$ lb, $F_y = -30$ lb, and $F_z = 60$ lb. Determine its magnitude *F* and the angles $\theta_x$, $\theta_y$, and $\theta_z$ it forms with the coordinate axes.

**Solution**    You can obtain the magnitude of **F** from formula (2.18):

$$F = \sqrt{F_x^2 + F_y^2 + F_z^2}$$
$$= \sqrt{(20 \text{ lb})^2 + (-30 \text{ lb})^2 + (60 \text{ lb})^2}$$
$$= \sqrt{4900 \text{ lb}} = 70 \text{ lb}$$

*(continued)*

Substituting the values of the components and magnitude of **F** into Eqs. (2.25), the direction cosines are

$$\cos\theta_x = \frac{F_x}{F} = \frac{20\text{ lb}}{70\text{ lb}} \qquad \cos\theta_y = \frac{F_y}{F} = \frac{-30\text{ lb}}{70\text{ lb}} \qquad \cos\theta_z = \frac{F_z}{F} = \frac{60\text{ lb}}{70\text{ lb}}$$

Calculating each quotient and its arc cosine gives you

$$\theta_x = 73.4° \qquad \theta_y = 115.4° \qquad \theta_x = 31.0°$$

These computations can be carried out easily with a calculator.

## 2.4B  Force Defined by Its Magnitude and Two Points on Its Line of Action

In many applications, the direction of a force **F** is defined by the coordinates of two points, $M(x_1, y_1, z_1)$ and $N(x_2, y_2, z_2)$, located on its line of action (Fig. 2.29).

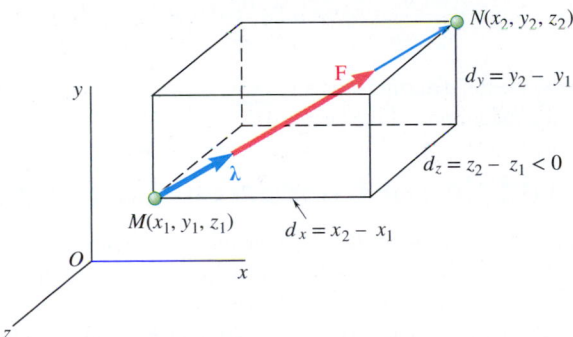

**Fig. 2.29** A case where the line of action of force **F** is determined by the two points $M$ and $N$. We can calculate the components of **F** and its direction cosines from the vector $\overrightarrow{MN}$.

Consider the vector $\overrightarrow{MN}$ joining $M$ and $N$ and of the same sense as a force **F**. Denoting its scalar components by $d_x$, $d_y$, and $d_z$, respectively, we write

$$\overrightarrow{MN} = d_x\mathbf{i} + d_y\mathbf{j} + d_z\mathbf{k} \tag{2.26}$$

We can obtain a unit vector $\lambda$ along the line of action of **F** (i.e., along the line $MN$) by dividing the vector $\overrightarrow{MN}$ by its magnitude $MN$. Substituting for $\overrightarrow{MN}$ from Eq. (2.26) and observing that $MN$ is equal to the distance $d$ from $M$ to $N$, we have

$$\lambda = \frac{\overrightarrow{MN}}{MN} = \frac{1}{d}(d_x\mathbf{i} + d_y\mathbf{j} + d_z\mathbf{k}) \tag{2.27}$$

Recalling that **F** is equal to the product of $F$ and $\lambda$, we have

$$\mathbf{F} = F\lambda = \frac{F}{d}(d_x\mathbf{i} + d_y\mathbf{j} + d_z\mathbf{k}) \tag{2.28}$$

It follows that the scalar components of **F** are, respectively,

**Scalar components
of force F**

$$F_x = \frac{Fd_x}{d} \qquad F_y = \frac{Fd_y}{d} \qquad F_z = \frac{Fd_z}{d} \qquad \text{(2.29)}$$

The relations in Eq. (2.29) considerably simplify the determination of the components of a force **F** of given magnitude $F$ when the line of action of **F** is defined by two points $M$ and $N$. The calculation consists of first subtracting the coordinates of $M$ from those of $N$, and then determining the components of the vector $\overrightarrow{MN}$ and the distance $d$ from $M$ to $N$. Thus,

$$d_x = x_2 - x_1 \qquad d_y = y_2 - y_1 \qquad d_z = z_2 - z_1$$
$$d = \sqrt{d_x^2 + d_y^2 + d_z^2}$$

Substituting for $F$ and for $d_x$, $d_y$, $d_z$, and $d$ into the relations in Eq. (2.29), we obtain the components $F_x$, $F_y$, and $F_z$ of the force.

We can then obtain the angles $\theta_x$, $\theta_y$, and $\theta_z$ that **F** forms with the coordinate axes from Eqs. (2.25). Comparing Eqs. (2.22) and (2.27), we can write

**Direction cosines
of force F**

$$\cos \theta_x = \frac{d_x}{d} \qquad \cos \theta_y = \frac{d_y}{d} \qquad \cos \theta_z = \frac{d_z}{d} \qquad \text{(2.30)}$$

In other words, we can determine the angles $\theta_x$, $\theta_y$, and $\theta_z$ directly from the components and the magnitude of the vector $\overrightarrow{MN}$.

## 2.4C    Addition of Concurrent Forces in Space

We can determine the resultant **R** of two or more forces in space by summing their rectangular components. Graphical or trigonometric methods are generally not practical in the case of forces in space.

The method followed here is similar to that used in Sec. 2.2B with coplanar forces. Setting

$$\mathbf{R} = \Sigma \mathbf{F}$$

we resolve each force into its rectangular components:

$$R_x\mathbf{i} + R_y\mathbf{j} + R_z\mathbf{k} = \Sigma(F_x\mathbf{i} + F_y\mathbf{j} + F_z\mathbf{k})$$
$$= (\Sigma F_x)\mathbf{i} + (\Sigma F_y)\mathbf{j} + (\Sigma F_z)\mathbf{k}$$

From this equation, it follows that

**Rectangular components
of the resultant**

$$R_x = \Sigma F_x \qquad R_y = \Sigma F_y \qquad R_z = \Sigma F_z \qquad \text{(2.31)}$$

The magnitude of the resultant and the angles $\theta_x$, $\theta_y$, and $\theta_z$ that the resultant forms with the coordinate axes are obtained using the method discussed earlier in this section. We end up with

**Resultant of concurrent
forces in space**

$$R = \sqrt{R_x^2 + R_y^2 + R_z^2} \qquad \text{(2.32)}$$

$$\cos \theta_x = \frac{R_x}{R} \qquad \cos \theta_y = \frac{R_y}{R} \qquad \cos \theta_z = \frac{R_z}{R} \qquad \text{(2.33)}$$

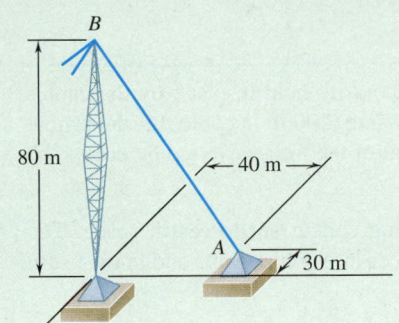

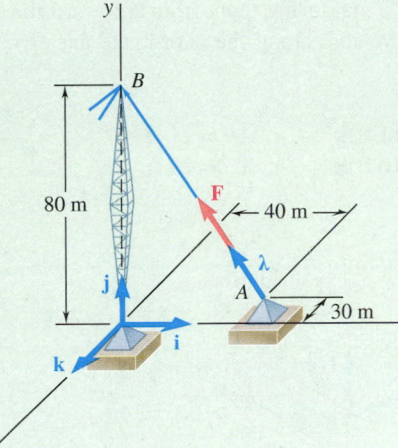

**Fig. 1** Cable force acting on bolt at A, and its unit vector.

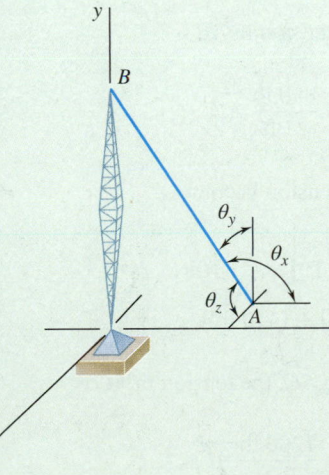

**Fig. 2** Direction angles for cable AB.

## Sample Problem 2.7

A tower guy wire is anchored by means of a bolt at A. The tension in the wire is 2500 N. Determine (*a*) the components $F_x$, $F_y$, and $F_z$ of the force acting on the bolt and (*b*) the angles $\theta_x$, $\theta_y$, and $\theta_z$ defining the direction of the force.

**STRATEGY:** From the given distances, we can determine the length of the wire and the direction of a unit vector along it. From that, we can find the components of the tension and the angles defining its direction.

**MODELING and ANALYSIS:**

**a. Components of the Force.** The line of action of the force acting on the bolt passes through points A and B, and the force is directed from A to B. The components of the vector $\overrightarrow{AB}$, which has the same direction as the force, are

$$d_x = -40 \text{ m} \qquad d_y = +80 \text{ m} \qquad d_z = +30 \text{ m}$$

The total distance from A to B is

$$AB = d = \sqrt{d_x^2 + d_y^2 + d_z^2} = 94.3 \text{ m}$$

Denoting the unit vectors along the coordinate axes by **i**, **j**, and **k**, you have

$$\overrightarrow{AB} = -(40 \text{ m})\mathbf{i} + (80 \text{ m})\mathbf{j} + (30 \text{ m})\mathbf{k}$$

Introducing the unit vector $\boldsymbol{\lambda} = \overrightarrow{AB}/AB$ (Fig. 1), you can express **F** in terms of $\overrightarrow{AB}$ as

$$\mathbf{F} = F\boldsymbol{\lambda} = F\frac{\overrightarrow{AB}}{AB} = \frac{2500 \text{ N}}{94.3 \text{ m}}\overrightarrow{AB}$$

Substituting the expression for $\overrightarrow{AB}$ gives you

$$\mathbf{F} = \frac{2500 \text{ N}}{94.3 \text{ m}}[-(40 \text{ m})\mathbf{i} + (80 \text{ m})\mathbf{j} + (30 \text{ m})\mathbf{k}]$$
$$= -(1060 \text{ N})\mathbf{i} + (2120 \text{ N})\mathbf{j} + (795 \text{ N})\mathbf{k}$$

The components of **F**, therefore, are

$$F_x = -1060 \text{ N} \qquad F_y = +2120 \text{ N} \qquad F_z = +795 \text{ N} \blacktriangleleft$$

**b. Direction of the Force.** Using Eqs. (2.25), you can write the direction cosines directly (Fig. 2):

$$\cos \theta_x = \frac{F_x}{F} = \frac{-1060 \text{ N}}{2500 \text{ N}} \qquad \cos \theta_y = \frac{F_y}{F} = \frac{+2120 \text{ N}}{2500 \text{ N}}$$
$$\cos \theta_z = \frac{F_z}{F} = \frac{+795 \text{ N}}{2500 \text{ N}}$$

Calculating each quotient and its arc cosine, you obtain

$$\theta_x = 115.1° \qquad \theta_y = 32.0° \qquad \theta_z = 71.5° \blacktriangleleft$$

(*Note:* You could have obtained this same result by using the components and magnitude of the vector $\overrightarrow{AB}$ rather than those of the force **F**.)

**REFLECT and THINK:** It makes sense that, for a given geometry, only a certain set of components and angles characterize a given resultant force. The methods in this section allow you to translate back and forth between forces and geometry.

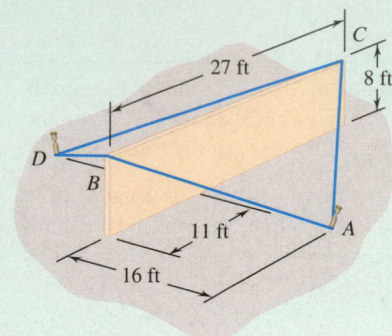

## Sample Problem 2.8

A wall section of precast concrete is temporarily held in place by the cables shown. If the tension is 840 lb in cable $AB$ and 1200 lb in cable $AC$, determine the magnitude and direction of the resultant of the forces exerted by cables $AB$ and $AC$ on stake $A$.

**STRATEGY:** This is a problem in adding concurrent forces in space. The simplest approach is to first resolve the forces into components and to then sum the components and find the resultant.

**MODELING and ANALYSIS:**

**Components of the Forces.** First resolve the force exerted by each cable on stake $A$ into $x$, $y$, and $z$ components. To do this, determine the components and magnitude of the vectors $\overrightarrow{AB}$ and $\overrightarrow{AC}$, measuring them from $A$ toward the wall section (Fig. 1). Denoting the unit vectors along the coordinate axes by $\mathbf{i}$, $\mathbf{j}$, and $\mathbf{k}$, these vectors are

$$\overrightarrow{AB} = -(16 \text{ ft})\mathbf{i} + (8 \text{ ft})\mathbf{j} + (11 \text{ ft})\mathbf{k} \qquad AB = 21 \text{ ft}$$
$$\overrightarrow{AC} = -(16 \text{ ft})\mathbf{i} + (8 \text{ ft})\mathbf{j} - (16 \text{ ft})\mathbf{k} \qquad AC = 24 \text{ ft}$$

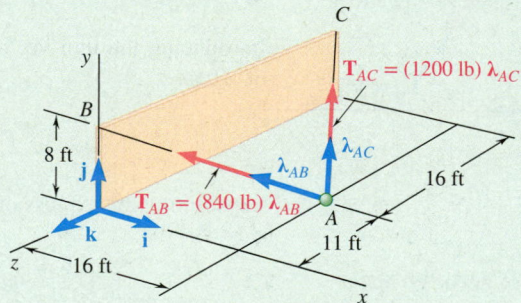

**Fig. 1** Cable forces acting on stake at $A$, and their unit vectors.

Denoting by $\boldsymbol{\lambda}_{AB}$ the unit vector along $AB$, the tension in $AB$ is

$$\mathbf{T}_{AB} = T_{AB}\boldsymbol{\lambda}_{AB} = T_{AB}\frac{\overrightarrow{AB}}{AB} = \frac{840 \text{ lb}}{21 \text{ ft}}\overrightarrow{AB}$$

Substituting the expression found for $\overrightarrow{AB}$, the tension becomes

$$\mathbf{T}_{AB} = \frac{840 \text{ lb}}{21 \text{ ft}}[-(16 \text{ ft})\mathbf{i} + (8 \text{ ft})\mathbf{j} + (11 \text{ ft})\mathbf{k}]$$
$$\mathbf{T}_{AB} = -(640 \text{ lb})\mathbf{i} + (320 \text{ lb})\mathbf{j} + (440 \text{ lb})\mathbf{k}$$

Similarly, denoting by $\boldsymbol{\lambda}_{AC}$ the unit vector along $AC$, the tension in $AC$ is

$$\mathbf{T}_{AC} = T_{AC}\boldsymbol{\lambda}_{AC} = T_{AC}\frac{\overrightarrow{AC}}{AC} = \frac{1200 \text{ lb}}{24 \text{ ft}}\overrightarrow{AC}$$
$$\mathbf{T}_{AC} = -(800 \text{ lb})\mathbf{i} + (400 \text{ lb})\mathbf{j} - (800 \text{ lb})\mathbf{k}$$

*(continued)*

**Resultant of the Forces.**   The resultant **R** of the forces exerted by the two cables is

$$\mathbf{R} = \mathbf{T}_{AB} + \mathbf{T}_{AC} = -(1440 \text{ lb})\mathbf{i} + (720 \text{ lb})\mathbf{j} - (360 \text{ lb})\mathbf{k}$$

You can now determine the magnitude and direction of the resultant as

$$R = \sqrt{R_x^2 + R_y^2 + R_z^2} = \sqrt{(-1440)^2 + (720)^2 + (-300)^2}$$

$$R = 1650 \text{ lb} \quad \blacktriangleleft$$

The direction cosines come from Eqs. (2.33):

$$\cos \theta_x = \frac{R_x}{R} = \frac{-1440 \text{ lb}}{1650 \text{ lb}} \qquad \cos \theta_y = \frac{R_y}{R} = \frac{+720 \text{ lb}}{1650 \text{ lb}}$$

$$\cos \theta_z = \frac{R_z}{R} = \frac{-360 \text{ lb}}{1650 \text{ lb}}$$

Calculating each quotient and its arc cosine, the angles are

$$\theta_x = 150.8° \qquad \theta_y = 64.1° \qquad \theta_z = 102.6° \quad \blacktriangleleft$$

**REFLECT and THINK:**   Based on visual examination of the cable forces, you might have anticipated that $\theta_x$ for the resultant should be obtuse and $\theta_y$ should be acute. The outcome of $\theta_z$ was not as apparent.

# SOLVING PROBLEMS ON YOUR OWN

In this section, we saw that we can define a **force in space** by its magnitude and direction or by the three rectangular components $F_x$, $F_y$, and $F_z$.

**A. When a force is defined by its magnitude and direction,** you can find its rectangular components $F_x$, $F_y$, and $F_z$ as follows.

**Case 1.** If the direction of the force **F** is defined by the angles $\theta_y$ and $\phi$ shown in Fig. 2.25, projections of **F** through these angles or their complements will yield the components of **F** [Eqs. (2.17)]. Note that to find the $x$ and $z$ components of **F**, first project **F** onto the horizontal plane; the projection $\mathbf{F}_h$ obtained in this way is then resolved into the components $\mathbf{F}_x$ and $\mathbf{F}_z$ (Fig. 2.25c).

**Case 2.** If the direction of the force **F** is defined by the angles $\theta_x$, $\theta_y$, and $\theta_z$ that **F** forms with the coordinate axes, you can obtain each component by multiplying the magnitude $F$ of the force by the cosine of the corresponding angle (Concept Application 2.4):

$$F_x = F \cos \theta_x \qquad F_y = F \cos \theta_y \qquad F_z = F \cos \theta_z$$

**Case 3.** If the direction of the force **F** is defined by two points $M$ and $N$ located on its line of action (Fig. 2.29), first express the vector $\overrightarrow{MN}$ drawn from $M$ to $N$ in terms of its components $d_x$, $d_y$, and $d_z$ and the unit vectors **i**, **j**, and **k**:

$$\overrightarrow{MN} = d_x\mathbf{i} + d_y\mathbf{j} + d_z\mathbf{k}$$

Then, determine the unit vector $\boldsymbol{\lambda}$ along the line of action of **F** by dividing the vector $\overrightarrow{MN}$ by its magnitude $MN$. Multiplying $\boldsymbol{\lambda}$ by the magnitude of **F** gives you the desired expression for **F** in terms of its rectangular components (Sample Prob. 2.7):

$$\mathbf{F} = F\boldsymbol{\lambda} = \frac{F}{d}(d_x\mathbf{i} + d_y\mathbf{j} + d_z\mathbf{k})$$

When writing vector expressions such as this, it is often helpful to use a consistent and meaningful system of notation when determining the rectangular components of a force. The method used in this text is illustrated in Sample Prob. 2.8, where the force $\mathbf{T}_{AB}$ acts from stake $A$ toward point $B$. Note that the subscripts have been ordered to agree with the direction of the force. We recommend that you adopt the same notation, because it will help you identify point 1 (the first subscript) and point 2 (the second subscript). This practice is not so important for scalar analysis methods (such as typically used in two-dimensional problems) as long as the diagrams that are drawn to support the analysis clearly establish the directions of the vectors involved. It is also not necessary for certain forces where sign conventions are used, such as in the case of truss members (Chap. 6) and beams (Chap. 7).

*(continued)*

When calculating the vector defining the line of action of a force, you might think of its scalar components as the number of steps you must take in each coordinate direction to go from point 1 to point 2. It is essential that you always remember to assign the correct sign to each of the components.

**B. When a force is defined by its rectangular components $F_x$, $F_y$, and $F_z$,** you can obtain its magnitude $F$ from

$$F = \sqrt{F_x^2 + F_y^2 + F_z^2}$$

You can determine the direction cosines of the line of action of **F** by dividing the components of the force by $F$:

$$\cos \theta_x = \frac{F_x}{F} \qquad \cos \theta_y = \frac{F_y}{F} \qquad \cos \theta_z = \frac{F_z}{F}$$

From the direction cosines, you can obtain the angles $\theta_x$, $\theta_y$, and $\theta_z$ that **F** forms with the coordinate axes (Concept Application 2.5).

**C. To determine the resultant $R$ of two or more forces** in three-dimensional space, first determine the rectangular components of each force by one of the procedures described previously. Adding these components will yield the components $R_x$, $R_y$, and $R_z$ of the resultant. You can then obtain the magnitude and direction of the resultant as indicated previously for a force **F** (Sample Prob. 2.8).

# Problems

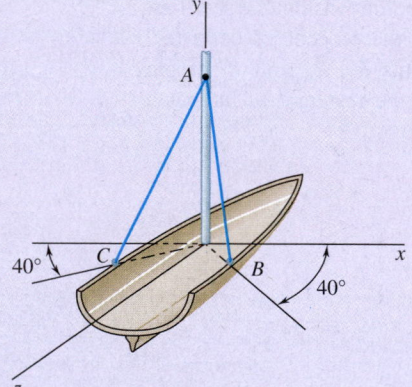

Fig. P2.71 and P2.72

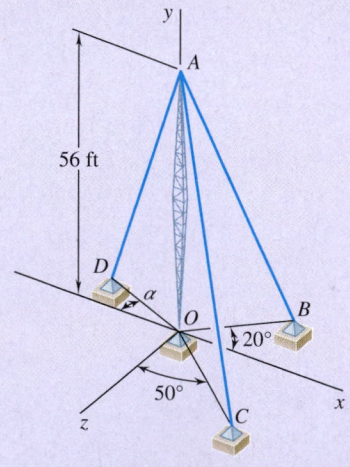

Fig. P2.75 and P2.76

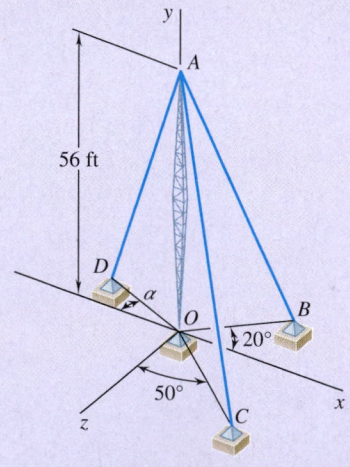

Fig. P2.77 and P2.78

**2.71** Determine (*a*) the *x*, *y*, and *z* components of the 500-N force, (*b*) the angles $\theta_x$, $\theta_y$, and $\theta_z$ that the force forms with the coordinate axes.

**2.72** Determine (*a*) the *x*, *y*, and *z* components of the 800-N force, (*b*) the angles $\theta_x$, $\theta_y$, and $\theta_z$ that the force forms with the coordinate axes.

**2.73** A gun is aimed at a point *A* located 35° east of north. Knowing that the barrel of the gun forms an angle of 40° with the horizontal and that the maximum recoil force is 400 N, determine (*a*) the *x*, *y*, and *z* components of that force, (*b*) the values of the angles $\theta_x$, $\theta_y$, and $\theta_z$ defining the direction of the recoil force. (Assume that the *x*, *y*, and *z* axes are directed, respectively, east, up, and south.)

**2.74** Solve Prob. 2.73 assuming that point *A* is located 15° north of west and that the barrel of the gun forms an angle of 25° with the horizontal.

**2.75** The angle between the guy wire *AB* and the mast is 20°. Knowing that the tension in *AB* is 300 lb, determine (*a*) the *x*, *y*, and *z* components of the force exerted on the boat at *B*, (*b*) the angles $\theta_x$, $\theta_y$, and $\theta_z$ defining the direction of the force exerted at *B*.

**2.76** The angle between the guy wire *AC* and the mast is 20°. Knowing that the tension in *AC* is 300 lb, determine (*a*) the *x*, *y*, and *z* components of the force exerted on the boat at *C*, (*b*) the angles $\theta_x$, $\theta_y$, and $\theta_z$ defining the direction of the force exerted at *C*.

**2.77** Cable *AB* is 65 ft long, and the tension in that cable is 3900 lb. Determine (*a*) the *x*, *y*, and *z* components of the force exerted by the cable on the anchor *B*, (*b*) the angles $\theta_x$, $\theta_y$, and $\theta_z$ defining the direction of that force.

**2.78** Cable *AC* is 70 ft long, and the tension in that cable is 5250 lb. Determine (*a*) the *x*, *y*, and *z* components of the force exerted by the cable on the anchor *C*, (*b*) the angles $\theta_x$, $\theta_y$, and $\theta_z$ defining the direction of that force.

**2.79** Determine the magnitude and direction of the force $\mathbf{F} = (260\ \text{N})\mathbf{i} - (320\ \text{N})\mathbf{j} + (800\ \text{N})\mathbf{k}$.

**2.80** Determine the magnitude and direction of the force $\mathbf{F} = (700\ \text{N})\mathbf{i} - (820\ \text{N})\mathbf{j} + (960\ \text{N})\mathbf{k}$.

**2.81** A force $\mathbf{F}$ of magnitude 250 lb acts at the origin of a coordinate system. Knowing that $\theta_x = 65°$, $\theta_y = 40°$, and $F_z > 0$, determine (*a*) the components of the force, (*b*) the angle $\theta_z$.

**2.82** A force acts at the origin of a coordinate system in a direction defined by the angles $\theta_x = 70.9°$ and $\theta_y = 144.9°$. Knowing that the $z$ component of the force is −52.0 lb, determine (a) the angle $\theta_z$, (b) the other components and the magnitude of the force.

**2.83** A force **F** of magnitude 210 N acts at the origin of a coordinate system. Knowing that $F_x = 80$ N, $\theta_z = 151.2°$, and $F_y < 0$, determine (a) the components $F_y$ and $F_z$, (b) the angles $\theta_x$ and $\theta_y$.

**2.84** A force acts at the origin of a coordinate system in a direction defined by the angles $\theta_y = 120°$ and $\theta_z = 75°$. Knowing that the $x$ component of the force is +40 N, determine (a) the angle $\theta_x$, (b) the magnitude of the force.

**2.85** Two cables *BG* and *BH* are attached to frame *ACD* as shown. Knowing that the tension in cable *BG* is 540 N, determine the components of the force exerted by cable *BG* on the frame at *B*.

**2.86** Two cables *BG* and *BH* are attached to frame *ACD* as shown. Knowing that the tension in cable *BH* is 750 N, determine the components of the force exerted by cable *BH* on the frame at *B*.

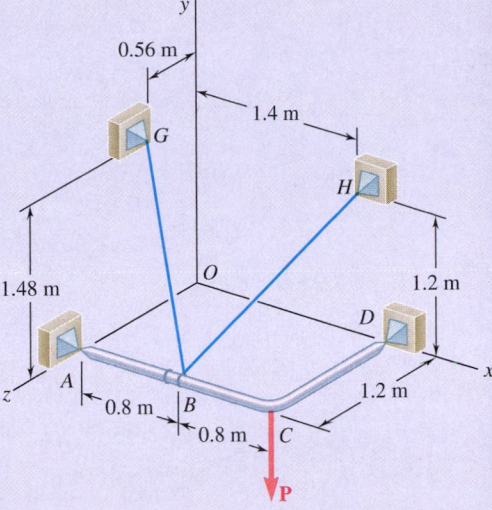

**Fig. P2.85 and *P2.86***

**2.87** In order to move a wrecked truck, two cables are attached at *A* and pulled by winches *B* and *C* as shown. Knowing that the tension in cable *AB* is 2 kips, determine the components of the force exerted at *A* by the cable.

**2.88** In order to move a wrecked truck, two cables are attached at *A* and pulled by winches *B* and *C* as shown. Knowing that the tension in cable *AC* is 1.5 kips, determine the components of the force exerted at *A* by the cable.

**2.89** A rectangular plate is supported by three cables as shown. Knowing that the tension in cable *AB* is 408 N, determine the components of the force exerted on the plate at *B*.

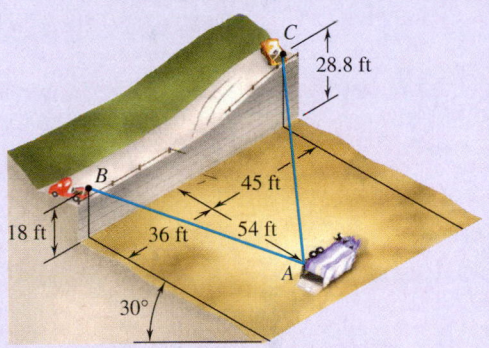

**Fig. P2.87 and P2.88**

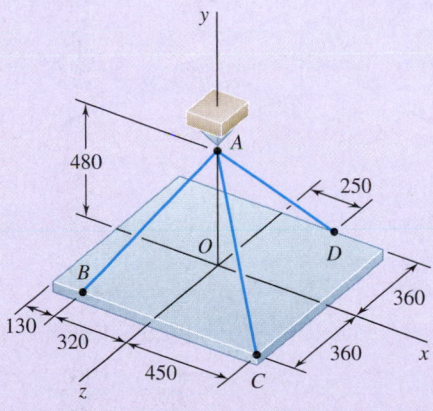

**Fig. P2.89 and *P2.90***

**2.90** A rectangular plate is supported by three cables as shown. Knowing that the tension in cable *AD* is 429 N, determine the components of the force exerted on the plate at *D*.

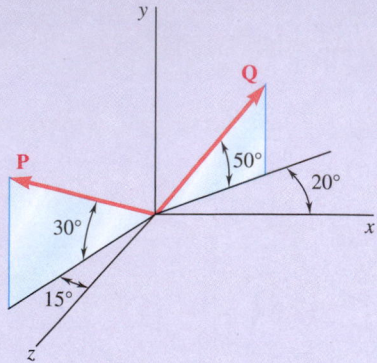

**Fig. P2.91 and P2.92**

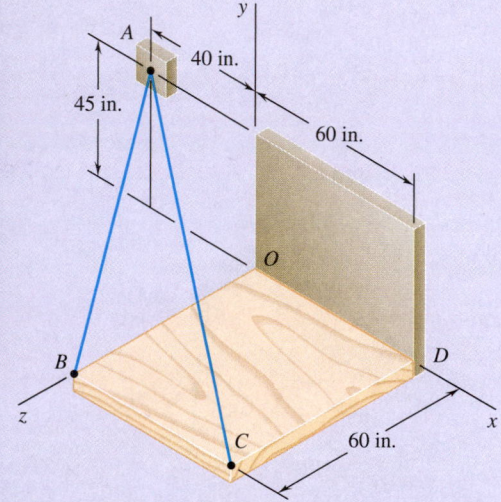

**Fig. *P2.93* and P2.94**

**2.91** Find the magnitude and direction of the resultant of the two forces shown, knowing that $P = 300$ N and $Q = 400$ N.

**2.92** Find the magnitude and direction of the resultant of the two forces shown, knowing that $P = 400$ N and $Q = 300$ N.

**2.93** Knowing that the tension is 425 lb in cable $AB$ and 510 lb in cable $AC$, determine the magnitude and direction of the resultant of the forces exerted at $A$ by the two cables.

**2.94** Knowing that the tension is 510 lb in cable $AB$ and 425 lb in cable $AC$, determine the magnitude and direction of the resultant of the forces exerted at $A$ by the two cables.

**2.95** For the frame of Prob. 2.85, determine the magnitude and direction of the resultant of the forces exerted by the cables at $B$ knowing that the tension is 540 N in cable $BG$ and 750 N in cable $BH$.

**2.96** For the plate of Prob. 2.89, determine the tensions in cables $AB$ and $AD$ knowing that the tension in cable $AC$ is 54 N and that the resultant of the forces exerted by the three cables at $A$ must be vertical.

**2.97** The boom $OA$ carries a load **P** and is supported by two cables as shown. Knowing that the tension in cable $AB$ is 183 lb and that the resultant of the load **P** and of the forces exerted at $A$ by the two cables must be directed along $OA$, determine the tension in cable $AC$.

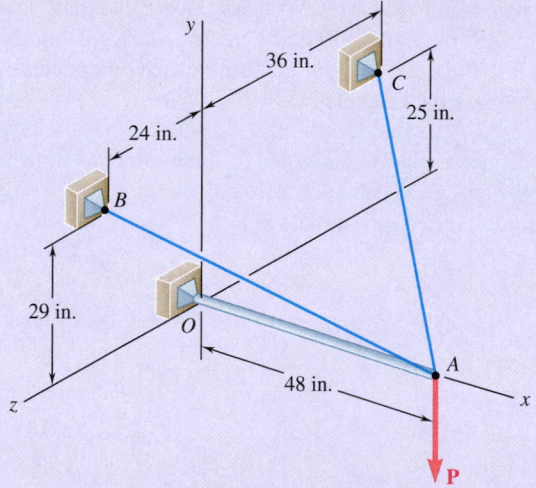

**Fig. P2.97**

**2.98** For the boom and loading of Prob. 2.97, determine the magnitude of the load **P**.

# 2.5   FORCES AND EQUILIBRIUM IN SPACE

According to the definition given in Sec. 2.3, a particle $A$ is in equilibrium if the resultant of all the forces acting on $A$ is zero. The components $R_x$, $R_y$, and $R_z$ of the resultant of forces in space are given by Eqs. (2.31); when the components of the resultant are zero, we have

$$\Sigma F_x = 0 \qquad \Sigma F_y = 0 \qquad \Sigma F_z = 0 \tag{2.34}$$

Eqs. (2.34) represent the necessary and sufficient conditions for the equilibrium of a particle in space. We can use them to solve problems dealing with the equilibrium of a particle involving no more than three unknowns.

The first step in solving three-dimensional equilibrium problems is to draw a free-body diagram showing the particle in equilibrium and *all* of the forces acting on it. You can then write the equations of equilibrium (2.34) and solve them for three unknowns. In the more common types of problems, these unknowns will represent (1) the three components of a single force or (2) the magnitude of three forces, each of known direction.

**Photo 2.4**   Although we cannot determine the tension in the four cables supporting the car by using the three equations (2.34), we can obtain a relation among the tensions by analyzing the equilibrium of the hook.

©WIN-Initiative/Neleman/Getty Images

## Sample Problem 2.9

A 200-kg cylinder is hung by means of two cables $AB$ and $AC$ that are attached to the top of a vertical wall. A horizontal force $\mathbf{P}$ perpendicular to the wall holds the cylinder in the position shown. Determine the magnitude of $\mathbf{P}$ and the tension in each cable.

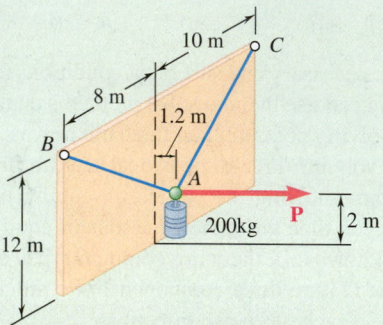

**STRATEGY:**  Connection point $A$ is acted upon by four forces, including the weight of the cylinder. You can use the given geometry to express the force components of the cables and then apply equilibrium conditions to calculate the tensions.

**MODELING and ANALYSIS:**

**Free-Body Diagram.**  Choose point $A$ as a free body; this point is subjected to four forces, three of which are of unknown magnitude. Introducing the unit vectors $\mathbf{i}$, $\mathbf{j}$, and $\mathbf{k}$, resolve each force into rectangular components (Fig. 1):

$$\mathbf{P} = P\mathbf{i}$$
$$\mathbf{W} = -mg\mathbf{j} = -(200 \text{ kg})(9.81 \text{ m/s}^2)\mathbf{j} = -(1962 \text{ N})\mathbf{j} \qquad \textbf{(1)}$$

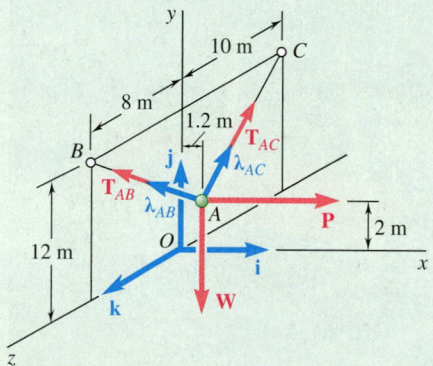

**Fig. 1**  Free-body diagram of particle $A$.

*(continued)*

For $\mathbf{T}_{AB}$ and $\mathbf{T}_{AC}$, it is first necessary to determine the components and magnitudes of the vectors $\overrightarrow{AB}$ and $\overrightarrow{AC}$. Denoting the unit vector along $AB$ by $\lambda_{AB}$, you can write $\mathbf{T}_{AB}$ as

$$\overrightarrow{AB} = -(1.2 \text{ m})\mathbf{i} + (10 \text{ m})\mathbf{j} + (8 \text{ m})\mathbf{k} \qquad AB = 12.862 \text{ m}$$

$$\lambda_{AB} = \frac{\overrightarrow{AB}}{12.862 \text{ m}} = -0.09330\mathbf{i} + 0.7775\mathbf{j} + 0.6220\mathbf{k}$$

$$\mathbf{T}_{AB} = T_{AB}\lambda_{AB} = -0.09330T_{AB}\mathbf{i} + 0.7775T_{AB}\mathbf{j} + 0.6220T_{AB}\mathbf{k} \qquad \textbf{(2)}$$

Similarly, denoting the unit vector along $AC$ by $\lambda_{AC}$, you have for $\mathbf{T}_{AC}$

$$\overrightarrow{AC} = -(1.2 \text{ m})\mathbf{i} + (10 \text{ m})\mathbf{j} - (10 \text{ m})\mathbf{k} \qquad AC = 14.193 \text{ m}$$

$$\lambda_{AC} = \frac{\overrightarrow{AC}}{14.193 \text{ m}} = -0.08455\mathbf{i} + 0.7046\mathbf{j} - 0.7046\mathbf{k}$$

$$\mathbf{T}_{AC} = T_{AC}\lambda_{AC} = -0.08455T_{AC}\mathbf{i} + 0.7046T_{AC}\mathbf{j} - 0.7046T_{AC}\mathbf{k} \qquad \textbf{(3)}$$

**Equilibrium Condition.**   Because $A$ is in equilibrium, you must have

$$\Sigma\mathbf{F} = 0: \qquad \mathbf{T}_{AB} + \mathbf{T}_{AC} + \mathbf{P} + \mathbf{W} = 0$$

or substituting from Eqs. (1), (2), and (3) for the forces and factoring $\mathbf{i}$, $\mathbf{j}$, and $\mathbf{k}$, you have

$$(-0.09330T_{AB} - 0.08455T_{AC} + P)\mathbf{i}$$
$$+ (0.7775T_{AB} + 0.7046T_{AC} - 1962 \text{ N})\mathbf{j}$$
$$+ (0.6220T_{AB} - 0.7046T_{AC})\mathbf{k} = 0$$

Setting the coefficients of $\mathbf{i}$, $\mathbf{j}$, and $\mathbf{k}$ equal to zero, you can write three scalar equations, which express that the sums of the $x$, $y$, and $z$ components of the forces are respectively equal to zero.

$$(\Sigma F_x = 0:) \qquad -0.09330T_{AB} - 0.08455T_{AC} + P = 0$$
$$(\Sigma F_y = 0:) \qquad +0.7775T_{AB} + 0.7046T_{AC} - 1962 \text{ N} = 0$$
$$(\Sigma F_z = 0:) \qquad +0.6220T_{AB} - 0.7046T_{AC} = 0$$

Solving these equations, you obtain

$$P = 235 \text{ N} \qquad T_{AB} = 1402 \text{ N} \qquad T_{AC} = 1238 \text{ N} \blacktriangleleft$$

**REFLECT and THINK:**   The solution of the three unknown forces yielded positive results, which is completely consistent with the physical situation of this problem. Conversely, if one of the cable force results had been negative, thereby reflecting compression instead of tension, you should recognize that the solution is in error.

# SOLVING PROBLEMS
# ON YOUR OWN

We saw earlier that when a particle is in **equilibrium**, the resultant of the forces acting on the particle must be zero. In the case of the equilibrium of a particle in three-dimensional space, this equilibrium condition provides you with three relations among the forces acting on the particle. These relations may be used to determine three unknowns—usually the magnitudes of three forces.

The solution usually consists of the following steps:

**1. Draw a free-body diagram of the particle.** This diagram shows the particle and all the forces acting on it. Indicate on the diagram the magnitudes of known forces, as well as any angles or dimensions that define the direction of a force. Any unknown magnitude or angle should be denoted by an appropriate symbol. Nothing else should be included in the free-body diagram.

**2. Resolve each force into rectangular components.** Following the method used earlier, determine for each force $\mathbf{F}$ the unit vector $\lambda$ defining the direction of that force, and express $\mathbf{F}$ as the product of its magnitude $F$ and $\lambda$. You will obtain an expression of the form

$$\mathbf{F} = F\lambda = \frac{F}{d}(d_x\mathbf{i} + d_y\mathbf{j} + d_z\mathbf{k})$$

where $d$, $d_x$, $d_y$, and $d_z$ are dimensions obtained from the free-body diagram of the particle. If you know the magnitude as well as the direction of the force, then $F$ is known and the expression obtained for $\mathbf{F}$ is well defined; otherwise, $F$ is one of the three unknowns that should be determined.

**3. Set the resultant, or sum, of the forces exerted on the particle equal to zero.** You will obtain a vector equation consisting of terms containing the unit vectors $\mathbf{i}$, $\mathbf{j}$, or $\mathbf{k}$. Group the terms containing the same unit vector and then factor that vector. For the vector equation to be satisfied, you must set the coefficient of each of the unit vectors equal to zero. This yields three scalar equations that you can solve for no more than three unknowns (Sample Prob. 2.9).

# Problems

## FREE-BODY PRACTICE PROBLEMS

**2.F5** Three cables are used to tether a balloon as shown. Knowing that the tension in cable *AC* is 444 N, draw the free-body diagram needed to determine the vertical force **P** exerted by the balloon at *A*.

**2.F6** A container of mass *m* = 120 kg is supported by three cables as shown. Draw the free-body diagram needed to determine the tension in each cable.

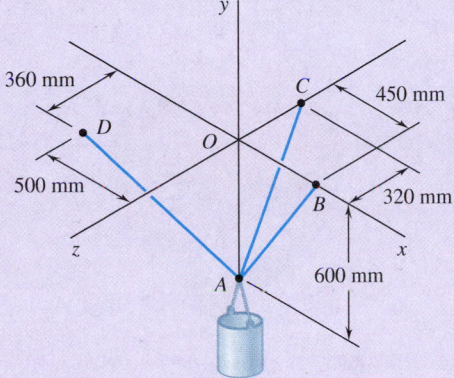

**Fig. P2.F6**

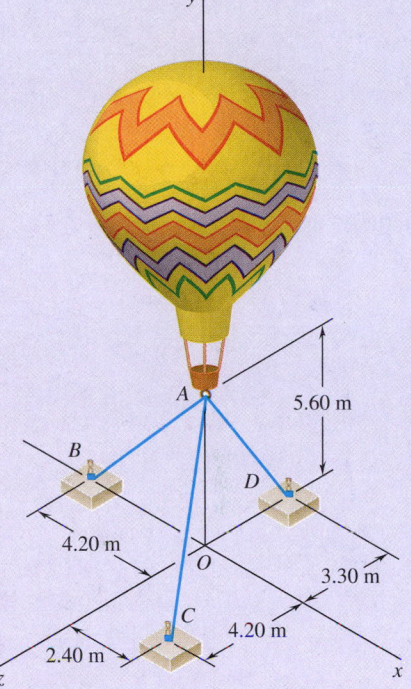

**Fig. P2.F5**

**2.F7** A 150-lb cylinder is supported by two cables *AC* and *BC* that are attached to the top of vertical posts. A horizontal force **P**, which is perpendicular to the plane containing the posts, holds the cylinder in the position shown. Draw the free-body diagram needed to determine the magnitude of **P** and the force in each cable.

**2.F8** A transmission tower is held by three guy wires attached to a pin at *A* and anchored by bolts at *B, C,* and *D*. Knowing that the tension in wire *AB* is 630 lb, draw the free-body diagram needed to determine the vertical force **P** exerted by the tower on the pin at *A*.

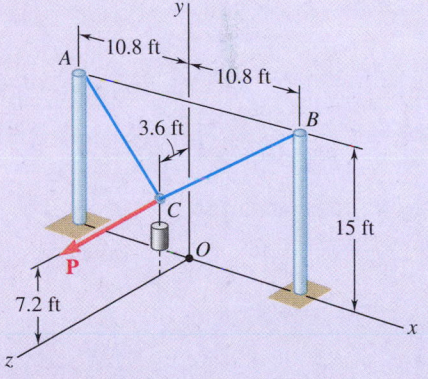

**Fig. P2.F7**

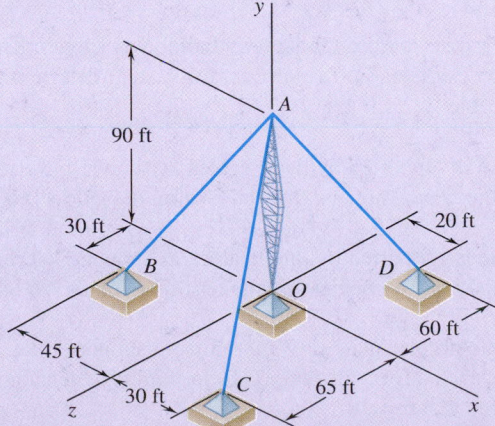

**Fig. P2.F8**

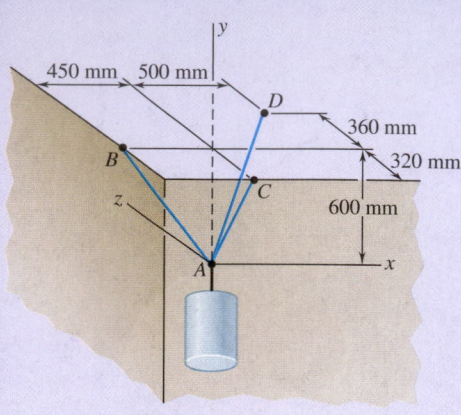

450 mm 500 mm

360 mm

320 mm

600 mm

**Fig. P2.99 and *P2.100***

**END-OF-SECTION PROBLEMS**

**2.99** A container is supported by three cables that are attached to a ceiling as shown. Determine the weight $W$ of the container, knowing that the tension in cable $AB$ is 6 kN.

**2.100** A container is supported by three cables that are attached to a ceiling as shown. Determine the weight $W$ of the container, knowing that the tension in cable $AD$ is 4.3 kN.

**2.101** Three cables are used to tether a balloon as shown. Determine the vertical force **P** exerted by the balloon at $A$ knowing that the tension in cable $AD$ is 481 N.

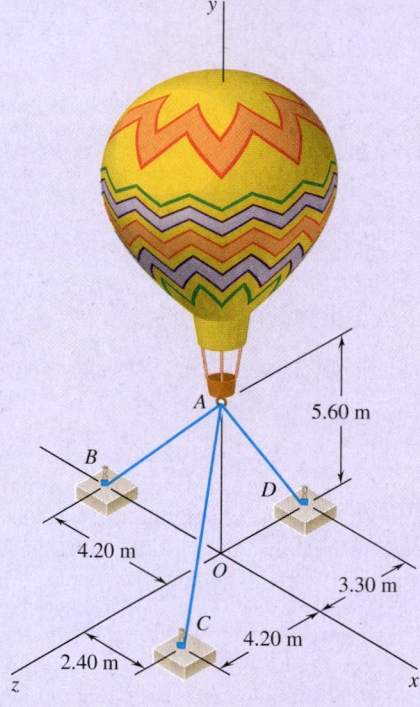

5.60 m

4.20 m

3.30 m

2.40 m

4.20 m

**Fig. P2.101 and *P2.102***

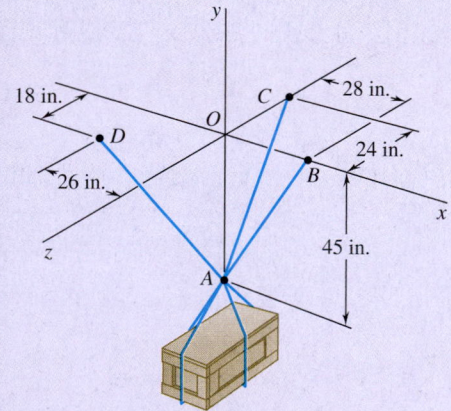

18 in.

28 in.

24 in.

26 in.

45 in.

**Fig. P2.103 and P2.104**

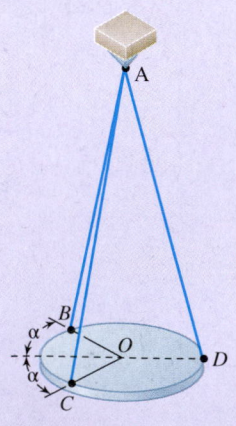

**Fig. *P2.105***

**2.102** Three cables are used to tether a balloon as shown. Knowing that the balloon exerts an 800-N vertical force at $A$, determine the tension in each cable.

**2.103** A crate is supported by three cables as shown. Determine the weight $W$ of the crate, knowing that the tension in cable $AD$ is 924 lb.

**2.104** A crate is supported by three cables as shown. Determine the weight $W$ of the crate, knowing that the tension in cable $AB$ is 1378 lb.

**2.105** A 12-lb circular plate of 7-in. radius is supported as shown by three wires, each of 25-in. length. Determine the tension in each wire, knowing that $\alpha = 30°$.

**2.106** Solve Prob. 2.105, knowing that $\alpha = 45°$.

**2.107** Three cables are connected at $A$, where the forces **P** and **Q** are applied as shown. Knowing that $Q = 0$, find the value of $P$ for which the tension in cable $AD$ is 305 N.

**2.108** Three cables are connected at $A$, where the forces **P** and **Q** are applied as shown. Knowing that $P = 1200$ N, determine the values of $Q$ for which cable $AD$ is taut.

**2.109** A rectangular plate is supported by three cables as shown. Knowing that the tension in cable $AC$ is 60 N, determine the weight of the plate.

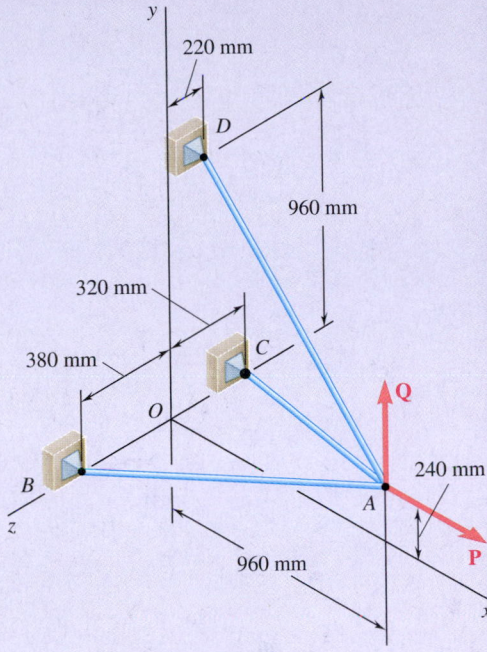

Fig. P2.107 and P2.108

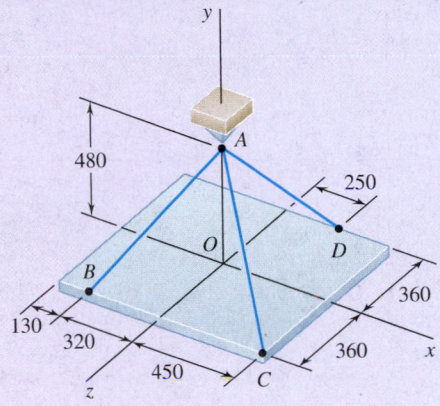

Dimensions in mm

**Fig. P2.109 and P2.110**

**2.110** A rectangular plate is supported by three cables as shown. Knowing that the tension in cable $AD$ is 520 N, determine the weight of the plate.

**2.111** A transmission tower is held by three guy wires attached to a pin at $A$ and anchored by bolts at $B$, $C$, and $D$. If the tension in wire $AB$ is 840 lb, determine the vertical force **P** exerted by the tower on the pin at $A$.

**2.112** A transmission tower is held by three guy wires attached to a pin at $A$ and anchored by bolts at $B$, $C$, and $D$. If the tension in wire $AC$ is 590 lb, determine the vertical force **P** exerted by the tower on the pin at $A$.

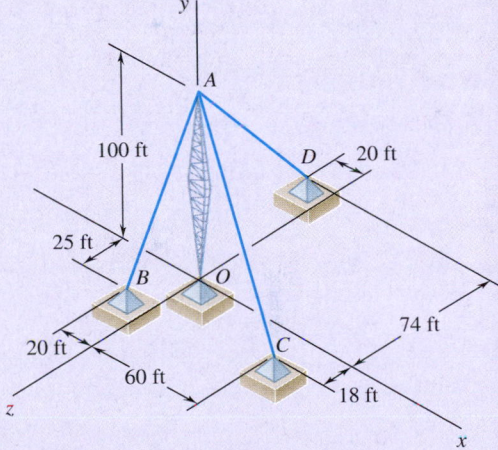

Fig. *P2.111* and P2.112

**2.113** In trying to move across a slippery icy surface, a 175-lb man uses two ropes, $AB$ and $AC$. Knowing that the force exerted on the man by the icy surface is perpendicular to that surface, determine the tension in each rope.

**2.114** Solve Prob. 2.113 assuming that a friend is helping the man at $A$ by pulling on him with a force $\mathbf{P} = -(45 \text{ lb})\mathbf{k}$.

**2.115** For the rectangular plate of Probs. 2.109 and 2.110, determine the tension in each of the three cables knowing that the weight of the plate is 792 N.

**2.116** For the cable system of Probs. 2.107 and 2.108, determine the tension in each cable knowing that $P = 2880$ N and $Q = 0$.

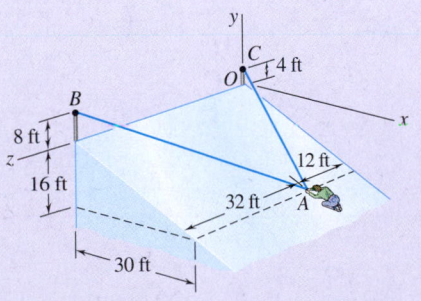

Fig. P2.113

**2.117** For the cable system of Probs. 2.107 and 2.108, determine the tension in each cable knowing that $P = 2880$ N and $Q = 576$ N.

**2.118** Three cables are connected at $D$, where an upward force of 30 kN is applied. Determine the tension in each cable.

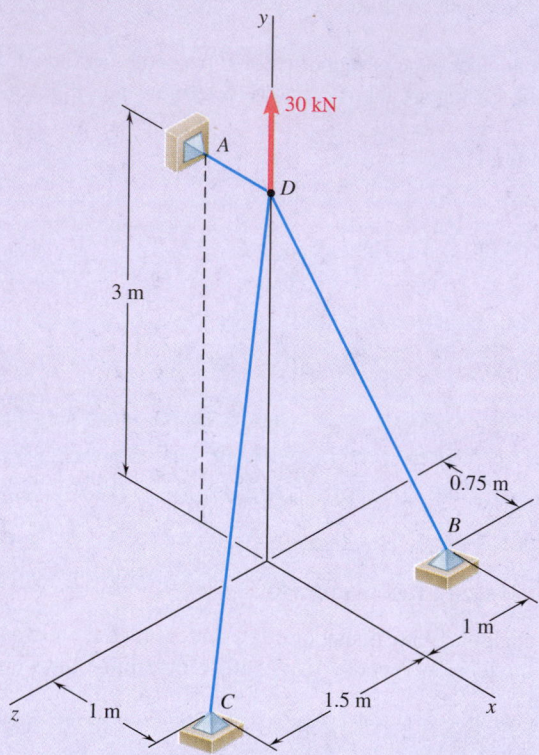

**2.119** For the transmission tower of Probs. 2.111 and 2.112, determine the tension in each guy wire knowing that the tower exerts on the pin at $A$ an upward vertical force of 1800 lb.

**2.120** Three wires are connected at point $D$, which is located 18 in. below the T-shaped pipe support $ABC$. Determine the tension in each wire when a 180-lb cylinder is suspended from point $D$ as shown.

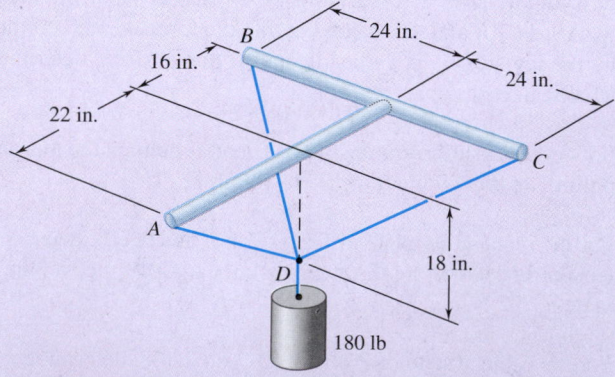

**Fig. P2.120**

**2.121** A container of weight $W$ is suspended from ring $A$, to which cables $AC$ and $AE$ are attached. A force $\mathbf{P}$ is applied to the end $F$ of a third cable that passes over a pulley at $B$ and through ring $A$ and that is attached to a support at $D$. Knowing that $W = 1000$ N, determine the magnitude of $\mathbf{P}$. (*Hint:* The tension is the same in all portions of cable $FBAD$.)

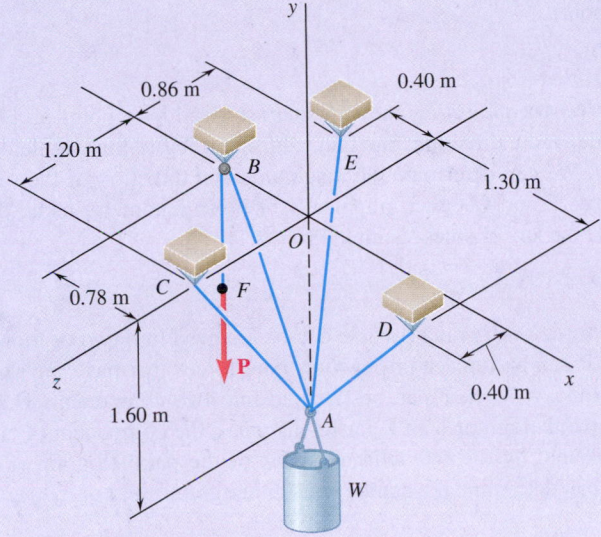

Fig. P2.121

**2.122** Knowing that the tension in cable $AC$ of the system described in Prob. 2.121 is 150 N, determine (*a*) the magnitude of the force $\mathbf{P}$, (*b*) the weight $W$ of the container.

**2.123** A container of weight $W$ is suspended from ring $A$. Cable $BAC$ passes through the ring and is attached to fixed supports at $B$ and $C$. Two forces $\mathbf{P} = P\mathbf{i}$ and $\mathbf{Q} = Q\mathbf{k}$ are applied to the ring to maintain the container in the position shown. Knowing that $W = 270$ lb, determine $P$ and $Q$. (*Hint:* The tension is the same in both portions of cable $BAC$.)

**2.124** For the system of Prob. 2.123, determine $W$ and $P$ knowing that $Q = 36$ lb.

**2.125** Collars $A$ and $B$ are connected by a 525-mm-long wire and can slide freely on frictionless rods. If a force $\mathbf{P} = (341 \text{ N})\mathbf{j}$ is applied to collar $A$, determine (*a*) the tension in the wire when $y = 155$ mm, (*b*) the magnitude of the force $\mathbf{Q}$ required to maintain the equilibrium of the system.

**2.126** Solve Prob. 2.125 assuming that $y = 275$ mm.

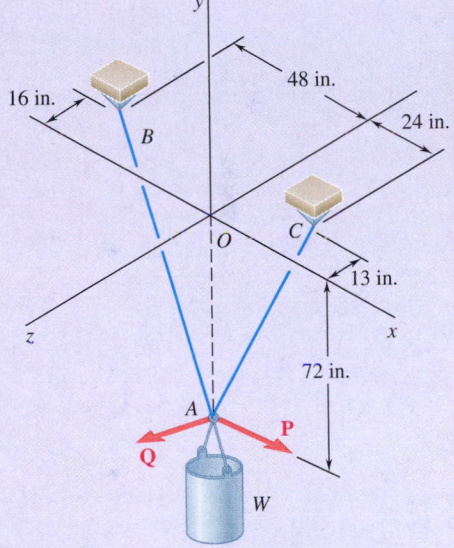

Fig. P2.123

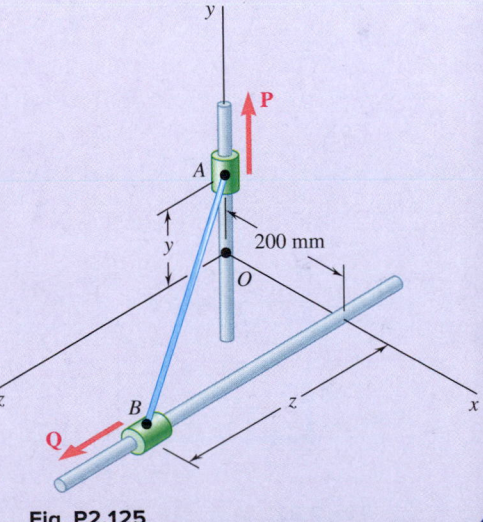

Fig. P2.125

# Review and Summary

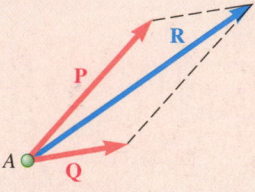

**Fig. 2.30**

In this chapter, we have studied the effect of forces on particles, i.e., on bodies of such shape and size that we may assume all forces acting on them apply at the same point.

## Resultant of Two Forces

Forces are *vector quantities;* they are characterized by a point of application, a magnitude, and a direction, and they add according to the parallelogram law (Fig. 2.30). We can determine the magnitude and direction of the resultant **R** of two forces **P** and **Q** either graphically or by trigonometry using the law of cosines and the law of sines (Sample Prob. 2.1).

## Components of a Force

Any given force acting on a particle can be resolved into two or more components, i.e., it can be replaced by two or more forces that have the same effect on the particle. A force **F** can be resolved into two components **P** and **Q** by drawing a parallelogram with **F** for its diagonal; the components **P** and **Q** are then represented by the two adjacent sides of the parallelogram (Fig. 2.31). Again, we can determine the components either graphically or by trigonometry (Sec. 2.1E).

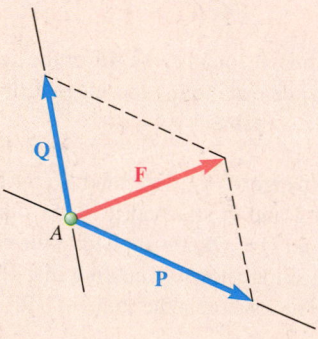

**Fig. 2.31**

## Rectangular Components; Unit Vectors

A force **F** is resolved into two rectangular components if its components $\mathbf{F}_x$ and $\mathbf{F}_y$ are perpendicular to each other and are directed along the coordinate axes (Fig. 2.32). Introducing the unit vectors **i** and **j** along the $x$ and $y$ axes, respectively, we can write the components and the vector as (Sec. 2.2A)

$$\mathbf{F}_x = F_x\mathbf{i} \qquad \mathbf{F}_y = F_y\mathbf{j} \tag{2.6}$$

and

$$\mathbf{F} = F_x\mathbf{i} + F_y\mathbf{j} \tag{2.7}$$

where $F_x$ and $F_y$ are the *scalar components* of **F**. These components, which can be positive or negative, are defined by the relations

$$F_x = F\cos\theta \qquad F_y = F\sin\theta \tag{2.8}$$

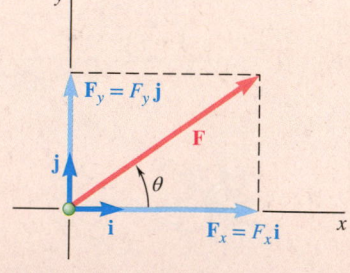

**Fig. 2.32**

When the rectangular components $F_x$ and $F_y$ of a force **F** are given, we can obtain the angle $\theta$ defining the direction of the force from

$$\tan \theta = \frac{F_y}{F_x} \qquad (2.9)$$

We can obtain the magnitude $F$ of the force by solving one of the Eqs. (2.8) for $F$ or by applying the Pythagorean theorem:

$$F = \sqrt{F_x^2 + F_y^2} \qquad (2.10)$$

## Resultant of Several Coplanar Forces

When three or more coplanar forces act on a particle, we can obtain the rectangular components of their resultant **R** by adding the corresponding components of the given forces algebraically (Sec. 2.2B):

$$R_x = \Sigma F_x \qquad R_y = \Sigma F_y \qquad (2.13)$$

The magnitude and direction of **R** then can be determined from relations similar to Eqs. (2.9) and (2.10) (Sample Prob. 2.3).

## Forces in Space

A force **F** in three-dimensional space can be resolved into rectangular components $\mathbf{F}_x$, $\mathbf{F}_y$, and $\mathbf{F}_z$ (Sec. 2.4A). Denoting by $\theta_x$, $\theta_y$, and $\theta_z$, respectively, the angles that **F** forms with the $x$, $y$, and $z$ axes (Fig. 2.33), we have

$$F_x = F \cos \theta_x \qquad F_y = F \cos \theta_y \qquad F_z = F \cos \theta_z \qquad (2.19)$$

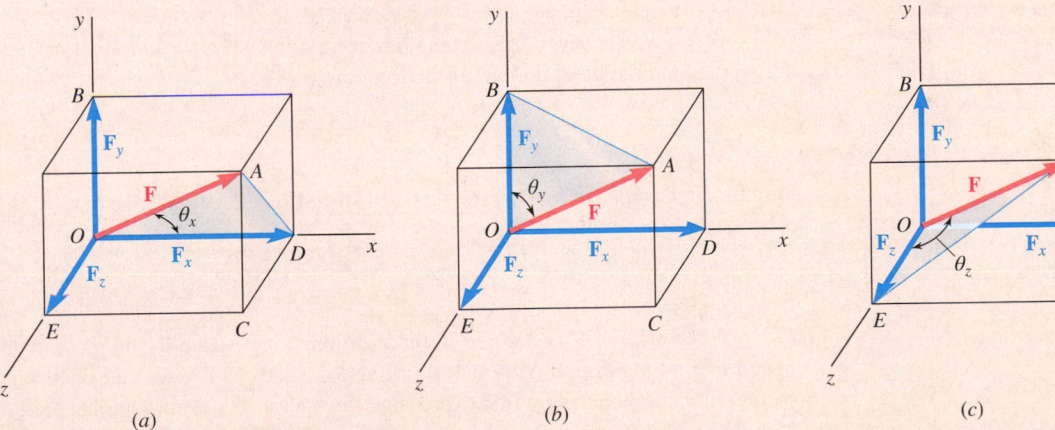

(a)            (b)            (c)

**Fig. 2.33**

## Direction Cosines

The cosines of $\theta_x$, $\theta_y$, and $\theta_z$ are known as the *direction cosines* of the force **F**. Introducing the unit vectors **i**, **j**, and **k** along the coordinate axes, we can write **F** as

$$\mathbf{F} = F_x \mathbf{i} + F_y \mathbf{j} + F_z \mathbf{k} \qquad (2.20)$$

or

$$\mathbf{F} = F(\cos \theta_x \mathbf{i} + \cos \theta_y \mathbf{j} + \cos \theta_z \mathbf{k}) \qquad (2.21)$$

This last equation shows (Fig. 2.34) that **F** is the product of its magnitude $F$ and the unit vector expressed by

$$\boldsymbol{\lambda} = \cos\theta_x \mathbf{i} + \cos\theta_y \mathbf{j} + \cos\theta_z \mathbf{k}$$

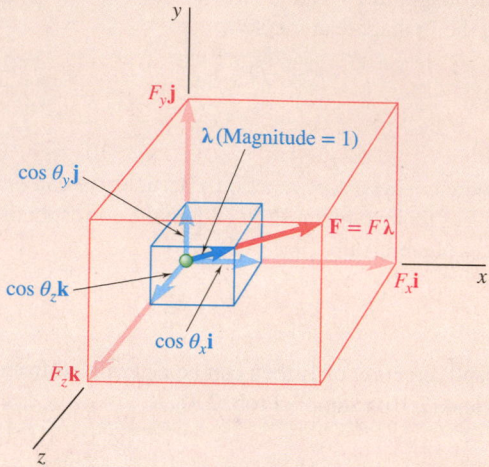

**Fig. 2.34**

Because the magnitude of $\boldsymbol{\lambda}$ is equal to unity, we must have

$$\cos^2\theta_x + \cos^2\theta_y + \cos^2\theta_z = 1 \qquad \textbf{(2.24)}$$

When we are given the rectangular components $F_x$, $F_y$, and $F_z$ of a force **F**, we can find the magnitude **F** of the force by

$$F = \sqrt{F_x^2 + F_y^2 + F_z^2} \qquad \textbf{(2.18)}$$

and the direction cosines of **F** are obtained from Eqs. (2.19). We have

$$\cos\theta_x = \frac{F_x}{F} \qquad \cos\theta_y = \frac{F_y}{F} \qquad \cos\theta_z = \frac{F_z}{F} \qquad \textbf{(2.25)}$$

When a force **F** is defined in three-dimensional space by its magnitude $F$ and two points $M$ and $N$ on its line of action (Sec. 2.4B), we can obtain its rectangular components by first expressing the vector $\overrightarrow{MN}$ joining points $M$ and $N$ in terms of its components $d_x$, $d_y$, and $d_z$ (Fig. 2.35):

$$\overrightarrow{MN} = d_x \mathbf{i} + d_y \mathbf{j} + d_z \mathbf{k} \qquad \textbf{(2.26)}$$

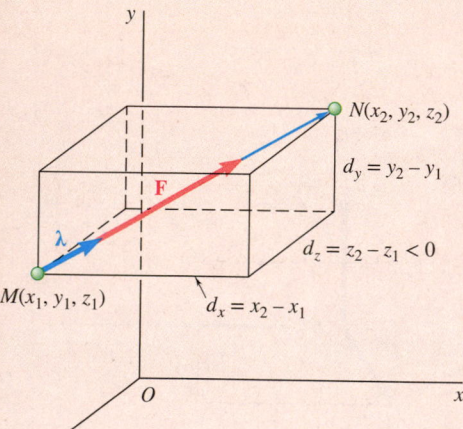

**Fig. 2.35**

We next determine the unit vector $\boldsymbol{\lambda}$ along the line of action of **F** by dividing $\overrightarrow{MN}$ by its magnitude $MN = d$:

$$\boldsymbol{\lambda} = \frac{\overrightarrow{MN}}{MN} = \frac{1}{d}(d_x \mathbf{i} + d_y \mathbf{j} + d_z \mathbf{k}) \qquad \textbf{(2.27)}$$

Recalling that **F** is equal to the product of $F$ and $\boldsymbol{\lambda}$, we have

$$\mathbf{F} = F\boldsymbol{\lambda} = \frac{F}{d}(d_x \mathbf{i} + d_y \mathbf{j} + d_z \mathbf{k}) \qquad \textbf{(2.28)}$$

From this equation it follows (Sample Probs. 2.7 and 2.8) that the scalar components of $\mathbf{F}$ are, respectively,

$$F_x = \frac{F d_x}{d} \qquad F_y = \frac{F d_y}{d} \qquad F_z = \frac{F d_z}{d} \qquad \text{(2.29)}$$

## Resultant of Forces in Space

When two or more forces act on a particle in three-dimensional space, we can obtain the rectangular components of their resultant $\mathbf{R}$ by adding the corresponding components of the given forces algebraically (Sec. 2.4C). We have

$$R_x = \Sigma F_x \qquad R_y = \Sigma F_y \qquad R_z = \Sigma F_z \qquad \text{(2.31)}$$

We can then determine the magnitude and direction of $\mathbf{R}$ from relations similar to Eqs. (2.18) and (2.25) (Sample Prob. 2.8).

## Equilibrium of a Particle

A particle is said to be in equilibrium when the resultant of all the forces acting on it is zero (Sec. 2.3A). The particle remains at rest (if originally at rest) or moves with constant speed in a straight line (if originally in motion) (Sec. 2.3B).

## Free-Body Diagram

To solve a problem involving a particle in equilibrium, first draw a free-body diagram of the particle showing all of the forces acting on it (Sec. 2.3C). If only three coplanar forces act on the particle, you can draw a force triangle to express that the particle is in equilibrium. Using graphical methods of trigonometry, you can solve this triangle for no more than two unknowns (Sample Prob. 2.4). If more than three coplanar forces are involved, you should use the equations of equilibrium:

$$\Sigma F_x = 0 \qquad \Sigma F_y = 0 \qquad \text{(2.15)}$$

These equations can be solved for no more than two unknowns (Sample Prob. 2.6).

## Equilibrium in Space

When a particle is in equilibrium in three-dimensional space (Sec. 2.5), use the three equations of equilibrium:

$$\Sigma F_x = 0 \qquad \Sigma F_y = 0 \qquad \Sigma F_z = 0 \qquad \text{(2.34)}$$

These equations can be solved for no more than three unknowns (Sample Prob. 2.9).

# Review Problems

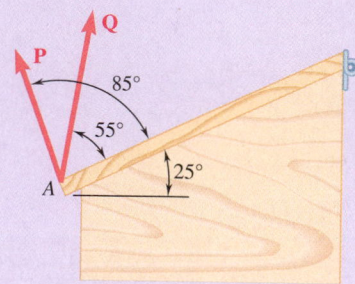

**Fig. P2.127**

**2.127** Two forces **P** and **Q** are applied to the lid of a storage bin as shown. Knowing that $P = 48$ N and $Q = 60$ N, determine by trigonometry the magnitude and direction of the resultant of the two forces.

**2.128** Determine the $x$ and $y$ components of each of the forces shown.

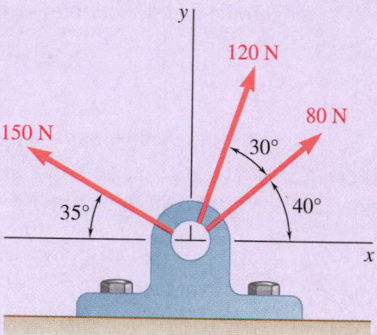

**Fig. P2.128**

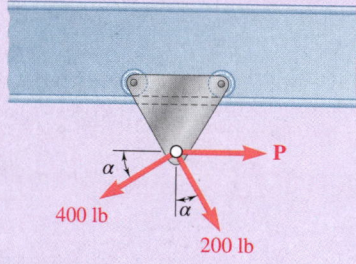

**Fig. P2.129**

**2.129** A hoist trolley is subjected to the three forces shown. Knowing that $\alpha = 40°$, determine (*a*) the required magnitude of the force **P** if the resultant of the three forces is to be vertical, (*b*) the corresponding magnitude of the resultant.

**2.130** Knowing that $\alpha = 55°$ and that boom $AC$ exerts on pin $C$ a force directed along line $AC$, determine (*a*) the magnitude of that force, (*b*) the tension in cable $BC$.

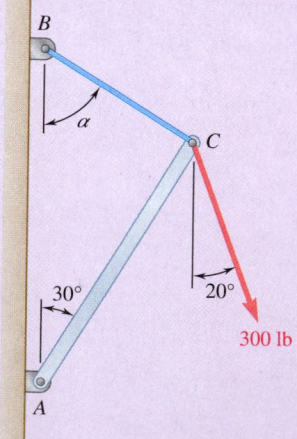

**Fig. P2.130**

**2.131** Two cables are tied together at $C$ and loaded as shown. Knowing that $P = 360$ N, determine the tension $(a)$ in cable $AC$, $(b)$ in cable $BC$.

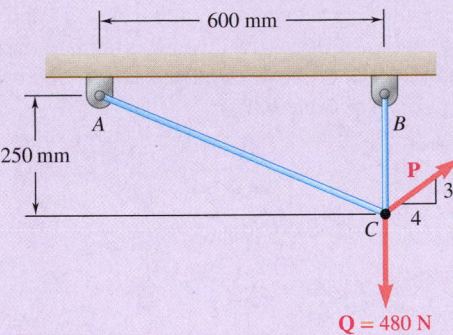

**Fig. P2.131**

**2.132** Two cables tied together at $C$ are loaded as shown. Knowing that the maximum allowable tension in each cable is 800 N, determine $(a)$ the magnitude of the largest force $\mathbf{P}$ that can be applied at $C$, $(b)$ the corresponding value of $\alpha$.

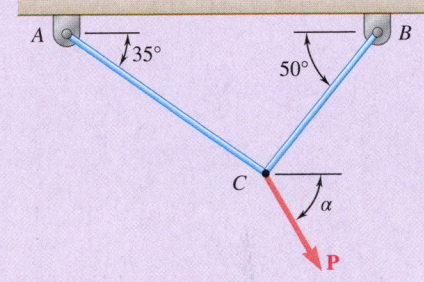

**Fig. P2.132**

**2.133** The end of the coaxial cable $AE$ is attached to the pole $AB$, which is strengthened by the guy wires $AC$ and $AD$. Knowing that the tension in wire $AC$ is 120 lb, determine $(a)$ the components of the force exerted by this wire on the pole, $(b)$ the angles $\theta_x$, $\theta_y$. and $\theta_z$ that the force forms with the coordinate axes.

**2.134** Knowing that the tension in cable $AC$ is 2130 N, determine the components of the force exerted on the plate at $C$.

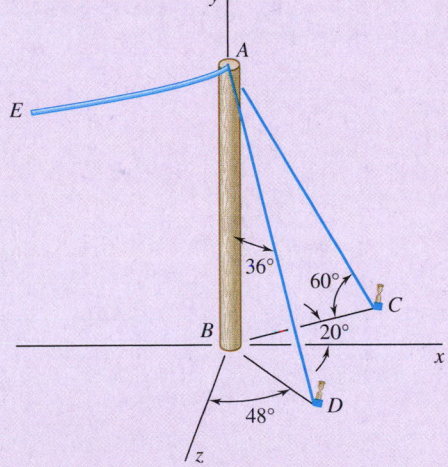

**Fig. P2.133**

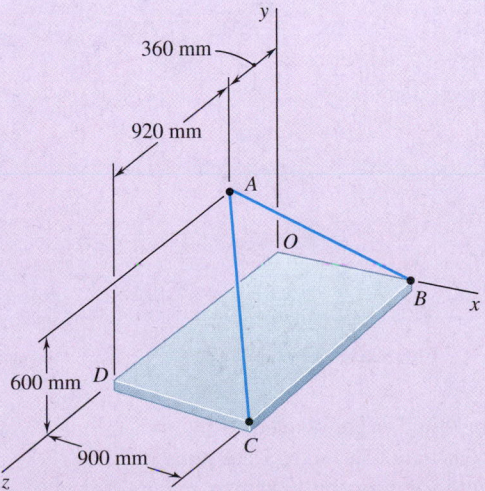

**Fig. P2.134**

81

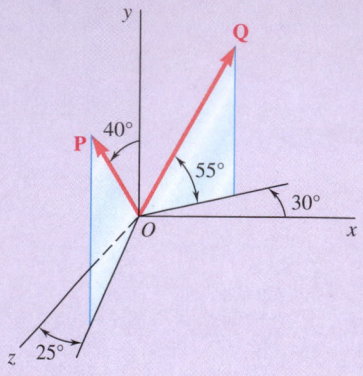

**Fig. P2.135**

**2.135** Find the magnitude and direction of the resultant of the two forces shown, knowing that $P = 600$ N and $Q = 450$ N.

**2.136** Cable *BAC* passes through a frictionless ring *A* and is attached to fixed supports at *B* and *C*, while cables *AD* and *AE* are both tied to the ring and are attached, respectively, to supports at *D* and *E*. Knowing that a 200-lb vertical load **P** is applied to ring *A*, determine the tension in each of the three cables. (*Hint:* The tension is the same in both portions of cable *BAC*.)

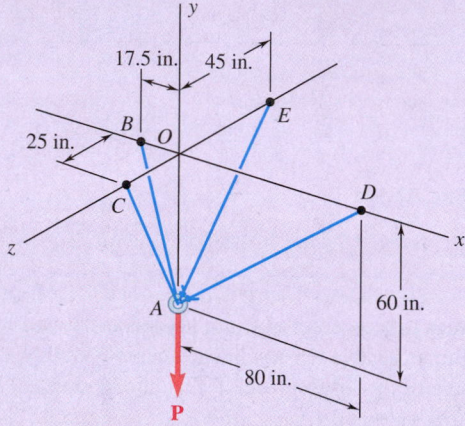

**Fig. P2.136**

**2.137** Collars *A* and *B* are connected by a 25-in.-long wire and can slide freely on frictionless rods. If a 60-lb force **Q** is applied to collar *B* as shown, determine (*a*) the tension in the wire when $x = 9$ in., (*b*) the corresponding magnitude of the force **P** required to maintain the equilibrium of the system.

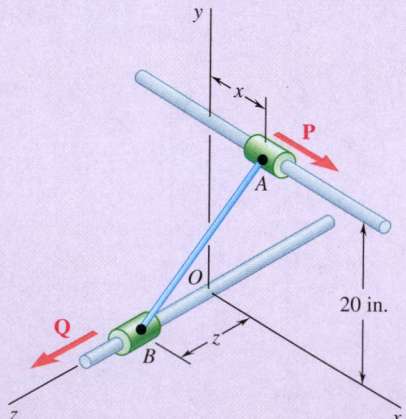

**Fig. P2.137 and *P2.138***

**2.138** Collars *A* and *B* are connected by a 25-in.-long wire and can slide freely on frictionless rods. Determine the distances *x* and *z* for which the equilibrium of the system is maintained when $P = 120$ lb and $Q = 60$ lb.

# 3

# Rigid Bodies: Equivalent Systems of Forces

Four tugboats work together to free the oil tanker *Coastal Eagle Point* that ran aground while attempting to navigate a channel in Tampa Bay. It will be shown in this chapter that the forces exerted on the ship by the tugboats could be replaced by an equivalent force exerted by a single, more powerful, tugboat.

# Objectives

- **Discuss** the principle of transmissibility that enables a force to be treated as a sliding vector.
- **Define** the moment of a force about a point.
- **Examine** vector and scalar products, useful in analysis involving moments.
- **Apply** Varignon's theorem to simplify certain moment analyses.
- **Define** the mixed triple product and use it to determine the moment of a force about an axis.
- **Define** the moment of a couple, and consider the particular properties of couples.
- **Resolve** a given force into an equivalent force-couple system at another point.
- **Reduce** a system of forces into an equivalent force-couple system.
- **Examine** circumstances where a system of forces can be reduced to a single force.
- **Define** a wrench and consider how any general system of forces can be reduced to a wrench.

# Introduction

In Chap. 2, we assumed that each of the bodies considered could be treated as a single particle. Such a view, however, is not always possible. In general, a body should be treated as a combination of a large number of particles. In this case, we need to consider the size of the body as well as the fact that forces act on different parts of the body and thus have different points of application.

Most of the bodies considered in elementary mechanics are assumed to be rigid. We define a **rigid body** as one that does not deform. Actual structures and machines are never absolutely rigid and deform under the loads to which they are subjected. However, these deformations are usually small and do not appreciably affect the conditions of equilibrium or the motion of the structure under consideration. They are important, though, as far as the resistance of the structure to failure is concerned and are considered in the study of mechanics of materials.

In this chapter, you will study the effect of forces exerted on a rigid body, and you will learn how to replace a given system of forces by a simpler equivalent system. This analysis rests on the fundamental assumption that the effect of a given force on a rigid body remains unchanged if that force is moved along its line of action (*principle of transmissibility*). It follows that forces acting on a rigid body can be represented by *sliding vectors*, as indicated earlier in Sec. 2.1B.

Two important concepts associated with the effect of a force on a rigid body are the *moment of a force about a point* (Sec. 3.1E) and the *moment of a force about an axis* (Sec. 3.2C). The determination of these quantities involves computing vector products and scalar products of two vectors, so in this chapter, we introduce the fundamentals of vector algebra and apply them to the solution of problems involving forces acting on rigid bodies.

Another concept introduced in this chapter is that of a *couple*, i.e., the combination of two forces that have the same magnitude, parallel lines of action, and opposite sense (Sec. 3.3A). As you will see, we can replace any system of forces acting on a rigid body by an equivalent system consisting of one force acting at a given point and one couple. This basic combination is called a *force-couple system*. In the case of concurrent, coplanar, or parallel forces, we can further reduce the equivalent force-couple system to a single force, called the *resultant* of the system, or to a single couple, called the *resultant couple* of the system.

# 3.1  FORCES AND MOMENTS

The basic definition of a force does not change if the force acts on a point or on a rigid body. However, the effects of the force can be very different, depending on factors such as the point of application or line of action of that force. As a result, calculations involving forces acting on a rigid body are generally more complicated than situations involving forces acting on a point. We begin by examining some general classifications of forces acting on rigid bodies.

## 3.1A  External and Internal Forces

Forces acting on rigid bodies can be separated into two groups: (1) *external forces* and (2) *internal forces.*

1. **External forces** are exerted by other bodies on the rigid body under consideration. They are entirely responsible for the external behavior of the rigid body, either causing it to move or ensuring that it remains at rest. We shall be concerned only with external forces in this chapter and in Chaps. 4 and 5.
2. **Internal forces** hold together the particles forming the rigid body. If the rigid body is structurally composed of several parts, the forces holding the component parts together are also defined as internal forces. We will consider internal forces in Chaps. 6 and 7.

As an example of external forces, consider the forces acting on a disabled truck that three people are pulling forward by means of a rope attached to the front bumper (Fig. 3.1*a*). The external forces acting on the truck are shown in a **free-body diagram** (Fig. 3.1*b*). Note that this free-body diagram shows the entire object, not just a particle representing the object. Let us first consider the **weight** of the truck. Although it embodies the effect of the earth's pull on each of the particles forming the truck, the weight can be represented by the single force **W**. The **point of application** of this force—that is, the point at which the force acts—is defined as the **center of gravity** of the truck. (In Chap. 5, we will show how to determine the location of centers of gravity.) The weight **W** tends to make the truck move vertically downward. In fact, it would actually cause the truck to move downward, i.e., to fall, if it were not for the presence of the ground. The ground opposes the downward motion of the truck by means of the reactions $\mathbf{R}_1$ and $\mathbf{R}_2$. These forces are exerted *by* the ground *on* the truck and must therefore be included among the external forces acting on the truck.

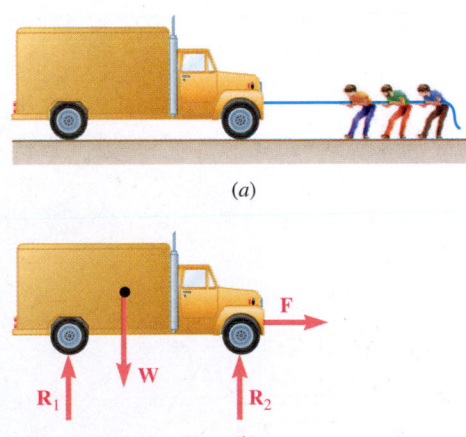

(a)

(b)

**Fig. 3.1** (*a*) Three people pulling on a truck with a rope; (*b*) free-body diagram of the truck, shown as a rigid body instead of a particle.

The people pulling on the rope exert the force **F**. The point of application of **F** is on the front bumper. The force **F** tends to make the truck move forward in a straight line and does actually make it move, because no external force opposes this motion. (We are ignoring rolling resistance here for simplicity.) This forward motion of the truck, during which each straight line keeps its original orientation (the floor of the truck remains horizontal, and the walls remain vertical), is known as a **translation**. Other forces might cause the truck to move differently. For example, the force exerted by a jack placed under the front axle would cause the truck to pivot about its rear axle. Such a motion is a **rotation**. We conclude, therefore, that each *external force* acting on a *rigid body* can, if unopposed, impart to the rigid body a motion of translation or rotation, or both.

## 3.1B Principle of Transmissibility: Equivalent Forces

The **principle of transmissibility** states that the conditions of equilibrium or motion of a rigid body remain unchanged if a force **F** acting at a given point of the rigid body is replaced by a force **F′** of the same magnitude and same direction, but acting at a different point, *provided that the two forces have the same line of action* (Fig. 3.2). The two forces **F** and **F′** have the same effect on the rigid body and are said to be **equivalent forces**. This principle, which states that the action of a force may be *transmitted* along its line of action, is based on experimental evidence. It *cannot* be derived from the properties established so far in this text and therefore must be accepted as an experimental law. (You will see in Sec. 16.1D that we *can* derive the principle of transmissibility from the study of the dynamics of rigid bodies, but this study requires the use of Newton's second and third laws and of several other concepts as well.) Therefore, our study of the statics of rigid bodies is based on the three principles introduced so far: the parallelogram law of vector addition, Newton's first law, and the principle of transmissibility.

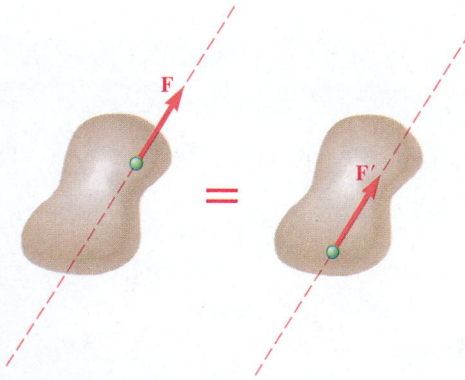

**Fig. 3.2** Two forces **F** and **F′** are equivalent if they have the same magnitude and direction and the same line of action, even if they act at different points.

We indicated in Chap. 2 that we could represent the forces acting on a particle by vectors. These vectors had a well-defined point of application—namely, the particle itself—and were therefore fixed, or bound, vectors. In the case of forces acting on a rigid body, however, the point of application of the force does not matter, as long as the line of action remains unchanged. Thus, forces acting on a rigid body must be represented by a different kind of vector, known as a **sliding vector**, because forces are allowed to slide along their lines of action. Note that all of the properties we derive in the following sections for the forces acting on a rigid body are valid more generally for any system of sliding vectors. In order to keep our presentation more intuitive, however, we will carry it out in terms of physical forces rather than in terms of mathematical sliding vectors.

Returning to the example of the truck, we first observe that the line of action of the force **F** is a horizontal line passing through both the front and rear bumpers of the truck (Fig. 3.3). Using the principle of transmissibility, we can therefore replace **F** by an *equivalent force* **F′** acting on the rear bumper. In other words, the conditions of motion are unaffected, and all of the other external forces acting on the truck (**W**, $\mathbf{R}_1$, $\mathbf{R}_2$) remain unchanged if the people push on the rear bumper instead of pulling on the front bumper.

The principle of transmissibility and the concept of equivalent forces have limitations. Consider, for example, a short bar *AB* acted upon by equal and opposite axial forces $\mathbf{P}_1$ and $\mathbf{P}_2$, as shown in Fig. 3.4a. According to the principle of

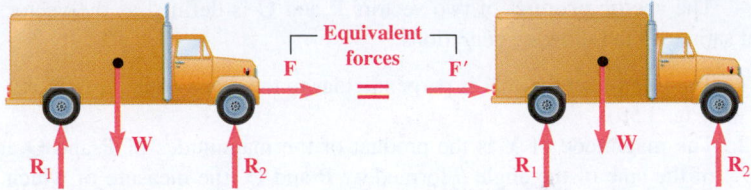

**Fig. 3.3** Force **F**′ is equivalent to force **F**, so the motion of the truck is the same whether you pull it or push it.

transmissibility, we can replace force $\mathbf{P}_2$ by a force $\mathbf{P}_2'$ having the same magnitude, the same direction, and the same line of action but acting at $A$ instead of $B$ (Fig. 3.4*b*). The forces $\mathbf{P}_1$ and $\mathbf{P}_2'$ acting on the same particle can be added according to the rules of Chap. 2, and because these forces are equal and opposite, their sum is equal to zero. Thus, in terms of the external behavior of the bar, the original system of forces shown in Fig. 3.4*a* is equivalent to no force at all (Fig. 3.4*c*).

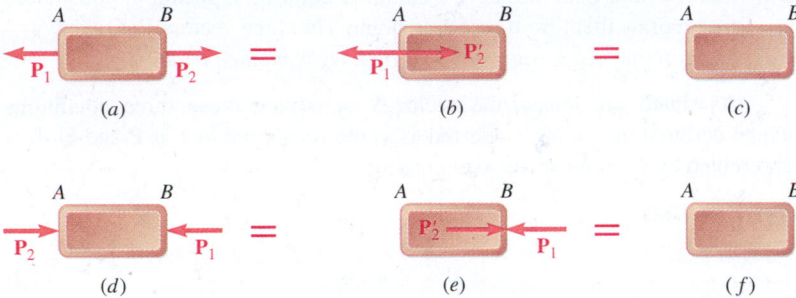

**Fig. 3.4** (*a*–*c*) A set of equivalent forces acting on bar *AB*; (*d*–*f*) another set of equivalent forces acting on bar *AB*. Both sets produce the same external effect (equilibrium in this case) but different internal forces and deformations.

Consider now the two equal and opposite forces $\mathbf{P}_1$ and $\mathbf{P}_2$ acting on the bar *AB* as shown in Fig. 3.4*d*. We can replace the force $\mathbf{P}_2$ by a force $\mathbf{P}_2'$ having the same magnitude, the same direction, and the same line of action but acting at *B* instead of at *A* (Fig. 3.4*e*). We can add forces $\mathbf{P}_1$ and $\mathbf{P}_2'$, and their sum is again zero (Fig. 3.4*f*). From the point of view of the mechanics of rigid bodies, the systems shown in Fig. 3.4*a* and *d* are thus equivalent. However, the *internal forces* and *deformations* produced by the two systems are distinctly different. The bar of Fig. 3.4*a* is in *tension* and, if not absolutely rigid, increases in length slightly; the bar of Fig. 3.4*d* is in *compression* and, if not absolutely rigid, decreases in length slightly. Thus, although we can use the principle of transmissibility to determine the conditions of motion or equilibrium of rigid bodies and to compute the external forces acting on these bodies, it should be avoided, or at least used with care, in determining internal forces and deformations.

## 3.1C Vector Products

In order to gain a better understanding of the effect of a force on a rigid body, we need to introduce a new concept, the *moment of a force about a point*. However, this concept is more clearly understood and is applied more effectively if we first add to the mathematical tools at our disposal the vector product of two vectors.

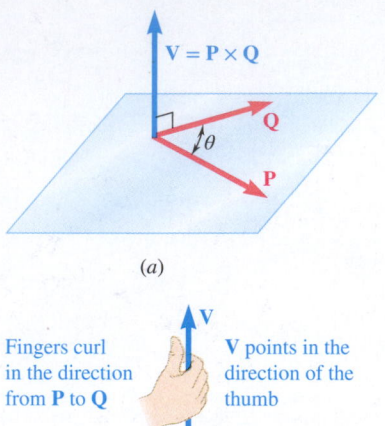

(a)

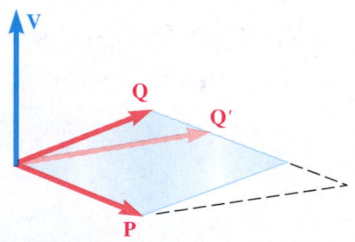

Fingers curl in the direction from **P** to **Q**

**V** points in the direction of the thumb

(b)

**Fig. 3.5** (a) The vector product **V** has the magnitude $PQ \sin \theta$ and is perpendicular to the plane of **P** and **Q**; (b) you can determine the direction of **V** by using the right-hand rule.

**Fig. 3.6** The magnitude of the vector product **V** equals the area of the parallelogram formed by **P** and **Q**. If you change **Q** to **Q'** in such a way that the parallelogram changes shape but **P** and the area are still the same, then the magnitude of **V** remains the same.

The **vector product** of two vectors **P** and **Q** is defined as the vector **V** that satisfies the following conditions.

1. The line of action of **V** is perpendicular to the plane containing **P** and **Q** (Fig. 3.5a).
2. The magnitude of **V** is the product of the magnitudes of **P** and **Q** and of the sine of the angle $\theta$ formed by **P** and **Q** (the measure of which is always 180° or less). We thus have

**Magnitude of a vector product**

$$V = PQ \sin \theta \tag{3.1}$$

3. The direction of **V** is obtained from the **right-hand rule**. Close your right hand and hold it so that your fingers are curled in the same sense as the rotation through $\theta$ that brings the vector **P** in line with the vector **Q**. Your thumb then indicates the direction of the vector **V** (Fig. 3.5b). Note that if **P** and **Q** do not have a common point of application, you should first redraw them from the same point. The three vectors **P**, **Q**, and **V**—taken in that order—are said to form a *right-handed triad*.[†]

As stated previously, the vector **V** satisfying these three conditions (which define it uniquely) is referred to as the *vector product* of **P** and **Q**. It is represented by the mathematical expression

**Vector product**

$$\mathbf{V} = \mathbf{P} \times \mathbf{Q} \tag{3.2}$$

Because of this notation, the vector product of two vectors **P** and **Q** is also referred to as the *cross product* of **P** and **Q**.

It follows from Eq. (3.1) that if the vectors **P** and **Q** have either the same direction or opposite directions, their vector product is zero. In the general case when the angle $\theta$ formed by the two vectors is neither 0° nor 180°, Eq. (3.1) has a simple geometric interpretation: The magnitude $V$ of the vector product of **P** and **Q** is equal to the area of the parallelogram that has **P** and **Q** for sides (Fig. 3.6). The vector product **P** × **Q** is therefore unchanged if we replace **Q** by a vector **Q'** that is coplanar with **P** and **Q** such that the line joining the tips of **Q** and **Q'** is parallel to **P**:

$$\mathbf{V} = \mathbf{P} \times \mathbf{Q} = \mathbf{P} \times \mathbf{Q'} \tag{3.3}$$

From the third condition used to define the vector product **V** of **P** and **Q**—namely, that **P**, **Q**, and **V** must form a right-handed triad—it follows that vector products *are not commutative*; i.e., **Q** × **P** is not equal to **P** × **Q**. Indeed, we can check that **Q** × **P** is represented by the vector −**V**, which is equal and opposite to **V**:

$$\mathbf{Q} \times \mathbf{P} = -(\mathbf{P} \times \mathbf{Q}) \tag{3.4}$$

We saw that the commutative property does not apply to vector products. However, it can be demonstrated that the *distributive* property

$$\mathbf{P} \times (\mathbf{Q}_1 + \mathbf{Q}_2) = \mathbf{P} \times \mathbf{Q}_1 + \mathbf{P} \times \mathbf{Q}_2 \tag{3.5}$$

does hold.

---

[†]Note that the *x, y,* and *z* axes used in Chap. 2 form a right-handed system of orthogonal axes and that the unit vectors **i, j,** and **k** defined in Sec. 2.4A form a right-handed orthogonal triad.

## Concept Application 3.1

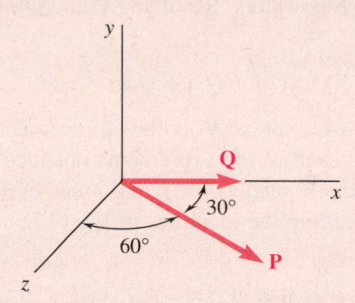

Let us compute the vector product $V = P \times Q$, where the vector $P$ is of magnitude 6 and lies in the $zx$ plane at an angle of 30° with the $x$ axis, and where the vector $Q$ is of magnitude 4 and lies along the $x$ axis (Fig. 3.7).

### Solution

It follows immediately from the definition of the vector product that the vector $V$ must lie along the $y$ axis, directed upward, with the magnitude

$$V = PQ \sin \theta = (6)(4) \sin 30° = 12 \quad \blacksquare$$

**Fig. 3.7** Two vectors $P$ and $Q$ with a 30° angle between them.

A third property, the associative property, does not apply to vector products; we have in general

$$(P \times Q) \times S \neq P \times (Q \times S) \tag{3.6}$$

## 3.1D Rectangular Components of Vector Products

Before we turn back to forces acting on rigid bodes, let's look at a more convenient way to express vector products using rectangular components. To do this, we use the unit vectors $i$, $j$, and $k$ that were defined in Chap. 2.

Consider first the vector product $i \times j$ (Fig. 3.8a). Because both vectors have a magnitude equal to 1 and because they are at a right angle to each other, their vector product is also a unit vector. This unit vector must be $k$, because the vectors $i$, $j$, and $k$ are mutually perpendicular and form a right-handed triad. Similarly, it follows from the right-hand rule given in Sec. 3.1C that the product $j \times i$ is equal to $-k$ (Fig. 3.8b). Finally, note that the vector product of a unit vector with itself, such as $i \times i$, is equal to zero, because both vectors have the same direction. Thus, we can list the vector products of all the various possible pairs of unit vectors:

$$
\begin{array}{lll}
i \times i = 0 & j \times i = -k & k \times i = j \\
i \times j = k & j \times j = 0 & k \times j = -i \\
i \times k = -j & j \times k = i & k \times k = 0
\end{array}
\tag{3.7}
$$

We can determine the sign of the vector product of two unit vectors simply by arranging them in a circle and reading them in the order of the multiplication (Fig. 3.9). The product is positive if they follow each other in counterclockwise order and is negative if they follow each other in clockwise order.

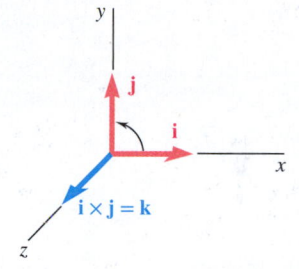

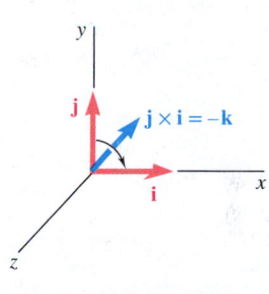

**Fig. 3.8** (a) The vector product of the $i$ and $j$ unit vectors is the $k$ unit vector; (b) the vector product of the $j$ and $i$ unit vectors is the $-k$ unit vector.

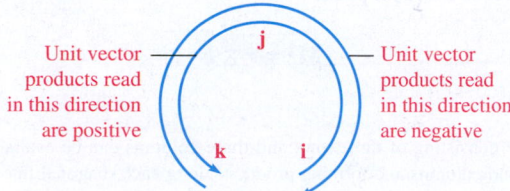

Unit vector products read in this direction are positive

Unit vector products read in this direction are negative

**Fig. 3.9** Arrange the three letters $i$, $j$, and $k$ in a counterclockwise circle. You can use the order of letters for the three unit vectors in a vector product to determine its sign.

We can now express the vector product **V** of two given vectors **P** and **Q** in terms of the rectangular components of these vectors. Resolving **P** and **Q** into components, we first write

$$\mathbf{V} = \mathbf{P} \times \mathbf{Q} = (P_x\mathbf{i} + P_y\mathbf{j} + P_z\mathbf{k}) \times (Q_x\mathbf{i} + Q_y\mathbf{j} + Q_z\mathbf{k})$$

Making use of the distributive property, we express **V** as the sum of vector products, such as $P_x\mathbf{i} \times Q_y\mathbf{j}$. We find that each of the expressions obtained is equal to the vector product of two unit vectors, such as $\mathbf{i} \times \mathbf{j}$, multiplied by the product of two scalars, such as $P_x Q_y$. Recalling the identities of Eq. (3.7) and factoring out **i**, **j**, and **k**, we obtain

$$\mathbf{V} = (P_y Q_z - P_z Q_y)\mathbf{i} + (P_z Q_x - P_x Q_z)\mathbf{j} + (P_x Q_y - P_y Q_x)\mathbf{k} \qquad (3.8)$$

Thus, the rectangular components of the vector product **V** are

**Rectangular components
of a vector product**

$$\begin{aligned} V_x &= P_y Q_z - P_z Q_y \\ V_y &= P_z Q_x - P_x Q_z \\ V_z &= P_x Q_y - P_y Q_x \end{aligned} \qquad (3.9)$$

Returning to Eq. (3.8), notice that the right-hand side represents the expansion of a determinant. Thus, we can express the vector product **V** in the following form, which is more easily memorized:[†]

**Rectangular components
of a vector product (determinant form)**

$$\mathbf{V} = \begin{vmatrix} \mathbf{i} & \mathbf{j} & \mathbf{k} \\ P_x & P_y & P_z \\ Q_x & Q_y & Q_z \end{vmatrix} \qquad (3.10)$$

## 3.1E   Moment of a Force about a Point

We are now ready to consider a force **F** acting on a rigid body (Fig. 3.10*a*). As we know, the force **F** is represented by a vector that defines its magnitude and direction. However, the effect of the force on the rigid body depends also upon its point of application *A*. The position of *A* can be conveniently defined by the vector **r** that joins the fixed reference point *O* with *A*; this vector is known as the *position vector* of *A*. The position vector **r** and the force **F** define the plane shown in Fig. 3.10*a*.

We define the **moment of F about O** as the vector product of **r** and **F**:

**Moment of a force about a point *O***

$$\mathbf{M}_O = \mathbf{r} \times \mathbf{F} \qquad (3.11)$$

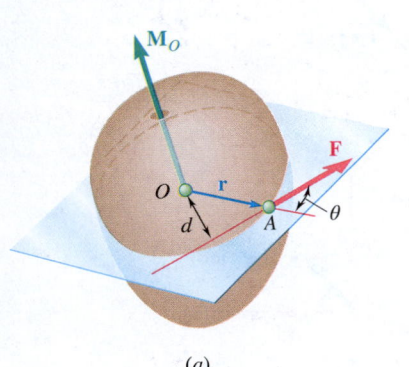

(a)

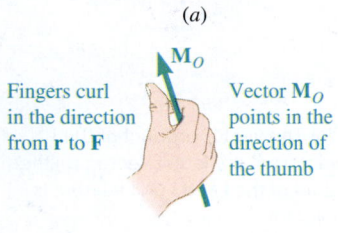

Fingers curl
in the direction
from **r** to **F**

Vector **M**$_O$
points in the
direction of
the thumb

(b)

**Fig. 3.10**  Moment of a force about a point. (*a*) The moment **M**$_O$ is the vector product of the position vector **r** and the force **F**; (*b*) a right-hand rule indicates the sense of **M**$_O$.

[†]Any determinant consisting of three rows and three columns can be evaluated by repeating the first and second columns and forming products along each diagonal line. The sum of the products obtained along the red lines is then subtracted from the sum of the products obtained along the black lines.

According to the definition of the vector product given in Sec. 3.1C, the moment $\mathbf{M}_O$ must be perpendicular to the plane containing $O$ and force $\mathbf{F}$. The sense of $\mathbf{M}_O$ is defined by the sense of the rotation that will bring vector $\mathbf{r}$ in line with vector $\mathbf{F}$; this rotation is observed as *counterclockwise* by an observer located at the tip of $\mathbf{M}_O$. Another way of defining the sense of $\mathbf{M}_O$ is furnished by a variation of the right-hand rule: Close your right hand and hold it so that your fingers curl in the sense of the rotation that $\mathbf{F}$ would impart to the rigid body about a fixed axis directed along the line of action of $\mathbf{M}_O$. This way, your thumb indicates the sense of the moment $\mathbf{M}_O$ (Fig. 3.10*b*).

Finally, denoting by $\theta$ the angle between the lines of action of the position vector $\mathbf{r}$ and the force $\mathbf{F}$, we find that the magnitude of the moment of $\mathbf{F}$ about $O$ is

**Magnitude of the moment of a force**

$$M_O = rF \sin \theta = Fd \qquad \textbf{(3.12)}$$

where $d$ represents the perpendicular distance from $O$ to the line of action of $\mathbf{F}$ (see Fig. 3.10). Experimentally, the tendency of a force $\mathbf{F}$ to make a rigid body rotate about a fixed axis perpendicular to the force depends upon the distance of $\mathbf{F}$ from that axis, as well as upon the magnitude of $\mathbf{F}$. For example, a child's breath can exert enough force to make a toy propeller spin (Fig. 3.11*a*), but a wind turbine requires the force of a substantial wind to rotate the blades and generate electrical power (Fig. 3.11*b*). However, the perpendicular distance between the rotation point and the line of action of the force (often called the *moment arm*) is just as important. If you want to apply a small moment to turn a nut on a pipe without breaking it, you might use a small pipe wrench that gives you a small moment arm (Fig. 3.11*c*). But if you need a larger moment, you could use a large wrench with a long moment arm (Fig. 3.11*d*). Therefore,

> *The magnitude of $\mathbf{M}_O$ measures the tendency of the force $\mathbf{F}$ to make the rigid body rotate about a fixed axis directed along $\mathbf{M}_O$.*

In the SI system of units, where a force is expressed in newtons (N) and a distance in meters (m), the moment of a force is expressed in newton-meters (N·m). In the U.S. customary system of units, where a force is expressed in pounds and a distance in feet or inches, the moment of a force is expressed in lb·ft or lb·in.

Note that although the moment $\mathbf{M}_O$ of a force about a point depends upon the magnitude, the line of action, and the sense of the force, it does *not* depend upon the actual position of the point of application of the force along its line of action. Conversely, the moment $\mathbf{M}_O$ of a force $\mathbf{F}$ does not characterize the position of the point of application of $\mathbf{F}$.

However, as we will see shortly, the moment $\mathbf{M}_O$ of a force $\mathbf{F}$ of a given magnitude and direction *completely defines the line of action of $\mathbf{F}$*. Indeed, the line of action of $\mathbf{F}$ must lie in a plane through $O$ perpendicular to the moment $\mathbf{M}_O$; its distance $d$ from $O$ must be equal to the quotient $M_O/F$ of the magnitudes of $\mathbf{M}_O$ and $\mathbf{F}$; and the sense of $\mathbf{M}_O$ determines whether the line of action of $\mathbf{F}$ occurs on one side or the other of the point $O$.

Recall from Sec. 3.1B that the principle of transmissibility states that two forces $\mathbf{F}$ and $\mathbf{F}'$ are equivalent (i.e., have the same effect on a rigid body) if they have the same magnitude, same direction, and same line of action. We can now restate this principle:

> *Two forces $\mathbf{F}$ and $\mathbf{F}'$ are equivalent if, and only if, they are equal (i.e., have the same magnitude and same direction) and have equal moments about a given point $O$.*

(*a*) Small force

(*b*) Large force

(*c*) Small moment arm

(*d*) Large moment arm

**Fig. 3.11**    (*a, b*) The moment of a force depends on the magnitude of the force; (*c, d*) it also depends on the length of the moment arm. (*a*) ©darko64/123rf.com; (*b*) ©Image Source/Getty Images RF; (*c*) ©Valery Voennyy/Alamy RF; (*d*) ©Monty Rakusen/Getty Images RF

The necessary and sufficient conditions for two forces **F** and **F′** to be equivalent are thus

$$\mathbf{F} = \mathbf{F'} \qquad \text{and} \qquad \mathbf{M}_O = \mathbf{M}_O' \qquad\qquad \textbf{(3.13)}$$

We should observe that if the relations of Eqs. (3.13) hold for a given point *O*, then they hold for any other point.

**Two-Dimensional Problems.**    Many applications in statics deal with two-dimensional structures. Such structures have length and breadth but only negligible depth. Often, they are subjected to forces contained in the plane of the structure. We can easily represent two-dimensional structures and the forces acting on them on a sheet of paper or on a blackboard. Their analysis is therefore considerably simpler than that of three-dimensional structures and forces.

Consider, for example, a rigid slab acted upon by a force **F** in the plane of the slab (Fig. 3.12). The moment of **F** about a point $O$, which is chosen in the plane of the figure, is represented by a vector $\mathbf{M}_O$ perpendicular to that plane and of magnitude $Fd$. In the case of Fig. 3.12a, the vector $\mathbf{M}_O$ points *out of* the page, whereas in the case of Fig. 3.12b, it points *into* the page. As we look at the figure, we observe in the first case that **F** tends to rotate the slab counterclockwise and in the second case that it tends to rotate the slab clockwise. Therefore, it is natural to refer to the sense of the moment of **F** about $O$ in Fig. 3.12a as counterclockwise ↺, and in Fig. 3.12b as clockwise ↻.

Because the moment of a force **F** acting in the plane of the figure must be perpendicular to that plane, we need only specify the *magnitude* and the *sense* of the moment of **F** about $O$. We do this by assigning to the magnitude $M_O$ of the moment a positive or negative sign according to whether the vector $\mathbf{M}_O$ points out of or into the page.

## 3.1F Rectangular Components of the Moment of a Force

We can use the distributive property of vector products to determine the moment of the resultant of several *concurrent forces*. If several forces $\mathbf{F}_1$, $\mathbf{F}_2$, . . . are applied at the same point $A$ (Fig. 3.13) and if we denote by **r** the position vector of $A$, it follows immediately from Eq. (3.5) that

$$\mathbf{r} \times (\mathbf{F}_1 + \mathbf{F}_2 + \cdots) = \mathbf{r} \times \mathbf{F}_1 + \mathbf{r} \times \mathbf{F}_2 + \cdots \qquad (3.14)$$

In words,

> ***The moment about a given point O of the resultant of several concurrent forces is equal to the sum of the moments of the various forces about the same point O.***

This property, which was originally established by the French mathematician Pierre Varignon (1654–1722) long before the introduction of vector algebra, is known as **Varignon's theorem**.

The relation in Eq. (3.14) makes it possible to replace the direct determination of the moment of a force **F** by determining the moments of two or more component forces. As you will see shortly, **F** is generally resolved into components parallel to the coordinate axes. However, it may be more expeditious in some instances to resolve **F** into components that are not parallel to the coordinate axes (see Sample Prob. 3.3).

In general, determining the moment of a force in space is considerably simplified if the force and the position vector of its point of application are resolved into rectangular $x$, $y$, and $z$ components. Consider, for example, the moment $\mathbf{M}_O$ about $O$ of a force **F** whose components are $F_x$, $F_y$, and $F_z$ and that is applied at a point $A$ with coordinates $x$, $y$, and $z$ (Fig. 3.14). Because the components of the position vector **r** are respectively equal to the coordinates $x$, $y$, and $z$ of the point $A$, we can write **r** and **F** as

$$\mathbf{r} = x\mathbf{i} + y\mathbf{j} + z\mathbf{k} \qquad (3.15)$$
$$\mathbf{F} = F_x\mathbf{i} + F_y\mathbf{j} + F_z\mathbf{k} \qquad (3.16)$$

Substituting for **r** and **F** from Eqs. (3.15) and (3.16) into

$$\mathbf{M}_O = \mathbf{r} \times \mathbf{F} \qquad (3.11)$$

and recalling Eqs. (3.8) and (3.9), we can write the moment $\mathbf{M}_O$ of **F** about $O$ in the form

$$\mathbf{M}_O = M_x\mathbf{i} + M_y\mathbf{j} + M_z\mathbf{k} \qquad (3.17)$$

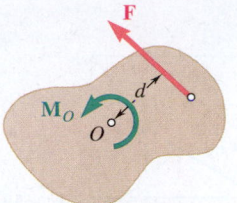

(a) $M_O = + Fd$

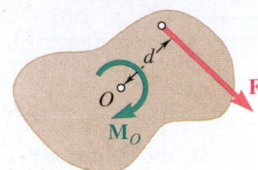

(b) $M_O = - Fd$

**Fig. 3.12** (a) A moment that tends to produce a counterclockwise rotation is positive; (b) a moment that tends to produce a clockwise rotation is negative.

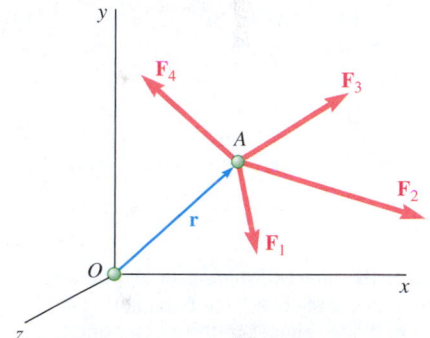

**Fig. 3.13** Varignon's theorem says that the moment about point $O$ of the resultant of these four forces equals the sum of the moments about point $O$ of the individual forces.

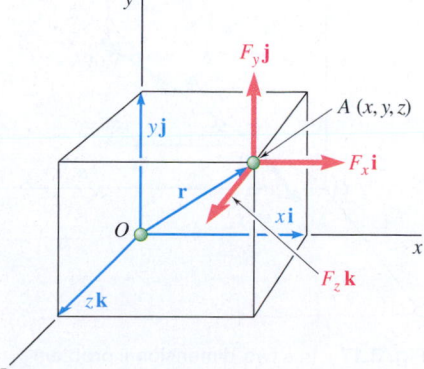

**Fig. 3.14** The moment $\mathbf{M}_O$ about point $O$ of a force **F** applied at point $A$ is the vector product of the position vector **r** and the force **F**, which can both be expressed in rectangular components.

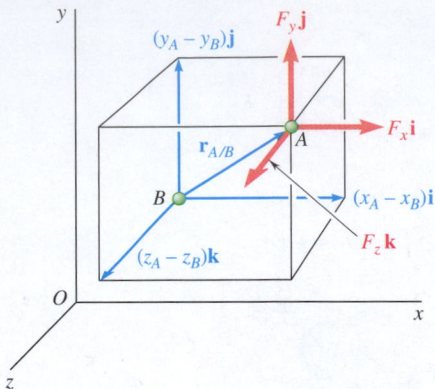

**Fig. 3.15** The moment $\mathbf{M}_B$ about the point $B$ of a force $\mathbf{F}$ applied at point $A$ is the vector product of the position vector $\mathbf{r}_{A/B}$ and force $\mathbf{F}$.

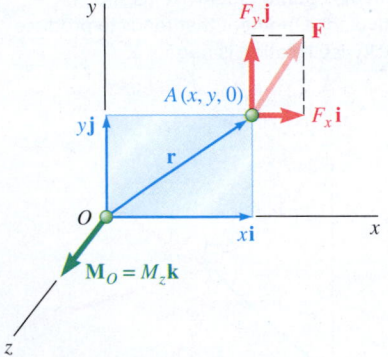

**Fig. 3.16** In a two-dimensional problem, the moment $\mathbf{M}_O$ of a force $\mathbf{F}$ applied at $A$ in the $xy$ plane reduces to the $z$ component of the vector product of $\mathbf{r}$ with $\mathbf{F}$.

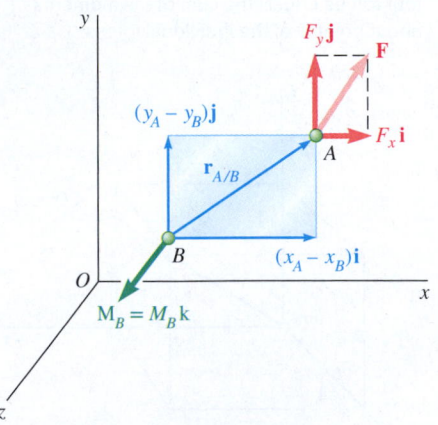

**Fig. 3.17** In a two-dimensional problem, the moment $\mathbf{M}_B$ about a point $B$ of a force $\mathbf{F}$ applied at $A$ in the $xy$ plane reduces to the $z$ component of the vector product of $\mathbf{r}_{A/B}$ with $\mathbf{F}$.

where the components $M_x$, $M_y$, and $M_z$ are defined by the relations

**Rectangular components of a moment**

$$
\begin{aligned}
M_x &= yF_z - zF_y \\
M_y &= zF_x - xF_z \\
M_z &= xF_y - yF_x
\end{aligned} \tag{3.18}
$$

As you will see in Sec.3.2C, the scalar components $M_x$, $M_y$, and $M_z$ of the moment $\mathbf{M}_O$ measure the tendency of the force $\mathbf{F}$ to impart to a rigid body a rotation about the $x$, $y$, and $z$ axes, respectively. Substituting from Eq. (3.18) into Eq. (3.17), we can also write $\mathbf{M}_O$ in the form of the determinant, as

$$
\mathbf{M}_O = \begin{vmatrix} \mathbf{i} & \mathbf{j} & \mathbf{k} \\ x & y & z \\ F_x & F_y & F_z \end{vmatrix} \tag{3.19}
$$

To compute the moment $\mathbf{M}_B$ about an arbitrary point $B$ of a force $\mathbf{F}$ applied at $A$ (Fig. 3.15), we must replace the position vector $\mathbf{r}$ in Eq. (3.11) by a vector drawn from $B$ to $A$. This vector is the *position vector of A relative to B*, denoted by $\mathbf{r}_{A/B}$. Observing that $\mathbf{r}_{A/B}$ can be obtained by subtracting $\mathbf{r}_B$ from $\mathbf{r}_A$, we write

$$
\mathbf{M}_B = \mathbf{r}_{A/B} \times \mathbf{F} = (\mathbf{r}_A - \mathbf{r}_B) \times \mathbf{F} \tag{3.20}
$$

or using the determinant form,

$$
\mathbf{M}_B = \begin{vmatrix} \mathbf{i} & \mathbf{j} & \mathbf{k} \\ x_{A/B} & y_{A/B} & z_{A/B} \\ F_x & F_y & F_z \end{vmatrix} \tag{3.21}
$$

where $x_{A/B}$, $y_{A/B}$, and $z_{A/B}$ denote the components of the vector $\mathbf{r}_{A/B}$:

$$
x_{A/B} = x_A - x_B \qquad y_{A/B} = y_A - y_B \qquad z_{A/B} = z_A - z_B
$$

In the case of two-dimensional problems, we can assume without loss of generality that the force $\mathbf{F}$ lies in the $xy$ plane (Fig. 3.16). Setting $z = 0$ and $F_z = 0$ in Eq. (3.19), we obtain

$$
\mathbf{M}_O = \left( xF_y - yF_x \right) \mathbf{k}
$$

We can verify that the moment of $\mathbf{F}$ about $O$ is perpendicular to the plane of the figure and that it is completely defined by the scalar

$$
M_O = M_z = xF_y - yF_x \tag{3.22}
$$

As noted earlier, a positive value for $M_O$ indicates that the vector $\mathbf{M}_O$ points out of the paper (the force $\mathbf{F}$ tends to rotate the body counterclockwise about $O$), and a negative value indicates that the vector $\mathbf{M}_O$ points into the paper (the force $\mathbf{F}$ tends to rotate the body clockwise about $O$).

To compute the moment about $B(x_B, y_B)$ of a force lying in the $xy$ plane and applied at $A(x_A, y_A)$ (Fig. 3.17), we set $z_{A/B} = 0$ and $F_z = 0$ in Eq. (3.21) and note that the vector $\mathbf{M}_B$ is perpendicular to the $xy$ plane and is defined in magnitude and sense by the scalar

$$
M_B = (x_A - x_B)F_y - (y_A - y_B)F_x \tag{3.23}
$$

## Sample Problem 3.1

A 100-lb vertical force is applied to the end of a lever, which is attached to a shaft at $O$. Determine (a) the moment of the 100-lb force about $O$; (b) the horizontal force applied at $A$ that creates the same moment about $O$; (c) the smallest force applied at $A$ that creates the same moment about $O$; (d) how far from the shaft a 240-lb vertical force must act to create the same moment about $O$; (e) whether any one of the forces obtained in parts $b$, $c$, or $d$ is equivalent to the original force.

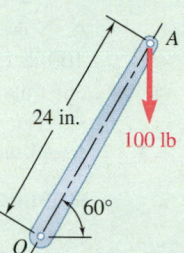

**STRATEGY:**   The calculations asked for all involve variations on the basic defining equation of a moment, $M_O = Fd$.

### MODELING and ANALYSIS:

**a. Moment about O.**   The perpendicular distance from $O$ to the line of action of the 100-lb force (Fig. 1) is

$$d = (24 \text{ in.})\cos 60° = 12 \text{ in.}$$

The magnitude of the moment about $O$ of the 100-lb force is

$$M_O = Fd = (100 \text{ lb})(12 \text{ in.}) = 1200 \text{ lb·in.}$$

Because the force tends to rotate the lever clockwise about $O$, represent the moment by a vector $\mathbf{M}_O$ perpendicular to the plane of the figure and pointing *into* the paper. You can express this fact with the notation

$$\mathbf{M}_O = 1200 \text{ lb·in.} \circlearrowright \quad \blacktriangleleft$$

**b. Horizontal Force.**   In this case, you have (Fig. 2)

$$d = (24 \text{ in.}) \sin 60° = 20.8 \text{ in.}$$

Because the moment about $O$ must be 1200 lb·in., you obtain

$$M_O = Fd$$
$$1200 \text{ lb·in.} = F(20.8 \text{ in.})$$
$$F = 57.7 \text{ lb} \qquad \mathbf{F} = 57.7 \text{ lb} \rightarrow \quad \blacktriangleleft$$

**c. Smallest Force.**   Because $M_O = Fd$, the smallest value of $F$ occurs when $d$ is maximum. Choose the force perpendicular to $OA$ and note that $d = 24$ in. (Fig. 3); thus,

$$M_O = Fd$$
$$1200 \text{ lb·in.} = F(24 \text{ in.})$$
$$F = 50 \text{ lb} \qquad \mathbf{F} = 50 \text{ lb} \; \measuredangle 30° \quad \blacktriangleleft$$

*(continued)*

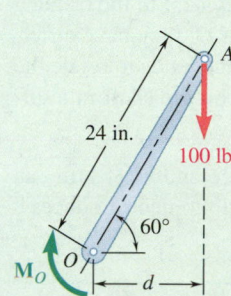

**Fig. 1**   Determination of the moment of the 100-lb force about $O$ using perpendicular distance $d$.

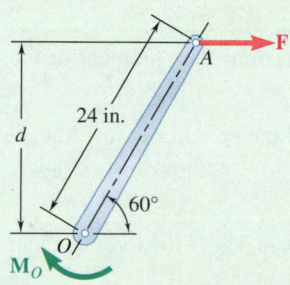

**Fig. 2**   Determination of horizontal force at $A$ that creates same moment about $O$.

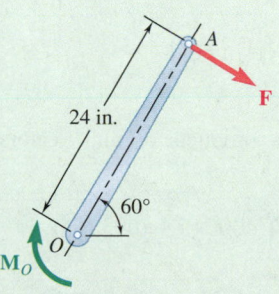

**Fig. 3**   Determination of smallest force at $A$ that creates same moment about $O$.

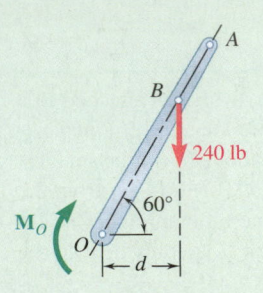

**Fig. 4** Position of vertical 240-lb force that creates same moment about $O$.

**d. 240-lb Vertical Force.** In this case (Fig. 4), $M_O = Fd$ yields

$$1200 \text{ lb·in.} = (240 \text{ lb})d \qquad d = 5 \text{ in.}$$

but

$$OB \cos 60° = d$$

so

$$OB = 10 \text{ in.} \ \blacktriangleleft$$

**e.** None of the forces considered in parts *b*, *c*, or *d* is equivalent to the original 100-lb force. Although they have the same moment about $O$, they have different *x* and *y* components. In other words, although each force tends to rotate the shaft in the same direction, each causes the lever to pull on the shaft in a different way.

**REFLECT and THINK:** Various combinations of force and lever arm can produce equivalent moments, but the system of force and moment produces a different overall effect in each case.

---

## Sample Problem 3.2

A force of 800 N acts on a bracket as shown. Determine the moment of the force about $B$.

**STRATEGY:** You can resolve both the force and the position vector from $B$ to $A$ into rectangular components and then use a vector approach to complete the solution.

**MODELING and ANALYSIS:** Obtain the moment $\mathbf{M}_B$ of the force $\mathbf{F}$ about $B$ by forming the vector product

$$\mathbf{M}_B = \mathbf{r}_{A/B} \times \mathbf{F}$$

where $\mathbf{r}_{A/B}$ is the vector drawn from $B$ to $A$ (Fig. 1). Resolving $\mathbf{r}_{A/B}$ and $\mathbf{F}$ into rectangular components, you have

$$\mathbf{r}_{A/B} = -(0.2 \text{ m})\mathbf{i} + (0.16 \text{ m})\mathbf{j}$$
$$\mathbf{F} = (800 \text{ N}) \cos 60°\mathbf{i} + (800 \text{ N}) \sin 60°\mathbf{j}$$
$$= (400 \text{ N})\mathbf{i} + (693 \text{ N})\mathbf{j}$$

Recalling the relations in Eq. (3.7) for the cross products of unit vectors (Sec. 3.1D), you obtain

$$\mathbf{M}_B = \mathbf{r}_{A/B} \times \mathbf{F} = [-(0.2 \text{ m})\mathbf{i} + (0.16 \text{ m})\mathbf{j}] \times [(400 \text{ N})\mathbf{i} + (693 \text{ N})\mathbf{j}]$$
$$= -(138.6 \text{ N·m})\mathbf{k} - (64.0 \text{ N·m})\mathbf{k}$$
$$= -(202.6 \text{ N·m})\mathbf{k} \qquad \mathbf{M}_B = 203 \text{ N·m} \ \circlearrowright \ \blacktriangleleft$$

The moment $\mathbf{M}_B$ is a vector perpendicular to the plane of the figure and pointing *into* the page.

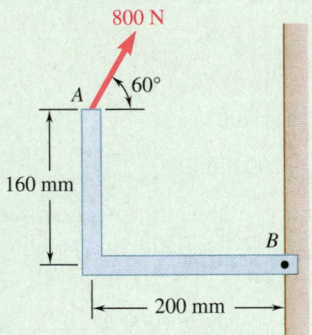

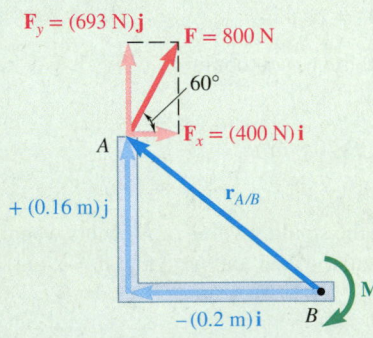

**Fig. 1** The moment $\mathbf{M}_B$ is determined from the vector product of position vector $\mathbf{r}_{A/B}$ and force vector $\mathbf{F}$.

*(continued)*

**REFLECT and THINK:** We can also use a scalar approach to solve this problem using the components for the force **F** and the position vector $\mathbf{r}_{A/B}$. Following the right-hand rule for assigning signs, we have

$$+\circlearrowleft M_B = \Sigma M_B = \Sigma Fd = -(400\text{ N})(0.16\text{ m}) - (693\text{ N})(0.2\text{ m}) = -202.6\text{ N·m}$$

$$\mathbf{M}_B = 203\text{ N·m} \circlearrowright \quad \triangleleft$$

## Sample Problem 3.3

A 30-lb force acts on the end of the 3-ft lever as shown. Determine the moment of the force about $O$.

**STRATEGY:** Resolving the force into components that are perpendicular and parallel to the axis of the lever greatly simplifies the moment calculation.

**MODELING and ANALYSIS:** Replace the force by two components: one component **P** in the direction of $OA$ and one component **Q** perpendicular to $OA$ (Fig. 1). Because $O$ is on the line of action of **P**, the moment of **P** about $O$ is zero. Thus, the moment of the 30-lb force reduces to the moment of **Q**, which is clockwise and can be represented by a negative scalar.

$$Q = (30\text{ lb})\sin 20° = 10.26\text{ lb}$$
$$M_O = -Q(3\text{ ft}) = -(10.26\text{ lb})(3\text{ ft}) = -30.8\text{ lb·ft}$$

Because the value obtained for the scalar $M_O$ is negative, the moment $\mathbf{M}_O$ points *into* the page. You can write it as

$$\mathbf{M}_O = 30.8\text{ lb·ft} \circlearrowright \quad \triangleleft$$

**REFLECT and THINK:** Always be alert for simplifications that can reduce the amount of computation.

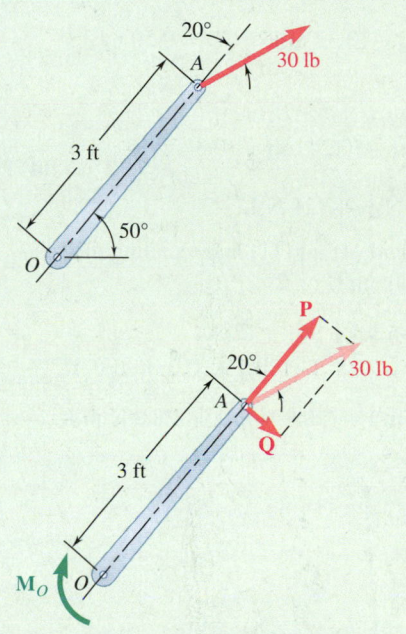

**Fig. 1** 30-lb force at $A$ resolved into components **P** and **Q** to simplify the determination of the moment $\mathbf{M}_O$.

## Sample Problem 3.4

A rectangular plate is supported by brackets at $A$ and $B$ and by a wire $CD$. If the tension in the wire is 200 N, determine the moment about $A$ of the force exerted by the wire on point $C$.

**STRATEGY:** The solution requires resolving the tension in the wire and the position vector from $A$ to $C$ into rectangular components. You will need a unit vector approach to determine the force components.

**MODELING and ANALYSIS:** Obtain the moment $\mathbf{M}_A$ about $A$ of the force **F** exerted by the wire on point $C$ by forming the vector product

$$\mathbf{M}_A = \mathbf{r}_{C/A} \times \mathbf{F} \quad (1)$$

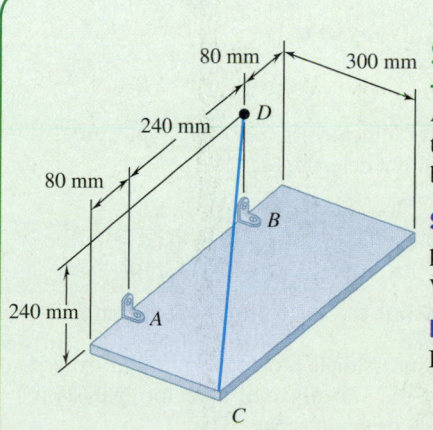

*(continued)*

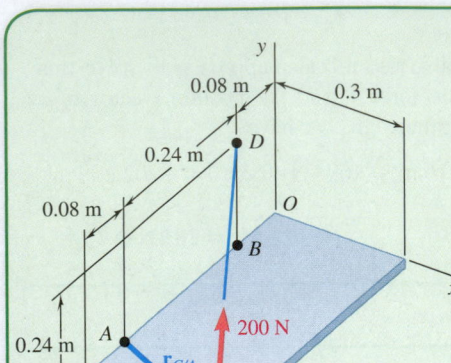

**Fig. 1** The moment $\mathbf{M}_A$ is determined from position vector $\mathbf{r}_{C/A}$ and force vector $\mathbf{F}$.

where $\mathbf{r}_{C/A}$ is the vector from $A$ to $C$

$$\mathbf{r}_{C/A} = \overrightarrow{AC} = (0.3 \text{ m})\mathbf{i} + (0.08 \text{ m})\mathbf{k} \qquad (2)$$

and $\mathbf{F}$ is the 200-N force directed along $CD$ (Fig. 1). Introducing the unit vector

$$\lambda = \overrightarrow{CD}/CD,$$

you can express $\mathbf{F}$ as

$$\mathbf{F} = F\lambda = (200 \text{ N})\frac{\overrightarrow{CD}}{CD} \qquad (3)$$

Resolving the vector $\overrightarrow{CD}$ into rectangular components, you have

$$\overrightarrow{CD} = -(0.3 \text{ m})\mathbf{i} + (0.24 \text{ m})\mathbf{j} - (0.32 \text{ m})\mathbf{k} \quad CD = 0.50 \text{ m}$$

Substituting into (3) gives you

$$\mathbf{F} = \frac{200 \text{ N}}{0.50 \text{ m}} [-(0.3 \text{ m})\mathbf{i} + (0.24 \text{ m})\mathbf{j} - (0.32 \text{ m})\mathbf{k}] \qquad (4)$$

$$= -(120 \text{ N})\mathbf{i} + (96 \text{ N})\mathbf{j} - (128 \text{ N})\mathbf{k}$$

Substituting for $\mathbf{r}_{C/A}$ and $\mathbf{F}$ from (2) and (4) into (1) and recalling the relations in Eq. (3.7) of Sec. 3.1D, you obtain (Fig. 2)

$$\mathbf{M}_A = \mathbf{r}_{C/A} \times \mathbf{F} = (0.3\mathbf{i} + 0.08\mathbf{k}) \times (-120\mathbf{i} + 96\mathbf{j} - 128\mathbf{k})$$

$$= (0.3)(96)\mathbf{k} + (0.3)(-128)(-\mathbf{j}) + (0.08)(-120)\mathbf{j} + (0.08)(96)(-\mathbf{i})$$

$$\mathbf{M}_A = -(7.68 \text{ N·m})\mathbf{i} + (28.8 \text{ N·m})\mathbf{j} + (28.8 \text{ N·m})\mathbf{k} \quad \blacktriangleleft$$

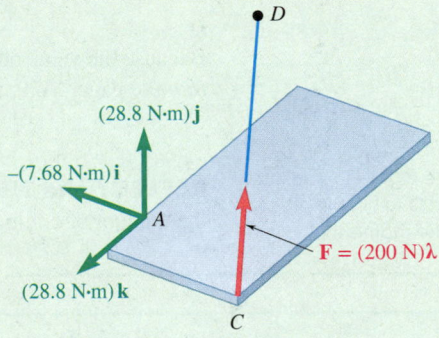

**Fig. 2** Components of moment $\mathbf{M}_A$ applied at $A$.

**ALTERNATIVE SOLUTION.** As indicated in Sec. 3.1F, you can also express the moment $\mathbf{M}_A$ in the form of a determinant:

$$\mathbf{M}_A = \begin{vmatrix} \mathbf{i} & \mathbf{j} & \mathbf{k} \\ x_C - x_A & y_C - y_A & z_C - z_A \\ F_x & F_y & F_z \end{vmatrix} = \begin{vmatrix} \mathbf{i} & \mathbf{j} & \mathbf{k} \\ 0.3 & 0 & 0.08 \\ -120 & 96 & -128 \end{vmatrix}$$

$$\mathbf{M}_A = -(7.68 \text{ N·m})\mathbf{i} + (28.8 \text{ N·m})\mathbf{j} + (28.8 \text{ N·m})\mathbf{k} \quad \blacktriangleleft$$

**REFLECT and THINK:** Two-dimensional problems often are solved easily using a scalar approach, but the versatility of a vector analysis is quite apparent in a three-dimensional problem such as this.

# SOLVING PROBLEMS ON YOUR OWN

In this section, we introduced the *vector product* or *cross product* of two vectors. In the following problems, you will use the vector product to compute the *moment of a force about a point* and also to determine the *perpendicular distance* from a point to a line.

We defined the moment of the force $\mathbf{F}$ about the point $O$ of a rigid body as

$$\mathbf{M}_O = \mathbf{r} \times \mathbf{F} \qquad (3.11)$$

where $\mathbf{r}$ is the position vector *from $O$ to any point* on the line of action of $\mathbf{F}$. Because the vector product is not commutative, it is absolutely necessary when computing such a product that you place the vectors in the proper order and that each vector have the correct sense. The moment $\mathbf{M}_O$ is important because its magnitude is a measure of the tendency of the force $\mathbf{F}$ to cause the rigid body to rotate about an axis directed along $\mathbf{M}_O$.

**1. Computing the moment $\mathbf{M}_O$ of a force in two dimensions.** You can use one of the following procedures:

   **a.** Use Eq. (3.12), $M_O = Fd$, which expresses the magnitude of the moment as the product of the magnitude of $\mathbf{F}$ and the *perpendicular distance $d$* from $O$ to the line of action of $\mathbf{F}$ (Sample Prob. 3.1).

   **b.** Express $\mathbf{r}$ and $\mathbf{F}$ in component form and formally evaluate the vector product $\mathbf{M}_O = \mathbf{r} \times \mathbf{F}$ (Sample Prob. 3.2).

   **c.** Resolve $\mathbf{F}$ into components respectively parallel and perpendicular to the position vector $\mathbf{r}$. Only the perpendicular component contributes to the moment of $\mathbf{F}$ (Sample Prob. 3.3).

   **d.** Use Eq. (3.22), $M_O = M_z = xF_y - yF_x$. When applying this method, the simplest approach is to treat the scalar components of $\mathbf{r}$ and $\mathbf{F}$ as positive and then to assign, by observation, the proper sign to the moment produced by each force component (Sample Prob. 3.2).

**2. Computing the moment $\mathbf{M}_O$ of a force $\mathbf{F}$ in three dimensions.** Following the method of Sample Prob. 3.4, the first step in the calculation is to select the most convenient (simplest) position vector $\mathbf{r}$. You should next express $\mathbf{F}$ in terms of its rectangular components. The final step is to evaluate the vector product $\mathbf{r} \times \mathbf{F}$ to determine the moment. In most three-dimensional problems, you will find it easiest to calculate the vector product using a determinant.

**3. Determining the perpendicular distance $d$ from a point $A$ to a given line.** First, assume that a force $\mathbf{F}$ of known magnitude $F$ lies along the given line. Next, determine its moment about $A$ by forming the vector product $\mathbf{M}_A = \mathbf{r} \times \mathbf{F}$, and calculate this product as indicated earlier. Then, compute its magnitude $M_A$. Finally, substitute the values of $F$ and $M_A$ into the equation $M_A = Fd$ and solve for $d$.

# Problems

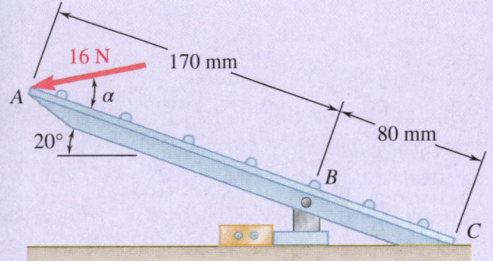

**Fig. P3.1 and P3.2**

**3.1** A foot valve for a pneumatic system is hinged at *B*. Knowing that $\alpha = 28°$, determine the moment of the 16-N force about point *B* by resolving the force into horizontal and vertical components.

**3.2** A foot valve for a pneumatic system is hinged at *B*. Knowing that $\alpha = 28°$, determine the moment of the 16-N force about point *B* by resolving the force into components along *ABC* and in a direction perpendicular to *ABC*.

**3.3** It is known that a vertical force of 200 lb is required to remove the nail at *C* from the board. As the nail first starts moving, determine (*a*) the moment about *B* of the force exerted on the nail, (*b*) the magnitude of the force **P** that creates the same moment about *B* if $\alpha = 10°$, (*c*) the smallest force **P** that creates the same moment about *B*.

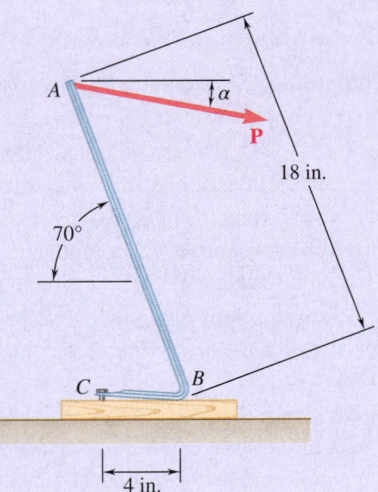

**Fig. *P3.3***

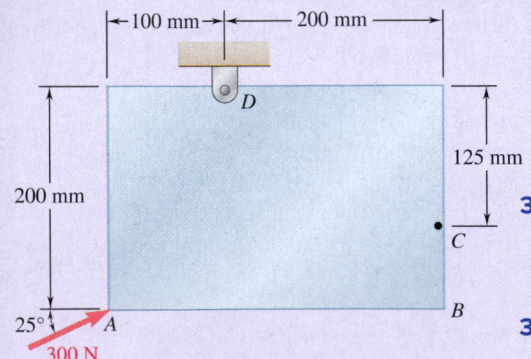

**Fig. P3.4 and P3.5**

**3.4** A 300-N force is applied at *A* as shown. Determine (*a*) the moment of the 300-N force about *D*, (*b*) the smallest force applied at *B* that creates the same moment about *D*.

**3.5** A 300-N force is applied at *A* as shown. Determine (*a*) the moment of the 300-N force about *D*, (*b*) the magnitude and sense of the horizontal force applied at *C* that creates the same moment about *D*, (*c*) the smallest force applied at *C* that creates the same moment about *D*.

**3.6** An 8-lb force **P** is applied to a shift lever. Determine the moment of **P** about B when α is equal to 25°.

**3.7** For the shift lever shown, determine the magnitude and the direction of the smallest force **P** that has a 210-lb·in. clockwise moment about B.

**3.8** An 11-lb force **P** is applied to a shift lever. The moment of **P** about B is clockwise and has a magnitude of 250 lb·in. Determine the value of α.

**3.9** Rod AB is held in place by the cord AC. Knowing that the tension in the cord is 1350 N and that c = 360 mm, determine the moment about B of the force exerted by the cord at point A by resolving that force into horizontal and vertical components applied (a) at point A, (b) at point C.

**3.10** Rod AB is held in place by the cord AC. Knowing that c = 840 mm and that the moment about B of the force exerted by the cord at point A is 756 N·m, determine the tension in the cord.

**3.11 and 3.12** The tailgate of a car is supported by the hydraulic lift BC. If the lift exerts a 125-lb force directed along its centerline on the ball and socket at B, determine the moment of the force about A.

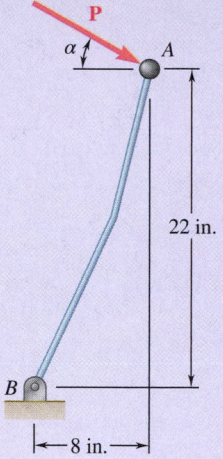

**Fig. P3.6, P3.7, and *P3.8***

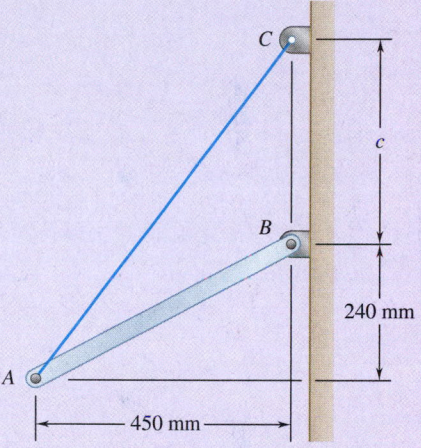

**Fig. P3.9 and *P3.10***

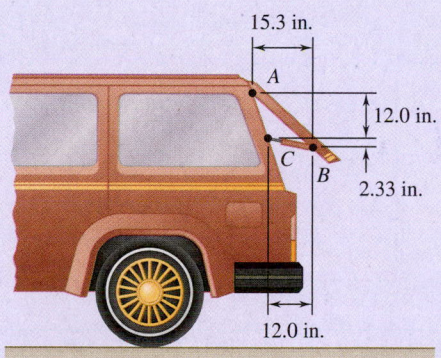

**Fig. P3.11**

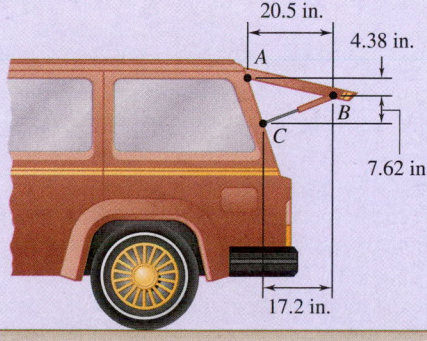

**Fig. P3.12**

**3.13 and 3.14**  It is known that the connecting rod AB exerts on the crank BC a 2.5-kN force directed down and to the left along the centerline of AB. Determine the moment of the force about C.

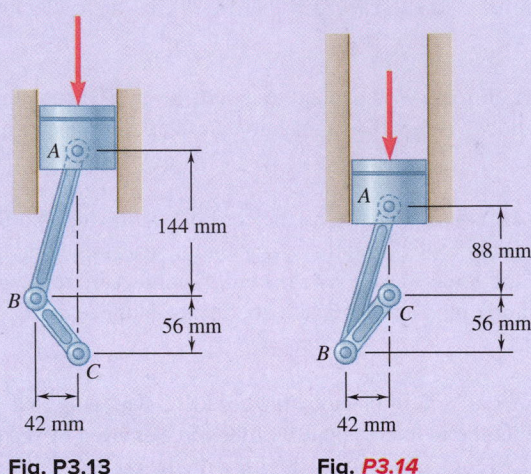

**Fig. P3.13**          **Fig. P3.14**

**3.15**  Form the vector product $\mathbf{P}_1 \times \mathbf{P}_2$ and use the result obtained to prove the identity

$$\sin(\theta_1 - \theta_2) = \sin\theta_1\cos\theta_2 - \cos\theta_1\sin\theta_2$$

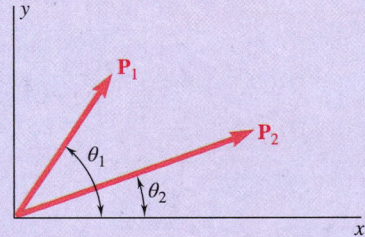

**Fig. P3.15**

**3.16**  The vectors **P** and **Q** are two adjacent sides of a parallelogram. Determine the area of the parallelogram when (a) $\mathbf{P} = -2\mathbf{i} + 2\mathbf{j} - 6\mathbf{k}$ and $\mathbf{Q} = 8\mathbf{i} + 2\mathbf{j} + 4\mathbf{k}$, (b) $\mathbf{P} = 3\mathbf{i} - 9\mathbf{j} - 7\mathbf{k}$ and $\mathbf{Q} = -4\mathbf{i} + 2\mathbf{j} - 5\mathbf{k}$.

**3.17**  A plane contains the vectors **A** and **B**. Determine the unit vector normal to the plane when **A** and **B** are equal to, respectively, (a) $7\mathbf{i} + 8\mathbf{j} - 2\mathbf{k}$ and $9\mathbf{i} - 4\mathbf{j} - 5\mathbf{k}$, (b) $6\mathbf{i} - 3\mathbf{j} + 9\mathbf{k}$ and $-5\mathbf{i} + 4\mathbf{j} - 3\mathbf{k}$.

**3.18**  A line passes through the points (–4 m, –3 m) and (2 m, 7 m). Determine the perpendicular distance d from the line to the origin O of the system of coordinates.

**3.19**  Determine the moment about the origin O of the force $\mathbf{F} = -2\mathbf{i} - 3\mathbf{j} + 5\mathbf{k}$ that acts at a point A. Assume that the position vector of A is (a) $\mathbf{r} = \mathbf{i} + \mathbf{j} + \mathbf{k}$, (b) $\mathbf{r} = 4\mathbf{i} + 6\mathbf{j} - 10\mathbf{k}$, (c) $\mathbf{r} = 4\mathbf{i} + 3\mathbf{j} - 5\mathbf{k}$.

**3.20** Determine the moment about the origin $O$ of the force $\mathbf{F} = 4\mathbf{i} + 10\mathbf{j} + 6\mathbf{k}$ that acts at a point $A$. Assume that the position vector of $A$ is (a) $\mathbf{r} = 2\mathbf{i} - 3\mathbf{j} + 4\mathbf{k}$, (b) $\mathbf{r} = 2\mathbf{i} + 6\mathbf{j} + 3\mathbf{k}$, (c) $\mathbf{r} = 2\mathbf{i} + 5\mathbf{j} + 6\mathbf{k}$.

**3.21** Before the trunk of a large tree is felled, cables $AB$ and $BC$ are attached as shown. Knowing that the tensions in cables $AB$ and $BC$ are 555 N and 660 N, respectively, determine the moment about $O$ of the resultant force exerted on the tree by the cables at $B$.

**3.22** The 12-ft boom $AB$ has a fixed end $A$. A steel cable is stretched from the free end $B$ of the boom to a point $C$ located on the vertical wall. If the tension in the cable is 380 lb, determine the moment about $A$ of the force exerted by the cable at $B$.

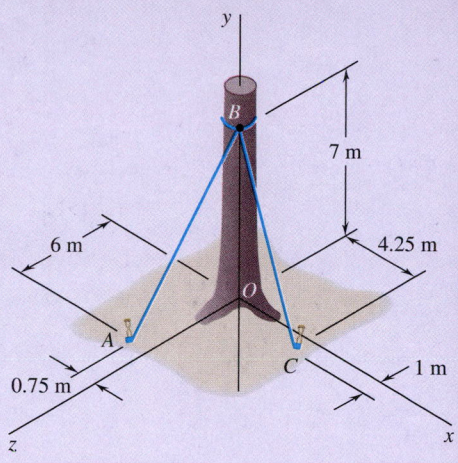

**Fig. P3.21**

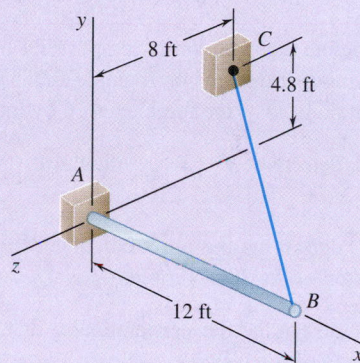

**Fig. P3.22**

**3.23** A 200-N force is applied as shown to the bracket $ABC$. Determine the moment of the force about $A$.

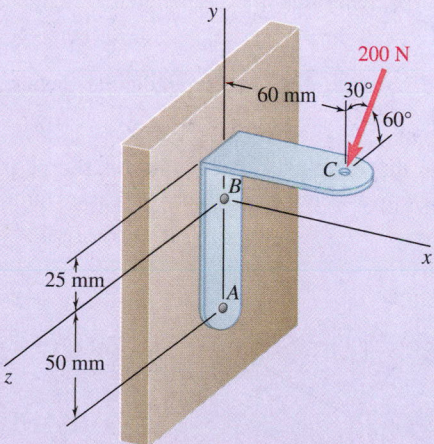

**Fig. P3.23**

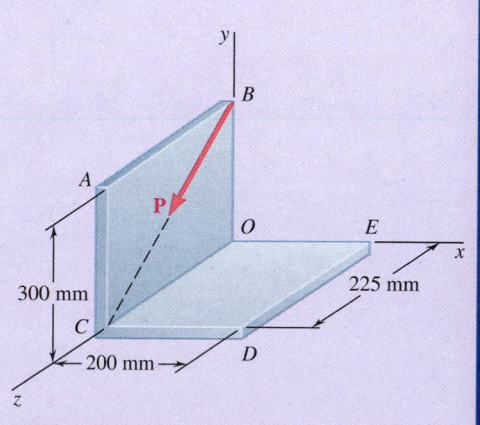

**Fig. P3.24**

**3.24** A force $\mathbf{P}$ of magnitude 200 N acts along the diagonal $BC$ of the bent plate shown. Determine the moment of $\mathbf{P}$ about point $E$.

**3.25** A 6-ft-long fishing rod *AB* is securely anchored in the sand of a beach. After a fish takes the bait, the resulting force in the line is 6 lb. Determine the moment about *A* of the force exerted by the line at *B*.

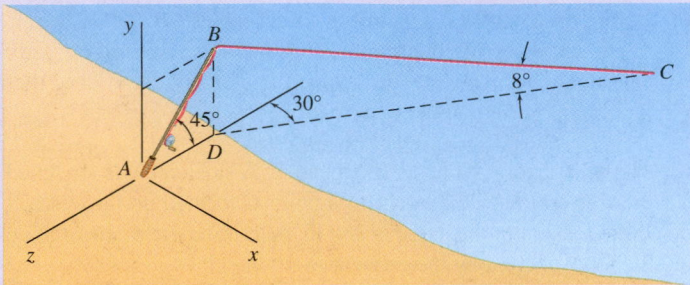

Fig. P3.25

**3.26** A precast concrete wall section is temporarily held by two cables as shown. Knowing that the tension in cable *BD* is 900 N, determine the moment about point *O* of the force exerted by the cable at *B*.

**3.27** In Prob. 3.22, determine the perpendicular distance from point *A* to cable *BC*.

**3.28** In Prob. 3.23, determine the perpendicular distance from point *A* to the line of action of the 200-N force.

**3.29** In Prob. 3.24, determine the perpendicular distance from point *E* to the line of action of force **P**.

**3.30** In Prob. 3.25, determine the perpendicular distance from point *A* to a line drawn through points *B* and *C*.

**3.31** In Prob. 3.25, determine the perpendicular distance from point *D* to a line drawn through points *B* and *C*.

**3.32** In Prob. 3.26, determine the perpendicular distance from point *O* to cable *BD*.

**3.33** In Prob. 3.26, determine the perpendicular distance from point *C* to cable *BD*.

**3.34** Determine the value of *a* that minimizes the perpendicular distance from point *C* to a section of pipeline that passes through points *A* and *B*.

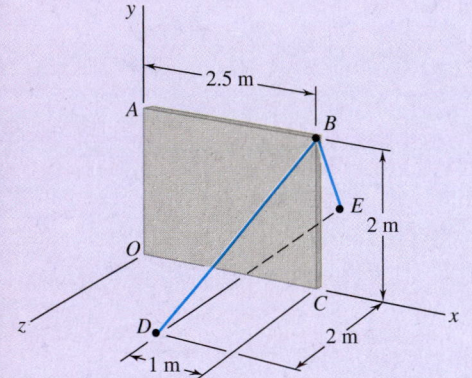

Fig. P3.26

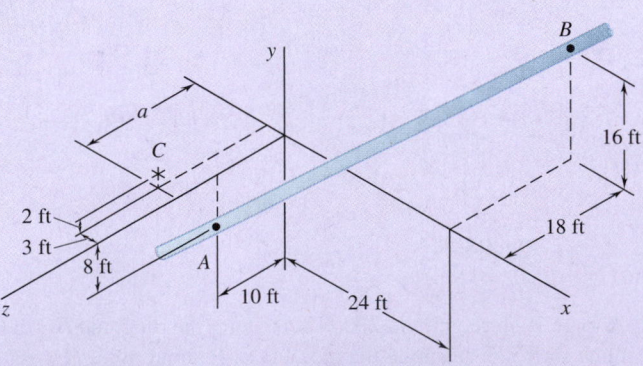

Fig. P3.34

# 3.2 MOMENT OF A FORCE ABOUT AN AXIS

We want to extend the idea of the moment about a point to the often useful concept of the moment about an axis. However, first we need to introduce another tool of vector mathematics. We have seen that the vector product multiplies two vectors together and produces a new vector. Here we examine the scalar product, which multiplies two vectors together and produces a scalar quantity.

## 3.2A Scalar Products

The **scalar product** of two vectors **P** and **Q** is defined as the product of the magnitudes of **P** and **Q** and of the cosine of the angle $\theta$ formed between them (Fig. 3.18). The scalar product of **P** and **Q** is denoted by $\mathbf{P} \cdot \mathbf{Q}$.

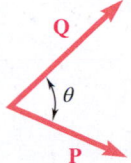

**Fig. 3.18** Two vectors **P** and **Q** and the angle $\theta$ between them.

**Scalar product**
$$\mathbf{P} \cdot \mathbf{Q} = PQ \cos \theta \qquad (3.24)$$

Note that this expression is not a vector but a *scalar*, which explains the name *scalar product*. Because of the notation used, $\mathbf{P} \cdot \mathbf{Q}$ is also referred to as the *dot product* of the vectors **P** and **Q**.

It follows from its very definition that the scalar product of two vectors is *commutative*, i.e., that

$$\mathbf{P} \cdot \mathbf{Q} = \mathbf{Q} \cdot \mathbf{P} \qquad (3.25)$$

It can also be proven that the scalar product is *distributive*, as shown by

$$\mathbf{P} \cdot (\mathbf{Q}_1 + \mathbf{Q}_2) = \mathbf{P} \cdot \mathbf{Q}_1 + \mathbf{P} \cdot \mathbf{Q}_2 \qquad (3.26)$$

As far as the associative property is concerned, this property cannot apply to scalar products. Indeed, $(\mathbf{P} \cdot \mathbf{Q}) \cdot \mathbf{S}$ has no meaning, because $\mathbf{P} \cdot \mathbf{Q}$ is not a vector but a scalar.

We can also express the scalar product of two vectors **P** and **Q** in terms of their rectangular components. Resolving **P** and **Q** into components, we first write

$$\mathbf{P} \cdot \mathbf{Q} = (P_x\mathbf{i} + P_y\mathbf{j} + P_z\mathbf{k}) \cdot (Q_x\mathbf{i} + Q_y\mathbf{j} + Q_z\mathbf{k})$$

Making use of the distributive property, we express $\mathbf{P} \cdot \mathbf{Q}$ as the sum of scalar products, such as $P_x\mathbf{i} \cdot Q_x\mathbf{i}$ and $P_x\mathbf{i} \cdot Q_y\mathbf{j}$. However, from the definition of the scalar product, it follows that the scalar products of the unit vectors are either zero or one.

$$\begin{array}{lll} \mathbf{i} \cdot \mathbf{i} = 1 & \mathbf{j} \cdot \mathbf{j} = 1 & \mathbf{k} \cdot \mathbf{k} = 1 \\ \mathbf{i} \cdot \mathbf{j} = 0 & \mathbf{j} \cdot \mathbf{k} = 0 & \mathbf{k} \cdot \mathbf{i} = 0 \end{array} \qquad (3.27)$$

Thus, the expression for $\mathbf{P} \cdot \mathbf{Q}$ reduces to scalar product

**Scalar product**
$$\mathbf{P} \cdot \mathbf{Q} = P_xQ_x + P_yQ_y + P_zQ_z \qquad (3.28)$$

In the particular case when **P** and **Q** are equal, we note that

$$\mathbf{P} \cdot \mathbf{P} = P_x^2 + P_y^2 + P_z^2 = P^2 \qquad (3.29)$$

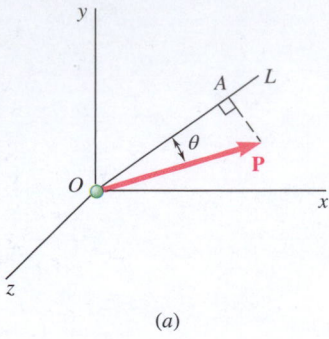

(a)

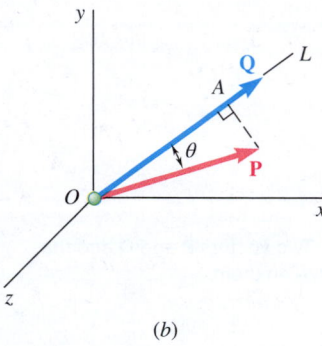

(b)

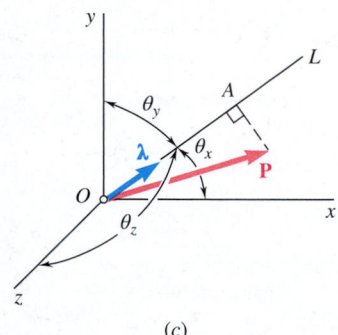

(c)

**Fig. 3.19** (a) The projection of vector **P** at an angle $\theta$ to a line *OL*; (b) the projection of **P** and a vector **Q** along *OL*; (c) the projection of **P**, a unit vector $\lambda$ along *OL*, and the angles of *OL* with the coordinate axes.

## Applications of the Scalar Product

1. **Angle formed by two given vectors.** Let two vectors be given in terms of their components:

$$\mathbf{P} = P_x\mathbf{i} + P_y\mathbf{j} + P_z\mathbf{k}$$

$$\mathbf{Q} = Q_x\mathbf{i} + Q_y\mathbf{j} + Q_z\mathbf{k}$$

To determine the angle formed by the two vectors, we equate the expressions obtained in Eqs. (3.24) and (3.28) for their scalar product,

$$PQ \cos \theta = P_xQ_x + P_yQ_y + P_zQ_z$$

Solving for $\cos \theta$, we have

$$\cos \theta = \frac{P_xQ_x + P_yQ_y + P_zQ_z}{PQ} \tag{3.30}$$

2. **Projection of a vector on a given axis.** Consider a vector **P** forming an angle $\theta$ with an axis, or directed line, *OL* (Fig. 3.19a). We define the *projection of **P** on the axis OL* as the scalar

$$P_{OL} = P \cos \theta \tag{3.31}$$

The projection $P_{OL}$ is equal in absolute value to the length of the segment *OA*. It is positive if *OA* has the same sense as the axis *OL*—that is, if $\theta$ is acute—and negative otherwise. If **P** and *OL* are at a right angle, the projection of **P** on *OL* is zero.

Now consider a vector **Q** directed along *OL* and of the same sense as *OL* (Fig. 3.19b). We can express the scalar product of **P** and **Q** as

$$\mathbf{P} \cdot \mathbf{Q} = PQ \cos \theta = P_{OL}Q \tag{3.32}$$

from which it follows that

$$P_{OL} = \frac{\mathbf{P} \cdot \mathbf{Q}}{Q} = \frac{P_xQ_x + P_yQ_y + P_zQ_z}{Q} \tag{3.33}$$

In the particular case when the vector selected along *OL* is the unit vector $\lambda$ (Fig. 3.19c), we have

$$P_{OL} = \mathbf{P} \cdot \lambda \tag{3.34}$$

Recall from Sec. 2.4A that the components of $\lambda$ along the coordinate axes are respectively equal to the direction cosines of *OL*. Resolving **P** and $\lambda$ into rectangular components, we can express the projection of **P** on *OL* as

$$P_{OL} = P_x \cos \theta_x + P_y \cos \theta_y + P_z \cos \theta_z \tag{3.35}$$

where $\theta_x$, $\theta_y$, and $\theta_z$ denote the angles that the axis *OL* forms with the coordinate axes.

# 3.2B   Mixed Triple Products

We have now seen both forms of multiplying two vectors together: the vector product and the scalar product. Here we define the **mixed triple product** of the three vectors **S**, **P**, and **Q** as the scalar expression

**Mixed triple product**

$$\mathbf{S} \cdot (\mathbf{P} \times \mathbf{Q}) \tag{3.36}$$

This is obtained by forming the scalar product of **S** with the vector product of **P** and **Q**. [In Chap. 15, we will introduce another kind of triple product, called the vector triple product: $\mathbf{S} \times (\mathbf{P} \times \mathbf{Q})$.]

The mixed triple product of **S**, **P**, and **Q** has a simple geometrical interpretation (Fig. 3.20*a*). Recall from Sec. 3.4 that the vector **P** × **Q** is perpendicular to the plane containing **P** and **Q** and that its magnitude is equal to the area of the parallelogram that has **P** and **Q** for sides. Also, Eq. (3.32) indicates that we can obtain the scalar product of **S** and **P** × **Q** by multiplying the magnitude of **P** × **Q** (i.e., the area of the parallelogram defined by **P** and **Q**) by the projection of **S** on the vector **P** × **Q** (i.e., by the projection of **S** on the normal to the plane containing the parallelogram). The mixed triple product is thus equal, in absolute value, to the volume of the parallelepiped having the vectors **S**, **P**, and **Q** for sides (Fig. 3.20*b*). The sign of the mixed triple product is positive if **S**, **P**, and **Q** form a right-handed triad and negative if they form a left-handed triad. [That is, **S** · (**P** × **Q**) is negative if the rotation that brings **P** into line with **Q** is observed as clockwise from the tip of **S**.] The mixed triple product is zero if **S**, **P**, and **Q** are coplanar.

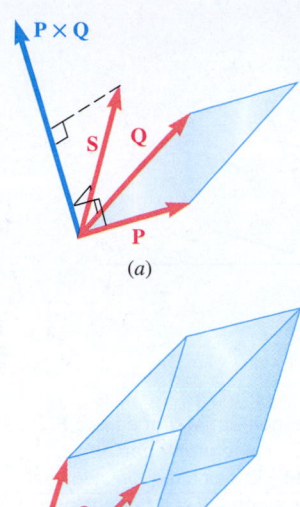

(*a*)

(*b*)

**Fig. 3.20**  (*a*) The mixed triple product is equal to the magnitude of the cross product of two vectors multiplied by the projection of the third vector onto that cross product; (*b*) the result equals the volume of the parallelepiped formed by the three vectors.

Because the parallelepiped defined in this way is independent of the order in which the three vectors are taken, the six mixed triple products that can be formed with **S**, **P**, and **Q** all have the same absolute value, although not the same sign. It can be shown that

$$\mathbf{S} \cdot (\mathbf{P} \times \mathbf{Q}) = \mathbf{P} \cdot (\mathbf{Q} \times \mathbf{S}) \quad = \mathbf{Q} \cdot (\mathbf{S} \times \mathbf{P})$$
$$= -\mathbf{S} \cdot (\mathbf{Q} \times \mathbf{P}) = -\mathbf{P} \cdot (\mathbf{S} \times \mathbf{Q}) = -\mathbf{Q} \cdot (\mathbf{P} \times \mathbf{S}) \quad (3.37)$$

Arranging the letters representing the three vectors counterclockwise in a circle (Fig. 3.21), we observe that the sign of the mixed triple product remains unchanged if the vectors are permuted in such a way that they still read in counterclockwise order. Such a permutation is said to be a *circular permutation*. It also follows from Eq. (3.37) and from the commutative property of scalar products that the mixed triple product of **S**, **P**, and **Q** can be defined equally well as **S** · (**P** × **Q**) or (**S** × **P**) · **Q**.

We can also express the mixed triple product of the vectors **S**, **P**, and **Q** in terms of the rectangular components of these vectors. Denoting **P** × **Q** by **V** and using formula (3.28) to express the scalar product of **S** and **V**, we have

$$\mathbf{S} \cdot (\mathbf{P} \times \mathbf{Q}) = \mathbf{S} \cdot \mathbf{V} = S_x V_x + S_y V_y + S_z V_z$$

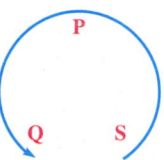

**Fig. 3.21**  Counterclockwise arrangement for determining the sign of the mixed triple product of three vectors: **P**, **Q**, and **S**.

Substituting from the relations in Eq. (3.9) for the components of **V**, we obtain

$$\mathbf{S} \cdot (\mathbf{P} \times \mathbf{Q}) = S_x(P_y Q_z - P_z Q_y) + S_y(P_z Q_x - P_x Q_z) + S_z(P_x Q_y - P_y Q_x) \quad (3.38)$$

We can write this expression in a more compact form if we observe that it represents the expansion of a determinant:

**Mixed triple product, determinant form**

$$\mathbf{S} \cdot (\mathbf{P} \times \mathbf{Q}) = \begin{vmatrix} S_x & S_y & S_z \\ P_x & P_y & P_z \\ Q_x & Q_y & Q_z \end{vmatrix} \quad (3.39)$$

By applying the rules governing the permutation of rows in a determinant, we could verify the relations in Eq. (3.37), which we derived earlier from geometrical considerations.

## 3.2C   Moment of a Force about a Given Axis

Now that we have the necessary mathematical tools, we can introduce the concept of moment of a force about an axis. Consider again a force **F** acting on a rigid body and the moment $\mathbf{M}_O$ of that force about *O* (Fig. 3.22). Let *OL* be an axis through *O*.

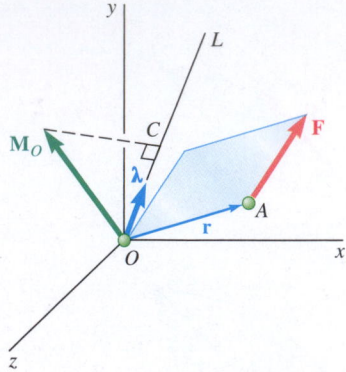

**Fig. 3.22** The moment $\mathbf{M}_{OL}$ of a force $\mathbf{F}$ about the axis $OL$ is the projection on $OL$ of the moment $\mathbf{M}_O$. The calculation involves the unit vector $\boldsymbol{\lambda}$ along $OL$ and the position vector $\mathbf{r}$ from $O$ to $A$, the point upon which the force $\mathbf{F}$ acts.

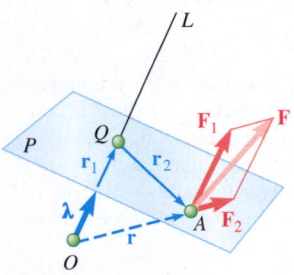

**Fig. 3.23** By resolving the force $\mathbf{F}$ into components parallel to the axis $OL$ and in a plane perpendicular to the axis, we can show that the moment $\mathbf{M}_{OL}$ of $\mathbf{F}$ about $OL$ measures the tendency of $\mathbf{F}$ to rotate the rigid body about the axis.

*We define the moment $M_{OL}$ of $\mathbf{F}$ about OL as the projection OC of the moment $M_O$ onto the axis OL.*

Suppose we denote the unit vector along $OL$ by $\boldsymbol{\lambda}$ and recall the expressions (3.34) and (3.11) for the projection of a vector on a given axis and for the moment $\mathbf{M}_O$ of a force $\mathbf{F}$. Then, we can express $M_{OL}$ as

**Moment about an axis through the origin**

$$M_{OL} = \boldsymbol{\lambda} \cdot \mathbf{M}_O = \boldsymbol{\lambda} \cdot (\mathbf{r} \times \mathbf{F}) \tag{3.40}$$

This shows that the moment $M_{OL}$ of $\mathbf{F}$ about the axis $OL$ is the scalar obtained by forming the mixed triple product of $\boldsymbol{\lambda}$, $\mathbf{r}$, and $\mathbf{F}$. We can also express $M_{OL}$ in the form of a determinant,

$$M_{OL} = \begin{vmatrix} \lambda_x & \lambda_y & \lambda_z \\ x & y & z \\ F_x & F_y & F_z \end{vmatrix} \tag{3.41}$$

where $\lambda_x, \lambda_y, \lambda_z$ = direction cosines of axis $OL$
$x, y, z$ = coordinates of point of application of $\mathbf{F}$
$F_x, F_y, F_z$ = components of force $\mathbf{F}$

The physical significance of the moment $M_{OL}$ of a force $\mathbf{F}$ about a fixed axis $OL$ becomes more apparent if we resolve $\mathbf{F}$ into two rectangular components $\mathbf{F}_1$ and $\mathbf{F}_2$, with $\mathbf{F}_1$ parallel to $OL$ and $\mathbf{F}_2$ lying in a plane $P$ perpendicular to $OL$ (Fig. 3.23). Resolving $\mathbf{r}$ similarly into two components $\mathbf{r}_1$ and $\mathbf{r}_2$ and substituting for $\mathbf{F}$ and $\mathbf{r}$ into Eq. (3.40), we get

$$M_{OL} = \boldsymbol{\lambda} \cdot [(\mathbf{r}_1 + \mathbf{r}_2) \times (\mathbf{F}_1 + \mathbf{F}_2)]$$
$$= \boldsymbol{\lambda} \cdot (\mathbf{r}_1 \times \mathbf{F}_1) + \boldsymbol{\lambda} \cdot (\mathbf{r}_1 \times \mathbf{F}_2) + \boldsymbol{\lambda} \cdot (\mathbf{r}_2 \times \mathbf{F}_1) + \boldsymbol{\lambda} \cdot (\mathbf{r}_2 \times \mathbf{F}_2)$$

Note that all of the mixed triple products except the last one are equal to zero because they involve vectors that are coplanar when drawn from a common origin (Sec. 3.2B). Therefore, this expression reduces to

$$M_{OL} = \boldsymbol{\lambda} \cdot (\mathbf{r}_2 \times \mathbf{F}_2) \tag{3.42}$$

The vector product $\mathbf{r}_2 \times \mathbf{F}_2$ is perpendicular to the plane $P$ and represents the moment of the component $\mathbf{F}_2$ of $\mathbf{F}$ about the point $Q$ where $OL$ intersects $P$. Therefore, the scalar $M_{OL}$, which is positive if $\mathbf{r}_2 \times \mathbf{F}_2$ and $OL$ have the same sense and is negative otherwise, measures the tendency of $\mathbf{F}_2$ to make the rigid body rotate about the fixed axis $OL$. The other component $\mathbf{F}_1$ of $\mathbf{F}$ does not tend to make the body rotate about $OL$, because $\mathbf{F}_1$ and $OL$ are parallel. Therefore, we conclude that

*The moment $M_{OL}$ of $\mathbf{F}$ about OL measures the tendency of the force $\mathbf{F}$ to impart to the rigid body a rotation about the fixed axis OL.*

From the definition of the moment of a force about an axis, it follows that the moment of $\mathbf{F}$ about a coordinate axis is equal to the component of $\mathbf{M}_O$ along that axis. If we substitute each of the unit vectors $\mathbf{i}$, $\mathbf{j}$, and $\mathbf{k}$ for $\boldsymbol{\lambda}$ in Eq. (3.40), we obtain expressions for the *moments of $\mathbf{F}$ about the coordinate axes.* These expressions are respectively equal to those obtained earlier for the components of the moment $\mathbf{M}_O$ of $\mathbf{F}$ about $O$:

$$\begin{aligned} M_x &= yF_z - zF_y \\ M_y &= zF_x - xF_z \\ M_z &= xF_y - yF_x \end{aligned} \tag{3.18}$$

Just as the components $F_x$, $F_y$, and $F_z$ of a force $\mathbf{F}$ acting on a rigid body measure, respectively, the tendency of $\mathbf{F}$ to move the rigid body in the $x$, $y$, and $z$ directions, the moments $M_x$, $M_y$, and $M_z$ of $\mathbf{F}$ about the coordinate axes measure the tendency of $\mathbf{F}$ to impart to the rigid body a rotation about the $x$, $y$, and $z$ axes, respectively.

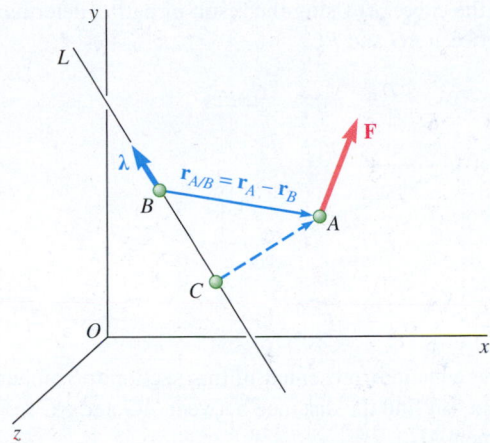

**Fig. 3.24**   The moment of a force about an axis or line $L$ can be found by evaluating the mixed triple product at a point $B$ on the line. The choice of $B$ is arbitrary, because using any other point on the line, such as $C$, yields the same result.

More generally, we can obtain the moment of a force $\mathbf{F}$ applied at $A$ about an axis that does not pass through the origin by choosing an arbitrary point $B$ on the axis (Fig. 3.24) and determining the projection on the axis $BL$ of the moment $\mathbf{M}_B$ of $\mathbf{F}$ about $B$. The equation for this projection is given next.

**Moment about an arbitrary axis**

$$M_{BL} = \boldsymbol{\lambda} \cdot \mathbf{M}_B = \boldsymbol{\lambda} \cdot (\mathbf{r}_{A/B} \times \mathbf{F}) \tag{3.43}$$

where $\mathbf{r}_{A/B} = \mathbf{r}_A - \mathbf{r}_B$ represents the vector drawn from $B$ to $A$. Expressing $M_{BL}$ in the form of a determinant, we have

$$M_{BL} = \begin{vmatrix} \lambda_x & \lambda_y & \lambda_z \\ x_{A/B} & y_{A/B} & z_{A/B} \\ F_x & F_y & F_z \end{vmatrix} \tag{3.44}$$

where   $\lambda_x$, $\lambda_y$, $\lambda_z$ = direction cosines of axis $BL$

$$x_{A/B} = x_A - x_B \quad y_{A/B} = y_A - y_B \quad z_{A/B} = z_A - z_B$$

$F_x, F_y, F_z$ = components of force $\mathbf{F}$

Note that this result is independent of the choice of the point $B$ on the given axis. Indeed, denoting by $M_{CL}$ the moment obtained with a different point $C$, we have

$$M_{CL} = \boldsymbol{\lambda} \cdot [(\mathbf{r}_A - \mathbf{r}_C) \times \mathbf{F}]$$
$$= \boldsymbol{\lambda} \cdot [(\mathbf{r}_A - \mathbf{r}_B) \times \mathbf{F}] + \boldsymbol{\lambda} \cdot [(\mathbf{r}_B - \mathbf{r}_C) \times \mathbf{F}]$$

However, because the vectors $\boldsymbol{\lambda}$ and $\mathbf{r}_B - \mathbf{r}_C$ lie along the same line, the volume of the parallelepiped having the vectors $\boldsymbol{\lambda}$, $\mathbf{r}_B - \mathbf{r}_C$, and $\mathbf{F}$ for sides is zero, as is the mixed triple product of these three vectors (Sec. 3.2B). The expression obtained for $M_{CL}$ thus reduces to its first term, which is the expression used earlier to define $M_{BL}$. In addition, it follows from Sec. 3.1E that, when computing the moment of $\mathbf{F}$ about the given axis, $A$ can be any point on the line of action of $\mathbf{F}$.

## Sample Problem 3.5

A cube of side $a$ is acted upon by a force **P** along the diagonal of a face, as shown. Determine the moment of **P** (*a*) about *A*, (*b*) about the edge *AB*, (*c*) about the diagonal *AG* of the cube. (*d*) Using the result of part *c*, determine the perpendicular distance between *AG* and *FC*.

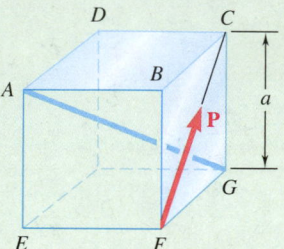

**STRATEGY:**   Use the equations presented in this section to compute the moments asked for. You can find the distance between *AG* and *FC* from the expression for the moment $M_{AG}$.

**MODELING and ANALYSIS:**

**a. Moment about *A*.**   Choosing $x$, $y$, and $z$ axes as shown (Fig. 1), resolve into rectangular components the force **P** and the vector $\mathbf{r}_{F/A} = \overrightarrow{AF}$ drawn from *A* to the point of application *F* of **P**.

$$\mathbf{r}_{F/A} = a\mathbf{i} - a\mathbf{j} = a(\mathbf{i} - \mathbf{j})$$

$$\mathbf{P} = (P/\sqrt{2})\mathbf{j} - (P/\sqrt{2})\mathbf{k} = (P/\sqrt{2})(\mathbf{j} - \mathbf{k})$$

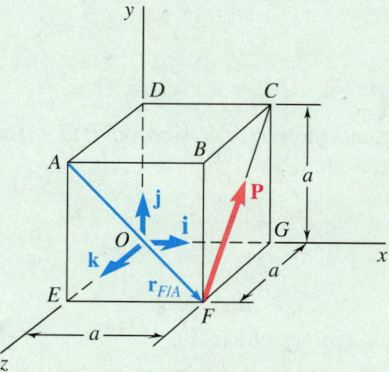

**Fig. 1**   Position vector $\mathbf{r}_{F/A}$ and force vector **P** relative to chosen coordinate system.

The moment of **P** about *A* is the vector product of these two vectors:

$$\mathbf{M}_A = \mathbf{r}_{F/A} \times \mathbf{P} = a(\mathbf{i} - \mathbf{j}) \times (P/\sqrt{2})(\mathbf{j} - \mathbf{k})$$

$$\mathbf{M}_A = (aP/\sqrt{2})(\mathbf{i} + \mathbf{j} + \mathbf{k}) \quad \blacktriangleleft$$

**b. Moment about *AB*.**   You want the projection of $\mathbf{M}_A$ on *AB*:

$$M_{AB} = \mathbf{i} \cdot \mathbf{M}_A = \mathbf{i} \cdot (aP/\sqrt{2})(\mathbf{i} + \mathbf{j} + \mathbf{k})$$

$$M_{AB} = aP/\sqrt{2} \quad \blacktriangleleft$$

*(continued)*

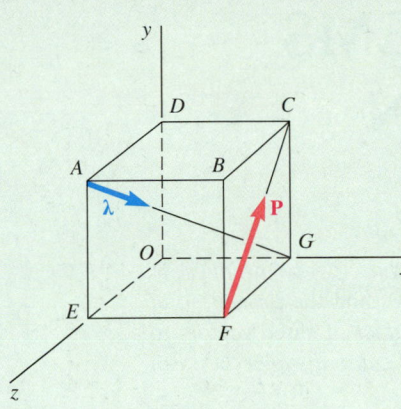

**Fig. 2** Unit vector λ used to determine moment of **P** about *AG*.

You can verify that because *AB* is parallel to the *x* axis, $M_{AB}$ is also the *x* component of the moment **M**$_A$.

**c. Moment about Diagonal *AG*.** You obtain the moment of **P** about *AG* by projecting **M**$_A$ on *AG*. If you denote the unit vector along *AG* by λ (Fig. 2), the calculation looks like this:

$$\lambda = \frac{\overrightarrow{AG}}{AG} = \frac{a\mathbf{i} - a\mathbf{j} - a\mathbf{k}}{a\sqrt{3}} = (1/\sqrt{3})(\mathbf{i} - \mathbf{j} - \mathbf{k})$$

$$M_{AG} = \lambda \cdot \mathbf{M}_A = (1/\sqrt{3})(\mathbf{i} - \mathbf{j} - \mathbf{k}) \cdot (aP/\sqrt{2})(\mathbf{i} + \mathbf{j} + \mathbf{k})$$

$$M_{AG} = (aP/\sqrt{6})(1 - 1 - 1) \qquad M_{AG} = -aP/\sqrt{6}$$

**Alternative Method.** You can also calculate the moment of **P** about *AG* from the determinant form:

$$M_{A/G} = \begin{vmatrix} \lambda_x & \lambda_y & \lambda_z \\ x_{F/A} & y_{F/A} & z_{F/A} \\ F_x & F_y & F_z \end{vmatrix} = \begin{vmatrix} 1/\sqrt{3} & -1/\sqrt{3} & -1/\sqrt{3} \\ a & -a & 0 \\ 0 & P/\sqrt{2} & -P/\sqrt{2} \end{vmatrix} = -aP/\sqrt{6}$$

**d. Perpendicular Distance between *AG* and *FC*.** First note that **P** is perpendicular to the diagonal *AG*. You can check this by forming the scalar product **P** · λ and verifying that it is zero:

$$\mathbf{P} \cdot \lambda = (P/\sqrt{2})(\mathbf{j} - \mathbf{k}) \cdot (1/\sqrt{3})(\mathbf{i} - \mathbf{j} - \mathbf{k}) = (P\sqrt{6})(0 - 1 + 1) = 0$$

You can then express the moment $M_{AG}$ as $-Pd$, where *d* is the perpendicular distance from *AG* to *FC* (Fig. 3). (The negative sign is needed because the rotation imparted to the cube by **P** appears as clockwise to an observer at *G*.) Using the value found for $M_{AG}$ in part *c*,

$$M_{AG} = -Pd = -aP/\sqrt{6} \qquad d = a/\sqrt{6}$$

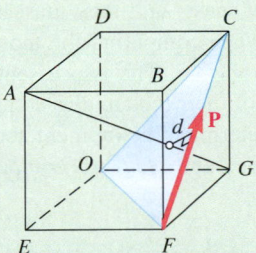

**Fig. 3** Perpendicular distance *d* from *AG* to *FC*.

**REFLECT and THINK:** In a problem like this, it is important to visualize the forces and moments in three dimensions so you can choose the appropriate equations for finding them and also recognize the geometric relationships between them.

# SOLVING PROBLEMS ON YOUR OWN

In the problems for this section, you will apply the *scalar product* (or *dot product*) of two vectors to determine the *angle formed by two given vectors* and the *projection of a force on a given axis*. You will also use the *mixed triple product* of three vectors to find the *moment of a force about a given axis* and the *perpendicular distance between two lines*.

**1. Calculating the angle formed by two given vectors.** First express the vectors in terms of their components and determine the magnitudes of the two vectors. Then, find the cosine of the desired angle by dividing the scalar product of the two vectors by the product of their magnitudes [Eq. (3.30)].

**2. Computing the projection of a vector P on a given axis *OL*.** In general, begin by expressing **P** and the unit vector $\lambda$, which defines the direction of the axis, in component form. Take care that $\lambda$ has the correct sense (i.e., $\lambda$ is directed from $O$ to $L$). The required projection is then equal to the scalar product **P · $\lambda$**. However, if you know the angle $\theta$ formed by **P** and $\lambda$, the projection is also given by $P \cos \theta$.

**3. Determining the moment M$_{OL}$ of a force about a given axis *OL*.** We defined $M_{OL}$ as

$$M_{OL} = \lambda \cdot \mathbf{M}_O = \lambda \cdot (\mathbf{r} \times \mathbf{F}) \qquad \qquad (3.40)$$

where $\lambda$ is the unit vector along *OL* and **r** is a position vector *from any point* on the line *OL to any point* on the line of action of **F**. As was the case for the moment of a force about a point, choosing the most convenient position vector will simplify your calculations. Also, recall the warning of the preceding section: The vectors **r** and **F** must have the correct sense, and they must be placed in the proper order. The procedure you should follow when computing the moment of a force about an axis is illustrated in part *c* of Sample Prob. 3.5. The two essential steps in this procedure are (1) express $\lambda$, **r**, and **F** in terms of their rectangular components and (2) evaluate the mixed triple product $\lambda \cdot (\mathbf{r} \times \mathbf{F})$ to determine the moment about the axis. In most three-dimensional problems, the most convenient way to compute the mixed triple product is by using a determinant.

As noted in the text, when $\lambda$ is directed along one of the coordinate axes, $M_{OL}$ is equal to the scalar component of **M**$_O$ along that axis.

**4. Determining the perpendicular distance between two lines.** Remember that it is the perpendicular component **F**$_2$ of the force **F** that tends to make a body rotate about a given axis *OL* (Fig. 3.23). It then follows that

$$M_{OL} = F_2 d$$

where $M_{OL}$ is the moment of **F** about axis *OL* and $d$ is the perpendicular distance between *OL* and the line of action of **F**. This last equation provides a simple technique for

*(continued)*

determining $d$. First, assume that a force $\mathbf{F}$ of known magnitude $F$ lies along one of the given lines and that the unit vector $\boldsymbol{\lambda}$ lies along the other line. Next, compute the moment $M_{OL}$ of the force $\mathbf{F}$ about the second line using the method discussed earlier. The magnitude of the parallel component, $F_1$, of $\mathbf{F}$ is obtained using the scalar product:

$$F_1 = \mathbf{F} \cdot \boldsymbol{\lambda}$$

The value of $F_2$ is then determined from

$$F_2 = \sqrt{F^2 - F_1^2}$$

Finally, substitute the values of $M_{OL}$ and $F_2$ into the equation $M_{OL} = F_2 d$ and solve for $d$.

You should now realize that the calculation of the perpendicular distance in part $d$ of Sample Prob. 3.5 was simplified by $\mathbf{P}$ being perpendicular to the diagonal $AG$. In general, the two given lines will not be perpendicular, so you will have to use the technique just outlined when determining the perpendicular distance between them.

# Problems

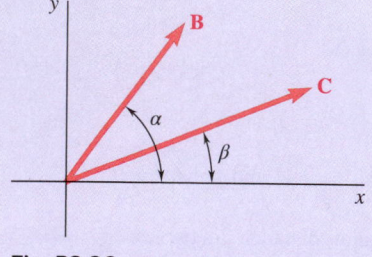

**Fig. P3.36**

**3.35** Given the vectors $\mathbf{P} = 2\mathbf{i} + \mathbf{j} + 2\mathbf{k}$, $\mathbf{Q} = 3\mathbf{i} + 4\mathbf{j} - 5\mathbf{k}$, and $\mathbf{S} = -4\mathbf{i} + \mathbf{j} - 2\mathbf{k}$, compute the scalar products $\mathbf{P} \cdot \mathbf{Q}$, $\mathbf{P} \cdot \mathbf{S}$, and $\mathbf{Q} \cdot \mathbf{S}$.

**3.36** Form the scalar product $\mathbf{B} \cdot \mathbf{C}$ and use the result obtained to prove the identity

$$\cos(\alpha - \beta) = \cos \alpha \cos \beta + \sin \alpha \sin \beta$$

**3.37** Three cables are attached to the top of the tower at $A$. Determine the angle formed by cables $AB$ and $AC$.

**3.38** Three cables are attached to the top of the tower at $A$. Determine the angle formed by cables $AD$ and $AB$.

**3.39** Knowing that the tension in cable $AC$ is 280 lb, determine (a) the angle between cable $AC$ and the boom $AB$, (b) the projection on $AB$ of the force exerted by cable $AC$ at point $A$.

**3.40** Knowing that the tension in cable $AD$ is 180 lb, determine (a) the angle between cable $AD$ and the boom $AB$, (b) the projection on $AB$ of the force exerted by cable $AD$ at point $A$.

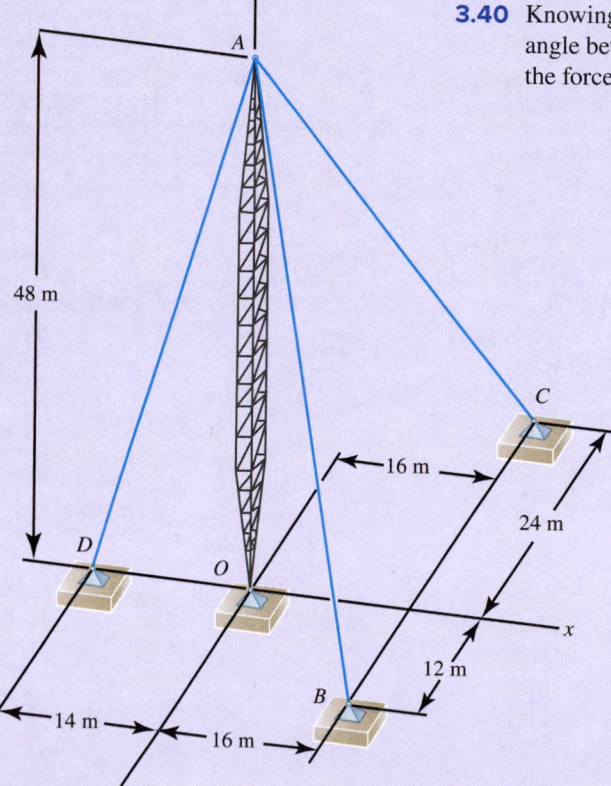

**Fig. P3.37 and P3.38**

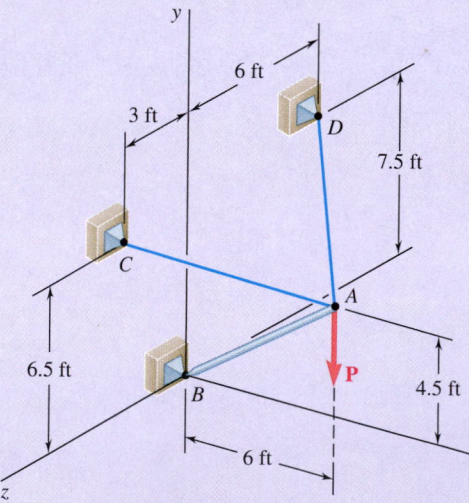

**Fig. P3.39 and P3.40**

**3.41** Ropes *AB* and *BC* are two of the ropes used to support a tent. The two ropes are attached to a stake at *B*. If the tension in rope *AB* is 540 N, determine (*a*) the angle between rope *AB* and the stake, (*b*) the projection on the stake of the force exerted by rope *AB* at point *B*.

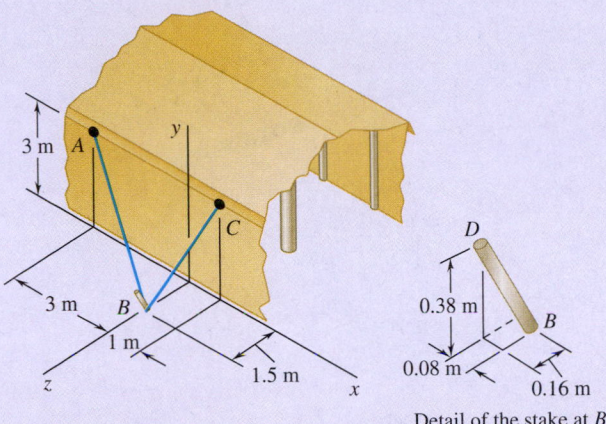

Detail of the stake at *B*

**Fig. P3.41 and *P3.42***

**3.42** Ropes *AB* and *BC* are two of the ropes used to support a tent. The two ropes are attached to a stake at *B*. If the tension in rope *BC* is 490 N, determine (*a*) the angle between rope *BC* and the stake, (*b*) the projection on the stake of the force exerted by rope *BC* at point *B*.

**3.43** The 20-in. tube *AB* can slide along a horizontal rod. The ends *A* and *B* of the tube are connected by elastic cords to the fixed point *C*. For the position corresponding to $x = 11$ in., determine the angle formed by the two cords, (*a*) using Eq. (3.30), (*b*) applying the law of cosines to triangle *ABC*.

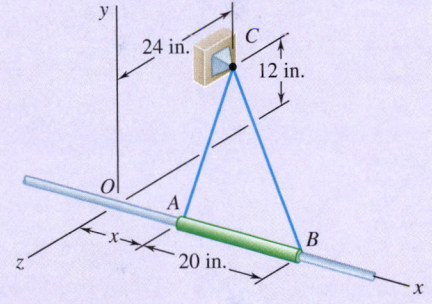

**Fig. P3.43**

**3.44** Solve Prob. 3.43 for the position corresponding to $x = 4$ in.

**3.45** Determine the volume of the parallelepiped of Fig. 3.20*b* when
(*a*) $\mathbf{P} = 2\mathbf{i} - 4\mathbf{j} + 3\mathbf{k}$, $\mathbf{Q} = -5\mathbf{i} - \mathbf{j} + 2\mathbf{k}$, and $\mathbf{S} = \mathbf{i} - 7\mathbf{j} - \mathbf{k}$
(*b*) $\mathbf{P} = 6\mathbf{i} - 5\mathbf{j} - 2\mathbf{k}$, $\mathbf{Q} = \mathbf{i} + 2\mathbf{j} + 3\mathbf{k}$, and $\mathbf{S} = -4\mathbf{i} - 3\mathbf{j} + 2\mathbf{k}$

**3.46** Given the vectors $\mathbf{P} = 3\mathbf{i} + 2\mathbf{j} + \mathbf{k}$, $\mathbf{Q} = 5\mathbf{i} + \mathbf{j} - 2\mathbf{k}$, and $\mathbf{S} = \mathbf{i} + 3\mathbf{j} + S_z\mathbf{k}$, determine the value of $S_z$ for which the three vectors are coplanar.

**3.47** A crane is oriented so that the end of the 25-m boom *AO* lies in the *yz* plane. At the instant shown, the tension in cable *AB* is 4 kN. Determine the moment about each of the coordinate axes of the force exerted on *A* by cable *AB*.

**3.48** The 25-m crane boom *AO* lies in the *yz* plane. Determine the maximum permissible tension in cable *AB* if the absolute value of moments about the coordinate axes of the force exerted on *A* by cable *AB* must be

$$|M_x| \leq 60 \text{ kN·m}, |M_y| \leq 12 \text{ kN·m}, |M_z| \leq 8 \text{ kN·m}$$

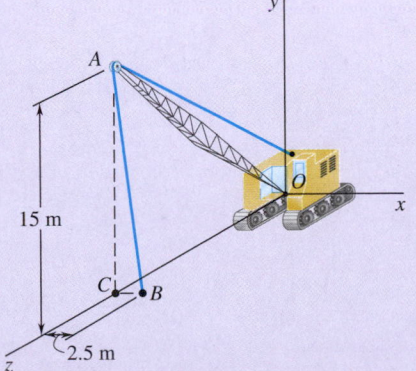

**Fig. P3.47 and P3.48**

**3.49** To loosen a frozen valve, a force **F** with a magnitude of 70 lb is applied to the handle of the valve. Knowing that $\theta = 25°$, $M_x = -61$ lb·ft, and $M_z = -43$ lb·ft, determine $\phi$ and $d$.

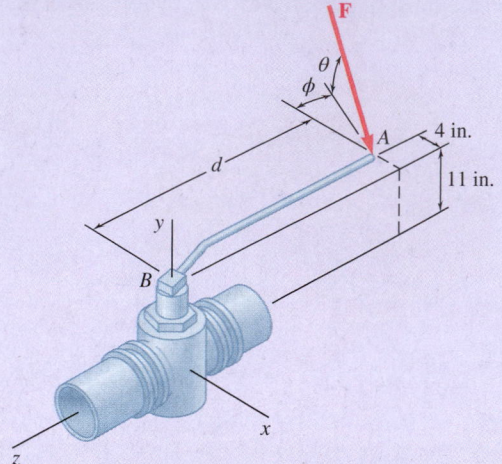

**Fig. P3.49 and *P3.50***

**3.50** When a force **F** is applied to the handle of the valve shown, its moments about the $x$ and $z$ axes are $M_x = -77$ lb·ft and $M_z = -81$ lb·ft, respectively. For $d = 27$ in., determine the moment $M_y$ of **F** about the $y$ axis.

**3.51** The $0.61 \times 1.00$-m lid *ABCD* of a storage bin is hinged along side *AB* and is held open by looping cord *DEC* over a frictionless hook at *E*. If the tension in the cord is 66 N, determine the moment about each of the coordinate axes of the force exerted by the cord at *D*.

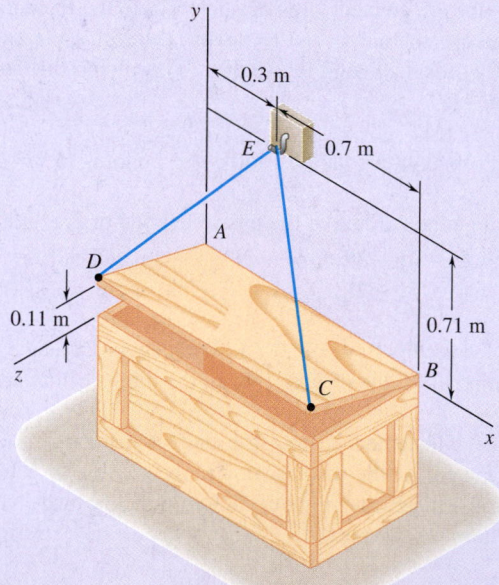

**Fig. P3.51 and P3.52**

**3.52** The $0.61 \times 1.00$-m lid *ABCD* of a storage bin is hinged along side *AB* and is held open by looping cord *DEC* over a frictionless hook at *E*. If the tension in the cord is 66 N, determine the moment about each of the coordinate axes of the force exerted by the cord at *C*.

**3.53** A farmer uses cables and winch pullers $B$ and $E$ to plumb one side of a small barn. If it is known that the sum of the moments about the $x$ axis of the forces exerted by the cables on the barn at points $A$ and $D$ is equal to 4728 lb·ft, determine the magnitude of $\mathbf{T}_{DE}$ when $T_{AB} = 255$ lb.

**3.54** Solve Prob. 3.53 when the tension in cable $AB$ is 306 lb.

**3.55** A force $\mathbf{P}$ of magnitude 520 lb acts on the frame shown at point $E$. Determine the moment of $\mathbf{P}$ about a line joining points $O$ and $D$.

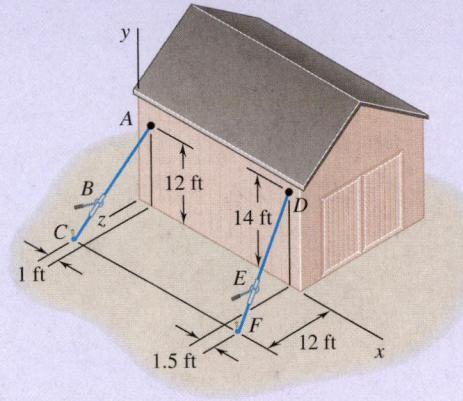

**Fig. P3.53**

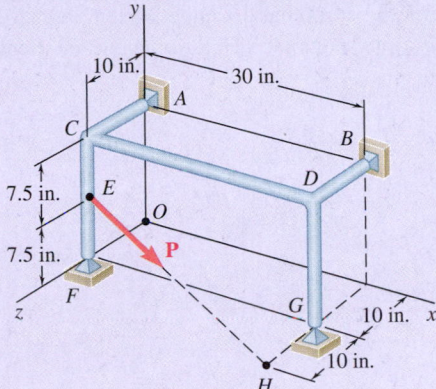

**Fig. P3.55 and P3.56**

**3.56** A force $\mathbf{P}$ acts on the frame shown at point $E$. Knowing that the absolute value of the moment of $\mathbf{P}$ about a line joining points $F$ and $B$ is 300 lb·ft, determine the magnitude of the force $\mathbf{P}$.

**3.57** The frame $ACD$ is hinged at $A$ and $D$ and is supported by a cable that passes through a ring at $B$ and is attached to hooks at $G$ and $H$. Knowing that the tension in the cable is 450 N, determine the moment about the diagonal $AD$ of the force exerted on the frame by portion $BH$ of the cable.

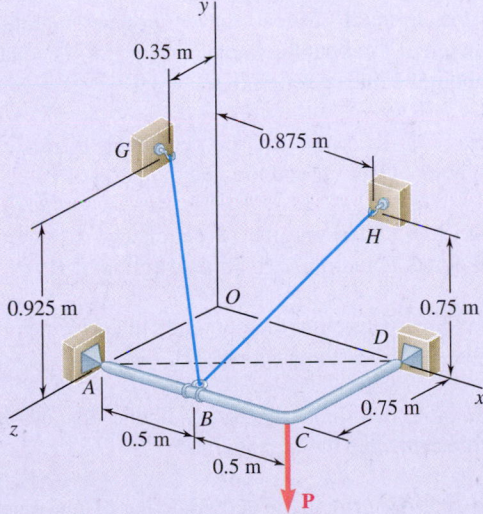

**Fig. P3.57**

**3.58** In Prob. 3.57, determine the moment about the diagonal $AD$ of the force exerted on the frame by portion $BG$ of the cable.

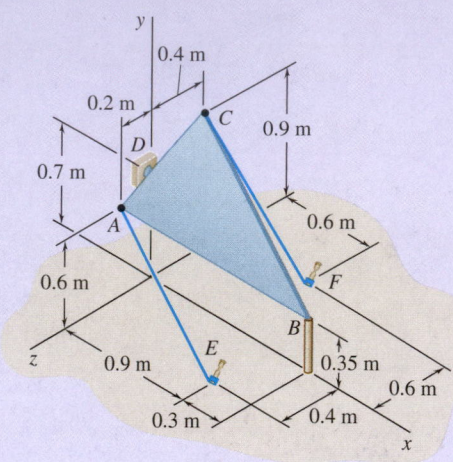

**Fig. P3.59 and P3.60**

**3.59** The triangular plate $ABC$ is supported by ball-and-socket joints at $B$ and $D$ and is held in the position shown by cables $AE$ and $CF$. If the force exerted by cable $AE$ at $A$ is 55 N, determine the moment of that force about the line joining points $D$ and $B$.

**3.60** The triangular plate $ABC$ is supported by ball-and-socket joints at $B$ and $D$ and is held in the position shown by cables $AE$ and $CF$. If the force exerted by cable $CF$ at $C$ is 33 N, determine the moment of that force about the line joining points $D$ and $B$.

**3.61** A regular tetrahedron has six edges of length $a$. A force $\mathbf{P}$ is directed as shown along edge $BC$. Determine the moment of $\mathbf{P}$ about edge $OA$.

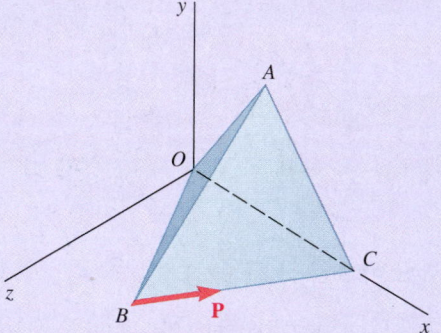

**Fig. P3.61 and *P3.62***

**3.62** A regular tetrahedron has six edges of length $a$. (*a*) Show that two opposite edges, such as $OA$ and $BC$, are perpendicular to each other. (*b*) Use this property and the result obtained in Prob. 3.61 to determine the perpendicular distance between edges $OA$ and $BC$.

**3.63** Two forces $\mathbf{F}_1$ and $\mathbf{F}_2$ in space have the same magnitude $F$. Prove that the moment of $\mathbf{F}_1$ about the line of action of $\mathbf{F}_2$ is equal to the moment of $\mathbf{F}_2$ about the line of action of $\mathbf{F}_1$.

**\*3.64** In Prob. 3.55, determine the perpendicular distance between a line joining points $O$ and $D$ and the line of action of $\mathbf{P}$.

**\*3.65** In Prob. 3.56, determine the perpendicular distance between a line joining points $F$ and $B$ and the line of action of $\mathbf{P}$.

**\*3.66** In Prob. 3.57, determine the perpendicular distance between portion $BH$ of the cable and the diagonal $AD$.

**\*3.67** In Prob. 3.58, determine the perpendicular distance between portion $BG$ of the cable and the diagonal $AD$.

**\*3.68** In Prob. 3.59, determine the perpendicular distance between cable $AE$ and the line joining points $D$ and $B$.

**\*3.69** In Prob. 3.60, determine the perpendicular distance between cable $CF$ and the line joining points $D$ and $B$.

# 3.3   COUPLES AND FORCE-COUPLE SYSTEMS

Now that we have studied the effects of forces and moments on a rigid body, we can ask if it is possible to simplify a system of forces and moments without changing these effects. It turns out that we *can* replace a system of forces and moments with a simpler and equivalent system. One of the key ideas used in such a transformation is called a couple.

## 3.3A   Moment of a Couple

*Two forces* **F** *and* **–F**, *having the same magnitude, parallel lines of action, and opposite sense, are said to form a* **couple** (Fig. 3.25). The sum of the components of the two forces in any direction is zero. The sum of the moments of the two forces about a given point, however, is not zero. The two forces do not cause the body on which they act to move along a line (translation), but they do tend to make it rotate.

Let us denote the position vectors of the points of application of **F** and **–F** by $\mathbf{r}_A$ and $\mathbf{r}_B$, respectively (Fig. 3.26). The sum of the moments of the two forces about $O$ is

$$\mathbf{r}_A \times \mathbf{F} + \mathbf{r}_B \times (-\mathbf{F}) = (\mathbf{r}_A - \mathbf{r}_B) \times \mathbf{F}$$

Setting $\mathbf{r}_A - \mathbf{r}_B = \mathbf{r}$, where **r** is the vector joining the points of application of the two forces, we conclude that the sum of the moments of **F** and **–F** about $O$ is represented by the vector

$$\mathbf{M} = \mathbf{r} \times \mathbf{F} \qquad (3.45)$$

The vector **M** is called the *moment of the couple*. It is perpendicular to the plane containing the two forces, and its magnitude is

$$M = rF \sin \theta = Fd \qquad (3.46)$$

where $d$ is the perpendicular distance between the lines of action of **F** and **–F**, and $\theta$ is the angle between **F** (or **–F**) and **r**. The sense of **M** is defined by the right-hand rule.

Note that the vector **r** in Eq. (3.45) is independent of the choice of the origin $O$ of the coordinate axes. Therefore, we would obtain the same result if the moments of **F** and **–F** had been computed about a different point $O'$. Thus, the moment **M** of a couple is a *free vector* (Sec. 2.1B), which can be applied at any point (Fig. 3.27).

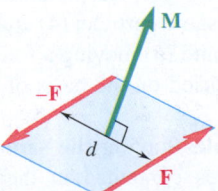

**Fig. 3.27**   The moment **M** of a couple equals the product of **F** and $d$, is perpendicular to the plane of the couple, and may be applied at any point of that plane.

From the definition of the moment of a couple, it also follows that two couples—one consisting of the forces $\mathbf{F}_1$ and $-\mathbf{F}_1$, the other of the forces $\mathbf{F}_2$ and $-\mathbf{F}_2$ (Fig. 3.28)—have equal moments if

$$F_1 d_1 = F_2 d_2 \qquad (3.47)$$

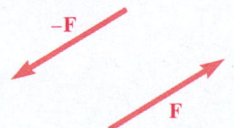

**Fig. 3.25**   A couple consists of two forces with equal magnitude, parallel lines of action, and opposite sense.

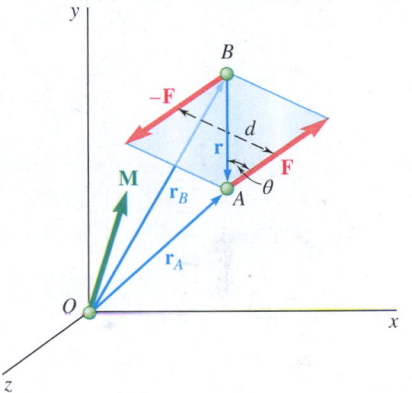

**Fig. 3.26**   The moment **M** of the couple about $O$ is the sum of the moments of **F** and of **–F** about $O$.

**Photo 3.1**   The parallel upward and downward forces of equal magnitude exerted on the arms of the lug nut wrench are an example of a couple. ©McGraw-Hill Education/Lucinda Dowell

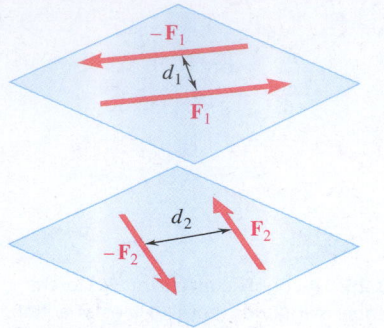

**Fig. 3.28** Two couples have the same moment if they lie in parallel planes, have the same sense, and if $F_1d_1 = F_2d_2$.

provided that the two couples lie in parallel planes (or in the same plane) and have the same sense (i.e., clockwise or counterclockwise).

## 3.3B Equivalent Couples

Imagine that three couples act successively on the same rectangular box (Fig. 3.29). As we have just seen, the only motion a couple can impart to a rigid body is a rotation. Because each of the three couples shown has the same moment **M** (same direction and same magnitude $M = 120$ lb·in.), we can expect each couple to have the same effect on the box.

As reasonable as this conclusion appears, we should not accept it hastily. Although intuition is of great help in the study of mechanics, it should not be accepted as a substitute for logical reasoning. Before stating that two systems (or groups) of forces have the same effect on a rigid body, we should prove that

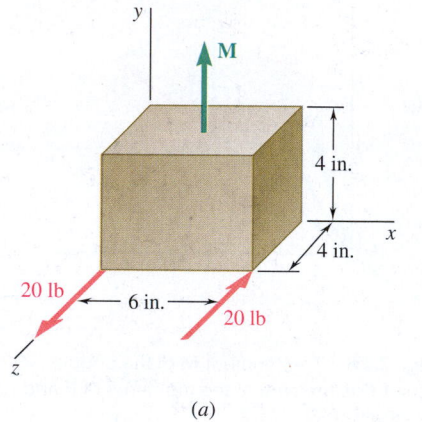
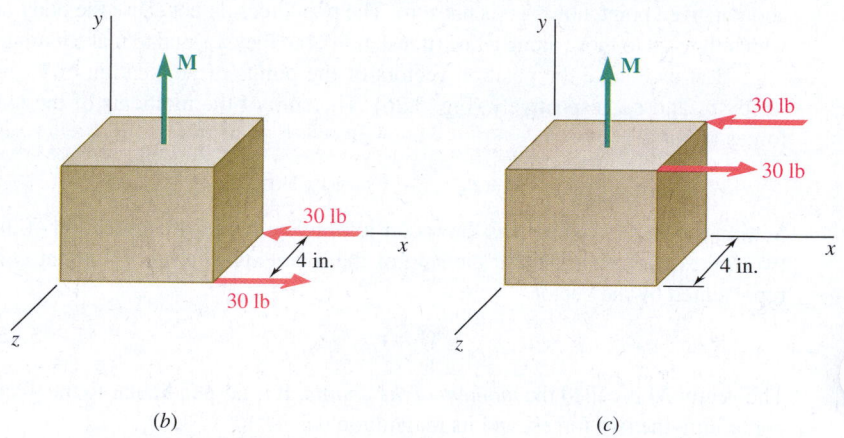

$(a)$ $(b)$ $(c)$

**Fig. 3.29** Three equivalent couples. ($a$) A couple acting on the bottom of the box, acting counterclockwise viewed from above; ($b$) a couple in the same plane and with the same sense but larger forces than in ($a$); ($c$) a couple acting in a different plane but with the same sense.

fact on the basis of the experimental evidence introduced so far. This evidence consists of the parallelogram law for the addition of two forces (Sec. 2.1A) and the principle of transmissibility (Sec. 3.1B). Therefore, we state that **two systems of forces are equivalent** (i.e., they have the same effect on a rigid body) **if we can transform one of them into the other by means of one or several of the following operations:** (1) replacing two forces acting on the same particle by their resultant; (2) resolving a force into two components; (3) canceling two equal and opposite forces acting on the same particle; (4) attaching to the same particle two equal and opposite forces; and (5) moving a force along its line of action. Each of these operations is justified on the basis of the parallelogram law or the principle of transmissibility.

Let us now prove that **two couples having the same moment M are equivalent**. First, consider two couples contained in the same plane, and assume that this plane coincides with the plane of the figure (Fig. 3.30). The first couple consists of the forces $\mathbf{F}_1$ and $-\mathbf{F}_1$ of magnitude $F_1$, located at a distance $d_1$ from each other (Fig. 3.30$a$). The second couple consists of the forces $\mathbf{F}_2$ and $-\mathbf{F}_2$ of magnitude $F_2$, located at a distance $d_2$ from each other (Fig. 3.30$d$). Because the two couples have the same moment **M**, which is perpendicular to the plane of the figure, they must have the same sense (assumed here to be counterclockwise), and the relation

$$F_1d_1 = F_2d_2 \tag{3.47}$$

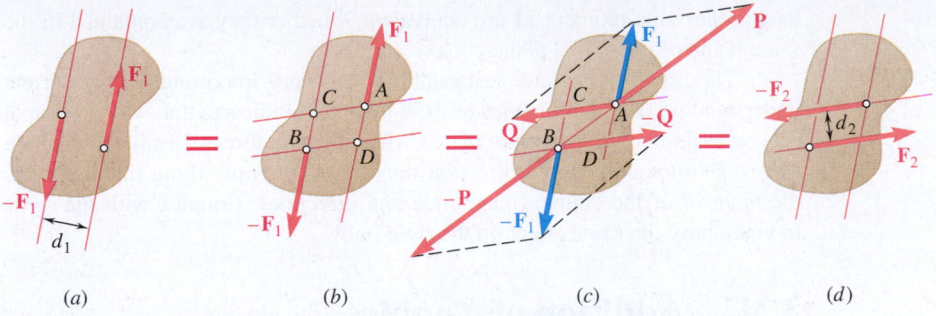

(a)  (b)  (c)  (d)

**Fig. 3.30** Four steps in transforming one couple to another couple in the same plane by using simple operations. (a) Starting couple; (b) label points of intersection of lines of action of the two couples; (c) resolve forces from first couple into components; (d) final couple.

must be satisfied. To prove that they are equivalent, we shall show that the first couple can be transformed into the second by means of the operations listed previously.

Let us denote by $A$, $B$, $C$, and $D$ the points of intersection of the lines of action of the two couples. We first slide the forces $\mathbf{F}_1$ and $-\mathbf{F}_1$ until they are attached, respectively, at $A$ and $B$, as shown in Fig. 3.30b. We then resolve force $\mathbf{F}_1$ into a component $\mathbf{P}$ along line $AB$ and a component $\mathbf{Q}$ along $AC$ (Fig. 3.30c). Similarly, we resolve force $-\mathbf{F}_1$ into $-\mathbf{P}$ along $AB$ and $-\mathbf{Q}$ along $BD$. The forces $\mathbf{P}$ and $-\mathbf{P}$ have the same magnitude, the same line of action, and opposite sense; we can move them along their common line of action until they are applied at the same point and may then be canceled. Thus, the couple formed by $\mathbf{F}_1$ and $-\mathbf{F}_1$ reduces to a couple consisting of $\mathbf{Q}$ and $-\mathbf{Q}$.

We now show that the forces $\mathbf{Q}$ and $-\mathbf{Q}$ are respectively equal to the forces $-\mathbf{F}_2$ and $\mathbf{F}_2$. We obtain the moment of the couple formed by $\mathbf{Q}$ and $-\mathbf{Q}$ by computing the moment of $\mathbf{Q}$ about $B$. Similarly, the moment of the couple formed by $\mathbf{F}_1$ and $-\mathbf{F}_1$ is the moment of $\mathbf{F}_1$ about $B$. However, by Varignon's theorem, the moment of $\mathbf{F}_1$ is equal to the sum of the moments of its components $\mathbf{P}$ and $\mathbf{Q}$. Because the moment of $\mathbf{P}$ about $B$ is zero, the moment of the couple formed by $\mathbf{Q}$ and $-\mathbf{Q}$ must be equal to the moment of the couple formed by $\mathbf{F}_1$ and $-\mathbf{F}_1$. Recalling Eq. (3.47), we have

$$Qd_2 = F_1d_1 = F_2d_2 \qquad \text{and} \qquad Q = F_2$$

Thus, the forces $\mathbf{Q}$ and $-\mathbf{Q}$ are respectively equal to the forces $-\mathbf{F}_2$ and $\mathbf{F}_2$, and the couple of Fig. 3.30a is equivalent to the couple of Fig. 3.30d.

Now consider two couples contained in parallel planes $P_1$ and $P_2$. We prove that they are equivalent if they have the same moment. In view of the preceding discussion, we can assume that the couples consist of forces of the same magnitude $F$ acting along parallel lines (Fig. 3.31a and d). We propose to show that the couple contained in plane $P_1$ can be transformed into the couple contained in plane $P_2$ by means of the standard operations listed previously.

Let us consider the two diagonal planes defined respectively by the lines of action of $\mathbf{F}_1$ and $-\mathbf{F}_2$ and by those of $-\mathbf{F}_1$ and $\mathbf{F}_2$ (Fig. 3.31b). At a point on their line of intersection, we attach two forces $\mathbf{F}_3$ and $-\mathbf{F}_3$, which are respectively equal to $\mathbf{F}_1$ and $-\mathbf{F}_1$. The couple formed by $\mathbf{F}_1$ and $-\mathbf{F}_3$ can be replaced by a couple consisting of $\mathbf{F}_3$ and $-\mathbf{F}_2$ (Fig. 3.31c), because both couples clearly have the same moment and are contained in the same diagonal plane. Similarly, the couple formed by $-\mathbf{F}_1$ and $\mathbf{F}_3$ can be replaced by a couple consisting of $-\mathbf{F}_3$ and $\mathbf{F}_2$. Canceling the two equal and opposite forces $\mathbf{F}_3$ and $-\mathbf{F}_3$, we obtain the desired couple in plane $P_2$ (Fig. 3.31d). Thus, we conclude that two couples

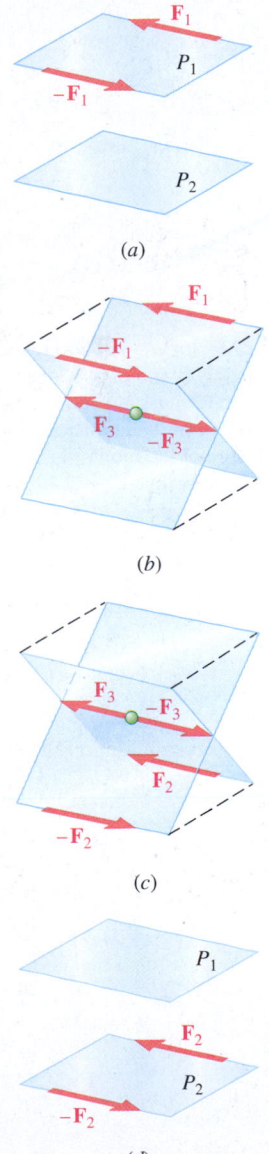

(a)

(b)

(c)

(d)

**Fig. 3.31** Four steps in transforming one couple to another couple in a parallel plane by using simple operations. (a) Initial couple; (b) add a force pair along the line of intersection of two diagonal planes; (c) replace two couples with equivalent couples in the same planes; (d) final couple.

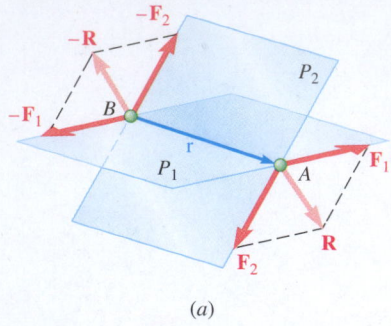

(a)

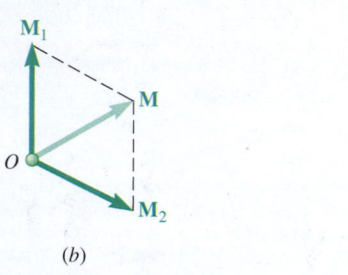

(b)

**Fig. 3.32**    (a) We can add two couples, each acting in one of two intersecting planes, to form a new couple. (b) The moment of the resultant couple is the vector sum of the moments of the component couples.

having the same moment **M** are equivalent, whether they are contained in the same plane or in parallel planes.

The property we have just established is very important for the correct understanding of the mechanics of rigid bodies. It indicates that when a couple acts on a rigid body, it does not matter where the two forces forming the couple act or what magnitude and direction they have. The only thing that counts is the *moment* of the couple (magnitude and direction). Couples with the same moment have the same effect on the rigid body.

## 3.3C    Addition of Couples

Consider two intersecting planes $P_1$ and $P_2$ and two couples acting respectively in $P_1$ and $P_2$. Recall that each couple is a free vector in its respective plane and can be represented within this plane by any combination of equal, opposite, and parallel forces and of perpendicular distance of separation that provides the same sense and magnitude for this couple. Thus, we can assume, without any loss of generality, that the couple in $P_1$ consists of two forces $\mathbf{F}_1$ and $-\mathbf{F}_1$ perpendicular to the line of intersection of the two planes and acting respectively at $A$ and $B$ (Fig. 3.32a). Similarly, we can assume that the couple in $P_2$ consists of two forces $\mathbf{F}_2$ and $-\mathbf{F}_2$ perpendicular to $AB$ and acting respectively at $A$ and $B$. The resultant $\mathbf{R}$ of $\mathbf{F}_1$ and $\mathbf{F}_2$ and the resultant $-\mathbf{R}$ of $-\mathbf{F}_1$ and $-\mathbf{F}_2$ form a couple. Denoting the vector joining $B$ to $A$ by $\mathbf{r}$ and recalling the definition of the moment of a couple (Sec. 3.3A), we express the moment **M** of the resulting couple as

$$\mathbf{M} = \mathbf{r} \times \mathbf{R} = \mathbf{r} \times (\mathbf{F}_1 + \mathbf{F}_2)$$

By Varignon's theorem, we can expand this expression as

$$\mathbf{M} = \mathbf{r} \times \mathbf{F}_1 + \mathbf{r} \times \mathbf{F}_2$$

The first term in this expression represents the moment $\mathbf{M}_1$ of the couple in $P_1$, and the second term represents the moment $\mathbf{M}_2$ of the couple in $P_2$. Therefore, we have

$$\mathbf{M} = \mathbf{M}_1 + \mathbf{M}_2 \qquad\qquad \textbf{(3.48)}$$

We conclude that the sum of two couples of moments $\mathbf{M}_1$ and $\mathbf{M}_2$ is a couple of moment **M** equal to the vector sum of $\mathbf{M}_1$ and $\mathbf{M}_2$ (Fig. 3.32b). We can extend this conclusion to state that any number of couples can be added to produce one resultant couple, as

$$\mathbf{M} = \Sigma\mathbf{M} = \Sigma(\mathbf{r} \times \mathbf{F})$$

## 3.3D    Couple Vectors

We have seen that couples with the same moment, whether they act in the same plane or in parallel planes, are equivalent. Therefore, we have no need to draw the actual forces forming a given couple in order to define its effect on a rigid body (Fig. 3.33a). It is sufficient to draw an arrow equal in magnitude and direction to the moment **M** of the couple (Fig. 3.33b). We have also seen that the sum of two couples is itself a couple and that we can obtain the moment **M** of the resultant couple by forming the vector sum of the moments $\mathbf{M}_1$ and $\mathbf{M}_2$ of the given couples. Thus, couples obey the law of addition of vectors, so the arrow used in Fig. 3.33b to represent the couple defined in Fig. 3.33a truly can be considered a vector.

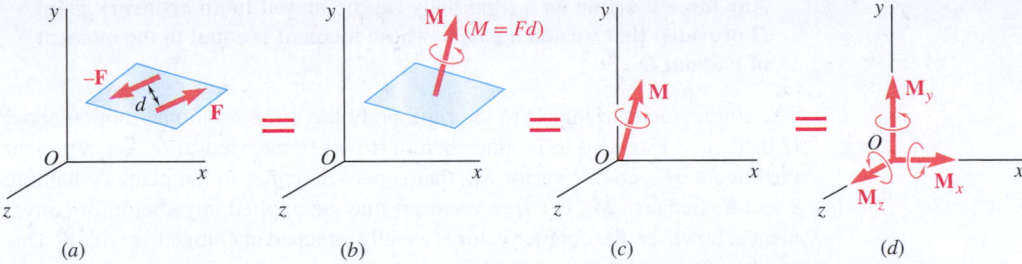

**Fig. 3.33** (a) A couple formed by two forces can be represented by (b) a couple vector, oriented perpendicular to the plane of the couple. (c) The couple vector is a free vector and can be moved to other points of application, such as the origin. (d) A couple vector can be resolved into components along the coordinate axes.

The vector representing a couple is called a **couple vector**. Note that, in Fig. 3.33, we use a red arrow to distinguish the couple vector, *which represents the couple itself*, from the *moment* of the couple, which was represented by a green arrow in earlier figures. Also note that we added the symbol ↺ to this red arrow to avoid any confusion with vectors representing forces. A couple vector, like the moment of a couple, is a free vector. Therefore, we can choose its point of application at the origin of the system of coordinates, if so desired (Fig. 3.33c). Furthermore, we can resolve the couple vector $\mathbf{M}$ into component vectors $\mathbf{M}_x$, $\mathbf{M}_y$, and $\mathbf{M}_z$ that are directed along the coordinate axes (Fig. 3.33d). These component vectors represent couples acting, respectively, in the $yz$, $zx$, and $xy$ planes.

## 3.3E Resolution of a Given Force into a Force at $O$ and a Couple

Consider a force $\mathbf{F}$ acting on a rigid body at a point $A$ defined by the position vector $\mathbf{r}$ (Fig. 3.34a). Suppose that for some reason it would simplify the analysis to have the force act at point $O$ instead. Although we can move $\mathbf{F}$ along its line of action (principle of transmissibility), we cannot move it to a point $O$ that does not lie on the original line of action without modifying the action of $\mathbf{F}$ on the rigid body.

We can, however, attach two forces at point $O$, one equal to $\mathbf{F}$ and the other equal to $-\mathbf{F}$, without modifying the action of the original force on the rigid body (Fig. 3.34b). As a result of this transformation, we now have a force $\mathbf{F}$ applied at $O$; the other two forces form a couple of moment $\mathbf{M}_O = \mathbf{r} \times \mathbf{F}$. Thus,

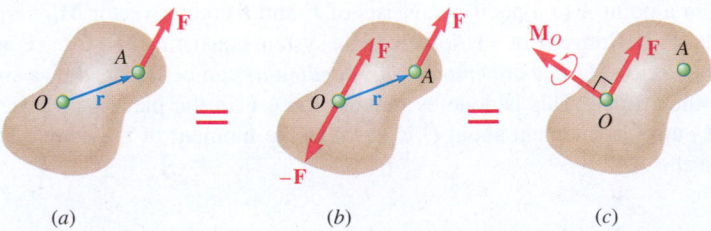

**Fig. 3.34** Replacing a force with a force and a couple. (a) Initial force $\mathbf{F}$ acting at point $A$; (b) attaching equal and opposite forces at $O$; (c) force $\mathbf{F}$ acting at point $O$ and a couple.

**Any force F acting on a rigid body can be moved to an arbitrary point O provided that we add a couple whose moment is equal to the moment of F about O.**

The couple tends to impart to the rigid body the same rotational motion about O that force **F** tended to produce before it was transferred to O. We represent the couple by a couple vector $\mathbf{M}_O$ that is perpendicular to the plane containing **r** and **F**. Because $\mathbf{M}_O$ is a free vector, it may be applied anywhere; for convenience, however, the couple vector is usually attached at O together with **F**. This combination is referred to as a **force-couple system** (Fig. 3.34c).

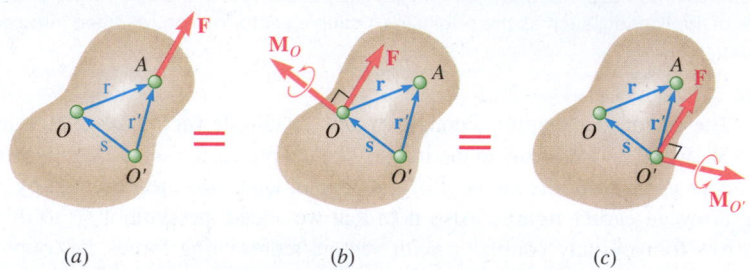

(a)                           (b)                           (c)

**Fig. 3.35** Moving a force to different points. (a) Initial force **F** acting at A; (b) force **F** acting at O and a couple; (c) force **F** acting at O′ and a different couple.

If we move force **F** from A to a different point O′ (Fig. 3.35a and c), we have to compute the moment $\mathbf{M}_{O'} = \mathbf{r}' \times \mathbf{F}$ of **F** about O′ and add a new force-couple system consisting of **F** and the couple vector $\mathbf{M}_{O'}$ at O′. We can obtain the relation between the moments of **F** about O and O′ as

$$\mathbf{M}_{O'} = \mathbf{r}' \times \mathbf{F} = (\mathbf{r} + \mathbf{s}) \times \mathbf{F} = \mathbf{r} \times \mathbf{F} + \mathbf{s} \times \mathbf{F}$$

$$\mathbf{M}_{O'} = \mathbf{M}_O + \mathbf{s} \times \mathbf{F} \tag{3.49}$$

where **s** is the vector joining O′ to O. Thus, we obtain the moment $\mathbf{M}_{O'}$ of **F** about O′ by adding to the moment $\mathbf{M}_O$ of **F** about O the vector product $\mathbf{s} \times \mathbf{F}$, representing the moment about O′ of the force **F** applied at O.

We also could have established this result by observing that, in order to transfer to O′ the force-couple system attached at O (Fig. 3.35b and c), we could freely move the couple vector $\mathbf{M}_O$ to O′. However, to move force **F** from O to O′, we need to add to **F** a couple vector whose moment is equal to the moment about O′ of force **F** applied at O. Thus, the couple vector $\mathbf{M}_{O'}$ must be the sum of $\mathbf{M}_O$ and the vector $\mathbf{s} \times \mathbf{F}$.

As noted here, the force-couple system obtained by transferring a force **F** from a point A to a point O consists of **F** and a couple vector $\mathbf{M}_O$ perpendicular to **F**. Conversely, any force-couple system consisting of a force **F** and a couple vector $\mathbf{M}_O$ that are *mutually perpendicular* can be replaced by a single equivalent force. This is done by moving force **F** in the plane perpendicular to $\mathbf{M}_O$ until its moment about O is equal to the moment of the couple being replaced.

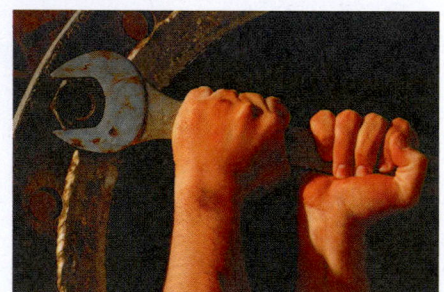

**Photo 3.2** The force exerted by each hand on the wrench could be replaced with an equivalent force-couple system acting on the nut. ©Steve Hix

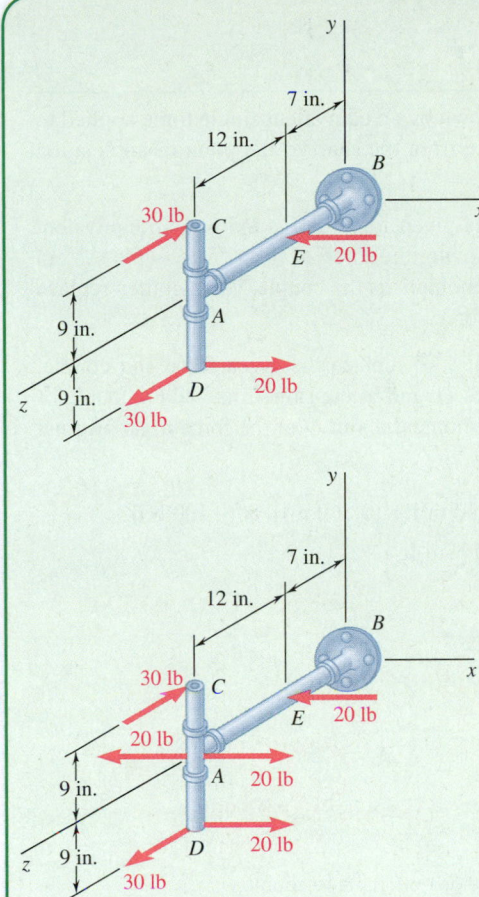

**Fig. 1** Placing two equal and opposite 20-lb forces at *A* to simplify calculations.

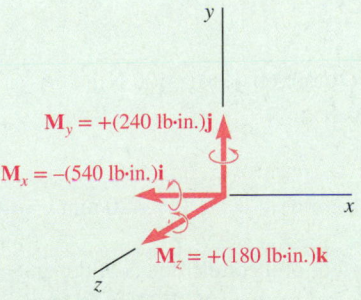

$M_y = +(240 \text{ lb·in.})\mathbf{j}$

$M_x = -(540 \text{ lb·in.})\mathbf{i}$

$M_z = +(180 \text{ lb·in.})\mathbf{k}$

**Fig. 2** The three couples represented as couple vectors.

# Sample Problem 3.6

Determine the components of the single-couple equivalent to the two couples shown.

**STRATEGY:** Look for ways to add equal and opposite forces to the diagram that, along with already known perpendicular distances, will produce new couples with moments along the coordinate axes. These can be combined into a single equivalent couple.

**MODELING:** You can simplify the computations by attaching two equal and opposite 20-lb forces at *A* (Fig. 1). This enables you to replace the original 20-lb-force couple by two new 20-lb-force couples: one lying in the *zx* plane and the other in a plane parallel to the *xy* plane.

**ANALYSIS:** You can represent these three couples by three couple vectors $\mathbf{M}_x$, $\mathbf{M}_y$, and $\mathbf{M}_z$ directed along the coordinate axes (Fig. 2). The corresponding moments are

$$M_x = -(30 \text{ lb})(18 \text{ in.}) = -540 \text{ lb·in.}$$
$$M_y = +(20 \text{ lb})(12 \text{ in.}) = +240 \text{ lb·in.}$$
$$M_z = +(20 \text{ lb})(9 \text{ in.}) = +180 \text{ lb·in.}$$

These three moments represent the components of the single couple **M** equivalent to the two given couples. You can write **M** as

$$\mathbf{M} = -(540 \text{ lb·in.})\mathbf{i} + (240 \text{ lb·in.})\mathbf{j} + (180 \text{ lb·in.})\mathbf{k} \quad \blacktriangleleft$$

**REFLECT and THINK:** You can also obtain the components of the equivalent single couple **M** by computing the sum of the moments of the four given forces about an arbitrary point. Selecting point *D*, the moment is (Fig. 3)

$$\mathbf{M} = \mathbf{M}_D = (18 \text{ in.})\mathbf{j} \times (-30 \text{ lb})\mathbf{k} + [(9 \text{ in.})\mathbf{j} - (12 \text{ in.})\mathbf{k}] \times (-20 \text{ lb})\mathbf{i}$$

After computing the various cross products, you get the same result, as

$$\mathbf{M} = -(540 \text{ lb·in.})\mathbf{i} + (240 \text{ lb·in.})\mathbf{j} + (180 \text{ lb·in.})\mathbf{k} \quad \blacktriangleleft$$

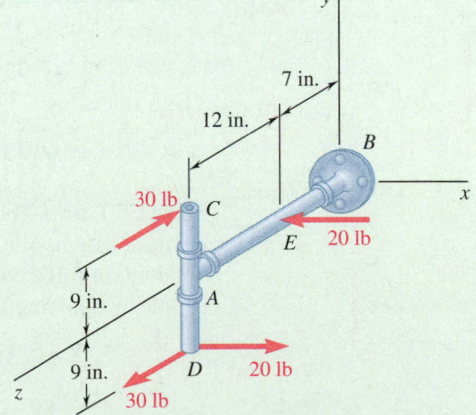

**Fig. 3** Using the given force system, the equivalent single couple can also be determined from the sum of moments of the forces about any point, such as point *D*.

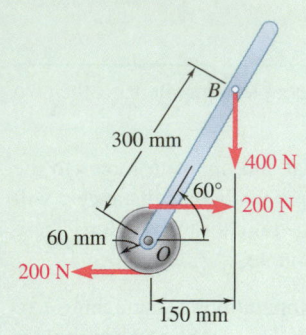

# Sample Problem 3.7

Replace the couple and force shown by an equivalent single force applied to the lever. Determine the distance from the shaft to the point of application of this equivalent force.

**STRATEGY:** First, replace the given force and couple by an equivalent force-couple system at $O$. By moving the force of this force-couple system a distance that creates the same moment as the couple, you can then replace the system with one equivalent force.

**MODELING and ANALYSIS:** To replace the given force and couple, move the force $\mathbf{F} = -(400 \text{ N})\mathbf{j}$ to $O$, and at the same time, add a couple of moment $\mathbf{M}_O$ that is equal to the moment about $O$ of the force in its original position (Fig. 1). Thus,

$$\mathbf{M}_O = \overrightarrow{OB} \times \mathbf{F} = [(0.150 \text{ m})\mathbf{i} + (0.260 \text{ m})\mathbf{j}] \times (-400 \text{ N})\mathbf{j}$$
$$= -(60 \text{ N·m})\mathbf{k}$$

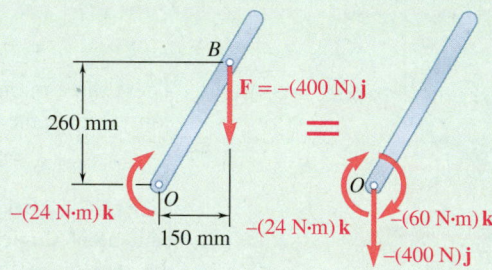

**Fig. 1** Replacing the given force and couple with an equivalent force-couple at $O$.

When you add this new couple to the couple of moment $-(24 \text{ N·m})\mathbf{k}$ formed by the two 200-N forces, you obtain a couple of moment $-(84 \text{ N·m})\mathbf{k}$ (Fig. 2). You can replace this last couple by applying $\mathbf{F}$ at a point $C$ chosen in such a way that

$$-(84 \text{ N·m})\mathbf{k} = \overrightarrow{OC} \times \mathbf{F}$$
$$= [(OC)\cos 60°\mathbf{i} + (OC)\sin 60°\mathbf{j}] \times (-400 \text{ N})\mathbf{j}$$
$$= -(OC)\cos 60°(400 \text{ N})\mathbf{k}$$

The result is

$$(OC) \cos 60° = 0.210 \text{ m} = 210 \text{ mm} \qquad OC = 420 \text{ mm} \blacktriangleleft$$

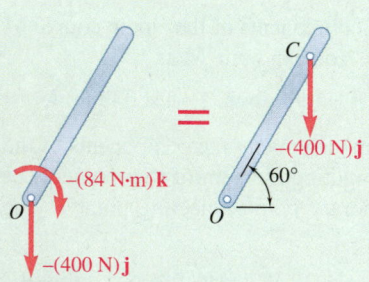

**Fig. 2** Resultant couple eliminated by moving force $\mathbf{F}$.

**REFLECT and THINK:** Because the effect of a couple does not depend on its location, you can move the couple of moment $-(24 \text{ N·m})\mathbf{k}$ to $B$, obtaining a force-couple system at $B$ (Fig. 3). Now you can eliminate this couple by applying $\mathbf{F}$ at a point $C$ chosen in such a way that

$$-(24 \text{ N·m})\mathbf{k} = \overrightarrow{BC} \times \mathbf{F}$$
$$= -(BC)\cos 60°(400 \text{ N})\mathbf{k}$$

The conclusion is

$$(BC) \cos 60° = 0.060 \text{ m} = 60 \text{ mm} \qquad BC = 120 \text{ mm}$$
$$OC = OB + BC = 300 \text{ mm} + 120 \text{ mm} \qquad OC = 420 \text{ mm} \blacktriangleleft$$

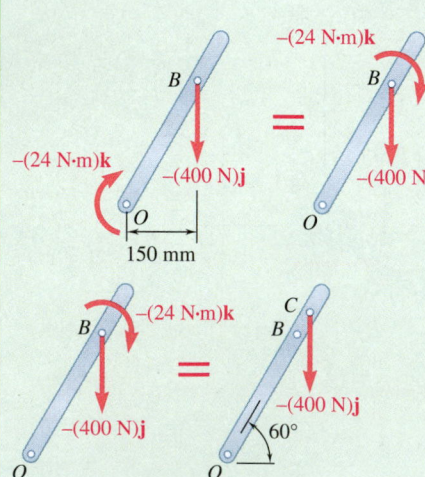

**Fig. 3** Couple can be moved to $B$ with no change in effect. This couple can then be eliminated by moving force $\mathbf{F}$.

## Case Study 3.1

The Vlooybergtoren tower in Tielt-Winge, Belgium, was constructed to provide a distinctive and unique platform for visitors to view the Kabouterbos "fairytale forest" (CS Photo 3.1). The staircase and observation deck is supported by a structural steel frame (CS Photo 3.2) that is clad in weathering steel (which oxidizes to produce the reddish-orange hue shown, forming a protective layer that inhibits further corrosion). Overall, this cantilever structure rises 11.3 m above the ground and weighs approximately 130 kN.* The base of the tower is supported against overturning by the anchor points shown in CS Photo 3.2. Considering only the self-weight of the tower, let's estimate the resulting equivalent force-couple applied at the support that prevents uplift (i.e., the anchor toward the rear of the tower). We will then use this equivalent force-couple to determine the total uplift force acting on the support.

**CS Photo 3.1**  Vlooybergtoren Tower in Tielt-Winge, Belgium.

Left: ©Kris Van den Bosch Right: ©Yves Willem

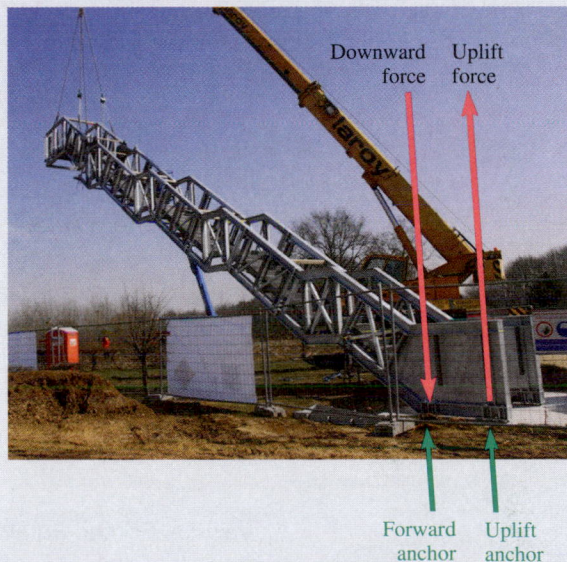

**CS Photo 3.2**  Tower under construction, showing steel frame and anchor supports. ©Yves Willem

### STRATEGY:

Use a two-dimensional model to represent the tower, and simple supports (a pin and a roller) to model the actual base conditions. Replace the self-weight load with an equivalent force-couple at the uplift anchor support. Then, by replacing the moment of the couple with vertical forces applied at the two support locations, the overall uplift force can be determined.

### MODELING:

CS Fig. 3.1 provides the geometry assumed for the tower, with supports *A* and *B* located at the tower's base and below the ground surface as shown. (Support *A* reflects the anchorage subject to uplift.) The 135-kN dead load is applied at the structure's *center of gravity,* a concept that will be examined in detail in Chap. 5.

* Source: "What's Cool in Steel?" *Modern Steel Construction,* Chicago, IL: The American Institute of Steel Construction, August 2016, pp. 42–43.

*(continued)*

(For demonstration purposes, we will assume a center of gravity $G$ approximated as shown in CS Fig. 3.1. It has been positioned closer to the left end than the right because the supporting structure becomes increasingly heavier toward the base of the tower.)

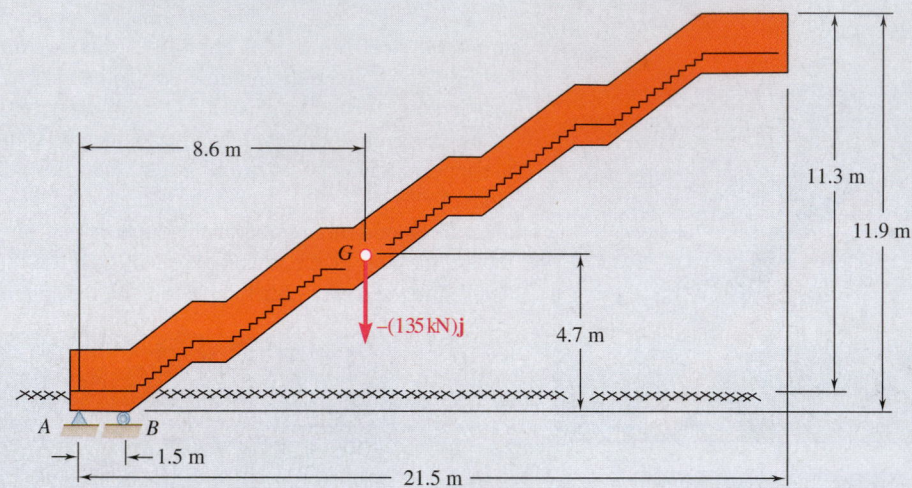

**CS Fig. 3.1**    Tower model.

**ANALYSIS:**

**a. Force-Couple System at A.**    To replace the given 135-kN force, move the force $\mathbf{F} = -(135 \text{ kN})\mathbf{j}$ to $A$, and at the same time, add a couple of moment $\mathbf{M}_A$ that is equal to the moment about $A$ of the force in its original position (CS Fig. 3.2a):

$$\mathbf{M}_A = \overrightarrow{AG} \times \mathbf{F} = [(8.6 \text{ m})\mathbf{i} + (4.7\text{m})\mathbf{j}] \times (-135 \text{ kN})\mathbf{j}$$
$$= -(1161 \text{ kN} \cdot \text{m})\mathbf{k}$$

**b. Vertical Forces Equivalent to Moment of Couple M$_A$.**    To replace the moment of couple $\mathbf{M}_A$ with two equal and opposite vertical forces at support locations $A$ and $B$, separated by perpendicular distance $d = 1.5$ m, divide $M_A$ by this perpendicular distance. The resulting magnitude of each force is

$$F = \frac{M_A}{d} = \frac{1161 \text{ kN} \cdot \text{m}}{1.5 \text{ m}} = 774 \text{ kN}$$

These forces are directed as shown in CS Fig. 3.2b.

**c. Uplift Force at Anchor A.**    Combining the forces acting at $A$ in CS Fig. 3.2b, the result shown in CS Fig. 3.2c is obtained. Being a complete system that is equivalent to the original load, the pair of forces in CS Fig. 3.2c represents the total load exerted by the tower's self-weight on these supports. Thus, the total uplift exerted on the anchorage at $A$ is the equivalent force at this point, or

639 kN ↑  ◄

(Because the structural frame consists of two equal sides, this total uplift force would be divided equally over both sides.)

*(continued)*

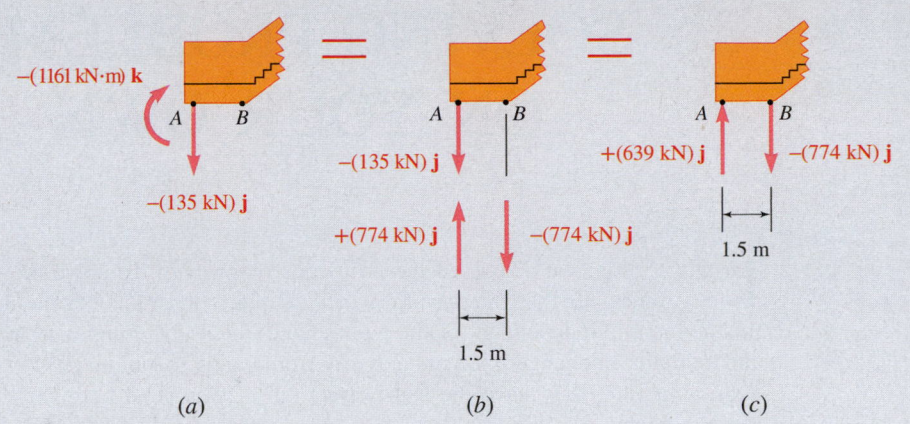

(a)                          (b)                          (c)

**CS Fig. 3.2**   (a) Equivalent force-couple at A, (b) moment of the couple replaced by two vertical forces at A and B, (c) the overall equivalent system of vertical forces applied at supports A and B.

**REFLECT and THINK:**

The equivalent forces exerted on the supports as shown in CS Fig. 3.2c are equal and opposite to the support reactions acting on the structure at these points. Such reactions can be determined more directly by the principles of rigid-body equilibrium that we will examine in Chap. 4.

# SOLVING PROBLEMS ON YOUR OWN

In this section, we discussed the properties of *couples*. To solve the following problems, remember that the net effect of a couple is to produce a moment **M**. Because this moment is independent of the point about which it is computed, **M** is a *free vector* and remains unchanged if you move it from point to point. Also, two couples are *equivalent* (i.e., they have the same effect on a given rigid body) if they produce the same moment.

When determining the moment of a couple, all previous techniques for computing moments apply. Also, because the moment of a couple is a free vector, you should compute its value relative to the most convenient point.

Because the only effect of a couple is to produce a moment, it is possible to represent a couple with a vector, called the *couple vector*, that is equal to the moment of the couple. The couple vector is a free vector and is represented by a special symbol, ⤸, to distinguish it from force vectors.

In solving the problems in this section, you will be called upon to perform the following operations:

1. **Adding two or more couples.** This results in a new couple, the moment of which is obtained by adding vectorially the moments of the given couples (Sample Prob. 3.6).

2. **Replacing a force with an equivalent force-couple system at a specified point.** As explained in Sec. 3.3E, the force of a force-couple system is equal to the original force, whereas the required couple vector is equal to the moment of the original force about the given point. In addition, it is important to note that the force and the couple vector are perpendicular to each other. Conversely, it follows that a force-couple system can be reduced to a single force only if the force and couple vector are mutually perpendicular (see the next paragraph).

3. **Replacing a force-couple system (with F perpendicular to M) with a single equivalent force.** The requirement that **F** and **M** be mutually perpendicular is satisfied in all two-dimensional problems. The single equivalent force is equal to **F** and is applied in such a way that its moment about the original point of application is equal to **M** (Sample Prob. 3.7).

# Problems

**3.70** Two 80-N forces are applied as shown to the corners *B* and *D* of a rectangular plate. (*a*) Determine the moment of the couple formed by the two forces by resolving each force into horizontal and vertical components and adding the moments of the two resulting couples. (*b*) Use the result obtained to determine the perpendicular distance between lines *BE* and *DF*.

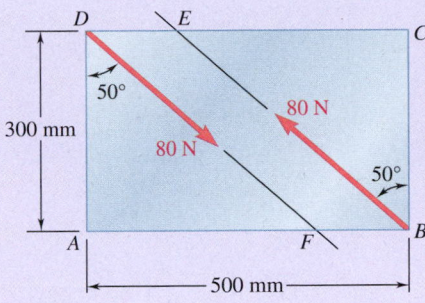

**Fig. P3.70**

**3.71** Two parallel 60-N forces are applied as shown to the corners *A* and *C* of a 200-mm square plate. Determine the moment of the couple formed by the two forces (*a*) by multiplying their magnitude by their perpendicular distance, (*b*) by resolving each force into horizontal and vertical components and adding the moments of the two resulting couples.

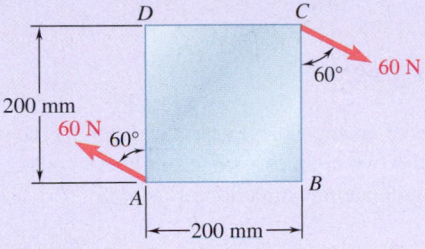

**Fig. P3.71**

**3.72** A multiple-drilling machine is used to drill simultaneously six holes in the steel plate shown. Each drill exerts a clockwise couple of magnitude 40 lb·in. on the plate. Determine an equivalent couple formed by the smallest possible forces acting (*a*) at *A* and *C*, (*b*) at *A* and *D*, (*c*) on the plate.

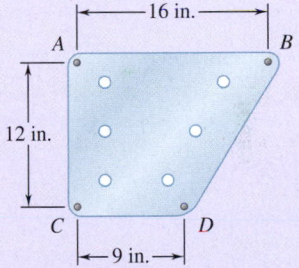

**Fig. *P3.72***

**3.73** Four pegs of the same diameter are attached to a board as shown. Two strings are passed around the pegs and pulled with the forces indicated. Determine the diameter of the pegs knowing that the resultant couple applied to the board is 1132.5 lb·in. counterclockwise.

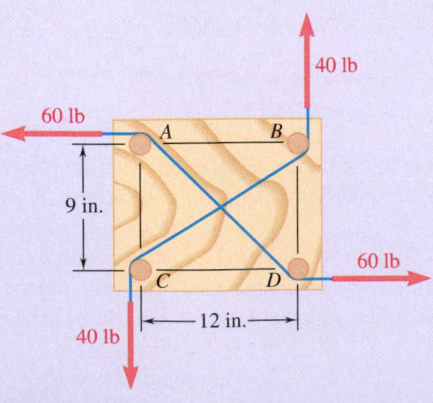

**Fig. P3.73**

**3.74** A piece of plywood in which several holes are being drilled successively has been secured to a workbench by means of two nails. Knowing that the drill exerts a 12-N·m couple on the piece of plywood, determine the magnitude of the resulting forces applied to the nails if they are located (*a*) at *A* and *B*, (*b*) at *B* and *C*, (*c*) at *A* and *C*.

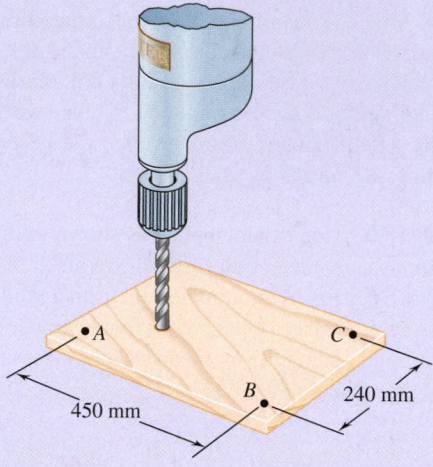

Fig. P3.74

**3.75** The shafts of an angle drive are acted upon by the two couples shown. Replace the two couples with a single equivalent couple, specifying its magnitude and the direction of its axis.

**3.76** If *P* = 0 in the figure, replace the two remaining couples with a single equivalent couple, specifying its magnitude and the direction of its axis.

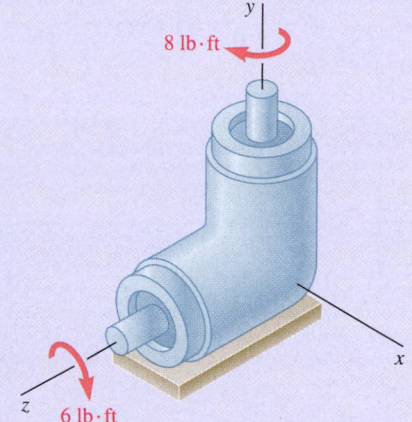

Fig. *P3.75*

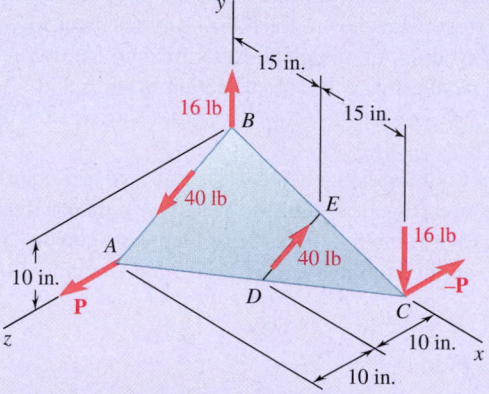

Fig. P3.76 and P3.77

**3.77** If *P* = 20 lb in the figure, replace the three couples with a single equivalent couple, specifying its magnitude and the direction of its axis.

**3.78** The two couples shown are to be replaced with a single equivalent couple. Determine (*a*) the couple vector representing the equivalent couple, (*b*) the two forces acting at *B* and *C* that can be used to form that couple.

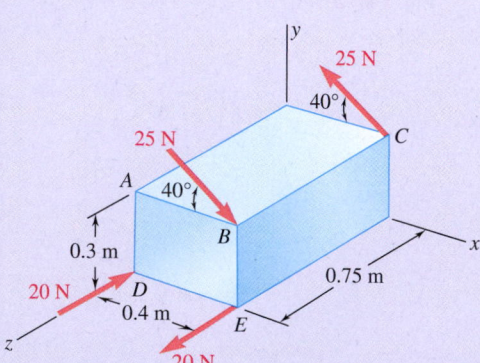

Fig. P3.78

**3.79** Solve part *a* of Prob. 3.78, assuming that two 15-N vertical forces have been added, one acting upward at *A* and the other downward at *C*.

**3.80** Shafts $A$ and $B$ connect the gear box to the wheel assemblies of a tractor, and shaft $C$ connects it to the engine. Shafts $A$ and $B$ lie in the vertical $yz$ plane, while shaft $C$ is directed along the $x$ axis. Replace the couples applied to the shafts by a single equivalent couple, specifying its magnitude and the direction of its axis.

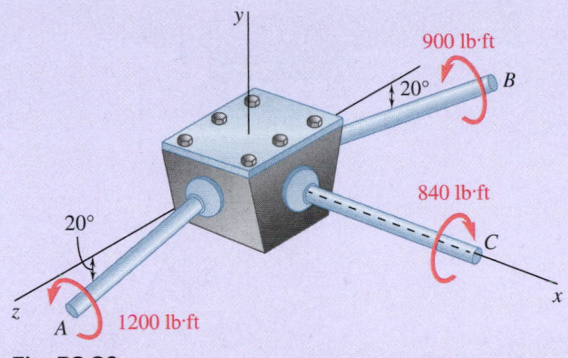

**Fig. P3.80**

**3.81** A 500-N force is applied to a bent plate as shown. Determine (*a*) an equivalent force-couple system at $B$, (*b*) an equivalent system formed by a vertical force at $A$ and a force at $B$.

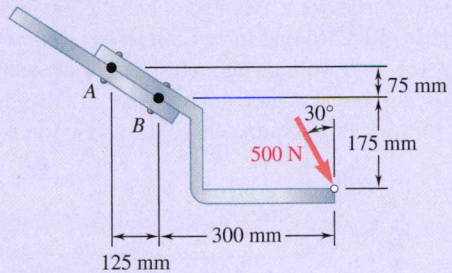

**Fig. *P3.81***

**3.82** A crane column supports a 16-kip load as shown. Replace the load with an equivalent system consisting of an axial force along $AB$ and a couple.

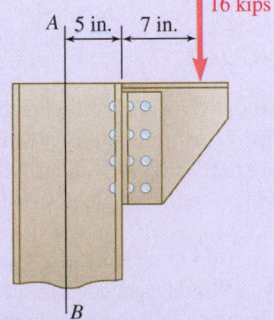

**Fig. P3.82**

**3.83** A dirigible is tethered by a cable attached to its cabin at *B*. If the tension in the cable is 1040 N, replace the force exerted by the cable at *B* with an equivalent system formed by two parallel forces applied at *A* and *C*.

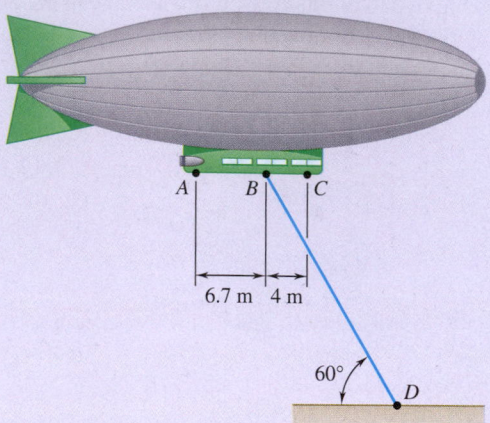

**Fig. P3.83**

**3.84** A 30-lb vertical force **P** is applied at *A* to the bracket shown, which is held by screws at *B* and *C*. (*a*) Replace **P** with an equivalent force-couple system at *B*. (*b*) Find the two horizontal forces at *B* and *C* that are equivalent to the couple obtained in part *a*.

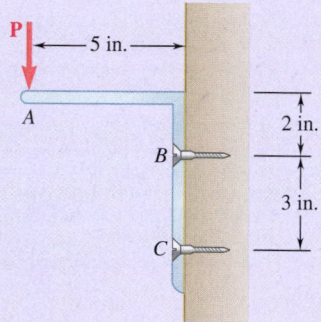

**Fig. P3.84**

**3.85** A worker tries to move a rock by applying a 360-N force to a steel bar as shown. (*a*) Replace that force with an equivalent force-couple system at *D*. (*b*) Two workers attempt to move the same rock by applying a vertical force at *A* and another force at *D*. Determine these two forces if they are to be equivalent to the single force of part *a*.

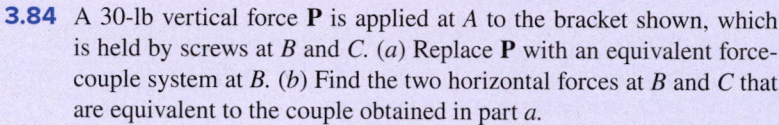

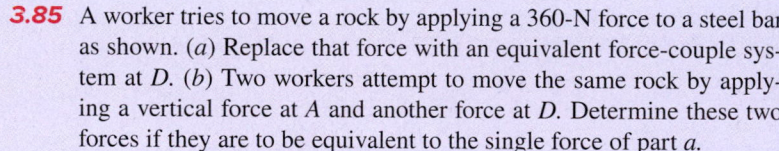

**Fig. *P3.85* and P3.86**

**3.86** A worker tries to move a rock by applying a 360-N force to a steel bar as shown. If two workers attempt to move the same rock by applying a force at *A* and a parallel force at *C*, determine these two forces so that they will be equivalent to the single 360-N force shown in the figure.

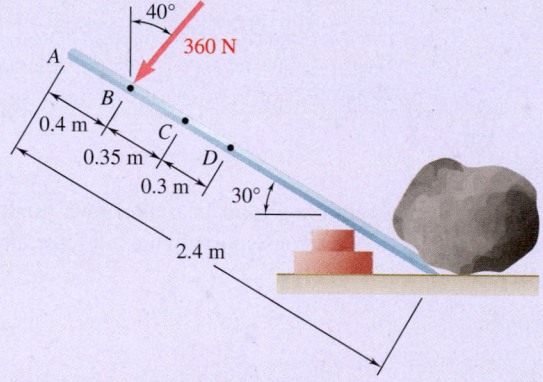

**3.87** The shearing forces exerted on the cross section of a steel channel can be represented by a 900-N vertical force and two 250-N horizontal forces as shown. Replace this force and couple with a single force **F** applied at point *C*, and determine the distance *x* from *C* to line *BD*. (Point *C* is defined as the *shear center* of the section.)

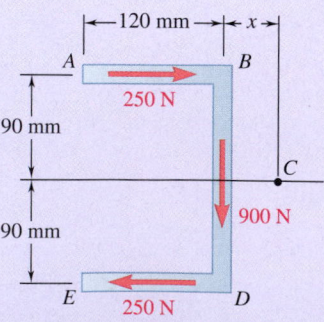

**Fig. P3.87**

**3.88** Knowing that $\alpha = 60°$, replace the force and couple shown with a single force applied at a point located (a) on line $AB$, (b) on line $CD$. In each case, determine the distance from the center $O$ to the point of application of the force.

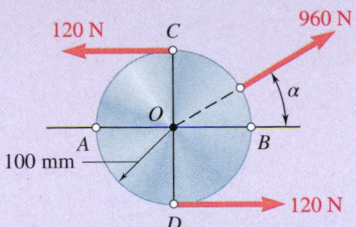

**3.89** Three control rods attached to a lever $ABC$ exert on it the forces shown. (a) Replace the three forces with an equivalent force-couple system at $B$. (b) Determine the single force that is equivalent to the force-couple system obtained in part a, and specify its point of application on the lever.

**Fig. P3.88**

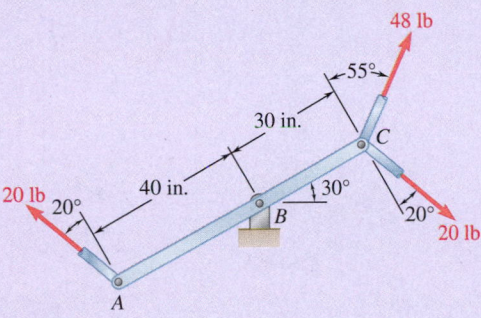

**Fig. P3.89**

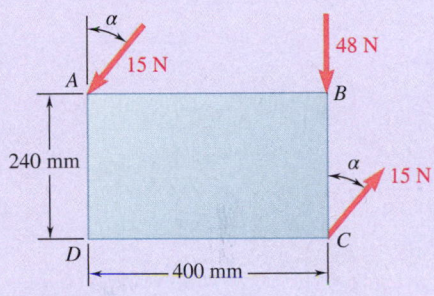

**Fig. P3.90**

**3.90** A rectangular plate is acted upon by the force and couple shown. This system is to be replaced with a single equivalent force. (a) For $\alpha = 40°$, specify the magnitude and line of action of the equivalent force. (b) Specify the value of $\alpha$ if the line of action of the equivalent force is to intersect line $CD$ 300 mm to the right of $D$.

**3.91** While tapping a hole, a machinist applies the horizontal forces shown to the handle of the tap wrench. Show that these forces are equivalent to a single force, and specify, if possible, the point of application of the single force on the handle.

**3.92** A hexagonal plate is acted upon by the force **P** and the couple shown. Determine the magnitude and the direction of the smallest force **P** for which this system can be replaced with a single force at $E$.

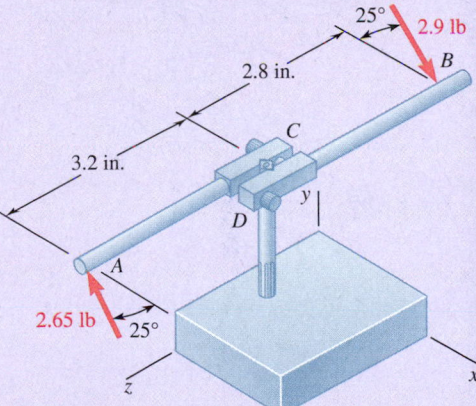

**Fig. P3.91**

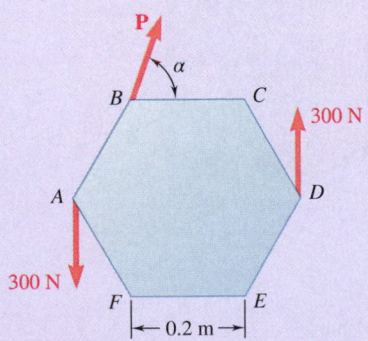

**Fig. P3.92**

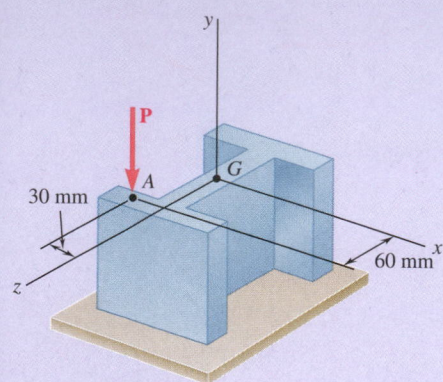

**Fig. P3.93**

**3.93** Replace the 250-kN force **P** with an equivalent force-couple system at *G*.

**3.94** A 2.6-kip force is applied at point *D* of the cast-iron post shown. Replace that force with an equivalent force-couple system at the center *A* of the base section.

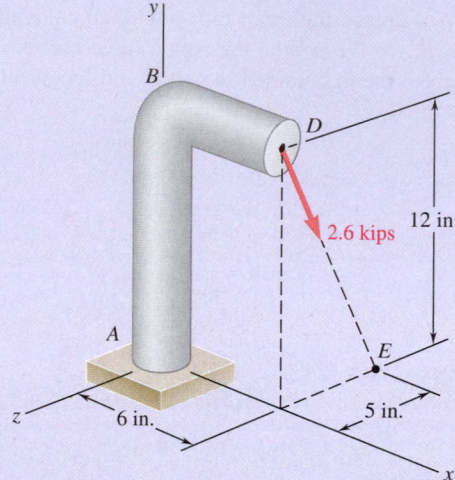

**Fig. *P3.94***

**3.95** Replace the 150-N force with an equivalent force-couple system at *A*.

**3.96** To keep a door closed, a wooden stick is wedged between the floor and the doorknob. The stick exerts at *B* a 175-N force directed along line *AB*. Replace that force with an equivalent force-couple system at *C*.

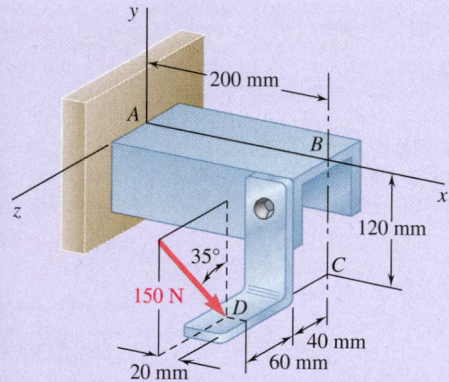

**Fig. P3.95**

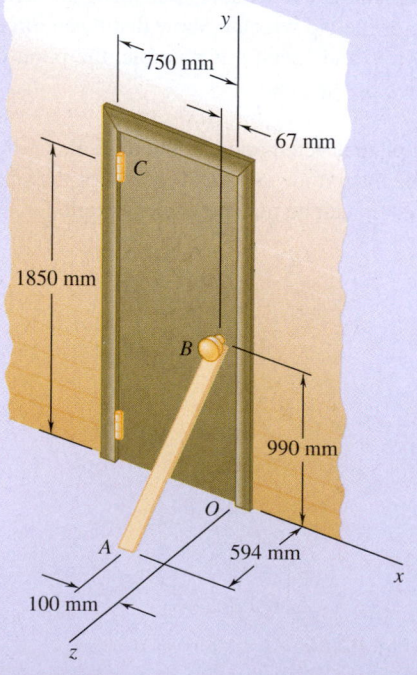

**Fig. P3.96**

**3.97** A 46-lb force **F** and a 2120-lb·in. couple **M** are applied to corner *A* of the block shown. Replace the given force-couple system with an equivalent force-couple system at corner *H*.

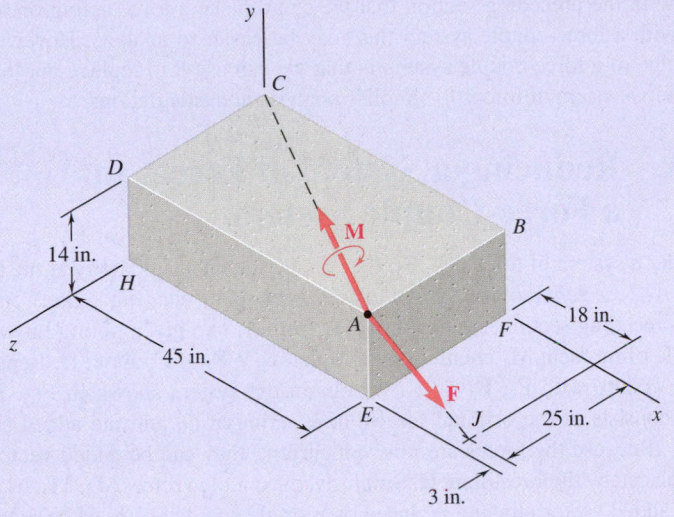

**Fig. P3.97**

**3.98** A 110-N force acting in a vertical plane parallel to the *yz* plane is applied to the 220-mm-long horizontal handle *AB* of a socket wrench. Replace the force with an equivalent force-couple system at the origin *O* of the coordinate system.

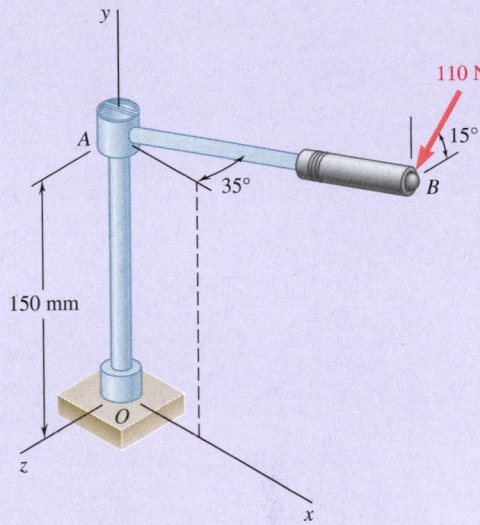

**Fig. P3.98**

**3.99** The 12-ft boom *AB* has a fixed end *A*, and the tension in cable *BC* is 570 lb. Replace the force that the cable exerts at *B* with an equivalent force-couple system at *A*.

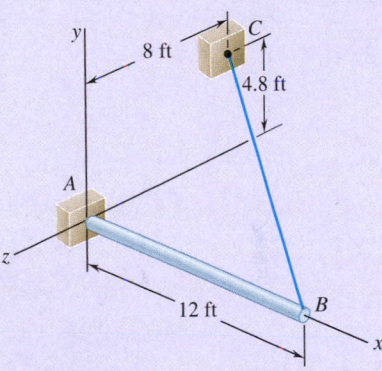

**Fig. P3.99**

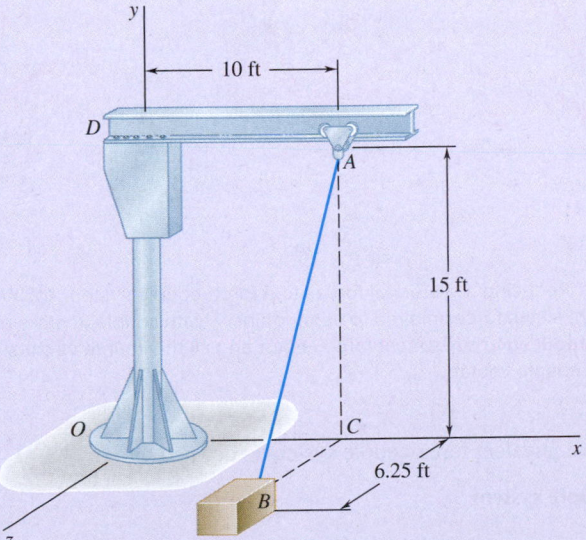

**Fig. P3.100**

**3.100** The jib crane shown is oriented so that its boom *AD* is parallel to the *x* axis and is used to move a heavy crate. Knowing that the tension in cable *AB* is 2.6 kips, replace the force exerted by the cable at *A* with an equivalent force-couple system at the center *O* of the base of the crane.

## 3.4   SIMPLIFYING SYSTEMS OF FORCES

We saw in the preceding section that we can replace a force acting on a rigid body with a force-couple system that may be easier to analyze. However, the true value of a force-couple system is that we can use it to replace not just one force but a system of forces to simplify analysis and calculations.

### 3.4A   Reducing a System of Forces to a Force-Couple System

Consider a system of forces $\mathbf{F}_1, \mathbf{F}_2, \mathbf{F}_3, \ldots$, acting on a rigid body at the points $A_1, A_2, A_3, \ldots$, *defined by the position vectors* $\mathbf{r}_1, \mathbf{r}_2, \mathbf{r}_3$, etc. (Fig. 3.36*a*). As seen in the preceding section, we can move $\mathbf{F}_1$ from $A_1$ to a given point $O$ if we add a couple of moment $\mathbf{M}_1$ equal to the moment $\mathbf{r}_1 \times \mathbf{F}_1$ of $\mathbf{F}_1$ about $O$. Repeating this procedure with $\mathbf{F}_2, \mathbf{F}_3, \ldots$, we obtain the system shown in Fig. 3.36*b*, which consists of the original forces, now acting at $O$, and the added couple vectors. Because the forces are now concurrent, they can be added vectorially and replaced by their resultant $\mathbf{R}$. Similarly, the couple vectors $\mathbf{M}_1, \mathbf{M}_2, \mathbf{M}_3, \ldots$, can be added vectorially and replaced by a single couple vector $\mathbf{M}_O^R$. Thus,

> **We can reduce any system of forces, however complex, to an equivalent force-couple system acting at a given point $O$.**

Note that, although each of the couple vectors $\mathbf{M}_1, \mathbf{M}_2, \mathbf{M}_3, \ldots$ in Fig. 3.36*b* is perpendicular to its corresponding force, the resultant force $\mathbf{R}$ and the resultant couple vector $\mathbf{M}_O^R$ shown in Fig. 3.36*c* are not, in general, perpendicular to each other.

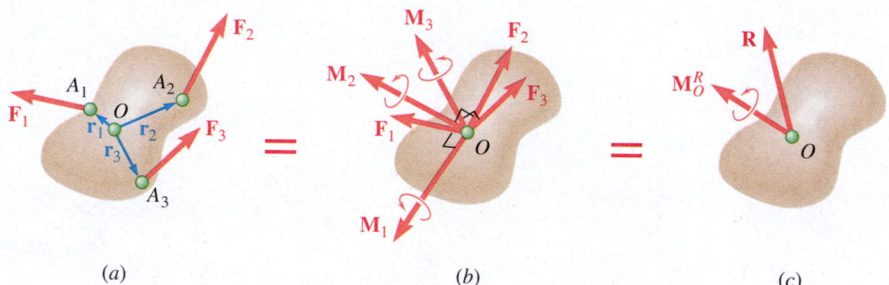

(*a*)                    (*b*)                    (*c*)

**Fig. 3.36**   Reducing a system of forces to a force-couple system. (*a*) Initial system of forces; (*b*) all the forces moved to act at point *O*, with couple vectors added; (*c*) all the forces reduced to a resultant force vector and all the couple vectors reduced to a resultant couple vector.

The equivalent force-couple system is defined by

**Force-couple system**

$$\mathbf{R} = \Sigma \mathbf{F} \qquad \mathbf{M}_O^R = \Sigma \mathbf{M}_O = \Sigma (\mathbf{r} \times \mathbf{F}) \tag{3.50}$$

These equations state that we obtain force $\mathbf{R}$ by adding all of the forces of the system, whereas we obtain the moment of the resultant couple vector $\mathbf{M}_O^R$, called the **moment resultant** of the system, by adding the moments about $O$ of all the forces of the system.

Once we have reduced a given system of forces to a force and a couple at a point $O$, we can replace it with a force and a couple at another point $O'$. The resultant force $\mathbf{R}$ will remain unchanged, whereas the new moment resultant $\mathbf{M}_{O'}^R$ will be equal to the sum of $\mathbf{M}_O^R$ and the moment about $O'$ of force $\mathbf{R}$ attached at $O$ (Fig. 3.37). We have

$$\mathbf{M}_{O'}^R = \mathbf{M}_O^R + \mathbf{s} \times \mathbf{R} \tag{3.51}$$

In practice, the reduction of a given system of forces to a single force $\mathbf{R}$ at $O$ and a couple vector $\mathbf{M}_O^R$ is carried out in terms of components. Resolving each position vector $\mathbf{r}$ and each force $\mathbf{F}$ of the system into rectangular components, we have

$$\mathbf{r} = x\mathbf{i} + y\mathbf{j} + z\mathbf{k} \tag{3.52}$$
$$\mathbf{F} = F_x\mathbf{i} + F_y\mathbf{j} + F_z\mathbf{k} \tag{3.53}$$

Substituting for $\mathbf{r}$ and $\mathbf{F}$ in Eq. (3.50) and factoring out the unit vectors $\mathbf{i}$, $\mathbf{j}$, and $\mathbf{k}$, we obtain $\mathbf{R}$ and $\mathbf{M}_O^R$ in the form

$$\mathbf{R} = R_x\mathbf{i} + R_y\mathbf{j} + R_z\mathbf{k} \qquad \mathbf{M}_O^R = M_x^R\mathbf{i} + M_y^R\mathbf{j} + M_z^R\mathbf{k} \tag{3.54}$$

The components $R_x$, $R_y$, and $R_z$ represent, respectively, the sums of the $x$, $y$, and $z$ components of the given forces and measure the tendency of the system to impart to the rigid body a translation in the $x$, $y$, or $z$ direction. Similarly, the components $M_x^R$, $M_y^R$, and $M_z^R$ represent, respectively, the sum of the moments of the given forces about the $x$, $y$, and $z$ axes and measure the tendency of the system to impart to the rigid body a rotation about the $x$, $y$, or $z$ axis.

If we need to know the magnitude and direction of force $\mathbf{R}$, we can obtain them from the components $R_x$, $R_y$, and $R_z$ by means of the relations in Eqs. (2.18) and (2.19) of Sec. 2.4A. Similar computations yield the magnitude and direction of the couple vector $\mathbf{M}_O^R$.

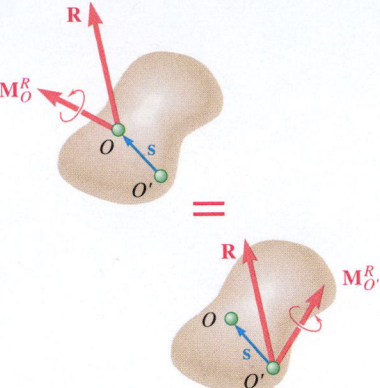

**Fig. 3.37** Once a system of forces has been reduced to a force-couple system at one point, we can replace it with an equivalent force-couple system at another point. The force resultant stays the same, but we have to add the moment of the resultant force about the new point to the resultant couple vector.

## 3.4B Equivalent and Equipollent Systems of Forces

We have just seen that any system of forces acting on a rigid body can be reduced to a force-couple system at a given point $O$. This equivalent force-couple system characterizes completely the effect of the given force system on the rigid body.

> **Two systems of forces are equivalent if they can be reduced to the same force-couple system at a given point $O$.**

Recall that the force-couple system at $O$ is defined by the relations in Eq. (3.50). Therefore, we can state that

> **Two systems of forces, $F_1$, $F_2$, $F_3$, . . . , and $F_1'$, $F_2'$, $F_3'$, . . . , that act on the same rigid body are equivalent if, and only if, the sums of the forces and the sums of the moments about a given point $O$ of the forces of the two systems are, respectively, equal.**

Mathematically, the necessary and sufficient conditions for the two systems of forces to be equivalent are

**Conditions for equivalent systems of forces**

$$\Sigma\mathbf{F} = \Sigma\mathbf{F}' \qquad \text{and} \qquad \Sigma\mathbf{M}_O = \Sigma\mathbf{M}_O' \tag{3.55}$$

**Photo 3.3** The forces exerted by the children upon the wagon can be replaced with an equivalent force-couple system when analyzing the motion of the wagon.
©Ingram Publishing/Getty Images RF

Note that to prove that two systems of forces are equivalent, we must establish the second of the relations in Eq. (3.55) with respect to *only one point O*. It will hold, however, with respect to *any point* if the two systems are equivalent.

Resolving the forces and moments in Eqs. (3.55) into their rectangular components, we can express the necessary and sufficient conditions for the equivalence of two systems of forces acting on a rigid body as

$$\Sigma F_x = \Sigma F_x' \qquad \Sigma F_y = \Sigma F_y' \qquad \Sigma F_z = \Sigma F_z'$$
$$\Sigma M_x = \Sigma M_x' \qquad \Sigma M_y = \Sigma M_y' \qquad \Sigma M_z = \Sigma M_z' \tag{3.56}$$

These equations have a simple physical significance. They express that

**Two systems of forces are equivalent if they tend to impart to the rigid body (1) the same translation in the *x*, *y*, and *z* directions, respectively, and (2) the same rotation about the *x*, *y*, and *z* axes, respectively.**

In general, when two systems of vectors satisfy Eqs. (3.55) or (3.56), i.e., when their resultants and their moment resultants about an arbitrary point *O* are respectively equal, the two systems are said to be **equipollent**. The result just established can thus be restated as

**If two systems of forces acting on a rigid body are equipollent, they are also equivalent.**

It is important to note that this statement does not apply to *any* system of vectors. Consider, for example, a system of forces acting on a set of independent particles that do *not* form a rigid body. A different system of forces acting on the same particles may happen to be equipollent to the first one; i.e., it may have the same resultant and the same moment resultant. Yet, because different forces now act on the various particles, their effects on these particles are different; the two systems of forces, while equipollent, are *not equivalent*.

## 3.4C  Further Reduction of a System of Forces

We have now seen that any given system of forces acting on a rigid body can be reduced to an equivalent force-couple system at *O*, consisting of a force **R** equal to the sum of the forces of the system, and a couple vector $\mathbf{M}_O^R$ of moment equal to the moment resultant of the system.

When $\mathbf{R} = 0$, the force-couple system reduces to the couple vector $\mathbf{M}_O^R$. The given system of forces then can be reduced to a single couple called the **resultant couple** of the system.

What are the conditions under which a given system of forces can be reduced to a single force? It follows from the preceding section that we can replace the force-couple system at *O* by a single force **R** acting along a new line of action if **R** and $\mathbf{M}_O^R$ are mutually perpendicular. The systems of forces that can be reduced to a single force, or *resultant*, are therefore the systems for which force **R** and the couple vector $\mathbf{M}_O^R$ are mutually perpendicular. This condition *is generally not satisfied* by systems of forces in space, but it *is satisfied* by systems consisting of (1) concurrent forces, (2) coplanar forces, or (3) parallel forces. Let's look at each case separately.

1. **Concurrent forces** act at the same point; therefore, we can add them directly to obtain their resultant **R**. Thus, they always reduce to a single force. Concurrent forces were discussed in detail in Chap. 2.

2. **Coplanar forces** act in the same plane, which we assume to be the plane of the figure (Fig. 3.38a). The sum **R** of the forces of the system also lies in the plane of the figure, whereas the moment of each force about $O$ and thus the moment resultant $\mathbf{M}_O^R$ are perpendicular to that plane. The force-couple system at $O$ consists, therefore, of a force **R** and a couple vector $\mathbf{M}_O^R$ that are mutually perpendicular (Fig. 3.38b).[†] We can reduce them to a single force **R** by moving **R** in the plane of the figure until its moment about $O$ becomes equal to $\mathbf{M}_O^R$. The distance from $O$ to the line of action of **R** is $d = M_O^R/R$ (Fig. 3.38c).

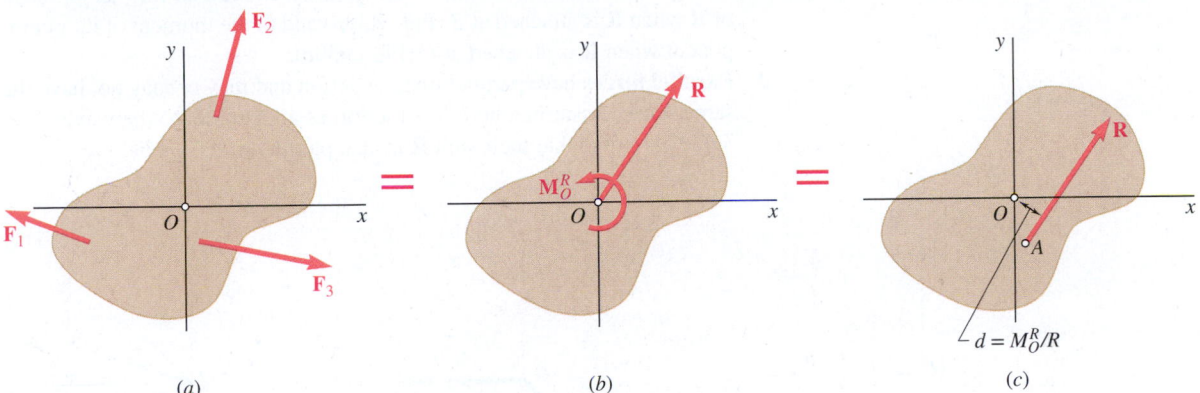

**Fig. 3.38**   Reducing a system of coplanar forces. (a) Initial system of forces; (b) equivalent force-couple system at O; (c) moving the resultant force to a point A such that the moment of **R** about O equals the couple vector.

As noted earlier, the reduction of a system of forces is considerably simplified if we resolve the forces into rectangular components. The force-couple system at $O$ is then characterized by the components (Fig. 3.39a)

$$R_x = \Sigma F_x \qquad R_y = \Sigma F_y \qquad M_z^R = M_O^R = \Sigma M_O \qquad (3.57)$$

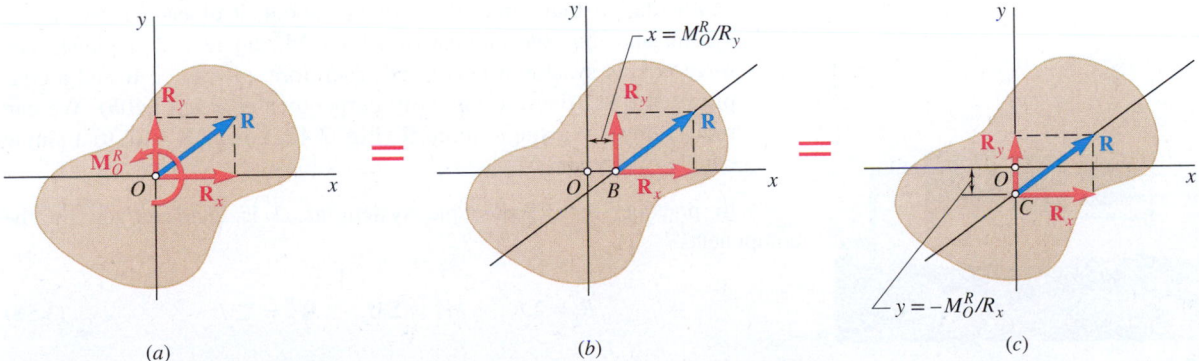

**Fig. 3.39**   Reducing a system of coplanar forces by using rectangular components. (a) From Fig. 3.38(b), resolve the resultant into components along the x and y axes; (b) determining the x intercept of the final line of action of the resultant; (c) determining the y intercept of the final line of action of the resultant.

[†]Because the couple vector $\mathbf{M}_O^R$ is perpendicular to the plane of the figure, we represent it by the symbol ↺. A counterclockwise couple ↺ represents a vector pointing out of the page and a clockwise couple ↻ represents a vector pointing into the page.

To reduce the system to a single force **R**, the moment of **R** about $O$ must be equal to $\mathbf{M}_O^R$. If we denote the coordinates of the point of application of the resultant by $x$ and $y$ and apply Eq. (3.22) of Sec. 3.1F, we have

$$xR_y - yR_x = M_O^R$$

This represents the equation of the line of action of **R**. We can also determine the $x$ and $y$ intercepts of the line of action of the resultant directly by noting that $\mathbf{M}_O^R$ must be equal to the moment about $O$ of the $y$ component of **R** when **R** is attached at $B$ (Fig. 3.39$b$) and to the moment of its $x$ component when **R** is attached at $C$ (Fig. 3.39$c$).

3. **Parallel forces** have parallel lines of action and may or may not have the same sense. Assuming here that the forces are parallel to the $y$ axis (Fig. 3.40$a$), we note that their sum **R** is also parallel to the $y$ axis.

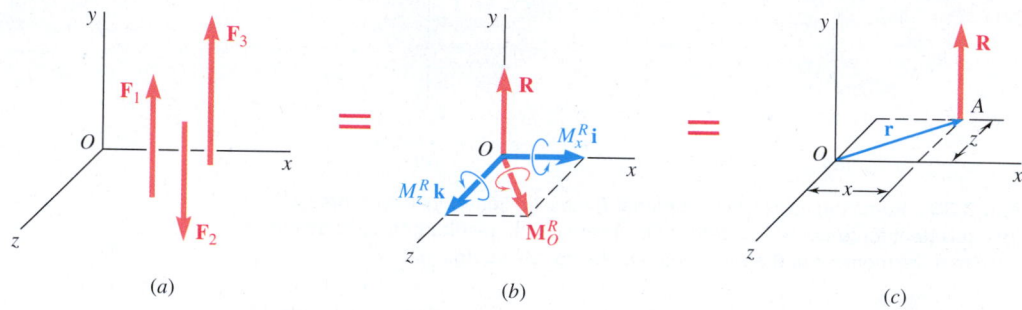

**Fig. 3.40** Reducing a system of parallel forces. ($a$) Initial system of forces; ($b$) equivalent force-couple system at $O$, resolved into components; ($c$) moving **R** to point $A$, chosen so that the moment of **R** about $O$ equals the resultant moment about $O$.

**Photo 3.4** The parallel wind forces acting on the highway signs can be reduced to a single equivalent force. Determining this force can simplify the calculation of the forces acting on the supports of the frame to which the signs are attached. ©Images-USA/Alamy RF

On the other hand, because the moment of a given force must be perpendicular to that force, the moment about $O$ of each force of the system and thus the moment resultant $\mathbf{M}_O^R$ lie in the $zx$ plane. The force-couple system at $O$ consists, therefore, of a force **R** and a couple vector $\mathbf{M}_O^R$ that are mutually perpendicular (Fig. 3.40$b$). We can reduce them to a single force **R** (Fig. 3.40$c$) or, if $\mathbf{R} = 0$, to a single couple of moment $\mathbf{M}_O^R$.

In practice, the force-couple system at $O$ is characterized by the components

$$R_y = \Sigma F_y \qquad M_x^R = \Sigma M_x \qquad M_z^R = \Sigma M_z \qquad (3.58)$$

The reduction of the system to a single force can be carried out by moving **R** to a new point of application $A(x, 0, z)$, which is chosen so that the moment of **R** about $O$ is equal to $\mathbf{M}_O^R$.

$$\mathbf{r} \times \mathbf{R} = \mathbf{M}_O^R$$
$$(x\mathbf{i} + z\mathbf{k}) \times R_y\mathbf{j} = M_x^R\mathbf{i} + M_z^R\mathbf{k}$$

By computing the vector products and equating the coefficients of the corresponding unit vectors in both sides of the equation, we obtain two scalar equations that define the coordinates of $A$:

$$-zR_y = M_x^R \quad \text{and} \quad xR_y = M_z^R$$

These equations express the fact that the moments of $\mathbf{R}$ about the $x$ and $z$ axes must be equal, respectively, to $M_x^R$ and $M_z^R$.

## *3.4D    Reduction of a System of Forces to a Wrench

In the general case of a system of forces in space, the equivalent force-couple system at $O$ consists of a force $\mathbf{R}$ and a couple vector $\mathbf{M}_O^R$ that are not perpendicular and where neither is zero (Fig. 3.41$a$). This system of forces *cannot* be reduced to a single force or to a single couple. However, we still have a way of simplifying this system further.

The simplification method consists of first replacing the couple vector by two other couple vectors that are obtained by resolving $\mathbf{M}_O^R$ into a component $\mathbf{M}_1$ along $\mathbf{R}$ and a component $\mathbf{M}_2$ in a plane perpendicular to $\mathbf{R}$ (Fig. 3.41$b$). Then, we can replace the couple vector $\mathbf{M}_2$ and force $\mathbf{R}$ by a single force $\mathbf{R}$ acting along a new line of action. The original system of forces thus reduces to $\mathbf{R}$ and to the couple vector $\mathbf{M}_1$ (Fig. 3.41$c$), i.e., to $\mathbf{R}$ and a couple acting in the plane perpendicular to $\mathbf{R}$.

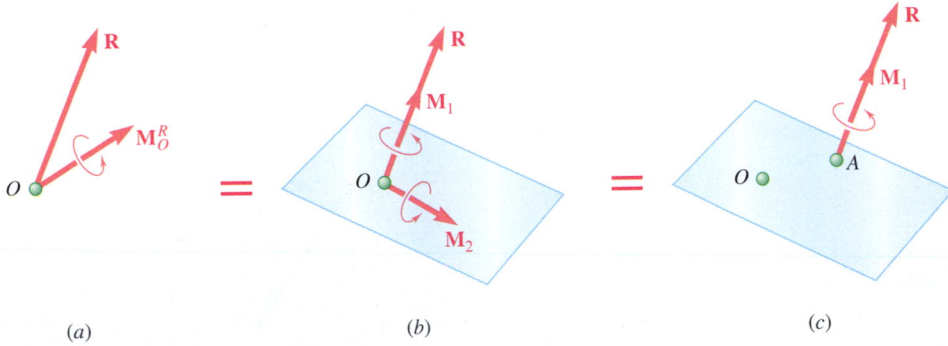

(a)                    (b)                    (c)

**Fig. 3.41**  Reducing a system of forces to a wrench. (a) General force system reduced to a single force and a couple vector, not perpendicular to each other; (b) resolving the couple vector into components along the line of action of the force and perpendicular to it; (c) moving the force and collinear couple vector (the wrench) to eliminate the couple vector perpendicular to the force.

This particular force-couple system is called a **wrench** because the resulting combination of push and twist is the same as that caused by an actual wrench. The line of action of $\mathbf{R}$ is known as the *axis of the wrench*, and the ratio $p = M_1/R$ is called the *pitch of the wrench*. A wrench therefore consists of two collinear vectors: a force $\mathbf{R}$ and a couple vector

$$\mathbf{M}_1 = p\mathbf{R} \tag{3.59}$$

Recall the expression in Eq. (3.33) for the projection of a vector on the line of action of another vector. Using this equation, we note that the projection of $\mathbf{M}_O^R$ on the line of action of $\mathbf{R}$ is

$$M_1 = \frac{\mathbf{R} \cdot \mathbf{M}_O^R}{R}$$

Thus, we can express the pitch of the wrench as[†]

$$p = \frac{M_1}{R} = \frac{\mathbf{R} \cdot \mathbf{M}_O^R}{R^2} \qquad (3.60)$$

To define the axis of the wrench, we can write a relation involving the position vector $\mathbf{r}$ of an arbitrary point $P$ located on that axis. We first attach the resultant force $\mathbf{R}$ and couple vector $\mathbf{M}_1$ at $P$ (Fig. 3.42). Then, because the moment about $O$ of this force-couple system must be equal to the moment resultant $\mathbf{M}_O^R$ of the original force system, we have

$$\mathbf{M}_1 + \mathbf{r} \times \mathbf{R} = \mathbf{M}_O^R \qquad (3.61)$$

Alternatively, using Eq. (3.59), we have

$$p\mathbf{R} + \mathbf{r} \times \mathbf{R} = \mathbf{M}_O^R \qquad (3.62)$$

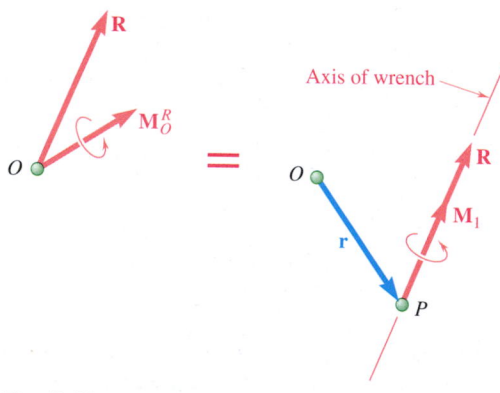

**Fig. 3.42** By finding the position vector $\mathbf{r}$ that locates any arbitrary point on the axis of the wrench, you can define the axis.

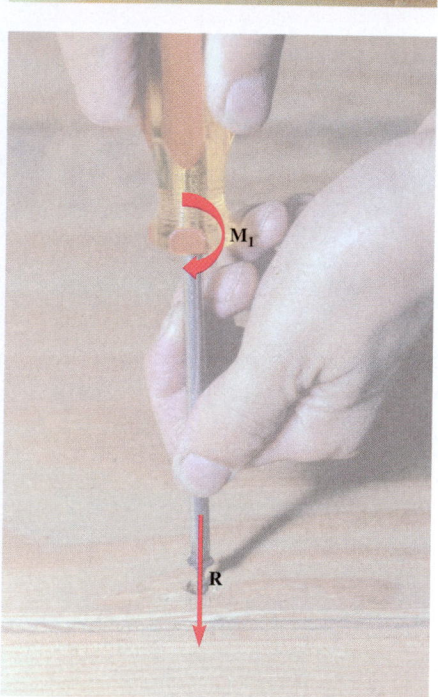

**Photo 3.5** The pushing-turning action associated with the tightening of a screw illustrates the collinear lines of action of the force and couple vector that constitute a wrench. ©Dana White/PhotoEdit.

[†]The expressions obtained for the projection of the couple vector on the line of action of $\mathbf{R}$ and for the pitch of the wrench are independent of the choice of point $O$. Using the relation (3.51) of Sec. 3.4A, we note that if a different point $O'$ had been used, the numerator in (3.60) would have been

$$\mathbf{R} \cdot \mathbf{M}_{O'}^R = \mathbf{R} \cdot (\mathbf{M}_O^R + \mathbf{s} \times \mathbf{R}) = \mathbf{R} \cdot \mathbf{M}_O^R + \mathbf{R} \cdot (\mathbf{s} \times \mathbf{R})$$

Because the mixed triple product $\mathbf{R} \cdot (\mathbf{s} \times \mathbf{R})$ is identically equal to zero, we have

$$\mathbf{R} \cdot \mathbf{M}_{O'}^R = \mathbf{R} \cdot \mathbf{M}_O^R$$

Thus, the scalar product $\mathbf{R} \cdot \mathbf{M}_O^R$ is independent of the choice of point $O$.

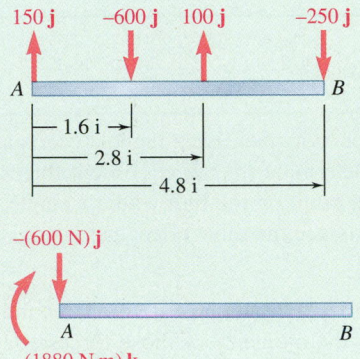

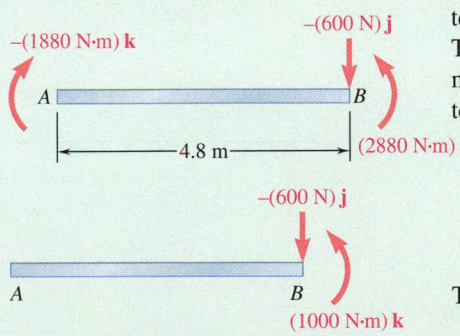

**Fig. 1**  Force-couple system at *A* that is equivalent to the given system of forces.

**Fig. 2**  Finding the force-couple system at *B* that is equivalent to that determined in part *a*.

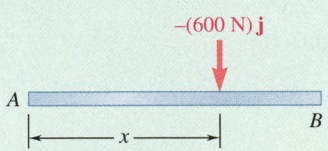

**Fig. 3**  Single force that is equivalent to the given system of forces.

## Sample Problem 3.8

A 4.80-m-long beam is subjected to the forces shown. Reduce the given system of forces to (*a*) an equivalent force-couple system at *A*, (*b*) an equivalent force-couple system at *B*, (*c*) a single force or resultant. *Note*: Because the reactions at the supports are not included in the given system of forces, the given system will not maintain the beam in equilibrium.

**STRATEGY:**  The *force* part of an equivalent force-couple system is simply the sum of the forces involved. The *couple* part is the sum of the moments caused by each force relative to the point of interest. Once you find the equivalent force-couple at one point, you can transfer it to any other point by a moment calculation.

**MODELING and ANALYSIS:**

**a. Force-Couple System at *A*.**  The force-couple system at *A* equivalent to the given system of forces consists of a force **R** and a couple $\mathbf{M}_A^R$ defined as (Fig. 1)

$$\mathbf{R} = \Sigma \mathbf{F}$$
$$= (150\text{ N})\mathbf{j} - (600\text{ N})\mathbf{j} + (100\text{ N})\mathbf{j} - (250\text{ N})\mathbf{j} = -(600\text{ N})\mathbf{j}$$
$$\mathbf{M}_A^R = \Sigma (\mathbf{r} \times \mathbf{F})$$
$$= (1.6\mathbf{i}) \times (-600\mathbf{j}) + (2.8\mathbf{i}) \times (100\mathbf{j}) + (4.8\mathbf{i}) \times (-250\mathbf{j})$$
$$= -(1880\text{ N·m})\mathbf{k}$$

The equivalent force-couple system at *A* is thus

$$\mathbf{R} = 600\text{ N} \downarrow \qquad \mathbf{M}_A^R = 1880\text{ N·m} \circlearrowleft \quad \blacktriangleleft$$

**b. Force-Couple System at *B*.**  You want to find a force-couple system at *B* equivalent to the force-couple system at *A* determined in part *a*. The force **R** is unchanged, but you must determine a new couple $\mathbf{M}_B^R$, the moment of which is equal to the moment about *B* of the force-couple system determined in part *a* (Fig. 2). You have

$$\mathbf{M}_B^R = \mathbf{M}_A^R + \overrightarrow{BA} \times \mathbf{R}$$
$$= -(1880\text{ N·m})\mathbf{k} + (-4.8\text{ m})\mathbf{i} \times (-600\text{ N})\mathbf{j}$$
$$= -(1880\text{ N·m})\mathbf{k} + (2880\text{ N·m})\mathbf{k} = +(1000\text{ N·m})\mathbf{k}$$

The equivalent force-couple system at *B* is thus

$$\mathbf{R} = 600\text{ N} \downarrow \quad \mathbf{M}_B^R = 1000\text{ N·m} \circlearrowleft \quad \blacktriangleleft$$

**c. Single Force or Resultant.**  The resultant of the given system of forces is equal to **R**, and its point of application must be such that the moment of **R** about *A* is equal to $\mathbf{M}_A^R$ (Fig. 3). This equality of moments leads to

$$\mathbf{r} \times \mathbf{R} = \mathbf{M}_A^R$$
$$x\mathbf{i} \times (-600\text{ N})\mathbf{j} = -(1880\text{ N·m})\mathbf{k}$$
$$-x(600\text{ N})\mathbf{k} = -(1880\text{ N·m})\mathbf{k}$$

*(continued)*

Solving for $x$, you get $x = 3.13$ m. Thus, the single force equivalent to the given system is defined as

$$R = 600\ N \downarrow \qquad x = 3.13\ m$$

**REFLECT and THINK:** This reduction of a given system of forces to a single equivalent force uses the same principles you will use later for finding centers of gravity and centers of mass, which are important parameters in engineering mechanics.

## Sample Problem 3.9

Four tugboats are bringing an ocean liner to its pier. Each tugboat exerts a 5000-lb force in the direction shown. Determine (a) the equivalent force-couple system at the foremast $O$, (b) the point on the hull where a single, more powerful tugboat should push to produce the same effect as the original four tugboats.

**STRATEGY:** The equivalent force-couple system is defined by the sum of the given forces and the sum of the moments of those forces at a particular point. A single tugboat could produce this system by exerting the resultant force at a point of application that produces an equivalent moment.

**MODELING and ANALYSIS:**

**a. Force-Couple System at O.** Resolve each of the given forces into components, as in Fig. 1 (kip units are used). The force-couple system at $O$ equivalent to the given system of forces consists of a force $\mathbf{R}$ and a couple $\mathbf{M}_O^R$ defined as

$$\begin{aligned}
\mathbf{R} &= \Sigma \mathbf{F} \\
&= (2.50\mathbf{i} - 4.33\mathbf{j}) + (3.00\mathbf{i} - 4.00\mathbf{j}) + (-5.00\mathbf{j}) + (3.54\mathbf{i} + 3.54\mathbf{j}) \\
&= 9.04\mathbf{i} - 9.79\mathbf{j}
\end{aligned}$$

$$\begin{aligned}
\mathbf{M}_O^R &= \Sigma (\mathbf{r} \times \mathbf{F}) \\
&= (-90\mathbf{i} + 50\mathbf{j}) \times (2.50\mathbf{i} - 4.33\mathbf{j}) \\
&\quad + (100\mathbf{i} + 70\mathbf{j}) \times (3.00\mathbf{i} - 4.00\mathbf{j}) \\
&\quad + (400\mathbf{i} + 70\mathbf{j}) \times (-5.00\mathbf{j}) \\
&\quad + (300\mathbf{i} - 70\mathbf{j}) \times (3.54\mathbf{i} + 3.54\mathbf{j}) \\
&= (390 - 125 - 400 - 210 - 2000 + 1062 + 248)\mathbf{k} \\
&= -1035\mathbf{k}
\end{aligned}$$

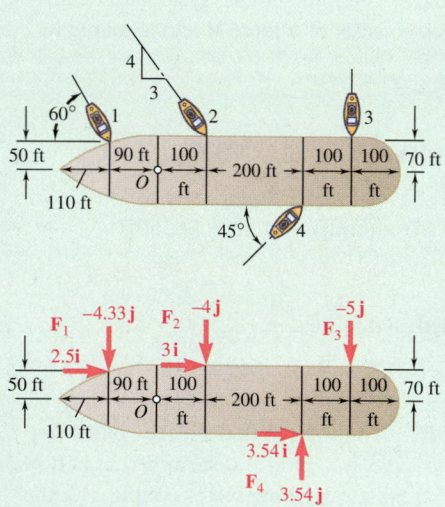

**Fig. 1** The given forces resolved into components.

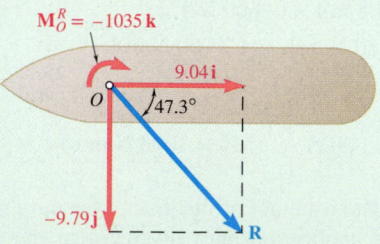

**Fig. 2** The equivalent force-couple system at $O$.

The equivalent force-couple system at $O$ is thus (Fig. 2)

$$\mathbf{R} = (9.04\ kips)\mathbf{i} - (9.79\ kips)\mathbf{j} \qquad \mathbf{M}_O^R = -(1035\ kip{\cdot}ft)\mathbf{k}$$

or

$$\mathbf{R} = 13.33\ kips \ ⦨47.3° \qquad \mathbf{M}_O^R = 1035\ kip{\cdot}ft \ \circlearrowleft$$

*(continued)*

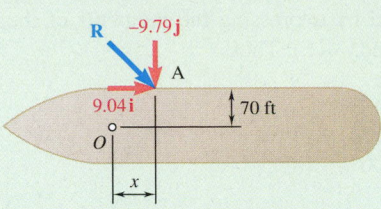

**Fig. 3** The point of application of a single tugboat to create the same effect as the given force system.

*Remark*: Because all the forces are contained in the plane of the figure, you would expect the sum of their moments to be perpendicular to that plane. Note that you could obtain the moment of each force component directly from the diagram by first forming the product of its magnitude and perpendicular distance to $O$ and then assigning to this product a positive or a negative sign, depending upon the sense of the moment.

**b. Single Tugboat.**   The force exerted by a single tugboat must be equal to $\mathbf{R}$, and its point of application $A$ must be such that the moment of $\mathbf{R}$ about $O$ is equal to $\mathbf{M}_O^R$ (Fig. 3). Observing that the position vector of $A$ is

$$\mathbf{r} = x\mathbf{i} + 70\mathbf{j}$$

you have

$$\mathbf{r} \times \mathbf{R} = \mathbf{M}_O^R$$
$$(x\mathbf{i} + 70\mathbf{j}) \times (9.04\mathbf{i} - 9.79\mathbf{j}) = -1035\mathbf{k}$$
$$-x(9.79)\mathbf{k} - 633\mathbf{k} = -1035\mathbf{k} \qquad x = 41.1 \text{ ft} \quad \triangleleft$$

**REFLECT and THINK:**   Reducing the given situation to that of a single force makes it easier to visualize the overall effect of the tugboats in maneuvering the ocean liner. But in practical terms, having four boats applying force allows for greater control in slowing and turning a large ship in a crowded harbor.

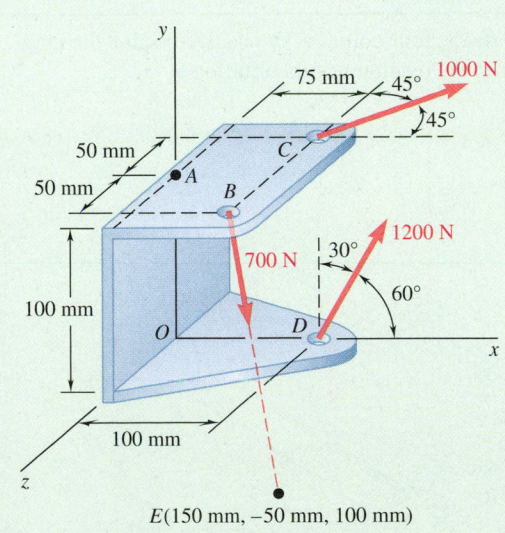

## Sample Problem 3.10

Three cables are attached to a bracket as shown. Replace the forces exerted by the cables with an equivalent force-couple system at $A$.

**STRATEGY:**   First determine the relative position vectors drawn from point $A$ to the points of application of the various forces and resolve the forces into rectangular components. Then, sum the forces and moments.

**MODELING and ANALYSIS:**   Note that $\mathbf{F}_B = (700 \text{ N})\boldsymbol{\lambda}_{BE}$, where

$$\boldsymbol{\lambda}_{BE} = \frac{\overrightarrow{BE}}{BE} = \frac{75\mathbf{i} - 150\mathbf{j} + 50\mathbf{k}}{175}$$

Using meters and newtons, the position and force vectors are

$$\mathbf{r}_{B/A} = \overrightarrow{AB} = 0.075\mathbf{i} + 0.050\mathbf{k} \qquad \mathbf{F}_B = 300\mathbf{i} - 600\mathbf{j} + 200\mathbf{k}$$
$$\mathbf{r}_{C/A} = \overrightarrow{AC} = 0.075\mathbf{i} - 0.050\mathbf{k} \qquad \mathbf{F}_C = 707\mathbf{i} - 707\mathbf{k}$$
$$\mathbf{r}_{D/A} = \overrightarrow{AD} = 0.100\mathbf{i} - 0.100\mathbf{j} \qquad \mathbf{F}_D = 600\mathbf{i} + 1039\mathbf{j}$$

The force-couple system at $A$ equivalent to the given forces consists of a force $\mathbf{R} = \Sigma\mathbf{F}$ and a couple $\mathbf{M}_A^R = \Sigma(\mathbf{r} \times \mathbf{F})$. Obtain the force $\mathbf{R}$ by adding respectively the $x$, $y$, and $z$ components of the forces:

$$\mathbf{R} = \Sigma\mathbf{F} = (1607 \text{ N})\mathbf{i} + (439 \text{ N})\mathbf{j} - (507 \text{ N})\mathbf{k} \quad \triangleleft$$

*(continued)*

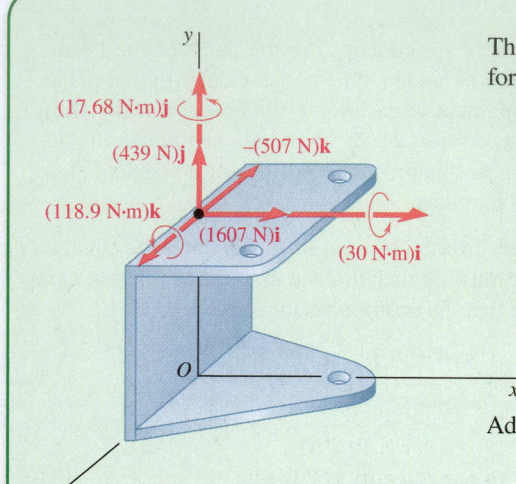

**Fig. 1** The rectangular components of the equivalent force-couple system at *A*.

The computation of $\mathbf{M}_A^R$ is facilitated by expressing the moments of the forces in the form of determinants (Sec. 3.1F). Thus,

$$\mathbf{r}_{B/A} \times \mathbf{F}_B = \begin{vmatrix} \mathbf{i} & \mathbf{j} & \mathbf{k} \\ 0.075 & 0 & 0.050 \\ 300 & -600 & 200 \end{vmatrix} = 30\mathbf{i} \qquad -45\mathbf{k}$$

$$\mathbf{r}_{C/A} \times \mathbf{F}_C = \begin{vmatrix} \mathbf{i} & \mathbf{j} & \mathbf{k} \\ 0.075 & 0 & -0.050 \\ 707 & 0 & -707 \end{vmatrix} = \qquad 17.68\mathbf{j}$$

$$\mathbf{r}_{D/A} \times \mathbf{F}_D = \begin{vmatrix} \mathbf{i} & \mathbf{j} & \mathbf{k} \\ 0.100 & -0.100 & 0 \\ 600 & 1039 & 0 \end{vmatrix} = \qquad 163.9\mathbf{k}$$

Adding these expressions, you have

$$\mathbf{M}_A^R = \Sigma(\mathbf{r} \times \mathbf{F}) = (30 \text{ N·m})\mathbf{i} + (17.68 \text{ N·m})\mathbf{j} + (118.9 \text{ N·m})\mathbf{k} \quad \blacktriangleleft$$

Fig. 1 shows the rectangular components of the force **R** and the couple $\mathbf{M}_A^R$.

**REFLECT and THINK:** The determinant approach to calculating moments shows its advantages in a general three-dimensional problem such as this.

## Sample Problem 3.11

A square foundation mat supports the four columns shown. Determine the magnitude and point of application of the resultant of the four loads.

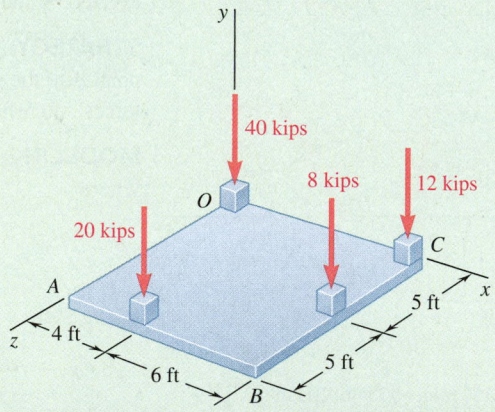

**STRATEGY:** Start by reducing the given system of forces to a force-couple system at the origin *O* of the coordinate system. Then, reduce the system further to a single force applied at a point with coordinates *x*, *z*.

**MODELING:** The force-couple system consists of a force **R** and a couple vector $\mathbf{M}_O^R$ defined as

$$\mathbf{R} = \Sigma\mathbf{F} \qquad \mathbf{M}_O^R = \Sigma(\mathbf{r} \times \mathbf{F})$$

*(continued)*

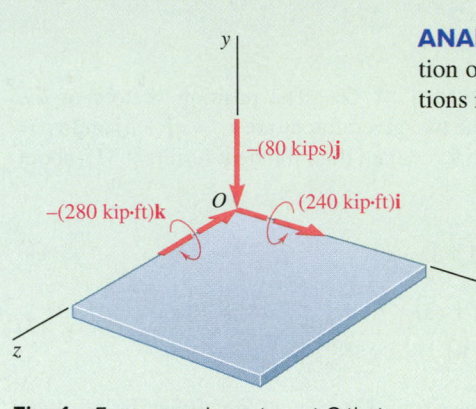

**Fig. 1** Force-couple system at $O$ that is equivalent to given force system.

**ANALYSIS:** After determining the position vectors of the points of application of the various forces, you may find it convenient to arrange the computations in tabular form. The results are shown in Fig. 1.

| r, ft | F, kips | r × F, kip·ft |
|---|---|---|
| 0 | $-40\mathbf{j}$ | 0 |
| $10\mathbf{i}$ | $-12\mathbf{j}$ | $-120\mathbf{k}$ |
| $10\mathbf{i} + 5\mathbf{k}$ | $-8\mathbf{j}$ | $40\mathbf{i} - 80\mathbf{k}$ |
| $4\mathbf{i} + 10\mathbf{k}$ | $-20\mathbf{j}$ | $200\mathbf{i} - 80\mathbf{k}$ |
| | $\mathbf{R} = -80\mathbf{j}$ | $\mathbf{M}_O^R = 240\mathbf{i} - 280\mathbf{k}$ |

The force $\mathbf{R}$ and the couple vector $\mathbf{M}_O^R$ are mutually perpendicular, so you can reduce the force-couple system further to a single force $\mathbf{R}$. Select the new point of application of $\mathbf{R}$ in the plane of the mat and in such a way that the moment of $\mathbf{R}$ about $O$ is equal to $\mathbf{M}_O^R$. Denote the position vector of the desired point of application by $\mathbf{r}$ and its coordinates by $x$ and $z$ (Fig. 2). Then

$$\mathbf{r} \times \mathbf{R} = \mathbf{M}_O^R$$
$$(x\mathbf{i} + z\mathbf{k}) \times (-80\mathbf{j}) = 240\mathbf{i} - 280\mathbf{k}$$
$$-80x\mathbf{k} + 80z\mathbf{i}) = 240\mathbf{i} - 280\mathbf{k}$$

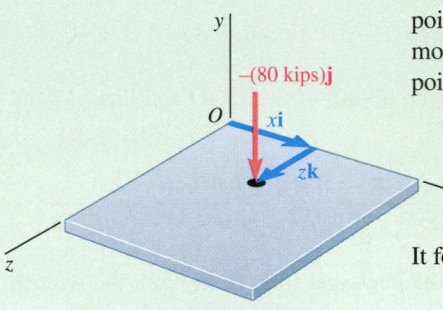

**Fig. 2** The single force that is equivalent to the given force system.

It follows that

$$-80x = -280 \qquad 80z = 240$$
$$x = 3.50 \text{ ft} \qquad z = 3.00 \text{ ft}$$

The resultant of the given system of forces is

$$\mathbf{R} = 80 \text{ kips} \downarrow \qquad \text{at } x = 3.50 \text{ ft}, z = 3.00 \text{ ft} \quad \blacktriangleleft$$

**REFLECT and THINK:** The fact that the given forces are all parallel simplifies the calculations, so the final step becomes just a two-dimensional analysis.

## Sample Problem 3.12

Two forces of the same magnitude $P$ act on a cube of side $a$ as shown. Replace the two forces by an equivalent wrench, and determine (*a*) the magnitude and direction of the resultant force $\mathbf{R}$, (*b*) the pitch of the wrench, (*c*) the point where the axis of the wrench intersects the *yz* plane.

**STRATEGY:** The first step is to determine the equivalent force-couple system at the origin $O$. Then, you can reduce this system to a wrench and determine its properties.

*(continued)*

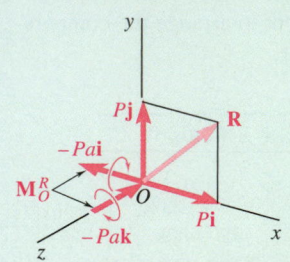

**Fig. 1** The force-couple system at $O$ that is equivalent to the given force system.

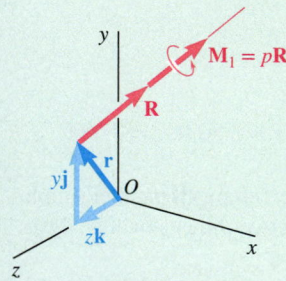

**Fig. 2** The wrench that is equivalent to the given force system.

**MODELING and ANALYSIS:**

**Equivalent Force-Couple System at O.** The position vectors of the points of application $E$ and $D$ of the two given forces are $\mathbf{r}_E = a\mathbf{i} + a\mathbf{j}$ and $\mathbf{r}_D = a\mathbf{j} + a\mathbf{k}$. The resultant $\mathbf{R}$ of the two forces and their moment resultant $\mathbf{M}_O^R$ about $O$ are (Fig. 1)

$$\mathbf{R} = \mathbf{F}_1 + \mathbf{F}_2 = P\mathbf{i} + P\mathbf{j} = P(\mathbf{i} + \mathbf{j}) \tag{1}$$

$$\begin{aligned}\mathbf{M}_O^R &= \mathbf{r}_E \times \mathbf{F}_1 + \mathbf{r}_D \times \mathbf{F}_2 = (a\mathbf{i} + a\mathbf{j}) \times P\mathbf{i} + (a\mathbf{j} + a\mathbf{k}) \times P\mathbf{j} \\ &= -Pa\mathbf{k} - Pa\mathbf{i} = -Pa(\mathbf{i} + \mathbf{k})\end{aligned} \tag{2}$$

**a. Resultant Force R.** It follows from Eq. (1) and Fig. 1 that the resultant force $\mathbf{R}$ has a magnitude of $R = P\sqrt{2}$, lies in the $xy$ plane, and forms angles of $45°$ with the $x$ and $y$ axes. Thus,

$$R = P\sqrt{2} \quad \theta_x = \theta_y = 45° \quad \theta_z = 90° \quad \blacktriangleleft$$

**b. Pitch of the Wrench.** Using Eq. (3.60) of Sec. 3.4D and Eqs. (1) and (2) above, the pitch $p$ of the wrench is

$$p = \frac{\mathbf{R} \cdot \mathbf{M}_O^R}{R^2} = \frac{P(\mathbf{i} + \mathbf{j}) \cdot (-Pa)(\mathbf{i} + \mathbf{k})}{(P\sqrt{2})2} = \frac{-P^2a(1 + 0 + 0)}{2P^2} \quad p = -\frac{a}{2} \quad \blacktriangleleft$$

**c. Axis of the Wrench.** From the pitch and from Eq. (3.59), the wrench consists of the force $\mathbf{R}$ found in Eq. (1) and the couple vector

$$\mathbf{M}_1 = p\mathbf{R} = -\frac{a}{2}P(\mathbf{i} + \mathbf{j}) = -\frac{Pa}{2}(\mathbf{i} + \mathbf{j}) \tag{3}$$

To find the point where the axis of the wrench intersects the $yz$ plane, set the moment of the wrench about $O$ equal to the moment resultant $\mathbf{M}_O^R$ of the original system:

$$\mathbf{M}_1 + \mathbf{r} \times \mathbf{R} = \mathbf{M}_O^R$$

Alternatively, noting that $\mathbf{r} = y\mathbf{j} + z\mathbf{k}$ (Fig. 2) and substituting for $\mathbf{R}$, $\mathbf{M}_O^R$, and $\mathbf{M}_1$ from Eqs. (1), (2), and (3), we have

$$-\frac{Pa}{2}(\mathbf{i} + \mathbf{j}) + (y\mathbf{j} + z\mathbf{k}) \times P(\mathbf{i} + \mathbf{j}) = -Pa(\mathbf{i} + \mathbf{k})$$

$$-\frac{Pa}{2}\mathbf{i} - \frac{Pa}{2}\mathbf{j} - Py\mathbf{k} + Pz\mathbf{j} - Pz\mathbf{i} = -Pa\mathbf{i} - Pa\mathbf{k}$$

Equating the coefficients of $\mathbf{k}$ and then the coefficients of $\mathbf{j}$, the final result is

$$y = a \quad z = a/2 \quad \blacktriangleleft$$

**REFLECT and THINK:** Conceptually, reducing a system of forces to a wrench is simply an additional application of finding an equivalent force-couple system.

# SOLVING PROBLEMS
# ON YOUR OWN

In this section, you studied the reduction and simplification of force systems. In solving the problems that follow, you will be asked to perform the following operations.

**1. Reducing a force system to a force and a couple at a given point A.** The force is the *resultant* **R** of the system that is obtained by adding the various forces. The moment of the couple is the *moment resultant* of the system that is obtained by adding the moments about $A$ of the various forces. We have

$$\mathbf{R} = \Sigma \mathbf{F} \qquad \mathbf{M}_A^R = \Sigma(\mathbf{r} \times \mathbf{F})$$

where the position vector **r** is drawn from $A$ to *any point* on the line of action of **F**.

**2. Moving a force-couple system from point A to point B.** If you wish to reduce a given force system to a force-couple system at point $B$, you need not recompute the moments of the forces about $B$ after you have reduced it to a force-couple system at point $A$. The resultant **R** remains unchanged, and you can obtain the new moment resultant $\mathbf{M}_B^R$ by adding the moment about $B$ of the force **R** applied at $A$ to $\mathbf{M}_A^R$ (Sample Prob. 3.8). Denoting the vector drawn from $B$ to $A$ as **s**, you have

$$\mathbf{M}_B^R = \mathbf{M}_A^R + \mathbf{s} \times \mathbf{R}$$

**3. Checking whether two force systems are equivalent.** First, reduce each force system to a force-couple system *at the same, but arbitrary, point A* (as explained in the first operation). The two force systems are equivalent (i.e., they have the same effect on the given rigid body) if the two reduced force-couple systems are identical; that is, if

$$\Sigma \mathbf{F} = \Sigma \mathbf{F}' \qquad \text{and} \qquad \Sigma \mathbf{M}_A = \Sigma \mathbf{M}_A'$$

You should recognize that if the first of these equations is not satisfied—that is, if the two systems do not have the same resultant **R**—the two systems cannot be equivalent, and there is no need to check whether or not the second equation is satisfied.

**4. Reducing a given force system to a single force.** First, reduce the given system to a force-couple system consisting of the resultant **R** and the couple vector $\mathbf{M}_A^R$ at some convenient point $A$ (as explained in the first operation). Recall from Sec. 3.4 that further reduction to a single force is possible *only if the force **R** and the couple vector $\mathbf{M}_A^R$ are mutually perpendicular.* This will certainly be the case for systems of forces that are either *concurrent, coplanar,* or *parallel.* You can then obtain the required single force by moving **R** until its moment about $A$ is equal to $\mathbf{M}_A^R$, as you did in several problems in Sec. 3.4. More formally, the position vector **r** drawn from $A$ to any point on the line of action of the single force **R** must satisfy the equation

$$\mathbf{r} \times \mathbf{R} = \mathbf{M}_A^R$$

*(continued)*

151

This procedure was illustrated in Sample Probs. 3.8, 3.9, and 3.11.

**5. Reducing a given force system to a wrench.** If the given system includes forces that are not concurrent, coplanar, or parallel, the equivalent force-couple system at a point $A$ will consist of a force $\mathbf{R}$ and a couple vector $\mathbf{M}_A^R$ that, in general, *are not mutually perpendicular.* (To check whether $\mathbf{R}$ and $\mathbf{M}_A^R$ are mutually perpendicular, form their scalar product. If this product is zero, they are mutually perpendicular; otherwise, they are not.) If $\mathbf{R}$ and $\mathbf{M}_A^R$ are not mutually perpendicular, the force-couple system (and thus the given system of forces) *cannot be reduced to a single force.* However, the system can be reduced to a *wrench*—the combination of a force $\mathbf{R}$ and a couple vector $\mathbf{M}_1$ directed along a common line of action called the *axis of the wrench* (Fig. 3.42). The ratio $p = M_1/R$ is called the *pitch* of the wrench.

To reduce a given force system to a wrench, you should follow these steps:

a. Reduce the given system to an equivalent force-couple system ($\mathbf{R}$, $\mathbf{M}_O^R$), typically located at the origin $O$.

b. Determine the pitch $p$ from Eq. (3.60),

$$p = \frac{M_1}{R} = \frac{\mathbf{R} \cdot \mathbf{M}_O^R}{R^2}$$

and the couple vector from $\mathbf{M}_1 = p\mathbf{R}$.

c. Set the moment about $O$ of the wrench equal to the moment resultant $\mathbf{M}_O^R$ of the force-couple system at $O$:

$$\mathbf{M}_1 + \mathbf{r} \times \mathbf{R} = \mathbf{M}_O^R \qquad (3.61)$$

This equation allows you to determine the point where the line of action of the wrench intersects a specified plane, because the position vector $\mathbf{r}$ is directed from $O$ to that point. These steps are illustrated in Sample Prob. 3.12. Although determining a wrench and the point where its axis intersects a plane may appear difficult, the process is simply the application of several of the ideas and techniques developed in this chapter. Once you have mastered the wrench, you can feel confident that you understand much of Chap. 3.

# Problems

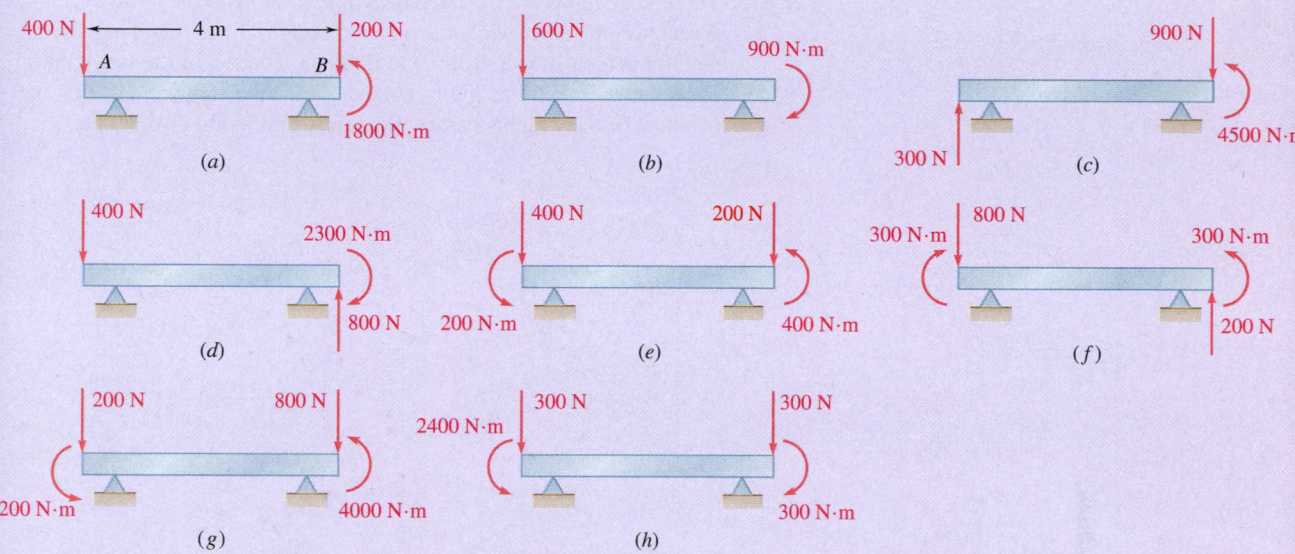

Fig. P3.101

**3.101** A 4-m-long beam is subjected to a variety of loadings. (*a*) Replace each loading with an equivalent force-couple system at end *A* of the beam. (*b*) Which of the loadings are equivalent?

**3.102** A 4-m-long beam is loaded as shown. Determine the loading of Prob. 3.101 that is equivalent to this loading.

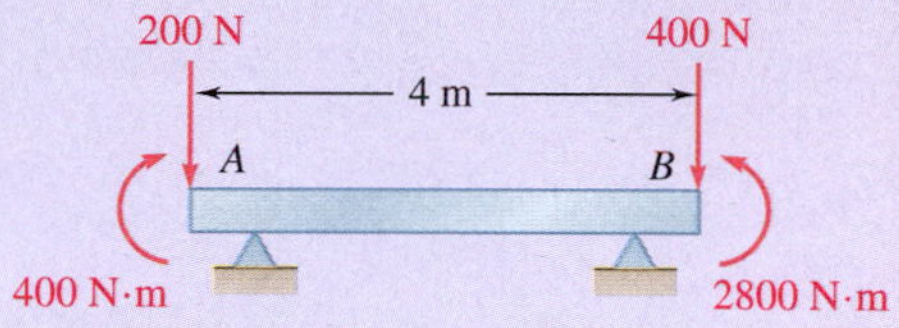

**3.103** Determine the single equivalent force and the distance from point *A* to its line of action for the beam and loading of (*a*) Prob. 3.101*b*, (*b*) Prob. 3.101*d*, (*c*) Prob. 3.101*e*.

Fig. P3.102

**3.104** Five separate force-couple systems act at the corners of a piece of sheet metal that has been bent into the shape shown. Determine which of these systems is equivalent to a force $\mathbf{F} = (10\ \text{lb})\mathbf{i}$ and a couple of moment $\mathbf{M} = (15\ \text{lb·ft})\mathbf{j} + (15\ \text{lb·ft})\mathbf{k}$ located at the origin.

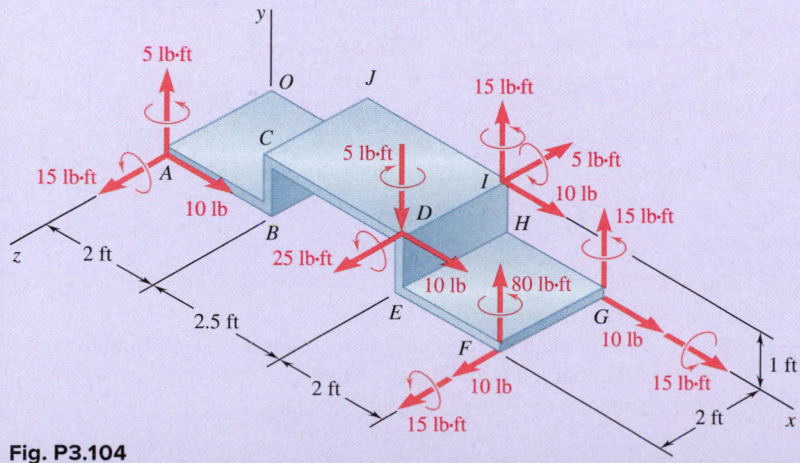

Fig. P3.104

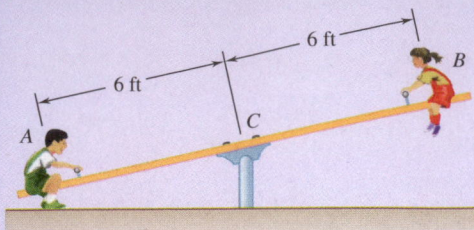

**Fig. P3.105**

**3.105** The weights of two children sitting at ends $A$ and $B$ of a seesaw are 84 lb and 64 lb, respectively. Where should a third child sit so that the resultant of the weights of the three children will pass through $C$ if she weighs (a) 60 lb, (b) 52 lb?

**3.106** Three stage lights are mounted on a pipe as shown. The lights at $A$ and $B$ each weigh 4.1 lb, while the one at $C$ weighs 3.5 lb. (a) If $d = 25$ in., determine the distance from $D$ to the line of action of the resultant of the weights of the three lights. (b) Determine the value of $d$ so that the resultant of the weights passes through the midpoint of the pipe.

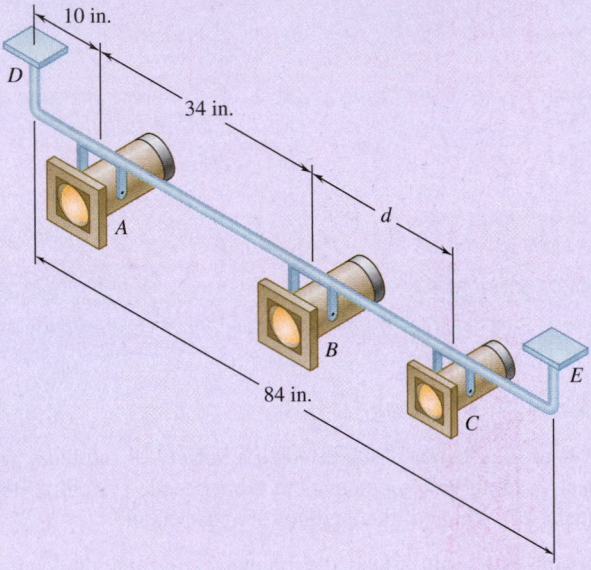

**Fig. P3.106**

**3.107** A beam supports three loads of given magnitude and a fourth load whose magnitude is a function of position. If $b = 1.5$ m and the loads are to be replaced with a single equivalent force, determine (a) the value of $a$ so that the distance from support $A$ to the line of action of the equivalent force is maximum, (b) the magnitude of the equivalent force and its point of application on the beam.

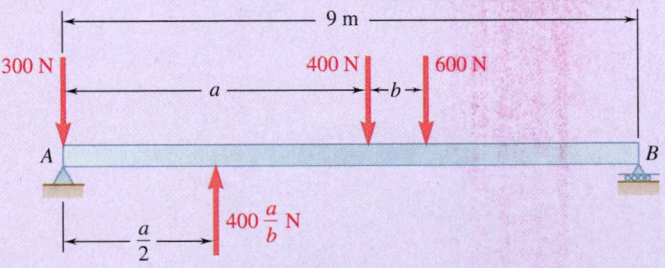

**Fig. *P3.107***

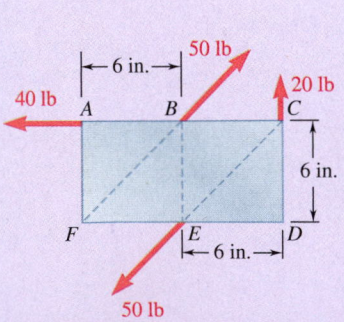

**Fig. P3.108**

**3.108** A $6 \times 12$-in. plate is subjected to four loads as shown. Find the resultant of the four loads and the two points at which the line of action of the resultant intersects the edge of the plate.

**3.109** Gear *C* is rigidly attached to arm *AB*. If the forces and couple shown can be reduced to a single equivalent force at *A*, determine the equivalent force and the magnitude of the couple **M**.

**3.110** To test the strength of a 625 × 500-mm suitcase, forces are applied as shown. If *P* = 88 N, (*a*) determine the resultant of the applied forces, (*b*) locate the two points where the line of action of the resultant intersects the edge of the suitcase.

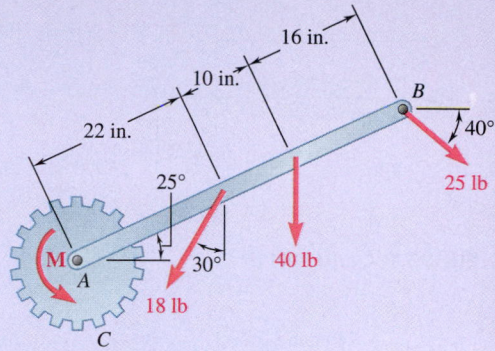

Fig. *P3.109*

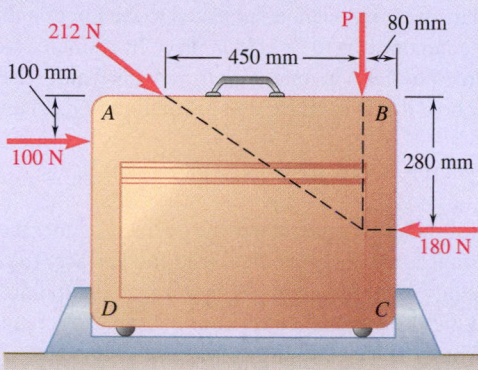

Fig. P3.110

**3.111** Two cables exert forces of 90 kN each on a truss of weight *W* = 200 kN. Find the resultant of these three forces acting on the truss and the point of intersection of its line of action with line *AB*.

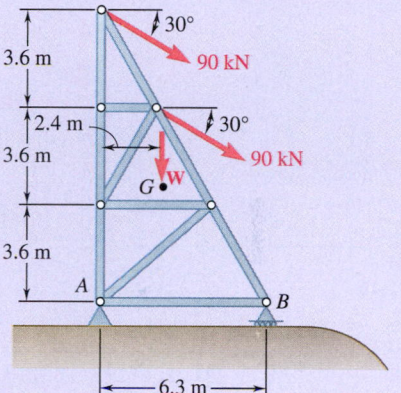

Fig. P3.111

**3.112** Pulleys *A* and *B* are mounted on bracket *CDEF*. The tension on each side of the two belts is as shown. Replace the four forces with a single equivalent force, and determine where its line of action intersects the bottom edge of the bracket.

**3.113** The roof of a building frame is subjected to the wind loading shown. Determine (*a*) the equivalent force-couple system at *D*, (*b*) the resultant of the loading and its line of action.

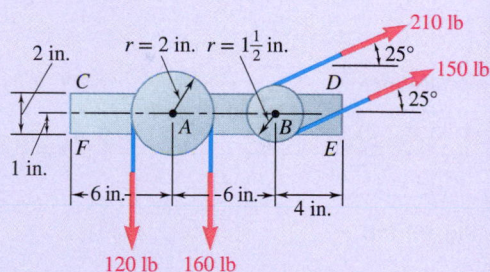

Fig. *P3.112*

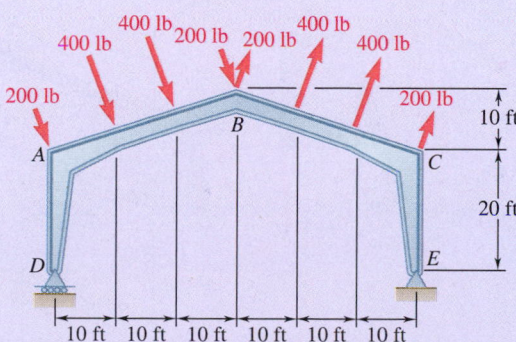

Fig. P3.113

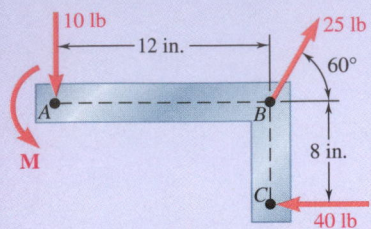

**Fig. P3.114 and P3.115**

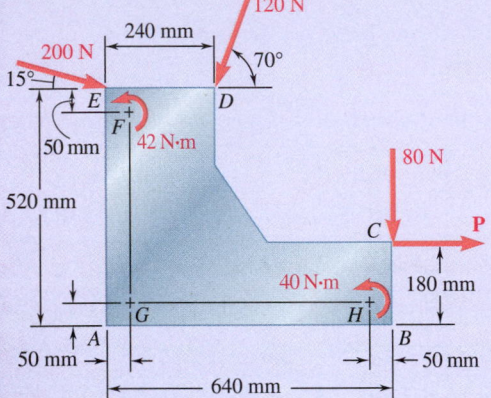

**Fig. P3.116**

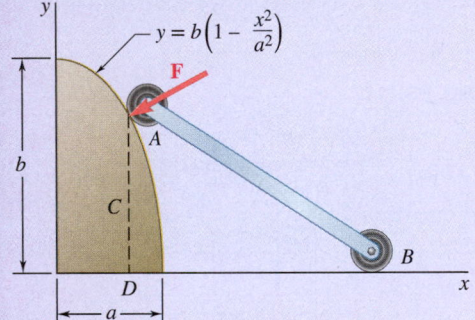

**Fig. P3.118**

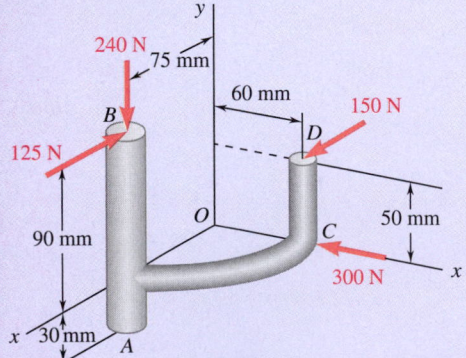

**Fig. P3.119**

**3.114** A couple of magnitude $M = 80$ lb·in. and the three forces shown are applied to an angle bracket. (*a*) Find the resultant of this system of forces. (*b*) Locate the points where the line of action of the resultant intersects line *AB* and line *BC*.

**3.115** A couple **M** and the three forces shown are applied to an angle bracket. Find the moment of the couple if the line of action of the resultant of the force system is to pass through (*a*) point *A*, (*b*) point *B*, (*c*) point *C*.

**3.116** A machine component is subjected to the forces and couples shown. The component is to be held in place by a single rivet that can resist a force but not a couple. For $P = 0$, determine the location of the rivet hole if it is to be located (*a*) on line *FG*, (*b*) on line *GH*.

**3.117** Solve Prob. 3.116, assuming that $P = 60$ N.

**3.118** As follower *AB* rolls along the surface of member *C*, it exerts a constant force **F** perpendicular to the surface. (*a*) Replace **F** with an equivalent force-couple system at the point *D* obtained by drawing the perpendicular from the point of contact to the *x* axis. (*b*) For $a = 1$ m and $b = 2$ m, determine the value of *x* for which the moment of the equivalent force-couple system at *D* is maximum.

**3.119** A machine component is subjected to the forces shown, each of which is parallel to one of the coordinate axes. Replace these forces with an equivalent force-couple system at *A*.

**3.120** Two 150-mm-diameter pulleys are mounted on line shaft *AD*. The belts at *B* and *C* lie in vertical planes parallel to the *yz* plane. Replace the belt forces shown with an equivalent force-couple system at *A*.

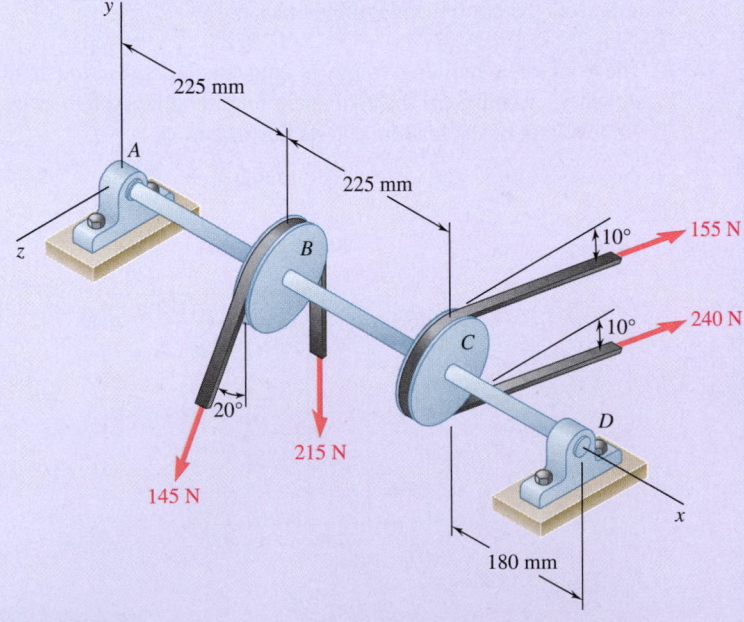

**Fig. P3.120**

**3.121** As an adjustable brace *BC* is used to bring a wall into plumb, the force-couple system shown is exerted on the wall. Replace this force-couple system with an equivalent force-couple system at *A* if *R* = 21.2 lb and *M* = 13.25 lb·ft.

**3.122** In order to unscrew the tapped faucet *A*, a plumber uses two pipe wrenches as shown. By exerting a 40-lb force on each wrench at a distance of 10 in. from the axis of the pipe and in a direction perpendicular to the pipe and to the wrench, he prevents the pipe from rotating, and thus he avoids loosening or further tightening the joint between the pipe and the tapped elbow *C*. Determine (*a*) the angle *θ* that the wrench at *A* should form with the vertical if elbow *C* is not to rotate about the vertical, (*b*) the force-couple system at *C* equivalent to the two 40-lb forces when this condition is satisfied.

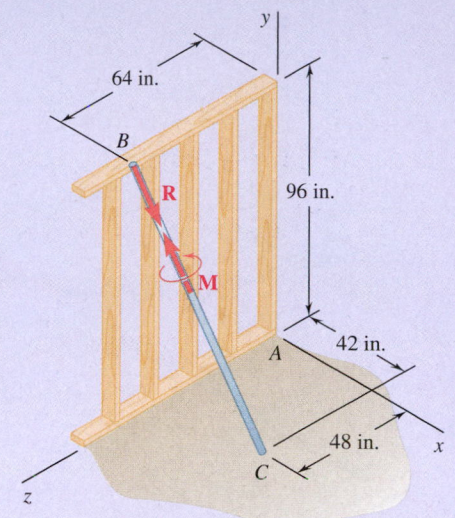

Fig. *P3.121*

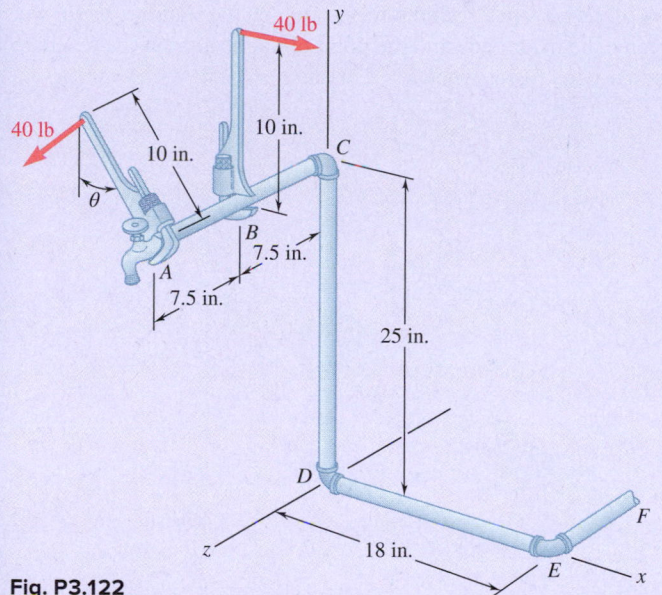

Fig. P3.122

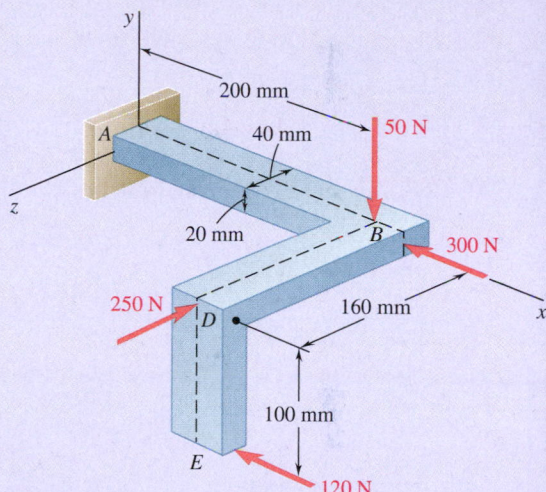

Fig. P3.124

**3.123** Assuming *θ* = 60° in Prob. 3.122, replace the two 40-lb forces with an equivalent force-couple system at *D* and determine whether the plumber's action tends to tighten or loosen the joint between (*a*) pipe *CD* and elbow *D*, (*b*) elbow *D* and pipe *DE*. Assume all threads to be right-handed.

**3.124** Four forces are applied to the machine component *ABDE* as shown. Replace these forces with an equivalent force-couple system at *A*.

**3.125** A blade held in a brace is used to tighten a screw at *A*. (*a*) Determine the forces exerted at *B* and *C*, knowing that these forces are equivalent to a force-couple system at *A* consisting of $\mathbf{R} = -(25\ \text{N})\mathbf{i} + R_y\mathbf{j} + R_z\mathbf{k}$ and $\mathbf{M}_A^R = -(13.5\ \text{N·m})\mathbf{i}$. (*b*) Find the corresponding values of $R_y$ and $R_z$. (*c*) What is the orientation of the slot in the head of the screw for which the blade is least likely to slip when the brace is in the position shown?

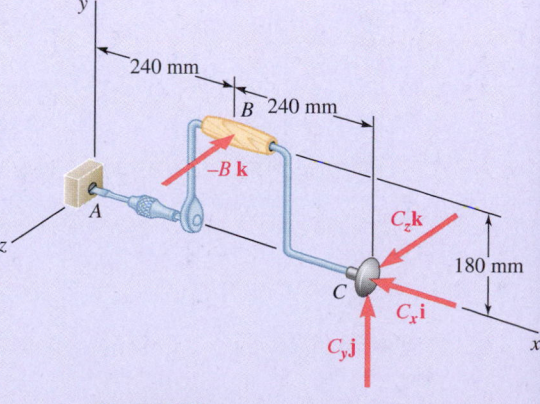

Fig. P3.125

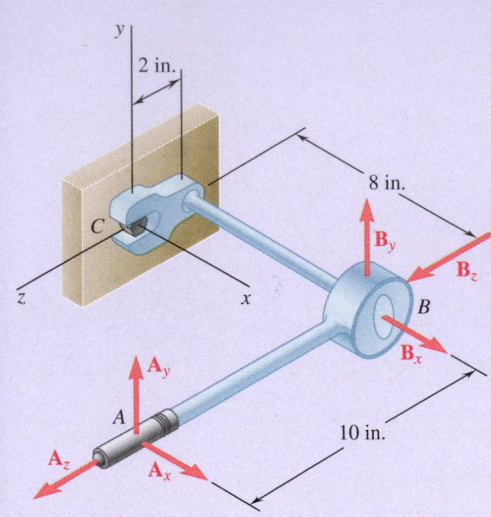

Fig. P3.126

**3.126** A mechanic uses a crowfoot wrench to loosen a bolt at $C$. The mechanic holds the socket wrench handle at points $A$ and $B$ and applies forces at these points. Knowing that these forces are equivalent to a force-couple system at $C$ consisting of the force $\mathbf{C} = -(8\text{ lb})\mathbf{i} + (4\text{ lb})\mathbf{k}$ and the couple $\mathbf{M}_C = (360\text{ lb·in.})\mathbf{i}$, determine the forces applied at $A$ and at $B$ when $A_z = 2$ lb.

**3.127** Four horizontal forces act on a vertical quarter-circular plate of radius 250 mm. Determine the magnitude and point of application of the resultant of the four forces if $P = 40$ N.

**3.128** Determine the magnitude of the force $\mathbf{P}$ for which the resultant of the four forces acts on the rim of the plate.

**3.129** Four signs are mounted on a frame spanning a highway, and the magnitudes of the horizontal wind forces acting on the signs are as shown. Determine the magnitude and the point of application of the resultant of the four wind forces when $a = 1$ ft and $b = 12$ ft.

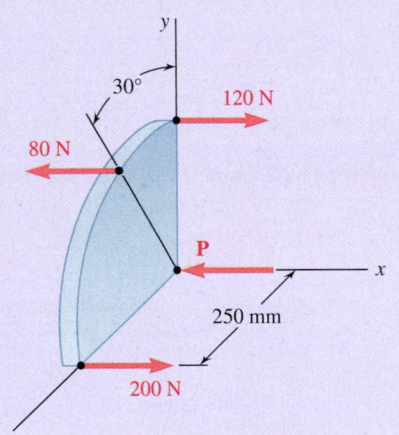

Fig. P3.127 and P3.128

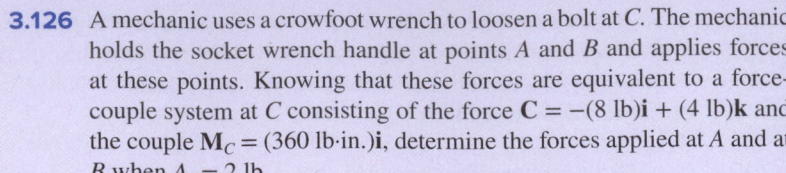

Fig. P3.129 and P3.130

**3.130** Four signs are mounted on a frame spanning a highway, and the magnitudes of the horizontal wind forces acting on the signs are as shown. Determine $a$ and $b$ so that the point of application of the resultant of the four forces is at $G$.

**3.131** A concrete foundation mat of 5-m radius supports four equally spaced columns, each of which is located 4 m from the center of the mat. Determine the magnitude and the point of application of the resultant of the four loads.

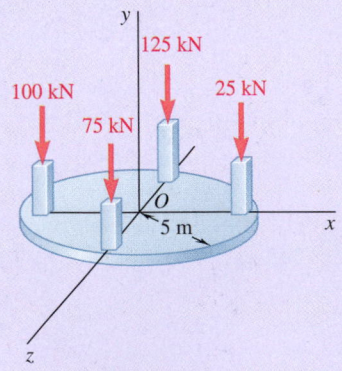

Fig. *P3.131*

**3.132** Determine the magnitude and the point of application of the smallest additional load that must be applied to the foundation mat of Prob. 3.131 if the resultant of the five loads is to pass through the center of the mat.

**\*3.133** Three forces of the same magnitude $P$ act on a cube of side $a$ as shown. Replace the three forces with an equivalent wrench and determine (a) the magnitude and direction of the resultant force **R**, (b) the pitch of the wrench, (c) the axis of the wrench.

**\*3.134** A piece of sheet metal is bent into the shape shown and is acted upon by three forces. If the forces have the same magnitude $P$, replace them with an equivalent wrench and determine (a) the magnitude and the direction of the resultant force **R**, (b) the pitch of the wrench, (c) the axis of the wrench.

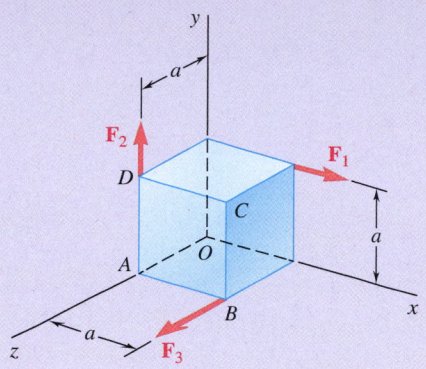

**Fig. P3.133**

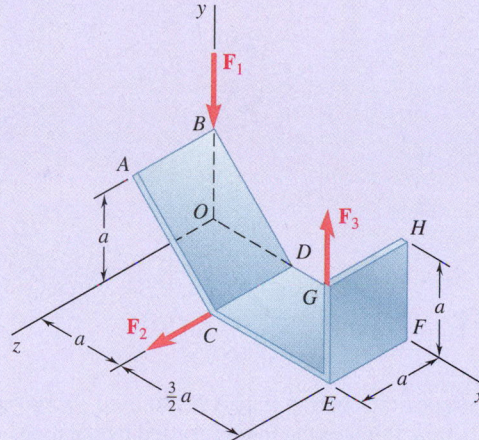

**Fig. P3.134**

**\*3.135 and \*3.136** The forces and couples shown are applied to two screws as a piece of sheet metal is fastened to a block of wood. Reduce the forces and the couples to an equivalent wrench and determine (a) the resultant force **R**, (b) the pitch of the wrench, (c) the point where the axis of the wrench intersects the $xz$ plane.

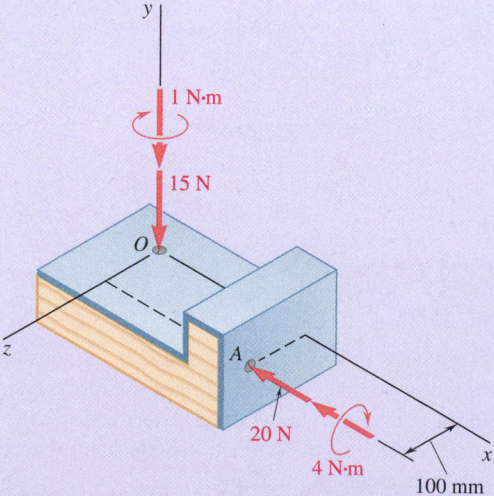

**Fig. P3.135**

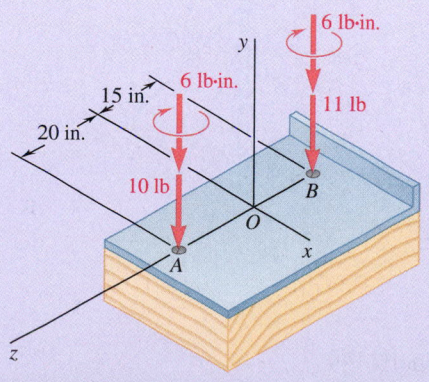

**Fig. P3.136**

*3.137 and *3.138  Two bolts at *A* and *B* are tightened by applying the forces and couples shown. Replace the two wrenches with a single equivalent wrench and determine (*a*) the resultant **R**, (*b*) the pitch of the single equivalent wrench, (*c*) the point where the axis of the wrench intersects the *xz* plane.

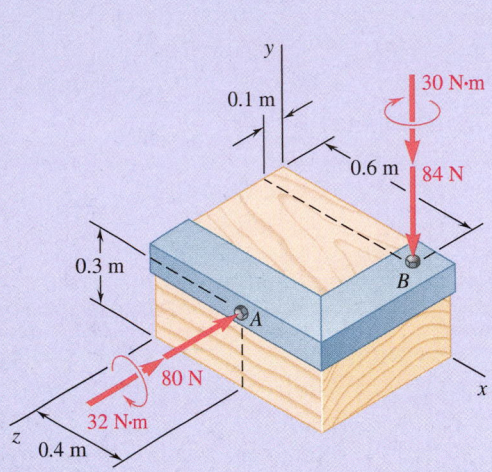

Fig. P3.137

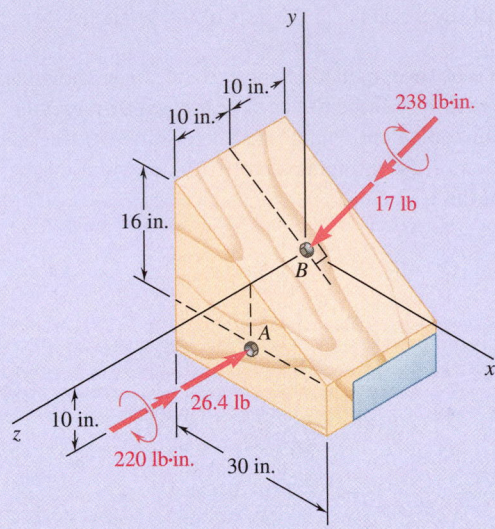

Fig. *P3.138*

*3.139  Two ropes attached at *A* and *B* are used to move the trunk of a fallen tree. Replace the forces exerted by the ropes with an equivalent wrench and determine (*a*) the resultant force **R**, (*b*) the pitch of the wrench, (*c*) the point where the axis of the wrench intersects the *yz* plane.

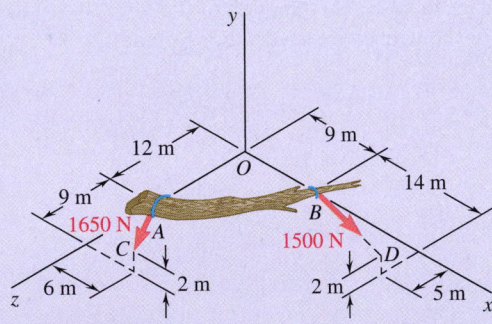

Fig. *P3.139*

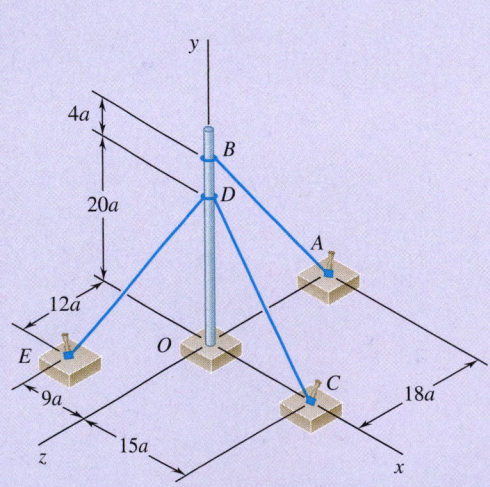

Fig. P3.140

*3.140  A flagpole is guyed by three cables. If the tensions in the cables have the same magnitude *P*, replace the forces exerted on the pole with an equivalent wrench and determine (*a*) the resultant force **R**, (*b*) the pitch of the wrench, (*c*) the point where the axis of the wrench intersects the *xz* plane.

**\*3.141 and \*3.142**  Determine whether the force-and-couple system shown can be reduced to a single equivalent force **R**. If it can, determine **R** and the point where the line of action of **R** intersects the *yz* plane. If it cannot be reduced, replace the given system with an equivalent wrench and determine its resultant, its pitch, and the point where its axis intersects the *yz* plane.

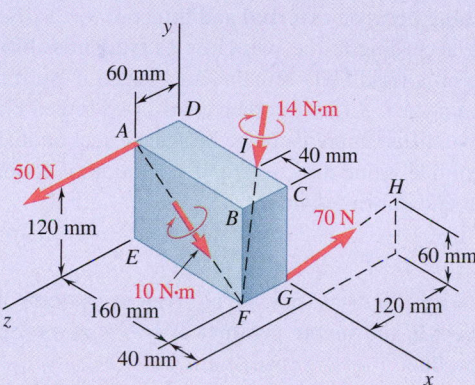

Fig. P3.141

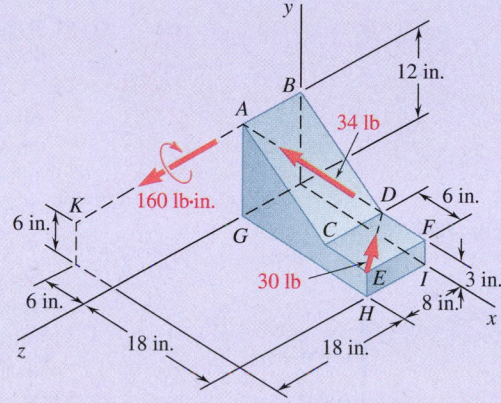

Fig. *P3.142*

**\*3.143**  Replace the wrench shown with an equivalent system consisting of two forces perpendicular to the *y* axis and applied respectively at *A* and *B*.

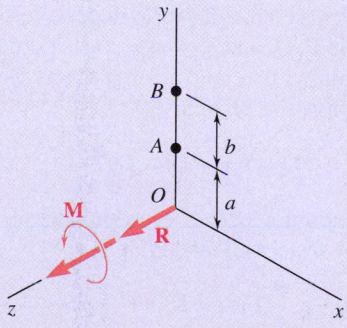

Fig. P3.143

**\*3.144**  Show that, in general, a wrench can be replaced with two forces chosen in such a way that one force passes through a given point while the other force lies in a given plane.

**\*3.145**  Show that a wrench can be replaced with two perpendicular forces, one of which is applied at a given point.

**\*3.146**  Show that a wrench can be replaced with two forces, one of which has a prescribed line of action.

# Review and Summary

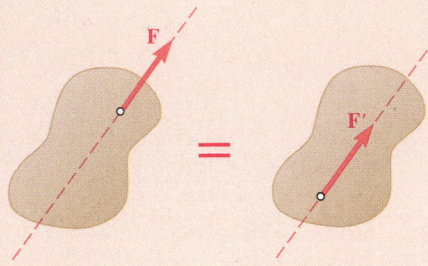

**Fig. 3.43**

## Principle of Transmissibility

In this chapter, we presented the effects of forces exerted on a rigid body. We began by distinguishing between **external** and **internal** forces (Sec. 3.1A). We then explained that, according to the **principle of transmissibility**, the effect of an external force on a rigid body remains unchanged if we move that force along its line of action (Sec. 3.1B). In other words, two forces **F** and **F′** acting on a rigid body at two different points have the same effect on that body if they have the same magnitude, same direction, and same line of action (Fig. 3.43). Two such forces are said to be **equivalent**.

## Vector Product

Before proceeding with the discussion of **equivalent systems of forces**, we introduced the concept of the **vector product of two vectors** (Sec. 3.1C). We defined the vector product

$$\mathbf{V} = \mathbf{P} \times \mathbf{Q}$$

of the vectors **P** and **Q** as a vector perpendicular to the plane containing **P** and **Q** (Fig. 3.44) with a magnitude of

$$V = PQ \sin \theta \tag{3.1}$$

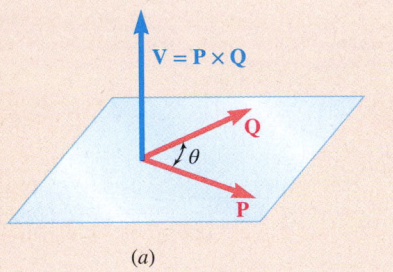

(a)

(b)

**Fig. 3.44**

and directed in such a way that a person located at the tip of **V** will observe the rotation to be counterclockwise through $\theta$, bringing the vector **P** in line with the vector **Q**. The three vectors **P**, **Q**, and **V**—taken in that order—are said to form a *right-handed triad*. It follows that the vector products **Q** × **P** and **P** × **Q** are represented by equal and opposite vectors:

$$\mathbf{Q} \times \mathbf{P} = -(\mathbf{P} \times \mathbf{Q}) \tag{3.4}$$

It also follows from the definition of the vector product of two vectors that the vector products of the unit vectors **i**, **j**, and **k** are

$$\mathbf{i} \times \mathbf{i} = 0 \qquad \mathbf{i} \times \mathbf{j} = \mathbf{k} \qquad \mathbf{j} \times \mathbf{i} = -\mathbf{k}$$

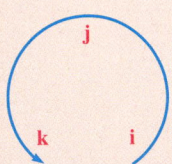

**Fig. 3.45**

and so on. You can determine the sign of the vector product of two unit vectors by arranging in a circle and in counterclockwise order the three letters representing the unit vectors (Fig. 3.45): The vector product of two unit vectors is positive if they follow each other in counterclockwise order and negative if they follow each other in clockwise order.

## Rectangular Components of Vector Product

The **rectangular components of the vector product V** of two vectors **P** and **Q** are expressed (Sec. 3.1D) as

$$\begin{aligned} V_x &= P_y Q_z - P_z Q_y \\ V_y &= P_z Q_x - P_x Q_z \\ V_z &= P_x Q_y - P_y Q_x \end{aligned} \tag{3.9}$$

We can also express the components of a vector product as a determinant:

$$\mathbf{V} = \begin{vmatrix} \mathbf{i} & \mathbf{j} & \mathbf{k} \\ P_x & P_y & P_z \\ Q_x & Q_y & Q_z \end{vmatrix} \qquad (3.10)$$

## Moment of a Force about a Point

We defined the **moment of a force F about a point** $O$ (Sec. 3.1E) as the vector product

$$\mathbf{M}_O = \mathbf{r} \times \mathbf{F} \qquad (3.11)$$

where **r** is the *position vector* drawn from $O$ to the point of application $A$ of the force **F** (Fig. 3.46). Denoting the angle between the lines of action of **r** and **F** as $\theta$, we found that the magnitude of the moment of **F** about $O$ is

$$M_O = rF \sin \theta = Fd \qquad (3.12)$$

where $d$ represents the perpendicular distance from $O$ to the line of action of **F**.

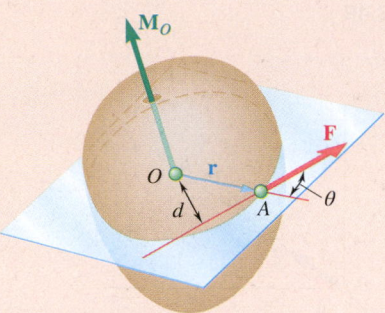

**Fig. 3.46**

## Rectangular Components of Moment

The **rectangular components of the moment $\mathbf{M}_O$ of a force F** (Sec. 3.1F) are

$$\begin{aligned} M_x &= yF_z - zF_y \\ M_y &= zF_x - xF_z \\ M_z &= xF_y - yF_x \end{aligned} \qquad (3.18)$$

where $x$, $y$, and $z$ are the components of the position vector **r** (Fig. 3.47). Using a determinant form, we also wrote

$$\mathbf{M}_O = \begin{vmatrix} \mathbf{i} & \mathbf{j} & \mathbf{k} \\ x & y & z \\ F_x & F_y & F_z \end{vmatrix} \qquad (3.19)$$

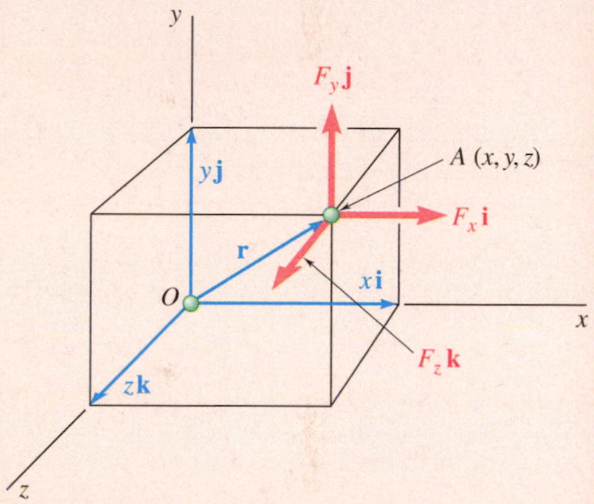

In the more general case of the moment about an arbitrary point $B$ of a force **F** applied at $A$, we had

$$\mathbf{M}_B = \begin{vmatrix} \mathbf{i} & \mathbf{j} & \mathbf{k} \\ x_{A/B} & y_{A/B} & z_{A/B} \\ F_x & F_y & F_z \end{vmatrix} \qquad (3.21)$$

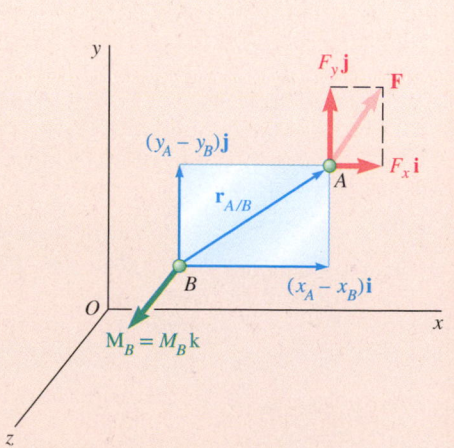

**Fig. 3.47**

where $x_{A/B}$, $y_{A/B}$, and $z_{A/B}$ denote the components of the vector $\mathbf{r}_{A/B}$:

$$x_{A/B} = x_A - x_B \quad y_{A/B} = y_A - y_B \quad z_{A/B} = z_A - z_B$$

In the case of *problems involving only two dimensions*, we can assume the force **F** lies in the $xy$ plane. Its moment $\mathbf{M}_B$ about a point $B$ in the same plane is perpendicular to that plane (Fig. 3.48) and is completely defined by the scalar

$$M_B = (x_A - x_B)F_y - (y_A - y_B)F_x \qquad (3.23)$$

Various methods for computing the moment of a force about a point were illustrated in Sample Probs. 3.1, 3.2, 3.3, and 3.4.

## Scalar Product of Two Vectors

The **scalar product** of two vectors **P** and **Q** (Sec. 3.2A), denoted by $\mathbf{P} \cdot \mathbf{Q}$, is defined as the scalar quantity

$$\mathbf{P} \cdot \mathbf{Q} = PQ \cos \theta \qquad (3.24)$$

**Fig. 3.48**

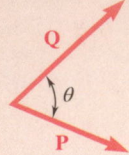

**Fig. 3.49**

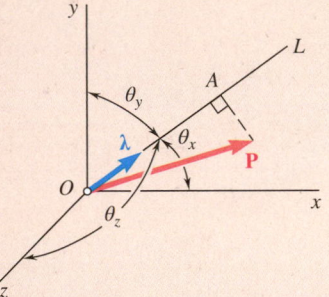

**Fig. 3.50**

where $\theta$ is the angle between **P** and **Q** (Fig. 3.49). By expressing the scalar product of **P** and **Q** in terms of the rectangular components of the two vectors, we determined that

$$\mathbf{P} \cdot \mathbf{Q} = P_x Q_x + P_y Q_y + P_z Q_z \tag{3.28}$$

### Projection of a Vector on an Axis

We obtain the **projection of a vector P on an axis** $OL$ (Fig. 3.50) by forming the scalar product of **P** and the unit vector $\boldsymbol{\lambda}$ along $OL$. We have

$$P_{OL} = \mathbf{P} \cdot \boldsymbol{\lambda} \tag{3.34}$$

Using rectangular components, this becomes

$$P_{OL} = P_x \cos \theta_x + P_y \cos \theta_y + P_z \cos \theta_z \tag{3.35}$$

where $\theta_x$, $\theta_y$, and $\theta_z$ denote the angles that the axis $OL$ forms with the coordinate axes.

### Mixed Triple Product of Three Vectors

We defined the **mixed triple product** of the three vectors **S**, **P**, and **Q** as the scalar expression

$$\mathbf{S} \cdot (\mathbf{P} \times \mathbf{Q}) \tag{3.36}$$

obtained by forming the scalar product of **S** with the vector product of **P** and **Q** (Sec. 3.2B). We showed that

$$\mathbf{S} \cdot (\mathbf{P} \times \mathbf{Q}) = \begin{vmatrix} S_x & S_y & S_z \\ P_x & P_y & P_z \\ Q_x & Q_y & Q_z \end{vmatrix} \tag{3.39}$$

where the elements of the determinant are the rectangular components of the three vectors.

### Moment of a Force about an Axis

We defined the **moment of a force F about an axis** $OL$ (Sec. 3.2C) as the projection $OC$ on $OL$ of the moment $\mathbf{M}_O$ of the force **F** (Fig. 3.51), i.e., as the mixed triple product of the unit vector $\boldsymbol{\lambda}$, the position vector **r**, and the force **F**:

$$M_{OL} = \boldsymbol{\lambda} \cdot \mathbf{M}_o = \boldsymbol{\lambda} \cdot (\mathbf{r} \times \mathbf{F}) \tag{3.40}$$

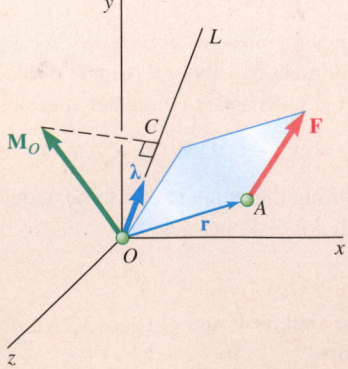

**Fig. 3.51**

The determinant form for the mixed triple product is

$$M_{OL} = \begin{vmatrix} \lambda_x & \lambda_y & \lambda_z \\ x & y & z \\ F_x & F_y & F_z \end{vmatrix} \qquad (3.41)$$

where

$\lambda_x, \lambda_y, \lambda_z$ = direction cosines of axis $OL$

$x, y, z$ = components of $\mathbf{r}$

$F_x, F_y, F_z$ = components of $\mathbf{F}$

An example of determining the moment of a force about a skew axis appears in Sample Prob. 3.5.

## Couples

*Two forces $\mathbf{F}$ and $-\mathbf{F}$ having the same magnitude, parallel lines of action, and opposite sense are said to form a* **couple** (Sec. 3.3A). The moment of a couple is independent of the point about which it is computed; it is a vector $\mathbf{M}$ perpendicular to the plane of the couple and equal in magnitude to the product of the common magnitude $F$ of the forces and the perpendicular distance $d$ between their lines of action (Fig. 3.52).

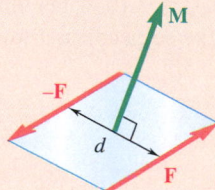

**Fig. 3.52**

Two couples having the same moment $\mathbf{M}$ are *equivalent*, i.e., they have the same effect on a given rigid body (Sec. 3.3B). The sum of two couples is itself a couple (Sec. 3.3C), and we can obtain the moment $\mathbf{M}$ of the resultant couple by adding vectorially the moments $\mathbf{M}_1$ and $\mathbf{M}_2$ of the original couples (Sample Prob. 3.6). It follows that we can represent a couple by a vector, called a **couple vector**, equal in magnitude and direction to the moment $\mathbf{M}$ of the couple (Sec. 3.3D). A couple vector is a *free vector* that can be attached to the origin $O$ if so desired and resolved into components (Fig. 3.53).

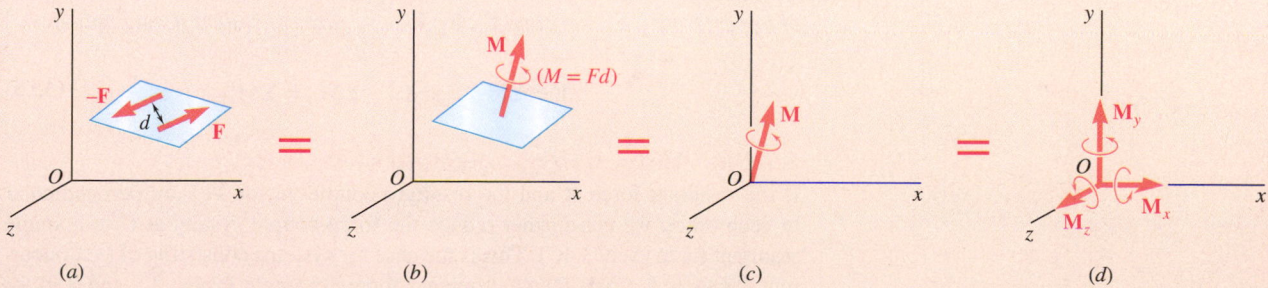

**Fig. 3.53**

## Force-Couple System

Any force $\mathbf{F}$ acting at a point $A$ of a rigid body can be replaced by a **force-couple system** at an arbitrary point $O$ consisting of the force $\mathbf{F}$ applied at $O$ and a couple of moment $\mathbf{M}_O$, which is equal to the moment about $O$ of the force $\mathbf{F}$ in its original position (Sec. 3.3E). Note that the force $\mathbf{F}$ and the couple vector $\mathbf{M}_O$ are always perpendicular to each other (Fig. 3.54).

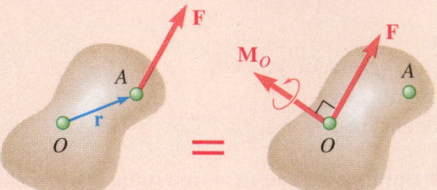

**Fig. 3.54**

### Reduction of a System of Forces to a Force-Couple System

It follows (Sec. 3.4A) that *any system of forces can be reduced to a force-couple system at a given point O* by first replacing each of the forces of the system by an equivalent force-couple system at $O$ (Fig. 3.55) and then adding all of the forces and all of the couples to obtain a resultant force $\mathbf{R}$ and a resultant couple vector $\mathbf{M}_O^R$ (Sample Probs. 3.8, 3.9, 3.10, 3.11). In general, the resultant $\mathbf{R}$ and the couple vector $\mathbf{M}_O^R$ will not be perpendicular to each other.

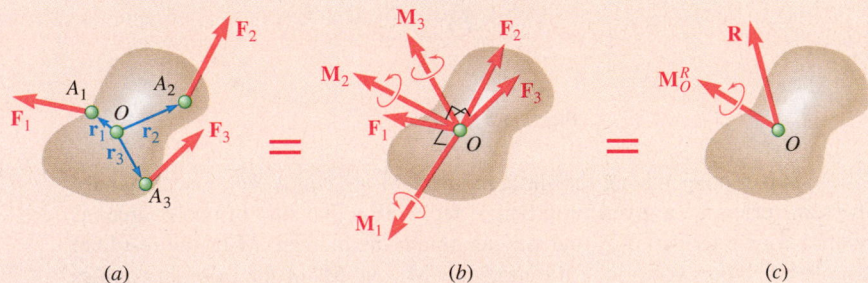

(a)                                    (b)                                    (c)

**Fig. 3.55**

### Equivalent Systems of Forces

We concluded (Sec. 3.4B) that, as far as rigid bodies are concerned, *two systems of forces, $\mathbf{F}_1$, $\mathbf{F}_2$, $\mathbf{F}_3$, . . . , and $\mathbf{F}'_1$, $\mathbf{F}'_2$, $\mathbf{F}'_3$, . . . , are equivalent if, and only if,*

$$\Sigma\mathbf{F} = \Sigma\mathbf{F}' \quad \text{and} \quad \Sigma\mathbf{M}_O = \Sigma\mathbf{M}'_O \tag{3.55}$$

### Further Reduction of a System of Forces

If the resultant force $\mathbf{R}$ and the resultant couple vector $\mathbf{M}_O^R$ are perpendicular to each other, we can further reduce the force-couple system at $O$ to a single resultant force (Sec. 3.4C). This is the case for systems consisting of (a) concurrent forces (c.f. Chap. 2), (b) coplanar forces (Sample Probs. 3.8 and 3.9), or (c) parallel forces (Sample Prob. 3.11). If the resultant $\mathbf{R}$ and the couple vector $\mathbf{M}_O^R$ are *not* perpendicular to each other, the system *cannot* be reduced to a single force. We can, however, reduce it to a special type of force-couple system called a *wrench*, consisting of the resultant $\mathbf{R}$ and a couple vector $\mathbf{M}_1$ directed along $\mathbf{R}$ (Sec. 3.4D and Sample Prob. 3.12).

# Review Problems

**3.147** A 300-N force **P** is applied at point *A* of the bell crank shown. (*a*) Compute the moment of the force **P** about *O* by resolving it into horizontal and vertical components. (*b*) Using the result of part *a*, determine the perpendicular distance from *O* to the line of action of **P**.

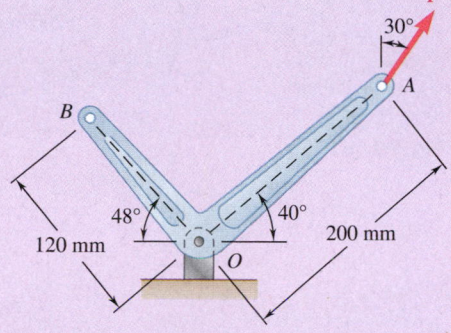

**Fig. P3.147**

**3.148** A winch puller *AB* is used to straighten a fence post. Knowing that the tension in cable *BC* is 1040 N and length *d* is 1.90 m, determine the moment about *D* of the force exerted by the cable at *C* by resolving that force into horizontal and vertical components applied (*a*) at point *C*, (*b*) at point *E*.

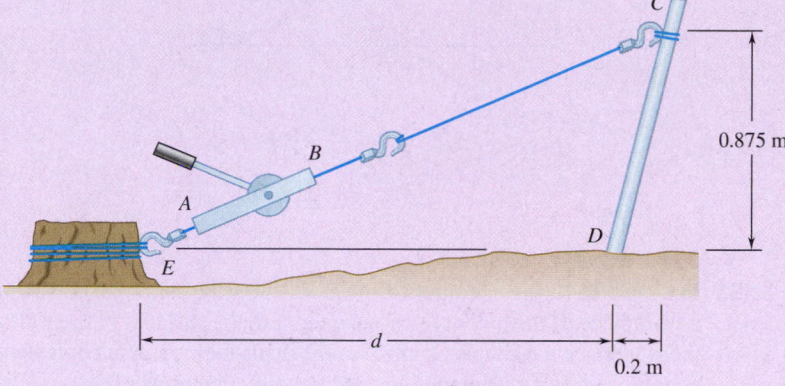

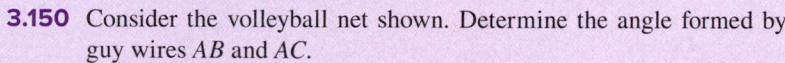

**Fig. P3.148**

**3.149** A small boat hangs from two davits, one of which is shown in the figure. The tension in line *ABAD* is 82 lb. Determine the moment about *C* of the resultant force **R**$_A$ exerted on the davit at *A*.

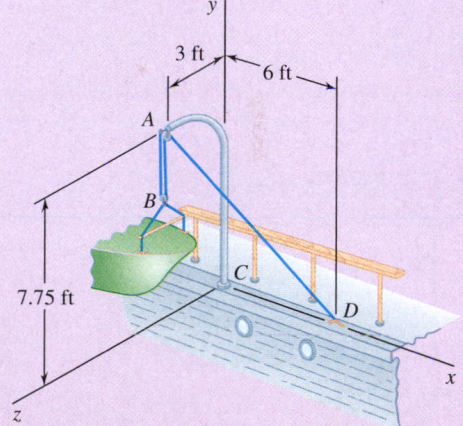

**Fig. *P3.149***

**3.150** Consider the volleyball net shown. Determine the angle formed by guy wires *AB* and *AC*.

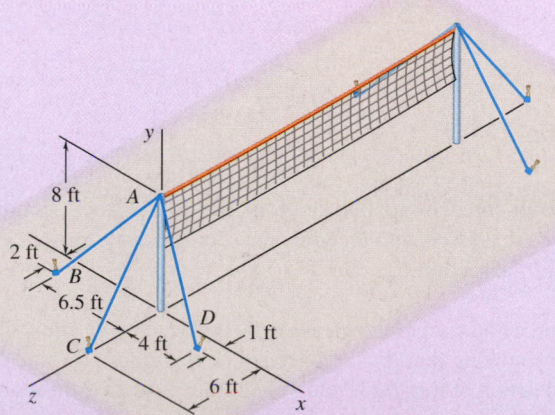

**Fig. P3.150**

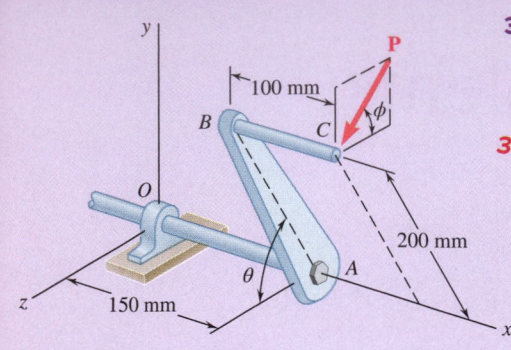

**Fig. P3.151**

**3.151** A single force **P** acts at $C$ in a direction perpendicular to the handle $BC$ of the crank shown. Determine the moment $M_x$ of **P** about the $x$ axis when $\theta = 65°$, knowing that $M_y = -15$ N·m and $M_z = -36$ N·m.

**3.152** The 23-in. vertical rod $CD$ is welded to the midpoint $C$ of the 50-in. rod $AB$. Determine the moment about $AB$ of the 174-lb force **Q**.

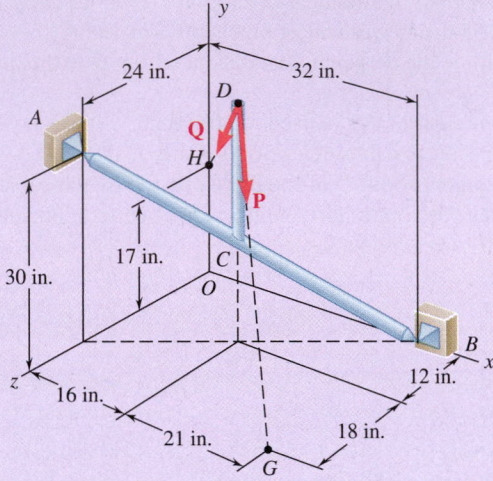

**Fig. P3.152**

**3.153** In a manufacturing operation, three holes are drilled simultaneously in a workpiece. If the holes are perpendicular to the surfaces of the workpiece, replace the couples applied to the drills with a single equivalent couple, specifying its magnitude and the direction of its axis.

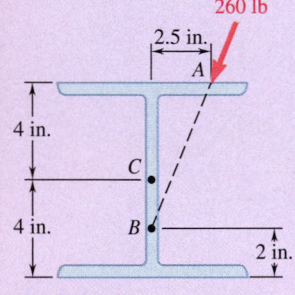

**Fig. P3.154**

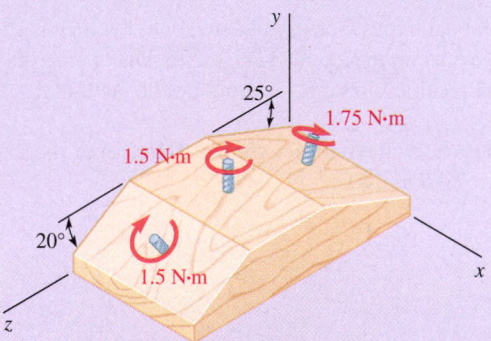

**Fig. P3.153**

**3.154** A 260-lb force is applied at $A$ to the rolled-steel section shown. Replace that force with an equivalent force-couple system at the center $C$ of the section.

**3.155** The force and couple shown are to be replaced by an equivalent single force. Knowing that $P = 2Q$, determine the required value of $\alpha$ if the line of action of the single equivalent force is to pass through (*a*) point $A$, (*b*) point $C$.

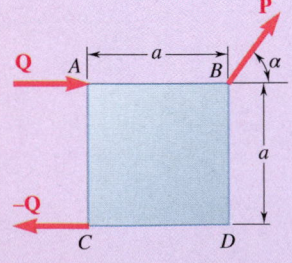

**Fig. P3.155**

**3.156** A 77-N force $\mathbf{F}_1$ and a 31-N·m couple $\mathbf{M}_1$ are applied to corner $E$ of the bent plate shown. If $\mathbf{F}_1$ and $\mathbf{M}_1$ are to be replaced with an equivalent force-couple system $(\mathbf{F}_2, \mathbf{M}_2)$ at corner $B$ and if $(M_2)_z = 0$, determine (*a*) the distance $d$, (*b*) $\mathbf{F}_2$ and $\mathbf{M}_2$.

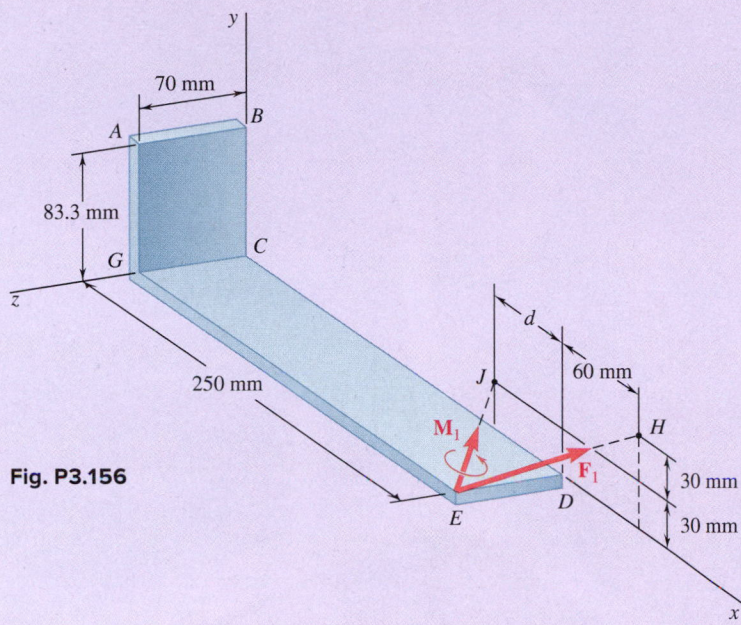

Fig. P3.156

**3.157** Three horizontal forces are applied as shown to a vertical cast-iron arm. Determine the resultant of the forces and the distance from the ground to its line of action when (*a*) $P = 200$ N, (*b*) $P = 2400$ N, (*c*) $P = 1000$ N.

**3.158** While using a pencil sharpener, a student applies the forces and couple shown. (*a*) Determine the forces exerted at $B$ and $C$ knowing that these forces and the couple are equivalent to a force-couple system at $A$ consisting of the force $\mathbf{R} = (2.6\ \text{lb})\mathbf{i} + R_y\mathbf{j} - (0.7\ \text{lb})\mathbf{k}$ and the couple $\mathbf{M}_A^R = M_x\mathbf{i} + (1.0\ \text{lb·ft})\mathbf{j} - (0.72\ \text{lb·ft})\mathbf{k}$. (*b*) Find the corresponding values of $R_y$ and $M_x$.

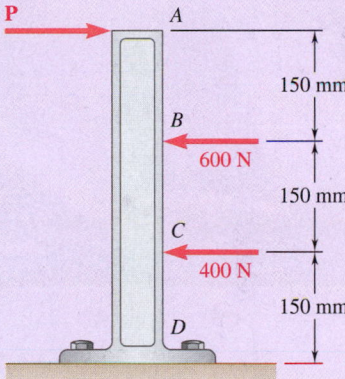

Fig. *P3.157*

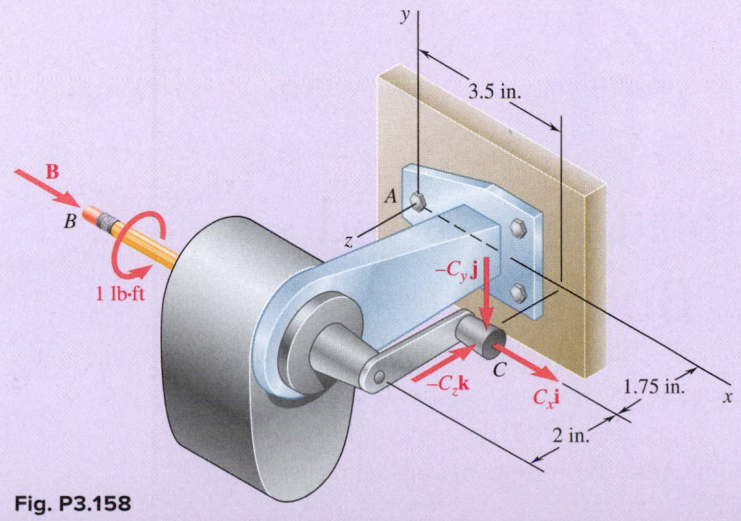

Fig. P3.158

169

©View Stock/Getty Images RF

# 4

# Equilibrium of Rigid Bodies

The Tianjin Eye is a Ferris wheel that straddles a bridge over the Hai River in China. The structure is designed so that the support reactions at the wheel bearings, as well as those at the base of the frame, maintain equilibrium under the effects of vertical gravity and horizontal wind forces.

# Objectives

- **Analyze** the static equilibrium of rigid bodies in two and three dimensions.
- **Consider** the attributes of a properly drawn free-body diagram, an essential tool for the equilibrium analysis of rigid bodies.
- **Examine** rigid bodies supported by statically indeterminate reactions and partial constraints.
- **Study** two cases of particular interest: the equilibrium of two-force and three-force bodies.

## Introduction

We saw in Chap. 3 how to reduce the external forces acting on a rigid body to a force-couple system at some arbitrary point $O$. When the force and the couple are both equal to zero, the external forces form a system equivalent to zero, and the rigid body is said to be in **equilibrium**.

We can obtain the necessary and sufficient conditions for the equilibrium of a rigid body by setting $\mathbf{R}$ and $\mathbf{M}_O^R$ equal to zero in the relations of Eq. (3.50) of Sec. 3.4A:

$$\Sigma\mathbf{F} = 0 \qquad \Sigma\mathbf{M}_O = \Sigma(\mathbf{r} \times \mathbf{F}) = 0 \qquad (4.1)$$

Resolving each force and each moment into its rectangular components, we can replace these vector equations for the equilibrium of a rigid body with the following six scalar equations:

$$\Sigma F_x = 0 \qquad \Sigma F_y = 0 \qquad \Sigma F_z = 0 \qquad (4.2)$$
$$\Sigma M_x = 0 \qquad \Sigma M_y = 0 \qquad \Sigma M_z = 0 \qquad (4.3)$$

We can use these equations to determine unknown forces applied to the rigid body or unknown reactions exerted on it by its supports. Note that Eqs. (4.2) express the fact that the components of the external forces in the $x$, $y$, and $z$ directions are balanced; Eqs. (4.3) express the fact that the moments of the external forces about the $x$, $y$, and $z$ axes are balanced. Therefore, for a rigid body in equilibrium, the system of external forces imparts no translational or rotational motion to the body.

In order to write the equations of equilibrium for a rigid body, we must first identify all of the forces acting on that body and then draw the corresponding **free-body diagram**. In this chapter, we first consider the equilibrium of *two-dimensional structures* subjected to forces contained in their planes and study how to draw their free-body diagrams. In addition to the forces *applied* to a structure, we must also consider the *reactions* exerted on the structure by its supports. A specific reaction is associated with each type of support. You will see how to determine whether the structure is properly supported, so that you can know in advance whether you can solve the equations of equilibrium for the unknown forces and reactions.

Later in this chapter, we consider the equilibrium of three-dimensional structures, and we provide the same kind of analysis to these structures and their supports.

# Free-Body Diagrams

In solving a problem concerning a rigid body in equilibrium, it is essential to consider *all* of the forces acting on the body. It is equally important to exclude any force that is *not* directly applied to the body. Omitting a force or adding an extraneous one would destroy the conditions of equilibrium. Therefore, the first step in solving the problem is to draw a **free-body diagram** of the rigid body under consideration.

We have already used free-body diagrams on many occasions in Chap. 2. However, in view of their importance to the solution of equilibrium problems, we summarize here the steps you must follow in drawing a correct free-body diagram.

1. Start with a clear decision regarding the choice of the free body to be analyzed. Mentally, you need to detach this body from the ground and separate it from all other bodies. Then, you can sketch the contour of this isolated body.

2. Indicate all external forces on the free-body diagram. These forces represent the actions exerted *on* the free body *by* the ground and *by* the bodies that have been detached. In the diagram, apply these forces at the various points where the free body was supported by the ground or was connected to the other bodies. Generally, you should include the *weight* of the free body among the external forces, because it represents the attraction exerted by the earth on the various particles forming the free body. You will see in Chap. 5 that you should draw the weight so it acts at the center of gravity of the body. If the free body is made of several parts, do *not* include the forces the various parts exert on each other among the

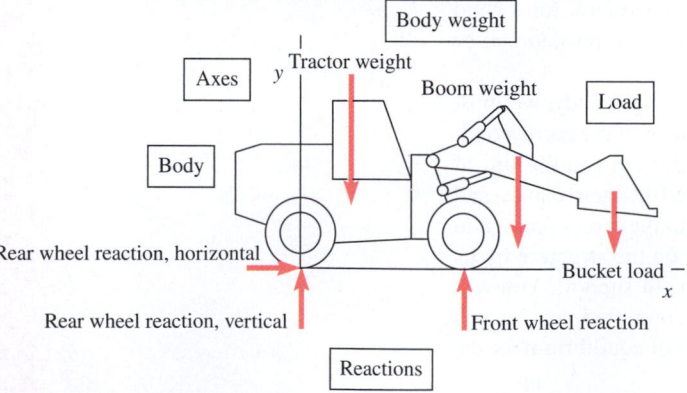

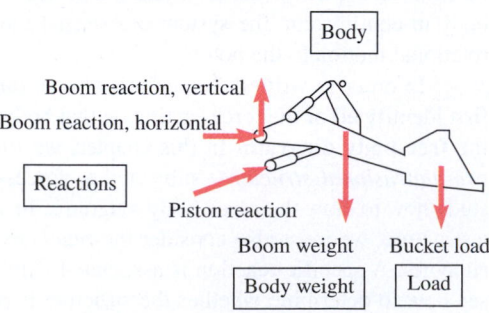

**Photo 4.1** A tractor supporting a bucket load. As shown, its free-body diagram should include all external forces acting on the tractor.
©McGraw-Hill Education/Photos by Lucinda Dowell

**Photo 4.2** Tractor bucket and boom. In Chap. 6, we will see how to determine the internal forces associated with interconnected members such as these using free-body diagrams like the one shown.
©McGraw-Hill Education/Photos by Lucinda Dowell

external forces. These forces are internal forces as far as the free body is concerned.

3. Clearly mark the magnitudes and directions of the *known external forces* on the free-body diagram. Recall that when indicating the directions of these forces, the forces are those exerted *on,* and not *by,* the free body. Known external forces generally include the *weight* of the free body and *forces applied* for a given purpose.

4. *Unknown external forces* usually consist of the **reactions** through which the ground and other bodies oppose a possible motion of the free body. The reactions constrain the free body to remain in the same position; for that reason, they are sometimes called *constraining forces.* Reactions are exerted at the points where the free body is *supported by* or *connected to* other bodies; you should clearly indicate these points. Reactions are discussed in detail in Secs. 4.1 and 4.3.

5. The free-body diagram should also include dimensions, because these may be needed for computing moments of forces. Any other detail, however, should be omitted.

# 4.1   EQUILIBRIUM IN TWO DIMENSIONS

In the first part of this chapter, we consider the equilibrium of two-dimensional structures; i.e., we assume that the structure being analyzed and the forces applied to it are contained in the same plane. The reactions needed to maintain the structure in the same position are also contained in this plane.

## 4.1A   Reactions for a Two-Dimensional Structure

The reactions exerted on a two-dimensional structure fall into three categories that correspond to three types of **supports** or **connections**.

1. **Reactions Equivalent to a Force with a Known Line of Action.** Supports and connections causing reactions of this type include *rollers, rockers, frictionless surfaces, short links and cables, collars on frictionless rods,* and *frictionless pins in slots.* Each of these supports and connections can prevent motion in one direction only. Fig. 4.1 shows these supports and connections together with the reactions they produce. Each reaction involves *one unknown*—specifically, the magnitude of the reaction. In problem solving, you should denote this magnitude by an appropriate letter. The line of action of the reaction is known and should be indicated clearly in the free-body diagram.

   The sense of the reaction must be as shown in Fig. 4.1 for cases of a frictionless surface (toward the free body) or a cable (away from the free body). The reaction can be directed either way in the cases of double-track rollers, links, collars on rods, or pins in slots. Generally, we assume that single-track rollers and rockers are reversible, so the corresponding reactions can be directed either way.

2. **Reactions Equivalent to a Force of Unknown Direction and Magnitude.** Supports and connections causing reactions of this type include *frictionless pins in fitted holes, hinges,* and *rough surfaces.* They can prevent translation of the free body in all directions, but they cannot prevent the body from rotating about the connection. Reactions of this group involve

| Support or Connection | Reaction | Number of Unknowns |
|---|---|---|
| 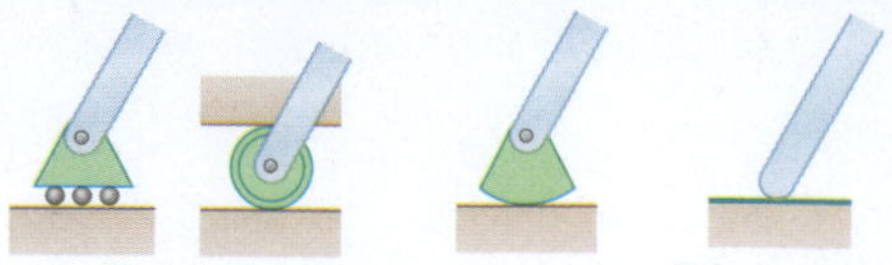 Rollers    Rocker    Frictionless surface |  Force with known line of action perpendicular to surface | 1 |
| 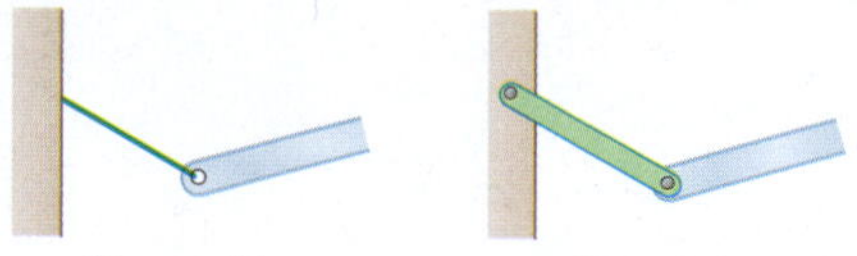 Short cable    Short link |  Force with known line of action along cable or link | 1 |
| 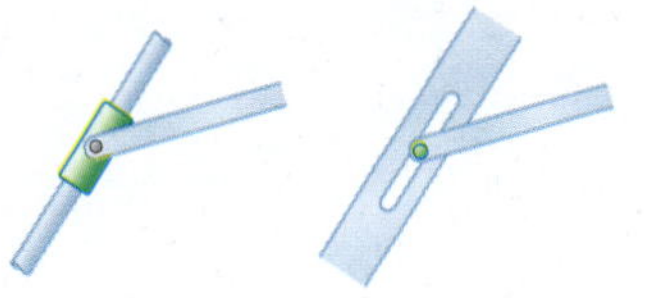 Collar on frictionless rod    Frictionless pin in slot | 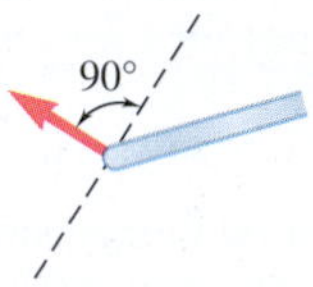 90°  Force with known line of action perpendicular to rod or slot | 1 |
| Frictionless pin or hinge    Rough surface | 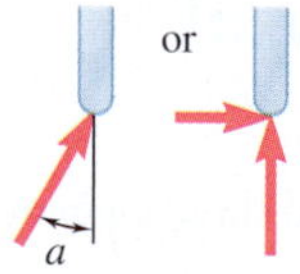 or  Force of unknown direction | 2 |
| 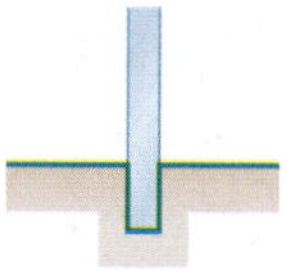 Fixed support | 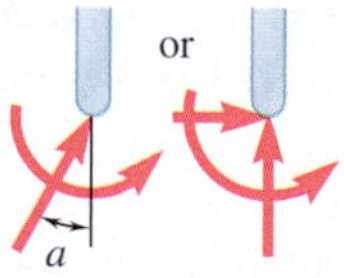 or  Force and couple | 3 |

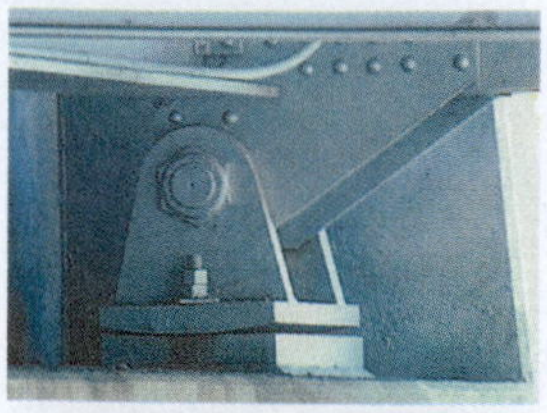

This rocker bearing supports the weight of a bridge. The convex surface of the rocker allows the bridge to move slightly horizontally.

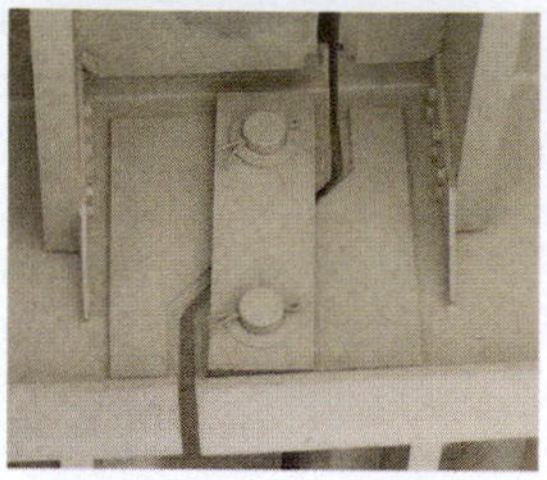

Links are often used to support suspended spans of highway bridges.

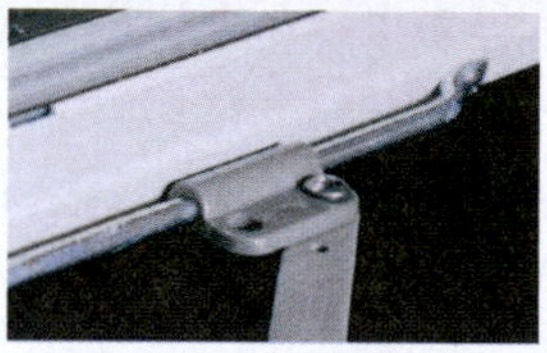

Force applied to the slider exerts a normal force on the rod, causing the window to open.

Pin supports are common on bridges and overpasses.

This cantilever support is fixed at one end and extends out into space at the other end.

**Fig. 4.1** Reactions of supports and connections in two dimensions. (Rocker bearing): Courtesy Godden Collection. National Information Service for Earthquake Engineering, University of California, Berkeley (Links): Courtesy Michigan Department of Transportation (Slider and rod): ©McGraw-Hill Education/Photo by Lucinda Dowell (Pin support): Courtesy Michigan Department of Transportation (Cantilever support): ©Richard Ellis/Alamy

two unknowns and are usually represented by their x and y components. In the case of a rough surface, the component normal to the surface must be directed away from the surface.

3. **Reactions Equivalent to a Force and a Couple.** These reactions are caused by *fixed supports* that oppose any motion of the free body and

thus constrain it completely. Fixed supports actually produce forces over the entire surface of contact; these forces, however, form a system that can be reduced to a force and a couple. Reactions of this group involve *three unknowns* usually consisting of the two components of the force and the moment of the couple.

When the sense of an unknown force or couple is not readily apparent, do not attempt to determine it. Instead, arbitrarily assume the sense of the force or couple; the sign of the answer will indicate whether the assumption is correct or not. (A positive answer means the assumption is correct, while a negative answer means the assumption is incorrect.)

## 4.1B    Rigid-Body Equilibrium in Two Dimensions

The conditions stated in Sec. 4.1A for the equilibrium of a rigid body become considerably simpler for the case of a two-dimensional structure. Choosing the $x$ and $y$ axes to be in the plane of the structure, we have

$$F_z = 0 \qquad M_x = M_y = 0 \qquad M_z = M_O$$

for each of the forces applied to the structure. Thus, the six equations of equilibrium stated in Sec. 4.1 reduce to three equations:

$$\Sigma F_x = 0 \qquad \Sigma F_y = 0 \qquad \Sigma M_O = 0 \qquad \textbf{(4.4)}$$

Because $\Sigma M_O = 0$ must be satisfied regardless of the choice of the origin $O$, we can write the equations of equilibrium for a two-dimensional structure in the more general form

**Equations of equilibrium in two dimensions**

$$\Sigma F_x = 0 \qquad \Sigma F_y = 0 \qquad \Sigma M_A = 0 \qquad \textbf{(4.5)}$$

where $A$ is any point in the plane of the structure. These three equations can be solved for no more than *three unknowns*.

You have just seen that unknown forces include reactions and that the number of unknowns corresponding to a given reaction depends upon the type of support or connection causing that reaction. Referring to Fig. 4.1, note that you can use the equilibrium Eqs. (4.5) to determine the reactions associated with two rollers and one cable, or one fixed support, or one roller and one pin in a fitted hole, etc.

For example, consider Fig. 4.2a, in which the truss shown is in equilibrium and is subjected to the given forces **P**, **Q**, and **S**. The truss is held in place by a pin at $A$ and a roller at $B$. The pin prevents point $A$ from moving by exerting a force on the truss that can be resolved into the components $\mathbf{A}_x$ and $\mathbf{A}_y$. The roller keeps the truss from rotating about $A$ by exerting the vertical force **B**. The free-body diagram of the truss is shown in Fig. 4.2b; it includes the reactions $\mathbf{A}_x$, $\mathbf{A}_y$, and **B**, as well as the applied forces **P**, **Q**, and **S** (in $x$ and $y$ component form) and the weight **W** of the truss.

Because the truss is in equilibrium, the sum of the moments about $A$ of all of the forces shown in Fig. 4.2b is zero, or $\Sigma M_A = 0$. We can use this equation to determine the magnitude $B$ because the equation does not contain $A_x$ or $A_y$. Then, because the sum of the $x$ components and the sum of the $y$ components of the forces are zero, we write the equations $\Sigma F_x = 0$ and $\Sigma F_y = 0$. From these equations, we can obtain the components $A_x$ and $A_y$, respectively.

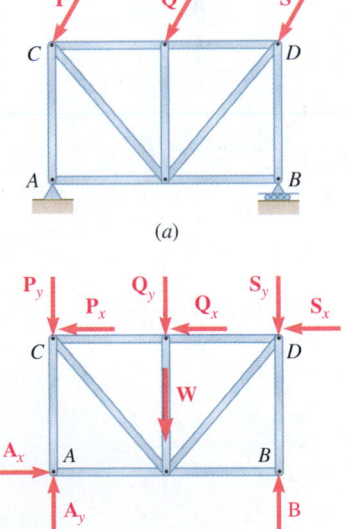

**Fig. 4.2** (*a*) A truss supported by a pin and a roller; (*b*) free-body diagram of the truss.

We could obtain an additional equation by noting that the sum of the moments of the external forces about a point other than $A$ is zero. We could write, for instance, $\Sigma M_B = 0$. This equation, however, does not contain any new information, because we have already established that the system of forces shown in Fig. 4.2$b$ is equivalent to zero. The additional equation *is not independent* and cannot be used to determine a fourth unknown. It can be useful, however, for checking the solution obtained from the original three equations of equilibrium.

Although the three equations of equilibrium cannot be *augmented* by additional equations, any of them can be *replaced* by another equation. Properly chosen, the new system of equations still describes the equilibrium conditions but may be easier to work with. For example, an alternative system of equations for equilibrium is

$$\Sigma F_x = 0 \qquad \Sigma M_A = 0 \qquad \Sigma M_B = 0 \qquad \textbf{(4.6)}$$

Here the second point about which the moments are summed (in this case, point $B$) cannot lie on the line parallel to the $y$ axis that passes through point $A$ (Fig. 4.2$b$). These equations are sufficient conditions for the equilibrium of the truss. The first two equations indicate that the external forces must reduce to a single vertical force at $A$. Because the third equation requires that the moment of this force be zero about a point $B$ that is not on its line of action, the force must be zero, and the rigid body is in equilibrium.

A third possible set of equilibrium equations is

$$\Sigma M_A = 0 \qquad \Sigma M_B = 0 \qquad \Sigma M_C = 0 \qquad \textbf{(4.7)}$$

where the points $A$, $B$, and $C$ do not lie in a straight line (Fig. 4.2$b$). The first equation requires that the external forces reduce to a single force at $A$; the second equation requires that this force pass through $B$; and the third equation requires that it pass through $C$. Because the points $A$, $B$, $C$ do not lie in a straight line, the force must be zero, and the rigid body is in equilibrium.

Notice that the equation $\Sigma M_A = 0$, stating that the sum of the moments of the forces about pin $A$ is zero, possesses a more definite physical meaning than either of the other two Eqs. (4.7). These two equations express a similar idea of balance, but with respect to points about which the rigid body is not actually hinged. They are, however, as useful as the first equation. The choice of equilibrium equations should not be unduly influenced by their physical meaning. Indeed, in practice, it is desirable to choose equations of equilibrium containing only one unknown, because this eliminates the necessity of solving simultaneous equations. You can obtain equations containing only one unknown by summing moments about the point of intersection of the lines of action of two unknown forces or, if these forces are parallel, by summing force components in a direction perpendicular to their common direction.

For example, in Fig. 4.3, in which the truss shown is held by rollers at $A$ and $B$ and a short link at $D$, we can eliminate the reactions at $A$ and $B$ by summing $x$ components. We can eliminate the reactions at $A$ and $D$ by summing moments about $C$, and the reactions at $B$ and $D$ by summing moments about $D$. The resulting equations are

$$\Sigma F_x = 0 \qquad \Sigma M_C = 0 \qquad \Sigma M_D = 0$$

Each of these equations contains only one unknown.

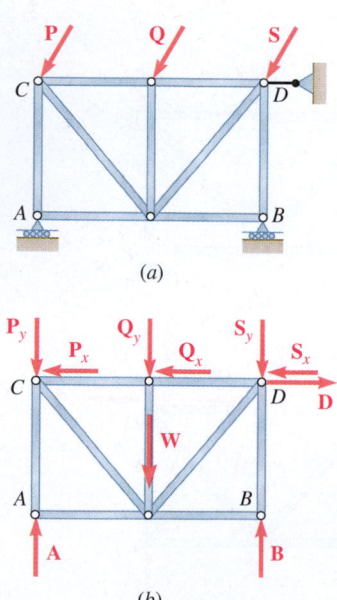

(a)

(b)

**Fig. 4.3** (*a*) A truss supported by two rollers and a short link; (*b*) free-body diagram of the truss.

## 4.1C  Statically Indeterminate Reactions and Partial Constraints

In the two examples considered in Figs. 4.2 and 4.3, the types of supports used were such that the rigid body could not possibly move under the given loads or under any other loading conditions. In such cases, the rigid body is said to be **completely constrained**. Recall that the reactions corresponding to these supports involved *three unknowns* and could be determined by solving the three equations of equilibrium. When such a situation exists, the reactions are said to be **statically determinate**.

Consider Fig. 4.4*a*, in which the truss shown is held by pins at *A* and *B*. These supports provide more constraints than are necessary to keep the truss from moving under the given loads or under any other loading conditions. Note from the free-body diagram of Fig. 4.4*b* that the corresponding reactions involve *four unknowns*. We pointed out in Sec. 4.1B that only three independent equilibrium equations are available; therefore, in this case, we have *more unknowns than equations*. As a result, we cannot determine all of the unknowns. The equations $\Sigma M_A = 0$ and $\Sigma M_B = 0$ yield the vertical components $B_y$ and $A_y$, respectively, but the equation $\Sigma F_x = 0$ gives only the sum $A_x + B_x$ of the horizontal components of the reactions at *A* and *B*. The components $A_x$ and $B_x$ are **statically indeterminate**. We could determine their magnitudes by considering the deformations produced in the truss by the given loading, but this method is beyond the scope of statics and belongs to the study of mechanics of materials.

Let's consider the opposite situation. The supports holding the truss shown in Fig. 4.5*a* consist of rollers at *A* and *B*. The constraints provided by these supports are not sufficient to keep the truss from moving. Although they prevent any vertical motion, the truss is free to move horizontally. The truss is said to be **partially constrained**.[†] From the free-body diagram in Fig. 4.5*b*, note that the reactions at *A* and *B* involve only *two unknowns*. Because three equations of equilibrium must still be satisfied, we have *fewer unknowns than equations*. In such a case, one of the equilibrium equations will not be satisfied in general. The equations $\Sigma M_A = 0$ and $\Sigma M_B = 0$ can be satisfied by a proper choice of reactions at *A* and *B*, but the equation $\Sigma F_x = 0$ is not satisfied unless the sum of the horizontal components of the applied forces happens to be zero. We thus observe that the equilibrium of the truss of Fig. 4.5 cannot be maintained under general loading conditions.

From these examples, it would appear that, if a rigid body is to be completely constrained and if the reactions at its supports are to be statically determinate, **there must be as many unknowns as there are equations of equilibrium**. When this condition is *not* satisfied, we can be certain that either the rigid body is not completely constrained or that the reactions at its supports are not statically determinate. It is also possible that the rigid body is not completely constrained *and* that the reactions are statically indeterminate.

You should note, however, that, although this condition is *necessary,* it is *not sufficient*. In other words, the fact that the number of unknowns is equal to the number of equations is no guarantee that a body is completely constrained or

---

[†]Partially constrained bodies are often referred to as *unstable.* However, to avoid confusion between this type of instability, due to insufficient constraints, and the type of instability considered in Chap. 10, which relates to the behavior of a rigid body when its equilibrium is disturbed, we shall restrict the use of the words *stable* and *unstable* to the latter case.

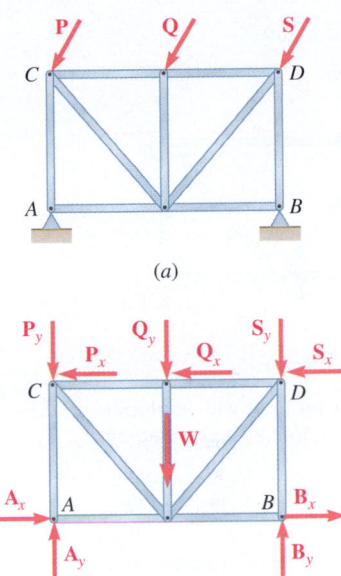

**Fig. 4.4**  (*a*) Truss with statically indeterminate reactions; (*b*) free-body diagram.

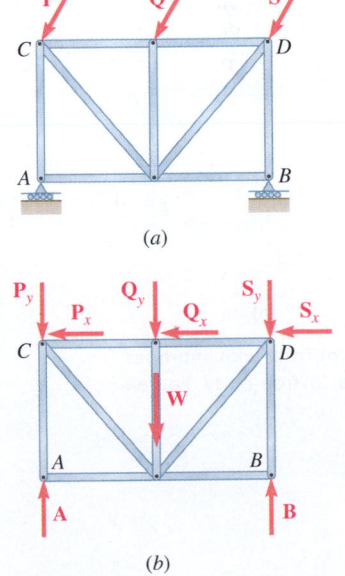

**Fig. 4.5**  (*a*) Truss with partial constraints; (*b*) free-body diagram.

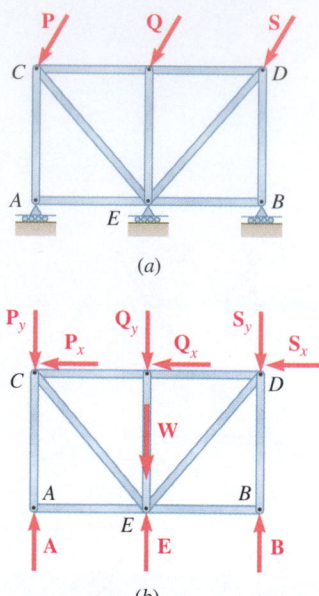

**Fig. 4.6** (*a*) Truss with improper constraints; (*b*) free-body diagram.

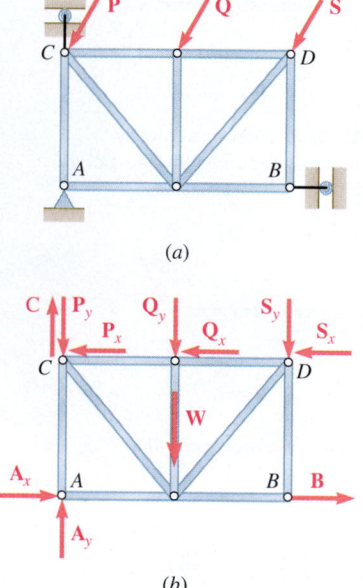

**Fig. 4.7** (*a*) Truss with improper constraints; (*b*) free-body diagram.

that the reactions at its supports are statically determinate. Consider Fig. 4.6*a*, which shows a truss held by rollers at *A, B,* and *E*. We have three unknown reactions of **A**, **B**, and **E** (Fig. 4.6*b*), but the equation $\Sigma F_x = 0$ is not satisfied unless the sum of the horizontal components of the applied forces happens to be zero. Although there are a sufficient number of constraints, these constraints are not properly arranged, so the truss is free to move horizontally. We say that the truss is **improperly constrained**. Because only two equilibrium equations are left for determining three unknowns, the reactions are statically indeterminate. Thus, improper constraints also produce static indeterminacy.

The truss shown in Fig. 4.7 is another example of improper constraints—and of static indeterminacy. This truss is held by a pin at *A* and by rollers at *B* and *C,* which altogether involve four unknowns. Because only three independent equilibrium equations are available, the reactions at the supports are statically indeterminate. On the other hand, we note that the equation $\Sigma M_A = 0$ cannot be satisfied under general loading conditions, because the lines of action of the reactions **B** and **C** pass through *A*. We conclude that the truss can rotate about *A* and that it is improperly constrained.[†]

The examples of Figs. 4.6 and 4.7 lead us to conclude that

> **A rigid body is improperly constrained whenever the supports (even though they may provide a sufficient number of reactions) are arranged in such a way that the reactions must be either concurrent or parallel.[‡]**

In summary, to be sure that a two-dimensional rigid body is completely constrained and that the reactions at its supports are statically determinate, you should verify that the reactions involve three—and only three—unknowns and that the supports are arranged in such a way that they do not require the reactions to be either concurrent or parallel.

Supports involving statically indeterminate reactions should be used with care in the design of structures and only with a full knowledge of the problems they may cause. On the other hand, the analysis of structures possessing statically indeterminate reactions often can be partially carried out by the methods of statics. In the case of the truss of Fig. 4.4, for example, we can determine the vertical components of the reactions at *A* and *B* from the equilibrium equations.

For obvious reasons, supports producing partial or improper constraints should be avoided in the design of stationary structures. However, a partially or improperly constrained structure will not necessarily collapse; under particular loading conditions, equilibrium can be maintained. For example, the trusses of Figs. 4.5 and 4.6 will be in equilibrium if the applied forces **P**, **Q**, and **S** are vertical. Besides, structures designed to move *should* be only partially constrained. A railroad car, for instance, would be of little use if it were completely constrained by having its brakes applied permanently.

---

[†]Rotation of the truss about *A* requires some "play" in the supports at *B* and *C*. In practice, such play will always exist. In addition, we note that if the play is kept small, the displacements of the rollers *B* and *C* and, thus, the distances from *A* to the lines of action of the reactions **B** and **C** will also be small. The equation $\Sigma M_A = 0$ then requires that **B** and **C** be very large, a situation which can result in the failure of the supports at *B* and *C*.

[‡]Because this situation arises from an inadequate arrangement or *geometry* of the supports, it is often referred to as *geometric instability*.

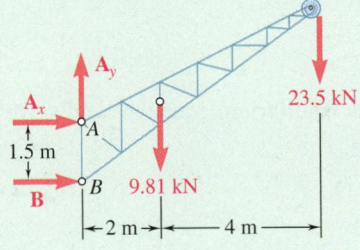

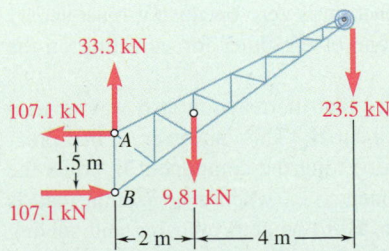

**Fig. 1** Free-body diagram of crane.

**Fig. 2** Free-body diagram of crane with solved reactions.

# Sample Problem 4.1

A fixed crane has a mass of 1000 kg and is used to lift a 2400-kg crate. It is held in place by a pin at $A$ and a rocker at $B$. The center of gravity of the crane is located at $G$. Determine the components of the reactions at $A$ and $B$.

**STRATEGY:** Draw a free-body diagram to show all of the forces acting on the crane, then use the equilibrium equations to calculate the values of the unknown forces.

**MODELING:**

**Free-Body Diagram.** By multiplying the masses of the crane and of the crate by $g = 9.81$ m/s$^2$, you obtain the corresponding weights—that is, 9810 N or 9.81 kN, and 23 500 N or 23.5 kN (Fig. 1). The reaction at pin $A$ is a force of unknown direction; you can represent it by components $A_x$ and $A_y$. The reaction at the rocker $B$ is perpendicular to the rocker surface; thus, it is horizontal. Assume that $A_x$, $A_y$, and $B$ act in the directions shown.

**ANALYSIS:**

**Determination of B.** The sum of the moments of all external forces about point $A$ is zero. The equation for this sum contains neither $A_x$ nor $A_y$, because the moments of $A_x$ and $A_y$ about $A$ are zero. Multiplying the magnitude of each force by its perpendicular distance from $A$, you have

$$+\circlearrowleft \Sigma M_A = 0: \qquad +B(1.5 \text{ m}) - (9.81 \text{ kN})(2 \text{ m}) - (23.5 \text{ kN})(6 \text{ m}) = 0$$
$$B = +107.1 \text{ kN} \qquad \mathbf{B} = 107.1 \text{ kN} \rightarrow \quad \blacktriangleleft$$

Because the result is positive, the reaction is directed as assumed.

**Determination of $A_x$.** Determine the magnitude of $A_x$ by setting the sum of the horizontal components of all external forces to zero.

$$\overset{+}{\rightarrow}\Sigma F_x = 0: \qquad A_x + B = 0$$
$$A_x + 107.1 \text{ kN} = 0$$
$$A_x = -107.1 \text{ kN} \qquad \mathbf{A}_x = 107.1 \text{ kN} \leftarrow \quad \blacktriangleleft$$

Because the result is negative, the sense of $A_x$ is opposite to that assumed originally.

**Determination of $A_y$.** The sum of the vertical components must also equal zero. Therefore,

$$+\uparrow\Sigma F_y = 0: \qquad A_y - 9.81 \text{ kN} - 23.5 \text{ kN} = 0$$
$$A_y = +33.3 \text{ kN} \qquad \mathbf{A}_y = 33.3 \text{ kN}\uparrow \quad \blacktriangleleft$$

Adding the components $A_x$ and $A_y$ vectorially, you can find that the reaction at $A$ is 112.2 kN ⦨17.3°.

**REFLECT and THINK:** You can check the values obtained for the reactions by recalling that the sum of the moments of all the external forces about any point must be zero. For example, considering point $B$ (Fig. 2), you can show that

$$+\circlearrowleft\Sigma M_B = -(9.81 \text{ kN})(2 \text{ m}) - (23.5 \text{ kN})(6 \text{ m}) + (107.1 \text{ kN})(1.5 \text{ m}) = 0$$

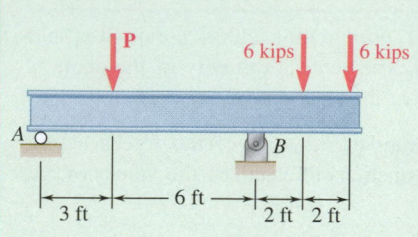

# Sample Problem 4.2

Three loads are applied to a beam as shown. The beam is supported by a roller at $A$ and by a pin at $B$. Neglecting the weight of the beam, determine the reactions at $A$ and $B$ when $P = 15$ kips.

**STRATEGY:**   Draw a free-body diagram of the beam, then write the equilibrium equations, first summing forces in the $x$ direction and then summing moments at $A$ and at $B$.

## MODELING:

**Free-Body Diagram.**   The reaction at $A$ is vertical and is denoted by **A** (Fig. 1). Represent the reaction at $B$ by components $\mathbf{B}_x$ and $\mathbf{B}_y$. Assume that each component acts in the direction shown.

## ANALYSIS:

**Equilibrium Equations.**   Write the three equilibrium equations and solve for the reactions indicated:

$$\xrightarrow{+}\ \Sigma F_x = 0: \qquad\qquad B_x = 0 \qquad\qquad \mathbf{B}_x = 0 \ \blacktriangleleft$$

$$+\circlearrowleft\ \Sigma M_A = 0:$$
$$-(15\ \text{kips})(3\ \text{ft}) + B_y(9\ \text{ft}) - (6\ \text{kips})(11\ \text{ft}) - (6\ \text{kips})(13\ \text{ft}) = 0$$
$$B_y = +21.0\ \text{kips} \quad \mathbf{B}_y = 21.0\ \text{kips} \uparrow \ \blacktriangleleft$$

$$+\circlearrowleft\ \Sigma M_B = 0:$$
$$-A(9\ \text{ft}) + (15\ \text{kips})(6\ \text{ft}) - (6\ \text{kips})(2\ \text{ft}) - (6\ \text{kips})(4\ \text{ft}) = 0$$
$$A = +6.00\ \text{kips} \quad \mathbf{A} = 6.00\ \text{kips} \uparrow \ \blacktriangleleft$$

**Fig. 1**   Free-body diagram of beam.

**REFLECT and THINK:**   Check the results by adding the vertical components of all of the external forces:

$$+\uparrow \Sigma F_y = +6.00\ \text{kips} - 15\ \text{kips} + 21.0\ \text{kips} - 6\ \text{kips} - 6\ \text{kips} = 0$$

**REMARK.**   In this problem, the reactions at both $A$ and $B$ are vertical; however, these reactions are vertical for different reasons. At $A$, the beam is supported by a roller; hence, the reaction cannot have any horizontal component. At $B$, the horizontal component of the reaction is zero because it must satisfy the equilibrium equation $\Sigma F_x = 0$ and none of the other forces acting on the beam has a horizontal component.

You might have noticed at first glance that the reaction at $B$ was vertical and dispensed with the horizontal component $\mathbf{B}_x$. This, however, is bad practice. In following it, you run the risk of forgetting the component $\mathbf{B}_x$ when the loading conditions require such a component (i.e., when a horizontal load is included). Also, you found the component $\mathbf{B}_x$ to be zero by using and solving an equilibrium equation, $\Sigma F_x = 0$. By setting $\mathbf{B}_x$ equal to zero immediately, you might not realize that you actually made use of this equation. Thus, you might lose track of the number of equations available for solving the problem.

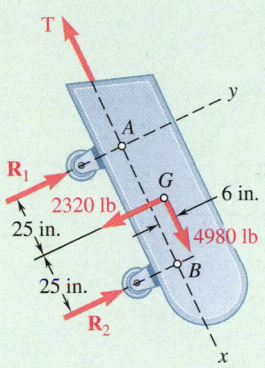

# Sample Problem 4.3

A loading car is at rest on a track forming an angle of 25° with the vertical. The gross weight of the car and its load is 5500 lb, and it acts at a point 30 in. from the track, halfway between the two axles. The car is held by a cable attached 24 in. from the track. Determine the tension in the cable and the reaction at each pair of wheels.

**STRATEGY:**   Draw a free-body diagram of the car to determine the unknown forces, and write equilibrium equations to find their values, summing moments at $A$ and $B$ and then summing forces.

**MODELING:**

**Free-Body Diagram.**   The reaction at each wheel is perpendicular to the track, and the tension force $\mathbf{T}$ is parallel to the track. Therefore, for convenience, choose the $x$ axis parallel to the track and the $y$ axis perpendicular to the track (Fig. 1). Then, resolve the 5500-lb weight into $x$ and $y$ components.

$$W_x = +(5500 \text{ lb}) \cos 25° = +4980 \text{ lb}$$
$$W_y = -(5500 \text{ lb}) \sin 25° = -2320 \text{ lb}$$

**ANALYSIS:**

**Equilibrium Equations.**   Take moments about $A$ to eliminate $\mathbf{T}$ and $\mathbf{R}_1$ from the computation.

$$+\circlearrowleft \Sigma M_A = 0: \qquad -(2320 \text{ lb})(25 \text{ in.}) - (4980 \text{ lb})(6 \text{ in.}) + R_2(50 \text{ in.}) = 0$$
$$R_2 = +1758 \text{ lb} \qquad\qquad \mathbf{R_2 = 1758 \text{ lb}} \nearrow \quad \blacktriangleleft$$

Then, take moments about $B$ to eliminate $\mathbf{T}$ and $\mathbf{R}_2$ from the computation.

$$+\circlearrowleft \Sigma M_B = 0: \qquad (2320 \text{ lb})(25 \text{ in.}) - (4980 \text{ lb})(6 \text{ in.}) - R_1(50 \text{ in.}) = 0$$
$$R_1 = +562 \text{ lb} \qquad\qquad \mathbf{R_1 = +562 \text{ lb}} \nearrow \quad \blacktriangleleft$$

Determine the value of $T$ by summing forces in the $x$ direction.

$$\searrow+\Sigma F_x = 0: \qquad +4980 \text{ lb} - T = 0$$
$$T = +4980 \text{ lb} \qquad\qquad \mathbf{T = 4980 \text{ lb}} \nwarrow \quad \blacktriangleleft$$

Fig. 2 shows the computed values of the reactions.

**REFLECT and THINK:**   You can verify the computations by summing forces in the $y$ direction.

$$\nearrow+\Sigma F_y = +562 \text{ lb} + 1758 \text{ lb} - 2320 \text{ lb} = 0$$

You could also check the solution by computing moments about any point other than $A$ or $B$.

**Fig. 1**  Free-body diagram of car.

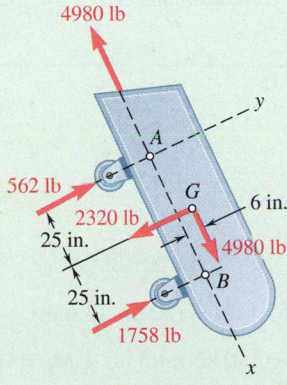

**Fig. 2**  Free-body diagram of car with solved reactions.

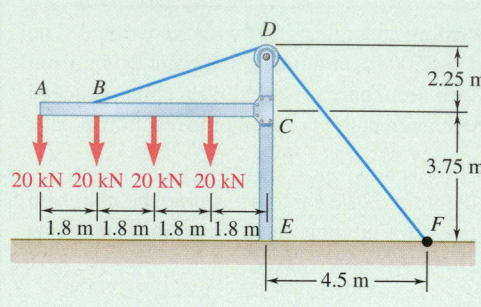

## Sample Problem 4.4

The frame shown supports part of the roof of a small building. Knowing that the tension in the cable is 150 kN, determine the reaction at the fixed end $E$.

**STRATEGY:** Draw a free-body diagram of the frame and of the cable $BDF$. The support at $E$ is fixed, so the reactions here include a moment. To determine its value, sum moments about point $E$.

**MODELING:**

**Free-Body Diagram.** Represent the reaction at the fixed end $E$ by the force components $\mathbf{E}_x$ and $\mathbf{E}_y$ and the couple $\mathbf{M}_E$ (Fig. 1). The other forces acting on the free body are the four 20-kN loads and the 150-kN force exerted at end $F$ of the cable.

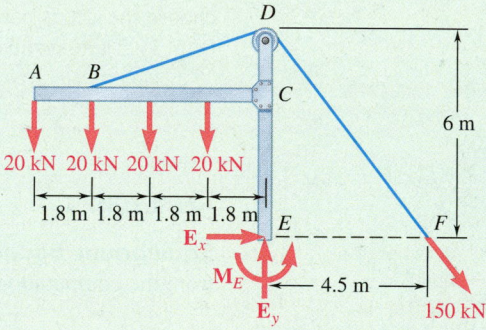

**Fig. 1** Free-body diagram of frame.

**ANALYSIS:**

**Equilibrium Equations.** First note that

$$DF = \sqrt{(4.5 \text{ m})^2 + (6 \text{ m})^2} = 7.5 \text{ m}$$

Then, you can write the three equilibrium equations and solve for the reactions at E.

$\xrightarrow{+} \Sigma F_x = 0:$     $E_x + \dfrac{4.5}{7.5}(150 \text{ kN}) = 0$

$E_x = -90.0 \text{ kN}$     $\mathbf{E}_x = 90.0 \text{ kN} \leftarrow$ ◄

$+\uparrow \Sigma F_y = 0:$     $E_y - 4(20 \text{ kN}) - \dfrac{6}{7.5}(150 \text{ kN}) = 0$

$E_y = +200 \text{ kN}$     $\mathbf{E}_y = 200 \text{ kN} \uparrow$ ◄

$+\circlearrowleft \Sigma M_E = 0:$     $(20 \text{ kN})(7.2 \text{ m}) + (20 \text{ kN})(5.4 \text{ m}) + (20 \text{ kN})(3.6 \text{ m})$

$+(20 \text{ kN})(1.8 \text{ m}) - \dfrac{6}{7.5}(150 \text{ kN})(4.5 \text{ m}) + M_E = 0$

$M_E = +180.0 \text{ kN·m}$     $\mathbf{M}_E = 180.0 \text{ kN·m} \circlearrowleft$ ◄

**REFLECT and THINK:** The cable provides a fourth constraint, making this situation statically indeterminate. This problem therefore gave us the value of the cable tension, which would have been determined by means other than statics. We could then use the three available independent static equilibrium equations to solve for the remaining three reactions.

## Sample Problem 4.5

A 400-lb weight is attached at $A$ to the lever shown. The constant of the spring $BC$ is $k = 250$ lb/in., and the spring is unstretched when $\theta = 0$. Determine the position of equilibrium.

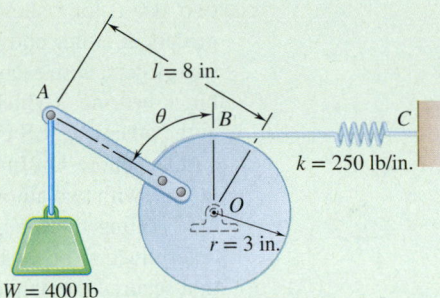

**STRATEGY:**   Draw a free-body diagram of the lever and cylinder to show all forces acting on the body (Fig. 1), then sum moments about $O$. Your final answer should be the angle $\theta$.

**MODELING:**

**Free-Body Diagram.**   Denote by $s$ the deflection of the spring from its unstretched position and note that $s = r\theta$. Then, $F = ks = kr\theta$.

**ANALYSIS:**

**Equilibrium Equation.**   Sum the moments of **W** and **F** about $O$ to eliminate the reactions supporting the cylinder. The result is

$$+\circlearrowleft \Sigma M_O = 0: \qquad Wl \sin\theta - r(kr\theta) = 0 \qquad \sin\theta = \frac{kr^2}{Wl}\theta$$

Substituting the given data yields

$$\sin\theta = \frac{(250 \text{ lb/in.})(3 \text{ in.})^2}{(400 \text{ lb})(8 \text{ in.})}\theta \quad \sin\theta = 0.703\,\theta$$

Solving by trial and error, the angle is $\qquad\qquad \theta = 0 \qquad\qquad \theta = 80.3° \blacktriangleleft$

**REFLECT and THINK:**   The weight could represent any vertical force acting on the lever. The key to the problem is to express the spring force as a function of the angle $\theta$.

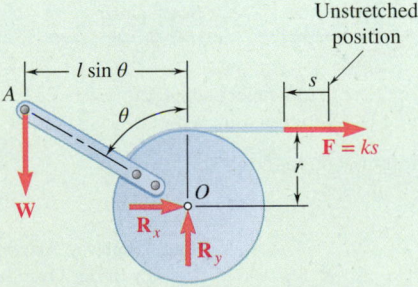

**Fig. 1**   Free-body diagram of the lever and cylinder.

# Case Study 4.1

The Mianus River Bridge in Greenwich, Connecticut, is a 24-span highway structure completed in 1958 that carries Interstate 95 over the Mianus River. Using separate northbound and southbound roadways, each direction includes two 100-ft-long skewed suspended spans that are supported by cantilevered girders at either end (CS Fig. 4.1). The suspended spans themselves contain two girders, with each attached to the cantilevered support girders using pillow-block bearings (which function as pin supports) at one end and twin hangers at the other end. CS Fig. 4.2 depicts the hanger connection, which functions as a link support. On June 28, 1983, one of the northbound suspended spans collapsed, with two automobiles and two trucks plunging into the void (CS Photo 4.1), killing three people and seriously injuring three more. The cause was determined to be corrosion-induced lateral displacement of the lower pin cap that secured the hangers onto the pin supporting the south girder, causing one of the hangers to work itself off the pin and transferring all load at this corner to the remaining hanger. The resulting increase in loading on the two pins eventually caused the upper pin to fracture, leading to the collapse of the entire span. The corrosion was accelerated by water and deicing agents draining through the deck expansion joint and regularly wetting the hanger connection. Compounding the situation was an inadequate routine inspection program that resulted in the severely compromised hanger condition remaining undetected before failing.*

Among the loads that the bridge was designed to support is a live load consisting of a standard truck as shown in CS Fig. 4.3, placed in each of the three

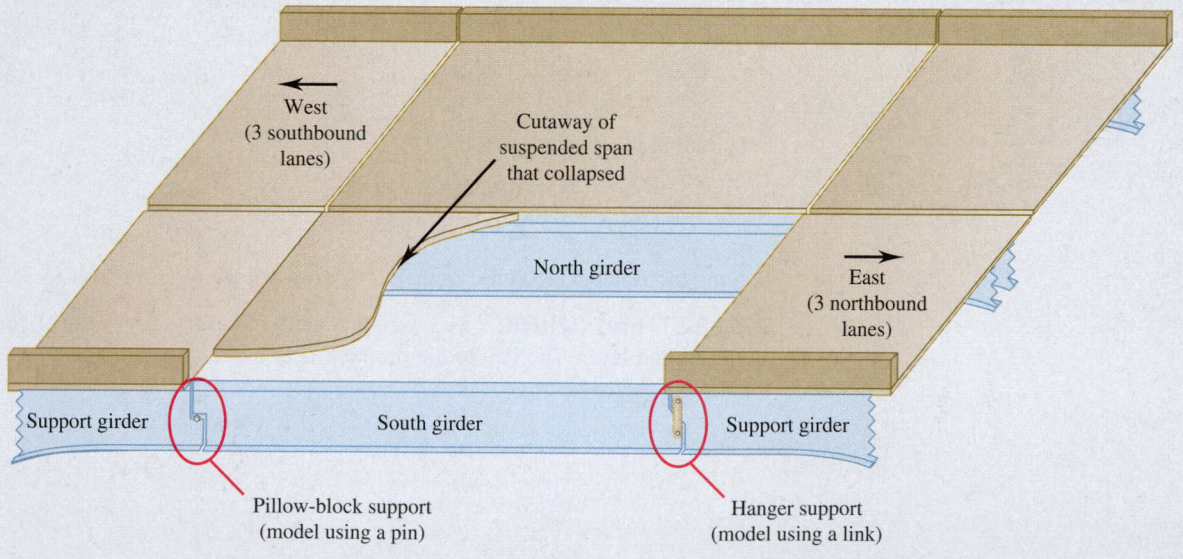

**CS Fig. 4.1**   Diagram of suspended span.

*Ref: "Highway Accident Report – Collapse of a Suspended Span of Interstate Route 95 Highway Bridge Over the Mianus River, Greenwich, Connecticut, June 28, 1983," Report No. NTSB/HAR-84/03, National Transportation Safety Board, July 19, 1984.

*(continued)*

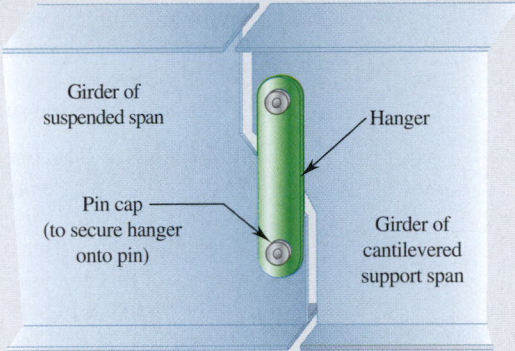

**CS Fig. 4.2**  Hanger support of suspended span.

**CS Photo 4.1**  View of Mianus River Bridge collapse.
©Hank Morgan/Getty Images

32 kips    32 kips    8 kips

14 ft    14 ft

**CS Fig. 4.3**  Standard design truck.

lanes of travel. Considering this live load (and disregarding any effects of the skewed deck), let's determine the maximum value of the resulting support reaction at the failed hanger connection.

**STRATEGY:**  Disregarding the effects of the skewed deck and positioning the three trucks so that they are aligned with each other as they travel over the three lanes, we will assume that the live load is equally distributed to the two girders. The live load carried by the south suspended girder will then be equivalent to the axle loads of 1.5 trucks, as shown in CS Fig. 4.4. It can be demonstrated

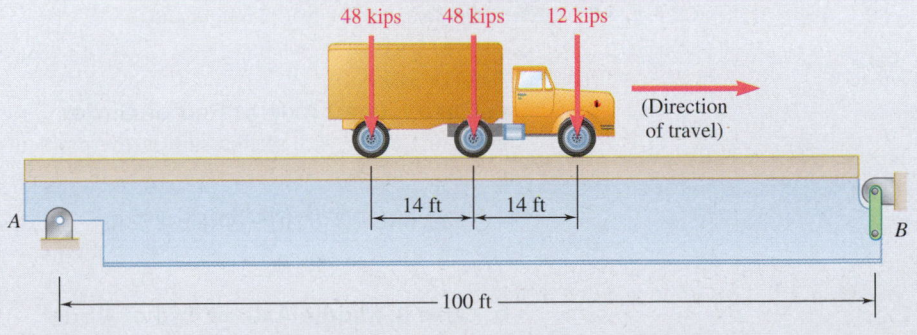

48 kips    48 kips    12 kips

(Direction of travel)

14 ft    14 ft

A    B

100 ft

**CS Fig. 4.4**  South suspended girder subject to live load of 1.5 trucks.

*(continued)*

that the maximum hanger reaction (at support *B*) will occur when one of the axle loads is positioned at this end of the girder. Considering the three possible cases, this reaction can then be determined by drawing the free-body diagram of the south girder and summing moments about end *A*.

**MODELING:** The free-body diagram for Case I is shown in CS Fig. 4.5, where the 12-kip axle is positioned at the end of the girder. Note that there is a short offset between this point and the hanger at *B*. Relative to the 100-ft span, the length of this offset is very small, and will therefore be disregarded for the purposes of this analysis. In the same manner, the free-body diagram for Case II is shown in CS Fig. 4.6, where the middle axle is now positioned at the end of the girder. Case III (not shown) would have the last axle placed at the end of the girder.

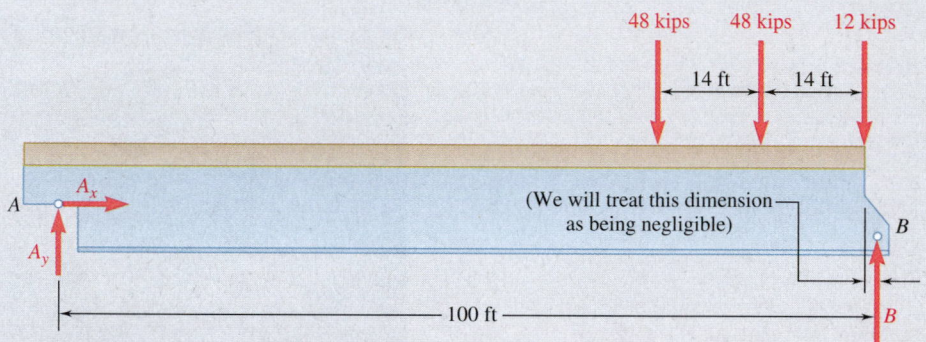

**CS Fig. 4.5**   Case I: 12-kip axle at far right end.

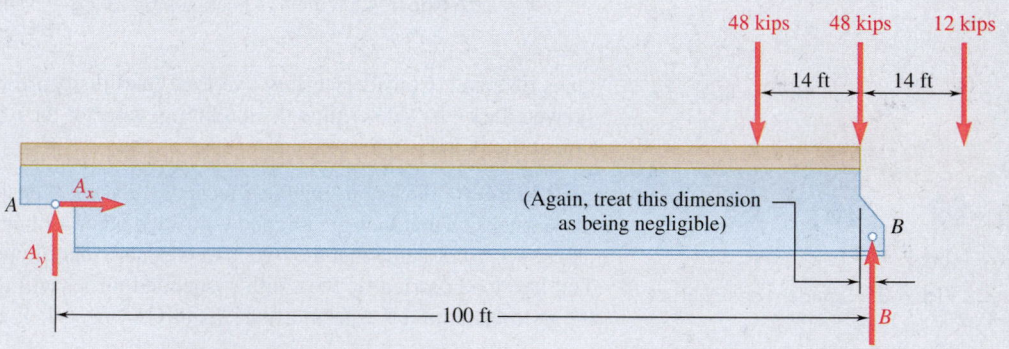

**CS Fig. 4.6**   Case II: middle axle (48 kips) at far right end.

**ANALYSIS:**

**a. Case I: Lead Axle at End of Girder.**   Referring to the free-body diagram of CS Fig. 4.5, set the sum of the moments of all external forces about point *A* equal to zero:

$$+\circlearrowleft \Sigma M_A = 0: \ +B(100 \text{ ft}) - (48 \text{ kips})(72 \text{ ft}) - (48 \text{ kips})(86 \text{ ft}) - (12 \text{ kips})(100 \text{ ft}) = 0$$
$$B = 87.8 \text{ kips} \qquad \mathbf{B} = 87.8 \text{ kips} \uparrow \ \blacktriangleleft$$

**b. Case II: Middle Axle at End of Girder.**   As shown in CS Fig. 4.6, the lead axle no longer acts on the suspended girder, *and should therefore not be*

*(continued)*

*included in its equilibrium analysis.* (It would thus be appropriate to not show this force at all with the girder's free-body diagram.) Setting the sum of the moments about point $A$ equal to zero of only those external forces acting on the girder:

$$+\circlearrowleft \Sigma M_A = 0: \qquad +B(100 \text{ ft}) - (48 \text{ kips})(86 \text{ ft}) - (48 \text{ kips})(100 \text{ ft}) = 0$$
$$B = 89.3 \text{ kips} \qquad \qquad \mathbf{B} = 89.3 \text{ kips} \uparrow \quad \blacktriangleleft$$

**c. Case III: Last Axle at End of Girder.**   Here, the only axle remaining on the girder is the trailing 48-kip axle. With it positioned at the end of the girder, by inspection it is apparent that the reaction at $B$ is equal to this force.

$$\mathbf{B} = 48 \text{ kips} \uparrow \quad \blacktriangleleft$$

Comparing the three cases, we conclude that Case II governs.

$$\mathbf{B}_{max} = 89.3 \text{ kips} \uparrow \quad \blacktriangleleft$$

**REFLECT and THINK:**   In addition to the live load considered here, this hanger is subject to other loads as well. Among these are the dead load (i.e., self-weight of the suspended span) and an impact load that is a code-prescribed percentage of the live load. Also note that if it were possible for the truck to be reversed and travel backward, we would obtain an even larger maximum live-load hanger reaction (97.9 kips) when the now-leading 48-kip axle is positioned at the hanger end of the girder, and with the other two axles trailing behind.

# SOLVING PROBLEMS
# ON YOUR OWN

You saw that, for a rigid body in equilibrium, the system of external forces is equivalent to zero. To solve an equilibrium problem, your first task is to draw a neat, reasonably large **free-body diagram** on which you show all external forces. You should include both known and unknown forces.

**For a two-dimensional rigid body,** the reactions at the supports can involve one, two, or three unknowns, depending on the type of support (Fig. 4.1). A correct free-body diagram is essential for the successful solution of a problem. Never proceed with the solution of a problem until you are sure your free-body diagram includes all loads, all reactions, and the weight of the body (if appropriate).

**1. You can write three equilibrium equations** and solve them for *three unknowns*. The three equations might be

$$\Sigma F_x = 0 \qquad \Sigma F_y = 0 \qquad \Sigma M_O = 0$$

However, usually several alternative sets of equations are possible, such as

$$\Sigma F_x = 0 \qquad \Sigma M_A = 0 \qquad \Sigma M_B = 0$$

where point $B$ is chosen in such a way that the line $AB$ is not parallel to the $y$ axis, or

$$\Sigma M_A = 0 \qquad \Sigma M_B = 0 \qquad \Sigma M_C = 0$$

where the points $A$, $B$, and $C$ do not lie along a straight line.

**2. To simplify your solution,** it may be helpful to use one of the following solution techniques.
   **a. By summing moments about the point of intersection** of the lines of action of two unknown forces, you obtain an equation in a single unknown.
   **b. By summing components in a direction perpendicular to two unknown parallel forces,** you also obtain an equation in a single unknown.

**3. After drawing your free-body diagram,** you may find that one of the following special situations arises.
   **a. The reactions involve fewer than three unknowns.** The body is said to be **partially constrained** and motion of the body is possible.
   **b. The reactions involve more than three unknowns.** The reactions are said to be **statically indeterminate**. Although you may be able to calculate one or two reactions, you cannot determine all of them.
   **c. The reactions pass through a single point or are parallel.** The body is said to be **improperly constrained** and motion can occur under a general loading condition.

# Problems

**FREE-BODY PRACTICE PROBLEMS**

**4.F1** Two crates, each of mass 350 kg, are placed as shown in the bed of a 1400-kg pickup truck. Draw the free-body diagram needed to determine the reactions at each of the two rear wheels A and front wheels B.

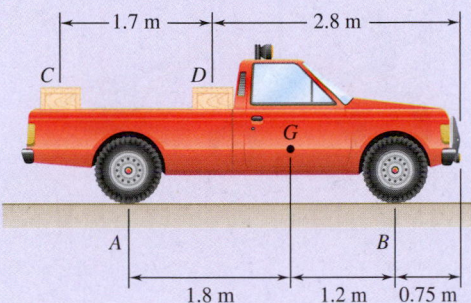

**Fig. P4.F1**

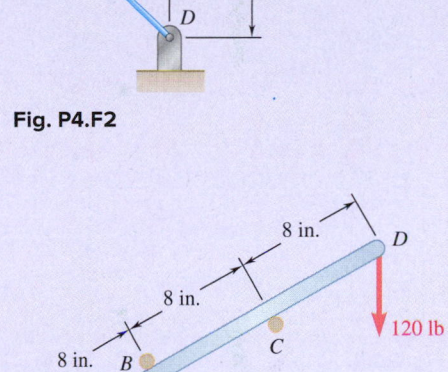

**Fig. P4.F2**

**4.F2** A lever AB is hinged at C and attached to a control cable at A. If the lever is subjected to a 75-lb vertical force at B, draw the free-body diagram needed to determine the tension in the cable and the reaction at C.

**4.F3** A light rod AD is supported by frictionless pegs at B and C and rests against a frictionless wall at A. A vertical 120-lb force is applied at D. Draw the free-body diagram needed to determine the reactions at A, B, and C.

**4.F4** A tension of 20 N is maintained in a tape as it passes through the support system shown. Knowing that the radius of each pulley is 10 mm, draw the free-body diagram needed to determine the reaction at C.

**Fig. P4.F3**

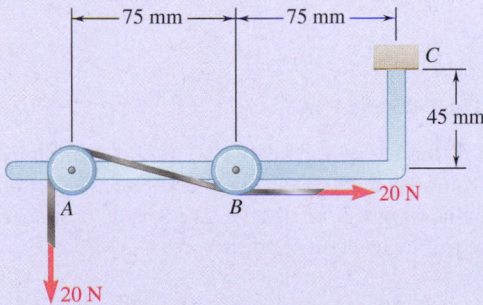

**Fig. P4.F4**

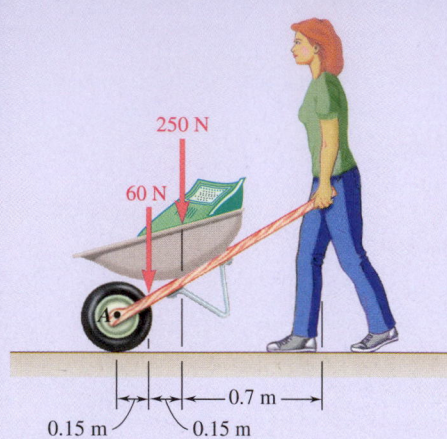

250 N

60 N

0.7 m

0.15 m  0.15 m

**Fig. P4.1**

**4.1** A gardener uses a 60-N wheelbarrow to transport a 250-N bag of fertilizer. What force must she exert on each handle?

**4.2** The gardener of Prob. 4.1 wishes to transport a second 250-N bag of fertilizer at the same time as the first one. Determine the maximum allowable horizontal distance from the axle $A$ of the wheelbarrow to the center of gravity of the second bag if she can hold only 75 N with each arm.

**4.3** A 2100-lb tractor is used to lift 900 lb of gravel. Determine the reaction at each of the two (*a*) rear wheels $A$, (*b*) front wheels $B$.

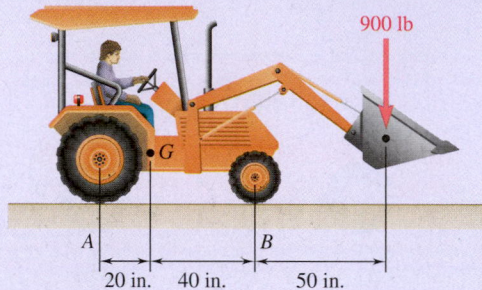

900 lb

$G$

$A$     $B$

20 in.   40 in.   50 in.

**Fig. *P4.3***

**4.4** For the beam and loading shown, determine (*a*) the reaction at $A$, (*b*) the tension in cable $BC$.

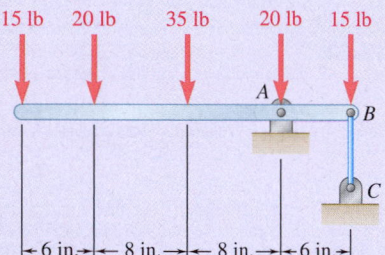

15 lb   20 lb   35 lb   20 lb   15 lb

$A$

$B$

$C$

←6 in.→←8 in.→←8 in.→←6 in.→

**Fig. P4.4**

**4.5** A load of lumber of weight $W = 25$ kN is being raised by a mobile crane. The weight of boom $ABC$ and the combined weight of the truck and driver are as shown. Determine the reaction at each of the two (*a*) front wheels $H$, (*b*) rear wheels $K$.

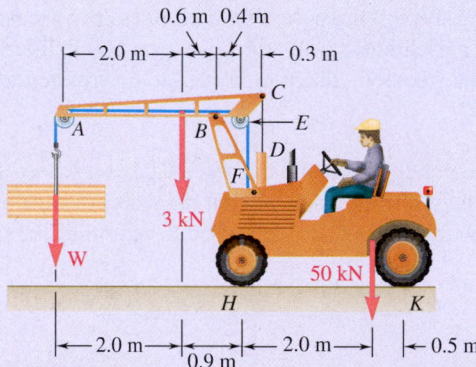

0.6 m  0.4 m

←2.0 m→   ←0.3 m

$A$   $B$   $C$   $E$

$D$

$F$

3 kN

$W$

50 kN

$H$   $K$

←2.0 m→←2.0 m→←0.5 m
0.9 m

**Fig. P4.5 and P4.6**

**4.6** A load of lumber of weight $W = 25$ kN is being raised by a mobile crane. Knowing that the tension is 25 kN in all portions of cable $AEF$ and that the weight of boom $ABC$ is 3 kN, determine (*a*) the tension in rod $CD$, (*b*) the reaction at pin $B$.

**4.7** A hand truck is used to move a compressed-air cylinder. Knowing that the combined weight of the truck and cylinder is 180 lb, determine (*a*) the vertical force **P** that should be applied to the handle to maintain the cylinder in the position shown, (*b*) the corresponding reaction at each of the two wheels.

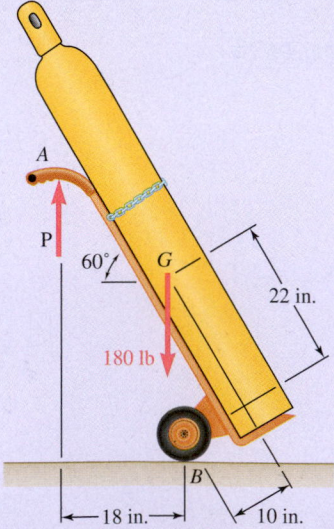

$A$

$P$

60°   $G$

22 in.

180 lb

$B$

←18 in.→   10 in.

**Fig. P4.7**

**4.8** Two external shafts of a gearbox are subject to the torques (or moments) shown. Determine the vertical components of the forces that must be exerted by the bolts at $A$ and $B$ to maintain the gearbox in equilibrium.

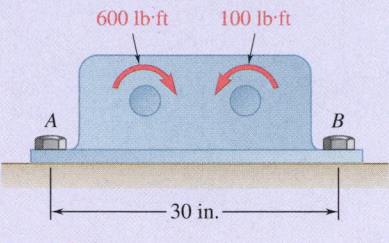

600 lb·ft    100 lb·ft

A            B

30 in.

**Fig. P4.8**

**4.9** Three loads are applied as shown to a light beam supported by cables attached at $B$ and $D$. Neglecting the weight of the beam, determine the range of values of $Q$ for which neither cable becomes slack when $P = 0$.

**4.10** The 10-m beam $AB$ rests upon, but is not attached to, supports at $C$ and $D$. Neglecting the weight of the beam, determine the range of values of $P$ for which the beam will remain in equilibrium.

**4.11** The maximum allowable value of each of the reactions is 50 kN, and each reaction must be directed upward. Neglecting the weight of the beam, determine the range of values of $P$ for which the beam is safe.

**4.12** For the beam of Sample Prob. 4.2, determine the range of values of $P$ for which the beam will be safe, knowing that the maximum allowable value of each of the reactions is 25 kips and that the reaction at $A$ must be directed upward.

**4.13** The maximum allowable value of each of the reactions is 180 N. Neglecting the weight of the beam, determine the range of the distance $d$ for which the beam is safe.

**4.14** For the beam and loading shown, determine the range of the distance $a$ for which the reaction at $B$ does not exceed 100 lb downward or 200 lb upward.

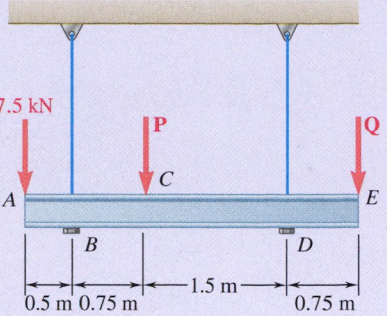

7.5 kN    P    Q

A    C    E
    B    D

0.5 m 0.75 m    1.5 m    0.75 m

**Fig. P4.9**

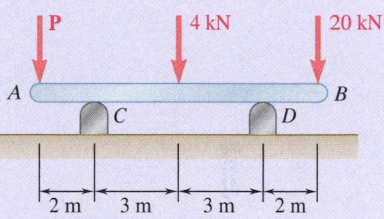

P    4 kN    20 kN

A    C    D    B

2 m    3 m    3 m    2 m

**Fig. P4.10 and P4.11**

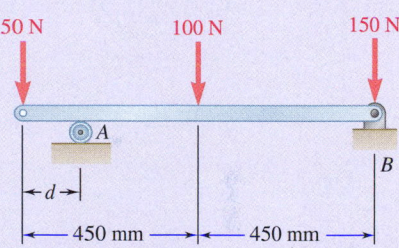

50 N    100 N    150 N

A    B

d

450 mm    450 mm

**Fig. P4.13**

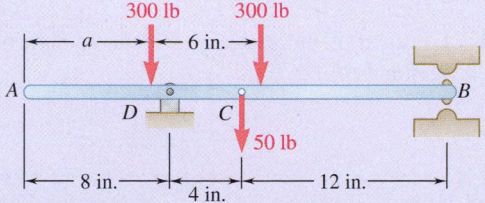

300 lb    300 lb

a    6 in.

A    B
D    C

50 lb

8 in.    4 in.    12 in.

**Fig. P4.14**

**4.15** The required tension in cable $AB$ is 1200 N. Determine (*a*) the vertical force **P** that must be applied to the pedal, (*b*) the corresponding reaction at $C$.

**4.16** Determine the maximum tension that can be developed in cable $AB$ if the maximum allowable value of the reaction at $C$ is 2.6 kN.

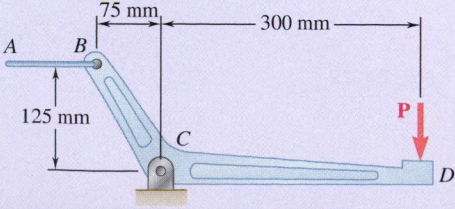

75 mm    300 mm

A    B

125 mm

C

P

D

**Fig. P4.15 and P4.16**

**4.17** Two links *AB* and *DE* are connected by a bell crank as shown. Knowing that the tension in link *AB* is 150 lb, determine (*a*) the tension in link *DE*, (*b*) the reaction at *C*.

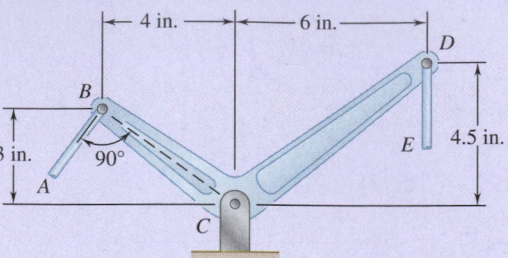

**Fig. P4.17 and P4.18**

**4.18** Two links *AB* and *DE* are connected by a bell crank as shown. Determine the maximum force that can be safely exerted by link *AB* on the bell crank if the maximum allowable value for the reaction at *C* is 400 lb.

**4.19** The bracket *BCD* is hinged at *C* and attached to a control cable at *B*. For the loading shown, determine (*a*) the tension in the cable, (*b*) the reaction at *C*.

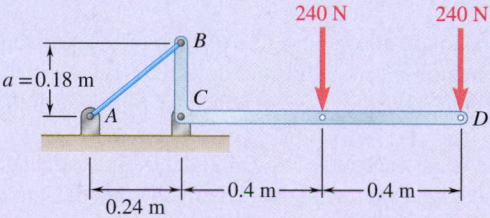

**Fig. P4.19**

**4.20** Solve Prob. 4.19, assuming that *a* = 0.32 m.

**4.21** The ladder *AB*, of length *L* and weight *W*, can be raised by cable *BC*. Determine the tension *T* required to raise end *B* just off the floor (*a*) in terms of *W* and $\theta$, (*b*) if *h* = 8 ft, *L* = 10 ft, and *W* = 35 lb.

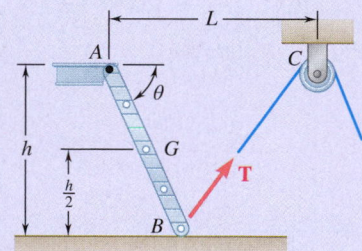

**Fig. P4.21**

**4.22** A lever *AB* is hinged at *C* and attached to a control cable at *A*. If the lever is subjected to a 500-N horizontal force at *B*, determine (*a*) the tension in the cable, (*b*) the reaction at *C*.

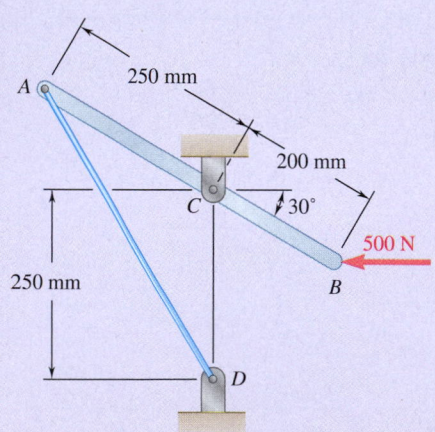

**Fig. P4.22**

**4.23 and 4.24** For each of the plates and loadings shown, determine the reactions at *A* and *B*.

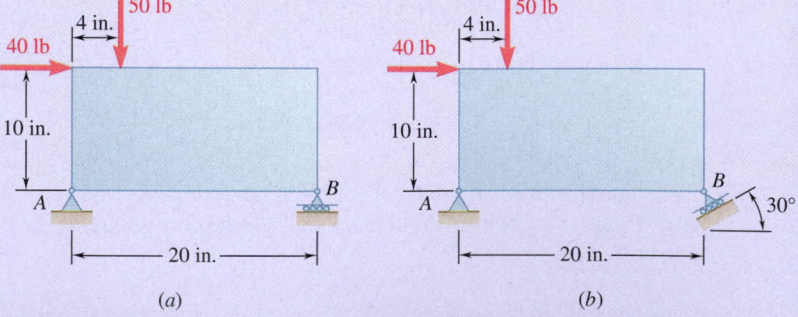

(*a*)          (*b*)

**Fig. P4.23**

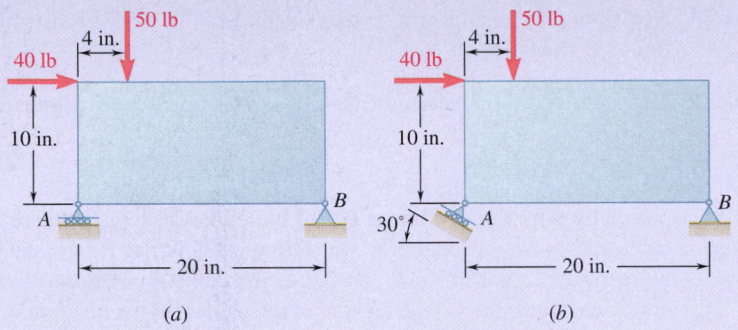

**Fig. P4.24**

**4.25** A rod *AB*, hinged at *A* and attached at *B* to cable *BD*, supports the loads shown. Knowing that *d* = 200 mm, determine (*a*) the tension in cable *BD*, (*b*) the reaction at *A*.

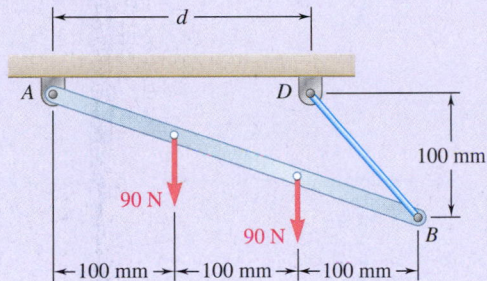

**Fig. P4.25 and P4.26**

**4.26** A rod *AB*, hinged at *A* and attached at *B* to cable *BD*, supports the loads shown. Knowing that *d* = 150 mm, determine (*a*) the tension in cable *BD*, (*b*) the reaction at *A*.

**4.27** For the frame and loading shown, determine the reactions at *A* and *E* when (*a*) $\alpha = 30°$, (*b*) $\alpha = 45°$.

**4.28** Determine the reactions at *A* and *C* when (*a*) $\alpha = 0$, (*b*) $\alpha = 30°$.

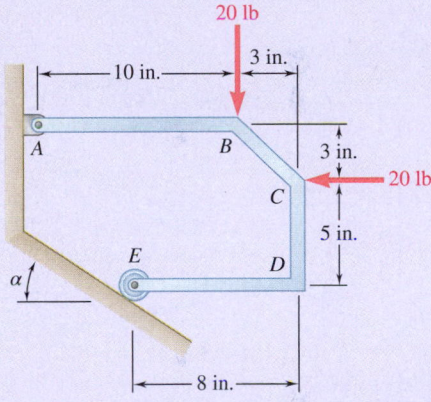

**Fig. P4.27**

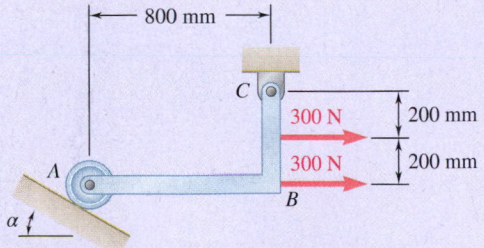

**Fig. P4.28**

**4.29** The spanner shown is used to rotate a shaft. A pin fits in a hole at *A*, while a flat, frictionless surface rests against the shaft at *B*. If a 300-N force **P** is exerted on the spanner at *D*, find (*a*) the reaction at *B*, (*b*) the component of the reaction at *A* in a direction perpendicular to *AC*.

**4.30** The spanner shown is used to rotate a shaft. A pin fits in a hole at *A*, while a flat, frictionless surface rests against the shaft at *B*. If the moment about *C* of the force exerted on the shaft at *A* is to be 90 N·m, find (*a*) the force **P** that should be exerted on the spanner at *D*, (*b*) the corresponding value of the force exerted on the spanner at *B*.

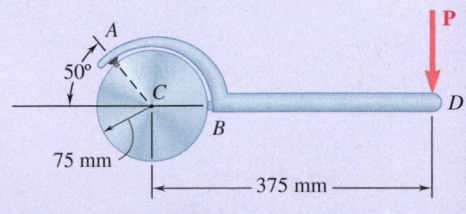

**Fig. P4.29 and P4.30**

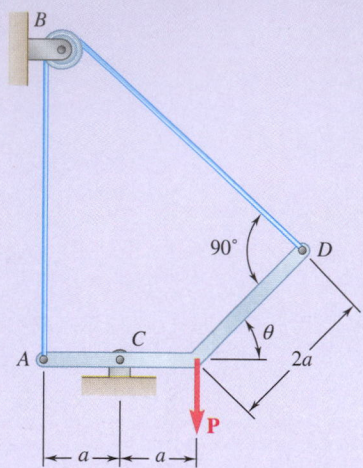

**Fig. P4.31 and P4.32**

**4.31** Neglecting friction, determine the tension in cable *ABD* and the reaction at *C* when $\theta = 60°$.

**4.32** Neglecting friction, determine the tension in cable *ABD* and the reaction at *C* when $\theta = 45°$.

**4.33** A force **P** of magnitude 90 lb is applied to member *ACDE* that is supported by a frictionless pin at *D* and by the cable *ABE*. Because the cable passes over a small pulley at *B*, the tension may be assumed to be the same in portions *AB* and *BE* of the cable. For the case when $a = 3$ in., determine (*a*) the tension in the cable, (*b*) the reaction at *D*.

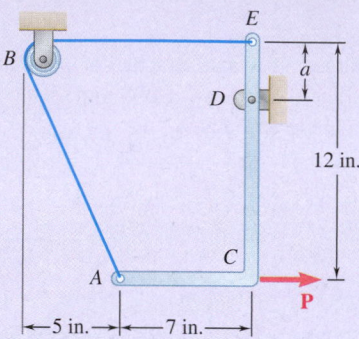

**Fig. P4.33**

**4.34** Solve Prob. 4.33 for $a = 6$ in.

**4.35** Bar *AC* supports two 400-N loads as shown. Rollers at *A* and *C* rest against frictionless surfaces and a cable *BD* is attached at *B*. Determine (*a*) the tension in cable *BD*, (*b*) the reaction at *A*, (*c*) the reaction at *C*.

**4.36** A light bar *AD* is suspended from a cable *BE* and supports a 20-kg block at *C*. The ends *A* and *D* of the bar are in contact with frictionless vertical walls. Determine the tension in cable *BE* and the reactions at *A* and *D*.

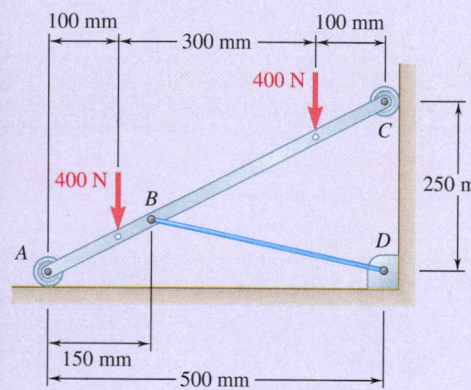

**Fig. P4.35**

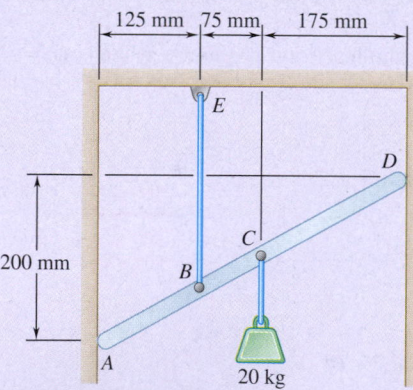

**Fig. P4.36**

**4.37** A 160-lb overhead garage door consists of a uniform rectangular panel *AC*, 84 in. long, supported by the cable *AE* attached at the middle of the upper edge of the door and by two sets of frictionless rollers at *A* and *B*. Each set consists of two rollers located on either side of the door. The rollers *A* are free to move in horizontal channels, while the rollers *B* are guided by vertical channels. If the door is held in the position for which *BD* = 42 in., determine (*a*) the tension in cable *AE*, (*b*) the reaction at each of the four rollers.

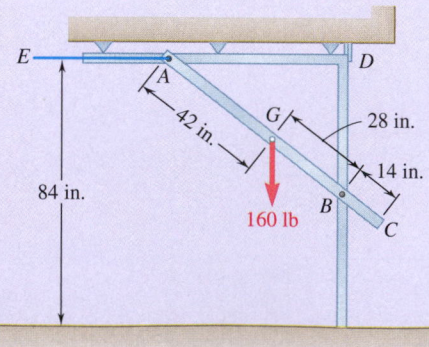

**Fig. P4.37**

**4.38** In Prob. 4.37, determine the distance *BD* for which the tension in cable *AE* is equal to 600 lb.

**4.39** A movable bracket is held at rest by a cable attached at *E* and by frictionless rollers. Knowing that the width of post *FG* is slightly less than the distance between the rollers, determine the force exerted on the post by each roller when $\alpha = 20°$.

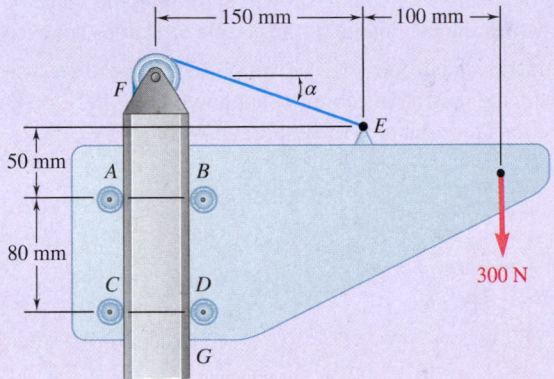

Fig. P4.39

**4.40** Solve Prob. 4.39 when $\alpha = 30°$.

**4.41** The semicircular rod *ABCD* is maintained in equilibrium by the small wheel at *D* and the rollers at *B* and *C*. Knowing that $\alpha = 45°$, determine the reactions at *B*, *C*, and *D*.

**4.42** Determine the range of values of $\alpha$ for which the semicircular rod shown can be maintained in equilibrium by the small wheel at *D* and the rollers at *B* and *C*.

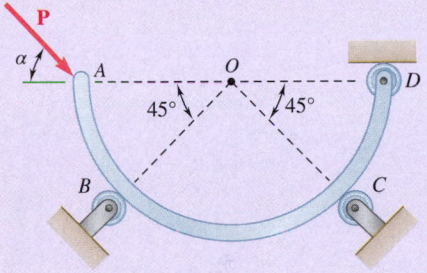

Fig. P4.41 and P4.42

**4.43** The rig shown consists of a 1200-lb horizontal member *ABC* and a vertical member *DBE* welded together at *B*. The rig is being used to raise a 3600-lb crate at a distance $x = 12$ ft from the vertical member *DBE*. If the tension in the cable is 4 kips, determine the reaction at *E*, assuming that the cable is (*a*) anchored at *F* as shown in the figure, (*b*) attached to the vertical member at a point located 1 ft above *E*.

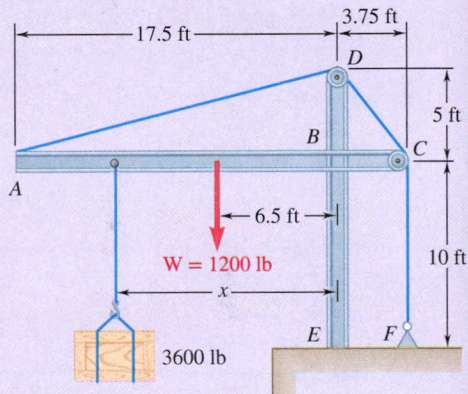

Fig. P4.43

**4.44** For the rig and crate of Prob. 4.43 and assuming that cable is anchored at *F* as shown, determine (*a*) the required tension in cable *ADCF* if the maximum value of the couple at *E* as *x* varies from 1.5 to 17.5 ft is to be as small as possible, (*b*) the corresponding maximum value of the couple.

**4.45** A 175-kg utility pole is used to support at *C* the end of an electric wire. The tension in the wire is 600 N, and the wire forms an angle of 15° with the horizontal at *C*. Determine the largest and smallest allowable tensions in the guy cable *BD* if the magnitude of the couple at *A* may not exceed 500 N·m.

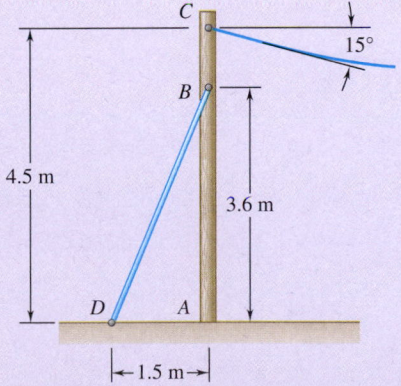

Fig. P4.45

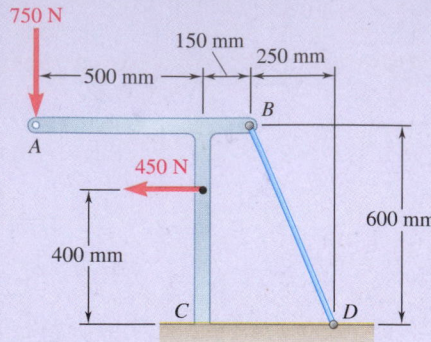

Fig. P4.46 and P4.47

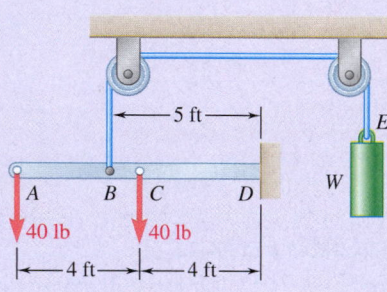

Fig. P4.48 and P4.49

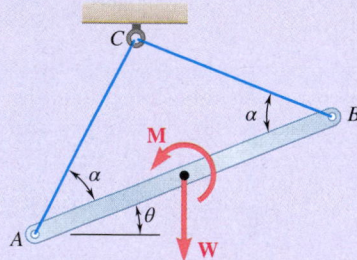

Fig. P4.51

**4.46** Knowing that the tension in wire BD is 1300 N, determine the reaction at the fixed support C of the frame shown.

**4.47** Determine the range of allowable values of the tension in wire BD if the magnitude of the couple at the fixed support C is not to exceed 100 N·m.

**4.48** Beam AD carries the two 40-lb loads shown. The beam is held by a fixed support at D and by the cable BE that is attached to the counterweight W. Determine the reaction at D when (a) W = 100 lb, (b) W = 90 lb.

**4.49** For the beam and loading shown, determine the range of values of W for which the magnitude of the couple at D does not exceed 40 lb·ft.

**4.50** A traffic-signal pole may be supported in the three ways shown; in part c, the tension in cable BC is known to be 1950 N. Determine the reactions for each type of support shown.

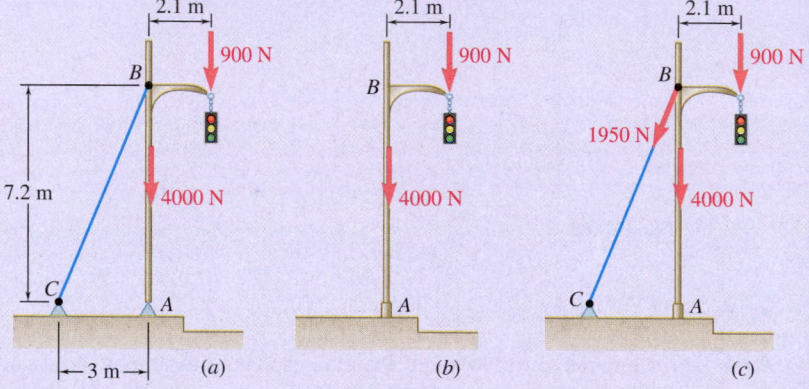

Fig. P4.50

**4.51** A uniform rod AB with a length of l and weight of W is suspended from two cords AC and BC of equal length. Determine the angle θ corresponding to the equilibrium position when a couple **M** is applied to the rod.

**4.52** Rod AD is acted upon by a vertical force **P** at end A and by two equal and opposite horizontal forces of magnitude Q at points B and C. Neglecting the weight of the rod, express the angle θ corresponding to the equilibrium position in terms of P and Q.

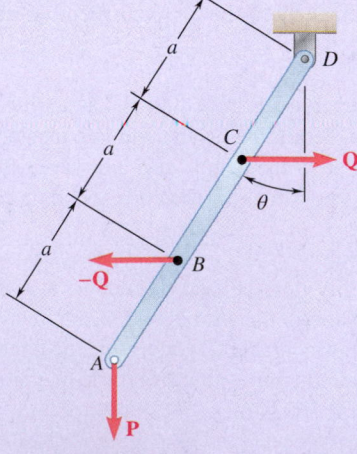

Fig. P4.52

**4.53**  A slender rod $AB$ with a weight of $W$ is attached to blocks $A$ and $B$ that move freely in the guides shown. The blocks are connected by an elastic cord that passes over a pulley at $C$. ($a$) Express the tension in the cord in terms of $W$ and $\theta$. ($b$) Determine the value of $\theta$ for which the tension in the cord is equal to $3W$.

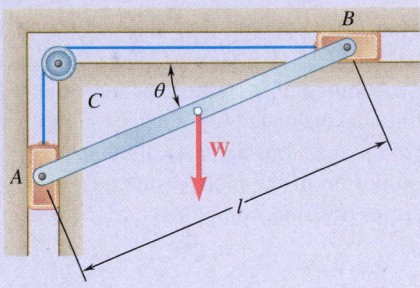

**Fig. P4.53**

**4.54 and 4.55**  A vertical load $\mathbf{P}$ is applied at end $B$ of rod $BC$. ($a$) Neglecting the weight of the rod, express the angle $\theta$ corresponding to the equilibrium position in terms of $P$, $l$, and the counterweight $W$. ($b$) Determine the value of $\theta$ corresponding to equilibrium if $P = 2W$.

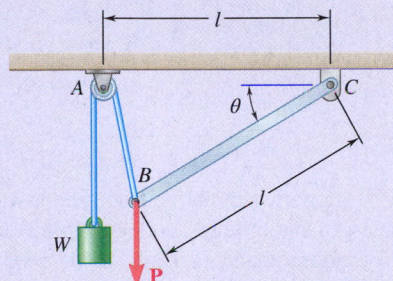

**Fig. P4.54**

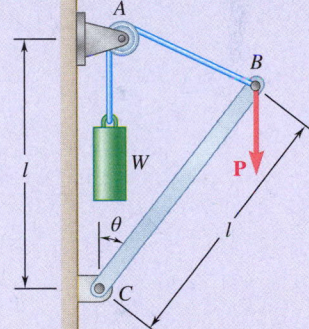

**Fig. P4.55**

**4.56**  A collar $B$ with a weight of $W$ can move freely along the vertical rod shown. The constant of the spring is $k$, and the spring is unstretched when $\theta = 0$. ($a$) Derive an equation in $\theta$, $W$, $k$, and $l$ that must be satisfied when the collar is in equilibrium. ($b$) Knowing that $W = 300$ N, $l = 500$ mm, and $k = 800$ N/m, determine the value of $\theta$ corresponding to equilibrium.

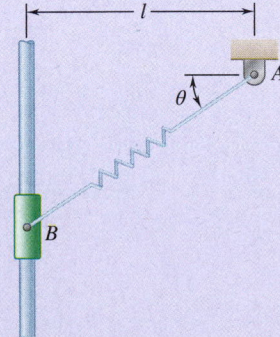

**Fig. P4.56**

**4.57**  Solve Sample Prob. 4.5, assuming that the spring is unstretched when $\theta = 90°$.

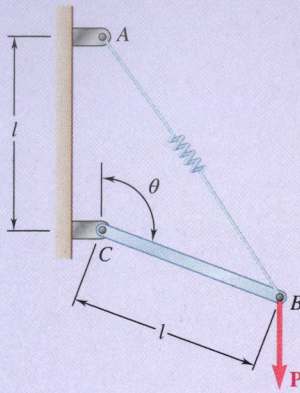

Fig. *P4.58*

**4.58** A vertical load **P** is applied at end *B* of rod *BC*. The constant of the spring is *k*, and the spring is unstretched when $\theta = 90°$. (*a*) Neglecting the weight of the rod, express the angle $\theta$ corresponding to the equilibrium position in terms of *P*, *k*, and *l*. (*b*) Determine the value of $\theta$ corresponding to equilibrium if $P = \frac{1}{4}kl$.

**4.59** Eight identical $500 \times 750$-mm rectangular plates, each of mass $m = 40$ kg, are held in a vertical plane as shown. All connections consist of frictionless pins, rollers, or short links. In each case, determine whether (*a*) the plate is completely, partially, or improperly constrained, (*b*) the reactions are statically determinate or indeterminate, (*c*) the equilibrium of the plate is maintained in the position shown. Also, wherever possible, compute the reactions.

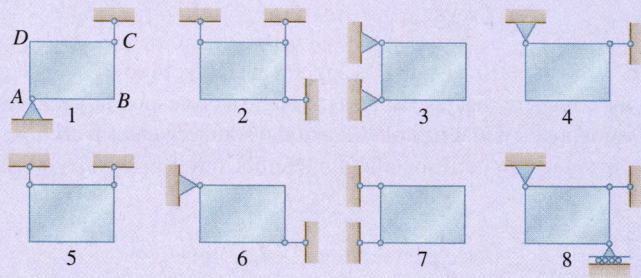

Fig. **P4.59**

**4.60** A truss can be supported in the eight different ways shown. All connections consist of smooth pins, rollers, or short links. For each case, answer the questions listed in Prob. 4.59, and, wherever possible, compute the reactions, assuming that the magnitude of the force **P** is 12 kips.

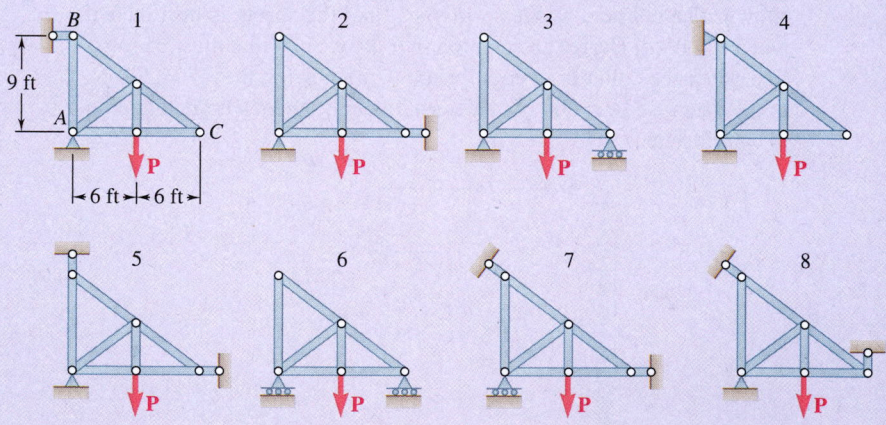

Fig. *P4.60*

# 4.2 TWO SPECIAL CASES

In practice, some simple cases of equilibrium occur quite often, either as part of a more complicated analysis or as the complete models of a situation. By understanding the characteristics of these cases, you can often simplify the overall analysis.

## 4.2A Equilibrium of a Two-Force Body

A particular case of equilibrium of considerable interest in practical applications is that of a rigid body subjected to two forces. Such a body is commonly called a **two-force body**. We show here that, **if a two-force body is in equilibrium, the two forces must have the same magnitude, the same line of action, and opposite sense**.

Consider a corner plate subjected to two forces $\mathbf{F}_1$ and $\mathbf{F}_2$ acting at $A$ and $B$, respectively (Fig. 4.8*a*). If the plate is in equilibrium, the sum of the moments of $\mathbf{F}_1$ and $\mathbf{F}_2$ about any axis must be zero. First, we sum moments about $A$. Because the moment of $\mathbf{F}_1$ is zero, the moment of $\mathbf{F}_2$ also must be zero and the line of action of $\mathbf{F}_2$ must pass through $A$ (Fig. 4.8*b*). Similarly, summing moments about $B$, we can show that the line of action of $\mathbf{F}_1$ must pass through $B$ (Fig. 4.8*c*). Therefore, both forces have the same line of action (line $AB$). You can see from either of the equations $\Sigma F_x = 0$ and $\Sigma F_y = 0$ that they must also have the same magnitude but opposite sense.

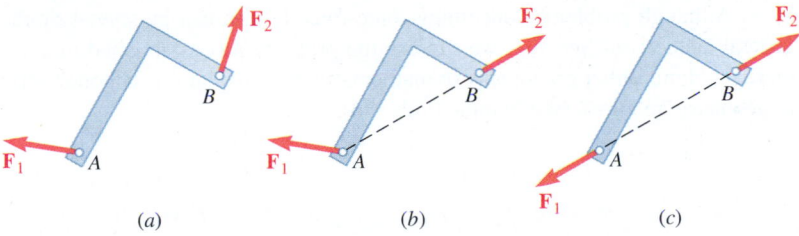

(a)          (b)          (c)

**Fig. 4.8** A two-force body in equilibrium. (*a*) Forces act at two points of the body; (*b*) summing moments about point *A* shows that the line of action of $\mathbf{F}_2$ must pass through *A*; (*c*) summing moments about point *B* shows that the line of action of $\mathbf{F}_1$ must pass through *B*.

If several forces act at two points $A$ and $B$, the forces acting at $A$ can be replaced by their resultant $\mathbf{F}_1$, and those acting at $B$ can be replaced by their resultant $\mathbf{F}_2$. Thus, a two-force body can be more generally defined as **a rigid body subjected to forces acting at only two points**. The resultants $\mathbf{F}_1$ and $\mathbf{F}_2$ then must have the same line of action, the same magnitude, and opposite sense (Fig. 4.8).

Later, in the study of structures, frames, and machines, you will see how the recognition of two-force bodies simplifies the solution of certain problems.

## 4.2B Equilibrium of a Three-Force Body

Another case of equilibrium that is of great practical interest is that of a **three-force body**, i.e., a rigid body subjected to three forces or, more generally, **a rigid body subjected to forces acting at only three points**. Consider a rigid body subjected to a system of forces that can be reduced to three forces $\mathbf{F}_1$, $\mathbf{F}_2$, and $\mathbf{F}_3$ acting at $A$, $B$, and $C$, respectively (Fig. 4.9*a*). We show that if the body is in equilibrium, **the lines of action of the three forces must be either concurrent or parallel**.

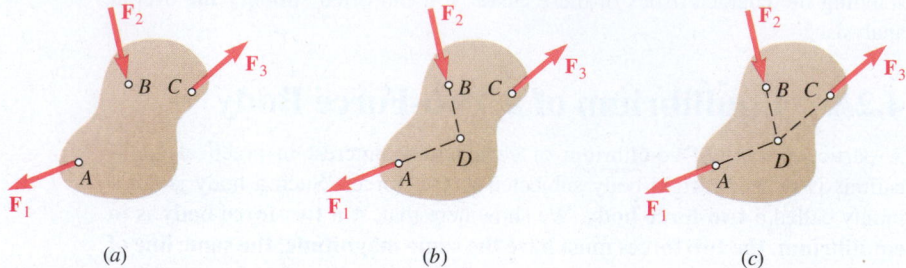

(a)                    (b)                    (c)

**Fig. 4.9**    A three-force body in equilibrium. Figures (*a*–*c*) demonstrate that the lines of action of the three forces must be either concurrent or parallel.

Because the rigid body is in equilibrium, the sum of the moments of $\mathbf{F}_1$, $\mathbf{F}_2$, and $\mathbf{F}_3$ about any axis must be zero. Assuming that the lines of action of $\mathbf{F}_1$ and $\mathbf{F}_2$ intersect and denoting their point of intersection by $D$, we sum moments about $D$ (Fig. 4.9*b*). Because the moments of $\mathbf{F}_1$ and $\mathbf{F}_2$ about $D$ are zero, the moment of $\mathbf{F}_3$ about $D$ also must be zero, and the line of action of $\mathbf{F}_3$ must pass through $D$ (Fig. 4.9*c*). Therefore, the three lines of action are concurrent. The only exception occurs when none of the lines intersect; in this case, the lines of action are parallel.

Although problems concerning three-force bodies can be solved by the general methods of Sec. 4.1, we can use the property just established to solve these problems either graphically or mathematically using simple trigonometric or geometric relations (see Sample Prob. 4.6).

## Sample Problem 4.6

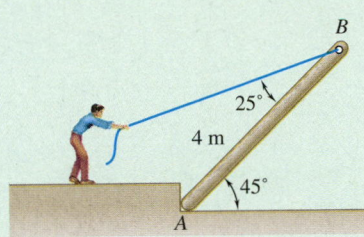

A man raises a 10-kg joist with a length of 4 m by pulling on a rope. Find the tension $T$ in the rope and the reaction at $A$.

**STRATEGY:** The joist is acted upon by three forces: its weight **W**, the force **T** exerted by the rope, and the reaction **R** of the ground at $A$. Therefore, it is a three-force body, and you can compute the forces by using a force triangle.

**MODELING:** First note that

$$W = mg = (10\ \text{kg})(9.81\ \text{m/s}^2) = 98.1\ \text{N}$$

Because the joist is a three-force body, the forces acting on it must be concurrent. The reaction **R** therefore must pass through the point of intersection $C$ of the lines of action of the weight **W** and the tension force **T**, as shown in the free-body diagram (Fig. 1). You can use this fact to determine the angle $\alpha$ that **R** forms with the horizontal.

**ANALYSIS:** Draw the vertical line $BF$ through $B$ and the horizontal line $CD$ through $C$ (Fig. 2). Then

$$AF = BF = (AB)\cos 45° = (4\ \text{m})\cos 45° = 2.828\ \text{m}$$

$$CD = EF = AE = \frac{1}{2}(AF) = 1.414\ \text{m}$$

$$BD = (CD)\cot(45° + 25°) = (1.414\ \text{m})\tan 20° = 0.515\ \text{m}$$
$$CE = DF = BF - BD = 2.828\ \text{m} - 0.515\ \text{m} = 2.313\ \text{m}$$

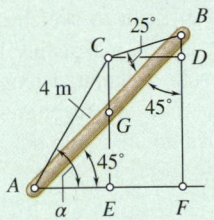

**Fig. 2** Geometry analysis of the lines of action for the three forces acting on the joist, concurrent at point C.

From these calculations, you can determine the angle $\alpha$ as

$$\tan \alpha = \frac{CE}{AE} = \frac{2.313\ \text{m}}{1.414\ \text{m}} = 1.636$$

$$\alpha = 58.6° \blacktriangleleft$$

You now know the directions of all the forces acting on the joist.

**Force Triangle.** Draw a force triangle as shown (Fig. 3) with its interior angles computed from the known directions of the forces. You can then use the law of sines to find the unknown forces.

$$\frac{T}{\sin 31.4°} = \frac{R}{\sin 110°} = \frac{98.1\ \text{N}}{\sin 38.6°}$$

$$T = 81.9\ \text{N} \blacktriangleleft$$
$$R = 147.8\ \text{N} \measuredangle 58.6° \blacktriangleleft$$

**REFLECT and THINK:** In practice, three-force members occur often, so learning this method of analysis is useful in many situations.

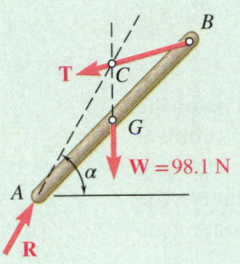

**Fig. 1** Free-body diagram of the joist.

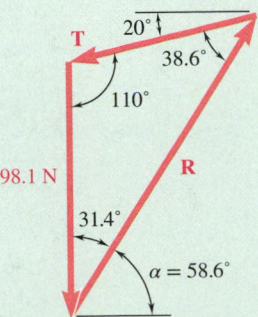

**Fig. 3** Force triangle.

# SOLVING PROBLEMS
# ON YOUR OWN

This section covered two particular cases of equilibrium of a rigid body.

**1. A two-force body is subjected to forces at only two points.** The resultants of the forces acting at each of these points must have the *same magnitude, the same line of action, and opposite sense*. This property allows you to simplify the solutions of some problems by replacing the two unknown components of a reaction by a single force of unknown magnitude but of *known direction*.

**2. A three-force body is subjected to forces at only three points.** The resultants of the forces acting at each of these points must be *concurrent or parallel*. To solve a problem involving a three-force body with concurrent forces, draw the free-body diagram showing that these three forces pass through the same point. You may be able to complete the solution by using simple geometry, such as a force triangle and the law of sines (see Sample Prob. 4.6).

This method for solving problems involving three-force bodies is not difficult to understand, but in practice, it can be difficult to sketch the necessary geometric constructions. If you encounter difficulty, first draw a reasonably large free-body diagram and then seek a relation between known or easily calculated lengths and a dimension that involves an unknown. Sample Prob. 4.6 illustrates this technique, where we used the easily calculated dimensions $AE$ and $CE$ to determine the angle $\alpha$.

# Problems

**4.61** A 500-lb cylindrical tank, 8 ft in diameter, is to be raised over a 2-ft obstruction. A cable is wrapped around the tank and pulled horizontally as shown. Knowing that the corner of the obstruction at *A* is rough, find the required tension in the cable and the reaction at *A*.

**4.62** Determine the reactions at *A* and *E* when $\alpha = 0$.

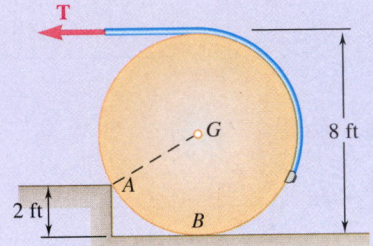

**Fig. P4.61**

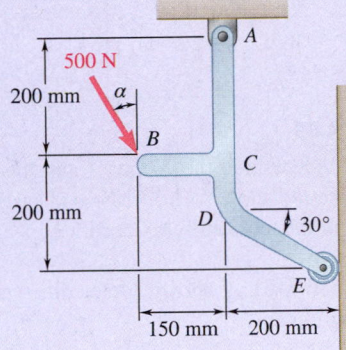

**Fig. P4.62 and P4.63**

**4.63** Determine (*a*) the value of $\alpha$ for which the reaction at *A* is vertical, (*b*) the corresponding reactions at *A* and *E*.

**4.64** A 12-ft ladder, weighing 40 lb, leans against a frictionless vertical wall. The lower end of the ladder rests on rough ground, 4 ft away from the wall. Determine the reactions at both ends.

**4.65** Determine the reactions at *B* and *C* when $a = 30$ mm.

**4.66** Determine the reactions at *A* and *E*.

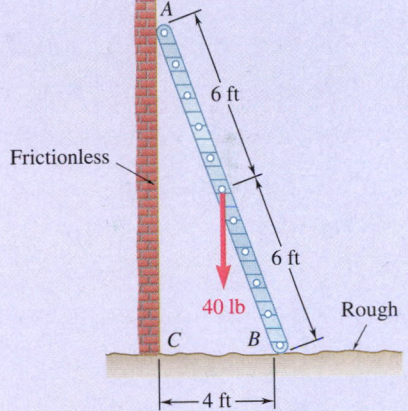

**Fig. P4.64**

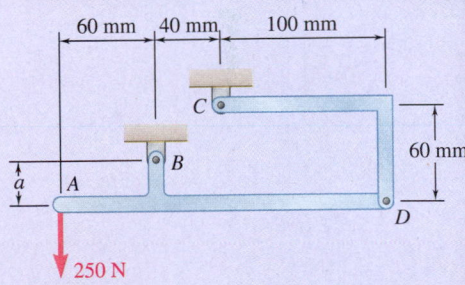

**Fig. P4.65**

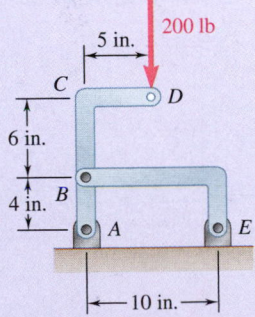

**Fig. P4.66**

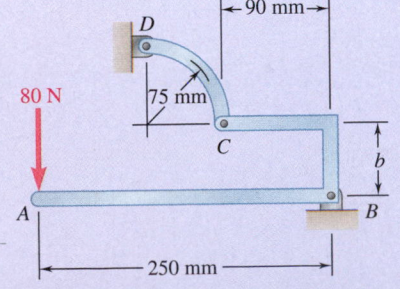

**Fig. P4.67**

**4.67** Determine the reactions at *B* and *D* when $b = 60$ mm.

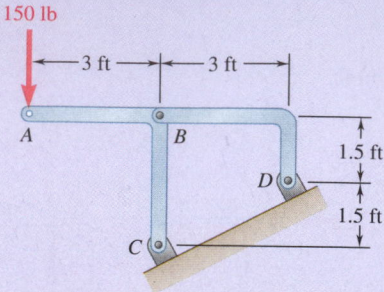

**Fig. P4.68**

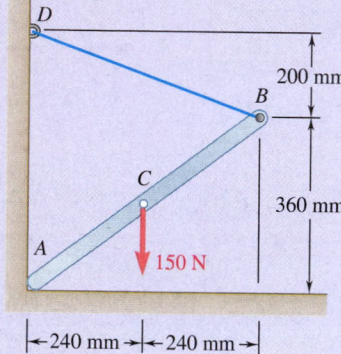

**Fig. P4.70**

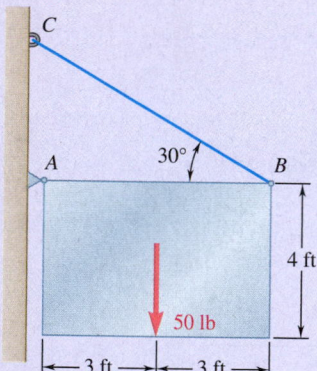

**Fig. P4.72**

**4.68** For the frame and loading shown, determine the reactions at *C* and *D*.

**4.69** A 50-kg crate is attached to the trolley-beam system shown. Knowing that $a = 1.5$ m, determine (*a*) the tension in cable *CD*, (*b*) the reaction at *B*.

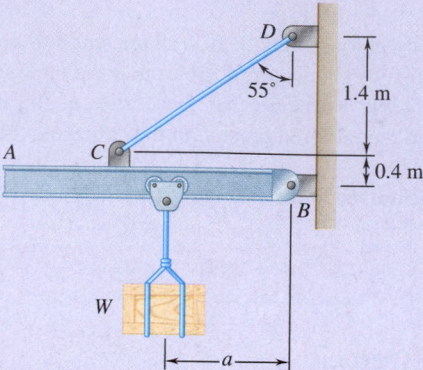

**Fig. P4.69**

**4.70** One end of rod *AB* rests in the corner *A* and the other end is attached to cord *BD*. If the rod supports a 150-N load at its midpoint *C*, find the reaction at *A* and the tension in the cord.

**4.71** For the boom and loading shown, determine (*a*) the tension in cord *BD*, (*b*) the reaction at *C*.

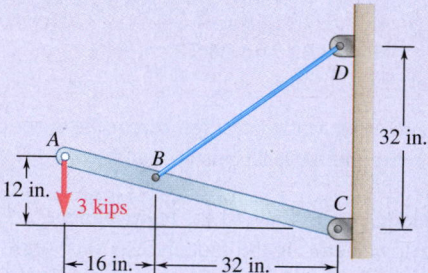

**Fig. P4.71**

**4.72** A 50-lb sign is supported by a pin and bracket at *A* and by a cable *BC*. Determine the reaction at *A* and the tension in the cable.

**4.73** Determine the reactions at *A* and *D* when $\beta = 30°$.

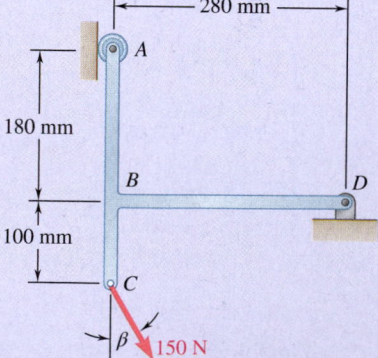

**Fig. P4.73 and P4.74**

**4.74** Determine the reactions at *A* and *D* when $\beta = 60°$.

204

**4.75** Rod *AB* is supported by a pin and bracket at *A* and rests against a frictionless peg at *C*. Determine the reactions at *A* and *C* when a 170-N vertical force is applied at *B*.

**4.76** Solve Prob. 4.75, assuming that the 170-N force applied at *B* is horizontal and directed to the left.

**4.77** The L-shaped member *ACB* is supported by a pin and bracket at *C* and by an inextensible cord attached at *A* and *B* and passing over a frictionless pulley at *D*. The tension may be assumed to be the same in portions *AD* and *BD* of the cord. If the magnitudes of the forces applied at *A* and *B* are, respectively, *P* = 25 lb and *Q* = 0, determine (*a*) the tension in the cord, (*b*) the reaction at *C*.

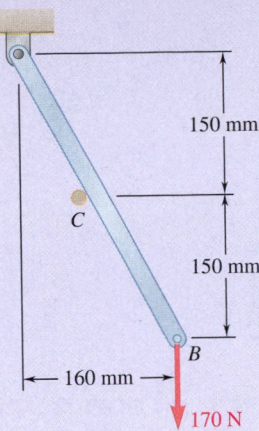

**Fig. P4.75**

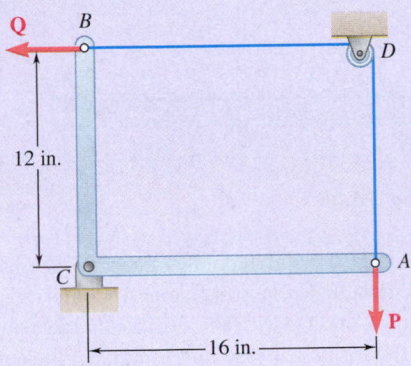

**Fig. P4.77**

**4.78** Using the method of Sec. 4.2B, solve Prob. 4.22.

**4.79** Knowing that $\theta = 30°$, determine the reaction (*a*) at *B*, (*b*) at *C*.

**4.80** Knowing that $\theta = 60°$, determine the reaction (*a*) at *B*, (*b*) at *C*.

**4.81** Determine the reactions at *A* and *B* when $\beta = 50°$.

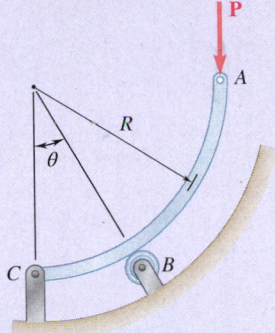

**Fig. P4.79 and P4.80**

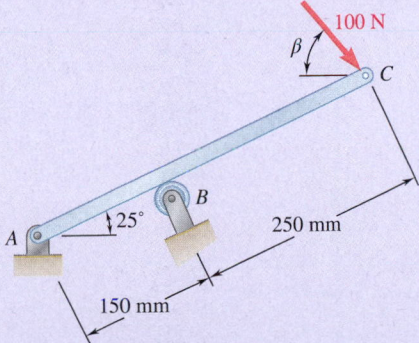

**Fig. P4.81 and *P4.82***

**4.82** Determine the reactions at *A* and *B* when $\beta = 80°$.

**4.83** Rod *AB* is bent into the shape of an arc of circle and is lodged between two pegs *D* and *E*. It supports a load **P** at end *B*. Neglecting friction and the weight of the rod, determine the distance *c* corresponding to equilibrium when *a* = 20 mm and *R* = 100 mm.

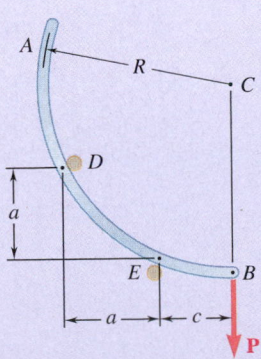

**Fig. P4.83**

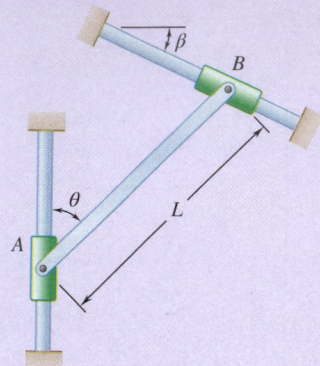

**Fig. P4.84 and P4.85**

**4.84** A slender rod of length $L$ is attached to collars that can slide freely along the guides shown. Knowing that the rod is in equilibrium, derive an expression for angle $\theta$ in terms of angle $\beta$.

**4.85** An 8-kg slender rod of length $L$ is attached to collars that can slide freely along the guides shown. Knowing that the rod is in equilibrium and that $\beta = 30°$, determine (a) the angle $\theta$ that the rod forms with the vertical, (b) the reactions at $A$ and $B$.

**4.86** A uniform plate girder weighing 6000 lb is held in a horizontal position by two crane cables. Determine the angle $\alpha$ and the tension in each cable.

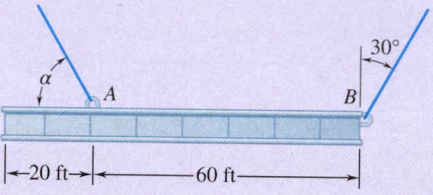

**Fig. P4.86**

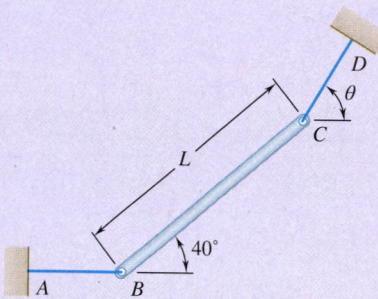

**Fig. P4.87**

**4.87** A slender rod $BC$ with a length of $L$ and weight $W$ is held by two cables as shown. Knowing that cable $AB$ is horizontal and that the rod forms an angle of 40° with the horizontal, determine (a) the angle $\theta$ that cable $CD$ forms with the horizontal, (b) the tension in each cable.

**4.88** A thin ring with a mass of 2 kg and radius $r = 140$ mm is held against a frictionless wall by a 125-mm string $AB$. Determine (a) the distance $d$, (b) the tension in the string, (c) the reaction at $C$.

**4.89** A slender rod with a length of $L$ and weight $W$ is attached to a collar at $A$ and is fitted with a small wheel at $B$. Knowing that the wheel rolls freely along a cylindrical surface of radius $R$, and neglecting friction, derive an equation in $\theta$, $L$, and $R$ that must be satisfied when the rod is in equilibrium.

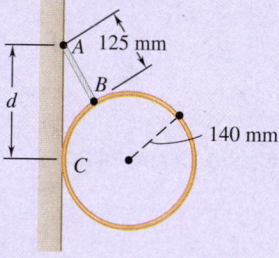

**Fig. P4.88**

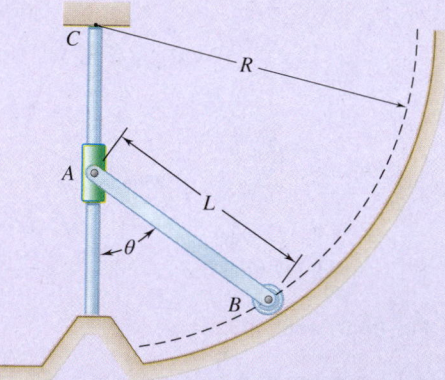

**Fig. P4.89**

**4.90** Knowing that for the rod of Prob. 4.89, $L = 15$ in., $R = 20$ in., and $W = 10$ lb, determine (a) the angle $\theta$ corresponding to equilibrium, (b) the reactions at $A$ and $B$.

# 4.3   EQUILIBRIUM IN THREE DIMENSIONS

The most general situation of rigid-body equilibrium occurs in three dimensions. The approach to modeling and analyzing these situations is the same as in two dimensions: Draw a free-body diagram and then write and solve the equilibrium equations. However, you now have more equations and more variables to deal with. In addition, reactions at supports and connections can be more varied, having as many as three force components and three couples acting at one support. As you will see in the Sample Problems, you need to visualize clearly in three dimensions and recall the vector analysis from Chaps. 2 and 3.

## 4.3A   Rigid-Body Equilibrium in Three Dimensions

We saw in Sec. 4.1 that six scalar equations are required to express the conditions for the equilibrium of a rigid body in the general three-dimensional case:

$$\Sigma F_x = 0 \qquad \Sigma F_y = 0 \qquad \Sigma F_z = 0 \qquad \textbf{(4.2)}$$
$$\Sigma M_x = 0 \qquad \Sigma M_y = 0 \qquad \Sigma M_z = 0 \qquad \textbf{(4.3)}$$

We can solve these equations for no more than *six unknowns,* which generally represent reactions at supports or connections.

In most problems, we can obtain the scalar Eqs. (4.2) and (4.3) more conveniently if we first write the conditions for the equilibrium of the rigid body considered in vector form:

$$\Sigma \mathbf{F} = 0 \qquad \Sigma \mathbf{M}_O = \Sigma(\mathbf{r} \times \mathbf{F}) = 0 \qquad \textbf{(4.1)}$$

Then, we can express the forces $\mathbf{F}$ and position vectors $\mathbf{r}$ in terms of scalar components and unit vectors. This enables us to compute all vector products either by direct calculation or by means of determinants (see Sec. 3.1F). Note that we can eliminate as many as three unknown reaction components from these computations through a judicious choice of the point $O$. By equating to zero the coefficients of the unit vectors in each of the two relations in Eq. (4.1), we obtain the desired scalar equations.[†]

Some equilibrium problems and their associated free-body diagrams might involve individual couples $\mathbf{M}_i$ either as applied loads or as support reactions. In such situations, you can accommodate these couples by expressing the second part of Eq. (4.1) as

$$\Sigma \mathbf{M}_O = \Sigma(\mathbf{r} \times \mathbf{F}) + \Sigma \mathbf{M}_i = 0 \qquad \textbf{(4.1′)}$$

## 4.3B   Reactions for a Three-Dimensional Structure

The reactions on a three-dimensional structure range from a single force of known direction exerted by a frictionless surface to a force-couple system exerted by a fixed support. Consequently, in problems involving the equilibrium

---

[†]In some problems, it may be convenient to eliminate from the solution the reactions at two points $A$ and $B$ by writing the equilibrium equation $\Sigma M_{AB} = 0$. This involves determining the moments of the forces about the axis $AB$ joining points $A$ and $B$ (see Sample Prob. 4.10).

**Photo 4.3** Universal joints, seen on the drive shafts of rear-wheel-drive cars and trucks, allow rotational motion to be transferred between two noncollinear shafts.
©McGraw-Hill Education/Lucinda Dowell

of a three-dimensional structure, between one and six unknowns may be associated with the reaction at each support or connection.

Fig. 4.10 shows various types of supports and connections with their corresponding reactions. A simple way of determining the type of reaction corresponding to a given support or connection and the number of unknowns involved is to find which of the six fundamental motions (translation in the *x, y,* and *z* directions and rotation about the *x, y,* and *z* axes) are allowed and which motions are prevented. The number of motions prevented equals the number of reactions.

Ball supports, frictionless surfaces, and cables, for example, prevent translation in one direction only and thus exert a single force whose line of action is known. Therefore, each of these supports involves one unknown—namely, the magnitude of the reaction. Rollers on rough surfaces and wheels on rails prevent translation in two directions; the corresponding reactions consist of two unknown force components. Rough surfaces in direct contact and ball-and-socket supports prevent translation in three directions while still allowing rotation; these supports involve three unknown force components.

Some supports and connections can prevent rotation as well as translation; the corresponding reactions include couples as well as forces. For example, the reaction at a fixed support, which prevents any motion (rotation as well as translation) consists of three unknown forces and three unknown couples. A universal joint, which is designed to allow rotation about two axes, exerts a reaction consisting of three unknown force components and one unknown couple.

Other supports and connections are primarily intended to prevent translation; their design, however, is such that they also prevent some rotations. The corresponding reactions consist essentially of force components but *may* also include couples. One group of supports of this type includes hinges and bearings designed to support radial loads only (e.g., journal bearings or roller bearings). The corresponding reactions consist of two force components but may also include two couples. Another group includes pin-and-bracket supports, hinges, and bearings designed to support an axial thrust as well as a radial load (e.g., ball bearings). The corresponding reactions consist of three force components but may include two couples. However, these supports do not exert any appreciable couples under normal conditions of use. Therefore, *only* force components should be included in their analysis *unless* it is clear that couples are necessary to maintain the equilibrium of the rigid body or unless the support is known to have been specifically designed to exert a couple (see Probs. 4.119, 4.120, 4.121, 4.122).

If the reactions involve more than six unknowns, you have more unknowns than equations, and some of the reactions are **statically indeterminate**. If the reactions involve fewer than six unknowns, you have more equations than unknowns, and some of the equations of equilibrium cannot be satisfied under general loading conditions. In this case, the rigid body is only **partially constrained**. Under the particular loading conditions corresponding to a given problem, however, the extra equations often reduce to trivial identities, such as $0 = 0$, and can be disregarded; although only partially constrained, the rigid body remains in equilibrium (see Sample Probs. 4.7 and 4.8). Even with six or more unknowns, it is possible that some equations of equilibrium are not satisfied. This can occur when the reactions associated with the given supports either are parallel or intersect the same line; the rigid body is then **improperly constrained**.

**Photo 4.4** This pillow block bearing supports the shaft of a fan used in an industrial facility. Courtesy of SKF, Limited

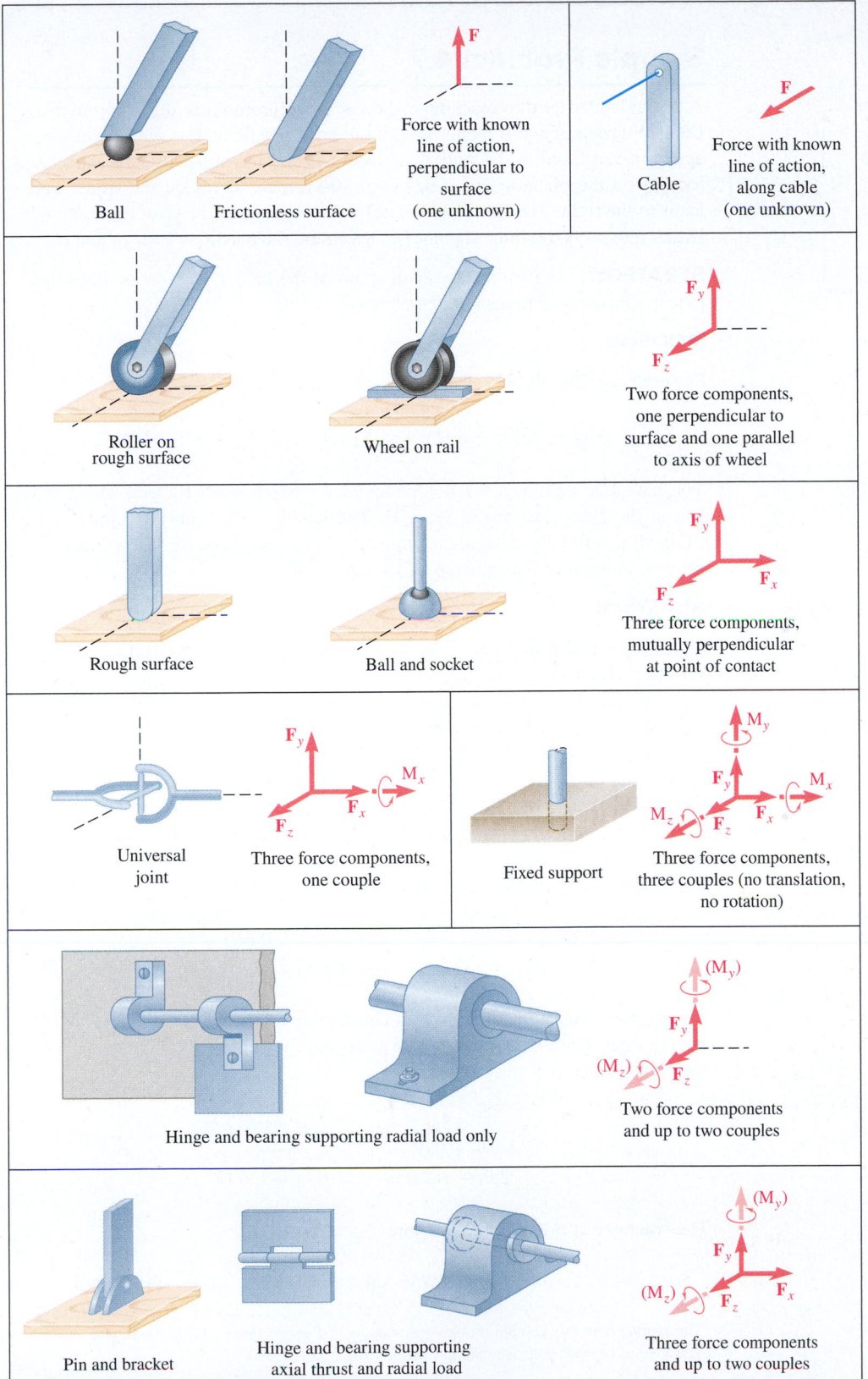

**Fig. 4.10** Reactions at supports and connections in three dimensions.

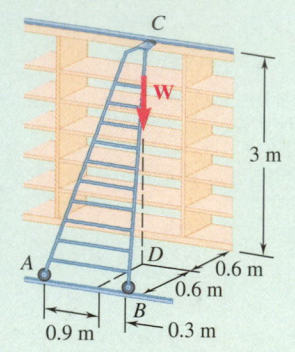

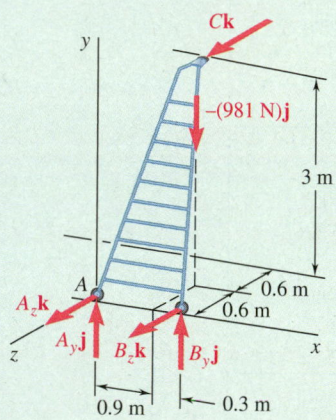

**Fig. 1**   Free-body diagram of the ladder.

# Sample Problem 4.7

A 20-kg ladder used to reach high shelves in a storeroom is supported by two flanged wheels $A$ and $B$ mounted on a rail and by a flangeless wheel $C$ resting against a rail fixed to the wall. (The ladder itself is symmetric, with wheel $C$ located on the plane of symmetry.) An 80-kg man stands on the ladder and leans to the right. The line of action of the combined weight $W$ of the man and ladder intersects the floor at point $D$. Determine the reactions at $A$, $B$, and $C$.

**STRATEGY:**   Draw a free-body diagram of the ladder, then write and solve the equilibrium equations in three dimensions.

**MODELING:**

**Free-Body Diagram.**   The combined weight of the man and ladder is

$$W = -mg\mathbf{j} = -(80\text{ kg} + 20\text{ kg})(9.81\text{ m/s}^2)\mathbf{j} = -(981\text{ N})\mathbf{j}$$

You have five unknown reaction components: two at each flanged wheel and one at the flangeless wheel (Fig. 1). The ladder is thus only partially constrained; it is free to roll along the rails. It is, however, in equilibrium under the given load because the equation $\Sigma F_x = 0$ is satisfied.

**ANALYSIS:**

**Equilibrium Equations.**   The forces acting on the ladder form a system equivalent to zero:

$$\Sigma \mathbf{F} = 0:\qquad A_y\mathbf{j} + A_z\mathbf{k} + B_y\mathbf{j} + B_z\mathbf{k} - (981\text{ N})\mathbf{j} + C\mathbf{k} = 0$$
$$(A_y + B_y - 981\text{ N})\mathbf{j} + (A_z + B_z + C)\mathbf{k} = 0 \qquad (1)$$

$$\Sigma \mathbf{M}_A = \Sigma(\mathbf{r} \times \mathbf{F}) = 0:\qquad 1.2\mathbf{j} \times (B_y\mathbf{j} + B_z\mathbf{k}) + (0.9\mathbf{i} - 0.6\mathbf{k}) \times (-981\mathbf{j})$$
$$+ (0.6\mathbf{i} + 3\mathbf{j} - 1.2\mathbf{k}) \times C\mathbf{k} = 0$$

Computing the vector products gives you[†]

$$1.2B_y\mathbf{k} - 1.2B_z\mathbf{j} - 882.9\mathbf{k} - 588.6\mathbf{i} - 0.6C\mathbf{j} + 3C\mathbf{i} = 0$$
$$(3C - 588.6)\mathbf{i} - (1.2B_z + 0.6C)\mathbf{j} + (1.2B_y - 882.9)\mathbf{k} = 0 \qquad (2)$$

Setting the coefficients of $\mathbf{i}$, $\mathbf{j}$, and $\mathbf{k}$ equal to zero in Eq. (2) produces the following three scalar equations, which state that the sum of the moments about each coordinate axis must be zero:

$$\begin{aligned} 3C - 588.6 &= 0 & C &= +196.2\text{ N} \\ 1.2B_z + 0.6C &= 0 & B_z &= -98.1\text{ N} \\ 1.2B_y - 882.9 &= 0 & B_y &= +736\text{ N} \end{aligned}$$

The reactions at $B$ and $C$ are therefore

$$\mathbf{B} = +(736\text{ N})\mathbf{j} - (98.1\text{ N})\mathbf{k} \qquad \mathbf{C} = +(196.2\text{ N})\mathbf{k}$$

[†]The moments in this sample problem, as well as in Sample Probs. 4.8 and 4.9, also can be expressed as determinants (see Sample Prob. 3.10).

*(continued)*

Setting the coefficients of **j** and **k** equal to zero in Eq. (1), you obtain two scalar equations stating that the sums of the components in the $y$ and $z$ directions are zero. Substitute the values on the previous page for $B_y$, $B_z$, and $C$ to get

$$A_y + B_y - 981 = 0 \qquad A_y + 736 - 981 = 0 \qquad A_y = +245 \text{ N}$$
$$A_z + B_z + C = 0 \qquad A_z - 98.1 + 196.2 = 0 \qquad A_z = -98.1 \text{ N}$$

Therefore, the reaction at $A$ is

$$\mathbf{A} = + (245 \text{ N})\mathbf{j} - (98.1 \text{ N})\mathbf{k} \quad \blacktriangleleft$$

**REFLECT and THINK:**   You summed moments about $A$ as part of the analysis. As a check, you could now use these results and demonstrate that the sum of moments about any other point, such as point $B$, is also zero.

## Sample Problem 4.8

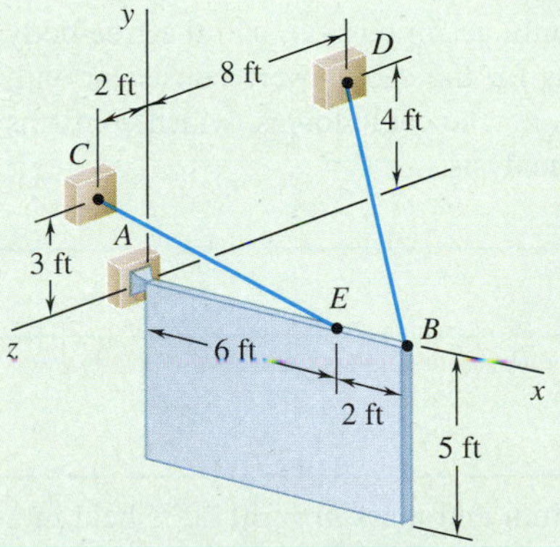

A $5 \times 8$-ft sign of uniform density weighs 270 lb and is supported by a ball-and-socket joint at $A$ and by two cables. Determine the tension in each cable and the reaction at $A$.

**STRATEGY:**   Draw a free-body diagram of the sign, and express the unknown cable tensions as Cartesian vectors. Then, determine the cable tensions and the reaction at $A$ by writing and solving the equilibrium equations.

**MODELING:**

**Free-Body Diagram.**   The forces acting on the sign are its weight $\mathbf{W} = -(270 \text{ lb})\mathbf{j}$ and the reactions at $A$, $B$, and $E$ (Fig. 1). The reaction at $A$ is a force of unknown direction represented by three unknown components. Because the directions of the forces exerted by the cables are known, these forces involve only one unknown each: specifically, the magnitudes $T_{BD}$ and $T_{EC}$. The total of five unknowns means that the sign is partially constrained. It can rotate freely about the $x$ axis; it is, however, in equilibrium under the given loading, because the equation $\Sigma M_x = 0$ is satisfied.

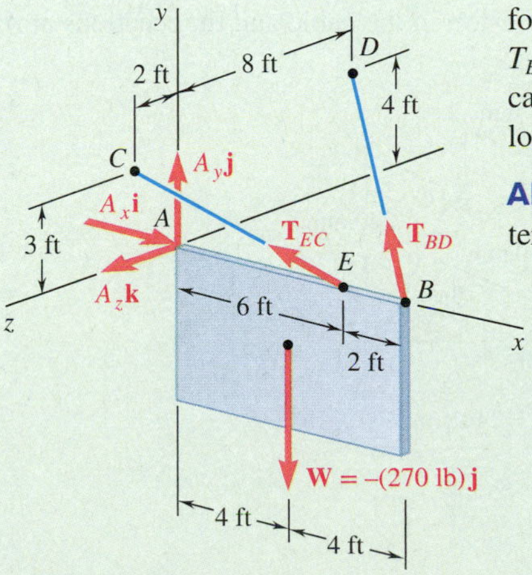

**ANALYSIS:**   You can express the components of the forces $\mathbf{T}_{BD}$ and $\mathbf{T}_{EC}$ in terms of the unknown magnitudes $T_{BD}$ and $T_{EC}$ as follows:

$$\overrightarrow{BD} = -(8 \text{ ft})\mathbf{i} + (4 \text{ ft})\mathbf{j} - (8 \text{ ft})\mathbf{k} \qquad BD = 12 \text{ ft}$$
$$\overrightarrow{EC} = -(6 \text{ ft})\mathbf{i} + (3 \text{ ft})\mathbf{j} + (2 \text{ ft})\mathbf{k} \qquad EC = 7 \text{ ft}$$

$$\mathbf{T}_{BD} = T_{BD}\left(\frac{\overrightarrow{BD}}{BD}\right) = T_{BD}(-\tfrac{2}{3}\mathbf{i} + \tfrac{1}{3}\mathbf{j} - \tfrac{2}{3}\mathbf{k})$$

$$\mathbf{T}_{EC} = T_{EC}\left(\frac{\overrightarrow{EC}}{EC}\right) = T_{EC}(-\tfrac{6}{7}\mathbf{i} + \tfrac{3}{7}\mathbf{j} + \tfrac{2}{7}\mathbf{k})$$

**Fig. 1**   Free-body diagram of the sign.

*(continued)*

**Equilibrium Equations.** The forces acting on the sign form a system equivalent to zero:

$$\Sigma\mathbf{F} = 0: \qquad A_x\mathbf{i} + A_y\mathbf{j} + A_z\mathbf{k} + \mathbf{T}_{BD} + \mathbf{T}_{EC} - (270\text{ lb})\mathbf{j} = 0$$

$$(A_x - \tfrac{2}{3}T_{BD} - \tfrac{6}{7}T_{EC})\mathbf{i} + (A_y + \tfrac{1}{3}T_{BD} + \tfrac{3}{7}T_{EC} - 270\text{ lb})\mathbf{j}$$
$$+ (A_z - \tfrac{2}{3}T_{BD} + \tfrac{2}{7}T_{EC})\mathbf{k} = 0 \quad \textbf{(1)}$$

$$\Sigma\mathbf{M}_A = \Sigma(\mathbf{r}\times\mathbf{F}) = 0:$$
$$(8\text{ ft})\mathbf{i}\times T_{BD}(-\tfrac{2}{3}\mathbf{i}+\tfrac{1}{3}\mathbf{j}-\tfrac{2}{3}\mathbf{k}) + (6\text{ ft})\mathbf{i}\times T_{EC}(-\tfrac{6}{7}\mathbf{i}+\tfrac{3}{7}\mathbf{j}+\tfrac{2}{7}\mathbf{k})$$
$$+ (4\text{ ft})\mathbf{i}\times(-270\text{ lb})\mathbf{j} = 0$$
$$(2.667T_{BD} + 2.571T_{EC} - 1080\text{ lb})\mathbf{k} + (5.333T_{BD} - 1.714T_{EC})\mathbf{j} = 0 \quad \textbf{(2)}$$

Setting the coefficients of $\mathbf{j}$ and $\mathbf{k}$ equal to zero in Eq. (2) yields two scalar equations that can be solved for $T_{BD}$ and $T_{EC}$:

$$T_{BD} = 101.3\text{ lb} \qquad T_{EC} = 315\text{ lb} \blacktriangleleft$$

Setting the coefficients of $\mathbf{i}$, $\mathbf{j}$, and $\mathbf{k}$ equal to zero in Eq. (1) produces three more equations, which yield the components of $\mathbf{A}$.

$$\mathbf{A} = +(338\text{ lb})\mathbf{i} + (101.2\text{ lb})\mathbf{j} - (22.5\text{ lb})\mathbf{k} \blacktriangleleft$$

**REFLECT and THINK:** Cables can only act in tension, and the free-body diagram and Cartesian vector expressions for the cables were consistent with this. The solution yielded positive results for the cable forces, which confirms that they are in tension and validates the analysis.

## Sample Problem 4.9

A uniform pipe cover of radius $r = 240$ mm and mass $m = 30$ kg is held in a horizontal position by the cable $CD$. Assuming that the bearing at $B$ does not exert any axial thrust, determine the tension in the cable and the reactions at $A$ and $B$.

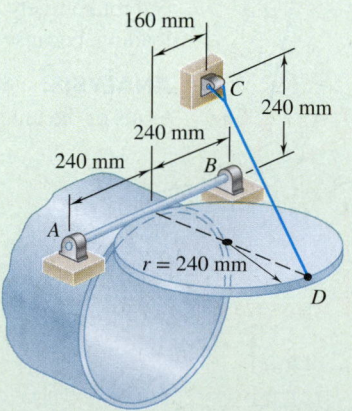

*(continued)*

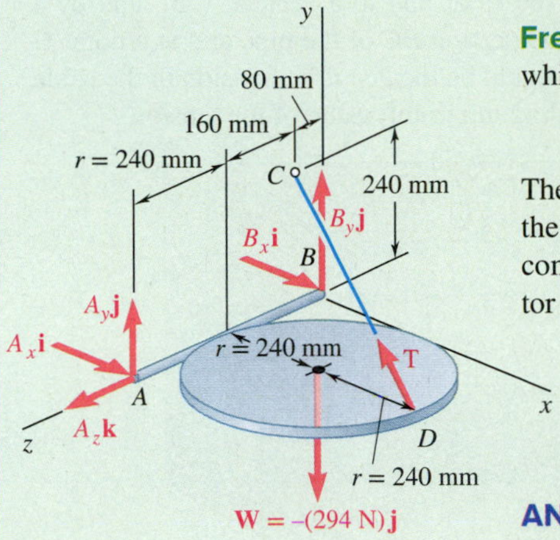

**STRATEGY:** Draw a free-body diagram with the coordinate axes shown (Fig. 1) and express the unknown cable tension as a Cartesian vector. Then, apply the equilibrium equations to determine this tension and the support reactions.

**MODELING:**

**Free-Body Diagram.**   The forces acting on the free body include its weight, which is

$$\mathbf{W} = -mg\mathbf{j} = -(30\text{ kg})(9.81\text{ m/s}^2)\mathbf{j} = -(294\text{ N})\mathbf{j}$$

The reactions involve six unknowns: the magnitude of the force $\mathbf{T}$ exerted by the cable, three force components at hinge $A$, and two at hinge $B$. Express the components of $\mathbf{T}$ in terms of the unknown magnitude $T$ by resolving the vector $\overrightarrow{DC}$ into rectangular components:

$$\overrightarrow{DC} = -(480\text{ mm})\mathbf{i} + (240\text{ mm})\mathbf{j} - (160\text{ mm})\mathbf{k} \qquad DC = 560\text{ mm}$$

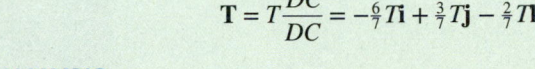

$$\mathbf{T} = T\frac{\overrightarrow{DC}}{DC} = -\tfrac{6}{7}T\mathbf{i} + \tfrac{3}{7}T\mathbf{j} - \tfrac{2}{7}T\mathbf{k}$$

**ANALYSIS:**

**Equilibrium Equations.**   The forces acting on the pipe cover form a system equivalent to zero. Thus,

$$\Sigma\mathbf{F} = 0: \qquad A_x\mathbf{i} + A_y\mathbf{j} + A_z\mathbf{k} + B_x\mathbf{i} + B_y\mathbf{j} + \mathbf{T} - (294\text{ N})\mathbf{j} = 0$$
$$(A_x + B_x - \tfrac{6}{7}T)\mathbf{i} + (A_y + B_y + \tfrac{3}{7}T - 294\text{ N})\mathbf{j} + (A_z - \tfrac{2}{7}T)\mathbf{k} = 0 \qquad (1)$$

$$\Sigma\mathbf{M}_B = \Sigma(\mathbf{r} \times \mathbf{F}) = 0:$$
$$2r\mathbf{k} \times (A_x\mathbf{i} + A_y\mathbf{j} + A_z\mathbf{k})$$
$$+ (2r\mathbf{i} + r\mathbf{k}) \times (-\tfrac{6}{7}T\mathbf{i} + \tfrac{3}{7}T\mathbf{j} - \tfrac{2}{7}T\mathbf{k})$$
$$+ (r\mathbf{i} + r\mathbf{k}) \times (-294\text{ N})\mathbf{j} = 0$$
$$(-2A_y - \tfrac{3}{7}T + 294\text{ N})r\mathbf{i} + (2A_x - \tfrac{2}{7}T)r\mathbf{j} + (\tfrac{6}{7}T - 294\text{ N})r\mathbf{k} = 0 \qquad (2)$$

Setting the coefficients of the unit vectors equal to zero in Eq. (2) gives three scalar equations, which yield

$$A_x = +49.0\text{ N} \qquad A_y = +73.5\text{ N} \qquad T = 343\text{ N} \ \blacktriangleleft$$

Setting the coefficients of the unit vectors equal to zero in Eq. (1) produces three more scalar equations. After substituting the values of $T$, $A_x$, and $A_y$ into these equations, you obtain

$$A_z = +98.0\text{ N} \qquad B_x = +245\text{ N} \qquad B_y = +73.5\text{ N}$$

The reactions at $A$ and $B$ are therefore

$$\mathbf{A} = +(49.0\text{ N})\mathbf{i} + (73.5\text{ N})\mathbf{j} + (98.0\text{ N})\mathbf{k} \ \blacktriangleleft$$

$$\mathbf{B} = +(245\text{ N})\mathbf{i} + (73.5\text{ N})\mathbf{j} \ \blacktriangleleft$$

**REFLECT and THINK:**   As a check, you can determine the tension in the cable using a scalar analysis. Assigning signs by the right-hand rule (rhr), we have

$$(+\text{rhr}) \qquad \Sigma M_z = 0: \qquad \tfrac{3}{7}T(0.48\text{ m}) - (294\text{ N})(0.24\text{ m}) = 0 \qquad T = 343\text{ N} \ \blacktriangleleft$$

**Fig. 1**   Free-body diagram of the pipe cover.

## Sample Problem 4.10

A 450-lb load hangs from the corner $C$ of a rigid piece of pipe $ABCD$ that has been bent as shown. The pipe is supported by ball-and-socket joints $A$ and $D$, which are fastened, respectively, to the floor and to a vertical wall, and by a cable attached at the midpoint $E$ of the portion $BC$ of the pipe and at a point $G$ on the wall. Determine (a) where $G$ should be located if the tension in the cable is to be minimum, (b) the corresponding minimum value of the tension.

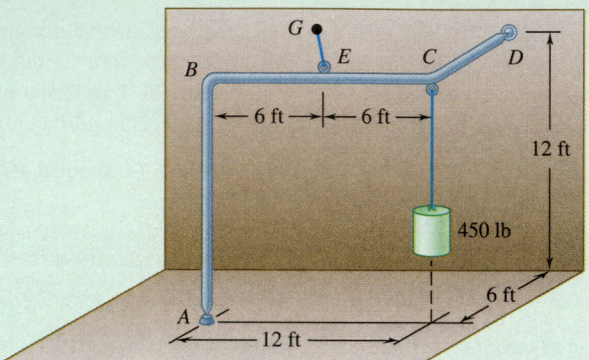

**STRATEGY:** Draw the free-body diagram of the pipe showing the reactions at $A$ and $D$. Isolate the unknown tension $\mathbf{T}$ and the known weight $\mathbf{W}$ by summing moments about the diagonal line $AD$, and compute values from the equilibrium equations.

**MODELING and ANALYSIS:**

**Free-Body Diagram.** The free-body diagram of the pipe includes the load $\mathbf{W} = (-450 \text{ lb})\mathbf{j}$, the reactions at $A$ and $D$, and the force $\mathbf{T}$ exerted by the cable (Fig. 1). To eliminate the reactions at $A$ and $D$ from the computations, take the sum of the moments of the forces about the line $AD$ and set it equal to zero. Denote the unit vector along $AD$ by $\boldsymbol{\lambda}$, which enables you to write

$$\Sigma M_{AD} = 0: \qquad \boldsymbol{\lambda} \cdot (\overrightarrow{AE} \times \mathbf{T}) + \boldsymbol{\lambda} \cdot (\overrightarrow{AC} \times \mathbf{W}) = 0 \qquad \textbf{(1)}$$

**Fig. 1** Free-body diagram of the pipe.

*(continued)*

You can compute the second term in Eq. (1) as follows:

$$\vec{AC} \times \mathbf{W} = (12\mathbf{i} + 12\mathbf{j}) \times (-450\mathbf{j}) = -5400\mathbf{k}$$

$$\lambda = \frac{\vec{AD}}{AD} = \frac{12\mathbf{i} + 12\mathbf{j} - 6\mathbf{k}}{18} = \tfrac{2}{3}\mathbf{i} + \tfrac{2}{3}\mathbf{j} - \tfrac{1}{3}\mathbf{k}$$

$$\lambda \cdot (\vec{AC} \times \mathbf{W}) = (\tfrac{2}{3}\mathbf{i} + \tfrac{2}{3}\mathbf{j} - \tfrac{1}{3}\mathbf{k}) \cdot (-5400\mathbf{k}) = +1800$$

Substituting this value into Eq. (1) gives

$$\lambda \cdot (\vec{AE} \times \mathbf{T}) = -1800 \text{ lb·ft} \qquad (2)$$

**Minimum Value of Tension.** Recalling the commutative property for mixed triple products, you can rewrite Eq. (2) in the form

$$\mathbf{T} \cdot (\lambda \times \vec{AE}) = -1800 \text{ lb·ft} \qquad (3)$$

This shows that the projection of $\mathbf{T}$ on the vector $\lambda \times \vec{AE}$ is a constant. It follows that $\mathbf{T}$ is minimum when it is parallel to the vector

$$\lambda \times \vec{AE} = (\tfrac{2}{3}\mathbf{i} + \tfrac{2}{3}\mathbf{j} - \tfrac{1}{3}\mathbf{k}) \times (6\mathbf{i} + 12\mathbf{j}) = 4\mathbf{i} - 2\mathbf{j} + 4\mathbf{k}$$

The corresponding unit vector is $\tfrac{2}{3}\mathbf{i} - \tfrac{1}{3}\mathbf{j} + \tfrac{2}{3}\mathbf{k}$, which gives

$$\mathbf{T}_{min} = T(\tfrac{2}{3}\mathbf{i} - \tfrac{1}{3}\mathbf{j} + \tfrac{2}{3}\mathbf{k}) \qquad (4)$$

Substituting for $\mathbf{T}$ and $\lambda \times \vec{AE}$ in Eq. (3) and computing the dot products yields $6T = -1800$ and, thus, $T = -300$. Carrying this value into Eq. (4) gives you

$$\mathbf{T}_{min} = -200\mathbf{i} + 100\mathbf{j} - 200\mathbf{k} \qquad T_{min} = 300 \text{ lb} \ \blacktriangleleft$$

**Location of G.** Because the vector $\vec{EG}$ and the force $\mathbf{T}_{min}$ have the same direction, their components must be proportional. Denoting the coordinates of $G$ by $x$, $y$, and 0 (Fig. 2), you get

$$\frac{x-6}{-200} = \frac{y-12}{+100} = \frac{0-6}{-200} \qquad x = 0 \qquad y = 15 \text{ ft} \ \blacktriangleleft$$

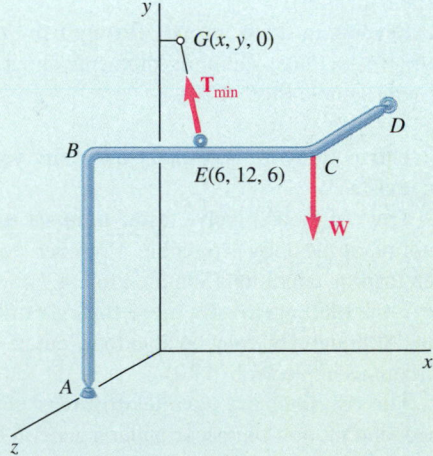

**Fig. 2** Location of point G for minimum tension in the cable.

**REFLECT and THINK:** Sometimes you have to rely on the vector analysis presented in Chaps. 2 and 3 as much as on the conditions for equilibrium described in this chapter.

# SOLVING PROBLEMS
# ON YOUR OWN

In this section, you considered the equilibrium of a *three-dimensional body*. It is again most important that you draw a complete *free-body diagram* as the first step of your solution.

**1. Pay particular attention to the reactions at the supports** as you draw the free-body diagram. The number of unknowns at a support can range from one to six (Fig. 4.10). To decide whether an unknown reaction or reaction component exists at a support, ask yourself whether the support prevents motion of the body in a certain direction or about a certain axis.

    **a. If motion is prevented in a certain direction,** include in your free-body diagram an unknown *reaction* or *reaction component* that acts in the *same direction.*

    **b. If a support prevents rotation about a certain axis,** include in your free-body diagram a *couple* of unknown magnitude that acts about the *same axis.*

**2. The external forces acting on a three-dimensional body form a system equivalent to zero.** Writing $\Sigma \mathbf{F} = 0$ and $\Sigma \mathbf{M}_A = 0$ about an appropriate point $A$ and setting the coefficients of $\mathbf{i}$, $\mathbf{j}$, and $\mathbf{k}$ in both equations equal to zero provides you with six scalar equations. In general, these equations contain six unknowns and may be solved for these unknowns.

**3. After completing your free-body diagram, you may want to seek equations involving as few unknowns as possible.** The following strategies may help you.

    **a. By summing moments about a ball-and-socket support or a hinge,** you obtain equations from which three unknown reaction components have been eliminated (Sample Probs. 4.8 and 4.9).

    **b. If you can draw an axis through the points of application of all but one of the unknown reactions,** summing moments about that axis will yield an equation in a single unknown (Sample Prob. 4.10).

**4. After drawing your free-body diagram, you may find that one of the following situations exists.**

    **a. The reactions involve fewer than six unknowns.** The body is partially constrained and motion of the body is possible. However, you may be able to determine the reactions for a given loading condition (Sample Prob. 4.7).

    **b. The reactions involve more than six unknowns.** The reactions are statically indeterminate. Although you may be able to calculate one or two reactions, you cannot determine all of them (Sample Prob. 4.10).

    **c. The reactions are parallel or intersect the same line.** The body is improperly constrained, and motion can occur under a general loading condition.

# Problems

**FREE-BODY PRACTICE PROBLEMS**

**4.F5** Two tape spools are attached to an axle supported by bearings at $A$ and $D$. The radius of spool $B$ is 1.5 in. and the radius of spool $C$ is 2 in. Knowing that $T_B = 20$ lb and that the system rotates at a constant rate, draw the free-body diagram needed to determine the reactions at $A$ and $D$. Assume that the bearing at $A$ does not exert any axial thrust and neglect the weights of the spools and axle.

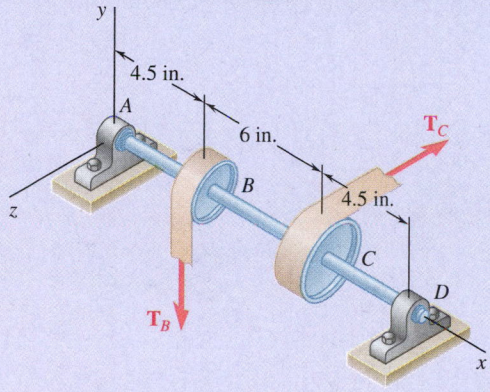

**Fig. P4.F5**

**4.F6** A 12-m pole supports a horizontal cable $CD$ and is held by a ball and socket at $A$ and two cables $BE$ and $BF$. Knowing that the tension in cable $CD$ is 14 kN and assuming that $CD$ is parallel to the $x$ axis ($\phi = 0$), draw the free-body diagram needed to determine the tension in cables $BE$ and $BF$ and the reaction at $A$.

**4.F7** A 20-kg cover for a roof opening is hinged at corners $A$ and $B$. The roof forms an angle of 30° with the horizontal, and the cover is maintained in a horizontal position by the brace $CE$. Draw the free-body diagram needed to determine the magnitude of the force exerted by the brace and the reactions at the hinges. Assume that the hinge at $A$ does not exert any axial thrust.

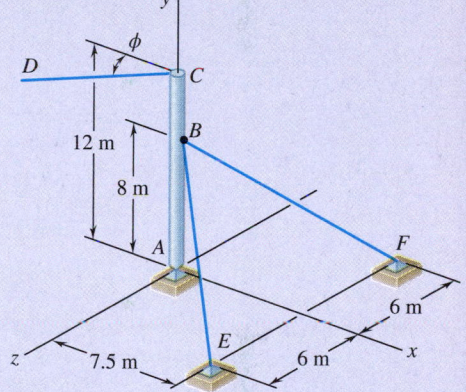

**Fig. P4.F6**

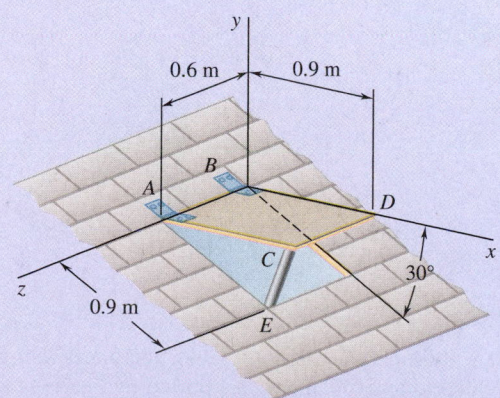

**Fig. P4.F7**

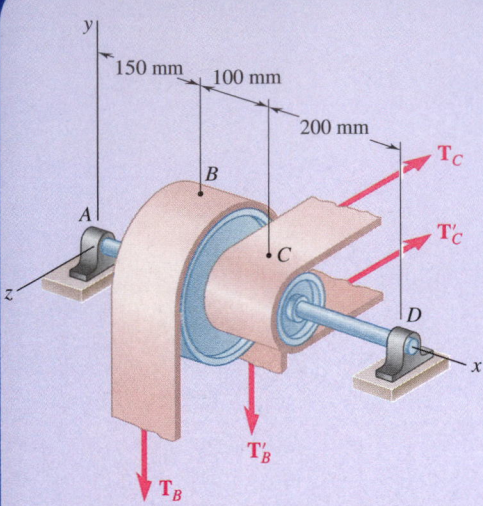

**Fig. P4.91**

**4.91** Two transmission belts pass over a double-sheaved pulley that is attached to an axle supported by bearings at $A$ and $D$. The radius of the inner sheave is 125 mm and the radius of the outer sheave is 250 mm. Knowing that when the system is at rest, the tension is 90 N in both portions of belt $B$ and 150 N in both portions of belt $C$, determine the reactions at $A$ and $D$. Assume that the bearing at $D$ does not exert any axial thrust.

**4.92** Solve Prob. 4.91, assuming that the pulley rotates at a constant rate and that $T_B = 104$ N, $T_B' = 84$ N, and $T_C = 175$ N.

**4.93** A small winch is used to raise a 120-lb load. Find (*a*) the magnitude of the vertical force $\mathbf{P}$ that should be applied at $C$ to maintain equilibrium in the position shown, (*b*) the reactions at $A$ and $B$, assuming that the bearing at $B$ does not exert any axial thrust.

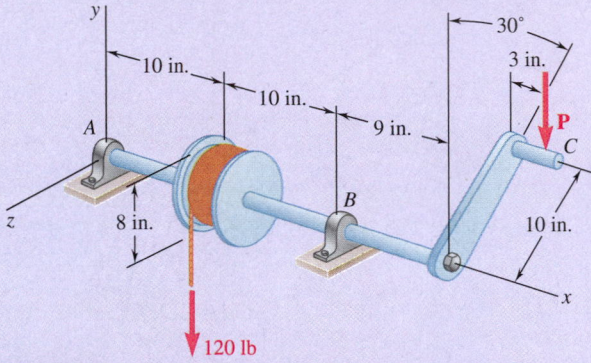

**Fig. P4.93**

**4.94** Two transmission belts pass over sheaves welded to an axle supported by bearings at $B$ and $D$. The sheave at $A$ has a radius of 2.5 in., and the sheave at $C$ has a radius of 2 in. Knowing that the system rotates at a constant rate, determine (*a*) the tension $T$, (*b*) the reactions at $B$ and $D$. Assume that the bearing at $D$ does not exert any axial thrust and neglect the weights of the sheaves and axle.

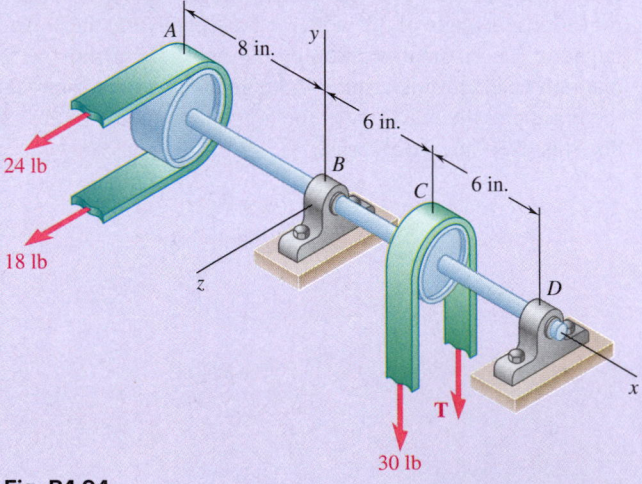

**Fig. P4.94**

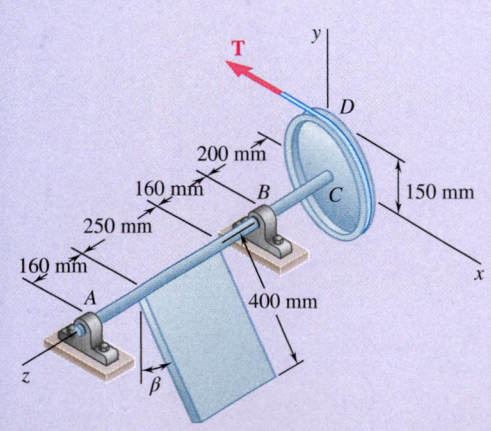

**Fig. P4.95**

**4.95** A $250 \times 400$-mm plate of mass 12 kg and a 300-mm-diameter pulley are welded to axle $AC$ that is supported by bearings at $A$ and $B$. For $\beta = 30°$, determine (*a*) the tension in the cable, (*b*) the reactions at $A$ and $B$. Assume that the bearing at $B$ does not exert any axial thrust.

**4.96** Solve Prob. 4.95 for $\beta = 60°$.

**4.97** The rectangular plate shown weighs 60 lb and is supported by three vertical wires. Determine the tension in each wire.

**4.98** A load $W$ is to be placed on the 60-lb plate of Prob. 4.97. Determine the magnitude of $W$ and the point where it should be placed if the tension is to be 50 lb in each of the three wires.

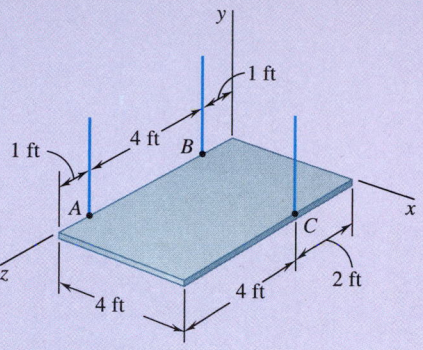

**Fig. P4.97**

**4.99** An opening in a floor is covered by a $1 \times 1.2$-m sheet of plywood with a mass of 18 kg. The sheet is hinged at $A$ and $B$ and is maintained in a position slightly above the floor by a small block $C$. Determine the vertical component of the reaction (a) at $A$, (b) at $B$, (c) at $C$.

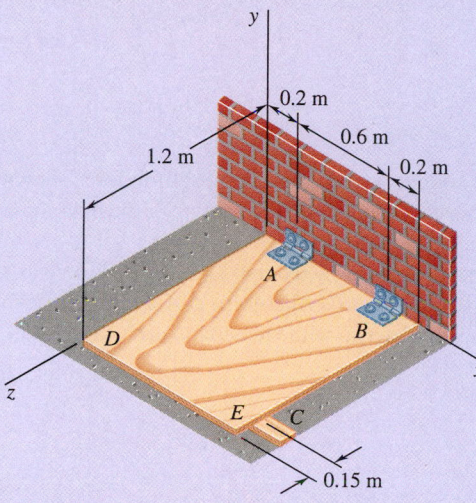

**Fig. P4.99**

**4.100** Solve Prob. 4.99, assuming that the small block $C$ is moved and placed under edge $DE$ at a point 0.15 m from corner $E$.

**4.101** Two steel pipes $AB$ and $BC$, each having a mass per unit length of 8 kg/m, are welded together at $B$ and supported by three vertical wires. Knowing that $a = 0.4$ m, determine the tension in each wire.

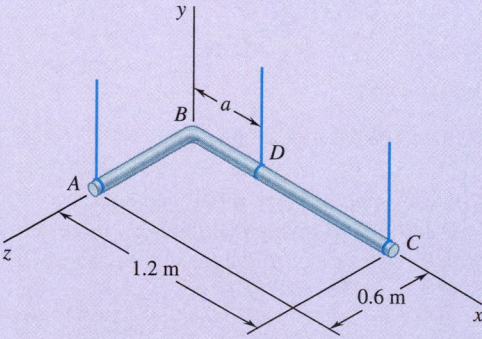

**Fig. P4.101**

**4.102** For the pipe assembly of Prob. 4.101, determine (a) the largest permissible value of $a$ if the assembly is not to tip, (b) the corresponding tension in each wire.

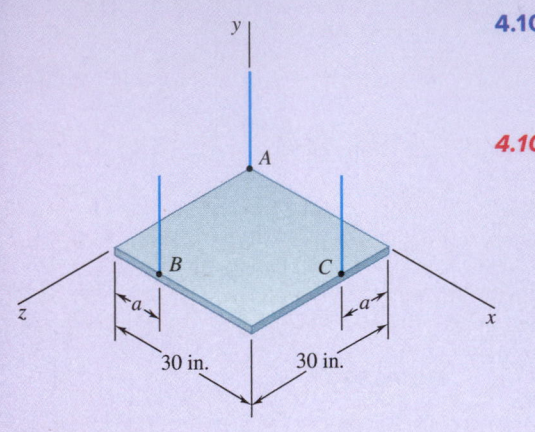

Fig. P4.103

**4.103** The 24-lb square plate shown is supported by three vertical wires. Determine (a) the tension in each wire when $a = 10$ in., (b) the value of $a$ for which the tension in each wire is 8 lb.

**4.104** The table shown weighs 30 lb and has a diameter of 4 ft. It is supported by three legs equally spaced around the edge. A vertical load **P** with a magnitude of 100 lb is applied to the top of the table at $D$. Determine the maximum value of $a$ if the table is not to tip over. Show, on a sketch, the area of the table over which **P** can act without tipping the table.

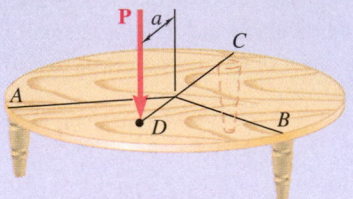

Fig. *P4.104*

**4.105** A 10-ft boom is acted upon by the 840-lb force shown. Determine the tension in each cable and the reaction at the ball-and-socket joint at $A$.

**4.106** The 6-m pole $ABC$ is acted upon by a 455-N force as shown. The pole is held by a ball-and-socket joint at $A$ and by two cables $BD$ and $BE$. For $a = 3$ m, determine the tension in each cable and the reaction at $A$.

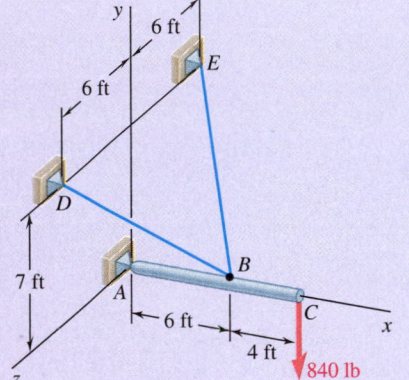

Fig. P4.105

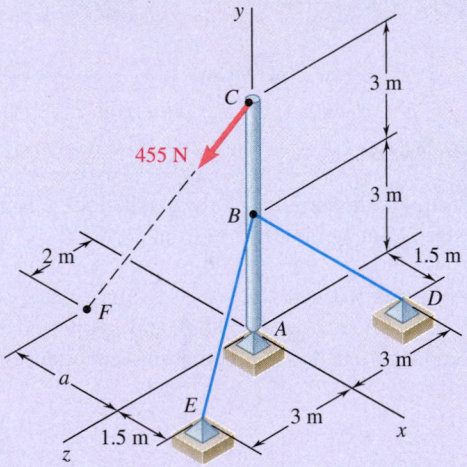

Fig. P4.106

**4.107** Solve Prob. 4.106 for $a = 1.5$ m.

**4.108** A 3-m pole is supported by a ball-and-socket joint at $A$ and by the cables $CD$ and $CE$. Knowing that the 5-kN force acts vertically downward ($\phi = 0$), determine (a) the tension in cables $CD$ and $CE$, (b) the reaction at $A$.

**4.109** A 3-m pole is supported by a ball-and-socket joint at $A$ and by the cables $CD$ and $CE$. Knowing that the line of action of the 5-kN force forms an angle $\phi = 30°$ with the vertical $xy$ plane (and is parallel to the $yz$ plane), determine (a) the tension in cables $CD$ and $CE$, (b) the reaction at $A$.

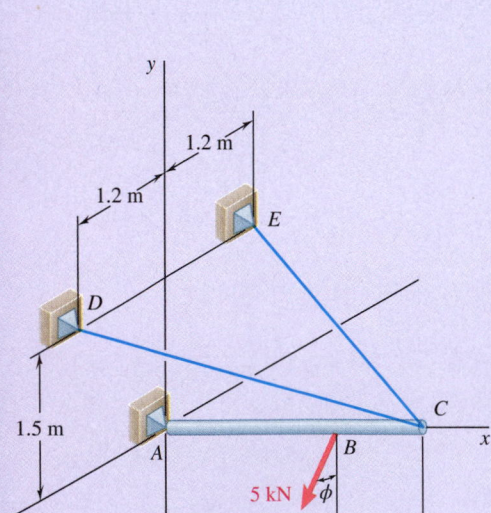

Fig. P4.108 and P4.109

**4.110** A 7-ft boom is held by a ball and socket at *A* and by two cables *EBF* and *DC;* cable *EBF* passes around a frictionless pulley at *B*. Determine the tension in each cable.

**4.111** A 48-in. boom is held by a ball-and-socket joint at *C* and by two cables *BF* and *DAE;* cable *DAE* passes around a frictionless pulley at *A*. For the loading shown, determine the tension in each cable and the reaction at *C*.

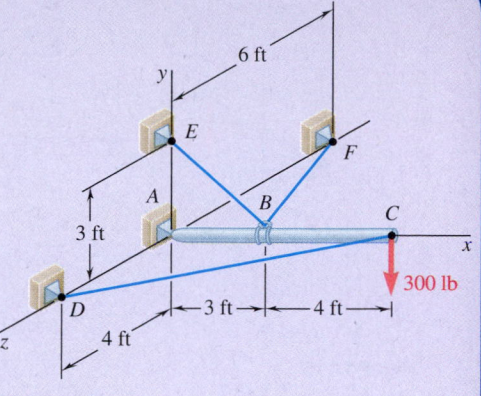

**Fig. P4.110**

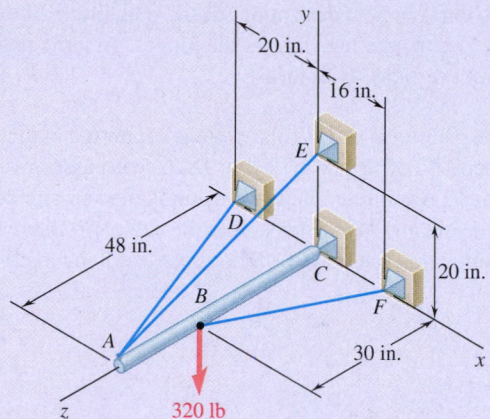

**Fig. *P4.111***

**4.112** Solve Prob. 4.111, assuming that the 320-lb load is applied at *A*.

**4.113** A 10-kg storm window measuring 900 × 1500 mm is held by hinges at *A* and *B*. In the position shown, it is held away from the side of the house by a 600-mm stick *CD*. Assuming that the hinge at *A* does not exert any axial thrust, determine the magnitude of the force exerted by the stick and the components of the reactions at *A* and *B*.

**4.114** The bent rod *ABEF* is supported by bearings at *C* and *D* and by wire *AH*. Knowing that portion *AB* of the rod is 250 mm long, determine (*a*) the tension in wire *AH*, (*b*) the reactions at *C* and *D*. Assume that the bearing at *D* does not exert any axial thrust.

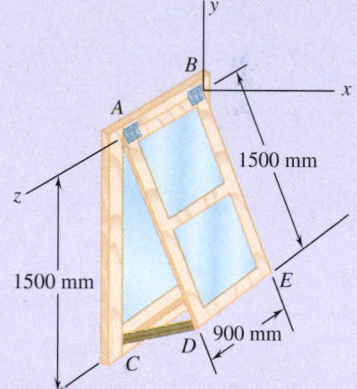

**Fig. P4.113**

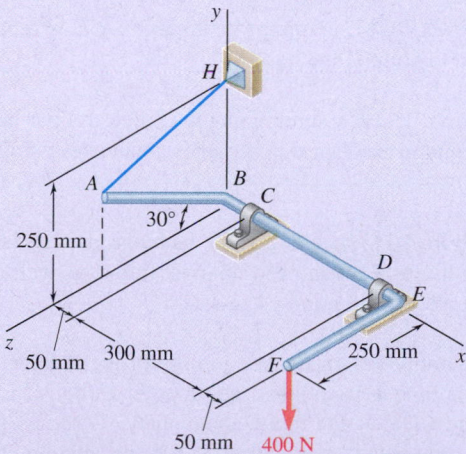

**Fig. *P4.114***

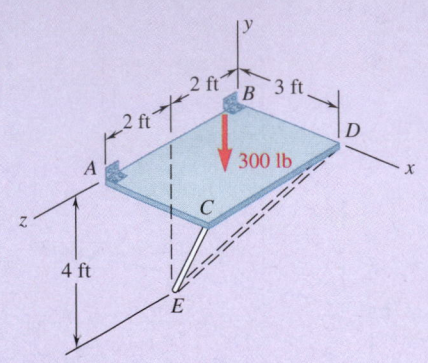

Fig. P4.115

**4.115** The horizontal platform *ABCD* weighs 60 lb and supports a 240-lb load at its center. The platform is normally held in position by hinges at *A* and *B* and by braces *CE* and *DE*. If brace *DE* is removed, determine the reactions at the hinges and the force exerted by the remaining brace *CE*. The hinge at *A* does not exert any axial thrust.

**4.116** The lid of a roof scuttle weighs 75 lb. It is hinged at corners *A* and *B* and maintained in the desired position by a rod *CD* pivoted at *C*. A pin at end *D* of the rod fits into one of several holes drilled in the edge of the lid. For $\alpha = 50°$, determine (*a*) the magnitude of the force exerted by rod *CD*, (*b*) the reactions at the hinges. Assume that the hinge at *B* does not exert any axial thrust.

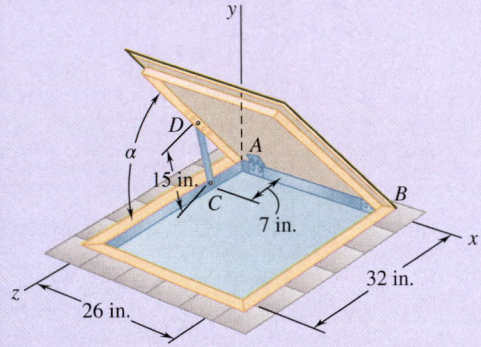

Fig. P4.116

**4.117** A 100-kg uniform rectangular plate is supported in the position shown by hinges *A* and *B* and by cable *DCE* that passes over a frictionless hook at *C*. Assuming that the tension is the same in both parts of the cable, determine (*a*) the tension in the cable, (*b*) the reactions at *A* and *B*. Assume that the hinge at *B* does not exert any axial thrust.

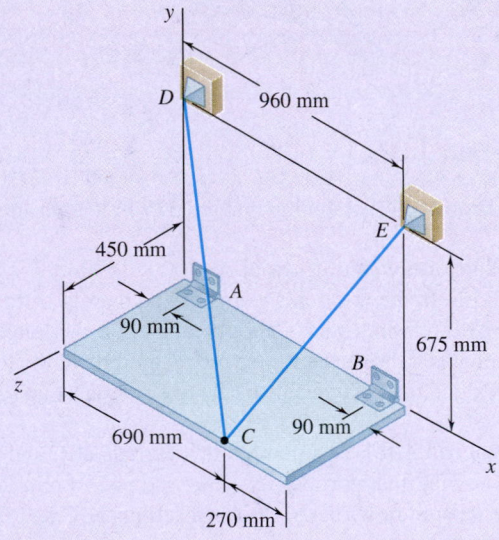

Fig. P4.117

**4.118** Solve Prob. 4.117, assuming that cable *DCE* is replaced by a cable attached to point *E* and hook *C*.

**4.119** Solve Prob. 4.113, assuming that the hinge at *A* has been removed and that the hinge at *B* can exert couples about axes parallel to the *x* and *y* axes.

**4.120** Solve Prob. 4.115, assuming that the hinge at *B* has been removed and that the hinge at *A* can exert an axial thrust, as well as couples about axes parallel to the *x* and *y* axes.

**4.121** The assembly shown is used to control the tension *T* in a tape that passes around a frictionless spool at *E*. Collar *C* is welded to rods *ABC* and *CDE*. It can rotate about shaft *FG* but its motion along the shaft is prevented by a washer *S*. For the loading shown, determine (*a*) the tension *T* in the tape, (*b*) the reaction at *C*.

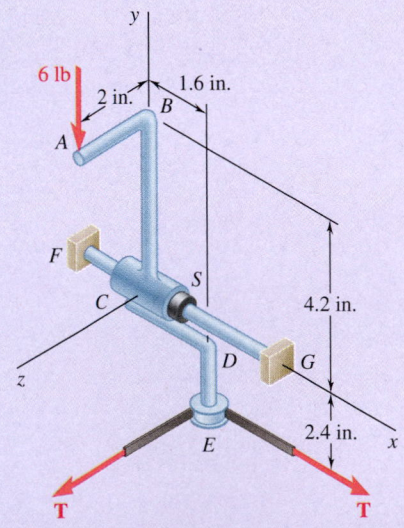

Fig. P4.121

**4.122** The assembly shown is welded to collar A that fits on the vertical pin shown. The pin can exert couples about the x and z axes but does not prevent motion about or along the y axis. For the loading shown, determine the tension in each cable and the reaction at A.

**4.123** The rigid L-shaped member ABC is supported by a ball-and-socket joint at A and by three cables. Determine the tension in each cable and the reaction at A caused by the 5-kN load applied at G.

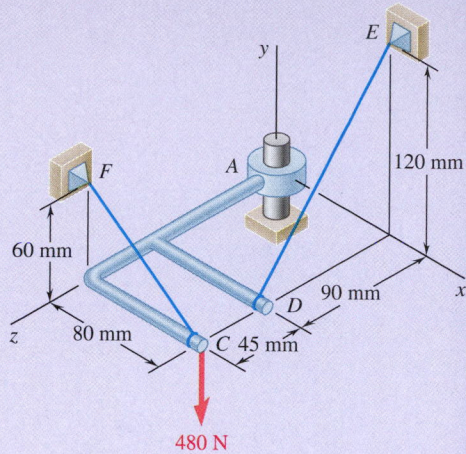

Fig. P4.122

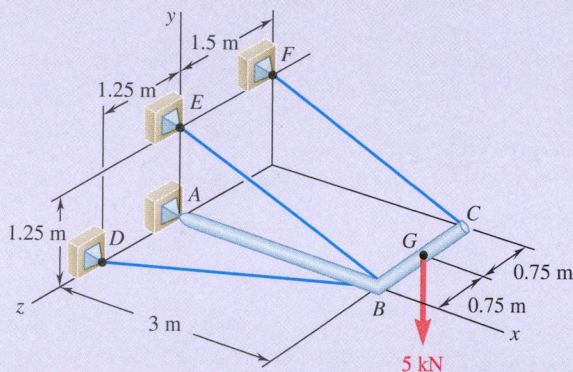

Fig. P4.123

**4.124** Solve Prob. 4.123, assuming that cable BD is removed and replaced by a cable joining points E and C.

**4.125** The rigid L-shaped member ABF is supported by a ball-and-socket joint at A and by three cables. For the loading shown, determine the tension in each cable and the reaction at A.

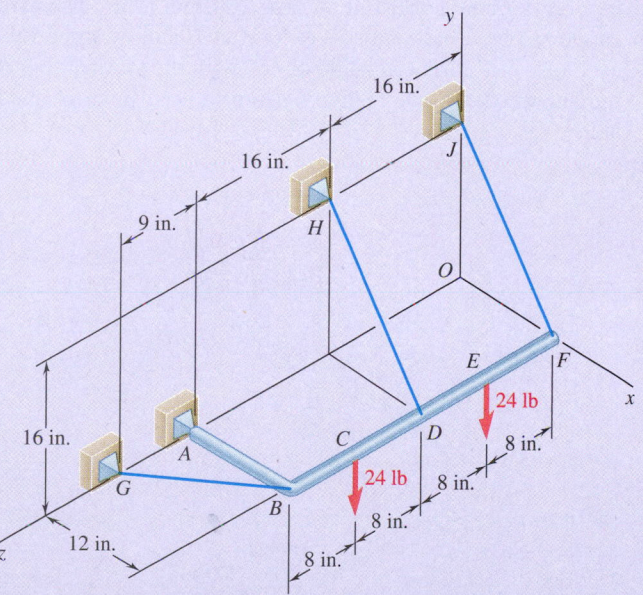

Fig. P4.125

**4.126** Solve Prob. 4.125, assuming that the load at C has been removed.

**4.127** Three rods are welded together to form a "corner" that is supported by three eyebolts. Neglecting friction, determine the reactions at A, B, and C when $P = 240$ lb, $a = 12$ in., $b = 8$ in., and $c = 10$ in.

**4.128** Solve Prob. 4.127, assuming that the force **P** is removed and is replaced by a couple $\mathbf{M} = +(600 \text{ lb·in.})\mathbf{j}$ acting at B.

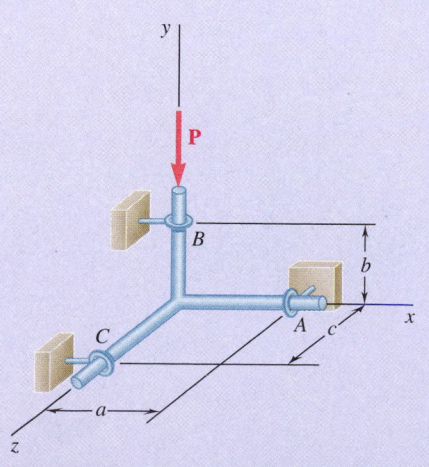

Fig. P4.127

**4.129** Frame *ABCD* is supported by a ball-and-socket joint at *A* and by three cables. For *a* = 150 mm, determine the tension in each cable and the reaction at *A*.

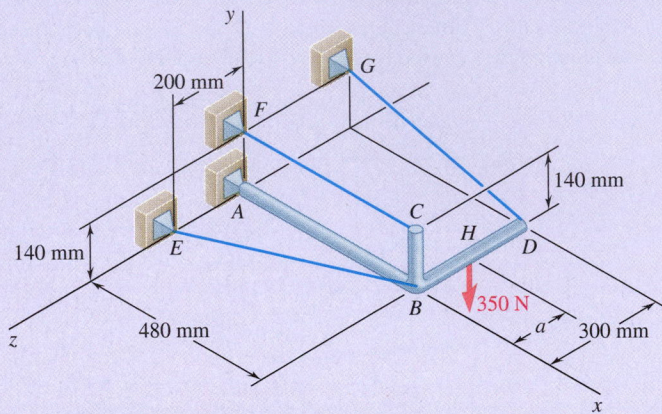

**Fig. P4.129 and *P4.130***

**4.130** Frame *ABCD* is supported by a ball-and-socket joint at *A* and by three cables. Knowing that the 350-N load is applied at *D* (*a* = 300 mm), determine the tension in each cable and the reaction at *A*.

**4.131** The assembly shown consists of an 80-mm rod *AF* that is welded to a cross frame consisting of four 200-mm arms. The assembly is supported by a ball-and-socket joint at *F* and by three short links, each of which forms an angle of 45° with the vertical. For the loading shown, determine (*a*) the tension in each link, (*b*) the reaction at *F*.

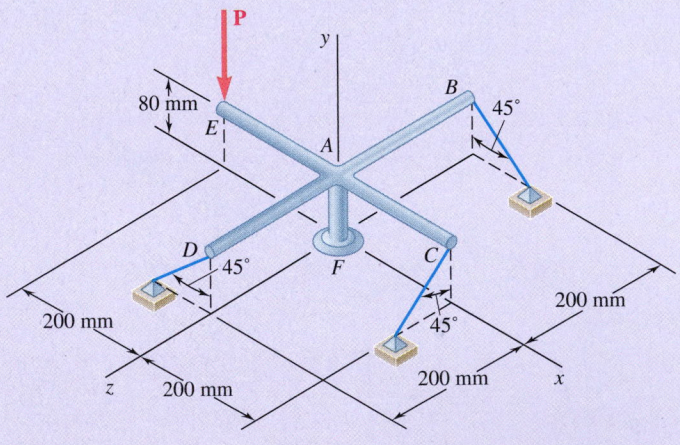

**Fig. P4.131**

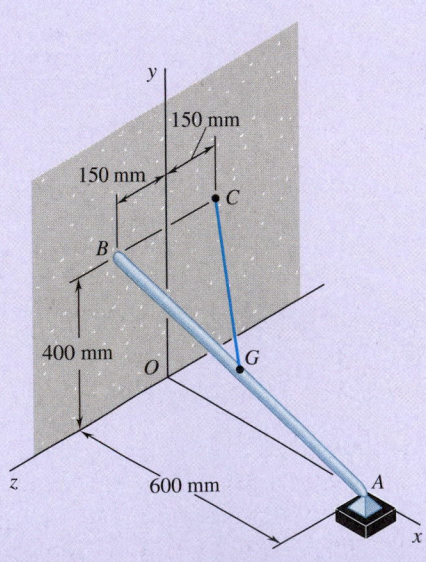

**Fig. *P4.132***

**4.132** The uniform 10-kg rod *AB* is supported by a ball-and-socket joint at *A* and by the cord *CG* that is attached to the midpoint *G* of the rod. Knowing that the rod leans against a frictionless vertical wall at *B*, determine (*a*) the tension in the cord, (*b*) the reactions at *A* and *B*.

**4.133** The frame *ACD* is supported by ball-and-socket joints at *A* and *D* and by a cable that passes through a ring at *B* and is attached to hooks at *G* and *H*. Knowing that the frame supports at point *C* a load of magnitude $P = 268$ N, determine the tension in the cable.

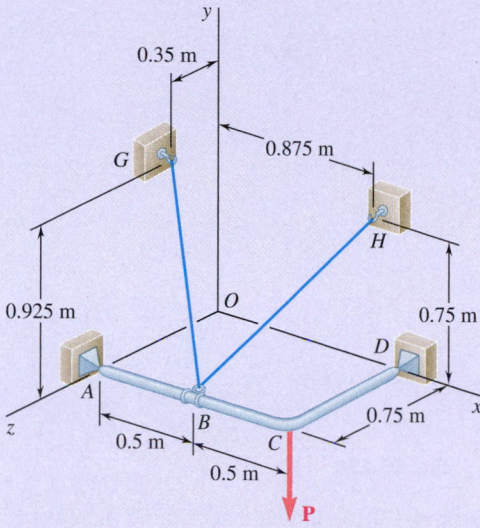

**Fig. P4.133**

**4.134** Solve Prob. 4.133, assuming that cable *GBH* is replaced by a cable *GB* attached at *G* and *B*.

**4.135** The 8-ft rod *AB* and the 6-ft rod *BC* are hinged at *B* and supported by cable *DE* and by ball-and-socket joints at *A* and *C*. Knowing that $h = 3$ ft, determine the tension in the cable for the loading shown.

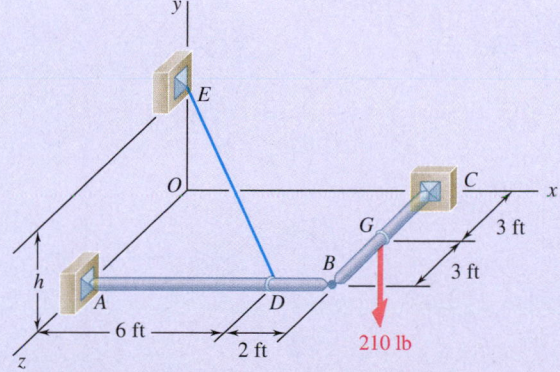

**Fig. P4.135**

**4.136** Solve Prob. 4.135 when $h = 10.5$ ft.

**4.137** Two rectangular plates are welded together to form the assembly shown. The assembly is supported by ball-and-socket joints at *B* and *D* and by a ball on a horizontal surface at *C*. For the loading shown, determine the reaction at *C*.

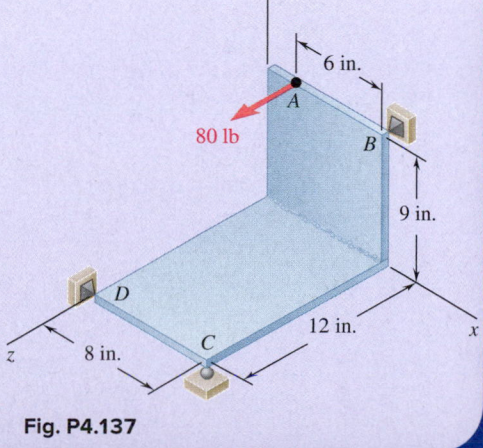

**Fig. P4.137**

**4.138** The pipe *ACDE* is supported by ball-and-socket joints at *A* and *E* and by the wire *DF*. Determine the tension in the wire when a 640-N load is applied at *B* as shown.

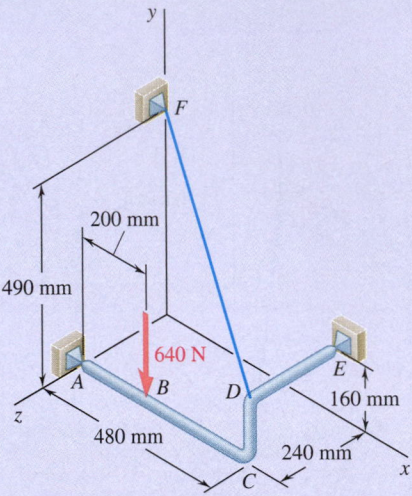

**Fig. P4.138**

**4.139** Solve Prob. 4.138, assuming that wire *DF* is replaced by a wire connecting *C* and *F*.

**4.140** Two 2 × 4-ft plywood panels, each with a weight of 12 lb, are nailed together as shown. The panels are supported by ball-and-socket joints at *A* and *F* and by the wire *BH*. Determine (*a*) the location of *H* in the *xy* plane if the tension in the wire is to be minimum, (*b*) the corresponding minimum tension.

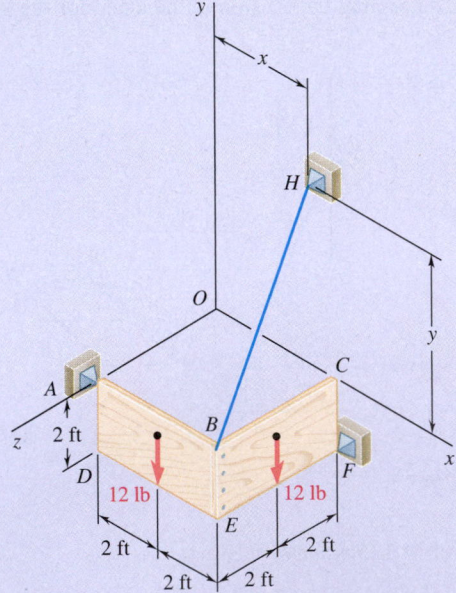

**Fig. P4.140**

**4.141** Solve Prob. 4.140, subject to the restriction that *H* must lie on the *y* axis.

# Review and Summary

## Equilibrium Equations

This chapter was devoted to the study of the **equilibrium of rigid bodies**, i.e., to the situation when the external forces acting on a rigid body *form a system equivalent to zero* (Introduction). We then have

$$\Sigma \mathbf{F} = 0 \qquad \Sigma \mathbf{M}_O = \Sigma(\mathbf{r} \times \mathbf{F}) = 0 \qquad \text{(4.1)}$$

Resolving each force and each moment into its rectangular components, we can express the necessary and sufficient conditions for the equilibrium of a rigid body with the following six scalar equations:

$$\Sigma F_x = 0 \qquad \Sigma F_y = 0 \qquad \Sigma F_z = 0 \qquad \text{(4.2)}$$
$$\Sigma M_x = 0 \qquad \Sigma M_y = 0 \qquad \Sigma M_z = 0 \qquad \text{(4.3)}$$

We can use these equations to determine unknown forces applied to the rigid body or unknown reactions exerted by its supports.

## Free-Body Diagram

When solving a problem involving the equilibrium of a rigid body, it is essential to consider *all* of the forces acting on the body. Therefore, the first step in the solution of the problem should be to draw a **free-body diagram** showing the body under consideration and all of the unknown as well as known forces acting on it.

## Equilibrium of a Two-Dimensional Structure

In the first part of this chapter, we considered the **equilibrium of a two-dimensional structure**; i.e., we assumed that the structure considered and the forces applied to it were contained in the same plane. We saw that each of the reactions exerted on the structure by its supports could involve one, two, or three unknowns, depending upon the type of support (Sec. 4.1A).

In the case of a two-dimensional structure, the equations given previously reduce to *three equilibrium equations:*

$$\Sigma F_x = 0 \qquad \Sigma F_y = 0 \qquad \Sigma M_A = 0 \qquad \text{(4.5)}$$

where $A$ is an arbitrary point in the plane of the structure (Sec. 4.1B). We can use these equations to solve for three unknowns. Although the three equilibrium Eqs. (4.5) cannot be *augmented* with additional equations, any of them can be *replaced* by another equation. Therefore, we can write alternative sets of equilibrium equations, such as

$$\Sigma F_x = 0 \qquad \Sigma M_A = 0 \qquad \Sigma M_B = 0 \qquad \text{(4.6)}$$

where point $B$ is chosen in such a way that the line $AB$ is not parallel to the $y$ axis, or

$$\Sigma M_A = 0 \qquad \Sigma M_B = 0 \qquad \Sigma M_C = 0 \qquad \text{(4.7)}$$

where the points $A$, $B$, and $C$ do not lie in a straight line.

## Static Indeterminacy, Partial Constraints, Improper Constraints

Because any set of equilibrium equations can be solved for only three unknowns, the reactions at the supports of a rigid two-dimensional structure cannot be completely determined if they involve *more than three unknowns;* they are said to be *statically indeterminate* (Sec. 4.1C). On the other hand, if the reactions involve *fewer than three unknowns,* equilibrium is not maintained under general loading conditions; the structure is said to be *partially constrained.* The fact that the reactions involve exactly three unknowns is no guarantee that you can solve the equilibrium equations for all three unknowns. If the supports are arranged in such a way that the reactions are *either concurrent or parallel,* the reactions are statically indeterminate, and the structure is said to be *improperly constrained.*

## Two-Force Body, Three-Force Body

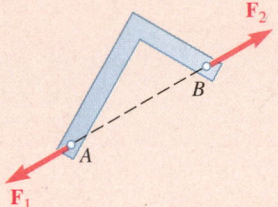

**Fig. 4.11**

We gave special attention in Sec. 4.2 to two particular cases of equilibrium of a rigid body. We defined a **two-force body** as a rigid body subjected to forces at only two points, and we showed that the resultants $F_1$ and $F_2$ of these forces must have the *same magnitude, the same line of action, and opposite sense* (Fig. 4.11), which is a property that simplifies the solution of certain problems in later chapters. We defined a **three-force body** as a rigid body subjected to forces at only three points, and we demonstrated that the resultants $F_1$, $F_2$, and $F_3$ of these forces must be *either concurrent* (Fig. 4.12) *or parallel.* This property provides us with an alternative approach to the solution of problems involving a three-force body (Sample Prob. 4.6).

## Equilibrium of a Three-Dimensional Body

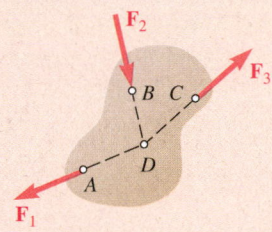

**Fig. 4.12**

In the second part of this chapter, we considered the *equilibrium of a three-dimensional body.* We saw that each of the reactions exerted on the body by its supports could involve between one and six unknowns, depending upon the type of support (Sec. 4.3A).

In the general case of the equilibrium of a three-dimensional body, all six of the scalar equilibrium Eqs. (4.2) and (4.3) should be used and solved for *six unknowns* (Sec. 4.3B). In most problems, however, we can obtain these equations more conveniently if we start from

$$\Sigma F = 0 \qquad \Sigma M_O = \Sigma(r \times F) = 0 \qquad (4.1)$$

and then express the forces $F$ and position vectors $r$ in terms of scalar components and unit vectors. We can compute the vector products either directly or by means of determinants, and obtain the desired scalar equations by equating to zero the coefficients of the unit vectors (Sample Probs. 4.7, 4.8, 4.9).

We noted that we may eliminate as many as three unknown reaction components from the computation of $\Sigma M_O$ in the second of the relations (4.1) through a judicious choice of point $O$. Also, we can eliminate the reactions at two points $A$ and $B$ from the solution of some problems by writing the equation $\Sigma M_{AB} = 0$, which involves the computation of the moments of the forces about an axis $AB$ joining points $A$ and $B$ (Sample Prob. 4.10).

We observed that when a body is subjected to individual couples $M_i$, either as applied loads or as support reactions, we can include these couples by expressing the second part of Eq. (4.1) as

$$\Sigma M_O = \Sigma(r \times F) + \Sigma M_i = 0 \qquad (4.1')$$

If the reactions involve more than six unknowns, some of the reactions are *statically indeterminate;* if they involve fewer than six unknowns, the rigid body is only *partially constrained.* Even with six or more unknowns, the rigid body is *improperly constrained* if the reactions associated with the given supports are either parallel or intersect the same line.

# Review Problems

**4.142** A 3200-lb forklift truck is used to lift a 1700-lb crate. Determine the reaction at each of the two (*a*) front wheels *A*, (*b*) rear wheels *B*.

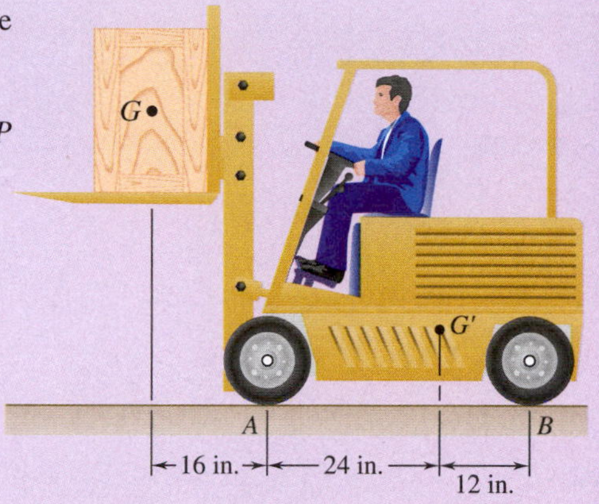

**4.143** The lever *BCD* is hinged at *C* and attached to a control rod at *B*. If *P* = 100 lb, determine (*a*) the tension in rod *AB*, (*b*) the reaction at *C*.

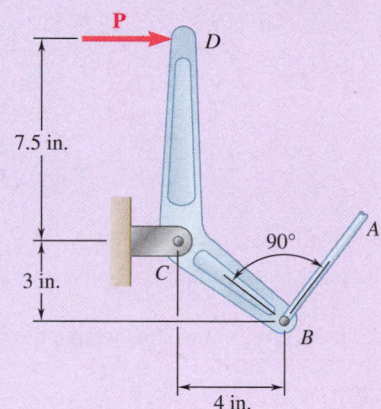

**Fig. P4.143**

**Fig. P4.142**

**4.144** Determine the reactions at *A* and *B* when (*a*) *h* = 0, (*b*) *h* = 200 mm.

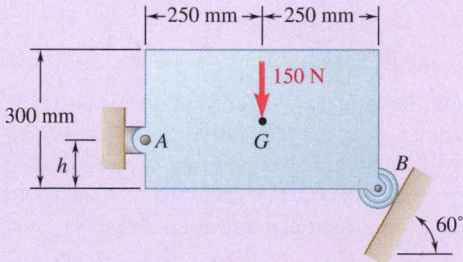

**Fig. P4.144**

**4.145** Neglecting friction and the radius of the pulley, determine (*a*) the tension in cable *ADB*, (*b*) the reaction at *C*.

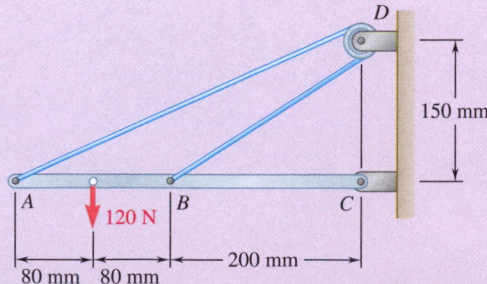

**Fig. P4.145**

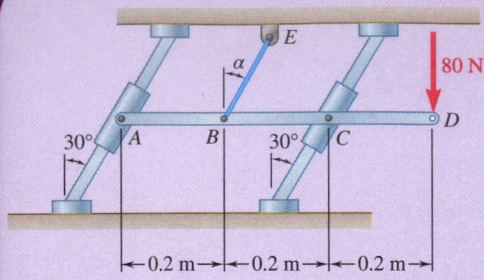

**Fig. P4.146**

**4.146** Bar $AD$ is attached at $A$ and $C$ to collars that can move freely on the rods shown. If the cord $BE$ is vertical ($\alpha = 0$), determine the tension in the cord and the reactions at $A$ and $C$.

**4.147** A slender rod $AB$, with a weight of $W$, is attached to blocks $A$ and $B$ that move freely in the guides shown. The constant of the spring is $k$, and the spring is unstretched when $\theta = 0$. (*a*) Neglecting the weight of the blocks, derive an equation in $W$, $k$, $l$, and $\theta$ that must be satisfied when the rod is in equilibrium. (*b*) Determine the value of $\theta$ when $W = 75$ lb, $l = 30$ in., and $k = 3$ lb/in.

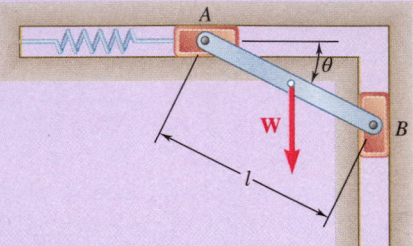

**Fig. P4.147**

**4.148** Determine the reactions at $A$ and $B$ when $a = 150$ mm.

**4.149** For the frame and loading shown, determine the reactions at $A$ and $C$.

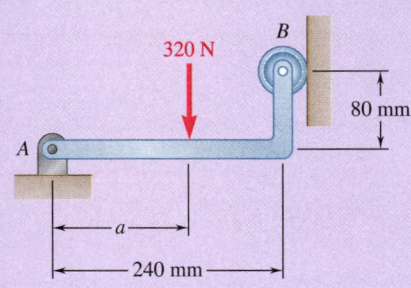

**Fig. P4.148**

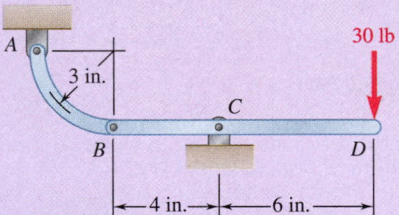

**Fig. P4.149**

**4.150** A 200-mm lever and a 240-mm-diameter pulley are welded to the axle $BE$ that is supported by bearings at $C$ and $D$. If a 720-N vertical load is applied at $A$ when the lever is horizontal, determine (*a*) the tension in the cord, (*b*) the reactions at $C$ and $D$. Assume that the bearing at $D$ does not exert any axial thrust.

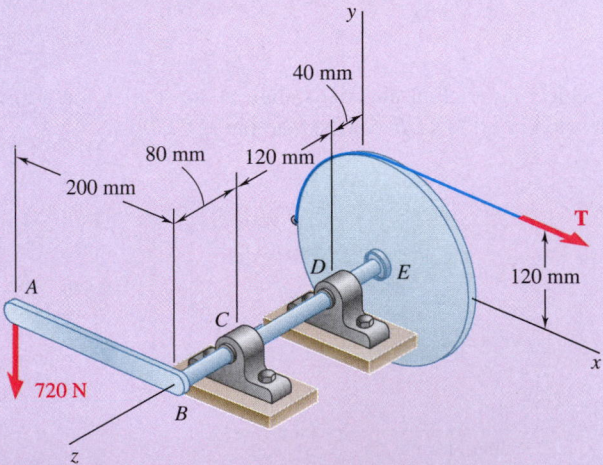

**Fig. P4.150**

**4.151** The 45-lb square plate shown is supported by three vertical wires. Determine the tension in each wire.

**4.152** The rectangular plate shown weighs 75 lb and is held in the position shown by hinges at *A* and *B* and by cable *EF*. Assuming that the hinge at *B* does not exert any axial thrust, determine (*a*) the tension in the cable, (*b*) the reactions at *A* and *B*.

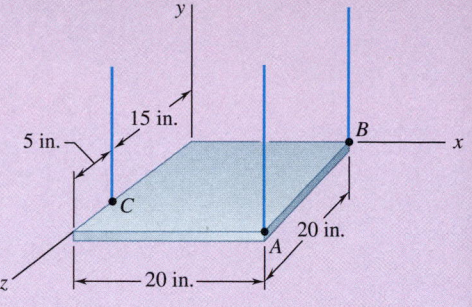

**Fig. P4.151**

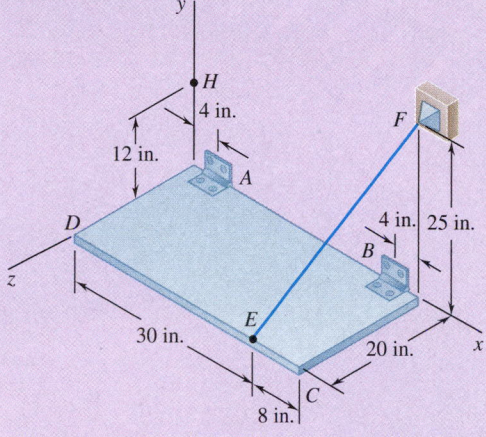

**Fig. P4.152**

**4.153** A force **P** is applied to a bent rod *ABC*, which may be supported in four different ways as shown. In each case, if possible, determine the reactions at the supports.

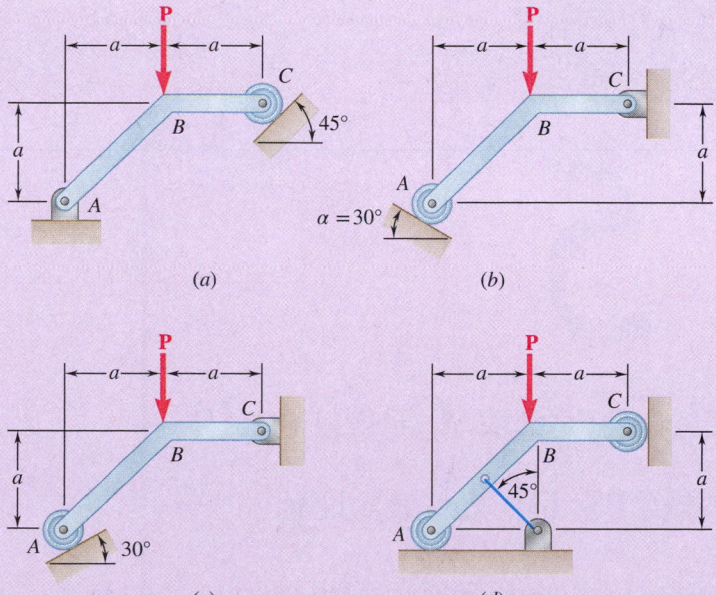

(*a*)          (*b*)

(*c*)          (*d*)

**Fig. P4.153**

# 5

# Distributed Forces: Centroids and Centers of Gravity

Loads on dams include three types of distributed forces: the weights of its constituent elements, the pressure forces exerted by the water on its submerged face, and the pressure forces exerted by the ground on its base.

# Objectives

- **Describe** the centers of gravity of two- and three-dimensional bodies.
- **Define** the centroids of lines, areas, and volumes.
- **Consider** the first moments of lines and areas, and examine their properties.
- **Determine** centroids of composite lines, areas, and volumes by summation methods.
- **Determine** centroids of composite lines, areas, and volumes by integration.
- **Apply** the theorems of Pappus-Guldinus to analyze surfaces and bodies of revolution.
- **Analyze** distributed loads on beams and forces on submerged surfaces.

# Introduction

We have assumed so far that we could represent the attraction exerted by the earth on a rigid body by a single force **W**. This force, called the force due to gravity or the weight of the body, is applied at the **center of gravity** of the body (Sec. 3.1A). Actually, the earth exerts a force on each of the particles forming the body, so we should represent the attraction of the earth on a rigid body by a large number of small forces distributed over the entire body. You will see in this chapter, however, that all of these small forces can be replaced by a single equivalent force **W**. You will also see how to determine the center of gravity—i.e., the point of application of the resultant **W**—for bodies of various shapes.

In the first part of this chapter, we study two-dimensional bodies, such as flat plates and wires contained in a given plane. We introduce two concepts closely associated with determining the center of gravity of a plate or a wire: the **centroid** of an area or a line and the **first moment** of an area or a line with respect to a given axis. Computing the area of a surface of revolution or the volume of a body of revolution is directly related to determining the centroid of the line or area used to generate that surface or body of revolution (theorems of Pappus-Guldinus). Also, as we show in Sec. 5.3, the determination of the centroid of an area simplifies the analysis of beams subjected to distributed loads and the computation of the forces exerted on submerged rectangular surfaces, such as hydraulic gates and portions of dams.

In the last part of this chapter, you will see how to determine the center of gravity of a three-dimensional body, as well as how to calculate the centroid of a volume and the first moments of that volume with respect to the coordinate planes.

**Photo 5.1** The precise balancing of the components of a mobile requires an understanding of centers of gravity and centroids, the main topics of this chapter.

©Christie's Images Ltd./SuperStock

# 5.1 PLANAR CENTERS OF GRAVITY AND CENTROIDS

In Chap. 4, we showed how the locations of the lines of action of forces affect the replacement of a system of forces with an equivalent system of forces and couples. In this section, we extend this idea to show how a distributed system of forces (in particular, the elements of an object's weight) can be replaced by a single resultant force acting at a specific point on an object. The specific point is called the object's center of gravity.

## 5.1A Center of Gravity of a Two-Dimensional Body

Let us first consider a flat horizontal plate (Fig. 5.1). We can divide the plate into $n$ small elements. We denote the coordinates of the first element by $x_1$ and $y_1$, those of the second element by $x_2$ and $y_2$, etc. The forces exerted by the earth on the elements of the plate are denoted, respectively, by $\Delta \mathbf{W}_1, \Delta \mathbf{W}_2, \ldots, \Delta \mathbf{W}_n$. These forces or weights are directed toward the center of the earth; however, for all practical purposes, we can assume them to be parallel. Their resultant is therefore a single force in the same direction. The magnitude $W$ of this force is obtained by adding the magnitudes of the elemental weights.

$$\Sigma F_z: \qquad W = \Delta W_1 + \Delta W_2 + \cdots + \Delta W_n$$

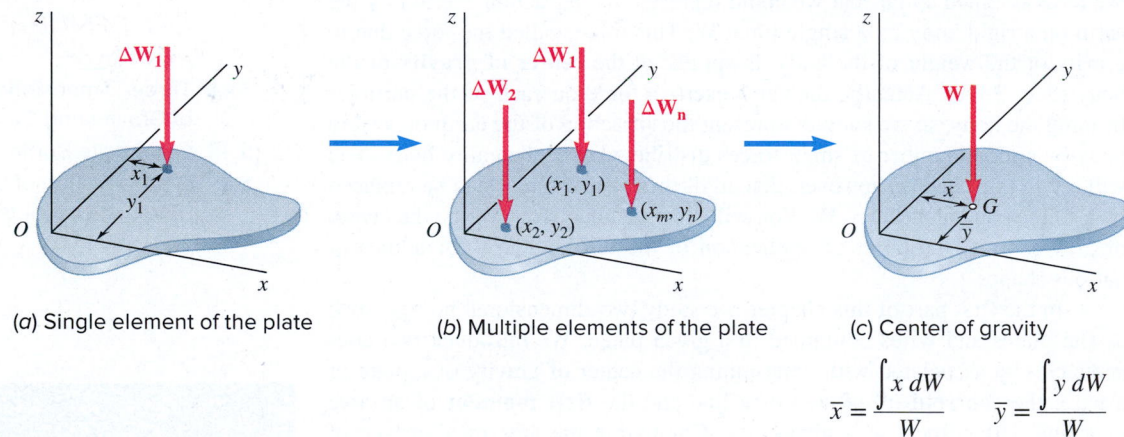

(a) Single element of the plate

(b) Multiple elements of the plate

(c) Center of gravity

$$\bar{x} = \frac{\int x \, dW}{W} \qquad \bar{y} = \frac{\int y \, dW}{W}$$

**Fig. 5.1** The center of gravity of a plate is the point where the resultant weight of the plate acts. It is the weighted average of all the elements of weight that make up the plate.

To obtain the coordinates $\bar{x}$ and $\bar{y}$ of point $G$ where the resultant $\mathbf{W}$ should be applied, we note that the moments of $\mathbf{W}$ about the $y$ and $x$ axes are equal to the sum of the corresponding moments of the elemental weights:

$$\begin{aligned} \Sigma M_y: & \qquad \bar{x}W = x_1 \Delta W_1 + x_2 \Delta W_2 + \cdots + x_n \Delta W_n \\ \Sigma M_x: & \qquad \bar{y}W = y_1 \Delta W_1 + y_2 \Delta W_2 + \cdots + y_n \Delta W_n \end{aligned} \qquad \textbf{(5.1)}$$

Solving these equations for $\bar{x}$ and $\bar{y}$ gives us

$$\bar{x} = \frac{x_1 \Delta W_1 + x_2 \Delta W_2 + \cdots + x_n \Delta W_n}{W}$$

$$\bar{y} = \frac{y_1 \Delta W_1 + y_2 \Delta W_2 + \cdots + y_n \Delta W_n}{W}$$

We could use these equations in this form to find the center of gravity of a collection of $n$ objects, each with a weight of $W_i$.

If we now increase the number of elements into which we divide the plate and simultaneously decrease the size of each element, in the limit of infinitely many elements of infinitesimal size, we obtain the expressions

**Photo 5.2** The center of gravity of a boomerang is not located on the object itself. ©C Squared Studios/Getty Images RF

**Weight, center of gravity of a flat plate**

$$W = \int dW \qquad \overline{x}\,W = \int x\,dW \qquad \overline{y}\,W = \int y\,dW \tag{5.2}$$

Or, solving for $\overline{x}$ and $\overline{y}$, we have

$$W = \int dW \qquad \overline{x} = \frac{\int x\,dW}{W} \qquad \overline{y} = \frac{\int y\,dW}{W} \tag{5.2'}$$

These equations define the weight $\mathbf{W}$ and the coordinates $\overline{x}$ and $\overline{y}$ of the **center of gravity** $G$ of a flat plate. The same equations can be derived for a wire lying in the $xy$ plane (Fig. 5.2). Note that the center of gravity $G$ of a wire is usually not located on the wire.

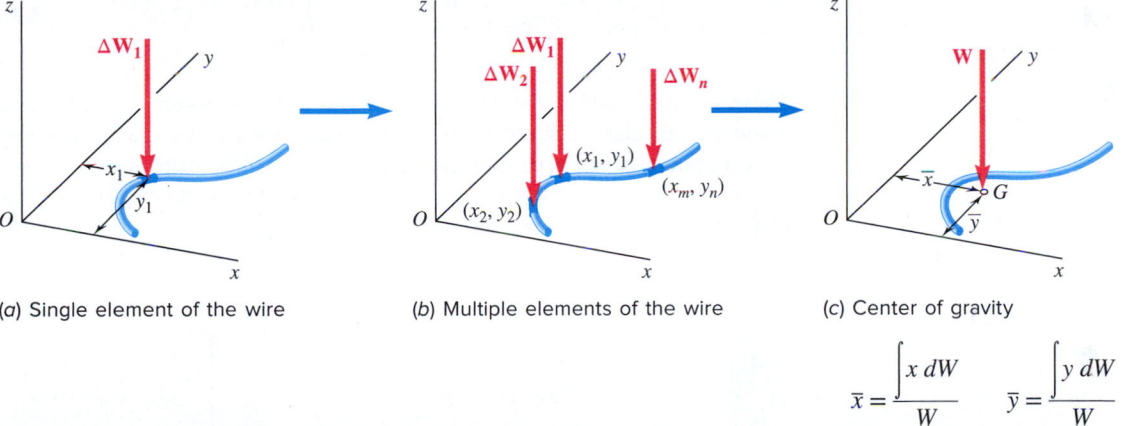

(a) Single element of the wire  (b) Multiple elements of the wire  (c) Center of gravity

$$\overline{x} = \frac{\int x\,dW}{W} \qquad \overline{y} = \frac{\int y\,dW}{W}$$

**Fig. 5.2** The center of gravity of a wire is the point where the resultant weight of the wire acts. The center of gravity may not actually be located on the wire.

## 5.1B Centroids of Areas and Lines

In the case of a flat homogeneous plate of uniform thickness, we can express the magnitude $\Delta W$ of the weight of an element of the plate as

$$\Delta W = \gamma t\,\Delta A$$

where $\gamma =$ specific weight (weight per unit volume) of the material
$t =$ thickness of the plate
$\Delta A =$ area of the element

Similarly, we can express the magnitude $W$ of the weight of the entire plate as

$$W = \gamma t A$$

where $A$ is the total area of the plate.

If U.S. customary units are used, the specific weight $\gamma$ should be expressed in $lb/ft^3$, the thickness $t$ in feet, and the areas $\Delta A$ and $A$ in square feet. Then, $\Delta W$ and $W$ are expressed in pounds. If SI units are used, $\gamma$ should be expressed in $N/m^3$, $t$ in meters, and the areas $\Delta A$ and $A$ in square meters; the weights $\Delta W$ and $W$ are then expressed in newtons.[†]

Substituting for $\Delta W$ and $W$ in the moment Eqs. (5.1) and dividing throughout by $\gamma t$, we obtain

$$\Sigma M_y: \quad \bar{x}A = x_1 \Delta A_1 + x_2 \Delta A_2 + \cdots + x_n \Delta A_n$$
$$\Sigma M_x: \quad \bar{y}A = y_1 \Delta A_1 + y_2 \Delta A_2 + \cdots + y_n \Delta A_n$$

If we increase the number of elements into which the area $A$ is divided and simultaneously decrease the size of each element, in the limit we obtain

**Centroid of an area $A$**

$$\bar{x}A = \int x \, dA \qquad \bar{y}A = \int y \, dA \qquad (5.3)$$

Or, solving for $\bar{x}$ and $\bar{y}$, we obtain

$$\bar{x} = \frac{\int x \, dA}{A} \qquad \bar{y} = \frac{\int y \, dA}{A} \qquad (5.3')$$

These equations define the coordinates $\bar{x}$ and $\bar{y}$ of the center of gravity of a homogeneous plate. The point whose coordinates are $\bar{x}$ and $\bar{y}$ is also known as the **centroid $C$ of the area** $A$ of the plate (Fig. 5.3). If the plate is not homogeneous, you cannot use these equations to determine the center of gravity of the plate; they still define, however, the centroid of the area.

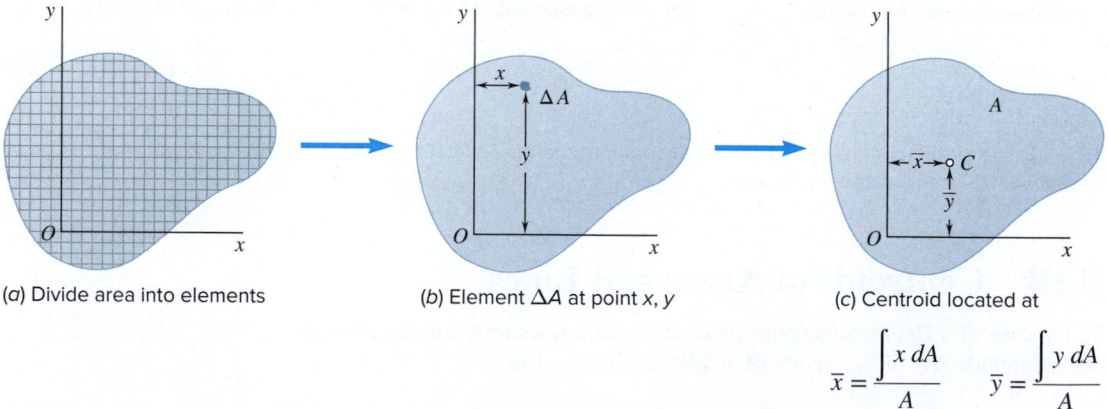

(a) Divide area into elements     (b) Element $\Delta A$ at point $x, y$     (c) Centroid located at

$$\bar{x} = \frac{\int x \, dA}{A} \qquad \bar{y} = \frac{\int y \, dA}{A}$$

**Fig. 5.3** The centroid of an area is the point where a homogeneous plate of uniform thickness would balance.

[†]We should note that in the SI system of units, a given material is generally characterized by its density $\rho$ (mass per unit volume) rather than by its specific weight $\gamma$. You can obtain the specific weight of the material from the relation

$$\gamma = \rho g$$

where $g = 9.81$ m/s$^2$. Note that because $\rho$ is expressed in kg/m$^3$, the units of $\gamma$ are (kg/m$^3$)(m/s$^2$), or N/m$^3$.

In the case of a homogeneous wire of uniform cross section, we can express the magnitude $\Delta W$ of the weight of an element of wire as

$$\Delta W = \gamma a\,\Delta L$$

where $\gamma$ = specific weight of the material
$a$ = cross-sectional area of the wire
$\Delta L$ = length of the element

The center of gravity of the wire then coincides with the **centroid** $C$ **of the line** $L$ defining the shape of the wire (Fig. 5.4). We can obtain the coordinates $\bar{x}$ and $\bar{y}$ of the centroid of line $L$ from the equations

**Centroid of a line $L$**

$$\bar{x}L = \int x\,dL \qquad \bar{y}L = \int y\,dL \tag{5.4}$$

Solving for $\bar{x}$ and $\bar{y}$ gives us

$$\bar{x} = \frac{\int x\,dL}{L} \qquad \bar{y} = \frac{\int y\,dL}{L} \tag{5.4'}$$

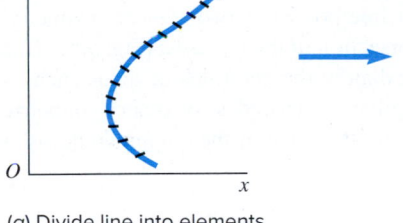

(a) Divide line into elements

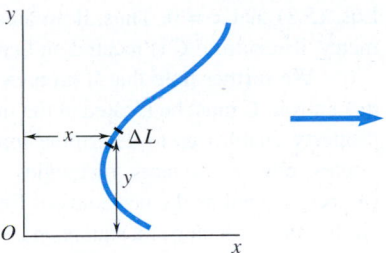

(b) Element $\Delta L$ at point $x$, $y$

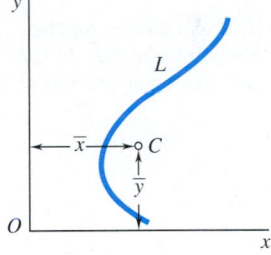

(c) Centroid located at

$$\bar{x} = \frac{\int x\,dL}{L} \qquad \bar{y} = \frac{\int y\,dL}{L}$$

**Fig. 5.4** The centroid of a line is the point where a homogeneous wire of uniform cross section would balance.

## 5.1C First Moments of Areas and Lines

The integral $\int x\,dA$ in Eqs. (5.3) is known as the **first moment of the area $A$ with respect to the $y$ axis** and is denoted by $Q_y$. Similarly, the integral $\int y\,dA$ defines the **first moment of $A$ with respect to the $x$ axis** and is denoted by $Q_x$. That is,

**First moments of area $A$**

$$Q_y = \int x\,dA \qquad Q_x = \int y\,dA \tag{5.5}$$

Comparing Eqs. (5.3) with Eqs. (5.5), we note that we can express the first moments of the area $A$ as the products of the area and the coordinates of its centroid:

$$Q_y = \bar{x}A \qquad Q_x = \bar{y}A \tag{5.6}$$

It follows from Eqs. (5.6) that we can obtain the coordinates of the centroid of an area by dividing the first moments of that area by the area itself. The

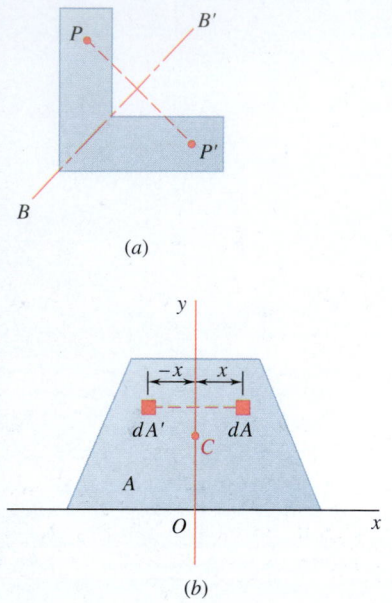

(a)

(b)

**Fig. 5.5** Symmetry about an axis. (a) The area is symmetric about the axis $BB'$. (b) The centroid of the area is located on the axis of symmetry.

first moments of the area are also useful in mechanics of materials for determining the shearing stresses in beams under transverse loadings. Finally, we observe from Eqs. (5.6) that, if the centroid of an area is located on a coordinate axis, the first moment of the area with respect to that axis is zero. Conversely, if the first moment of an area with respect to a coordinate axis is zero, the centroid of the area is located on that axis.

We can use equations similar to Eqs. (5.5) and (5.6) to define the first moments of a line with respect to the coordinate axes and to express these moments as the products of the length $L$ of the line and the coordinates $\bar{x}$ and $\bar{y}$ of its centroid.

An area $A$ is said to be **symmetric with respect to an axis** $BB'$ if for every point $P$ of the area there exists a point $P'$ of the same area such that the line $PP'$ is perpendicular to $BB'$ and is divided into two equal parts by that axis (Fig. 5.5a). The axis $BB'$ is called an **axis of symmetry**. A line $L$ is said to be symmetric with respect to an axis $BB'$ if it satisfies similar conditions. When an area $A$ or a line $L$ possesses an axis of symmetry $BB'$, its first moment with respect to $BB'$ is zero, and its centroid is located on that axis. For example, note that, for the area $A$ of Fig. 5.5b, which is symmetric with respect to the $y$ axis, every element of area $dA$ with abscissa $x$ corresponds to an element $dA'$ of equal area and with abscissa $-x$. It follows that the integral in the first of Eqs. (5.5) is zero and, thus, that $Q_y = 0$. It also follows from the first of the relations in Eqs. (5.3) that $\bar{x} = 0$. Thus, if an area $A$ or a line $L$ possesses an axis of symmetry, its centroid $C$ is located on that axis.

We further note that if an area or line possesses two axes of symmetry, its centroid $C$ must be located at the intersection of the two axes (Fig. 5.6). This property enables us to determine immediately the centroids of areas such as circles, ellipses, squares, rectangles, equilateral triangles, or other symmetric figures, as well as the centroids of lines in the shape of the circumference of a circle, the perimeter of a square, etc.

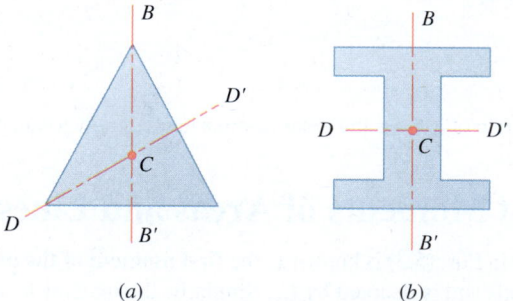

(a)                    (b)

**Fig. 5.6** If an area has two axes of symmetry, the centroid is located at their intersection. (a) An area with two axes of symmetry but no center of symmetry; (b) an area with two axes of symmetry and a center of symmetry.

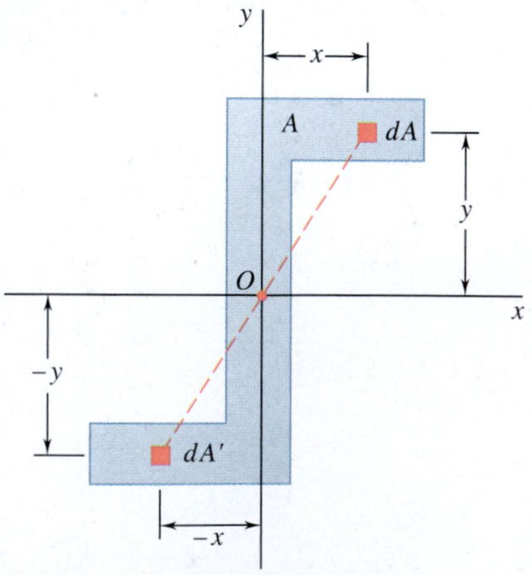

We say that an area $A$ is **symmetric with respect to a center** $O$ if, for every element of area $dA$ of coordinates $x$ and $y$, there exists an element $dA'$ of equal area with coordinates $-x$ and $-y$ (Fig. 5.7). It then follows that the integrals in Eqs. (5.5) are both zero and that $Q_x = Q_y = 0$. It also follows from Eqs. (5.3) that $\bar{x} = \bar{y} = 0$; that is, that the centroid of the area coincides with its center of symmetry $O$. Similarly, if a line possesses a center of symmetry $O$, the centroid of the line coincides with the center $O$.

Note that a figure possessing a center of symmetry does not necessarily possess an axis of symmetry (Fig. 5.7), whereas a figure possessing two axes

**Fig. 5.7** An area may have a center of symmetry but no axis of symmetry.

of symmetry does not necessarily possess a center of symmetry (Fig. 5.6*a*). However, if a figure possesses two axes of symmetry at right angles to each other, the point of intersection of these axes is a center of symmetry (Fig. 5.6*b*).

Determining the centroids of unsymmetrical areas and lines and of areas and lines possessing only one axis of symmetry will be discussed in the next section. Centroids of common shapes of areas and lines are shown in Fig. 5.8A and B.

| Shape | | $\overline{x}$ | $\overline{y}$ | Area |
|---|---|---|---|---|
| Triangular area | 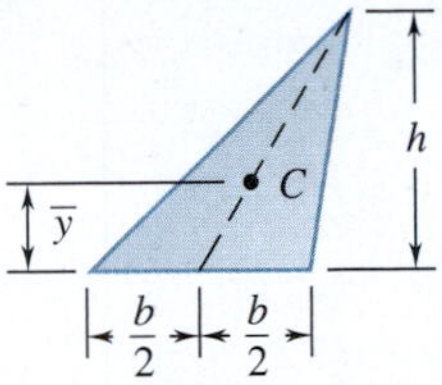 | | $\dfrac{h}{3}$ | $\dfrac{bh}{2}$ |
| Quarter-circular area | 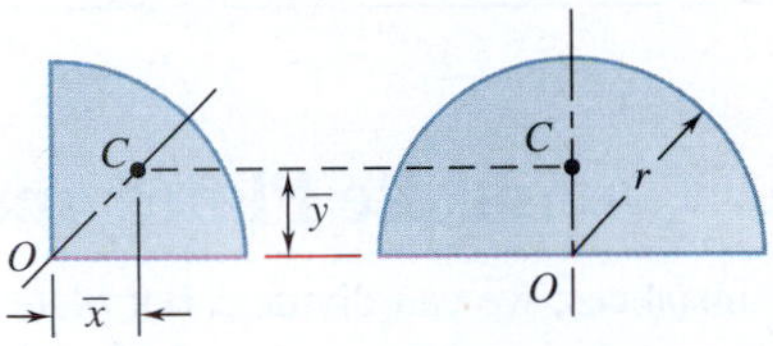 | $\dfrac{4r}{3\pi}$ | $\dfrac{4r}{3\pi}$ | $\dfrac{\pi r^2}{4}$ |
| Semicircular area | | $0$ | $\dfrac{4r}{3\pi}$ | $\dfrac{\pi r^2}{2}$ |
| Quarter-elliptical area | 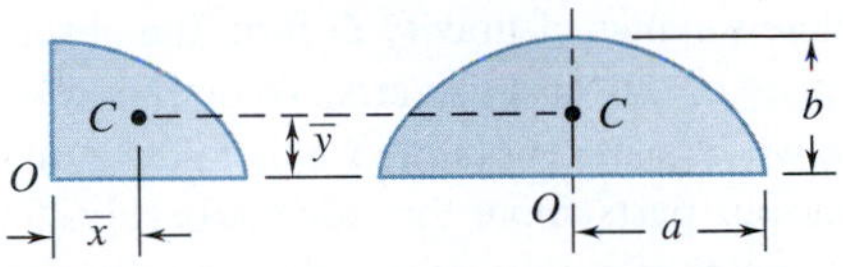 | $\dfrac{4a}{3\pi}$ | $\dfrac{4b}{3\pi}$ | $\dfrac{\pi ab}{4}$ |
| Semielliptical area | | $0$ | $\dfrac{4b}{3\pi}$ | $\dfrac{\pi ab}{2}$ |
| Semiparabolic area | 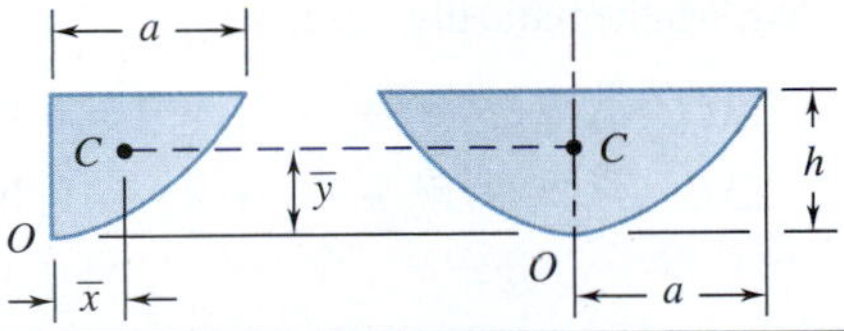 | $\dfrac{3a}{8}$ | $\dfrac{3h}{5}$ | $\dfrac{2ah}{3}$ |
| Parabolic area | | $0$ | $\dfrac{3h}{5}$ | $\dfrac{4ah}{3}$ |
| Parabolic spandrel | 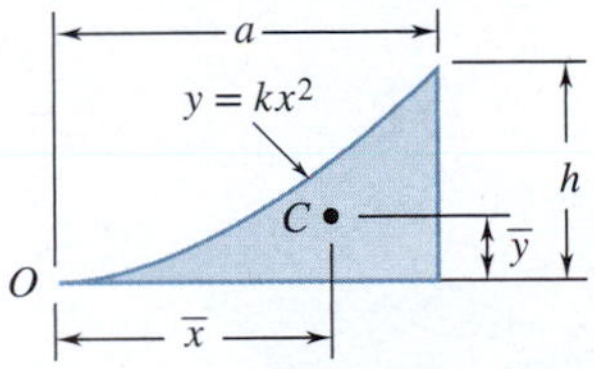 | $\dfrac{3a}{4}$ | $\dfrac{3h}{10}$ | $\dfrac{ah}{3}$ |
| General spandrel | 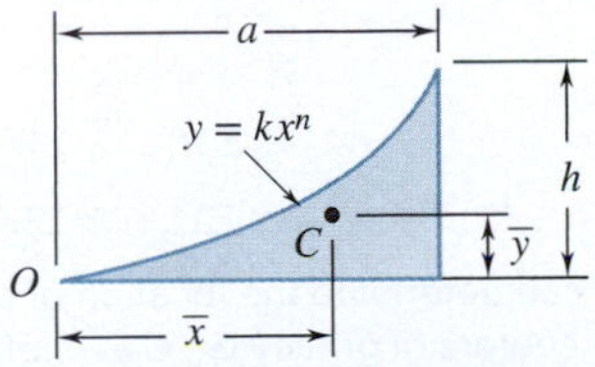 | $\dfrac{n+1}{n+2}a$ | $\dfrac{n+1}{4n+2}h$ | $\dfrac{ah}{n+1}$ |
| Circular sector | 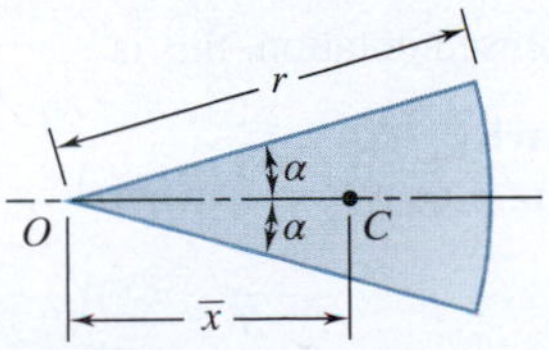 | $\dfrac{2r\sin\alpha}{3\alpha}$ | $0$ | $\alpha r^2$ |

**Fig. 5.8A**  Centroids of common shapes of areas.

| Shape | | $\bar{x}$ | $\bar{y}$ | Length |
|---|---|---|---|---|
| Quarter-circular arc | 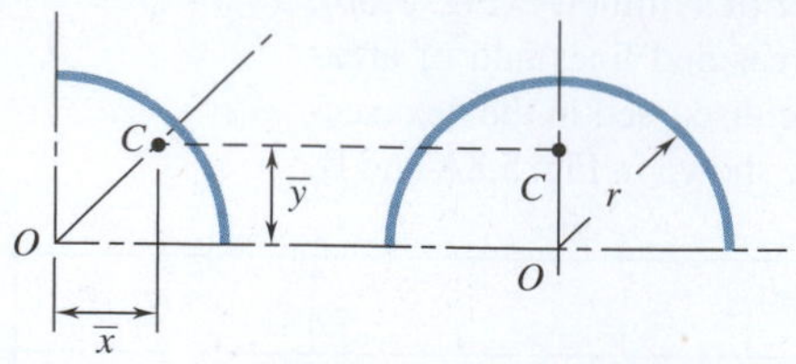 | $\dfrac{2r}{\pi}$ | $\dfrac{2r}{\pi}$ | $\dfrac{\pi r}{2}$ |
| Semicircular arc | | $0$ | $\dfrac{2r}{\pi}$ | $\pi r$ |
| Arc of circle | 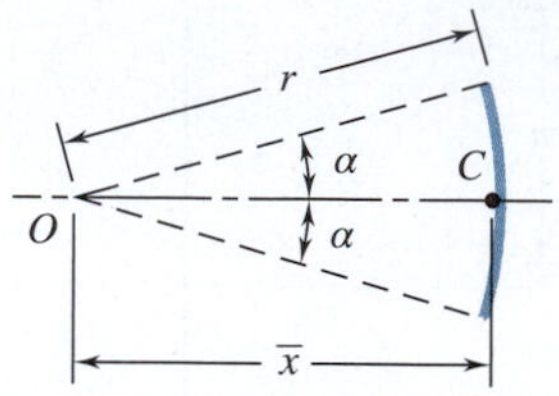 | $\dfrac{r \sin \alpha}{\alpha}$ | $0$ | $2\alpha r$ |

**Fig. 5.8B** Centroids of common shapes of lines.

## 5.1D Composite Plates and Wires

In many instances, we can divide a flat plate into rectangles, triangles, or the other common shapes shown in Fig. 5.8A. We can determine the abscissa $\bar{X}$ of the plate's center of gravity $G$ from the abscissas $\bar{x}_1, \bar{x}_2, \ldots, \bar{x}_n$ of the centers of gravity of the various parts. To do this, we equate the moment of the weight of the whole plate about the $y$ axis to the sum of the moments of the weights of the various parts about the same axis (Fig. 5.9). We can obtain the ordinate $\bar{Y}$ of the center of gravity of the plate in a similar way by equating moments about the $x$ axis. Mathematically, we have

$$\Sigma M_y: \quad \bar{X}(W_1 + W_2 + \cdots + W_n) = \bar{x}_1 W_1 + \bar{x}_2 W_2 + \cdots + \bar{x}_n W_n$$

$$\Sigma M_x: \quad \bar{Y}(W_1 + W_2 + \cdots + W_n) = \bar{y}_1 W_1 + \bar{y}_2 W_2 + \cdots + \bar{y}_n W_n$$

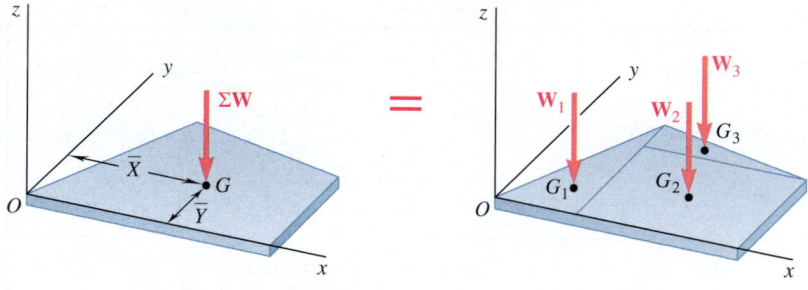

$$\Sigma M_y: \quad \bar{X}\, \Sigma W = \Sigma \bar{x}\, W$$

$$\Sigma M_x: \quad \bar{Y}\, \Sigma W = \Sigma \bar{y}\, W$$

**Fig. 5.9** We can determine the location of the center of gravity $G$ of a composite plate from the centers of gravity $G_1, G_2, \ldots$ of the component plates.

In more condensed notation, this is

**Center of gravity
of a composite plate**

$$\bar{X} = \frac{\Sigma \bar{x} W}{W} \qquad \bar{Y} = \frac{\Sigma \bar{y} W}{W}$$

(5.7)

We can use these equations to find the coordinates $\overline{X}$ and $\overline{Y}$ of the center of gravity of the plate from the centers of gravity of its component parts.

If the plate is homogeneous and of uniform thickness, the center of gravity coincides with the centroid $C$ of its area. We can determine the abscissa $\overline{X}$ of the centroid of the area by noting that we can express the first moment $Q_y$ of the composite area with respect to the $y$ axis as (1) the product of $\overline{X}$ and the total area and (2) as the sum of the first moments of the elementary areas with respect to the $y$ axis (Fig. 5.10). We obtain the ordinate $\overline{Y}$ of

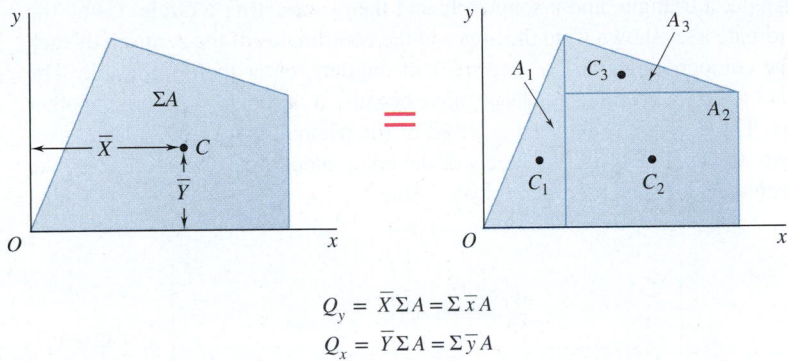

$$Q_y = \overline{X}\Sigma A = \Sigma \overline{x} A$$
$$Q_x = \overline{Y}\Sigma A = \Sigma \overline{y} A$$

**Fig. 5.10** We can find the location of the centroid of a composite area from the centroids of the component areas.

the centroid in a similar way by considering the first moment $Q_x$ of the composite area. We have

$$Q_y = \overline{X}(A_1 + A_2 + \cdots + A_n) = \overline{x}_1 A_1 + \overline{x}_2 A_2 + \cdots + \overline{x}_n A_n$$
$$Q_x = \overline{Y}(A_1 + A_2 + \cdots + A_n) = \overline{y}_1 A_1 + \overline{y}_2 A_2 + \cdots + \overline{y}_n A_n$$

Again, in shorter form,

**Centroid of a composite area**

$$Q_y = \overline{X}\Sigma A = \Sigma \overline{x} A \qquad Q_x = \overline{Y}\Sigma A = \Sigma \overline{y} A \qquad \textbf{(5.8)}$$

These equations yield the first moments of the composite area, or we can use them to obtain the coordinates $\overline{X}$ and $\overline{Y}$ of its centroid.

First moments of areas, like moments of forces, can be positive or negative. Thus, you need to take care to assign the appropriate sign to the moment of each area. For example, an area whose centroid is located to the left of the $y$ axis has a negative first moment with respect to that axis. Also, the area of a hole should be assigned a negative sign (Fig. 5.11).

Similarly, it is possible in many cases to determine the center of gravity of a composite wire or the centroid of a composite line by dividing the wire or line into simpler elements (see Sample Prob. 5.2).

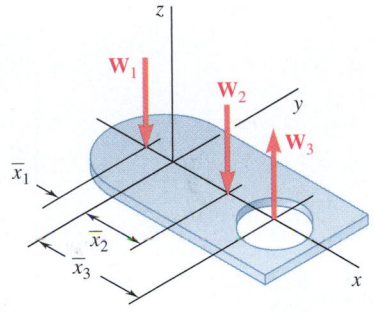

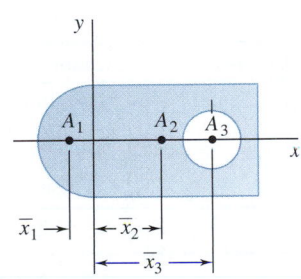

|  | $\overline{x}$ | $A$ | $\overline{x}A$ |
|---|---|---|---|
| $A_1$ Semicircle | − | + | − |
| $A_2$ Full rectangle | + | + | + |
| $A_3$ Circular hole | + | − | − |

**Fig. 5.11** When calculating the centroid of a composite area, note that if the centroid of a component area has a negative coordinate distance relative to the origin, or if the area represents a hole, then the first moment is negative.

# Sample Problem 5.1

For the plane area shown, determine (a) the first moments with respect to the x and y axes; (b) the location of the centroid.

**STRATEGY:** Break up the given area into simple components, find the centroid of each component, and then find the overall first moments and centroid.

**MODELING:** As shown in Fig. 1, you obtain the given area by adding a rectangle, a triangle, and a semicircle and then subtracting a circle. Using the coordinate axes shown, find the area and the coordinates of the centroid of each of the component areas. To keep track of the data, enter them in a table. The area of the circle is indicated as negative because it is subtracted from the other areas. The coordinate $\bar{y}$ of the centroid of the triangle is negative for the axes shown. Compute the first moments of the component areas with respect to the coordinate axes and enter them in your table.

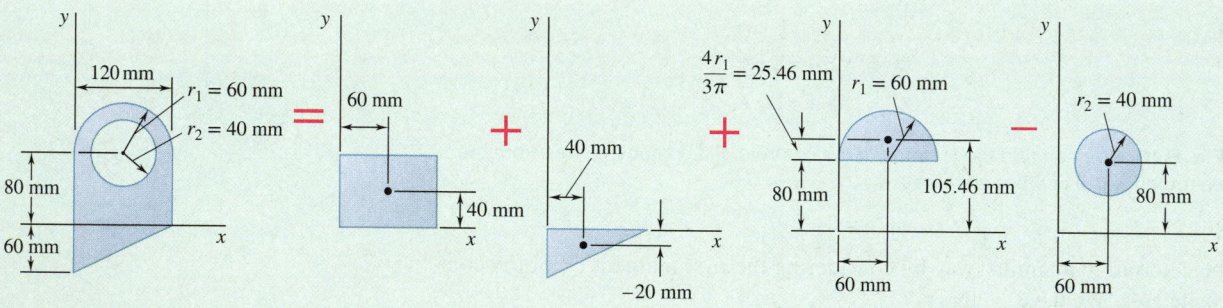

| Component | $A$, mm$^2$ | $\bar{x}$, mm | $\bar{y}$, mm | $\bar{x}A$, mm$^3$ | $\bar{y}A$, mm$^3$ |
|---|---|---|---|---|---|
| Rectangle | $(120)(80) = 9.6 \times 10^3$ | 60 | 40 | $+576 \times 10^3$ | $+384 \times 10^3$ |
| Triangle | $\frac{1}{2}(120)(60) = 3.6 \times 10^3$ | 40 | $-20$ | $+144 \times 10^3$ | $-72 \times 10^3$ |
| Semicircle | $\frac{1}{2}\pi(60)^2 = 5.655 \times 10^3$ | 60 | 105.46 | $+339.3 \times 10^3$ | $+596.4 \times 10^3$ |
| Circle | $-\pi(40)^2 = -5.027 \times 10^3$ | 60 | 80 | $-301.6 \times 10^3$ | $-402.2 \times 10^3$ |
| | $\Sigma A = 13.828 \times 10^3$ | | | $\Sigma \bar{x}A = +757.7 \times 10^3$ | $\Sigma \bar{y}A = +506.2 \times 10^3$ |

**Fig. 1** Given area modeled as the combination of simple geometric shapes.

## ANALYSIS:

### a. First Moments of the Area.
Using Eqs. (5.8), you obtain

$$Q_x = \Sigma \bar{y}A = 506.2 \times 10^3 \text{ mm}^3 \qquad Q_x = 506 \times 10^3 \text{ mm}^3 \blacktriangleleft$$
$$Q_y = \Sigma \bar{x}A = 757.7 \times 10^3 \text{ mm}^3 \qquad Q_y = 758 \times 10^3 \text{ mm}^3 \blacktriangleleft$$

### b. Location of Centroid.
Substituting the values given in the table into the equations defining the centroid of a composite area yields (Fig. 2)

$$\bar{X}\Sigma A = \Sigma \bar{x}A: \quad \bar{X}(13.828 \times 10^3 \text{ mm}^2) = 757.7 \times 10^3 \text{ mm}^3$$
$$\bar{X} = 54.8 \text{ mm} \blacktriangleleft$$
$$\bar{Y}\Sigma A = \Sigma \bar{y}A: \quad \bar{Y}(13.828 \times 10^3 \text{ mm}^2) = 506.2 \times 10^3 \text{ mm}^3$$
$$\bar{Y} = 36.6 \text{ mm} \blacktriangleleft$$

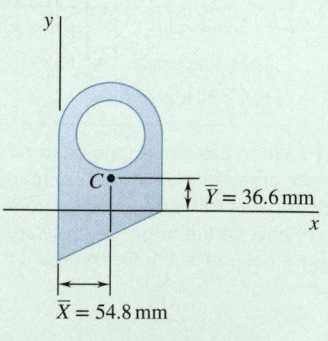

**Fig. 2** Centroid of composite area.

*(continued)*

**REFLECT AND THINK:**   Given that the lower portion of the shape has more area to the left and that the upper portion has a hole, the location of the centroid seems reasonable upon visual inspection.

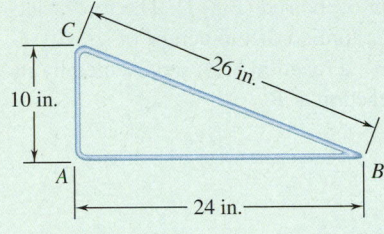

## Sample Problem 5.2

The figure shown is made from a piece of thin, homogeneous wire. Determine the location of its center of gravity.

**STRATEGY:**   Because the figure is formed of homogeneous wire, its center of gravity coincides with the centroid of the corresponding line. Therefore, you can simply determine that centroid.

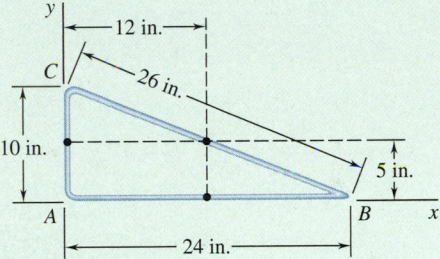

**Fig. 1**   Location of each line segment's centroid.

**MODELING:**   Choosing the coordinate axes shown in Fig. 1 with the origin at $A$, determine the coordinates of the centroid of each line segment and compute the first moments with respect to the coordinate axes. You may find it convenient to list the data in a table.

| Segment | $L$, in. | $\bar{x}$, in. | $\bar{y}$, in. | $\bar{x}L$, in$^2$ | $\bar{y}L$, in$^2$ |
|---------|----------|----------------|----------------|--------------------|--------------------|
| $AB$    | 24       | 12             | 0              | 288                | 0                  |
| $BC$    | 26       | 12             | 5              | 312                | 130                |
| $CA$    | 10       | 0              | 5              | 0                  | 50                 |
|         | $\Sigma L = 60$ |         |                | $\Sigma\bar{x}L = 600$ | $\Sigma\bar{y}L = 180$ |

**ANALYSIS:**   Substituting the values obtained from the table into the equations defining the centroid of a composite line gives

$$\bar{X}\Sigma L = \Sigma\bar{x}L: \qquad \bar{X}(60 \text{ in.}) = 600 \text{ in}^2 \qquad\qquad \bar{X} = 10 \text{ in.} \blacktriangleleft$$
$$\bar{Y}\Sigma L = \Sigma\bar{y}L: \qquad \bar{Y}(60 \text{ in.}) = 180 \text{ in}^2 \qquad\qquad \bar{Y} = 3 \text{ in.} \blacktriangleleft$$

**REFLECT and THINK:**   The centroid is not on the wire itself, but it is within the area enclosed by the wire.

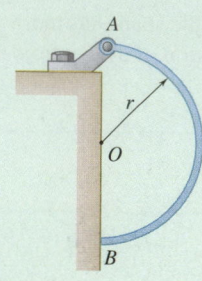

## Sample Problem 5.3

A uniform semicircular rod of weight $W$ and radius $r$ is attached to a pin at $A$ and rests against a frictionless surface at $B$. Determine the reactions at $A$ and $B$.

**STRATEGY:** The key to solving the problem is finding where the weight $W$ of the rod acts. Because the rod is a simple geometrical shape, you can look in Fig. 5.8 for the location of the wire's centroid.

**MODELING:** Draw a free-body diagram of the rod (Fig. 1). The forces acting on the rod are its weight $\mathbf{W}$, which is applied at the center of gravity $G$ (whose position is obtained from Fig. 5.8B); a reaction at $A$, represented by its components $\mathbf{A}_x$ and $\mathbf{A}_y$; and a horizontal reaction at $B$.

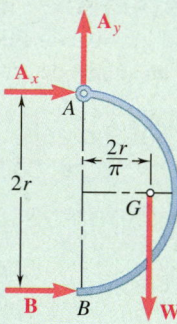

**Fig. 1** Free-body diagram of the rod.

## ANALYSIS:

$+\circlearrowleft \Sigma M_A = 0:$ $\qquad B(2r) - W\left(\dfrac{2r}{\pi}\right) = 0$

$$B = +\frac{W}{\pi} \qquad\qquad \mathbf{B} = \frac{W}{\pi} \rightarrow$$

$\xrightarrow{+} \Sigma F_x = 0:$ $\qquad A_x + B = 0$

$$A_x = -B = -\frac{W}{\pi} \qquad \mathbf{A}_x = \frac{W}{\pi} \leftarrow$$

$+\uparrow \Sigma F_y = 0:$ $\qquad A_y - W = 0 \qquad\qquad \mathbf{A}_y = W\uparrow$

Adding the two components of the reaction at $A$ (Fig. 2), we have

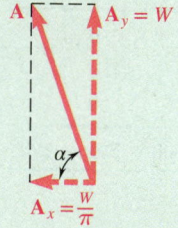

**Fig. 2** Reaction at $A$.

$$A = \left[ W^2 + \left(\frac{W}{\pi}\right)^2\right]^{1/2} \qquad \mathbf{A} = W\left(1 + \frac{1}{\pi^2}\right)^{1/2}$$

$$\tan\alpha = \frac{W}{W/\pi} = \pi \qquad\qquad \alpha = \tan^{-1}\pi$$

The answers can also be expressed as

$$\mathbf{A} = 1.049W \measuredangle 72.3° \qquad \mathbf{B} = 0.318W\rightarrow$$

**REFLECT and THINK:** Once you know the location of the rod's center of gravity, the problem is a straightforward application of the concepts in Chap. 4.

# SOLVING PROBLEMS
# ON YOUR OWN

In this section, we developed the general equations for locating the centers of gravity of two-dimensional bodies and wires [Eqs. (5.2)] and the centroids of plane areas [Eqs. (5.3)] and lines [Eqs. (5.4)]. In the following problems, you will have to locate the centroids of composite areas and lines or determine the first moments of the area for composite plates [Eqs. (5.8)].

**1. Locating the centroids of composite areas and lines.** Sample Probs. 5.1 and 5.2 illustrate the procedure you should follow when solving problems of this type. However, several points are worth emphasizing.

   **a.** The first step in your solution should be to decide how to construct the given area or line from the common shapes of Fig. 5.8. You should recognize that for plane areas it is often possible to construct a particular shape in more than one way. Also, showing the different components (as is done in Sample Prob. 5.1) can help you correctly establish their centroids and areas or lengths. Do not forget that you can subtract areas as well as add them to obtain a desired shape.

   **b.** We strongly recommend that, for each problem, you construct a table listing the areas or lengths and the respective coordinates of the centroids. Remember, any areas that are "removed" (such as holes) are treated as negative. Also, the sign of negative coordinates must be included. Therefore, you should always carefully note the location of the origin of the coordinate axes.

   **c.** When possible, use symmetry (Sec. 5.1C) to help you determine the location of a centroid.

   **d.** In the formulas for the circular sector and for the arc of a circle in Fig. 5.8, the angle $\alpha$ must always be expressed in radians.

**2. Calculating the first moments of an area.** The procedures for locating the centroid of an area and for determining the first moments of an area are similar; however, it is not necessary to compute the total area for finding first moments. Also, as noted in Sec. 5.1C, you should recognize that the first moment of an area relative to a centroidal axis is zero.

**3. Solving problems involving the center of gravity.** The bodies considered in the following problems are homogeneous; thus, their centers of gravity and centroids coincide. In addition, when a body that is suspended from a single pin is in equilibrium, the pin and the body's center of gravity must lie on the same vertical line.

It may appear that many of the problems in this section have little to do with the study of mechanics. However, being able to locate the centroid of composite shapes will be essential in several topics that you will study later in this course.

# Problems

**5.1 through 5.9** Locate the centroid of the plane area shown.

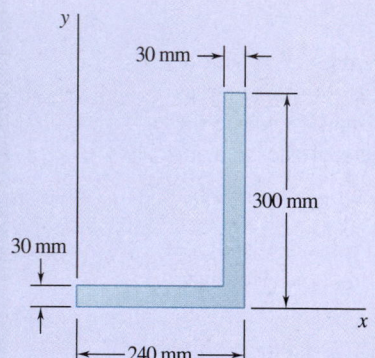

**Fig. P5.1**

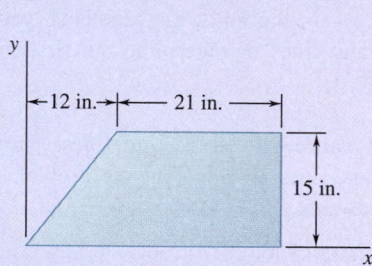

**Fig. P5.2**

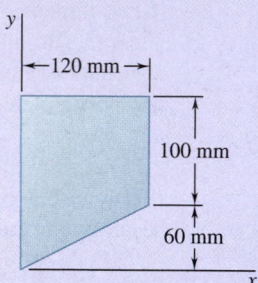

**Fig. P5.3**

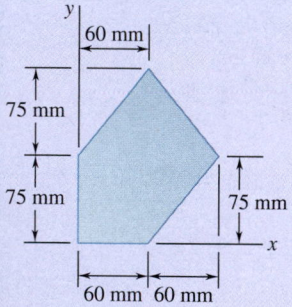

**Fig. P5.4**

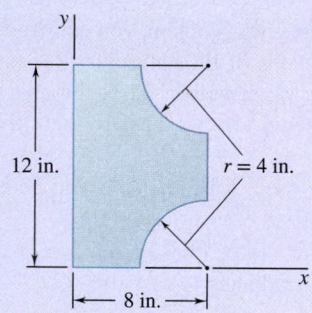

**Fig. P5.5**

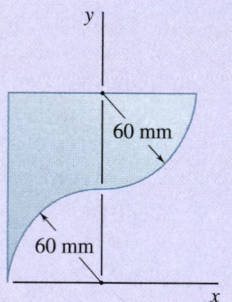

**Fig. P5.6**

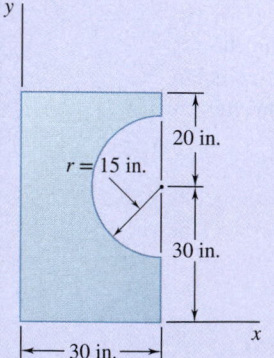

**Fig. P5.7**

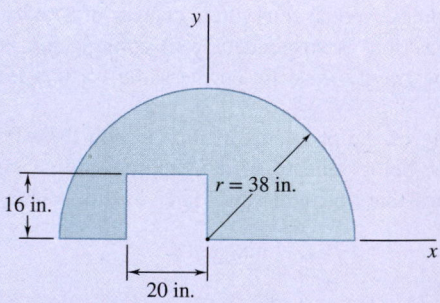

**Fig. P5.8**

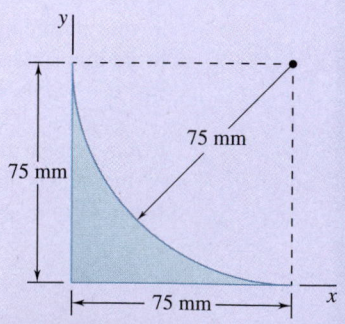

**Fig. P5.9**

**5.10 through 5.15** Locate the centroid of the plane area shown.

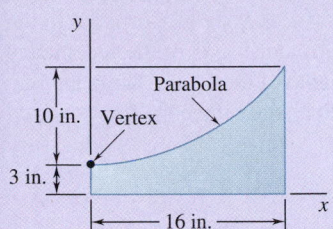

Fig. P5.10

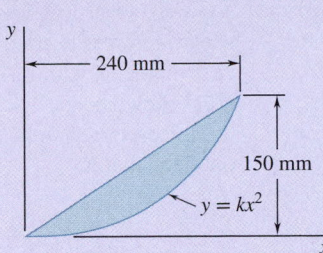

Fig. P5.11

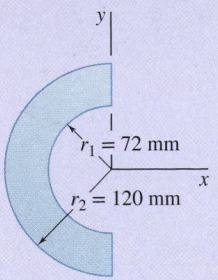

Fig. *P5.12*

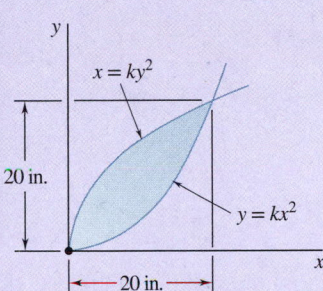

Fig. P5.13

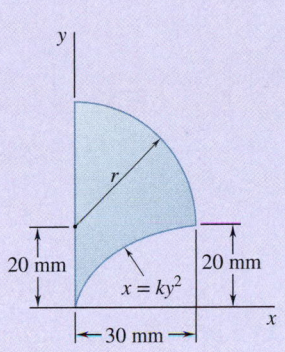

Fig. P5.14

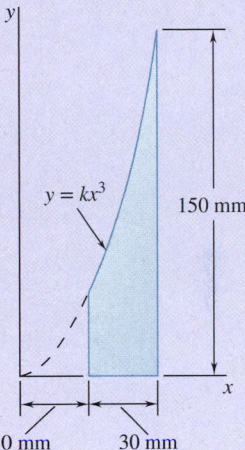

Fig. *P5.15*

**5.16** Determine the $y$ coordinate of the centroid of the shaded area in terms of $r_1$, $r_2$, and $\alpha$.

**5.17** Show that as $r_1$ approaches $r_2$, the location of the centroid approaches that for an arc of circle of radius $(r_1 + r_2)/2$.

**5.18** For the area shown, determine the ratio $a/b$ for which $\bar{x} = \bar{y}$.

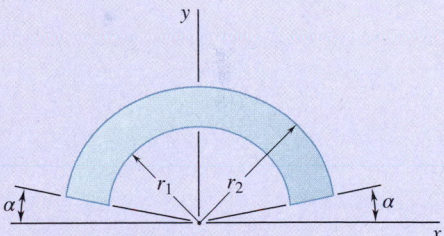

Fig. P5.16 and P5.17

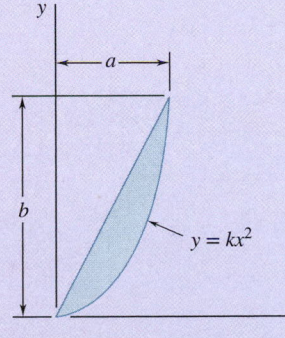

Fig. *P5.18*

**5.19** For the semiannular area of Prob. 5.12, determine the ratio $r_1$ to $r_2$ so that the centroid of the area is located at $x = -\frac{1}{2}r_2$ and $y = 0$.

**5.20** A built-up beam is constructed by nailing seven boards together as shown. The nails are equally spaced along the beam, and the beam supports a vertical load. As proved in mechanics of materials, the shearing forces exerted on the nails at $A$ and $B$ are proportional to the first moments with respect to the centroidal $x$ axis of the red shaded areas shown, respectively, in parts $a$ and $b$ of the figure. Knowing that the force exerted on the nail at $A$ is 120 N, determine the force exerted on the nail at $B$.

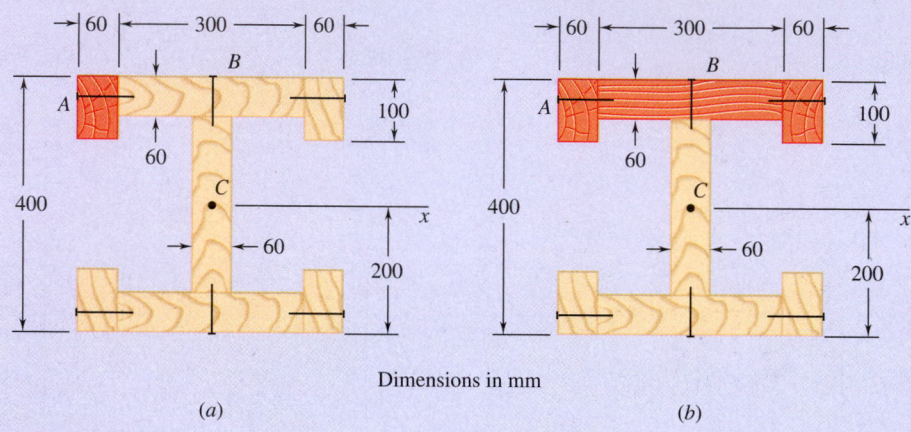

Dimensions in mm

(a)                                    (b)

**Fig. P5.20**

**5.21 and 5.22** The horizontal $x$ axis is drawn through the centroid $C$ of the area shown, and it divides the area into two component areas $A_1$ and $A_2$. Determine the first moment of each component area with respect to the $x$ axis, and explain the results obtained.

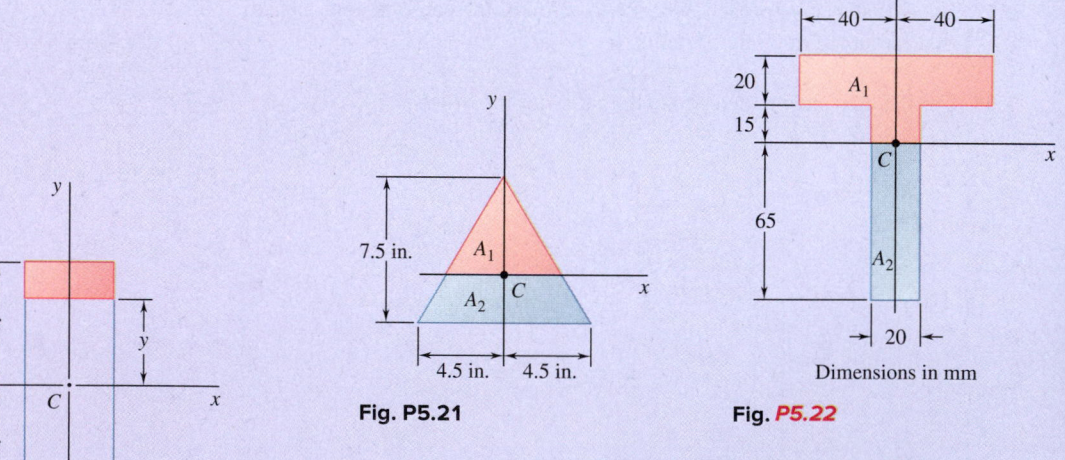

**Fig. P5.23**

**Fig. P5.21**

**Fig. P5.22**

Dimensions in mm

**5.23** The first moment of the shaded area with respect to the $x$ axis is denoted by $Q_x$. (a) Express $Q_x$ in terms of $b$, $c$, and the distance $y$ from the base of the shaded area to the $x$ axis. (b) For what value of $y$ is $Q_x$ maximum, and what is that maximum value?

**5.24 through 5.27**  A thin, homogeneous wire is bent to form the perimeter of the figure indicated. Locate the center of gravity of the wire figure thus formed.

  **5.24**  Fig. P5.1.
  **5.25**  Fig. P5.3.
  **5.26**  Fig. P5.5.
  **5.27**  Fig. P5.8.

**5.28**  The homogeneous wire *ABC* is bent into a semicircular arc and a straight section as shown and is attached to a hinge at *A*. Determine the value of $\theta$ for which the wire is in equilibrium for the indicated position.

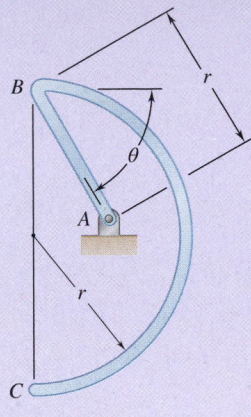

**Fig. P5.28**

**5.29**  The frame for a sign is fabricated from thin, flat steel bar stock of mass per unit length 4.73 kg/m. The frame is supported by a pin at *C* and by a cable *AB*. Determine (*a*) the tension in the cable, (*b*) the reaction at *C*.

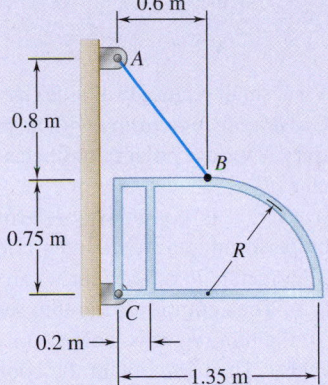

**Fig. P5.29**

**5.30**  The homogeneous wire *ABCD* is bent as shown and is attached to a hinge at *C*. Determine the length *L* for which portion *BCD* of the wire is horizontal.

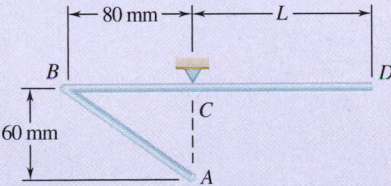

**Fig. P5.30 and P5.31**

**5.31**  The homogeneous wire *ABCD* is bent as shown and is attached to a hinge at *C*. Determine the length *L* for which portion *AB* of the wire is horizontal.

**5.32**  Determine the distance *h* for which the centroid of the shaded area is as far above line *BB′* as possible when (*a*) $k = 0.10$, (*b*) $k = 0.80$.

**5.33**  Knowing that the distance *h* has been selected to maximize the distance $\bar{y}$ from line *BB′* to the centroid of the shaded area, show that $\bar{y} = 2h/3$.

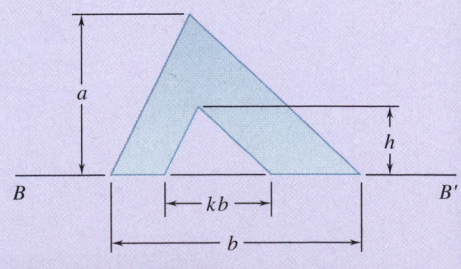

**Fig. P5.32 and P5.33**

# 5.2   FURTHER CONSIDERATIONS OF CENTROIDS

The objects we analyzed in Sec. 5.1 were composites of basic geometric shapes like rectangles, triangles, and circles. The same idea of locating a center of gravity or centroid applies for an object with a more complicated shape, but the mathematical techniques for finding the location are a little more difficult.

## 5.2A   Determination of Centroids by Integration

For an area bounded by analytical curves (i.e., curves defined by algebraic equations), we usually determine the centroid by evaluating the integrals in Eqs. (5.3'):

$$\bar{x} = \frac{\int x\,dA}{A} \qquad \bar{y} = \frac{\int y\,dA}{A} \tag{5.3'}$$

If the element of area $dA$ is a small rectangle of sides $dx$ and $dy$, evaluating each of these integrals requires a *double integration* with respect to $x$ and $y$. A double integration is also necessary if we use polar coordinates for which $dA$ is a small element with sides $dr$ and $r\,d\theta$.

In most cases, however, it is possible to determine the coordinates of the centroid of an area by performing a single integration. We can achieve this by choosing $dA$ to be a thin rectangle or strip, or it can be a thin sector or pie-shaped element (Fig. 5.12). The centroid of the thin rectangle is located at its center, and the centroid of the thin sector is located at a distance $(2/3)r$ from its vertex (as it is for a triangle). Then, we obtain the coordinates of the centroid of the area under consideration by setting the first moment of the entire area with respect to each of the coordinate axes equal to the sum (or integral) of the

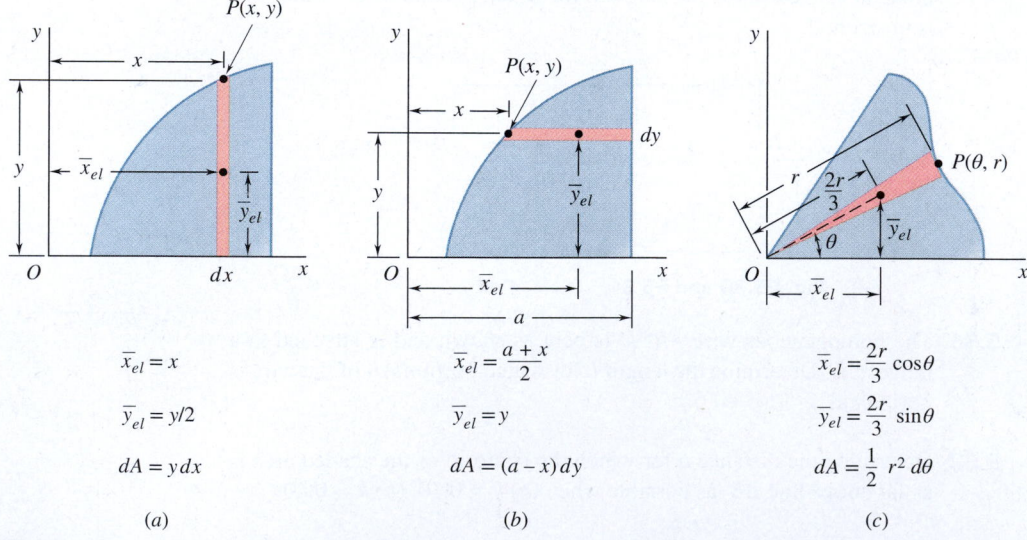

$$\bar{x}_{el} = x$$

$$\bar{y}_{el} = y/2$$

$$dA = y\,dx$$

(a)

$$\bar{x}_{el} = \frac{a + x}{2}$$

$$\bar{y}_{el} = y$$

$$dA = (a - x)\,dy$$

(b)

$$\bar{x}_{el} = \frac{2r}{3}\cos\theta$$

$$\bar{y}_{el} = \frac{2r}{3}\sin\theta$$

$$dA = \frac{1}{2}r^2\,d\theta$$

(c)

**Fig. 5.12**   Centroids and areas of differential elements. (*a*) Vertical rectangular strip; (*b*) horizontal rectangular strip; (*c*) triangular sector.

corresponding moments of the elements of the area. Denoting the coordinates of the centroid of the element $dA$ by $\bar{x}_{el}$ and $\bar{y}_{el}$ we have

**First moments of area**

$$Q_y = \bar{x}A = \int \bar{x}_{el}\, dA$$
$$Q_x = \bar{y}A = \int \bar{y}_{el}\, dA$$

$$(5.9)$$

If we do not already know the area $A$, we can also compute it from these elements.

In order to carry out the integration, we need to express the coordinates $\bar{x}_{el}$ and $\bar{y}_{el}$ of the centroid of the element of area $dA$ in terms of the coordinates of a point located on the curve bounding the area under consideration. Also, we should express the area of the element $dA$ in terms of the coordinates of that point and the appropriate differentials. This has been done in Fig. 5.12 for three common types of elements; the pie-shaped element of part (c) should be used when the equation of the curve bounding the area is given in polar coordinates. You can substitute the appropriate expressions into formulas (5.9), and then use the equation of the bounding curve to express one of the coordinates in terms of the other. This process reduces the double integration to a single integration. Once you have determined the area and evaluated the integrals in Eqs. (5.9), you can solve these equations for the coordinates $\bar{x}$ and $\bar{y}$ of the centroid of the area.

When a line is defined by an algebraic equation, you can determine its centroid by evaluating the integrals in Eqs. (5.4′):

$$\bar{x} = \frac{\int x\, dL}{L} \qquad \bar{y} = \frac{\int y\, dL}{L}$$

$$(5.4')$$

You can replace the differential length $dL$ with one of the following expressions, depending upon which coordinate, $x$, $y$, or $\theta$, is chosen as the independent variable in the equation used to define the line (these expressions can be derived using the Pythagorean theorem):

$$dL = \sqrt{1 + \left(\frac{dy}{dx}\right)^2}\, dx \qquad dL = \sqrt{1 + \left(\frac{dx}{dy}\right)^2}\, dy$$

$$dL = \sqrt{r^2 + \left(\frac{dr}{d\theta}\right)^2}\, d\theta$$

After you have used the equation of the line to express one of the coordinates in terms of the other, you can perform the integration and solve Eqs. (5.4) for the coordinates $\bar{x}$ and $\bar{y}$ of the centroid of the line.

## 5.2B   Theorems of Pappus-Guldinus

These two theorems, which were first formulated by the Greek geometer Pappus during the third century C.E. and later restated by the Swiss mathematician Guldinus or Guldin (1577–1643), deal with surfaces and bodies of revolution. A **surface of revolution** is a surface that can be generated by rotating a plane curve about a fixed axis. For example, we can obtain the surface of a sphere by rotating a semicircular arc $ABC$ about the diameter $AC$ (Fig. 5.13). Similarly, rotating a straight line $AB$ about an axis $AC$ produces the surface of a cone, and rotating the circumference of a circle about a nonintersecting axis generates the surface of a torus or ring. A **body of revolution** is a body that can be generated

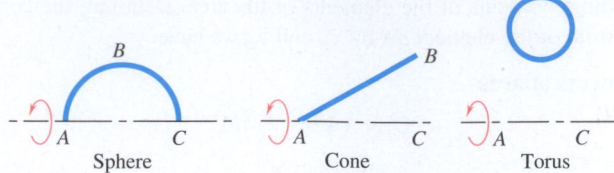

**Fig. 5.13** Rotating plane curves about an axis generates surfaces of revolution.

by rotating a plane area about a fixed axis. As shown in Fig. 5.14, we can generate a sphere, a cone, and a torus by rotating the appropriate shape about the indicated axis.

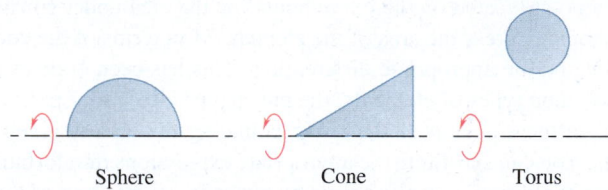

**Fig. 5.14** Rotating plane areas about an axis generates volumes of revolution.

**Theorem I.** *The area of a surface of revolution is equal to the length of the generating curve times the distance traveled by the centroid of the curve while the surface is being generated.*

**Proof.** Consider an element $dL$ of the line $L$ (Fig. 5.15) that is revolved about the $x$ axis. The circular strip generated by the element $dL$ has an area $dA$ equal to $2\pi y\,dL$. Thus, the entire area generated by $L$ is $A = \int 2\pi y\,dL$. Recall our earlier result that the integral $\int y\,dL$ is equal to $\bar{y}L$. Therefore, we have

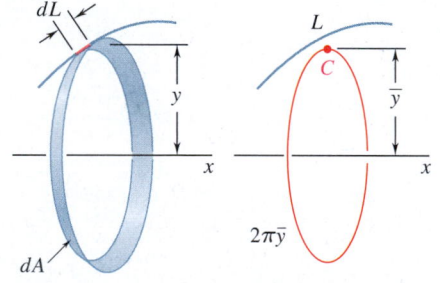

**Fig. 5.15** An element of length $dL$ rotated about the $x$ axis generates a circular strip of area $dA$. The area of the entire surface of revolution equals the length of the line $L$ multiplied by the distance traveled by the centroid $C$ of the line during one revolution.

$$A = 2\pi\bar{y}L \tag{5.10}$$

Here $2\pi\bar{y}$ is the distance traveled by the centroid $C$ of $L$ (Fig. 5.15).  □

Note that the generating curve must not cross the axis about which it is rotated; if it did, the two sections on either side of the axis would generate areas having opposite signs, and the theorem would not apply.

**Theorem II.** *The volume of a body of revolution is equal to the generating area times the distance traveled by the centroid of the area while the body is being generated.*

**Proof.** Consider an element $dA$ of the area $A$ that is revolved about the $x$ axis (Fig. 5.16). The circular ring generated by the element $dA$ has a volume $dV$ equal to $2\pi y\,dA$. Thus, the entire volume generated by $A$ is $V = \int 2\pi y\,dA$, and because we showed earlier that the integral $\int y\,dA$ is equal to $\bar{y}A$, we have

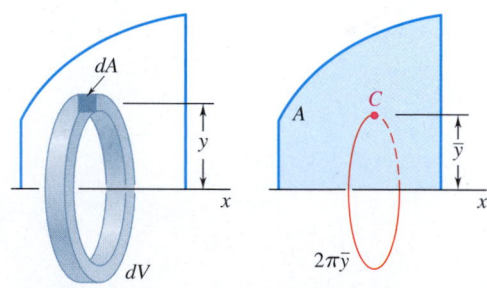

**Fig. 5.16** An element of area $dA$ rotated about the $x$ axis generates a circular ring of volume $dV$. The volume of the entire body of revolution equals the area of the region $A$ multiplied by the distance traveled by the centroid $C$ of the region during one revolution.

$$V = 2\pi\bar{y}A \tag{5.11}$$

Here $2\pi\bar{y}$ is the distance traveled by the centroid of $A$.  □

Again, note that the theorem does not apply if the axis of rotation intersects the generating area.

The theorems of Pappus-Guldinus offer a simple way to compute the areas of surfaces of revolution and the volumes of bodies of revolution. Conversely, they also can be used to determine the centroid of a plane curve if you know the area of the surface generated by the curve or to determine the centroid of a plane area if you know the volume of the body generated by the area (see Sample Prob. 5.8).

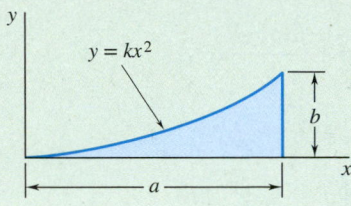

## Sample Problem 5.4

Determine the location of the centroid of a parabolic spandrel by direct integration.

**STRATEGY:**   First express the parabolic curve using the parameters $a$ and $b$. Then, choose a differential element of area and express its area in terms of $a$, $b$, $x$, and $y$. We illustrate the solution first with a vertical element and then a horizontal element.

**MODELING:**

**Determination of the Constant $k$.**   Determine the value of $k$ by substituting $x = a$ and $y = b$ into the given equation. We have $b = ka^2$ or $k = b/a^2$. The equation of the curve is thus

$$y = \frac{b}{a^2}x^2 \qquad \text{or} \qquad x = \frac{a}{b^{1/2}}y^{1/2}$$

**ANALYSIS:**

**Vertical Differential Element.**   Choosing the differential element shown in Fig. 1, the total area of the region is

$$A = \int dA = \int y\,dx = \int_0^a \frac{b}{a^2}x^2\,dx = \left[\frac{b}{a^2}\frac{x^3}{3}\right]_0^a = \frac{ab}{3}$$

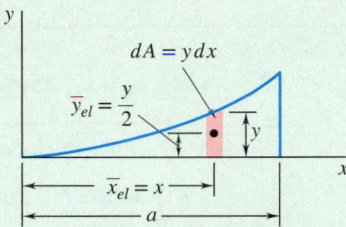

**Fig. 1**   Vertical differential element used to determine the centroid.

The first moment of the differential element with respect to the $y$ axis is $\bar{x}_{el}\,dA$; hence, the first moment of the entire area with respect to this axis is

$$Q_y = \int \bar{x}_{el}\,dA = \int xy\,dx = \int_0^a x\left(\frac{b}{a^2}x^2\right)dx = \left[\frac{b}{a^2}\frac{x^4}{4}\right]_0^a = \frac{a^2 b}{4}$$

Because $Q_y = \bar{x}A$, you have

$$\bar{x}A = \int \bar{x}_{el}\,dA \qquad \bar{x}\frac{ab}{3} = \frac{a^2 b}{4} \qquad \bar{x} = \tfrac{3}{4}a \blacktriangleleft$$

Likewise, the first moment of the differential element with respect to the $x$ axis is $\bar{y}_{el}\,dA$, so the first moment of the entire area about the $x$ axis is

$$Q_x = \int \bar{y}_{el}\,dA = \int \frac{y}{2}y\,dx = \int_0^a \frac{1}{2}\left(\frac{b}{a^2}x^2\right)^2 dx = \left[\frac{b^2}{2a^4}\frac{x^5}{5}\right]_0^a = \frac{ab^2}{10}$$

Because $Q_x = \bar{y}A$, you get

$$\bar{y}A = \int \bar{y}_{el}\,dA \qquad \bar{y}\frac{ab}{3} = \frac{a b^2}{10} \qquad \bar{y} = \tfrac{3}{10}b \blacktriangleleft$$

*(continued)*

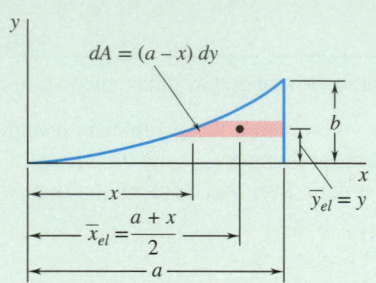

**Fig. 2** Horizontal differential element used to determine the centroid.

**Horizontal Differential Element.** You obtain the same results by considering a horizontal element (Fig. 2). The first moments of the area are

$$Q_y = \int \bar{x}_{el}\, dA = \int \frac{a+x}{2}(a-x)\, dy = \int_0^b \frac{a^2 - x^2}{2}\, dy$$

$$= \frac{1}{2}\int_0^b \left( a^2 - \frac{a^2}{b}y \right) dy = \frac{a^2 b}{4}$$

$$Q_x = \int \bar{y}_{el}\, dA = \int y(a-x)\, dy = \int y\left( a - \frac{a}{b^{1/2}}y^{1/2} \right) dy$$

$$= \int_0^b \left( ay - \frac{a}{b^{1/2}}y^{3/2} \right) dy = \frac{a b^2}{10}$$

To determine $\bar{x}$ and $\bar{y}$, again substitute these expressions into the equations defining the centroid of the area.

**REFLECT and THINK:** You obtain the same results whether you choose a vertical or a horizontal element of area, as you should. You can use both methods as a check against making a mistake in your calculations.

## Sample Problem 5.5

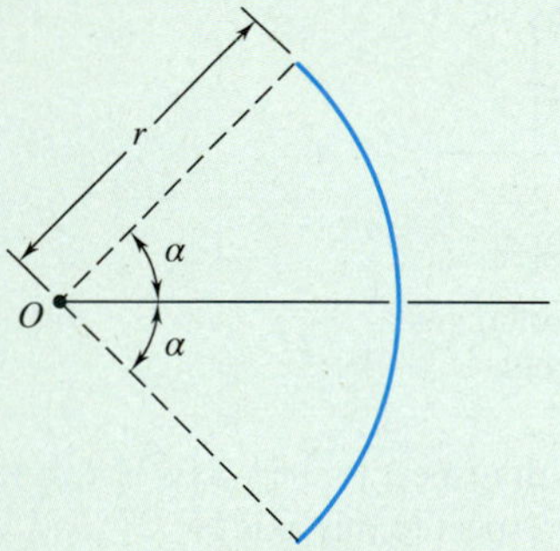

Determine the location of the centroid of the circular arc shown.

**STRATEGY:** For a simple figure with circular geometry, you should use polar coordinates.

**MODELING:** The arc is symmetrical with respect to the $x$ axis, so $\bar{y} = 0$. Choose a differential element, as shown in Fig. 1.

**ANALYSIS:** Determine the length of the arc by integration.

$$L = \int dL = \int_{-\alpha}^{\alpha} r\, d\theta = r\int_{-\alpha}^{\alpha} d\theta = 2r\alpha$$

The first moment of the arc with respect to the $y$ axis is

$$Q_y = \int x\, dL = \int_{-\alpha}^{\alpha}(r\cos\theta)(r\, d\theta) = r^2\int_{-\alpha}^{\alpha}\cos\theta\, d\theta$$

$$= r^2[\sin\theta]_{-\alpha}^{\alpha} = 2r^2\sin\alpha$$

Because $Q_y = \bar{x}L$, you obtain

$$\bar{x}(2r\alpha) = 2r^2\sin\alpha \qquad \bar{x} = \frac{r\sin\alpha}{\alpha} \blacktriangleleft$$

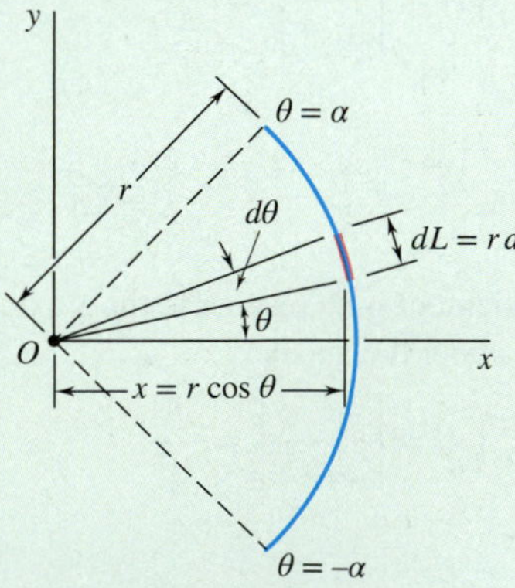

**Fig. 1** Differential element used to determine the centroid.

**REFLECT and THINK:** Observe that this result matches that given for this case in Fig. 5.8B.

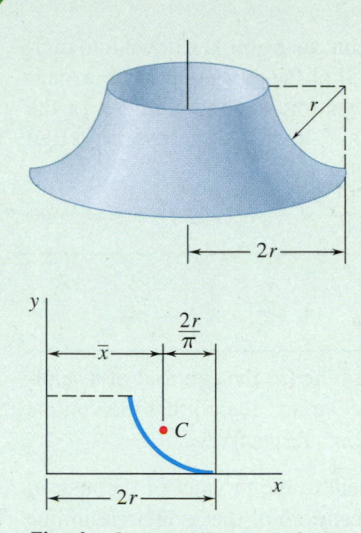

**Fig. 1** Centroid location of arc.

## Sample Problem 5.6

Determine the area of the surface of revolution shown that is obtained by rotating a quarter-circular arc about a vertical axis.

**STRATEGY:** According to the first Pappus-Guldinus theorem, the area of the surface of revolution is equal to the product of the length of the arc and the distance traveled by its centroid.

**MODELING and ANALYSIS:** Referring to Fig. 5.8B and Fig. 1, you have

$$\bar{x} = 2r - \frac{2r}{\pi} = 2r\left(1 - \frac{1}{\pi}\right)$$

$$A = 2\pi\bar{x}L = 2\pi\left[2r\left(1 - \frac{1}{\pi}\right)\right]\left(\frac{\pi r}{2}\right)$$

$$A = 2\pi r^2(\pi - 1) \blacktriangleleft$$

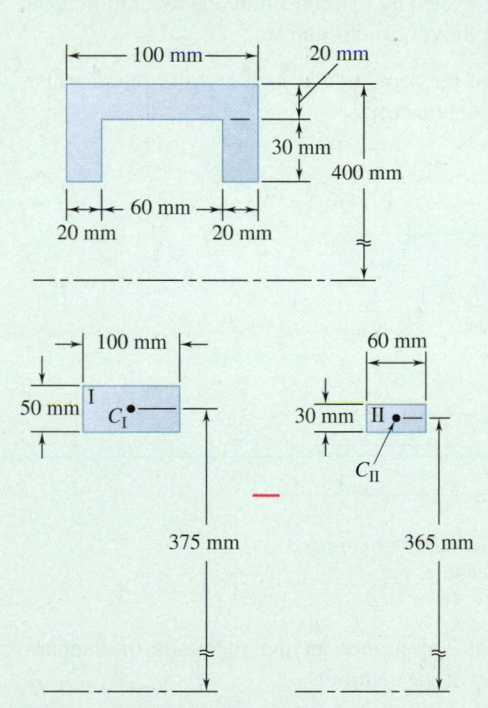

**Fig. 1** Modeling the given area by subtracting area II from area I.

## Sample Problem 5.7

The outside diameter of a pulley is 0.8 m, and the cross section of its rim is as shown. Knowing that the pulley is made of steel and that the density of steel is $\rho = 7.85 \times 10^3$ kg/m$^3$, determine the mass and weight of the rim.

**STRATEGY:** You can determine the volume of the rim by applying the second Pappus-Guldinus theorem, which states that the volume equals the product of the given cross-sectional area and the distance traveled by its centroid in one complete revolution. However, you can find the volume more easily by observing that the cross section can be formed from rectangle I with a positive area and from rectangle II with a negative area (Fig. 1).

**MODELING:** Use a table to keep track of the data, as you did in Sec. 5.1.

| | Area, mm² | $\bar{y}$, mm | Distance Traveled by $C$, mm | Volume, mm³ |
|---|---|---|---|---|
| I | +5000 | 375 | $2\pi(375) = 2356$ | $(5000)(2356) = 11.78 \times 10^6$ |
| II | -1800 | 365 | $2\pi(365) = 2293$ | $(-1800)(2293) = -4.13 \times 10^6$ |
| | | | | Volume of rim = $7.65 \times 10^6$ |

**ANALYSIS:** Because 1 mm = $10^{-3}$ m, you have 1 mm$^{-3}$ = $(10^{-3}$ m$)^3$ = $10^{-9}$ m$^3$. Thus, you obtain $V = 7.65 \times 10^6$ mm$^3$ = $(7.65 \times 10^6)(10^{-9}$ m$^3)$ = $7.65 \times 10^{-3}$ m$^3$.

$$m = \rho V = (7.85 \times 10^3 \text{ kg/m}^3)(7.65 \times 10^{-3} \text{ m}^3) \qquad m = 60.0 \text{ kg} \blacktriangleleft$$

$$W = mg = (60.0 \text{ kg})(9.81 \text{ m/s}^2) = 589 \text{ kg·m/s}^2 \qquad W = 589 \text{ N} \blacktriangleleft$$

*(continued)*

**REFLECT and THINK:** When a cross section can be broken down into multiple common shapes, you can apply Theorem II of Pappus-Guldinus in a manner that involves finding the products of the centroid ($\bar{y}$) and area ($A$), or the first moments of area ($\bar{y}A$), for each shape. Thus, it was not necessary to find the centroid or the area of the overall cross section.

## Sample Problem 5.8

Using the theorems of Pappus-Guldinus, determine (*a*) the centroid of a semicircular area and (*b*) the centroid of a semicircular arc. Recall that the volume and the surface area of a sphere are $\frac{4}{3}\pi r^3$ and $4\pi r^2$, respectively.

**STRATEGY:** The volume of a sphere is equal to the product of the area of a semicircle and the distance traveled by the centroid of the semicircle in one revolution about the *x* axis. Given the volume, you can determine the distance traveled by the centroid and thus the distance of the centroid from the axis. Similarly, the area of a sphere is equal to the product of the length of the generating semicircle and the distance traveled by its centroid in one revolution. You can use this to find the location of the centroid of the arc.

**MODELING:** Draw diagrams of the semicircular area and the semicircular arc (Fig. 1) and label the important geometries.

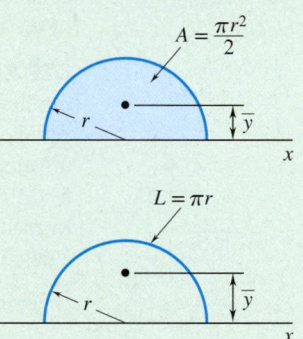

**Fig. 1** Semicircular area and semicircular arc.

**ANALYSIS:** Set up the equalities described in the theorems of Pappus-Guldinus and solve for the location of the centroid.

$$V = 2\pi\bar{y}A \qquad \frac{4}{3}\pi r^3 = 2\pi\bar{y}\left(\frac{1}{2}\pi r^2\right) \qquad \bar{y} = \frac{4r}{3\pi} \blacktriangleleft$$

$$A = 2\pi\bar{y}L \qquad 4\pi r^2 = 2\pi\bar{y}(\pi r) \qquad \bar{y} = \frac{2r}{\pi} \blacktriangleleft$$

**REFLECT and THINK:** Observe that these results match those given for these cases in Fig. 5.8.

# SOLVING PROBLEMS
# ON YOUR OWN

In the problems for this section, you will use the equations

$$\bar{x} = \frac{\int x\,dA}{A} \qquad \bar{y} = \frac{\int y\,dA}{A} \qquad \textbf{(5.3')}$$

$$\bar{x} = \frac{\int x\,dL}{L} \qquad \bar{y} = \frac{\int y\,dL}{L} \qquad \textbf{(5.4')}$$

to locate the centroids of plane areas and lines, respectively. You will also apply the theorems of Pappus-Guldinus to determine the areas of surfaces of revolution and the volumes of bodies of revolution.

**1. Determining the centroids of areas and lines by direct integration.** When solving problems of this type, you should follow the method of solution shown in Sample Probs. 5.4 and 5.5. To compute $A$ or $L$, determine the first moments of the area or the line, and solve Eqs. (5.3') or (5.4') for the coordinates of the centroid. In addition, you should pay particular attention to the following points.

   **a.** Begin your solution by carefully defining or determining each term in the applicable integral formulas. We strongly encourage you to show on your sketch of the given area or line your choice for $dA$ or $dL$ and the distances to its centroid.

   **b.** As explained in Sec. 5.2A, $x$ and $y$ in Eqs. (5.3') and (5.4') represent the *coordinates of the centroid* of the differential elements $dA$ and $dL$. It is important to recognize that the coordinates of the centroid of $dA$ are not equal to the coordinates of a point located on the curve bounding the area under consideration. You should carefully study Fig. 5.12 until you fully understand this important point.

   **c.** To possibly simplify or minimize your computations, always examine the shape of the given area or line before defining the differential element you will use. For example, sometimes it may be preferable to use horizontal rectangular elements instead of vertical ones. Also, it is usually advantageous to use polar coordinates when a line or an area has circular symmetry.

   **d.** Although most of the integrations in this section are straightforward, at times it may be necessary to use more advanced techniques, such as trigonometric substitution or integration by parts. Using a table of integrals is often the fastest method to evaluate difficult integrals.

**2. Applying the theorems of Pappus-Guldinus.** As shown in Sample Probs. 5.6, 5.7, and 5.8, these simple, yet very useful theorems allow you to apply your knowledge of centroids to the computation of areas and volumes. Although the theorems refer to the distance traveled by the centroid and to the length of the generating curve or to the generating area, the resulting equations [Eqs. (5.10) and (5.11)] contain the products of these quantities, which are simply the first moments of a line ($\bar{y}L$) and an area ($\bar{y}A$), respectively. Thus, for those problems for which the generating line or area consists of more than one common shape, you need only determine $\bar{y}L$ or $\bar{y}A$; you do not have to calculate the length of the generating curve or the generating area.

# Problems

**5.34 through 5.36** Determine by direct integration the centroid of the area shown.

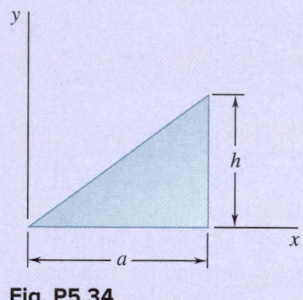

**Fig. P5.34**

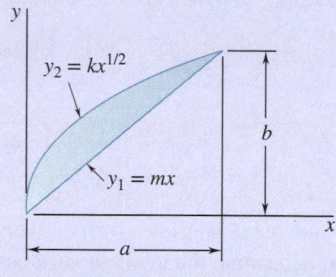

**Fig. P5.35**

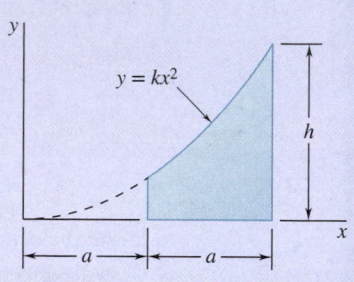

**Fig. P5.36**

**5.37 through 5.39** Determine by direct integration the centroid of the area shown.

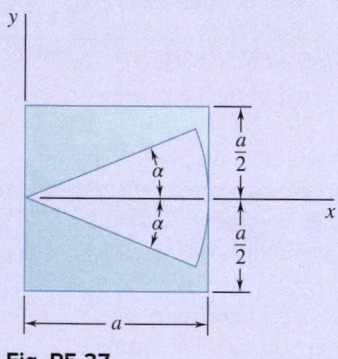

**Fig. P5.37**

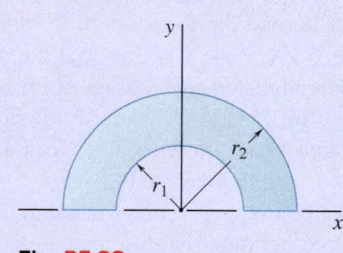

**Fig. P5.38**

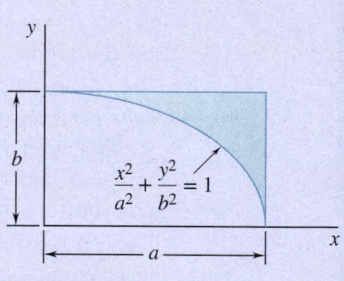

**Fig. P5.39**

**5.40 and 5.41** Determine by direct integration the centroid of the area shown. Express your answer in terms of $a$ and $b$.

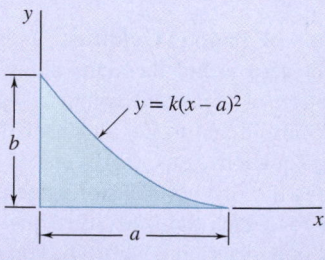

**Fig. P5.40**

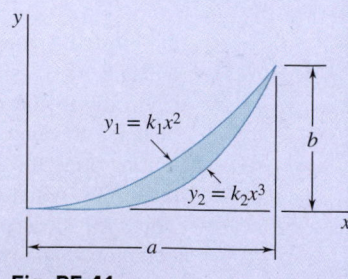

**Fig. P5.41**

**5.42** Determine by direct integration the centroid of the area shown.

**5.43 and 5.44** Determine by direct integration the centroid of the area shown. Express your answer in terms of $a$ and $b$.

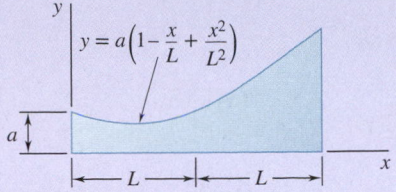

Fig. *P5.42*

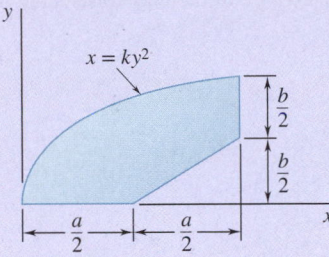

Fig. P5.43

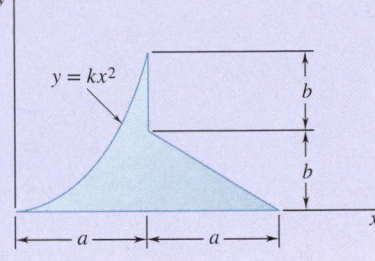

Fig. P5.44

**5.45 and 5.46** A homogeneous wire is bent into the shape shown. Determine by direct integration the $x$ coordinate of its centroid.

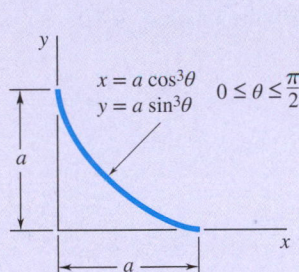

Fig. P5.45

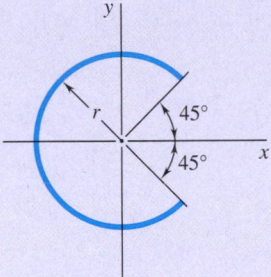

Fig. P5.46

***5.47** A homogeneous wire is bent into the shape shown. Determine by direct integration the $x$ coordinate of its centroid. Express your answer in terms of $a$.

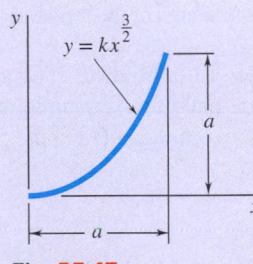

Fig. *P5.47*

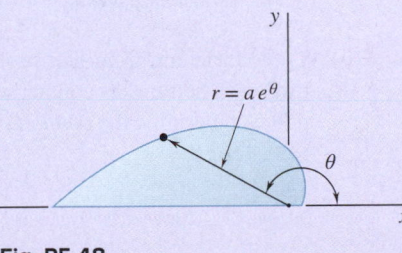

Fig. P5.48

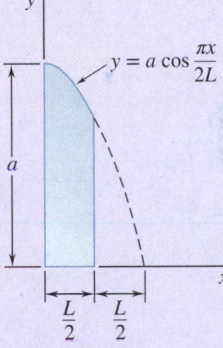

Fig. P5.49

***5.48 and *5.49** Determine by direct integration the centroid of the area shown.

**5.50** Determine the centroid of the area shown in terms of $a$.

**5.51** Determine the centroid of the area shown when $a = 4$ in.

**5.52** Determine the volume and the surface area of the solid obtained by rotating the area of Prob. 5.1 about (a) the line $x = 240$ mm, (b) the $y$ axis.

**5.53** Determine the volume and the surface area of the solid obtained by rotating the area of Prob. 5.7 about (a) the $x$ axis, (b) the $y$ axis.

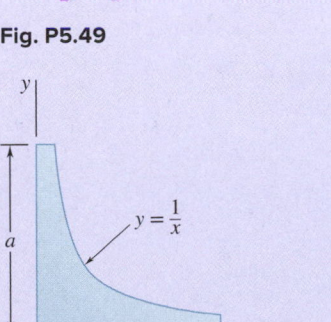

Fig. *P5.50* and P5.51

**5.54** Determine the volume and the surface area of the solid obtained by rotating the area of Prob. 5.6 about (*a*) the line $x = -60$ mm, (*b*) the line $y = 120$ mm.

**5.55** Determine the volume and the surface area of the half-torus shown.

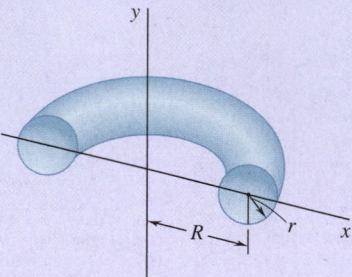

Fig. P5.55

**5.56** Determine the volume of the solid generated by rotating the semiparabolic area shown about (*a*) the *y* axis, (*b*) the *x* axis.

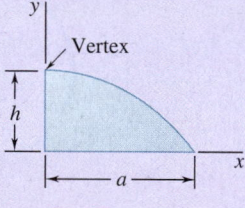

Fig. *P5.56*

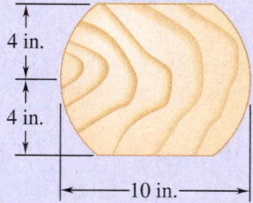

Fig. P5.58

**5.57** Verify that the expressions for the volumes of the first four shapes in Fig. 5.21 are correct.

**5.58** Knowing that two equal caps have been removed from a 10-in.-diameter wooden sphere, determine the total surface area of the remaining portion.

**5.59** Three different drive belt profiles are to be studied. If at any given time each belt makes contact with one-half of the circumference of its pulley, determine the *contact area* between the belt and the pulley for each design.

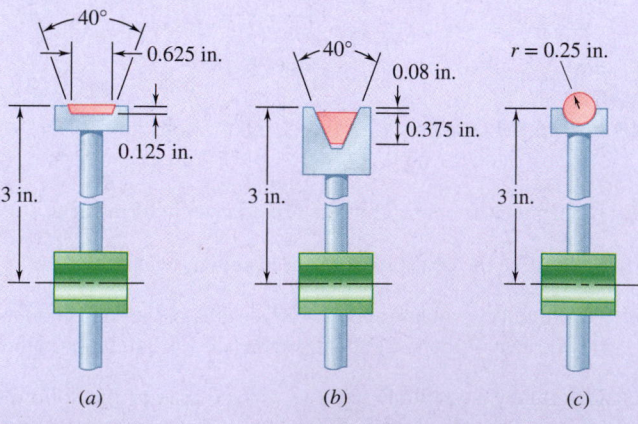

Fig. *P5.59*

**5.60** Determine the capacity, in liters, of the punch bowl shown if $R = 250$ mm.

**Fig. P5.60**

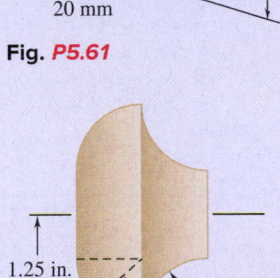

**Fig. P5.61**

**5.61** Determine the volume and total surface area of the bushing shown.

**5.62** Determine the volume and weight of the solid brass knob shown, knowing that the specific weight of brass is 0.306 lb/in³.

**5.63** Determine the total surface area of the solid brass knob shown.

**5.64** Determine the volume of the brass collar obtained by rotating the shaded area shown about the vertical axis $AA'$.

**Fig. P5.62 and P5.63**

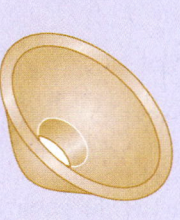

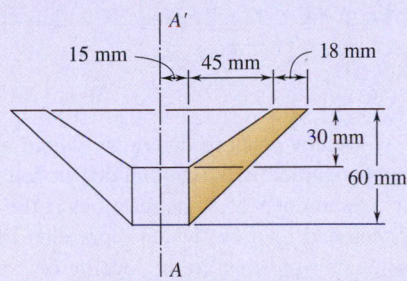

**Fig. P5.64**

*****5.65** The shade for a wall-mounted light is formed from a thin sheet of translucent plastic. Determine the surface area of the outside of the shade, knowing it has the parabolic cross section shown.

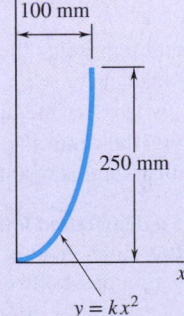

$y = kx^2$

**Fig. P5.65**

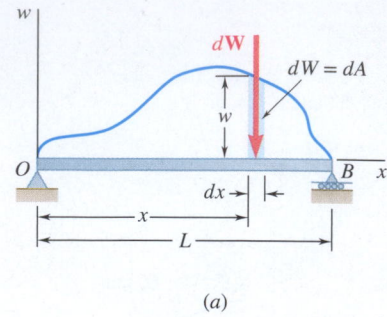

(a)

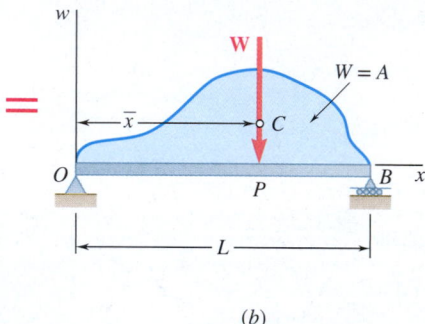

(b)

**Fig. 5.17** (*a*) A load curve representing the distribution of load forces along a horizontal beam, with an element of length *dx*; (*b*) the resultant load *W* has a magnitude equal to the area *A* under the load curve and acts through the centroid of the area.

# 5.3 ADDITIONAL APPLICATIONS OF CENTROIDS

We can use the concept of the center of gravity or the centroid of an area to solve other problems besides those dealing with the weights of flat plates. The same techniques allow us to deal with other kinds of distributed loads on objects, such as the forces on a straight beam (a bridge girder or the main carrying beam of a house floor) or a flat plate under water (the side of a dam or a window in an aquarium tank).

## 5.3A Distributed Loads on Beams

Consider a beam supporting a **distributed load**; this load may consist of the weight of materials supported directly or indirectly by the beam, or it may be caused by wind or hydrostatic pressure. We can represent the distributed load by plotting the load $w$ supported per unit length (Fig. 5.17); this load is expressed in N/m or in lb/ft. The magnitude of the force exerted on an element of the beam with length $dx$ is $dW = w\,dx$, and the total load supported by the beam is

$$W = \int_0^L w\,dx$$

Note that the product $w\,dx$ is equal in magnitude to the element of area $dA$ shown in Fig. 5.17*a*. The load $W$ is thus equal in magnitude to the total area $A$ under the load curve, as

$$W = \int dA = A$$

We now want to determine where a *single concentrated load* $\mathbf{W}$, of the same magnitude $W$ as the total distributed load, should be applied on the beam if it is to produce the same reactions at the supports (Fig. 5.17*b*). However, this concentrated load $\mathbf{W}$, which represents the resultant of the given distributed loading, is equivalent to the loading only when considering the free-body diagram of the entire beam. We obtain the point of application $P$ of the equivalent concentrated load $\mathbf{W}$ by setting the moment of $\mathbf{W}$ about point $O$ equal to the sum of the moments of the elemental loads $d\mathbf{W}$ about $O$. Thus,

$$(OP)W = \int x\,dW$$

Then, because $dW = w\,dx = dA$ and $W = A$, we have

$$(OP)A = \int_0^L x\,dA \qquad \textbf{(5.12)}$$

Because this integral represents the first moment with respect to the $w$ axis of the area under the load curve, we can replace it with the product $\bar{x}A$. We therefore have $OP = \bar{x}$, where $\bar{x}$ is the distance from the $w$ axis to the centroid $C$ of the area $A$ (this is *not* the centroid of the beam).

We can summarize this result:

> **We can replace a distributed load on a beam by a concentrated load; the magnitude of this single load is equal to the area under the load curve, and its line of action passes through the centroid of that area.**

Note, however, that the concentrated load is equivalent to the given loading only so far as external forces are concerned. It can be used to determine reactions, but should not be used to compute internal forces and deflections.

**Photo 5.4** The roof of the building shown must be able to support not only the total weight of the snow but also the nonsymmetric distributed loads resulting from drifting of the snow. ©maurice joseph/Alamy

## *5.3B    Forces on Submerged Surfaces

The approach used for distributed loads on beams works in other applications as well. Here, we use it to determine the resultant of the hydrostatic pressure forces exerted on a *rectangular surface* submerged in a liquid. We can use these methods to determine the resultant of the hydrostatic forces exerted on the surfaces of dams, rectangular gates, and vanes. In Chap. 9, we discuss the resultants of forces on submerged surfaces of variable width.

Consider a rectangular plate with a length of $L$ and width of $b$, where $b$ is measured perpendicular to the plane of the figure (Fig. 5.18). As for the case of distributed loads on a beam, the load exerted on an element of the plate with a length of $dx$ is $w\,dx$, where $w$ is the load per unit length and $x$ is the distance along the length. However, this load also can be expressed as $p\,dA = pb\,dx$, where $p$ is the gage pressure in the liquid[†] and $b$ is the width of the plate; thus, $w = bp$. Because the gage pressure in a liquid is $p = \gamma h$, where $\gamma$ is the specific weight of the liquid and $h$ is the vertical distance from the free surface, it follows that

$$w = bp = b\gamma h \qquad (5.13)$$

This equation shows that the load per unit length $w$ is proportional to $h$ and, thus, varies linearly with $x$.

From the results of Sec. 5.3A, the resultant **R** of the hydrostatic forces exerted on one side of the plate is equal in magnitude to the trapezoidal area under the load curve, and its line of action passes through the centroid $C$ of that area. The point $P$ of the plate where **R** is applied is known as the *center of pressure*.[‡]

Now consider the forces exerted by a liquid on a curved surface of constant width (Fig. 5.19a). Because determining the resultant **R** of these forces by direct integration would not be easy, we consider the free body obtained by detaching the volume of liquid $ABD$ bounded by the curved surface $AB$ and by the two plane surfaces $AD$ and $DB$ shown in Fig. 5.19b. The forces acting on the free body $ABD$ are the weight **W** of the detached volume of liquid, the resultant $\mathbf{R}_1$ of the forces exerted on $AD$, the resultant $\mathbf{R}_2$ of the forces exerted on $BD$, and the resultant $-\mathbf{R}$ of the forces exerted *by the curved surface on the liquid*. The resultant $-\mathbf{R}$ is both equal and opposite to and has the same line of action as the resultant **R** of the forces exerted *by the liquid on the curved surface*. We can determine the forces **W**, $\mathbf{R}_1$, and $\mathbf{R}_2$ by standard methods. After their values have been found, we obtain the force $-\mathbf{R}$ by solving the equations of equilibrium for the free body of Fig. 5.19b. The resultant **R** of the hydrostatic forces exerted on the curved surface is just the reverse of $-\mathbf{R}$.

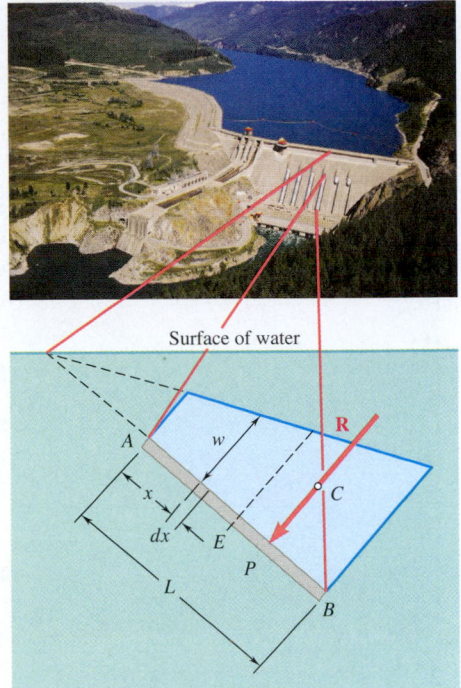

**Fig. 5.18**  The waterside face of a hydroelectric dam can be modeled as a rectangular plate submerged under water. Shown is a side view of the plate. ©North Light Images/agefotostock

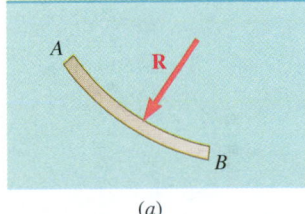

(a)

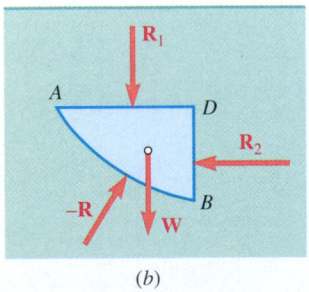

(b)

**Fig. 5.19**  (a) Force $R$ exerted by a liquid on a submerged curved surface of constant width; (b) free-body diagram of the volume of liquid $ABD$.

[†]The pressure $p$, which represents a load per unit area, is measured in N/m² or in lb/ft². The derived SI unit N/m² is called a pascal (Pa).

[‡]The area under the load curve is equal to $w_E L$, where $w_E$ is the load per unit length at the center $E$ of the plate. Then from Eq. (5.13), we have

$$R = w_E L = (bp_E)\,L = p_E\,(bL) = p_E A$$

where $A$ denotes the area of the plate. Thus, we can obtain the magnitude of **R** by multiplying the area of the plate by the pressure at its center $E$. Note, however, that the resultant **R** *should be applied at P, not at E.*

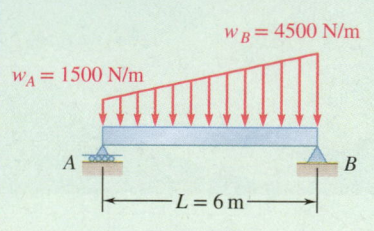

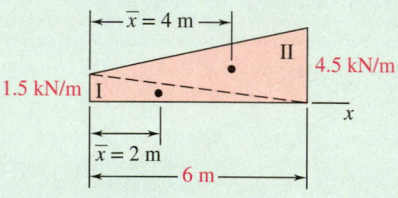

**Fig. 1**    The load modeled as two triangular areas.

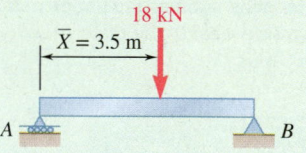

**Fig. 2**    Equivalent concentrated load.

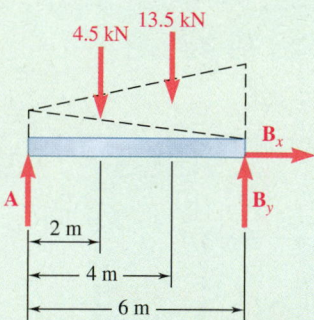

**Fig. 3**    Free-body diagram of the beam.

## Sample Problem 5.9

A beam supports a distributed load as shown. (*a*) Determine the equivalent concentrated load. (*b*) Determine the reactions at the supports.

**STRATEGY:**    The magnitude of the resultant of the load is equal to the area under the load curve, and the line of action of the resultant passes through the centroid of the same area. Break down the area into pieces for easier calculation, and determine the resultant load. Then, use the calculated forces or their resultant to determine the reactions.

**MODELING and ANALYSIS:**

**a. Equivalent Concentrated Load.**    Divide the area under the load curve into two triangles (Fig. 1), and construct the table below. To simplify the computations and tabulation, the given loads per unit length have been converted into kN/m.

| Component | $A$, kN | $\bar{x}$, m | $\bar{x}A$, kN·m |
|---|---|---|---|
| Triangle I | 4.5 | 2 | 9 |
| Triangle II | 13.5 | 4 | 54 |
| | $\Sigma A = 18.0$ | | $\Sigma \bar{x}A = 63$ |

Thus, $\bar{X}\Sigma A = \Sigma \bar{x}A$:    $\bar{X}(18 \text{ kN}) = 63 \text{ kN·m}$    $\bar{X} = 3.5 \text{ m}$

The equivalent concentrated load (Fig. 2) is

$$\mathbf{W} = 18 \text{ kN} \downarrow \quad \blacktriangleleft$$

Its line of action is located at a distance

$$\bar{X} = 3.5 \text{ m to the right of } A \quad \blacktriangleleft$$

**b. Reactions.**    The reaction at $A$ is vertical and is denoted by **A**. Represent the reaction at $B$ by its components $\mathbf{B}_x$ and $\mathbf{B}_y$. Consider the given load to be the sum of two triangular loads (see the free-body diagram, Fig. 3). The resultant of each triangular load is equal to the area of the triangle and acts at its centroid.

Write the following equilibrium equations from the free-body diagram:

$$\xrightarrow{+} \Sigma F_x = 0: \qquad\qquad\qquad\qquad\qquad B_x = 0 \quad \blacktriangleleft$$

$$+\circlearrowleft \Sigma M_A = 0: \quad -(4.5 \text{ kN})(2 \text{ m}) - (13.5 \text{ kN})(4 \text{ m}) + B_y(6 \text{ m}) = 0$$

$$B_y = 10.5 \text{ kN} \uparrow \quad \blacktriangleleft$$

$$+\circlearrowleft \Sigma M_B = 0: \quad +(4.5 \text{ kN})(4 \text{ m}) + (13.5 \text{ kN})(2 \text{ m}) - A(6 \text{ m}) = 0$$

$$A = 7.5 \text{ kN} \uparrow \quad \blacktriangleleft$$

**REFLECT and THINK:**    You can replace the given distributed load by its resultant, which you found in part *a*. Then, you can determine the reactions from the equilibrium equations $\Sigma F_x = 0$, $\Sigma M_A = 0$, and $\Sigma M_B = 0$. Again, the results are

$$\mathbf{B}_x = 0 \qquad \mathbf{B}_y = 10.5 \text{ kN} \uparrow \qquad \mathbf{A} = 7.5 \text{ kN} \uparrow \quad \blacktriangleleft$$

# Sample Problem 5.10

The cross section of a concrete dam is shown. Consider a 1-ft-thick section of the dam, and determine (*a*) the resultant of the reaction forces exerted by the ground on the base *AB* of the dam, (*b*) the resultant of the pressure forces exerted by the water on the face *BC* of the dam. The specific weights of concrete and water are 150 lb/ft³ and 62.4 lb/ft³, respectively.

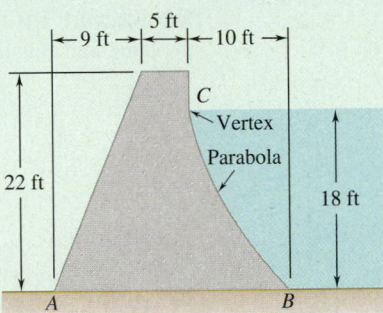

**STRATEGY:**  Draw a free-body diagram of the section of the dam, breaking it into parts to simplify the calculations. Model the resultant of the reactions as a force-couple system at *A*. Use the method described in Sec. 5.3B to find the force exerted by the dam on the water and reverse it to find the force exerted by the water on face *BC*.

## MODELING and ANALYSIS:

**a. Ground Reaction.**  Choose as a free body the 1-ft-thick section *AEFCDB* of the dam and water (Fig. 1). The reaction forces exerted by the ground on the base *AB* are represented by an equivalent force-couple system at *A*. Other forces acting on the free body are the weight of the dam represented by the weights of its components $W_1$, $W_2$, and $W_3$; the weight of the water $W_4$; and the resultant **P** of the pressure forces exerted on section *BD* by the water to the right of section *BD*.

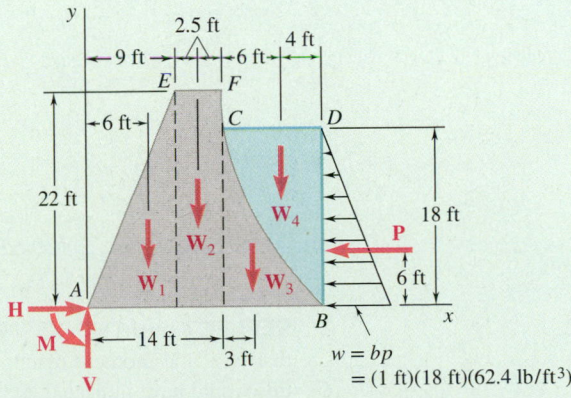

**Fig. 1**  Free-body diagram of the dam and water.

*(continued)*

Calculate each of the forces that appear in the free-body diagram (Fig. 1):

$$W_1 = \tfrac{1}{2}(9 \text{ ft})(22 \text{ ft})(1 \text{ ft})(150 \text{ lb/ft}^3) = 14{,}850 \text{ lb}$$
$$W_2 = (5 \text{ ft})(22 \text{ ft})(1 \text{ ft})(150 \text{ lb/ft}^3) = 16{,}500 \text{ lb}$$
$$W_3 = \tfrac{1}{3}(10 \text{ ft})(18 \text{ ft})(1 \text{ ft})(150 \text{ lb/ft}^3) = 9000 \text{ lb}$$
$$W_4 = \tfrac{2}{3}(10 \text{ ft})(18 \text{ ft})(1 \text{ ft})(62.4 \text{ lb/ft}^3) = 7488 \text{ lb}$$
$$P = \tfrac{1}{2}(18 \text{ ft})(1 \text{ ft})(18 \text{ ft})(62.4 \text{ lb/ft}^3) = 10{,}109 \text{ lb}$$

**Equilibrium Equations**    Write the equilibrium equations for the section of the dam, and calculate the forces and moment labeled at $A$ in Fig. 1.

$$\xrightarrow{+} \Sigma F_x = 0: \qquad H - 10{,}109 \text{ lb} = 0 \qquad\qquad\qquad \mathbf{H = 10{,}110 \text{ lb}} \rightarrow \;\blacktriangleleft$$

$$+\uparrow \Sigma F_y = 0: \qquad V - 14{,}850 \text{ lb} - 16{,}500 \text{ lb} - 9000 \text{ lb} - 7488 \text{ lb} = 0$$

$$\mathbf{V = 47{,}840 \text{ lb}} \uparrow \;\blacktriangleleft$$

$$+\circlearrowleft \Sigma M_A = 0: \qquad -(14{,}850 \text{ lb})(6 \text{ ft}) - (16{,}500 \text{ lb})(11.5 \text{ ft})$$
$$- (9000 \text{ lb})(17 \text{ ft}) - (7488 \text{ lb})(20 \text{ ft}) + (10{,}109 \text{ lb})(6 \text{ ft}) + M = 0$$

$$\mathbf{M = 520{,}960 \text{ lb·ft}} \circlearrowleft \;\blacktriangleleft$$

You can replace the force-couple system by a single force acting at a distance $d$ to the right of $A$, where

$$d = \frac{520{,}960 \text{ lb·ft}}{47{,}840 \text{ lb}} = 10.89 \text{ ft}$$

**b. Resultant R of Water Forces.**    Draw a free-body diagram for the parabolic section of water $BCD$ (Fig. 2). The forces involved are the resultant $-\mathbf{R}$ of the forces exerted by the dam on the water, the weight $\mathbf{W}_4$, and the force $\mathbf{P}$. Because these forces must be concurrent, $-\mathbf{R}$ passes through the point of intersection $G$ of $\mathbf{W}_4$ and $\mathbf{P}$. Draw a force triangle to determine the magnitude and direction of $-\mathbf{R}$. The resultant $\mathbf{R}$ of the forces exerted by the water on the face $BC$ is equal and opposite. Hence,

$$\mathbf{R = 12{,}580 \text{ lb}} \;\nwarrow\; 36.5° \;\blacktriangleleft$$

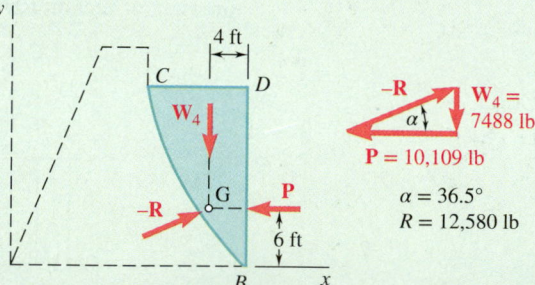

$$\alpha = 36.5°$$
$$R = 12{,}580 \text{ lb}$$

**Fig. 2**    Free-body diagram of the parabolic section of water $BCD$.

**REFLECT and THINK:**    Note that if you found the distance $d$ to be negative— that is, if the moment reaction at $A$ had been acting in the opposite direction— this would have indicated an instability condition of the dam. In this situation, the effects of the water pressure would overcome the weight of the dam, causing it to tip about $A$.

# Case Study 5.1

For structures consisting of multi-level roofs, snow on the lower roofs often tends to drift toward the upper building structure, such as is shown in CS Photo 5.1. ASCE Standard 7*, used by structural engineers in the U.S., addresses such drifting in its requirements governing snow loads. For example, let's consider an enclosed parking structure located in Duluth, Minnesota, that has a 10-m-long flat roof attached to a taller building. The roof is supported by beams equally spaced at 1.2 m. Following the provisions of ASCE 7 as a guide, we will assume a combination of building geometry, wind exposure, roof thermal characteristics, and function of structure that result in the drifting snow load shown in CS Fig. 5.1. For this distributed loading, let's determine the magnitude and location of the resultant load that acts on one of the beams. Then, treating the beam's connection at $B$ as a roller and the connection at $C$ as a pin, we'll determine the support reactions at these locations.

**CS Photo 5.1**  Drifting snow on a lower roof, with depth increasing toward the upper structure.

©Jessica Puckett

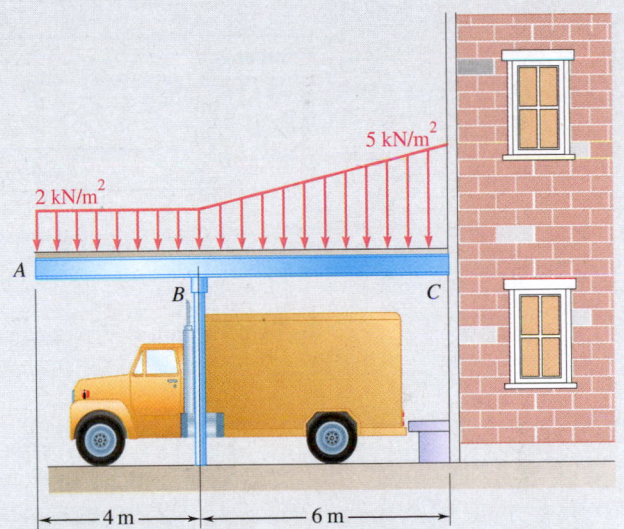

**CS Fig. 5.1**  Enclosed parking structure with the design snowdrift roof load.

## STRATEGY:

Because the beams are equally spaced, the load applied along the length of one beam will be the given roof load multiplied by the 1.2-m beam spacing. (In other words, this is the portion of the total roof load that acts on one of the beams.) CS Fig. 5.2 shows the resulting load, along with the support conditions to be assumed for the beam. The magnitude of this distributed load's resultant is the area under the load curve, and its location is the centroid of the load area. The total load area can be broken up into smaller areas of simple geometry and used

*Source: "ASCE/SEI 7-05," *Minimum Design Loads for Buildings and Other Structures*, Reston, VA: American Society of Civil Engineers, 2005, Ch. 7.

*(continued)*

to calculate the magnitude and location of the resultant. This resultant can then be applied to a free-body diagram of the beam to determine the support reactions.

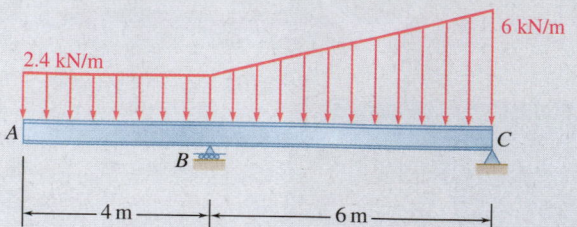

**CS Fig. 5.2**   Single roof beam showing its distributed loading and given support conditions.

## MODELING and ANALYSIS:

### a. Equivalent Concentrated Load.
Divide the area under the load curve into a rectangle and a triangle (CS Fig. 5.3), and construct the table below.

| Component | $A$, kN | $\bar{x}$, m | $\bar{x}A$, kN·m |
|---|---|---|---|
| Area I | 24 | 5 | 120 |
| Area II | 10.8 | 8 | 86.4 |
| | $\Sigma A = 34.8$ | | $\Sigma \bar{x}A = 206.4$ |

The equivalent concentrated load (CS Fig. 5.4) is

$$\mathbf{W} = 34.8 \text{ kN} \blacktriangleleft$$

Its line of action is located at a distance

$$\overline{X} = 5.93 \text{ m to the right of } A \blacktriangleleft$$

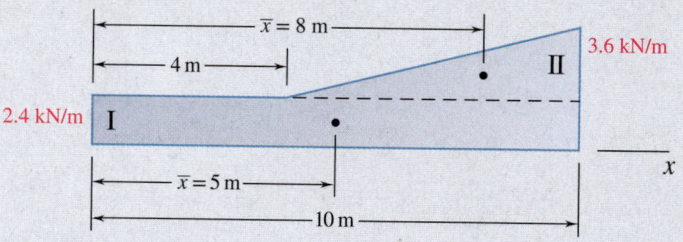

**CS Fig. 5.3**   The load modeled as a triangular and a rectangular area.

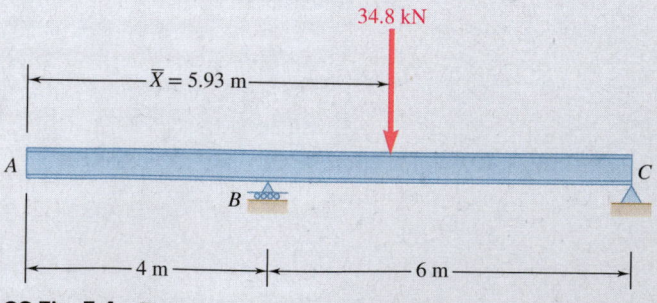

**CS Fig. 5.4**   Equivalent concentrated load.

*(continued)*

**b. Reactions.**  The reaction at $B$ is vertical and is denoted by $\mathbf{B}$. The reaction at $C$ is represented by the components $\mathbf{C}_x$ and $\mathbf{C}_y$. Referring to a free-body diagram of the beam with the applied resultant load (CS Fig. 5.5), write the following equilibrium equations:

$+\rightarrow \Sigma F_x = 0:$ $\qquad\qquad\qquad\qquad\qquad\qquad\qquad\qquad\qquad$ $\mathbf{C}_x = 0$ ◄

$+\circlearrowleft\Sigma M_C = 0: \; +(34.8 \text{ kN})(10 \text{ m} - 5.93 \text{ m}) - B(6 \text{ m}) = 0$ $\quad$ $\mathbf{B} = 23.6 \text{ kN} \uparrow$ ◄

$+\uparrow\Sigma F_y = 0: \; -(34.8 \text{ kN}) + (23.6 \text{ kN}) + C_y = 0$ $\qquad$ $\mathbf{C}_y = 11.2 \text{ kN} \uparrow$ ◄

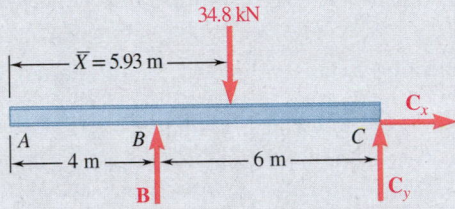

**CS Fig. 5.5**  Free-body diagram of the beam with equivalent concentrated load.

**REFLECT and THINK:**

Both natural and human mechanisms can result in the partial removal of snow loadings, often leading to worsened effects for certain parameters in a structure. Therefore, ASCE 7 also requires the consideration of partial snow loadings. We will revisit this scenario in Case Study 7.1 to study the effects of the loading examined here, as well as the potential effects of partial loadings.

# SOLVING PROBLEMS ON YOUR OWN

The problems in this section involve two common and very important types of loading: distributed loads on beams and forces on submerged surfaces of constant width. As we discussed in Sec. 5.3 and illustrated in Sample Probs. 5.9 and 5.10, determining the single equivalent force for each of these loadings requires a knowledge of centroids.

**1. Analyzing beams subjected to distributed loads.** In Sec. 5.3A, we showed that a distributed load on a beam can be replaced by a single equivalent force. The magnitude of this force is equal to the area under the distributed load curve, and its line of action passes through the centroid of that area. Thus, you should begin solving this kind of problem by replacing the various distributed loads on a given beam by their respective single equivalent forces. You can then determine the reactions at the supports of the beam by using the methods of Chap. 4.

When possible, divide complex distributed loads into the common-shape areas shown in Fig. 5.8A (Sample Prob. 5.9). You can replace each of these areas under the loading curve by a single equivalent force. If required, you can further reduce the system of equivalent forces to a single equivalent force. As you study Sample Prob. 5.9, note how we used the analogy between force and area under the loading curve and applied the techniques for locating the centroid of a composite area to analyze a beam subjected to a distributed load.

**2. Solving problems involving forces on submerged bodies.** Remember the following points and techniques when solving problems of this type.

   **a.** The pressure $p$ at a depth $h$ below the free surface of a liquid is equal to $\gamma h$ or $\rho g h$, where $\gamma$ and $\rho$ are the specific weight and the density of the liquid, respectively. The load per unit length $w$ acting on a submerged surface of constant width $b$ is then

$$w = bp = b\gamma h = b\rho g h$$

   **b.** The line of action of the resultant force **R** acting on a submerged plane surface is perpendicular to the surface.

   **c.** For a vertical or inclined plane rectangular surface with a width of $b$, you can represent the loading on the surface using a linearly distributed load that is trapezoidal in shape (Fig. 5.18). The magnitude of the resultant **R** is given by

$$R = \gamma h_E A$$

where $h_E$ is the vertical distance to the center of the surface and $A$ is the area of the surface.

   **d.** The load curve is triangular (rather than trapezoidal) when the top edge of a plane rectangular surface coincides with the free surface of the liquid, because the pressure of the liquid at the free surface is zero. For this case, it is straightforward to determine the line of action of **R**, because it passes through the centroid of a *triangular* distributed load.

**e.** For the general case, rather than analyzing a trapezoid, we suggest you use the method indicated in part *b* of Sample Prob. 5.9. First, divide the trapezoidal distributed load into two triangles, and then compute the magnitude of the resultant of each triangular load. (The magnitude is equal to the area of the triangle times the width of the plate.) Note that the line of action of each resultant force passes through the centroid of the corresponding triangle and that the sum of these forces is equivalent to **R**. Thus, rather than using **R**, you can use the two equivalent resultant forces whose points of application are easily calculated. You should use the equation given for *R* here in paragraph **c** when you need only the magnitude of **R**.

**f.** When the submerged surface of a constant width is curved, you can obtain the resultant force acting on the surface by considering the equilibrium of the volume of liquid bounded by the curved surface and by using horizontal and vertical planes (Fig. 5.19). Observe that the force **R**₁ of Fig. 5.19 is equal to the weight of the liquid lying above the plane *AD*. The method of solution for problems involving curved surfaces is shown in part *b* of Sample Prob. 5.10.

In subsequent mechanics courses (in particular, mechanics of materials and fluid mechanics), you will have ample opportunity to use the ideas introduced in this section.

# Problems

**5.66 and 5.67**  For the beam and loading shown, determine (*a*) the magnitude and location of the resultant of the distributed load, (*b*) the reactions at the beam supports.

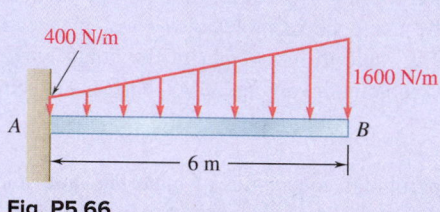

**Fig. P5.66**

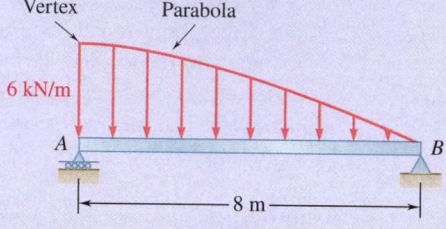

**Fig. P5.67**

**5.68 through 5.73**  Determine the reactions at the beam supports for the given loading.

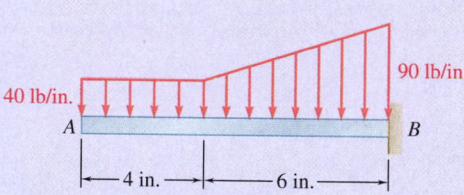

**Fig. P5.68**

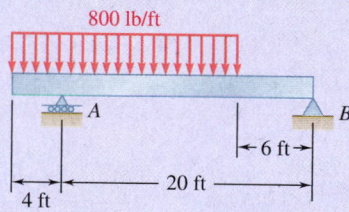

**Fig. P5.69**

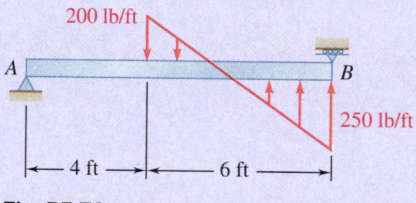

**Fig. P5.70**

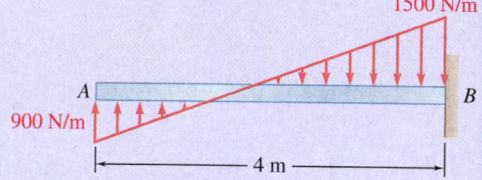

**Fig. P5.71**

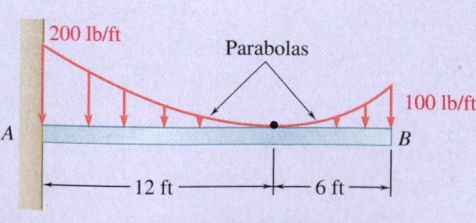

**Fig. P5.72**

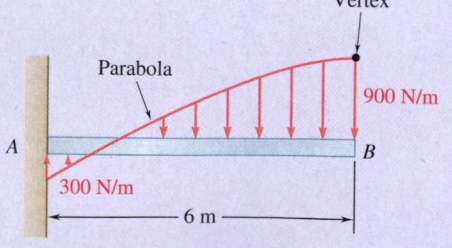

**Fig. P5.73**

**5.74** Determine (a) the distance a so that the vertical reactions at supports A and B are equal, (b) the corresponding reactions at the supports.

**5.75** Determine (a) the distance a so that the reaction at support B is minimum, (b) the corresponding reactions at the supports.

**5.76** Determine the reactions at the beam supports for the given loading when $w_0 = 150$ lb/ft.

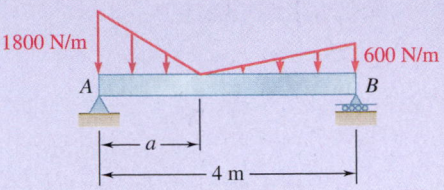

**Fig. P5.74 and *P5.75***

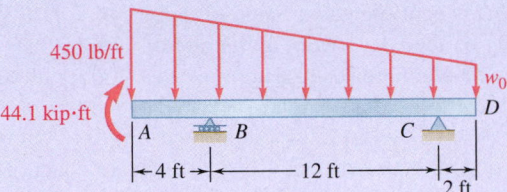

**Fig. P5.76 and P5.77**

**5.77** Determine (a) the distributed load $w_0$ at the end D of the beam ABCD for which the reaction at B is zero, (b) the corresponding reaction at C.

**5.78** The beam AB supports two concentrated loads and rests on soil that exerts a linearly distributed upward load as shown. Determine the values of $w_A$ and $w_B$ corresponding to equilibrium.

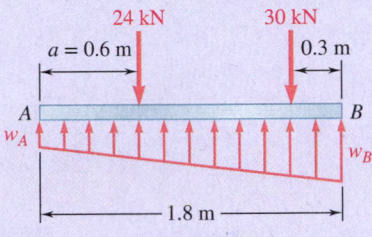

**Fig. P5.78**

**5.79** For the beam and loading of Prob. 5.78, determine (a) the distance a for which $w_A = 20$ kN/m, (b) the corresponding value of $w_B$.

In the following problems, use $\gamma = 62.4$ lb/ft³ for the specific weight of fresh water and $\gamma_c = 150$ lb/ft³ for the specific weight of concrete if U.S. customary units are used. With SI units, use $\rho = 10^3$ kg/m³ for the density of fresh water and $\rho_c = 2.40 \times 10^3$ kg/m³ for the density of concrete. (See the **footnote** on page 236 for how to determine the specific weight of a material given its density.)

**5.80** The cross section of a concrete dam is as shown. For a 1-ft-wide dam section, determine (a) the resultant of the reaction forces exerted by the ground on the base AB of the dam, (b) the point of application of the resultant of part a, (c) the resultant of the pressure forces exerted by the water on the face BC of the dam.

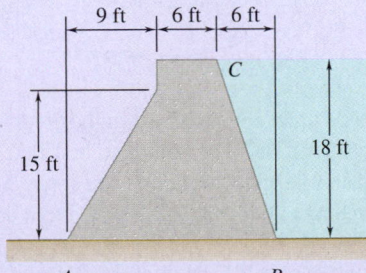

**Fig. P5.80**

**5.81** The cross section of a concrete dam is as shown. For a 1-m-wide dam section determine (a) the resultant of the reaction forces exerted by the ground on the base AB of the dam, (b) the point of application of the resultant of part a, (c) the resultant of the pressure forces exerted by the water on the face BC of the dam.

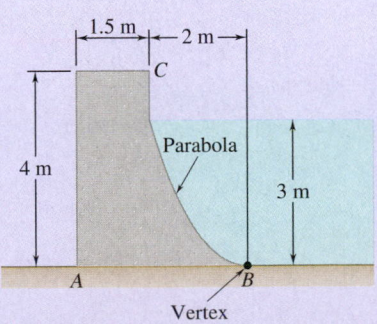

**Fig. P5.81**

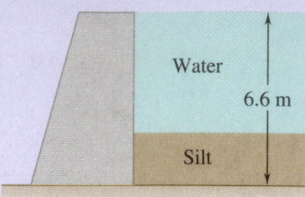

**Fig. P5.82 and *P5.83***

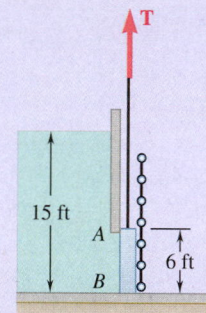

**Fig. P5.84**

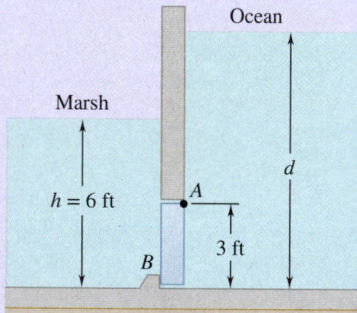

**Fig. P5.85**

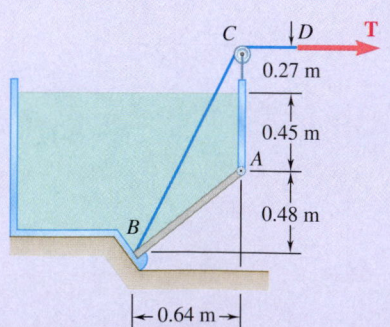

**Fig. P5.88 and P5.89**

**5.82** The dam for a lake is designed to withstand the additional force caused by silt that has settled on the lake bottom. Assuming that silt is equivalent to a liquid of density $\rho_s = 1.76 \times 10^3$ kg/m$^3$ and considering a 1-m-wide section of dam, determine the percentage increase in the force acting on the dam face for a silt accumulation of depth 2 m.

**5.83** The base of a dam for a lake is designed to resist up to 120 percent of the horizontal force of the water. After construction, it is found that silt (that is equivalent to a liquid of density $\rho_s = 1.76 \times 10^3$ kg/m$^3$) is settling on the lake bottom at the rate of 12 mm/year. Considering a 1-m-wide section of dam, determine the number of years of service until the dam becomes unsafe.

**5.84** The friction force between a 6 × 6-ft square sluice gate *AB* and its guides is equal to 10 percent of the resultant of the pressure forces exerted by the water on the face of the gate. Determine the initial force needed to lift the gate if it weighs 1000 lb.

**5.85** A freshwater marsh is drained to the ocean through an automatic tide gate that is 4 ft wide and 3 ft high. The gate is held by hinges located along its top edge at *A* and bears on a sill at *B*. If the water level in the marsh is *h* = 6 ft, determine the ocean level *d* for which the gate will open. (Specific weight of salt water = 64 lb/ft$^3$.)

**5.86** The 3 × 4-m side *AB* of a tank is hinged at its bottom *A* and is held in place by a thin rod *BC*. The maximum tensile force the rod can withstand without breaking is 200 kN, and the design specifications require the force in the rod not to exceed 20 percent of this value. If the tank is slowly filled with water, determine the maximum allowable depth of water *d* in the tank.

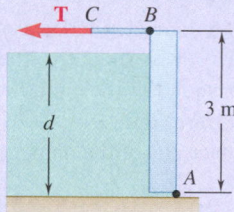

**Fig. *P5.86* and P5.87**

**5.87** The 3 × 4-m side of an open tank is hinged at its bottom *A* and is held in place by a thin rod *BC*. The tank is to be filled with glycerine with a density of 1263 kg/m$^3$. Determine the force **T** in the rod and the reaction at the hinge after the tank is filled to a depth of 2.9 m.

**5.88** A 0.5 × 0.8-m gate *AB* is located at the bottom of a tank filled with water. The gate is hinged along its top edge *A* and rests on a frictionless stop at *B*. Determine the reactions at *A* and *B* when cable *BCD* is slack.

**5.89** A 0.5 × 0.8-m gate *AB* is located at the bottom of a tank filled with water. The gate is hinged along its top edge *A* and rests on a frictionless stop at *B*. Determine the minimum tension required in cable *BCD* to open the gate.

**5.90** A 4 × 2-ft gate is hinged at $A$ and is held in position by rod $CD$. End $D$ rests against a spring whose constant is 828 lb/ft. The spring is undeformed when the gate is vertical. Assuming that the force exerted by rod $CD$ on the gate remains horizontal, determine the minimum depth of water $d$ for which the bottom $B$ of the gate will move to the end of the cylindrical portion of the floor.

**5.91** Solve Prob. 5.90 if the gate weighs 1000 lb.

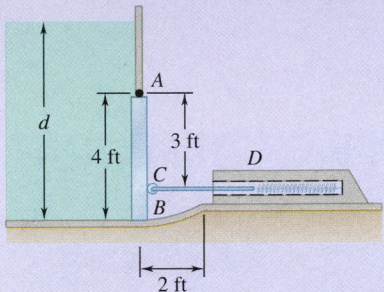

**Fig. P5.90**

**5.92** A prismatically shaped gate placed at the end of a freshwater channel is supported by a pin and bracket at $A$ and rests on a frictionless support at $B$. The pin is located at a distance $h = 0.10$ m below the center of gravity $C$ of the gate. Determine the depth of water $d$ for which the gate will open.

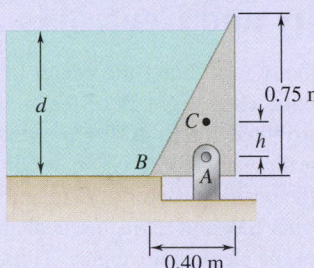

**Fig. P5.92 and P5.93**

**5.93** A prismatically shaped gate placed at the end of a freshwater channel is supported by a pin and bracket at $A$ and rests on a frictionless support at $B$. The pin is located at a distance $h$ below the center of gravity $C$ of the gate. Determine the distance $h$ if the gate is to open when $d = 0.75$ m.

**5.94** A long trough is supported by a continuous hinge along its lower edge and by a series of horizontal cables attached to its upper edge. Determine the tension in each of the cables at a time when the trough is completely full of water.

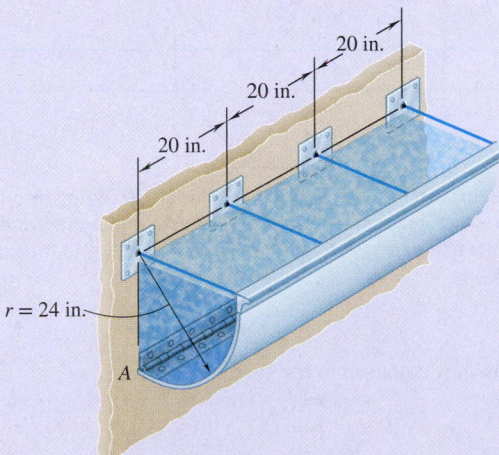

**Fig. P5.94**

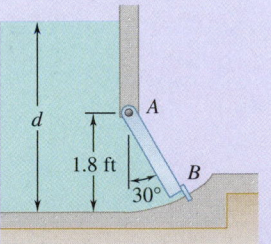

**Fig. P5.95**

**5.95** The square gate $AB$ is held in the position shown by hinges along its top edge $A$ and by a shear pin at $B$. For a depth of water $d = 3.5$ ft, determine the force exerted on the gate by the shear pin.

275

# 5.4 CENTERS OF GRAVITY AND CENTROIDS OF VOLUMES

So far in this chapter, we have dealt with finding centers of gravity and centroids of two-dimensional areas and objects such as flat plates and plane surfaces. However, the same ideas apply to three-dimensional objects as well. The most general situations require the use of multiple integration for analysis, but we can often use symmetry considerations to simplify the calculations. In this section, we show how to do this.

## 5.4A Three-Dimensional Centers of Gravity and Centroids

For a three-dimensional body, we obtain the center of gravity $G$ by dividing the body into small elements. The weight $\mathbf{W}$ of the body acting at $G$ is equivalent to the system of distributed forces $\Delta\mathbf{W}$ representing the weights of the small elements. Choosing the $y$ axis to be vertical with positive sense upward (Fig. 5.20) and denoting the position vector of $G$ to be $\bar{\mathbf{r}}$, we set $\mathbf{W}$ equal to the sum of the elemental weights $\Delta\mathbf{W}$ and set its moment about $O$ equal to the sum of the moments about $O$ of the elemental weights. Thus,

$$\Sigma\mathbf{F}: \qquad\qquad -W\mathbf{j} = \Sigma(-\Delta W\mathbf{j})$$

$$\Sigma\mathbf{M}_O: \qquad\qquad \bar{\mathbf{r}} \times (-W\mathbf{j}) = \Sigma[\mathbf{r} \times (-\Delta W\mathbf{j})] \qquad\qquad \textbf{(5.14)}$$

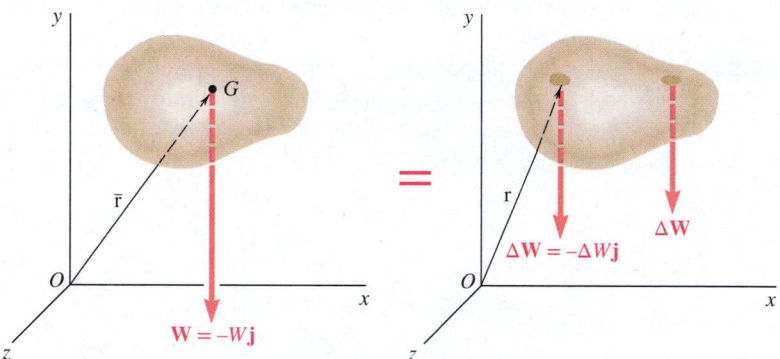

**Fig. 5.20** For a three-dimensional body, the weight $\mathbf{W}$ acting through the center of gravity $G$ and its moment about $O$ is equivalent to the system of distributed weights acting on all the elements of the body and the sum of their moments about $O$.

We can rewrite the last equation in the form

$$\bar{\mathbf{r}}W \times (-\mathbf{j}) = (\Sigma\mathbf{r}\,\Delta W) \times (-\mathbf{j}) \qquad\qquad \textbf{(5.15)}$$

From these equations, we can see that the weight $\mathbf{W}$ of the body is equivalent to the system of the elemental weights $\Delta\mathbf{W}$ if the following conditions are satisfied:

$$W = \Sigma\,\Delta W \qquad\qquad \bar{\mathbf{r}}W = \Sigma\mathbf{r}\,\Delta W$$

Increasing the number of elements and simultaneously decreasing the size of each element, we obtain in the limit as

**Weight, center of gravity of a three-dimensional body**

$$W = \int dW \qquad \bar{\mathbf{r}}W = \int \mathbf{r}\, dW \qquad (5.16)$$

Note that these relations are independent of the orientation of the body. For example, if the body and the coordinate axes were rotated so that the $z$ axis pointed upward, the unit vector $-\mathbf{j}$ would be replaced by $-\mathbf{k}$ in Eqs. (5.14) and (5.15), but the relations in Eqs. (5.16) would remain unchanged.

Resolving the vectors $\bar{\mathbf{r}}$ and $\mathbf{r}$ into rectangular components, we note that the second of the relations in Eqs. (5.16) is equivalent to the three scalar equations

$$\bar{x}W = \int x\, dW \qquad \bar{y}W = \int y\, dW \qquad \bar{z}W = \int z\, dW \qquad (5.17)$$

or

$$\bar{x} = \frac{\int x\, dW}{W} \qquad \bar{y} = \frac{\int y\, dW}{W} \qquad \bar{z} = \frac{\int z\, dW}{W} \qquad (5.17')$$

If the body is made of a homogeneous material of specific weight $\gamma$, we can express the magnitude $dW$ of the weight of an infinitesimal element in terms of the volume $dV$ of the element and express the magnitude $W$ of the total weight in terms of the total volume $V$. We obtain

$$dW = \gamma\, dV \qquad W = \gamma V$$

Substituting for $dW$ and $W$ in the second of the relations in Eqs. (5.16), we have

$$\bar{\mathbf{r}}V = \int \mathbf{r}\, dV \qquad (5.18)$$

In scalar form, this becomes

**Centroid of a volume $V$**

$$\bar{x}V = \int x\, dV \qquad \bar{y}V = \int y\, dV \qquad \bar{z}V = \int z\, dV \qquad (5.19)$$

or

$$\bar{x} = \frac{\int x\, dV}{V} \qquad \bar{y} = \frac{\int y\, dV}{V} \qquad \bar{z} = \frac{\int z\, dV}{V} \qquad (5.19')$$

The center of gravity of a homogeneous body whose coordinates are $\bar{x}, \bar{y}, \bar{z}$ is also known as the **centroid $C$ of the volume $V$** of the body. If the body is not homogeneous, we cannot use Eqs. (5.19) to determine the center of gravity of the body; however, Eqs. (5.19) still define the centroid of the volume.

The integral $\int x\, dV$ is known as the **first moment of the volume with respect to the $yz$ plane**. Similarly, the integrals $\int y\, dV$ and $\int z\, dV$ define the first moments of the volume with respect to the $zx$ plane and the $xy$ plane, respectively. You can see from Eqs. (5.19) that if the centroid of a volume is located in a coordinate plane, the first moment of the volume with respect to that plane is zero.

A volume is said to be symmetrical with respect to a given plane if, for every point $P$ of the volume, there exists a point $P'$ of the same volume such that the line $PP'$ is perpendicular to the given plane and is bisected by that plane. We say the plane is a **plane of symmetry** for the given volume. When a volume $V$ possesses a plane of symmetry, the first moment of $V$ with respect to that plane is zero, and the centroid of the volume is located in the plane of symmetry. If a volume possesses two planes of symmetry, the centroid of the volume is located on the line of intersection of the two planes. Finally, if a volume possesses three planes of symmetry that intersect at a well-defined point (i.e., not along a common line), the point of intersection of the three planes coincides with the centroid of the volume. This property enables us to determine immediately the locations of the centroids of spheres, ellipsoids, cubes, rectangular parallelepipeds, etc.

For unsymmetrical volumes or volumes possessing only one or two planes of symmetry, we can determine the location of the centroid by integration (Sec. 5.4C). The centroids of several common volumes are shown in Fig. 5.21. Note that, in general, the centroid of a volume of revolution *does not coincide* with the centroid of its cross section. Thus, the centroid of a hemisphere is different from that of a semicircular area, and the centroid of a cone is different from that of a triangle.

## 5.4B  Composite Bodies

If a body can be divided into several of the common shapes shown in Fig. 5.21, we can determine its center of gravity $G$ by setting the moment about $O$ of its total weight equal to the sum of the moments about $O$ of the weights of the various component parts. Proceeding in this way, we obtain the following equations defining the coordinates $\overline{X}, \overline{Y}, \overline{Z}$ of the center of gravity $G$ as

**Center of gravity of a body with weight $W$**

$$\overline{X}\Sigma W = \Sigma \overline{x}W \qquad \overline{Y}\Sigma W = \Sigma \overline{y}W \qquad \overline{Z}\Sigma W = \Sigma \overline{z}W \qquad \text{(5.20)}$$

or

$$\overline{X} = \frac{\Sigma \overline{x}W}{\Sigma W} \qquad \overline{Y} = \frac{\Sigma \overline{y}W}{\Sigma W} \qquad \overline{Z} = \frac{\Sigma \overline{z}W}{\Sigma W} \qquad \text{(5.20')}$$

If the body is made of a homogeneous material, its center of gravity coincides with the centroid of its volume, and we obtain

**Centroid of a volume $V$**

$$\overline{X}\Sigma V = \Sigma \overline{x}V \qquad \overline{Y}\Sigma V = \Sigma \overline{y}V \qquad \overline{Z}\Sigma V = \Sigma \overline{z}V \qquad \text{(5.21)}$$

or

$$\overline{X} = \frac{\Sigma \overline{x}V}{\Sigma V} \qquad \overline{Y} = \frac{\Sigma \overline{y}V}{\Sigma V} \qquad \overline{Z} = \frac{\Sigma \overline{z}V}{\Sigma V} \qquad \text{(5.21')}$$

| Shape | | $\bar{x}$ | Volume |
|---|---|---|---|
| Hemisphere | 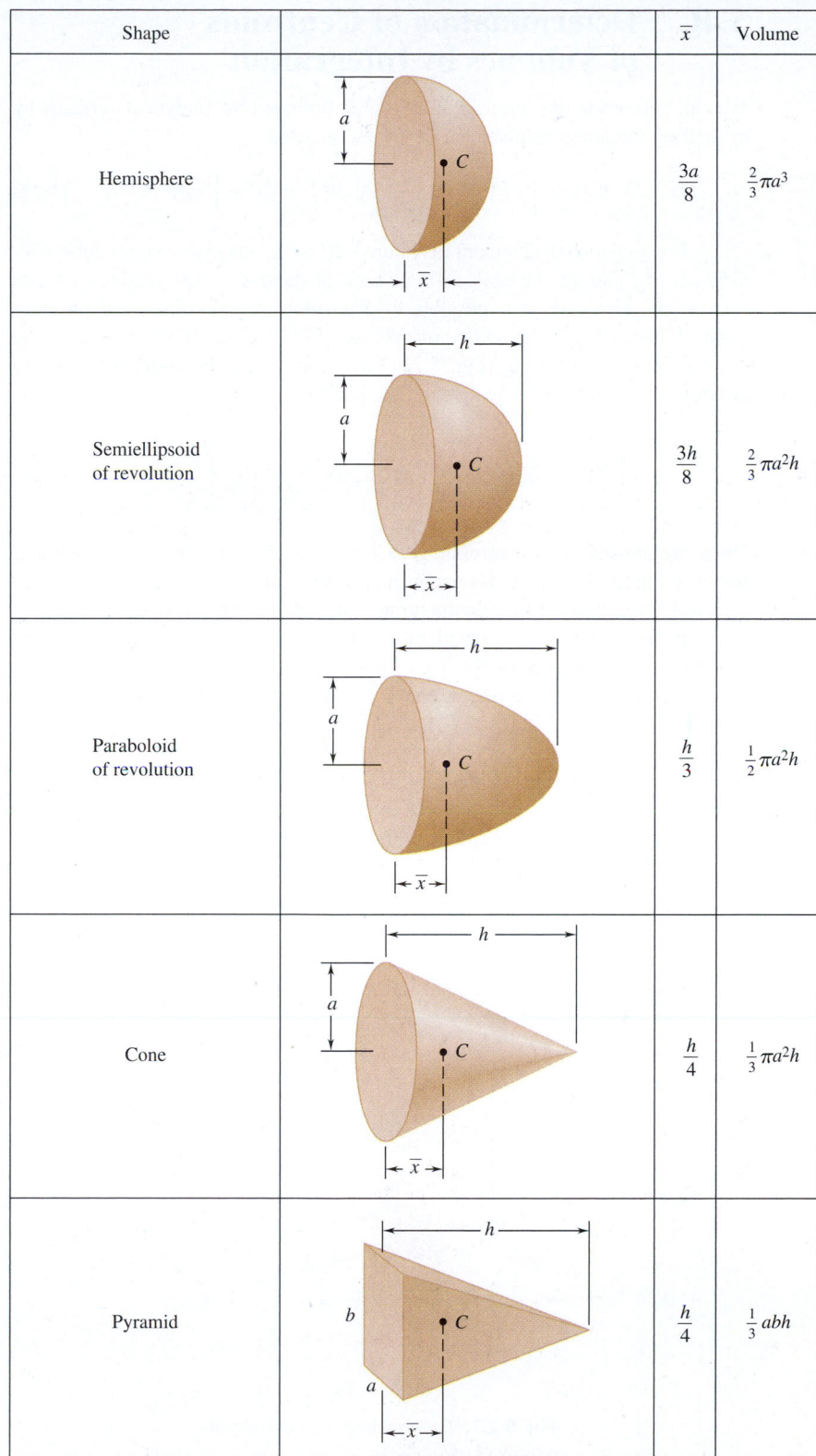 | $\dfrac{3a}{8}$ | $\dfrac{2}{3}\pi a^3$ |
| Semiellipsoid of revolution | | $\dfrac{3h}{8}$ | $\dfrac{2}{3}\pi a^2 h$ |
| Paraboloid of revolution | | $\dfrac{h}{3}$ | $\dfrac{1}{2}\pi a^2 h$ |
| Cone | | $\dfrac{h}{4}$ | $\dfrac{1}{3}\pi a^2 h$ |
| Pyramid | | $\dfrac{h}{4}$ | $\dfrac{1}{3}abh$ |

**Fig. 5.21** Centroids of common shapes and volumes.

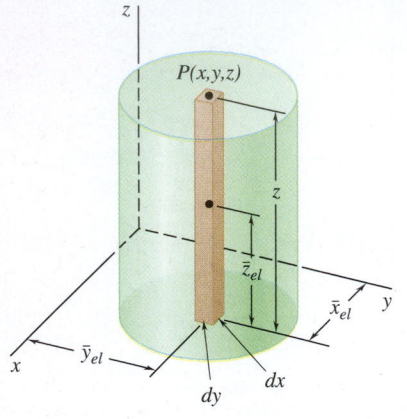

$\bar{x}_{el} = x, \ \bar{y}_{el} = y, \ \bar{z}_{el} = \frac{z}{2}$

$dV = z \, dx \, dy$

**Fig. 5.22** Determining the centroid of a volume by double integration.

# 5.4C Determination of Centroids of Volumes by Integration

We can determine the centroid of a volume bounded by analytical surfaces by evaluating the integrals given earlier in this section:

$$\bar{x}V = \int x \, dV \qquad \bar{y}V = \int y \, dV \qquad \bar{z}V = \int z \, dV \qquad \textbf{(5.22)}$$

If we choose the element of volume $dV$ to be equal to a small cube with sides $dx$, $dy$, and $dz$, the evaluation of each of these integrals requires a *triple integration*. However, it is possible to determine the coordinates of the centroid of most volumes by *double integration* if we choose $dV$ to be equal to the volume of a thin filament (Fig. 5.22). We then obtain the coordinates of the centroid of the volume by rewriting Eqs. (5.22) as

$$\bar{x}V = \int \bar{x}_{el} \, dV \qquad \bar{y}V = \int \bar{y}_{el} \, dV \qquad \bar{z}V = \int \bar{z}_{el} \, dV \qquad \textbf{(5.23)}$$

Then, we substitute the expressions given in Fig. 5.22 for the volume $dV$ and the coordinates $\bar{x}_{el}$, $\bar{y}_{el}$, $\bar{z}_{el}$. By using the equation of the surface to express $z$ in terms of $x$ and $y$, we reduce the integration to a double integration in $x$ and $y$.

If the volume under consideration possesses *two planes of symmetry*, its centroid must be located on the line of intersection of the two planes. Choosing the $x$ axis to lie along this line, we have

$$\bar{y} = \bar{z} = 0$$

and the only coordinate to determine is $\bar{x}$. This can be done with a *single integration* by dividing the given volume into thin slabs parallel to the $yz$ plane and expressing $dV$ in terms of $x$ and $dx$ in the equation

$$\bar{x}V = \int \bar{x}_{el} \, dV \qquad \textbf{(5.24)}$$

For a body of revolution, the slabs are circular, and their volume is given in Fig. 5.23.

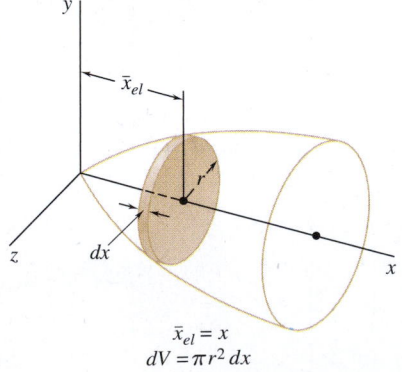

$\bar{x}_{el} = x$

$dV = \pi r^2 \, dx$

**Fig. 5.23** Determining the centroid of a body of revolution.

## Sample Problem 5.11

Determine the location of the center of gravity of the homogeneous body of revolution shown that was obtained by joining a hemisphere and a cylinder and carving out a cone.

**STRATEGY:**   The body is homogeneous, so the center of gravity coincides with the centroid. Because the body was formed from a composite of three simple shapes, you can find the centroid of each shape and combine them using Eq. (5.21).

**MODELING:**   Because of symmetry, the center of gravity lies on the $x$ axis. As shown in Fig. 1, the body is formed by adding a hemisphere to a cylinder and then subtracting a cone. Find the volume and the abscissa of the centroid of each of these components from Fig. 5.21 and enter them in a table (below). Then, you can determine the total volume of the body and the first moment of its volume with respect to the $yz$ plane.

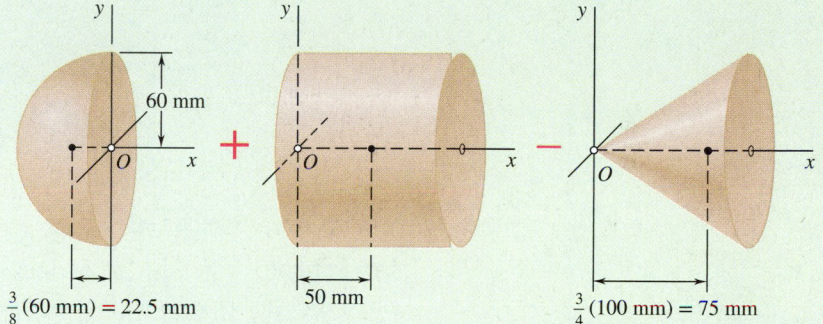

**Fig. 1**   The given body modeled as the combination of simple geometric shapes.

**ANALYSIS:**   Note that the location of the centroid of the hemisphere is negative because it lies to the left of the origin.

| Component | Volume, mm³ | | $\bar{x}$, mm | $\bar{x}V$, mm⁴ |
|---|---|---|---|---|
| Hemisphere | $\dfrac{1}{2}\dfrac{4\pi}{3}(60)^3 =$ | $0.4524 \times 10^6$ | $-22.5$ | $-10.18 \times 10^6$ |
| Cylinder | $\pi(60)^2(100) =$ | $1.1310 \times 10^6$ | $+50$ | $+56.55 \times 10^6$ |
| Cone | $-\dfrac{\pi}{3}(60)^2(100) =$ | $-0.3770 \times 10^6$ | $+75$ | $-28.28 \times 10^6$ |
| | $\Sigma V =$ | $1.206 \times 10^6$ | | $\Sigma\bar{x}V = +18.09 \times 10^6$ |

Thus,

$$\bar{X}\Sigma V = \Sigma \bar{x}V: \qquad \bar{X}(1.206 \times 10^6 \text{ mm}^3) = 18.09 \times 10^6 \text{ mm}^4$$

$$\bar{X} = 15 \text{ mm} \;\blacktriangleleft$$

**REFLECT and THINK:**   Adding the hemisphere and subtracting the cone have the effect of shifting the centroid of the composite shape to the left of that for the cylinder (50 mm). However, because the first moment of volume for

*(continued)*

the cylinder is larger than for the hemisphere and cone combined, you should expect the centroid for the composite to still be in the positive $x$ domain. Thus, as a rough visual check, the result of +15 mm is reasonable.

## Sample Problem 5.12

Locate the center of gravity of the steel machine part shown. The diameter of each hole is 1 in.

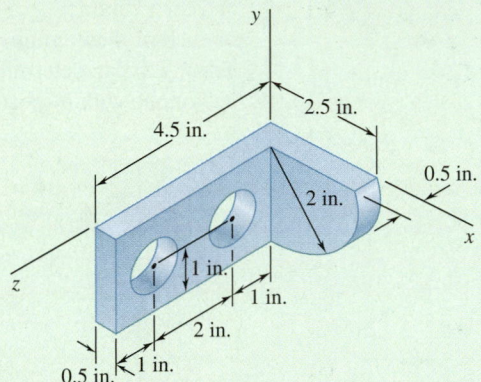

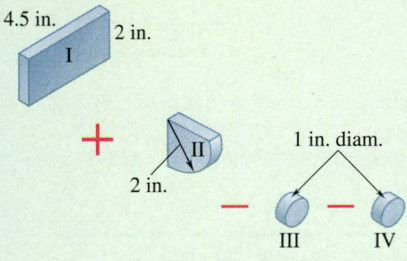

**Fig. 1** The given body modeled as the combination of simple geometric shapes.

**STRATEGY:** This part can be broken down into the sum of two volumes minus two smaller volumes (holes). Find the volume and centroid of each volume and combine them using Eq. (5.21) to find the overall centroid.

**MODELING:** As shown in Fig. 1, the machine part can be obtained by adding a rectangular parallelepiped (I) to a quarter cylinder (II) and then subtracting two 1-in.-diameter cylinders (III and IV). Determine the volume and the coordinates of the centroid of each component and enter them in a table (on the next page). Using the data in the table, determine the total volume and the moments of the volume with respect to each of the coordinate planes.

**ANALYSIS:** You can treat each component volume as a planar shape using Fig. 5.8A to find the volumes and centroids, but the right-angle joining of components I and II requires calculations in three dimensions. You may find it helpful to draw more detailed sketches of components with the centroids carefully labeled (Fig. 2).

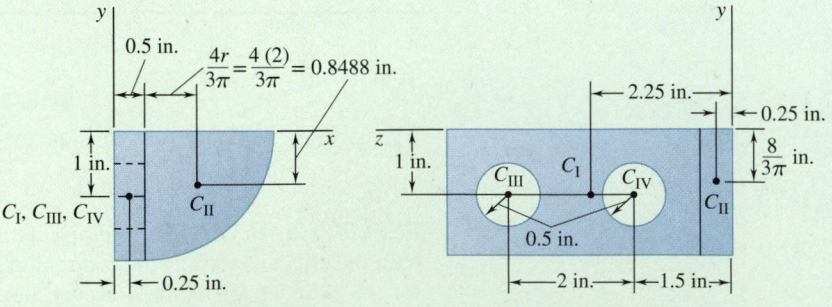

**Fig. 2** Centroids of components.

*(continued)*

| | $V$, in$^3$ | $\bar{x}$, in. | $\bar{y}$, in. | $\bar{z}$, in. | $\bar{x}V$, in$^4$ | $\bar{y}V$, in$^4$ | $\bar{z}V$, in$^4$ |
|---|---|---|---|---|---|---|---|
| I | $(4.5)(2)(0.5) = 4.5$ | 0.25 | $-1$ | 2.25 | 1.125 | $-4.5$ | 10.125 |
| II | $\frac{1}{4}\pi(2)^2(0.5) = 1.571$ | 1.3488 | $-0.8488$ | 0.25 | 2.119 | $-1.333$ | 0.393 |
| III | $-\pi(0.5)^2(0.5) = -0.3927$ | 0.25 | $-1$ | 3.5 | $-0.098$ | 0.393 | $-1.374$ |
| IV | $-\pi(0.5)^2(0.5) = -0.3927$ | 0.25 | $-1$ | 1.5 | $-0.098$ | 0.393 | $-0.589$ |
| | $\Sigma V = 5.286$ | | | | $\Sigma\bar{x}V = 3.048$ | $\Sigma\bar{y}V = -5.047$ | $\Sigma\bar{z}V = 8.555$ |

Thus,

$$\bar{X}\Sigma V = \Sigma\bar{x}V: \qquad \bar{X}(5.286 \text{ in}^3) = 3.048 \text{ in}^4 \qquad \bar{X} = \quad 0.577 \text{ in.} \quad \blacktriangleleft$$

$$\bar{Y}\Sigma V = \Sigma\bar{y}V: \qquad \bar{Y}(5.286 \text{ in}^3) = -5.047 \text{ in}^4 \qquad \bar{Y} = -0.955 \text{ in.} \quad \blacktriangleleft$$

$$\bar{Z}\Sigma V = \Sigma\bar{z}V: \qquad \bar{Z}(5.286 \text{ in}^3) = 8.555 \text{ in}^4 \qquad \bar{Z} = \quad 1.618 \text{ in.} \quad \blacktriangleleft$$

**REFLECT and THINK:**   By inspection, you should expect $\bar{X}$ and $\bar{Z}$ to be considerably less than (1/2)(2.5 in.) and (1/2)(4.5 in.), respectively, and $\bar{Y}$ to be slightly less in magnitude than (1/2)(2 in.). Thus, as a rough visual check, the results obtained are as expected.

## Sample Problem 5.13

Determine the location of the centroid of the half right circular cone shown.

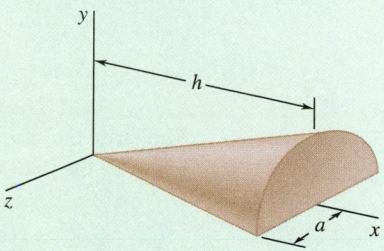

**STRATEGY:**   This is not one of the shapes in Fig. 5.21, so you have to determine the centroid by using integration.

**MODELING:**   Because the $xy$ plane is a plane of symmetry, the centroid lies in this plane, and $\bar{z} = 0$. Choose a slab of thickness $dx$ as a differential element. The volume of this element is

$$dV = \frac{1}{2}\pi r^2\, dx$$

Obtain the coordinates $\bar{x}_{el}$ and $\bar{y}_{el}$ of the centroid of the element from Fig. 5.8 (semicircular area):

$$\bar{x}_{el} = x \qquad \bar{y}_{el} = \frac{4r}{3\pi}$$

*(continued)*

Noting that $r$ is proportional to $x$, use similar triangles (Fig. 1) to write

$$\frac{r}{x} = \frac{a}{h} \qquad r = \frac{a}{h}x$$

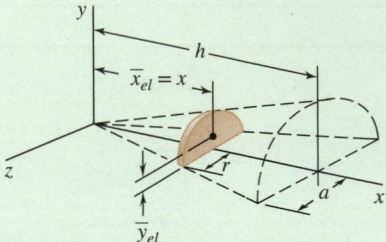

**Fig. 1** Geometry of the differential element.

**ANALYSIS:** The volume of the body is

$$V = \int dV = \int_0^h \tfrac{1}{2}\pi r^2\, dx = \int_0^h \tfrac{1}{2}\pi \left(\frac{a}{h}x\right)^2 dx = \frac{\pi a^2 h}{6}$$

The moment of the differential element with respect to the $yz$ plane is $\overline{x}_{el}dV$; the total moment of the body with respect to this plane is

$$\int \overline{x}_{el}\, dV = \int_0^h x(\tfrac{1}{2}\pi r^2)\, dx = \int_0^h x(\tfrac{1}{2}\pi)\left(\frac{a}{h}x\right)^2 dx = \frac{\pi a^2 h^2}{8}$$

Thus,

$$\overline{x}V = \int \overline{x}_{el}\, dV \qquad \overline{x}\,\frac{\pi a^2 h}{6} = \frac{\pi a^2 h^2}{8} \qquad\qquad \overline{x} = \tfrac{3}{4}h \quad \blacktriangleleft$$

Similarly, the moment of the differential element with respect to the $zx$ plane is $\overline{y}_{el}\, dV$; the total moment is

$$\int \overline{y}_{el}\, dV = \int_0^h \frac{4r}{3\pi}(\tfrac{1}{2}\pi r^2)\, dx = \frac{2}{3}\int_0^h \left(\frac{a}{h}x\right)^3 dx = \frac{a^3 h}{6}$$

Thus,

$$\overline{y}V = \int \overline{y}_{el}\, dV \qquad \overline{y}\,\frac{\pi a^2 h}{6} = \frac{a^3 h}{6} \qquad\qquad \overline{y} = \frac{a}{\pi} \quad \blacktriangleleft$$

**REFLECT and THINK:** Because a full right circular cone is a body of revolution, its $\overline{x}$ is unchanged for any portion of the cone bounded by planes intersecting along the $x$ axis. The same centroid location in the $x$ direction was therefore obtained for the half cone that Fig. 5.21 shows for the full cone. Similarly, the same $\overline{x}$ result would be obtained for a quarter cone.

# SOLVING PROBLEMS
# ON YOUR OWN

In the problems for this section, you will be asked to locate the centers of gravity of three-dimensional bodies or the centroids of their volumes. All of the techniques we previously discussed for two-dimensional bodies—using symmetry, dividing the body into common shapes, choosing the most efficient differential element, etc.—also may be applied to the general three-dimensional case.

**1. Locating the centers of gravity of composite bodies.** In general, you must use Eqs. (5.20):

$$\overline{X}\Sigma W = \Sigma \overline{x} W \qquad \overline{Y}\Sigma W = \Sigma \overline{y} W \qquad \overline{Z}\Sigma W = \Sigma \overline{z} W \qquad \textbf{(5.20)}$$

However, for the case of a *homogeneous body,* the center of gravity of the body coincides with the *centroid of its volume.* Therefore, for this special case, you can also use Eqs. (5.21) to locate the center of gravity of the body:

$$\overline{X}\Sigma V = \Sigma \overline{x} V \qquad \overline{Y}\Sigma V = \Sigma \overline{y} V \qquad \overline{Z}\Sigma V = \Sigma \overline{z} V \qquad \textbf{(5.21)}$$

Note that these equations are simply an extension of the equations used for the two-dimensional problems considered earlier in the chapter. As the solutions of Sample Probs. 5.11 and 5.12 illustrate, the methods of solution for two- and three-dimensional problems are identical. Thus, we once again strongly encourage you to construct appropriate diagrams and tables when analyzing composite bodies. Also, as you study Sample Prob. 5.12, observe how we obtained the $x$ and $y$ coordinates of the centroid of the quarter cylinder using the equations for the centroid of a quarter circle.

*Two special cases* of interest occur when the given body consists of either uniform wires or uniform plates made of the same material.

**a.** For a body made of *several wire elements* of the *same uniform cross section,* the cross-sectional area $A$ of the wire elements factors out of Eqs. (5.21) when $V$ is replaced with the product $AL$, where $L$ is the length of a given element. Eqs. (5.21) thus reduce in this case to

$$\overline{X}\Sigma L = \Sigma \overline{x} L \qquad \overline{Y}\Sigma L = \Sigma \overline{y} L \qquad \overline{Z}\Sigma L = \Sigma \overline{z} L$$

**b.** For a body made of *several plates* of the *same uniform thickness,* the thickness $t$ of the plates factors out of Eqs. (5.21) when $V$ is replaced with the product $tA$, where $A$ is the area of a given plate. Eqs. (5.21) thus reduce in this case to

$$\overline{X}\Sigma A = \Sigma \overline{x} A \qquad \overline{Y}\Sigma A = \Sigma \overline{y} A \qquad \overline{Z}\Sigma A = \Sigma \overline{z} A$$

**2. Locating the centroids of volumes by direct integration.** As explained in Sec. 5.4C, you can simplify evaluating the integrals of Eqs. (5.22) by choosing either a thin filament (Fig. 5.22) or a thin slab (Fig. 5.23) for the element of volume $dV$. Thus, you should begin your solution by identifying, if possible, the $dV$ that produces the single or double integrals that are easiest to compute. For bodies of revolution, this may be a thin slab (as in Sample Prob. 5.13) or a thin cylindrical shell. However, it is important to remember that the relationship you establish among the variables (like the relationship between $r$ and $x$ in Sample Prob. 5.13) directly affects the complexity of the integrals you have to compute. Finally, we again remind you that $\overline{x}_{el}$, $\overline{y}_{el}$, and $\overline{z}_{el}$ in Eqs. (5.23) are the coordinates of the centroid of $dV$.

# Problems

**5.96** Consider the composite body shown. Determine (*a*) the value of $\bar{x}$ when $h = L/2$, (*b*) the ratio $h/L$ for which $\bar{x} = L$.

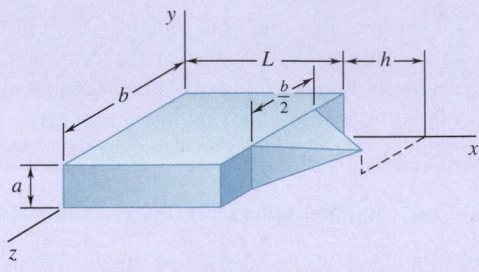

**Fig. P5.96**

**5.97** A cone and a cylinder of the same radius $a$ and height $h$ are attached as shown. Determine the location of the centroid of the composite body.

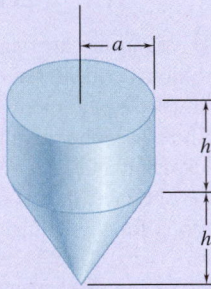

**Fig. P5.97**

**5.98** Determine the location of the center of gravity of the parabolic reflector shown, which is formed by machining a rectangular block so that the curved surface is a paraboloid of revolution of base radius $a$ and height $h$.

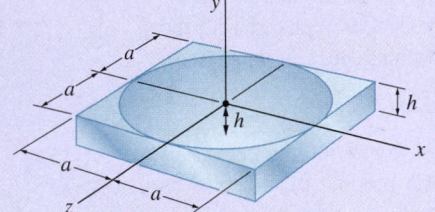

**Fig. P5.98**

**5.99** Locate the centroid of the frustum of a right circular cone when $r_1 = 40$ mm, $r_2 = 50$ mm, and $h = 60$ mm.

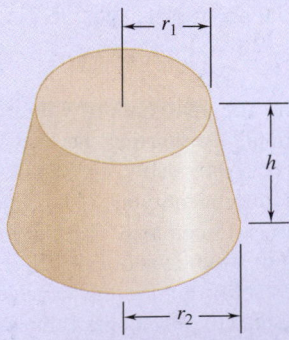

**Fig. P5.99**

**5.100** For the stop bracket shown, locate the *x* coordinate of the center of gravity.

**5.101** For the stop bracket shown, locate the *z* coordinate of the center of gravity.

**5.102** For the machine element shown, locate the *x* coordinate of the center of gravity.

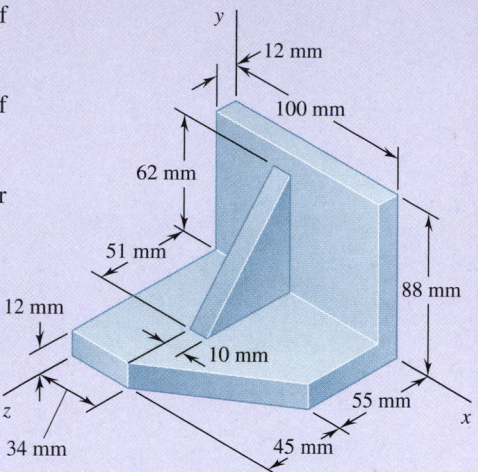

Fig. P5.100 and *P5.101*

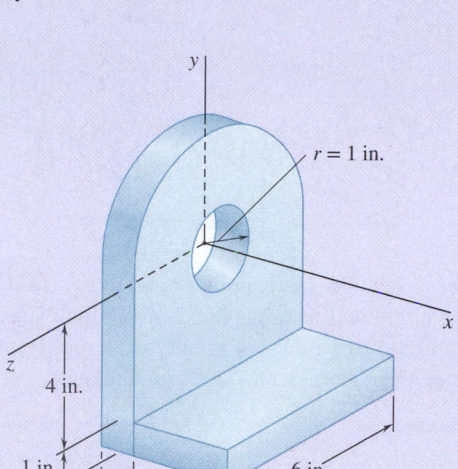

Fig. P5.102 and P5.103

**5.103** For the machine element shown, locate the *y* coordinate of the center of gravity.

**5.104** For the machine element shown, locate the *y* coordinate of the center of gravity.

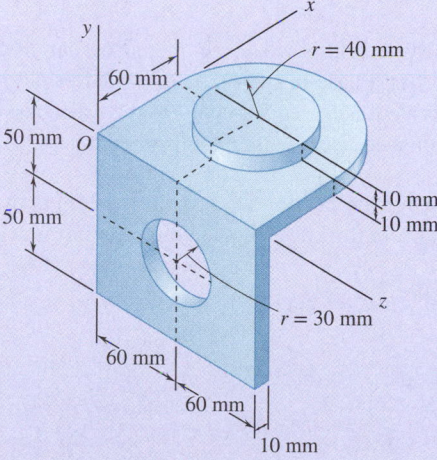

Fig. P5.104 and *P5.105*

**5.105** For the machine element shown, locate the *x* coordinate of the center of gravity.

**5.106 and 5.107** Locate the center of gravity of the sheet-metal form shown.

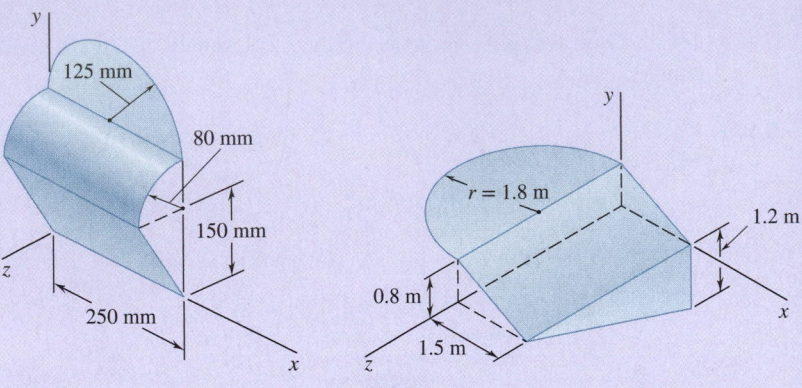

Fig. P5.106                    Fig. P5.107

**5.108** A corner reflector for tracking by radar has two sides in the shape of a quarter circle with a radius of 15 in. and one side in the shape of a triangle. Locate the center of gravity of the reflector, knowing that it is made of sheet metal with a uniform thickness.

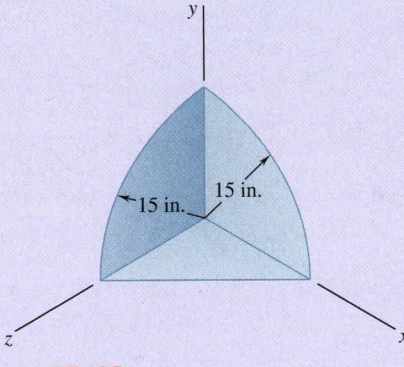

Fig. *P5.108*

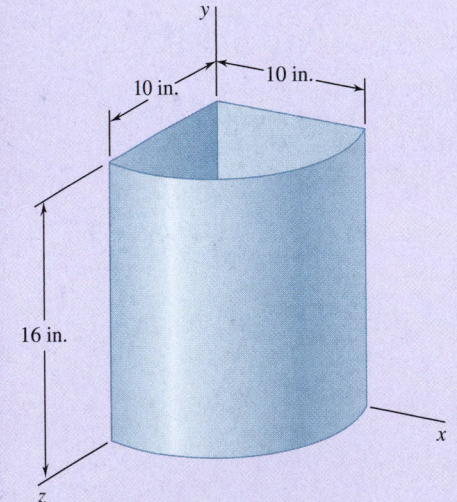

Fig. P5.109

**5.109** A wastebasket, designed to fit in the corner of a room, is 16 in. high and has a base in the shape of a quarter circle with a radius of 10 in. Locate the center of gravity of the wastebasket, knowing that it is made of sheet metal with a uniform thickness.

**5.110** An elbow for the duct of a ventilating system is made of sheet metal with a uniform thickness. Locate the center of gravity of the elbow.

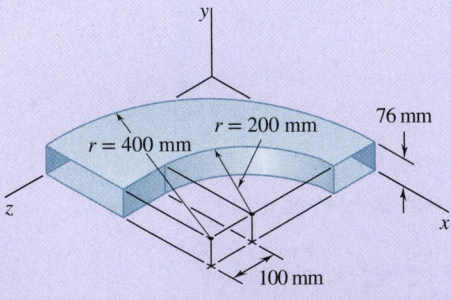

Fig. P5.110

**5.111** A window awning is fabricated from sheet metal with a uniform thickness. Locate the center of gravity of the awning.

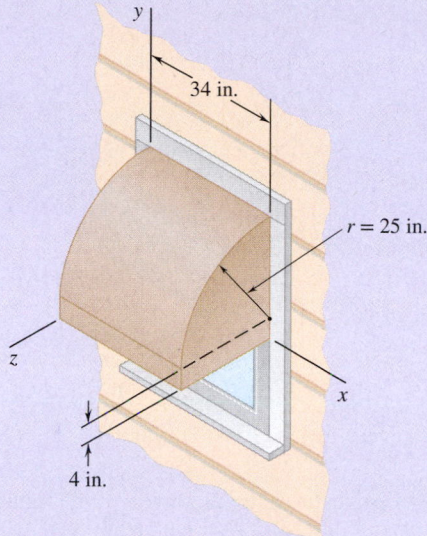

**Fig. P5.111**

**5.112 and 5.113** Locate the center of gravity of the sheet-metal form shown.

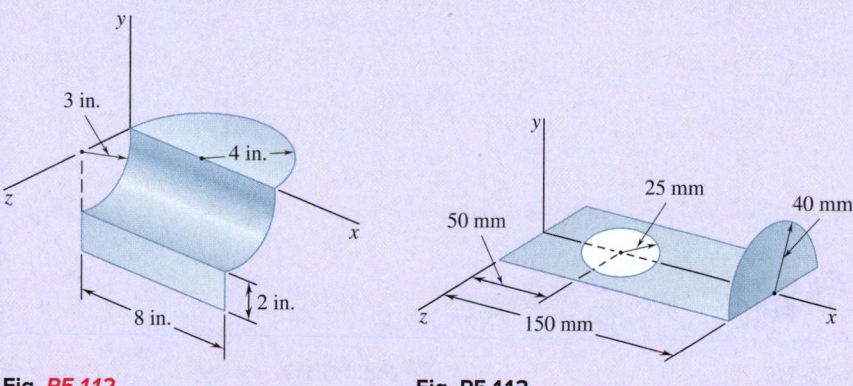

**Fig. P5.112**          **Fig. P5.113**

**5.114** A thin steel wire with a uniform cross section is bent into the shape shown. Locate its center of gravity.

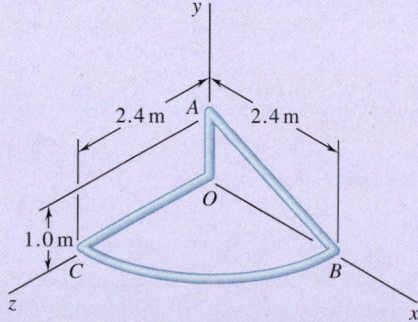

**Fig. P5.114**

**5.115** The frame of a greenhouse is constructed from uniform aluminum channels. Locate the center of gravity of the portion of the frame shown.

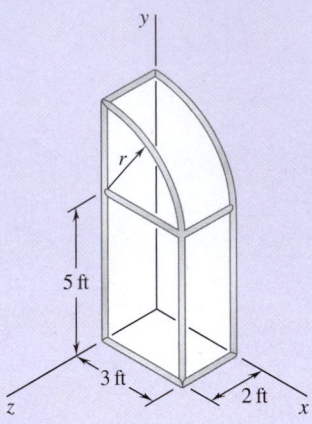

Fig. *P5.115*

**5.116 and 5.117** Locate the center of gravity of the figure shown, knowing that it is made of thin brass rods with a uniform diameter.

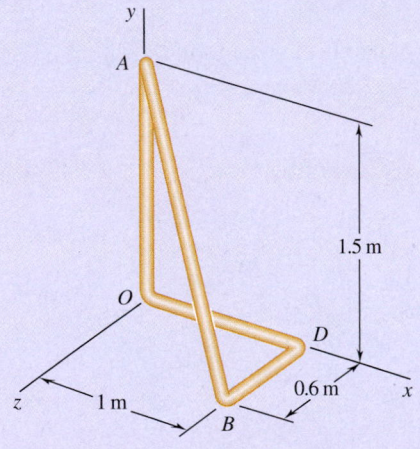

Fig. P5.116

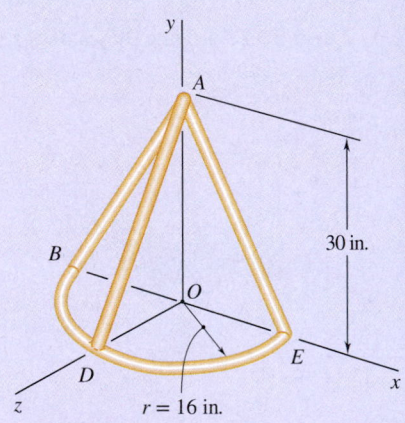

Fig. P5.117

**5.118** A scratch awl has a plastic handle and a steel blade and shank. Knowing that the density of plastic is 1030 kg/m$^3$ and of steel is 7860 kg/m$^3$, locate the center of gravity of the awl.

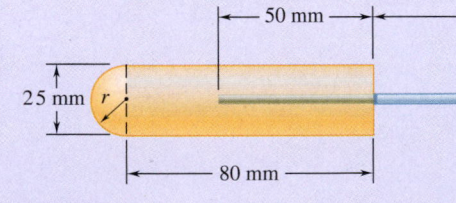

Fig. P5.118

**5.119** A bronze bushing is mounted inside a steel sleeve. Knowing that the specific weight of bronze is 0.318 lb/in$^3$ and of steel is 0.284 lb/in$^3$, determine the location of the center of gravity of the assembly.

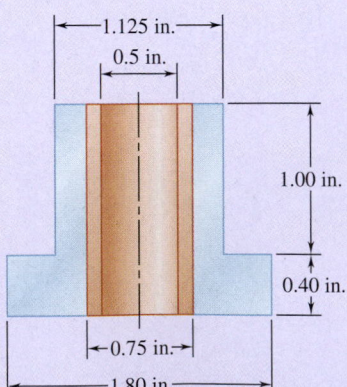

Fig. P5.119

**5.120** A brass collar with a length of 2.5 in. is mounted on an aluminum rod with a length of 4 in. Locate the center of gravity of the composite body. (Specific weights: brass = 0.306 lb/in³, aluminum = 0.101 lb/in³.)

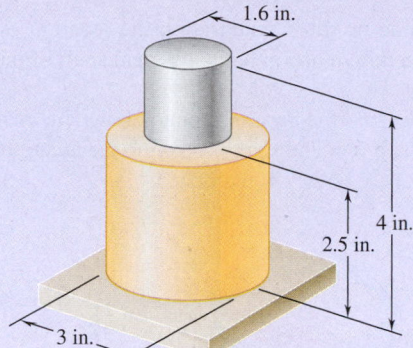

Fig. *P5.120*

**5.121** The three legs of a small glass-topped table are equally spaced and are made of steel tubing that has an outside diameter of 24 mm and a cross-sectional area of 150 mm². The diameter and the thickness of the table top are 600 mm and 10 mm, respectively. Knowing that the density of steel is 7860 kg/m³ and of glass is 2190 kg/m³, locate the center of gravity of the table.

**5.122 through 5.124** Determine by direct integration the values of $\bar{x}$ for the two volumes obtained by passing a vertical cutting plane through the given shape of Fig. 5.21. The cutting plane is parallel to the base of the given shape and divides the shape into two volumes of equal height.

    **5.122** A hemisphere.
    **5.123** A semiellipsoid of revolution.
    **5.124** A paraboloid of revolution.

**5.125 and 5.126** Locate the centroid of the volume obtained by rotating the shaded area about the *x* axis.

**5.127** Locate the centroid of the volume obtained by rotating the shaded area about the line $x = h$.

**\*5.128** Locate the centroid of the volume generated by revolving the portion of the sine curve shown about the *x* axis.

**\*5.129** Locate the centroid of the volume generated by revolving the portion of the sine curve shown about the *y* axis. (*Hint:* Use a thin cylindrical shell of radius *r* and thickness *dr* as the element of volume.)

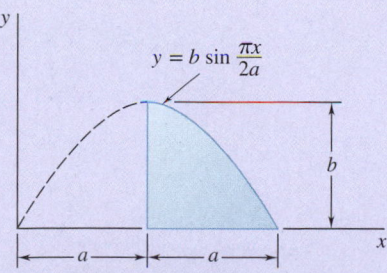

Fig. P5.128 and P5.129

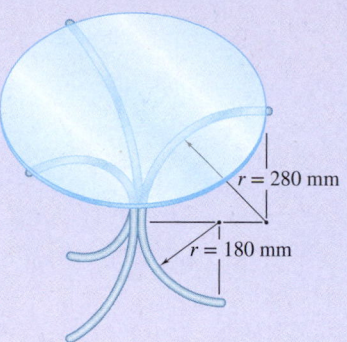

Fig. P5.121

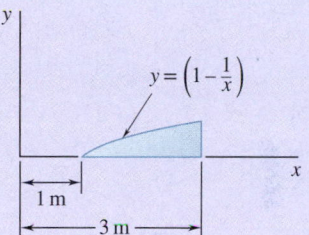

Fig. P5.125

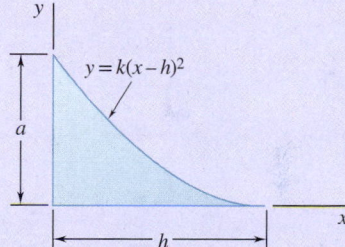

Fig. *P5.126*

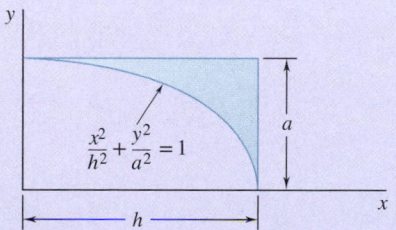

Fig. *P5.127*

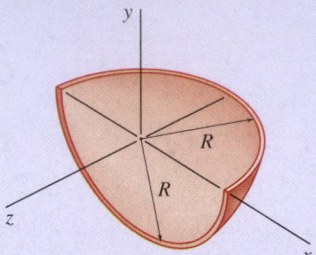

Fig. *P5.131*

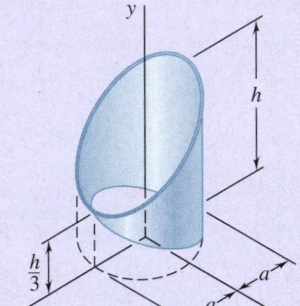

Fig. *P5.133*

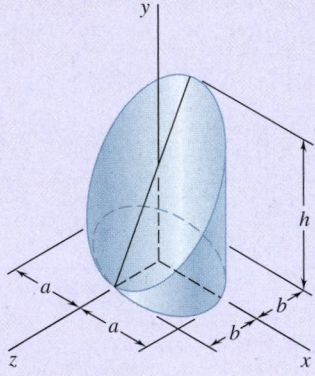

Fig. P5.134

**\*5.130** Show that for a regular pyramid of height $h$ and $n$ sides ($n = 3, 4, \ldots$) the centroid of the volume of the pyramid is located at a distance $h/4$ above the base.

**5.131** Determine by direct integration the location of the centroid of one-half of a thin, uniform hemispherical shell of radius $R$.

**5.132** The sides and the base of a punch bowl are of uniform thickness $t$. If $t << R$ and $R = 250$ mm, determine the location of the center of gravity of ($a$) the bowl, ($b$) the punch.

Fig. P5.132

**5.133** Locate the centroid of the section shown, which was cut from a thin circular pipe by two oblique planes.

**\*5.134** Locate the centroid of the section shown, which was cut from an elliptical cylinder by an oblique plane.

**5.135** Determine by direct integration the location of the centroid of the volume between the $xz$ plane and the portion shown of the surface $y = 16h(ax - x^2)(bz - z^2)/a^2b^2$.

**5.136** After grading a lot, a builder places four stakes to designate the corners of the slab for a house. To provide a firm, level base for the slab, the builder places a minimum of 3 in. of gravel beneath the slab. Determine the volume of gravel needed and the $x$ coordinate of the centroid of the volume of the gravel. (*Hint:* The bottom surface of the gravel is an oblique plane, which can be represented by the equation $y = a + bx + cz$.)

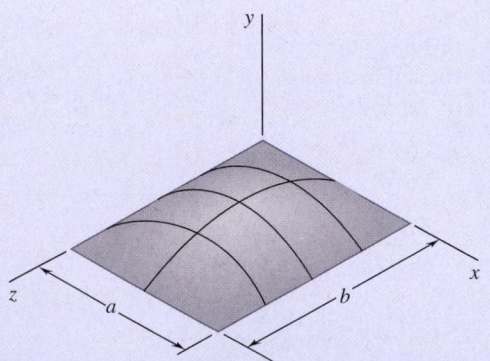

Fig. P5.135

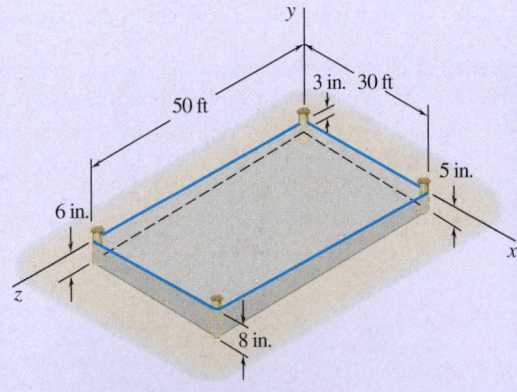

Fig. P5.136

292

# Review and Summary

This chapter was devoted chiefly to determining the **center of gravity** of a rigid body, i.e., to determining the point $G$ where we can apply a single force $\mathbf{W}$—the *weight* of the body—to represent the effect of earth's attraction on the body.

## Center of Gravity of a Two-Dimensional Body

In the first part of this chapter, we considered *two-dimensional bodies,* such as flat plates and wires contained in the $xy$ plane. By adding force components in the vertical $z$ direction and moments about the horizontal $y$ and $x$ axes (Sec. 5.1A), we derived the relations

$$W = \int dW \qquad \overline{x}W = \int x \, dW \qquad \overline{y}W = \int y \, dW \qquad \textbf{(5.2)}$$

These equations define the weight of the body and the coordinates $\overline{x}$ and $\overline{y}$ of its center of gravity.

## Centroid of an Area or Line

In the case of a *homogeneous flat* plate of uniform thickness (Sec. 5.1B), the center of gravity $G$ of the plate coincides with the **centroid $C$ of the area** $A$ of the plate. The coordinates are defined by the relations

$$\overline{x}A = \int x \, dA \qquad \overline{y}A = \int y \, dA \qquad \textbf{(5.3)}$$

Similarly, determining the center of gravity of a *homogeneous wire of uniform cross section* contained in a plane reduces to determining the **centroid $C$ of the line** $L$ representing the wire; we have

$$\overline{x}L = \int x \, dL \qquad \overline{y}L = \int y \, dL \qquad \textbf{(5.4)}$$

## First Moments

The integrals in Eqs. (5.3) are referred to as the **first moments** of the area $A$ with respect to the $y$ and $x$ axes and are denoted by $Q_y$ and $Q_x$, respectively (Sec. 5.1C). We have

$$Q_y = \overline{x}A \qquad Q_x = \overline{y}A \qquad \textbf{(5.6)}$$

The first moments of a line can be defined in a similar way.

## Properties of Symmetry

Determining the centroid $C$ of an area or line is simplified when the area or line possesses certain properties of symmetry. If the area or line is symmetric with respect to an axis, its centroid $C$ lies on that axis; if it is symmetric with respect to two axes, $C$ is located at the intersection of the two axes; if it is symmetric with respect to a center $O$, $C$ coincides with $O$.

## Center of Gravity of a Composite Body

The areas and the centroids of various common shapes are tabulated in Fig. 5.8A. When a flat plate can be divided into several of these shapes, the coordinates $\overline{X}$ and $\overline{Y}$ of its center of gravity $G$ can be determined from the coordinates $\overline{x}_1, \overline{x}_2, \ldots$ and $\overline{y}_1, \overline{y}_2, \ldots$ of the centers of gravity $G_1, G_2, \ldots$ of the various parts (Sec. 5.1D). Equating moments about the $y$ and $x$ axes, respectively (Fig. 5.24), we have

$$\overline{X}\Sigma W = \Sigma \overline{x} W \qquad \overline{Y}\Sigma W = \Sigma \overline{y} W \tag{5.7}$$

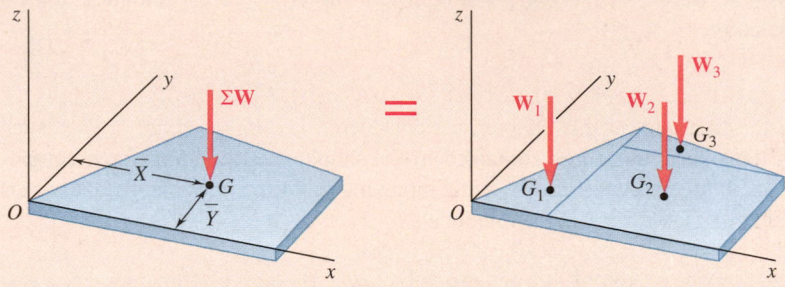

**Fig. 5.24**

If the plate is homogeneous and of uniform thickness, its center of gravity coincides with the centroid $C$ of the area of the plate, and Eqs. (5.7) reduce to

$$Q_y = \overline{X}\Sigma A = \Sigma \overline{x} A \qquad Q_x = \overline{Y}\Sigma A = \Sigma \overline{y} A \tag{5.8}$$

These equations yield the first moments of the composite area, or they can be solved for the coordinates $\overline{X}$ and $\overline{Y}$ of its centroid (Sample Prob. 5.1). Determining the center of gravity of a composite wire is carried out in a similar fashion (Sample Prob. 5.2).

## Determining a Centroid by Integration

When an area is bounded by analytical curves, you can determine the coordinates of its centroid by *integration* (Sec. 5.2A). This can be done by evaluating either the double integrals in Eqs. (5.3) or a single integral that uses one of the thin rectangular or pie-shaped elements of area shown in Fig. 5.12. Denoting by $\overline{x}_{el}$ and $\overline{y}_{el}$ the coordinates of the centroid of the element $dA$, we have

$$Q_y = \overline{x} A = \int \overline{x}_{el} \, dA \qquad Q_x = \overline{y} A = \int \overline{y}_{el} \, dA \tag{5.9}$$

It is advantageous to use the same element of area to compute both of the first moments $Q_y$ and $Q_x$; we can also use the same element to determine the area $A$ (Sample Prob. 5.4).

## Theorems of Pappus-Guldinus

The **theorems of Pappus-Guldinus** relate the area of a surface of revolution or the volume of a body of revolution to the centroid of the generating curve or area (Sec. 5.2B). The area $A$ of the surface generated by rotating a curve of length $L$ about a fixed axis (Fig. 5.25a) is

$$A = 2\pi \overline{y} L \tag{5.10}$$

where $\overline{y}$ represents the distance from the centroid $C$ of the curve to the fixed axis. Similarly, the volume $V$ of the body generated by rotating an area $A$ about a fixed axis (Fig. 5.25b) is

$$V = 2\pi \overline{y} A \tag{5.11}$$

where $\overline{y}$ represents the distance from the centroid $C$ of the area to the fixed axis.

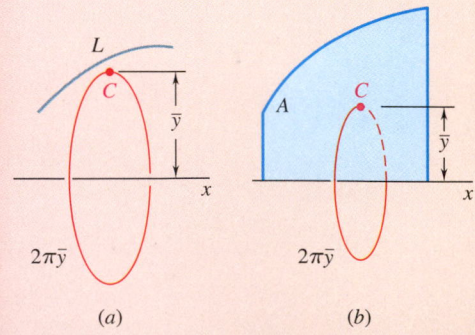

**Fig. 5.25**

## Distributed Loads

The concept of the centroid of an area also can be used to solve problems other than those dealing with the weight of flat plates. For example, to determine the reactions at the supports of a beam (Sec. 5.3A), we can replace a **distributed load** $w$ by a concentrated load **W** equal in magnitude to the area $A$ under the load curve and passing through the centroid $C$ of that area (Fig. 5.26). We can use this same approach to determine the resultant of the hydrostatic forces exerted on a **rectangular plate submerged in a liquid** (Sec. 5.3B).

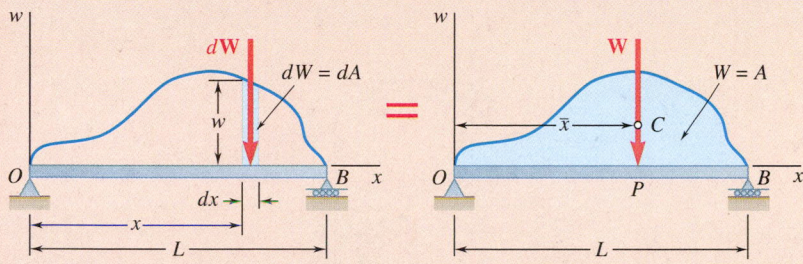

**Fig. 5.26**

## Center of Gravity of a Three-Dimensional Body

The last part of this chapter was devoted to determining the center of gravity $G$ of a three-dimensional body. We defined the coordinates $\bar{x}, \bar{y}, \bar{z}$ of $G$ by the relations

$$\bar{x}W = \int x \, dW \qquad \bar{y}W = \int y \, dW \qquad \bar{z}W = \int z \, dW \qquad \textbf{(5.17)}$$

## Centroid of a Volume

In the case of a homogeneous body, the center of gravity $G$ coincides with the centroid $C$ of the volume $V$ of the body. The coordinates of $C$ are defined by the relations

$$\bar{x}V = \int x \, dV \qquad \bar{y}V = \int y \, dV \qquad \bar{z}V = \int z \, dV \qquad \textbf{(5.19)}$$

If the volume possesses a *plane of symmetry,* its centroid $C$ lies in that plane; if it possesses two planes of symmetry, $C$ is located on the line of intersection of the two planes; if it possesses three planes of symmetry that intersect at only one point, $C$ coincides with that point (Sec. 5.4A).

## Center of Gravity of a Composite Body

The volumes and centroids of various common three-dimensional shapes are tabulated in Fig. 5.21. When a body can be divided into several of these shapes, we can determine the coordinates $\bar{X}, \bar{Y}, \bar{Z}$ of its center of gravity $G$ from the corresponding coordinates of the centers of gravity of its various parts (Sec. 5.4B). We have

$$\bar{X}\Sigma W = \Sigma \bar{x}W \qquad \bar{Y}\Sigma W = \Sigma \bar{y}W \qquad \bar{Z}\Sigma W = \Sigma \bar{z}W \qquad \textbf{(5.20)}$$

If the body is made of a homogeneous material, its center of gravity coincides with the centroid $C$ of its volume, and we have (Sample Probs. 5.11 and 5.12)

$$\bar{X}\Sigma V = \Sigma \bar{x}V \qquad \bar{Y}\Sigma V = \Sigma \bar{y}V \qquad \bar{Z}\Sigma V = \Sigma \bar{z}V \qquad \textbf{(5.21)}$$

### Determining a Centroid by Integration

When a volume is bounded by analytical surfaces, we can find the coordinates of its centroid by *integration* (Sec. 5.4C). To avoid the computation of triple integrals in Eqs. (5.19), we can use elements of volume in the shape of thin filaments, as shown in Fig. 5.27. Denoting the coordinates of the centroid of the element $dV$ as $\bar{x}_{el}$, $\bar{y}_{el}$, $\bar{z}_{el}$, we rewrite Eqs. (5.19) as

$$\bar{x}V = \int \bar{x}_{el}\, dV \qquad \bar{y}V = \int \bar{y}_{el}\, dV \qquad \bar{z}V = \int \bar{z}_{el}\, dV \qquad (5.23)$$

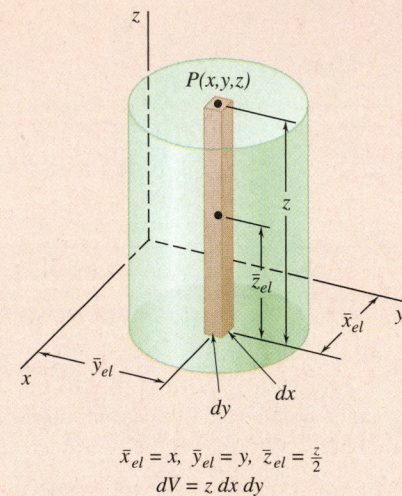

$$\bar{x}_{el} = x, \ \bar{y}_{el} = y, \ \bar{z}_{el} = \frac{z}{2}$$
$$dV = z\, dx\, dy$$

**Fig. 5.27**

that involve only double integrals. If the volume possesses *two planes of symmetry,* its centroid $C$ is located on their line of intersection. Choosing the $x$ axis to lie along that line and dividing the volume into thin slabs parallel to the $yz$ plane, we can determine $C$ from the relation

$$\bar{x}V = \int \bar{x}_{el}\, dV \qquad (5.24)$$

with a *single integration* (Sample Prob. 5.13). For a body of revolution, these slabs are circular and their volume is given in Fig. 5.28.

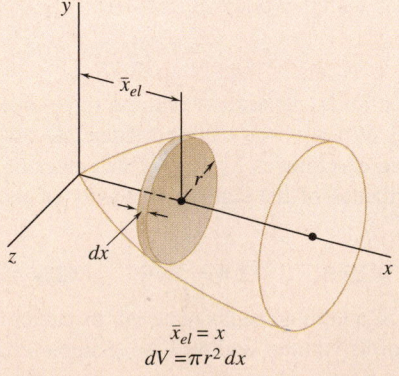

$$\bar{x}_{el} = x$$
$$dV = \pi r^2\, dx$$

**Fig. 5.28**

# Review Problems

**5.137 and 5.138**  Locate the centroid of the plane area shown.

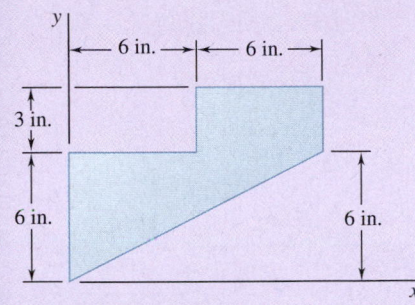

**Fig. P5.137**

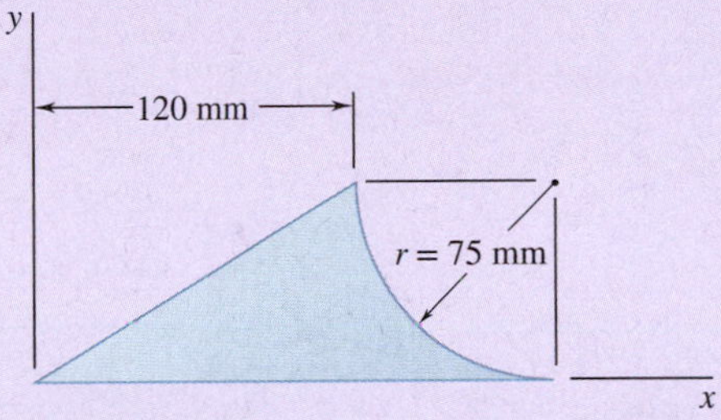

y axis diagram with dimensions 120 mm and r = 75 mm

**Fig. P5.138**

**5.139**  A uniform circular rod with a weight of 8 lb and radius of 10 in. is attached to a pin at $C$ and to the cable $AB$. Determine ($a$) the tension in the cable, ($b$) the reaction at $C$.

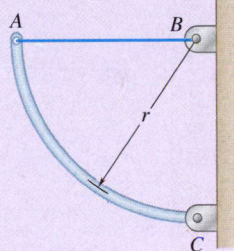

**Fig. P5.139**

**5.140**  Determine by direct integration the centroid of the area shown. Express your answer in terms of $a$ and $h$.

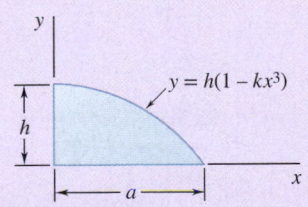

$y = h(1 - kx^3)$

**Fig. P5.140**

**5.141**  Determine by direct integration the centroid of the area shown.

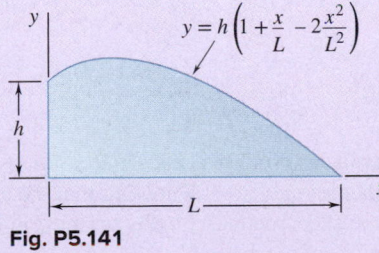

$y = h\left(1 + \dfrac{x}{L} - 2\dfrac{x^2}{L^2}\right)$

**Fig. P5.141**

**5.142**  The escutcheon (a decorative plate placed on a pipe where the pipe exits from a wall) shown is cast from brass. Knowing that the density of brass is 8470 kg/m$^3$, determine the mass of the escutcheon.

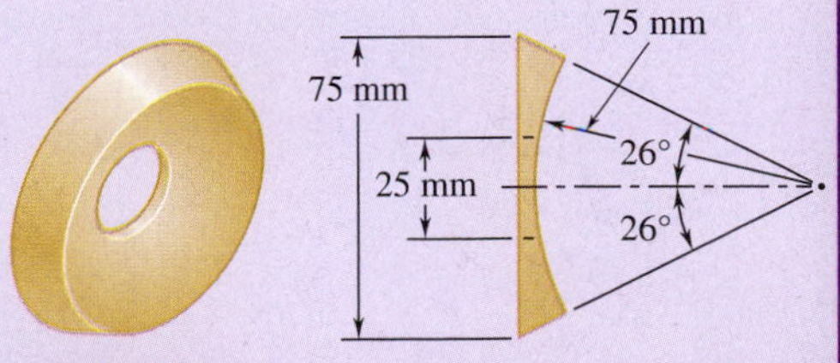

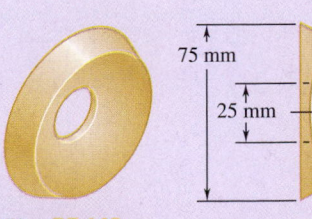

**Fig. P5.142**

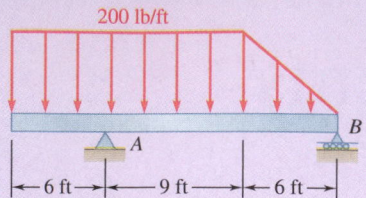

200 lb/ft

A

B

|← 6 ft →|← 9 ft →|← 6 ft →|

**Fig. P5.143**

**5.143** Determine the reactions at the beam supports for the given loading.

**5.144** A beam is subjected to a linearly distributed downward load and rests on two wide supports $BC$ and $DE$ that exert uniformly distributed upward loads as shown. Determine the values of $w_{BC}$ and $w_{DE}$ corresponding to equilibrium when $w_A = 600$ N/m.

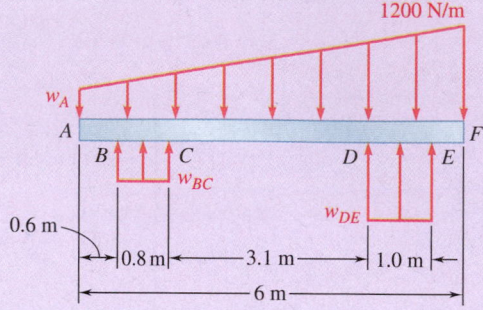

**Fig. P5.144**

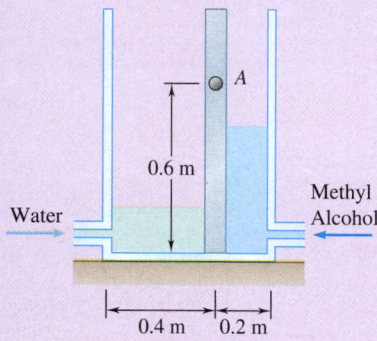

Water

Methyl Alcohol

A

0.6 m

|← 0.4 m →|← 0.2 m →|

**Fig. P5.145**

**5.145** A tank is divided into two sections by a 1 × 1-m square gate that is hinged at $A$. A couple with a magnitude of 490 N·m is required for the gate to rotate. If one side of the tank is filled with water at the rate of 0.1 m³/min and the other side is filled simultaneously with methyl alcohol (density $\rho_{ma} = 789$ kg/m³) at the rate of 0.2 m³/min, determine at what time and in which direction the gate will rotate.

**5.146** Determine the $y$ coordinate of the centroid of the body shown.

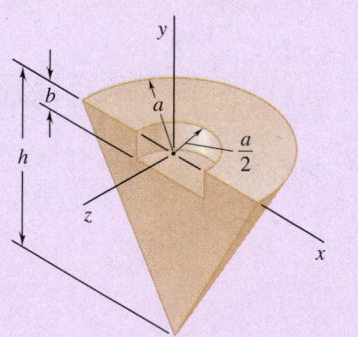

**Fig. P5.146**

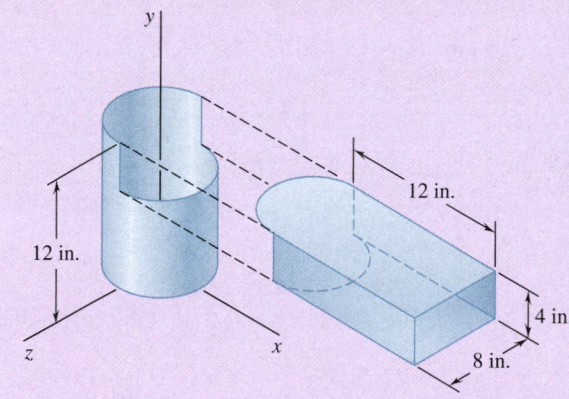

**Fig. P5.147**

**5.147** An 8-in.-diameter cylindrical duct and a 4 × 8-in. rectangular duct are to be joined as indicated. Knowing that the ducts were fabricated from the same sheet metal, which is of uniform thickness, locate the center of gravity of the assembly.

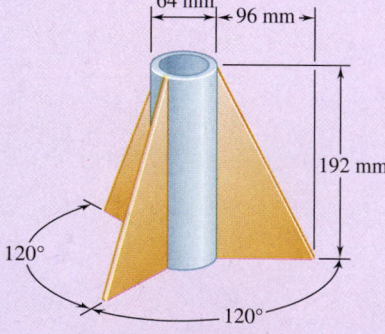

64 mm

96 mm

192 mm

120°

120°

**Fig. P5.148**

**5.148** Three brass plates are brazed to a steel pipe to form the flagpole base shown. Knowing that the pipe has a wall thickness of 8 mm and that each plate is 6 mm thick, determine the location of the center of gravity of the base. (Densities: brass = 8470 kg/m³; steel = 7860 kg/m³.)

©Lee Rentz/Avalon

# 6

# Analysis of Structures

Trusses, such as this cantilever arch bridge over Deception Pass in Washington state, provide both a practical and an economical solution to many engineering problems.

# Objectives

- **Define** an ideal truss and consider the attributes of simple trusses.
- **Analyze** plane and space trusses by the method of joints.
- **Simplify** certain truss analyses by recognizing special loading and geometry conditions.
- **Analyze** trusses by the method of sections.
- **Consider** the characteristics of compound trusses.
- **Analyze** structures containing multi-force members, such as frames and machines.

## Introduction

In the preceding chapters, we studied the equilibrium of a single rigid body, where all forces involved were external to the rigid body. We now consider the equilibrium of structures made of several connected parts. This situation calls for determining not only the external forces acting on the structure, but also the forces that hold together the various parts of the structure. From the point of view of the structure as a whole, these forces are **internal forces**.

Consider, for example, the crane shown in Fig. 6.1*a* that supports a load *W*. The crane consists of three beams *AD*, *CF,* and *BE* connected by frictionless pins; it is supported by a pin at *A* and by a cable *DG*. The free-body diagram of the crane is drawn in Fig. 6.1*b*. The external forces shown in the diagram include the weight **W**, the two components $\mathbf{A}_x$ and $\mathbf{A}_y$ of the reaction at *A,* and the force **T** exerted by the cable at *D*. The internal forces holding the various parts of the crane together do not appear in the free-body diagram. If, however, we dismember the crane and draw a free-body diagram for each of its component parts, we can see the forces holding the three beams together because these forces are external forces from the point of view of each component part (Fig. 6.1*c*).

Note that we represent the force exerted at *B* by member *BE* on member *AD* as equal and opposite to the force exerted at the same point by member *AD* on member *BE*. Similarly, the force exerted at *E* by *BE* on *CF* is shown equal and opposite to the force exerted by *CF* on *BE*, and the components of the force exerted at *C* by *CF* on *AD* are shown equal and opposite to the components

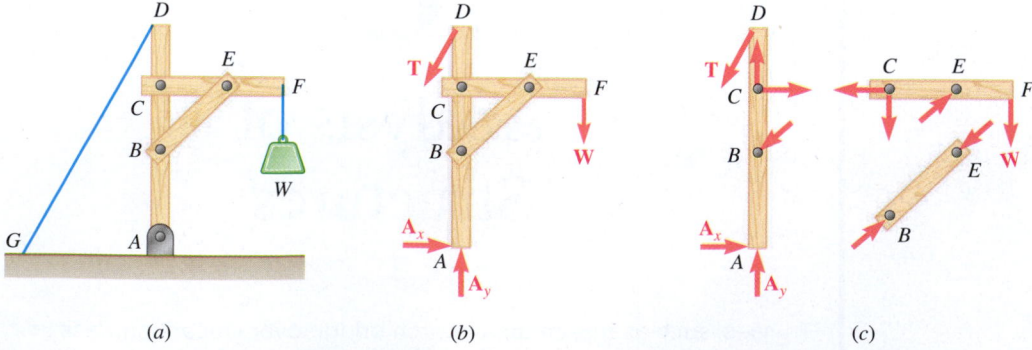

(a)                      (b)                      (c)

**Fig. 6.1**   A structure in equilibrium. (*a*) Diagram of a crane supporting a load; (*b*) free-body diagram of the crane; (*c*) free-body diagrams of the components of the crane.

of the force exerted by *AD* on *CF*. These representations agree with Newton's third law, which states that

> **The forces of action and reaction between two bodies in contact have the same magnitude, same line of action, and opposite sense.**

We pointed out in Chap. 1 that this law, which is based on experimental evidence, is one of the six fundamental principles of elementary mechanics. Its application is essential for solving problems involving connected bodies.

In this chapter, we consider three broad categories of engineering structures:

1. **Trusses**, which are designed to support loads and are usually stationary, fully constrained structures. Trusses consist exclusively of straight members connected at joints located at the ends of each member. Members of a truss, therefore, are **two-force members**, i.e., members acted upon by two equal and opposite forces directed along the member.
2. **Frames**, which are also designed to support loads and are also usually stationary, fully constrained structures. However, like the crane of Fig. 6.1, frames always contain at least one **multi-force member**, i.e., a member acted upon by three or more forces that, in general, are not directed along the member.
3. **Machines**, which are designed to transmit and modify forces and are structures containing moving parts. Machines, like frames, always contain at least one multi-force member.

**Two-force member**

**Multi-force member**

**Multi-force member**

(*a*) A truss bridge          (*b*) A bicycle frame          (*c*) A hydraulic machine arm

**Photo 6.1**   The structures you see around you to support loads or transmit forces are generally trusses, frames, or machines. (a) ©Datacraft Co Ltd/Getty Images RF; (b) ©Fuse/Getty Images RF; (c) ©Design Pics/Ken Welsh RF

# 6.1 ANALYSIS OF TRUSSES

The truss is one of the major types of engineering structures. It provides a practical and economical solution to many engineering situations, especially in the design of bridges and buildings. In this section, we describe the basic elements of a truss and study a common method for analyzing the forces acting in a truss.

## 6.1A Simple Trusses

A truss consists of straight members connected at joints, as shown in Fig. 6.2*a*. Truss members are connected at their extremities only; no member is continuous through a joint. In Fig. 6.2*a*, for example, there is no member *AB*; instead

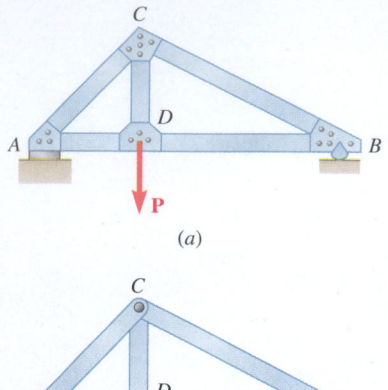

**Fig. 6.2** (*a*) A typical truss consists of straight members connected at joints; (*b*) we can model a truss as two-force members connected by pins.

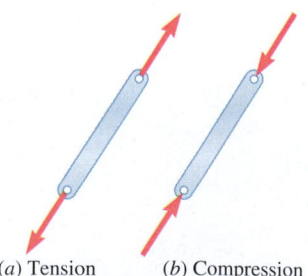

**Photo 6.2** Shown is a pin-jointed connection on the approach span to the San Francisco–Oakland Bay Bridge.

(*a*) Tension   (*b*) Compression

**Fig. 6.4** A two-force member of a truss can be in tension or compression.

we have two distinct members *AD* and *DB*. Most actual structures are made of several trusses joined together to form a space framework. Each truss is designed to carry those loads that act in its plane and thus may be treated as a two-dimensional structure.

In general, the members of a truss are slender and can support little lateral load; all loads, therefore, must be applied at the various joints and not to the members themselves. When a concentrated load is to be applied between two joints or when the truss must support a distributed load, as in the case of a bridge truss, a floor system must be provided. The floor transmits the load to the joints through the use of stringers and floor beams (Fig. 6.3).

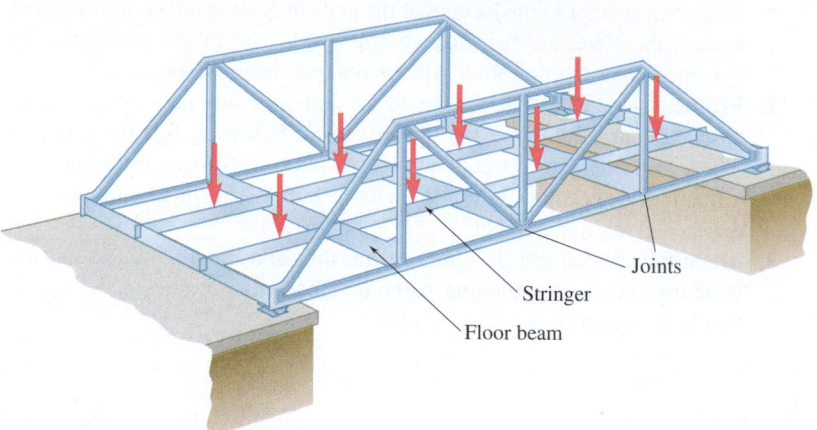

**Fig. 6.3** A floor system of a truss uses stringers and floor beams to transmit an applied load to the joints of the truss.

We assume that the weights of the truss members can be applied to the joints, with half of the weight of each member applied to each of the two joints the member connects. Although the members are actually joined together by means of welded, bolted, or riveted connections, it is customary to assume that the members are pinned together; therefore, the forces acting at each end of a member reduce to a single force and no couple. This enables us to model the forces applied to a truss member as a single force at each end of the member. We can then treat each member as a two-force member, and we can consider the entire truss as a group of pins and two-force members (Fig. 6.2*b*). An individual member can be acted upon as shown in either of the two sketches of Fig. 6.4. In Fig. 6.4*a*, the forces tend to pull the member apart, and the member is in tension; in Fig. 6.4*b*, the forces tend to push the member together, and the member is in compression. Some typical trusses are shown in Fig. 6.5.

Consider the truss of Fig. 6.6*a*, which is made of four members connected by pins at *A, B, C,* and *D*. If we apply a load at *B*, the truss will greatly deform, completely losing its original shape. In contrast, the truss of Fig. 6.6*b*, which is made of three members connected by pins at *A, B,* and *C*, will deform only slightly under a load applied at *B*. The only possible deformation for this truss is one involving small changes in the length of its members. The truss of Fig. 6.6*b* is said to be a **rigid truss**, the term "rigid" being used here to indicate that the truss *will not collapse.*

As shown in Fig. 6.6*c*, we can obtain a larger rigid truss by adding two members *BD* and *CD* to the basic triangular truss of Fig. 6.6*b*. We can repeat

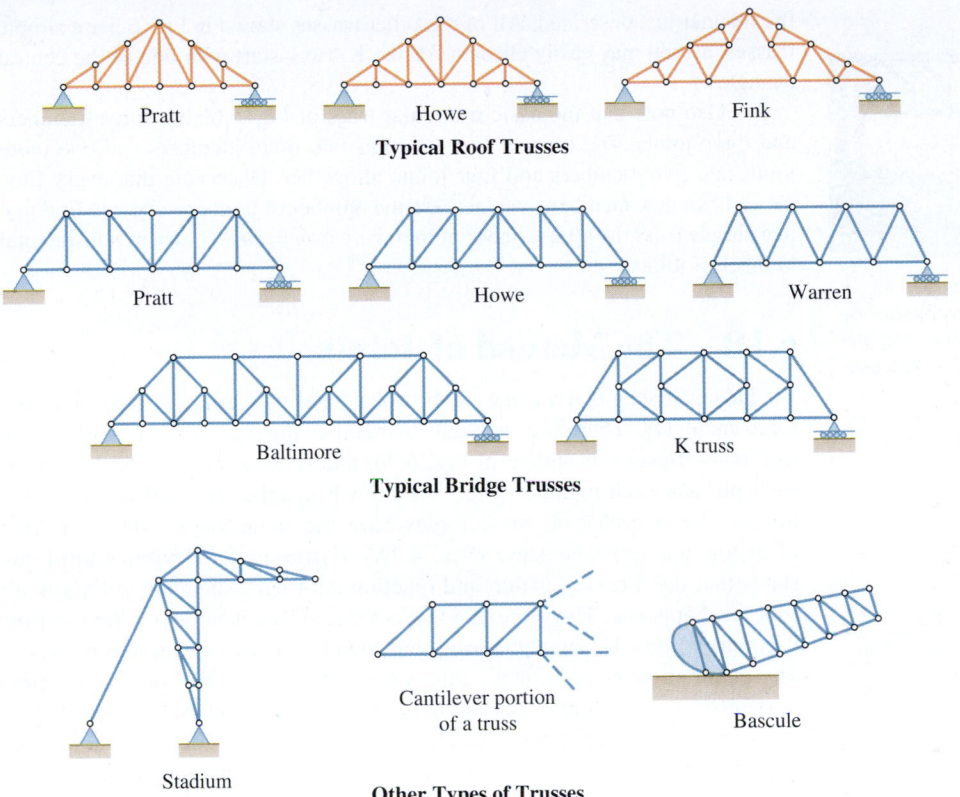

**Fig. 6.5**   You can often see trusses in the design of a building roof, a bridge, or other other larger structures.

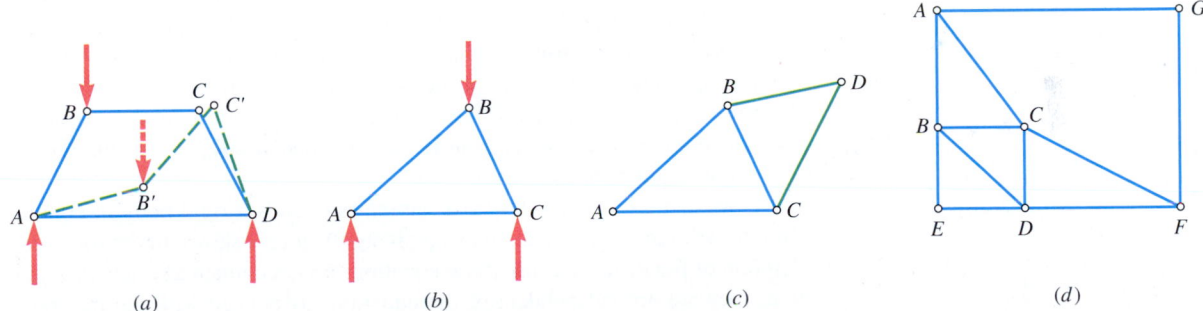

**Fig. 6.6**   (a) A poorly designed truss that cannot support a load; (b) the most elementary rigid truss consists of a simple triangle; (c) a larger rigid truss built up from the triangle in (b); (d) a rigid truss not made up of triangles alone.

this procedure as many times as we like, and the resulting truss will be rigid if each time we add two new members they are attached to two existing joints and connected at a new joint. (The three joints must not be in a straight line.) A truss that can be constructed in this manner is called a **simple truss**.

Note that a simple truss is not necessarily made only of triangles. The truss of Fig. 6.6d, for example, is a simple truss that we constructed from triangle *ABC* by adding successively the joints *D, E, F,* and *G*. On the other hand, rigid trusses are not always simple trusses, even when they appear to be made of triangles. The Fink and Baltimore trusses shown in Fig. 6.5, for instance, are not simple trusses, because they cannot be constructed from a single triangle in

**Photo 6.3** Two K trusses were used as the main components of the movable bridge shown, which moved above a large stockpile of ore. The bucket below the trusses picked up ore and redeposited it until the ore was thoroughly mixed. The ore was then sent to the mill for processing into steel.

Courtesy of Ferdinand Beer

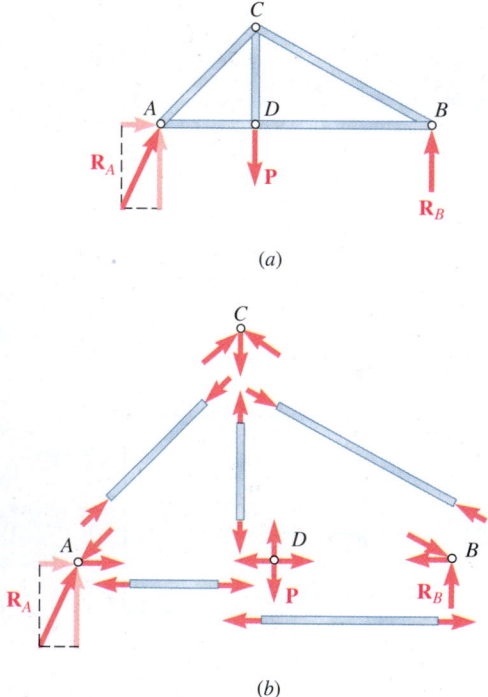

**Fig. 6.7** (a) Free-body diagram of the truss as a rigid body; (b) free-body diagrams of the five members and four pins that make up the truss.

the manner just described. All of the other trusses shown in Fig. 6.5 are simple trusses, as you may easily check. (For the K truss, start with one of the central triangles.)

Also note that the basic triangular truss of Fig. 6.6b has three members and three joints. The truss of Fig. 6.6c has two more members and one more joint; i.e., five members and four joints altogether. Observing that every time we add two new members, we increase the number of joints by one, we find that in a simple truss the total number of members is $m = 2n - 3$, where $n$ is the total number of joints.

## 6.1B The Method of Joints

We have just seen that a truss can be considered as a group of pins and two-force members. Therefore, we can dismember the truss of Fig. 6.2, whose free-body diagram is shown in Fig. 6.7a, and draw a free-body diagram for each pin and each member (Fig. 6.7b). Each member is acted upon by two forces, one at each end; these forces have the same magnitude, same line of action, and opposite sense (Sec. 4.2A). Furthermore, Newton's third law states that the forces of action and reaction between a member and a pin are equal and opposite. Therefore, the forces exerted by a member on the two pins it connects must be directed along that member and be equal and opposite. The common magnitude of the forces exerted by a member on the two pins it connects is commonly referred to as the *force in the member*, even though this quantity is actually a scalar. Because we know the lines of action of all the internal forces in a truss, the analysis of a truss reduces to computing the forces in its various members and determining whether each of its members is in tension or compression.

Because the entire truss is in equilibrium, each pin must be in equilibrium. We can use the fact that a pin is in equilibrium to draw its free-body diagram and write two equilibrium equations (Sec. 2.3A). Thus, if the truss contains $n$ pins, we have $2n$ equations available, which can be solved for $2n$ unknowns. In the case of a simple truss, we have $m = 2n - 3$; that is, $2n = m + 3$, and the number of unknowns that we can determine from the free-body diagrams of the pins is $m + 3$. This means that we can find the forces in all the members, the two components of the reaction $\mathbf{R}_A$, and the reaction $\mathbf{R}_B$ by considering the free-body diagrams of the pins.

We can also use the fact that the entire truss is a rigid body in equilibrium to write three more equations involving the forces shown in the free-body diagram of Fig. 6.7a. Because these equations do not contain any new information, they are not independent of the equations associated with the free-body diagrams of the pins. Nevertheless, we can use them to determine the components of the reactions at the supports. The arrangement of pins and members in a simple truss is such that it is always possible to find a joint involving only two unknown forces. We can determine these forces by using the methods of Sec. 2.3C and then transferring their values to the adjacent joints, treating them as known quantities at these joints. We repeat this procedure until we have determined all unknown forces.

As an example, let's analyze the truss of Fig. 6.7 by considering the equilibrium of each pin successively, starting with a joint at which only two forces are unknown. In this truss, all pins are subjected to at least three unknown forces. Therefore, we must first determine the reactions at the supports by considering the entire truss as a free body and using the equations of equilibrium of a rigid body. In this way, we find that $\mathbf{R}_A$ is vertical, and we determine the magnitudes of $\mathbf{R}_A$ and $\mathbf{R}_B$.

This reduces the number of unknown forces at joint *A* to two, and we can determine these forces by considering the equilibrium of pin *A*. The reaction $\mathbf{R}_A$ and the forces $\mathbf{F}_{AC}$ and $\mathbf{F}_{AD}$ exerted on pin *A* by members *AC* and *AD*, respectively, must form a force triangle. First we draw $\mathbf{R}_A$ (Fig. 6.8); noting that $\mathbf{F}_{AC}$ and $\mathbf{F}_{AD}$ are directed along *AC* and *AD*, respectively, we complete the triangle and determine the magnitude and sense of $\mathbf{F}_{AC}$ and $\mathbf{F}_{AD}$. The magnitudes $F_{AC}$ and $F_{AD}$ represent the forces in members *AC* and *AD*. Because $\mathbf{F}_{AC}$ is directed down and to the left—that is, *toward* joint *A*—member *AC* pushes on pin *A* and is in compression. (From Newton's third law, pin *A* pushes *on* member AC.) Because $\mathbf{F}_{AD}$ is directed *away* from joint *A*, member *AD* pulls on pin *A* and is in tension. (From Newton's third law, pin *A* pulls *away* from member *AD*.)

We can now proceed to joint *D*, where only two forces, $\mathbf{F}_{DC}$ and $\mathbf{F}_{DB}$, are still unknown. The other forces are the load $\mathbf{P}$, which is given, and the force $\mathbf{F}_{DA}$ exerted on the pin by member *AD*. As indicated previously, this force is equal and opposite to the force $\mathbf{F}_{AD}$ exerted by the same member on pin *A*. We can draw the force polygon corresponding to joint *D*, as shown in Fig. 6.8, and determine the forces $\mathbf{F}_{DC}$ and $\mathbf{F}_{DB}$ from that polygon. However, when more than

**Photo 6.4**  Because roof trusses, such as those shown, require support only at their ends, it is possible to construct buildings with large, unobstructed interiors. ©McGraw-Hill Education/Sabina Dowell

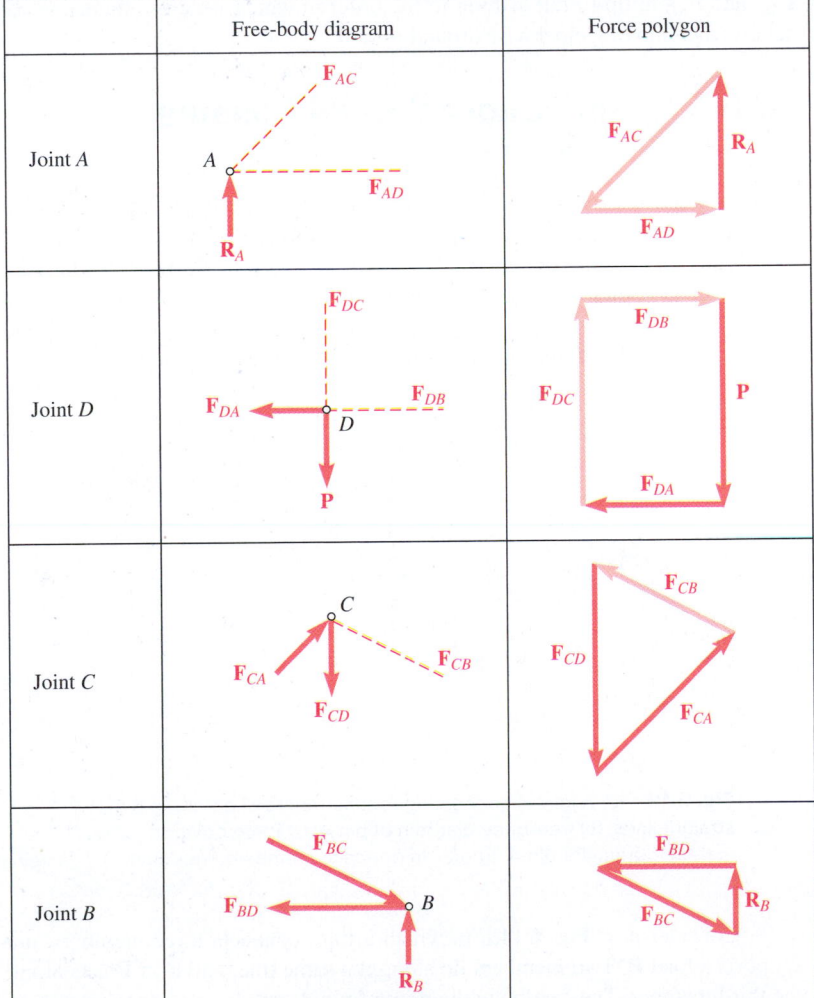

**Fig. 6.8**  Free-body diagrams and force polygons used to determine the forces on the pins and in the members of the truss in Fig. 6.7.

three forces are involved, it is usually more convenient to solve the equations of equilibrium $\Sigma F_x = 0$ and $\Sigma F_y = 0$ for the two unknown forces. Because both of these forces are directed away from joint $D$, members $DC$ and $DB$ pull on the pin and are in tension.

Next, we consider joint $C$; its free-body diagram is shown in Fig. 6.8. Both $\mathbf{F}_{CD}$ and $\mathbf{F}_{CA}$ are known from the analysis of the preceding joints, so only $\mathbf{F}_{CB}$ is unknown. Because the equilibrium of each pin provides sufficient information to determine two unknowns, we can check our analysis at this joint. We draw the force triangle and determine the magnitude and sense of $\mathbf{F}_{CB}$. Because $\mathbf{F}_{CB}$ is directed toward joint $C$, member $CB$ pushes on pin $C$ and is in compression. The check is obtained by verifying that the force $\mathbf{F}_{CB}$ and member $CB$ are parallel.

Finally, at joint $B$, we know all of the forces. Because the corresponding pin is in equilibrium, the force triangle must close, giving us an additional check of the analysis.

Note that the force polygons shown in Fig. 6.8 are not unique; we could replace each of them by an alternative configuration. For example, the force triangle corresponding to joint $A$ could be drawn as shown in Fig. 6.9. We obtained the triangle actually shown in Fig. 6.8 by drawing the three forces $\mathbf{R}_A$, $\mathbf{F}_{AC}$, and $\mathbf{F}_{AD}$ in tip-to-tail fashion in the order in which we cross their lines of action when moving clockwise around joint $A$.

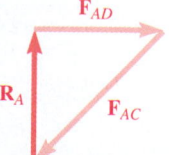

**Fig. 6.9** Alternative force polygon for joint $A$ in Fig. 6.8.

## *6.1C Joints Under Special Loading Conditions

Some geometric arrangements of members in a truss are particularly simple to analyze by observation. For example, Fig. 6.10*a* shows a joint connecting four members lying along two intersecting straight lines. The free-body diagram of Fig. 6.10*b* shows that pin $A$ is subjected to two pairs of directly opposite forces. The corresponding force polygon, therefore, must be a parallelogram (Fig. 6.10*c*), and **the forces in opposite members must be equal**.

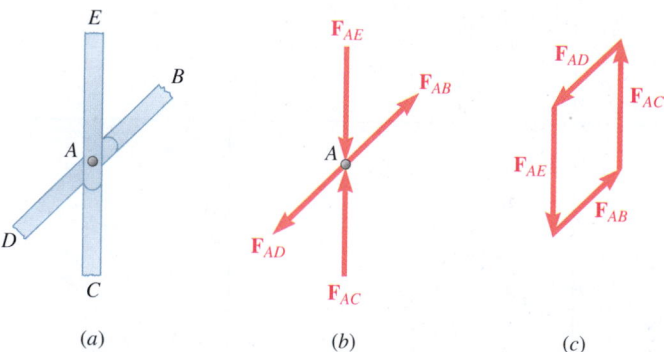

**Fig. 6.10** (*a*) A joint $A$ connecting four members of a truss in two straight lines; (*b*) free-body diagram of pin $A$; (*c*) force polygon (parallelogram) for pin $A$. Forces in opposite members are equal.

Consider next Fig. 6.11*a*, in which a joint connects three members and supports a load $\mathbf{P}$. Two members lie along the same line, and load $\mathbf{P}$ acts along the third member. The free-body diagram of pin $A$ and the corresponding force polygon are the same as in Fig. 6.10*b* and *c*, with $\mathbf{F}_{AE}$ replaced by load $\mathbf{P}$. Thus, **the forces in the two opposite members must be equal, and the force in** the other member must equal *P.* Fig. 6.11*b* shows a particular case of special interest. Because, in this case, no external load is applied to the joint, we have $P = 0$, and the force in member *AC* is zero. Member *AC* is said to be a **zero-force member**.

Now consider a joint connecting two members only. From Sec. 2.3A, we know that a particle acted upon by two forces is in equilibrium if the two forces have the same magnitude, same line of action, and opposite sense. In the case of the joint of Fig. 6.12*a*, which connects two members *AB* and *AD* lying along the same line, the forces in the two members must be equal for pin *A* to be in equilibrium. In the case of the joint of Fig. 6.12*b*, pin *A* cannot be in equilibrium unless the forces in both members are zero. Members connected as shown in Fig. 6.12*b*, therefore, must be **zero-force members**.

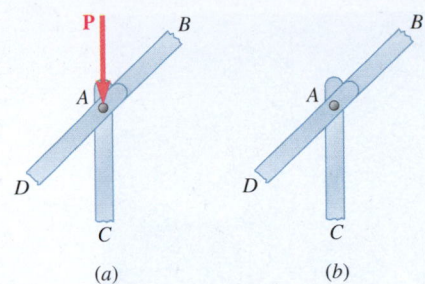

**Fig. 6.11** (*a*) Joint *A* in a truss connects three members, two in a straight line and the third along the line of a load. Force in the third member equals the load. (*b*) If the load is zero, the third member is a zero-force member.

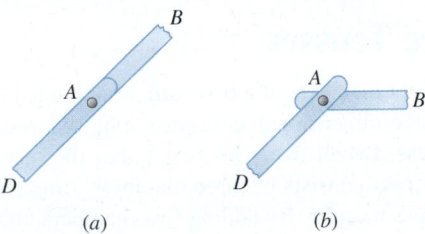

**Fig. 6.12** (*a*) A joint in a truss connecting two members in a straight line. Forces in the members are equal. (*b*) If the two members are not in a straight line, they must be zero-force members.

Spotting joints that are under the special loading conditions just described will expedite the analysis of a truss. Consider, for example, a Howe truss loaded as shown in Fig. 6.13. We can recognize all of the members represented by green lines as zero-force members. Joint *C* connects three members, two of which lie in the same line, and is not subjected to any external load; member *BC* is thus a zero-force member. Applying the same reasoning to joint *K,* we find that member *JK* is also a zero-force member. But joint *J* is now in the same situation as joints *C* and *K,* so member *IJ* also must be a zero-force member. Examining joints *C, J,* and *K* also shows that the forces in members *AC* and *CE* are equal, that the forces in members *HJ* and *JL* are equal, and that the forces in members *IK* and *KL* are equal. Turning our attention to joint *I,* where the 20-kN load and member *HI* are collinear, we note that the force in member *HI* is 20 kN (tension) and that the forces in members *GI* and *IK* are equal. Hence, the forces in members *GI, IK,* and *KL* are equal.

Note that the conditions described here do not apply to joints *B* and *D* in Fig. 6.13, so it is wrong to assume that the force in member *DE* is 25 kN or that the forces in members *AB* and *BD* are equal. To determine the forces in these members and in all remaining members, you need to carry out the analysis

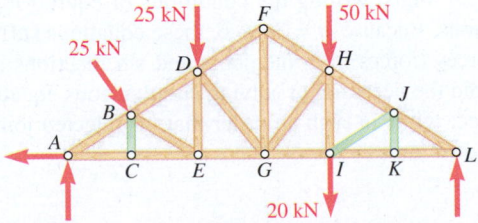

**Fig. 6.13** An example of loading on a Howe truss; identifying special loading conditions.

**Photo 6.5** Three-dimensional or space trusses are used for broadcast and power transmission line towers, roof framing, and spacecraft applications, such as components of the *International Space Station*. ©James Hardy/PhotoAlto RF

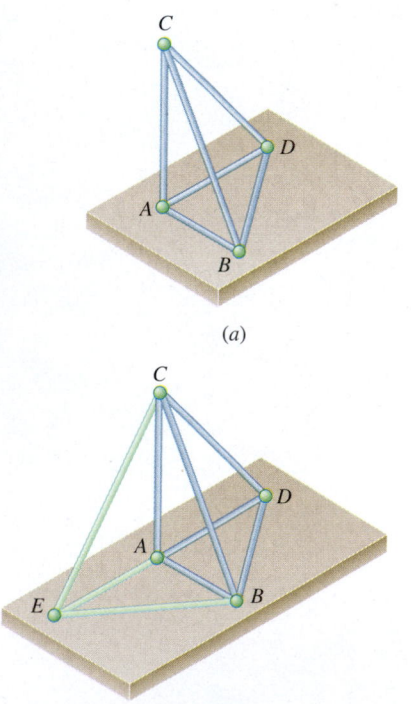

**Fig. 6.14** (*a*) The most elementary space truss consists of six members joined at their ends to form a tetrahedron. (*b*) We can add three members at a time to three joints of an existing space truss, connecting the new members at a new joint, to build a larger simple space truss.

of joints *A, B, D, E, F, G, H,* and *L* in the usual manner. Thus, until you have become thoroughly familiar with the conditions under which you can apply the rules described in this section, you should draw the free-body diagrams of all pins and write the corresponding equilibrium equations (or draw the corresponding force polygons) whether or not the joints being considered are under one of these special loading conditions.

A final remark concerning zero-force members: these members are not useless. For example, although the zero-force members of Fig. 6.13 do not carry any loads under the loading conditions shown, the same members would probably carry loads if the loading conditions were changed. Besides, even in the case considered, these members are needed to support the weight of the truss and to maintain the truss in the desired shape.

## *6.1D Space Trusses

When several straight members of a truss are joined together at their extremities to form a three-dimensional configuration, the resulting structure is called a **space truss**. Recall from Sec. 6.1A that the most elementary two-dimensional rigid truss consists of three members joined at their extremities to form the sides of a triangle. By adding two members at a time to this basic configuration and connecting them at a new joint, we could obtain a larger rigid structure that we defined as a simple truss. Similarly, the most elementary rigid space truss consists of six members joined at their extremities to form the edges of a tetrahedron *ABCD* (Fig. 6.14*a*). By adding three members at a time to this basic configuration, such as *AE, BE,* and *CE* (Fig. 6.14*b*), attaching them to three existing joints, and connecting them at a new joint, we can obtain a larger rigid structure that we define as a **simple space truss**. (The four joints must not lie in a plane.) Note that the basic tetrahedron has six members and four joints, and every time we add three members, the number of joints increases by one. Therefore, we conclude that in a simple space truss the total number of members is $m = 3n - 6$, where *n* is the total number of joints.

If a space truss is to be completely constrained and if the reactions at its supports are to be statically determinate, the supports should consist of a combination of balls, rollers, and balls and sockets, providing six unknown reactions (see Sec. 4.3B). We can determine these unknown reactions by solving the six equations expressing that the three-dimensional truss is in equilibrium.

Although the members of a space truss are actually joined together by means of bolted or welded connections, we assume for analysis purposes that each joint consists of a ball-and-socket connection. Thus, no couple is applied to the members of the truss, and we can treat each member as a two-force member. The conditions of equilibrium for each joint are expressed by the three equations $\Sigma F_x = 0$, $\Sigma F_y = 0$, and $\Sigma F_z = 0$. Thus, in the case of a simple space truss containing *n* joints, writing the conditions of equilibrium for each joint yields 3*n* equations. Because $m = 3n - 6$, these equations suffice to determine all unknown forces (forces in *m* members and six reactions at the supports). However, to avoid the necessity of solving simultaneous equations, you should take care to select joints in such an order that no selected joint involves more than three unknown forces.

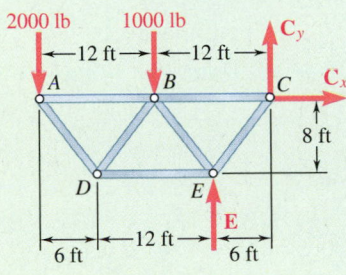

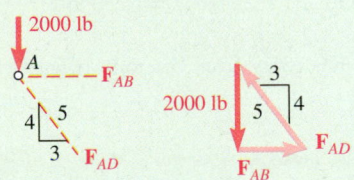

**Fig. 1** Free-body diagram of the entire truss.

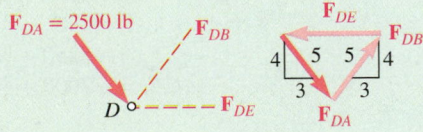

**Fig. 2** Free-body diagram of joint A.

# Sample Problem 6.1

Using the method of joints, determine the force in each member of the truss shown.

**STRATEGY:** To use the method of joints, you start with an analysis of the free-body diagram of the entire truss. Then, look for a joint connecting only two members as a starting point for the calculations. In this example, we start at joint A and proceed through joints D, B, E, and C, but you could also start at joint C and proceed through joints E, B, D, and A.

**MODELING and ANALYSIS:** You can combine these steps for each joint of the truss in turn. Draw a free-body diagram; draw a force polygon or write the equilibrium equations; and solve for the unknown forces.

**Entire Truss.** Draw a free-body diagram of the entire truss (Fig. 1); external forces acting on this free body are the applied loads and the reactions at C and E. Write the equilibrium equations, taking moments about C.

$$+\circlearrowright\Sigma M_C = 0: \quad (2000 \text{ lb})(24 \text{ ft}) + (1000 \text{ lb})(12 \text{ ft}) - E(6 \text{ ft}) = 0$$
$$E = +10{,}000 \text{ lb} \qquad\qquad \mathbf{E} = 10{,}000 \text{ lb} \uparrow$$

$$\overset{+}{\rightarrow}\Sigma F_x = 0: \qquad\qquad\qquad\qquad\qquad\qquad \mathbf{C}_x = 0$$
$$+\uparrow\Sigma F_y = 0: \quad -2000 \text{ lb} - 1000 \text{ lb} + 10{,}000 \text{ lb} + C_y = 0$$
$$C_y = -7000 \text{ lb} \qquad\qquad \mathbf{C}_y = 7000 \text{ lb} \downarrow$$

**Joint A.** This joint is subject to only two unknown forces: the forces exerted by AB and those by AD. Use a force triangle to determine $\mathbf{F}_{AB}$ and $\mathbf{F}_{AD}$ (Fig. 2). Note that member AB pulls on the joint so AB is in tension, and member AD pushes on the joint so AD is in compression. Obtain the magnitudes of the two forces from the proportion

$$\frac{2000 \text{ lb}}{4} = \frac{F_{AB}}{3} = \frac{F_{AD}}{5}$$

$$F_{AB} = 1500 \text{ lb } T \blacktriangleleft$$
$$F_{AD} = 2500 \text{ lb } C \blacktriangleleft$$

As an alternative to the force triangle approach, remember that a more general analytic solution can also be used. This alternate method is especially conducive to joint equilibrium problems that involve more than three forces, and is illustrated later in this sample problem for the analysis of joints B, E, and C.

**Joint D.** Because you have already determined the force exerted by member AD, only two unknown forces are now involved at this joint. Again, use a force triangle to determine the unknown forces in members DB and DE (Fig. 3).

**Fig. 3** Free-body diagram of joint D.

*(continued)*

$$F_{DB} = F_{DA} \qquad\qquad F_{DB} = 2500 \text{ lb } T \blacktriangleleft$$
$$F_{DE} = 2(\tfrac{3}{5})F_{DA} \qquad\qquad F_{DE} = 3000 \text{ lb } C \blacktriangleleft$$

**Joint B.** Because more than three forces act at this joint (Fig. 4), determine the two unknown forces $\mathbf{F}_{BC}$ and $\mathbf{F}_{BE}$ by solving the equilibrium equations $\Sigma F_x = 0$ and $\Sigma F_y = 0$. Suppose you arbitrarily assume that both unknown forces act away from the joint, i.e., that the members are in tension. The positive value obtained for $F_{BC}$ indicates that this assumption is correct; member $BC$ is in tension. The negative value of $F_{BE}$ indicates that the second assumption is wrong; member $BE$ is in compression.

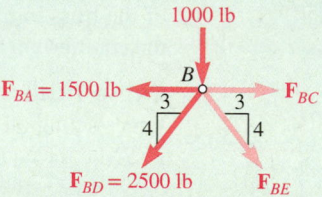

**Fig. 4** Free-body diagram of joint $B$.

$$+\uparrow\Sigma F_y = 0: \qquad -1000 - \tfrac{4}{5}(2500) - \tfrac{4}{5}F_{BE} = 0$$
$$F_{BE} = -3750 \text{ lb} \qquad\qquad F_{BE} = 3750 \text{ lb } C \blacktriangleleft$$

$$\overset{+}{\rightarrow}\Sigma F_x = 0: \qquad F_{BC} - 1500 - \tfrac{3}{5}(2500) - \tfrac{3}{5}(3750) = 0$$
$$F_{BC} = +5250 \text{ lb} \qquad\qquad F_{BC} = 5250 \text{ lb } T \blacktriangleleft$$

**Joint E.** Assume the unknown force $\mathbf{F}_{EC}$ acts away from the joint (Fig. 5). Summing $x$ components, you obtain

$$\overset{+}{\rightarrow}\Sigma F_x = 0: \qquad \tfrac{3}{5}F_{EC} + 3000 + \tfrac{3}{5}(3750) = 0$$
$$F_{EC} = -8750 \text{ lb} \qquad\qquad F_{EC} = 8750 \text{ lb } C \blacktriangleleft$$

Summing $y$ components, you obtain a check of your computations:

$$+\uparrow\Sigma F_y = 10{,}000 - \tfrac{4}{5}(3750) - \tfrac{4}{5}(8750)$$
$$= 10{,}000 - 3000 - 7000 = 0 \qquad\qquad \text{(checks)}$$

**REFLECT and THINK:** Using the computed values of $\mathbf{F}_{CB}$ and $\mathbf{F}_{CE}$, you can determine the reactions $\mathbf{C}_x$ and $\mathbf{C}_y$ by considering the equilibrium of joint $C$ (Fig. 6). Because these reactions have already been determined from the equilibrium of the entire truss, this provides two checks of your computations. You can also simply use the computed values of all forces acting on the joint (forces in members and reactions) and check that the joint is in equilibrium:

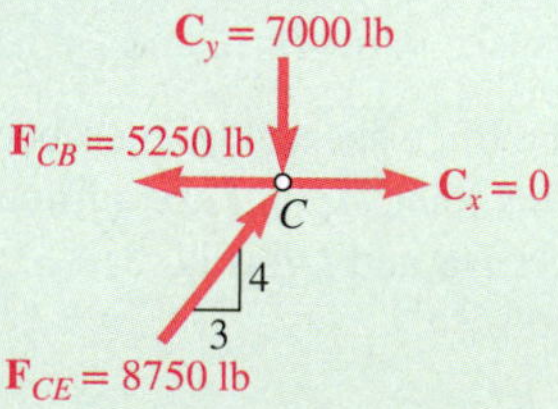

$$\overset{+}{\rightarrow}\Sigma F_x = -5250 + \tfrac{3}{5}(8750) = -5250 + 5250 = 0 \qquad\qquad \text{(checks)}$$
$$+\uparrow\Sigma F_y = -7000 + \tfrac{4}{5}(8750) = -7000 + 7000 = 0 \qquad\qquad \text{(checks)}$$

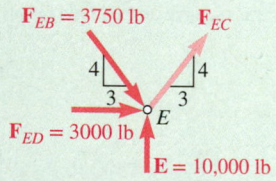

**Fig. 5** Free-body diagram of joint $E$.

**Fig. 6** Free-body diagram of joint $C$.

# SOLVING PROBLEMS
# ON YOUR OWN

In this section, you learned to use the **method of joints** to determine the forces in the members of a **simple truss**; that is, a truss that can be constructed from a basic triangular truss by adding to it two new members at a time and connecting them at a new joint.

The method consists of the following steps:

**1. Draw a free-body diagram of the entire truss, and use this diagram to determine the reactions at the supports.**

**2. Locate a joint connecting only two members, and draw the free-body diagram of its pin.** Use this free-body diagram to determine the unknown force in each of the two members. If only three forces are involved (the two unknown forces and a known one), you will probably find it more convenient to draw and solve the corresponding force triangle. If more than three forces are involved, you should write and solve the equilibrium equations for the pin, $\Sigma F_x = 0$ and $\Sigma F_y = 0$, assuming that the members are in tension. A positive answer means that the member is in tension, whereas a negative answer means that the member is in compression. Once you have found the forces, enter their values on a sketch of the truss with $T$ for tension and $C$ for compression.

**3. Next, locate a joint where the forces in only two of the connected members are still unknown.** Draw the free-body diagram of the pin and use it as indicated in Step 2 to determine the two unknown forces.

**4. Repeat this procedure until you have found the forces in all the members of the truss.** Because you previously used the three equilibrium equations associated with the free-body diagram of the entire truss to determine the reactions at the supports, you will end up with three extra equations. These equations can be used to check your computations.

**5. Note that the choice of the first joint is not unique.** Once you have determined the reactions at the supports of the truss, you can choose either of two joints as a starting point for your analysis. In Sample Prob. 6.1, we started at joint $A$ and proceeded through joints $D$, $B$, $E$, and $C$, but we could also have started at joint $C$ and proceeded through joints $E$, $B$, $D$, and $A$. On the other hand, having selected a first joint, you may in some cases reach a point in your analysis beyond which you cannot proceed. You must then start again from another joint to complete your solution.

Keep in mind that the analysis of a simple truss always can be carried out by the method of joints. Also remember that it is helpful to outline your solution *before* starting any computations.

# Problems

6.1 through 6.8 Using the method of joints, determine the force in each member of the truss shown. State whether each member is in tension ($T$) or compression ($C$).

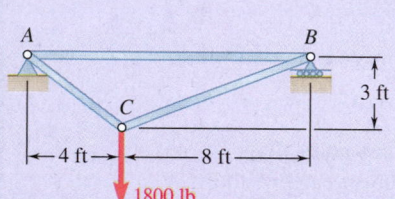

**Fig. P6.1**

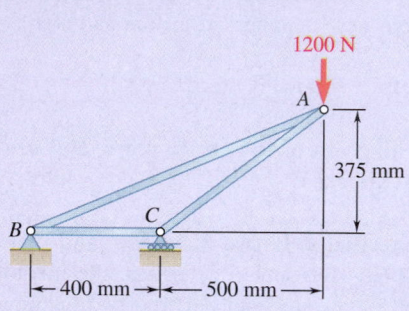

**Fig. P6.2**

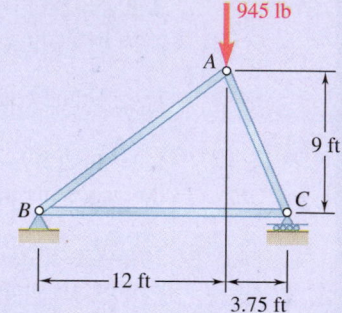

**Fig. P6.3**

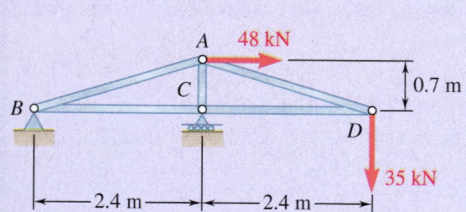

**Fig. P6.4**

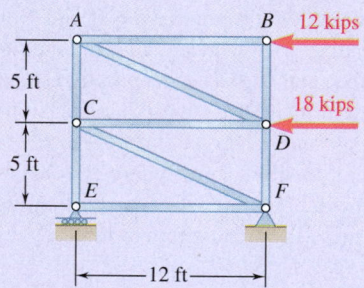

**Fig. *P6.5***

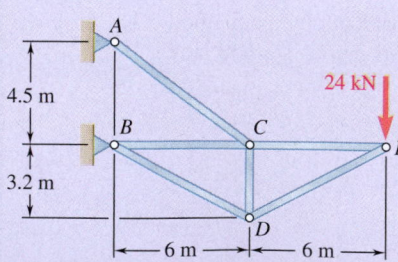

**Fig. P6.6**

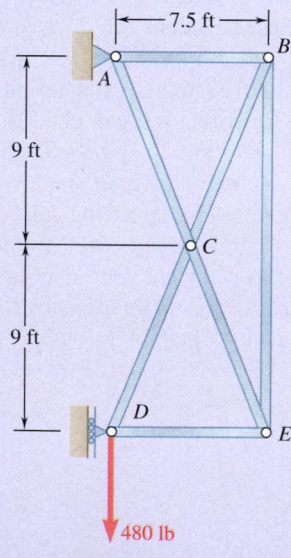

**Fig. *P6.7***

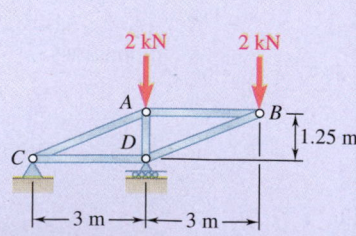

**Fig. P6.8**

**6.9 and 6.10** Determine the force in each member of the truss shown. State whether each member is in tension (T) or compression (C).

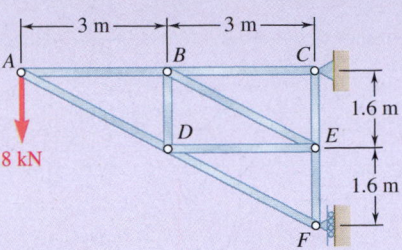

Fig. P6.9

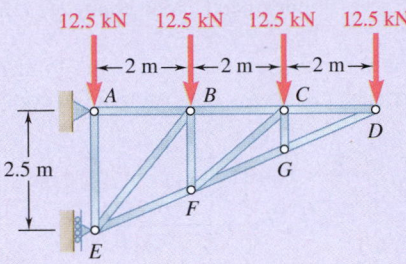

Fig. P6.10

**6.11** Determine the force in each member of the Gambrel roof truss shown. State whether each member is in tension (T) or compression (C).

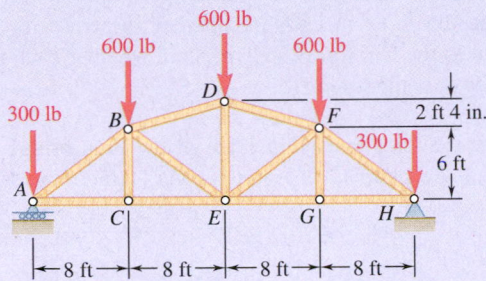

Fig. P6.11

**6.12** Determine the force in each member of the Howe roof truss shown. State whether each member is in tension (T) or compression (C).

**6.13** Determine the force in each member of the roof truss shown. State whether each member is in tension (T) or compression (C).

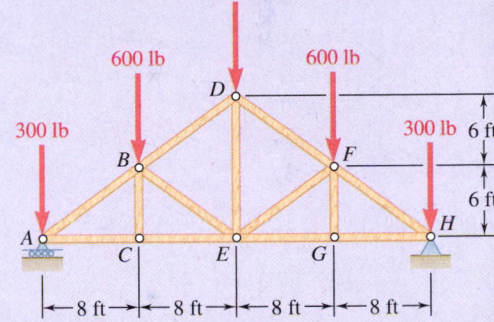

Fig. P6.12

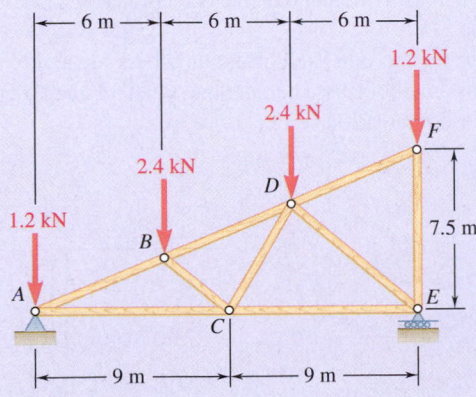

Fig. P6.13

**6.14** Determine the force in each member of the fan roof truss shown. State whether each member is in tension (T) or compression (C).

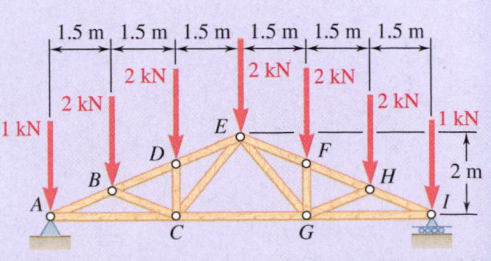

Fig. P6.14

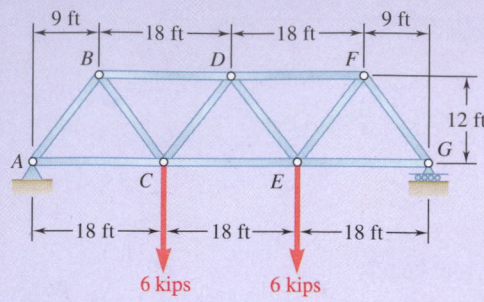

**Fig. P6.15**

**6.15** Determine the force in each member of the Warren bridge truss shown. State whether each member is in tension (*T*) or compression (*C*).

**6.16** Solve Prob. 6.15 assuming that the load applied at *E* has been removed.

**6.17** Determine the force in each member of the Pratt roof truss shown. State whether each member is in tension (*T*) or compression (*C*).

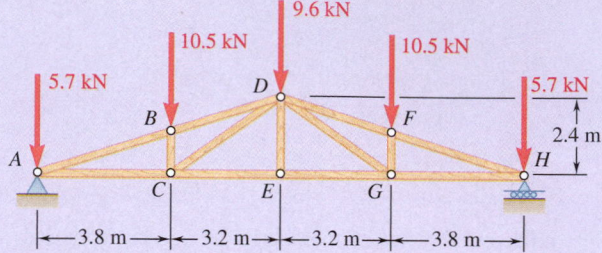

**Fig. P6.17**

**6.18** The truss shown is one of several supporting an advertising panel. Determine the force in each member of the truss for a wind load equivalent to the two forces shown. State whether each member is in tension (*T*) or compression (*C*).

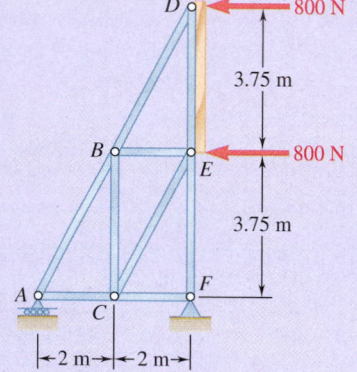

**Fig. P6.18**

**6.19** Determine the force in each member of the Pratt bridge truss shown. State whether each member is in tension (*T*) or compression (*C*).

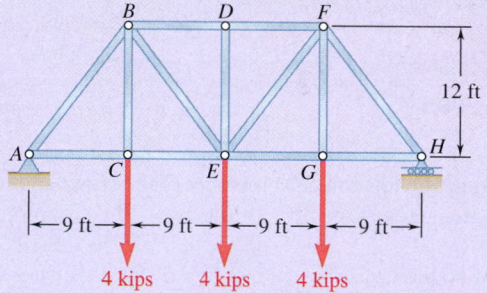

**Fig. P6.19**

**6.20** Solve Prob. 6.19 assuming that the load applied at *G* has been removed.

**6.21** Determine the force in each of the members located to the left of *FG* for the scissors roof truss shown. State whether each member is in tension (*T*) or compression (*C*).

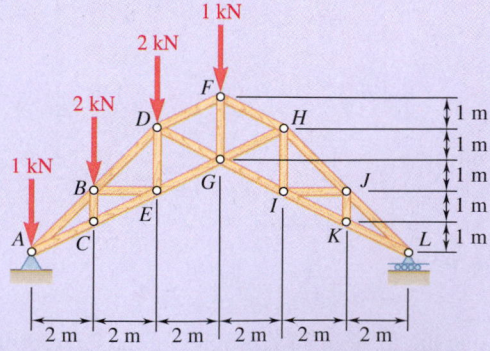

**Fig. P6.21**

**6.22** Determine the force in member *DE* and in each of the members located to the left of *DE* for the inverted Howe roof truss shown. State whether each member is in tension (*T*) or compression (*C*).

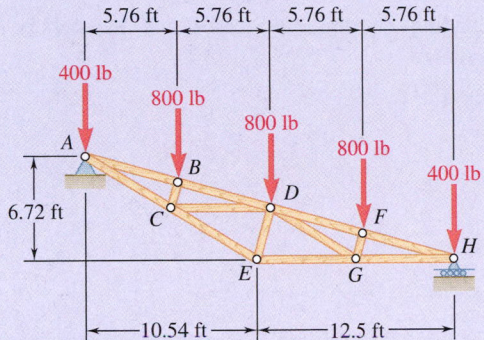

**Fig. P6.22 and P6.23**

**6.23** Determine the force in each of the members located to the right of *DE* for the inverted Howe roof truss shown. State whether each member is in tension (*T*) or compression (*C*).

**6.24** The portion of truss shown represents the upper part of a power transmission line tower. For the given loading, determine the force in each of the members located above *HJ*. State whether each member is in tension (*T*) or compression (*C*).

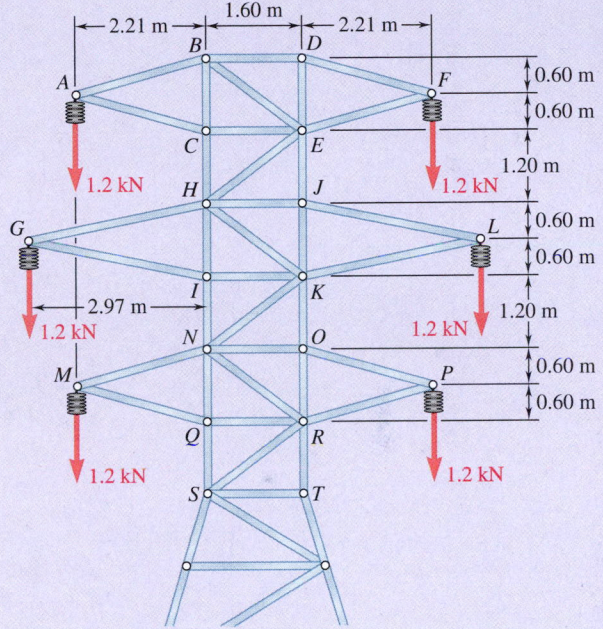

**Fig. P6.24**

**6.25** For the tower and loading of Prob. 6.24 and knowing that $F_{CH} = F_{EJ} = 1.2$ kN C and $F_{EH} = 0$, determine the force in member *HJ* and in each of the members located between *HJ* and *NO*. State whether each member is in tension (*T*) or compression (*C*).

**6.26** Solve Prob. 6.24 assuming that the cables hanging from the right side of the tower have fallen to the ground.

**6.27 and 6.28** Determine the force in each member of the truss shown. State whether each member is in tension (*T*) or compression (*C*).

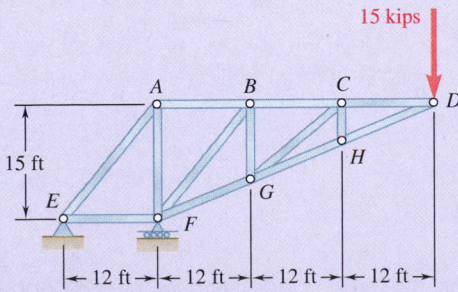

**Fig. P6.27**

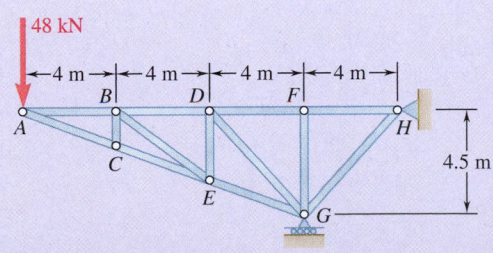

**Fig. P6.28**

**6.29** Determine whether the trusses of Probs. 6.31*a*, 6.32*a*, and 6.33*a* are simple trusses.

**6.30** Determine whether the trusses of Probs. 6.31*b*, 6.32*b*, and 6.33*b* are simple trusses.

**6.31** For the given loading, determine the zero-force members in each of the two trusses shown.

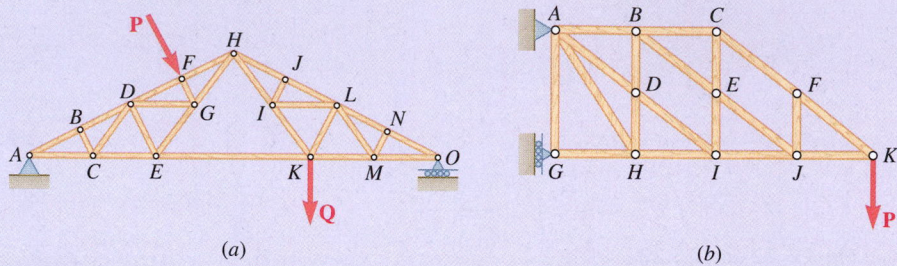

(*a*)                                         (*b*)

**Fig. *P6.31***

**6.32** For the given loading, determine the zero-force members in each of the two trusses shown.

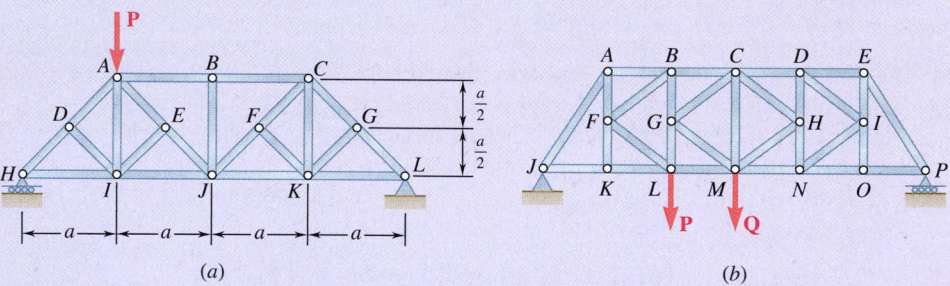

(*a*)                                         (*b*)

**Fig. P6.32**

**6.33** For the given loading, determine the zero-force members in each of the two trusses shown.

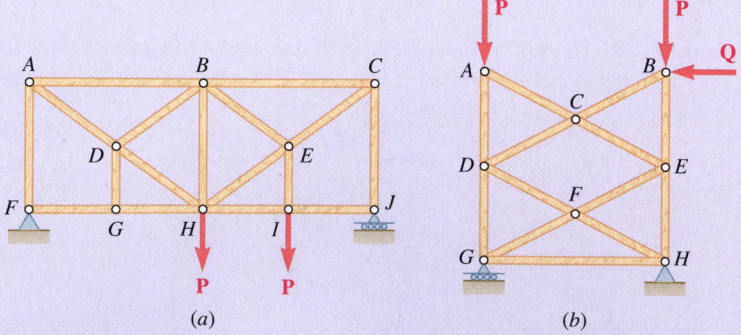

(*a*)                                         (*b*)

**Fig. *P6.33***

**6.34** Determine the zero-force members in the truss of (*a*) Prob. 6.21, (*b*) Prob. 6.27.

316

*6.35 The truss shown consists of six members and is supported by a short link at *A*, two short links at *B*, and a ball-and-socket at *D*. Determine the force in each of the members for the given loading.

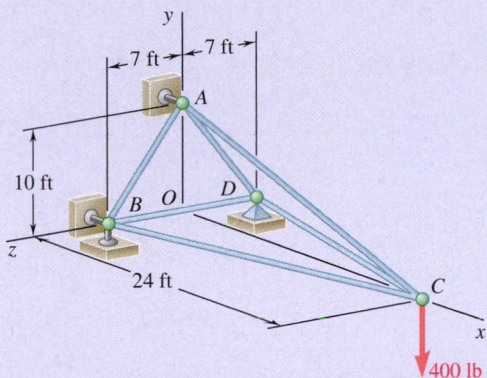

**Fig. P6.35**

*6.36 The truss shown consists of six members and is supported by a ball-and-socket at *B*, a short link at *C*, and two short links at *D*. Determine the force in each of the members for $\mathbf{P} = (-2184 \text{ N})\mathbf{j}$ and $\mathbf{Q} = 0$.

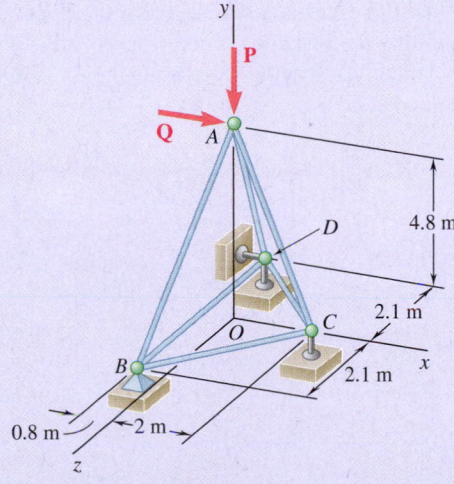

**Fig. P6.36 and P6.37**

*6.37 The truss shown consists of six members and is supported by a ball-and-socket at *B*, a short link at *C*, and two short links at *D*. Determine the force in each of the members for $\mathbf{P} = 0$ and $\mathbf{Q} = (2968 \text{ N})\mathbf{i}$.

*6.38 The truss shown consists of nine members and is supported by a ball-and-socket at *A*, two short links at *B*, and a short link at *C*. Determine the force in each of the members for the given loading.

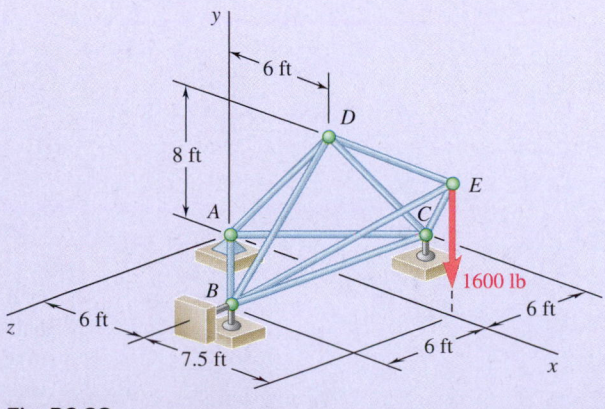

**Fig. P6.38**

*6.39 The truss shown consists of nine members and is supported by a ball-and-socket at $B$, a short link at $C$, and two short links at $D$. ($a$) Check that this truss is a simple truss, that it is completely constrained, and that the reactions at its supports are statically determinate. ($b$) Determine the force in each member for $\mathbf{P} = (-1200 \text{ N})\mathbf{j}$ and $\mathbf{Q} = 0$.

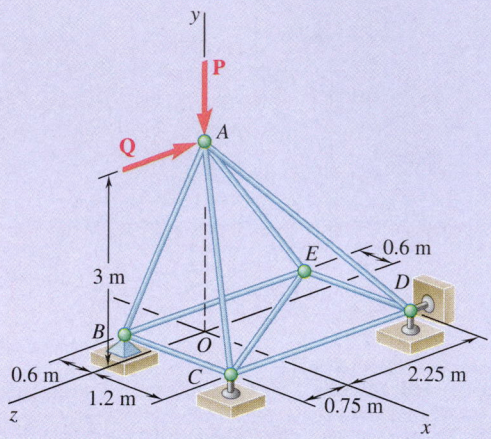

**Fig. P6.39**

*6.40 Solve Prob. 6.39 for $\mathbf{P} = 0$ and $\mathbf{Q} = (-900 \text{ N})\mathbf{k}$.

*6.41 The truss shown consists of 18 members and is supported by a ball-and-socket at $A$, two short links at $B$, and one short link at $G$. ($a$) Check that this truss is a simple truss, that it is completely constrained, and that the reactions at its supports are statically determinate. ($b$) For the given loading, determine the force in each of the six members joined at $E$.

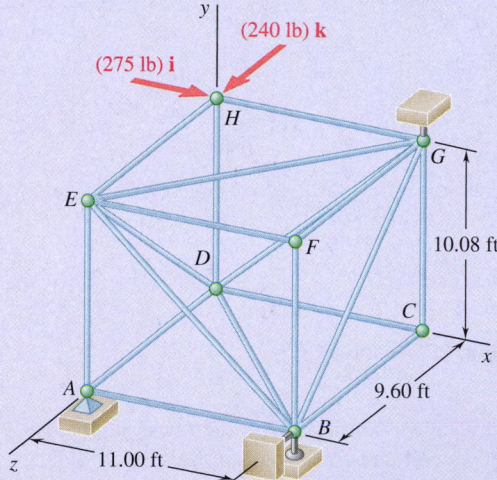

**Fig. P6.41 and P6.42**

*6.42 The truss shown consists of 18 members and is supported by a ball-and-socket at $A$, two short links at $B$, and one short link at $G$. ($a$) Check that this truss is a simple truss, that it is completely constrained, and that the reactions at its supports are statically determinate. ($b$) For the given loading, determine the force in each of the six members joined at $G$.

# 6.2   OTHER TRUSS ANALYSES

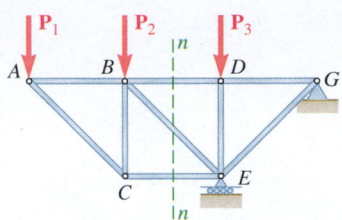

The method of joints is most effective when we want to determine the forces in all the members of a truss. If, however, we need to determine the force in only one member or in a very few members, the method of sections is more efficient.

## 6.2A   The Method of Sections

Assume, for example, that we want to determine the force in member *BD* of the truss shown in Fig. 6.15*a*. To do this, we must determine the force with which member *BD* acts on either joint *B* or joint *D*. If we were to use the method of joints, we would choose either joint *B* or joint *D* as a free body. However, we can also choose a larger portion of the truss that is composed of several joints and members, provided that the force we want to find is one of the external forces acting on that portion. If, in addition, we choose the portion of the truss as a free body where a total of only three unknown forces act upon it, we can obtain the desired force by solving the equations of equilibrium for this portion of the truss. In practice, we isolate a portion of the truss by *passing a section* through three members of the truss, one of which is the desired member. That is, we draw a line that divides the truss into two completely separate parts but does not intersect more than three members. We can then use as a free body either of the two portions of the truss obtained after the intersected members have been removed.[†]

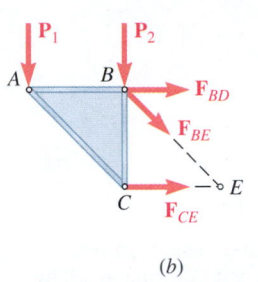

In Fig. 6.15*a*, we have passed the section *nn* through members *BD*, *BE*, and *CE*, and we have chosen the portion *ABC* of the truss as the free body (Fig. 6.15*b*). The forces acting on this free body are the loads $\mathbf{P}_1$ and $\mathbf{P}_2$ at points *A* and *B* and the three unknown forces $\mathbf{F}_{BD}$, $\mathbf{F}_{BE}$, and $\mathbf{F}_{CE}$. Because we do not know whether the members removed are in tension or compression, we have arbitrarily drawn the three forces away from the free body as if the members are in tension.

**Fig. 6.15**   (*a*) We can pass a section *nn* through the truss, dividing the three members *BD*, *BE*, and *CE*. (*b*) Free-body diagram of portion *ABC* of the truss. We assume that members *BD*, *BE*, and *CE* are in tension.

We use the fact that the rigid body *ABC* is in equilibrium to write three equations that we can solve for the three unknown forces. If we want to determine only force $\mathbf{F}_{BD}$, say, we need write only one equation, provided that the equation does not contain the other unknowns. Thus, the equation $\Sigma M_E = 0$ yields the value of the magnitude $F_{BD}$ (Fig. 6.15*b*). A positive sign in the answer will indicate that our original assumption regarding the sense of $\mathbf{F}_{BD}$ was correct and that member *BD* is in tension; a negative sign will indicate that our assumption was incorrect and that *BD* is in compression.

On the other hand, if we want to determine only force $\mathbf{F}_{CE}$, we need to write an equation that does not involve $\mathbf{F}_{BD}$ or $\mathbf{F}_{BE}$; the appropriate equation is $\Sigma M_B = 0$. Again, a positive sign for the magnitude $F_{CE}$ of the desired force indicates a correct assumption—that is, tension; and a negative sign indicates an incorrect assumption—that is, compression.

If we want to determine only force $\mathbf{F}_{BE}$, the appropriate equation is $\Sigma F_y = 0$. Whether the member is in tension or compression is again determined from the sign of the answer.

If we determine the force in only one member, no independent check of the computation is available. However, if we calculate all of the unknown forces acting on the free body, we can check the computations by writing an additional equation. For instance, if we determine $\mathbf{F}_{BD}$, $\mathbf{F}_{BE}$, and $\mathbf{F}_{CE}$ as indicated previously, we can check the work by verifying that $\Sigma F_x = 0$.

---

[†]In the analysis of some trusses, we can pass sections through more than three members, provided we can write equilibrium equations involving only one unknown that we can use to determine the forces in one, or possibly two, of the intersected members. See Probs. 6.61, 6.62, 6.63, and 6.64.

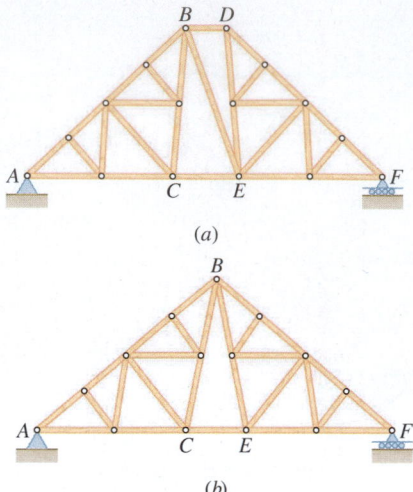

(a)

(b)

**Fig. 6.16** Compound trusses. (a) Two simple trusses ABC and DEF connected by three bars. (b) Two simple trusses ABC and DEF connected by one joint and one bar (a Fink truss).

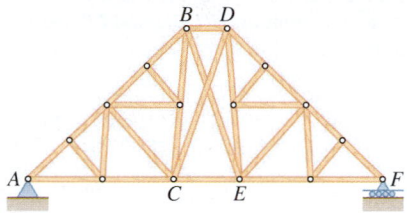

**Fig. 6.17** A statically indeterminate, overrigid compound truss, due to a redundant member.

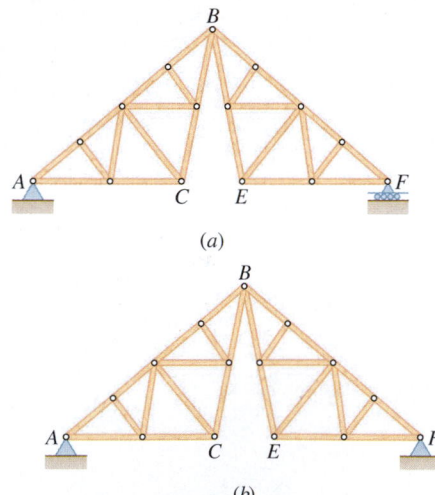

(a)

(b)

**Fig. 6.18** Two simple trusses joined by a pin. (a) Supported by a pin and a roller, the truss will collapse under its own weight. (b) Supported by two pins, the truss becomes rigid and does not collapse.

# 6.2B    Trusses Made of Several Simple Trusses

Consider two simple trusses ABC and DEF. If we connect them by three bars BD, BE, and CE as shown in Fig. 6.16a, together they form a rigid truss ABDF. We can also combine trusses ABC and DEF into a single rigid truss by joining joints B and D at a single joint B and connecting joints C and E by a bar CE (Fig. 6.16b). This is known as a *Fink truss*. The trusses of Fig. 6.16a and b are *not* simple trusses; you cannot construct them from a triangular truss by adding successive pairs of members as described in Sec. 6.1A. They are rigid trusses, however, as you can check by comparing the systems of connections used to hold the simple trusses ABC and DEF together (three bars in Fig. 6.16a, one pin and one bar in Fig. 6.16b) with the systems of supports discussed in Sec. 4.1. Trusses made of several simple trusses rigidly connected are known as **compound trusses**.

In a compound truss, the number of members $m$ and the number of joints $n$ are still related by the formula $m = 2n - 3$. You can verify this by observing that if a compound truss is supported by a frictionless pin and a roller (involving three unknown reactions), the total number of unknowns is $m + 3$, and this number must be equal to the number $2n$ of equations obtained by expressing that the $n$ pins are in equilibrium. It follows that $m = 2n - 3$.

Compound trusses supported by a pin and a roller or by an equivalent system of supports are *statically determinate, rigid,* and *completely constrained.* This means that we can determine all of the unknown reactions and the forces in all of the members by using the methods of statics, and the truss will neither collapse nor move. However, the only way to determine all of the forces in the members using the method of joints requires solving a large number of simultaneous equations. In the case of the compound truss of Fig. 6.16a, for example, it is more efficient to pass a section through members BD, BE, and CE to determine the forces in these members.

Suppose, now, that the simple trusses ABC and DEF are connected by *four* bars: BD, BE, CD, and CE (Fig. 6.17). The number of members $m$ is now larger than $2n - 3$. This truss is said to be **overrigid**, and one of the four members BD, BE, CD, or CE is **redundant**. If the truss is supported by a pin at A and a roller at F, the total number of unknowns is $m + 3$. Because $m > 2n - 3$, the number $m + 3$ of unknowns is now larger than the number $2n$ of available independent equations; the truss is *statically indeterminate.*

Finally, let us assume that the two simple trusses ABC and DEF are joined by a single pin, as shown in Fig. 6.18a. The number of members, $m$, is now smaller than $2n - 3$. If the truss is supported by a pin at A and a roller at F, the total number of unknowns is $m + 3$. Because $m < 2n - 3$, the number $m + 3$ of unknowns is now smaller than the number $2n$ of equilibrium equations that need to be satisfied. This truss is **nonrigid** and will collapse under its own weight. However, if two pins are used to support it, the truss becomes *rigid* and will not collapse (Fig. 6.18b). Note that the total number of unknowns is now $m + 4$ and is equal to the number $2n$ of equations.

More generally, if the reactions at the supports involve $r$ unknowns, the condition for a compound truss to be statically determinate, rigid, and completely constrained is $m + r = 2n$. However, although this condition is necessary, it is not sufficient for the equilibrium of a structure that ceases to be rigid when detached from its supports (see Sec. 6.3B).

## Sample Problem 6.2

Determine the forces in members *EF* and *GI* of the truss shown.

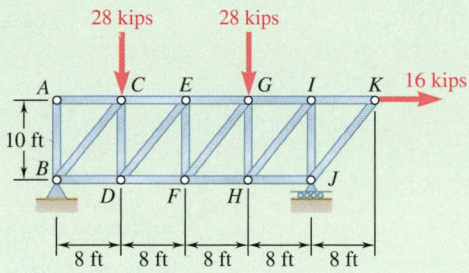

**STRATEGY:** You are asked to determine the forces in only two of the members in this truss, so the method of sections is more appropriate than the method of joints. You can use a free-body diagram of the entire truss to help determine the reactions, and then pass sections through the truss to isolate parts of it for calculating the desired forces.

**MODELING and ANALYSIS:** You can go through the steps that follow for the determination of the support reactions, and then for the analysis of portions of the truss.

**Free-Body: Entire Truss.** Draw a free-body diagram of the entire truss. External forces acting on this free body consist of the applied loads and the reactions at *B* and *J* (Fig. 1). Write and solve the following equilibrium equations.

$+\circlearrowleft \Sigma M_B = 0$:

$$-(28 \text{ kips})(8 \text{ ft}) - (28 \text{ kips})(24 \text{ ft}) - (16 \text{ kips})(10 \text{ ft}) + J(32 \text{ ft}) = 0$$
$$J = +33 \text{ kips} \qquad \mathbf{J} = 33 \text{ kips} \uparrow$$

$\xrightarrow{+} \Sigma F_x = 0$: $\qquad B_x + 16 \text{ kips} = 0$

$$B_x = -16 \text{ kips} \qquad \mathbf{B}_x = 16 \text{ kips} \leftarrow$$

$+\circlearrowleft \Sigma M_J = 0$:

$$(28 \text{ kips})(24 \text{ ft}) + (28 \text{ kips})(8 \text{ ft}) - (16 \text{ kips})(10 \text{ ft}) - B_y(32 \text{ ft}) = 0$$
$$B_y = +23 \text{ kips} \qquad \mathbf{B}_y = 23 \text{ kips} \uparrow$$

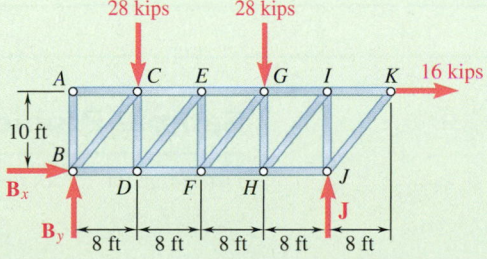

**Fig. 1** Free-body diagram of the entire truss.

**Force in Member *EF*.** Pass section *nn* through the truss diagonally so that it intersects member *EF* and only two additional members (Fig. 2). Remove the intersected members and choose the left-hand portion of the truss as a free body

*(continued)*

(Fig. 3). Three unknowns are involved; to eliminate the two horizontal forces, we write

$$+\uparrow\Sigma F_y = 0: \qquad +23 \text{ kips} - 28 \text{ kips} - F_{EF} = 0$$
$$F_{EF} = -5 \text{ kips}$$

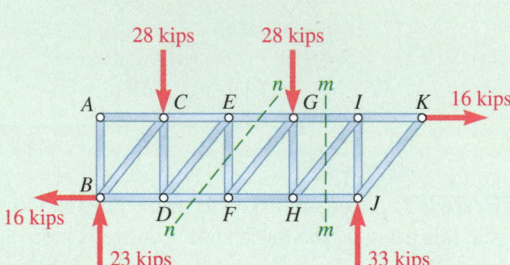

**Fig. 2** Sections *nn* and *mm* that will be used to analyze members *EF* and *GI*.

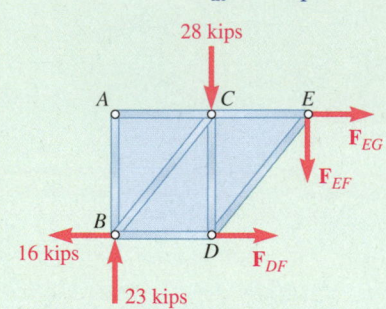

$F_{EF} = 5 \text{ kips } C$ ◄

**Fig. 3** Free-body diagram to analyze member *EF*.

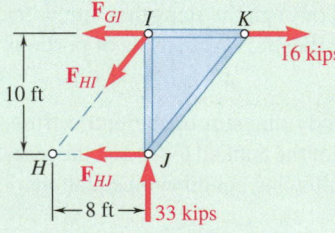

**Fig. 4** Free-body diagram to analyze member *GI*.

The sense of $\mathbf{F}_{EF}$ was chosen assuming member *EF* to be in tension; the negative sign indicates that the member is in compression.

**Force in Member *GI*.** Pass section *mm* through the truss vertically so that it intersects member *GI* and only two additional members (Fig. 2). Remove the intersected members and choose the right-hand portion of the truss as a free body (Fig. 4). Again, three unknown forces are involved; to eliminate the two forces passing through point *H*, sum the moments about that point.

$$+\circlearrowleft\Sigma M_H = 0: \quad (33 \text{ kips})(8 \text{ ft}) - (16 \text{ kips})(10 \text{ ft}) + F_{GI}(10 \text{ ft}) = 0$$
$$F_{GI} = -10.4 \text{ kips} \qquad F_{GI} = 10.4 \text{ kips } C \blacktriangleleft$$

**REFLECT and THINK:** Note that a section passed through a truss does not have to be vertical or horizontal; it can be diagonal as well. Choose the orientation that cuts through no more than three members of unknown force and also gives you the simplest part of the truss for which you can write equilibrium equations and determine the unknowns.

## Sample Problem 6.3

Determine the forces in members *FH*, *GH*, and *GI* of the roof truss shown.

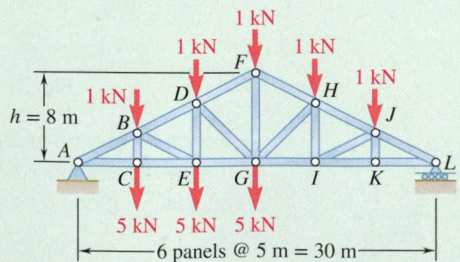

*(continued)*

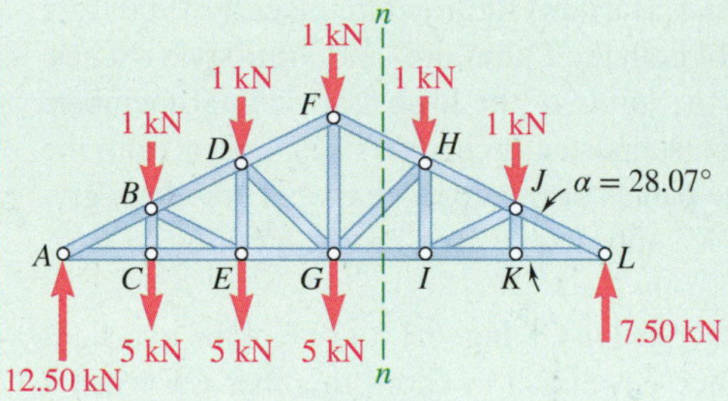

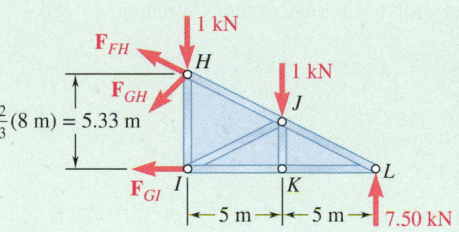

**Fig. 1**   Free-body diagram of the entire truss.

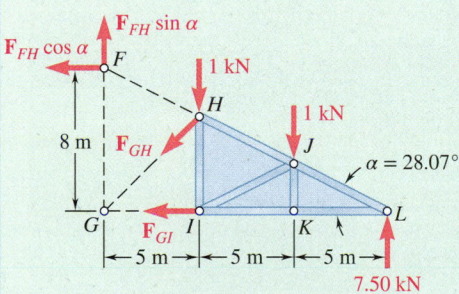

**Fig. 2**   Free-body diagram to analyze member *GI*.

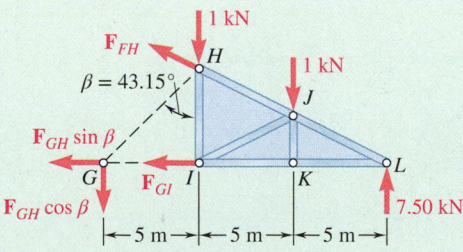

**Fig. 3**   Simplifying the analysis of member *FH* by first sliding its force to point *F*.

**Fig. 4**   Simplifying the analysis of member *GH* by first sliding its force to point *G*.

**STRATEGY:**   You are asked to determine the forces in only three members of the truss, so use the method of sections. Determine the reactions by treating the entire truss as a free body and then isolate part of it for analysis. In this case, you can use the same smaller part of the truss to determine all three desired forces.

**MODELING and ANALYSIS:**   Your reasoning and computation should go something like the sequence given here.

**Free Body: Entire Truss.**   From the free-body diagram of the entire truss (Fig. 1), find the reactions at *A* and *L*:

$$\mathbf{A} = 12.50 \text{ kN}\uparrow \qquad \mathbf{L} = 7.50 \text{ kN}\uparrow$$

Note that

$$\tan \alpha = \frac{FG}{GL} = \frac{8 \text{ m}}{15 \text{ m}} = 0.5333 \qquad \alpha = 28.07°$$

**Force in Member *GI*.**   Pass section *nn* vertically through the truss (Fig. 1). Using the portion *HLI* of the truss as a free body (Fig. 2), obtain the value of $F_{GI}$:

$$+\circlearrowleft \Sigma M_H = 0: \quad (7.50 \text{ kN})(10 \text{ m}) - (1 \text{ kN})(5 \text{ m}) - F_{GI}(5.33 \text{ m}) = 0$$
$$F_{GI} = +13.13 \text{ kN} \qquad F_{GI} = 13.13 \text{ kN } T \blacktriangleleft$$

**Force in Member *FH*.**   Determine the value of $F_{FH}$ from the equation $\Sigma M_G = 0$. To do this, move $\mathbf{F}_{FH}$ along its line of action until it acts at point *F*, where you can resolve it into its *x* and *y* components (Fig. 3). The moment of $\mathbf{F}_{FH}$ with respect to point *G* is now $(F_{FH} \cos \alpha)(8 \text{ m})$.

$$+\circlearrowleft \Sigma M_G = 0:$$
$$(7.50 \text{ kN})(15 \text{ m}) - (1 \text{ kN})(10 \text{ m}) - (1 \text{ kN})(5 \text{ m}) + (F_{FH} \cos \alpha)(8 \text{ m}) = 0$$
$$F_{FH} = -13.81 \text{ kN} \qquad F_{FH} = 13.81 \text{ kN } C \blacktriangleleft$$

**Force in Member *GH*.**   First note that

$$\tan \beta = \frac{GI}{HI} = \frac{5 \text{ m}}{\frac{2}{3}(8 \text{ m})} = 0.9375 \qquad \beta = 43.15°$$

Then, determine the value of $F_{GH}$ by resolving the force $\mathbf{F}_{GH}$ into *x* and *y* components at point *G* (Fig. 4) and solving the equation $\Sigma M_L = 0$.

$$+\circlearrowleft \Sigma M_L = 0: \quad (1 \text{ kN})(10 \text{ m}) + (1 \text{ kN})(5 \text{ m}) + (F_{GH} \cos \beta)(15 \text{ m}) = 0$$
$$F_{GH} = -1.371 \text{ kN} \qquad F_{GH} = 1.371 \text{ kN } C \blacktriangleleft$$

**REFLECT and THINK:**   Sometimes you should resolve a force into components to include it in the equilibrium equations. By first sliding this force along its line of action to a more strategic point, you might eliminate one of its components from a moment equilibrium equation.

## Case Study 6.1

CS Photo 6.1 shows a pin-connected truss that is part of the railroad bridge spanning the Connecticut River at Warehouse Point in East Windsor, Connecticut. Built by the New York, New Haven & Hartford Railroad in the early 1900s, the seven-panel structure has features of both the Baltimore- and Pratt-style trusses. In addition to being connected to the joints of the lowest horizontal members (or *chords*), the floor system is also supported by hangers suspended from the mid-panel points. The central three panels also employ diagonal *counters,* giving these panels a characteristic "X" appearance (see Probs. 6.65, 6.66, 6.67, and 6.68).

If we assume the geometry illustrated in CS Fig. 6.1 for one of the two sides of the truss, let's determine the force developed in chord member *CF* when a unit 1-kip load is applied at *D* as shown, and then repeat the analysis with the unit load moved to joint *G*. For clarity, we will omit the mid-panel hangers, because they are zero-force members for both load cases. (What other zero-force members are present?)

**CS Photo 6.1** Pin-connected truss bridge at Warehouse Point, CT.

Courtesy of Jeffrey C. Mazurek

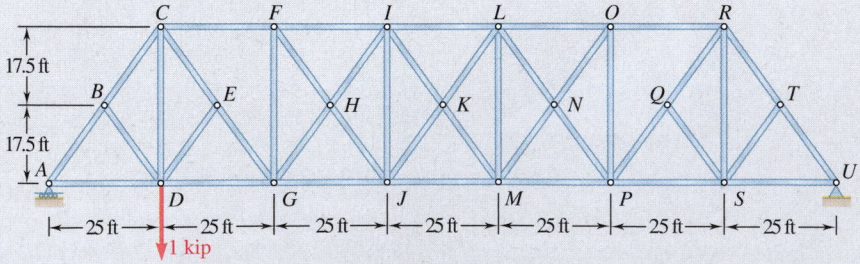

**CS Fig. 6.1** Assumed truss geometry and loading.

*(continued)*

**STRATEGY:**

For each load case, determine the reactions by treating the entire truss as a free body. Then, using the method of sections, cut through the second panel to expose the force in member *CF* and apply equilibrium to determine this force.

**MODELING and ANALYSIS:**

**Case I: Unit 1-kip Load at *D*.** For this load case, CS Fig. 6.2 shows the free-body diagram of the entire truss. From this diagram, find the reaction at *A*:

$$+\circlearrowleft \Sigma M_U = 0: \qquad +(1\text{ kip})(150\text{ ft}) - A(175\text{ ft}) = 0 \quad \mathbf{A} = 0.8571 \text{ kips} \uparrow$$

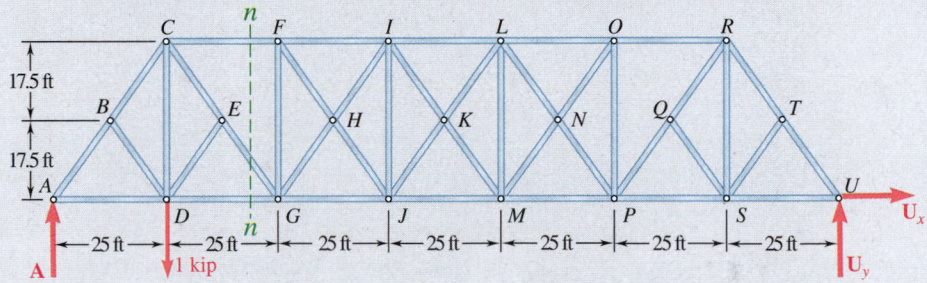

**CS Fig. 6.2** Free-body diagram of entire truss for Case I.

Now pass a section *nn* vertically through the truss (CS Fig. 6.2) and choose the left-hand portion as a free body (CS Fig. 6.3).

Applying equilibrium,

$$+\circlearrowleft \Sigma M_G = 0: \quad +(0.8571\text{ kips})(50\text{ ft}) - (1\text{ kip})(25\text{ ft}) + F_{CF}(35\text{ ft}) = 0$$
$$F_{CF} = -0.5101 \text{ kips} \qquad F_{CF} = 0.5101 \text{ kips } C$$

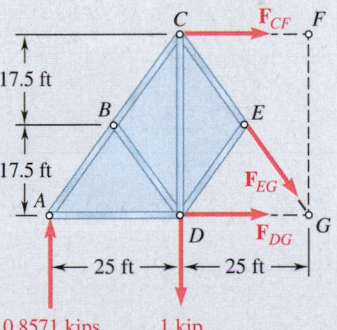

**CS Fig. 6.3** Free-body diagram for Case I: unit load at *D*.

**Case II: Unit 1-kip Load at *G*.** Moving the unit load from point *D* to point *G* and repeating the process for finding the reaction in Case I:

$$\mathbf{A} = 0.7143 \text{ kips} \uparrow$$

Again, pass a section *nn* vertically through the truss (CS Fig. 6.2) and choose the left-hand portion as a free body (CS Fig. 6.4). Because the unit load is no

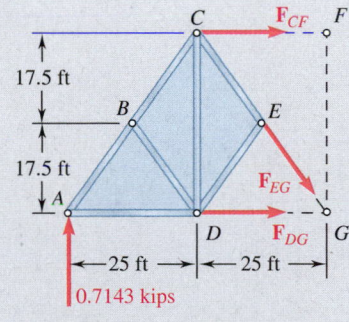

**CS Fig. 6.4** Free-body diagram for Case II: unit load at *G*.

*(continued)*

longer acting directly on the chosen free body, it is not shown in this figure. Applying equilibrium,

$$+\circlearrowleft \Sigma M_G = 0: \quad +(0.7143 \text{ kips})(50 \text{ ft}) + F_{CF}(35 \text{ ft}) = 0$$
$$F_{CF} = -1.0204 \text{ kips} \qquad F_{CF} = 1.0204 \text{ kips } C \blacktriangleleft$$

**REFLECT and THINK:**   Repeating the process to consider the force developed in member $CF$ due to the unit 1-kip load applied in order to each of the joints along the lower chord, the following table can be compiled:

| Location of 1-kip Load | $F_{CF}$, kips |
|:---:|:---:|
| $A$ | 0 |
| $D$ | −0.5101 |
| $G$ | −1.0204 |
| $J$ | −0.8163 |
| $M$ | −0.6123 |
| $P$ | −0.4082 |
| $S$ | −0.2041 |
| $U$ | 0 |

These results can also be plotted as shown in CS Fig. 6.5. This plot is known as an *influence line,* and it displays the force developed in member $F_{CF}$ as a unit load traverses the deck. A valuable tool for bridge design, such influence lines allow engineers to study the structural effects of moving loads. And while they are developed using a single unit force as has been done here, techniques exist that allow these plots to be readily applied in the evaluation of multiple-axle loads, such as the train shown in CS Photo 6.1.

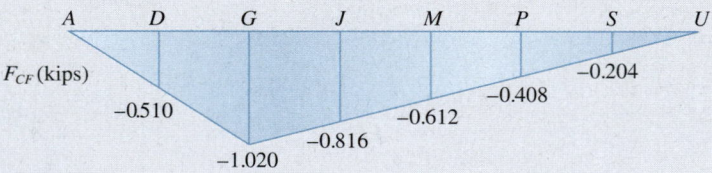

**CS Fig. 6.5**   Influence line for member *CF.*

# SOLVING PROBLEMS
# ON YOUR OWN

The **method of joints** that you studied in Sec. 6.1 is usually the best method to use when you need to find the forces *in all of the members* of a simple truss. However, the **method of sections**, which was covered in this section, is more efficient when you need to find the force *in only one member* or the forces *in a very few members* of a simple truss. The method of sections also must be used when the truss *is not a simple truss*.

**A. To determine the force in a given truss member by the method of sections,** follow these steps:

**1. Draw a free-body diagram of the entire truss,** and use this diagram to determine the reactions at the supports.

**2. Pass a section through three members of the truss,** one of which is the member whose force you want to find. After you cut through these members, you will have two separate portions of truss.

**3. Select one of these two portions of truss and draw its free-body diagram.** This diagram should include the external forces applied to the selected portion, as well as the forces exerted on it by the intersected members that were removed.

**4. You can now write three equilibrium equations** that can be solved for the forces in the three intersected members.

**5. An alternative approach is to write a single equation** that can be solved for the force in the desired member. To do so, first observe whether the forces exerted by the other two members on the free body are parallel or whether their lines of action intersect.

    **a. If these forces are parallel,** you can eliminate them by writing an equilibrium equation involving *components in a direction perpendicular* to these two forces.

    **b. If their lines of action intersect at a point $H$,** you can eliminate them by writing an equilibrium equation involving *moments about H*.

**6. Keep in mind that the section you use must intersect three members only.** The reason is that the equilibrium equations in Step 4 can be solved for only three unknowns. However, you can pass a section through more than three members to find the force in one of those members if you can write an equilibrium equation containing only that force as an unknown. Such special situations are found in Probs. 6.61, 6.62, 6.63, and 6.64.

*(continued)*

**B. About completely constrained and determinate trusses:**

**1. Any simple truss that is simply supported** is a completely constrained and determinate truss.

**2. To determine whether any other truss is or is not completely constrained and determinate,** count the number $m$ of its members, the number $n$ of its joints, and the number $r$ of the reaction components at its supports. Compare the sum $m + r$ representing the number of unknowns and the product $2n$ representing the number of available independent equilibrium equations.

    **a. If $m + r < 2n$,** there are fewer unknowns than equations. Thus, some of the equations cannot be satisfied, and the truss is only *partially constrained*.

    **b. If $m + r > 2n$,** there are more unknowns than equations. Thus, some of the unknowns cannot be determined, and the truss is *indeterminate*.

    **c. If $m + r = 2n$,** there are as many unknowns as there are equations. This, however, does not mean that all of the unknowns can be determined and that all of the equations can be satisfied. To find out whether the truss is *completely* or *improperly constrained*, try to determine the reactions at its supports and the forces in its members. If you can find all of them, the truss is *completely constrained and determinate*.

# Problems

**6.43** Determine the force in members *BD* and *DE* of the truss shown.

**6.44** Determine the force in members *DG* and *EG* of the truss shown.

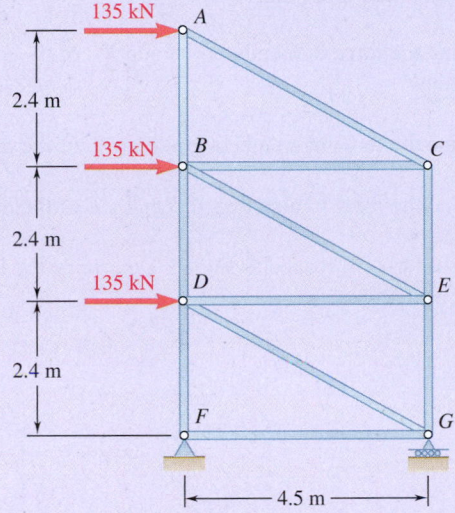

**Fig. P6.43 and P6.44**

**6.45** Determine the force in members *BD* and *CD* of the truss shown.

**6.46** Determine the force in members *DF* and *DG* of the truss shown.

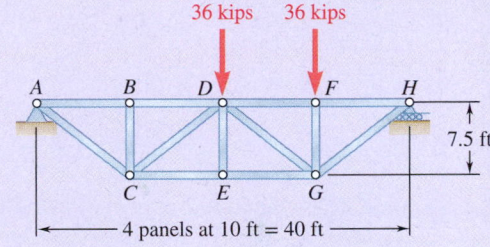

**Fig. P6.45 and P6.46**

**6.47** A floor truss is loaded as shown. Determine the force in members *CF*, *EF*, and *EG*.

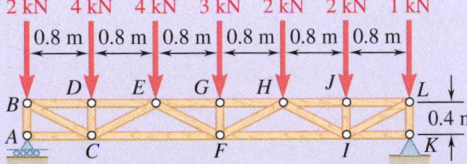

**Fig. P6.47 and P6.48**

**6.48** A floor truss is loaded as shown. Determine the force in members *FI*, *HI*, and *HJ*.

**6.49** Determine the force in members *CD* and *DF* of the truss shown.

**6.50** Determine the force in members *CE* and *EF* of the truss shown.

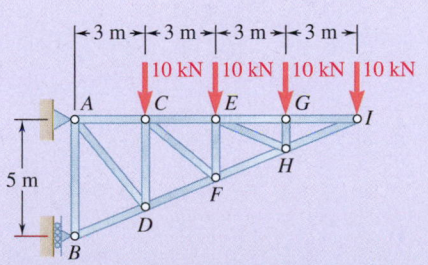

**Fig. P6.49 and P6.50**

**6.51** Determine the force in members *DE* and *DF* of the truss shown when *P* = 20 kips.

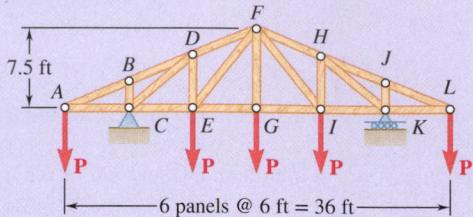

**Fig. P6.51 and P6.52**

**6.52** Determine the force in members *EG* and *EF* of the truss shown when *P* = 20 kips.

**6.53** Determine the force in members *DF* and *DE* of the truss shown.

**6.54** Determine the force in members *CD* and *CE* of the truss shown.

**6.55** A Pratt roof truss is loaded as shown. Determine the force in members *CE*, *DE*, and *DF*.

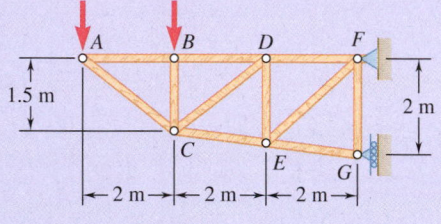

**Fig. P6.53 and P6.54**

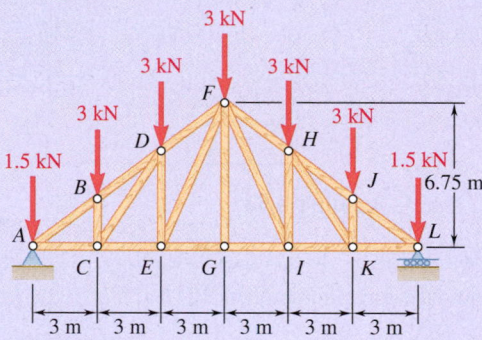

**Fig. P6.55 and P6.56**

**6.56** A Pratt roof truss is loaded as shown. Determine the force in members *FH*, *FI*, and *GI*.

**6.57** A Howe scissors roof truss is loaded as shown. Determine the force in members *DF*, *DG*, and *EG*.

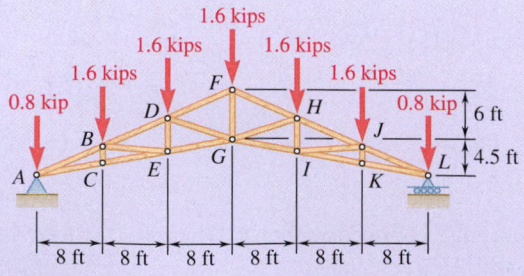

**Fig. P6.57 and P6.58**

**6.58** A Howe scissors roof truss is loaded as shown. Determine the force in members *GI*, *HI*, and *HJ*.

**6.59** Determine the force in members *AD*, *CD*, and *CE* of the truss shown.

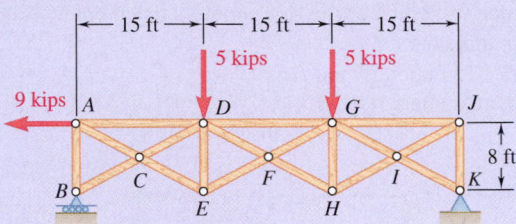

Fig. P6.59 and P6.60

**6.60** Determine the force in members *DG*, *FG*, and *FH* of the truss shown.

**6.61** Determine the force in member *GJ* of the truss shown. (*Hint:* Use section *aa.*)

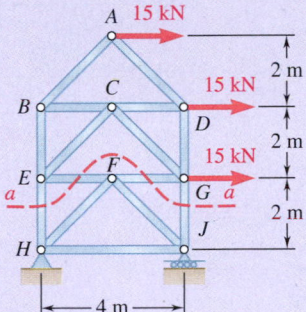

Fig. P6.61

**6.62** Determine the force in members *DG* and *FH* of the truss shown. (*Hint:* Use section *aa.*)

**6.63** Determine the force in members *CD* and *JK* of the truss shown. (*Hint:* Use section *aa.*)

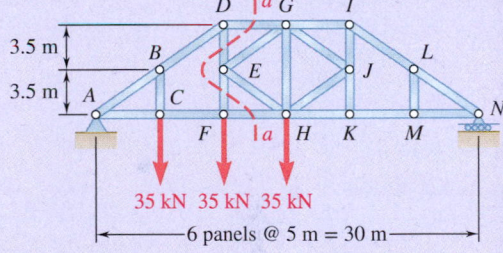

Fig. P6.62

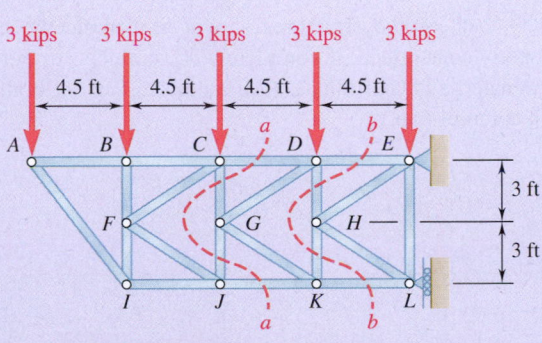

Fig. *P6.63* and *P6.64*

**6.64** Determine the force in members *DE* and *KL* of the truss shown. (*Hint:* Use section *bb.*)

**6.65 and 6.66** The diagonal members in the center panels of the power transmission line tower shown are very slender and can act only in tension; such members are known as *counters*. For the given loading, determine (*a*) which of the two counters listed below is acting, (*b*) the force in that counter.

**6.65** Counters *CJ* and *HE*.

**6.66** Counters *IO* and *KN*.

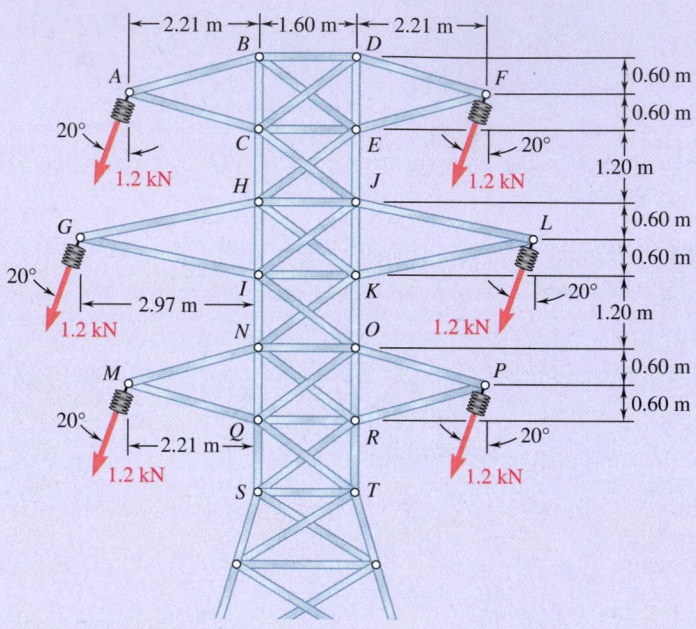

Fig. P6.65 and P6.66

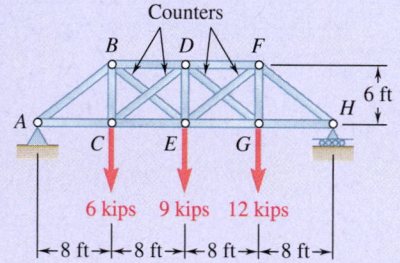

Fig. P6.67

**6.67** The diagonal members in the center panels of the truss shown are very slender and can act only in tension; such members are known as *counters*. Determine the force in member *DE* and in the counters that are acting under the given loading.

**6.68** Solve Prob. 6.67 assuming that the 9-kip load has been removed.

**6.69** Classify each of the structures shown as completely, partially, or improperly constrained; if completely constrained, further classify as determinate or indeterminate. (All members can act both in tension and in compression.)

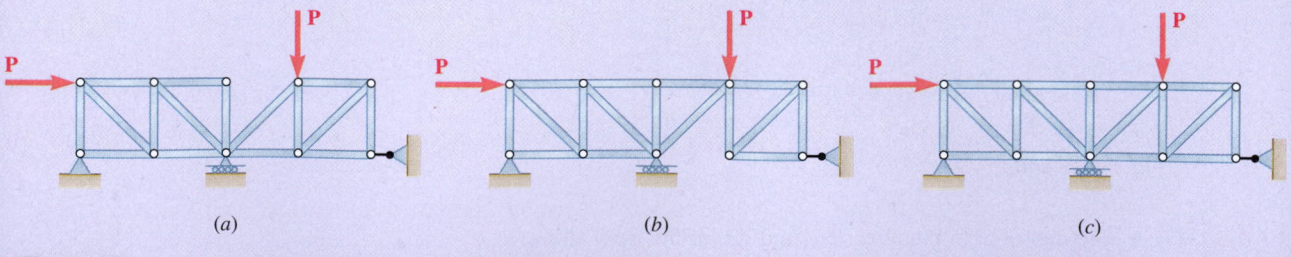

Fig. P6.69

**6.70 through 6.74** Classify each of the structures shown as completely, partially, or improperly constrained; if completely constrained, further classify as determinate or indeterminate. (All members can act both in tension and in compression.)

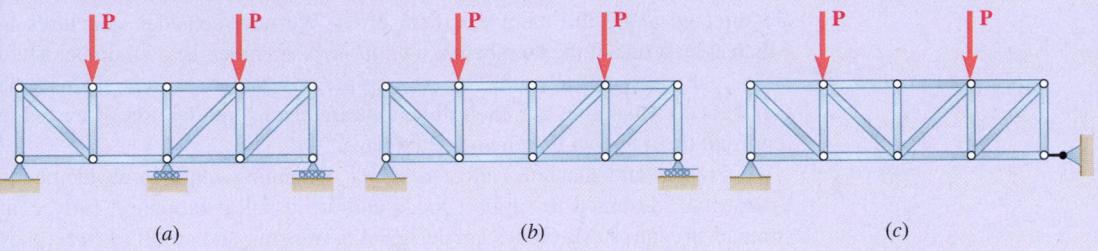

(a)           (b)           (c)

**Fig. P6.70**

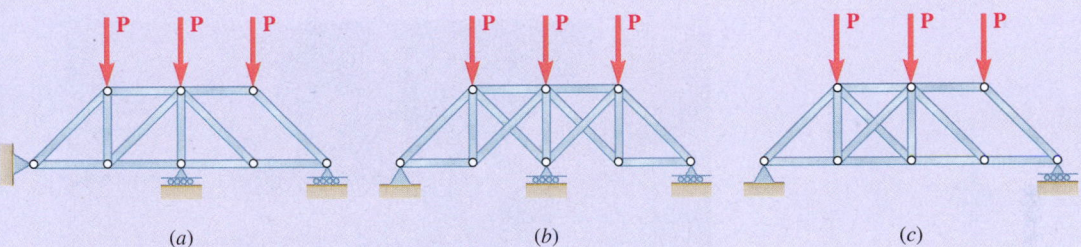

(a)           (b)           (c)

**Fig. P6.71**

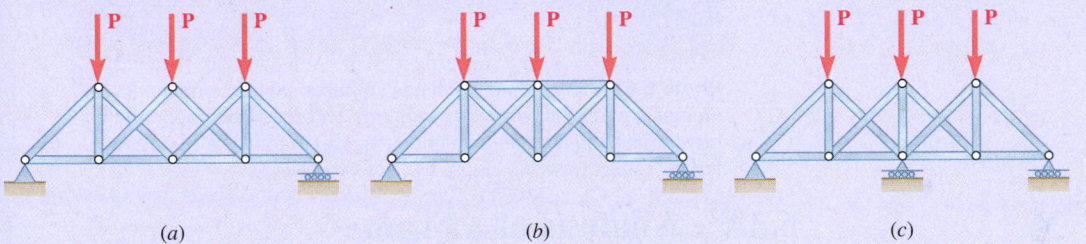

(a)           (b)           (c)

**Fig. P6.72**

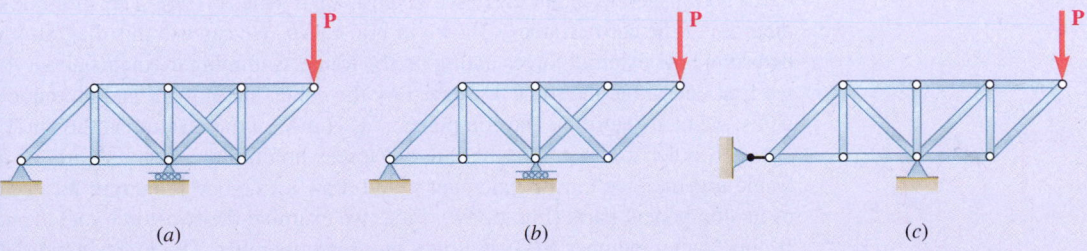

(a)           (b)           (c)

**Fig. P6.73**

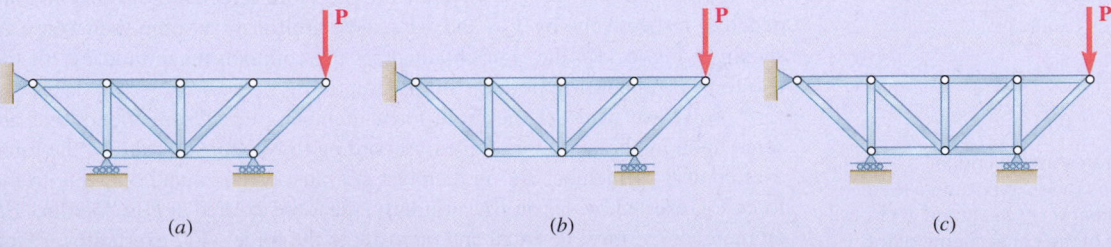

(a)           (b)           (c)

**Fig. P6.74**

# 6.3 FRAMES

When we study trusses, we are looking at structures consisting entirely of pins and straight two-force members. The forces acting on the two-force members are directed along the members themselves. We now consider structures in which at least one of the members is a *multi-force* member, i.e., a member acted upon by three or more forces. These forces are generally not directed along the members on which they act; their directions are unknown; therefore, we need to represent them by two unknown components.

Frames and machines are structures containing multi-force members. **Frames** are designed to support loads and are usually stationary, fully constrained structures. **Machines** are designed to transmit and modify forces; they may or may not be stationary and always contain moving parts.

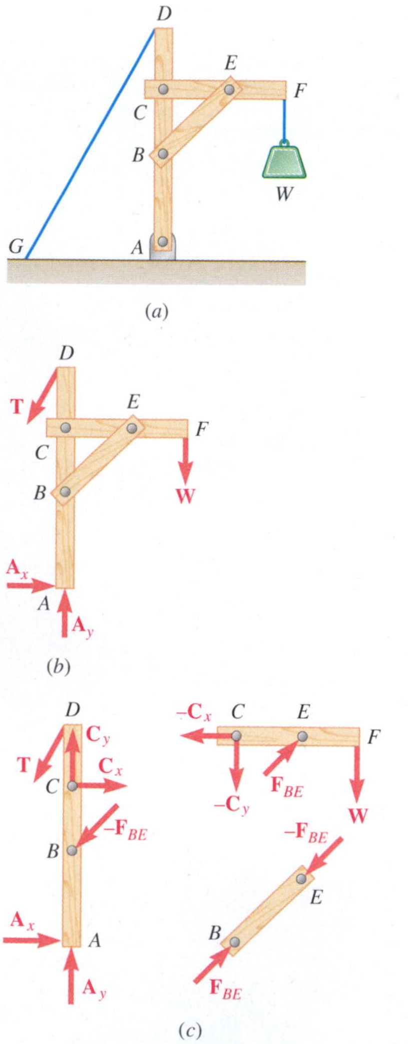

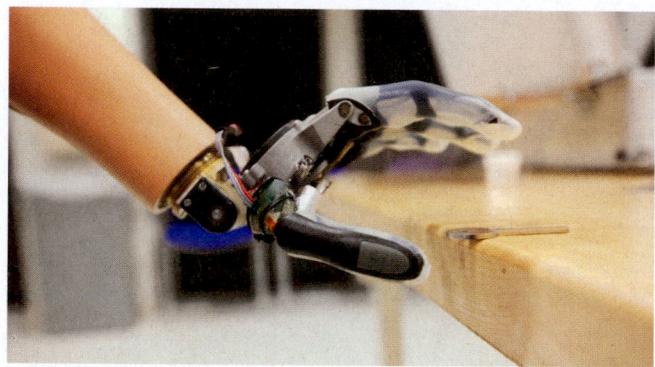

**Photo 6.6** Frames and machines contain multi-force members. Frames are fully constrained structures, whereas machines like this prosthetic hand are movable and designed to transmit or modify forces. ©Mark Thiessen/National Geographic Society/Corbis

## 6.3A Analysis of a Frame

As the first example of analysis of a frame, we consider again the crane described in Sec. 6.1 that carries a given load $W$ (Fig. 6.19a). The free-body diagram of the entire frame is shown in Fig. 6.19b. We can use this diagram to determine the external forces acting on the frame. Summing moments about $A$, we first determine the force **T** exerted by the cable; summing $x$ and $y$ components, we then determine the components $\mathbf{A}_x$ and $\mathbf{A}_y$ of the reaction at the pin $A$.

In order to determine the internal forces holding the various parts of a frame together, we must dismember it and draw a free-body diagram for each of its component parts (Fig. 6.19c). First, we examine the two-force members. In this frame, member $BE$ is the only two-force member. The forces acting at each end of this member must have the same magnitude, same line of action, and opposite sense (Sec. 4.2A). They are therefore directed along $BE$ and are denoted, respectively, by $\mathbf{F}_{BE}$ and $-\mathbf{F}_{BE}$. We arbitrarily assume their sense as shown in Fig. 6.19c; the sign obtained for the common magnitude $F_{BE}$ of the two forces will confirm or deny this assumption.

Next, we consider the multi-force members, i.e., the members that are acted upon by three or more forces. According to Newton's third law, the force exerted at $B$ by member $BE$ on member $AD$ must be equal and opposite to the force $\mathbf{F}_{BE}$ exerted by $AD$ on $BE$. Similarly, the force exerted at $E$ by member $BE$ on member $CF$ must be equal and opposite to the force $-\mathbf{F}_{BE}$ exerted by $CF$ on $BE$. Thus, the forces that the two-force member $BE$ exerts on $AD$ and $CF$ are,

**Fig. 6.19** A frame in equilibrium. (a) Diagram of a crane supporting a load; (b) free-body diagram of the crane; (c) free-body diagrams of the components of the crane.

respectively, equal to $-\mathbf{F}_{BE}$ and $\mathbf{F}_{BE}$; they have the same magnitude $F_{BE}$, opposite sense, and should be directed as shown in Fig. 6.19c.

Joint $C$ connects two multi-force members. Because neither the direction nor the magnitude of the forces acting at $C$ are known, we represent these forces by their $x$ and $y$ components. The components $\mathbf{C}_x$ and $\mathbf{C}_y$ of the force acting on member $AD$ are arbitrarily directed to the right and upward. Because, according to Newton's third law, the forces exerted by member $CF$ on $AD$ and by member $AD$ on $CF$ are equal and opposite, the components of the force acting on member $CF$ *must* be directed to the left and downward; we denote them, respectively, by $-\mathbf{C}_x$ and $-\mathbf{C}_y$. Whether the force $\mathbf{C}_x$ is actually directed to the right and the force $-\mathbf{C}_x$ is actually directed to the left will be determined later from the sign of their common magnitude $C_x$ with a plus sign indicating that the assumption was correct and a minus sign that it was wrong. We complete the free-body diagrams of the multi-force members by showing the external forces acting at $A$, $D$, and $F$.[†]

We can now determine the internal forces by considering the free-body diagram of either of the two multi-force members. Choosing the free-body diagram of $CF$, for example, we write the equations $\Sigma M_C = 0$, $\Sigma M_E = 0$, and $\Sigma F_x = 0$, which yield the values of the magnitudes $F_{BE}$, $C_y$, and $C_x$, respectively. We can check these values by verifying that member $AD$ is also in equilibrium.

Note that we assume the pins in Fig. 6.19 form an integral part of one of the two members they connected, so it is not necessary to show their free-body diagrams. We can always use this assumption to simplify the analysis of frames and machines. However, when a pin connects three or more members, connects a support and two or more members, or when a load is applied to a pin, we must make a clear decision in choosing the member to which we assume the pin belongs. (If multi-force members are involved, the pin should be attached to one of these members.) We then need to identify clearly the various forces exerted on the pin. This is illustrated in Sample Prob. 6.6.

## 6.3B Frames that Collapse Without Supports

The crane we just analyzed was constructed so it could keep the same shape without the help of its supports; we therefore considered it to be a rigid body. Many frames, however, will collapse if detached from their supports; such frames cannot be considered rigid bodies. Consider, for example, the frame shown in Fig. 6.20a that consists of two members $AC$ and $CB$ carrying loads $\mathbf{P}$ and $\mathbf{Q}$ at their midpoints. The members are supported by pins at $A$ and $B$ and are connected by a pin at $C$. If we detach this frame from its supports, it will not maintain its shape. Therefore, we should consider it to be made of *two distinct rigid parts AC and CB*.

The equations $\Sigma F_x = 0$, $\Sigma F_y = 0$, and $\Sigma M = 0$ (about any given point) express the conditions for the *equilibrium of a rigid body* (Chap. 4); we should use them, therefore, in connection with the free-body diagrams of members $AC$ and $CB$ (Fig. 6.20b). Because these members are multi-force members and because pins are used at the supports and at the connection, we represent each of the reactions at $A$ and $B$ and the forces at $C$ by two components. In accordance with Newton's third law, we represent the components of the force exerted by

---

[†]It is not strictly necessary to use a minus sign to distinguish the force exerted by one member on another from the equal and opposite force exerted by the second member on the first, since the two forces belong to different free-body diagrams and thus are not easily confused. In the Sample Problems, we use the same symbol to represent equal and opposite forces that are applied to different free bodies. Note that, under these conditions, the sign obtained for a given force component does not directly relate the sense of that component to the sense of the corresponding coordinate axis. Rather, a positive sign indicates that *the sense assumed for that component in the free-body diagram* is correct, and a negative sign indicates that it is wrong.

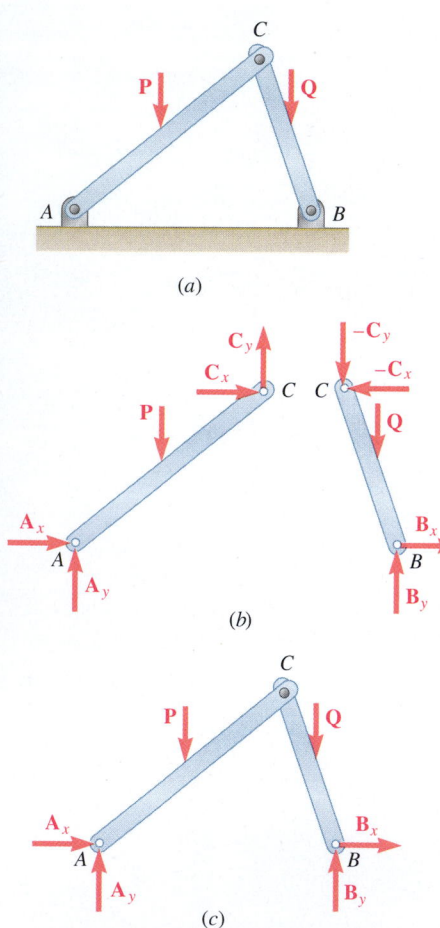

*CB* on *AC* and the components of the force exerted by *AC* on *CB* by vectors of the same magnitude and opposite sense. Thus, if the first pair of components consists of $C_x$ and $C_y$, the second pair is represented by $-C_x$ and $-C_y$.

Note that four unknown force components act on free body *AC*, whereas we need only three independent equations to express that the body is in equilibrium. Similarly, four unknowns, but only three equations, are associated with *CB*. However, only six different unknowns are involved in the analysis of the two members, and altogether, six equations are available to express that the members are in equilibrium. Setting $\Sigma M_A = 0$ for free body *AC* and $\Sigma M_B = 0$ for *CB*, we obtain two simultaneous equations that we can solve for the common magnitude $C_x$ of the components $C_x$ and $-C_x$ and for the common magnitude $C_y$ of the components $C_y$ and $-C_y$. We then have $\Sigma F_x = 0$ and $\Sigma F_y = 0$ for each of the two free bodies, successively obtaining the magnitudes $A_x$, $A_y$, $B_x$, and $B_y$.

Observe that, because the equations of equilibrium $\Sigma F_x = 0$, $\Sigma F_y = 0$, and $\Sigma M = 0$ (about any given point) are satisfied by the forces acting on free body *AC* and because they are also satisfied by the forces acting on free body *CB*, they must be satisfied when the forces acting on the two free bodies are considered simultaneously. Because the internal forces at *C* cancel each other, we find that the equations of equilibrium must be satisfied by the external forces shown on the free-body diagram of the frame *ACB* itself (Fig. 6.20c), even though the frame is not a rigid body. We can use these equations to determine some of the components of the reactions at *A* and *B*. We will find, however, that **the reactions cannot be completely determined from the free-body diagram of the whole frame**. It is thus necessary to dismember the frame and consider the free-body diagrams of its component parts (Fig. 6.20b), even when we are interested in determining external reactions only. The reason is that the equilibrium equations obtained for free body *ACB are necessary conditions* for the equilibrium of a nonrigid structure, *but these are not sufficient conditions.*

The method of solution outlined here involved simultaneous equations. We now present a more efficient method that utilizes the free body *ACB*, as well as the free bodies *AC* and *CB*. Writing $\Sigma M_A = 0$ and $\Sigma M_B = 0$ for free body *ACB*, we obtain $B_y$ and $A_y$. From $\Sigma M_C = 0$, $\Sigma F_x = 0$, and $\Sigma F_y = 0$ for free body *AC*, we successively obtain $A_x$, $C_x$, and $C_y$. Finally, setting $\Sigma F_x = 0$ for *ACB* gives us $B_x$.

We noted previously that the analysis of the frame in Fig. 6.20 involves six unknown force components and six independent equilibrium equations. (The equilibrium equations for the whole frame were obtained from the original six equations and, therefore, are not independent.) Moreover, we checked that all unknowns could be actually determined and that all equations could be satisfied. This frame is **statically determinate and rigid**. (We use the word "rigid" here to indicate that the frame maintains its shape as long as it remains attached to its supports.) In general, to determine whether a structure is statically determinate and rigid, you should draw a free-body diagram for each of its component parts and count the reactions and internal forces involved. You should then determine the number of independent equilibrium equations (excluding equations expressing the equilibrium of the whole structure or of groups of component parts already analyzed). If you have more unknowns than equations, the structure is *statically indeterminate*. If you have fewer unknowns than equations, the structure is *nonrigid*. If you have as many unknowns as equations *and if all unknowns can be determined and all equations satisfied* under general loading conditions, the structure is statically determinate and rigid. If, however, due to an improper arrangement of members and supports, all unknowns cannot be determined and all equations cannot be satisfied, the structure is **statically indeterminate and nonrigid**.

**Fig. 6.20** (*a*) A frame of two members supported by two pins and joined together by a third pin. Without the supports, the frame would collapse and is therefore not a rigid body. (*b*) Free-body diagrams of the two members. (*c*) Free-body diagram of the whole frame.

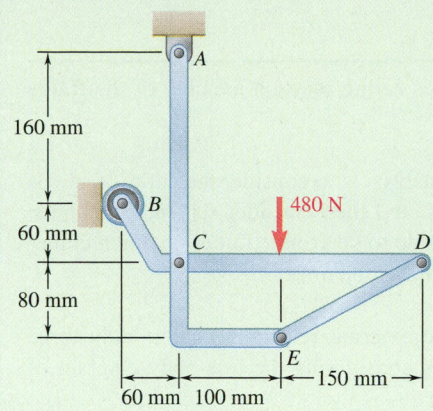

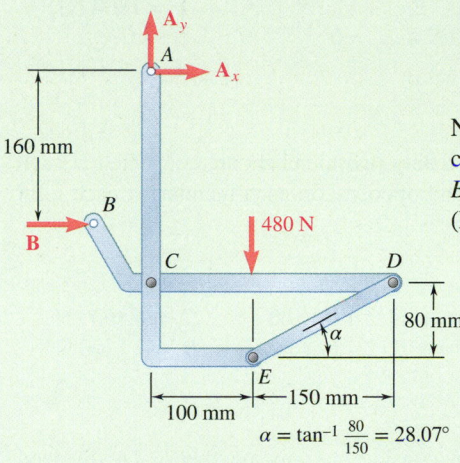

**Fig. 1** Free-body diagram of the entire frame.

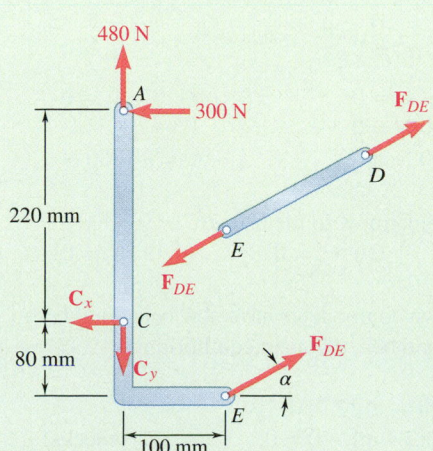

**Fig. 3** Free-body diagrams of members *ACE* and *DE*.

## Sample Problem 6.4

In the frame shown, members *ACE* and *BCD* are connected by a pin at *C* and by the link *DE*. For the loading shown, determine the force in link *DE* and the components of the force exerted at *C* on member *BCD*.

**STRATEGY:** Follow the general procedure discussed in this section. First treat the entire frame as a free body, which will enable you to find the reactions at *A* and *B*. Then, dismember the frame and treat each member as a free body, which will give you the equations needed to find the force at *C*.

**MODELING and ANALYSIS:** Because the external reactions involve only three unknowns, compute the reactions by considering the free-body diagram of the entire frame (Fig. 1).

$$+\uparrow\Sigma F_y = 0: \quad A_y - 480\text{ N} = 0 \qquad A_y = +480\text{ N} \qquad \mathbf{A}_y = 480\text{ N}\uparrow$$
$$+\circlearrowleft\Sigma M_A = 0: \quad -(480\text{ N})(100\text{ mm}) + B(160\text{ mm}) = 0$$
$$B = +300\text{ N} \qquad \mathbf{B} = 300\text{ N}\rightarrow$$
$$\xrightarrow{+}\Sigma F_x = 0: \quad B + A_x = 0$$
$$300\text{ N} + A_x = 0 \qquad A_x = -300\text{ N} \qquad \mathbf{A}_x = 300\text{ N}\leftarrow$$

Now dismember the frame (Fig. 2 and Fig. 3). Because only two members are connected at *C*, the components of the unknown forces acting on *ACE* and *BCD* are, respectively, equal and opposite. Assume that link *DE* is in tension (Fig. 3) and exerts equal and opposite forces at *D* and *E*, directed as shown.

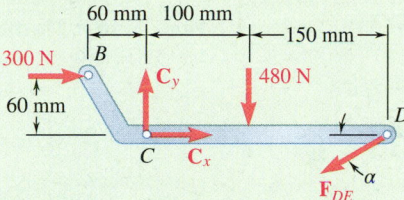

**Fig. 2** Free-body diagram of member *BCD*.

**Free Body: Member BCD.** Using the free body *BCD* (Fig. 2), you can write and solve three equilibrium equations:

$$+\circlearrowright\Sigma M_C = 0:$$
$$(F_{DE}\sin\alpha)(250\text{ mm}) + (300\text{ N})(60\text{ mm}) + (480\text{ N})(100\text{ mm}) = 0$$
$$F_{DE} = -561\text{ N} \qquad F_{DE} = 561\text{ N } C \blacktriangleleft$$
$$\xrightarrow{+}\Sigma F_x = 0: \quad C_x - F_{DE}\cos\alpha + 300\text{ N} = 0$$
$$C_x - (-561\text{ N})\cos 28.07° + 300\text{ N} = 0 \qquad C_x = -795\text{ N}$$
$$+\uparrow\Sigma F_y = 0: \quad C_y - F_{DE}\sin\alpha - 480\text{ N} = 0$$
$$C_y - (-561\text{ N})\sin 28.07° - 480\text{ N} = 0 \qquad C_y = +216\text{ N}$$

From the signs obtained for $C_x$ and $C_y$, the force components $\mathbf{C}_x$ and $\mathbf{C}_y$ exerted on member *BCD* are directed to the left and up, respectively. Thus, you have

$$\mathbf{C}_x = 795\text{ N}\leftarrow, \ \mathbf{C}_y = 216\text{ N}\uparrow \blacktriangleleft$$

**REFLECT and THINK:** Check the computations by considering the free body *ACE* (Fig. 3). For example,

$$+\circlearrowleft\Sigma M_A = (F_{DE}\cos\alpha)(300\text{ mm}) + (F_{DE}\sin\alpha)(100\text{ mm}) - C_x(220\text{ mm})$$
$$= (-561\cos\alpha)(300) + (-561\sin\alpha)(100) - (-795)(220) = 0$$

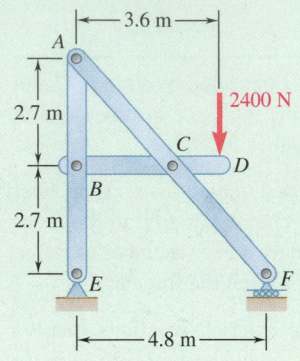

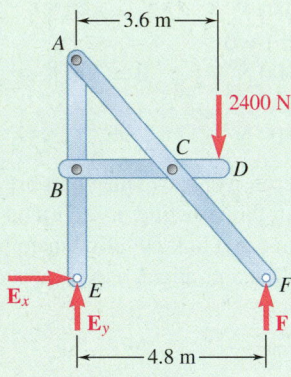

**Fig. 1** Free-body diagram of the entire frame.

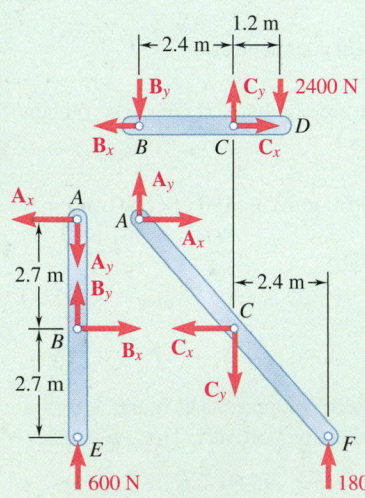

**Fig. 2** Free-body diagrams of the individual members.

## Sample Problem 6.5

Determine the components of the forces acting on each member of the frame shown.

**STRATEGY:** The approach to this analysis is to consider the entire frame as a free body to determine the reactions, and then consider separate members. However, in this case, you will not be able to determine forces on one member without analyzing a second member at the same time.

**MODELING and ANALYSIS:** The external reactions involve only three unknowns, so compute the reactions by considering the free-body diagram of the entire frame (Fig. 1).

$$+\circlearrowleft \Sigma M_E = 0: \qquad -(2400 \text{ N})(3.6 \text{ m}) + F(4.8 \text{ m}) = 0$$
$$F = +1800 \text{ N} \qquad\qquad \mathbf{F} = 1800 \text{ N}\uparrow \blacktriangleleft$$
$$+\uparrow \Sigma F_y = 0: \qquad -2400 \text{ N} + 1800 \text{ N} + E_y = 0$$
$$E_y = +600 \text{ N} \qquad\qquad \mathbf{E}_y = 600 \text{ N}\uparrow \blacktriangleleft$$
$$\xrightarrow{+} \Sigma F_x = 0: \qquad\qquad\qquad\qquad \mathbf{E}_x = 0 \blacktriangleleft$$

Now dismember the frame. Because only two members are connected at each joint, force components are equal and opposite on each member at each joint (Fig. 2).

**Free Body: Member *BCD*.**

$$+\circlearrowleft \Sigma M_B = 0: \qquad -(2400 \text{ N})(3.6 \text{ m}) + C_y(2.4 \text{ m}) = 0 \qquad C_y = +3600 \text{ N} \blacktriangleleft$$
$$+\circlearrowleft \Sigma M_C = 0: \qquad -(2400 \text{ N})(1.2 \text{ m}) + B_y(2.4 \text{ m}) = 0 \qquad B_y = +1200 \text{ N} \blacktriangleleft$$
$$\xrightarrow{+} \Sigma F_x = 0: \qquad -B_x + C_x = 0$$

Neither $B_x$ nor $C_x$ can be obtained by considering only member *BCD*; you need to look at member *ABE*. The positive values obtained for $B_y$ and $C_y$ indicate that the force components $\mathbf{B}_y$ and $\mathbf{C}_y$ are directed as assumed.

**Free Body: Member *ABE*.**

$$+\circlearrowleft \Sigma M_A = 0: \qquad\qquad B_x(2.7 \text{ m}) = 0 \qquad\qquad B_x = 0 \blacktriangleleft$$
$$\xrightarrow{+} \Sigma F_x = 0: \qquad\qquad +B_x - A_x = 0 \qquad\qquad A_x = 0 \blacktriangleleft$$
$$+\uparrow \Sigma F_y = 0: \qquad -A_y + B_y + 600 \text{ N} = 0$$
$$-A_y + 1200 \text{ N} + 600 \text{ N} = 0 \qquad A_y = +1800 \text{ N} \blacktriangleleft$$

**Free Body: Member *BCD*.** Returning now to member *BCD*, you have

$$\xrightarrow{+} \Sigma F_x = 0: \qquad -B_x + C_x = 0 \qquad 0 + C_x = 0 \qquad C_x = 0 \blacktriangleleft$$

**REFLECT and THINK:** All unknown components have now been found. To check the results, you can verify that member *ACF* is in equilibrium.

$$+\circlearrowleft \Sigma M_C = (1800 \text{ N})(2.4 \text{ m}) - A_y(2.4 \text{ m}) - A_x(2.7 \text{ m})$$
$$= (1800 \text{ N})(2.4 \text{ m}) - (1800 \text{ N})(2.4 \text{ m}) - 0 = 0 \qquad \text{(checks)}$$

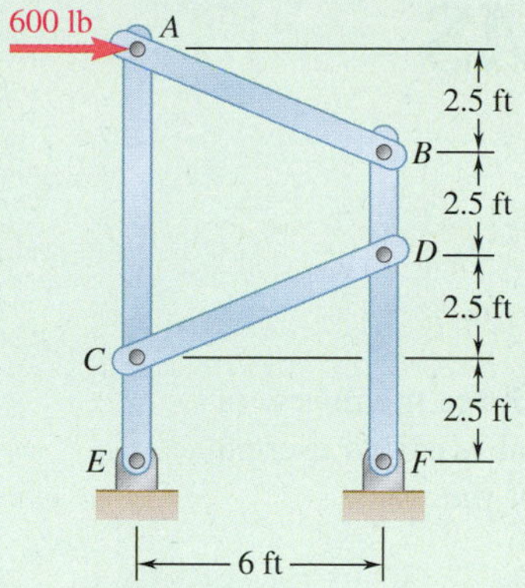

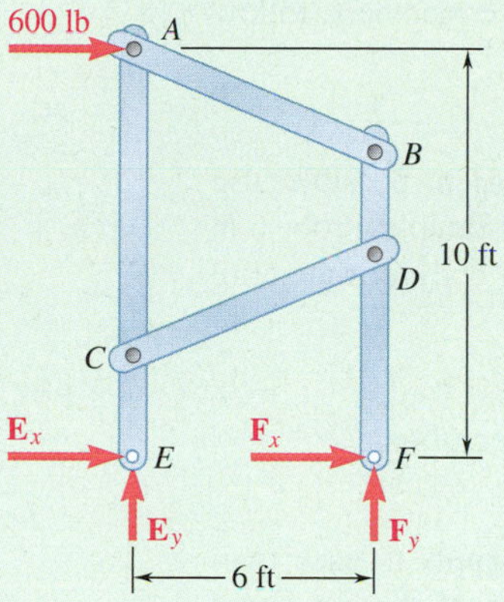

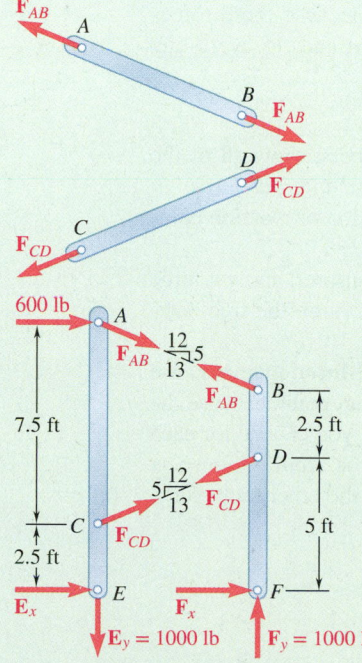

**Fig. 1**  Free-body diagram of the entire frame.

**Fig. 2**  Free-body diagrams of the individual members.

# Sample Problem 6.6

A 600-lb horizontal force is applied to pin $A$ of the frame shown. Determine the forces acting on the two vertical members of the frame.

**STRATEGY:**   Begin as usual with a free-body diagram of the entire frame, but this time you will not be able to determine all of the reactions. You will have to analyze a separate member and then return to the entire frame analysis in order to determine the remaining reaction forces.

**MODELING and ANALYSIS:**   Choosing the entire frame as a free body (Fig. 1), you can write equilibrium equations to determine the two force components $\mathbf{E}_y$ and $\mathbf{F}_y$. However, these equations are not sufficient to determine $\mathbf{E}_x$ and $\mathbf{F}_x$.

$$+\circlearrowleft \Sigma M_E = 0: \qquad -(600\text{ lb})(10\text{ ft}) + F_y(6\text{ ft}) = 0$$
$$F_y = +1000\text{ lb} \qquad\qquad \mathbf{F}_y = 1000\text{ lb}\uparrow \ \blacktriangleleft$$
$$+\uparrow \Sigma F_y = 0: \qquad E_y + F_y = 0$$
$$E_y = -1000\text{ lb} \qquad\qquad \mathbf{E}_y = 1000\text{ lb}\downarrow \ \blacktriangleleft$$

To proceed with the solution, now consider the free-body diagrams of the various members (Fig. 2). In dismembering the frame, assume that pin $A$ is attached to the multi-force member $ACE$ so that the 600-lb force is applied to that member. Note that $AB$ and $CD$ are two-force members.

**Free Body: Member *ACE*.**
$$+\uparrow\Sigma F_y = 0: \qquad -\tfrac{5}{13}F_{AB} + \tfrac{5}{13}F_{CD} - 1000\text{ lb} = 0$$
$$+\circlearrowleft\Sigma M_E = 0: \qquad -(600\text{ lb})(10\text{ ft}) - (\tfrac{12}{13}F_{AB})(10\text{ ft}) - (\tfrac{12}{13}F_{CD})(2.5\text{ ft}) = 0$$

Solving these equations simultaneously gives you

$$F_{AB} = -1040\text{ lb} \qquad\qquad F_{CD} = +1560\text{ lb} \ \blacktriangleleft$$

The signs indicate that the sense assumed for $F_{CD}$ was correct and the sense for $F_{AB}$ was incorrect. Now summing $x$ components, you have

$$\xrightarrow{+}\Sigma F_x = 0: \qquad 600\text{ lb} + \tfrac{12}{13}(-1040\text{ lb}) + \tfrac{12}{13}(+1560\text{ lb}) + E_x = 0$$
$$E_x = -1080\text{ lb} \qquad\qquad \mathbf{E}_x = 1080\text{ lb}\leftarrow \ \blacktriangleleft$$

**Free Body: Entire Frame.**   Now that $\mathbf{E}_x$ is determined, you can return to the free-body diagram of the entire frame.

$$\xrightarrow{+}\Sigma F_x = 0: \qquad 600\text{ lb} - 1080\text{ lb} + F_x = 0$$
$$F_x = +480\text{ lb} \qquad\qquad \mathbf{F}_x = 480\text{ lb}\rightarrow \ \blacktriangleleft$$

**REFLECT and THINK:**   Check your computations by verifying that the equation $\Sigma M_B = 0$ is satisfied by the forces acting on member $BDF$.

$$+\circlearrowleft\Sigma M_B = -(\tfrac{12}{13}F_{CD})(2.5\text{ ft}) + (F_x)(7.5\text{ ft})$$
$$= -\tfrac{12}{13}(1560\text{ lb})(2.5\text{ ft}) + (480\text{ lb})(7.5\text{ ft})$$
$$= -3600\text{ lb·ft} + 3600\text{ lb·ft} = 0 \qquad \text{(checks)}$$

# SOLVING PROBLEMS
# ON YOUR OWN

In this section, we analyzed **frames containing one or more multi-force members**. In the problems that follow, you will be asked to determine the external reactions exerted on the frame and the internal forces that hold together the members of the frame.

In solving problems involving frames containing one or more multi-force members, follow these steps.

**1. Draw a free-body diagram of the entire frame.** To the greatest extent possible, use this free-body diagram to calculate the reactions at the supports. (In Sample Prob. 6.6, only two of the four reaction components could be found from the free body of the entire frame.)

**2. Dismember the frame, and draw a free-body diagram of each member.**

**3. First consider the two-force members.** Equal and opposite forces apply to each two-force member at the points where it is connected to another member. If the two-force member is straight, these forces are directed along the axis of the member. If you cannot tell at this point whether the member is in tension or compression, *assume* that the member is in tension and *direct both of the forces away from the member.* Because these forces have the same unknown magnitude, give them both the *same name* and, to avoid any confusion later, *do not use a plus sign or a minus sign.*

**4. Next consider the multi-force members.** For each of these members, show all of the forces acting on the member, including *applied loads, reactions, and internal forces at connections.* Clearly indicate the magnitude and direction of any reaction or reaction component found earlier from the free-body diagram of the entire frame.

   **a. Where a multi-force member is connected to a two-force member,** apply a force to the multi-force member that is *equal and opposite* to the force drawn on the free-body diagram of the two-force member, *giving it the same name.*

   **b. Where a multi-force member is connected to another multi-force member,** use *horizontal and vertical components* to represent the internal forces at that point, because the directions and magnitudes of these forces are unknown. The direction you choose for each of the two force components exerted on the first multi-force member is arbitrary, but *you must apply equal and opposite force components of the same name* to the other multi-force member. Again, *do not use a plus sign or a minus sign.*

*(continued)*

**5. Now determine the internal forces as well as any reactions** you have not already found.

    **a. The free-body diagram** of each multi-force member can provide you with *three equilibrium equations.*

    **b. To simplify your solution,** seek a way to write an equation involving a single unknown. If you can locate *a point where all but one of the unknown force components intersect,* you can obtain an equation in a single unknown by summing moments about that point. *If all unknown forces except one are parallel,* you can obtain an equation in a single unknown by summing force components in a direction perpendicular to the parallel forces.

    **c. Because you arbitrarily chose the direction of each of the unknown forces,** you cannot determine whether your guess was correct until the solution is complete. To do that, consider the *sign* of the value found for each of the unknowns: a *positive* sign means the direction you selected was *correct;* a *negative* sign means the direction is *opposite* to the direction you assumed.

**6. To be more effective and efficient** as you proceed through your solution, observe the following rules.

    **a. If you can find an equation involving only one unknown,** write that equation and *solve it for that unknown.* Immediately replace that unknown wherever it appears on other free-body diagrams by the value you have found. Repeat this process by seeking equilibrium equations involving only one unknown until you have found all of the internal forces and unknown reactions.

    **b. If you cannot find an equation involving only one unknown,** you may have to *solve a pair of simultaneous equations.* Before doing so, check that you have included the values of all of the reactions you obtained from the free-body diagram of the entire frame.

    **c. The total number of equations** of equilibrium for the entire frame and for the individual members *will be larger than the number of unknown forces and reactions.* After you have found all of the reactions and all of the internal forces, you can use the remaining equations to check the accuracy of your computations.

# Problems

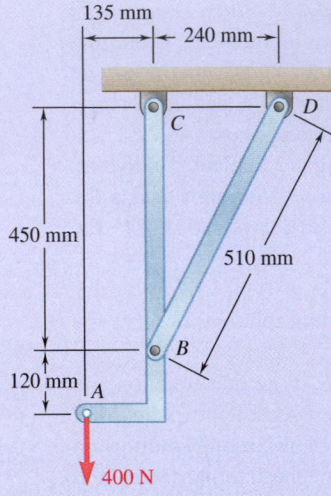

Fig. P6.F1

**6.F1** For the frame and loading shown, draw the free-body diagram(s) needed to determine the force in member *BD* and the components of the reaction at *C*.

**6.F2** For the frame and loading shown, draw the free-body diagram(s) needed to determine the components of all forces acting on member *ABC*.

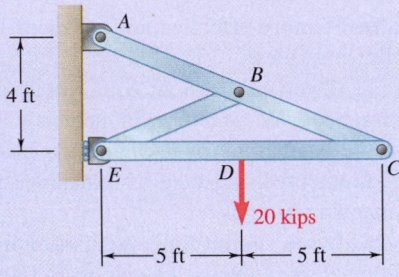

Fig. P6.F2

**6.F3** Draw the free-body diagram(s) needed to determine all the forces exerted on member *AI* if the frame is loaded by a clockwise couple of magnitude 1200 lb·in. applied at point *D*.

**6.F4** Knowing that the pulley has a radius of 0.5 m, draw the free-body diagram(s) needed to determine the components of the reactions at *A* and *E*.

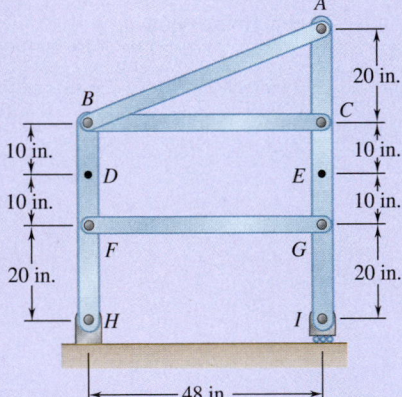

Fig. P6.F3

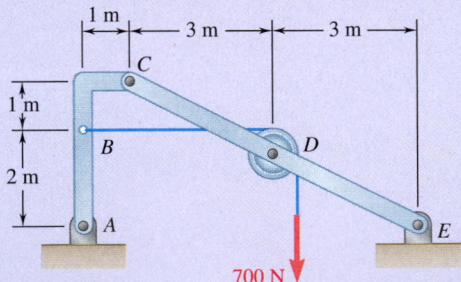

Fig. P6.F4

**6.75 and 6.76** Determine the force in member *BD* and the components of the reaction at *C*.

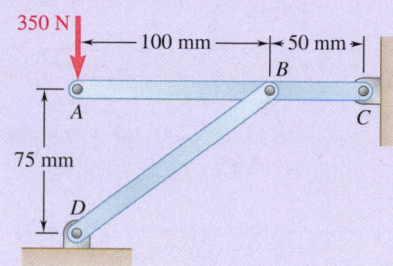

Fig. P6.75

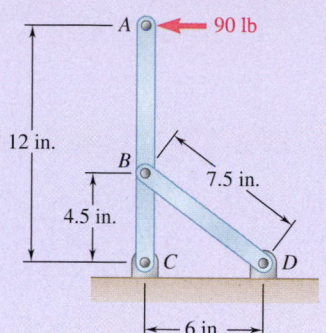

Fig. P6.76

**6.77** For the frame and loading shown, determine the force acting on member *ABC* (*a*) at *B*, (*b*) at *C*.

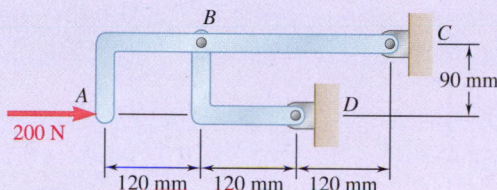

Fig. P6.77

**6.78** Determine the components of all forces acting on member *ABCD* of the assembly shown.

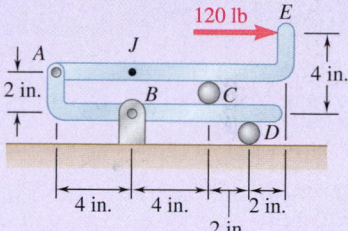

Fig. **P6.78**

**6.79** The hydraulic cylinder *CF*, which partially controls the position of rod *DE*, has been locked in the position shown. Knowing that $\theta = 60°$, determine (*a*) the force **P** for which the tension in link *AB* is 410 N, (*b*) the corresponding force exerted on member *BCD* at point *C*.

**6.80** The hydraulic cylinder *CF*, which partially controls the position of rod *DE*, has been locked in the position shown. Knowing that $P = 400$ N and $\theta = 75°$, determine (*a*) the force in link *AB*, (*b*) the corresponding force exerted on member *BCD* at point *C*.

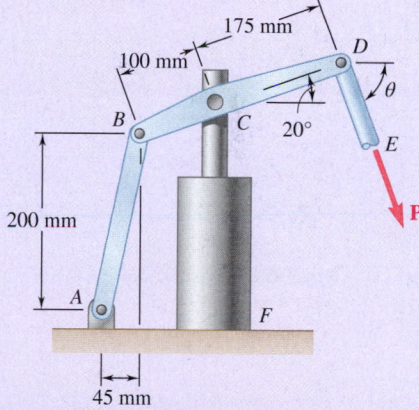

Fig. P6.79 and P6.80

**6.81** Determine the components of all forces acting on member *ABCD* when $\theta = 0$.

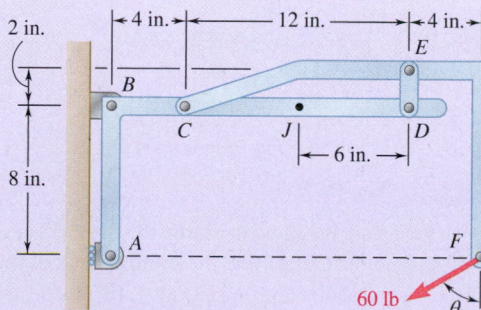

Fig. P6.81 and *P6.82*

**6.82** Determine the components of all forces acting on member *ABCD* when $\theta = 90°$.

343

**6.83** Determine the components of the reactions at $A$ and $E$, ($a$) if the 800-N load is applied as shown, ($b$) if the 800-N load is moved along its line of action and is applied at point $D$.

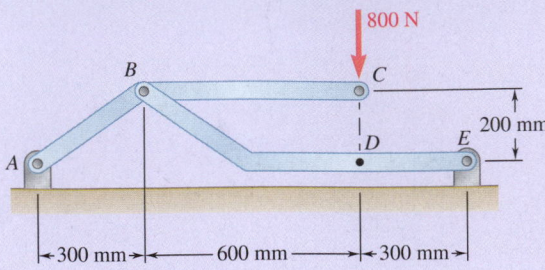

**Fig. P6.83**

**6.84** Determine the components of the reactions at $D$ and $E$ if the frame is loaded by a clockwise couple of magnitude 150 N·m applied ($a$) at $A$, ($b$) at $B$.

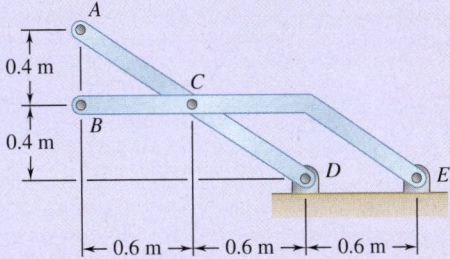

**Fig. P6.84**

**6.85** Determine the components of the reactions at $A$ and $E$ if a 750-N force directed vertically downward is applied ($a$) at $B$, ($b$) at $D$.

**6.86** Determine the components of the reactions at $A$ and $E$ if the frame is loaded by a clockwise couple with a magnitude of 36 N·m applied ($a$) at $B$, ($b$) at $D$.

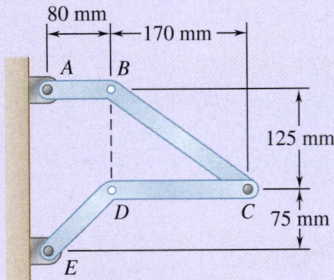

**Fig. P6.85 and P6.86**

**6.87** Determine the components of the reactions at $A$ and $B$, ($a$) if the 60-lb load is applied as shown, ($b$) if the 60-lb load is moved along its line of action and is applied at point $E$.

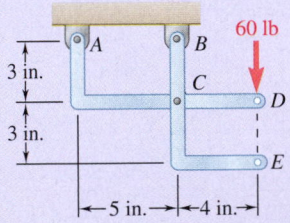

**Fig. P6.87**

**6.88** The 48-lb load can be moved along the line of action shown and applied at $A$, $D$, or $E$. Determine the components of the reactions at $B$ and $F$ if the 48-lb load is applied ($a$) at $A$, ($b$) at $D$, ($c$) at $E$.

**6.89** The 48-lb load is removed and a 288-lb·in. clockwise couple is applied successively at $A$, $D$, and $E$. Determine the components of the reactions at $B$ and $F$ if the couple is applied ($a$) at $A$, ($b$) at $D$, ($c$) at $E$.

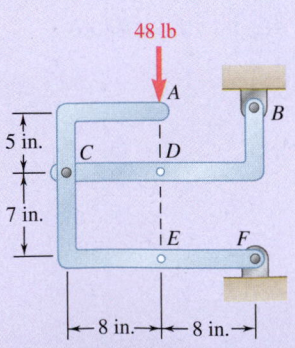

**Fig. P6.88 and P6.89**

**6.90** (*a*) Show that, when a frame supports a pulley at *A*, an equivalent loading of the frame and of each of its component parts can be obtained by removing the pulley and applying at *A* two forces equal and parallel to the forces that the cable exerted on the pulley. (*b*) Show that, if one end of the cable is attached to the frame at a point *B*, a force of magnitude equal to the tension in the cable should also be applied at *B*.

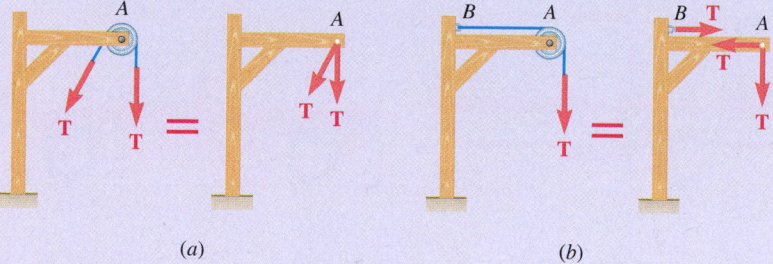

(*a*)                    (*b*)

**Fig. P6.90**

**6.91** Knowing that each pulley has a radius of 250 mm, determine the components of the reactions at *D* and *E*.

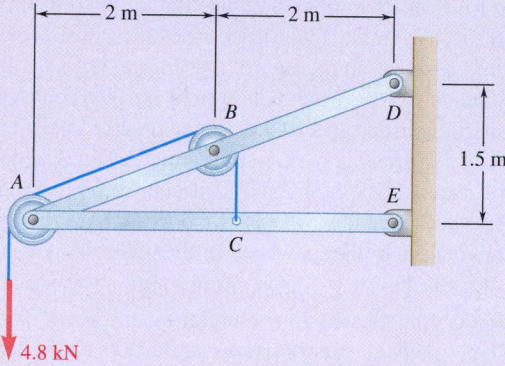

**Fig. P6.91**

**6.92** Knowing that the pulley has a radius of 75 mm, determine the components of the reactions at *A* and *B*.

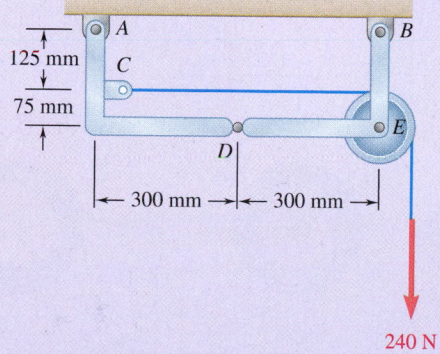

**Fig. P6.92**

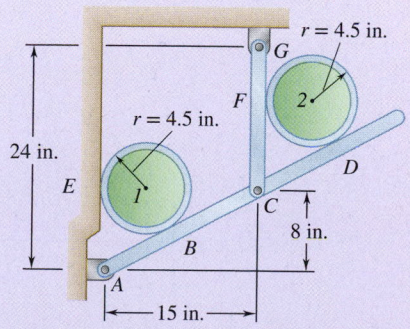

**Fig. P6.93**

**6.93** Two 9-in.-diameter pipes (pipe *1* and pipe *2*) are supported every 7.5 ft by a small frame like that shown. Knowing that the combined weight of each pipe and its contents is 30 lb/ft and assuming frictionless surfaces, determine the components of the reactions at *A* and *G*.

**6.94** Solve Prob. 6.93 assuming that pipe *1* is removed and that only pipe *2* is supported by the frames.

**6.95** A trailer weighing 2400 lb is attached to a 2900-lb pickup truck by a ball-and-socket truck hitch at *D*. Determine (*a*) the reactions at each of the six wheels when the truck and trailer are at rest, (*b*) the additional load on each of the truck wheels due to the trailer.

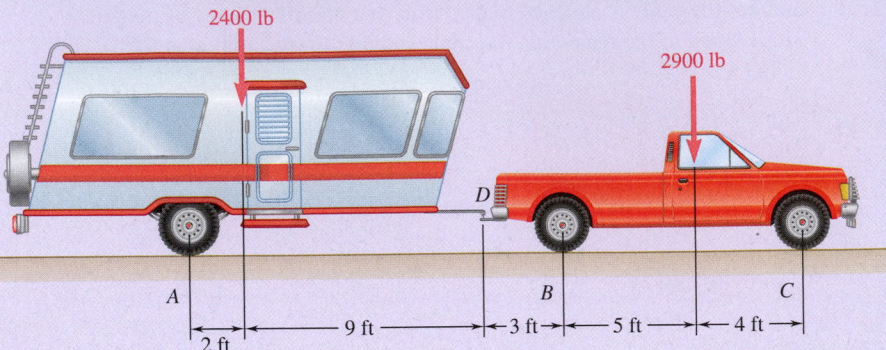

Fig. P6.95

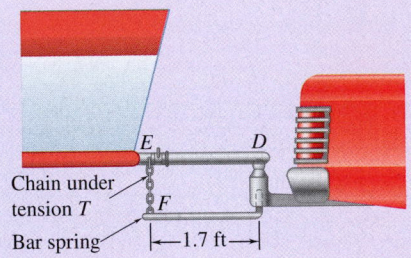

Fig. P6.96

**6.96** In order to obtain a better weight distribution over the four wheels of the pickup truck of Prob. 6.95, a compensating hitch of the type shown is used to attach the trailer to the truck. The hitch consists of two bar springs (only one is shown in the figure) that fit into bearings inside a support rigidly attached to the truck. The springs are also connected by chains to the trailer frame, and specially designed hooks make it possible to place both chains in tension. (*a*) Determine the tension *T* required in each of the two chains if the additional load due to the trailer is to be evenly distributed over the four wheels of the truck. (*b*) What are the resulting reactions at each of the six wheels of the trailer-truck combination?

**6.97** The cab and motor units of the front-end loader shown are connected by a vertical pin located 2 m behind the cab wheels. The distance from *C* to *D* is 1 m. The center of gravity of the 300-kN motor unit is located at $G_m$, while the centers of gravity of the 100-kN cab and 75-kN load are located, respectively, at $G_c$ and $G_l$. Knowing that the machine is at rest with its brakes released, determine (*a*) the reactions at each of the four wheels, (*b*) the forces exerted on the motor unit at *C* and *D*.

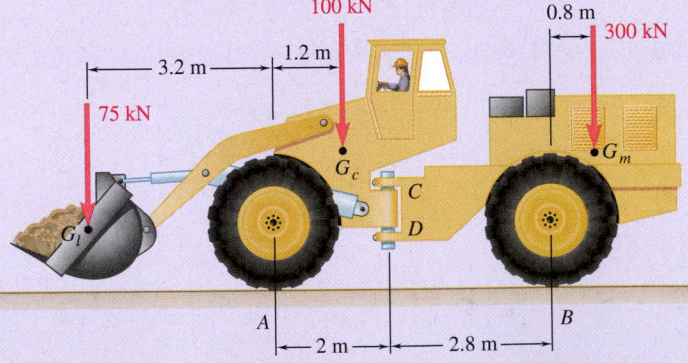

Fig. *P6.97*

**6.98** Solve Prob. 6.97 assuming that the 75-kN load has been removed.

**6.99** Knowing that *P* = 90 lb and *Q* = 60 lb, determine the components of all forces acting on member *BCDE* of the assembly shown.

**6.100** Knowing that *P* = 60 lb and *Q* = 90 lb, determine the components of all forces acting on member *BCDE* of the assembly shown.

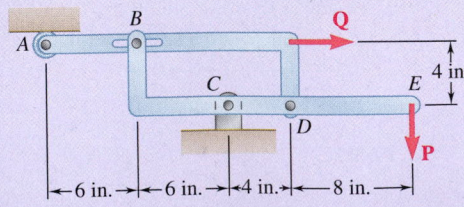

Fig. P6.99 and P6.100

**6.101 and 6.102** For the frame and loading shown, determine the components of all forces acting on member *ABE*.

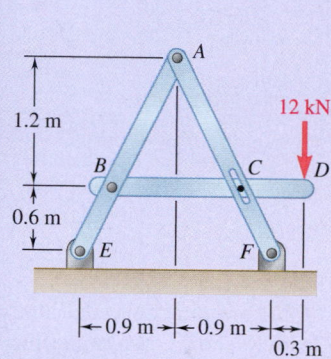

**Fig. P6.101**

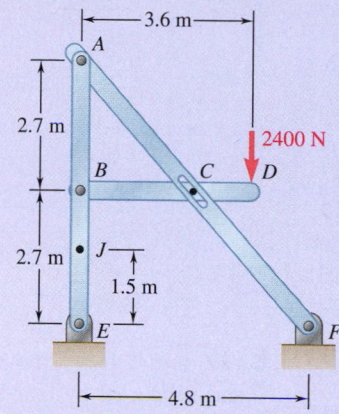

**Fig. P6.102**

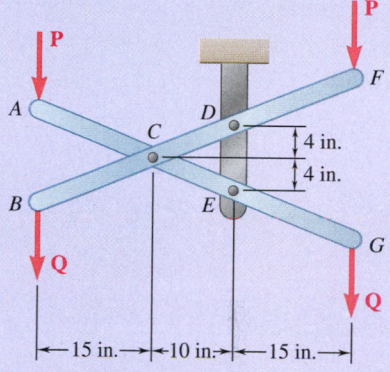

**Fig. P6.103 and P6.104**

**6.103** Knowing that $P = 15$ lb and $Q = 65$ lb, determine the components of the forces exerted (*a*) on member *BCDF* at *C* and *D*, (*b*) on member *ACEG* at *E*.

**6.104** Knowing that $P = 25$ lb and $Q = 55$ lb, determine the components of the forces exerted (*a*) on member *BCDF* at *C* and *D*, (*b*) on member *ACEG* at *E*.

**6.105** For the frame and loading shown, determine the components of the forces acting on member *DABC* at *B* and *D*.

**6.106** Solve Prob. 6.105 assuming that the 6-kN load has been removed.

**6.107** The axis of the three-hinge arch *ABC* is a parabola with vertex at *B*. Knowing that $P = 112$ kN and $Q = 140$ kN, determine (*a*) the components of the reaction at *A*, (*b*) the components of the force exerted at *B* on segment *AB*.

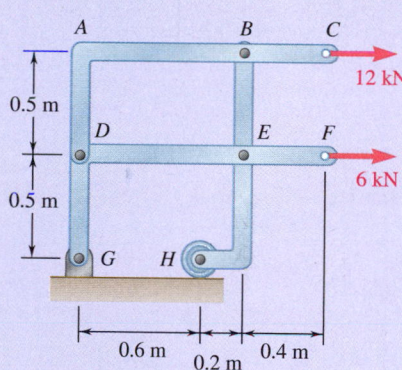

**Fig. P6.105**

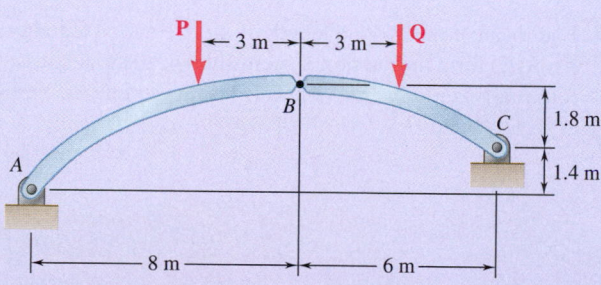

**Fig. P6.107 and P6.108**

**6.108** The axis of the three-hinge arch *ABC* is a parabola with vertex at *B*. Knowing that $P = 140$ kN and $Q = 112$ kN, determine (*a*) the components of the reaction at *A*, (*b*) the components of the force exerted at *B* on segment *AB*.

**6.109 and 6.110**  Neglecting the effect of friction at the horizontal and vertical surfaces, determine the forces exerted at $B$ and $C$ on member $BCE$.

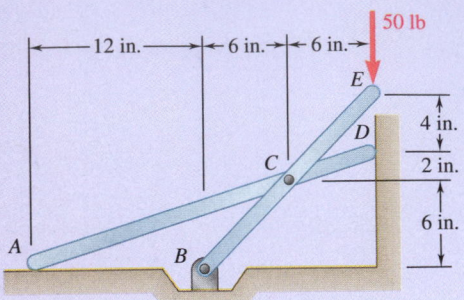

Fig. P6.109

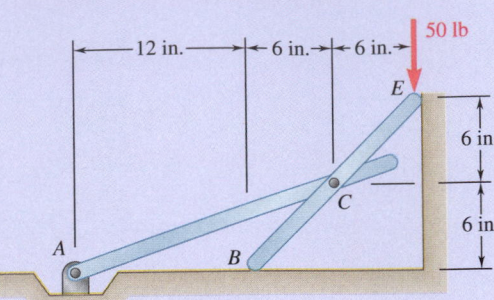

Fig. P6.110

**6.111, 6.112, and 6.113**  Members $ABC$ and $CDE$ are pin-connected at $C$ and supported by four links. For the loading shown, determine the force in each link.

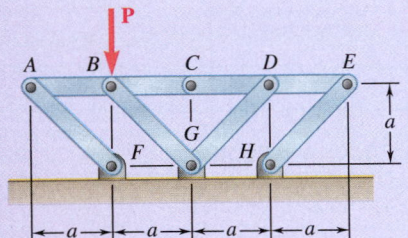

Fig. *P6.111*

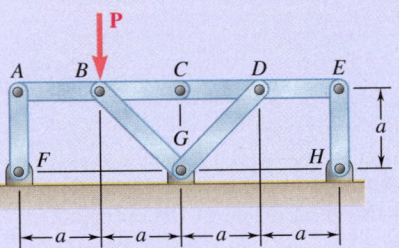

Fig. P6.112

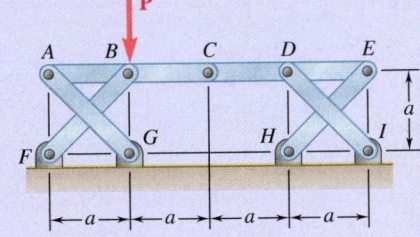

Fig. P6.113

**6.114**  Members $ABC$ and $CDE$ are pin-connected at $C$ and supported by the four links $AF$, $BG$, $DG$, and $EH$. For the loading shown, determine the force in each link.

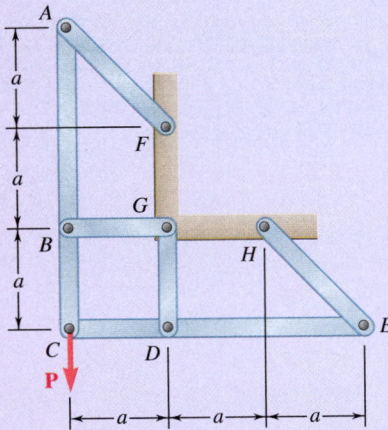

Fig. *P6.114*

**6.115**  Solve Prob. 6.112 assuming that the force **P** is replaced by a clockwise couple of moment $\mathbf{M}_0$ applied to member $CDE$ at $D$.

**6.116**  Solve Prob. 6.114 assuming that the force **P** is replaced by a clockwise couple of moment $\mathbf{M}_0$ applied to member $CDE$ at $D$.

**6.117**  Four beams, each with a length of $2a$, are nailed together at their midpoints to form the support system shown. Assuming that only vertical forces are exerted at the connections, determine the vertical reactions at $A$, $D$, $E$, and $H$.

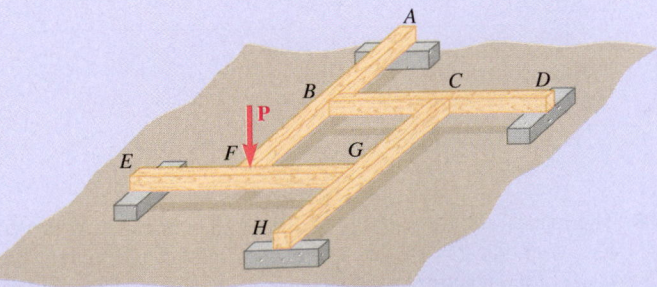

Fig. P6.117

**6.118** Four beams, each with a length of $3a$, are held together by single nails at $A$, $B$, $C$, and $D$. Each beam is attached to a support located at a distance $a$ from an end of the beam as shown. Assuming that only vertical forces are exerted at the connections, determine the vertical reactions at $E$, $F$, $G$, and $H$.

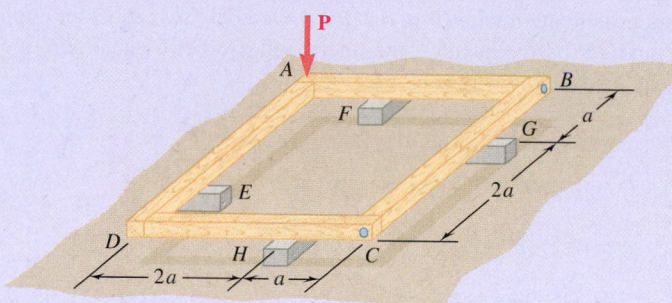

**Fig. P6.118**

**6.119 through 6.121** Each of the frames shown consists of two L-shaped members connected by two rigid links. For each frame, determine the reactions at the supports and indicate whether the frame is rigid.

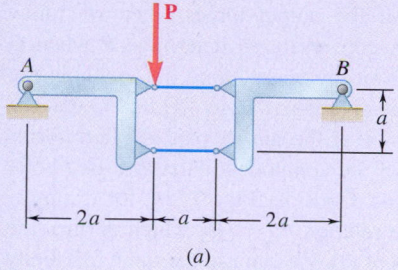

(a)

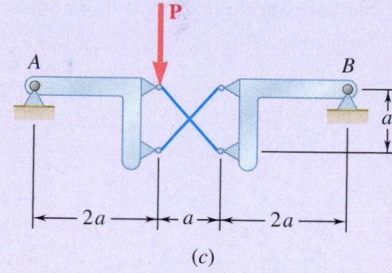

(b)

(c)

**Fig. P6.119**

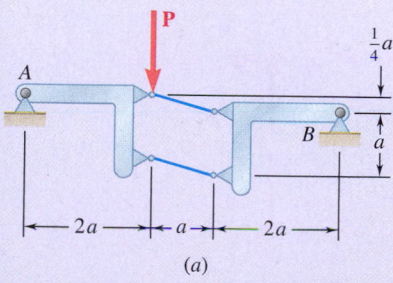

(a)

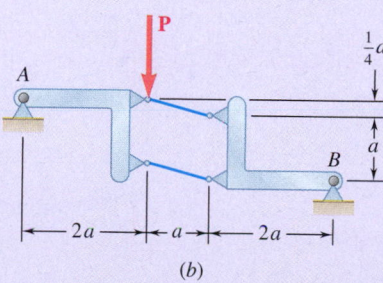

(b)

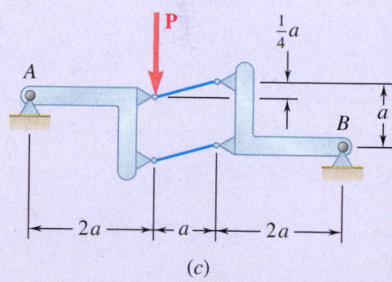

(c)

**Fig. P6.120**

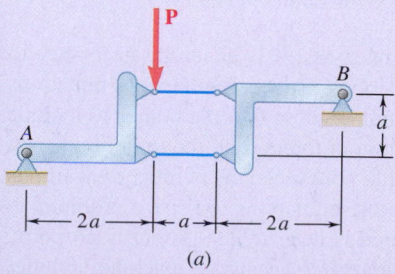

(a)

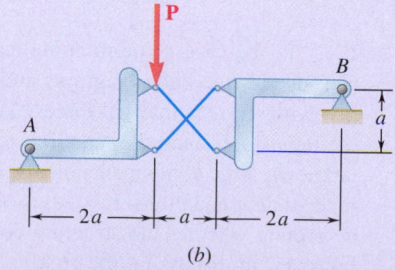

(b)

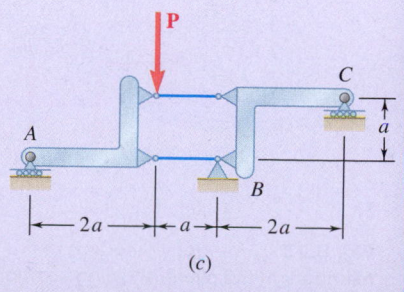

(c)

**Fig. P6.121**

## 6.4   MACHINES

Machines are structures designed to transmit and modify forces. Whether they are simple tools or include complicated mechanisms, their main purpose is to transform **input forces** into **output forces**. Consider, for example, a pair of cutting pliers used to cut a wire (Fig. 6.21*a*). If we apply two equal and opposite forces **P** and −**P** on the handles, the pliers will exert two equal and opposite forces **Q** and −**Q** on the wire (Fig. 6.21*b*).

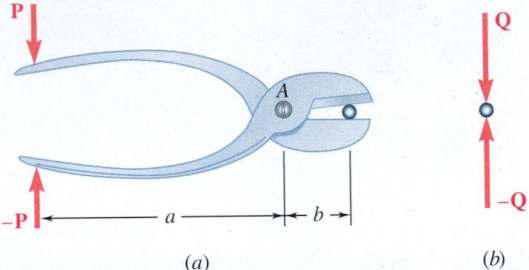

**Fig. 6.21** (*a*) Input forces on the handles of a pair of cutting pliers; (*b*) output forces cut a wire.

To determine the magnitude $Q$ of the output forces when we know the magnitude $P$ of the input forces (or, conversely, to determine $P$ when $Q$ is known), we draw a free-body diagram of the pliers *alone* (i.e., without the wire), showing the input forces **P** and −**P** and the *reactions* −**Q** and **Q** that the wire exerts on the pliers (Fig. 6.22). However, because a pair of pliers forms a nonrigid structure, we must treat one of the component parts as a free body in order to determine the unknown forces. Consider Fig. 6.23*a*, for example. Taking moments about $A$, we obtain the relation $Pa = Qb$, which defines the magnitude $Q$ in terms of $P$ (or $P$ in terms of $Q$). We can use the same free-body diagram to determine the components of the internal force at $A$; we find $A_x = 0$ and $A_y = P + Q$.

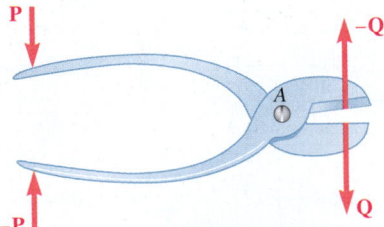

**Fig. 6.22** To show a free-body diagram of the pliers in equilibrium, we include the input forces and the reactions to the output forces.

**Fig. 6.23** Free-body diagrams of the members of the pliers, showing components of the internal force at joint *A*.

In the case of more complicated machines, it is generally necessary to use several free-body diagrams and, possibly, to solve simultaneous equations involving various internal forces. You should choose the free bodies to include the input forces and the reactions to the output forces, and the total number of unknown force components involved should not exceed the number of available independent equations. It is advisable, before attempting to solve a problem, to determine whether the structure considered is determinate. There is no point, however, in discussing the rigidity of a machine, because a machine includes moving parts and thus *must* be nonrigid.

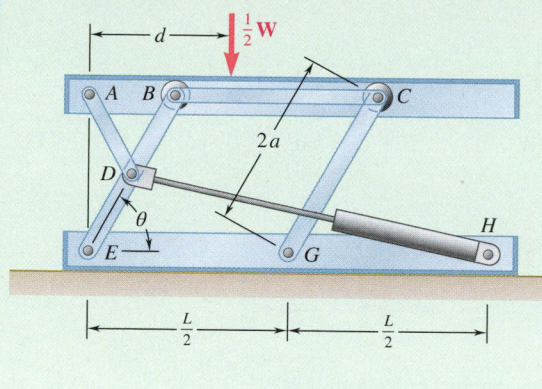

# Sample Problem 6.7

A hydraulic-lift table is used to raise a 1000-kg crate. The table consists of a platform and two identical linkages on which hydraulic cylinders exert equal forces. (Only one linkage and one cylinder are shown.) Members *EDB* and *CG* are each of length 2*a*, and member *AD* is pinned to the midpoint of *EDB*. If the crate is placed on the table so that half of its weight is supported by the system shown, determine the force exerted by each cylinder in raising the crate for $\theta = 60°$, $a = 0.70$ m, and $L = 3.20$ m. Show that the result is independent of the distance *d*.

**STRATEGY:**   The free-body diagram of the entire frame will involve more than three unknowns, so it alone can not be used to solve this problem. Instead, draw free-body diagrams of each component of the machine and work from them.

**MODELING:**   The machine consists of the platform and the linkage. Its free-body diagram (Fig. 1) includes an input force $\mathbf{F}_{DH}$ exerted by the cylinder; the weight **W**/2, which is equal and opposite to the output force; and reactions at *E* and *G*, which are assumed to be directed as shown. Dismember the mechanism and draw a free-body diagram for each of its component parts (Fig. 2). Note that *AD, BC,* and *CG* are two-force members. Member *CG* has already been assumed to be in compression; now assume that *AD* and *BC* are in tension and direct the forces exerted on them as shown. Use equal and opposite vectors to represent the forces exerted by the two-force members on the platform, on member *BDE,* and on roller *C.*

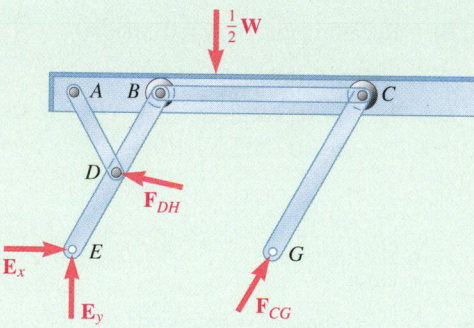

**Fig. 1**   Free-body diagram of the machine.

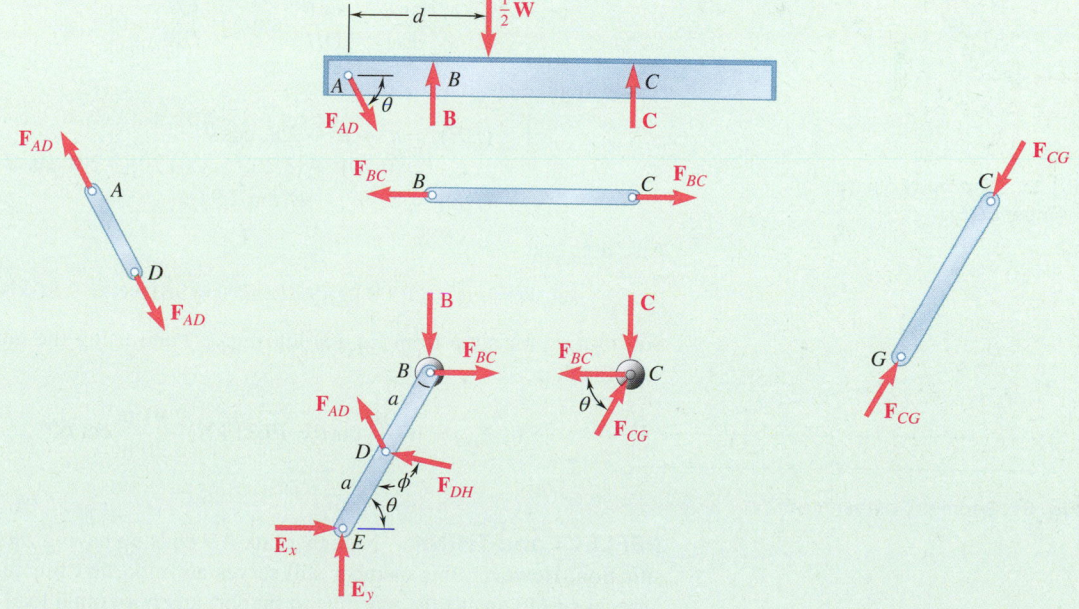

**Fig. 2**   Free-body diagram of each component part.

*(continued)*

## ANALYSIS:

### Free Body: Platform *ABC* (Fig. 3).

$$\xrightarrow{+}\Sigma F_x = 0: \qquad F_{AD}\cos\theta = 0 \qquad F_{AD} = 0$$
$$+\uparrow\Sigma F_y = 0: \qquad B + C - \tfrac{1}{2}W = 0 \qquad B + C = \tfrac{1}{2}W \tag{1}$$

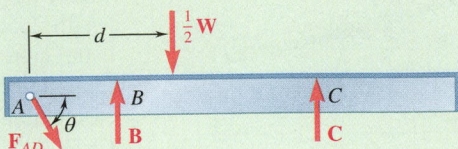

**Fig. 3** Free-body diagram of platform *ABC*.

### Free Body: Roller *C* (Fig. 4).
Draw a force triangle and obtain $F_{BC} = C\cot\theta$.

### Free Body: Member *BDE* (Fig. 5).
Recalling that $F_{AD}, = 0$, you have

$$+\circlearrowleft\Sigma M_E = 0: \quad F_{DH}\cos(\phi - 90°)a - B(2a\cos\theta) - F_{BC}(2a\sin\theta) = 0$$
$$F_{DH}\,a\sin\phi - B(2a\cos\theta) - (C\cot\theta)(2a\sin\theta) = 0$$
$$F_{DH}\sin\phi - 2(B + C)\cos\theta = 0$$

From Eq. (1), you obtain

$$F_{DH} = W\frac{\cos\theta}{\sin\phi} \tag{2}$$

Note that *the result obtained is independent of d.* ◄

Applying first the law of sines to triangle *EDH* (Fig. 6), you have

$$\frac{\sin\phi}{EH} = \frac{\sin\theta}{DH} \quad \sin\phi = \frac{EH}{DH}\sin\theta \tag{3}$$

Now using the law of cosines, you get

$$(DH)^2 = a^2 + L^2 - 2aL\cos\theta$$
$$= (0.70)^2 + (3.20)^2 - 2(0.70)(3.20)\cos 60°$$
$$(DH)^2 = 8.49 \qquad DH = 2.91\text{ m}$$

Also note that

$$W = mg = (1000\text{ kg})(9.81\text{ m/s}^2) = 9810\text{ N} = 9.81\text{ kN}$$

Substituting for $\sin\phi$ from Eq. (3) into Eq. (2) and using the numerical data, your result is

$$F_{DH} = W\frac{DH}{EH}\cot\theta = (9.81\text{ kN})\frac{2.91\text{ m}}{3.20\text{ m}}\cot 60°$$

$$F_{DH} = 5.15\text{ kN} ◄$$

**REFLECT and THINK:** Note that link *AD* ends up having zero force in this situation. However, this member still serves an important function, because it is necessary to enable the machine to support any horizontal load that might be exerted on the platform.

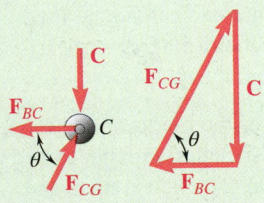

**Fig. 4** Free-body diagram of roller *C* and its force triangle.

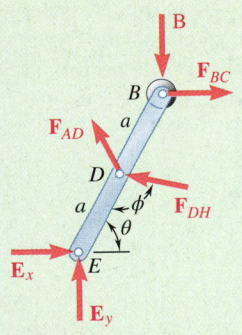

**Fig. 5** Free-body diagram of member *BDE*.

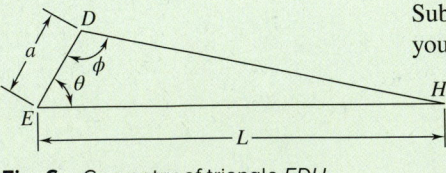

**Fig. 6** Geometry of triangle *EDH*.

# SOLVING PROBLEMS ON YOUR OWN

This section dealt with the analysis of *machines*. Because machines are designed to transmit or modify forces, they always contain moving parts. However, the machines considered here are always at rest, and you will be working with the set of *forces required to maintain the equilibrium of the machine.*

Known forces that act on a machine are called *input forces. A machine transforms the input forces into output forces,* such as the cutting forces applied by the pliers of Fig. 6.21. You will determine the output forces by finding the equal and opposite forces that should be applied to the machine to maintain its equilibrium.

In Sec. 6.3, you analyzed frames; you will use almost the same procedure to analyze machines by following these steps.

**1. Draw a free-body diagram of the whole machine,** and use it to determine as many as possible of the unknown forces exerted on the machine.

**2. Dismember the machine and draw a free-body diagram of each member.**

**3. First consider the two-force members.** Apply equal and opposite forces to each two-force member at the points where it is connected to another member. If you cannot tell at this point whether the member is in tension or in compression, *assume* that the member is in tension and *direct both of the forces away from the member.* Because these forces have the same unknown magnitude, *give them both the same name.*

**4. Next consider the multi-force members.** For each of these members, show all of the forces acting on it, including applied loads and forces, reactions, and internal forces at connections.

    **a. Where a multi-force member is connected to a two-force member,** apply to the multi-force member a force that is *equal and opposite* to the force drawn on the free-body diagram of the two-force member, *giving it the same name.*

    **b. Where a multi-force member is connected to another multi-force member,** use *horizontal and vertical components* to represent the internal forces at that point. The directions you choose for each of the two force components exerted on the first multi-force member are arbitrary, but *you must apply equal and opposite force components of the same name* to the other multi-force member.

**5. Write equilibrium equations** after you have completed the various free-body diagrams.

    **a. To simplify your solution,** you should, whenever possible, write and solve equilibrium equations involving single unknowns.

    **b. Because you arbitrarily chose the direction of each of the unknown forces,** you must determine at the end of the solution whether your guess was correct. To that effect, *consider the sign* of the value found for each of the unknowns. A *positive* sign indicates your guess was correct, whereas a *negative* sign indicates it was not.

**6. Finally, check your solution** by substituting the results obtained into an equilibrium equation you have not previously used.

# Problems

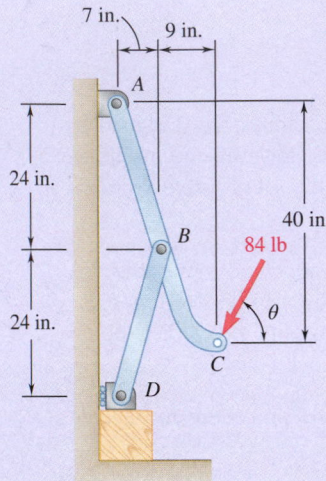

**Fig. P6.F5**

**6.F5** An 84-lb force is applied to the toggle vise at *C*. Knowing that $\theta = 90°$, draw the free-body diagram(s) needed to determine the vertical force exerted on the block at *D*.

**6.F6** For the system and loading shown, draw the free-body diagram(s) needed to determine the force **P** required for equilibrium.

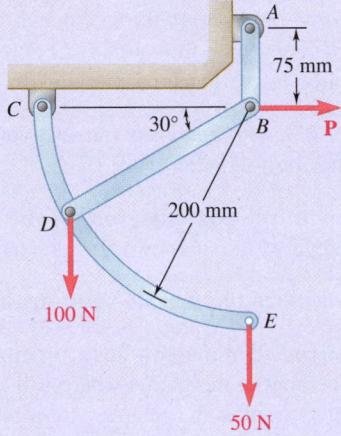

**Fig. P6.F6**

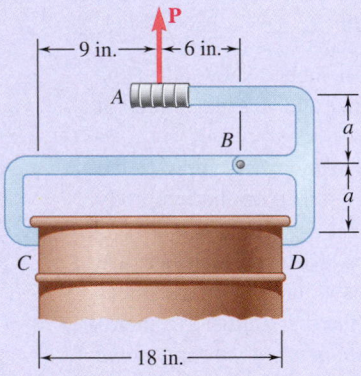

**Fig. P6.F7**

**6.F7** A small barrel weighing 60 lb is lifted by a pair of tongs as shown. Knowing that *a* = 5 in., draw the free-body diagram(s) needed to determine the forces exerted at *B* and *D* on tong *ABD*.

**6.F8** The position of member *ABC* is controlled by the hydraulic cylinder *CD*. Knowing that $\theta = 30°$, draw the free-body diagram(s) needed to determine the force exerted by the hydraulic cylinder on pin *C*, and the reaction at *B*.

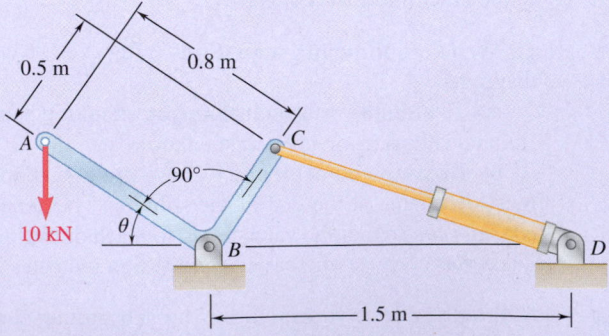

**Fig. P6.F8**

## END-OF-SECTION PROBLEMS

**6.122** The shear shown is used to cut and trim electronic-circuit-board laminates. For the position shown, determine (*a*) the vertical component of the force exerted on the shearing blade at *D*, (*b*) the reaction at *C*.

**6.123** A 100-lb force directed vertically downward is applied to the toggle vise at *C*. Knowing that link *BD* is 6 in. long and that *a* = 4 in., determine the horizontal force exerted on block *E*.

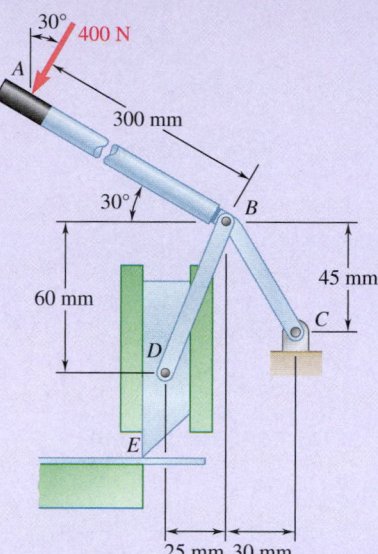

**Fig. P6.122**

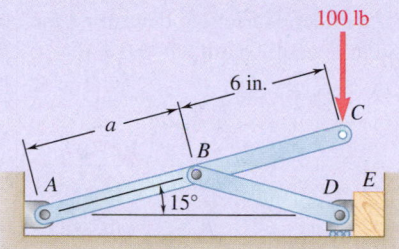

**Fig. P6.123 and P6.124**

**6.124** A 100-lb force directed vertically downward is applied to the toggle vise at *C*. Knowing that link *BD* is 6 in. long and that *a* = 8 in., determine the horizontal force exerted on block *E*.

**6.125** The control rod *CE* passes through a horizontal hole in the body of the toggle system shown. Knowing that link *BD* is 250 mm long, determine the force **Q** required to hold the system in equilibrium when *β* = 20°.

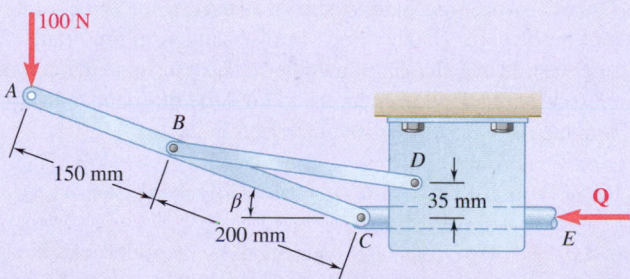

**Fig. P6.125**

**6.126** Solve Prob. 6.125 when (*a*) *β* = 0, (*b*) *β* = 6°.

**6.127** The press shown is used to emboss a small seal at *E*. Knowing that *P* = 250 N, determine (*a*) the vertical component of the force exerted on the seal, (*b*) the reaction at *A*.

**6.128** The press shown is used to emboss a small seal at *E*. Knowing that the vertical component of the force exerted on the seal must be 900 N, determine (*a*) the required vertical force **P**, (*b*) the corresponding reaction at *A*.

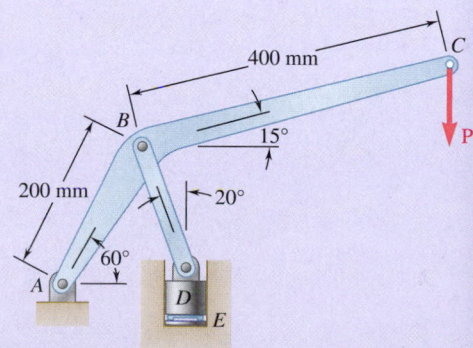

**Fig. P6.127 and *P6.128***

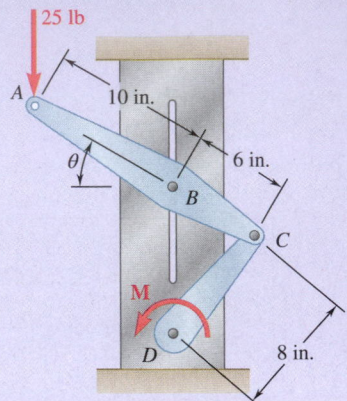

**Fig. P6.129 and P6.130**

**6.129** The pin at *B* is attached to member *ABC* and can slide freely along the slot cut in the fixed plate. Neglecting the effect of friction, determine the couple **M** required to hold the system in equilibrium when $\theta = 30°$.

**6.130** The pin at *B* is attached to member *ABC* and can slide freely along the slot cut in the fixed plate. Neglecting the effect of friction, determine the couple **M** required to hold the system in equilibrium when $\theta = 60°$.

**6.131** Arm *ABC* is connected by pins to a collar at *B* and to crank *CD* at *C*. Neglecting the effect of friction, determine the couple **M** required to hold the system in equilibrium when $\theta = 0$.

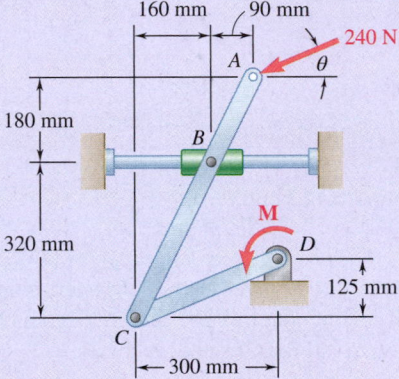

**Fig. P6.131 and P6.132**

**6.132** Arm *ABC* is connected by pins to a collar at *B* and to crank *CD* at *C*. Neglecting the effect of friction, determine the couple **M** required to hold the system in equilibrium when $\theta = 90°$.

**6.133** The Whitworth mechanism shown is used to produce a quick-return motion of point *D*. The block at *B* is pinned to the crank *AB* and is free to slide in a slot cut in member *CD*. Determine the couple **M** that must be applied to the crank *AB* to hold the mechanism in equilibrium when (*a*) $\alpha = 0$, (*b*) $\alpha = 30°$.

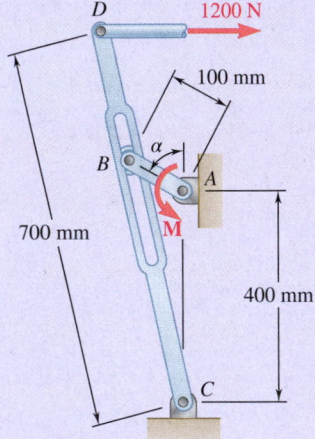

**Fig. P6.133**

**6.134** Solve Prob. 6.133 when (*a*) $\alpha = 60°$, (*b*) $\alpha = 90°$.

**6.135 and 6.136** Two rods are connected by a slider block as shown. Neglecting the effect of friction, determine the couple **M**$_A$ required to hold the system in equilibrium.

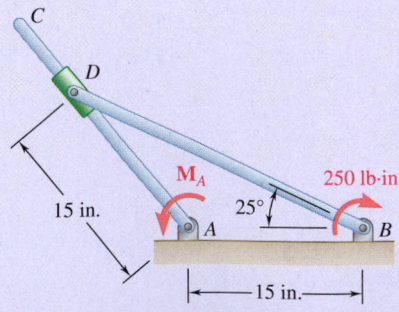

**Fig. P6.135**

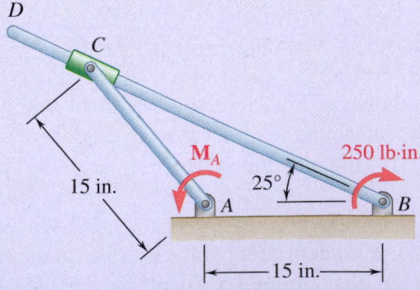

**Fig. P6.136**

**6.137 and 6.138** Rod *CD* is attached to the collar *D* and passes through a collar welded to end *B* of lever *AB*. Neglecting the effect of friction, determine the couple **M** required to hold the system in equilibrium when $\theta = 30°$.

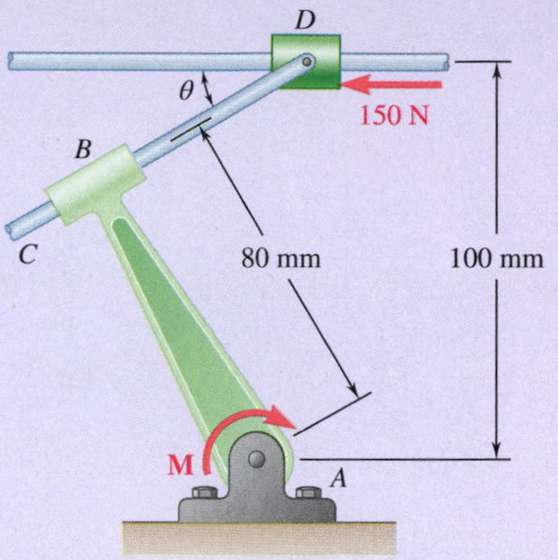

**6.139** Two hydraulic cylinders control the position of the robotic arm *ABC*. Knowing that in the position shown the cylinders are parallel, determine the force exerted by each cylinder when $P = 160$ N and $Q = 80$ N.

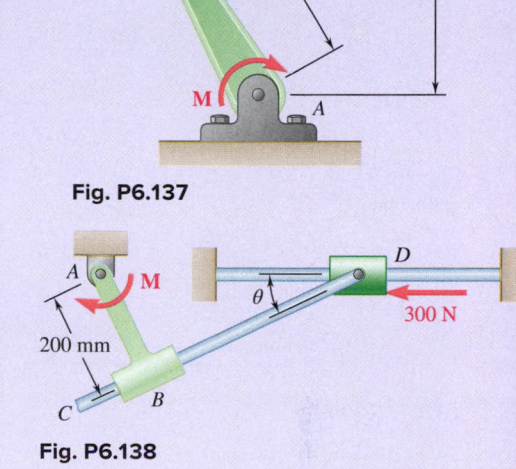

Fig. P6.137

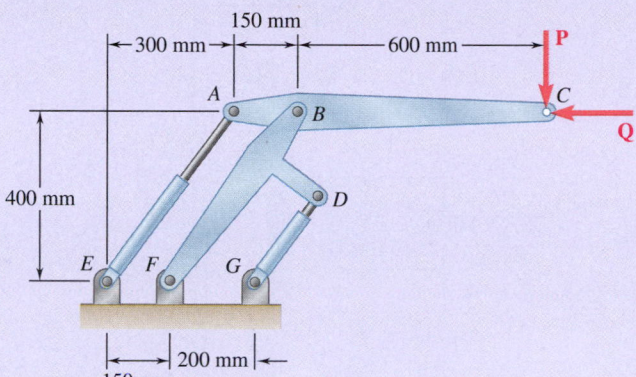

Fig. P6.139 and P6.140

Fig. P6.138

**6.140** Two hydraulic cylinders control the position of the robotic arm *ABC*. In the position shown, the cylinders are parallel and both are in tension. Knowing that $F_{AE} = 600$ N and $F_{DG} = 50$ N, determine the forces **P** and **Q** applied at *C* to arm *ABC*.

**6.141** A steel ingot weighing 8000 lb is lifted by a pair of tongs as shown. Determine the forces exerted at *C* and *E* on the tong *BCE*.

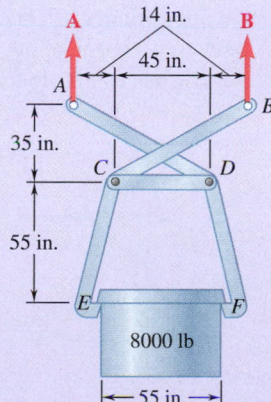

Fig. P6.141

**6.142** If the toggle shown is added to the tongs of Prob. 6.141 and the load is lifted by applying a single force at *G*, determine the forces exerted at *C* and *E* on the tong *BCE*.

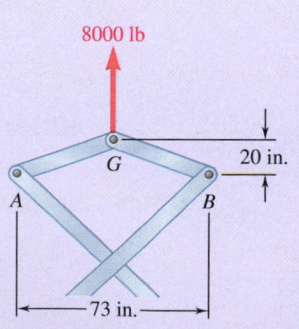

Fig. P6.142

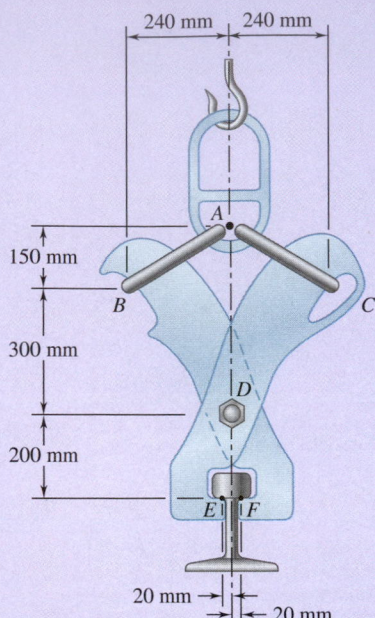

**Fig. P6.143**

**6.143** A 9-m length of railroad rail of mass 40 kg/m is lifted by the tongs shown. Determine the forces exerted at *D* and *F* on tong *BDF*.

**6.144** The gear-pulling assembly shown consists of a crosshead *CF*, two grip arms *ABC* and *FGH*, two links *BD* and *EG*, and a threaded center rod *JK*. Knowing that the center rod *JK* must exert a 4800-N force on the vertical shaft *KL* in order to start the removal of the gear, determine all forces acting on grip arm *ABC*. Assume that the rounded ends of the crosshead are smooth and exert horizontal forces only on the grip arms.

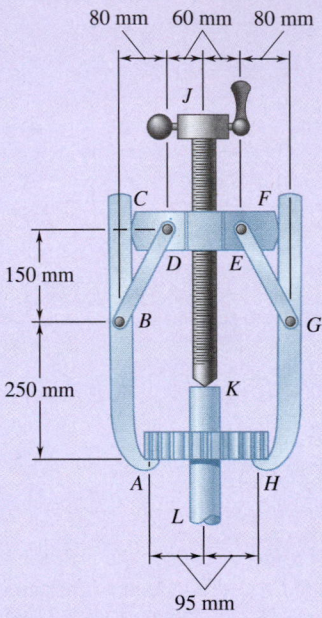

**Fig. P6.144**

**6.145** The pliers shown are used to grip a 0.3-in.-diameter rod. Knowing that two 60-lb forces are applied to the handles, determine (*a*) the magnitude of the forces exerted on the rod, (*b*) the force exerted by the pin at *A* on portion *AB* of the pliers.

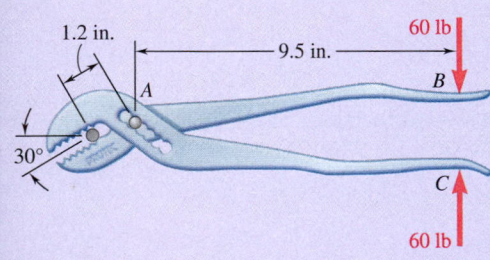

**Fig. P6.145**

**6.146** A hand-operated hydraulic cylinder has been designed for use where space is severely limited. Determine the magnitude of the force exerted on the piston at *D* when two 90-lb forces are applied as shown.

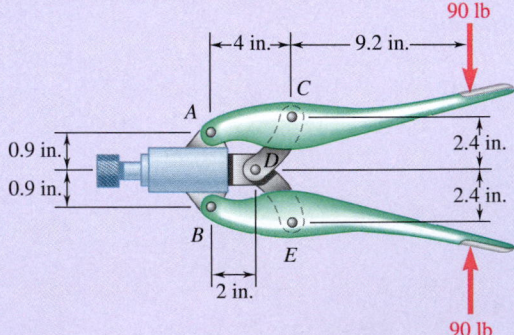

**Fig. P6.146**

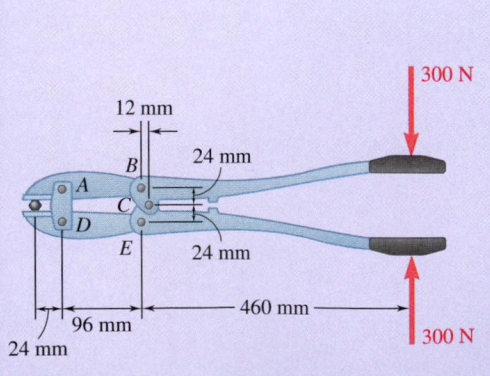

**Fig. P6.147**

**6.147** In using the bolt cutter shown, a worker applies two 300-N forces to the handles. Determine the magnitude of the forces exerted by the cutter on the bolt.

**6.148** The upper blade and lower handle of the compound-lever shears are pin-connected to the main element *ABE* at *A* and *B*, respectively, and to the short link *CD* at *C* and *D*, respectively. Determine the forces exerted on a twig when two 120-N forces are applied to the handles.

**6.149 and 6.150** Determine the force **P** that must be applied to the toggle *CDE* to maintain bracket *ABC* in the position shown.

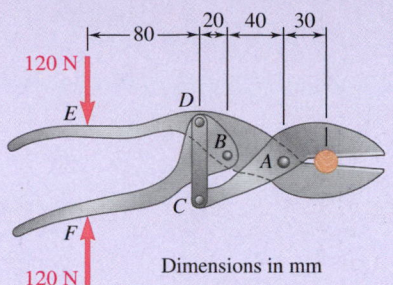

**Fig. P6.148**

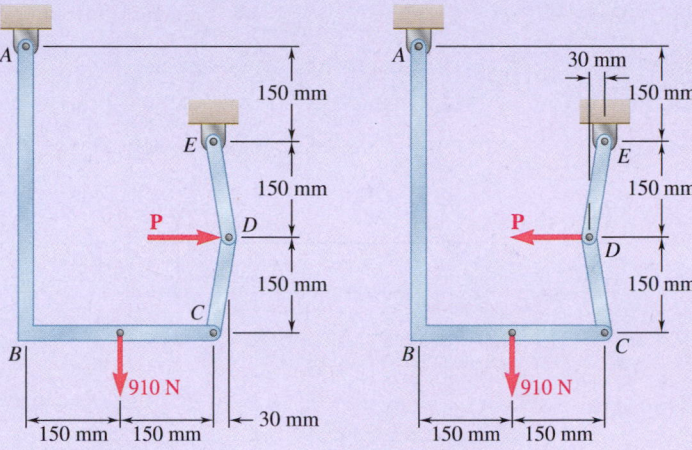

**Fig. P6.149**        **Fig. P6.150**

**6.151** Because the brace shown must remain in position even when the magnitude of **P** is very small, a single safety spring is attached at *D* and *E*. The spring *DE* has a constant of 50 lb/in. and an unstretched length of 7 in. Knowing that *l* = 10 in. and that the magnitude of **P** is 800 lb, determine the force **Q** required to release the brace.

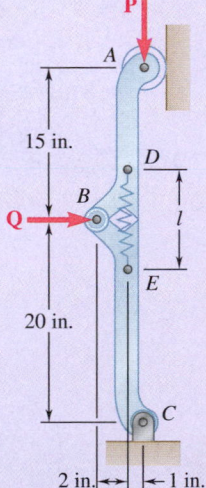

**Fig. P6.151**

**6.152** The specialized plumbing wrench shown is used in confined areas (e.g., under a basin or sink). It consists essentially of a jaw *BC* pinned at *B* to a long rod. Knowing that the forces exerted on the nut are equivalent to a clockwise (when viewed from above) couple with a magnitude of 135 lb·in., determine (*a*) the magnitude of the force exerted by pin *B* on jaw *BC*, (*b*) the couple **M**$_0$ that is applied to the wrench.

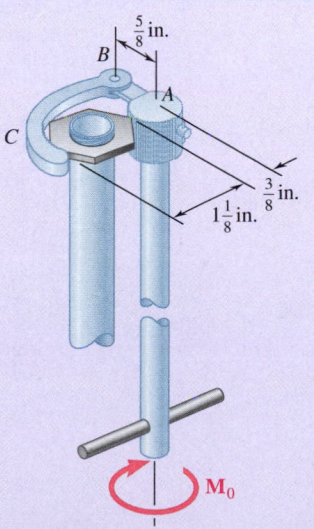

**Fig. P6.152**

**6.153** The elevation of the platform is controlled by two identical mechanisms, only one of which is shown. A load of 1200 lb is applied to the mechanism shown. Knowing that the pin at $C$ can transmit only a horizontal force, determine (a) the force in link $BE$, (b) the components of the force exerted by the hydraulic cylinder on pin $H$.

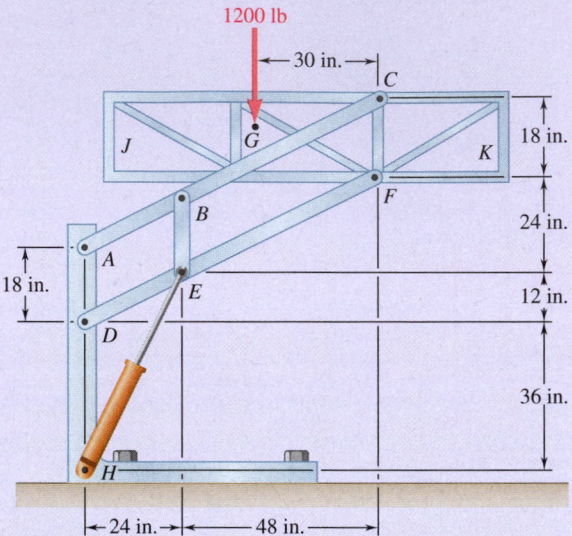

Fig. P6.153

**6.154** The action of the backhoe bucket is controlled by the three hydraulic cylinders shown. Determine the force exerted by each cylinder in supporting the 3000-lb load shown.

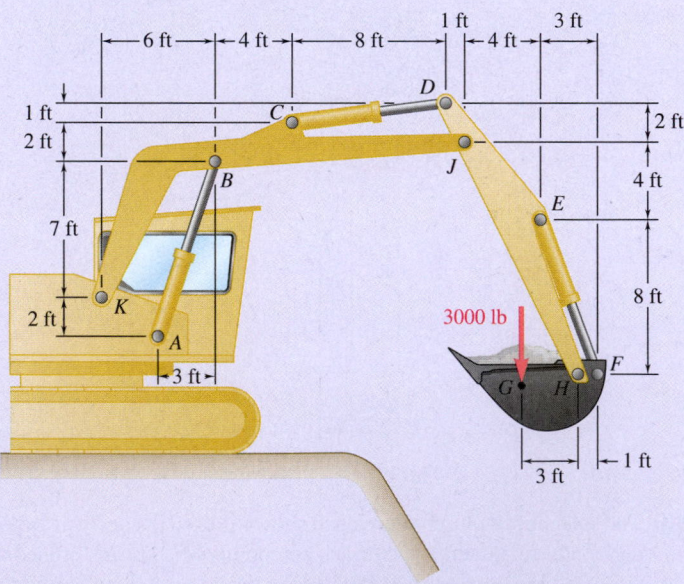

Fig. P6.154

**6.155** The telescoping arm *ABC* is used to provide an elevated platform for construction workers. The workers and the platform together have a mass of 200 kg and a combined center of gravity located directly above *C*. For the position when $\theta = 20°$, determine (*a*) the force exerted at *B* by the single hydraulic cylinder *BD*, (*b*) the force exerted on the supporting carriage at *A*.

**6.156** The telescoping arm *ABC* of Prob. 6.155 can be lowered until end *C* is close to the ground, so that workers can easily board the platform. For the position when $\theta = -20°$, determine (*a*) the force exerted at *B* by the single hydraulic cylinder *BD*, (*b*) the force exerted on the supporting carriage at *A*.

**6.157** The motion of the backhoe bucket shown is controlled by the hydraulic cylinders *AD, CG,* and *EF*. As a result of an attempt to dislodge a portion of a slab, a 2-kip force **P** is exerted on the bucket teeth at *J*. Knowing that $\theta = 45°$, determine the force exerted by each cylinder.

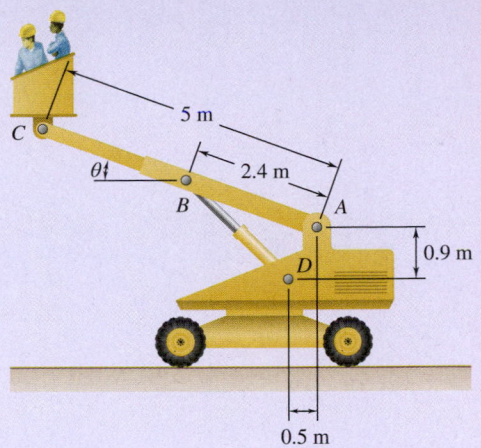

**Fig. P6.155**

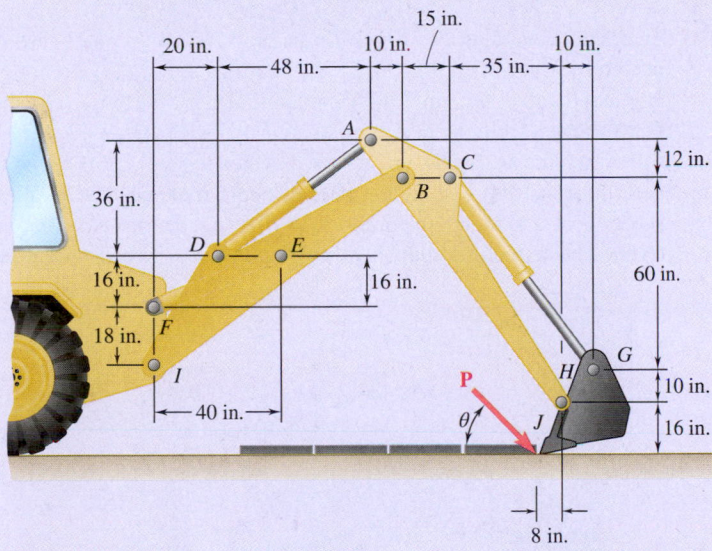

**Fig. P6.157**

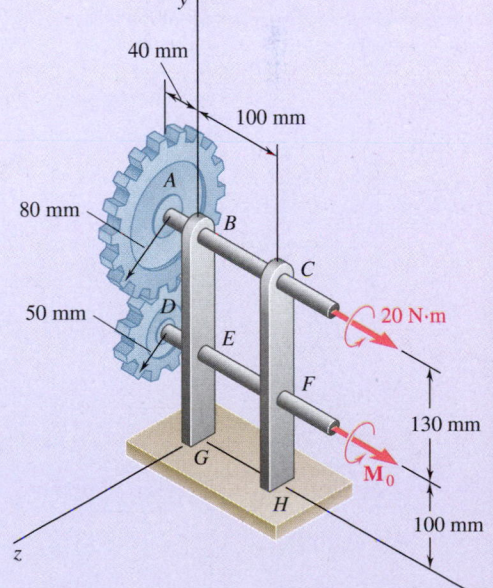

**6.158** Solve Prob. 6.157 assuming that the 2-kip force **P** acts horizontally to the right ($\theta = 0$).

**6.159** The gears *A* and *D* are rigidly attached to horizontal shafts that are held by frictionless bearings. Determine (*a*) the couple **M**$_0$ that must be applied to shaft *DEF* to maintain equilibrium, (*b*) the reactions at *G* and *H*.

**Fig. P6.159**

**6.160** In the planetary gear system shown, the radius of the central gear $A$ is $a = 18$ mm, the radius of each planetary gear is $b$, and the radius of the outer gear $E$ is $(a + 2b)$. A clockwise couple with a magnitude of $M_A = 10$ N·m is applied to the central gear $A$, and a counterclockwise couple with a magnitude of $M_S = 50$ N·m is applied to the spider $BCD$. If the system is to be in equilibrium, determine (a) the required radius $b$ of the planetary gears, (b) the magnitude $M_E$ of the couple that must be applied to the outer gear $E$.

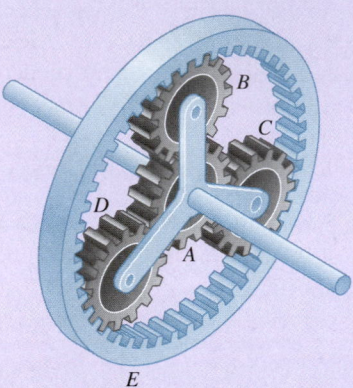

Fig. P6.160

**\*6.161** Two shafts $AC$ and $CF$, which lie in the vertical $xy$ plane, are connected by a universal joint at $C$. The bearings at $B$ and $D$ do not exert any axial force. A couple with a magnitude of 500 lb·in. (clockwise when viewed from the positive $x$ axis) is applied to shaft $CF$ at $F$. At a time when the arm of the crosspiece attached to shaft $CF$ is horizontal, determine (a) the magnitude of the couple that must be applied to shaft $AC$ at $A$ to maintain equilibrium, (b) the reactions at $B$, $D$, and $E$. (*Hint:* The sum of the couples exerted on the crosspiece must be zero.)

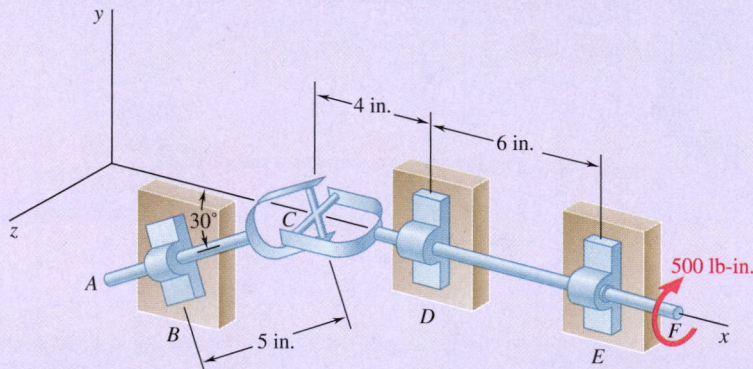

Fig. P6.161

**\*6.162** Solve Prob. 6.161 assuming that the arm of the crosspiece attached to shaft $CF$ is vertical.

**\*6.163** The large mechanical tongs shown are used to grab and lift a thick 7500-kg steel slab $HJ$. Knowing that slipping does not occur between the tong grips and the slab at $H$ and $J$, determine the components of all forces acting on member $EFH$. (*Hint:* Consider the symmetry of the tongs to establish relationships between the components of the force acting at $E$ on $EFH$ and the components of the force acting at $D$ on $DGJ$.)

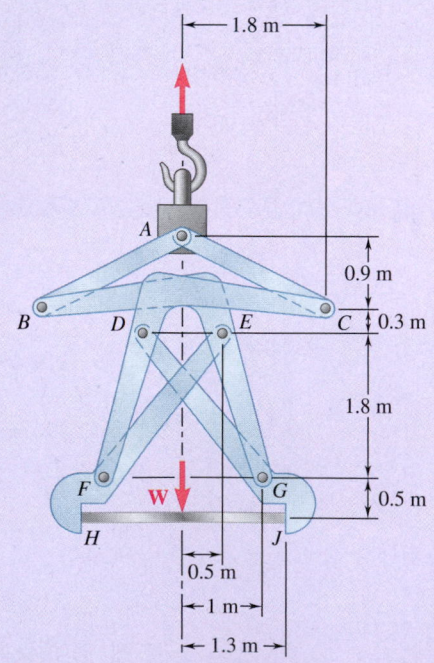

Fig. P6.163

# Review and Summary

In this chapter, you studied ways to determine the **internal forces** holding together the various parts of a structure.

## Analysis of Trusses

The first half of the chapter presented the analysis of **trusses**, i.e., structures consisting of *straight members connected at their extremities only*. Because the members are slender and unable to support lateral loads, all of the loads must be applied at the joints; thus, we can assume that a truss consists of *pins and two-force members* (Sec. 6.1A).

## Simple Trusses

A truss is **rigid** if it is designed in such a way that it does not greatly deform or collapse under a small load. A triangular truss consisting of three members connected at three joints is a rigid truss (Fig. 6.24a). The truss obtained by adding two new members to the first one and connecting them at a new joint (Fig. 6.24b) is also rigid. Trusses obtained by repeating this procedure are called **simple trusses**. We may check that, in a simple truss, the total number of members is $m = 2n - 3$, where $n$ is the total number of joints (Sec. 6.1A).

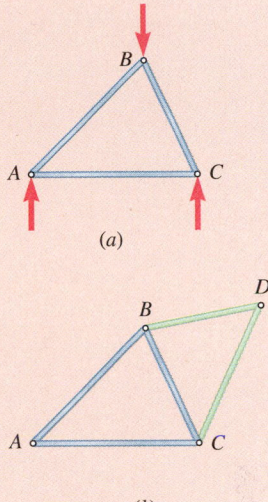

**Fig. 6.24**

## Method of Joints

We can determine the forces in the various members of a simple truss by using the **method of joints** (Sec. 6.1B). First, we obtain the reactions at the supports by considering the entire truss as a free body. Then, we draw the free-body diagram of each pin, showing the forces exerted on the pin by the members or supports it connects. Because the members are straight two-force members, the force exerted by a member on the pin is directed along that member, and only the magnitude of the force is unknown. In the case of a simple truss, it is always possible to draw the free-body diagrams of the pins in such an order that only two unknown forces are included in each diagram. We obtain these forces from the corresponding two equilibrium equations or—if only three forces are involved—from the corresponding force triangle. If the force exerted by a member on a pin is directed toward that pin, the member is in **compression**; if it is directed away from the pin, the member is in **tension** (Sample Prob. 6.1). The analysis of a truss is sometimes expedited by first recognizing **joints under special loading conditions** (Sec. 6.1C). The method of joints also can be extended for the analysis of three-dimensional or **space trusses** (Sec. 6.1D).

## Method of Sections

The **method of sections** is usually preferable to the method of joints when we want to determine the force in only one member—or very few members—of a truss (Sec. 6.2A). To determine the force in member *BD* of the truss of Fig. 6.25a, for example, we *pass a section* through members *BD*, *BE*, and *CE*; remove these members; and use the portion *ABC* of the truss as a free body (Fig. 6.25b). Setting $\Sigma M_E = 0$, we determine the magnitude of force $\mathbf{F}_{BD}$ that represents the force in member *BD*. A positive sign indicates that the member is in *tension*; a negative sign indicates that it is in *compression* (Sample Probs. 6.2 and 6.3).

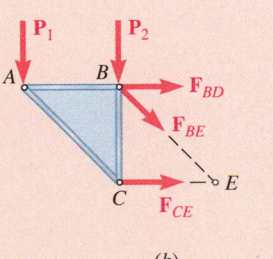

**Fig. 6.25**

## Compound Trusses

The method of sections is particularly useful in the analysis of **compound trusses**, i.e., trusses that cannot be constructed from the basic triangular truss of Fig. 6.24a but are built by rigidly connecting several simple trusses (Sec. 6.2B). If the component trusses are properly connected (e.g., one pin and one link, or three non-concurrent and unparallel links) and if the resulting structure is properly supported (e.g., one pin and one roller), the compound truss is **statically determinate, rigid, and completely constrained**. The following necessary—but not sufficient—condition is then satisfied: $m + r = 2n$, where $m$ is the number of members, $r$ is the number of unknowns representing the reactions at the supports, and $n$ is the number of joints.

## Frames and Machines

In the second part of the chapter, we analyzed **frames** and **machines**. These structures contain *multi-force members,* i.e., members acted upon by three or more forces. Frames are designed to support loads and are usually stationary, fully constrained structures. Machines are designed to transmit or modify forces and always contain moving parts (Sec. 6.3).

## Analysis of a Frame

To analyze a frame, we first consider the entire frame to be a free body and write three equilibrium equations (Sec. 6.3A). If the frame remains rigid when detached from its supports, the reactions involve only three unknowns and may be determined from these equations (Sample Probs. 6.4 and 6.5). On the other hand, if the frame ceases to be rigid when detached from its supports, the reactions involve more than three unknowns, and we cannot determine them completely from the equilibrium equations of the frame (Sec. 6.3B; Sample Prob. 6.6).

## Multi-force Members

We then dismember the frame and identify the various members as either two-force members or multi-force members; we assume pins form an integral part of one of the members they connect. We draw the free-body diagram of each of the multi-force members, noting that, when two multi-force members are connected to the same two-force member, they are acted upon by that member with *equal and opposite forces of unknown magnitude but known direction.* When two multi-force members are connected by a pin, they exert on each other *equal and opposite forces of unknown direction* that should be represented by *two unknown components.* We can then solve the equilibrium equations obtained from the free-body diagrams of the multi-force members for the various internal forces (Sample Probs. 6.4 and 6.5). We also can use the equilibrium equations to complete the determination of the reactions at the supports (Sample Prob. 6.6). Actually, if the frame is *statically determinate and rigid,* the free-body diagrams of the multi-force members could provide as many equations as there are unknown forces (including the reactions) (Sec. 6.3B). However, as suggested previously, it is advisable to first consider the free-body diagram of the entire frame to minimize the number of equations that must be solved simultaneously.

## Analysis of a Machine

To analyze a machine, we dismember it and, following the same procedure as for a frame, draw the free-body diagram of each multi-force member. The corresponding equilibrium equations yield the **output forces** exerted by the machine in terms of the **input forces** applied to it, as well as the **internal forces** at the various connections (Sec. 6.4; Sample Prob. 6.7).

# Review Problems

**6.164** Using the method of joints, determine the force in each member of the truss shown. State whether each member is in tension or compression.

**6.165** Using the method of joints, determine the force in each member of the double-pitch roof truss shown. State whether each member is in tension or compression.

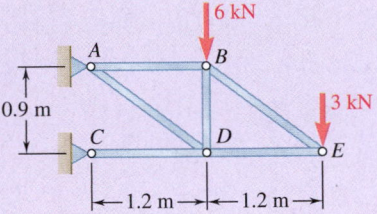

**Fig. P6.164**

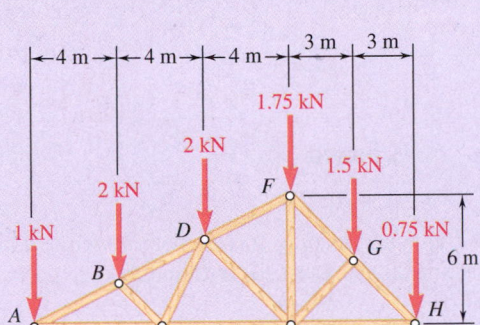

**Fig. P6.165**

**6.166** A stadium roof truss is loaded as shown. Determine the force in members *AB*, *AG*, and *FG*.

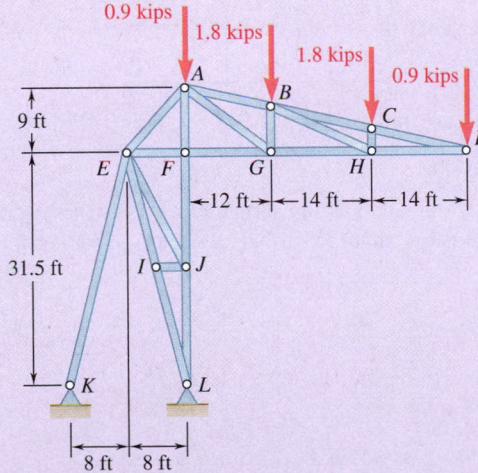

**Fig. P6.166 and *P6.167***

**6.167** A stadium roof truss is loaded as shown. Determine the force in members *AE*, *EF*, and *FJ*.

**6.168** Determine the components of all forces acting on member *ABD* of the frame shown.

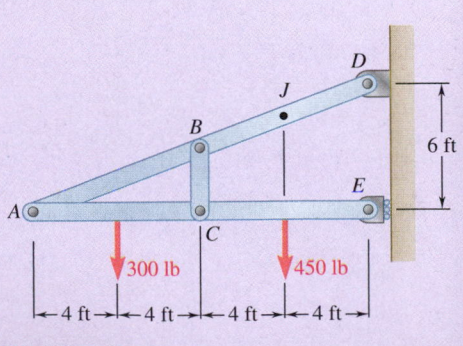

**Fig. P6.168**

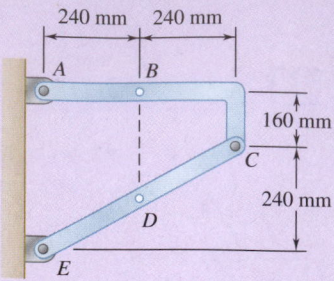

**Fig. P6.169**

**6.169** Determine the components of the reactions at *A* and *E* if the frame is loaded by a clockwise couple of magnitude 36 N·m applied (*a*) at *B*, (*b*) at *D*.

**6.170** Knowing that the pulley has a radius of 50 mm, determine the components of the reactions at *B* and *E*.

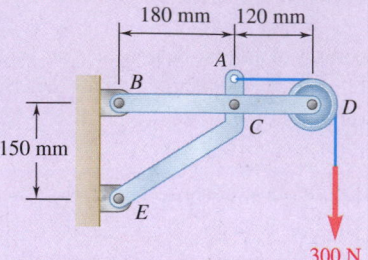

**Fig. P6.170**

**6.171** For the frame and loading shown, determine the components of the forces acting on member *CFE* at *C* and *F*.

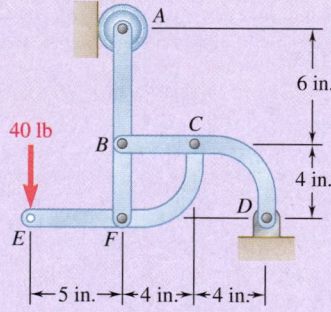

**Fig. P6.171**

**6.172** For the frame and loading shown, determine the reactions at *A*, *B*, *D*, and *E*. Assume that the surface at each support is frictionless.

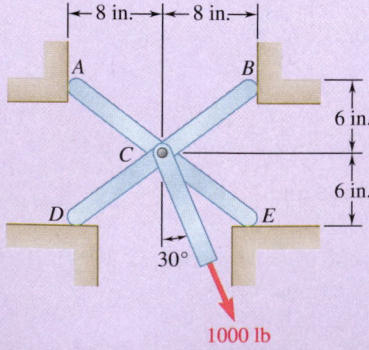

**Fig. P6.172**

**6.173** Water pressure in the supply system exerts a downward force of 135 N on the vertical plug at *A*. Determine the tension in the fusible link *DE* and the force exerted on member *BCE* at *B*.

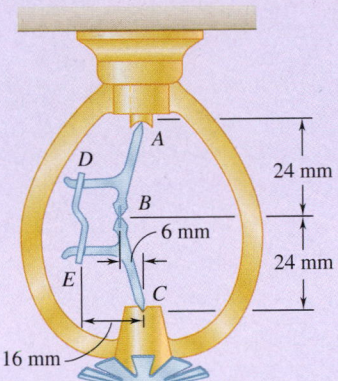

**Fig. P6.173**

**6.174** A couple **M** with a magnitude of 1.5 kN·m is applied to the crank of the engine system shown. For each of the two positions shown, determine the force **P** required to hold the system in equilibrium.

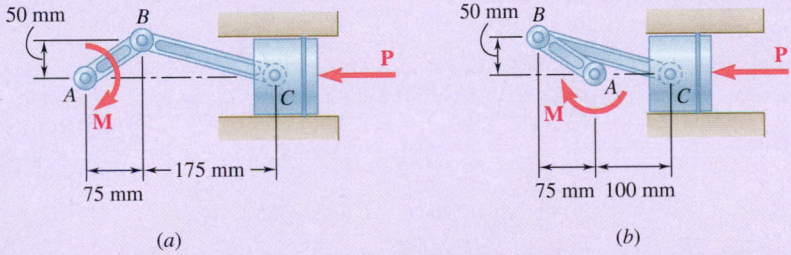

(a)                              (b)

**Fig. P6.174**

**6.175** The compound-lever pruning shears shown can be adjusted by placing pin *A* at various ratchet positions on blade *ACE*. Knowing that 300-lb vertical forces are required to complete the pruning of a small branch, determine the magnitude *P* of the forces that must be applied to the handles when the shears are adjusted as shown.

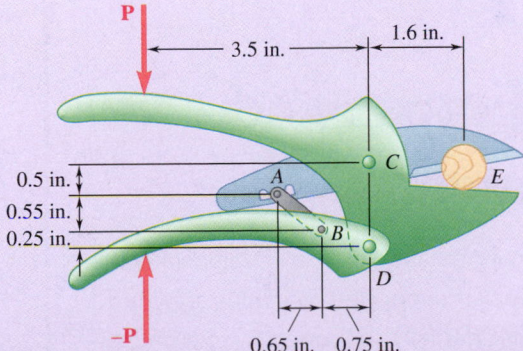

**Fig. P6.175**

# 7

# Internal Forces and Moments

The Assut de l'Or Bridge in the City of Arts and Science in Valencia, Spain, is cable-stayed, where the bridge deck is supported by cables attached to the curved tower. The tower itself is partially supported by four anchor cables. The deck of the bridge consists of a system of beams that support the roadway.

# Objectives

- **Consider** the general state of internal member forces, which includes axial force, shearing force, and bending moment.
- **Apply** equilibrium analysis methods to obtain specific values, general expressions, and diagrams for shear and bending moment in beams.
- **Examine** relations among load, shear, and bending moment, and use these to obtain shear and bending-moment diagrams for beams.
- **Analyze** the tension forces in cables subjected to concentrated loads, loads uniformly distributed along the horizontal, and loads uniformly distributed along the cable itself.

## Introduction

In previous chapters, we considered two basic problems involving structures: (1) determining the external forces acting on a structure (Chap. 4) and (2) determining the internal forces that hold together the various members forming a structure (Chap. 6). Now we consider the problem of determining the internal forces that hold together the parts of a given individual member.

We will first analyze the internal forces in the members of a frame, such as the crane considered in Fig. 6.1. Note that, whereas the internal forces in a straight two-force member can produce only **tension** or **compression** in that member, the internal forces in any other type of member usually produce **shear** and **bending** as well.

Most of this chapter is devoted to the analysis of the internal forces in two important types of engineering elements:

1. **Beams**, which are usually long, straight prismatic members designed to support loads applied at various points along their length.
2. **Cables**, which are flexible members capable of withstanding only tension and are designed to support either concentrated or distributed loads. Cables are used in many engineering applications, such as suspension bridges and power transmission lines.

# 7.1   INTERNAL FORCES IN MEMBERS

Consider a straight two-force member *AB* (Fig. 7.1*a*). From Sec. 4.2A, we know that the forces **F** and −**F** acting at *A* and *B*, respectively, must be directed along *AB* in opposite sense and have the same magnitude *F*. Suppose we cut the member at *C*. To maintain equilibrium of the resulting free bodies *AC* and *CB*, we must apply to *AC* a force −**F** equal and opposite to **F**, and to *CB* a force **F** equal and opposite to −**F** (Fig. 7.1*b*). These new forces are directed along *AB* in opposite sense and have the same magnitude *F*. Because the two parts *AC* and *CB* were in equilibrium before the member was cut, **internal forces** equivalent to

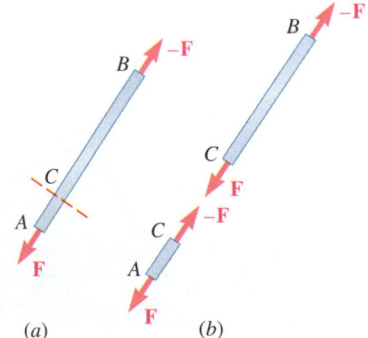

**Fig. 7.1**   A straight two-force member in tension. (*a*) External forces act at the ends of the member; (*b*) internal axial forces do not depend on the location of section *C*.

these new forces must have existed in the member itself. We conclude that, in the case of a straight two-force member, the internal forces that the two portions of the member exert on each other are equivalent to **axial forces**. The common magnitude $F$ of these forces does not depend upon the location of the section $C$ and is referred to as the *force in member AB*. In the case shown in Fig. 7.1, the member is in tension and elongates under the action of the internal forces. In the case represented in Fig. 7.2, the member is in compression and decreases in length under the action of the internal forces.

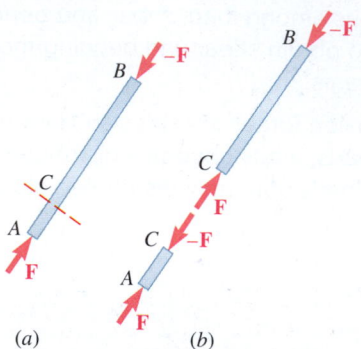

**Fig. 7.2**   A straight two-force member in compression. (*a*) External forces act at the ends; (*b*) internal axial forces are independent of the location of section *C*.

Next, consider a **multi-force member**. Take, for instance, member $AD$ of the crane analyzed in Sec. 6.3A. This crane is shown again in Fig. 7.3*a*, and we drew the free-body diagram of member $AD$ in Fig. 7.3*b*. Suppose we cut member $AD$ at $J$ and draw a free-body diagram for each of the portions $JD$ and $AJ$ (Fig. 7.3*c* and *d*). Considering the free body $JD$, we find that, to maintain its equilibrium, we need to apply at $J$ a force $\mathbf{F}$ to balance the vertical component of $\mathbf{T}$, a force $\mathbf{V}$ to balance the horizontal component of $\mathbf{T}$, and a couple $\mathbf{M}$ to balance the moment of $\mathbf{T}$ about $J$. Again, we conclude that internal forces must have existed at $J$ before member $AD$ was cut, which is equivalent to the force-couple system shown in Fig. 7.3*c*.

According to Newton's third law, the internal forces acting on $AJ$ must be equivalent to an equal and opposite force-couple system, as shown in Fig. 7.3*d*. The action of the internal forces in member $AD$ *is not limited to producing tension or compression*, as in the case of straight two-force members;

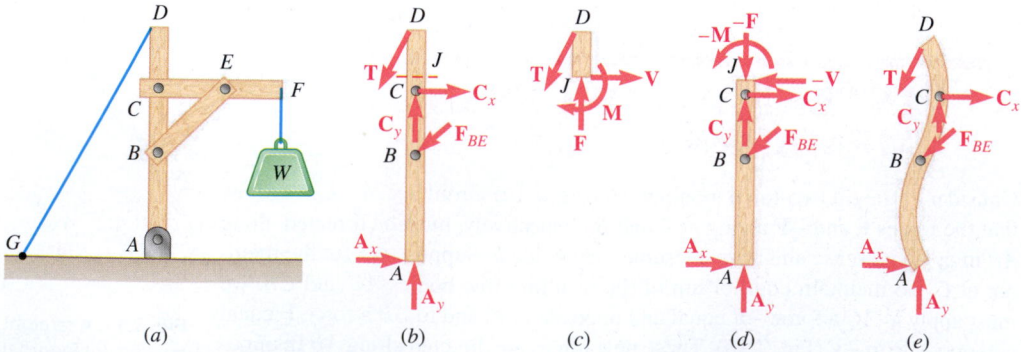

**Fig. 7.3**   (*a*) Crane from Chapter 6; (*b*) free-body diagram of multi-force member *AD*; (*c, d*) free-body diagrams of sections of member *AD* showing internal force-couple systems; (*e*) deformation of member *AD*.

**Photo 7.1**   The design of the shaft of a circular saw must account for the internal forces resulting from the forces applied to the teeth of the blade. At a given point in the shaft, these internal forces are equivalent to a force-couple system consisting of axial and shearing forces and couples representing the bending and torsional moments. ©McGraw-Hill Education/Sabina Dowell

the internal forces *also produce shear and bending.* The force **F** is an **axial force**; the force **V** is called a **shearing force**; and the moment **M** of the couple is known as the **bending moment at** *J*. Note that, when determining internal forces in a member, you should clearly indicate on which portion of the member the forces are supposed to act. The deformation that occurs in member *AD* is sketched in Fig. 7.3*e*. The actual analysis of such a deformation is part of the study of mechanics of materials.

Also note that, in a **two-force member that is not straight**, the internal forces are also equivalent to a force-couple system. This is shown in Fig. 7.4, where the two-force member *ABC* has been cut at *D*.

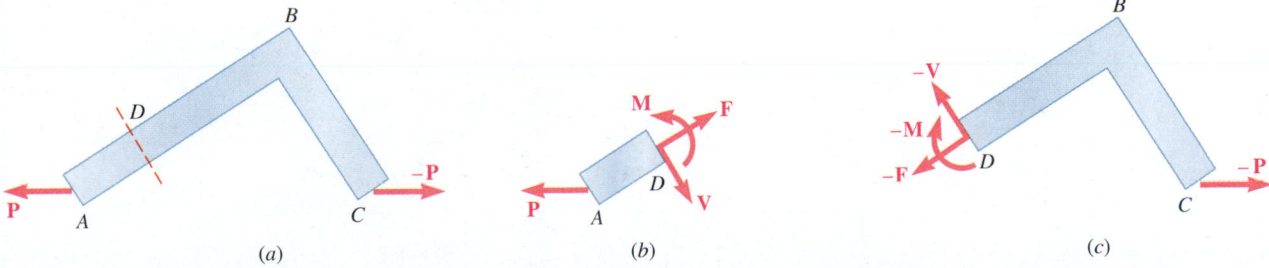

*(a)*                *(b)*                *(c)*

**Fig. 7.4**   (*a*) Free-body diagram of a two-force member that is not straight; (*b,c*) free-body diagrams of sections of member *ABC* showing internal force-couple systems.

## Sample Problem 7.1

In the frame shown, determine the internal forces (*a*) in member *ACF* at point *J*, (*b*) in member *BCD* at point *K*. This frame was previously analyzed in Sample Prob. 6.5.

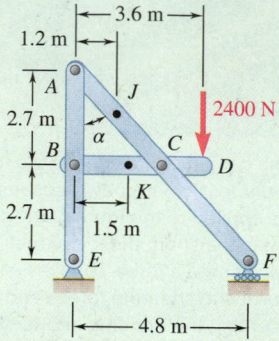

**STRATEGY:** After isolating each member, you can cut it at the given point and treat the resulting parts as objects in equilibrium. Analysis of the equilibrium equations, as we did before in Sample Prob. 6.5, will determine the internal force-couple system.

**MODELING:** The reactions and the connection forces acting on each member of the frame were determined previously in Sample Prob. 6.5. The results are repeated in Fig. 1.

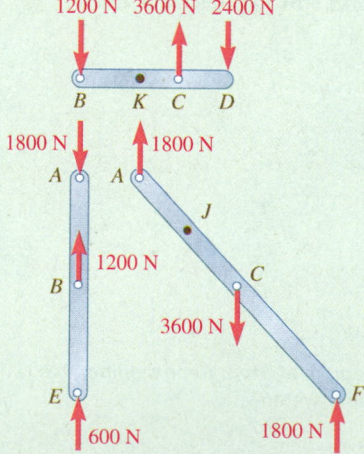

**Fig. 1** Reactions and connection forces acting on each member of the frame.

## ANALYSIS:

**a. Internal Forces at J.** Cut member *ACF* at point *J*, obtaining the two parts shown in Fig. 2. Represent the internal forces at *J* by an equivalent force-couple system, which can be determined by considering the equilibrium of either part. Considering the free body *AJ*, you have

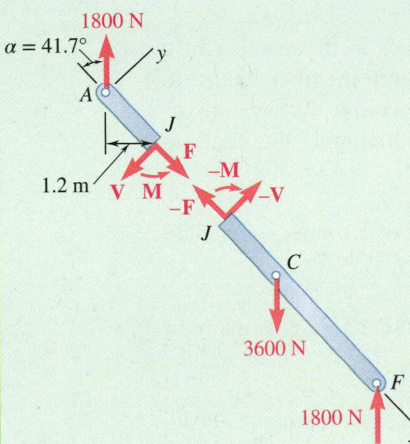

$+\circlearrowleft \Sigma M_J = 0:$     $-(1800 \text{ N})(1.2 \text{ m}) + M = 0$
$M = +2160 \text{ N·m}$     **M = 2160 N·m $\circlearrowleft$** ◀

$+\searrow \Sigma F_x = 0:$     $F - (1800 \text{ N}) \cos 41.7° = 0$
$F = +1344 \text{ N}$     **F = 1344 N $\searrow$** ◀

$+\nearrow \Sigma F_y = 0:$     $-V + (1800 \text{ N}) \sin 41.7° = 0$
$V = +1197 \text{ N}$     **V = 1197 N $\nearrow$** ◀

The internal forces at *J* are therefore equivalent to a couple **M**, an axial force **F**, and a shearing force **V**. The internal force-couple system acting on part *JCF* is equal and opposite.

**Fig. 2** Free-body diagrams of portions *AJ* and *FJ* of member *ACF*.

**b. Internal Forces at K.** Cut member *BCD* at *K*, obtaining the two parts shown in Fig. 3. Considering the free body *BK*, you obtain

$+\circlearrowleft \Sigma M_K = 0:$     $(1200 \text{ N})(1.5 \text{ m}) + M = 0$
$M = -1800 \text{ N·m}$     **M = 1800 N·m $\circlearrowleft$** ◀

$\xrightarrow{+} \Sigma F_x = 0:$     $F = 0$     **F = 0** ◀

$+\uparrow \Sigma F_y = 0:$     $-1200 \text{ N} - V = 0$
$V = -1200 \text{ N}$     **V = 1200 N↑** ◀

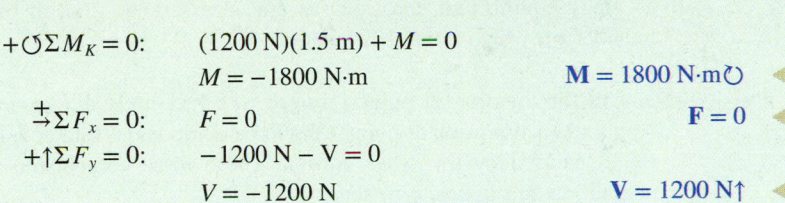

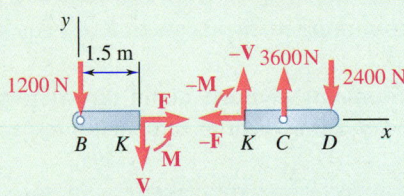

**Fig. 3** Free-body diagrams of portions *BK* and *DK* of member *BCD*.

**REFLECT and THINK:** The mathematical techniques involved in solving a problem of this type are not new; they are simply applications of concepts presented in earlier chapters. However, the physical interpretation is new: We are now determining the internal forces and moments within a structural member. These are of central importance in the study of mechanics of materials.

# SOLVING PROBLEMS
# ON YOUR OWN

In this section, we discussed how to determine the internal forces in the member of a frame. The internal forces at a given point in a **straight two-force member** reduce to an axial force, but in all other cases, they are equivalent to a **force-couple system** consisting of an **axial force F**, a **shearing force V**, and a couple **M** representing the **bending moment** at that point.

To determine the internal forces at a given point $J$ of the member of a frame, you should take the following steps.

**1. Draw a free-body diagram of the entire frame,** and use it to determine as many of the reactions at the supports as you can.

**2. Dismember the frame and draw a free-body diagram of each of its members.** Write as many equilibrium equations as are necessary to find all of the forces acting on the member on which point $J$ is located.

**3. Cut the member at point $J$ and draw a free-body diagram of each resulting portion.** Apply to each portion at point $J$ the force components and couple representing the internal forces exerted by the other portion. These force components and couples are equal in magnitude and opposite in sense.

**4. Select one of the two free-body diagrams** you have drawn and use it to write three equilibrium equations for the corresponding portion of the member.

    **a. Summing moments about $J$** and equating them to zero yields the bending moment at point $J$.

    **b. Summing components in directions parallel and perpendicular** to the member at $J$ and equating them to zero yields, respectively, the axial and shearing forces.

**5. When recording your answers, be sure to specify the portion of the member** you have used, because the forces and couples acting on the two portions have opposite senses.

The solutions of the problems in this section require you to determine the forces exerted on each other by the various members of a frame, so be sure to review the methods used in Chap. 6 to solve this type of problem. When frames involve pulleys and cables, for instance, remember that the forces exerted by a pulley on the member of the frame to which it is attached have the same magnitude and direction as the forces exerted by the cable on the pulley (Prob. 6.90).

# Problems

**7.1 and 7.2** Determine the internal forces (axial force, shearing force, and bending moment) at point *J* of the structure indicated.

    **7.1** Frame and loading of Prob. 6.78.

    **7.2** Frame and loading of Prob. 6.81.

**7.3** Determine the internal forces at point *J* when $\alpha = 90°$.

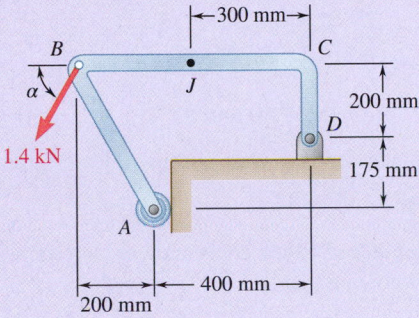

**Fig. P7.3 and P7.4**

**7.4** Determine the internal forces at point *J* when $\alpha = 0$.

**7.5** Determine the internal forces at point *J* when $\alpha = 90°$.

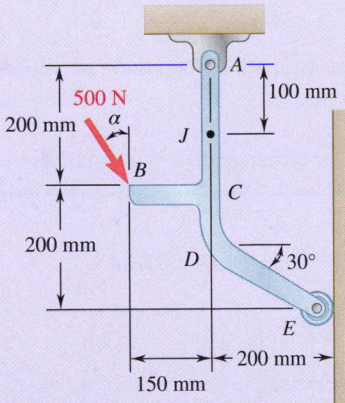

**Fig. *P7.5* and *P7.6***

**7.6** Determine the internal forces at point *J* when $\alpha = 0$.

**7.7** An archer aiming at a target is pulling with a 45-lb force on the bow-string. Assuming that the shape of the bow can be approximated by a parabola, determine the internal forces at point *J*.

**7.8** For the bow of Prob. 7.7, determine the magnitude and location of the maximum (*a*) axial force, (*b*) shearing force, (*c*) bending moment.

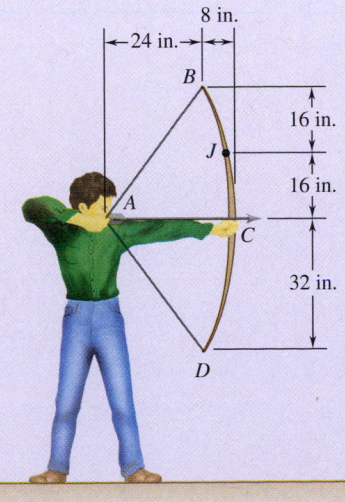

**Fig. P7.7**

**7.9** A semicircular rod is loaded as shown. Determine the internal forces at point $J$.

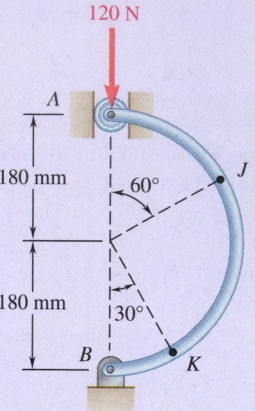

Fig. P7.9 and P7.10

**7.10** A semicircular rod is loaded as shown. Determine the internal forces at point $K$.

**7.11** A semicircular rod is loaded as shown. Determine the internal forces at point $J$ knowing that $\theta = 30°$.

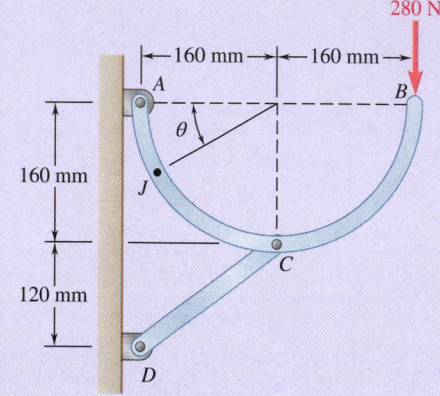

Fig. P7.11 and P7.12

**7.12** A semicircular rod is loaded as shown. Determine the magnitude and location of the maximum bending moment in the rod.

**7.13** The axis of the curved member $AB$ is a parabola with vertex at $A$. If a vertical load **P** of magnitude 450 lb is applied at $A$, determine the internal forces at $J$ when $h = 12$ in., $L = 40$ in., and $a = 24$ in.

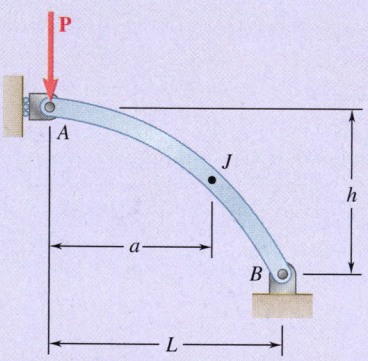

Fig. **P7.13** and **P7.14**

**7.14** Knowing that the axis of the curved member $AB$ is a parabola with vertex at $A$, determine the magnitude and location of the maximum bending moment.

**7.15** Knowing that the radius of each pulley is 120 mm and neglecting friction, determine the internal forces at (*a*) point $C$, (*b*) point $J$ that is 100 mm to the left of $C$.

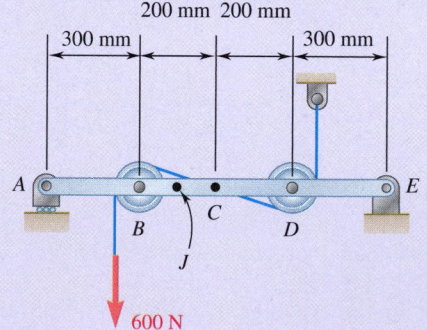

Fig. P7.15 and P7.16

**7.16** Knowing that the radius of each pulley is 100 mm and neglecting friction, determine the internal forces at (*a*) point $C$, (*b*) point $J$ that is 100 mm to the left of $C$.

**7.17** A 5-in.-diameter pipe is supported every 9 ft by a small frame consisting of two members as shown. Knowing that the combined weight of the pipe and its contents is 10 lb/ft and neglecting the effect of friction, determine the magnitude and location of the maximum bending moment in member *AC*.

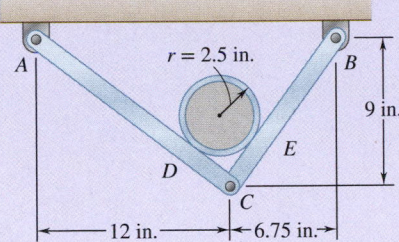

**Fig. P7.17**

**7.18** For the frame of Prob. 7.17, determine the magnitude and location of the maximum bending moment in member *BC*.

**7.19** Knowing that the radius of each pulley is 200 mm and neglecting friction, determine the internal forces at point *J* of the frame shown.

**7.20** Knowing that the radius of each pulley is 200 mm and neglecting friction, determine the internal forces at point *K* of the frame shown.

**7.21 and 7.22** A force **P** is applied to a bent rod that is supported by a roller and a pin and bracket. For each of the three cases shown, determine the internal forces at point *J*.

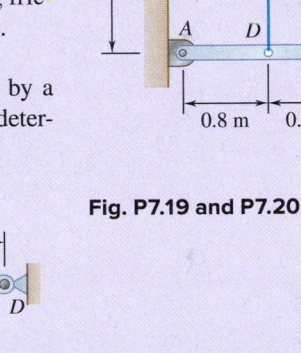

**Fig. P7.19 and P7.20**

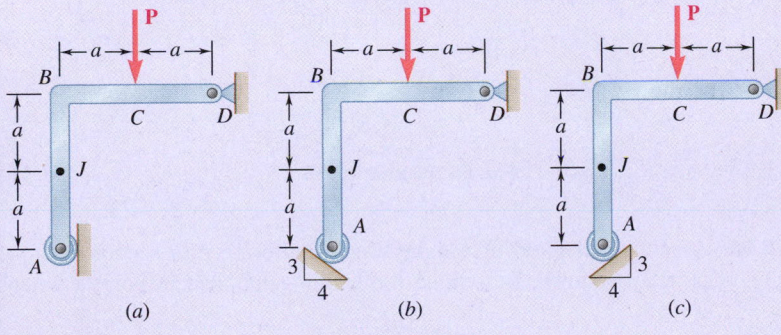

**Fig. P7.21**

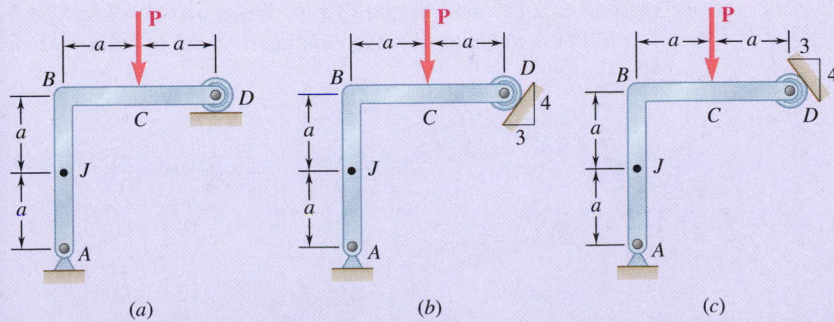

**Fig. P7.22**

**7.23** A quarter-circular rod of weight $W$ and uniform cross section is supported as shown. Determine the bending moment at point $J$ when $\theta = 30°$.

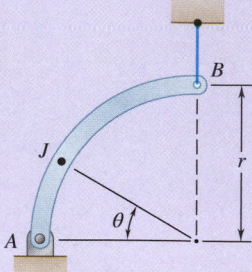

**Fig. P7.23**

**7.24** For the rod of Prob. 7.23, determine the magnitude and location of the maximum bending moment.

**7.25** A semicircular rod of weight $W$ and uniform cross section is supported as shown. Determine the bending moment at point $J$ when $\theta = 60°$.

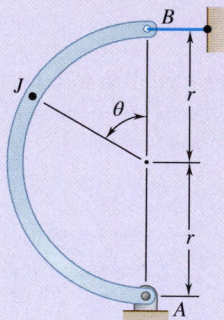

**Fig. P7.25 and P7.26**

**7.26** A semicircular rod of weight $W$ and uniform cross section is supported as shown. Determine the bending moment at point $J$ when $\theta = 150°$.

**7.27 and 7.28** A half section of pipe rests on a frictionless horizontal surface as shown. If the half section of pipe has a mass of 9 kg and a diameter of 300 mm, determine the bending moment at point $J$ when $\theta = 90°$.

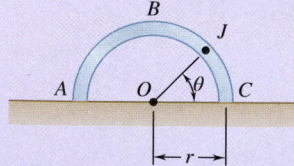

**Fig. P7.27**

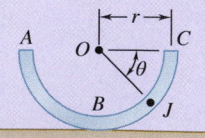

**Fig. P7.28**

# 7.2 BEAMS

A structural member designed to support loads applied at various points along the member is known as a **beam**. In most cases, the loads are perpendicular to the axis of the beam and cause only shear and bending in the beam. When the loads are not at a right angle to the beam, they also produce axial forces in the beam.

Beams are usually long, straight prismatic bars. Designing a beam for the most effective support of the applied loads is a two-part process: (1) determine the shearing forces and bending moments produced by the loads and (2) select the cross section best suited to resist these shearing forces and bending moments. Here we are concerned with the first part of the problem of beam design. The second part belongs to the study of the mechanics of materials.

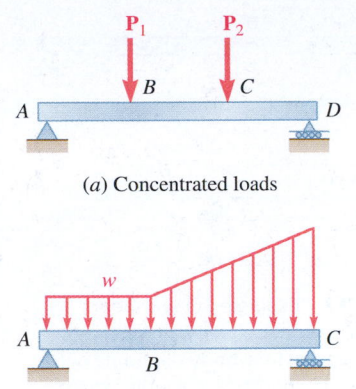

(a) Concentrated loads

(b) Distributed load

**Fig. 7.5** A beam may be subjected to (a) concentrated loads or (b) distributed loads, or a combination of both.

## 7.2A Various Types of Loading and Support

A beam can be subjected to **concentrated loads** $P_1$, $P_2$, ... that are expressed in newtons, pounds, or their multiples, kilonewtons and kips (Fig. 7.5a). We can also subject a beam to a **distributed load** $w$, expressed in N/m, kN/m, lb/ft, or kips/ft (Fig. 7.5b). In many cases, a beam is subjected to a combination of both types of load. When the load $w$ per unit length has a constant value over part of the beam (as between $A$ and $B$ in Fig. 7.5b), the load is said to be **uniformly distributed** over that part of the beam. Determining the reactions at the supports is considerably simplified if we replace distributed loads by equivalent concentrated loads, as explained in Sec. 5.3A. However, you should not do this substitution, or you should at least perform it with care, when calculating internal forces (see Sample Prob. 7.3).

Beams are classified according to the way in which they are supported. Fig. 7.6 shows several types of beams used frequently. The distance $L$ between

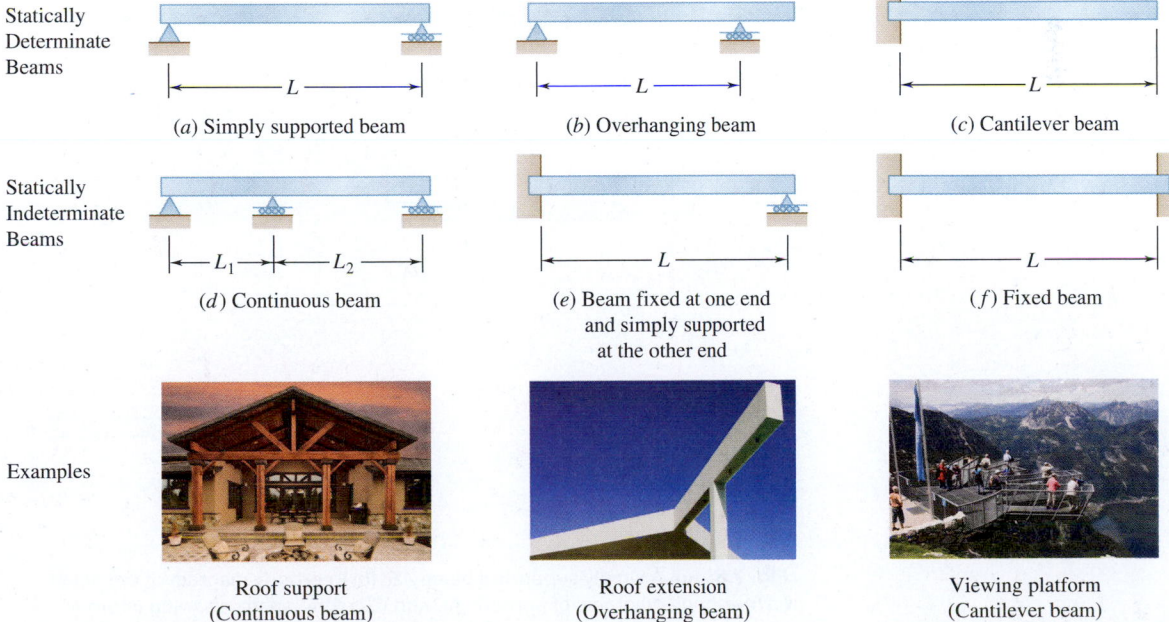

| Statically Determinate Beams | | |
|---|---|---|
| (a) Simply supported beam | (b) Overhanging beam | (c) Cantilever beam |

| Statically Indeterminate Beams | | |
|---|---|---|
| (d) Continuous beam | (e) Beam fixed at one end and simply supported at the other end | (f) Fixed beam |

Examples

Roof support (Continuous beam)

Roof extension (Overhanging beam)

Viewing platform (Cantilever beam)

**Fig. 7.6** Some common types of beams and their supports. Continuous beam: ©Ross Chandler/Getty Images RF; overhanging beam: ©Goodshoot/Getty Images RF; cantilever beam: ©Ange/Alamy

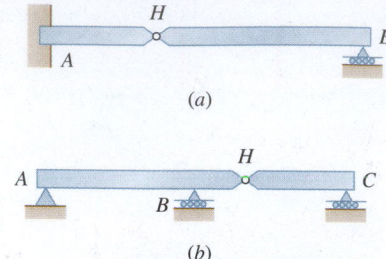

**Fig. 7.7**   Examples of two-beam systems connected by a hinge. In both cases, free-body diagrams of each individual beam enable you to determine the support reactions.

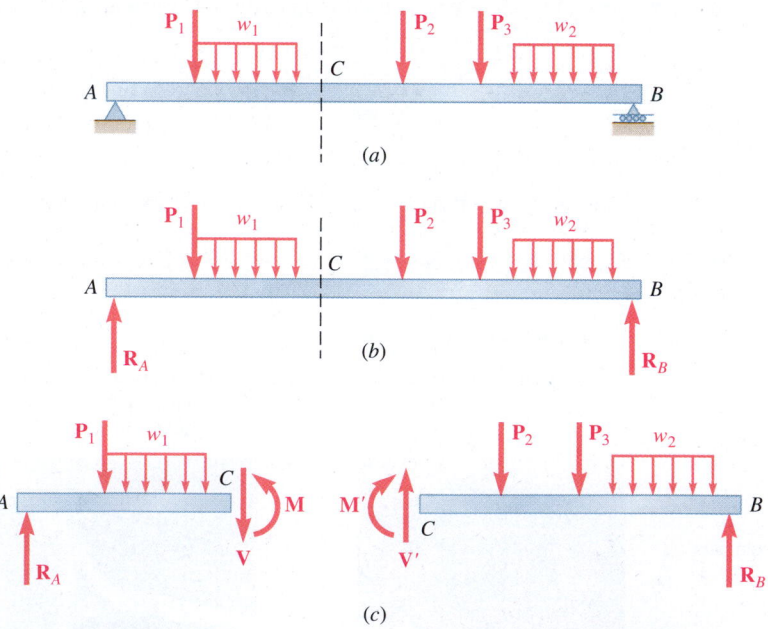

**Photo 7.2**   As a truck crosses a highway overpass, the internal forces vary in the beams of the overpass. ©Alan Thornton/Getty Images

supports is called the **span**. Note that the reactions are determinate if the supports involve only three unknowns. If more unknowns are involved, the reactions are statically indeterminate, and the methods of statics are not sufficient to determine the reactions. In such a case, we must take into account the properties of the beam with regard to its resistance to bending. Beams supported by only two rollers are not shown here; they are partially constrained and move under certain types of loadings.

Sometimes two or more beams are connected by hinges to form a single continuous structure. Two examples of beams hinged at a point $H$ are shown in Fig. 7.7. Here, the reactions at the supports involve four unknowns and cannot be determined from the free-body diagram of the two-beam system. However, we can determine the reactions by considering the free-body diagram of each beam separately. Analysis of this situation involves six unknowns (including two force components at the hinge), and six equations are available.

## 7.2B   Shear and Bending Moment in a Beam

Consider a beam $AB$ subjected to various concentrated and distributed loads (Fig. 7.8a). We propose to determine the shearing force and bending moment at any point of the beam. In the example considered here, the beam is simply supported, but the method used could be applied to any type of statically determinate beam.

First, we determine the reactions at $A$ and $B$ by choosing the entire beam as a free body (Fig. 7.8b). Setting $\Sigma M_A = 0$ and $\Sigma M_B = 0$, we obtain, respectively, $\mathbf{R}_B$ and $\mathbf{R}_A$.

**Fig. 7.8**   (a) A simply supported beam $AB$; (b) free-body diagram of the beam; (c) free-body diagrams of portions $AC$ and $CB$ of the beam, showing internal shearing forces and couples.

To determine the internal forces at an arbitrary point *C,* we cut the beam at *C* and draw the free-body diagrams of the portions *AC* and *CB* (Fig. 7.8*c*). Using the free-body diagram of *AC,* we can determine the shearing force **V** at *C* by equating the sum of the vertical components of all forces acting on *AC* to zero. Similarly, we can find the bending moment **M** at *C* by equating the sum of the moments about *C* of all forces and couples acting on *AC* to zero. Alternatively, we could use the free-body diagram of CB[†] and determine the shearing force **V′** and the bending moment **M′** by equating the sum of the vertical components and the sum of the moments about *C* of all forces and couples acting on *CB* to zero. Although this choice of free bodies may make the computation of the numerical values of the shearing force and bending moment easier, it requires us to indicate on which portion of the beam the internal forces considered are acting. If we want to calculate and efficiently record the shearing force and bending moment at every point of the beam, we must devise a way to avoid having to specify which portion of the beam is used as a free body every time. Therefore, we shall adopt the following conventions.

In determining the shearing force in a beam, *we always assume* that the internal forces **V** and **V′** are directed as shown in Fig. 7.8*c*. A positive value obtained for their common magnitude *V* indicates that this assumption is correct and that the shearing forces are actually directed as shown. A negative value obtained for *V* indicates that the assumption is wrong and the shearing forces are directed in the opposite way. Thus, to define completely the shearing forces at a given point of the beam, we only need to record the magnitude *V,* together with a plus or minus sign. The scalar *V* is commonly referred to as the **shear** at the given point of the beam.

Similarly, *we always assume* that the internal couples **M** and **M′** are directed as shown in Fig. 7.8*c*. A positive value obtained for their magnitude *M,* commonly referred to as the **bending moment,** indicates that this assumption is correct, whereas a negative value indicates it is wrong.

Summarizing these sign conventions, we state:

> *The shear V and the bending moment M at a given point of a beam are said to be positive when the internal forces and couples acting on each portion of the beam are directed as shown in Fig. 7.9a.*

You may be able to remember these conventions more easily by noting that:

1. *The shear at C is positive when the* **external** *forces (loads and reactions) acting on the beam tend to shear off the beam at C as indicated in Fig. 7.9b.*
2. *The bending moment at C is positive when the* **external** *forces acting on the beam tend to bend the beam at C in a concave-up fashion as indicated in Fig. 7.9c.*

It may also help to note that the situation described in Fig. 7.9, in which the values of both the shear and the bending moment are positive, is precisely the situation that occurs in the left half of a simply supported beam carrying a single concentrated load at its midpoint. This particular example is fully discussed in the following section.

---

[†]We now designate the force and couple representing the internal forces acting on *CB* by **V′** and **M′**, rather than by −**V** and −**M** as done earlier. The reason is to avoid confusion when applying the sign convention we are about to introduce.

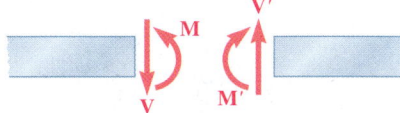

(*a*) Internal forces at section
(positive shear and positive bending moment)

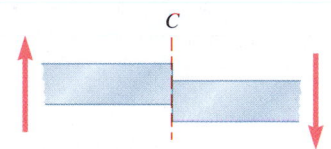

(*b*) Effect of external forces
(positive shear)

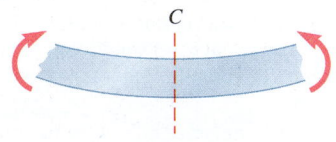

(*c*) Effect of external forces
(positive bending moment)

**Fig. 7.9** Figure for remembering the signs of shear and bending moment.

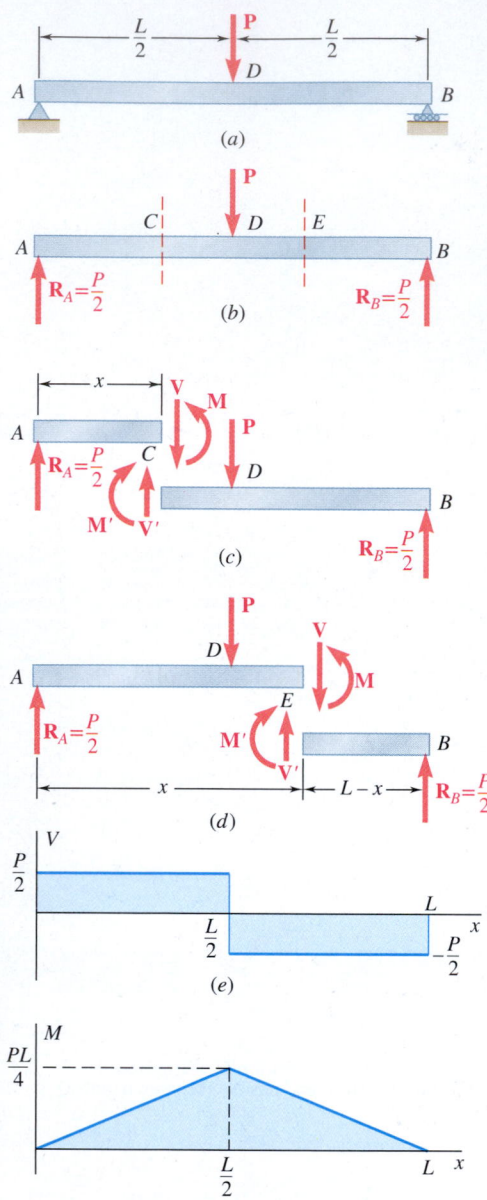

**Fig. 7.10** (a) A beam supporting a single concentrated load at its midpoint; (b) free-body diagram of the beam; (c) free-body diagrams of parts of the beam after a cut at C; (d) free-body diagrams of parts of the beam after a cut at E; (e) shear diagram of the beam; (f) bending-moment diagram of the beam.

## 7.2C  Shear and Bending-Moment Diagrams

Now that we have clearly defined the shear and bending moment in sense as well as in magnitude, we can easily record their values at any point along a beam by plotting these values against the distance $x$ measured from one end of the beam. The graphs obtained in this way are called, respectively, the **shear diagram** and the **bending-moment diagram**.

As an example, consider a simply supported beam $AB$ of span $L$ subjected to a single concentrated load **P** applied at its midpoint $D$ (Fig. 7.10a). We first determine the reactions at the supports from the free-body diagram of the entire beam (Fig. 7.10b); we find that the magnitude of each reaction is equal to $P/2$.

Next, we cut the beam at a point $C$ between $A$ and $D$ and draw the free-body diagrams of $AC$ and $CB$ (Fig. 7.10c). *Assuming that the shear and bending moment are positive,* we direct the internal forces **V** and **V′** and the internal couples **M** and **M′** as indicated in Fig. 7.9a. Considering the free body $AC$, we set the sum of the vertical components and the sum of the moments about $C$ of the forces acting on the free body to zero. From this, we find $V = +P/2$ and $M = +Px/2$. Therefore, both the shear and bending moment are positive. (You can check this by observing that the reaction at $A$ tends to shear off and to bend the beam at $C$ as indicated in Fig. 7.9b and c.) Now let's plot $V$ and $M$ between $A$ and $D$ (Fig. 7.10e and f). The shear has a constant value $V = P/2$, whereas the bending moment increases linearly from $M = 0$ at $x = 0$ to $M = PL/4$ at $x = L/2$.

Proceeding along the beam, we cut it at a point $E$ between $D$ and $B$ and consider the free body $EB$ (Fig. 7.10d). As before, the sum of the vertical components and the sum of the moments about $E$ of the forces acting on the free body are zero. We obtain $V = -P/2$ and $M = P(L - x)/2$. The shear is therefore negative and the bending moment is positive. (Again, you can check this by observing that the reaction at $B$ bends the beam at $E$ as indicated in Fig. 7.9c, but tends to shear it off in a manner opposite to that shown in Fig. 7.9b.) We can now complete the shear and bending-moment diagrams of Fig. 7.10e and f. The shear has a constant value $V = -P/2$ between $D$ and $B$, whereas the bending moment decreases linearly from $M = PL/4$ at $x = L/2$ to $M = 0$ at $x = L$.

Note that when a beam is subjected to concentrated loads only, the shear is of constant value between loads and the bending moment varies linearly between loads. However, when a beam is subjected to distributed loads, the shear and bending moment vary quite differently (see Sample Prob. 7.3).

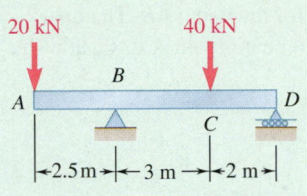

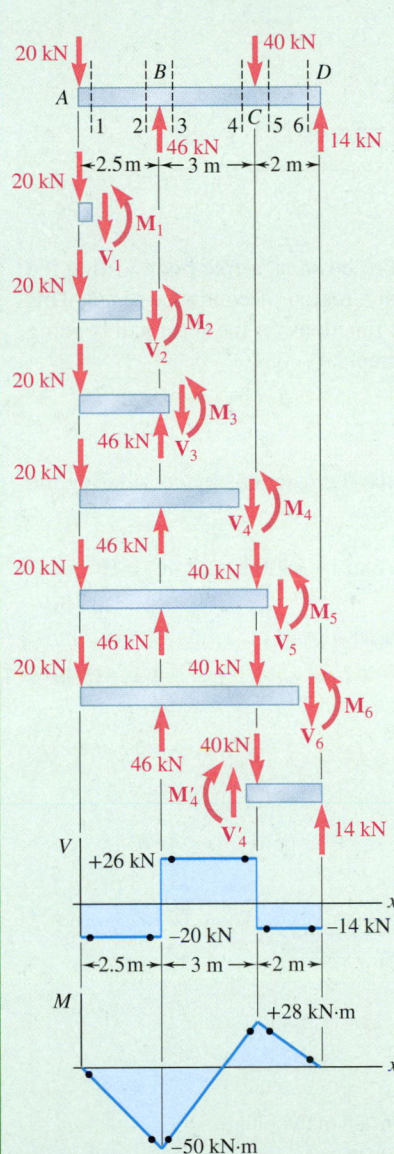

**Fig. 1** Free-body diagrams of the beam sections, and the resulting shear and bending-moment diagrams.

## Sample Problem 7.2

Draw the shear and bending-moment diagrams for the beam and loading shown.

**STRATEGY:** Treat the entire beam as a free body to determine the reactions, and then cut the beam just before and just after each external concentrated force (Fig. 1) to see how the shear and bending moment change along the length of the beam.

**MODELING and ANALYSIS:**

**Free-Body, Entire Beam.** From the free-body diagram of the entire beam, find the reactions at $B$ and $D$:

$$\mathbf{R}_B = 46 \text{ kN} \uparrow \qquad \mathbf{R}_D = 14 \text{ kN} \uparrow$$

**Shear and Bending Moment.** First, determine the internal forces just to the right of the 20-kN load at $A$. Consider the stub of the beam to the left of point 1 as a free body, and assume $V$ and $M$ are positive (according to the standard convention). Then, you have

$$+\uparrow \Sigma F_y = 0: \qquad -20 \text{ kN} - V_1 = 0 \qquad V_1 = -20 \text{ kN}$$
$$+\circlearrowleft \Sigma M_1 = 0: \qquad (20 \text{ kN})(0 \text{ m}) + M_1 = 0 \qquad M_1 = 0$$

Next, consider the portion of the beam to the left of point 2 as a free body:

$$+\uparrow \Sigma F_y = 0: \qquad -20 \text{ kN} - V_2 = 0 \qquad V_2 = -20 \text{ kN}$$
$$+\circlearrowleft \Sigma M_2 = 0: \qquad (20 \text{ kN})(2.5 \text{ m}) + M_2 = 0 \qquad M_2 = -50 \text{ kN·m}$$

Determine the shear and bending moment at sections 3, 4, 5, and 6 in a similar way from the free-body diagrams. The results are

$$V_3 = +26 \text{ kN} \qquad M_3 = -50 \text{ kN·m}$$
$$V_4 = +26 \text{ kN} \qquad M_4 = +28 \text{ kN·m}$$
$$V_5 = -14 \text{ kN} \qquad M_5 = +28 \text{ kN·m}$$
$$V_6 = -14 \text{ kN} \qquad M_6 = 0$$

For several of the later cuts, the results are easier to obtain by considering as a free body the portion of the beam to the right of the cut. For example, consider the portion of the beam to the right of point 4. You have

$$+\uparrow \Sigma F_y = 0: \qquad V_4 - 40 \text{ kN} + 14 \text{ kN} = 0 \qquad V_4 = +26 \text{ kN}$$
$$+\circlearrowleft \Sigma M_4 = 0: \qquad -M_4 + (14 \text{ kN})(2 \text{ m}) = 0 \qquad M_4 = +28 \text{ kN·m}$$

**Shear and Bending-Moment Diagrams.** Now plot the six points shown on the shear and bending-moment diagrams. As indicated in Sec. 7.2C, the shear is of constant value between concentrated loads, and the bending moment varies linearly. You therefore obtain the shear and bending-moment diagrams shown in Fig. 1.

**REFLECT and THINK:** The calculations are pretty similar for each new choice of free body. However, moving along the beam, the shear changes magnitude whenever you pass a transverse force, and the graph of the bending moment changes slope at these points.

*(continued)*

## Sample Problem 7.3

Draw the shear and bending-moment diagrams for the beam *AB*. The distributed load of 40 lb/in. extends over 12 in. of the beam from *A* to *C*, and the 400-lb load is applied at *E*.

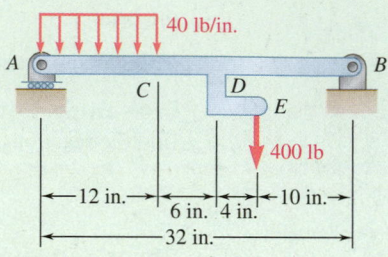

**STRATEGY:** Again, consider the entire beam as a free body to find the reactions. Then, cut the beam within each region of continuous load. This will enable you to determine continuous functions for the shear and bending moment, which you can then plot on a graph.

**MODELING and ANALYSIS:**

**Free-Body, Entire Beam.** Determine the reactions by considering the entire beam as a free body (Fig. 1).

$+\circlearrowleft \Sigma M_A = 0$:   $B_y(32 \text{ in.}) - (480 \text{ lb})(6 \text{ in.}) - (400 \text{ lb})(22 \text{ in.}) = 0$
$B_y = +365 \text{ lb}$    $\mathbf{B}_y = 365 \text{ lb} \uparrow$

$+\circlearrowleft \Sigma M_B = 0$:   $(480 \text{ lb})(26 \text{ in.}) + (400 \text{ lb})(10 \text{ in.}) - A(32 \text{ in.}) = 0$
$A = +515 \text{ lb}$    $\mathbf{A} = 515 \text{ lb} \uparrow$

$\xrightarrow{+} \Sigma F_x = 0$:   $B_x = 0$    $\mathbf{B}_x = 0$

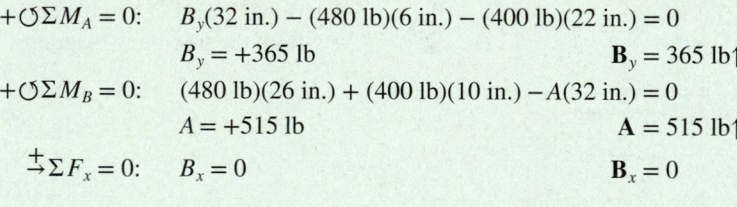

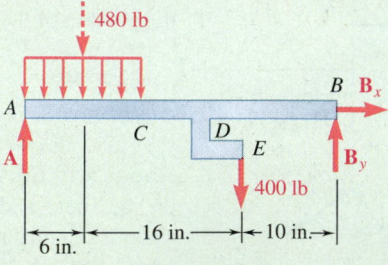

**Fig. 1** Free-body diagram of the entire beam.

Now, replace the 400-lb load by an equivalent force-couple system acting on the beam at point *D* and cut the beam at several points (Fig. 2).

**Fig. 2** Free-body diagrams of the beam sections, and the resulting shear and bending-moment diagrams.

**Shear and Bending Moment.**    *From A to C.* Determine the internal forces at a distance $x$ from point $A$ by considering the portion of the beam to the left of point 1. Replace that part of the distributed load acting on the free body by its resultant. You get

$$+\uparrow \Sigma F_y = 0: \qquad 515 - 40x - V = 0 \qquad\qquad V = 515 - 40x$$

$$+\circlearrowleft \Sigma M_1 = 0: \qquad -515x + 40x(\tfrac{1}{2}x) + m = 0 \qquad M = 515x - 20x^2$$

Note that $V$ and $M$ are not numerical values, but they are expressed as functions of $x$. The free-body diagram shown can be used for all values of $x$ smaller than 12 in., so the expressions obtained for $V$ and $M$ are valid throughout the region $0 < x < 12$ in.

*From C to D.*    Consider the portion of the beam to the left of point 2. Again replacing the distributed load by its resultant, you have

$$+\uparrow \Sigma F_y = 0: \qquad 515 - 480 - V = 0 \qquad\qquad V = 35 \text{ lb}$$

$$+\circlearrowleft \Sigma M_2 = 0: \qquad -515x + 480(x - 6) + M = 0 \qquad M = (2880 + 35x) \text{ lb·in.}$$

These expressions are valid in the region 12 in. $< x <$ 18 in.

*From D to B.*    Use the portion of the beam to the left of point 3 for the region 18 in. $< x <$ 32 in. Thus,

$$+\uparrow \Sigma F_y = 0: \qquad 515 - 480 - 400 - V = 0 \qquad V = -365 \text{ lb}$$

$$+\circlearrowleft \Sigma M_3 = 0: \qquad -515x + 480(x - 6) - 1600 + 400(x - 18) + M = 0$$

$$M = (11{,}680 - 365x) \text{ lb·in.}$$

**Shear and Bending-Moment Diagrams.**    Plot the shear and bending-moment diagrams for the entire beam. Note that the couple of moment 1600 lb·in. applied at point $D$ introduces a discontinuity into the bending-moment diagram. Also note that the bending-moment diagram under the distributed load is not straight but is slightly curved.

**REFLECT and THINK:**    Shear and bending-moment diagrams typically feature various kinds of curves and discontinuities. In such cases, it is often useful to express $V$ and $M$ as functions of location $x$, as well as to determine certain numerical values.

# SOLVING PROBLEMS
# ON YOUR OWN

In this section, you saw how to determine the **shear** $V$ and the **bending moment** $M$ at any point in a beam. You also learned to draw the **shear diagram** and the **bending-moment diagram** for the beam by plotting, respectively, $V$ and $M$ against the distance $x$ measured along the beam.

**A. Determining the shear and bending moment in a beam.** To determine the shear $V$ and the bending moment $M$ at a given point $C$ of a beam, take the following steps.

**1. Draw a free-body diagram of the entire beam,** and use it to determine the reactions at the beam supports.

**2. Cut the beam at point $C$,** and using the original loading, select one of the two resulting portions of the beam.

**3. Draw the free-body diagram of the portion of the beam you have selected.** Show:

   **a. The loads and the reactions** exerted on that portion of the beam, replacing each distributed load by an equivalent concentrated load, as explained in Sec. 5.3A.

   **b. The shearing force and the bending moment representing the internal forces at $C$.** To facilitate recording the shear $V$ and the bending moment $M$ after determining them, follow the convention indicated in Figs. 7.8 and 7.9. Thus, if you are using the portion of the beam located to the *left of C*, apply at $C$ a *shearing force* **V** *directed downward* and a *bending moment* **M** *directed counterclockwise.* If you are using the portion of the beam located to the *right of C*, apply at $C$ a *shearing force* **V′** *directed upward* and a *bending moment* **M′** *directed clockwise* (Sample Prob. 7.2).

**4. Write the equilibrium equations for the portion of the beam you have selected.** Solve the equation $\Sigma F_y = 0$ for $V$ and the equation $\Sigma M_C = 0$ for $M$.

**5. Record the values of V and M with the sign obtained for each of them.** A positive sign for $V$ means that the shearing forces exerted at $C$ on each of the two portions of the beam are directed as shown in Figs. 7.8 and 7.9; a negative sign means they have the opposite sense. Similarly, a positive sign for $M$ means that the bending couples at $C$ are directed as shown in these figures, and a negative sign means they have the opposite sense. In addition, a positive sign for $M$ means that the concavity of the beam at $C$ is directed upward, and a negative sign means it is directed downward.

**B. Drawing the shear and bending-moment diagrams for a beam.** Obtain these diagrams by plotting, respectively, $V$ and $M$ against the distance $x$ measured along the beam. However, in most cases, you need to compute the values of $V$ and $M$ at only a few points.

**1. For a beam supporting only concentrated loads,** note (Sample Prob. 7.2) that

   **a. The shear diagram consists of segments of horizontal lines.** Thus, to draw the shear diagram of the beam, you need to compute $V$ only just to the left or just to the right of the points where the loads or reactions are applied.

**b. The bending-moment diagram consists of segments of oblique straight lines.** Thus, to draw the bending-moment diagram of the beam, you need to compute $M$ only at the points where the loads or reactions are applied.

**2. For a beam supporting uniformly distributed loads,** note (Sample Prob. 7.3) that under each of the distributed loads:

**a. The shear diagram consists of a segment of an oblique straight line.** Thus, you need to compute $V$ only where the distributed load begins and where it ends.

**b. The bending-moment diagram consists of an arc of parabola.** In most cases, you need to compute $M$ only where the distributed load begins and where it ends.

**3. For a beam with a more complicated loading,** you need to consider the free-body diagram of a portion of the beam of arbitrary length $x$ and determine $V$ and $M$ as functions of $x$. This procedure may have to be repeated several times, because $V$ and $M$ are often represented by different functions in various parts of the beam (Sample Prob. 7.3).

**4. When a couple is applied to a beam,** the shear has the same value on both sides of the point of application of the couple, but the bending-moment diagram shows a discontinuity at that point, rising or falling by an amount equal to the magnitude of the couple. Note that a couple can either be applied directly to the beam or result from the application of a load on a member rigidly attached to the beam (Sample Prob. 7.3).

# Problems

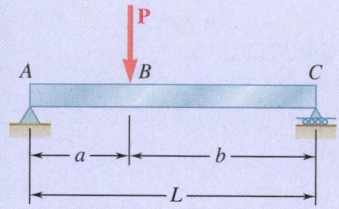

Fig. P7.29

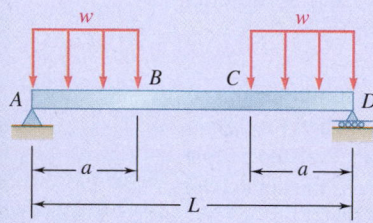

Fig. P7.30

**7.29 through 7.32** For the beam and loading shown, (a) draw the shear and bending-moment diagrams, (b) determine the maximum absolute value of the bending moment.

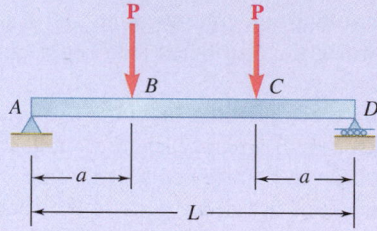

Fig. P7.31

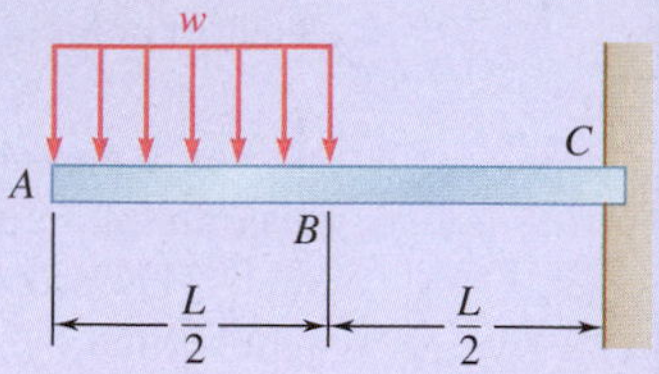

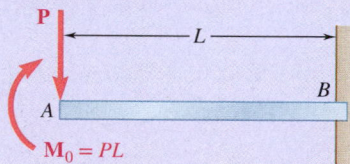

Fig. P7.32

**7.33 and 7.34** For the beam and loading shown, (a) draw the shear and bending-moment diagrams, (b) determine the maximum absolute values of the shear and bending moment.

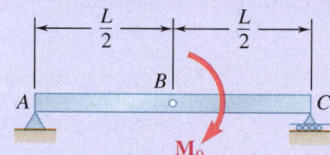

Fig. P7.33

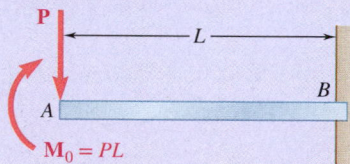

Fig. P7.34

**7.35 and 7.36** For the beam and loading shown, (a) draw the shear and bending-moment diagrams, (b) determine the maximum absolute values of the shear and bending moment.

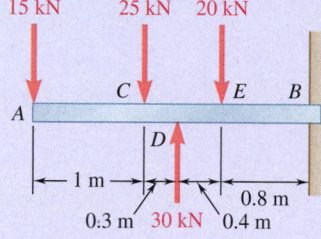

Fig. P7.35

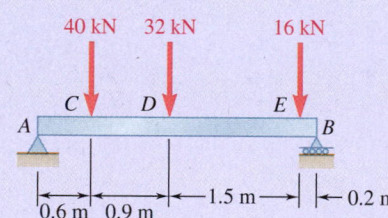

Fig. P7.36

**7.37 and 7.38** For the beam and loading shown, (a) draw the shear and bending-moment diagrams, (b) determine the maximum absolute values of the shear and bending moment.

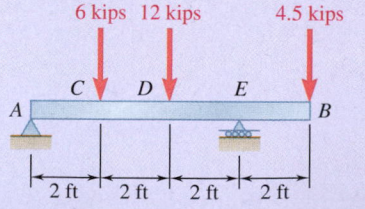

Fig. P7.37

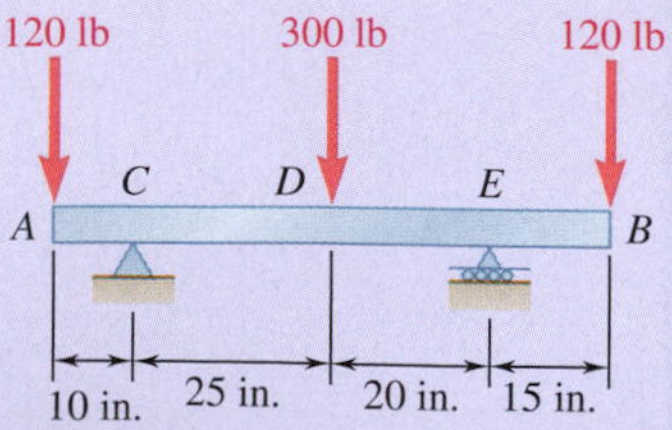

Fig. P7.38

388

**7.39 through 7.42** For the beam and loading shown, (a) draw the shear and bending-moment diagrams, (b) determine the maximum absolute values of the shear and bending moment.

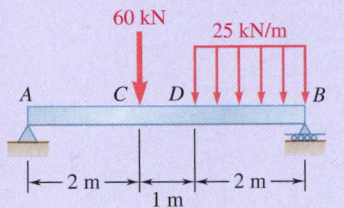

Fig. P7.39

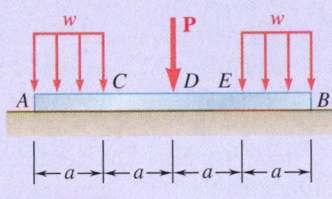

Fig. P7.40

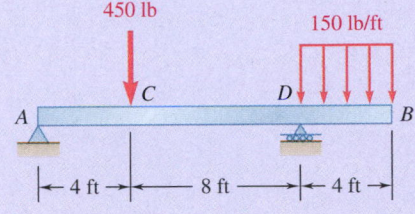

Fig. P7.41

**7.43** Assuming the upward reaction of the ground on beam AB to be uniformly distributed and knowing that $P = wa$, (a) draw the shear and bending-moment diagrams, (b) determine the maximum absolute values of the shear and bending moment.

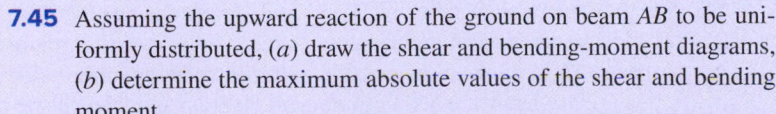
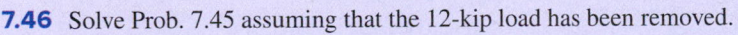

Fig. P7.43

**7.44** Solve Prob. 7.43 knowing that $P = 3wa$.

**7.45** Assuming the upward reaction of the ground on beam AB to be uniformly distributed, (a) draw the shear and bending-moment diagrams, (b) determine the maximum absolute values of the shear and bending moment.

**7.46** Solve Prob. 7.45 assuming that the 12-kip load has been removed.

**7.47 and 7.48** Assuming the upward reaction of the ground on beam AB to be uniformly distributed, (a) draw the shear and bending-moment diagrams, (b) determine the maximum absolute values of the shear and bending moment.

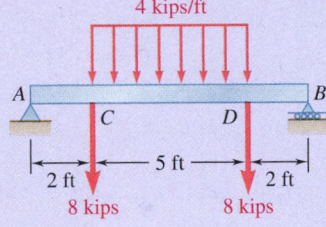

Fig. P7.42

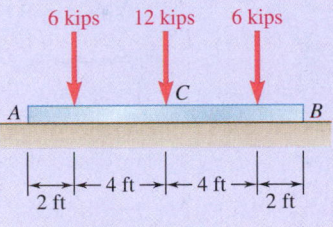

Fig. P7.45

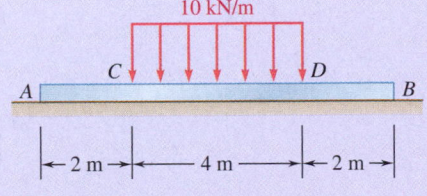

Fig. P7.47

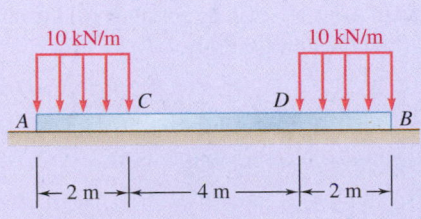

Fig. P7.48

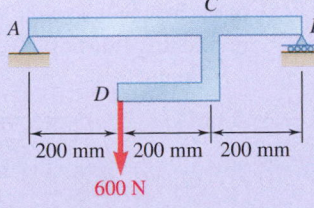

Fig. P7.49

**7.49** Draw the shear and bending-moment diagrams for the beam AB, and determine the shear and bending moment (a) just to the left of C, (b) just to the right of C.

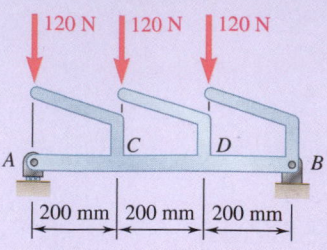

120 N  120 N  120 N

A  C  D  B

200 mm  200 mm  200 mm

**Fig. P7.50**

**7.50 through 7.52** Draw the shear and bending-moment diagrams for the beam *AB*, and determine the maximum absolute values of the shear and bending moment.

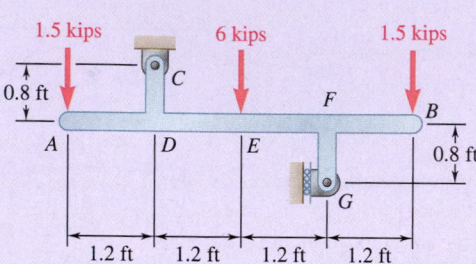

1.5 kips  6 kips  1.5 kips

0.8 ft

C

A  D  E  F  B

0.8 ft

G

1.2 ft  1.2 ft  1.2 ft  1.2 ft

**Fig. P7.51**

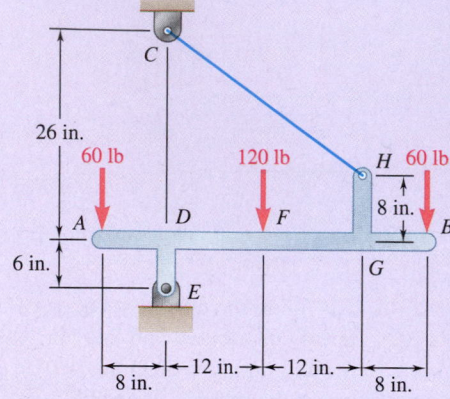

26 in.

C

60 lb  120 lb  H  60 lb

8 in.

A  D  F  B

6 in.

E

G

8 in.  12 in.  12 in.  8 in.

**Fig. P7.52**

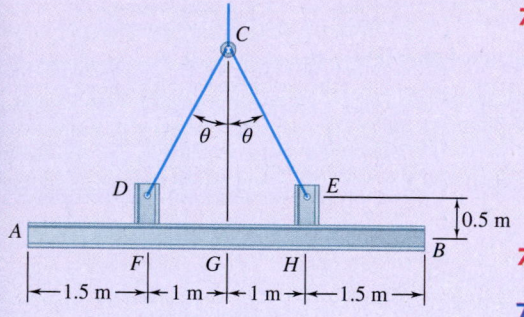

C

θ  θ

D  E

A  0.5 m

F  G  H  B

1.5 m  1 m  1 m  1.5 m

**Fig. P7.53**

**7.53** Two small channel sections *DF* and *EH* have been welded to the uniform beam *AB* of weight $W = 3$ kN to form the rigid structural member shown. This member is being lifted by two cables attached at *D* and *E*. Knowing that $\theta = 30°$ and neglecting the weight of the channel sections, (*a*) draw the shear and bending-moment diagrams for beam *AB*, (*b*) determine the maximum absolute values of the shear and bending moment in the beam.

**7.54** Solve Prob. 7.53 when $\theta = 60°$.

**7.55** For the structural member of Prob. 7.53, determine (*a*) the angle $\theta$ for which the maximum absolute value of the bending moment in beam *AB* is as small as possible, (*b*) the corresponding value of $|M|_{max}$. (*Hint:* Draw the bending-moment diagram and then equate the absolute values of the largest positive and negative bending moments obtained.)

**7.56** For the beam of Prob. 7.43, determine (*a*) the ratio $k = P/wa$ for which the maximum absolute value of the bending moment in the beam is as small as possible, (*b*) the corresponding value of $|M|_{max}$. (See the hint for Prob. 7.55.)

**7.57** Determine (*a*) the distance *a* for which the maximum absolute value of the bending moment in beam *AB* is as small as possible, (*b*) the corresponding value of $|M|_{max}$. (See the hint for Prob. 7.55.)

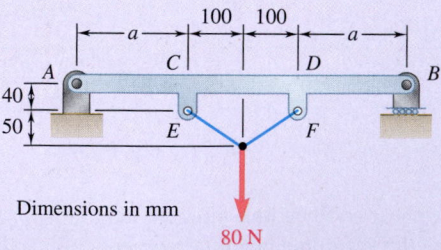

100  100

a  a

A  C  D  B

40

50

E  F

Dimensions in mm

80 N

**Fig. P7.57**

**7.58** For the beam and loading shown, determine (*a*) the distance *a* for which the maximum absolute value of the bending moment in the beam is as small as possible, (*b*) the corresponding value of $|M|_{max}$. (See the hint for Prob. 7.55.)

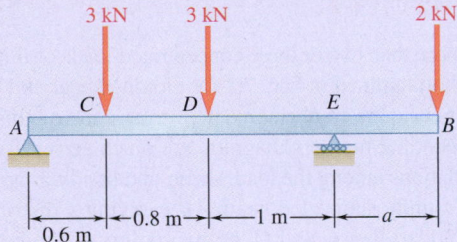

**Fig. P7.58**

**7.59** A uniform beam is to be picked up by crane cables attached at *A* and *B*. Determine the distance *a* from the ends of the beam to the points where the cables should be attached if the maximum absolute value of the bending moment in the beam is to be as small as possible. (*Hint:* Draw the bending-moment diagram in terms of *a*, *L*, and the weight per unit length *w*, and then equate the absolute values of the largest positive and negative bending moments obtained.)

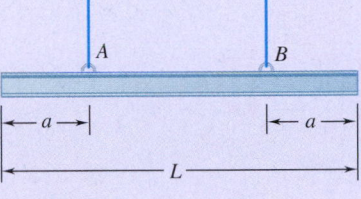

**Fig. P7.59**

**7.60** Knowing that $P = Q = 150$ lb, determine (*a*) the distance *a* for which the maximum absolute value of the bending moment in beam *AB* is as small as possible, (*b*) the corresponding value of $|M|_{max}$. (See the hint for Prob. 7.55.)

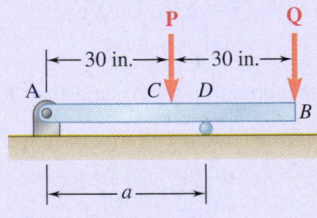

**Fig. P7.60**

**7.61** Solve Prob. 7.60 assuming that $P = 300$ lb and $Q = 150$ lb.

**\*7.62** In order to reduce the bending moment in the cantilever beam *AB*, a cable and counterweight are permanently attached at end *B*. Determine the magnitude of the counterweight for which the maximum absolute value of the bending moment in the beam is as small as possible, and find the corresponding value of $|M|_{max}$. Consider (*a*) the case when the distributed load is permanently applied to the beam, (*b*) the more general case when the distributed load may either be applied or removed.

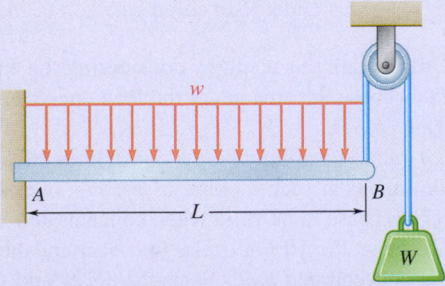

**Fig. P7.62**

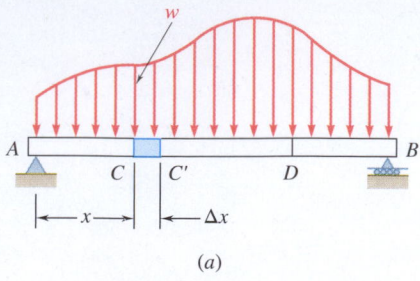

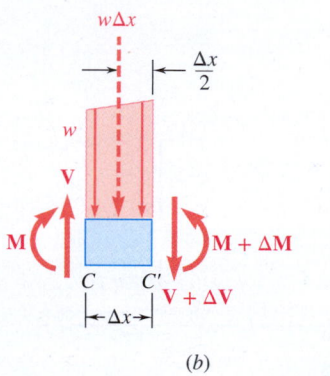

**Fig. 7.11** (*a*) A simply supported beam carrying a distributed load; (*b*) free-body diagram of a portion *CC'* of the beam.

# 7.3 RELATIONS AMONG LOAD, SHEAR, AND BENDING MOMENT

If a beam carries more than two or three concentrated loads or if it carries a distributed load, the method outlined in Sec. 7.2 for plotting shear and bending-moment diagrams is likely to be quite cumbersome. However, constructing a shear diagram and, especially, a bending-moment diagram, are much easier if we take into consideration some relations among the load, shear, and bending moment.

Consider a simply supported beam *AB* carrying a distributed load *w* per unit length (Fig. 7.11*a*). Let *C* and *C'* be two points of the beam at a distance $\Delta x$ from each other. We denote the shear and bending moment at *C* by *V* and *M*, respectively, and we assume they are positive. We denote the shear and bending moment at *C'* by $V + \Delta V$ and $M + \Delta M$.

Let us now detach the portion of beam *CC'* and draw its free-body diagram (Fig. 7.11*b*). The forces exerted on the free body include a load with a magnitude of $w\Delta x$ (indicated by a dashed arrow to distinguish it from the original distributed load from which it is derived) and internal forces and couples at *C* and *C'*. Because we assumed both the shear and bending moment are positive, the forces and couples are directed as shown in the figure.

**Relations Between Load and Shear.**  Because the free body *CC'* is in equilibrium, we set the sum of the vertical components of the forces acting on it to zero:

$$V - (V + \Delta V) - w\Delta x = 0$$
$$\Delta V = -w\Delta x$$

Dividing both sides of this equation by $\Delta x$ and then letting $\Delta x$ approach zero, we obtain

$$\frac{dV}{dx} = -w \tag{7.1}$$

Eq. (7.1) indicates that, for a beam loaded as shown in Fig. 7.11*a*, the slope *dV/dx* of the shear curve is negative and the numerical value of the slope at any point is equal to the load per unit length at that point.

Integrating (7.1) between arbitrary points *C* and *D*, we have

$$V_D - V_C = -\int_{x_C}^{x_D} w\, dx \tag{7.2}$$

or

$$V_D - V_C = -(\text{area under load curve between } C \text{ and } D) \tag{7.2'}$$

Note that we could also obtain this result by considering the equilibrium of the portion of beam *CD*, because the area under the load curve represents the total load applied between *C* and *D*.

Eq. (7.1) is *not* valid at a point where a concentrated load is applied; the shear curve is discontinuous at such a point, as we saw in Sec. 7.2. Similarly, formulas (7.2) and (7.2') cease to be valid when concentrated loads are applied between *C* and *D*, because they do not take into account the sudden change in shear caused by a concentrated load. Formulas (7.2) and (7.2'), therefore, should be applied only between successive concentrated loads.

**Relations Between Shear and Bending Moment.**   Returning to the free-body diagram of Fig. 7.11*b*, we can set the sum of the moments about $C'$ to be zero, obtaining

$$(M + \Delta M) - M - V\Delta x + w\Delta x \frac{\Delta x}{2} = 0$$

$$\Delta M = V\Delta x - \tfrac{1}{2}w(\Delta x)^2$$

Dividing both sides of this equation by $\Delta x$ and then letting $\Delta x$ approach zero, we have

$$\frac{dM}{dx} = V \tag{7.3}$$

Eq. (7.3) indicates that the slope $dM/dx$ of the bending-moment curve is equal to the value of the shear. This is true at any point where the shear has a well-defined value, i.e., at any point where no concentrated load is applied. Formula (7.3) also shows that the shear is zero at points where the bending moment is maximum. This property simplifies the determination of points where the beam is likely to fail under bending.

Integrating Eq. (7.3) between arbitrary points $C$ and $D$, we obtain

$$M_D - M_C = \int_{x_C}^{x_D} V dx \tag{7.4}$$

$$M_D - M_C = \text{area under shear curve between } C \text{ and } D \tag{7.4'}$$

Note that the area under the shear curve should be considered positive where the shear is positive and should be negative where the shear is negative. Formulas (7.4) and (7.4′) are valid even when concentrated loads are applied between $C$ and $D$, as long as the shear curve has been drawn correctly. The formulas cease to be valid, however, if a *couple* is applied at a point between $C$ and $D$, because they do not take into account the sudden change in the bending moment caused by a couple (see Sample Prob. 7.7).

In most engineering applications, you need to know the value of the bending moment at only a few specific points. Once you have drawn the shear diagram and determined $M$ at one end of the beam, you can obtain the value of the bending moment at any given point by computing the area under the shear curve and using formula (7.4′). For instance, because $M_A = 0$ for the beam of Fig. 7.12, you can determine the maximum value of the bending moment for that beam simply by measuring the area of the shaded triangle in the shear diagram as

$$M_{\text{max}} = \frac{1}{2}\frac{L}{2}\frac{wL}{2} = \frac{wL^2}{8}$$

In this example, the load curve is a horizontal straight line, the shear curve is an oblique straight line, and the bending-moment curve is a parabola. If the load curve had been an oblique straight line (first degree), the shear curve would have been a parabola (second degree), and the bending-moment curve would have been a cubic (third degree). The equations of the shear and bending-moment curves are always, respectively, one and two degrees higher than the equation of the load curve. Thus, once you have computed a few values of the shear and bending moment, you should be able to sketch the shear and bending-moment diagrams without actually determining the functions $V(x)$ and $M(x)$. The sketches will be more accurate if you make use of the fact that, at any point where the curves are continuous, the slope of the shear curve is equal to $-w$ and the slope of the bending-moment curve is equal to $V$.

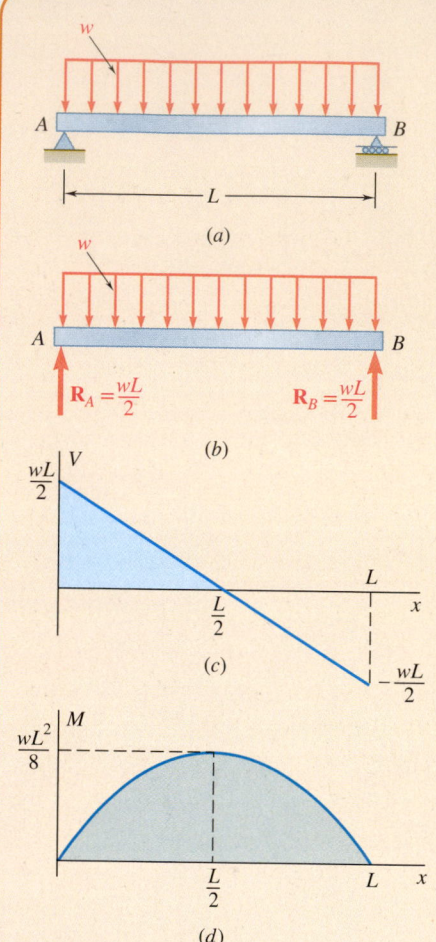

$w$

$A$ — — — $B$

|←——— $L$ ———→|

(a)

$w$

$A$ ———————— $B$

$R_A = \dfrac{wL}{2}$    $R_B = \dfrac{wL}{2}$

(b)

$\dfrac{wL}{2}$ $V$

$\dfrac{L}{2}$    $L$    $x$

$-\dfrac{wL}{2}$

(c)

$\dfrac{wL^2}{8}$ $M$

$\dfrac{L}{2}$    $L$    $x$

(d)

**Fig. 7.12** (a) A simply supported beam carrying a uniformly distributed load; (b) free-body diagram of the beam to determine the reactions at the supports; (c) the shear curve is an oblique straight line; (d) the bending-moment diagram is a parabola.

## Concept Application 7.1

Consider a simply supported beam $AB$ with a span of $L$ carrying a uniformly distributed load $w$ (Fig. 7.12a). From the free-body diagram of the entire beam, we determine the magnitude of the reactions at the supports: $R_A = R_B = wL/2$ (Fig. 7.12b). Then, we draw the shear diagram. Close to end $A$ of the beam, the shear is equal to $R_A$; that is, to $wL/2$, which we can check by considering a very small portion of the beam as a free body. Using formula (7.2), we can then determine the shear $V$ at any distance $x$ from $A$ as

$$V - V_A = -\int_0^x w\,dx = -wx$$

$$V = V_A - wx = \frac{wL}{2} - wx = w\left(\frac{L}{2} - x\right)$$

The shear curve is thus an oblique straight line that crosses the $x$ axis at $x = L/2$ (Fig. 7.12c). Now consider the bending moment. We first observe that $M_A = 0$. The value $M$ of the bending moment at any distance $x$ from $A$, then, can be obtained from Eq. (7.4) as

$$M - M_A = \int_0^x V\,dx$$

$$M = \int_0^x w\left(\frac{L}{2} - x\right)dx = \frac{w}{2}(Lx - x^2)$$

The bending-moment curve is a parabola. The maximum value of the bending moment occurs when $x = L/2$, because $V$ (and thus $dM/dx$) is zero for that value of $x$. Substituting $x = L/2$ in the last equation, we obtain $M_{max} = wL^2/8$.

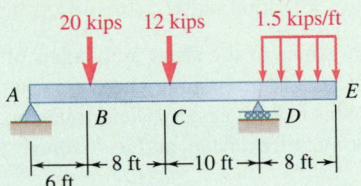

## Sample Problem 7.4

Draw the shear and bending-moment diagrams for the beam and loading shown.

**STRATEGY:**   The beam supports two concentrated loads and one distributed load. You can use the equations in this section between these loads and under the distributed load, but you should expect certain changes in the diagrams at the load points.

**MODELING and ANALYSIS:**

**Free-Body, Entire Beam.**   Consider the entire beam as a free body and determine the reactions (Fig. 1):

$+\circlearrowleft \Sigma M_A = 0$:

$$D(24 \text{ ft}) - (20 \text{ kips})(6 \text{ ft}) - (12 \text{ kips})(14 \text{ ft}) - (12 \text{ kips})(28 \text{ ft}) = 0$$

$$D = +26 \text{ kips} \qquad \mathbf{D} = 26 \text{ kips} \uparrow$$

$+\uparrow \Sigma F_y = 0$:   $\quad A_y - 20 \text{ kips} - 12 \text{ kips} + 26 \text{ kips} - 12 \text{ kips} = 0$

$$A_y = +18 \text{ kips} \qquad \mathbf{A}_y = 18 \text{ kips} \uparrow$$

$\xrightarrow{+} \Sigma F_x = 0$:   $\quad A_x = 0 \qquad\qquad \mathbf{A}_x = 0$

Note that the bending moment is zero at both $A$ and $E$; thus, you know two points (indicated by small circles) on the bending-moment diagram.

**Shear Diagram.**   Because $dV/dx = -w$, the slope of the shear diagram is zero (i.e., the shear is constant between concentrated loads and reactions). To find the shear at any point, divide the beam into two parts and consider either part as a free body. For example, using the portion of the beam to the left of point 1 (Fig. 1), you can obtain the shear between $B$ and $C$:

$+\uparrow \Sigma F_y = 0$:   $\quad + 18 \text{ kips} - 20 \text{ kips} - V = 0 \qquad V = -2 \text{ kips}$

You can also find that the shear is $+12$ kips just to the right of $D$ and zero at end $E$. Because the slope $dV/dx = -w$ is constant between $D$ and $E$, the shear diagram between these two points is a straight line.

**Bending-Moment Diagram.**   Recall that the area under the shear curve between two points is equal to the change in the bending moment between the same two points. For convenience, compute the area of each portion of the shear diagram and indicate it on the diagram (Fig. 1). Because you know the bending moment $M_A$ at the left end is zero, you have

| | |
|---|---|
| $M_B - M_A = +108$ | $M_B = +108 \text{ kip·ft}$ |
| $M_C - M_B = -16$ | $M_C = +92 \text{ kip·ft}$ |
| $M_D - M_C = -140$ | $M_D = -48 \text{ kip·ft}$ |
| $M_E - M_D = +48$ | $M_E = 0$ |

Because you know $M_E$ is zero, this gives you a check of the calculations.

Between the concentrated loads and reactions, the shear is constant; thus, the slope $dM/dx$ is constant. Therefore, you can draw the bending-moment diagram by connecting the known points with straight lines. Between $D$ and

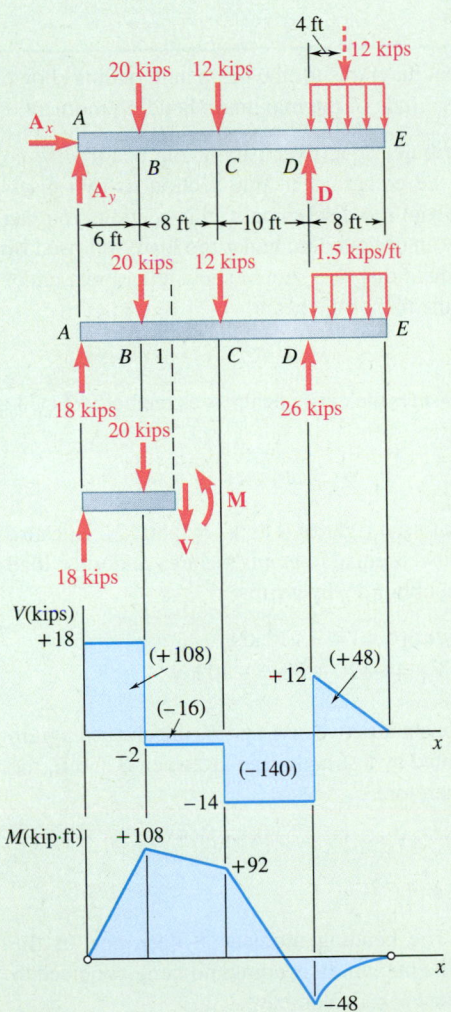

**Fig. 1**   Free-body diagrams of the beam; free-body diagram of the section to the left of the cut; shear diagram; bending-moment diagram.

*(continued)*

$E$, where the shear diagram is an oblique straight line, the bending-moment diagram is a parabola.

From the $V$ and $M$ diagrams, note that $V_{max} = 18$ kips and $M_{max} = 108$ kip·ft.

**REFLECT and THINK:** As expected, the values of the shear and slopes of the bending-moment curves show abrupt changes at the points where concentrated loads act. Useful for design, these diagrams make it easier to determine the maximum values of the shear and bending moment for a beam and its loading.

## Sample Problem 7.5

Draw the shear and bending-moment diagrams for the beam and loading shown and determine the location and magnitude of the maximum bending moment.

**STRATEGY:** The load is a distributed load over part of the beam with no concentrated loads. You can use the equations in this section in two parts: for the load and no load regions. From the discussion in this section, you can expect the shear diagram will show an oblique line under the load, followed by a horizontal line. The bending-moment diagram should show a parabola under the load and an oblique line under the rest of the beam.

**MODELING and ANALYSIS:**

**Free-Body, Entire Beam.** Consider the entire beam as a free body (Fig. 1) to obtain the reactions

$$\mathbf{R}_A = 80 \text{ kN}\uparrow \qquad \mathbf{R}_C = 40 \text{ kN}\uparrow$$

**Shear Diagram.** The shear just to the right of $A$ is $V_A = +80$ kN. Because the change in shear between two points is equal to *minus* the area under the load curve between these points, you can obtain $V_B$ by writing

$$V_B - V_A = -(20 \text{ kN/m})(6 \text{ m}) = -120 \text{ kN}$$
$$V_B = -120 + V_A = -120 + 80 = -40 \text{ kN}$$

Because the slope $dV/dx = -w$ is constant between $A$ and $B$, the shear diagram between these two points is represented by a straight line. Between $B$ and $C$, the area under the load curve is zero; therefore,

$$V_C - V_B = 0 \qquad V_C = V_B = -40 \text{ kN}$$

and the shear is constant between $B$ and $C$ (Fig. 1).

**Bending-Moment Diagram.** The bending moment at each end of the beam is zero. In order to determine the maximum bending moment, you need to locate the section $D$ of the beam where $V = 0$. You have

$$V_D - V_A = -wx$$
$$0 - 80 \text{ kN} = -(20 \text{ kN/m})x$$

Solving for $x$:

$$x = 4 \text{ m} \blacktriangleleft$$

*(continued)*

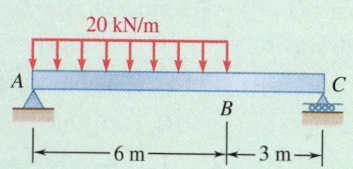

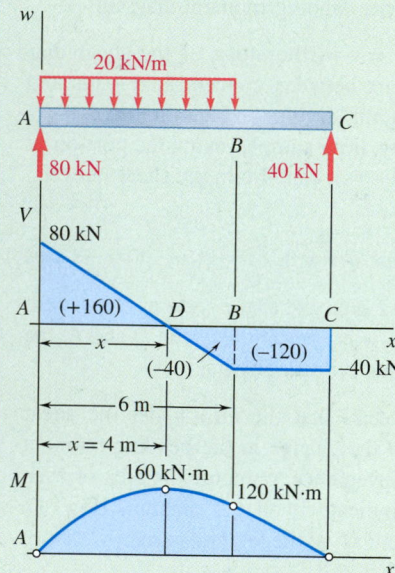

**Fig. 1** Free-body diagram of the beam; shear diagram; bending-moment diagram.

The maximum bending moment occurs at point $D$, where we have $dM/dx = V = 0$. Calculate the areas of the various portions of the shear diagram and mark them (in parentheses) on the diagram (Fig. 1). Because the area of the shear diagram between two points is equal to the change in bending moment between those points, you can write

$$M_D - M_A = +160 \text{ kN·m} \qquad M_D = +160 \text{ kN·m}$$
$$M_B - M_D = -40 \text{ kN·m} \qquad M_B = +120 \text{ kN·m}$$
$$M_C - M_B = -120 \text{ kN·m} \qquad M_C = 0$$

The bending-moment diagram consists of an arc of parabola followed by a segment of straight line; the slope of the parabola at $A$ is equal to the value of $V$ at that point.

The maximum bending moment is

$$M_{\max} = M_D = +160 \text{ kN·m} \quad \blacktriangleleft$$

**REFLECT and THINK:** The analysis conforms to our initial expectations. It is often useful to predict what the results of analysis will be as a way of checking against large-scale errors. However, final results can only depend on detailed modeling and analysis.

## Sample Problem 7.6

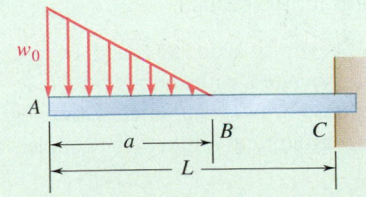

Sketch the shear and bending-moment diagrams for the cantilever beam shown.

**STRATEGY:** Because no support reactions appear until the right end of the beam, you can rely on the equations from this section without needing to use free-body diagrams and equilibrium equations. Due to the nonuniform load, you should expect the results to involve equations of higher degree with a parabolic curve in the shear diagram and a cubic curve in the bending-moment diagram.

**MODELING and ANALYSIS:**

**Shear Diagram.** At the free end of the beam, $V_A = 0$. Between $A$ and $B$, the area under the load curve is $\frac{1}{2} w_0 a$; we find $V_B$ by writing

$$V_B - V_A = -\tfrac{1}{2} w_0 a \qquad V_B = -\tfrac{1}{2} w_0 a$$

Between $B$ and $C$, the beam is not loaded; thus, $V_C = V_B$. At $A$, we have $w = w_0$, and according to Eq. (7.1), the slope of the shear curve is $dV/dx = -w_0$.

*(continued)*

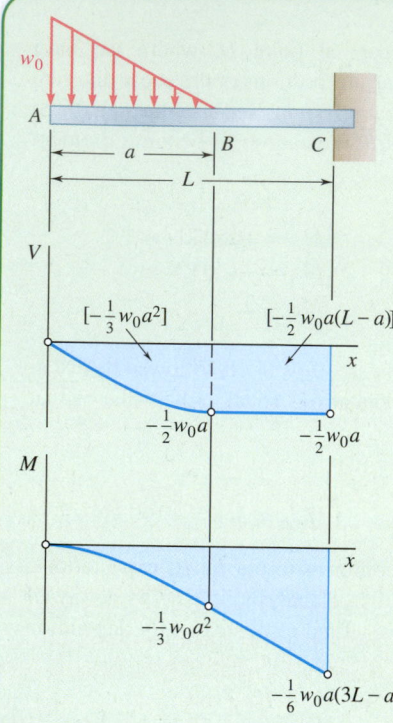

**Fig. 1** Beam with load; shear diagram; bending-moment diagram.

At $B$, the slope is $dV/dx = 0$. Between $A$ and $B$, the loading decreases linearly, and the shear diagram is parabolic (Fig. 1). Between $B$ and $C$, $w = 0$ and the shear diagram is a horizontal line.

**Bending-Moment Diagram.**   Note that $M_A = 0$ at the free end of the beam. You can compute the area under the shear curve, obtaining

$$M_B - M_A = -\tfrac{1}{3}w_0a^2 \qquad M_B = -\tfrac{1}{3}w_0a^2$$
$$M_C - M_B = -\tfrac{1}{2}w_0a(L - a)$$
$$M_C = -\tfrac{1}{6}w_0a(L - a)$$

You can complete the sketch of the bending-moment diagram by recalling that $dM/dx = V$. The result is that between $A$ and $B$ the diagram is represented by a cubic curve with zero slope at $A$, and between $B$ and $C$ the diagram is represented by a straight line.

**REFLECT and THINK:**   Although not strictly required for the solution of this problem, determining the support reactions would serve as an excellent check of the final values of the shear and bending-moment diagrams.

# Sample Problem 7.7

The simple beam $AC$ is loaded by a couple of magnitude $T$ applied at point $B$. Draw the shear and bending-moment diagrams for the beam.

**STRATEGY:**   The load supported by the beam is a concentrated couple. Because the only vertical forces are those associated with the support reactions, you should expect the shear diagram to be of constant value. However, the bending-moment diagram will have a discontinuity at $B$ due to the couple.

**MODELING and ANALYSIS:**

**Free-Body, Entire Beam.**   Consider the entire beam as a free body and determine the reactions:

$$\mathbf{R}_A = \frac{T}{L}\uparrow \qquad \mathbf{R}_B = \frac{T}{L}\downarrow$$

**Shear and Bending-Moment Diagrams (Fig. 1).**   The shear at any section is constant and equal to $T/L$. Because a couple is applied at $B$, the bending-moment diagram is discontinuous at $B$; because the couple is counterclockwise, the bending moment *decreases* suddenly by an amount equal to $T$. You can demonstrate this by taking a section to the immediate right of $B$ and applying equilibrium to solve for the bending moment at this location.

**REFLECT and THINK:**   You can generalize the effect of a couple applied to a beam. At the point where the couple is applied, the bending-moment diagram increases by the value of the couple if it is clockwise and decreases by the value of the couple if it is counterclockwise.

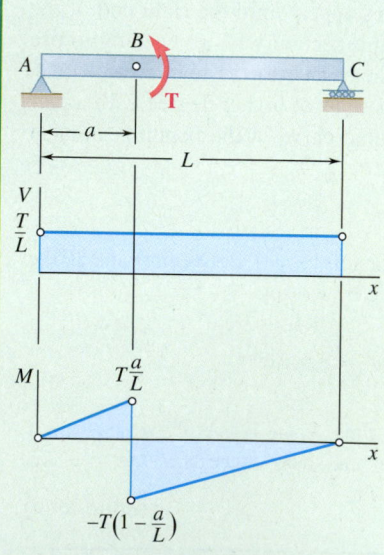

**Fig. 1** Beam with load; shear diagram; bending-moment diagram.

# Case Study 7.1

Using the provisions of Standard ASCE 7 as a guide, Case Study 5.1 modeled a drifting snow load on a flat-roof parking enclosure as shown in CS Fig. 7.1. The roof frame consists of equally spaced beams, and each is supported as shown in CS Fig. 7.2. This figure also shows the corresponding distributed load as obtained in Case Study 5.1, and represents the portion of the total roof snow load acting on any particular beam. Let's develop the shear and bending-moment diagrams for this beam and loading.

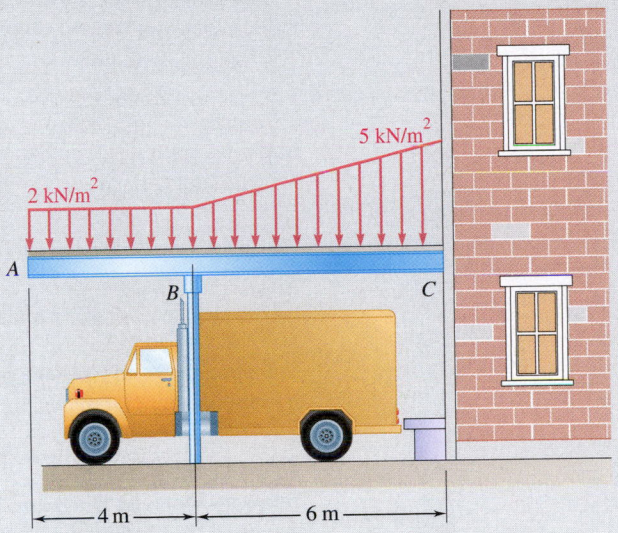

**CS Fig. 7.1**   Enclosed parking structure with the design snowdrift roof load.

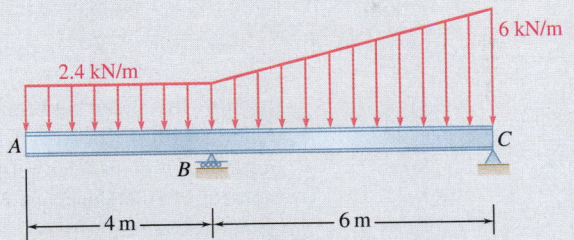

**CS Fig. 7.2**   Single roof beam showing its distributed loading and support conditions.

**STRATEGY:**   The usual first step would be to draw the free-body diagram of the beam and apply equilibrium to obtain the support reactions; this was already completed in Case Study 5.1. You can then use the equations of this section within each of the two regions of continuous load, *AB* and *BC,* to determine the distribution of shear and moment within each region.

**MODELING and ANALYSIS:**

**Free-Body, Entire Beam.** Consider the entire beam as a free body (CS Fig. 7.3) to determine the reactions. This was completed in Case Study 5.1, resulting in

$$\mathbf{R}_B = 23.6 \text{ kN} \uparrow \qquad\qquad \mathbf{R}_C = 11.2 \text{ kN} \uparrow$$

**Shear Diagram.** The shear at $A$ is zero. Because the change in shear between two points is equal to *minus* the area under the load curve between these points, you can obtain $V_B$ by

$$V_B - V_A = -(2.4 \text{ kN/m})(4 \text{ m}) = -9.6 \text{ kN}$$

$$V_B = -9.6 + V_A = -9.6 + 0 = -9.6 \text{ kN}$$

Because the slope $dV/dx = -w$ is constant between $A$ and $B$, the shear diagram is linear between these points. Taking a section to the immediate right of $B$, you can sum forces in the $y$ direction on a free body of portion $AB$ to obtain $V_B = +14$ kN. Using an origin at $A$ as shown in CS Fig. 7.3, the load function between points $B$ and $C$ is

$$w = (0.6x) \text{ kN/m}$$

Applying Eq. (7.2), you can determine the shear $V$ at any distance $x$ between points $B$ and $C$ from

$$V - V_B = -\int_{x_B}^{x} w \, dx = -\int_{4}^{x} (0.6x) dx = -0.3x^2 + 0.3(4)^2 = \left(-0.3x^2 + 4.8\right) \text{ kN}$$

$$V = V_B - 0.3x^2 + 4.8 = 14 - 0.3x^2 + 4.8 = \left(-0.3x^2 + 18.8\right) \text{ kN}$$

The shear curve in this region is thus a parabola, as shown in CS Fig. 7.3. To determine where it crosses the $x$ axis, set this shear expression equal to zero:

$$V = \left(-0.3\,x^2 + 18.8\right) \text{ kN} = 0 \qquad\qquad x = 7.92 \text{ m}$$

**Bending-Moment Diagram.** The bending moment at $A$ is zero. Because the change in the bending moment between two points is equal to the area under the shear curve between these points, you can obtain $M_B$ by

$$M_B - M_A = \tfrac{1}{2}(-9.6 \text{ kN})(4 \text{ m}) = -19.2 \text{ kN·m}$$

$$M_B = -19.2 + M_A = -19.2 + 0 = -19.2 \text{ kN·m}$$

Because the shear between $A$ and $B$ is linear, the bending moment is parabolic (CS Fig. 7.3). Also, because the shear is zero at $A$, the slope of the bending moment is also zero at this point. Using Eq. (7.4), you can determine the moment $M$ at any distance $x$ between points $B$ and $C$ from

$$M - M_B = \int_{x_B}^{x} V \, dx = \int_{4}^{x} (-0.3x^2 + 18.8) \, dx$$

$$= [-0.1x^3 + 18.8x + 0.1(4)^3 - 18.8(4)] \text{ kN·m}$$

$$M = M_B - 0.1x^3 + 18.8x - 68.8 = \left(-19.2 - 0.1x^3 + 18.8x - 68.8\right) \text{ kN·m}$$

$$M = \left(-0.1\,x^3 + 18.8x - 88.0\right) \text{ kN·m}$$

The moment curve in this region is thus a cubic function, as shown in CS Fig. 7.3. The maximum positive value occurs at $x = 7.92$ m, because the shear is zero at this point.

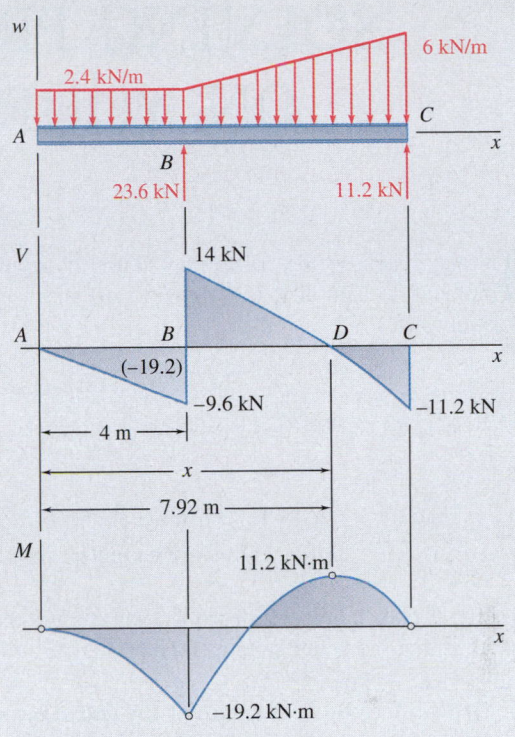

**CS Fig. 7.3** Free-body diagram of the beam; shear diagram; bending-moment diagram.

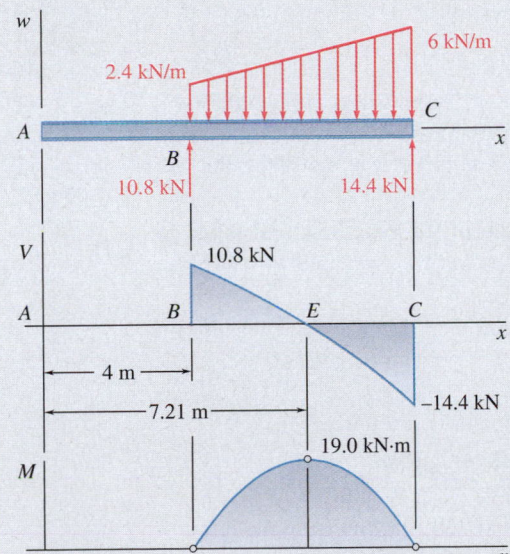

**CS Fig. 7.4** Free-body diagram of the beam; shear diagram; bending-moment diagram for partial loading case.

**REFLECT and THINK:** As noted in Case Study 5.1, Standard ASCE 7 requires the consideration of partial loadings because these could lead to worsened effects for certain parameters in a structure. For instance, in the full design snowdrift load just considered, we see that the maximum absolute shear and bending moment occur at *B*, and are 14 kN and 19.2 kN·m, respectively. In an attempt to protect the roof, suppose snow removal operations are performed, beginning with the snow on the overhang *AB*. For the case where all the snow is removed from portion *AB*, you can use the same methods applied earlier for the original loading to obtain the shear and bending-moment diagrams shown in CS Fig. 7.4. We see that this partial loading results in a larger overall shear (14.4 kN at *C*), and a maximum bending moment (19.0 kN·m within *BC*) that is nearly as great in magnitude as that in the original loading. In addition, the wall reaction at *C* is significantly larger for this second load case, and the beam connection at this location would need to be designed accordingly.

# SOLVING PROBLEMS
## ON YOUR OWN

In this section, we described how to use the relations among the load, shear, and bending moment to simplify the drawing of shear and bending-moment diagrams. These relations are

$$\frac{dV}{dx} = -w \tag{7.1}$$

$$\frac{dM}{dx} = V \tag{7.3}$$

$$V_D - V_C = -(\text{area under load curve between } C \text{ and } D) \tag{7.2'}$$
$$M_D - M_C = -(\text{area under shear curve between } C \text{ and } D) \tag{7.4'}$$

Taking these relations into account, you can use the following procedure to draw the shear and bending-moment diagrams for a beam.

**1. Draw a free-body diagram of the entire beam,** and use it to determine the reactions at the beam supports.

**2. Draw the shear diagram.** This can be done as in the preceding section by cutting the beam at various points and considering the free-body diagram of one of the two resulting portions of the beam (Sample Prob. 7.3). You can, however, consider one of the following alternative procedures.

　　**a. The shear $V$ at any point of the beam is the sum of the reactions and loads to the left of that point;** an upward force is counted as positive, whereas a downward force is counted as negative.

　　**b. For a beam carrying a distributed load,** you can start from a point where you know $V$ and use Eq. (7.2') repeatedly to find $V$ at all other points of interest.

**3. Draw the bending-moment diagram,** using the following procedure.

　　**a. Compute the area under each portion of the shear curve,** assigning a positive sign to areas above the $x$ axis and a negative sign to areas below the $x$ axis.

　　**b. Apply Eq. (7.4') repeatedly** (Sample Probs. 7.4 and 7.5), starting from the left end of the beam, where $M = 0$ (except if a couple is applied at that end, or if the beam is a cantilever beam with a fixed left end).

　　**c. Where a couple is applied to the beam,** be careful to show a discontinuity in the bending-moment diagram by *increasing* the value of $M$ at that point by an amount equal to the magnitude of the couple if the couple is *clockwise,* or *decreasing* the value of $M$ by that amount if the couple is *counterclockwise* (Sample Prob. 7.7).

**4. Determine the location and magnitude of $|M|_{max}$.** The maximum absolute value of the bending moment occurs at one of the points where $dM/dx = 0$ [according to Eq. (7.3), that is at a point where $V$ is equal to zero or changes sign]. You should

    **a. Determine from the shear diagram the value of $|M|$ where $V$ changes sign;** this will occur under a concentrated load (Sample Prob. 7.4).

    **b. Determine the points where $V = 0$ and the corresponding values of $|M|$;** this will occur under a distributed load. To find the distance $x$ between point $C$, where the distributed load starts, and point $D$, where the shear is zero, use Eq. (7.2'). For $V_C$, use the known value of the shear at point $C$; for $V_D$, use zero and express the area under the load curve as a function of $x$ (Sample Prob. 7.5).

**5. You can improve the quality of your drawings** by keeping in mind that, at any given point according to Eqs. (7.1) and (7.3), the slope of the $V$ curve is equal to $-w$ and the slope of the $M$ curve is equal to $V$.

**6. Finally, for beams supporting a distributed load expressed as a function $w(x)$,** remember that you can obtain the shear $V$ by integrating the function $-w(x)$, and you can obtain the bending moment $M$ by integrating $V(x)$ [Eqs. (7.2) and (7.4)].

# Problems

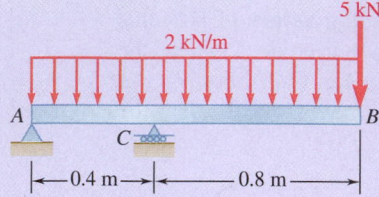

Fig. P7.69

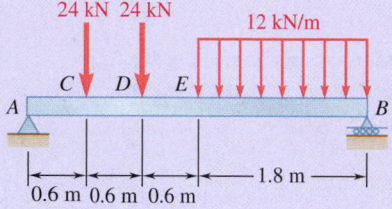

Fig. P7.70

**7.63** Using the method of Sec. 7.3, solve Prob. 7.29.

**7.64** Using the method of Sec. 7.3, solve Prob. 7.30.

**7.65** Using the method of Sec. 7.3, solve Prob. 7.31.

**7.66** Using the method of Sec. 7.3, solve Prob. 7.32.

**7.67** Using the method of Sec. 7.3, solve Prob. 7.33.

**7.68** Using the method of Sec. 7.3, solve Prob. 7.34.

**7.69 and 7.70** For the beam and loading shown, (*a*) draw the shear and bending-moment diagrams, (*b*) determine the maximum absolute values of the shear and bending moment.

**7.71** Using the method of Sec. 7.3, solve Prob. 7.39.

**7.72** Using the method of Sec. 7.3, solve Prob. 7.40.

**7.73** Using the method of Sec. 7.3, solve Prob. 7.41.

**7.74** Using the method of Sec. 7.3, solve Prob. 7.42.

**7.75 and 7.76** For the beam and loading shown, (*a*) draw the shear and bending-moment diagrams, (*b*) determine the maximum absolute values of the shear and bending moment.

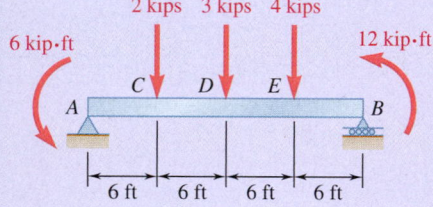

Fig. *P7.75*

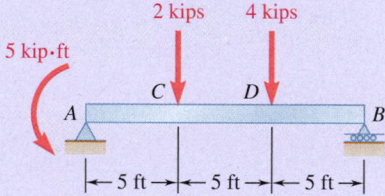

Fig. *P7.76*

**7.77 and 7.78** For the beam and loading shown, (*a*) draw the shear and bending-moment diagrams, (*b*) determine the magnitude and location of the maximum absolute value of the bending moment.

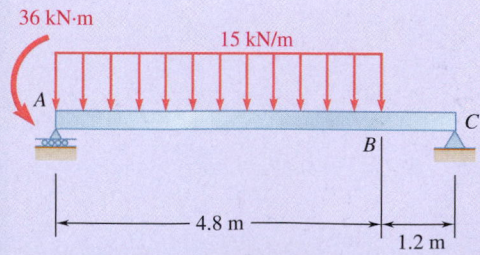

Fig. P7.77

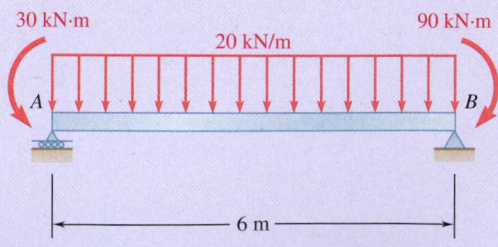

Fig. P7.78

**7.79 and 7.80** For the beam and loading shown, (*a*) draw the shear and bending-moment diagrams, (*b*) determine the magnitude and location of the maximum absolute value of the bending moment.

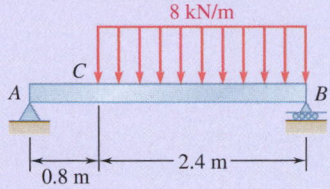

Fig. P7.79

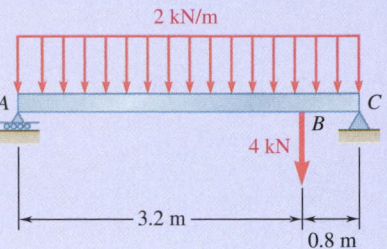

Fig. P7.80

**7.81 and 7.82** For the beam and loading shown, (*a*) draw the shear and bending-moment diagrams, (*b*) determine the magnitude and location of the maximum absolute value of the bending moment.

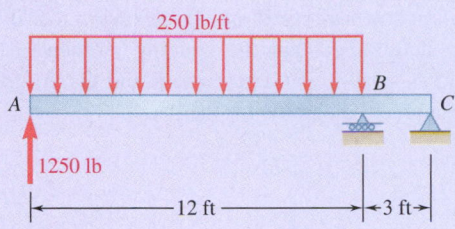

Fig. P7.81

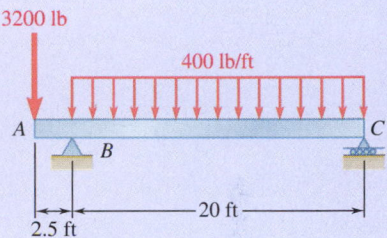

Fig. P7.82

**7.83** (*a*) Draw the shear and bending-moment diagrams for beam *AB*, (*b*) determine the magnitude and location of the maximum absolute value of the bending moment.

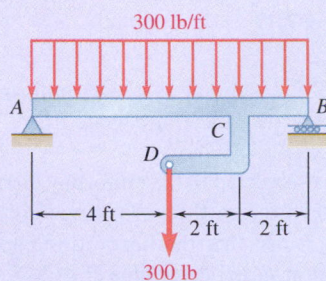

Fig. *P7.83*

**7.84** Solve Prob. 7.83 assuming that the 300-lb force applied at *D* is directed upward.

**7.85 and 7.86** For the beam and loading shown, (*a*) write the equations of the shear and bending-moment curves, (*b*) determine the magnitude and location of the maximum bending moment.

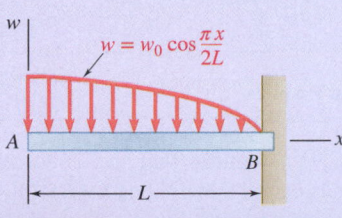

Fig. *P7.85*

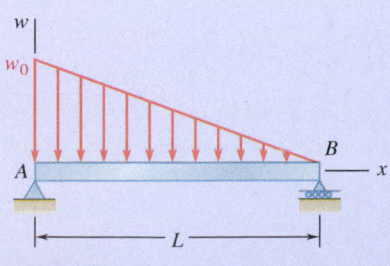

Fig. P7.86

**7.87 and *7.88*** For the beam and loading shown, (*a*) write the equations of the shear and bending-moment curves, (*b*) determine the magnitude and location of the maximum bending moment.

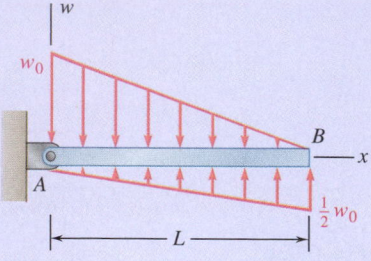

Fig. P7.87

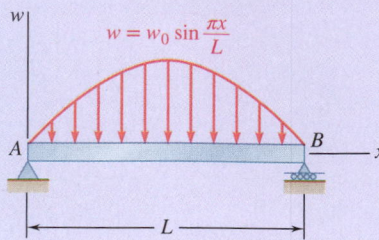

Fig. *P7.88*

**7.89** The beam *AB* supports the uniformly distributed load of 1000 N/m and two unknown forces **P** and **Q**. Knowing that it has been experimentally determined that the bending moment is −395 N·m at *A* and −215 N·m at *C*, (*a*) determine **P** and **Q**, (*b*) draw the shear and bending-moment diagrams for the beam.

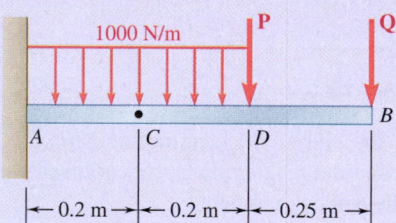

Fig. P7.89

**7.90** Solve Prob. 7.89 assuming that the uniformly distributed load of 1000 N/m extends over the entire beam *AB*.

***7.91** The beam *AB* is subjected to the uniformly distributed load of 2 kips/ft and two unknown forces **P** and **Q**. Knowing that it has been experimentally determined that the bending moment is +172 kip·ft at *D* and +235 kip·ft at *E*, (*a*) determine **P** and **Q**, (*b*) draw the shear and bending-moment diagrams for the beam.

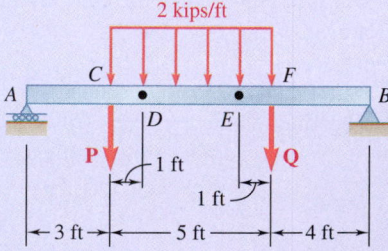

Fig. P7.91

***7.92** Solve Prob. 7.91 assuming that the uniformly distributed load of 2 kips/ft extends over the entire beam *AB*.

# *7.4 CABLES

Cables are used in many engineering applications, such as suspension bridges, transmission lines, aerial tramways, guy wires for high towers, etc. Cables may be divided into two categories, according to their loading: (1) supporting concentrated loads and (2) supporting distributed loads.

## 7.4A Cables with Concentrated Loads

Consider a cable attached to two fixed points $A$ and $B$ and supporting $n$ vertical concentrated loads $\mathbf{P}_1, \mathbf{P}_2, \ldots, \mathbf{P}_n$ (Fig. 7.13a). We assume that the cable is *flexible*, i.e., that its resistance to bending is small and can be neglected. We further assume that the *weight of the cable is negligible* compared with the loads supported by the cable. We can therefore approximate any portion of cable between successive loads as a two-force member. Thus, the internal forces at any point in the cable reduce to a *force of tension directed along the cable*.

We assume that each of the loads lies in a given vertical line, i.e., that the horizontal distance from support $A$ to each of the loads is known. We also assume that we know the horizontal and vertical distances between the supports. With these assumptions, we want to determine the shape of the cable (i.e., the vertical distance from support $A$ to each of the points $C_1, C_2, \ldots, C_n$) and also the tension $T$ in each portion of the cable.

We first draw the free-body diagram of the entire cable (Fig. 7.13b). Because we do not know the slopes of the portions of cable attached at $A$ and $B$, we represent the reactions at $A$ and $B$ by two components each. Thus, four unknowns are involved, and the three equations of equilibrium are not sufficient to determine the reactions. (A cable is not a rigid body; thus, the equilibrium equations represent *necessary but not sufficient conditions*. See Sec. 6.3B.) We must therefore obtain an additional equation by considering the equilibrium of a portion of the cable. This is possible if we know the coordinates $x$ and $y$ of a point $D$ of the cable.

We draw the free-body diagram of the portion of cable $AD$ (Fig. 7.14a). From the equilibrium condition $\Sigma M_D = 0$, we obtain an additional relation between the scalar components $A_x$ and $A_y$ and can determine the reactions at $A$ and $B$. However, the problem remains indeterminate if we do not know the coordinates of $D$ unless we are given some other relation between $A_x$ and $A_y$

**Photo 7.3** The weight of the chairlift cables is negligible compared to the weights of the chairs and skiers, so we can use the methods of this section to determine the force at any point in the cable. ©Bob Edme/AP Photo

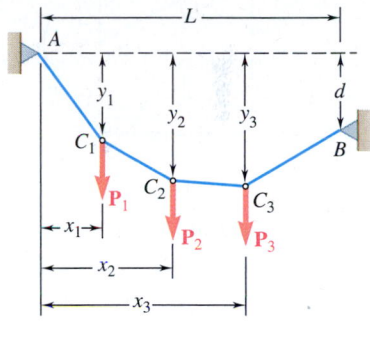
(a)

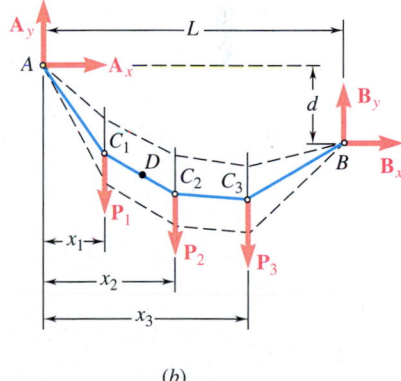
(b)

**Fig. 7.13** (a) A cable supporting vertical concentrated loads; (b) free-body diagram of the entire cable.

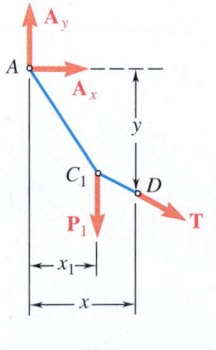
(a)

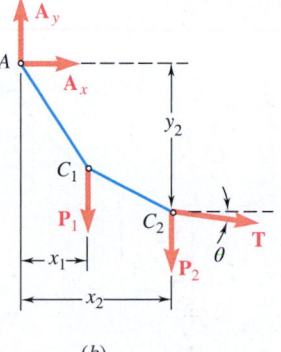
(b)

**Fig. 7.14** (a) Free-body diagram of the portion of cable $AD$; (b) free-body diagram of the portion of cable $AC_2$.

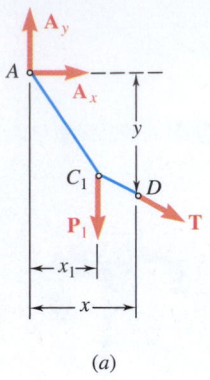

(a)

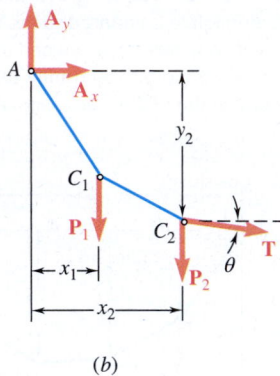

(b)

**Fig. 7.14** (repeated)

(or between $B_x$ and $B_y$). The cable might hang in any of various possible ways, as indicated by the dashed lines in Fig. 7.13b.

Once we have determined $A_x$ and $A_y$, we can find the vertical distance from $A$ to any point of the cable. Considering point $C_2$, for example, we draw the free-body diagram of the portion of cable $AC_2$ (Fig. 7.14b). From $\Sigma F_{C2} = 0$, we obtain an equation that we can solve for $y_2$. From $\Sigma F_x = 0$ and $\Sigma F_y = 0$, we obtain the components of force $\mathbf{T}$ representing the tension in the portion of cable to the right of $C_2$. Note that $T\cos\theta = -A_x$; that is, *the horizontal component of the tension force is the same at any point of the cable*. It follows that the tension $T$ is maximum when $\cos\theta$ is minimum, i.e., in the portion of cable that has the largest angle of inclination $\theta$. This portion of cable must be adjacent to one of the two supports of the cable.

## 7.4B Cables with Distributed Loads

Consider a cable attached to two fixed points $A$ and $B$ and carrying a *distributed load* (Fig. 7.15a). We just saw that for a cable supporting concentrated loads, the internal force at any point is a force of tension directed along the cable. By contrast, in the case of a cable carrying a distributed load, the cable hangs in the shape of a curve, and the internal force at a point $D$ is a force of tension $\mathbf{T}$ *directed along the tangent to the curve*. Here, we examine how to determine the tension at any point of a cable supporting a given distributed load. In the following sections, we will determine the shape of the cable for two common types of distributed loads.

Considering the most general case of distributed load, we draw the free-body diagram of the portion of cable extending from the lowest point $C$ to a given point $D$ of the cable (Fig. 7.15b). The three forces acting on the free body are the tension force $\mathbf{T}_0$ at $C$, which is horizontal; the tension force $\mathbf{T}$ at $D$, which is directed along the tangent to the cable at $D$; and the resultant $\mathbf{W}$ of the distributed load supported by the portion of cable $CD$. Drawing the corresponding force triangle (Fig. 7.15c), we obtain the relations

$$T\cos\theta = T_0 \qquad T\sin\theta = w \qquad (7.5)$$

$$T = \sqrt{T_0^2 + w^2} \qquad \tan\theta = \frac{w}{T_0} \qquad (7.6)$$

From the relations in Eqs. (7.5), we see that the horizontal component of the tension force $\mathbf{T}$ is the same at any point. Furthermore, the vertical component of $\mathbf{T}$ at any point is equal to the magnitude $W$ of the load when measured from the lowest point ($C$) to the point in question ($D$). Relations in Eqs. (7.6) show that the tension $T$ is minimum at the lowest point and maximum at one of the two support points.

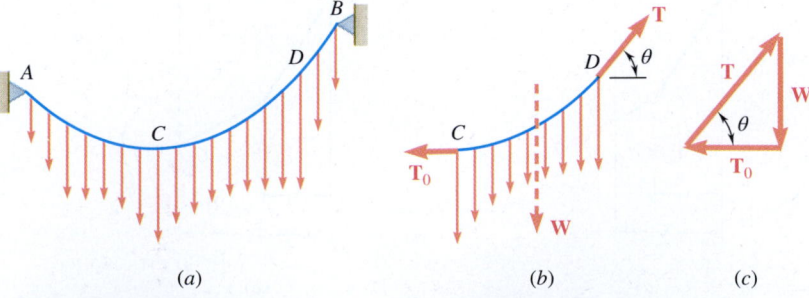

(a)                    (b)              (c)

**Fig. 7.15** (a) A cable carrying a distributed load; (b) free-body diagram of the portion of the cable $CD$; (c) force triangle for the free-body diagram in part (b).

## 7.4C Parabolic Cables

Now suppose that cable *AB* carries a load *uniformly distributed along the horizontal* (Fig. 7.16*a*). We can approximate the load on the cables of a suspension bridge in this way, because the weight of the cables is small compared with the uniform weight of the roadway. We denote the load per unit length by *w* (*measured horizontally*) and express it in N/m or lb/ft. Choosing coordinate axes with the origin at the lowest point *C* of the cable, we find that the magnitude *W* of the total load carried by the portion of cable extending from *C* to the point *D* with coordinates *x* and *y* is $W = wx$. The relations in Eqs. (7.6) defining the magnitude and direction of the tension force at *D* become

$$T = \sqrt{T_0^2 + w^2 x^2} \qquad \tan \theta = \frac{wx}{T_0} \qquad \textbf{(7.7)}$$

Moreover, the distance from *D* to the line of action of the resultant **W** is equal to half of the horizontal distance from *C* to *D* (Fig. 7.16*b*). Summing moments about *D,* we have

$$+\circlearrowleft \Sigma M_D = 0: \quad wx\frac{x}{2} - T_0 y = 0$$

Solving for *y,* we have

**Equation of parabolic cable**

$$y = \frac{wx^2}{2T_0} \qquad \textbf{(7.8)}$$

This is the equation of a *parabola* with a vertical axis and its vertex at the origin of coordinates. Thus, the curve formed by cables loaded uniformly along the horizontal is a parabola.[‡]

When the supports *A* and *B* of the cable have the same elevation, the distance *L* between the supports is called the **span** of the cable and the vertical distance *h* from the supports to the lowest point is called the **sag** of the cable (Fig. 7.17*a*). If you know the span and sag of a cable and if the load *w* per unit horizontal length is given, you can find the minimum tension $T_0$ by substituting $x = L/2$ and $y = h$ in Eq. (7.8). Eqs. (7.7) then yield the tension and the slope at any point of the cable, and Eq. (7.8) defines the shape of the cable.

When the supports have different elevations, the position of the lowest point of the cable is not known, and we must determine the coordinates $x_A$, $y_A$ and $x_B$, $y_B$ of the supports. To do this, we note that the coordinates of *A* and *B* satisfy Eq. (7.8) and that

$$x_B - x_A = L \quad \text{and} \quad y_B - y_A = d$$

where *L* and *d* denote, respectively, the horizontal and vertical distances between the two supports (Fig. 7.17*b* and *c*).

We can obtain the length of the cable from its lowest point *C* to its support *B* from the formula

$$s_B = \int_0^{x_B} \sqrt{1 + \left(\frac{dy}{dx}\right)^2}\, dx \qquad \textbf{(7.9)}$$

**Photo 7.4** The main cables of suspension bridges, like the Golden Gate Bridge above, may be assumed to carry a loading that is uniformly distributed along the horizontal. ©Thinkstock/Getty Images RF

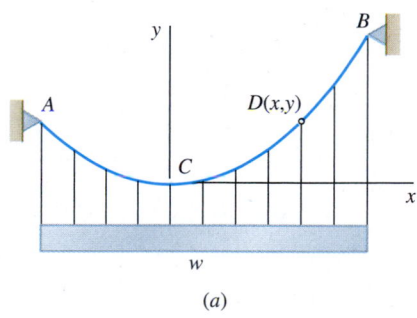

(*a*)

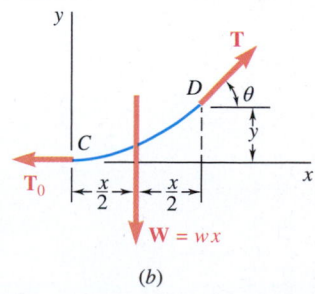

(*b*)

**Fig. 7.16** (*a*) A cable carrying a uniformly distributed load along the horizontal; (*b*) free-body diagram of the portion of cable *CD*.

---

[‡]Cables hanging under their own weight are not loaded uniformly along the horizontal and do not form parabolas. However, the error introduced by assuming a parabolic shape for cables hanging under their own weight is small when the cable is sufficiently taut. In the next section, we give a complete discussion of cables hanging under their own weight.

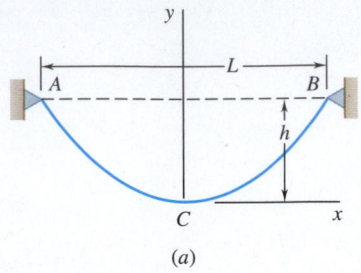
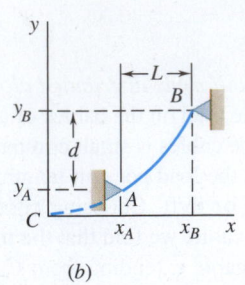
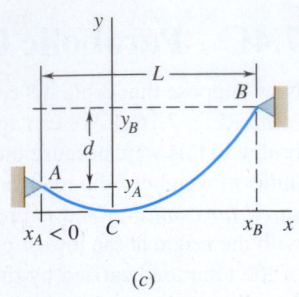

(a)  (b)  (c)

**Fig. 7.17**  (a) The shape of a parabolic cable is determined by its span $L$ and sag $h$; (b,c) the span and vertical distance between supports for cables with supports at different elevations.

Differentiating Eq. (7.8), we obtain the derivative $dy/dx = wx/T_0$. Substituting this into Eq. (7.9) and using the binomial theorem to expand the radical in an infinite series, we have

$$s_B = \int_0^{x_B} \sqrt{1 + \frac{w^2 x^2}{T_0^2}}\, dx = \int_0^{x_B} \left(1 + \frac{w^2 x^2}{2T_0^2} - \frac{w^4 x^4}{8T_0^4} + \cdots\right) dx$$

$$s_B = x_B \left(1 + \frac{w^2 x_B^2}{6T_0^2} - \frac{w^4 x_B^4}{40T_0^4} + \cdots\right)$$

Then, because $wx_B^2/2T_0 = y_B$, we obtain

$$s_B = x_B \left[1 + \frac{2}{3}\left(\frac{y_B}{x_B}\right)^2 - \frac{2}{5}\left(\frac{y_B}{x_B}\right)^4 + \cdots\right] \tag{7.10}$$

This series converges for values of the ratio $y_B/x_B$ less than 0.5. In most cases, this ratio is much smaller, and only the first two terms of the series need be computed.

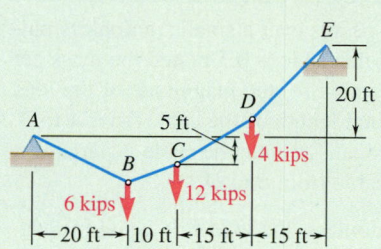

# Sample Problem 7.8

The cable $AE$ supports three vertical loads from the points indicated. If point $C$ is 5 ft below the left support, determine (a) the elevation of points $B$ and $D$, (b) the maximum slope and the maximum tension in the cable.

**STRATEGY:** To solve for the support reactions at $A$, consider a free-body diagram of the entire cable, as well as one that takes a section at $C$, because you know the coordinates of this point. Taking subsequent sections at $B$ and $D$ will then enable you to determine their elevations. The resulting cable geometry establishes the maximum slope, which is where the maximum tension in the cable occurs.

## MODELING and ANALYSIS:

**Free Body, Entire Cable.** Determine the reaction components $\mathbf{A}_x$ and $\mathbf{A}_y$ as

$+\circlearrowleft \Sigma M_E = 0$:

$A_x(20\text{ ft}) - A_y(60\text{ ft}) + (6\text{ kips})(40\text{ ft}) + (12\text{ kips})(30\text{ ft}) + (4\text{ kips})(15\text{ ft}) = 0$

$$20A_x - 60A_y + 660 = 0$$

**Free Body, ABC.** Consider the portion $ABC$ of the cable as a free body (Fig. 1). Then you have

$+\circlearrowleft \Sigma M_C = 0$: $\qquad -A_x(5\text{ ft}) - A_y(30\text{ ft}) + (6\text{ kips})(10\text{ ft}) = 0$

$$-5A_x - 30A_y + 60 = 0$$

Solving the two equations simultaneously, you obtain

$$A_x = -18\text{ kips} \qquad \mathbf{A}_x = 18\text{ kips} \leftarrow$$
$$A_y = +5\text{ kips} \qquad \mathbf{A}_y = 5\text{ kips} \uparrow$$

### a. Elevation of Points B and D:

**Free Body, AB.** Considering the portion of cable $AB$ as a free body, you obtain

$+\circlearrowleft \Sigma M_B = 0$: $\qquad (18\text{ kips})y_B - (5\text{ kips})(20\text{ ft}) = 0$

$$y_B = 5.56\text{ ft below } A \quad \blacktriangleleft$$

**Free Body, ABCD.** Using the portion of cable $ABCD$ as a free body gives you

$+\circlearrowleft \Sigma M_D = 0$:

$-(18\text{ kips})y_D - (5\text{ kips})(45\text{ ft}) + (6\text{ kips})(25\text{ ft}) + (12\text{ kips})(15\text{ ft}) = 0$

$$y_D = 5.83\text{ ft above } A \quad \blacktriangleleft$$

### b. Maximum Slope and Maximum Tension.

Note that the maximum slope occurs in portion $DE$. Because the horizontal component of the tension is constant and equal to 18 kips, you have

$$\tan\theta = \frac{14.17}{15\text{ ft}} \qquad\qquad \theta = 43.4° \quad \blacktriangleleft$$

$$T_{max} = \frac{18\text{ kips}}{\cos\theta} \qquad\qquad T_{max} = 24.8\text{ kips} \quad \blacktriangleleft$$

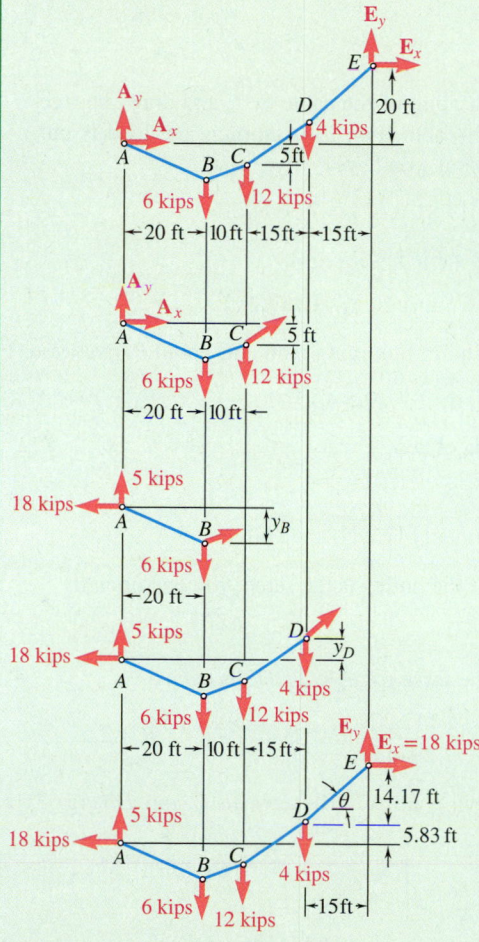

**Fig. 1** Free-body diagrams of the cable system.

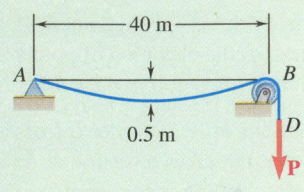

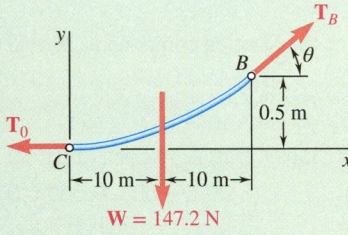

**Fig. 1** Free-body diagram of cable portion *CB*.

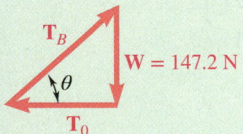

**Fig. 2** Force triangle for cable portion *CB*.

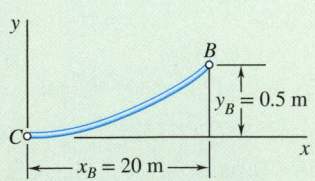

**Fig. 3** Dimensions used to determine the length of the cable.

## Sample Problem 7.9

A light cable is attached to a support at *A*, passes over a small frictionless pulley at *B*, and supports a load **P**. The sag of the cable is 0.5 m and the mass per unit length of the cable is 0.75 kg/m. Determine (*a*) the magnitude of the load **P**, (*b*) the slope of the cable at *B*, (*c*) the total length of the cable from *A* to *B*. Because the ratio of the sag to the span is small, assume the cable is parabolic. Also, neglect the weight of the portion of cable from *B* to *D*.

**STRATEGY:** Because the pulley is frictionless, the load **P** is equal in magnitude to the tension in the cable at *B*. You can determine the tension using the methods of this section and then use that value to determine the slope and length of the cable.

**MODELING and ANALYSIS:**

**a. Load P.** Denote the lowest point of the cable by *C* and draw the free-body diagram of the portion *CB* of cable (Fig. 1). Assuming the load is uniformly distributed along the horizontal, you have

$$w = (0.75 \text{ kg/m})(9.81 \text{ m/s}^2) = 7.36 \text{ N/m}$$

The total load for the portion *CB* of cable is

$$W = wx_B = (7.36 \text{ N/m})(20 \text{ m}) = 147.2 \text{ N}$$

This load acts halfway between *C* and *B*. Summing moments about *B* gives you

$$+\circlearrowleft \Sigma M_B = 0: \qquad (147.2 \text{ N})(10 \text{ m}) - T_0(0.5 \text{ m}) = 0 \qquad T_0 = 2944 \text{ N}$$

From the force triangle (Fig. 2), you obtain

$$T_B = \sqrt{T_0^2 + W^2}$$
$$= \sqrt{(2944 \text{ N})^2 + (147.2 \text{ N})^2} = 2948 \text{ N}$$

Because the tension on each side of the pulley is the same, you end up with

$$P = T_B = 2948 \text{ N} \quad \blacktriangleleft$$

**b. Slope of Cable at *B*.** The force triangle also tells us that

$$\tan \theta = \frac{W}{T_0} = \frac{147.2 \text{ N}}{2944 \text{ N}} = 0.05 \qquad \theta = 2.9° \quad \blacktriangleleft$$

**c. Length of Cable.** Applying Eq. (7.10) between *C* and *B* (Fig. 3) gives you

$$s_B = x_B \left[ 1 + \frac{2}{3}\left(\frac{y_B}{x_B}\right)^2 + \cdots \right]$$
$$= (20 \text{ m}) \left[ 1 + \frac{2}{3}\left(\frac{0.5 \text{ m}}{20 \text{ m}}\right)^2 + \cdots \right] = 20.00833 \text{ m}$$

The total length of the cable between *A* and *B* is twice this value. Thus,

$$\text{Length} = 2s_B = 40.0167 \text{ m} \quad \blacktriangleleft$$

**REFLECT and THINK:** Notice that the length of the cable is only very slightly more than the length of the span between *A* and *B*. This means that the cable must be very taut, which is consistent with the relatively large value of load **P** (compared to the weight of the cable).

# SOLVING PROBLEMS
# ON YOUR OWN

In the problems of this section, you will apply the equations of equilibrium to *cables that lie in a vertical plane*. We assume that a cable cannot resist bending, so the force of tension in the cable is always directed along the cable.

**A.** In the first part of this lesson, we considered **cables subjected to concentrated loads.** Because we assume the weight of the cable is negligible, the cable is straight between loads.

Your solution will consist of the following steps.

**1. Draw a free-body diagram of the entire cable** showing the loads and the horizontal and vertical components of the reaction at each support. Use this free-body diagram to write the corresponding equilibrium equations.

**2. You will have four unknown components and only three equations of equilibrium** (see Fig. 7.13). You must therefore find an additional piece of information, such as the *position* of a point on the cable or the *slope* of the cable at a given point.

**3. After you have identified the point of the cable where the additional information exists,** cut the cable at that point, and draw a free-body diagram of one of the two resulting portions of the cable.

    **a. If you know the position** of the point where you have cut the cable, set $\Sigma M = 0$ about that point for the new free body. This will yield the additional equation required to solve for the four unknown components of the reactions (Sample Prob. 7.8).

    **b. If you know the slope** of the portion of the cable you have cut, set $\Sigma F_x = 0$ and $\Sigma F_y = 0$ for the new free body. This will yield two equilibrium equations that, together with the original three, you can solve for the four reaction components and for the tension in the cable where it has been cut.

**4. To find the elevation of a given point of the cable and the slope and tension at that point** once you have found the reactions at the supports, you should cut the cable at that point and draw a free-body diagram of one of the two resulting portions of the cable. Setting $\Sigma M = 0$ about the given point yields its elevation. Writing $\Sigma F_x = 0$ and $\Sigma F_y = 0$ yields the components of the tension force from which you can find its magnitude and direction.

**5. For a cable supporting vertical loads only,** *the horizontal component of the tension force is the same at any point.* It follows that, for such a cable, the *maximum tension occurs in the steepest portion of the cable.*

*(continued)*

**B.** In the second portion of this section, we considered **cables carrying a load that is uniformly distributed along the horizontal.** The shape of the cable is then parabolic.

Your solution will use one or more of the following concepts.

**1. Place the origin of coordinates at the lowest point of the cable** and direct the $x$ and $y$ axes to the right and upward, respectively. Then *the equation of the parabola* is

$$y = \frac{wx^2}{2T_0} \tag{7.8}$$

The minimum cable tension occurs at the origin, where the cable is horizontal. The maximum tension is at the support where the slope is maximum.

**2. If the supports of the cable have the same elevation,** the sag $h$ of the cable is the vertical distance from the lowest point of the cable to the horizontal line joining the supports. To solve a problem involving such a parabolic cable, use Eq. (7.8) for one of the supports; this equation can be solved for one unknown.

**3. If the supports of the cable have different elevations,** you will have to write Eq. (7.8) for each of the supports (see Fig. 7.17).

**4. To find the length of the cable** from the lowest point to one of the supports, you can use Eq. (7.10). In most cases, you will need to compute only the first two terms of the series.

# Problems

**7.93** Three loads are suspended as shown from the cable *ABCDE*. Knowing that $d_C = 4$ m, determine (*a*) the components of the reaction at *E*, (*b*) the maximum tension in the cable.

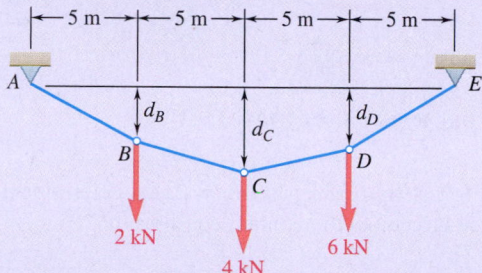

**Fig. P7.93 and P7.94**

**7.94** Knowing that the maximum tension in cable *ABCDE* is 25 kN, determine the distance $d_C$.

**7.95** If $d_A = 8$ ft and $d_C = 10$ ft, determine the components of the reaction at *E*.

**7.96** If $d_A = d_C = 6$ ft, determine (*a*) the components of the reaction at *E*, (*b*) the maximum tension in the cable.

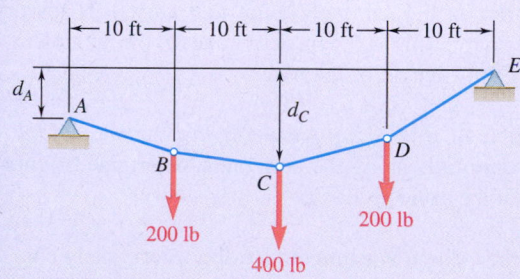

**Fig. P7.95 and P7.96**

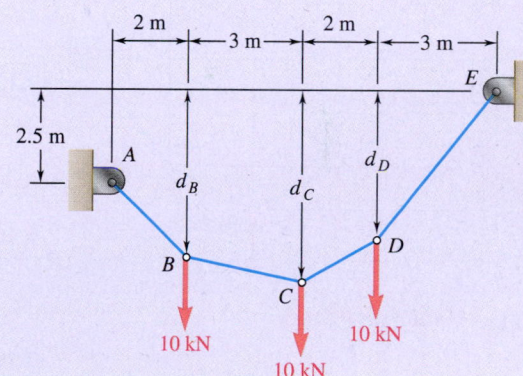

**Fig. P7.97 and P7.98**

**7.97** Knowing that $d_C = 5$ m, determine (*a*) the distances $d_B$ and $d_D$, (*b*) the maximum tension in the cable.

**7.98** Determine (*a*) distance $d_C$ for which portion *BC* of the cable is horizontal, (*b*) the corresponding components of the reaction at *E*.

**7.99** Knowing that $d_C = 9$ ft, determine (*a*) the distances $d_B$ and $d_D$, (*b*) the reaction at *E*.

**7.100** Determine (*a*) the distance $d_C$ for which portion *DE* of the cable is horizontal, (*b*) the corresponding reactions at *A* and *E*.

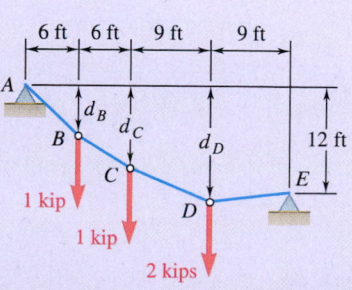

**Fig. P7.99 and P7.100**

415

**7.101** Knowing that $m_B = 70$ kg and $m_C = 25$ kg, determine the magnitude of the force **P** required to maintain equilibrium.

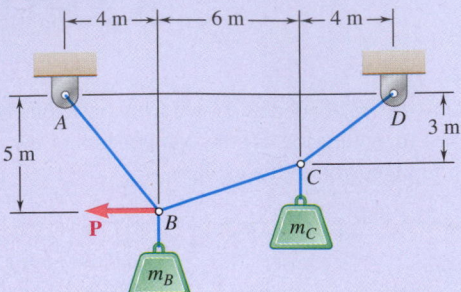

Fig. P7.101 and P7.102

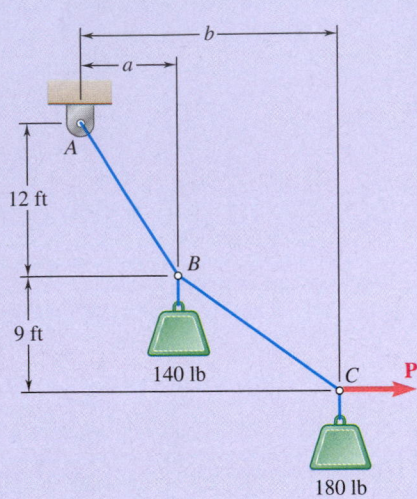

Fig. P7.103 and P7.104

**7.102** Knowing that $m_B = 18$ kg and $m_c = 10$ kg, determine the magnitude of the force **P** required to maintain equilibrium.

**7.103** Cable *ABC* supports two loads as shown. Knowing that $b = 21$ ft, determine (*a*) the required magnitude of the horizontal force **P**, (*b*) the corresponding distance *a*.

**7.104** Cable *ABC* supports two loads as shown. Determine the distances *a* and *b* when a horizontal force **P** of magnitude 200 lb is applied at *C*.

**7.105** If $a = 3$ m, determine the magnitudes of **P** and **Q** required to maintain the cable in the shape shown.

**7.106** If $a = 4$ m, determine the magnitudes of **P** and **Q** required to maintain the cable in the shape shown.

**7.107** An electric wire having a mass per unit length of 0.6 kg/m is strung between two insulators at the same elevation that are 60 m apart. Knowing that the sag of the wire is 1.5 m, determine (*a*) the maximum tension in the wire, (*b*) the length of the wire.

**7.108** The total mass of cable *ACB* is 20 kg. Assuming that the mass of the cable is distributed uniformly along the horizontal, determine (*a*) the sag *h*, (*b*) the slope of the cable at *A*.

**7.109** The center span of the George Washington Bridge, as originally constructed, consisted of a uniform roadway suspended from four cables. The uniform load supported by each cable was $w = 9.75$ kips/ft along the horizontal. Knowing that the span *L* is 3500 ft and that the sag *h* is 316 ft, determine for the original configuration (*a*) the maximum tension in each cable, (*b*) the length of each cable.

**7.110** The center span of the Verrazano-Narrows Bridge consists of two uniform roadways suspended from four cables. The design of the bridge allows for the effect of extreme temperature changes that cause the sag of the center span to vary from $h_w = 386$ ft in winter to $h_s = 394$ ft in summer. Knowing that the span is $L = 4260$ ft, determine the change in length of the cables due to extreme temperature changes.

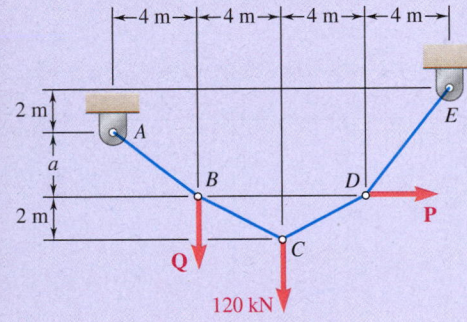

Fig. P7.105 and P7.106

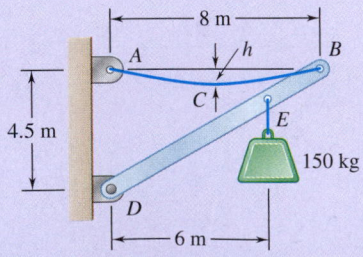

Fig. P7.108

**7.111** Each cable of the Golden Gate Bridge supports a load $w = 11.1$ kips/ft along the horizontal. Knowing that the span $L$ is 4150 ft and that the sag $h$ is 464 ft, determine (*a*) the maximum tension in each cable, (*b*) the length of each cable.

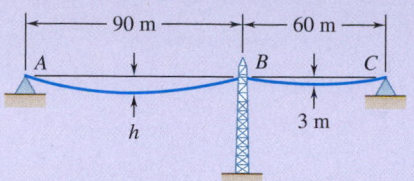

**7.112** Two cables of the same gauge are attached to a transmission tower at $B$. Because the tower is slender, the horizontal component of the resultant of the forces exerted by the cables at $B$ is to be zero. Knowing that the mass per unit length of the cables is 0.4 kg/m, determine (*a*) the required sag $h$, (*b*) the maximum tension in each cable.

**Fig. P7.112**

**7.113** A 76-m length of wire having a mass per unit length of 2.2 kg/m is used to span a horizontal distance of 75 m. Determine (*a*) the approximate sag of the wire, (*b*) the maximum tension in the wire. (*Hint:* Use only the first two terms of Eq. (7.10).)

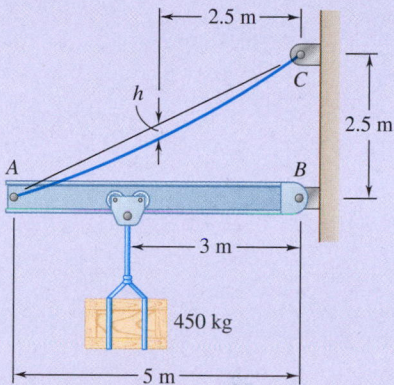

**7.114** A cable of length $L + \Delta$ is suspended between two points that are at the same elevation and a distance $L$ apart. (*a*) Assuming that $\Delta$ is small compared to $L$ and that the cable is parabolic, determine the approximate sag in terms of $L$ and $\Delta$. (*b*) If $L = 100$ ft and $\Delta = 4$ ft, determine the approximate sag. (*Hint:* Use only the first two terms of Eq. (7.10).)

**7.115** The total mass of cable $AC$ is 25 kg. Assuming that the mass of the cable is distributed uniformly along the horizontal, determine the sag $h$ and the slope of the cable at $A$ and $C$.

**Fig. P7.115**

**7.116** Cable $ACB$ supports a load uniformly distributed along the horizontal as shown. The lowest point $C$ is located 9 m to the right of $A$. Determine (*a*) the vertical distance $a$, (*b*) the length of the cable, (*c*) the components of the reaction at $A$.

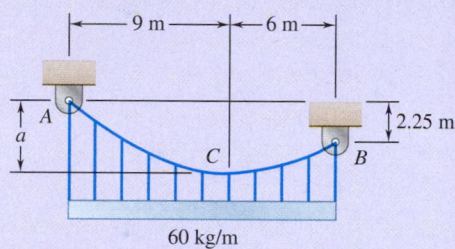

**7.117** Each cable of the side spans of the Golden Gate Bridge supports a load $w = 10.2$ kips/ft along the horizontal. Knowing that for the side spans the maximum vertical distance $h$ from each cable to the chord $AB$ is 30 ft and occurs at midspan, determine (*a*) the maximum tension in each cable, (*b*) the slope at $B$.

**Fig. P7.116**

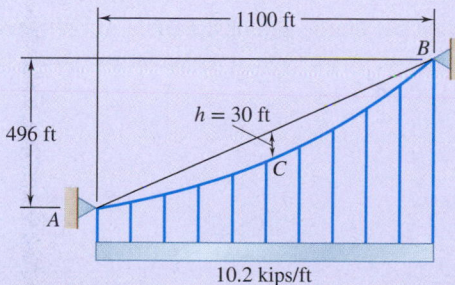

**Fig. P7.117**

**7.118** A steam pipe weighing 45 lb/ft that passes between two buildings 40 ft apart is supported by a system of cables as shown. Assuming that the weight of the cable system is equivalent to a uniformly distributed loading of 5 lb/ft, determine (*a*) the location of the lowest point $C$ of the cable, (*b*) the maximum tension in the cable.

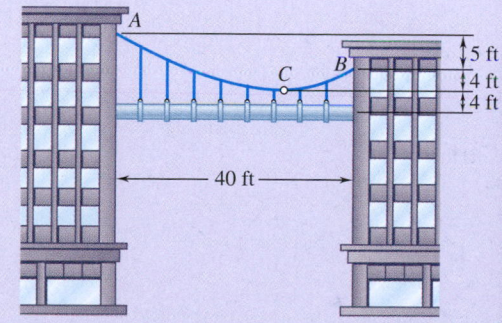

**Fig. P7.118**

*7.119 A cable $AB$ of span $L$ and a simple beam $A'B'$ of the same span are subjected to identical vertical loadings as shown. Show that the magnitude of the bending moment at a point $C'$ in the beam is equal to the product $T_0h$, where $T_0$ is the magnitude of the horizontal component of the tension force in the cable and $h$ is the vertical distance between point $C$ and the chord joining the points of support $A$ and $B$.

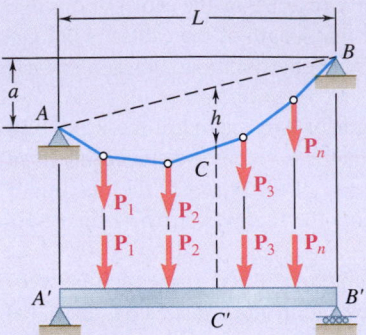

Fig. P7.119

7.120 through 7.123 Making use of the property established in Prob. 7.119, solve the problem indicated by first solving the corresponding beam problem.

7.120   Prob. 7.94.

7.121   Prob. 7.97a.

7.122   Prob. 7.99a.

7.123   Prob. 7.100a.

*7.124 Show that the curve assumed by a cable that carries a distributed load $w(x)$ is defined by the differential equation $d^2y/dx^2 = w(x)/T_0$, where $T_0$ is the tension at the lowest point.

*7.125 Using the property indicated in Prob. 7.124, determine the curve assumed by a cable of span $L$ and sag $h$ carrying a distributed load $w = w_0 \cos(\pi x/L)$, where $x$ is measured from midspan. Also determine the maximum and minimum values of the tension in the cable.

*7.126 If the weight per unit length of the cable $AB$ is $w_0/\cos^2\theta$, prove that the curve formed by the cable is a circular arc. (Hint: Use the property indicated in Prob. 7.124.)

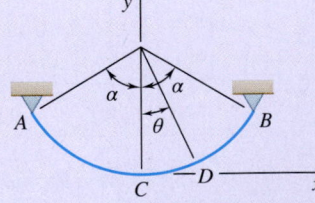

Fig. P7.126

# *7.5   CATENARY CABLES

Let us now consider a cable *AB* carrying a load that is **uniformly distributed along the cable itself** (Fig. 7.18*a*). Cables hanging under their own weight are loaded in this way. We denote the load per unit length by *w* (*measured along the cable*) and express it in N/m or lb/ft. The magnitude *W* of the total load carried by a portion of cable with a length of *s*, extending from the lowest point *C*

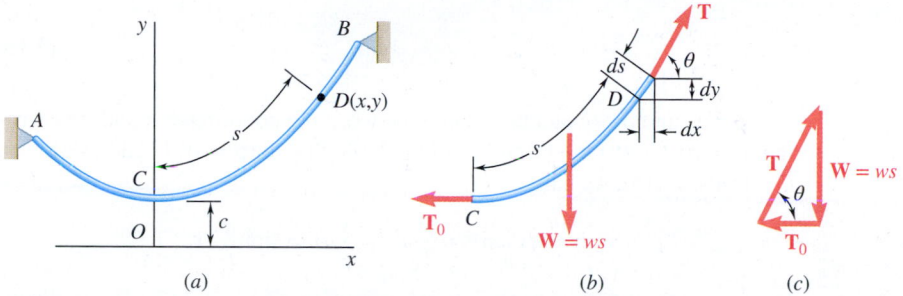

**Fig. 7.18**   (*a*) A cable carrying a load uniformly distributed along the cable; (*b*) free-body diagram of a portion of the cable *CD*; (*c*) force triangle for part (*b*).

to some point *D*, is $W = ws$. Substituting this value for *W* in formula (7.6), we obtain the tension at *D*, as

$$T = \sqrt{T_0^2 + w^2 s^2}$$

In order to simplify the subsequent computations, we introduce the constant $c = T_0/w$. This gives us

$$T_0 = wc \qquad W = ws \qquad T = w\sqrt{c^2 + s^2} \qquad \textbf{(7.11)}$$

The free-body diagram of the portion of cable *CD* is shown in Fig. 7.18*b*. However, we cannot use this diagram directly to obtain the equation of the curve assumed by the cable, because we do not know the horizontal distance from *D* to the line of action of the resultant **W** of the load. To obtain this equation, we note that the horizontal projection of a small element of cable of length *ds* is $dx = ds \cos\theta$. Observing from Fig. 7.18*c* that $\cos\theta = T_0/T$ and using Eq. (7.11), we have

$$dx = ds \cos\theta = \frac{T_0}{T} ds = \frac{wc\, ds}{w\sqrt{c^2 + s^2}} = \frac{ds}{\sqrt{1 + s^2/c^2}}$$

(*a*) High-voltage power lines

(*b*) A spider's web

(*c*) The Gateway Arch

**Photo 7.5**   Catenary cables occur in nature, as well as in engineered structures. (*a*) High-voltage power lines, common all across the country and in much of the world, support only their own weight. (*b*) Catenary cables can be as delicate as the silk threads of a spider's web. (*c*) The Gateway to the West Arch in St. Louis is an inverted catenary arch cast in concrete (which is in compression instead of tension). (a) ©Ingram Publishing/Newscom; (b) ©Karl Weatherly/Getty Images RF; (c) ©Eric Nathan/Alamy

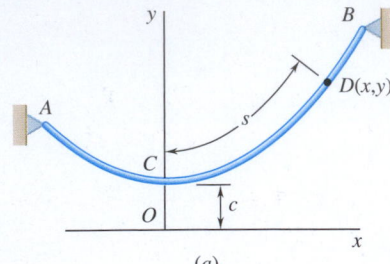

(a)

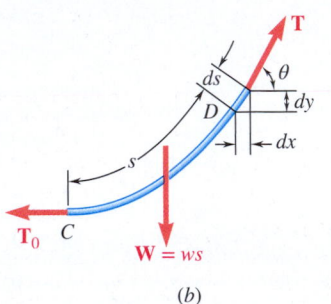

(b)

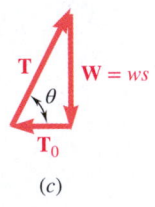

(c)

**Fig. 7.18**   *(repeated)*

Selecting the origin $O$ of the coordinates at a distance $c$ directly below $C$ (Fig. 7.18a) and integrating from $C(0, c)$ to $D(x, y)$, we obtain[†]

$$x = \int_0^s \frac{ds}{\sqrt{1 + s^2/c^2}} = c\left[\sinh^{-1}\frac{s}{c}\right]_0^s = c\,\sinh^{-1}\frac{s}{c}$$

This equation, which relates the length $s$ of the portion of cable $CD$ and the horizontal distance $x$, can be written in the form

**Length of catenary cable**

$$s = c\,\sinh\frac{x}{c} \tag{7.15}$$

We can now obtain the relation between the coordinates $x$ and $y$ by writing $dy = dx\,\tan\theta$. Observing from Fig. 7.18c that $\tan\theta = W/T_0$ and using Eqs. (7.11) and (7.15), we have

$$dy = dx\,\tan\theta = \frac{W}{T_0}dx = \frac{s}{c}dx = \sinh\frac{x}{c}dx$$

Integrating from $C(0, c)$ to $D(x, y)$ and using Eqs. (7.12) and (7.13), we obtain

$$y - c = \int_0^x \sinh\frac{x}{c}dx = c\left[\cosh\frac{x}{c}\right]_0^x = c\left(\cosh\frac{x}{c} - 1\right)$$

$$y - c = \cosh\frac{x}{c} - c$$

which reduces to

**Equation of catenary cable**

$$y = c\,\cosh\frac{x}{c} \tag{7.16}$$

This is the equation of a **catenary** with a vertical axis. The ordinate $c$ of the lowest point $C$ is called the *parameter* of the catenary. By squaring both sides of Eqs. (7.15) and (7.16), subtracting, and taking Eq. (7.14) into account, we obtain the following relation between $y$ and $s$:

$$y^2 - s^2 = c^2 \tag{7.17}$$

---

[†]This integral appears in all standard integral tables. The function

$$z = \sinh^{-1}u$$

(read "arc hyperbolic sine $u$") is the inverse of the function $u = \sinh z$ (read "hyperbolic sine $z$"). This function and the function $v = \cosh z$ (read "hyperbolic cosine $z$") are defined as

$$u = \sinh z = \frac{1}{2}(e^z - e^{-z}) \qquad v = \cosh z = \frac{1}{2}(e^z + e^{-z})$$

Numerical values of the functions $\sinh z$ and $\cosh z$ are listed in tables of hyperbolic functions and also may be computed on most calculators, either directly or from the definitions. Refer to any calculus text for a complete description of the properties of these functions. In this section, we use only the following properties, which are easy to derive from the definitions:

$$\frac{d\sinh z}{dz} = \cosh z \qquad \frac{d\cosh z}{dz} = \sinh z \tag{7.12}$$

$$\sinh 0 = 0 \qquad \cosh 0 = 1 \tag{7.13}$$

$$\cosh^2 z - \sinh^2 z = 1 \tag{7.14}$$

Solving Eq. (7.17) for $s^2$ and carrying into the last of the relations in Eqs. (7.11), we write these relations as

$$T_0 = wc \qquad W = ws \qquad T = wy \qquad (7.18)$$

The last relation indicates that the tension at any point $D$ of the cable is proportional to the vertical distance from $D$ to the horizontal line representing the $x$ axis.

When the supports $A$ and $B$ of the cable have the same elevation, the distance $L$ between the supports is called the *span* of the cable and the vertical distance $h$ from the supports to the lowest point $C$ is called the *sag* of the cable. These definitions are the same as those given for parabolic cables; note that, because of our choice of coordinate axes, the sag $h$ is now

$$h = y_A - c \qquad (7.19)$$

Also note that some catenary problems involve transcendental equations, which must be solved by successive approximations (see Sample Prob. 7.10). When the cable is fairly taut, however, we can assume that the load is uniformly distributed *along the horizontal* and replace the catenary by a parabola. This greatly simplifies the solution of the problem, and the error introduced is small.

When the supports $A$ and $B$ have different elevations, the position of the lowest point of the cable is not known. We can then solve the problem in a manner similar to that indicated for parabolic cables by noting that the cable must pass through the supports and that $x_B - x_A = L$ and $y_B - y_A = d$, where $L$ and $d$ denote, respectively, the horizontal and vertical distances between the two supports.

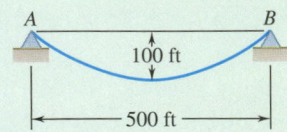

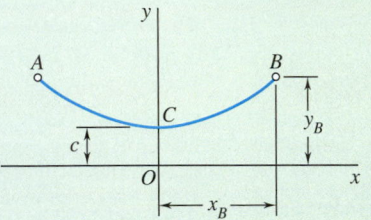

**Fig. 1** Cable geometry.

## Sample Problem 7.10

A uniform cable weighing 3 lb/ft is suspended between two points $A$ and $B$ as shown. Determine (*a*) the maximum and minimum values of the tension in the cable, (*b*) the length of the cable.

**STRATEGY:** This is a cable carrying only its own weight that is supported by its ends at the same elevation. You can use the analysis in this section to solve the problem.

**MODELING and ANALYSIS:**

**Equation of Cable.** Place the origin of coordinates at a distance $c$ below the lowest point of the cable (Fig. 1). The equation of the cable is given by Eq. (7.16), as

$$y = c \cosh \frac{x}{c}$$

The coordinates of point $B$ are

$$x_B = 250 \text{ ft} \qquad y_B = 100 + c$$

Substituting these coordinates into the equation of the cable, you obtain

$$100 + c = c \cosh \frac{250}{c}$$

$$\frac{100}{c} + 1 = \cosh \frac{250}{c}$$

Determine the value of $c$ by substituting successive trial values, as shown in the following table.

| $c$ | $\dfrac{250}{c}$ | $\dfrac{100}{c}$ | $\dfrac{100}{c} + 1$ | $\cosh \dfrac{250}{c}$ |
|-----|------------------|------------------|----------------------|------------------------|
| 300 | 0.833 | 0.333 | 1.333 | 1.367 |
| 350 | 0.714 | 0.286 | 1.286 | 1.266 |
| 330 | 0.758 | 0.303 | 1.303 | 1.301 |
| 328 | 0.762 | 0.305 | 1.305 | 1.305 |

Taking $c = 328$, you have

$$y_B = 100 + c = 428 \text{ ft}$$

**a. Maximum and Minimum Values of the Tension.** Using Eqs. (7.18), you obtain

$$T_{\min} = T_0 = wc = (3 \text{ lb/ft})(328 \text{ ft}) \qquad T_{\min} = 984 \text{ lb} \blacktriangleleft$$
$$T_{\max} = T_B = wy_B = (3 \text{ lb/ft})(428 \text{ ft}) \qquad T_{\max} = 1284 \text{ lb} \blacktriangleleft$$

**b. Length of Cable.** You can find one-half of the length of the cable by solving Eq. (7.17). Hence,

$$y_B^2 - s_{CB}^2 = c^2 \quad s_{CB}^2 = y_B^2 - c^2 = (428)^2 - (328)^2 \qquad s_{CB} = 275 \text{ ft}$$

The total length of the cable is therefore

$$s_{AB} = 2s_{CB} = 2(275 \text{ ft}) \qquad s_{AB} = 550 \text{ ft} \blacktriangleleft$$

**REFLECT and THINK:** The sag in the cable is one-fifth of the cable's span, so it is not very taut. The weight of the cable is $ws = (3 \text{ lb/ft})(550 \text{ ft}) = 1650 \text{ lb}$, while its maximum tension is only 1284 lb. This demonstrates that the total weight of a cable can exceed its maximum tension.

# SOLVING PROBLEMS
# ON YOUR OWN

In the last section of this chapter, we described how to solve problems involving a *cable carrying a load uniformly distributed along the cable.* The shape assumed by the cable is a catenary and is defined by

$$y = c \cosh \frac{x}{c} \qquad (7.16)$$

**1. Keep in mind that the origin of coordinates for a catenary is located at a distance $c$ directly below its lowest point.** The length of the cable from the origin to any point is expressed as

$$s = c \sinh \frac{x}{c} \qquad (7.15)$$

**2. You should first identify all of the known and unknown quantities.** Then, consider each of the equations listed in the text [Eqs. (7.15), (7.16), (7.17), (7.18), and (7.19)] and solve an equation that contains only one unknown. Substitute the value found into another equation, and solve that equation for another unknown.

**3. If the sag $h$ is given,** use Eq. (7.19) to replace $y$ by $h + c$ in Eq. (7.16) if $x$ is known (Sample Prob. 7.10) or in Eq. (7.17) if $s$ is known, and solve the resulting equation for the constant $c$.

**4. Many of the problems you will encounter will involve the solution by trial and error** of an equation involving a hyperbolic sine or cosine. You can make your work easier by keeping track of your calculations in a table, as in Sample Prob. 7.10, or by applying a numerical methods approach using a computer or calculator.

# Problems

**7.127** A 25-ft chain with a weight of 30 lb is suspended between two points at the same elevation. Knowing that the sag is 10 ft, determine (*a*) the distance between the supports, (*b*) the maximum tension in the chain.

**7.128** A 500-ft-long aerial tramway cable having a weight per unit length of 2.8 lb/ft is suspended between two points at the same elevation. Knowing that the sag is 125 ft, find (*a*) the horizontal distance between the supports, (*b*) the maximum tension in the cable.

**7.129** A 40-m cable is strung as shown between two buildings. The maximum tension is found to be 350 N, and the lowest point of the cable is observed to be 6 m above the ground. Determine (*a*) the horizontal distance between the buildings, (*b*) the total mass of the cable.

**7.130** A 50-m steel surveying tape has a mass of 1.6 kg. If the tape is stretched between two points at the same elevation and pulled until the tension at each end is 60 N, determine the horizontal distance between the ends of the tape. Neglect the elongation of the tape due to the tension.

**7.131** A 20-m length of wire having a mass per unit length of 0.2 kg/m is attached to a fixed support at *A* and to a collar at *B*. Neglecting the effect of friction, determine (*a*) the force **P** for which $h = 8$ m, (*b*) the corresponding span *L*.

**7.132** A 20-m length of wire having a mass per unit length of 0.2 kg/m is attached to a fixed support at *A* and to a collar at *B*. Knowing that the magnitude of the horizontal force applied to the collar is $P = 20$ N, determine (*a*) the sag *h*, (*b*) the span *L*.

**7.133** A 20-m length of wire having a mass per unit length of 0.2 kg/m is attached to a fixed support at *A* and to a collar at *B*. Neglecting the effect of friction, determine (*a*) the sag *h* for which $L = 15$ m, (*b*) the corresponding force **P**.

**7.134** Determine the sag of a 30-ft chain that is attached to two points at the same elevation that are 20 ft apart.

**7.135** A counterweight *D* is attached to a cable that passes over a small pulley at *A* and is attached to a support at *B*. Knowing that $L = 45$ ft and $h = 15$ ft, determine (*a*) the length of the cable from *A* to *B*, (*b*) the weight per unit length of the cable. Neglect the weight of the cable from *A* to *D*.

**7.136** A 90-m wire is suspended between two points at the same elevation that are 60 m apart. Knowing that the maximum tension is 300 N, determine (*a*) the sag of the wire, (*b*) the total mass of the wire.

**7.137** A cable weighing 2 lb/ft is suspended between two points at the same elevation that are 160 ft apart. Determine the smallest allowable sag of the cable if the maximum tension is not to exceed 400 lb.

**7.138** A uniform cord 50 in. long passes over a pulley at *B* and is attached to a pin support at *A*. Knowing that $L = 20$ in. and neglecting the effect of friction, determine the smaller of the two values of *h* for which the cord is in equilibrium.

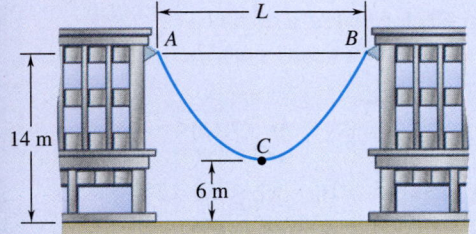

**Fig. P7.129**

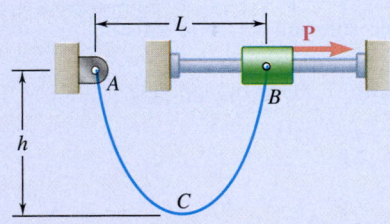

**Fig. P7.131, P7.132, and P7.133**

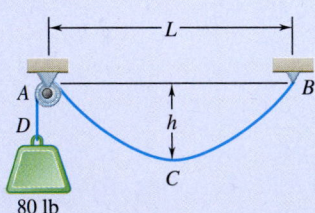

80 lb

**Fig. P7.135**

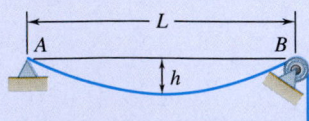

**Fig. P7.138**

**7.139** A motor $M$ is used to slowly reel in the cable shown. Knowing that the mass per unit length of the cable is 0.4 kg/m, determine the maximum tension in the cable when $h = 5$ m.

**7.140** A motor $M$ is used to slowly reel in the cable shown. Knowing that the mass per unit length of the cable is 0.4 kg/m, determine the maximum tension in the cable when $h = 3$ m.

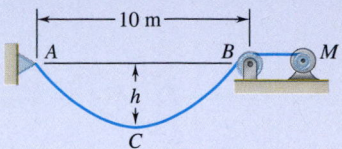

**Fig. P7.139 and P7.140**

**7.141** The cable $ACB$ has a mass per unit length of 0.45 kg/m. Knowing that the lowest point of the cable is located at a distance $a = 0.6$ m below the support $A$, determine (a) the location of the lowest point $C$, (b) the maximum tension in the cable.

**7.142** The cable $ACB$ has a mass per unit length of 0.45 kg/m. Knowing that the lowest point of the cable is located at a distance $a = 2$ m below the support $A$, determine (a) the location of the lowest point $C$, (b) the maximum tension in the cable.

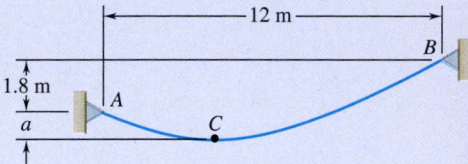

**Fig. P7.141 and P7.142**

**7.143** A uniform cable weighing 3 lb/ft is held in the position shown by a horizontal force $\mathbf{P}$ applied at $B$. Knowing that $P = 180$ lb and $\theta_A = 60°$, determine (a) the location of point $B$, (b) the length of the cable.

**7.144** A uniform cable weighing 3 lb/ft is held in the position shown by a horizontal force $\mathbf{P}$ applied at $B$. Knowing that $P = 150$ lb and $\theta_A = 60°$, determine (a) the location of point $B$, (b) the length of the cable.

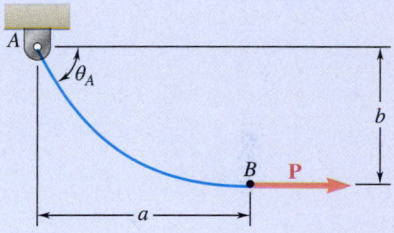

**Fig. P7.143 and P7.144**

**7.145** To the left of point $B$, the long cable $ABDE$ rests on the rough horizontal surface shown. Knowing that the mass per unit length of the cable is 2 kg/m, determine the force $\mathbf{F}$ when $a = 3.6$ m.

**7.146** To the left of point $B$, the long cable $ABDE$ rests on the rough horizontal surface shown. Knowing that the mass per unit length of the cable is 2 kg/m, determine the force $\mathbf{F}$ when $a = 6$ m.

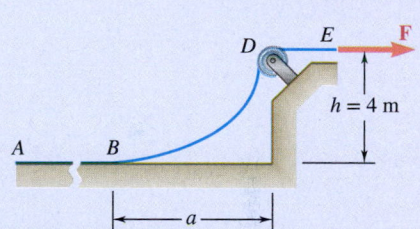

**Fig. P7.145 and P7.146**

**\*7.147** The 10-ft cable $AB$ is attached to two collars as shown. The collar at $A$ can slide freely along the rod; a stop attached to the rod prevents the collar at $B$ from moving on the rod. Neglecting the effect of friction and the weight of the collars, determine the distance $a$.

**\*7.148** Solve Prob. 7.147 assuming that the angle $\theta$ formed by the rod and the horizontal is 45°.

**7.149** Denoting the angle formed by a uniform cable and the horizontal by $\theta$, show that at any point (a) $s = c \tan \theta$, (b) $y = c \sec \theta$.

**\*7.150** (a) Determine the maximum allowable horizontal span for a uniform cable with a weight per unit length of $w$ if the tension in the cable is not to exceed a given value $T_m$. (b) Using the result of part $a$, determine the maximum span of a steel wire for which $w = 0.25$ lb/ft and $T_m = 8000$ lb.

**\*7.151** A cable has a mass per unit length of 3 kg/m and is supported as shown. Knowing that the span $L$ is 6 m, determine the *two* values of the sag $h$ for which the maximum tension is 350 N.

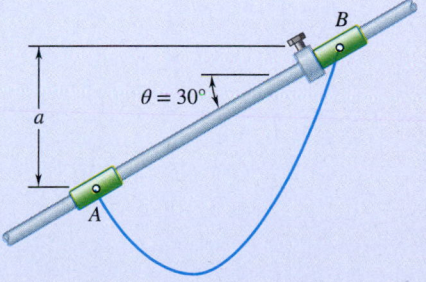

**Fig. P7.147**

**\*7.152** Determine the sag-to-span ratio for which the maximum tension in the cable is equal to the total weight of the entire cable $AB$.

**\*7.153** A cable with a weight per unit length of $w$ is suspended between two points at the same elevation that are a distance $L$ apart. Determine (a) the sag-to-span ratio for which the maximum tension is as small as possible, (b) the corresponding values of $\theta_B$ and $T_m$.

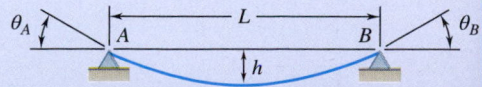

**Fig. P7.151, P7.152, and P7.153**

# Review and Summary

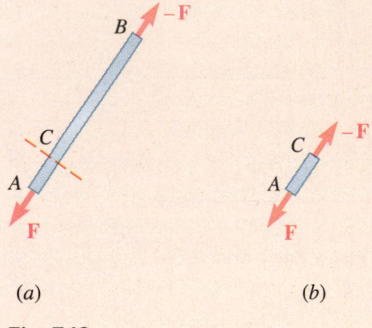

**Fig. 7.19**

In this chapter, you learned to determine the internal forces that hold together the various parts of a given member in a structure.

## Forces in Straight Two-Force Members

Considering first a **straight two-force member** $AB$ (Sec. 7.1), recall that such a member is subjected at $A$ and $B$ to equal and opposite forces $\mathbf{F}$ and $-\mathbf{F}$ directed along $AB$ (Fig. 7.19$a$). Cutting member $AB$ at $C$ and drawing the free-body diagram of portion $AC$, we concluded that the internal forces existing at $C$ in member $AB$ are equivalent to an **axial force** $-\mathbf{F}$ equal and opposite to $\mathbf{F}$ (Fig. 7.19$b$). Note that, in the case of a two-force member that is not straight, the internal forces reduce to a force-couple system and not to a single force.

## Forces in Multi-Force Members

Consider next a **multi-force member** $AD$ (Fig. 7.20$a$). Cutting it at $J$ and drawing the free-body diagram of portion $JD$, we concluded that the internal forces at $J$ are equivalent to a force-couple system consisting of the **axial force** $\mathbf{F}$, the **shearing force** $\mathbf{V}$, and a couple $\mathbf{M}$ (Fig. 7.20$b$). The magnitude of the shearing force measures the **shear** at point $J$, and the moment of the couple is referred to as the **bending moment** at $J$. Because an equal and opposite force-couple system is obtained by considering the free-body diagram of portion $AJ$, it is necessary to specify which portion of member $AD$ is used when recording the answers (Sample Prob. 7.1).

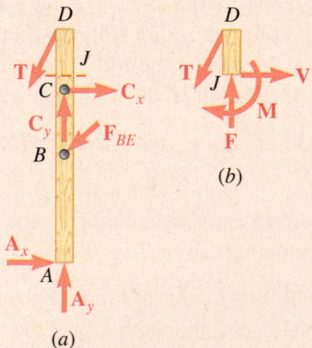

**Fig. 7.20**

## Forces in Beams

Most of the chapter was devoted to the analysis of the internal forces in two important types of engineering structures: beams and cables. **Beams** are usually long, straight prismatic members designed to support loads applied at various points along the member. In general, the loads are perpendicular to the axis of the beam and produce only shear and bending in the beam. The loads may be either **concentrated** at specific points or **distributed** along the entire length or a portion of the beam. The beam itself may be supported in various ways; because only statically determinate beams are considered in this text, we limited our analysis to that of simply supported beams, overhanging beams, and cantilever beams (Sec. 7.2).

## Shear and Bending Moment in a Beam

To obtain the shear $V$ and bending moment $M$ at a given point $C$ of a beam, we first determine the reactions at the supports by considering the entire beam as a free body. We then cut the beam at $C$ and use the free-body diagram of one of the two resulting portions to determine $V$ and $M$. In order to avoid any confusion regarding the sense of the shearing force $\mathbf{V}$ and couple $\mathbf{M}$ (which act in opposite directions on the two portions of the beam), we adopted the sign convention illustrated in Fig. 7.21 (Sec. 7.2B). Once we have determined the values of the shear and bending moment at a few selected points of the beam, it is usually possible to draw a **shear diagram** and a **bending-moment diagram** representing, respectively, the shear and bending moment at any point of the beam (Sec. 7.2C). When a beam is subjected to concentrated loads only, the shear is of constant value between loads, and the bending moment varies linearly between loads (Sample Prob. 7.2). When a beam is subjected to distributed loads, the shear and bending moment vary quite differently (Sample Prob. 7.3).

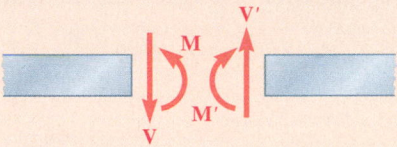

Internal forces at section
(positive shear and positive bending moment)

**Fig. 7.21**

## Relations among Load, Shear, and Bending Moment

Construction of the shear and bending-moment diagrams is simplified by taking into account the following relations. Denoting the distributed load per unit length by $w$ (assumed positive if directed downward), we have (Sec. 7.3):

$$\frac{dV}{dx} = -w \tag{7.1}$$

$$\frac{dM}{dx} = V \tag{7.3}$$

In integrated form, these equations become

$$V_D - V_C = -(\text{area under load curve between } C \text{ and } D) \tag{7.2'}$$

$$M_D - M_C = \text{area under shear curve between } C \text{ and } D \tag{7.4'}$$

Eq. (7.2′) makes it possible to draw the shear diagram of a beam from the curve representing the distributed load on that beam and the value of $V$ at one end of the beam. Similarly, Eq. (7.4′) makes it possible to draw the bending-moment diagram from the shear diagram and the value of $M$ at one end of the beam. However, discontinuities are introduced in the shear diagram by concentrated loads and in the bending-moment diagram by concentrated couples, none of which are accounted for in these equations (Sample Probs. 7.4 and 7.7). Finally, we note from Eq. (7.3) that the points of the beam where the bending moment is maximum or minimum are also the points where the shear is zero (Sample Prob. 7.5).

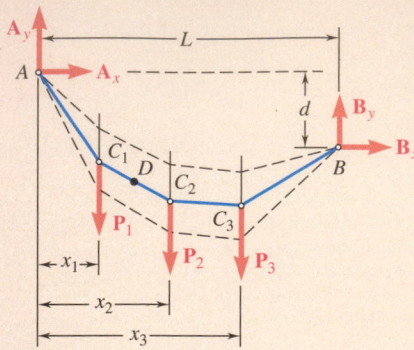
**Fig. 7.22**

## Cables with Concentrated Loads

The second half of the chapter was devoted to the analysis of **flexible cables**. We first considered a cable of negligible weight supporting **concentrated loads** (Sec. 7.4A). Using the entire cable $AB$ as a free body (Fig. 7.22), we noted that the three available equilibrium equations were not sufficient to determine the four unknowns representing the reactions at supports $A$ and $B$. However, if the coordinates of a point $D$ of the cable are known, we can obtain an additional equation by considering the free-body diagram of portion $AD$ or $DB$ of the cable. Once we have determined the reactions at the supports, we can find the elevation of any point of the cable and the tension in any portion of the cable from the appropriate free-body diagram (Sample Prob. 7.8). We noted that the horizontal component of the force $\mathbf{T}$ representing the tension is the same at any point of the cable.

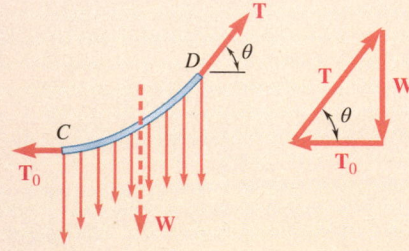
**Fig. 7.23**

## Cables with Distributed Loads

We next considered cables carrying **distributed loads** (Sec. 7.4B). Using as a free body a portion of cable $CD$ extending from the lowest point $C$ to an arbitrary point $D$ of the cable (Fig. 7.23), we observed that the horizontal component of the tension force $\mathbf{T}$ at $D$ is constant and equal to the tension $T_0$ at $C$, whereas its vertical component is equal to the weight $W$ of the portion of cable $CD$. The magnitude and direction of $\mathbf{T}$ were obtained from the force triangle:

$$T = \sqrt{T_0^2 + W^2} \quad \tan\theta = \frac{W}{T_0} \tag{7.6}$$

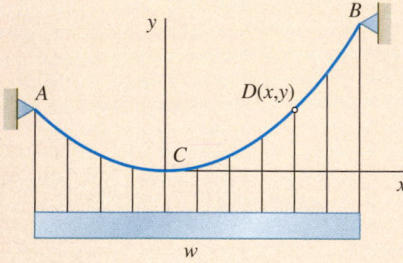
**Fig. 7.24**

## Parabolic Cable

In the case of a load uniformly distributed along the horizontal—as in a suspension bridge (Fig. 7.24)—the load supported by portion $CD$ is $W = wx$, where $w$ is the constant load per unit horizontal length (Sec. 7.4C). We also found that the curve formed by the cable is a **parabola** with equation

$$y = \frac{wx^2}{2T_0} \tag{7.8}$$

and that the length of the cable can be found by using the expansion in series given in Eq. (7.10) (Sample Prob. 7.9).

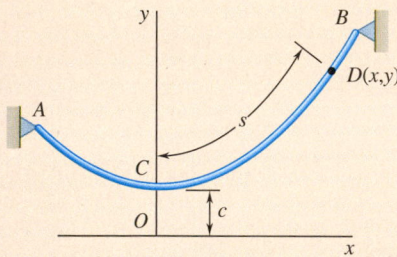
**Fig. 7.25**

## Catenary

In the case of a load uniformly distributed along the cable itself—e.g., a cable hanging under its own weight (Fig. 7.25)—the load supported by portion $CD$ is $W = ws$, where $s$ is the length measured along the cable and $w$ is the constant load per unit length (Sec. 7.5). Choosing the origin $O$ of the coordinate axes at a distance $c = T_0/w$ below $C$, we derived the relations

$$s = c \sinh\frac{x}{c} \tag{7.15}$$

$$y = c \cosh\frac{x}{c} \tag{7.16}$$

$$y^2 - s^2 = c^2 \tag{7.17}$$

$$T_0 = wc \qquad W = ws \qquad T = wy \tag{7.18}$$

These equations can be used to solve problems involving cables hanging under their own weight (Sample Prob. 7.10). Eq. (7.16), which defines the shape of the cable, is the equation of a **catenary**.

# Review Problems

**7.154 and 7.155** Knowing that the turnbuckle has been tightened until the tension in wire $AD$ is 850 N, determine the internal forces at the point indicated:

    **7.154** Point $J$

    **7.155** Point $K$

**7.156** Two members, each consisting of a straight and a quarter-circular portion of rod, are connected as shown and support a 75-lb load at $A$. Determine the internal forces at point $J$.

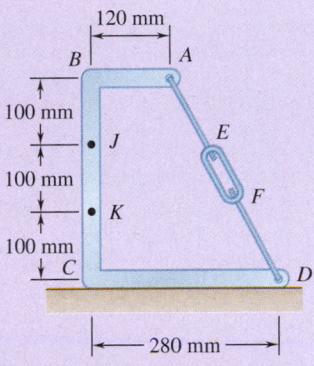

**Fig. P7.154 and P7.155**

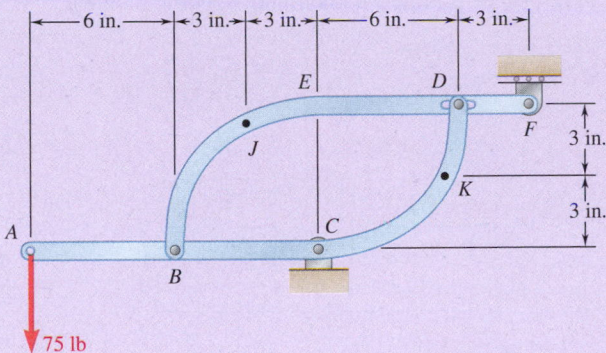

**Fig. P7.156**

**7.157** Knowing that the radius of each pulley is 150 mm, that $\alpha = 20°$, and neglecting friction, determine the internal forces at (a) point $J$, (b) point $K$.

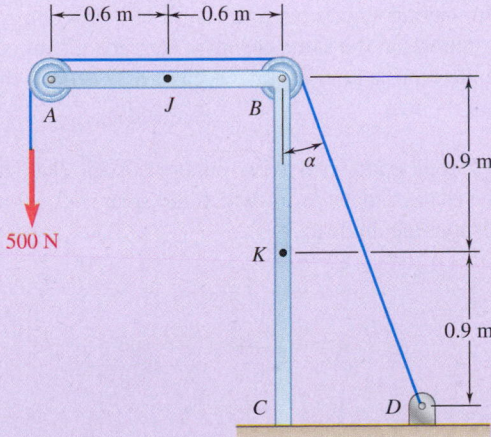

**Fig. P7.157**

**7.158** For the beam shown, determine (a) the magnitude $P$ of the two upward forces for which the maximum absolute value of the bending moment in the beam is as small as possible, (b) the corresponding value of $|M|_{max}$.

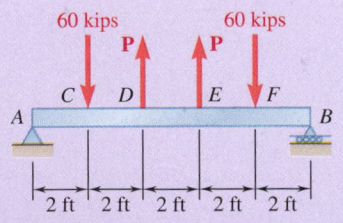

**Fig. P7.158**

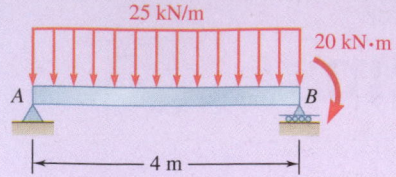

**Fig. P7.159**

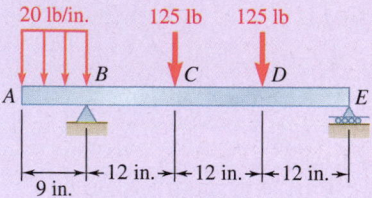

**Fig. P7.160**

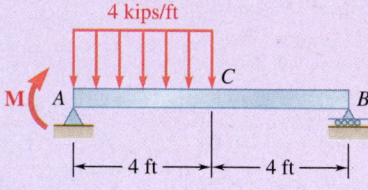

**Fig. P7.161**

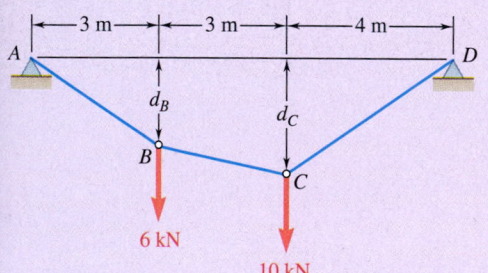

**Fig. P7.163**

**7.159** For the beam and loading shown, (*a*) draw the shear and bending-moment diagrams, (*b*) determine the magnitude and location of the maximum absolute value of the bending moment.

**7.160** For the beam and loading shown, (*a*) draw the shear and bending-moment diagrams, (*b*) determine the maximum absolute values of the shear and bending moment.

**7.161** For the beam shown, draw the shear and bending-moment diagrams, and determine the magnitude and location of the maximum absolute value of the bending moment, knowing that (*a*) *M* = 0, (*b*) *M* = 24 kip·ft.

**7.162** The beam *AB,* which lies on the ground, supports the parabolic load shown. Assuming the upward reaction of the ground to be uniformly distributed, (*a*) write the equations of the shear and bending-moment curves, (*b*) determine the maximum bending moment.

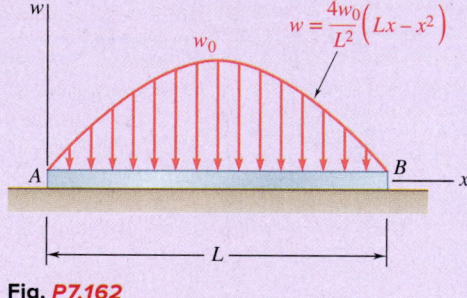

**Fig. P7.162**

**7.163** Two loads are suspended as shown from the cable *ABCD.* Knowing that $d_B = 1.8$ m, determine (*a*) the distance $d_C$, (*b*) the components of the reaction at *D,* (*c*) the maximum tension in the cable.

**7.164** A wire having a mass per unit length of 0.65 kg/m is suspended from two supports at the same elevation that are 120 m apart. If the sag is 30 m, determine (*a*) the total length of the wire, (*b*) the maximum tension in the wire.

**7.165** A 10-ft rope is attached to two supports *A* and *B* as shown. Determine (*a*) the span of the rope for which the span is equal to the sag, (*a*) the corresponding angle $\theta_B$.

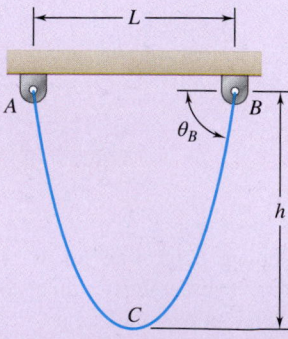

**Fig. P7.165**

430

# 8

## Friction

The tractive force that a railroad locomotive can develop depends upon the frictional resistance between the drive wheels and the rails. When the potential exists for wheel slip to occur, such as when a train travels upgrade over wet rails, sand is deposited on top of the railhead to increase this friction.

## Introduction

# Objectives

- **Examine** the laws of dry friction and the associated coefficients and angles of friction.
- **Consider** the equilibrium of rigid bodies where dry friction at contact surfaces is modeled.
- **Apply** the laws of friction to analyze problems involving wedges and square-threaded screws.
- **Study** engineering applications of the laws of friction, such as in modeling axle, disk, wheel, and belt friction.

## Introduction

In the previous chapters, we assumed that surfaces in contact are either *frictionless* or *rough*. If they are frictionless, the force each surface exerts on the other is normal to the surfaces, and the two surfaces can move freely with respect to each other. If they are rough, tangential forces can develop that prevent the motion of one surface with respect to the other.

This view is a simplified one. Actually, no perfectly frictionless surface exists. When two surfaces are in contact, tangential forces, called **friction forces**, always develop if you attempt to move one surface with respect to the other. However, these friction forces are limited in magnitude and do not prevent motion if you apply sufficiently large forces. Thus, the distinction between frictionless and rough surfaces is a matter of degree. You will see this more clearly in this chapter, which is devoted to the study of friction and its applications to common engineering situations.

There are two types of friction: **dry friction**, sometimes called *Coulomb friction,* and **fluid friction** or *viscosity.* Fluid friction develops between layers

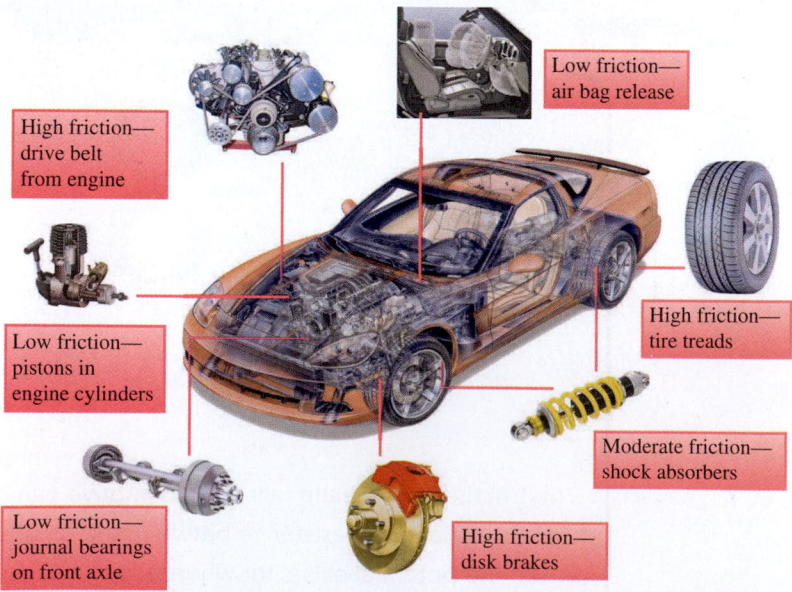

**Photo 8.1**   Examples of friction in an automobile. Depending upon the application, the degree of friction is controlled by design engineers.

of fluid moving at different velocities. This is of great importance in analyzing problems involving the flow of fluids through pipes and orifices or dealing with bodies immersed in moving fluids. It is also basic for the analysis of the motion of *lubricated mechanisms*. Such problems are considered in texts on fluid mechanics. The present study is limited to dry friction, i.e., to situations involving rigid bodies that are in contact along *unlubricated* surfaces.

In the first section of this chapter, we examine the equilibrium of various rigid bodies and structures, assuming dry friction at the surfaces of contact. Afterward, we consider several specific engineering applications where dry friction plays an important role: wedges, square-threaded screws, journal bearings, thrust bearings, rolling resistance, and belt friction.

# 8.1   THE LAWS OF DRY FRICTION

We can illustrate the laws of dry friction by the following experiment. Place a block of weight **W** on a horizontal plane surface (Fig. 8.1*a*). The forces acting on the block are its weight **W** and the reaction of the surface. Because the weight has no horizontal component, the reaction of the surface also has no horizontal component; the reaction is therefore *normal* to the surface and is represented by **N** in Fig. 8.1*a*. Now suppose that you apply a horizontal force **P** to the block (Fig. 8.1*b*). If **P** is small, the block does not move; some other horizontal force must therefore exist, which balances **P**. This other force is the **static-friction force F**, which is actually the resultant of a great number of forces acting over the entire surface of contact between the block and the plane. The nature of these forces is not known exactly, but we generally assume that these forces are due to the irregularities of the surfaces in contact and, to a certain extent, to molecular attraction.

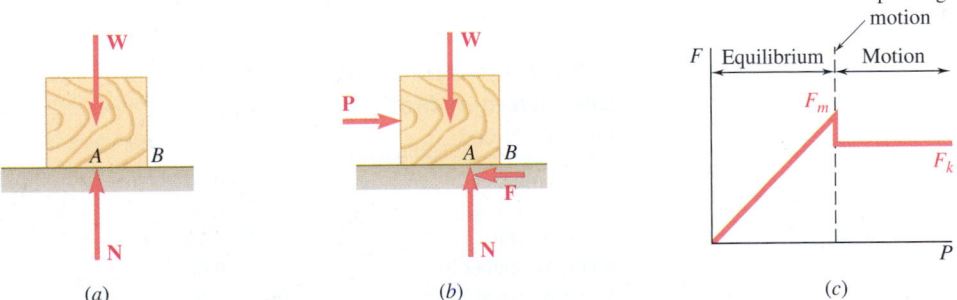

**Fig. 8.1**   (*a*) Block on a horizontal plane; friction force is zero; (*b*) a horizontally applied force **P** produces an opposing friction force **F**; (*c*) graph of **F** with increasing **P**.

If you increase the force **P**, the friction force **F** also increases, continuing to oppose **P**, until its magnitude reaches a certain *maximum value* $F_m$ (Fig. 8.1*c*). If **P** is further increased, the friction force cannot balance it anymore, and the block starts sliding. As soon as the block has started in motion, the magnitude of **F** drops from $F_m$ to a lower value $F_k$. This happens because less interpenetration occurs between the irregularities of the surfaces in contact when these surfaces move with respect to each other. From then on, the block keeps sliding with increasing velocity while the friction force, denoted by $\mathbf{F}_k$ and called the **kinetic-friction force**, remains approximately constant.

Note that, as the magnitude $F$ of the friction force increases from 0 to $F_m$, the point of application $A$ of the resultant $N$ of the normal forces of contact moves to the right. In this way, the couples formed by $P$ and $F$ and by $W$ and $N$, respectively, remain balanced. If $N$ reaches $B$ before $F$ reaches its maximum value $F_m$, the block starts to tip about $B$ before it can start sliding (see Sample Prob. 8.4).

## 8.1A    Coefficients of Friction

Experimental evidence shows that the maximum value $F_m$ of the static-friction force is proportional to the normal component $N$ of the reaction of the surface. We have

**Static friction**

$$F_m = \mu_s N \qquad (8.1)$$

where $\mu_s$ is a constant called the **coefficient of static friction**. Similarly, we can express the magnitude $F_k$ of the kinetic-friction force in the form

**Kinetic friction**

$$F_k = \mu_k N \qquad (8.2)$$

where $\mu_k$ is a constant called the **coefficient of kinetic friction**. The coefficients of friction $\mu_s$ and $\mu_k$ do not depend upon the area of the surfaces in contact. Both coefficients, however, depend strongly on the *nature* of the surfaces in contact. Because they also depend upon the exact condition of the surfaces, their value is seldom known with an accuracy greater than 5%. Approximate values of coefficients of static friction for various combinations of dry surfaces are listed in Table 8.1. The corresponding values of the coefficient of kinetic friction are about 25% smaller. Because coefficients of friction are dimensionless quantities, the values given in Table 8.1 can be used with both SI units and U.S. customary units.

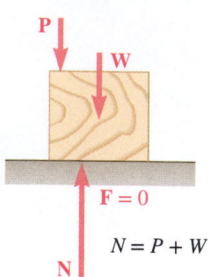

(a) No friction ($P_x = 0$)

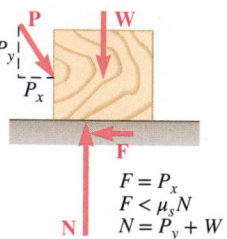

$F = P_x$
$F < \mu_s N$
$N = P_y + W$

(b) No motion ($P_x < F_m$)

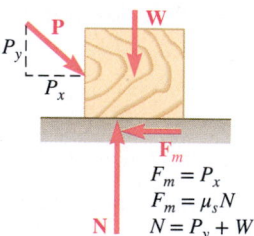

$F_m = P_x$
$F_m = \mu_s N$
$N = P_y + W$

(c) Motion impending $\longrightarrow$ ($P_x = F_m$)

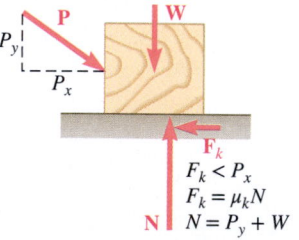

$F_k < P_x$
$F_k = \mu_k N$
$N = P_y + W$

(d) Motion $\longrightarrow$ ($P_x > F_k$)

**Fig. 8.2** (a) Applied force is vertical, friction force is zero; (b) horizontal component of applied force is less than $F_m$, no motion occurs; (c) horizontal component of applied force equals $F_m$, motion is impending; (d) horizontal component of applied force is greater than $F_k$, forces are unbalanced and motion continues.

**Table 8.1   Approximate Values of Coefficient of Static Friction for Dry Surfaces**

| | |
|---|---|
| Metal on metal | 0.15–0.60 |
| Metal on wood | 0.20–0.60 |
| Metal on stone | 0.30–0.70 |
| Metal on leather | 0.30–0.60 |
| Wood on wood | 0.25–0.50 |
| Wood on leather | 0.25–0.50 |
| Stone on stone | 0.40–0.70 |
| Earth on earth | 0.20–1.00 |
| Rubber on concrete | 0.60–0.90 |

From this discussion, it appears that four different situations can occur when a rigid body is in contact with a horizontal surface:

1. The forces applied to the body do not tend to move it along the surface of contact; there is no friction force (Fig. 8.2a).
2. The applied forces tend to move the body along the surface of contact, but are not large enough to set it in motion. We can find the static-friction force $F$ that has developed by solving the equations of equilibrium for the body.

Because there is no evidence that **F** has reached its maximum value, the equation $F_m = \mu_s N$ *cannot be used* to determine the friction force (Fig. 8.2*b*).

3. The applied forces are such that the body is just about to slide. We say that *motion is impending*. The friction force **F** has reached its maximum value $F_m$ and, together with the normal force **N**, balances the applied forces. Both the equations of equilibrium and the equation $F_m = \mu_s N$ *can be used*. Note that the friction force has a sense opposite to the sense of impending motion (Fig. 8.2*c*).

4. The body is sliding under the action of the applied forces, and the equations of equilibrium no longer apply. However, **F** is now equal to $\mathbf{F}_k$, and we can use the equation $F_k = \mu_k N$. The sense of $\mathbf{F}_k$ is opposite to the sense of motion (Fig. 8.2*d*).

# 8.1B Angles of Friction

It is sometimes convenient to replace the normal force **N** and the friction force **F** by their resultant **R**. Let's see what happens when we do that.

Consider again a block of weight **W** resting on a horizontal plane surface. If no horizontal force is applied to the block, the resultant **R** reduces to the normal force **N** (Fig. 8.3*a*). However, if the applied force **P** has a horizontal component $\mathbf{P}_x$ that tends to move the block, force **R** has a horizontal component **F** and, thus, forms an angle $\phi$ with the normal to the surface (Fig. 8.3*b*). If you increase $\mathbf{P}_x$ until motion becomes impending, the angle between **R** and the vertical grows and reaches a maximum value (Fig. 8.3*c*). This value is called the **angle of static friction** and is denoted by $\phi_s$. From the geometry of Fig. 8.3*c*, we note that

**Angle of static friction**

$$\tan \phi_s = \frac{F_m}{N} = \frac{\mu_s N}{N}$$

$$\tan \phi_s = \mu_s \qquad \textbf{(8.3)}$$

If motion actually takes place, the magnitude of the friction force drops to $F_k$; similarly, the angle between **R** and **N** drops to a lower value $\phi_k$, which is called the **angle of kinetic friction** (Fig. 8.3*d*). From the geometry of Fig. 8.3*d*, we have

**Angle of kinetic friction**

$$\tan \phi_k = \frac{F_k}{N} = \frac{\mu_k N}{N}$$

$$\tan \phi_k = \mu_k \qquad \textbf{(8.4)}$$

Another example shows how the angle of friction can be used to advantage in the analysis of certain types of problems. Consider a block resting on a board and subjected to no other force than its weight **W** and the reaction **R** of the board. The board can be given any desired inclination. If the board is horizontal, the force **R** exerted by the board on the block is perpendicular to the board and balances the weight **W** (Fig. 8.4*a*). If the board is given a small angle of inclination $\theta$, force **R** deviates from the perpendicular to the board by angle $\theta$ and continues to balance **W** (Fig. 8.4*b*). The reaction **R** now has a normal component **N** with a magnitude of $N = W \cos \theta$ and a tangential component **F** with a magnitude of $F = W \sin \theta$.

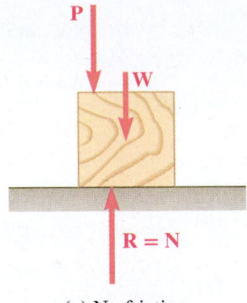

(*a*) No friction

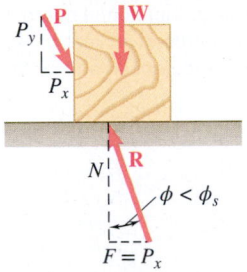

(*b*) No motion

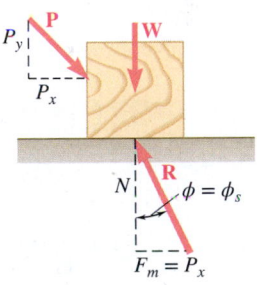

(*c*) Motion impending ⟶

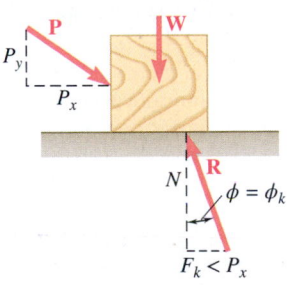

(*d*) Motion ⟶

**Fig. 8.3** (*a*) Applied force is vertical, friction force is zero; (*b*) applied force is at an angle, its horizontal component balanced by the horizontal component of the surface resultant; (*c*) impending motion, the horizontal component of the applied force equals the maximum horizontal component of the resultant; (*d*) motion, the horizontal component of the resultant is less than the horizontal component of the applied force.

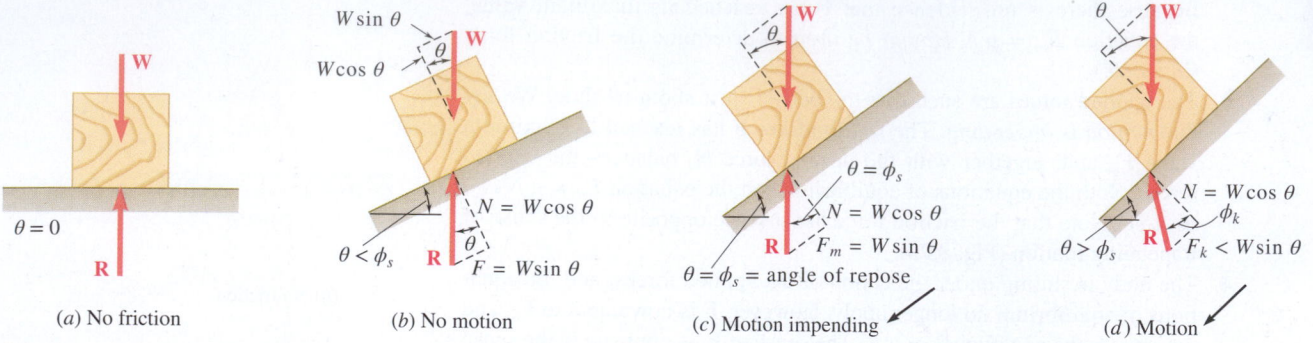

**Fig. 8.4** (*a*) Block on horizontal board, friction force is zero; (*b*) board's angle of inclination is less than angle of static friction, no motion; (*c*) board's angle of inclination equals angle of friction, motion is impending; (*d*) angle of inclination is greater than angle of friction, forces are unbalanced and motion occurs.

If we keep increasing the angle of inclination, motion soon becomes impending. At that time, the angle between **R** and the normal reaches its maximum value $\theta = \phi_s$ (Fig. 8.4*c*). The value of the angle of inclination corresponding to impending motion is called the **angle of repose**, and as demonstrated here, the angle of repose is equal to the angle of static friction $\phi_s$. If we further increase the angle of inclination $\theta$, motion starts and the angle between **R** and the normal drops to the lower value $\phi_k$ (Fig. 8.4*d*). The reaction **R** is not vertical anymore, and the forces acting on the block are unbalanced.

## 8.1C Problems Involving Dry Friction

Many engineering applications involve dry friction. Some are simple situations, such as variations on the block sliding on a plane just described. Others involve more complicated situations, as in Sample Prob. 8.3. Many problems deal with the stability of rigid bodies in accelerated motion and will be studied in dynamics. Also, several common machines and mechanisms can be analyzed by applying the laws of dry friction, including wedges, screws, journal and thrust bearings, and belt transmissions. We will study these applications in the following sections.

The methods used to solve problems involving dry friction are the same that we used in the preceding chapters. If a problem involves only a motion of translation with no possible rotation, we can usually treat the body under consideration as a particle and use the methods of Chap. 2. If the problem involves a possible rotation, we must treat the body as a rigid body and use the methods of Chap. 4. If the structure considered is made of several parts, we must apply the principle of action and reaction, as we did in Chap. 6.

If the body being considered is acted upon by more than three forces (including the reactions at the surfaces of contact), the reaction at each surface is represented by its components **N** and **F**, and we solve the problem using the equations of equilibrium. If only three forces act on the body under consideration, it may be more convenient to represent each reaction by the single force **R** and solve the problem by using a force triangle.

Most problems involving friction fall into one of the following three groups.

1. All applied forces are given, and we know the coefficients of friction; we are to determine whether the body being considered remains at rest or slides. The friction force **F** *required to maintain equilibrium* is unknown

**Photo 8.2** The coefficient of static friction between a crate and the inclined conveyer belt must be sufficiently large to enable the crate to be transported without slipping. ©Tomohiro Ohsumi/Bloomberg/Getty Images

(its magnitude is *not* equal to $\mu_s N$) and needs to be determined, together with the normal force **N**, by drawing a free-body diagram and solving the equations of equilibrium (Fig. 8.5a). We then compare the value found for the magnitude $F$ of the friction force with the maximum value $F_m = \mu_s N$. If $F$ is smaller than or equal to $F_m$, the body remains at rest. If the value found for $F$ is larger than $F_m$, equilibrium cannot be maintained and motion takes place; the actual magnitude of the friction force is then $F_k = \mu_k N$.

2. All applied forces are given, and we know the motion is impending; we are to determine the value of the coefficient of static friction. Here again, we determine the friction force and the normal force by drawing a free-body diagram and solving the equations of equilibrium (Fig. 8.5b). Because we know that the value found for $F$ is the maximum value $F_m$, we determine the coefficient of friction by solving the equation $F_m = \mu_s N$.

3. The coefficient of static friction is given, and we know that the motion is impending in a given direction; we are to determine the magnitude or the direction of one of the applied forces. The friction force should be shown in the free-body diagram with a *sense opposite to that of the impending motion* and with a magnitude $F_m = \mu_s N$ (Fig. 8.5c). We can then write the equations of equilibrium and determine the desired force.

As noted previously, when only three forces are involved, it may be more convenient to represent the reaction of the surface by a single force **R** and to solve the problem by drawing a force triangle. Such a solution is used in Sample Prob. 8.2.

When two bodies $A$ and $B$ are in contact (Fig. 8.6a), the forces of friction exerted, respectively, by $A$ on $B$ and by $B$ on $A$ are equal and opposite (Newton's third law). In drawing the free-body diagram of one of these bodies, it is important to include the appropriate friction force with its correct sense. Observe the following rule: *The sense of the friction force acting on A is opposite to that of the motion (or impending motion) of A as observed from B* (Fig. 8.6b). (It is therefore the same as the motion of $B$ as observed from $A$.) The sense of the friction force acting on $B$ is determined in a similar way (Fig. 8.6c). Note that the motion of $A$ as observed from $B$ is a *relative motion*. For example, if body $A$ is fixed and body $B$ moves, body $A$ has a relative motion with respect to $B$. Also, if both $B$ and $A$ are moving down but $B$ is moving faster than $A$, then body $A$ is observed, from $B$, to be moving up.

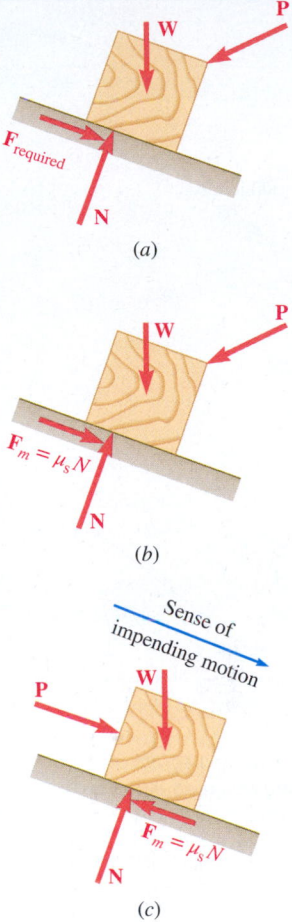

(a)

(b)

(c)

**Fig. 8.5** Three types of friction problems: (a) Given the forces and coefficient of friction, will the block slide or stay? (b) Given the forces and that motion is pending, determine the coefficient of friction; (c) given the coefficient of friction and that motion is impending, determine the applied force.

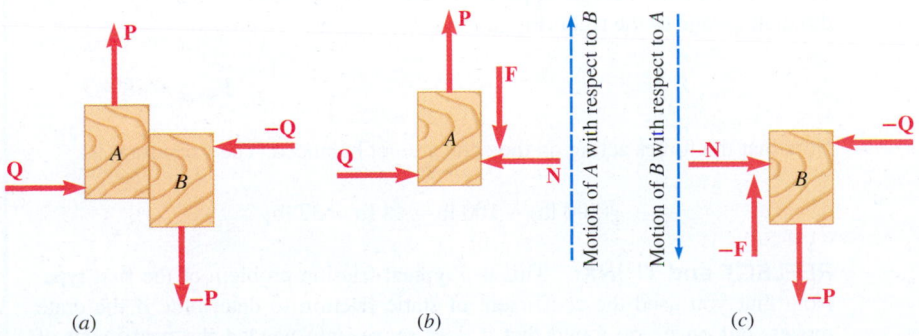

(a)            (b)            (c)

**Fig. 8.6** (a) Two blocks held in contact by forces; (b) free-body diagram for block A, including direction of friction force; (c) free-body diagram for block B, including direction of friction force.

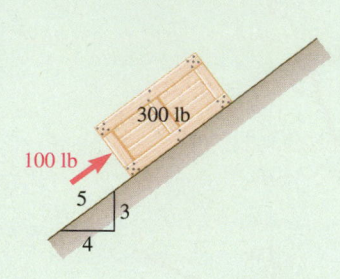

## Sample Problem 8.1

A 100-lb force acts as shown on a 300-lb crate placed on an inclined plane. The coefficients of friction between the crate and the plane are $\mu_s = 0.25$ and $\mu_k = 0.20$. Determine whether the crate is in equilibrium, and find the value of the friction force.

**STRATEGY:**    This is a friction problem of the first type: You know the forces and the friction coefficients and want to determine if the crate moves. You also want to find the friction force.

**MODELING and ANALYSIS:**

**Force Required for Equilibrium.**    First determine the value of the friction force *required to maintain equilibrium*. Assuming that **F** is directed down and to the left, draw the free-body diagram of the crate (Fig. 1) and solve the equilibrium equations:

$$+ \nearrow \Sigma F_x = 0: \qquad 100 \text{ lb} - \tfrac{3}{5}(300 \text{ lb}) - F = 0$$
$$F = -80 \text{ lb} \quad \mathbf{F} = 80 \text{ lb} \nearrow$$

$$+ \nwarrow \Sigma F_y = 0: \qquad N - \tfrac{4}{5}(300 \text{ lb}) = 0$$
$$N = +240 \text{ lb} \quad \mathbf{N} = 240 \text{ lb} \nwarrow$$

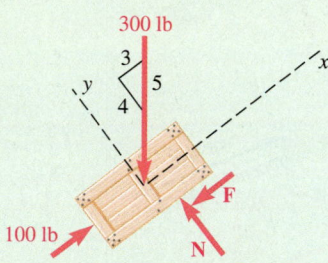

**Fig. 1**  Free-body diagram of crate showing assumed direction of friction force.

The force **F** required to maintain equilibrium is an 80-lb force directed up and to the right; the tendency of the crate is thus to move down the plane.

**Maximum Friction Force.**    The magnitude of the maximum friction force that may be developed between the crate and the plane is

$$F_m = \mu_s N \qquad F_m = 0.25(240 \text{ lb}) = 60 \text{ lb}$$

Because the value of the force required to maintain equilibrium (80 lb) is larger than the maximum value that may be obtained (60 lb), equilibrium is not maintained and *the crate will slide down the plane*.

**Actual Value of Friction Force.**    The magnitude of the actual friction force is

$$F_{\text{actual}} = F_k = \mu_k N = 0.20(240 \text{ lb}) = 48 \text{ lb}$$

The sense of this force is opposite to the sense of motion; the force is thus directed up and to the right (Fig. 2):

$$\mathbf{F}_{\text{actual}} = 48 \text{ lb} \nearrow \quad \blacktriangleleft$$

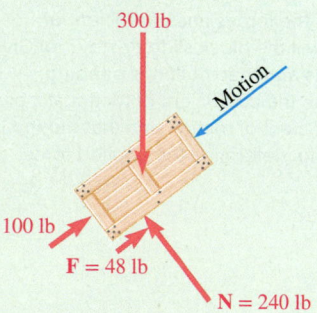

**Fig. 2**  Free-body diagram of crate showing actual friction force.

Note that the forces acting on the crate are not balanced. Their resultant is

$$\tfrac{3}{5}(300 \text{ lb}) - 100 \text{ lb} - 48 \text{ lb} = 32 \text{ lb} \swarrow$$

**REFLECT and THINK:**    This is a typical friction problem of the first type. Note that you used the coefficient of static friction to determine if the crate moves, but once you found that it does move, you needed the coefficient of kinetic friction to determine the friction force.

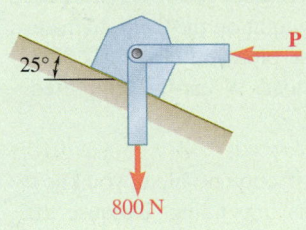

# Sample Problem 8.2

A support block is acted upon by two forces as shown. Knowing that the coefficients of friction between the block and the incline are $\mu_s = 0.35$ and $\mu_k = 0.25$, determine the force **P** required to (a) start the block moving up the incline, (b) keep it moving up, (c) prevent it from sliding down.

**STRATEGY:**   This problem involves practical variations of the third type of friction problem. You can approach the solutions through the concept of the angles of friction.

**MODELING:**

**Free-Body Diagram.**   For each part of the problem, draw a free-body diagram of the block and a force triangle including the 800-N vertical force, the horizontal force **P**, and the force **R** exerted on the block by the incline. You must determine the direction of **R** in each separate case. Note that, because **P** is perpendicular to the 800-N force, the force triangle is a right triangle, which easily can be solved for **P**. In most other problems, however, the force triangle will be an oblique triangle and should be solved by applying the law of sines.

**ANALYSIS:**

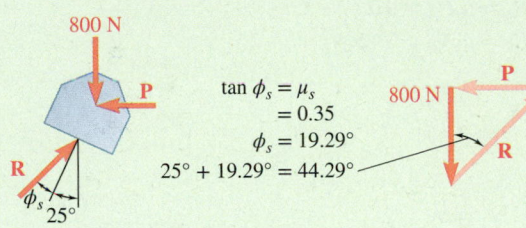

$$\tan \phi_s = \mu_s$$
$$= 0.35$$
$$\phi_s = 19.29°$$
$$25° + 19.29° = 44.29°$$

**Fig. 1**   Free-body diagram of block and its force triangle—motion impending up the incline.

**a. Force *P* to Start Block Moving Up.**   In this case, motion is impending up the incline, so the resultant is directed at the angle of static friction (Fig. 1). Note that the resultant is oriented to the left of the normal such that its friction component (not shown) is directed opposite the direction of impending motion.

$$P = (800 \text{ N}) \tan 44.29° \qquad \mathbf{P} = 780 \text{ N} \leftarrow$$

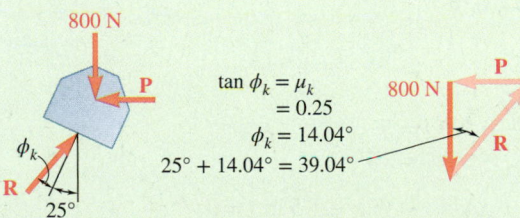

$$\tan \phi_k = \mu_k$$
$$= 0.25$$
$$\phi_k = 14.04°$$
$$25° + 14.04° = 39.04°$$

**Fig. 2**   Free-body diagram of block and its force triangle—motion continuing up the incline.

**b. Force *P* to Keep Block Moving Up.**   Motion is continuing, so the resultant is directed at the angle of kinetic friction (Fig. 2). Again, the resultant is oriented to the left of the normal such that its friction component is directed opposite the direction of motion.

$$P = (800 \text{ N}) \tan 39.04° \qquad \mathbf{P} = 649 \text{ N} \leftarrow$$

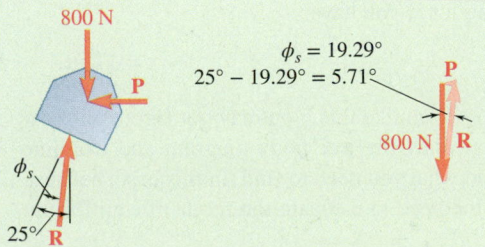

$$\phi_s = 19.29°$$
$$25° - 19.29° = 5.71°$$

**Fig. 3**   Free-body diagram of block and its force triangle—motion prevented down the slope.

**c. Force *P* to Prevent Block from Sliding Down.**   Here, motion is impending down the incline, so the resultant is directed at the angle of static friction (Fig. 3). Note that the resultant is oriented to the right of the normal such that its friction component is directed opposite the direction of impending motion.

$$P = (800 \text{ N}) \tan 5.71° \qquad \mathbf{P} = 80.0 \text{ N} \leftarrow$$

**REFLECT and THINK:**   As expected, considerably more force is required to begin moving the block up the slope than is necessary to restrain it from sliding down the slope.

## Sample Problem 8.3

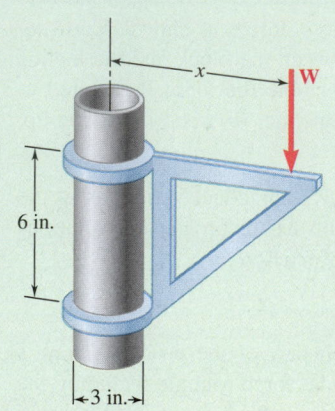

The movable bracket shown may be placed at any height on the 3-in.-diameter pipe. If the coefficient of static friction between the pipe and bracket is 0.25, determine the minimum distance $x$ at which the load **W** can be supported. Neglect the weight of the bracket.

**STRATEGY:**   In this variation of the third type of friction problem, you know the coefficient of static friction and that motion is impending. Because the problem involves consideration of resistance to rotation, you should apply both moment equilibrium and force equilibrium.

**MODELING:**

**Free-Body Diagram.**   Draw the free-body diagram of the bracket (Fig. 1). When **W** is placed at the minimum distance $x$ from the axis of the pipe, the bracket is just about to slip, and the forces of friction at $A$ and $B$ have reached their maximum values:

$$F_A = \mu_s N_A = 0.25\, N_A$$
$$F_B = \mu_s N_B = 0.25\, N_B$$

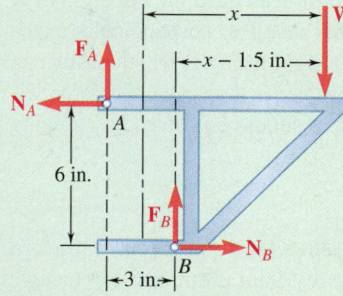

**Fig. 1**   Free-body diagram of the bracket.

**ANALYSIS:**

**Equilibrium Equations.**

$\xrightarrow{+}\Sigma F_x = 0:$ 
$$N_B - N_A = 0$$
$$N_B = N_A$$

$+\uparrow\Sigma F_y = 0:$ 
$$F_A + F_B - W = 0$$
$$0.25 N_A + 0.25 N_B = W$$

Because $N_B$ is equal to $N_A$,

$$0.50 N_A = W$$
$$N_A = 2W$$

$+\circlearrowleft\Sigma M_B = 0:$    $N_A(6\text{ in.}) - F_A(3\text{ in.}) - W(x - 1.5\text{ in.}) = 0$
$$6N_A - 3(0.25 N_A) - Wx + 1.5W = 0$$
$$6(2W) - 0.75(2W) - Wx + 1.5W = 0$$

Dividing through by $W$ and solving for $x$, you have

$$x = 12 \text{ in.}\ \blacktriangleleft$$

**REFLECT and THINK:**   In a problem like this, you may not figure out how to approach the solution until you draw the free-body diagram and examine what information you are given and what you need to find. In this case, because you are asked to find a distance, the need to evaluate the moment equilibrium should be clear.

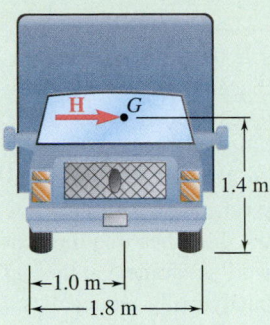

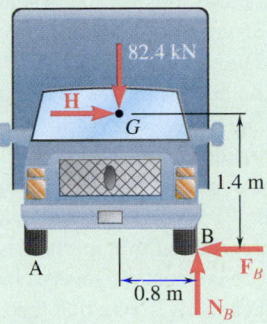

**Fig. 1** Free-body diagram of the truck.

## Sample Problem 8.4

An 8400-kg truck is traveling on a level horizontal curve, resulting in an effective lateral force **H** (applied at the center of gravity $G$ of the truck). Treating the truck as a rigid system with the center of gravity shown, and knowing that the distance between the outer edges of the tires is 1.8 m, determine (a) the maximum force **H** before tipping of the truck occurs, (b) the minimum coefficient of static friction between the tires and roadway such that slipping does not occur before tipping.

**STRATEGY:** For the direction of **H** shown, the truck would tip about the outer edge of the tire shown on the right in the diagram. At the verge of tip, the normal force and friction force are zero at the tire on the left, and the normal force at the right tire is at the outer edge. You can apply equilibrium to determine the value of **H** necessary for tip and the required friction force such that slipping does not occur.

**MODELING:** Draw the free-body diagram of the truck (Fig. 1), which reflects impending tip about point $B$. Obtain the weight of the truck by multiplying its mass of 8400 kg by $g = 9.81$ m/s²; that is, $W = 82\,400$ N or 82.4 kN.

**ANALYSIS:**

**Free Body: Truck (Fig. 1).**

$$+\circlearrowleft \Sigma M_B = 0: \quad (82.4 \text{ kN})(0.8 \text{ m}) - H(1.4 \text{ m}) = 0$$
$$H = +47.1 \text{ kN} \qquad\qquad \mathbf{H = 47.1 \text{ kN}} \rightarrow \quad \blacktriangleleft$$

$$\xrightarrow{+} \Sigma F_x = 0: \quad 47.1 \text{ kN} - F_B = 0$$
$$F_B = +47.1 \text{ kN}$$

$$+\uparrow \Sigma F_y = 0: \quad N_B - 82.4 \text{ kN} = 0$$
$$N_B = +82.4 \text{ kN}$$

**Minimum Coefficient of Static Friction.** The magnitude of the maximum friction force that can be developed is

$$F_m = \mu_s N_B = \mu_s (82.4 \text{ kN})$$

Setting this equal to the friction force required, $F_B = 47.1$ kN, gives

$$\mu_s(82.4 \text{ kN}) = 47.1 \text{ kN} \qquad\qquad \mu_s = 0.572 \quad \blacktriangleleft$$

**REFLECT and THINK:** Recall from physics that **H** represents the force due to the centripetal acceleration of the truck (of mass $m$), and its magnitude is

$$H = m(v^2/\rho)$$

where

$\quad v =$ velocity of the truck
$\quad \rho =$ radius of curvature

In this problem, if the truck were traveling around a curve of 100-m radius (measured to $G$), the velocity at which it would begin to tip would be 23.7 m/s (or 85.2 km/h). You will learn more about this aspect in your study of dynamics.

# Case Study 8.1

A common detail in structural steel building frames is the simple shear connection. Circled in CS Photo 8.1 is an example of such a connection, showing a beam attached to a column using a pair of framing angles welded to either side of the beam web and bolted to the column flange. CS Fig. 8.1 further illustrates the details of the connection. Because the flanges of an I-shaped beam primarily resist bending moment and the web primarily resists shear, and because only the web is connected in a simple shear connection, very little bending moment is transmitted through the joint. For this reason, the bending moment at the end of the beam is assumed to be zero, and the joint is analytically modeled as a pin connection. The American Institute of Steel Construction (AISC) publishes the *Steel Construction Manual,** which contains numerous aids for the design of steel buildings, as well as the *Specification for Structural Steel Buildings* (AISC 360-10) that governs their design. In accordance with AISC 360-10, one way that the bolts of a simple shear connection can be designed is as being *slip-critical,* where the friction of the clamped interface is relied upon to support the end-shear of the beam. If the connection considered in CS Photo 8.1 was designed as slip-critical using ¾-in. AISC Group A bolts, the design aids in Part 10 of the AISC Manual indicate its capacity to be 75.9 kips, provided that standard bolt holes are used and that the surface of the steel at the interface is unpainted clean mill scale. Assuming that friction of the connection governs its design, let's perform an analysis to confirm this rated capacity. (Note that there could be other factors that govern the overall capacity of the connection, such as the strength of the framing angles.)

**CS Photo 8.1**    Bolted double-angle simple shear connection (circled). American Institute of Steel Construction "Teaching Aids," Photo "BeamToColumn23," on "For Faculty & Students" channel at www.aisc.org

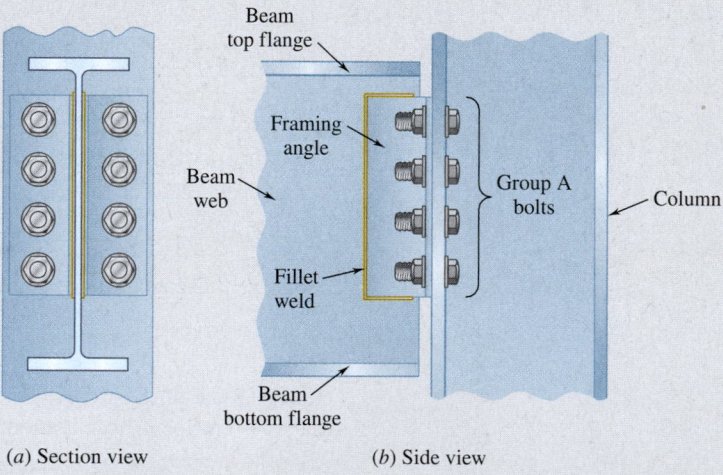

(*a*) Section view                  (*b*) Side view

**CS Fig. 8.1**    Connection details.

**STRATEGY:**    The friction capacity of the connection can be determined using a suitable static coefficient of friction along with the normal force acting on the interface, where this normal force is the clamping force developed by the tensioned bolts. Using the provisions of AISC 360-10, accepted values for the

*Ref: *Steel Construction Manual,* American Institute of Steel Construction, 14e, 2011.

*(continued)*

coefficient of friction and the minimum tension for properly installed bolts can be obtained.

**MODELING:** Treat one of the framing angles as a free body, cutting through the bolts and weld (CS Fig. 8.2). Because there are two framing angles that support the end of the beam, one half of the beam end-shear is shown. The tension in each bolt can be obtained from AISC 360-10, where the minimum tension in a properly installed ¾-in. Group A bolt is listed as 28 kips. Because the average bolt tension in a proper installation can be expected to be somewhat higher than this minimum, the specification allows an increase of 13%. For unpainted clean mill scale steel surfaces, AISC 360-10 specifies a static friction coefficient $\mu = 0.30$ that can be used to determine the friction force.

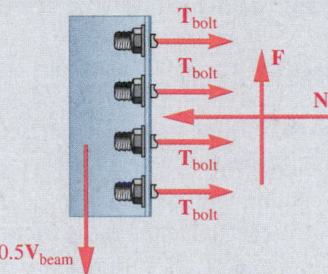

**CS Fig. 8.2** Free-body diagram of the framing angle.

**ANALYSIS:** **Normal Force.** Each bolt force is the minimum prescribed tension, increased by 13%, or

$$T_{bolt} = 1.13(28 \text{ kips}) = 31.64 \text{ kips}$$

Applying equilibrium (CS Fig. 8.2):

$$+\rightarrow \Sigma F_x = 0: \qquad 4T_{bolt} - N = 4(31.64 \text{ kips}) - N = 0 \qquad \mathbf{N = 126.56 \text{ kips}}\leftarrow$$

**Maximum Friction Force.** The magnitude of the maximum friction force that can be developed is

$$F_m = \mu N = (0.30)(126.56 \text{ kips}) = 37.968 \text{ kips}$$

**Capacity of Connection.** The capacity of the connection reflects the beam end-shear $V_{beam}$ that can be supported. Applying equilibrium (CS Fig. 8.2):

$$+\uparrow \Sigma F_y = 0: \qquad -0.5V_{beam} + F_m = -0.5V_{beam} + 37.968 \text{ kips} = 0$$
$$V_{beam} = 75.9 \text{ kips} \qquad \text{(checks)}$$

**REFLECT and THINK:** A slip-critical bolted connection is intended not to slip under the maximum anticipated design loads. Should an overload situation occur that does cause the connecting elements to slip, the connection won't actually fail unless the bolts shear off, or unless the parts that the bolts bear against fail in bearing. Often this involves loads much greater than those necessary to cause slip, thereby adding to the overall margin of safety.

# SOLVING PROBLEMS
# ON YOUR OWN

In this section, you studied and applied the **laws of dry friction**. Previously, you had encountered only (*a*) frictionless surfaces that could move freely with respect to each other or (*b*) rough surfaces that allowed no motion relative to each other.

**A. In solving problems involving dry friction,** keep the following ideas in mind.

**1. The reaction R exerted by a surface on a free body** can be resolved into a normal component **N** and a tangential component **F**. The tangential component is known as the **friction force**. When a body is in contact with a fixed surface, the direction of the friction force **F** is opposite to that of the actual or impending motion of the body.

   **a. No motion will occur** as long as $F$ does not exceed the maximum value $F_m = \mu_s N$, where $\mu_s$ is the **coefficient of static friction**.

   **b. Motion will occur** if a value of $F$ larger than $F_m$ is required to maintain equilibrium. As motion takes place, the actual value of $F$ drops to $F_k = \mu_k N$, where $\mu_k$ is the **coefficient of kinetic friction** (Sample Prob. 8.1).

   **c. Motion may also occur** at a value of $F$ smaller than $F_m$ if tipping of the rigid body is a possibility (Sample Prob. 8.4)

**2. When only three forces are involved,** you might prefer an alternative approach to the analysis of friction (Sample Prob. 8.2). The reaction **R** is defined by its magnitude $R$ and the angle $\phi$ it forms with the normal to the surface. No motion occurs as long as $\phi$ does not exceed the maximum value $\phi_s$, where $\tan \phi_s = \mu_s$. Motion does occur if a value of $\phi$ larger than $\phi_s$ is required to maintain equilibrium, and the actual value of $\phi$ drops to $\phi_k$, where $\tan \phi_k = \mu_k$.

**3. When two bodies are in contact,** you must determine the sense of the actual or impending relative motion at the point of contact. On each of the two bodies, a friction force **F** is in a direction opposite to that of the actual or impending motion of the body as seen from the other body (see Fig. 8.6).

**B. Methods of solution.** The first step in your solution is to draw a free-body diagram of the body under consideration, resolving the force exerted on each surface where friction exists into a normal component **N** and a friction force **F**. If several bodies are involved, draw a free-body diagram for each of them, labeling and directing the forces at each surface of contact, as described for analyzing frames in Chap. 6.

The problem you have to solve may fall into one of the following three categories.

**1. You know all the applied forces and the coefficients of friction, and you must determine whether equilibrium is maintained.** In this situation, the friction force is unknown and *cannot be assumed to be equal* to $\mu_s N$.

　　**a. Write the equations of equilibrium** to determine $N$ and $F$.

　　**b. Calculate the maximum allowable friction force,** $F_m = \mu_s N$. If $F \leq F_m$, equilibrium is maintained. If $F \geq F_m$, motion occurs, and the magnitude of the friction force is $F_k = \mu_k N$ (Sample Prob. 8.1).

**2. You know all the applied forces, and you must find the smallest allowable value of $\mu_s$ for which equilibrium is maintained.** Assume that motion is impending, and determine the corresponding value of $\mu_s$.

　　**a. Write the equations of equilibrium** to determine $N$ and $F$.

　　**b. Because motion is impending,** $F = F_m$. Substitute the values found for $N$ and $F$ into the equation $F_m = \mu_s N$ and solve for $\mu_s$ (Sample Prob. 8.4).

**3. The motion of the body is impending and $\mu_s$ is known; you must find some unknown quantity,** such as a distance, an angle, the magnitude of a force, or the direction of a force.

　　**a. Assume a possible motion of the body** and, on the free-body diagram, draw the friction force in a direction opposite to that of the assumed motion.

　　**b. Because motion is impending,** $F = F_m = \mu_s N$. Substituting the known value for $\mu_s$, you can express $F$ in terms of $N$ on the free-body diagram, thus eliminating one unknown.

　　**c. Write and solve the equilibrium equations** for the unknown you seek (Sample Prob. 8.3).

# Problems

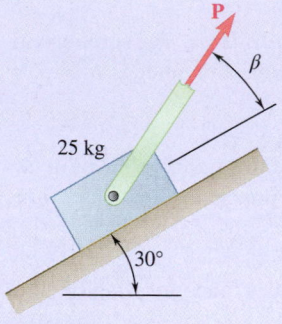

**Fig. P8.F1**

**FREE-BODY PRACTICE PROBLEMS**

**8.F1** Knowing that the coefficient of friction between the 25-kg block and the incline is $\mu_s = 0.25$, draw the free-body diagram needed to determine both the smallest value of $P$ required to start the block moving up the incline and the corresponding value of $\beta$.

**8.F2** Two blocks $A$ and $B$ are connected by a cable as shown. Knowing that the coefficient of static friction at all surfaces of contact is 0.30 and neglecting the friction of the pulleys, draw the free-body diagrams needed to determine the smallest force $P$ required to move the blocks.

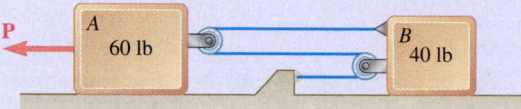

**Fig. P8.F2**

**8.F3** A cord is attached to and partially wound around a cylinder with a weight of $W$ and radius $r$ that rests on an incline as shown. Knowing that $\theta = 30°$, draw the free-body diagram needed to determine both the tension in the cord and the smallest allowable value of the coefficient of static friction between the cylinder and the incline for which equilibrium is maintained.

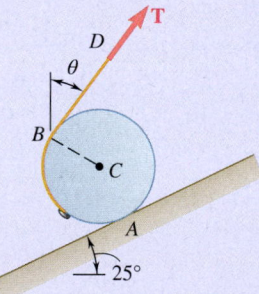

**Fig. P8.F3**

**8.F4** A 40-kg packing crate must be moved to the left along the floor without tipping. Knowing that the coefficient of static friction between the crate and the floor is 0.35, draw the free-body diagram needed to determine both the largest allowable value of $\alpha$ and the corresponding magnitude of the force $P$.

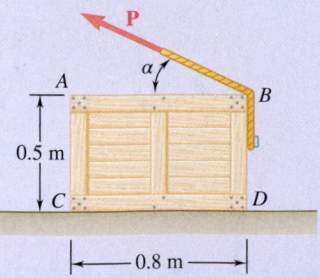

**Fig. P8.F4**

**8.1** Determine whether the block shown is in equilibrium and find the magnitude and direction of the friction force when $\theta = 40°$ and $P = 400$ N.

**8.2** Determine whether the block shown is in equilibrium and find the magnitude and direction of the friction force when $\theta = 35°$ and $P = 200$ N.

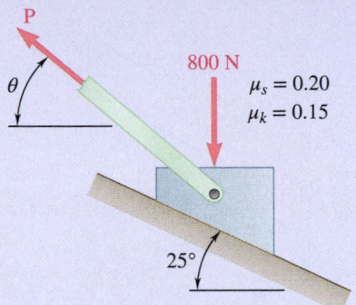

**Fig. P8.1 and P8.2**

**8.3** Determine whether the block shown is in equilibrium and find the magnitude and direction of the friction force when $\theta = 30°$ and $P = 50$ lb.

**8.4** Determine whether the block shown is in equilibrium and find the magnitude and direction of the friction force when $\theta = 35°$ and $P = 100$ lb.

**8.5** Knowing that $\theta = 45°$ in Prob. 8.1, determine the range of values of $P$ for which equilibrium of the block shown is maintained.

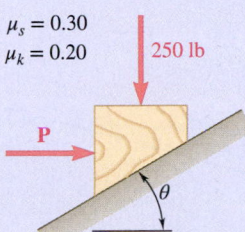

**Fig. P8.3 and P8.4**

**8.6** The 20-lb block $A$ hangs from a cable as shown. Pulley $C$ is connected by a short link to block $E$, which rests on a horizontal rail. Knowing that the coefficient of static friction between block $E$ and the rail is $\mu_s = 0.35$ and neglecting the weight of block $E$ and the friction in the pulleys, determine the maximum allowable value of $\theta$ if the system is to remain in equilibrium.

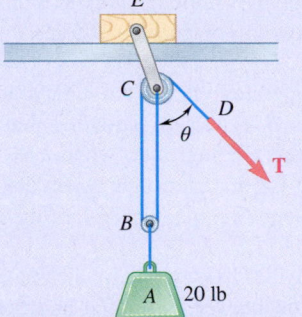

**Fig. P8.6**

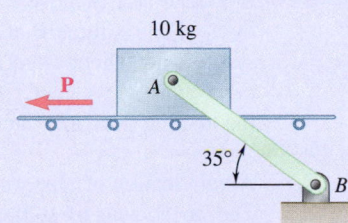

**Fig. P8.7**

**8.7** The 10-kg block is attached to link $AB$ and rests on a moving belt. Knowing that $\mu_s = 0.30$ and $\mu_k = 0.25$ and neglecting the weight of the link, determine the magnitude of the horizontal force **P** that should be applied to the belt to maintain its motion (*a*) to the left as shown, (*b*) to the right.

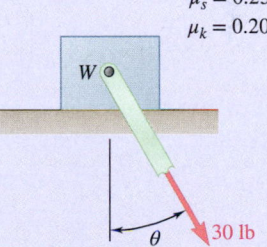

**Fig. P8.8**

**8.8** Considering only values of $\theta$ less than 90°, determine the smallest value of $\theta$ required to start the block moving to the right when (*a*) $W = 75$ lb, (*b*) $W = 100$ lb.

**8.9** The coefficients of friction between the block and the rail are $\mu_s = 0.25$ and $\mu_k = 0.20$. Knowing that $\theta = 60°$, determine the smallest value of $P$ required (*a*) to start the block moving up the rail, (*b*) to keep it moving up, (*c*) to prevent it from moving down.

**8.10** The coefficients of friction between the block and the rail are $\mu_s = 0.25$ and $\mu_k = 0.20$. Find the magnitude and direction of the smallest force **P** required (*a*) to start the block moving up the rail, (*b*) to keep the block from moving down.

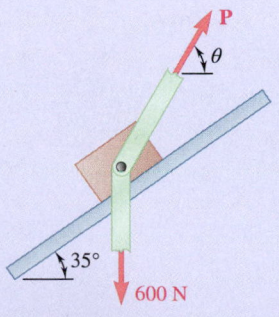

**Fig. P8.9 and P8.10**

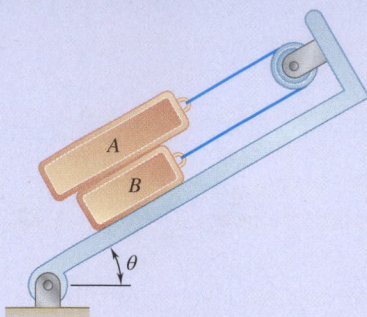

**Fig. P8.11 and P8.12**

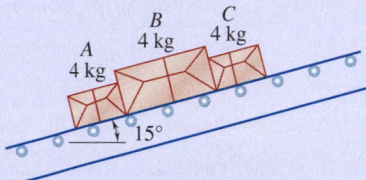

**Fig. P8.13**

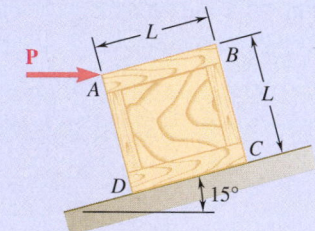

**Fig. P8.15**

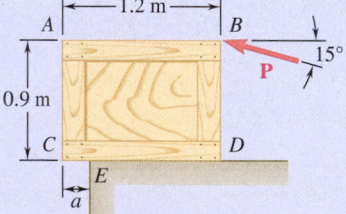

**Fig. P8.16**

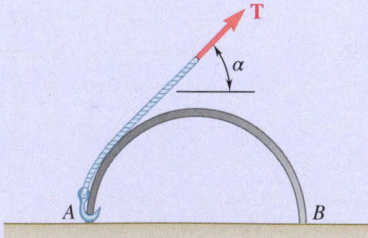

**Fig. P8.17**

**8.11** The 50-lb block $A$ and the 25-lb block $B$ are supported by an incline that is held in the position shown. Knowing that the coefficient of static friction is 0.15 between the two blocks and zero between block $B$ and the incline, determine the value of $\theta$ for which motion is impending.

**8.12** The 50-lb block $A$ and the 25-lb block $B$ are supported by an incline that is held in the position shown. Knowing that the coefficient of static friction is 0.15 between all surfaces of contact, determine the value of $\theta$ for which motion is impending.

**8.13** Three 4-kg packages $A$, $B$, and $C$ are placed on a conveyor belt that is at rest. Between the belt and both packages $A$ and $C$, the coefficients of friction are $\mu_s = 0.30$ and $\mu_k = 0.20$; between package $B$ and the belt, the coefficients are $\mu_s = 0.10$ and $\mu_k = 0.08$. The packages are placed on the belt so that they are in contact with each other and at rest. Determine which, if any, of the packages will move and the friction force acting on each package.

**8.14** Solve Prob. 8.13 assuming that package $B$ is placed to the right of both packages $A$ and $C$.

**8.15** A uniform crate with a mass of 30 kg must be moved up along the 15° incline without tipping. Knowing that force $\mathbf{P}$ is horizontal, determine (*a*) the largest allowable coefficient of static friction between the crate and the incline, (*b*) the corresponding magnitude of force $\mathbf{P}$.

**8.16** A worker slowly moves a 50-kg crate to the left along a loading dock by applying a force $\mathbf{P}$ at corner $B$ as shown. Knowing that the crate starts to tip about edge $E$ of the loading dock when $a = 200$ mm, determine (*a*) the coefficient of kinetic friction between the crate and the loading dock, (*b*) the corresponding magnitude $P$ of the force.

**8.17** A half-section of pipe weighing 200 lb is pulled by a cable as shown. The coefficient of static friction between the pipe and the floor is 0.40. If $\alpha = 30°$, determine (*a*) the tension $T$ required to move the pipe, (*b*) whether the pipe will slide or tip.

**8.18** A 200-lb sliding door is mounted on a horizontal rail as shown. The coefficients of static friction between the rail and the door at $A$ and $B$ are 0.15 and 0.25, respectively. Determine the horizontal force that must be applied to the handle $C$ in order to move the door to the right.

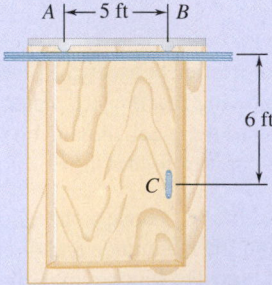

**Fig. P8.18**

**8.19** Wire is being drawn at a constant rate from a spool by applying a vertical force **P** to the wire as shown. The spool and the wire wrapped on the spool have a combined weight of 20 lb. Knowing that the coefficients of friction at both $A$ and $B$ are $\mu_s = 0.40$ and $\mu_k = 0.30$, determine the required magnitude of force **P**.

**8.20** Solve Prob. 8.19 assuming that the coefficients of friction at $B$ are zero.

**8.21** The cylinder shown has a weight $W$ and radius $r$. Express in terms of $W$ and $r$ the magnitude of the largest couple **M** that can be applied to the cylinder if it is not to rotate, assuming the coefficient of static friction to be (*a*) zero at $A$ and 0.30 at $B$, (*b*) 0.25 at $A$ and 0.30 at $B$.

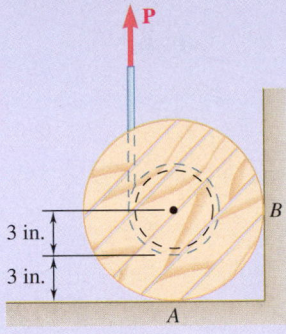

**Fig. P8.19**

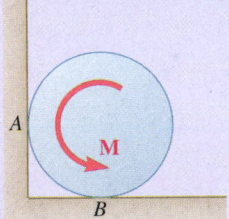

**Fig. P8.21 and P.22**

**8.22** The cylinder shown has a weight $W$ and radius $r$, and the coefficient of static friction $\mu_s$ is the same at $A$ and $B$. Determine the magnitude of the largest couple **M** that can be applied to the cylinder if it is not to rotate.

**8.23** The 10-lb uniform rod $AB$ is held in the position shown by the force **P**. Knowing that the coefficient of static friction is 0.20 at $A$ and $B$, determine the smallest value of $P$ for which equilibrium is maintained.

**8.24** In Prob. 8.23, determine the largest value of **P** for which equilibrium is maintained.

**8.25** A 6.5-m ladder $AB$ leans against a wall as shown. Assuming that the coefficient of static friction $\mu_s$ is zero at $B$, determine the smallest value of $\mu_s$ at $A$ for which equilibrium is maintained.

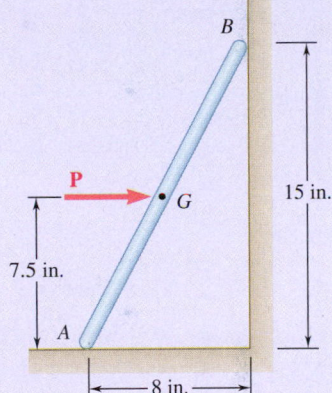

**Fig. P8.23**

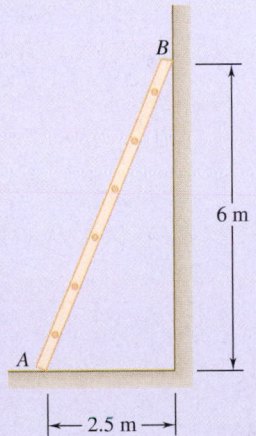

**Fig. P8.25 and P8.26**

**8.26** A 6.5-m ladder $AB$ leans against a wall as shown. Assuming that the coefficient of static friction $\mu_s$ is the same at $A$ and $B$, determine the smallest value of $\mu_s$ for which equilibrium is maintained.

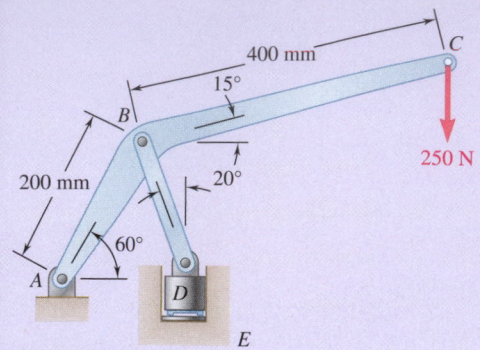

**Fig. P8.27**

**8.27** The press shown is used to emboss a small seal at *E*. Knowing that the coefficient of static friction between the vertical guide and the embossing die *D* is 0.30, determine the force exerted by the die on the seal.

**8.28** The machine base shown has a mass of 75 kg and is fitted with skids at *A* and *B*. The coefficient of static friction between the skids and the floor is 0.30. If a force **P** with a magnitude of 500 N is applied at corner *C*, determine the range of values of $\theta$ for which the base will not move.

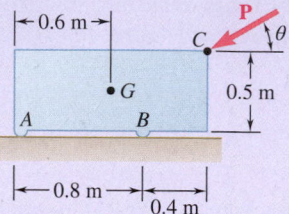

**Fig. P8.28**

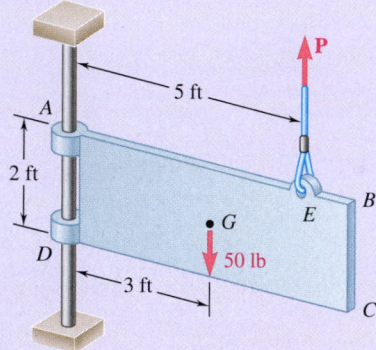

**Fig. P8.29**

**8.29** The 50-lb plate *ABCD* is attached at *A* and *D* to collars that can slide on the vertical rod. Knowing that the coefficient of static friction is 0.40 between both collars and the rod, determine whether the plate is in equilibrium in the position shown when the magnitude of the vertical force applied at *E* is (*a*) $P = 0$, (*b*) $P = 20$ lb.

**8.30** In Prob. 8.29, determine the range of values of the magnitude *P* of the vertical force applied at *E* for which the plate will move downward.

**8.31** A window sash weighing 10 lb is normally supported by two 5-lb sash weights. Knowing that the window remains open after one sash cord has broken, determine the smallest possible value of the coefficient of static friction. (Assume that the sash is slightly smaller than the frame and will bind only at points *A* and *D*.)

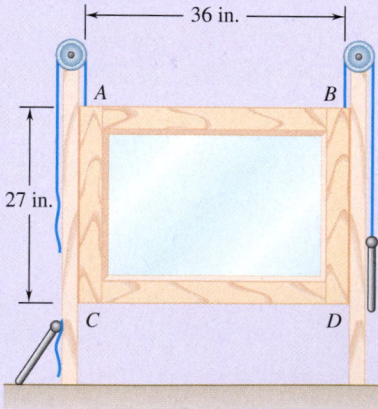

**Fig. P8.31**

**8.32** A 500-N concrete block is to be lifted by the pair of tongs shown. Determine the smallest allowable value of the coefficient of static friction between the block and the tongs at *F* and *G*.

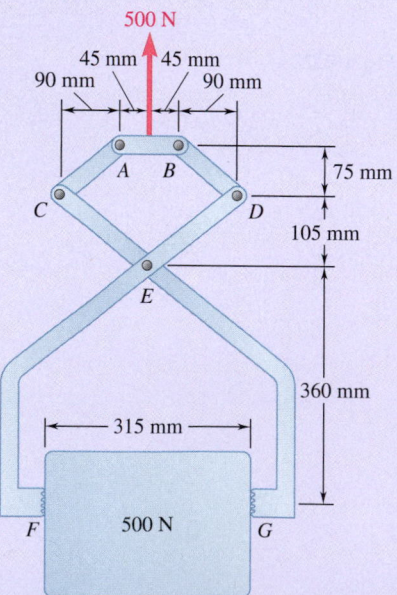

**Fig. P8.32**

**8.33** A pipe with a diameter of 60 mm is gripped by the stillson wrench shown. Portions $AB$ and $DE$ of the wrench are rigidly attached to each other, and portion $CF$ is connected by a pin at $D$. If the wrench is to grip the pipe and be self-locking, determine the required minimum coefficients of friction at $A$ and $C$.

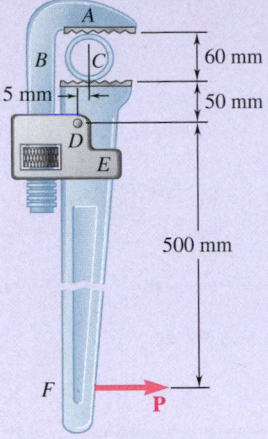

**Fig. P8.33**

**8.34** A driver starts the engine of an automobile that is stopped with its front wheels resting against a curb and tries to drive over the curb. Knowing that the radius of the wheels is 12 in., that the coefficient of static friction between the tires and the pavement is 0.90, and that 60 percent of the weight of the automobile is distributed over its front wheels and 40 percent over its rear wheels, determine the largest curb height $h$ that the automobile can negotiate, assuming (a) front-wheel drive, (b) rear-wheel drive.

**Fig. P8.34**

**8.35** Solve Prob. 8.34, assuming that the weight of the car is equally distributed over its front and rear wheels.

**8.36** Two uniform rods each of weight $W$ and length $L$ are maintained in the position shown by a couple $M_0$ applied to rod $CD$. Knowing that the coefficient of static friction between the rods is 0.40, determine the range of values of $M_0$ for which equilibrium is maintained.

**8.37** A 1.2-m plank with a mass of 3 kg rests on two joists. Knowing that the coefficient of static friction between the plank and the joists is 0.30, determine the magnitude of the horizontal force required to move the plank when (a) $a = 750$ mm, (b) $a = 900$ mm.

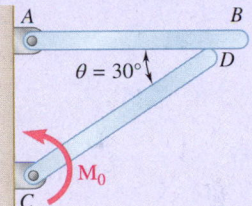

**Fig. P8.36**

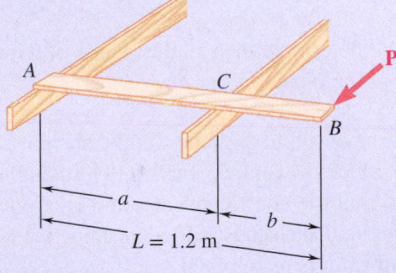

**Fig. P8.37**

**8.38** Two identical uniform boards, each with a weight of 40 lb, are temporarily leaned against each other as shown. Knowing that the coefficient of static friction between all surfaces is 0.40, determine (a) the largest magnitude of the force $P$ for which equilibrium will be maintained, (b) the surface at which motion will impend.

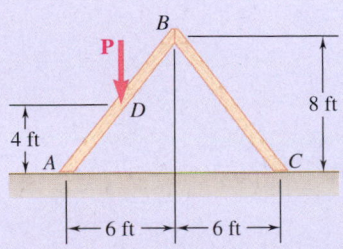

**Fig. P8.38**

451

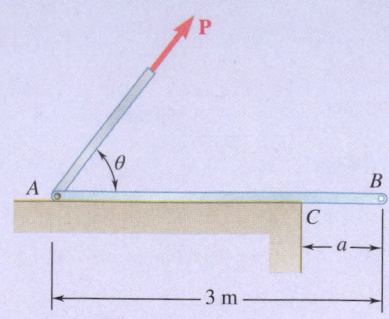

**Fig. P8.39 and P8.40**

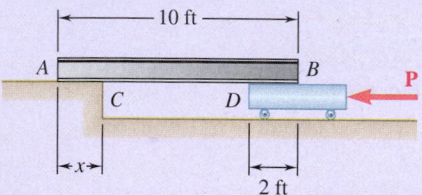

**Fig. P8.41**

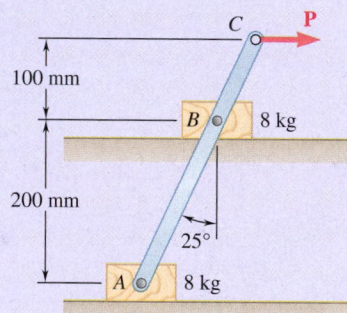

**Fig. P8.43**

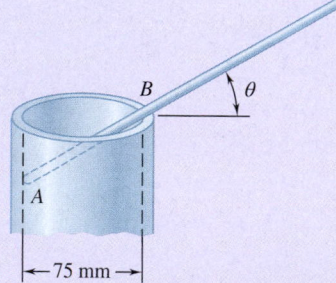

**Fig. P8.44**

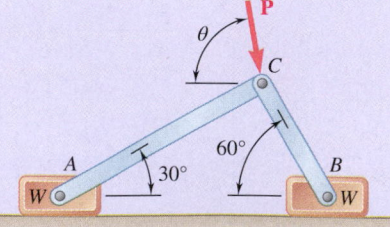

**Fig. P8.46 and P8.47**

**8.39** A uniform 20-kg tube resting on a loading dock will be moved by means of a cable attached at end $A$. Knowing that the coefficient of static friction between the tube and the dock is 0.30, determine the largest angle $\theta$ for which the tube will slide horizontally to the right and the corresponding magnitude of the force $\mathbf{P}$ when (a) $a = 0$, (b) $a = 0.75$ m.

**8.40** A uniform 20-kg tube rests on a loading dock with its end $B$ located at a distance $a = 0.25$ m from the edge $C$ of the dock. A cable attached at end $A$ forming an angle $\theta = 60°$ with the tube will be used to move the tube. Knowing that the coefficient of static friction between the tube and the dock is 0.30, determine (a) the smallest value of $P$ for which motion of the tube impends, (b) whether the tube tends to slide or to rotate about the edge $C$ of the dock.

**8.41** A 10-ft beam, weighing 1200 lb, is to be moved to the left onto the platform as shown. A horizontal force $\mathbf{P}$ is applied to the dolly, which is mounted on frictionless wheels. The coefficients of friction between all surfaces are $\mu_s = 0.30$ and $\mu_s = 0.25$, and initially, $x = 2$ ft. Knowing that the top surface of the dolly is slightly higher than the platform, determine the force $\mathbf{P}$ required to start moving the beam. (*Hint:* The beam is supported at $A$ and $D$.)

**8.42** (a) Show that the beam of Prob. 8.41 *cannot* be moved if the top surface of the dolly is slightly *lower* than the platform. (b) Show that the beam *can* be moved if two 175-lb workers stand on the beam at $B$, and determine how far to the left the beam can be moved.

**8.43** Two 8-kg blocks $A$ and $B$ resting on shelves are connected by a rod of negligible mass. Knowing that the magnitude of a horizontal force $\mathbf{P}$ applied at $C$ is slowly increased from zero, determine the value of $P$ for which motion occurs and what that motion is when the coefficient of static friction between all surfaces is (a) $\mu_s = 0.40$, (b) $\mu_s = 0.50$.

**8.44** A slender steel rod with a length of 225 mm is placed inside a pipe as shown. Knowing that the coefficient of static friction between the rod and the pipe is 0.20, determine the largest value of $\theta$ for which the rod will not fall into the pipe.

**8.45** In Prob. 8.44, determine the smallest value of $\theta$ for which the rod will not fall out of the pipe.

**8.46** Two slender rods of negligible weight are pin-connected at $C$ and attached to blocks $A$ and $B$, each with a weight $W$. Knowing that $\theta = 80°$ and that the coefficient of static friction between the blocks and the horizontal surface is 0.30, determine the largest value of $P$ for which equilibrium is maintained.

**8.47** Two slender rods of negligible weight are pin-connected at $C$ and attached to blocks $A$ and $B$, each with a weight $W$. Knowing that $P = 1.260W$ and that the coefficient of static friction between the blocks and the horizontal surface is 0.30, determine the range of values of $\theta$ between 0 and 180° for which equilibrium is maintained.

# 8.2 WEDGES AND SCREWS

Friction is a key element in analyzing the function and operation of several types of simple machines. Here we examine the wedge and the screw, which are both extensions of the inclined plane we analyzed in Sec. 8.1.

## 8.2A Wedges

Wedges are simple machines used to raise large stone blocks and other heavy loads. These loads are raised by applying to the wedge a force usually considerably smaller than the weight of the load. In addition, because of the friction between the surfaces in contact, a properly shaped wedge remains in place after being forced under the load. In this way, you can use a wedge advantageously to make small adjustments in the position of heavy pieces of machinery.

Consider the block $A$ shown in Fig. 8.7a. This block rests against a vertical wall $B$, and we want to raise it slightly by forcing a wedge $C$ between block $A$ and a second wedge $D$. We want to find the minimum value of the force $\mathbf{P}$ that we must apply to wedge $C$ to move the block. We assume that we know the weight $\mathbf{W}$ of the block, which is either given in pounds or determined in newtons from the mass of the block expressed in kilograms.

We have drawn the free-body diagrams of block $A$ and wedge $C$ in Fig. 8.7b and c. The forces acting on the block include its weight and the normal and friction forces at the surfaces of contact with wall $B$ and wedge $C$. The magnitudes of the friction forces $\mathbf{F}_1$ and $\mathbf{F}_2$ are equal, respectively, to $\mu_s N_1$ and $\mu_s N_2$, because the motion of the block must be started. It is important to show the friction forces with their correct sense. Because the block will move upward, the force $\mathbf{F}_1$ exerted by the wall on the block must be directed downward. On the other hand, because wedge $C$ moves to the right, the relative motion of $A$ with respect to $C$ is to the left, and the force $\mathbf{F}_2$ exerted by $C$ on $A$ must be directed to the right.

Now consider the free body $C$ in Fig. 8.7c. The forces acting on $C$ include the applied force $\mathbf{P}$ and the normal and friction forces at the surfaces of contact with $A$ and $D$. The weight of the wedge is small compared with the other forces involved and can be neglected. The forces exerted by $A$ on $C$ are equal and opposite to the forces $\mathbf{N}_2$ and $\mathbf{F}_2$ exerted by $C$ on $A$, so we denote them, respectively, by $-\mathbf{N}_2$ and $-\mathbf{F}_2$; the friction force $-\mathbf{F}_2$ therefore must be directed to the left. We check that the force $\mathbf{F}_3$ exerted by $D$ is also directed to the left.

We can reduce the total number of unknowns involved in the two free-body diagrams to four if we express the friction forces in terms of the normal forces. Then, because block $A$ and wedge $C$ are in equilibrium, we obtain four equations that we can solve to obtain the magnitude of $\mathbf{P}$. Note that, in the example considered here, it is more convenient to replace each pair of normal and friction forces by their resultant. Each free body is then subjected to only three forces, and we can solve the problem by drawing the corresponding force triangles (see Sample Prob. 8.5).

## 8.2B Square-Threaded Screws

Square-threaded screws are frequently part of jacks, presses, and other mechanisms. Their analysis is similar to the analysis of a block sliding along an inclined plane. (Screws are also commonly used as fasteners, but the threads on these screws are shaped differently.)

**Photo 8.3** Wedges are used as shown to split tree trunks because the normal forces exerted by a wedge on the wood are much larger than the force required to insert the wedge. ©Leslie Miller/agefotostock

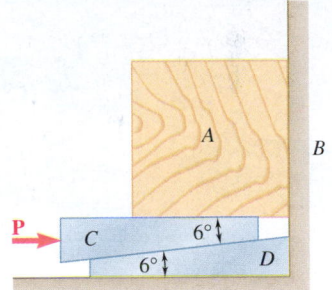

(a)

(b)

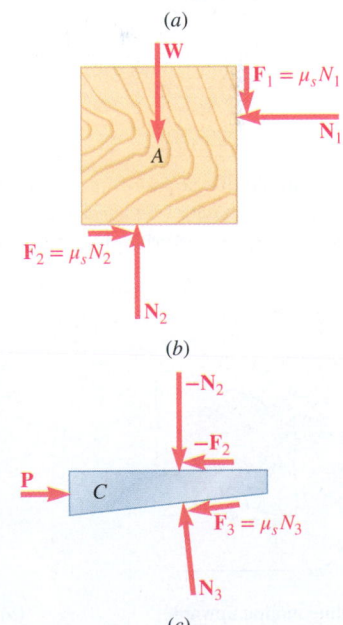

(c)

**Fig. 8.7** (a) A wedge $C$ used to raise a block $A$; (b) free-body diagram of block $A$; (c) free-body diagram of wedge $C$. Note the directions of the friction forces.

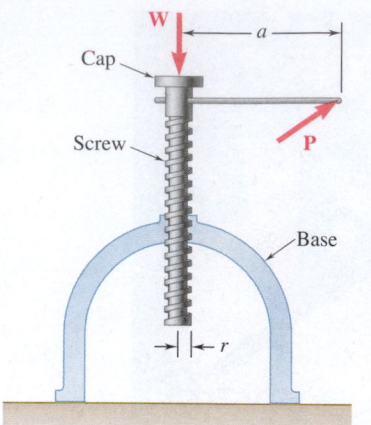

**Fig. 8.8** A screw as part of a jack carrying a load **W**.

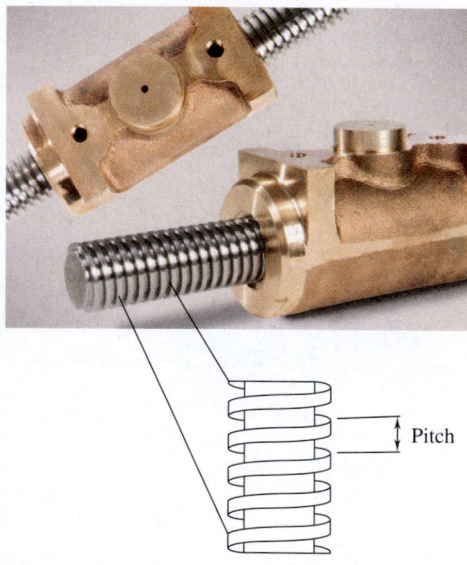

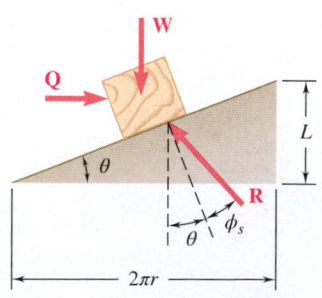

Pitch

**Photo 8.4** An example of a square-threaded screw, fitted to a sleeve, as might be used in an industrial application. Courtesy of REMPCO Inc.

Consider the jack shown in Fig. 8.8. The screw carries a load **W** and is supported by the base of the jack. Contact between the screw and the base takes place along a portion of their threads. By applying a force **P** on the handle, the screw can be made to turn and to raise the load **W**.

In Fig. 8.9a, we have unwrapped the thread of the base and shown it as a straight line. We obtained the correct slope by horizontally drawing the product $2\pi r$, where $r$ is the mean radius of the thread, and vertically drawing the **lead** $L$ of the screw, i.e., the distance through which the screw advances in one turn. The angle $\theta$ this line forms with the horizontal is the **lead angle**. Because the force of friction between two surfaces in contact does not depend upon the area of contact, we can assume a much smaller than actual area of contact between the two threads, which allows us to represent the screw as the block shown in Fig. 8.9a. Note that, in this analysis of the jack, we neglect the small friction force between cap and screw.

The free-body diagram of the block includes the load **W**, the reaction **R** of the base thread, and a horizontal force **Q**, which has the same effect as the force **P** exerted on the handle. The force **Q** should have the same moment as **P** about the axis of the screw, so its magnitude should be $Q = Pa/r$. We can obtain the value of force **Q**, and thus that of force **P** required to raise load **W**, from the free-body diagram shown in Fig. 8.9a. The friction angle is taken to be equal to $\phi_s$, because presumably the load is raised through a succession of short strokes. In mechanisms providing for the continuous rotation of a screw, it may be desirable to distinguish between the force required to start motion (using $\phi_s$) and that required to maintain motion (using $\phi_k$).

If the friction angle $\phi_s$ is larger than the lead angle $\theta$, the screw is said to be *self-locking;* it will remain in place under the load. To lower the load, we must then apply the force shown in Fig. 8.9b. If $\phi_s$ is smaller than $\theta$, the screw will unwind under the load; it is then necessary to apply the force shown in Fig. 8.9c to maintain equilibrium.

The lead of a screw should not be confused with its **pitch**. The *lead* is defined as the distance through which the screw advances in one turn; the *pitch* is the distance measured between two consecutive threads. Lead and pitch are equal in the case of *single-threaded* screws, but they are different in the case of *multiple-threaded* screws, i.e., screws having several independent threads. It is easily verified that for double-threaded screws the lead is twice as large as the pitch; for triple-threaded screws, it is three times as large as the pitch; etc.

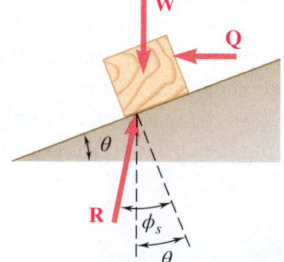

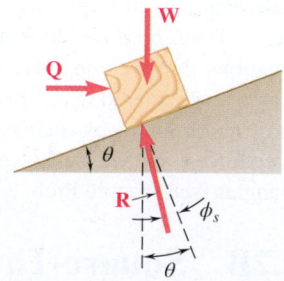

(a) Impending motion upward      (b) Impending motion downward with $\phi_s > \theta$      (c) Impending motion downward with $\phi_s < \theta$

**Fig. 8.9** Block-and-incline analysis of a screw. We can represent the screw as a block, because the force of friction does not depend on the area of contact between two surfaces.

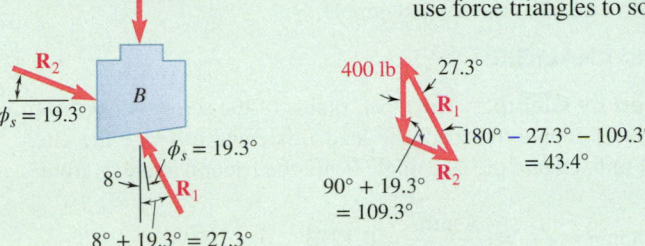

# Sample Problem 8.5

The position of the machine block $B$ is adjusted by moving the wedge $A$. Knowing that the coefficient of static friction is 0.35 between all surfaces of contact, determine the force $\mathbf{P}$ required to (a) raise block $B$, (b) lower block $B$.

**STRATEGY:** For both parts of the problem, normal forces and friction forces act between the wedge and the block. In part (a), you also have normal and friction forces at the left surface of the block; for part (b), they are on the right surface of the block. If you combine the normal and friction forces at each surface into resultants, you have a total of three forces acting on each body and can use force triangles to solve.

**MODELING:** For each part, draw the free-body diagrams of block $B$ and wedge $A$ together with the corresponding force triangles. Then, use the law of sines to find the desired forces. Note that, because $\mu_s = 0.35$, the angle of friction is

$$\phi_s = \tan^{-1} 0.35 = 19.3°$$

**ANALYSIS:** **a. Force P to Raise Block.**

**Free Body: Block $B$ (Fig. 1).** The friction force on block $B$ due to wedge $A$ is to the left, so the resultant $\mathbf{R}_1$ is at an angle equal to the slope of the wedge plus the angle of friction.

$$\frac{R_1}{\sin 109.3°} = \frac{400 \text{ lb}}{\sin 43.4°} \qquad R_1 = 549 \text{ lb}$$

**Fig. 1** Free-body diagram of the block and its force triangle—block being raised.

**Free Body: Wedge $A$ (Fig. 2).** The friction forces on wedge $A$ are to the right.

$$\frac{P}{\sin 46.6°} = \frac{549 \text{ lb}}{\sin 70.7°} \qquad \mathbf{P} = 423 \text{ lb} \leftarrow \ \blacktriangleleft$$

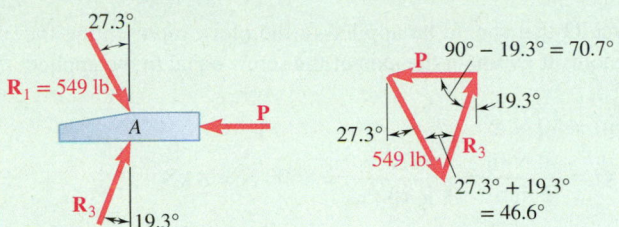

**Fig. 2** Free-body diagram of the wedge and its force triangle—block being raised.

**b. Force P to Lower Block.**

**Free Body: Block $B$ (Fig. 3).** Now the friction force on block $B$ due to wedge $A$ is to the right, so the resultant $\mathbf{R}_1$ is at an angle equal to the angle of friction minus the slope of the wedge.

$$\frac{R_1}{\sin 70.7°} = \frac{400 \text{ lb}}{\sin 98.0°} \qquad R_1 = 381 \text{ lb}$$

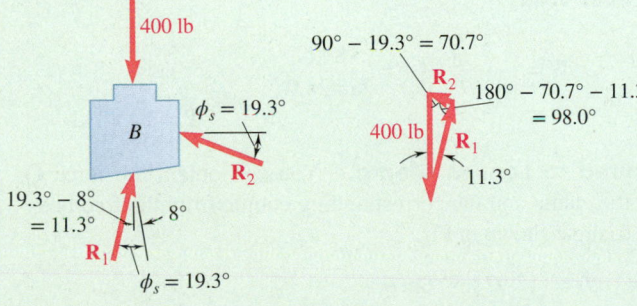

**Fig. 3** Free-body diagram of the block and its force triangle—block being lowered.

**Free Body: Wedge $A$ (Fig. 4).** The friction forces on wedge $A$ are to the left.

$$\frac{P}{\sin 30.6°} = \frac{381 \text{ lb}}{\sin 70.7°} \qquad \mathbf{P} = 206 \text{ lb} \rightarrow \ \blacktriangleleft$$

**REFLECT and THINK:** The force needed to lower the block is much less than the force needed to raise the block, which makes sense.

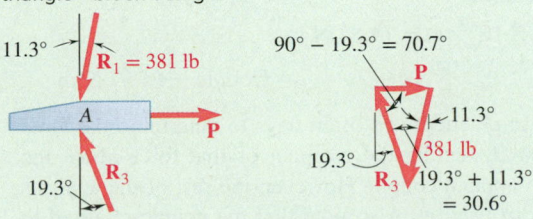

**Fig. 4** Free-body diagram of the wedge and its force triangle—block being lowered.

# Sample Problem 8.6

A clamp is used to hold two pieces of wood together as shown. The clamp has a double square thread with a mean diameter of 10 mm and a pitch of 2 mm. The coefficient of friction between threads is $\mu_s = 0.30$. If a maximum couple of 40 N·m is applied in tightening the clamp, determine (a) the force exerted on the pieces of wood, (b) the couple required to loosen the clamp.

**STRATEGY:**    If you represent the screw by a block, as in the analysis of this section, you can determine the incline of the screw from the geometry given in the problem, and you can find the force applied to the block by setting the moment of that force equal to the applied couple.

## MODELING and ANALYSIS:

**a. Force Exerted by Clamp.**    The mean radius of the screw is $r = 5$ mm. Because the screw is double-threaded, the lead $L$ is equal to twice the pitch: $L = 2(2$ mm$) = 4$ mm. Obtain the lead angle $\theta$ and the friction angle $\phi_s$ from

$$\tan \theta = \frac{L}{2\pi r} = \frac{4 \text{ mm}}{10\pi \text{ mm}} = 0.1273 \quad \theta = 7.3°$$

$$\tan \phi_s = \mu_s = 0.30 \qquad\qquad \phi_s = 16.7°$$

You can find the force **Q** that should be applied to the block representing the screw by setting its moment $Qr$ about the axis of the screw equal to the applied couple.

$$Q(5 \text{ mm}) = 40 \text{ N·m}$$

$$Q = \frac{40 \text{ N·m}}{5 \text{ mm}} = \frac{40 \text{ N·m}}{5 \times 10^{-3}\text{m}} = 8000 \text{ N} = 8 \text{ kN}$$

Now you can draw the free-body diagram and the corresponding force triangle for the block (Fig. 1). Solve the triangle to find the magnitude of the force **W** exerted on the pieces of wood.

$$W = \frac{Q}{\tan(\theta + \phi_s)} = \frac{8 \text{ kN}}{\tan 24.0°}$$

$$W = 17.97 \text{ kN} \quad \blacktriangleleft$$

**b. Couple Required to Loosen Clamp.**    You can obtain the force **Q** required to loosen the clamp and the corresponding couple from the free-body diagram and force triangle shown in Fig. 2.

$$Q = W \tan(\phi_s - \theta) = (17.97 \text{ kN}) \tan 9.4°$$
$$= 2.975 \text{ kN}$$
$$\text{Couple} = Qr = (2.975 \text{ kN})(5 \text{ mm})$$
$$= (2.975 \times 10^3 \text{ N})(5 \times 10^{-3} \text{ m}) = 14.87 \text{ N·m}$$

$$\text{Couple} = 14.87 \text{ N·m} \quad \blacktriangleleft$$

**REFLECT and THINK:**    In practice, you often have to determine the force effectively acting on a screw by setting the moment of that force about the axis of the screw equal to an applied couple. However, the rest of the analysis is mostly an application of dry friction. Also note that the couple required to loosen a screw is not the same as the couple required to tighten it.

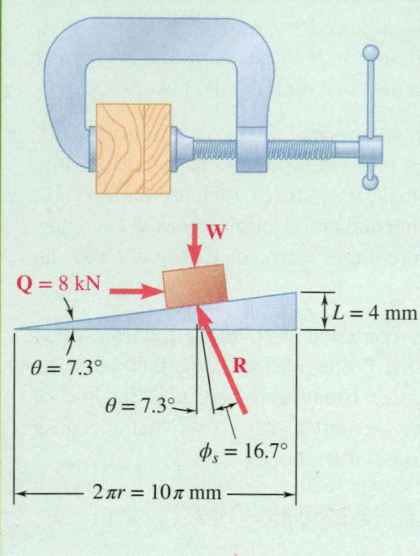

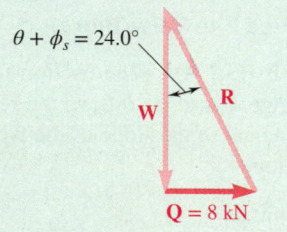

**Fig. 1** Free-body diagram of the block and its force triangle—clamp being tightened.

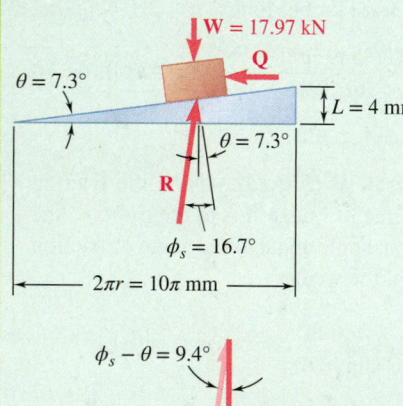

**Fig. 2** Free-body diagram of the block and its force triangle—clamp being loosened.

# SOLVING PROBLEMS
# ON YOUR OWN

In this section, you saw how to apply the laws of friction to the solution of problems involving **wedges** and **square-threaded screws**.

**1. Wedges.** Keep the following steps in mind when solving a problem involving a wedge.

    **a. First draw a free-body diagram of the wedge and of each of the other bodies involved.** Carefully note the sense of the relative motion of all surfaces of contact and show each friction force acting in *a direction opposite* to the direction of that relative motion.

    **b. Show the maximum static friction force $F_m$** at each surface if the wedge is to be inserted or removed, because motion will be impending in each of these cases.

    **c. The reaction R and the angle of friction,** rather than the normal force and the friction force, are most useful in many applications. You can then draw one or more force triangles and determine the unknown quantities either graphically or by trigonometry (Sample Prob. 8.5).

**2. Square-threaded screws.** The analysis of a square-threaded screw is equivalent to the analysis of a block sliding on an incline. To draw the appropriate incline, you need to unwrap the thread of the screw and represent it as a straight line (Sample Prob. 8.6). When solving a problem involving a square-threaded screw, keep the following steps in mind.

    **a. Do not confuse the pitch of a screw with the lead of a screw.** The *pitch* of a screw is the distance between two consecutive threads, whereas the *lead* of a screw is the distance the screw advances in one full turn. The lead and the pitch are equal only in single-threaded screws. In a double-threaded screw, the lead is twice the pitch.

    **b. The couple required to tighten a screw is different from the couple required to loosen it.** Also, screws used in jacks and clamps are usually *self-locking;* that is, the screw will remain stationary as long as no couple is applied to it, and a couple must be applied to the screw to loosen it (Sample Prob. 8.6).

# Problems

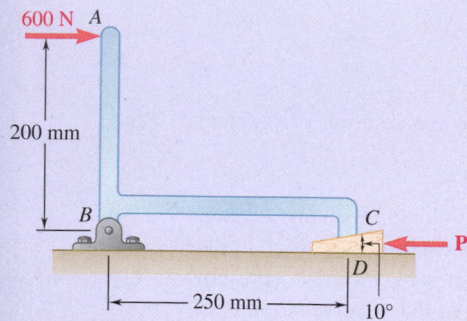

600 N  A

200 mm

B

C

D

250 mm

10°

P

**Fig. P8.48**

**8.48** The machine part *ABC* is supported by a frictionless hinge at *B* and a 10° wedge at *C*. Knowing that the coefficient of static friction is 0.20 at both surfaces of the wedge, determine (*a*) the force **P** required to move the wedge to the left, (*b*) the components of the corresponding reaction at *B*.

**8.49** Solve Prob. 8.48 assuming that the wedge is moved to the right.

**8.50 and 8.51**  Two 6° wedges of negligible weight are used to move and position the 900-kg block. Knowing that the coefficient of static friction is 0.30 at all surfaces of contact, determine the smallest force **P** that should be applied as shown to one of the wedges.

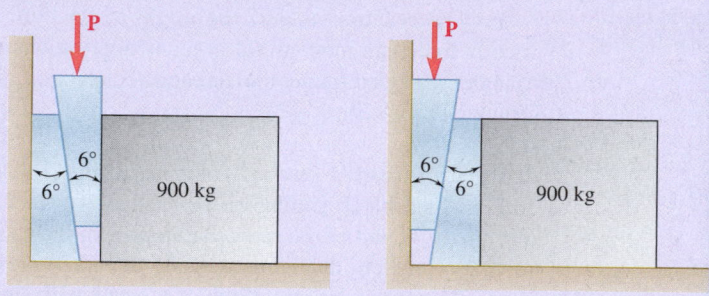

P

6°

6°

900 kg

**Fig. P8.50**

P

6°

6°

900 kg

**Fig. P8.51**

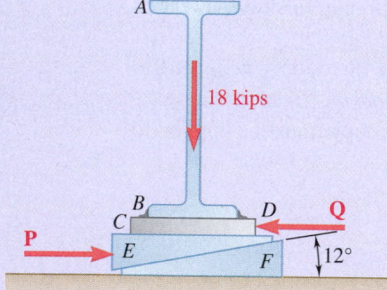

A

18 kips

B

C

D

Q

P

E

F

12°

**Fig. P8.52**

**8.52** The elevation of the end of the steel beam supported by a concrete floor is adjusted by means of the steel wedges *E* and *F*. The base plate *CD* has been welded to the lower flange of the beam, and the end reaction of the beam is known to be 18 kips. The coefficient of static friction is 0.30 between two steel surfaces and 0.60 between steel and concrete. If horizontal motion of the beam is prevented by the force **Q**, determine (*a*) the force **P** required to raise the beam, (*b*) the corresponding force **Q**.

**8.53** Solve Prob. 8.52 assuming that the end of the beam is to be lowered.

**8.54** Block *A* supports a pipe column and rests as shown on wedge *B*. Knowing that the coefficient of static friction at all surfaces of contact is 0.25 and that $\theta = 45°$, determine the smallest force **P** required to raise block *A*.

**8.55** Block *A* supports a pipe column and rests as shown on wedge *B*. Knowing that the coefficient of static friction at all surfaces of contact is 0.25 and that $\theta = 45°$, determine the smallest force **P** for which equilibrium is maintained.

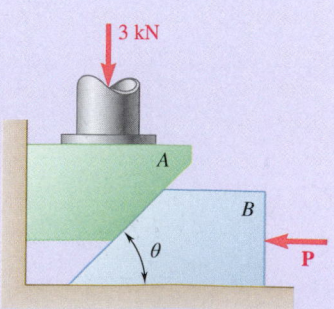

3 kN

A

B

θ

P

**Fig. P8.54, P8.55, and P8.56**

**8.56** Block *A* supports a pipe column and rests as shown on wedge *B*. The coefficient of static friction at all surfaces of contact is 0.25. If **P** = 0, determine (*a*) the angle $\theta$ for which sliding is impending, (*b*) the corresponding force exerted on the block by the vertical wall.

**8.57** A 200-lb block rests as shown on a wedge of negligible weight. Knowing that the coefficient of static friction is 0.25 at all surfaces of contact, determine the angle $\theta$ for which sliding is impending and compute the corresponding value of the normal force exerted on the block by the vertical wall.

**8.58** A 15° wedge is forced into a saw cut to prevent binding of the circular saw. The coefficient of static friction between the wedge and the wood is 0.25. Knowing that a horizontal force **P** with a magnitude of 30 lb was required to insert the wedge, determine the magnitude of the forces exerted on the board by the wedge after insertion.

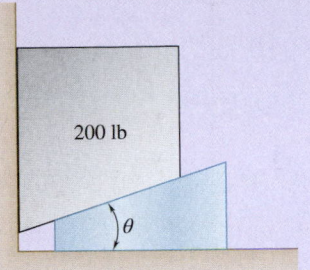

**Fig. P8.57**

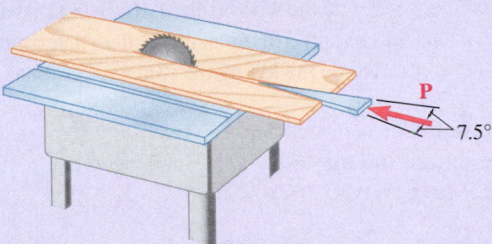

**Fig. P8.58**

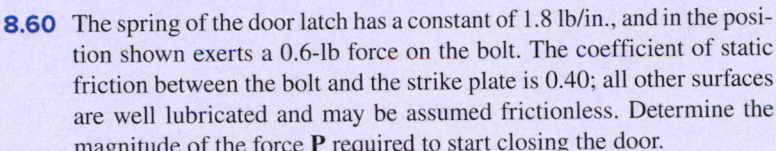

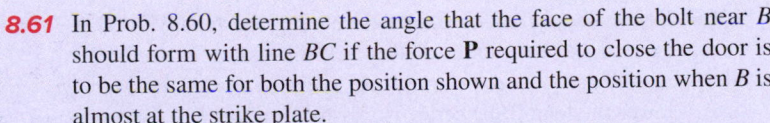

**8.59** A 12° wedge is used to spread a split ring. The coefficient of static friction between the wedge and the ring is 0.30. Knowing that a force **P** with a magnitude of 120 N was required to insert the wedge, determine the magnitude of the forces exerted on the ring by the wedge after insertion.

**8.60** The spring of the door latch has a constant of 1.8 lb/in., and in the position shown exerts a 0.6-lb force on the bolt. The coefficient of static friction between the bolt and the strike plate is 0.40; all other surfaces are well lubricated and may be assumed frictionless. Determine the magnitude of the force **P** required to start closing the door.

**8.61** In Prob. 8.60, determine the angle that the face of the bolt near $B$ should form with line $BC$ if the force **P** required to close the door is to be the same for both the position shown and the position when $B$ is almost at the strike plate.

**8.62** An 8° wedge is to be forced under a machine base at $B$. Knowing that the coefficient of static friction at all surfaces is 0.15, (a) determine the force **P** required to move the wedge, (b) indicate whether the machine base will slide on the floor.

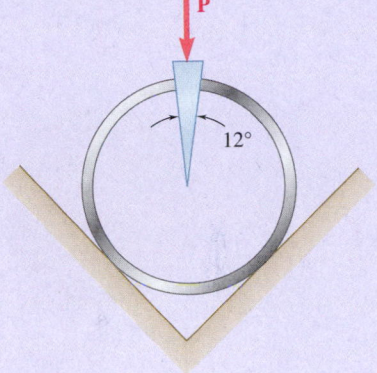

**Fig. P8.59**

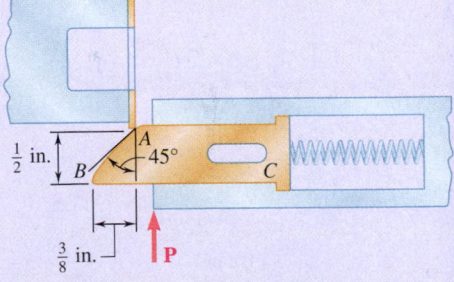

**Fig. P8.60**

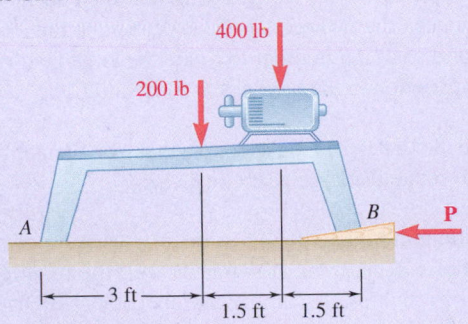

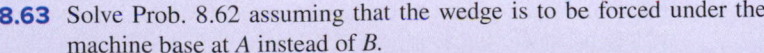

**Fig. P8.62**

**8.63** Solve Prob. 8.62 assuming that the wedge is to be forced under the machine base at $A$ instead of $B$.

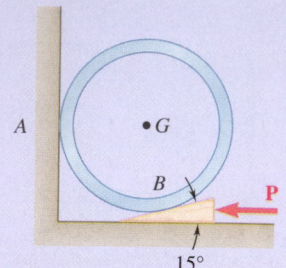

Fig. P8.64 and P8.65

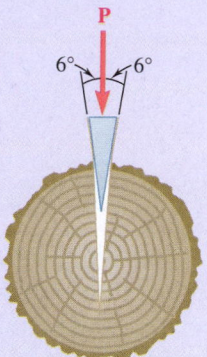

Fig. P8.66

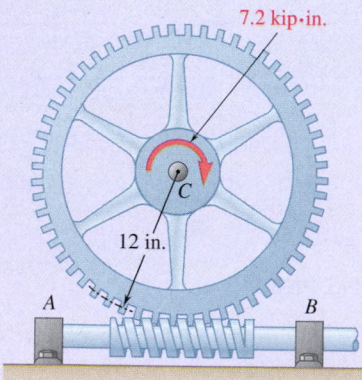

Fig. P8.69

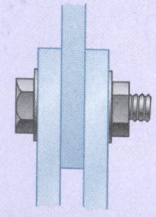

Fig. P8.71

**8.64** A 15° wedge is forced under a 50-kg pipe as shown. The coefficient of static friction at all surfaces is 0.20. (*a*) Show that slipping will occur between the pipe and the vertical wall. (*b*) Determine the force **P** required to move the wedge.

**8.65** A 15° wedge is forced under a 50-kg pipe as shown. Knowing that the coefficient of static friction at both surfaces of the wedge is 0.20, determine the largest coefficient of static friction between the pipe and the vertical wall for which slipping will occur at *A*.

**8.66** A 12° wedge is used to split a log. The coefficient of static friction between the wedge and the wood is 0.40. Knowing that a force **P** of magnitude 3.2 kN was required to insert the wedge, determine the magnitude of the forces exerted on the log by the wedge after it has been inserted.

**8.67** A conical wedge is placed between two horizontal plates that are then slowly moved toward each other. Indicate what will happen to the wedge if (*a*) $\mu_s = 0.20$, (*b*) $\mu_s = 0.30$.

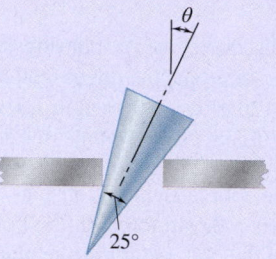

Fig. P8.67

**8.68** Derive the following formulas relating the load **W** and the force **P** exerted on the handle of the jack discussed in Sec. 8.2B. (*a*) $P = (Wr/a) \tan (\theta + \phi_s)$ to raise the load; (*b*) $P = (Wr/a) \tan (\phi_s - \theta)$ to lower the load if the screw is self-locking; (*c*) $P = (Wr/a) \tan (\theta - \phi_s)$ to hold the load if the screw is not self-locking.

**8.69** The square-threaded worm gear shown has a mean radius of 1.5 in. and a lead of 0.375 in. The large gear is subjected to a constant clockwise couple of 7.2 kip·in. Knowing that the coefficient of static friction between the two gears is 0.12, determine the couple that must be applied to shaft *AB* in order to rotate the large gear counterclockwise. Neglect friction in the bearings at *A*, *B*, and *C*.

**8.70** In Prob. 8.69, determine the couple that must be applied to shaft *AB* in order to rotate the large gear clockwise.

**8.71** High-strength bolts are used in the construction of many steel structures. For a 1-in.-nominal-diameter bolt, the required minimum bolt tension is 51 kips. Assuming the coefficient of friction to be 0.30, determine the required couple that should be applied to the bolt and nut. The mean diameter of the thread is 0.94 in., and the lead is 0.125 in. Neglect friction between the nut and washer, and assume the bolt to be square-threaded.

**8.72** The position of the automobile jack shown is controlled by a screw *ABC* that is single-threaded at each end (right-handed thread at *A*, left-handed thread at *C*). Each thread has a pitch of 2.5 mm and a mean diameter of 9 mm. If the coefficient of static friction is 0.15, determine the magnitude of the couple **M** that must be applied to raise the automobile.

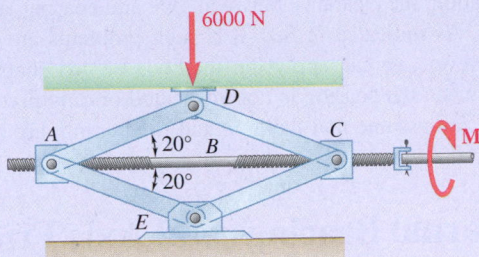

**Fig. P8.72**

**8.73** For the jack of Prob. 8.72, determine the magnitude of the couple **M** that must be applied to lower the automobile.

**8.74** The vise shown consists of two members connected by two double-threaded screws with a mean radius of 0.25 in. and pitch of 0.08 in. The lower member is threaded at *A* and *B* ($\mu_s = 0.35$), but the upper member is not threaded. It is desired to apply two equal and opposite forces of 120 lb on the blocks held between the jaws. (*a*) What screw should be adjusted first? (*b*) What is the maximum couple applied in tightening the second screw?

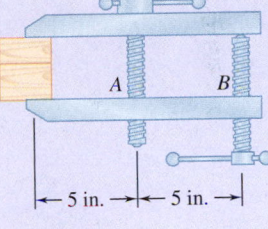

**Fig. P8.74**

**8.75** In the vise shown, the screw is single-threaded in the upper member; it passes through the lower member and is held by a frictionless washer. The pitch of the screw is 3 mm, its mean radius is 12 mm, and the coefficient of static friction is 0.15. Determine the magnitude *P* of the forces exerted by the jaws when a 60-N·m couple is applied to the screw.

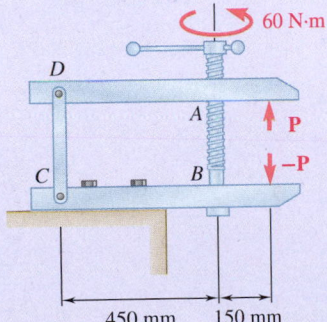

**Fig. P8.75**

**8.76** Solve Prob. 8.75, assuming that the screw is single-threaded at both *A* and *B* (right-handed thread at *A* and left-handed thread at *B*).

## *8.3 FRICTION ON AXLES, DISKS, AND WHEELS

**Journal bearings** are used to provide lateral support to rotating shafts and axles. **Thrust bearings** are used to provide axial support to shafts and axles. If the journal bearing is fully lubricated, the frictional resistance depends upon the speed of rotation, the clearance between axle and bearing, and the viscosity of the lubricant. As indicated in Sec. 8.1, such problems are studied in fluid mechanics. However, we can apply the methods of this chapter to the study of axle friction when the bearing is not lubricated or only partially lubricated. In this case, we can assume that the axle and the bearing are in direct contact along a single straight line.

### 8.3A Journal Bearings and Axle Friction

Consider two wheels, each with a weight of **W**, rigidly mounted on an axle supported symmetrically by two journal bearings (Fig. 8.10a). If the wheels rotate, we find that, to keep them rotating at constant speed, it is necessary to apply a couple **M** to each of them. The free-body diagram in Fig. 8.10c represents one of the wheels and the corresponding half axle in projection on a plane perpendicular to the axle. The forces acting on the free body include the weight **W** of the wheel,

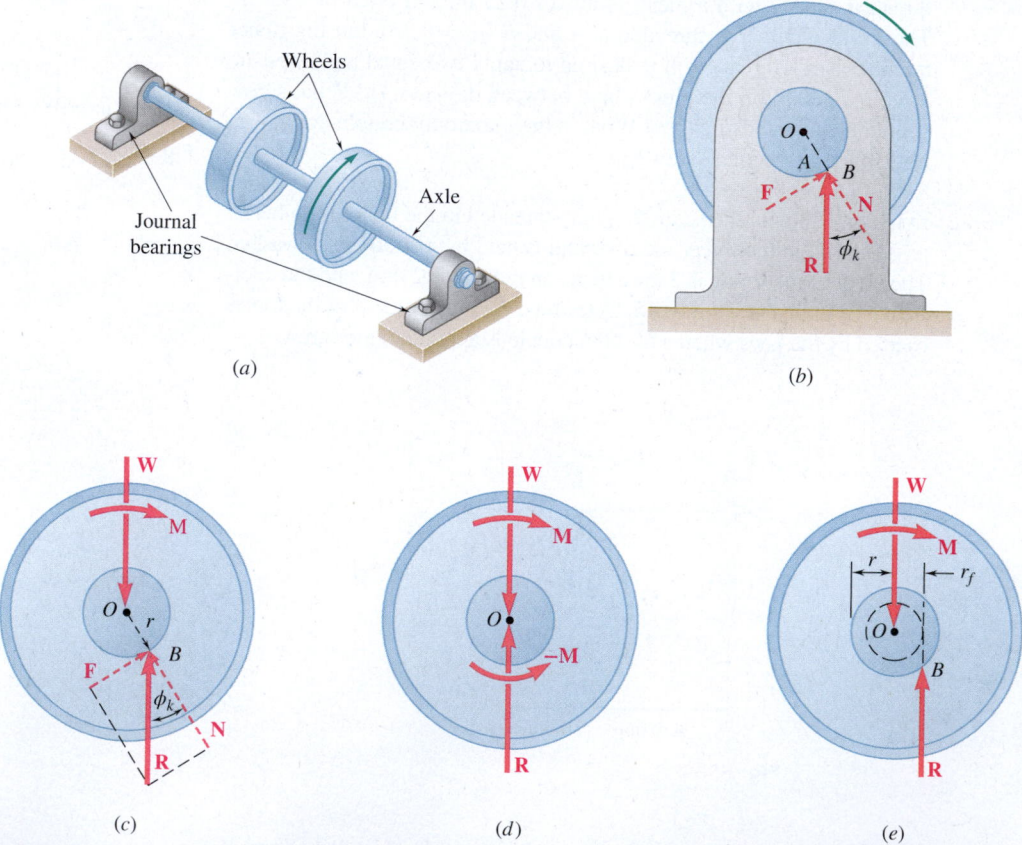

**Fig. 8.10** (a) Two wheels supported by two journal bearings; (b) point of contact when the axle is rotating; (c) free-body diagram of one wheel and corresponding half axle; (d) frictional resistance produces a couple that opposes the couple maintaining the axle in motion; (e) graphical analysis with circle of friction.

the couple **M** required to maintain its motion, and a force **R** representing the reaction of the bearing. This force is vertical, equal, and opposite to **W**, but it does not pass through the center $O$ of the axle; **R** is located to the right of $O$ at a distance such that its moment about $O$ balances the moment **M** of the couple. Therefore, when the axle rotates, contact between the axle and bearing does not take place at the lowest point $A$. Instead, contact takes place at point $B$ (Fig. 8.10b) or, rather, along a straight line intersecting the plane of the figure at $B$.

Physically, the location of contact is explained by the fact that, when the wheels are set in motion, the axle "climbs" in the bearings until slippage occurs. After sliding back slightly, the axle settles more or less in the position shown. This position is such that the angle between the reaction **R** and the normal to the surface of the bearing is equal to the angle of kinetic friction $\phi_k$. The distance from $O$ to the line of action of **R** is thus $r \sin \phi_k$, where $r$ is the radius of the axle. Setting $\Sigma M_O = 0$ for the forces acting on the free body (the wheel), we obtain the magnitude of the couple **M** required to overcome the frictional resistance of one of the bearings:

$$M = Rr \sin \phi_k \qquad (8.5)$$

For small values of the angle of friction, we can replace $\sin \phi_k$ by $\tan \phi_k$; that is, by $\mu_k$. This gives us the approximate formula

$$M \approx Rr\mu_k \qquad (8.6)$$

In the solution of certain problems, it may be more convenient to let the line of action of **R** pass through $O$, as it does when the axle does not rotate. In such a case, you need to add a couple $-\textbf{M}$, with the same magnitude as the couple **M** but of opposite sense, to the reaction **R** (Fig. 8.10d). This couple represents the frictional resistance of the bearing.

If a graphical solution is preferred, you can readily draw the line of action of **R** (Fig. 8.10e) if you note that it must be tangent to a circle centered at $O$ and with a radius

$$r_f = r \sin \phi_k \approx r\mu_k \qquad (8.7)$$

This circle is called the **circle of friction** of the axle and bearing, and it is independent of the loading conditions of the axle.

## 8.3B  Thrust Bearings and Disk Friction

Two types of thrust bearings are commonly used to provide axial support to rotating shafts and axles: (1) **end bearings** and (2) **collar bearings** (Fig. 8.11). In the case of collar bearings, friction forces develop between the two ring-shaped areas in contact. In the case of end bearings, friction takes place over full circular areas or over ring-shaped areas when the end of the shaft is hollow. Friction between circular areas, called **disk friction**, also occurs in other mechanisms, such as disk clutches.

To obtain a formula for the most general case of disk friction, let us consider a rotating hollow shaft. A couple **M** keeps the shaft rotating at constant speed, while an axial force **P** maintains it in contact with a fixed bearing (Fig. 8.12). Contact between the shaft and the bearing takes place over a ring-shaped area with an inner radius of $R_1$ and an outer radius of $R_2$. Assuming that the pressure between the two surfaces in contact is uniform, we find that the magnitude of the normal force $\Delta \textbf{N}$ exerted on an element of area $\Delta A$ is $\Delta N = P \, \Delta A/A$,

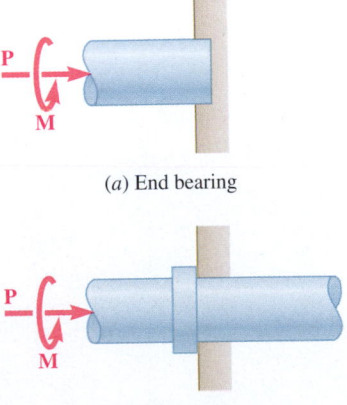

(a) End bearing

(b) Collar bearing

**Fig. 8.11** In thrust bearings, an axial force keeps the rotating axle in contact with the support bearing.

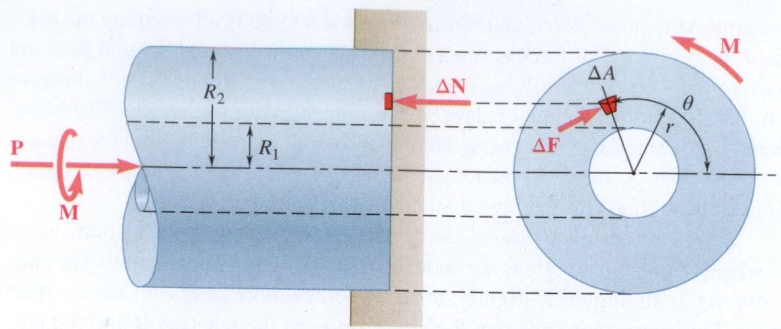

**Fig. 8.12** Geometry of the frictional contact surface in a thrust bearing.

where $A = \pi(R_2^2 - R_1^2)$, and that the magnitude of the friction force $\Delta \mathbf{F}$ acting on $\Delta A$ is $\Delta F = \mu_k \, \Delta N$. Let's use $r$ to denote the distance from the axis of the shaft to the element of area $\Delta A$. Then, the magnitude $\Delta M$ of the moment of $\Delta \mathbf{F}$ about the axis of the shaft is

$$\Delta M = r \, \Delta F = \frac{r \mu_k P \, \Delta A}{\pi(R_2^2 - R_1^2)}$$

Equilibrium of the shaft requires that the moment $\mathbf{M}$ of the couple applied to the shaft be equal in magnitude to the sum of the moments of the friction forces $\Delta \mathbf{F}$ opposing the motion of the shaft. Replacing $\Delta A$ by the infinitesimal element $dA = r \, d\theta \, dr$ used with polar coordinates and integrating over the area of contact, the expression for the magnitude of the couple $\mathbf{M}$ required to overcome the frictional resistance of the bearing is

$$M = \frac{\mu_k P}{\pi(R_2^2 - R_1^2)} \int_0^{2\pi} \int_{R_1}^{R_2} r^2 \, dr \, d\theta$$

$$= \frac{\mu_k P}{\pi(R_2^2 - R_1^2)} \int_0^{2\pi} \tfrac{1}{3}(R_2^3 - R_1^3) d\theta$$

$$M = \tfrac{2}{3}\mu_k P \frac{R_2^3 - R_1^3}{R_2^2 - R_1^2} \tag{8.8}$$

When contact takes place over a full circle with a radius of $R$, formula (8.8) reduces to

$$M = \tfrac{2}{3}\mu_k PR \tag{8.9}$$

This value of $M$ is the same value we would obtain if contact between the shaft and bearing took place at a single point located at a distance $2R/3$ from the axis of the shaft.

The largest couple that can be transmitted by a disk clutch without causing slippage is given by a formula similar to Eq. (8.9), where $\mu_k$ has been replaced by the coefficient of static friction $\mu_s$.

## 8.3C   Wheel Friction and Rolling Resistance

The wheel is one of the most important inventions of our civilization. Among many other uses, with a wheel we can move heavy loads with relatively little effort. Because the point where the wheel is in contact with the ground at any given instant has no relative motion with respect to the ground, use of the wheel avoids the large friction forces that would arise if the load were in direct contact with the ground. However, some resistance to the wheel's motion does occur. This resistance has two distinct causes. It is due to (1) a combined effect of axle friction and friction at the rim and (2) the fact that the wheel and the ground deform, causing contact between the wheel and ground to take place over an area rather than at a single point.

To understand better the first cause of resistance to the motion of a wheel, consider a railroad car supported by eight wheels mounted on axles and bearings. We assume the car is moving to the right at constant speed along a straight horizontal track. The free-body diagram of one of the wheels is shown in Fig. 8.13*a*. The forces acting on the free body include the load **W** supported by the wheel and the normal reaction **N** of the track. Because **W** passes through the center *O* of the axle, we represent the frictional resistance of the bearing by a counterclockwise couple **M** (see Sec. 8.3A). Then, to keep the free body in equilibrium, we must add two equal and opposite forces **P** and **F**, forming a clockwise couple of moment −**M**. The force **F** is the friction force exerted by the track on the wheel, and **P** represents the force that should be applied to the wheel to keep it rolling at constant speed. Note that the forces **P** and **F** would not exist if there were no friction between the wheel and the track. The couple **M** representing the axle friction would then be zero; the wheel would slide on the track without turning in its bearing.

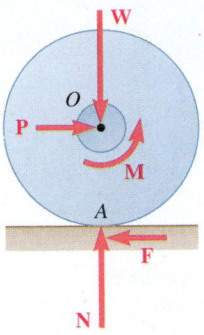

(*a*) Effect of axle friction

The couple **M** and the forces **P** and **F** also reduce to zero when there is no axle friction. For example, a wheel that is not held in bearings but rolls freely and at constant speed on horizontal ground (Fig. 8.13*b*) is subjected to only two forces: its own weight **W** and the normal reaction **N** of the ground. No friction force acts on the wheel regardless of the value of the coefficient of friction between the wheel and ground. Thus, a wheel rolling freely on horizontal ground should keep rolling indefinitely.

Experience, however, indicates that a free wheel does slow down and eventually come to rest. This is due to the second type of resistance mentioned at the beginning of this section, known as **rolling resistance**. Under the load **W**, both the wheel and the ground deform slightly, causing the contact between wheel and ground to take place over a certain area. Experimental evidence shows that the resultant of the forces exerted by the ground on the wheel over this area is a force **R** applied at a point *B*, which is not located directly under the center *O* of the wheel but slightly in front of it (Fig. 8.13*c*). To balance the moment of **W** about *B* and to keep the wheel rolling at constant speed, it is necessary to apply a horizontal force **P** at the center of the wheel. Setting $\Sigma M_B = 0$, we obtain

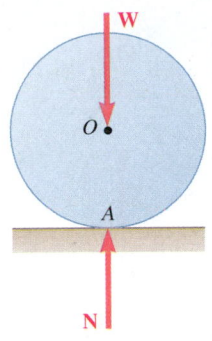

(*b*) Free wheel

$$Pr = Wb \qquad \textbf{(8.10)}$$

where *r* = radius of wheel
  *b* = horizontal distance between *O* and *B*

The distance *b* is commonly called the **coefficient of rolling resistance**. Note that *b* is not a dimensionless coefficient, because it represents a length; *b* is usually expressed in inches or in millimeters. The value of *b* depends upon several parameters in a manner that has not yet been clearly established. Values of the coefficient of rolling resistance vary from about 0.01 in. or 0.25 mm for a steel wheel on a steel rail to 5.0 in. or 125 mm for the same wheel on soft ground.

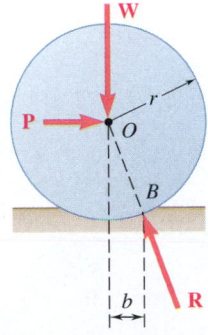

(*c*) Rolling resistance

**Fig. 8.13**   (*a*) Free-body diagram of a rolling wheel, showing the effect of axle friction; (*b*) free-body diagram of a free wheel, not connected to an axle; (*c*) free-body diagram of a rolling wheel, showing the effect of rolling resistance.

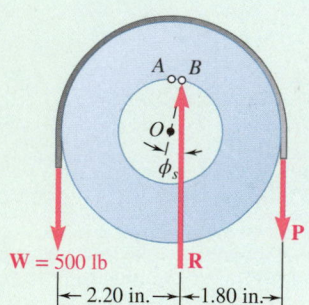

**Fig. 1** Free-body diagram of the pulley—smallest vertical force to raise the load.

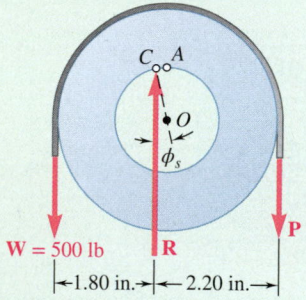

**Fig. 2** Free-body diagram of the pulley—smallest vertical force to hold load.

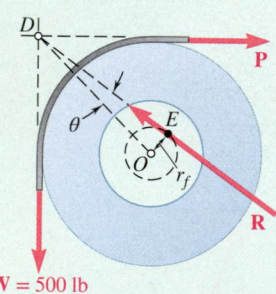

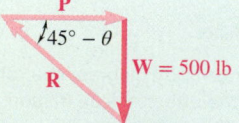

**Fig. 3** Free-body diagram of the pulley and the force triangle—smallest horizontal force to raise load.

## Sample Problem 8.7

A pulley with a diameter of 4 in. can rotate about a fixed shaft with a diameter of 2 in. The coefficient of static friction between the pulley and shaft is 0.20. Determine (*a*) the smallest vertical force **P** required to start raising a 500-lb load, (*b*) the smallest vertical force **P** required to hold the load, (*c*) the smallest horizontal force **P** required to start raising the same load.

**STRATEGY:** You can use the radius of the circle of friction to position the reaction of the pulley in each scenario and then apply the principles of equilibrium.

**MODELING and ANALYSIS:**

**a. Vertical Force P Required to Start Raising the Load.** When the forces in both parts of the rope are equal, contact between the pulley and shaft takes place at *A* (Fig. 1). When **P** is increased, the pulley rolls around the shaft slightly and contact takes place at *B*. Draw the free-body diagram of the pulley when motion is impending. The perpendicular distance from the center *O* of the pulley to the line of action of **R** is

$$r_f = r \sin \phi_s \approx r\mu_s \qquad r_f \approx (1 \text{ in.})0.20 = 0.20 \text{ in.}$$

Summing moments about *B*, you obtain

$$+\circlearrowleft \Sigma M_B = 0: \qquad (2.20 \text{ in.})(500 \text{ lb}) - (1.80 \text{ in.})P = 0$$
$$P = 611 \text{ lb} \qquad\qquad \mathbf{P = 611 \text{ lb}} \downarrow \ \blacktriangleleft$$

**b. Vertical Force P to Hold the Load.** As the force **P** is decreased, the pulley rolls around the shaft, and contact takes place at *C* (Fig. 2). Considering the pulley as a free body and summing moments about *C*, you find

$$+\circlearrowleft \Sigma M_C = 0: \qquad (1.80 \text{ in.})(500 \text{ lb}) - (2.20 \text{ in.})P = 0$$
$$P = 409 \text{ lb} \qquad\qquad \mathbf{P = 409 \text{ lb}} \downarrow \ \blacktriangleleft$$

**c. Horizontal Force P to Start Raising the Load.** Because the three forces **W**, **P**, and **R** are not parallel, they must be concurrent (Fig. 3). The direction of **R** is thus determined from the fact that its line of action must pass through the point of intersection *D* of **W** and **P** and must be tangent to the circle of friction. Recall that the radius of the circle of friction is $r_f = 0.20$ in., so you can calculate the angle marked $\theta$ in Fig. 3 as

$$\sin \theta = \frac{OE}{OD} = \frac{0.20 \text{ in.}}{(2 \text{ in.})\sqrt{2}} = 0.0707 \qquad \theta = 4.1°$$

From the force triangle, you can determine

$$P = W \cot (45° - \theta) = (500 \text{ lb}) \cot 40.9°$$
$$= 577 \text{ lb} \qquad\qquad \mathbf{P = 577 \text{ lb}} \rightarrow \ \blacktriangleleft$$

**REFLECT and THINK:** Many elementary physics problems treat pulleys as frictionless, but when you do take friction into account, the results can be quite different, depending on the direction of motion, the directions of the forces involved, and especially the coefficient of friction.

# SOLVING PROBLEMS
# ON YOUR OWN

In this section, we described several additional engineering applications of the laws of friction.

**1. Journal bearings and axle friction.** In journal bearings, the reaction does not pass through the center of the shaft or axle that is being supported. The distance from the center of the shaft or axle to the line of action of the reaction (Fig. 8.10) is defined by

$$r_f = r \sin \phi_k \approx r\mu_k$$

if motion is actually taking place.

It is defined by

$$r_f = r \sin \phi_s \approx r\mu_s$$

if motion is impending.

Once you have determined the line of action of the reaction, you can draw a free-body diagram and use the corresponding equations of equilibrium to complete the solution (Sample Prob. 8.7). In some problems, it is useful to observe that the line of action of the reaction must be tangent to a circle with a radius of $r_f \approx r\mu_k$ or $r_f \approx r\mu_s$, which is known as the **circle of friction** (Sample Prob. 8.7, part *c*).

**2. Thrust bearings and disk friction.** In a thrust bearing, the magnitude of the couple required to overcome frictional resistance is equal to the sum of the moments of the *kinetic* friction forces exerted on the end of the shaft [Eqs. (8.8) and (8.9)].

An example of disk friction is the **disk clutch**. It is analyzed in the same way as a thrust bearing, except that to determine the largest couple that can be transmitted you must compute the sum of the moments of the maximum *static* friction forces exerted on the disk.

**3. Wheel friction and rolling resistance.** The rolling resistance of a wheel is caused by deformations of both the wheel and the ground. The line of action of the reaction **R** of the ground on the wheel intersects the ground at a horizontal distance $b$ from the center of the wheel. The distance $b$ is known as the **coefficient of rolling resistance** and is expressed in inches or millimeters.

**4. In problems involving both rolling resistance and axle friction,** the free-body diagram should show that the line of action of the reaction **R** of the ground on the wheel is tangent to the friction circle of the axle and intersects the ground at a horizontal distance from the center of the wheel equal to the coefficient of rolling resistance.

# Problems

**8.77** A lever of negligible weight is loosely fitted onto a 75-mm-diameter fixed shaft. It is observed that the lever will just start rotating if a 3-kg mass is added at *C*. Determine the coefficient of static friction between the shaft and the lever.

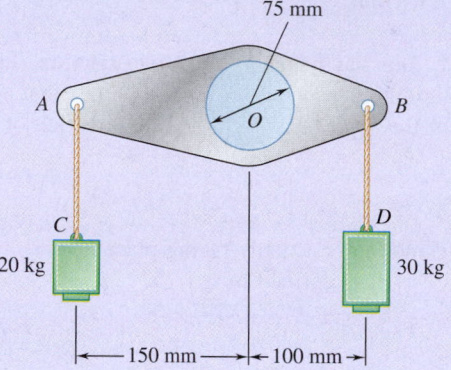

**Fig. P8.77**

**8.78** A 6-in.-radius pulley of weight 5 lb is attached to a 1.5-in.-radius shaft that fits loosely in a fixed bearing. It is observed that the pulley will just start rotating if a 0.5-lb weight is added to block *A*. Determine the coefficient of static friction between the shaft and the bearing.

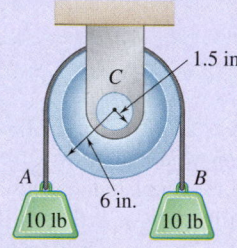

**Fig. P8.78**

**8.79 and 8.80** The double pulley shown is attached to a 10-mm-radius shaft that fits loosely in a fixed bearing. Knowing that the coefficient of static friction between the shaft and the poorly lubricated bearing is 0.40, determine the magnitude of the force **P** required to start raising the load.

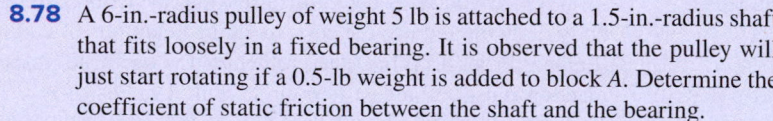

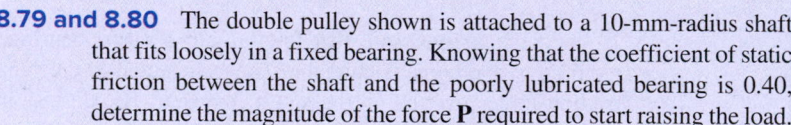

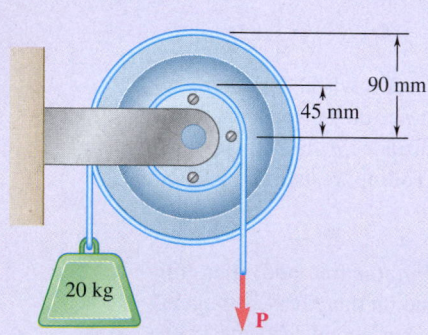

**Fig. P8.79 and P8.81**

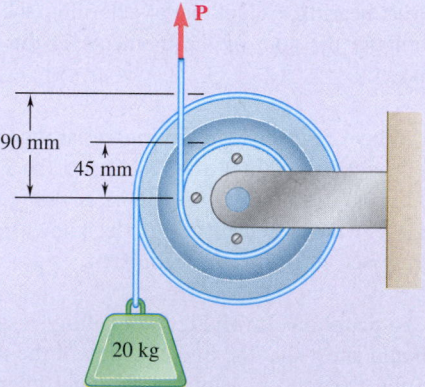

**Fig. P8.80 and P8.82**

**8.81 and 8.82** The double pulley shown is attached to a 10-mm-radius shaft that fits loosely in a fixed bearing. Knowing that the coefficient of static friction between the shaft and the poorly lubricated bearing is 0.40, determine the magnitude of the smallest force **P** required to maintain equilibrium.

**8.83** The block and tackle shown are used to raise a 150-lb load. Each of the 3-in.-diameter pulleys rotates on a 0.5-in.-diameter axle. Knowing that the coefficient of static friction is 0.20, determine the tension in each portion of the rope as the load is slowly raised.

**8.84** The block and tackle shown are used to lower a 150-lb load. Each of the 3-in.-diameter pulleys rotates on a 0.5-in.-diameter axle. Knowing that the coefficient of static friction is 0.20, determine the tension in each portion of the rope as the load is slowly lowered.

**8.85** A scooter is to be designed to roll down a 2 percent slope at a constant speed. Assuming that the coefficient of kinetic friction between the 25-mm-diameter axles and the bearings is 0.10, determine the required diameter of the wheels. Neglect the rolling resistance between the wheels and the ground.

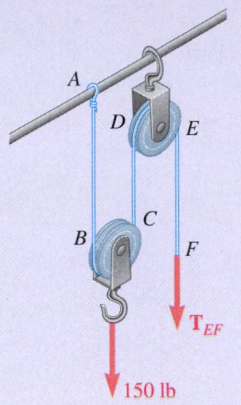

**Fig. P8.83** and **P8.84**

**8.86** The link arrangement shown is frequently used in highway bridge construction to allow for expansion due to changes in temperature. At each of the 60-mm-diameter pins $A$ and $B$, the coefficient of static friction is 0.20. Knowing that the vertical component of the force exerted by $BC$ on the link is 200 kN, determine (a) the horizontal force that should be exerted on beam $BC$ to just move the link, (b) the angle that the resulting force exerted by beam $BC$ on the link will form with the vertical.

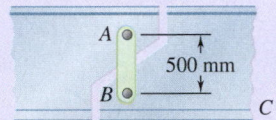

**Fig. P8.86**

**8.87 and 8.88** A lever $AB$ of negligible weight is loosely fitted onto a 2.5-in.-diameter fixed shaft. Knowing that the coefficient of static friction between the fixed shaft and the lever is 0.15, determine the force **P** required to start the lever rotating counterclockwise.

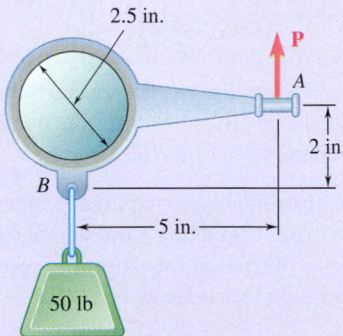

**Fig. P8.87** and P8.89

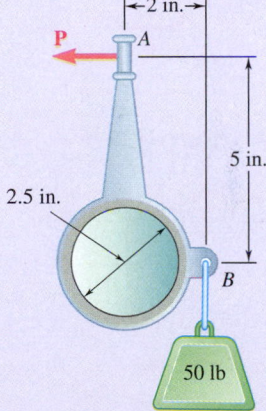

**Fig. P8.88 and P8.90**

**8.89 and 8.90** A lever $AB$ of negligible weight is loosely fitted onto a 2.5-in.-diameter fixed shaft. Knowing that the coefficient of static friction between the fixed shaft and the lever is 0.15, determine the force **P** required to start the lever rotating clockwise.

**8.91** A loaded railroad car has a mass of 30 Mg and is supported by eight 800-mm-diameter wheels with 125-mm-diameter axles. Knowing that the coefficients of friction are $\mu_s = 0.020$ and $\mu_k = 0.015$, determine the horizontal force required (a) to start the car moving, (b) to keep the car moving at a constant speed. Neglect rolling resistance between the wheels and the rails.

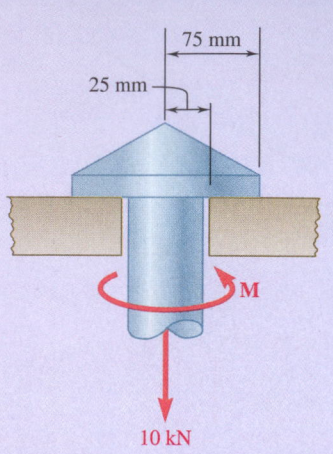

Fig. P8.92

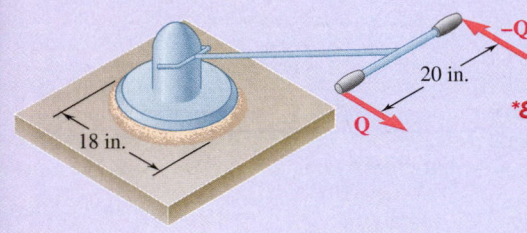

Fig. P8.93

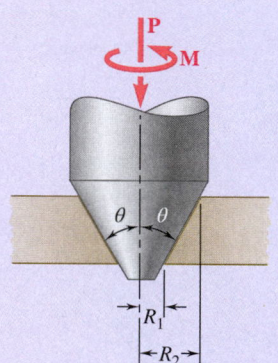

Fig. P8.96

Fig. P8.100

**8.92** Knowing that a couple of magnitude 15 N·m is required to start the vertical shaft rotating, determine the coefficient of static friction between the annular surfaces of contact.

**8.93** A 50-lb electric floor polisher is operated on a surface for which the coefficient of kinetic friction is 0.25. Assuming that the normal force per unit area between the disk and the floor is uniformly distributed, determine the magnitude $Q$ of the horizontal forces required to prevent motion of the machine.

**\*8.94** The frictional resistance of a thrust bearing decreases as the shaft and bearing surfaces wear out. It is generally assumed that the wear is directly proportional to the distance traveled by any given point of the shaft and thus to the distance $r$ from the point to the axis of the shaft. Assuming then that the normal force per unit area is inversely proportional to $r$, show that the magnitude $M$ of the couple required to overcome the frictional resistance of a worn-out end bearing (with contact over the full circular area) is equal to 75 percent of the value given by Eq. (8.9) for a new bearing.

**\*8.95** Assuming that bearings wear out as indicated in Prob. 8.94, show that the magnitude $M$ of the couple required to overcome the frictional resistance of a worn-out collar bearing is

$$M = \frac{1}{2}\mu_k P(R_1 + R_2)$$

where $P$ = magnitude of the total axial force
$R_1, R_2$ = inner and outer radii of collar

**\*8.96** Assuming that the pressure between the surfaces of contact is uniform, show that the magnitude $M$ of the couple required to overcome frictional resistance for the conical bearing shown is

$$M = \frac{2}{3}\frac{\mu_k P}{\sin\theta}\frac{R_2^3 - R_1^3}{R_2^2 - R_1^2}$$

**8.97** Solve Prob. 8.93 assuming that the normal force per unit area between the disk and the floor varies linearly from a maximum at the center to zero at the circumference of the disk.

**8.98** Determine the horizontal force required to move a 2500-lb automobile with 23-in.-diameter tires along a horizontal road at a constant speed. Neglect all forms of friction except rolling resistance, and assume the coefficient of rolling resistance to be 0.05 in.

**8.99** Knowing that a 6-in.-diameter disk rolls at a constant velocity down a 2 percent incline, determine the coefficient of rolling resistance between the disk and the incline.

**8.100** A 900-kg machine base is rolled along a concrete floor using a series of steel pipes with outside diameters of 100 mm. Knowing that the coefficient of rolling resistance is 0.5 mm between the pipes and the base and 1.25 mm between the pipes and the concrete floor, determine the magnitude of the force **P** required to slowly move the base along the floor.

**8.101** Solve Prob. 8.85 including the effect of a coefficient of rolling resistance of 1.75 mm.

**8.102** Solve Prob. 8.91 including the effect of a coefficient of rolling resistance of 0.5 mm.

# 8.4 BELT FRICTION

Another common application of dry friction concerns belts, which serve many different purposes in engineering, such as transmitting a torque from a lawn mower engine to its wheels. Some of the same analysis affects the design of band brakes and the operation of ropes and pulleys.

Consider a flat belt passing over a fixed cylindrical drum (Fig. 8.14*a*). We want to determine the relation between the values $T_1$ and $T_2$ of the tension in the two parts of the belt when the belt is just about to slide toward the right.

First, we detach from the belt a small element $PP'$ subtending an angle $\Delta\theta$. Denoting the tension at $P$ by $T$ and the tension at $P'$ by $T + \Delta T$, we draw the free-body diagram of the element of the belt (Fig. 8.14*b*). Besides the two forces of tension, the forces acting on the free body are the normal component $\Delta\mathbf{N}$ of the reaction of the drum and the friction force $\Delta\mathbf{F}$. Because we assume motion is impending, we have $\Delta F = \mu_s\,\Delta N$. Note that if $\Delta\theta$ approaches zero, the magnitudes $\Delta N$ and $\Delta F$ and the *difference* $\Delta T$ between the tension at $P$ and the tension at $P'$ also approach zero; the value $T$ of the tension at $P$, however, remains unchanged. This observation helps in understanding our choice of notation.

Choosing the coordinate axes shown in Fig. 8.14*b*, we can write the equations of equilibrium for the element $PP'$ as

$$\Sigma F_x = 0: \quad (T + \Delta T) \cos\frac{\Delta\theta}{2} - T \cos\frac{\Delta\theta}{2} - \mu_s\Delta N = 0 \tag{8.11}$$

$$\Sigma F_y = 0: \quad \Delta N - (T + \Delta T) \sin\frac{\Delta\theta}{2} - T \sin\frac{\Delta\theta}{2} = 0 \tag{8.12}$$

Solving Eq. (8.12) for $\Delta N$ and substituting into Eq. (8.11), we obtain after reductions

$$\Delta T \cos\frac{\Delta\theta}{2} - \mu_s(2T + \Delta T) \sin\frac{\Delta\theta}{2} = 0$$

Now we divide both terms by $\Delta\theta$. For the first term, we do this simply by dividing $\Delta T$ by $\Delta\theta$. We carry out the division of the second term by dividing the terms in parentheses by 2 and the sine by $\Delta\theta/2$. The result is

$$\frac{\Delta T}{\Delta\theta} \cos\frac{\Delta\theta}{2} - \mu_s\left(T + \frac{\Delta T}{2}\right) \frac{\sin(\Delta\theta/2)}{\Delta\theta/2} = 0$$

If we now let $\Delta\theta$ approach zero, the cosine approaches one and $\Delta T/2$ approaches zero, as noted above. The quotient of $\sin(\Delta\theta/2)$ over $\Delta\theta/2$ approaches one, according to a lemma derived in all calculus textbooks. Because the limit as $\Delta\theta$ approaches zero of $\Delta T/\Delta\theta$ is equal to the derivative $dT/d\theta$ by definition, we get

$$\frac{dT}{d\theta} - \mu_s T = 0 \qquad \frac{dT}{T} = \mu_s d\theta$$

Now we integrate both members of the last equation from $P_1$ to $P_2$ (see Fig. 8.14*a*). At $P_1$, we have $\theta = 0$ and $T = T_1$; at $P_2$, we have $\theta = \beta$ and $T = T_2$. Integrating between these limits, we have

$$\int_{T_1}^{T_2} \frac{dT}{T} = \int_0^\beta \mu_s d\theta$$

$$\ln T_2 - \ln T_1 = \mu_s\beta$$

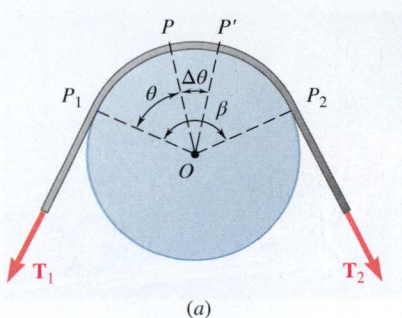

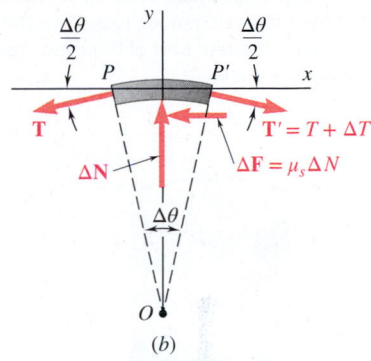

**Fig. 8.14** (*a*) Tensions at the ends of a belt passing over a drum; (*b*) free-body diagram of an element of the belt, indicating the condition that the belt is about to slip to the right.

**Photo 8.5** A sailor wraps a rope around the smooth post (called a *bollard*) in order to control the rope using much less force than the tension in the taut part of the rope. ©Bart Sadowski/Getty Images RF

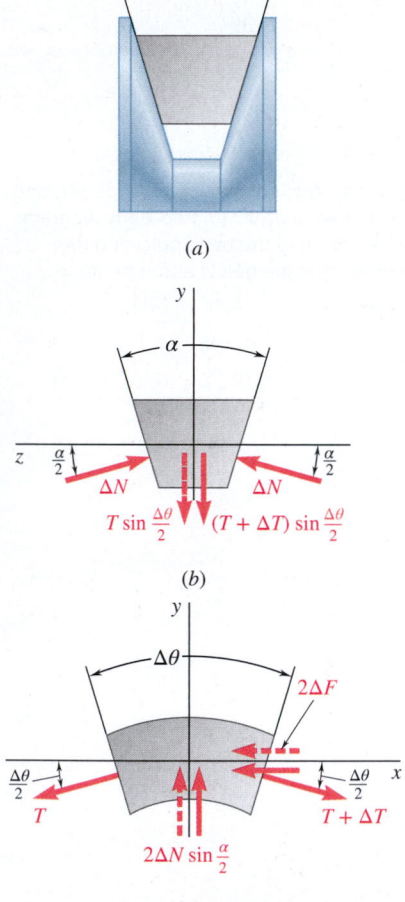

**Fig. 8.15** (*a*) A V belt lying in the groove of a pulley; (*b*) free-body diagram of a cross-sectional element of the belt; (*c*) free-body diagram of a short length of belt.

Noting that the left-hand side is equal to the natural logarithm of the quotient of $T_2$ and $T_1$, this reduces to

$$\ln \frac{T_2}{T_1} = \mu_s \beta \tag{8.13}$$

We can also write this relation in the form

**Belt friction, impending slip**

$$\frac{T_2}{T_1} = e^{\mu_s \beta} \tag{8.14}$$

The formulas we have derived apply equally well to problems involving flat belts passing over fixed cylindrical drums and to problems involving ropes wrapped around a post or capstan. They also can be used to solve problems involving band brakes. (In this situation, the drum is about to rotate, but the band remains fixed.) The formulas also can be applied to problems involving belt drives. In these problems, both the pulley and the belt rotate; our concern is then to find whether the belt will slip; i.e., whether it will move with respect to the pulley.

Formulas (8.13) and (8.14) should be used only if the belt, rope, or brake is *about to slip*. Generally, it is easier to use Eq. (8.14) if you need to find $T_1$ or $T_2$; it is preferable to use Eq. (8.13) if you need to find either $\mu_s$ or the angle of contact $\beta$. Note that $T_2$ is always larger than $T_1$. $T_2$ therefore represents the tension in that part of the belt or rope that *pulls*, whereas $T_1$ is the tension in the part that *resists*. Also observe that the angle of contact $\beta$ must be expressed in *radians*. The angle of contact $\beta$ may be larger than $2\pi$; for example, if a rope is wrapped $n$ times around a post, $\beta$ is equal to $2\pi n$.

If the belt, rope, or brake is actually slipping, you should use formulas similar to Eqs. (8.13) and (8.14) involving the coefficient of kinetic friction $\mu_k$ to find the difference in forces. If the belt, rope, or brake is *not* slipping and is *not* about to slip, none of these formulas can be used.

The belts used in belt drives are often V-shaped. In the V belt shown in Fig. 8.15*a*, contact between belt and pulley takes place along the sides of the groove. Again, we can obtain the relation between the values $T_1$ and $T_2$ of the tension in the two parts of the belt when the belt is just about to slip by drawing the free-body diagram of an element of the belt (Fig. 8.15*b* and *c*). Formulas similar to Eqs. (8.11) and (8.12) are derived, but the magnitude of the total friction force acting on the element is now $2\Delta F$, and the sum of the $y$ components of the normal forces is $2\Delta N \sin(\alpha/2)$. Proceeding as previously, we obtain

$$\ln \frac{T_2}{T_1} = \frac{\mu_s \beta}{\sin(\alpha/2)} \tag{8.15}$$

or

$$\frac{T_2}{T_1} = e^{\mu_s \beta / \sin(\alpha/2)} \tag{8.16}$$

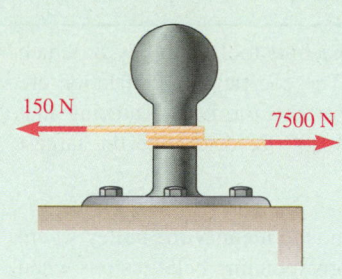

# Sample Problem 8.8

A hawser (a thick docking rope) thrown from a ship to a pier is wrapped two full turns around a bollard. The tension in the hawser is 7500 N; by exerting a force of 150 N on its free end, a dockworker can just keep the hawser from slipping. (*a*) Determine the coefficient of friction between the hawser and the bollard. (*b*) Determine the tension in the hawser that could be resisted by the 150-N force if the hawser were wrapped three full turns around the bollard.

**STRATEGY:**  You are given the difference in forces and the angle of contact through which the friction acts. You can insert these data in the equations of belt friction to determine the coefficient of friction, and then you can use the result to determine the ratio of forces in the second situation.

**MODELING and ANALYSIS:**

**a. Coefficient of Friction.**  Because slipping of the hawser is impending, we use Eq. (8.13):

$$\ln\frac{T_2}{T_1} = \mu_s\beta$$

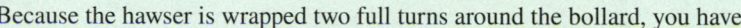

Because the hawser is wrapped two full turns around the bollard, you have

$$\beta = 2(2\pi \text{ rad}) = 12.57 \text{ rad}$$
$$T_1 = 150 \text{ N} \qquad T_2 = 7500 \text{ N}$$

Therefore,

$$\mu_s\beta = \ln\frac{T_2}{T_1}$$
$$\mu_s(12.57 \text{ rad}) = \ln\frac{7500 \text{ N}}{150 \text{ N}} = \ln 50 = 3.91$$
$$\mu_s = 0.311$$

$\mu_s = 0.311$ ◄

**Photo 8.6**  Dockworker mooring a ship using a hawser wrapped around a bollard. ©Fuse/Getty Images RF

**b. Hawser Wrapped Three Turns Around Bollard.**  Using the value of $\mu_s$ obtained in part *a*, you now have (Fig. 1)

$$\beta = 3(2\pi \text{ rad}) = 18.85 \text{ rad}$$
$$T_1 = 150 \text{ N} \qquad \mu_s = 0.311$$

Substituting these values into Eq. (8.14), you obtain

$$\frac{T_2}{T_1} = e^{\mu_s\beta}$$
$$\frac{T_2}{150 \text{ N}} = e^{(0.311)(18.85)} = e^{5.862} = 351.5$$
$$T_2 = 52\,725 \text{ N}$$

$T_2 = 52.7 \text{ kN}$ ◄

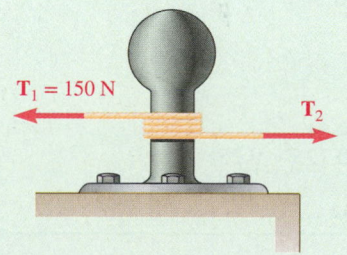

**Fig. 1**  Hawser wrapped three turns around a bollard.

**REFLECT and THINK:**  You can see how the use of a simple post or pulley can have an enormous effect of the magnitude of a force. This is why such systems are commonly used to control, load, and unload container ships in a harbor.

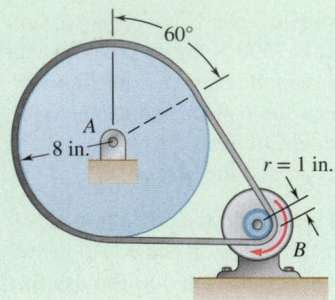

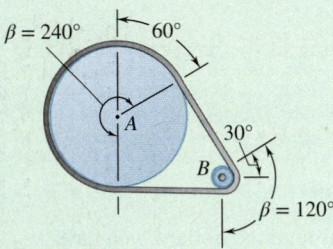

**Fig. 1** Angles of contact for the pulleys.

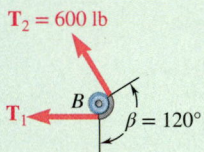

**Fig. 2** Belt tensions at pulley B.

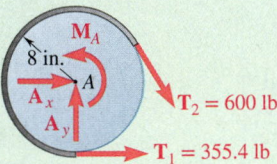

**Fig. 3** Free-body diagram of pulley A.

## Sample Problem 8.9

A flat belt connects pulley A, which drives a machine tool, to pulley B, which is attached to the shaft of an electric motor. The coefficients of friction are $\mu_s = 0.25$ and $\mu_k = 0.20$ between both pulleys and the belt. Knowing that the maximum allowable tension in the belt is 600 lb, determine the largest torque that the belt can exert on pulley A.

**STRATEGY:** The key to solving this problem is to identify the pulley where slippage would first occur, and then find the corresponding belt tensions when slippage is impending. The resistance to slippage depends upon the angle of contact $\beta$ between pulley and belt, as well as upon the coefficient of static friction $\mu_s$. Because $\mu_s$ is the same for both pulleys, slippage occurs first on pulley B, for which $\beta$ is smaller (Fig. 1).

**MODELING and ANALYSIS:**

**Pulley B.** Using Eq. (8.14) with $T_2 = 600$ lb, $\mu_s = 0.25$, and $\beta = 120° = 2\pi/3$ rad (Fig. 2), you obtain

$$\frac{T_2}{T_1} = e^{\mu_s \beta} \qquad \frac{600 \text{ lb}}{T_1} = e^{0.25(2\pi/3)} = 1.688$$

$$T_1 = \frac{600 \text{ lb}}{1.688} = 355.4 \text{ lb}$$

**Pulley A.** Draw the free-body diagram of pulley A (Fig. 3). The couple $\mathbf{M}_A$ is applied to the pulley using the machine tool to which it is attached and is equal and opposite to the torque exerted by the belt. Setting the sum of the moments equal to zero gives

$$+\circlearrowleft \Sigma M_A = 0: \quad M_A - (600 \text{ lb})(8 \text{ in.}) + (355.4 \text{ lb})(8 \text{ in.}) = 0$$
$$M_A = 1957 \text{ lb·in.} \qquad \qquad M_A = 163.1 \text{ lb·ft} \blacktriangleleft$$

**REFLECT and THINK:** You may check that the belt does not slip on pulley A by computing the value of $\mu_s$ required to prevent slipping at A and verify that it is smaller than the actual value of $\mu_s$. From Eq. (8.13), you have

$$\mu_s \beta = \ln \frac{T_2}{T_1} = \ln \frac{600 \text{ lb}}{355.4 \text{ lb}} = 0.524$$

Because $\beta = 240° = 4\pi/3$ rad,

$$\frac{4\pi}{3} \mu_s = 0.524 \qquad \mu_s = 0.125 < 0.25$$

# SOLVING PROBLEMS
# ON YOUR OWN

In the preceding section, you studied **belt friction**. The problems you will solve include belts passing over fixed drums, band brakes in which the drum rotates when the band remains fixed, and belt drives.

**1. Problems involving belt friction** fall into one of the following two categories.

   **a. Problems in which slipping is impending.** You can use one of the following formulas involving the *coefficient of static friction* $\mu_s$.

$$\ln \frac{T_2}{T_1} = \mu_s \beta \tag{8.13}$$

or

$$\frac{T_2}{T_1} = e^{\mu_s \beta} \tag{8.14}$$

   **b. Problems in which slipping is occurring.** You can obtain the formulas to be used from Eqs. (8.13) and (8.14) by replacing $\mu_s$ with the *coefficient of kinetic friction* $\mu_k$.

**2. As you start solving a belt-friction problem,** remember these conventions:

   **a. The angle $\beta$ must be expressed in radians.** In a belt-and-drum problem, this is the angle subtending the arc of the drum on which the belt is wrapped.

   **b. The larger tension is always denoted by $T_2$** and the smaller tension is denoted by $T_1$.

   **c. The larger tension occurs at the end of the belt that is in the direction of the motion,** or impending motion, of the belt relative to the drum.

**3. In each of the problems you will be asked to solve,** three of the four quantities $T_1$, $T_2$, $\beta$, and $\mu_s$ (or $\mu_k$) will either be given or readily found, and you will then solve the appropriate equation for the fourth quantity. You will encounter two kinds of problems.

   **a. Find $\mu_s$ between the belt and drum, knowing that slipping is impending.** From the given data, determine $T_1$, $T_2$, and $\beta$; substitute these values into Eq. (8.13) and solve for $\mu_s$ (Sample Prob. 8.8, part *a*). Follow the same procedure to find the smallest value of $\mu_s$ for which slipping will not occur.

   **b. Find the magnitude of a force or couple applied to the belt or drum, knowing that slipping is impending.** The given data should include $\mu_s$ and $\beta$. If it also includes $T_1$ or $T_2$, use Eq. (8.14) to find the other tension. If neither $T_1$ nor $T_2$ is known but some other data is given, use the free-body diagram of the belt-drum system to write an equilibrium equation that you can solve simultaneously with Eq. (8.14) for $T_1$ and $T_2$. You then will be able to find the magnitude of the specified force or couple from the free-body diagram of the system. Follow the same procedure to determine the largest value of a force or couple that can be applied to the belt or drum if no slipping is to occur (Sample Prob. 8.9).

# Problems

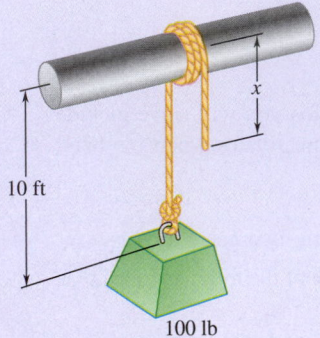

Fig. P8.103

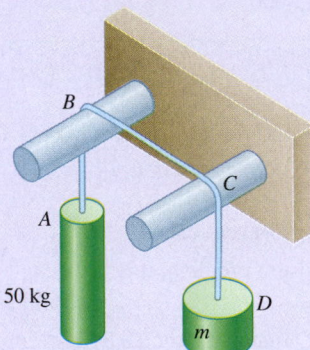

Fig. P8.105 and P8.106

**8.103** A rope having a weight per unit length of 0.4 lb/ft is wound $2\frac{1}{2}$ times around a horizontal rod. Knowing that the coefficient of static friction between the rope and the rod is 0.30, determine the minimum length $x$ of rope that should be left hanging if a 100-lb load is to be supported.

**8.104** A hawser is wrapped two full turns around a capstan head. By exerting a 160-lb force on the free end of the hawser, a sailor can resist a force of 10,000 lb on the other end of the hawser. Determine (a) the coefficient of static friction between the hawser and the capstan, (b) the number of times the hawser should be wrapped around the capstan if a 40,000-lb force is to be resisted by the same 160-lb force.

**8.105** Two cylinders are connected by a rope that passes over two fixed rods as shown. Knowing that the coefficient of static friction between the rope and the rods is 0.40, determine the range of the mass $m$ of cylinder $D$ for which equilibrium is maintained.

**8.106** Two cylinders are connected by a rope that passes over two fixed rods as shown. Knowing that for cylinder $D$ upward motion impends when $m = 20$ kg, determine (a) the coefficient of static friction between the rope and the rods, (b) the corresponding tension in portion $BC$ of the rope.

**8.107** The coefficient of static friction between block $B$ and the horizontal surface and between the rope and support $C$ is 0.40. Knowing that $m_A = 12$ kg, determine the smallest mass of block $B$ for which equilibrium is maintained.

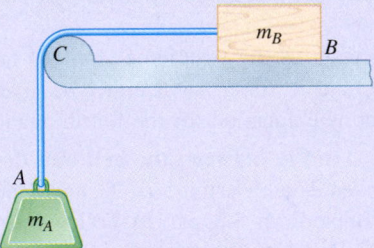
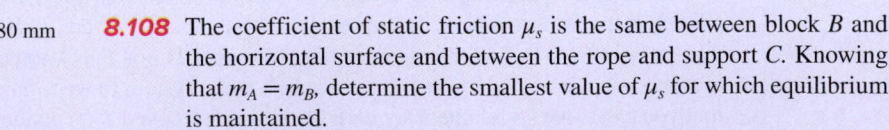

Fig. *P8.107* and *P8.108*

**8.108** The coefficient of static friction $\mu_s$ is the same between block $B$ and the horizontal surface and between the rope and support $C$. Knowing that $m_A = m_B$, determine the smallest value of $\mu_s$ for which equilibrium is maintained.

**8.109** A band belt is used to control the speed of a flywheel as shown. Determine the magnitude of the couple being applied to the flywheel, knowing that the coefficient of kinetic friction between the belt and the flywheel is 0.25 and that the flywheel is rotating clockwise at a constant speed. Show that the same result is obtained if the flywheel rotates counterclockwise.

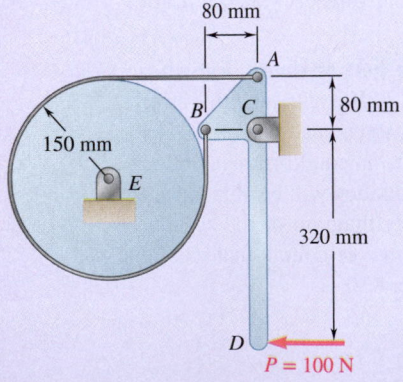

Fig. P8.109

**8.110** The setup shown is used to measure the output of a small turbine. When the flywheel is at rest, the reading of each spring scale is 14 lb. If a 105-lb·in. couple must be applied to the flywheel to keep it rotating clockwise at a constant speed, determine (*a*) the reading of each scale at that time, (*b*) the coefficient of kinetic friction. Assume that the length of the belt does not change.

**8.111** The setup shown is used to measure the output of a small turbine. The coefficient of kinetic friction is 0.20, and the reading of each spring scale is 16 lb when the flywheel is at rest. Determine (*a*) the reading of each scale when the flywheel is rotating clockwise at a constant speed, (*b*) the couple that must be applied to the flywheel. Assume that the length of the belt does not change.

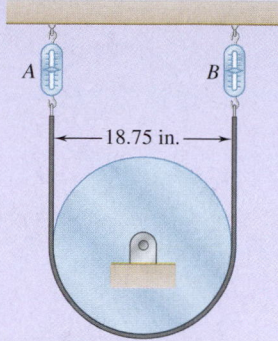

**Fig. P8.110 and P8.111**

**8.112** A flat belt is used to transmit a couple from drum *B* to drum *A*. Knowing that the coefficient of static friction is 0.40 and that the allowable belt tension is 450 N, determine the largest couple that can be exerted on drum *A*.

**8.113** A flat belt is used to transmit a couple from pulley *A* to pulley *B*. The radius of each pulley is 60 mm, and a force of magnitude $P = 900$ N is applied as shown to the axle of pulley *A*. Knowing that the coefficient of static friction is 0.35, determine (*a*) the largest couple that can be transmitted, (*b*) the corresponding maximum value of the tension in the belt.

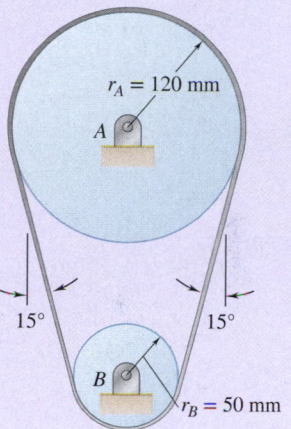

**Fig. P8.112**

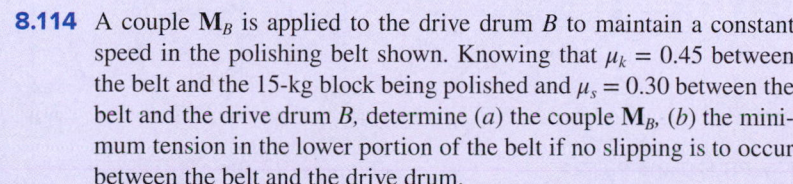

**Fig. P8.113**

**8.114** A couple $\mathbf{M}_B$ is applied to the drive drum *B* to maintain a constant speed in the polishing belt shown. Knowing that $\mu_k = 0.45$ between the belt and the 15-kg block being polished and $\mu_s = 0.30$ between the belt and the drive drum *B*, determine (*a*) the couple $\mathbf{M}_B$, (*b*) the minimum tension in the lower portion of the belt if no slipping is to occur between the belt and the drive drum.

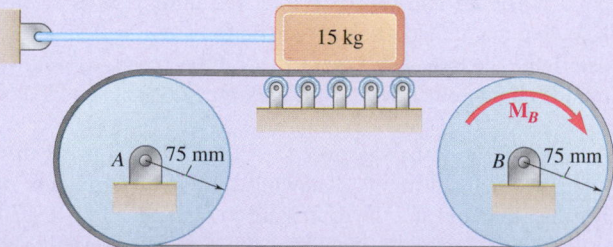

**Fig. P8.114**

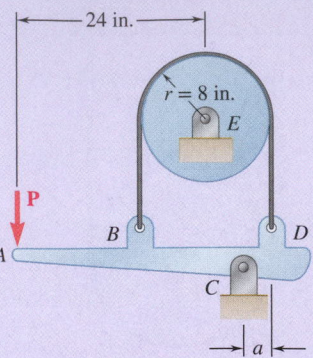

Fig. *P8.115, P8.116,* and P8.117

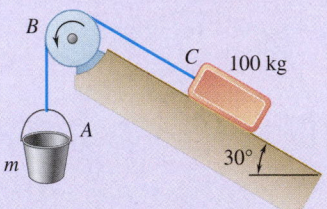

Fig. P8.118

**8.115** The speed of the brake drum shown is controlled by a belt attached to the control bar *AD*. A force **P** with a magnitude of 25 lb is applied to the control bar at *A*. Determine the magnitude of the couple being applied to the drum knowing that the coefficient of kinetic friction between the belt and the drum is 0.25, that $a = 4$ in., and that the drum is rotating at a constant speed (*a*) counterclockwise, (*b*) clockwise.

**8.116** The speed of the brake drum shown is controlled by a belt attached to the control bar *AD*. Knowing that $a = 4$ in., determine the maximum value of the coefficient of static friction for which the brake is not self-locking when the drum rotates counterclockwise.

**8.117** The speed of the brake drum shown is controlled by a belt attached to the control bar *AD*. Knowing that the coefficient of static friction is 0.30 and that the brake drum is rotating counterclockwise, determine the minimum value of *a* for which the brake is not self-locking.

**8.118** Bucket *A* and block *C* are connected by a cable that passes over drum *B*. Knowing that drum *B* rotates slowly counterclockwise and that the coefficients of friction at all surfaces are $\mu_s = 0.35$ and $\mu_k = 0.25$, determine the smallest combined mass *m* of the bucket and its contents for which block *C* will (*a*) remain at rest, (*b*) start moving up the incline, (*c*) continue moving up the incline at a constant speed.

**8.119** Solve Prob. 8.118 assuming that drum *B* is frozen and cannot rotate.

**8.120 and 8.122** A cable is placed around three parallel pipes. Knowing that the coefficients of friction are $\mu_s = 0.25$ and $\mu_k = 0.20$, determine (*a*) the smallest weight *W* for which equilibrium is maintained, (*b*) the largest weight *W* that can be raised if pipe *B* is slowly rotated counterclockwise while pipes *A* and *C* remain fixed.

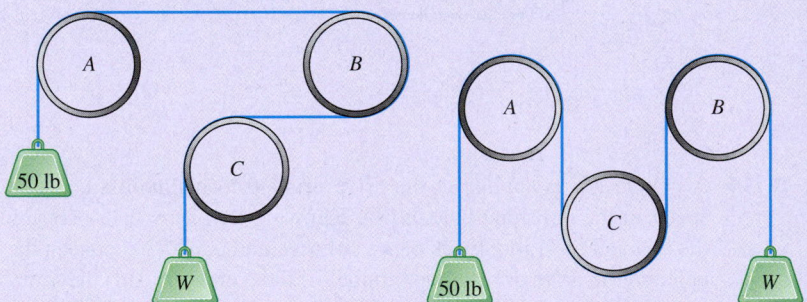

Fig. P8.120 and P8.121          Fig. *P8.122* and *P8.123*

**8.121 and 8.123** A cable is placed around three parallel pipes. Two of the pipes are fixed and do not rotate; the third pipe is slowly rotated. Knowing that the coefficients of friction are $\mu_s = 0.25$ and $\mu_k = 0.20$, determine the largest weight *W* that can be raised (*a*) if only pipe *A* is rotated counterclockwise, (*b*) if only pipe *C* is rotated clockwise.

**8.124** A recording tape passes over the 20-mm-radius drive drum *B* and under the idler drum *C*. Knowing that the coefficients of friction between the tape and the drums are $\mu_s = 0.40$ and $\mu_k = 0.30$ and that drum *C* is free to rotate, determine the smallest allowable value of *P* if slipping of the tape on drum *B* is not to occur.

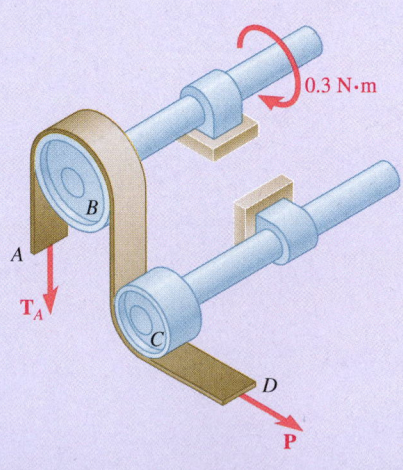

Fig. P8.124

**8.125** Solve Prob. 8.124 assuming that the idler drum *C* is frozen and cannot rotate.

**8.126** The strap wrench shown is used to grip the pipe firmly without marring the external surface of the pipe. Knowing that the coefficient of static friction is the same for all surfaces of contact, determine the smallest value of $\mu_s$ for which the wrench will be self-locking when $a = 200$ mm, $r = 30$ mm, and $\theta = 65°$.

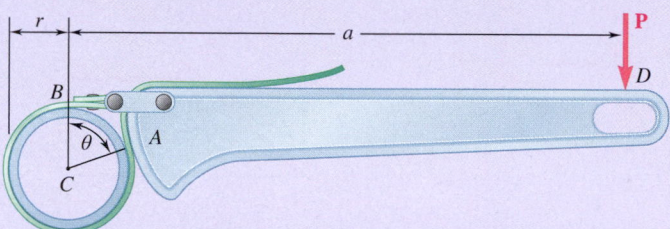

Fig. P8.126

**8.127** The axle of the pulley is frozen and cannot rotate with respect to the block. Knowing that the coefficient of static friction between cable *ABCD* and the pulley is 0.30, determine (*a*) the maximum allowable value of $\theta$ if the system is to remain in equilibrium, (*b*) the corresponding reactions at *A* and *D*. (Assume that the straight portions of the cable meet at point *E*.)

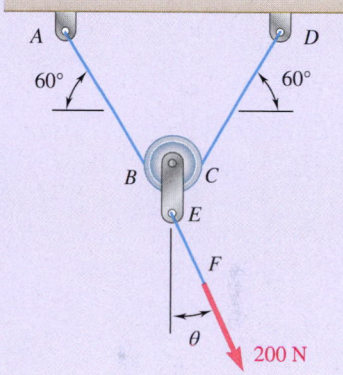

Fig. P8.127

**8.128** The 10-lb bar *AE* is suspended by a cable that passes over a 5-in.-radius drum. Vertical motion of end *E* of the bar is prevented by the two stops shown. Knowing that $\mu_s = 0.30$ between the cable and the drum, determine (*a*) the largest counterclockwise couple $\mathbf{M}_0$ that can be applied to the drum if slipping is not to occur, (*b*) the corresponding force exerted on end *E* of the bar.

**8.129** Solve Prob. 8.128 assuming that a clockwise couple $\mathbf{M}_0$ is applied to the drum.

**8.130** Prove that Eqs. (8.13) and (8.14) are valid for any shape of surface provided that the coefficient of friction is the same at all points of contact.

**8.131** Complete the derivation of Eq. (8.15), which relates the tension in both parts of a V belt.

**8.132** Solve Prob. 8.112 assuming that the flat belt and drums are replaced by a V belt and V pulleys with $\alpha = 36°$. (The angle $\alpha$ is as shown in Fig. 8.15*a*.)

**8.133** Solve Prob. 8.113 assuming that the flat belt and pulleys are replaced by a V belt and V pulleys with $\alpha = 36°$. (The angle $\alpha$ is as shown in Fig. 8.15*a*.)

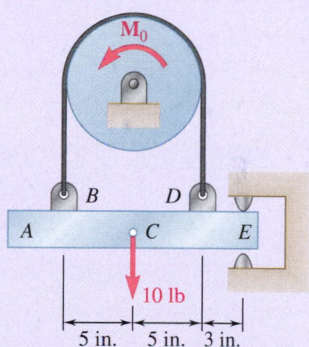

Fig. P8.128

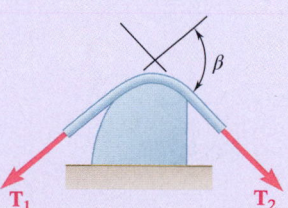

Fig. P8.130

# Review and Summary

This chapter was devoted to the study of **dry friction**, i.e., to problems involving rigid bodies in contact along unlubricated surfaces.

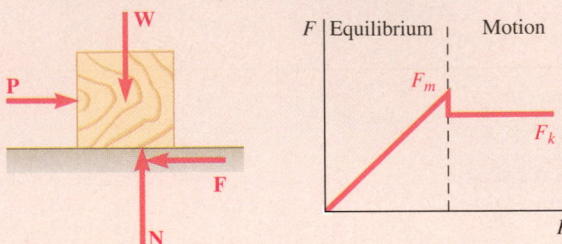

**Fig. 8.16**

## Static and Kinetic Friction

If we apply a horizontal force **P** to a block resting on a horizontal surface (Sec. 8.1), we note that at first the block does not move. This shows that a **friction force F** must have developed to balance **P** (Fig. 8.16). As the magnitude of **P** increases, the magnitude of **F** also increases until it reaches a maximum value $F_m$. If **P** is further increased, the block starts sliding, and the magnitude of **F** drops from $F_m$ to a lower value $F_k$. Experimental evidence shows that $F_m$ and $F_k$ are proportional to the normal component $N$ of the reaction of the surface. We have

$$F_m = \mu_s N \tag{8.1}$$
$$F_k = \mu_k N \tag{8.2}$$

where $\mu_s$ and $\mu_k$ are called, respectively, the **coefficient of static friction** and the **coefficient of kinetic friction**. These coefficients depend on the nature and the condition of the surfaces in contact. Approximate values of the coefficients of static friction are given in Table 8.1.

## Angles of Friction

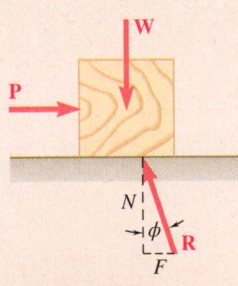

**Fig. 8.17**

It is sometimes convenient to replace the normal force **N** and the friction force **F** by their resultant **R** (Fig. 8.17). As the friction force increases and reaches its maximum value $F_m = \mu_s N$, the angle $\phi$ that **R** forms with the normal to the surface increases and reaches a maximum value $\phi_s$, which is called the **angle of static friction**. If motion actually takes place, the magnitude of **F** drops to $F_k$; similarly, the angle $\phi$ drops to a lower value $\phi_k$, which is called the **angle of kinetic friction**. As shown in Sec. 8.1B, we have

$$\tan \phi_s = \mu_s \tag{8.3}$$
$$\tan \phi_k = \mu_k \tag{8.4}$$

## Problems Involving Friction

When solving equilibrium problems involving friction, you should keep in mind that the magnitude $F$ of the friction force is equal to $F_m = \mu_s N$ only if *the body is about to slide* (Sec. 8.1C). *If motion is not impending,* you should treat $F$ and $N$ as independent unknowns to be determined from the equilibrium equations (Fig. 8.18a). You should also check that the value of $F$ required to maintain equilibrium is not larger than $F_m$; if it were, the body would move, and the magnitude of the friction force would be $F_k = \mu_k N$ (Sample Prob. 8.1). On the other hand, *if motion is known to be impending*, $F$ has reached its maximum value $F_m = \mu_s N$ (Fig. 8.18b), and you should substitute this expression for $F$ in the equilibrium equations (Sample Prob. 8.3). When only three forces are involved in a free-body diagram, including the reaction **R** of the surface in contact with the body, it is usually more convenient to solve the problem by drawing a force triangle (Sample Prob. 8.2). In some problems, impending motion can be due to tipping instead of slipping; the assessment of this condition requires a moment equilibrium analysis of the body (Sample Prob. 8.4).

When a problem involves the analysis of the forces exerted on each other by *two bodies A and B*, it is important to show the friction forces with their correct sense. The correct sense for the friction force exerted by $B$ on $A$, for instance, is opposite to that of the *relative motion* (or impending motion) of $A$ with respect to $B$ (Fig. 8.6).

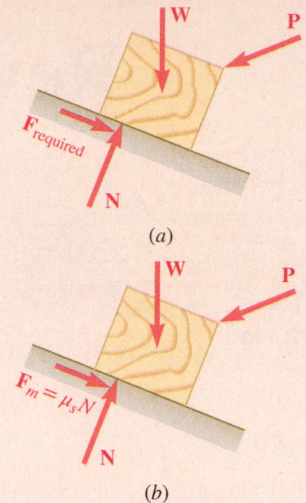

**Fig. 8.18**

## Wedges and Screws

In the later sections of this chapter, we considered several specific engineering applications where dry friction plays an important role. In the case of **wedges**, which are simple machines used to raise heavy loads (Sec. 8.2A), we must draw two or more free-body diagrams, taking care to show each friction force with its correct sense (Sample Prob. 8.5). The analysis of **square-threaded screws**, which are frequently used in jacks, presses, and other mechanisms, is reduced to the analysis of a block sliding on an incline by unwrapping the thread of the screw and showing it as a straight line (Sec. 8.2B). This is shown again in Fig. 8.19, where $r$ denotes the *mean radius* of the thread, $L$ is the *lead* of the screw (i.e., the distance through which the screw advances in one turn), **W** is the load, and $Qr$ is equal to the couple exerted on the screw. We noted in the case of multiple-threaded screws that the lead $L$ of the screw is *not* equal to its pitch, which is the distance measured between two consecutive threads.

Other engineering applications considered in this chapter were **journal bearings** and **axle friction** (Sec. 8.3A), **thrust bearings** and **disk friction** (Sec. 8.3B), **wheel friction** and **rolling resistance** (Sec. 8.3C), and **belt friction** (Sec. 8.4).

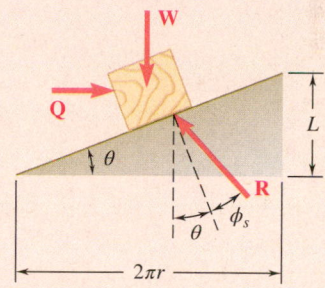

**Fig. 8.19**

## Belt Friction

In solving a problem involving a flat belt passing over a fixed cylinder, it is important to first determine the direction in which the belt slips or is about to slip. If the drum is rotating, the motion or impending motion of the belt should be determined *relative* to the rotating drum. For instance, if the belt shown in Fig. 8.20 is about to slip to the right relative to the drum, the friction forces exerted by the drum on the belt are directed to the left, and the tension is larger in the right-hand portion of the belt than in the left-hand portion. Denoting the larger tension by $T_2$, the smaller tension by $T_1$, the coefficient of static friction by $\mu_s$, and the angle (in radians) subtended by the belt by $\beta$, we derived in Sec. 8.4 the formulas

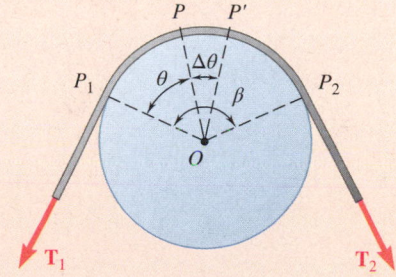

**Fig. 8.20**

$$\ln \frac{T_2}{T_1} = \mu_s \beta \qquad (8.13)$$

$$\frac{T_2}{T_1} = e^{\mu_s \beta} \qquad (8.14)$$

that we used in solving Sample Probs. 8.8 and 8.9. If the belt actually slips on the drum, the coefficient of static friction $\mu_s$ should be replaced by the coefficient of kinetic friction $\mu_k$ in both of these formulas.

# Review Problems

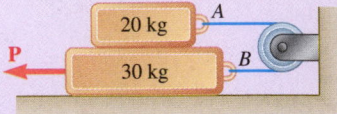

**Fig. P8.134**

**8.134 and 8.135** The coefficients of friction are $\mu_s = 0.40$ and $\mu_k = 0.30$ between all surfaces of contact. Determine the smallest force **P** required to start the 30-kg block moving if cable $AB$ (*a*) is attached as shown, (*b*) is removed.

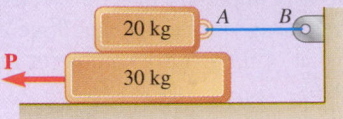

**Fig. P8.135**

**8.136** A 120-lb cabinet is mounted on casters that can be locked to prevent their rotation. The coefficient of static friction between the floor and each caster is 0.30. If $h = 32$ in., determine the magnitude of the force **P** required to move the cabinet to the right (*a*) if all casters are locked, (*b*) if the casters at $B$ are locked and the casters at $A$ are free to rotate, (*c*) if the casters at $A$ are locked and the casters at $B$ are free to rotate.

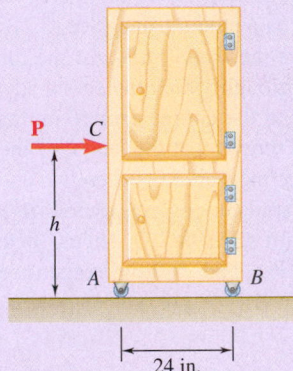

**Fig. P8.136**

**8.137** A slender rod with a length of $L$ is lodged between peg $C$ and the vertical wall, and supports a load **P** at end $A$. Knowing that the coefficient of static friction between the peg and the rod is 0.15 and neglecting friction at the roller, determine the range of values of the ratio $L/a$ for which equilibrium is maintained.

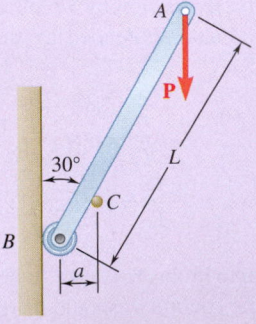

**Fig. P8.137**

**8.138** The hydraulic cylinder shown exerts a force of 3 kN directed to the right on point $B$ and to the left on point $E$. Determine the magnitude of the couple **M** required to rotate the drum clockwise at a constant speed.

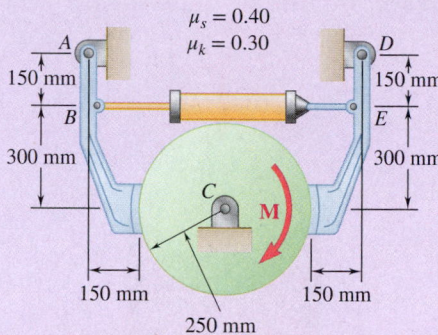

**Fig. P8.138**

**8.139** A rod $DE$ and a small cylinder are placed between two guides as shown. The rod is not to slip downward, no matter how large the force **P** may be; i.e., the arrangement is said to be self-locking. Neglecting the weight of the cylinder, determine the minimum allowable coefficients of static friction at $A$, $B$, and $C$.

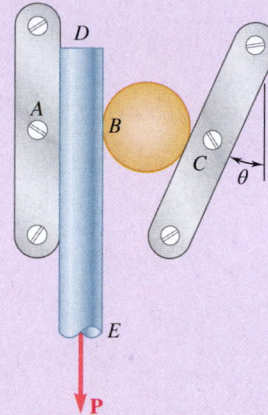

**Fig. *P8.139***

**8.140** Bar $AB$ is attached to collars that can slide on the inclined rods shown. A force **P** is applied at point $D$ located at a distance $a$ from end $A$. Knowing that the coefficient of static friction $\mu_s$ between each collar and the rod upon which it slides is 0.30 and neglecting the weights of the bar and of the collars, determine the smallest value of the ratio $a/L$ for which equilibrium is maintained.

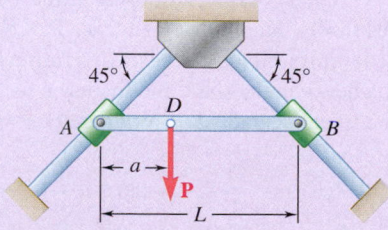

**Fig. P8.140**

**8.141** Two $10°$ wedges of negligible weight are used to move and position the 400-lb block. Knowing that the coefficient of static friction is 0.25 at all surfaces of contact, determine the smallest force **P** that should be applied as shown to one of the wedges.

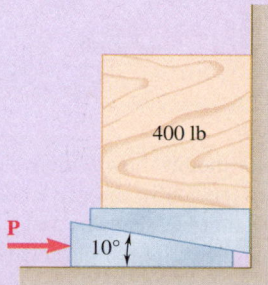

**Fig. P8.141**

**8.142** A 10° wedge is used to split a section of a log. The coefficient of static friction between the wedge and the log is 0.35. Knowing that a force **P** with a magnitude of 600 lb was required to insert the wedge, determine the magnitude of the forces exerted on the wood by the wedge after insertion.

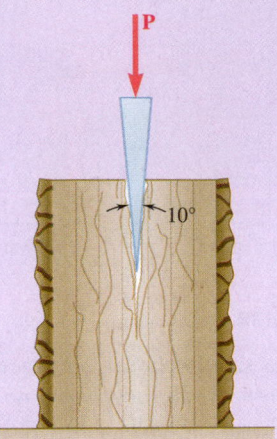

Fig. *P8.142*

**8.143** In the gear-pulling assembly shown, the square-threaded screw *AB* has a mean radius of 15 mm and a lead of 4 mm. Knowing that the coefficient of static friction is 0.10, determine the couple that must be applied to the screw in order to produce a force of 3 kN on the gear. Neglect friction at end *A* of the screw.

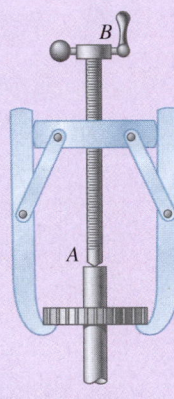

Fig. P8.143

**8.144** A lever of negligible weight is loosely fitted onto a 30-mm-radius fixed shaft as shown. Knowing that a force **P** of magnitude 275 N will just start the lever rotating clockwise, determine (*a*) the coefficient of static friction between the shaft and the lever, (*b*) the smallest force **P** for which the lever does not start rotating counterclockwise.

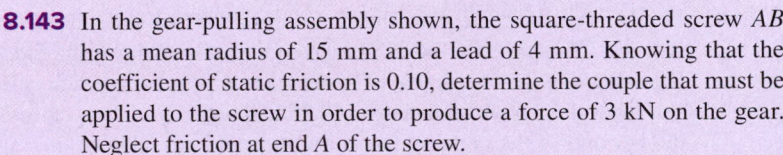
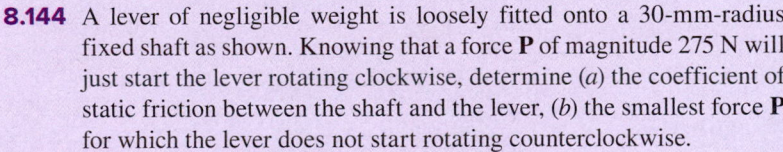

Fig. P8.144

**8.145** In the pivoted motor mount shown, the weight **W** of the 175-lb motor is used to maintain tension in the drive belt. Knowing that the coefficient of static friction between the flat belt and drums *A* and *B* is 0.40 and neglecting the weight of platform *CD*, determine the largest couple that can be transmitted to drum *B* when the drive drum *A* is rotating clockwise.

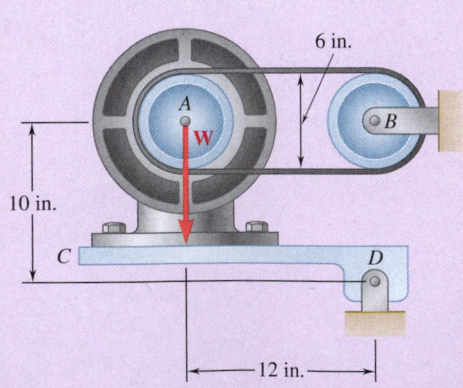

Fig. *P8.145*

# 9

# Distributed Forces: Moments of Inertia

The strength of structural members used in the construction of buildings depends to a large extent on the properties of their cross sections. This includes the second moments of area, or moments of inertia, of these cross sections.

# Objectives

- **Describe** the second moment, or moment of inertia, of an area.

- **Determine** the rectangular and polar moments of inertia of areas and their corresponding radii of gyration by integration.

- **Develop** the parallel-axis theorem and apply it to determine the moments of inertia of composite areas.

- **Introduce** the product of inertia and apply it to analyze the transformation of moments of inertia when coordinate axes are rotated.

- **Describe** the moment of inertia of a mass with respect to an axis.

- **Apply** the parallel-axis theorem to facilitate mass moment of inertia computations.

- **Analyze** the transformation of mass moments of inertia when coordinate axes are rotated.

## Introduction

In Chap. 5, we analyzed various systems of forces distributed over an area or volume. The three main types of forces considered were (1) weights of homogeneous plates of uniform thickness (Secs. 5.1 and 5.2); (2) distributed loads on beams and submerged surfaces (Sec. 5.3); and (3) weights of homogeneous three-dimensional bodies (Sec. 5.4). In all of these cases, the distributed forces were proportional to the elemental areas or volumes associated with them. Therefore, we could obtain the resultant of these forces by summing the corresponding areas or volumes, and we determined the moment of the resultant about any given axis by computing the first moments of the areas or volumes about that axis.

In the first part of this chapter, we consider distributed forces $\Delta\mathbf{F}$ where the magnitudes depend not only upon the elements of area $\Delta A$ on which these forces act but also upon the distance from $\Delta A$ to some given axis. More precisely, we assume the magnitude of the force per unit area $\Delta F/\Delta A$ varies linearly with the distance to the axis. Forces of this type arise in the study of the bending of beams and in problems involving submerged nonrectangular surfaces.

Starting with the assumption that the elemental forces involved are distributed over an area $A$ and vary linearly with the distance $y$ to the $x$ axis, we will show that the magnitude of their resultant $\mathbf{R}$ depends upon the first moment $Q_x$ of the area $A$. However, the location of the point where $\mathbf{R}$ is applied depends upon the *second moment*, or *moment of inertia*, $I_x$ of the same area with respect to the $x$ axis. You will see how to compute the moments of inertia of various areas with respect to given $x$ and $y$ axes. We also introduce the *polar moment of inertia* $J_O$ of an area. To facilitate these computations, we establish a relation between the moment of inertia $I_x$ of an area $A$ with respect to a given $x$ axis and the moment of inertia $I_{x'}$ of the same area with respect to the parallel centroidal $x'$ axis (a relation known as the parallel-axis theorem). You will also study the transformation of the moments of inertia of a given area when the coordinate axes are rotated.

In the second part of this chapter, we will explain how to determine the moments of inertia of various *masses* with respect to a given axis. Moments of inertia of masses are common in dynamics problems involving the rotation of a rigid body about an axis. To facilitate the computation of mass moments of inertia, we introduce another version of the parallel-axis theorem. Finally, we will analyze the transformation of moments of inertia of masses when the coordinate axes are rotated.

# 9.1 MOMENTS OF INERTIA OF AREAS

In the first part of this chapter, we consider distributed forces $\Delta\mathbf{F}$ whose magnitudes $\Delta F$ are proportional to the elements of area $\Delta A$ on which the forces act and, at the same time, vary linearly with the distance from $\Delta A$ to a given axis.

## 9.1A Second Moment, or Moment of Inertia, of an Area

Consider a beam with a uniform cross section that is subjected to two equal and opposite couples: one applied at each end of the beam. Such a beam is said to be in **pure bending**. The internal forces in any section of the beam are distributed forces whose magnitudes $\Delta F = ky\,\Delta A$ vary linearly with the distance $y$ between the element of area $\Delta A$ and an axis passing through the centroid of the section. (This statement can be derived in a course on mechanics of materials.) This axis, represented by the $x$ axis in Fig. 9.1, is known as the **neutral axis** of the section. The forces on one side of the neutral axis are forces of compression, whereas those on the other side are forces of tension. On the neutral axis itself, the forces are zero.

The magnitude of the resultant $\mathbf{R}$ of the elemental forces $\Delta\mathbf{F}$ that act over the entire section is

$$R = \int ky\,dA = k\int y\,dA$$

You might recognize this last integral as the **first moment** $Q_x$ of the section about the $x$ axis; it is equal to $\bar{y}A$ and is thus equal to zero, because the centroid of the section is located on the $x$ axis. The system of forces $\Delta\mathbf{F}$ thus reduces to a couple. The magnitude $M$ of this couple (bending moment) must be equal to the sum of the moments $\Delta M_x = y\,\Delta F = ky^2\,\Delta A$ of the elemental forces. Integrating over the entire section, we obtain

$$M = \int ky^2\,dA = k\int y^2\,dA$$

This last integral is known as the **second moment**, or **moment of inertia**,[†] of the beam section with respect to the $x$ axis and is denoted by $I_x$. We obtain it by multiplying each element of area $dA$ by the *square of its distance* from the $x$ axis and integrating over the beam section. Because each product $y^2\,dA$ is positive, regardless of the sign of $y$, or zero (if $y$ is zero), the integral $I_x$ is always positive.

Another example of a second moment, or moment of inertia, of an area is provided by the following problem from hydrostatics. A vertical circular gate used to close the outlet of a large reservoir is submerged under water as shown in Fig. 9.2. What is the resultant of the forces exerted by the water on the gate,

[†]The term *second moment* is more proper than the term *moment of inertia*, which logically should be used only to denote integrals of mass (see Sec. 9.5). In engineering practice, however, moment of inertia is used in connection with areas as well as masses.

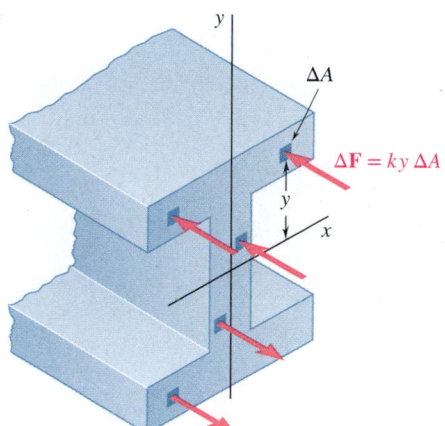
**Fig. 9.1** Representative forces on a cross section of a beam subjected to equal and opposite couples at each end.

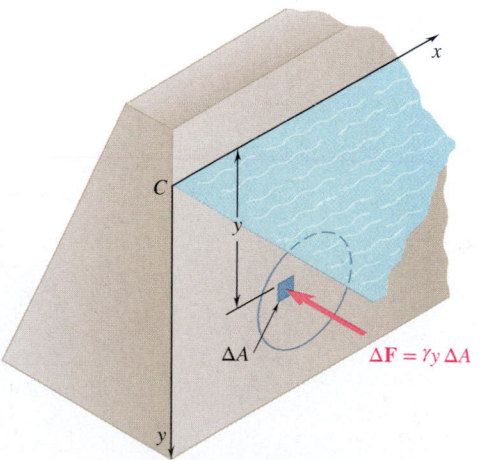
**Fig. 9.2** Vertical circular gate, submerged under water, used to close the outlet of a reservoir.

and what is the moment of the resultant about the line of intersection of the plane of the gate and the water surface ($x$ axis)?

If the gate were rectangular, we could determine the resultant of the forces due to water pressure from the pressure curve, as we did in Sec. 5.3B. Because the gate is circular, however, we need to use a more general method. Denoting the depth of an element of area $\Delta A$ by $y$ and the specific weight of water by $\gamma$, the pressure at an element is $p = \gamma y$, and the magnitude of the elemental force exerted on $\Delta A$ is $\Delta F = p\,\Delta A = \gamma y\,\Delta A$. The magnitude of the resultant of the elemental forces is thus

$$R = \int \gamma y\,dA = \gamma \int y\,dA$$

We can obtain this by computing the first moment $Q_x = \int y\,dA$ of the area of the gate with respect to the $x$ axis. The moment $M_x$ of the resultant must be equal to the sum of the moments $\Delta M_x = y\,\Delta F = \gamma\,y^2\,\Delta A$ of the elemental forces. Integrating over the area of the gate, we have

$$M_x = \int \gamma y^2\,dA = \gamma \int y^2\,dA$$

Here again, the last integral represents the second moment, or moment of inertia, $I_x$ of the area with respect to the $x$ axis.

## 9.1B   Determining the Moment of Inertia of an Area by Integration

We just defined the second moment, or moment of inertia, $I_x$ of an area $A$ with respect to the $x$ axis. In a similar way, we can also define the moment of inertia $I_y$ of the area $A$ with respect to the $y$ axis (Fig. 9.3a):

**Moments of inertia of an area**

$$I_x = \int y^2\,dA \quad I_y = \int x^2\,dA \tag{9.1}$$

We can evaluate these integrals, which are known as the **rectangular moments of inertia** of the area $A$, more easily if we choose $dA$ to be a thin strip parallel to one of the coordinate axes. To compute $I_x$, we choose the strip parallel to the $x$ axis, so that all points of the strip are at the same distance $y$ from the $x$ axis (Fig. 9.3b). We obtain the moment of inertia $dI_x$ of the strip by multiplying the area $dA$ of

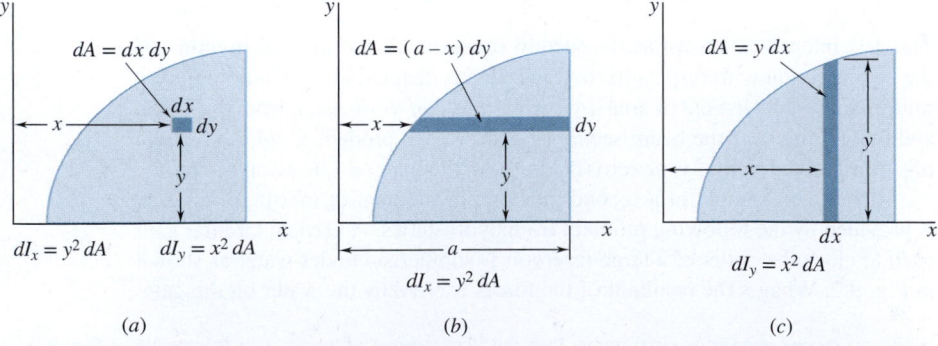

**Fig. 9.3**   (a) Rectangular moments of inertia $dI_x$ and $dI_y$ of an area $dA$; (b) calculating $I_x$ with a horizontal strip; (c) calculating $I_y$ with a vertical strip.

the strip by $y^2$. To compute $I_y$, we choose the strip parallel to the $y$ axis, so that all points of the strip are at the same distance $x$ from the $y$ axis (Fig. 9.3$c$). Then, the moment of inertia $dI_y$ of the strip is $x^2\,dA$.

**Moment of Inertia of a Rectangular Area.**   As an example, let us determine the moment of inertia of a rectangle with respect to its base (Fig. 9.4). Dividing the rectangle into strips parallel to the $x$ axis, we have

$$dA = b\,dy \qquad dI_x = y^2 b\,dy$$

$$I_x = \int_0^h by^2\,dy = \frac{1}{3}bh^3 \tag{9.2}$$

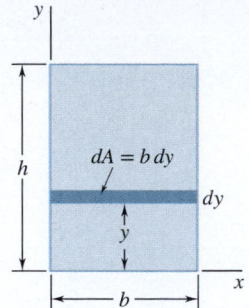

**Fig. 9.4**   Calculating the moment of inertia of a rectangular area with respect to its base.

**Computing $I_x$ and $I_y$ Using the Same Elemental Strips.**   We can use Eq. (9.2) to determine the moment of inertia $dI_x$ with respect to the $x$ axis of a rectangular strip that is parallel to the $y$ axis, such as the strip shown in Fig. 9.3$c$. Setting $b = dx$ and $h = y$ in formula (9.2), we obtain

$$dI_x = \frac{1}{3}y^3\,dx$$

We also have

$$dI_y = x^2\,dA = x^2 y\,dx$$

Thus, we can use the same element to compute the moments of inertia $I_x$ and $I_y$ of a given area (Fig. 9.5).

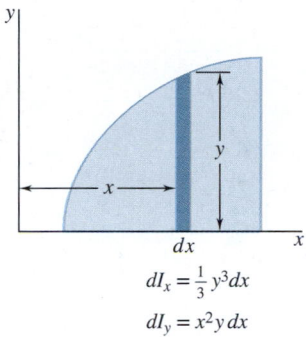

$$dI_x = \tfrac{1}{3}y^3 dx$$

$$dI_y = x^2 y\,dx$$

**Fig. 9.5**   Using the same strip element of a given area to calculate $I_x$ and $I_y$.

## 9.1C  Polar Moment of Inertia

An integral of great importance in problems concerning the torsion of cylindrical shafts and in problems dealing with the rotation of slabs is

**Polar moment of inertia**

$$J_O = \int r^2 dA \tag{9.3}$$

where $r$ is the distance from $O$ to the element of area $dA$ (Fig. 9.6). This integral is called the **polar moment of inertia** of the area $A$ with respect to the "pole" $O$.

We can compute the polar moment of inertia of a given area from the rectangular moments of inertia $I_x$ and $I_y$ of the area if these quantities are already known. Indeed, noting that $r^2 = x^2 + y^2$, we have

$$J_O = \int r^2 dA = \int (x^2 + y^2)dA = \int y^2 dA + \int x^2 dA$$

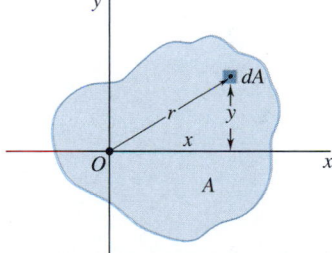

**Fig. 9.6**   Distance $r$ used to evaluate the polar moment of inertia of area $A$.

That is,

$$J_O = I_x + I_y \tag{9.4}$$

## 9.1D  Radius of Gyration of an Area

Consider an area $A$ that has a moment of inertia $I_x$ with respect to the $x$ axis (Fig. 9.7$a$). Imagine that we concentrate this area into a thin strip parallel to the $x$ axis (Fig. 9.7$b$). If the concentrated area $A$ is to have the same moment of inertia with respect to the $x$ axis, the strip should be placed at a distance $k_x$ from the $x$ axis, where $k_x$ is defined by the relation

$$I_x = k_x^2 A$$

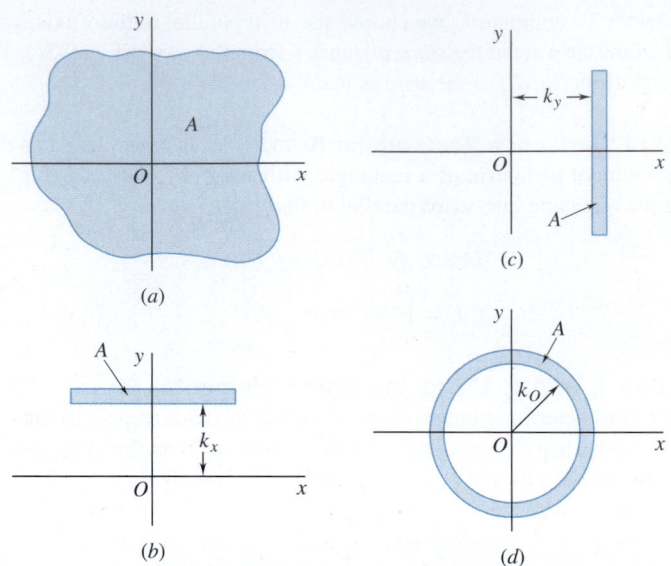

**Fig. 9.7** (a) Area A with given moment of inertia $I_x$; (b) compressing the area to a horizontal strip with radius of gyration $k_x$; (c) compressing the area to a vertical strip with radius of gyration $k_y$; (d) compressing the area to a circular ring with polar radius of gyration $k_O$.

Solving for $k_x$, we have

**Radius of gyration**

$$k_x = \sqrt{\frac{I_x}{A}} \qquad (9.5)$$

The distance $k_x$ is referred to as the **radius of gyration** of the area with respect to the $x$ axis. In a similar way, we can define the radii of gyration $k_y$ and $k_O$ (Fig. 9.7c and d); we have

$$I_y = k_y^2 A \qquad k_y = \sqrt{\frac{I_y}{A}} \qquad (9.6)$$

$$J_O = k_O^2 A \qquad k_O = \sqrt{\frac{J_O}{A}} \qquad (9.7)$$

If we rewrite Eq. (9.4) in terms of the radii of gyration, we find that

$$k_O^2 = k_x^2 + k_y^2 \qquad (9.8)$$

## Concept Application 9.1

For the rectangle shown in Fig. 9.8, compute the radius of gyration $k_x$ with respect to its base. Using formulas (9.5) and (9.2), you have

$$k_x^2 = \frac{I_x}{A} = \frac{\frac{1}{3}bh^3}{bh} = \frac{h^2}{3} \qquad k_x = \frac{h}{\sqrt{3}}$$

The radius of gyration $k_x$ of the rectangle is shown in Fig. 9.8. Do not confuse it with the ordinate $\bar{y} = h/2$ of the centroid of the area. The radius of gyration $k_x$ depends upon the *second moment* of the area, whereas the ordinate $\bar{y}$ is related to the *first moment* of the area.

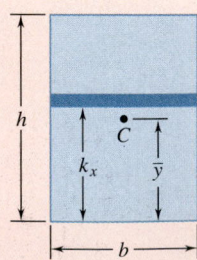

**Fig. 9.8** Radius of gyration of a rectangle with respect to its base.

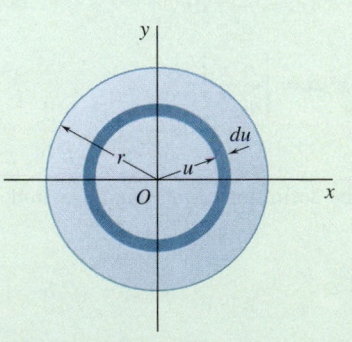

**Fig. 1** Triangle with differential strip element parallel to its base.

## Sample Problem 9.1

Determine the moment of inertia of a triangle with respect to its base.

**STRATEGY:**  To find the moment of inertia with respect to the base, it is expedient to use a differential strip of area parallel to the base. Use the geometry of the situation to carry out the integration.

**MODELING:**  Draw a triangle with a base $b$ and height $h$, choosing the $x$ axis to coincide with the base (Fig. 1). Choose a differential strip parallel to the $x$ axis to be $dA$. Because all portions of the strip are at the same distance from the $x$ axis, you have

$$dI_x = y^2 dA \qquad dA = l\,dy$$

**ANALYSIS:**  Using similar triangles, you have

$$\frac{l}{b} = \frac{h - y}{h} \qquad l = b\frac{h - y}{h} \qquad dA = b\frac{h - y}{h}dy$$

Integrating $dI_x$ from $y = 0$ to $y = h$, you obtain

$$I_x = \int y^2\,dA = \int_0^h y^2 b\frac{h - y}{h}\,dy = \frac{b}{h}\int_0^h (hy^2 - y^3)dy = \frac{b}{h}\left[h\frac{y^3}{3} - \frac{y^4}{4}\right]_0^h \qquad I_x = \frac{bh^3}{12} \blacktriangleleft$$

**REFLECT and THINK:**  This problem also could have been solved using a differential strip perpendicular to the base by applying Eq. (9.2) to express the moment of inertia of this strip. However, because of the geometry of this triangle, you would need two integrals to complete the solution.

## Sample Problem 9.2

(*a*) Determine the centroidal polar moment of inertia of a circular area by direct integration. (*b*) Using the result of part *a*, determine the moment of inertia of a circular area with respect to a diameter.

**STRATEGY:**  Because the area is circular, you can evaluate part *a* by using an annular differential area. For part *b*, you can use symmetry and Eq. (9.4) to solve for the moment of inertia with respect to a diameter.

**MODELING and ANALYSIS:**

**a. Polar Moment of Inertia.**  Choose an annular differential element of area to be $dA$ (Fig. 1). Because all portions of the differential area are at the same distance from the origin, you have

$$dJ_O = u^2 dA \quad dA = 2\pi u\,du$$

$$J_O = \int dJ_O = \int_0^r u^2(2\pi u\,du) = 2\pi\int_0^r u^3\,du$$

$$J_O = \frac{\pi}{2}r^4 \blacktriangleleft$$

**Fig. 1** Circular area with an annular differential element.

**b. Moment of Inertia with Respect to a Diameter.**  Because of the symmetry of the circular area, $I_x = I_y$. Then, from Eq. (9.4), you have

$$J_O = I_x + I_y = 2I_x \qquad \frac{\pi}{2}r^4 = 2I_x \qquad I_{\text{diameter}} = I_x = \frac{\pi}{4}r^4 \blacktriangleleft$$

**REFLECT and THINK:**  Always look for ways to simplify a problem by the use of symmetry. This is especially true for situations involving circles or spheres.

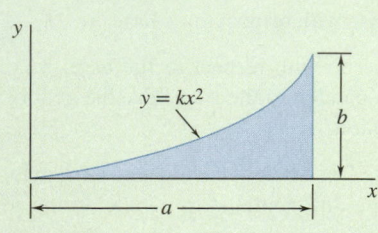

$y = kx^2$

## Sample Problem 9.3

(a) Determine the moment of inertia of the shaded region shown with respect to each of the coordinate axes. (Properties of this region were considered in Sample Prob. 5.4.) (b) Using the results of part a, determine the radius of gyration of the shaded area with respect to each of the coordinate axes.

**STRATEGY:**  You can determine the moments of inertia by using a single differential strip of area; a vertical strip will be more convenient. You can calculate the radii of gyration from the moments of inertia and the area of the region.

**MODELING:**   Referring to Sample Prob. 5.4, you can find the equation of the curve and the total area using

$$y = \frac{b}{a^2}x^2 \qquad A = \frac{1}{3}ab$$

**ANALYSIS:**

### a. Moments of Inertia.

**Moment of Inertia $I_x$.**   Choose a vertical differential element of area for $dA$ (Fig. 1). Because all portions of this element are *not* at the same distance from the $x$ axis, you must treat the element as a thin rectangle. The moment of inertia of the element with respect to the $x$ axis is then

$$dI_x = \frac{1}{3}y^3\,dx = \frac{1}{3}\left(\frac{b}{a^2}x^2\right)^3 dx = \frac{1}{3}\frac{b^3}{a^6}x^6\,dx$$

$$I_x = \int dI_x = \int_0^a \frac{1}{3}\frac{b^3}{a^6}x^6\,dx = \left[\frac{1}{3}\frac{b^3}{a^6}\frac{x^7}{7}\right]_0^a$$

$$I_x = \frac{ab^3}{21} \quad \blacktriangleleft$$

**Fig. 1**  Subject area with vertical differential strip element.

**Moment of Inertia $I_y$.**   Use the same vertical differential element of area. Because all portions of the element are at the same distance from the $y$ axis, you have

$$dI_y = x^2\,dA = x^2(y\,dx) = x^2\left(\frac{b}{a^2}x^2\right)dx = \frac{b}{a^2}x^4 dx$$

$$I_y = \int dI_y = \int_0^a \frac{b}{a^2}x^4 dx = \left[\frac{b}{a^2}\frac{x^5}{5}\right]_0^a$$

$$I_y = \frac{a^3b}{5} \quad \blacktriangleleft$$

### b. Radii of Gyration $k_x$ and $k_y$.   From the definition of radius of gyration, you have

$$k_x^2 = \frac{I_x}{A} = \frac{ab^3/21}{ab/3} = \frac{b^2}{7} \qquad\qquad k_x = \sqrt{\tfrac{1}{7}}\,b \quad \blacktriangleleft$$

and

$$k_y^2 = \frac{I_y}{A} = \frac{a^3b/5}{ab/3} = \tfrac{3}{5}a^2 \qquad\qquad k_y = \sqrt{\tfrac{3}{5}}\,a \quad \blacktriangleleft$$

**REFLECT and THINK:**   This problem demonstrates how you can calculate $I_x$ and $I_y$ using the same strip element. However, the general mathematical approach in each case is distinctly different.

# SOLVING PROBLEMS
# ON YOUR OWN

In this section, we introduced the **rectangular and polar moments of inertia of areas** and the corresponding **radii of gyration**. Although the problems you are about to solve may appear more appropriate for a calculus class than for one in mechanics, we hope that our introductory comments have convinced you of the relevance of moments of inertia to your study of a variety of engineering topics.

**1. Calculating the rectangular moments of inertia $I_x$ and $I_y$.** We defined these quantities as

$$I_x = \int y^2 \, dA \qquad I_y = \int x^2 \, dA \qquad\qquad (9.1)$$

where $dA$ is a differential element of area $dx \, dy$. The moments of inertia are the **second moments of the area**; it is for that reason that $I_x$, for example, depends on the perpendicular distance $y$ to the area $dA$. As you study Sec. 9.1, you should recognize the importance of carefully defining the shape and the orientation of $dA$. Furthermore, you should note the following points.

    **a. You can obtain the moments of inertia of most areas by means of a single integration.** You can use the expressions given in Figs. 9.3b, 9.3c, and 9.5 to calculate $I_x$ and $I_y$. Regardless of whether you use a single or a double integration, be sure to show the element $dA$ that you have chosen on your sketch.

    **b. The moment of inertia of an area is always positive,** regardless of the location of the area with respect to the coordinate axes. The reason is that the moment of inertia is obtained by integrating the product of $dA$ and the *square* of distance. (Note how this differs from the first moment of the area.) Only when an area is *removed* (as in the case for a hole) does its moment of inertia enter in your computations with a minus sign.

    **c. As a partial check of your work,** observe that the moments of inertia are equal to an area times the square of a length. Thus, every term in an expression for a moment of inertia must be a length to the fourth power.

**2. Computing the polar moment of inertia $J_O$.** We defined $J_O$ as

$$J_O = \int r^2 \, dA \qquad\qquad (9.3)$$

where $r^2 = x^2 + y^2$. If the given area has circular symmetry (as in Sample Prob. 9.2), it is possible to express $dA$ as a function of $r$ and to compute $J_O$ with a single integration. When the area lacks circular symmetry, it is usually easier first to calculate $I_x$ and $I_y$ and then to determine $J_O$ from

$$J_O = I_x + I_y \qquad\qquad (9.4)$$

Lastly, if the equation of the curve that bounds the given area is expressed in polar coordinates, then $dA = r \, dr \, d\theta$, and you need to perform a double integration to compute the integral for $J_O$ (see Prob. 9.27).

    **3. Determining the radii of gyration $k_x$ and $k_y$ and the polar radius of gyration $k_O$.** These quantities are defined in Sec. 9.1D. You should realize that they can be determined only after you have computed the area and the appropriate moments of inertia. It is important to remember that $k_x$ is measured in the $y$ direction, whereas $k_y$ is measured in the $x$ direction; you should carefully study Sec. 9.1D until you understand this point.

# Problems

9.1 through 9.4    Determine by direct integration the moment of inertia of the shaded area with respect to the $y$ axis.

9.5 through 9.8    Determine by direct integration the moment of inertia of the shaded area with respect to the $x$ axis.

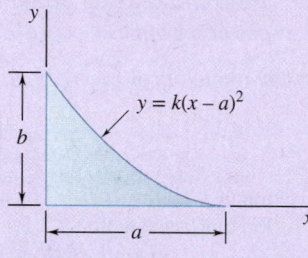

Fig. P9.1 and *P9.5*

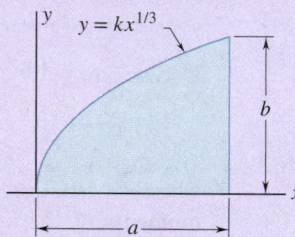

Fig. P9.2 and P9.6

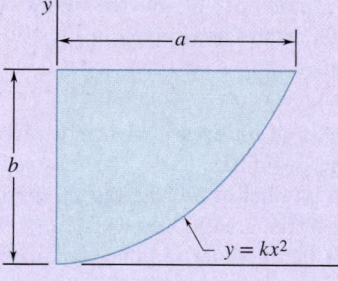

Fig. P9.3 and *P9.7*

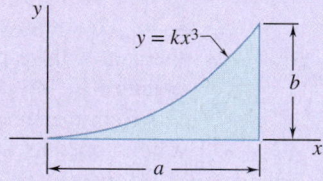

Fig. P9.4 and P9.8

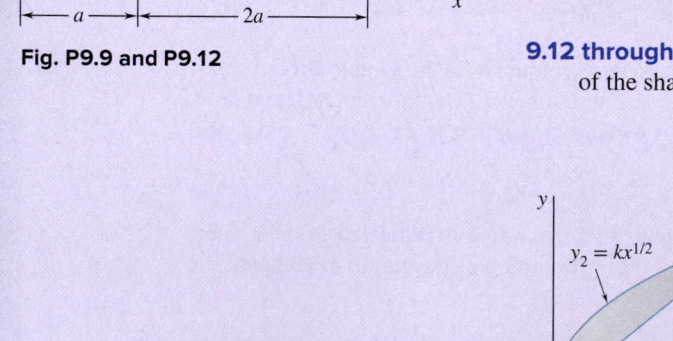

Fig. P9.9 and P9.12

9.9 through 9.11    Determine by direct integration the moment of inertia of the shaded area with respect to the $x$ axis.

9.12 through 9.14    Determine by direct integration the moment of inertia of the shaded area with respect to the $y$ axis.

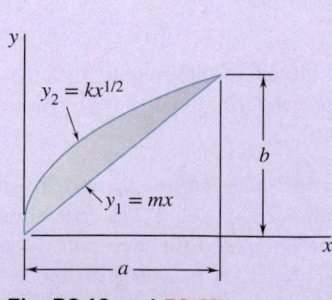

Fig. P9.10 and *P9.13*

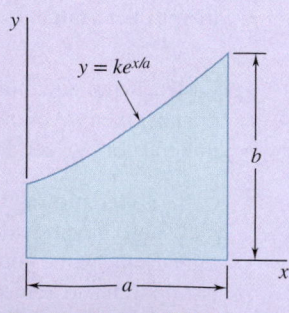

Fig. P9.11 and *P9.14*

**9.15 and 9.16**  Determine the moment of inertia and the radius of gyration of the shaded area shown with respect to the $x$ axis.

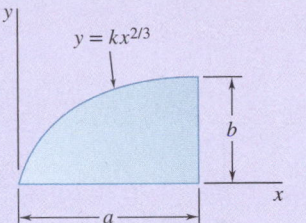

Fig. P9.15 and P9.17          Fig. P9.16 and P9.18

**9.17 and 9.18**  Determine the moment of inertia and the radius of gyration of the shaded area shown with respect to the $y$ axis.

**9.19**  Determine the moment of inertia and the radius of gyration of the shaded area shown with respect to the $x$ axis.

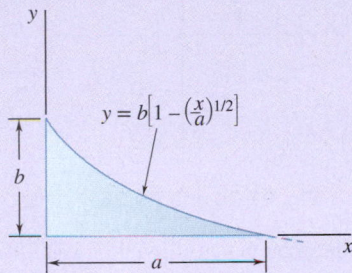

Fig. *P9.19 and P9.20*

**9.20**  Determine the moment of inertia and the radius of gyration of the shaded area shown with respect to the $y$ axis.

**9.21**  Determine the polar moment of inertia and the polar radius of gyration of the rectangle shown with respect to the midpoint of one of its (*a*) longer sides, (*b*) shorter sides.

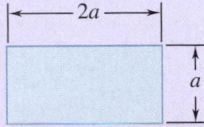

Fig. P9.21

**9.22**  Determine the polar moment of inertia and the polar radius of gyration of the trapezoid shown with respect to point $P$.

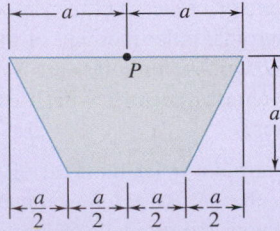

Fig. P9.22

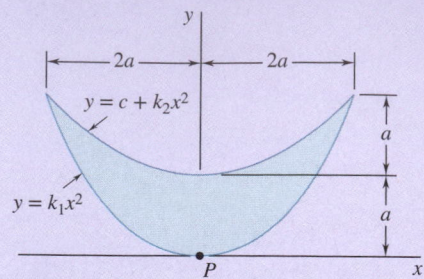

**Fig. P9.23**

**9.23 and 9.24** Determine the polar moment of inertia and the polar radius of gyration of the shaded area shown with respect to point $P$.

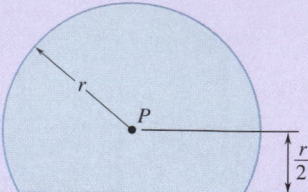

**Fig. P9.24**

**9.25** (*a*) Determine by direct integration the polar moment of inertia of the annular area shown with respect to point $O$. (*b*) Using the result of part *a*, determine the moment of inertia of the given area with respect to the *x* axis.

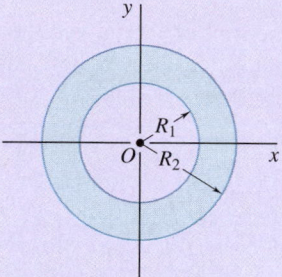

**Fig. P9.25 and P9.26**

**9.26** (*a*) Show that the polar radius of gyration $k_O$ of the annular area shown is approximately equal to the mean radius $R_m = (R_1 + R_2)/2$ for small values of the thickness $t = R_2 - R_1$. (*b*) Determine the percentage error introduced by using $R_m$ in place of $k_O$ for the following values of $t/R_m$: $1$, $\frac{1}{2}$, and $\frac{1}{10}$.

**9.27** Determine the polar moment of inertia and the polar radius of gyration of the shaded area shown with respect to point $O$.

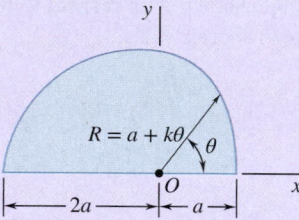

**Fig. P9.27**

**9.28** Determine the polar moment of inertia and the polar radius of gyration of the isosceles triangle shown with respect to point $O$.

**\*9.29** Using the polar moment of inertia of the isosceles triangle of Prob. 9.28, show that the centroidal polar moment of inertia of a circular area of radius $r$ is $\pi r^4/2$. (*Hint:* As a circular area is divided into an increasing number of equal circular sectors, what is the approximate shape of each circular sector?)

**\*9.30** Prove that the centroidal polar moment of inertia of a given area $A$ cannot be smaller than $A^2/2\pi$. (*Hint:* Compare the moment of inertia of the given area with the moment of inertia of a circle that has the same area and the same centroid.)

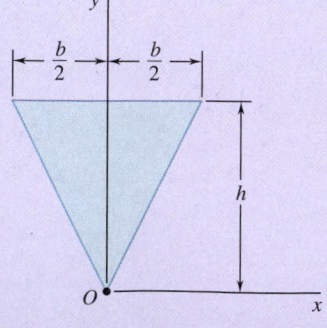

**Fig. P9.28**

# 9.2  PARALLEL-AXIS THEOREM AND COMPOSITE AREAS

In practice, we often need to determine the moment of inertia of a complicated area that can be broken down into a sum of simple areas. However, in doing these calculations, we have to determine the moment of inertia of each simple area with respect to the same axis. In this section, we first derive a formula for computing the moment of inertia of an area with respect to a centroidal axis parallel to a given axis. Then, we show how you can use this formula for finding the moment of inertia of a composite area.

## 9.2A  The Parallel-Axis Theorem

Consider the moment of inertia $I$ of an area $A$ with respect to an axis $AA'$ (Fig. 9.9). We denote the distance from an element of area $dA$ to $AA'$ by $y$. This gives us

$$I = \int y^2 \, dA$$

Let us now draw through the centroid $C$ of the area an axis $BB'$ parallel to $AA'$; this axis is called a *centroidal axis*. Denoting the distance from the element $dA$ to $BB'$ by $y'$, we have $y = y' + d$, where $d$ is the distance between the axes $AA'$ and $BB'$. Substituting for $y$ in the previous integral, we obtain

$$I = \int y^2 \, dA = \int (y' + d)^2 \, dA$$
$$= \int y'^2 \, dA + 2d \int y' \, dA + d^2 \int dA$$

The first integral represents the moment of inertia $\bar{I}$ of the area with respect to the centroidal axis $BB'$. The second integral represents the first moment of the area with respect to $BB'$, but because the centroid $C$ of the area is located on this axis, the second integral must be zero. The last integral is equal to the total area $A$. Therefore, we have

**Parallel-axis theorem**

$$I = \bar{I} + Ad^2 \qquad (9.9)$$

This formula states that the moment of inertia $I$ of an area with respect to any given axis $AA'$ is equal to the moment of inertia $\bar{I}$ of the area with respect to a centroidal axis $BB'$ parallel to $AA'$ *plus* the product of the area $A$ and the square of the distance $d$ between the two axes. This theorem is known as the **parallel-axis theorem**. Substituting $k^2A$ for $I$ and $\bar{k}^2A$ for $\bar{I}$, we can also express this theorem as

$$k^2 = \bar{k}^2 + d^2 \qquad (9.10)$$

A similar theorem relates the polar moment of inertia $J_O$ of an area about a point $O$ to the polar moment of inertia $\bar{J}_C$ of the same area about its centroid $C$. Denoting the distance between $O$ and $C$ by $d$, we have

$$J_O = \bar{J}_C + Ad^2 \quad \text{or} \quad k_O^2 = \bar{k}_C^2 + d^2 \qquad (9.11)$$

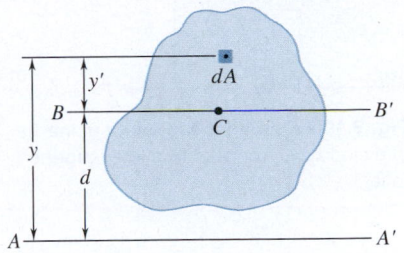

**Fig. 9.9**  The moment of inertia of an area $A$ with respect to an axis $AA'$ can be determined from its moment of inertia with respect to the centroidal axis $BB'$ by a calculation involving the distance $d$ between the axes.

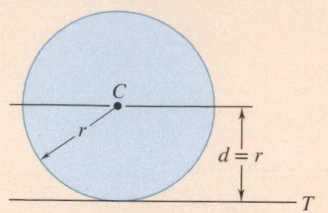

**Fig. 9.10** Finding the moment of inertia of a circle with respect to a line tangent to it.

## Concept Application 9.2

As an application of the parallel-axis theorem, let us determine the moment of inertia $I_T$ of a circular area with respect to a line tangent to the circle (Fig. 9.10). We found in Sample Prob. 9.2 that the moment of inertia of a circular area about a centroidal axis is $\bar{I} = \frac{1}{4}\pi r^4$. Therefore, we have

$$I_T = \bar{I} + Ad^2 = \frac{1}{4}\pi r^4 + (\pi r^2)r^2 = \frac{5}{4}\pi r^4$$

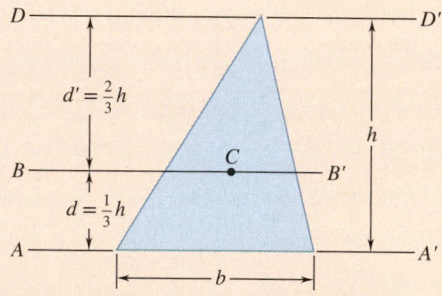

**Fig. 9.11** Finding the centroidal moment of inertia of a triangle from the moment of inertia about a parallel axis.

## Concept Application 9.3

We can also use the parallel-axis theorem to determine the centroidal moment of inertia of an area when we know the moment of inertia of the area with respect to a parallel axis. Consider, for instance, a triangular area (Fig. 9.11). We found in Sample Prob. 9.1 that the moment of inertia of a triangle with respect to its base $AA'$ is equal to $\frac{1}{12}bh^3$. Using the parallel-axis theorem, we have

$$I_{AA'} = \bar{I}_{BB'} + Ad^2$$
$$\bar{I}_{BB'} = I_{AA'} - Ad^2 = \frac{1}{12}bh^3 - \frac{1}{2}bh\left(\frac{1}{3}h\right)^2 = \frac{1}{36}bh^3$$

Note that we *subtracted* the product $Ad^2$ from the given moment of inertia in order to obtain the centroidal moment of inertia of the triangle. That is, this product is *added* when transferring *from* a centroidal axis to a parallel axis, but it is *subtracted* when transferring *to* a centroidal axis. In other words, the moment of inertia of an area is always smaller with respect to a centroidal axis than with respect to any parallel axis.

Returning to Fig. 9.11, we can obtain the moment of inertia of the triangle with respect to the line $DD'$ (which is drawn through a vertex) by writing

$$I_{DD'} = \bar{I}_{BB'} + Ad'^2 = \frac{1}{36}bh^3 + \frac{1}{2}bh\left(\frac{2}{3}h\right)^2 = \frac{1}{4}bh^3$$

Note that we could not have obtained $I_{DD'}$ directly from $I_{AA'}$. We can apply the parallel-axis theorem only if one of the two parallel axes passes through the centroid of the area.

## 9.2B Moments of Inertia of Composite Areas

Consider a composite area $A$ made of several component areas $A_1, A_2, A_3, \ldots$. The integral representing the moment of inertia of $A$ can be subdivided into integrals evaluated over $A_1, A_2, A_3, \ldots$. Therefore, we can obtain the moment of inertia of $A$ with respect to a given axis by adding the moments of inertia of the areas $A_1, A_2, A_3, \ldots$ with respect to the same axis.

Figure 9.12 shows several common geometric shapes along with formulas for the moments of inertia of each one. Before adding the moments of inertia of the component areas, however, you may have to use the parallel-axis theorem to transfer each moment of inertia to the desired axis. Sample Probs. 9.4 and 9.5 illustrate the technique.

Properties of the cross sections of various structural shapes are given in Figs. 9.13A and 9.13B. As we noted in Sec. 9.1A, the moment of inertia

**Photo 9.1** Fig. 9.13 tabulates data for a small sample of the rolled-steel shapes that are readily available. Shown above are examples of wide-flange shapes that are commonly used in the construction of buildings. ©Barry Willis/Getty Images

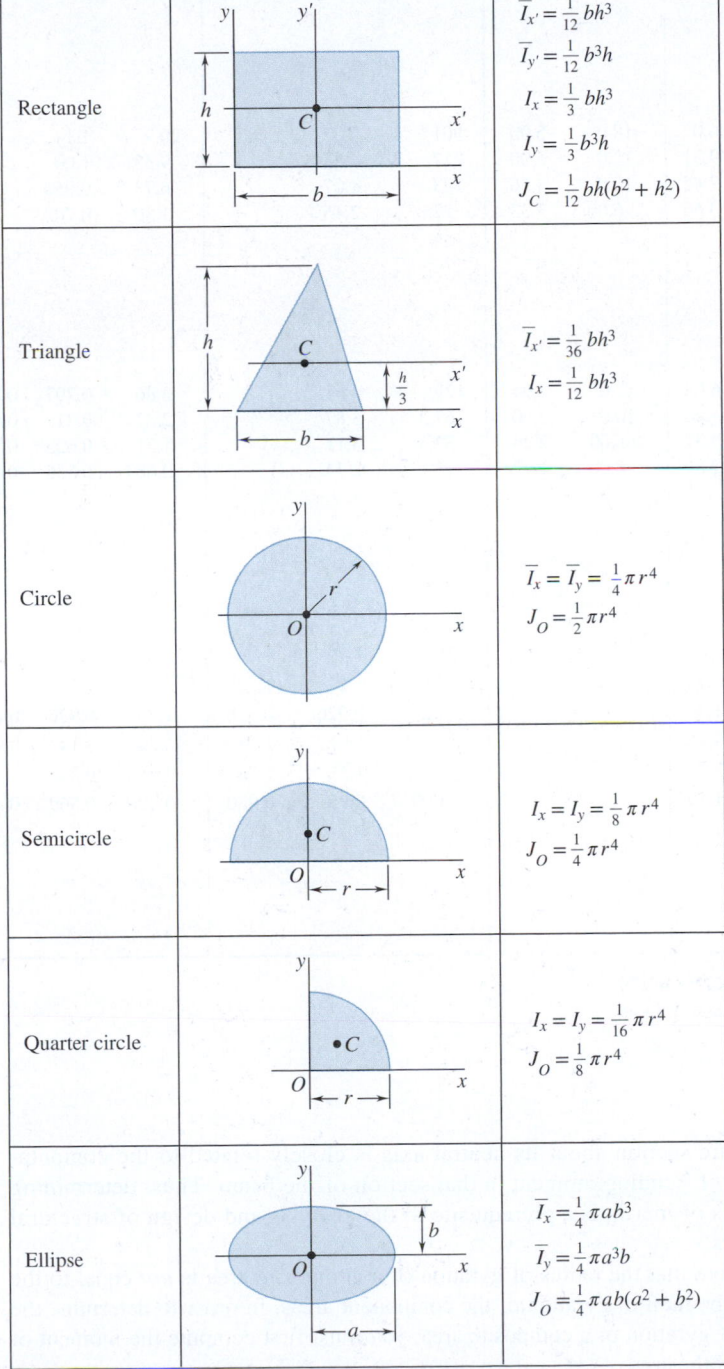

| | | |
|---|---|---|
| Rectangle | | $\overline{I}_{x'} = \frac{1}{12} bh^3$ $\overline{I}_{y'} = \frac{1}{12} b^3 h$ $I_x = \frac{1}{3} bh^3$ $I_y = \frac{1}{3} b^3 h$ $J_C = \frac{1}{12} bh(b^2 + h^2)$ |
| Triangle | | $\overline{I}_{x'} = \frac{1}{36} bh^3$ $I_x = \frac{1}{12} bh^3$ |
| Circle | | $\overline{I}_x = \overline{I}_y = \frac{1}{4}\pi r^4$ $J_O = \frac{1}{2}\pi r^4$ |
| Semicircle | | $I_x = I_y = \frac{1}{8}\pi r^4$ $J_O = \frac{1}{4}\pi r^4$ |
| Quarter circle | | $I_x = I_y = \frac{1}{16}\pi r^4$ $J_O = \frac{1}{8}\pi r^4$ |
| Ellipse | | $\overline{I}_x = \frac{1}{4}\pi ab^3$ $\overline{I}_y = \frac{1}{4}\pi a^3 b$ $J_O = \frac{1}{4}\pi ab(a^2 + b^2)$ |

**Fig. 9.12** Moments of inertia of common geometric shapes.

| | Designation | Area in² | Depth in. | Width in. | Axis X–X | | | Axis Y–Y | | |
|---|---|---|---|---|---|---|---|---|---|---|
| | | | | | $\bar{I}_x$, in⁴ | $\bar{k}_x$, in. | $\bar{y}$, in. | $\bar{I}_y$, in⁴ | $\bar{k}_y$, in. | $\bar{x}$, in. |
| W Shapes (Wide-Flange Shapes) | 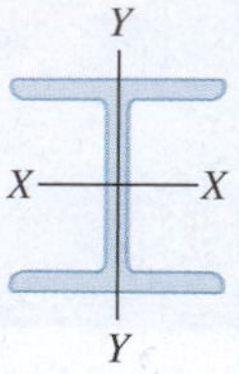 W18 × 76† | 22.3 | 18.2 | 11.0 | 1330 | 7.73 | | 152 | 2.61 | |
| | W16 × 57 | 16.8 | 16.4 | 7.12 | 758 | 6.72 | | 43.1 | 1.60 | |
| | W14 × 38 | 11.2 | 14.1 | 6.77 | 385 | 5.87 | | 26.7 | 1.55 | |
| | W8 × 31 | 9.12 | 8.00 | 8.00 | 110 | 3.47 | | 37.1 | 2.02 | |
| S Shapes (American Standard Shapes) | 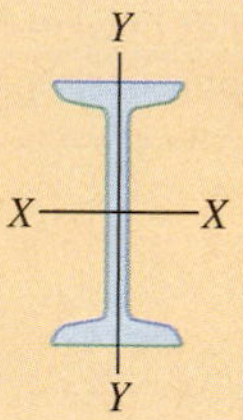 S18 × 54.7† | 16.0 | 18.0 | 6.00 | 801 | 7.07 | | 20.7 | 1.14 | |
| | S12 × 31.8 | 9.31 | 12.0 | 5.00 | 217 | 4.83 | | 9.33 | 1.00 | |
| | S10 × 25.4 | 7.45 | 10.0 | 4.66 | 123 | 4.07 | | 6.73 | 0.950 | |
| | S6 × 12.5 | 3.66 | 6.00 | 3.33 | 22.0 | 2.45 | | 1.80 | 0.702 | |
| C Shapes (American Standard Channels) | 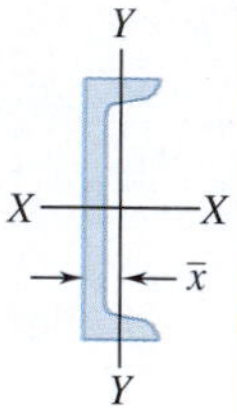 C12 × 20.7† | 6.08 | 12.0 | 2.94 | 129 | 4.61 | | 3.86 | 0.797 | 0.698 |
| | C10 × 15.3 | 4.48 | 10.0 | 2.60 | 67.3 | 3.87 | | 2.27 | 0.711 | 0.634 |
| | C8 × 11.5 | 3.37 | 8.00 | 2.26 | 32.5 | 3.11 | | 1.31 | 0.623 | 0.572 |
| | C6 × 8.2 | 2.39 | 6.00 | 1.92 | 13.1 | 2.34 | | 0.687 | 0.536 | 0.512 |
| Angles | 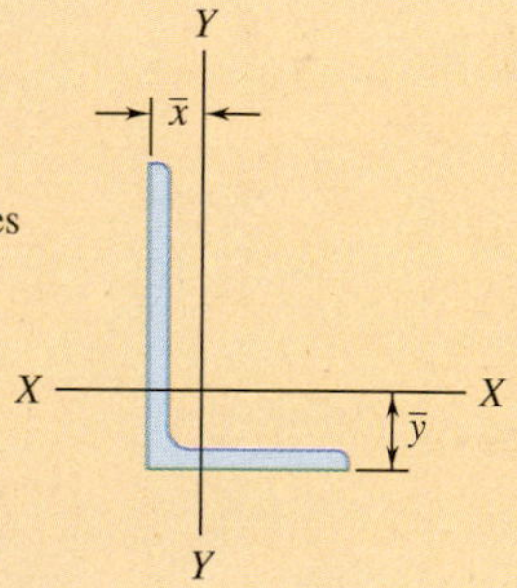 L6 × 6 × 1‡ | 11.0 | | | 35.4 | 1.79 | 1.86 | 35.4 | 1.79 | 1.86 |
| | L4 × 4 × ½ | 3.75 | | | 5.52 | 1.21 | 1.18 | 5.52 | 1.21 | 1.18 |
| | L3 × 3 × ¼ | 1.44 | | | 1.23 | 0.926 | 0.836 | 1.23 | 0.926 | 0.836 |
| | L6 × 4 × ½ | 4.75 | | | 17.3 | 1.91 | 1.98 | 6.22 | 1.14 | 0.981 |
| | L5 × 3 × ½ | 3.75 | | | 9.43 | 1.58 | 1.74 | 2.55 | 0.824 | 0.746 |
| | L3 × 2 × ¼ | 1.19 | | | 1.09 | 0.953 | 0.980 | 0.390 | 0.569 | 0.487 |

**Fig. 9.13A**  Properties of rolled-steel shapes (U.S. customary units).*

*Courtesy of the American Institute of Steel Construction, Chicago, Illinois

†Nominal depth in inches and weight in pounds per foot

‡Depth, width, and thickness in inches

of a beam section about its neutral axis is closely related to the computation of the bending moment in that section of the beam. Thus, determining moments of inertia is a prerequisite to the analysis and design of structural members.

Note that the radius of gyration of a composite area is *not* equal to the sum of the radii of gyration of the component areas. In order to determine the radius of gyration of a composite area, you must first compute the moment of inertia of the area.

| | Designation | Area $mm^2$ | Depth mm | Width mm | Axis $X$–$X$ | | | Axis $Y$–$Y$ | | |
|---|---|---|---|---|---|---|---|---|---|---|
| | | | | | $\bar{I}_x$ $10^6\ mm^4$ | $\bar{k}_x$ mm | $\bar{y}$ mm | $\bar{I}_y$ $10^6\ mm^4$ | $\bar{k}_y$ mm | $\bar{x}$ mm |
| W Shapes (Wide-Flange Shapes) | 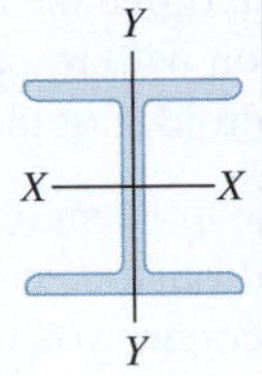 W460 × 113† | 14 400 | 462 | 279 | 554 | 196 | | 63.3 | 66.3 | |
| | W410 × 85 | 10 800 | 417 | 181 | 316 | 171 | | 17.9 | 40.6 | |
| | W360 × 57.8 | 7230 | 358 | 172 | 160 | 149 | | 11.1 | 39.4 | |
| | W200 × 46.1 | 5880 | 203 | 203 | 45.8 | 88.1 | | 15.4 | 51.3 | |
| S Shapes (American Standard Shapes) | 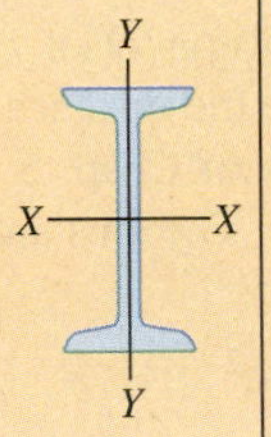 S460 × 81.4† | 10 300 | 457 | 152 | 333 | 180 | | 8.62 | 29.0 | |
| | S310 × 47.3 | 6010 | 305 | 127 | 90.3 | 123 | | 3.88 | 25.4 | |
| | S250 × 37.8 | 4810 | 254 | 118 | 51.2 | 103 | | 2.80 | 24.1 | |
| | S150 × 18.6 | 2360 | 152 | 84.6 | 9.16 | 62.2 | | 0.749 | 17.8 | |
| C Shapes (American Standard Channels) | 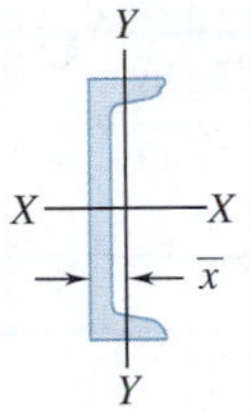 C310 × 30.8† | 3920 | 305 | 74.7 | 53.7 | 117 | | 1.61 | 20.2 | 17.7 |
| | C250 × 22.8 | 2890 | 254 | 66.0 | 28.0 | 98.3 | | 0.945 | 18.1 | 16.1 |
| | C200 × 17.1 | 2170 | 203 | 57.4 | 13.5 | 79.0 | | 0.545 | 15.8 | 14.5 |
| | C150 × 12.2 | 1540 | 152 | 48.8 | 5.45 | 59.4 | | 0.286 | 13.6 | 13.0 |
| Angles | 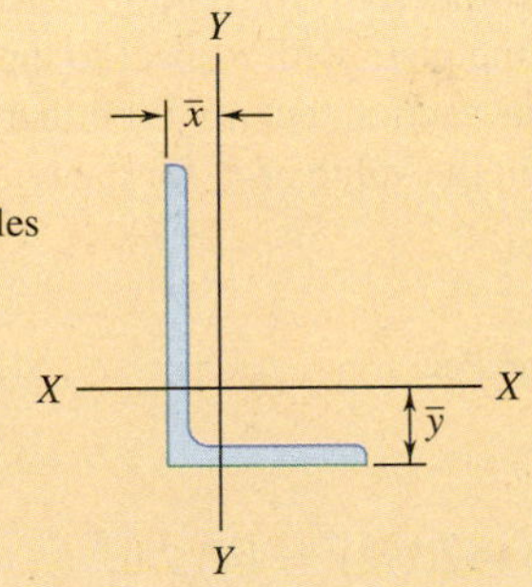 L152 × 152 × 25.4‡ | 7100 | | | 14.7 | 45.5 | 47.2 | 14.7 | 45.5 | 47.2 |
| | L102 × 102 × 12.7 | 2420 | | | 2.30 | 30.7 | 30.0 | 2.30 | 30.7 | 30.0 |
| | L76 × 76 × 6.4 | 929 | | | 0.512 | 23.5 | 21.2 | 0.512 | 23.5 | 21.2 |
| | L152 × 102 × 12.7 | 3060 | | | 7.20 | 48.5 | 50.3 | 2.59 | 29.0 | 24.9 |
| | L127 × 76 × 12.7 | 2420 | | | 3.93 | 40.1 | 44.2 | 1.06 | 20.9 | 18.9 |
| | L76 × 51 × 6.4 | 768 | | | 0.454 | 24.2 | 24.9 | 0.162 | 14.5 | 12.4 |

**Fig. 9.13B**   Properties of rolled-steel shapes (SI units).

†Nominal depth in millimeters and mass in kilograms per meter

‡Depth, width, and thickness in millimeters

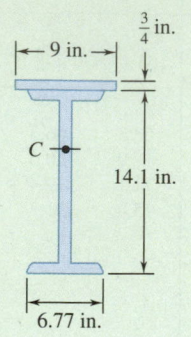

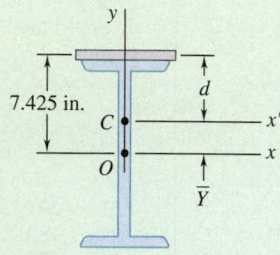

**Fig. 1** Origin of coordinates placed at centroid of wide-flange shape.

## Sample Problem 9.4

The strength of a W14 × 38 rolled-steel beam is increased by attaching a 9 × 3/4-in. plate to its upper flange as shown. Determine the moment of inertia and the radius of gyration of the composite section with respect to an axis that is parallel to the plate and passes through the centroid $C$ of the section.

**STRATEGY:** This problem involves finding the moment of inertia of a composite area with respect to its centroid. You should first determine the location of this centroid. Then, using the parallel-axis theorem, you can determine the moment of inertia relative to this centroid for the overall section from the centroidal moment of inertia for each component part.

**MODELING and ANALYSIS:** Place the origin $O$ of coordinates at the centroid of the wide-flange shape, and compute the distance $\overline{Y}$ to the centroid of the composite section by using the methods of Chap. 5 (Fig. 1). Refer to Fig. 9.13A for the area of the wide-flange shape. The area and the $y$ coordinate of the centroid of the plate are

$$A = (9 \text{ in.})(0.75 \text{ in.}) = 6.75 \text{ in}^2$$

$$\overline{y} = \tfrac{1}{2}(14.1 \text{ in.}) + \tfrac{1}{2}(0.75 \text{ in.}) = 7.425 \text{ in.}$$

| Section | Area, in² | $\overline{y}$, in. | $\overline{y}A$, in³ |
|---|---|---|---|
| Plate | 6.75 | 7.425 | 50.12 |
| Wide-flange shape | 11.2 | 0 | 0 |
| | $\Sigma A = 17.95$ | | $\Sigma\overline{y}A = 50.12$ |

$$\overline{Y}\Sigma A = \Sigma\overline{y}A \qquad \overline{Y}(17.95) = 50.12 \qquad \overline{Y} = 2.792 \text{ in.}$$

**Moment of Inertia.** Use the parallel-axis theorem to determine the moments of inertia of the wide-flange shape and the plate with respect to the $x'$ axis. This axis is a centroidal axis for the composite section, but *not* for either of the elements considered separately. You can obtain the value of $\overline{I}_x$ for the wide-flange shape from Fig. 9.13A.

For the wide-flange shape,

$$I_{x'} = \overline{I}_x + A\overline{Y}^2 = 385 + (11.2)(2.792)^2 = 472.3 \text{ in}^4$$

For the plate,

$$I_{x'} = \overline{I}_x + Ad^2 = (\tfrac{1}{12})(9)(\tfrac{3}{4})^3 + (6.75)(7.425 - 2.792)^2 = 145.2 \text{ in}^4$$

For the composite area,

$$I_{x'} = 472.3 + 145.2 = 617.5 \text{ in}^4 \qquad I_{x'} = 618 \text{ in}^4 \;\blacktriangleleft$$

**Radius of Gyration.** From the moment of inertia and area just calculated, you obtain

$$k_{x'}^2 = \frac{I_{x'}}{A} = \frac{617.5 \text{ in}^4}{17.95 \text{ in}^2} \qquad k_{x'} = 5.87 \text{ in.} \;\blacktriangleleft$$

**REFLECT and THINK:** This is a common type of calculation for many different situations. It is often helpful to list data in a table to keep track of the numbers and identify which data you need.

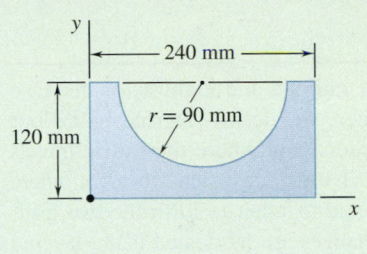

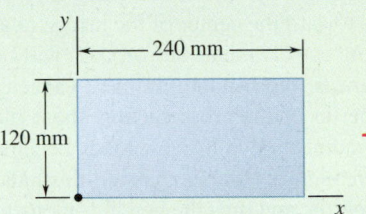

**Fig. 1** Modeling given area by subtracting a half circle from a rectangle.

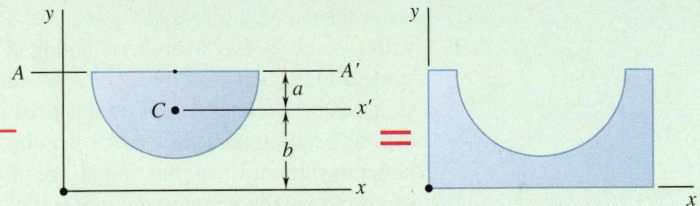

# Sample Problem 9.5

Determine the moment of inertia of the shaded area with respect to the $x$ axis.

**STRATEGY:** You can obtain the given area by subtracting a half circle from a rectangle (Fig. 1). Then, compute the moments of inertia of the rectangle and the half circle separately.

## MODELING and ANALYSIS:

**Moment of Inertia of Rectangle.** Referring to Fig. 9.12, you have

$$I_x = \tfrac{1}{3}bh^3 = \tfrac{1}{3}(240 \text{ mm})(120 \text{ mm})^3 = 138.2 \times 10^6 \text{ mm}^4$$

**Moment of Inertia of Half Circle.** Refer to Fig. 5.8 and determine the location of the centroid $C$ of the half circle with respect to diameter $AA'$. As shown in Fig. 2, you have

$$a = \frac{4r}{3\pi} = \frac{(4)(90 \text{ mm})}{3\pi} = 38.2 \text{ mm}$$

The distance $b$ from the centroid $C$ to the $x$ axis is

$$b = 120 \text{ mm} - a = 120 \text{ mm} - 38.2 \text{ mm} = 81.8 \text{ mm}$$

Referring now to Fig. 9.12, compute the moment of inertia of the half circle with respect to diameter $AA'$ and then compute the area of the half circle.

$$I_{AA'} = \tfrac{1}{8}\pi r^4 = \tfrac{1}{8}\pi(90 \text{ mm})^4 = 25.76 \times 10^6 \text{ mm}^4$$

$$A = \tfrac{1}{2}\pi r^2 = \tfrac{1}{2}\pi(90 \text{ mm})^2 = 12.72 \times 10^3 \text{ mm}^2$$

Next, using the parallel-axis theorem, obtain the value of $\overline{I}_{x'}$ as

$$I_{AA'} = \overline{I}_{x'} + Aa^2$$

$$25.76 \times 10^6 \text{ mm}^4 = \overline{I}_{x'} + (12.72 \times 10^3 \text{ mm}^2)(38.2 \text{ mm})^2$$

$$\overline{I}_{x'} = 7.20 \times 10^6 \text{ mm}^4$$

Again using the parallel-axis theorem, obtain the value of $I_x$ as

$$I_x = \overline{I}_{x'} + Ab^2 = 7.20 \times 10^6 \text{ mm}^4 + (12.72 \times 10^3 \text{ mm}^2)(81.8 \text{ mm})^2$$
$$= 92.3 \times 10^6 \text{ mm}^4$$

**Moment of Inertia of Given Area.** Subtracting the moment of inertia of the half circle from that of the rectangle, you obtain

$$I_x = 138.2 \times 10^6 \text{ mm}^4 - 92.3 \times 10^6 \text{ mm}^4$$

$$I_x = 45.9 \times 10^6 \text{ mm}^4 \blacktriangleleft$$

**REFLECT and THINK:** Figs. 5.8A, 5.8B, and 9.12 are useful references for locating centroids and moments of inertia of common areas; don't forget to use them.

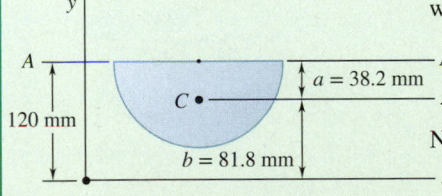

**Fig. 2** Centroid location of the half circle.

## Case Study 9.1

Twin-girder steel bridges are often used for railroad and highway spans. One style is the *through plate girder* bridge, where the deck is carried by a floor system supported along its sides by the girders, and where the traffic travels between or *through* the girders. CS Photo 9.1 illustrates such a bridge, where steel plates and angle shapes have been riveted together to form the two girders. This photo also shows how the girder flanges are fabricated using layered cover plates, with increasingly more layers toward the center of the bridge (and with the successive layers terminating at the cut-off points shown). As will be studied in a course on *mechanics of materials*, the bending-moment capacity of a beam is related to the moment of inertia of the cross section about the axis of bending. Because girder bending moments due to deck loads become larger toward midspan, the flanges are often reinforced in the manner illustrated here to gradually increase the moment of inertia, and thus the bending-moment capacity, in this region.

**CS Photo 9.1**    Riveted through plate girder bridge. ©Martin Matlack

Based on a standard 60-ft steel through plate girder railroad bridge design used by the Harriman Lines,[†] CS Fig. 9.1 shows the primary characteristics for one half of a girder (with the other half being symmetrical). Considering section *n-n*, CS Fig. 9.2 shows the approximate dimensions of the cross section at this location, where three $16 \times \frac{1}{2}$-in. cover plates are attached to two $L6 \times 6 \times \frac{5}{8}$ angles to form the top and bottom flanges. The web is a $76 \times \frac{3}{8}$-in. plate. Let's find the moment of inertia of this composite section with respect to its bending axis (i.e., the centroidal *x* axis).

[†] Ref: Ketchum, M. S., *Structural Engineers Handbook*, 3e, McGraw-Hill, 1924, p. 238.

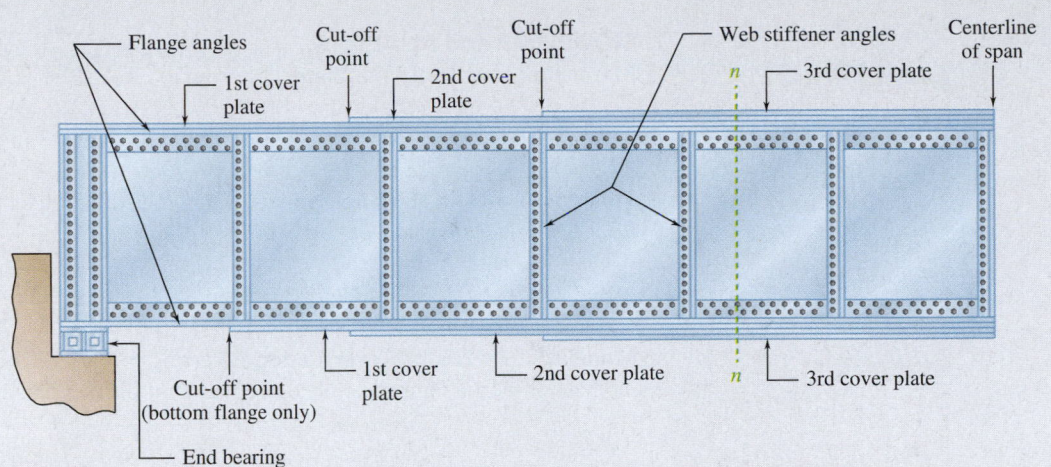

**CS Fig. 9.1** Characteristics of Harriman Lines 60-ft through plate girder railroad bridge (showing half a span).

Source: Ketchum, M. S., *Structural Engineers Handbook*, 3e, New York: McGraw-Hill, 1924, p. 238.

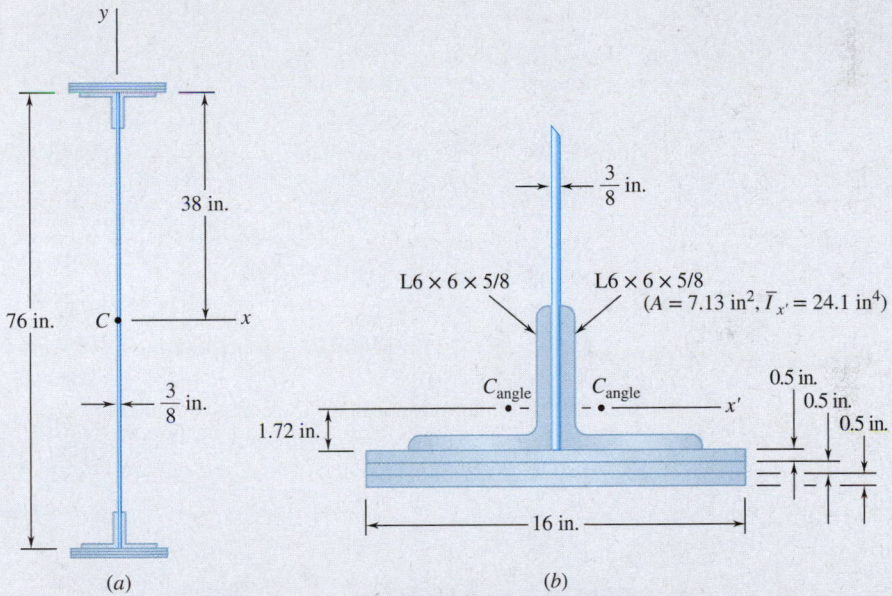

**CS Fig. 9.2** (a) Girder cross section at n-n. (b) Detail of bottom flange (top flange similar).

## STRATEGY:

To determine the moment of inertia of the composite area with respect to one of its centroidal axes, the first step would normally be to determine the location of the centroid. Due to symmetry, this centroid can be established by inspection as shown in CS Fig. 9.2(a). Using the parallel-axis theorem, the moment of inertia of each component part relative to the centroid of the composite area can be found. The moment of inertia for the composite area can then be determined by summing the inertias of the component parts.

*(continued)*

**MODELING and ANALYSIS:**

**Moment of Inertia of Web Plate.** Referring to Fig. 9.12,

$$I_x = \tfrac{1}{12}bh^3 = \tfrac{1}{12}(0.375 \text{ in.})(76 \text{ in.})^3 = 13{,}718 \text{ in}^4$$

**Moment of Inertia of Flange Angles.** Using the parallel-axis theorem and the L6 × 6 × $\tfrac{5}{8}$ data shown in CS Fig. 9.2($b$),

$$I_x = \Sigma(\bar{I}_{x'} + Ad^2) = 4\big[24.1 \text{ in}^4 + (7.13 \text{ in}^2)(38 \text{ in.} - 1.72 \text{ in.})^2\big] = 37{,}636 \text{ in}^4$$

**Moment of Inertia of Cover Plates.** Referring to Fig. 9.12, the area and the moment of inertia for a cover plate about its centroidal $x'$ axis is

$$A = (16 \text{ in.})(0.5 \text{ in.}) = 8 \text{ in}^2$$

$$\bar{I}_{x'} = \tfrac{1}{12}bh^3 = \tfrac{1}{12}(16 \text{ in.})(0.5 \text{ in.})^3 = 0.1667 \text{ in}^4$$

Using the parallel-axis theorem for the first pair of cover plates (i.e., top and bottom flanges),

$$I_x = \bar{I}_{x'} + Ad^2 = 2[0.1667 \text{ in}^4 + (8 \text{ in}^2)(38 \text{ in.} + 0.25 \text{ in.})^2] = 23{,}409 \text{ in}^4$$

For the second pair of cover plates,

$$I_x = \bar{I}_{x'} + Ad^2 = 2[0.1667 \text{ in}^4 + (8 \text{ in}^2)(38 \text{ in.} + 0.75 \text{ in.})^2] = 24{,}025 \text{ in}^4$$

For the third pair of cover plates,

$$I_x = \bar{I}_{x'} + Ad^2 = 2[0.1667 \text{ in}^4 + (8 \text{ in}^2)(38 \text{ in.} + 1.25 \text{ in.})^2] = 24{,}649 \text{ in}^4$$

Thus, the moment of inertia for the composite area is

$$I_x = 13{,}718 + 37{,}636 + 23{,}409 + 24{,}025 + 24{,}649 = 123{,}437 \text{ in}^4$$

$$I_x = 123{,}400 \text{ in}^4 \quad \blacktriangleleft$$

**REFLECT and THINK:**

I-shapes like the plate girder considered here are an efficient means to distribute material to resist bending. For comparison, consider a hypothetical beam made up of the girder's three pairs of cover plates fastened together to form the cross section shown in CS Fig. 9.3. The moment of inertia about its bending axis (i.e., centroidal $x$ axis) can be determined by applying the parallel-axis theorem to each of the six cover plates:

$$\begin{aligned}
I_x = \Sigma(\bar{I}_{x'} + Ad^2) &= 2[0.1667 \text{ in}^4 + (8 \text{ in}^2)(0.25 \text{ in.})^2] \\
&\quad + 2[0.1667 \text{ in}^4 + (8 \text{ in}^2)(0.75 \text{ in.})^2] \\
&\quad + 2[0.1667 \text{ in}^4 + (8 \text{ in}^2)(1.25 \text{ in.})^2] = 36 \text{ in}^4
\end{aligned}$$

The same result can be obtained more directly by referring to Fig. 9.12 and evaluating the resulting composite $16 \times 3$-in. rectangular section as a whole:

$$I_x = \tfrac{1}{12}bh^3 = \tfrac{1}{12}(16 \text{ in.})(3 \text{ in.})^3 = 36 \text{ in}^4$$

For the plate girder of CS Fig. 9.2, the total moment of inertia attributed to these same cover plates was found to be

$$I_x = 23{,}409 + 24{,}025 + 24{,}649 = 72{,}083 \text{ in}^4$$

Thus, by separating the cover plates to form the girder's flanges, the moment of inertia associated with these plates relative to the axis of bending is over 2000 times larger! Considering each individual plate, note that this increase is due to the second term of the parallel-axis theorem $(Ad^2)$, and that the first term, representing the moment of inertia with respect to the centroidal axis of each plate $(\bar{I}_{x'} = 0.1667 \text{ in}^4)$, is negligible in comparison.

There are additional factors that contribute to the overall capacity of a plate girder as well. Nonetheless, strategies such as this to increase moment of inertia can greatly enhance bending strength.

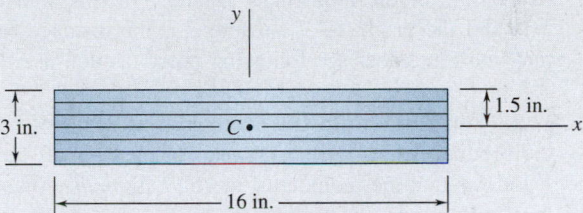

**CS Fig. 9.3**  Cross section of a hypothetical beam consisting of the three pairs of cover plates.

# SOLVING PROBLEMS
# ON YOUR OWN

In this section, we introduced the **parallel-axis theorem** and showed how to use it to simplify the computation of moments and polar moments of inertia of composite areas. The areas that you will consider in the following problems will consist of common shapes and rolled-steel shapes. You will also use the parallel-axis theorem to locate the point of application (the center of pressure) of the resultant of the hydrostatic forces acting on a submerged plane area.

**1. Applying the parallel-axis theorem.** In Sec. 9.2, we derived the parallel-axis theorem

$$I = \bar{I} + Ad^2 \tag{9.9}$$

which states that the moment of inertia $I$ of an area $A$ with respect to a given axis is equal to the sum of the moment of inertia $\bar{I}$ of that area with respect to a *parallel centroidal axis* and the product $Ad^2$, where $d$ is the distance between the two axes. It is important that you remember the following points as you use the parallel-axis theorem.

   **a. You can obtain the centroidal moment of inertia $\bar{I}$ of an area $A$** by subtracting the product $Ad^2$ from the moment of inertia $I$ of the area with respect to a parallel axis. It follows that the moment of inertia $\bar{I}$ is *smaller* than the moment of inertia $I$ of the same area with respect to any parallel axis.

   **b. You can apply the parallel-axis theorem only if one of the two axes involved is a centroidal axis.** Therefore, as we noted in Concept Application 9.3, to compute the moment of inertia of an area with respect to a *noncentroidal axis* when the moment of inertia of the area is known with respect to *another noncentroidal axis,* it is necessary to first compute the moment of inertia of the area with respect to a centroidal axis parallel to the two given axes.

**2. Computing the moments and polar moments of inertia of composite areas.** Sample Probs. 9.4 and 9.5 illustrate the steps you should follow to solve problems of this type. As with all composite-area problems, you should show on your sketch the common shapes or rolled-steel shapes that constitute the various elements of the given area, as well as the distances between the centroidal axes of the elements and the axes about which the moments of inertia are to be computed. In addition, it is important to note the following points.

   **a. The moment of inertia of an area is always positive,** regardless of the location of the axis with respect to which it is computed. As pointed out in the comments for the preceding section, only when an area is *removed* (as in the case of a hole) should you enter its moment of inertia in your computations with a minus sign.

   **b. The moments of inertia of a semiellipse and a quarter ellipse** can be determined by dividing the moment of inertia of an ellipse by 2 and 4, respectively. Note, however, that the moments of inertia obtained in this manner are *with respect to the axes of* *symmetry of the ellipse*. To obtain the *centroidal* moments of inertia of these shapes, use the parallel-axis theorem. This remark also applies to a semicircle and to a quarter circle. Also note that the expressions given for these shapes in Fig. 9.12 are *not* centroidal moments of inertia.

**c. To calculate the polar moment of inertia** of a composite area, you can use either the expressions given in Fig. 9.12 for $J_O$ or the relationship

$$J_O = I_x + I_y \tag{9.4}$$

depending on the shape of the given area.

**d. Before computing the centroidal moments of inertia** of a given area, you may find it necessary to first locate the centroid of the area using the methods of Chap. 5.

**3. Locating the point of application of the resultant of a system of hydrostatic forces.** In Sec. 9.1, we found that

$$R = \gamma \int y \, dA = \gamma \overline{y} A$$

$$M_x = \gamma \int y^2 \, dA = \gamma I_x$$

where $\overline{y}$ is the distance from the $x$ axis to the centroid of the submerged plane area. Because **R** is equivalent to the system of elemental hydrostatic forces, it follows that

$$\Sigma M_x: \qquad y_P R = M_x$$

where $y_P$ is the depth of the point of application of **R**. Then

$$y_P(\gamma \overline{y} A) = \gamma I_x \quad \text{or} \quad y_P = \frac{I_x}{\overline{y} A}$$

In closing, we encourage you to carefully study the notation used in Figs. 9.13A and 9.13B for the rolled-steel shapes, because you will likely encounter it again in subsequent engineering courses.

# Problems

**9.31 and 9.32** Determine the moment of inertia and the radius of gyration of the shaded area with respect to the $x$ axis.

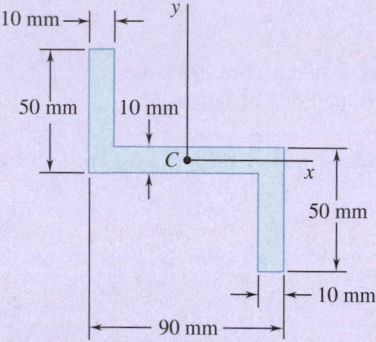

**Fig. P9.31 and P9.33**

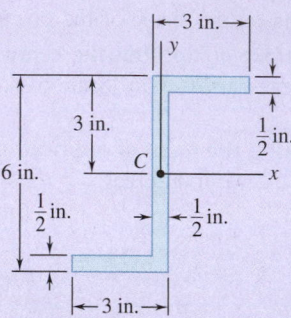

**Fig. P9.32 and P9.34**

**9.33 and 9.34** Determine the moment of inertia and the radius of gyration of the shaded area with respect to the $y$ axis.

**9.35** Determine the moments of inertia of the shaded area shown with respect to the $x$ and $y$ axes when $a = 20$ mm.

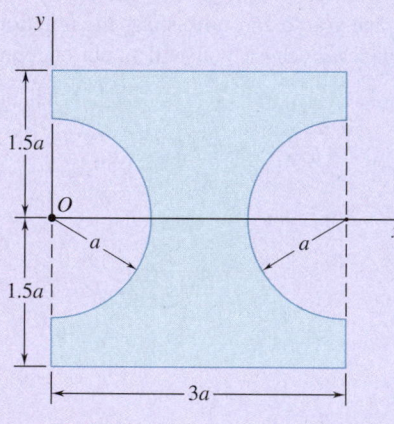

**Fig. P9.35**

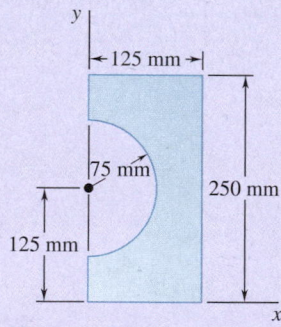

**Fig. P9.36**

**9.36** Determine the moments of inertia of the shaded area shown with respect to the $x$ and $y$ axes.

**9.37** Determine the shaded area and its moment of inertia with respect to the centroidal axis parallel to $AA'$, knowing that its moments of inertia with respect to $AA'$ and $BB'$ are 2000 in$^4$ and 4000 in$^4$, respectively, and that $d_1 = 8$ in. and $d_2 = 4$ in.

**9.38** Knowing that the shaded area is equal to 25 in$^2$ and that its moment of inertia with respect to $AA'$ is 800 in$^4$, determine its moment of inertia with respect to $BB'$, for $d_1 = 5$ in. and $d_2 = 2$ in.

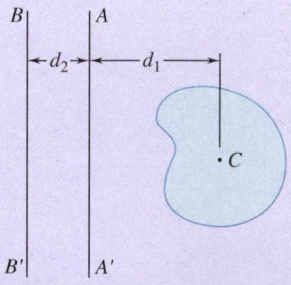

**Fig. P9.37 and P9.38**

**9.39** The shaded area is equal to 5000 mm². Determine its centroidal moments of inertia $\bar{I}_x$ and $\bar{I}_y$, knowing that $2\bar{I}_x = \bar{I}_y$ and that the polar moment of inertia of the area about point $A$ is $J_A = 22.5 \times 10^6$ mm⁴.

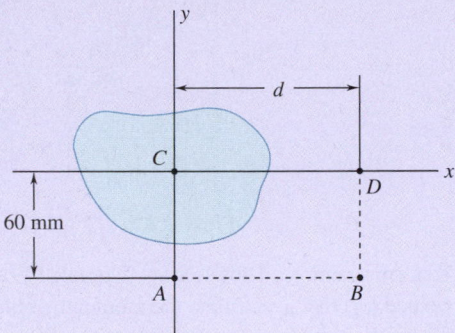

**Fig. P9.39 and P9.40**

**9.40** The polar moments of inertia of the shaded area with respect to points $A$, $B$, and $D$ are, respectively, $J_A = 28.8 \times 10^6$ mm⁴, $J_B = 67.2 \times 10^6$ mm⁴, and $J_D = 45.6 \times 10^6$ mm⁴. Determine the shaded area, its centroidal polar moment of inertia $\bar{J}_C$, and the distance $d$ from $C$ to $D$.

**9.41 through 9.44** Determine the moments of inertia $\bar{I}_x$ and $\bar{I}_y$ of the area shown with respect to centroidal axes respectively parallel and perpendicular to side $AB$.

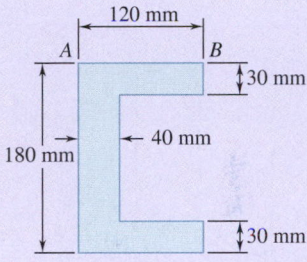

**Fig. P9.41**

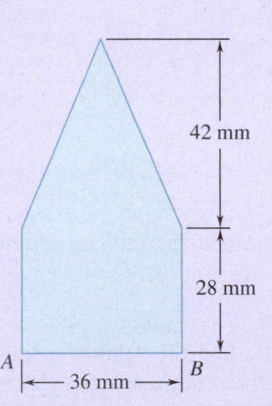

**Fig. P9.42**

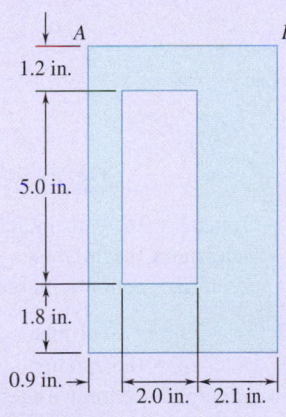

**Fig. P9.43**

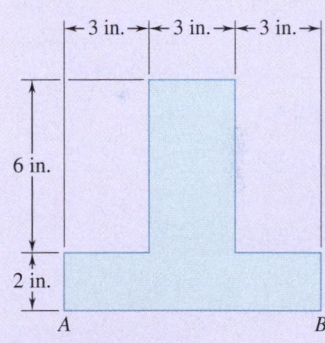

**Fig. P9.44**

**9.45 and 9.46** Determine the polar moment of inertia of the area shown with respect to (*a*) point $O$, (*b*) the centroid of the area.

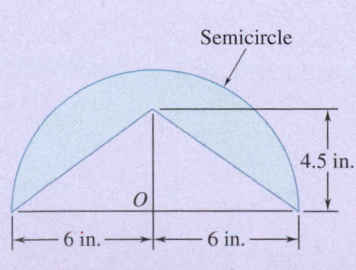

**Fig. P9.45**

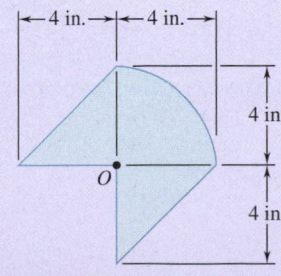

**Fig. P9.46**

**9.47 and 9.48** Determine the polar moment of inertia of the area shown with respect to (a) point $O$, (b) the centroid of the area.

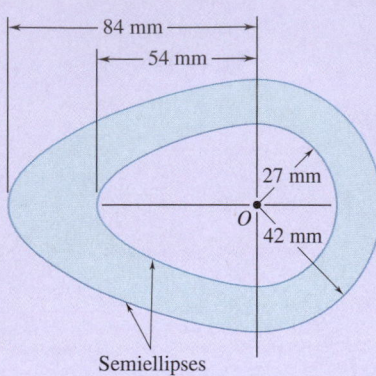

Semiellipses

**Fig. P9.48**

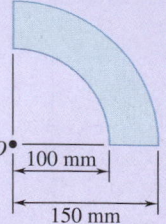

**Fig. P9.47**

**9.49** To form a reinforced box section, two rolled W sections and two plates are welded together. Determine the moments of inertia and the radii of gyration of the combined section with respect to the centroidal axes shown.

**9.50** Two channels are welded to a $d \times 12$-in. steel plate as shown. Determine the width $d$ for which the ratio $\overline{I}_x / \overline{I}_y$ of the centroidal moments of inertia of the section is 16.

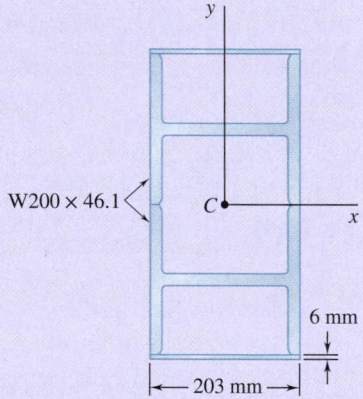

W200 × 46.1

6 mm

203 mm

**Fig. P9.49**

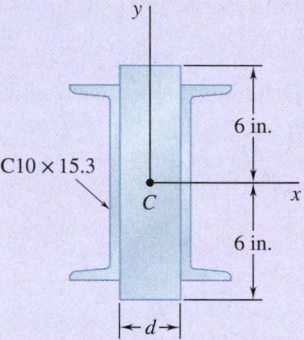

C10 × 15.3

6 in.

6 in.

$d$

**Fig. P9.50**

**9.51** Four L3 × 3 × $\frac{1}{4}$-in. angles are welded to a rolled W section as shown. Determine the moments of inertia and the radii of gyration of the combined section with respect to the centroidal $x$ and $y$ axes.

**9.52** Two 20-mm steel plates are welded to a rolled S section as shown. Determine the moments of inertia and the radii of gyration of the combined section with respect to the centroidal $x$ and $y$ axes.

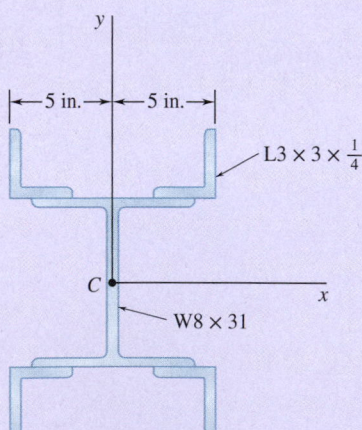

5 in.    5 in.

L3 × 3 × $\frac{1}{4}$

W8 × 31

**Fig. P9.51**

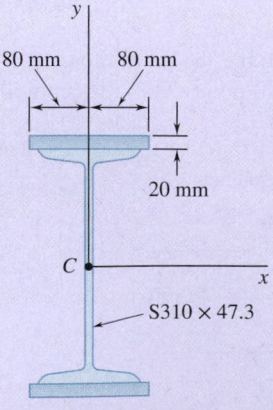

80 mm    80 mm

20 mm

S310 × 47.3

**Fig. P9.52**

**9.53** A channel and a plate are welded together as shown to form a section that is symmetrical with respect to the *y* axis. Determine the moments of inertia of the combined section with respect to its centroidal *x* and *y* axes.

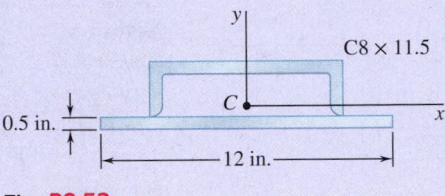

Fig. **P9.53**

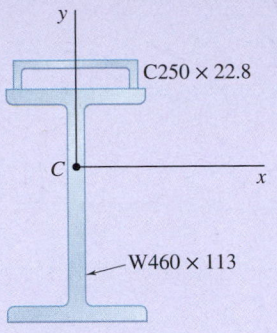

Fig. **P9.54**

**9.54** The strength of the rolled W section shown is increased by welding a channel to its upper flange. Determine the moments of inertia of the combined section with respect to its centroidal *x* and *y* axes.

**9.55** Two L76 × 76 × 6.4-mm angles are welded to a C250 × 22.8 channel. Determine the moments of inertia of the combined section with respect to centroidal axes respectively parallel and perpendicular to the web of the channel.

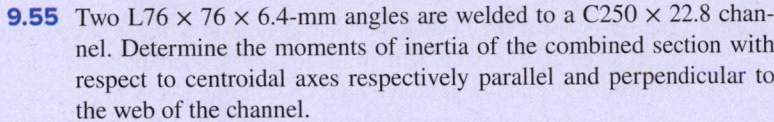

Fig. **P9.55**

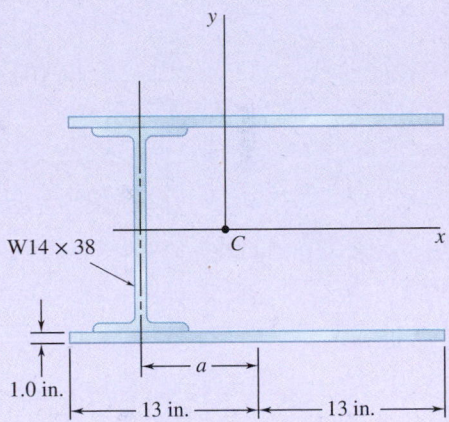

Fig. **P9.56**

**9.56** Two steel plates are welded to a rolled W section as indicated. Knowing that the centroidal moments of inertia $\bar{I}_x$ and $\bar{I}_y$ of the combined section are equal, determine (*a*) the distance *a*, (*b*) the moments of inertia with respect to the centroidal *x* and *y* axes.

**9.57 and 9.58** The panel shown forms the end of a trough that is filled with water to the line *AA'*. Referring to Sec. 9.1A, determine the depth of the point of application of the resultant of the hydrostatic forces acting on the panel (the center of pressure).

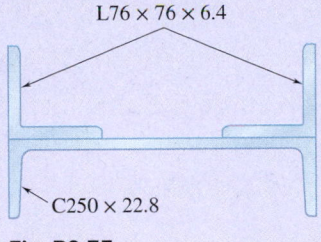

Fig. P9.57

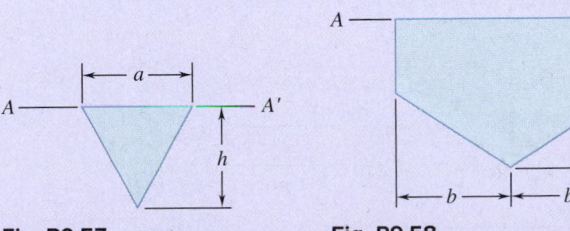

Fig. P9.58

**9.59 and 9.60** The panel shown forms the end of a trough that is filled with water to the line $AA'$. Referring to Sec. 9.1A, determine the depth of the point of application of the resultant of the hydrostatic forces acting on the panel (the center of pressure).

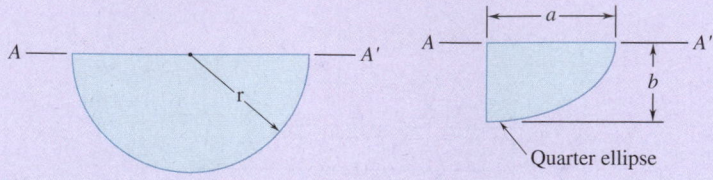

Fig. P9.59                     Quarter ellipse

Fig. P9.60

**9.61** A vertical trapezoidal gate that is used as an automatic valve is held shut by two springs attached to hinges located along edge $AB$. Knowing that each spring exerts a couple of magnitude 1470 N·m, determine the depth $d$ of water for which the gate will open.

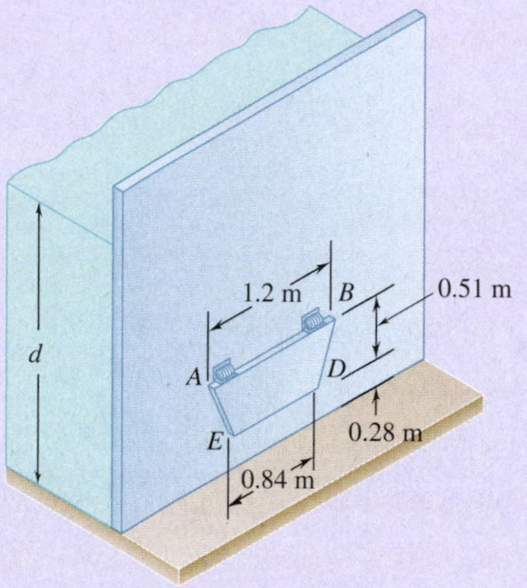

**9.62** The cover for a 0.5-m-diameter access hole in a water storage tank is attached to the tank with four equally spaced bolts as shown. Determine the additional force on each bolt due to the water pressure when the center of the cover is located 1.4 m below the water surface.

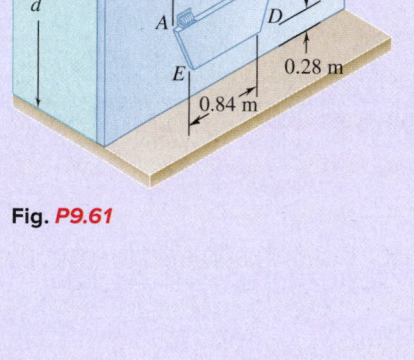

Fig. *P9.61*

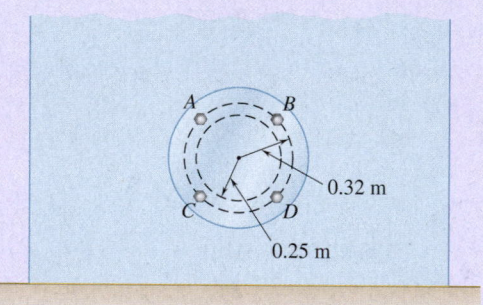

Fig. *P9.62*

**\*9.63** Determine the $x$ coordinate of the centroid of the volume shown. (*Hint*: The height $y$ of the volume is proportional to the $x$ coordinate; consider an analogy between this height and the water pressure on a submerged surface.)

**\*9.64** Determine the $x$ coordinate of the centroid of the volume shown; this volume was obtained by intersecting an elliptic cylinder with an oblique plane. (See the hint for Prob. 9.63.)

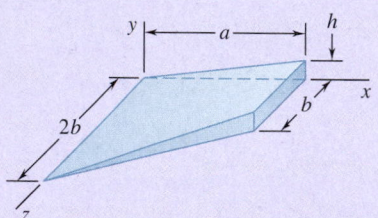

Fig. P9.63

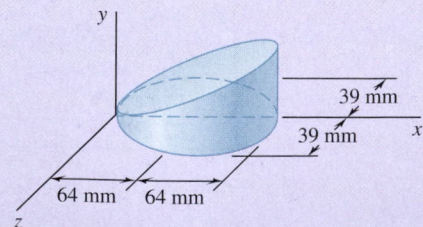

Fig. P9.64

*9.65 Show that the system of hydrostatic forces acting on a submerged plane area $A$ can be reduced to a force $\mathbf{P}$ at the centroid $C$ of the area and two couples. The force $\mathbf{P}$ is perpendicular to the area and has a magnitude of $P = \gamma A \bar{y} \sin \theta$, where $\gamma$ is the specific weight of the liquid. The couples are $\mathbf{M}_{x'} = (\gamma \bar{I}_{x'} \sin \theta)\mathbf{i}$ and $\mathbf{M}_{y'} = (\gamma \bar{I}_{x'y'} \sin \theta)\mathbf{j}$, where $\bar{I}_{x'y'} = \int x'y' \, dA$ (see Sec. 9.3). Note that the couples are independent of the depth at which the area is submerged.

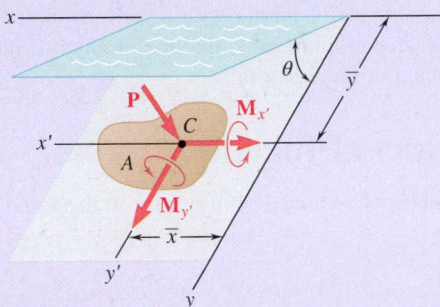

Fig. P9.65

*9.66 Show that the resultant of the hydrostatic forces acting on a submerged plane area $A$ is a force $\mathbf{P}$ perpendicular to the area and of magnitude $P = \gamma A \bar{y} \sin \theta = \bar{p}A$, where $\gamma$ is the specific weight of the liquid and $\bar{p}$ is the pressure at the centroid $C$ of the area. Show that $\mathbf{P}$ is applied at a point $C_P$, called the center of pressure, whose coordinates are $x_p = I_{xy}/A\bar{y}$ and $y_p = I_x/A\bar{y}$, where $I_{xy} = \int xy \, dA$ (see Sec. 9.3). Show also that the difference of ordinates $y_p - \bar{y}$ is equal to $\bar{k}_{x'}^2/\bar{y}$ and thus depends upon the depth at which the area is submerged.

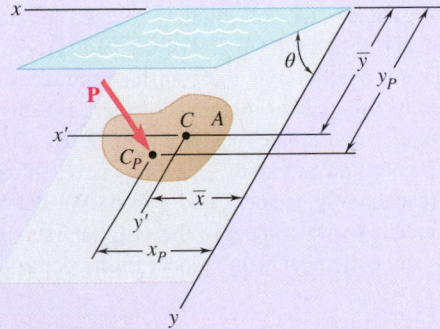

Fig. P9.66

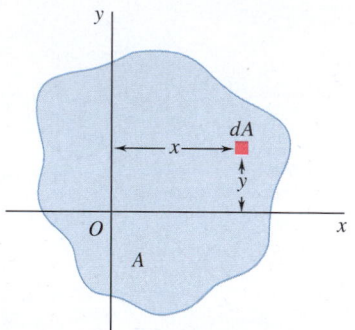

**Fig. 9.14** An element of area $dA$ with coordinates $x$ and $y$.

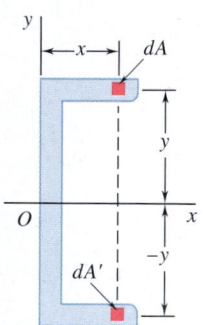

**Fig. 9.15** If an area has an axis of symmetry, its product of inertia is zero.

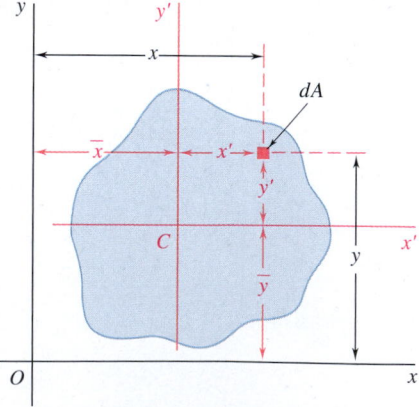

**Fig. 9.16** An element of area $dA$ with respect to $x$ and $y$ axes and the centroidal axes $x'$ and $y'$ for area $A$.

# *9.3 TRANSFORMATION OF MOMENTS OF INERTIA

The moments of inertia of an area can have different values depending on what axes we use to calculate them. It turns out that it is often important to determine the maximum and minimum values of the moments of inertia, which means finding the particular orientation of axes that produce these values. The first step in calculating moments of inertia with regard to rotated axes is to determine a new kind of second moment, called the product of inertia. In this section, we illustrate the procedures for this.

## 9.3A Product of Inertia

The **product of inertia** of an area $A$ with respect to the $x$ and $y$ axes is defined by the integral

**Product of inertia**

$$I_{xy} = \int xy\,dA \qquad (9.12)$$

We calculate it by multiplying each element $dA$ of an area $A$ by its coordinates $x$ and $y$ and integrating over the area (Fig. 9.14). Unlike the moments of inertia $I_x$ and $I_y$, the product of inertia $I_{xy}$ can be positive, negative, or zero. We will see shortly that the product of inertia is necessary for transforming moments of inertia with respect to a different set of axes; in a course on mechanics of materials, you will find other applications of this quantity.

When one or both of the $x$ and $y$ axes are axes of symmetry for the area $A$, the product of inertia $I_{xy}$ is zero. Consider, for example, the channel section shown in Fig. 9.15. Because this section is symmetrical with respect to the $x$ axis, we can associate with each element $dA$ of coordinates $x$ and $y$ an element $dA'$ of coordinates $x$ and $-y$. Clearly, the contributions to $I_{xy}$ of any pair of elements chosen in this way cancel out, and the integral of Eq. (9.12) reduces to zero.

We can derive a parallel-axis theorem for products of inertia similar to the one established in Sec. 9.2 for moments of inertia. Consider an area $A$ and a system of rectangular coordinates $x$ and $y$ (Fig. 9.16). Through the centroid $C$ of the area, with coordinates $\bar{x}$ and $\bar{y}$, we draw two centroidal axes $x'$ and $y'$ that are parallel, respectively, to the $x$ and $y$ axes. We denote the coordinates of an element of area $dA$ with respect to the original axes by $x$ and $y$, and the coordinates of the same element with respect to the centroidal axes by $x'$ and $y'$. This gives us

$$x = x' + \bar{x} \quad \text{and} \quad y = y' + \bar{y}$$

Substituting into Eq. (9.12), we obtain the expression for the product of inertia $I_{xy}$ as

$$I_{xy} = \int xy\,dA = \int (x' + \bar{x})(y' + \bar{y})\,dA$$

$$= \int x'y'\,dA + \bar{y}\int x'\,dA + \bar{x}\int y'\,dA + \bar{x}\,\bar{y}\int dA$$

The first integral represents the product of inertia $\bar{I}_{xy}$ of the area $A$ with respect to the centroidal axes $x'$ and $y'$. The next two integrals represent first moments of the area with respect to the centroidal axes; they reduce to zero, because the centroid $C$ is located on these axes. The last integral is equal to the total area $A$. Therefore, we have

**Parallel-axis theorem for products of inertia**

$$I_{xy} = \overline{I}_{x'y'} + \overline{x}\,\overline{y}A \qquad (9.13)$$

## 9.3B   Principal Axes and Principal Moments of Inertia

Consider an area $A$ with coordinate axes $x$ and $y$ (Fig. 9.17) and assume that we know the moments and product of inertia of the area $A$. We have

$$I_x = \int y^2\, dA \qquad I_y = \int x^2\, dA \qquad I_{xy} = \int xy\, dA \qquad (9.14)$$

We propose to determine the moments and product of inertia $I_{x'}$, $I_{y'}$, and $I_{x'y'}$ of $A$ with respect to new axes $x'$ and $y'$ that we obtain by rotating the original axes about the origin through an angle $\theta$.

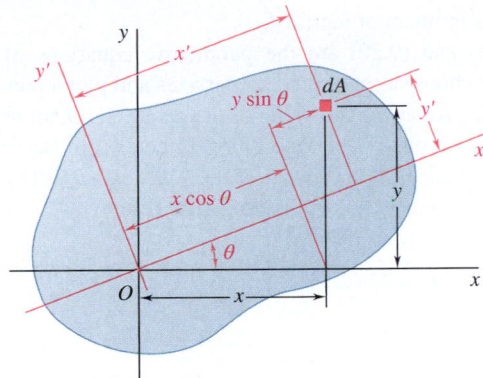

**Fig. 9.17**   An element of area $dA$ with respect to the $x$ and $y$ axes and a set of $x'$ and $y'$ axes rotated about the origin through an angle $\theta$.

We first note that the relations between the coordinates $x'$, $y'$ and $x$, $y$ of an element of area $dA$ are

$$x' = x\cos\theta + y\sin\theta \qquad y' = y\cos\theta - x\sin\theta$$

Substituting for $y'$ in the expression for $I_{x'}$, we obtain

$$I_{x'} = \int (y')^2 dA = \int (y\cos\theta - x\sin\theta)^2 dA$$

$$= \cos^2\theta \int y^2 dA - 2\sin\theta\cos\theta \int xy\, dA + \sin^2\theta \int x^2\, dA$$

Using the relations in Eq. (9.14), we have

$$I_{x'} = I_x \cos^2\theta - 2I_{xy}\sin\theta\cos\theta + I_y \sin^2\theta \qquad (9.15)$$

Similarly, we obtain for $I_{y'}$ and $I_{x'y'}$ the expressions

$$I_{y'} = I_x \sin^2\theta + 2I_{xy}\sin\theta\cos\theta + I_y \cos^2\theta \qquad (9.16)$$

$$I_{x'y'} = (I_x - I_y)\sin\theta\cos\theta + I_{xy}(\cos^2\theta - \sin^2\theta) \qquad (9.17)$$

Recalling the trigonometric relations

$$\sin 2\theta = 2\sin\theta\cos\theta \qquad \cos 2\theta = \cos^2\theta - \sin^2\theta$$

and

$$\cos^2\theta = \frac{1+\cos 2\theta}{2} \qquad \sin^2\theta = \frac{1-\cos 2\theta}{2}$$

we can write Eqs. (9.15), (9.16), and (9.17) as

$$I_{x'} = \frac{I_x+I_y}{2} + \frac{I_x-I_y}{2}\cos 2\theta - I_{xy}\sin 2\theta \qquad \text{(9.18)}$$

$$I_{y'} = \frac{I_x+I_y}{2} + \frac{I_x-I_y}{2}\cos 2\theta + I_{xy}\sin 2\theta \qquad \text{(9.19)}$$

$$I_{x'y'} = \frac{I_x-I_y}{2}\sin 2\theta + I_{xy}\cos 2\theta \qquad \text{(9.20)}$$

Now, adding Eqs. (9.18) and (9.19), we observe that

$$I_{x'} + I_{y'} = I_x + I_y \qquad \text{(9.21)}$$

We could have anticipated this result, because both members of Eq. (9.21) are equal to the polar moment of inertia $J_O$.

Eqs. (9.18) and (9.20) are the parametric equations of a circle. This means that, if we choose a set of rectangular axes and plot a point $M$ of abscissa $I_{x'}$ and ordinate $I_{x'y'}$ for any given value of the parameter $\theta$, all of the points will lie on a circle. To establish this property algebraically, we can eliminate $\theta$ from Eqs. (9.18) and (9.20) by transposing $(I_x + I_y)/2$ in Eq. (9.18), squaring both sides of Eqs. (9.18) and (9.20), and adding. The result is

$$\left(I_{x'} - \frac{I_x+I_y}{2}\right)^2 + I_{x'y'}^2 = \left(\frac{I_x-I_y}{2}\right)^2 + I_{xy}^2 \qquad \text{(9.22)}$$

Setting

$$I_{\text{ave}} = \frac{I_x+I_y}{2} \quad \text{and} \quad R = \sqrt{\left(\frac{I_x-I_y}{2}\right)^2 + I_{xy}^2} \qquad \text{(9.23)}$$

we can write the identity equation (9.22) in the form

$$(I_{x'} - I_{\text{ave}})^2 + I_{x'y'}^2 = R^2 \qquad \text{(9.24)}$$

This is the equation of a circle of radius $R$ centered at the point $C$ whose $x$ and $y$ coordinates are $I_{\text{ave}}$ and 0, respectively (Fig. 9.18a).

Note that Eqs. (9.19) and (9.20) are parametric equations of the same circle. Furthermore, because of the symmetry of the circle about the horizontal axis, we would obtain the same result if we plot a point $N$ of coordinates $I_{y'}$ and $-I_{x'y'}$ (Fig. 9.18b) instead of plotting $M$. We will use this property in Sec. 9.4.

The two points $A$ and $B$ where this circle intersects the horizontal axis (Fig. 9.18a) are of special interest: Point $A$ corresponds to the maximum value of the moment of inertia $I_{x'}$, whereas point $B$ corresponds to its minimum value. In addition, both points correspond to a zero value of the product of inertia $I_{x'y'}$. Thus, we can obtain the values $\theta_m$ of the parameter $\theta$ corresponding to the points $A$ and $B$ by setting $I_{x'y'} = 0$ in Eq. (9.20). The result is[†]

$$\tan 2\theta_m = \frac{2I_{xy}}{I_x - I_y} \qquad \text{(9.25)}$$

[†]We can also obtain this relation by differentiating $I_{x'}$ in Eq. (9.18) and setting $dI_{x'}/d\theta = 0$.

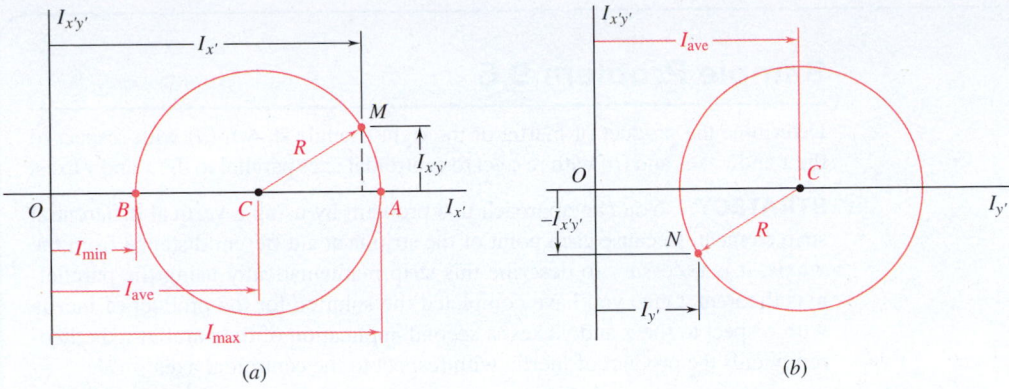

**Fig. 9.18** Plots of $I_{x'y'}$ versus (a) $I_{x'}$ and (b) $I_{y'}$ for different values of the parameter $\theta$ are identical circles. The circle in part (a) indicates the average, maximum, and minimum values of the moment of inertia.

This equation defines two values ($2\theta_m$) that are 180° apart and thus two values ($\theta_m$) that are 90° apart. One of these values corresponds to point $A$ in Fig. 9.18a and to an axis through $O$ in Fig. 9.17 with respect to which the moment of inertia of the given area is maximum. The other value corresponds to point $B$ and to an axis through $O$ with respect to which the moment of inertia of the area is minimum. These two perpendicular axes are called the **principal axes of the area about** $O$. The corresponding values $I_{max}$ and $I_{min}$ of the moment of inertia are called the **principal moments of inertia of the area about** $O$. Because we obtained the two values $\theta_m$ defined by Eq. (9.25) by setting $I_{x'y'} = 0$ in Eq. (9.20), it is clear that the product of inertia of the given area with respect to its principal axes is zero.

Note from Fig. 9.18a that

$$I_{max} = I_{ave} + R \qquad I_{min} = I_{ave} - R \qquad \textbf{(9.26)}$$

Using the values for $I_{ave}$ and $R$ from formulas (9.23), we obtain

$$I_{max, min} = \frac{I_x + I_y}{2} \pm \sqrt{\left(\frac{I_x - I_y}{2}\right)^2 + I_{xy}^2} \qquad \textbf{(9.27)}$$

Unless you can tell by inspection which of the two principal axes corresponds to $I_{max}$ and which corresponds to $I_{min}$, you must substitute one of the values of $\theta_m$ into Eq. (9.18) in order to determine which of the two corresponds to the maximum value of the moment of inertia of the area about $O$.

Referring to Sec. 9.3A, note that, if an area possesses an axis of symmetry through a point $O$, this axis must be a principal axis of the area about $O$. On the other hand, a principal axis does not need to be an axis of symmetry; whether or not an area possesses any axes of symmetry, it will always have two principal axes of inertia about any point $O$.

The properties we have established hold for any point $O$ located inside or outside the given area. If we choose the point $O$ to coincide with the centroid of the area, any axis through $O$ is a centroidal axis; the two principal axes of the area about its centroid are referred to as the **principal centroidal axes of the area**.

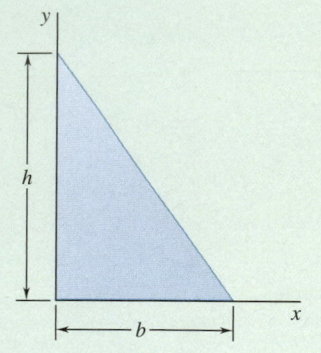

# Sample Problem 9.6

Determine the product of inertia of the right triangle shown (*a*) with respect to the *x* and *y* axes and (*b*) with respect to centroidal axes parallel to the *x* and *y* axes.

**STRATEGY:** You can approach this problem by using a vertical differential strip element. Because each point of the strip is at a different distance from the *x* axis, it is necessary to describe this strip mathematically using the parallel-axis theorem. Once you have completed the solution for the product of inertia with respect to the *x* and *y* axes, a second application of the parallel-axis theorem yields the product of inertia with respect to the centroidal axes.

**MODELING and ANALYSIS:**

**a. Product of Inertia $I_{xy}$.** Choose a vertical rectangular strip as the differential element of area (Fig. 1). Using a differential version of the parallel-axis theorem, you have

$$dI_{xy} = dI_{x'y'} + \bar{x}_{el}\bar{y}_{el}\,dA$$

The element is symmetrical with respect to the *x'* and *y'* axes, so $dI_{x'y'} = 0$. From the geometry of the triangle, you can express the variables in terms of *x* and *y*.

$$y = h\left(1 - \frac{x}{b}\right) \qquad dA = y\,dx = h\left(1 - \frac{x}{b}\right)dx$$

$$\bar{x}_{el} = x \qquad \bar{y}_{el} = \frac{1}{2}y = \frac{1}{2}h\left(1 - \frac{x}{b}\right)$$

Integrating $dI_{xy}$ from $x = 0$ to $x = b$ gives you $I_{xy}$:

$$I_{xy} = \int dI_{xy} = \int \bar{x}_{el}\bar{y}_{el}\,dA = \int_0^b x(\tfrac{1}{2})h^2\left(1 - \frac{x}{b}\right)^2 dx$$

$$= h^2\int_0^b\left(\frac{x}{2} - \frac{x^2}{b} + \frac{x^3}{2b^2}\right)dx = h^2\left[\frac{x^2}{4} - \frac{x^3}{3b} + \frac{x^4}{8b^2}\right]_0^b$$

$$I_{xy} = \tfrac{1}{24}b^2h^2 \quad \blacktriangleleft$$

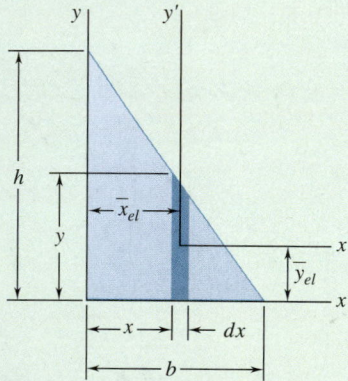

**Fig. 1** Using a vertical rectangular strip as the differential element.

**b. Product of Inertia $\bar{I}_{x''y''}$.** The coordinates of the centroid of the triangle relative to the *x* and *y* axes are (Fig. 2 and Fig. 5.8A)

$$\bar{x} = \frac{1}{3}b \qquad \bar{y} = \frac{1}{3}h$$

Using the expression for $I_{xy}$ obtained in part *a*, apply the parallel-axis theorem again:

$$I_{xy} = \bar{I}_{x''y''} + \bar{x}\,\bar{y}A$$

$$\frac{1}{24}b^2h^2 = \bar{I}_{x''y''} + (\tfrac{1}{3}b)(\tfrac{1}{3}h)(\tfrac{1}{2}bh)$$

$$\bar{I}_{x''y''} = \tfrac{1}{24}b^2h^2 - \tfrac{1}{18}b^2h^2$$

$$\bar{I}_{x''y''} = -\tfrac{1}{72}b^2h^2 \quad \blacktriangleleft$$

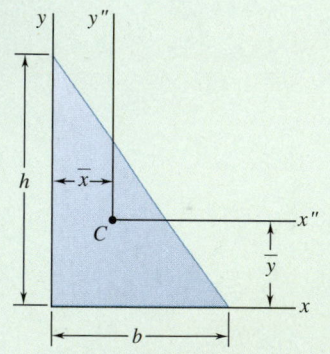

**Fig. 2** Centroid of the triangular area.

**REFLECT and THINK:** An equally effective alternative strategy would be to use a horizontal strip element. Again, you would need to use the parallel-axis theorem to describe this strip, because each point in the strip would be a different distance from the *y* axis.

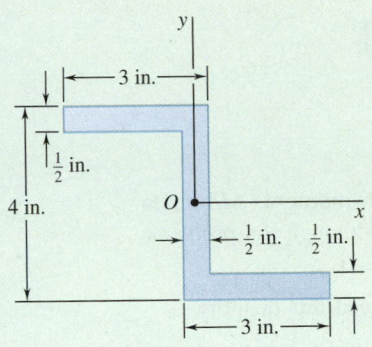

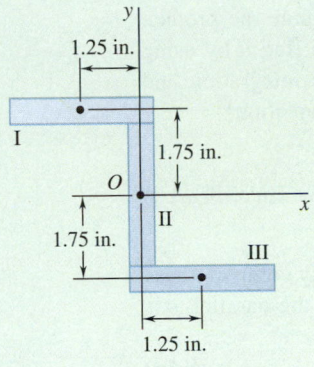

**Fig. 1** Modeling the given area as three rectangles.

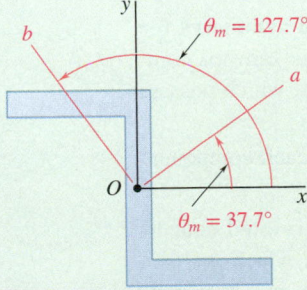

**Fig. 2** Orientation of principal axes.

# Sample Problem 9.7

For the section shown, the moments of inertia with respect to the $x$ and $y$ axes have been computed and are known to be

$$I_x = 10.38 \text{ in}^4 \qquad I_y = 6.97 \text{ in}^4$$

Determine (a) the orientation of the principal axes of the section about $O$, (b) the values of the principal moments of inertia of the section about $O$.

**STRATEGY:** The first step is to compute the product of inertia with respect to the $x$ and $y$ axes, treating the section as a composite area of three rectangles. Then, you can use Eq. (9.25) to find the principal axes and Eq. (9.27) to find the principal moments of inertia.

**MODELING and ANALYSIS:** Divide the area into three rectangles as shown (Fig. 1). Note that the product of inertia $I_{x'y'}$ with respect to centroidal axes parallel to the $x$ and $y$ axes is zero for each rectangle. Thus, using the parallel-axis theorem

$$I_{xy} = I_{x'y'} + \bar{x}\,\bar{y}\,A$$

you find that $I_{xy}$ reduces to $\bar{x}\,\bar{y}\,A$ for each rectangle.

| Rectangle | Area, in$^2$ | $\bar{x}$, in. | $\bar{y}$, in. | $\bar{x}\,\bar{y}A$, in$^4$ |
|---|---|---|---|---|
| I | 1.5 | −1.25 | +1.75 | −3.28 |
| II | 1.5 | 0 | 0 | 0 |
| III | 1.5 | +1.25 | −1.75 | −3.28 |
| | | | | $\Sigma \bar{x}\bar{y}A = -6.56$ |

$$I_{xy} = \Sigma \bar{x}\,\bar{y}A = -6.56 \text{ in}^4$$

**a. Principal Axes.** Because you know the magnitudes of $I_x$, $I_y$, and $I_{xy}$, you can use Eq. (9.25) to determine the values of $\theta_m$ (Fig. 2):

$$\tan 2\theta_m = -\frac{2I_{xy}}{I_x - I_y} = \frac{2(-6.56)}{10.38 - 6.97} = +3.85$$

$$2\theta_m = 75.4° \text{ and } 255.4°$$

$$\theta_m = 37.7° \text{ and } \theta_m = 127.7° \quad \blacktriangleleft$$

**b. Principal Moments of Inertia.** Using Eq. (9.27), you have

$$I_{\text{max, min}} = \frac{I_x + I_y}{2} \pm \sqrt{\left(\frac{I_x - I_y}{2}\right)^2 + I_{xy}^2}$$

$$= \frac{10.38 + 6.97}{2} \pm \sqrt{\left(\frac{10.38 - 6.97}{2}\right)^2 + (-6.56)^2}$$

$$I_{\text{max}} = 15.45 \text{ in}^4 \quad I_{\text{min}} = 1.897 \text{ in}^4 \quad \blacktriangleleft$$

**REFLECT and THINK:** Note that the elements of the area of the section are more closely distributed about the $b$ axis than about the $a$ axis. Therefore, you can conclude that $I_a = I_{\text{max}} = 15.45 \text{ in}^4$ and $I_b = I_{\text{min}} = 1.897 \text{ in}^4$. You can verify this conclusion by substituting $\theta = 37.7°$ into Eqs. (9.18) and (9.19).

# SOLVING PROBLEMS
# ON YOUR OWN

In the problems for this section, you will continue your work with **moments of inertia** and use various techniques for computing **products of inertia**. Although the problems are generally straightforward, several items are worth noting.

**1. Calculating the product of inertia $I_{xy}$ by integration.** We defined this quantity as

$$I_{xy} = \int xy\, dA \tag{9.12}$$

and stated that its value can be positive, negative, or zero. You can compute the product of inertia directly from this equation using double integration, or you can find it by using single integration as shown in Sample Prob. 9.6. When applying single integration and using the parallel-axis theorem, it is important to remember that in the equation

$$dI_{xy} = dI_{x'y'} + \bar{x}_{el}\bar{y}_{el}\, dA$$

$\bar{x}_{el}$ and $\bar{y}_{el}$ are the coordinates of the centroid of the element of area $dA$. Thus, if $dA$ is not in the first quadrant, one or both of these coordinates is negative.

**2. Calculating the products of inertia of composite areas.** You can easily compute these quantities from the products of inertia of their component parts by using the parallel-axis theorem, as

$$I_{xy} = \bar{I}_{x'y'} + \bar{x}\,\bar{y}A \tag{9.13}$$

The proper technique to use for problems of this type is illustrated in Sample Probs. 9.6 and 9.7. In addition to the usual rules for composite-area problems, it is essential that you remember the following points.

   **a. If either of the centroidal axes of a component area is an axis of symmetry for that area, the product of inertia $\bar{I}_{x'y'}$ for that area is zero.** Thus, $\bar{I}_{x'y'}$ is zero for component areas such as circles, semicircles, rectangles, and isosceles triangles, which possess an axis of symmetry parallel to one of the coordinate axes.

   **b. Pay careful attention to the signs of the coordinates $\bar{x}$ and $\bar{y}$** of each component area when you use the parallel-axis theorem (Sample Prob. 9.7).

**3. Determining the moments of inertia and the product of inertia for rotated coordinate axes.** In Sec. 9.3B, we derived Eqs. (9.18), (9.19), and (9.20) from which you can compute the moments of inertia and the product of inertia for coordinate axes that have been rotated about the origin $O$. To apply these equations, you must know a set of values $I_x$, $I_y$, and $I_{xy}$ for a given orientation of the axes, and you must remember that $\theta$ is positive for counterclockwise rotations of the axes and negative for clockwise rotations of the axes.

**4. Computing the principal moments of inertia.** We showed in Sec. 9.3B that a particular orientation of the coordinate axes exists for which the moments of inertia attain their maximum and minimum values, $I_{max}$ and $I_{min}$, and for which the product of inertia is zero. Eq. (9.27) can be used to compute these values that are known as the **principal moments of inertia** of the area about $O$. The corresponding axes are referred to as the **principal axes** of the area about $O$, and their orientation is defined by Eq. (9.25). To determine which of the principal axes corresponds to $I_{max}$ and which corresponds to $I_{min}$, you can either follow the procedure outlined in the text after Eq. (9.27) or observe about which of the two principal axes the area is more closely distributed; that axis corresponds to $I_{min}$ (Sample Prob. 9.7).

# Problems

**9.67 through 9.70** Determine by direct integration the product of inertia of the given area with respect to the $x$ and $y$ axes.

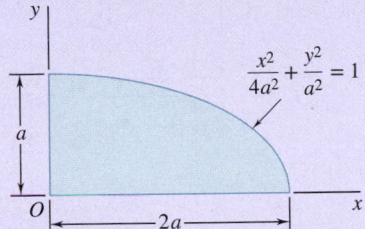

$$\frac{x^2}{4a^2} + \frac{y^2}{a^2} = 1$$

**Fig. P9.67**

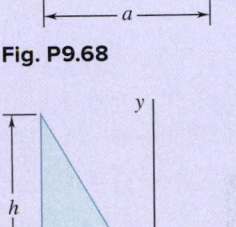

**Fig. P9.68**

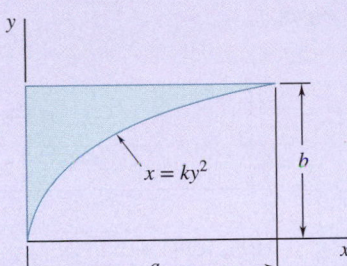

$x = ky^2$

**Fig. P9.69**

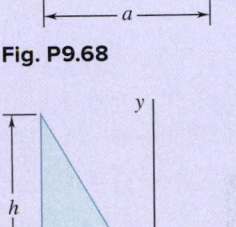

**Fig. *P9.70***

**9.71 through 9.74** Using the parallel-axis theorem, determine the product of inertia of the area shown with respect to the centroidal $x$ and $y$ axes.

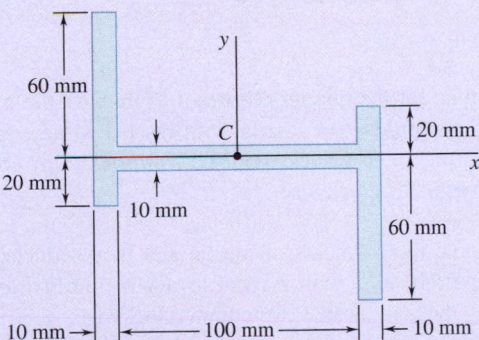

**Fig. P9.71**

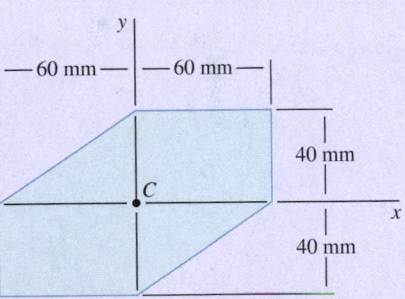

**Fig. P9.72**

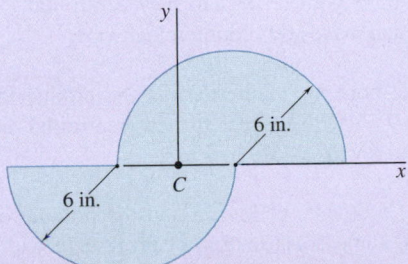

**Fig. *P9.73***

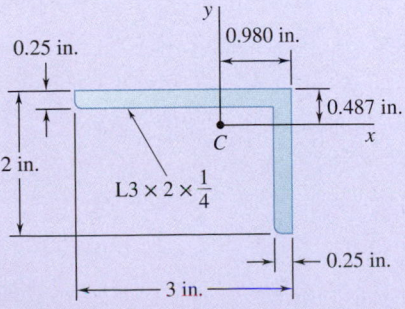

$L3 \times 2 \times \frac{1}{4}$

**Fig. P9.74**

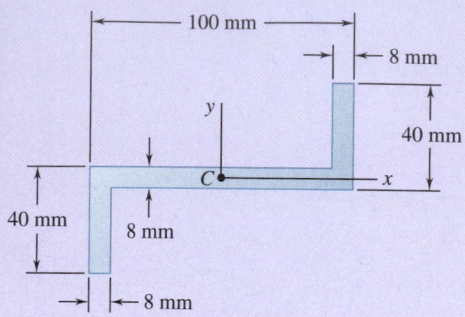

**Fig. P9.75**

**9.75 through 9.78** Using the parallel-axis theorem, determine the product of inertia of the area shown with respect to the centroidal x and y axes.

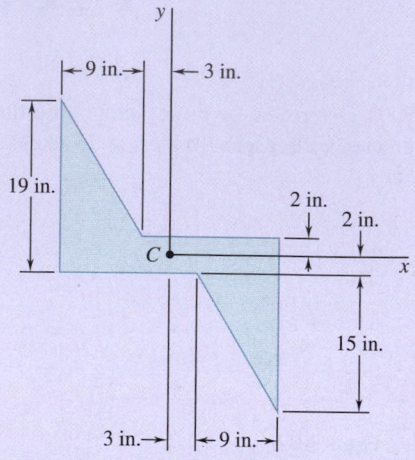

**Fig. P9.76**

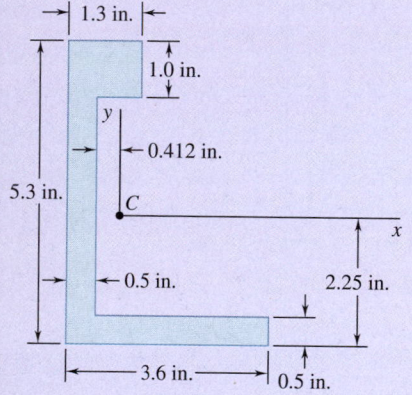

**Fig. *P9.77***

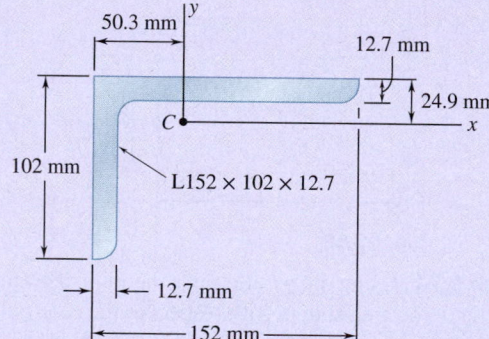

**Fig. P9.78**

**9.79** Determine for the quarter ellipse of Prob. 9.67 the moments of inertia and the product of inertia with respect to new axes obtained by rotating the x and y axes about O (a) through 45° counterclockwise, (b) through 30° clockwise.

**9.80** Determine the moments of inertia and the product of inertia of the area of Prob. 9.72 with respect to new centroidal axes obtained by rotating the x and y axes 30° counterclockwise.

**9.81** Determine the moments of inertia and the product of inertia of the area of Prob. 9.73 with respect to new centroidal axes obtained by rotating the x and y axes 60° counterclockwise.

**9.82** Determine the moments of inertia and the product of inertia of the area of Prob. 9.75 with respect to new centroidal axes obtained by rotating the x and y axes 45° clockwise.

**9.83** Determine the moments of inertia and the product of inertia of the L3 × 2 × $\frac{1}{4}$-in. angle cross section of Prob. 9.74 with respect to new centroidal axes obtained by rotating the x and y axes 30° clockwise.

**9.84** Determine the moments of inertia and the product of inertia of the L152 × 102 × 12.7-mm angle cross section of Prob. 9.78 with respect to new centroidal axes obtained by rotating the $x$ and $y$ axes 30° clockwise.

**9.85** For the quarter ellipse of Prob. 9.67, determine the orientation of the principal axes at the origin, as well as the corresponding values of the moments of inertia.

**9.86 through 9.88** For the area indicated, determine the orientation of the principal axes at the origin, as well as the corresponding values of the moments of inertia.

**9.86** Area of Prob. 9.72
**9.87** Area of Prob. 9.73
**9.88** Area of Prob. 9.75

**9.89 and 9.90** For the angle cross section indicated, determine the orientation of the principal axes at the origin, as well as the corresponding values of the moments of inertia.

**9.89** The L3 × 2 × $\frac{1}{4}$-in. angle cross section of Prob. 9.74
**9.90** The L152 × 102 × 12.7-mm angle cross section of Prob. 9.78

# *9.4 MOHR'S CIRCLE FOR MOMENTS OF INERTIA

The circle introduced in the preceding section to illustrate the relations between the moments and products of inertia of a given area with respect to axes passing through a fixed point $O$ was first introduced by the German engineer Otto Mohr (1835–1918) and is known as **Mohr's circle**. Here, we show that, if we know the moments and product of inertia of an area $A$ with respect to two rectangular $x$ and $y$ axes that pass through a point $O$, we can use Mohr's circle to graphically determine (a) the principal axes and principal moments of inertia of the area about $O$ and (b) the moments and product of inertia of the area with respect to any other pair of rectangular axes $x'$ and $y'$ through $O$.

Consider a given area $A$ and two rectangular coordinate axes $x$ and $y$ (Fig. 9.19a). Assuming that we know the moments of inertia $I_x$ and $I_y$ and the product of inertia $I_{xy}$, we can represent them on a diagram by plotting a point $x$ with coordinates $I_x$ and $I_{xy}$ and a point $Y$ with coordinates $I_y$ and $-I_{xy}$ (Fig. 9.19b). If $I_{xy}$ is positive, as assumed in Fig. 9.19a, then point $x$ is located above the horizontal axis and point $Y$ is located below, as shown in Fig. 9.19b. If $I_{xy}$ is negative, $X$ is located below the horizontal axis and $Y$ is located above. Joining $x$ and $y$ with a straight line, we denote the point of intersection of line $XY$ with the horizontal axis by $C$. Then, we draw the circle of center $C$ and diameter $XY$. Noting that the abscissa of $C$ and the radius of the circle are respectively equal to the quantities $I_{ave}$ and $R$ defined by formula (9.23), we conclude that the circle obtained is Mohr's circle for the given area about point $O$. Thus, the abscissas of the points $A$ and $B$ where the circle intersects the horizontal axis represent, respectively, the principal moments of inertia $I_{max}$ and $I_{min}$ of the area.

Also note that, because $\tan(XCA) = 2I_{xy}/(I_x - I_y)$, the angle $XCA$ is equal in magnitude to one of the angles $2\theta_m$ that satisfy Eq. (9.25). Thus, the angle $\theta_m$, which in Fig. 9.19a defines the principal axis $Oa$ corresponding to point $A$ in Fig. 9.19b, is equal to half of the angle $XCA$ of Mohr's circle. In addition, if $I_x > I_y$ and $I_{xy} > 0$, as in the case considered here, the rotation that brings $CX$ into $CA$ is clockwise. Also, under these conditions, the angle $\theta_m$ obtained from Eq. (9.25) is negative; thus, the rotation that brings $Ox$ into $Oa$ is also clockwise. We conclude that the senses of rotation in both parts of Fig. 9.19 are the same. If a clockwise rotation through $2\theta_m$ is required to bring $CX$ into $CA$ on Mohr's circle, a clockwise rotation through $\theta_m$ will bring $Ox$ into the corresponding principal axis $Oa$ in Fig. 9.19a.

Because Mohr's circle is uniquely defined, we can obtain the same circle by considering the moments and product of inertia of the area $A$ with respect to the rectangular axes $x'$ and $y'$ (Fig. 9.19a). The point $X'$ with coordinates $I_{x'}$ and $-I_{x'y'}$ and the point $Y'$ with coordinates $I_{y'}$ and $-I_{x'y'}$ are thus located on Mohr's circle, and the angle $X'CA$ in Fig. 9.19b must be equal to twice the angle $x'Oa$ in Fig. 9.19a. Because, as noted previously, the angle $XCA$ is twice the angle $xOa$, it follows that the angle $XCX'$ in Fig. 9.19b is twice the angle $xOx'$ in Fig. 9.19a. The diameter $X'Y'$, which defines the moments and product of inertia $I_{x'}$, $I_{y'}$, and $I_{x'y'}$ of the given area with respect to rectangular axes $x'$ and $y'$ forming an angle $\theta$ with the $x$ and $y$ axes, can be obtained by rotating through an angle $2\theta$ the diameter $XY$, which corresponds to the moments and product of inertia $I_x$, $I_y$, and $I_{xy}$. Note that the rotation that brings the diameter $XY$ into the diameter $X'Y'$ in Fig. 9.19b has the same

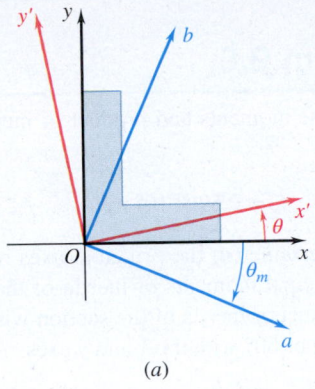

(a)

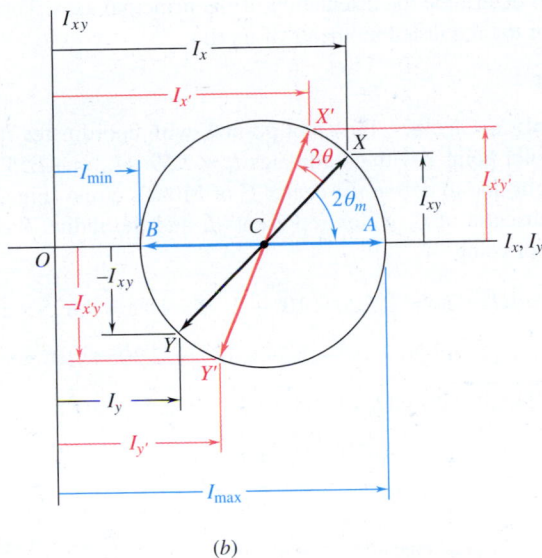

(b)

**Fig. 9.19**   (a) An area A with principal axes Oa and Ob and axes Ox' and Oy' obtained by rotation through an angle θ; (b) Mohr's circle used to calculate angles and moments of inertia.

sense as the rotation that brings the $x$ and $y$ axes into the $x'$ and $y'$ axes in Fig. 9.19a.

Finally, also note that the use of Mohr's circle is not limited to graphical solutions; that is, to solutions based on the careful drawing and measuring of the various parameters involved. By merely sketching Mohr's circle and using trigonometry, you can easily derive the various relations required for a numerical solution of a given problem (see Sample Prob. 9.8).

## Sample Problem 9.8

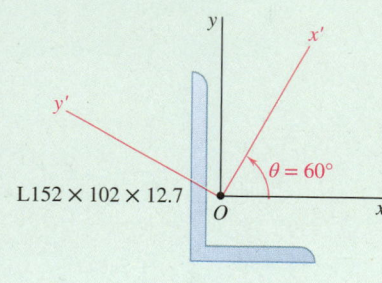

L152 × 102 × 12.7

For the section shown, the moments and product of inertia with respect to the $x$ and $y$ axes are

$$I_x = 7.20 \times 10^6 \text{ mm}^4 \qquad I_y = 2.59 \times 10^6 \text{ mm}^4 \qquad I_{xy} = -2.54 \times 10^6 \text{ mm}^4$$

Using Mohr's circle, determine (*a*) the principal axes of the section about $O$, (*b*) the values of the principal moments of inertia of the section about $O$, and (*c*) the moments and product of inertia of the section with respect to the $x'$ and $y'$ axes that form an angle of 60° with the $x$ and $y$ axes.

**STRATEGY:** You should carefully draw Mohr's circle and use the geometry of the circle to determine the orientation of the principal axes. Then, complete the analysis for the requested moments of inertia.

**MODELING:**

**Drawing Mohr's Circle.** First plot point $X$ with coordinates $I_x = 7.20$, $I_{xy} = -2.54$, and plot point $Y$ with coordinates $I_y = 2.59$, $-I_{xy} = +2.54$. Join $X$ and $Y$ with a straight line to define the center $C$ of Mohr's circle (Fig. 1). You can measure the abscissa of $C$, which represents $I$, and the radius $R$ of the circle either directly or using

$$I_{\text{ave}} = OC = \tfrac{1}{2}(I_x + I_y) = \tfrac{1}{2}(7.20 \times 10^6 + 2.59 \times 10^6) = 4.895 \times 10^6 \text{ mm}^4$$

$$CD = \tfrac{1}{2}(I_x - I_y) = \tfrac{1}{2}(7.20 \times 10^6 - 2.59 \times 10^6) = 2.305 \times 10^6 \text{ mm}^4$$

$$R = \sqrt{(CD)^2 + (DX)^2} = \sqrt{(2.305 \times 10^6)^2 + (2.54 \times 10^6)^2}$$

$$= 3.430 \times 10^6 \text{ mm}^4$$

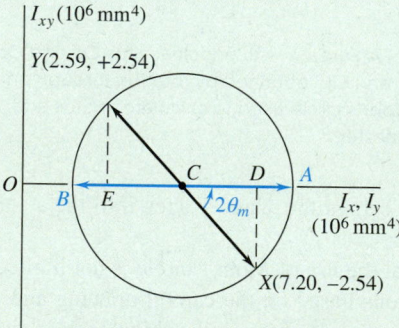

**Fig. 1** Mohr's circle.

**ANALYSIS:**

**a. Principal Axes.** The principal axes of the section correspond to points $A$ and $B$ on Mohr's circle, and the angle through which you should rotate $CX$ to bring it into $CA$ defines $2\theta_m$. You obtain

$$\tan 2\theta_m = \frac{DX}{CD} = \frac{2.54}{2.305} = 1.102 \quad 2\theta_m = 47.8° \circlearrowleft \qquad \theta_m = 23.9° \circlearrowleft \quad \blacktriangleleft$$

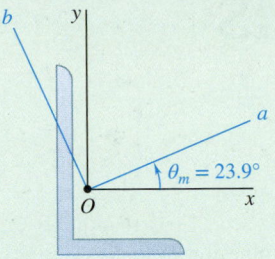

**Fig. 2** Orientation of the principal axes.

Thus, the principal axis $Oa$ corresponding to the maximum value of the moment of inertia is obtained by rotating the $x$ axis through 23.9° counterclockwise; the principal axis $Ob$ corresponding to the minimum value of the moment of inertia can be obtained by rotating the $y$ axis through the same angle (Fig. 2).

**b. Principal Moments of Inertia.** The principal moments of inertia are represented by the abscissas of A and B. The results are

$$I_{max} = OA = OC + CA = I_{ave} + R = (4.895 + 3.430)10^6 \text{ mm}^4$$

$$I_{max} = 8.33 \times 10^6 \text{ mm}^4 \quad \blacktriangleleft$$

$$I_{min} = OB = OC - BC = I_{ave} - R = (4.895 - 3.430)10^6 \text{ mm}^4$$

$$I_{min} = 1.47 \times 10^6 \text{ mm}^4 \quad \blacktriangleleft$$

**c. Moments and Product of Inertia with Respect to the $x'$ and $y'$ Axes.** On Mohr's circle, you obtain the points $X'$ and $Y'$, which correspond to the $x'$ and $y'$ axes, by rotating $CX$ and $CY$ through an angle $2\theta = 2(60°) = 120°$ counterclockwise (Fig. 3). The coordinates of $X'$ and $Y'$ yield the desired moments and product of inertia. Noting that the angle that $CX'$ forms with the horizontal axis is $\phi = 120° - 47.8° = 72.2°$, you have

$$I_{x'} = OF = OC + CF = 4.895 \times 10^6 \text{ mm}^4 + (3.430 \times 10^6 \text{ mm}^4) \cos 72.2°$$

$$I_{x'} = 5.94 \times 10^6 \text{ mm}^4 \quad \blacktriangleleft$$

$$I_{y'} = OG = OC - GC = 4.895 \times 10^6 \text{ mm}^4 - (3.430 \times 10^6 \text{ mm}^4) \cos 72.2°$$

$$I_{y'} = 3.85 \times 10^6 \text{ mm}^4 \quad \blacktriangleleft$$

$$I_{x'y'} = FX' = (3.430 \times 10^6 \text{ mm}^4) \sin 72.2°$$

$$I_{x'y'} = 3.27 \times 10^6 \text{ mm}^4 \quad \blacktriangleleft$$

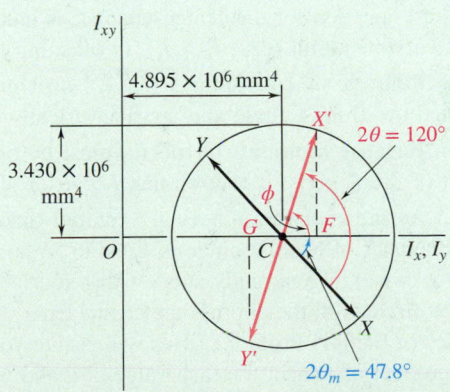

**Fig. 3** Using Mohr's circle to determine the moments and product of inertia with respect to $x'$ and $y'$ axes.

**REFLECT and THINK:** This problem illustrates typical calculations with Mohr's circle. The technique is a useful one to learn and remember.

# SOLVING PROBLEMS
# ON YOUR OWN

In the problems for this section, you will use **Mohr's circle** to determine the moments and products of inertia of a given area for different orientations of the coordinate axes. Although in some cases using Mohr's circle may not be as direct as substituting into the appropriate equations [Eqs. (9.18), (9.19), and (9.20)], this method of solution has the advantage of providing a visual representation of the relationships among the variables involved. Also, Mohr's circle shows all of the values of the moments and products of inertia that are possible for a given problem.

**Using Mohr's circle.** We presented the underlying theory in Sec. 9.3B, and we discussed the application of this method in Sec. 9.4 and in Sample Prob. 9.8. In the same problem, we presented the steps you should follow to determine the **principal axes**, the **principal moments of inertia**, and the **moments and product of inertia with respect to a specified orientation of the coordinates axes**. When you use Mohr's circle to solve problems, it is important that you remember the following points.

   **a. Mohr's circle is completely defined by the quantities $R$ and $I_{ave}$,** which represent, respectively, the radius of the circle and the distance from the origin $O$ to the center $C$ of the circle. You can obtain these quantities from Eqs. (9.23) if you know the moments and product of inertia for a given orientation of the axes. However, Mohr's circle can be defined by other combinations of known values (Probs. 9.103, 9.106, and 9.107). For these cases, it may be necessary to first make one or more assumptions, such as choosing an arbitrary location for the center when $I_{ave}$ is unknown, assigning relative magnitudes to the moments of inertia (e.g., $I_x > I_y$), or selecting the sign of the product of inertia.

   **b. Point X of coordinates $(I_x, I_{xy})$ and point Y of coordinates $(I_y, -I_{xy})$** are both located on Mohr's circle and are diametrically opposite.

   **c. Because moments of inertia must be positive,** all of Mohr's circle must lie to the right of the $I_{xy}$ axis; it follows that $I > R$ for all cases.

   **d. As the coordinate axes are rotated through an angle $\theta$,** the associated rotation of the diameter of Mohr's circle is equal to $2\theta$ and is in the same sense (clockwise or counterclockwise). We strongly suggest that you label the known points on the circumference of the circle with the appropriate capital letter, as was done in Fig. 9.19b and for the Mohr circles of Sample Prob. 9.8. This will enable you to determine the sign of the corresponding product of inertia for each value of $\theta$ and which moment of inertia is associated with each of the coordinate axes (Sample Prob. 9.8, parts a and c).

Although we have introduced Mohr's circle within the specific context of the study of moments and products of inertia, the Mohr circle technique also applies to the solution of analogous but physically different problems in the mechanics of materials. This multiple use of a specific technique is not unique, and as you pursue your engineering studies, you will encounter several methods of solution that can be applied to a variety of problems.

# Problems

**9.91** Using Mohr's circle, determine for the quarter ellipse of Prob. 9.67 the moments of inertia and the product of inertia with respect to new axes obtained by rotating the $x$ and $y$ axes about $O$ (*a*) through 45° counterclockwise, (*b*) through 30° clockwise.

**9.92** Using Mohr's circle, determine the moments of inertia and the product of inertia of the area of Prob. 9.72 with respect to new centroidal axes obtained by rotating the $x$ and $y$ axes 30° counterclockwise.

**9.93** Using Mohr's circle, determine the moments of inertia and the product of inertia of the area of Prob. 9.73 with respect to new centroidal axes obtained by rotating the $x$ and $y$ axes 60° counterclockwise.

*9.94* Using Mohr's circle, determine the moments of inertia and the product of inertia of the area of Prob. 9.75 with respect to new centroidal axes obtained by rotating the $x$ and $y$ axes 45° clockwise.

**9.95** Using Mohr's circle, determine the moments of inertia and the product of inertia of the L3 $\times$ 2 $\times$ $\frac{1}{4}$-in. angle cross section of Prob. 9.74 with respect to new centroidal axes obtained by rotating the $x$ and $y$ axes 30° clockwise.

*9.96* Using Mohr's circle, determine the moments of inertia and the product of inertia of the L152 $\times$ 102 $\times$ 12.7-mm angle cross section of Prob. 9.78 with respect to new centroidal axes obtained by rotating the $x$ and $y$ axes 30° clockwise.

**9.97** For the quarter ellipse of Prob. 9.67, use Mohr's circle to determine the orientation of the principal axes at the origin and the corresponding values of the moments of inertia.

**9.98 through *9.102*** Using Mohr's circle, determine for the area indicated the orientation of the principal centroidal axes and the corresponding values of the moments of inertia.

  **9.98** Area of Prob. 9.72
  **9.99** Area of Prob. 9.76
  **9.100** Area of Prob. 9.73
  *9.101* Area of Prob. 9.74
  *9.102* Area of Prob. 9.77

**9.103** The moments and product of inertia of an L4 $\times$ 3 $\times$ $\frac{1}{4}$-in. angle cross section with respect to two rectangular axes $x$ and $y$ through $C$ are, respectively, $\bar{I}_x = 1.33$ in$^4$, $\bar{I}_y = 2.75$ in$^4$, and $\bar{I}_{xy} < 0$, with the minimum value of the moment of inertia of the area with respect to any axis through $C$ being $\bar{I}_{min} = 0.692$ in$^4$. Using Mohr's circle, determine (*a*) the product of inertia $\bar{I}_{xy}$ of the area, (*b*) the orientation of the principal axes, (*c*) the value of $\bar{I}_{max}$.

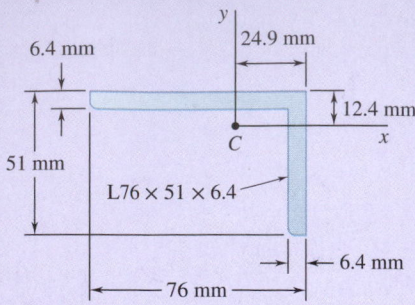

6.4 mm

24.9 mm

12.4 mm

C

x

51 mm

L76 × 51 × 6.4

6.4 mm

76 mm

**Fig. P9.104**

**9.104 and 9.105** Using Mohr's circle, determine the orientation of the principal centroidal axes and the corresponding values of the moments of inertia for the cross section of the rolled-steel angle shown. (Properties of the cross sections are given in Fig. 9.13.)

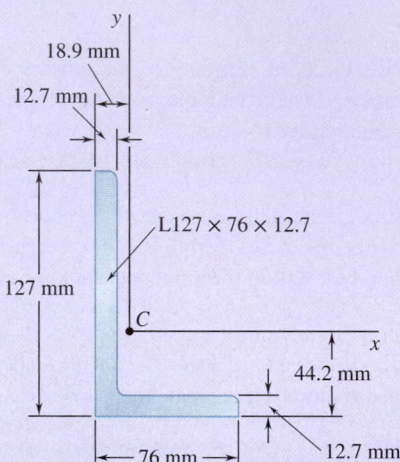

y

18.9 mm

12.7 mm

L127 × 76 × 12.7

127 mm

C

x

44.2 mm

76 mm

12.7 mm

**Fig. P9.105**

**\*9.106** For a given area, the moments of inertia with respect to two rectangular centroidal $x$ and $y$ axes are $\bar{I}_x = 1200$ in⁴ and $\bar{I}_y = 300$ in⁴, respectively. Knowing that, after rotating the $x$ and $y$ axes about the centroid $30°$ counterclockwise, the moment of inertia relative to the rotated $x$ axis is $1450$ in⁴, use Mohr's circle to determine (*a*) the orientation of the principal axes, (*b*) the principal centroidal moments of inertia.

**9.107** It is known that for a given area $\bar{I}_y = 48 × 10^6$ mm⁴ and $\bar{I}_{xy} = -20 × 10^6$ mm⁴, where the $x$ and $y$ axes are rectangular centroidal axes. If the axis corresponding to the maximum product of inertia is obtained by rotating the $x$ axis $67.5°$ counterclockwise about $C$, use Mohr's circle to determine (*a*) the moment of inertia $\bar{I}_x$ of the area, (*b*) the principal centroidal moments of inertia.

**9.108** Using Mohr's circle, show that for any regular polygon (such as a pentagon) (*a*) the moment of inertia with respect to every axis through the centroid is the same, (*b*) the product of inertia with respect to every pair of rectangular axes through the centroid is zero.

**9.109** Using Mohr's circle, prove that the expression $I_{x'}I_{y'} - I_{x'y'}^2$ is independent of the orientation of the $x'$ and $y'$ axes, where $I_{x'}$, $I_{y'}$, and $I_{x'y'}$ represent the moments and product of inertia, respectively, of a given area with respect to a pair of rectangular axes $x'$ and $y'$ through a given point $O$. Also show that the given expression is equal to the square of the length of the tangent drawn from the origin of the coordinate system to Mohr's circle.

**9.110** Using the invariance property established in the preceding problem, express the product of inertia $I_{xy}$ of an area $A$ with respect to a pair of rectangular axes through $O$ in terms of the moments of inertia $I_x$ and $I_y$ of $A$ and the principal moments of inertia $I_{min}$ and $I_{max}$ of $A$ about $O$. Use the formula obtained to calculate the product of inertia $I_{xy}$ of the L3 × 2 × $\frac{1}{4}$-in. angle cross section shown in Fig. 9.13A, knowing that its maximum moment of inertia is $1.257$ in⁴.

# 9.5 MASS MOMENTS OF INERTIA

So far in this chapter, we have examined moments of inertia of areas. In the rest of this chapter, we consider moments of inertia associated with the masses of bodies. This will be an important concept in dynamics when studying the rotational motion of a rigid body about an axis.

## 9.5A Moment of Inertia of a Simple Mass

Consider a small mass $\Delta m$ mounted on a rod of negligible mass that can rotate freely about an axis $AA'$ (Fig. 9.20a). If we apply a couple to the system, the rod and mass (assumed to be initially at rest) start rotating about $AA'$. We will study the details of this motion later in dynamics. At present, we wish to indicate only that the time required for the system to reach a given speed of rotation is proportional to the mass $\Delta m$ and to the square of the distance $r$. The product $r^2 \Delta m$ thus provides a measure of the **inertia** of the system; that is, a measure of the resistance the system offers when we try to set it in motion. For this reason, the product $r^2 \Delta m$ is called the **moment of inertia** of the mass $\Delta m$ with respect to axis $AA'$.

Now suppose a body of mass $m$ is to be rotated about an axis $AA'$ (Fig. 9.20b). Dividing the body into elements of mass $\Delta m_1$, $\Delta m_2$, etc., we find that the body's resistance to being rotated is measured by the sum $r_1^2 \Delta m_1 + r_2^2 \Delta m_2 + \ldots$. This sum defines the moment of inertia of the body with respect to axis $AA'$. Increasing the number of elements, we find that the moment of inertia is equal, in the limit, to the integral

**Moment of inertia of a mass**

$$I = \int r^2 \, dm \qquad (9.28)$$

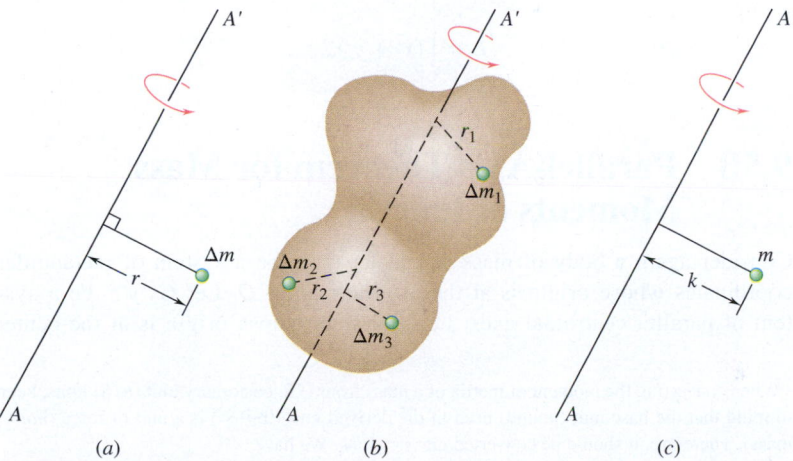

(a)          (b)          (c)

**Fig. 9.20** (a) An element of mass $\Delta m$ at a distance $r$ from an axis $AA'$; (b) the moment of inertia of a rigid body is the sum of the moments of inertia of many small masses; (c) the moment of inertia is unchanged if all the mass is concentrated at a point at a distance from the axis equal to the radius of gyration.

We define the **radius of gyration** $k$ of the body with respect to axis $AA'$ by the relation

**Radius of gyration
of a mass**

$$I = k^2 m \qquad \text{or} \qquad k = \sqrt{\frac{I}{m}} \tag{9.29}$$

The radius of gyration $k$ represents the distance at which the entire mass of the body should be concentrated if its moment of inertia with respect to $AA'$ is to remain unchanged (Fig. 9.20$c$). Whether it stays in its original shape (Fig. 9.20$b$) or is concentrated as shown in Fig. 9.20$c$, the mass $m$ reacts in the same way to a rotation (or *gyration*) about $AA'$.

If SI units are used, the radius of gyration $k$ is expressed in meters and the mass $m$ in kilograms, so the unit for the moment of inertia of a mass is kg·m$^2$. If U.S. customary units are used, the radius of gyration is expressed in feet and the mass in slugs (i.e., in lb·s$^2$/ft), so the derived unit for the moment of inertia of a mass is lb·ft·s$^{2}$.[†]

We can express the moment of inertia of a body with respect to a coordinate axis in terms of the coordinates $x$, $y$, $z$ of the element of mass $dm$ (Fig. 9.21). Noting, for example, that the square of the distance $r$ from the element $dm$ to the $y$ axis is $z^2 + x^2$, the moment of inertia of the body with respect to the $y$ axis is

$$I_y = \int r^2 dm = \int (z^2 + x^2) dm$$

We obtain similar expressions for the moments of inertia with respect to the $x$ and $z$ axes.

**Moments of inertia with
respect to coordinate axes**

$$\begin{aligned} I_x &= \int (y^2 + z^2) dm \\ I_y &= \int (z^2 + x^2) dm \\ I_z &= \int (x^2 + y^2) dm \end{aligned} \tag{9.30}$$

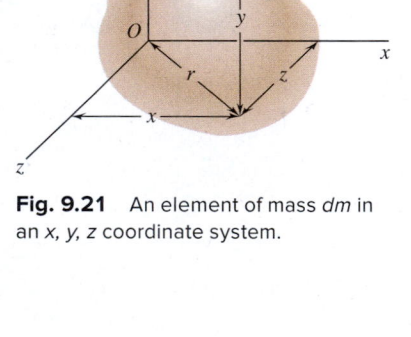

**Fig. 9.21**  An element of mass $dm$ in an $x$, $y$, $z$ coordinate system.

## 9.5B   Parallel-Axis Theorem for Mass Moments of Inertia

Consider again a body of mass $m$ and let $Oxyz$ be a system of rectangular coordinates whose origin is at the arbitrary point $O$. Let $Gx'y'z'$ be a system of parallel centroidal axes; i.e., a system whose origin is at the center

**Photo 9.2**  The rotational behavior of this crankshaft depends upon its mass moment of inertia with respect to its axis of rotation, as you will see in a dynamics course. ©loraks/ Getty Images RF

---

[†] When converting the moment of inertia of a mass from U.S. customary units to SI units, keep in mind that the base unit (pound) used in the derived unit (lb·ft·s$^2$) is a unit of force (*not* of mass). Therefore, it should be converted into newtons. We have

$$1 \text{ lb} \cdot \text{ft} \cdot \text{s}^2 = (4.45 \text{ N}) (0.3048 \text{ m}) (1 \text{ s})^2 = 1.356 \text{ N} \cdot \text{m} \cdot \text{s}^2$$

or because $1 \text{ N} = 1 \text{ kg} \cdot \text{m/s}^2$

$$1 \text{ lb} \cdot \text{ft} \cdot \text{s}^2 = 1.356 \text{ kg} \cdot \text{m}^2$$

of gravity $G$ of the body and whose axes $x'$, $y'$, $z'$ are parallel to the $x$, $y$, and $z$ axes, respectively (Fig. 9.22). (Note that we use the term centroidal here to define axes passing through the center of gravity $G$ of the body, regardless of whether or not $G$ coincides with the centroid of the volume of the body.) We denote by $\bar{x}$, $\bar{y}$, $\bar{z}$ the coordinates of $G$ with respect to $Oxyz$. Then, we have the following relations between the coordinates $x$, $y$, $z$ of the element $dM$ with respect to $Oxyz$ and its coordinates $x'$, $y'$, $z'$ with respect to the centroidal axes $Gx'y'z'$:

$$x = x' + \bar{x} \quad y = y' + \bar{y} \quad z = z' + \bar{z} \tag{9.31}$$

Referring to Eqs. (9.30), we can express the moment of inertia of the body with respect to the $x$ axis as

$$I_x = \int (y^2 + z^2)\, dm = \int [(y' + \bar{y})^2 + (z' + \bar{z})^2]dm$$

$$= \int (y'^2 + z'^2)\, dm + 2\bar{y}\int y'\, dm + 2\bar{z}\int z'\, dm + (\bar{y}^2 + \bar{z}^2)\int dm$$

The first integral in this expression represents the moment of inertia $\bar{I}_{x'}$ of the body with respect to the centroidal axis $x'$. The second and third integrals represent the first moment of the body with respect to the $z'x'$ and $x'y'$ planes, respectively, and because both planes contain $G$, these two integrals are zero. The last integral is equal to the total mass $m$ of the body. Therefore, we have

$$I_x = \bar{I}_{x'} + m(\bar{y}^2 + \bar{z}^2) \tag{9.32}$$

Similarly,

$$I_y = \bar{I}_{y'} + m(\bar{z}^2 + \bar{x}^2) \quad I_z = \bar{I}_{z'} + m(\bar{x}^2 + \bar{y}^2) \tag{9.32'}$$

We easily verify from Fig. 9.22 that the sum $\bar{z}^2 + \bar{x}^2$ represents the square of the distance $OB$ between the $y$ and $y'$ axes. Similarly, $\bar{y}^2 + \bar{z}^2$ and $\bar{x}^2 + \bar{y}^2$ represent the squares of the distance between the $x$ and $x'$ axes and the $z$ and $z'$ axes, respectively. We denote the distance between an arbitrary axis $AA'$ and a parallel centroidal axis $BB'$ by $d$ (Fig. 9.23). Then, the general relation between the moment of inertia $I$ of the body with respect to $AA'$ and its moment of inertia $\bar{I}$ with respect to $BB'$, known as the parallel-axis theorem for mass moments of inertia, is

**Parallel-axis theorem for
mass moments of inertia**

$$I = \bar{I} + md^2 \tag{9.33}$$

Expressing the moments of inertia in terms of the corresponding radii of gyration, we can also write

$$k^2 + \bar{k}^2 + d^2 \tag{9.34}$$

where $k$ and $\bar{k}$ represent the radii of gyration of the body about $AA'$ and $BB'$, respectively.

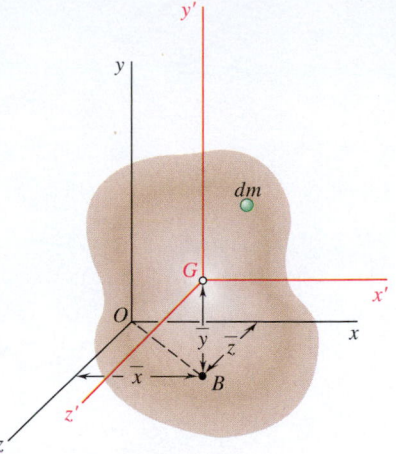

**Fig. 9.22**    A body of mass $m$ with an arbitrary rectangular coordinate system at $O$ and a parallel centroidal coordinate system at $G$. Also shown is an element of mass $dm$.

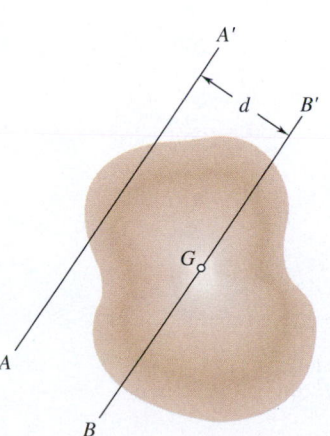

**Fig. 9.23**    We use $d$ to denote the distance between an arbitrary axis $AA'$ and a parallel centroidal axis $BB'$.

## 9.5C    Moments of Inertia of Thin Plates

Now imagine a thin plate of uniform thickness $t$, made of a homogeneous material of density $\rho$ (density = mass per unit volume). The mass moment of inertia of the plate with respect to an axis $AA'$ *contained in the plane* of the plate (Fig. 9.24a) is

$$I_{AA', \text{ mass}} = \int r^2 \, dm$$

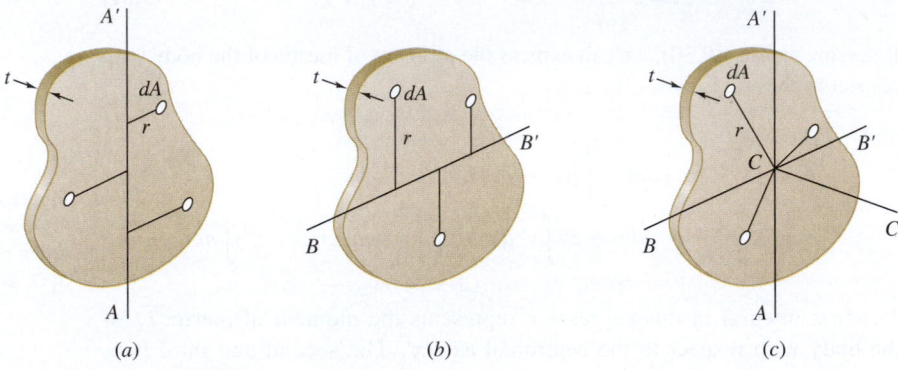

**Fig. 9.24**  (a) A thin plate with an axis $AA'$ in the plane of the plate; (b) an axis $BB'$ in the plane of the plate and perpendicular to $AA'$; (c) an axis $CC'$ perpendicular to the plate and passing through the intersection of $AA'$ and $BB'$.

Because $dm = \rho t \, dA$, we have

$$I_{AA', \text{ mass}} = \rho t \int r^2 \, dA$$

However, $r$ represents the distance of the element of area $dA$ to the axis $AA'$. Therefore, the integral is equal to the moment of inertia of the area of the plate with respect to $AA'$.

$$I_{AA', \text{ mass}} = \rho t I_{AA', \text{ area}} \tag{9.35}$$

Similarly, for an axis $BB'$ that is contained in the plane of the plate and is perpendicular to $AA'$ (Fig. 9.24b), we have

$$I_{BB', \text{ mass}} = \rho t I_{BB', \text{ area}} \tag{9.36}$$

Consider now the axis $CC'$, which is *perpendicular* to the plate and passes through the point of intersection $C$ of $AA'$ and $BB'$ (Fig. 9.24c). This time we have

$$I_{CC', \text{ mass}} = \rho t J_{C, \text{ area}} \tag{9.37}$$

where $J_C$ is the polar moment of inertia of the area of the plate with respect to point $C$.

Recall the relation $J_C = I_{AA'} + I_{BB'}$ between the polar and rectangular moments of inertia of an area. We can use this to write the relation between the mass moments of inertia of a thin plate as

$$I_{CC'} = I_{AA'} + I_{BB'} \tag{9.38}$$

**Rectangular Plate.**   In the case of a rectangular plate of sides $a$ and $b$ (Fig. 9.25), we obtain the mass moments of inertia with respect to axes through the center of gravity of the plate as

$$I_{AA', \text{mass}} = \rho t I_{AA', \text{area}} = \rho t \left(\tfrac{1}{12}a^3 b\right)$$
$$I_{BB', \text{mass}} = \rho t I_{BB', \text{area}} = \rho t \left(\tfrac{1}{12}ab^3\right)$$

Because the product $\rho abt$ is equal to the mass $m$ of the plate, we can also write the mass moments of inertia of a thin rectangular plate as

$$I_{AA'} = \tfrac{1}{12}ma^2 \qquad I_{BB'} = \tfrac{1}{12}mb^2 \qquad \text{(9.39)}$$
$$I_{CC'} = I_{AA'} + I_{BB'} = \tfrac{1}{12}m(a^2+b^2) \qquad \text{(9.40)}$$

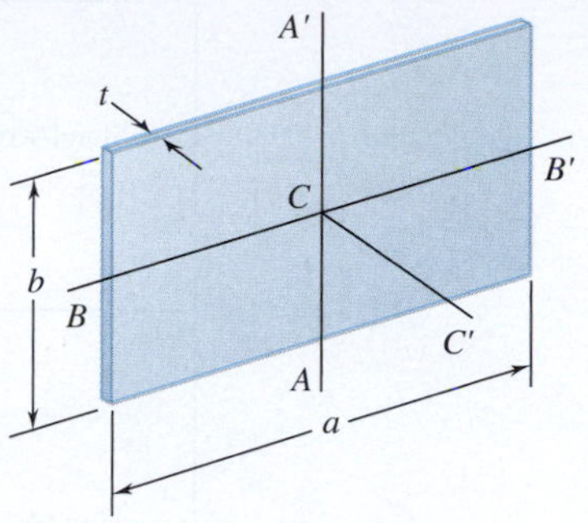

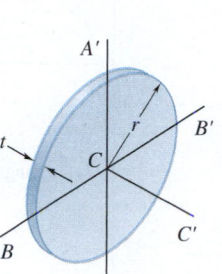

**Fig. 9.25**   A thin rectangular plate of sides $a$ and $b$.

**Circular Plate.**   In the case of a circular plate, or disk, of radius $r$ (Fig. 9.26), Eq. (9.35) becomes

$$I_{AA', \text{mass}} = \rho t I_{AA', \text{area}} = \rho t \left(\tfrac{1}{4}\pi r^4\right)$$

In this case, the product $\rho \pi r^2 t$ is equal to the mass $m$ of the plate, and $I_{AA'} = I_{BB'}$. Therefore, we can write the mass moments of inertia of a circular plate as

$$I_{AA'} = I_{BB'} = \tfrac{1}{4}mr^2 \qquad \text{(9.41)}$$
$$I_{CC'} = I_{AA'} + I_{BB'} = \tfrac{1}{2}mr^2 \qquad \text{(9.42)}$$

**Fig. 9.26**   A thin circular plate of radius $r$.

## 9.5D   Determining the Moment of Inertia of a Three-Dimensional Body by Integration

We obtain the moment of inertia of a three-dimensional body by evaluating the integral $I = \int r^2\, dm$. If the body is made of a homogeneous material with a density $\rho$, the element of mass $dm$ is equal to $\rho\, dV$, and we have $I = \rho \int r^2\, dV$. This integral depends only upon the shape of the body. Thus, in order to compute the moment of inertia of a three-dimensional body, it is generally necessary to perform a triple, or at least a double, integration.

However, if the body possesses two planes of symmetry, it is usually possible to determine the body's moment of inertia with a single integration. We do this by choosing as the element of mass $dm$ a thin slab that is perpendicular to the planes of symmetry. In the case of bodies of revolution, for example, the element of mass is a thin disk (Fig. 9.27). Using formula (9.42), we can express the moment of inertia of the disk with respect to the axis of revolution as indicated in Fig. 9.27. Its moment of inertia with respect to each of the other two coordinate axes is obtained by using formula (9.41) and the parallel-axis theorem. Integration of these expressions yields the desired moment of inertia of the body.

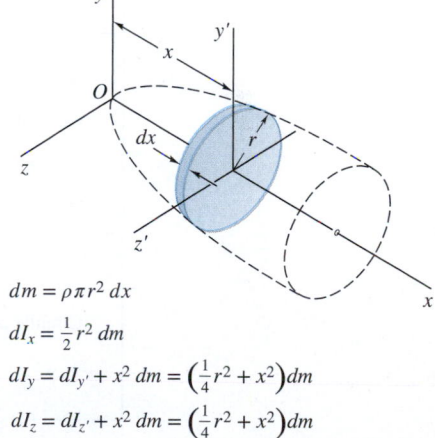

$$dm = \rho\pi r^2\, dx$$
$$dI_x = \tfrac{1}{2}r^2\, dm$$
$$dI_y = dI_{y'} + x^2\, dm = \left(\tfrac{1}{4}r^2 + x^2\right)dm$$
$$dI_z = dI_{z'} + x^2\, dm = \left(\tfrac{1}{4}r^2 + x^2\right)dm$$

**Fig. 9.27**   Using a thin disk to determine the moment of inertia of a body of revolution.

## 9.5E   Moments of Inertia of Composite Bodies

Fig. 9.28 lists the moments of inertia of a few common shapes. For a body consisting of several of these simple shapes in combination, you can obtain the moment of inertia of the body with respect to a given axis by first computing the moments of inertia of its component parts about the desired axis and then adding them together. As was the case for areas, the radius of gyration of a composite body *cannot* be obtained by adding the radii of gyration of its component parts.

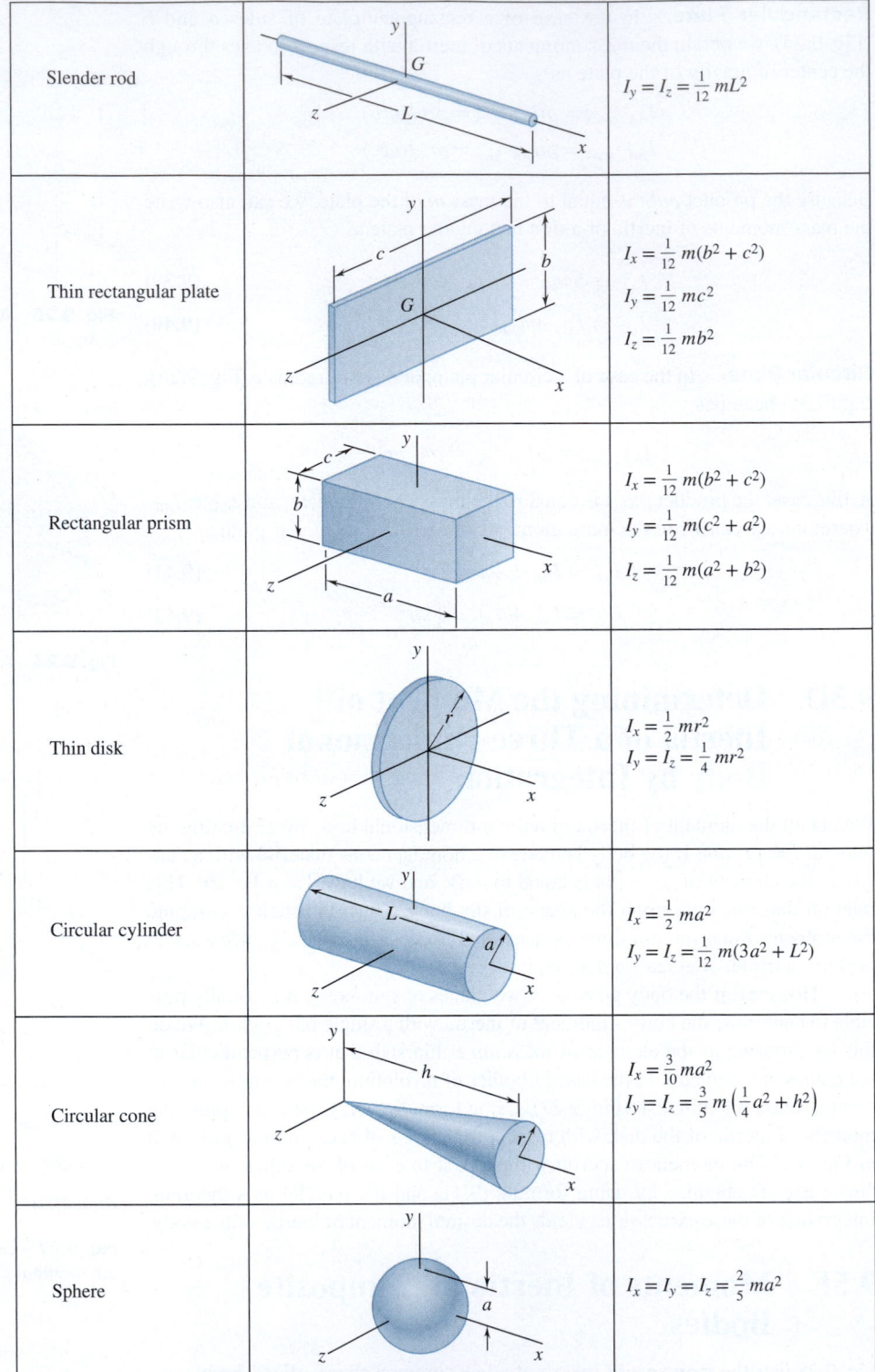

| | | |
|---|---|---|
| Slender rod | | $I_y = I_z = \frac{1}{12} mL^2$ |
| Thin rectangular plate | | $I_x = \frac{1}{12} m(b^2 + c^2)$ <br> $I_y = \frac{1}{12} mc^2$ <br> $I_z = \frac{1}{12} mb^2$ |
| Rectangular prism | | $I_x = \frac{1}{12} m(b^2 + c^2)$ <br> $I_y = \frac{1}{12} m(c^2 + a^2)$ <br> $I_z = \frac{1}{12} m(a^2 + b^2)$ |
| Thin disk | | $I_x = \frac{1}{2} mr^2$ <br> $I_y = I_z = \frac{1}{4} mr^2$ |
| Circular cylinder | | $I_x = \frac{1}{2} ma^2$ <br> $I_y = I_z = \frac{1}{12} m(3a^2 + L^2)$ |
| Circular cone | | $I_x = \frac{3}{10} ma^2$ <br> $I_y = I_z = \frac{3}{5} m\left(\frac{1}{4}a^2 + h^2\right)$ |
| Sphere | | $I_x = I_y = I_z = \frac{2}{5} ma^2$ |

**Fig. 9.28** Mass moments of inertia of common geometric shapes.

## Sample Problem 9.9

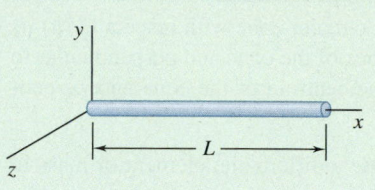

Determine the moment of inertia of a slender rod of length $L$ and mass $m$ with respect to an axis that is perpendicular to the rod and passes through one end.

**STRATEGY:**   Approximating the rod as a one-dimensional body enables you to solve the problem by a single integration.

**MODELING and ANALYSIS:**   Choose the differential element of mass shown in Fig. 1 and express it as a mass per unit length.

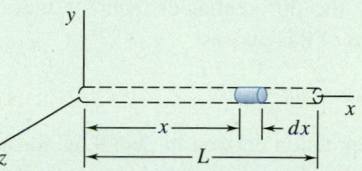

$$dm = \frac{m}{L}dx$$

$$I_y = \int x^2\,dm = \int_0^L x^2\frac{m}{L}dx = \left[\frac{m}{L}\frac{x^3}{3}\right]_0^L \qquad I_y = \frac{1}{3}mL^2 \blacktriangleleft$$

**Fig. 1**   Differential element of mass.

**REFLECT and THINK:**   This problem could also have been solved by starting with the moment of inertia for a slender rod with respect to its centroid, as given in Fig. 9.28, and using the parallel-axis theorem to obtain the moment of inertia with respect to an end of the rod.

## Sample Problem 9.10

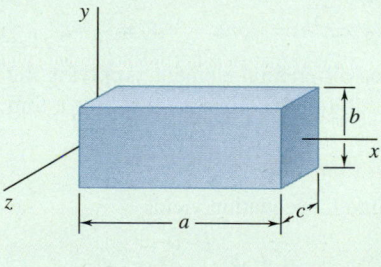

For the homogeneous rectangular prism shown, determine the moment of inertia with respect to the $z$ axis.

**STRATEGY:**   You can approach this problem by choosing a differential element of mass perpendicular to the long axis of the prism; find its moment of inertia with respect to a centroidal axis parallel to the $z$ axis; and then apply the parallel-axis theorem.

**MODELING and ANALYSIS:**   Choose as the differential element of mass the thin slab shown in Fig. 1. Then

$$dm = \rho bc\,dx$$

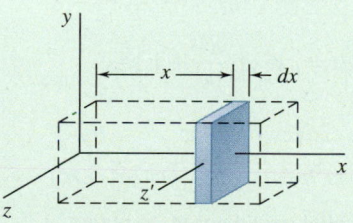

Referring to Sec. 9.5C, the moment of inertia of the element with respect to the $z'$ axis is

$$dI_{z'} = \tfrac{1}{12}b^2\,dm$$

**Fig. 1**   Differential element of mass.

Applying the parallel-axis theorem, you can obtain the mass moment of inertia of the slab with respect to the $z$ axis.

$$dI_z = dI_{z'} + x^2\,dm = \tfrac{1}{12}b^2\,dm + x^2\,dm = (\tfrac{1}{12}b^2 + x^2)\rho bc\,dx$$

Integrating from $x = 0$ to $x = a$ gives you

$$I_z = \int dI_z = \int_0^a (\tfrac{1}{12}b^2 + x^2)\rho bc\,dx = \rho abc(\tfrac{1}{12}b^2 + \tfrac{1}{3}a^2)$$

Because the total mass of the prism is $m = \rho abc$, you can write

$$I_z = m(\tfrac{1}{12}b^2 + \tfrac{1}{3}a^2) \qquad I_z = \tfrac{1}{12}m(4a^2 + b^2) \blacktriangleleft$$

**REFLECT AND THINK:**   Note that if the prism is thin, $b$ is small compared to $a$, and the expression for $I_z$ reduces to $\tfrac{1}{3}ma^2$, which is the result obtained in Sample Prob. 9.9 when $L = a$.

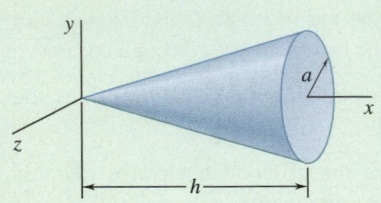

## Sample Problem 9.11

Determine the moment of inertia of a right circular cone with respect to (a) its longitudinal axis, (b) an axis through the apex of the cone and perpendicular to its longitudinal axis, (c) an axis through the centroid of the cone and perpendicular to its longitudinal axis.

**STRATEGY:** For parts (a) and (b), choose a differential element of mass in the form of a thin circular disk perpendicular to the longitudinal axis of the cone. You can solve part (c) by an application of the parallel-axis theorem.

**MODELING and ANALYSIS:** Choose the differential element of mass shown in Fig. 1. Express the radius and mass of this disk as

$$r = a\frac{x}{h} \qquad dm = \rho\pi r^2 dx = \rho\pi\frac{a^2}{h^2}x^2 dx$$

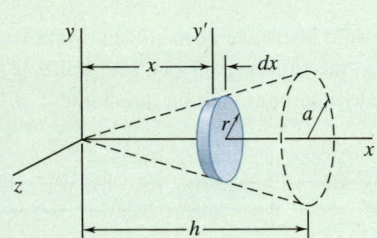

**Fig. 1** Differential element of mass.

**a. Moment of Inertia $I_x$.** Using the expression derived in Sec. 9.5C for a thin disk, compute the mass moment of inertia of the differential element with respect to the $x$ axis.

$$dI_x = \tfrac{1}{2}r^2 dm = \tfrac{1}{2}\left(a\frac{x}{h}\right)^2\left(\rho\pi\frac{a^2}{h^2}x^2 dx\right) = \tfrac{1}{2}\rho\pi\frac{a^4}{h^4}x^4 dx$$

Integrating from $x = 0$ to $x = h$ gives you

$$I_x = \int dI_x = \int_0^h \tfrac{1}{2}\rho\pi\frac{a^4}{h^4}x^4\,dx = \tfrac{1}{2}\rho\pi\frac{a^4}{h^4}\frac{h^5}{5} = \tfrac{1}{10}\rho\pi a^4 h$$

Because the total mass of the cone is $m = \tfrac{1}{3}\rho\pi a^2 h$, you can write this as

$$I_x = \tfrac{1}{10}\rho\pi a^4 h = \tfrac{3}{10}a^2(\tfrac{1}{3}\rho\pi a^2 h) = \tfrac{3}{10}ma^2 \qquad I_x = \tfrac{3}{10}ma^2 \quad \blacktriangleleft$$

**b. Moment of Inertia $I_y$.** Use the same differential element. Applying the parallel-axis theorem and using the expression derived in Sec. 9.5C for a thin disk, you have

$$dI_y = dI_{y'} + x^2 dm = \tfrac{1}{4}r^2 dm + x^2 dm = (\tfrac{1}{4}r^2 + x^2)\,dm$$

Substituting the expressions for $r$ and $dm$ into this equation yields

$$dI_y = \left(\frac{1}{4}\frac{a^2}{h^2}x^2 + x^2\right)\left(\rho\pi\frac{a^2}{h^2}x^2 dx\right) = \rho\pi\frac{a^2}{h^2}\left(\frac{a^2}{4h^2} + 1\right)x^4 dx$$

$$I_y = \int dI_y = \int_0^h \rho\pi\frac{a^2}{h^2}\left(\frac{a^2}{4h^2} + 1\right)x^4 dx = \rho\pi\frac{a^2}{h^2}\left(\frac{a^2}{4h^2} + 1\right)\frac{h^5}{5}$$

Introducing the total mass of the cone $m$, you can rewrite $I_y$ as

$$I_y = \tfrac{3}{5}(\tfrac{1}{4}a^2 + h^2)\tfrac{1}{3}\rho\pi a^2 h \qquad I_y = \tfrac{3}{5}m(\tfrac{1}{4}a^2 + h^2) \quad \blacktriangleleft$$

**c. Moment of Inertia $\bar{I}_{y''}$.** Apply the parallel-axis theorem to obtain

$$I_y = \bar{I}_{y''} + m\bar{x}^2$$

Solve for $\bar{I}_{y''}$ and recall from Fig. 5.21 that $\bar{x} = \tfrac{3}{4}h$ (Fig. 2). The result is

$$\bar{I}_{y''} = I_y - m\bar{x}^2 = \tfrac{3}{5}m(\tfrac{1}{4}a^2 + h^2) - m(\tfrac{3}{4}h)^2$$

$$\bar{I}_{y''} = \tfrac{3}{20}m(a^2 + \tfrac{1}{4}h^2) \quad \blacktriangleleft$$

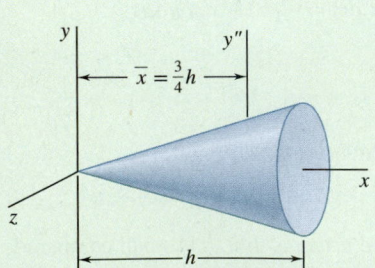

**Fig. 2** Centroid of a right circular cone.

**REFLECT and THINK:** The parallel-axis theorem for masses can be just as useful as the version for areas. Don't forget to use the reference figures for centroids of volumes when needed.

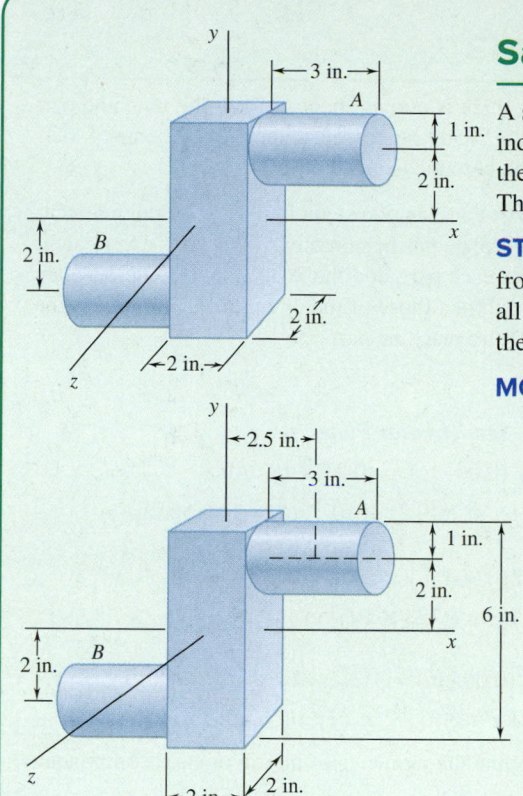

**Fig. 1**   Geometry of each component.

## Sample Problem 9.12

A steel forging consists of a $6 \times 2 \times 2$-in. rectangular prism and two cylinders with a diameter of 2 in. and length of 3 in. as shown. Determine the moments of inertia of the forging with respect to the coordinate axes. The specific weight of steel is 490 lb/ft$^3$.

**STRATEGY:**   Compute the moments of inertia of each component from Fig. 9.28 using the parallel-axis theorem when necessary. Note that all lengths should be expressed in feet to be consistent with the units for the given specific weight.

**MODELING and ANALYSIS:**

### Computation of Masses.

**Prism.**

$$V = (2 \text{ in.})(2 \text{ in.})(6 \text{ in.}) = 24 \text{ in}^3$$

$$W = \frac{(24 \text{ in}^3)(490 \text{ lb/ft}^3)}{1728 \text{ in}^3/\text{ft}^3} = 6.81 \text{ lb}$$

$$m = \frac{6.81 \text{ lb}}{32.2 \text{ ft/s}^2} = 0.211 \text{ lb·s}^2/\text{ft}$$

**Each Cylinder.**

$$V = \pi(1 \text{ in.})^2(3 \text{ in.}) = 9.42 \text{ in}^3$$

$$W = \frac{(9.42 \text{ in}^3)(490 \text{ lb/ft}^3)}{1728 \text{ in}^3/\text{ft}^3} = 2.67 \text{ lb}$$

$$m = \frac{2.67 \text{ lb}}{32.2 \text{ ft/s}^2} = 0.0829 \text{ lb·s}^2/\text{ft}$$

### Moments of Inertia (Fig. 1).

**Prism.**

$$I_x = I_z = \tfrac{1}{12}(0.211 \text{ lb·s}^2/\text{ft})[(\tfrac{6}{12}\text{ft})^2 + (\tfrac{2}{12} \text{ ft})^2] = 4.88 \times 10^{-3} \text{ lb·ft·s}^2$$

$$I_y = \tfrac{1}{12}(0.211 \text{ lb·s}^2/\text{ft})[(\tfrac{2}{12}\text{ft})^2 + (\tfrac{2}{12} \text{ ft})^2] = 0.977 \times 10^{-3} \text{ lb·ft·s}^2$$

**Each Cylinder.**

$$I_x = \tfrac{1}{2}ma^2 + m\bar{y}^2 = \tfrac{1}{2}(0.0829 \text{ lb·s}^2/\text{ft})(\tfrac{1}{12}\text{ft})^2$$
$$+ (0.0829 \text{ lb·s}^2/\text{ft})(\tfrac{2}{12}\text{ft})^2 = 2.59 \times 10^{-3} \text{ lb·ft·s}^2$$

$$I_y = \tfrac{1}{12}m(3a^2 + L^2) = m\bar{x}^2 = \tfrac{1}{12}(0.0829 \text{ lb·s}^2/\text{ft})[3(\tfrac{1}{12}\text{ft})^2 + (\tfrac{3}{12}\text{ft})^2]$$
$$+ (0.0829 \text{ lb·s}^2/\text{ft})(\tfrac{2.5}{12}\text{ft})^2 = 4.17 \times 10^{-3} \text{ lb·ft·s}^2$$

$$I_z = \frac{1}{12}m(3a^2 + L^2) + m(\bar{x}^2 + \bar{y}^2) = \tfrac{1}{12}(0.0829 \text{ lb·s}^2/\text{ft})[3(\tfrac{1}{12}\text{ft})^2 + (\tfrac{3}{12}\text{ft})^2]$$
$$+ (0.0829 \text{ lb·s}^2/\text{ft})[(\tfrac{2.5}{12}\text{ft})^2 + (\tfrac{2}{12}\text{ft})^2] = 6.48 \times 10^{-3} \text{ lb·ft·s}^2$$

**Entire Body.**   Adding the values obtained for the prism and two cylinders, you have

$$I_x = 4.88 \times 10^{-3} + 2(2.59 \times 10^{-3}) \qquad \boxed{I_x = 10.06 \times 10^{-3} \text{ lb·ft·s}^2} \blacktriangleleft$$
$$I_y = 0.977 \times 10^{-3} + 2(4.17 \times 10^{-3}) \qquad \boxed{I_y = 9.32 \times 10^{-3} \text{ lb·ft·s}^2} \blacktriangleleft$$
$$I_z = 4.88 \times 10^{-3} + 2(6.48 \times 10^{-3}) \qquad \boxed{I_z = 17.84 \times 10^{-3} \text{ lb·ft·s}^2} \blacktriangleleft$$

**REFLECT and THINK:**   The results indicate this forging has more resistance to rotation about the $z$ axis (largest moment of inertia) than about the $x$ or $y$ axes. This makes intuitive sense, because more of the mass is farther from the $z$ axis than from the $x$ or $y$ axes.

## Sample Problem 9.13

A thin steel plate that is 4 mm thick is cut and bent to form the machine part shown. The density of the steel is 7850 kg/m$^3$. Determine the moments of inertia of the machine part with respect to the coordinate axes.

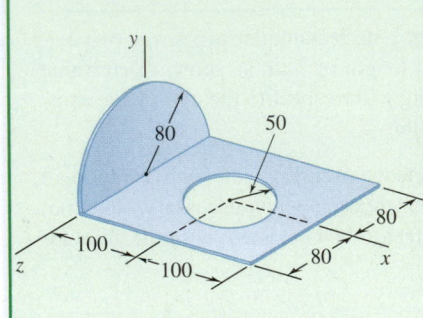

Dimensions in mm

**STRATEGY:**   The machine part consists of a semicircular plate and a rectangular plate from which a circular plate has been removed (Fig. 1). After calculating the moments of inertia for each part, add those of the semicircular plate and the rectangular plate, then subtract those of the circular plate to determine the moments of inertia for the entire machine part.

### MODELING and ANALYSIS:

**Computation of Masses.**   *Semicircular Plate.*

$$V_1 = \tfrac{1}{2}\pi r^2 t = \tfrac{1}{2}\pi(0.08 \text{ m})^2(0.004 \text{ m}) = 40.21 \times 10^{-6} \text{m}^3$$

$$m_1 = \rho V_1 = (7.85 \times 10^3 \text{ kg/m}^3)(40.21 \times 10^{-6} \text{ m}^3) = 0.3156 \text{ kg}$$

**Rectangular Plate.**

$$V_2 = (0.200 \text{ m})(0.160 \text{ m})(0.004 \text{ m}) = 128 \times 10^{-6}\text{m}^3$$

$$m_2 = \rho V_2 = (7.85 \times 10^3 \text{ kg/m}^3)(128 \times 10^{-6} \text{ m}^3) = 1.005 \text{ kg}$$

**Circular Plate.**

$$V_3 = \pi a^2 t = \pi(0.050 \text{ m})^2(0.004 \text{ m}) = 31.42 \times 10^{-6} \text{ m}^3$$

$$m_3 = \rho V_3 = (7.85 \times 10^3 \text{ kg/m}^3)(31.42 \times 10^{-6} \text{ m}^3) = 0.2466 \text{ kg}$$

**Moments of Inertia.**   Compute the moments of inertia of each component, using the method presented in Sec. 9.5C.

**Semicircular Plate.**   Observe from Fig. 9.28 that, for a circular plate of mass $m$ and radius $r$,

$$I_x = \tfrac{1}{2}mr^2 \qquad I_y = I_z = \tfrac{1}{4}mr^2$$

Because of symmetry, halve these values for a semicircular plate. Thus,

$$I_x = \tfrac{1}{2}(\tfrac{1}{2}mr^2) \qquad I_y = I_z = \tfrac{1}{2}(\tfrac{1}{4}mr^2)$$

Because the mass of the semicircular plate is $m_1 = \tfrac{1}{2}m$, you have

$$I_x = \tfrac{1}{2}m_1 r^2 = \tfrac{1}{2}(0.3156 \text{ kg})(0.08 \text{ m})^2 = 1.010 \times 10^{-3} \text{ kg·m}^2$$

$$I_y = I_z = \tfrac{1}{4}\left(\tfrac{1}{2}mr^2\right) = \tfrac{1}{4}m_1 r^2 = \tfrac{1}{4}(0.3156 \text{ kg})(0.08 \text{ m})^2 = 0.505 \times 10^{-3} \text{ kg·m}^2$$

**Rectangular Plate.**

$$I_x = \tfrac{1}{12}m_2 c^2 = \tfrac{1}{12}(1.005 \text{ kg})(0.16 \text{ m})^2 = 2.144 \times 10^{-3} \text{ kg·m}^2$$

$$I_z = \tfrac{1}{3}m_2 b^2 = \tfrac{1}{3}(1.005 \text{ kg})(0.2 \text{ m})^2 = 13.400 \times 10^{-3} \text{ kg·m}^2$$

$$I_y = I_x + I_z = (2.144 + 13.400)(10^{-3}) = 15.544 \times 10^{-3} \text{ kg·m}^2$$

**Circular Plate.**

$$I_x = \tfrac{1}{4}m_3 a^2 = \tfrac{1}{4}(0.2466 \text{ kg})(0.05 \text{ m})^2 = 0.154 \times 10^{-3} \text{ kg·m}^2$$

$$I_y = \tfrac{1}{2}m_3 a^2 + m_3 d^2$$
$$= \tfrac{1}{2}(0.2466 \text{ kg})(0.05 \text{ m})^2 + (0.2466 \text{ kg})(0.1 \text{ m})^2 = 2.774 \times 10^{-3} \text{ kg·m}^2$$

$$I_z = \tfrac{1}{4}m_3 a^2 + m_3 d^2 = \tfrac{1}{4}(0.2466 \text{ kg})(0.05 \text{ m})^2 + (0.2466 \text{ kg})(0.1 \text{ m})^2$$
$$= 2.620 \times 10^{-3} \text{ kg·m}^2$$

**Entire Machine Part.**

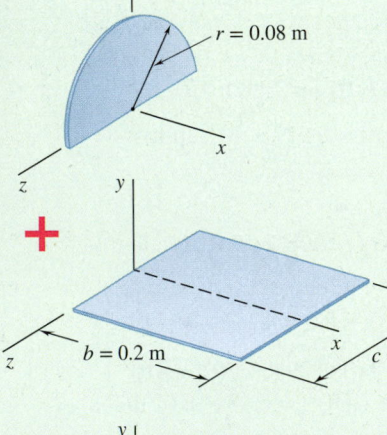

**Fig. 1**   Modeling the machine part as a combination of simple geometric shapes.

$$I_x = (1.010 + 2.144 - 0.154)(10^{-3}) \text{ kg·m}^2 \qquad \boxed{I_x = 3.00 \times 10^{-3} \text{ kg·m}^2} \blacktriangleleft$$

$$I_y = (0.505 + 15.544 - 2.774)(10^{-3}) \text{ kg·m}^2 \qquad \boxed{I_y = 13.28 \times 10^{-3} \text{ kg·m}^2} \blacktriangleleft$$

$$I_z = (0.505 + 13.400 - 2.620)(10^{-3}) \text{ kg·m}^2 \qquad \boxed{I_z = 11.29 \times 10^{-3} \text{ kg·m}^2} \blacktriangleleft$$

# SOLVING PROBLEMS ON YOUR OWN

In this section, we introduced the **mass moment of inertia** and the **radius of gyration** of a three-dimensional body with respect to a given axis [Eqs. (9.28) and (9.29)]. We also derived a **parallel-axis theorem** for use with mass moments of inertia and discussed the computation of the mass moments of inertia of thin plates and three-dimensional bodies.

**1. Computing mass moments of inertia.** You can calculate the mass moment of inertia $I$ of a body with respect to a given axis directly from the definition given in Eq. (9.28) for simple shapes (Sample Prob. 9.9). In most cases, however, it is necessary to divide the body into thin slabs, compute the moment of inertia of a typical slab with respect to the given axis—using the parallel-axis theorem if necessary—and integrate the resulting expression.

**2. Applying the parallel-axis theorem.** In Sec. 9.5B, we derived the parallel-axis theorem for mass moments of inertia as

$$I = \bar{I} + md^2 \tag{9.33}$$

This theorem states that the moment of inertia $I$ of a body of mass $m$ with respect to a given axis is equal to the sum of the moment of inertia $I$ of that body with respect to a parallel centroidal axis and the product $md^2$, where $d$ is the distance between the two axes. When you calculate the moment of inertia of a three-dimensional body with respect to one of the coordinate axes, you can replace $d^2$ by the sum of the squares of distances measured along the other two coordinate axes [Eqs. (9.32) and (9.32′)].

**3. Avoiding unit-related errors.** To avoid errors, you must be consistent in your use of units. Thus, all lengths should be expressed in meters or feet, as appropriate, and for problems using U.S. customary units, masses should be given in lb·s²/ft. In addition, we strongly recommend that you include units as you perform your calculations (Sample Probs. 9.12 and 9.13).

**4. Calculating the mass moment of inertia of thin plates.** We showed in Sec. 9.5C that you can obtain the mass moment of inertia of a thin plate with respect to a given axis by multiplying the corresponding moment of inertia of the area of the plate by the density $\rho$ and the thickness $t$ of the plate [Eqs. (9.35), (9.36), and (9.37)]. Note that, because the axis $CC'$ in Fig. 9.24c is perpendicular to the plate, $I_{CC',\text{mass}}$ is associated with the *polar* moment of inertia $J_{C,\text{area}}$.

Instead of calculating the moment of inertia of a thin plate with respect to a specified axis directly, you may sometimes find it convenient to first compute its moment of inertia with respect to an axis parallel to the specified axis and to then apply the parallel-axis theorem. Furthermore, to determine the moment of inertia of a thin plate with respect to an axis perpendicular to the plate, you may wish to first determine its moments of inertia with respect to two perpendicular in-plane axes and to then use Eq. (9.38). Finally, remember that the mass of a plate consists of area $A$, thickness $t$, and density $\rho$, because $m = \rho t A$.

*(continued)*

**5. Determining the moment of inertia of a body by direct single integration.** We discussed in Sec. 9.5D and illustrated in Sample Probs. 9.10 and 9.11 how you can use single integration to compute the moment of inertia of a body that can be divided into a series of thin, parallel slabs. For such cases, you will often need to express the mass of the body in terms of the body's density and dimensions. Assuming that the body has been divided, as in the sample problems, into thin slabs perpendicular to the *x* axis, you will need to express the dimensions of each slab as functions of the variable *x*.

  **a. In the special case of a body of revolution,** the elemental slab is a thin disk, and you can use the equations given in Fig. 9.27 to determine the moments of inertia of the body (Sample Prob. 9.11).

  **b. In the general case, when the body is not a solid of revolution,** the differential element is not a disk but a thin slab of a different shape. You cannot use the equations of Fig. 9.27 in this case. See, for example, Sample Prob. 9.10, where the element was a thin, rectangular slab. For more complex configurations, you may want to use one or more of the following equations, which are based on Eqs. (9.32) and (9.32′) of Sec. 9.5B.

$$dI_x = dI_{x'} + (\bar{y}_{el}^2 + \bar{z}_{el}^2)dm$$

$$dI_y = dI_{y'} + (\bar{z}_{el}^2 + \bar{x}_{el}^2)dm$$

$$dI_z = dI_{z'} + (\bar{x}_{el}^2 + \bar{y}_{el}^2)dm$$

Here, the primes denote the centroidal axes of each elemental slab and $\bar{x}_{el}$, $\bar{y}_{el}$, and $\bar{z}_{el}$ represent the coordinates of its centroid. Determine the centroidal moments of inertia of the slab in the manner described earlier for a thin plate: Refer to Fig. 9.12, calculate the corresponding moments of inertia of the area of the slab, and multiply the result by the density $\rho$ and the thickness $t$ of the slab. Also, assuming that the body has been divided into thin slabs perpendicular to the *x* axis, remember that you can obtain $dI_{x'}$ by adding $dI_{y'}$ and $dI_{z'}$ instead of computing it directly. Finally, using the geometry of the body, express the result obtained in terms of the single variable *x,* and integrate in *x*.

**6. Computing the moment of inertia of a composite body.** As stated in Sec. 9.5E, the moment of inertia of a composite body with respect to a specified axis is equal to the sum of the moments of its components with respect to that axis. Sample Probs. 9.12 and 9.13 illustrate the appropriate method of solution. Also remember that the moment of inertia of a component is negative only if the component is *removed* (as in the case of a hole).

Although the composite-body problems in this section are relatively straightforward, you will have to work carefully to avoid computational errors. In addition, if some of the moments of inertia that you need are not given in Fig. 9.28, you will have to derive your own formulas, using the techniques described in this section.

# Problems

**9.111** A thin plate with a mass $m$ is cut in the shape of an equilateral triangle of side $a$. Determine the mass moment of inertia of the plate with respect to ($a$) the centroidal axes $AA'$ and $BB'$, ($b$) the centroid axis $CC'$ that is perpendicular to the plate.

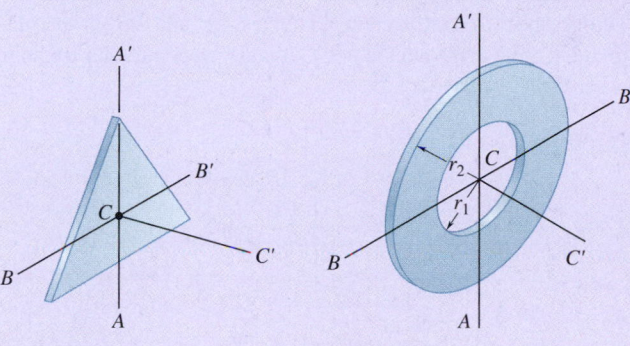

**Fig. P9.111**        **Fig. P9.112**

**9.112** A ring with a mass $m$ is cut from a thin uniform plate. Determine the mass moment of inertia of the ring with respect to ($a$) the axis $AA'$, ($b$) the centroidal axis $CC'$ that is perpendicular to the plane of the ring.

**9.113** A thin elliptical plate has a mass $m$. Determine the mass moment of inertia of the plate with respect to ($a$) the centroidal axes $AA'$ and $BB'$, ($b$) the centroidal axis $CC'$ that is perpendicular to the plate.

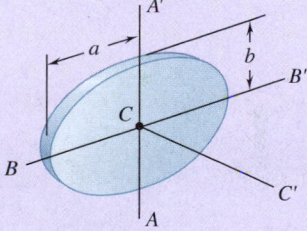

**Fig. P9.113**

**9.114** The parabolic spandrel shown was cut from a thin, uniform plate. Denoting the mass of the spandrel by $m$, determine its mass moment of inertia with respect to ($a$) the axis $BB'$, ($b$) the axis $DD'$ that is perpendicular to the spandrel. (*Hint:* See Sample Prob. 9.3.)

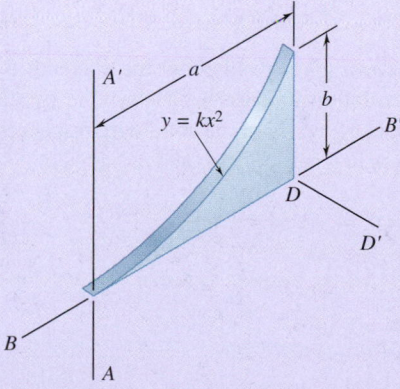

**Fig. P9.114**

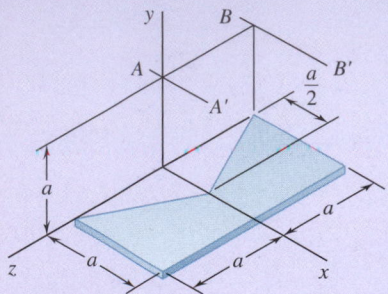

**Fig. P9.115 and P9.116**

**9.115** A piece of thin, uniform sheet metal is cut to form the machine component shown. Denoting the mass of the component by $m$, determine its mass moment of inertia with respect to (*a*) the $x$ axis, (*b*) the $y$ axis.

**9.116** A piece of thin, uniform sheet metal is cut to form the machine component shown. Denoting the mass of the component by $m$, determine its mass moment of inertia with respect to (*a*) the axis $AA'$, (*b*) the axis $BB'$, where the $AA'$ and $BB'$ axes are parallel to the $x$ axis and lie in a plane parallel to and at a distance $a$ above the $xz$ plane.

**9.117** A thin plate of mass $m$ is cut in the shape of an isosceles triangle of base $b$ and height $h$. Determine the mass moment of inertia of the plate with respect to (*a*) the centroidal axes $AA'$ and $BB'$ in the plane of the plate, (*b*) the centroidal axis $CC'$ that is perpendicular to the plate.

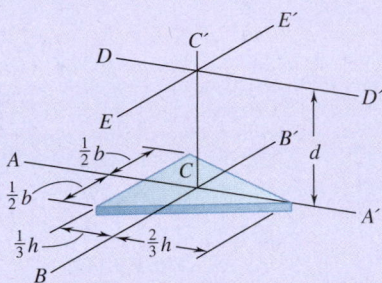

**Fig. P9.117 and P9.118**

**9.118** A thin plate of mass $m$ is cut in the shape of an isosceles triangle of base $b$ and height $h$. Determine the mass moments of inertia of the plate with respect to axes $DD'$ and $EE'$ that are parallel to the centroidal axes $AA'$ and $BB'$ and located at a distance $d$ from the plane of the plate.

**9.119** Determine by direct integration the mass moment of inertia with respect to the $y$ axis of the right circular cylinder shown, assuming that it has a uniform density and a mass $m$.

**9.120** The area shown is revolved about the $x$ axis to form a homogeneous solid of revolution of mass $m$. Using direct integration, express the mass moment of inertia of the solid with respect to the $x$ axis in terms of $m$ and $h$.

**9.121** The area shown is revolved about the $x$ axis to form a homogeneous solid of revolution of mass $m$. Determine by direct integration the mass moment of inertia of the solid with respect to the $x$ axis. Express your answers in terms of $m$, $a$, and $n$.

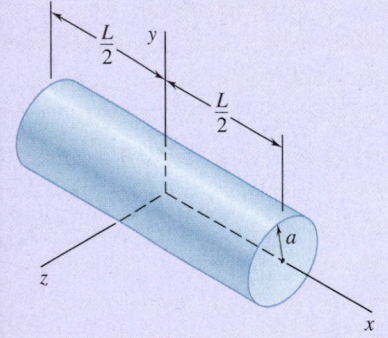

**Fig. P9.119**

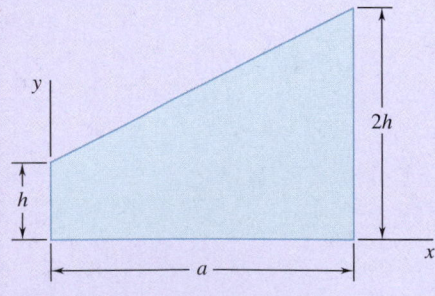

**Fig. P9.120**

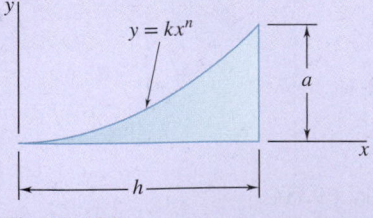

**Fig. P9.121**

**9.122** Determine by direct integration the mass moment of inertia with respect to the $x$ axis of the pyramid shown, assuming that it has a uniform density and a mass $m$.

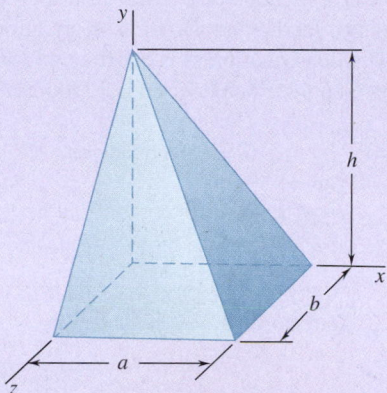

**Fig. P9.122 and *P9.123***

**9.123** Determine by direct integration the mass moment of inertia with respect to the $y$ axis of the pyramid shown, assuming that it has a uniform density and a mass $m$.

**9.124** Determine by direct integration the mass moment of inertia and the radius of gyration with respect to the $x$ axis of the paraboloid shown, assuming that it has a uniform density and a mass $m$.

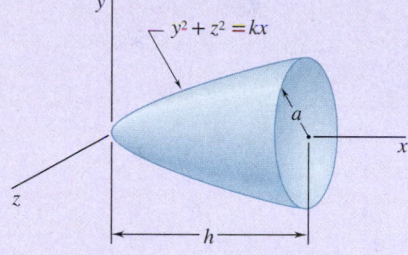

**Fig. P9.124**

**9.125** A thin triangular plate of mass $m$ is welded along its base $AB$ to a block as shown. Knowing that the plate forms an angle $\theta$ with the $y$ axis, determine by direct integration the mass moment of inertia of the plate with respect to (a) the $x$ axis, (b) the $y$ axis, (c) the $z$ axis.

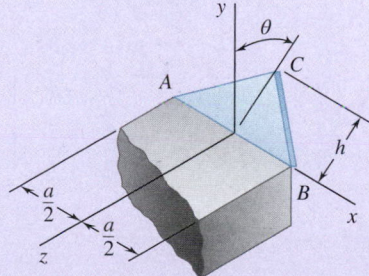

**Fig. *P9.125***

**\*9.126** A thin steel wire is bent into the shape shown. Denoting the mass per unit length of the wire by $m'$, determine by direct integration the mass moment of inertia of the wire with respect to each of the coordinate axes.

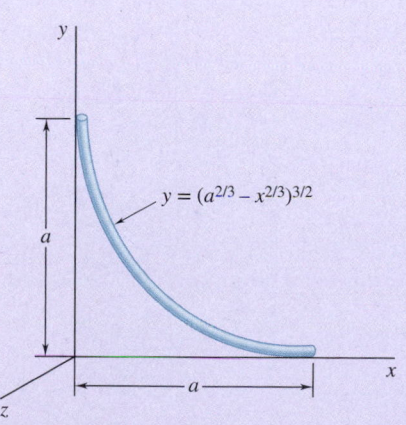

**Fig. P9.126**

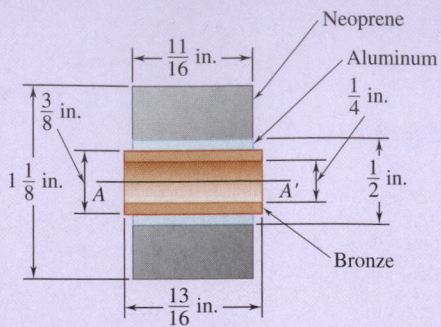

**Fig. P9.127**

**9.127** Shown is the cross section of an idler roller. Determine its mass moment of inertia and its radius of gyration with respect to the axis $AA'$. (The specific weight of bronze is 0.310 lb/in³; of aluminum, 0.100 lb/in³; and of neoprene, 0.0452 lb/in³.)

**9.128** Shown is the cross section of a molded flat-belt pulley. Determine its mass moment of inertia and its radius of gyration with respect to the axis $AA'$. (The density of brass is 8650 kg/m³, and the density of the fiber-reinforced polycarbonate used is 1250 kg/m³.)

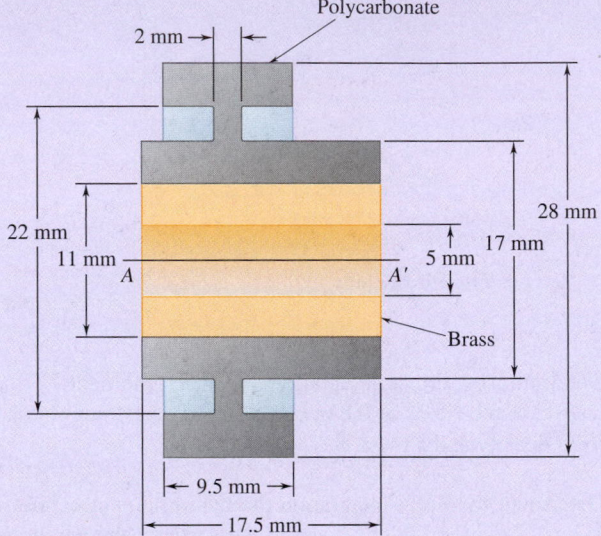

**Fig. P9.128**

**9.129** The machine part shown is formed by machining a conical surface into a circular cylinder. For $b = \frac{1}{2}h$, determine the mass moment of inertia and the radius of gyration of the machine part with respect to the $y$ axis.

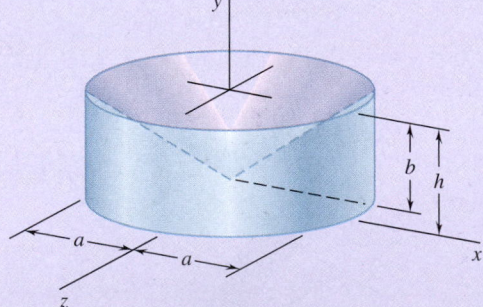

**Fig. *P9.129***

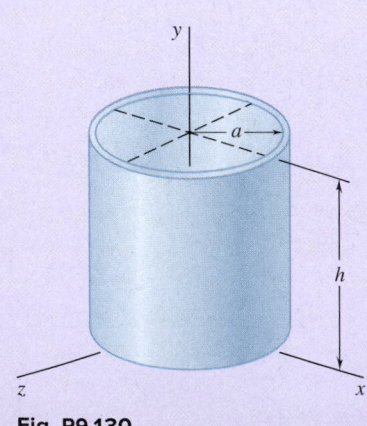

**Fig. P9.130**

**9.130** Knowing that the thin cylindrical shell shown has a mass $m$, thickness $t$, and height $h$, determine the mass moment of inertia of the shell with respect to the $x$ axis. (*Hint:* Consider the shell as formed by removing a cylinder of radius $a$ and height $h$ from a cylinder of radius $a + t$ and height $h$; then neglect the terms containing $t^2$ and $t^3$ and keep those terms containing $t$.)

**9.131** A circular hole of radius $r$ is to be drilled through the center of a rectangular steel plate to form the machine component shown. Denoting the density of steel by $\rho$, determine (a) the mass moment of inertia of the component with respect to the axis $BB'$, (b) the value of $r$ for which, given $a$ and $h$, $I_{BB'}$ is minimum, (c) the corresponding value of $I_{BB'}$ and radius of gyration $k_{BB'}$.

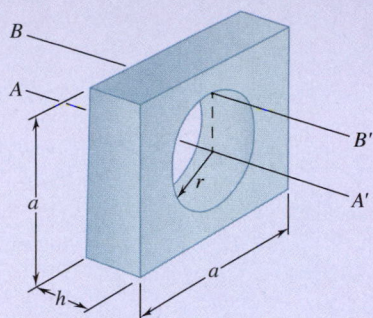

**Fig. P9.131**

**9.132** The cups and the arms of an anemometer are fabricated from a material of density $\rho$. Knowing that the mass moment of inertia of a thin, hemispherical shell with a mass $m$ and thickness $t$ with respect to its centroidal axis $GG'$ is $5ma^2/12$, determine (a) the mass moment of inertia of the anemometer with respect to the axis $AA'$, (b) the ratio of $a$ to $l$ for which the centroidal moment of inertia of the cups is equal to 1 percent of the moment of inertia of the cups with respect to the axis $AA'$.

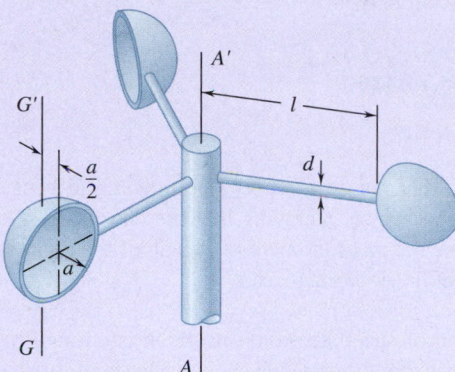

**Fig. P9.132**

**9.133** After a period of use, one of the blades of a shredder has been worn to the shape shown and is of mass 0.18 kg. Knowing that the mass moments of inertia of the blade with respect to the $AA'$ and $BB'$ axes are 0.320 g·m² and 0.680 g·m², respectively, determine (a) the location of the centroidal axis $GG'$, (b) the radius of gyration with respect to axis $GG'$.

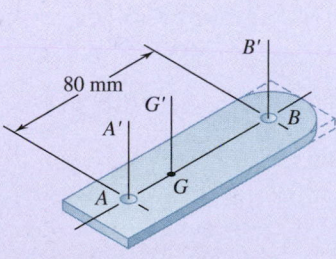

**Fig. P9.133**

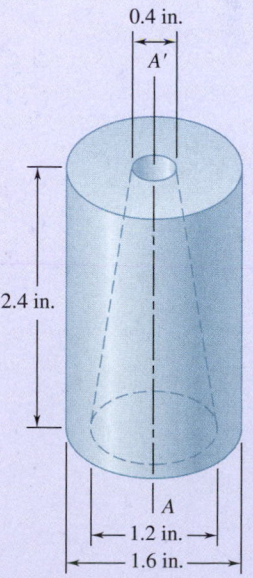

**Fig. P9.134**

**9.134** Determine the mass moment of inertia of the 0.9-lb machine component shown with respect to the axis $AA'$.

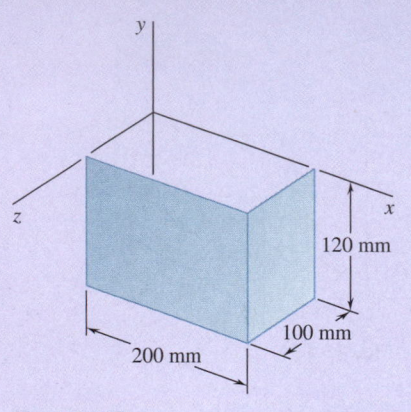

**Fig. P9.135**

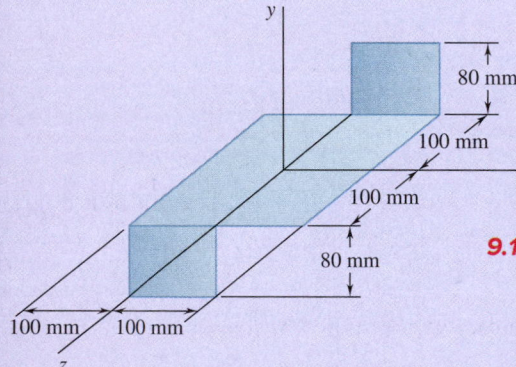

**Fig. *P9.137***

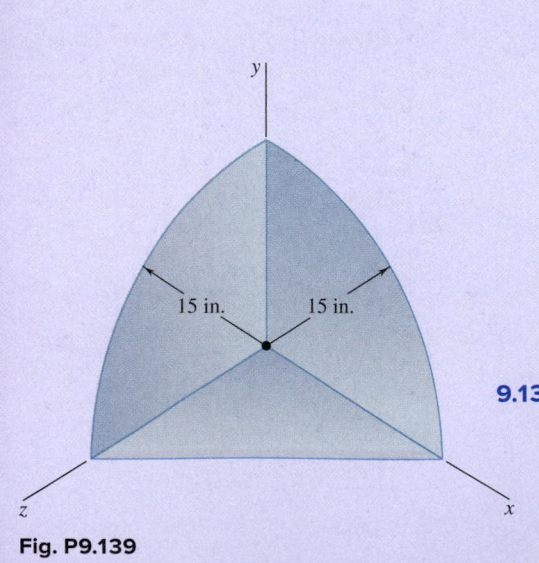

**Fig. P9.139**

**9.135 and 9.136** A 2-mm thick piece of sheet steel is cut and bent into the machine component shown. Knowing that the density of steel is 7850 kg/m³, determine the mass moment of inertia of the component with respect to each of the coordinate axes.

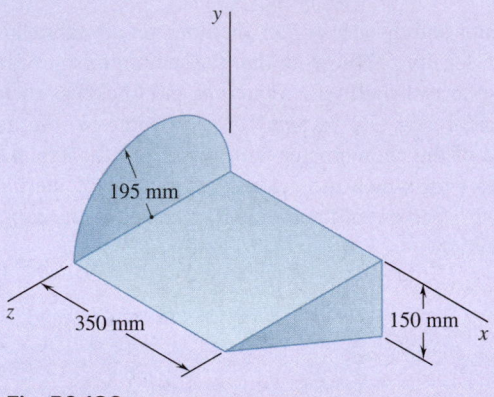

**Fig. P9.136**

**9.137** A 2-mm thick piece of sheet steel is cut and bent into the machine component shown. Knowing that the density of steel is 7850 kg/m³, determine the mass moment of inertia of the component with respect to each of the coordinate axes.

**9.138** A section of sheet steel 0.03 in. thick is cut and bent into the sheet metal machine component shown. Determine the mass moment of inertia of the component with respect to each of the coordinate axes. (The specific weight of steel is 490 lb/ft³.)

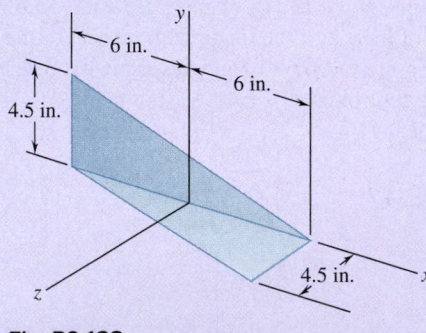

**Fig. P9.138**

**9.139** A corner reflector for tracking by radar has two sides in the shape of a quarter circle of radius 15 in. and one side in the shape of a triangle. Each part of the reflector is formed from aluminum plate of uniform 0.05-in. thickness. Knowing that the specific weight of the aluminum used is 170 lb/ft³, determine the mass moment of inertia of the reflector with respect to each of the coordinate axes.

**\*9.140** A farmer constructs a trough by welding a rectangular piece of 2-mm-thick sheet steel to half of a steel drum. Knowing that the density of steel is 7850 kg/m³ and that the thickness of the walls of the drum is 1.8 mm, determine the mass moment of inertia of the trough with respect to each of the coordinate axes. Neglect the mass of the welds.

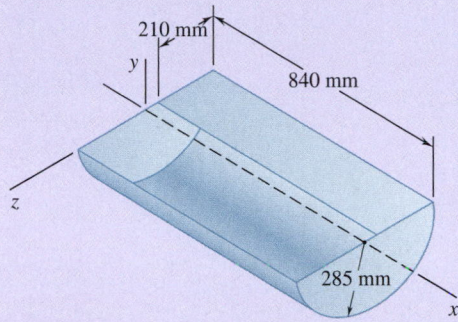

**Fig. P9.140**

**9.141** The machine element shown is fabricated from steel. Determine the mass moment of inertia of the assembly with respect to (*a*) the *x* axis, (*b*) the *y* axis, (*c*) the *z* axis. (The density of steel is 7850 kg/m³.)

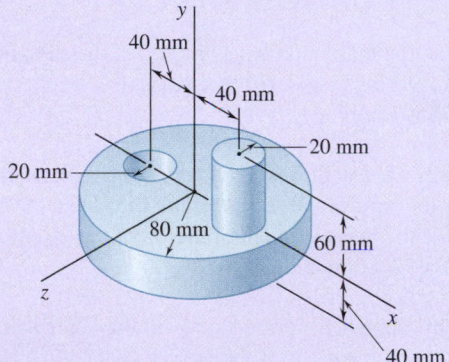

**Fig. P9.141**

**9.142** Determine the mass moments of inertia and the radii of gyration of the steel machine element shown with respect to the *x* and *y* axes. (The density of steel is 7850 kg/m³.)

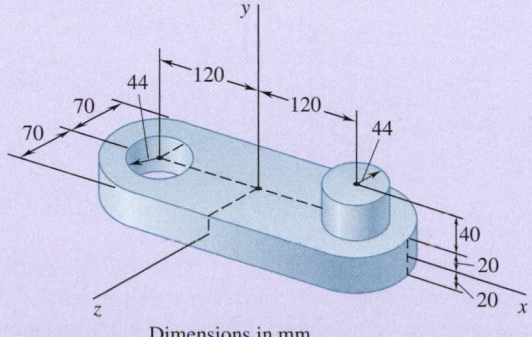

Dimensions in mm

**Fig. P9.142**

**9.143** Determine the mass moment of inertia of the steel machine element shown with respect to the *x* axis. (The specific weight of steel is 490 lb/ft$^3$.)

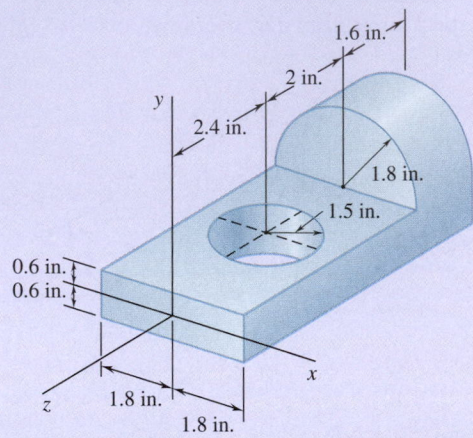

**Fig. P9.143 and P9.144**

**9.144** Determine the mass moment of inertia of the steel machine element shown with respect to the *y* axis. (The specific weight of steel is 490 lb/ft$^3$.)

**9.145** Determine the mass moment of inertia of the steel fixture shown with respect to (*a*) the *x* axis, (*b*) the *y* axis, (*c*) the *z* axis. (The density of steel is 7850 kg/m$^3$.)

**9.146** Aluminum wire with a weight per unit length of 0.033 lb/ft is used to form the circle and the straight members of the figure shown. Determine the mass moment of inertia of the assembly with respect to each of the coordinate axes.

**9.147** The figure shown is formed of $\frac{1}{8}$-in.-diameter steel wire. Knowing that the specific weight of the steel is 490 lb/ft$^3$, determine the mass moment of inertia of the wire with respect to each of the coordinate axes.

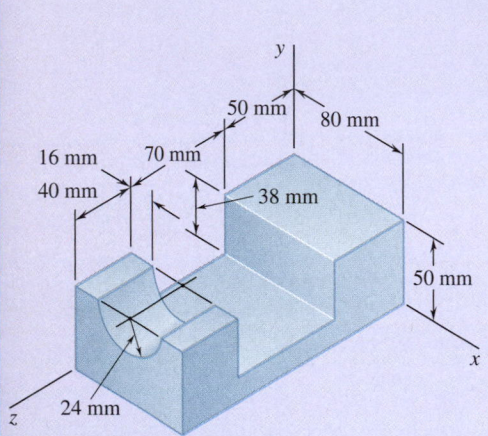

**Fig. P9.145**

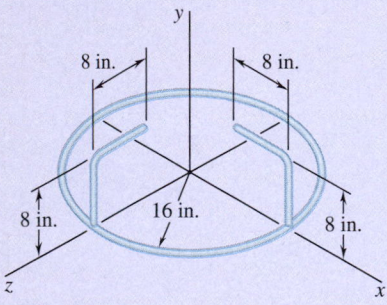

**Fig. P9.146**

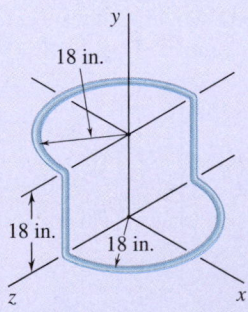

**Fig. P9.147**

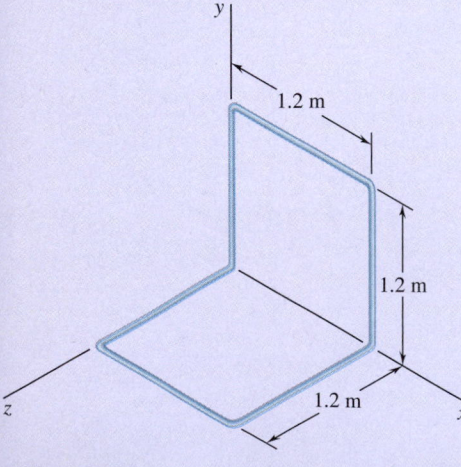

**Fig. P9.148**

**9.148** A homogeneous wire with a mass per unit length of 0.056 kg/m is used to form the figure shown. Determine the mass moment of inertia of the wire with respect to each of the coordinate axes.

# *9.6 ADDITIONAL CONCEPTS OF MASS MOMENTS OF INERTIA

In this final section of the chapter, we present several concepts involving mass moments of inertia that are analogous to material presented in Sec. 9.4 involving moments of inertia of areas. These ideas include mass products of inertia, principal axes of inertia, and principal moments of inertia for masses, which are necessary for the study of the dynamics of rigid bodies in three dimensions.

## 9.6A Mass Products of Inertia

In this section, you will see how to determine the moment of inertia of a body with respect to an arbitrary axis $OL$ through the origin (Fig. 9.29) if its moments of inertia with respect to the three coordinate axes, as well as certain other quantities defined here, have already been determined.

The moment of inertia $I_{OL}$ of the body with respect to $OL$ is equal to $\int p^2 dm$, where $p$ denotes the perpendicular distance from the element of mass $dm$ to the axis $OL$. If we denote the unit vector along $OL$ by $\boldsymbol{\lambda}$ and the position vector of the element $dm$ by $\mathbf{r}$, the perpendicular distance $p$ is equal to $r \sin \theta$, which is the magnitude of the vector product $\boldsymbol{\lambda} \times \mathbf{r}$. We therefore have

$$I_{OL} = \int p^2 dm = \int |\boldsymbol{\lambda} \times \mathbf{r}|^2 dm \qquad \textbf{(9.43)}$$

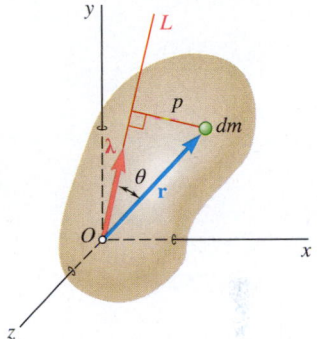

**Fig. 9.29** An element of mass $dm$ of a body and its perpendicular distance to an arbitrary axis $OL$ through the origin.

Expressing $|\boldsymbol{\lambda} \times \mathbf{r}|^2$ in terms of the rectangular components of the vector product, we have

$$I_{OL} = \int \left[ (\lambda_x y - \lambda_y x)^2 + (\lambda_y z - \lambda_z y)^2 + (\lambda_z x - \lambda_x z)^2 \right] dm$$

Here, the components $\lambda_x$, $\lambda_y$, $\lambda_z$ of the unit vector $\boldsymbol{\lambda}$ represent the direction cosines of the axis $OL$, and the components $x$, $y$, $z$ of $\mathbf{r}$ represent the coordinates of the element of mass $dm$. Expanding the squares and rearranging the terms, we obtain

$$I_{OL} = \lambda_x^2 \int (y^2 + z^2) dm + \lambda_y^2 \int (z^2 + x^2) dm + \lambda_z^2 \int (x^2 + y^2) dm$$
$$- 2\lambda_x \lambda_y \int xy\, dm - 2\lambda_y \lambda_z \int yz\, dm - 2\lambda_z \lambda_x \int zx\, dm \qquad \textbf{(9.44)}$$

Referring to Eqs. (9.30), note that the first three integrals in Eq. (9.44) represent, respectively, the moments of inertia $I_x$, $I_y$, and $I_z$ of the body with respect to the coordinate axes. The last three integrals in Eq. (9.44), which involve products of coordinates, are called the **products of inertia** of the body with respect to the $x$ and $y$ axes, the $y$ and $z$ axes, and the $z$ and $x$ axes, respectively.

**Mass products of inertia**

$$I_{xy} = \int xy\, dm \qquad I_{yz} = \int yz\, dm \qquad I_{zx} = \int zx\, dm \qquad \textbf{(9.45)}$$

Rewriting Eq. (9.44) in terms of the integrals defined in Eqs. (9.30) and (9.45), we have

$$I_{OL} = I_x \lambda_x^2 + I_y \lambda_y^2 + I_z \lambda_z^2 - 2I_{xy} \lambda_x \lambda_y - 2I_{yz} \lambda_y \lambda_z - 2I_{zx} \lambda_z \lambda_x \qquad \textbf{(9.46)}$$

The definition of the products of inertia of a mass given in Eqs. (9.45) is an extension of the definition of the product of inertia of an area (Sec. 9.3).

Mass products of inertia reduce to zero under the same conditions of symmetry as do products of inertia of areas, and the parallel-axis theorem for mass products of inertia is expressed by relations similar to the formula derived for the product of inertia of an area. Substituting the expressions for $x$, $y$, and $z$ given in Eqs. (9.31) into Eqs. (9.45), we find that

**Parallel-axis theorem for mass products of inertia**

$$
\begin{aligned}
I_{xy} &= \bar{I}_{x'y'} + m\bar{x}\,\bar{y} \\
I_{yz} &= \bar{I}_{y'z'} + m\bar{y}\,\bar{z} \\
I_{zx} &= \bar{I}_{z'x'} + m\bar{z}\,\bar{x}
\end{aligned}
\qquad\qquad (9.47)
$$

Here, $\bar{x}$, $\bar{y}$, $\bar{z}$ are the coordinates of the center of gravity $G$ of the body and $\bar{I}_{x'y'}$, $\bar{I}_{y'z'}$, $\bar{I}_{z'x'}$ denote the products of inertia of the body with respect to the centroidal axes $x'$, $y'$, and $z'$ (see Fig. 9.22).

## 9.6B   Principal Axes and Principal Moments of Inertia

Let us assume that we have determined the moment of inertia of the body considered in the preceding section with respect to a large number of axes $OL$ through the fixed point $O$. Suppose that we plot a point $Q$ on each axis $OL$ at a distance $OQ = 1/\sqrt{I_{OL}}$ from $O$. The locus of the points $Q$ forms a surface (Fig. 9.30). We can obtain the equation of that surface by substituting $1/(OQ)^2$ for $I_{OL}$ in Eq. (9.46) and then multiplying both sides of the equation by $(OQ)^2$. Observing that

$$
(OQ)\lambda_x = x \qquad (OQ)\lambda_y = y \qquad (OQ)\lambda_z = z
$$

where $x$, $y$, $z$ denote the rectangular coordinates of $Q$, we have

$$
I_x x^2 + I_y y^2 + I_z z^2 - 2I_{xy}xy - 2I_{yz}yz - 2I_{zx}zx = 1 \qquad (9.48)
$$

This is the equation of a *quadric surface*. Because the moment of inertia $I_{OL}$ is different from zero for every axis $OL$, no point $Q$ can be at an infinite distance from $O$. Thus, the quadric surface obtained is an *ellipsoid*. This ellipsoid, which defines the moment of inertia of the body with respect to any axis through $O$, is known as the **ellipsoid of inertia** of the body at $O$.

Observe that, if we rotate the axes in Fig. 9.30, the coefficients of the equation defining the ellipsoid change, because they are equal to the moments and products of inertia of the body with respect to the rotated coordinate axes. However, the *ellipsoid itself remains unaffected*, because its shape depends only upon the distribution of mass in the given body. Suppose that we choose as coordinate axes the principal axes $x'$, $y'$, and $z'$ of the ellipsoid of inertia (Fig. 9.31). The equation of the ellipsoid with respect to these coordinate axes is known to be of the form

$$
I_{x'}x'^2 + I_{y'}y'^2 + I_{z'}z'^2 = 1 \qquad (9.49)
$$

which does not contain any products of the coordinates. Comparing Eqs. (9.48) and (9.49), we observe that the products of inertia of the body with respect to the $x'$, $y'$, and $z'$ axes must be zero. The $x'$, $y'$, and $z'$ axes are known as the **principal axes of inertia** of the body at $O$, and the coefficients $I_{x'}$, $I_{y'}$, and $I_{z'}$ are referred to as the **principal moments of inertia** of the body at $O$. Note that, given a body of arbitrary shape and a point $O$, it is always possible to find

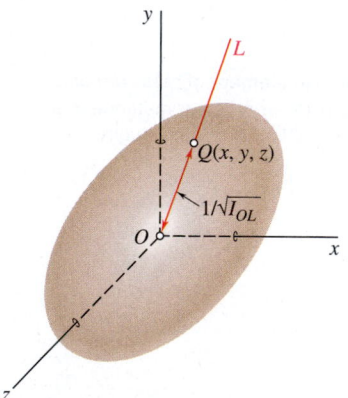

**Fig. 9.30**   The ellipsoid of inertia defines the moment of inertia of a body with respect to any axis through $O$.

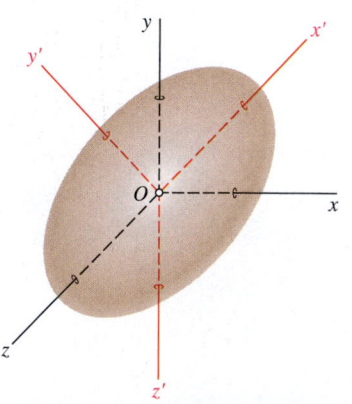

**Fig. 9.31**   Principal axes of inertia $x'$, $y'$, $z'$ of the body at $O$.

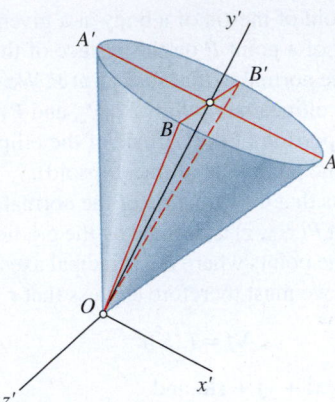

**Fig. 9.32**   A homogeneous cone with elliptical base has two mutually perpendicular planes of symmetry.

principal axes of inertia of the body at $O$; that is, axes with respect to which the products of inertia of the body are zero. Indeed, whatever the shape of the body, the moments and products of inertia of the body with respect to the $x$, $y$, and $z$ axes through $O$ define an ellipsoid, and this ellipsoid has principal axes that, by definition, are the principal axes of inertia of the body at $O$.

If the principal axes of inertia $x'$, $y'$, and $z'$ are used as coordinate axes, the expression in Eq. (9.46) for the moment of inertia of a body with respect to an arbitrary axis through $O$ reduces to

$$I_{OL} = I_{x'}\lambda_{x'}^2 + I_{y'}\lambda_{y'}^2 + I_{z'}\lambda_{z'}^2 \qquad \textbf{(9.50)}$$

The determination of the principal axes of inertia of a body of arbitrary shape is somewhat involved and is discussed in the next section. In many cases, however, these axes can be spotted immediately. Consider, for instance, the homogeneous cone of elliptical base shown in Fig. 9.32; this cone possesses two mutually perpendicular planes of symmetry $OAA'$ and $OBB'$. From the definition of Eq. (9.45), we observe that if we choose the $x'y'$ and $y'z'$ planes to coincide with the two planes of symmetry, all of the products of inertia are zero. The $x'$, $y'$, and $z'$ axes selected in this way are therefore the principal axes of inertia of the cone at $O$. In the case of the homogeneous regular tetrahedron $OABC$ shown in Fig. 9.33, the line joining the corner $O$ to the center $D$ of the opposite face is a principal axis of inertia at $O$, and any line through $O$ perpendicular to $OD$ is also a principal axis of inertia at $O$. This property is apparent if we observe that rotating the tetrahedron through 120° about $OD$ leaves its shape and mass distribution unchanged. It follows that the ellipsoid of inertia at $O$ also remains unchanged under this rotation. The ellipsoid, therefore, is a body of revolution whose axis of revolution is $OD$, and the line $OD$, as well as any perpendicular line through $O$, must be a principal axis of the ellipsoid.

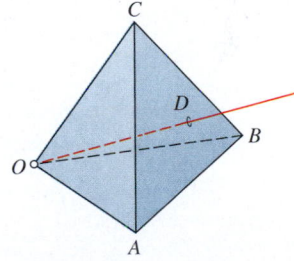

**Fig. 9.33**   A line drawn from a corner to the center of the opposite face of a homogeneous regular tetrahedron is a principal axis, because each 120° rotation of the body about this axis leaves its shape and mass distribution unchanged.

## 9.6C   Principal Axes and Moments of Inertia for a Body of Arbitrary Shape

The method of analysis described in this section extends the analysis in the preceding section. However, generally speaking, you should use it only when the body under consideration has no obvious property of symmetry.

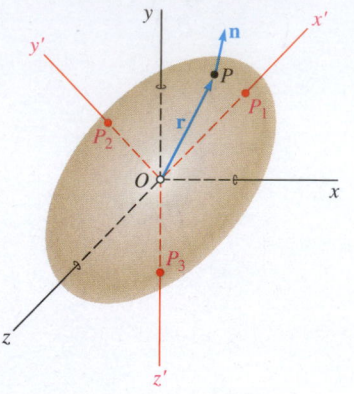

**Fig. 9.34** The principal axes intersect an ellipsoid of inertia at points where the radius vectors are collinear with the unit normal vectors to the surface.

Consider the ellipsoid of inertia of a body at a given point $O$ (Fig. 9.34). Let $\mathbf{r}$ be the radius vector of a point $P$ on the surface of the ellipsoid, and let $\mathbf{n}$ be the unit vector along the normal to that surface at $P$. We observe that the only points where $\mathbf{r}$ and $\mathbf{n}$ are collinear are points $P_1$, $P_2$, and $P_3$, where the principal axes intersect the visible portion of the surface of the ellipsoid (along with the corresponding points on the other side of the ellipsoid).

Recall from calculus that the direction of the normal to a surface of equation $f(x, y, z) = 0$ at a point $P(x, y, z)$ is defined by the gradient $\nabla f$ of the function $f$ at that point. To obtain the points where the principal axes intersect the surface of the ellipsoid of inertia, we must therefore express that $\mathbf{r}$ and $\nabla f$ are collinear,

$$\nabla f = (2K)\mathbf{r} \tag{9.51}$$

where $K$ is a constant, $\mathbf{r} = x\mathbf{i} + y\mathbf{j} + z\mathbf{k}$, and

$$\nabla f = \frac{\partial f}{\partial x}\mathbf{i} + \frac{\partial f}{\partial y}\mathbf{j} + \frac{\partial f}{\partial z}\mathbf{k}$$

Recalling Eq. (9.48), we note that the function $f(x, y, z)$ corresponding to the ellipsoid of inertia is

$$f(x, y, z) = I_x x^2 + I_y y^2 + I_z z^2 - 2I_{xy} xy - 2I_{yz} yz - 2I_{zx} zx - 1$$

Substituting for $\mathbf{r}$ and $\nabla f$ into Eq. (9.51) and equating the coefficients of the unit vectors, we obtain

$$\begin{aligned}
I_x x - I_{xy} y - I_{zx} z &= Kx \\
-I_{xy} x + I_y y - I_{yz} z &= Ky \\
-I_{zx} x - I_{yz} y + I_z z &= Kz
\end{aligned} \tag{9.52}$$

Dividing each term by the distance $r$ from $O$ to $P$, we obtain similar equations involving the direction cosines $\lambda_x$, $\lambda_y$, and $\lambda_z$:

$$\begin{aligned}
I_x \lambda_x - I_{xy} \lambda_y - I_{zx} \lambda_z &= K\lambda_x \\
-I_{xy} \lambda_x + I_y \lambda_y - I_{yz} \lambda_z &= K\lambda_y \\
-I_{zx} \lambda_x - I_{yz} \lambda_y + I_z \lambda_z &= K\lambda_z
\end{aligned} \tag{9.53}$$

Transposing the right-hand members leads to the homogeneous linear equations, as

$$\begin{aligned}
(I_x - K)\lambda_x - I_{xy} \lambda_y - I_{zx} \lambda_z &= 0 \\
-I_{xy} \lambda_x + (I_y - K)\lambda_y - I_{yz} \lambda_z &= 0 \\
-I_{zx} \lambda_x - I_{yz} \lambda_y + (I_z - K)\lambda_z &= 0
\end{aligned} \tag{9.54}$$

For this system of equations to have a solution different from $\lambda_x = \lambda_y = \lambda_z = 0$, its discriminant must be zero. Thus,

$$\begin{vmatrix}
I_x - K & -I_{xy} & -I_{zx} \\
-I_{xy} & I_y - K & -I_{yz} \\
-I_{zx} & -I_{yz} & I_z - K
\end{vmatrix} = 0 \tag{9.55}$$

Expanding this determinant and changing signs, we have

$$\begin{aligned}
K^3 - (I_x + I_y + I_z)K^2 + (I_x I_y + I_y I_z + I_z I_x - I_{xy}^2 - I_{yz}^2 - I_{zx}^2)K \\
- (I_x I_y I_z - I_x I_{yz}^2 - I_y I_{zx}^2 - I_z I_{xy}^2 - 2I_{xy} I_{yz} I_{zx}) = 0
\end{aligned} \tag{9.56}$$

This is a cubic equation in $K$, which yields three real, positive roots: $K_1$, $K_2$, and $K_3$.

To obtain the direction cosines of the principal axis corresponding to the root $K_1$, we substitute $K_1$ for $K$ in Eqs. (9.54). Because these equations are now linearly dependent, only two of them may be used to determine $\lambda_x$, $\lambda_y$, and $\lambda_z$. We can obtain an additional equation, however, by recalling from Sec. 2.4A that the direction cosines must satisfy the relation

$$\lambda_x^2 + \lambda_y^2 + \lambda_z^2 = 1 \tag{9.57}$$

Repeating this procedure with $K_2$ and $K_3$, we obtain the direction cosines of the other two principal axes.

We now show that *the roots $K_1$, $K_2$, and $K_3$ of* Eq. (9.56) *are the principal moments of inertia of the given body.* Let us substitute for $K$ in Eqs. (9.53) the root $K_1$, and for $\lambda_x$, $\lambda_y$, and $\lambda_z$ the corresponding values $(\lambda_x)_1$, $(\lambda_y)_1$, and $(\lambda_z)_1$ of the direction cosines; the three equations are satisfied. We now multiply by $(\lambda_x)_1$, $(\lambda_y)_1$, and $(\lambda_z)_1$, respectively, each term in the first, second, and third equation and add the equations obtained in this way. The result is

$$I_x^2(\lambda_x)_1^2 + I_y^2(\lambda_y)_1^2 + I_z^2(\lambda_z)_1^2 - 2I_{xy}(\lambda_x)_1(\lambda_y)_1$$
$$- 2I_{yz}(\lambda_y)_1(\lambda_z)_1 - 2I_{zx}(\lambda_z)_1(\lambda_x)_1 = K_1[(\lambda_x)_1^2 + (\lambda_y)_1^2 + (\lambda_z)_1^2]$$

Recalling Eq. (9.46), we observe that the left-hand side of this equation represents the moment of inertia of the body with respect to the principal axis corresponding to $K_1$; it is thus the principal moment of inertia corresponding to that root. On the other hand, recalling Eq. (9.57), we note that the right-hand member reduces to $K_1$. Thus, $K_1$ itself is the principal moment of inertia. In the same fashion, we can show that $K_2$ and $K_3$ are the other two principal moments of inertia of the body.

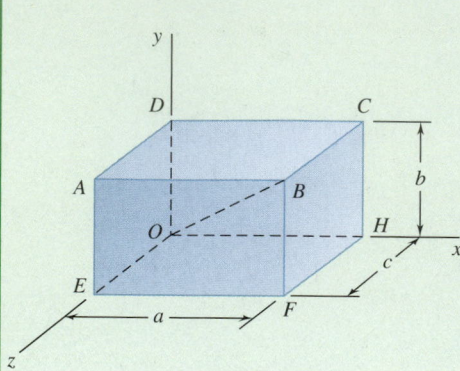

**Fig. 1** Centroidal axes for the rectangular prism.

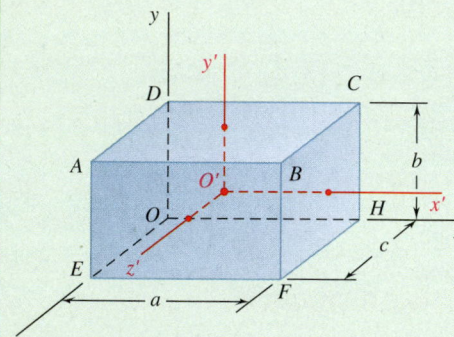

**Fig. 2** Direction angles for *OB*.

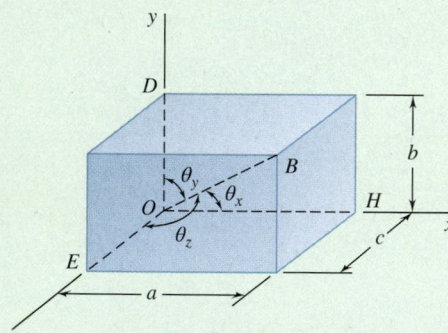

**Fig. 3** Line *OB* passes through the centroid *O'*.

## Sample Problem 9.14

Consider a rectangular prism with a mass of $m$ and sides $a$, $b$, and $c$. Determine (a) the moments and products of inertia of the prism with respect to the coordinate axes shown, (b) its moment of inertia with respect to the diagonal $OB$.

**STRATEGY:** For part (a), you can introduce centroidal axes and apply the parallel-axis theorem. For part (b), determine the direction cosines of line $OB$ from the given geometry and use either Eq. (9.46) or (9.50).

**MODELING and ANALYSIS:**

### a. Moments and Products of Inertia with Respect to the Coordinate Axes.

*Moments of Inertia.* Introduce the centroidal axes $x'$, $y'$, and $z'$ with respect to which the moments of inertia are given in Fig. 9.28, and then apply the parallel-axis theorem (Fig. 1). Thus,

$$I_x = \overline{I}_{x'} + m(\overline{y}^2 + \overline{z}^2) = \tfrac{1}{12}m(b^2 + c^2) + m(\tfrac{1}{4}b^2 + \tfrac{1}{4}c^2)$$

$$I_x = \tfrac{1}{3}m(b^2 + c^2) \quad ◄$$

Similarly,

$$I_y = \tfrac{1}{3}m(c^2 + a^2) \qquad I_z = \tfrac{1}{3}m(a^2 + b^2) \quad ◄$$

*Products of Inertia.* Because of symmetry, the products of inertia with respect to the centroidal axes $x'$, $y'$, and $z'$ are zero, and these axes are principal axes of inertia. Using the parallel-axis theorem, you have

$$I_{xy} = \overline{I}_{x'y'} + m\overline{x}\,\overline{y} = 0 + m(\tfrac{1}{2}a)(\tfrac{1}{2}b) \qquad I_{xy} = \tfrac{1}{4}mab \quad ◄$$

Similarly,

$$I_{yz} = \tfrac{1}{4}mbc \quad I_{zx} = \tfrac{1}{4}mca \quad ◄$$

### b. Moment of Inertia with Respect to *OB*. Recall Eq. (9.46):

$$I_{OB} = I_x\lambda_x^2 + I_y\lambda_y^2 + I_z\lambda_z^2 - 2I_{xy}\lambda_x\lambda_y - 2I_{yz}\lambda_y\lambda_z - 2I_{zx}\lambda_z\lambda_x$$

where the direction cosines of $OB$ are (Fig. 2)

$$\lambda_x = \cos\theta_x = \frac{OH}{OB} = \frac{a}{(a^2 + b^2 + c^2)^{1/2}}$$

$$\lambda_y = \frac{b}{(a^2 + b^2 + c^2)^{1/2}} \qquad \lambda_z = \frac{c}{(a^2 + b^2 + c^2)^{1/2}}$$

Substituting the values obtained in part (a) for the moments and products of inertia and for the direction cosines into the equation for $I_{OB}$, you obtain

$$I_{OB} = \frac{1}{a^2 + b^2 + c^2}\Big[\tfrac{1}{3}m(b^2 + c^2)a^2 + \tfrac{1}{3}m(c^2 + a^2)b^2 + \tfrac{1}{3}m(a^2 + b^2)c^2$$
$$- \tfrac{1}{2}ma^2b^2 - \tfrac{1}{2}mb^2c^2 - \tfrac{1}{2}mc^2a^2\Big]$$

$$I_{OB} = \frac{m}{6}\frac{a^2b^2 + b^2c^2 + c^2a^2}{a^2 + b^2 + c^2} \quad ◄$$

**REFLECT and THINK:** You can also obtain the moment of inertia $I_{OB}$ directly from the principal moments of inertia $\overline{I}_{x'}$, $\overline{I}_{y'}$, and $\overline{I}_{z'}$, because the line $OB$ passes through the centroid $O'$. Because the $x'$, $y'$, and $z'$ axes are principal axes of inertia (Fig. 3), use Eq. (9.50) to write

$$I_{OB} = \overline{I}_{x'}\lambda_x^2 + \overline{I}_{y'}\lambda_y^2 + \overline{I}_{z'}\lambda_z^2$$

$$= \frac{1}{a^2 + b^2 + c^2}\Big[\frac{m}{12}(b^2 + c^2)a^2 + \frac{m}{12}(c^2 + a^2)b^2 + \frac{m}{12}(a^2 + b^2)c^2\Big]$$

$$I_{OB} = \frac{m}{6}\frac{a^2b^2 + b^2c^2 + c^2a^2}{a^2 + b^2 + c^2} \quad ◄$$

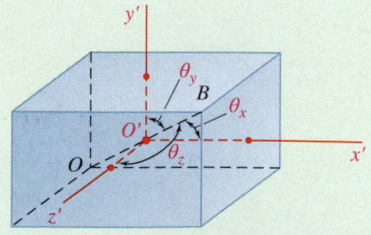

## Sample Problem 9.15

If $a = 3c$ and $b = 2c$ for the rectangular prism of Sample Prob. 9.14, determine (a) the principal moments of inertia at the origin $O$, (b) the principal axes of inertia at $O$.

**STRATEGY:** Substituting the data into the results from Sample Prob. 9.14 gives you values you can use with Eq. (9.56) to determine the principal moments of inertia. You can then use these values to set up a system of equations for finding the direction cosines of the principal axes.

**MODELING and ANALYSIS:**

### a. Principal Moments of Inertia at the Origin O.
Substituting $a = 3c$ and $b = 2c$ into the solution to Sample Prob. 9.14 gives you

$$I_x = \tfrac{5}{3}mc^2 \qquad I_y = \tfrac{10}{3}mc^2 \qquad I_z = \tfrac{13}{3}mc^2$$
$$I_{xy} = \tfrac{3}{2}mc^2 \qquad I_{yz} = \tfrac{1}{2}mc^2 \qquad I_{zx} = \tfrac{3}{4}mc^2$$

Substituting the values of the moments and products of inertia into Eq. (9.56) and collecting terms yields

$$K^3 - (\tfrac{28}{3}mc^2)K^2 + (\tfrac{3479}{144}m^2c^4)K - \tfrac{589}{54}m^3c^6 = 0$$

Now solve for the roots of this equation; from the discussion in Sec. 9.6C, it follows that these roots are the principal moments of inertia of the body at the origin.

$$K_1 = 0.568867mc^2 \qquad K_2 = 4.20885mc^2 \qquad K_3 = 4.55562mc^2$$

$$K_1 = 0.569mc^2 \qquad K_2 = 4.21mc^2 \qquad K_3 = 4.56mc^2 \quad \blacktriangleleft$$

### b. Principal Axes of Inertia at O.
To determine the direction of a principal axis of inertia, first substitute the corresponding value of $K$ into two of the equations (9.54). The resulting equations, together with Eq. (9.57), constitute a system of three equations from which you can determine the direction cosines of the corresponding principal axis. Thus, for the first principal moment of inertia $K_1$, you have

$$(\tfrac{5}{3} - 0.568867)mc^2(\lambda_x)_1 - \tfrac{3}{2}mc^2(\lambda_y)_1 - \tfrac{3}{4}mc^2(\lambda_z)_1 = 0$$
$$-\tfrac{3}{2}mc^2(\lambda_x)_1 + (\tfrac{10}{3} - 0.568867)mc^2(\lambda_y)_1 - \tfrac{1}{2}mc^2(\lambda_z)_1 = 0$$
$$(\lambda_x)_1^2 + (\lambda_y)_1^2 + (\lambda_z)_1^2 = 1$$

Solving yields

$$(\lambda_x)_1 = 0.836600 \qquad (\lambda_y)_1 = 0.496001 \qquad (\lambda_z)_1 = 0.232557$$

The angles that the first principal axis of inertia forms with the coordinate axes are then

$$(\theta_x)_1 = 33.2° \qquad (\theta_y)_1 = 60.3° \qquad (\theta_z)_1 = 76.6° \quad \blacktriangleleft$$

Using the same set of equations successively with $K_2$ and $K_3$, you can find that the angles associated with the second and third principal moments of inertia at the origin are, respectively,

$$(\theta_x)_2 = 57.8° \qquad (\theta_y)_2 = 146.6° \qquad (\theta_z)_2 = 98.0° \quad \blacktriangleleft$$

and

$$(\theta_x)_3 = 82.8° \qquad (\theta_y)_3 = 76.1° \qquad (\theta_z)_3 = 164.3° \quad \blacktriangleleft$$

# SOLVING PROBLEMS
# ON YOUR OWN

In this section, we defined the **mass products of inertia** $I_{xy}$, $I_{yz}$, and $I_{zx}$ of a body and showed you how to determine the moments of inertia of that body with respect to an arbitrary axis passing through the origin $O$. You also saw how to determine at the origin $O$ the **principal axes of inertia** of a body and the corresponding **principal moments of inertia**.

**1. Determining the mass products of inertia of a composite body.** You can express the mass products of inertia of a composite body with respect to the coordinate axes as the sums of the products of inertia of its component parts with respect to those axes. For each component part, use the parallel-axis theorem to write Eqs. (9.47)

$$I_{xy} = \bar{I}_{x'y'} + m\bar{x}\,\bar{y} \qquad I_{yz} = \bar{I}_{y'z'} + m\bar{y}\,\bar{z} \qquad I_{zx} = \bar{I}_{z'x'} + m\bar{z}\,\bar{x}$$

Here, the primes denote the centroidal axes of each component part, and $\bar{x}$, $\bar{y}$, and $\bar{z}$ represent the coordinates of its center of gravity. Keep in mind that the mass products of inertia can be positive, negative, or zero, and be sure to take into account the signs of $\bar{x}$, $\bar{y}$, and $\bar{z}$.

 **a. From the properties of symmetry of a component part,** you can deduce that two or all three of its centroidal mass products of inertia are zero. For instance, you can verify for a thin plate parallel to the $xy$ plane, a wire lying in a plane parallel to the $xy$ plane, a body with a plane of symmetry parallel to the $xy$ plane, and a body with an axis of symmetry parallel to the $z$ axis that the products of inertia $\bar{I}_{y'z'}$ and $\bar{I}_{z'x'}$ are zero.

 For rectangular, circular, or semicircular plates with axes of symmetry parallel to the coordinate axes, straight wires parallel to a coordinate axis, circular and semicircular wires with axes of symmetry parallel to the coordinate axes, and rectangular prisms with axes of symmetry parallel to the coordinate axes, the products of inertia $\bar{I}_{x'y'}$, $\bar{I}_{y'z'}$, and $\bar{I}_{z'x'}$ are all zero.

 **b. Mass products of inertia that are different from zero** can be computed from Eqs. (9.45). Although, in general, you need a triple integration to determine a mass product of inertia, you can use a single integration if you can divide the given body into a series of thin, parallel slabs. The computations are then similar to those discussed in the preceding section for moments of inertia.

**2. Computing the moment of inertia of a body with respect to an arbitrary axis $OL$.** In Sec. 9.6A, we derived an expression for the moment of inertia $I_{OL}$ that was given in Eq. (9.46). Before computing $I_{OL}$, you must first determine the mass moments and products of inertia of the body with respect to the given coordinate axes, as well as the direction cosines of the unit vector $\boldsymbol{\lambda}$ along $OL$.

**3. Calculating the principal moments of inertia of a body and determining its principal axes of inertia.** You saw in Sec. 9.6B that it is always possible to find an orientation of the coordinate axes for which the mass products of inertia are zero. These axes are referred to as the **principal axes of inertia**, and the corresponding moments of inertia are known as the **principal moments of inertia** of the body. In many cases, you can determine the principal axes of inertia of a body from its properties of symmetry. The procedure required to determine the principal moments and principal axes of inertia of a body with no obvious property of symmetry was discussed in Sec. 9.6C and was illustrated in Sample Prob. 9.15. It consists of the following steps.

**a. Expand the determinant in Eq. (9.55) and solve the resulting cubic equation.** You can obtain the solution by trial and error or (preferably) with an advanced scientific calculator or appropriate computer software. The roots $K_1$, $K_2$, and $K_3$ of this equation are the principal moments of inertia of the body.

**b. To determine the direction of the principal axis corresponding to $K_1$,** substitute this value for $K$ in two of the equations (9.54) and solve these equations, together with Eq. (9.57), for the direction cosines of the principal axis corresponding to $K_1$.

**c. Repeat this procedure with $K_2$ and $K_3$** to determine the directions of the other two principal axes. As a check of your computations, you may wish to verify that the scalar product of any two of the unit vectors along the three axes you have obtained is zero and, thus, that these axes are perpendicular to each other.

# Problems

**9.149** Determine the mass products of inertia $I_{xy}$, $I_{yz}$, and $I_{zx}$ of the steel fixture shown. (The density of steel is 7850 kg/m$^3$.)

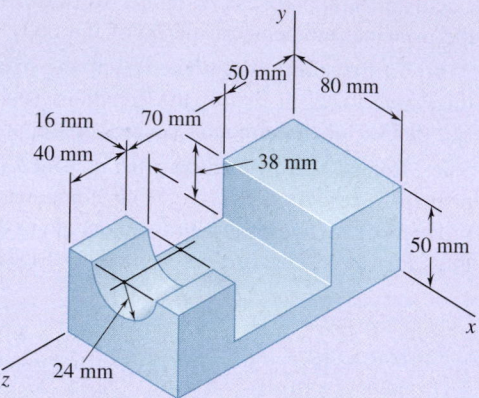

**Fig. P9.149**

**9.150** Determine the mass products of inertia $I_{xy}$, $I_{yz}$, and $I_{zx}$ of the steel machine element shown. (The density of steel is 7850 kg/m$^3$.)

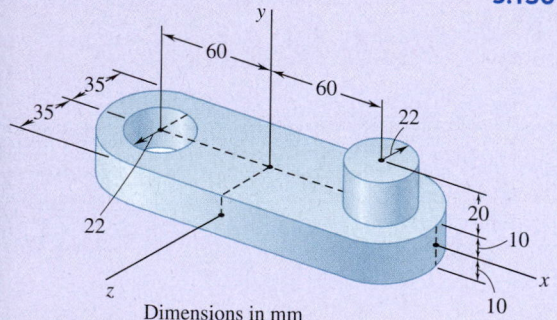

Dimensions in mm

**Fig. P9.150**

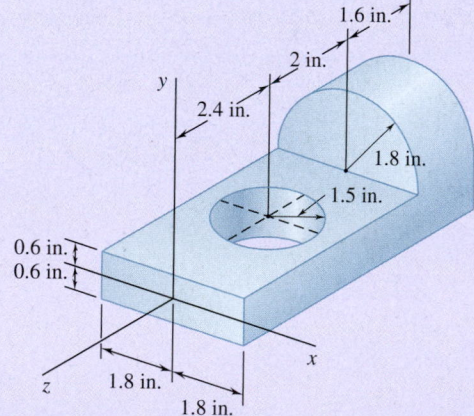

**Fig. P9.151**

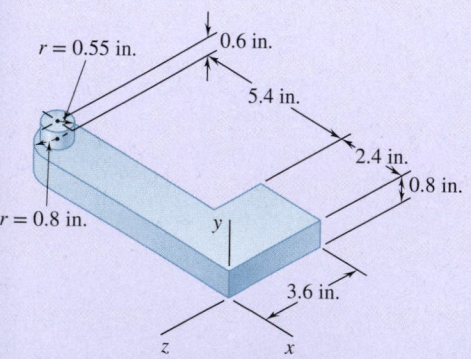

**Fig. P9.152**

**9.151** Determine the mass products of inertia $I_{xy}$, $I_{yz}$, and $I_{zx}$ of the steel machine component shown. (The specific weight of steel is 490 lb/ft$^3$.)

**9.152** Determine the mass products of inertia $I_{xy}$, $I_{yz}$, and $I_{zx}$ of the cast aluminum machine component shown. (The specific weight of aluminum is 0.100 lb/in$^3$.)

**9.153 through 9.156** A section of sheet steel 2 mm thick is cut and bent into the machine component shown. Knowing that the density of steel is 7850 kg/m³, determine the mass products of inertia $I_{xy}$, $I_{yz}$, and $I_{zx}$ of the component.

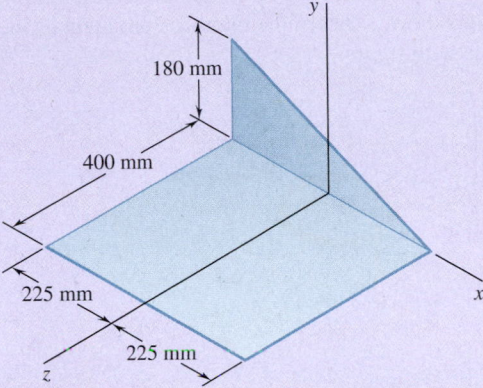

Fig. *P9.153*

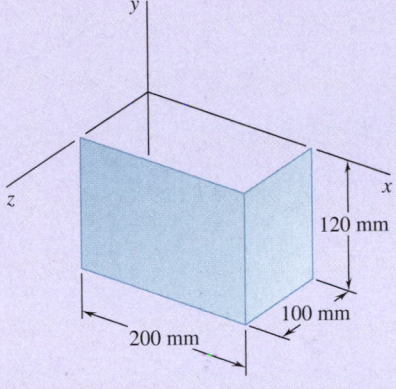

Fig. *P9.154*

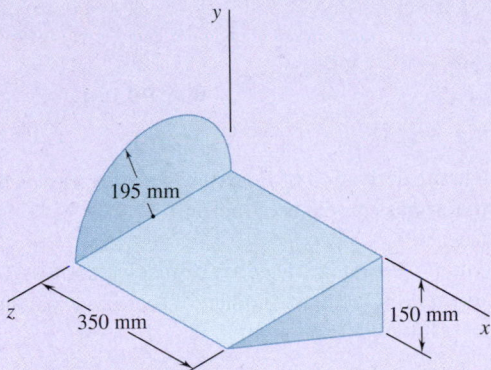

Fig. P9.155

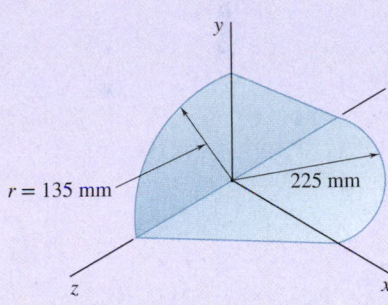

Fig. P9.156

**9.157** The figure shown is formed of 1.5-mm-diameter aluminum wire. Knowing that the density of aluminum is 2800 kg/m³, determine the mass products of inertia $I_{xy}$, $I_{yz}$, and $I_{zx}$ of the wire figure.

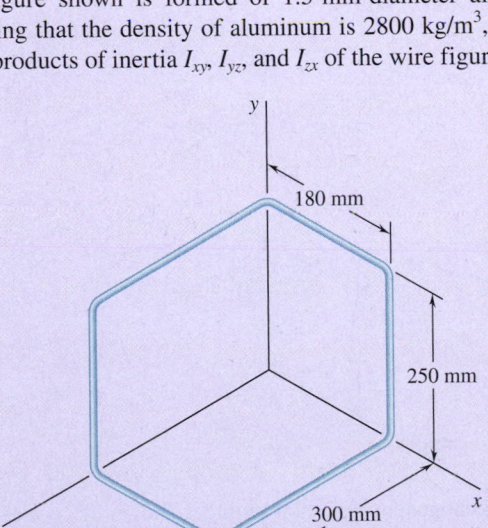

Fig. P9.157

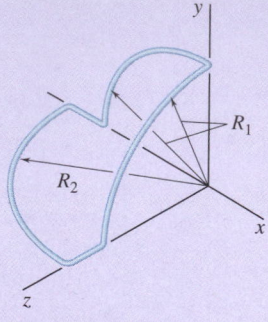

**Fig. P9.158**

**9.158** Thin aluminum wire of uniform diameter is used to form the figure shown. Denoting the mass per unit length of the wire by $m'$, determine the mass products of inertia $I_{xy}$, $I_{yz}$, and $I_{zx}$ of the wire figure.

**9.159 and 9.160** Brass wire with a weight per unit length $w$ is used to form the figure shown. Determine the mass products of inertia $I_{xy}$, $I_{yz}$, and $I_{zx}$ of the wire figure.

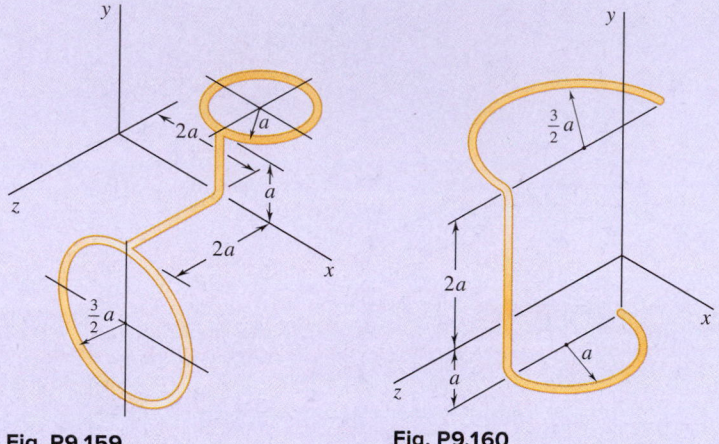

**Fig. P9.159**          **Fig. P9.160**

**9.161** Complete the derivation of Eqs. (9.47) that expresses the parallel-axis theorem for mass products of inertia.

**9.162** For the homogeneous tetrahedron of mass $m$ shown, (a) determine by direct integration the mass product of inertia $I_{zx}$, (b) deduce $I_{yz}$ and $I_{xy}$ from the result obtained in part $a$.

**9.163** The homogeneous circular cone shown has a mass $m$. Determine the mass moment of inertia of the cone with respect to the line joining the origin $O$ and point $A$.

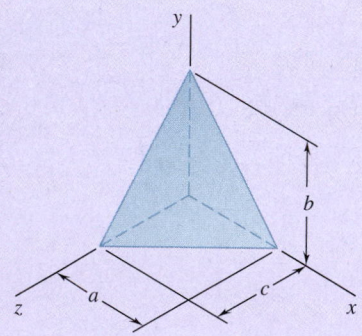

**Fig. P9.162**

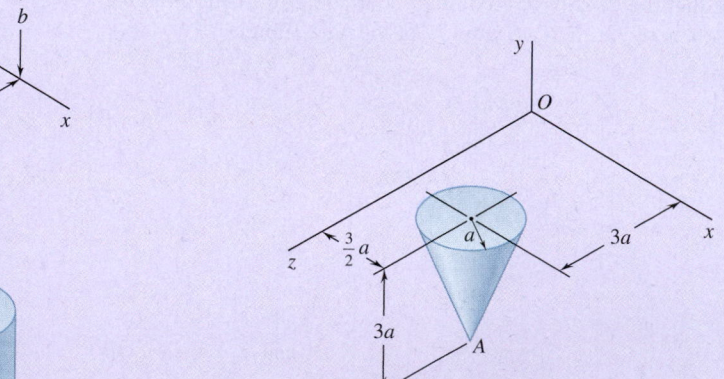

**Fig. P9.163**

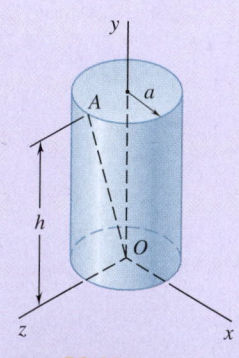

**Fig. P9.164**

**9.164** The homogeneous circular cylinder shown has a mass $m$. Determine the mass moment of inertia of the cylinder with respect to the line joining the origin $O$ and point $A$ that is located on the perimeter of the top surface of the cylinder.

**9.165** Shown is the machine element of Prob. 9.141. Determine its mass moment of inertia with respect to the line joining the origin $O$ and point $A$.

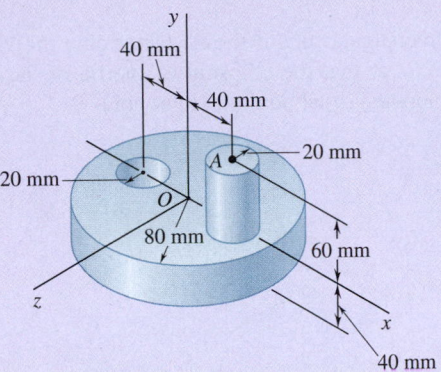

**Fig. P9.165**

**9.166** Determine the mass moment of inertia of the steel fixture of Probs. 9.145 and 9.149 with respect to the axis through the origin that forms equal angles with the $x$, $y$, and $z$ axes.

**9.167** The thin, bent plate shown is of uniform density and weight $W$. Determine its mass moment of inertia with respect to the line joining the origin $O$ and point $A$.

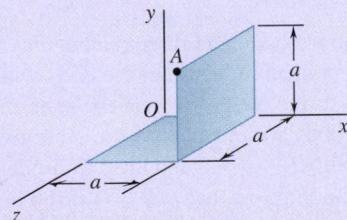

**Fig. P9.167**

**9.168** A piece of sheet steel with thickness $t$ and specific weight $\gamma$ is cut and bent into the machine component shown. Determine the mass moment of inertia of the component with respect to the line joining the origin $O$ and point $A$.

**9.169** Determine the mass moment of inertia of the machine component of Probs. 9.136 and 9.155 with respect to the axis through the origin characterized by the unit vector $\lambda = (-4\mathbf{i} + 8\mathbf{j} + \mathbf{k})/9$.

**9.170 through 9.172** For the wire figure of the problem indicated, determine the mass moment of inertia of the figure with respect to the axis through the origin characterized by the unit vector $\lambda = (-3\mathbf{i} - 6\mathbf{j} + 2\mathbf{k})/7$.

    **9.170** Prob. 9.148
    **9.171** Prob. 9.147
    **9.172** Prob. 9.146

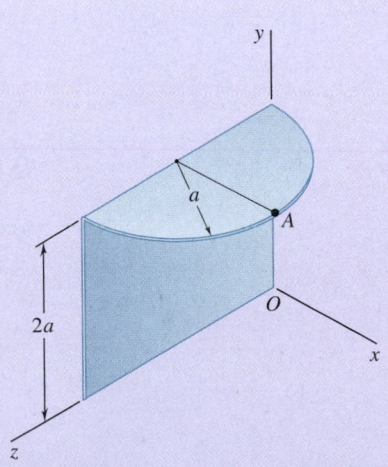

**Fig. P9.168**

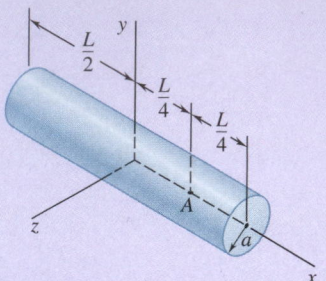

**Fig. P9.173**

**9.173** For the homogeneous circular cylinder shown with radius $a$ and length $L$, determine the value of the ratio $a/L$ for which the ellipsoid of inertia of the cylinder is a sphere when computed ($a$) at the centroid of the cylinder, ($b$) at point $A$.

**9.174** For the rectangular prism shown, determine the values of the ratios $b/a$ and $c/a$ so that the ellipsoid of inertia of the prism is a sphere when computed ($a$) at point $A$, ($b$) at point $B$.

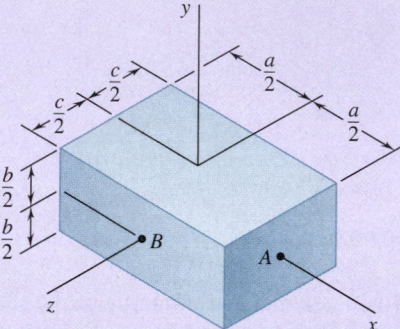

**Fig. P9.174**

**9.175** For the right circular cone of Sample Prob. 9.11, determine the value of the ratio $a/h$ for which the ellipsoid of inertia of the cone is a sphere when computed ($a$) at the apex of the cone, ($b$) at the center of the base of the cone.

**9.176** Given an arbitrary body and three rectangular axes $x$, $y$, and $z$, prove that the mass moment of inertia of the body with respect to any one of the three axes cannot be larger than the sum of the mass moments of inertia of the body with respect to the other two axes. That is, prove that the inequality $I_x \leq I_y + I_z$ and the two similar inequalities are satisfied. Furthermore, prove that $I_y \geq \frac{1}{2} I_x$ if the body is a homogeneous solid of revolution, where $x$ is the axis of revolution and $y$ is a transverse axis.

**9.177** Consider a cube with mass $m$ and side $a$. ($a$) Show that the ellipsoid of inertia at the center of the cube is a sphere, and use this property to determine the moment of inertia of the cube with respect to one of its diagonals. ($b$) Show that the ellipsoid of inertia at one of the corners of the cube is an ellipsoid of revolution, and determine the principal moments of inertia of the cube at that point.

**9.178** Given a homogeneous body of mass $m$ and of arbitrary shape and three rectangular axes $x$, $y$, and $z$ with origin at $O$, prove that the sum $I_x + I_y + I_z$ of the mass moments of inertia of the body cannot be smaller than the similar sum computed for a sphere of the same mass and the same material centered at $O$. Furthermore, using the result of Prob. 9.176, prove that, if the body is a solid of revolution, where $x$ is the axis of revolution, its mass moment of inertia $I_y$ about a transverse axis $y$ cannot be smaller than $3ma^2/10$, where $a$ is the radius of the sphere of the same mass and the same material.

*9.179 The homogeneous circular cylinder shown has a mass $m$, and the diameter $OB$ of its top surface forms 45° angles with the $x$ and $z$ axes. (a) Determine the principal mass moments of inertia of the cylinder at the origin $O$. (b) Compute the angles that the principal axes of inertia at $O$ form with the coordinate axes. (c) Sketch the cylinder, and show the orientation of the principal axes of inertia relative to the $x$, $y$, and $z$ axes.

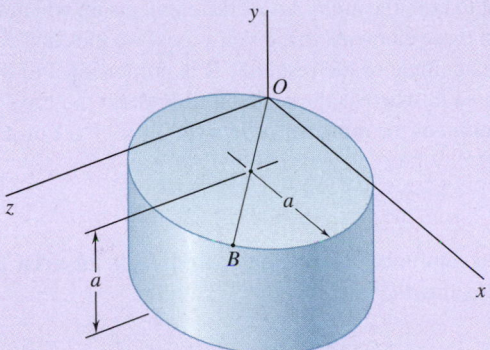

Fig. P9.179

9.180 through 9.184 For the component described in the problem indicated, determine (a) the principal mass moments of inertia at the origin, (b) the principal axes of inertia at the origin. Sketch the body and show the orientation of the principal axes of inertia relative to the $x$, $y$, and $z$ axes.

*9.180 Prob. 9.165
*9.181 Probs. 9.145 and 9.149
*9.182 Prob. 9.167
*9.183 Prob. 9.168
*9.184 Probs. 9.148 and 9.170

# Review and Summary

In the first half of this chapter, we discussed how to determine the resultant $\mathbf{R}$ of forces $\Delta\mathbf{F}$ distributed over a plane area $A$ when the magnitudes of these forces are proportional to both the areas $\Delta A$ of the elements on which they act and the distances $y$ from these elements to a given $x$ axis; we thus had $\Delta F = ky\,\Delta A$. We found that the magnitude of the resultant $\mathbf{R}$ is proportional to the first moment $Q_x = \int y\,dA$ of area $A$, whereas the moment of $\mathbf{R}$ about the $x$ axis is proportional to the **second moment**, or **moment of inertia**, $I_x = \int y^2 dA$ of $A$ with respect to the same axis (Sec. 9.1A).

## Rectangular Moments of Inertia

The **rectangular moments of inertia** $I_x$ and $I_y$ of an area (Sec. 9.1B) are obtained by evaluating the integrals

$$I_x = \int y^2 dA \qquad I_y = \int x^2 dA \qquad (9.1)$$

We can reduce these computations to single integrations by choosing $dA$ to be a thin strip parallel to one of the coordinate axes. We also recall that it is possible to compute $I_x$ and $I_y$ from the same elemental strip (Fig. 9.35) using the formula for the moment of inertia of a rectangular area (Sample Prob. 9.3).

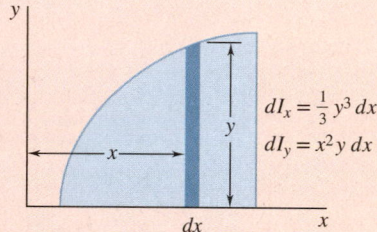

$$dI_x = \tfrac{1}{3}y^3\,dx$$
$$dI_y = x^2 y\,dx$$

**Fig. 9.35**

## Polar Moment of Inertia

We defined the **polar moment of inertia of an area** $A$ with respect to the pole $O$ (Sec. 9.1C) as

$$J_O = \int r^2 dA \qquad (9.3)$$

where $r$ is the distance from $O$ to the element of area $dA$ (Fig. 9.36). Observing that $r^2 = x^2 + y^2$, we established the relation

$$J_O = I_x + I_y \qquad (9.4)$$

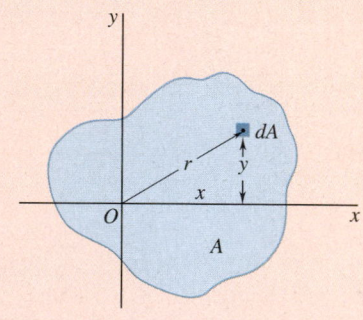

**Fig. 9.36**

## Radius of Gyration

We defined the **radius of gyration of an area** $A$ with respect to the $x$ axis (Sec. 9.1D) as the distance $k_x$, where $I_x = k_x^2 A$. With similar definitions for the radii of gyration of $A$ with respect to the $y$ axis and with respect to $O$, we have

$$k_x = \sqrt{\frac{I_x}{A}} \qquad k_y = \sqrt{\frac{I_y}{A}} \qquad k_o = \sqrt{\frac{J_O}{A}} \qquad (9.5\text{–}9.7)$$

## Parallel-Axis Theorem

The **parallel-axis theorem**, presented in Sec. 9.2A, states that the moment of inertia $I$ of an area with respect to any given axis $AA'$ (Fig. 9.37) is equal to the moment of inertia $\bar{I}$ of the area with respect to the centroidal axis $BB'$ that is parallel to $AA'$ *plus* the product of the area $A$ and the square of the distance $d$ between the two axes:

$$I = \bar{I} + Ad^2 \qquad (9.9)$$

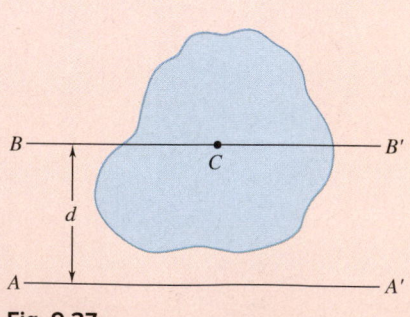

**Fig. 9.37**

You can use this formula to determine the moment of inertia $\bar{I}$ of an area with respect to a centroidal axis $BB'$ if you know its moment of inertia $I$ with respect to a parallel axis $AA'$. In this case, however, the product $Ad^2$ should be *subtracted* from the known moment of inertia $I$.

A similar relation holds between the polar moment of inertia $J_O$ of an area about a point $O$ and the polar moment of inertia $\bar{J}_C$ of the same area about its centroid $C$. Letting $d$ be the distance between $O$ and $C$, we have

$$J_C = \bar{J}_C + Ad^2 \tag{9.11}$$

## Composite Areas

The parallel-axis theorem can be used very effectively to compute the **moment of inertia of a composite area** with respect to a given axis (Sec. 9.2B). Considering each component area separately, we first compute the moment of inertia of each area with respect to its centroidal axis, using the data provided in Figs. 9.12 and 9.13 whenever possible. Then, apply the parallel-axis theorem to determine the moment of inertia of each component area with respect to the desired axis, and add the values (Sample Probs. 9.4 and 9.5).

## Product of Inertia

Sec. 9.3 was devoted to the transformation of the moments of inertia of an area under a rotation of the coordinate axes. First, we defined the **product of inertia of an area** $A$ as

$$I_{xy} = \int xy \, dA \tag{9.12}$$

and showed that $I_{xy} = 0$ if the area $A$ is symmetrical with respect to either or both of the coordinate axes. We also derived the **parallel-axis theorem for products of inertia** as

$$I_{xy} = \bar{I}_{x'y'} + \bar{x}\bar{y}A \tag{9.13}$$

where $\bar{I}_{x'y'}$ is the product of inertia of the area with respect to the centroidal axes $x'$ and $y'$ that are parallel to the $x$ and $y$ axes and $\bar{x}$ and $\bar{y}$ are the coordinates of the centroid of the area (Sec. 9.3A).

## Rotation of Axes

In Sec. 9.3B, we determined the moments and product of inertia $I_{x'}$, $I_{y'}$, and $I_{x'y'}$ of an area with respect to the $x'$ and $y'$ axes obtained by rotating the original $x$ and $y$ coordinate axes counterclockwise through an angle $\theta$ (Fig. 9.38). We expressed $I_{x'}$, $I_{y'}$, and $I_{x'y'}$ in terms of the moments and product of inertia $I_x$, $I_y$, and $I_{xy}$ computed with respect to the original $x$ and $y$ axes.

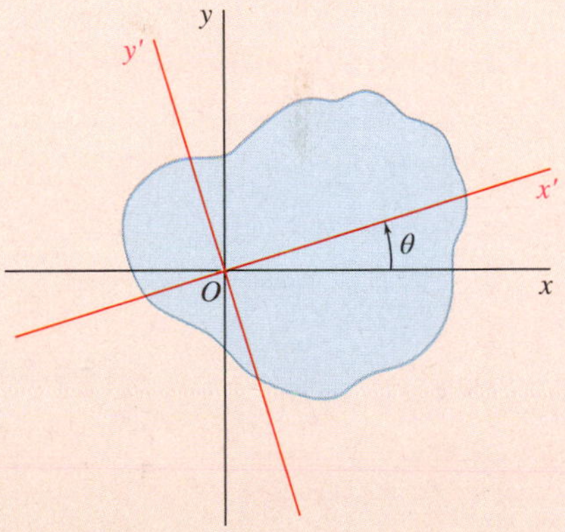

$$I_{x'} = \frac{I_x + I_y}{2} + \frac{I_x - I_y}{2}\cos 2\theta - I_{xy}\sin 2\theta \tag{9.18}$$

$$I_{y'} = \frac{I_x + I_y}{2} - \frac{I_x - I_y}{2}\cos 2\theta + I_{xy}\sin 2\theta \tag{9.19}$$

$$I_{x'y'} = \frac{I_x - I_y}{2}\sin 2\theta + I_{xy}\cos 2\theta \tag{9.20}$$

**Fig. 9.38**

## Principal Axes

We defined the **principal axes of the area about** $O$ as the two axes perpendicular to each other with respect to which the moments of inertia of the area are maximum and minimum. The corresponding values of $\theta$, denoted by $\theta_m$, were obtained from

$$\tan 2\theta_m = -\frac{2I_{xy}}{I_x - I_y} \tag{9.25}$$

## Principal Moments of Inertia

The corresponding maximum and minimum values of $I$ are called the **principal moments of inertia** of the area about $O$:

$$I_{max, min} = \frac{I_x + I_y}{2} \pm \sqrt{\left(\frac{I_x - I_y}{2}\right)^2 + I_{xy}^2} \qquad (9.27)$$

We also noted that the corresponding value of the product of inertia is zero.

## Mohr's Circle

The transformation of the moments and product of inertia of an area under a rotation of axes can be represented graphically by drawing **Mohr's circle** (Sec. 9.4). Given the moments and product of inertia $I_x$, $I_y$, and $I_{xy}$ of the area with respect to the $x$ and $y$ coordinate axes, we plot points $X$ $(I_x, I_{xy})$ and $Y$ $(I_y, -I_{xy})$ and draw the line joining these two points (Fig. 9.39). This line is a diameter of Mohr's circle and thus defines this circle. As the coordinate axes are rotated through $\theta$, the diameter rotates through *twice that angle*, and the coordinates of $X'$ and $Y'$ yield the new values $I_{x'}$, $I_{y'}$, and $I_{x'y'}$ of the moments and product of inertia of the area. Also, the angle $\theta_m$ and the coordinates of points $A$ and $B$ define the principal axes $a$ and $b$ and the principal moments of inertia of the area (Sample Prob. 9.8).

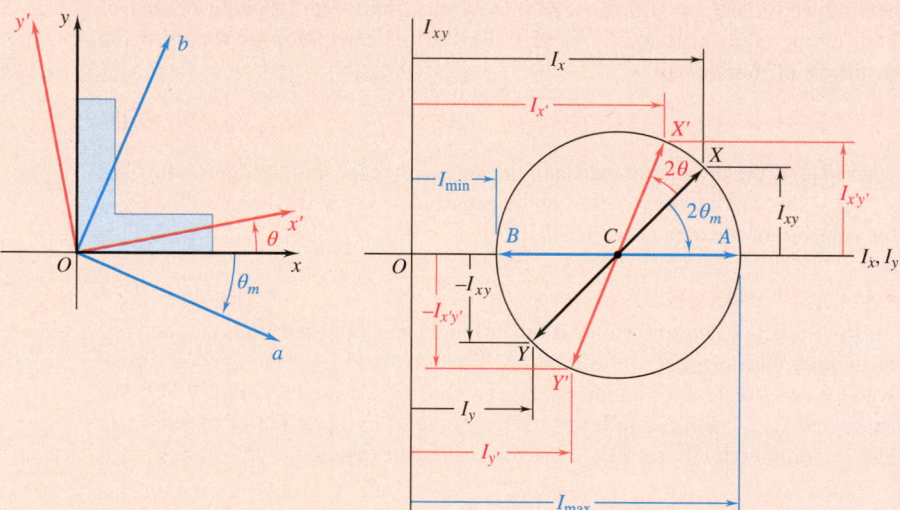

**Fig. 9.39**

## Moments of Inertia of Masses

The second half of the chapter was devoted to determining **moments of inertia of masses**, which are encountered in dynamics problems involving the rotation of a rigid body about an axis. We defined the mass moment of inertia of a body with respect to an axis $AA'$ (Fig. 9.40) as

$$I = \int r^2 \, dm \qquad (9.28)$$

where $r$ is the distance from $AA'$ to the element of mass (Sec. 9.5A). We defined the **radius of gyration** of the body as

$$k = \sqrt{\frac{I}{m}} \qquad (9.29)$$

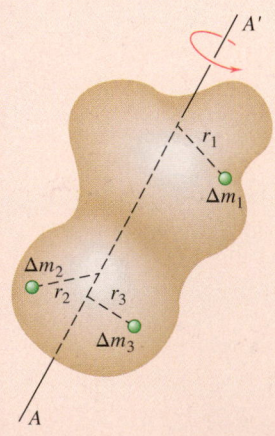

**Fig. 9.40**

The moments of inertia of a body with respect to the coordinate axes were expressed as

$$I_x = \int (y^2 + z^2)\, dm$$

$$I_y = \int (z^2 + x^2)\, dm \qquad \text{(9.30)}$$

$$I_z = \int (x^2 + y^2)\, dm$$

## Parallel-Axis Theorem

We saw that the **parallel-axis theorem** also applies to mass moments of inertia (Sec. 9.5B). Thus, the moment of inertia $I$ of a body with respect to an arbitrary axis $AA'$ (Fig. 9.41) can be expressed as

$$I = \bar{I} + md^2 \qquad \text{(9.33)}$$

where $\bar{I}$ is the moment of inertia of the body with respect to the centroidal axis $BB'$ that is parallel to the axis $AA'$, $m$ is the mass of the body, and $d$ is the distance between the two axes.

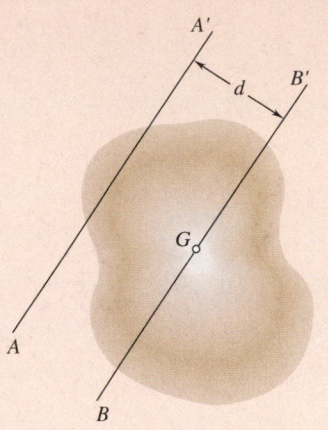

**Fig. 9.41**

## Moments of Inertia of Thin Plates

We can readily obtain the moments of inertia of thin plates from the moments of inertia of their areas (Sec. 9.5C). We found that for a rectangular plate the moments of inertia with respect to the axes shown (Fig. 9.42) are

$$I_{AA'} = \tfrac{1}{12}ma^2 \qquad I_{BB'} = \tfrac{1}{12}mb^2 \qquad \text{(9.39)}$$

$$I_{CC'} = I_{AA'} + I_{BB'} = \tfrac{1}{12}m(a^2 + b^2) \qquad \text{(9.40)}$$

whereas for a circular plate (Fig. 9.43) they are

$$I_{AA'} = I_{BB'} = \tfrac{1}{4}mr^2 \qquad \text{(9.41)}$$

$$I_{CC'} = I_{AA'} + I_{BB'} = \tfrac{1}{2}mr^2 \qquad \text{(9.42)}$$

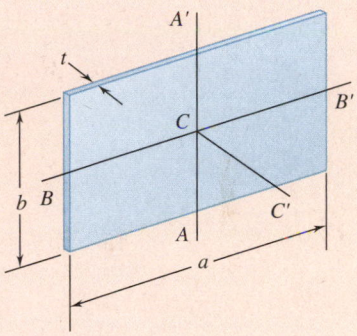

**Fig. 9.42**

## Composite Bodies

When a body possesses two planes of symmetry, it is usually possible to use a single integration to determine its moment of inertia with respect to a given axis by selecting the element of mass $dm$ to be a thin plate (Sample Probs. 9.10 and 9.11). On the other hand, when a body consists of several common geometric shapes, we can obtain its moment of inertia with respect to a given axis by using the formulas given in Fig. 9.28 together with the parallel-axis theorem (Sample Probs. 9.12 and 9.13).

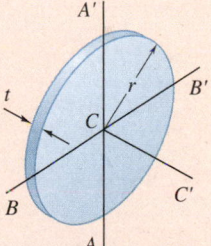

**Fig. 9.43**

## Moment of Inertia with Respect to an Arbitrary Axis

In the last section of the chapter, we described how to determine the moment of inertia of a body with respect to an arbitrary axis $OL$ that is drawn through the origin $O$ (Sec. 9.6A). We denoted the components of the unit vector $\lambda$ along $OL$ by $\lambda_x$, $\lambda_y$, and $\lambda_z$ (Fig. 9.44) and introduced the **products of inertia** as

$$I_{xy} = \int xy\, dm \qquad I_{yz} = \int yz\, dm \qquad I_{zx} = \int zx\, dm \qquad \text{(9.45)}$$

We found that the moment of inertia of the body with respect to $OL$ could be expressed as

$$I_{OL} = I_x \lambda_x^2 + I_y \lambda_y^2 + I_z \lambda_z^2 - 2I_{xy}\lambda_x\lambda_y - 2I_{yz}\lambda_y\lambda_z - 2I_{zx}\lambda_z\lambda_x \qquad \text{(9.46)}$$

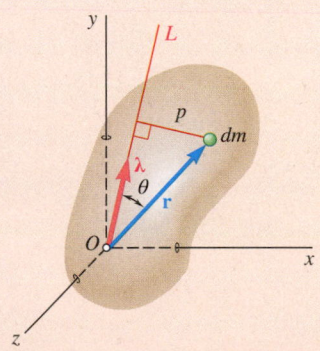

**Fig. 9.44**

### Ellipsoid of Inertia

By plotting a point $Q$ along each axis $OL$ at a distance $OQ = 1/\sqrt{I_{OL}}$ from $O$ (Sec. 9.6B), we obtained the surface of an ellipsoid, known as the **ellipsoid of inertia** of the body at point $O$.

### Principal Axes and Principal Moments of Inertia

The principal axes $x'$, $y'$, and $z'$ of this ellipsoid (Fig. 9.45) are the **principal axes of inertia** of the body; that is, the products of inertia $I_{x'y'}$, $I_{y'z'}$, and $I_{z'x'}$ of the body with respect to these axes are all zero. In many situations, you can deduce the principal axes of inertia of a body from its properties of symmetry. Choosing these axes to be the coordinate axes, we can then express $I_{OL}$ as

$$I_{OL} = I_{x'}\lambda_{x'}^2 + I_{y'}\lambda_{y'}^2 + I_{z'}\lambda_{z'}^2 \qquad \textbf{(9.50)}$$

where $I_{x'}$, $I_{y'}$, and $I_{z'}$ are the **principal moments of inertia** of the body at $O$.

   When the principal axes of inertia cannot be obtained by observation (Sec. 9.6B), it is necessary to solve the cubic equation

$$K^3 - (I_x + I_y + I_z)K^2 + (I_xI_y + I_yI_z + I_zI_x - I_{xy}^2 - I_{yz}^2 - I_{zx}^2)K$$
$$- (I_xI_yI_z - I_xI_{yz}^2 - I_yI_{zx}^2 - I_zI_{xy}^2 - 2I_{xy}I_{yz}I_{zx}) = 0 \qquad \textbf{(9.56)}$$

We found (Sec. 9.6C) that the roots $K_1$, $K_2$, and $K_3$ of this equation are the principal moments of inertia of the given body. The direction cosines $(\lambda_x)_1$, $(\lambda_y)_1$, and $(\lambda_z)_1$ of the principal axis corresponding to the principal moment of inertia $K_1$ are then determined by substituting $K_1$ into Eqs. (9.54) and by solving two of these equations and Eq. (9.57) simultaneously. The same procedure is then repeated using $K_2$ and $K_3$ to determine the direction cosines of the other two principal axes (Sample Prob. 9.15).

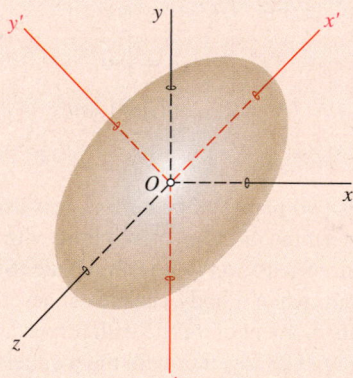

**Fig. 9.45**

# Review Problems

**9.185** Determine by direct integration the moments of inertia of the shaded area with respect to the $x$ and $y$ axes.

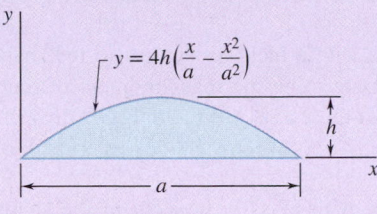

**Fig. P9.185**

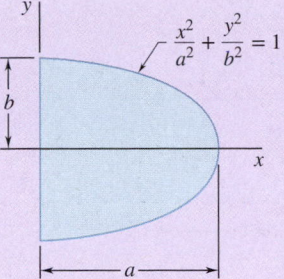

**Fig. P9.186**

**9.186** Determine the moment of inertia and the radius of gyration of the shaded area shown with respect to the $y$ axis.

**9.187** Determine the moment of inertia and the radius of gyration of the shaded area shown with respect to the $x$ axis.

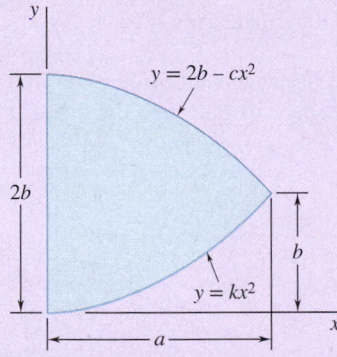

**Fig. P9.187**

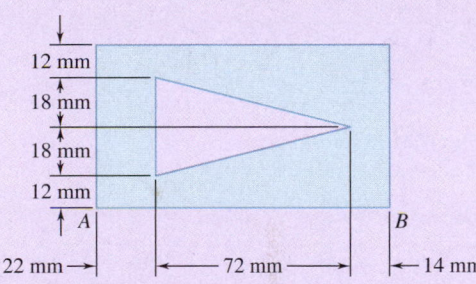

**Fig. P9.188**

**9.188** Determine the moments of inertia $\overline{I}_x$ and $\overline{I}_y$ of the area shown with respect to centroidal axes respectively parallel and perpendicular to side $AB$.

**9.189** Determine the polar moment of inertia of the area shown with respect to (*a*) point $O$, (*b*) the centroid of the area.

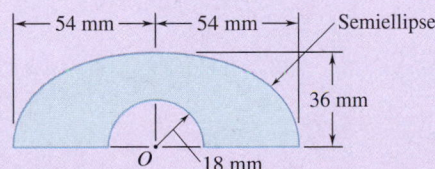

**Fig. P9.189**

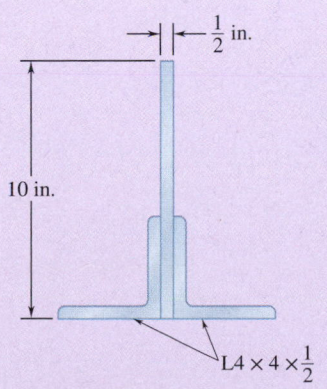

**Fig. P9.190**

**9.190** Two L4 × 4 × $\frac{1}{2}$-in. angles are welded to a steel plate as shown. Determine the moments of inertia of the combined section with respect to centroidal axes respectively parallel and perpendicular to the plate.

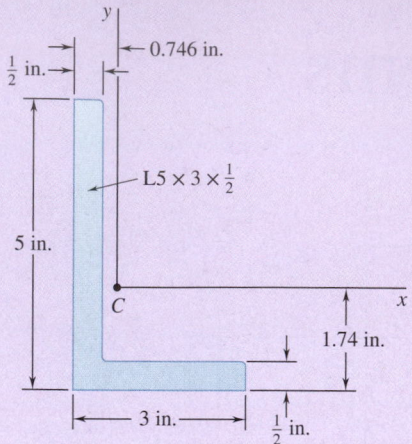

Fig. P9.191 and *P9.192*

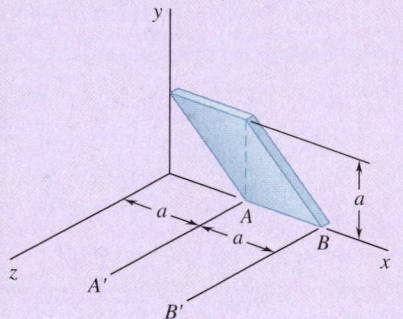

Fig. P9.193 and *P9.194*

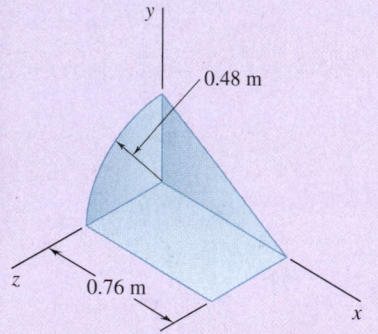

Fig. P9.195

**9.191** Using the parallel-axis theorem, determine the product of inertia of the L5 × 3 × $\frac{1}{2}$-in. angle cross section shown with respect to the centroidal $x$ and $y$ axes.

**9.192** For the L5 × 3 × $\frac{1}{2}$-in. angle cross section shown, use Mohr's circle to determine (*a*) the moments of inertia and the product of inertia with respect to new centroidal axes obtained by rotating the $x$ and $y$ axes 30° clockwise, (*b*) the orientation of the principal axes through the centroid and the corresponding values of the moments of inertia.

**9.193** A thin plate with a mass $m$ was cut in the shape of a parallelogram as shown. Determine the mass moment of inertia of the plate with respect to (*a*) the $x$ axis, (*b*) the axis $BB'$ that is perpendicular to the plate.

**9.194** A thin plate with mass $m$ was cut in the shape of a parallelogram as shown. Determine the mass moment of inertia of the plate with respect to (*a*) the $y$ axis, (*b*) the axis $AA'$ that is perpendicular to the plate.

**9.195** A 2-mm-thick piece of sheet steel is cut and bent into the machine component shown. Knowing that the density of steel is 7850 kg/m$^3$, determine the mass moment of inertia of the component with respect to each of the coordinate axes.

**9.196** Determine the mass moment of inertia of the steel machine element shown with respect to the $z$ axis. (The specific weight of steel is 490 lb/ft$^3$.)

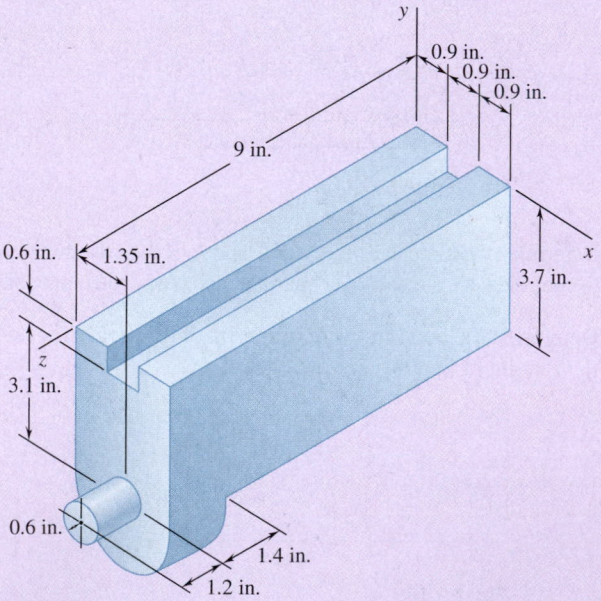

Fig. P9.196

# 10

# Method of Virtual Work

The method of virtual work is particularly effective when a simple relation can be found among the displacements of the points of application of the various forces involved. This is the case for the scissor lift platform being used by a worker to gain access to the structural frame of a building under construction.

# Objectives

- **Define** the work of a force, and consider the circumstances when a force does no work.

- **Examine** the principle of virtual work, and apply it to analyze the equilibrium of machines and mechanisms.

- **Apply** the concept of potential energy to determine the equilibrium position of a rigid body or a system of rigid bodies.

- **Evaluate** the mechanical efficiency of machines, and consider the stability of equilibrium.

## *Introduction

In the preceding chapters, we solved problems involving the equilibrium of rigid bodies by expressing the balance of external forces acting on the bodies. We wrote the equations of equilibrium $\Sigma F_x = 0$, $\Sigma F_y = 0$, and $\Sigma M_A = 0$ and solved them for the desired unknowns. We now consider a different method, which turns out to be more effective for solving certain types of equilibrium problems. This method, based on the **principle of virtual work**, was first formally used by the Swiss mathematician Jean Bernoulli in the eighteenth century.

As you will see in Sec. 10.1B, the principle of virtual work considers a particle or rigid body (or more generally, a system of connected rigid bodies) that is in equilibrium under various external forces. The principle states that, if the body is given an arbitrary displacement from that position of equilibrium, the total work done by the external forces during the displacement is zero. This principle is particularly effective when applied to the solution of problems involving the equilibrium of machines or mechanisms consisting of several connected members.

In the second part of this chapter, we apply the method of virtual work in an alternative form based on the concept of **potential energy**. We will show in Sec. 10.2 that, if a particle, rigid body, or system of rigid bodies is in equilibrium, the derivative of its potential energy with respect to a variable defining its position must be zero.

You will also learn in this chapter to evaluate the mechanical efficiency of a machine (Sec. 10.1D) and to determine whether a given position of equilibrium is stable, unstable, or neutral (Sec. 10.2D).

## *10.1   THE BASIC METHOD

The first step in explaining the method of virtual work is to define the terms *displacement* and *work* as they are used in mechanics. Then, we can state the principle of virtual work and show how to apply it in practical situations. We also take the opportunity to define mechanical efficiency, which is a useful and important parameter for the design of real machines.

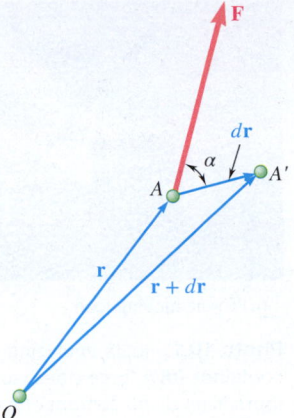

## 10.1A    Work of a Force

Consider a particle that moves from a point $A$ to a neighboring point $A'$ (Fig. 10.1). If $\mathbf{r}$ denotes the position vector corresponding to point $A$, we denote the small vector joining $A$ and $A'$ by the differential $d\mathbf{r}$; we call the vector $d\mathbf{r}$ the **displacement** of the particle.

Now let us assume that a force $\mathbf{F}$ is acting on the particle. The **work** $dU$ **of force F corresponding to the displacement** $d\mathbf{r}$ is defined as the quantity

**Definition of work**

$$dU = \mathbf{F} \cdot d\mathbf{r} \qquad \textbf{(10.1)}$$

That is, $dU$ is the scalar product of the force $\mathbf{F}$ and the displacement $d\mathbf{r}$. Suppose we denote the magnitudes of the force by $F$, the displacement by $ds$, and the angle formed by $\mathbf{F}$ and $d\mathbf{r}$ by $\alpha$. Then, recalling the definition of the scalar product of two vectors (Sec. 3.2A), we have

**Fig. 10.1**   The work of a force acting on a particle is the scalar product of the force and the particle's displacement.

$$dU = F\, ds \cos \alpha \qquad \textbf{(10.1')}$$

Work is a scalar quantity, so it has a magnitude and a sign, but no direction. Note that work should be expressed in units obtained by multiplying units of length by units of force. Thus, if we use U.S. customary units, we should express work in ft·lb or in·lb. If we use SI units, we express work in N · m. This unit of work is called a **joule** (J).[†]

It follows from (10.1') that work $dU$ is positive if the angle $\alpha$ is acute, and negative if $\alpha$ is obtuse. Three particular cases are of special interest.

- If the force $\mathbf{F}$ has the same direction as $d\mathbf{r}$, the work $dU$ reduces to $F\, ds$.
- If $\mathbf{F}$ has a direction opposite to that of $d\mathbf{r}$, the work is $dU = -F\, ds$.
- Finally, if $\mathbf{F}$ is perpendicular to $d\mathbf{r}$, the work $dU$ is zero.

We can also consider the work $dU$ of a force $\mathbf{F}$ during a displacement $d\mathbf{r}$ to be the product of $F$ and the component $ds \cos \alpha$ of the displacement $d\mathbf{r}$ along $\mathbf{F}$ (Fig. 10.2a). This view is particularly useful in computing the work done by the weight $\mathbf{W}$ of a body (Fig. 10.2b). The work of $\mathbf{W}$ is equal to the product of $W$ and the vertical displacement $dy$ of the center of gravity $G$ of the body. If the displacement is downward, the work is positive; if the displacement is upward, the work is negative.

Some forces frequently encountered in statics do no work, such as forces applied to fixed points ($ds = 0$) or acting in a direction perpendicular to the displacement ($\cos \alpha = 0$). Among these forces are the reaction at a frictionless pin when the body supported rotates about the pin; the reaction at a frictionless surface when the body in contact moves along the surface; the reaction at a roller moving along its track; the weight of a body when its center of gravity moves horizontally; and the friction force acting on a wheel rolling without slipping (because at any instant the point of contact does not move). Examples of forces that do work are the weight of a body (except in the case considered previously), the friction force acting on a body sliding on a rough surface, and most forces applied on a moving body.

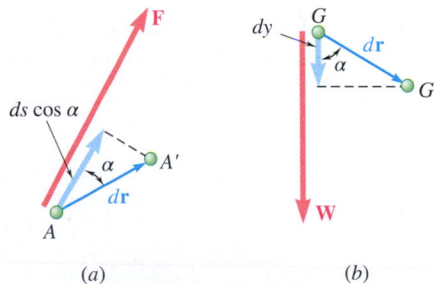

**Fig. 10.2**   (a) You can think of work as the product of a force and the component of displacement in the direction of the force. (b) This is useful for computing the work done by an object's weight.

---

[†]The joule is the SI unit of *energy,* whether in mechanical form (work, potential energy, kinetic energy) or in chemical, electrical, or thermal form. Note that even though 1 N·m = 1 J, we must express the moment of a force in N · m, and not in joules, because the moment of a force is not a form of energy.

(a) Crane moving load

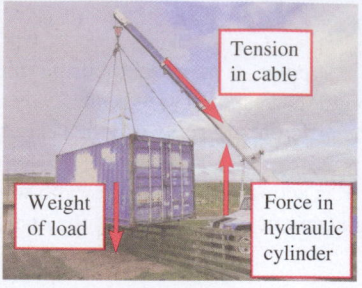
(b) Forces that do work

(c) Forces that do no work

**Photo 10.1**   (a) In analyzing a crane, we might consider displacements associated with vertical movement of a container. (b) A force does work if it has a component in the direction of a displacement. (c) A force does no work if there is no displacement or if the force is perpendicular to a displacement. Courtesy Brian Miller

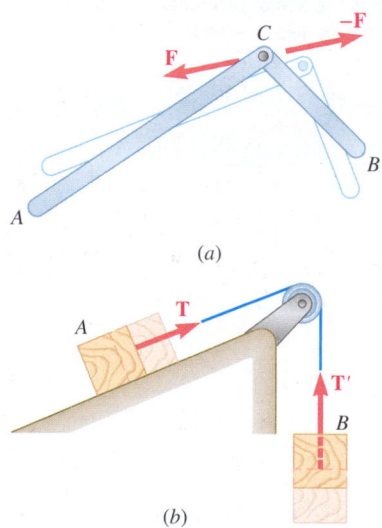

**Fig. 10.3**   For (a) a frictionless pin or (b) a cord that is not extensible, the total work done by the pairs of internal forces is zero.

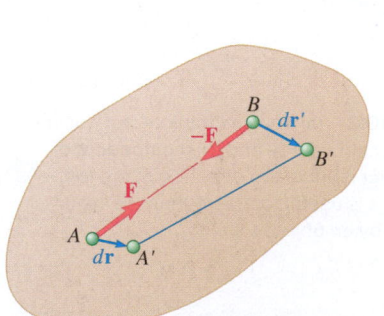

**Fig. 10.4**   As demonstrated here for an arbitrary pair of particles, the total work of the internal forces holding a rigid body together is zero.

In certain cases, the sum of the work done by several forces is zero. Consider, for example, two rigid bodies $AC$ and $BC$ that are connected at $C$ by a *frictionless pin* (Fig. 10.3a). Among the forces acting on $AC$ is the force $\mathbf{F}$ exerted at $C$ by $BC$. In general, the work of this force is not zero, but it is equal in magnitude and opposite in sign to the work of the force $-\mathbf{F}$ exerted by $AC$ on $BC$, because these forces are equal and opposite and are applied to the same particle. Thus, when the total work done by all the forces acting on $AB$ and $BC$ is considered, the work of the two internal forces at $C$ cancels out. We obtain a similar result if we consider a system consisting of two blocks connected by a *cord AB* that is not extensible (Fig. 10.3b). The work of the tension force $\mathbf{T}$ at $A$ is equal in magnitude to the work of the tension force $\mathbf{T}'$ at $B$, because these forces have the same magnitude and the points $A$ and $B$ move through the same distance. However, in one case, the work is positive, and in the other, it is negative. Thus, the work of the internal forces again cancels out.

We can show that the total work of the internal forces holding together the particles of a rigid body is zero. Consider two particles $A$ and $B$ of a rigid body and the two equal and opposite forces $\mathbf{F}$ and $-\mathbf{F}$ they exert on each other (Fig. 10.4). Although, in general, small displacements $d\mathbf{r}$ and $d\mathbf{r}'$ of the two particles are different, the components of these displacements along $AB$ must be equal; otherwise, the particles would not remain at the same distance from each other, so the body would not be rigid. Therefore, the work of $\mathbf{F}$ is equal in magnitude and opposite in sign to the work of $-\mathbf{F}$, and their sum is zero.

In computing the work of the external forces acting on a rigid body, it is often convenient to determine the work of a couple without considering separately the work of each of the two forces forming the couple. Consider the two forces $\mathbf{F}$ and $-\mathbf{F}$ forming a couple of moment $\mathbf{M}$ and acting on a rigid body (Fig. 10.5). Any small displacement of the rigid body bringing $A$ and $B$, respectively, into $A'$ and $B''$ can be divided into two parts: one in which points $A$ and $B$ undergo equal displacements $d\mathbf{r}_1$, the other in which $A'$ remains fixed while $B'$ moves into $B''$ through a displacement $d\mathbf{r}_2$ with a magnitude of $ds_2 = r\,d\theta$. In the first part of the motion, the work of $\mathbf{F}$ is equal in magnitude and opposite in sign to the work of $-\mathbf{F}$, and their sum is zero. In the second part of the motion, only force $\mathbf{F}$ works, and its work is $dU = F\,ds_2 = Fr\,d\theta$. But the product $Fr$ is equal to the magnitude $M$ of the moment of the couple. Thus, the work of a couple of moment $\mathbf{M}$ acting on a rigid body is

**Work of a couple**

$$dU = M\,d\theta \qquad\qquad (10.2)$$

where $d\theta$ is the small angle (expressed in radians) through which the body rotates. We again note that work should be expressed in units obtained by multiplying units of force by units of length.

## 10.1B  The Principle of Virtual Work

Consider a particle acted upon by several forces $\mathbf{F}_1, \mathbf{F}_2, \ldots, \mathbf{F}_n$ (Fig. 10.6). We can imagine that the particle undergoes a small displacement from $A$ to $A'$. This displacement is possible, but it does not necessarily take place. The forces may be balanced and the particle remains at rest, or the particle may move under the action of the given forces in a direction different from that of $AA'$. Because the considered displacement does not actually occur, it is called a **virtual displacement**, which is denoted by $\delta\mathbf{r}$. The symbol $\delta\mathbf{r}$ represents a differential of the first order; it is used to distinguish the virtual displacement from the displacement $d\mathbf{r}$ that would take place under actual motion. As you will see, we can use virtual displacements to determine whether the conditions of equilibrium of a particle are satisfied.

The work of each of the forces $\mathbf{F}_1, \mathbf{F}_2, \ldots, \mathbf{F}_n$ during the virtual displacement $\delta\mathbf{r}$ is called **virtual work**. The virtual work of all the forces acting on the particle of Fig. 10.6 is

$$\delta U = \mathbf{F}_1 \cdot \delta\mathbf{r} + \mathbf{F}_2 \cdot \delta\mathbf{r} + \cdots + \mathbf{F}_n \cdot \delta\mathbf{r}$$
$$= (\mathbf{F}_1 + \mathbf{F}_2 + \cdots + \mathbf{F}_n) \cdot \delta\mathbf{r}$$

or

$$\delta U = \mathbf{R} \cdot \delta\mathbf{r} \qquad (10.3)$$

where $\mathbf{R}$ is the resultant of the given forces. Thus, the total virtual work of the forces $\mathbf{F}_1, \mathbf{F}_2, \ldots, \mathbf{F}_n$ is equal to the virtual work of their resultant $\mathbf{R}$.

The principle of virtual work for a particle states:

> **If a particle is in equilibrium, the total virtual work of the forces acting on the particle is zero for any virtual displacement of the particle.**

This condition is necessary: If the particle is in equilibrium, the resultant $\mathbf{R}$ of the forces is zero, and it follows from Eq. (10.3) that the total virtual work $\delta U$ is zero. The condition is also sufficient: If the total virtual work $\delta U$ is zero for any virtual displacement, the scalar product $\mathbf{R} \cdot \delta\mathbf{r}$ is zero for any $\delta\mathbf{r}$, and the resultant $\mathbf{R}$ must be zero.

In the case of a rigid body, the principle of virtual work states:

> **If a rigid body is in equilibrium, the total virtual work of the external forces acting on the rigid body is zero for any virtual displacement of the body.**

The condition is necessary: If the body is in equilibrium, all the particles forming the body are in equilibrium and the total virtual work of the forces acting on all the particles must be zero. However, we have seen in the preceding section that the total work of the internal forces is zero; therefore, the total work of the external forces also must be zero. The condition can also be proven to be sufficient.

The principle of virtual work can be extended to the case of a **system of connected rigid bodies**. If the system remains connected during the virtual displacement, **only the work of the forces external to the system need be considered**, because the total work of the internal forces at the various connections is zero.

## 10.1C  Applying the Principle of Virtual Work

The principle of virtual work is particularly effective when applied to the solution of problems involving machines or mechanisms consisting of several connected rigid bodies. Consider, for instance, the toggle vise $ACB$ of Fig. 10.7a

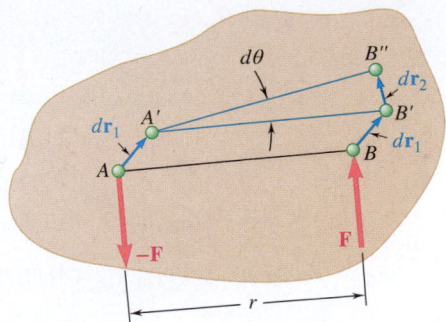

**Fig. 10.5**  The work of a couple acting on a rigid body is the moment of the couple times the angular rotation.

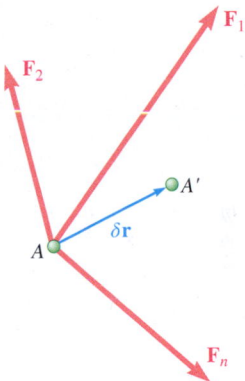

**Fig. 10.6**  Forces acting on a particle that goes through a virtual displacement.

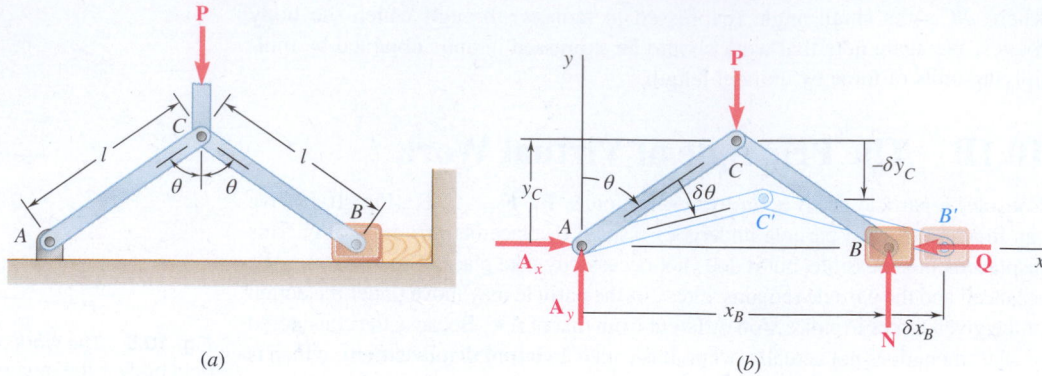

**Fig. 10.7** (*a*) A toggle vise used to compress a wooden block, assuming no friction; (*b*) a virtual displacement of the vise.

used to compress a wooden block. Suppose we wish to determine the force exerted by the vise on the block when a given force **P** is applied at *C*, assuming there is no friction. Denoting the reaction of the block on the vise by **Q**, we draw the free-body diagram of the vise and consider the virtual displacement obtained by giving a positive increment $\delta\theta$ to angle $\theta$ (Fig. 10.7*b*). Choosing a system of coordinate axes with origin at *A*, we note that $x_B$ increases as $y_C$ decreases. This is indicated in the figure, where we show a positive increment $\delta x_B$ and a negative increment $-\delta y_C$. The reactions $\mathbf{A}_x$, $\mathbf{A}_y$, and **N** do no work during the virtual displacement considered, so we need only compute the work done by **P** and **Q**. Because **Q** and $\delta x_B$ have opposite senses, the virtual work of **Q** is $\delta U_Q = -Q\,\delta x_B$. Because **P** and the increment shown ($-\delta y_C$) have the same sense, the virtual work of **P** is $\delta U_P = +P(-\delta y_C) = -P\,\delta y_C$. (We could have predicted the minus signs by simply noting that the forces **Q** and **P** are directed opposite to the positive *x* and *y* axes, respectively.) Expressing the coordinates $x_B$ and $y_C$ in terms of the angle $\theta$ and differentiating, we obtain

$$x_B = 2l\sin\theta \qquad y_C = l\cos\theta$$
$$\delta x_B = 2l\cos\theta\,\delta\theta \qquad \delta y_C = -l\sin\theta\,\delta\theta \tag{10.4}$$

The total virtual work of the forces **Q** and **P** is thus

$$\delta U = \delta U_Q + \delta U_P = -Q\,\delta x_B - P\,\delta y_C$$
$$= -2Ql\cos\theta\,\delta\theta + Pl\sin\theta\,\delta\theta$$

Setting $\delta U = 0$, we obtain

$$2Ql\cos\theta\,\delta\theta = Pl\sin\theta\,\delta\theta \tag{10.5}$$

and

$$Q = \frac{1}{2}P\tan\theta \tag{10.6}$$

The superiority of the method of virtual work over the conventional equilibrium equations in the problem considered here is clear: By using the method of virtual work, we were able to eliminate all unknown reactions, whereas the equation $\Sigma M_A = 0$ would have eliminated only two of the unknown reactions. This property of the method of virtual work can be used in solving many problems involving machines and mechanisms.

**If the virtual displacement considered is consistent with the constraints imposed by the supports and connections, all reactions and internal forces are eliminated and only the work of the loads, applied forces, and friction forces need be considered.**

**Photo 10.2** The method of virtual work is useful for determining the forces exerted by the hydraulic cylinders positioning the bucket lift. The reason is that a simple relation exists among the displacements of the points of application of the forces acting on the members of the lift. Courtesy of Altec, Inc.

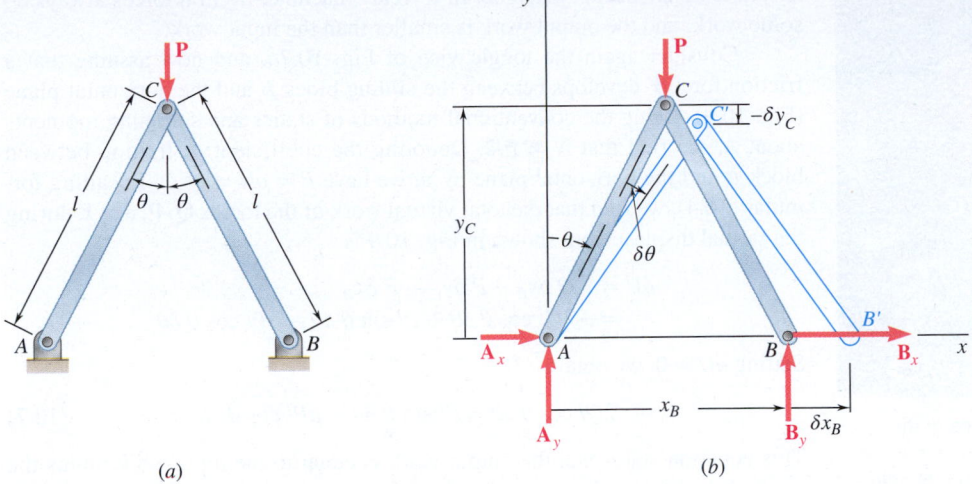

**Fig. 10.8** (*a*) A completely constrained frame *ACB*; (*b*) a virtual displacement of the frame in order to determine $B_x$, keeping *A* fixed.

We can also use the method of virtual work to solve problems involving completely constrained structures, although the virtual displacements considered never actually take place. Consider, for example, the frame *ACB* shown in Fig. 10.8*a*. If point *A* is kept fixed while point *B* is given a horizontal virtual displacement (Fig. 10.8*b*), we need consider only the work of **P** and **B**$_x$. We can thus determine the reaction component **B**$_x$ in the same way as the force **Q** of the preceding example (Fig. 10.7*b*); we have

$$B_x = \frac{1}{2} P \tan \theta$$

By keeping *B* fixed and giving a horizontal virtual displacement to *A*, we can similarly determine the reaction component **A**$_x$. Then, we can determine the components **A**$_y$ and **B**$_y$ by rotating the frame *ACB* as a rigid body about *B* and *A*, respectively.

We can also use the method of virtual work to determine the configuration of a system in equilibrium under given forces. For example, we can obtain the value of the angle $\theta$ for which the linkage of Fig. 10.7 is in equilibrium under two given forces **P** and **Q** by solving Eq. (10.6) for tan $\theta$.

Note, however, that the attractiveness of the method of virtual work depends to a large extent upon the existence of simple geometric relations between the various virtual displacements involved in the solution of a given problem. When no such simple relations exist, it is usually advisable to revert to the conventional method of Chap. 6.

## 10.1D Mechanical Efficiency of Real Machines

In analyzing the toggle vise of Fig. 10.7, we assumed that no friction forces were involved. Thus, the virtual work consisted only of the work of the applied force **P** and of the reaction **Q**. However, the work of reaction **Q** is equal in magnitude and opposite in sign to the work of the force exerted by the vise on the block. Therefore, Eq. (10.5) states that the **output work** $2Ql \cos \theta \, \delta\theta$ is equal to the **input work** $Pl \sin \theta \, \delta\theta$. A machine in which input and output work are equal

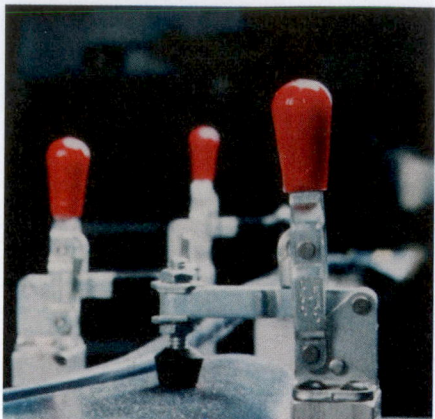

**Photo 10.3** The clamping force of the toggle clamp shown can be expressed as a function of the force applied to the handle by first establishing the geometric relations among the members of the clamp and then applying the method of virtual work. Courtesy of DE-STA-CO

is said to be an "ideal" machine. In a "real" machine, friction forces always do some work, and the output work is smaller than the input work.

Consider again the toggle vise of Fig. 10.7a, and now assume that a friction force **F** develops between the sliding block $B$ and the horizontal plane (Fig. 10.9). Using the conventional methods of statics and summing moments about $A$, we find that $N = P/2$. Denoting the coefficient of friction between block $B$ and the horizontal plane by $\mu$, we have $F = \mu N = \mu P/2$. Recalling formulas (10.4), we find that the total virtual work of the forces **Q**, **P**, and **F** during the virtual displacement shown in Fig. 10.9 is

$$\delta U = -Q\,\delta x_B - P\,\delta y_C - F\,\delta x_B$$
$$= -2Ql\cos\theta\,\delta\theta + Pl\sin\theta\,\delta\theta - \mu Pl\cos\theta\,\delta\theta$$

Setting $\delta U = 0$, we obtain

$$2Ql\cos\theta\,\delta\theta = Pl\sin\theta\,\delta\theta - \mu Pl\cos\theta\,\delta\theta \qquad (10.7)$$

This equation states that the output work is equal to the input work minus the work of the friction force. Solving for $Q$, we have

$$Q = \frac{1}{2}P\,(\tan\theta - \mu) \qquad (10.8)$$

Note that $Q = 0$ when $\tan\theta = \mu$; that is, when $\theta$ is equal to the angle of friction $\phi$, and that $Q < 0$ when $\theta < \phi$. Thus, we can use the toggle vise only for values of $\theta$ larger than the angle of friction.

We define the **mechanical efficiency** $\eta$ of a machine as the ratio

**Mechanical efficiency**

$$\eta = \frac{\text{output work}}{\text{input work}} \qquad (10.9)$$

Thus, the mechanical efficiency of an ideal machine is $\eta = 1$ when input and output work are equal, whereas the mechanical efficiency of a real machine is always less than 1.

In the case of the toggle vise we have just analyzed, we have

$$\eta = \frac{\text{output work}}{\text{input work}} = \frac{2Ql\cos\theta\,\delta\theta}{Pl\sin\theta\,\delta\theta} \qquad (10.10)$$

We can check that, in the absence of friction forces, we would have $\mu = 0$ and $\eta = 1$. In the general case when $\mu$ is different from zero, the efficiency $\eta$ becomes zero for $\mu\cot\theta = 1$; that is, for $\tan\theta = \mu$ or $\theta = \tan^{-1}\mu = \phi$. We note again that the toggle vise can be used only for values of $\theta$ larger than the angle of friction $\phi$.

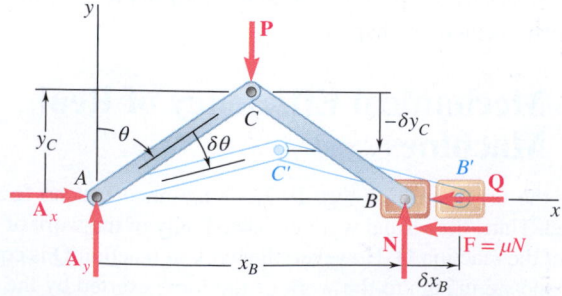

**Fig. 10.9** A virtual displacement of the toggle vise with friction.

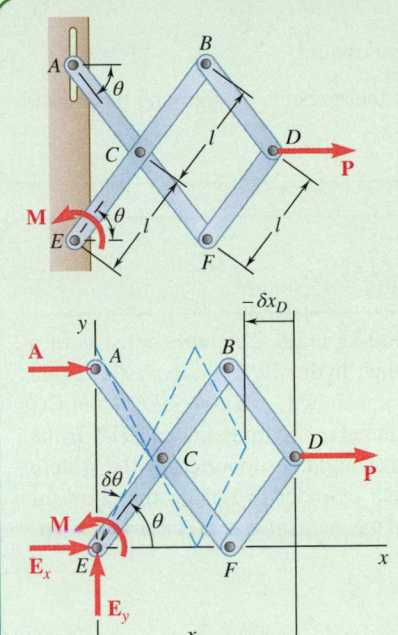

## Sample Problem 10.1

Using the method of virtual work, determine the magnitude of the couple **M** required to maintain the equilibrium of the mechanism shown.

**STRATEGY:** For a virtual displacement consistent with the constraints, the reactions do no work, so you can focus solely on the force **P** and the moment **M**. You can solve for **M** in terms of **P** and the geometric parameters.

**MODELING:** Choose a coordinate system with origin at $E$ (Fig. 1). Then

$$x_D = 3l \cos \theta \qquad\qquad \delta x_D = -3l \sin \theta \, \delta\theta$$

**ANALYSIS: Principle of Virtual Work.** Because the reactions **A**, $\mathbf{E}_x$, and $\mathbf{E}_y$ do no work during the virtual displacement, the total virtual work done by **M** and **P** must be zero. Notice that **P** acts in the positive $x$ direction and **M** acts in the positive $\theta$ direction. You obtain

$$\delta U - 0: \qquad\qquad +M \, \delta\theta + P \, \delta x_D = 0$$
$$+M \, \delta\theta + P(-3l \sin \theta \, \delta\theta) = 0$$
$$M = 3Pl \sin \theta \quad \blacktriangleleft$$

**REFLECT and THINK:** This problem illustrates that the principle of virtual work can help determine a moment as well as a force in a straightforward computation.

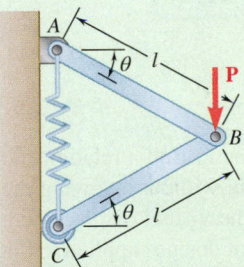

**Fig. 1** Free-body diagram of mechanism showing a virtual displacement.

## Sample Problem 10.2

Determine the expressions for $\theta$ and for the tension in the spring that correspond to the equilibrium position of the mechanism. The unstretched length of the spring is $h$, and the spring constant is $k$. Neglect the weight of the mechanism.

**STRATEGY:** The tension in the spring is a force **F** exerted at $C$. Applying the principle of virtual work, you can obtain a relationship between **F** and the applied force **P**.

**MODELING:** With the coordinate system shown in Fig. 1,

$$y_B = l \sin \theta \qquad\qquad y_C = 2l \sin \theta$$
$$\delta y_B = l \cos \theta \, \delta\theta \qquad\qquad \delta y_C = 2l \cos \theta \, \delta\theta$$

The elongation of the spring is $s = y_C - h = 2l \sin \theta - h$. The magnitude of the force exerted at $C$ by the spring is

$$F = ks = k(2l \sin \theta - h) \qquad\qquad (1)$$

**ANALYSIS: Principle of Virtual Work.** Because the reactions $\mathbf{A}_x$, $\mathbf{A}_y$, and $C$ do no work, the total virtual work done by **P** and **F** must be zero.

$$\delta U = 0: \qquad P \, \delta y_B - F \, \delta y_C = 0$$
$$P(l \cos \theta \, \delta\theta) - k(2l \sin \theta - h)(2l \cos \theta \, \delta\theta) = 0$$
$$\sin \theta = \frac{P + 2kh}{4kl} \quad \blacktriangleleft$$

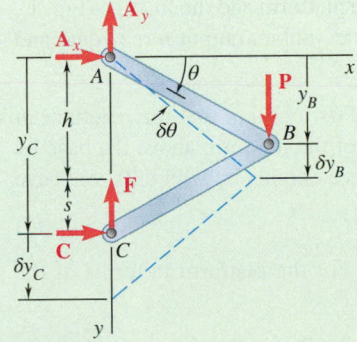

**Fig. 1** Free-body diagram of mechanism showing a virtual displacement.

*(continued)*

Substituting this expression into Eq. (1), you obtain

$$F = \tfrac{1}{2}P \quad \blacktriangleleft$$

**REFLECT and THINK:** You can verify these results by applying the appropriate equations of equilibrium.

## Sample Problem 10.3

A hydraulic-lift table is used to raise a 1000-kg crate. The table consists of a platform and two identical linkages on which hydraulic cylinders exert equal forces. (Only one linkage and one cylinder are shown.) Members *EDB* and *CG* are each of length 2*a*, and member *AD* is pinned to the midpoint of *EDB*. If the crate is placed on the table so that half of its weight is supported by the system shown, determine the force exerted by each cylinder in raising the crate for $\theta = 60°$, $a = 0.70$ m, and $L = 3.20$ m. (This mechanism was previously considered in Sample Prob. 6.7.)

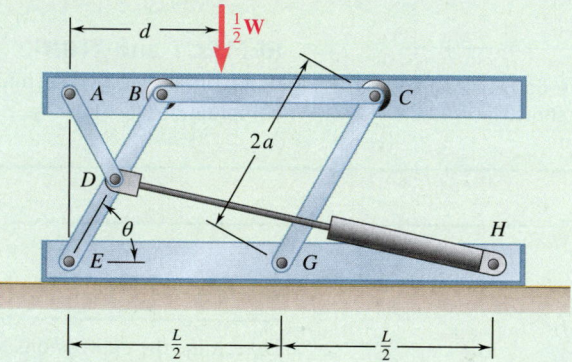

**STRATEGY:** The principle of virtual work allows you to find a relationship between the force applied by the cylinder and the weight without involving the reactions. However, you need a relationship between the virtual displacement and the change in angle $\theta$, which is found from the law of cosines applied to the given geometry.

**MODELING:** The free body consists of the platform and the linkage (Fig. 1), with an input force $\mathbf{F}_{DH}$ exerted by the cylinder and an output force equal and opposite to $\tfrac{1}{2}\mathbf{W}$.

**ANALYSIS:** **Principle of Virtual Work.** First observe that the reactions at *E* and *G* do no work. Denoting the elevation of the platform above the base by *y* and the length *DH* of the cylinder-and-piston assembly by *s* (Fig. 2), you have

$$\delta U = 0: \qquad -\tfrac{1}{2}W\,\delta y + F_{DH}\,\delta s = 0 \qquad (1)$$

You can express the vertical displacement $\delta y$ of the platform in terms of the angular displacement $\delta\theta$ of *EDB* as

$$y = (EB)\sin\theta = 2a\sin\theta$$
$$\delta y = 2a\cos\theta\,\delta\theta$$

*(continued)*

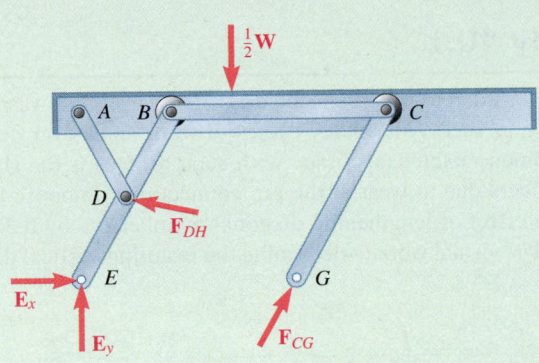

**Fig. 1** Free-body diagram of the platform and linkage.

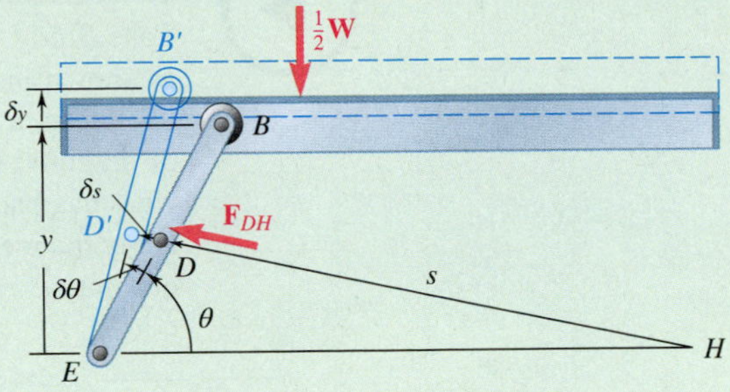

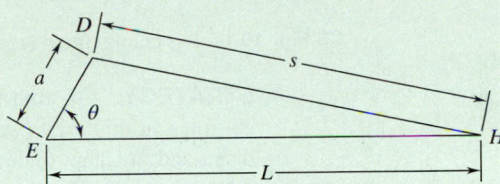

**Fig. 2** Virtual displacement of the machine.

To express $\delta s$ similarly in terms of $\delta\theta$, first note that by the law of cosines (Fig. 3),

$$s^2 = a^2 + L^2 - 2aL\cos\theta$$

**Fig. 3** Geometry associated with the cylinder-and-piston assembly.

Differentiating,

$$2s\,\delta s = -2aL(-\sin\theta)\,\delta\theta$$

$$\delta s = \frac{aL\sin\theta}{s}\,\delta\theta$$

Substituting for $\delta y$ and $\delta s$ into Eq. (1), you have

$$(-\tfrac{1}{2}W)2a\cos\theta\,\delta\theta + F_{DH}\frac{aL\sin\theta}{s}\,\delta\theta = 0$$

$$F_{DH} = W\frac{s}{L}\cot\theta$$

With the given numerical data, you obtain

$$W = mg = (1000\text{ kg})(9.81\text{ m/s}^2) = 9810\text{ N} = 9.81\text{ kN}$$
$$s^2 = a^2 + L^2 - 2aL\cos\theta$$
$$= (0.70)^2 + (3.20)^2 - 2(0.70)(3.20)\cos 60° = 8.49$$
$$s = 2.91\text{ m}$$
$$F_{DH} = W\frac{s}{L}\cot\theta = (9.81\text{ kN})\frac{2.91\text{ m}}{3.20\text{ m}}\cot 60°$$

$$F_{DH} = 5.15\text{ kN} \blacktriangleleft$$

**REFLECT and THINK:** The principle of virtual work gives you a relationship between forces, but sometimes you need to review the geometry carefully to find a relationship between the displacements.

## Case Study 10.1

Case Study 6.1 considered a pin-connected railroad truss that was modeled as shown in CS Fig. 10.1. After many cycles of train loads over decades of operation, a common maintenance issue with such bridges is the slackening of diagonal members due to wear at the pin connections. Suppose that such wear has had the effect of lengthening diagonal member *EG* by 0.3 in. Let's apply the method of virtual work to determine the resulting vertical deflection of joint *G*.

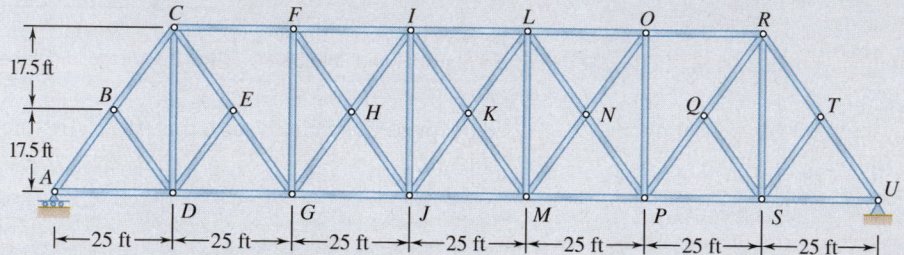

**CS Fig. 10.1**    Pin-connected railroad truss model.

**STRATEGY:**    To find the vertical deflection at *G*, apply an arbitrary verti-cal force at this joint (we will use a unit force). Using the method of sections discussed in Chap. 6, determine the resulting force exerted by member *EG* on joints *E* and *G*. Then, apply the method of virtual work for a virtual displace-ment equal to the specified 0.3-in. increase in the length of member *EG*.

**MODELING and ANALYSIS:**

**Force in Member *EG*.**    CS Fig. 10.2 shows the free-body diagram of the entire truss, with a vertical unit 1-kip force placed at joint *G*. From this dia-gram, find the reaction at *A*:

$$+\circlearrowleft \Sigma M_U = 0: \qquad + (1 \text{ kip})(125 \text{ ft}) - A(175 \text{ ft}) = 0 \qquad \mathbf{A} = 0.7143 \text{ kips} \uparrow$$

Now pass a section *nn* vertically through the truss (CS Fig. 10.2) and choose the left-hand portion as a free body (CS Fig. 10.3). Note that

$$\tan \alpha = \frac{FG}{CF} = \frac{35 \text{ ft}}{25 \text{ ft}} = 1.4 \qquad \alpha = 54.46°$$

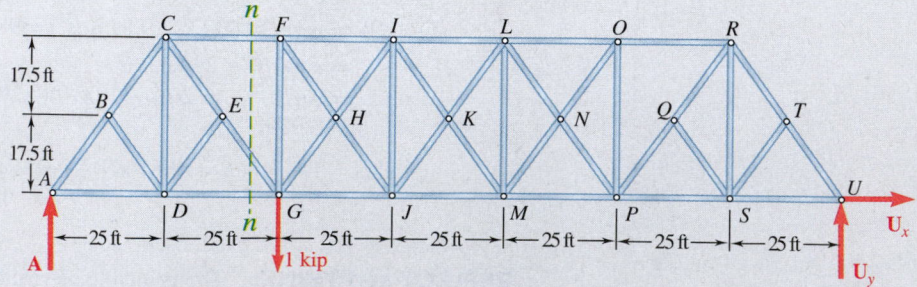

**CS Fig. 10.2**    Free-body diagram of entire truss, with unit load placed at joint *G*.

*(continued)*

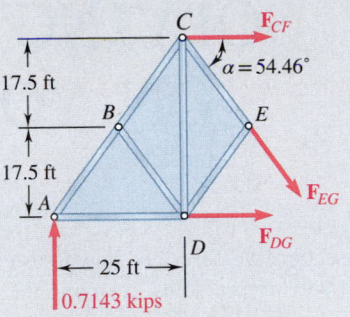

**CS Fig. 10.3**  Free-body diagram to analyze member *EG*.

Applying equilibrium,

$$+\uparrow \Sigma F_y = 0:  + (0.7143 \text{ kips}) - F_{EG} \sin 54.46° = 0$$
$$F_{EG} = + 0.8778 \text{ kips}$$   $F_{EG} = 0.8778 \text{ kips } T \blacktriangleleft$

**Vertical Deflection at G.**   Remove member *EG* and replace it with forces $F_{EG}$ applied at *E* and *G* as shown in CS Fig. 10.4. The vertical deflection at *G* will be denoted by $\delta y_G$, and the increase in length of member *EG* will be represented by a virtual displacement of joint *E* relative to joint *G*, $\delta_{EG}$. These displacements are also shown in CS Fig. 10.4. Applying the method of virtual work, where the unit load and $\delta y_G$ have the same direction, and $F_{EG}$ and $\delta_{EG}$ have opposite directions, you have

$$\delta U = 0:  + (1 \text{ kip})\delta y_G - F_{EG}\delta_{EG} = 0$$
$$+ (1 \text{ kip})\delta y_G - (0.8778 \text{ kips})(0.3 \text{ in.}) = 0$$
$$\delta y_G = +0.263 \text{ in.}$$   $\delta y_G = 0.263 \text{ in.}\downarrow \blacktriangleleft$

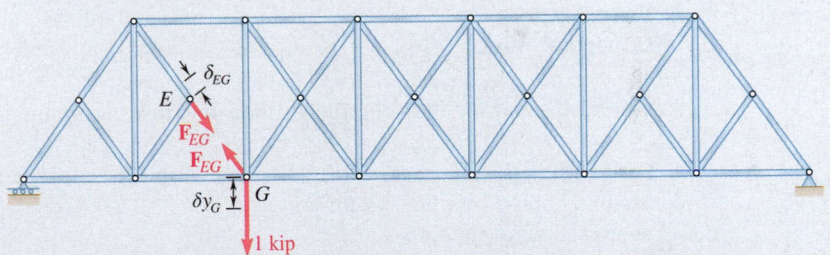

**CS Fig. 10.4**  Forces and displacements for virtual work analysis.

**REFLECT and THINK:**   The method of virtual work was used here to determine the deflection of a truss joint due to a prescribed change in length of one of its members, and was facilitated through the application of an *imaginary* unit force of the same direction and location as the desired deflection. Of course, the actual service loads acting on the structure, such as train loads, will also cause the bridge to deflect as well. In addition to these *real* loads, displacements depend upon the material and section properties of the individual members as well as the geometry of the overall structure, and rely on the principles of deformable solids developed in a mechanics of materials course.

# SOLVING PROBLEMS
# ON YOUR OWN

In this section, we described how to use the **method of virtual work**, which is a different way of solving problems involving the equilibrium of rigid bodies.

The work done by a force during a displacement of its point of application or by a couple during a rotation is found, respectively, by using:

$$dU = F \, ds \cos \alpha \qquad \textbf{(10.1)}$$
$$dU = M \, d\theta \qquad \textbf{(10.2)}$$

**Principle of virtual work.** In its more general and more useful form, this principle can be stated as:

> **If a system of connected rigid bodies is in equilibrium, the total virtual work of the external forces applied to the system is zero for any virtual displacement of the system.**

As you apply the principle of virtual work, keep in mind the following points.

**1. Virtual displacement.** A machine or mechanism in equilibrium has no tendency to move. However, we can cause—or imagine—a small displacement. Because it does not actually occur, such a displacement is called a **virtual displacement**.

**2. Virtual work.** The work done by a force or couple during a virtual displacement is called **virtual work**.

**3. You need consider only the forces that do work** during the virtual displacement.

**4. Forces that do no work** during a virtual displacement that are consistent with the constraints imposed on the system are
   **a.** Reactions at supports
   **b.** Internal forces at connections
   **c.** Forces exerted by inextensible cords and cables

None of these forces need to be considered when you use the method of virtual work.

**5. Be sure to express the various virtual displacements** involved in your computations in terms of a single virtual displacement. This is done in each of the three preceding sample problems, where the virtual displacements are all expressed in terms of $\delta\theta$.

**6. Remember that the method of virtual work is effective only in those cases** where the geometry of the system makes it relatively easy to relate the displacements involved.

# Problems

**10.1** Determine the vertical force **P** that must be applied at *C* to maintain the equilibrium of the linkage.

**10.2** Determine the horizontal force **P** that must be applied at *A* to maintain the equilibrium of the linkage.

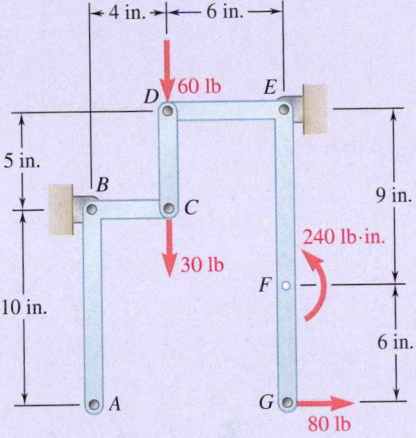

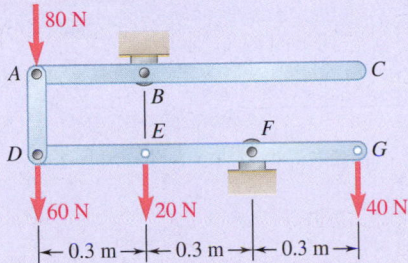

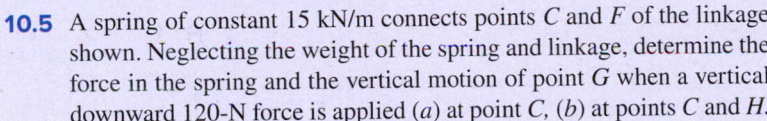

**Fig. P10.1 and P10.3**

**10.3 and 10.4** Determine the couple **M** that must be applied to member *ABC* to maintain the equilibrium of the linkage.

**Fig. P10.2 and P10.4**

**10.5** A spring of constant 15 kN/m connects points *C* and *F* of the linkage shown. Neglecting the weight of the spring and linkage, determine the force in the spring and the vertical motion of point *G* when a vertical downward 120-N force is applied (*a*) at point *C*, (*b*) at points *C* and *H*.

**10.6** A spring of constant 15 kN/m connects points *C* and *F* of the linkage shown. Neglecting the weight of the spring and linkage, determine the force in the spring and the vertical motion of point *G* when a vertical downward 120-N force is applied (*a*) at point *E*, (*b*) at points *E* and *F*.

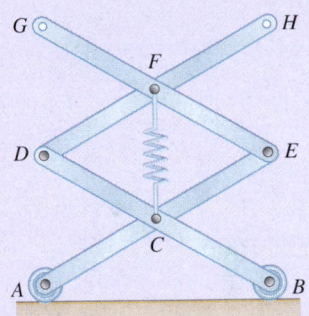

**Fig. P10.5 and P10.6**

**10.7** The two-bar linkage shown is supported by a pin and bracket at *B* and a collar at *D* that slides freely on a vertical rod. Determine the force **P** required to maintain the equilibrium of the linkage.

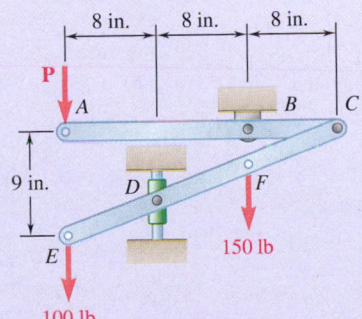

**Fig. P10.7**

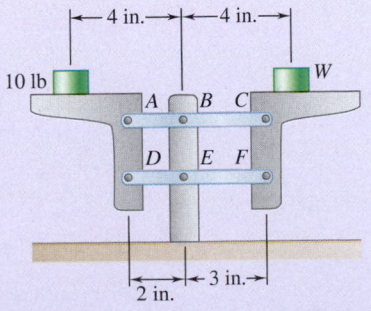

**Fig. P10.8**

**10.8** Determine the weight *W* that balances the 10-lb load placed on the linkage shown.

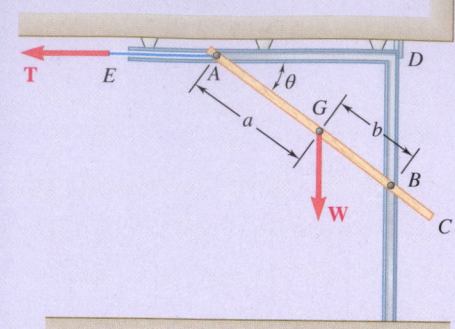

Fig. P10.9

**10.9** An overhead garage door of weight $W$ consists of a uniform rectangular panel $AC$ supported by a cable $AE$ attached at the middle of the upper edge of the door and by two sets of frictionless rollers $A$ and $B$ that can slide in horizontal and vertical channels. Express the tension $T$ in cable $AE$ in terms of $W$, $a$, $b$, and $\theta$.

**10.10** The slender rod $AB$ is attached to a collar $B$ and rests on a small wheel at $C$. Neglecting the radius of the wheel and the effect of friction, derive an expression for the magnitude of the force $\mathbf{Q}$ required to maintain the equilibrium of the rod.

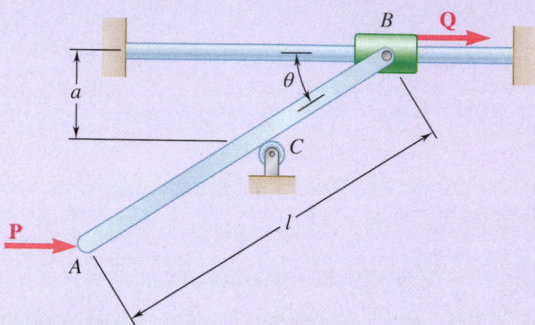

Fig. P10.10

**10.11** Solve Prob. 10.10, assuming that the force $\mathbf{P}$ applied at $A$ is vertical and directed downward.

**10.12** Knowing that the line of action of the force $\mathbf{Q}$ passes through point $C$, derive an expression for the magnitude of $\mathbf{Q}$ required to maintain equilibrium.

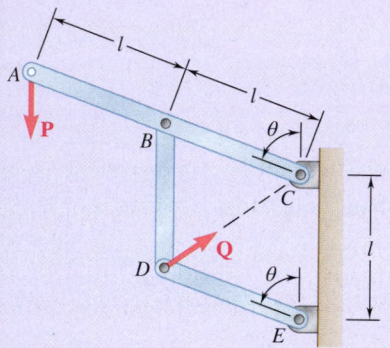

Fig. P10.12

**10.13** Solve Prob. 10.12 assuming that the force $\mathbf{P}$ applied at point $A$ acts horizontally to the left.

**10.14** The mechanism shown is acted upon by the force $\mathbf{P}$. Derive an expression for the magnitude of the force $\mathbf{Q}$ required to maintain equilibrium.

**10.15 and 10.16** Derive an expression for the magnitude of the couple $\mathbf{M}$ required to maintain the equilibrium of the linkage shown.

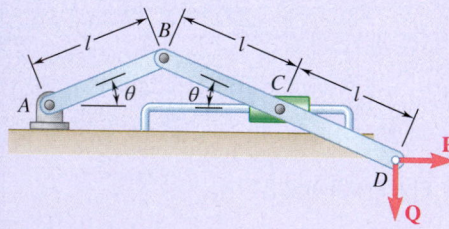

Fig. P10.14

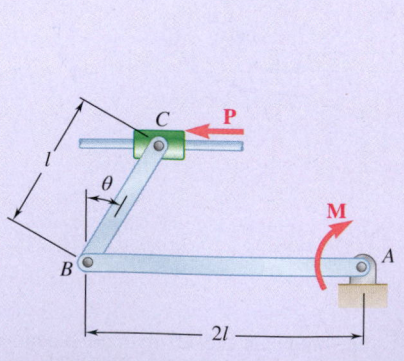

Fig. P10.15

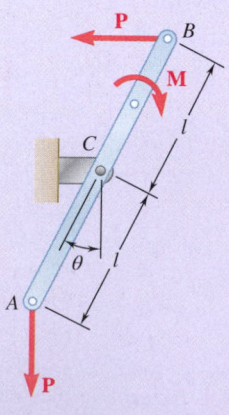

Fig. P10.16

**10.17** A uniform rod $AB$ with length $l$ and weight $W$ is suspended from two cords $AC$ and $BC$ of equal length. Derive an expression for the magnitude of the couple $\mathbf{M}$ required to maintain equilibrium of the rod in the position shown.

**10.18** The pin at $C$ is attached to member $BCD$ and can slide along a slot cut in the fixed plate shown. Neglecting the effect of friction, derive an expression for the magnitude of the couple $\mathbf{M}$ required to maintain equilibrium when the force $\mathbf{P}$ that acts at $D$ is directed ($a$) as shown, ($b$) vertically downward, ($c$) horizontally to the right.

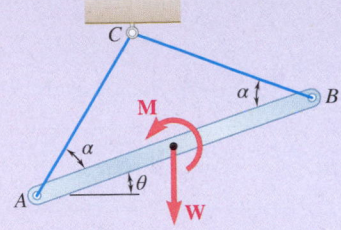

**Fig. P10.17**

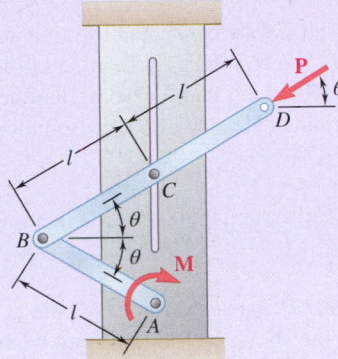

**Fig. P10.18**

**10.19** For the linkage shown, determine the couple $\mathbf{M}$ required for equilibrium when $l = 1.8$ ft, $Q = 40$ lb, and $\theta = 65°$.

**10.20** For the linkage shown, determine the force $\mathbf{Q}$ required for equilibrium when $l = 18$ in., $M = 600$ lb·in., and $\theta = 70°$.

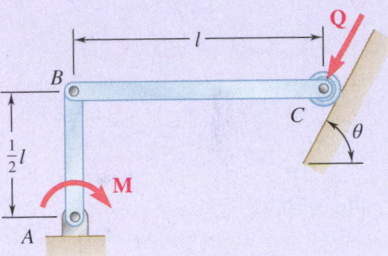

**Fig. P10.19 and P10.20**

**10.21** A 4-kN force $\mathbf{P}$ is applied as shown to the piston of the engine system. Knowing that $AB = 50$ mm and $BC = 200$ mm, determine the couple $\mathbf{M}$ required to maintain the equilibrium of the system when ($a$) $\theta = 30°$, ($b$) $\theta = 150°$.

**10.22** A couple $\mathbf{M}$ with a magnitude of 100 N·m is applied as shown to the crank of the engine system. Knowing that $AB = 50$ mm and $BC = 200$ mm, determine the force $\mathbf{P}$ required to maintain the equilibrium of the system when ($a$) $\theta = 60°$, ($b$) $\theta = 120°$.

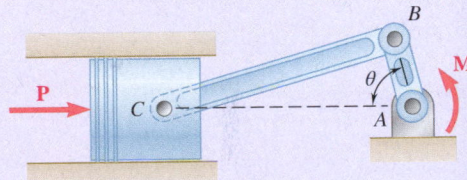

**Fig. P10.21 and P10.22**

**10.23** Rod $AB$ is attached to a block at $A$ that can slide freely in the vertical slot shown. Neglecting the effect of friction and the weights of the rods, determine the value of $\theta$ corresponding to equilibrium.

**10.24** Solve Prob. 10.23, assuming that the 800-N force is replaced by a 24-N·m clockwise couple applied at $D$.

**10.25** In Prob. 10.9, knowing that $a = 42$ in., $b = 28$ in., and $W = 160$ lb, determine the tension $T$ in cable $AE$ when the door is held in the position for which $BD = 42$ in.

**10.26** Determine the value of $\theta$ corresponding to the equilibrium position of the mechanism of Prob. 10.14 when $P = 50$ lb and $Q = 75$ lb.

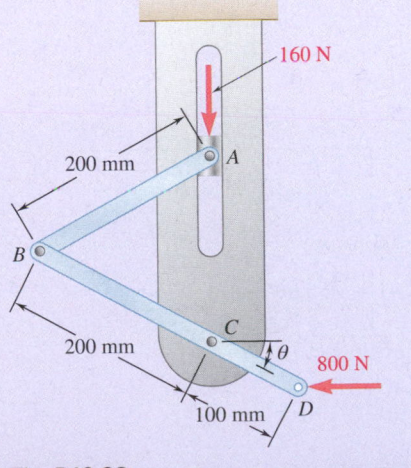

**Fig. P10.23**

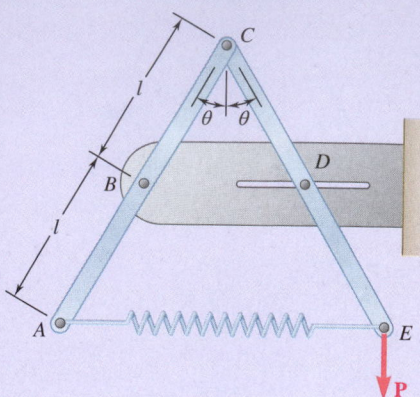

**Fig. P10.29 and P10.30**

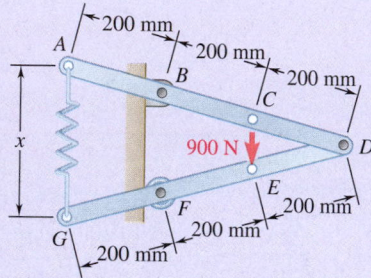

**Fig. P10.32**

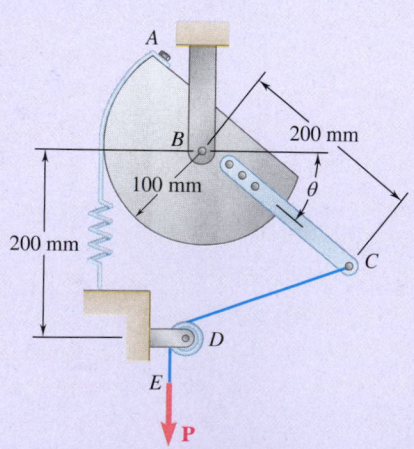

**Fig. P10.35**

**10.27** Determine the value of $\theta$ corresponding to the equilibrium position of the mechanism of Prob. 10.12 when $P = 80$ N and $Q = 100$ N.

**10.28** Determine the value of $\theta$ corresponding to the equilibrium position of the mechanism of Prob. 10.10 when $P = 250$ N, $Q = 500$ N, $l = 400$ mm, and $a = 80$ mm.

**10.29** Two rods $AC$ and $CE$ are connected by a pin at $C$ and by a spring $AE$. The constant of the spring is $k$, and the spring is unstretched when $\theta = 30°$. For the loading shown, derive an equation in $P$, $\theta$, $l$, and $k$ that must be satisfied when the system is in equilibrium.

**10.30** Two rods $AC$ and $CE$ are connected by a pin at $C$ and by a spring $AE$. The constant of the spring is 1.5 lb/in., and the spring is unstretched when $\theta = 30°$. Knowing that $l = 10$ in. and neglecting the weight of the rods, determine the value of $\theta$ corresponding to equilibrium when $P = 40$ lb.

**10.31** Solve Prob. 10.30 assuming that force **P** is moved to $C$ and acts vertically downward.

**10.32** Two bars $AD$ and $DG$ are connected by a pin at $D$ and by a spring $AG$. Knowing that the spring is 300 mm long when unstretched and that the constant of the spring is 5 kN/m, determine the value of $x$ corresponding to equilibrium when a 900-N load is applied at $E$ as shown.

**10.33** Solve Prob. 10.32 assuming that the 900-N vertical force is applied at $C$ instead of $E$.

**10.34** Two 5-kg bars $AB$ and $BC$ are connected by a pin at $B$ and by a spring $DE$. Knowing that the spring is 150 mm long when unstretched and that the constant of the spring is 1 kN/m, determine the value of $x$ corresponding to equilibrium.

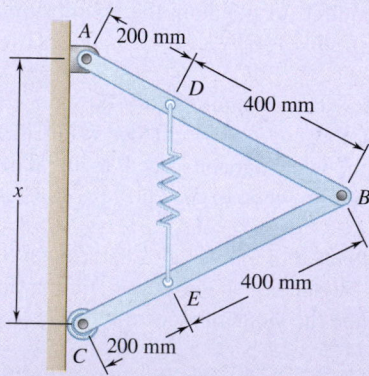

**Fig. P10.34**

**10.35** A vertical force **P** with a magnitude of 150 N is applied to end $E$ of cable $CDE$ that passes over a small pulley $D$ and is attached to the mechanism at $C$. The constant of the spring is $k = 4$ kN/m, and the spring is unstretched when $\theta = 0$. Neglecting the weight of the mechanism and the radius of the pulley, determine the value of $\theta$ corresponding to equilibrium.

**10.36** A load **W** with a magnitude of 72 lb is applied to the mechanism at *C*. Neglecting the weight of the mechanism, determine the value of $\theta$ corresponding to equilibrium. The constant of the spring is $k = 20$ lb/in., and the spring is unstretched when $\theta = 0$.

**10.37 and 10.38** Knowing that the constant of spring *CD* is *k* and that the spring is unstretched when rod *ABC* is horizontal, determine the value of $\theta$ corresponding to equilibrium for the data indicated.

  **10.37** $P = 300$ N, $l = 400$ mm, and $k = 5$ kN/m
  **10.38** $P = 75$ lb, $l = 15$ in., and $k = 20$ lb/in.

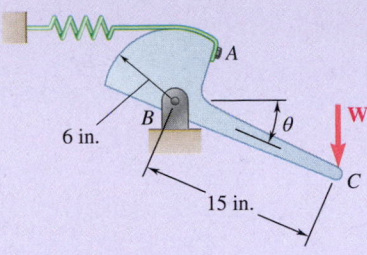

**Fig. P10.36**

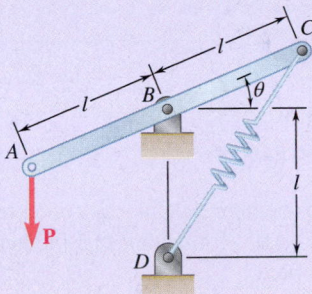

**Fig. P10.37 and P10.38**

**10.39** The lever *AB* is attached to the horizontal shaft *BC* that passes through a bearing and is welded to a fixed support at *C*. The torsional spring constant of the shaft *BC* is *K*; that is, a couple of magnitude *K* is required to rotate end *B* through 1 rad. Knowing that the shaft is untwisted when *AB* is horizontal, determine the value of $\theta$ corresponding to the position of equilibrium when $P = 100$ N, $l = 250$ mm, and $K = 12.5$ N·m/rad.

**10.40** Solve Prob. 10.39, assuming that $P = 350$ N, $l = 250$ mm, and $K = 12.5$ N·m/rad. Obtain answers in each of the following quadrants: $0 < \theta < 90°$, $270° < \theta < 360°$, and $360° < \theta < 450°$.

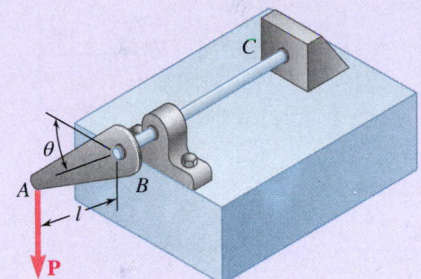

**Fig. P10.39**

**10.41** The position of boom *ABC* is controlled by the hydraulic cylinder *BD*. For the loading shown, determine the force exerted by the hydraulic cylinder on pin *B* when $\theta = 70°$.

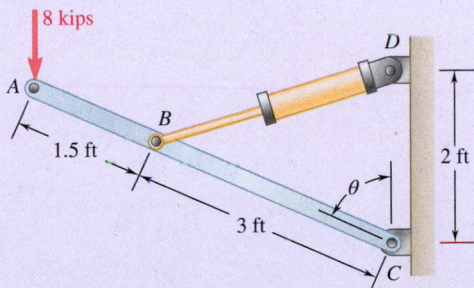

**Fig. P10.41 and P10.42**

**10.42** The position of boom *ABC* is controlled by the hydraulic cylinder *BD*. For the loading shown, determine the largest allowable value of the angle $\theta$ if the maximum force that the cylinder can exert on pin *B* is 25 kips.

**593**

**10.43** The position of member *ABC* is controlled by the hydraulic cylinder *CD*. For the loading shown, determine the force exerted by the hydraulic cylinder on pin *C* when $\theta = 55°$.

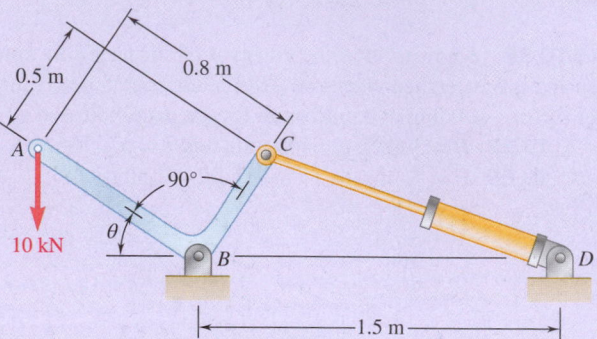

**Fig. P10.43 and P10.44**

**10.44** The position of member *ABC* is controlled by the hydraulic cylinder *CD*. Determine angle $\theta$, knowing that the hydraulic cylinder exerts a 15-kN force on pin *C*.

**10.45** The telescoping arm *ABC* is used to provide an elevated platform for construction workers. The workers and the platform together weigh 500 lb, and their combined center of gravity is located directly above *C*. For the position when $\theta = 20°$, determine the force exerted on pin *B* by the single hydraulic cylinder *BD*.

**10.46** Solve Prob. 10.45, assuming that the workers are lowered to a point near the ground so that $\theta = -20°$.

**10.47** Denoting the coefficient of static friction between collar *C* and the vertical rod by $\mu_s$, derive an expression for the magnitude of the largest couple **M** for which equilibrium is maintained in the position shown. Explain what happens if $\mu_s \geq \tan \theta$.

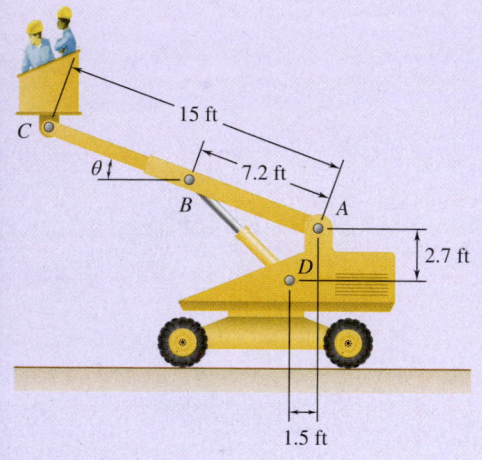

**Fig. P10.45**

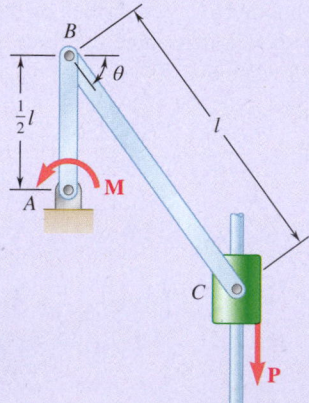

**Fig. *P10.47* and P10.48**

**10.48** Knowing that the coefficient of static friction between collar *C* and the vertical rod is 0.40, determine the magnitude of the largest and smallest couple **M** for which equilibrium is maintained in the position shown when $\theta = 35°$, $l = 600$ mm, and $P = 300$ N.

**10.49** A block with weight $W$ is pulled up a plane forming an angle $\alpha$ with the horizontal by a force **P** directed along the plane. If $\mu$ is the coefficient of friction between the block and the plane, derive an expression for the mechanical efficiency of the system. Show that the mechanical efficiency cannot exceed $\frac{1}{2}$ if the block is to remain in place when the force **P** is removed.

**10.50** Derive an expression for the mechanical efficiency of the jack discussed in Sec. 8.2B. Show that if the jack is to be self-locking, the mechanical efficiency cannot exceed $\frac{1}{2}$.

**10.51** Denoting the coefficient of static friction between the block attached to rod $ACE$ and the horizontal surface by $\mu_s$, derive expressions in terms of $P$, $\mu_s$, and $\theta$ for the largest and smallest magnitude of the force **Q** for which equilibrium is maintained.

**10.52** Knowing that the coefficient of static friction between the block attached to rod $ACE$ and the horizontal surface is 0.15, determine the magnitude of the largest and smallest force **Q** for which equilibrium is maintained when $\theta = 30°$, $l = 0.2$ m, and $P = 40$ N.

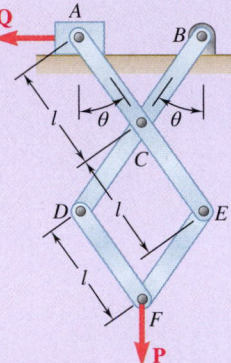

**Fig. P10.51 and P10.52**

**10.53** Using the method of virtual work, determine separately the force and couple representing the reaction at $A$.

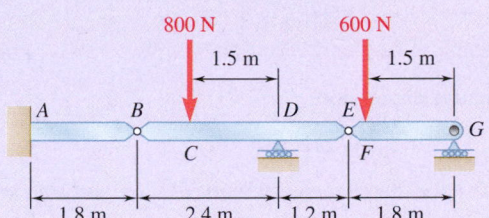

800 N    600 N
1.5 m    1.5 m

1.8 m    2.4 m    1.2 m    1.8 m

**Fig. P10.53 and P10.54**

**10.54** Using the method of virtual work, determine the reaction at $D$.

**10.55** Referring to Prob. 10.43 and using the value found for the force exerted by the hydraulic cylinder $CD$, determine the change in the length of $CD$ required to raise the 10-kN load by 15 mm.

**10.56** Referring to Prob. 10.45 and using the value found for the force exerted by the hydraulic cylinder $BD$, determine the change in the length of $BD$ required to raise the platform attached at $C$ by 2.5 in.

**10.57** Determine the vertical movement of joint $C$ if the length of member $FG$ is increased by 1.5 in. (*Hint:* Apply a vertical load at joint $C$, and using the methods of Chap. 6, compute the force exerted by member $FG$ on joints $F$ and $G$. Then, apply the method of virtual work for a virtual displacement resulting in the specified increase in length of member $FG$. This method should be used only for small changes in the lengths of members.)

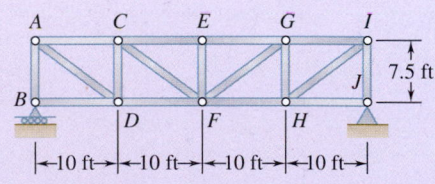

7.5 ft

←10 ft→←10 ft→←10 ft→←10 ft→

**Fig. P10.57**

**10.58** Determine the horizontal movement of joint $C$ if the length of member $FG$ is increased by 1.5 in. (See the hint for Prob. 10.57.)

## *10.2 WORK, POTENTIAL ENERGY, AND STABILITY

The concept of virtual work has another important connection with equilibrium, leading to criteria for conditions of stable, unstable, and neutral equilibrium. However, to explain this connection, we first need to introduce expressions for the work of a force during a finite displacement and then to define the concept of potential energy.

### 10.2A Work of a Force During a Finite Displacement

Consider a force $\mathbf{F}$ acting on a particle. In Sec. 10.1A, we defined the work of $\mathbf{F}$ corresponding to an infinitesimal displacement $d\mathbf{r}$ of the particle as

$$dU = \mathbf{F} \cdot d\mathbf{r} \tag{10.1}$$

We obtain the work of $\mathbf{F}$ corresponding to a finite displacement of the particle from $A_1$ to $A_2$ (Fig. 10.10$a$) that is denoted by $U_{1\rightarrow2}$ by integrating Eq. (10.1) along the curve described by the particle. Thus,

**Work during a finite displacement**

$$U_{1\rightarrow2} = \int_{A_1}^{A_2} \mathbf{F} \cdot d\mathbf{r} \tag{10.11}$$

Using the alternative expression

$$dU = F \, ds \cos \alpha \tag{10.1'}$$

given in Sec. 10.1 for the elementary work $dU$, we can also express the work $U_{1\rightarrow2}$ as

$$U_{1\rightarrow2} = \int_{s_1}^{s_2} (F \cos \alpha) \, ds \tag{10.11'}$$

Here, the variable of integration $s$ measures the distance along the path traveled by the particle. We can represent the work $U_{1\rightarrow2}$ by the area under the curve obtained by plotting $F \cos \alpha$ against $s$ (Fig. 10.10$b$). In the case of a force $\mathbf{F}$ of constant magnitude acting in the direction of motion, formula (10.11$'$) yields $U_{1\rightarrow2} = F(s_2 - s_1)$.

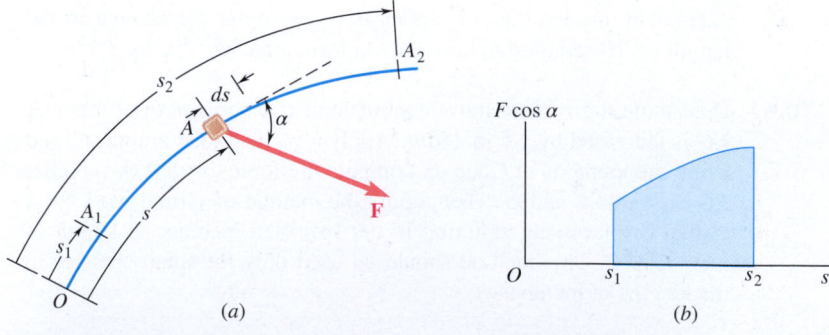

**Fig. 10.10** ($a$) A force acting on a particle moving along a path from $A_1$ to $A_2$; ($b$) the work done by the force in ($a$) equals the area under the graph of $F \cos \alpha$ versus $s$.

Recall from Sec. 10.1 that the work of a couple of moment **M** during an infinitesimal rotation $d\theta$ of a rigid body is

$$dU = M \, d\theta \qquad (10.2)$$

Therefore, we can express the work of the couple during a finite rotation of the body as

**Work during a finite rotation**

$$U_{1\rightarrow2} = \int_{\theta_1}^{\theta_2} M \, d\theta \qquad (10.12)$$

In the case of a constant couple, formula (10.12) yields

$$U_{1\rightarrow2} = M(\theta_2 - \theta_1)$$

**Work of a Weight.** We stated in Sec. 10.1 that the work of a body's weight **W** during an infinitesimal displacement of the body is equal to the product of $W$ and the vertical displacement of the body's center of gravity. With the $y$ axis pointing upward, we obtain the work of **W** during a finite displacement of the body (Fig. 10.11) from

$$dU = -W \, dy$$

Integrating from $A_1$ to $A_2$, we have

$$U_{1\rightarrow2} = -\int_{y_1}^{y_2} W \, dy = Wy_1 - Wy_2 \qquad (10.13)$$

or

$$U_{1\rightarrow2} = -W(y_2 - y_1) = -W \, \Delta y \qquad (10.13')$$

where $\Delta y$ is the vertical displacement from $A_1$ to $A_2$. The work of the weight **W** is thus equal to **the product of** $W$ **and the vertical displacement of the center of gravity of the body.** The work is *positive* when $\Delta y < 0$; that is, when the body moves down.

**Work of the Force Exerted by a Spring.** Consider a body A attached to a fixed point $B$ by a spring. We assume that the spring is undeformed when the body is at $A_0$ (Fig. 10.12a). Experimental evidence shows that the magnitude of the force **F** exerted by the spring on a body A is proportional to the deflection $x$ of the spring measured from position $A_0$. We have

$$F = kx \qquad (10.14)$$

where $k$ is the **spring constant** expressed in SI units of N/m or U.S. customary units of lb/ft or lb/in. The work of force **F** exerted by the spring during a finite displacement of the body from $A_1(x = x_1)$ to $A_2(x = x_2)$ is obtained from

$$dU = -F \, dx = -kx \, dx$$

$$U_{1\rightarrow2} = -\int_{x_1}^{x_2} kx \, dx = \frac{1}{2}kx_1^2 - \frac{1}{2}kx_2^2 \qquad (10.15)$$

You should take care to express $k$ and $x$ in consistent units. For example, if you use U.S. customary units, $k$ should be expressed in lb/ft and $x$ expressed in feet, or $k$ is given in lb/in. and $x$ in inches. In the first case, the work is obtained in ft·lb; in the second case, it is in in·lb. We note that the work of the force **F**

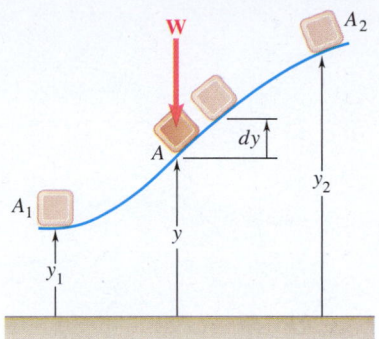

**Fig. 10.11** The work done by the weight of a body equals the magnitude of the weight times the vertical displacement of its center of gravity.

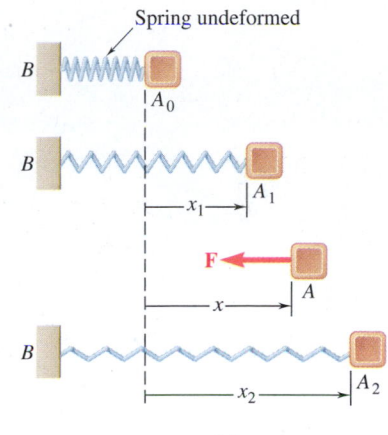

(a)

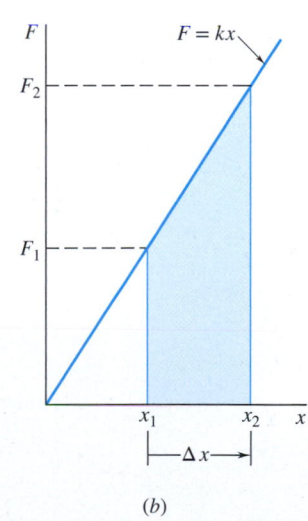

(b)

**Fig. 10.12** (a) When a body is attached to a fixed point by a spring, the force on it is the product of the spring constant and the displacement from the undeformed position; (b) the work of the force equals the area under the graph of $F$ versus $x$ between $x_1$ and $x_2$.

exerted by the spring on the body is *positive* when $x_2 < x_1$; that is, *when the spring is returning to its undeformed position.*

Because Eq. (10.14) is the equation of a straight line of slope $k$ passing through the origin, we can obtain the work $U_{1 \to 2}$ of **F** during the displacement from $A_1$ to $A_2$ by evaluating the area of the trapezoid shown in Fig. 10.12*b*. This is done by computing the values $F_1$ and $F_2$ and multiplying the base $\Delta x$ of the trapezoid by its mean height as $\frac{1}{2}(F_1 + F_2)$. Because the work of the force **F** exerted by the spring is positive for a negative value of $\Delta x$, we have

$$U_{1 \to 2} = -\frac{1}{2}(F_1 + F_2)\,\Delta x \qquad (10.16)$$

Eq. (10.16) is usually more convenient to use than Eq. (10.15) and affords fewer chances of confusing the units involved.

## 10.2B Potential Energy

Let's consider again the body of Fig. 10.11. Using Eq. (10.13), we obtain the work of weight **W** during a finite displacement by subtracting the value of the function $Wy$ corresponding to the second position of the body from its value corresponding to the first position. Thus, the work of **W** is independent of the actual path followed; it depends only upon the initial and final values of the function $Wy$. This function is called the **potential energy** of the body with respect to the force due to gravity **W** and is denoted by $V_g$. Thus,

$$U_{1 \to 2} = (V_g)_1 - (V_g)_2 \qquad \text{with } V_g = Wy \qquad (10.17)$$

Note that if $(V_g)_2 > (V_g)_1$, that is, *if the potential energy increases* during the displacement (as in the case considered here), *the work $U_{1 \to 2}$ is negative.* If, on the other hand, the work of **W** is positive, the potential energy decreases. Therefore, the potential energy $V_g$ of the body provides a measure of *the work that can be done* by its weight **W**. Because only the *change* in potential energy—not the actual value of $V_g$—is involved in formula (10.17), we can add an arbitrary constant to the expression obtained for $V_g$. In other words, the level from which the elevation $y$ is measured can be chosen arbitrarily. Note that potential energy is expressed in the same units as work, that is, in joules (J) if SI units are used[†] and in ft·lb or in·lb if U.S. customary units are used.

Now consider the body of Fig. 10.12*a*. Using Eq. (10.15), we obtain the work of the elastic force **F** by subtracting the value of the function $\frac{1}{2}kx^2$ corresponding to the second position of the body from its value corresponding to the first position. This function, denoted by $V_e$, is called the **potential energy** of the body with respect to the **elastic force F**. We have

$$U_{1 \to 2} = (V_e)_1 - (V_e)_2 \qquad \text{with } V_e = \frac{1}{2}kx^2 \qquad (10.18)$$

Note that during the displacement considered, the work of force **F** exerted by the spring on the body is negative and the potential energy $V_e$ increases. Also note that the expression obtained for $V_e$ is valid only if the deflection of the spring is measured from its undeformed position.

We can use the concept of potential energy when forces other than gravity forces and elastic forces are involved. It remains valid as long as the elementary

---

[†]The joule is the SI unit of *energy,* whether in mechanical form (work, potential energy, kinetic energy) or in chemical, electrical, or thermal form. Note that even though 1 N·m = 1 J, we must express the moment of a force in N·m, and not in joules, because the moment of a force is not a form of energy.

work $dU$ of the force considered is an *exact differential*. It is then possible to find a function $V$, called potential energy, such that

$$dU = -dV \qquad (10.19)$$

Integrating Eq. (10.19) over a finite displacement, we obtain

**Potential energy,
general formulation**

$$U_{1\to2} = V_1 - V_2 \qquad (10.20)$$

This equation says that **the work of the force is independent of the path followed and is equal to minus the change in potential energy**. A force that satisfies Eq. (10.20) is called a **conservative force**.[†]

## 10.2C   Potential Energy and Equilibrium

Applying the principle of virtual work is considerably simplified if we know the potential energy of a system. In the case of a virtual displacement, formula (10.19) becomes $\delta U = -\delta V$. Moreover, if the position of the system is defined by a single independent variable $\theta$, we can write $\delta V = (dV/d\theta)\,\delta\theta$. Because $\delta\theta$ must be different from zero, the condition $\delta U = 0$ for the equilibrium of the system becomes

**Equilibrium condition**
$$\frac{dV}{d\theta} = 0 \qquad (10.21)$$

In terms of potential energy, therefore, the principle of virtual work states:

> **If a system is in equilibrium, the derivative of its total potential energy is zero.**

If the position of the system depends upon several independent variables (the system is then said to possess *several degrees of freedom*), the partial derivatives of $V$ with respect to each of the independent variables must be zero.

Consider, for example, a structure made of two members $AC$ and $CB$ and carrying a load $W$ at $C$. The structure is supported by a pin at $A$ and a roller at $B$, and a spring $BD$ connects $B$ to a fixed point $D$ (Fig. 10.13a). The constant of the spring is $k$, and we assume that the natural length of the spring is equal to $AD$, so that the spring is undeformed when $B$ coincides with $A$. Neglecting friction forces and the weights of the members, we find that the only forces that do work during a virtual displacement of the structure are the weight $\mathbf{W}$ and the force $\mathbf{F}$ exerted by the spring at point $B$ (Fig. 10.13b). Therefore, we can obtain the total potential energy of the system by adding the potential energy $V_g$ corresponding to the gravity force $\mathbf{W}$ and the potential energy $V_e$ corresponding to the elastic force $\mathbf{F}$.

Choosing a coordinate system with the origin at $A$ and noting that the deflection of the spring measured from its undeformed position is $AB = x_B$, we have

$$V_e = \frac{1}{2}kx_B^2 \quad \text{and} \quad V_g = Wy_C$$

Expressing the coordinates $x_B$ and $y_C$ in terms of the angle $\theta$, we have

$$x_B = 2l\sin\theta \qquad y_C = l\cos\theta$$

$$V_e = \frac{1}{2}k(2l\sin\theta)^2 \qquad V_g = W(l\cos\theta)$$

$$V = V_e + V_g = 2kl^2\sin^2\theta + Wl\cos\theta \qquad (10.22)$$

[†]A detailed discussion of conservative forces is given in Sec. 13.2B of *Dynamics*.

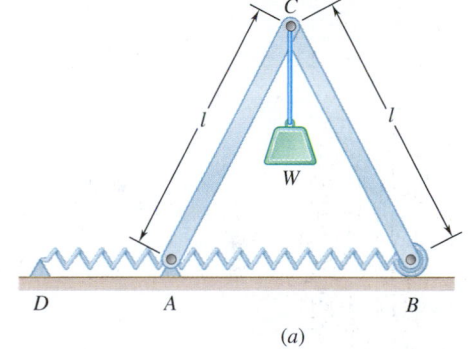

(a)

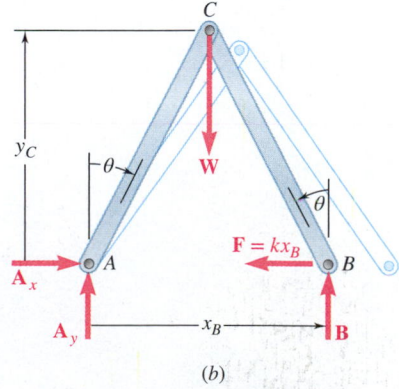

(b)

**Fig. 10.13**   (a) Structure carrying a load at $C$ with a spring from $B$ to $D$; (b) free-body diagram of the structure and a virtual displacement.

We obtain the positions of equilibrium of the system by setting the derivative of the potential energy $V$ to zero, as

$$\frac{dV}{d\theta} = 4kl^2 \sin\theta \cos\theta - Wl \sin\theta = 0$$

or, factoring out $l \sin\theta$, as

$$\frac{dV}{d\theta} = l \sin\theta(4kl \cos\theta - W) = 0$$

There are therefore two positions of equilibrium corresponding to the values $\theta = 0$ and $\theta = \cos^{-1}(W/4kl)$, respectively.[†]

## 10.2D  Stability of Equilibrium

Consider the three uniform rods with a length of $2a$ and weight $\mathbf{W}$ shown in Fig. 10.14. Although each rod is in equilibrium, there is an important difference between the three cases considered. Suppose that each rod is slightly disturbed from its position of equilibrium and then released. Rod $a$ moves back toward its original position; rod $b$ keeps moving away from its original position; and rod $c$ remains in its new position. In case $a$, the equilibrium of the rod is said to be **stable**; in case $b$, it is **unstable**; and in case $c$, it is **neutral**.

Recall from Sec. 10.2B that the potential energy $V_g$ with respect to gravity is equal to $Wy$, where $y$ is the elevation of the point of application of $\mathbf{W}$ measured from an arbitrary level. We observe that the potential energy of rod $a$ is minimum in the position of equilibrium considered, that the potential energy of rod $b$ is maximum, and that the potential energy of rod $c$ is constant. Equilibrium is thus *stable, unstable,* or *neutral* according to whether the potential energy is *minimum, maximum,* or *constant* (Fig. 10.15).

This result is quite general, as we now show. We first observe that a force always tends to do positive work and thus to decrease the potential energy of the system on which it is applied. Therefore, when a system is disturbed from its position of equilibrium, the forces acting on the system tend to bring it back to its original position if $V$ is minimum (Fig. 10.15$a$) and to move it farther away if $V$ is maximum (Fig. 10.15$b$). If $V$ is constant (Fig. 10.15$c$), the forces do not tend to move the system either way.

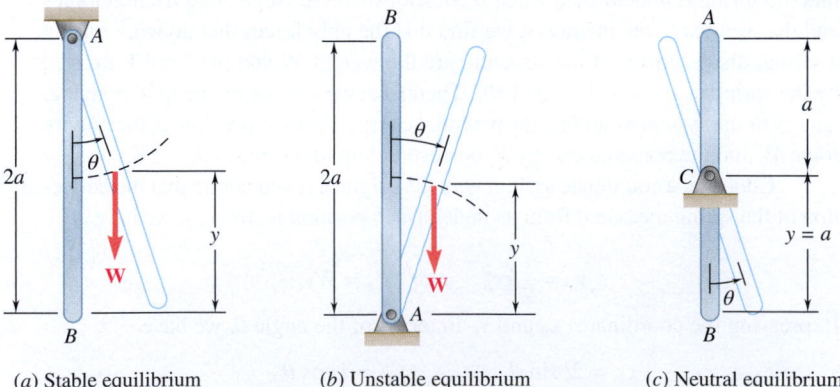

(a) Stable equilibrium        (b) Unstable equilibrium        (c) Neutral equilibrium

**Fig. 10.14**  (*a*) Rod supported from above, stable equilibrium; (*b*) rod supported from below, unstable equilibrium; (*c*) rod supported at its midpoint, neutral equilibrium.

[†]The second position does not exist if $W > 4kl$.

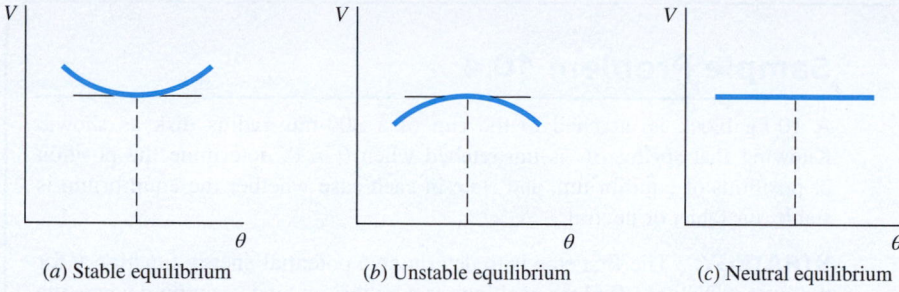

(*a*) Stable equilibrium          (*b*) Unstable equilibrium          (*c*) Neutral equilibrium

**Fig. 10.15**   Stable, unstable, and neutral equilibria correspond to potential energy values that are minimum, maximum, or constant, respectively.

Recall from calculus that a function is minimum or maximum according to whether its second derivative is positive or negative. Therefore, we can summarize the conditions for the equilibrium of a system with one degree of freedom (i.e., a system for which the position is defined by a single independent variable $\theta$) as

$$\frac{dV}{d\theta} = 0 \quad \frac{d^2V}{d\theta^2} > 0 : \text{stable equilibrium}$$

$$\frac{dV}{d\theta} = 0 \quad \frac{d^2V}{d\theta^2} < 0 : \text{unstable equilibrium}$$
(10.23)

If both the first and the second derivatives of $V$ are zero, it is necessary to examine derivatives of a higher order to determine whether the equilibrium is stable, unstable, or neutral. The equilibrium is neutral if all derivatives are zero, because the potential energy $V$ is then a constant. The equilibrium is stable if the first derivative found to be different from zero is of even order and positive. In all other cases, the equilibrium is unstable.

If the system of interest possesses *several degrees of freedom,* the potential energy $V$ depends upon several variables. Thus, it becomes necessary to apply the theory of functions of several variables to determine whether $V$ is minimum. It can be verified that a system with two degrees of freedom is stable, and the corresponding potential energy $V(\theta_1, \theta_2)$ is minimum, if the following relations are satisfied simultaneously:

$$\frac{\partial V}{\partial \theta_1} = \frac{\partial V}{\partial \theta_2} = 0$$

$$\left(\frac{\partial^2 V}{\partial \theta_1 \partial \theta_2}\right)^2 - \frac{\partial^2 V \partial^2 V}{\partial \theta_1^2 \partial \theta_2^2} < 0$$
(10.24)

$$\frac{\partial^2 V}{\partial \theta_1^2} > 0 \quad \text{or} \quad \frac{\partial^2 V}{\partial \theta_2^2} > 0$$

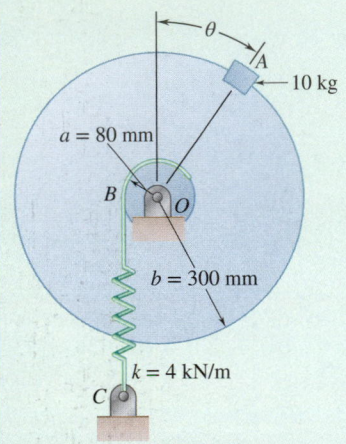

$a = 80$ mm

$B$

$O$

$b = 300$ mm

$k = 4$ kN/m

$C$

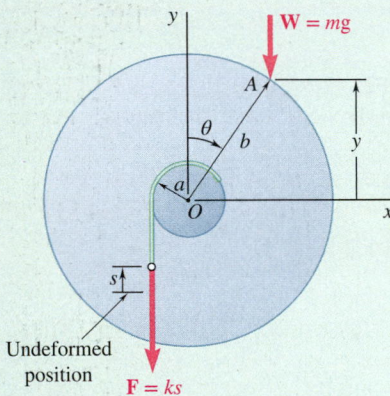

Undeformed position

$\mathbf{F} = ks$

**Fig. 1**   Free-body diagram of rotated disk, showing only those forces that do work.

## Sample Problem 10.4

A 10-kg block is attached to the rim of a 300-mm-radius disk as shown. Knowing that spring $BC$ is unstretched when $\theta = 0$, determine the position or positions of equilibrium, and state in each case whether the equilibrium is stable, unstable, or neutral.

**STRATEGY:**   The first step is to determine a potential energy function $V$ for the system. You can find the positions of equilibrium by determining where the derivative of $V$ is zero. You can find the types of stability by finding where $V$ is maximum or minimum.

**MODELING and ANALYSIS:**

**Potential Energy.**   Denote the deflection of the spring from its undeformed position by $s$, and place the origin of coordinates at $O$ (Fig. 1). You obtain

$$V_e = \tfrac{1}{2}ks^2 \quad V_g = Wy = mgy$$

Measuring $\theta$ in radians, you have

$$s = a\theta \qquad y = b\cos\theta$$

Substituting for $s$ and $y$ in the expressions for $V_e$ and $V_g$ gives you

$$V_e = \tfrac{1}{2}ka^2\theta^2 \quad V_g = mgb\cos\theta$$
$$V = V_e + V_g = \tfrac{1}{2}ka^2\theta^2 + mgb\cos\theta$$

**Positions of Equilibrium.**   Setting $dV/d\theta = 0$, you obtain

$$\frac{dV}{d\theta} = ka^2\theta - mgb\sin\theta = 0$$
$$\sin\theta = \frac{ka^2}{mgb}\theta$$

Now substitute $a = 0.08$ m, $b = 0.3$ m, $k = 4$ kN/m, and $m = 10$ kg. The result is

$$\sin\theta = \frac{(4\text{ kN/m})(0.08\text{ m})^2}{(10\text{ kg})(9.81\text{ m/s}^2)(0.3\text{ m})}\theta$$
$$\sin\theta = 0.8699\,\theta$$

where $\theta$ is expressed in radians. Solving by trial and error for $\theta$, you find

$$\theta = 0 \quad \text{and} \quad \theta = 0.902 \text{ rad}$$
$$\theta = 0 \quad \text{and} \quad \theta = 51.7° \blacktriangleleft$$

**Stability of Equilibrium.**   The second derivative of the potential energy $V$ with respect to $\theta$ is

$$\frac{d^2V}{d\theta^2} = ka^2 - mgb\cos\theta$$
$$= (4\text{ kN/m})(0.08\text{ m})^2 - (10\text{ kg})(9.81\text{ m/s}^2)(0.3\text{ m})\cos\theta$$
$$= 25.6 - 29.43\cos\theta$$

For $\theta = 0$,   $\dfrac{d^2V}{d\theta^2} = 25.6 - 29.43\cos 0° = -3.83 < 0$

The equilibrium is unstable for $\theta = 0$. $\blacktriangleleft$

For $\theta = 51.7°$,   $\dfrac{d^2V}{d\theta^2} = 25.6 - 29.43\cos 51.7° = +7.36 > 0$

The equilibrium is stable for $\theta = 51.7°$. $\blacktriangleleft$

**REFLECT and THINK:**   If you just let the block-and-disk system fall on its own, it will come to rest at $\theta = 51.7°$. If you balance the system at $\theta = 0$, the slightest touch will put it in motion.

# SOLVING PROBLEMS
# ON YOUR OWN

In this section, we defined the **work of a force during a finite displacement** and the **potential energy** of a rigid body or a system of rigid bodies. You saw how to use the concept of potential energy to determine the **equilibrium position** of a rigid body or a system of rigid bodies.

**1. The potential energy $V$ of a system** is the sum of the potential energies associated with the various forces acting on the system that do work as the system moves. In the problems of this section, you will determine the following energies.

    **a. Potential energy of a weight.** This is the potential energy due to *gravity*, $V_g = Wy$, where $y$ is the elevation of the weight $W$ measured from some arbitrary reference level. You can use the potential energy $V_g$ with any vertical force $\mathbf{P}$ of constant magnitude directed downward; we write $V_g = Py$.

    **b. Potential energy of a spring.** This is the potential energy due to the *elastic* force exerted by a spring, $V_e = \frac{1}{2}kx^2$, where $k$ is the constant of the spring and $x$ is the deformation of the spring measured from its unstretched position.

Reactions at fixed supports, internal forces at connections, forces exerted by inextensible cords and cables, and other forces that do no work do not contribute to the potential energy of the system.

**2. Express all distances and angles in terms of a single variable,** such as an angle $\theta$, when computing the potential energy $V$ of a system. This is necessary because determining the equilibrium position of the system requires computing the derivative $dV/d\theta$.

**3. When a system is in equilibrium, the first derivative of its potential energy is zero.** Therefore,

    **a. To determine a position of equilibrium of a system,** first express its potential energy $V$ in terms of the single variable $\theta$, and then compute its derivative and solve the equation $dV/d\theta = 0$ for $\theta$.

    **b. To determine the force or couple required to maintain a system in a given position of equilibrium,** substitute the known value of $\theta$ in the equation $dV/d\theta = 0$, and solve this equation for the desired force or couple.

**4. Stability of equilibrium.** The following rules generally apply:

    **a. Stable equilibrium** occurs when the potential energy of the system is *minimum*; that is, when $dV/d\theta = 0$ and $d^2V/d\theta^2 > 0$ (Figs. 10.14*a* and 10.15*a*).

    **b. Unstable equilibrium** occurs when the potential energy of the system is *maximum*; that is, when $dV/d\theta = 0$ and $d^2V/d\theta^2 < 0$ (Fig. 10.14*b* and Fig. 10.15*b*).

    **c. Neutral equilibrium** occurs when the potential energy of the system is *constant*; $dV/d\theta$, $d^2V/d\theta^2$, and all the successive derivatives of $V$ are then equal to zero (Fig. 10.14*c* and Fig. 10.15*c*).

See the latter part of Section 10.2D for a discussion of the case when $dV/d\theta$, $d^2V/d\theta^2$, but *not all* of the successive derivatives of $V$ are equal to zero.

# Problems

**10.59** Using the method of Sec. 10.2C, solve Prob. 10.29.

**10.60** Using the method of Sec. 10.2C, solve Prob. 10.30.

**10.61** Using the method of Sec. 10.2C, solve Prob. 10.31.

**10.62** Using the method of Sec. 10.2C, solve Prob. 10.32.

**10.63** Using the method of Sec. 10.2C, solve Prob. 10.34.

**10.64** Using the method of Sec. 10.2C, solve Prob. 10.35.

**10.65** Using the method of Sec. 10.2C, solve Prob. 10.37.

**10.66** Using the method of Sec. 10.2C, solve Prob. 10.38.

**10.67** Show that equilibrium is neutral in Prob. 10.1.

**10.68** Show that equilibrium is neutral in Prob. 10.7.

**10.69** Two uniform rods, each with a mass *m,* are attached to gears of equal radii as shown. Determine the positions of equilibrium of the system and state in each case whether the equilibrium is stable, unstable, or neutral.

**10.70** Two uniform rods, *AB* and *CD,* are attached to gears of equal radii as shown. Knowing that $W_{AB} = 8$ lb and $W_{CD} = 4$ lb, determine the positions of equilibrium of the system and state in each case whether the equilibrium is stable, unstable, or neutral.

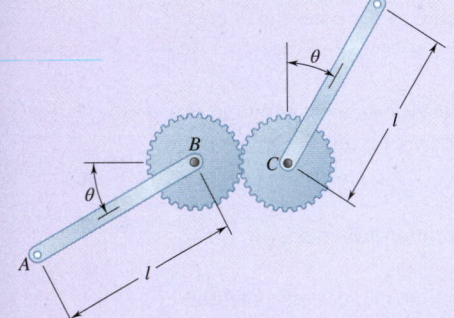

**Fig. P10.69 and P10.70**

**10.71** Two uniform rods *AB* and *CD,* of the same length *l,* are attached to gears as shown. Knowing that rod *AB* weighs 3 lb and that rod *CD* weighs 2 lb, determine the positions of equilibrium of the system and state in each case whether the equilibrium is stable, unstable, or neutral.

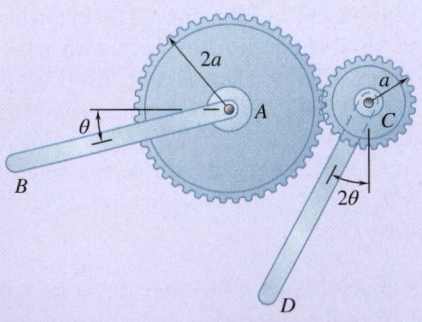

**Fig. P10.71**

**10.72** Two uniform rods, each of mass $m$ and length $l$, are attached to drums that are connected by a belt as shown. Assuming that no slipping occurs between the belt and the drums, determine the positions of equilibrium of the system and state in each case whether the equilibrium is stable, unstable, or neutral.

**10.73** Using the method of Sec. 10.2C, solve Prob. 10.39. Determine whether the equilibrium is stable, unstable, or neutral. (*Hint:* The potential energy corresponding to the couple exerted by a torsion spring is $\frac{1}{2}K\theta^2$, where $K$ is the torsional spring constant and $\theta$ is the angle of twist.)

**10.74** In Prob. 10.40, determine whether each of the positions of equilibrium is stable, unstable, or neutral. (See the hint for Prob. 10.73.)

**10.75** A load **W** of magnitude 144 lb is applied to the mechanism at $C$. Knowing that the constant of the spring is $k = 20$ lb/in., and that the spring is unstretched when $\theta = 0$, determine that value of $\theta$ corresponding to equilibrium and check that the equilibrium is stable.

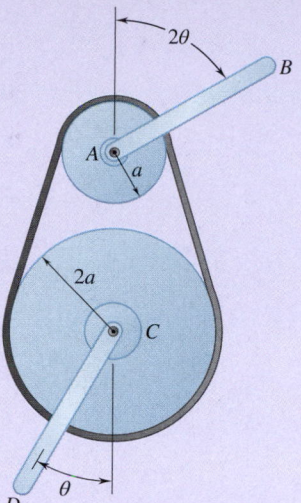

**Fig. P10.72**

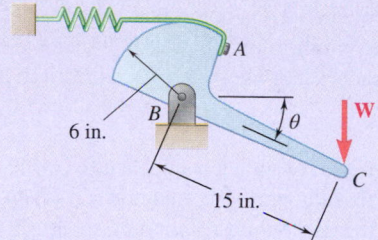

**Fig. P10.75**

**10.76** Solve Prob. 10.75, assuming that the spring is unstretched when $\theta = 30°$.

**10.77** Bar $ABC$ is attached to collars $A$ and $B$ that can move freely on the rods shown. The constant of the spring is $k$ and the spring is unstretched when $\theta = 0$. (*a*) Neglecting the weight of bar $ABC$, derive an equation in $\theta$, $m$, $k$, and $l$ that must be satisfied when bar $ABC$ is in equilibrium. (*b*) Determine the value of $\theta$ corresponding to equilibrium when $m = 5$ kg, $k = 800$ N/m, and $l = 250$ mm, and check that the equilibrium is stable.

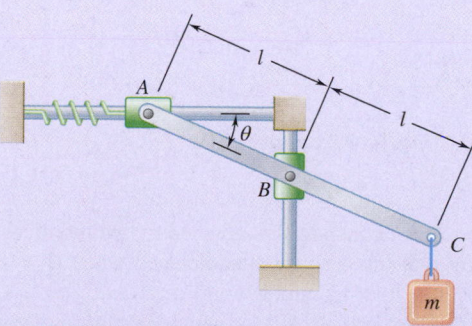

**Fig. P10.77**

**10.78** Solve Prob. 10.77, assuming that the spring is unstretched when $\theta = 30°$.

**10.79** A slender rod $AB$ with a weight $W$ is attached to two blocks $A$ and $B$ that can move freely in the guides shown. The constant of the spring is $k$, and the spring is unstretched when $AB$ is horizontal. Neglecting the weight of the blocks, derive an equation in $\theta$, $W$, $l$, and $k$ that must be satisfied when the rod is in equilibrium.

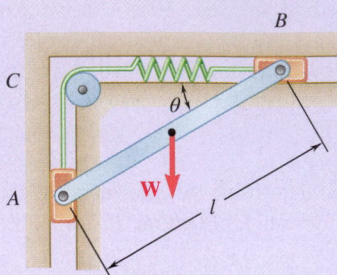

**Fig. *P10.79* and P10.80**

**10.80** A slender rod $AB$ with a weight $W$ is attached to two blocks $A$ and $B$ that can move freely in the guides shown. Knowing that the spring is unstretched when $AB$ is horizontal, determine three values of $\theta$ corresponding to equilibrium when $W = 300$ lb, $l = 16$ in., and $k = 75$ lb/in. State in each case whether the equilibrium is stable, unstable, or neutral.

**10.81** A spring $AB$ of constant $k$ is attached to two identical gears as shown. Knowing that the spring is undeformed when $\theta = 0$, determine two values of the angle $\theta$ corresponding to equilibrium when $P = 30$ lb, $a = 4$ in., $b = 3$ in., $r = 6$ in., and $k = 5$ lb/in. State in each case whether the equilibrium is stable, unstable, or neutral.

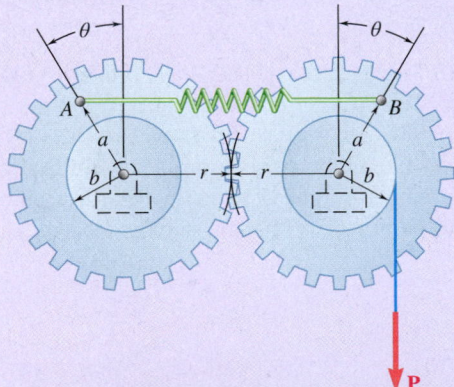

**Fig. P10.81 and *P10.82***

**10.82** A spring $AB$ of constant $k$ is attached to two identical gears as shown. Knowing that the spring is undeformed when $\theta = 0$, and given that $a = 60$ mm, $b = 45$ mm, $r = 90$ mm, and $k = 6$ kN/m, determine (*a*) the range of values of $P$ for which a position of equilibrium exists, (*b*) two values of $\theta$ corresponding to equilibrium if the value of $P$ is equal to half the upper limit of the range found in part *a*.

**10.83** A slender rod $AB$ is attached to two collars $A$ and $B$ that can move freely along the guide rods shown. Knowing that $\beta = 30°$ and $P = Q = 400$ N, determine the value of the angle $\theta$ corresponding to equilibrium.

**10.84** A slender rod $AB$ is attached to two collars $A$ and $B$ that can move freely along the guide rods shown. Knowing that $\beta = 30°$, $P = 100$ N, and $Q = 25$ N, determine the value of the angle $\theta$ corresponding to equilibrium.

**10.85 and 10.86** Cart $B$, which weighs 75 kN, rolls along a sloping track that forms an angle $\beta$ with the horizontal. The spring constant is 5 kN/m, and the spring is unstretched when $x = 0$. Determine the distance $x$ corresponding to equilibrium for the angle $\beta$ indicated.

> **10.85** Angle $\beta = 30°$
> **10.86** Angle $\beta = 60°$

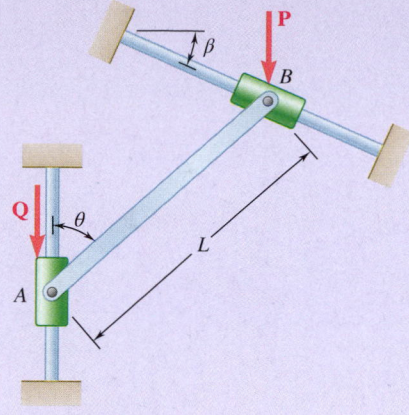

Fig. P10.83 and *P10.84*

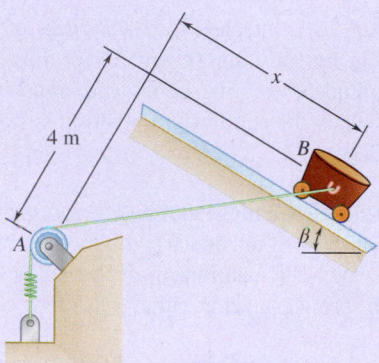

Fig. *P10.85* and P10.86

**10.87 and 10.88** Collar $A$ can slide freely on the semicircular rod shown. Knowing that the constant of the spring is $k$ and that the unstretched length of the spring is equal to the radius $r$, determine the value of $\theta$ corresponding to equilibrium when $W = 50$ lb, $r = 9$ in., and $k = 15$ lb/in.

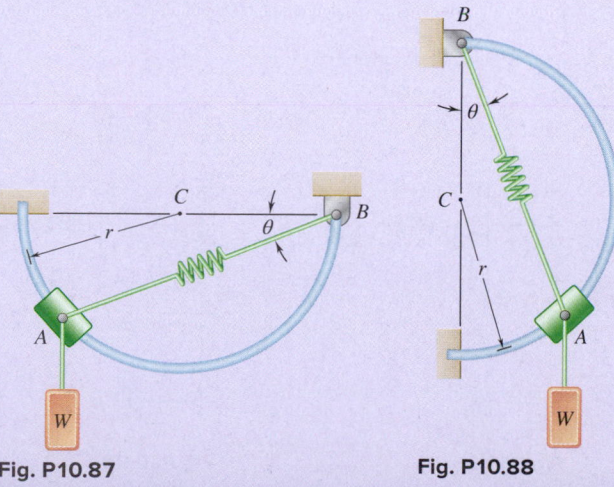

Fig. P10.87

Fig. P10.88

**10.89** Two bars $AB$ and $BC$ of negligible weight are attached to a single spring of constant $k$ that is unstretched when the bars are horizontal. Determine the range of values of the magnitude $P$ of two equal and opposite forces **P** and $-$**P** for which the equilibrium of the system is stable in the position shown.

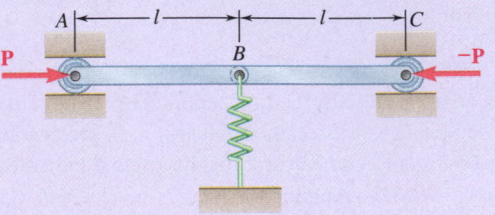

**Fig. P10.89**

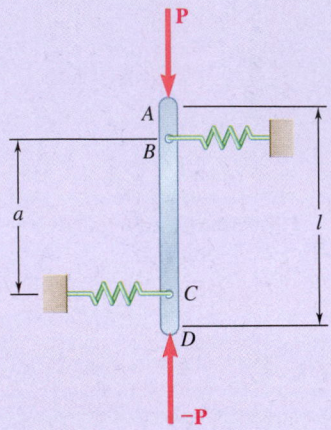

**Fig. *P10.90***

**10.90** A vertical bar $AD$ is attached to two springs of constant $k$ and is in equilibrium in the position shown. Determine the range of values of the magnitude $P$ of two equal and opposite vertical forces **P** and $-$**P** for which the equilibrium position is stable if (*a*) $AB = CD$, (*b*) $AB = 2CD$.

**10.91** Rod $AB$ is attached to a hinge at $A$ and to two springs, each of constant $k$. If $h = 25$ in., $d = 12$ in., and $W = 80$ lb, determine the range of values of $k$ for which the equilibrium of the rod is stable in the position shown. Each spring can act in either tension or compression.

**10.92** Rod $AB$ is attached to a hinge at $A$ and to two springs, each of constant $k$. If $h = 45$ in., $k = 6$ lb/in., and $W = 60$ lb, determine the smallest distance $d$ for which the equilibrium of the rod is stable in the position shown. Each spring can act in either tension or compression.

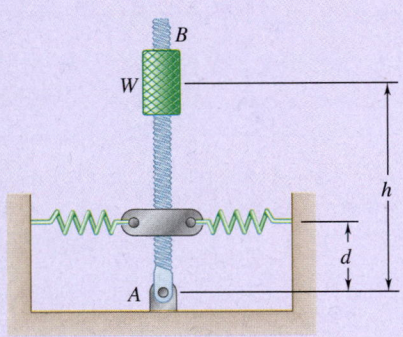

**Fig. P10.91 and P10.92**

**10.93 and 10.94** Two bars are attached to a single spring of constant $k$ that is unstretched when the bars are vertical. Determine the range of values of $P$ for which the equilibrium of the system is stable in the position shown.

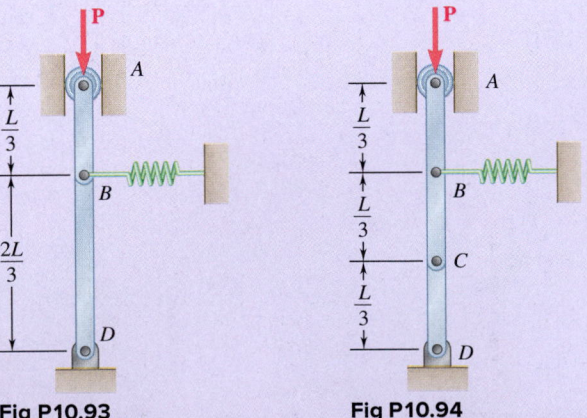

Fig P10.93          Fig P10.94

**10.95** The horizontal bar *BEH* is connected to three vertical bars. The collar at *E* can slide freely on bar *DF*. Determine the range of values of *Q* for which the equilibrium of the system is stable in the position shown when *a* = 24 in., *b* = 20 in., and *P* = 150 lb.

**10.96** The horizontal bar *BEH* is connected to three vertical bars. The collar at *E* can slide freely on bar *DF*. Determine the range of values of *P* for which the equilibrium of the system is stable in the position shown when *a* = 150 mm, *b* = 200 mm, and *Q* = 45 N.

**\*10.97** Bars *AB* and *BC*, each with a length *l* and of negligible weight, are attached to two springs, each of constant *k*. The springs are undeformed, and the system is in equilibrium when $\theta_1 = \theta_2 = 0$. Determine the range of values of *P* for which the equilibrium position is stable.

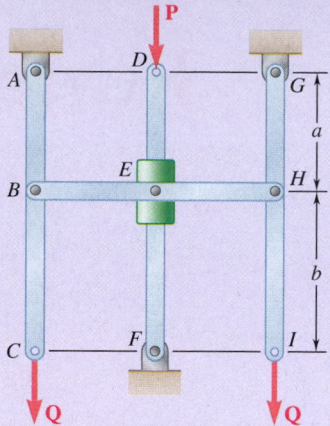

**Fig. P10.95 and P10.96**

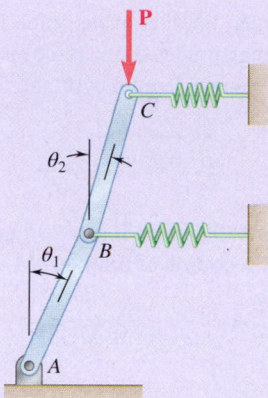

**Fig. P10.97**

**\*10.98** Solve Prob. 10.97 knowing that *l* = 30 in. and *k* = 50 lb/in.

**\*10.99** Bars *AB* and *CD*, each of length *l* and of negligible weight, are attached to a spring of constant *k*. The spring is undeformed and the system is in equilibrium when $\theta_1 = \theta_2 = 0$. Determine the range of values of *P* for which the equilibrium position is stable.

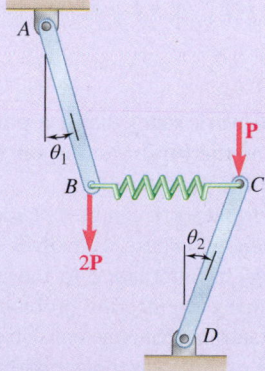

**Fig. P10.99**

**\*10.100** Solve Prob. 10.99, assuming that the vertical force applied at *B* is increased to 5**P**.

# Review and Summary

### Work of a Force

The first section of this chapter was devoted to the **principle of virtual work** and to its direct application to the solution of equilibrium problems. We first defined the **work of a force F corresponding to the small displacement** $d\mathbf{r}$ (Sec. 10.1A) as the quantity

$$dU = \mathbf{F} \cdot d\mathbf{r} \tag{10.1}$$

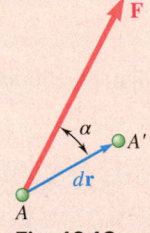

**Fig. 10.16**

obtained by forming the scalar product of the force $\mathbf{F}$ and the displacement $d\mathbf{r}$ (Fig. 10.16). Denoting the magnitudes of the force and of the displacement by $F$ and $ds$, respectively, and the angle formed by $\mathbf{F}$ and $d\mathbf{r}$ by $\alpha$, we have

$$dU = F\, ds \cos \alpha \tag{10.19}$$

The work $dU$ is positive if $\alpha < 90°$, zero if $\alpha = 90°$, and negative if $\alpha > 90°$. We also found that the **work of a couple of moment M** acting on a rigid body is

$$dU = M\, d\theta \tag{10.2}$$

where $d\theta$ is the small angle expressed in radians through which the body rotates.

### Virtual Displacement

Considering a particle located at $A$ and acted upon by several forces $\mathbf{F}_1$, $\mathbf{F}_2$, . . . , $\mathbf{F}_n$ (Sec. 10.1B), we imagined that the particle moved to a new position $A'$ (Fig. 10.17). Because this displacement does not actually take place, we refer to it to as a **virtual displacement** denoted by $\delta\mathbf{r}$. The corresponding work of the forces is called **virtual work** and is denoted by $\delta U$. We have

$$\delta U = \mathbf{F}_1 \cdot \delta\mathbf{r} + \mathbf{F}_2 \cdot \delta\mathbf{r} + \ldots + \mathbf{F}_n \cdot \delta\mathbf{r}$$

### Principle of Virtual Work

The **principle of virtual work** states that **if a particle is in equilibrium, the total virtual work $\delta U$ of the forces acting on the particle is zero for any virtual displacement of the particle**.

The principle of virtual work can be extended to the case of rigid bodies and systems of rigid bodies. Because it involves only forces that do work, its application provides a useful alternative to the use of the equilibrium equations in the solution of many engineering problems. It is particularly effective in the case of machines and mechanisms consisting of connected rigid bodies, because the work of the reactions at the supports is zero and the work of the internal forces at the pin connections cancels out (Sec. 10.1C; Sample Probs. 10.1, 10.2, and 10.3).

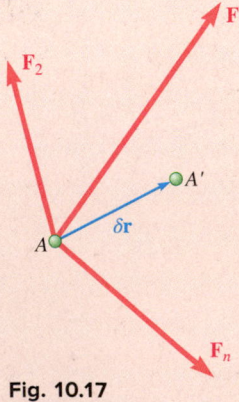

**Fig. 10.17**

## Mechanical Efficiency

In the case of real machines, however (Sec. 10.1D), the work of the friction forces should be taken into account with the result that the **output work is less than the input work**. We defined the **mechanical efficiency** of a machine as the ratio

$$\eta = \frac{\text{output work}}{\text{input work}} \qquad (10.9)$$

We noted that, for an ideal machine (no friction), $\eta = 1$, whereas for a real machine, $\eta < 1$.

## Work of a Force over a Finite Displacement

In the second section of this chapter, we considered the work of forces corresponding to finite displacements of their points of application. We obtained the work $U_{1 \to 2}$ of the force $\mathbf{F}$ corresponding to a displacement of the particle $A$ from $A_1$ to $A_2$ (Fig. 10.18) by integrating the right-hand side of Eqs. (10.1) or (10.1') along the curve described by the particle (Sec. 10.2A). Thus,

$$U_{1 \to 2} = \int_{A_1}^{A_2} \mathbf{F} \cdot d\mathbf{r} \qquad (10.11)$$

or

$$U_{1 \to 2} = \int_{s_1}^{s_2} (F \cos \alpha)\, ds \qquad (10.11')$$

Similarly, we expressed the work of a couple of moment $\mathbf{M}$ corresponding to a finite rotation from $\theta_1$ to $\theta_2$ of a rigid body as

$$U_{1 \to 2} = \int_{\theta_1}^{\theta_2} M\, d\theta \qquad (10.12)$$

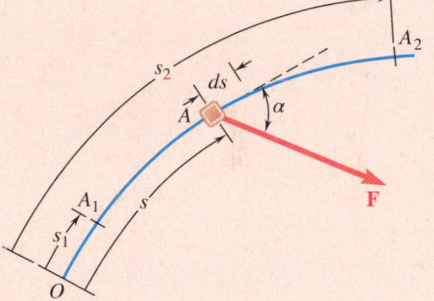

**Fig. 10.18**

## Work of a Weight

We obtained the **work of the weight W of a body** as its center of gravity moves from the elevation $y_1$ to $y_2$ (Fig. 10.19) by setting $F = W$ and $\alpha = 180°$ in Eq. (10.11') as

$$U_{1 \to 2} = -\int_{y_1}^{y_2} W\, dy = W y_1 - W y_2 \qquad (10.13)$$

The work of $\mathbf{W}$ is therefore positive when the elevation $y$ decreases.

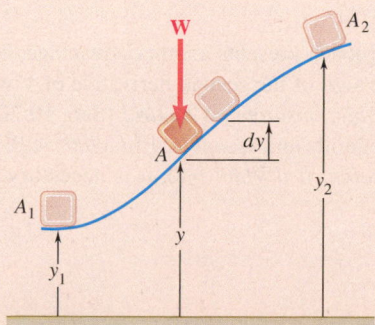

**Fig. 10.19**

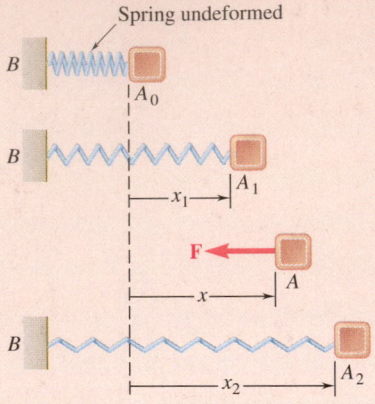

Spring undeformed

**Fig. 10.20**

## Work of the Force Exerted by a Spring

The **work of the force F exerted by a spring** on a body $A$ as the spring is stretched from $x_1$ to $x_2$ (Fig. 10.20) can be obtained by setting $F = kx$, where $k$ is the constant of the spring, and $\alpha = 180°$ in Eq. (10.11'). Hence,

$$U_{1\to2} = -\int_{x_1}^{x_2} kx\,dx = \frac{1}{2}kx_1^2 - \frac{1}{2}kx_2^2 \qquad (10.15)$$

The work of **F** is therefore positive when the spring is returning to its undeformed position.

## Potential Energy

When the work of a force **F** is independent of the path actually followed between $A_1$ and $A_2$, the force is said to be a **conservative force**, and we can express its work as

$$U_{1\to2} = V_1 - V_2 \qquad (10.20)$$

Here, $V$ is the **potential energy** associated with **F**, and $V_1$ and $V_2$ represent the values of $V$ at $A_1$ and $A_2$, respectively (Sec. 10.2B). We found the potential energies associated, respectively, with the force of gravity **W** and the elastic force **F** exerted by a spring to be

$$V_g = Wy \qquad \text{and} \qquad V_e = \frac{1}{2}kx^2 \qquad (10.17, 10.18)$$

## Alternative Expression for the Principle of Virtual Work

When the position of a mechanical system depends upon a single independent variable $\theta$, the potential energy of the system is a function $V(\theta)$ of that variable, and it follows from Eq. (10.20) that $\delta U = -\delta V = -(dV/d\delta)\,\delta\theta$. The condition $\delta U = 0$ required by the principle of virtual work for the equilibrium of the system thus can be replaced by the condition

$$\frac{dV}{d\theta} = 0 \qquad (10.21)$$

When all the forces involved are conservative, it may be preferable to use Eq. (10.21) rather than apply the principle of virtual work directly (Sec. 10.2C; Sample Prob. 10.4).

## Stability of Equilibrium

This alternative approach presents another advantage, because it is possible to determine from the sign of the second derivative of $V$ whether the equilibrium of the system is *stable, unstable,* or *neutral* (Sec. 10.2D). If $d^2V/d\theta^2 > 0$, $V$ is *minimum* and the equilibrium is *stable;* if $d^2V/d\theta^2 < 0$, $V$ is *maximum* and the equilibrium is *unstable;* if $d^2V/d\theta^2 = 0$, it is necessary to examine derivatives of a higher order.

# Review Problems

**10.101** Determine the vertical force **P** that must be applied at $G$ to maintain the equilibrium of the linkage.

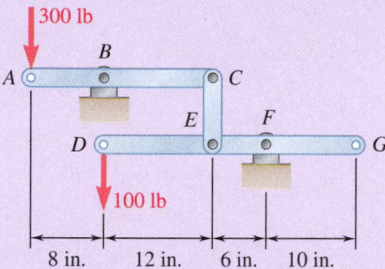

**Fig. P10.101 and P10.102**

**10.102** Determine the couple **M** that must be applied to member *DEFG* to maintain the equilibrium of the linkage.

**10.103** Determine the force **P** required to maintain the equilibrium of the linkage shown. All members are of the same length, and the wheels at $A$ and $B$ roll freely on the horizontal rod.

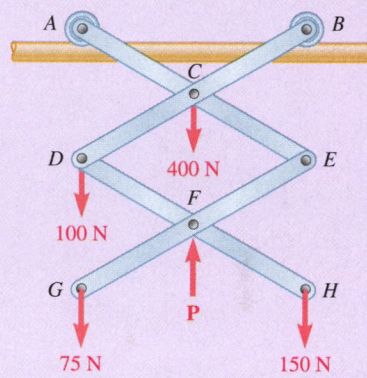

**Fig. P10.103**

**10.104** Derive an expression for the magnitude of the force **Q** required to maintain the equilibrium of the mechanism shown.

**10.105** Derive an expression for the magnitude of the couple **M** required to maintain the equilibrium of the linkage shown.

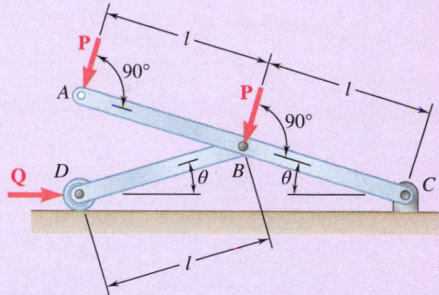

**Fig. P10.104**

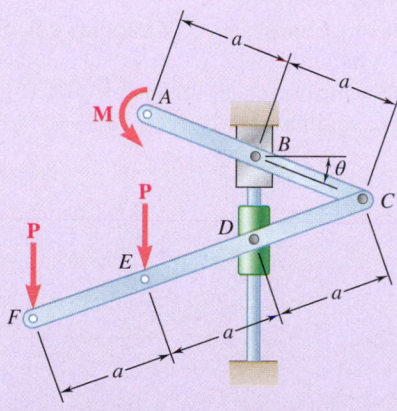

**Fig. P10.105**

**10.106** A vertical load **W** is applied to the linkage at $B$. The constant of the spring is $k$, and the spring is unstretched when $AB$ and $BC$ are horizontal. Neglecting the weight of the linkage, derive an equation in $\theta$, $W$, $l$, and $k$ that must be satisfied when the linkage is in equilibrium.

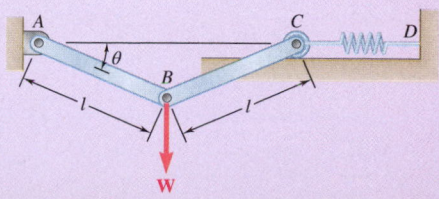

**Fig. P10.106**

**10.107** A force **P** with a magnitude of 240 N is applied to end $E$ of cable $CDE$, which passes under pulley $D$ and is attached to the mechanism at $C$. Neglecting the weight of the mechanism and the radius of the pulley, determine the value of $\theta$ corresponding to equilibrium. The constant of the spring is $k = 4$ kN/m, and the spring is unstretched when $\theta = 90°$.

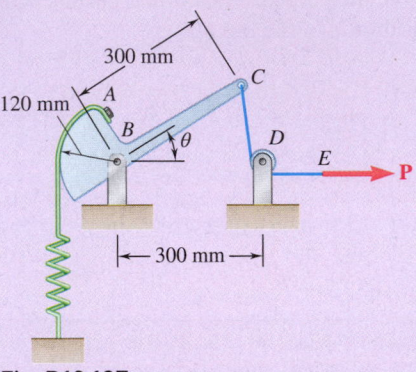

**Fig. P10.107**

**10.108** Two identical rods $ABC$ and $DBE$ are connected by a pin at $B$ and by a spring $CE$. Knowing that the spring is 4 in. long when unstretched and that the constant of the spring is 8 lb/in., determine the distance $x$ corresponding to equilibrium when a 24-lb load is applied at $E$ as shown.

**10.109** Solve Prob. 10.108 assuming that the 24-lb load is applied at $C$ instead of $E$.

**10.110** Two uniform rods each with a mass $m$ and length $l$ are attached to gears as shown. For the range $0 \leq \theta \leq 180°$, determine the positions of equilibrium of the system, and state in each case whether the equilibrium is stable, unstable, or neutral.

**10.111** A homogeneous hemisphere with a radius $r$ is placed on an incline as shown. Assuming that friction is sufficient to prevent slipping between the hemisphere and the incline, determine the angle $\theta$ corresponding to equilibrium when $\beta = 10°$.

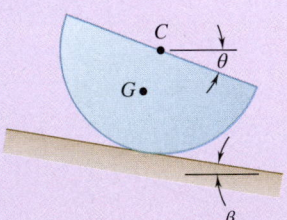

**Fig. P10.111 and P10.112**

**10.112** A homogeneous hemisphere with a radius $r$ is placed on an incline as shown. Assuming that friction is sufficient to prevent slipping between the hemisphere and the incline, determine (a) the largest angle $\beta$ for which a position of equilibrium exists, (b) the angle $\theta$ corresponding to equilibrium when the angle $\beta$ is equal to half the value found in part $a$.

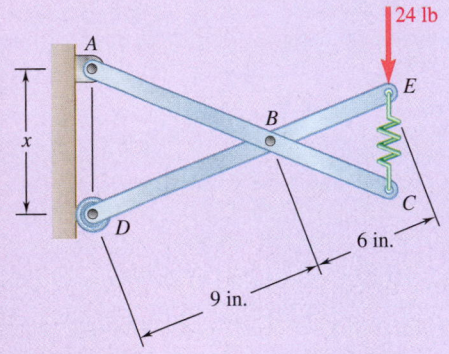

**Fig. P10.108**

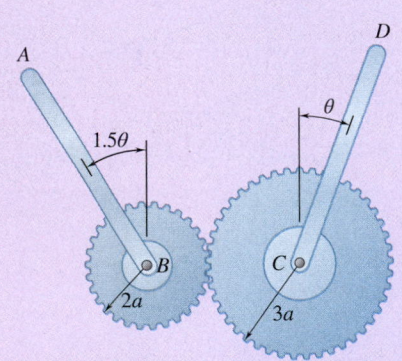

**Fig. P10.110**

# 11

# Kinematics of Particles

The motion of the paraglider can be described in terms of its *position, velocity,* and *acceleration.* When landing, the pilot of the paraglider needs to consider the wind velocity and the *relative motion* of the glider with respect to the wind. The study of motion is known as *kinematics* and is the subject of this chapter.

# Objectives

- **Describe** the basic kinematic relationships between position, velocity, acceleration, and time.
- **Solve** problems using these basic kinematic relationships and calculus or graphical methods.
- **Define** position, velocity, and acceleration in terms of Cartesian, tangential and normal, and radial and transverse coordinates.
- **Analyze** the relative motion of multiple particles by using a translating coordinate system.
- **Determine** the motion of a particle that depends on the motion of another particle.
- **Determine** which coordinate system is most appropriate for solving a curvilinear kinematics problem.
- **Calculate** the position, velocity, and acceleration of a particle undergoing curvilinear motion using Cartesian, tangential and normal, and radial and transverse coordinates.

## Introduction

Chaps. 1 to 10 were devoted to **statics**; that is, to the analysis of bodies at rest. We now begin the study of **dynamics**, which is the part of mechanics that deals with the analysis of bodies in motion.

Although the study of statics goes back to the time of the Greek philosophers, the first significant contribution to dynamics was made by Galileo (1564–1642). Galileo's experiments on uniformly accelerated bodies led Newton (1642–1727) to formulate his fundamental laws of motion.

Dynamics includes two broad areas of study:

1. **Kinematics**, which is the study of the geometry of motion. The principles of kinematics relate the displacement, velocity, acceleration, and time of a body's motion, without reference to the cause of the motion.
2. **Kinetics**, which is the study of the relation between the forces acting on a body, the mass of the body, and the motion of the body. We use kinetics to predict the motion caused by given forces or to determine the forces required to produce a given motion.

Chaps. 11 through 14 describe the **dynamics of particles**; in Chap. 11, we consider the **kinematics of particles**. The use of the word *particles* does not mean that our study is restricted to small objects; rather, it indicates that in these first chapters we study the motion of bodies—possibly as large as cars, rockets, or airplanes—without regard to their size or shape. By saying that we analyze the bodies as particles, we mean that we consider only their motion as an entire unit; we neglect any rotation about their own centers of mass. In some cases, however, such a rotation is not negligible, and we cannot treat the bodies as particles. Such motions are analyzed in later chapters dealing with the **dynamics of rigid bodies**.

In the first part of Chap. 11, we describe the rectilinear motion of a particle; that is, we determine the position, velocity, and acceleration of a particle at every instant as it moves along a straight line. We first use general methods of analysis to study the motion of a particle; we then consider two important particular cases, namely, the uniform motion and the uniformly accelerated motion of a particle (Sec. 11.2). We then discuss the simultaneous motion of several particles and introduce the concept of the relative motion of one particle with respect to another. The first part of this chapter concludes with a study of graphical methods of analysis and their application to the solution of problems involving the rectilinear motion of particles.

In the second part of this chapter, we analyze the motion of a particle as it moves along a curved path. We define the position, velocity, and acceleration of a particle as vector quantities and introduce the derivative of a vector function to add to our mathematical tools. We consider applications in which we define the motion of a particle by the rectangular components of its velocity and acceleration; at this point, we analyze the motion of a projectile (Sec. 11.4C). Then, we examine the motion of a particle relative to a reference frame in translation. Finally, we analyze the curvilinear motion of a particle in terms of components other than rectangular. In Sec. 11.5, we introduce the tangential and normal components of an object's velocity and acceleration, and then examine the radial and transverse components of an object's motion.

# 11.1  RECTILINEAR MOTION OF PARTICLES

A particle moving along a straight line is said to be in **rectilinear motion**. The only variables we need to describe this motion are the time, $t$, and the distance along the line, $x$, as a function of time. With these variables, we can define the particle's position, velocity, and acceleration, which completely describe the particle's motion. When we study the motion of a particle moving in a plane (two dimensions) or in space (three dimensions), we will use a more general position vector rather than simply the distance along a line.

## 11.1A  Position, Velocity, and Acceleration

At any given instant $t$, a particle in rectilinear motion occupies some position on the straight line. To define the particle's position $P$, we choose a fixed origin $O$ on the straight line and a positive direction along the line. We measure the distance $x$ from $O$ to $P$ and record it with a plus or minus sign, according to whether we reach $P$ from $O$ by moving along the line in the positive or negative direction. The distance $x$, with the appropriate sign, completely defines the position of the particle; it is called the **position coordinate** of the particle. For example, the position coordinate corresponding to $P$ in Fig. 11.1a is $x = +5$ m; the coordinate corresponding to $P'$ in Fig. 11.1b is $x' = -2$ m.

When we know the position coordinate $x$ of a particle for every value of time $t$, we say that the motion of the particle is known. We can provide a "timetable" of the motion in the form of an equation in $x$ and $t$, such as $x = 6t^2 - t^3$, or in the form of a graph of $x$ versus $t$, as shown in Fig. 11.6. The units most often used to measure the position coordinate $x$ are the meter (m) in the SI system of units[†] and the foot (ft) in the U.S. customary system of units. Time $t$ is usually measured in seconds (s).

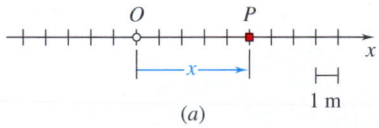

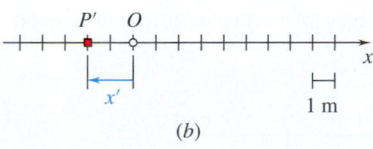

**Fig. 11.1**  Position is measured from a fixed origin. (a) A positive position coordinate; (b) a negative position coordinate.

[†]See Sec. 1.3.

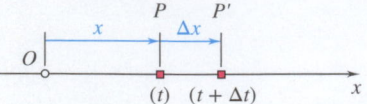

**Fig. 11.2** A small displacement $\Delta x$ from time $t$ to time $t + \Delta t$.

**Photo 11.1** The motion of this solar car can be described by its position, velocity, and acceleration.

Source: Stefano Paltera/NREL

Now consider the position $P$ occupied by the particle at time $t$ and the corresponding coordinate $x$ (Fig. 11.2). Consider also the position $P'$ occupied by the particle at a later time $t + \Delta t$. We can obtain the position coordinate of $P'$ by adding the small displacement $\Delta x$ to the coordinate $x$ of $P$. This displacement is positive or negative according to whether $P'$ is to the right or to the left of $P$. We define the **average velocity** of the particle over the time interval $\Delta t$ as the quotient of the displacement $\Delta x$ and the time interval $\Delta t$ as

$$\text{Average velocity} = \frac{\Delta x}{\Delta t}$$

If we use SI units, $\Delta x$ is expressed in meters and $\Delta t$ in seconds; the average velocity is then expressed in meters per second (m/s). If we use U.S. customary units, $\Delta x$ is expressed in feet and $\Delta t$ in seconds; the average velocity is then expressed in feet per second (ft/s).

We can determine the **instantaneous velocity** $v$ of a particle at the instant $t$ by allowing the time interval $\Delta t$ to become infinitesimally small. Thus,

$$\text{Instantaneous velocity} = v = \lim_{\Delta t \to 0} \frac{\Delta x}{\Delta t}$$

The instantaneous velocity is also expressed in m/s or ft/s. Observing that the limit of the quotient is equal, by definition, to the derivative of $x$ with respect to $t$, we have

**Velocity of a particle along a line**

$$v = \frac{dx}{dt} \tag{11.1}$$

We represent the velocity $v$ by an algebraic number that can be positive or negative.[‡] A positive value of $v$ indicates that $x$ increases, i.e., that the particle moves in the positive direction (Fig. 11.3a). A negative value of $v$ indicates that $x$ decreases; that is, that the particle moves in the negative direction (Fig. 11.3b). The magnitude of $v$ is known as the **speed** of the particle.

Consider the velocity $v$ of the particle at time $t$ and also its velocity $v + \Delta v$ at a later time $t + \Delta t$ (Fig. 11.4). We define the **average acceleration** of the particle over the time interval $\Delta t$ as the quotient of $\Delta v$ and $\Delta t$ as

$$\text{Average acceleration} = \frac{\Delta v}{\Delta t}$$

If we use SI units, $\Delta v$ is expressed in m/s and $\Delta t$ in seconds; the average acceleration is then expressed in m/s$^2$. If we use U.S. customary units, $\Delta v$ is expressed in ft/s and $\Delta t$ in seconds; the average acceleration is then expressed in ft/s$^2$.

We obtain the **instantaneous acceleration** $a$ of the particle at the instant $t$ by again allowing the time interval $\Delta t$ to approach zero. Thus,

$$\text{Instantaneous acceleration} = a = \lim_{\Delta t \to 0} \frac{\Delta v}{\Delta t}$$

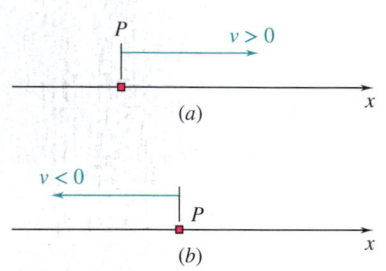

**Fig. 11.3** In rectilinear motion, velocity can be only (a) positive or (b) negative along the line.

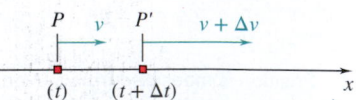

**Fig. 11.4** A change in velocity from $v$ to $v + \Delta v$ corresponding to a change in time from $t$ to $t + \Delta t$.

[‡]As you will see in Sec. 11.4A, velocity is actually a vector quantity. However, because we are considering here the rectilinear motion of a particle where the velocity has a known and fixed direction, we need only specify its sense and magnitude. We can do this conveniently by using a scalar quantity with a plus or minus sign. This is also true of the acceleration of a particle in rectilinear motion.

The instantaneous acceleration is also expressed in m/s² or ft/s². The limit of the quotient, which is by definition the derivative of $v$ with respect to $t$, measures the rate of change of the velocity. We have

**Acceleration of a particle along a line**

$$a = \frac{dv}{dt} \qquad\qquad (11.2)$$

or substituting for $v$ from Eq. (11.1),

$$a = \frac{d^2x}{dt^2} \qquad\qquad (11.3)$$

We represent the acceleration $a$ by an algebraic number that can be positive or negative (see the footnote on the preceding page). A positive value of $a$ indicates that the velocity (i.e., the algebraic number $v$) increases. This may mean that the particle is moving faster in the positive direction (Fig. 11.5a) or that it is moving more slowly in the negative direction (Fig. 11.5b); in both cases, $\Delta v$ is positive. A negative value of $a$ indicates that the velocity decreases; either the particle is moving more slowly in the positive direction (Fig. 11.5c), or it is moving faster in the negative direction (Fig. 11.5d).

Sometimes we use the term *deceleration* to refer to $a$ when the speed of the particle (i.e., the magnitude of $v$) decreases; the particle is then moving more slowly. For example, the particle of Fig. 11.5 is decelerating in parts $b$ and $c$; it is truly accelerating (i.e., moving faster) in parts $a$ and $d$.

We can obtain another expression for the acceleration by eliminating the differential $dt$ in Eqs. (11.1) and (11.2). Solving Eq. (11.1) for $dt$, we have $dt = dx/v;$ substituting into Eq. (11.2) gives us

$$a = v\frac{dv}{dx} \qquad\qquad (11.4)$$

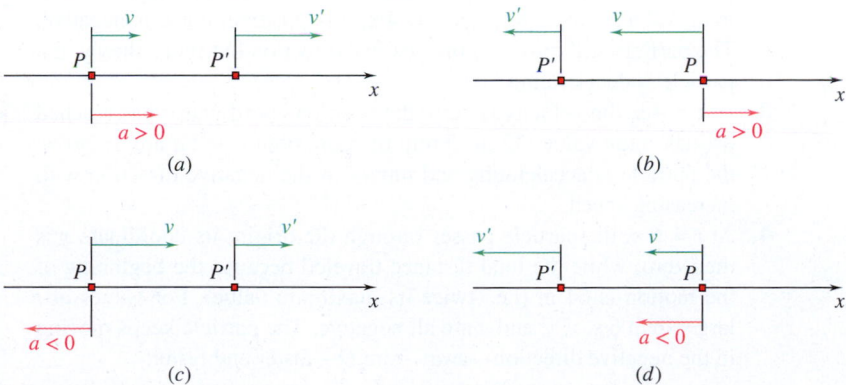

**Fig. 11.5** Velocity and acceleration can be in the same or different directions. (a, d) When $a$ and $v$ are in the same direction, the particle speeds up; (b, c) when $a$ and $v$ are in opposite directions, the particle slows down.

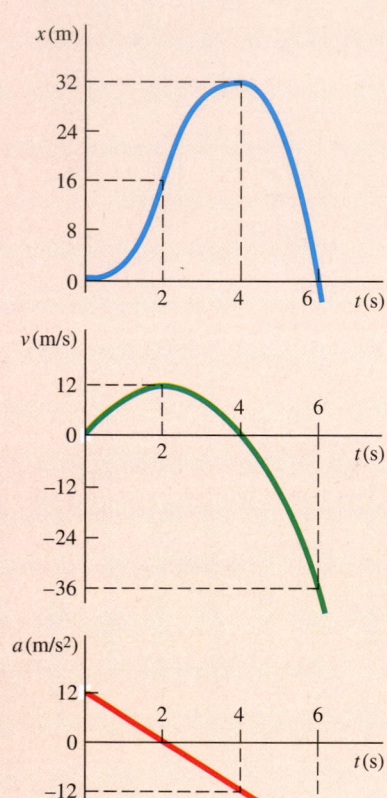

**Fig. 11.6** Graphs of position, velocity, and acceleration as functions of time for Concept Application 11.1.

## Concept Application 11.1

Consider a particle moving in a straight line, and assume that its position is defined by

$$x = 6t^2 - t^3$$

where $t$ is in seconds and $x$ in meters. We can obtain the velocity $v$ at any time $t$ by differentiating $x$ with respect to $t$ as

$$v = \frac{dx}{dt} = 12t - 3t^2$$

We can obtain the acceleration $a$ by differentiating again with respect to $t$. Hence,

$$a = \frac{dv}{dt} = 12 - 6t$$

In Fig. 11.6, we have plotted the position coordinate, the velocity, and the acceleration. These curves are known as *motion curves*. Keep in mind, however, that the particle does not move along any of these curves; the particle moves in a straight line.

Because the derivative of a function measures the slope of the corresponding curve, the slope of the $x$–$t$ curve at any given time is equal to the value of $v$ at that time. Similarly, the slope of the $v$–$t$ curve is equal to the value of $a$. Because $a = 0$ at $t = 2$ s, the slope of the $v$–$t$ curve must be zero at $t = 2$ s; the velocity reaches a maximum at this instant. Also, because $v = 0$ at $t = 0$ and at $t = 4$ s, the tangent to the $x$–$t$ curve must be horizontal for both of these values of $t$.

A study of the three motion curves of Fig. 11.6 shows that the motion of the particle from $t = 0$ to $t = \infty$ can be divided into four phases:

1. The particle starts from the origin, $x = 0$, with no velocity but with a positive acceleration. Under this acceleration, the particle gains a positive velocity and moves in the positive direction. From $t = 0$ to $t = 2$ s, $x$, $v$, and $a$ are all positive.
2. At $t = 2$ s, the acceleration is zero; the velocity has reached its maximum value. From $t = 2$ s to $t = 4$ s, $v$ is positive, but $a$ is negative. The particle still moves in the positive direction but more slowly; the particle is decelerating.
3. At $t = 4$ s, the velocity is zero; the position coordinate $x$ has reached its maximum value (32 m). From then on, both $v$ and $a$ are negative; the particle is accelerating and moves in the negative direction with increasing speed.
4. At $t = 6$ s, the particle passes through the origin; its coordinate $x$ is then zero, while the total distance traveled because the beginning of the motion is 64 m (i.e., twice its maximum value). For values of $t$ larger than 6 s, $x$, $v$, and $a$ are all negative. The particle keeps moving in the negative direction—away from $O$—faster and faster.

# 11.1B Determining the Motion of a Particle

We have just seen that the motion of a particle is said to be known if we know its position for every value of the time $t$. In practice, however, a motion is seldom defined by a relation between $x$ and $t$. More often, the conditions of the motion are specified by the type of acceleration that the particle possesses. For example, a freely falling body has a constant acceleration that is directed downward and equal to 9.81 m/s$^2$ or 32.2 ft/s$^2$, a mass attached to a stretched spring has an acceleration proportional to the instantaneous elongation of the spring measured from its equilibrium position, etc. In general, we can express the acceleration of the particle as a function of one or more of the variables $x$, $v$, and $t$. Thus, in order to determine the position coordinate $x$ in terms of $t$, we need to perform two successive integrations.

Let us consider three common classes of motion.

1. $a = f(t)$. **The Acceleration Is a Given Function of $t$.** Solving Eq. (11.2) for $dv$ and substituting $f(t)$ for $a$, we have

$$dv = a\, dt$$
$$dv = f(t)dt$$

Integrating both sides of the equation, we obtain

$$\int dv = \int f(t)\, dt$$

This equation defines $v$ in terms of $t$. Note, however, that an arbitrary constant is introduced after the integration is performed. This is due to the fact that many motions correspond to the given acceleration $a = f(t)$. In order to define the motion of the particle uniquely, it is necessary to specify the **initial conditions** of the motion; that is, the value $v_0$ of the velocity and the value $x_0$ of the position coordinate at $t = 0$. Rather than use an arbitrary constant that is determined by the initial conditions, it is often more convenient to replace the indefinite integrals with **definite integrals**. Definite integrals have lower limits corresponding to the initial conditions $t = 0$ and $v = v_0$ and upper limits corresponding to $t = t$ and $v = v$. This gives us

$$\int_{v_0}^{v} dv = \int_{0}^{t} f(t)\, dt$$

$$v - v_0 = \int_{0}^{t} f(t)\, dt$$

which yields $v$ in terms of $t$.

We can now solve Eq. (11.1) for $dx$ as

$$dx = v\, dt$$

and substitute the expression obtained from the first integration for $v$. Then, we integrate both sides of this equation via the left-hand side with respect to $x$ from $x = x_0$ to $x = x$ and the right-hand side with respect to $t$ from $t = 0$ to $t = t$. In this way, we obtain the position coordinate $x$ in terms of $t$; the motion is completely determined.

We will study two important cases in greater detail in Sec. 11.2: the case when $a = 0$, corresponding to a *uniform motion*, and the case when $a = $ constant, corresponding to a *uniformly accelerated motion*.

2. $a = f(x)$. **The Acceleration Is a Given Function of $x$.** Rearranging Eq. (11.4) and substituting $f(x)$ for $a$, we have

$$v\, dv = a\, dx$$
$$v\, dv = f(x)dx$$

Because each side contains only one variable, we can integrate the equation. Denoting again the initial values of the velocity and of the position coordinate by $v_0$ and $x_0$, respectively, we obtain

$$\int_{v_0}^{v} v\, dv = \int_{x_0}^{x} f(x)\, dx$$

$$\frac{1}{2}v^2 - \frac{1}{2}v_0^2 = \int_{x_0}^{x} f(x)\, dx$$

which yields $v$ in terms of $x$. We now solve Eq. (11.1) for $dt$, giving us

$$dt = \frac{dx}{v}$$

and substitute for $v$ the expression just obtained. We can then integrate both sides to obtain the desired relation between $x$ and $t$. However, in most cases, this last integration cannot be performed analytically, and we must resort to a numerical method of integration.

3. $a = f(v)$. **The Acceleration Is a Given Function of $v$.** We can now substitute $f(v)$ for $a$ in either Eqs. (11.2) or (11.4) to obtain either

$$f(v) = \frac{dv}{dt} \qquad\qquad f(v) = v\frac{dv}{dx}$$

$$dt = \frac{dv}{f(v)} \qquad\qquad dx = \frac{v\, dv}{f(v)}$$

Integration of the first equation yields a relation between $v$ and $t$; integration of the second equation yields a relation between $v$ and $x$. Either of these relations can be used in conjunction with Eq. (11.1) to obtain the relation between $x$ and $t$ that characterizes the motion of the particle.

## Sample Problem 11.1

The position of a particle moving along a straight line is defined by the relation $x = t^3 - 6t^2 - 15t + 40$, where $x$ is expressed in feet and $t$ in seconds. Determine (a) the time at which the velocity is zero, (b) the position and distance traveled by the particle at that time, (c) the acceleration of the particle at that time, (d) the distance traveled by the particle from $t = 4$ s to $t = 6$ s.

**STRATEGY:** You need to use the basic kinematic relationships between position, velocity, and acceleration. Because the position is given as a function of time, you can differentiate it to find equations for the velocity and acceleration. Once you have these equations, you can solve the problem.

**MODELING and ANALYSIS:** Taking the derivative of position, you obtain

$$x = t^3 - 6t^2 - 15t + 40 \tag{1}$$

$$v = \frac{dx}{dt} = 3t^2 - 12t - 15 \tag{2}$$

$$a = \frac{dv}{dt} = 6t - 12 \tag{3}$$

These equations are graphed in Fig. 1.

**a. Time When $v = 0$.** Set $v = 0$ in Eq. (2) for

$$3t^2 - 12t - 15 = 0 \quad t = -1 \text{ s} \quad \text{and} \quad t = +5 \text{ s} \blacktriangleleft$$

Only the root $t = +5$ s corresponds to a time after the motion has begun: for $t < 5$ s, $v < 0$ and the particle moves in the negative direction; for $t > 5$ s, $v > 0$ and the particle moves in the positive direction.

**b. Position and Distance Traveled When $v = 0$.** Substitute $t = +5$ s into Eq. (1), yielding

$$x_5 = (5)^3 - 6(5)^2 - 15(5) + 40 \qquad x_5 = -60 \text{ ft} \blacktriangleleft$$

The initial position at $t = 0$ was $x_0 = +40$ ft. Because $v \neq 0$ during the interval $t = 0$ to $t = 5$ s, you have

$$\text{Distance traveled} = x_5 - x_0 = -60 \text{ ft} - 40 \text{ ft} = -100 \text{ ft}$$

Distance traveled = 100 ft in the negative direction ◀

**c. Acceleration When $v = 0$.** Substitute $t = +5$ s into Eq. (3) for

$$a_5 = 6(5) - 12 \qquad a_5 = +18 \text{ ft/s}^2 \blacktriangleleft$$

**d. Distance Traveled from $t = 4$ s to $t = 6$ s.** The particle moves in the negative direction from $t = 4$ s to $t = 5$ s and in the positive direction from $t = 5$ s to $t = 6$ s; therefore, the distance traveled during each of these time intervals must be computed separately.

From $t = 4$ s to $t = 5$ s: $\quad x_5 = -60$ ft

$$x_4 = (4)^3 - 6(4)^2 - 15(4) + 40 = -52 \text{ ft}$$

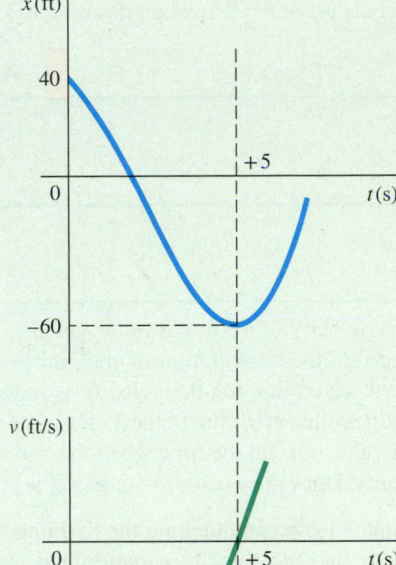

**Fig. 1** Motion curves for the particle.

*(continued)*

Distance traveled $= x_5 - x_4 = -60$ ft $- (-52$ ft$) = -8$ ft
$= 8$ ft in the negative direction

From $t = 5$ s to $t = 6$ s:     $x_5 = -60$ ft
$x_6 = (6)^3 - 6(6)^2 - 15(6) + 40 = -50$ ft
Distance traveled $= x_6 - x_5 = -50$ ft $- (-60$ ft$) = +10$ ft
$= 10$ ft in the positive direction

*Total distance traveled* from $t = 4$ s to $t = 6$ s is 8 ft + 10 ft     $= 18$ ft

**REFLECT and THINK:**   The total distance traveled by the particle in the two-second interval is 18 ft, but because one distance is positive and one is negative, the net change in position is only 2 ft (in the positive direction). This illustrates the difference between total distance traveled and net change in position. Note that the maximum displacement occurs at $t = 5$ s, when the velocity is zero.

## Sample Problem 11.2

You throw a ball vertically upward with a velocity of 10 m/s from a window located 20 m above the ground. Knowing that the acceleration of the ball is constant and equal to 9.81 m/s$^2$ downward, determine (*a*) the velocity $v$ and elevation $y$ of the ball above the ground at any time $t$, (*b*) the highest elevation reached by the ball and the corresponding value of $t$, (*c*) the time when the ball hits the ground and the corresponding velocity. Draw the $v$–$t$ and $y$–$t$ curves.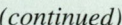

**STRATEGY:**   The acceleration is constant, so you can integrate the defining kinematic equation for acceleration once to find the velocity equation and a second time to find the position relationship. Once you have these equations, you can solve the problem.

**MODELING and ANALYSIS:**   Model the ball as a particle with negligible drag.

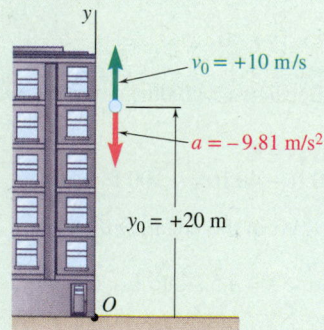

**Fig. 1**  Acceleration, initial velocity, and initial position of the ball.

   **a. Velocity and Elevation.**   Choose the $y$ axis measuring the position coordinate (or elevation) with its origin $O$ on the ground and its positive sense upward. The value of the acceleration and the initial values of $v$ and $y$ are as indicated in Fig. 1. Substituting for $a$ in $a = dv/dt$ and noting that when $t = 0$, $v_0 = +10$ m/s, you have

$$\frac{dv}{dt} = a = -9.81 \text{ m/s}^2$$

$$\int_{v_0=10}^{v} dv = -\int_{0}^{t} 9.81 \, dt$$

$$[v]_{10}^{v} = -[9.81t]_{0}^{t}$$
$$v - 10 = -9.81t$$

$$v = 10 - 9.81t \quad (1) \blacktriangleleft$$

*(continued)*

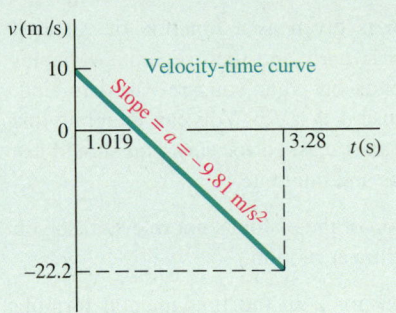

**Fig. 2** Velocity of the ball as a function of time.

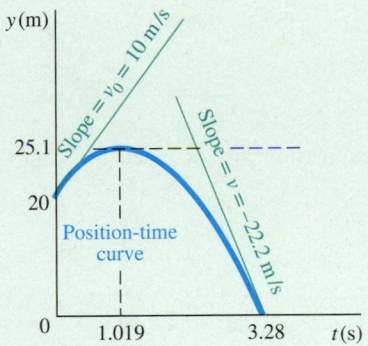

**Fig. 3** Height of the ball as a function of time.

Substituting for $v$ in $v = dy/dt$ and noting that when $t = 0$, $y_0 = 20$ m, you have

$$\frac{dy}{dt} = v = 10 - 9.81t$$

$$\int_{y_0=20}^{y} dy = \int_0^t (10 - 9.81t)\, dt$$

$$[y]_{20}^{y} = [10t - 4.905t^2]_0^t$$

$$y - 20 = 10t - 4.905\,t^2$$

$$y = 20 + 10t - 4.905\,t^2 \quad \textbf{(2)} \;\blacktriangleleft$$

Graphs of these equations are shown in Figs. 2 and 3.

   **b. Highest Elevation.**   The ball reaches its highest elevation when $v = 0$. Substituting into Eq. (1), you obtain

$$10 - 9.81t = 0 \qquad\qquad t = 1.019 \text{ s} \;\blacktriangleleft$$

Substituting $t = 1.019$ s into Eq. (2), you find

$$y = 20 + 10(1.019) - 4.905(1.019)^2 \qquad y = 25.1 \text{ m} \;\blacktriangleleft$$

   **c. Ball Hits the Ground.**   The ball hits the ground when $y = 0$. Substituting into Eq. (2), you obtain

$$20 + 10t - 4.905t^2 = 0 \quad t = -1.243 \text{ s} \quad \text{and} \quad t = +3.28 \text{ s} \;\blacktriangleleft$$

Only the root $t = +3.28$ s corresponds to a time after the motion has begun. Carrying this value of $t$ into Eq. (1), you find

$$v = 10 - 9.81(3.28) = -22.2 \text{ m/s} \qquad v = 22.2 \text{ m/s} \downarrow \;\blacktriangleleft$$

**REFLECT and THINK:**   When the acceleration is constant, the velocity changes linearly, and the position is a quadratic function of time. You will see in Sec. 11.2 that the motion in this problem is an example of free fall, where the acceleration in the vertical direction is constant and equal to $-g$.

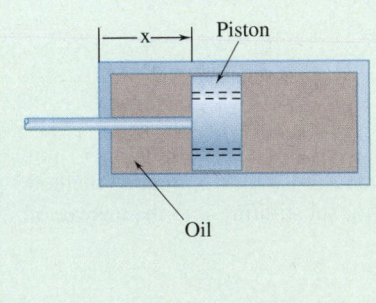

## Sample Problem 11.3

Many mountain bike shocks utilize a piston that travels in an oil-filled cylinder to provide shock absorption; this system is shown schematically. When the front tire goes over a bump, the cylinder is given an initial velocity $v_0$. The piston, which is attached to the fork, then moves with respect to the cylinder, and oil is forced through orifices in the piston. This causes the piston to decelerate at a rate proportional to the velocity at $a = -kv$. At time $t = 0$, the position of the piston is $x = 0$. Express (a) the velocity $v$ in terms of $t$, (b) the position $x$ in terms of $t$, (c) the velocity $v$ in terms of $x$. Draw the corresponding motion curves.

*(continued)*

**STRATEGY:** Because the acceleration is given as a function of velocity, you need to use either $a = dv/dt$ or $a = v\,dv/dx$ and then separate variables and integrate. Which one you use depends on what you are asked to find. Because part $a$ asks for $v$ in terms of $t$, use $a = dv/dt$. You can integrate this again using $v = dx/dt$ for part $b$. Because part $c$ asked for $v(x)$, you should use $a = v\,dv/dx$ and then separate the variables and integrate.

**MODELING and ANALYSIS:** Rotation of the piston is not relevant, so you can model it as a particle undergoing rectilinear motion.

**a. v in Terms of t.** Substitute $-kv$ for $a$ in the fundamental formula defining acceleration, $a = dv/dt$. You obtain

$$-kv = \frac{dv}{dt} \qquad \frac{dv}{v} = -k\,dt \qquad \int_{v_0}^{v} \frac{dv}{v} = -k\int_{0}^{t} dt$$

$$\ln\frac{v}{v_0} = -kt \qquad\qquad v = v_0 e^{-kt} \blacktriangleleft$$

**b. x in Terms of t.** Substitute the expression just obtained for $v$ into $v = dx/dt$. You get

$$v_0 e^{-kt} = \frac{dx}{dt}$$

$$\int_{0}^{x} dx = v_0 \int_{0}^{t} e^{-kt}dt$$

$$x = -\frac{v_0}{k}[e^{-kt}]_0^t = -\frac{v_0}{k}(e^{-kt} - 1)$$

$$x = \frac{v_0}{k}(1 - e^{-kt}) \blacktriangleleft$$

**c. v in Terms of x.** Substitute $-kv$ for $a$ in $a = v\,dv/dx$. You have

$$-kv = v\frac{dv}{dx}$$

$$dv = -k\,dx$$

$$\int_{v_0}^{v} dv = -k\int_{0}^{x} dx$$

$$v - v_0 = -kx \qquad\qquad v = v_0 - kx \blacktriangleleft$$

The motion curves are shown in Fig. 1.

**REFLECT and THINK:** You could have solved part $c$ by eliminating $t$ from the answers obtained for parts $a$ and $b$. You could use this alternative method as a check. From part $a$, you obtain $e^{-kt} = v/v_0$; substituting into the answer of part $b$, you have

$$x = \frac{v_0}{k}(1 - e^{-kt}) = \frac{v_0}{k}\left(1 - \frac{v}{v_0}\right) \qquad v = v_0 - kx \qquad \text{(checks)}$$

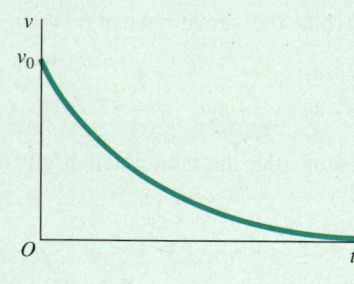

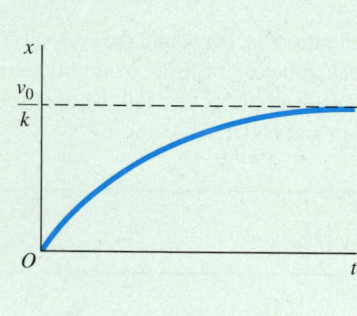

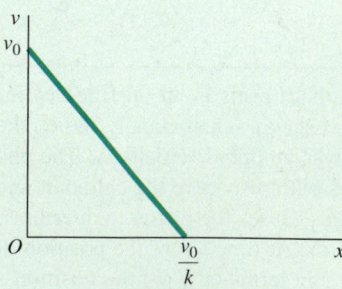

**Fig. 1** Motion curves for the piston.

## Sample Problem 11.4

An uncontrolled automobile traveling at 45 mph strikes a highway crash barrier square on. After initially hitting the barrier, the automobile decelerates at a rate proportional to the distance $x$ the automobile has moved into the barrier; specifically, $a = -60\sqrt{x}$, where $a$ and $x$ are expressed in ft/s$^2$ and ft, respectively. Determine the distance the automobile will move into the barrier before it comes to rest.

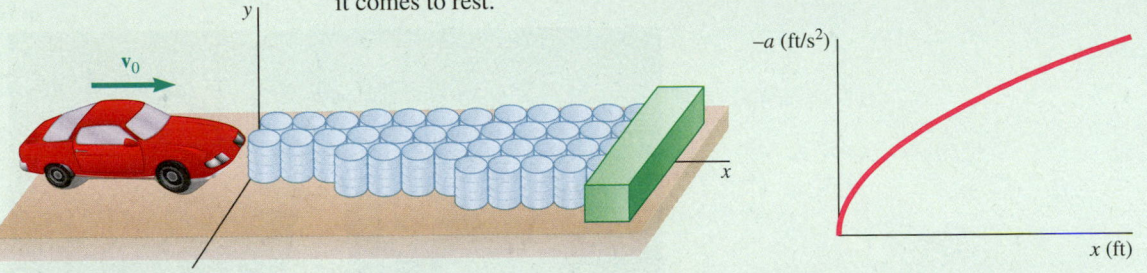

**STRATEGY:** Because you are given the deceleration as a function of displacement, you should start with the basic kinematic relationship $a = v\, dv/dx$.

**MODELING and ANALYSIS:** Model the car as a particle. First find the initial speed in ft/s,

$$v_0 = \left(45\frac{\text{mi}}{\text{hr}}\right)\left(\frac{1\ \text{hr}}{3600\ \text{s}}\right)\left(\frac{5280\ \text{ft}}{\text{mi}}\right) = 66\frac{\text{ft}}{\text{s}}$$

Substituting $a = -60\sqrt{x}$ into $a = v\, dv/dx$ gives

$$a = -60\sqrt{x} = \frac{v\, dv}{dx}$$

Separating variables and integrating gives

$$v\, dv = -60\sqrt{x}\, dx \rightarrow \int_{v_0}^{0} v\, dv = -\int_{0}^{x} 60\sqrt{x}\, dx$$

$$\frac{1}{2}v^2 - \frac{1}{2}v_0^2 = -40x^{3/2} \rightarrow x = \left(\frac{1}{80}(v_0^2 - v^2)\right)^{2/3} \qquad \textbf{(1)}$$

Substituting $v = 0$, $v_0 = 45$ ft/s gives

$$d = 14.37\ \text{ft} \blacktriangleleft$$

**REFLECT and THINK:** A distance of 14 ft seems reasonable for a barrier of this type. If you substitute $d$ into the equation for $a$, you find a maximum deceleration of about 7 g's. Note that this problem would have been much harder to solve if you had been asked to find the time for the automobile to stop. In this case, you would need to determine $v(t)$ from Eq. (1). This gives $v = \sqrt{v_0^2 - 80x^{3/2}}$. Using the basic kinematic relationship $v = dx/dt$, you can easily show that

$$\int_{0}^{t} dt = \int_{0}^{x} \frac{dx}{\sqrt{v_0^2 - 80x^{3/2}}}$$

Unfortunately, there is no closed-form solution to this integral, so you would need to solve it numerically.

## Case Study 11.1

People with mobility impairments often have difficulty participating in athletic and recreational activities, missing out on the social and physiological benefits such recreation has to offer. Some senior engineering students decided to design and build an adapted dart launcher to allow athletes with disabilities to play this fun game with their friends.

**CS Photo 11.1**    Adapted dart launcher.
©Katherine Mavrommati

The device they designed, shown in CS Photo 11.1, is similar to an air cannon. The athlete fills a pressure tank by moving a pneumatic cylinder back and forth; the displacement necessary is less than 150 mm and the required force is less than 20 N. Each pump supplies approximately 5 kN/m$^2$ pressure to the reservoir.

Assuming an adiabatic system, the acceleration $a$ of the dart while in the launch tube can be expressed as

$$a = \frac{1}{m}\left[\frac{AP_0 V_0}{V_0 + Ax} - f\right] \tag{1}$$

where $A$ is the cross-sectional area of the launch tube, $P_0$ is the stored pressure in the reservoir, $V_0$ is the volume of the reservoir, $x$ is the distance along the tube, and $f$ is the friction force between the dart launch carriage and the tube wall.

During testing, the darts stick best when they hit perpendicular to the board. You can experiment with different launch angles (20°, 30°, 40°, and 45°) and the number of pumps to determine the best combination when trying to hit the bull's eye. CS Fig. 11.1 shows the location of the dart board with respect to the launcher.

The cannon has the following parameters: cannon inner radius $r = 20$ mm, cannon tube length $L = 330$ mm, volume of reservoir $V_0 = 0.0005$ m$^3$, mass of dart and piston $m = 0.2$ kg, friction force along walls $f = 0.1$ N, and launch height $y_0 = 1.0$ m.

**STRATEGY:**    You are given acceleration as a function of position, so you should start with the basic relationship $a = v\, dv/dx$. You will then need to use projectile motion to analyze the flight of the dart toward the board.

*(continued)*

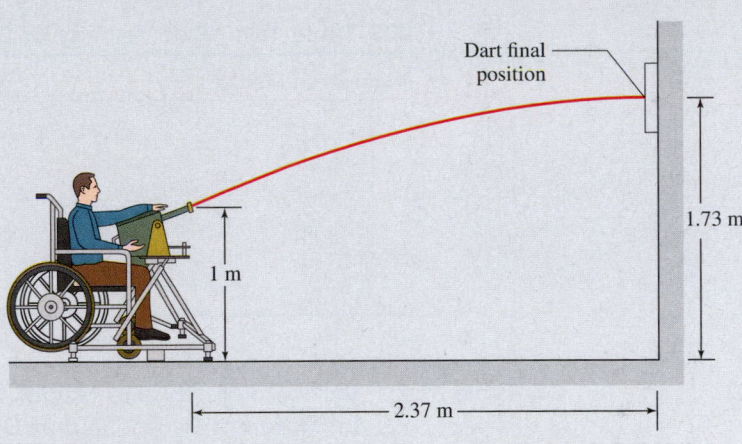

**CS Fig. 11.1** Dart launch schematic—distance and height to bull's eye.

**MODELING:** Treat the air cannon as an adiabatic system (you will learn more about this in your thermodynamics class) and the dart as a particle, with drag considered negligible during flight.

**ANALYSIS:** Substituting the expression for $a$ in Eq. (1) into $a = v\, dv/dx$, and then separating variables and integrating gives you:

$$\int_0^L \frac{1}{m}\left[\frac{AP_0V_0}{V_0 + Ax} - f\right]dx = \int_0^{v_0} v\,dv$$

Performing these integrations and solving for the exit velocity $v_0$ as a function of the system parameters gives

$$v_0 = \left[\frac{2}{m}\left(P_0V_0 \ln\left(\frac{V_0 + AL}{V_0}\right) - fL\right)\right]^{1/2} \tag{2}$$

The pressure $P_0$ is equal to the number of pumps $N$ times the pressure per pump, $5000\ \text{N/m}^2$. You can calculate $v_0$ of the dart for a given number of pumps, and then use this to solve for the projectile motion of the dart after it leaves the cannon. For the $x$ direction, you get

$$x = x_0 + (v_0)_x t, \text{ where } x_0 = 0 \text{ and } (v_0)_x = v_0 \cos(\theta)$$

Solving for the time gives

$$t = \frac{d}{v_0 \cos\theta}$$

Using this time, you can determine the vertical component of velocity $(v_{\text{hit}})_y$ using

$$(v_{\text{hit}})_y = v_0 \sin\theta - gt$$

and the height $y$ where the dart hits the board

$$y = y_0 + v_0 \sin\theta\, t - \frac{1}{2}gt^2$$

The exit velocity $v_0$ for different pumps is shown in CS Table 11.1, and the $y$ component of velocity (which we want close to zero to ensure it sticks) and the height of the dart when it hits the board are shown in CS Table 11.2 for

*(continued)*

**CS Table 11.1:   Exit Velocity of Dart**

| Number of Pumps | Velocity (m/s) |
|---|---|
| 2 | 5.47 |
| 3 | 6.71 |
| 4 | 7.75 |
| 5 | 8.67 |
| 6 | 9.50 |
| 7 | 10.27 |
| 8 | 10.98 |

**CS Table 11.2:   Vertical Velocity and Height When the Dart Hits the Board for Different Pumps and at Different Launch Angles**

| Angle | Velocity and Position of Hit | Number of Pumps | | | | | | |
|---|---|---|---|---|---|---|---|---|
| | | 2 | 3 | 4 | 5 | 6 | 7 | 8 |
| 20° | $(v_{hit})_y$ (m/s) | −2.66 | −1.396 | −0.542 | 0.1115 | 0.645 | 1.100 | 1.500 |
| | $y$ (m) | 0.818 | 1.169 | 1.343 | 1.448 | 1.517 | 1.567 | 1.604 |
| 30° | $(v_{hit})_y$ (m/s) | −2.18 | −0.651 | 0.411 | 1.238 | 1.925 | 2.52 | 3.04 |
| | $y$ (m) | 1.138 | 1.550 | 1.750 | 1.880 | 1.961 | 2.02 | 2.06 |
| 40° | $(v_{hit})_y$ (m/s) | −2.04 | −0.216 | 1.066 | 2.0722 | 2.91 | 3.64 | 4.29 |
| | $y$ (m) | 1.417 | 1.945 | 2.21 | 2.36 | 2.47 | 2.54 | 2.60 |

launch angles of 20, 30, and 40 degrees. From this table, you can see that a launch angle at 30° with four pumps, which results in a vertical velocity component of 0.411 m/s and a height of 1.75 m, is the best combination.

**REFLECT and THINK:**   Wheelchair dart competitions place the bull's-eye at a height of 1.37 m. How would this change the calculations performed above? A number of different variables have to be considered when designing the system, including cannon length, reservoir size, displacement and force required to pump the air cylinders, aesthetics, and athlete engagement. Athletes with mobility impairments may have limited range and force production to actuate the pump cylinder, and the design should provide them with the opportunity to get some exercise and to control how the dart is "thrown." Additionally, there are a number of safety considerations that have to be taken into account.

# SOLVING PROBLEMS
## ON YOUR OWN

In the problems for this section, you will be asked to determine the **position, velocity, and/or acceleration** of a particle in **rectilinear motion**. As you read each problem, it is important to identify both the independent variable (typically $t$ or $x$) and what is required (e.g., the need to express $v$ as a function of $x$). You may find it helpful to start each problem by writing down both the given information and a simple statement of what is to be determined.

**1. Determining $v(t)$ and $a(t)$ for a given $x(t)$.** As explained in Sec. 11.1A, the first and second derivatives of $x$ with respect to $t$ are equal to the velocity and the acceleration, respectively, of the particle [Eqs. (11.1) and (11.2)]. If the velocity and acceleration have opposite signs, the particle can come to rest and then move in the opposite direction (Sample Prob. 11.1). Thus, when computing the total distance traveled by a particle, you should first determine if the particle comes to rest during the specified interval of time. Constructing a diagram similar to that of Sample Prob. 11.1, which shows the position and the velocity of the particle at each critical instant ($v = v_{max}$, $v = 0$, etc.), will help you to visualize the motion.

**2. Determining $v(t)$ and $x(t)$ for a given $a(t)$.** We discussed the solution of problems of this type in the first part of Sec. 11.1B. We used the initial conditions, $t = 0$ and $v = v_0$, for the lower limits of the integrals in $t$ and $v$, but any other known state (e.g., $t = t_1$ and $v = v_1$) could be used instead. Also, if the given function $a(t)$ contains an unknown constant (e.g., the constant $k$ if $a = kt$), you will first have to determine that constant by substituting a set of known values of $t$ and $a$ in the equation defining $a(t)$.

**3. Determining $v(x)$ and $x(t)$ for a given $a(x)$.** This is the second case considered in Sec. 11.1B and is illustrated in Sample Prob. 11.4. We again note that the lower limits of integration can be any known state (e.g., $x = x_1$ and $v = v_1$). In addition, because $v = v_{max}$ when $a = 0$, you can determine the positions where the maximum or minimum values of the velocity occur by setting $a(x) = 0$ and solving for $x$.

**4. Determining $v(x)$, $v(t)$, and $x(t)$ for a given $a(v)$.** This is the last case treated in Sec. 11.1B; the appropriate solution techniques for problems of this type are illustrated in Sample Prob. 11.3. All of the general comments for the preceding cases once again apply. Note that Sample Prob. 11.3 provides a summary of how and when to use the equations $v = dx/dt$, $a = dv/dt$, and $a = v\, dv/dx$.

*(continued)*

We can summarize these relationships in Table 11.1.

**Table 11.1**

| If. . . | Kinematic Relationship | Integrate |
|---------|------------------------|-----------|
| $a = a(t)$ | $\dfrac{dv}{dt} = a(t)$ | $\displaystyle\int_{v_0}^{v} dv = \int_{0}^{t} a(t)\,dt$ |
| $a = a(x)$ | $v\dfrac{dv}{dx} = a(x)$ | $\displaystyle\int_{v_0}^{v} v\,dv = \int_{x_0}^{x} a(x)\,dx$ |
| $a = a(v)$ | $\dfrac{dv}{dt} = a(v)$ | $\displaystyle\int_{v_0}^{v} \dfrac{dv}{a(v)} = \int_{0}^{t} dt$ |
|  | $v\dfrac{dv}{dx} = a(v)$ | $\displaystyle\int_{x_0}^{x} dx = \int_{v_0}^{v} \dfrac{v\,dv}{a(v)}$ |

# Problems[†]

**11.CQ1** A bus travels the 100 miles between $A$ and $B$ at 50 mi/h and then another 100 miles between $B$ and $C$ at 70 mi/h. The average speed of the bus for the entire 200-mile trip is:
  **a.** More than 60 mi/h.
  **b.** Equal to 60 mi/h.
  **c.** Less than 60 mi/h.

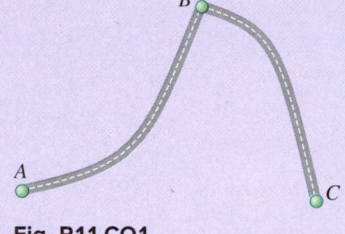

**Fig. P11.CQ1**

**11.CQ2** Two cars $A$ and $B$ race each other down a straight road. The position of each car as a function of time is shown. Which of the following statements are true? (More than one answer can be correct.)
  **a.** At time $t_2$, both cars have traveled the same distance.
  **b.** At time $t_1$, both cars have the same speed.
  **c.** Both cars have the same speed at some time $t < t_1$.
  **d.** Both cars have the same acceleration at some time $t < t_1$.
  **e.** Both cars have the same acceleration at some time $t_1 < t < t_2$.

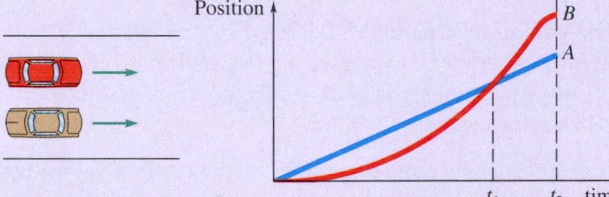

**Fig. P11.CQ2**

## END-OF-SECTION PROBLEMS

**11.1** A snowboarder starts from rest at the top of a double black diamond hill. As she rides down the slope, GPS coordinates are used to determine her displacement as a function of time: $x = 0.5t^3 + t^2 + 2t$, where $x$ and $t$ are expressed in feet and seconds, respectively. Determine the position, velocity, and acceleration of the boarder when $t = 5$ seconds.

**11.2** The motion of a particle is defined by the relation $x = t^3 - 12t^2 + 36t + 30$, where $x$ and $t$ are expressed in feet and seconds, respectively. Determine the time, the position, and the acceleration of the particle when $v = 0$.

**11.3** The vertical motion of mass $A$ is defined by the relation $x = \cos(10t) - 0.1\sin(10t)$, where $x$ and $t$ are expressed in mm and seconds, respectively. Determine (*a*) the position, velocity, and acceleration of $A$ when $t = 0.4$ s, (*b*) the maximum velocity and acceleration of $A$.

**Fig. P11.3**

**11.4** A loaded railroad car is rolling at a constant velocity when it couples with a spring and dashpot bumper system. After the coupling, the motion of the car is defined by the relation $x = 60e^{-4.8t} \sin 16t$, where $x$ and $t$ are expressed in millimeters and seconds, respectively. Determine the position, the velocity, and the acceleration of the railroad car when (*a*) $t = 0$, (*b*) $t = 0.3$ s.

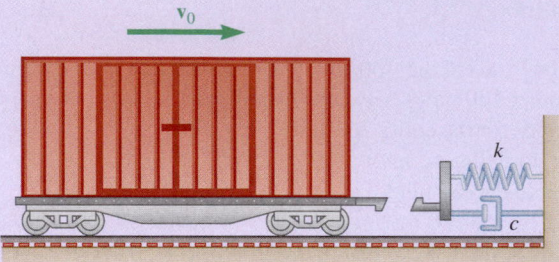

**Fig. P11.4**

**11.5** A group of hikers uses a GPS while doing a 40-mile trek in Colorado. A curve fit to the data shows that their altitude can be approximated by the function $y(t) = 0.12t^5 - 6.75t^4 + 135t^3 - 1120t^2 + 3200t + 9070$, where $y$ and $t$ are expressed in feet and hours, respectively. During the 18-hour hike, determine (*a*) the maximum altitude that the hikers reach, (*b*) the total feet they ascend, (*c*) the total feet they descend. *Hint:* You will need to use a calculator or computer to solve for the roots of a fourth-order polynomial.

**11.6** The motion of a particle is defined by the relation $x = t^3 - 6t^2 + 9t + 5$, where $x$ is expressed in feet and $t$ in seconds. Determine (*a*) when the velocity is zero, (*b*) the position, acceleration, and total distance traveled when $t = 5$ s.

**11.7** A girl operates a radio-controlled model car in a vacant parking lot. The girl's position is at the origin of the $xy$ coordinate axes, and the surface of the parking lot lies in the $x$–$y$ plane. She drives the car in a straight line so that the $x$ coordinate is defined by the relation $x(t) = 0.5t^3 - 3t^2 + 3t + 2$, where $x$ and $t$ are expressed in meters and seconds, respectively. Determine (*a*) when the velocity is zero, (*b*) the position and total distance traveled when the acceleration is zero.

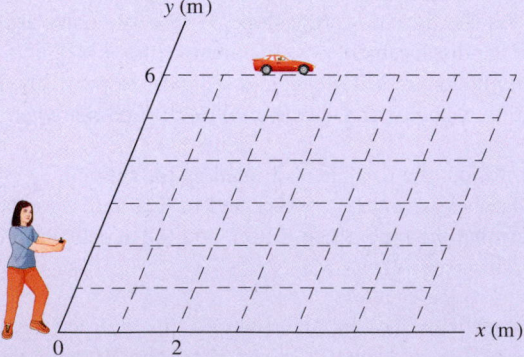

**Fig. P11.7**

**11.8** The motion of a particle is defined by the relation $x = t^2 - (t - 2)^3$, where $x$ and $t$ are expressed in feet and seconds, respectively. Determine (*a*) the two positions at which the velocity is zero, (*b*) the total distance traveled by the particle from $t = 0$ to $t = 4$ s.

**11.9** The brakes of a car are applied, causing it to slow down at a rate of 10 ft/s². Knowing that the car stops in 300 ft, determine (a) how fast the car was traveling immediately before the brakes were applied, (b) the time required for the car to stop.

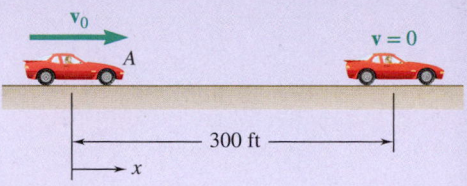

Fig. P11.9

**11.10** The acceleration of a particle is defined by the relation $a = 3e^{-0.2t}$, where $a$ and $t$ are expressed in ft/s² and seconds, respectively. Knowing that $x = 0$ and $v = 0$ at $t = 0$, determine the velocity and position of the particle when $t = 0.5$ s.

**11.11** The acceleration of a particle is defined by the relation $a = 9 - 3t^2$, where $a$ and $t$ are expressed in ft/s² and seconds, respectively. The particle starts at $t = 0$ with $v = 0$ and $x = 5$ ft. Determine (a) the time when the velocity is again zero, (b) the position and velocity when $t = 4$ s, (c) the total distance traveled by the particle from $t = 0$ to $t = 4$ s.

**11.12** Many car companies are performing research on collision avoidance systems. A small prototype applies engine braking that decelerates the vehicle according to the relationship $a = -k\sqrt{t}$, where $a$ and $t$ are expressed in m/s² and seconds, respectively. The vehicle is traveling at 20 m/s when its radar sensors detect a stationary obstacle. Knowing that it takes the prototype vehicle 4 seconds to stop, determine (a) expressions for its velocity and position as a function of time, (b) how far the vehicle traveled before it stopped.

**11.13** A Scotch yoke is a mechanism that transforms the circular motion of a crank into the reciprocating motion of a shaft (or vice versa). It has been used in a number of different internal combustion engines and in control valves. In the Scotch yoke shown, the acceleration of point $A$ is defined by the relation $a = -1.8 \sin kt$, where $a$ and $t$ are expressed in m/s² and seconds, respectively, and $k = 3$ rad/s. Knowing that $x = 0$ and $v = 0.6$ m/s when $t = 0$, determine the velocity and position of point $A$ when $t = 0.5$ s.

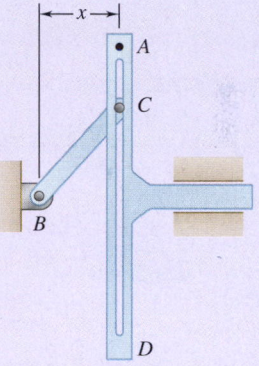

Fig. *P11.13* and *P11.14*

**11.14** For the Scotch yoke mechanism shown, the acceleration of point $A$ is defined by the relation $a = -1.08 \sin kt - 1.44 \cos kt$, where $a$ and $t$ are expressed in m/s² and seconds, respectively, and $k = 3$ rad/s. Knowing that $x = 0.16$ m and $v = 0.36$ m/s when $t = 0$, determine the velocity and position of point $A$ when $t = 0.5$ s.

Fig. P11.15

**11.15** A piece of electronic equipment that is surrounded by packing material is dropped so that it hits the ground with a speed of 4 m/s. After contact, the equipment experiences an acceleration of $a = -kx$, where $k$ is a constant and $x$ is the compression of the packing material. If the packing material experiences a maximum compression of 15 mm, determine the maximum acceleration of the equipment.

**11.16** A projectile enters a resisting medium at $x = 0$ with an initial velocity $v_0 = 1000$ ft/s and travels 3 in. before coming to rest. Knowing that the velocity of the projectile is defined by the relation $v = v_0 - kx$, where $v$ is expressed in ft/s and $x$ is in feet, determine (a) the initial acceleration of the projectile, (b) the time required for the projectile to penetrate 2.5 in. into the resisting medium.

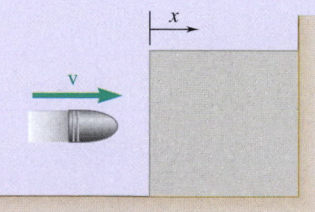

Fig. P11.16

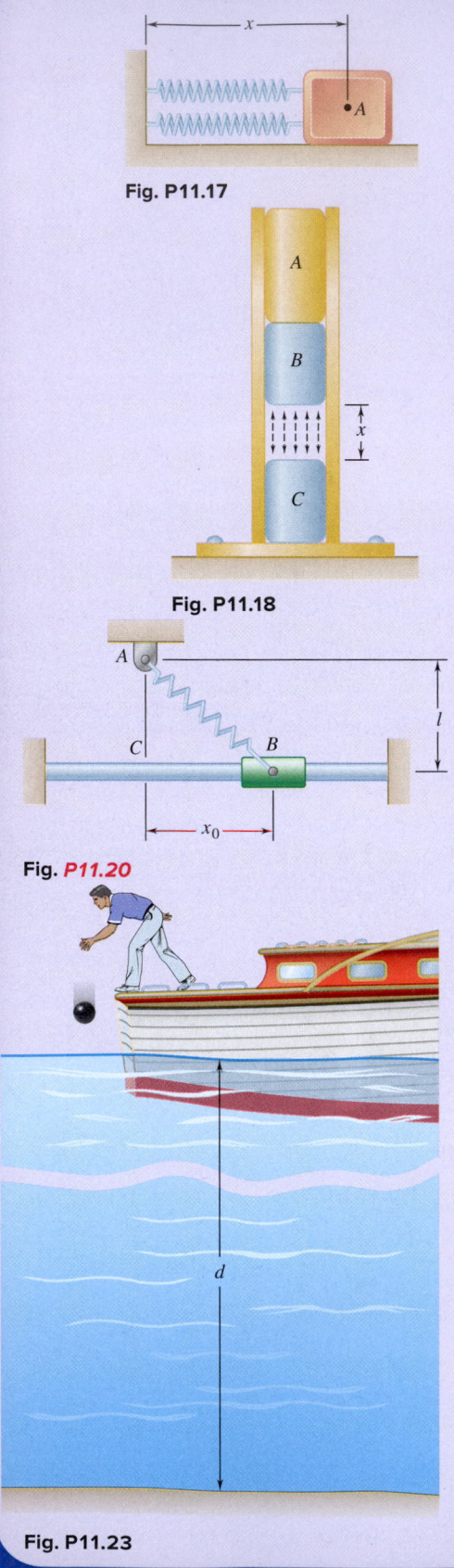

Fig. P11.17

Fig. P11.18

Fig. P11.20

Fig. P11.23

**11.17** Point $A$ oscillates with an acceleration $a = 100(0.25 - x)$, where $a$ and $x$ are expressed in m/s$^2$ and meters, respectively. Knowing that the system starts at time $t = 0$ with $v = 0$ and $x = 0.2$ m, determine the position and the velocity of $A$ when $t = 0.2$ s.

**11.18** A brass (nonmagnetic) block $A$ and a steel magnet $B$ are in equilibrium in a brass tube under the magnetic repelling force of another steel magnet $C$ located at a distance $x = 0.004$ m from $B$. The force is inversely proportional to the square of the distance between $B$ and $C$. If block $A$ is suddenly removed, the acceleration of block $B$ is $a = -9.81 + k/x^2$, where $a$ and $x$ are expressed in m/s$^2$ and meters, respectively, and $k = 4 \times 10^{-4}$ m$^3$/s$^2$. Determine the maximum velocity and acceleration of $B$.

**11.19** Based on experimental observations, the acceleration of a particle is defined by the relation $a = -(0.1 + \sin x/b)$, where $a$ and $x$ are expressed in m/s$^2$ and meters, respectively. Knowing that $b = 0.8$ m and that $v = 1$ m/s when $x = 0$, determine (a) the velocity of the particle when $x = -1$ m, (b) the position where the velocity is maximum, (c) the maximum velocity.

**11.20** A spring $AB$ is attached to a support at $A$ and to a collar. The unstretched length of the spring is $l$. Knowing that the collar is released from rest at $x = x_0$ and has an acceleration defined by the relation $a = -100(x - lx/\sqrt{l^2 + x^2})$, determine the velocity of the collar as it passes through point $C$.

**11.21** The acceleration of a particle is defined by the relation $a = k(1 - e^{-x})$, where $k$ is a constant. Knowing that the velocity of the particle is $v = +9$ m/s when $x = -3$ m and that the particle comes to rest at the origin, determine (a) the value of $k$, (b) the velocity of the particle when $x = -2$ m.

**11.22** Starting from $x = 0$ with no initial velocity, a particle is given an acceleration $a = 0.8\sqrt{v^2 + 49}$, where $a$ and $v$ are expressed in ft/s$^2$ and ft/s, respectively. Determine (a) the position of the particle when $v = 24$ ft/s, (b) the speed and acceleration of the particle when $x = 40$ ft.

**11.23** A ball is dropped from a boat so that it strikes the surface of a lake with a speed of 16.5 ft/s. While in the water the ball experiences an acceleration of $a = 10 - 0.8v$, where $a$ and $v$ are expressed in ft/s$^2$ and ft/s, respectively. Knowing that the ball takes 3 s to reach the bottom of the lake, determine (a) the depth of the lake, (b) the speed of the ball when it hits the bottom of the lake.

**11.24** The acceleration of a particle is defined by the relation $a = -k\sqrt{v}$, where $k$ is a constant. Knowing that $x = 0$ and $v = 81$ m/s at $t = 0$, and that $v = 36$ m/s when $x = 18$ m, determine (a) the velocity of the particle when $x = 20$ m, (b) the time required for the particle to come to rest.

**11.25** The acceleration of a particle is defined by the relation $a = -kv^{2.5}$, where $k$ is a constant. The particle starts at $x = 0$ with a velocity of 16 mm/s, and when $x = 6$ mm, the velocity is observed to be 4 mm/s. Determine (a) the velocity of the particle when $x = 5$ mm, (b) the time at which the velocity of the particle is 9 mm/s.

**11.26** A human-powered vehicle (HPV) team wants to model the acceleration during the 260-m sprint race (the first 60 m is called a flying start) using $a = A - Cv^2$, where $a$ is the acceleration in m/s$^2$ and $v$ is the velocity in m/s. From wind tunnel testing, they found that $C = 0.0012$ m$^{-1}$. Knowing that the cyclist starts from rest and is going 100 km/h at the 260-meter mark, what is the value of $A$?

**11.27** Experimental data indicate that in a region downstream of a given louvered supply vent the velocity of the emitted air is defined by $v = 0.18v_0/x$, where $v$ and $x$ are expressed in m/s and meters, respectively, and $v_0$ is the initial discharge velocity of the air. For $v_0 = 3.6$ m/s, determine (a) the acceleration of the air at $x = 2$ m, (b) the time required for the air to flow from $x = 1$ to $x = 3$ m.

**11.28** Based on observations, the speed of a jogger can be approximated by the relation $v = 7.5(1 - 0.04x)^{0.3}$, where $v$ and $x$ are expressed in km/h and kilometers, respectively. Knowing that $x = 0$ at $t = 0$, determine (a) the distance the jogger has run when $t = 1$ h, (b) the jogger's acceleration in m/s$^2$ at $t = 0$, (c) the time required for the jogger to run 6 km.

**11.29** The acceleration due to gravity at an altitude $y$ above the surface of the earth can be expressed as

$$a = \frac{-32.2}{\left[1 + \left(y/20.9 \times 10^6\right)\right]^2}$$

where $a$ and $y$ are expressed in ft/s$^2$ and feet, respectively. Using this expression, compute the height reached by a projectile fired vertically upward from the surface of the earth if its initial velocity is (a) 1800 ft/s, (b) 3000 ft/s, (c) 36,700 ft/s.

**11.30** The acceleration due to gravity of a particle falling toward the earth is $a = -gR^2/r^2$, where $r$ is the distance from the *center* of the earth to the particle, $R$ is the radius of the earth, and $g$ is the acceleration due to gravity at the surface of the earth. If $R = 3960$ mi, calculate the *escape velocity*; that is, the minimum velocity with which a particle must be projected vertically upward from the surface of the earth if it is not to return to the earth. (*Hint:* $v = 0$ for $r = \infty$.)

**11.31** The velocity of a particle is $v = v_0[1 - \sin(\pi t/T)]$. Knowing that the particle starts from the origin with an initial velocity $v_0$, determine (a) its position and its acceleration at $t = 3T$, (b) its average velocity during the interval $t = 0$ to $t = T$.

**11.32** An eccentric circular cam, which serves a similar function as the Scotch yoke mechanism in Problem 11.13, is used in conjunction with a flat face follower to control motion in pumps and in steam engine valves. Knowing that the eccentricity is denoted by $e$, the maximum range of the displacement of the follower is $d_{max}$, and the maximum velocity of the follower is $v_{max}$, determine the displacement, velocity, and acceleration of the follower.

**Fig. P11.26**
Couresty of Phillip Cornwell

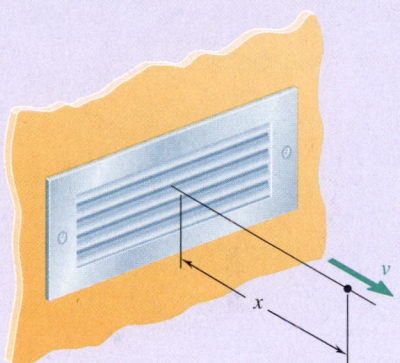

**Fig. P11.27**

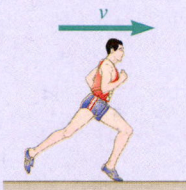

**Fig. P11.28**

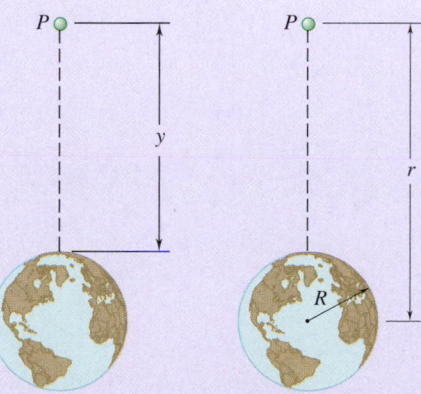

**Fig. P11.29**          **Fig. P11.30**

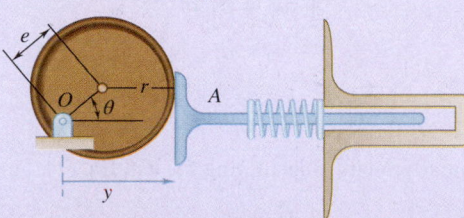

**Fig. P11.32**

## 11.2    SPECIAL CASES AND RELATIVE MOTION

In this section, we derive the equations that describe uniform rectilinear motion and uniformly accelerated rectilinear motion. We also introduce the concept of relative motion, which is of fundamental importance whenever we consider the motion of more than one particle at the same time.

### 11.2A    Uniform Rectilinear Motion

Uniform rectilinear motion is a type of straight-line motion that is frequently encountered in practical applications. In this motion, the acceleration $a$ of the particle is zero for every value of $t$. The velocity $v$ is therefore constant, and Eq. (11.1) becomes

$$\frac{dx}{dt} = v = \text{constant}$$

We can obtain the position coordinate $x$ by integrating this equation. Denoting the initial value of $x$ by $x_0$, we have

**Distance in uniform rectilinear motion**

$$\int_{x_0}^{x} dx = v \int_{0}^{t} dt$$
$$x - x_0 = vt$$

$$\boxed{x = x_0 + vt} \tag{11.5}$$

This equation can be used *only if the velocity of the particle is known to be constant.* For example, this would be true for an airplane in steady flight or a car cruising along a highway at a constant speed.

### 11.2B    Uniformly Accelerated Rectilinear Motion

Uniformly accelerated rectilinear motion is another common type of motion. In this case, the acceleration $a$ of the particle is constant, and Eq. (11.2) becomes

$$\frac{dv}{dt} = a = \text{constant}$$

We obtain the velocity $v$ of the particle by integrating this equation as

$$\int_{v_0}^{v} dv = a \int_{0}^{t} dt$$
$$v - v_0 = at$$

$$\boxed{v = v_0 + at} \tag{11.6}$$

where $v_0$ is the initial velocity. Substituting for $v$ in Eq. (11.1), we have

$$\frac{dx}{dt} = v_0 + at$$

Denoting the initial value of $x$ by $x_0$ and integrating, we have

$$\int_{x_0}^{x} dx = \int_{0}^{t} (v_0 + at)dt$$

$$x - x_0 = v_0 t + \tfrac{1}{2}at^2$$

$$\boxed{x = x_0 + v_0 t + \tfrac{1}{2}at^2} \tag{11.7}$$

We can also use Eq. (11.4) and write

$$v\frac{dv}{dx} = a = \text{constant}$$

$$v\,dv = a\,dx$$

Integrating both sides, we obtain

$$\int_{v_0}^{v} v\,dv = a\int_{x_0}^{x} dx$$

$$\tfrac{1}{2}(v^2 - v_0^2) = a(x - x_0)$$

$$\boxed{v^2 = v_0^2 + 2a(x - x_0)} \tag{11.8}$$

The three equations we have derived provide useful relations among position, velocity, and time in the case of constant acceleration, once you have provided appropriate values for $a$, $v_0$, and $x_0$. You first need to define the origin $O$ of the $x$ axis and choose a positive direction along the axis; this direction determines the signs of $a$, $v_0$, and $x_0$. Eq. (11.6) relates $v$ and $t$ and should be used when the value of $v$ corresponding to a given value of $t$ is desired, or vice versa, Eq. (11.7) relates $x$ and $t$; Eq. (11.8) relates $v$ and $x$. An important application of uniformly accelerated motion is the motion of a body in **free fall**. The acceleration of a body in free fall (usually denoted by $g$) is equal to 9.81 m/s² or 32.2 ft/s² (we ignore air resistance in this case).

It is important to keep in mind that the three equations can be used *only when the acceleration of the particle is known to be constant.* If the acceleration of the particle is variable, you need to determine its motion from the fundamental Eqs. (11.1), (11.2), (11.3), and (11.4) according to the methods outlined in Sec. 11.1B.

## 11.2C   Motion of Several Particles

When several particles move independently along the same line, you can write independent equations of motion for each particle. Whenever possible, you should record time from the same initial instant for all particles and measure displacements from the same origin and in the same direction. In other words, use a single clock and a single measuring tape.

**Relative Motion of Two Particles.** Consider two particles $A$ and $B$ moving along the same straight line (Fig. 11.7). If we measure the position coordinates $x_A$ and $x_B$ from the same origin, the difference $x_B - x_A$ defines the **relative position coordinate of $B$ with respect to $A$**, which is denoted by $x_{B/A}$. We have

**Relative position of two particles**

$$x_{B/A} = x_B - x_A \quad \text{or} \quad \boxed{x_B = x_A + x_{B/A}} \tag{11.9}$$

Regardless of the positions of $A$ and $B$ with respect to the origin, a positive sign for $x_{B/A}$ means that $B$ is to the right of $A$, and a negative sign means that $B$ is to the left of $A$.

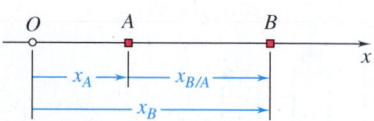

**Fig. 11.7** Two particles $A$ and $B$ in motion along the same straight line.

The rate of change of $x_{B/A}$ is known as the **relative velocity of B with respect to A** and is denoted by $v_{B/A}$. Differentiating Eq. (11.9), we obtain

**Relative velocity of two particles**
$$v_{B/A} = v_B - v_A \quad \text{or} \quad \boxed{v_B = v_A + v_{B/A}} \quad \textbf{(11.10)}$$

A positive sign for $v_{B/A}$ means that $B$ is *observed from A* to move in the positive direction; a negative sign means that it is observed to move in the negative direction.

The rate of change of $v_{B/A}$ is known as the **relative acceleration of B with respect to A** and is denoted by $a_{B/A}$. Differentiating Eq. (11.10), we obtain[†]

**Relative acceleration of two particles**
$$a_{B/A} = a_B - a_A \quad \text{or} \quad \boxed{a_B = a_A + a_{B/A}} \quad \textbf{(11.11)}$$

**Dependent Motion of Particles.** Sometimes, the position of a particle depends upon the position of another particle or of several other particles. These motions are called **dependent**. For example, the position of block $B$ in Fig. 11.8 depends upon the position of block $A$. Because the rope $ACDEFG$ is of constant length, and because the lengths of the portions of rope $CD$ and $EF$ wrapped around the pulleys remain constant, it follows that the sum of the lengths of the segments $AC$, $DE$, and $FG$ is constant. Observing that the length of the segment $AC$ differs from $x_A$ only by a constant and that, similarly, the lengths of the segments $DE$ and $FG$ differ from $x_B$ only by a constant, we have

$$x_A + 2x_B = \text{constant}$$

Because only one of the two coordinates $x_A$ and $x_B$ can be chosen arbitrarily, we say that the system shown in Fig. 11.8 has **one degree of freedom**. From the relation between the position coordinates $x_A$ and $x_B$, it follows that if $x_A$ is given an increment $\Delta x_A$—that is, if block $A$ is lowered by an amount $\Delta x_A$—the coordinate $x_B$ receives an increment $\Delta x_B = -\frac{1}{2}\Delta x_A$. In other words, block $B$ rises by half the same amount. You can check this directly from Fig. 11.8.

In the case of the three blocks of Fig. 11.9, we can again observe that the length of the rope that passes over the pulleys is constant. Thus, the following relation must be satisfied by the position coordinates of the three blocks:

$$2x_A + 2x_B + x_C = \text{constant}$$

Because two of the coordinates can be chosen arbitrarily, we say that the system shown in Fig. 11.9 has **two degrees of freedom**.

When the relation existing between the position coordinates of several particles is *linear*, a similar relation holds between the velocities and between the accelerations of the particles. In the case of the blocks of Fig. 11.9, for instance, we can differentiate the position equation twice and obtain

$$2\frac{dx_A}{dt} + 2\frac{dx_B}{dt} + \frac{dx_C}{dt} = 0 \quad \text{or} \quad 2v_A + 2v_B + v_C = 0$$

$$2\frac{dv_A}{dt} + 2\frac{dv_B}{dt} + \frac{dv_C}{dt} = 0 \quad \text{or} \quad 2a_A + 2a_B + a_C = 0$$

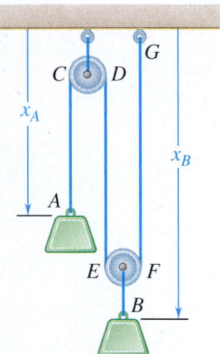

**Fig. 11.8** A system of blocks and pulleys with one degree of freedom.

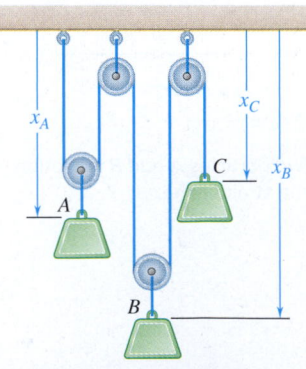

**Fig. 11.9** A system of blocks and pulleys with two degrees of freedom.

[†]Note that the product of the subscripts $A$ and $B/A$ used in the right-hand sides of Eqs. (11.9), (11.10), and (11.11) is equal to the subscript $B$ that appears in the left-hand sides. This may help you remember the correct order of subscripts in various situations.

## Sample Problem 11.5

In an elevator shaft, a ball is thrown vertically upward with an initial velocity of 18 m/s from a height of 12 m above ground. At the same instant, an open-platform elevator passes the 5-m level, moving upward with a constant velocity of 2 m/s. Determine (*a*) when and where the ball hits the elevator, (*b*) the relative velocity of the ball with respect to the elevator when the ball hits the elevator. ▶

**STRATEGY:** The ball has a constant acceleration, so its motion is *uniformly accelerated*. The elevator has a constant velocity, so its motion is *uniform*. You can write equations to describe each motion and then set the position coordinates equal to each other to find when the particles meet. The relative velocity is determined from the calculated motion of each particle.

### MODELING and ANALYSIS:

**Motion of Ball.** Place the origin $O$ of the $y$ axis at ground level and choose its positive direction upward (Fig. 1). Then, the initial position of the ball is $y_0 = +12$ m, its initial velocity is $v_0 = +18$ m/s, and its acceleration is $a = -9.81$ m/s². Substituting these values in the equations for uniformly accelerated motion, you get

$$v_B = v_0 + at \qquad v_B = 18 - 9.81t \qquad (1)$$

$$y_B = y_0 + v_0 t + \frac{1}{2}at^2 \qquad y_B = 12 + 18t - 4.905t^2 \qquad (2)$$

**Motion of Elevator.** Again place the origin $O$ at ground level and choose the positive direction upward (Fig. 2). Noting that $y_0 = +5$ m, you have

$$v_E = +2 \text{ m/s} \qquad (3)$$

$$y_E = y_0 + v_E t \qquad y_E = 5 + 2t \qquad (4)$$

**Ball Hits Elevator.** First note that you used the same time $t$ and the same origin $O$ in writing the equations of motion for both the ball and the elevator. From Fig. 3, when the ball hits the elevator,

$$y_E = y_B \qquad (5)$$

Substituting for $y_E$ and $y_B$ from Eqs. (2) and (4) into Eq. (5), you have

$$5 + 2t = 12 + 18t - 4.905\,t^2$$
$$t = -0.39 \text{ s} \quad \text{and} \qquad t = 3.65 \text{ s} \quad \blacktriangleleft$$

Only the root $t = 3.65$ s corresponds to a time after the motion has begun. Substituting this value into Eq. (4), you obtain

$$y_E = 5 + 2(3.65) = 12.30 \text{ m}$$
$$\text{Elevation from ground} = 12.30 \text{ m} \quad \blacktriangleleft$$

**Relative Velocity.** The relative velocity of the ball with respect to the elevator is

$$v_{B/E} = v_B - v_E = (18 - 9.81t) - 2 = 16 - 9.81t$$

When the ball hits the elevator at time $t = 3.65$ s, you have

$$v_{B/E} = 16 - 9.81(3.65) \qquad v_{B/E} = -19.81 \text{ m/s} \quad \blacktriangleleft$$

*(continued)*

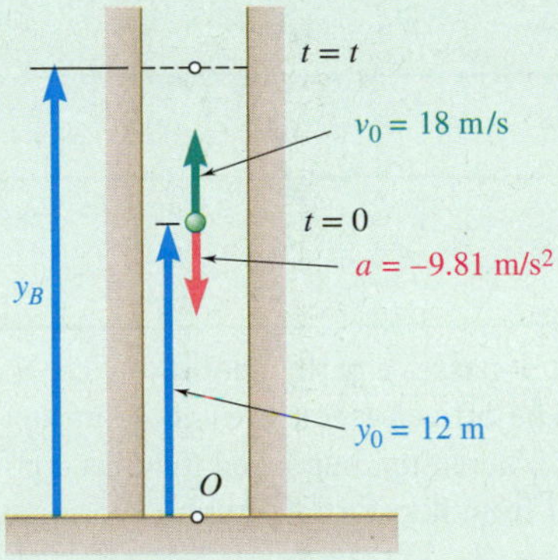

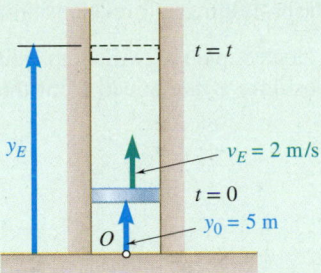

**Fig. 1** Acceleration, initial velocity, and initial position of the ball.

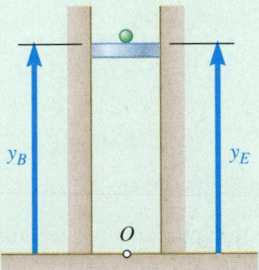

**Fig. 2** Initial velocity and initial position of the elevator.

**Fig. 3** Position of ball and elevator at time $t$.

The negative sign means that if you are riding on the elevator, it will appear as if the ball is moving downward.

**REFLECT and THINK:**   The key insight is that, when two particles collide, their position coordinates must be equal. Also, although you can use the basic kinematic relationships in this problem, you may find it easier to use the equations relating $a$, $v$, $x$, and $t$ when the acceleration is constant or zero.

## Sample Problem 11.6

Car $A$ is traveling at a constant 90 mi/h when it passes a parked police officer $B$, who gives chase when the car passes her. The officer accelerates at a constant rate until she reaches the speed of 105 mi/h. Thereafter, her speed remains constant. The police officer catches the car 3 mi from her starting point. Determine the initial acceleration of the police officer.

**STRATEGY:**   One car is traveling at a constant speed and the other has a constant acceleration, so you can start with the algebraic relationships found in Sec. 11.2 rather than separating and integrating the basic kinematic relationships.

**MODELING and ANALYSIS:**   A clearly labeled picture will help you understand the problem better (Fig. 1). The position, $x$, is defined from the point the car passes the officer.

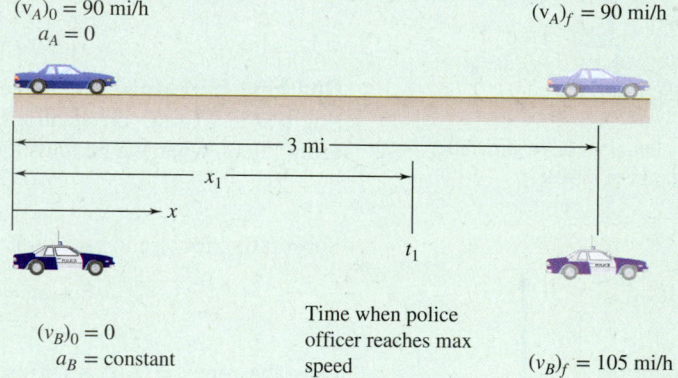

**Fig. 1**   Velocities and accelerations of the cars at various times.

**Unit Conversions.**   First you should convert everything to units of feet and seconds. Use the subscript $A$ for the car and the subscript $B$ for the officer:

$$v_A = \left(90\frac{\text{mi}}{\text{hr}}\right)\left(\frac{1 \text{ hr}}{3600 \text{ s}}\right)\left(\frac{5280 \text{ ft}}{\text{mi}}\right) = 132\frac{\text{ft}}{\text{s}}$$

$$v_B = \left(105\frac{\text{mi}}{\text{hr}}\right)\left(\frac{1 \text{ hr}}{3600 \text{ s}}\right)\left(\frac{5280 \text{ ft}}{\text{mi}}\right) = 154\frac{\text{ft}}{\text{s}}$$

**Motion of the Speeding Car $A$.**   Because the car has a constant speed,

$$x_A = v_A t = 132t \tag{1}$$

*(continued)*

**Motion of the Officer B.** The officer has a constant acceleration until she reaches a final speed of 105 mph. This time is labeled $t_1$ in Fig. 1. Therefore, from time $0 < t < t_1$, the officer has a velocity of

$$v_B = a_B t \quad \text{for } 0 < t < t_1$$

or at time $t = t_1$, it is

$$154 = a_B t_1 \tag{2}$$

The distance the officer travels is going to be the distance from 0 to $t_1$ and then from $t_1$ to $t_f$. Hence,

$$x_B = \frac{1}{2} a_B t_1^2 + v_B (t - t_1) \text{ for } t > t_1 \tag{3}$$

The officer catches the speeder when $x_A = x_B = 3$ mi $= 15{,}840$ ft. From Eq. (1), you can solve for the time $t_f = (15{,}840 \text{ ft})/(132 \text{ ft/s}) = 120$ s. Therefore, you have two equations: Eq. (2) and

$$15{,}840 = \frac{1}{2} a_B t_1^2 + 154(120 - t_1) \tag{4}$$

Substituting Eq. (2) into Eq. (4) allows you to solve for $t_1$:

$$t_1 = 34.39 \text{ s}$$

Substituting this into Eq. (2) gives

$$a_B = 4.49 \text{ ft/s} \quad \blacktriangleleft$$

**REFLECT and THINK:** It is important to use the same origin for the position of both vehicles. The time to accelerate from 0 to 105 mph seems reasonable, although it is perhaps longer than you would expect. A high-performance sports car can go from 0 to 60 mph in less than 5 seconds. It is very likely that the officer could have accelerated to 105 mph in less time if she had wanted to, but perhaps she had to consider the safety of other motorists.

## Sample Problem 11.7

Collar A and block B are connected by a cable passing over three pulleys, labeled C, D, and E as shown. Pulleys C and E are fixed, but D is attached to a collar that is pulled downward with a constant velocity of 3 in./s. At $t = 0$, collar A starts moving downward from position K with a constant acceleration and no initial velocity. Knowing that the velocity of collar A is 12 in./s as it passes through point L, determine the change in elevation, the velocity, and the acceleration of block B when collar A passes through L. ▶

**STRATEGY:** You have multiple objects connected by cables, so this is a problem with *dependent motion*. Use the given data to write a single equation relating the changes in position coordinates of collar A, pulley D, and block B. Based on the given information, you will also need to use the algebraic relationships we found for uniformly accelerated motion.

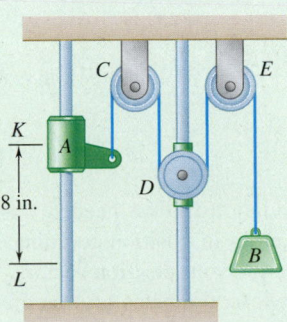

*(continued)*

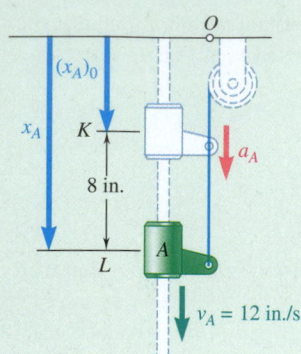

**Fig. 1** Position, velocity, and acceleration of collar A.

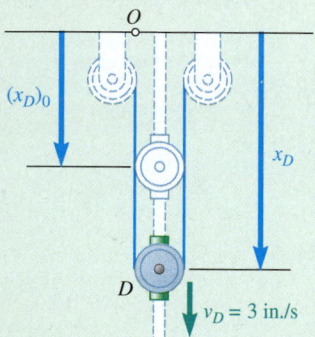

**Fig. 2** Position and velocity of pulley D.

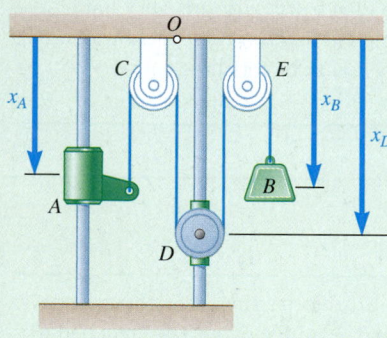

**Fig. 3** Position of A, B, and D.

## MODELING and ANALYSIS:

**Motion of Collar A.**   Place the origin $O$ at the upper horizontal surface and choose the positive direction downward. Then when $t = 0$, collar $A$ is at position $K$ and $(v_A)_0 = 0$ (Fig. 1). Because $v_A = 12$ in./s and $x_A - (x_A)_0 = 8$ in. when the collar passes through $L$, you have

$$v_A^2 = (v_A)_0^2 + 2a_A[x_A - (x_A)_0] \qquad (12)^2 = 0 + 2a_A(8)$$
$$a_A = 9 \text{ in./s}^2$$

To find the time at which collar $A$ reaches point $L$, use the equation for velocity as a function of time with uniform acceleration. Thus,

$$v_A = (v_A)_0 + a_A t \qquad 12 = 0 + 9t \qquad t = 1.333 \text{ s}$$

**Motion of Pulley D.**   Because the positive direction is downward, you have (Fig. 2)

$$a_D = 0 \qquad v_D = 3 \text{ in./s} \qquad x_D = (x_D)_0 + v_D t = (x_D)_0 + 3t$$

When collar $A$ reaches $L$ at $t = 1.333$ s, the position of pulley $D$ is

$$x_D = (x_D)_0 + 3(1.333) = (x_D)_0 + 4$$

Thus, $\qquad\qquad\qquad\qquad x_D - (x_D)_0 = 4 \text{ in.}$

**Motion of Block B.**   From Fig. 3, note that the total length of cable $ACDEB$ differs from the quantity $(x_A + 2x_D + x_B)$ only by a constant. Because the cable length is constant during the motion, this quantity must also remain constant. Thus, considering the times $t = 0$ and $t = 1.333$ s, you can write

$$x_A + 2x_D + x_B = (x_A)_0 + 2(x_D)_0 + (x_B)_0 \qquad \textbf{(1)}$$
$$[x_A - (x_A)_0] + 2[x_D - (x_D)_0] + [x_B - (x_B)_0] = 0 \qquad \textbf{(2)}$$

But you know that $x_A - (x_A)_0 = 8$ in. and $x_D - (x_D)_0 = 4$ in. Substituting these values in Eq. (2), you find

$$8 + 2(4) + [x_B - (x_B)_0] = 0 \qquad x_B - (x_B)_0 = -16 \text{ in.}$$

Thus, $\qquad\qquad\qquad$ **Change in elevation of $B$ = 16 in. ↑** ◄

Differentiating Eq. (1) twice, you obtain equations relating the velocities and the accelerations of $A$, $B$, and $D$. Substituting for the velocities and accelerations of $A$ and $D$ at $t = 1.333$ s, you have

$v_A + 2v_D + v_B = 0$:    $12 + 2(3) + v_B = 0$
$$v_B = -18 \text{ in./s} \qquad\qquad \textbf{\textit{v}}_B = 18 \text{ in./s ↑} ◄$$

$a_A + 2a_D + a_B = 0$:    $9 + 2(0) + a_B = 0$
$$a_B = -9 \text{ in./s}^2 \qquad\qquad \textbf{\textit{a}}_B = 9 \text{ in./s}^2 ↑ ◄$$

**REFLECT and THINK:**   In this case, the relationship we needed was not between position coordinates, but between changes in position coordinates at two different times. The key step is to clearly define your position vectors. This is a two-degree-of-freedom system, because two coordinates are required to completely describe it.

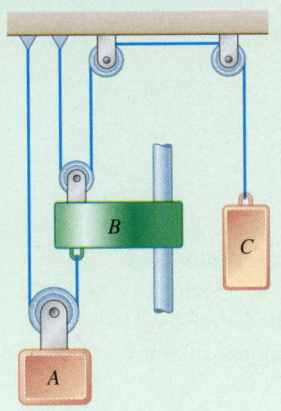

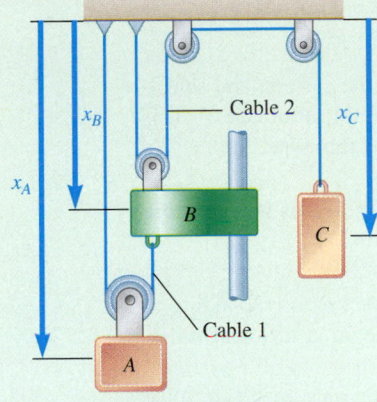

**Fig. 1**   Position of A, B, and C.

## Sample Problem 11.8

Block $C$ starts from rest and moves down with a constant acceleration. Knowing that after block $A$ has moved 1.5 ft its velocity is 0.6 ft/s, determine (a) the acceleration of $A$ and $C$, (b) the change in velocity and the change in position of block $B$ after 2.5 seconds.

**STRATEGY:**   Because you have blocks connected by cables, this is a dependent-motion problem. You should define coordinates for each mass and write constraint equations for both cables.

**MODELING and ANALYSIS:**   Define position vectors as shown in Fig. 1, where positive is defined to be down.

**Constraint Equations.**   Assuming the cables are inextensible, you can write the lengths in terms of the defined coordinates and then differentiate.

**Cable 1:**
$$x_A + (x_A - x_B) = \text{constant}$$

Differentiating this, you find

$$2v_A = v_B \quad \text{and} \quad 2a_A = a_B \tag{1}$$

**Cable 2:**
$$2x_B + x_C = \text{constant}$$

Differentiating this, you find

$$v_C = -2v_B \quad \text{and} \quad a_C = -2a_B \tag{2}$$

Substituting Eq. (1) into Eq. (2) gives

$$v_C = -4v_A \quad \text{and} \quad a_C = -4a_A \tag{3}$$

**Motion of A.**   You can use the constant-acceleration equations for block $A$:

$$v_A^2 - v_{A_0}^2 = 2a_A[x_A - (x_A)_0] \quad \text{or} \quad a_A = \frac{v_A^2 - (v_A)_0^2}{2[x_A - (x_A)_0]} \tag{4}$$

**a. Acceleration of A and C.**   You know $v_C$ and $a_C$ are down, so from Eq. (3), you also know $v_A$ and $a_A$ are up. Substituting the given values into Eq. (4), you find

$$a_A = \frac{(0.6 \text{ ft/s})^2 - 0}{2(-1.5 \text{ ft})} = -0.12 \text{ ft/s}^2 \qquad \mathbf{a}_A = 0.120 \text{ ft/s}^2 \uparrow \ \blacktriangleleft$$

Substituting this value into $a_C = -4a_A$, you obtain

$$\mathbf{a}_C = 0.480 \text{ ft/s}^2 \downarrow \ \blacktriangleleft$$

**b. Velocity and change in position of B after 2.5 s.**   Substituting $a_A$ into $a_B = 2a_A$ gives

$$a_B = 2(-0.12 \text{ ft/s}^2) = -0.24 \text{ ft/s}^2$$

You can use the equations of constant acceleration to find

$$\Delta v_B = a_B t = (-0.24 \text{ ft/s}^2)(2.5 \text{ s}) = -0.600 \text{ ft/s} \qquad \Delta v_B = 0.600 \text{ ft/s} \uparrow \ \blacktriangleleft$$
$$\Delta x_B = \tfrac{1}{2}a_B t^2 = \tfrac{1}{2}(-0.24 \text{ ft/s}^2)(2.5 \text{ s})^2 = -0.750 \text{ ft} \qquad \Delta x_B = 0.750 \text{ ft} \uparrow \ \blacktriangleleft$$

**REFLECT and THINK:**   One of the keys to solving this problem is recognizing that because there are two cables, you need to write two constraint equations. The directions of the answers also make sense. If block $C$ is accelerating downward, you would expect $A$ and $B$ to accelerate upward.

# SOLVING PROBLEMS
# ON YOUR OWN

In this section, we derived the equations that describe **uniform rectilinear motion** (constant velocity) and **uniformly accelerated rectilinear motion** (constant acceleration). We also introduced the concept of **relative motion**. We can apply the equations for relative motion [Eqs. (11.9), (11.10), and (11.11)] to the independent or dependent motions of any two particles moving along the same straight line.

**A. Independent motion of one or more particles.** Organize the solution of problems of this type as follows.

**1. Begin your solution** by listing the given information, sketching the system, and selecting the origin and the positive direction of the coordinate axis (Sample Prob. 11.5). It is always advantageous to have a visual representation of problems of this type.

**2. Write the equations** that describe the motions of the various particles, as well as those that describe how these motions are related [Eq. (5) of Sample Prob. 11.5].

**3. Define the initial conditions;** that is, specify the state of the system corresponding to $t = 0$. This is especially important if the motions of the particles begin at different times. In such cases, either of two approaches can be used.
    **a.** Let $t = 0$ be the time when the last particle begins to move. You must then determine the initial position $x_0$ and the initial velocity $v_0$ of each of the other particles.
    **b.** Let $t = 0$ be the time when the first particle begins to move. You must then, in each of the equations describing the motion of another particle, replace $t$ with $t - t_0$, where $t_0$ is the time at which that specific particle begins to move. It is important to recognize that the equations obtained in this way are valid only for $t \geq t_0$.

**B. Dependent motion of two or more particles.** In problems of this type, the particles of the system are connected to each other, typically by ropes or cables. The method of solution of these problems is similar to that of the preceding group of problems, except that it is now necessary to describe the *physical connections* between the particles. In the following problems, the connection is provided by one or more cables. For each cable, you will have to write equations similar to the last three equations of Sec. 11.2C. We suggest that you use the following procedure.

**1. Draw a sketch of the system** and select a coordinate system, indicating clearly a positive sense for each of the coordinate axes. For example, in Sample Probs. 11.7 and 11.8, we measured lengths downward from the upper horizontal support. It thus follows that those displacements, velocities, and accelerations that have positive values are directed downward.

**2. Write the equation describing the constraint** imposed by each cable on the motion of the particles involved. Differentiating this equation twice, you obtain the corresponding relations among velocities and accelerations.

**3. If several directions of motion are involved,** you must select a coordinate axis and a positive sense for each of these directions. You should also try to locate the origins of your coordinate axes so that the equations of constraints are as simple as possible. For example, in Sample Prob. 11.7, it is easier to define the various coordinates by measuring them downward from the upper support than by measuring them upward from the bottom support.

**Finally, keep in mind** that the method of analysis described in this section and the corresponding equations can be used only for particles moving with *uniform* or *uniformly accelerated rectilinear motion.*

# Problems

**11.33** An airplane begins its take-off run at $A$ with zero velocity and a constant acceleration $a$. Knowing that it becomes airborne 30 s later at $B$ with a take-off velocity of 270 km/h, determine ($a$) the acceleration $a$, ($b$) distance $AB$.

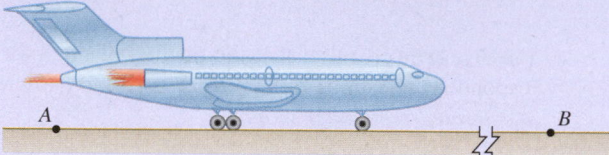

**Fig. P11.33**

**11.34** A minivan is tested for acceleration and braking. In the street-start acceleration test, the elapsed time is 8.2 s for a velocity increase from 10 km/h to 100 km/h. In the braking test, the distance traveled is 44 m during braking to a stop from 100 km/h. Assuming constant values of acceleration and deceleration, determine ($a$) the acceleration during the street-start test, ($b$) the deceleration during the braking test.

**Fig. P11.34**

**11.35** Steep safety ramps are built beside mountain highways to enable vehicles with defective brakes to stop safely. A truck enters a 750-ft ramp at a high speed $v_0$ and travels 540 ft in 6 s at constant deceleration before its speed is reduced to $v_0/2$. Assuming the same constant deceleration, determine ($a$) the additional time required for the truck to stop, ($b$) the additional distance traveled by the truck.

**Fig. P11.35**

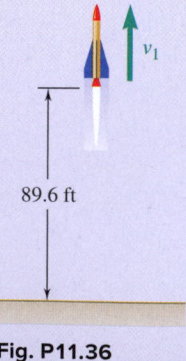

**Fig. P11.36**

**11.36** A group of students launches a model rocket in the vertical direction. Based on tracking data, they determine that the altitude of the rocket was 89.6 ft at the end of the powered portion of the flight and that the rocket landed 16 s later. Knowing that the descent parachute failed to deploy so that the rocket fell freely to the ground after reaching its maximum altitude, and assuming that $g = 32.2$ ft/s$^2$, determine ($a$) the speed $v_1$ of the rocket at the end of powered flight, ($b$) the maximum altitude reached by the rocket.

**11.37** A small package is released from rest at $A$ and moves along the skate wheel conveyor *ABCD*. The package has a uniform acceleration of 4.8 m/s$^2$ as it moves down sections *AB* and *CD*, and its velocity is constant between $B$ and $C$. If the velocity of the package at $D$ is 7.2 m/s, determine (*a*) the distance $d$ between $C$ and $D$, (*b*) the time required for the package to reach $D$.

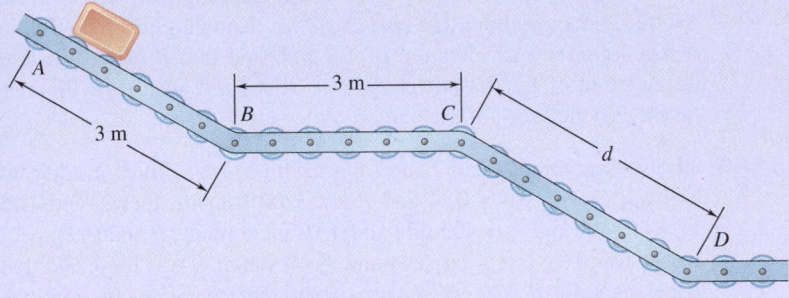

Fig. *P11.37*

**11.38** A sprinter in a 100-m race accelerates uniformly for the first 35 m and then runs with constant velocity. If the sprinter's time for the first 35 m is 5.4 s, determine (*a*) his acceleration, (*b*) his final velocity, (*c*) his time for the race.

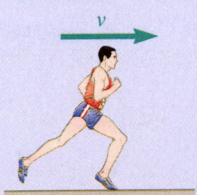

Fig. *P11.38*

**11.39** Automobile $A$ starts from $O$ and accelerates at the constant rate of 0.75 m/s$^2$. A short time later it is passed by bus $B$, which is traveling in the opposite direction at a constant speed of 6 m/s. Knowing that bus $B$ passes point $O$ 20 s after automobile $A$ started from there, determine when and where the vehicles passed each other.

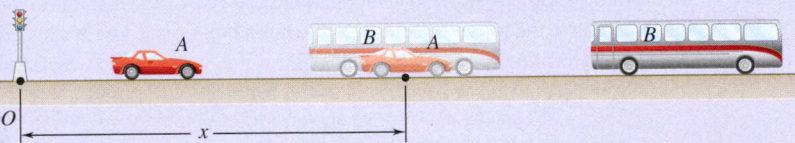

Fig. P11.39

**11.40** In a boat race, boat $A$ is leading boat $B$ by 50 m and both boats are traveling at a constant speed of 180 km/h. At $t = 0$, the boats accelerate at constant rates. Knowing that when $B$ passes $A$, $t = 8$ s and $v_A = 225$ km/h, determine (*a*) the acceleration of $A$, (*b*) the acceleration of $B$.

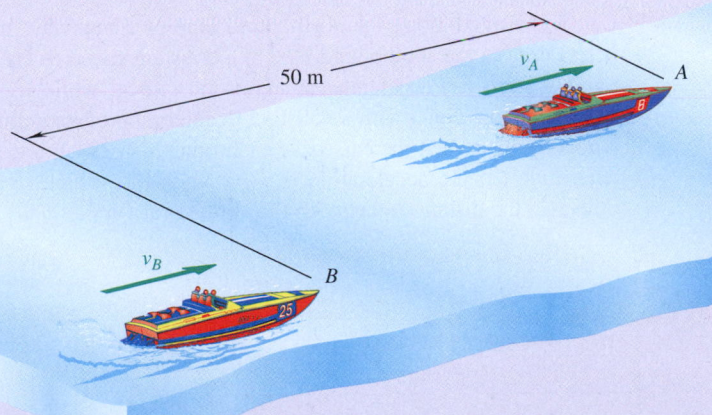

Fig. P11.40

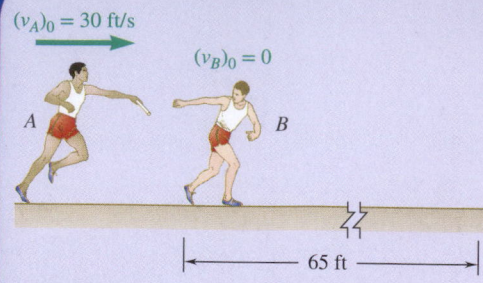

$(v_A)_0 = 30$ ft/s

$(v_B)_0 = 0$

A

B

|← 65 ft →|

**Fig. P11.41**

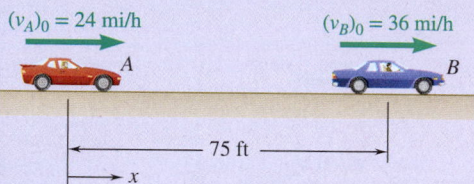

$(v_A)_0 = 24$ mi/h

A

$(v_B)_0 = 36$ mi/h

B

|← 75 ft →|

x

**Fig. P11.42**

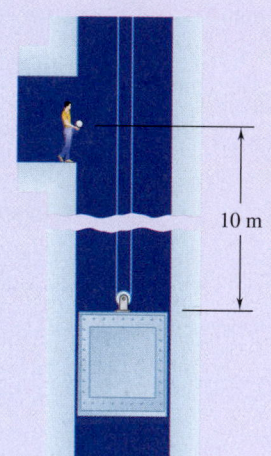

10 m

**Fig. P11.44**

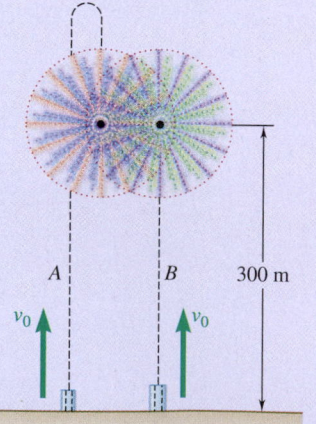

A    B    300 m

$v_0$    $v_0$

**Fig. P11.45**

**11.41** As relay runner A enters the 65-ft-long exchange zone with a speed of 30 ft/s, he begins to slow down. He hands the baton to runner B 2.5 s later as they leave the exchange zone with the same velocity. Determine (a) the uniform acceleration of each of the runners, (b) when runner B should begin to run.

**11.42** Automobiles A and B are traveling in adjacent highway lanes and at t = 0 have the positions and speeds shown. Knowing that automobile A has a constant acceleration of 1.8 ft/s² and that B has a constant deceleration of 1.2 ft/s², determine (a) when and where A will overtake B, (b) the speed of each automobile at that time.

**11.43** Two automobiles A and B are approaching each other in adjacent highway lanes. At t = 0, A and B are 3200 ft apart, their speeds are $v_A = 65$ mi/h and $v_B = 40$ mi/h, and they are at points P and Q, respectively. Knowing that A passes point Q 40 s after B was there and that B passes point P 42 s after A was there, determine (a) the uniform accelerations of A and B, (b) when the vehicles pass each other, (c) the speed of B at that time.

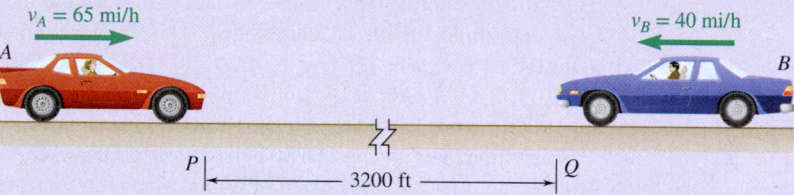

$v_A = 65$ mi/h

A

$v_B = 40$ mi/h

B

P |← 3200 ft →| Q

**Fig. P11.43**

**11.44** An elevator is moving upward at a constant speed of 4 m/s. A man standing 10 m above the top of the elevator throws a ball upward with a speed of 3 m/s. Determine (a) when the ball will hit the elevator, (b) where the ball will hit the elevator with respect to the location of the man.

**11.45** Two rockets are launched at a fireworks display. Rocket A is launched with an initial velocity $v_0 = 100$ m/s and rocket B is launched $t_1$ seconds later with the same initial velocity. The two rockets are timed to explode simultaneously at a height of 300 m as A is falling and B is rising. Assuming a constant acceleration g = 9.81 m/s², determine (a) the time $t_1$, (b) the velocity of B relative to A at the time of the explosion.

**11.46** Car A is parked along the northbound lane of a highway, and car B is traveling in the southbound lane at a constant speed of 60 mi/h. At t = 0, A starts and accelerates at a constant rate $a_A$, while at t = 5 s, B begins to slow down with a constant deceleration of magnitude $a_A/6$. Knowing that when the cars pass each other x = 294 ft and $v_A = v_B$, determine (a) the acceleration $a_A$, (b) when the vehicles pass each other, (c) the distance d between the vehicles at t = 0.

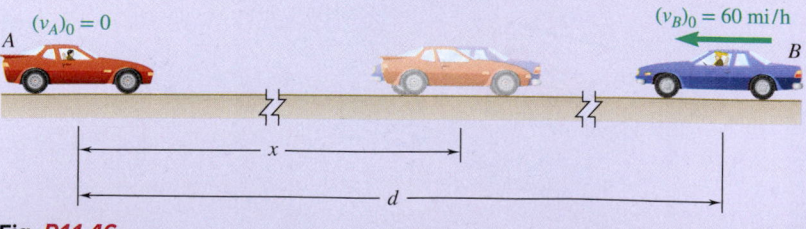

$(v_A)_0 = 0$

A

$(v_B)_0 = 60$ mi/h

B

|← x →|

|← d →|

**Fig. P11.46**

**11.47** The elevator $E$ shown in the figure moves downward with a constant velocity of 4 m/s. Determine (a) the velocity of the cable $C$, (b) the velocity of the counterweight $W$, (c) the relative velocity of the cable $C$ with respect to the elevator, (d) the relative velocity of the counterweight $W$ with respect to the elevator.

**11.48** The elevator $E$ shown starts from rest and moves upward with a constant acceleration. If the counterweight $W$ moves through 30 ft in 5 s, determine (a) the acceleration of the elevator and the cable $C$, (b) the velocity of the elevator after 5 s.

**11.49** An athlete pulls handle $A$ to the left with a constant velocity of 0.5 m/s. Determine (a) the velocity of the weight $B$, (b) the relative velocity of weight $B$ with respect to the handle $A$.

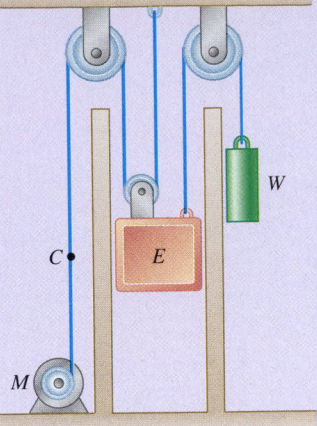

**Fig. P11.47 and P11.48**

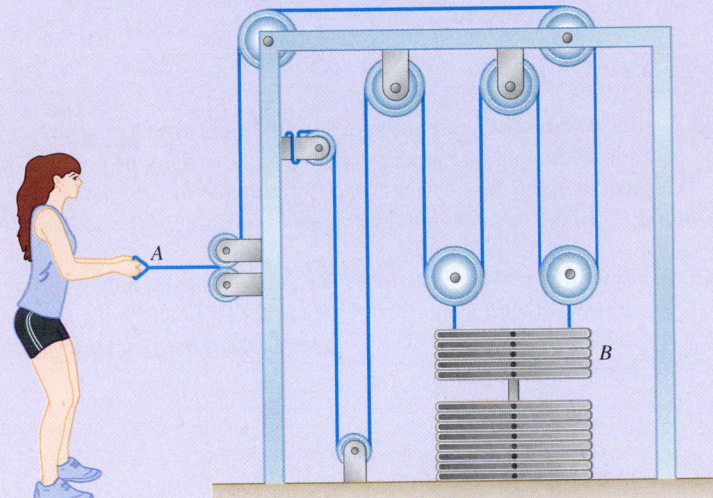

**Fig. P11.49**

**11.50** An athlete pulls handle $A$ to the left with a constant acceleration. Knowing that after the weight $B$ has been lifted 4 in. its velocity is 2 ft/s, determine (a) the accelerations of handle $A$ and weight $B$, (b) the velocity and change in position of handle $A$ after 0.5 s.

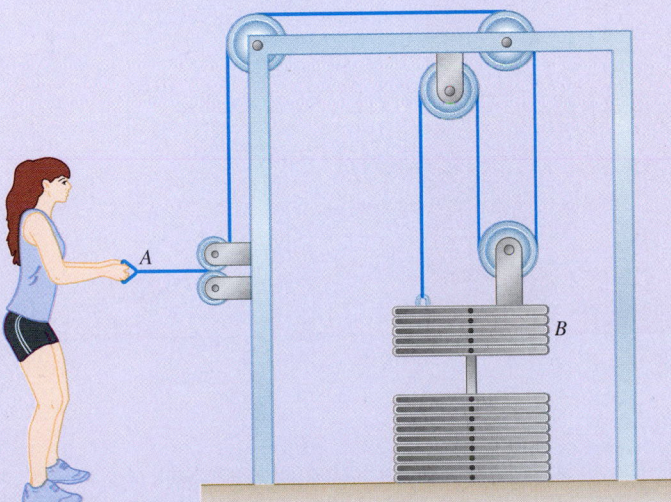

**Fig. P11.50**

**11.51** In the position shown, collar $B$ moves to the left with a constant velocity of 300 mm/s. Determine (a) the velocity of collar $A$, (b) the velocity of portion $C$ of the cable, (c) the relative velocity of portion $C$ of the cable with respect to collar $B$.

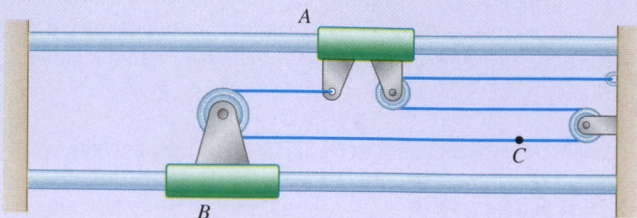

Fig. P11.51 and P11.52

**11.52** Collar $A$ starts from rest and moves to the right with a constant acceleration. Knowing that after 8 s the relative velocity of collar $B$ with respect to collar $A$ is 610 mm/s, determine (a) the accelerations of $A$ and $B$, (b) the velocity and the change in position of $B$ after 6 s.

**11.53** A farmer lifts his hay bales into the top loft of his barn by walking his horse forward with a constant velocity of 1 ft/s. Determine the velocity and acceleration of the hay bale when the horse is 10 ft away from the barn.

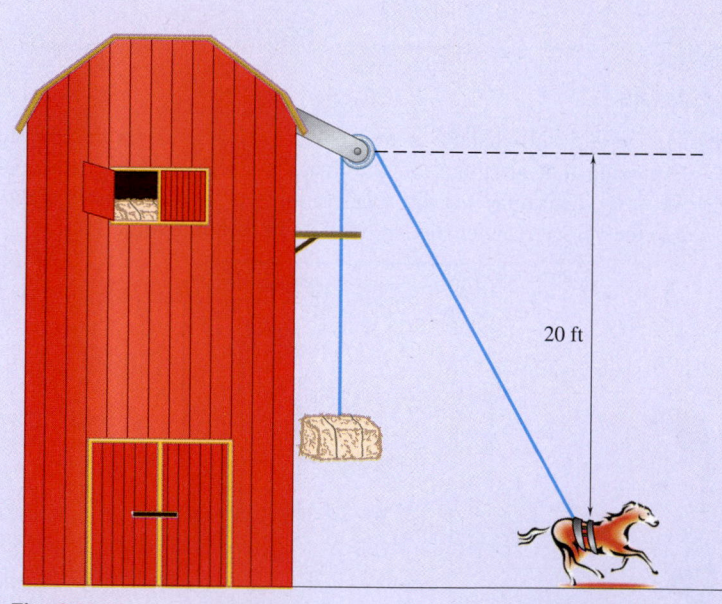

Fig. P11.53

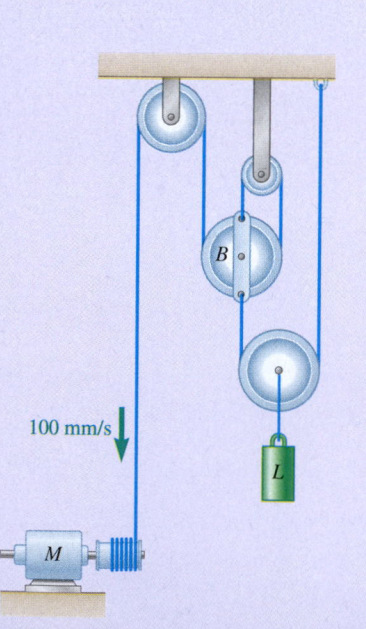

Fig. P11.54

**11.54** The motor $M$ reels in the cable at a constant rate of 100 mm/s. Determine (a) the velocity of load $L$, (b) the velocity of pulley $B$ with respect to load $L$.

**11.55** Collar $A$ starts from rest at $t = 0$ and moves upward with a constant acceleration of 3.6 in./s². Knowing that collar $B$ moves downward with a constant velocity of 18 in./s, determine (a) the time at which the velocity of block $C$ is zero, (b) the corresponding position of block $C$.

**11.56** Collars $A$ and $B$ start from rest, and collar $A$ moves upward with an acceleration of $3t^2$ mm/s². Knowing that collar $B$ moves downward with a constant acceleration and that its velocity is 150 mm/s after moving 700 mm, determine (a) the acceleration of block $C$ as a function of time, (b) the distance through which block $C$ will have moved after 3 s.

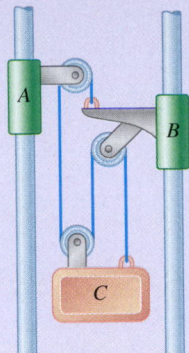

**Fig. P11.55**

**11.57** Block $B$ starts from rest, block $A$ moves with a constant acceleration, and slider block $C$ moves to the right with a constant acceleration of 75 mm/s². Knowing that at $t = 2$ s the velocities of $B$ and $C$ are 480 mm/s downward and 280 mm/s to the right, respectively, determine (a) the accelerations of $A$ and $B$, (b) the initial velocities of $A$ and $C$, (c) the change in position of slider block $C$ after 3 s.

**11.58** Block $B$ moves downward with a constant velocity of 20 mm/s. At $t = 0$, block $A$ is moving upward with a constant acceleration, and its velocity is 30 mm/s. Knowing that at $t = 3$ s slider block $C$ has moved 57 mm to the right, determine (a) the velocity of slider block $C$ at $t = 0$, (b) the accelerations of $A$ and $C$, (c) the change in position of block $A$ after 5 s.

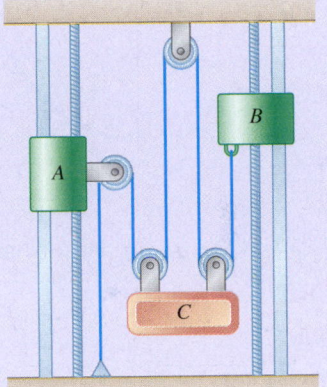

**Fig. P11.56**

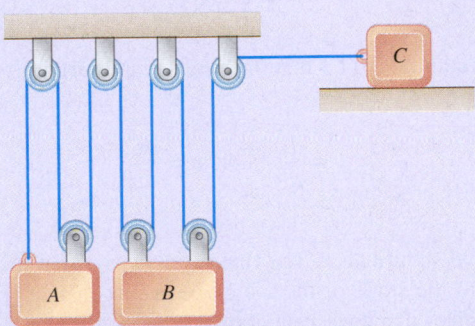

**Fig. P11.57 and P11.58**

**11.59** The system shown starts from rest, and each component moves with a constant acceleration. If the relative acceleration of block $C$ with respect to collar $B$ is 60 mm/s² upward and the relative acceleration of block $D$ with respect to block $A$ is 110 mm/s² downward, determine (a) the velocity of block $C$ after 3 s, (b) the change in position of block $D$ after 5 s.

**\*11.60** The system shown starts from rest, and the length of the upper cord is adjusted so that $A$, $B$, and $C$ are initially at the same level. Each component moves with a constant acceleration, and after 2 s the relative change in position of block $C$ with respect to block $A$ is 280 mm upward. Knowing that when the relative velocity of collar $B$ with respect to block $A$ is 80 mm/s downward, the displacements of $A$ and $B$ are 160 mm downward and 320 mm downward, respectively, determine (a) the accelerations of $A$ and $B$ if $a_B > 10$ mm/s², (b) the change in position of block $D$ when the velocity of block $C$ is 600 mm/s upward.

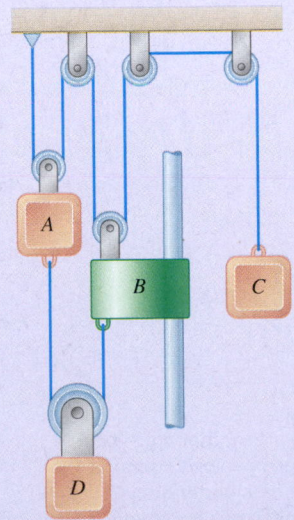

**Fig. P11.59 and P11.60**

# *11.3    GRAPHICAL SOLUTIONS

In analyzing problems in rectilinear motion, it is often useful to draw graphs of position, velocity, or acceleration versus time. Sometimes these graphs can provide insight into the situation by indicating when quantities increase, decrease, or stay the same. In other cases, the graphs can provide numerical solutions when analytical methods are not available. In many experimental situations, data are collected as a function of time, and the methods of this section are very useful for the analysis.

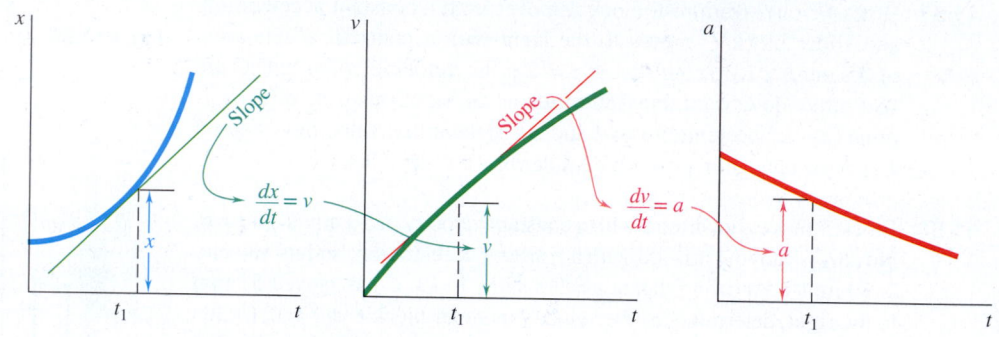

**Fig. 11.10**    The slope of an $x$–$t$ curve at time $t_1$ equals the velocity $v$ at that time; the slope of the $v$–$t$ curve at time $t_1$ equals the acceleration $a$ at that time.

We observed in Sec. 11.1 that the fundamental formulas

$$v = \frac{dx}{dt} \quad \text{and} \quad a = \frac{dv}{dt}$$

have a geometrical significance. The first formula says that the velocity at any instant is equal to the slope of the $x$–$t$ curve at that instant (Fig. 11.10). The second formula states that the acceleration is equal to the slope of the $v$–$t$ curve. We can use these two properties to determine graphically the $v$–$t$ and $a$–$t$ curves of a motion when the $x$–$t$ curve is known.

Integrating the two fundamental formulas from a time $t_1$ to a time $t_2$, we have

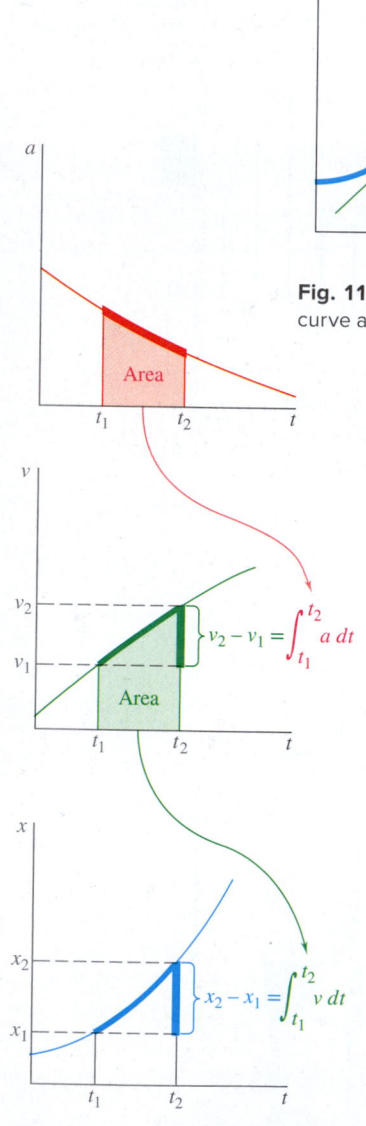

**Fig. 11.11**    The area under an $a$–$t$ curve equals the change in velocity during that time interval; the area under the $v$–$t$ curve equals the change in position during that time interval.

$$x_2 - x_1 = \int_{t_1}^{t_2} v\, dt \quad \text{and} \quad v_2 - v_1 = \int_{t_1}^{t_2} a\, dt \qquad \textbf{(11.12)}$$

The first formula says that the area measured under the $v$–$t$ curve from $t_1$ to $t_2$ is equal to the change in $x$ during that time interval (Fig. 11.11). Similarly, the second formula states that the area measured under the $a$–$t$ curve from $t_1$ to $t_2$ is equal to the change in $v$ during that time interval. We can use these two properties to determine graphically the $x$–$t$ curve of a motion when its $v$–$t$ curve or its $a$–$t$ curve is known (see Sample Prob. 11.9).

Graphical solutions are particularly useful when the motion considered is defined from experimental data and when $x$, $v$, and $a$ are not analytical functions of $t$. They also can be used to advantage when the motion consists of distinct parts and when its analysis requires writing a different equation for each of its parts. When using a graphical solution, however, be careful to note that (1) the area under the $v$–$t$ curve measures the *change in x*—not $x$ itself—and similarly, that the area under the $a$–$t$ curve measures the *change in v;* (2) an area above the $t$ axis corresponds to an *increase* in $x$ or $v$, whereas an area located below the $t$ axis measures a *decrease* in $x$ or $v$.

In drawing motion curves, it is useful to remember that, if the velocity is constant, it is represented by a horizontal straight line; the position coordinate $x$ is then a linear function of $t$ and is represented by an oblique straight line. If the acceleration is constant and different from zero, it is represented by a horizontal straight line; $v$ is then a linear function of $t$ and is represented by an oblique straight line, and $x$ is a second-degree polynomial in $t$ and is represented by a parabola. If the acceleration is a linear function of $t$, the velocity and the position coordinate are equal to second-degree and third-degree polynomials, respectively; $a$ is then represented by an oblique straight line, $v$ by a parabola, and $x$ by a cubic. In general, if the acceleration is a polynomial of degree $n$ in $t$, the velocity is a polynomial of degree $n + 1$, and the position coordinate is a polynomial of degree $n + 2$. These polynomials are represented by motion curves of a corresponding degree.

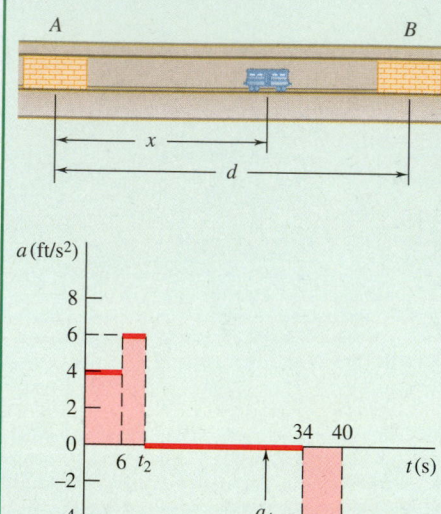

**Fig. 1** Acceleration of the subway car as a function of time.

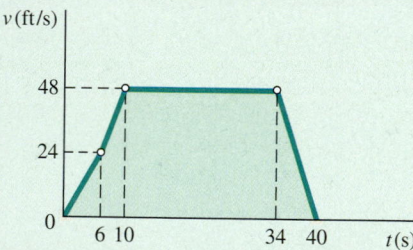

**Fig. 2** Velocity of the subway car as a function of time.

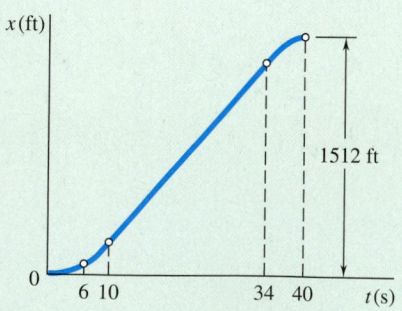

**Fig. 3** Position of the subway car as a function of time.

## Sample Problem 11.9

A subway car leaves station $A$; it gains speed at the rate of 4 ft/s$^2$ for 6 s and then at the rate of 6 ft/s$^2$ until it has reached the speed of 48 ft/s. The car maintains the same speed until it approaches station $B$; then the driver applies the brakes, giving the car a constant deceleration and bringing it to a stop in 6 s. The total running time from $A$ to $B$ is 40 s. Draw the $a$–$t$, $v$–$t$, and $x$–$t$ curves, and determine the distance between stations $A$ and $B$.

**STRATEGY:** You are given acceleration data, so first draw the graph of $a$ versus $t$. You can calculate areas under the curve to determine the $v$–$t$ curve and calculate areas under the $v$–$t$ curve to determine the $x$–$t$ curve.

**MODELING and ANALYSIS:** You can model the subway car as a particle without drag.

**Acceleration–Time Curve.** Because the acceleration is either constant or zero, the $a$–$t$ curve consists of horizontal straight-line segments. Determine the values of $t_2$ and $a_4$ as

$0 < t < 6$:    Change in $v$ = area under $a$–$t$ curve
$$v_6 - 0 = (6 \text{ s})(4 \text{ ft/s}^2) = 24 \text{ ft/s}$$

$6 < t < t_2$:   Because the velocity increases from 24 to 48 ft/s
Change in $v$ = area under $a$–$t$ curve
$$48 \text{ ft/s} - 24 \text{ ft/s} = (t_2 - 6)(6 \text{ ft/s}^2) \quad t_2 = 10 \text{ s}$$

$t_2 < t < 34$:  Because the velocity is constant, the acceleration is zero

$34 < t < 40$: Change in $v$ = area under $a$–$t$ curve
$$0 - 48 \text{ ft/s} = (6 \text{ s})a_4 \quad a_4 = -8 \text{ ft/s}^2$$

The acceleration is negative, so the corresponding area is below the $t$ axis; this area represents a decrease in velocity (Fig. 1).

**Velocity–Time Curve.** Because the acceleration is either constant or zero, the $v$–$t$ curve consists of straight-line segments connecting the points determined previously (Fig. 2).

                               Change in $x$ area under $v$–$t$ curve

$0 < t < 6$:         $x_6 - 0 = \frac{1}{2}(6)(24) = 72 \text{ ft}$

$6 < t < 10$:      $x_{10} - x_6 = \frac{1}{2}(4)(24 + 48) = 144 \text{ ft}$

$10 < t < 34$:    $x_{34} - x_{10} = (24)(48) = 1152 \text{ ft}$

$34 < t < 40$:    $x_{40} - x_{34} = \frac{1}{2}(6)(48) = 144 \text{ ft}$

Adding the changes in $x$ gives you the distance from $A$ to $B$:

$$d = x_{40} - 0 = 1512 \text{ ft}$$

$$d = 1512 \text{ ft} \blacktriangleleft$$

**Position–Time Curve.** The points determined previously should be joined by three parabolic arcs and one straight-line segment (Fig. 3). In constructing the $x$–$t$ curve, keep in mind that for any value of $t$, the slope of the tangent to the $x$–$t$ curve is equal to the value of $v$ at that instant.

**REFLECT and THINK:** This problem also could have been solved using the uniform motion equations for each interval of time that has a different acceleration, but it would have been much more difficult and time consuming. For a real subway car, the acceleration does not instantaneously change from one value to another.

# SOLVING PROBLEMS
# ON YOUR OWN

In this section, we reviewed and developed several **graphical techniques** for the solution of problems involving rectilinear motion. These techniques can be used to solve problems directly or to complement analytical methods of solution by providing a visual description, and thus a better understanding, of the motion of a given body. We suggest that you sketch one or more motion curves for several of the problems in this section, even if these problems are not part of your homework assignment.

**1. Drawing $x$–$t$, $v$–$t$, and $a$–$t$ curves and applying graphical methods.** We described the following properties in Sec. 11.3, and they should be kept in mind as you use a graphical method of solution.

    **a.** **The slopes of the $x$–$t$ and $v$–$t$ curves** at a time $t_1$ are equal to the velocity and the acceleration at time $t_1$, respectively.

    **b.** **The areas under the $a$–$t$ and $v$–$t$ curves** between the times $t_1$ and $t_2$ are equal to the change $\Delta v$ in the velocity and to the change $\Delta x$ in the position coordinate, respectively, during that time interval.

    **c.** **If you know one of the motion curves,** the fundamental properties we have summarized in paragraphs $a$ and $b$ will enable you to construct the other two curves. However, when using the properties of paragraph $b$, you must know the velocity and the position coordinate at time $t_1$ in order to determine the velocity and the position coordinate at time $t_2$. Thus, in Sample Prob. 11.9, knowing that the initial value of the velocity was zero allowed us to find the velocity at $t = 6$ s: $v_6 = v_0 + \Delta v = 0 + 24$ ft/s $= 24$ ft/s.

If you have studied the shear and bending-moment diagrams for a beam previously, you should recognize the analogy between the three motion curves and the three diagrams representing, respectively, the distributed load, the shear, and the bending moment in the beam. Thus, any techniques that you have learned regarding the construction of these diagrams can be applied when drawing the motion curves.

**2. Using approximate methods.** When the $a$–$t$ and $v$–$t$ curves are not represented by analytical functions or when they are based on experimental data, it is often necessary to use approximate methods to calculate the areas under these curves. In those cases, the given area is approximated by a series of rectangles of width $\Delta t$. The smaller the value of $\Delta t$, the smaller is the error introduced by the approximation. You can obtain the velocity and the position coordinate from

$$v = v_0 + \Sigma a_{\mathrm{ave}}\, \Delta t \qquad x = x_0 + \Sigma v_{\mathrm{ave}}\, \Delta t$$

where $a_{\mathrm{ave}}$ and $v_{\mathrm{ave}}$ are the heights of an acceleration rectangle and a velocity rectangle, respectively.

# Problems

**11.61** A particle moves in a straight line with a constant acceleration of $-4$ ft/s$^2$ for 6 s, zero acceleration for the next 4 s, and a constant acceleration of $+4$ ft/s$^2$ for the next 4 s. Knowing that the particle starts from the origin and that its velocity is $-8$ ft/s during the zero acceleration time interval, (a) construct the $v$–$t$ and $x$–$t$ curves for $0 \leq t \leq 14$ s, (b) determine the position and the velocity of the particle and the total distance traveled when $t = 14$ s.

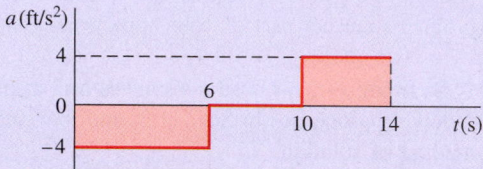

**Fig. P11.61 and P11.62**

**11.62** A particle moves in a straight line with a constant acceleration of $-4$ ft/s$^2$ for 6 s, zero acceleration for the next 4 s, and a constant acceleration of $+4$ ft/s$^2$ for the next 4 s. Knowing that the particle starts from the origin with $v_0 = 16$ ft/s, (a) construct the $v$–$t$ and $x$–$t$ curves for $0 \leq t \leq 14$ s, (b) determine the amount of time during which the particle is further than 16 ft from the origin.

**11.63** A particle moves in a straight line with the velocity shown in the figure. Knowing that $x = -540$ m at $t = 0$, (a) construct the $a$–$t$ and $x$–$t$ curves for $0 < t < 50$ s, and determine (b) the total distance traveled by the particle when $t = 50$ s, (c) the two times at which $x = 0$.

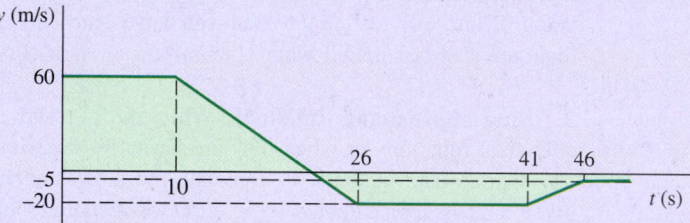

**Fig. P11.63 and P11.64**

**11.64** A particle moves in a straight line with the velocity shown in the figure. Knowing that $x = -540$ m at $t = 0$, (a) construct the $a$–$t$ and $x$–$t$ curves for $0 < t < 50$ s, and determine (b) the maximum value of the position coordinate of the particle, (c) the values of $t$ for which the particle is at $x = 100$ m.

**11.65** A particle moves in a straight line with the velocity shown in the figure. Knowing that $x = -48$ ft at $t = 0$, draw the $a$–$t$ and $x$–$t$ curves for $0 < t < 40$ s and determine (a) the maximum value of the position coordinate of the particle, (b) the values of $t$ for which the particle is at a distance of 108 ft from the origin.

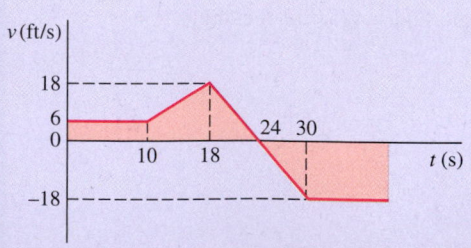

**Fig. P11.65**

**11.66** A parachutist is in free fall at a rate of 200 km/h when he opens his parachute at an altitude of 600 m. Following a rapid and constant deceleration, he then descends at a constant rate of 50 km/h from 586 m to 30 m, where he maneuvers the parachute into the wind to further slow his descent. Knowing that the parachutist lands with a negligible downward velocity, determine (*a*) the time required for the parachutist to land after opening his parachute, (*b*) the initial deceleration.

Fig. P11.66

**11.67** A commuter train traveling at 40 mi/h is 3 mi from a station. The train then decelerates so that its speed is 20 mi/h when it is 0.5 mi from the station. Knowing that the train arrives at the station 7.5 min after beginning to decelerate and assuming constant decelerations, determine (*a*) the time required for the train to travel the first 2.5 mi, (*b*) the speed of the train as it arrives at the station, (*c*) the final constant deceleration of the train.

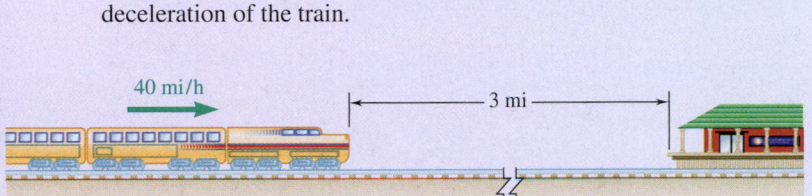

Fig. *P11.67*

**11.68** A temperature sensor is attached to slider *AB,* which moves back and forth through 60 in. The maximum velocities of the slider are 12 in./s to the right and 30 in./s to the left. When the slider is moving to the right, it accelerates and decelerates at a constant rate of 6 in./s²; when moving to the left, the slider accelerates and decelerates at a constant rate of 20 in./s². Determine the time required for the slider to complete a full cycle, and construct the *v–t* and *x–t* curves of its motion.

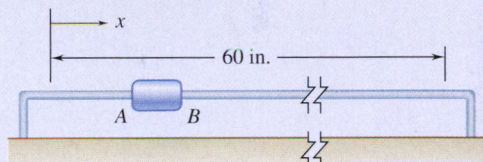

Fig. *P11.68*

**11.69** In a water-tank test involving the launching of a small model boat, the model's initial horizontal velocity is 6 m/s and its horizontal acceleration varies linearly from −12 m/s² at *t* = 0 to −2 m/s² at *t* = $t_1$ and then remains equal to −2 m/s² until *t* = 1.4 s. Knowing that *v* = 1.8 m/s when *t* = $t_1$, determine (*a*) the value of $t_1$, (*b*) the velocity and the position of the model at *t* = 1.4 s.

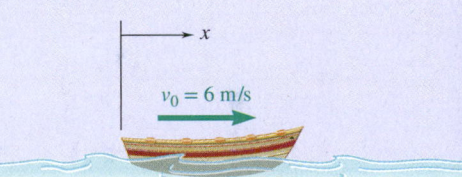

Fig. P11.69

**11.70** The acceleration record shown was obtained for a small airplane traveling along a straight course. Knowing that *x* = 0 and *v* = 60 m/s when *t* = 0, determine (*a*) the velocity and position of the plane at *t* = 20 s, (*b*) its average velocity during the interval 6 s < *t* < 14 s.

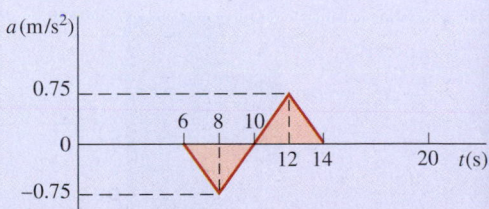

Fig. P11.70

**11.71** In a 400-m race, runner *A* reaches her maximum velocity $v_A$ in 4 s with constant acceleration and maintains that velocity until she reaches the halfway point with a split time of 25 s. Runner *B* reaches her maximum velocity $v_B$ in 5 s with constant acceleration and maintains that velocity until she reaches the halfway point with a split time of 25.2 s. Both runners then run the second half of the race with the same constant deceleration of 0.1 m/s². Determine (*a*) the race times for both runners, (*b*) the position of the winner relative to the loser when the winner reaches the finish line.

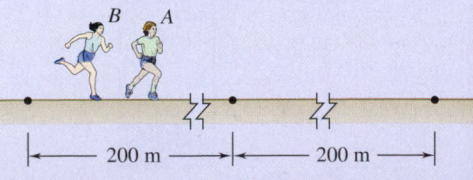

Fig. P11.71

**11.72** A car and a truck are both traveling at the constant speed of 35 mi/h; the car is 40 ft behind the truck. The driver of the car wants to pass the truck; that is, he wishes to place his car at $B$, 40 ft in front of the truck, and then resume the speed of 35 mi/h. The maximum acceleration of the car is 5 ft/s$^2$ and the maximum deceleration obtained by applying the brakes is 20 ft/s$^2$. What is the shortest time in which the driver of the car can complete the passing operation if he does not at any time exceed a speed of 50 mi/h? Draw the $v$–$t$ curve.

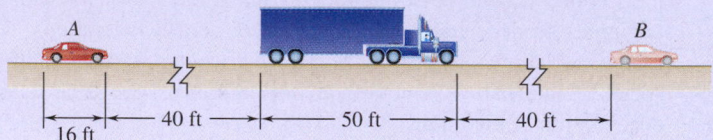

$A$　　　$B$

16 ft ← 40 ft → ← 50 ft → ← 40 ft →

**Fig. P11.72**

**11.73** Solve Prob. 11.72, assuming that the driver of the car does not pay any attention to the speed limit while passing and concentrates on reaching position $B$ and resuming a speed of 35 mi/h in the shortest possible time. What is the maximum speed reached? Draw the $v$–$t$ curve.

**11.74** Car $A$ is traveling on a highway at a constant speed $(v_A)_0 = 60$ mi/h and is 380 ft from the entrance of an access ramp when car $B$ enters the acceleration lane at that point at a speed $(v_B)_0 = 15$ mi/h. Car $B$ accelerates uniformly and enters the main traffic lane after traveling 200 ft in 5 s. It then continues to accelerate at the same rate until it reaches a speed of 60 mi/h, which it then maintains. Determine the final distance between the two cars.

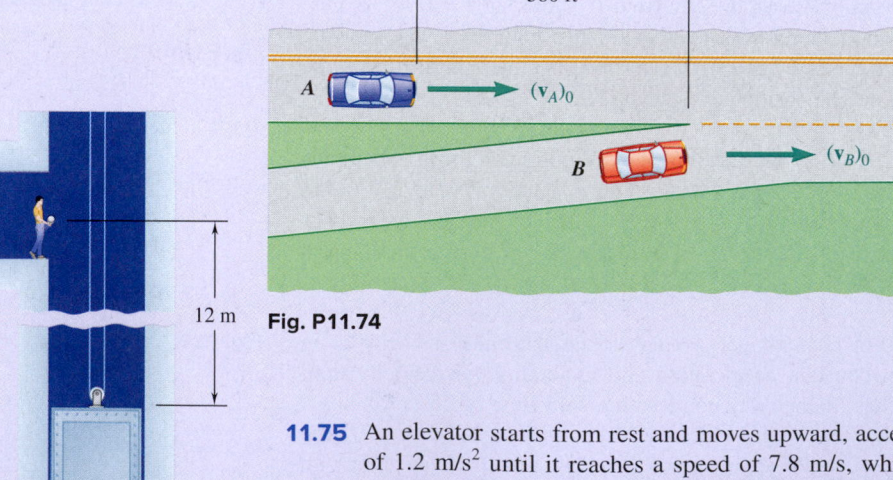

← 380 ft →

$A$ 　$(v_A)_0$

$B$ 　$(v_B)_0$

12 m　**Fig. P11.74**

**11.75** An elevator starts from rest and moves upward, accelerating at a rate of 1.2 m/s$^2$ until it reaches a speed of 7.8 m/s, which it then maintains. Two seconds after the elevator begins to move, a man standing 12 m above the initial position of the top of the elevator throws a ball upward with an initial velocity of 20 m/s. Determine when the ball will hit the elevator.

**Fig. P11.75**

**11.76** Car A is traveling at 40 mi/h when it enters a 30 mi/h speed zone. The driver of car A decelerates at a rate of 16 ft/s² until reaching a speed of 30 mi/h, which she then maintains. When car B, which was initially 60 ft behind car A and traveling at a constant speed of 45 mi/h, enters the speed zone, its driver decelerates at a rate of 20 ft/s² until reaching a speed of 28 mi/h. Knowing that the driver of car B maintains a speed of 28 mi/h, determine (a) the closest that car B comes to car A, (b) the time at which car A is 70 ft in front of car B.

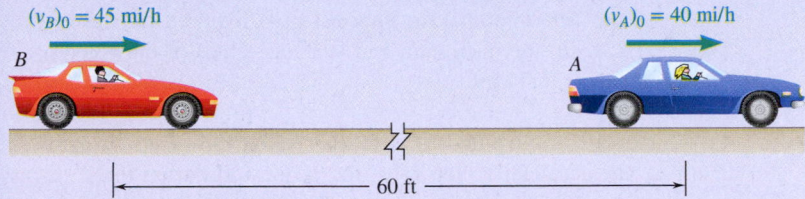

$(v_B)_0 = 45$ mi/h

$(v_A)_0 = 40$ mi/h

B

A

60 ft

**Fig. P11.76**

**11.77** An accelerometer record for the motion of a given part of a mechanism is approximated by an arc of a parabola for 0.2 s and a straight line for the next 0.2 s, as shown in the figure. Knowing that $v = 0$ when $t = 0$ and $x = 0.8$ ft when $t = 0.4$ s, (a) construct the $v$–$t$ curve for $0 \leq t \leq 0.4$ s, (b) determine the position of the part at $t = 0.3$ s and $t = 0.2$ s.

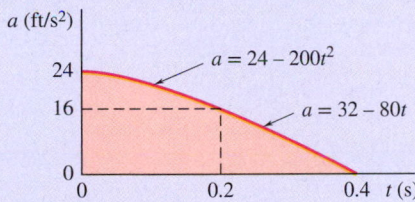

$a$ (ft/s²)

24

16

0

0                0.2                0.4   $t$ (s)

$a = 24 - 200t^2$

$a = 32 - 80t$

**Fig. P11.77**

**11.78** A car is traveling at a constant speed of 54 km/h when its driver sees a child run into the road. The driver applies her brakes until the child returns to the sidewalk and then accelerates to resume her original speed of 54 km/h; the acceleration record of the car is shown in the figure. Assuming $x = 0$ when $t = 0$, determine (a) the time $t_1$ at which the velocity is again 54 km/h, (b) the position of the car at that time, (c) the average velocity of the car during the interval $1$ s $\leq t \leq t_1$.

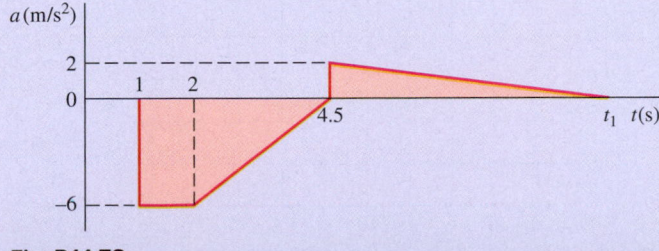

$a$ (m/s²)

2

0

1    2

4.5

$t_1$   $t$ (s)

−6

**Fig. P11.78**

**11.79** An airport shuttle train travels between two terminals that are 1.6 mi apart. To maintain passenger comfort, the acceleration of the train is limited to $\pm 4$ ft/s$^2$, and the jerk, or rate of change of acceleration, is limited to $\pm 0.8$ ft/s$^2$ per second. If the shuttle has a maximum speed of 20 mi/h, determine (*a*) the shortest time for the shuttle to travel between the two terminals, (*b*) the corresponding average velocity of the shuttle.

**11.80** During a manufacturing process, a conveyor belt starts from rest and travels a total of 1.2 ft before temporarily coming to rest. Knowing that the jerk, or rate of change of acceleration, is limited to $\pm 4.8$ ft/s$^2$ per second, determine (*a*) the shortest time required for the belt to move 1.2 ft, (*b*) the maximum and average values of the velocity of the belt during that time.

**11.81** Two seconds are required to bring the piston rod of an air cylinder to rest; the acceleration record of the piston rod during the 2 s is as shown. Determine by approximate means (*a*) the initial velocity of the piston rod, (*b*) the distance traveled by the piston rod as it is brought to rest.

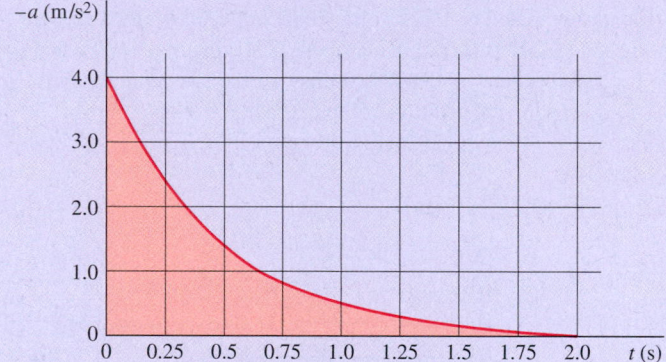

Fig. *P11.81*

**11.82** The acceleration record shown was obtained during the speed trials of a sports car. Knowing that the car starts from rest, determine by approximate means (*a*) the velocity of the car at $t = 8$ s, (*b*) the distance the car has traveled at $t = 20$ s.

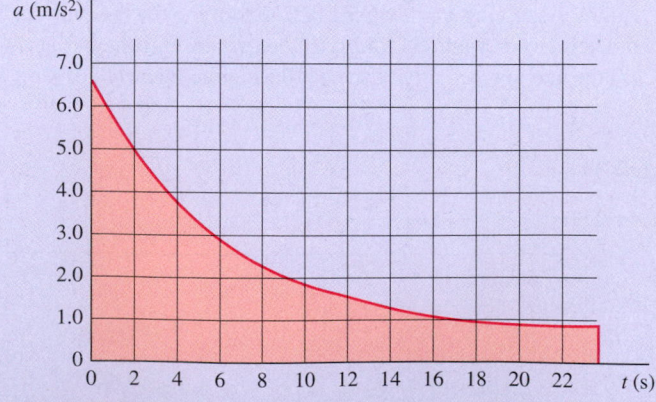

Fig. *P11.82*

**11.83**  A training airplane has a velocity of 126 ft/s when it lands on an aircraft carrier. As the arresting gear of the carrier brings the airplane to rest, the velocity and the acceleration of the airplane are recorded; the results are shown (solid curve) in the figure. Determine by approximate means (*a*) the time required for the airplane to come to rest, (*b*) the distance traveled in that time.

**Fig. P11.83**

**11.84**  Shown in the figure is a portion of the experimentally determined *v–x* curve for a shuttle cart. Determine by approximate means the acceleration of the cart when (*a*) *x* = 10 in., (*b*) *v* = 80 in./s.

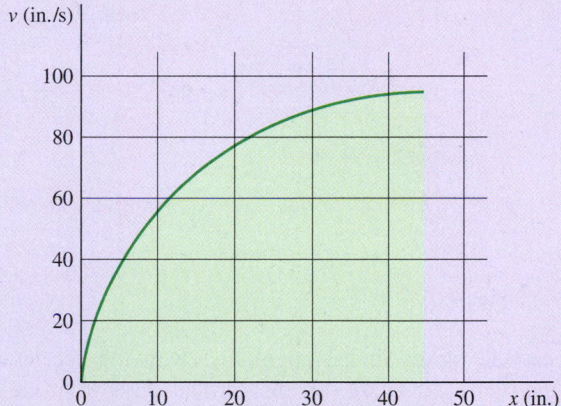

**Fig. P11.84**

**11.85**  An elevator starts from rest and rises 40 m to its maximum velocity in *T* s with the acceleration record shown in the figure. Determine (*a*) the required time *T*, (*b*) the maximum velocity, (*c*) the velocity and position of the elevator at *t* = *T*/2.

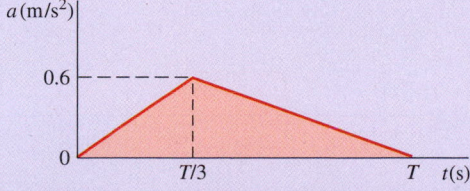

**Fig. P11.85**

**11.86** Two road rally checkpoints $A$ and $B$ are located on the same highway and are 8 mi apart. The speed limits for the first 5 mi and the last 3 mi are 60 mi/h and 35 mi/h, respectively. Drivers must stop at each checkpoint, and the specified time between points $A$ and $B$ is 10 min 20 s. Knowing that the driver accelerates and decelerates at the same constant rate, determine the magnitude of her acceleration if she travels at the speed limit as much as possible.

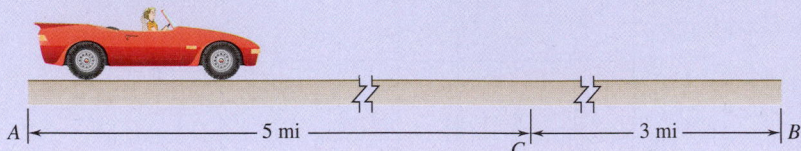

**Fig. P11.86**

**11.87** As shown in the figure, from $t = 0$ to $t = 4$ s, the acceleration of a given particle is represented by a parabola. Knowing that $x = 0$ and $v = 8$ m/s when $t = 0$, (a) construct the $v$–$t$ and $x$–$t$ curves for $0 < t < 4$ s, (b) determine the position of the particle at $t = 3$ s. (*Hint:* Use the table Centroids of Common Shapes of Areas and Lines in the Appendix.)

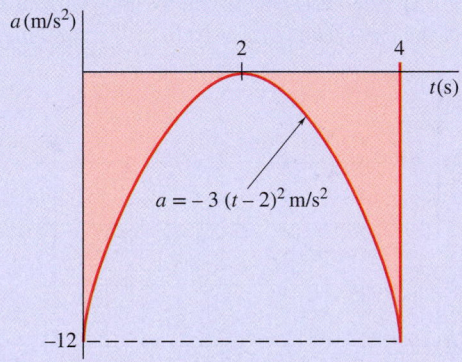

**Fig. P11.87**

**11.88** A particle moves in a straight line with the acceleration shown in the figure. Knowing that the particle starts from the origin with $v_0 = -2$ m/s, (a) construct the $v$–$t$ and $x$–$t$ curves for $0 < t < 18$ s, (b) determine the position and the velocity of the particle and the total distance traveled when $t = 18$ s.

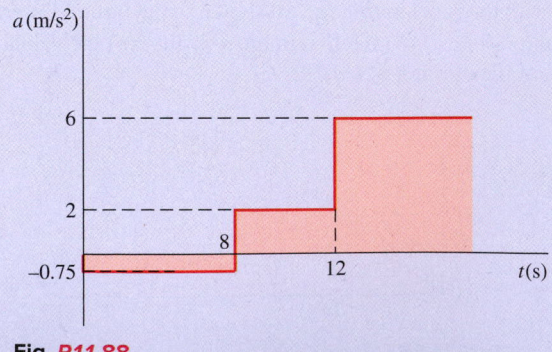

**Fig. P11.88**

# 11.4 CURVILINEAR MOTION OF PARTICLES

When a particle moves along a curve other than a straight line, we say that the particle is in **curvilinear motion**. We can use position, velocity, and acceleration to describe the motion, but now we must treat these quantities as vectors because they can have directions in two or three dimensions.

## 11.4A Position, Velocity, and Acceleration Vectors

To define the position $P$ occupied by a particle in curvilinear motion at a given time $t$, we select a fixed reference system, such as the $x$, $y$, $z$ axes shown in Fig. 11.12$a$, and draw the vector $\mathbf{r}$ joining the origin $O$ and point $P$. The vector $\mathbf{r}$ is characterized by its magnitude $r$ and its direction with respect to the reference axes, so it completely defines the position of the particle with respect to those axes. We refer to vector $\mathbf{r}$ as the **position vector** of the particle at time $t$.

Consider now the vector $\mathbf{r}'$ defining the position $P'$ occupied by the same particle at a later time $t + \Delta t$. The vector $\Delta \mathbf{r}$ joining $P$ and $P'$ represents the change in the position vector during the time interval $\Delta t$ and is called the **displacement vector**. We can check this directly from Fig. 11.12$a$, where we obtain the vector $\mathbf{r}'$ by adding the vectors $\mathbf{r}$ and $\Delta \mathbf{r}$ according to the triangle rule. Note that $\Delta \mathbf{r}$ represents a change in *direction* as well as a change in *magnitude* of the position vector $\mathbf{r}$.

We define the **average velocity** of the particle over the time interval $\Delta t$ as the quotient of $\Delta \mathbf{r}$ and $\Delta t$. Because $\Delta \mathbf{r}$ is a vector and $\Delta t$ is a scalar, the quotient $\Delta \mathbf{r}/\Delta t$ is a vector attached at $P$ with the same direction as $\Delta \mathbf{r}$ and a magnitude equal to the magnitude of $\Delta \mathbf{r}$ divided by $\Delta t$ (Fig. 11.12$b$).

We obtain the **instantaneous velocity** of the particle at time $t$ by taking the limit as the time interval $\Delta t$ approaches zero. The instantaneous velocity is thus represented by the vector

$$\mathbf{v} = \lim_{\Delta t \to 0} \frac{\Delta \mathbf{r}}{\Delta t} \qquad (11.13)$$

As $\Delta t$ and $\Delta \mathbf{r}$ become shorter, the points $P$ and $P'$ get closer together. Thus, the vector $\mathbf{v}$ obtained in the limit must be tangent to the path of the particle (Fig. 11.12$c$).

Because the position vector $\mathbf{r}$ depends upon the time $t$, we can refer to it as a **vector function** of the scalar variable $t$ and denote it by $\mathbf{r}(t)$. Extending the concept of the derivative of a scalar function introduced in elementary calculus, we refer to the limit of the quotient $\Delta \mathbf{r}/\Delta t$ as the **derivative** of the vector function $\mathbf{r}(t)$. We have

**Velocity vector**
$$\mathbf{v} = \frac{d\mathbf{r}}{dt} \qquad (11.14)$$

The magnitude $v$ of the vector $\mathbf{v}$ is called the **speed** of the particle. We can obtain the speed by substituting the magnitude of this vector, which is represented by the straight-line segment $PP'$, for the vector $\Delta \mathbf{r}$ in formula (11.13). However, the length of segment $PP'$ approaches the length $\Delta s$ of arc $PP'$ as $\Delta t$ decreases (Fig. 11.12$a$). Therefore, we can write

$$v = \lim_{\Delta t \to 0} \frac{PP'}{\Delta t} = \lim_{\Delta t \to 0} \frac{\Delta s}{\Delta t} \qquad \boxed{v = \frac{ds}{dt}} \qquad (11.15)$$

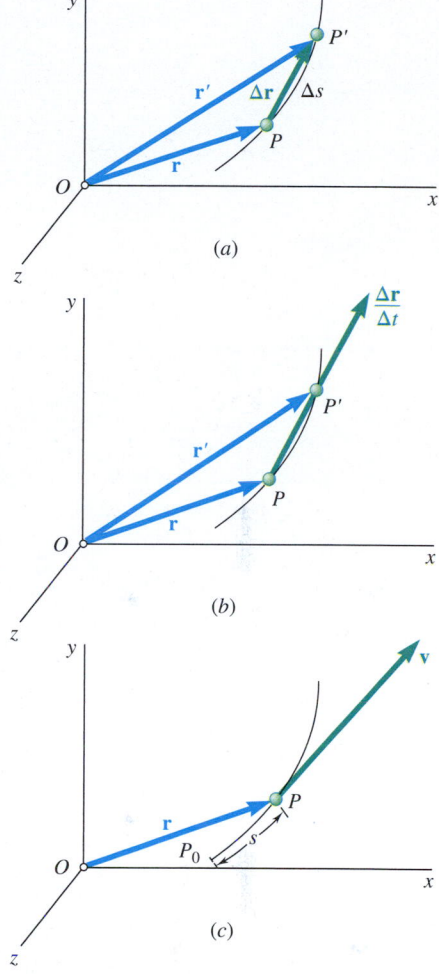

*(a)*

*(b)*

*(c)*

**Fig. 11.12** (*a*) Position vectors for a particle moving along a curve from $P$ to $P'$; (*b*) the average velocity vector is the quotient of the change in position to the elapsed time interval; (*c*) the instantaneous velocity vector is tangent to the particle's path.

Thus, we obtain the speed $v$ by finding the length $s$ of the arc described by the particle and differentiating it with respect to $t$.

Now let's consider the velocity **v** of the particle at time $t$ and its velocity **v′** at a later time $t + \Delta t$ (Fig. 11.13*a*). Let us draw both vectors **v** and **v′** from the same origin $O'$ (Fig. 11.13*b*). The vector $\Delta$**v** joining $Q$ and $Q'$ represents the change in the velocity of the particle during the time interval $\Delta t$, because we can obtain the vector **v′** by adding the vectors **v** and $\Delta$**v**. Again, note that $\Delta$**v** represents a change in the *direction* of the velocity as well as a change in *speed*. We define the **average acceleration** of the particle over the time interval $\Delta t$ as the quotient of $\Delta$**v** and $\Delta t$. Because $\Delta$**v** is a vector and $\Delta t$ is a scalar, the quotient $\Delta$**v**$/\Delta t$ is a vector in the same direction as $\Delta$**v**.

We obtain the **instantaneous acceleration** of the particle at time $t$ by choosing increasingly smaller values for $\Delta t$ and $\Delta$**v**. The instantaneous acceleration is thus represented by the vector

$$\mathbf{a} = \lim_{\Delta t \to 0} \frac{\Delta \mathbf{v}}{\Delta t} \tag{11.16}$$

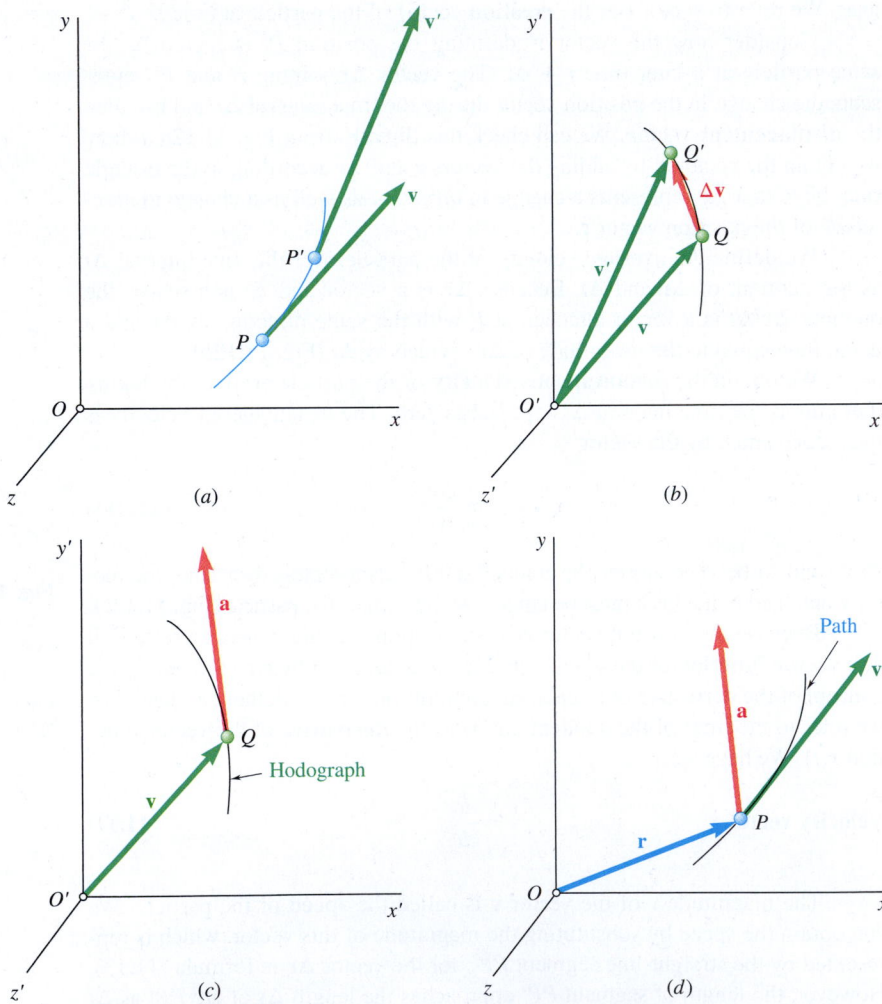

**Fig. 11.13**   *(a)* Velocities $v$ and $v'$ of a particle at two different times; *(b)* the vector change in the particle's velocity during the time interval; *(c)* the instantaneous acceleration vector is tangent to the hodograph; *(d)* in general, the acceleration vector is not tangent to the particle's path.

Noting that the velocity **v** is a vector function **v**(*t*) of the time *t,* we can refer to the limit of the quotient $\Delta\mathbf{v}/\Delta t$ as the derivative of **v** with respect to *t.* We have

**Acceleration vector**     $$\mathbf{a} = \frac{d\mathbf{v}}{dt}$$     **(11.17)**

Observe that the acceleration **a** is tangent to the curve described by the tip *Q* of the vector **v** when we draw **v** from a fixed origin *O'* (Fig. 11.13*c*). However, in general, the acceleration is *not* tangent to the path of the particle (Fig. 11.13*d*). The curve described by the tip of **v** and shown in Fig. 11.13*c* is called the *hodograph* of the motion.

## 11.4B   Derivatives of Vector Functions

We have just seen that we can represent the velocity **v** of a particle in curvilinear motion by the derivative of the vector function **r**(*t*) characterizing the position of the particle. Similarly, we can represent the acceleration **a** of the particle by the derivative of the vector function **v**(*t*). Here we give a formal definition of the derivative of a vector function and establish a few rules governing the differentiation of sums and products of vector functions.

Let **P**(*u*) be a vector function of the scalar variable *u.* By that, we mean that the scalar *u* completely defines the magnitude and direction of the vector **P**. If the vector **P** is drawn from a fixed origin *O* and the scalar *u* is allowed to vary, the tip of **P** describes a given curve in space. Consider the vectors **P** corresponding, respectively, to the values *u* and *u* + $\Delta u$ of the scalar variable (Fig. 11.14*a*). Let $\Delta\mathbf{P}$ be the vector joining the tips of the two given vectors. Then we have

$$\Delta\mathbf{P} = \mathbf{P}(u + \Delta u) - \mathbf{P}(u)$$

Dividing through by $\Delta u$ and letting $\Delta u$ approach zero, we define the derivative of the vector function **P**(*u*) as

$$\frac{d\mathbf{P}}{du} = \lim_{\Delta u \to 0} \frac{\Delta\mathbf{P}}{\Delta u} = \lim_{\Delta u \to 0} \frac{\mathbf{P}(u + \Delta u) - \mathbf{P}(u)}{\Delta u}$$     **(11.18)**

As $\Delta u$ approaches zero, the line of action of $\Delta\mathbf{P}$ becomes tangent to the curve of Fig. 11.14*a*. Thus, the derivative *d***P**/*du* of the vector function **P**(*u*) *is tangent to the curve described by the tip of* **P**(*u*) (Fig. 11.14*b*).

The standard rules for the differentiation of the sums and products of scalar functions extend to vector functions. Consider first the **sum of two vector functions P**(*u*) and **Q**(*u*) of the same scalar variable *u.* According to the definition given in Eq. (11.18), the derivative of the vector **P** + **Q** is

$$\frac{d(\mathbf{P} + \mathbf{Q})}{du} = \lim_{\Delta u \to 0} \frac{\Delta(\mathbf{P} + \mathbf{Q})}{\Delta u} = \lim_{\Delta u \to 0} \left( \frac{\Delta\mathbf{P}}{\Delta u} + \frac{\Delta\mathbf{Q}}{\Delta u} \right)$$

or because the limit of a sum is equal to the sum of the limits of its terms,

$$\frac{d(\mathbf{P} + \mathbf{Q})}{du} = \lim_{\Delta u \to 0} \frac{\Delta\mathbf{P}}{\Delta u} + \lim_{\Delta u \to 0} \frac{\Delta\mathbf{Q}}{\Delta u}$$

$$\frac{d(\mathbf{P} + \mathbf{Q})}{du} = \frac{d\mathbf{P}}{du} + \frac{d\mathbf{Q}}{du}$$     **(11.19)**

That is, the derivative of a sum of vector functions equals the sum of the derivative of each function separately.

We now consider the **product of a scalar function** *f*(*u*) **and a vector function P**(*u*) of the same scalar variable *u.* The derivative of the vector *f* **P** is

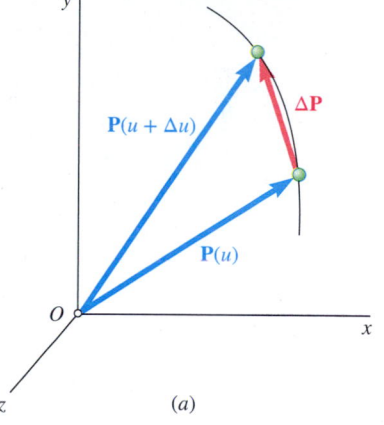

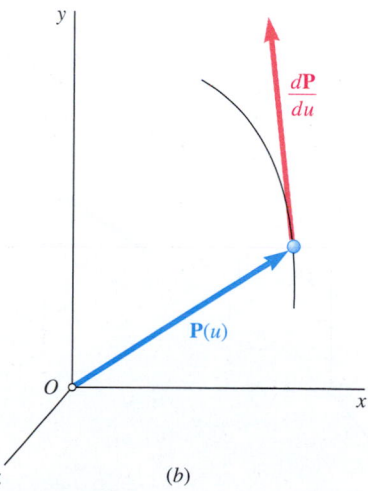

**Fig. 11.14** (*a*) The change in vector function for a particle moving along a curvilinear path; (*b*) the derivative of the vector function is tangent to the path described by the tip of the function.

$$\frac{d(f\mathbf{P})}{du} = \lim_{\Delta u \to 0} \frac{(f + \Delta f)(\mathbf{P} + \Delta \mathbf{P}) - f\mathbf{P}}{\Delta u} = \lim_{\Delta u \to 0} \left( \frac{\Delta f}{\Delta u}\mathbf{P} + f\frac{\Delta \mathbf{P}}{\Delta u} \right)$$

or recalling the properties of the limits of sums and products,

$$\frac{d(f\mathbf{P})}{du} = \frac{df}{du}\mathbf{P} + f\frac{d\mathbf{P}}{du} \tag{11.20}$$

In a similar way, we can obtain the derivatives of the **scalar product** and the **vector product** of two vector functions $\mathbf{P}(u)$ and $\mathbf{Q}(u)$. Thus,

$$\frac{d(\mathbf{P} \cdot \mathbf{Q})}{du} = \frac{d\mathbf{P}}{du} \cdot \mathbf{Q} + \mathbf{P} \cdot \frac{d\mathbf{Q}}{du} \tag{11.21}$$

$$\frac{d(\mathbf{P} \times \mathbf{Q})}{du} = \frac{d\mathbf{P}}{du} \times \mathbf{Q} + \mathbf{P} \times \frac{d\mathbf{Q}}{du} \tag{11.22}†$$

We can use the properties just established to determine the **rectangular components of the derivative of a vector function** $\mathbf{P}(u)$. Resolving $\mathbf{P}$ into components along fixed rectangular axes $x$, $y$, and $z$, we have

$$\mathbf{P} = P_x\mathbf{i} + P_y\mathbf{j} + P_z\mathbf{k} \tag{11.23}$$

where $P_x$, $P_y$, and $P_z$ are the rectangular scalar components of the vector $\mathbf{P}$, and $\mathbf{i}$, $\mathbf{j}$, and $\mathbf{k}$ are the unit vectors corresponding, respectively, to the $x$, $y$, and $z$ axes. From Eq. (11.19), the derivative of $\mathbf{P}$ is equal to the sum of the derivatives of the terms on the right-hand side. Because each of these terms is the product of a scalar and a vector function, we should use Eq. (11.20). However, the unit vectors $\mathbf{i}$, $\mathbf{j}$, and $\mathbf{k}$ have a constant magnitude (equal to 1) and fixed directions. Their derivatives are therefore zero, and we obtain

$$\frac{d\mathbf{P}}{du} = \frac{dP_x}{du}\mathbf{i} + \frac{dP_y}{du}\mathbf{j} + \frac{dP_z}{du}\mathbf{k} \tag{11.24}$$

Note that the coefficients of the unit vectors are, by definition, the scalar components of the vector $d\mathbf{P}/du$. We conclude that we can obtain the rectangular scalar components of the derivative $d\mathbf{P}/du$ of the vector function $\mathbf{P}(u)$ by differentiating the corresponding scalar components of $\mathbf{P}$.

**Rate of Change of a Vector.** When the vector $\mathbf{P}$ is a function of the time $t$, its derivative $d\mathbf{P}/dt$ represents the **rate of change** of $\mathbf{P}$ with respect to the frame $Oxyz$. Resolving $\mathbf{P}$ into rectangular components and using Eq. (11.24), we have

$$\frac{d\mathbf{P}}{dt} = \frac{dP_x}{dt}\mathbf{i} + \frac{dP_y}{dt}\mathbf{j} + \frac{dP_z}{dt}\mathbf{k}$$

Alternatively, using dots to indicate differentiation with respect to $t$ gives

$$\dot{\mathbf{P}} = \dot{P}_x\mathbf{i} + \dot{P}_y\mathbf{j} + \dot{P}_z\mathbf{k} \tag{11.24'}$$

As you will see in Sec. 15.5, the rate of change of a vector as observed from a *moving frame of reference* is, in general, different from its rate of change as observed from a fixed frame of reference. However, if the moving frame $O'x'y'z'$ is in *translation*—that is, if its axes remain parallel to the corresponding axes of the fixed frame $Oxyz$ (Fig. 11.15)—we can use the same unit vectors $\mathbf{i}$, $\mathbf{j}$,

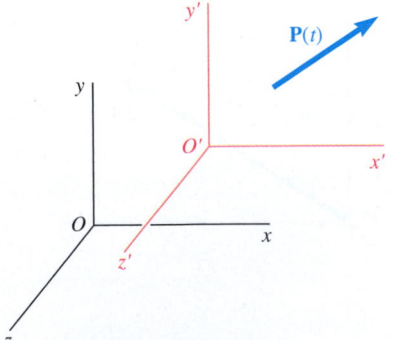

**Fig. 11.15** The rate of change of a vector is the same with respect to a fixed frame of reference and with respect to a frame in translation.

---

†Because the vector product is not commutative (see Sec. 3.4), the order of the factors in Eq. (11.22) must be maintained.

and **k** in both frames, and at any given instant, the vector **P** has the same components $P_x$, $P_y$, and $P_z$ in both frames. It follows from Eq. (11.24′) that the rate of change **P** is the same with respect to the frames $Oxyz$ and $O'x'y'z'$. Therefore,

**The rate of change of a vector is the same with respect to a fixed frame and with respect to a frame in translation.**

This property will greatly simplify our work, because we will be concerned mainly with frames in translation.

## 11.4C  Rectangular Components of Velocity and Acceleration

Suppose the position of a particle $P$ is defined at any instant by its rectangular coordinates $x$, $y$, and $z$. In this case, it is often convenient to resolve the velocity **v** and the acceleration **a** of the particle into rectangular components (Fig. 11.16).

To resolve the position vector **r** of the particle into rectangular components, we write

$$\mathbf{r} = x\mathbf{i} + y\mathbf{j} + z\mathbf{k} \qquad (11.25)$$

Here the coordinates $x$, $y$, and $z$ are functions of $t$. Differentiating twice, we obtain

**Velocity and acceleration in rectangular components**

$$\mathbf{v} = \frac{d\mathbf{r}}{dt} = \dot{x}\mathbf{i} + \dot{y}\mathbf{j} + \dot{z}\mathbf{k} \qquad (11.26)$$

$$\mathbf{a} = \frac{d\mathbf{v}}{dt} = \ddot{x}\mathbf{i} + \ddot{y}\mathbf{j} + \ddot{z}\mathbf{k} \qquad (11.27)$$

where $\dot{x}$, $\dot{y}$, and $\dot{z}$ and $\ddot{x}$, $\ddot{y}$, and $\ddot{z}$ represent, respectively, the first and second derivatives of $x$, $y$, and $z$ with respect to $t$. It follows from Eqs. (11.26) and (11.27) that the scalar components of the velocity and acceleration are

$$v_x = \dot{x} \qquad v_y = \dot{y} \qquad v_z = \dot{z} \qquad (11.28)$$

$$a_x = \ddot{x} \qquad a_y = \ddot{y} \qquad a_z = \ddot{z} \qquad (11.29)$$

A positive value for $v_x$ indicates that the vector component $\mathbf{v}_x$ is directed to the right, and a negative value indicates that it is directed to the left. The sense of each of the other vector components is determined in a similar way from the sign of the corresponding scalar component. If desired, we can obtain the magnitudes and directions of the velocity and acceleration from their scalar components using the methods of Secs. 2.2A and 2.4A (or Appendix A).

The use of rectangular components to describe the position, velocity, and acceleration of a particle is particularly effective when the component $a_x$ of the acceleration depends only upon $t$, $x$, and/or $v_x$, and similarly when $a_y$ depends only upon $t$, $y$, and/or $v_y$, and when $a_z$ depends upon $t$, $z$, and/or $v_z$. In this case, we can integrate Eqs. (11.28) and (11.29) independently. In other words, the motion of the particle in the $x$ direction, its motion in the $y$ direction, and its motion in the $z$ direction can be treated separately.

In the case of the **motion of a projectile**, we can show (see Sec. 12.1D) that the components of the acceleration are ▶

$$a_x = \ddot{x} = 0 \qquad a_y = \ddot{y} = -g \qquad a_z = \ddot{z} = 0$$

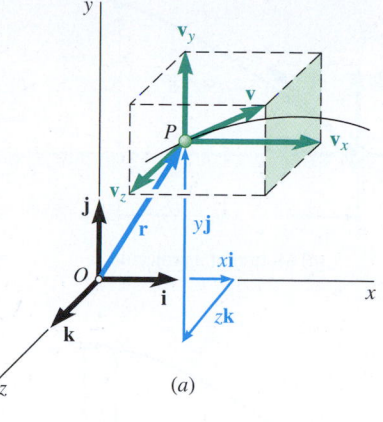

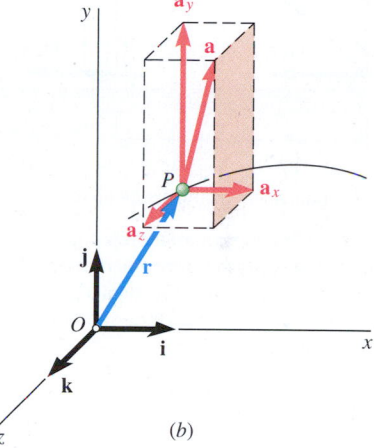

**Fig. 11.16** (*a*) Rectangular components of position and velocity for a particle *P*; (*b*) rectangular components of acceleration for particle *P*.

**Photo 11.3** The motion of this snowboarder in the air is a parabola, assuming we can neglect air resistance.
©Purestock/SuperStock RF

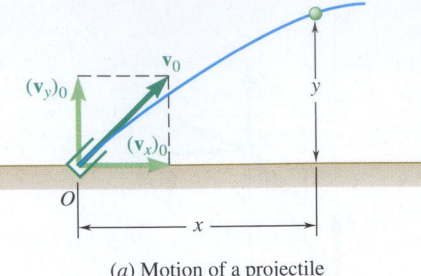

(a) Motion of a projectile

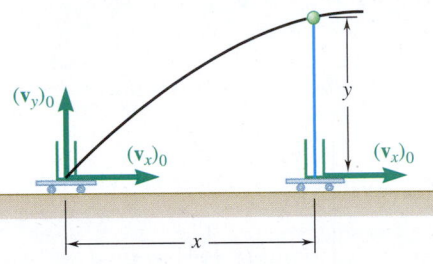

(b) Equivalent rectilinear motions

**Fig. 11.17** The motion of a projectile (a) consists of uniform horizontal motion and uniformly accelerated vertical motion and (b) is equivalent to two independent rectilinear motions.

if the resistance of the air is neglected. Denoting initial coordinates by $x_0$, $y_0$, and $z_0$ and the components of the initial velocity $\mathbf{v}_0$ of the projectile by $(v_x)_0$, $(v_y)_0$, and $(v_z)_0$, we can integrate twice in $t$ and obtain

$$v_x = \dot{x} = (v_x)_0 \qquad v_y = \dot{y} = (v_y)_0 - gt \qquad v_z = \dot{z} = (v_z)_0$$
$$x = x_0 + (v_x)_0 t \qquad y = y_0 + (v_y)_0 t - \tfrac{1}{2}gt^2 \qquad z = z_0 + (v_z)_0 t$$

If the projectile is fired in the $xy$ plane from the origin $O$, we have $x_0 = y_0 = z_0 = 0$ and $(v_z)_0 = 0$, so the equations of motion reduce to

$$v_x = (v_x)_0 \qquad v_y = (v_y)_0 - gt \qquad v_z = 0$$
$$x = (v_x)_0 t \qquad y = (v_y)_0 t - \tfrac{1}{2}gt^2 \qquad z = 0$$

These equations show that the projectile remains in the $xy$ plane, that its motion in the horizontal direction is uniform, and that its motion in the vertical direction is uniformly accelerated. Thus, we can replace the motion of a projectile by two independent rectilinear motions, which are easily visualized if we assume that the projectile is fired vertically with an initial velocity $(\mathbf{v}_y)_0$ from a platform moving with a constant horizontal velocity $(\mathbf{v}_x)_0$ (Fig. 11.17). The coordinate $x$ of the projectile is equal at any instant to the distance traveled by the platform, and we can compute its coordinate $y$ as if the projectile were moving along a vertical line. Additionally, because the $(\mathbf{v}_x)_0$ values are the same, the projectile will land on the platform regardless of the value of $(\mathbf{v}_y)_0$.

Note that the equations defining the coordinates $x$ and $y$ of a projectile at any instant are the parametric equations of a parabola. Thus, the trajectory of a projectile is *parabolic*. This result, however, ceases to be valid if we take into account the resistance of the air or the variation with altitude of the acceleration due to gravity.

## 11.4D Motion Relative to a Frame in Translation

We have just seen how to describe the motion of a particle by using a single frame of reference. In most cases, this frame was attached to the earth and was considered to be fixed. Now we want to analyze situations in which it is convenient to use several frames of reference simultaneously. If one of the frames

**Photo 11.4** The pilot of a helicopter landing on a moving carrier must take into account the relative motion of the ship. ©Digital Vision/Getty Images RF

is attached to the earth, it is called a **fixed frame of reference**, and the other frames are referred to as **moving frames of reference**. You should recognize, however, that the selection of a fixed frame of reference is purely arbitrary. Any frame can be designated as "fixed"; all other frames not rigidly attached to this frame are then described as "moving."

Consider two particles $A$ and $B$ moving in space (Fig. 11.18). The vectors $\mathbf{r}_A$ and $\mathbf{r}_B$ define their positions at any given instant with respect to the fixed frame of reference $Oxyz$. Consider now a system of axes $x'$, $y'$, and $z'$ centered at $A$ and parallel to the $x$, $y$, and $z$ axes. Suppose that, while the origin of these axes moves, their orientation remains the same; then the frame of reference $Ax'y'z'$ is in *translation* with respect to $Oxyz$. The vector $\mathbf{r}_{B/A}$ joining $A$ and $B$ defines **the position of $B$ relative to the moving frame** $Ax'y'z'$ (or for short, **the position of $B$ relative to $A$**).

Fig. 11.18 shows that the position vector $\mathbf{r}_B$ of particle $B$ is the sum of the position vector $\mathbf{r}_A$ of particle $A$ and of the position vector $\mathbf{r}_{B/A}$ of $B$ relative to $A$; that is,

**Relative position**

$$\mathbf{r}_B = \mathbf{r}_A + \mathbf{r}_{B/A} \qquad \textbf{(11.30)}$$

Differentiating Eq. (11.30) with respect to $t$ within the fixed frame of reference, and using dots to indicate time derivatives, we have

$$\dot{\mathbf{r}}_B = \dot{\mathbf{r}}_A + \dot{\mathbf{r}}_{B/A} \qquad \textbf{(11.31)}$$

The derivatives $\dot{\mathbf{r}}_A$ and $\dot{\mathbf{r}}_B$ represent, respectively, the velocities $\mathbf{v}_A$ and $\mathbf{v}_B$ of the particles $A$ and $B$. Because $Ax'y'z'$ is in translation, the derivative $\dot{\mathbf{r}}_{B/A}$ represents the rate of change of $\mathbf{r}_{B/A}$ with respect to the frame $Ax'y'z'$, as well as with respect to the fixed frame (Sec. 11.4B). This derivative, therefore, defines **the velocity $\mathbf{v}_{B/A}$ of $B$ relative to the frame** $Ax'y'z'$ (or for short, **the velocity $\mathbf{v}_{B/A}$ of $B$ relative to $A$**). We have

**Relative velocity**

$$\mathbf{v}_B = \mathbf{v}_A + \mathbf{v}_{B/A} \qquad \textbf{(11.32)}$$

Differentiating Eq. (11.32) with respect to $t$, and using the derivative $\dot{\mathbf{v}}_{B/A}$ to define **the acceleration $\mathbf{a}_{B/A}$ of $B$ relative to the frame** $Ax'y'z'$ (or for short, **the acceleration $\mathbf{a}_{B/A}$ of $B$ relative to $A$**), we obtain

**Relative acceleration**

$$\mathbf{a}_B = \mathbf{a}_A + \mathbf{a}_{B/A} \qquad \textbf{(11.33)}$$

We refer to the motion of $B$ with respect to the fixed frame $Oxyz$ as the **absolute motion of** $B$. The equations derived in this section show that **we can obtain the absolute motion of $B$ by combining the motion of $A$ and the relative motion of $B$ with respect to the moving frame attached to $A$.** Eq. (11.32), for example, expresses that the absolute velocity $\mathbf{v}_B$ of particle $B$ can be obtained by vectorially adding the velocity of $A$ and the velocity of $B$ relative to the frame $Ax'y'z'$. Eq. (11.33) expresses a similar property in terms of the accelerations. (Note that the product of the subscripts $A$ and $B/A$ used in the right-hand sides of Eqs. (11.30), (11.31), (11.32), and (11.33) is equal to the subscript $B$ used in their left-hand sides.) Keep in mind, however, that the frame $Ax'y'z'$ is in *translation*; that is, while it moves with $A$, it maintains the same orientation. As you will see later (Sec. 15.7), you must use different relations in the case of a rotating frame of reference.

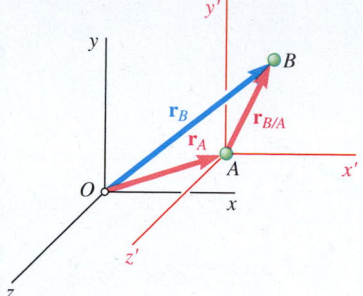

**Fig. 11.18** The vector $\mathbf{r}_{B/A}$ defines the position of $B$ with respect to moving frame $A$.

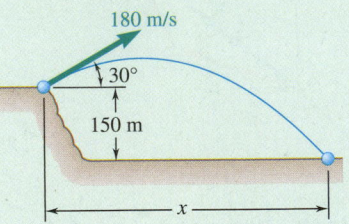

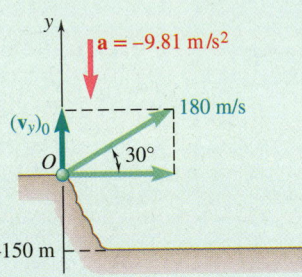

**Fig. 1** Acceleration and initial velocity of the projectile in the y direction.

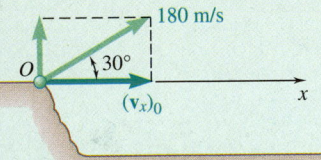

**Fig. 2** Initial velocity of the projectile in the x direction.

## Sample Problem 11.10

A projectile is fired from the edge of a 150-m cliff with an initial velocity of 180 m/s at an angle of 30° with the horizontal. Neglecting air resistance, find (a) the horizontal distance from the cannon to the point where the projectile strikes the ground, (b) the greatest elevation above the ground reached by the projectile.

**STRATEGY:** This is a projectile motion problem, so you can consider the vertical and horizontal motions separately. First determine the equations governing each direction, and then use them to find the distances.

**MODELING and ANALYSIS:** Model the projectile as a particle and neglect the effects of air resistance. The vertical motion has a constant acceleration. Choosing the positive sense of the y axis upward and placing the origin O at the cannon (Fig. 1), you have

$$(v_y)_0 = (180 \text{ m/s}) \sin 30° = +90 \text{ m/s}$$
$$a = -9.81 \text{ m/s}^2$$

Substitute these values into the equations for motion with constant acceleration. Thus,

$$v_y = (v_y)_0 + at \qquad v_y = 90 - 9.81t \qquad \textbf{(1)}$$
$$y = (v_y)_0 t + \tfrac{1}{2}at^2 \qquad y = 90t - 4.90t^2 \qquad \textbf{(2)}$$
$$v_y^2 = (v_y)_0^2 + 2ay \qquad v_y^2 = 8100 - 19.62y \qquad \textbf{(3)}$$

The horizontal motion has zero acceleration. Choose the positive sense of the x axis to the right (Fig. 2), which gives you

$$(v_x)_0 = (180 \text{ m/s}) \cos 30° = +155.9 \text{ m/s}$$

Substituting into the equation for constant acceleration, you obtain

$$x = (v_x)_0 t \qquad x = 155.9t \qquad \textbf{(4)}$$

**a. Horizontal Distance.** When the projectile strikes the ground,

$$y = -150 \text{ m}$$

Substituting this value into Eq. (2) for the vertical motion, you have

$$-150 = 90t - 4.90t^2 \qquad t^2 - 18.37t - 30.6 = 0 \qquad t = 19.91 \text{ s}$$

Substituting $t = 19.91$ s into Eq. (4) for the horizontal motion, you obtain

$$x = 155.9(19.91) \qquad x = 3100 \text{ m} \blacktriangleleft$$

**b. Greatest Elevation.** When the projectile reaches its greatest elevation, $v_y = 0$; substituting this value into Eq. (3) for the vertical motion, you have

$$0 = 8100 - 19.62y \qquad y = 413 \text{ m}$$

Greatest elevation above ground $= 150 \text{ m} + 413 \text{ m} = 563 \text{ m} \blacktriangleleft$

**REFLECT and THINK:** Because there is no air resistance, you can treat the vertical and horizontal motions separately and can immediately write down the algebraic equations of motion. If you did want to include air resistance, you must know the acceleration as a function of the speed (you will see how to derive this in Chap. 12), and then you need to use the basic kinematic relationships, separate variables, and integrate.

## Sample Problem 11.11

A projectile is fired with an initial velocity of 800 ft/s at a target $B$ located 2000 ft above the cannon $A$ and at a horizontal distance of 12,000 ft. Neglecting air resistance, determine the value of the firing angle $\alpha$ needed to hit the target.

**STRATEGY:** This is a projectile motion problem, so you can consider the vertical and horizontal motions separately. First determine the equations governing the motion in each direction, and then use them to find the firing angle.

**MODELING and ANALYSIS:**

**Horizontal Motion.** Place the origin of the coordinate axes at the cannon (Fig. 1). Then

$$(v_x)_0 = 800 \cos \alpha$$

Substituting into the equation of uniform horizontal motion, you obtain

$$x = (v_x)_0 t \qquad x = (800 \cos \alpha)t$$

Obtain the time required for the projectile to move through a horizontal distance of 12,000 ft by setting $x$ equal to 12,000 ft.

$$12,000 = (800 \cos \alpha)t$$

$$t = \frac{12,000}{800 \cos \alpha} = \frac{15}{\cos \alpha}$$

**Vertical Motion.** Again, place the origin at the cannon (Fig. 2).

$$(v_y)_0 = 800 \sin \alpha \qquad a = -32.2 \text{ ft/s}^2$$

Substituting into the equation for constant acceleration in the vertical direction, you obtain

$$y = (v_y)_0 t + \tfrac{1}{2}at^2 \qquad y = (800 \sin \alpha)t - 16.1t^2$$

**Projectile Hits Target.** When $x = 12,000$ ft, you want $y = 2000$ ft. Substituting for $y$ and setting $t$ equal to the value found previously, you have

$$2000 = 800 \sin \alpha \frac{15}{\cos \alpha} - 16.1 \left( \frac{15}{\cos \alpha} \right)^2 \qquad \textbf{(1)}$$

Because $1/\cos^2 \alpha = \sec^2 \alpha = 1 + \tan^2 \alpha$, you have

$$2000 = 800(15) \tan \alpha - 16.1(15^2)(1 + \tan^2 \alpha)$$
$$3622 \tan^2 \alpha - 12,000 \tan \alpha + 5622 = 0$$

Solving this quadratic equation for $\tan \alpha$ gives you

$$\tan \alpha = 0.565 \qquad \text{and} \qquad \tan \alpha = 2.75$$
$$\alpha = 29.5° \qquad \text{and} \qquad \alpha = 70.0° \quad \blacktriangleleft$$

The target will be hit if either of these two firing angles is used (Fig. 3).

**REFLECT and THINK:** It is a well-known characteristic of projectile motion that you can hit the same target by using either of two firing angles. We used trigonometry to write the equation in terms of $\tan \alpha$, but most calculators or computer programs like Maple, Matlab, or Mathematica also can be used to solve Eq. (1) for $\alpha$. You must be careful when using these tools, however, to make sure that you find both angles.

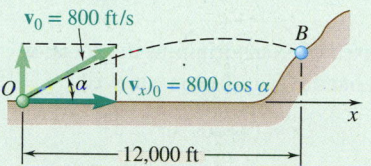

**Fig. 1** Initial velocity of the projectile in the $x$ direction.

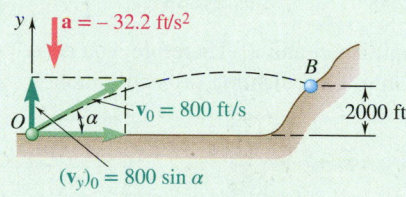

**Fig. 2** Acceleration and initial velocity of the projectile in the $y$ direction.

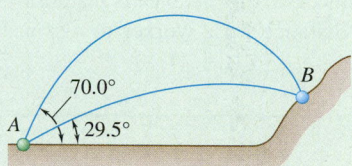

**Fig. 3** Firing angles that will hit target $B$.

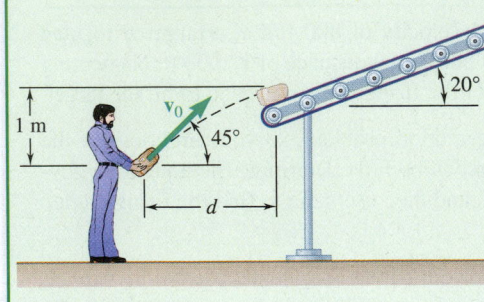

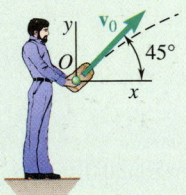

**Fig. 1** Initial velocity of the package.

## Sample Problem 11.12

A conveyor belt at an angle of 20° with the horizontal is used to transfer small packages to other parts of an industrial plant. A worker tosses a package with an initial velocity $\mathbf{v}_0$ at an angle of 45° so that its velocity is parallel to the belt as it lands 1 m above the release point. Determine (*a*) the magnitude of $\mathbf{v}_0$, (*b*) the horizontal distance *d*.

**STRATEGY:**   This is a projectile motion problem, so you can consider the vertical and the horizontal motions separately. First determine the equations governing the motion in each direction, then use them to determine the unknown quantities.

**MODELING and ANALYSIS:**

**Horizontal Motion.**   Placing the axes of your origin at the location where the package leaves the worker's hands (Fig. 1), you can write

**Horizontal:** $v_x = v_0 \cos 45°$   and   $x = (v_0 \cos 45°)t$

**Vertical:** $v_y = v_0 \sin 45° - gt$   and   $y = (v_0 \sin 45°)t - \frac{1}{2}gt^2$

**Landing on the Belt.**   The problem statement indicates that when the package lands on the belt, its velocity vector will be in the same direction as the belt is moving. If this happens when $t = t_1$, you can write

$$\frac{v_y}{v_x} = \tan 20° = \frac{v_0 \sin 45° - gt_1}{v_0 \cos 45°} = 1 - \frac{gt_1}{v_0 \cos 45°} \tag{1}$$

This equation has two unknown quantities: $t_1$ and $v_0$. Therefore, you need more equations. Substituting $t = t_1$ into the remaining projectile motion equations gives

$$d = (v_0 \cos 45°)t \tag{2}$$

$$1 \text{ m} = (v_0 \sin 45°)t_1 - \frac{1}{2}gt_1^2 \tag{3}$$

You now have three equations (1), (2), and (3), and three unknowns $t_1$, $v_0$, and *d*. Using $g = 9.81 \text{ m/s}^2$ and solving these three equations give $t_1 = 0.3083$ s and

$$v_0 = 6.73 \text{ m/s} \blacktriangleleft$$

$$d = 1.466 \text{ m} \blacktriangleleft$$

**REFLECT and THINK:**   All of these projectile problems are similar. You write down the governing equations for motion in the horizontal and vertical directions and then use additional information in the problem statement to solve the problem. In this case, the distance is just less than 1.5 meters, which is a reasonable distance for a worker to toss a package.

## Sample Problem 11.13

Airplane $B$, which is traveling at a constant 560 km/h, is pursuing airplane $A$, which is traveling northeast at a constant 800 km/hr. At time $t = 0$, airplane $A$ is 640 km east of airplane $B$. Determine (a) the direction of the course airplane $B$ should follow (measured from the east) to intercept plane $A$, (b) the rate at which the distance between the airplanes is decreasing, (c) how long it takes for airplane $B$ to catch airplane $A$.

**STRATEGY:**   To find when $B$ intercepts $A$, you just need to find out when the two planes are at the same location. The rate at which the distance is decreasing is the magnitude of $v_{B/A}$, so you can use the relative velocity equation for part (b).

**MODELING and ANALYSIS:**   Choose $x$ to be east, $y$ to be north, and place the origin of your coordinate system at $B$ (Fig. 1).

**Positions of the Planes.**   You know that each plane has a constant speed, so you can write a position vector for each plane. Thus,

$$\mathbf{r}_A = [(v_A \cos 45°)t + 640 \text{ km}]\mathbf{i} + [(v_A \sin 45°)t]\mathbf{j} \quad (1)$$
$$\mathbf{r}_B = [(v_B \cos \theta)t]\mathbf{i} + [(v_B \sin \theta)t]\mathbf{j} \quad (2)$$

**a. Direction of $B$.**   Plane $B$ will catch up when they are at the same location; that is, when $\mathbf{r}_A = \mathbf{r}_B$. You can equate components in the $\mathbf{j}$ direction to find

$$v_A \sin 45° \, t_1 = v_B \sin \theta \, t_1$$

After you substitute in values,

$$\sin \theta = \frac{(v_A \sin 45°)t_1}{v_B t_1} = \frac{(560 \text{ km/hr}) \sin 45°}{800 \text{ km/hr}} = 0.4950$$
$$\theta = \sin^{-1} 0.4950 = 29.668° \qquad \qquad \boxed{\theta = 29.7°} \blacktriangleleft$$

**b. Rate.**   The rate at which the distance is decreasing is the magnitude of $\mathbf{v}_{B/A}$, so

$$\mathbf{v}_{B/A} = \mathbf{v}_B - \mathbf{v}_A = (v_B \cos \theta \, \mathbf{i} + v_B \sin \theta \, \mathbf{j}) - (v_A \cos 45° \, \mathbf{i} + v_A \sin 45° \, \mathbf{j})$$
$$= [(800 \text{ km/h}) \cos 29.668° - (560 \text{ km/h}) \cos 45°]\mathbf{i}$$
$$\quad + [(800 \text{ km/h}) \sin 29.668° - (560 \text{ km/h}) \sin 45°]\mathbf{j}$$
$$= 299.15 \text{ km/h } \mathbf{i} \qquad\qquad \boxed{|\mathbf{v}_{B/A}| = 299 \text{ km/h}} \blacktriangleleft$$

**c. Time for $B$ to Catch Up with $A$.**   To find the time, you equate the $\mathbf{i}$ components of each position vector, giving

$$(v_A \cos 45°)t_1 + 640 \text{ km} = (v_B \cos \theta)t_1$$

Solve this for $t_1$. Thus,

$$t_1 = \frac{640 \text{ km}}{v_B \cos \theta - v_A \cos 45°}$$

$$= \frac{640 \text{ km}}{(800 \text{ km/h}) \cos 29.67° - (560 \text{ km/h}) \cos 45°} = 2.139\text{h}$$

$$\boxed{t_1 = 2.14 \text{ h}} \blacktriangleleft$$

**REFLECT and THINK:**   The relative velocity is only in the horizontal (eastern) direction. This makes sense, because the vertical (northern) components have to be equal in order for the two planes to intersect.

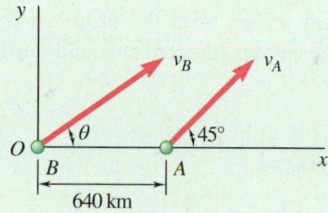

**Fig. 1**   Initial velocity of airplanes $A$ and $B$.

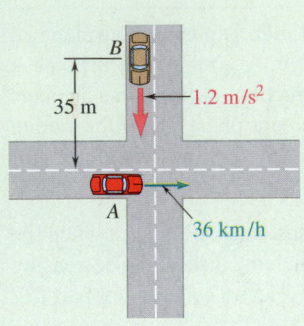

## Sample Problem 11.14

Automobile $A$ is traveling east at the constant speed of 36 km/h. As automobile $A$ crosses the intersection shown, automobile $B$ starts from rest 35 m north of the intersection and moves south with a constant acceleration of 1.2 m/s². Determine the position, velocity, and acceleration of $B$ relative to $A$ 5 s after $A$ crosses the intersection. ⏵

**STRATEGY:** This is a relative motion problem. Determine the motion of each vehicle independently, and then use the definition of relative motion to determine the desired quantities.

**MODELING and ANALYSIS:**

**Motion of Automobile A.** Choose $x$ and $y$ axes with the origin at the intersection of the two streets and with positive senses directed east and north, respectively. First express the speed in m/s, as

$$v_A = \left(36 \frac{km}{h}\right)\left(\frac{1000 \ m}{1 \ km}\right)\left(\frac{1 \ h}{3600 \ s}\right) = 10 \ m/s$$

The motion of $A$ is uniform, so for any time $t$

$$a_A = 0$$
$$v_A = +10 \ m/s$$
$$x_A = (x_A)_0 + v_A t = 0 + 10t$$

For $t = 5$ s, you have (Fig. 1)

$$a_A = 0 \qquad\qquad \mathbf{a}_A = 0$$
$$v_A = +10 \ m/s \qquad\qquad \mathbf{v}_A = 10 \ m/s \rightarrow$$
$$x_A = +(10 \ m/s)(5 \ s) = +50 \ m \qquad \mathbf{r}_A = 50 \ m \rightarrow$$

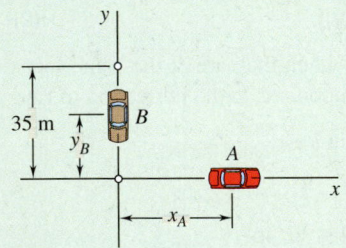

**Fig. 1** Initial positions of cars $A$ and $B$.

**Motion of Automobile B.** The motion of $B$ is uniformly accelerated, so

$$a_B = -1.2 \ m/s^2$$
$$v_B = (v_B)_0 + at = 0 - 1.2t$$
$$y_B = (y_B)_0 + (v_B)_0 t + \tfrac{1}{2}a_B t^2 = 35 + 0 - \tfrac{1}{2}(1.2)t^2$$

For $t = 5$ s, you have (Fig. 1)

$$a_B = -1.2 \ m/s^2 \qquad\qquad \mathbf{a}_B = 1.2 \ m/s^2 \downarrow$$
$$v_B = -(1.2 \ m/s^2)(5 \ s) = -6 \ m/s \qquad \mathbf{v}_B = 6 \ m/s \downarrow$$
$$y_B = 35 - \tfrac{1}{2}(1.2 \ m/s^2)(5 \ s)^2 = +20 \ m \qquad \mathbf{r}_B = 20 \ m \uparrow$$

**Motion of B Relative to A.** Draw the triangle corresponding to the vector equation $\mathbf{r}_B = \mathbf{r}_A + \mathbf{r}_{B/A}$ (Fig. 2) and obtain the magnitude and direction of the position vector of $B$ relative to $A$.

$$r_{B/A} = 53.9 \ m \qquad \alpha = 21.8° \qquad \mathbf{r}_{B/A} = 53.9 \ m \ ⦨ \ 21.8° ◄$$

Proceeding in a similar fashion (Fig. 2), find the velocity and acceleration of $B$ relative to $A$. Hence,

$$\mathbf{v}_B = \mathbf{v}_A + \mathbf{v}_{B/A}$$
$$v_{B/A} = 11.66 \ m/s \qquad \beta = 31.0° \qquad \mathbf{v}_{B/A} = 11.66 \ m/s \ ⦩ \ 31.0° ◄$$
$$\mathbf{a}_B = \mathbf{a}_A + \mathbf{a}_{B/A} \qquad\qquad\qquad\qquad \mathbf{a}_{B/A} = 1.2 \ m/s^2 \downarrow ◄$$

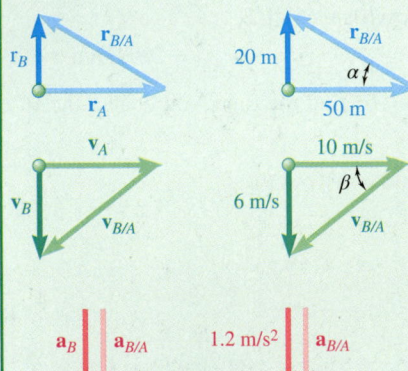

**Fig. 2** Vector triangles for position, velocity, and acceleration.

*(continued)*

**REFLECT and THINK:** Note that the relative position and velocity of $B$ relative to $A$ change with time; the values given here are only for the moment $t = 5$ s. Rather than drawing triangles, you could have also used vector algebra. When the vectors are at right angles, as in this problem, drawing vector triangles is usually easiest.

## Sample Problem 11.15

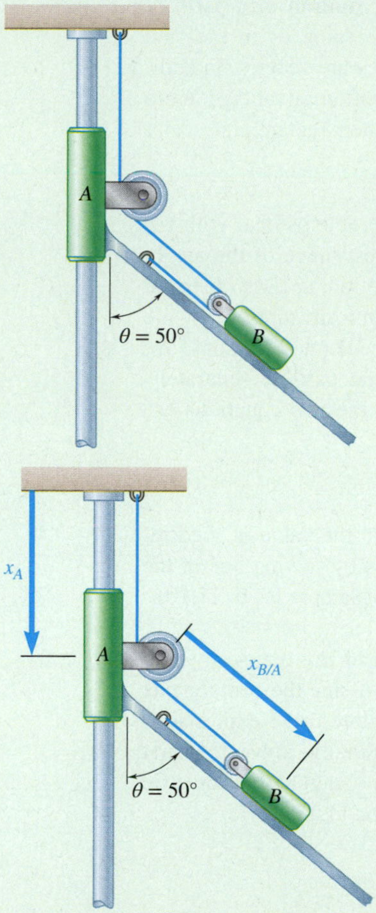

Knowing that at the instant shown cylinder/ramp $A$ has a velocity of 8 in./s directed down, determine the velocity of block $B$.

**STRATEGY:** You have objects connected by cables, so this is a dependent-motion problem. You should define coordinates for each object and write a constraint equation for the cable. You will also need to use relative motion, because $B$ slides on $A$.

**MODELING and ANALYSIS:** Define position vectors, as shown in Fig. 1.

**Constraint Equations.** Assuming the cable is inextensible, you can write the length in terms of the coordinates and then differentiate.

The constraint equation for the cable is

$$x_A + 2x_{B/A} = \text{constant}$$

Differentiating this gives

$$v_A = -2v_{B/A} \tag{1}$$

Substituting for $v_A$ gives $v_{B/A} = -4$ in./s or 4 in./s up the incline.

**Dependent Motion.** You know that the direction of $v_{B/A}$ is directed up the incline. Therefore, the relative motion equation relating the velocities of blocks $A$ and $B$ is $\mathbf{v}_B = \mathbf{v}_A + \mathbf{v}_{B/A}$. You could either draw a vector triangle or use vector algebra. Let's use vector algebra. Using the coordinate system shown in Fig. 2 and substituting in the magnitudes gives

$$(v_B)_x \mathbf{i} + (v_B)_y \mathbf{j} = (-8 \text{ in./s})\mathbf{j} + (-4 \text{ in./s}) \sin 50° \, \mathbf{i} + (4 \text{ in./s}) \cos 50° \, \mathbf{j}$$

Equating components gives

**i:** $(v_B)_x = -(4 \text{ in./s})\sin 50°$ $\qquad \rightarrow v_{B_x} = -3.064$ in./s
**j:** $(v_B)_y = (-8 \text{ in./s}) + (4 \text{ in./s})\cos 50°$ $\qquad \rightarrow v_{B_y} = -5.429$ in./s

Finding the magnitude and direction gives

$$\mathbf{v}_B = 6.23 \text{ in./s} \; \measuredangle \; 60.6° \; \blacktriangleleft$$

**REFLECT and THINK:** Rather than using vector algebra, you could have also drawn a vector triangle, as shown in Fig. 3. To use this vector triangle, you need to use the law of cosines and the law of sines. Looking at the mechanism, block $B$ should move up the incline if block $A$ moves downward; our mathematical result is consistent with this. It is also interesting to note that, even though $B$ moves up the incline relative to $A$, block $B$ is actually moving down and to the left, as shown in the calculation here. This occurs because block $A$ is also moving down.

**Fig. 1** Position vectors to $A$ and $B$.

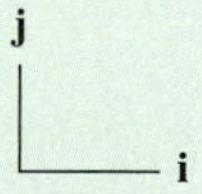

**Fig. 2** Coordinates for vector algebra.

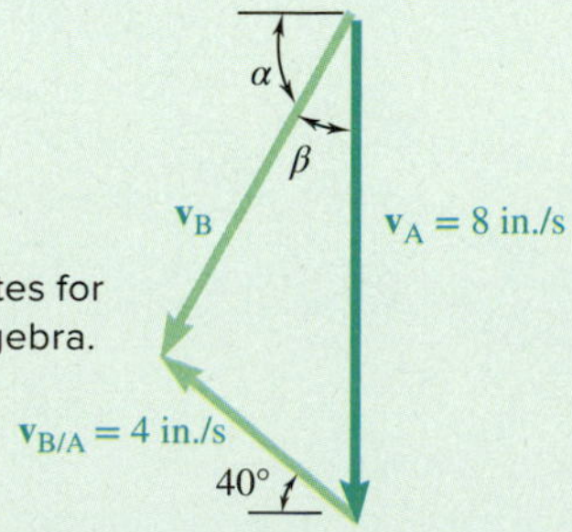

**Fig. 3** Vector triangle for the velocity of blocks $A$ and $B$.

# SOLVING PROBLEMS
# ON YOUR OWN

In the problems for this section, you will analyze the **curvilinear motion** of a particle. The physical interpretations of velocity and acceleration are the same as in the first sections of the chapter, but you should remember that these quantities are vectors. In addition, recall from your experience with vectors in statics that it is often advantageous to express position vectors, velocities, and accelerations in terms of their rectangular scalar components [Eqs. (11.25), (11.26), and (11.27)].

**A. Analyzing the motion of a projectile.** Many of the following problems deal with the two-dimensional motion of a projectile where we can neglect the resistance of the air. In Sec. 11.4C, we developed the equations that describe this type of motion, and we observed that the horizontal component of the velocity remains constant (uniform motion), while the vertical component of the acceleration is constant (uniformly accelerated motion). We are able to consider the horizontal and the vertical motions of the particle separately. Assuming that the projectile is fired from the origin, we can write the two equations as

$$x = (v_x)_0 t \qquad y = (v_y)_0 t - \tfrac{1}{2}gt^2$$

**1. If you know the initial velocity and firing angle,** you can obtain the value of $y$ corresponding to any given value of $x$ (or the value of $x$ for any value of $y$) by solving one of the previous equations for $t$ and substituting for $t$ into the other equation (Sample Prob. 11.10).

**2. If you know the initial velocity and the coordinates of a point of the trajectory** and you wish to determine the firing angle $\alpha$, begin your solution by expressing the components $(v_x)_0$ and $(v_y)_0$ of the initial velocity as functions of $\alpha$. Then, substitute these expressions and the known values of $x$ and $y$ into the previous equations. Finally, solve the first equation for $t$ and substitute that value of $t$ into the second equation to obtain a trigonometric equation in $\alpha$, which you can solve for that unknown (Sample Prob. 11.11).

**B. Solving translational two-dimensional relative-motion problems.** You saw in Sec. 11.4D that you can obtain the absolute motion of a particle $B$ by combining the motion of a particle $A$ and the **relative motion** of $B$ with respect to a frame attached to $A$ that is in *translation* (Sample Probs. 11.12, 11.3 and 11.14). You can then express the velocity and acceleration of $B$ as shown in Eqs. (11.32) and (11.33), respectively.

**1. To visualize the relative motion of $B$ with respect to $A$,** imagine that you are attached to particle $A$ as you observe the motion of particle $B$. For example, to a passenger in automobile $A$ of Sample Prob. 11.14, automobile $B$ appears to be heading in a southwesterly direction (*south* should be obvious; *west* is due to the fact that automobile $A$ is moving to the east—automobile $B$ then appears to travel to the west). Note that this conclusion is consistent with the direction of $\mathbf{v}_{B/A}$.

**2. To solve a relative-motion problem,** first write the vector equations (11.30), (11.32), and (11.33), which relate the motions of particles $A$ and $B$. You may then use either of the following methods.

    **a. Construct the corresponding vector triangles** and solve them for the desired position vector, velocity, and acceleration (Sample Prob. 11.14).

    **b. Express all vectors in terms of their rectangular components** and solve the resulting two independent sets of scalar equations (Sample Prob. 11.15). If you choose this approach, be sure to select the same positive direction for the displacement, velocity, and acceleration of each particle.

# Problems

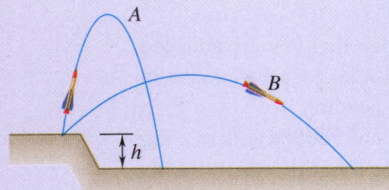

Fig. P11.CQ3

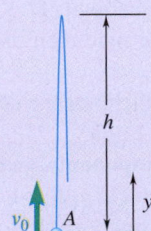

Fig. P11.CQ4

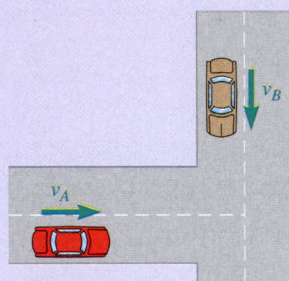

Fig. P11.CQ6

**CONCEPT QUESTIONS**

**11.CQ3** Two model rockets are fired simultaneously from a ledge and follow the trajectories shown. Neglecting air resistance, which of the rockets will hit the ground first?

**a.** $A$

**b.** $B$

**c.** They hit at the same time.

**d.** The answer depends on $h$.

**11.CQ4** Ball $A$ is thrown straight up. Which of the following statements about the ball are true at the highest point in its path?

**a.** The velocity and acceleration are both zero.

**b.** The velocity is zero, but the acceleration is not zero.

**c.** The velocity is not zero, but the acceleration is zero.

**d.** Neither the velocity nor the acceleration is zero.

**11.CQ5** Ball $A$ is thrown straight up with an initial speed $v_0$ and reaches a maximum elevation $h$ before falling back down. When $A$ reaches its maximum elevation, a second ball is thrown straight upward with the same initial speed $v_0$. At what height, $y$, will the balls cross paths?

**a.** $y = h$

**b.** $y > h/2$

**c.** $y = h/2$

**d.** $y < h/2$

**e.** $y = 0$

**11.CQ6** Two cars are approaching an intersection at constant speeds as shown. What velocity will car $B$ appear to have to an observer in car $A$?

**a.** $\rightarrow$    **b.** $\searrow$    **c.** $\nwarrow$    **d.** $\nearrow$    **e.** $\swarrow$

**11.CQ7** Blocks $A$ and $B$ are released from rest in the positions shown. Neglecting friction between all surfaces, which figure best indicates the direction $\alpha$ of the acceleration of block $B$?

**a.** 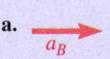    **b.**     **c.**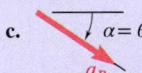

**d.** 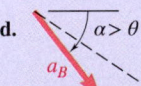    **e.**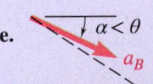

Fig. P11.CQ7

**END-OF-SECTION PROBLEMS**

**11.89** A ball is thrown so that the motion is defined by the equations $x = 5t$ and $y = 2 + 6t - 4.9t^2$, where $x$ and $y$ are expressed in meters and $t$ is expressed in seconds. Determine (a) the velocity at $t = 1$ s, (b) the horizontal distance the ball travels before hitting the ground.

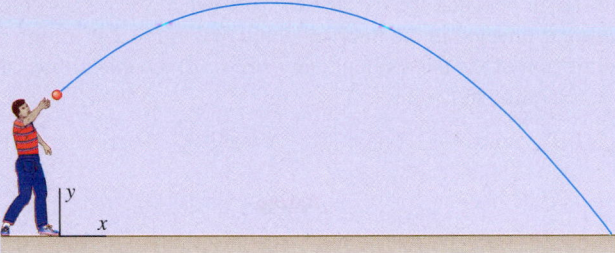

**Fig. P11.89**

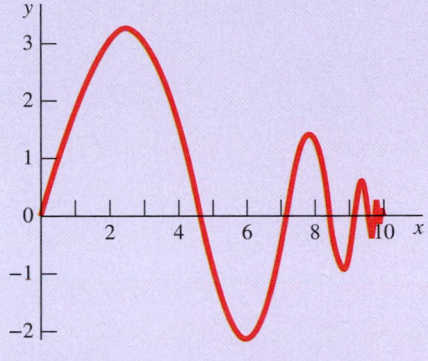

**Fig. P11.90**

**11.90** The motion of a vibrating particle is defined by the position vector $\mathbf{r} = 10(1 - e^{-3t})\mathbf{i} + (4e^{-2t} \sin 15t)\mathbf{j}$, where $\mathbf{r}$ and $t$ are expressed in millimeters and seconds, respectively. Determine the velocity and acceleration when (a) $t = 0$, (b) $t = 0.5$ s.

**11.91** The motion of a particle is defined by the equations $x = \dfrac{(4 \cos \pi t - 2)}{(2 - \cos \pi t)}$ and $\dfrac{y = 3 \sin \pi t}{(2 - \cos \pi t)}$, where $x$ and $y$ are expressed in feet and $t$ is expressed in seconds. Show that the path of the particle is part of the ellipse shown, and determine the velocity when (a) $t = 0$, (b) $t = 1/3$ s, (c) $t = 1$ s.

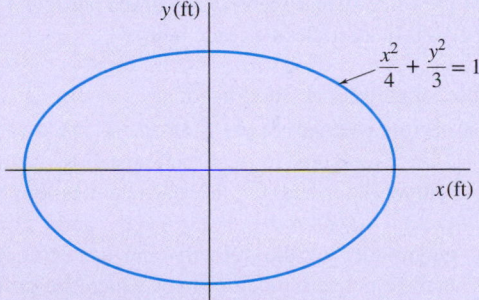

**Fig. P11.91**

**11.92** The motion of a particle is defined by the equations $x = 10t - 5 \sin t$ and $y = 10 - 5 \cos t$, where $x$ and $y$ are expressed in feet and $t$ is expressed in seconds. Sketch the path of the particle for the time interval $0 \leq t \leq 2\pi$, and determine (a) the magnitudes of the smallest and largest velocities reached by the particle, (b) the corresponding times, positions, and directions of the velocities.

**11.93** Engineers are examining how shock absorber designs affect the displacement of a mountain biker's hip after she lands from a jump. Immediately after the jump, the $x$ and $y$ locations of her hip can be described by $x(t) = 20t + 0.2 \sin\left(\dfrac{4\pi}{3}t\right)$ and $y(t) = 3.5 + 0.1\,e^{-0.3t} \sin\left(\dfrac{4\pi}{3}t\right)$, where $x$ and $y$ are expressed in feet and $t$ is expressed in seconds.

Plot the path of the hip for the time interval $0 \leq t \leq 3$ s and determine the hip's vertical position and the components of the hip velocity at $t = 2$ s.

**11.94** A girl operates a radio-controlled model car in a vacant parking lot. The girl's position is at the origin of the $xy$ coordinate axes, and the surface of the parking lot lies in the $x$–$y$ plane. The motion of the car is defined by the position vector $\mathbf{r} = (2 + 2t^2)\mathbf{i} + (6 + t^3)\mathbf{j}$, where $\mathbf{r}$ and $t$ are expressed in meters and seconds, respectively. Determine (a) the distance between the car and the girl when $t = 2$ s, (b) the distance the car traveled in the interval from $t = 0$ to $t = 2$ s, (c) the speed and direction of the car's velocity at $t = 2$ s, (d) the magnitude of the car's acceleration at $t = 2$ s.

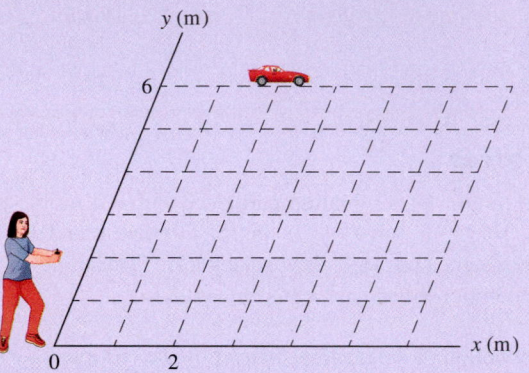

**Fig. P11.94**

**11.95** The three-dimensional motion of a particle is defined by the position vector $\mathbf{r} = (Rt \cos \omega_n t)\mathbf{i} + ct\mathbf{j} + (Rt \sin \omega_n t)\mathbf{k}$. Determine the magnitudes of the velocity and acceleration of the particle. (The space curve described by the particle is a conic helix.)

**\*11.96** The three-dimensional motion of a particle is defined by the position vector, $\mathbf{r} = (At \cos t)\mathbf{i} + (A\sqrt{t^2 + 1})\mathbf{j} + (Bt \sin t)\mathbf{k}$, where $r$ and $t$ are expressed in feet and seconds, respectively. Show that the curve described by the particle lies on the hyperboloid $(y/A)^2 - (x/A)^2 - (z/B)^2 = 1$. For $A = 3$ and $B = 1$, determine (a) the magnitudes of the velocity and acceleration when $t = 0$, (b) the smallest nonzero value of $t$ for which the position vector and the velocity are perpendicular to each other.

**11.97** An airplane used to drop water on brushfires is flying horizontally in a straight line at 130 knots (1 knot = 1.15 mph) at an altitude of 175 ft. Determine the distance $d$ at which the pilot should release the water so that it will hit the fire at $B$.

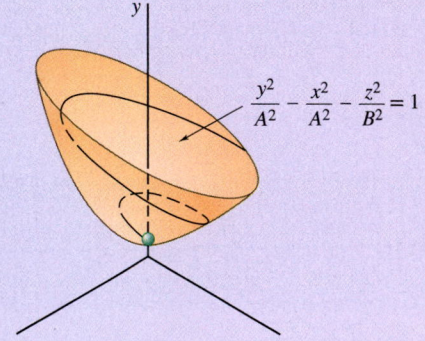

$$\frac{y^2}{A^2} - \frac{x^2}{A^2} - \frac{z^2}{B^2} = 1$$

**Fig. P11.96**

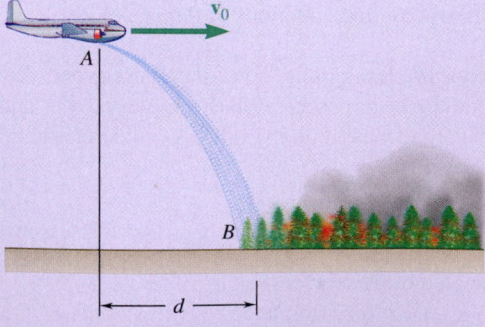

**Fig. P11.97**

**11.98** A ski jumper starts with a horizontal take-off velocity of 25 m/s and lands on a straight landing hill inclined at 30°. Determine (*a*) the time between take-off and landing, (*b*) the length *d* of the jump, (*c*) the maximum vertical distance between the jumper and the landing hill.

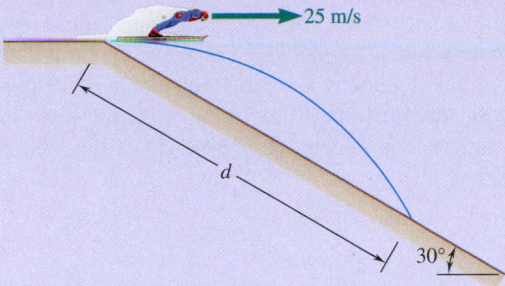

**Fig. P11.98**

**11.99** A baseball pitching machine "throws" baseballs with a horizontal velocity $v_0$. Knowing that height *h* varies between 788 mm and 1068 mm, determine (*a*) the range of values of $v_0$, (*b*) the values of $\alpha$ corresponding to *h* = 788 mm and *h* = 1068 mm.

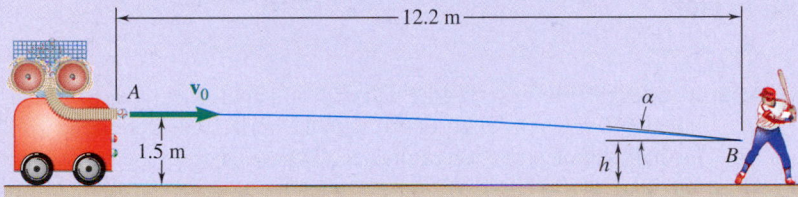

**Fig. P11.99**

**11.100** While delivering newspapers, a girl throws a newspaper with a horizontal velocity $v_0$. Determine the range of values of $v_0$ if the newspaper is to land between points *B* and *C*.

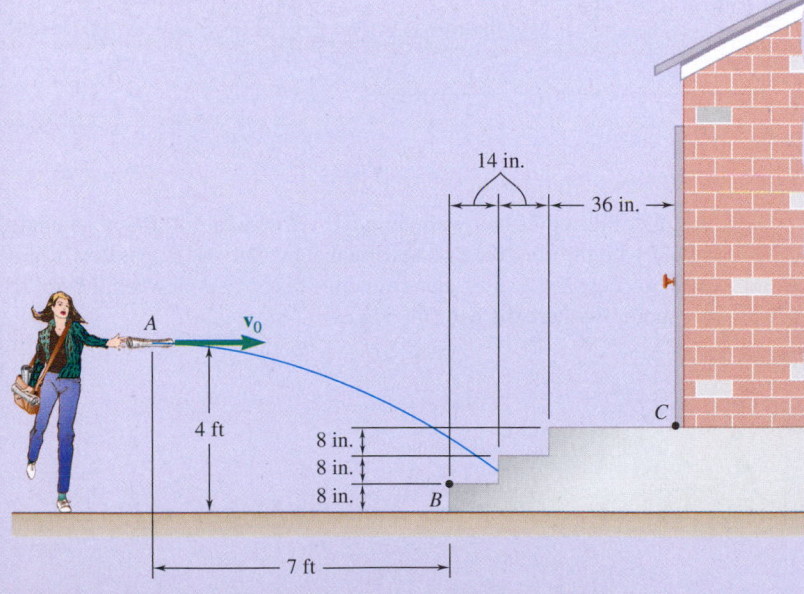

**Fig. P11.100**

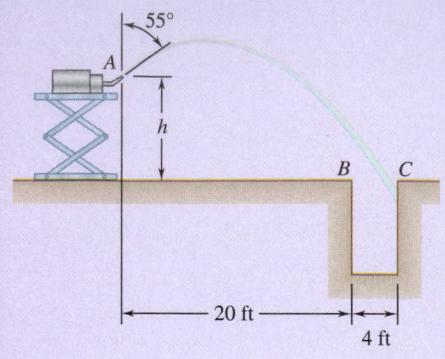

**Fig. P11.101**

**11.101** A pump is located near the edge of the horizontal platform shown. The nozzle at A discharges water with an initial velocity of 25 ft/s at an angle of 55° with the vertical. Determine the range of values of the height h for which the water enters the opening BC.

**11.102** In slow pitch softball, the underhand pitch must reach a maximum height of between 1.8 m and 3.7 m above the ground. A pitch is made with an initial velocity $v_0$ with a magnitude of 13 m/s at an angle of 33° with the horizontal. Determine (a) if the pitch meets the maximum height requirement, (b) the height of the ball as it reaches the batter.

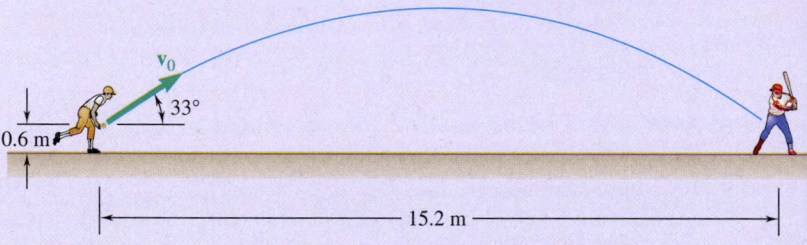

**Fig. P11.102**

**11.103** A volleyball player serves the ball with an initial velocity $v_0$ of magnitude 13.40 m/s at an angle of 20° with the horizontal. Determine (a) if the ball will clear the top of the net, (b) how far from the net the ball will land.

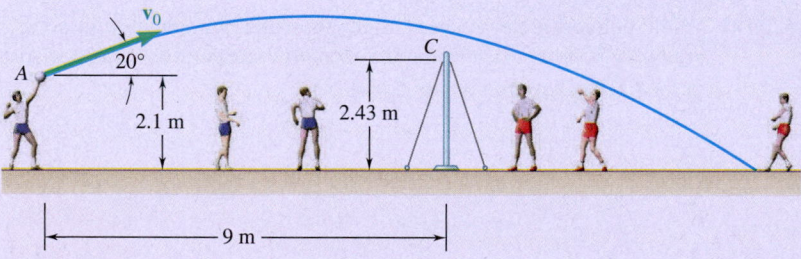

**Fig. P11.103**

**11.104** A golfer hits a golf ball with an initial velocity of 140 ft/s at an angle of 25° with the horizontal. Knowing that the fairway slopes downward at an average angle of 5°, determine the distance d between the golfer and point B where the ball first lands.

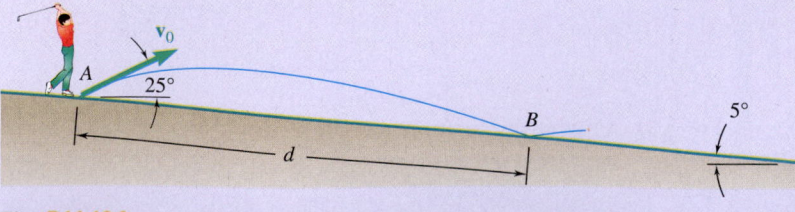

**Fig. P11.104**

**11.105** A homeowner uses a snowblower to clear his driveway. Knowing that the snow is discharged at an average angle of 40° with the horizontal, determine the initial velocity $v_0$ of the snow.

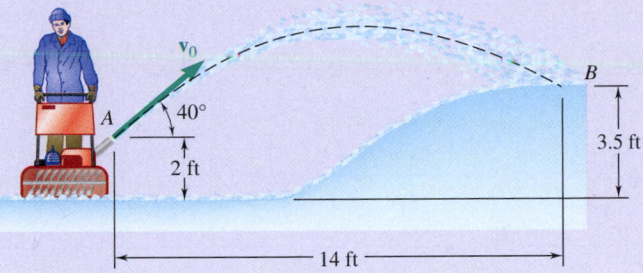

**Fig. P11.105**

**11.106** At halftime of a football game, souvenir balls are thrown to the spectators with a velocity $v_0$. Determine the range of values of $v_0$ if the balls are to land between points $B$ and $C$.

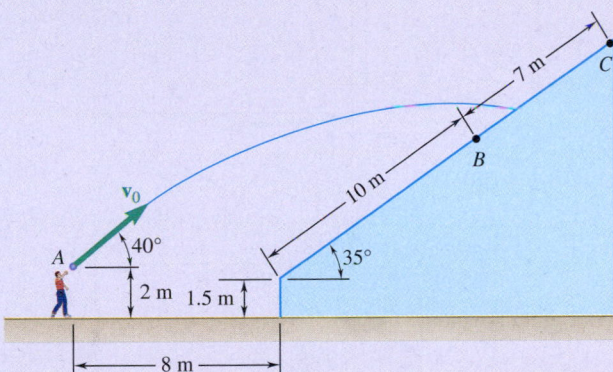

**Fig. P11.106**

**11.107** A basketball player shoots when she is 16 ft from the backboard. Knowing that the ball has an initial velocity $v_0$ at an angle of 30° with the horizontal, determine the value of $v_0$ when $d$ is equal to (*a*) 9 in., (*b*) 17 in.

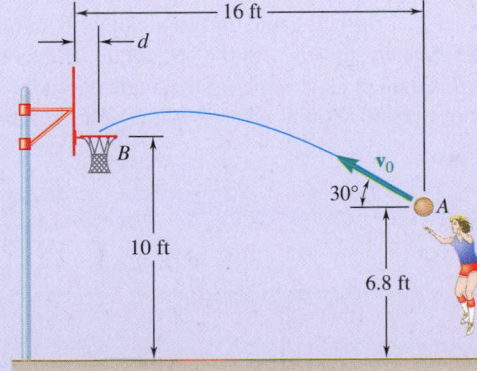

**Fig. P11.107**

**11.108** A tennis player serves the ball at a height $h = 2.5$ m with an initial velocity of $\mathbf{v}_0$ at an angle of 5° with the horizontal. Determine the range of $v_0$ for which the ball will land in the service area that extends to 6.4 m beyond the net.

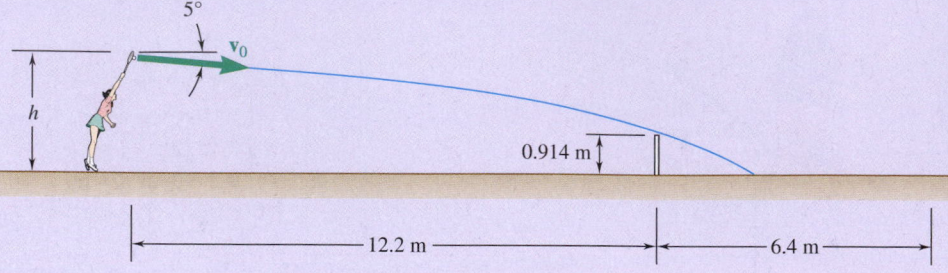

Fig. P11.108

**11.109** The nozzle at $A$ discharges cooling water with an initial velocity $\mathbf{v}_0$ at an angle of 6° with the horizontal onto a grinding wheel 350 mm in diameter. Determine the range of values of the initial velocity for which the water will land on the grinding wheel between points $B$ and $C$.

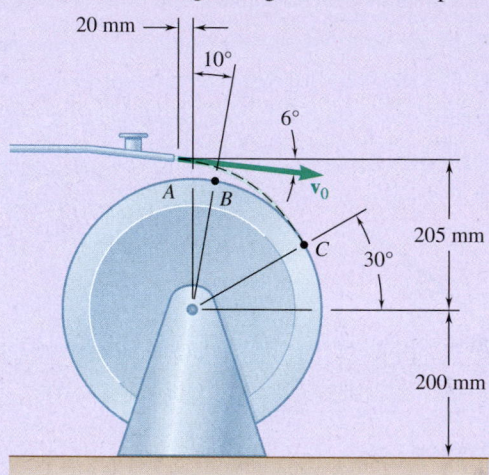

Fig. *P11.109*

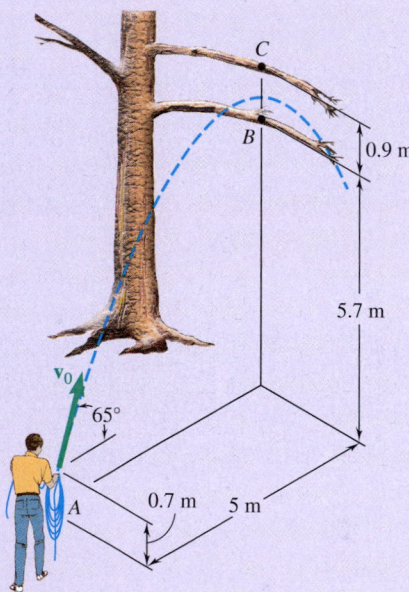

Fig. *P11.110*

**11.110** While holding one of its ends, a worker lobs a coil of rope over the lowest limb of a tree. If he throws the rope with an initial velocity $\mathbf{v}_0$ at an angle of 65° with the horizontal, determine the range of values of $v_0$ for which the rope will go over only the lowest limb.

**11.111** The pitcher in a softball game throws a ball with an initial velocity $\mathbf{v}_0$ of 108 km/h at an angle $\alpha$ with the horizontal. If the height of the ball at point $B$ is 0.68 m, determine (a) the angle $\alpha$, (b) the angle $\theta$ that the velocity of the ball at point $B$ forms with the horizontal.

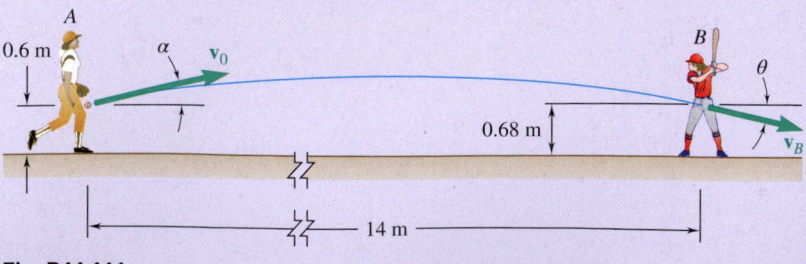

Fig. P11.111

**11.112** A model rocket is launched from point $A$ with an initial velocity $\mathbf{v}_0$ of 75 m/s. If the rocket's descent parachute does not deploy and the rocket lands a distance $d = 100$ m from $A$, determine ($a$) the angle $\alpha$ that $\mathbf{v}_0$ forms with the vertical, ($b$) the maximum height above point $A$ reached by the rocket, ($c$) the duration of the flight.

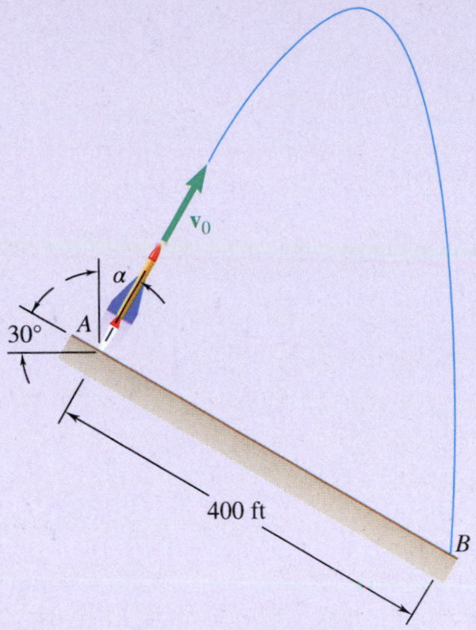

**11.113** The initial velocity $\mathbf{v}_0$ of a hockey puck is 90 mi/h. Determine ($a$) the largest value (less than 45°) of the angle $\alpha$ for which the puck will enter the net, ($b$) the corresponding time required for the puck to reach the net.

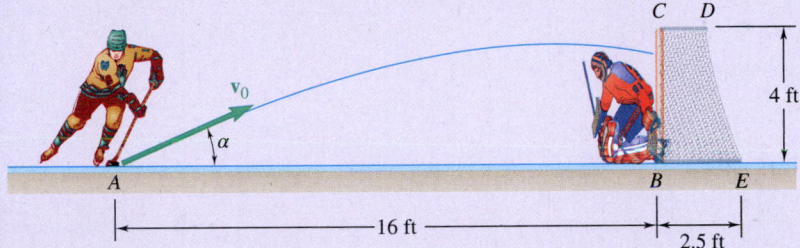

Fig. P11.112

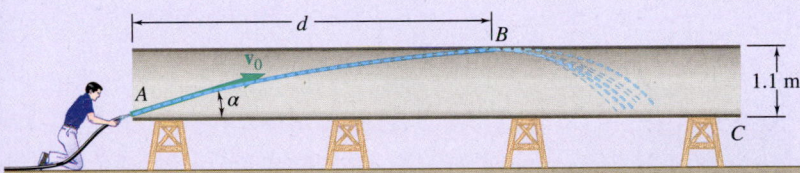

Fig. P11.113

**11.114** A worker uses high-pressure water to clean the inside of a long drainpipe. If the water is discharged with an initial velocity $\mathbf{v}_0$ of 11.5 m/s, determine ($a$) the distance $d$ to the farthest point $B$ on the top of the pipe that the worker can wash from his position at $A$, ($b$) the corresponding angle $\alpha$.

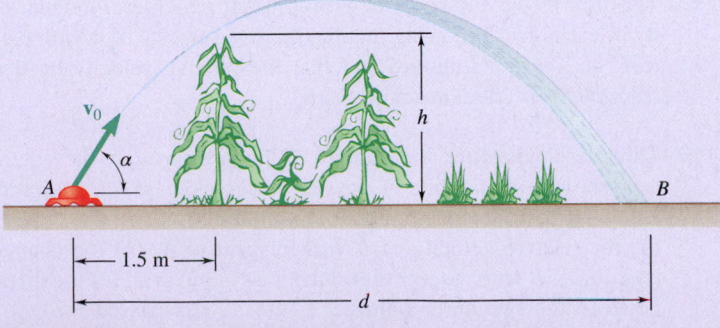

Fig. P11.114

**11.115** An oscillating garden sprinkler which discharges water with an initial velocity $\mathbf{v}_0$ of 8 m/s is used to water a vegetable garden. Determine the distance $d$ to the farthest point $B$ that will be watered and the corresponding angle $\alpha$ when ($a$) the vegetables are just beginning to grow, ($b$) the height $h$ of the corn is 1.8 m.

Fig. *P11.115*

**\*11.116** A nozzle at *A* discharges water with an initial velocity of 36 ft/s at an angle *α* with the horizontal. Determine (*a*) the distance *d* to the farthest point *B* on the roof that the water can reach, (*b*) the corresponding angle *α*. Check that the stream will clear the edge of the roof.

Fig. *P11.116*

**11.117** The velocities of skiers *A* and *B* are as shown. Determine the velocity of *A* with respect to *B*.

Fig. P11.117

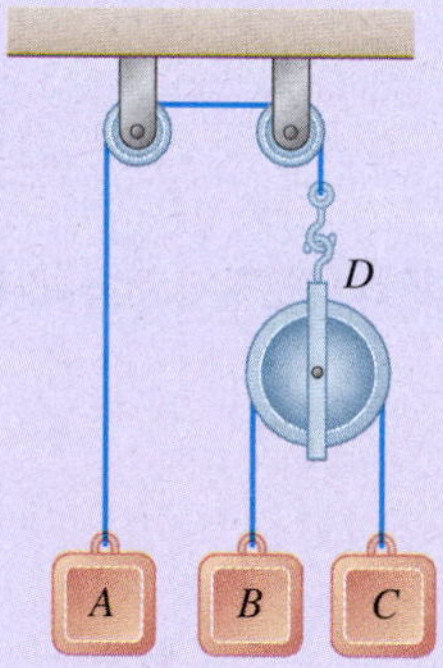

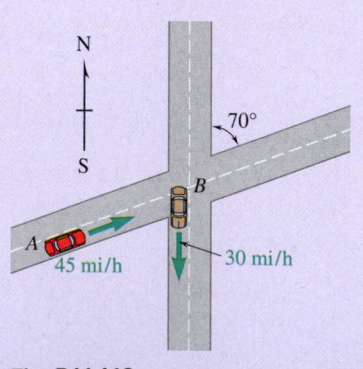

Fig. P11.118

Fig. P11.119

**11.118** The three blocks shown move with constant velocities. Find the velocity of each block, knowing that the relative velocity of *A* with respect to *C* is 300 mm/s upward and that the relative velocity of *B* with respect to *A* is 200 mm/s downward.

**11.119** Three seconds after automobile *B* passes through the intersection shown, automobile *A* passes through the same intersection. Knowing that the speed of each automobile is constant, determine (*a*) the relative velocity of *B* with respect to *A*, (*b*) the change in position of *B* with respect to *A* during a 4-s interval, (*c*) the distance between the two automobiles 2 s after *A* has passed through the intersection.

**11.120** Shore-based radar indicates that a ferry leaves its slip with a velocity $\mathbf{v} = 18$ km/h ⦨ 70°, while instruments aboard the ferry indicate a speed of 18.4 km/h and a heading of 30° west of south relative to the river. Determine the velocity of the river.

**11.121** Airplanes $A$ and $B$ are flying at the same altitude and are tracking the eye of hurricane $C$. The relative velocity of $C$ with respect to $A$ is $\mathbf{v}_{C/A} = 350$ km/h ⦨ 75°, and the relative velocity of $C$ with respect to $B$ is $\mathbf{v}_{C/B} = 400$ km/h ⦧ 40°. Determine (a) the relative velocity of $B$ with respect to $A$, (b) the velocity of $A$ if ground-based radar indicates that the hurricane is moving at a speed of 30 km/h due north, (c) the change in position of $C$ with respect to $B$ during a 15-min interval.

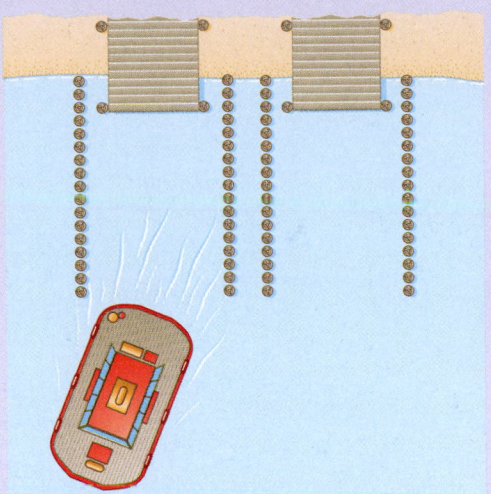

**Fig. P11.120**

**Fig. P11.121**

**11.122** Instruments in an airplane indicate that with respect to the air, the plane is moving north at a speed of 500 km/h. At the same time, ground-based radar indicates that the plane is moving at a speed of 530 km/h in a direction 5° east of north. Determine the magnitude and direction of the velocity of the air.

**11.123** Knowing that at the instant shown block $B$ has a velocity of 2 ft/s to the right and an acceleration of 3 ft/s² to the left, determine (a) the velocity of block $A$, (b) the acceleration of block $A$.

**11.124** Knowing that at the instant shown block $A$ has a velocity of 8 in./s and an acceleration of 6 in./s², both directed down the incline, determine (a) the velocity of block $B$, (b) the acceleration of block $B$.

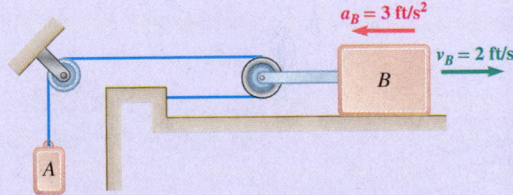

**Fig. P11.123**

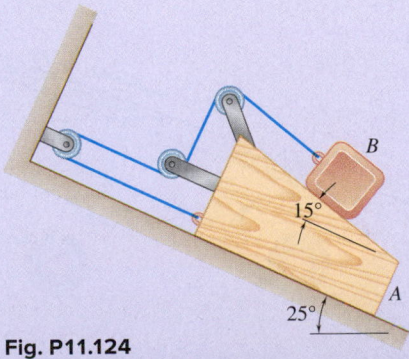

**Fig. P11.124**

**11.125** A boat is moving to the right with a constant deceleration of 0.3 m/s$^2$ when a boy standing on the deck $D$ throws a ball with an initial velocity relative to the deck which is vertical. The ball rises to a maximum height of 8 m above the release point and the boy must step forward a distance $d$ to catch it at the same height as the release point. Determine ($a$) the distance $d$, ($b$) the relative velocity of the ball with respect to the deck when the ball is caught.

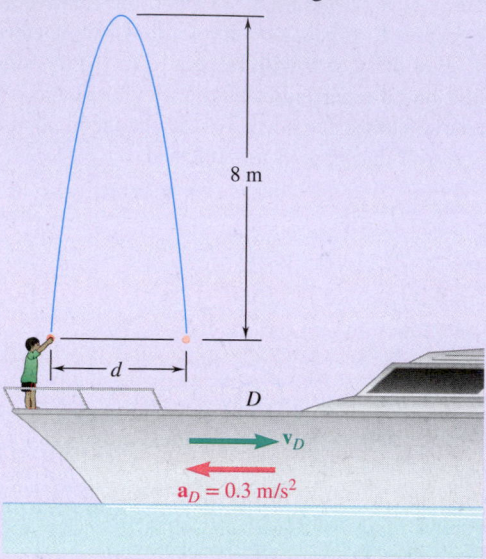

Fig. P11.125

**11.126** The assembly of rod $A$ and wedge $B$ starts from rest and moves to the right with a constant acceleration of 2 mm/s$^2$. Determine ($a$) the acceleration of wedge $C$, ($b$) the velocity of wedge $C$ when $t = 10$ s.

**11.127** Coal discharged from a dump truck with an initial velocity $(v_c)_0 = 5$ ft/s ↘ 50° falls onto conveyor belt $B$. Determine the required velocity $\mathbf{v}_B$ of the belt if the relative velocity with which the coal hits the belt is to be ($a$) vertical, ($b$) as small as possible.

**11.128** Conveyor belt $A$, which forms a 20° angle with the horizontal, moves at a constant speed of 4 ft/s and is used to load an airplane. Knowing that a worker tosses duffel bag $B$ with an initial velocity of 2.5 ft/s at an angle of 30° with the horizontal, determine the velocity of the bag relative to the belt as it lands on the belt.

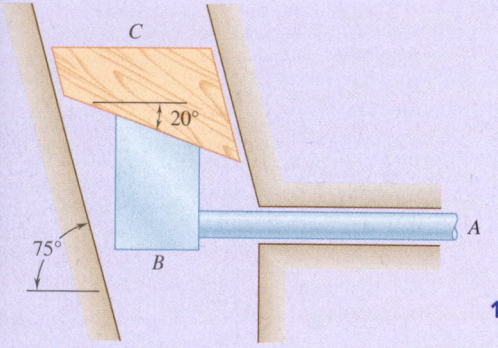

Fig. P11.126

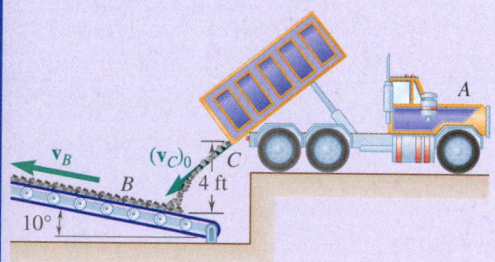

Fig. P11.127

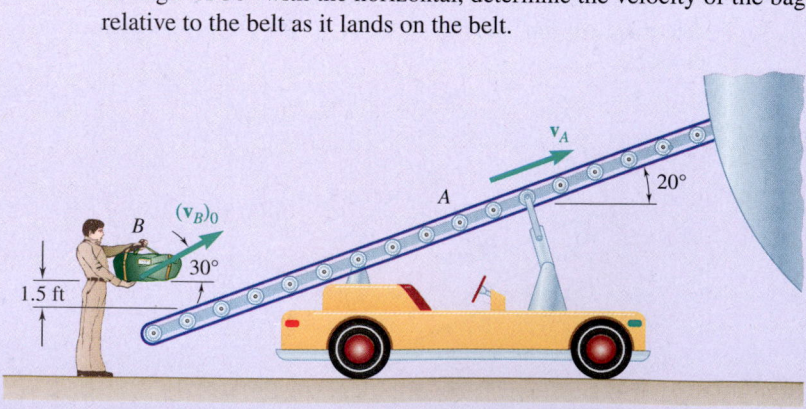

Fig. P11.128

**11.129** During a rainstorm, the paths of the raindrops appear to form an angle of 30° with the vertical and to be directed to the left when observed from a side window of a train moving at a speed of 15 km/h. A short time later, after the speed of the train has increased to 24 km/h, the angle between the vertical and the paths of the drops appears to be 45°. If the train were stopped, at what angle and with what velocity would the drops be observed to fall?

**11.130** Instruments in airplane A indicate that, with respect to the air, the plane is headed 30° north of east with an air speed of 300 mi/h. At the same time, radar on ship B indicates that the relative velocity of the plane with respect to the ship is 280 mi/h in the direction 33° north of east. Knowing that the ship is steaming due south at 12 mi/h, determine (a) the velocity of the airplane, (b) the wind speed and direction.

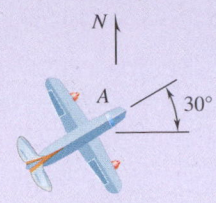

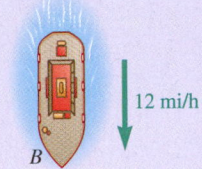

Fig. *P11.130*

**11.131** When a small boat travels north at 15 km/h, a flag mounted on its stern forms an angle θ = 50° with the centerline of the boat as shown. A short time later, when the boat travels east at 25 km/h, angle θ is again 50°. Determine the speed and the direction of the wind.

**11.132** As part of a department store display, a model train D runs on a slight incline between the store's up and down escalators. When the train and shoppers pass point A, the train appears to a shopper on the up escalator B to move downward at an angle of 22° with the horizontal, and to a shopper on the down escalator C to move upward at an angle of 23° with the horizontal and to travel to the left. Knowing that the speed of the escalators is 3 ft/s, determine the speed and the direction of the train.

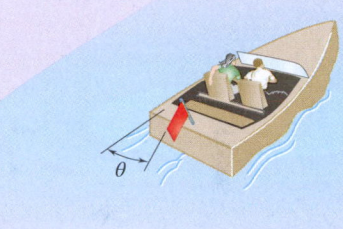

Fig. **P11.131**

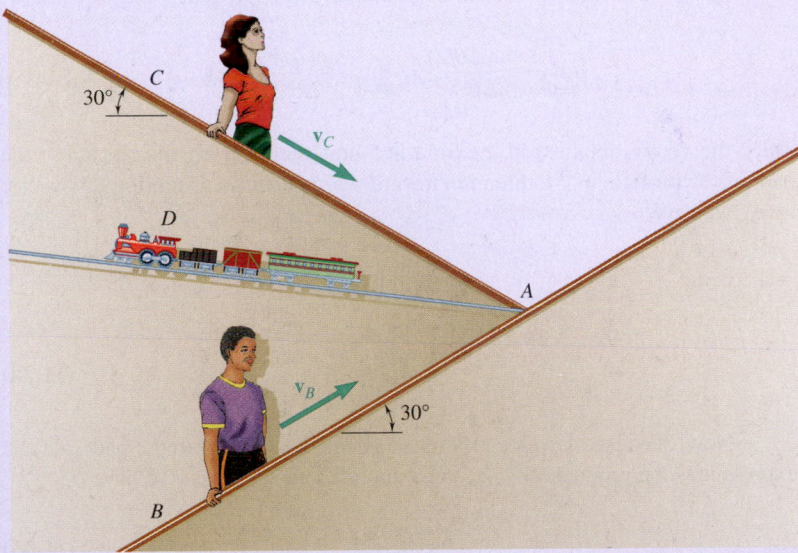

Fig. *P11.132*

## 11.5    NON-RECTANGULAR COMPONENTS

Sometimes it is useful to analyze the motion of a particle in a coordinate system that is not rectangular. In this section, we introduce two common and important systems. The first system is based on the path of the particle; the second system is based on the radial distance and angular displacement of the particle.

### 11.5A    Tangential and Normal Components

We saw in Sec. 11.4 that the velocity of a particle is a vector tangent to the path of the particle, but in general, the acceleration is not tangent to the path. It is sometimes convenient to resolve the acceleration into components directed, respectively, along the tangent and the normal to the path of the particle. We will refer to this reference frame as tangential and normal coordinates, which are sometimes called path coordinates.

**Planar Motion of a Particle.**    First we consider a particle that moves along a curve contained in a plane. Let $P$ be the position of the particle at a given instant. We attach at $P$ a unit vector $\mathbf{e}_t$ tangent to the path of the particle and pointing in the direction of motion (Fig. 11.19$a$). Let $\mathbf{e}_t'$ be the unit vector corresponding to the position $P'$ of the particle at a later instant. Drawing both vectors from the same origin $O'$, we define the vector $\Delta\mathbf{e}_t = \mathbf{e}_t' - \mathbf{e}_t$ (Fig. 11.19$b$). Because $\mathbf{e}_t$ and $\mathbf{e}_t'$ are of unit length, their tips lie on a circle with a radius of 1. Denote the angle formed by $\mathbf{e}_t$ and $\mathbf{e}_t'$ by $\Delta\theta$. Then, the magnitude of $\Delta\mathbf{e}_t$ is $2\sin(\Delta\theta/2)$. Considering now the vector $\Delta\mathbf{e}_t/\Delta\theta$, we note that, as $\Delta\theta$ approaches zero, this vector becomes tangent to the unit circle of Fig. 11.19$b$—that is, perpendicular to $\mathbf{e}_t$—and that its magnitude approaches

$$\lim_{\Delta\theta\to 0}\frac{2\sin(\Delta\theta/2)}{\Delta\theta} = \lim_{\Delta\theta\to 0}\frac{\sin(\Delta\theta/2)}{\Delta\theta/2} = 1$$

Thus, the vector obtained in the limit is a unit vector along the normal to the path of the particle in the direction toward which $\mathbf{e}_t$ turns. Denoting this vector by $\mathbf{e}_n$, we have

$$\mathbf{e}_n = \lim_{\Delta\theta\to 0}\frac{\Delta\mathbf{e}_t}{\Delta\theta}$$

$$\mathbf{e}_n = \frac{d\mathbf{e}_t}{d\theta} \tag{11.34}$$

Now, because the velocity $\mathbf{v}$ of the particle is tangent to the path, we can express it as the product of the scalar $v$ and the unit vector $\mathbf{e}_t$. We have

$$\mathbf{v} = v\mathbf{e}_t \tag{11.35}$$

To obtain the acceleration of the particle, we differentiate Eq. (11.35) with respect to $t$. Applying the rule for the differentiation of the product of a scalar and a vector function (Sec. 11.4B), we have

$$\mathbf{a} = \frac{d\mathbf{v}}{dt} = \frac{dv}{dt}\mathbf{e}_t + v\frac{d\mathbf{e}_t}{dt} \tag{11.36}$$

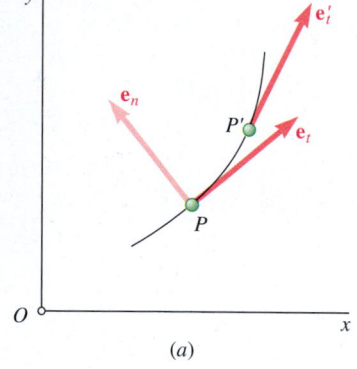

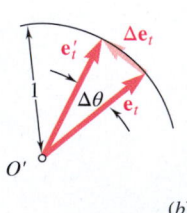

**Fig. 11.19**    ($a$) Unit tangent vectors for two positions of particle $P$; ($b$) the angle between the unit tangent vectors and their difference $\Delta\mathbf{e}_t$.

However,

$$\frac{d\mathbf{e}_t}{dt} = \frac{d\mathbf{e}_t}{d\theta}\frac{d\theta}{ds}\frac{ds}{dt}$$

Recall from Eq. (11.15) that $ds/dt = v$, from Eq. (11.34) that $d\mathbf{e}_t/d\theta = \mathbf{e}_n$, and from elementary calculus that $d\theta/ds$ is equal to $1/\rho$, where $\rho$ is the radius of curvature of the path at $P$ (Fig. 11.20). Then, we have

$$\frac{d\mathbf{e}_t}{dt} = \frac{v}{\rho}\mathbf{e}_n \qquad (11.37)$$

Substituting into Eq. (11.36), we obtain

**Acceleration in normal and tangential components**

$$\mathbf{a} = \frac{dv}{dt}\mathbf{e}_t + \frac{v^2}{\rho}\mathbf{e}_n \qquad (11.38)$$

Thus, the scalar components of the acceleration are

$$a_t = \frac{dv}{dt} \qquad a_n = \frac{v^2}{\rho} \qquad (11.39)$$

These relations state that the **tangential component** of the acceleration is equal to the **rate of change of the speed of the particle**, whereas the **normal component** is equal to the **square of the speed divided by the radius of curvature of the path at $P$**. For a given speed, the normal acceleration increases as the radius of curvature decreases. If the particle travels in a straight line, then $\rho$ is infinite, and the normal acceleration is zero. If the speed of the particle increases, $a_t$ is positive, and the vector component $\mathbf{a}_t$ points in the direction of motion. If the speed of the particle decreases, $a_t$ is negative, and $\mathbf{a}_t$ points against the direction of motion. The vector component $\mathbf{a}_n$, on the other hand, **is always directed toward the center of curvature $C$ of the path** (Fig. 11.21).

We conclude from this discussion that the tangential component of the acceleration reflects a change in the speed of the particle, whereas its normal component reflects a change in the direction of motion of the particle. The acceleration of a particle is zero only if both of its components are zero. Thus, the acceleration of a particle moving with constant speed along a curve is not zero unless the particle happens to pass through a point of inflection of the curve (where the radius of curvature is infinite) or unless the curve is a straight line. ▶

The fact that the normal component of acceleration depends upon the radius of curvature of the particle's path is taken into account in the design of structures or mechanisms as widely different as airplane wings, railroad tracks, and cams. In order to avoid sudden changes in the acceleration of the air particles flowing past a wing, wing profiles are designed without any sudden change in curvature. Similar care is taken in designing railroad curves to avoid sudden changes in the acceleration of the cars (which would be hard on the equipment and unpleasant for the passengers). A straight section of track, for instance, is never directly followed by a circular section. Special transition sections are used to help pass smoothly from the infinite radius of curvature of the straight section to the finite radius of the circular track. Likewise, in the design of high-speed cams (that can be used to transform rotary motion into translational motion), abrupt changes in acceleration are avoided by using transition curves that produce a continuous change in acceleration.

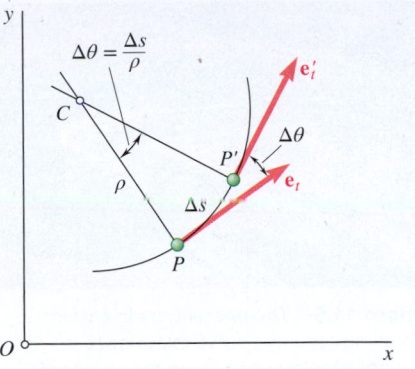

**Fig. 11.20** Relationship among $\Delta\theta$, $\Delta s$, and $\rho$. Recall that for a circle, the arc length is equal to the radius multiplied by the angle.

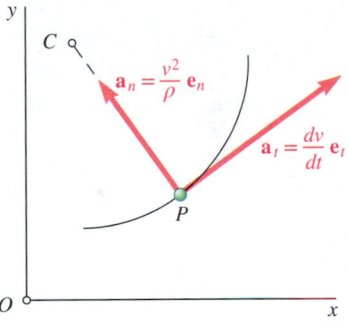

**Fig. 11.21** Acceleration components in normal and tangential coordinates; the normal component always points toward the center of curvature of the path.

**Photo 11.5**   The passengers in a train traveling around a curve experience a normal acceleration toward the center of curvature of the path. ©Tony Hertz/Alamy

**Motion of a Particle in Space.**   The relations in Eqs. (11.38) and (11.39) still hold in the case of a particle moving along a space curve. However, because an infinite number of straight lines are perpendicular to the tangent at a given point $P$ of a space curve, it is necessary to define more precisely the direction of the unit vector $\mathbf{e}_n$.

Let us consider again the unit vectors $\mathbf{e}_t$ and $\mathbf{e}'_t$ tangent to the path of the particle at two neighboring points $P$ and $P'$ (Fig. 11.22a). Again the vector $\Delta\mathbf{e}_t$ represents the difference between $\mathbf{e}_t$ and $\mathbf{e}'_t$ (Fig. 11.22b). Let us now imagine a plane through $P$ (Fig. 11.22c) parallel to the plane defined by the vectors $\mathbf{e}_t$, $\mathbf{e}'_t$, and $\Delta\mathbf{e}_t$ (Fig. 11.22b). This plane contains the tangent to the curve at $P$ and is parallel to the tangent at $P'$. If we let $P'$ approach $P$, we obtain in the limit the plane that fits the curve most closely in the neighborhood of $P$. This plane is called the **osculating plane** at $P$ (from the Latin *osculari*, to kiss). It follows from this definition that the osculating plane contains the unit vector $\mathbf{e}_n$, because this vector represents the limit of the vector $\Delta\mathbf{e}_t/\Delta\theta$. The normal defined by $\mathbf{e}_n$ is thus contained in the osculating plane; it is called the **principal normal** at $P$. The unit vector $\mathbf{e}_b = \mathbf{e}_t \times \mathbf{e}_n$ that completes the right-handed triad $\mathbf{e}_t$, $\mathbf{e}_n$, and $\mathbf{e}_b$ (Fig. 11.22c) defines the **binormal** at $P$. The binormal is thus perpendicular to the osculating plane. We conclude that the acceleration of the particle at $P$ can be resolved into two components: one along the tangent and the other along the principal normal at $P$, as indicated in Eq. (11.38). Note that the acceleration has no component along the binormal.

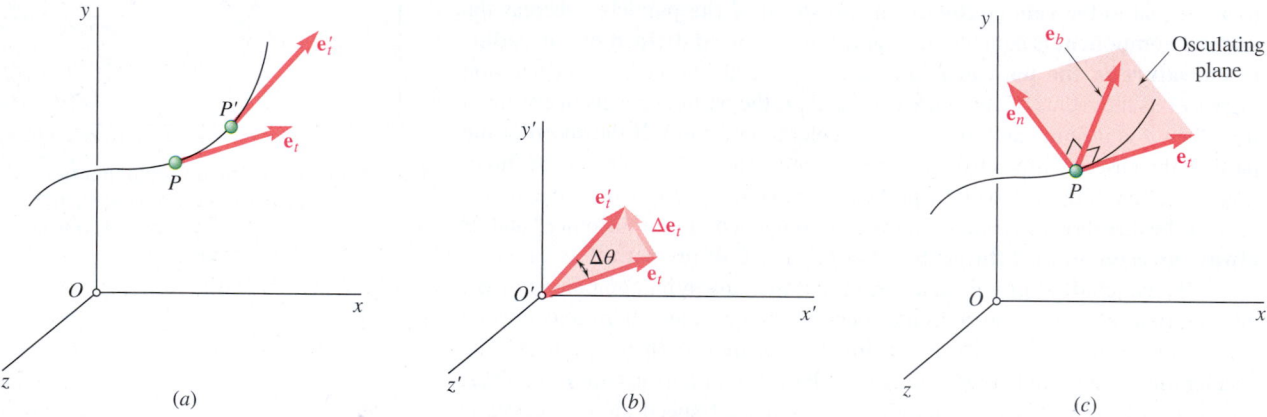

$(a)$          $(b)$          $(c)$

**Fig. 11.22**   $(a)$ Unit tangent vectors for a particle moving in space; $(b)$ the plane defined by the unit vectors and the vector difference $\Delta\mathbf{e}_t$; $(c)$ the osculating plane contains the unit tangent and principal normal vectors and is perpendicular to the unit binormal vector.

## 11.5B   Radial and Transverse Components

In some situations in planar motion, the position of particle $P$ is defined by its polar coordinates $r$ and $\theta$ (Fig. 11.23a). It is then convenient to resolve the velocity and acceleration of the particle into components parallel and perpendicular to the radial line $OP$. These components are called **radial and transverse components**.

We attach two unit vectors, $\mathbf{e}_r$ and $\mathbf{e}_\theta$, at $P$ (Fig. 11.23b). The vector $\mathbf{e}_r$ is directed along $OP$ and the vector $\mathbf{e}_\theta$ is obtained by rotating $\mathbf{e}_r$ through 90° counterclockwise. The unit vector $\mathbf{e}_r$ defines the **radial** direction; that is, the direction in which $P$ would move if $r$ were increased and $\theta$ were kept constant. The unit vector $\mathbf{e}_\theta$ defines the **transverse** direction; that is, the direction in which $P$ would move if $\theta$ were increased and $r$ were kept constant. *A* derivation similar

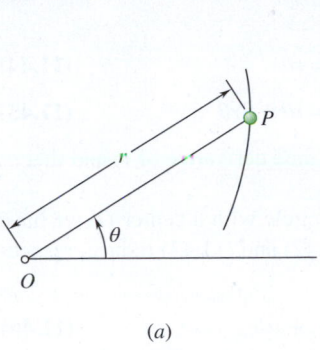
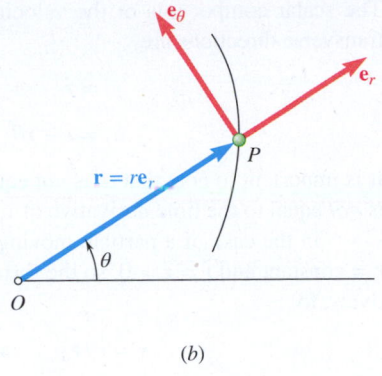
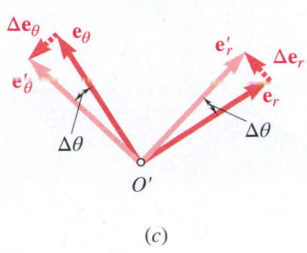

(a)    (b)    (c)

**Fig. 11.23** (a) Polar coordinates $r$ and $\theta$ of a particle at $P$; (b) radial and transverse unit vectors; (c) changes of the radial and transverse unit vectors resulting from a change in angle $\Delta\theta$.

to the one we used in the preceding section to determine the unit vector $\mathbf{e}_t$ leads to the relations

$$\frac{d\mathbf{e}_r}{d\theta} = \mathbf{e}_\theta \qquad \frac{d\mathbf{e}_\theta}{d\theta} = -\mathbf{e}_r \qquad \textbf{(11.40)}$$

Here $-\mathbf{e}_r$ denotes a unit vector with a sense opposite to that of $\mathbf{e}_r$ (Fig. 11.23c). Using the chain rule of differentiation, we express the time derivatives of the unit vectors $\mathbf{e}_r$ and $\mathbf{e}_\theta$ as

$$\frac{d\mathbf{e}_r}{dt} = \frac{d\mathbf{e}_r}{d\theta}\frac{d\theta}{dt} = \mathbf{e}_\theta\frac{d\theta}{dt} \qquad \frac{d\mathbf{e}_\theta}{dt} = \frac{d\mathbf{e}_\theta}{d\theta}\frac{d\theta}{dt} = -\mathbf{e}_r\frac{d\theta}{dt}$$

or using dots to indicate differentiation with respect to $t$ as

$$\dot{\mathbf{e}}_r = \dot{\theta}\mathbf{e}_\theta \qquad \dot{\mathbf{e}}_\theta = -\dot{\theta}\mathbf{e}_r \qquad \textbf{(11.41)}$$

To obtain the velocity $\mathbf{v}$ of particle $P$, we express the position vector $\mathbf{r}$ of $P$ as the product of the scalar $r$ and the unit vector $\mathbf{e}_r$ and then differentiate with respect to $t$ for

$$\mathbf{v} = \frac{d}{dt}(r\mathbf{e}_r) = \dot{r}\mathbf{e}_r + r\dot{\mathbf{e}}_r$$

Using the first of the relations of Eqs. (11.41), we can rewrite this as

**Velocity in radial and transverse components**

$$\mathbf{v} = \dot{r}\mathbf{e}_r + r\dot{\theta}\mathbf{e}_\theta \qquad \textbf{(11.42)}$$

Differentiating again with respect to $t$ to obtain the acceleration, we have

$$\mathbf{a} = \frac{d\mathbf{v}}{dt} = \ddot{r}\mathbf{e}_r + \dot{r}\dot{\mathbf{e}}_r + \dot{r}\dot{\theta}\mathbf{e}_\theta + r\ddot{\theta}\mathbf{e}_\theta + r\dot{\theta}\dot{\mathbf{e}}_\theta$$

Substituting for $\dot{\mathbf{e}}_r$ and $\dot{\mathbf{e}}_\theta$ from Eqs. (11.41) and factoring $\mathbf{e}_r$ and $\mathbf{e}_\theta$, we obtain

**Acceleration in radial and transverse components**

$$\mathbf{a} = (\ddot{r} - r\dot{\theta}^2)\mathbf{e}_r + (r\ddot{\theta} + 2\dot{r}\dot{\theta})\mathbf{e}_\theta \qquad \textbf{(11.43)} \; \blacktriangleright$$

**Photo 11.6** The foot pedals on an elliptical trainer undergo curvilinear motion. ©Fuse/Getty Images RF

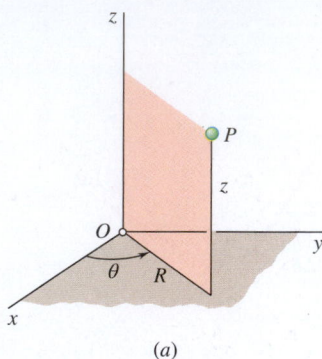

(a)

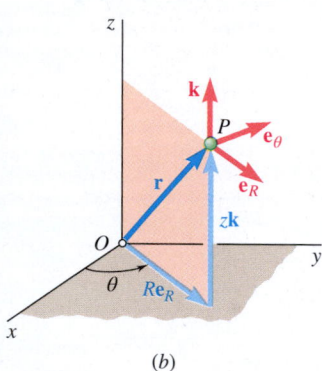

(b)

**Fig. 11.24** (a) Cylindrical coordinates $R$, $\theta$, and $z$; (b) unit vectors in cylindrical coordinates for a particle in space.

The scalar components of the velocity and the acceleration in the radial and transverse directions are

$$v_r = \dot{r} \qquad\qquad v_\theta = r\dot{\theta} \qquad\qquad \textbf{(11.44)}$$

$$a_r = \ddot{r} - r\dot{\theta}^2 \qquad a_\theta = r\ddot{\theta} + 2\dot{r}\dot{\theta} \qquad \textbf{(11.45)}$$

It is important to note that $a_r$ is *not* equal to the time derivative of $v_r$ and that $a_\theta$ is *not* equal to the time derivative of $v_\theta$.

In the case of a particle moving along a circle with a center $O$, we have $r = $ constant and $\dot{r} = \ddot{r} = 0$, so the formulas (11.42) and (11.43) reduce, respectively, to

$$\mathbf{v} = r\dot{\theta}\mathbf{e}_\theta \qquad \mathbf{a} = -r\dot{\theta}^2\mathbf{e}_r + r\ddot{\theta}\mathbf{e}_\theta \qquad \textbf{(11.46)}$$

Compare this to using tangential and normal coordinates for a particle in a circular path. In this case, the radius of curvature $\rho$ is equal to the radius of the circle $r$, and we have $\mathbf{v} = v\mathbf{e}_t$ and $\mathbf{a} = \dot{v}\mathbf{e}_t + (v^2/r)\mathbf{e}_n$. Note that $\mathbf{e}_r$ and $\mathbf{e}_n$ point in opposite directions ($\mathbf{e}_n$ inward and $\mathbf{e}_r$ outward).

**Extension to the Motion of a Particle in Space: Cylindrical Coordinates.** Sometimes it is convenient to define the position of a particle $P$ in space by its cylindrical coordinates $R$, $\theta$, and $z$ (Fig. 11.24a). We can then use the unit vectors $\mathbf{e}_R$, $\mathbf{e}_\theta$, and $\mathbf{k}$ shown in Fig. 11.24b. Resolving the position vector $\mathbf{r}$ of particle $P$ into components along the unit vectors, we have

$$\mathbf{r} = R\mathbf{e}_R + z\mathbf{k} \qquad \textbf{(11.47)}$$

Observe that $\mathbf{e}_R$ and $\mathbf{e}_\theta$ define the radial and transverse directions in the horizontal $xy$ plane, respectively, and that the vector $\mathbf{k}$, which defines the **axial** direction, is constant in direction as well as in magnitude. Then, we can verify that

$$\mathbf{v} = \frac{d\mathbf{r}}{dt} = \dot{R}\mathbf{e}_R + R\dot{\theta}\mathbf{e}_\theta + \dot{z}\mathbf{k} \qquad \textbf{(11.48)}$$

$$\mathbf{a} = \frac{d\mathbf{v}}{dt} = (\ddot{R} - R\dot{\theta}^2)\mathbf{e}_R + (R\ddot{\theta} + 2\dot{R}\dot{\theta})\mathbf{e}_\theta + \ddot{z}\mathbf{k} \qquad \textbf{(11.49)}$$

## Sample Problem 11.16

A motorist is traveling on a curved section of highway with a radius of 2500 ft at a speed of 60 mi/h. The motorist suddenly applies the brakes, causing the automobile to slow down at a constant rate. If the speed has been reduced to 45 mi/h after 8 s, determine the acceleration of the automobile immediately after the brakes have been applied.

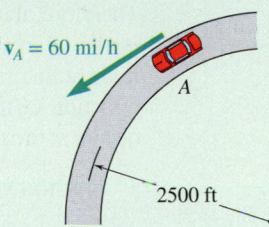

**STRATEGY:**  You know the path of the motion, and that the forward speed of the vehicle defines the direction of $\mathbf{e}_t$. Therefore, you can use tangential and normal components.

**MODELING and ANALYSIS:**
**Tangential Component of Acceleration.**  First express the speeds in ft/s.

$$60 \text{ mi/h} = \left(60\frac{\text{mi}}{\text{h}}\right)\left(\frac{5280 \text{ ft}}{1 \text{ min}}\right)\left(\frac{1 \text{ h}}{3600 \text{ s}}\right) = 88 \text{ ft/s}$$

$$45 \text{ mi/h} = 66 \text{ ft/s}$$

Because the automobile slows down at a constant rate, you have the tangential acceleration of

$$a_t = \text{average } a_t = \frac{\Delta v}{\Delta t} = \frac{66 \text{ ft/s} - 88 \text{ ft/s}}{8 \text{ s}} = -2.75 \text{ ft/s}^2$$

**Normal Component of Acceleration.**  Immediately after the brakes have been applied, the speed is still 88 ft/s. Therefore, you have

$$a_n = \frac{v^2}{\rho} = \frac{(88 \text{ ft/s})^2}{2500 \text{ ft}} = 3.10 \text{ ft/s}^2$$

**Magnitude and Direction of Acceleration.**  The magnitude and direction of the resultant $\mathbf{a}$ of the components $\mathbf{a}_n$ and $\mathbf{a}_t$ are (Fig. 1)

$$\tan \alpha = \frac{a_n}{a_t} = \frac{3.10 \text{ ft/s}^2}{2.75 \text{ ft/s}^2} \qquad \alpha = 48.4° \blacktriangleleft$$

$$a = \frac{a_n}{\sin \alpha} = \frac{3.10 \text{ ft/s}^2}{\sin 48.4°} \qquad \mathbf{a} = 4.14 \text{ ft/s}^2 \blacktriangleleft$$

**REFLECT and THINK:**  The tangential component of acceleration is opposite the direction of motion, and the normal component of acceleration points to the center of curvature, which is what you would expect for slowing down on a curved path. Attempting to do this problem in Cartesian coordinates is quite difficult.

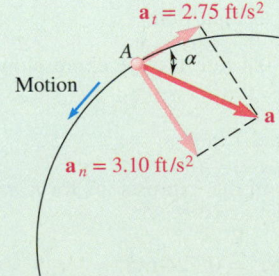

**Fig. 1**  Acceleration of the car.

## Sample Problem 11.17

Determine the minimum radius of curvature of the trajectory described by the projectile considered in Sample Prob. 11.10.

**STRATEGY:** You are asked to find the radius of curvature, so you should use normal and tangential coordinates.

**MODELING and ANALYSIS:** Because $a_n = v^2/\rho$, you have $\rho = v^2/a_n$. Therefore, the radius is small when $v$ is small or when $a_n$ is large. The speed $v$ is minimum at the top of the trajectory, because $v_y = 0$ at that point; $a_n$ is maximum at that same point, because the direction of the vertical coincides with the direction of the normal (Fig. 1). Therefore, the minimum radius of curvature occurs at the top of the trajectory. At this point, you have

$$v = v_x = 155.9 \text{ m/s} \qquad a_n = a = 9.81 \text{ m/s}^2$$

$$\rho = \frac{v^2}{a_n} = \frac{(155.9 \text{ m/s})^2}{9.81 \text{ m/s}^2} \qquad \rho = 2480 \text{ m} \blacktriangleleft$$

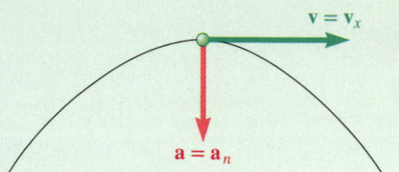

**Fig. 1** Acceleration and velocity of the projectile.

**REFLECT and THINK:** The top of the trajectory is the easiest point to determine the radius of curvature. At any other point in the trajectory, you need to find the normal component of acceleration. You can do this easily at the top, because you know that the total acceleration is pointed vertically downward and the normal component is simply the component perpendicular to the tangent to the path. Once you have the normal acceleration, it is straightforward to find the radius of curvature if you know the speed.

## Sample Problem 11.18

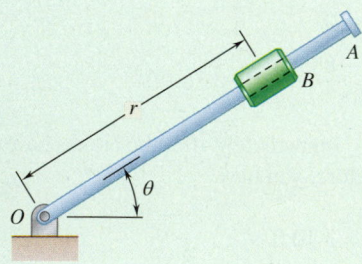

The rotation of the 0.9-m arm $OA$ about $O$ is defined by the relation $\theta = 0.15t^2$, where $\theta$ is expressed in radians and $t$ in seconds. Collar $B$ slides along the arm in such a way that its distance from $O$ is $r = 0.9 - 0.12t^2$, where $r$ is expressed in meters and $t$ in seconds. After the arm $OA$ has rotated through 30°, determine (a) the total velocity of the collar, (b) the total acceleration of the collar, (c) the relative acceleration of the collar with respect to the arm. ▶

**STRATEGY:** You are given information in terms of $r$ and $\theta$, so you should use polar coordinates.

**MODELING and ANALYSIS:** Model the collar as a particle.

**Time $t$ When $\theta = 30°$.** Substitute $\theta = 30° = 0.524$ rad into the expression for $\theta$. You obtain

$$\theta = 0.15t^2 \qquad 0.524 = 0.15t^2 \qquad t = 1.869 \text{ s}$$

**Equations of Motion.** Substituting $t = 1.869$ s in the expressions for $r$, $\theta$, and their first and second derivatives, you have

$$r = 0.9 - 0.12t^2 = 0.481 \text{ m} \qquad \theta = 0.15t^2 = 0.524 \text{ rad}$$
$$\dot{r} = -0.24t = -0.449 \text{ m/s} \qquad \dot{\theta} = 0.30t = 0.561 \text{ rad/s}$$
$$\ddot{r} = -0.24 = -0.240 \text{ m/s}^2 \qquad \ddot{\theta} = 0.30 = 0.300 \text{ rad/s}^2$$

*(continued)*

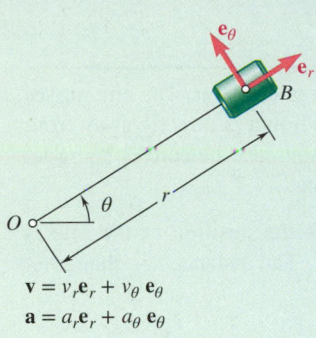

$$\mathbf{v} = v_r\mathbf{e}_r + v_\theta\,\mathbf{e}_\theta$$
$$\mathbf{a} = a_r\mathbf{e}_r + a_\theta\,\mathbf{e}_\theta$$

**Fig. 1** Radial and transverse coordinates for collar B.

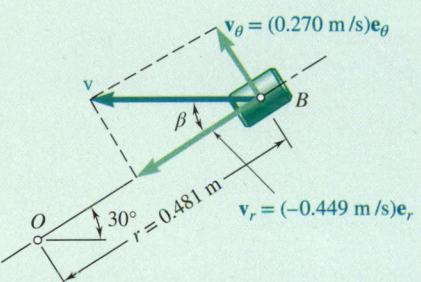

**Fig. 2** Velocity of collar B.

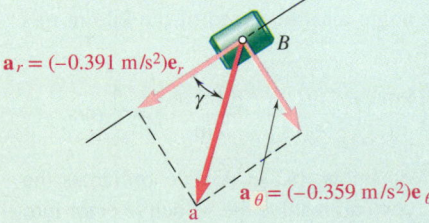

**Fig. 3** Acceleration of collar B.

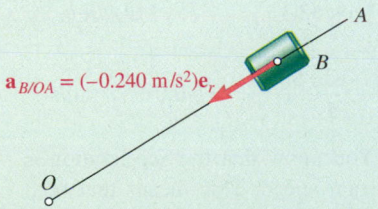

**Fig. 4** Acceleration of collar B with respect to arm OA.

**a. Velocity of B.** Using Eqs. (11.44), you can obtain the values of $v_r$ and $v_\theta$ when $t = 1.869$ s (Fig. 1).

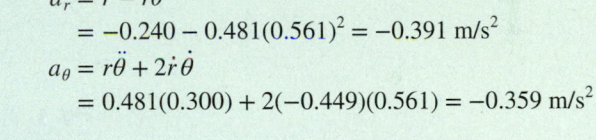

$$v_r = \dot{r} = -0.449 \text{ m/s}$$
$$v_\theta = r\dot{\theta} = 0.481(0.561) = 0.270 \text{ m/s}$$

Solve the right triangle shown in Fig. 2 to obtain the magnitude and direction of the velocity,

$$v = 0.524 \text{ m/s} \quad \beta = 31.0° \quad \blacktriangleleft$$

**b. Acceleration of B.** Using Eqs. (11.45), you obtain (Fig. 3)

$$a_r = \ddot{r} - r\dot{\theta}^2$$
$$= -0.240 - 0.481(0.561)^2 = -0.391 \text{ m/s}^2$$
$$a_\theta = r\ddot{\theta} + 2\dot{r}\dot{\theta}$$
$$= 0.481(0.300) + 2(-0.449)(0.561) = -0.359 \text{ m/s}^2$$

$$a = 0.531 \text{ m/s}^2 \quad \gamma = 42.6° \quad \blacktriangleleft$$

**c. Acceleration of B with Respect to Arm OA.** Note that the motion of the collar with respect to the arm is rectilinear and defined by the coordinate $r$ (Fig. 4). You have

$$a_{B/OA} = \ddot{r} = -0.240 \text{ m/s}^2$$
$$a_{B/OA} = 0.240 \text{ m/s}^2 \text{ toward } O. \quad \blacktriangleleft$$

**REFLECT and THINK:** You should consider polar coordinates for any kind of rotational motion. They turn this problem into a straightforward solution, whereas any other coordinate system would make this problem much more difficult. One way to make this problem harder would be to ask you to find the radius of curvature in addition to the velocity and acceleration. To do this, you would have to find the normal component of the acceleration—that is, the component of acceleration that is perpendicular to the tangential direction defined by the velocity vector.

## Sample Problem 11.19

A boy is flying a kite that is 60 m high with 75 m of cord out. The kite moves horizontally from this position at a constant 6 km/h that is directly away from the boy. Ignoring the sag in the cord, determine how fast the cord is being let out at this instant and how fast this rate is increasing.

**STRATEGY:** The most natural way to describe the position of the kite is using a radial vector and angle, as shown in Fig. 1. The distance $r$ is changing, so use polar coordinates.

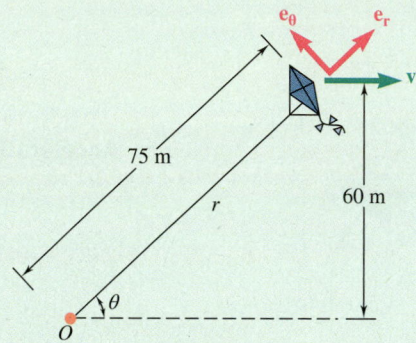

**Fig. 1** Radial and transverse coordinates for the kite.

**MODELING and ANALYSIS:** The angle and the speed of the kite in m/s are found by

$$\theta = \sin^{-1}\left(\frac{60}{75}\right) = 53.13° \text{ and } v = 6\left(\frac{\text{km}}{\text{hr}}\right)\left(\frac{\text{hr}}{3600 \text{ s}}\right)\left(\frac{1000 \text{ m}}{\text{km}}\right) = \frac{5}{3} \text{ m/s}$$

**Velocity in Polar Coordinates.** You know that in polar coordinates the velocity is $\mathbf{v} = \dot{r}\mathbf{e}_r + r\dot{\theta}\mathbf{e}_\theta$. Using Fig. 1, you can resolve the velocity vector into polar coordinates, giving

$$\dot{r} = v\cos\theta = \left(\frac{5}{3}\text{m/s}\right)\cos 53.13° \qquad \dot{r} = 1.000 \text{ m/s} \blacktriangleleft$$

$$r\dot{\theta} = -v\sin\theta \qquad \dot{\theta} = -\frac{v\sin\theta}{r} = -\frac{(5/3 \text{ m/s})\sin 53.13°}{75} = -0.01778 \text{ rad/s}$$

**Acceleration in Polar Coordinates.** You know that the acceleration is zero, because the kite is traveling at a constant speed. This means that both components of the acceleration need to be zero. You know the radial component is $a_r = \ddot{r} - r\dot{\theta}^2 = 0$. So,

$$\ddot{r} = r\dot{\theta}^2 = (75 \text{ m})(-0.01778 \text{ rad/s})^2 \qquad \ddot{r} = 0.0237 \text{ m/s}^2 \blacktriangleleft$$

**REFLECT and THINK:** When the angle is 90°, then $\dot{r}$ will be zero. When the angle is very small—that is, when the kite is far away—you would expect the cord to increase at a rate of 6 m/s, which is the speed of the kite. Our answer is reasonable because it is between these two limits.

## Sample Problem 11.20

At the instant shown, the length of the boom $AB$ is being *decreased* at the constant rate of 0.2 m/s, and the boom is being lowered at the constant rate of 0.08 rad/s. Determine (*a*) the velocity of point $B$, (*b*) the acceleration of point $B$.

**STRATEGY:**   Use polar coordinates, because that is the most natural way to describe the position of point $B$.

**MODELING and ANALYSIS:**   From the problem statement, you know that

$$\dot{r} = -0.2 \text{ m/s} \qquad \ddot{r} = 0 \qquad \dot{\theta} = -0.08 \text{ rad/s} \qquad \ddot{\theta} = 0$$

**a. Velocity of B.**   Using Eqs. (11.44), you can determine the values of $\mathbf{v}_r$ and $\mathbf{v}_\theta$ at this instant to be

$$v_r = \dot{r} = -0.2 \text{ m/s}$$
$$v_\theta = r\dot{\theta} = (6 \text{ m})(-0.08 \text{ rad/s}) = -0.48 \text{ m/s}$$

Therefore, you can write the velocity vector as

$$\mathbf{v} = (-0.200 \text{ m/s})\mathbf{e}_r + (-0.480 \text{ m/s})\mathbf{e}_\theta \quad \blacktriangleleft$$

**b. Acceleration of B.**   Using Eqs. (11.45), you find that

$$a_r = \ddot{r} - r\dot{\theta}^2 = 0 - (6 \text{ m})(-0.08 \text{ rad/s})^2 = -0.0384 \text{ m/s}^2$$
$$a_\theta = r\ddot{\theta} + 2\dot{r}\dot{\theta} = 0 + 2(-0.02 \text{ m/s})(-0.08 \text{ rad/s}) = 0.00320 \text{ m/s}^2$$

or

$$\mathbf{a} = (-0.0384 \text{ m/s}^2)\mathbf{e}_r + (0.00320 \text{ m/s}^2)\mathbf{e}_\theta \quad \blacktriangleleft$$

**REFLECT and THINK:**   Once you identify what you are given in the problem statement, this problem is quite straightforward. Sometimes you will be asked to express your answer in terms of a magnitude and direction. The easiest way is to first determine the $x$ and $y$ components and then to find the magnitude and direction. From Fig. 1,

$$\overset{+}{\rightarrow}: (v_B)_x = 0.48 \cos 60° - 0.2 \cos 30° = 0.06680 \text{ m/s}$$
$$+\uparrow: (v_B)_y = -0.48 \sin 60° - 0.2 \sin 30° = -0.5157 \text{ m/s}$$

So, the magnitude and direction are

$$v_B = \sqrt{0.06680^2 + 0.5157^2}$$
$$= 0.520 \text{ m/s} \qquad \tan\beta = \frac{0.51569}{0.06680}, \beta = 82.6°$$

So, an alternative way of expressing the velocity of $B$ is $\mathbf{v}_B = 0.520$ m/s ⦨82.6°

You could also find the magnitude and direction of the acceleration if you needed it expressed in this way. It is important to note that no matter what coordinate system we choose, the resultant velocity vector is the same. You can choose to express this vector in whatever coordinate system is most useful. Fig. 2 shows the velocity vector $\mathbf{v}_B$ resolved into $x$ and $y$ components and $r$ and $\theta$ coordinates.

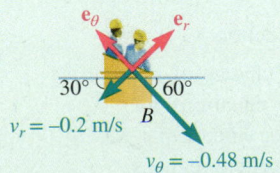

**Fig. 1**   Velocity of $B$.

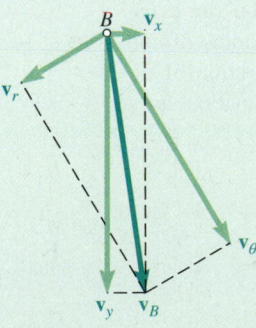

**Fig. 2**   Resultant velocity of collar $B$ in Cartesian and in radial and transverse coordinates.

# SOLVING PROBLEMS
# ON YOUR OWN

In the following problems, you will be asked to express the velocity and the acceleration of particles in terms of either their **tangential and normal components** or their **radial and transverse components**. Although these components may not be as familiar to you as rectangular components, you will find that they can simplify the solution of many problems and that certain types of motion are more easily described when they are used.

**1. Using tangential and normal components.** These components are most often used when the particle of interest travels along a known curvilinear path or when the radius of curvature of the path is to be determined (Sample Probs. 11.16 and 11.17). Remember that the unit vector $\mathbf{e}_t$ is tangent to the path of the particle (and thus aligned with the velocity), whereas the unit vector $\mathbf{e}_n$ is directed along the normal to the path and always points toward its center of curvature. It follows that the directions of the two unit vectors are constantly changing as the particle moves.

**2. Acceleration in terms of tangential and normal components.** We derived in Sec. 11.5A the following equation, which is applicable to both the two-dimensional and the three-dimensional motion of a particle:

$$\mathbf{a} = \frac{dv}{dt}\mathbf{e}_t + \frac{v^2}{\rho}\mathbf{e}_n \qquad (11.38)$$

The following observations may help you in solving the problems of this section.

  **a. The tangential component** of the acceleration measures the rate of change of the speed as $a_t = dv/dt$. It follows that, when $a_t$ is constant, you can use the equations for uniformly accelerated motion with the acceleration equal to $a_t$. Furthermore, when a particle moves at a constant speed, we have $a_t = 0$, and the acceleration of the particle reduces to its normal component.

  **b. The normal component** of the acceleration is always directed toward the center of curvature of the path of the particle, and its magnitude is $a_n = v^2/\rho$. Thus, you can determine the normal component if you know the speed of the particle and the radius of curvature $\rho$ of the path. Conversely, if you know the speed and normal acceleration of the particle, you can find the radius of curvature of the path by solving this equation for $\rho$ (Sample Prob. 11.17).

**3. Using radial and transverse components.** These components are used to analyze the planar motion of a particle $P$ when the position of $P$ is defined by its polar coordinates $r$ and $\theta$. As shown in Fig. 11.23, the unit vector $\mathbf{e}_r$, which defines the **radial** direction, is attached to $P$ and points away from the fixed point $O$, whereas the unit vector $\mathbf{e}_\theta$, which defines the **transverse** direction, is obtained by rotating $\mathbf{e}_r$ *counterclockwise* through $90°$. The velocity and acceleration of a particle are expressed in terms of their radial and transverse components in Eqs. (11.42) and (11.43), respectively. Note that the expressions obtained contain the first and second derivatives with respect to $t$ of both coordinates $r$ and $\theta$.

In the problems of this section, you will encounter the following types of problems involving radial and transverse components.

   **a. Both $r$ and $\theta$ are known functions of $t$.** In this case, you compute the first and second derivatives of $r$ and $\theta$ and substitute the resulting expressions into Eqs. (11.42) and (11.43).

   **b. A certain relationship exists between $r$ and $\theta$.** First, you should determine this relationship from the geometry of the given system and use it to express $r$ as a function of $\theta$. Once you know the function $r = f(\theta)$, you can apply the chain rule to determine $\dot{r}$ in terms of $\theta$ and $\dot{\theta}$, and $\ddot{r}$ in terms of $\theta$, $\dot{\theta}$, and $\ddot{\theta}$:

$$\dot{r} = f'(\theta)\dot{\theta}$$
$$\ddot{r} = f''(\theta)\dot{\theta}^2 + f'(\theta)\ddot{\theta}$$

You can then substitute these expressions into Eqs. (11.42) and (11.43).

   **c. The three-dimensional motion of a particle,** as indicated at the end of Sec. 11.5B, often can be described effectively in terms of the **cylindrical coordinates** $R$, $\theta$, and $z$ (Fig. 11.24). The unit vectors then should consist of $\mathbf{e}_R$, $\mathbf{e}_\theta$, and $\mathbf{k}$. The corresponding components of the velocity and the acceleration are given in Eqs. (11.48) and (11.49). Note that the radial distance $R$ is always measured in a plane parallel to the $xy$ plane, and be careful not to confuse the position vector $\mathbf{r}$ with its radial component $R\mathbf{e}_R$.

# Problem

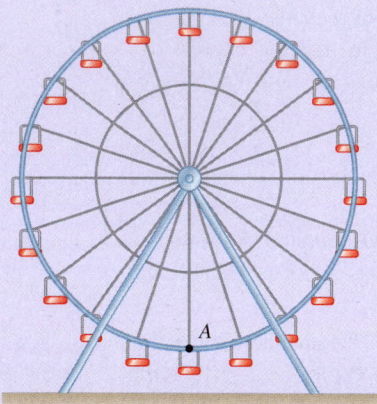

Fig. P11.CQ8

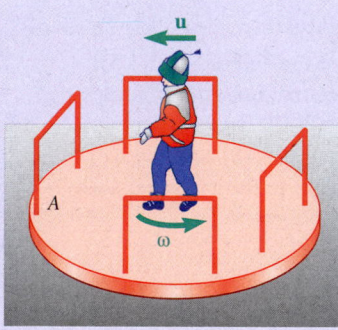

Fig. P11.CQ10

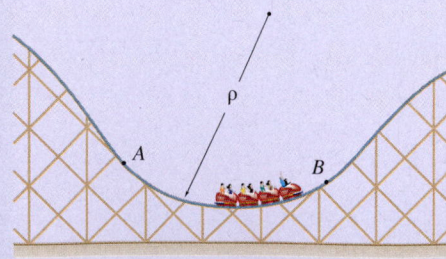

Fig. P11.134

## CONCEPT QUESTIONS

**11.CQ8** The Ferris wheel is rotating with a constant angular velocity $\omega$. What is the direction of the acceleration of point $A$?
**a.** → **b.** ↑ **c.** ↓ **d.** ← **e.** The acceleration is zero.

**11.CQ9** A race car travels around the track shown at a constant speed. At which point will the race car have the largest acceleration?
**a.** $A$ **b.** $B$ **c.** $C$ **d.** $D$
**e.** The acceleration will be zero at all the points.

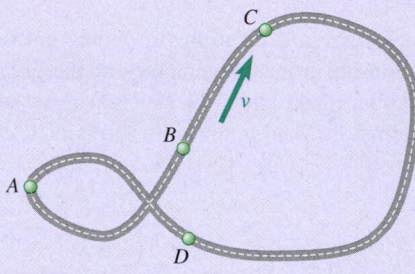

Fig. P11.CQ9

**11.CQ10** A child walks across merry-go-round $A$ with a constant speed $u$ relative to $A$. The merry-go-round undergoes fixed-axis rotation about its center with a constant angular velocity $\omega$ counterclockwise. When the child is at the center of $A$, as shown, what is the direction of his acceleration when viewed from above?
**a.** → **b.** ← **c.** ↑ **d.** ↓ **e.** The acceleration is zero.

## END-OF-SECTION PROBLEMS

**11.133** Determine the normal component of acceleration of a car traveling at 72 km/h on an exit ramp of radius $\rho = 150$ m.

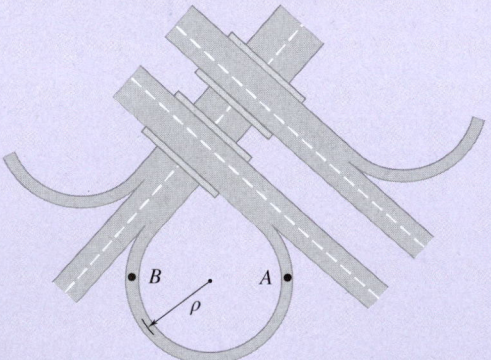

Fig. P11.133

**11.134** Determine the maximum speed that the cars of the roller coaster can reach along the circular portion $AB$ of the track if $\rho = 25$ m and the normal component of their acceleration cannot exceed 3g.

**11.135** Human centrifuges are often used to simulate different acceleration levels for pilots and astronauts. Pilots typically face inward toward the center of the gondola in order to experience a simulated forward acceleration. Knowing that the pilots sits 5 m from the axis of rotation and experiences 5 g's inward, determine her velocity.

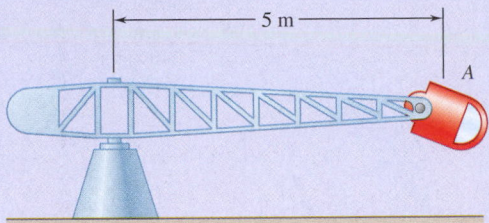

**Fig. P11.135**

**11.136** The diameter of the eye of a stationary hurricane is 20 mi and the maximum wind speed is 100 mi/h at the eye wall with $r = 10$ mi. Assuming that the wind speed is constant for constant $r$ and decreases uniformly with increasing $r$ to 40 mi/h at $r = 110$ mi, determine the magnitude of the acceleration of the air at (a) $r = 10$ mi, (b) $r = 60$ mi, (c) $r = 110$ mi.

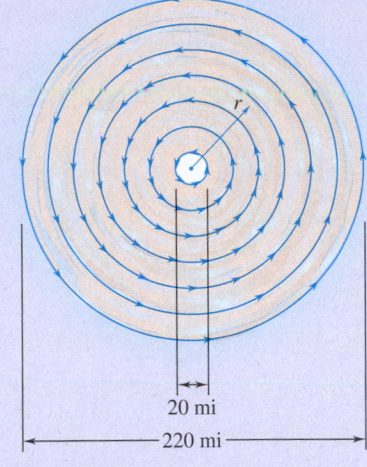

**Fig. P11.136**

**11.137** The peripheral speed of the tooth of a 10-in.-diameter circular saw blade is 150 ft/s when the power to the saw is turned off. The speed of the tooth decreases at a constant rate, and the blade comes to rest in 9 s. Determine the time at which the total acceleration of the tooth is 130 ft/s$^2$.

**11.138** A robot arm moves so that $P$ travels in a circle about point $B$, which is not moving. Knowing that $P$ starts from rest, and its speed increases at a constant rate of 10 mm/s$^2$, determine (a) the magnitude of the acceleration when $t = 4$ s, (b) the time for the magnitude of the acceleration to be 80 mm/s$^2$.

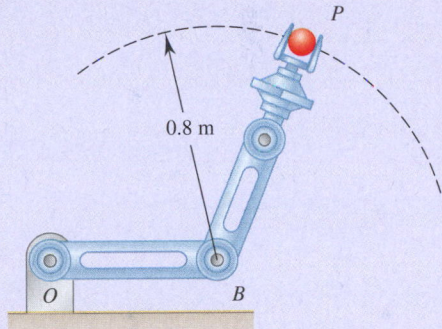

**Fig. P11.138**

**11.139** A monorail train starts from rest on a curve of radius 400 m and accelerates at the constant rate $a_t$. If the maximum total acceleration of the train must not exceed 1.5 m/s$^2$, determine (a) the shortest distance in which the train can reach a speed of 72 km/h, (b) the corresponding constant rate of acceleration $a_t$.

**11.140** A motorist starts from rest at point $A$ on a circular entrance ramp when $t = 0$, increases the speed of her automobile at a constant rate, and enters the highway at point $B$. Knowing that her speed continues to increase at the same rate until it reaches 100 km/h at point $C$, determine (a) the speed at point $B$, (b) the magnitude of the total acceleration when $t = 20$ s.

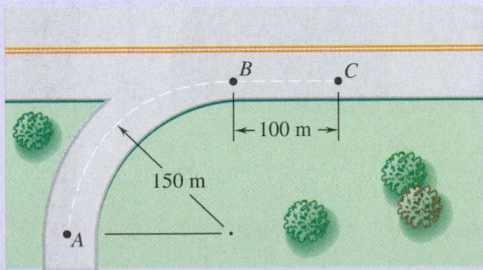

**Fig. *P11.140***

**11.141** Race car $A$ is traveling on a straight portion of the track while race car $B$ is traveling on a circular portion of the track. At the instant shown, the speed of $A$ is increasing at the rate of 10 m/s$^2$, and the speed of $B$ is decreasing at the rate of 6 m/s$^2$. For the position shown, determine (*a*) the velocity of $B$ relative to $A$, (*b*) the acceleration of $B$ relative to $A$.

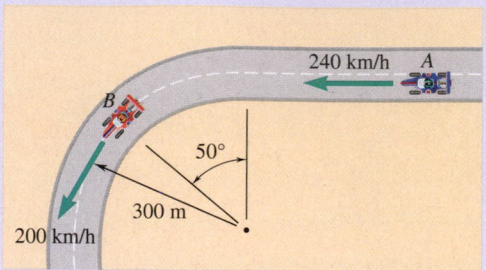

**Fig. P11.141**

**11.142** At a given instant in an airplane race, airplane $A$ is flying horizontally in a straight line, and its speed is being increased at the rate of 8 m/s$^2$. Airplane $B$ is flying at the same altitude as airplane $A$ and, as it rounds a pylon, is following a circular path of 300-m radius. Knowing that at the given instant the speed of $B$ is being decreased at the rate of 3 m/s$^2$, determine, for the positions shown, (*a*) the velocity of $B$ relative to $A$, (*b*) the acceleration of $B$ relative to $A$.

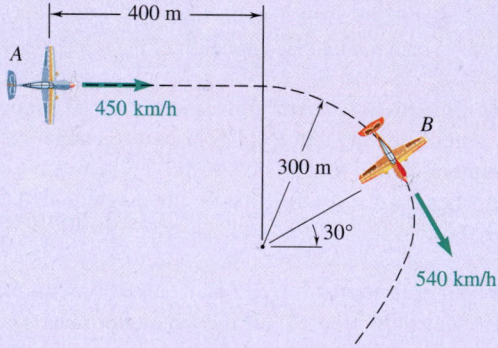

**Fig. *P11.142***

**11.143** A race car enters the circular portion of a track that has a radius of 70 m. When the car enters the curve at point $P$, it is traveling with a speed of 120 km/h that is increasing at 5 m/s$^2$. Three seconds later, determine the $x$ and $y$ components of velocity and acceleration of the car.

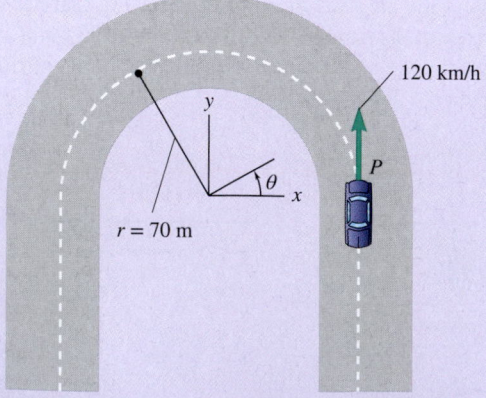

**Fig. P11.143**

**11.144** Pin $A$, which is attached to link $AB$, is constrained to move in the circular slot $CD$. Knowing that at $t = 0$ the pin starts from rest and moves so that its speed increases at a constant rate of 0.8 in/s$^2$, determine the magnitude of its total acceleration when (a) $t = 0$, (b) $t = 2$ s.

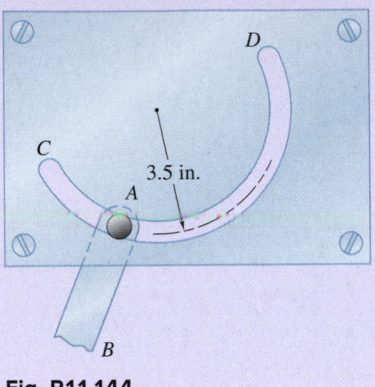

**Fig. P11.144**

**11.145** A golfer hits a golf ball from point $A$ with an initial velocity of 50 m/s at an angle of 25° with the horizontal. Determine the radius of curvature of the trajectory described by the ball (a) at point $A$, (b) at the highest point of the trajectory.

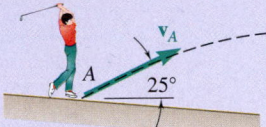

**Fig. P11.145**

**11.146** A nozzle discharges a stream of water in the direction shown with an initial velocity of 24 ft/s. Determine the radius of curvature of the stream (a) as it leaves the nozzle, (b) at the maximum height of the stream.

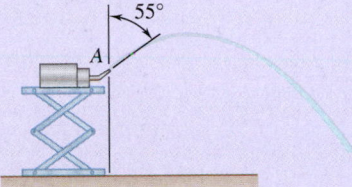

**Fig. P11.146**

**11.147** Coal is discharged from the tailgate $A$ of a dump truck with an initial velocity $\mathbf{v}_A = 2$ m/s $\searrow 50°$. Determine the radius of curvature of the trajectory described by the coal (a) at point $A$, (b) at the point of the trajectory 1 m below point $A$.

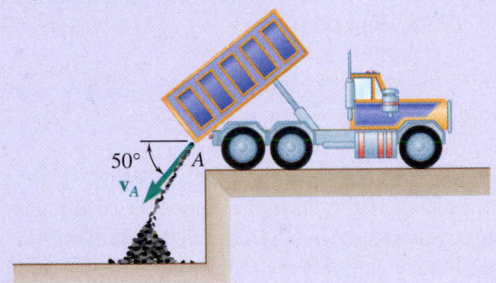

**Fig. P11.147**

**11.148** From measurements of a photograph, it has been found that as the stream of water shown left the nozzle at $A$, it had a radius of curvature of 25 m. Determine (a) the initial velocity $\mathbf{v}_A$ of the stream, (b) the radius of curvature of the stream as it reaches its maximum height at $B$.

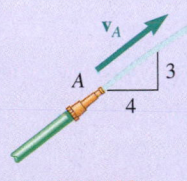

**Fig. P11.148**

**11.149** A child throws a ball from point $A$ with an initial velocity $\mathbf{v}_0$ at an angle of 3° with the horizontal. Knowing that the ball hits a wall at point $B$, determine (a) the magnitude of the initial velocity, (b) the minimum radius of curvature of the trajectory.

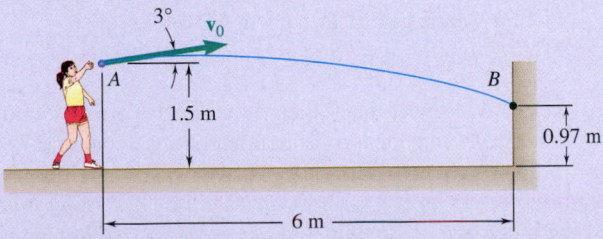

**Fig. P11.149**

707

**11.150** A projectile is fired from point $A$ with an initial velocity $\mathbf{v}_0$. ($a$) Show that the radius of curvature of the trajectory of the projectile reaches its minimum value at the highest point $B$ of the trajectory. ($b$) Denoting by $\theta$ the angle formed by the trajectory and the horizontal at a given point $C$, show that the radius of curvature of the trajectory at $C$ is $\rho = \rho_{min}/\cos^3\theta$.

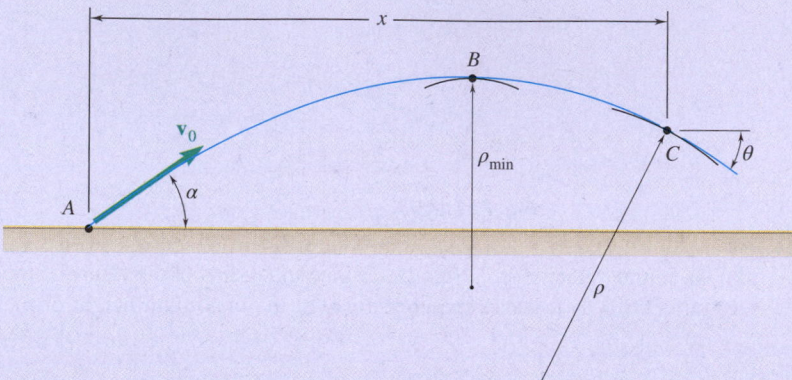

Fig. *P11.150*

**\*11.151** Determine the radius of curvature of the path described by the particle of Prob. 11.95 when $t = 0$.

**\*11.152** Determine the radius of curvature of the path described by the particle of Prob. 11.96 when $t = 0$, $A = 3$, and $B = 1$.

**11.153 and 11.154** A satellite will travel indefinitely in a circular orbit around a planet if the normal component of the acceleration of the satellite is equal to $g(R/r)^2$, where $g$ is the acceleration of gravity at the surface of the planet, $R$ is the radius of the planet, and $r$ is the distance from the center of the planet to the satellite. Knowing that the diameter of the sun is 1.39 Gm and that the acceleration of gravity at its surface is 274 m/s², determine the radius of the orbit of the indicated planet around the sun assuming that the orbit is circular.

    **11.153** Earth: $(v_{mean})_{orbit} = 107$ Mm/h.
    **11.154** Saturn: $(v_{mean})_{orbit} = 34.7$ Mm/h.

**11.155 through *11.157*** Determine the speed of a satellite relative to the indicated planet if the satellite is to travel indefinitely in a circular orbit 100 mi above the surface of the planet. (See the information given in Probs. 11.153 and 11.154.)

    **11.155** Venus: $g = 29.20$ ft/s², $R = 3761$ mi.
    **11.156** Mars: $g = 12.17$ ft/s², $R = 2102$ mi.
    **11.157** Jupiter: $g = 75.35$ ft/s², $R = 44,432$ mi.

**11.158** A satellite will travel indefinitely in a circular orbit around the earth if the normal component of its acceleration is equal to $g(R/r)^2$, where $g = 9.81$ m/s², $R = $ radius of the earth $= 6370$ km, and $r = $ distance from the center of the earth to the satellite. Assuming that the orbit of the moon is a circle with a radius of $384 \times 10^3$ km, determine the speed of the moon relative to the earth.

**11.159** Knowing that the radius of the earth is 6370 km, determine the time of one orbit of the Hubble Space Telescope if the telescope travels in a circular orbit 590 km above the surface of the earth. (See the information given in Probs. 11.153 and 11.154.)

**11.160** Satellites $A$ and $B$ are traveling in the same plane in circular orbits around the earth at altitudes of 120 and 200 mi, respectively. If at $t = 0$ the satellites are aligned as shown and knowing that the radius of the earth is $R = 3960$ mi, determine when the satellites will next be radially aligned. (See the information given in Probs. 11.153 and 11.154.)

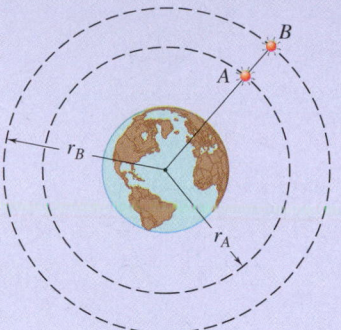

Fig. *P11.160*

**11.161** Robots with telescoping arms are sometimes used to perform tasks (e.g., welding or placing screws) where access may be difficult for other robotic types. During a test run, a robot arm is programmed to extend according to the relationship $r = 3 + 0.5 \cos(4\theta)$ and the arm rotates according to the relationship $\theta = -\dfrac{\pi}{4}t^2 + \pi t$, where $r$ is in feet, $\theta$ is in radians, and $t$ is in seconds. Determine (a) the velocity and acceleration of the robot tip $A$ at $t = 3$ s and (b) use a computer program to plot the path of tip $A$ in $x$ and $y$ coordinates for $0 \le t \le 4$s.

**11.162** The angular displacement of the robotic arm is programmed according to the relationship $\theta = (1/\pi)(\sin \pi t)$, where $\theta$ and $t$ are expressed in radians and seconds, respectively. Simultaneously, the arm is programmed to extend so that the distance to $A$ follows the relationship $r = 4(1 + e^{-2t})$, where $r$ and $t$ are expressed in feet and seconds, respectively. When $t = 1.5$ s, determine (a) the velocity of point $A$, (b) the acceleration of point $A$.

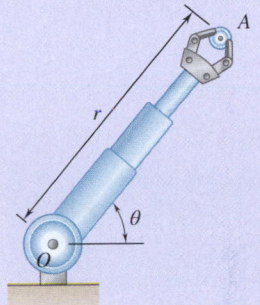

Fig. P11.161 and P11.162

**11.163** During a parasailing ride, the boat is traveling at a constant 30 km/hr with a 200-m long tow line. At the instant shown, the angle between the line and the water is 30° and is increasing at a constant rate of 2°/s. Determine the velocity and acceleration of the parasailer at this instant.

Fig. P11.163 and P11.164

**11.164** Some parasailing systems use a winch to pull the rider back to the boat. During the interval when $\theta$ is between 20° and 40° (where $t = 0$ at $\theta = 20°$), the angle increases at the constant rate of 2°/s. During this time, the length of the rope is defined by the relationship $r = 600 - \frac{1}{8}t^{5/2}$, where $r$ and $t$ are expressed in feet and seconds, respectively. Knowing that the boat is traveling at a constant rate of 15 knots (where 1 knot = 1.15 mi/h), (a) plot the magnitude of the velocity of the parasailer as a function of time, (b) determine the magnitude of the acceleration of the parasailer when $t = 5$ s.

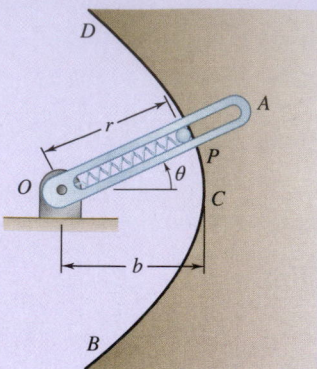

**Fig. P11.165**

**11.165** As rod $OA$ rotates, pin $P$ moves along the parabola $BCD$. Knowing that the equation of this parabola is $r = 2b/(1 + \cos \theta)$ and that $\theta = kt$, determine the velocity and acceleration of $P$ when (a) $\theta = 0$, (b) $\theta = 90°$.

**11.166** The pin at $B$ is free to slide along the circular slot $DE$ and along the rotating rod $OC$. Assuming that the rod $OC$ rotates at a constant rate $\dot{\theta}$, (a) show that the acceleration of pin $B$ is of constant magnitude, (b) determine the direction of the acceleration of pin $B$.

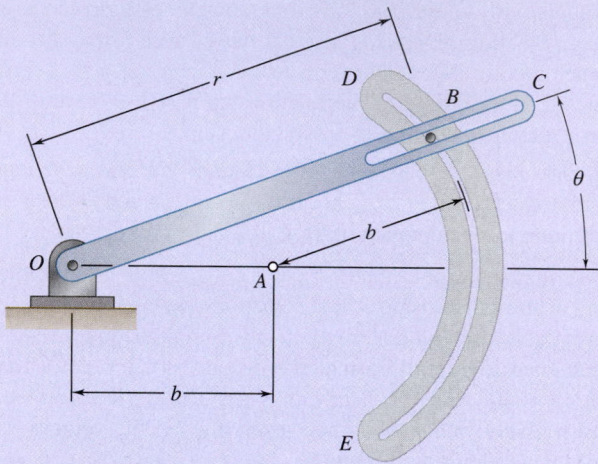

**Fig. P11.166**

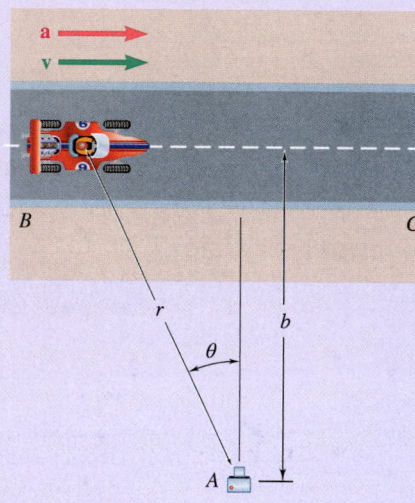

**11.167** To study the performance of a racecar, a high-speed camera is positioned at point $A$. The camera is mounted on a mechanism which permits it to record the motion of the car as the car travels on straightway $BC$. Determine (a) the speed of the car in terms of $b$, $\theta$, and $\dot{\theta}$, (b) the magnitude of the acceleration in terms of $b$, $\theta$, $\dot{\theta}$, and $\ddot{\theta}$.

**11.168** After taking off, a helicopter climbs in a straight line at a constant angle $\beta$. Its flight is tracked by radar from point $A$. Determine the speed of the helicopter in terms of $d$, $\beta$, $\theta$, and $\dot{\theta}$.

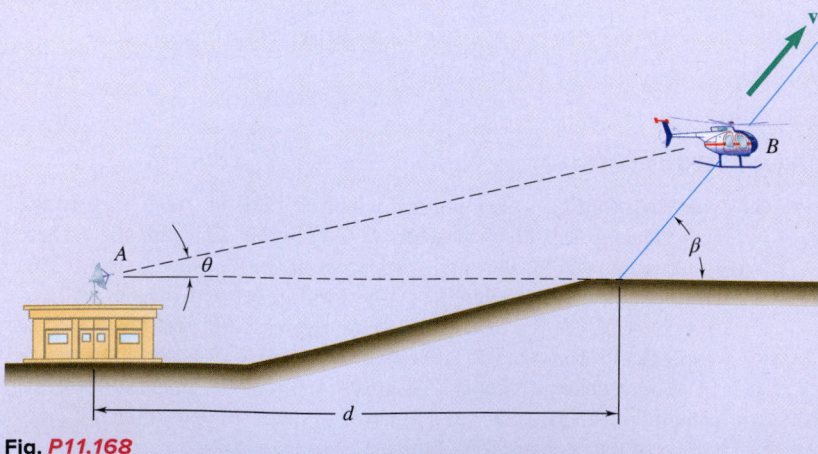

**Fig. P11.168**

**11.169** At the bottom of a loop in the vertical plane, an airplane has a horizontal velocity of 315 mi/h and is speeding up at a rate of 10 ft/s². The radius of curvature of the loop is 1 mi. The plane is being tracked by radar at $O$. What are the recorded values of $\dot{r}$, $\ddot{r}$, $\dot{\theta}$, and $\ddot{\theta}$ for this instant?

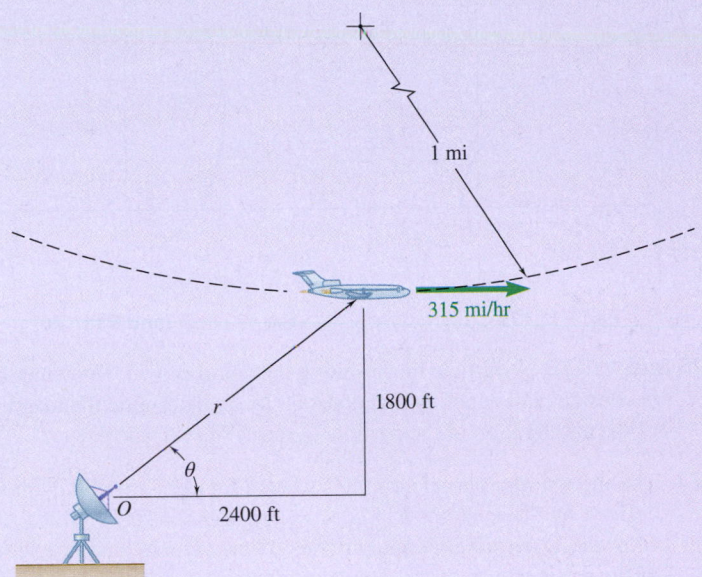

1 mi

315 mi/hr

1800 ft

$\theta$

$O$      2400 ft

**Fig. P11.169**

**11.170** An airplane passes over a radar tracking station at $A$ and continues to fly due east. When the plane is at $P$, the distance and angle of elevation of the plane are, respectively, $r = 12{,}600$ ft and $\theta = 31.2°$. Two seconds later, the radar station sights the plane at $r = 13{,}600$ ft and $\theta = 28.3°$. Determine approximately the speed and the angle of dive $\alpha$ of the plane during the 2-s interval.

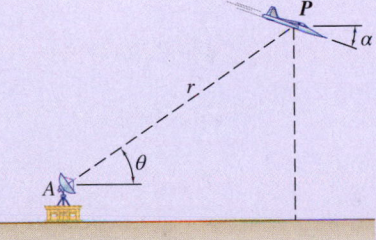

$P$

$\alpha$

$r$

$\theta$

$A$

**Fig. P11.170**

**11.171** For the race car of Prob. 11.167, it was found that it took 0.4 s for the car to travel from the position $\theta = 60°$ to the position $\theta = 35°$. Knowing that $b = 25$ m, determine the average speed of the car during the 0.4-s interval.

**11.172** For the helicopter of Prob. 11.168, it was found that when the helicopter was at $B$, the distance and the angle of elevation of the helicopter were $r = 3000$ ft and $\theta = 20°$, respectively. Four seconds later, the radar station sighted the helicopter at $r = 3320$ ft and $\theta = 23.1°$. Determine the average speed and the angle of climb $\beta$ of the helicopter during the 4-s interval.

**11.173 and 11.174** A particle moves along the spiral shown. Determine the magnitude of the velocity of the particle in terms of $b$, $\theta$, and $\dot{\theta}$.

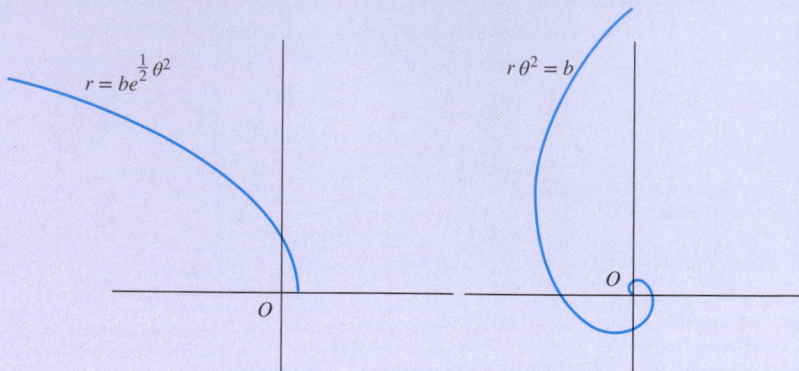

$r = be^{\frac{1}{2}\theta^2}$

$r\theta^2 = b$

Fig. *P11.173* and P11.175          Fig. *P11.174* and P11.176

**11.175 and 11.176** A particle moves along the spiral shown. Knowing that $\dot{\theta}$ is constant and denoting this constant by $\omega$, determine the magnitude of the acceleration of the particle in terms of $b$, $\theta$, and $\dot{\theta}$.

**11.177** The motion of a particle on the surface of a right circular cylinder is defined by the relations $R = A$, $\theta = 2\pi t$, and $z = At^2/4$, where $A$ is a constant. Determine the magnitudes of the velocity and acceleration of the particle at any time $t$.

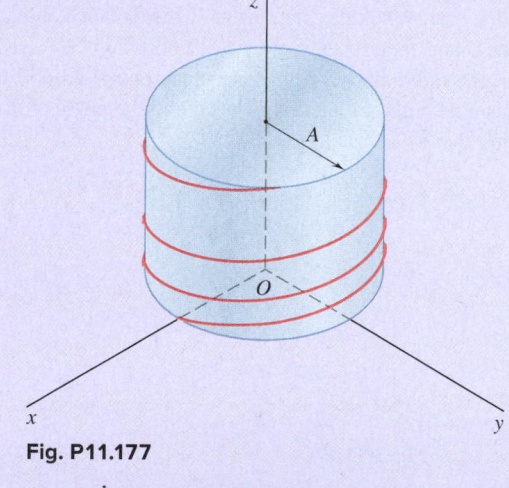

Fig. P11.177

**11.178** Show that $\dot{r} = h\dot{\phi}\sin\theta$, knowing that at the instant shown, step $AB$ of the step exerciser is rotating counterclockwise at a constant rate $\dot{\phi}$.

**11.179** The three-dimensional motion of a particle is defined by the cylindrical coordinates $R = A/(t + 1)$, $\theta = Bt$, and $z = Ct/(t + 1)$. Determine the magnitudes of the velocity and acceleration when (a) $t = 0$, (b) $t = \infty$.

**\*11.180** For the conic helix of Prob. 11.95, determine the angle that the osculating plane forms with the $y$ axis.

**\*11.181** Determine the direction of the binormal of the path described by the particle of Prob. 11.96 when (a) $t = 0$, (b) $t = \pi/2$ s.

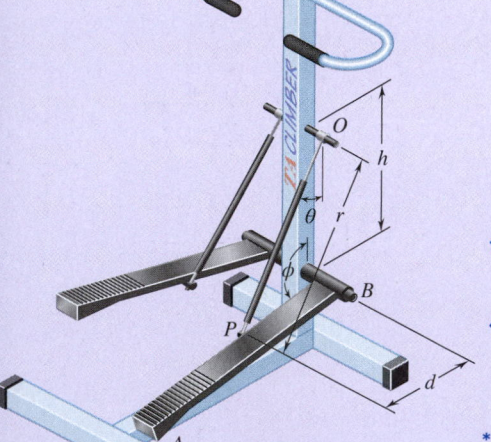

Fig. P11.178

# Review and Summary

## Position Coordinate of a Particle in Rectilinear Motion

In the first half of this chapter, we analyzed the **rectilinear motion of a particle**; that is, the motion of a particle along a straight line. To define the position $P$ of the particle on that line, we chose a fixed origin $O$ and a positive direction (Fig. 11.25). The distance $x$ from $O$ to $P$, with the appropriate sign, completely defines the position of the particle on the line and is called the **position coordinate** of the particle (Sec. 11.1A).

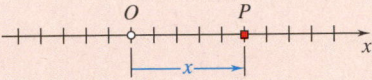

**Fig. 11.25**

## Velocity and Acceleration in Rectilinear Motion

The **velocity** $v$ of the particle was shown to be equal to the time derivative of the position coordinate $x$, so

$$v = \frac{dx}{dt} \tag{11.1}$$

And we obtained the **acceleration** $a$ by differentiating $v$ with respect to $t$, as

$$a = \frac{dv}{dt} \tag{11.2}$$

or

$$a = \frac{d^2x}{dt^2} \tag{11.3}$$

We also noted that $a$ could be expressed as

$$a = v\frac{dv}{dx} \tag{11.4}$$

We observed that the velocity $v$ and the acceleration $a$ are represented by algebraic numbers that can be positive or negative. A positive value for $v$ indicates that the particle moves in the positive direction, and a negative value shows that it moves in the negative direction. A positive value for $a$, however, may mean that the particle is truly accelerated (i.e., moves faster) in the positive direction or that it is decelerated (i.e., moves more slowly) in the negative direction. A negative value for $a$ is subject to a similar interpretation.

## Determination of the Velocity and Acceleration by Integration

In most problems, the conditions of motion of a particle are defined by the type of acceleration that the particle possesses and by the initial conditions (Sec. 11.1B). Then, we can obtain the velocity and position of the particle by integrating two of the equations (11.1), (11.2), (11.3), and (11.4). The selection of these equations depends upon the type of acceleration involved (Sample Probs. 11.2, 11.3, and 11.4).

### Uniform Rectilinear Motion

Two types of motion are frequently encountered. **Uniform rectilinear motion** (Sec. 11.2A), in which the velocity $v$ of the particle is constant, is described by

$$x = x_0 + vt \tag{11.5}$$

### Uniformly Accelerated Rectilinear Motion

**Uniformly accelerated rectilinear motion** (Sec. 11.2B), in which the acceleration $a$ of the particle is constant, is described by

$$v = v_0 + at \tag{11.6}$$

$$x = x_0 + v_0 t + \tfrac{1}{2}at^2 \tag{11.7}$$

$$v^2 = v_0^2 + 2a(x - x_0) \tag{11.8}$$

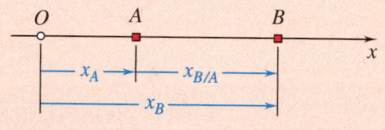

**Fig. 11.26**

### Relative Motion of Two Particles

When two particles $A$ and $B$ (such as two aircraft) move, we may wish to consider the **relative motion** of $B$ with respect to $A$ (Sec. 11.2C). Denoting the **relative position coordinate** of $B$ with respect to $A$ by $x_{B/A}$ (Fig. 11.26), we have

$$x_B = x_A + x_{B/A} \tag{11.9}$$

Differentiating Eq. (11.9) twice with respect to $t$, we obtained successively

$$v_B = v_A + v_{B/A} \tag{11.10}$$

and

$$a_B = a_A + a_{B/A} \tag{11.11}$$

where $v_{B/A}$ and $a_{B/A}$ represent, respectively, the **relative velocity** and the **relative acceleration** of $B$ with respect to $A$.

### Dependent Motion

When several blocks are **connected by inextensible cords**, it is possible to write a linear relation between their position coordinates. We can then write similar relations between their velocities and between their accelerations, which we can use to analyze their motion (Sample Probs. 11.7 and 11.8).

### Graphical Solutions

It is sometimes convenient to use a **graphical solution** for problems involving the rectilinear motion of a particle (Sec. 11.3). The graphical solution most commonly used involves the $x$–$t$, $v$–$t$, and $a$–$t$ curves (Sample Prob. 11.9). It was shown at any given time $t$ that

$$v = \text{slope of } x\text{–}t \text{ curve}$$
$$a = \text{slope of } v\text{–}t \text{ curve}$$

Also, over any given time interval from $t_1$ to $t_2$, we have

$$v_2 - v_1 = \text{area under } a\text{–}t \text{ curve}$$
$$x_2 - x_1 = \text{area under } v\text{–}t \text{ curve}$$

### Position Vector and Velocity in Curvilinear Motion

In the second half of this chapter, we analyzed the **curvilinear motion of a particle**; that is, the motion of a particle along a curved path. We defined the position $P$ of the particle at a given time (Sec. 11.4A) by the **position vector r** joining the $O$ of the coordinates and point $P$ (Fig. 11.27). We defined the **velocity v** of the particle by the relation

$$\mathbf{v} = \frac{d\mathbf{r}}{dt} \qquad (11.14)$$

The velocity is a **vector tangent to the path of the particle** with a magnitude $v$ (called the **speed** of the particle) equal to the time derivative of the length $s$ of the arc described by the particle. Thus,

$$v = \frac{ds}{dt} \qquad (11.15)$$

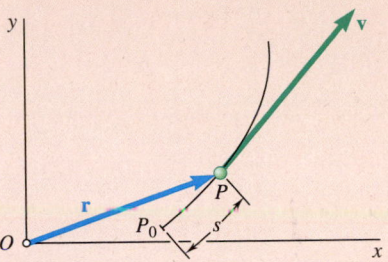

**Fig. 11.27**

## Acceleration in Curvilinear Motion

We defined the **acceleration a** of the particle by the relation

$$\mathbf{a} = \frac{d\mathbf{v}}{dt} \qquad (11.17)$$

and we noted that, in general, *the acceleration is not tangent to the path of the particle.*

## Derivative of a Vector Function

Before proceeding to the consideration of the components of velocity and acceleration, we reviewed the formal definition of the derivative of a vector function and established a few rules governing the differentiation of sums and products of vector functions. We then showed that the rate of change of a vector is the same with respect both to a fixed frame and to a frame in translation (Sec. 11.4B).

## Rectangular Components of Velocity and Acceleration

Denoting the rectangular coordinates of a particle $P$ by $x$, $y$, and $z$, we found that the rectangular components of the velocity and acceleration of $P$ equal, respectively, the first and second derivatives with respect to $t$ of the corresponding coordinates. Thus,

$$v_x = \dot{x} \qquad v_y = \dot{y} \qquad v_z = \dot{z} \qquad (11.28)$$
$$a_x = \ddot{x} \qquad a_y = \ddot{y} \qquad a_z = \ddot{z} \qquad (11.29)$$

## Component Motions

When the component $a_x$ of the acceleration depends only upon $t$, $x$, and/or $v_x$; when, similarly, $a_y$ depends only upon $t$, $y$, and/or $v_y$; and $a_z$ upon $t$, $z$, and/or $v_z$, Eq. (11.29) can be integrated independently. The analysis of the given curvilinear motion then reduces to the analysis of three independent rectilinear component motions (Sec. 11.4C). This approach is particularly effective in the study of the motion of projectiles (Sample Probs. 11.10, 11.11, and 11.12).

## Relative Motion of Two Particles

For two particles $A$ and $B$ moving in space (Fig. 11.28), we considered the relative motion of $B$ with respect to $A$, or more precisely, with respect to a moving frame attached to $A$ and in translation with $A$ (Sec. 11.4D). Denoting the **relative position vector** of $B$ with respect to $A$ by $\mathbf{r}_{B/A}$ (Fig. 11.28), we have

$$\mathbf{r}_B = \mathbf{r}_A + \mathbf{r}_{B/A} \qquad (11.30)$$

Denoting the **relative velocity** and the **relative acceleration** of $B$ with respect to $A$ by $\mathbf{v}_{B/A}$ and $\mathbf{a}_{B/A}$, respectively, we also showed that

$$\mathbf{v}_B = \mathbf{v}_A + \mathbf{v}_{B/A} \qquad (11.32)$$

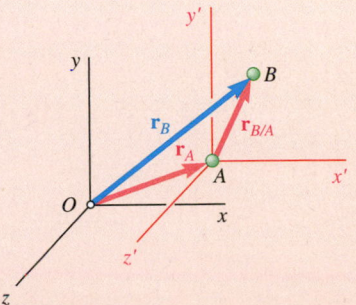

**Fig. 11.28**

and

$$\mathbf{a}_B = \mathbf{a}_A + \mathbf{a}_{B/A} \qquad (11.33)$$

## Tangential and Normal Components

It is sometimes convenient to resolve the velocity and acceleration of a particle $P$ into components other than the rectangular $x$, $y$, and $z$ components. For a particle $P$ moving along a path contained in a plane, we attached to $P$ unit vectors $\mathbf{e}_t$ tangent to the path and $\mathbf{e}_n$ normal to the path and directed toward the center of curvature of the path (Sec. 11.5A). We then express the velocity and acceleration of the particle in terms of tangential and normal components. We have

$$\mathbf{v} = v\mathbf{e}_t \qquad (11.35)$$

and

$$\mathbf{a} = \frac{dv}{dt}\mathbf{e}_t + \frac{v^2}{\rho}\mathbf{e}_n \qquad (11.38)$$

where $v$ is the speed of the particle and $\rho$ is the radius of curvature of its path (Sample Probs. 11.16 and 11.17). We observed that, while the velocity $\mathbf{v}$ is directed along the tangent to the path, the acceleration $\mathbf{a}$ consists of a component $\mathbf{a}_t$ directed along the tangent to the path and a component $\mathbf{a}_n$ directed toward the center of curvature of the path (Fig. 11.29).

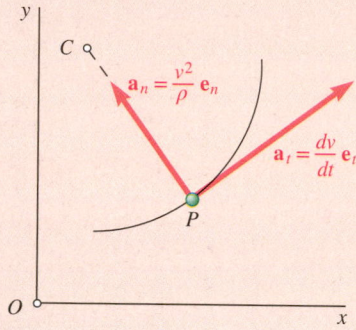

**Fig. 11.29**

## Motion Along a Space Curve

For a particle $P$ moving along a space curve, we defined the plane that most closely fits the curve in the neighborhood of $P$ as the **osculating plane**. This plane contains the unit vectors $\mathbf{e}_t$ and $\mathbf{e}_n$ that define the tangent and principal normal to the curve, respectively. The unit vector $\mathbf{e}_b$, which is perpendicular to the osculating plane, defines the **binormal**.

## Radial and Transverse Components

When the position of a particle $P$ moving in a plane is defined by its polar coordinates $r$ and $\theta$, it is convenient to use radial and transverse components directed, respectively, along the position vector $\mathbf{r}$ of the particle and in the direction obtained by rotating $\mathbf{r}$ through 90° counterclockwise (Sec. 11.5B). We attached to $P$ unit vectors $\mathbf{e}_r$ and $\mathbf{e}_\theta$ directed in the radial and transverse directions, respectively (Fig. 11.30). We then expressed the velocity and acceleration of the particle in terms of radial and transverse components as

$$\mathbf{v} = \dot{r}\mathbf{e}_r + r\dot{\theta}\mathbf{e}_\theta \qquad (11.42)$$

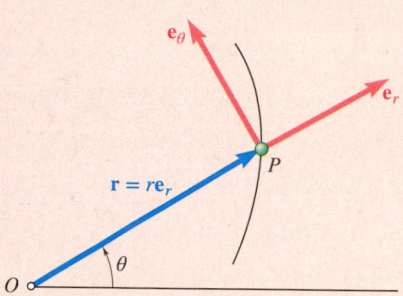

**Fig. 11.30**

and

$$\mathbf{a} = (\ddot{r} - r\dot{\theta}^2)\mathbf{e}_r + (r\ddot{\theta} + 2\dot{r}\dot{\theta})\mathbf{e}_\theta \qquad (11.43)$$

where dots are used to indicate differentiation with respect to time. The scalar components of the velocity and acceleration in the radial and transverse directions are therefore

$$v_r = \dot{r} \qquad v_\theta = r\dot{\theta} \qquad (11.44)$$

and

$$a_r = \ddot{r} - r\dot{\theta}^2 \qquad a_\theta = r\ddot{\theta} + 2\dot{r}\dot{\theta} \qquad (11.45)$$

It is important to note that $a_r$ is *not* equal to the time derivative of $v_r$ and that $a_\theta$ is *not* equal to the time derivative of $v_\theta$ (Sample Probs. 11.18, 11.19, and 11.20).

This chapter ended with a discussion of the use of cylindrical coordinates to define the position and motion of a particle in space.

# Review Problems

**11.182** Students are testing their new drone to see if it can safely deliver packages to different departments on campus. Position data can be approximated using the expressions $x(t) = -0.0000225t^4 + 0.003t^3 + 0.01t^2$ and $y(t) = 300\left[1 - \cos\left(\dfrac{\pi}{40}t\right)\right]$, where $x$ and $y$ are expressed in meters and $t$ is expressed in seconds. Knowing that the take-off and landing altitudes are the same, plot the path of the drone and determine (*a*) the duration of the flight, (*b*) its maximum speed in the $x$ direction, (*c*) its maximum altitude and the horizontal distance traveled during the flight.

**11.183** A drag racing car starts from rest and moves down the racetrack with an acceleration defined by $a = 50 - 10t$, where $a$ and $t$ are in m/s² and seconds, respectively. After reaching a speed of 125 m/s, a parachute is deployed to help slow down the dragster. Knowing that this deceleration is defined by the relationship $a = -0.02v^2$, where $v$ is the velocity in m/s, determine (*a*) the total time from the beginning of the race until the car slows back down to 10 m/s, (*b*) the total distance the car travels during this time.

**11.184** A driver is traveling at a speed of 72 km/h in car $A$ when he looks down to text a friend that he is running late. Just before he looks down, he is 80 m from an intersection, and a bicyclist $B$, traveling at a constant speed of 8 m/s, is 32 m from that same intersection. The light turns red during the 3 s text, and when the driver looks up, he hits the brakes, causing a constant deceleration of 5 m/s². Determine (*a*) the distance between the car and the bike when the bike reaches the center of the intersection, (*b*) the velocity that the car appears to have to the cyclist at this time, (*c*) where the car stops.

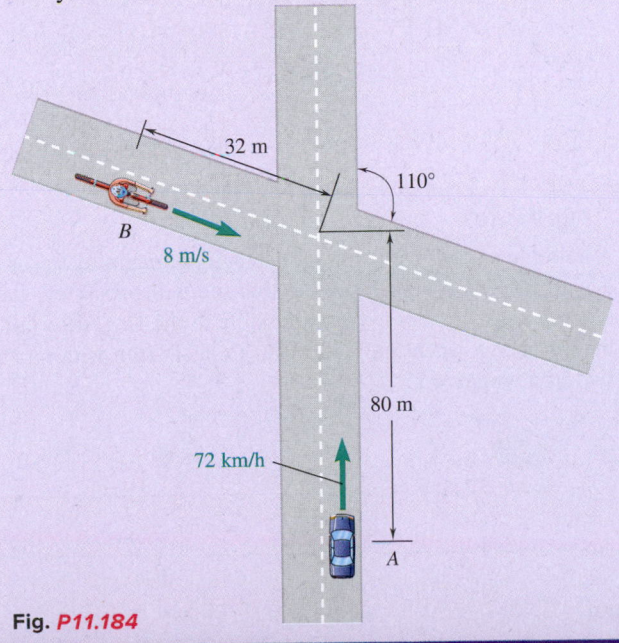

Fig. *P11.184*

**11.185** The velocities of commuter trains $A$ and $B$ are as shown. Knowing that the speed of each train is constant and that $B$ reaches the crossing 10 min after $A$ passed through the same crossing, determine ($a$) the relative velocity of $B$ with respect to $A$, ($b$) the distance between the fronts of the engines 3 min after $A$ passed through the crossing.

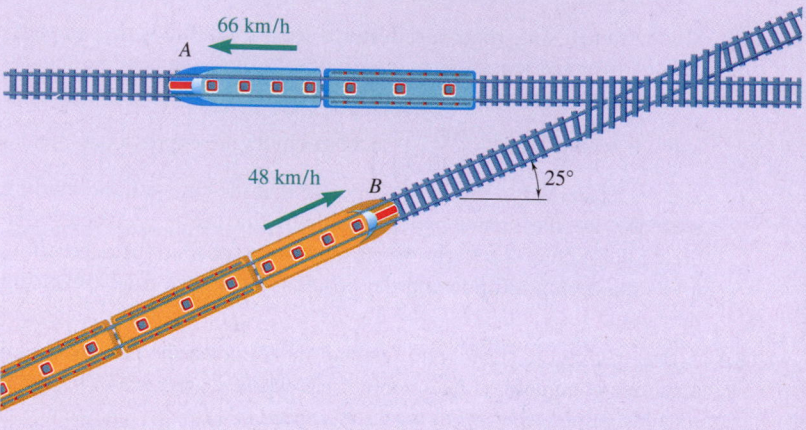

Fig. P11.185

**11.186** Knowing that slider block $A$ starts from rest and moves to the left with a constant acceleration of 1 ft/s$^2$, determine ($a$) the relative acceleration of block $A$ with respect to block $B$, ($b$) the velocity of block $B$ after 2 s.

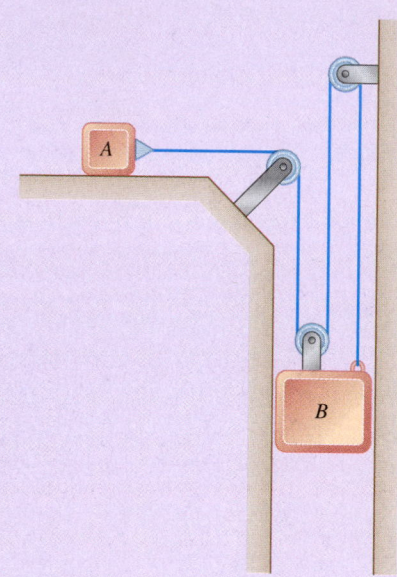

Fig. P11.186

**11.187** A roller-coaster car is traveling at a speed of 20 m/s when it passes through point $B$. At that point, it enters a concave down circular section of the track that has a radius of curvature of 60 m. After it reaches point $B$, the car's speed increases at a rate of 4 m/s$^2$. Determine ($a$) the time it takes the car to reach point $D$, ($b$) the $x$ and $y$ components of the overall acceleration of the car when it reaches point D.

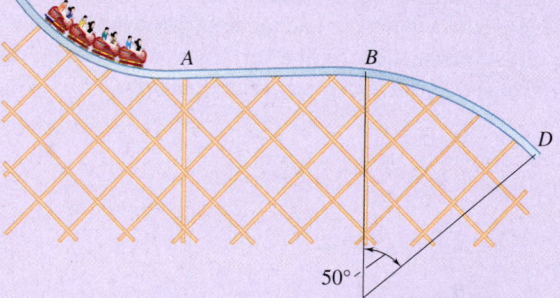

Fig. P11.187

**11.188** A golfer hits a ball with an initial velocity of magnitude $v_0$ at an angle $\alpha$ with the horizontal. Knowing that the ball must clear the tops of two trees and land as close as possible to the flag, determine $v_0$ and the distance $d$ when the golfer uses ($a$) a six-iron with $\alpha = 31°$, ($b$) a five-iron with $\alpha = 27°$.

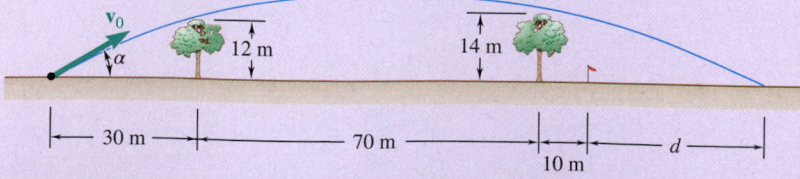

Fig. P11.188

**11.189** As the truck shown begins to back up with a constant acceleration of 4 ft/s², the outer section $B$ of its boom starts to retract with a constant acceleration of 1.6 ft/s² relative to the truck. Determine (*a*) the acceleration of section $B$, (*b*) the velocity of section $B$ when $t = 2$ s.

**Fig. P11.189**

**11.190** A velodrome is a specially designed track used in bicycle racing that has constant radius curves at each end. Knowing that a rider starts from rest and that $a_t = (11.46 - 0.01878v^2)$ m/s², determine her acceleration at point $B$.

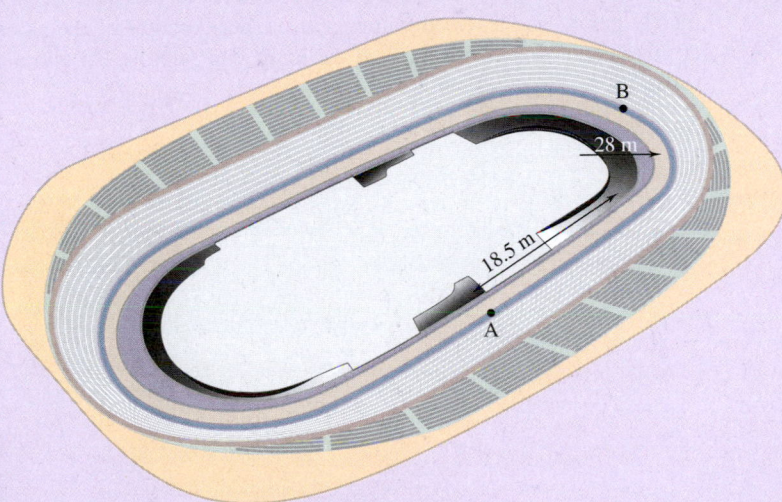

**Fig. P11.190**

**11.191** Sand is discharged at $A$ from a conveyor belt and falls onto the top of a stockpile at $B$. Knowing that the conveyor belt forms an angle $\alpha = 25°$ with the horizontal, determine (*a*) the speed $v_0$ of the belt, (*b*) the radius of curvature of the trajectory described by the sand at point $B$.

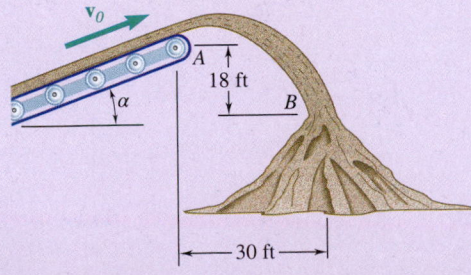

**Fig. P11.191**

**11.192** The end point $B$ of a boom is originally 5 m from fixed point $A$ when the driver starts to retract the boom with a constant radial acceleration of $\ddot{r} = -1.0$ m/s² and lower it with a constant angular acceleration $\ddot{\theta} = -0.5$ rad/s². At $t = 2$ s, determine (a) the velocity of point $B$, (b) the acceleration of point $B$, (c) the radius of curvature of the path.

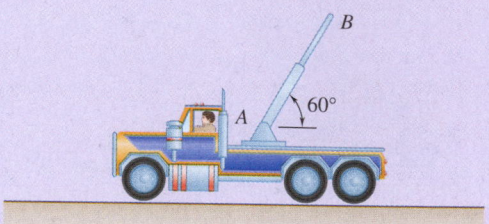

**Fig. P11.192**

**11.193** A telemetry system is used to quantify kinematic values of a ski jumper immediately before she leaves the ramp. According to the system $r = 500$ ft, $\dot{r} = -105$ ft/s, $\ddot{r} = -10$ ft/s², $\theta = 25°$, $\dot{\theta} = 0.07$ rad/s, and $\ddot{\theta} = 0.06$ rad/s². Determine (a) the velocity of the skier immediately before she leaves the jump, (b) the acceleration of the skier at this instant, (c) the distance of the jump $d$ neglecting lift and air resistance.

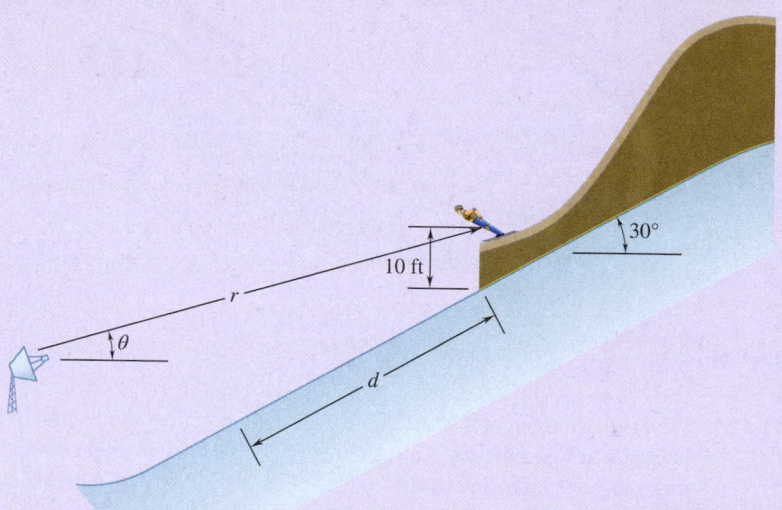

**Fig. P11.193**

# 12

# Kinetics of Particles: Newton's Second Law

The forces experienced by the passengers on a roller coaster will depend on whether the roller-coaster car is traveling up a hill or down a hill, in a straight line, or along a horizontal or vertical curved path. The relation existing among force, mass, and acceleration will be studied in this chapter.

# Objectives

- **Explain** the relationships between mass, force, and acceleration.
- **Model** physical systems by drawing complete free-body diagrams and kinetic diagrams.
- **Apply** Newton's second law of motion to solve particle kinetics problems using different coordinate systems.
- **Analyze** central force motion problems using principles of angular momentum and Newton's law of gravitation.

## Introduction

In statics, we used Newton's first and third laws of motion extensively to study bodies at rest and the forces acting upon them. We also use these two laws in dynamics; in fact, they are sufficient for analyzing the motion of bodies that have no acceleration. However, when a body is accelerated—that is, when the magnitude or the direction of its velocity changes—it is necessary to use Newton's second law of motion to relate the motion of the body to the forces acting on it.

In this chapter, we discuss Newton's second law and apply it to analyzing the motion of particles. According to the second law, if the resultant of the forces acting on a particle is not zero, the particle has an acceleration proportional to the magnitude of the resultant and in the direction of this resultant force. Moreover, we use the ratio of the magnitudes of the resultant force and of the acceleration to define the *mass* of the particle. In Sec. 12.1B, we define the *linear momentum* of a particle as the product $\mathbf{L} = m\mathbf{v}$ of the mass $m$ and velocity $\mathbf{v}$ of the particle. Then, we can express Newton's second law in an alternative form, relating the rate of change of the linear momentum to the resultant of the forces acting on that particle.

In the Sample Problems, we apply Newton's second law to the solution of engineering problems using either rectangular components, tangential and normal components, or radial and transverse coordinates of the forces and accelerations involved. Recall that we can consider an actual body—including bodies as large as a car, rocket, or airplane—as a particle for the purpose of analyzing its motion, as long as the effect of a rotation of the body about its center of mass can be ignored. We stress the need for consistent units in solving these problems, briefly reviewing the International System of Units (SI units) and the system of U.S. customary units.

The second part of this chapter is devoted to the solution of the motion of a particle under a central force. We define the *angular momentum* $\mathbf{H}_O$ of a particle about a point $O$ as the moment about $O$ of the linear momentum of the particle: $\mathbf{H}_O = \mathbf{r} \times m\mathbf{v}$. It then follows from Newton's second law that the rate of change of the angular momentum $\mathbf{H}_O$ of a particle is equal to the sum of the moments about $O$ of the forces acting on that particle.

We can use this form of the second law to deal with the motion of a particle under a *central force*; that is, under a force directed toward or away from a fixed point $O$. Because such a force has zero moment about $O$, it follows that the angular momentum of the particle about $O$ is conserved. This property greatly simplifies the analysis of the motion, as we show by solving problems involving the orbital motion of bodies under gravitational attraction.

In Sec. 12.3, which is optional, we present a more extensive discussion of orbital motion, including several problems related to space mechanics.

# 12.1  NEWTON'S SECOND LAW AND LINEAR MOMENTUM

In statics, we dealt with forces acting on particles that led to a state of equilibrium. Now we study forces acting on particles that lead to a state of motion. The key relationship connecting force and motion is Newton's second law.

## 12.1A  Newton's Second Law of Motion

We can state Newton's second law as follows:

**If the resultant force acting on a particle is not zero, the particle has an acceleration proportional to the magnitude of the resultant and in the direction of this resultant force.**

Newton's second law of motion is best understood by imagining the following experiment: A particle is subjected to a force $F_1$ of constant direction and constant magnitude $F_1$. Under the action of that force, the particle moves in a straight line and *in the direction of the force* (Fig. 12.1a). By determining the position of the particle at various instants, we find that its acceleration has a constant magnitude $a_1$. If we repeat the experiment with forces $F_2$, $F_3$, . . . of a different magnitude or direction (Fig. 12.1b and c), we find each time that the particle moves in the direction of the force acting on it and that the magnitudes $a_1$, $a_2$, $a_3$, . . . of the accelerations are proportional to the magnitudes $F_1$, $F_2$, $F_3$, . . . of the corresponding forces. Thus,

$$\frac{F_1}{a_1} = \frac{F_2}{a_2} = \frac{F_3}{a_3} = \cdots = \text{constant}$$

The constant value obtained for the ratio of the magnitudes of the forces and accelerations is a characteristic of the particle under consideration; it is called the **mass** of the particle and is denoted by $m$. When a particle of mass $m$ is acted upon by a force $F$, the force $F$ and the acceleration $a$ of the particle must therefore satisfy the relation

**Newton's second law**           $$F = ma \qquad (12.1)$$

This relation provides a complete formulation of Newton's second law; it states not only that the magnitudes of $F$ and $a$ are proportional, but also (because $m$ is a positive scalar) that the vectors $F$ and $a$ have the same direction (Fig. 12.2). Note that Eq. (12.1) still holds when $F$ is not constant, but varies with time in magnitude or direction. The magnitudes of $F$ and $a$ remain proportional, and the two vectors have the same direction at any given instant. However, they are not, in general, tangent to the path of the particle.

When a particle is subjected simultaneously to several forces, Eq. (12.1) should be replaced by

**Newton's second law, multiple forces:**

$$\Sigma F = ma \qquad (12.2)$$

where $\Sigma F$ represents the sum or resultant of all the forces acting on the particle.

Note that the system of axes with respect to which we determine the acceleration $a$ is not arbitrary. These axes must have a constant orientation with respect to the stars, and their origin either must be attached to the sun

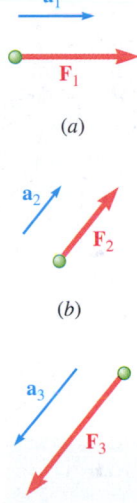

**Fig. 12.1** Experiments show that a force applied to a particle gives the particle an acceleration proportional to the magnitude of the force and in the same direction as the force.

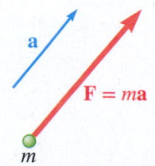

**Fig. 12.2** By Newton's second law, the proportionality constant between an applied force and the resulting acceleration is the particle's mass $m$.

(more accurately, to the center of mass of the solar system) or move with a constant velocity with respect to the sun. Such a system of axes is called a **newtonian frame of reference**.[†] A system of axes attached to the earth does *not* constitute a newtonian frame of reference, because the earth rotates with respect to the stars and is accelerated with respect to the sun. However, in most engineering applications, we can determine the acceleration **a** with respect to axes attached to the earth and use Eqs. (12.1) and (12.2) without any appreciable error. However, these equations do not hold if **a** represents a relative acceleration measured with respect to moving axes, such as axes attached to an accelerated car or to a rotating piece of machinery.

If the resultant $\Sigma\mathbf{F}$ of the forces acting on the particle is zero, it follows from Eq. (12.2) that the acceleration **a** of the particle is also zero. If the particle is initially at rest ($\mathbf{v}_0 = 0$) with respect to the newtonian frame of reference used, it will thus remain at rest ($\mathbf{v} = 0$). If originally moving with a velocity $\mathbf{v}_0$, the particle will maintain a constant velocity $\mathbf{v} = \mathbf{v}_0$; that is, it will move with the constant speed $v_0$ in a straight line. This, we recall, is the statement of Newton's first law (Sec. 2.3B); thus, Newton's first law is a particular case of Newton's second law.

## 12.1B Linear Momentum of a Particle and Its Rate of Change

Suppose we replace the acceleration **a** in Eq. (12.2) by the derivative $d\mathbf{v}/dt$. We have

$$\Sigma\mathbf{F} = m\frac{d\mathbf{v}}{dt}$$

Because the mass $m$ of the particle is constant, we can write this as

$$\Sigma\mathbf{F} = \frac{d}{dt}(m\mathbf{v}) \tag{12.3}$$

The product $m\mathbf{v}$ is called the **linear momentum**, or simply the **momentum**, of the particle. It has the same direction as the velocity of the particle, and its magnitude is equal to the product of the mass $m$ and the speed $v$ of the particle (Fig. 12.3). Eq. (12.3) says:

> **The resultant of the forces acting on the particle is equal to the rate of change of the linear momentum of the particle.**

The second law of motion was originally stated by Newton in this form. Denoting the linear momentum of the particle by **L**, we have

**Linear momentum** $\qquad\qquad \boxed{\mathbf{L} = m\mathbf{v}} \qquad\qquad$ **(12.4)**

If we denote its derivative with respect to $t$ as $\dot{\mathbf{L}}$, we can write Eq. (12.3) in the alternative form as

**Newton's second law, momentum form**

$$\boxed{\Sigma\mathbf{F} = \dot{\mathbf{L}}} \tag{12.5}$$

We assumed that the mass $m$ of the particle is constant in Eqs. (12.3), (12.4), and (12.5). Therefore, you should not use Eq. (12.3) or (12.5) to solve

**Photo 12.1** When the racecar accelerates forward, the rear tires have a friction force acting on them in the direction the car is moving. ©Glow Images RF

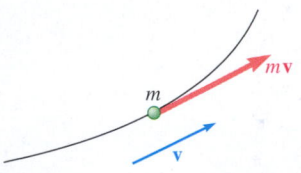

**Fig. 12.3** Linear momentum is the product of the mass $m$ and the velocity **v** of a particle. It is a vector in the same direction as the velocity.

[†]Stars are not actually fixed, so a more rigorous definition of a newtonian frame of reference (also called an *inertial system*) is one with respect to which Eq. (12.2) holds.

problems involving the motion of bodies, such as rockets, that gain or lose mass. We will consider problems of that type in Sec. 14.3B.[†]

It follows from Eq. (12.3) that the rate of change of the linear momentum $m\mathbf{v}$ is zero when $\Sigma\mathbf{F} = 0$. Thus, we have the statement:

**If the resultant force acting on a particle is zero, the linear momentum of the particle remains constant in both magnitude and direction.**

This is the principle of **conservation of linear momentum** for a particle.

## 12.1C  Systems of Units

In using the fundamental equation $\mathbf{F} = m\mathbf{a}$, the units of force, mass, length, and time cannot be chosen arbitrarily. If they are, the magnitude of the force $\mathbf{F}$ required to give an acceleration $\mathbf{a}$ to the mass $m$ will *not* be numerically equal to the product $ma$; it will only be proportional to this product. Thus, we can choose three of the four units arbitrarily, but we must choose the fourth unit so that the equation $\mathbf{F} = m\mathbf{a}$ is satisfied. The units are then said to form a system of consistent kinetic units.

Two systems of consistent kinetic units are currently used by American engineers: the International System of Units (SI units[‡]) and the system of U.S. customary units. Both systems were discussed in detail in Sec. 1.3, so we describe them only briefly in this section.

**International System of Units (SI Units).**  In this system, the base units are the units of length, mass, and time and are called, respectively, the *meter* (m), the *kilogram* (kg), and the *second* (s). All three are arbitrarily defined (Sec. 1.3). The unit of force is a derived unit. It is called the *newton* (N) and is defined as the force that gives an acceleration of $1 \text{ m/s}^2$ to a mass of 1 kg (Fig. 12.4). From Eq. (12.1), we have

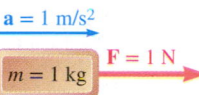

$$1 \text{ N} = (1 \text{ kg})(1 \text{ m/s}^2) = 1 \text{ kg·m/s}^2$$

**Fig. 12.4**  A force of 1 newton gives a 1-kilogram mass an acceleration of $1 \text{ m/s}^2$.

The SI units are said to form an *absolute* system of units. This means that the three base units chosen are independent of the location where measurements are made. The meter, the kilogram, and the second may be used anywhere on the earth; they may even be used on another planet. They always have the same meaning.

The *weight* **W** of a body, or the *force of gravity* exerted on that body, should, like any other force, be expressed in newtons. A body subjected only to its own weight acquires an acceleration equal to the acceleration due to gravity $g$. (Be careful using the term *acceleration due to gravity*, because the only time an object accelerates with a magnitude $g$ is during free-fall in the absence of drag.) It follows from Newton's second law that the magnitude $W$ of the weight of a body of mass $m$ is

$$W = mg \qquad\qquad (12.6)$$

Recall that $g = 9.81 \text{ m/s}^2$, so the weight of a body of mass 1 kg (Fig. 12.5) is

$$W = (1 \text{ kg})(9.81 \text{ m/s}^2) = 9.81 \text{ N}$$

This value would be much less on the moon, where the acceleration due to gravity is $1.6249 \text{ m/s}^2$.

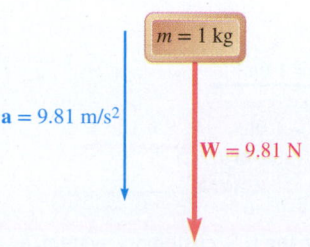

**Fig. 12.5**  In the SI system, a block with mass 1 kg has a weight of 9.81 N.

[†]Note that Eqs. (12.3) and (12.5) do hold in relativistic mechanics, where the mass $m$ of the particle is assumed to vary with the speed of the particle.

[‡]SI stands for *Système International d'Unités* (French).

Multiples and submultiples of the units of length, mass, and force are frequently used in engineering practice. They are, respectively, the *kilometer* (km) and the *millimeter* (mm); the *megagram* (Mg, which is also called the metric ton) and the *gram* (g); and the *kilonewton* (kN). By definition,

$$\begin{array}{ll} 1 \text{ km} = 1000 \text{ m} & 1 \text{ mm} = 0.001 \text{ m} \\ 1 \text{ Mg} = 1000 \text{ kg} & 1 \text{ g} = 0.001 \text{ kg} \\ & 1 \text{ kN} = 1000 \text{ N} \end{array}$$

You can convert these units to meters, kilograms, and newtons, respectively, simply by moving the decimal point three places to the right or to the left.

Units other than those of mass, length, and time all can be expressed in terms of these three base units. For example, we can obtain the unit of linear momentum by recalling the definition and writing

$$mv = (\text{kg})(\text{m/s}) = \text{kg} \cdot \text{m/s}$$

**U.S. Customary Units.**   Most practicing American engineers still commonly use a system in which the base units are those of length, force, and time. These units are, respectively, the *foot* (ft), the *pound* (lb), and the *second* (s). The second is the same as the corresponding SI unit. The foot is equal to 0.3048 m. The pound is defined as the *weight* of a platinum standard, called the *standard pound,* which is kept at the National Institute of Standards and Technology outside Washington, DC. The mass of this standard is 0.453 592 43 kg. Because the weight of a body depends upon the gravitational attraction of the earth, which varies with location, the standard pound should be placed at sea level and at a latitude of 45° to properly define a force of 1 lb. Clearly, the U.S. customary units do not form an absolute system of units. Because of their dependence upon the earth's gravitational attraction, they are said to form a *gravitational* system of units.

Although the standard pound also serves as the unit of mass in commercial transactions in the United States, it cannot be used that way in engineering computations because such a unit would not be consistent with the base units defined in this system. Indeed, when acted upon by a force of 1 lb—that is, when subjected to its own weight—the standard pound receives the acceleration of gravity, $g = 32.2 \text{ ft/s}^2$ (Fig. 12.6), and not the unit acceleration required by Eq. (12.1). The unit of mass consistent with the foot, the pound, and the second is the mass that receives an acceleration of $1 \text{ ft/s}^2$ when a force of 1 lb is applied to it (Fig. 12.7). This unit, sometimes called a *slug,* can be derived from the equation $F = ma$ after substituting 1 lb and $1 \text{ ft/s}^2$ for $F$ and $a$, respectively. We have

$$F = ma \qquad 1 \text{ lb} = (1 \text{ slug})(1 \text{ ft/s}^2)$$

From this, we obtain

$$1 \text{ slug} = \frac{1 \text{ lb}}{1 \text{ ft/s}^2} = 1 \text{ lb} \cdot \text{s}^2/\text{ft}$$

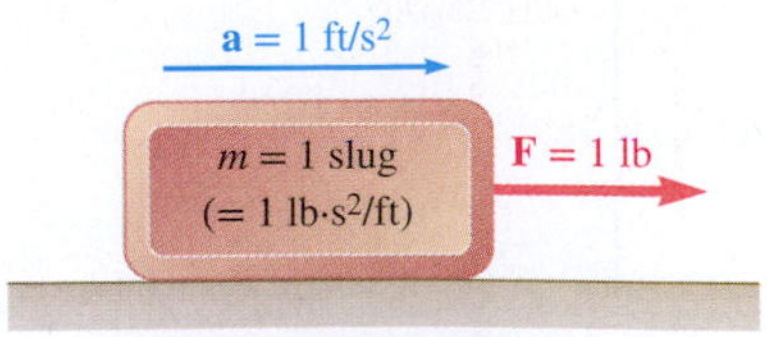

Comparing Figs. 12.6 and 12.7, we conclude that the slug is a mass 32.2 times larger than the mass of the standard pound. (On a horizontal surface, when acted on by a force of 1 pound, the motion of the larger mass is relatively "sluggish.")

The fact that bodies are characterized in the U.S. customary system of units by their weight in pounds rather than by their mass in slugs was convenient in the study of statics, where we were dealing (for the most part) with weights and other forces and only seldom with masses. However, in the study of kinetics, which involves forces, masses, and accelerations, we will often have

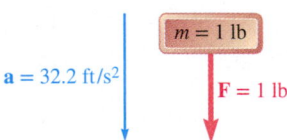

$\mathbf{a} = 32.2 \text{ ft/s}^2$

$m = 1$ lb

$\mathbf{F} = 1$ lb

**Fig. 12.6**   In the U.S. customary system, a block with a weight of 1 lb in free fall has an acceleration of 32.2 ft/s².

$\mathbf{a} = 1 \text{ ft/s}^2$

$m = 1$ slug
$(= 1 \text{ lb} \cdot \text{s}^2/\text{ft})$

$\mathbf{F} = 1$ lb

**Fig. 12.7**   In the U.S. customary system, a force of 1 lb applied to a block with a mass of 1 slug produces an acceleration of 1 ft/s².

to express the mass $m$ of a body in slugs, the weight $W$ of which is given in pounds. Recalling Eq. (12.6), we have

$$m = \frac{W}{g} \qquad \textbf{(12.7)}$$

where $g$ is the acceleration due to gravity ($g = 32.2$ ft/s$^2$).

Units other than the units of force, length, and time all can be expressed in terms of these three base units. For example, we can obtain the unit of linear momentum from its definition as

$$mv = (\text{slug})(\text{ft/s}) = (\text{lb·s}^2/\text{ft})(\text{ft/s}) = \text{lb·s}$$

**Conversion from One System of Units to Another.** The conversion from U.S. customary units to SI units, and vice versa, was discussed in Sec. 1.4. Recall that the conversion factors obtained for the units of length, force, and mass are, respectively,

Length:                 1 ft = 0.3048 m
Force:                  1 lb = 4.448 N
Mass:                1 slug = 1 lb·s$^2$/ft = 14.59 kg

Thermodynamicists often use a unit called the pound-mass (lbm); this is not a unit that is consistent with Newton's second law, and whenever we use pounds in dynamics, it will refer to pounds-force (lbf).

Although it cannot be used as a consistent unit of mass, the mass of the standard pound is, by definition,

$$1 \text{ pound-mass} = 0.4536 \text{ kg}$$

This constant can be used to determine the *mass* in SI units (kilograms) of a body that has been characterized by its *weight* in U.S. customary units (pounds).

## 12.1D    Equations of Motion

Consider a particle of mass $m$ acted upon by several forces. Recall that we can express Newton's second law by the equation

$$\Sigma \mathbf{F} = m\mathbf{a} \qquad \textbf{(12.2)}$$

which relates the forces acting on the particle to the vector $m\mathbf{a}$ (Fig. 12.8).[†] Two of the most important tools you will use in solving dynamics problems, particularly those involving Newton's second law, are the free-body diagram (FBD) and the kinetic diagram (KD). These diagrams will help you to model dynamic systems and apply appropriate equations of motion. The free-body diagram shown on the left side of Fig. 12.9 is no different from what you did in statics in Chapter 4 and consists of the following steps:

**Body:** Define your system by isolating the body (or bodies) of interest. If a problem has multiple bodies (such as in Sample Probs. 12.3, 12.4, and 12.5), you may have to draw multiple free-body diagrams and kinetic diagrams.

**Fig. 12.8** The sum of forces applied to a particle of mass $m$ produces a vector $m\mathbf{a}$ in the direction of the resultant force.

[†]In the 1700s, Jean-Baptiste le Rond d'Alembert expressed Newton's second law as $\Sigma \mathbf{F} - m\mathbf{a} = 0$ so he could solve dynamics problems using the principles of statics. The $-m\mathbf{a}$ term has been called a fictitious *inertial force,* but it is important for you to realize that there is no such thing as inertial forces (or centrifugal forces that "push" you outward when going around a curve). D'Alembert's principle (also called dynamic equilibrium) is seldom used in modern engineering.

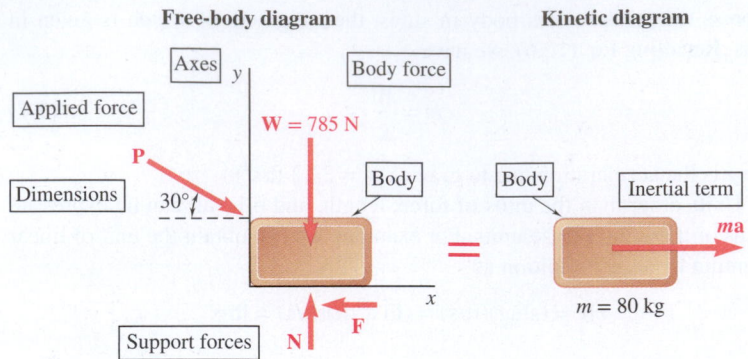

**Fig. 12.9** Steps in drawing a free-body diagram and a kinetic diagram for solving dynamics problems.

**Axes:** Draw an appropriate coordinate system (e.g., Cartesian, normal and tangential, or radial and transverse).

**Support Forces:** Replace supports or constraints with appropriate forces (e.g., two perpendicular forces for a pin, normal forces, friction forces).

**Applied Forces and Body Forces:** Draw any applied forces and body forces (also sometimes called field forces) on your diagram (e.g., weight, magnetic forces, a known pulling force).

**Dimensions:** Add any angles or distances that are important for solving the problem.

In statics problems, we deal with bodies in equilibrium, and the inertial term in Newton's second law is zero. For dynamics problems, this is not the case. We utilize the kinetic diagram to visualize this term.

**Body:** This is the same body as in the free-body diagram; place this beside the free-body diagram.

**Inertial Terms:** Draw the $m\mathbf{a}$ term to be consistent with the coordinate system. Generally, draw this term in different components (e.g., $ma_x$ and $ma_y$ or $ma_n$ and $ma_t$). If they are unknown quantities, it is best to draw them in the positive directions as defined by your coordinates.

Drawing these two diagrams clarifies how to develop your equations of motion. The free-body diagram is a visual representation of the $\Sigma\mathbf{F}$ term, and the kinetic diagram is a visual representation of the $m\mathbf{a}$ term. Because Newton's second law is a vector equation, you can use the free-body diagram and kinetic diagram to write $\Sigma\mathbf{F} = m\mathbf{a}$ directly in component form. Examples of using these diagrams to help you write your equations of motion are shown in the Sample Problems, and you can get extra practice by solving the Free-Body Problems 12.F1 through 12.F12.

As mentioned, it is usually more convenient to replace Eq. (12.2) with equivalent equations involving scalar quantities. As we saw in Chap. 11, we can resolve these vectors into components using several different coordinate systems (e.g., Cartesian, tangential and normal, or radial and transverse), depending on the type of problem we are solving.

**Rectangular Components.**   Resolving each force $\mathbf{F}$ and the acceleration $\mathbf{a}$ into rectangular components, we have

$$\Sigma(F_x\mathbf{i} + F_y\mathbf{j} + F_z\mathbf{k}) = m(a_x\mathbf{i} + a_y\mathbf{j} + a_z\mathbf{k})$$

**Photo 12.2**   Biomechanics researchers use video analysis and force plate measurements in Cartesian coordinates to analyze human motion.
©Chris Ryan/agefotostock RF

It follows from this equation that

$$\Sigma F_x = ma_x \qquad \Sigma F_y = ma_y \qquad \Sigma F_z = ma_z \qquad \textbf{(12.8)}$$

Recall from Sec. 11.4C that the components of the acceleration are equal to the second derivatives of the coordinates of the particle. This gives us

$$\Sigma F_x = m\ddot{x} \qquad \Sigma F_y = m\ddot{y} \qquad \Sigma F_z = m\ddot{z} \qquad \textbf{(12.8′)}$$

Consider, as an example, the motion of a projectile. If we neglect air resistance, the only force acting on the projectile after it has been fired is its weight $\mathbf{W} = -W\mathbf{j}$. The equations defining the motion of the projectile are therefore

$$m\ddot{x} = 0 \qquad m\ddot{y} = -W \qquad m\ddot{z} = 0$$

and the components of the acceleration of the projectile are

$$\ddot{x} = 0 \qquad \ddot{y} = -\frac{W}{m} = -g \qquad \ddot{z} = 0$$

where $g$ is 9.81 m/s$^2$ or 32.2 ft/s$^2$. You can integrate these equations independently, as shown in Sec. 11.4C, to obtain the velocity and displacement of the projectile at any instant.

When a problem involves two or more bodies, you should write equations of motion for each of the bodies (see Sample Probs. 12.3, 12.4, and 12.5). Recall from Sec. 12.1A that all accelerations should be measured with respect to a newtonian frame of reference. In most engineering applications, you can determine accelerations with respect to axes attached to the earth, but relative accelerations measured with respect to moving axes, such as axes attached to an accelerated body, cannot be substituted for $\mathbf{a}$ in the equations of motion.

**Photo 12.3** A fighter jet making a sharp turn has a large normal component of acceleration, often equal to several *g*. As a result, the pilot experiences a large normal force, which in extreme cases, can cause blackouts. ©Purestock/SuperStock RF

**Tangential and Normal Components.** We can also resolve the forces and the acceleration of the particle into components along the tangent to the path (in the direction of motion) and the normal direction (toward the inside of the path) (Fig. 12.10). Substituting into Eq. (12.2), we obtain the two scalar equations of

$$\Sigma F_t = ma_t \qquad \Sigma F_n = ma_n \qquad \text{(12.9)}$$

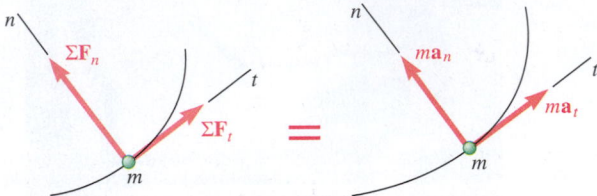

**Fig. 12.10** The net force acting on a particle moving in a curvilinear path can be resolved into components tangent to the path and normal to the path, producing tangential and normal components of acceleration.

Now substituting for $a_t$ and $a_n$ from Eqs. (11.39), we have

$$\Sigma F_t = m\frac{dv}{dt} \qquad \Sigma F_n = m\frac{v^2}{\rho} \qquad \text{(12.9')}$$

We can solve these equations for two unknowns.

**Radial and Transverse Components.** Consider a particle *P*, with polar coordinates *r* and *θ*, that moves in a plane under the action of several forces. Resolving the forces and the acceleration of the particle into radial and transverse components (Fig. 12.11) and substituting into Eq. (12.2), we obtain the two scalar equations of

$$\Sigma F_r = ma_r \qquad \Sigma F_\theta = ma_\theta \qquad \text{(12.10)}$$

Substituting for $a_r$ and $a_\theta$ from Eqs. (11.45), we have

$$\Sigma F_r = m(\ddot{r} - r\dot{\theta}^2) \qquad \text{(12.11)}$$
$$\Sigma F_\theta = m(r\ddot{\theta} + 2\dot{r}\dot{\theta}) \qquad \text{(12.12)}$$

We can solve these equations for two unknowns.

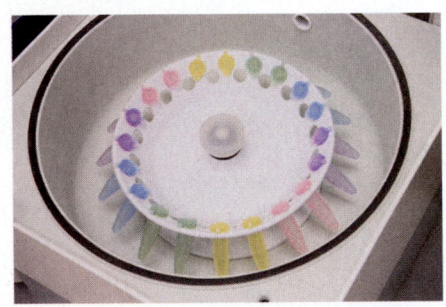

**Photo 12.4** The forces on the specimens used in a high-speed centrifuge can be described in terms of radial and transverse components. ©Russell Illig/Getty Images RF

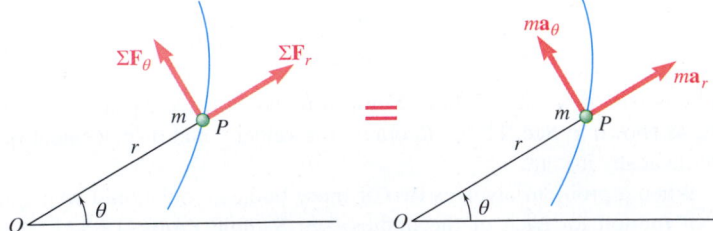

**Fig. 12.11** Pictorial representation of Newton's second law in radial and transverse components.

## Sample Problem 12.1

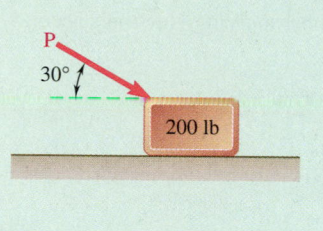

A 200-lb block rests on a horizontal plane. Find the magnitude of the force **P** required to give the block an acceleration of 10 ft/s² to the right. The coefficient of kinetic friction between the block and the plane is $\mu_k = 0.25$. ▶

**STRATEGY:**   You are given an acceleration and want to find the applied force. Therefore, you need to use Newton's second law.

**MODELING:**   Pick the block as your system and model it as a particle. Drawing its free-body and kinetic diagrams, you obtain Fig. 1.

**ANALYSIS:**   Before using Fig. 1, it is convenient to determine the mass of the object.

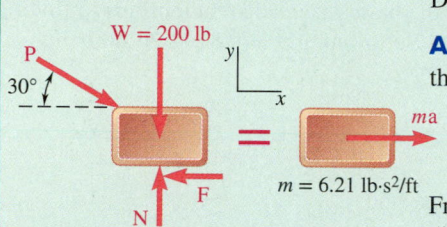

**Fig. 1**   Free-body diagram and kinetic diagram for the block.

$$m = \frac{W}{g} = \frac{200 \text{ lb}}{32.2 \text{ ft/s}^2} = 6.21 \text{ lb·s}^2/\text{ft}$$

From Fig. 1, it is clear that the forces acting on the block shown in the free-body diagram need to be equal to the vector $m\mathbf{a}$, as shown in the kinetic diagram. Using these diagrams and recognizing that $F = \mu_k N$, you can write

$\xrightarrow{+} \Sigma F_x = ma:$     $P \cos 30° - 0.25N = (6.21 \text{ lb·s}^2/\text{ft})(10 \text{ ft/s}^2)$
$P \cos 30° - 0.25N = 62.1 \text{ lb}$     **(1)**

$+\uparrow \Sigma F_y = 0:$     $N - P \sin 30° - 200 \text{ lb} = 0$     **(2)**

Solving Eq. (2) for $N$ and substituting the result into Eq. (1), you obtain

$$N = P \sin 30° + 200 \text{ lb}$$

$P \cos 30° - 0.25(P \sin 30° + 200 \text{ lb}) = 62.1 \text{ lb}$     $P = 151 \text{ lb}$ ◀

**REFLECT and THINK:**   When you begin pushing on an object, you first have to overcome the static friction force ($F = \mu_s N$) before the object will move. Also note that the downward component of force **P** increases the normal force **N**, which in turn increases the friction force **F** that you must overcome.

## Sample Problem 12.2

A 0.5-kg fragile glass vase is dropped onto a thick pad that has a force-deflection relationship as shown. Knowing that the vase has a speed of 3 m/s when it first contacts the pad, determine the maximum downward displacement of the vase.

**STRATEGY:**   Use Newton's second law to find the acceleration of the vase and then integrate it to find the displacement.

**MODELING:**   Choose the vase to be your system and model it as a particle. Because the force is a linear function of displacement, you can write the force acting on the vase as

$$F_P = \frac{200 \text{ N}}{0.02 \text{ m}} x = (10\,000 \text{ N/m})x$$

*(continued)*

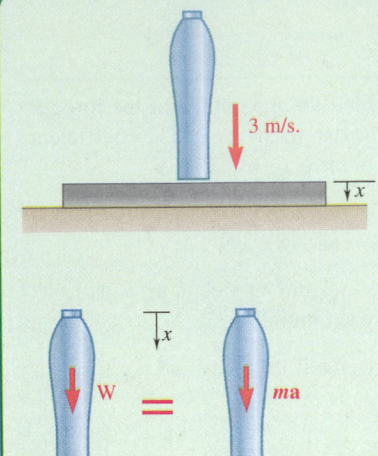

**Fig. 1** Free-body diagram and kinetic diagram for the vase.

Draw its free-body diagram and kinetic diagram (Fig. 1).

**ANALYSIS:**    You can obtain a scalar equation by applying Newton's second law in the vertical direction. Thus,

$$+\downarrow \Sigma F_x = ma \qquad W - (10\,000)x = ma$$

Substituting in values and solving for $a$ gives

$$a = 9.81 - 20\,000x$$

**Maximum Displacement.**    Now that you have the acceleration as a function of displacement, you need to use the basic kinematic relationships to find the maximum compression of the pad. Substituting $a = 9.81 - 20\,000x$ into $a = v\, dv/dx$ gives

$$a = 9.81 - 20\,000x = \frac{v\,dv}{dx}$$

Separating variables and integrating, you find

$$v\,dv = (9.81 - 20\,000x)dx \rightarrow \int_{v_0}^{0} v\,dv = \int_{0}^{x_{max}} (9.81 - 20\,000x)dx$$

$$0 - \frac{1}{2}v_0^2 = 9.81 x_{max} - 10\,000 x_{max}^2 \tag{1}$$

Substituting $v_0 = 3$ m/s into Eq. (1) and solving for $x_{max}$ using the quadratic formula gives $x_{max} = 0.0217$ m.

$$x_{max} = 21.7 \text{ mm} \blacktriangleleft$$

**REFLECT and THINK:**    A distance of 21.7 mm indicates that the pad must be relatively thick. For a real pad, the assumption that it acts as a linear spring may not be an accurate model. For the numbers given in this problem, the maximum acceleration the vase experiences is

$$a = 9.81 - (20\,000)(0.0217) = -207.3 \text{ m/s}^2 \text{ or about 21 g's}$$

## Sample Problem 12.3

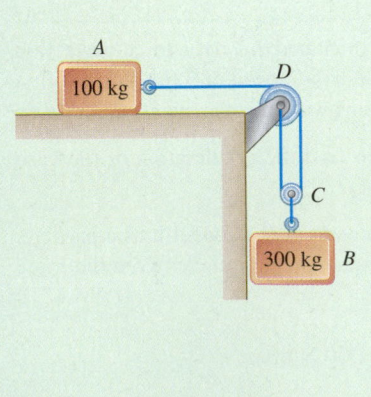

The two blocks shown start from rest. The horizontal plane and the pulley are frictionless, and the pulley is assumed to be of negligible mass. Determine the acceleration of each block and the tension in each cord.

**STRATEGY:**    You are interested in finding the tension in the rope and the acceleration of the two blocks, so use Newton's second law. The two blocks are connected by a cable, indicating that you need to relate their accelerations using the techniques discussed in Chap. 11 for objects with dependent motion.

**MODELING:**    Treat both blocks as particles and assume that the pulley is massless and frictionless. Because there are two masses, you need two systems: block A by itself and block B by itself. The free-body and kinetic diagrams for these objects are shown in Figs. 1 and 2, To help determine the forces acting on block B, you can also isolate the massless pulley C as a system (Fig. 3).

*(continued)*

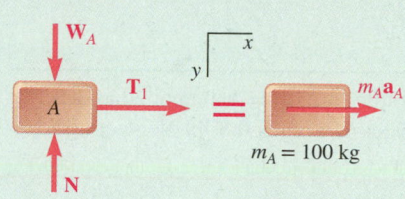

**Fig. 1** Free-body diagram and kinetic diagram for A.

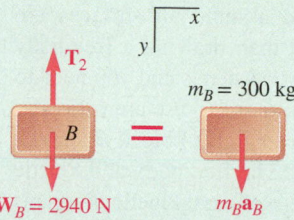

**Fig. 2** Free-body diagram and kinetic diagram for B.

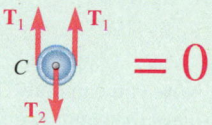

**Fig. 3** Free-body diagram and kinetic diagram for the pulley.

**ANALYSIS:** You can start with either kinetics or kinematics. The key is to make sure you keep track of your equations and unknowns.

**Kinetics.** Apply Newton's second law successively to block A, block B, and pulley C.

**Block A.** Denote the tension in cord *ACD* by $T_1$ (Fig. 1). Then you have

$$\xrightarrow{+}\Sigma F_x = m_A a_A: \qquad T_1 = 100a_A \qquad (1)$$

**Block B.** Observe that the weight of block B is

$$W_B = m_B g = (300 \text{ kg})(9.81 \text{ m/s}^2) = 2940 \text{ N}$$

Denote the tension in cord *BC* by $T_2$ (Fig. 2). Then

$$+\downarrow \Sigma F_y = m_B a_B: \qquad 2940 - T_2 = 300a_B \qquad (2)$$

**Pulley C.** Assuming $m_C$ is zero, you have (Fig. 3)

$$+\downarrow \Sigma F_y = m_C a_C = 0: \qquad T_2 - 2T_1 = 0 \qquad (3)$$

At this point, you have three equations, (1), (2), and (3), and four unknowns, $T_1$, $T_2$, $a_B$, and $a_A$. Therefore, you need one more equation, which you can get from kinematics.

**Kinematics.** It is important to make sure that the directions you assumed for the kinetic diagrams are consistent with the kinematic analysis. Note that if block A moves through a distance $x_A$ to the right, block B moves down through a distance

$$x_B = \frac{1}{2}x_A$$

Differentiating twice with respect to $t$, you have

$$a_B = \frac{1}{2}a_A \qquad (4)$$

You now have four equations and four unknowns, so you can solve this problem. You can do this using a computer, a calculator, or by hand. To solve these equations by hand, you can substitute for $a_B$ from Eq. (4) into Eq. (2).

$$2940 - T_2 = 300(\tfrac{1}{2}a_A)$$
$$T_2 = 2940 - 150a_A \qquad (5)$$

Now substitute for $T_1$ and $T_2$ from Eqs. (1) and (5), respectively, into Eq. (3).

$$2940 - 150a_A - 2(100a_A) = 0$$
$$2940 - 350a_A = 0 \qquad a_A = 8.40 \text{ m/s}^2 \quad \blacktriangleleft$$

Then substitute the value obtained for $a_A$ into Eqs. (4) and (1).

$$a_B = \tfrac{1}{2}a_A = \tfrac{1}{2}(8.40 \text{ m/s}^2) \qquad a_B = 4.20 \text{ m/s}^2 \quad \blacktriangleleft$$

$$T_1 = 100a_A = (100 \text{ kg})(8.40 \text{ m/s}^2) \qquad T_1 = 840 \text{ N} \quad \blacktriangleleft$$

Recalling Eq. (3), you have

$$T_2 = 2T_1 \qquad T_2 = 2(840 \text{ N}) \qquad T_2 = 1680 \text{ N} \quad \blacktriangleleft$$

**REFLECT and THINK:** Note that the value obtained for $T_2$ is *not* equal to the weight of block B. Rather than choosing B and the pulley as separate systems, you could have chosen the system to be B *and* the pulley. In this case, $T_2$ would have been an internal force.

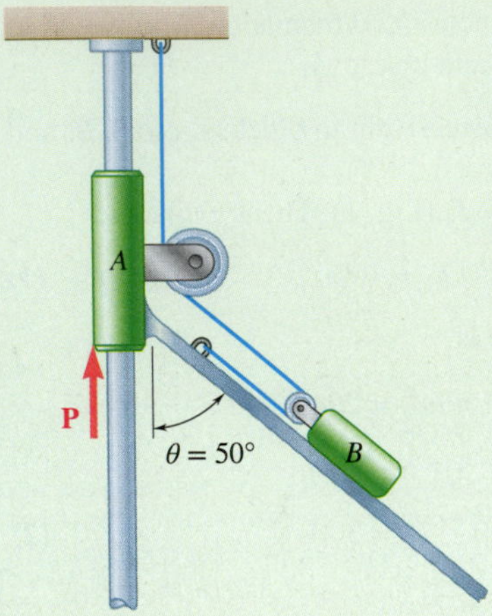

## Sample Problem 12.4

Collar $A$ has a ramp that is welded to it and a force $P = 5$ lb applied as shown. Collar $A$ and the ramp weigh 3 lb, and block $B$ weighs 0.8 lb. Neglecting friction, determine the tension in the cable.

**STRATEGY:** The principle you need to use is Newton's second law. Because a block is sliding down an incline and a cable is connecting $A$ and $B$, you also need to use relative motion and dependent motion.

**MODELING:** Model $A$ and $B$ as particles and assume all surfaces are smooth. As usual, start by choosing a system and then drawing a free-body diagram and a kinetic diagram. This problem has two systems, and you need to be careful with how you define them. The easiest systems to use are (a) collar $A$ with its pulley and the ramp welded to it (system 1) and (b) block $B$ and the pulley attached to it (system 2), as shown in Fig. 1. The free-body and kinetic diagrams for system 1 are shown in Fig. 2. The free-body and kinetic diagrams for $B$ are a little trickier, because you don't know the direction of the acceleration of $B$.

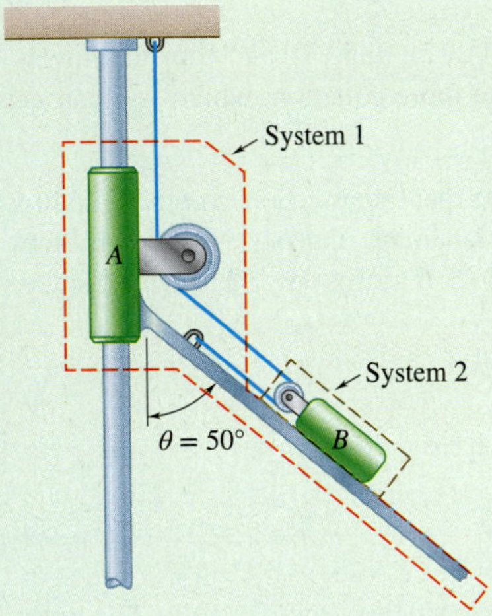

**Kinematics for Block $B$.** Express the acceleration $\mathbf{a}_B$ of block $B$ as the sum of the acceleration of $A$ and the acceleration of $B$ relative to $A$. Hence,

$$\mathbf{a}_B = \mathbf{a}_A + \mathbf{a}_{B/A}$$

Here $\mathbf{a}_{B/A}$ is directed along the inclined surface of the wedge. Now you can draw the appropriate diagrams (Fig. 3). Note that you do not need to use the same $x$–$y$ coordinate system for each mass, because these directions are simply used for obtaining the scalar equations.

**Fig. 1.** System boundaries.

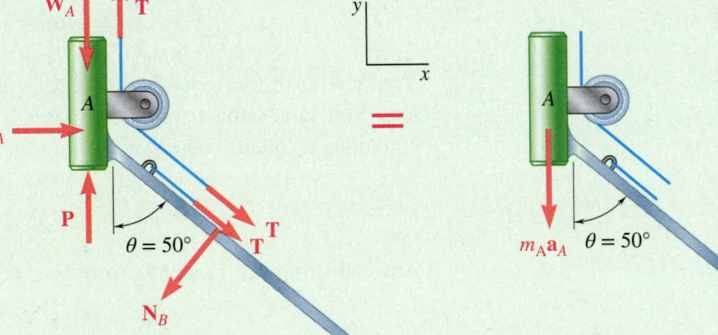

**Fig. 2.** Free-body diagram and kinetic diagram for system 1.

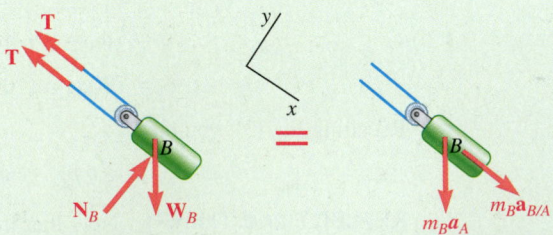

**Fig. 3.** Free-body diagram and kinetic diagram for $B$.

*(continued)*

**ANALYSIS:**  You can obtain a scalar equation by applying Newton's second law to each of these systems.

### a. System 1:

$$\xrightarrow{+}\Sigma F_x = m_A a_{A_x} \qquad N_A - N_B \cos 50° + 2T \cos 40° = 0 \tag{1}$$

$$+\uparrow \Sigma F_y = m_A a_{A_y} \qquad -W_A + P + T - 2T \sin 40° - N_B \sin 50° = -m_A a_A \tag{2}$$

### b. Block *B*:

$$+\searrow \Sigma F_x = m_B a_{B_x} \qquad -2T + W_B \sin 40° = m_B a_{B/A} + m_B a_A \sin 40° \tag{3}$$

$$+\nearrow \Sigma F_y = m_B a_{B_y} \qquad N_B - W_B \cos 40° = -m_B a_A \cos 40° \tag{4}$$

You now have four equations and five unknowns ($T$, $N_A$, $N_B$, $a_A$, and $a_{B/A}$), so you need one more equation. The motions of $A$ and $B$ are related because they are connected by a cable.

**Constraint Equations.**  Define position vectors as shown in Fig. 4. Note that the positive directions for the position vectors for $A$ and $B$ are defined from the kinetic diagrams in Figs. 2 and 3. Assuming the cable is inextensible, you can write the lengths in terms of the coordinates and then differentiate.

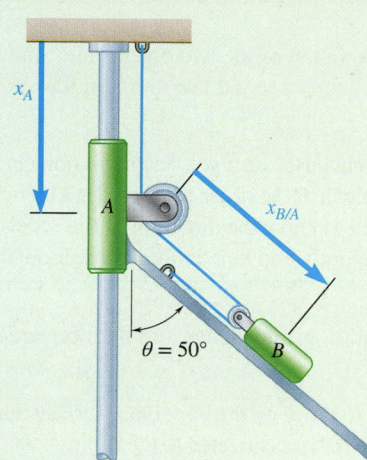

**Fig. 4.**  Position vectors for dependent motion.

Constraint equation for the cable: $x_A + 2x_{B/A} = $ constant

Differentiating this twice gives

$$a_A = -2a_{B/A} \tag{5}$$

You now have five equations and five unknowns, so all that remains is to substitute the known values and solve for the unknowns. The results are $N_A = -0.1281$ lb, $N_B = 0.869$ lb, $T = 0.281$ lb, $a_A = -13.46$ ft/s$^2$, and $a_{B/A} = 6.73$ ft/s$^2$.

$$T = 0.281 \text{ lb} \qquad \blacktriangleleft$$

**REFLECT and THINK:**  In this problem, we focused on the problem formulation and assumed that you can solve the resulting equations either by hand or by using a calculator/computer. It is important to note that you are given the weights of $A$ and $B$, so you need to calculate the masses in slugs using $m = W/g$. The solution required multiple systems and multiple concepts, including Newton's second law, relative motion, and dependent motion. If friction occurred between $B$ and the ramp, you would first need to determine whether or not the system would move under the applied force by assuming that it does not move and calculating the friction force. Then, you would compare this force to the maximum allowable force $\mu_s N$.

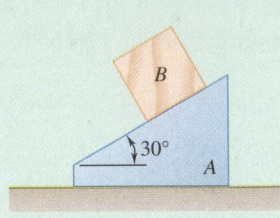

## Sample Problem 12.5

The 12-lb block $B$ starts from rest and slides on the 30-lb wedge $A$, which is supported by a horizontal surface. Neglecting friction, determine (a) the acceleration of the wedge, (b) the acceleration of the block relative to the wedge.

**STRATEGY:**    You are given the forces (weights) of the two objects and want to find their accelerations. You can use Newton's second law, but you have to take into account relative motion as well.

**MODELING:**    Treat both objects as particles. Because you have two objects, you will need two systems: wedge $A$ and block $B$. In order to draw the kinetic diagrams for each of these systems, you need to know the direction of the accelerations. Therefore, before drawing the free-body and kinetic diagrams, look at the kinematics.

**Kinematics.**    First examine the acceleration of the wedge and the acceleration of the block.

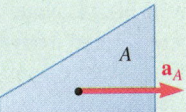

**Fig. 1**    Acceleration of $A$.

*Wedge A.* Because the wedge is constrained to move on the horizontal surface, its acceleration $\mathbf{a}_A$ is horizontal (Fig. 1). Assume that it is directed to the right.

*Block B.* You can express the acceleration $\mathbf{a}_B$ of block $B$ as the sum of the acceleration of $A$ and the acceleration of $B$ relative to $A$ (Fig. 2), so

$$\mathbf{a}_B = \mathbf{a}_A + \mathbf{a}_{B/A}$$

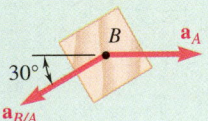

**Fig. 2**    Acceleration of $B$.

Here $\mathbf{a}_{B/A}$ is directed along the inclined surface of the wedge. Now you can draw the appropriate diagrams. The free-body diagrams and kinetic diagrams for $A$ and $B$ are shown in Figs. 3 and 4, respectively. The forces exerted by the block and the horizontal surface on wedge $A$ are represented by $\mathbf{N}_1$ and $\mathbf{N}_2$, respectively.

**ANALYSIS:**
**Kinetics.**    Recall that Figs. 3 and 4 are visual representations of Newton's second law. Therefore, you can use them to obtain scalar equations.

*Wedge A.* For Wedge $A$, the positive $x$ direction is defined to be to the right. Applying Newton's second law in the $x$ direction gives

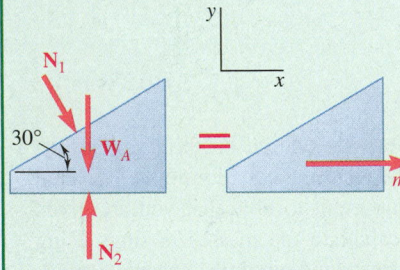

**Fig. 3**    Free-body diagram and kinetic diagram for $A$.

$$\xrightarrow{+} \Sigma F_x = m_A a_A: \qquad N_1 \sin 30° = m_A a_A$$
$$0.5 N_1 = (W_A/g) a_A \qquad \textbf{(1)}$$

*Block B.* Using the coordinate axes shown in Fig. 4 and resolving $\mathbf{a}_B$ into its components $\mathbf{a}_A$ and $\mathbf{a}_{B/A}$, you have

$$+\nearrow \Sigma F_x = m_B a_x: \qquad -W_B \sin 30° = m_B a_A \cos 30° - m_B a_{B/A}$$
$$-W_B \sin 30° = (W_B/g)(a_A \cos 30° - a_{B/A})$$
$$a_{B/A} = a_A \cos 30° + g \sin 30° \qquad \textbf{(2)}$$
$$+\nwarrow \Sigma F_y = m_B a_y: \quad N_1 - W_B \cos 30° = -m_B a_A \sin 30°$$
$$N_1 - W_B \cos 30° = -(W_B/g) a_A \sin 30° \qquad \textbf{(3)}$$

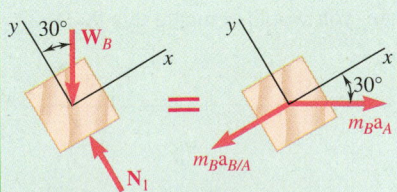

**Fig. 4**    Free-body diagram and kinetic diagram for $B$.

You now have three equations, (1), (2), and (3), and three unknowns, $N_1$, $a_A$, and $a_{B/A}$, so you can solve these with your calculator or by hand as shown here.

**a. Acceleration of Wedge A.**    Substitute for $N_1$ from Eq. (1) into Eq. (3).

$$2(W_A/g) a_A - W_B \cos 30° = -(W_B/g) a_A \sin 30°$$

Then, solve for $a_A$ and substitute the numerical data.

$$a_A = \frac{W_B \cos 30°}{2W_A + W_B \sin 30°} g = \frac{(12 \text{ lb}) \cos 30°}{2(30 \text{ lb}) + (12 \text{ lb}) \sin 30°} (32.2 \text{ ft/s}^2)$$

$$a_A = +5.07 \text{ ft/s}^2 \qquad \mathbf{a}_A = 5.07 \text{ ft/s}^2 \rightarrow \quad \blacktriangleleft$$

### b. Acceleration of Block *B* Relative to *A*.

Now substitute the value obtained for $a_A$ into Eq. (2).

$$a_{B/A} = (5.07 \text{ ft/s}^2) \cos 30° + (32.2 \text{ ft/s}^2) \sin 30°$$

$$a_{B/A} = +20.5 \text{ ft/s}^2 \qquad \mathbf{a}_{B/A} = 20.5 \text{ ft/s}^2 \; \searrow \; 30° \quad \blacktriangleleft$$

**REFLECT and THINK:** Many students are tempted to draw the acceleration of block *B* down the incline in the kinetic diagram. It is important to recognize that this is the direction of the *relative* acceleration. Rather than the kinetic diagram you used for block *B*, you could have simply put unknown accelerations in the *x* and *y* directions and then used your relative motion equation to obtain more scalar equations.

## Sample Problem 12.6

The bob of a 2-m pendulum describes an arc of a circle in a vertical plane. If the tension in the cord is 2.5 times the weight of the bob for the position shown, find the velocity and the acceleration of the bob in that position. ⏵

**STRATEGY:** The most direct approach is to use Newton's law with tangential and normal components.

**MODELING:** Choose the bob as your system; if its radius is small, you can model it as a particle. Draw the free-body and kinetic diagrams for the bob knowing that the weight of the bob is $W = mg$; the tension in the cord is $2.5mg$. The normal acceleration $\mathbf{a}_n$ is directed toward *O*, and you can assume that $\mathbf{a}_t$ is in the direction shown in Fig. 1.

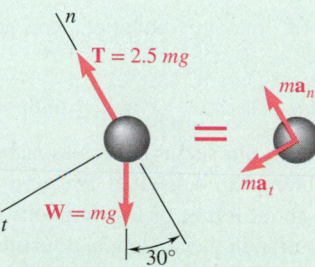

**Fig. 1** Free-body diagram and kinetic diagram for the bob.

**ANALYSIS:** You can obtain scalar equations by applying Newton's second law in the normal and tangential directions. Hence,

$$+\nearrow \Sigma F_t = ma_t: \qquad mg \sin 30° = ma_t$$

$$a_t = g \sin 30° = +4.90 \text{ m/s}^2 \qquad \mathbf{a}_t = 4.90 \text{ m/s}^2 \nearrow \quad \blacktriangleleft$$

$$+\nwarrow \Sigma F_n = ma_n: \qquad 2.5 \, mg - mg \cos 30° = ma_n$$

$$a_n = 1.634g = +16.03 \text{ m/s}^2 \qquad \mathbf{a}_n = 16.03 \text{ m/s}^2 \nwarrow \quad \blacktriangleleft$$

Because $a_n = v^2/\rho$, you have $v^2 = \rho a_n = (2 \text{ m})(16.03 \text{ m/s}^2)$. Thus,

$$v = \pm 5.66 \text{ m/s} \qquad \mathbf{v} = 5.66 \text{ m/s} \nearrow \text{ (up or down)} \quad \blacktriangleleft$$

**REFLECT and THINK:** If you look at these equations for an angle of zero instead of 30°, you will see that when the bob is straight below point *O*, the tangential acceleration is zero, and the velocity is a maximum. The normal acceleration is not zero because the bob has a velocity at this point.

## Sample Problem 12.7

Determine the rated speed of a highway curve with a radius of $\rho = 400$ ft banked through an angle $\theta = 18°$. The *rated speed* of a banked highway curve is the speed at which a car should travel to have no lateral friction force exerted on its wheels.

**STRATEGY:** You are given information about the lateral friction force—that is, it is equal to zero—so use Newton's second law. Use normal and tangential components, because the car is traveling in a curved path and the problem involves speed and a radius of curvature.

**MODELING:** Choose the car to be the system. Assuming you can neglect the rotation of the car about its center of mass, treat it as a particle. The car travels in a *horizontal* circular path with a radius of $\rho$. The normal component $\mathbf{a}_n$ of the acceleration is directed toward the center of the path, as shown in the kinetic diagram (Fig. 1); its magnitude is $a_n = v^2/\rho$, where $v$ is the speed of the car in ft/s. The mass $m$ of the car is $W/g$, where $W$ is the weight of the car. Because no lateral friction force is exerted on the car, the reaction $\mathbf{R}$ of the road is perpendicular to the roadway, as shown in the free-body diagram (Fig. 1).

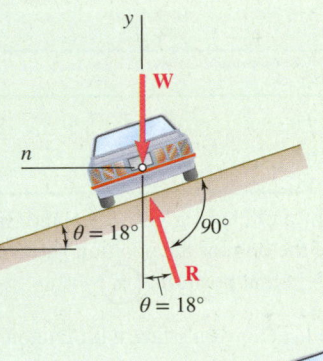

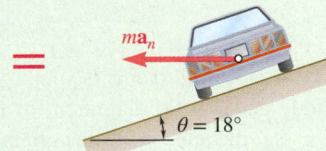

**Fig. 1** Free-body diagram and kinetic diagram of the car.

**ANALYSIS:** You can obtain scalar equations by applying Newton's second law in the vertical and normal directions. Thus,

$$+\uparrow \Sigma F_y = 0: \qquad R\cos\theta - W = 0 \qquad R = \frac{W}{\cos\theta} \qquad (1)$$

$$\overset{+}{\leftarrow} \Sigma F_n = ma_n: \qquad R\sin\theta = \frac{W}{g}a_n \qquad (2)$$

Substituting $R$ from Eq. (1) into Eq. (2), and recalling that $a_n = v^2/\rho$, you obtain

$$\frac{W}{\cos\theta}\sin\theta = \frac{W}{g}\frac{v^2}{\rho} \qquad v^2 = g\rho\tan\theta$$

Finally, substituting $\rho = 400$ ft and $\theta = 18°$ into this equation, you get $v^2 = (32.2 \text{ ft/s}^2)(400 \text{ ft})\tan 18°$. Hence,

$$v = 64.7 \text{ ft/s} \qquad\qquad v = 44.1 \text{ mi/h} \blacktriangleleft$$

**REFLECT and THINK:** For a highway curve, this seems like a reasonable speed for avoiding a spin-out. For this problem, the tangential direction is into the page; because you were not asked about forces or accelerations in this direction, you did not need to analyze motion in the tangential direction. If the roadway were banked at a larger angle, would the rated speed be larger or smaller than this calculated value?

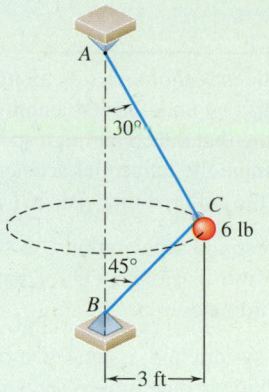

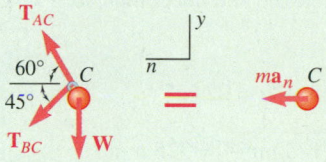

**Fig. 1** Free-body diagram and kinetic diagram for the sphere.

## Sample Problem 12.8

Two wires $AC$ and $BC$ are tied at $C$ to a sphere that revolves at the constant speed $v$ in the horizontal circle shown. Knowing that the wires will break if their tension exceeds 15 lb, determine the range of values of $v$ for which both wires remain taut and the wires do not break.

**STRATEGY:**    You are given information about the forces in the wires, so use Newton's second law. The sphere is moving along a curved path, so use normal and tangential coordinates.

**MODELING:**    Choose the sphere for the system and assume you can treat it as a particle. Draw the free-body and kinetic diagrams as shown in Fig. 1. The tensions act in the direction of the wires, and the normal direction is toward the center of the circular path.

**ANALYSIS:**    You can obtain scalar equations by applying Newton's second law in the normal and vertical directions. Thus,

$$+\!\leftarrow \Sigma F_n = ma_n \qquad T_{AC} \cos 60° + T_{BC} \cos 45° = ma_n = m\frac{v^2}{\rho} \qquad \textbf{(1)}$$

$$+\!\uparrow \Sigma F_y = ma_y \qquad -W + T_{AC} \sin 60° - T_{BC} \sin 45° = 0 \qquad \textbf{(2)}$$

where $m = W/g = 6 \text{ lb}/(32.2 \text{ ft/s}^2) = 0.1863 \text{ lb·s}^2/\text{ft}$ and $\rho = 3$ ft. In these two equations, you have three unknowns, $T_{AC}$, $T_{BC}$, and $v$, so you need a third equation. The problem statement indicates that you want the range of speeds when both wires remain taut (that is, the tension is positive) and that this tension must be less than 15 lb. To find this range, first set each tension equal to zero and solve the resulting set of equations.

For $T_{AC} = 0$, you find $v = 9.83$ ft/s and $T_{BC} = -8.485$ lb, which is impossible for a wire.

For $T_{BC} = 0$, you find $v = 7.468$ ft/s and $T_{AC} = 6.928$ lb.

Thus, the minimum speed is 7.47 ft/s. Now set the tensions equal to 15 lb to find the maximum speed.

For $T_{AC} = 15$ lb, you find $v = 15.29$ ft/s and $T_{BC} = 9.886$ lb.

For $T_{BC} = 15$ lb, you find $v = 18.03$ ft/s and $T_{AC} = 19.18$ lb.

Therefore, the maximum speed is 15.29 ft/s. Combining these results gives you

$$7.47 \text{ ft/s} \leq v \leq 15.29 \text{ ft/s} \quad \blacktriangleleft$$

**REFLECT and THINK:**    In this problem, you needed to use the information in the problem statement to obtain additional equations so that you could determine the range of speeds. Another way to look at the solution is to solve Eqs. (1) and (2) for $T_{AC}$ and $T_{BC}$ in terms of $v$ and to plot these as shown in Fig. 2. It is easy to see from this graph that $T_{AC}$ determines the maximum speed and $T_{BC}$ determines the minimum speed if both wires are to remain taut and also have tensions less than 15 lb.

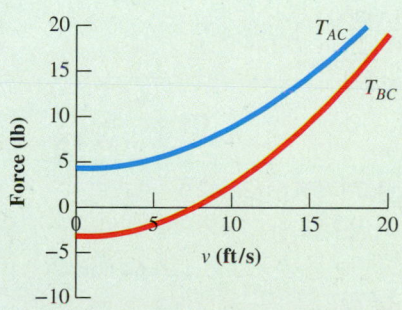

**Fig. 2** Tension in cables as a function of speed.

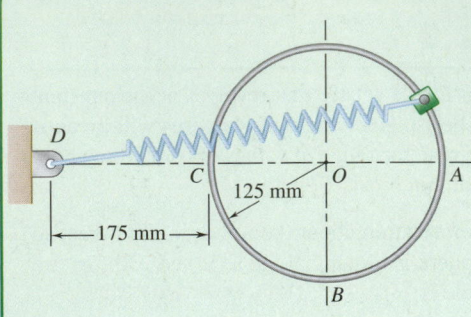

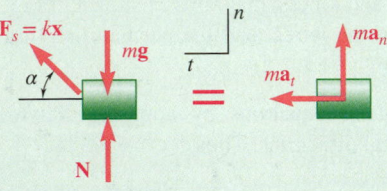

**Fig. 1** Free-body diagram and kinetic diagram for the collar.

## Sample Problem 12.9

A 0.5-kg collar is attached to a spring and slides without friction along a circular rod in a *vertical* plane. The spring has an undeformed length of 150 mm and a constant $k = 200$ N/m. Knowing that the collar has a speed of 3 m/s as it passes through point $B$, determine the tangential acceleration of the collar and the force of the rod on the collar at this instant. ▶

**STRATEGY:** This problem deals with forces and accelerations, so you need to use Newton's second law. The collar moves along a curved path, so you should use normal and tangential coordinates.

**MODELING:** Choose the collar as your system and assume you can treat it as a particle. Draw the free-body and kinetic diagrams as shown in Fig. 1. The spring force acts in the direction of the spring, and the force is drawn assuming that the spring is stretched and not compressed. Check this using geometry.

$$\tan \alpha = \frac{125 \text{ mm}}{300 \text{ mm}} = 0.4167 \rightarrow \alpha = 22.62°$$

$$L_{BD} = \sqrt{(300 \text{ mm})^2 + (125 \text{ mm})^2} = 325 \text{ mm}$$

Thus, when the collar is at $B$, the spring is extended as $x = L_{BD} - L_0 = 325 \text{ mm} - 150 \text{ mm} = 175 \text{ mm}$.

**ANALYSIS:** You can obtain scalar equations by applying Newton's second law in the normal and tangential directions. Hence,

$$+\uparrow \Sigma F_n = ma_n \qquad kx \sin \alpha + N - mg = ma_n = m\frac{v^2}{\rho} \qquad \textbf{(1)}$$

$$\underset{\rightarrow}{+}\Sigma F_t = ma_t \qquad Fx \cos \alpha = ma_t \qquad \textbf{(2)}$$

You now have two equations, (1) and (2), and two unknowns, $a_t$ and $N$. You can solve for these by hand or using your calculator/computer. You can solve for the normal force in Eq. (1) as

$$N = mg + m\frac{v^2}{\rho} - kx \sin \alpha$$

Substituting values gives

$$N = (0.5 \text{ kg})(9.81 \text{ m/s}^2) + (0.5 \text{ kg})\frac{(3 \text{ m/s})^2}{0.125 \text{ m}}$$
$$- (200 \text{ N/m})(0.175 \text{ m}) \sin (22.62°)$$

$$N = 27.4 \text{ N} \blacktriangleleft$$

$$a_t = \frac{Fx \cos \alpha}{m} = \frac{(200 \text{ N/m})(0.175) \cos (22.62°)}{0.5 \text{ kg}}$$

$$a_t = 64.6 \text{ m/s}^2 \blacktriangleleft$$

**REFLECT and THINK:** How would this problem have changed if you had been told friction was acting between the rod and the collar? You would have had one additional term in your free-body diagram, $\mu_k N$, in the direction opposite to the velocity. Thus, you would need to be told the direction the collar was moving as well as the coefficient of kinetic friction.

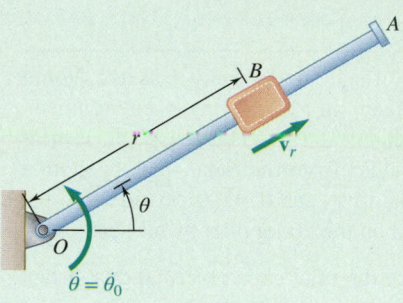

# Sample Problem 12.10

A block $B$ with a mass $m$ can slide freely on a frictionless arm $OA$ that rotates in a horizontal plane at a constant rate $\dot{\theta}_0$. Knowing that $B$ is released at a distance $r_0$ from $O$, express as a function of $r$ (a) the component $v_r$ of the velocity of $B$ along $OA$, (b) the magnitude of the horizontal force $\mathbf{F}$ exerted on $B$ by the arm $OA$.

**STRATEGY:** You want to find a force, so use Newton's second law. The radial distance $r$ of the mass is changing, as is the angular displacement $\theta$, so use radial and transverse coordinates.

**MODELING:** Choose block $B$ as your system and assume you can model it as a particle. Because all other forces are perpendicular to the plane of the figure, the only force shown acting on $B$ is the force $\mathbf{F}$ perpendicular to $OA$. Draw free-body and kinetic diagrams for block B as shown in Fig. 1.

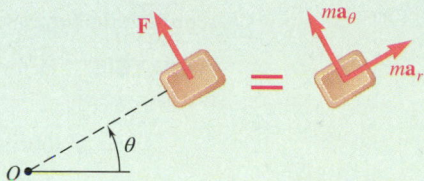

**Fig. 1**   Free-body diagram and kinetic diagram for the block.

**ANALYSIS:**

**Equations of Motion.** You can obtain scalar equations by applying Newton's second law in the radial and transverse directions. Hence,

$$+\nearrow \Sigma F_r = ma_r: \qquad\qquad 0 = m(\ddot{r} - r\dot{\theta}^2) \qquad\qquad (1)$$
$$+\nwarrow \Sigma F_\theta = ma_\theta: \qquad\qquad F = m(r\ddot{\theta} + 2\dot{r}\dot{\theta}) \qquad\qquad (2)$$

**a. Component $v_r$ of Velocity.** Because $v_r = \dot{r}$, you have

$$\ddot{r} = \dot{v}_r = \frac{dv_r}{dt} = \frac{dv_r}{dr}\frac{dr}{dt} = v_r\frac{dv_r}{dr}$$

After using Eq. (1) to obtain $\ddot{r} = r\dot{\theta}^2$ and recalling that $\dot{\theta} = \dot{\theta}_0$, you can separate the variables to obtain

$$v_r\, dv_r = \dot{\theta}_0^2\, r\, dr$$

Multiply by 2 and integrate from 0 to $v_r$ and from $r_0$ to $r$. The result is

$$v_r^2 = \dot{\theta}_0^2(r^2 - r_0^2) \qquad\qquad v_r = \dot{\theta}_0(r^2 - r_0^2)^{1/2} \quad \blacktriangleleft$$

**b. Horizontal Force $F$.** Set $\dot{\theta} = \dot{\theta}_0$, $\ddot{\theta} = 0$ and $\dot{r} = v_r$ in Eq. (2). Then, substitute for $v_r$ the expression obtained in part a. The result is

$$F = 2m\,\dot{\theta}_0(r^2 - r_0^2)^{1/2}\dot{\theta}_0 \qquad F = 2m\,\dot{\theta}_0^2(r^2 - r_0^2)^{1/2} \quad \blacktriangleleft$$

**REFLECT and THINK:** Introducing radial and transverse components of force and acceleration also involves using components of velocity in the computation. This is still much simpler and more direct than trying to use other coordinate systems. Even though the radial acceleration is zero, the block accelerates relative to the rod with acceleration $\ddot{r}$.

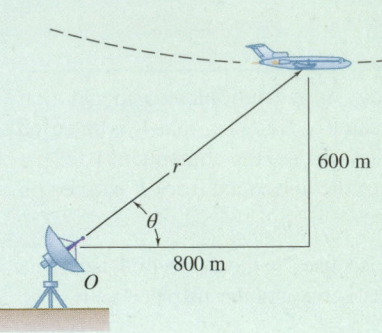

## Sample Problem 12.11

NASA flies a reduced-gravity aircraft (affectionately known as the Vomit Comet) in an elliptic flight to train astronauts in a microgravity environment. The plane is being tracked by radar located at $O$. When the plane is near the bottom of its trajectory, as shown, values from the radar tracking station are $\dot{r} = 120$ m/s, $\dot{\theta} = -0.090$ rad/s, $\ddot{r} = 34.8$ m/s², and $\ddot{\theta} = 0.0156$ rad/s². At the instant shown, determine the force exerted on the 80-kg pilot by his seat.

**STRATEGY:** You want to find the force the pilot experiences at this instant and you can calculate the accelerations, so you should use Newton's second law. Because you know that the radial distance and the angle are changing with time, use radial and transverse components.

**MODELING:** Choosing the pilot as the system, draw the free-body and kinetic diagrams as shown in Fig. 1. You could choose to put the forces and the pilot in the $r$ and $\theta$ direction or the $x$ and $y$ direction (we chose $F_x$ and $F_y$ to represent the forces from the seat back and bottom, respectively).

**ANALYSIS:** Before you apply Newton's second law, determine $r$ and $\theta$ from the geometry.

$$r = \sqrt{800^2 + 600^2} = 1000 \text{ m} \qquad \theta = \tan^{-1}(600/800) = 36.87°$$

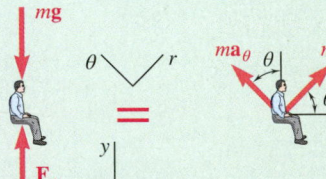

**Kinematics.** Determine the components of the accelerations as

$$a_r = \ddot{r} - r\dot{\theta}^2 = 34.8 \text{ m/s}^2 - (1000 \text{ m})(-0.090 \text{ rad/s})^2 = 26.7 \text{ m/s}^2$$
$$a_\theta = r\ddot{\theta} + 2\dot{r}\dot{\theta} = (1000 \text{ m})(0.0156 \text{ rad/s}^2) + 2(120 \text{ m/s})(-0.090 \text{ rad/s})$$
$$= -6.00 \text{ m/s}^2$$

**Fig. 1**

**Kinetics.** Obtain scalar equations by applying Newton's second law in the horizontal and vertical directions. Thus,

$$\xrightarrow{+} \Sigma F_x = ma_x \qquad\qquad F_x = ma_r \cos\theta - ma_\theta \sin\theta \qquad\qquad (1)$$

$$+\uparrow \Sigma F_y = ma_y \qquad\qquad F_y - mg = ma_r \sin\theta + ma_\theta \cos\theta \qquad (2)$$

You have two equations, (1) and (2), and two unknowns, $F_x$ and $F_y$. Substituting the known values into Eqs. (1) and (2) gives

$$F_x = (80\text{kg})(26.7 \text{ m/s}^2) \cos 36.87° - (80 \text{ kg})(-6.00 \text{ m/s}^2) \sin 36.87°$$
$$F_x = 1997 \text{ N} \rightarrow \quad \blacktriangleleft$$

$$F_y = (80 \text{ kg})(9.81 \text{ m/s}^2) + (80 \text{ kg})(26.7 \text{ m/s}^2) \sin 36.87° +$$
$$(80 \text{ kg})(-6.00 \text{ m/s}^2) \cos 36.87°$$

$$F_y = 1682 \text{ N} \uparrow \quad \blacktriangleleft$$

**REFLECT and THINK:** These forces correspond to a forward acceleration of 2.54 $g$ and a vertical acceleration of 2.14 $g$. Although this is a bit high for a passenger aircraft, it is within the flight characteristics for the Vomit Comet. If you had been asked to determine whether the plane was speeding up or slowing down, you would need to find the component of the acceleration in the tangential direction, which is defined by the direction of the velocity vector.

## Case Study 12.1

In 2001, drivers practicing for the Firestone Firehawk 600 CART series race (see racecar in CS Photo 12.1) complained of being dizzy and disoriented. Due to concerns over driver safety, the race was canceled (resulting in a huge loss of revenue and some ensuing lawsuits).

The turns at the Texas Motor Speedway have a radius of 750 feet, as shown in CS Fig. 12.1, and are banked at 24 degrees. This bank angle is much higher than that found at Indianapolis (9 degrees) or at Michigan (18 degrees), and allowed the cars to reach speeds of 230 mph (337.3 ft/s). To investigate why the drivers were feeling disoriented, determine (*a*) the normal acceleration of the driver, (*b*) the normal force from the seat, assuming that the driver weighs 150 lbs and that the seat harness applies a sideways force on the driver.

**CS Photo 12.1** CART racecar. ©Donald Miralle/Allsport/Getty Images

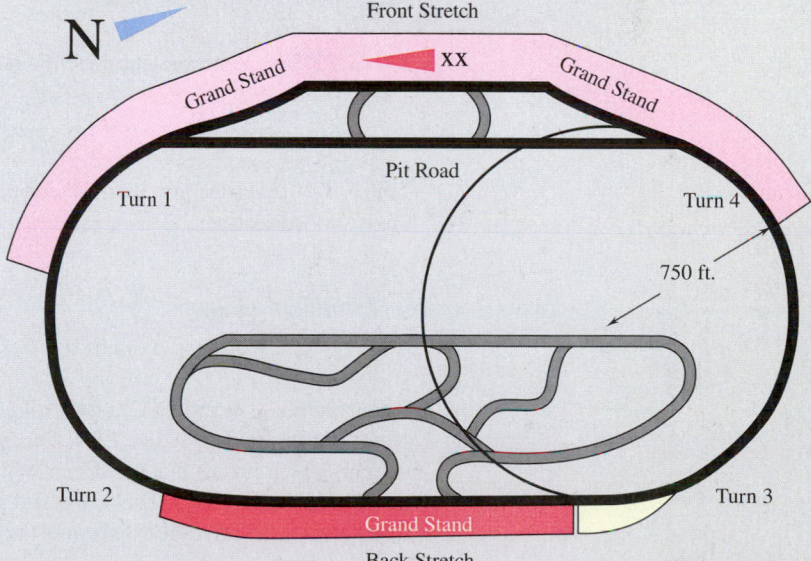

**CS Fig. 12.1** Texas Motor Speedway.

*(continued)*

Additionally, knowing that the weight of the car and driver is 1750 lbs, determine the minimum coefficient of sideways friction necessary to keep the car from skidding up the track. We can estimate that the downforce caused by the spoilers of the car is 2.5 times the car's weight at these high speeds.

**STRATEGY:** For the first part of the problem, we are analyzing the driver. Using normal and tangential coordinates, we can determine the normal acceleration of the driver. Then, we will analyze the forces on the driver by using Newton's second law. For the second portion of the problem, we will analyze the forces on the car by again using Newton's second law, but now will isolate the racecar for our system.

**MODELING (THE DRIVER):** Draw the free-body and kinetic diagrams of the driver; this includes the normal force $N$ from the seat, the weight $mg$ of the driver, and the side force $F_s$ due to the harness belt. Note that we are only looking at the forces in the vertical and normal direction and ignoring the forces in the tangential direction.

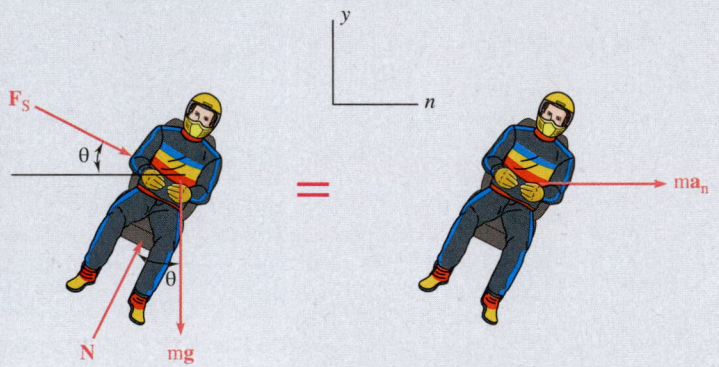

**ANALYSIS:** You can determine the normal acceleration using

$$a_n = \frac{v^2}{R} = \frac{(337.33 \text{ ft/s})^2}{750 \text{ ft}} = 151.7 \text{ ft/s}^2$$

This is 4.71 times the gravitational constant, 32.2 ft/s². Now, summing forces in the $n$ and $y$ directions:

$$\xrightarrow{+} \Sigma F_n = ma_n = m\frac{v^2}{R} \qquad F_s \cos \theta + N \sin \theta = ma_n = m(151.7 \text{ ft/s}^2) \qquad \textbf{(1)}$$

$$+\uparrow \Sigma F_y = 0 \qquad N \cos \theta - F_s \sin \theta - mg = 0 \qquad \textbf{(2)}$$

Where $m = W/g = 150 \text{ lb}/32.2 \text{ ft/s}^2 = 4.658 \text{ lbs}^2/\text{ft}$. Solving (1) and (2), you find $N = 424.5$ lbs and $F_s = 584.7$ lbs. This normal force is 2.83 times the person's weight, which means that the driver will be pulling almost 3 g's for prolonged periods during the race. This can result in blood pooling in the lower extremities, causing the driver the discomfort reported.

*(continued)*

**MODELING (THE CAR):**   Now you can analyze the car. Drawing the free-body and kinetic diagrams yields:

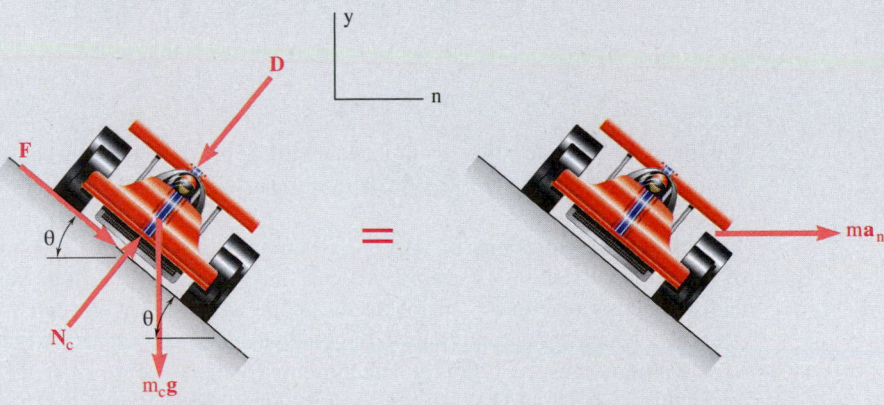

**ANALYSIS:**   You can use the diagrams to set up your equations of motion.

$$\xrightarrow{+}\Sigma F_n = ma_n = m\frac{v^2}{R} \qquad F\cos\theta + (N_C - D)\sin\theta = ma_n = m(151.7 \text{ ft/s}^2) \quad \textbf{(4)}$$

$$+\uparrow\Sigma F_y = 0 \qquad\qquad (N_C - D)\cos\theta - F\sin\theta - mg = 0 \qquad\qquad\qquad \textbf{(5)}$$

Solving (4) and (5) for $F$ and $N_C$ gives $F = 7146$ lb and $N_C = 8597$ lb. The minimum friction coefficient to allow this to occur is $\mu = F/N_C = 0.831$.

**REFLECT and THINK:**   You could also solve for the rated speed of the curve, exactly like what was done in Sample Prob. 12.7 (you get an answer of 70.7 mph when you have no friction and no downforce). The Texas Motor Speedway was originally designed for NASCAR racing, and these cars typically reach speeds of 190 mph during qualifying. The turbocharged engines and large downforce of the CART racers allowed much greater speeds, which put the racers at risk. If you were designing a new track for the CART racecars, what is the minimum radius of curvature that you should use to keep the normal force on the drivers under two times their weight?

# SOLVING PROBLEMS
# ON YOUR OWN

In the problems for this section, you will apply **Newton's second law of motion**, $\Sigma\mathbf{F} = m\mathbf{a}$, to relate the forces acting on a particle to its motion.

**1. Writing the equations of motion.** When applying Newton's second law to the types of motion discussed in this section, you will find it most convenient to express the vectors **F** and **a** in terms of either their rectangular components, their tangential and normal components, or their radial and transverse components.

**a. When using rectangular components** (Sample Probs. 12.1, 12.2, 12.3, 12.4, and 12.5), recall from Sec. 11.4C the expressions found for $a_x$, $a_y$, and $a_z$. Then you can write

$$\Sigma F_x = m\ddot{x} \qquad \Sigma F_y = m\ddot{y} \qquad \Sigma F_z = m\ddot{z}$$

**b. When using tangential and normal components** (Sample Probs. 12.6 through 12.9), recall from Sec. 11.5A the expressions found for $a_t$ and $a_n$. Then you can write

$$\Sigma F_t = m\frac{dv}{dt} \qquad \Sigma F_n = m\frac{v^2}{\rho}$$

**c. When using radial and transverse components** (Sample Probs. 12.10 and 12.11), recall from Sec. 11.5B the expressions found for $a_r$ and $a_\theta$. Then you can write

$$\Sigma F_r = m(\ddot{r} - r\dot{\theta}^2) \qquad \Sigma F_\theta = m(r\ddot{\theta} + 2\dot{r}\dot{\theta})$$

**2. Drawing a free-body diagram and a kinetic diagram.** Drawing a free-body diagram showing the applied forces and a kinetic diagram showing the vector $m\mathbf{a}$ or its components will provide you with a pictorial representation of Newton's second law (Sample Probs. 12.1 through 12.11). These diagrams will be of great help to you when writing the equations of motion. Note that when a problem involves two or more bodies, it is usually best to consider each body separately.

**3. Applying Newton's second law.** As we observed in Sec. 12.1A, the acceleration used in the equation $\Sigma\mathbf{F} = m\mathbf{a}$ always should be the absolute acceleration of the particle (i.e., it should be measured with respect to a newtonian frame of reference). Also, if the sense of the acceleration **a** is unknown or is not easily deduced, assume an arbitrary sense for **a** (usually the positive direction of a coordinate axis), and then let the solution provide the correct sense. Finally, note how the solutions of Sample Probs. 12.3, 12.4, and 12.5 were divided into a *kinematics* portion and a *kinetics* portion, and how in Sample Probs. 12.4 and 12.5 we used two systems of coordinate axes to simplify the equations of motion.

**4. When a problem involves dry friction,** be sure to review the relevant section of *Statics* (Sec. 8.1) before attempting to solve that problem. In particular, you should know when to use each of the equations $F = \mu_s N$ and $F = \mu_k N$. You should also recognize that if the motion of a system is not specified, it is necessary first to assume a possible motion and then to check the validity of that assumption. For example, you can assume that the motion is impending, then check to see if the friction force is greater than $\mu_s N$ (if it is, then your assumption was wrong and the particle is moving).

**5. Solving problems involving relative motion.** When a body $B$ moves with respect to a body $A$, as in Sample Probs. 12.4 and 12.5, it is often convenient to express the acceleration of $B$ as

$$\mathbf{a}_B = \mathbf{a}_A + \mathbf{a}_{B/A}$$

where $\mathbf{a}_{B/A}$ is the acceleration of $B$ relative to $A$; that is, the acceleration of $B$ as observed from a frame of reference attached to $A$ and in translation. If $B$ is observed to move in a straight line, $\mathbf{a}_{B/A}$ is directed along that line. On the other hand, if $B$ is observed to move along a circular path, you should resolve the relative acceleration $\mathbf{a}_{B/A}$ into components tangential and normal to that path.

**6. Finally, always consider the implications of any assumption you make.** Thus, in a problem involving two cords, if you assume that the tension in one of the cords is equal to its maximum allowable value, check whether any requirements set for the other cord will be satisfied. For instance, will the tension $T$ in that cord satisfy the relation $0 \leq T \leq T_{max}$? That is, will the cord remain taut and will its tension be less than its maximum allowable value?

# Problems

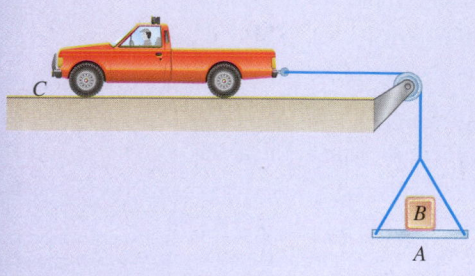

Fig. P12.CQ1

## CONCEPT QUESTIONS

**12.CQ1** A 1000-lb boulder $B$ is resting on a 200-lb platform $A$ when truck $C$ accelerates to the left with a constant acceleration. Which of the following statements are true? (More than one may be true.)
- **a.** The tension in the cord connected to the truck is 200 lb.
- **b.** The tension in the cord connected to the truck is 1200 lb.
- **c.** The tension in the cord connected to the truck is greater than 1200 lb.
- **d.** The normal force between $A$ and $B$ is 1000 lb.
- **e.** The normal force between $A$ and $B$ is 1200 lb.
- **f.** None of the above is true.

**12.CQ2** Marble $A$ is placed in a hollow tube, and the tube is swung in a horizontal plane causing the marble to be thrown out. As viewed from the top, which of the following choices best describes the path of the marble after leaving the tube?
- **a.** 1  **b.** 2  **c.** 3  **d.** 4  **e.** 5

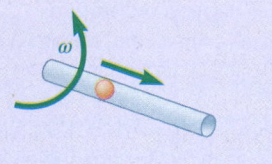

Top View

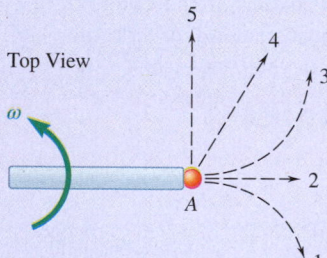

Fig. P12.CQ2

**12.CQ3** The two systems shown start from rest. On the left, two 40-lb weights are connected by an inextensible cord, and on the right, a constant 40-lb force pulls on the cord. Neglecting all frictional forces, which of the following statements is true?
- **a.** Blocks $A$ and $C$ will have the same acceleration.
- **b.** Block $C$ will have a larger acceleration than block $A$.
- **c.** Block $A$ will have a larger acceleration than block $C$.
- **d.** Block $A$ will not move.
- **e.** None of the above is true.

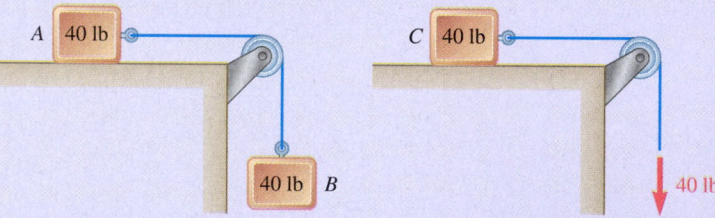

Fig. P12.CQ3

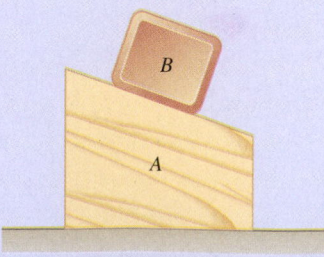

Fig. P12.CQ4

**12.CQ4** Blocks $A$ and $B$ are released from rest in the position shown. Neglecting friction, the normal force between block $A$ and the ground is:
- **a.** Less than the weight of $A$ plus the weight of $B$.
- **b.** Equal to the weight of $A$ plus the weight of $B$.
- **c.** Greater than the weight of $A$ plus the weight of $B$.

**12.CQ5** People sit on a Ferris wheel at points $A$, $B$, $C$, and $D$. The Ferris wheel travels at a constant angular velocity. At the instant shown, which person experiences the largest force from his or her chair (back and seat)? Assume you can neglect the size of the chairs—that is, the people are located the same distance from the axis of rotation.

   **a.** $A$
   **b.** $B$
   **c.** $C$
   **d.** $D$
   **e.** The force is the same for all the passengers.

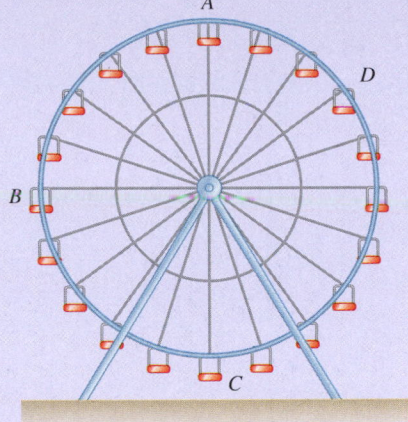

**Fig. P12.CQ5**

## FREE-BODY PRACTICE PROBLEMS

**12.F1** Crate $A$ is gently placed with zero initial velocity onto a moving conveyor belt. The coefficient of kinetic friction between the crate and the belt is $\mu_k$. Draw the free-body diagram (FBD) and kinetic diagram (KD) for $A$ immediately after it contacts the belt.

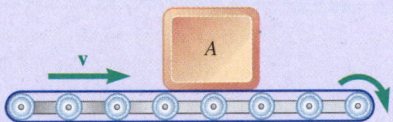

**Fig. P12.F1**

**12.F2** Two blocks weighing $W_A$ and $W_B$ are at rest on a conveyor that is initially at rest. The belt is suddenly started in an upward direction so that slipping occurs between the belt and the boxes. Assuming the coefficient of friction between the boxes and the belt is $\mu_k$, draw the FBDs and KDs for blocks $A$ and $B$. How would you determine if $A$ and $B$ remain in contact?

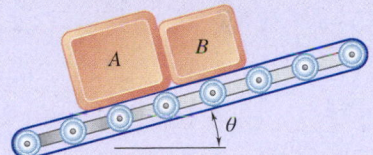

**Fig. P12.F2**

**12.F3** Objects $A$, $B$, and $C$ have masses $m_A$, $m_B$, and $m_C$, respectively. The coefficient of kinetic friction between $A$ and $B$ is $\mu_k$, and the friction between $A$ and the ground is negligible and the pulleys are massless and frictionless. Assuming $B$ slides on $A$, draw the FBD and KD for each of the three masses $A$, $B$, and $C$.

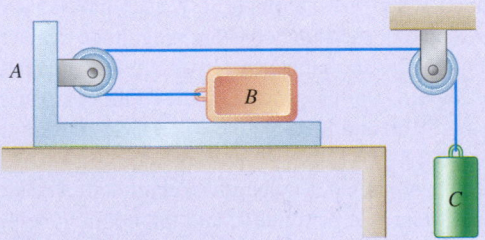

**Fig. P12.F3**

**12.F4** Blocks $A$ and $B$ have masses $m_A$ and $m_B$, respectively. Neglecting friction between all surfaces, draw the FBD and KD for each mass.

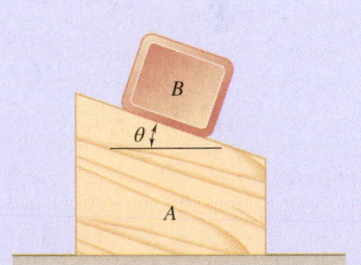

**Fig. P12.F4**

**12.F5** Blocks $A$ and $B$ have masses $m_A$ and $m_B$, respectively. Neglecting friction between all surfaces, draw the FBD and KD for the two systems shown.

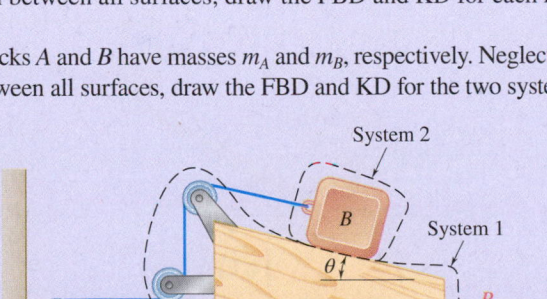

**Fig. P12.F5**

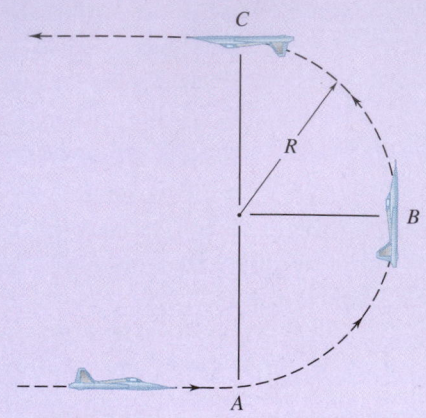

**Fig. P12.F6**

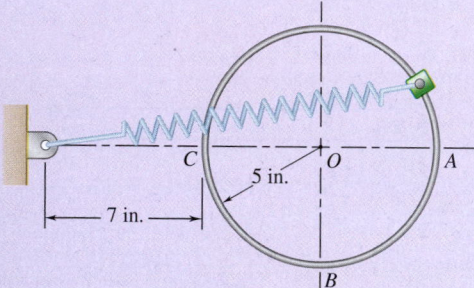

**Fig. P12.F8**

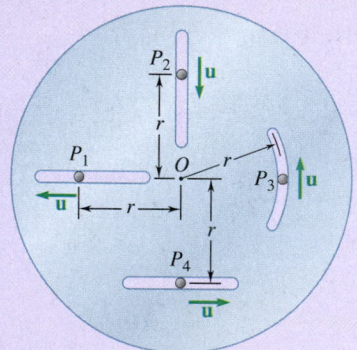

**Fig. P12.F9**

**12.F6** A pilot of mass $m$ flies a jet in a half-vertical loop of radius $R$ so that the speed of the jet, $v$, remains constant. Draw a FBD and KD of the pilot at points $A$, $B$, and $C$.

**12.F7** Wires $AC$ and $BC$ are attached to a sphere that revolves at a constant speed $v$ in the horizontal circle of radius $r$ as shown. Draw a FBD and KD of $C$.

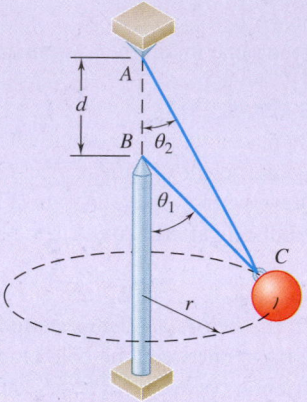

**Fig. P12.F7**

**12.F8** A collar of mass $m$ is attached to a spring and slides without friction along a circular rod in a vertical plane. The spring has an undeformed length of 5 in. and a constant $k$. Knowing that the collar has a speed $v$ at point $C$, draw the FBD and KD of the collar at this point.

**12.F9** Four pins slide in four separate slots cut in a horizontal circular plate as shown. When the plate is at rest, each pin has a velocity directed as shown and of the same constant magnitude $u$. Each pin has a mass $m$ and maintains the same velocity relative to the plate when the plate rotates about $O$ with a constant counterclockwise angular velocity $\omega$. Draw the FBDs and KDs to determine the forces on pins $P_1$ and $P_2$.

**12.F10** At the instant shown, the length of the boom $AB$ is being *decreased* at the constant rate of 0.2 m/s, and the boom is being lowered at the constant rate of 0.08 rad/s. If the mass of the men and lift connected to the boom at point $B$ is $m$, draw the FBD and KD that could be used to determine the horizontal and vertical forces at $B$.

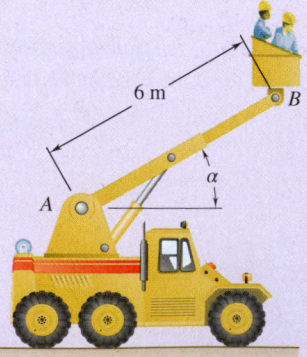

**Fig. P12.F10**

**12.F11** Disk $A$ rotates in a horizontal plane about a vertical axis at the constant rate $\dot{\theta}_0$. Slider $B$ has a mass $m$ and moves in a frictionless slot cut in the disk. The slider is attached to a spring of constant $k$, which is undeformed when $r = 0$. Knowing that the slider is released with no radial velocity in the position $r = r_0$, draw a FBD and KD at an arbitrary distance $r$ from $O$.

**12.F12** Pin $B$ has a mass $m$ and slides along the slot in the rotating arm $OC$ and along the slot $DE$ which is cut in a fixed horizontal plate. Neglecting friction and knowing that rod $OC$ rotates at the constant rate $\dot{\theta}_0$, draw a FBD and KD that can be used to determine the forces $\mathbf{P}$ and $\mathbf{Q}$ exerted on pin $B$ by rod $OC$ and the wall of slot $DE$, respectively.

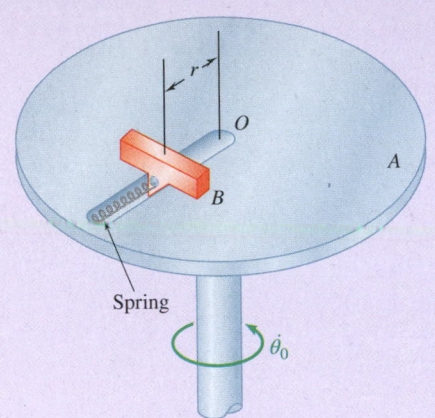

Fig. P12.F11

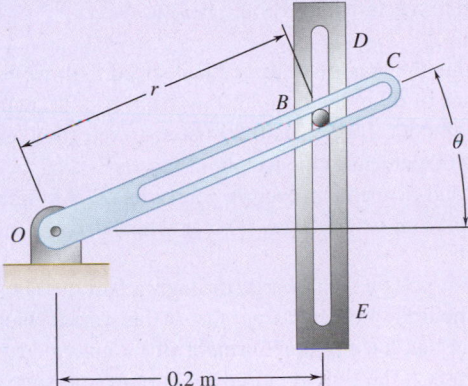

Fig. P12.F12

### END-OF-SECTION PROBLEMS

**12.1** The acceleration due to gravity on Mars is 3.75 m/s². Knowing that the mass of a silver bar has been officially designated as 10 kg, determine its weight in newtons on Mars.

**12.2** The value of $g$ at any latitude $\phi$ may be obtained from the formula

$$g = 32.09(1 + 0.0053 \sin^2 \phi)\text{ft/s}^2$$

which takes into account the effect of the rotation of the earth, as well as the fact that the earth is not truly spherical. Knowing that the weight of a silver bar has been officially designated as 5 lb, determine to four significant figures (a) the mass in slugs, (b) the weight in pounds at latitudes of 0°, 30°, and 50°.

**12.3** A Global Positioning System (GPS) satellite is in a circular orbit 12,580 mi above the surface of the earth and completes one orbit every 12 h. Knowing that the magnitude of the linear momentum of the satellite is $750 \times 10^3$ lb·s and the radius of the earth is 3960 mi, determine (a) the mass of the satellite, (b) the weight of the satellite before it was launched from earth.

**12.4** A spring scale $A$ and a lever scale $B$ having equal lever arms are fastened to the roof of an elevator, and identical packages are attached to the scales as shown. Knowing that when the elevator moves downward with an acceleration of 1 m/s² the spring scale indicates a load of 60 N, determine (a) the weight of the packages, (b) the load indicated by the spring scale and the mass needed to balance the lever scale when the elevator moves upward with an acceleration of 1 m/s².

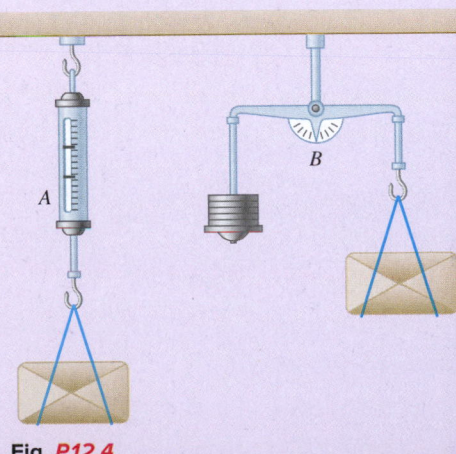

Fig. P12.4

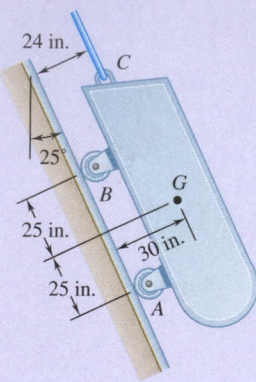

**Fig. P12.5**

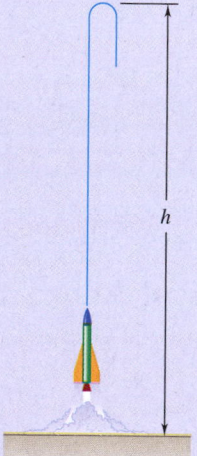

**Fig. P12.6**

**12.5** A loading car is at rest on a track forming an angle of 25° with the vertical when a force is applied to the cable attached at C. The gross weight of the car and its load is 5500 lb, and it acts at point G. Knowing the tension in the cable connected at C is 5000 lb, determine (a) the acceleration of the car, (b) the distance the car moves in 20 s, (c) the time it takes for the car to return to its original position if the cable breaks after 20 s.

**12.6** A 0.5-oz model rocket is launched vertically from rest at time $t = 0$ with a constant thrust of 0.9 lb for 0.3 s and no thrust for $t > 0.3$ s. Neglecting air resistance and the decrease in mass of the rocket, determine (a) the maximum height $h$ reached by the rocket, (b) the time required to reach this maximum height.

**12.7** Determine the maximum theoretical speed that may be achieved over a distance of 60 m by a car starting from rest, knowing that the coefficient of static friction is 0.80 between the tires and the pavement and that 60 percent of the weight of the car is distributed over its front wheels and 40 percent over its rear wheels. Assume (a) four-wheel drive, (b) front-wheel drive, (c) rear-wheel drive.

**12.8** A tugboat pulls a small barge through a harbor. The propeller thrust minus the drag produces a net thrust that varies linearly with speed. Knowing that the combined weight of the tug and barge is 3600 kN, determine (a) the time required to increase the speed from an initial value $v_1 = 1.0$ m/s to a final value $v_2 = 2.5$ m/s, (b) the distance traveled during this time interval.

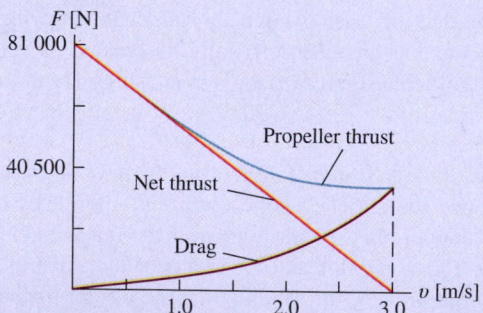

**Fig. P12.8**

**12.9** If an automobile's braking distance from 108 km/h is 75 m on level pavement, determine the automobile's braking distance from 108 km/h when it is (a) going up a 5° incline, (b) going down a 3-percent incline. Assume the braking force is independent of grade.

**12.10** A 4-kg package is released from rest at point $A$ and travels down the conveyor shown. Portions $AB$ and $CD$ are parallel to each other. Neglecting friction and any other energy loss, determine (*a*) the acceleration of the package at $A$, (*b*) the acceleration of the package during the horizontal portion of the track, (*c*) the speed of the package at point $D$.

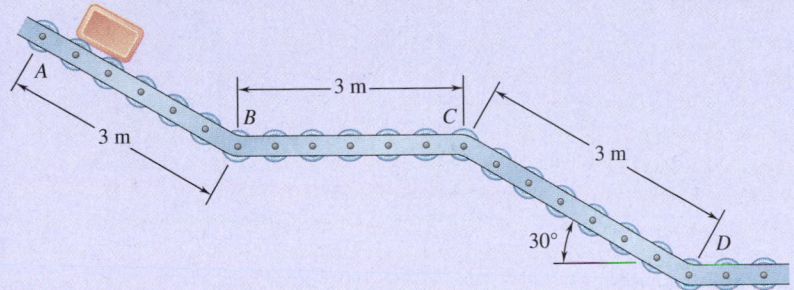

**Fig. P12.10**

**12.11** The coefficients of friction between the load and the flatbed trailer shown are $\mu_s = 0.40$ and $\mu_k = 0.30$. Knowing that the speed of the rig is 72 km/h, determine the shortest distance in which the rig can be brought to a stop if the load is not to shift.

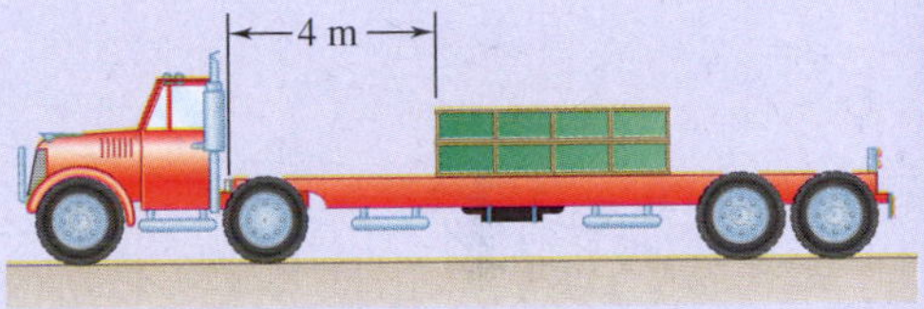

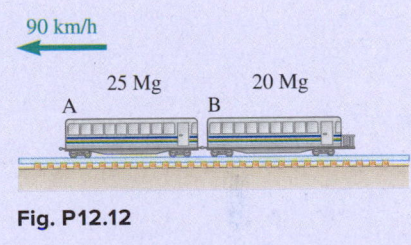

**Fig. P12.11**

**12.12** A light train made up of two cars is traveling at 90 km/h when the brakes are applied to both cars. Knowing that car $A$ has a mass of 25 Mg and car $B$ a mass of 20 Mg, and that the braking force is 30 kN on each car, determine (*a*) the distance traveled by the train before it comes to a stop, (*b*) the force in the coupling between the cars while the train is slowing down.

**Fig. P12.12**

**12.13** The two blocks shown are originally at rest. Neglecting the masses of the pulleys and the effect of friction in the pulleys and between the blocks and the incline, determine (*a*) the acceleration of each block, (*b*) the tension in the cable.

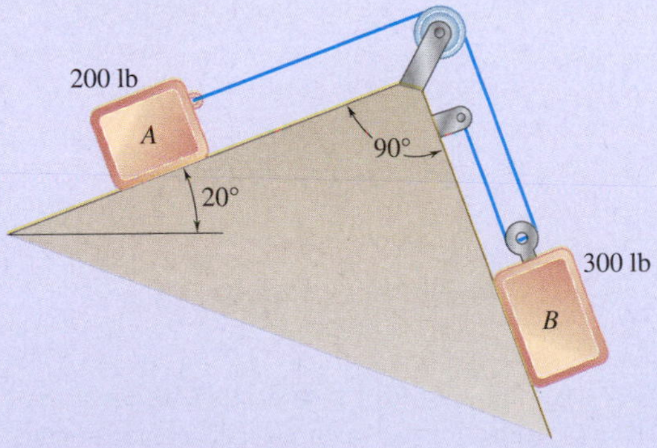

**Fig. P12.13 and P12.14**

**12.14** The two blocks shown are originally at rest. Neglecting the masses of the pulleys and the effect of friction in the pulleys and knowing that the coefficients of friction between the blocks and the inclines are $\mu_s = 0.25$ and $\mu_k = 0.20$, determine (*a*) the acceleration of each block, (*b*) the tension in the cable.

753

**12.15** Each of the systems shown is initially at rest. Neglecting axle friction and the masses of the pulleys, determine for each system (*a*) the acceleration of block *A*, (*b*) the velocity of block *A* after it has moved through 10 ft, (*c*) the time required for block *A* to reach a velocity of 20 ft/s.

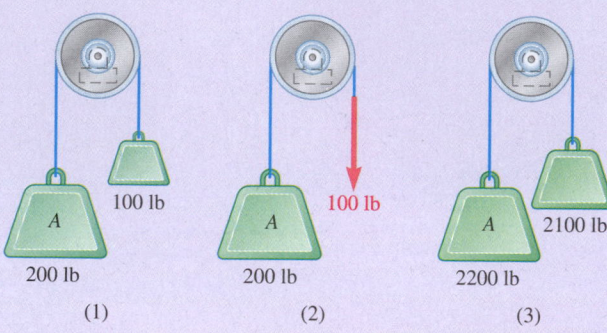

(1)　　　　(2)　　　　(3)

**Fig. P12.15**

**12.16** Boxes *A* and *B* are at rest on a conveyor belt that is initially at rest. The belt is suddenly started in an upward direction so that slipping occurs between the belt and the boxes. Knowing that the coefficients of kinetic friction between the belt and the boxes are $(\mu_k)_A = 0.30$ and $(\mu_k)_B = 0.32$, determine the initial acceleration of each box.

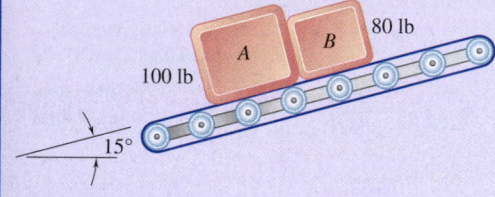

**Fig. P12.16**

**12.17** A 5000-lb truck is being used to lift a 1000-lb boulder *B* that is on a 200-lb pallet *A*. Knowing the acceleration of the truck is 1 ft/s², determine (*a*) the horizontal force between the tires and the ground, (*b*) the force between the boulder and the pallet.

**Fig. P12.17**

**12.18** Block *A* has a mass of 40 kg, and block *B* has a mass of 8 kg. The coefficients of friction between all surfaces of contact are $\mu_s = 0.20$ and $\mu_k = 0.15$. If $P = 0$, determine (*a*) the acceleration of block *B*, (*b*) the tension in the cord.

**12.19** Block *A* has a mass of 40 kg, and block *B* has a mass of 8 kg. The coefficients of friction between all surfaces of contact are $\mu_s = 0.20$ and $\mu_k = 0.15$. If $P = 40$ N, determine (*a*) the acceleration of block *B*, (*b*) the tension in the cord.

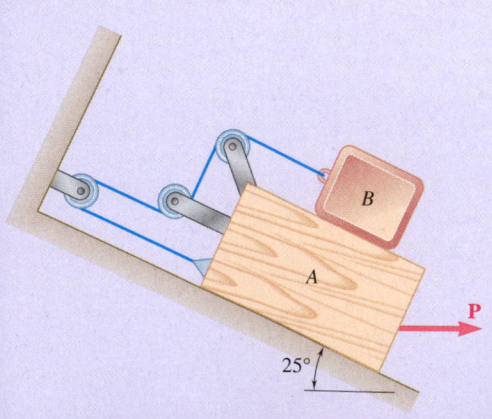

**Fig. P12.18 and P12.19**

**12.20** The flat-bed trailer carries two 1500-kg beams with the upper beam secured by a cable. The coefficients of static friction between the two beams and between the lower beam and the bed of the trailer are 0.25 and 0.30, respectively. Knowing that the load does not shift, determine (a) the maximum acceleration of the trailer and the corresponding tension in the cable, (b) the maximum deceleration of the trailer.

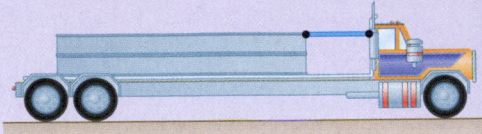

Fig. **P12.20**

**12.21** The position of the 10-lb machine block B is adjusted by moving the 5-lb wedge A. Neglect friction between all surfaces of contact. Knowing that **P** = 10 lb, determine the (a) acceleration of B, (b) force between A and B.

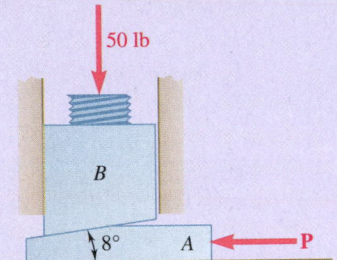

Fig. **P12.21**

**12.22** To unload a bound stack of plywood from a truck, the driver first tilts the bed of the truck and then accelerates from rest. Knowing that the coefficients of friction between the bottom sheet of plywood and the bed are $\mu_s = 0.40$ and $\mu_k = 0.30$, determine (a) the smallest acceleration of the truck which will cause the stack of plywood to slide, (b) the acceleration of the truck which causes corner A of the stack to reach the end of the bed in 0.9 s.

Fig. **P12.22**

**12.23** To transport a series of bundles of shingles A to a roof, a contractor uses a motor-driven lift consisting of a horizontal platform BC which rides on rails attached to the sides of a ladder. The lift starts from rest and initially moves with a constant acceleration **a**₁ as shown. The lift then decelerates at a constant rate **a**₂ and comes to rest at D, near the top of the ladder. Knowing that the coefficient of static friction between a bundle of shingles and the horizontal platform is 0.30, determine the largest allowable acceleration **a**₁ and the largest allowable deceleration **a**₂ if the bundle is not to slide on the platform.

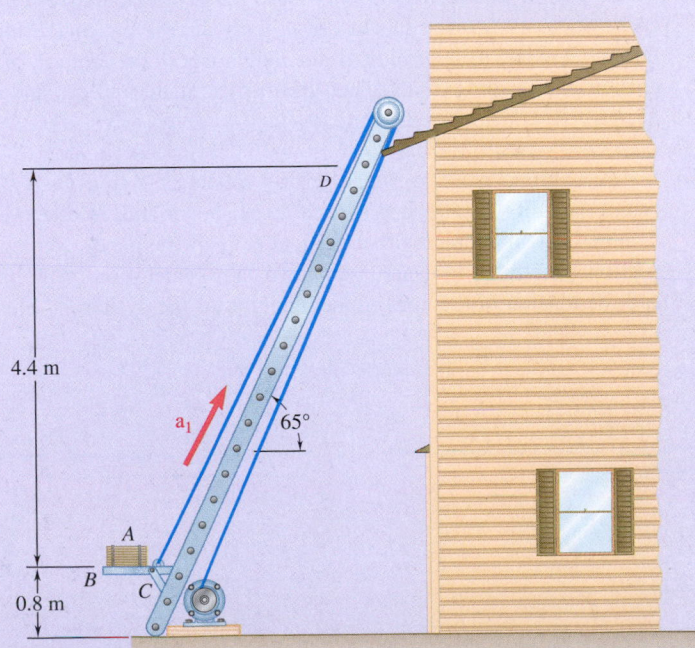

Fig. **P12.23**

**12.24** An airplane has a mass of 25 Mg and its engines develop a total thrust of 40 kN during take-off. If the drag **D** exerted on the plane has a magnitude $D = 2.25v^2$, where $v$ is expressed in meters per second and $D$ in newtons, and if the plane becomes airborne at a speed of 240 km/h, determine (a) the length of runway required for the plane to take off, (b) the time required to take off.

**12.25** Determine the maximum theoretical speed that a 1225 kg automobile starting from rest can reach after traveling 400 m if air resistance is considered. Assume that the coefficient of static friction between the tires and the pavement is 0.70, that the automobile has front-wheel drive, that the front wheels support 62 percent of the automobile's weight, and that the aerodynamic drag **D** has a magnitude $D = 0.575v^2$, where $D$ and $v$ are expressed in newtons and m/s, respectively.

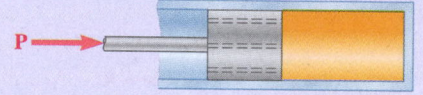

Fig. *P12.26*

**12.26** A constant force **P** is applied to a piston and rod of total mass $m$ to make them move in a cylinder filled with oil. As the piston moves, the oil is forced through orifices in the piston and exerts on the piston a force of magnitude $kv$ in a direction opposite to the motion of the piston. Knowing that the piston starts from rest at $t = 0$ and $x = 0$, show that the equation relating $x$, $v$, and $t$, where $x$ is the distance traveled by the piston and $v$ is the speed of the piston, is linear in each of these variables.

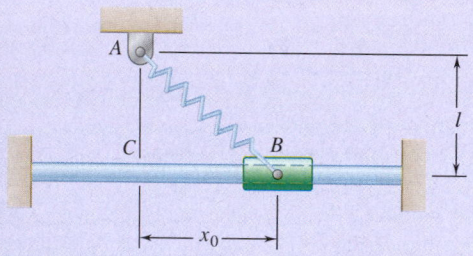

Fig. **P12.27**

**12.27** A spring $AB$ of constant $k$ is attached to a support at $A$ and to a collar of mass $m$. The unstretched length of the spring is $l$. Knowing that the collar is released from rest at $x = x_0$ and neglecting friction between the collar and the horizontal rod, determine the magnitude of the velocity of the collar as it passes through point $C$.

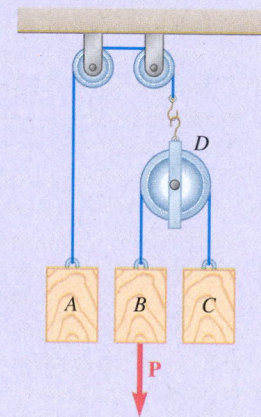

Fig. **P12.28**

**12.28** Block $A$ has a mass of 10 kg, and blocks $B$ and $C$ have masses of 5 kg each. Knowing that the blocks are initially at rest and that $B$ moves through 3 m in 2 s, determine (a) the magnitude of the force **P**, (b) the tension in the cord $AD$. Neglect the masses of the pulleys and axle friction.

**12.29** A 40-lb sliding panel is supported by rollers at $B$ and $C$. A 25-lb counterweight $A$ is attached to a cable as shown and, in cases $a$ and $c$, is initially in contact with a vertical edge of the panel. Neglecting friction, determine in each case shown the acceleration of the panel and the tension in the cord immediately after the system is released from rest.

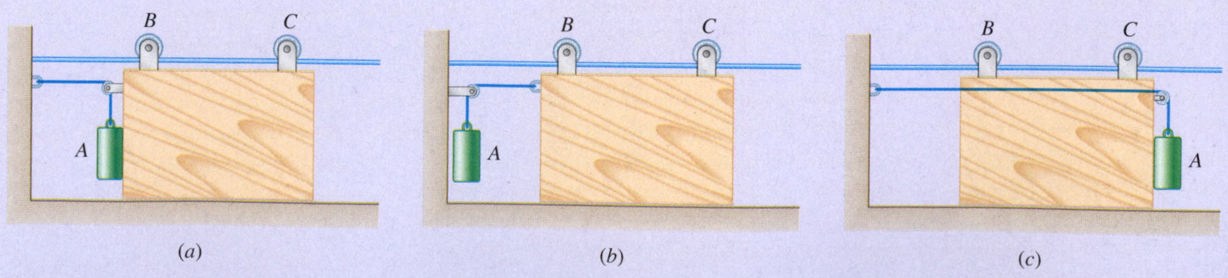

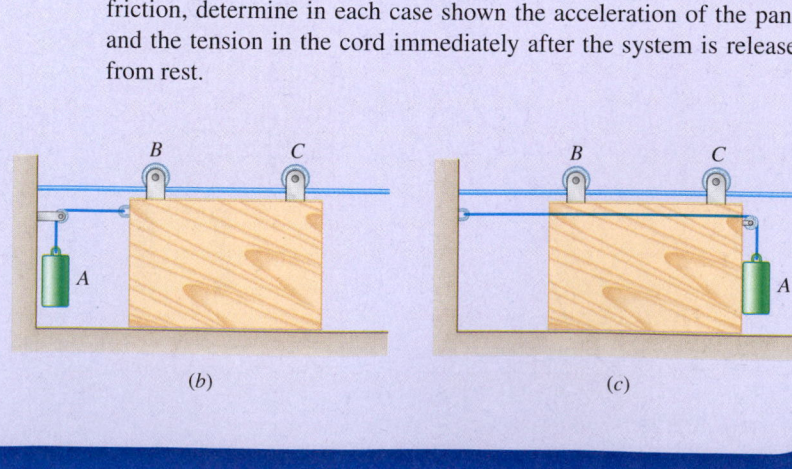

(a)                    (b)                    (c)

Fig. **P12.29**

**12.30** An athlete pulls handle $A$ to the left with a constant force of $P = 100$ N. Knowing that after the handle $A$ has been pulled 30 cm its velocity is 3 m/s, determine the mass of the weight stack $B$.

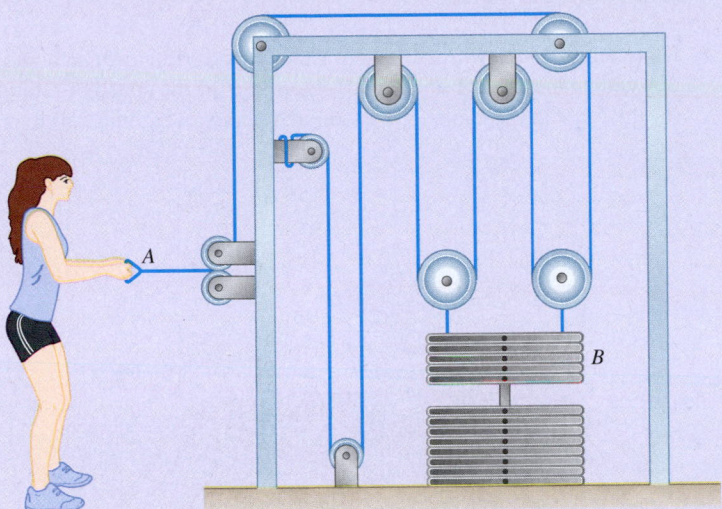

**Fig. P12.30**

**12.31** A 10-lb block $B$ rests as shown on a 20-lb bracket $A$. The coefficients of friction are $\mu_s = 0.30$ and $\mu_k = 0.25$ between block $B$ and bracket $A$, and there is no friction in the pulley or between the bracket and the horizontal surface. (a) Determine the maximum weight of block $C$ if block $B$ is not to slide on bracket $A$. (b) If the weight of block $C$ is 10 percent larger than the answer found in $a$, determine the accelerations of $A$, $B$, and $C$.

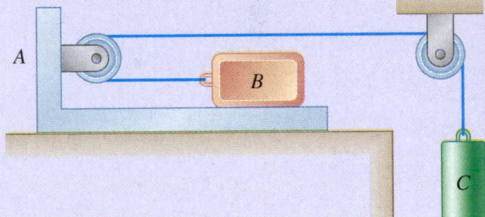

**Fig. P12.31**

**12.32** Knowing that $\mu_k = 0.30$, determine the acceleration of each block when $m_A = m_B = m_C$.

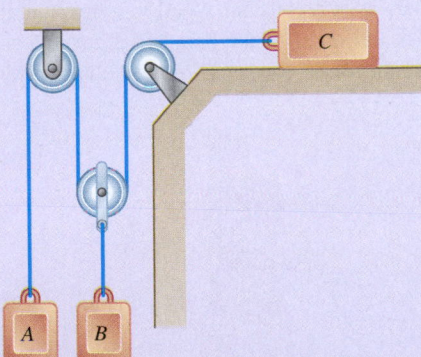

**Fig. P12.32 and P12.33**

**12.33** Knowing that $\mu_k = 0.30$, determine the acceleration of each block when $m_A = 5$ kg, $m_B = 30$ kg, and $m_C = 15$ kg.

**12.34** The 30-lb block $B$ is supported by the 55-lb block $A$ and is attached to a cord to which a 50-lb horizontal force is applied as shown. Neglecting friction, determine (a) the acceleration of block $A$, (b) the acceleration of block $B$ relative to $A$.

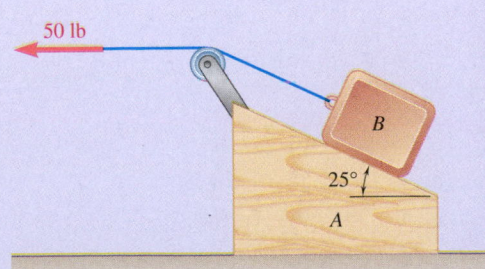

**Fig. P12.34**

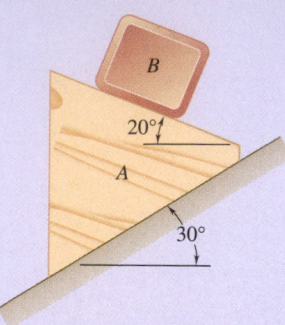

**Fig. P12.35**

**12.35** Block *B* of mass 10 kg rests as shown on the upper surface of a 22-kg wedge *A*. Knowing that the system is released from rest and neglecting friction, determine (*a*) the acceleration of *B*, (*b*) the velocity of *B* relative to *A* at *t* = 0.5 s.

**12.36** Knowing that the swings of an amusement park ride form an angle of 40° with respect to the horizontal, determine (*a*) the speed of rotation, (*b*) the force in the cable for a swing and person weighing 250 lb.

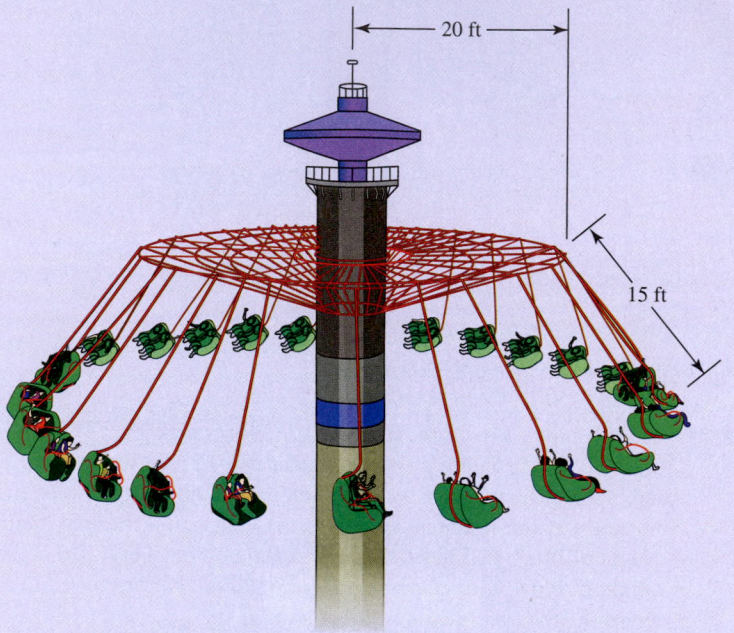

**Fig. P12.36**

**12.37** During a hammer thrower's practice swings, the 7.1-kg head *A* of the hammer revolves at a constant speed in a horizontal circle as shown. Knowing that the speed of the hammer is 2.5 m/s and $\theta = 60°$, determine (*a*) the tension in wire *BC*, (*b*) the radius of the circle, $\rho$.

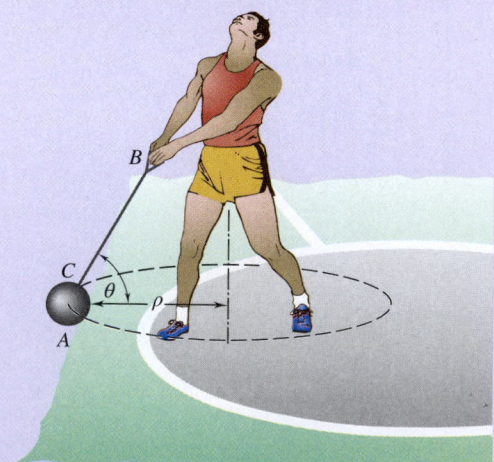

**Fig. P12.37**

**12.38** Human centrifuges are often used to simulate different acceleration levels for pilots. When aerospace physiologists say that a pilot is pulling 9 g's, they mean that the resultant normal force on the pilot from the bottom of the seat is nine times their weight. Knowing that the centrifuge starts from rest and has a constant angular acceleration of 1.5 RPM per second until the pilot is pulling 9 g's and then continues with a constant angular velocity, determine (a) how long it will take for the pilot to reach 9 g's (b) the angle θ of the normal force once the pilot reaches 9 g's.

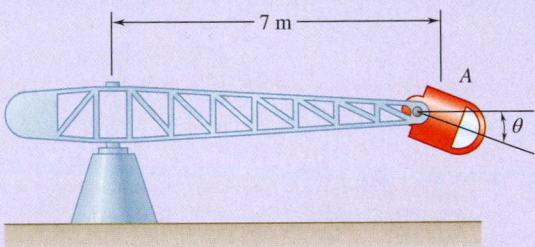

**Fig. P12.38**

**12.39** A single wire ACB passes through a ring at C attached to a sphere which revolves at a constant speed v in the horizontal circle shown. Knowing that the tension is the same in both portions of the wire, determine the speed v.

**\*12.40** Two wires AC and BC are tied at C to a sphere that revolves at a constant speed v in the horizontal circle shown. Determine the range of the allowable values of v if both wires are to remain taut and if the tension in either of the wires is not to exceed 60 N.

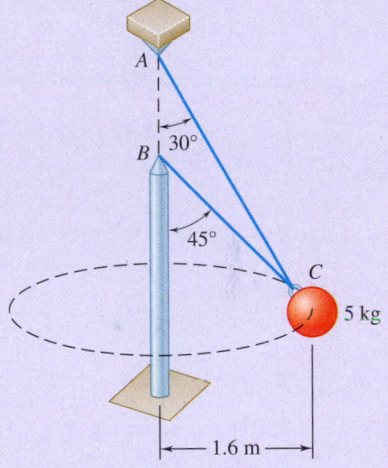

**Fig. P12.39 and P12.40**

**12.41** A 2-lb sphere is at rest relative to a parabolic dish which rotates at a constant rate about a vertical axis. Neglecting friction and knowing that r = 3 ft, determine (a) the velocity v of the sphere, (b) the magnitude of the normal force exerted by the sphere on the inclined surface of the dish.

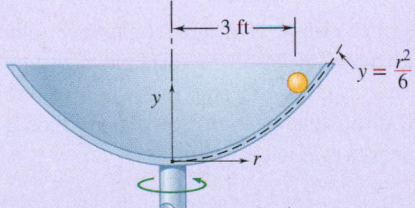

**Fig. P12.41**

**\*12.42** The 0.5-kg flyballs of a centrifugal governor revolve at a constant speed v in the horizontal circle of 150-mm radius shown. Neglecting the mass of links AB, BC, AD, and DE, and requiring that the links support only tensile forces, determine the range of the allowable values of v so that the magnitudes of the forces in the links do not exceed 75 N.

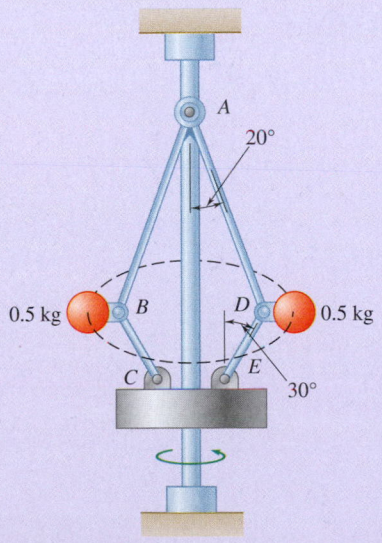

**Fig. P12.42**

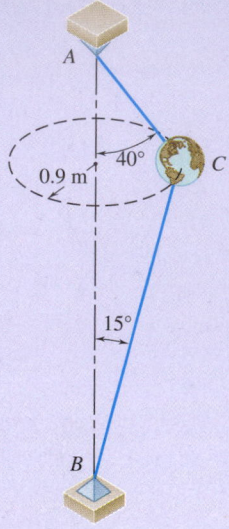

Fig. P12.43

**\*12.43** As part of an outdoor display, a 5-kg model $C$ of the earth is attached to wires $AC$ and $BC$ and revolves at a constant speed $v$ in the horizontal circle shown. Determine the range of the allowable values of $v$ if both wires are to remain taut and if the tension in either of the wires is not to exceed 116 N.

**12.44** A 130-lb wrecking ball $B$ is attached to a 45-ft-long steel cable $AB$ and swings in the vertical arc shown. Determine the tension in the cable (*a*) at the top $C$ of the swing, (*b*) at the bottom $D$ of the swing, where the speed of $B$ is 13.2 ft/s.

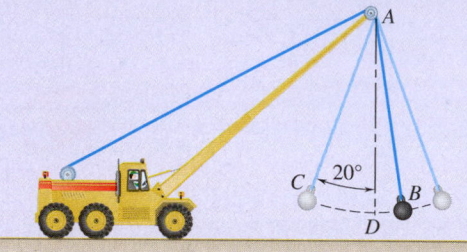

Fig. P12.44

**12.45** During a high-speed chase, a 2400-lb sports car traveling at a speed of 100 mi/h just loses contact with the road as it reaches the crest $A$ of a hill. (*a*) Determine the radius of curvature $\rho$ of the vertical profile of the road at $A$. (*b*) Using the value of $\rho$ found in part *a*, determine the force exerted on a 160-lb driver by the seat of his 3100-lb car as the car, traveling at a constant speed of 50 mi/h, passes through $A$.

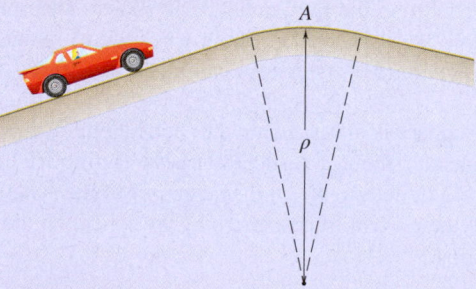

Fig. P12.45

**12.46** An airline pilot climbs to a new flight level along the path shown. Knowing that the speed of the airplane decreases at a constant rate from 180 m/s at point $A$ to 160 m/s at point $C$, determine the magnitude of the abrupt change in the force exerted on a 90-kg passenger as the airplane passes point $B$.

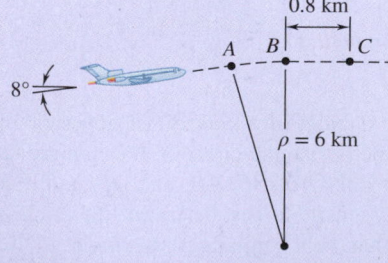

Fig. P12.46

**12.47** The roller-coaster track shown is contained in a vertical plane. The portion of track between $A$ and $B$ is straight and horizontal, while the portions to the left of $A$ and to the right of $B$ have radii of curvature as indicated. A car is traveling at a speed of 72 km/h when the brakes are suddenly applied, causing the wheels of the car to slide on the track ($\mu_k = 0.20$). Determine the initial deceleration of the car if the brakes are applied as the car (a) has almost reached $A$, (b) is traveling between $A$ and $B$, (c) has just passed $B$.

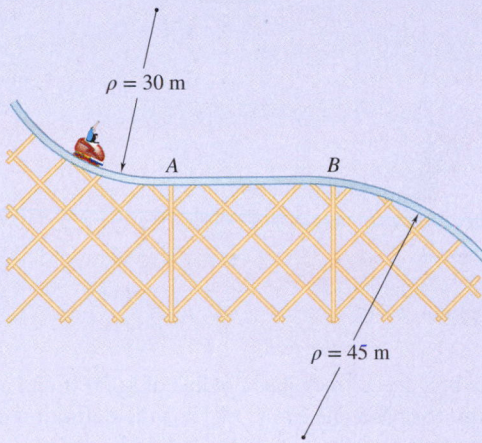

**Fig. P12.47**

**12.48** A spherical-cap governor is fixed to a vertical shaft that rotates with angular velocity $\omega$. When the string-supported clapper of mass $m$ touches the cap, a cutoff switch is operated electrically to reduce the speed of the shaft. Knowing that the radius of the clapper is small relative to the cap, determine the minimum angular speed at which the cutoff switch operates.

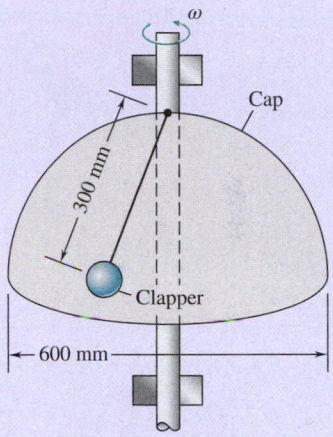

**Fig. P12.48**

**12.49** A series of small packages, each with a mass of 0.5 kg, are discharged from a conveyor belt as shown. Knowing that the coefficient of static friction between each package and the conveyor belt is 0.4, determine (a) the force exerted by the belt on the package just after it has passed point $A$, (b) the angle $\theta$ defining the point $B$ where the packages first *slip* relative to the belt.

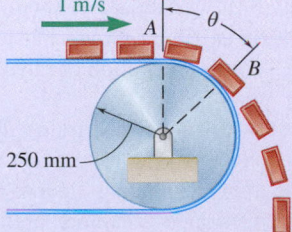

**Fig. P12.49**

**12.50** A 55-kg pilot flies a jet trainer in a half vertical loop of 1200-m radius so that the speed of the trainer decreases at a constant rate. Knowing that the plane has a speed of 550 km/h at point $A$, and the pilot experiences weightlessness at point $C$ (i.e., the normal force from the seat bottom is zero), determine (a) the deceleration of the plane, (b) the force exerted on her by the seat of the trainer when the trainer is at point $B$.

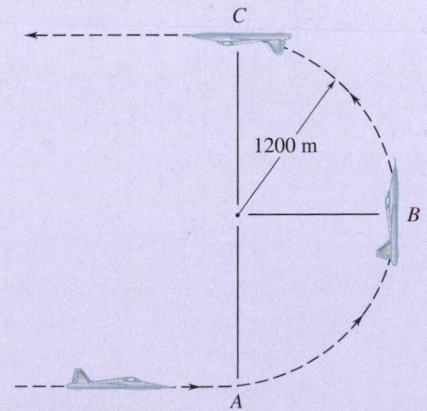

**Fig. P12.50**

**12.51** A carnival ride is designed to allow the general public to experience high-acceleration motion. The ride rotates about point $O$ in a horizontal circle such that the rider has a speed $v_0$. The rider reclines on a platform $A$ which rides on rollers such that friction is negligible. A mechanical stop prevents the platform from rolling down the incline. Determine (a) the speed $v_0$ at which the platform $A$ begins to roll upward, (b) the normal force experienced by an 80-kg rider at this speed.

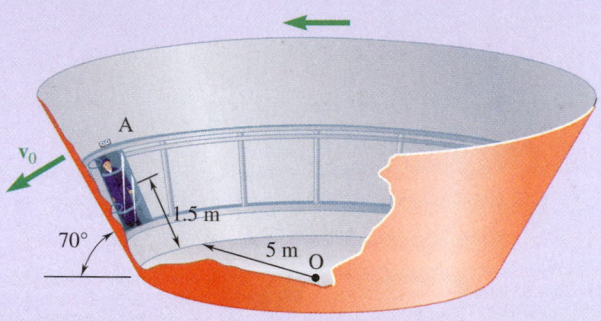

Fig. P12.51

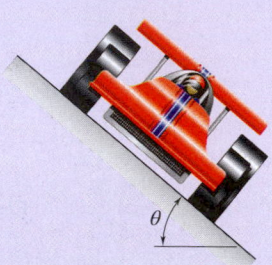

Fig. *P12.52*

**12.52** A curve in a speed track has a radius of 1000 ft and a rated speed of 120 mi/h. (See Sample Prob. 12.7 for the definition of rated speed.) Knowing that a racing car starts skidding on the curve when traveling at a speed of 180 mi/h, determine (a) the banking angle $\theta$, (b) the coefficient of static friction between the tires and the track under the prevailing conditions, (c) the minimum speed at which the same car could negotiate the curve.

**12.53** Tilting trains, such as the Acela Express that serves the Northeast Corridor in the Northeast United States, are designed to travel safely at high speeds on curved sections of track which were built for slower, conventional trains. As it enters a curve, each car is tilted by hydraulic actuators mounted on its trucks. The tilting feature of the cars also increases passenger comfort by eliminating or greatly reducing the side force $\mathbf{F}_s$ (parallel to the floor of the car) to which passengers feel subjected. For a train traveling at 100 mi/h on a curved section of track banked through an angle $\theta = 6°$ and with a rated speed of 60 mi/h, determine (a) the magnitude of the side force felt by a passenger of weight $W$ in a standard car with no tilt ($\phi = 0$), (b) the required angle of tilt $\phi$ if the passenger is to feel no side force. (See Sample Prob. 12.7 for the definition of rated speed.)

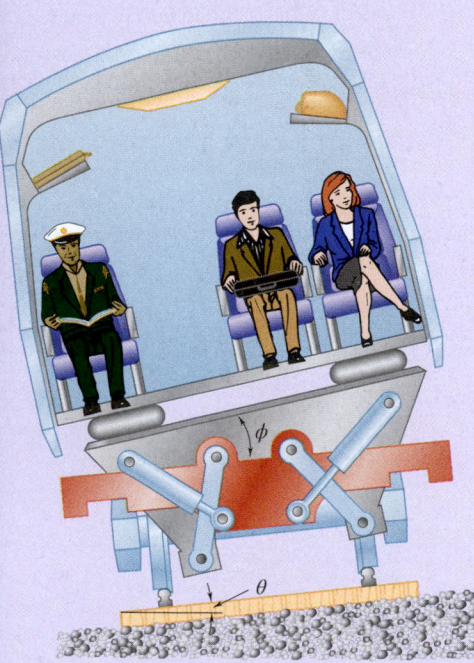

Fig. P12.53 and *P12.54*

**12.54** Tests carried out with the tilting trains described in Prob. 12.53 revealed that passengers feel queasy when they see through the car windows that the train is rounding a curve at high speed, yet do not feel any side force. Designers, therefore, prefer to reduce, but not eliminate that force. For the train of Prob. 12.53, determine the required angle of tilt $\phi$ if passengers are to feel side forces equal to 10 percent of their weights.

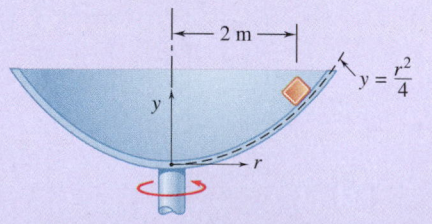

Fig. P12.55

**12.55** A 3-kg block is at rest relative to a parabolic dish which rotates at a constant rate about a vertical axis. Knowing that the coefficient of static friction is 0.5 and that $r = 2$ m, determine the maximum allowable velocity $v$ of the block.

**12.56** A polisher is started so that the fleece along the circumference undergoes a constant tangential acceleration of 4 m/s². Three seconds after it is started, small tufts of fleece from along the circumference of the 225-mm-diameter polishing pad are observed to fly free of the pad. At this instant, determine (a) the speed $v$ of a tuft as it leaves the pad, (b) the magnitude of the force required to free a tuft if the average mass of a tuft is 1.6 mg.

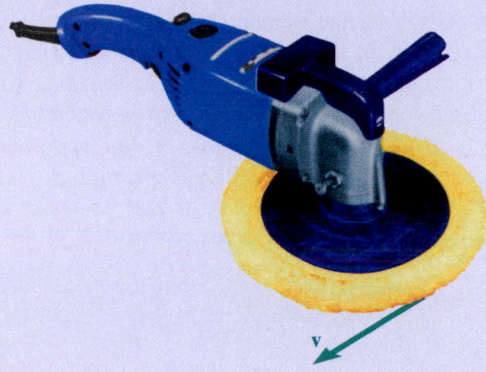

Fig. P12.56

**12.57** A turntable $A$ is built into a stage for use in a theatrical production. It is observed during a rehearsal that a trunk $B$ starts to slide on the turntable 10 s after the turntable begins to rotate. Knowing that the trunk undergoes a constant tangential acceleration of 0.24 m/s², determine the coefficient of static friction between the trunk and the turntable.

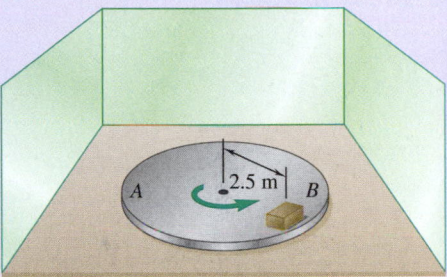

Fig. P12.57

**12.58** The carnival ride from Prob. 12.51 is modified so that the 80-kg riders can move up and down the inclined wall as the speed of the ride increases. Assuming that the friction between the wall and the carriage is negligible, determine the position $h$ of the rider if the speed $v_0 = 13$ m/s.

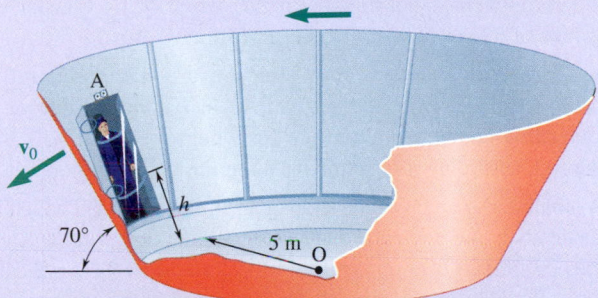

Fig. P12.58 and *P12.59*

**12.59** The carnival ride from Prob. 12.51 is modified so that the 80-kg riders can move up and down the inclined wall as the speed of the ride increases. Knowing that the coefficient of static friction between the wall and the platform is 0.2, determine the range of values of the constant speed $v_0$ for which the platform will remain at $h = 1.5$ m.

**12.60** A small 8-oz collar $D$ can slide on portion $AB$ of a rod which is bent as shown. Knowing that the rod rotates about the vertical $AC$ at a constant rate and that $\alpha = 40°$ and $r = 24$ in., determine the range of values of the speed $v$ for which the collar will not slide on the rod if the coefficient of static friction between the rod and the collar is 0.35.

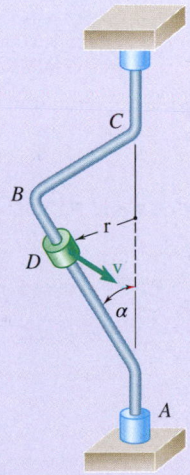

Fig. *P12.60*

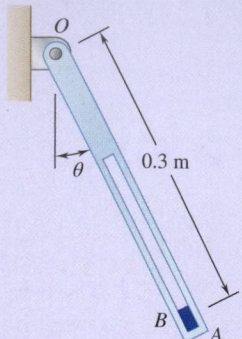

Fig. P12.61

**12.61** A small block $B$ fits inside a slot cut in arm $OA$ that rotates in a vertical plane at a constant rate. The block remains in contact with the end of the slot closest to $A$ and its speed is 1.4 m/s for $0 \le \theta \le 150°$. Knowing that the block begins to slide when $\theta = 150°$, determine the coefficient of static friction between the block and the slot.

**12.62** The parallel-link mechanism $ABCD$ is used to transport a component $I$ between manufacturing processes at stations $E$, $F$, and $G$ by picking it up at a station when $\theta = 0$ and depositing it at the next station when $\theta = 180°$. Knowing that member $BC$ remains horizontal throughout its motion and that links $AB$ and $CD$ rotate at a constant rate in a vertical plane in such a way that $v_B = 2.2$ ft/s, determine (a) the minimum value of the coefficient of static friction between the component and $BC$ if the component is not to slide on $BC$ while being transferred, (b) the values of $\theta$ for which sliding is impending.

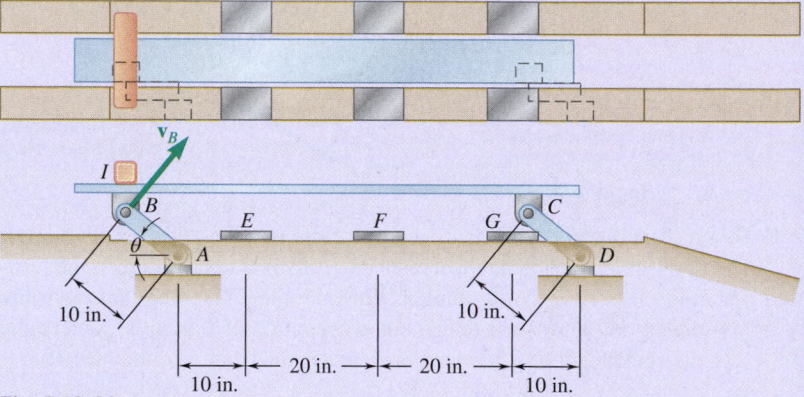

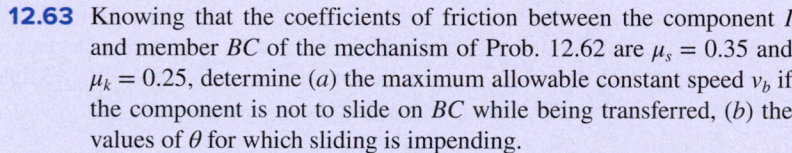

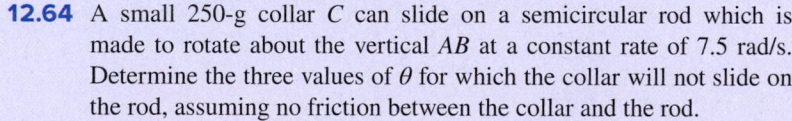

Fig. P12.62

**12.63** Knowing that the coefficients of friction between the component $I$ and member $BC$ of the mechanism of Prob. 12.62 are $\mu_s = 0.35$ and $\mu_k = 0.25$, determine (a) the maximum allowable constant speed $v_b$ if the component is not to slide on $BC$ while being transferred, (b) the values of $\theta$ for which sliding is impending.

**12.64** A small 250-g collar $C$ can slide on a semicircular rod which is made to rotate about the vertical $AB$ at a constant rate of 7.5 rad/s. Determine the three values of $\theta$ for which the collar will not slide on the rod, assuming no friction between the collar and the rod.

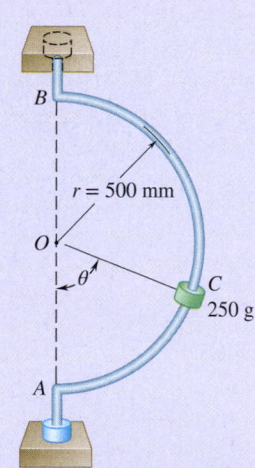

Fig. P12.64 and P12.65

**12.65** A small 250-g collar $C$ can slide on a semicircular rod which is made to rotate about the vertical $AB$ at a constant rate of 7.5 rad/s. Knowing that the coefficients of friction are $\mu_s = 0.25$ and $\mu_k = 0.20$, indicate whether the collar will slide on the rod if it is released in the position corresponding to (a) $\theta = 75°$, (b) $\theta = 40°$. Also, determine the magnitude and direction of the friction force exerted on the collar immediately after release.

**12.66** An advanced spatial disorientation trainer allows the cab to rotate around multiple axes, as well as to extend inward and outward. It can be used to simulate driving, fixed-wing aircraft flying, and helicopter maneuvering. In one training scenario, the trainer rotates and translates in the horizontal plane, where the location of the pilot is defined by the relationships $r = 10 + 2 \cos(\frac{\pi}{5}t)$ and $\theta = 0.1(2t^2 - t)$, where $r$, $\theta$, and $t$ are expressed in feet, radians, and seconds, respectively. Knowing that the pilot has a weight of 175 lbs, (a) determine the magnitude of the resulting force acting on the pilot at $t = 5$ s, (b) plot the magnitudes of the radial and transverse components of the force exerted on the pilot from 0 to 10 seconds.

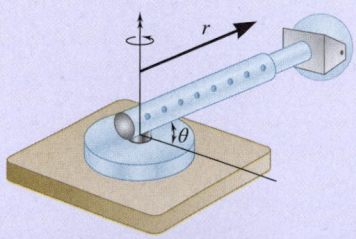

Fig. P12.66 and P12.67

**12.67** An advanced spatial disorientation trainer is programmed to only rotate and translate in the horizontal plane. The pilot's location is defined by the relationships $r = 8(1 - e^{-t})$ and $\theta = 2/\pi(\sin \frac{\pi}{2} t)$, where $r$, $\theta$, and $t$ are expressed in feet, radians, and seconds, respectively. Determine the radial and transverse components of the force exerted on the 175-lb pilot at $t = 3$ s.

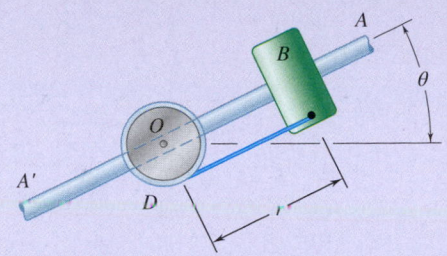

**Fig. P12.68**

**12.68** The 3-kg collar $B$ slides on the frictionless arm $AA'$. The arm is attached to drum $D$ and rotates about $O$ in a horizontal plane at the rate $\dot{\theta} = 0.75t$, where $\dot{\theta}$ and $t$ are expressed in rad/s and seconds, respectively. As the arm-drum assembly rotates, a mechanism within the drum releases cord so that the collar moves outward from $O$ with a constant speed of 0.5 m/s. Knowing that at $t = 0$, $r = 0$, determine the time at which the tension in the cord is equal to the magnitude of the horizontal force exerted on $B$ by arm $AA'$.

**12.69** A 0.5-kg block $B$ slides without friction inside a slot cut in arm $OA$ that rotates in a vertical plane. The rod has a constant angular acceleration $\ddot{\theta} = 10$ rad/s². Knowing that when $\theta = 45°$ and $r = 0.8$ m the velocity of the block is zero, determine at this instant, (a) the force exerted on the block by the arm, (b) the relative acceleration of the block with respect to the arm.

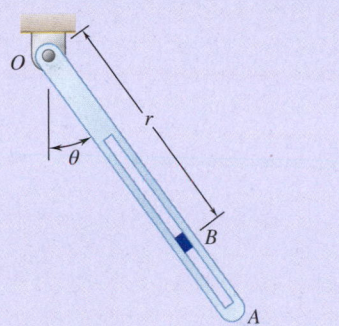

**Fig. P12.69**

**12.70** Pin $B$ weighs 4 oz and is free to slide in a horizontal plane along the rotating arm $OC$ and along the fixed circular slot $DE$ of radius $b = 20$ in. Neglecting friction and assuming that $\dot{\theta} = 15$ rad/s and $\ddot{\theta} = 250$ rad/s² for the position $\theta = 20°$, determine for that position (a) the radial and transverse components of the resultant force exerted on pin $B$, (b) the forces **P** and **Q** exerted on pin $B$, respectively, by rod $OC$ and the wall of slot $DE$.

**12.71** The parasailing system shown uses a winch to let rope out at a constant rate so that the 70-kg rider moves away from the boat, which is traveling with a constant velocity. At the instant shown, the rope has a length of 30 m, it is increasing in length at a constant 1 m/s, the angle is increasing at a rate of 0.05 rad/s, and $\ddot{\theta}$ is −0.01 rad/s². Knowing that when the rope makes a 30° angle with respect to the water, the tension in the rope is 10 kN, determine the magnitude and direction of the force of the parasail on the parasailor.

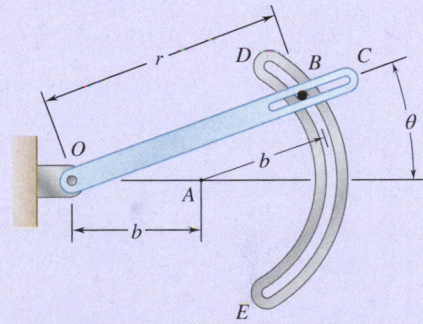

**Fig. P12.70**

**Fig. P12.71**

**12.72** A 700-kg horse $A$ lifts a 50-kg hay bale $B$ as shown. At the instant when $r = 8$ m and $\theta = 60°$, the velocity and acceleration of the horse are 2 m/s to the right and 0.5 m/s² to the left, respectively. Neglecting the mass of the pulley, determine at that instant (*a*) the tension in the cable, (*b*) the average horizontal force between the ground and the horse's feet.

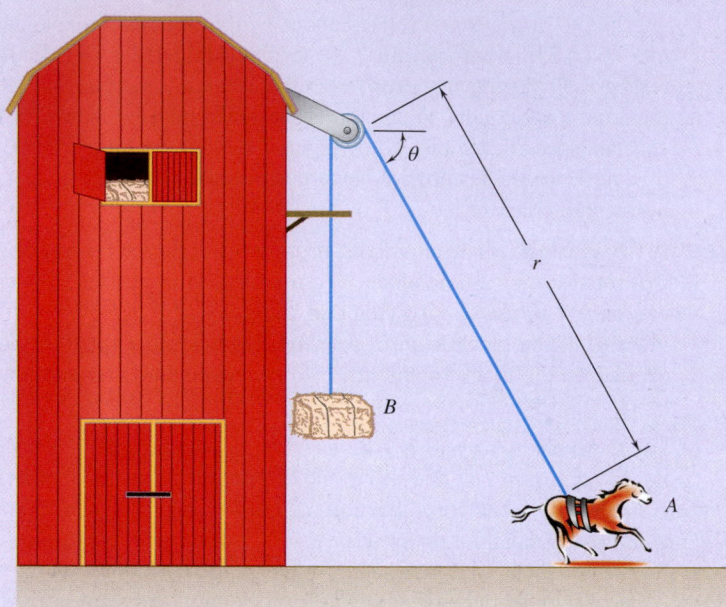

Fig. P12.72

**\*12.73** Slider $C$ has a weight of 0.5 lb and may move in a slot cut in arm $AB$, which rotates at the constant rate $\dot{\theta}_0 = 10$ rad/s in a horizontal plane. The slider is attached to a spring of constant $k = 2.5$ lb/ft, which is unstretched when $r = 0$. Knowing that the slider is released from rest with no radial velocity in the position $r = 18$ in. and neglecting friction, determine for the position $r = 12$ in. (*a*) the radial and transverse components of the velocity of the slider, (*b*) the radial and transverse components of its acceleration, (*c*) the horizontal force exerted on the slider by arm $AB$.

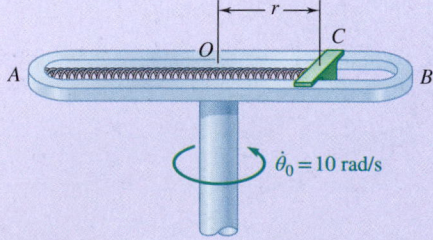

Fig. *P12.73*

# 12.2   ANGULAR MOMENTUM AND ORBITAL MOTION

In Sec. 12.1, we introduced the idea of linear momentum and showed how Newton's second law could be expressed as the rate of change of linear momentum. Angular momentum, or the moment of linear momentum, is another useful quantity. In this section, we define angular momentum for a particle and discuss the motion of a particle under a central force, which is applicable to many types of orbital motion.

## 12.2A   Angular Momentum of a Particle and Its Rate of Change

Consider a particle $P$ with a mass $m$ moving with respect to a newtonian frame of reference $Oxyz$. As we saw in Sec. 12.1B, the linear momentum of the particle at a given instant is defined as the vector $m\mathbf{v}$ that is obtained by multiplying the velocity $\mathbf{v}$ of the particle by its mass $m$. The moment about $O$ of the vector $m\mathbf{v}$ is called the *moment of momentum,* or the **angular momentum**, of the particle about $O$ at that instant and is denoted by $\mathbf{H}_O$. Recall the definition of the moment of a vector (Sec. 3.1E) and denote the position vector of $P$ by $\mathbf{r}$. Then we have

**Angular momentum of a particle**
$$\mathbf{H}_O = \mathbf{r} \times m\mathbf{v} \qquad (12.13)$$

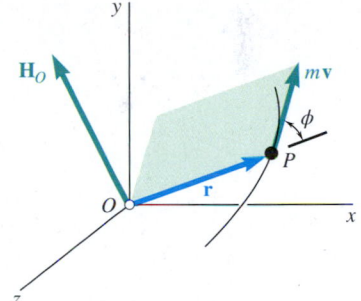

Note that $\mathbf{H}_O$ is a vector perpendicular to the plane containing $\mathbf{r}$ and $m\mathbf{v}$ and has a magnitude

$$\mathbf{H}_O = rmv \sin \phi \qquad (12.14)$$

**Fig. 12.12**   The angular momentum vector of a particle is the vector product of the position vector $\mathbf{r}$ and the linear momentum vector $m\mathbf{v}$.

where $\phi$ is the angle between $\mathbf{r}$ and $m\mathbf{v}$ (Fig. 12.12). We can determine the sense of $\mathbf{H}_O$ from the sense of $m\mathbf{v}$ by applying the right-hand rule. The unit of angular momentum is obtained by multiplying the units of length and of linear momentum (Sec. 12.1C). In SI units, we have

$$(\text{m})(\text{kg·m/s}) = \text{kg·m}^2/\text{s}$$

In U.S. customary units, we have

$$(\text{ft})(\text{slug})(\text{ft/s}) = (\text{ft})(\text{lb·s}) = \text{ft·lb·s}$$

Resolving the vectors $\mathbf{r}$ and $m\mathbf{v}$ into components and applying formula (3.10), we obtain

$$\mathbf{H}_O = \begin{vmatrix} \mathbf{i} & \mathbf{j} & \mathbf{k} \\ x & y & z \\ mv_x & mv_y & mv_z \end{vmatrix} \qquad (12.15)$$

The components of $\mathbf{H}_O$, which also represent the moments of the linear momentum $m\mathbf{v}$ about the coordinate axes, can be obtained by expanding the determinant in Eq. (12.15). The results are

$$\begin{aligned} H_x &= m(yv_z - zv_y) \\ H_y &= m(zv_x - xv_z) \\ H_z &= m(xv_y - yv_x) \end{aligned} \qquad (12.16)$$

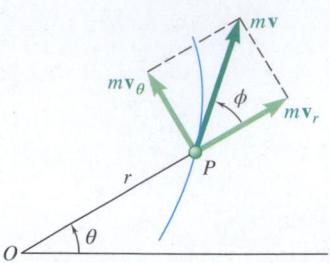

**Fig. 12.13** In polar coordinates, angular momentum of a particle is the product of the position $r$ and the transverse component of linear momentum.

In the case of a particle moving in the $xy$ plane, we have $z = v_z = 0$ and the components $H_x$ and $H_y$ reduce to zero. The angular momentum is thus perpendicular to the $xy$ plane; it is then completely defined by the scalar

$$H_O = H_z = m(xv_y - yv_x) \tag{12.17}$$

This value can be positive or negative, according to the sense in which the particle is observed to move about $O$. If we use polar coordinates, we resolve the linear momentum of the particle into radial and transverse components (Fig. 12.13), which gives us

$$H_O = rmv \sin \phi = rmv_\theta \tag{12.18}$$

Alternatively, recalling from Eq. (11.44) that $v_\theta = r\dot{\theta}$, we have

**Angular momentum
in polar coordinates**

$$H_O = m r^2 \dot{\theta} \tag{12.19}$$

Let us now compute the derivative with respect to $t$ of the angular momentum $\mathbf{H}_O$ of a particle $P$ moving in space. Differentiating both sides of Eq. (12.13) and recalling the rule for the differentiation of a vector product (Sec. 11.4B), we have

$$\dot{\mathbf{H}}_O = \dot{\mathbf{r}} \times m\mathbf{v} + \mathbf{r} \times m\dot{\mathbf{v}} = \mathbf{v} \times m\mathbf{v} + \mathbf{r} \times m\mathbf{a}$$

Because the vectors $\mathbf{v}$ and $m\mathbf{v}$ are collinear, the first term of this expression is zero; by Newton's second law, $m\mathbf{a}$ is equal to the sum $\Sigma\mathbf{F}$ of the forces acting on $P$. Noting that $\mathbf{r} \times \Sigma\mathbf{F}$ represents the sum $\Sigma\mathbf{M}_O$ of the moments about $O$ of these forces, we obtain

$$\Sigma\mathbf{M}_O = \dot{\mathbf{H}}_O \tag{12.20}$$

Eq. (12.20), which results directly from Newton's second law, states:

**The sum of the moments about $O$ of the forces acting on the particle is equal to the rate of change of angular momentum (or moment of momentum) of the particle about $O$.**

## 12.2B Motion Under a Central Force and Conservation of Angular Momentum

When the only force acting on a particle $P$ is a force $\mathbf{F}$ directed toward or away from a fixed point $O$, the particle is said to be moving under a **central force**, and the point $O$ is referred to as the **center of force** (Fig. 12.14). Because the line of action of $\mathbf{F}$ passes through $O$, we must have $\Sigma\mathbf{M}_O = 0$ at any given instant. Substituting into Eq. (12.20), we obtain

$$\dot{\mathbf{H}}_O = 0$$

for all values of $t$ and, integrating in $t$,

$$\mathbf{H}_O = \text{constant} \tag{12.21}$$

We thus conclude that

**The angular momentum of a particle moving under a central force is constant in both magnitude and direction.**

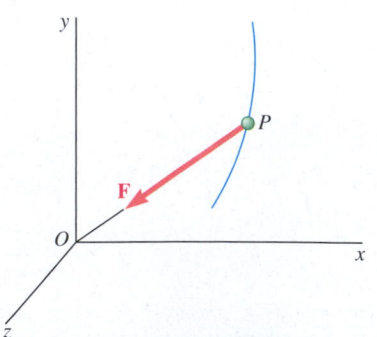

**Fig. 12.14** The central force $\mathbf{F}$ acts toward the center of force $O$.

Recall the definition of the angular momentum of a particle (Sec. 12.2A). From that, we have

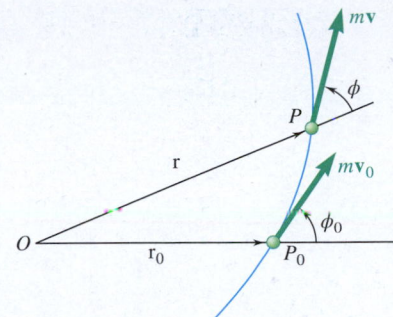

**Fig. 12.15** Angular momentum of a particle moving in a fixed plane under the action of a central force.

**Conservation of angular momentum**

$$\mathbf{r} \times m\mathbf{v} = \mathbf{H}_O = \text{constant} \tag{12.22}$$

It follows that the position vector $\mathbf{r}$ of the particle $P$ must be perpendicular to the constant vector $\mathbf{H}_O$. Thus, a particle under a central force moves in a fixed plane perpendicular to $\mathbf{H}_O$. The vector $\mathbf{H}_O$ and the fixed plane are defined by the initial position vector $\mathbf{r}_0$ and the initial velocity $\mathbf{v}_0$ of the particle. For convenience, let us assume that the plane of the figure coincides with the fixed plane of motion (Fig. 12.15).

Because the magnitude $H_O$ of the angular momentum of the particle $P$ is constant, the right-hand side in Eq. (12.14) must be constant. Therefore, we have

$$rmv \sin \phi = r_0 mv_0 \sin \phi_0 \tag{12.23}$$

This is another way to express the conservation of angular momentum; this relation applies to the motion of any particle under a central force. Because the gravitational force exerted by the sun on a planet is a central force directed toward the center of the sun, Eq. (12.23) is fundamental to the study of planetary motion. For a similar reason, it is also fundamental to studying the motion of space vehicles in orbit about the earth.

Alternatively, from Eq. (12.19), we can express the fact that the magnitude $H_O$ of the angular momentum of the particle $P$ is constant by writing

$$mr^2\dot{\theta} = H_O = \text{constant} \tag{12.24}$$

Dividing by $m$ and using $h$ to denote the angular momentum per unit mass $H_O/m$, we have

$$r^2\dot{\theta} = h \tag{12.25}$$

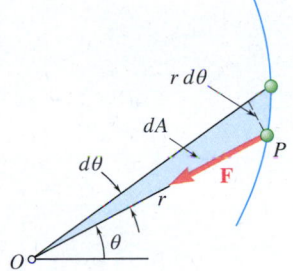

**Fig. 12.16** When a particle moves under a central force, its areal velocity is constant.

Eq. (12.25) has an interesting geometric interpretation. Note from Fig. 12.16 that the radius vector $OP$ sweeps across an infinitesimal area $dA = \frac{1}{2}r^2 d\theta$ as it rotates through an angle $d\theta$. Then, defining the **areal velocity** of the particle as the quotient $dA/dt$, we see that the left-hand side of Eq. (12.25) represents twice the areal velocity of the particle. We thus conclude that

> When a particle moves under a central force, its areal velocity is constant.

## 12.2C   Newton's Law of Gravitation

As you saw in the preceding section, the gravitational force exerted by the sun on a planet or by the earth on an orbiting satellite is an important example of a central force. In this section, you will learn how to determine the magnitude of a gravitational force.

Newton's **law of universal gravitation** states that two particles of masses $M$ and $m$ at a distance $r$ from each other have a mutual attraction of equal and opposite forces $\mathbf{F}$ and $-\mathbf{F}$ directed along the line joining the particles (Fig. 12.17). The common magnitude $F$ of the two forces is

**Newton's law of universal gravitation**

$$F = G\frac{Mm}{r^2} \tag{12.26}$$

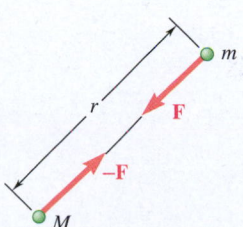

**Fig. 12.17** By Newton's law of gravitation, two masses attract each other with equal force.

where $G$ is a universal constant, called the **constant of gravitation**. Experiments show that the value of $G$ is $(66.73 \pm 0.03) \times 10^{-12}$ m$^3$/kg·s$^2$ in SI units or approximately $34.4 \times 10^{-9}$ ft$^4$/lb·s$^4$ in U.S. customary units. Gravitational forces exist between any pair of bodies, but their effect is appreciable only when one of the bodies has a very large mass. The effect of gravitational forces is apparent in the cases of the motion of a planet about the sun, of satellites orbiting about the earth, or of bodies falling on the surface of the earth.

Because the force exerted by the earth on a body of mass $m$ located on or near its surface is defined as the weight **W** of the body, we can substitute the magnitude $W = mg$ of the weight for $F$, and the earth's radius $R$ for $r$ in Eq. (12.26). We obtain

$$W = mg = \frac{GM}{R^2}m \quad \text{or} \quad g = \frac{GM}{R^2} \tag{12.27}$$

where $M$ is the mass of the earth. Because the earth is not truly spherical, the distance $R$ from the center of the earth depends upon the point selected on its surface. Thus, the values of $W$ and $g$ vary with the altitude and latitude of the point considered. Another reason for the variation of $W$ and $g$ with latitude is that a system of axes attached to the earth does not constitute a newtonian frame of reference (see Sec. 12.1A). A more accurate definition of the weight of a body should therefore include a component representing the effects of the centripetal acceleration due to the earth's rotation. Values of $g$ at sea level vary from 9.781 m/s$^2$ (or 32.09 ft/s$^2$) at the equator to 9.833 m/s$^2$ (or 32.26 ft/s$^2$) at the poles.[†]

We can use Eq. (12.26) to find the force exerted by the earth on a body of mass $m$ located in space at a distance $r$ from its center. The computations are somewhat simplified by noting that, according to Eq. (12.27), we can express the product of the constant of gravitation $G$ and the mass $M$ of the earth as

$$GM = gR^2 \tag{12.28}$$

Here we give $g$ and the earth's radius $R$ as their average values $g = 9.81$ m/s$^2$ and $R = 6.37 \times 10^6$ m in SI units[‡] and $g = 32.2$ ft/s$^2$ and $R = (3960 \text{ mi})(5280 \text{ ft/mi})$ in U.S. customary units.

The discovery of the law of universal gravitation often has been attributed to the belief that, after observing an apple falling from a tree, Newton realized that the earth must attract an apple in much the same way as the moon. It is doubtful that this incident actually took place, but we can say that Newton would not have formulated his law if he had not first perceived that the acceleration of a falling body must have the same cause as the acceleration that keeps the moon in its orbit.

---

[†]A formula expressing $g$ in terms of the latitude $\phi$ was given in Prob. 12.2.
[‡]You can find the value of $R$ simply by relating the earth's circumference to its radius as $2\pi r = 40 \times 10^6$ m.

## Sample Problem 12.12

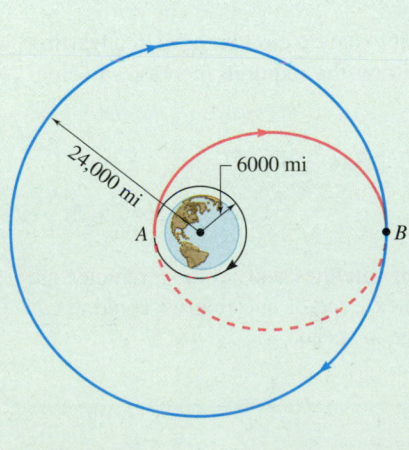

18,820 mi/h

Earth

*B*                    *A*

2340 mi

240 mi

$\phi$

*mv*

*mv*$_A$

*B*      $r_B$      *O*   $r_A$   *A*

*mv*$_B$

**Fig. 1**   The satellite at various positions.

A satellite is launched in a direction parallel to the surface of the earth with a velocity of 18,820 mi/h from an altitude of 240 mi. Determine the velocity of the satellite as it reaches its maximum altitude of 2340 mi. Recall that the earth's radius is 3960 mi.

**STRATEGY:**   The satellite is acted on by a central force, so angular momentum is conserved. You can use the principle of conservation of angular momentum to determine the velocity of the satellite.

**MODELING and ANALYSIS:**   Because the satellite is moving under a central force directed toward the center *O* of the earth, its angular momentum $\mathbf{H}_O$ is constant. From Eq. (12.14), you have

$$rmv \sin \phi = H_O = \text{constant}$$

This equation shows that *v* is at a minimum at *B*, where both *r* and $\sin \phi$ are maximum. Expressing the conservation of angular momentum between *A* and *B*, we have

$$r_A m v_A = r_B m v_B$$

Hence,

$$v_B = v_A \frac{r_A}{r_B} = (18{,}820 \text{ mi/h}) \frac{3960 \text{ mi} + 240 \text{ mi}}{3960 \text{ mi} + 2340 \text{ mi}}$$

$$v_B = 12{,}550 \text{ mi/h} \ \blacktriangleleft$$

**REFLECT and THINK:**   Note that in order to increase velocity, you could choose to apply thrusters pushing the spacecraft closer to the earth. Because this is a central force, the spacecraft's angular momentum remains constant. Therefore, its velocity *v* increases as the radial distance *r* decreases.

## Sample Problem 12.13

24,000 mi

6000 mi

*A*

*B*

A space tug travels a circular orbit with a 6000-mi radius around the earth. In order to transfer it to a larger orbit with a 24,000-mi radius, the tug is first placed on an elliptical path *AB* by firing its engines as it passes through *A*, thus increasing its velocity by 3810 mi/h. Determine how much the tug's velocity should be increased as it reaches *B* to insert it into the larger circular orbit.

**STRATEGY:**   Use Newton's second law and conservation of angular momentum.

**MODELING:**   Choose the space tug as the system, and assume you can treat it as a particle. Draw free-body and kinetic diagrams of the system at *A* as shown in Fig. 1.

*(continued)*

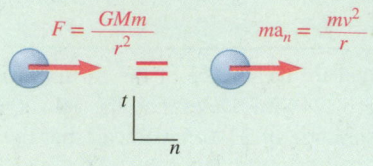

**Fig. 1** Free-body diagram and kinetic diagram of satellite at point $A$.

**ANALYSIS:**

**Circular Orbit through A.** Applying Newton's second law in the normal direction when the tug is at $A$ gives

$$\xrightarrow{+}\Sigma F_n = ma_n \qquad\qquad \frac{GMm}{r_A^2} = \frac{mv_A^2}{r_A} \qquad\qquad (1)$$

Solve Eq. (1) for $v_A$, use Eq. 12.28, and substitute in numbers to find

$$v_A = \sqrt{\frac{GM}{r_A}} = \sqrt{\frac{gR^2}{r_A}} = \sqrt{\frac{(32.2 \text{ ft/s}^2)((3960 \text{ mi})(5280 \text{ ft/mi}))^2}{(600 \text{ mi})(5280 \text{ ft/mi})}} = 21{,}080 \text{ ft/s}$$

Converting this to mi/h gives $v_A = 14{,}370$ mi/h. Thus, the increased velocity to put the space tug into an elliptic orbit is $(v_A)_{\text{ell}} = 14{,}370$ mi/h $+ 3810$ mi/h $= 18{,}180$ mi/h.

**Elliptical Path AB.** To find the velocity at $B$, use the conservation of angular momentum between $A$ and $B$. The velocity is perpendicular to $r$ at both $A$ and $B$, so you have

$$H_O = r_A m v_A = r_B m v_B \qquad\qquad (2)$$

Solving Eq. (2) for $v_B$ and substituting in numbers give

$$(v_B)_{\text{ell}} = \frac{r_A}{r_B}(v_A)_{\text{ell}} = \frac{6000 \text{ mi}}{24{,}000 \text{ mi}}(18{,}180 \text{ mi/h}) = 4545 \text{ mi/h}$$

**Circular Orbit through B.** Applying Newton's second law in the normal direction when the tug is at $B$ gives

$$\xleftarrow{+}\Sigma F_n = ma_n \qquad\qquad \frac{GMm}{r_B^2} = \frac{mv_B^2}{r_B} \qquad\qquad (3)$$

By solving Eq. (3) for $v_B$ and substituting in numbers, you find

$$v_B = \sqrt{\frac{GM}{r_B}} = \sqrt{\frac{gR^2}{r_B}} = \sqrt{\frac{(32.2 \text{ ft/s}^2)((3960 \text{ mi})(5280 \text{ ft/mi}))^2}{(24{,}000 \text{ mi})(5280 \text{ ft/mi})}} = 10{,}540 \text{ ft/s}$$

This is the speed of the space tug at $B$ for it to have a circular orbit. Converting this to mi/h gives $v_B = 7186$ mi/h. Therefore, the required increase in velocity is

$$\Delta v_B = 7186 \text{ mi/h} - 4545 \text{ mi/h}$$

$$\Delta v_B = 2640 \text{ mi/h} \blacktriangleleft$$

**REFLECT and THINK:** The speeds of satellites and orbiting vehicles are quite large, as seen in this problem. The next type of question we could ask is what force is required to impart this change in speed.

# SOLVING PROBLEMS
# ON YOUR OWN

In this section, we introduced the *angular momentum* or the *moment of the momentum*, $\mathbf{H}_O$, of a particle about $O$ as

$$\mathbf{H}_O = r \times m\mathbf{v} \qquad (12.13)$$

and found that $\mathbf{H}_O$ is constant when the particle moves under a **central force** with its center located at $O$.

**1. Solving problems involving the motion of a particle under a central force.** In problems of this type, the angular momentum $\mathbf{H}_O$ of the particle about the center of force $O$ is conserved. Therefore, we can express the conservation of angular momentum of particle $P$ about $O$ by $rmv \sin \phi = r_0 mv_0 \sin \phi_0$.

**2. In space mechanics problems** involving the orbital motion of a planet about the sun or of a satellite about the earth, the moon, or some other planet, the central force $\mathbf{F}$ is the force of gravitational attraction. This force is directed *toward* the center of force $O$ and has the magnitude

$$F = G\frac{Mm}{r^2} \qquad (12.26)$$

Note that in the particular case of the gravitational force exerted by the earth, the product $GM$ can be replaced by $gR^2$, where $R$ is the earth's radius [Eq. (12.28)].

The following two cases of orbital motion are frequently encountered:

    **a. For a satellite in a circular orbit,** the force $\mathbf{F}$ is normal to the orbit and you can write $F = ma_n$ (Sample Prob. 12.13). Substituting for $F$ from Eq. (12.26) and observing that $a_n = v^2/\rho = v^2/r$, you obtain

$$G\frac{Mm}{r^2} = m\frac{v^2}{r} \quad \text{or} \quad v^2 = \frac{GM}{r}$$

    **b. For a satellite in an elliptical orbit,** the radius vector $\mathbf{r}$ and the velocity $\mathbf{v}$ of the satellite are perpendicular to each other at points $A$ and $B$, which are closest and farthest to the center of force $O$, respectively (Sample Prob. 12.13). Thus, the conservation of angular momentum of the satellite between these two points can be expressed as

$$r_A mv_A = r_B mv_B$$

# Problems

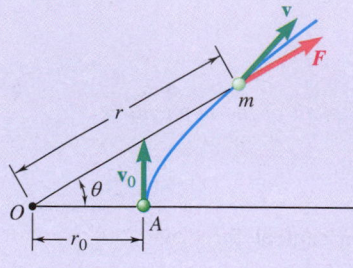

Fig. P12.74

**12.74** A particle of mass $m$ is projected from point $A$ with an initial velocity $\mathbf{v}_0$ perpendicular to line $OA$ and moves under a central force $\mathbf{F}$ directed away from the center of force $O$. Knowing that the particle follows a path defined by the equation $r = r_0/\sqrt{\cos 2\theta}$ and using Eq. (12.25), express the radial and transverse components of the velocity $\mathbf{v}$ of the particle as functions of $\theta$.

**12.75** For the particle of Prob. 12.74, show ($a$) that the velocity of the particle and the central force $\mathbf{F}$ are proportional to the distance $r$ from the particle to the center of force $O$, ($b$) that the radius of curvature of the path is proportional to $r^3$.

**12.76** A particle of mass $m$ is projected from point $A$ with an initial velocity $\mathbf{v}_0$ perpendicular to line $OA$ and moves under a central force $\mathbf{F}$ along a semicircular path of diameter $OA$. Observing that $r = r_0 \cos \theta$ and using Eq. (12.25), show that the speed of the particle is $v = v_0/\cos^2 \theta$.

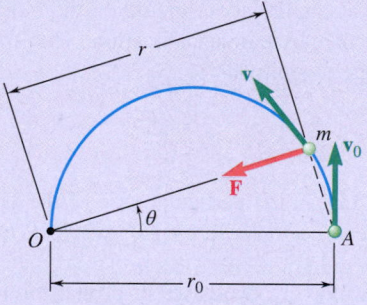

Fig. P12.76

**12.77** For the particle of Prob. 12.76, determine the tangential component $F_t$ of the central force $\mathbf{F}$ along the tangent to the path of the particle for ($a$) $\theta = 0$, ($b$) $\theta = 45°$.

**12.78** Determine the mass of the earth knowing that the mean radius of the moon's orbit about the earth is 238,910 mi and that the moon requires 27.32 days to complete one full revolution about the earth.

**12.79** Show that the radius $r$ of the moon's orbit can be determined from the radius $R$ of the earth, the acceleration of gravity $g$ at the surface of the earth, and the time $\tau$ required for the moon to complete one full revolution about the earth. Compute $r$ knowing that $\tau = 27.3$ days, giving the answer in both SI and U.S. customary units.

**12.80** Communication satellites are placed in a geosynchronous orbit—that is, in a circular orbit such that they complete one full revolution about the earth in one sidereal day (23.934 h), and thus appear stationary with respect to the ground. Determine ($a$) the altitude of these satellites above the surface of the earth, ($b$) the velocity with which they describe their orbit. Give the answers in both SI and U.S. customary units.

**12.81** Show that the radius $r$ of the orbit of a moon of a given planet can be determined from the radius $R$ of the planet, the acceleration of gravity at the surface of the planet, and the time $\tau$ required by the moon to complete one full revolution about the planet. Determine the acceleration of gravity at the surface of the planet Jupiter knowing that $R = 71\ 492$ km and that $\tau = 3.551$ days and $r = 670.9 \times 10^3$ km for its moon Europa.

**12.82** The orbit of the planet Venus is nearly circular with an orbital velocity of $126.5 \times 10^3$ km/h. Knowing that the mean distance from the center of the sun to the center of Venus is $108 \times 10^6$ km and that the radius of the sun is $695.5 \times 10^3$ km, determine (*a*) the mass of the sun, (*b*) the acceleration of gravity at the surface of the sun.

**12.83** A satellite is placed into a circular orbit about the planet Saturn at an altitude of 2100 mi. The satellite describes its orbit with a velocity of $54.7 \times 10^3$ mi/h. Knowing that the radius of the orbit about Saturn and the periodic time of Atlas, one of Saturn's moons, are $85.54 \times 10^3$ mi and 0.6017 days, respectively, determine (*a*) the radius of Saturn, (*b*) the mass of Saturn. (The *periodic time* of a satellite is the time it requires to complete one full revolution about the planet.)

**12.84** The periodic time (see Prob. 12.83) of an earth satellite in a circular polar orbit is 120 minutes. Determine (*a*) the altitude $h$ of the satellite, (*b*) the time during which the satellite is above the horizon for an observer located at the north pole.

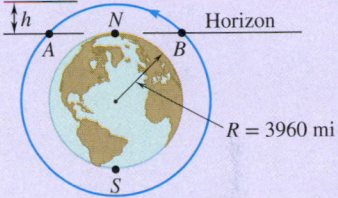

**Fig. P12.84**

**12.85** A 500-kg spacecraft first is placed into a circular orbit about the earth at an altitude of 4500 km and then is transferred to a circular orbit about the moon. Knowing that the mass of the moon is 0.01230 times the mass of the earth and that the radius of the moon is 1737 km, determine (*a*) the gravitational force exerted on the spacecraft as it was orbiting the earth, (*b*) the required radius of the orbit of the spacecraft about the moon if the periodic times (see Prob. 12.83) of the two orbits are to be equal, (*c*) the acceleration of gravity at the surface of the moon.

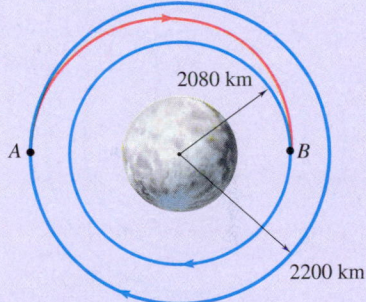

**Fig. P12.86**

**12.86** A space vehicle is in a circular orbit of 2200-km radius around the moon. To transfer it to a smaller circular orbit of 2080-km radius, the vehicle is first placed on an elliptic path $AB$ by reducing its speed by 26.3 m/s as it passes through $A$. Knowing that the mass of the moon is $73.49 \times 10^{21}$ kg, determine (*a*) the speed of the vehicle as it approaches $B$ on the elliptic path, (*b*) the amount by which its speed should be reduced as it approaches $B$ to insert it into the smaller circular orbit.

**12.87** As a first approximation to the analysis of a space flight from the earth to the planet Mars, assume the orbits of the earth and Mars are circular and coplanar. The mean distances from the sun to the earth and to Mars are $149.6 \times 10^6$ km and $227.8 \times 10^6$ km, respectively. To place the spacecraft into an elliptical transfer orbit at point $A$, its speed is increased over a short interval of time to $v_A$, which is 2.94 km/s faster than the earth's orbital speed. When the spacecraft reaches point $B$ on the elliptical transfer orbit, its speed $v_B$ is increased to the orbital speed of Mars. Knowing that the mass of the sun is $332.8 \times 10^3$ times the mass of the earth, determine the increase in speed required at $B$.

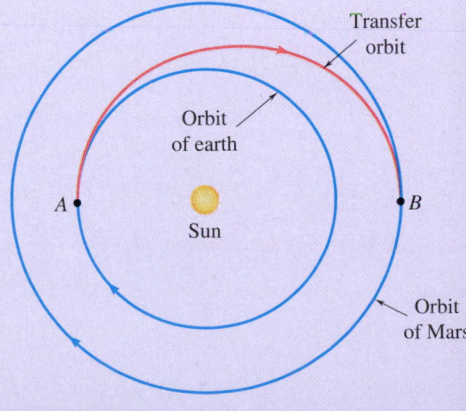

**Fig. P12.87**

**12.88** To place a communications satellite into a geosynchronous orbit (see Prob. 12.80) at an altitude of 22,240 mi above the surface of the earth, the satellite first is released from a space shuttle, which is in a circular orbit at an altitude of 185 mi, and then is propelled by an upper-stage booster to its final altitude. As the satellite passes through A, the booster's motor is fired to insert the satellite into an elliptical transfer orbit. The booster is again fired at B to insert the satellite into a geosynchronous orbit. Knowing that the second firing increases the speed of the satellite by 4810 ft/s, determine (a) the speed of the satellite as it approaches B on the elliptic transfer orbit, (b) the increase in speed resulting from the first firing at A.

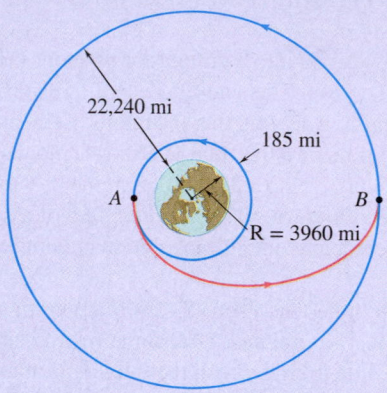

**Fig. P12.88**

**12.89** A space vehicle is in a circular orbit with a 1400-mi radius around the moon. To transfer to a smaller orbit with a 1300-mi radius, the vehicle is first placed in an elliptic path AB by reducing its speed by 86 ft/s as it passes through A. Knowing that the mass of the moon is $5.03 \times 10^{21}$ lb·s$^2$/ft, determine (a) the speed of the vehicle as it approaches B on the elliptic path, (b) the amount by which its speed should be reduced as it approaches B to insert it into the smaller circular orbit.

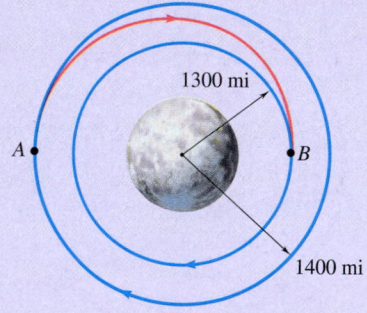

**Fig. P12.89**

**12.90** A 1-kg collar can slide on a horizontal rod that is free to rotate about a vertical shaft. The collar is initially held at A by a cord attached to the shaft. A spring of constant 30 N/m is attached to the collar and to the shaft and is undeformed when the collar is at A. As the rod rotates at the rate $\dot{\theta} = 16$ rad/s, the cord is cut and the collar moves out along the rod. Neglecting friction and the mass of the rod, determine (a) the radial and transverse components of the acceleration of the collar at A, (b) the acceleration of the collar relative to the rod at A, (c) the transverse component of the velocity of the collar at B.

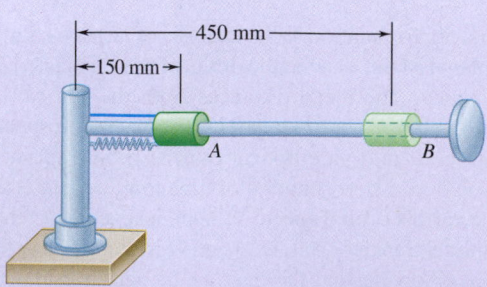

**Fig. P12.90**

**12.91** A 1-lb ball $A$ and a 2-lb ball $B$ are mounted on a horizontal rod that rotates freely about a vertical shaft. The balls are held in the positions shown by pins. The pin holding $B$ is suddenly removed and the ball moves to position $C$ as the rod rotates. Neglecting friction and the mass of the rod and knowing that the initial speed of $A$ is $v_A = 8$ ft/s, determine ($a$) the radial and transverse components of the acceleration of ball $B$ immediately after the pin is removed, ($b$) the acceleration of ball $B$ relative to the rod at that instant, ($c$) the speed of ball $A$ after ball $B$ has reached the stop at $C$.

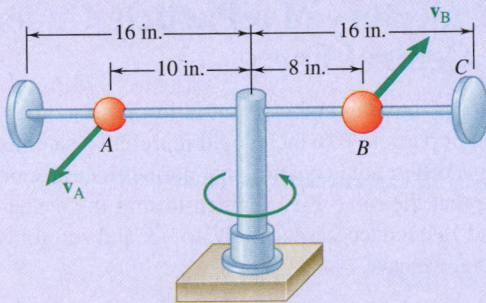

**Fig. P12.91**

**12.92** Two 2.6-lb collars $A$ and $B$ can slide without friction on a frame, consisting of the horizontal rod $OE$ and the vertical rod $CD$, which is free to rotate about $CD$. The two collars are connected by a cord running over a pulley that is attached to the frame at $O$, and a stop prevents collar $B$ from moving. The frame is rotating at the rate $\dot{\theta} = 12$ rad/s and $r = 0.6$ ft when the stop is removed, allowing collar $A$ to move out along rod $OE$. Neglecting friction and the mass of the frame, determine, for the position $r = 1.2$ ft, ($a$) the transverse component of the velocity of collar $A$, ($b$) the tension in the cord and the acceleration of collar $A$ relative to the rod $OE$.

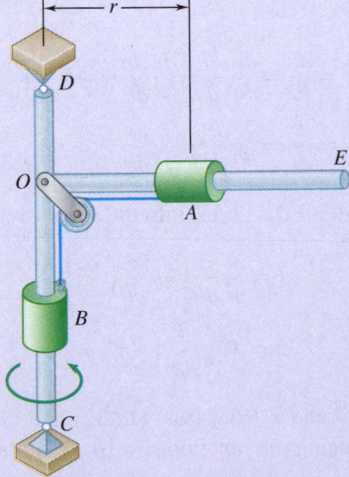

**Fig. P12.92**

**12.93** A small ball swings in a horizontal circle at the end of a cord of length $l_1$, which forms an angle $\theta_1$ with the vertical. The cord is then slowly drawn through the support at $O$ until the length of the free end is $l_2$. ($a$) Derive a relation among $l_1$, $l_2$, $\theta_1$, and $\theta_2$. ($b$) If the ball is set in motion so that initially $l_1 = 0.8$ m and $\theta_1 = 35°$, determine the angle $\theta_2$ when $l_2 = 0.6$ m.

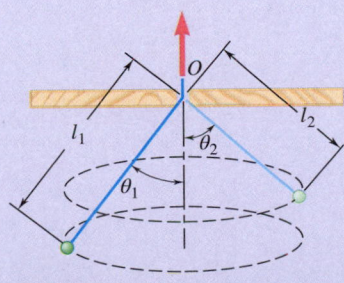

**Fig. P12.93**

# *12.3 APPLICATIONS OF CENTRAL-FORCE MOTION

The most important examples of a particle moving under the action of a central force occur in space mechanics, where gravity is the central force. In this section, we examine some of the basic ideas of this motion, concentrating on the motions of satellites around the earth and on planets around a star.

## 12.3A Trajectory of a Particle Under a Central Force

Consider a particle $P$ moving under a central force $\mathbf{F}$. In order to fully characterize the motion of particle $P$ (which could represent a satellite, a moon, etc.), we must develop a differential equation that defines its trajectory.

Assuming that the force $\mathbf{F}$ is directed toward the center of force $O$, we note that $\Sigma F_r$ and $\Sigma F_\theta$ reduce, respectively, to $-F$ and zero in Eqs. (12.11) and (12.12). Therefore, we have

$$m(\ddot{r} - r\dot{\theta}^2) = -F \qquad (12.29)$$
$$m(r\ddot{\theta} + 2\dot{r}\dot{\theta}) = 0 \qquad (12.30)$$

These equations define the motion of $P$. We can also use Eq. (12.25) to analyze the motion of $P$, obtaining

$$r^2\dot{\theta} = h \quad \text{or} \quad r^2\frac{d\theta}{dt} = h \qquad (12.31)$$

We can use Eq. (12.31) to eliminate the independent variable $t$ from Eq. (12.29). Solving Eq. (12.31) for $\dot{\theta}$ or $d\theta/dt$, we have

$$\dot{\theta} = \frac{d\theta}{dt} = \frac{h}{r^2} \qquad (12.32)$$

It follows that

$$\dot{r} = \frac{dr}{dt} = \frac{dr}{d\theta}\frac{d\theta}{dt} = \frac{h}{r^2}\frac{dr}{d\theta} = -h\frac{d}{d\theta}\left(\frac{1}{r}\right) \qquad (12.33)$$

$$\ddot{r} = \frac{d\dot{r}}{dt} = \frac{d\dot{r}}{d\theta}\frac{d\theta}{dt} = \frac{h}{r^2}\frac{d\dot{r}}{d\theta}$$

If we substitute for $\dot{r}$ from Eq. (12.33) into the expression for $\ddot{r}$, we obtain

$$\ddot{r} = \frac{h}{r^2}\frac{d}{d\theta}\left[-h\frac{d}{d\theta}\left(\frac{1}{r}\right)\right]$$

$$\ddot{r} = -\frac{h^2}{r^2}\frac{d^2}{d\theta^2}\left(\frac{1}{r}\right) \qquad (12.34)$$

Now, substituting for $\dot{\theta}$ and $\ddot{r}$ from Eqs. (12.32) and (12.34), respectively, in Eq. (12.29) and introducing the function $u = 1/r$, we obtain, after reductions,

$$\frac{d^2u}{d\theta^2} + u = \frac{F}{mh^2u^2} \qquad (12.35)$$

In deriving Eq. (12.35), we assumed force $\mathbf{F}$ to be directed toward $O$. The magnitude $F$ therefore should be positive if $\mathbf{F}$ is actually directed toward $O$ (attractive force) and negative if $\mathbf{F}$ is directed away from $O$ (repulsive force).

If $F$ is a known function of $r$ and thus of $u$, Eq. (12.35) is a differential equation in $u$ and $\theta$. This differential equation defines the trajectory followed by the particle under the central force **F**. We can obtain the equation of the trajectory by solving the differential equation (12.35) for $u$ as a function of $\theta$ and determining the constants of integration from the initial conditions.

## *12.3B  Application to Space Mechanics

After the last stages of their launching rockets have burned out, earth satellites and other space vehicles are subject to only the gravitational pull of the earth. We can therefore determine their motion from Eqs. (12.31) and (12.35), which govern the motion of a particle under a central force, after replacing $F$ by the expression for the force of gravitational attraction.[†] We set $F$ in Eq. (12.35) as

$$F = \frac{GMm}{r^2} = GMmu^2$$

where $M$ = mass of the earth
$m$ = mass of space vehicle
$r$ = distance from center of the earth to vehicle
$u = 1/r$

Then we obtain the differential equation

$$\frac{d^2u}{d\theta^2} + u = \frac{GM}{h^2} \qquad \textbf{(12.36)}$$

**Photo 12.5**  The Hubble telescope was carried into orbit by the space shuttle in 1990. Source: NASA/JSC

Note that the right-hand side is a constant.

To solve the differential equation (12.36), we add the particular solution $u = GM/h^2$ to the general solution $u = C \cos(\theta - \theta_0)$ of the corresponding homogeneous equation (i.e., the equation obtained by setting the right-hand side equal to zero). Choosing the polar axis so that $\theta_0 = 0$, we have

$$\frac{1}{r} = u = \frac{GM}{h^2} + C \cos \theta \qquad \textbf{(12.37)}$$

Eq. (12.37) is the equation of a *conic section* (ellipse, parabola, or hyperbola) in the polar coordinates $r$ and $\theta$. The origin $O$ of the coordinates, which is located at the center of the earth, is a *focus* of this conic section, and the polar axis is one of its axes of symmetry (Fig. 12.18).

The ratio of the constants $C$ and $GM/h^2$ defines the **eccentricity** $\varepsilon$ of the conic section. If we set

$$\varepsilon = \frac{C}{GM/h^2} = \frac{Ch^2}{GM} \qquad \textbf{(12.38)}$$

we can write Eq. (12.37) in the form

$$\frac{1}{r} = \frac{GM}{h^2}(1 + \varepsilon \cos \theta) \qquad \textbf{(12.37′)}$$

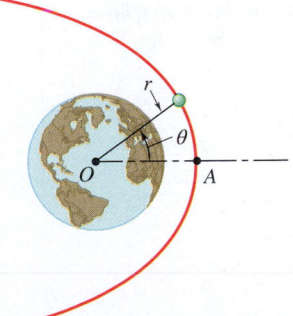

**Fig. 12.18**  The trajectory of an earth satellite is a conic section with the center of the earth as one of its foci.

This equation represents three possible trajectories.

1.  $\varepsilon > 1$, or $C > GM/h^2$: There are two values $\theta_1$ and $-\theta_1$ of the polar angle, defined by $\cos \theta_1 = -GM/Ch^2$, for which the right-hand side of

---

[†]We assume that the space vehicles considered here are attracted by the earth only and that their masses are negligible compared to the mass of the earth. If a vehicle travels very far from the earth, its path may be affected by the gravitational attraction of the sun, the moon, or another planet.

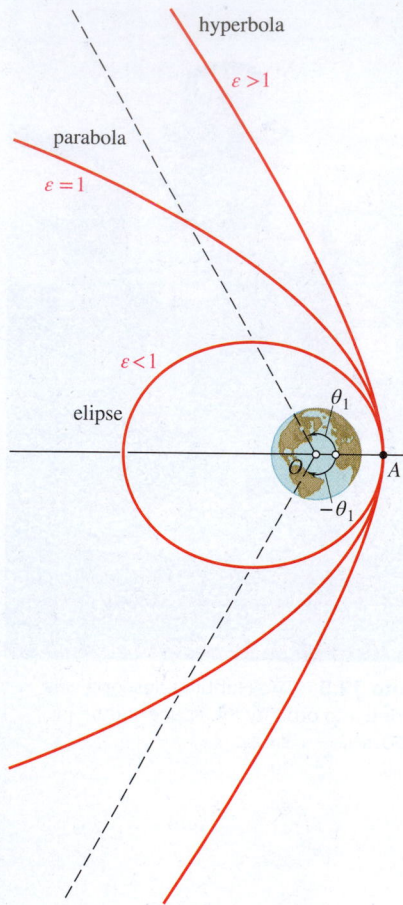

**Fig. 12.19** Depending on the eccentricity, the orbit of an earth satellite can be a hyperbola, a parabola, or an ellipse.

Eq. (12.37) becomes zero. For both these values, the radius vector $r$ becomes infinite; the conic section is a *hyperbola* (Fig. 12.19).

2. $\varepsilon = 1$, or $C = GM/h^2$: The radius vector becomes infinite for $\theta = 180°$; the conic section is a *parabola*.

3. $\varepsilon < 1$, or $C < GM/h^2$: The radius vector remains finite for every value of $\theta$; the conic section is an *ellipse*. In the particular case when $\varepsilon = C = 0$, the length of the radius vector is constant; the conic section is a circle.

Let's now see how we can determine the constants $C$ and $GM/h^2$, which characterize the trajectory of a space vehicle, from the vehicle's position and velocity at the beginning of its free flight. We assume that, as is generally the case, the powered phase of its flight has been programmed in such a way that as the last stage of the launching rocket burns out, the vehicle has a velocity parallel to the surface of the earth (Fig. 12.20). In other words, we assume that the space vehicle begins its free flight at the vertex $A$ of its trajectory. (In Sec. 13.2D, we consider problems involving oblique launchings.)

Denoting the radius and speed of the vehicle at the beginning of its free flight by $r_0$ and $v_0$, respectively, we observe that the velocity reduces to its transverse component. Thus, $v_0 = r_0\dot{\theta}_0$. Recalling Eq. (12.25), we express the angular momentum per unit mass $h$ as

$$h = r_0^2\dot{\theta}_0 = r_0 v_0 \tag{12.39}$$

The value obtained for $h$ can be used to determine the constant $GM/h^2$. We also note that the computation of this constant is simplified if we use the relation obtained in Sec. 12.2C.

$$GM = gR^2 \tag{12.28}$$

where $R$ is the radius of the earth ($R = 6.37 \times 10^6$ m or 3960 mi) and $g$ is the acceleration due to gravity at the earth's surface.

We obtain the constant $C$ by setting $\theta = 0$, $r = r_0$ in Eq. (12.37). Hence,

$$C = \frac{1}{r_0} - \frac{GM}{h^2} \tag{12.40}$$

Substituting for $h$ from Eq. (12.39), we can easily express $C$ in terms of $r_0$ and $v_0$.

**Initial Conditions.** Now we can determine the initial conditions corresponding to each of the three fundamental trajectories indicated. Considering first the parabolic trajectory, we set $C$ equal to $GM/h^2$ in Eq. (12.40) and eliminate $h$ between Eqs. (12.39) and (12.40). Solving for $v_0$, we obtain

$$v_0 = \sqrt{\frac{2GM}{r_0}}$$

We can check that a larger value of the initial velocity corresponds to a hyperbolic trajectory and a smaller value corresponds to an elliptic orbit. Because the value of $v_0$ obtained for the parabolic trajectory is the smallest value for which the space vehicle does not return to its starting point, it is called the **escape velocity**. Therefore, making use of Eq. (12.28), we have

$$v_{esc} = \sqrt{\frac{2GM}{r_0}} \quad \text{or} \quad v_{esc} = \sqrt{\frac{2gR^2}{r_0}} \tag{12.41}$$

Note that the trajectory is (1) hyperbolic if $v_0 > v_{esc}$, (2) parabolic if $v_0 = v_{esc}$, and (3) elliptic if $v_0 < v_{esc}$.

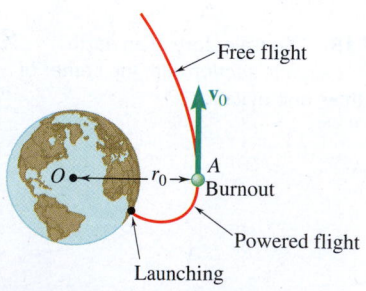

**Fig. 12.20** Typically, a space vehicle has a velocity parallel to the surface of the earth after the powered portion of its flight.

Among the various possible elliptic orbits, the one obtained when $C = 0$, the *circular orbit,* is of special interest. Taking into account Eq. (12.28), the value of the initial velocity corresponding to a circular orbit is

$$v_{circ} = \sqrt{\frac{GM}{r_0}} \quad \text{or} \quad v_{circ} = \sqrt{\frac{gR^2}{r_0}} \qquad (12.42)$$

Note from Fig. 12.21 that, for values of $v_0$ larger than $v_{circ}$ but smaller than $v_{esc}$, point $A$ is the point of the orbit closest to the earth where free flight begins. This point is called the *perigee,* whereas point $A'$, which is farthest away from the earth, is known as the *apogee.* For values of $v_0$ smaller than $v_{circ}$, point $A$ is the apogee and point $A''$, which is on the other side of the orbit, is the perigee. For values of $v_0$ much smaller than $v_{circ}$, the trajectory of the space vehicle intersects the surface of the earth; in such a case, the vehicle does not go into orbit.

Ballistic missiles, which were designed to hit the surface of the earth, also travel along elliptic trajectories. In fact, you should now realize that any object projected in vacuum with an initial velocity $v_0$ smaller than $v_{esc}$ moves along an elliptic path. Only when the distances involved are small enough that we can assume the gravitational field of the earth is uniform can we approximate the elliptic path by a parabolic path, as we did earlier (Sec. 11.4C) in the case of conventional projectiles.

**Periodic Time.** An important characteristic of the motion of an earth satellite is the time required by the satellite to travel through one complete orbit. This time, which is known as the satellite's **periodic time**, is denoted by $\tau$. We first observe, in view of the definition of areal velocity (Sec. 12.2B), that we can obtain $\tau$ by dividing the area inside the orbit by the areal velocity. The area of an ellipse is equal to $\pi ab$, where $a$ and $b$ denote the semimajor and semiminor axes, respectively. Because the areal velocity is equal to $h/2$, we have

$$\tau = \frac{2\pi ab}{h} \qquad (12.43)$$

Although we can readily determine $h$ from $r_0$ and $v_0$ in the case of a satellite launched in a direction parallel to the earth's surface, the semiaxes $a$ and $b$ are not directly related to the initial conditions. However, the values $r_0$ and $r_1$ of $r$ corresponding to the perigee and apogee of the orbit can be determined from Eq. (12.37), so we can express the semiaxes $a$ and $b$ in terms of $r_0$ and $r_1$.

Consider the elliptic orbit shown in Fig. 12.22. The earth's center is located at $O$ and coincides with one of the two foci of the ellipse, and the points $A$ and $A'$ represent, respectively, the perigee and apogee of the orbit. We easily check that

$$r_0 + r_1 = 2a$$

and thus

$$a = \tfrac{1}{2}(r_0 + r_1) \qquad (12.44)$$

Recall that the sum of the distances from each of the foci to any point of the ellipse is constant, so we have

$$O'B + BO = O'A + OA = 2a \quad \text{or} \quad BO = a$$

On the other hand, we have $CO = a - r_0$. We can therefore write

$$b^2 = (BC)^2 = (BO)^2 - (CO)^2 = a^2 - (a - r_0)^2$$
$$b^2 = r_0(2a - r_0) = r_0 r_1$$

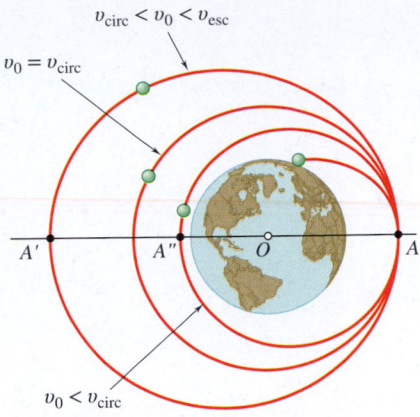

**Fig. 12.21** Various elliptic orbits are possible for earth satellites, depending on the initial velocity.

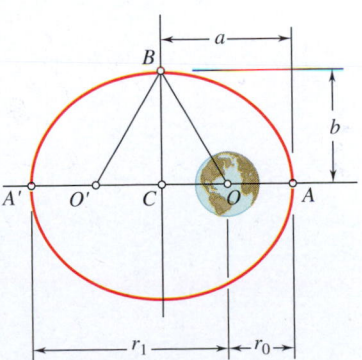

**Fig. 12.22** For an elliptic orbit, the distances to apogee ($A'$) and perigee ($A$) are related to the semimajor and semiminor axes.

and thus

$$b = \sqrt{r_0 r_1} \tag{12.45}$$

Formulas (12.44) and (12.45) indicate that the semimajor and semiminor axes of the orbit are equal, respectively, to the arithmetic and geometric means of the maximum and minimum values of the radius vector. Once you have determined $r_0$ and $r_1$, you can compute the lengths of the semiaxes and substitute for $a$ and $b$ in Eq. (12.43).

## *12.3C   Kepler's Laws of Planetary Motion

We can use the equations governing the motion of an earth satellite to describe the motion of the moon around the earth. In that case, however, the mass of the moon is not negligible compared with the earth's mass, and the results are not entirely accurate.

We can also apply the theory developed in the preceding sections to the study of the motion of the planets around the sun. Although another error is introduced by neglecting the forces exerted by the planets on one another, the approximation obtained is excellent. Indeed, even before Newton had formulated his fundamental theory, the properties expressed by Eq. (12.37), where $M$ now represents the mass of the sun, and by Eq. (12.31) had been discovered by the German astronomer Johannes Kepler (1571–1630) from astronomical observations of the motion of the planets.

Kepler's three **laws of planetary motion** can be stated as follows.

1.  The path of each planet describes an ellipse, with the sun located at one of its foci.
2.  The radius vector drawn from the sun to a planet sweeps equal areas in equal times.
3.  The squares of the periodic times of the planets are proportional to the cubes of the semimajor axes of their orbits.

The first law states a particular case of the result established in Sec. 12.3B, and the second law expresses that the areal velocity of each planet is constant (see Sec. 12.2B). Kepler's third law also can be derived from the results obtained in Sec. 12.3B. (See also Prob. 12.120.)

# Sample Problem 12.14

A satellite is launched in a direction parallel to the earth's surface with a velocity of 36 900 km/h from an altitude of 500 km. Determine (a) the maximum altitude reached by the satellite, (b) the periodic time of the satellite.

**STRATEGY:** After the satellite is launched, it is subjected to the earth's gravitational attraction only and undergoes central-force motion. Knowing this, you can determine the satellite's trajectory, maximum altitude, and periodic time.

**MODELING and ANALYSIS:** The satellite can be modeled as a particle.

**a. Maximum Altitude.** After the satellite is launched, it is subject only to the earth's gravitational attraction. Thus, its motion is governed by Eq. (12.37), so

$$\frac{1}{r} = \frac{GM}{h^2} + C \cos \theta \qquad (1)$$

Because the radial component of the velocity is zero at the point of launching $A$, you have $h = r_0 v_0$. Recalling that for the earth, $R = 6370$ km, you can compute

$$r_0 = 6370 \text{ km} + 500 \text{ km} = 6870 \text{ km} = 6.87 \times 10^6 \text{ m}$$

$$v_0 = 36\,900 \text{ km/h} = \frac{36.9 \times 10^6 \text{ m}}{3.6 \times 10^3 \text{ s}} = 10.25 \times 10^3 \text{ m/s}$$

$$h = r_0 v_0 = (6.87 \times 10^6 \text{ m})(10.25 \times 10^3 \text{ m/s}) = 70.4 \times 10^9 \text{ m}^2/\text{s}$$
$$h^2 = 4.96 \times 10^{21} \text{ m}^4/\text{s}^2$$

Because $GM = gR^2$, where $R$ is the radius of the earth, you also have

$$GM = gR^2 = (9.81 \text{ m/s}^2)(6.37 \times 10^6 \text{ m})^2 = 398 \times 10^{12} \text{ m}^3/\text{s}^2$$

$$\frac{GM}{h^2} = \frac{398 \times 10^{12} \text{ m}^3/\text{s}^2}{4.96 \times 10^{21} \text{ m}^4/\text{s}^2} = 80.3 \times 10^{-9} \text{ m}^{-1}$$

Substituting this value into Eq. (1) gives

$$\frac{1}{r} = 80.3 \times 10^{-9} \text{ m}^{-1} + C \cos \theta \qquad (2)$$

Note that at point $A$, $\theta = 0$ and $r = r_0 = 6.87 \times 10^6$ m (Fig. 1). From this, you can compute the constant $C$ as

$$\frac{1}{6.87 \times 10^6 \text{ m}} = 80.3 \times 10^{-9} \text{ m}^{-1} + C \cos 0° \qquad C = 65.3 \times 10^{-9} \text{ m}^{-1}$$

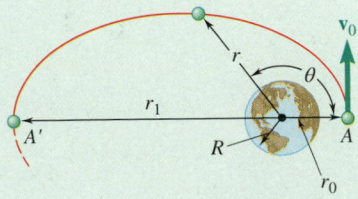

**Fig. 1** Satellite orbit after launch velocity $v_0$.

(continued)

At $A'$, which is the point on the orbit farthest from the earth, you have $\theta = 180°$ (Fig. 1). Using Eq. (2), you can compute the corresponding distance $r_1$ to be

$$\frac{1}{r_1} = 80.3 \times 10^{-9}\ \text{m}^{-1} + (65.3 \times 10^{-9}\ \text{m}^{-1})\cos 180°$$

$$r_1 = 66.7 \times 10^6\ \text{m} = 66\ 700\ \text{km}$$

*Maximum altitude* $= 66\ 700\ \text{km} - 6370\ \text{km} = 60\ 300\ \text{km}$  ◄

**b. Periodic Time.**   Because $A$ and $A'$ are the perigee and apogee, respectively, of the elliptical orbit, use Eqs. (12.44) and (12.45) to compute the semimajor and semiminor axes of the orbit (Fig. 2):

$$a = \tfrac{1}{2}(r_0 + r_1) = \tfrac{1}{2}(6.87 + 66.7)(10^6)\ \text{m} = 36.8 \times 10^6\ \text{m}$$

$$b = \sqrt{r_0 r_1} = \sqrt{(6.87)(66.7)} \times 10^6\ \text{m} = 21.4 \times 10^6\ \text{m}$$

$$\tau = \frac{2\pi ab}{h} = \frac{2\pi(36.8 \times 10^6\ \text{m})(21.4 \times 10^6\ \text{m})}{70.4 \times 10^9\ \text{m}^2/\text{s}}$$

$$\tau = 70.3 \times 10^3\ \text{s} = 1171\ \text{min} = 19\ \text{h}\ 31\ \text{min}\ \ ◄$$

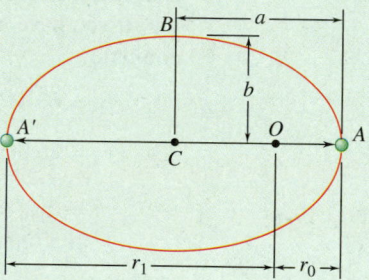

**Fig. 2**   Semimajor and semiminor axes of the orbit.

**REFLECT and THINK:**   The satellite takes less than one day to travel over 60 000 km around the earth. In this problem, you started with Eq. 12.37, but it is important to remember that this formula was the solution to a differential equation that was derived using Newton's second law.

# SOLVING PROBLEMS
# ON YOUR OWN

In this section, we continued our study of the motion of a particle under a central force and applied the results to problems in space mechanics. We found that the trajectory of a particle under a central force is defined by the differential equation

$$\frac{d^2u}{d\theta^2} + u = \frac{F}{mh^2u^2} \tag{12.35}$$

where $u$ is the reciprocal of the distance $r$ of the particle from the center of force ($u = 1/r$), $F$ is the magnitude of the central force $\mathbf{F}$, and $h$ is a constant equal to the angular momentum per unit mass of the particle. In space mechanics problems, $\mathbf{F}$ is the force of gravitational attraction exerted on the satellite or spacecraft by the sun, the earth, or other planet about which it travels. Substituting $F = GMm/r^2 = GMmu^2$ into Eq. (12.35), we obtain for that case

$$\frac{d^2u}{d\theta^2} + u = \frac{GM}{h^2} \tag{12.36}$$

where the right-hand side is a constant.

**1. Analyzing the motion of satellites and spacecraft.** The solution of the differential equation (12.36) defines the trajectory of a satellite or spacecraft. We obtained it in Sec. 12.3B in the alternative forms

$$\frac{1}{r} = \frac{GM}{h^2} + C \cos \theta \qquad \text{or} \qquad \frac{1}{r} = \frac{GM}{h^2}(1 + \varepsilon \cos \theta) \tag{12.37, 12.37'}$$

Remember when applying these equations that $\theta = 0$ always corresponds to the perigee (the point of closest approach) of the trajectory (Fig. 12.18) and that $h$ is a constant for a given trajectory. Depending on the value of the eccentricity $\varepsilon$, the trajectory is either a hyperbola, a parabola, or an ellipse.

   **a. $\varepsilon > 1$: The trajectory is a hyperbola.** For this case, the spacecraft never returns to its starting point.

   **b. $\varepsilon = 1$: The trajectory is a parabola.** This is the limiting case between open (hyperbolic) and closed (elliptic) trajectories. We had observed for this case that the velocity $v_0$ at the perigee is equal to the escape velocity $v_{\text{esc}}$. Hence,

$$v_0 = v_{\text{esc}} = \sqrt{\frac{2GM}{r_0}} \tag{12.41}$$

Note that the escape velocity is the smallest velocity for which the spacecraft does not return to its starting point.

   **c. $\varepsilon < 1$: The trajectory is an elliptical orbit.** For problems involving elliptical orbits, you may find that the relation derived in Prob. 12.102

$$\frac{1}{r_0} + \frac{1}{r_1} = \frac{2GM}{h^2}$$

is useful in the solution of subsequent problems. When you apply this equation, remember that $r_0$ and $r_1$ are the distances from the center of force to the perigee ($\theta = 0$) and apogee ($\theta = 180°$), respectively; that $h = r_0 v_0 = r_1 v_1$; and that, for a satellite orbiting the earth, $GM_{\text{earth}} = gR^2$, where $R$ is the radius of the earth. Also recall that the trajectory is a circle when $\varepsilon = 0$.

**2. Determining the point of impact of a descending spacecraft.** For problems of this type, you may assume that the trajectory is elliptic and that the initial point of the descent trajectory is the apogee of the path (Fig. 12.21). Note that at the point of impact, the distance $r$ in Eqs. (12.37) and (12.37′) is equal to the radius $R$ of the body on which the spacecraft lands or crashes. In addition, we have $h = R v_1 \sin \phi_1$, where $v_1$ is the speed of the spacecraft at impact and $\phi_1$ is the angle that its path forms with the vertical at the point of impact.

**3. Calculating the time to travel between two points on a trajectory.** For central-force motion, you can determine the time $t$ required for a particle to travel along a portion of its trajectory by recalling from Sec. 12.2B that the rate at which area is swept per unit time by the position vector $\mathbf{r}$ is equal to one-half of the angular momentum per unit mass $h$ of the particle: $dA/dt = h/2$. Because $h$ is a constant for a given trajectory, it follows that

$$t = \frac{2A}{h}$$

where $A$ is the total area swept in the time $t$.

 **a. In the case of an elliptic trajectory,** the time required to complete one orbit is called the **periodic time** and is expressed as

$$\tau = \frac{2(\pi a b)}{h} \tag{12.43}$$

where $a$ and $b$ are the semimajor and semiminor axes, respectively, of the ellipse and are related to the distances $r_0$ and $r_1$ by

$$a = \tfrac{1}{2}(r_0 + r_1) \qquad \text{and} \qquad b = \sqrt{r_0 r_1} \tag{12.44, 12.45}$$

 **b. Kepler's third law** provides a convenient relation between the periodic times of two satellites describing elliptic orbits about the same body (Sec. 12.3C). Denoting the semimajor axes of the two orbits by $a_1$ and $a_2$, respectively, and the corresponding periodic times by $\tau_1$ and $\tau_2$, we have

$$\frac{\tau_1^2}{\tau_2^2} = \frac{a_1^3}{a_2^3}$$

 **c. In the case of a parabolic trajectory,** you may be able to use the expression given on the inside of the front cover of this book for a parabolic or a semiparabolic area to calculate the time required to travel between two points of the trajectory.

# Problems

## CONCEPT QUESTIONS

**12.CQ6** A uniform crate $C$ with mass $m$ is being transported to the left by a forklift with a constant speed $v_1$. What is the magnitude of the angular momentum of the crate about point $D$; that is, the upper left corner of the crate?

   **a.** 0
   **b.** $mv_1a$
   **c.** $mv_1b$
   **d.** $mv_1\sqrt{a^2+b^2}$

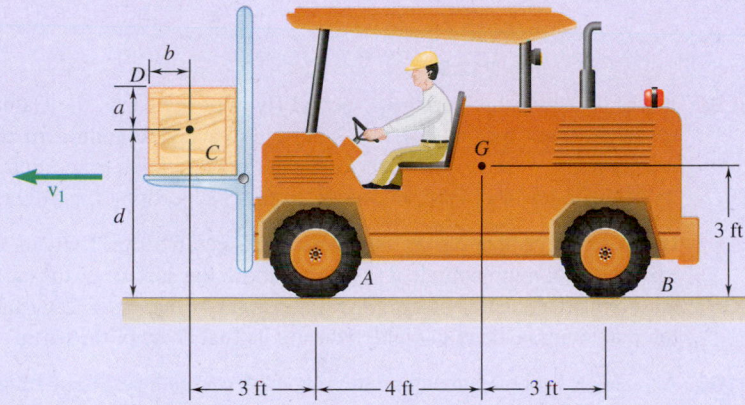

**Fig. P12.CQ6 and P12.CQ7**

**12.CQ7** A uniform crate $C$ with mass $m$ is being transported to the left by a forklift with a constant speed $v_1$. What is the magnitude of the angular momentum of the crate about point $A$; that is, the point of contact between the front tire of the forklift and the ground?

   **a.** 0
   **b.** $mv_1d$
   **c.** $3mv_1$
   **d.** $mv_1\sqrt{3^2+d^2}$

## END-OF-SECTION PROBLEMS

**12.94** A particle of mass $m$ is projected from point $A$ with an initial velocity $\mathbf{v}_0$ perpendicular to $OA$ and moves under a central force $\mathbf{F}$ along an elliptic path defined by the equation $r = r_0/(2 - \cos\theta)$. Using Eq. (12.35), show that $\mathbf{F}$ is inversely proportional to the square of the distance $r$ from the particle to the center of force $O$.

**12.95** A particle of mass $m$ describes the logarithmic spiral $r = r_0\,e^{b\theta}$ under a central force $\mathbf{F}$ directed toward the center of force $O$. Using Eq. (12.35), show that $\mathbf{F}$ is inversely proportional to the cube of the distance $r$ from the particle to $O$.

**12.96** A particle with a mass $m$ describes the path defined by the equation $r = r_0/(6\cos\theta - 5)$ under a central force $\mathbf{F}$ directed away from the center of force $O$. Using Eq. (12.35), show that $\mathbf{F}$ is inversely proportional to the square of the distance $r$ from the particle to $O$.

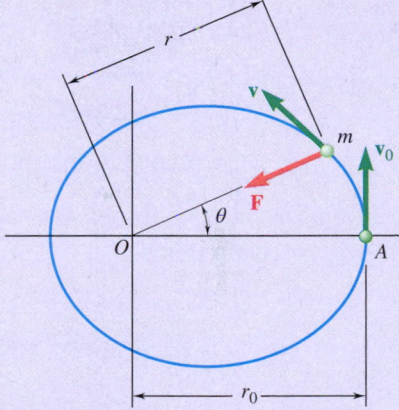

**Fig. P12.94**

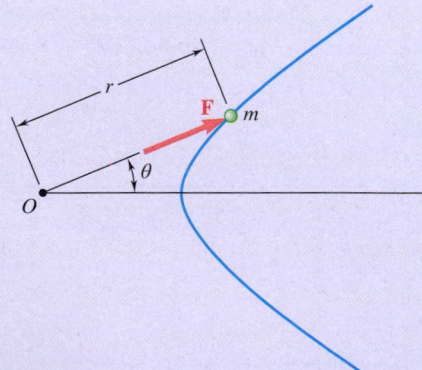

**Fig. P12.96**

**12.97** A particle of mass $m$ describes the parabola $y = x^2/4r_0$ under a central force $\mathbf{F}$ directed toward the center of force $C$. Using Eq. (12.35) and Eq. (12.37') with $\varepsilon = 1$, show that $\mathbf{F}$ is inversely proportional to the square of the distance $r$ from the particle to the center of force and that the angular momentum per unit mass $h = \sqrt{2GMr_0}$.

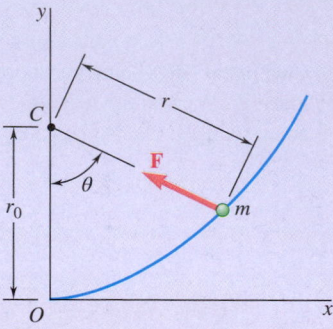

Fig. P12.97

**12.98** It was observed that during its second flyby of the earth, the Galileo spacecraft had a velocity of 14.1 km/s as it reached its minimum altitude of 303 km above the surface of the earth. Determine the eccentricity of the trajectory of the spacecraft during this portion of its flight.

**12.99** It was observed that during the Galileo spacecraft's first flyby of the earth, its minimum altitude was 600 mi above the surface of the earth. Assuming that the trajectory of the spacecraft was parabolic, determine the maximum velocity of Galileo during its first flyby of the earth.

**12.100** As a space probe approaching the planet Venus on a parabolic trajectory reaches point $A$ closest to the planet, its velocity is decreased to insert it into a circular orbit. Knowing that the mass and the radius of Venus are $4.87 \times 10^{24}$ kg and 6052 km, respectively, determine (a) the velocity of the probe as it approaches $A$, (b) the decrease in velocity required to insert it into the circular orbit.

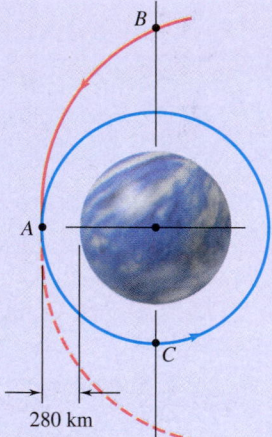

280 km

Fig. P12.100

**12.101** It was observed that as the Voyager I spacecraft reached the point of its trajectory closest to the planet Saturn, it was at a distance of $185 \times 10^3$ km from the center of the planet and had a velocity of 21.0 km/s. Knowing that Tethys, one of Saturn's moons, describes a circular orbit of radius $295 \times 10^3$ km at a speed of 11.35 km/s, determine the eccentricity of the trajectory of Voyager I on its approach to Saturn.

**12.102** A satellite describes an elliptic orbit about a planet of mass $M$. Denoting by $r_0$ and $r_1$, respectively, the minimum and maximum values of the distance $r$ from the satellite to the center of the planet, derive the relation

$$\frac{1}{r_0} + \frac{1}{r_1} = \frac{2GM}{h^2}$$

where $h$ is the angular momentum per unit mass of the satellite.

**12.103** A space probe is describing a circular orbit about a planet of radius $R$. The altitude of the probe above the surface of the planet is $\alpha R$ and its speed is $v_0$. To place the probe in an elliptic orbit which will bring it closer to the planet, its speed is reduced from $v_0$ to $\beta v_0$, where $\beta < 1$, by firing its engine for a short interval of time. Determine the smallest permissible value of $\beta$ if the probe is not to crash on the surface of the planet.

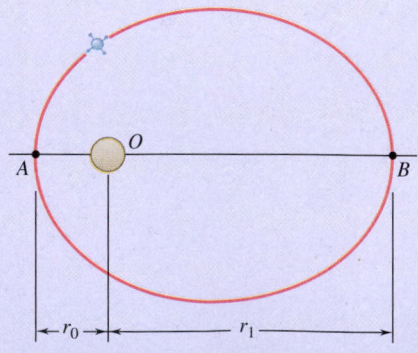

Fig. P12.102

**12.104** A satellite describes a circular orbit at an altitude of 19 110 km above the surface of the earth. Determine (*a*) the increase in speed required at point *A* for the satellite to achieve the escape velocity and enter a parabolic orbit, (*b*) the decrease in speed required at point *A* for the satellite to enter an elliptic orbit with a minimum altitude of 6370 km, (*c*) the eccentricity $\varepsilon$ of the elliptic orbit.

**12.105** A space probe is to be placed in a circular orbit of 5600-mi radius about the planet Venus in a specified plane. As the probe reaches *A*, the point of its original trajectory closest to Venus, it is inserted in a first elliptic transfer orbit by reducing its speed by $\Delta v_A$. This orbit brings it to point *B* with a much reduced velocity. There the probe is inserted in a second transfer orbit located in the specified plane by changing the direction of its velocity and further reducing its speed by $\Delta v_B$. Finally, as the probe reaches point *C*, it is inserted in the desired circular orbit by reducing its speed by $\Delta v_C$. Knowing that the mass of Venus is 0.82 times the mass of the earth, that $r_A = 9.3 \times 10^3$ mi and $r_B = 190 \times 10^3$ mi, and that the probe approaches *A* on a parabolic trajectory, determine by how much the velocity of the probe should be reduced (*a*) at *A*, (*b*) at *B*, (*c*) at *C*.

**12.106** For the space probe of Prob. 12.105, it is known that $r_A = 9.3 \times 10^3$ mi and that the velocity of the probe is reduced to 20,000 ft/s as it passes through *A*. Determine (*a*) the distance from the center of Venus to point *B*, (*b*) the amounts by which the velocity of the probe should be reduced at *B* and *C*, respectively.

**12.107** As it describes an elliptic orbit about the sun, a spacecraft reaches a maximum distance of $202 \times 10^6$ mi from the center of the sun at point *A* (called the aphelion) and a minimum distance of $92 \times 10^6$ mi at point *B* (called the perihelion). To place the spacecraft in a smaller elliptic orbit with aphelion at *A'* and perihelion at *B'*, where *A'* and *B'* are located $164.5 \times 10^6$ mi and $85.5 \times 10^6$ mi, respectively, from the center of the sun, the speed of the spacecraft is first reduced as it passes through *A* and then is further reduced as it passes through *B'*. Knowing that the mass of the sun is $332.8 \times 10^3$ times the mass of the earth, determine (*a*) the speed of the spacecraft at *A*, (*b*) the amounts by which the speed of the spacecraft should be reduced at *A* and *B'* to insert it into the desired elliptic orbit.

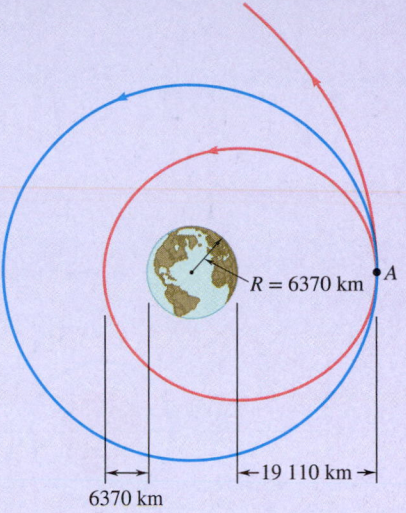

**Fig. P12.104**

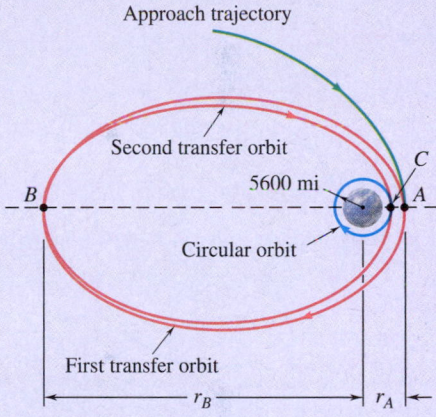

**Fig. P12.105**

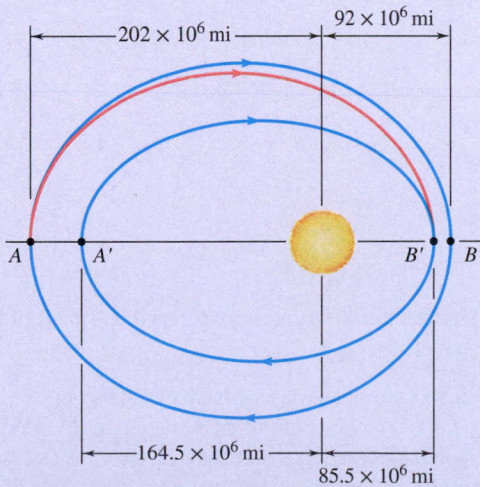

**Fig. P12.107**

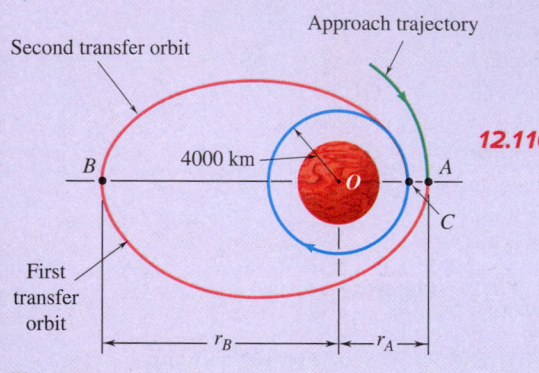

Fig. *P12.110*

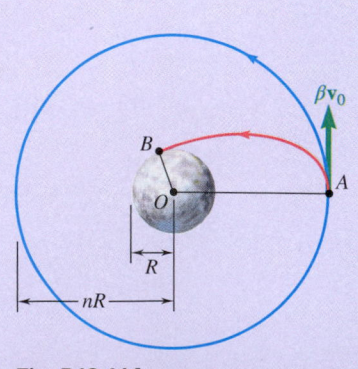

Fig. *P12.111*

**12.108** Halley's comet travels in an elongated elliptic orbit for which the minimum distance from the sun is approximately $\frac{1}{2}r_E$, where $r_E = 150 \times 10^6$ km is the mean distance from the sun to the earth. Knowing that the periodic time of Halley's comet is about 76 years, determine the maximum distance from the sun reached by the comet.

**12.109** Based on observations made during the 1996 sighting of comet Hyakutake, it was concluded that the trajectory of the comet is a highly elongated ellipse for which the eccentricity is approximately $\varepsilon = 0.999887$. Knowing that for the 1996 sighting the minimum distance between the comet and the sun was $0.230R_E$, where $R_E$ is the mean distance from the sun to the earth, determine the periodic time of the comet.

**12.110** A space probe is to be placed in a circular orbit of radius 4000 km about the planet Mars. As the probe reaches $A$, the point of its original trajectory closest to Mars, it is inserted into a first elliptic transfer orbit by reducing its speed. This orbit brings it to point $B$ with a much-reduced velocity. There the probe is inserted into a second transfer orbit by further reducing its speed. Knowing that the mass of Mars is 0.1074 times the mass of the earth, that $r_A = 9000$ km and $r_B = 180\,000$ km, and that the probe approaches $A$ on a parabolic trajectory, determine the time needed for the space probe to travel from $A$ to $B$ on its first transfer orbit.

**12.111** A spacecraft and a satellite are at diametrically opposite positions in the same circular orbit of altitude 500 km above the earth. As it passes through point $A$, the spacecraft fires its engine for a short interval of time to increase its speed and enter an elliptic orbit. Knowing that the spacecraft returns to $A$ at the same time the satellite reaches $A$ after completing one and a half orbits, determine (*a*) the increase in speed required, (*b*) the periodic time for the elliptic orbit.

**12.112** The Clementine spacecraft described an elliptic orbit of minimum altitude $h_A = 400$ km and maximum altitude $h_B = 2940$ km above the surface of the moon. Knowing that the radius of the moon is 1737 km and that the mass of the moon is 0.01230 times the mass of the earth, determine the periodic time of the spacecraft.

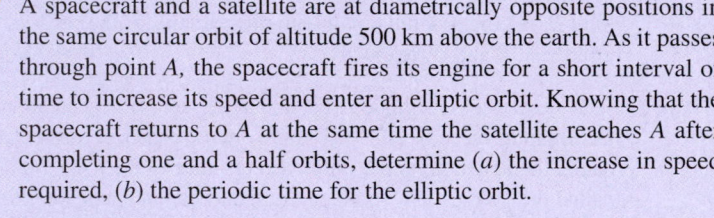

Fig. P12.112

**12.113** Determine the time needed for the space probe of Prob. 12.100 to travel from $B$ to $C$.

**12.114** A space probe is describing a circular orbit of radius $nR$ with a velocity $v_0$ about a planet of radius $R$ and center $O$. As the probe passes through point $A$, its velocity is reduced from $v_0$ to $\beta v_0$, where $\beta < 1$, to place the probe on a crash trajectory. Express in terms of $n$ and $\beta$ the angle $AOB$, where $B$ denotes the point of impact of the probe on the planet.

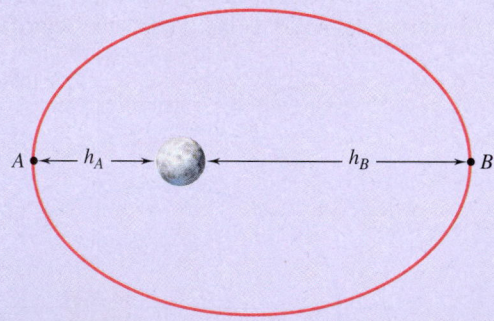

Fig. *P12.114*

**12.115** A long-range ballistic trajectory between points $A$ and $B$ on the earth's surface consists of a portion of an ellipse with the apogee at point $C$. Knowing that point $C$ is 1500 km above the surface of the earth and the range $R\phi$ of the trajectory is 6000 km, determine (a) the velocity of the projectile at $C$, (b) the eccentricity $\varepsilon$ of the trajectory.

**12.116** A space shuttle is describing a circular orbit at an altitude of 563 km above the surface of the earth. As it passes through point $A$, it fires its engine for a short interval of time to reduce its speed by 152 m/s and begin its descent toward the earth. Determine the angle $AOB$ so that the altitude of the shuttle at point $B$ is 121 km. (*Hint:* Point $A$ is the apogee of the elliptic descent trajectory.)

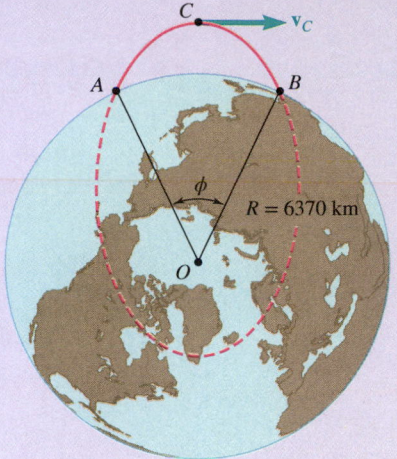

**Fig. P12.115**

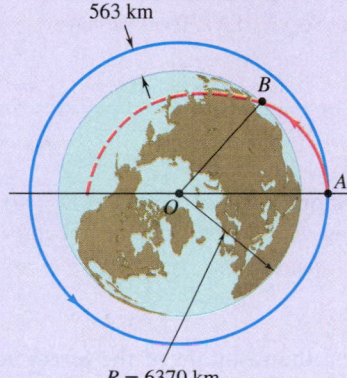

**Fig. P12.116**

**12.117** As a spacecraft approaches the planet Jupiter, it releases a probe which is to enter the planet's atmosphere at point $B$ at an altitude of 280 mi above the surface of the planet. The trajectory of the probe is a hyperbola of eccentricity $\varepsilon = 1.031$. Knowing that the radius and the mass of Jupiter are 44,423 mi and $1.30 \times 10^{26}$ slug, respectively, and that the velocity $\mathbf{v}_B$ of the probe at $B$ forms an angle of 82.9° with the direction of $OA$, determine (a) the angle $AOB$, (b) the speed $v_B$ of the probe at $B$.

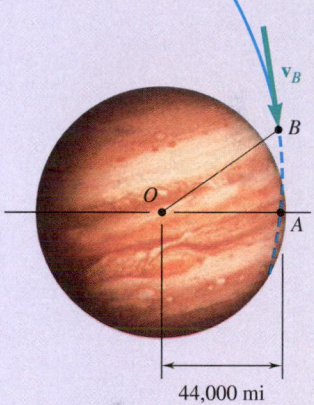

**Fig. P12.117**
©Edward Slater/Getty Images

**12.118** A satellite describes an elliptic orbit about a planet. Denoting by $r_0$ and $r_1$ the distances corresponding, respectively, to the perigee and apogee of the orbit, show that the curvature of the orbit at each of these two points can be expressed as

$$\frac{1}{\rho} = \frac{1}{2}\left(\frac{1}{r_0} + \frac{1}{r_1}\right)$$

**12.119** (a) Express the eccentricity $\varepsilon$ of the elliptic orbit described by a satellite about a planet in terms of the distances $r_0$ and $r_1$ corresponding, respectively, to the perigee and apogee of the orbit. (b) Use the result obtained in part $a$ and the data given in Prob. 12.109, where $R_E = 149.6 \times 10^6$ km, to determine the approximate maximum distance from the sun reached by comet Hyakutake.

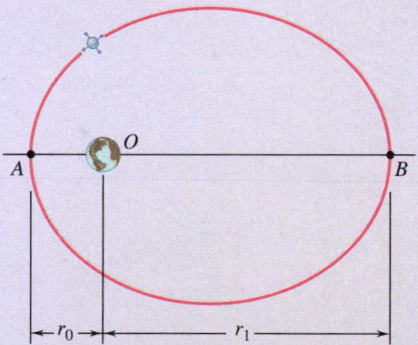

**Fig. P12.118 and P12.119**

**12.120** Derive Kepler's third law of planetary motion from Eqs. (12.37) and (12.43).

**12.121** Show that the angular momentum per unit mass $h$ of a satellite describing an elliptic orbit of semimajor axis $a$ and eccentricity $\varepsilon$ about a planet of mass $M$ can be expressed as

$$h = \sqrt{GMa(1 - \varepsilon^2)}$$

# Review and Summary

This chapter was devoted to Newton's second law and its application to analyzing the motion of particles.

## Newton's Second Law

Denote the mass of a particle by $m$, the sum (or resultant) of the forces acting on the particle by $\Sigma \mathbf{F}$, and the acceleration of the particle relative to a newtonian frame of reference by $\mathbf{a}$ (Sec. 12.1A). Then we have

$$\Sigma \mathbf{F} = m\mathbf{a} \tag{12.2}$$

## Linear Momentum

Introducing the **linear momentum** of a particle, $\mathbf{L} = m\mathbf{v}$ (Sec. 12.1B), we saw that Newton's second law also can be written in the form

$$\Sigma \mathbf{F} = \dot{\mathbf{L}} \tag{12.5}$$

This equation states that **the resultant of the forces acting on a particle is equal to the rate of change of the linear momentum of the particle**.

## Consistent Systems of Units

Eq. (12.2) holds only if we use a consistent system of units. With SI units, the forces should be expressed in newtons, the masses in kilograms, and the accelerations in $m/s^2$; with U.S. customary units, the forces should be expressed in pounds, the masses in $lb \cdot s^2/ft$ (also referred to as *slugs*), and the accelerations in $ft/s^2$ (Sec. 12.1C).

## Free-Body Diagram and Kinetic Diagram

A **free-body diagram** for a system shows the applied forces, while a **kinetic diagram** shows the vector $m\mathbf{a}$ or its components. These diagrams provide a pictorial representation of Newton's second law. Drawing them will be of great help to you when writing the equations of motion. Note that when a problem involves two or more bodies, it is usually best to consider each body separately.

## Equations of Motion for a Particle

To solve a problem involving the motion of a particle, we should first draw the free-body diagram and kinetic diagram for each particle in the system. Then, we can use these diagrams to help us write equations containing scalar quantities (Sec. 12.1D). Using **rectangular components** of $\mathbf{F}$ and $\mathbf{a}$, we have

$$\Sigma F_x = ma_x \qquad \Sigma F_y = ma_y \qquad \Sigma F_z = ma_z \tag{12.8}$$

Using **tangential and normal components**, we have

$$\Sigma F_t = m\frac{dv}{dt} \qquad \Sigma F_n = m\frac{v^2}{\rho} \tag{12.9'}$$

Using **radial and transverse components**, we have

$$\Sigma F_r = m(\ddot{r} - r\dot{\theta}^2) \tag{12.11}$$
$$\Sigma F_\theta = m(r\ddot{\theta} + 2\dot{r}\dot{\theta}) \tag{12.12}$$

Sample Probs. 12.1 through 12.5 used rectangular components, Sample Probs. 12.6 through 12.9 used tangential and normal coordinates, and Sample Probs. 12.10 and 12.11 used radial and transverse coordinates.

### Angular Momentum

In the second part of this chapter, we defined the **angular momentum $\mathbf{H}_O$** of a particle about a point $O$ as the moment about $O$ of the linear momentum $m\mathbf{v}$ of that particle (Sec. 12.2A). Thus,

$$\mathbf{H}_O = \mathbf{r} \times m\mathbf{v} \tag{12.13}$$

We noted that $\mathbf{H}_O$ is a vector perpendicular to the plane containing $\mathbf{r}$ and $m\mathbf{v}$ (Fig. 12.23) and has a magnitude of

$$H_O = rmv \sin \phi \tag{12.14}$$

Resolving the vectors $\mathbf{r}$ and $m\mathbf{v}$ into rectangular components, we expressed the angular momentum $\mathbf{H}_O$ in the determinant form

$$\mathbf{H}_O = \begin{vmatrix} \mathbf{i} & \mathbf{j} & \mathbf{k} \\ x & y & z \\ mv_x & mv_y & mv_z \end{vmatrix} \tag{12.15}$$

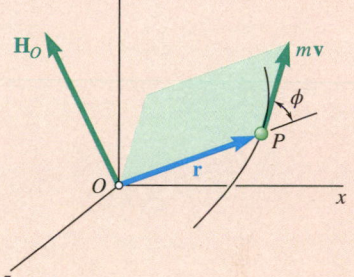

**Fig. 12.23**

In the case of a particle moving in the $xy$ plane, we have $z = v_z = 0$. The angular momentum is perpendicular to the $xy$ plane and is completely defined by its magnitude. We have

$$H_O = H_Z = m(xv_y - yv_x) \tag{12.17}$$

### Rate of Change of Angular Momentum

Computing the rate of change $\dot{\mathbf{H}}_O$ of the angular momentum $\mathbf{H}_O$ and applying Newton's second law, we obtain the equation

$$\Sigma \mathbf{M}_O = \dot{\mathbf{H}}_O \tag{12.20}$$

This equation states that **the sum of the moments about $O$ of the forces acting on a particle is equal to the rate of change of the angular momentum of the particle about $O$.**

### Motion Under a Central Force

When the only force acting on a particle $P$ is a force $\mathbf{F}$ directed toward or away from a fixed point $O$, the particle is said to be moving **under a central force** (Sec. 12.2B). Because $\Sigma \mathbf{M}_O = 0$ at any given instant, it follows from Eq. (12.20) that $\dot{\mathbf{H}}_O = 0$ for all values of $t$ and thus

$$\mathbf{H}_O = \text{constant} \tag{12.21}$$

We concluded that **the angular momentum of a particle moving under a central force is constant, both in magnitude and direction**, and that the particle moves in a plane perpendicular to the vector $\mathbf{H}_O$.

Recalling Eq. (12.14), we wrote the relation

$$rmv \sin \phi = r_0 mv_0 \sin \phi_0 \tag{12.23}$$

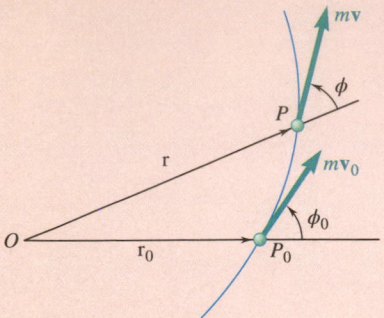

**Fig. 12.24**

for the motion of any particle under a central force (Fig. 12.24). Using polar coordinates and recalling Eq. (12.19), we also had

$$r^2 \dot{\theta} = h \qquad (12.25)$$

where $h$ is a constant representing the angular momentum per unit mass, $H_O/m$, of the particle. We observed (Fig. 12.25) that the infinitesimal area $dA$ swept by the radius vector $OP$ as it rotates through $d\theta$ is equal to $dA = \frac{1}{2} r^2 \, d\theta$ and, thus, that the left-hand side of Eq. (12.25) represents twice the **areal velocity** $dA/dt$ of the particle. Therefore, **the areal velocity of a particle moving under a central force is constant**.

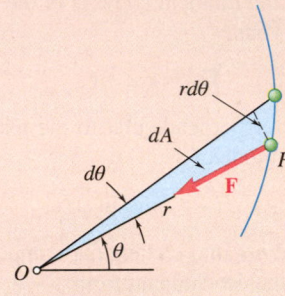

**Fig. 12.25**

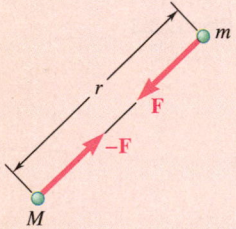

**Fig. 12.26**

### Newton's Law of Universal Gravitation

An important application of the motion under a central force is provided by the orbital motion of bodies under gravitational attraction (Sec. 12.2C). According to **Newton's law of universal gravitation**, two particles at a distance $r$ from each other and of masses $M$ and $m$, respectively, attract each other with equal and opposite forces $\mathbf{F}$ and $-\mathbf{F}$ directed along the line joining the particles (Fig. 12.26). The common magnitude $F$ of the two forces is

$$F = G \frac{Mm}{r^2} \qquad (12.26)$$

where $G$ is the **constant of gravitation**. In the case of a body of mass $m$ subjected to the gravitational attraction of the earth, we can express the product $GM$, where $M$ is the mass of the earth, as

$$GM = gR^2 \qquad (12.28)$$

where $g = 9.81 \text{ m/s}^2 = 32.2 \text{ ft/s}^2$ and $R$ is the radius of the earth.

### Orbital Motion

We showed in Sec. 12.3A that a particle moving under a central force describes a trajectory defined by the differential equation

$$\frac{d^2 u}{d\theta^2} + u = \frac{F}{mh^2 u^2} \qquad (12.35)$$

where $F > 0$ corresponds to an attractive force and $u = 1/r$. In the case of a particle moving under a force of gravitational attraction (Sec. 12.2C), we substituted for $F$ the expression given in Eq. (12.26). Measuring $\theta$ from the axis $OA$ joining the focus $O$ to the point $A$ of the trajectory closest to $O$ (Fig. 12.27), we found that the solution to Eq. (12.35) is

$$\frac{1}{r} = u = \frac{GM}{h^2} + C \cos \theta \qquad (12.37)$$

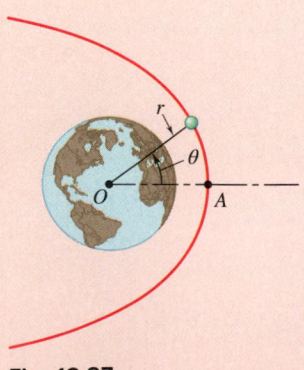

**Fig. 12.27**

This is the equation of a conic of eccentricity $\varepsilon = Ch^2/GM$. The conic is an **ellipse** if $\varepsilon < 1$, a **parabola** if $\varepsilon = 1$, and a **hyperbola** if $\varepsilon > 1$. We can determine the constants $C$ and $h$ from the initial conditions; if the particle is projected from point $A$ ($\theta = 0$, $r = r_0$) with an initial velocity $\mathbf{v}_0$ that is perpendicular to $OA$, we have $h = r_0 v_0$ (Sample Prob. 12.14).

## Escape Velocity

We also showed that the values of the initial velocity corresponding, respectively, to a parabolic and a circular trajectory are

$$v_{\text{esc}} = \sqrt{\frac{2GM}{r_0}} \qquad (12.41)$$

$$v_{\text{circ}} = \sqrt{\frac{GM}{r_0}} \qquad (12.42)$$

The first of these values, called the **escape velocity**, is the smallest value of $v_0$ for which the particle will not return to its starting point.

## Periodic Time

The **periodic time** $\tau$ of a planet or satellite is defined as the time required by that body to describe its orbit. We showed that

$$\tau = \frac{2\pi ab}{h} \qquad (12.43)$$

where $h = r_0 v_0$ and where $a$ and $b$ represent the semimajor and semiminor axes of the orbit. We further showed that these semiaxes are respectively equal to the arithmetic and geometric means of the maximum and minimum values of the radius $r$.

## Kepler's Laws

The last part of the chapter (Sec. 12.3C) presented **Kepler's laws of planetary motion** and showed that these empirical laws, obtained from early astronomical observations, confirm Newton's laws of motion, as well as his law of gravitation.

# Review Problems

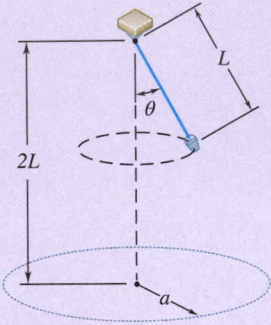

**Fig. P12.123**

**12.122** In the braking test of a sports car, its velocity is reduced from 70 mi/h to zero in a distance of 170 ft with slipping impending. Knowing that the coefficient of kinetic friction is 80 percent of the coefficient of static friction, determine (a) the coefficient of static friction, (b) the stopping distance for the same initial velocity if the car skids. Ignore air resistance and rolling resistance.

**12.123** A bucket is attached to a rope of length $L = 1.2$ m and is made to revolve in a horizontal circle. Drops of water leaking from the bucket fall and strike the floor along the perimeter of a circle of radius $a$. Determine the radius $a$ when $\theta = 30°$.

**12.124** Block $A$ has a weight of 40 lb, and block $B$ has a weight of 8 lb. The coefficient of kinetic friction between all surfaces of contact is $\mu_k = 0.15$. Knowing $\theta = 20°$ and $P = 50$ lb, determine (a) the acceleration of block $B$, (b) the tension in the cord.

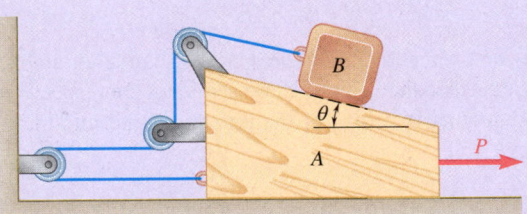

**Fig. P12.124**

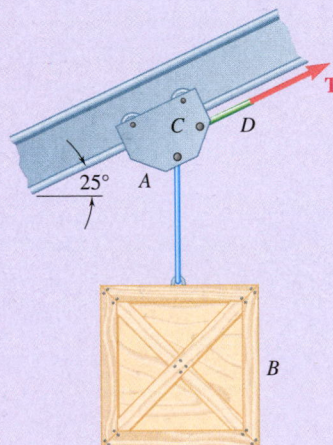

**Fig. P12.125**

**12.125** A 500-lb crate $B$ is suspended from a cable attached to a 40-lb trolley $A$ which rides on an inclined I-beam as shown. Knowing that at the instant shown the trolley has an acceleration of 1.2 ft/s² up and to the right, determine (a) the acceleration of $B$ relative to $A$, (b) the tension in cable $CD$.

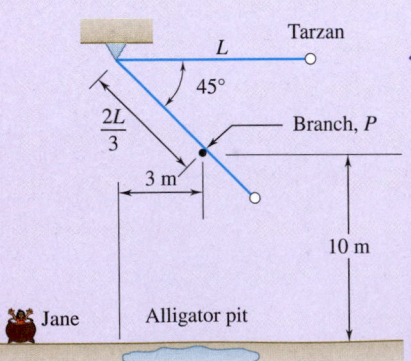

**Fig. P12.126**

**12.126** You have been hired to analyze a stunt for an upcoming Tarzan movie. In order to save Jane, Tarzan must swing over a pit of hungry alligators. He begins swinging when the vine is horizontal, but alas, he leaps before he looks, and the vine he is swinging on hits a branch as shown. Tarzan has a mass of 90 kg and is to be modeled as a point mass. The vine has a length of 9 m, and Tarzan's tangential acceleration is $a_t = g \cos \theta$, where $\theta$ is the angular position of the vine as measured from horizontal. Knowing that the rope can support three times Tarzan's weight, determine (a) Tarzan's speed before the rope hits the branch, (b) the tension in the rope before and after it hits the branch, (c) whether or not Tarzan clears the alligator pit and saves Jane. *Hint:* Of course he saves Jane, but you need to prove it.

**12.127** The parasailing system shown uses a winch to pull the rider in toward the boat, which is traveling with a constant velocity. During the interval when $\theta$ is between 20° and 40° (where $t = 0$ at $\theta = 20°$), the angle increases at the constant rate of 2°/s. During this time, the length of the rope is defined by the relationship $r = 125 - \frac{1}{3}t^{3/2}$, where $r$ and $t$ are expressed in meters and seconds, respectively. At the instant when the rope makes a 30° angle with the water, the tension in the rope is 18 kN. At this instant, what is the magnitude and direction of the force of the parasail on the 75-kg parasailer?

**Fig. P12.127**

**12.128** A robot arm moves in the vertical plane so that the 0.1-kg cylinder $P$ travels in a circle about point $B$, which is not moving. Knowing that arm $BP$ starts from rest in a horizontal position, and that the speed of $P$ increases at a constant rate of 200 mm/s², determine the force acting on the cylinder at $t = 3$ s.

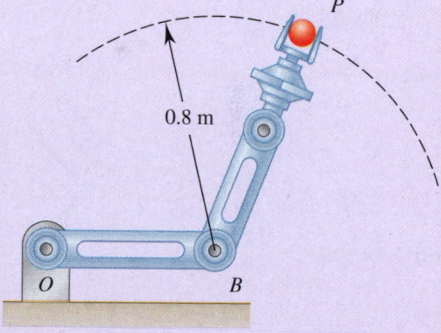

**Fig. P12.128**

**12.129** Telemetry technology is used to quantify kinematic values of a 200-kg roller-coaster cart as it passes overhead. According to the system, $r = 25$ m, $\dot{r} = -10$ m/s, $\ddot{r} = -2$ m/s², $\theta = 90°$, $\dot{\theta} = -0.4$ rad/s, and $\ddot{\theta} = -0.32$ rad/s². At this instant, determine (*a*) the normal force between the cart and the track, (*b*) the radius of curvature of the track.

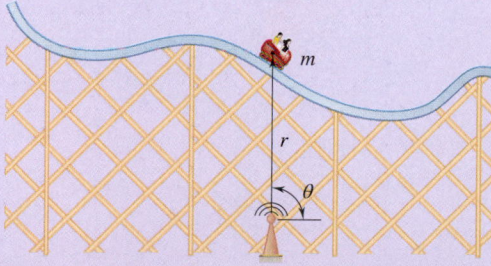

**Fig. P12.129**

**12.130** The radius of the orbit of a moon of a given planet is equal to twice the radius of that planet. Denoting by $\rho$ the mean density of the planet, show that the time required by the moon to complete one full revolution about the planet is $(24\pi/G\rho)^{1/2}$, where $G$ is the constant of gravitation.

**12.131** At main engine cutoff of its thirteenth flight, the space shuttle *Discovery* was in an elliptic orbit of minimum altitude 40.3 mi and maximum altitude 336 mi above the surface of the earth. Knowing that at point $A$ the shuttle had a velocity $\mathbf{v}_0$ parallel to the surface of the earth and that the shuttle was transferred to a circular orbit as it passed through point $B$, determine (*a*) the speed $v_0$ of the shuttle at $A$, (*b*) the increase in speed required at $B$ to insert the shuttle into the circular orbit.

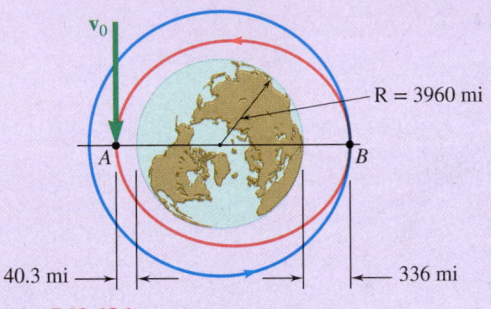

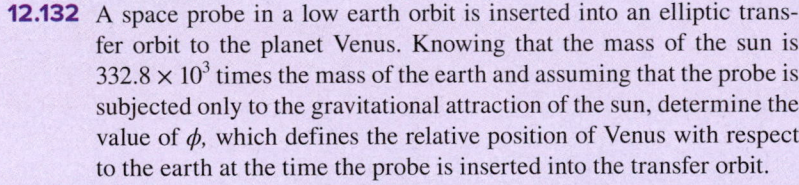

**Fig. *P12.131***

**12.132** A space probe in a low earth orbit is inserted into an elliptic transfer orbit to the planet Venus. Knowing that the mass of the sun is $332.8 \times 10^3$ times the mass of the earth and assuming that the probe is subjected only to the gravitational attraction of the sun, determine the value of $\phi$, which defines the relative position of Venus with respect to the earth at the time the probe is inserted into the transfer orbit.

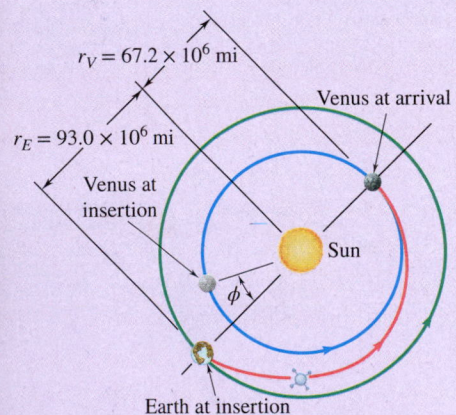

**Fig. P12.132**

**\*12.133** Disk $A$ rotates in a horizontal plane about a vertical axis at the constant rate $\dot{\theta}_0 = 10$ rad/s. Slider $B$ has mass 1 kg and moves in a frictionless slot cut in the disk. The slider is attached to a spring of constant $k$, which is undeformed when $r = 0$. Knowing that the slider is released with no radial velocity in the position $r = 500$ mm, determine the position of the slider and the horizontal force exerted on it by the disk at $t = 0.1$ s for (*a*) $k = 100$ N/m, (*b*) $k = 200$ N/m.

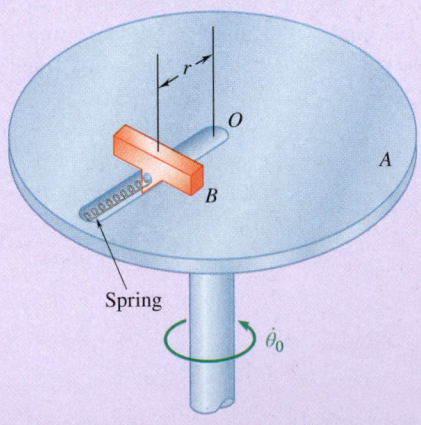

**Fig. P12.133**

# 13

# Kinetics of Particles: Energy and Momentum Methods

A golf ball will deform upon impact as shown by this high-speed photo. The maximum deformation will occur when the club head velocity and the ball velocity are the same. In this chapter, impacts will be analyzed using the coefficient of restitution and conservation of linear momentum. The kinetics of particles using energy and momentum methods is the subject of this chapter.

# Objectives

- **Calculate** the work done by a force.
- **Calculate** the kinetic energy of a particle.
- **Calculate** the gravitational and elastic potential energy of a system.
- **Solve** particle kinetics problems using the principle of work and energy.
- **Calculate** the power and efficiency of a mechanical system.
- **Solve** particle kinetics problems using conservation of energy.
- **Solve** particle kinetic problems involving conservative central forces.
- **Draw** complete and accurate impulse–momentum diagrams.
- **Solve** particle kinetics problems using the principle of impulse and momentum.
- **Solve** particle kinetics problems using conservation of linear momentum.
- **Solve** impact problems using the principle of impact and momentum and the coefficient of restitution.
- **Determine** the appropriate principle(s) to apply when solving a particle dynamics problem.
- **Solve** multi-step dynamics problems using multiple kinetics principles.

## Introduction

In the preceding chapter, we solved most problems dealing with the motion of particles through the use of the fundamental equation of motion $\mathbf{F} = m\mathbf{a}$. Given a particle acted upon by a force $\mathbf{F}$, we could solve this equation for the acceleration $\mathbf{a}$; then by applying the principles of kinematics, we could determine from $\mathbf{a}$ the velocity and position of the particle at any time.

However, using the general equation $\mathbf{F} = m\mathbf{a}$ together with kinematics allows us to obtain two additional concepts: the **principle of work and energy** and the **principle of impulse and momentum**. The advantage of these ideas lies in the fact that they make the determination of the acceleration unnecessary. Indeed, the principle of work and energy directly relates force, mass, velocity, and displacement, whereas the principle of impulse and momentum relates force, mass, velocity, and time.

We present work and energy first. In Sec. 13.1, we define the *work of a force* and the *kinetic energy of a particle*. Then, we apply the principle of work and energy to the solution of engineering problems. We also introduce the concepts of *power* and *efficiency* of a machine, which are important in engineering applications such as motors and hydraulic actuators.

In Sec. 13.2, we examine the concept of *potential energy* of a conservative force, and we apply the principle of conservation of energy to various problems of practical interest. In Sec. 13.2D, we use the principles of conservation of energy and of conservation of angular momentum jointly to solve problems of space mechanics.

The second part of this chapter deals with the principle of impulse and momentum and its application to the study of the motion of a particle. You will see in Sec. 13.3B that this principle is particularly effective in the study of the *impulsive motion* of a particle, where very large forces act for a very short time interval—like hitting a nail with a hammer.

We also consider the *central impact* of two bodies. We will show that a relation exists between the relative velocities of the two colliding bodies before and after impact. We can use this relation, together with the fact that the total momentum of the two bodies is conserved, to solve several types of practical problems.

Finally, we will discuss how to choose the best principle for solving a given problem from among Newton's second law, work and energy, or impulse and momentum. You may even need to apply multiple principles in order to solve some dynamics problems.

# 13.1 WORK AND ENERGY

Work and energy have very specific meanings in science and engineering. In everyday speech, you might say that holding up a concrete block is a lot of work, but in science, if the block doesn't move, you don't do any work at all while holding it. Similarly, people talk about energy all the time, from how you feel on a particular day ("I don't seem to have much energy today") to national and international policy ("The high cost of energy is affecting our trade balance with other countries."). In science and engineering, work and energy have very specific definitions that involve forces, displacements, masses, and velocities. These two concepts are of great value in analyzing a wide range of engineering problems.

## 13.1A Work of a Force

We first define the terms *displacement* and *work* as they are used in mechanics.[†] Consider a particle that moves from a point $A$ to a neighboring point $A'$ (Fig. 13.1). If **r** denotes the position vector corresponding to point $A$, we can denote the small vector joining $A$ and $A'$ by the differential $d\mathbf{r}$; the vector $d\mathbf{r}$ is called the **displacement** of the particle. Now, let us assume that a force **F** is acting on the particle. We define the **work of the force F corresponding to the displacement $d\mathbf{r}$** as the quantity

$$dU = \mathbf{F} \cdot d\mathbf{r} \tag{13.1}$$

We obtain $dU$ by taking the scalar product of the force **F** and the displacement $d\mathbf{r}$. We denote the magnitudes of the force and of the displacement by $F$ and $ds$, respectively, and the angle formed by **F** and $d\mathbf{r}$ by $\alpha$. Then, from the definition of the scalar product of two vectors (Sec. 3.2A), we have

$$dU = F\, ds \cos \alpha \tag{13.1'}$$

Using Eq. (3.30), we can also express the work $dU$ in terms of the rectangular components of the force and of the displacement:

$$dU = F_x\, dx + F_y\, dy + F_z\, dz \tag{13.1''}$$

Work is a scalar quantity, so it has a magnitude and a sign but no direction.

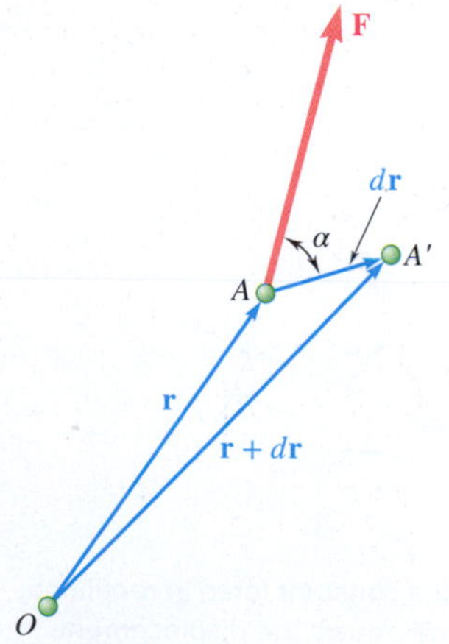

**Fig. 13.1** The work of a force acting on a particle is the scalar product of the force **F** and the displacement $d\mathbf{r}$ of the particle.

[†]We defined work in Sec. 10.1A and outlined its basic properties in Secs. 10.1A and 10.2A. For convenience, we repeat here the portions of this material that relate to the kinetics of particles.

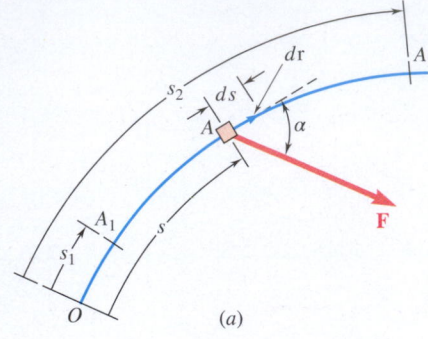

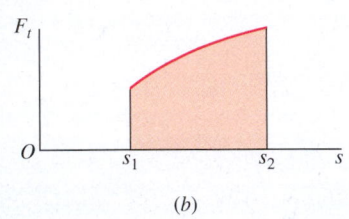

**Fig. 13.2** (*a*) The work of force **F** over a finite displacement is the integral of Eq. (13.1) from point $A_1$ to point $A_2$. (*b*) The work is represented by the area under the graph of $F_t$ versus $s$ from $s_1$ to $s_2$.

Note that work is expressed in units obtained by multiplying units of length by units of force. Thus, if we use U.S. customary units, work is expressed in ft·lb or in·lb. If we use SI units, work is expressed in N·m. The unit of work N·m is called a **joule** (J).[†] Recalling the conversion factors indicated in Sec. 12.1C, we have

$$1 \text{ ft·lb} = (1 \text{ ft})(1 \text{ lb}) = (0.3048 \text{ m})(4.448 \text{ N}) = 1.356 \text{ J}$$

It follows from Eq. (13.1′) that the work $dU$ is positive if angle $\alpha$ is acute and negative if $\alpha$ is obtuse. Three particular cases are of special interest. If the force **F** has the same direction as $d\mathbf{r}$, the work $dU$ reduces to $F \, ds$. If **F** has a direction opposite to that of $d\mathbf{r}$, the work is $dU = -F \, ds$. Finally, if **F** is perpendicular to $d\mathbf{r}$, the work $dU$ is zero.

We can obtain the work of **F** during a *finite* displacement of the particle from $A_1$ to $A_2$ (Fig. 13.2*a*) by integrating Eq. (13.1) along the path described by the particle. This work, denoted by $U_{1\rightarrow2}$, is

**Work of a force**

$$U_{1\rightarrow2} = \int_{A_1}^{A_2} \mathbf{F} \cdot d\mathbf{r} \qquad (13.2)$$

Using the alternative expression of Eq. (13.1′) for the elementary work $dU$, and observing that $F \cos \alpha$ represents the tangential component $F_t$ of the force, we can also express the work $U_{1\rightarrow2}$ as

$$U_{1\rightarrow2} = \int_{s_1}^{s_2} (F \cos \alpha) \, ds = \int_{s_1}^{s_2} F_t \, ds \qquad (13.2')$$

where the variable of integration $s$ measures the distance traveled by the particle along the path. The work $U_{1\rightarrow2}$ is represented by the area under the curve obtained by plotting $F_t = F \cos \alpha$ against $s$ (Fig. 13.2*b*).

When the force **F** is defined by its rectangular components, we can use the expression of Eq. (13.1″) for the elementary work. We have

$$U_{1\rightarrow2} = \int_{A_1}^{A_2} (F_x \, dx + F_y \, dy + F_z \, dz) \qquad (13.2'')$$

where the integration is performed along the path described by the particle.

We can use these equations to derive formulas for the work done by a force in several common and important situations, as we now show. These formulas can simplify the calculations needed to solve many common problems. For other situations, you can return to the basic equations (13.1) and (13.2) and their variants.

**Work of a Constant Force in Rectilinear Motion.** When a particle moving in a straight line is acted upon by a force **F** of constant magnitude and of constant direction (Fig. 13.3), formula (13.2′) yields

$$U_{1\rightarrow2} = (F \cos \alpha) \, \Delta x \qquad (13.3)$$

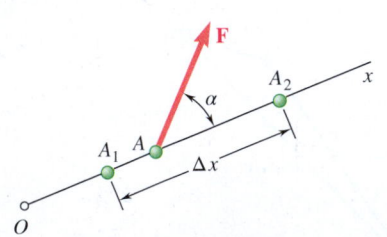

**Fig. 13.3** For a constant force in rectilinear motion, the work equals the displacement times the component of force in the direction of the displacement.

where $\alpha$ = angle the force forms with the direction of motion
      $\Delta x$ = displacement from $A_1$ to $A_2$

[†]The joule (J) is the SI unit of energy, whether in mechanical form (work, potential energy, or kinetic energy) or in chemical, electrical, or thermal form. Note that even though N·m = J, the moment of a force must be expressed in N·m and not in joules, because the moment of a force is not a form of energy.

**Work of the Force of Gravity.** We can obtain the work of the weight **W** of a body—that is, of the force of gravity exerted on that body—by substituting the components of **W** into Eqs. (13.1″) and (13.2″). Choosing the $y$ axis upward (Fig. 13.4), we have $F_x = 0$, $F_y = -W$, and $F_z = 0$. This gives us

$$dU = -W \, dy$$

$$U_{1\to2} = -\int_{y_1}^{y_2} W \, dy = Wy_1 - Wy_2 \qquad \textbf{(13.4)}$$

or

$$U_{1\to2} = -W(y_2 - y_1) = -W \, \Delta y \qquad \textbf{(13.4′)}$$

where $\Delta y$ is the vertical displacement from $A_1$ to $A_2$. The work of the weight **W** is thus equal to **the product of $W$ and the vertical displacement of the center of gravity of the body.** The work is positive when $\Delta y < 0$; that is, when the body moves down. When the body moves up (and $\Delta y > 0$), the force and displacement are in opposite directions, and the work is negative.

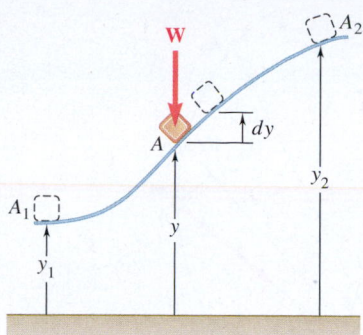

**Fig. 13.4** The work done by the force of gravity is the product of the weight and the vertical displacement of the object's center of gravity. If the object moves up, the work done by gravity is negative.

**Work of the Force Exerted by a Spring.** Consider a body $A$ attached to a fixed point $B$ by a spring; we assume that the spring is undeformed when the body is at $A_0$ (Fig. 13.5$a$). For a linear spring, the magnitude of the force **F** exerted by the spring on body $A$ is proportional to the deflection $x$ of the spring measured from the unstretched position $A_0$ (i.e., $x = L_{\text{stretched}} - L_{\text{unstretched}}$). We have

$$F = kx \qquad \textbf{(13.5)}$$

where $k$ is the **spring constant** expressed in N/m or kN/m if SI units are used and in lb/ft or lb/in. if U.S. customary units are used.[†]

We can obtain the work of force **F** exerted by the spring during a finite displacement of the body from $A_1(x = x_1)$ to $A_2(x = x_2)$ by writing

$$dU = -F \, dx = -kx \, dx$$

$$U_{1\to2} = -\int_{x_1}^{x_2} kx \, dx = \tfrac{1}{2}kx_1^2 - \tfrac{1}{2}kx_2^2 \qquad \textbf{(13.6)}$$

You need to be careful in expressing $k$ and $x$ in consistent units. For example, if you use U.S. customary units, $k$ should be expressed in lb/ft and $x$ in feet, or $k$ should be given in lb/in. and $x$ in inches. In the first case, the work will have units of ft·lb; in the second case, it will have units of in·lb. Note that the work of force **F** exerted by the spring on the body is positive when $x_2 < x_1$; that is, when the spring is returning to its undeformed position. When the body is moved from $x_1$ to $x_2$, the work done by the spring is negative, because the displacement and force are in opposite directions.

Because Eq. (13.5) is the equation of a straight line of slope $k$ passing through the origin, we can also obtain the work $U_{1\to2}$ of **F** during the displacement from $A_1$ to $A_2$ by evaluating the area of the trapezoid shown in Fig. 13.5$b$. We can do this by computing $F_1$ and $F_2$ and multiplying the base $\Delta x$ of the trapezoid by its mean height $\tfrac{1}{2}(F_1 + F_2)$. Because the work of the force **F** exerted by the spring is positive for a negative value of $\Delta x$, we have

$$U_{1\to2} = -\tfrac{1}{2}(F_1 + F_2) \, \Delta x \qquad \textbf{(13.6′)}$$

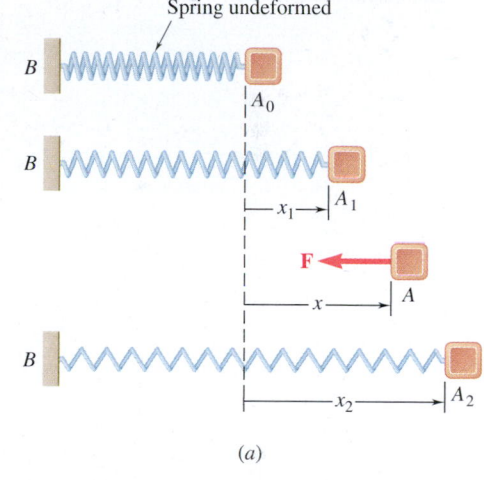

(a)

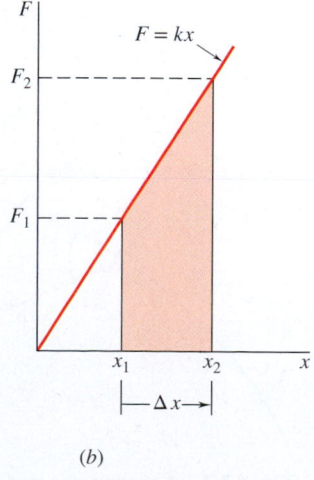

(b)

**Fig. 13.5** (a) The work of a force exerted by a spring depends on the spring constant and the initial and final positions of the spring. (b) The work is represented by the area under the graph of force versus position.

[†]The relation $F = kx$ is correct under static conditions only. Under dynamic conditions, Eq. (13.5) should be modified to take into account the inertia of the spring. However, the error introduced by using $F = kx$ in the solution of kinetics problems is small if the mass of the spring is small compared with the other masses in motion.

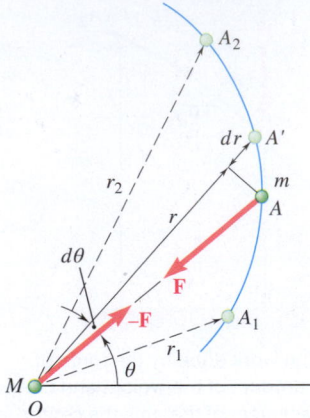

**Fig. 13.6** The work of a gravitational force depends on the gravitational constant, the masses of the interacting bodies, and the radial distance between them.

**Work of a Gravitational Force.** We saw in Sec. 12.2C that two particles of mass $M$ and $m$ separated by a distance $r$ attract each other with equal and opposite forces $\mathbf{F}$ and $-\mathbf{F}$, directed along the line joining the particles and of magnitude

$$F = G\frac{Mm}{r^2}$$

Let us assume that particle $M$ occupies a fixed position $O$, while particle $m$ moves along the path shown in Fig. 13.6. We can obtain the work of force $\mathbf{F}$ exerted on particle $m$ during an infinitesimal displacement of the particle from $A$ to $A'$ by multiplying the magnitude $F$ of the force by the radial component $dr$ of the displacement. Because $\mathbf{F}$ is directed toward $O$ and $dr$ is directed away from $O$, the work is negative, and we have

$$dU = -F\,dr = -G\frac{Mm}{r^2}dr$$

The work of the gravitational force $\mathbf{F}$ during a finite displacement from $A_1(r = r_1)$ to $A_2(r = r_2)$ is therefore

$$U_{1\to 2} = -\int_{r_1}^{r_2}\frac{GMm}{r^2}dr = \frac{GMm}{r_2} - \frac{GMm}{r_1} \tag{13.7}$$

where $M$ is the mass of the earth. We can use this formula to determine the work of the force exerted by the earth on a body of mass $m$ at a distance $r$ from the earth's center when $r$ is larger than the radius $R$ of the earth. Recalling the first of the relations in Eq. (12.27), we can replace the product $GMm$ in Eq. (13.7) by $WR^2$, where $R$ is the earth's radius ($R = 6.37 \times 10^6$ m or 3960 mi) and $W$ is the weight of the body at the earth's surface.

   Some forces frequently encountered in kinetics problems *do no work.* They are forces applied to fixed points ($ds = 0$) or acting in a direction perpendicular to the displacement ($\cos \alpha = 0$). Forces that do no work include the reaction at a frictionless pin when the body supported rotates about the pin; the normal force at a frictionless fixed surface when the body in contact moves along the surface; the reaction at a roller moving along its track; and the weight of a body when its center of gravity moves horizontally.

## 13.1B Principle of Work and Energy

Consider a particle of mass $m$ acted upon by a force $\mathbf{F}$ and moving along a path that is either rectilinear or curved (Fig. 13.7). Expressing Newton's second law in terms of the tangential components of the force and of the acceleration (see Sec. 12.1D), we have

$$F_t = ma_t \quad \text{or} \quad F_t = m\frac{dv}{dt}$$

where $v$ is the speed of the particle. Recalling from Sec. 11.4A that $v = ds/dt$, we obtain

$$F_t = m\frac{dv}{ds}\frac{ds}{dt} = mv\frac{dv}{ds}$$

$$F_t\,ds = mv\,dv$$

Integrating from $A_1$, where $s = s_1$ and $v = v_1$, to $A_2$, where $s = s_2$ and $v = v_2$, we have

$$\int_{s_1}^{s_2}F_t\,ds = m\int_{v_1}^{v_2}v\,dv = \tfrac{1}{2}mv_2^2 - \tfrac{1}{2}mv_1^2 \tag{13.8}$$

The left-hand side of Eq. (13.8) represents the work $U_{1\to 2}$ of the force $\mathbf{F}$ exerted on the particle during the displacement from $A_1$ to $A_2$; as indicated earlier, the

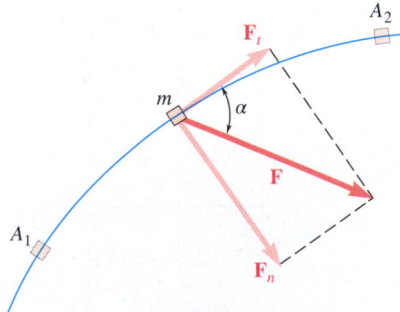

**Fig. 13.7** A particle $m$ acted upon by a force $\mathbf{F}$.

work $U_{1\rightarrow2}$ is a scalar quantity. Thus, the expression $\frac{1}{2}mv^2$ is also a scalar quantity. We define it as the kinetic energy of the particle, denoted by $T$. That is,

**Kinetic energy of a particle**

$$T = \frac{1}{2}mv^2 \qquad \text{(13.9)}$$

Substituting into Eq. (13.8), we have

**Principle of work and energy**

$$U_{1\rightarrow2} = T_2 - T_1 \qquad \text{(13.10)}$$

This equation states that when a particle moves from $A_1$ to $A_2$ under the action of a force **F**, **the work of the force F is equal to the change in kinetic energy of the particle**. This is known as the **principle of work and energy**. Rearranging the terms in Eq. (13.10) gives

$$T_1 + U_{1\rightarrow2} = T_2 \qquad \text{(13.11)}$$

Like Newton's second law from which it is derived, the principle of work and energy applies only with respect to a newtonian frame of reference (Sec. 12.1A). The speed $v$ used to determine the kinetic energy $T$ therefore should be measured with respect to a newtonian frame of reference.

Because both work and kinetic energy are scalar quantities, we can compute their sum as an ordinary algebraic sum with the work $U_{1\rightarrow2}$ being positive or negative according to the direction of **F**. When several forces act on the particle, the expression $U_{1\rightarrow2}$ represents the total work of the forces acting on the particle; it is obtained by adding algebraically the work of the various forces.

As just noted, the kinetic energy of a particle is a scalar quantity. It further appears from the definition $T = \frac{1}{2}mv^2$ that, regardless of the particle's direction of motion, the kinetic energy is always positive. Considering the particular case when $v_1 = 0$ and $v_2 = v$, and substituting $T_1 = 0$ and $T_2 = T$ into Eq. (13.10), we observe that the work done by the forces acting on the particle is equal to $T$. Thus, the kinetic energy of a particle moving with a speed $v$ represents the work that must be done to bring the particle from rest to the speed $v$. Substituting $T_1 = T$ and $T_2 = 0$ into Eq. (13.10), we also note that when a particle moving with a speed $v$ is brought to rest, the work done by the forces acting on the particle is $-T$. Assuming that no energy is dissipated into heat, we conclude that the work done by the forces exerted *by the particle* on the bodies that cause it to come to rest is equal to $T$. Thus, the kinetic energy of a particle also represents **the capacity to do work associated with the speed of the particle**.

The kinetic energy is measured in the same units as work; that is, in joules if we use SI units and in ft·lb if we use U.S. customary units. We check that, in SI units,

$$T = \tfrac{1}{2}mv^2 = \text{kg}(\text{m/s})^2 = (\text{kg}\cdot\text{m/s}^2)\text{m} = \text{N}\cdot\text{m} = \text{J}$$

whereas in customary units,

$$T = \tfrac{1}{2}mv^2 = (\text{slug})(\text{ft/s})^2 = (\text{lb}\cdot\text{s}^2/\text{ft})(\text{ft/s})^2 = \text{ft}\cdot\text{lb}$$

## 13.1C  Applications of the Principle of Work and Energy

Using the principle of work and energy greatly simplifies the solution of many problems involving forces, displacements, and velocities. Consider, for example, the pendulum $OA$ consisting of a bob $A$ of weight $W$ attached to a cord of length $l$ (Fig. 13.8$a$). The pendulum is released with no initial velocity from a

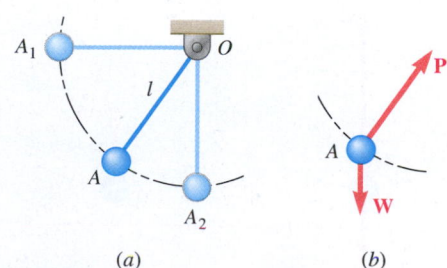

**Fig. 13.8** ($a$) A bob of weight $W$ swings from an initial position $A_1$ to a final position $A_2$; ($b$) free-body diagram of the bob at position $A$.

horizontal position $OA_1$ and allowed to swing in a vertical plane. We wish to determine the speed of the bob as it passes through $A_2$, directly under $O$.

We first determine the work done during the displacement from $A_1$ to $A_2$ by the forces acting on the bob. We draw a free-body diagram of the bob, showing all the *actual* forces acting on it; that is, the weight **W** and the force **P** exerted by the cord (Fig. 13.8*b*). (Recall that an inertia vector is not an actual force and *should not* be included in the free-body diagram.) Note that force **P** does no work, because it is normal to the path; the only force that does work is thus the weight **W**. We obtain the work of **W** by multiplying its magnitude $W$ by the vertical displacement $l$ (Sec. 13.1A); because the displacement is downward, the work is positive. We therefore have $U_{1\rightarrow2} = Wl$.

Now consider the kinetic energy of the bob. We have $T_1 = 0$ at $A_1$ and $T_2 = \frac{1}{2}(W/g)v_2^2$ at $A_2$. We can now apply the principle of work and energy. From Eq. (13.11), we have

$$T_1 + U_{1\rightarrow2} = T_2 \qquad 0 + Wl = \frac{1}{2}\frac{W}{g}v_2^2$$

Solving for $v_2$, we find $v_2 = \sqrt{2gl}$. Note that this speed is also that of a body falling freely from a height $l$.

This example illustrates the following advantages of the method of work and energy:

1. In order to find the speed at $A_2$, there is no need to determine the acceleration in an intermediate position $A$ and to integrate the acceleration expression from $A_1$ to $A_2$.
2. All quantities involved are scalars and can be added directly, without using $x$ and $y$ components.
3. Forces that do no work are eliminated from the solution of the problem.

What is an advantage in one problem, however, may be a disadvantage in another. It is evident, for instance, that the method of work and energy cannot be used to directly determine an acceleration. It is also evident that to determine a force that is normal to the path of the particle (i.e., a force that does no work) we must supplement the method of work and energy by the direct application of Newton's second law. Suppose, for example, that we wish to determine the tension in the cord of the pendulum of Fig. 13.8*a* as the bob passes through $A_2$. We draw a free-body diagram and kinetic diagram of the bob in that position (Fig. 13.9) and express Newton's second law in terms of tangential and normal components. The equations $\Sigma F_t = ma_t$ and $\Sigma F_n = ma_n$ yield, respectively, $a_t = 0$ and

$$P - W = ma_n = \frac{W}{g}\frac{v_2^2}{l}$$

But earlier, we determined the speed at $A_2$ by the method of work and energy. Substituting $v_2^2 = 2gl$ and solving for $P$, we have

$$P = W + \frac{W}{g}\frac{2gl}{l} = 3W$$

If we used only statics principles and designed the cord to hold the weight of the bob (or even twice the weight of the bob), the cord would have failed.

When a problem involves two particles or more, we can apply the principle of work and energy to each particle separately. Adding the kinetic energies of the various particles and considering the work of all the forces acting on them, we can also write a single equation of work and energy for all the particles involved. We have

$$T_1 + U_{1\rightarrow2} = T_2 \tag{13.11}$$

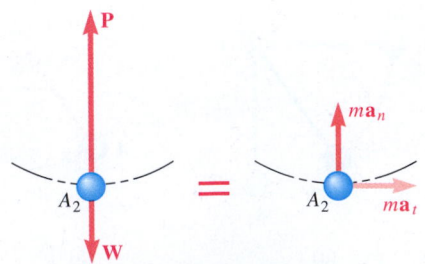

**Fig. 13.9** Free-body diagram and kinetic diagram for determining the force on a pendulum bob.

where $T_1$ represents the arithmetic sum of the kinetic energies of the particles involved at position 1, $T_2$ represents the arithmetic sum of the kinetic energies of the particles involved at position 2, and $U_{1\rightarrow2}$ is the work of all the forces acting on the particles, including the forces of action and reaction exerted by the particles on each other. In problems involving bodies connected by inextensible cords or links, however, the work of the forces exerted by a given cord or link on the two bodies it connects cancels out, because the points of application of these forces move through equal distances (see Sample Prob. 13.2). (In Chap. 14, we discuss how to apply the method of work and energy to a system of particles.)

In particle kinetics problems, typically the friction force acts in a direction opposite to that of the motion (e.g., a car braking or a box sliding down a ramp). In these cases, the work of friction represents energy dissipated into heat and results in a decrease in the kinetic energy of the body involved (see Sample Prob. 13.3).

## 13.1D  Power and Efficiency

We define **power** as the time rate at which work is done. In the selection of a motor or engine, power is a much more important criterion than is the actual amount of work to be performed. Either a small motor or a large power plant can be used to do a given amount of work, but the small motor may require a month to do the work done by the power plant in a matter of minutes. If $\Delta U$ is the work done during the time interval $\Delta t$, the average power during that time interval is

$$\text{Average power} = \frac{\Delta U}{\Delta t}$$

Letting $\Delta t$ approach zero, we obtain in the limit

**Power**

$$\text{Power} = \frac{dU}{dt} \qquad\qquad \textbf{(13.12)}$$

Substituting the scalar product $\mathbf{F}\cdot d\mathbf{r}$ for $dU$, we can also write

$$\text{Power} = \frac{dU}{dt} = \frac{\mathbf{F}\cdot d\mathbf{r}}{dt}$$

Then, recalling that $d\mathbf{r}/dt$ represents the velocity $\mathbf{v}$ of the point of application of $\mathbf{F}$, we have

$$\text{Power} = \mathbf{F}\cdot\mathbf{v} \qquad\qquad \textbf{(13.13)}$$

Because we defined power as the time rate at which work is done, we obtain its units by dividing units of work by the unit of time. Thus, if we use SI units, power is expressed in J/s; this unit is called a *watt* (W). We have

$$1\text{ W} = 1\text{ J/s} = 1\text{ N·m/s}$$

If we use U.S. customary units, power is expressed in ft·lb/s or in *horsepower* (hp), where one horsepower is defined as

$$1\text{ hp} = 550\text{ ft·lb/s}$$

Recall from Sec. 13.1A that 1 ft·lb = 1.356 J, so we can verify that

$$1\text{ ft·lb/s} = 1.356\text{ J/s} = 1.356\text{ W}$$

$$1\text{ hp} = 550(1.356\text{ W}) = 746\text{ W} = 0.746\text{ kW}$$

**Photo 13.1**  The power used to operate a chair lift at a ski resort is the product of the force applied and the speed of the lift.
©imagebroker.net/SuperStock RF

We defined the **mechanical efficiency** of a machine in Sec. 10.1D as the ratio of the output work to the input work:

$$\eta = \frac{\text{output work}}{\text{input work}} \tag{13.14}$$

This definition is based on the assumption that work is done at a constant rate. The ratio of the output to the input work is therefore equal to the ratio of the rates at which output and input work are done, and we have

**Mechanical efficiency**

$$\eta = \frac{\text{power output}}{\text{power input}} \tag{13.15}$$

Because of energy losses due to friction, the output work is always smaller than the input work, and consequently, the power output is always smaller than the power input. The mechanical efficiency of a machine is therefore always less than 1.

When we use a machine to transform mechanical energy into electrical energy or to change thermal energy into mechanical energy, we can obtain its *overall efficiency* from Eq. (13.15). The overall efficiency of a machine is always less than 1; it provides a measure of all the various energy losses involved (losses of electric or thermal energy, as well as frictional losses). Note that you have to express the power output and the power input in the same units before using Eq. (13.15).

# Sample Problem 13.1

An automobile weighing 4000 lb is driven down a 5° incline at a speed of 60 mi/h when the brakes are applied, causing a constant total braking force (applied by the road on the tires) of 1500 lb. Determine the distance traveled by the automobile as it comes to a stop.

**STRATEGY:** You are given the velocity of the car at two positions along the road and need to determine the distance $x$ between them (Fig. 1), so use the principle of work and energy.

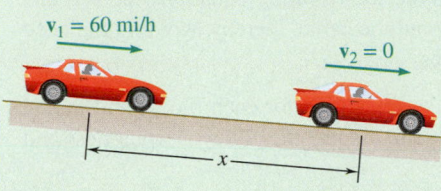

**MODELING:** Choose the car as the system and assume it can be modeled as a particle.

**ANALYSIS:** To apply the principle of work and energy, you find the kinetic energy at each position of the car. The difference between the kinetic energies will be equal to the work done by the braking force.

**Fig. 1** Car at the two positions of interest.

**Principle of Work and Energy.**

$$T_1 + U_{1 \to 2} = T_2 \tag{1}$$

Therefore, you need to calculate each term in this equation.

**Kinetic Energy.**

*Position 1.* $\quad v_1 = \left(60\frac{\text{mi}}{\text{h}}\right)\left(\frac{5280\text{ ft}}{1\text{ mi}}\right)\left(\frac{1\text{ h}}{3600\text{ s}}\right) = 88\text{ ft/s}$

$$T_1 = \tfrac{1}{2}mv_1^2 = \tfrac{1}{2}(4000/32.2)(88)^2 = 481{,}000\text{ ft·lb}$$

*Position 2.* $\qquad v_2 = 0 \qquad T_2 = 0$

**Work.** The best way to identify which forces do work is to draw a free-body diagram, as shown in Fig. 2. It is clear that the only external forces that do work are the total braking force and the weight. The normal force does no work because it is perpendicular to the motion. Using the definition of work gives

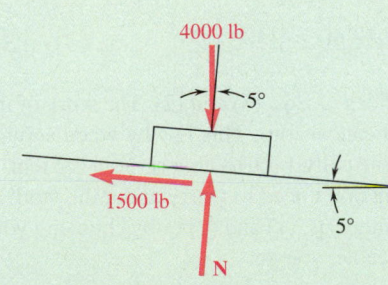

$$U_{1 \to 2} = -1500x + (4000\sin 5°)x = -1151x$$

Note that the work of the gravitational force is positive because the automobile is moving down. Substituting into Eq. (1) gives

$$481{,}000 - 1151x = 0 \qquad x = 418\text{ ft} \quad \blacktriangleleft$$

**Fig. 2** Free-body diagram for the car.

**REFLECT and THINK:** Solving this problem using Newton's second law would require determining the car's deceleration from the free-body diagram (Fig. 2) and then integrating this using the given velocity information. Using the principle of work and energy allows you to avoid that calculation.

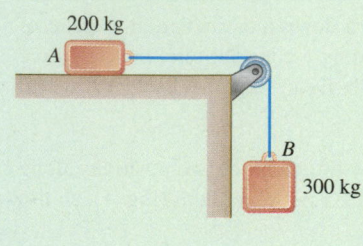

## Sample Problem 13.2

Two blocks are joined by an inextensible cable as shown. If the system is released from rest, determine the velocity of block A after it has moved 2 m. Assume that the coefficient of kinetic friction between block A and the plane is $\mu_k = 0.25$ and that the pulley is weightless and frictionless.

**STRATEGY:**   You are interested in determining the velocity and are given two locations in space, so use the principle of work and energy. You can apply this principle to each block and combine the resulting equations, or you can choose your system to be both blocks and the cable, thereby avoiding the need to determine the work of internal forces.

**MODELING:**   Define two separate systems, one for each block, and model them as particles. As stated in the problem, assume the pulley is weightless and frictionless.

**ANALYSIS:**

**Work and Energy for Block A.**   Denote the friction force by $\mathbf{F}_A$ and the force exerted by the cable by $\mathbf{F}_C$. Then you have (Fig. 1)

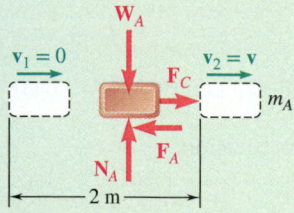

$$m_A = 200 \text{ kg} \qquad W_A = (200 \text{ kg})(9.81 \text{ m/s}^2) = 1962 \text{ N}$$

$$F_A = \mu_k N_A = \mu_k W_A = 0.25(1962 \text{ N}) = 490 \text{ N}$$

$$T_1 + U_{1\to2} = T_2: \qquad 0 + F_C(2 \text{ m}) - F_A(2 \text{ m}) = \tfrac{1}{2}m_A v^2$$

$$F_C(2 \text{ m}) - (490 \text{ N})(2 \text{ m}) = \tfrac{1}{2}(200 \text{ kg})v^2 \qquad \textbf{(1)}$$

**Fig. 1**   Free-body diagram and two positions for block A.

**Work and Energy for Block B.**   From the free-body diagram for block B (Fig. 2), you have

$$m_B = 300 \text{ kg} \qquad W_B = (300 \text{ kg})(9.81 \text{ m/s}^2) = 2940 \text{ N}$$

$$T_1 + U_{1\to2} = T_2: \qquad 0 + W_B(2 \text{ m}) - F_C(2 \text{ m}) = \tfrac{1}{2}m_B v^2$$

$$(2940 \text{ N})(2 \text{ m}) - F_C(2 \text{ m}) = \tfrac{1}{2}(300 \text{ kg})v^2 \qquad \textbf{(2)}$$

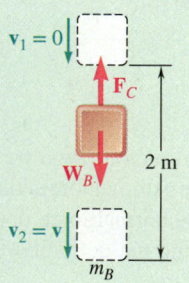

**Fig. 2**   Free-body diagram and two positions for block B.

Now add the left-hand and right-hand sides of Eqs. (1) and (2). The work of the forces exerted by the cable on A and B cancels out. This is why when solving problems using work and energy, it is usually best to choose your system to include all the objects of interest, so you don't need to worry about the work of internal forces. Therefore, after combining Eqs. (1) and (2) or by choosing your system to be block A, block B, and the cable, you get

$$(2940 \text{ N})(2 \text{ m}) - (490 \text{ N})(2 \text{ m}) = \tfrac{1}{2}(200 \text{ kg} + 300 \text{ kg})v^2$$

$$4900 \text{ J} = \tfrac{1}{2}(500 \text{ kg})v^2 \qquad v = 4.43 \text{ m/s} \quad \blacktriangleleft$$

**REFLECT and THINK:**   When using the principle of work and energy, it usually saves time to choose your system to be everything that moves. Now that you know the velocity of the block, you could use Eq. (1) to determine the force in the cable. Only when you need to determine an internal force would you need to isolate part of a system.

# Sample Problem 13.3

A spring is used to stop a 60-kg package that is sliding on a horizontal surface. The spring has a constant $k = 20$ kN/m and is held by cables so that it is initially compressed 120 mm. The package has a velocity of 2.5 m/s in the position shown, and the maximum additional deflection of the spring is 40 mm. Determine (*a*) the coefficient of kinetic friction between the package and the surface, (*b*) the velocity of the package as it passes again through the position shown.

**STRATEGY:**  You have velocity information and specific locations in space, so use the principle of work and energy. Break the motion into two segments: segment 1 is the initial position to the point where the spring has a maximum deflection (Fig. 1), and segment 2 is from the point the spring has a maximum deflection back to the original position.

**MODELING:**  The system is the crate, which you can model as a particle. A free-body diagram for the crate when it is not in contact with the spring is shown in Fig. 2. After it hits the spring, it has an additional force **P** acting on it due to the compression of the spring (Fig. 3).

**ANALYSIS:**  The principle of work and energy is

$$T_1 + U_{1\to2} = T_2 \tag{1}$$

Call the initial position of the package position 1 and the position where maximum spring deflection occurs position 2 (Fig. 1)

### a. Motion from Position 1 to Position 2

**Kinetic Energy.**  *Position 1.*  $v_1 = 2.5$ m/s

$$T_1 = \tfrac{1}{2}mv_1^2 = \tfrac{1}{2}(60 \text{ kg})(2.5 \text{ m/s})^2 = 187.5 \text{ N·m} = 187.5 \text{ J}$$

*Position 2.* (maximum spring deflection):  $v_2 = 0$     $T_2 = 0$

**Work.**  *Friction Force* **F.** You have (Fig. 2)

$$F = \mu_k N = \mu_k W = \mu_k mg = \mu_k(60 \text{ kg})(9.81 \text{ m/s}^2) = (588.6 \text{ N})\mu_k$$

The work of **F** is negative and equal to

$$(U_{1\to2})_f = -Fx = -(588.6 \text{ N})\mu_k(0.600 \text{ m} + 0.040 \text{ m}) = -(377 \text{ J})\mu_k$$

*Spring Force* **P.** The variable force **P** exerted by the spring does an amount of negative work equal to the area under the force-deflection curve of the spring force. You have

$$P_{min} = kx_0 = (20 \text{ kN/m})(120 \text{ mm}) = (20\,000 \text{ N/m})(0.120 \text{ m}) = 2400 \text{ N}$$
$$P_{max} = P_{min} + k\,\Delta x = 2400 \text{ N} + (20 \text{ kN/m})(40 \text{ mm}) = 3200 \text{ N}$$
$$(U_{1\to2})_e = -\tfrac{1}{2}(P_{min} + P_{max})\,\Delta x = -\tfrac{1}{2}(2400 \text{ N} + 3200 \text{ N})(0.040 \text{ m}) = -112.0 \text{ J}$$

The total work between positions 1 and 2 is thus

$$U_{1\to2} = (U_{1\to2})_f + (U_{1\to2})_e = -(377 \text{ J})\mu_k - 112.0 \text{ J}$$

**Principle of Work and Energy.**  You can determine the coefficient of kinetic friction from the expression for the principle of work and energy in this segment of the motion.

$$T_1 + U_{1\to2} = T_2: \quad 187.5 \text{ J} - (377 \text{ J})\mu_k - 112.0 \text{ J} = 0 \qquad \mu_k = 0.20 \ \blacktriangleleft$$

*(continued)*

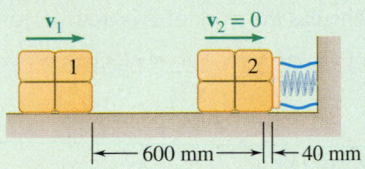

**Fig. 1**   The package at position 1 and position 2.

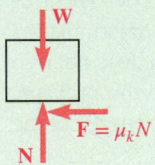

**Fig. 2**   Free-body diagram before spring is engaged.

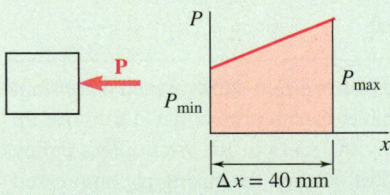

**Fig. 3**   Force *P* on the block after it hits the spring.

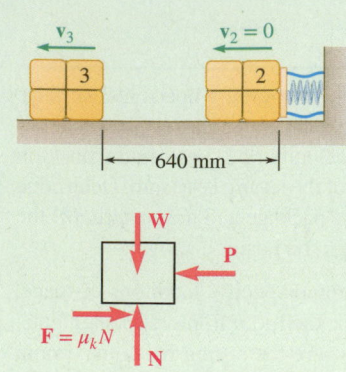

**Fig. 4** Free-body diagram when the package is moving to the left.

### b. Motion from Position 2 to Position 3.

Call the position where the package returns to its initial position as position 3 (Fig. 4).

**Kinetic Energy.** *Position 2.*

$$v_2 = 0 \qquad T_2 = 0$$

*Position 3.*

$$T_3 = \tfrac{1}{2}mv_3^2 = \tfrac{1}{2}(60 \text{ kg})v_3^2$$

**Work.** Because the distances involved are the same, the numerical values of the work of the friction force **F** and of the spring force **P** are the same as before. However, the work of **F** is still negative, whereas the work of **P** is now positive.

$$U_{2\rightarrow3} = -(377 \text{ J})\mu_k + 112.0 \text{ J} = -75.5 \text{ J} + 112.0 \text{ J} = +36.5 \text{ J}$$

**Principle of Work and Energy.**

$$T_2 + U_{2\rightarrow3} = T_3: \quad 0 + 36.5 \text{ J} = \tfrac{1}{2}(60 \text{ kg})v_3^2$$

$$v_3 = 1.103 \text{ m/s} \qquad \mathbf{v_3 = 1.103 \text{ m/s} \leftarrow}$$

**REFLECT and THINK:** You needed to break this problem into two segments. From the first segment, you were able to determine the coefficient of friction. Then, you could use the principle of work and energy to determine the velocity of the package at any other location. Note that the system does not lose any energy due to the spring; it returns all of its energy back to the package. You would need to design something that could absorb the kinetic energy of the package in order to bring it to rest.

---

## Sample Problem 13.4

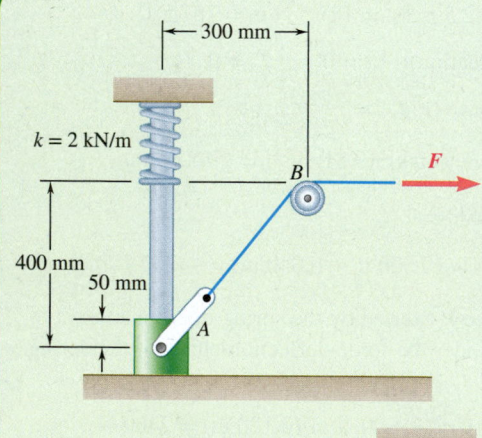

The 2-kg collar *A* starts from rest in the position shown when a constant force *F* = 100 N is applied to the cable, causing the collar *A* to move up the smooth vertical shaft. Neglecting the mass of the frictionless pulley and the spring, determine the speed of *A* when the spring is compressed 50 mm. ▶

**STRATEGY:** You have information about two positions and are asked to find a speed, so use the principle of work and energy.

**MODELING:** You have several choices of systems. Two possible systems are shown in Fig. 1.

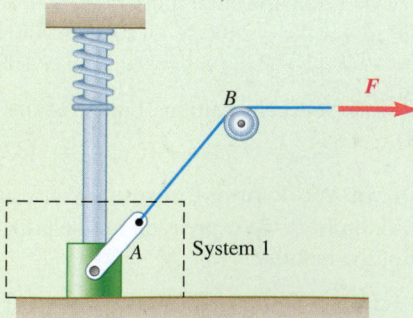

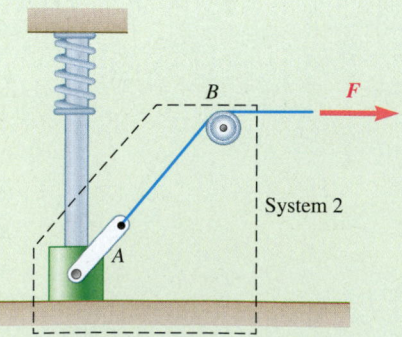

**Fig. 1** Possible systems for this problem.

Which one should you use? You can solve the problem using either one, but it turns out that some system choices make the problem easier to solve than others. For system 1, the tension in the rope is $F$, but only the component of $F$ in the direction of the motion does work. This component is continually changing, so calculating the work is difficult. For system 2, the work the force $F$ does is just the magnitude of $F$ (because it is constant) times the distance the force travels horizontally. Therefore, the problem is easiest to solve using system 2.

**ANALYSIS:**   The principle of work and energy is

$$T_1 + U_{1 \rightarrow 2} = T_2 \tag{1}$$

To start, draw the system in the two positions shown in Fig. 2. Because the figure will be very cluttered if you draw the two positions on the same figure, you should draw them side by side.

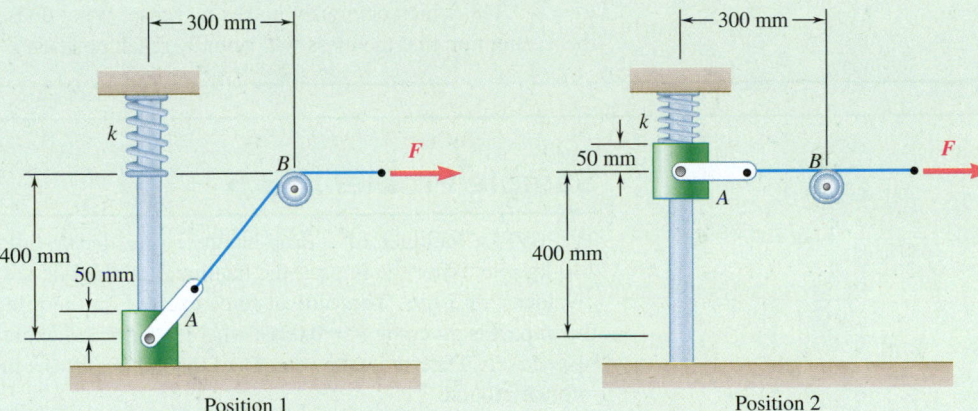

**Fig. 2**   System in the two positions of interest.

**Kinetic Energy.**   Because the collar is initially at rest, $T_1 = 0$. In position 2, when the upper spring is compressed 50 mm, the kinetic energy is

$$T_2 = \frac{1}{2}mv_2^2 = \frac{1}{2}(2 \text{ kg})v_2^2 = v_2^2$$

**Work.**   As the collar is raised from position 1 to where the spring is compressed 50 mm, the work done by the weight is

$$(U_{1 \rightarrow 2})_g = -mgy_2 = -(2 \text{ kg})(9.81 \text{ m/s}^2)(0.4 \text{ m}) = -7.848 \text{ J}$$

and the work of the spring force is

$$(U_{1 \rightarrow 2})_s = \frac{1}{2}kx_1^2 - \frac{1}{2}kx_2^2 = 0 - \frac{1}{2}(2000 \text{ N/m})(0.05 \text{ m})^2 = -2.50 \text{ J}$$

Finally, you must calculate the work of the 100-N force. In position 1, the length $AB$ is

$$(l_{AB})_1 = \sqrt{(0.4)^2 + (0.3)^2} = 0.5 \text{ m}$$

In position 2, the length $AB$ is $(l_{AB})_2 = 0.3$ m. The distance the 100-N force travels through is therefore

$$d = (l_{AB})_1 - (l_{AB})_2 = 0.5 \text{ m} - 0.3 \text{ m} = 0.2 \text{ m}$$

*(continued)*

The work done by the 100-N force $F$ is

$$(U_{1\rightarrow2})_F = Fd = (100 \text{ N})(0.5 \text{ m} - 0.3 \text{ m}) = 20 \text{ J}$$

Thus, the total work is

$$U_{1\rightarrow2} = (U_{1\rightarrow2})_g + (U_{1\rightarrow2})_s + (U_{1\rightarrow2})_F = -7.848 \text{ J} - 2.50 \text{ J} + 20 \text{ J} = 9.652 \text{ J}$$

Substituting these values in the principle of work and energy gives

$$T_1 + U_{1\rightarrow2} = T_2$$
$$0 + 9.652 = v_2^2$$

$$v_2 = 3.11 \text{ m/s} \blacktriangleleft$$

**REFLECT and THINK:** What if the force had been only 10 N instead of 100 N? The work would have been a factor of 10 smaller (i.e., 2 J), and you would have $v_2^2 = -8.348$, which obviously makes no sense. What does this mean? It means the assumption that the mass will actually reach position 2 is incorrect.

## Sample Problem 13.5

The 650-kg hammer of a drop-hammer pile driver falls onto the top of a 140-kg pile. After the impact, the hammer and the pile stick together and have a velocity of 3 m/s. The vertical force exerted on the pile by the ground after the impact is given by $F = 0.02x^2$, where $x$ and $F$ are expressed in mm and kN, respectively. Determine the velocity of the system after it has penetrated 80 mm into the ground.

**STRATEGY:** You are given a force as a function of displacement and are interested in two positions; therefore, use the principle of work and energy.

**MODELING:** The system is the hammer and the pile together after the impact. They can be modeled as a single particle. A free-body diagram for this system (Fig. 1) shows that the only two forces that do work are the weight and the force from the ground.

**ANALYSIS:** The principle of work and energy is

$$T_1 + U_{1\rightarrow2} = T_2 \tag{1}$$

**Kinetic Energy.** The two positions being considered are immediately after the impact and after the system has moved down 50 mm. Because the system is initially traveling at 3 m/s, the initial kinetic energy is

$$T_1 = \frac{1}{2}mv_1^2 = \frac{1}{2}(650 \text{ kg} + 140 \text{ kg})(3 \text{ m/s})^2 = 3555 \text{ J}$$

In position 2, the kinetic energy is

$$T_2 = \frac{1}{2}mv_2^2 = \frac{1}{2}(650 \text{ kg} + 140 \text{ kg})v_2^2 = 395\, v_2^2$$

**Work.** As the system moves into the ground, the weight and the resisting force, $F$, do work. The work the weight does is

$$(U_{1\rightarrow2})_g = mgy = (790 \text{ kg})(9.81 \text{ m/s}^2)(0.08 \text{ m}) = 620.0 \text{ J}$$

*(continued)*

**Fig. 1**  Free-body diagram after the impact.

The given equation for the force is such that $F$ is in kN when $x$ is expressed in mm. This means that the number in front (i.e., the 0.02) has to have the units of $kN/mm^2$ for the units to work out. The work of the resisting force is

$$(U_{1\to2})_F = \int_{x_1}^{x_2} F_x \, dx$$

$$= \int_0^{80} -(0.02 \text{ kN/mm}^2)x^2 \, dx = -\left(\frac{0.02}{3} \text{ kN/mm}^2\right)x^3 \Big|_0^{80}$$

$$= -3413 \text{ kN·mm} = -3413 \text{ J}$$

Thus, the total work is

$$U_{1\to2} = (U_{1\to2})_g + (U_{1\to2})_F = 620.0 \text{ J} - 3413 \text{ J} = -2793 \text{ J}$$

Substituting the kinetic energies and total work in the principle of work and energy gives

$$T_1 + U_{1\to2} = T_2$$
$$3555 - 2793 = 395v_2^2$$

$$v_2 = 1.389 \text{ m/s} \downarrow \quad \blacktriangleleft$$

**REFLECT and THINK:**  To determine how deep the system enters the ground before it stops, you need to set the final kinetic energy equal to zero and make the maximum depth, $x_m$, unknown. This gives

$$3555 + 790(9.81)x_m - \left(\frac{2 \times 10^7}{3} \text{ N/m}^2\right)x_m^3 = 0$$

Solving this, you find $x_m = 0.0859$ m or 85.9 mm.

## Sample Problem 13.6

A 2000-lb roller coaster car starts from rest at point 1 and moves without friction down the track shown. (*a*) Determine the force exerted by the track on the car at point 2, where the radius of curvature of the track is 20 ft. (*b*) Determine the minimum safe value of the radius of curvature at point 3.

**STRATEGY:**  Use the principle of work and energy to determine the speed of the car at any location along the track. To determine the force exerted by the track, you need to use Newton's second law. You will need to draw a free-body diagram and kinetic diagram of the car at each position.

**MODELING:**  Choose the car as the system and assume it can be modeled as a particle.

**ANALYSIS:**  Apply the principle of work and energy

$$T_1 + U_{1\to2} = T_2 \tag{1}$$

*(continued)*

### a. Force Exerted by the Track at Point 2.

Use the principle of work and energy to determine the velocity of the car as it passes through point 2.

**Kinetic Energy.** $\quad T_1 = 0 \qquad T_2 = \frac{1}{2}mv_2^2 = \frac{1}{2}\frac{W}{g}v_2^2$

**Work.** The only force that does work is the weight **W**. Because the vertical displacement from point 1 to point 2 is 40 ft downward, the work of the weight is

$$U_{1\rightarrow2} = +W(40\text{ ft})$$

**Principle of Work and Energy.** Substituting these values into Eq. (1) gives

$$T_1 + U_{1\rightarrow2} = T_2 \qquad 0 + W(40\text{ ft}) = \frac{1}{2}\frac{W}{g}v_2^2$$

$$v_2^2 = 80g = 80(32.2) \qquad v_2 = 50.8\text{ ft/s}$$

**Newton's Second Law at Point 2.** The acceleration $\mathbf{a}_n$ of the car at point 2 has a magnitude of $a_n = v_2^2/\rho$ and is directed upward. Because the external forces acting on the car are **W** and **N** (Fig. 1), you have

$$+\uparrow\Sigma F_n = ma_n: \qquad -W + N = ma_n$$

$$= \frac{W}{g}\frac{v_2^2}{\rho}$$

$$= \frac{W}{g}\frac{80g}{20}$$

$$N = 5W \qquad \mathbf{N} = 10{,}000\text{ lb }\uparrow \quad \blacktriangleleft$$

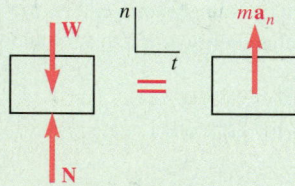

**Fig. 1** Free-body diagram and kinetic diagram at point 2.

### b. Minimum Value of $\rho$ at Point 3.

**Principle of Work and Energy.** Applying the principle of work and energy between point 1 and point 3, you obtain

$$T_1 + U_{1\rightarrow3} = T_3 \qquad 0 + W(25\text{ ft}) = \frac{1}{2}\frac{W}{g}v_3^2$$

$$v_3^2 = 50g = 50(32.2) \qquad v_3 = 40.1\text{ ft/s}$$

**Newton's Second Law at Point 3.** The minimum safe value of $\rho$ occurs when **N** = 0. In this case, the acceleration $\mathbf{a}_n$ with a magnitude of $a_n = v_3^2/\rho$, is directed downward (Fig. 2), and you have

$$+\downarrow\Sigma F_n = ma_n: \qquad W = \frac{W}{g}\frac{v_3^2}{\rho}$$

$$= \frac{W}{g}\frac{50g}{\rho} \qquad\qquad \rho = 50\text{ ft} \quad \blacktriangleleft$$

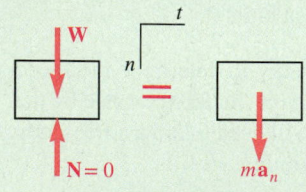

**Fig. 2** Free-body diagram and kinetic diagram at point 3.

**REFLECT and THINK:** This is an example where you need both Newton's second law and the principle of work and energy. Work–energy is used to determine the speed of the car, and Newton's second law is used to determine the normal force. A normal force of $5W$ is equivalent to a fighter pilot pulling 5 g's and should only be experienced for a very short time. For safety, you would also want to make sure your radius of curvature was quite a bit larger than 50 ft.

# Sample Problem 13.7

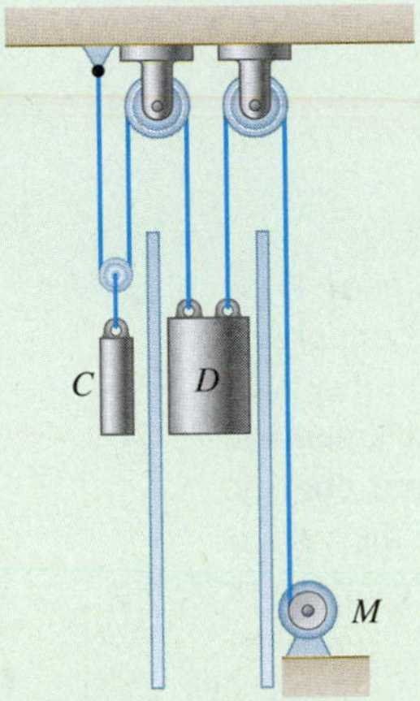

The dumbwaiter $D$ and its load have a combined weight of 600 lb, whereas the counterweight $C$ weighs 800 lb. Determine the power delivered by the electric motor $M$ when the dumbwaiter (a) is moving up at a constant speed of 8 ft/s, (b) has an instantaneous velocity of 8 ft/s and an acceleration of 2.5 ft/s², where both are directed upward.

**STRATEGY:**  This problem requires you to use the definition of power. You will need to use Newton's second law to determine the tensions in the two cables.

**MODELING:**  Define two separate systems, one for body $C$ and one for body $D$, and model them as particles. Assume the pulley is weightless and frictionless.

**ANALYSIS:**  The force **F** exerted by the motor cable has the same direction as the velocity $\mathbf{v}_D$ of the dumbwaiter, so the power is equal to $Fv_D$, where $v_D = 8$ ft/s. To obtain the power, you must first determine **F** in each of the two given situations.

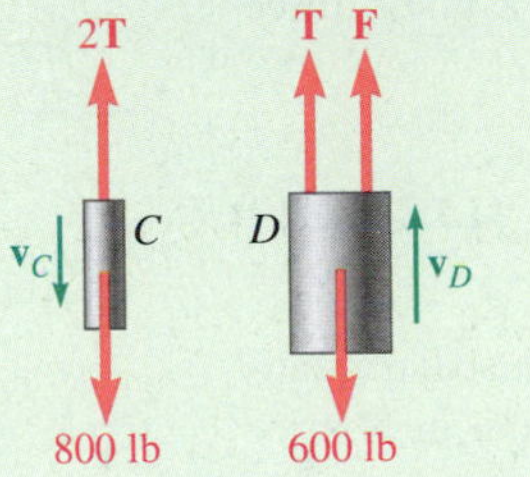

**a. Uniform Motion.**  You have $\mathbf{a}_C = \mathbf{a}_D = 0$; both bodies are in equilibrium (Fig. 1).

**Body C:** $\qquad +\uparrow\Sigma F_y = 0:\qquad 2T - 800\text{ lb} = 0 \qquad T = 400\text{ lb}$
**Body D:** $\qquad +\uparrow\Sigma F_y = 0:\qquad F + T - 600\text{ lb} = 0$

$$F = 600\text{ lb} - T = 600\text{ lb} - 400\text{ lb} = 200\text{ lb}$$
$$Fv_D = (200\text{ lb})(8\text{ ft/s}) = 1600\text{ ft·lb/s}$$
$$\text{Power} = (1600\text{ ft·lb/s})\frac{1\text{ hp}}{550\text{ ft·lb/s}} = 2.91\text{ hp} \blacktriangleleft$$

**b. Accelerated Motion.**  You have

$$\mathbf{a}_D = 2.5\text{ ft/s}^2\uparrow \qquad \mathbf{a}_C = -\tfrac{1}{2}\mathbf{a}_D = 1.25\text{ ft/s}^2\downarrow$$

The equations of motion are obtained using Figs. 2 and 3.

**Body C:** $\qquad +\downarrow\Sigma F_y = m_C a_C:\qquad 800 - 2T = \dfrac{800}{32.2}(1.25) \qquad T = 384.5\text{ lb}$

**Body D:** $\qquad +\uparrow\Sigma F_y = m_D a_D:\qquad F + T - 600 = \dfrac{600}{32.2}(2.5)$

$$F + 384.5 - 600 = 46.6 \qquad F = 262.1\text{ lb}$$
$$Fv_D = (262.1\text{ lb})(8\text{ ft/s}) = 2097\text{ ft·lb/s}$$
$$\text{Power} = (2097\text{ ft·lb/s})\frac{1\text{ hp}}{550\text{ ft·lb/s}} = 3.81\text{ hp} \blacktriangleleft$$

**REFLECT and THINK:**  As you might expect, the motor needs to deliver more power to produce accelerated motion than to produce motion at constant velocity.

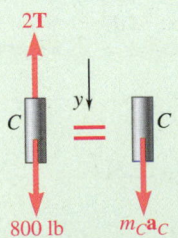

**Fig. 2**  Free-body diagram and kinetic diagram for C.

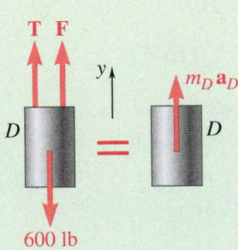

**Fig. 3**  Free-body diagram and kinetic diagram for D.

**Fig. 1**  Free-body diagrams for C and D.

# SOLVING PROBLEMS ON YOUR OWN

In the preceding chapter, you solved problems dealing with the motion of a particle by using the fundamental equation $\mathbf{F} = m\mathbf{a}$ to determine the acceleration $\mathbf{a}$. By applying the principles of kinematics, you could then use $\mathbf{a}$ to determine the velocity and displacement of the particle at any time. In this section, we combined $\mathbf{F} = m\mathbf{a}$ and kinematic relationships to obtain an additional principle called the **principle of work and energy**. This eliminates the need to calculate the acceleration and enables you to relate the velocities of the particle at two points along its path of motion. To solve a problem using work and energy, you need to follow these steps:

**1. Compute the work of each of the external forces.** The work $U_{1\rightarrow2}$ of a given force $\mathbf{F}$ during the finite displacement of a particle from $A_1$ to $A_2$ is defined as

$$U_{1\rightarrow2} = \int \mathbf{F}\cdot d\mathbf{r} \quad \text{or} \quad U_{1\rightarrow2} = \int (F\cos\alpha)\,ds \qquad \textbf{(13.2, 13.2}')$$

where $\alpha$ is the angle between $\mathbf{F}$ and the displacement $d\mathbf{r}$. The work $U_{1\rightarrow2}$ is a scalar quantity and is expressed in ft·lb or in·lb in the U.S. customary system of units and in N·m or joules (J) in the SI system of units. Note that the work done is zero for a force perpendicular to the displacement ($\alpha = 90°$). Negative work is done for $90° < \alpha < 180°$ and in particular for a friction force, which is generally opposite in direction to the displacement ($\alpha = 180°$).

The work $U_{1\rightarrow2}$ can be easily evaluated in the following cases that you will encounter.

**a. Work of a constant force in rectilinear motion (Sample Prob. 13.1)**

$$U_{1\rightarrow2} = (F\cos\alpha)\,\Delta x \qquad \textbf{(13.3)}$$

where $\alpha$ = angle the force forms with the direction of motion
$\Delta x$ = displacement from $A_1$ to $A_2$ (Fig. 13.3)

**b. Work of the force of gravity (Sample Probs. 13.1, 13.2, 13.4, 13.5, and 13.6)**

$$U_{1\rightarrow2} = -W\Delta y \qquad \textbf{(13.4}')$$

where $\Delta y$ is the vertical displacement of the center of gravity of the body of weight $W$. Note that the work is positive when $\Delta y$ is negative; that is, when the body moves down (Fig. 13.4).

**c. Work of the force exerted by a linear spring (Sample Probs. 13.3 and 13.4)**

$$U_{1\rightarrow2} = \tfrac{1}{2}kx_1^2 - \tfrac{1}{2}kx_2^2 \qquad \textbf{(13.6)}$$

where $k$ is the spring constant and $x_1$ and $x_2$ are the elongations of the spring corresponding to the positions $A_1$ and $A_2$ (Fig. 13.5).

### d. Work of a gravitational force

$$U_{1\to 2} = \frac{GMm}{r_2} - \frac{GMm}{r_1} \tag{13.7}$$

for a displacement of the body from $A_1(r = r_1)$ to $A_2(r = r_2)$ (Fig. 13.6).

**2. Calculate the kinetic energy at $A_1$ and $A_2$.** The kinetic energy $T$ is

$$T = \frac{1}{2}mv^2 \tag{13.9}$$

where $m$ is the mass of the particle and $v$ is the magnitude of its velocity. The units of kinetic energy are the same as the units of work; that is, ft·lb or in·lb if you use U.S. customary units and N·m or joules (J) if you use SI units.

**3. Substitute the values for the work done $U_{1\to 2}$ and the kinetic energies $T_1$ and $T_2$** into the equation

$$T_1 + U_{1\to 2} = T_2 \tag{13.11}$$

You will now have one scalar equation that you can solve for one unknown. Note that this equation does not yield the time of travel or the acceleration directly. However, if you know the radius of curvature $\rho$ of the path of the particle at a point where you have obtained the velocity $v$, you can express the normal component of the acceleration as $a_n = v^2/\rho$ and obtain the normal component of the force exerted on the particle by using Newton's second law.

**4. We introduced power in this section as the time rate at which work is done as $P = dU/dt$.** Power is measured in ft·lb/s or *horsepower* (hp) in U.S. customary units and in J/s or *watts* (W) in the SI system of units. To calculate the power, you can use the equivalent formula

$$P = \mathbf{F}\cdot\mathbf{v} \tag{13.13}$$

where $\mathbf{F}$ and $\mathbf{v}$ denote the force and the velocity, respectively, at a given time (Sample Prob. 13.7). In some problems (see, e.g., Prob. 13.47), you will be asked for the *average power*, which you can obtain by dividing the total work by the time interval during which the work is done.

# Problems

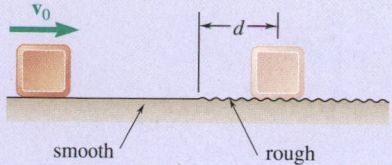

smooth      rough

Fig. P13.CQ1

## CONCEPT QUESTIONS

**13.CQ1** Block *A* is traveling with a speed $v_0$ on a smooth surface when the surface suddenly becomes rough with a coefficient of friction of $\mu$ causing the block to stop after a distance *d*. If block *A* were traveling twice as fast, that is, at a speed $2v_0$, how far will it travel on the rough surface before stopping?

**a.** *d*/2
**b.** *d*
**c.** $\sqrt{2}\,d$
**d.** 2*d*
**e.** 4*d*

## END-OF-SECTION PROBLEMS

**13.1** A 400-kg satellite is placed in a circular orbit 6394 km above the surface of the earth. At this elevation, the acceleration of gravity is 4.09 m/s². Knowing that its orbital speed is 20 000 km/h, determine the kinetic energy of the satellite.

**13.2** A 0.5-lb stone is dropped down the "bottomless pit" at Carlsbad Caverns and strikes the ground with a speed of 95 ft/s. Neglecting air resistance, determine (*a*) the kinetic energy of the stone as it strikes the ground and the height *h* from which it was dropped. (*b*) Solve part *a* assuming that the same stone is dropped down a hole on the moon. (Acceleration of gravity on the moon = 5.31 ft/s².)

Fig. P13.2

Fig. *P13.3*

**13.3** A baseball player hits a 5.1-oz baseball with an initial velocity of 130 ft/s at an angle of 40° with the horizontal as shown. Determine (*a*) the kinetic energy of the ball immediately after it is hit, (*b*) the kinetic energy of the ball when it reaches its maximum height, (*c*) the maximum height above the ground reached by the ball.

**13.4** A 500-kg communications satellite is in a circular geosynchronous orbit and completes one revolution about the earth in 23 h and 56 min at an altitude of 35 800 km above the surface of the earth. Knowing that the radius of the earth is 6370 km, determine the kinetic energy of the satellite.

**13.5** In an ore-mixing operation, a bucket full of ore is suspended from a traveling crane which moves along a stationary bridge. The bucket is to swing no more than 15° from vertical when the crane is brought to a sudden stop. Determine the maximum allowable speed *v* of the crane.

**13.6** In an ore-mixing operation, a bucket full of ore is suspended from a traveling crane which moves along a stationary bridge. The crane is traveling at a speed of 8 ft/s when it is brought to a sudden stop. Determine the maximum horizontal distance through which the bucket will swing.

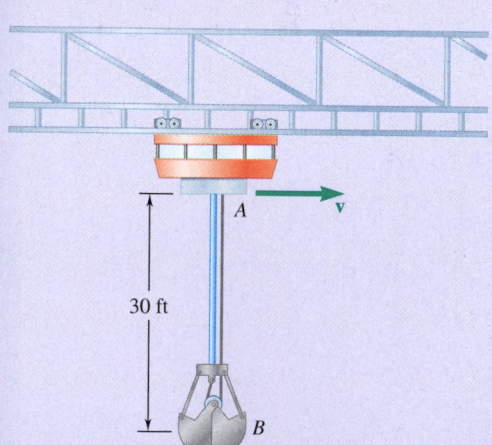

Fig. P13.5 and P13.6

**13.7** Determine the maximum theoretical speed that may be achieved over a distance of 100 m by a car starting from rest, knowing that the coefficient of static friction between the tires and pavement is 0.8 and 60 percent of the weight of the car is distributed over its front wheels and 40 percent over its rear wheels. Assume (a) front-wheel drive, (b) rear-wheel drive, (c) four-wheel drive.

**13.8** A 2000-kg automobile starts from rest at point A on a 6° incline and coasts through a distance of 150 m to point B. The brakes are then applied, causing the automobile to come to a stop at point C, which is 20 m from B. Knowing that slipping is impending during the braking period and neglecting air resistance and rolling resistance, determine (a) the speed of the automobile at point B, (b) the coefficient of static friction between the tires and the road.

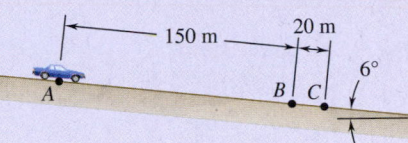

**Fig. P13.8**

**13.9** An athlete is holding 30 lb of weights at a height of 6 inches above the stack as shown. To lower the weights, she applies a constant force of 5 lb to the handle. Determine the velocity of the weights immediately before they hit the stack.

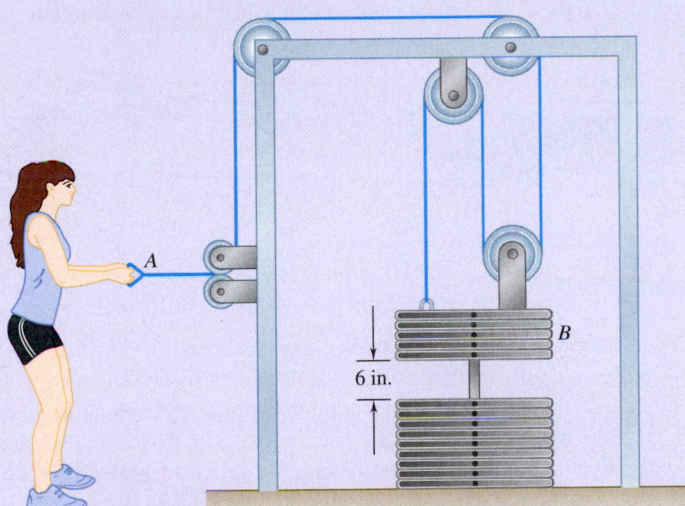

**Fig. P13.9**

**13.10** A 1.4-kg model rocket is launched vertically from rest with a constant thrust of 25 N until the rocket reaches an altitude of 15 m and the thrust ends. Neglecting air resistance, determine (a) the speed of the rocket when the thrust ends, (b) the maximum height reached by the rocket, (c) the speed of the rocket when it returns to the ground.

**13.11** Packages are thrown down an incline at A with a velocity of 1 m/s. The packages slide along the surface ABC to a conveyor belt which moves with a velocity of 2 m/s. Knowing that $\mu_k = 0.25$ between the packages and the surface ABC, determine the distance d if the packages are to arrive at C with a velocity of 2 m/s.

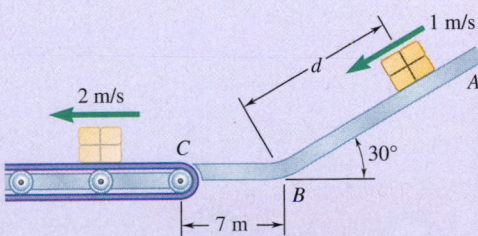

**Fig. P13.11 and P13.12**

**13.12** A package is thrown down an incline at A with a velocity of 1 m/s. The package slides along the surface ABC to a conveyor belt that moves with a velocity of 2 m/s. Knowing that $d = 6$ m and $\mu_k = 0.2$ between the package and all surfaces, determine (a) the speed of the package at C, (b) the distance the package will slide on the conveyor belt before it comes to rest relative to the belt.

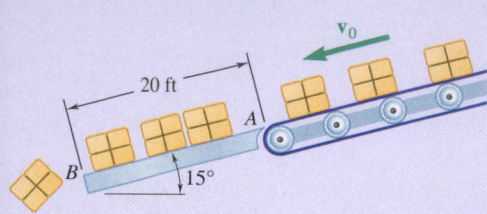

**Fig. P13.13 and P13.14**

**Fig. P13.15**

**13.13** Boxes are transported by a conveyor belt with a velocity $\mathbf{v}_0$ to a fixed incline at $A$ where they slide and eventually fall off at $B$. Knowing that $\mu_k = 0.40$, determine the velocity of the conveyor belt if the boxes leave the incline at $B$ with a velocity of 8 ft/s.

**13.14** Boxes are transported by a conveyor belt with a velocity $\mathbf{v}_0$ to a fixed incline at $A$ where they slide and eventually fall off at $B$. Knowing that $\mu_k = 0.40$, determine the velocity of the conveyor belt if the boxes are to have zero velocity at $B$.

**13.15** A 1200-kg trailer is hitched to a 1400-kg car. The car and trailer are traveling at 72 km/h when the driver applies the brakes on both the car and the trailer. Knowing that the braking forces exerted on the car and the trailer are 5000 N and 4000 N, respectively, determine (*a*) the distance traveled by the car and trailer before they come to a stop, (*b*) the horizontal component of the force exerted by the trailer hitch on the car.

**13.16** A trailer truck enters a 2 percent uphill grade traveling at 72 km/h and reaches a speed of 108 km/h in 300 m. The cab has a mass of 1800 kg and the trailer 5400 kg. Determine (*a*) the average force at the wheels of the cab, (*b*) the average force in the coupling between the cab and the trailer.

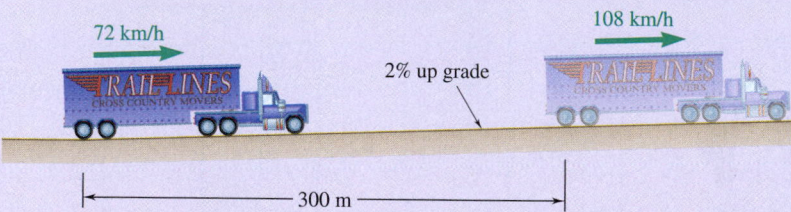

**Fig. P13.16**

**13.17** The subway train shown is traveling at a speed of 30 mi/h when the brakes are fully applied on the wheels of cars $B$ and $C$, causing them to slide on the track, but are not applied on the wheels of car $A$. Knowing that the coefficient of kinetic friction is 0.35 between the wheels and the track, determine (*a*) the distance required to bring the train to a stop, (*b*) the force in each coupling.

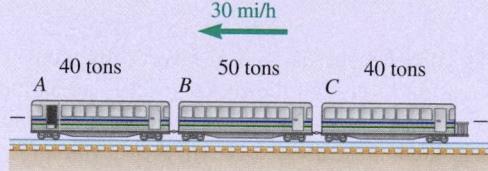

**Fig. P13.17 and P13.18**

**13.18** The subway train shown is traveling at a speed of 30 mi/h when the brakes are fully applied on the wheels of car $A$, causing it to slide on the track, but are not applied on the wheels of cars $B$ or $C$. Knowing that the coefficient of kinetic friction is 0.35 between the wheels and the track, determine (*a*) the distance required to bring the train to a stop, (*b*) the force in each coupling.

**13.19** A 5000-lb truck is being used to lift a 1000-lb boulder $B$ that is on a 200-lb pallet $A$. Knowing that the truck starts from rest and that the horizontal force between the tires and the ground is 700 lb, determine the velocity of the boulder after the truck has moved forward 6 ft.

**Fig. P13.19**

**13.20** The system shown is at rest when a constant 30-lb force is applied to collar *B*. (*a*) If the force acts through the entire motion, determine the speed of collar *B* as it strikes the support at *C*. (*b*) After what distance *d* should the 30-lb force be removed if the collar is to reach support *C* with zero velocity?

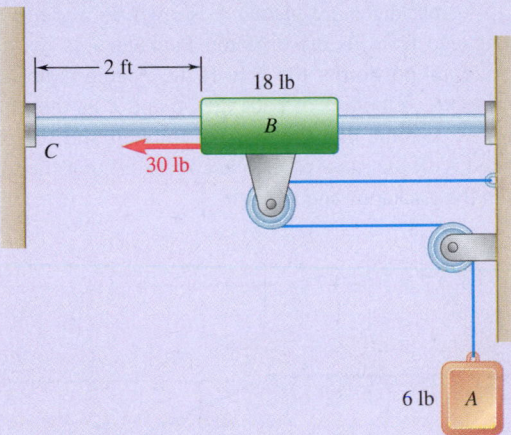

Fig. P13.20

**13.21** Car *B* is towing car *A* at a constant speed of 10 m/s on an uphill grade when the brakes of car *A* are fully applied causing all four wheels to skid. The driver of car *B* does not change the throttle setting or change gears. The masses of the cars *A* and *B* are 1400 kg and 1200 kg, respectively, and the coefficient of kinetic friction is 0.8. Neglecting air resistance and rolling resistance, determine (*a*) the distance traveled by the cars before they come to a stop, (*b*) the tension in the cable.

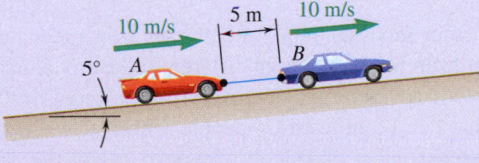

Fig. *P13.21*

**13.22** The motor applies a constant downward force $F = 1050$ lb to the cable used to raise the 4000-lb elevator *E* shown in the figure. The counterweight has a weight of 2000 lb and the elevator starts from rest. After the elevator travels 20 ft, determine (*a*) the velocity of the elevator, (*b*) the velocity of the counterweight.

**13.23** The motor applies a constant downward force *F* to the cable used to raise the 4000-lb elevator *E* shown in the figure. The counterweight has a weight of 2000 lb. Knowing that the elevator starts from rest and reaches a speed of 3 m/s after traveling 30 ft, determine *F*.

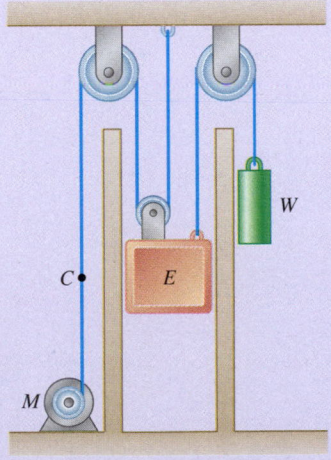

Fig. P13.22 and P13.23

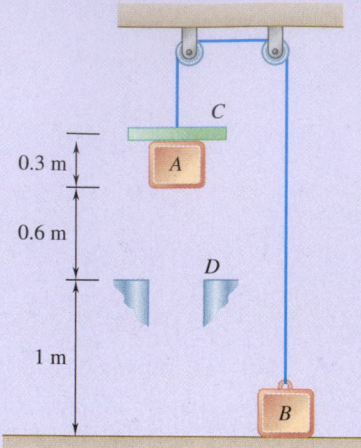

0.3 m

0.6 m

1 m

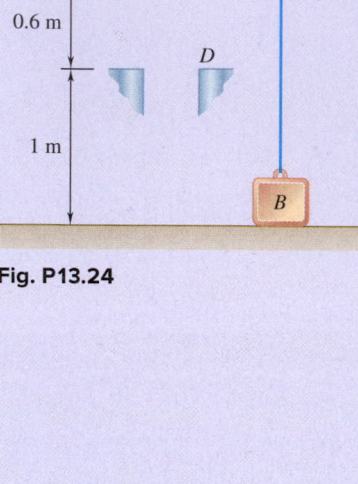

**Fig. P13.24**

**13.24** Two blocks $A$ and $B$, of mass 4 kg and 5 kg, respectively, are connected by a cord that passes over pulleys as shown. A 3-kg collar $C$ is placed on block $A$ and the system is released from rest. After the blocks have moved 0.9 m, collar $C$ is removed and blocks $A$ and $B$ continue to move. Determine the speed of block $A$ just before it strikes the ground.

**13.25** Four 15-kg packages are placed as shown on a conveyor belt which is disengaged from its drive motor. Package $1$ is just to the right of the horizontal portion of the belt. If the system is released from rest, determine the velocities of packages $1$ and $2$ as they fall off the belt at point $A$. Assume that the mass of the belt and the rollers is small compared with the mass of the packages and that there is no slipping between the packages and the belt.

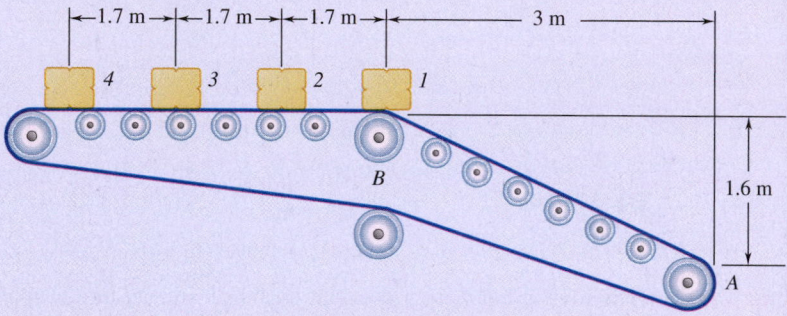

←1.7 m→←1.7 m→←1.7 m→← 3 m →

*4* *3* *2* *1*

$B$

1.6 m

$A$

**Fig. P13.25**

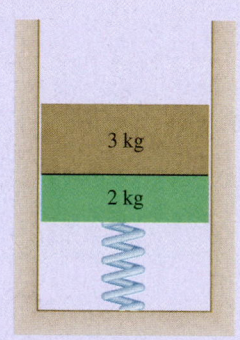

3 kg

2 kg

**Fig. P13.26**

**13.26** A 3-kg block rests on top of a 2-kg block supported by, but not attached to, a spring of constant 40 N/m. The upper block is suddenly removed. Determine (*a*) the maximum speed reached by the 2-kg block, (*b*) the maximum height reached by the 2-kg block.

**13.27** Solve Prob. 13.26, assuming that the 2-kg block is attached to the spring.

**13.28** People with mobility impairments can gain great health and social benefits from participating in different recreational activities. You are tasked with designing an adaptive spring-powered shuffleboard attachment that can be utilized by people who use wheelchairs. Knowing that the coefficient of kinetic friction between the 15-ounce puck $A$ and the wooden surface is 0.3, the maximum spring displacement you desire is 6 inches, and that you want the puck to travel at least 30 ft/s, determine (*a*) the spring constant $k$, (*b*) how far the athlete should pull back the spring to make the puck come to rest after 34 ft.

$k$

$A$

**Fig. P13.28**

**13.29** A 7.5-lb collar is released from rest in the position shown, slides down the inclined rod, and compresses the spring. The direction of motion is reversed and the collar slides up the rod. Knowing that the maximum deflection of the spring is 5 in., determine (a) the coefficient of kinetic friction between the collar and the rod, (b) the maximum speed of the collar.

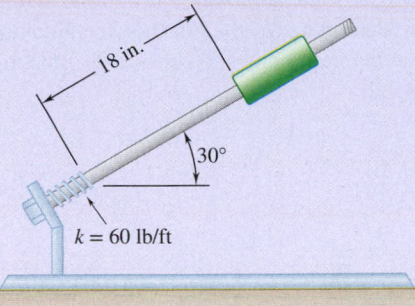

**Fig. P13.29**

**13.30** A 10-kg block is attached to spring A and connected to spring B by a cord and pulley. The block is held in the position shown with both springs unstretched when the support is removed and the block is released with no initial velocity. Knowing that the constant of each spring is 2 kN/m, determine (a) the velocity of the block after it has moved down 50 mm, (b) the maximum velocity achieved by the block.

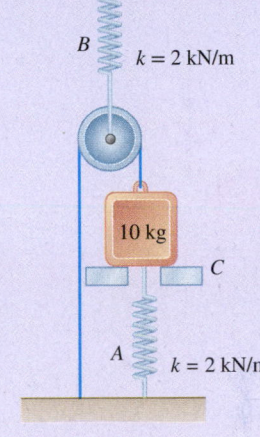

**Fig. P13.30**

**13.31** A 5-kg collar A is at rest on top of, but not attached to, a spring with stiffness $k_1 = 400$ N/m when a constant 150-N force is applied to the cable. Knowing A has a speed of 1 m/s when the upper spring is compressed 75 mm, determine the spring stiffness $k_2$. Ignore friction and the mass of the pulley.

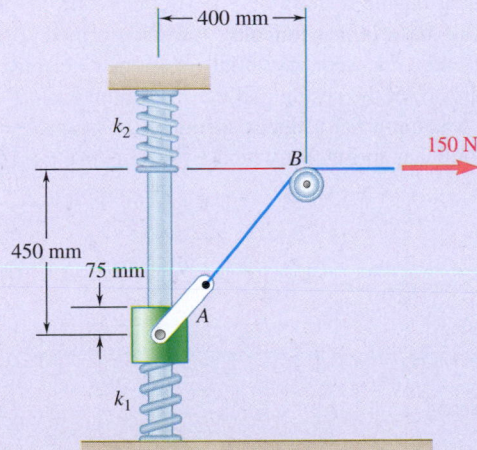

**Fig. P13.31**

**13.32** A 0.75-lb brass (nonmagnetic) block A and a 0.5-lb steel magnet B are in equilibrium in a brass tube under the magnetic repelling force of another steel magnet C located at a distance $x = 0.15$ in. from B. The force is inversely proportional to the square of the distance between B and C. If block A is suddenly removed, determine (a) the maximum velocity of B, (b) the maximum acceleration of B. Assume that air resistance and friction are negligible.

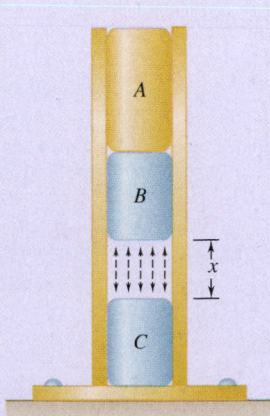

**Fig. P13.32**

**13.33** An uncontrolled automobile traveling at 65 mph strikes squarely a highway crash cushion of the type shown in which the automobile is brought to rest by successively crushing steel barrels. The magnitude $F$ of the force required to crush the barrels is shown as a function of the distance $x$ the automobile has moved into the cushion. Knowing that the weight of the automobile is 2250 lb and neglecting the effect of friction, determine (a) the distance the automobile will move into the cushion before it comes to rest, (b) the maximum deceleration of the automobile.

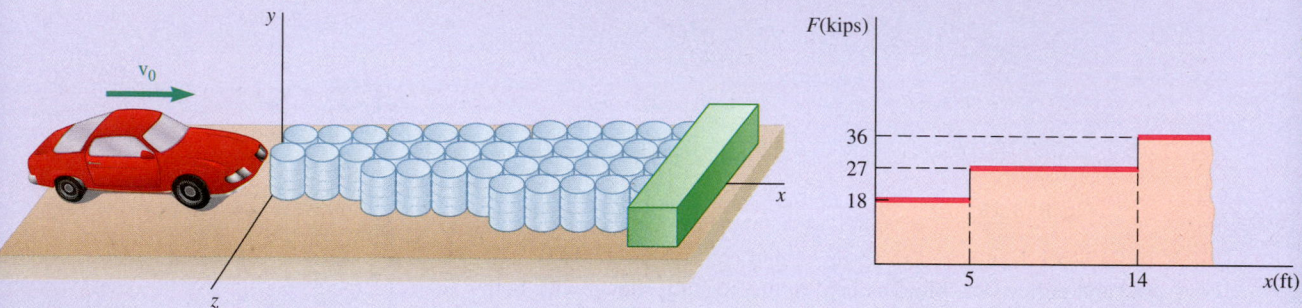

**Fig. P13.33**

**13.34** Two types of energy-absorbing fenders designed to be used on a pier are statically loaded. The force-deflection curve for each type of fender is given in the graph. Determine the maximum deflection of each fender when a 90-ton ship moving at 1 mi/h strikes the fender and is brought to rest.

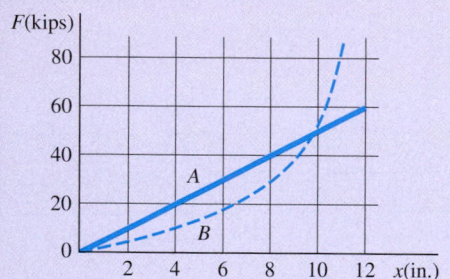

**Fig. P13.34**

**13.35** Nonlinear springs are classified as hard or soft, depending upon the curvature of their force-deflection curve (see figure). If a delicate instrument having a mass of 5 kg is placed on a spring of length $l$ so that its base is just touching the undeformed spring and then is inadvertently released from that position, determine the maximum deflection $x_m$ of the spring and the maximum force $F_m$ exerted by the spring, assuming (a) a linear spring of constant $k = 3$ kN/m, (b) a hard, nonlinear spring, for which $F = (3 \text{ kN/m})(x + 160x^3)$.

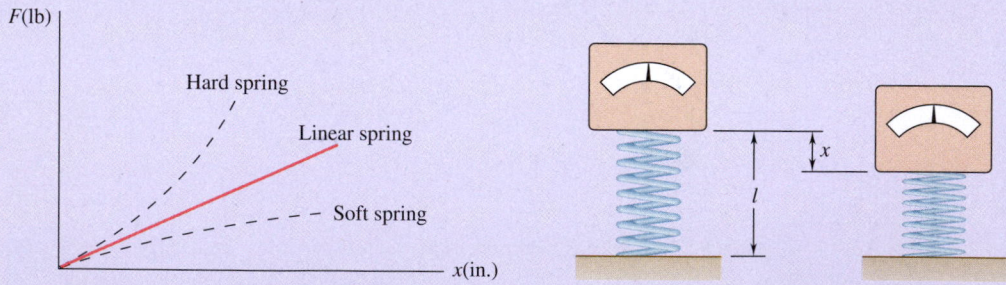

**Fig. P13.35**

**13.36** A meteor starts from rest at a very great distance from the earth. Knowing that the radius of the earth is 6370 km and neglecting all forces except the gravitational attraction of the earth, determine the speed of the meteor (a) when it enters the ionosphere at an altitude of 1000 km, (b) when it enters the stratosphere at an altitude of 50 km, (c) when it strikes the earth's surface.

**13.37** Express the acceleration of gravity $g_h$ at an altitude $h$ above the surface of the earth in terms of the acceleration of gravity $g_0$ at the surface of the earth, the altitude $h$, and the radius $R$ of the earth. Determine the percent error if the weight that an object has on the surface of the earth is used as its weight at an altitude of (a) 0.625 mi, (b) 625 mi.

**13.38** A golf ball struck on earth rises to a maximum height of 60 m and hits the ground 230 m away. How high will the same golf ball travel on the moon if the magnitude and direction of its velocity are the same as they were on earth immediately after the ball was hit? Assume that the ball is hit and lands at the same elevation in both cases and that the effect of the atmosphere on the earth is neglected, so that the trajectory in both cases is a parabola. The acceleration of gravity on the moon is 0.165 times that on earth.

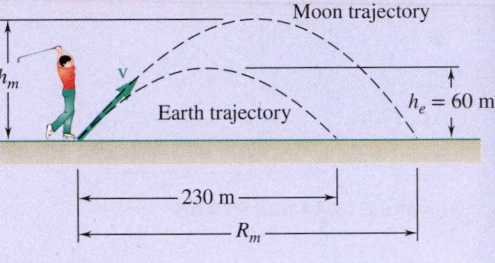

**Fig. P13.38**

**13.39** The sphere at $A$ is given a downward velocity $\mathbf{v}_0$ of magnitude 5 m/s and swings in a vertical plane at the end of a rope of length $l = 2$ m attached to a support at $O$. Determine the angle $\theta$ at which the rope will break, knowing that it can withstand a maximum tension equal to twice the weight of the sphere.

**13.40** The sphere at $A$ is given a downward velocity $\mathbf{v}_0$ and swings in a vertical circle of radius $l$ and center $O$. Determine the smallest velocity $\mathbf{v}_0$ for which the sphere will reach point $B$ as it swings about point $O$ (a) if $AO$ is a rope, (b) if $AO$ is a slender rod of negligible mass.

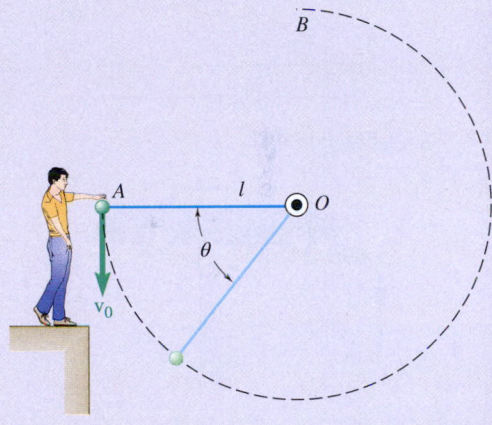

**13.41** A bag is gently pushed off the top of a wall at $A$ and swings in a vertical plane at the end of a rope of length $l$. Determine the angle $\theta$ for which the rope will break, knowing that it can withstand a maximum tension equal to twice the weight of the bag.

**Fig. P13.39 and P13.40**

**13.42** A roller coaster starts from rest at $A$, rolls down the track to $B$, describes a circular loop of 40-ft diameter, and moves up and down past point $E$. Knowing that $h = 60$ ft and assuming no energy loss due to friction, determine (a) the force exerted by his seat on a 160-lb rider at $B$ and $D$, (b) the minimum value of the radius of curvature at $E$ if the roller coaster is not to leave the track at that point.

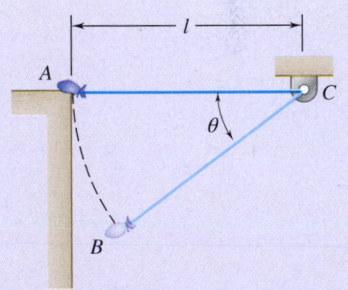

**Fig. P13.41**

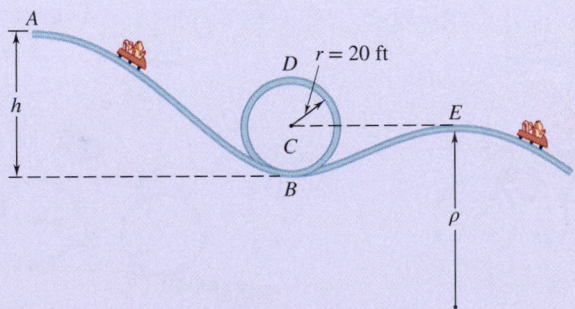

**Fig. P13.42**

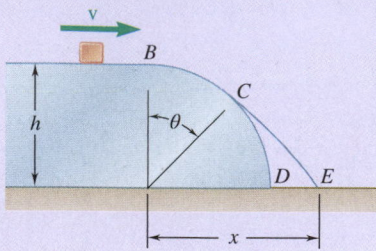

Fig. P13.44 and P13.45

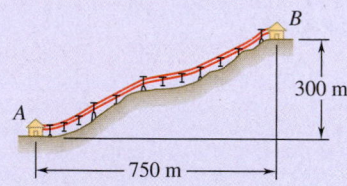

Fig. P13.46

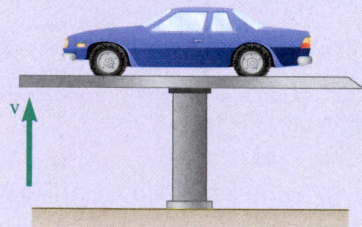

Fig. P13.47

**13.43** In Prob. 13.42, determine the range of values of $h$ for which the roller coaster will not leave the track at $D$ or $E$, knowing that the radius of curvature at $E$ is $\rho = 75$ ft. Assume no energy loss due to friction.

**13.44** A small block slides at a speed $v$ on a horizontal surface. Knowing that $h = 0.9$ m, determine the required speed of the block if it is to leave the cylindrical surface $BCD$ when $\theta = 30°$.

**13.45** A small block slides at a speed $v = 8$ ft/s on a horizontal surface at a height $h = 3$ ft above the ground. Determine ($a$) the angle $\theta$ at which it will leave the cylindrical surface $BCD$, ($b$) the distance $x$ at which it will hit the ground. Neglect friction and air resistance.

**13.46** A chairlift is designed to transport 1000 skiers per hour from the base $A$ to the summit $B$. The average mass of a skier is 70 kg and the average speed of the lift is 75 m/min. Determine ($a$) the average power required, ($b$) the required capacity of the motor if the mechanical efficiency is 85 percent and if a 300-percent overload is to be allowed.

**13.47** It takes 15 s to raise a 1200-kg car and the supporting 300-kg hydraulic car-lift platform to a height of 2.8 m. Determine ($a$) the average output power delivered by the hydraulic pump to lift the system, ($b$) the average electric power required, knowing that the overall conversion efficiency from electric to mechanical power for the system is 82 percent.

**13.48** The velocity of the lift of Prob. 13.47 increases uniformly from zero to its maximum value at mid-height in 7.5 s and then decreases uniformly to zero in 7.5 s. Knowing that the peak power output of the hydraulic pump is 6 kW when the velocity is maximum, determine the maximum lift force provided by the pump.

**13.49** ($a$) A 120-lb woman rides a 15-lb bicycle up a 3-percent slope at a constant speed of 5 ft/s. How much power must be developed by the woman? ($b$) A 180-lb man on an 18-lb bicycle starts down the same slope and maintains a constant speed of 20 ft/s by braking. How much power is dissipated by the brakes? Ignore air resistance and rolling resistance.

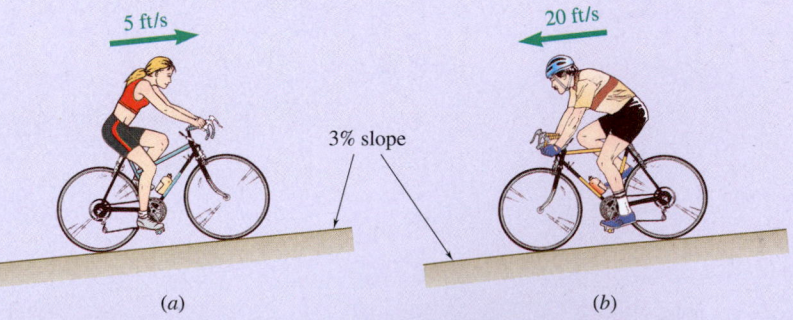

Fig. P13.49

**13.50** A power specification formula is to be derived for electric motors which drive conveyor belts moving solid material at different rates to different heights and distances. Denoting the efficiency of a motor by $\eta$ and neglecting the power needed to drive the belt itself, derive a formula (*a*) in the SI system of units for the power $P$ in kW, in terms of the mass flow rate $m$ in kg/h, the height $b$ and horizontal distance $l$ in meters and (*b*) in U.S. customary units, for the power in hp, in terms of the material flow rate $w$ in tons/h, and the height $b$ and horizontal distance $l$ in feet.

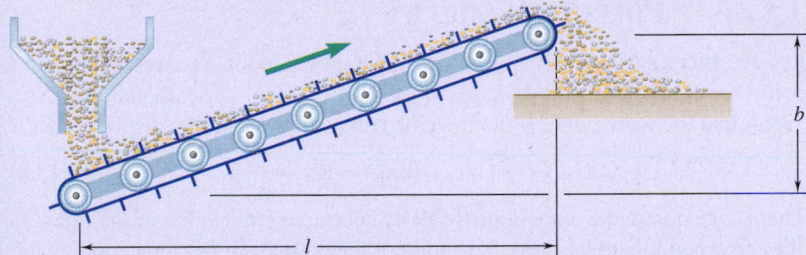

**Fig. P13.50**

**13.51** A 1400-kg automobile starts from rest and travels 400 m during a performance test. The motion of the automobile is defined by the relation $x = 4000 \ln(\cosh 0.03t)$, where $x$ and $t$ are expressed in meters and seconds, respectively. The magnitude of the aerodynamic drag is $D = 0.35v^2$, where $D$ and $v$ are expressed in newtons and m/s, respectively. Determine the power dissipated by the aerodynamic drag when (*a*) $t = 10$ s, (*b*) $t = 15$ s.

**Fig. P13.51**

**13.52** The frictional resistance of a ship is known to vary directly as the 1.75 power of the speed $v$ of the ship. A single tugboat at full power can tow the ship at a constant speed of 4.5 km/h by exerting a constant force of 300 kN. Determine (*a*) the power delivered by the tugboat, (*b*) the maximum speed at which two tugboats, each capable of delivering that same power, can tow the ship.

**13.53** The fluid transmission of a 15-Mg truck allows the engine to deliver an essentially constant power of 50 kW to the driving wheels. Determine the time required and the distance traveled as the speed of the truck is increased (*a*) from 36 km/h to 54 km/h, (*b*) from 54 km/h to 72 km/h.

**13.54** The elevator $E$ has a weight of 6600 lb when fully loaded and is connected as shown to a counterweight $W$ of weight of 2200 lb. Determine the power in hp delivered by the motor (*a*) when the elevator is moving down at a constant speed of 1 ft/s, (*b*) when it has an upward velocity of 1 ft/s and a deceleration of 0.18 ft/s$^2$.

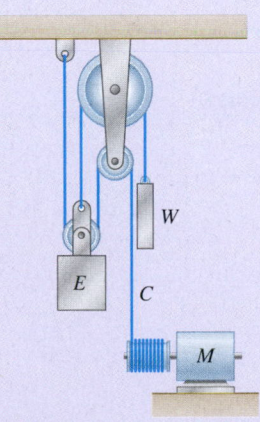

**Fig. P13.54**

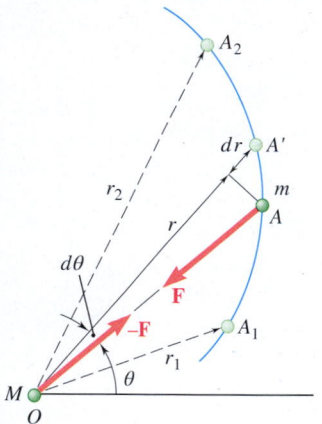

**Fig. 13.4** *(repeated)*

**Fig. 13.6** *(repeated)*

# 13.2 CONSERVATION OF ENERGY

The principle of work and energy is useful for solving many different types of engineering problems. However, in many engineering applications, the total mechanical energy remains constant, although it may be transformed from one form into another. This is known as the principle of conservation of energy. To formulate this principle, we must first define a quantity known as potential energy. (Some of the material in this section was considered in Sec. 10.2B.)

## 13.2A Potential Energy

Let's consider again a body of weight **W** that moves along a curved path from a point $A_1$ of elevation $y_1$ to a point $A_2$ of elevation $y_2$ (Fig. 13.4). Recall from Sec. 13.1A that the work done by the force of gravity **W** during this displacement is

$$U_{1\rightarrow2} = -(Wy_2 - Wy_1) = Wy_1 - Wy_2 \qquad (13.4)$$

That is, we obtain the work done by **W** by subtracting the value of the function $Wy$ corresponding to the second position of the body from its value corresponding to the first position. The work of **W** is independent of the actual path followed; it depends only upon the initial and final values of the function $Wy$. This function is called the **potential energy** of the body with respect to the **force of gravity W** and is denoted by $V_g$. We have

**Gravitational potential energy on earth**

$$U_{1\rightarrow2} = (V_g)_1 - (V_g)_2 \quad \text{where} \quad V_g = Wy \qquad (13.16)$$

where $y$ is measured from an arbitrary horizontal datum where the potential energy is zero by definition. Note that if $(V_g)_2 > (V_g)_1$, that is, **if the potential energy increases** during the displacement (as in the case considered here), **the work $U_{1\rightarrow2}$ is negative**. On the other hand, if the work of **W** is positive, the potential energy decreases. Therefore, the potential energy $V_g$ of the body provides a measure of the work that can be done by its weight **W**. Also note that the *change* in potential energy—not the actual value of $V_g$—is involved in formula (13.16). For this reason, the level, or datum, from which we measure the elevation $y$ can be chosen arbitrarily. Finally, note that potential energy is expressed in the same units as work; that is, in joules if we use SI units and in ft·lb or in·lb if we use U.S. customary units.

This expression for the potential energy of a body with respect to gravity is valid only as long as we can assume the weight **W** of the body remains constant; that is, as long as the displacements of the body are small compared with the radius of the earth. In the case of a space vehicle, however, we need to take into consideration the variation of the force of gravity with the distance $r$ from the center of the earth. Using the expression obtained in Sec. 13.1A for the work of a gravitational force, we have (Fig. 13.6)

$$U_{1\rightarrow2} = \frac{GMm}{r_2} - \frac{GMm}{r_1} \qquad (13.7)$$

Therefore, we can obtain the work of the force of gravity by subtracting the value of the function $-GMm/r$ corresponding to the second position of the body from its value corresponding to the first position. Thus, the expression that we use for the potential energy $V_g$ when the variation in the force of gravity cannot be neglected is

**Gravitational potential energy in space**

$$V_g = -\frac{GMm}{r} \qquad (13.17)$$

Taking the first of the relations of Eq. (12.27) into account, we can write $V_g$ in the alternative form

$$V_g = -\frac{WR^2}{r} \qquad (13.17')$$

where $R$ is the radius of the earth and $W$ is the value of the weight of the body at the surface of the earth. When using either of the relations in Eqs. (13.17) or (13.17′) to express $V_g$, the distance $r$ should, of course, be measured from the center of the earth.[†] Note that $V_g$ is always negative and that it approaches zero for very large values of $r$.

Consider now a body attached to a spring and moving from a position $A_1$, corresponding to a deflection $x_1$ of the spring, to a position $A_2$, corresponding to a deflection $x_2$ of the spring (Fig. 13.5). Recall from Sec. 13.1A that the work of the force $\mathbf{F}$ exerted by the spring on the body is

$$U_{1\rightarrow2} = \tfrac{1}{2}kx_1^2 - \tfrac{1}{2}kx_2^2 \qquad (13.6)$$

That is, we obtain the work of the elastic force by subtracting the value of the function $\tfrac{1}{2}kx^2$ corresponding to the second position of the body from its value corresponding to the first position. This function is denoted by $V_e$ and is called the **potential energy** of the body with respect to the **elastic force $\mathbf{F}$**. We have

**Elastic potential energy**

$$U_{1\rightarrow2} = (V_e)_1 - (V_e)_2 \quad \text{where} \quad V_e = \tfrac{1}{2}kx^2 \qquad (13.18)$$

where $x = L_{stretched} - L_{unstretched}$, or the deflection of the spring from its unde-formed position. Note that, during the displacement from $A_1$ to $A_2$, the work of the force $\mathbf{F}$ exerted by the spring on the body is negative and that the poten-tial energy $V_e$ increases. We can use formula (13.18) even when the spring is rotated about its fixed end (Fig. 13.10a). The work of the elastic force depends only upon the initial and final deflections of the spring (Fig. 13.10b).

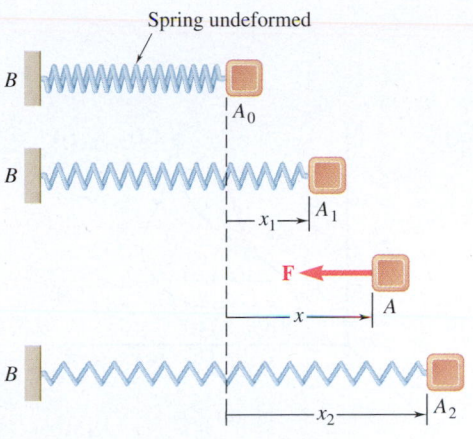

Spring undeformed

**Fig. 13.5** *(repeated)*

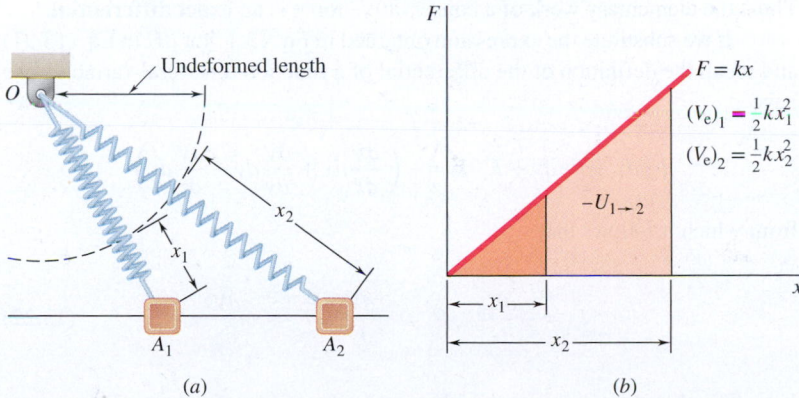

(a)            (b)

**Fig. 13.10** (a) The equation for potential energy of a spring force is valid if the spring stretches when rotated about a fixed end; (b) the work of the elastic force depends only on the initial and final deflections of the spring.

[†]The expressions for $V_g$ in Eqs. (13.17) and (13.17′) are valid only when $r \geq R$; that is, when the body considered is above the surface of the earth.

We can use the concept of potential energy when forces other than gravity forces and elastic forces are involved. Indeed, it remains valid as long as the work of the force considered is independent of the path followed by its point of application, as this point moves from a given position $A_1$ to a given position $A_2$. Such forces are said to be **conservative forces** or **path-independent forces**. We next consider their general properties.

## *13.2B   Conservative Forces

As indicated in the preceding section, a force **F** acting on a particle $A$ is said to be **conservative if its work $U_{1\rightarrow2}$ is independent of the path followed by the particle $A$ as it moves from $A_1$ to $A_2$** (Fig. 13.11$a$). We then have

$$U_{1\rightarrow2} = -[V(x_2, y_2, z_2) - V(x_1, y_1, z_1)] = V(x_1, y_1, z_1) - V(x_2, y_2, z_2) \quad \textbf{(13.19)}$$

or for short,

$$U_{1\rightarrow2} = V_1 - V_2 \quad \textbf{(13.19′)}$$

The function $V(x, y, z)$ is called the potential energy, or **potential function**, of **F**.

Note that if $A_2$ is chosen to coincide with $A_1$—that is, if the particle describes a closed path (Fig. 13.11$b$)—we have $V_1 = V_2$ and the work is zero. Thus, for any conservative force **F**, we can write

$$\oint \mathbf{F}\cdot d\mathbf{r} = 0 \quad \textbf{(13.20)}$$

where the circle on the integral sign indicates that the path is closed.

Let us now apply Eq. (13.19) between two neighboring points $A(x, y, z)$ and $A'(x + dx, y + dy, z + dz)$. The elementary work $dU$ corresponding to the displacement $d\mathbf{r}$ from $A$ to $A'$ is

$$dU = V(x, y, z) - V(x + dx, y + dy, z + dz)$$

or

$$dU = -dV(x, y, z) \quad \textbf{(13.21)}$$

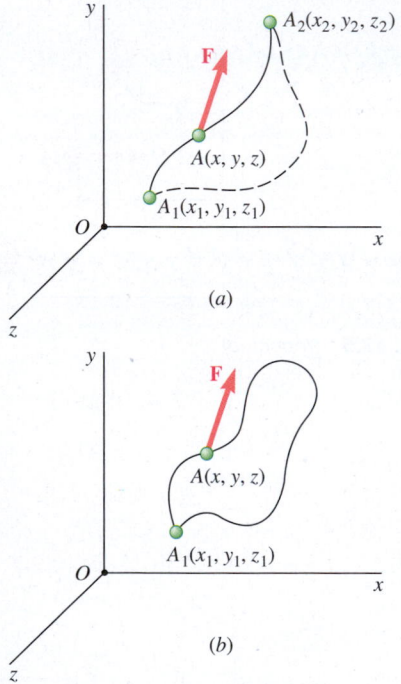

**Fig. 13.11** (*a*) The work of a conservative force acting on a particle is independent of the path of the particle; (*b*) if the particle travels a closed path, the work of a conservative force is zero.

Thus, the elementary work of a conservative force is an **exact differential**.

If we substitute the expression obtained in Eq. 13.1″ for $dU$ in Eq. (13.21) and recall the definition of the differential of a function of several variables, we have

$$F_x\,dx + F_y\,dy + F_z\,dz = -\left(\frac{\partial V}{\partial x}dx + \frac{\partial V}{\partial y}dy + \frac{\partial V}{\partial z}dz\right)$$

from which it follows that

$$F_x = -\frac{\partial V}{\partial x} \quad F_y = -\frac{\partial V}{\partial y} \quad F_z = -\frac{\partial V}{\partial z} \quad \textbf{(13.22)}$$

It is clear that the components of **F** must be functions of the coordinates $x$, $y$, and $z$. Thus, a necessary condition for a conservative force is that it depends only upon the position of its point of application. The relations in Eq. (13.22) can be expressed more concisely if we write

$$\mathbf{F} = F_x\mathbf{i} + F_y\mathbf{j} + F_z\mathbf{k} = -\left(\frac{\partial V}{\partial x}\mathbf{i} + \frac{\partial V}{\partial y}\mathbf{j} + \frac{\partial V}{\partial z}\mathbf{k}\right)$$

The vector in parentheses is known as the **gradient of the scalar function** $V$ and is denoted by **grad** $V$. We thus have for any conservative force

$$\mathbf{F} = -\mathbf{grad}\ V \qquad (13.23)$$

The relations in Eqs. (13.19), (13.20), (13.21), (13.22), and (13.23) are satisfied by any conservative force. It can also be shown that if a force $\mathbf{F}$ satisfies one of these relations, $\mathbf{F}$ must be a conservative force.

## 13.2C   The Principle of Conservation of Energy

We saw in the preceding two sections that we can express the work of a conservative force, such as the weight of a particle or the force exerted by a spring, as a change in potential energy. When a particle moves under the action of conservative forces, the principle of work and energy stated in Sec. 13.1B can be expressed in a modified form. Substituting for $U_{1 \rightarrow 2}$ from Eq. (13.19′) into Eq. (13.10), we have

$$V_1 - V_2 = T_2 - T_1$$

or

**Conservation of energy**

$$T_1 + V_1 = T_2 + V_2 \qquad (13.24)$$

Formula (13.24) indicates that when a particle moves under the action of conservative forces, **the sum of the kinetic energy and of the potential energy of the particle remains constant**. The sum $T + V$ is called the **total mechanical energy** of the particle and is denoted by $E$. So far, we have discussed two types of potential energy: gravitational potential energy, $V_g$, and elastic potential energy, $V_e$. Therefore, another way to write Eq. (13.24) is

$$T_1 + V_{g_1} + V_{e_1} = T_2 + V_{g_2} + V_{e_2} \qquad (13.24′)$$

Consider, for example, the pendulum analyzed in Sec. 13.1C that is released with no velocity from $A_1$ and allowed to swing in a vertical plane (Fig. 13.12). Measuring the potential energy from the level of $A_2$, that is, placing our datum at $A_2$, we have at $A_1$

$$T_1 = 0 \qquad V_1 = Wl \qquad T_1 + V_1 = Wl$$

Recalling that at $A_2$ the speed of the pendulum is $v_2 = \sqrt{2gl}$, we have

$$T_2 = \frac{1}{2}mv_2^2 = \frac{1}{2}\frac{W}{g}(2gl) = Wl \qquad V_2 = 0$$

$$T_2 + V_2 = Wl$$

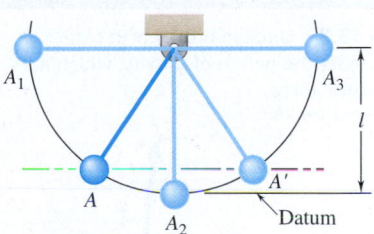

**Fig. 13.12**   The motion of a pendulum is easily analyzed using conservation of energy.

Thus, we can check that the total mechanical energy $E = T + V$ of the pendulum is the same at $A_1$ and $A_2$. Whereas the energy is entirely potential at $A_1$, it becomes entirely kinetic at $A_2$, and as the pendulum keeps swinging to the right past $A_2$, the kinetic energy is transformed back into potential energy. At $A_3$, $T_3 = 0$ and $V_3 = Wl$.

Because the total mechanical energy of the pendulum remains constant and its potential energy depends only upon its elevation, the kinetic energy of the pendulum must have the same value at any two points located at the same height. Thus, the speed of the pendulum is the same at $A$ and at $A′$ (Fig. 13.12).

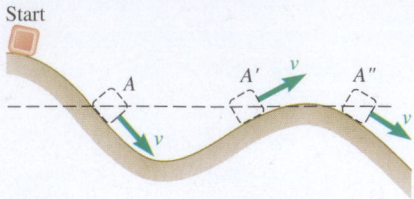

Start

**Fig. 13.13** A particle moving along a frictionless track has the same speed every time it passes the same elevation.

**Photo 13.2** The potential energy of the roller coaster car is converted into kinetic energy as it descends the track.
©Dynamic Graphics/SuperStock RF

**Photo 13.3** Once in orbit, Earth satellites move under the action of gravity, which acts as a central force.
©iLexx/Getty Images RF

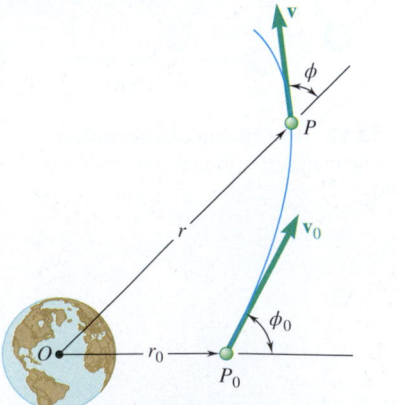

**Fig. 13.14** A space vehicle moving from $P_0$ to $P$ under the earth's gravitational force.

We can extend this result to the case of a particle moving along any given path, regardless of the shape of the path, as long as the only forces acting on the particle are its weight and the normal reaction of the path. The particle of Fig. 13.13, for example, which slides in a vertical plane along a frictionless track, has the same speed at $A$, $A'$, and $A''$.

The weight of a particle and the force exerted by a spring are conservative forces, but **friction forces are nonconservative, or *path-dependent*, forces**. In other words, *the work of a friction force cannot be expressed as a change in potential energy.* The work of a friction force depends upon the path followed by its point of application; and whereas the work $U_{1 \to 2}$ defined by Eq. (13.19) is positive or negative according to the sense of motion, the work of a friction force, as we noted in Sec. 13.1C, is almost always negative. It follows that when a mechanical system involves friction, its total mechanical energy does not remain constant but decreases. The energy of the system, however, is not lost; it is transformed into heat, and the sum of the *mechanical energy* and of the *thermal energy* of the system remains constant.

Other forms of energy also can be involved in a system. For instance, a generator converts mechanical energy into *electrical energy;* a gasoline engine converts *chemical energy* into mechanical energy; a nuclear reactor converts *mass* into thermal energy. If all forms of energy are considered, the energy of any system can be considered as constant, and the principle of conservation of energy remains valid under all conditions.

If we express the work done by non-conservative forces as $U_{1 \to 2}^{NC}$, we can express Eq. (13.2) as

$$T_1 + V_{g_1} + V_{e_1} + U_{1 \to 2}^{NC} = T_2 + V_{g_2} + V_{e_2} \qquad \textbf{(13.24'')}$$

Note that if $U_{1 \to 2}^{NC}$ is zero, then the expression reduces to the conservation of energy equation of Eq. (13.24').

## 13.2D   Application to Space Mechanics: Motion Under a Conservative Central Force

We saw in Sec. 12.2B that when a particle $P$ moves under a central force $\mathbf{F}$, the angular momentum $\mathbf{H}_O$ of the particle about the center of force $O$ is constant. If the force $\mathbf{F}$ is also conservative, there exists a potential energy $V$ associated with $\mathbf{F}$, and the total energy $E = T + V$ of the particle is constant. Thus, when a particle moves under a conservative central force, we can use both the principle of conservation of angular momentum and the principle of conservation of energy to study its motion.

Consider, for example, a space vehicle of mass $m$ moving under the earth's gravitational force. Let us assume that it begins its free flight at point $P_0$ at a distance $r_0$ from the center of the earth with a velocity $\mathbf{v}_0$ forming an angle $\phi_0$ with the radius vector $OP_0$ (Fig. 13.14). Let $P$ be a point of the trajectory described by the vehicle; we denote by $r$ the distance from $O$ to $P$, by $\mathbf{v}$ the velocity of the vehicle at $P$, and by $\phi$ the angle formed by $\mathbf{v}$ and the radius vector $OP$. Applying the principle of conservation of angular momentum about $O$ between $P_0$ and $P$ (Sec. 12.2B), we have

$$r_0 m v_0 \sin \phi_0 = r m v \sin \phi \qquad \textbf{(13.25)}$$

Recalling the expression in Eq. (13.17) for the potential energy due to a gravitational force, we apply the principle of conservation of energy between $P_0$ and $P$, obtaining

$$T_0 + V_0 = T + V$$

$$\frac{1}{2}mv_0^2 - \frac{GMm}{r_0} = \frac{1}{2}mv^2 - \frac{GMm}{r} \tag{13.26}$$

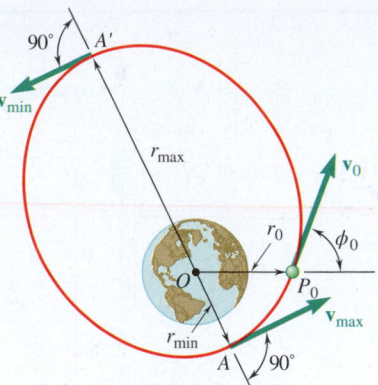

where $M$ is the mass of the earth.

We can solve Eq. (13.26) for the magnitude $v$ of the velocity of the vehicle at $P$ when we know the distance $r$ from $O$ to $P$. Then, we can use Eq. (13.25) to determine the angle $\phi$ that the velocity forms with the radius vector $OP$.

We can also use Eqs. (13.25) and (13.26) to determine the maximum and minimum values of $r$ in the case of a satellite launched from $P_0$ in a direction forming an angle $\phi_0$ with the vertical $OP_0$ (Fig. 13.15). We obtain the desired values of $r$ by making $\phi = 90°$ in Eq. (13.25) and eliminating $v$ between Eqs. (13.25) and (13.26).

**Fig. 13.15** A space vehicle launched from point $P_0$ into an orbit around the earth.

Note that applying the principles of conservation of energy and of conservation of angular momentum leads to a more fundamental formulation of the problems of space mechanics than does the method indicated in Sec. 12.3B. It also results in much simpler computations in all cases involving oblique launchings. Although you must use the method of Sec. 12.3B when the actual trajectory or the periodic time of a space vehicle is to be determined, the calculations will be simplified if you first use the conservation principles to compute the maximum and minimum values of the radius vector $r$.

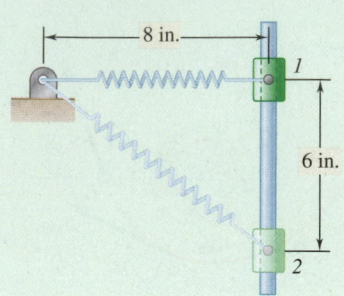

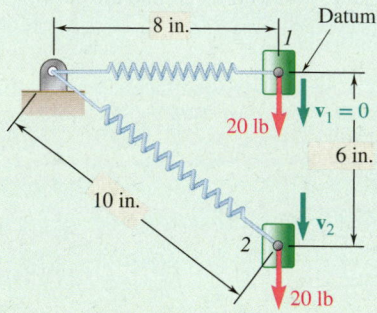

**Fig. 1** The system in position *1* and position *2*.

## Sample Problem 13.8

A 20-lb collar slides without friction along a vertical rod as shown. The spring attached to the collar has an undeformed length of 4 in. and a spring constant of 3 lb/in. If the collar is released from rest in position *1*, determine its velocity after it has moved 6 in. to position *2*.

**STRATEGY:**   You are given two positions and want to determine the velocity of the collar. No non-conservative forces are involved, so use the conservation of energy.

**MODELING:**   For your system, choose the collar and the spring. You can treat the collar as a particle.

**ANALYSIS:   Conservation of Energy.**   Applying the principle of conservation of energy between positions *1* and *2* gives

$$T_1 + V_{g_1} + V_{e_1} = T_2 + V_{g_2} + V_{e_2} \qquad (1)$$

You need to determine the kinetic and potential energy at these positions.

**Position 1.   *Potential Energy.*** The elongation of the linear spring (Fig. 1) is

$$x_1 = 8 \text{ in.} - 4 \text{ in.} = 4 \text{ in.}$$

This gives

$$V_{e_1} = \tfrac{1}{2}kx_1^2 = \tfrac{1}{2}(3 \text{ lb/in.})(4 \text{ in.})^2 = 24 \text{ in·lb} = 2 \text{ ft·lb}$$

Choosing the datum as shown, you have $V_{g_1} = 0$.

***Kinetic Energy.***   Because the velocity at position *1* is zero, $T_1 = 0$.

**Position 2.   *Potential Energy.*** The elongation of the spring is

$$x_2 = 10 \text{ in.} - 4 \text{ in.} = 6 \text{ in.}$$

so you have

$$V_{e_2} = \tfrac{1}{2}kx_2^2 = \tfrac{1}{2}(3 \text{ lb/in.})(6 \text{ in.})^2 = 54 \text{ in·lb} = 4.5 \text{ ft·lb}$$

$$V_{g_2} = Wy_2 = (20 \text{ lb})(-6 \text{ in.}) = -120 \text{ in·lb} = -10 \text{ ft·lb}$$

***Kinetic Energy.***

$$T_2 = \frac{1}{2}mv_2^2 = \frac{1}{2}\frac{20}{32.2}v_2^2 = 0.311\,v_2^2$$

**Conservation of Energy.**   Substituting into Eq. (1) gives

$$T_1 + V_{g_1} + V_{e_1} = T_2 + V_{g_2} + V_{e_2}$$
$$0 + 0 + 2 \text{ ft·lb} = 0.311\,v_2^2 + (-10 \text{ ft·lb}) + (4.5 \text{ ft·lb})$$
$$v_2 = \pm 4.91 \text{ ft/s}$$

$$\mathbf{v_2} = 4.91 \text{ ft/s} \downarrow \quad \blacktriangleleft$$

**REFLECT and THINK:**   If you had not included the spring in your system, you would have needed to treat it as an external force; therefore, you would have needed to determine the work. Similarly, if there was friction acting on the collar, you would have needed to use the more general work–energy principle to solve this problem. It turns out that the work done by friction is not very easy to calculate because the normal force depends on the spring force.

## Sample Problem 13.9

A 2.5-lb collar is attached to a spring and slides along a smooth circular rod in a vertical plane. The spring has an undeformed length of 4 in. and a spring constant $k$. The collar is at rest at point $C$ and is given a slight push to the right. Knowing that the maximum velocity of the collar is achieved as it passes through point $A$, determine (a) the spring constant $k$, (b) the force exerted by the rod on the collar at point $A$.

**STRATEGY:** Because you have two positions and are given information about the speed, use the conservation of energy. To find the force, you need to use Newton's second law.

**MODELING:** For the conservation of energy portion of the problem, model the collar as a particle and use it and the spring as your system. When using Newton's second law, use the collar as your system.

### ANALYSIS:

**Conservation of Energy.** Position 1 is when the collar is at point $C$, and position 2 is when it is at point $A$ (Fig. 1).

Applying conservation of energy between positions 1 and 2 gives

$$T_1 + V_{g_1} + V_{e_1} = T_2 + V_{g_2} + V_{e_2} \qquad (1)$$

**Position 1.** Because the system starts from rest, $T_1 = 0$, and because the spring has an unstretched length of 4 in., you know $V_{e_1} = 0$. Putting the datum at $A$ gives.

$$V_{g_1} = (2.5 \text{ lb})(7/12 \text{ ft}) = 1.4583 \text{ ft·lb}$$

**Position 2.** From geometry, the distance from the pin to $A$ is $\sqrt{(3 \text{ in.})^2 + (7 \text{ in.})^2} = 7.616$ in. Therefore, the elongation of the linear spring (Fig. 1) is $x_2 = 7.616$ in. $- 4$ in. $= 3.616$ in. $= 0.3013$ ft. You know $V_{g_2} = 0$ because the datum is at position 2. You also know

$$V_{e_2} = \tfrac{1}{2}kx_2^2 = \frac{1}{2}k(0.3013 \text{ ft})^2 = 0.04539k$$

$$T_2 = \tfrac{1}{2}mv_2^2 = \frac{1}{2}\left(\frac{2.5 \text{ lb}}{32.2 \text{ ft/s}^2}\right)v_2^2 = 0.03882v_2^2$$

Substituting these expressions into Eq. (1) gives

$$0 + 1.4585 + 0 = 0.03882v_2^2 + 0 + 0.04539k \qquad (2)$$

You have two unknowns in this equation, so you need another equation. In the problem statement, you are also given that the collar has a maximum velocity at point $A$. Therefore, the tangential acceleration must be zero at $A$, and you should use Newton's second law to get additional equations. The system now includes only the collar; the spring applies an external force to the system. A free-body diagram and kinetic diagram for the collar at position 2 are shown in Fig. 2. Applying Newton's second law in the $t$ direction gives

$$+\uparrow\Sigma F_t = 0 = kx_2 \sin\theta - W \quad \text{or} \quad k(0.3013 \text{ ft})(3/7.616) - 2 \text{ lb} = 0$$

Solving for $k$,

$$k = 16.85 \text{ lb/ft} \quad \blacktriangleleft$$

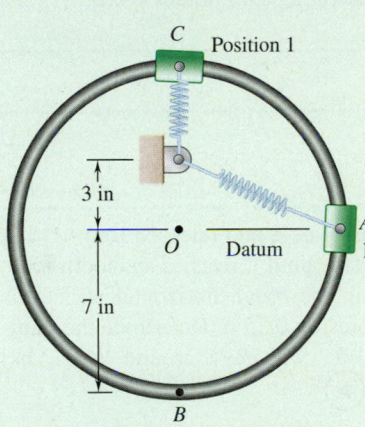

**Fig. 1** The system in the two positions of interest.

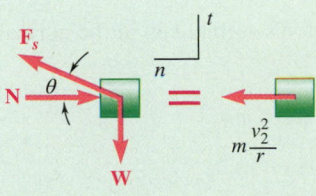

**Fig. 2** Free-body diagram and kinetic diagram for the collar at point $A$.

*(continued)*

**Force Exerted by the Rod.** Substituting this value of $k$ into Eq. (2) gives $v_2 = 3.597$ ft/s. Applying Newton's second law in the $n$ direction gives

$$\xleftarrow{+} \Sigma F_n = m\frac{v_2^2}{r} \qquad kx_2 \cos \theta - N = \frac{mv_2^2}{r}$$

Solving for $N$ and substituting in values provides

$$N = kx_2 \cos \theta - \frac{mv_2^2}{r}$$

$$= (16.85 \text{ lb/ft})(0.3013 \text{ ft})(7/7.616) - \frac{(2.5 \text{ lb}/32.2 \text{ ft/s}^2)(3.597 \text{ ft/s}^2)}{(7/12 \text{ ft})}$$

$$N = 4.19 \text{ lb} \blacktriangleleft$$

**REFLECT and THINK:** When the collar is pushed to the right, its speed increases until it reaches point $A$, and then it begins to decrease. The minimum speed occurs when the collar is at $B$, because the only forces are in the normal direction; that is, no forces act in the tangential direction. Therefore, the acceleration in the tangential direction is zero, indicating a minimum speed.

## Sample Problem 13.10

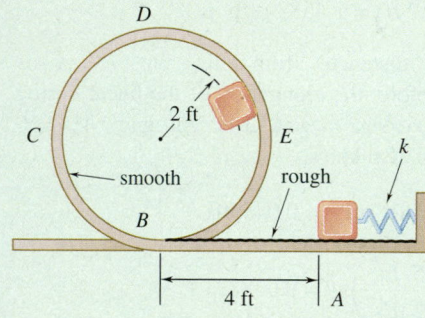

A 0.5-lb pellet is pushed against the spring at $A$ and released from rest. It moves 4 ft along a rough horizontal surface until it reaches a smooth loop. The coefficient of kinetic friction along the rough horizontal surface is $\mu_k = 0.3$, and the spring is initially compressed 0.25 ft. Determine the minimum spring constant $k$ for which the pellet will travel around $BCDE$ and always remain in contact with the loop. ▶

**STRATEGY:** You are given two positions and a non-conservative force is present, so use the work–energy principle. Also, for the pellet to remain in contact with the loop, the force $\mathbf{N}$ exerted on the pellet by the loop must be equal to or greater than zero. Therefore, you also need to use Newton's second law.

**MODELING:** Choose the pellet as your system and model it as a particle. A free-body diagram and kinetic diagram for the pellet when it is at point $D$ are shown in Fig. 1.

**ANALYSIS:**

**Newton's Second Law.** Applying Newton's second law in the normal direction and setting $N = 0$ gives you

$$+\downarrow \Sigma F_n = ma_n: \qquad W = ma_n \qquad mg = ma_n \qquad a_n = g$$

$$a_n = \frac{v_D^2}{r}: \qquad v_D^2 = ra_n = rg = (2 \text{ ft})(32.2 \text{ ft/s}^2) = 64.4 \text{ ft}^2/\text{s}^2$$

Note that $v_D$ is the minimum speed of the pellet at $D$ in order for it to remain in contact with the path.

*(continued)*

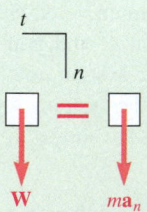

**Fig. 1** Free-body diagram and kinetic diagram for the pellet at point $D$.

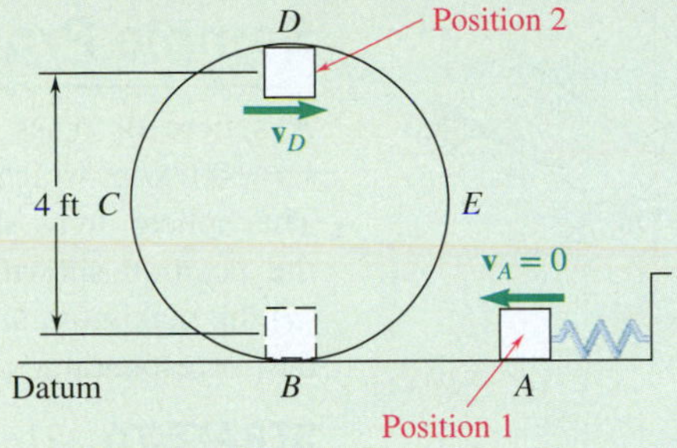

**Fig. 2** The system at the positions of interest.

**Work–Energy.** Choose the system to be the pellet and the spring. Apply the principle of work–energy between positions 1 and 2 (Fig. 2)

$$T_1 + V_{g_1} + V_{e_1} + U^{NC}_{1\to2} = T_2 + V_{g_2} + V_{e_2} \tag{1}$$

You need to determine the kinetic and potential energy at positions 1 and 2 and the work done by friction.

**Position 1.** *Potential Energy.* The elastic potential energy is

$$V_{e_1} = \tfrac{1}{2}kx^2 = \tfrac{1}{2}(k)(0.25 \text{ ft})^2 = 0.03125\,k$$

Choosing the datum at $A$, you have $V_{g_1} = 0$.

*Kinetic Energy.* Because the pellet is released from rest, $v_A = 0$ and $T_1 = 0$.

**Position 2.** *Potential Energy.* The spring is now undeformed; thus $V_{e_2} = 0$. Because the pellet is 4 ft above the datum, you have

$$V_{g_2} = Wy_2 = (0.5 \text{ lb})(4 \text{ ft}) = 2 \text{ ft·lb}$$

*Kinetic Energy.* Using the value of $v_D^2$ obtained you have

$$T_2 = \tfrac{1}{2}mv_D^2 = \frac{1}{2}\frac{0.5 \text{ lb}}{32.2 \text{ ft/s}^2}(64.4 \text{ ft}^2/\text{s}^2) = 0.5 \text{ ft·lb}$$

**Work.** Because the normal force is equal to the weight on a horizontal surface, you find the work that friction does to be

$$U^{NC}_{1\to2} = -\mu_k Nd = -0.3(0.5 \text{ lb})(4 \text{ ft}) = -0.6 \text{ ft·lb}$$

**Work–Energy.** Substituting these values into Eq. (1) gives you

$$T_1 + V_{g_1} + V_{e_1} + U^{NC}_{1\to2} = T_2 + V_{g_2} + V_{e_2}$$
$$0 + 0 + 0.3125k - 0.6 \text{ ft·lb} = 0.5 \text{ ft·lb} + 2 \text{ ft·lb} + 0$$

You can solve this for $k$.

$$k = 99.2 \text{ lb/ft} \quad \blacktriangleleft$$

**REFLECT and THINK:** A common misconception in problems like this is assuming that the speed of the particle is zero at the top of the loop, rather than that the normal force is equal to or greater than zero. If the pellet had a speed of zero at the top, it would clearly fall straight down, which is impossible.

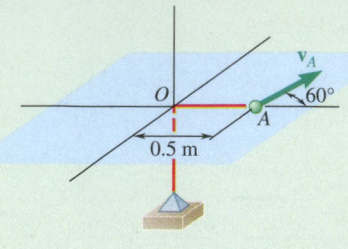

## Sample Problem 13.11

A sphere of mass $m = 0.6$ kg is attached to an elastic cord of constant $k = 100$ N/m, which is undeformed when the sphere is located at the origin $O$. The sphere may slide without friction on the horizontal surface and in the position shown its velocity $\mathbf{v}_A$ has a magnitude of 20 m/s. Determine (a) the maximum and minimum distances from the sphere to the origin $O$, (b) the corresponding values of its speed.

**STRATEGY:**   The force exerted by the cord on the sphere passes through the fixed point $O$, so use conservation of angular momentum. Also, you are interested in the speed at two locations, and no non-conservative forces act on the sphere. You can therefore use conservation of energy.

**MODELING:**   Choose the sphere, which can be modeled as a particle, as your system.

**ANALYSIS:**

### Conservation of Angular Momentum About O.
At point $B$, where the distance from $O$ is maximum (Fig. 1), the velocity of the sphere is perpendicular to $OB$ and the angular momentum is $r_m m v_m$. A similar property holds at point $C$, where the distance from $O$ is minimum. Expressing conservation of angular momentum between $A$ and $B$, you have

$$r_A m v_A \sin 60° = r_m m v_m$$
$$(0.5 \text{ m})(0.6 \text{ kg})(20 \text{ m/s}) \sin 60° = r_m (0.6 \text{ kg}) v_m$$
$$v_m = \frac{8.66}{r_m} \tag{1}$$

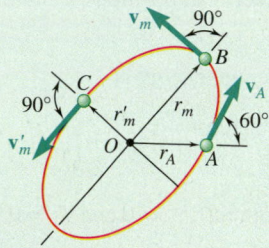

**Fig. 1**   The particle at locations $A$, $B$, and $C$.

You have one equation and two unknowns, $v_m$ and $r_m$. Therefore, you need to use conservation of energy to get a second equation.

### Conservation of Energy.

**At Point A.**   $T_A = \tfrac{1}{2} m v_A^2 = \tfrac{1}{2}(0.6 \text{ kg})(20 \text{ m/s})^2 = 120 \text{ J}$

$V_A = \tfrac{1}{2} k r_A^2 = \tfrac{1}{2}(100 \text{ N/m})(0.5 \text{ m})^2 = 12.5 \text{ J}$

**At Point B.**   $T_B = \tfrac{1}{2} m v_m^2 = \tfrac{1}{2}(0.6 \text{ kg}) v_m^2 = 0.3 v_m^2$

$V_B = \tfrac{1}{2} k r_m^2 = \tfrac{1}{2}(100 \text{ N/m}) r_m^2 = 50 r_m^2$

Apply the principle of conservation of energy between points $A$ and $B$:

$$T_A + V_A = T_B + V_B$$
$$120 + 12.5 = 0.3 v_m^2 + 50 r_m^2 \tag{2}$$

**a. Maximum and Minimum Values of Distance.**   Substituting for $v_m$ from Eq. (1) into Eq. (2) and solving for $r_m^2$, you obtain

$$r_m^2 = 2.468 \text{ or } 0.1824 \qquad r_m = 1.571 \text{ m}, \ r_m' = 0.427 \text{ m} \ \blacktriangleleft$$

**b. Corresponding Values of Speed.**   Substituting the values obtained for $r_m$ and $r_m'$ into Eq. (1), you have

$$v_m = \frac{8.66}{1.571} \qquad v_m = 5.51 \text{ m/s} \ \blacktriangleleft$$

$$v_m' = \frac{8.66}{0.427} \qquad v_m' = 20.3 \text{ m/s} \ \blacktriangleleft$$

*(continued)*

**REFLECT and THINK:** This problem is similar to problems dealing with space mechanics; instead of the gravitational central force acting on an orbiting body, you have the spring force acting on the sphere. It can be shown that the path of the sphere is an ellipse with center $O$.

# Sample Problem 13.12

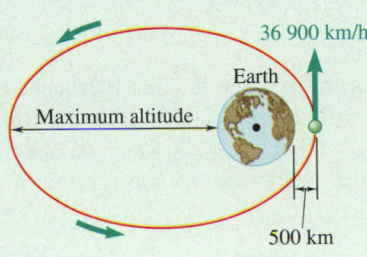

A satellite is launched from an altitude of 500 km in a direction parallel to the surface of the earth with a velocity of 36 900 km/h. Determine (*a*) the maximum altitude reached by the satellite, (*b*) the maximum allowable error in the direction of launching if the satellite is to go into orbit and come no closer than 200 km to the surface of the earth.

**STRATEGY:** Because the only force acting on the satellite is the force of gravity, which is a central force, and you are interested in two positions (the position of the satellite at launch and at its maximum altitude), you can use conservation of angular momentum and conservation of energy.

**MODELING:** Choose the satellite as your system and model it as a particle.

**ANALYSIS:**

**a. Maximum Altitude.** Denote the point of the orbit farthest from the earth by $A'$ and the corresponding distance from the center of the earth by $r_1$ (Fig. 1). Because the satellite is in free flight between $A$ and $A'$, you can apply the principle of conservation of energy as

$$T_A + V_A = T_{A'} + V_{A'}$$

$$\tfrac{1}{2}mv_0^2 - \frac{GMm}{r_0} = \tfrac{1}{2}mv_1^2 - \frac{GMm}{r_1} \qquad (1)$$

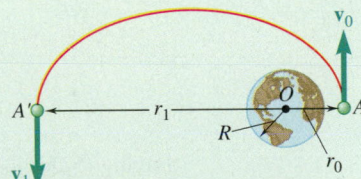

**Fig. 1** The system in the two positions of interest.

Now apply the principle of conservation of angular momentum of the satellite about $O$. Considering points $A$ and $A'$, you have

$$r_0 m v_0 = r_1 m v_1 \qquad v_1 = v_0 \frac{r_0}{r_1} \qquad (2)$$

*(continued)*

Substitute this expression for $v_1$ into Eq. (1), divide each term by the mass $m$, and rearrange the terms. The result is

$$\tfrac{1}{2}v_0^2\left(1 - \frac{r_0^2}{r_1^2}\right) = \frac{GM}{r_0}\left(1 - \frac{r_0}{r_1}\right) \qquad 1 + \frac{r_0}{r_1} = \frac{2GM}{r_0 v_0^2} \qquad (3)$$

Recall that the radius of the earth is $R = 6370$ km. This gives you

$$r_0 = 6370 \text{ km} + 500 \text{ km} = 6870 \text{ km} = 6.87 \times 10^6 \text{m}$$
$$v_0 = 36\,900 \text{ km/h} = (36.9 \times 10^6 \text{m})/(3.6 \times 10^3 \text{s}) = 10.25 \times 10^3 \text{m/s}$$
$$GM = gR^2 = (9.81 \text{ m/s}^2)(6.37 \times 10^6 \text{m})^2 = 398 \times 10^{12} \text{m}^3/\text{s}^2$$

Substituting these values into Eq. (3), you obtain $r_1 = 66.8 \times 10^6$ m.

Maximum altitude $= 66.8 \times 10^6$ m $- 6.37 \times 10^6$ m $= 60.4 \times 10^6$ m $=$

$$60\,400 \text{ km} \quad \blacktriangleleft$$

**b. Allowable Error in Direction of Launch.** The satellite is launched from $P_0$ in a direction forming an angle $\phi_0$ with the vertical $OP_0$ (Fig. 2). You obtain the value of $\phi_0$ corresponding to $r_{min} = 6370$ km $+ 200$ km $= 6570$ km by applying the principles of conservation of energy and of conservation of angular momentum between $P_0$ and $A$:

$$\tfrac{1}{2}mv_0^2 - \frac{GMm}{r_0} = \tfrac{1}{2}mv_{max}^2 - \frac{GMm}{r_{min}} \qquad (4)$$

$$r_0 m v_0 \sin\phi_0 = r_{min} m v_{max} \qquad (5)$$

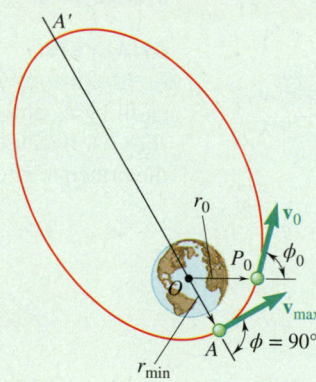

**Fig. 2** Two locations used to determine maximum allowable error in direction.

Solving (5) for $v_{max}$ and then substituting for $v_{max}$ into (4), you can solve (4) for $\sin\phi_0$. Finally, using the values of $v_0$ and $GM$ computed in part $a$ and noting that $r_0/r_{min} = 6870/6570 = 1.0457$, you find

$$\sin\phi_0 = 0.9801 \qquad \phi_0 = 90° \pm 11.5° \qquad \text{Allowable error} = \pm 11.5° \quad \blacktriangleleft$$

**REFLECT and THINK:** Space probes and other long-distance vehicles are designed with small rockets to allow for mid-course corrections. Satellites launched from the Space Station usually do not need this kind of fine-tuning.

# SOLVING PROBLEMS ON YOUR OWN

In this section you examined the work done by a force $\mathbf{F}$ acting on a particle $A$ as it moves from a given position $A_1$ to a given position $A_2$ (Fig. 13.11$a$). If the work is independent of the path followed by the particle, then we can define a function $V$, called **potential energy**, for the force $\mathbf{F}$.

$$U_{1\to2} = -[V(x_2, y_2, z_2) - V(x_1, y_1, z_1)] = V(x_1, y_1, z_1) - V(x_2, y_2, z_2) \quad \textbf{(13.9)}$$

or for short,

$$U_{1\to2} = V_1 - V_2 \quad \textbf{(13.19')}$$

The work is negative when the change in potential energy is positive; that is, when $V_2 > V_1$.

Substituting this expression into the equation for work and energy, you can write

$$T_1 + V_1 = T_2 + V_2 \quad \textbf{(13.24)}$$

or

$$T_1 + V_{g_1} + V_{e_1} = T_2 + V_{g_2} + V_{e_2} \quad \textbf{(13.24')}$$

This equation states that when a particle moves under the action of a conservative force, **the sum of the kinetic and potential energies of the particle remains constant**. We expanded this equation for cases when there are non-conservative forces present:

$$T_1 + V_{g_1} + V_{e_1} + U_{1\to2}^{NC} = T_2 + V_{g_2} + V_{e_2} \quad \textbf{(13.24'')}$$

Your solutions of problems using the above formulas will consist of the following steps.

**1. Determine whether all the forces involved are conservative.** If some of the forces are not conservative—for example, if friction is involved—you must use the second equation (13.24″), because the work done by such forces depends upon the path followed by the particle and a potential function does not exist for these non-conservative forces. You can then determine the work done by non-conservative forces as:

$$U_{1\to2}^{NC} = \int_1^2 \mathbf{F}^{NC} \cdot d\mathbf{s}$$

**2. Determine the kinetic energy $T = \frac{1}{2}mv^2$ at each end of the path.**

**3. Compute the potential energy for all the forces involved at each end of the path.** Recall the following expressions for potential energy derived in this section.

**a. The potential energy of a weight $W$** close to the surface of the earth and at a height $y$ above a given datum:

$$V_g = Wy \quad \textbf{(13.16)}$$

**b. The potential energy of a mass $m$ located at a distance $r$ from the center of the earth,** large enough so that the variation of the force of gravity must be taken into account:

$$V_g = -\frac{GMm}{r} \tag{13.17}$$

where the distance $r$ is measured from the center of the earth and $V_g$ is equal to zero at $r = \infty$.

**c. The potential energy of a body with respect to an elastic force $F = kx$:**

$$V_e = \tfrac{1}{2}kx^2 \tag{13.18}$$

where the distance $x$ is the deflection of the elastic spring measured from its *undeformed* position and $k$ is the spring constant. Note that $V_e$ depends only upon the deflection $x$ and not upon the path of the body attached to the spring. Also, $V_e$ is always positive, whether the spring is compressed or elongated.

**4. Substitute your expressions for the non-conservative work and the kinetic and potential energies** into Eq. (13.24″). You will be able to solve this equation for one unknown—for example, for a velocity (Sample Prob. 13.8). If more than one unknown is involved, you will have to search for another condition or equation, such as Newton's second law (Sample Prob. 13.10), the maximum speed (Sample Prob. 13.9), minimum speed (Sample Prob. 13.10), or the minimum potential energy of the particle. For problems involving a central force, you can obtain a second equation by using conservation of angular momentum (Sample Prob. 13.11). This is especially useful in space mechanics applications (Sec. 13.2D).

# Problems

## CONCEPT QUESTIONS

**13.CQ2** Two small balls $A$ and $B$ with masses $2m$ and $m$, respectively, are released from rest at a height $h$ above the ground. Neglecting air resistance, which of the following statements is true when the two balls hit the ground?
- **a.** The kinetic energy of $A$ is the same as the kinetic energy of $B$.
- **b.** The kinetic energy of $A$ is half the kinetic energy of $B$.
- **c.** The kinetic energy of $A$ is twice the kinetic energy of $B$.
- **d.** The kinetic energy of $A$ is four times the kinetic energy of $B$.

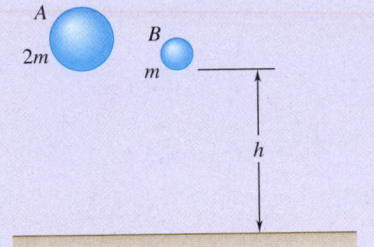

**Fig. P13.CQ2**

**13.CQ3** A small block $A$ is released from rest and slides down the frictionless ramp to the loop. The maximum height $h$ of the loop is the same as the initial height of the block. Will $A$ make it completely around the loop without losing contact with the track?
- **a.** Yes
- **b.** No
- **c.** Need more information

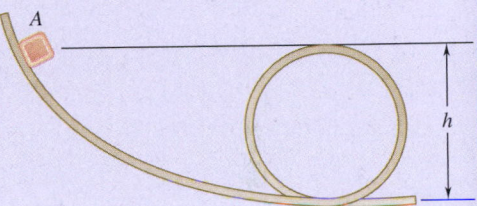

**Fig. P13.CQ3**

## END-OF-SECTION PROBLEMS

**13.55** A force **P** is slowly applied to a plate that is attached to two springs and causes a deflection $x_0$. In each of the two cases shown, derive an expression for the constant $k_e$, in terms of $k_1$ and $k_2$, of the single spring equivalent to the given system; that is, of the single spring which will undergo the same deflection $x_0$ when subjected to the same force **P**.

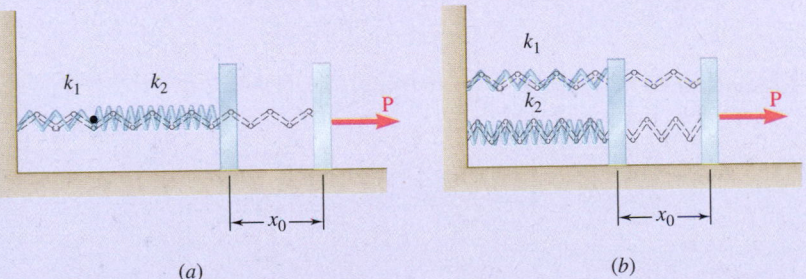

(a)                    (b)

**Fig. P13.55**

**13.56** A loaded railroad car of mass $m$ is rolling at a constant velocity $\mathbf{v}_0$ when it couples with a massless bumper system. Determine the maximum deflection of the bumper assuming the two springs are (a) in series (as shown), (b) in parallel.

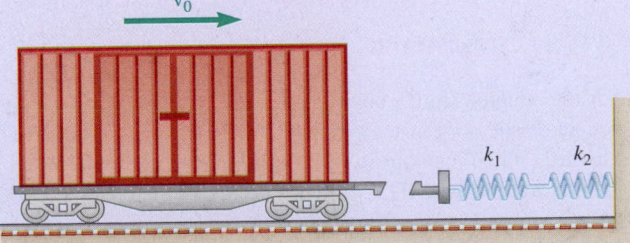

**Fig. P13.56**

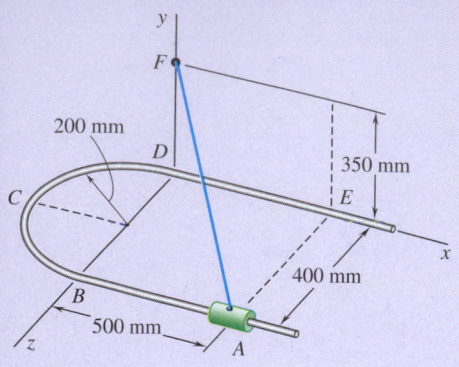

**Fig. P13.57**

**13.57** A 750-g collar can slide along the horizontal rod shown. It is attached to an elastic cord with an undeformed length of 300 mm and a spring constant of 150 N/m. Knowing that the collar is released from rest at A and neglecting friction, determine the speed of the collar (a) at B, (b) at E.

**13.58** A 2-lb collar C may slide without friction along a horizontal rod. It is attached to three springs, each of constant 30 lb/ft and 2-in. undeformed length. Knowing that the collar is released from rest in the position shown, determine the maximum speed it will reach in the ensuing motion.

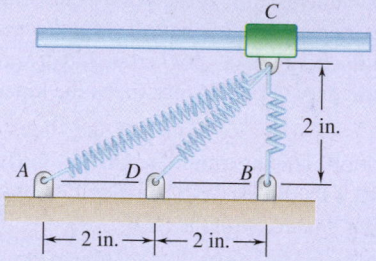

**Fig. P13.58**

**13.59** Solve Prob. 13.58 assuming the spring CD has been removed.

**13.60** A 500-g collar can slide without friction on the curved rod BC in a *horizontal* plane. Knowing that the undeformed length of the spring is 80 mm and that $k = 400$ kN/m, determine (a) the velocity that the collar should be given at A to reach B with zero velocity, (b) the velocity of the collar when it eventually reaches C.

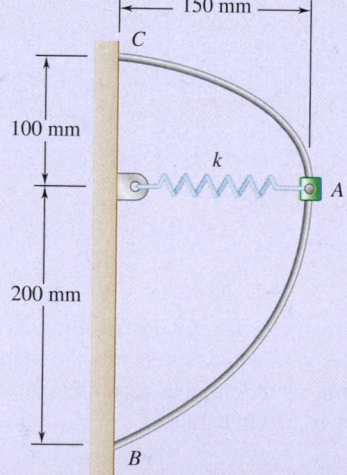

**Fig. P13.60**

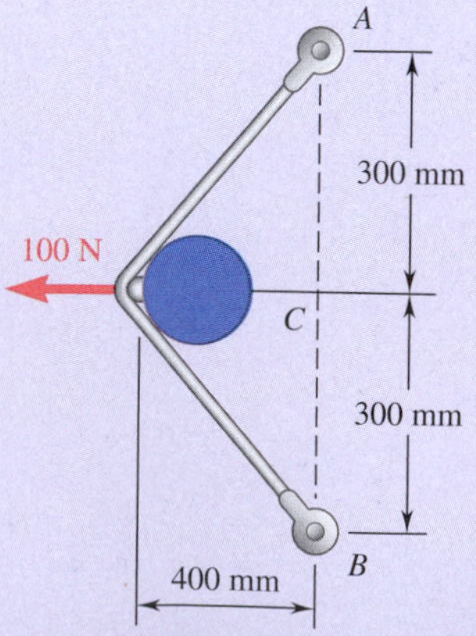

**13.61** For the adapted shuffleboard device in Prob 13.28, you decide to utilize an elastic cord instead of a compression spring to propel the puck forward. When the cord is stretched directly between points A and B, the tension is 20 N. The 425-gram puck is placed in the center and pulled back through a distance of 400 mm; a force of 100 N is required to hold it at this location. Knowing that the coefficient of friction is 0.3, determine how far the puck will travel.

**Fig. P13.61**

**13.62** An elastic cable is to be designed for bungee jumping from a tower 130 ft high. The specifications call for the cable to be 85 ft long when unstretched, and to stretch to a total length of 100 ft when a 600-lb weight is attached to it and dropped from the tower. Determine (*a*) the required spring constant *k* of the cable, (*b*) how close to the ground a 186-lb man will come if he uses this cable to jump from the tower.

**13.63** It is shown in mechanics of materials that the stiffness of an elastic cable is $k = AE/L$, where *A* is the cross-sectional area of the cable, *E* is the modulus of elasticity, and *L* is the length of the cable. A winch is lowering a 4000-lb piece of machinery using a constant speed of 3 ft/s when the winch suddenly stops. Knowing that the steel cable has a diameter of 0.4 in., $E = 29 \times 10^6$ lb/in$^2$, and when the winch stops $L = 30$ ft, determine the maximum downward displacement of the piece of machinery from the point it was when the winch stopped.

**Fig. P13.62**

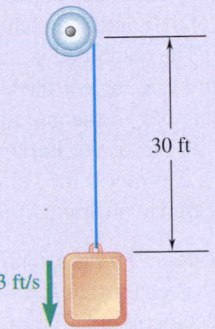

**Fig. P13.63**

**13.64** A 1.2-kg collar can slide along the rod shown. It is attached to an elastic cord anchored at *F*. The cord has an undeformed length of 300 mm and a spring constant of 70 N/m. Knowing that the collar is released from rest at *A* and neglecting friction, determine the speed of the collar (*a*) at *B*, (*b*) at *E*.

**13.65** A 500-g collar can slide without friction along the semicircular rod *BCD*. The spring is of constant 320 N/m and its undeformed length is 200 mm. Knowing that the collar is released from rest at *B*, determine (*a*) the speed of the collar as it passes through *C*, (*b*) the force exerted by the rod on the collar at *C*.

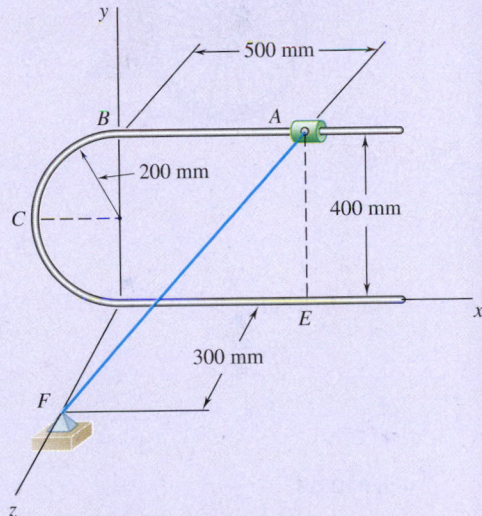

**Fig. P13.64**

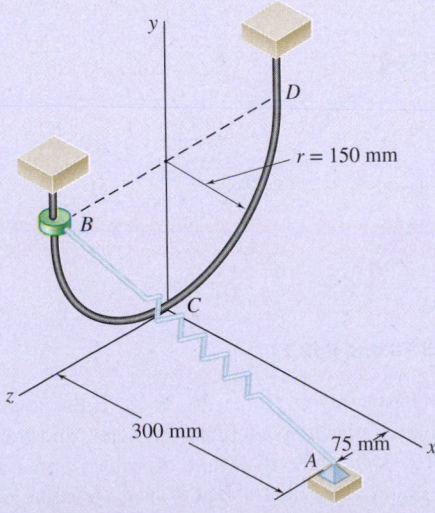

**Fig. P13.65**

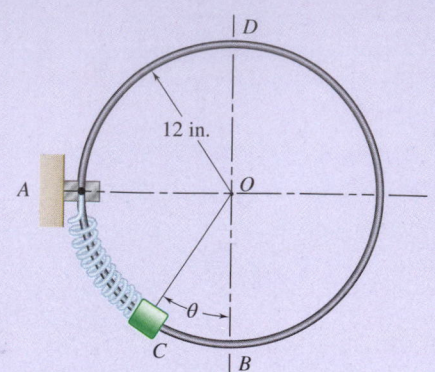

Fig. P13.66

**13.66** A thin circular rod is supported in a *vertical plane* by a bracket at A. Attached to the bracket and loosely wound around the rod is a spring of constant $k = 3$ lb/ft and undeformed length equal to the arc of circle AB. An 8-oz collar C, not attached to the spring, can slide without friction along the rod. Knowing that the collar is released from rest at an angle $\theta$ with the vertical, determine (a) the smallest value of $\theta$ for which the collar will pass through D and reach point A, (b) the velocity of the collar as it reaches point A.

**13.67** Cornhole is a game that requires you to toss beanbags through a hole in a wooden board. People with limited arm mobility often have difficulty enjoying this favorite tailgating activity. An adapted launching device attaches to a wheelchair so that points O and A are fixed. The device mimics an underhand throw by utilizing an elastic band to power the arm OC, which rotates about pin O. The elastic cord has an unstretched length of 1 ft and is attached to the fixed point A and to point B on the arm. The combined weight of the beanbag and holder at C is 4 lbs, and you can neglect the weight of the rod OB. A small clamp holds the beam bag in the cup until release. Knowing that the starting position is 30° from the horizontal, as shown in the figure, determine the spring constant if the velocity of the bean bag is 31 ft/s when the bag is released at an angle of $\theta = 45°$.

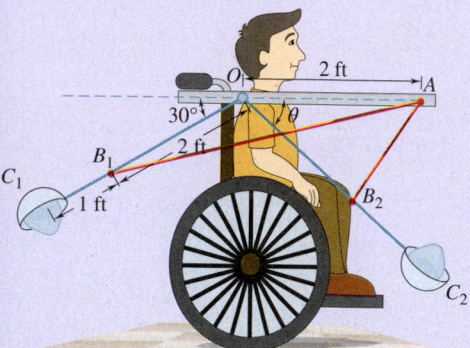

Fig. *P13.67*

**13.68** A spring is used to stop a 50-kg package that is moving down a 20° incline. The spring has a constant $k = 30$ kN/m and is held by cables so that it is initially compressed 50 mm. Knowing that the velocity of the package is 2 m/s when it is 8 m from the spring and neglecting friction, determine the maximum additional deformation of the spring in bringing the package to rest.

**13.69** Solve Prob. 13.68 assuming the coefficient of kinetic friction between the package and the incline is 0.15.

**13.70** A roller coaster starts from rest at A, rolls down the track to B, describes a circular loop of 12-m diameter, and travels up and down past point E. Knowing that $h = 20$ m and assuming no energy loss due to friction, determine the force exerted by the seat on a 50-kg rider at B, D, and E.

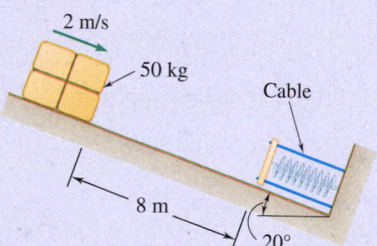

Fig. P13.68

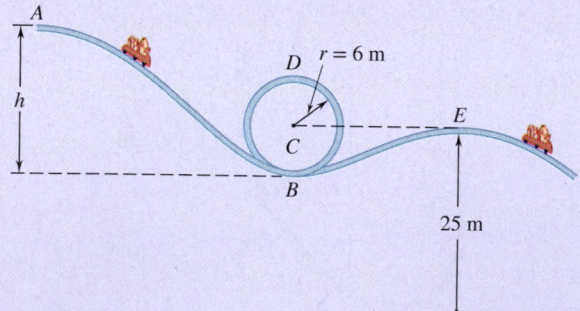

Fig. P13.70 and P13.71

**13.71** A roller coaster starts from rest at A, rolls down the track to B, describes a circular loop of 12-m diameter, and travels up and down past point E. Determine the range of values of $h$ for which the roller coaster will not leave the track at D or E. Assume no energy loss due to friction.

**13.72** A 1-lb collar is attached to a spring and slides without friction along a circular rod in a *vertical* plane. The spring has an undeformed length of 5 in. and a constant $k = 10$ lb/ft. Knowing that the collar is released from being held at $A$, determine the speed of the collar and the normal force between the collar and the rod as the collar passes through $B$.

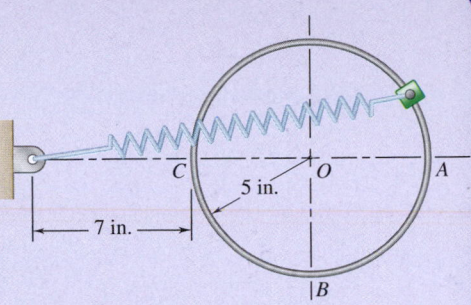

**Fig. P13.72**

**13.73** A 10-lb collar is attached to a spring and slides without friction along a fixed rod in a vertical plane. The spring has an undeformed length of 14 in. and a constant $k = 4$ lb/in. Knowing that the collar is released from rest in the position shown, determine the force exerted by the rod on the collar at (a) point $A$, (b) point $B$. Both these points are on the curved portion of the rod.

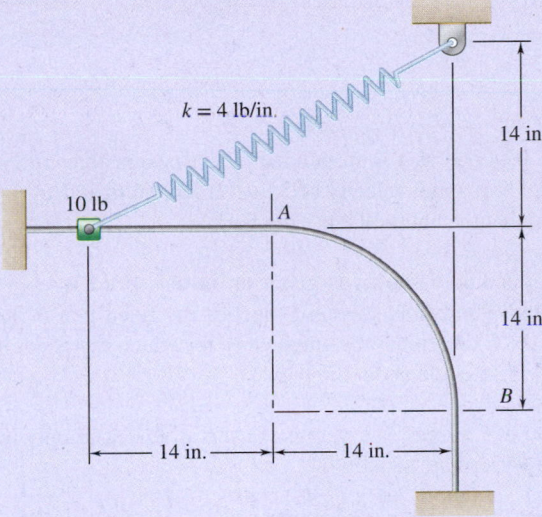

**Fig. P13.73**

**13.74** An 8-oz package is projected upward with a velocity $\mathbf{v}_0$ by a spring at $A$; it moves around a frictionless loop and is deposited at $C$. For each of the two loops shown, determine (a) the smallest velocity $\mathbf{v}_0$ for which the package will reach $C$, (b) the corresponding force exerted by the package on the loop just before the package leaves the loop at $C$.

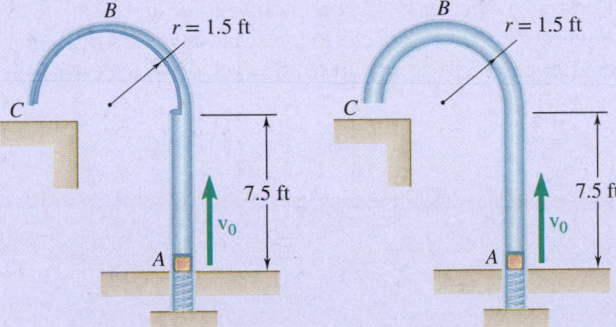

**Fig. P13.74 and P13.75**

**13.75** If the package of Prob. 13.74 is not to hit the horizontal surface at $C$ with a speed greater than 10 ft/s, (a) show that this requirement can be satisfied only by the second loop, (b) determine the largest allowable initial velocity $\mathbf{v}_0$ when the second loop is used.

**13.76** A small package of weight $W$ is projected into a vertical return loop at $A$ with a velocity $\mathbf{v}_0$. The package travels without friction along a circle of radius $r$ and is deposited on a horizontal surface at $C$. For each of the two loops shown, determine (a) the smallest velocity $\mathbf{v}_0$ for which the package will reach the horizontal surface at $C$, (b) the corresponding force exerted by the loop on the package as it passes point $B$.

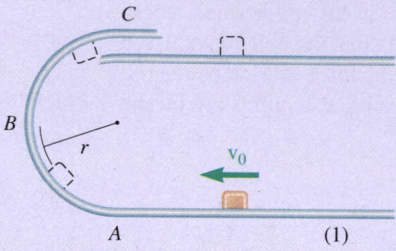

Fig. P13.76

**13.77** The 1-kg ball at $A$ is suspended by an inextensible cord and given an initial horizontal velocity of 5 m/s. If $l = 0.6$ m and $x_B = 0$, determine $y_B$ so that the ball will enter the basket.

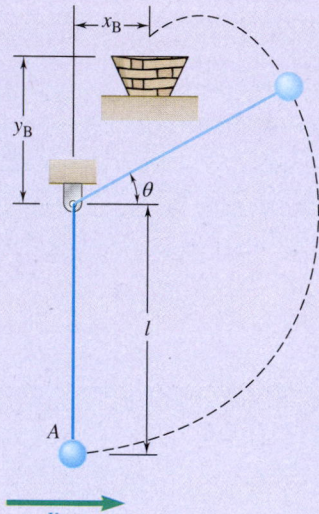

Fig. P13.77

**13.78** The pendulum shown is given an initial speed $v_0$ at $A$ and swings through $90°$ before the cord touches the fixed peg $B$. Knowing that $a = 0.4\,l$, determine the smallest $v_0$ for which the pendulum bob will describe a circle about the peg.

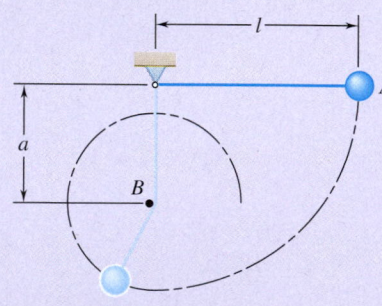

Fig. P13.78

*13.79 Prove that a force $F(x, y, z)$ is conservative if, and only if, the following relations are satisfied:

$$\frac{\partial F_x}{\partial y} = \frac{\partial F_y}{\partial x} \qquad \frac{\partial F_y}{\partial z} = \frac{\partial F_z}{\partial y} \qquad \frac{\partial F_z}{\partial x} = \frac{\partial F_x}{\partial z}$$

**13.80** The force $\mathbf{F} = (yz\mathbf{i} + zx\mathbf{j} + xy\mathbf{k})/xyz$ acts on the particle $P(x, y, z)$ which moves in space. (a) Using the relation derived in Prob. 13.79, show that this force is a conservative force. (b) Determine the potential function associated with $\mathbf{F}$.

*13.81 A force $\mathbf{F}$ acts on a particle $P(x, y)$ which moves in the $xy$ plane. Determine whether $\mathbf{F}$ is a conservative force and compute the work of $\mathbf{F}$ when $P$ describes in a clockwise sense the path $ABCA$, including the quarter circle $x^2 + y^2 = a^2$, if (a) $\mathbf{F} = ky\mathbf{i}$, (b) $\mathbf{F} = k\,(y\mathbf{i} + x\mathbf{j})$.

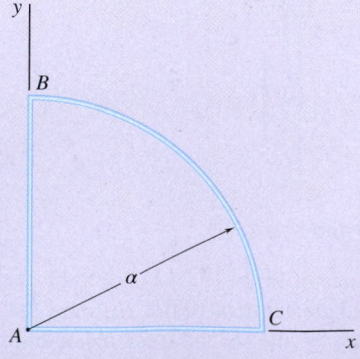

Fig. P13.81

**\*13.82** The potential function associated with a force **P** in space is known to be $V(x, y, z) = -(x^2 + y^2 + z^2)^{1/2}$. (*a*) Determine the *x*, *y*, and *z* components of **P**. (*b*) Calculate the work done by **P** from *O* to *D* by integrating along the path *OABD*, and show that it is equal to the negative of the change in potential from *O* to *D*.

**13.83** Certain springs are characterized by increasing stiffness with increasing deformation according to the relation $F = k_1 x + k_2 x^3$, where *F* is the force exerted by the spring, $k_1$ and $k_2$ are positive constants, and *x* is the deflection of the spring measured from its undeformed position. Determine (*a*) the potential energy $V_e$ as a function of *x*, (*b*) the maximum velocity of a particle of mass *m* attached to the spring and released from rest with $x = x_0$. Neglect friction.

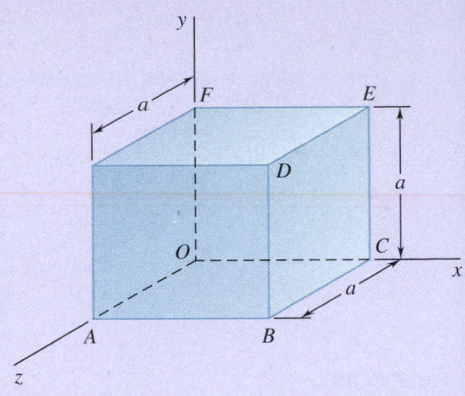

**Fig. P13.82**

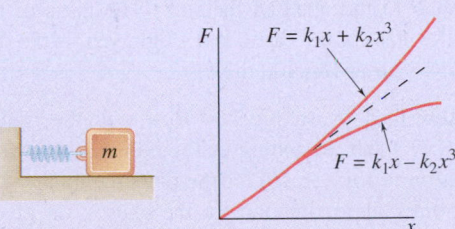

**Fig. *P13.83* and *P13.84***

**13.84** Certain springs are characterized by decreasing stiffness with increasing deformation according to the relation $F = k_1 x - k_2 x^3$, where *F* is the force exerted by the spring, $k_1$ and $k_2$ are positive constants, and *x* is the deflection of the spring measured from its undeformed position. Determine (*a*) the potential energy $V_e$ as a function of *x*, (*b*) the maximum velocity of a particle of mass *m* attached to the spring and released from rest with $x = x_0$. Neglect friction.

**13.85** (*a*) Determine the kinetic energy per unit mass that a missile must have after being fired from the surface of the earth if it is to reach an infinite distance from the earth. (*b*) What is the initial velocity of the missile (called the *escape velocity*)? Give your answers in SI units and show that the answer to part *b* is independent of the firing angle.

**13.86** A satellite describes an elliptic orbit of minimum altitude 606 km above the surface of the earth. The semimajor and semiminor axes are 17 440 km and 13 950 km, respectively. Knowing that the speed of the satellite at point *C* is 4.78 km/s, determine (*a*) the speed at point *A*, the perigee, (*b*) the speed at point *B*, the apogee.

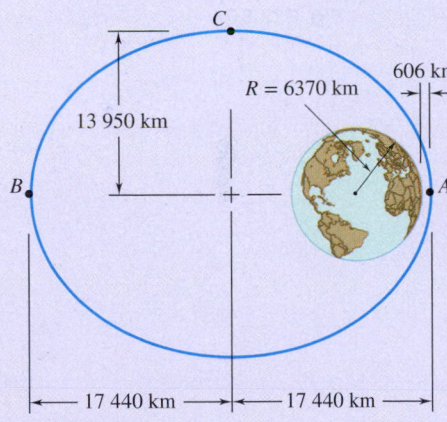

**Fig. P13.86**

**13.87** While describing a circular orbit 200 mi above the earth, a space vehicle launches a 6000-lb communications satellite. Determine (*a*) the additional energy required to place the satellite in a geosynchronous orbit at an altitude of 22,000 mi above the surface of the earth, (*b*) the energy required to place the satellite in the same orbit by launching it from the surface of the earth, excluding the energy needed to overcome air resistance. (A *geosynchronous orbit* is a circular orbit in which the satellite appears stationary with respect to the ground.)

**13.88** How much energy per pound should be imparted to a satellite in order to place it in a circular orbit at an altitude of (*a*) 400 mi, (*b*) 4000 mi?

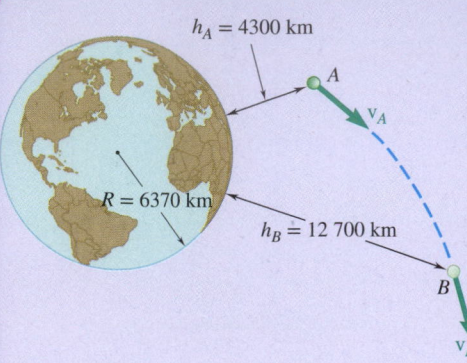

$h_A = 4300$ km

$A$

$v_A$

$R = 6370$ km

$h_B = 12\,700$ km

$B$

$v_B$

**Fig. P13.89**

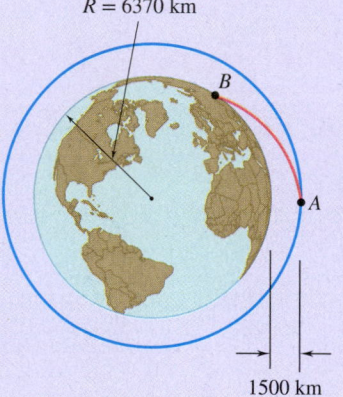

$R = 6370$ km

$B$

$A$

1500 km

**Fig. P13.90**

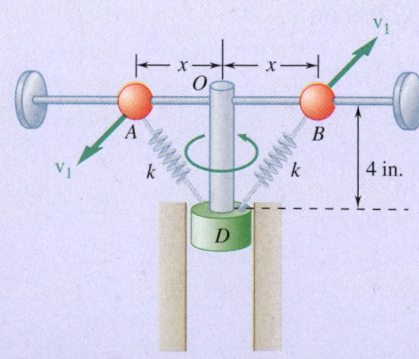

$v_1$

$x$  $O$  $x$

$A$  $B$

$v_1$

$k$  $k$  4 in.

$D$

**Fig. P13.95**

**13.89** Knowing that the velocity of an experimental space probe fired from the earth has a magnitude $v_A = 32.5$ Mm/h at point $A$, determine the speed of the probe as it passes through point $B$.

**13.90** A spacecraft is describing a circular orbit at an altitude of 1500 km above the surface of the earth. As it passes through point $A$, its speed is reduced by 40 percent and it enters an elliptic crash trajectory with the apogee at point $A$. Neglecting air resistance, determine the speed of the spacecraft when it reaches the earth's surface at point $B$.

**13.91** Observations show that a celestial body traveling at $1.2 \times 10^6$ mi/h appears to be describing about point $B$ a circle of radius equal to 60 light years. Point $B$ is suspected of being a very dense concentration of mass called a black hole. Determine the ratio $M_B/M_S$ of the mass at $B$ to the mass of the sun. (The mass of the sun is 330,000 times the mass of the earth, and a light year is the distance traveled by light in 1 year at 186,300 mi/s.)

**13.92** (a) Show that, by setting $r = R + y$ in the right-hand member of Eq. (13.17') and expanding that member in a power series in $y/R$, the expression in Eq. (13.16) for the potential energy $V_g$ due to gravity is a first-order approximation for the expression given in Eq. (13.17'). (b) Using the same expansion, derive a second-order approximation for $V_g$.

**13.93** Collar $A$ has a mass of 3 kg and is attached to a spring of constant 1200 N/m and of undeformed length equal to 0.5 m. The system is set in motion with $r = 0.3$ m, $v_\theta = 2$ m/s, and $v_r = 0$. Neglecting the mass of the rod and the effect of friction, determine the radial and transverse components of the velocity of the collar when $r = 0.6$ m.

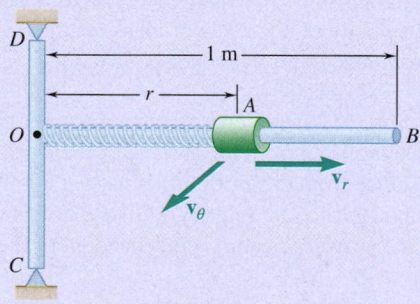

$D$

1 m

$r$

$A$

$O$

$B$

$v_r$

$v_\theta$

$C$

**Fig. P13.93 and P13.94**

**13.94** Collar $A$ has a mass of 3 kg and is attached to a spring of constant 1200 N/m and of undeformed length equal to 0.5 m. The system is set in motion with $r = 0.3$ m, $v_\theta = 2$ m/s, and $v_r = 0$. Neglecting the mass of the rod and the effect of friction, determine (a) the maximum distance between the origin and the collar, (b) the corresponding speed. (*Hint:* Solve the equation obtained for $r$ by trial and error.)

**13.95** A governor is designed so that the valve of negligible mass at $D$ will open once a vertical force greater than 20 lbs is exerted on it. In initial testing of the device, the two 1-lb masses are at $x = 1$ in. and are prevented from sliding along the rod by stops. Each mass is connected to the valve by 10 lb/in. springs that are both unstretched at $x = 1$ in. The governor rotates so that $v_1 = 30$ ft/s when the stops are removed. When the valve opens, determine the position and velocity of the masses.

**13.96** A 1.5-lb ball that can slide on a *horizontal* frictionless surface is attached to a fixed point $O$ by means of an elastic cord of constant $k = 1$ lb/in. and undeformed length 2 ft. The ball is placed at point $A$, 3 ft from $O$, and given an initial velocity $\mathbf{v}_0$ perpendicular to $OA$. Determine (a) the smallest allowable value of the initial speed $v_0$ if the cord is not to become slack, (b) the closest distance $d$ that the ball will come to point $O$ if it is given half the initial speed found in part a.

**13.97** A 1.5-lb ball that can slide on a *horizontal* frictionless surface is attached to a fixed point $O$ by means of an elastic cord of constant $k = 1$ lb/in. and undeformed length 2 ft. The ball is placed at point $A$, 3 ft from $O$, and given an initial velocity $\mathbf{v}_0$ perpendicular to $OA$, allowing the ball to come within a distance $d = 9$ in. of point $O$ after the cord has become slack. Determine (a) the initial speed $v_0$ of the ball, (b) its maximum speed.

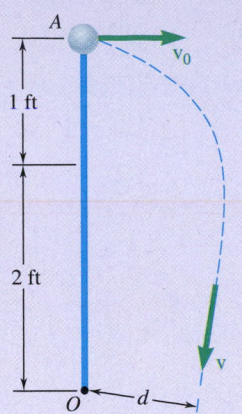

**Fig. P13.96 and P13.97**

**13.98** Using the principles of conservation of energy and conservation of angular momentum, solve part a of Sample Prob. 12.14.

**13.99** Solve Sample Prob. 13.11, assuming that the elastic cord is replaced by a central force $\mathbf{F}$ with a magnitude of $(80/r^2)$ N directed toward $O$.

**13.100** A spacecraft is describing an elliptic orbit of minimum altitude $h_A = 2400$ km and maximum altitude $h_B = 9600$ km above the surface of the earth. Determine the speed of the spacecraft at $A$.

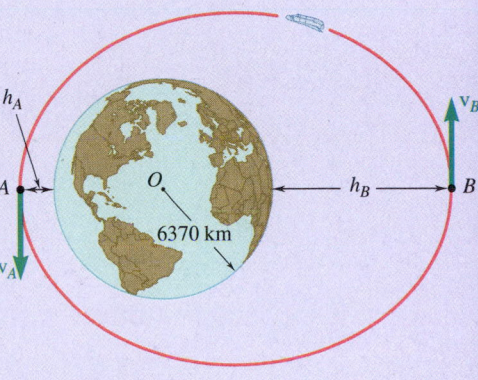

**Fig. P13.100**

**13.101** While describing a circular orbit, 185 mi above the surface of the earth, a space shuttle ejects at point $A$ an inertial upper stage (IUS) carrying a communications satellite to be placed in a geosynchronous orbit (see Prob. 13.87) at an altitude of 22,230 mi above the surface of the earth. Determine (a) the velocity of the IUS relative to the shuttle after its engine has been fired at $A$, (b) the increase in velocity required at $B$ to place the satellite in its final orbit.

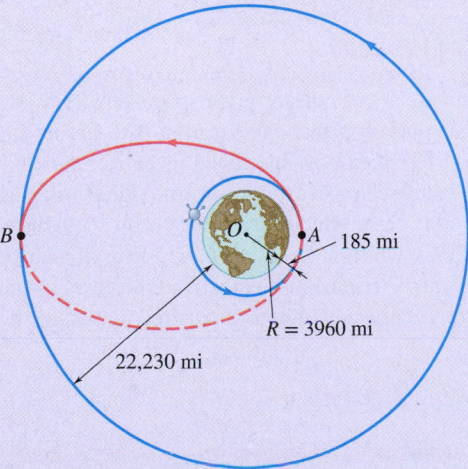

**Fig. P13.101**

**13.102** A spacecraft approaching the planet Saturn reaches point $A$ with a velocity $\mathbf{v}_A$ of magnitude $68.8 \times 10^3$ ft/s. It is to be placed in an elliptic orbit about Saturn so that it will be able to periodically examine Tethys, one of Saturn's moons. Tethys is in a circular orbit of radius $183 \times 10^3$ mi about the center of Saturn, traveling at a speed of $37.2 \times 10^3$ ft/s. Determine (a) the decrease in speed required by the spacecraft at $A$ to achieve the desired orbit, (b) the speed of the spacecraft when it reaches the orbit of Tethys at $B$.

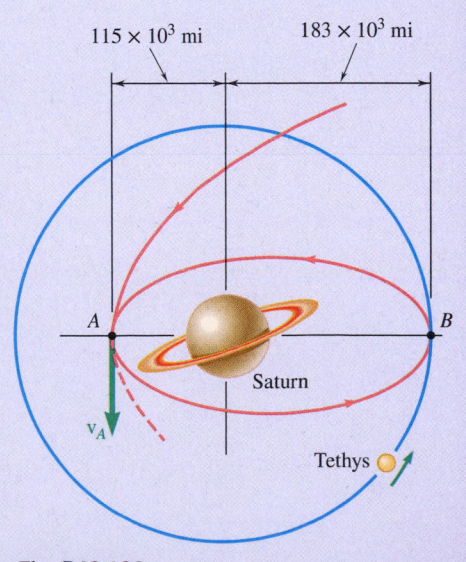

**Fig. P13.102**

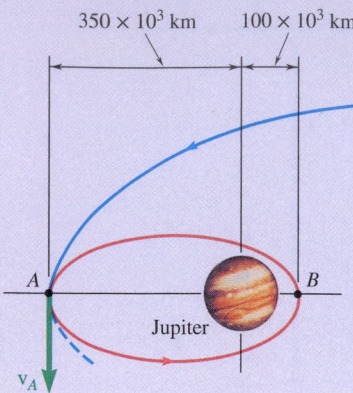

$350 \times 10^3$ km    $100 \times 10^3$ km

Jupiter

$v_A$

**Fig. P13.103**

**13.103** A spacecraft traveling along a parabolic path toward the planet Jupiter is expected to reach point $A$ with a velocity $\mathbf{v}_A$ of magnitude 26.9 km/s. Its engines will then be fired to slow it down, placing it into an elliptic orbit which will bring it to within $100 \times 10^3$ km of Jupiter. Determine the decrease in speed $\Delta v$ at point $A$ which will place the spacecraft into the required orbit. The mass of Jupiter is 319 times the mass of the earth.

**13.104** As a first approximation to the analysis of a space flight from the earth to Mars, it is assumed that the orbits of the earth and Mars are circular and coplanar. The mean distances from the sun to the earth and to Mars are $149.6 \times 10^6$ km and $227.8 \times 10^6$ km, respectively. To place the spacecraft into an elliptical transfer orbit at point $A$, its speed is increased over a short interval of time to $v_A$, which is faster than the earth's orbital speed. When the spacecraft reaches point $B$ on the elliptical transfer orbit, its speed $v_B$ is increased to the orbital speed of Mars. Knowing that the mass of the sun is $332.8 \times 10^3$ times the mass of the earth, determine the increase in velocity required $(a)$ at $A$, $(b)$ at $B$.

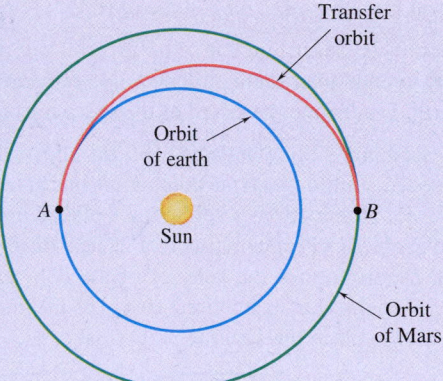

Transfer orbit

Orbit of earth

$A$

Sun

$B$

Orbit of Mars

**Fig. P13.104**

**13.105** The optimal way of transferring a space vehicle from an inner circular orbit to an outer coplanar circular orbit is to fire its engines as it passes through $A$ to increase its speed and place it in an elliptic transfer orbit. Another increase in speed as it passes through $B$ will place it in the desired circular orbit. For a vehicle in a circular orbit about the earth at an altitude $h_1 = 200$ mi, which is to be transferred to a circular orbit at an altitude $h_2 = 500$ mi, determine $(a)$ the required increases in speed at $A$ and at $B$, $(b)$ the total energy per unit mass required to execute the transfer.

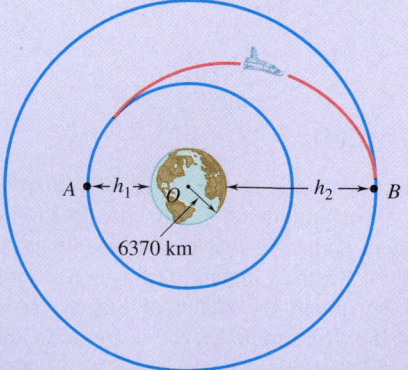

$A$    $h_1$    $O$    $h_2$    $B$

6370 km

**Fig. P13.105**

**13.106** During a flyby of the earth, the velocity of a spacecraft is 10.4 km/s as it reaches its minimum altitude of 990 km above the surface at point $A$. At point $B$, the spacecraft is observed to have an altitude of 8350 km. Determine (a) the magnitude of the velocity at point $B$, (b) the angle $\phi_B$.

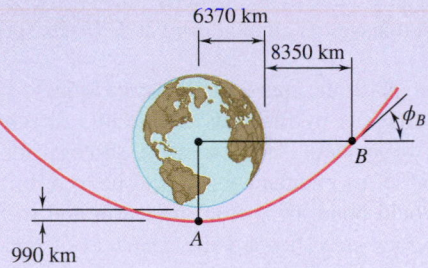

**Fig. P13.106**

**13.107** A space platform is in a circular orbit about the earth at an altitude of 300 km. As the platform passes through $A$, a rocket carrying a communications satellite is launched from the platform with a relative velocity of magnitude 3.44 km/s in a direction tangent to the orbit of the platform. This was intended to place the rocket in an elliptic transfer orbit bringing it to point $B$, where the rocket would again be fired to place the satellite in a geosynchronous orbit of radius 42 140 km. After launching, it was discovered that the relative velocity imparted to the rocket was too large. Determine the angle $\gamma$ at which the rocket will cross the intended orbit at point $C$.

**13.108** A satellite is projected into space with a velocity $\mathbf{v}_0$ at a distance $r_0$ from the center of the earth by the last stage of its launching rocket. The velocity $\mathbf{v}_0$ was designed to send the satellite into a circular orbit of radius $r_0$. However, owing to a malfunction of control, the satellite is not projected horizontally but at an angle $\alpha$ with the horizontal and, as a result, is propelled into an elliptic orbit. Determine the maximum and minimum values of the distance from the center of the earth to the satellite.

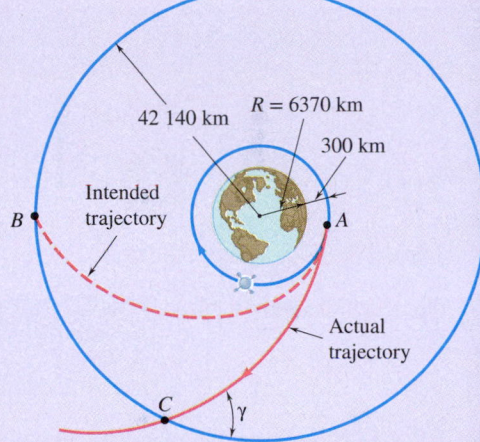

**Fig. P13.107**

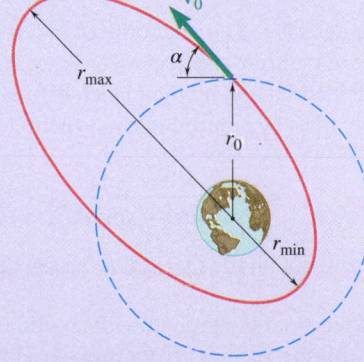

**Fig. P13.108**

**13.109** A space vehicle is to rendezvous with an orbiting laboratory that circles the earth at a constant altitude of 360 km. The vehicle has reached an altitude of 60 km when its engine is shut off, and its velocity $\mathbf{v}_0$ forms an angle $\phi_0 = 50°$ with the vertical $OB$ at that time. What magnitude should $\mathbf{v}_0$ have if the vehicle's trajectory is to be tangent at $A$ to the orbit of the laboratory?

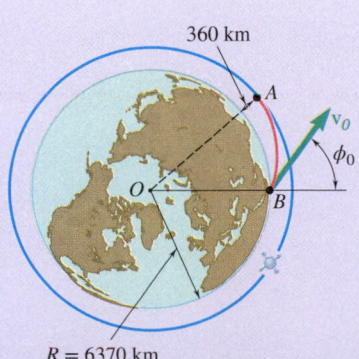

**Fig. P13.109**

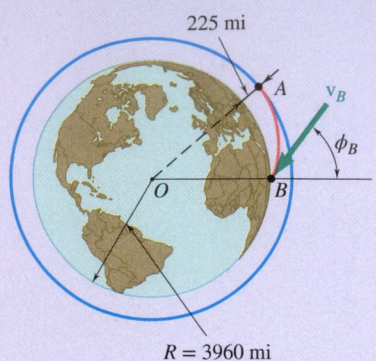

225 mi

$A$    $v_B$

$\phi_B$

$O$    $B$

$R = 3960$ mi

**Fig. P13.110**

**13.110** A space vehicle is in a circular orbit at an altitude of 225 mi above the earth. To return to earth, it decreases its speed as it passes through $A$ by firing its engine for a short interval of time in a direction opposite to the direction of its motion. Knowing that the velocity of the space vehicle should form an angle $\phi_B = 60°$ with the vertical as it reaches point $B$ at an altitude of 40 mi, determine (*a*) the required speed of the vehicle as it leaves its circular orbit at $A$, (*b*) its speed at point $B$.

**\*13.111** In Prob. 13.110, the speed of the space vehicle was decreased as it passed through $A$ by firing its engine in a direction opposite to the direction of motion. An alternative strategy for taking the space vehicle out of its circular orbit would be to turn it around so that its engine would point away from the earth and then give it an incremental velocity $\Delta \mathbf{v}_A$ toward the center $O$ of the earth. This would likely require a smaller expenditure of energy when firing the engine at $A$, but might result in too fast a descent at $B$. Assuming this strategy is used with only 50 percent of the energy expenditure used in Prob. 13.110, determine the resulting values of $\phi_B$ and $v_B$.

**13.112** Show that the values $v_A$ and $v_P$ of the speed of an earth satellite at the apogee $A$ and the perigee $P$ of an elliptic orbit are defined by the relations

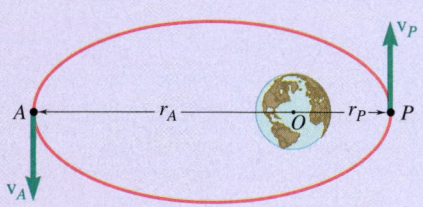

$A$    $r_A$    $O$    $r_P$    $P$    $v_P$

$v_A$

**Fig. P13.112 and P13.113**

$$v_A^2 = \left( \frac{2GM}{r_A + r_P} \frac{r_P}{r_A} \right) \qquad v_P^2 = \frac{2GM}{r_A + r_P} \frac{r_A}{r_P}$$

where $M$ is the mass of the earth, and $r_A$ and $r_P$ represent, respectively, the maximum and minimum distances of the orbit to the center of the earth.

**13.113** Show that the total energy $E$ of an earth satellite of mass $m$ describing an elliptic orbit is $E = -GMm/(r_A + r_P)$, where $M$ is the mass of the earth, and $r_A$ and $r_P$ represent, respectively, the maximum and minimum distances of the orbit to the center of the earth. (Recall that the gravitational potential energy of a satellite was defined as being zero at an infinite distance from the earth.)

**\*13.114** A space probe describes a circular orbit of radius $nR$ with a velocity $\mathbf{v}_0$ about a planet of radius $R$ and center $O$. Show that (*a*) in order for the probe to leave its orbit and hit the planet at an angle $\theta$ with the vertical, its velocity must be reduced to $\alpha v_0$, where

$$\alpha = \sin \theta \sqrt{\frac{2(n-1)}{n^2 - \sin^2 \theta}}$$

(*b*) the probe will not hit the planet if $\alpha$ is larger than $\sqrt{2/(1 + n)}$.

**13.115** A missile is fired from the ground with an initial velocity $\mathbf{v}_0$ forming an angle $\phi_0$ with the vertical. If the missile is to reach a maximum altitude equal to $\alpha R$, where $R$ is the radius of the earth, (*a*) show that the required angle $\phi_0$ is defined by the relation

$$\sin \phi_0 = (1 + \alpha) \sqrt{1 - \frac{\alpha}{1 + \alpha} \left( \frac{v_{\text{esc}}}{v_0} \right)^2}$$

where $v_{\text{esc}}$ is the escape velocity, (*b*) determine the range of allowable values of $v_0$.

**13.116** A spacecraft of mass $m$ describes a circular orbit of radius $r_1$ around the earth. (a) Show that the additional energy $\Delta E$ that must be imparted to the spacecraft to transfer it to a circular orbit of larger radius $r_2$ is

$$\Delta E = \frac{GMm(r_2 - r_1)}{2r_1 r_2}$$

where $M$ is the mass of the earth. (b) Further show that if the transfer from one circular orbit to the other is executed by placing the spacecraft on a transitional semielliptic path $AB$, the amounts of energy $\Delta E_A$ and $\Delta E_B$ which must be imparted at $A$ and $B$ are, respectively, proportional to $r_2$ and $r_1$:

$$\Delta E_A = \frac{r_2}{r_1 + r_2}\Delta E \qquad \Delta E_B = \frac{r_1}{r_1 + r_2}\Delta E$$

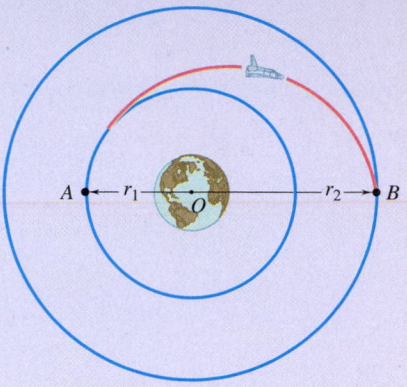

**Fig. P13.116**

**\*13.117** Using the answers obtained in Prob. 13.108, show that the intended circular orbit and the resulting elliptic orbit intersect at the ends of the minor axis of the elliptic orbit.

**\*13.118** (a) Express in terms of $r_{min}$ and $v_{max}$ the angular momentum per unit mass, $h$, and the total energy per unit mass, $E/m$, of a space vehicle moving under the gravitational attraction of a planet of mass $M$ (Fig. 13.15). (b) Eliminating $v_{max}$ between the equations obtained, derive the formula

$$\frac{1}{r_{min}} = \frac{GM}{h^2}\left[1 + \sqrt{1 + \frac{2E}{m}\left(\frac{h}{GM}\right)^2}\right]$$

(c) Show that the eccentricity $\varepsilon$ of the trajectory of the vehicle can be expressed as

$$\varepsilon = \sqrt{1 + \frac{2E}{m}\left(\frac{h}{GM}\right)^2}$$

(d) Further show that the trajectory of the vehicle is a hyperbola, an ellipse, or a parabola, depending on whether $E$ is positive, negative, or zero, respectively.

# 13.3    IMPULSE AND MOMENTUM

We now consider a third basic method for the solution of problems dealing with the motion of particles. This method is based on the principle of impulse and momentum and can be used to solve problems involving force, mass, velocity, and time. It is of particular interest in the solution of problems involving impulsive motion and problems involving impacts (Secs. 13.3B and 13.4).

## 13.3A    Principle of Impulse and Momentum

Consider a particle of mass $m$ acted upon by a force $\mathbf{F}$. As we saw in Sec. 12.1B, we can express Newton's second law in the form

$$\mathbf{F} = \frac{d}{dt}(m\mathbf{v}) \tag{13.27}$$

where $m\mathbf{v}$ is the linear momentum of the particle. Multiplying both sides of Eq. (13.27) by $dt$ and integrating from a time $t_1$ to a time $t_2$, we have

$$\mathbf{F}\,dt = d(m\mathbf{v})$$

$$\int_{t_1}^{t_2} \mathbf{F}\,dt = m\mathbf{v}_2 - m\mathbf{v}_1$$

Moving $m\mathbf{v}_1$ to the left side of this equation gives us

$$m\mathbf{v}_1 + \int_{t_1}^{t_2} \mathbf{F}\,dt = m\mathbf{v}_2 \tag{13.28}$$

The integral in Eq. (13.28) is a vector known as the **linear impulse**, or simply the **impulse**, of the force $\mathbf{F}$ during the interval of time considered. Resolving $\mathbf{F}$ into rectangular components, we have

$$\mathbf{Imp}_{1\rightarrow2} = \int_{t_1}^{t_2} \mathbf{F}\,dt$$

$$= \mathbf{i}\int_{t_1}^{t_2} F_x\,dt + \mathbf{j}\int_{t_1}^{t_2} F_y\,dt + \mathbf{k}\int_{t_1}^{t_2} F_z\,dt \tag{13.29}$$

Note that the components of the impulse of force $\mathbf{F}$ are, respectively, equal to the areas under the curves obtained by plotting the components $F_x$, $F_y$, and $F_z$ against $t$ (Fig. 13.16). In the case of a force $\mathbf{F}$ of constant magnitude and direction, the impulse is represented by the vector $\mathbf{F}(t_2 - t_1)$, which has the same direction as $\mathbf{F}$.

   If we use SI units, the magnitude of the impulse of a force is expressed in N·s. However, recalling the definition of the newton, we have

$$\text{N·s} = (\text{kg·m/s}^2)\text{·s} = \text{kg·m/s}$$

which is the unit obtained in Sec. 12.1C for the linear momentum of a particle. This verifies that Eq. (13.28) is dimensionally correct. If we use U.S. customary units, the impulse of a force is expressed in lb·s, which is also the unit obtained in Sec. 12.1C for the linear momentum of a particle.

   Eq. (13.28) states that when a particle is acted upon by a force $\mathbf{F}$ during a given time interval, **we can obtain the final momentum $m\mathbf{v}_2$ of the particle**

**Photo 13.4**   This impact test between an F-4 Phantom and a rigid reinforced target was to determine the impact force as a function of time. ©Sandia National Laboratories/Getty Images RF

**by adding vectorially its initial momentum $m\mathbf{v}_1$ and the impulse of the force F during the time interval considered.** This can be expressed as:

**Principle of impulse and momentum**

$$m\mathbf{v}_1 + \mathbf{Imp}_{1\to2} = m\mathbf{v}_2 \qquad \text{(13.30)}$$

Fig. 13.17 is a pictorial representation of this principle and is called an *impulse–momentum diagram.* To obtain an analytic solution, it is thus necessary to replace Eq. (13.30) with the corresponding component equations. Note that whereas kinetic energy and work are scalar quantities, momentum and impulse are vector quantities.

$$(mv_x)_1 + \int_{t_1}^{t_2} F_x \, dt = (mv_x)_2$$

$$(mv_y)_1 + \int_{t_1}^{t_2} F_y \, dt = (mv_y)_2 \qquad \text{(13.31)}$$

$$(mv_z)_1 + \int_{t_1}^{t_2} F_z \, dt = (mv_z)_2$$

When several forces act on a particle, we must consider the impulse of each of the forces. We have

$$m\mathbf{v}_1 + \Sigma\,\mathbf{Imp}_{1\to2} = m\mathbf{v}_2 \qquad \text{(13.32)}$$

Again, this equation represents a relation between vector quantities; in the actual solution of a problem, it should be replaced by the corresponding component equations.

When a problem involves two or more particles, we can consider each particle separately and write Eq. (13.32) for each particle. We can also add vectorially the momenta of all the particles and the impulses of all the forces involved. We then have

$$\Sigma m\mathbf{v}_1 + \Sigma\,\mathbf{Imp}_{1\to2} = \Sigma m\mathbf{v}_2 \qquad \text{(13.33)}$$

Because the forces of action and reaction exerted by the particles on each other form pairs of equal and opposite forces, and because the time interval from $t_1$ to $t_2$ is common to all of the forces involved, the impulses of the forces of action and reaction cancel out. Thus, we need consider only the impulses of the external forces.[†]

If no external force is exerted on the particles or, more generally, if the sum of the external forces is zero, the second term in Eq. (13.33) vanishes and the equation reduces to

**Conservation of linear momentum**

$$\Sigma m\mathbf{v}_1 = \Sigma m\mathbf{v}_2 \qquad \text{(13.34)}$$

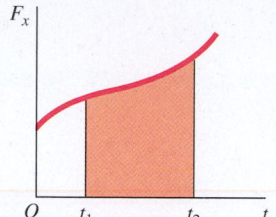

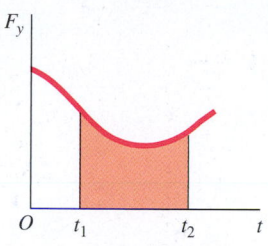

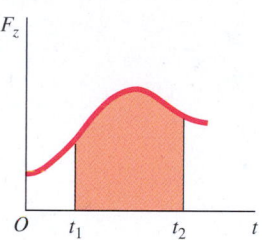

**Fig. 13.16**  Components of the impulse of a force **F** acting from times $t_1$ to $t_2$.

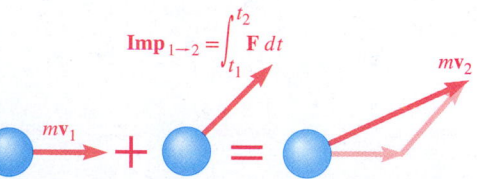

**Fig. 13.17**  Impulse–momentum diagram. Initial momentum plus impulse of a force **F** equals final momentum.

[†]Note the difference between this statement and the corresponding statement in Sec. 13.1C regarding the work of the forces of action and reaction between several particles. Although the sum of the impulses of these forces is always zero, the sum of their work is zero only under special circumstances, e.g., when the particles involved are connected by inextensible cords or links and thus are constrained to move through equal distances.

For two particles $A$ and $B$, this is

$$m_A\mathbf{v}_A + m_B\mathbf{v}_B = m_A\mathbf{v}'_A + m_B\mathbf{v}'_B \qquad \textbf{(13.34')}$$

where $\mathbf{v}'_A$ and $\mathbf{v}'_B$ represent the velocities of the bodies at the second time. This equation says that **the total momentum of the particles is conserved**. Consider, for example, two boats with masses of $m_A$ and $m_B$, initially at rest, that are being pulled together (Fig. 13.18). If we neglect the resistance of the water, the only external forces acting on the boats are their weights and the buoyant forces exerted on them. Because these forces are balanced, we have

$$\Sigma m\mathbf{v}_1 = \Sigma m\mathbf{v}_2$$
$$0 = m_A\mathbf{v}'_A + m_B\mathbf{v}'_B$$

where $\mathbf{v}'_A$ and $\mathbf{v}'_B$ represent the velocities of the boats after a finite interval of time. This equation indicates that the boats move in opposite directions (toward each other) with velocities inversely proportional to their masses.[†]

**Fig. 13.18** Neglecting the resistance of the water, linear momentum is conserved for two boats being pulled together.

## 13.3B Impulsive Motion

A force acting on a particle during a very short time interval but large enough to produce a definite change in momentum is called an **impulsive force**. The resulting motion is called an **impulsive motion**. For example, when a baseball is struck, the contact between bat and ball takes place during a very short time interval $\Delta t$. But the average value of the force $\mathbf{F}_{avg}$ exerted by the bat on the ball is very large, and the resulting impulse $\mathbf{F}_{avg}\Delta t$ is large enough to change the sense of motion of the ball (Fig. 13.19).

When impulsive forces act on a particle, Eq. (13.32) becomes

**Impulse–momentum principle for impulsive motion**

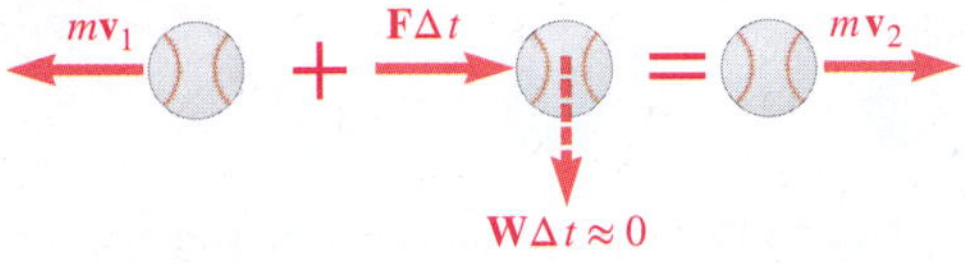

$$m\mathbf{v}_1 + \Sigma\mathbf{F}_{avg}\Delta t = m\mathbf{v}_2 \qquad \textbf{(13.35)}$$

We can neglect any force that is not an impulsive force because the corresponding impulse $\mathbf{F}_{avg}\Delta t$ is very small. Non-impulsive forces include the weight of the body, the force exerted by a spring, or any other force that is known to be small compared with an impulsive force. Unknown reactions may or may not be impulsive; their impulses therefore should be included in Eq. (13.35) as long as they have not been proved negligible. For example, we may neglect the impulse of the weight of the baseball considered previously. If we analyze the motion of the bat, we can neglect the impulse of the weight of the bat. The impulses of the

**Fig. 13.19** When an impulsive force (i.e., a large force that acts over a short time) acts on a system, we can often neglect non-impulsive forces, such as weight.

[†]We use blue equal signs in Fig. 13.18 and throughout the remainder of this chapter to indicate that two systems of vectors are *equipollent;* that is, that they have the same resultant and moment resultant (c.f. Sec. 3.4B). We continue to use red equal signs to indicate that two systems of vectors are *equivalent;* that is, they have the same effect. We will discuss this and the concept of the conservation of momentum for a system of particles in greater detail in Chap. 14.

reactions of the player's hands on the bat, however, should be included; these impulses are not negligible if the ball is incorrectly hit.

Note that the method of impulse and momentum is particularly effective in analyzing the impulsive motion of a particle, because it involves only the initial and final velocities of the particle and the impulses of the forces exerted on the particle. The direct application of Newton's second law, on the other hand, would require determining the forces as functions of time and integrating the equations of motion over the time interval $\Delta t$.

In the case of the impulsive motion of several particles, we can use Eq. (13.33). It reduces to

$$\Sigma m\mathbf{v}_1 + \Sigma \mathbf{F}_{\text{avg}}\, \Delta t = \Sigma m\mathbf{v}_2 \qquad \textbf{(13.36)}$$

where the second term involves only impulsive, external forces. If all of the external forces acting on the various particles are non-impulsive, the second term in Eq. (13.36) vanishes, and this equation reduces to Eq. (13.34):

$$\Sigma m\mathbf{v}_1 = \Sigma m\mathbf{v}_2 \qquad \textbf{(13.34)}$$

As before, for two particles, this reduces to

$$m_A\mathbf{v}_A + m_B\mathbf{v}_B = m_A\mathbf{v}_A' + m_B\mathbf{v}_B' \qquad \textbf{(13.34}')$$

In other words, the total momentum of the particles is conserved. This situation occurs, for example, when two freely moving particles collide with one another. We should note, however, that although the total momentum of the particles is conserved, their total energy is generally *not* conserved. Problems involving the collision or *impact* of two particles are discussed in detail in Sec. 13.4.

## Sample Problem 13.13

An automobile weighing 4000 lb is moving down a 5° incline at a speed of 60 mi/h when the brakes are applied, causing a constant total braking force (applied by the road on the tires) of 1500 lb. Determine the time required for the automobile to come to a stop.

**STRATEGY:**   Because you are given velocities at two different times, use the principle of impulse and momentum.

**MODELING:**   Choose the automobile to be your system and assume you can model it as a particle. The impulse–momentum diagram for this system is shown in Fig. 1.

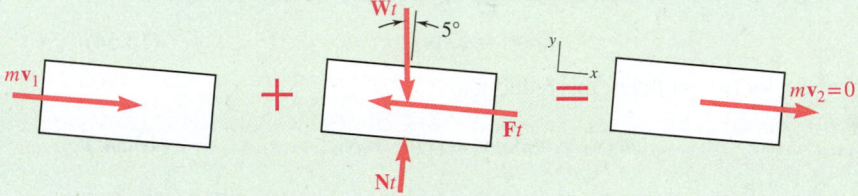

**Fig. 1**   Impulse–momentum diagram for the car.

**ANALYSIS:**   The general impulse–momentum principle is

$$m\mathbf{v}_1 + \Sigma\mathbf{Imp}_{1\to2} = m\mathbf{v}_2$$

This is a vector equation, and because the impulsive force is constant, the impulse is simply equal to the force multiplied by its time duration. You can obtain scalar equations by using Fig. 1. In the direction down the incline, you get

$+\searrow$ components:   $mv_1 + (W \sin 5°)t - Ft = 0$

$(4000/32.2)(88 \text{ ft/s}) + (4000 \sin 5°)t - 1500t = 0$      $t = 9.49 \text{ s}$ ◄

**REFLECT and THINK:**   You could use Newton's second law to solve this problem. First, you would determine the car's deceleration, separate variables, and then integrate $a = dv/dt$ to relate the velocity, deceleration, and time. You could not use conservation of energy to solve this problem, because this principle does not involve time.

## Sample Problem 13.14

In order to determine the weight of a freight train of 40 identical boxcars, an engineer attaches a dynamometer between the train and the locomotive. The train starts from rest, travels over a straight, level track, and reaches a speed of 30 mi/h after three minutes. During this time interval, the average reading of the dynamometer is 120 tons. Knowing that the effective coefficient of friction in the system is 0.03 and air resistance is negligible, determine (a) the weight of the train (in tons), (b) the coupling force between boxcars A and B.

(continued)

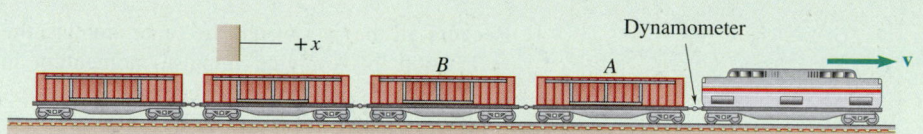

**STRATEGY:**   This problem could be solved using Newton's second law and kinematic relationships, but because you are given velocities at two times and asked to find the force, you can also use impulse and momentum.

**MODELING:**   Choose the system to be the 40 boxcars behind the engine. An impulse–momentum diagram for this system is shown in Fig. 1, where **F** is the dynamometer force.

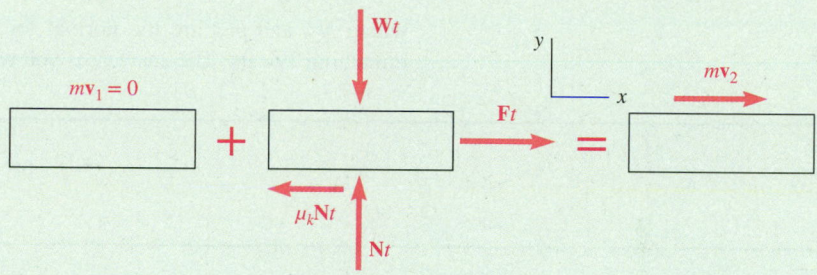

**Fig. 1**   Impulse–momentum diagram for the 40 boxcars.

**ANALYSIS:**   Apply the principle of impulse and momentum

$$m\mathbf{v}_1 + \Sigma\mathbf{Imp}_{1\to2} = m\mathbf{v}_2$$

You can obtain scalar equations by using Fig. 1 and looking at the $x$ and $y$ directions.

$+\uparrow y$ components:   $\qquad Nt - Wt = 0 \qquad N = W$

$\xrightarrow{+} x$ components:   $\qquad 0 + Ft - \mu_k Nt = mv_2$

$0 + (120 \text{ ton})(2000 \text{ lb/ton})(180 \text{ s}) - 0.03(W)(180 \text{ s})$

$$= \left(\frac{W}{32.2 \text{ ft/s}^2}\right)(30 \text{ mi/h})\left(\frac{1 \text{ h}}{3600 \text{ s}}\right)\left(\frac{5280 \text{ ft}}{\text{mi}}\right)$$

Solving for $W$, you obtain

$$W = 6.384 \times 10^6 \text{ lb} = 3190 \text{ tons} \quad \blacktriangleleft$$

**Coupling Force Between Cars *A* and *B*.**   You need to define a new system where the force of interest is an external force. Therefore, choose car *A* to be your system and define $F_A$ as the coupling force between cars *A* and *B*. The impulse–momentum diagram for this system is shown in Fig. 2.

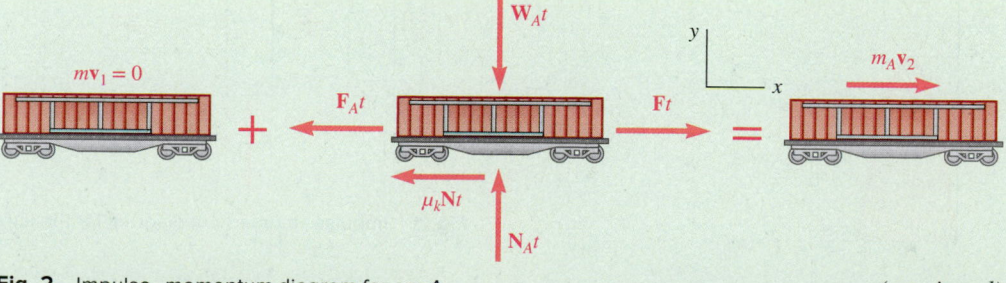

**Fig. 2**   Impulse–momentum diagram for car *A*.

*(continued)*

Because all the cars weigh the same amount, the weight of $A$ is $W_A = W/40$ = 159,600 lb. Applying impulse–momentum in the $y$ direction gives you $N_A = W_A$. Considering the $x$ direction,

$\xrightarrow{+}$ $x$ components:     $0 + Ft - \mu_k N_A t - F_A t = m_A v_2$

Substituting in numbers and solving for $F_A$ gives

$$F_A = 117.0 \text{ tons} \quad \blacktriangleleft$$

**REFLECT and THINK:**   Rather than using $A$ as your system, you could have chosen the remaining 39 cars to be your system. In this case, you would find

$$0 - \mu_k N_{39} t + F_A t = m_{39} v_2$$

where $N_{39}$ and $m_{39}$ are the normal force and the mass, respectively, for the remaining 39 cars. The answer, as you would expect, is the same.

## Sample Problem 13.15

A hammer and punch is used by a surgeon when inserting a hip implant. To better understand this process, an instrumented implant is inserted into a fixed replicate femur. The upward resisting force from the replicate femur on the hip implant can be neglected during the impact, and the impact force from the punch can be approximated by a half sine wave. Determine the speed of the 0.3-kg implant immediately after impact.

**STRATEGY:**   Because you are relating force, time, and velocities, you should use the principle of impulse and momentum.

**MODELING:**   Choose the system to be the implant. An impulse–momentum diagram for this system is shown in Fig. 1. The resisting force is left off Fig. 1 because it is assumed to be negligible.

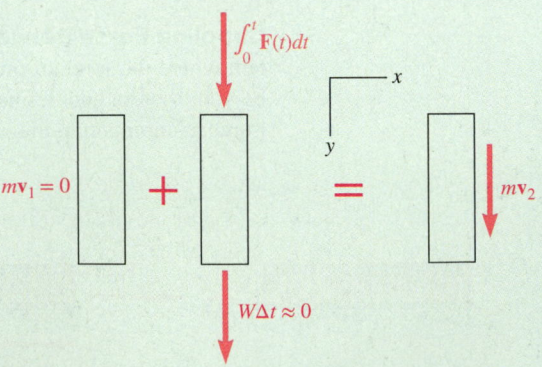

**Fig. 1**   Impulse–momentum diagram for the implant.

*(continued)*

**ANALYSIS:**    Apply the principle of impulse and momentum

$$m\mathbf{v}_1 + \Sigma \mathbf{Imp}_{1 \to 2} = m\mathbf{v}_2$$

You can obtain scalar equations by looking at the vertical components.

$\downarrow + y$ components:            $0 + \displaystyle\int_0^t F(t)\,dt = mv_2$            **(1)**

where

$$\int_0^t F(t)\,dt = \int_0^{0.002} 35\,000 \sin\left(\frac{2\pi}{0.004}t\right)dt = -35\,000 \frac{0.004}{2\pi}\cos\left(\frac{2\pi}{0.004}t\right)\Big|_0^{0.002}$$

$$= 45.56\ \text{N·s}$$

Substituting this into Eq. (1) and solving for $v_2$ gives

$$v_2 = \frac{\displaystyle\int_0^t F(t)\,dt}{m} = \frac{45.36\ \text{N·s}}{0.3\ \text{kg}}$$

$$v_2 = 148.5\ \text{m/s} \quad \blacktriangleleft$$

**REFLECT and THINK:**  This problem is similar to Sample Prob. 13.5, where the drop-hammer pile driver hits the pile, then the hammer and pile move down, and the earth resists the motion. In that problem, you analyzed the motion *after* the impact; in this problem, you are analyzing the motion *during* the impact. In reality, you would need to do some experimental measurements to determine if the resisting force really is negligible during the impact. If you knew the force relationship of the femur on the implant, you could solve this as a two-part problem to first find the velocity of the implant immediately after the impact using impulse and momentum, and then determine how far the implant moves down into the femur using work and energy.

## Sample Problem 13.16

A 4-oz baseball is pitched with a velocity of 80 ft/s toward a batter. After the ball is hit by the bat $B$, it has a velocity of 120 ft/s in the direction shown. If the bat and ball are in contact for 0.015 s, determine the average impulsive force exerted on the ball during the impact.

**STRATEGY:**  This situation features an impact, and therefore impulsive forces, so apply the principle of impulse and momentum to the ball.

**MODELING:**   Choose the ball as your system and treat it as a particle. The impulse–momentum diagram for this system is shown in Fig. 1. Because the weight of the ball is a non-impulsive force that is typically much smaller than the impulsive force, you can neglect it.

*(continued)*

**ANALYSIS:**    Apply the principle of impulse and momentum

$$m\mathbf{v}_1 + \Sigma\mathbf{Imp}_{1\to2} = m\mathbf{v}_2$$

Applying this in the $x$ and $y$ directions gives

$\xrightarrow{+}$ $x$ components:         $-mv_1 + F_x\Delta t = mv_2\cos 40°$

$$-\frac{\frac{4}{16}}{32.2}(80 \text{ ft/s}) + F_x(0.015 \text{ s}) = \frac{\frac{4}{16}}{32.2}(120 \text{ ft/s})\cos 40°$$

$$F_x = +89.0 \text{ lb}$$

$+\uparrow$ $y$ components:         $0 + F_y\Delta t = mv_2\sin 40°$

$$F_y(0.015 \text{ s}) = \frac{\frac{4}{16}}{32.2}(120 \text{ ft/s})\sin 40°$$

$$F_y = +39.9 \text{ lb}$$

From its components $F_x$ and $F_y$ you can determine the magnitude and direction of the average impulsive force $\mathbf{F}$ as

$$\mathbf{F} = 97.5 \text{ lb} \angle 24.2°$$

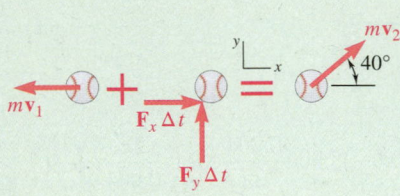

**Fig. 1**  Impulse–momentum diagram for the ball.

**REFLECT and THINK:**    In this problem, we neglected the impulse due to the weight. This would have had a magnitude of $(4/16 \text{ lb})(0.015 \text{ s}) = 0.00288 \text{ lb·s}$. This indeed is much smaller than the impulse exerted on the ball by the bat, which is $(97.4 \text{ lb})(0.015 \text{ s}) = 1.463 \text{ lb·s}$.

## Sample Problem 13.17

A 10-kg package drops from a chute into a 25-kg cart with a velocity of 3 m/s. The cart is initially at rest and can roll freely. Determine (*a*) the final velocity of the cart, (*b*) the impulse exerted by the cart on the package, (*c*) the fraction of the initial energy lost in the impact. ▶

**STRATEGY:**    Because you have an impact, and therefore impulsive forces, use the principle of impulse and momentum.

**MODELING:**    Choose the package and the cart to be your system, and assume that both can be treated as particles. The impulse–momentum diagram for this system is shown in Fig. 1. Note that a vertical impulse occurs between the cart and the ground, because the cart is constrained to move horizontally.

**ANALYSIS:**    Apply the principle of impulse and momentum

$$m\mathbf{v}_1 + \Sigma\mathbf{Imp}_{1\to2} = m\mathbf{v}_2$$

**Fig. 1**  Impulse–momentum diagram for the system.

*(continued)*

**a. Package and Cart.** Applying this principle in the $x$ direction gives

$\xrightarrow{+}$ $x$ components:
$$m_P v_1 \cos 30° + 0 = (m_P + m_C)v_2$$
$$(10 \text{ kg})(3 \text{ m/s}) \cos 30° = (10 \text{ kg} + 25 \text{ kg})v_2$$

$$\mathbf{v}_2 = 0.742 \text{ m/s} \rightarrow \quad \blacktriangleleft$$

In Fig. 1, the force between the package and the cart is not shown because it is internal to the defined system. To determine this force, you need a new system; that is, just the package by itself. The impulse–momentum diagram for the package alone is shown in Fig. 2.

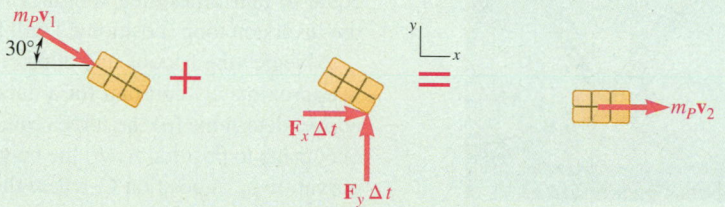

**Fig. 2** Impulse–momentum diagram for the package.

**b. Impulse–Momentum Principle: Package.** The package moves in both $x$ and $y$ directions, so write the conservation of momentum equation for each component of the motion.

$\xrightarrow{+}$ $x$ components:
$$-mv_1 + F_x \Delta t = mv_2$$
$$(10 \text{ kg})(3 \text{ m/s}) \cos 30° + F_x \Delta t = (10 \text{ kg})(0.742 \text{ m/s})$$
$$F_x \Delta t = -18.56 \text{ N·s}$$

$+\uparrow$ $y$ components:
$$-m_P v_1 \sin 30° + F_y \Delta t = 0$$
$$-(10 \text{ kg})(3 \text{ m/s}) \sin 30° + F_y \Delta t = 0$$
$$F_y \Delta t = +15 \text{ N·s}$$

The impulse exerted on the package is

$$\mathbf{F} \Delta t = 23.9 \text{ N·s} \measuredangle 38.9° \quad \blacktriangleleft$$

**c. Fraction of Energy Lost.** The initial and final energies are

$$T_1 = \tfrac{1}{2} m_P v_1^2 = \tfrac{1}{2}(10 \text{ kg})(3 \text{ m/s})^2 = 45 \text{ J}$$

$$T_2 = \tfrac{1}{2}(m_P + m_C) v_2^2 = \tfrac{1}{2}(10 \text{ kg} + 25 \text{ kg})(0.742 \text{ m/s})^2 = 9.63 \text{ J}$$

The fraction of energy lost is

$$\frac{T_1 - T_2}{T_1} = \frac{45 \text{ J} - 9.63 \text{ J}}{45 \text{ J}} = 0.786 \quad \blacktriangleleft$$

**REFLECT and THINK:** Except in the purely theoretical case of a "perfectly elastic" collision, mechanical energy is never conserved in a collision between two objects, even though linear momentum may be conserved. Note that, in this problem, momentum was conserved in the $x$ direction but was not conserved in the $y$ direction because of the vertical impulse on the wheels of the cart. Whenever you deal with an impact, you need to use impulse–momentum methods.

# Case Study 13.1

Amusement park operators are continually striving for more thrilling rides. Assume you have been hired to design a launched roller coaster to break the current total height and loop height records (an inversion loop is shown in CS Photo 13.1). The design uses a hydraulic system to launch the 6000-kg cars up a top hat feature (see CS Photo 13.2) that has a height $h_A = 200$ m above the launch platform. Park officials have requested the following specifications for the ride profile shown in CS Fig. 13.1: (1) at the crest of the top hat (position A), the rider will feel weightless (i.e., the normal force on them is zero), (2) at the top of the inversion loop (position B), the normal force on the rider will be equal to four times their weight, (3) the normal force on the rider as they exit the inversion loop (position C) will be the same as at the top of the loop.

Neglecting friction and other energy losses, assuming that the launch force is approximately constant for a duration of 3 s, and meeting the specifications above, determine (a) the force required to launch the 6000-kg coaster train and occupants to position A, (b) the height of point B if $\rho_B = 25$ m, (c) the radius of curvature $\rho_C$ at position C, where the track is 30° from the horizontal.

**CS Photo 13.1** Inversion loop on a roller coaster.
©Chaimdan/Getty Images RF

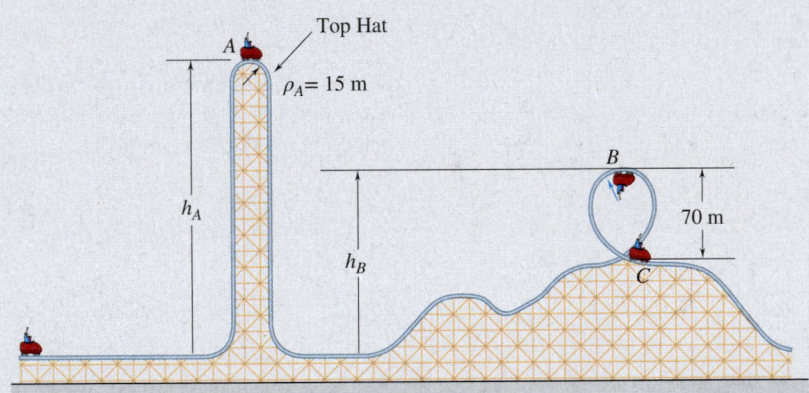

**CS Fig. 13.1** Profile of roller coaster ride.

**CS Photo 13.2** Top hat feature of a roller coaster.
©Stan Honda/AFP/Getty Images

**STRATEGY:** First, let's find the required velocity at point A using Newton's second law. From this velocity, you can use work and energy to find the velocity $v_2$ at the end of the launch pad, and then use impulse and momentum to determine the required hydraulic force to achieve this velocity. Then, you can use a combination of work and energy and Newton's second law to calculate parts (b) and (c).

**MODELING:** You should model the occupant as a particle at the crest of the top hat to determine the required velocity at point A, and draw the free-body and kinetic diagrams as shown in CS Fig. 13.2. Create similar diagrams for positions B and C.

Model the roller coaster cars as a particle, and draw the impulse–momentum diagram during the 3-second launch.

**ANALYSIS:** You can use your free-body and kinetic diagram CS Fig. 13.3 to find the velocity needed for the normal force to equal zero at point A (the ride will feel weightless at this point).

$$+\downarrow \Sigma F_n = ma_n: \quad N_A + mg = m\frac{v_A^2}{\rho_A}$$

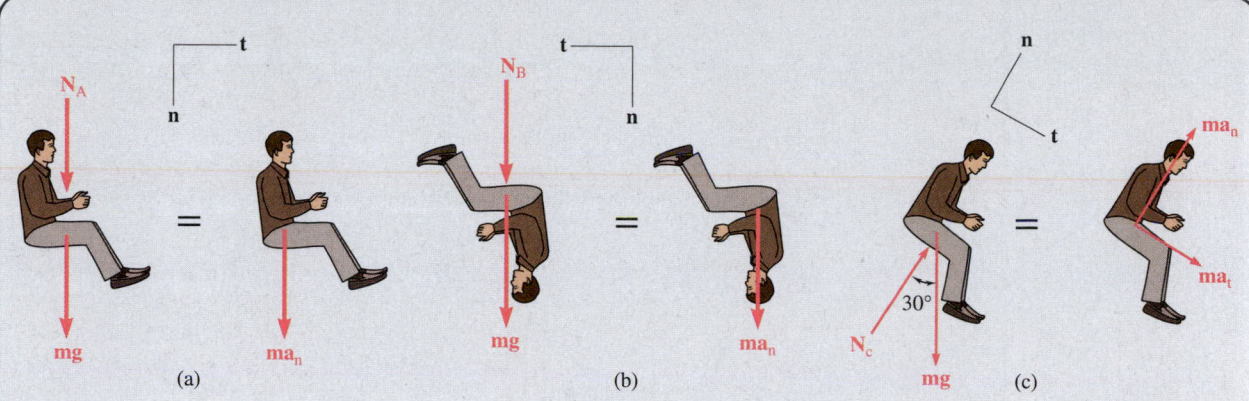

**CS Fig. 13.2**   Free-body and kinetic diagrams of a rider at points *A*, *B*, and *C*.

**CS Fig. 13.3**   Impulse–momentum diagram for the launch phase.

Setting $N_A$ equal to zero and solving for the velocity gives:

$$v_A = \sqrt{g\rho_A} = \sqrt{(9.81 \text{ m/s}^2)(15 \text{ m})} = 12.131 \text{ m/s}$$

Now determine the velocity $v_2$ at the end of the launch area using conservation of energy.

$$T_2 + V_2 = T_A + V_A: \quad \frac{1}{2}mv_2^2 + mgh_B = \frac{1}{2}mv_A^2 + mgh_A$$

Setting the datum at the launch height and substituting in values gives:

$$\frac{1}{2}(600 \text{ kg})v_2^2 = \frac{1}{2}(600 \text{ kg})(12.131 \text{ m/s}^2) + (600 \text{ kg})(9.81 \text{ m/s}^2)(200 \text{ m})$$

$$v_2 = 63.806 \text{ m/s}$$

You can use impulse and momentum and your diagram in CS Fig. 13.3 to calculate the required force to achieve this velocity:

$$m_{car}v_1 + \int_0^3 Fdt = m_{car}v_A: \quad 0 + F(3 \text{ s}) = 6000 \text{ kg}(63.806 \text{ m/s})$$

$$F \rightarrow = 127.61 \text{ kN}$$

To determine the maximum height for the inversion loop, use CS Fig. 13.2b and substitute in $N_B = 4mg$.

$$+\downarrow \Sigma F_n = ma_n: \quad N_B + mg = m\frac{v_B^2}{\rho_B} \rightarrow 5mg = m\frac{v_B^2}{\rho_B}$$

Solving for $v_B$ and substituting in values gives you

$$v_B = \sqrt{5g\rho_B} = \sqrt{5(9.81 \text{ m/s}^2)(25 \text{ m})} = 35.018 \text{ m/s}$$

*(continued)*

You can use work and energy to determine the height at the top of the inversion loop:

$$T_A + V_A = T_B + V_B: \quad \frac{1}{2}mv_A^2 + mgh_A = \frac{1}{2}mv_B^2 + mgh_B$$

Again, using the launch position as your datum and solving for $h_B$ gives you

$$h_B = \frac{1}{2g}(v_A^2 - v_B^2) + h_A = \frac{1}{2(9.81 \text{ m/s}^2)}[(12.131 \text{ m/s})^2 - (35.018 \text{ m/s})^2] + 200 \text{ m}$$

$$h_B = 145 \text{ m} \quad \blacktriangleleft$$

You can find the velocity at the bottom of the loop by again using work and energy:

$$T_B + V_B = T_C + V_C: \quad \frac{1}{2}mv_B^2 + mgh_B = \frac{1}{2}mv_C^2 + mgh_C$$

Recognizing that the masses cancel (which is good because it means velocity is independent of rider mass) and solving for $v_C$,

$$v_C = [(35.018 \text{ m/s})^2 + 2(9.81 \text{ m/s})^2(145 \text{ m}) - 2(9.81 \text{ m/s})^2(145 \text{ m} - 70 \text{ m})]^{1/2}$$

$$v_C = 50.987 \text{ m/s} \quad \blacktriangleleft$$

Finally, we can use CS Fig. 13.2 to determine the necessary radius of curvature at the bottom of the loop.

$$+\nearrow \Sigma F_n = ma_n: \quad N_C - mg\cos(30°) = m\frac{v_C^2}{\rho_C}$$

Solving for $\rho_C$, setting $N_C = 4mg$, and substituting in values gives you

$$\rho_C = \frac{v_C^2}{[4 - \cos(30°)]g} = \frac{(50.987 \text{ m/s})^2}{[4 - \cos(30°)](9.81 \text{ m/s}^2)} = 84.557 \text{ m}$$

$$\rho_C = 84.6 \text{ m} \quad \blacktriangleleft$$

**REFLECT and THINK:** Inversion loops on a roller coaster are rarely perfect circles, as shown in the current case study. Typically, they are in the shape of a clothoid loop that varies the radius of curvature to maintain a relatively constant g force on the rider. There are industry standards for the amount and duration of g forces that can be sustained during a coaster ride.

The calculations here assume no energy losses for factors such as friction, air resistance, vibration, and noise, so our results really place limits on the design of the coaster (e.g., you will need more than our calculated average thrust force, and the loop will need to be lower than $h_B$). How could you design a zero-g floater after you have gone through the loop?

# SOLVING PROBLEMS
# ON YOUR OWN

In this section, we integrated Newton's second law to derive the **principle of impulse and momentum** for a particle. Recalling that we defined the *linear momentum* of a particle as the product of its mass $m$ and its velocity $\mathbf{v}$ (Sec. 12.1B), we have

$$m\mathbf{v}_1 + \Sigma\mathbf{Imp}_{1\rightarrow2} = m\mathbf{v}_2 \tag{13.32}$$

This equation states that we can obtain the linear momentum $m\mathbf{v}_2$ of a particle at time $t_2$ by adding its linear momentum $m\mathbf{v}_1$ at time $t_1$ to the **impulses** of the forces exerted on the particle during the time interval $t_1$ to $t_2$. For computing purposes, we can express the momenta and impulses in terms of their rectangular components and replace Eq. (13.32) by the equivalent scalar equations. The units of momentum and impulse are N·s in the SI system of units and lb·s in U.S. customary units. To solve problems using this equation, you can follow these steps.

1. **Draw an impulse–momentum diagram** showing the particle, its momentum at $t_1$ and at $t_2$, and the impulses of the forces exerted on the particle during the time interval $t_1$ to $t_2$.

2. **Calculate the impulse of each force,** expressing it in terms of its rectangular components if more than one direction is involved. You may encounter the following cases:

   a. **The time interval is finite and the force is constant.**

   $$\mathbf{Imp}_{1\rightarrow2} = \mathbf{F}(t_2 - t_1)$$

   b. **The time interval is finite and the force is a function of $t$.**

   $$\mathbf{Imp}_{1\rightarrow2} = \int_{t_1}^{t_2}\mathbf{F}(t)dt$$

   c. **The time interval is very small and the force is very large.** The force is called an **impulsive force**, and its impulse over the time interval $t_2 - t_1 = \Delta t$ is

   $$\mathbf{Imp}_{1\rightarrow2} = \mathbf{F}_{\mathbf{avg}}\,\Delta t$$

Note that this impulse is assumed to be zero for a non-impulsive force such as the weight of a body, the force exerted by a spring, or any other force that is known to be small by comparison with the impulsive forces. However, we *cannot* assume unknown reactions are non-impulsive, and you should take their impulses into account.

3. **Substitute the values obtained for the impulses into Eq. (13.32)** or into the equivalent scalar equations. You will find that the forces and velocities in the problems of this section are contained in a plane. Therefore, you can write two scalar equations and solve these equations for two unknowns. These unknowns may be a *time* (Sample Prob. 13.13), a force (Sample Prob. 13.14), a *velocity* (Sample Prob. 13.15), an *average impulsive force* (Sample Prob. 13.16), or an *impulse* (Sample Prob. 13.17).

**4. When several particles are involved,** it is often necessary to draw a separate diagram for each particle showing the initial and final momentum of the particle, as well as the impulses of the forces exerted on the particle.

**a. It is usually convenient,** however, to first consider a system that includes all of the particles. This system leads to

$$\Sigma m\mathbf{v}_1 + \Sigma \mathbf{Imp}_{1\to 2} = \Sigma m\mathbf{v}_2 \tag{13.33}$$

where you need to consider the impulses of only the forces external to the system.

Therefore, the two equivalent scalar equations will not contain any of the impulses of the unknown internal forces.

**b. If the sum of the impulses of the external forces is zero,** Eq. (13.33) reduces to

$$\Sigma m\mathbf{v}_1 = \Sigma m\mathbf{v}_2 \tag{13.34}$$

or for two particles as

$$m_A\mathbf{v}_A + m_B\mathbf{v}_B = m_A\mathbf{v}_A' + m_B\mathbf{v}_B' \tag{13.34'}$$

which says that *the total linear momentum of the particles is conserved.* This occurs when the time interval is very short and the external forces are negligible compared to the impulsive forces. Keep in mind, however, that the total momentum may be conserved in one direction, but not in another (Sample Prob. 13.17).

# Problems

## CONCEPT QUESTIONS

**13.CQ4** A large insect impacts the front windshield of a sports car traveling down a road. Which of the following statements is true during the collision?
- **a.** The car exerts a greater force on the insect than the insect exerts on the car.
- **b.** The insect exerts a greater force on the car than the car exerts on the insect.
- **c.** The car exerts a force on the insect, but the insect does not exert a force on the car.
- **d.** The car exerts the same force on the insect as the insect exerts on the car.
- **e.** Neither exerts a force on the other; the insect gets smashed simply because it gets in the way of the car.

**13.CQ5** The expected damages associated with two types of perfectly plastic collisions are to be compared. In the first case, two identical cars traveling at the same speed impact each other head-on. In the second case, the car impacts a massive concrete wall. In which case would you expect the car to be more damaged?
- **a.** Case 1
- **b.** Case 2
- **c.** The same damage in each case

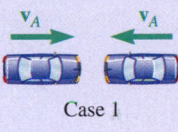

Case 1

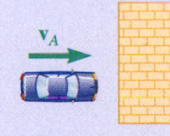

Case 2

**Fig. P13.CQ5**

## IMPULSE–MOMENTUM DIAGRAM PRACTICE PROBLEMS

**13.F1** The initial velocity of the block in position $A$ is 30 ft/s. The coefficient of kinetic friction between the block and the plane is $\mu_k = 0.30$. Draw the impulse–momentum diagram that can be used to determine the time it takes for the block to reach $B$ with zero velocity, if $\theta = 20°$.

**13.F2** A 4-lb collar which can slide on a frictionless vertical rod is acted upon by a force **P** which varies in magnitude as shown. Knowing that the collar is initially at rest, draw the impulse–momentum diagram that can be used to determine its velocity at $t = 3$ s.

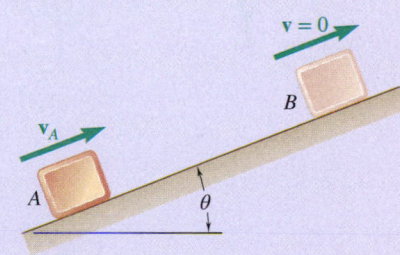

**Fig. P13.F1**

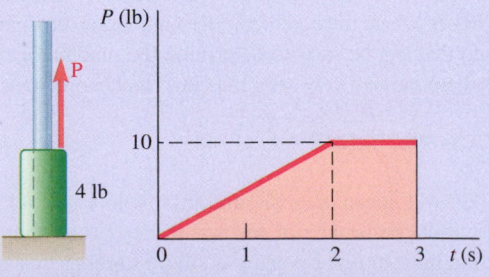

**Fig. P13.F2**

**13.F3** The 15-kg suitcase $A$ has been propped up against one end of a 40-kg luggage carrier $B$ and is prevented from sliding down by other luggage. When the luggage is unloaded and the last heavy trunk is removed from the carrier, the suitcase is free to slide down, causing the 40-kg carrier to move to the left with a velocity $v_B$ of magnitude 0.8 m/s. Neglecting friction, draw the impulse–momentum diagrams that can be used to determine (*a*) the velocity of $A$ as it rolls on the carrier, (*b*) the velocity of the carrier after the suitcase hits the right side of the carrier without bouncing back.

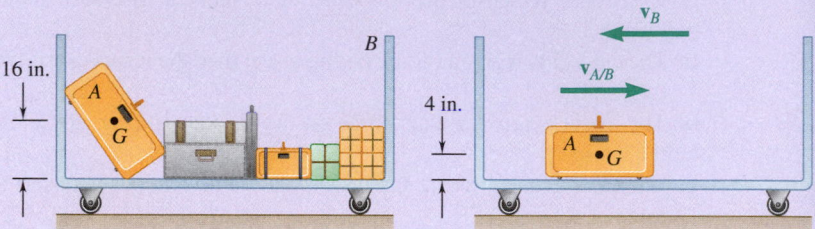

**Fig. P13.F3**

**13.F4** Car $A$ was traveling west at a speed of 15 m/s and car $B$ was traveling north at an unknown speed when they slammed into each other at an intersection. Upon investigation, it was found that after the crash the two cars got stuck and skidded off at an angle of 50° north of east. Knowing the masses of $A$ and $B$ are $m_A$ and $m_B$, respectively, draw the impulse–momentum diagram that can be used to determine the velocity of $B$ before impact.

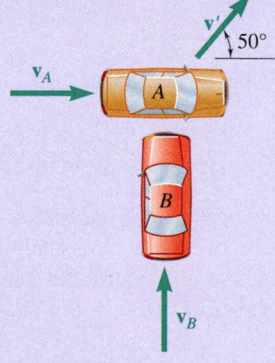

**Fig. P13.F4**

**13.F5** Two identical spheres $A$ and $B$, each of mass $m$, are attached to an inextensible inelastic cord of length $L$ and are resting at a distance $a$ from each other on a frictionless horizontal surface. Sphere $B$ is given a velocity $\mathbf{v}_0$ in a direction perpendicular to line $AB$ and moves it without friction until it reaches $B'$ where the cord becomes taut. Draw the impulse–momentum diagram that can be used to determine the magnitude of the velocity of each sphere immediately after the cord has become taut.

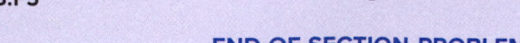

**Fig. P13.F5**

**END-OF-SECTION PROBLEMS**

**13.119** A 35 000-Mg ocean liner has an initial velocity of 4 km/h. Neglecting the frictional resistance of the water, determine the time required to bring the liner to rest by using a single tugboat that exerts a constant force of 150 kN.

874

**13.120** A 2500-lb automobile is moving at a speed of 60 mi/h when the brakes are fully applied, causing all four wheels to skid. Determine the time required to stop the automobile (a) on dry pavement ($\mu_k = 0.75$), (b) on an icy road ($\mu_k = 0.10$).

**13.121** A sailboat weighing 980 lb with its occupants is running downwind at 8 mi/h when its spinnaker is raised to increase its speed. Determine the net force provided by the spinnaker over the 10-s interval that it takes for the boat to reach a speed of 12 mi/h.

Fig. P13.121

**13.122** A truck is hauling a 300-kg log out of a ditch using a winch attached to the back of the truck. Knowing the winch applies a constant force of 2500 N and the coefficient of kinetic friction between the ground and the log is 0.45, determine the time for the log to reach a speed of 0.5 m/s.

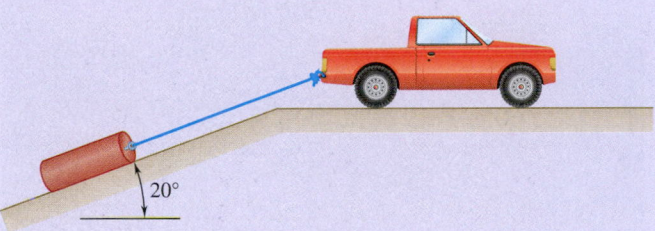

Fig. *P13.122*

**13.123** The coefficients of friction between the load and the flatbed trailer shown are $\mu_s = 0.40$ and $\mu_k = 0.35$. Knowing that the speed of the rig is 55 mi/h, determine the shortest time in which the rig can be brought to a stop if the load is not to shift.

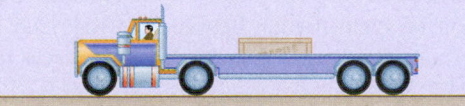

Fig. P13.123

**13.124** Steep safety ramps are built beside mountain highways to enable vehicles with defective brakes to stop. A 10-ton truck enters a 15° ramp at a high speed $v_0 = 108$ ft/s and travels for 6 s before its speed is reduced to 36 ft/s. Assuming constant deceleration, determine (a) the magnitude of the braking force, (b) the additional time required for the truck to stop. Neglect air resistance and rolling resistance.

Fig. P13.124

**13.125** Baggage on the floor of the baggage car of a high-speed train is not prevented from moving other than by friction. The train is traveling down a 5-percent grade when it decreases its speed at a constant rate from 120 mi/h to 60 mi/h in a time interval of 12 s. Determine the smallest allowable value of the coefficient of static friction between a trunk and the floor of the baggage car if the trunk is not to slide.

**13.126** The 18 000-kg F-35B uses thrust vectoring to allow it to take off vertically. In one maneuver, the pilot reaches the top of her static hover at 200 m. The combined thrust and lift force on the airplane applied at the end of the static hover can be expressed as $\mathbf{F} = (44t + 2500t^2)\mathbf{i} + (250t^2 + t + 176\,580)\mathbf{j}$, where $\mathbf{F}$ and $t$ are expressed in newtons and seconds, respectively. Determine (a) how long it will take the airplane to reach a cruising speed of 1000 km/hr (cruising speed is defined to be in the $x$ direction only), (b) the altitude of the plane at this time.

**Fig. P13.126**
Source: U.S. Air Force photo/Samuel King Jr.

**13.127** A truck is traveling down a road with a 4-percent grade at a speed of 60 mi/h when its brakes are applied to slow it down to 20 mi/h. An antiskid braking system limits the braking force to a value at which the wheels of the truck are just about to slide. Knowing that the coefficient of static friction between the road and the wheels is 0.60, determine the shortest time needed for the truck to slow down.

**13.128** In anticipation of a long 6° upgrade, a bus driver accelerates at a constant rate from 80 km/h to 100 km/h in 8 s while still on a level section of the highway. Knowing that the speed of the bus is 100 km/h as it begins to climb the grade at time $t = 0$ and that the driver does not change the setting of the throttle or shift gears, determine (a) the speed of the bus when $t = 10$ s, (b) the time when the speed is 60 km/h.

**13.129** The subway train shown is traveling at a speed of 30 mi/h when the brakes are fully applied on the wheels of cars $B$ and $C$, causing them to slide on the track. The brakes are not applied on the wheels of car $A$. Knowing that the coefficient of kinetic friction is 0.35 between the wheels and the track, determine (a) the time required to bring the train to a stop, (b) the force in each coupling.

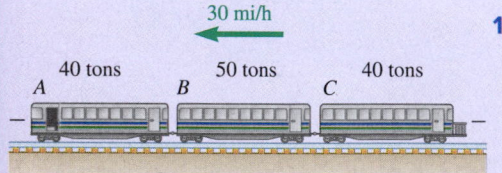

**Fig. P13.129 and P13.130**

**13.130** The subway train shown is traveling at a speed of 30 mi/h when the brakes are fully applied on the wheels of car $A$, causing it to slide on the track. The brakes are not applied on the wheels of cars $B$ or $C$. Knowing that the coefficient of kinetic friction is 0.35 between the wheels and the track, determine (a) the time required to bring the train to a stop, (b) the force in each coupling.

**13.131** A tractor-trailer rig with a 2000-kg tractor, a 4500-kg trailer, and a 3600-kg trailer is traveling on a level road at 90 km/h. The brakes on the rear trailer fail, and the antiskid system of the tractor and front trailer provide the largest possible force that will not cause the wheels to slide. Knowing that the coefficient of static friction is 0.75, determine (*a*) the shortest time for the rig to a come to a stop, (*b*) the force in the coupling between the two trailers during that time. Assume that the force exerted by the coupling on each of the two trailers is horizontal.

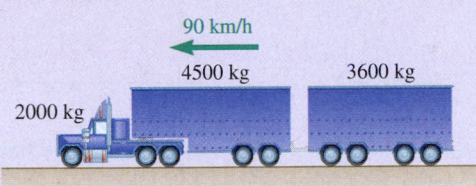

**Fig. P13.131**

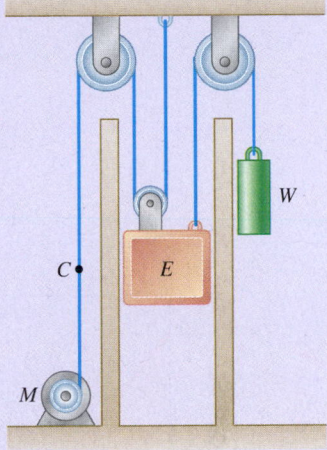

**Fig. P13.132**

**13.132** The motor applies a constant downward force $F = 550$ lb to the cable connected to the 4000-lb elevator $E$ shown in the figure. The counterweight has a weight of $W = 3000$ lb. Knowing that the elevator starts from rest, determine the time when the velocity of the elevator will be 3 m/s.

**13.133** An 8-kg cylinder $C$ rests on a 4-kg platform $A$ supported by a cord that passes over the pulleys $D$ and $E$ and is attached to a 4-kg block $B$. Knowing that the system is released from rest, determine (*a*) the velocity of block $B$ after 0.8 s, (*b*) the force exerted by the cylinder on the platform.

**13.134** An estimate of the expected load on over-the-shoulder seat belts is to be made before designing prototype belts that will be evaluated in automobile crash tests. Assuming that an automobile traveling at 45 mi/h is brought to a stop in 110 ms, determine (*a*) the average impulsive force exerted by a 200-lb man on the belt, (*b*) the maximum force $F_m$ exerted on the belt if the force–time diagram has the shape shown.

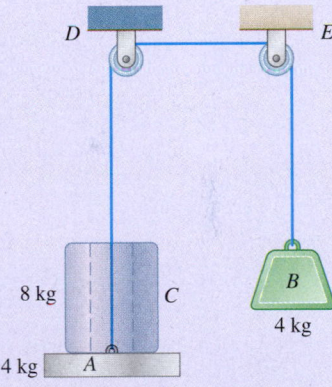

**Fig. P13.133**

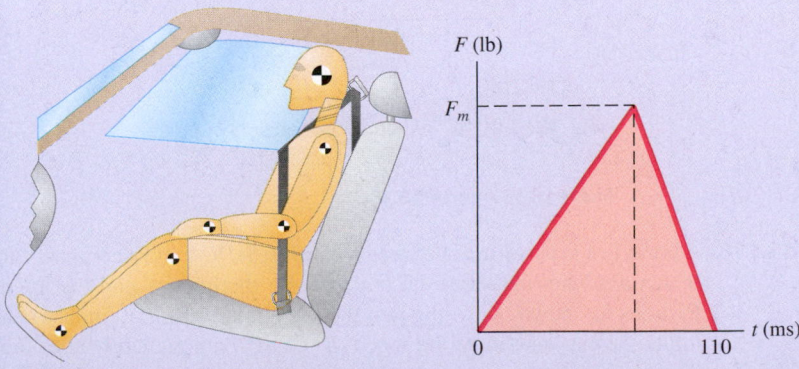

**Fig. P13.134**

**13.135** A 60-g model rocket is fired vertically. The engine applies a thrust **P** which varies in magnitude as shown. Neglecting air resistance and the change in mass of the rocket, determine (*a*) the maximum speed of the rocket as it goes up, (*b*) the time for the rocket to reach its maximum elevation.

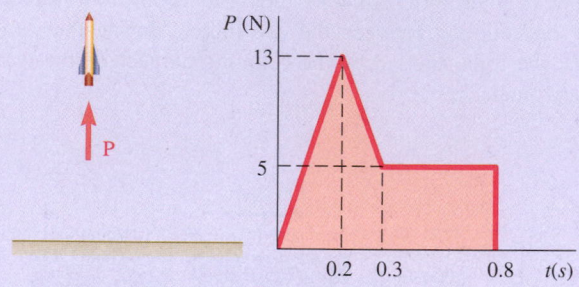

**Fig. *P13.135***

**13.136** A 12-lb block, which can slide on a frictionless inclined surface, is acted upon by a force **P** that varies in magnitude as shown. Knowing that the block is initially at rest, determine (*a*) the velocity of the block at $t = 5$ s, (*b*) the time at which the velocity of the block is zero.

**Fig. P13.136**

**13.137** A crash test is performed between an SUV *A* and a 2500-lb compact car *B*. The compact car is stationary before the impact and has its brakes applied. A transducer measures the force during the impact, and the force **P** varies as shown. Knowing that the coefficients of friction between the tires and road are $\mu_s = 0.9$ and $\mu_k = 0.7$, determine (*a*) the time at which the compact car will start moving, (*b*) the maximum speed of the car, (*c*) the time at which the car will come to a stop.

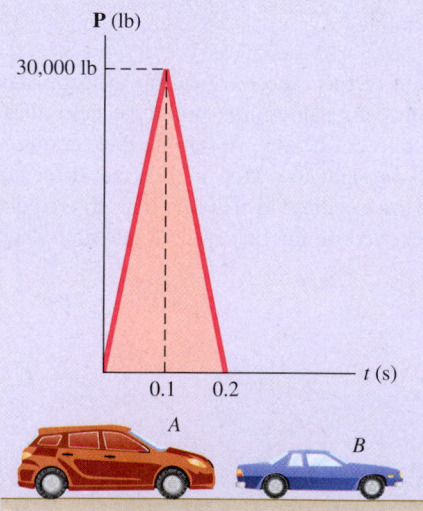

**Fig. *P13.137* and P13.138**

**13.138** A crash test is performed between a 4500-lb SUV *A* and a compact car *B*. A transducer measures the force during the impact, and the force **P** varies as shown. Knowing that the SUV is traveling 30 mph when it hits the car, determine the speed of the SUV immediately after the impact.

**13.139** A baseball player catching a ball can soften the impact by pulling his hand back. Assuming that a 5-oz ball reaches his glove at 90 mi/h and that the player pulls his hand back during the impact at an average speed of 30 ft/s over a distance of 6 in., bringing the ball to a stop, determine the average impulsive force exerted on the player's hand.

**13.140** A 1.62-oz golf ball is hit with a golf club and leaves it with a velocity of 100 mi/h. We assume that for $0 \leq t \leq t_0$, where $t_0$ is the duration of the impact, the magnitude $F$ of the force exerted on the ball can be expressed as $F = F_m \sin (\pi t/t_0)$. Knowing that $t_0 = 0.5$ ms, determine the maximum value $F_m$ of the force exerted on the ball.

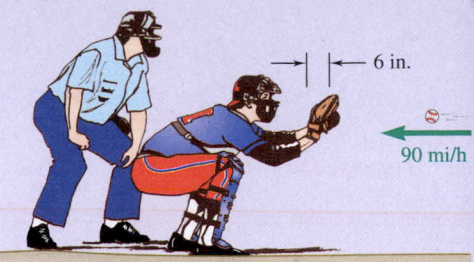

**Fig. P13.139**

**13.141** The triple jump is a track-and-field event in which an athlete gets a running start and tries to leap as far as he can with a hop, step, and jump. Shown in the figure is the initial hop of the athlete. Assuming that he approaches the takeoff line from the left with a horizontal velocity of 10 m/s, remains in contact with the ground for 0.18 s, and takes off at a 50° angle with a velocity of 12 m/s, determine the vertical component of the average impulsive force exerted by the ground on his foot. Give your answer in terms of the weight $W$ of the athlete.

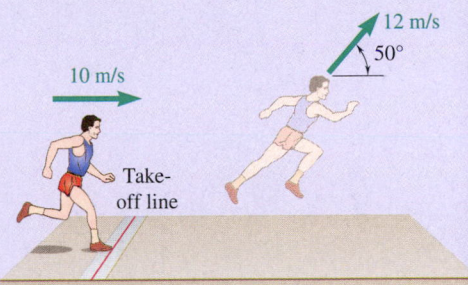

**13.142** The last segment of the triple jump track-and-field event is the jump, in which the athlete makes a final leap, landing in a sand-filled pit. Assuming that the velocity of an 80-kg athlete just before landing is 9 m/s at an angle of 35° with the horizontal and that the athlete comes to a complete stop in 0.22 s after landing, determine the horizontal component of the average impulsive force exerted on his feet during landing.

**Fig. P13.141**

**13.143** The design for a new cementless hip implant is to be studied using an instrumented implant and a fixed simulated femur. Assuming the punch applies an average force of 2 kN over a time of 2 ms to the 200-g implant, determine (a) the velocity of the implant immediately after impact, (b) the average resistance of the implant to penetration if the implant moves 1 mm before coming to rest.

**Fig. P13.142**

Landing pit

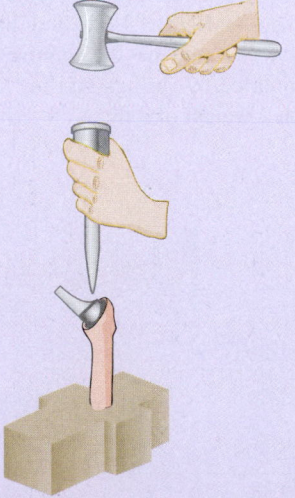

**Fig. P13.143**

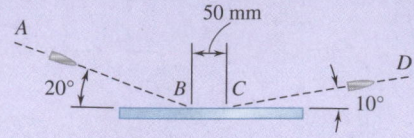

Fig. *P13.144*

**13.144** A 28-g steel-jacketed bullet is fired with a velocity of 650 m/s toward a steel plate and ricochets along path *CD* with a velocity of 500 m/s. Knowing that the bullet leaves a 50-mm scratch on the surface of the plate and assuming that it has an average speed of 600 m/s while in contact with the plate, determine the magnitude and direction of the impulsive force exerted by the plate on the bullet.

**13.145** A 120-ton tugboat is moving at 6 ft/s with a slack towing cable attached to a 100-ton barge that is at rest. The cable is being unwound from a drum on the tugboat at a constant rate of 5.4 ft/s and that rate is maintained after the cable becomes taut. Neglecting the resistance of the water, determine (*a*) the velocity of the tugboat after the cable becomes taut, (*b*) the impulse exerted on the barge as the cable becomes taut.

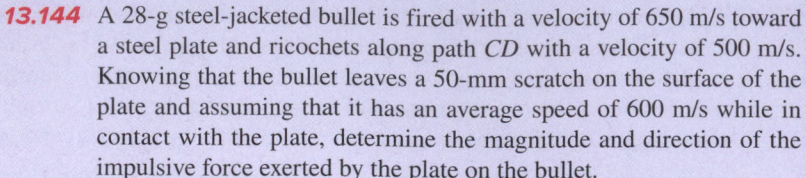

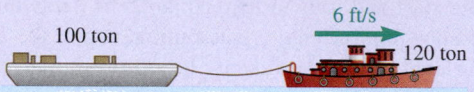

Fig. P13.145

**13.146** At an intersection, car *B* was traveling south and car *A* was traveling 30° north of east when they slammed into each other. Upon investigation, it was found that after the crash the two cars got stuck and skidded off at an angle of 10° north of east. Each driver claimed that he was going at the speed limit of 50 km/h and that he tried to slow down but couldn't avoid the crash because the other driver was going a lot faster. Knowing that the masses of cars *A* and *B* were 1500 kg and 1200 kg, respectively, determine (*a*) which car was going faster, (*b*) the speed of the faster of the two cars if the slower car was traveling at the speed limit.

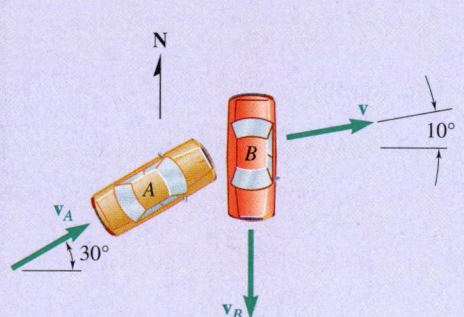

Fig. P13.146

**13.147** The 650-kg hammer of a drop-hammer pile driver falls from a height of 1.2 m onto the top of a 140-kg pile, driving it 110 mm into the ground. Assuming they stick together ($e = 0$), determine the average resistance of the ground to penetration.

**13.148** A small rivet connecting two pieces of sheet metal is being clinched by hammering. Determine the impulse exerted on the rivet and the energy absorbed by the rivet under each blow, knowing that the head of the hammer has a weight of 1.5 lb and that it strikes the rivet with a velocity of 20 ft/s. Assume that the hammer does not rebound and that the anvil is supported by springs and (*a*) has an infinite mass (rigid support), (*b*) has a weight of 9 lb.

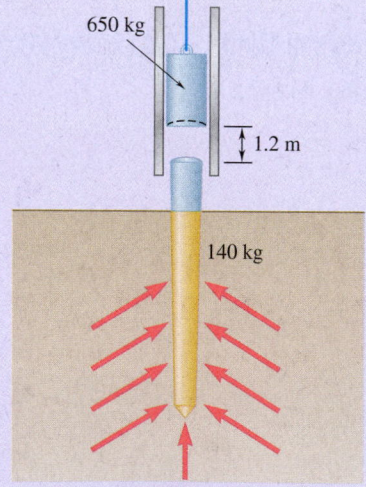

Fig. P13.147

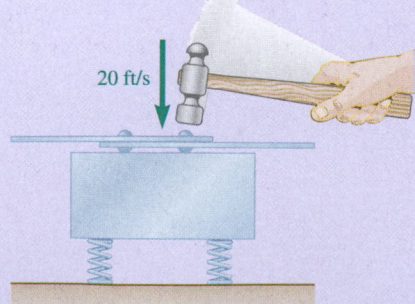

Fig. P13.148

**13.149** Bullet $B$ weighs 0.5 oz and blocks $A$ and $C$ both weigh 3 lb. The coefficient of friction between the blocks and the plane is $\mu_k = 0.25$. Initially, the bullet is moving at $v_0$ and blocks $A$ and $C$ are at rest (Fig. 1). After the bullet passes through $A$, it becomes embedded in block $C$ and all three objects come to stop in the positions shown (Fig. 2). Determine the initial speed of the bullet $v_0$.

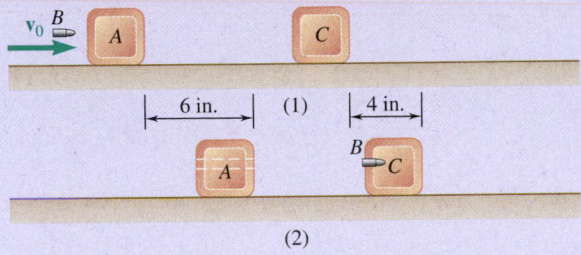

Fig. P13.149

**13.150** A 180-lb man and a 120-lb woman stand at opposite ends of a 300-lb boat, ready to dive, each with a 16-ft/s velocity relative to the boat. Determine the velocity of the boat after they have both dived, if (*a*) the woman dives first, (*b*) the man dives first.

Fig. P13.150

**13.151** A 75-g ball is projected from a height of 1.6 m with a horizontal velocity of 2 m/s and bounces from a 400-g smooth plate supported by springs. Knowing that the height of the rebound is 0.6 m, determine (*a*) the velocity of the plate immediately after the impact, (*b*) the energy lost due to the impact.

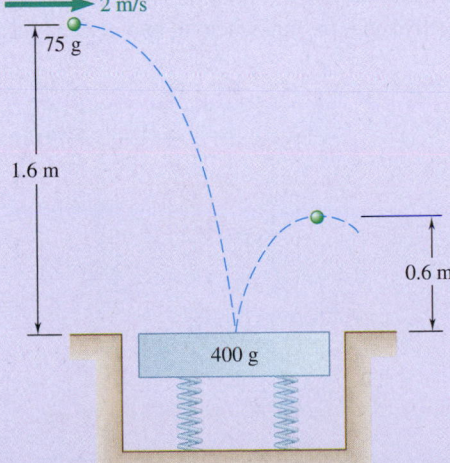

Fig. P13.151

**13.152** A ballistic pendulum is used to measure the speed of high-speed projectiles. A 6-g bullet $A$ is fired into a 1-kg wood block $B$ suspended by a cord with a length of $l = 2.2$ m. The block then swings through a maximum angle of $\theta = 60°$. Determine (a) the initial speed of the bullet $v_0$, (b) the impulse imparted by the bullet on the block, (c) the force on the cord immediately after the impact.

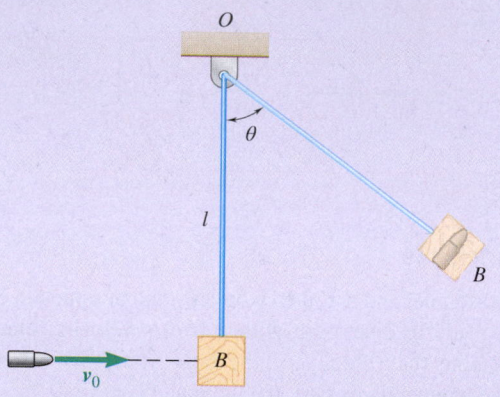

Fig. P13.152

**13.153** An old 3000-lb cannon fires an 18-lb shell with an initial velocity of 1500 ft/s at an angle of 30° (you may assume this is the absolute velocity of the shell). The cannon rests on a horizontal surface and is free to move horizontally. Assuming that the barrel of the cannon is rigidly attached to the frame (no recoil mechanism) and that the shell leaves the barrel 5 ms after firing, determine (a) the recoil velocity of the cannon, (b) the resultant of the vertical impulsive forces exerted by the ground on the cannon.

Fig. **P13.153**

**13.154** In order to test the resistance of a chain to impact, the chain is suspended from a 240-lb rigid beam supported by two columns. A rod attached to the last link is then hit by a 60-lb block dropped from a 5-ft height. Determine the initial impulse exerted on the chain and the energy absorbed by the chain, assuming that the block does not rebound from the rod and that the columns supporting the beam are (a) perfectly rigid, (b) equivalent to two perfectly elastic springs.

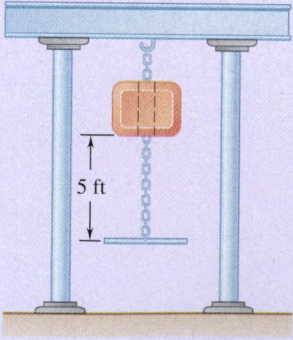

Fig. **P13.154**

# 13.4   IMPACTS

A collision between two bodies that occurs in a very small interval of time, and during which the two bodies exert relatively large forces on each other, is called an **impact**. The common normal to the surfaces in contact during the impact is called the **line of impact**. If the mass centers of the two colliding bodies are located on this line, the impact is called a **central impact**. Otherwise, the impact is said to be **eccentric**. Our present study is limited to the central impact of two particles. In Chap. 17, we consider the analysis of the eccentric impact of two rigid bodies.

 If the velocities of the two particles are directed along the line of impact, the impact is said to be a **direct impact** (Fig. 13.20a). If either or both particles move along a line other than the line of impact, the impact is said to be an **oblique impact** (Fig. 13.20b). ▶

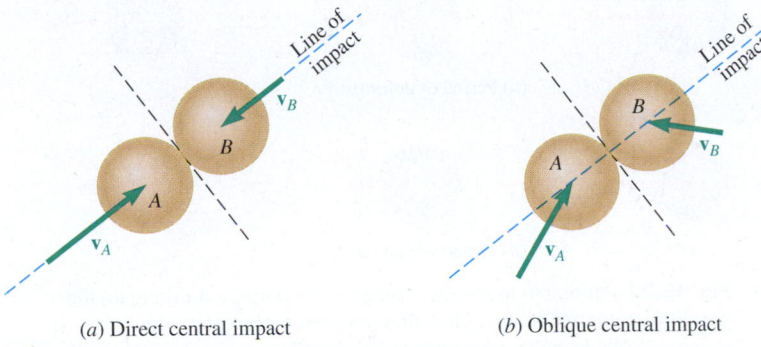

 (a) Direct central impact            (b) Oblique central impact

**Fig. 13.20**   Central impacts can be (a) direct (or "head-on") or (b) oblique.

## 13.4A   Direct Central Impact

Consider two particles A and B with mass $m_A$ and $m_B$ that are moving in the same straight line and to the right with known velocities $\mathbf{v}_A$ and $\mathbf{v}_B$ (Fig. 13.21a). If $\mathbf{v}_A$ is larger than $\mathbf{v}_B$, particle A eventually strikes particle B. Under the impact, the two particles **deform**, and at the end of the period of deformation, they have the same velocity $\mathbf{u}$ (Fig. 13.21b). A period of **restitution** then takes place. At the end of this period, depending upon the magnitude of the impact forces and upon the materials involved, the two particles either have regained their original shape or will stay permanently deformed. Our purpose here is to determine the velocities $\mathbf{v}'_A$ and $\mathbf{v}'_B$ of the particles at the end of the period of restitution (Fig. 13.21c).

 Considering first the two particles as a single system, we note that there is no impulsive, external force. Thus, the total linear momentum of the two particles is conserved, and we have

$$m_A\mathbf{v}_A + m_B\mathbf{v}_B = m_A\mathbf{v}'_A + m_B\mathbf{v}'_B \qquad (13.34')$$

Because all of the velocities considered are directed along the same axis, we can replace this equation by the following relation involving only scalar components, as

$$m_A v_A + m_B v_B = m_A v'_A + m_B v'_B \qquad (13.37)$$

A positive value for any of the scalar quantities $v_A$, $v_B$, $v'_A$, or $v'_B$ means that the corresponding vector is directed to the right; a negative value indicates that the corresponding vector is directed to the left.

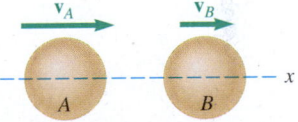

(a) Before impact

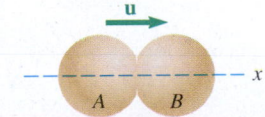

(b) At maximum deformation

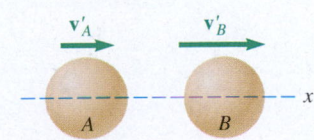

(c) After impact

**Fig. 13.21**   Every impact has three stages: (a) before the impact, (b) a maximum deformation when the particles have the same velocity, and (c) after the impact.

To obtain the velocities $\mathbf{v}_A'$ and $\mathbf{v}_B'$, it is necessary to establish a second relation between the scalars $\mathbf{v}_A'$ and $\mathbf{v}_B'$. For this purpose, let us now consider the motion of particle $A$ during the period of deformation and apply the principle of impulse and momentum. Because the only impulsive force acting on $A$ during this period is the force $\mathbf{P}$ exerted by $B$ (Fig. 13.22$a$), we have, again using scalar components,

$$m_A v_A - \int P\,dt = m_A u \tag{13.38}$$

where the integral extends over the period of deformation. Considering now the motion of $A$ during the period of restitution and denoting the force exerted by $B$ on $A$ during this period by $\mathbf{R}$ (Fig. 13.22$b$), we have

$$m_A u - \int R\,dt = m_A v_A' \tag{13.39}$$

where the integral extends over the period of restitution.

($a$) Period of deformation

($b$) Period of restitution

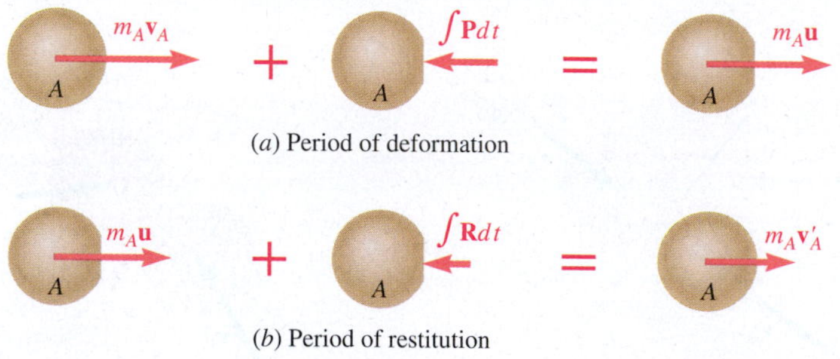

**Fig. 13.22** Impulse–momentum diagram for particle $A$ during ($a$) the period of deformation, and ($b$) during the period of restoration.

In general, the force $\mathbf{R}$ exerted on $A$ during the period of restitution differs from the force $\mathbf{P}$ exerted during the period of deformation, and the magnitude $\int R\,dt$ of its impulse is smaller than the magnitude $\int P\,dt$ of the impulse of $\mathbf{P}$. The ratio of the magnitudes of the impulses, corresponding, respectively, to the period of restitution and to the period of deformation, is called the **coefficient of restitution** and is denoted by $e$. We have

$$e = \frac{\int R\,dt}{\int P\,dt} \tag{13.40}$$

The value of the coefficient $e$ is always between 0 and 1. It depends to a large extent on the two materials involved, but it also varies considerably with the impact velocity and the shape and size of the two colliding bodies.

Solving Eqs. (13.38) and (13.39) for the two impulses and substituting into Eq. (13.40), we obtain

$$e = \frac{u - v_A'}{v_A - u} \tag{13.41}$$

A similar analysis of particle $B$ leads to the relation

$$e = \frac{v_B' - u}{u - v_B} \tag{13.42}$$

Because the quotients in Eqs. (13.41) and (13.42) are equal, they are also equal to the quotient obtained by adding, respectively, their numerators and their denominators. We therefore have

$$e = \frac{(u - v_A') + (v_B' - u)}{(v_A - u) + (u - v_B)} = \frac{v_B' - v_A'}{v_A - v_B}$$

and

**Coefficient of restitution**

$$v'_B - v'_A = e(v_A - v_B) \qquad (13.43)$$

Because $v'_B - v_A'$ represents the relative velocity of the two particles after impact and $v_A - v_B$ represents their relative velocity before impact, Eq. (13.43) says:

> We can obtain the relative velocity of the two particles after impact by multiplying their relative velocity before impact by the coefficient of restitution.

This property is used to determine experimentally the value of the coefficient of restitution of two given materials.

**Photo 13.5** The height the tennis ball bounces decreases after each impact because it has a coefficient of restitution less than one and energy is lost with each bounce.
©Terry Oakley/Alamy

We can now obtain the velocities of the two particles after impact by solving Eqs. (13.37) and (13.43) simultaneously for $v'_A$ and $v'_B$. Recall that the derivations of Eqs. (13.37) and (13.43) were based on the assumption that particle $B$ is located to the right of $A$ and that both particles are initially moving to the right. If particle $B$ is initially moving to the left, the scalar $v_B$ should be considered negative. The same sign convention holds for the velocities after impact: A positive sign for $v'_A$ indicates that particle $A$ moves to the right after impact, and a negative sign indicates that it moves to the left.

Two particular cases of impact are of special interest.

1. $e = 0$, **Perfectly Plastic Impact**. When $e = 0$, Eq. (13.43) yields $v'_B = v'_A$. There is no period of restitution, and both particles stay together after impact. Substituting $v'_B = v'_A = v'$ into Eq. (13.37), which expresses that the total momentum of the particles is conserved, we have

$$m_A v_A + m_B v_B = (m_A + m_B)v' \qquad (13.44)$$

We can solve this equation for the common velocity $v'$ of the two particles after impact.

2. $e = 1$, **Perfectly Elastic Impact**. When $e = 1$, Eq. (13.43) reduces to

$$v'_B - v'_A = v_A - v_B \qquad (13.45)$$

This equation says that the relative velocities before and after impact are equal. This means that the impulses received by each particle during the period of deformation and during the period of restitution are equal. We can obtain the velocities $v'_A$ and $v'_B$ by solving Eqs. (13.37) and (13.45) simultaneously.

It is worth noting that **in the idealized case of a perfectly elastic impact, the total energy of the two particles**, as well as their total momentum, **is conserved**. We can write Eqs. (13.37) and (13.45) as

$$m_A(v_A - v'_A) = m_B(v'_B - v_B) \qquad (13.37')$$
$$v_A + v'_A = v_B + v'_B \qquad (13.45')$$

Multiplying Eqs. (13.37′) and (13.45′) member by member, we have

$$m_A(v_A - v'_A)(v_A + v'_A) = m_B(v'_B - v_B)(v'_B + v_B)$$
$$m_A v_A^2 - m_A(v'_A)^2 = m_B(v'_B)^2 - m_B v_B^2$$

Rearranging the terms in this equation and multiplying by 1/2, we obtain

$$\frac{1}{2}m_A v_A^2 + \frac{1}{2}m_B v_B^2 = \frac{1}{2}m_A(v'_A)^2 + \frac{1}{2}m_B(v'_B)^2 \qquad (13.46)$$

This equation states that the kinetic energy of the particles is conserved. Note, however, that in the general case of impact, that is, when $e$ is not equal to 1,

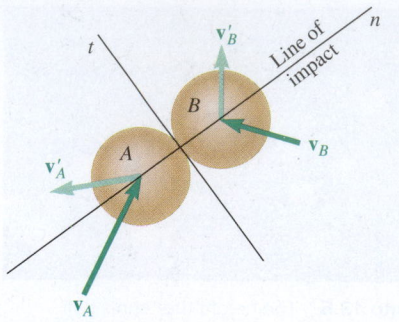

**Fig. 13.23** In an oblique central impact, the velocities of the colliding particles are not directed along the line of impact.

**Photo 13.6** When pool balls strike each other there is a transfer of momentum.
©Richard T. Nowitz/Corbis.

**the total mechanical energy of the particles is not conserved.** This can be shown in any given case by comparing the kinetic energies before and after impact. The lost kinetic energy may be transformed into other forms of energy, such as heat, sound, generation of elastic waves within the two colliding bodies, or permanent deformation of the bodies. ▶

## 13.4B Oblique Central Impact

Let us now consider the case when the velocities of the two colliding particles are *not* directed along the line of impact (Fig. 13.23). As mentioned earlier, the impact is said to be **oblique**. Because the velocities $\mathbf{v}_A'$ and $\mathbf{v}_B'$ of the particles after impact are unknown in direction as well as in magnitude, their determination requires the use of four independent equations.

We choose as coordinate axes the $n$ axis along the line of impact (i.e., along the common normal to the surfaces in contact) and the $t$ axis along their common tangent. In very special cases where we can assume that the particles are perfectly smooth and frictionless, we observe that the only impulses exerted on the particles during the impact are due to internal forces directed along the line of impact; that is, along the $n$ axis (Fig. 13.24). This leads to the following results.

1. The component along the $t$ axis of the momentum of each particle, considered separately, is conserved because no impulses act in the $t$ direction; hence the $t$ component of the velocity of each particle remains unchanged. We have

$$(v_A)_t = (v_A')_t \qquad (v_B)_t = (v_B')_t \qquad \textbf{(13.47)}$$

2. The component along the $n$ axis of the total momentum of the two particles is conserved because the two impulses are equal and opposite to one another. We have

$$m_A(v_A)_n + m_B(v_B)_n = m_A(v_A')_n + m_B(v_B')_n \qquad \textbf{(13.48)}$$

3. We obtain the component along the $n$ axis of the relative velocity of the two particles after impact by multiplying the $n$ component of their relative velocity before impact by the coefficient of restitution. Indeed, a derivation similar to that given in Sec. 13.4A for direct central impact yields

$$(v_B')_n - (v_A')_n = e[(v_A)_n - (v_B)_n] \qquad \textbf{(13.49)}$$

We have thus obtained four independent equations that can be solved for the components of the velocities of $A$ and $B$ after impact. This method of solution is illustrated in Sample Prob. 13.20.

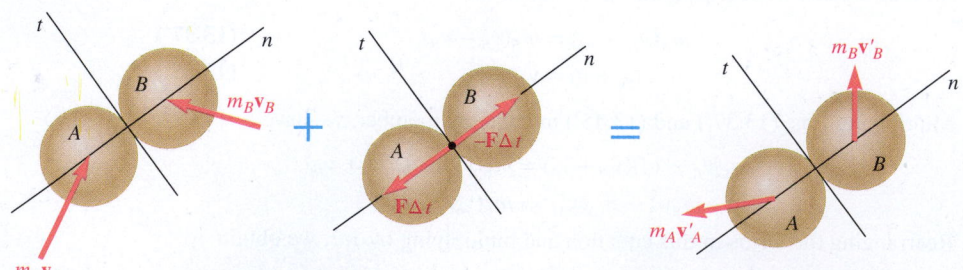

**Fig. 13.24** Impulse–momentum diagram for an oblique impact. By including the internal impulses as equal and opposite, you also have the impulse–momentum diagram for each individual object (just ignore the other object).

Our analysis of the oblique central impact of two particles has been based so far on the assumption that both particles move freely before and after the impact. Let us now examine the case when one or both of the colliding particles is constrained in its motion. Consider, for instance, the collision between block $A$, which is constrained to move on a horizontal surface, and ball $B$, which is free to move in the plane of the figure (Fig. 13.25). Assuming no friction between the block and the ball or between the block and the horizontal surface, we note that the impulses exerted on the system consist of the impulses of the internal forces $\mathbf{F}$ and $-\mathbf{F}$ directed along the line of impact, that is, along the $n$ axis, and of the impulse of the external force $\mathbf{F}_{ext}$ exerted by the horizontal surface on block $A$ and directed along the vertical, as shown in the impulse–momentum diagram (Fig. 13.26).

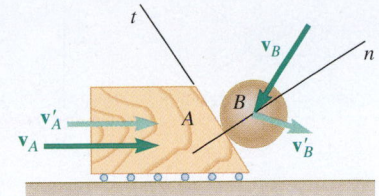

**Fig. 13.25** An impact between a block moving on a horizontal surface and a ball moving in the vertical plane is called a "constrained impact."

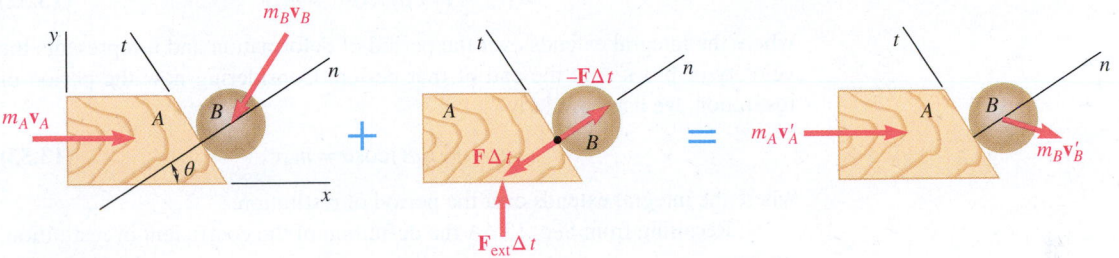

**Fig. 13.26** Impulse–momentum diagram for a constrained impact between block $A$ and ball $B$.

The velocities of block $A$ and ball $B$ immediately after the impact are represented by three unknowns: the magnitude of the velocity $v'_A$ of block $A$, which is known to be horizontal, and the magnitude and direction of the velocity $v'_B$ of ball $B$. We must therefore write three equations. We do this by using the impulse–momentum diagram and observing the following behavior.

1.  The component along the $t$ axis of the momentum of ball $B$ is conserved because no impulses act on the ball in the $t$ direction; hence, the $t$ component of the velocity of ball $B$ remains unchanged. We have

$$(v_B)_t = (v'_B)_t \tag{13.50}$$

2.  The component along the horizontal $x$ axis of the total momentum of block $A$ and ball $B$ is conserved because no external impulses act in the $x$-direction. We write this as

$$m_A v_A + m_B (v_B)_x = m_A v'_A + m_B (v'_B)_x \tag{13.51}$$

3.  We obtain the component along the $n$ axis of the relative velocity of block $A$ and ball $B$ after impact by multiplying the $n$ component of their relative velocity before impact by the coefficient of restitution. We again have

$$(v'_B)_n - (v'_A)_n = e[(v_A)_n - (v_B)_n] \tag{13.49}$$

Note, however, that in the case considered here, we cannot establish the validity of Eq. (13.49) through a mere extension of the derivation given in Sec. 13.4A for the direct central impact of two particles moving in a straight line. Indeed, these particles were not subjected to any external impulse, whereas block $A$ in the present analysis is subjected to the impulse exerted by the horizontal surface. To prove that Eq. (13.49) is still valid, we first apply the

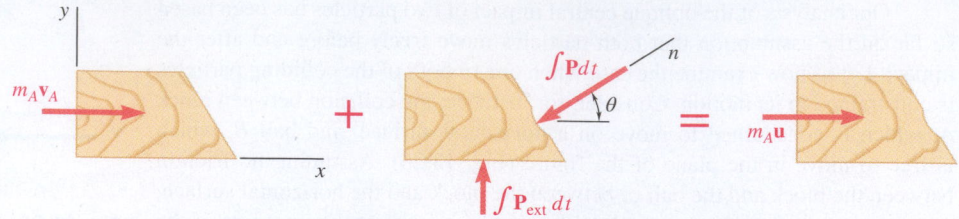

**Fig. 13.27**    Impulse–momentum diagram for block $A$ during the period of deformation.

principle of impulse and momentum to block $A$ over the period of deformation (Fig. 13.27). Considering only the horizontal components, we have

$$m_A v_A - \left( \int P dt \right) \cos\theta = m_A u \qquad (13.52)$$

where the integral extends over the period of deformation and $\mathbf{u}$ represents the velocity of block $A$ at the end of that period. Considering now the period of restitution, we have similarly

$$m_A u - \left( \int R dt \right) \cos\theta = m_A v_A' \qquad (13.53)$$

where the integral extends over the period of restitution.

Recalling from Sec. 13.4A the definition of the coefficient of restitution, we have

$$e = \frac{\int R dt}{\int P dt} \qquad (13.40)$$

Solving Eqs. (13.52) and (13.53) for the integrals $\int P dt$ and $\int R dt$ and substituting into Eq. (13.40), we have, after reductions,

$$e = \frac{u - v_A'}{v_A - u}$$

Then, multiplying all velocities by $\cos\theta$ to obtain their projections on the line of impact gives

$$e = \frac{u_n - (v_A')_n}{(v_A)_n - u_n} \qquad (13.54)$$

Note that Eq. (13.54) is identical to Eq. (13.41) except for the subscripts $n$ that we use here to indicate that we are considering velocity components along the line of impact. Because the motion of ball $B$ is unconstrained, we can complete the proof of Eq. (13.49) in the same manner as the derivation of Eq. (13.43). Thus, we conclude that the relation in Eq. (13.49) between the components along the line of impact of the relative velocities of two colliding particles remains valid when one of the particles is constrained in its motion. The validity of this relation is easily extended to the case when both particles are constrained in their motion.

## 13.4C    Problems Involving Multiple Principles

You now have at your disposal three different methods for the solution of kinetics problems.

- The direct application of Newton's second law, $\Sigma \mathbf{F} = m\mathbf{a}$.
- The method of work and energy, $T_1 + V_{g_1} + V_{e_1} + U_{1 \to 2}^{NC} = T_2 + V_{g_2} + V_{e_2}$, where $U_{1 \to 2}^{NC}$ is the work of external non-conservative forces such as friction.
- The method of impulse and momentum, $m\mathbf{v}_1 + \mathbf{Imp}_{1 \to 2} = m\mathbf{v}_2$.

To derive maximum benefit from these three methods, you should be able to choose the method best suited for the solution of a given problem. You also should be prepared to solve problems that require you to use multiple principles.

You have already seen that the method of work and energy is in many cases more expeditious than the direct application of Newton's second law. As indicated in Sec. 13.1C, however, the method of work and energy has limitations, and it must sometimes be supplemented by the use of $\Sigma F = ma$. This is the case, for example, when you wish to determine an acceleration or a normal force.

For the solution of problems involving no impulsive forces, usually the equation $\Sigma F = ma$ yields a solution just as fast as the method of impulse and momentum, and the method of work and energy (if it applies) is more rapid and more convenient. However, in problems involving impact, the method of impulse and momentum is the only practicable method. A solution based on the direct application of $\Sigma F = ma$ would be unwieldy, and the method of work and energy cannot be used, because impact (unless perfectly elastic) involves a loss of mechanical energy.

Many problems involve only conservative forces except for a short impact phase during which impulsive forces act. The solution of such problems can be divided into several parts. The part corresponding to the impact phase calls for the use of the method of impulse and momentum and of the relation between relative velocities. The other parts usually can be solved by using the method of work and energy. If the problem involves the determination of a normal force, however, the use of $\Sigma F = ma$ is necessary.

Consider, for example, a pendulum $A$, with a mass $m_A$ and a length $l$, that is released with no velocity from a position $A_1$ (Fig. 13.28a). The pendulum swings freely in a vertical plane and hits a second pendulum $B$, with a mass $m_B$ and the same length $l$, that is initially at rest. After the impact (with coefficient of restitution $e$), pendulum $B$ swings through an angle $\theta$ that we wish to determine.

The solution of the problem can be divided into three parts:

1. **Pendulum $A$ Swings from $A_1$ to $A_2$.** Use the principle of conservation of energy to determine the velocity $(v_A)_2$ of the pendulum at $A_2$ (Fig. 13.28b).

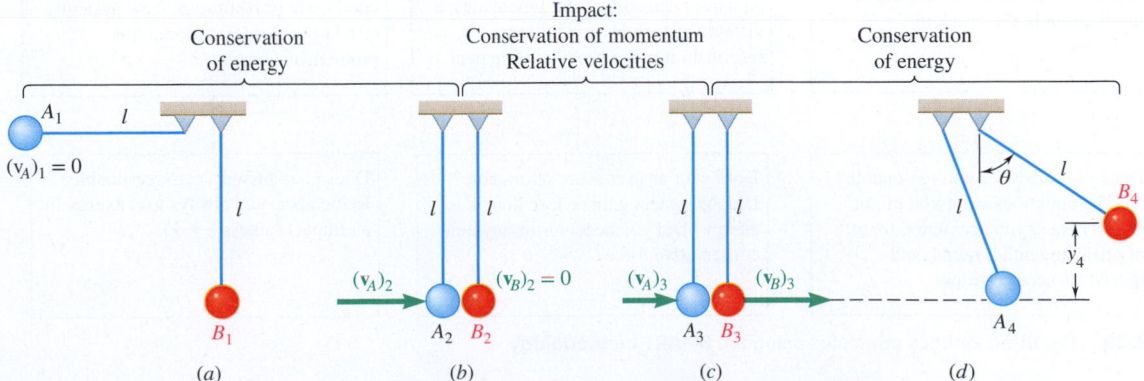

**Fig. 13.28** Analyzing an impact between two pendulum bobs by conservation of energy and conservation of momentum.

2. **Pendulum *A* Hits Pendulum *B*.** Use the fact that the total momentum of the two pendulums is conserved, and use the relation between their relative velocities—that is, the coefficient of restitution—to determine the velocities $(\mathbf{v}_A)_3$ and $(\mathbf{v}_B)_3$ of the two pendulums after impact (Fig. 13.28*c*).

3. **Pendulum *B* Swings from $B_3$ to $B_4$.** Apply the principle of conservation of energy to pendulum *B* to determine the maximum elevation $y_4$ reached by that pendulum (Fig. 13.28*d*). You can then determine the angle $\theta$ by trigonometry.

Note that if you need to determine the tensions in the cords holding the pendulums, the method of solution just described should be supplemented by the use of $\Sigma\mathbf{F} = m\mathbf{a}$. A summary of all the kinetics principles we have discussed so far and some clues as to when to apply them are shown in Fig. 13.29.

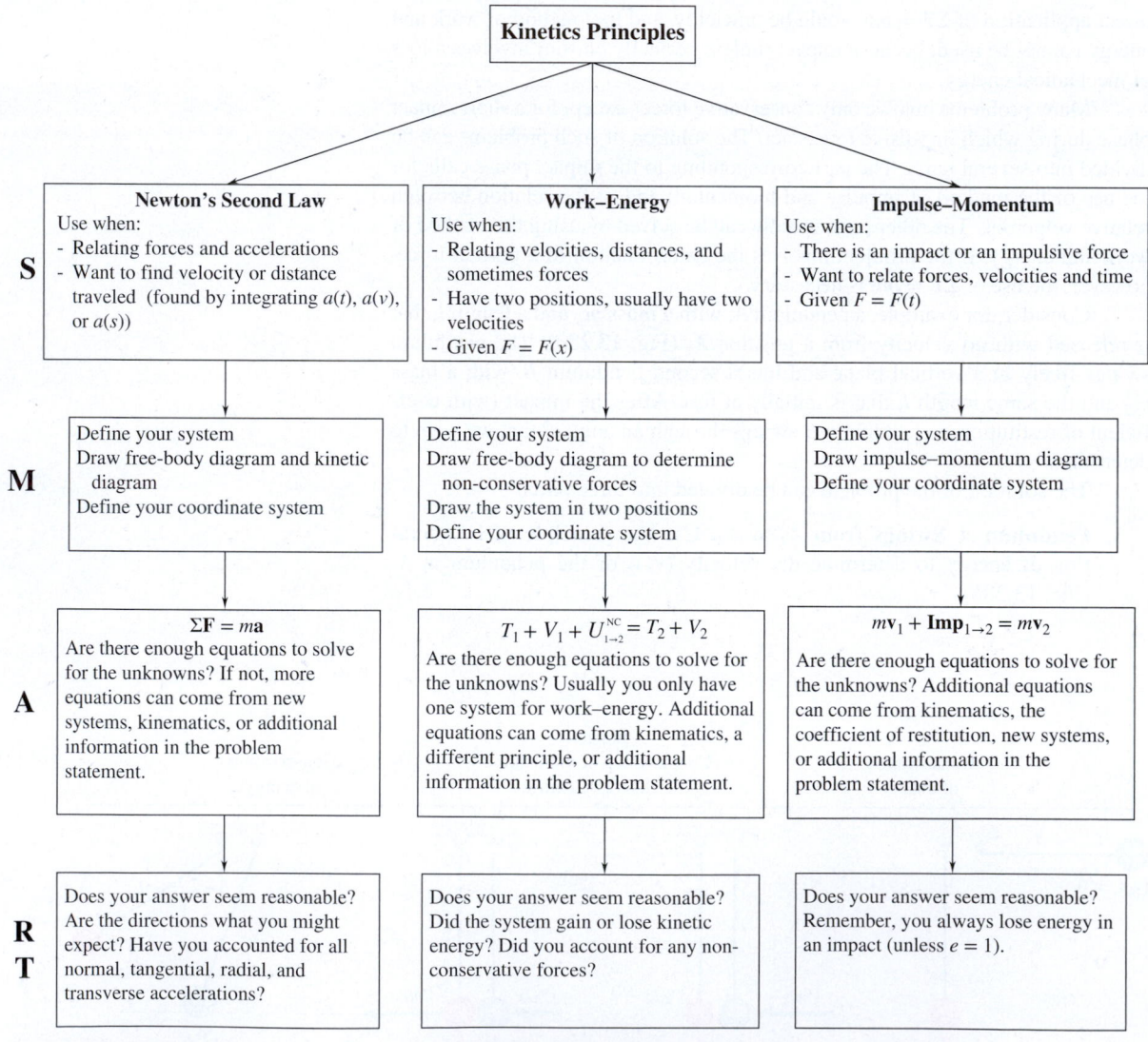

**Fig. 13.29** The three kinetics principles using the SMART methodology.

## Sample Problem 13.18

A 20-Mg railroad car moving at a speed of 0.5 m/s to the right collides with a 35-Mg car at rest. After the collision, the 35-Mg car moves to the right at a speed of 0.3 m/s. Determine the coefficient of restitution between the two cars.

**STRATEGY:**  Because there is an impact and no external impulses, use the conservation of linear momentum. You will also need to use the equation for the coefficient of restitution.

**MODELING:**  Choose your system to be both railroad cars and model them as particles. The impulse–momentum diagram for this system is shown in Fig. 1. There are no external impulses acting on this system.

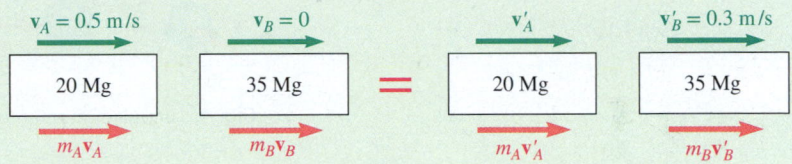

**Fig. 1**  Velocities and linear momenta of the cars before and after impact.

**ANALYSIS:**  The total momentum of the two cars is conserved, so

$$m_A \mathbf{v}_A + m_B \mathbf{v}_B = m_A \mathbf{v}'_A + m_B \mathbf{v}'_B$$

Substituting in the known values gives

$$(20 \text{ Mg})(+0.5 \text{ m/s}) + (35 \text{ Mg})(0) = (20 \text{ Mg})v'_A + (35 \text{ Mg})(+0.3 \text{ m/s})$$
$$v'_A = -0.025 \text{ m/s} \qquad v'_A = 0.025 \text{ m/s} \leftarrow$$

You can obtain the coefficient of restitution from its definition as

$$e = \frac{v'_B - v'_A}{v_A - v_B} = \frac{+0.3 - (-0.025)}{+0.5 - 0} = \frac{0.325}{0.5}$$

$$e = 0.65 \blacktriangleleft$$

**REFLECT and THINK:**  The railroad cars are constrained to move along the track, so this is a one-dimensional direct central impact. The interaction forces are large, but they last for only a very short time. Mechanical energy is lost during this impact, so you could not have used the conservation of energy.

## Sample Problem 13.19

A ball is thrown against a frictionless, vertical wall. Immediately before the ball strikes the wall, its velocity has a magnitude of $v$ and forms an angle of 30° with the horizontal. Knowing that $e = 0.90$, determine the magnitude and direction of the velocity of the ball as it rebounds from the wall.

**STRATEGY:** An impact occurs, and you are given the coefficient of restitution, so use conservation of momentum and the definition of the coefficient of restitution.

**MODELING:** Choose your system to be the ball, and model it as a particle. The impulse–momentum diagram for this system is shown in Fig. 1.

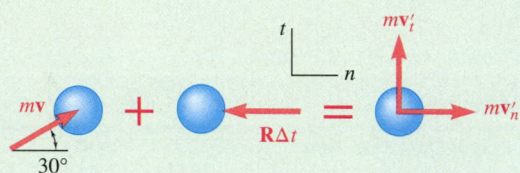

**Fig. 1** Impulse–momentum diagram for the ball.

**ANALYSIS:** Resolve the initial velocity of the ball into components perpendicular and parallel to the wall, as shown in Fig. 2.

$$v_n = v \cos 30° = 0.866v \qquad v_t = v \sin 30° = 0.500v$$

**Motion Parallel to the Wall.** Because the wall is frictionless, the impulse it exerts on the ball is perpendicular to the wall. Thus, the component of the momentum of the ball parallel to the wall is conserved. You have

$$\mathbf{v}_t' = \mathbf{v}_t = 0.500v\uparrow$$

**Motion Perpendicular to the Wall.** Because the mass of the wall (and of the earth) is essentially infinite, writing an equation for conservation of the total momentum of the ball and wall would yield no useful information. However, using the equation for coefficient of restitution, you have

$$0 - v_n' = e(v_n - 0)$$
$$v_n' = -0.90(0.866v) = -0.779v \qquad \mathbf{v}_n' = 0.779v \leftarrow$$

**Resultant Motion.** Adding vectorially the components $\mathbf{v}_n'$ and $\mathbf{v}_t'$ (Fig. 3), you find

$$\mathbf{v}' = 0.926v \ \measuredangle\ 32.7° \ \blacktriangleleft$$

**REFLECT and THINK:** Tests similar to this are done to make sure that sporting equipment—such as tennis balls, golf balls, and basketballs—are consistent and fall within certain specifications. Testing modern golf balls and clubs shows that the coefficient of restitution actually decreases with increasing club speed (from about 0.84 at a speed of 90 mph to about 0.80 at club speeds of 130 mph).

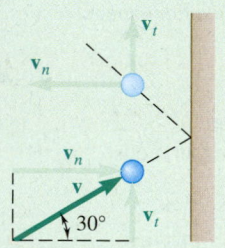

**Fig. 2** Components of the initial velocity.

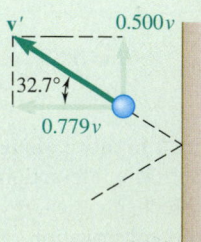

**Fig. 3** Finding the magnitude and direction for the final velocity.

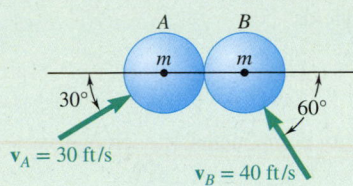

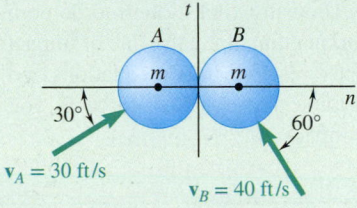

**Fig. 1** Initial velocities of *A* and *B* and the coordinate system to be used.

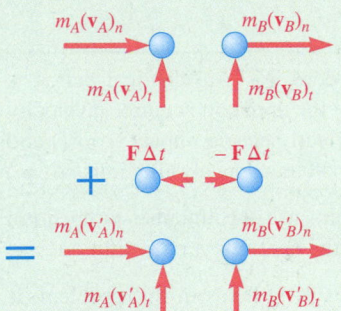

**Fig. 2** Impulse–momentum diagram for the system.

# Sample Problem 13.20

The magnitudes and directions of the velocities of two identical frictionless balls are shown before they strike each other. Assuming $e = 0.90$, determine the magnitude and direction of the velocity of each ball after the impact.

**STRATEGY:**  Because an impact occurs, use the principle of impulse and momentum. You also need the equation for the coefficient of restitution.

**MODELING:**  Choose your system to be both balls. Assuming they are small and do not rotate, you can model them as particles. Fig. 1 shows the normal and tangential directions; Fig. 2 shows the impulse–momentum diagram for this system. The impulsive forces that the balls exert on each other during the impact are directed along a line joining the centers of the balls (the line of impact). Therefore, it is best to resolve the velocities into components directed, respectively, along the line of impact and along the common tangent to the surfaces in contact. Thus,

$$(v_A)_n = v_A \cos 30° = +26.0 \text{ ft/s}$$
$$(v_A)_t = v_A \sin 30° = +15.0 \text{ ft/s}$$
$$(v_B)_n = -v_B \cos 60° = -20.0 \text{ ft/s}$$
$$(v_B)_t = v_B \sin 60° = +34.6 \text{ ft/s}$$

**ANALYSIS:**

**Motion Along the Common Tangent.**  Considering only the *t* components, apply the principle of impulse and momentum to each ball *separately*. Because the impulsive forces are directed along the line of impact, the *t* component of the momentum, and hence the *t* component of the velocity of each ball, is unchanged. You have

$$(\mathbf{v}_A')_t = 15.0 \text{ ft/s} \uparrow \qquad (\mathbf{v}_B')_t = 34.6 \text{ ft/s} \uparrow$$

**Motion Along the Line of Impact.**  In the *n* direction, consider the two balls as a single system. By Newton's third law, the internal impulses are, respectively, $\mathbf{F}\Delta t$ and $-\mathbf{F}\Delta t$, so they cancel. Thus, the total momentum of the balls is conserved as

$$m_A(v_A)_n + m_B(v_B)_n = m_A(v_A')_n + m_B(v_B')_n$$
$$m(26.0) + m(-20.0) = m(v_A')_n + m(v_B')_n$$
$$(v_A')_n + (v_B')_n = 6.0 \quad \textbf{(1)}$$

Using the equation for the coefficient of restitution relating the relative velocities, you have

$$(v_B')_n - (v_A')_n = e[(v_A)_n - (v_B)_n]$$

You can now substitute the known quantities into this equation. It is important to use the signs correctly when substituting into this equation; for example, $(\mathbf{v}_B)_n = -20$. This gives

$$(v_B')_n - (v_A')_n = (0.90)[26.0 - (-20.0)]$$
$$(v_B')_n - (v_A')_n = 41.4 \quad \textbf{(2)}$$

*(continued)*

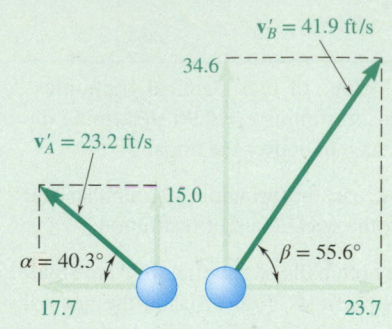

**Fig. 3** The velocity components can be resolved into their magnitudes and directions.

Solving Eqs. (1) and (2) simultaneously yields

$$(v_A')_n = -17.7 \qquad (v_B')_n = +23.7$$
$$(v_A')_n = 17.7 \text{ ft/s} \leftarrow \qquad (v_B')_n = 23.7 \text{ ft/s} \rightarrow$$

**Resultant Motion.**    Adding the velocity components of each ball vectorially (Fig. 3), you obtain

$$\mathbf{v}_A' = 23.2 \text{ ft/s} \measuredangle 40.3° \qquad \mathbf{v}_B' = 41.9 \text{ ft/s} \measuredangle 55.6° \blacktriangleleft$$

**REFLECT and THINK:**    Rather than choosing your system to be both balls, you could have applied impulse–momentum along the line of impact for each ball individually. This would have resulted in two equations and one additional unknown, $F\Delta t$. To determine the impulsive force $F$, you would need to be given the time for the impact, $\Delta t$.

## Sample Problem 13.21

Ball $B$ is hanging from an inextensible cord $BC$. An identical ball $A$ is released from rest when it is just touching the cord and acquires a velocity $\mathbf{v}_0$ before striking ball $B$. Assuming a perfectly elastic impact ($e = 1$) and no friction, determine the velocity of each ball immediately after impact. ▶

**STRATEGY:**    Because an impact occurs, use the impulse–momentum principle. You also need the equation for the coefficient of restitution.

**MODELING:**    You have several choices of systems in this problem. If you choose $A$ as your system, you obtain the impulse–momentum diagram shown in Fig. 1. Choosing the system to be both balls results in the impulse–momentum diagram shown in Fig. 2.

**ANALYSIS:**

**Impulse–Momentum Principle: Ball A.**    Applying the conservation of momentum of ball $A$ along the common tangent to balls $A$ and $B$ (Fig. 1) gives

$$m\mathbf{v}_A + \mathbf{F}\Delta t = m\mathbf{v}_A'$$
$$+\searrow t \text{ components:} \qquad mv_0 \sin 30° + 0 = m(v_A')_t$$
$$(v_A')_t = 0.5v_0 \tag{1}$$

**Impulse–Momentum Principle: Balls A and B.**    Because ball $B$ is constrained to move in a circle with center $C$, its velocity $\mathbf{v}_B$ after impact must be horizontal. Applying impulse and momentum (momentum is not conserved in the $x$ or $y$) for a system containing both balls (Fig. 2) gives

$$m\mathbf{v}_A + \mathbf{T}\Delta t = m\mathbf{v}_A' + m\mathbf{v}_B'$$
$$\xrightarrow{+} x \text{ components:} \qquad 0 = m(v_A')_t \cos 30° - m(v_A')_n \sin 30° - mv_B'$$

*(continued)*

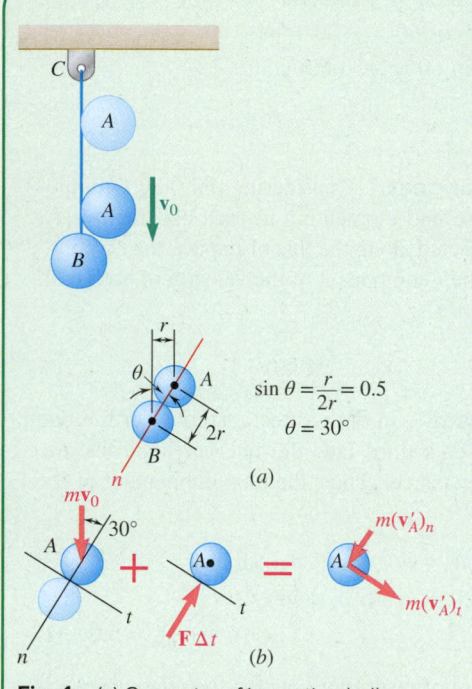

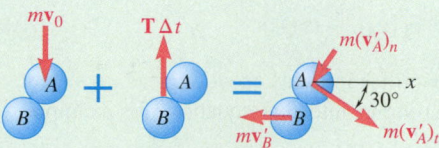

**Fig. 1** (a) Geometry of impacting balls, (b) impulse-momentum for ball a.

$$\sin \theta = \frac{r}{2r} = 0.5$$
$$\theta = 30°$$

**Fig. 2** Impulse–momentum diagram for both balls.

This equation expresses the conservation of total momentum in the $x$ direction. Substituting for $(v_A')_t$ from Eq. (1) and rearranging terms, you have

$$0.5(v_A')_n + v_B' = 0.433v_0 \qquad \textbf{(2)}$$

**Relative Velocities Along the Line of Impact.** Because $e = 1$, the equation for the coefficient of restitution gives

$$(v_B')_n - (v_A')_n = (v_A)_n - (v_B)_n$$

$$v_B' \sin 30° - (v_A')_n = v_0 \cos 30° - 0 \qquad \textbf{(3)}$$

$$0.5v_B' - (v_A')_n = 0.866v_0$$

It is important to note that the coefficient of restitution always uses the components of the velocities along the line of impact; that is, the $n$ direction. Solving Eqs. (2) and (3) simultaneously, you obtain

$$(v_A')_n = -0.520v_0 \qquad v_B' = 0.693v_0$$

$$\mathbf{v}_B' = 0.693v_0 \leftarrow \quad \blacktriangleleft$$

Recalling Eq. (1), draw a sketch (Fig. 3) and obtain by trigonometry

$$v_A' = 0.721v_0 \qquad \beta = 46.1° \qquad \alpha = 46.1° - 30° = 16.1°$$

$$\mathbf{v}_B' = 0.721v_0 \; \measuredangle \; 16.1° \quad \blacktriangleleft$$

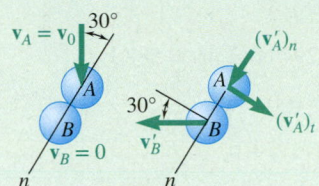

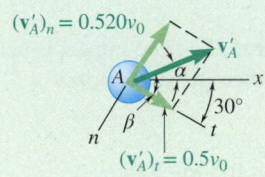

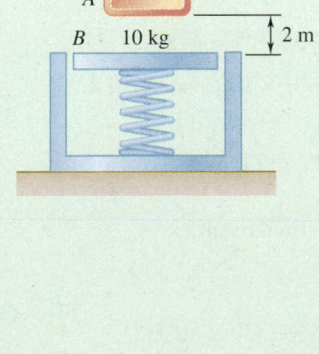

**Fig. 3** Diagram to find the magnitude and direction for the final velocity of $B$.

**REFLECT and THINK:** Because $e = 1$, the impact between $A$ and $B$ is perfectly elastic. Therefore, rather than using the coefficient of restitution, you could have used the conservation of energy as your final equation.

## Sample Problem 13.22

A 30-kg block is dropped from a height of 2 m onto the 10-kg pan of a spring scale. The constant of the spring is $k = 20$ kN/m. Assuming the impact to be perfectly plastic, determine the maximum deflection of the pan.

**STRATEGY:** This problem has three distinct phases, as shown in Fig. 1. In phase 1, $A$ falls (use the conservation of energy); in phase 2, $A$ hits $B$ (use the conservation of momentum); and in phase 3, $A$ and $B$ move down together (use the conservation of energy).

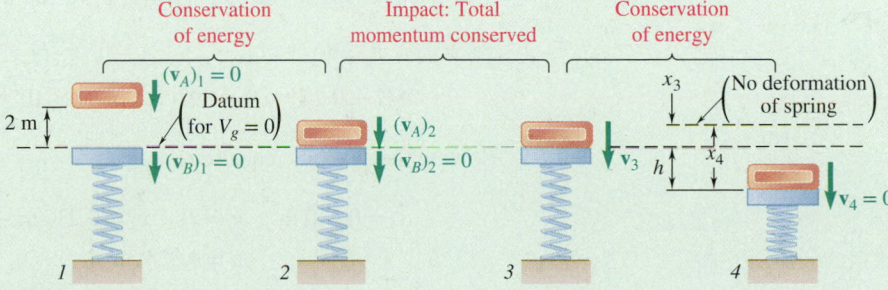

**Fig. 1** Three phases of the motion.

*(continued)*

**MODELING:** For each phase of the motion, define a different system. For phase 1, choose $A$ as your system, and for phase 2, define your system as $A$ and $B$ together. For phase 3, your system is $A$, $B$, and the spring.

**ANALYSIS:**

**Conservation of Energy A.**  Block $A$ weighs

$$W_A = (30\ \text{kg})(9.81\ \text{m/s}^2) = 294\ \text{N}$$

thus,

$$T_1 = \tfrac{1}{2}m_A(v_A)_1^2 = 0 \qquad V_1 = W_A y = (294\ \text{N})(2\ \text{m}) = 588\ \text{J}$$

$$T_2 = \tfrac{1}{2}m_A(v_A)_2^2 = \tfrac{1}{2}(30\ \text{kg})(v_A)_2^2 \qquad V_2 = 0$$

$$T_1 + V_1 = T_2 + V_2: \qquad 0 + 588\ \text{J} = \tfrac{1}{2}(30\ \text{kg})(v_A)_2^2 + 0$$

$$(v_A)_2 = +6.26\ \text{m/s} \qquad (\mathbf{v}_A)_2 = 6.26\ \text{m/s} \downarrow$$

**Impact: Conservation of Momentum for A and B.**  The impact is perfectly plastic, so $e = 0$; the block and pan move together after the impact.

$$m_A(v_A)_2 + m_B(v_B)_2 = (m_A + m_B)v_3$$
$$(30\ \text{kg})(6.26\ \text{m/s}) + 0 = (30\ \text{kg} + 10\ \text{kg})v_3$$
$$v_3 = +4.70\ \text{m/s} \qquad \mathbf{v}_3 = 4.70\ \text{m/s} \downarrow$$

**Conservation of Energy for A, B, and the Spring.**  Initially, the spring supports the weight $W_B$ of the pan; thus, the initial deflection of the spring is

$$x_3 = \frac{W_B}{k} = \frac{(10\ \text{kg})(9.81\ \text{m/s}^2)}{20 \times 10^3\ \text{N/m}} = \frac{98.1\ \text{N}}{20 \times 10^3\ \text{N/m}} = 4.91 \times 10^{-3}\ \text{m}$$

Denoting the total maximum deflection of the spring by $x_4$, you have

$$T_3 = \tfrac{1}{2}(m_A + m_B)v_3^2 = \tfrac{1}{2}(30\ \text{kg} + 10\ \text{kg})(4.70\ \text{m/s}^2) = 442\ \text{J}$$

$$V_3 = V_g + V_e = 0 + \tfrac{1}{2}kx_3^2 = \tfrac{1}{2}(20 \times 10^3)(4.91 \times 10^{-3})^2 = 0.241\ \text{J}$$

$$T_4 = 0$$

$$V_4 = V_g + V_e = (W_A + W_B)(-h) + \tfrac{1}{2}kx_4^2 = -(392)h + \tfrac{1}{2}(20 \times 10^3)x_4^2$$

The displacement of the pan is $h = x_4 - x_3$, so the final result is

$$T_3 + V_3 = T_4 + V_4:$$
$$442 + 0.241 = 0 - 392(x_4 - 4.91 \times 10^{-3}) + \tfrac{1}{2}(20 \times 10^3)x_4^2$$

Solving the quadratic equation, you get

$$x_4 = 0.230\ \text{m} \qquad h = x_4 - x_3 = 0.230\ \text{m} - 4.91 \times 10^{-3}\ \text{m}$$

$$h = 0.225\ \text{m} \qquad\qquad h = 225\ \text{mm} \blacktriangleleft$$

**REFLECT and THINK:**  The spring constant for this scale is pretty large, but the block is fairly massive and is dropped from a height of 2 m. From this perspective, the deflection seems reasonable.

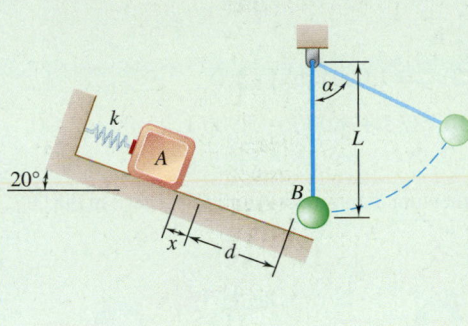

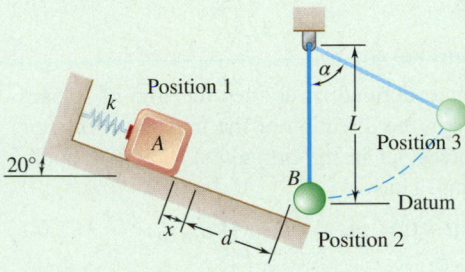

**Fig. 1** Three positions of interest for this problem.

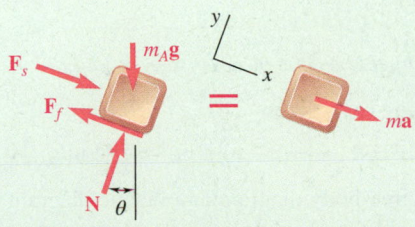

**Fig. 2** Free-body diagram and kinetic diagram for block *A*.

## Sample Problem 13.23

A 2-kg block *A* is pushed up against a spring, compressing it a distance $x = 0.1$ m. The block is then released from rest and slides down the 20° incline until it strikes a 1-kg sphere *B* that is suspended by a 1-m inextensible rope. The spring constant is $k = 800$ N/m, the coefficient of friction between *A* and the ground is 0.2, block *A* slides from the unstretched length of the spring a distance $d = 1.5$ m, and the coefficient of restitution between *A* and *B* is 0.8. When $\alpha = 40°$, determine (*a*) the speed of *B*, (*b*) the tension in the rope.

**STRATEGY:** A lot of things are going on in this problem, so you need to break the motion into steps.

**Step 1:** Block *A* slides down the incline, so there are two positions. Therefore, use the work–energy principle between position *1* and position *2* to find the velocity of *A* just before it strikes ball *B* (Fig. 1).

**Step 2:** Block *A* hits *B*, so an impact occurs. Therefore, use impulse–momentum and the equation for the coefficient of restitution.

**Step 3:** Ball *B* is swinging up, so you have two positions (position *2* and position *3* in Fig. 1). You are asked to find the speed at position *3*; therefore, use the conservation of energy.

**Step 4:** To find the tension when $\alpha = 40°$, use Newton's second law with normal and tangential coordinates.

**MODELING:** Each step requires a different system. For Step 1, your system is *A* and the spring. For Step 2, it is *A* and *B*. Finally, for Steps 3 and 4, it is *B*. We model *A* and *B* as particles and draw the appropriate figures in the analysis section.

**ANALYSIS:**

**Step 1. Block Slides Down the Incline.** The principle of work and energy between where the block is released to the point it strikes *B* is

$$T_1 + V_{g_1} + V_{e_1} + U^{\text{NC}}_{1 \to 2} = T_2 + V_{g_2} + V_{e_2} \tag{1}$$

**Work.** The only non-conservative force that does work is the friction force. A free-body diagram for *A* is shown in Fig. 2. Applying Newton's second law gives

$$+\nearrow \Sigma F_y = 0: \quad N - m_A g \cos \theta = 0 \quad \text{or}$$

$$N = m_A g \cos \theta = (2 \text{ kg})(9.81 \text{ m/s2}) \cos 20° = 18.437 \text{ N}$$

and the friction force is

$$F_f = \mu_k N = (0.2)(18.437 \text{ N}) = 3.687 \text{ N}$$

So the work is

$$U^{\text{NC}}_{1 \to 2} = -F_f(x + d) = -(3.687 \text{ N})(1.6 \text{ m}) = -5.900 \text{ J}$$

**Position 1.** Place your datum for $V_g$ at the impact point near *B* (see Fig. 1). Calculate the initial energy as

$$T_1 = 0 \qquad V_{e_1} = \tfrac{1}{2} k x_1^2 = \frac{1}{2}(800)(0.1)^2 = 4.00 \text{ J}$$

$$V_{g_1} = m_A g h_1 = m_A g(x + d) \sin \theta = (2)(9.81)(1.6) \sin 20° = 10.737 \text{ J}$$

*(continued)*

**Position 2.** The energy at position *2* is

$$T_2 = \tfrac{1}{2}m_A v_A^2 = \tfrac{1}{2}(2)v_A^2 = 1.000\,v_A^2 \quad V_2 = 0$$

Substituting into Eq. (1) gives $0 + 10.737\text{ J} + 4.00\text{ J} - 5.900\text{ J} = 1.00\,v_A^2 + 0$. Solving for $v_A$ gives $v_A = 2.973$ m/s.

**Step 2. Impact.**   The impulse–momentum diagram for *A* and *B* is shown in Fig. 3.

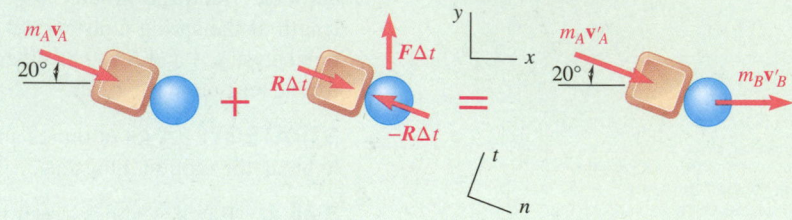

**Fig. 3**   Impulse–momentum diagram for *A* and *B*.

Note that two coordinate systems are defined: *n* defines the line of impact between the block and ball and *y* is in the direction of the impulsive force of the rope. Because no impulsive forces act in the horizontal direction, apply impulse–momentum in the *x* direction. Thus,

$\xrightarrow{+}$ *x* components:     $m_A v_A \cos\theta + 0 = m_A v_A' \cos\theta + m_B v_B'$     **(2)**

**Coefficient of Restitution.**

$$(v_B')_n - (v_A')_n = e[(v_A)_n - (v_B)_n] \text{ or } v_B\cos\theta - v_A' = e\,v_A \quad\quad \textbf{(3)}$$

In Eqs. (2) and (3) you can solve for two unknowns, $v_A'$ and $v_B'$. This gives

$$v_A' = 1.0382 \text{ m/s} \quad\quad v_B' = 3.6356 \text{ m/s}$$

**Step 3. Sphere *B* Rises.**   The tension does no work, so use the conservation of energy for *B* between positions *2* and *3*. Again, define the datum as shown in Fig 1.

$$T_2 + V_{g_2} + V_{e_2} = T_3 + V_{g_3} + V_{e_3} \quad\quad \textbf{(4)}$$

**Position 2.**

$$T_2 = \frac{1}{2}m_B(v_B')^2 \quad V_{g_2} = 0 \quad V_{e_2} = 0$$

**Position 3.**

$$T_3 = \frac{1}{2}m_B(v_B)_3 \quad V_{g_3} = m_B g L(1 - \cos\alpha) \quad V_{e_3} = 0$$

Substituting these into Eq. (4) and solving $(v_B)_3$ gives

$$v_{B_3} = 2.94 \text{ m/s} \quad\blacktriangleleft$$

**Step 4. Tension in the Rope.**   A free-body diagram and kinetic diagram for the sphere at position *3* are shown in Fig. 4. Applying Newton's second law in the normal direction gives

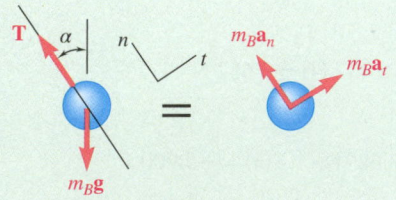

**Fig. 4**

$$+\nwarrow \Sigma F_n = m_B a_n: \quad\quad T - m_B g \cos\alpha = m_B a_n = m_B\frac{(v_B)_3}{L}$$

Solving for *T*, you find

$$T = 16.14 \text{ N} \quad\blacktriangleleft$$

**REFLECT and THINK:**   You cannot use work–energy from position *1* to position *3* because a loss of energy occurs when *A* hits *B*. If the coefficient of friction had been larger, say $\mu_k = 0.4$, you would find that after the impact, *B* has a speed of 2.10 m/s. Plugging this into Eq. (4) gives an imaginary number for the speed at $\alpha = 40°$, meaning sphere *B* does not reach this angle.

# SOLVING PROBLEMS
# ON YOUR OWN

This section deals with the **impact of two smooth bodies**; that is, with a collision occurring in a very small interval of time. You solved several impact problems by applying the principle of impulse and momentum and by expressing the relationship between the relative velocities of the two bodies before and after impact.

**1. As a first step in your solution,** you should select and draw two coordinate axes: the $t$ axis, which is tangent to the surfaces of contact of the two colliding bodies; and the $n$ axis, which is normal to the surfaces of contact and defines the line of impact. In all of the problems in this section, the line of impact passes through the mass centers of the colliding bodies, and the impact is referred to as a **central impact**.

**2. Next draw an impulse–momentum diagram** showing the momenta of the bodies before impact, the impulses exerted on the bodies during impact, and the final momenta of the bodies after impact (Fig. 13.24). Then, observe whether the impact is a **direct central impact** or an **oblique central impact**.

**3. Direct central impact** (Sample Prob. 13.18). This occurs when the velocities of bodies $A$ and $B$ are both directed along the line of impact before impact (Fig. 13.20*a*).

 **a. Conservation of momentum.** Because the impulsive forces are internal to the system, the total momentum of $A$ and $B$ is conserved as

$$m_A v_A + m_B v_B = m_A v_A' + m_B v_B' \qquad (13.37)$$

where $v_A$ and $v_B$ denote the velocities of bodies $A$ and $B$ before impact and $v_A'$ and $v_B'$ denote their velocities after impact.

 **b. Coefficient of restitution.** You can also write the relation between the relative velocities of the two bodies before and after impact as

$$v_B' - v_A' = e(v_A - v_B) \qquad (13.43)$$

where $e$ is the coefficient of restitution between the two bodies.

 Note that Eqs. (13.37) and (13.43) are scalar equations that you can solve for two unknowns. Also, be careful to adopt a consistent sign convention for all velocities.

**4. Oblique central impact** (Sample Prob. 13.20). This occurs when one or both of the initial velocities of the two bodies is not directed along the line of impact (Fig. 13.20*b*). Again, these solution steps are only applicable to problems where the impulsive forces in the tangential direction are negligible (e.g., you would not use these to solve Prob. 13.146). To solve problems of this type, you should first resolve the momenta and impulses shown in your diagram into components along the $t$ axis and the $n$ axis.

 **a. Conservation of momentum.** Because the impulsive forces act along the line of impact—that is, along the $n$ axis—the component along the $t$ axis of the momentum *of* *each body* is conserved. Therefore, for each body, you can write that the $t$ components of its velocity before and after impact are equal. So,

$$(v_A)_t = (v_A')_t \qquad (v_B)_t = (v_B')_t \tag{13.47}$$

Also, the component along the $n$ axis of the total momentum of the system is conserved as

$$m_A(v_A)_n + m_B(v_B)_n = m_A(v_A')_n + m_B(v_B')_n \tag{13.48}$$

**b. Coefficient of restitution.** The relation between the relative velocities of the two bodies before and after impact can be written in the $n$ direction only. Hence,

$$(v_B')_n - (v_A')_n = e[(v_A)_n - (v_B)_n] \tag{13.49}$$

You now have four equations that you can solve for four unknowns. Note that after finding all of the velocities, you can determine the impulse exerted by body $A$ on body $B$ by drawing an impulse–momentum diagram for $B$ alone and equating components in the $n$ direction.

**c. When the motion of one of the colliding bodies is constrained,** you must include the impulses of the external forces in your diagram (Sample Probs. 13.21 and 13.23). You will then observe that some of the previous relations do not hold. However, in the example shown in Fig. 13.26, the total momentum of the system is conserved in a direction perpendicular to the external impulse. Also note that, when a body $A$ bounces off a fixed surface $B$, the only conservation of momentum equation that you can use is the first of Eq. (13.47) (Sample Prob. 13.19).

**5. Remember that energy is lost during most impacts.** The only exception is for **perfectly elastic** impacts ($e = 1$), where energy is conserved. Thus, in the general case of impact where $e < 1$, mechanical energy is not conserved. Therefore, be careful *not to apply* the principle of conservation of energy through an impact situation. Instead, apply this principle separately to the motions preceding and following the impact (Sample Probs. 13.22 and 13.23).

# Problems

## CONCEPT QUESTIONS

**13.CQ6** A 5-kg ball $A$ strikes a 1-kg ball $B$ that is initially at rest. Is it possible that, after the impact, $A$ is not moving and $B$ has a speed of $5v$?
  **a.** Yes
  **b.** No
  Explain your answer.

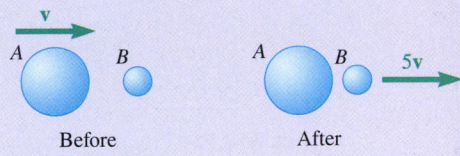

**Fig. P13.CQ6**

## IMPULSE–MOMENTUM DIAGRAM PRACTICE PROBLEMS

**13.F6** A sphere with a speed $v_0$ rebounds after striking a frictionless inclined plane as shown. Draw the impulse–momentum diagram that can be used to find the velocity of the sphere after the impact.

**13.F7** An 80-Mg railroad engine $A$ coasting at 6.5 km/h strikes a 20-Mg flatcar $C$ carrying a 30-Mg load $B$ which can slide along the floor of the car ($\mu_k = 0.25$). The flatcar was at rest with its brakes released. Instead of $A$ and $C$ coupling as expected, it is observed that $A$ rebounds with a speed of 2 km/h after the impact. Draw impulse–momentum diagrams that can be used to determine ($a$) the coefficient of restitution and the speed of the flatcar immediately after impact, ($b$) the time it takes the load to slide to a stop relative to the car.

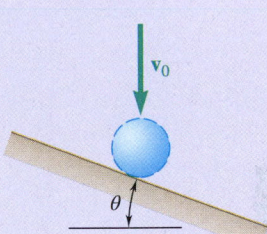

**Fig. P13.F6**

**Fig. P13.F7**

**13.F8** Two frictionless balls strike each other as shown. The coefficient of restitution between the balls is $e$. Draw the impulse–momentum diagram that could be used to find the velocities of $A$ and $B$ after the impact.

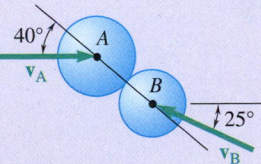

**Fig. P13.F8**

**13.F9** A 10-kg ball $A$ moving horizontally at 12 m/s strikes a 10-kg block $B$. The coefficient of restitution of the impact is 0.4 and the coefficient of kinetic friction between the block and the inclined surface is 0.5. Draw the impulse–momentum diagram that can be used to determine the speeds of $A$ and $B$ after the impact.

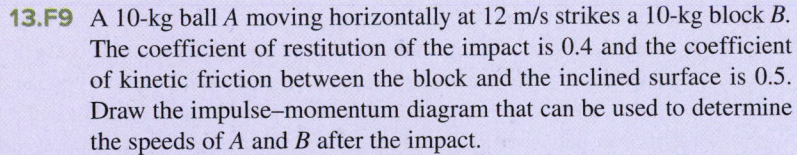

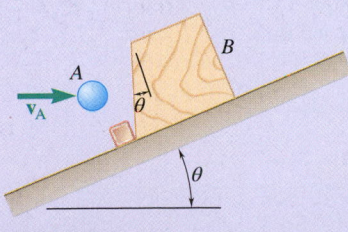

**Fig. P13.F9**

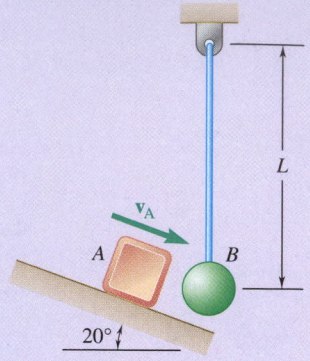

Fig. P13.F10

**13.F10** Block $A$ of mass $m_A$ strikes ball $B$ of mass $m_B$ with a speed of $v_A$ as shown. Draw the impulse–momentum diagram that can be used to determine the speeds of $A$ and $B$ after the impact and the impulse during the impact.

### END-OF-SECTION PROBLEMS

**13.155** Two steel blocks slide without friction on a horizontal surface; immediately before impact their velocities are as shown. Knowing that $e = 0.75$, determine (a) their velocities after impact, (b) the energy loss during impact.

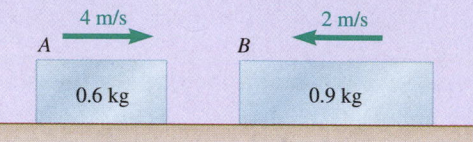

Fig. P13.155 and P13.156

**13.156** The velocities of two steel blocks before impact are as shown. Knowing that the velocity of block $B$ after the impact is observed to be 2.5 m/s to the right, determine the coefficient of restitution between the two blocks.

**13.157** One of the requirements for tennis balls to be used in official competition is that, when dropped onto a rigid surface from a height of 100 in., the height of the first bounce of the ball must be in the range 53 in. $\leq h \leq 58$ in. Determine the range of the coefficients of restitution of the tennis balls satisfying this requirement.

**13.158** A collar $A$ of mass $m$ is moving with a speed $v$ when it strikes an identical collar $B$ that is at rest. Knowing that the coefficient of restitution between the collars is $e$, determine the energy lost in the impact as a function of $m$, $e$, and $v$.

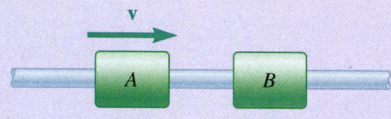

Fig. P13.158

**13.159** To apply shock loading to an artillery shell, a 20-kg pendulum $A$ is released from a known height and strikes impactor $B$ at a known velocity $\mathbf{v}_0$. Impactor $B$ then strikes the 1-kg artillery shell $C$. Knowing the coefficient of restitution between all objects is $e$, determine the mass of $B$ to maximize the impulse applied to the artillery shell $C$.

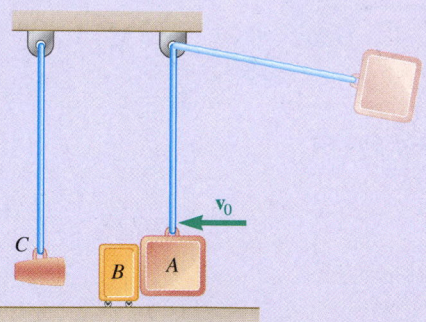

Fig. P13.159

**13.160** Packages in an automobile parts supply house are transported to the loading dock by pushing them along on a roller track with very little friction. At the instant shown, packages $B$ and $C$ are at rest, and package $A$ has a velocity of 2 m/s. Knowing that the coefficient of restitution between the packages is 0.3, determine (a) the velocity of package $C$ after $A$ hits $B$ and $B$ hits $C$, (b) the velocity of $A$ after it hits $B$ for the second time.

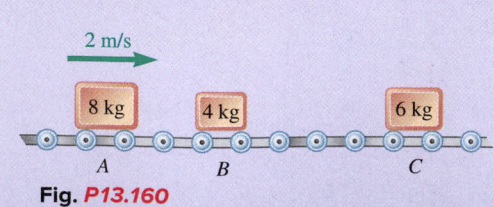

Fig. P13.160

**13.161** Three steel spheres of equal mass are suspended from the ceiling by cords of equal length which are spaced at a distance slightly greater than the diameter of the spheres. After being pulled back and released, sphere $A$ hits sphere $B$, which then hits sphere $C$. Denoting by $e$ the coefficient of restitution between the spheres and by $\mathbf{v}_0$ the velocity of $A$ just before it hits $B$, determine (a) the velocities of $A$ and $B$ immediately after the first collision, (b) the velocities of $B$ and $C$ immediately after the second collision. (c) If $n$ spheres are suspended from the ceiling and the first sphere is pulled back and released as described above, determine the velocity of the last sphere after it is hit for the first time. (d) Use the result of part $c$ to obtain the velocity of the last sphere when $n = 5$ and $e = 0.85$.

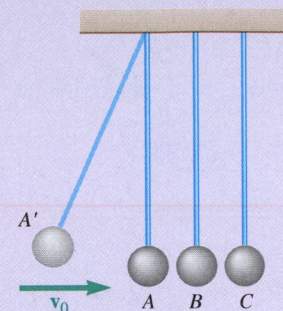

**Fig. P13.161**

**13.162** At an amusement park, there are 200-kg bumper cars $A$, $B$, and $C$ that have riders with masses of 40 kg, 60 kg, and 35 kg, respectively. Car $A$ is moving to the right with a velocity $\mathbf{v}_A = 2$ m/s and car $C$ has a velocity $\mathbf{v}_B = 1.5$ m/s to the left, but car $B$ is initially at rest. The coefficient of restitution between each car is 0.8. Determine the final velocity of each car, after all impacts, assuming (a) cars $A$ and $C$ hit car $B$ at the same time, (b) car $A$ hits car $B$ before car $C$ does.

**Fig. *P13.162* and P13.163**

**13.163** At an amusement park there are 200-kg bumper cars $A$, $B$, and $C$ that have riders with masses of 40 kg, 60 kg, and 35 kg, respectively. Car $A$ is moving to the right with a velocity $\mathbf{v}_A = 2$ m/s when it hits stationary car $B$. The coefficient of restitution between each car is 0.8. Determine the velocity of car $C$ so that after car $B$ collides with car $C$ the velocity of car $B$ is zero.

**13.164** Two identical billiard balls can move freely on a horizontal table. Ball $A$ has a velocity $\mathbf{v}_0$ as shown and hits ball $B$, which is at rest, at a point $C$ defined by $\theta = 45°$. Knowing that the coefficient of restitution between the two balls is $e = 0.8$ and assuming no friction, determine the velocity of each ball after impact.

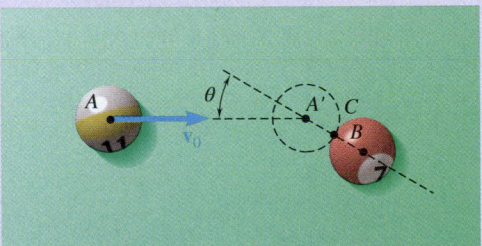

**Fig. P13.164**

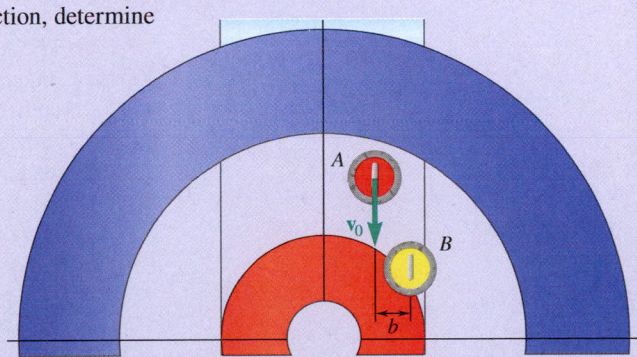

**Fig. P13.165**

**13.165** Two identical 40-lb curling stones have diameters of 11 in. and may move freely on a sheet of ice. Stone $B$ is at rest when stone $A$ strikes it with a speed of 0.5 m/s. (a) Knowing that $b = 5$ in and $e = 0.7$, determine the velocity of each stone after impact in terms of $\mathbf{v}_0$. (b) Show that if $e = 1$, the final velocities of the stones form a right angle for all values of $b$.

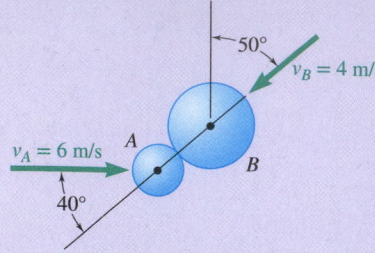

Fig. P13.166

**13.166** A 600-g ball $A$ is moving with a velocity of magnitude 6 m/s when it is hit as shown by a 1-kg ball $B$ that has a velocity of magnitude 4 m/s. Knowing that the coefficient of restitution is 0.8 and assuming no friction, determine the velocity of each ball after impact.

**13.167** Two identical hockey pucks are moving on a hockey rink at the same speed of 3 m/s and in perpendicular directions when they strike each other as shown. Assuming a coefficient of restitution $e = 0.9$, determine the magnitude and direction of the velocity of each puck after impact.

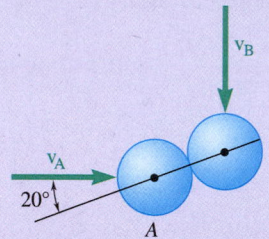

Fig. P13.167

**13.168** A billiard player wishes to have ball $A$ hit ball $B$ obliquely and then ball $C$ squarely. Assuming perfectly elastic impacts ($e = 1$), the radius of the balls is $r$, and the distance between the centers of $B$ and $C$ is $d$, determine ($a$) the angle $\theta$ defining point $D$ where ball $B$ should be hit, ($b$) the range of values of angles $ABC$ and $ACB$ for which this play is possible.

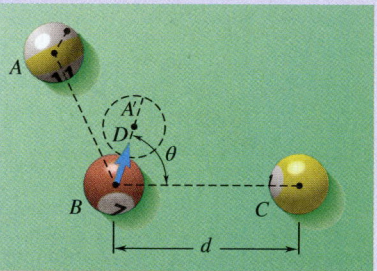

Fig. P13.168

**13.169** A boy located at point $A$ halfway between the center O of a semicircular wall and the wall itself throws a ball at the wall in a direction forming an angle of 45° with $OA$. Knowing that after hitting the wall the ball rebounds in a direction parallel to $OA$, determine the coefficient of restitution between the ball and the wall.

**13.170** The Mars Pathfinder spacecraft used large airbags to cushion its impact with the planet's surface when landing. Assuming the spacecraft had an impact velocity of 18.5 m/s at an angle of 45° with respect to the horizontal, the coefficient of restitution is 0.85 and neglecting friction, determine ($a$) the height of the first bounce, ($b$) the length of the first bounce. (Acceleration of gravity on Mars = 3.73 m/s$^2$.)

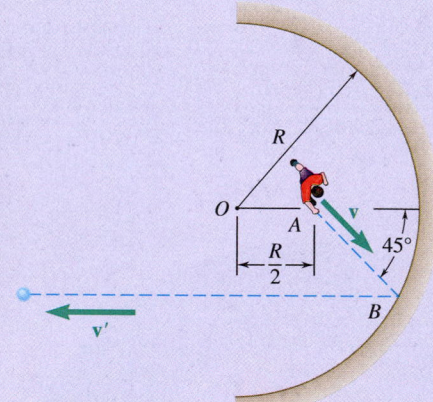

Fig. P13.169

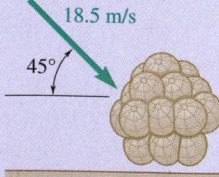

Fig. *P13.170*

**13.171** A girl throws a ball at an inclined wall from a height of 3 ft, hitting the wall at $A$ with a horizontal velocity $\mathbf{v}_0$ of magnitude 25 ft/s. Knowing that the coefficient of restitution between the ball and the wall is 0.9 and neglecting friction, determine the distance $d$ from the foot of the wall to the point $B$ where the ball will hit the ground after bouncing off the wall.

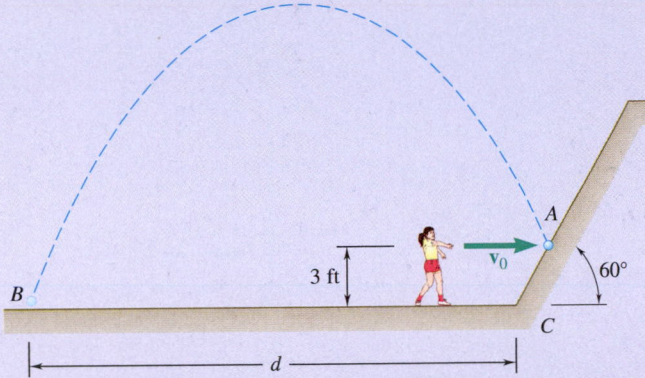

**Fig. P13.171**

**13.172** Rockfalls can cause major damage to roads and infrastructure. To design mitigation bridges and barriers, engineers use the coefficient of restitution to model the behavior of the rocks. Rock $A$ falls a distance of 20 m before striking an incline with a slope of $\alpha = 40°$. Knowing that the coefficient of restitution between rock $A$ and the incline is 0.2, determine the velocity of the rock after the impact.

**13.173** From experimental tests, smaller boulders tend to have a greater coefficient of restitution than larger boulders. Rock $A$ falls a distance of 20 meters before striking an incline with a slope of $\alpha = 45°$. Knowing that $h = 30$ m and $d = 20$ m, determine if a boulder will land on the road or beyond the road for a coefficient of restitution of (a) $e = 0.2$, (b) $e = 0.1$.

**13.174** Two cars of the same mass run head-on into each other at $C$. After the collision, the cars skid with their brakes locked and come to a stop in the positions shown in the lower part of the figure. Knowing that the speed of car $A$ just before impact was 5 mi/h and that the coefficient of kinetic friction between the pavement and the tires of both cars is 0.30, determine (a) the speed of car $B$ just before impact, (b) the effective coefficient of restitution between the two cars.

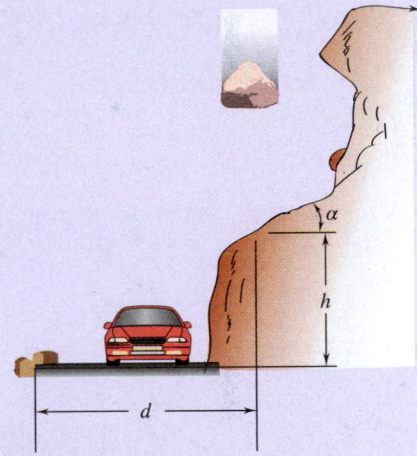

**Fig. P13.172 and P13.173**

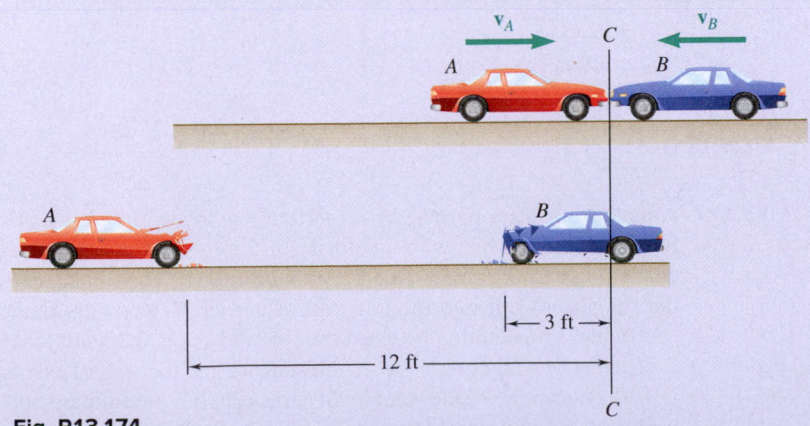

**Fig. P13.174**

**13.175** A 1-kg block $B$ is moving with a velocity $\mathbf{v}_0$ of magnitude $v_0 = 2$ m/s as it hits the 0.5-kg sphere $A$, which is at rest and hanging from a cord attached at $O$. Knowing that $\mu_k = 0.6$ between the block and the horizontal surface and $e = 0.8$ between the block and the sphere, determine after impact (a) the maximum height $h$ reached by the sphere, (b) the distance $x$ traveled by the block.

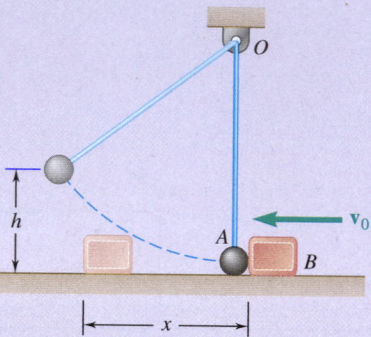

**Fig. P13.175**

**13.176** A 0.25-lb ball thrown with a horizontal velocity $\mathbf{v}_0$ strikes a 1.5-lb plate attached to a vertical wall at a height of 36 in. above the ground. It is observed that after rebounding, the ball hits the ground at a distance of 24 in. from the wall when the plate is rigidly attached to the wall (Fig. P13.176a) and at a distance of 10 in. when a foam-rubber mat is placed between the plate and the wall (Fig. P13.176b). Determine (a) the coefficient of restitution $e$ between the ball and the plate, (b) the initial velocity $\mathbf{v}_0$ of the ball.

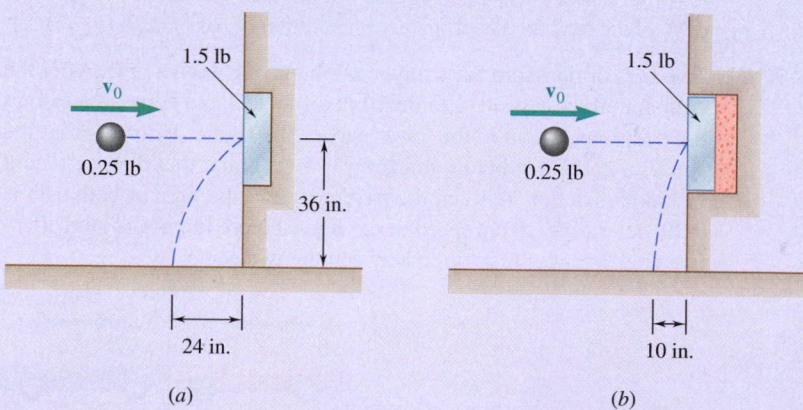

**Fig. P13.176**

**13.177** After having been pushed by an airline employee, an empty 40-kg luggage carrier $A$ hits with a velocity of 5 m/s an identical carrier $B$ containing a 15-kg suitcase equipped with rollers. The impact causes the suitcase to roll into the left wall of carrier $B$. Knowing that the coefficient of restitution between the two carriers is 0.80 and that the coefficient of restitution between the suitcase and the wall of carrier $B$ is 0.30, determine (a) the velocity of carrier $B$ after the suitcase hits its wall for the first time, (b) the total energy lost in that impact.

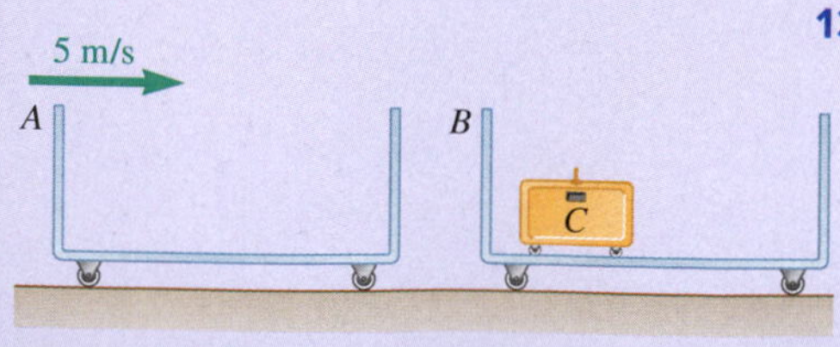

**Fig. P13.177**

**13.178** Blocks *A* and *B* each weigh 0.8 lb and block *C* weighs 2.4 lb. The coefficient of friction between the blocks and the plane is $\mu_k = 0.30$. Initially, block *A* is moving at a speed $v_0 = 15$ ft/s and blocks *B* and *C* are at rest (Fig. P13.178*a*). After *A* strikes *B* and *B* strikes *C*, all three blocks come to a stop in the positions shown (Fig. P13.178*b*). Determine (*a*) the coefficients of restitution between *A* and *B* and between *B* and *C*, (*b*) the displacement *x* of block *C*.

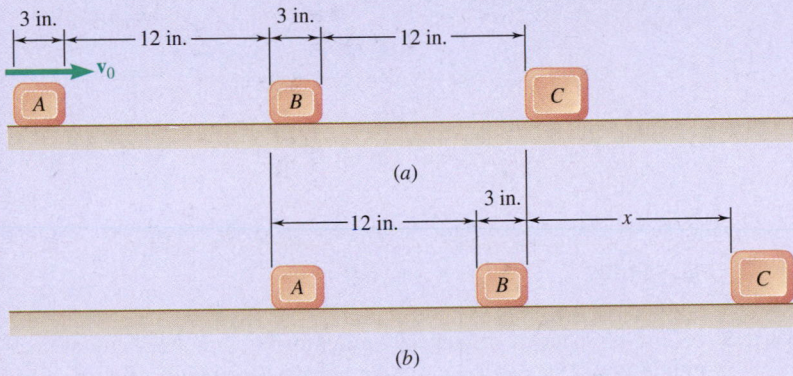

**Fig. P13.178**

**13.179** A 5-kg sphere is dropped from a height of $y = 2$ m to test newly designed spring floors used in gymnastics. The mass of the floor section is 10 kg, and the effective stiffness of the floor is $k = 120$ kN/m. Knowing that the coefficient of restitution between the ball and the platform is 0.6, determine (*a*) the height *h* reached by the sphere after rebound, (*b*) the maximum force in the springs.

**13.180** A 5-kg sphere is dropped from a height of $y = 3$ m to test a new spring floor used in gymnastics. The mass of floor section *B* is 12 kg, and the sphere bounces back upward a distance of 44 mm. Knowing that the maximum deflection of the floor section is 33 mm from its equilibrium position, determine (*a*) the coefficient of restitution between the sphere and the floor, (*b*) the effective spring constant *k* of the floor section.

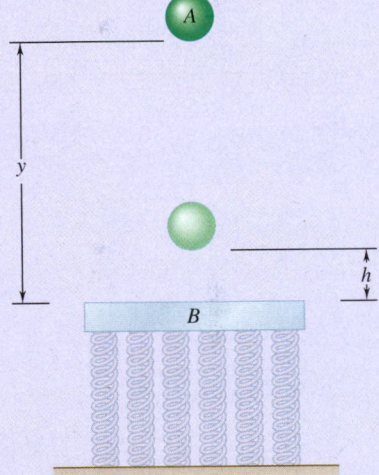

**Fig. P13.179 and P13.180**

**13.181** The three blocks shown are identical. Blocks *B* and *C* are at rest when block *B* is hit by block *A*, which is moving with a velocity $v_A$ of 3 ft/s. After the impact, which is assumed to be perfectly plastic ($e = 0$), the velocity of blocks *A* and *B* decreases due to friction, while block *C* picks up speed, until all three blocks are moving with the same velocity **v**. Knowing that the coefficient of kinetic friction between all surfaces is $\mu_k = 0.20$, determine (*a*) the time required for the three blocks to reach the same velocity, (*b*) the total distance traveled by each block during that time.

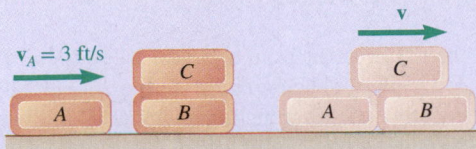

**Fig. P13.181**

**13.182** Block $A$ is released from rest and slides down the frictionless surface of $B$ until it hits a bumper on the right end of $B$. Block $A$ has a mass of 10 kg and object $B$ has a mass of 30 kg and $B$ can roll freely on the ground. Determine the velocities of $A$ and $B$ immediately after impact when ($a$) $e = 0$, ($b$) $e = 0.7$.

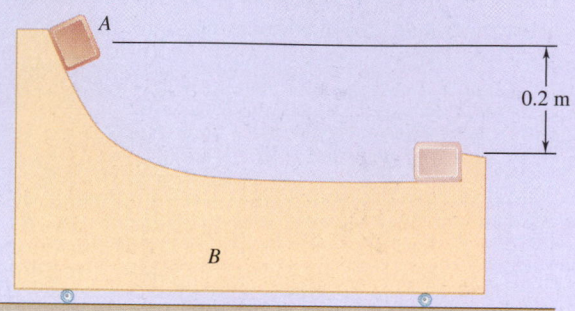

Fig. P13.182

**13.183** A 23.1-kg sphere $A$ of radius 90 mm moving with a velocity of magnitude $v_0 = 2$ m/s strikes a 2.1-kg sphere $B$ of radius 40 mm that is hanging from an inextensible cord and is initially at rest. Knowing that sphere $B$ swings to a maximum height $h = 0.25$ m, determine the coefficient of restitution between the two spheres.

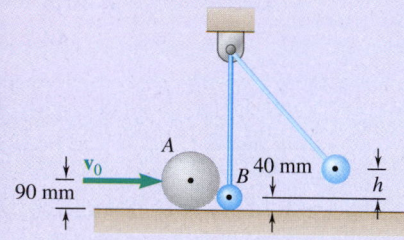

Fig. P13.183

**13.184** A test machine that kicks soccer balls has a 5-lb simulated foot attached to the end of a 6-ft long pendulum arm of negligible mass. Knowing that the arm is released from the horizontal position and that the coefficient of restitution between the foot and the 1-lb ball is 0.8, determine the exit velocity of the ball ($a$) if the ball is stationary, ($b$) if the ball is struck when it is rolling toward the foot with a velocity of 10 ft/s.

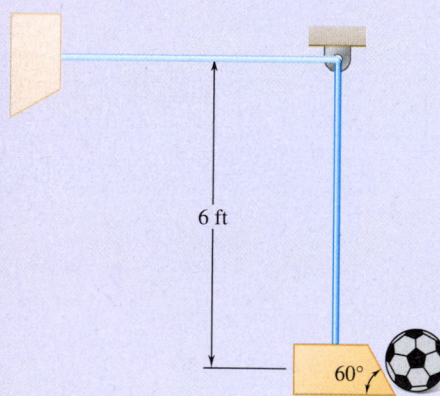

Fig. P13.184

**13.185** Ball $B$ is hanging from an inextensible cord. An identical ball $A$ is released from rest when it is just touching the cord and drops through the vertical distance $h_A = 8$ in. before striking ball $B$. Assuming $e = 0.9$ and no friction, determine the resulting maximum vertical displacement $h_B$ of the ball $B$.

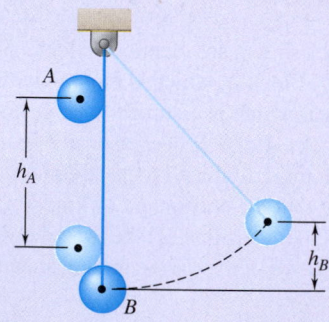

Fig. P13.185

**13.186** A 70-g ball $B$ dropped from a height $h_0 = 1.5$ m reaches a height $h_2 = 0.25$ m after bouncing twice from identical 210-g plates. Plate $A$ rests directly on hard ground, while plate $C$ rests on a foam-rubber mat. Determine ($a$) the coefficient of restitution between the ball and the plates, ($b$) the height $h_1$ of the ball's first bounce.

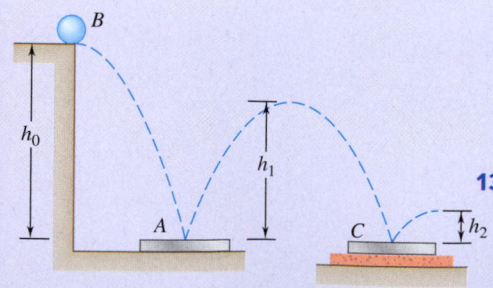

Fig. P13.186

908

**13.187** A 2-kg sphere moving to the right with a velocity of 5 m/s strikes at $A$, which is on the surface of a 9-kg quarter cylinder that is initially at rest and in contact with a spring with a constant of 20 kN/m. The spring is held by cables, so it is initially compressed 50 mm. Neglecting friction and knowing that the coefficient of restitution is 0.6, determine (*a*) the velocity of the sphere immediately after impact, (*b*) the maximum compressive force in the spring.

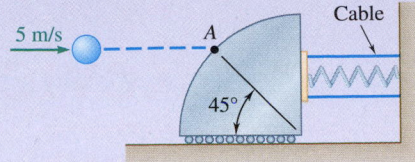

**Fig. P13.187**

**13.188** When the rope is at an angle of $\alpha = 30°$, the 1-lb sphere $A$ has a speed $v_0 = 4$ ft/s. The coefficient of restitution between $A$ and the 2-lb wedge $B$ is 0.7 and the length of rope $l = 2.6$ ft. The spring constant has a value of 2 lb/in. and $\theta = 20°$. Determine (*a*) the velocities of $A$ and $B$ immediately after the impact, (*b*) the maximum deflection of the spring, assuming $A$ does not strike $B$ again before this point.

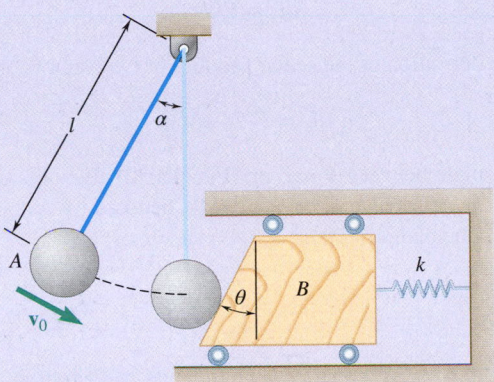

**Fig. P13.188 and *P13.189***

**13.189** When the rope is at an angle of $\alpha = 30°$, the 1-kg sphere $A$ has a speed $v_0 = 0.6$ m/s. The coefficient of restitution between $A$ and the 2-kg wedge $B$ is 0.8 and the length of rope $l = 0.9$ m. The spring constant has a value of 1500 N/m and $\theta = 20°$. Determine (*a*) the velocities of $A$ and $B$ immediately after the impact, (*b*) the maximum deflection of the spring, assuming $A$ does not strike $B$ again before this point.

# Review and Summary

This chapter was devoted to presenting the method of work and energy and the method of impulse and momentum. In the first half of the chapter, we studied the method of work and energy and its application to the analysis of the motion of particles.

## Work of a Force

We first considered a force $\mathbf{F}$ acting on a particle $A$ and defined the **work of F corresponding to the small displacement** $d\mathbf{r}$ (Sec. 13.1) as the quantity

$$dU = \mathbf{F} \cdot d\mathbf{r} \qquad (13.1)$$

or recalling the definition of the scalar product of two vectors, as

$$dU = F\, ds\, \cos\alpha \qquad (13.1')$$

where $\alpha$ is the angle between $\mathbf{F}$ and $d\mathbf{r}$ (Fig. 13.30). We obtained the work of $\mathbf{F}$ during a finite displacement from $A_1$ to $A_2$, denoted by $U_{1\to 2}$, by integrating Eq. (13.1) along the path described by the particle as

$$U_{1\to 2} = \int_{A_1}^{A_2} \mathbf{F} \cdot d\mathbf{r} \qquad (13.2)$$

For a force defined by its rectangular components, we wrote

$$U_{1\to 2} = \int_{A_1}^{A_2} (F_x\, dx + F_y\, dy + F_z\, dz) \qquad (13.2'')$$

## Work of a Weight

We obtain the work of the weight $\mathbf{W}$ of a body as its center of gravity moves from the elevation $y_1$ to $y_2$ (Fig. 13.31) by substituting $F_x = F_z = 0$ and $F_y = -W$ into Eq. (13.2'') and integrating. We found

$$U_{1\to 2} = -\int_{y_1}^{y_2} W\, dy = W y_1 - W y_2 \qquad (13.4)$$

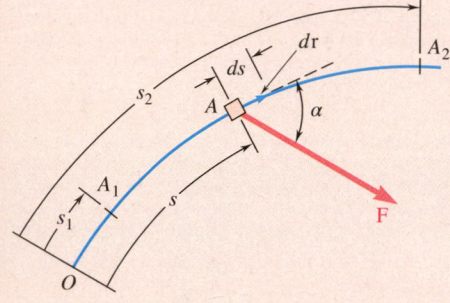

Fig. 13.30

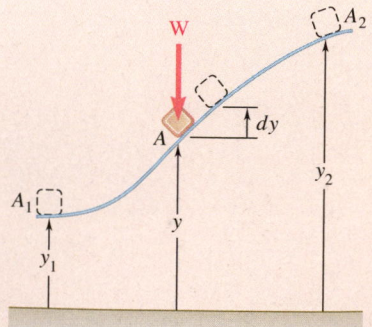

Fig. 13.31

## Work of the Force Exerted by a Spring

The work of a force $\mathbf{F}$ exerted by a spring on a body $A$ during a finite displacement of the body (Fig. 13.32) from $A_1(x = x_1)$ to $A_2(x = x_2)$ was obtained by

$$dU = -F\,dx = -kx\,dx$$

$$U_{1\to2} = -\int_{x_1}^{x_2} kx\,dx = \tfrac{1}{2}kx_1^2 - \tfrac{1}{2}kx_2^2 \qquad (13.6)$$

The work of $\mathbf{F}$ is therefore positive when the spring is returning to its undeformed position.

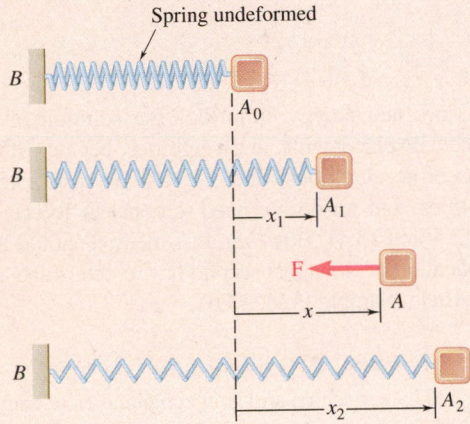

Spring undeformed

**Fig. 13.32**

## Work of the Gravitational Force

We obtained the **work of the gravitational force** $\mathbf{F}$ exerted by a particle of mass $M$ located at $O$ on a particle of mass $m$ as the latter moves from $A_1$ to $A_2$ (Fig. 13.33) by recalling from Sec. 12.2C the expression for the magnitude of $\mathbf{F}$ and writing

$$U_{1\to2} = -\int_{r_1}^{r_2} \frac{GMm}{r^2}\,dr = \frac{GMm}{r_2} - \frac{GMm}{r_1} \qquad (13.7)$$

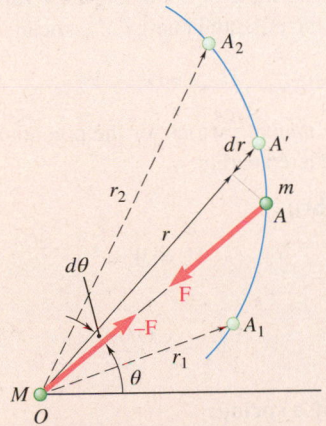

**Fig. 13.33**

## Kinetic Energy of a Particle

We defined the **kinetic energy of a particle** of mass $m$ moving with a velocity $\mathbf{v}$ (Sec. 13.1B) as the scalar quantity

$$T = \tfrac{1}{2}mv^2 \tag{13.9}$$

## Principle of Work and Energy

From Newton's second law, we derived the **principle of work and energy**, which states that we can obtain the kinetic energy of a particle at $A_2$ by adding its kinetic energy at $A_1$ to the work done during the displacement from $A_1$ to $A_2$ by the force $\mathbf{F}$ exerted on the particle as

$$T_1 + U_{1\to 2} = T_2 \tag{13.11}$$

## Method of Work and Energy

The method of work and energy simplifies the solution of many problems dealing with forces, displacements, and velocities, because it does not require the determination of accelerations (Sec. 13.1C). We also note that it involves only scalar quantities, and we do not need to consider forces that do no work (Sample Probs. 13.1 and 13.4). However, this method should be supplemented by the direct application of Newton's second law to determine a force normal to the path of the particle (Sample Prob. 13.6).

## Power and Mechanical Efficiency

The power developed by a machine and its mechanical efficiency were discussed in Sec. 13.1D. We defined power as the time rate at which work is done by

$$\text{Power} = \frac{dU}{dt} = \mathbf{F} \cdot \mathbf{v} \tag{13.12, 13.13}$$

where $\mathbf{F}$ is the force exerted on the particle and $\mathbf{v}$ is the velocity of the particle (Sample Prob. 13.7). The **mechanical efficiency**, denoted by $\eta$, was expressed as

$$\eta = \frac{\text{power output}}{\text{power input}} \tag{13.15}$$

## Conservative Force and Potential Energy

When the work of a force $\mathbf{F}$ is independent of the path followed (Secs. 13.2A and 13.2B), the force $\mathbf{F}$ is said to be a **conservative force**, and its work is equal to minus the change in the potential energy $V$ associated with $\mathbf{F}$

$$U_{1\to 2} = V_1 - V_2 \tag{13.19'}$$

We obtained the following expressions for the potential energy associated with each of the forces considered earlier.

**Force of gravity (weight):**

$$V_g = Wy \tag{13.16}$$

**Gravitational force:**

$$V_g = -\frac{GMm}{r} \tag{13.17}$$

**Elastic force exerted by a spring:**

$$V_e = \tfrac{1}{2}kx^2 \tag{13.18}$$

## Principle of Conservation of Energy

Substituting for $U_{1\to2}$ from Eq. (13.19′) into Eq. (13.11) and rearranging the terms (Sec. 13.2C), we obtained

$$T_1 + V_1 = T_2 + V_2 \qquad\qquad (13.24)$$

or

$$T_1 + V_{g_1} + V_{e_1} = T_2 + V_{g_2} + V_{e_2} \qquad\qquad (13.24')$$

This is the **principle of conservation of energy**, which states that, when a particle moves under the action of conservative forces, the sum of its kinetic and potential energies remains constant. The application of this principle facilitates the solution of problems involving only conservative forces (Sample Probs. 13.8 and 13.9).

## Alternative Expression for the Principle of Work and Energy

Rather than finding the work due to all external forces, you can write an alternative expression for the work–energy principle such that

$$T_1 + V_{g_1} + V_{e_1} + U_{1\to2}^{NC} = T_2 + V_{g_2} + V_{e_2} \qquad\qquad (13.24'')$$

where $U_{1\to2}^{NC}$ is the work of external non-conservative forces such as friction (Sample Prob. 13.10).

## Motion Under a Gravitational Force

Recalling from Sec. 12.2B that when a particle moves under a central force $\mathbf{F}$ its angular momentum about the center of force $O$ remains constant, we observed (Sec. 13.D) that, if the central force $\mathbf{F}$ is also conservative, the principles of conservation of angular momentum and of conservation of energy can be used jointly to analyze the motion of the particle (Sample Prob. 13.11). Because the gravitational force exerted by the earth on a space vehicle is both central and conservative, this approach was used to study the motion of such vehicles (Sample Prob. 13.12) and was found particularly effective in the case of an **oblique launching**. Considering the initial position $P_0$ and an arbitrary position $P$ of the vehicle (Fig. 13.34), we have

$$(H_O)_0 = H_O: \qquad r_0 m v_0 \sin \phi_0 = r m v \sin \phi \qquad (13.25)$$

$$T_0 + V_0 = T + V: \qquad \tfrac{1}{2}m v_0^2 - \frac{GMm}{r_0} = \tfrac{1}{2}m v^2 - \frac{GMm}{r} \qquad (13.26)$$

where $m$ is the mass of the vehicle and $M$ the mass of the earth.

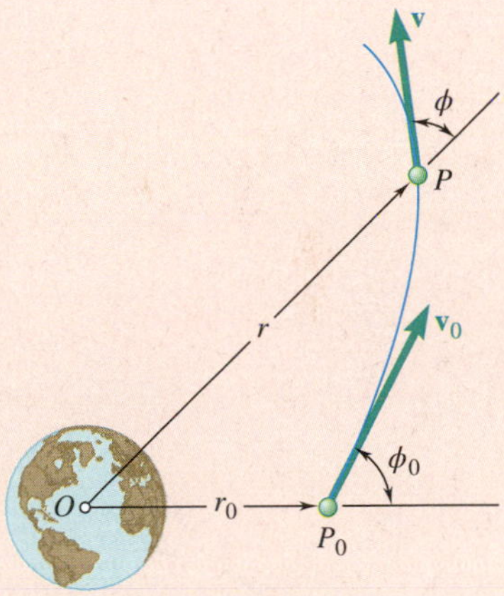

**Fig. 13.34**

## Principle of Impulse and Momentum for a Particle

The second half of this chapter was devoted to the method of impulse and momentum and to its application to the solution of various types of problems involving the motion of particles.

We defined the **linear momentum of a particle** (Sec. 13.3A) as the product $m\mathbf{v}$ of the mass $m$ of the particle and its velocity $\mathbf{v}$. From Newton's second law, $\mathbf{F} = m\mathbf{a}$, we derived the relation

$$m\mathbf{v}_1 + \int_{t_1}^{t_2} \mathbf{F}\, dt = m\mathbf{v}_2 \qquad\qquad (13.28)$$

where $m\mathbf{v}_1$ and $m\mathbf{v}_2$ represent the momentum of the particle at a time $t_1$ and a time $t_2$, respectively, and where the integral defines the **linear impulse of the force F** during the corresponding time interval. Therefore, we have

$$m\mathbf{v}_1 + \mathbf{Imp}_{1\to2} = m\mathbf{v}_2 \tag{13.30}$$

which expresses the principle of impulse and momentum for a particle.

When the particle considered is subjected to several forces, we need to use the sum of the impulses of these forces. So we have

$$m\mathbf{v}_1 + \Sigma\,\mathbf{Imp}_{1\to2} = m\mathbf{v}_2 \tag{13.32}$$

Because Eqs. (13.30) and (13.32) involve vector quantities, it is necessary to consider their $x$ and $y$ components separately when applying them to the solution of a given problem (Sample Probs. 13.13, 13.14, and 13.15).

### Impulsive Motion

The method of impulse and momentum is particularly effective in the study of the **impulsive motion** of a particle, when very large forces, called **impulsive forces**, are applied for a very short interval of time $\Delta t$, because this method involves the impulses $\mathbf{F}_{\text{avg}}\Delta t$ of the forces, rather than the forces themselves (Sec. 13.3B). Assuming that all non-impulsive forces (e.g., weight) are negligible, we wrote

$$m\mathbf{v}_1 + \Sigma\mathbf{F}_{\text{avg}}\,\Delta t = m\mathbf{v}_2 \tag{13.35}$$

In the case of the impulsive motion of several particles, we had

$$\Sigma m\mathbf{v}_1 + \Sigma\mathbf{F}_{\text{avg}}\,\Delta t = \Sigma m\mathbf{v}_2 \tag{13.36}$$

where the second term involves only impulsive, external forces (Sample Prob. 13.17).

In the particular case when the sum of the impulses of the external forces is zero, Eq. (13.36) reduces to $\Sigma m\mathbf{v}_1 = \Sigma m\mathbf{v}_2$; that is, the total momentum of the particles is conserved. For two particles, this reduces to

$$m_A\mathbf{v}_A + m_B\mathbf{v}_B = m_A\mathbf{v}'_A + m_B\mathbf{v}'_B \tag{13.34$'$}$$

### Direct Central Impact

In Sec. 13.4, we considered the **central impact** of two colliding bodies. In the case of a **direct central impact** (Sec. 13.4A), the two colliding bodies $A$ and $B$ were moving along the **line of impact** with velocities $\mathbf{v}_A$ and $\mathbf{v}_B$, respectively (Fig. 13.35). Two equations could be used to determine their velocities $\mathbf{v}'_A$ and $\mathbf{v}'_B$ after the impact. The first expressed conservation of the total momentum of the two bodies as

$$m_A v_A + m_B v_B = m_A v'_A + m_B v'_B \tag{13.37}$$

where a positive sign indicates that the corresponding velocity is directed to the right. The second equation, called the coefficient of restitution equation, related the *relative velocities* of the two bodies before and after the impact as

$$v'_B - v'_A = e(v_A - v_B) \tag{13.43}$$

The constant $e$ is known as the **coefficient of restitution**; its value lies between 0 and 1 and depends, in large measure, on the materials involved. When $e = 0$, the impact is said to be **perfectly plastic**; when $e = 1$, it is said to be **perfectly elastic**. Eq. (13.43) is only valid for direct central impact. (Sample Prob. 13.18).

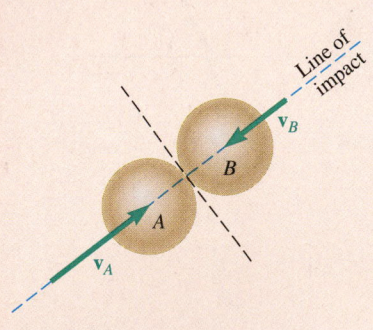

**Fig. 13.35**

## Oblique Central Impact

In the case of an **oblique central impact** (Sec. 13.4B), the velocities of the two colliding smooth bodies before and after the impact were resolved into $n$ components along the line of impact and $t$ components along the common tangent to the surfaces in contact (Fig. 13.36). We observed that the $t$ component of the velocity of each body remained unchanged, while the $n$ components satisfied equations similar to Eqs. (13.37) and (13.43) (Sample Probs. 13.19 and 13.20). We showed that, although this method was developed for bodies moving freely before and after the impact, it could be extended to the case when one or both of the colliding bodies is constrained in its motion (Sample Prob. 13.21). When the velocities are not along the line of impact, the coefficient of restitution equation uses the normal component,

$$(v_B')_n - (v_A')_n = e[(v_A)_n - (v_B)_n] \tag{13.49}$$

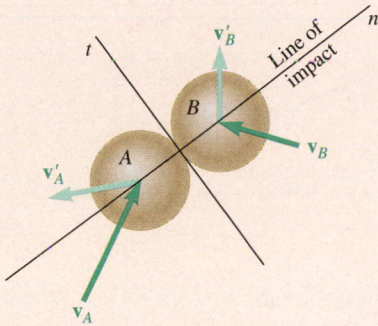

**Fig. 13.36**

## Using the Three Fundamental Methods of Kinetic Analysis

In Sec. 13.4C, we discussed the relative advantages of the three fundamental methods presented in this chapter and the preceding one, namely, Newton's second law, work and energy, and impulse and momentum. We noted that we can combine the method of work and energy and the method of impulse and momentum to solve problems involving a short impact phase during which impulsive forces must be taken into consideration (Sample Probs. 13.22 and 13.23).

# Review Problems

**13.190** A 34,000-lb airplane lands on an aircraft carrier and is caught by an arresting cable. The cable is inextensible and is paid out at $A$ and $B$ from mechanisms located below deck and consisting of pistons moving in long oil-filled cylinders. The piston–cylinder system is designed to maintain a constant tension in the cable. Knowing that the landing speed is 110 mi/h and the airplane travels a distance $d = 90$ ft after being caught, determine the tension in the cable.

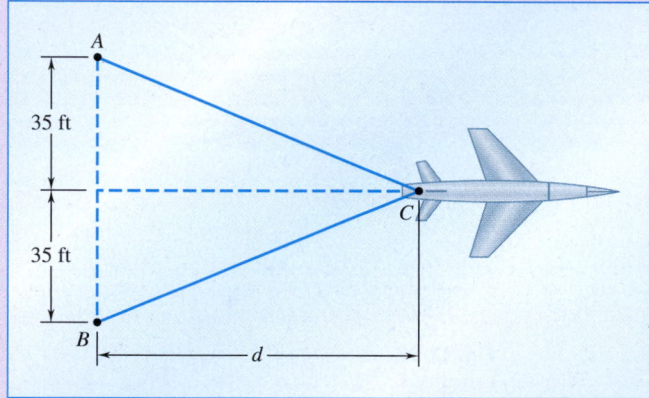

**Fig. P13.190**

**13.191** There has been renewed interest in pneumatic tube transportation systems (e.g., the Hyperloop®). These systems evacuate most of the air from sealed cylinders, and typically the passenger capsules float on air; both of these factors minimize drag. The passenger capsule starts from rest and climbs 1250 feet over a mountain pass. Knowing that the weight of a passenger capsule is 50,000 lb, determine the amount of energy the linear induction motors will need to supply if the capsule is traveling 300 mi/h at the top of the pass.

**Fig. P13.191**

**13.192** A satellite describes an elliptic orbit about a planet of mass $M$. The minimum and maximum values of the distance $r$ from the satellite to the center of the planet are, respectively, $r_0$ and $r_1$. Use the principles of conservation of energy and conservation of angular momentum to derive the relation

$$\frac{1}{r_0} + \frac{1}{r_1} = \frac{2GM}{h^2}$$

where $h$ is the angular momentum per unit mass of the satellite and $G$ is the constant of gravitation.

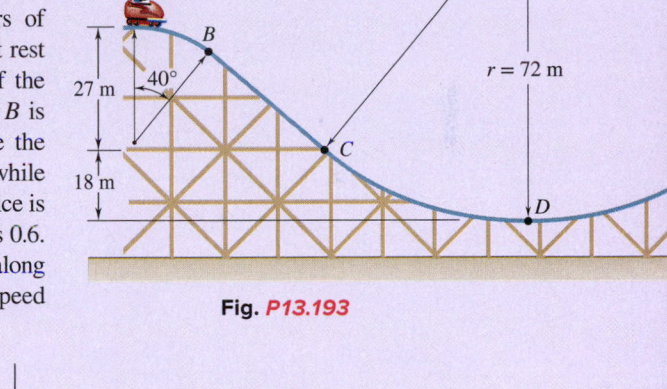

**Fig. P13.192**

**13.193** A section of track for a roller coaster consists of two circular arcs $AB$ and $CD$ joined by a straight portion $BC$. The radius of $AB$ is 27 m and the radius of $CD$ is 72 m. The car and its occupants, of total mass 250 kg, reach point $A$ with practically no velocity and then drop freely along the track. Determine the maximum and minimum values of the normal force exerted by the track on the car as the car travels from $A$ to $D$. Ignore air resistance and rolling resistance.

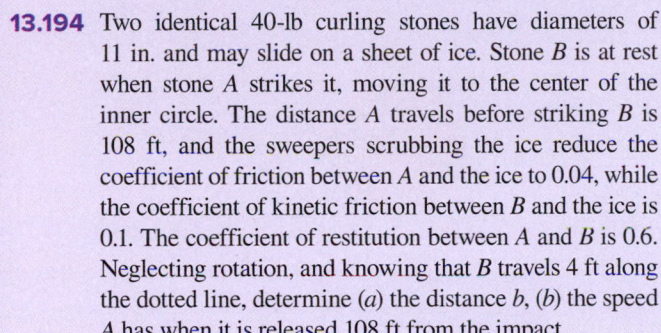

**13.194** Two identical 40-lb curling stones have diameters of 11 in. and may slide on a sheet of ice. Stone $B$ is at rest when stone $A$ strikes it, moving it to the center of the inner circle. The distance $A$ travels before striking $B$ is 108 ft, and the sweepers scrubbing the ice reduce the coefficient of friction between $A$ and the ice to 0.04, while the coefficient of kinetic friction between $B$ and the ice is 0.1. The coefficient of restitution between $A$ and $B$ is 0.6. Neglecting rotation, and knowing that $B$ travels 4 ft along the dotted line, determine (*a*) the distance $b$, (*b*) the speed $A$ has when it is released 108 ft from the impact.

**Fig. P13.193**

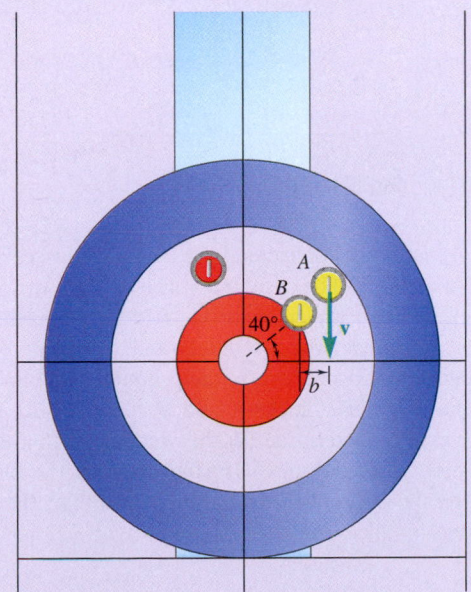

**Fig. P13.194**

**13.195** A 300-g block is released from rest after a spring of constant $k = 600$ N/m has been compressed 160 mm. Determine the force exerted by the loop $ABCD$ on the block as the block passes through (*a*) point $A$, (*b*) point $B$, (*c*) point $C$. Assume no friction.

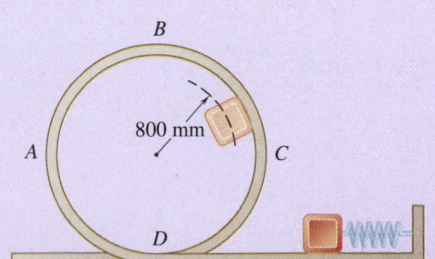

**Fig. P13.195**

917

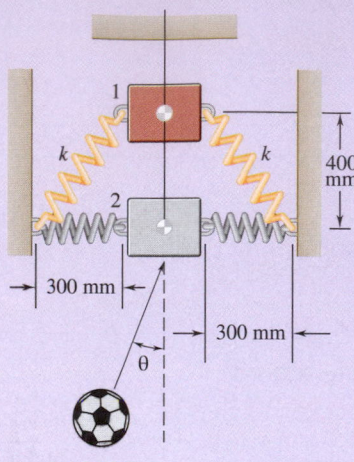

Fig. **P13.196**

**13.196** A kicking-simulation attachment goes on the front of a wheelchair, allowing athletes with mobility impairments to play soccer. The athletes load up the spring shown through a ratchet mechanism that pulls the 2-kg "foot" back to position 1. They then release the "foot" to impact the 0.45-kg soccer ball that is rolling toward the "foot" with a speed of 2 m/s at an angle $\theta = 30°$, as shown in the figure. The impact occurs with a coefficient of restitution $e = 0.75$ when the foot is at position 2, where the springs are unstretched. Knowing that the effective friction coefficient during rolling is $\mu_k = 0.1$, determine (*a*) the necessary spring coefficient to make the ball roll 30 m, (*b*) the direction the ball will travel after it is kicked.

**13.197** A 625-g basketball and a 58.5-g tennis ball are dropped from a height of $d = 1.5$ m onto the floor. The coefficient of restitution between the basketball and the ground is 0.85, and the coefficient of restitution between the tennis ball and the basketball is 0.9. Knowing that there is a small gap between the balls as they fall, determine the maximum height of the tennis ball after the impact.

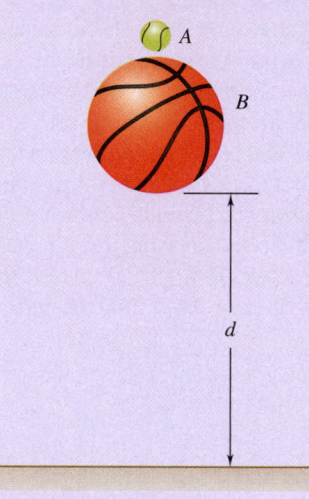

Fig. **P13.197**

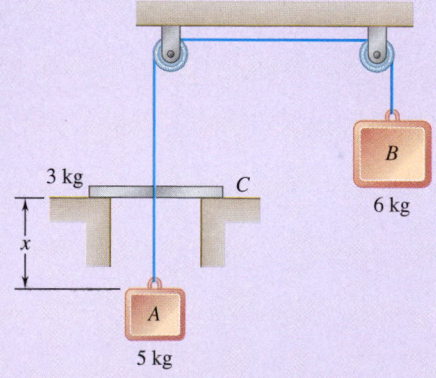

Fig. **P13.198**

**13.198** Blocks *A* and *B* are connected by a cord that passes over pulleys and through a collar *C*. The system is released from rest when $x = 1.7$ m. As block *A* rises, it strikes collar *C* with perfectly plastic impact ($e = 0$). After impact, the two blocks and the collar keep moving until they come to a stop and reverse their motion. As *A* and *C* move down, *C* hits the ledge and blocks *A* and *B* keep moving until they come to another stop. Determine (*a*) the velocity of the blocks and collar immediately after *A* hits *C*, (*b*) the distance the blocks and collar move after the impact before coming to a stop, (*c*) the value of *x* at the end of one complete cycle.

**13.199** A 2-kg ball $B$ is traveling horizontally at 10 m/s when it strikes 2-kg ball $A$. Ball $A$ is initially at rest and is attached to a spring with constant 100 N/m and an unstretched length of 1.2 m. Knowing the coefficient of restitution between $A$ and $B$ is 0.8 and friction between all surfaces is negligible, determine the normal force between $A$ and the ground when it is at the bottom of the hill.

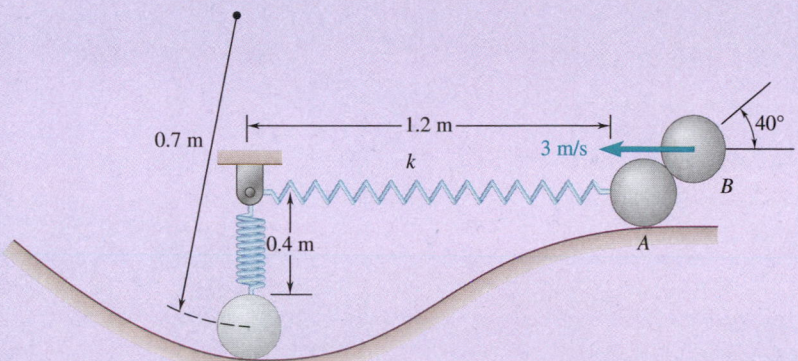

**Fig. P13.199**

**13.200** A 2-kg block $A$ is pushed up against a spring compressing it a distance $x$. The block is then released from rest and slides down the 20° incline until it strikes a 1-kg sphere $B$ that is suspended from a 1-m inextensible rope. The spring constant $k = 800$ N/m, the coefficient of friction between $A$ and the ground is 0.2, the distance $A$ slides from the unstretched length of the spring $d = 1.5$ m, and the coefficient of restitution between $A$ and $B$ is 0.8. Knowing the tension in the rope is 20 N when $\alpha = 30°$, determine the initial compression $x$ of the spring.

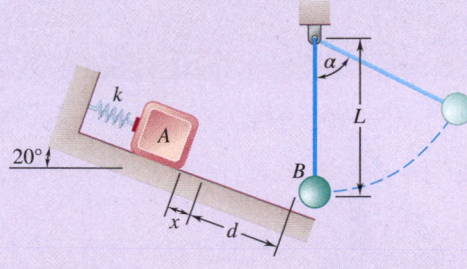

**Fig. P13.200**

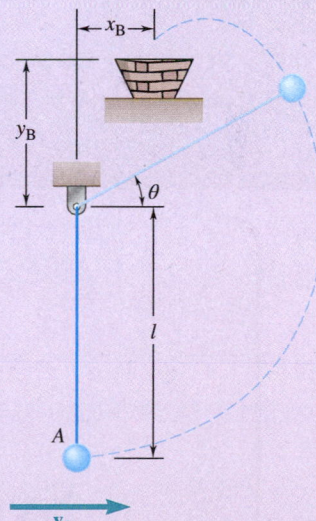

***13.201** The 2-lb ball at $A$ is suspended by an inextensible cord and given an initial horizontal velocity of $\mathbf{v}_0$. If $l = 2$ ft, $x_B = 0.3$ ft, and $y_B = 0.4$ ft, determine the initial velocity $\mathbf{v}_0$ so that the ball will enter the basket. (*Hint:* Use a computer to solve the resulting set of equations.)

**Fig. P13.201**

©XCOR Aerospace/Mike Massee

# 14

## Systems of Particles

The thrust for this XR-5M15 prototype engine is produced by gas particles being ejected at a high velocity. The determination of the forces on the test stand is based on the analysis of the motion of a *variable system of particles*; that is, the motion of a large number of air particles considered together rather than separately.

# Objectives

- **Apply** Newton's second law to a system of particles.
- **Calculate** the linear momentum and the angular momentum about a point of a system of particles.
- **Describe** the motion of the center of mass of a system of particles.
- **Determine** the kinetic energy of a system of particles.
- **Analyze** the motion of a system of particles by using the principle of work and energy and the principle of impulse and momentum.
- **Analyze** the motion of steady streams of particles.
- **Analyze** systems of particles gaining or losing mass.

# Introduction

In this chapter, you will study the motion of **systems of particles**; that is, the motion of a large number of particles considered together. In the first part of the chapter, we examine systems consisting of well-defined particles, like a set of billiard balls or a projectile that fragments into pieces. In the second part, we consider the motion of variable systems; these are systems that are continually gaining or losing particles or doing both at the same time. This could describe the motion of a stream of water or of a rocket during launch.

We start by applying Newton's second law to each particle of the system. We show that the *external forces* acting on the various particles form a system equipollent to the system of $m_i\mathbf{a}_i$ for the various particles. In other words, both systems have the same resultant and the same moment resultant about any given point. We further show that the resultant and moment resultant of the external forces are equal, respectively, to the rate of change of the total linear momentum and to the rate of change of the total angular momentum of the particles of the system.

We then define the *mass center* of a system of particles and describe the motion of that point, along with an analysis of the motion of the particles about their mass center. We discuss the conditions under which the linear momentum and the angular momentum of a system of particles are conserved and apply these results to the solution of various problems.

In Sec. 14.2, we apply the work–energy principle to a system of particles, and then we apply the impulse–momentum principle. We use these ideas to solve several problems of practical interest.

Note that although the derivations given in the first part of this chapter are carried out for a system of independent particles, they remain valid when the particles of the system are rigidly connected; that is, when they form a rigid body. In fact, these results form the foundation of our discussion of the kinetics of rigid bodies in Chaps. 16 through 18.

In Sec. 14.3, we consider steady streams of particles, such as a stream of water diverted by a fixed vane or the flow of air through a jet engine. We show how to determine the force exerted by the stream on the vane and the thrust developed by the engine. Finally, we analyze systems that gain mass by continually absorbing particles or lose mass by continually expelling particles. Among the various practical applications of this analysis is the determination of the thrust developed by a rocket engine.

# 14.1 APPLYING NEWTON'S SECOND LAW AND MOMENTUM PRINCIPLES TO SYSTEMS OF PARTICLES

In statics, we studied the effects of forces on particles and on rigid bodies in equilibrium. However, when you consider particles in motion, the situation of particles acting together but not forming a rigid body occurs in several important and practical applications. We analyze this kind of problem by applying Newton's laws to the system. The results are an interesting middle ground between the dynamics of particles and the dynamics of rigid bodies, which we will study next.

## 14.1A Newton's Second Law for a System of Particles

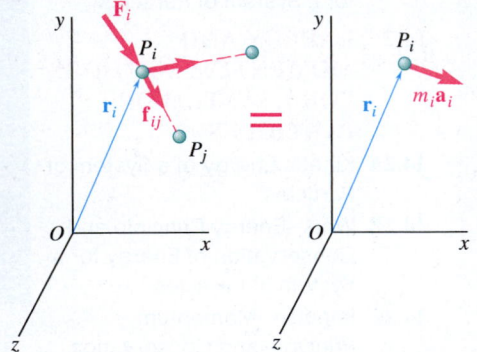

**Fig. 14.1** Newton's second law for the *i*th particle in a system of particles.

In order to derive the equations of motion for a system of $n$ particles, let us begin by writing Newton's second law for each individual particle of the system. Consider the particle $P_i$, where $1 \leq i \leq n$. Let $m_i$ be the mass of $P_i$ and let $\mathbf{a}_i$ be its acceleration with respect to the newtonian frame of reference $Oxyz$. The force exerted on $P_i$ by another particle $P_j$ of the system (Fig. 14.1), called an **internal force**, is denoted by $\mathbf{f}_{ij}$. The resultant of the internal forces exerted on $P_i$ by all the other particles of the system is thus $\sum_{j=1}^{n} \mathbf{f}_{ij}$ (where $\mathbf{f}_{ii}$ has no meaning and is assumed to be equal to zero). On the other hand, denoting the resultant of all the **external forces** acting on $P_i$ by $\mathbf{F}_i$, we write Newton's second law for the particle $P_i$ as

$$\mathbf{F}_i + \sum_{j=1}^{n} \mathbf{f}_{ij} = m_i \mathbf{a}_i \qquad (14.1)$$

Denoting the position vector of $P_i$ by $\mathbf{r}_i$ and taking the moments about $O$ of the various terms in Eq. (14.1), we also have

$$\mathbf{r}_i \times \mathbf{F}_i + \sum_{j=1}^{n} (\mathbf{r}_i \times \mathbf{f}_{ij}) = \mathbf{r}_i \times m_i \mathbf{a}_i \qquad (14.2)$$

Repeating this procedure for each particle $P_i$ of the system, we obtain $n$ equations of the type in Eq. (14.1) and $n$ equations of the type in Eq. (14.2), where $i$ takes successively the values $1, 2, \ldots, n$. Thus, these equations state that the external forces $\mathbf{F}_i$ and the internal forces $\mathbf{f}_{ij}$ acting on the various particles form a system equivalent to the system of the $m_i\mathbf{a}_i$ terms (i.e., one system may be replaced by the other) (Fig. 14.2).

Before proceeding further with our derivation, let us examine the internal forces $\mathbf{f}_{ij}$. These forces occur in pairs as $\mathbf{f}_{ij}$, $\mathbf{f}_{ji}$, where $\mathbf{f}_{ij}$ represents the force exerted by the particle $P_j$ on the particle $P_i$ and $\mathbf{f}_{ji}$ represents the force exerted by $P_i$ on $P_j$ (see Fig. 14.2). Now, according to Newton's third law (Sec. 6.1), as extended by Newton's law of gravitation to particles acting at a distance (Sec. 12.2C), the forces $\mathbf{f}_{ij}$ and $\mathbf{f}_{ji}$ are equal and opposite and have the

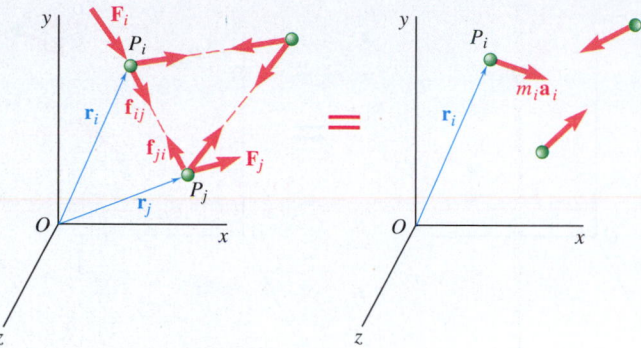

**Fig. 14.2** The sum of internal forces equals zero, and the sum of external forces equals the sum of the mass times acceleration for every particle in the system.

same line of action. Their sum is therefore $\mathbf{f}_{ij} + \mathbf{f}_{ji} = 0$, and the sum of their moments about $O$ is

$$\mathbf{r}_i \times \mathbf{f}_{ij} + \mathbf{r}_j \times \mathbf{f}_{ji} = \mathbf{r}_i \times (\mathbf{f}_{ij} + \mathbf{f}_{ji}) + (\mathbf{r}_j - \mathbf{r}_i) \times \mathbf{f}_{ji} = 0$$

because the vectors $\mathbf{r}_j - \mathbf{r}_i$ and $\mathbf{f}_{ji}$ in the last term are collinear. Adding all of the internal forces of the system and summing their moments about $O$, we obtain the equations

$$\sum_{i=1}^{n} \sum_{j=1}^{n} \mathbf{f}_{ij} = 0 \qquad \sum_{i=1}^{n} \sum_{j=1}^{n} (\mathbf{r}_i \times \mathbf{f}_{ij}) = 0 \qquad \textbf{(14.3)}$$

These equations state that the resultant and the moment resultant of the internal forces of the system are zero.

Returning now to the $n$ equations (14.1), where $i = 1, 2, \ldots, n$, we sum their left-hand sides and sum their right-hand sides. Taking into account the first of Eqs. (14.3), we obtain

$$\sum_{i=1}^{n} \mathbf{F}_i = \sum_{i=1}^{n} m_i \mathbf{a}_i \qquad \textbf{(14.4)}$$

Proceeding similarly with Eq. (14.2), and taking into account the second of Eqs. (14.3), we have

$$\sum_{i=1}^{n} (\mathbf{r}_i \times \mathbf{F}_i) = \sum_{i=1}^{n} (\mathbf{r}_i \times m_i \mathbf{a}_i) \qquad \textbf{(14.5)}$$

Eqs. (14.4) and (14.5) express the fact that the system of the external forces $\mathbf{F}_i$ and the system of $m_i \mathbf{a}_i$ have the same resultant and the same moment resultant. Referring to the definition given in *Statics* Sec. 3.4B for two equipollent systems of vectors, we can therefore state that **the system of the external forces acting on the particles and the system of the $m_i \mathbf{a}_i$ terms of the particles are equipollent** (Fig. 14.3). Fig. 14.3 basically shows that a free-body diagram for a system of particles is equal to its kinetic diagram.

Eqs. (14.3) state that the system of internal forces $\mathbf{f}_{ij}$ is equipollent to zero. Note, however, that it does *not* follow that the internal forces have no effect on the individual particles under consideration. Indeed, the gravitational forces that the sun and the planets exert on one another are internal to the solar system and are equipollent to zero. Yet these forces are responsible for the motion of the planets about the sun.

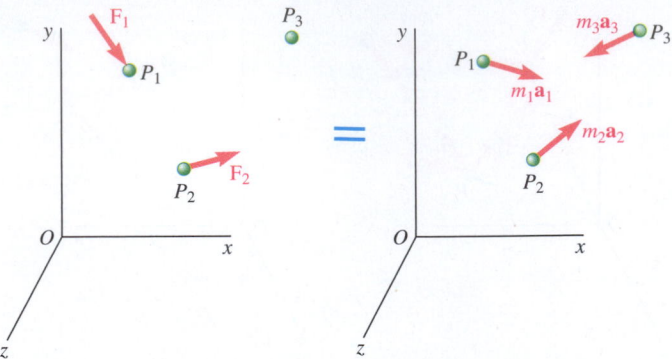

**Fig. 14.3** The free-body diagram for a system of particles is equal to the kinetic diagram for a system of particles.

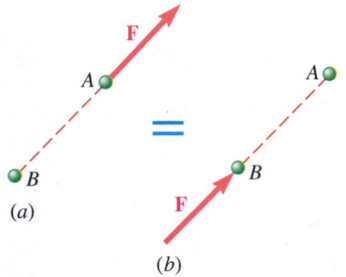

**Fig. 14.4** (*a*) A system of resultant force and moment applied to particle *A* is not equivalent to (*b*) the same force and moment applied to particle *B*.

Similarly, it does not follow from Eqs. (14.4) and (14.5) that two systems of external forces that have the same resultant and the same moment resultant will have the same effect on a given system of particles. Clearly, the systems shown in Figs. 14.4*a* and 14.4*b* have the same resultant and the same moment resultant; yet the first system accelerates particle *A* and leaves particle *B* unaffected, whereas the second system accelerates *B* and does not affect *A*. It is important to recall that when we stated in Sec. 3.4B that two equipollent systems of forces acting on a rigid body are also equivalent, we specifically noted that this property could *not* be extended to a system of forces acting on a set of independent particles such as those considered in this chapter.

In order to avoid any confusion, we use blue equal signs to connect equipollent systems of vectors, such as those shown in Figs. 14.3 and 14.4. These signs indicate that the two systems of vectors have the same resultant and the same moment resultant. We continue to use red equal signs to indicate that two systems of vectors are equivalent; that is, that one system can actually be replaced by the other (Fig. 14.2).

## 14.1B   Linear and Angular Momentum of a System of Particles

We can express Eqs. (14.4) and (14.5) in a more condensed form by introducing the linear and the angular momentum of the system of particles. We define the linear momentum **L** of the system of particles as the sum of the linear momenta of the various particles of the system (Sec. 12.1B). Then, we have

**Linear momentum, system of particles**

$$\mathbf{L} = \sum_{i=1}^{n} m_i \mathbf{v}_i \qquad \textbf{(14.6)}$$

Defining the angular momentum $\mathbf{H}_O$ about $O$ of the system of particles in a similar way (Sec. 12.2A) gives us

**Angular momentum, system of particles**

$$\mathbf{H}_O = \sum_{i=1}^{n} (\mathbf{r}_i \times m_i \mathbf{v}_i) \qquad \textbf{(14.7)}$$

Differentiating both sides of Eqs. (14.6) and (14.7) with respect to $t$, we have

$$\dot{\mathbf{L}} = \sum_{i=1}^{n} m_i \dot{\mathbf{v}}_i = \sum_{i=1}^{n} m_i \mathbf{a}_i \tag{14.8}$$

and

$$\dot{\mathbf{H}}_O = \sum_{i=1}^{n} (\dot{\mathbf{r}}_i \times m_i \mathbf{v}_i) + \sum_{i=1}^{n} (\mathbf{r}_i \times m_i \dot{\mathbf{v}}_i)$$
$$= \sum_{i=1}^{n} (\mathbf{v}_i \times m_i \mathbf{v}_i) + \sum_{i=1}^{n} (\mathbf{r}_i \times m_i \mathbf{a}_i)$$

Because the vectors $\mathbf{v}_i$ and $m_i \mathbf{v}_i$ are collinear, this last equation reduces to

$$\dot{\mathbf{H}}_O = \sum_{i=1}^{n} (\mathbf{r}_i \times m_i \mathbf{a}_i) \tag{14.9}$$

Note that the right-hand sides of Eqs. (14.8) and (14.9) are identical to the right-hand sides of Eqs. (14.4) and (14.5), respectively. It follows that the left-hand sides of these equations are also equal. Recall that the left-hand side of Eq. (14.5) represents the sum of the moments $\mathbf{M}_O$ about $O$ of the external forces acting on the particles of the system. So, omitting the subscript $i$ from the sums, we have

$$\Sigma \mathbf{F} = \dot{\mathbf{L}} \tag{14.10}$$
$$\Sigma \mathbf{M}_O = \dot{\mathbf{H}}_O \tag{14.11}$$

These equations state:

**The resultant and the moment resultant about the fixed point $O$ of the external forces are equal to the rates of change of the linear momentum and of the angular momentum about $O$, respectively, of the system of particles.**

## 14.1C   Motion of the Mass Center of a System of Particles

We can write Eq. (14.10) in an alternative form by considering the **mass center** of the system of particles. The mass center of the system is the point $G$ defined by the position vector $\bar{\mathbf{r}}$, which satisfies the relation

$$m\bar{\mathbf{r}} = \sum_{i=1}^{n} m_i \mathbf{r}_i \tag{14.12}$$

where $m$ represents the total mass $m = \sum_{i=1}^{n} m_i$ of the particles. Resolving the position vectors $\bar{\mathbf{r}}$ and $\mathbf{r}_i$ into rectangular components, we obtain the following three scalar equations, which we can use to determine the coordinates $\bar{x}, \bar{y}, \bar{z}$ of the mass center:

$$m\bar{x} = \sum_{i=1}^{n} m_i x_i \qquad m\bar{y} = \sum_{i=1}^{n} m_i y_i \qquad m\bar{z} = \sum_{i=1}^{n} m_i z_i \tag{14.12'}$$

Because $m_i g$ represents the weight of the particle $P_i$, and $mg$ is the total weight of the particles, $G$ is also the center of gravity of the system of particles. However, in order to avoid any confusion, we refer to $G$ as the *mass center* of the system of particles when we are discussing properties associated with the *mass* of the particles, and as the *center of gravity* of the system when we consider properties associated with the *weight* of the particles. Particles located outside the gravitational field of the earth, for example, have a mass but no weight. We can then properly refer to their mass center, but obviously not to their center of gravity.[†]

Differentiating both members of Eq. (14.12) with respect to $t$, we obtain

$$m\dot{\bar{\mathbf{r}}} = \sum_{i=1}^{n} m_i \dot{\mathbf{r}}_i$$

or

$$m\bar{\mathbf{v}} = \sum_{i=1}^{n} m_i \mathbf{v}_i \qquad (14.13)$$

where $\bar{\mathbf{v}}$ represents the velocity of the mass center $G$ of the system of particles. But the right-hand side of Eq. (14.13) is, by definition, the linear momentum $\mathbf{L}$ of the system [see Eq. (14.6)]. We therefore have

$$\mathbf{L} = m\bar{\mathbf{v}} \qquad (14.14)$$

and, differentiating both members with respect to $t$,

$$\dot{\mathbf{L}} = m\bar{\mathbf{a}} \qquad (14.15)$$

where $\bar{\mathbf{a}}$ represents the acceleration of the mass center $G$. Substituting for $\dot{\mathbf{L}}$ from Eq. (14.15) into Eq. (14.10), we obtain

$$\boxed{\Sigma\mathbf{F} = m\bar{\mathbf{a}}} \qquad (14.16)$$

which defines the motion of the mass center $G$ of the system of particles.

Note that Eq. (14.16) is identical to the equation we would obtain for a particle of mass $m$ equal to the total mass of the particles of the system, acted upon by all the external forces. We therefore state:

> **The mass center of a system of particles moves as if the entire mass of the system and all of the external forces were concentrated at that point.**

This principle is best illustrated by the motion of an exploding projectile. We know that if air resistance is neglected, we can assume that a projectile will travel along a parabolic path. After it has exploded, the mass center $G$ of the fragments of the projectile will continue to travel along the same path. Indeed, point $G$ must move as if the mass and the weight of all fragments were concentrated at $G;$ it must therefore move as if the projectile had not exploded.

Also note that the preceding derivation does not involve the moments of the external forces. Therefore, *it would be wrong to assume* that the external forces are equipollent to a vector $m\bar{\mathbf{a}}$ attached at the mass center $G$. In general, this is not the case because, as you will see next, the sum of the moments about $G$ of the external forces is not, in general, equal to zero.

## 14.1D  Angular Momentum of a System of Particles About Its Mass Center

In some applications (e.g., in analyzing the motion of a rigid body), it is convenient to consider the motion of the particles of the system with respect to a centroidal frame of reference $Gx'y'z'$ that translates with respect to the newtonian frame of reference $Oxyz$ (Fig. 14.5). Although a centroidal frame

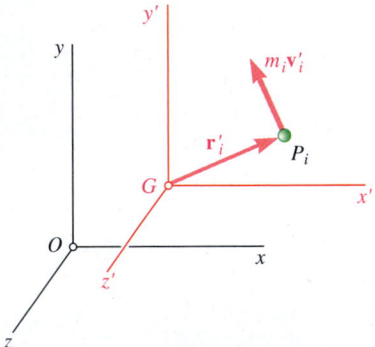

**Fig. 14.5** A centroidal frame of reference $Gx'y'z'$ moving in translation with respect to a newtonian frame of reference $Oxyz$.

---

[†]We should also point out that the mass center and the center of gravity of a system of particles do not exactly coincide, because the weights of the particles are directed toward the center of the earth and thus do not truly form a system of parallel forces. For particles on the earth, this difference is extremely small.

is not, in general, a newtonian frame of reference, we will show that the fundamental relation in Eq. (14.11) holds when the frame $Oxyz$ is replaced by $Gx'y'z'$.

Let's denote the position vector and the velocity of the particle $P_i$ relative to the moving frame of reference $Gx'y'z'$ by $\mathbf{r}'_i$ and $\mathbf{v}'_i$, respectively. We then define the **angular momentum $\mathbf{H}'_G$** of the system of particles **about the mass center $G$** as

$$\mathbf{H}'_G = \sum_{i=1}^{n} (\mathbf{r}'_i \times m_i \mathbf{v}'_i) \tag{14.17}$$

We now differentiate both members of Eq. (14.17) with respect to $t$. This operation is similar to that performed earlier on Eq. (14.7), so we can write immediately

$$\dot{\mathbf{H}}'_G = \sum_{i=1}^{n} (\mathbf{r}'_i \times m_i \mathbf{a}'_i) \tag{14.18}$$

where $\mathbf{a}'_i$ denotes the acceleration of $P_i$ relative to the moving frame of reference. Referring to Sec. 11.4D, we have

$$\mathbf{a}_i = \bar{\mathbf{a}} + \mathbf{a}'_i$$

where $\mathbf{a}_i$ and $\bar{\mathbf{a}}$ denote, respectively, the accelerations of $P_i$ and $G$ relative to the frame $Oxyz$. Solving for $\mathbf{a}'_i$ and substituting into Eq. (14.18), we have

$$\dot{\mathbf{H}}'_G = \sum_{i=1}^{n} (\mathbf{r}'_i \times m_i \mathbf{a}_i) - \left( \sum_{i=1}^{n} m_i \mathbf{r}'_i \right) \times \bar{\mathbf{a}} \tag{14.19}$$

However, by Eq. (14.12), the second sum in Eq. (14.19) is equal to $m\bar{\mathbf{r}}'$ and thus to zero, because the position vector $\bar{\mathbf{r}}'$ of $G$ relative to the frame $Gx'y'z'$ is clearly zero. On the other hand, because $\mathbf{a}_i$ represents the acceleration of $P_i$ relative to a newtonian frame, we can use Eq. (14.1) and replace $m_i\mathbf{a}_i$ by the sum of the internal forces $\mathbf{f}_{ij}$ and of the resultant $\mathbf{F}_i$ of the external forces acting on $P_i$. But a reasoning similar to that used in Sec. 14.1A shows that the moment resultant about $G$ of the internal forces $\mathbf{f}_{ij}$ of the entire system is zero. The first sum in Eq. (14.19) therefore reduces to the resultant moment about $G$ of the external forces acting on the particles of the system, and we have

$$\Sigma \mathbf{M}_G = \dot{\mathbf{H}}'_G \tag{14.20}$$

This equation states:

> **The resultant moment about $G$ of the external forces is equal to the rate of change of the angular momentum about $G$ of the system of particles.**

Note that in Eq. (14.17) we defined the angular momentum $\mathbf{H}'_G$ as the sum of the moments about $G$ of the momenta of the particles $m_i\mathbf{v}'_i$ *in their motion relative to the centroidal frame of reference $Gx'y'z'$*. We may sometimes want to compute the sum $\mathbf{H}_G$ of the moments about $G$ of the momenta of the particles $m_i\mathbf{v}_i$ *in their absolute motion*; that is, in their motion as observed from the newtonian frame of reference $Oxyz$ (Fig. 14.6):

$$\mathbf{H}_G = \sum_{i=1}^{n} (\mathbf{r}'_i \times m_i \mathbf{v}_i) \tag{14.21}$$

Remarkably, the angular momenta $\mathbf{H}'_G$ and $\mathbf{H}_G$ are identically equal. This can be verified by referring to Sec. 11.4D and writing

$$\mathbf{v}_i = \bar{\mathbf{v}} + \mathbf{v}'_i \tag{14.22}$$

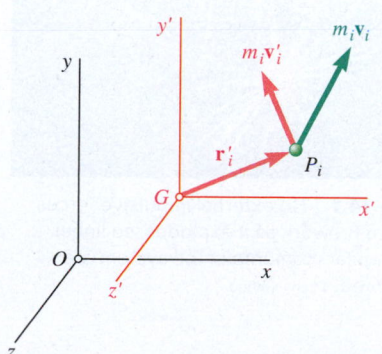

**Fig. 14.6**  The linear momentum of particle $P_i$ with respect to the centroidal frame ($m_i\mathbf{v}'_i$) and with respect to a newtonian frame ($m_i\mathbf{v}_i$).

Substituting for $\mathbf{v}_i$ from Eq. (14.22) into Eq. (14.21), we have

$$\mathbf{H}_G = \left(\sum_{i=1}^{n} m_i \mathbf{r}_i'\right) \times \bar{\mathbf{v}} + \sum_{i=1}^{n} (\mathbf{r}_i' \times m_i \mathbf{v}_i')$$

But, as observed earlier, the first sum is equal to zero. Thus, $\mathbf{H}_G$ reduces to the second sum, which by definition is equal to $\mathbf{H}_G'$.[†]

Taking advantage of the property we have just established, we simplify our notation by dropping the prime (') from Eq. (14.20) and writing

$$\Sigma \mathbf{M}_G = \dot{\mathbf{H}}_G \qquad (14.23)$$

Here we can compute the angular momentum $\mathbf{H}_G$ by taking the moments about $G$ of the momenta of the particles with respect to either the newtonian frame $Oxyz$ or the centroidal frame $Gx'y'z'$:

$$\mathbf{H}_G = \sum_{i=1}^{n} (\mathbf{r}_i' \times m_i \mathbf{v}_i) = \sum_{i=1}^{n} (\mathbf{r}_i' \times m_i \mathbf{v}_i') \qquad (14.24)$$

## 14.1E Conservation of Momentum for a System of Particles

If no external force acts on the particles of a system, the left-hand sides of Eqs. (14.10) and (14.11) are equal to zero. These equations then reduce to $\dot{\mathbf{L}} = 0$ and $\dot{\mathbf{H}}_O = 0$. We conclude that

$$\mathbf{L} = \text{constant} \qquad \mathbf{H}_O = \text{constant} \qquad (14.25)$$

These equations state that the linear momentum of the system of particles and its angular momentum about the fixed point $O$ are conserved.

In some applications, such as problems involving central forces, the moment about a fixed point $O$ of each of the external forces can be zero without any of the forces being zero. In such cases, the second of Eqs. (14.25) still holds; the angular momentum of the system of particles about $O$ is conserved.

We can also apply the concept of conservation of momentum to the analysis of the motion of the mass center $G$ of a system of particles and to the analysis of the motion of the system about $G$. For example, if the sum of the external forces is zero, the first of Eqs. (14.25) applies. Recalling Eq. (14.14), we have

$$\bar{\mathbf{v}} = \text{constant} \qquad (14.26)$$

This equation says that the mass center $G$ of the system moves in a straight line and at a constant speed. On the other hand, if the sum of the moments about $G$ of the external forces is zero, it follows from Eq. (14.23) that the angular momentum of the system about its mass center is conserved.

$$\mathbf{H}_O = \text{constant} \qquad (14.27)$$

**Photo 14.1** No external impulsive forces act on a firework as it explodes, so linear and angular momenta of the system are conserved. ©Lena Kofoed

---

[†]Note that this property is peculiar to the centroidal frame $Gx'y'z'$ and does not, in general, hold for other frames of reference (see Prob. 14.29).

## Sample Problem 14.1

A 200-kg space vehicle passes through the origin of a newtonian reference frame *Oxyz* at time $t = 0$ with velocity $\mathbf{v}_0 = (150 \text{ m/s})\mathbf{i}$ relative to the frame. Following the detonation of explosive charges, the vehicle separates into three parts *A*, *B*, and *C*, each with a mass of 100 kg, 60 kg, and 40 kg, respectively. At $t = 2.5$ s, the positions of parts *A* and *B* are observed to be $A(555, -180, 240)$ and $B(255, 0, -120)$, where the coordinates are expressed in meters. Determine the position of part *C* at that time.

**STRATEGY:**   There are no external forces, so the linear momentum of the system is conserved. Use kinematics to relate the motion of the center of mass of the spacecraft and the rectangular coordinates of its position.

**MODELING and ANALYSIS:**   The system is the space vehicle. After the explosion, the system is composed of all three parts: *A*, *B*, and *C*. The mass center *G* of the system moves with the constant velocity $\mathbf{v}_0 = (150 \text{ m/s})\mathbf{i}$. At $t = 2.5$ s, its position is

$$\bar{\mathbf{r}} = \mathbf{v}_0 t = (150 \text{ m/s})\mathbf{i}(2.5 \text{ s}) = (375 \text{ m})\mathbf{i}$$

Recalling Eq. (14.12), you have

$$m\bar{\mathbf{r}} = m_A \mathbf{r}_A + m_B \mathbf{r}_B + m_C \mathbf{r}_C$$

$$(200 \text{ kg})(375 \text{ m})\mathbf{i} = (100 \text{ kg})[(555 \text{ m})\mathbf{i} - (180 \text{ m})\mathbf{j} + (240 \text{ m})\mathbf{k}]$$

$$+ (60 \text{ kg})[(255 \text{ m})\mathbf{i} - (120 \text{ m})\mathbf{k}] + (40 \text{ kg})\mathbf{r}_C$$

$$\mathbf{r}_C = (105 \text{ m})\mathbf{i} + (450 \text{ m})\mathbf{j} - (420 \text{ m})\mathbf{k} \quad \blacktriangleleft$$

**REFLECT and THINK:**   This kind of calculation can serve as a model for any situation involving fragmentation of a projectile with no external forces present.

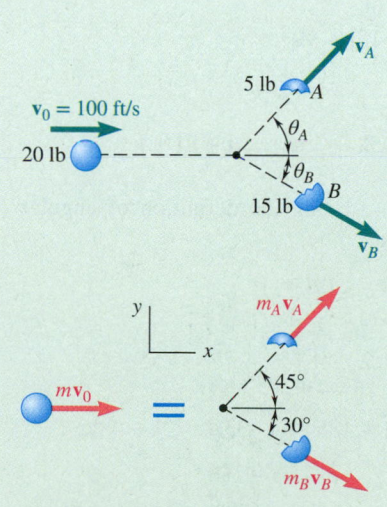

**Fig. 1**  Impulse–momentum diagram for the projectile.

## Sample Problem 14.2

A 20-lb projectile is moving with a velocity of 100 ft/s when it explodes into two fragments *A* and *B*, weighing 5 lb and 15 lb, respectively. Knowing that immediately after the explosion, fragments *A* and *B* travel in directions defined respectively by $\theta_A = 45°$ and $\theta_B = 30°$, determine the velocity of each fragment.

**STRATEGY:**   There are no external forces, so apply the conservation of linear momentum to the system.

**MODELING and ANALYSIS:**   The system is the projectile. After the explosion, the system is composed of the two fragments. The impulse–momentum diagram for this system is shown in Fig. 1. There are no external impulses acting on this system, so linear momentum is conserved and

$$m_A \mathbf{v}_A + m_B \mathbf{v}_B = m\mathbf{v}_0$$

$$(5/g)\mathbf{v}_A + (15/g)\mathbf{v}_B = (20/g)\mathbf{v}_0$$

*(continued)*

Applying this equation in the $x$ and $y$ directions gives you two scalar equations. Thus,

$\xrightarrow{+}$ $x$ components : $\qquad 5v_A \cos 45° + 15v_B \cos 30° = 20(100)$

$+\uparrow$ $y$ components : $\qquad 5v_A \sin 45° - 15v_B \sin 30° = 0$

Solving the two equations for $v_A$ and $v_B$ simultaneously gives

$$v_A = 207 \text{ ft/s} \qquad v_B = 97.6 \text{ ft/s}$$

$$\mathbf{v}_A = 207 \text{ ft/s} \measuredangle 45° \qquad \mathbf{v}_B = 97.6 \text{ ft/s} \measuredangle 30° \quad \blacktriangleleft$$

**REFLECT and THINK:** As you might have predicted, the less massive fragment winds up with a larger magnitude of velocity and departs the original trajectory at a larger angle.

## Sample Problem 14.3

A system consists of three particles $A$, $B$, and $C$, with masses $m_A = 1$ kg, $m_B = 2$ kg, and $m_C = 3$ kg. The velocities of the particles expressed in m/s are, respectively, $\mathbf{v}_A = 3\mathbf{i} - 2\mathbf{j} + 4\mathbf{k}$, $\mathbf{v}_B = 4\mathbf{i} + 3\mathbf{j}$, and $\mathbf{v}_C = 2\mathbf{i} + 5\mathbf{j} - 3\mathbf{k}$. Determine (*a*) the angular momentum $\mathbf{H}_O$ of the system about $O$, (*b*) the position vector $\bar{\mathbf{r}}$ of the mass center $G$ of the system, (*c*) the angular momentum $\mathbf{H}_G$ of the system about $G$.

**STRATEGY:** You have a system of particles, so use the definitions of angular momentum and center of mass.

**MODELING:** Choose the three particles as your system.

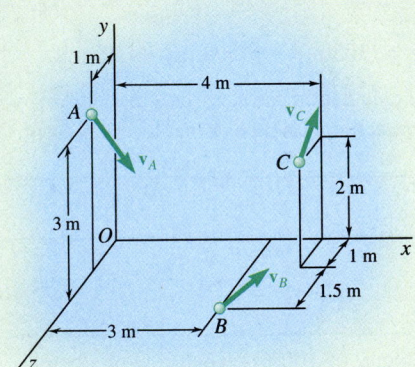

**ANALYSIS:** The linear momentum of each particle expressed in kg·m/s is

$$m_A\mathbf{v}_A = 3\mathbf{i} - 2\mathbf{j} + 4\mathbf{k}$$
$$m_B\mathbf{v}_B = 8\mathbf{i} + 6\mathbf{j}$$
$$m_C\mathbf{v}_C = 6\mathbf{i} + 15\mathbf{j} - 9\mathbf{k}$$

The position vectors (in meters) are

$$\mathbf{r}_A = 3\mathbf{j} + \mathbf{k} \qquad \mathbf{r}_B = 3\mathbf{i} + 2.5\mathbf{k} \qquad \mathbf{r}_C = 4\mathbf{i} + 2\mathbf{j} + \mathbf{k}$$

**a. Angular Momentum About O.** Using the definition of angular momentum about $O$ (in kg·m$^2$/s), you find

$$\mathbf{H}_O = \mathbf{r}_A \times (m_A\mathbf{v}_A) + \mathbf{r}_B \times (m_B\mathbf{v}_B) + \mathbf{r}_C \times (m_C\mathbf{v}_C)$$

$$= \begin{vmatrix} \mathbf{i} & \mathbf{j} & \mathbf{k} \\ 0 & 3 & 1 \\ 3 & -2 & 4 \end{vmatrix} + \begin{vmatrix} \mathbf{i} & \mathbf{j} & \mathbf{k} \\ 3 & 0 & 2.5 \\ 8 & 6 & 0 \end{vmatrix} + \begin{vmatrix} \mathbf{i} & \mathbf{j} & \mathbf{k} \\ 4 & 2 & 1 \\ 6 & 15 & -9 \end{vmatrix}$$

$$= (14\mathbf{i} + 3\mathbf{j} - 9\mathbf{k}) + (-15\mathbf{i} + 20\mathbf{j} + 18\mathbf{k}) + (-33\mathbf{i} + 42\mathbf{j} + 48\mathbf{k})$$

$$= 34\mathbf{i} + 65\mathbf{j} + 57\mathbf{k}$$

$$\mathbf{H}_O = -(34 \text{ kg·m}^2\text{/s})\mathbf{i} + (65 \text{ kg·m}^2\text{/s})\mathbf{j} + (57 \text{ kg·m}^2\text{/s})\mathbf{k} \quad \blacktriangleleft$$

*(continued)*

**b. Mass Center.**   Using the definition of mass center, you find

$$(m_A + m_B + m_C)\bar{\mathbf{r}} = m_A\mathbf{r}_A + m_B\mathbf{r}_B + m_C\mathbf{r}_C$$

$$6\bar{\mathbf{r}} = (1)(3\mathbf{j} + \mathbf{k}) + (2)(3\mathbf{i} + 2.5\mathbf{k}) + (3)(4\mathbf{i} + 2\mathbf{j} + \mathbf{k})$$

$$\bar{\mathbf{r}} = 3\mathbf{i} + 1.5\mathbf{j} + 1.5\mathbf{k}$$

$$\bar{\mathbf{r}} = (3.00 \text{ m})\mathbf{i} + (1.500 \text{ m})\mathbf{j} + (1.500 \text{ m})\mathbf{k} \quad \blacktriangleleft$$

**c. Angular Momentum About G.**   The angular momentum of the system about $G$ is

$$\mathbf{H}_G = \mathbf{r}'_A \times m_A\mathbf{v}_A + \mathbf{r}'_B \times m_B\mathbf{v}_B + \mathbf{r}'_C \times m_C\mathbf{v}_C$$

where $\mathbf{r}'_A$, $\mathbf{r}'_B$, and $\mathbf{r}'_C$ are the position vectors from the particles to the center of mass; that is,

$$\mathbf{r}'_A = \mathbf{r}_A - \bar{\mathbf{r}} = -3\mathbf{i} + 1.5\mathbf{j} - 0.5\mathbf{k}$$
$$\mathbf{r}'_B = \mathbf{r}_B - \bar{\mathbf{r}} = -1.5\mathbf{j} + \mathbf{k}$$
$$\mathbf{r}'_C = \mathbf{r}_C - \bar{\mathbf{r}} = \mathbf{i} + 0.5\mathbf{j} - 0.5\mathbf{k}$$

Therefore, you can calculate the angular momentum as

$$\mathbf{H}_G = \mathbf{r}'_A \times (m_A\mathbf{v}_A) + \mathbf{r}'_B \times (m_B\mathbf{v}_B) + \mathbf{r}'_C \times (m_C\mathbf{v}_C)$$

$$= \begin{vmatrix} \mathbf{i} & \mathbf{j} & \mathbf{k} \\ -3 & 1.5 & -0.5 \\ 3 & -2 & 4 \end{vmatrix} + \begin{vmatrix} \mathbf{i} & \mathbf{j} & \mathbf{k} \\ 0 & -1.5 & 1 \\ 8 & 6 & 0 \end{vmatrix} + \begin{vmatrix} \mathbf{i} & \mathbf{j} & \mathbf{k} \\ 1 & 0.5 & -0.5 \\ 6 & 15 & -9 \end{vmatrix}$$

$$= (5\mathbf{i} + 10.5\mathbf{j} + 1.5\mathbf{k}) + (-6\mathbf{i} + 8\mathbf{j} + 12\mathbf{k}) + (3\mathbf{i} + 6\mathbf{j} + 12\mathbf{k})$$

$$= 2\mathbf{i} + 24.5\mathbf{j} + 25.5\mathbf{k}$$

$$\mathbf{H}_G = (2.00 \text{ kg·m}^2\text{/s})\mathbf{i} + (24.5 \text{ kg·m}^2\text{/s})\mathbf{j} + (25.5 \text{ kg·m}^2\text{/s})\mathbf{k} \quad \blacktriangleleft$$

**REFLECT and THINK:**   You should be able to verify that the answers to this problem satisfy the equations given in Prob. 14.27; that is, $\mathbf{H}_O = \bar{\mathbf{r}} \times m\bar{\mathbf{v}} + \mathbf{H}_G$. Because no impulses act on the system, the linear momentum of the overall system is constant; the location of the center of mass of the system, however, changes with time.

# SOLVING PROBLEMS
# ON YOUR OWN

This chapter dealt with the motion of **systems of particles** where the motion of a large number of particles is considered together, rather than separately. In this first section, you learned to compute the **linear momentum** and the **angular momentum** of a system of particles. We defined the linear momentum $\mathbf{L}$ of a system of particles as the sum of the linear momenta of the particles, and we defined the angular momentum $\mathbf{H}_O$ of the system as the sum of the angular momenta of the particles about $O$:

$$\mathbf{L} = \sum_{i=1}^{n} m_i \mathbf{v}_i \qquad \mathbf{H}_O = \sum_{i=1}^{n} (\mathbf{r}_i \times m_i \mathbf{v}_i) \qquad \textbf{(14.6, 14.7)}$$

In this section, you will be asked to solve several problems of practical interest, either by observing that the linear momentum of a system of particles is conserved or by considering the motion of the mass center of a system of particles.

**1. Conservation of the linear momentum of a system of particles.** This occurs *when the resultant of the external forces acting on the particles of the system is zero.* You may encounter such a situation in the following types of problems.

**a. Problems involving the rectilinear motion** of objects, such as colliding automobiles and railroad cars. After you have checked that the resultant of the external forces is zero, equate the algebraic sums of the initial momenta and final momenta to obtain an equation that you can solve for one unknown.

**b. Problems involving the two-dimensional or three-dimensional motion** of objects, such as exploding shells or colliding aircraft, automobiles, or billiard balls. After you have checked that the resultant of the external forces is zero, add the initial momenta of the objects vectorially, add their final momenta vectorially, and equate the two sums to obtain a vector equation expressing that the linear momentum of the system is conserved.

In the case of two-dimensional motion, you can replace this equation with two scalar equations that you can solve for two unknowns. In the case of three-dimensional motion, you can replace the equation with three scalar equations that you can solve for three unknowns.

**2. Motion of the mass center of a system of particles.** You saw in Sec. 14.1C that *the mass center of a system of particles moves as if the entire mass of the system and all of the external forces were concentrated at that point.*

**a. In the case of a body exploding while in motion,** it follows that the mass center of the resulting fragments moves as the body itself would have moved if the explosion had not occurred. You can solve problems of this type by writing the equation of motion of the mass center of the system in vector form and expressing the position vector of the mass center in terms of the position vectors of the various fragments [Eq. (14.12) and Sample Prob. 14.1]. You can then rewrite the vector equation as two or three scalar equations and solve the equations for an equivalent number of unknowns.

**b. In the case of the collision of several moving bodies,** it follows that the motion of the mass center of the various bodies is unaffected by the collision. You can solve problems of this type by writing the equation of motion of the mass center of the system in vector form and expressing its position vector before and after the collision in terms of the position vectors of the relevant bodies [Eq.(14.12)]. You can then rewrite the vector equation as two or three scalar equations and solve these equations for an equivalent number of unknowns.

# Problems

**14.1** A 30-g bullet is fired with a horizontal velocity of 450 m/s and becomes embedded in block *B*, which has a mass of 3 kg. After the impact, block *B* slides on 30-kg carrier *C* until it impacts the end of the carrier. Knowing the impact between *B* and *C* is perfectly plastic and the coefficient of kinetic friction between *B* and *C* is 0.2, determine (*a*) the velocity of the bullet and *B* after the first impact, (*b*) the final velocity of the carrier.

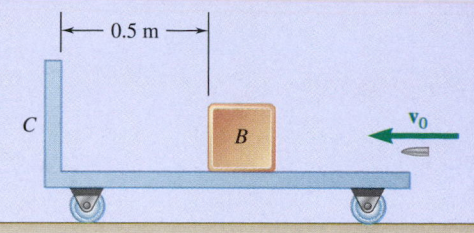

**Fig. P14.1**

**14.2** Two identical 1350-kg automobiles *A* and *B* are at rest with their brakes released when *B* is struck by a 5400-kg truck *C* that is moving to the left at 8 km/h. A second collision then occurs when *B* strikes *A*. Assuming the first collision is perfectly plastic and the second collision is perfectly elastic, determine the velocities of the three vehicles just after the second collision.

**Fig. P14.2**

**14.3** An airline employee tosses two suitcases in rapid succession, with a horizontal velocity of 7.2 ft/s, onto a 50-lb baggage carrier which is initially at rest. (*a*) Knowing that the final velocity of the baggage carrier is 3.6 ft/s and that the first suitcase the employee tosses onto the carrier has a weight of 30 lb, determine the weight of the other suitcase. (*b*) What would be the final velocity of the carrier if the employee reverses the order in which he tosses the suitcases?

**Fig. P14.3**

**14.4** Car *A* weighing 4000 lb and car *B* weighing 3700 lb are at rest on a 22-ton flatcar which is also at rest. Cars *A* and *B* then accelerate and quickly reach constant speeds relative to the flatcar of 7 ft/s and 3.5 ft/s, respectively, before decelerating to a stop at the opposite end of the flatcar. Neglecting friction and rolling resistance, determine the velocity of the flatcar when the cars are moving at constant speeds.

**Fig. P14.4**

**Fig. P14.5**

**14.5** Two swimmers $A$ and $B$, of weight 190 lb and 125 lb, respectively, are at diagonally opposite corners of a floating raft when they realize that the raft has broken away from its anchor. Swimmer $A$ immediately starts walking toward $B$ at a speed of 2 ft/s relative to the raft. Knowing that the raft weighs 300 lb, determine (*a*) the speed of the raft if $B$ does not move, (*b*) the speed with which $B$ must walk toward $A$ if the raft is not to move.

**14.6** A 180-lb man and a 120-lb woman stand side by side at the same end of a 300-lb boat, ready to dive, each with a 16-ft/s velocity relative to the boat. Determine the velocity of the boat after they have both dived, if (*a*) the woman dives first, (*b*) the man dives first.

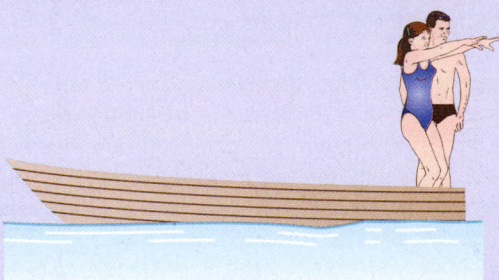

**Fig. P14.6**

**14.7** A 40-Mg boxcar $A$ is moving in a railroad switchyard with a velocity of 9 km/h toward cars $B$ and $C$, which are both at rest with their brakes off at a short distance from each other. Car $B$ is a 25-Mg flatcar supporting a 30-Mg container, and car $C$ is a 35-Mg boxcar. As the cars hit each other, they get automatically and tightly coupled. Determine the velocity of car $A$ immediately after each of the two couplings, assuming that the container (*a*) does not slide on the flatcar, (*b*) slides after the first coupling but hits a stop before the second coupling occurs, (*c*) slides and hits the stop only after the second coupling has occurred.

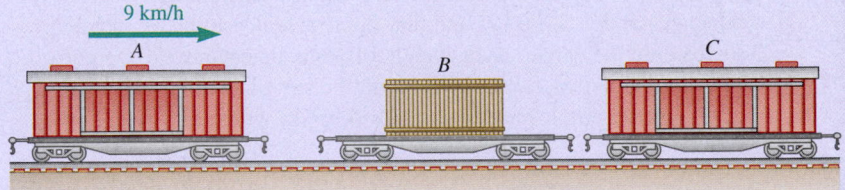

**Fig. P14.7**

**14.8** Two identical cars $A$ and $B$ are at rest on a loading dock with brakes released. Car $C$, of a slightly different style but of the same weight, has been pushed by dockworkers and hits car $B$ with a velocity of 1.5 m/s. Knowing that the coefficient of restitution is 0.8 between $B$ and $C$, and it is 0.5 between $A$ and $B$, determine the velocity of each car after all collisions have taken place.

**Fig. P14.8**

**14.9** A 20-kg base satellite deploys three sub-satellites, each of which has its own thrust capabilities, to perform research on tether propulsion. The masses of sub-satellites A, B, and C are 4 kg, 6 kg, and 8 kg, respectively, and their velocities expressed in m/s are given by $\mathbf{v}_A = 4\mathbf{i} - 2\mathbf{j} + 2\mathbf{k}$, $\mathbf{v}_B = \mathbf{i} + 4\mathbf{j}$, $\mathbf{v}_C = 2\mathbf{i} + 2\mathbf{j} + 4\mathbf{k}$. At the instant shown, what is the angular momentum $\mathbf{H}_O$ of the system about the base satellite?

**14.10** For the satellite system of Prob. 14.9, assuming that the velocity of the base satellite is zero, determine (a) the position vector $\bar{\mathbf{r}}$ of the mass center G of the system, (b) the linear momentum **L** of the system, (c) the angular momentum $\mathbf{H}_G$ of the system about G. Also, verify that the answers to this problem and to Prob. 14.9 satisfy the equation given in Prob. 14.27.

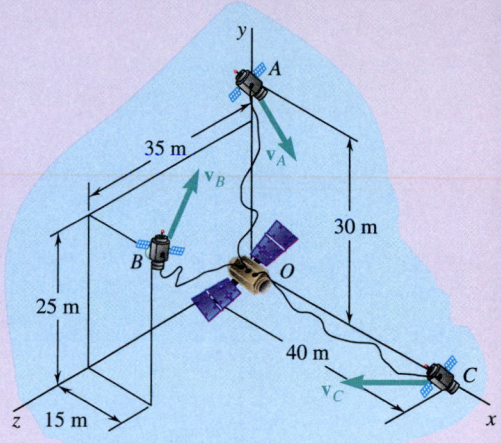

**Fig. P14.9 and P14.10**

**14.11** A system consists of three identical 19.32-lb particles A, B, and C. The velocities of the particles are, respectively, $\mathbf{v}_A = v_A\mathbf{j}$, $\mathbf{v}_B = v_B\mathbf{i}$, and $\mathbf{v}_C = v_C\mathbf{k}$. Knowing that the angular momentum of the system about O expressed in ft·lb·s is $\mathbf{H}_O = -1.2\mathbf{k}$, determine (a) the velocities of the particles, (b) the angular momentum of the system about its mass center G.

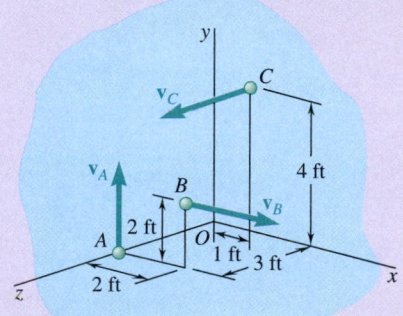

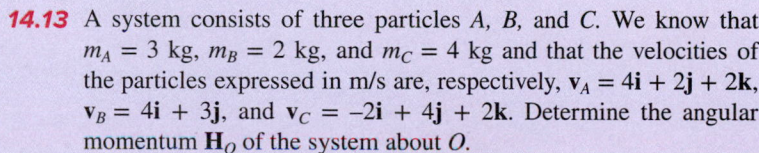

**Fig. P14.11 and P14.12**

**14.12** A system consists of three identical 19.32-lb particles A, B, and C. The velocities of the particles are, respectively, $\mathbf{v}_A = v_A\mathbf{j}$, $\mathbf{v}_B = v_B\mathbf{i}$, and $\mathbf{v}_C = v_C\mathbf{k}$, and the magnitude of the linear momentum **L** of the system is 9 lb·s. Knowing that $\mathbf{H}_G = \mathbf{H}_O$, where $\mathbf{H}_G$ is the angular momentum of the system about its mass center G and $\mathbf{H}_O$ is the angular momentum of the system about O, determine (a) the velocities of the particles, (b) the angular momentum of the system about O.

**14.13** A system consists of three particles A, B, and C. We know that $m_A = 3$ kg, $m_B = 2$ kg, and $m_C = 4$ kg and that the velocities of the particles expressed in m/s are, respectively, $\mathbf{v}_A = 4\mathbf{i} + 2\mathbf{j} + 2\mathbf{k}$, $\mathbf{v}_B = 4\mathbf{i} + 3\mathbf{j}$, and $\mathbf{v}_C = -2\mathbf{i} + 4\mathbf{j} + 2\mathbf{k}$. Determine the angular momentum $\mathbf{H}_O$ of the system about O.

**14.14** For the system of particles of Prob. 14.13, determine (a) the position vector $\bar{\mathbf{r}}$ of the mass center G of the system, (b) the linear momentum $m\bar{\mathbf{v}}$ of the system, (c) the angular momentum $\mathbf{H}_G$ of the system about G. Also verify that the answers to this problem and to Prob. 14.13 satisfy the equation given in Prob. 14.27.

**Fig. P14.13**

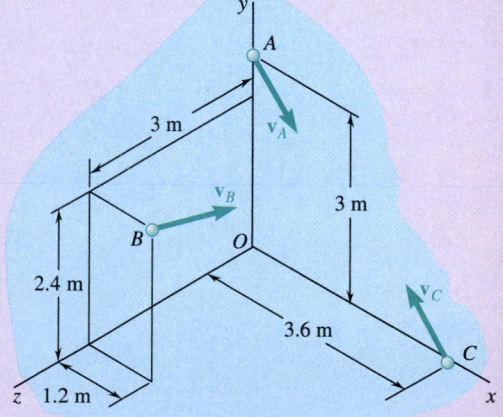

**14.15** A 13-kg projectile is passing through the origin $O$ with a velocity $\mathbf{v}_0 = (35 \text{ m/s})\mathbf{i}$ when it explodes into two fragments $A$ and $B$, of mass 5 kg and 8 kg, respectively. Knowing that 3 s later the position of fragment $A$ is (90 m, 7 m, −14 m), determine the position of fragment $B$ at the same instant. Assume $a_y = -g = -9.81 \text{ m/s}^2$ and neglect air resistance.

**14.16** A 300-kg space vehicle traveling with a velocity $\mathbf{v}_0 = (360 \text{ m/s})\mathbf{i}$ passes through the origin $O$ at $t = 0$. Explosive charges then separate the vehicle into three parts $A$, $B$, and $C$, with mass, respectively, 150 kg, 100 kg, and 50 kg. Knowing that at $t = 4$ s, the positions of parts $A$ and $B$ are observed to be $A$ (1170 m, −290 m, −585 m) and $B$ (1975 m, 365 m, 800 m), determine the corresponding position of part $C$. Neglect the effect of gravity.

**14.17** A 2-kg model rocket is launched vertically and reaches an altitude of 70 m with a speed of 30 m/s at the end of powered flight, time $t = 0$. As the rocket approaches its maximum altitude, it explodes into two parts of masses $m_A = 0.7$ kg and $m_B = 1.3$ kg. Part $A$ is observed to strike the ground 80 m west of the launch point at $t = 6$ s. Determine the position of part $B$ at that time.

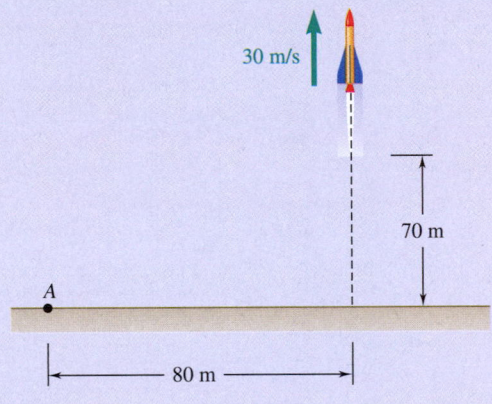

**Fig. P14.17**

**14.18** An 18-kg cannonball and a 12-kg cannonball are chained together and fired horizontally with a velocity of 165 m/s from the top of a 15-m wall. The chain breaks during the flight of the cannonballs and the 12-kg cannonball strikes the ground at $t = 1.5$ s, at a distance of 240 m from the foot of the wall, and 7 m to the right of the line of fire. Determine the position of the other cannonball at that instant. Neglect the resistance of the air.

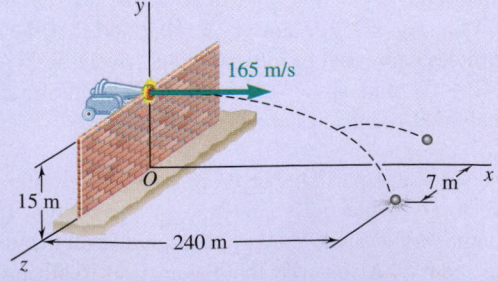

**Fig. P14.18**

**14.19 and 14.20** Cruiser $A$ was traveling east at 60 mi/h on a police emergency call when it was hit at an intersection by car $B$, which was traveling south at high speed. After sliding together on the wet pavement, the two cars hit cruiser $C$, which was traveling north at 45 mi/h and was 63 ft from the intersection at the time of the first collision. Stuck together, the three cars hit a wall and came to a stop at $D$. Knowing that car $B$ weighs 3600 lb and that each of the cruisers weighs 3000 lb, solve the problems below neglecting the forces exerted on the cars by the wet pavement and treating the cars as point masses.

**14.19** Knowing that the coordinates of point $D$ are $x_D = 42$ ft and $y_D = 46.5$ ft, determine (*a*) the time elapsed from the first collision to the stop at $D$, (*b*) the speed of car $B$.

**14.20** Knowing that the speed of car $B$ was 75 mi/h and that the time elapsed from the first collision to the stop at $D$ was 1.5 s, determine the coordinates of $D$.

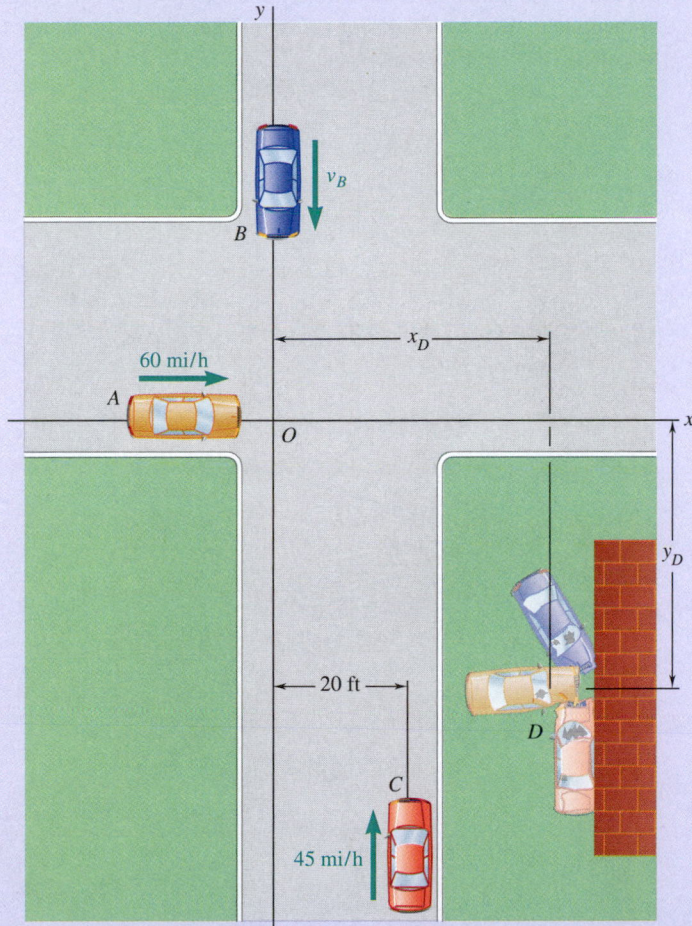

**Fig. P14.19 and P14.20**

**14.21** An expert archer demonstrates his ability by hitting tennis balls thrown by an assistant. A 2-oz tennis ball has a velocity of $(32 \text{ ft/s})\mathbf{i} - (7 \text{ ft/s})\mathbf{j}$ and is 33 ft above the ground when it is hit by a 1.2-oz arrow traveling with a velocity of $(165 \text{ ft/s})\mathbf{j} + (230 \text{ ft/s})\mathbf{k}$, where $\mathbf{j}$ is directed upward. Determine the position $P$ where the ball and arrow will hit the ground, relative to point $O$ located directly under the point of impact.

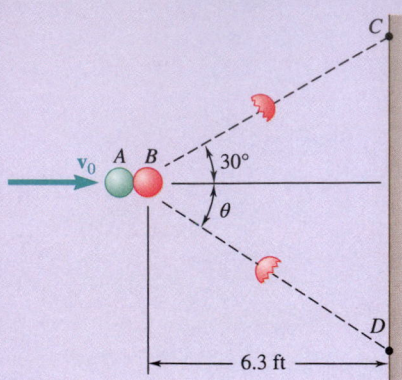

**Fig. P14.22**

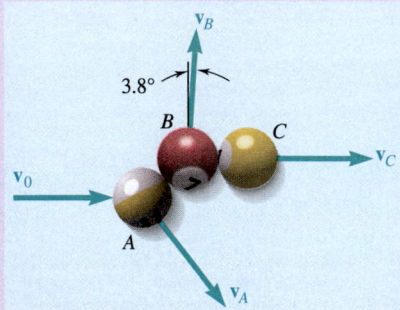

**Fig. P14.23**

**14.22** Two spheres, each of mass $m$, can slide freely on a frictionless, horizontal surface. Sphere $A$ is moving at a speed $v_0 = 16$ ft/s when it strikes sphere $B$, which is at rest, and the impact causes sphere $B$ to break into two pieces, each of mass $m/2$. Knowing that 0.7 s after the collision one piece reaches point $C$ and 0.9 s after the collision the other piece reaches point $D$, determine ($a$) the velocity of sphere $A$ after the collision, ($b$) the angle $\theta$ and the speeds of the two pieces after the collision.

**14.23** In a game of pool, ball $A$ is moving with a velocity $\mathbf{v}_0 = (10$ ft/s$)\mathbf{i}$ when it strikes balls $B$ and $C$, which are at rest side by side as shown. After the collision, $A$ is observed to move with the velocity $\mathbf{v}_A = (3.92$ ft/s$)\mathbf{i} - (4.56$ ft/s$)\mathbf{j}$, while $B$ and $C$ move in the directions shown. Determine the magnitudes of the velocities of $B$ and $C$.

**14.24** A 6-kg shell moving with a velocity $\mathbf{v}_0 = (12$ m/s$)\mathbf{i} - (9$ m/s$)\mathbf{j} - (360$ m/s$)\mathbf{k}$ explodes at point $D$ into three fragments $A$, $B$, and $C$ of mass, respectively, 3 kg, 2 kg, and 1 kg. Knowing that the fragments hit the vertical wall at the points indicated, determine the speed of each fragment immediately after the explosion. Assume that elevation changes due to gravity may be neglected.

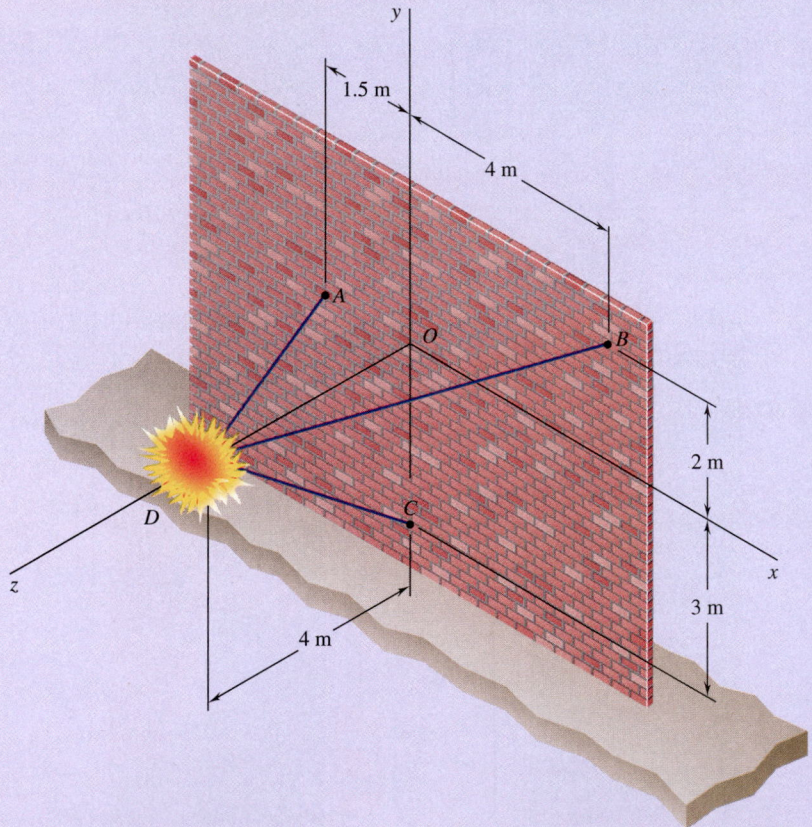

**Fig. P14.24 and P14.25**

**14.25** A 6-kg shell moving with a velocity $\mathbf{v}_0 = (12$ m/s$)\mathbf{i} - (9$ m/s$)\mathbf{j} - (360$ m/s$)\mathbf{k}$ explodes at point $D$ into three fragments $A$, $B$, and $C$ of mass, respectively, 2 kg, 1 kg, and 3 kg. Knowing that the fragments hit the vertical wall at the points indicated, determine the speed of each fragment immediately after the explosion. Assume that elevation changes due to gravity may be neglected.

**14.26** In a scattering experiment, an alpha particle $A$ is projected with the velocity $\mathbf{u}_0 = -(600 \text{ m/s})\mathbf{i} + (750 \text{ m/s})\mathbf{j} - (800 \text{ m/s})\mathbf{k}$ into a stream of oxygen nuclei moving with a common velocity $\mathbf{v}_0 = (600 \text{ m/s})\mathbf{j}$. After colliding successively with the nuclei $B$ and $C$, particle $A$ is observed to move along the path defined by the points $A_1$ (280, 240, 120) and $A_2$ (360, 320, 160), while nuclei $B$ and $C$ are observed to move along paths defined, respectively, by $B_1$ (147, 220, 130) and $B_2$ (114, 290, 120), and by $C_1$ (240, 232, 90) and $C_2$ (240, 280, 75). All paths are along straight lines and all coordinates are expressed in millimeters. Knowing that the mass of an oxygen nucleus is four times that of an alpha particle, determine the speed of each of the three particles after the collisions.

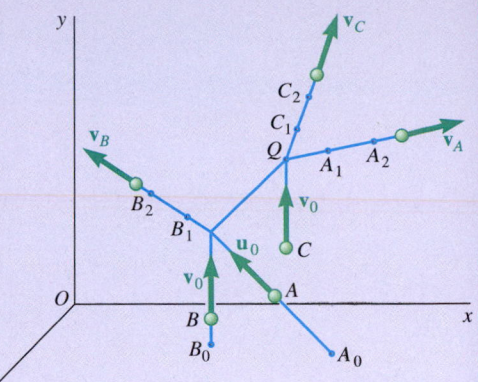

**Fig. P14.26**

**14.27** Derive the relation

$$\mathbf{H}_O = \bar{\mathbf{r}} \times m\bar{\mathbf{v}} + \mathbf{H}_G$$

between the angular momenta $\mathbf{H}_O$ and $\mathbf{H}_G$ defined in Eqs. (14.7) and (14.24), respectively. The vectors $\bar{\mathbf{r}}$ and $\bar{\mathbf{v}}$ define, respectively, the position and velocity of the mass center $G$ of the system of particles relative to the newtonian frame of reference $Oxyz$, and $m$ represents the total mass of the system.

**14.28** Show that Eq. (14.23) may be derived directly from Eq. (14.11) by substituting for $\mathbf{H}_O$ the expression given in Prob. 14.27.

**14.29** Consider the frame of reference $Ax'y'z'$ in translation with respect to the newtonian frame of reference $Oxyz$. We define the angular momentum $\mathbf{H}'_A$ of a system of $n$ particles about $A$ as the sum

$$\mathbf{H}'_A = \sum_{i=1}^{n} (\mathbf{r}'_i \times m_i \mathbf{v}'_i) \tag{1}$$

of the moments about $A$ of the momenta $m_i v'_i$ of the particles in their motion relative to the frame $Ax'y'z'$. Denoting by $\mathbf{H}_A$ the sum

$$\mathbf{H}_A = \sum_{i=1}^{n} (\mathbf{r}'_i \times m_i \mathbf{v}_i)$$

of the moments about $A$ of the momenta $m_i v_i$ of the particles in their motion relative to the newtonian frame $Oxyz$, show that $\mathbf{H}_A = \mathbf{H}'_A$ at a given instant if, and only if, one of the following conditions is satisfied at that instant: ($a$) $A$ has zero velocity with respect to the frame $Oxyz$, ($b$) $A$ coincides with the mass center $G$ of the system, ($c$) the velocity $\mathbf{v}_A$ relative to $Oxyz$ is directed along the line $AG$.

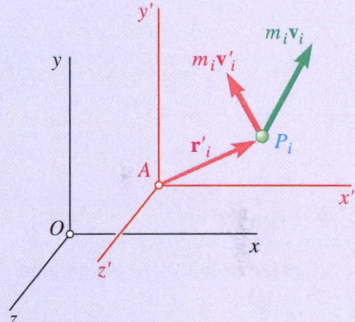

**Fig. P14.29**

**14.30** Show that the relation $\Sigma \mathbf{M}_A = \dot{\mathbf{H}}'_A$, where $\mathbf{H}'_A$ is defined by Eq. (1) of Prob. 14.29 and where $\Sigma \mathbf{M}_A$ represents the sum of the moments about $A$ of the external forces acting on the system of particles, is valid if, and only if, one of the following conditions is satisfied: ($a$) the frame $Ax'y'z'$ is itself a newtonian frame of reference, ($b$) $A$ coincides with the mass center $G$, ($c$) the acceleration $\mathbf{a}_A$ of $A$ relative to $Oxyz$ is directed along the line $AG$.

# 14.2 ENERGY AND MOMENTUM METHODS FOR A SYSTEM OF PARTICLES

Solving problems involving a system of particles is often made easier by applying energy and momentum methods, just as it was as for a single particle in Chap. 13. Definitions of terms and statements of the work–energy and impulse–momentum principles are very similar to the single-particle versions, especially when you take into account the mass center of the particles.

## 14.2A Kinetic Energy of a System of Particles

We define the kinetic energy $T$ of a system of particles as the sum of the kinetic energies of the various particles of the system. Referring to Sec. 13.1B, we have

**Kinetic energy, system of particles**

$$T = \frac{1}{2}\sum_{i=1}^{n} m_i v_i^2 \tag{14.28}$$

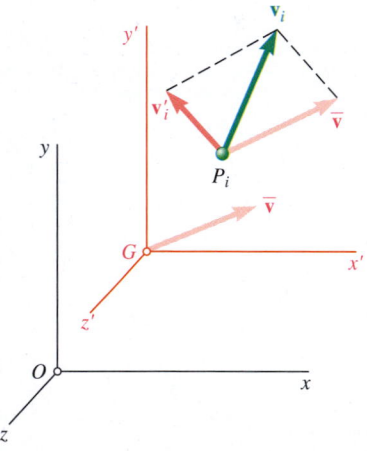

**Fig. 14.7** A centroidal frame of reference $Gx'y'z'$ moving in translation with velocity $\bar{\mathbf{v}}$ with respect to a newtonian reference frame $Oxyz$.

**Using a Centroidal Frame of Reference.** It is often convenient when computing the kinetic energy of a system comprised of a large number of particles (as in the case of a rigid body) to consider the motion of the mass center $G$ of the system and the motion of the system relative to a moving frame attached to $G$ separately.

Let $P_i$ be a particle of the system, $\mathbf{v}_i$ be its velocity relative to the newtonian frame of reference $Oxyz$, and $\mathbf{v}_i'$ be its velocity relative to the moving frame $Gx'y'z'$ that is in translation with respect to $Oxyz$ (Fig. 14.7). Recall from Sec. 14.1D that

$$\mathbf{v}_i = \bar{\mathbf{v}} + \mathbf{v}_i' \tag{14.22}$$

where $\bar{\mathbf{v}}$ denotes the velocity of the mass center $G$ relative to the newtonian frame $Oxyz$. Observing that $v_i^2$ is equal to the scalar product $\mathbf{v}_i \cdot \mathbf{v}_i$, we can express the kinetic energy $T$ of the system relative to the newtonian frame $Oxyz$ as

$$T = \frac{1}{2}\sum_{i=1}^{n} m_i v_i^2 = \frac{1}{2}\sum_{i=1}^{n}(m_i \mathbf{v}_i \cdot \mathbf{v}_i)$$

or, substituting for $\mathbf{v}_i$ from Eq. (14.22),

$$T = \frac{1}{2}\sum_{i=1}^{n}[m_i(\bar{\mathbf{v}} + \mathbf{v}_i') \cdot (\bar{\mathbf{v}} + \mathbf{v}_i')]$$

$$= \frac{1}{2}\left(\sum_{i=1}^{n} m_i\right)\bar{v}^2 + \bar{\mathbf{v}} \cdot \sum_{i=1}^{n} m_i \mathbf{v}_i' + \frac{1}{2}\sum_{i=1}^{n} m_i v_i'^2$$

In this equation, the first sum represents the total mass $m$ of the system. Recalling Eq. (14.13), we note that the second sum is equal to $m\bar{\mathbf{v}}'$ and thus to zero, because $\bar{\mathbf{v}}'$, which represents the velocity of $G$ relative to the frame $Gx'y'z'$, is clearly zero. We therefore have

$$T = \frac{1}{2}m\bar{v}^2 + \frac{1}{2}\sum_{i=1}^{n} m_i v_i'^2 \tag{14.29}$$

This equation states that we can obtain the kinetic energy $T$ of a system of particles **by adding the kinetic energy of the mass center $G$ and the kinetic energy of the system in its motion relative to the frame $Gx'y'z'$.**

## 14.2B   Work–Energy Principle and Conservation of Energy for a System of Particles

We can apply the principle of work and energy to each particle $P_i$ of a system of particles, obtaining for each particle $P_i$

$$(T_1)_i + (U_{1\to2})_i = (T_2)_i$$

where $(U_{1\to2})_i$ represents the work done by the internal forces $\mathbf{f}_{ij}$ and the resultant external force $\mathbf{F}_i$ acting on $P_i$. Adding the kinetic energies of the various particles of the system and considering the work of all the forces involved, we obtain an expression for the entire system as

**Work–energy principle,
system of particles**

$$T_1 + U_{1\to2} = T_2 \qquad\qquad (14.30)$$

The quantities $T_1$ and $T_2$ now represent the kinetic energy of the entire system and can be computed from either Eq. (14.28) or Eq. (14.29). The quantity $U_{1\to2}$ represents the work of all the forces acting on the particles of the system. Note that although the internal forces $\mathbf{f}_{ij}$ and $\mathbf{f}_{ji}$ are equal and opposite, the work of these forces does not, in general, cancel out, because the particles $P_i$ and $P_j$ on which they act generally undergo different displacements. Therefore, in computing $U_{1\to2}$, **we must consider the work of the internal forces** $\mathbf{f}_{ij}$**, as well as the work of the external forces** $\mathbf{F}_i$. An alternative way of writing Eq. (14.30) is

$$T_1 + V_{g_1} + V_{e_1} + U_{1\to2}^{NC} = T_2 + V_{g_2} + V_{e_2} \qquad\qquad (14.30')$$

where $V_g$ is the gravitational potential energy of the system, $V_e$ is the elastic potential energy, and $U_{1\to2}^{NC}$ is the work due to non-conservative forces.

If all of the forces acting on the particles of the system are conservative, we can replace Eq. (14.30) by

**Conservation of energy,
system of particles**

$$T_1 + V_1 = T_2 + V_2 \qquad\qquad (14.31)$$

where $V$ represents the potential energy associated with the internal and external forces acting on the particles of the system.

## 14.2C   Impulse–Momentum Principle and Conservation of Momentum for a System of Particles

Integrating Eqs. (14.10) and (14.11) with respect to $t$ from $t_1$ to $t_2$, we have

$$\sum \int_{t_1}^{t_2} \mathbf{F}\, dt = \mathbf{L}_2 - \mathbf{L}_1 \qquad\qquad (14.32)$$

$$\sum \int_{t_1}^{t_2} \mathbf{M}_O\, dt = (\mathbf{H}_O)_2 - (\mathbf{H}_O)_1 \qquad\qquad (14.33)$$

From the definition of the linear impulse of a force given in Sec. 13.3A, the integrals in Eq. (14.32) represent the linear impulses of the external forces acting on the particles of the system. In a similar way, we shall refer to the integrals in Eq. (14.33) as the **angular impulses** about $O$ of the external forces. Thus, Eq. (14.32) states that the sum of the linear impulses of the external forces acting

**Photo 14.2**   When a golf ball is hit out of a sand trap, some of the momentum of the club is transferred to the golf ball and any sand that is hit. ©Design Pics/Darren Greenwood RF

on the system is equal to the change in linear momentum of the system. Similarly, Eq. (14.33) says that the sum of the angular impulses about $O$ of the external forces is equal to the change in angular momentum about $O$ of the system.

To clarify the physical significance of Eqs. (14.32) and (14.33), we rearrange the terms in these equations, obtaining

$$\mathbf{L}_1 + \sum \int_{t_1}^{t_2} \mathbf{F}\, dt = \mathbf{L}_2 \qquad (14.34)$$

$$(\mathbf{H}_O)_1 + \sum \int_{t_1}^{t_2} \mathbf{M}_O\, dt = (\mathbf{H}_O)_2 \qquad (14.35)$$

In parts $a$ and $c$ of Fig. 14.8, we have sketched the momenta of the particles of the system at times $t_1$ and $t_2$, respectively. In part $b$, we show terms equal to the sum of the linear impulses of the external forces and the sum of the angular impulses about $O$ of the external forces. For simplicity, we have assumed the particles move in the plane of the figure, but the present discussion remains valid in the case of particles moving in space. Recall from Eq. (14.6) that $\mathbf{L}$, by definition, is the resultant of the momenta $m_i\mathbf{v}_i$. Then, Eq. (14.34) says that the resultant of the vectors shown in parts $a$ and $b$ of Fig. 14.8 is equal to the resultant of the vectors shown in part $c$. Recalling from Eq. (14.7) that $\mathbf{H}_O$ is the angular momentum, we note that Eq. (14.35) similarly says that the angular momentum of the vectors in part $a$ added to the angular impulses in part $b$ of Fig. 14.8 is equal to the angular momentum of the vectors in part $c$. Together, Eqs. (14.34) and (14.35) state that

> **The momenta of the particles at time $t_1$ and the impulses of the external forces from $t_1$ to $t_2$ form a system of vectors equipollent to the system of the momenta of the particles at time $t_2$.**

This is indicated in Fig. 14.8 by the use of blue plus and equal signs.

If no external force acts on the particles of the system, the integrals in Eqs. (14.34) and (14.35) are zero, and these equations yield

**Conservation of linear
and angular momentum**

$$\mathbf{L}_1 = \mathbf{L}_2 \qquad (14.36)$$
$$(\mathbf{H}_O)_1 = (\mathbf{H}_O)_2 \qquad (14.37)$$

We thus check the result obtained in Sec. 14.1E: If no external force acts on the particles of a system, the linear momentum and the angular momentum about $O$ of the system of particles are conserved. The system of the initial momenta is equipollent to the system of the final momenta, and it follows that the angular momentum of the system of particles about *any* fixed point is conserved.

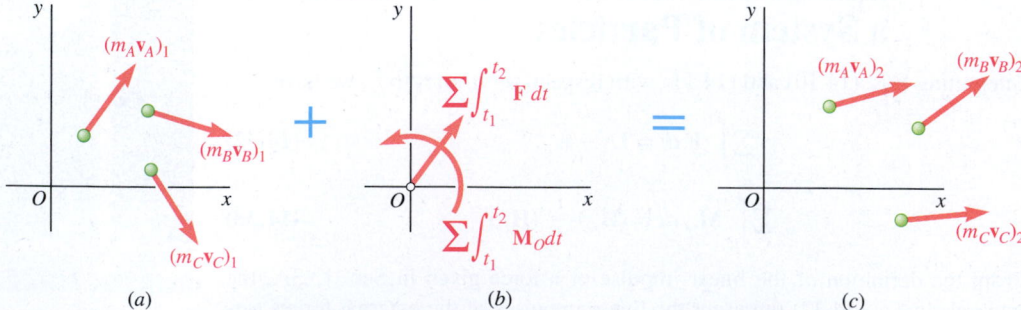

**Fig. 14.8**    The impulse–momentum diagram for a system of particles contains (*a*) momenta of particles at time $t_1$; (*b*) impulses of the external forces and moments about $O$; (*c*) momenta of the particles at time $t_2$.

## Sample Problem 14.4

For the 200-kg space vehicle of Sample Prob. 14.1, it is known that at $t = 2.5$ s, the velocity of part $A$ is $\mathbf{v}_A = (270 \text{ m/s})\mathbf{i} - (120 \text{ m/s})\mathbf{j} + (160 \text{ m/s})\mathbf{k}$, and the velocity of part $B$ is parallel to the $xz$ plane. Determine (a) the velocity of part $C$, (b) the energy gained during the detonation.

**STRATEGY:** Because there are no external forces, use the conservation of linear momentum. Although it is not immediately apparent, you will also need to use the conservation of angular momentum to solve this problem.

**MODELING and ANALYSIS:** Choose the space vehicle as your system. After the explosion, the system is composed of three parts: $A$, $B$, and $C$. Fig. 1 shows the momenta of the system before and after the explosion. From the conservation of linear momentum, you have

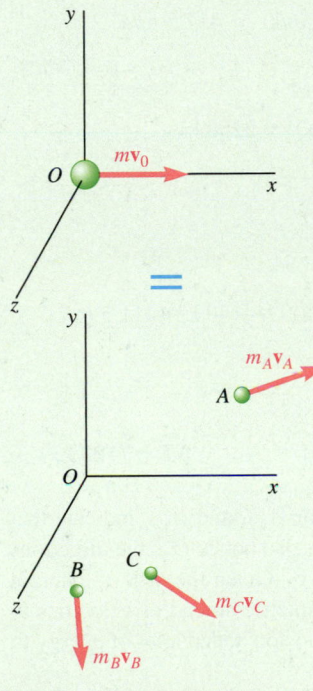

**Fig. 1** Impulse–momentum diagram for the system.

$$\mathbf{L}_1 = \mathbf{L}_2: \qquad m\mathbf{v}_0 = m_A\mathbf{v}_A + m_B\mathbf{v}_B + m_C\mathbf{v}_C \qquad (1)$$

From conservation of angular momentum about point $O$, you have

$$(\mathbf{H}_O)_1 = (\mathbf{H}_O)_2: \qquad 0 = \mathbf{r}_A \times m_A\mathbf{v}_A + \mathbf{r}_B \times m_B\mathbf{v}_B + \mathbf{r}_C \times m_C\mathbf{v}_C \qquad (2)$$

Recall from Sample Prob. 14.1 that $\mathbf{v}_0 = (150 \text{ m/s})\mathbf{i}$ and

$$m_A = 100 \text{ kg} \qquad m_B = 60 \text{ kg} \qquad m_C = 40 \text{ kg}$$
$$\mathbf{r}_A = (555 \text{ m})\mathbf{i} - (180 \text{ m})\mathbf{j} + (240 \text{ m})\mathbf{k}$$
$$\mathbf{r}_B = (255 \text{ m})\mathbf{i} - (120 \text{ m})\mathbf{k}$$
$$\mathbf{r}_C = (105 \text{ m})\mathbf{i} + (450 \text{ m})\mathbf{j} - (420 \text{ m})\mathbf{k}$$

Then, using the information given in the statement of this problem, rewrite Eqs. (1) and (2) as

$$200(150\mathbf{i}) = 100(270\mathbf{i} - 120\mathbf{j} + 160\mathbf{k}) + 60[(v_B)_x\mathbf{i} + (v_B)_z\mathbf{k}]$$
$$+ 40[(v_C)_x\mathbf{i} + (v_C)_y\mathbf{j} + (v_C)_z\mathbf{k}] \qquad (1')$$

$$0 = 100\begin{vmatrix} \mathbf{i} & \mathbf{j} & \mathbf{k} \\ 555 & -180 & 240 \\ 270 & -120 & 160 \end{vmatrix} + 60\begin{vmatrix} \mathbf{i} & \mathbf{j} & \mathbf{k} \\ 255 & 0 & -120 \\ (v_B)_x & 0 & (v_B)_z \end{vmatrix} + 40\begin{vmatrix} \mathbf{i} & \mathbf{j} & \mathbf{k} \\ 105 & 450 & -420 \\ (v_C)_x & (v_C)_y & (v_C)_z \end{vmatrix} \qquad (2')$$

Equate the coefficient of $\mathbf{j}$ in Eq. (1′) and the coefficients of $\mathbf{i}$ and $\mathbf{k}$ in Eq. (2′). After reductions, you obtain the following three scalar equations:

$$(v_C)_y - 300 = 0$$
$$450(v_C)_z + 420(v_C)_y = 0$$
$$105(v_C)_y - 450(v_C)_x - 45\,000 = 0$$

which yield, respectively,

$$(v_C)_y = 300 \qquad (v_C)_z = -280 \qquad (v_C)_x = -30$$

The velocity of part $C$ is thus

$$\mathbf{v}_C = -(30 \text{ m/s})\mathbf{i} + (300 \text{ m/s})\mathbf{j} - (280 \text{ m/s})\mathbf{k} \quad \blacktriangleleft$$

*(continued)*

Equating the coefficients of the **i** and **k** terms on each side of Eq. (1′) and solving for the unknown components of the velocity of $B$ gives

$$(v_B)_x = 70 \text{ m/s} \qquad (v_B)_z = -80 \text{ m/s}$$

So

$$v_A = \sqrt{(270 \text{ m/s})^2 + (-120 \text{ m/s})^2 + (160 \text{ m/s})^2} = 336.0 \text{ m/s}$$

$$v_B = \sqrt{(70 \text{ m/s})^2 + (0)^2 + (-80 \text{ m/s})^2} = 106.3 \text{ m/s}$$

$$v_C = \sqrt{(-30 \text{ m/s})^2 + (300)^2 + (-280 \text{ m/s})^2} = 411.5 \text{ m/s}$$

The initial kinetic energy is

$$T_1 = \frac{1}{2}mv_0^2 = \frac{1}{2}(200 \text{ kg})(150 \text{ m/s})^2 = 2250 \text{ kJ}$$

The final kinetic energy is

$$T_2 = \tfrac{1}{2}m_A v_A^2 + \tfrac{1}{2}m_A v_A^2 + \tfrac{1}{2}m_A v_A^2$$

$$= \frac{1}{2}(100 \text{ kg})(336.0 \text{ m/s})^2 + \frac{1}{2}(60 \text{ kg})(106.3 \text{ m/s})^2 + \frac{1}{2}(40 \text{ kg})(411.5 \text{ m/s})^2$$

$$= 9370 \text{ kJ}$$

So

$$\Delta T = T_2 - T_1 = 9370 \text{ kJ} - 2250 \text{ kJ} \qquad \Delta T = 7120 \text{ kJ} \;\blacktriangleleft$$

**REFLECT and THINK:**   The negative signs for $(v_C)_x$ and $(v_C)_z$ indicate that the velocity is not directed as shown in Fig. 1. We also notice that the directions of the components of $\mathbf{v}_C$ are opposite to those of $\mathbf{v}_A$. Given the lack of external forces, it seems reasonable to expect a more symmetric spread of velocities in all directions. You should also notice that the explosion added a lot of energy to the system.

## Sample Problem 14.5

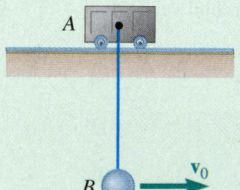

Ball $B$, with a mass of $m_B$, is suspended from a cord with a length $l$ attached to cart $A$, with a mass of $m_A$, that can roll freely on a frictionless horizontal track. If the ball is given an initial horizontal velocity $\mathbf{v}_0$ while the cart is at rest, determine (*a*) the velocity of $B$ as it reaches its maximum elevation, (*b*) the maximum vertical distance $h$ through which $B$ will rise. (Assume $v_0^2 < 2gl$.) ▶

**STRATEGY:**   You are asked about the velocity of the system at two different positions, so use the principle of work and energy for the cart–ball system. You will also use the impulse–momentum principle, because momentum is conserved in the $x$ direction.

*(continued)*

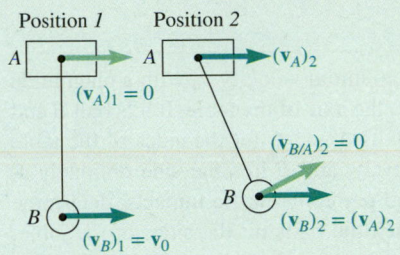

**Fig. 1** Velocity vectors at the two positions.

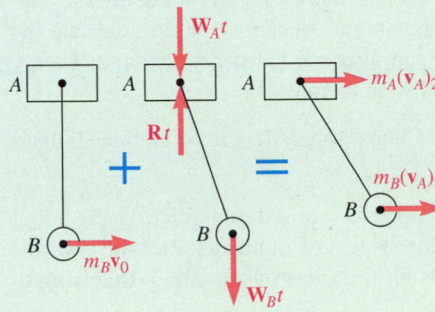

**Fig. 2** Impulse–momentum diagram for the system.

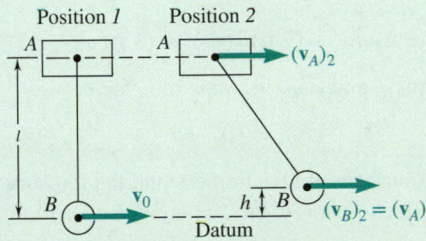

**Fig. 3** The system drawn in position *1* and position *2*.

**MODELING and ANALYSIS:** For your system, choose the ball and the cart and model them as particles.

**Velocities.**

**Position 1:**      $(\mathbf{v}_A)_1 = 0$      $(\mathbf{v}_B)_1 = \mathbf{v}_0$      **(1)**

**Position 2:** When ball $B$ reaches its maximum elevation, its velocity $(\mathbf{v}_{B/A})_2$ relative to its support $A$ is zero (Fig. 1). Thus, at that instant, its absolute velocity is

$$(\mathbf{v}_B)_2 = (\mathbf{v}_A)_2 + (\mathbf{v}_{B/A})_2 = (\mathbf{v}_A)_2 \qquad \textbf{(2)}$$

**Impulse–Momentum Principle.** The external impulses consist of $\mathbf{W}_A t$, $\mathbf{W}_B t$, and $\mathbf{R}t$, where $\mathbf{R}$ is the reaction of the track on the cart. Recalling Eqs. (1) and (2), draw the impulse–momentum diagram (Fig. 2) and write

$$\Sigma m\mathbf{v}_1 + \Sigma \mathbf{Ext\ Imp}_{1\to2} = \Sigma m\mathbf{v}_2$$

$\xrightarrow{+} x$ components:     $m_B v_0 = (m_A + m_B)(v_A)_2$

This expresses that the linear momentum of the system is conserved in the horizontal direction. Solving for $(v_A)_2$, you have

$$(v_A)_2 = \frac{m_B}{m_A + m_B} v_0 \qquad (\mathbf{v}_B)_2 = (\mathbf{v}_A)_2 = \frac{m_B}{m_A + m_B} v_0 \rightarrow \ \blacktriangleleft$$

**Conservation of Energy.** The system is shown in Fig. 3 in the two positions. Define your datum at the location of $B$ in position *1* (although you could also choose to place it at $A$). You can now calculate the kinetic and potential energies in the two positions:

**Position 1.**    *Potential Energy:*     $V_1 = m_A gl$
           *Kinetic Energy:*      $T_1 = \frac{1}{2} m_B v_0^2$
**Position 2.**    *Potential Energy:*     $V_2 = m_A gl + m_B gh$
           *Kinetic Energy:*      $T_2 = \frac{1}{2}(m_A + m_B)(v_A)_2^2$

Substituting these into the conservation of energy gives

$$T_1 + V_1 = T_2 + V_2: \qquad \tfrac{1}{2} m_B v_0^2 + m_A gl = \tfrac{1}{2}(m_A + m_B)(v_A)_2^2 + m_A gl + m_B gh$$

Solving for $h$, you have

$$h = \frac{v_0^2}{2g} - \frac{m_A + m_B}{m_B} \frac{(v_A)_2^2}{2g}$$

or substituting $(v_A)_2$ from above, you have

$$h = \frac{v_0^2}{2g} - \frac{m_B}{m_A + m_B} \frac{v_0^2}{2g} \qquad h = \frac{m_A}{m_A + m_B} \frac{v_0^2}{2g} \ \blacktriangleleft$$

**REFLECT and THINK:** Recalling that $v_0^2 < 2gl$, it follows from the last equation that $h < l$; this verifies that $B$ stays below $A$, as assumed in the solution. For $m_A \gg m_B$, the answers reduce to $(\mathbf{v}_B)_2 = (\mathbf{v}_A)_2 = 0$ and $h = v_0^2/2g$; $B$ oscillates as a simple pendulum with $A$ fixed. For $m_A \ll m_B$, they reduce to $(\mathbf{v}_B)_2 = (\mathbf{v}_A)_2 = \mathbf{v}_0$ and $h = 0$; $A$ and $B$ move with the same constant velocity $\mathbf{v}_0$.

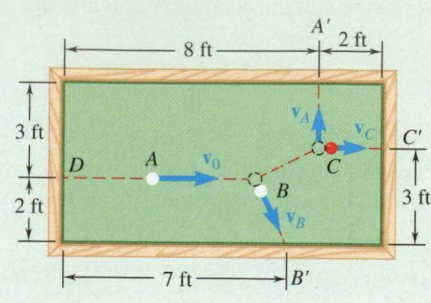

## Sample Problem 14.6

In a game of billiards, ball $A$ is given an initial velocity $\mathbf{v}_0$ with a magnitude of $v_0 = 10$ ft/s along line $DA$ parallel to the axis of the table. It hits ball $B$ and then ball $C$, which are both at rest. Balls $A$ and $C$ hit the sides of the table squarely at points $A'$ and $C'$, respectively, and $B$ hits the side obliquely at $B'$. Assuming frictionless surfaces and perfectly elastic impacts, determine the velocities $\mathbf{v}_A$, $\mathbf{v}_B$, and $\mathbf{v}_C$ with which the balls hit the sides of the table. (*Remark:* In this Sample Problem and in several of the problems that follow, we assume the billiard balls are particles moving freely in a horizontal plane, rather than the rolling and sliding spheres they actually are.)

**STRATEGY:** Because there are no externally applied forces, use the conservation of linear and angular momentum. Because you are told that the impacts are perfectly elastic, you can also use the conservation of energy (but note that in general, energy is lost in an impact).

**MODELING and ANALYSIS:** Choose the system to be all three billiard balls and model them as particles.

**Conservation of Momentum.** There is no external force, so the initial momentum $m\mathbf{v}_0$ is equipollent to the system of momenta after the two collisions (and before any of the balls hit the sides of the table). Referring to Fig. 1, you have

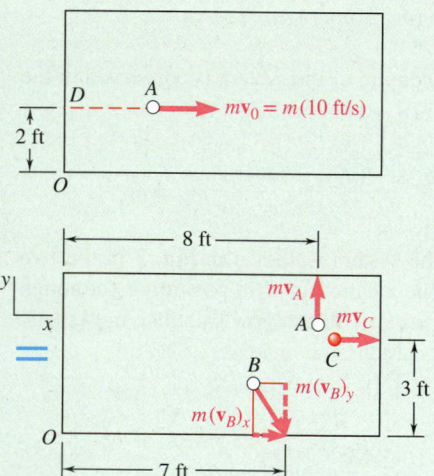

$\xrightarrow{+}$ $x$ components:
$$m(10 \text{ ft/s}) = m(v_B)_x + m v_c \tag{1}$$

$+\uparrow$ $y$ components:
$$0 = m v_A - m(v_B)_y \tag{2}$$

$+\circlearrowleft$ moments about $O$:
$$(-2 \text{ ft})m(10 \text{ ft/s}) = (8 \text{ ft})m v_A - (7 \text{ ft})m(v_B)_y - (3 \text{ ft})m v_C \tag{3}$$

**Fig. 1** Impulse-momentum diagram for the system.

Solving the three equations for $v_A$, $(v_B)_x$, and $(v_B)_y$ in terms of $v_C$ gives

$$v_A = (v_B)_y = 3v_C - 20 \qquad (v_B)_x = 10 - v_C \tag{4}$$

**Conservation of Energy.** The surfaces are frictionless and the impacts are perfectly elastic, so the initial kinetic energy $\frac{1}{2}mv_0^2$ is equal to the final kinetic energy of the system:

$$\tfrac{1}{2}mv_0^2 = \tfrac{1}{2}mv_A^2 + \tfrac{1}{2}mv_B^2 + \tfrac{1}{2}mv_C^2$$
$$v_A^2 + (v_B)_x^2 + (v_B)_y^2 + v_C^2 = (10 \text{ ft/s})^2 \tag{5}$$

Substituting for $v_A$, $(v_B)_x$, and $(v_B)_y$ from Eqs. (4) into Eq. (5), you have

$$2(3v_C - 20)^2 + (10 - v_C)^2 + v_C^2 = 100$$
$$20v_C^2 - 260v_C + 800 = 0$$

Solving for $v_C$, you find $v_C = 5$ ft/s and $v_C = 8$ ft/s. Because only the second root yields a positive value for $v_A$ after substitution into Eqs. (4), then $v_C = 8$ ft/s and

$$v_A = (v_B)_y = 3(8) - 20 = 4 \text{ ft/s} \qquad (v_B)_x = 10 - 8 = 2 \text{ ft/s}$$

$$\mathbf{v}_A = 4 \text{ ft/s} \uparrow \qquad \mathbf{v}_B = 4.47 \text{ ft/s} \searrow 63.4° \qquad \mathbf{v}_C = 8 \text{ ft/s} \rightarrow \blacktriangleleft$$

**REFLECT and THINK:** In a real situation, energy would not be conserved, and you would need to know the coefficient of restitution between the balls to solve this problem. We also neglected friction and the rotation of the balls in our analysis, which is often a poor assumption in pool or billiards. We discuss rigid-body impacts in Chap. 17.

# Case Study 14.1

Dynamics principles are used extensively in accident reconstruction. Police investigators measure skidmarks on pavement to determine the direction of motion, as well as to help determine the initial speeds of cars involved in an accident. Multiple car accidents can be especially difficult to analyze—a simplified case, shown in CS Fig. 14.1, is presented here.

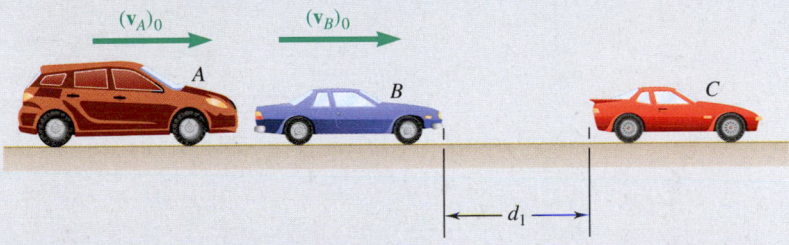

**CS Fig. 14.1**   Diagram of three-car collision.

Car $A$ is traveling with a speed of $(v_A)_0$ when it hits car $B$, which is slowing down to a speed of $(v_B)_0$ because of stationary vehicles in front of it. For this preliminary investigation, assume that the two cars stick together, skid a distance $d_1$, and strike car $C$. The three vehicles stick together, then slide an additional distance $d_2$ before coming to rest. By measuring the length of the skid marks, it is known that $d_1 = 30$ ft and $d_2 = 20$ ft.

The driver of car $B$ claims he was traveling 20 mph when he was hit from behind. You have been asked to do a preliminary analysis to determine if the driver of car $A$ was speeding, given that the speed limit at the site of the accident was 40 mph.

**STRATEGY:**   In accident reconstruction, you typically work "backward." When the cars are skidding, you can use work and energy to relate skid distances and vehicle velocities. You can use impulse and momentum to relate the velocities of the vehicles immediately before and after a collision.

**MODELING:**   You can model the cars as particles, and create your impulse–momentum diagrams for each collision, as shown in CS Fig. 14.2.

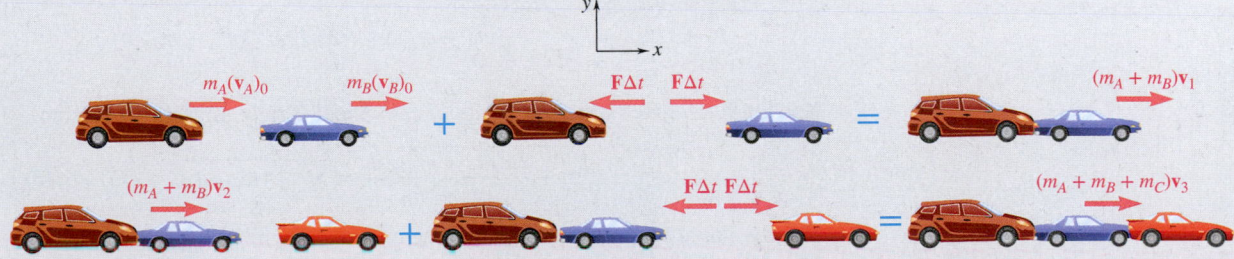

**CS Fig. 14.2**   Impulse–momentum diagram for two collisions; times 1, 2, and 3 are just after the first collision, just before the second collision, and just after the second collision, respectively.

*(continued)*

**ANALYSIS:** Defining position 4 as the point where the cars come to rest, applying work and energy gives you

$$T_3 + U_{3 \to 4} = T_4: \qquad \frac{1}{2}(m_A + m_B + m_C)\,v_3^2 - F_f d_2 = 0$$

Use CS Fig. 14.3 to help you determine the value for $F_f$. Summing forces in the $y$ direction gives you $F_f = \mu N = \mu(m_A + m_B + m_C)g$ (this would be different if the cars were on a slope). Solving for $v_3$ gives you

$$v_3 = \sqrt{\frac{2\mu(m_A + m_B + m_C)g\,d_2}{m_A + m_B + m_C}} = \sqrt{2\mu g\,d_2} \qquad \textbf{(1)}$$

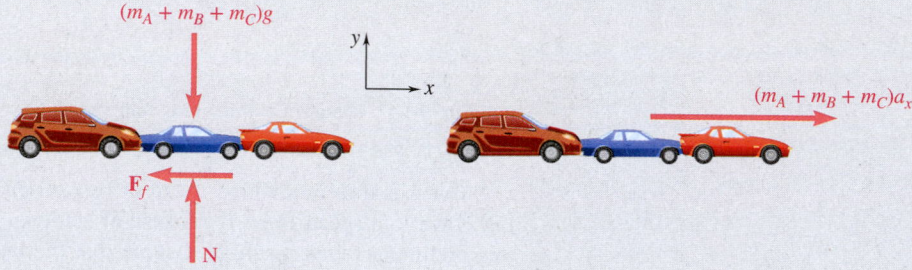

**CS Fig. 14.3** FBD and KD of cars skidding on asphalt.

Defining your system as cars $A$ and $B$ combined and car $C$, you can use CS Fig 14.2 and apply the conservation of linear momentum for the second impact, thus giving

$$(m_A + m_B)v_2 = (m_A + m_B + m_C)v_3$$

Solving for $v_2$, you get

$$v_2 = \frac{(m_A + m_B + m_C)}{(m_A + m_B)}v_3 \qquad \textbf{(2)}$$

Now you must determine how much energy is lost between position 1 (right after collision 1) and position 2 (right before collision 2). Using an approach analogous to Eq. (1) above, you get

$$T_1 + U_{1 \to 2} = T_2: \qquad \frac{1}{2}(m_A + m_B)\,v_1^2 - F_f d_1 = \frac{1}{2}(m_A + m_B)\,v_2^2$$

Solving for $v_1$ gives you the velocity of cars $A$ and $B$ right after their collision:

$$v_1 = \sqrt{v_2^2 + \frac{2\mu(m_A + m_B)g\,d_1}{m_A + m_B}} = \sqrt{v_2^2 + 2\mu g\,d_1} \qquad \textbf{(3)}$$

Finally, you can use impulse and momentum to determine the speed of car $A$ before the collision:

$$m_A(v_A)_0 + m_B(v_B)_0 = (m_A + m_B)v_1$$

*(continued)*

Solving for $(v_A)_0$ gives you:

$$(v_A)_0 = \frac{(m_A + m_B)(v_1) - m_B(v_B)_0}{m_A} \tag{4}$$

Knowing the makes of the cars, we can estimate the weights as $W_A = 4000$ lb, $W_B = 3000$ lb, and $W_C = 2700$ lb. From road testing, the kinetic coefficient of friction between rubber tires and dry concrete is $\mu = 0.70$. Using these data and the value for $d_1$ and $d_2$, and substituting Eq. (1) into Eq. (2), the resulting equation into Eq. (3), and finally this resulting equation into Eq. (4) gives you $(v_A)_0 = 75.2$ ft/s $= 51.3$ mph, which is well above the speed limit.

**REFLECT and THINK:**  Accident reconstruction experts rely on the principles of dynamics to help them determine what happened during an accident. In the current example, you had to rely on the speed estimate of the driver who was rear-ended. If this driver had been stationary [e.g., $(v_B)_0 = 0$], then car $A$ would have been going 66 mph. For car $A$ to be driving at the speed limit, then $(v_B)_0$ would have an initial speed of approximate 61 mph (or 20 miles over the speed limit). The investigator would most likely depend on other information as well, such as eyewitness accounts and damage done to the different automobiles.

# SOLVING PROBLEMS
## ON YOUR OWN

In Sec. 14.1, we defined the linear momentum and the angular momentum of a system of particles. In this section, we defined the **kinetic energy** $T$ of a system of particles as

$$T = \frac{1}{2}\sum_{i=1}^{n} m_i v_i^2 \qquad\qquad (14.28)$$

The solutions of the problems in Sec. 14.1 were based on the conservation of linear momentum of a system of particles or on the observation of the motion of the mass center of a system of particles. In this section, you will solve problems involving the following concepts.

**1. Computation of the kinetic energy lost in collisions.** You can compute the kinetic energy $T_1$ of the system of particles before the collisions and its kinetic energy $T_2$ after the collisions from Eq. (14.28) and subtract one from the other. Keep in mind that although linear momentum and angular momentum are vector quantities, kinetic energy is a *scalar* quantity.

**2. Conservation of linear momentum and conservation of energy.** As you saw in Sec. 14.1, when the resultant of the external forces acting on a system of particles is zero, the linear momentum of the system is conserved. In problems involving two-dimensional motion, expressing that the initial linear momentum and the final linear momentum of the system are equipollent yields two algebraic equations. Equating the initial total energy of the system of particles (including potential energy as well as kinetic energy) to its final total energy yields an additional equation. Thus, you can write three equations that you can solve for three unknowns (Sample Prob. 14.6). Note that if the resultant of the external forces is not zero but has a fixed direction, the component of the linear momentum in a direction perpendicular to the resultant is still conserved; the number of equations that you can use is then reduced to two (Sample Prob. 14.5).

**3. Conservation of linear and angular momentum.** When no external forces act on a system of particles, both the linear momentum of the system and its angular momentum about some arbitrary point are conserved. In the case of three-dimensional motion, this enables you to write as many as six equations, although you may need to solve only some of them to obtain the desired answers (Sample Prob. 14.4). In the case of two-dimensional motion, you will be able to write three equations that you can solve for three unknowns.

**4. Conservation of linear and angular momentum and conservation of energy.** In the case of the two-dimensional motion of a system of particles that is not subjected to any external forces, you can obtain two algebraic equations by expressing that the linear momentum of the system is conserved; one equation by writing that the angular momentum of the system about some arbitrary point is conserved; and a fourth equation by expressing that the total energy of the system is conserved. These equations can be solved for four unknowns.

# Problems

**14.31** Determine the energy lost due to friction and the impacts for Prob. 14.1.

**14.32** In Prob. 14.3, determine the energy lost (*a*) when the first suitcase hits the carrier, (*b*) when the second suitcase hits the carrier.

**14.33** In Prob. 14.6, determine the work done by the woman and by the man as each dives from the boat, assuming that the woman dives first.

**14.34** Determine the energy lost as a result of the series of collisions described in Prob. 14.8.

**14.35** Two automobiles *A* and *B*, of mass $m_A$ and $m_B$, respectively, are traveling in opposite directions when they collide head on. The impact is assumed perfectly plastic, and it is further assumed that the energy absorbed by each automobile is equal to its loss of kinetic energy with respect to a moving frame of reference attached to the mass center of the two-vehicle system. Denoting by $E_A$ and $E_B$, respectively, the energy absorbed by automobile *A* and by automobile *B*, (*a*) show that $E_A/E_B = m_B/m_A$—that is, the amount of energy absorbed by each vehicle is inversely proportional to its mass, (*b*) compute $E_A$ and $E_B$, knowing that $m_A = 1600$ kg and $m_B = 900$ kg and that the speeds of *A* and *B* are, respectively, 90 km/h and 60 km/h.

**Fig. P14.35**

**14.36** It is assumed that each of the two automobiles involved in the collision described in Prob. 14.35 had been designed to safely withstand a test in which it crashed into a solid, immovable wall at the speed $v_0$. The severity of the collision of Prob. 14.35 may then be measured for each vehicle by the ratio of the energy it absorbed in the collision to the energy it absorbed in the test. On that basis, show that the collision described in Prob. 14.35 is $(m_A/m_B)^2$ times more severe for automobile *B* than for automobile *A*.

**14.37** Solve Sample Prob. 14.5, assuming that cart *A* is given an initial horizontal velocity $\mathbf{v}_0$ while ball *B* is at rest.

**14.38** Ball *B* is suspended from a cord of length *l* attached to cart *A*, which can roll freely on a frictionless, horizontal track. The ball and the cart have the same mass *m*. If the cart is given an initial horizontal velocity $\mathbf{v}_0$ while the ball is at rest, describe the subsequent motion of the system, specifying the velocities of *A* and *B* for the following successive values of the angle $\theta$ (assume positive counterclockwise) that the cord will form with the vertical: (*a*) $\theta = \theta_{max}$, (*b*) $\theta = 0$, (*c*) $\theta = \theta_{min}$.

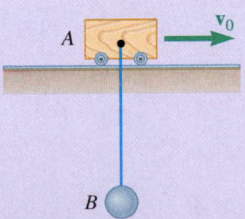

**Fig. P14.38**

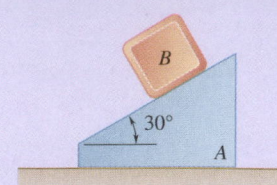

Fig. P14.39

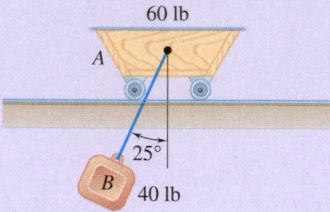

Fig. P14.40

**14.39** A 15-lb block $B$ starts from rest and slides on the 25-lb wedge $A$, which is supported by a horizontal surface. Neglecting friction, determine (*a*) the velocity of $B$ relative to $A$ after it has slid 3 ft down the inclined surface of the wedge, (*b*) the corresponding velocity of $A$.

**14.40** A 40-lb block $B$ is suspended from a 6-ft cord attached to a 60-lb cart $A$, which may roll freely on a frictionless, horizontal track. If the system is released from rest in the position shown, determine the velocities of $A$ and $B$ as $B$ passes directly under $A$.

**14.41 and 14.42** In a game of pool, ball $A$ is moving with a velocity $\mathbf{v}_0$ with a magnitude of $v_0 = 15$ ft/s when it strikes balls $B$ and $C$, which are at rest and aligned as shown. Knowing that after the collision the three balls move in the directions indicated and assuming frictionless surfaces and perfectly elastic impact (i.e., conservation of energy), determine the magnitudes of the velocities $\mathbf{v}_A$, $\mathbf{v}_B$, and $\mathbf{v}_C$.

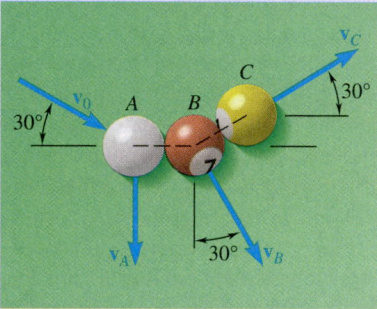

Fig. P14.41                    Fig. P14.42

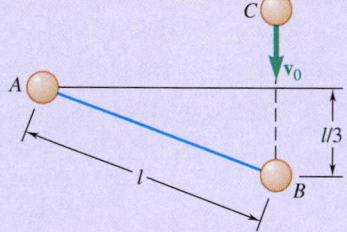

Fig. P14.43

**14.43** Three spheres, each with a mass of $m$, can slide freely on a frictionless, horizontal surface. Spheres $A$ and $B$ are attached to an inextensible, inelastic cord with a length $l$ and are at rest in the position shown when sphere $B$ is struck squarely by sphere $C$, which is moving with a velocity $\mathbf{v}_0$. Knowing that the cord is taut when sphere $B$ is struck by sphere $C$ and assuming perfectly elastic impact between $B$ and $C$, and thus the conservation of energy for the entire system, determine the velocity of each sphere immediately after impact.

**14.44** In a game of pool, ball $A$ is moving with the velocity $\mathbf{v}_0 = v_0\mathbf{i}$ when it strikes balls $B$ and $C$, which are at rest side by side. Assuming frictionless surfaces and perfectly elastic impact (i.e., conservation of energy), determine the final velocity of each ball, assuming that the path of $A$ is (*a*) perfectly centered and that $A$ strikes $B$ and $C$ simultaneously, (*b*) not perfectly centered and that $A$ strikes $B$ slightly before it strikes $C$.

Fig. P14.44

**14.45** The 2-kg sub-satellite $B$ has an initial velocity $\mathbf{v}_B = (3 \text{ m/s})\mathbf{j}$. It is connected to the 20-kg base satellite $A$ by a 500-m space tether. Determine the velocity of the base satellite and sub-satellite immediately after the tether becomes taut (assuming no rebound).

**14.46** A 900-lb space vehicle traveling with a velocity $\mathbf{v}_0 = (1500 \text{ ft/s})\mathbf{k}$ passes through the origin $O$. Explosive charges then separate the vehicle into three parts $A$, $B$, and $C$, with masses of 150 lb, 300 lb, and 450 lb, respectively. Knowing that shortly thereafter the positions of the three parts are, respectively, $A(250, 250, 2250)$, $B(600, 1300, 3200)$, and $C(-475, -950, 1900)$, where the coordinates are expressed in feet, that the velocity of $B$ is $\mathbf{v}_B = (500 \text{ ft/s})\mathbf{i} + (1100 \text{ ft/s})\mathbf{j} + (2100 \text{ ft/s})\mathbf{k}$, and that the $x$ component of the velocity of $C$ is $-400$ ft/s, determine the velocity of part $A$.

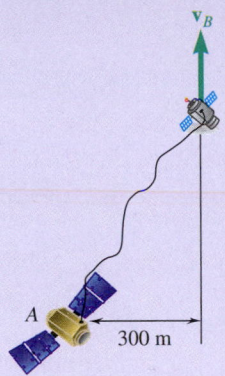

**Fig. P14.45**

**14.47** Four small disks $A$, $B$, $C$, and $D$ can slide freely on a frictionless horizontal surface. Disks $B$, $C$, and $D$ are connected by light rods and are at rest in the position shown when disk $B$ is struck squarely by disk $A$, which is moving to the right with a velocity $\mathbf{v}_0 = (38.5 \text{ ft/s})\mathbf{i}$. The weights of the disks are $W_A = W_B = W_C = 15$ lb, and $W_D = 30$ lb. Knowing that the velocities of the disks immediately after the impact are $\mathbf{v}_A = \mathbf{v}_B = (8.25 \text{ ft/s})\mathbf{i}$, $\mathbf{v}_C = v_C\mathbf{i}$, and $\mathbf{v}_D = v_D\mathbf{i}$, determine (a) the speeds $v_C$ and $v_D$, (b) the fraction of the initial kinetic energy of the system which is dissipated during the collision.

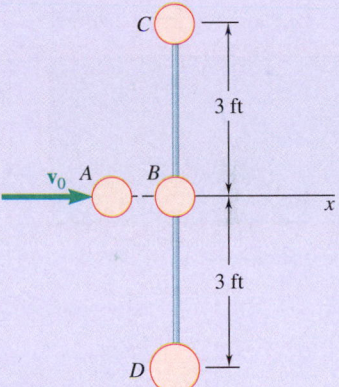

**Fig. P14.47**

**14.48** In the scattering experiment of Prob. 14.26, it is known that the alpha particle is projected from $A_0(300, 0, 300)$ and that it collides with the oxygen nucleus $C$ at $Q(240, 200, 100)$, where all coordinates are expressed in millimeters. Determine the coordinates of point $B_0$ where the original path of nucleus $B$ intersects the $zx$ plane. (*Hint:* Express that the angular momentum of the three particles about $Q$ is conserved.)

**14.49** Three identical small spheres, each weighing 2 lb, can slide freely on a horizontal frictionless surface. Spheres $B$ and $C$ are connected by a light rod and are at rest in the position shown when sphere $B$ is struck squarely by sphere $A$, which is moving to the right with a velocity $\mathbf{v}_0 = (6.5 \text{ ft/s})\mathbf{i}$. Knowing that $\theta = 30°$ and that the velocities of spheres $A$ and $B$ immediately after the impact are $\mathbf{v}_A = (0.5 \text{ ft/s})\mathbf{i}$ and $\mathbf{v}_B = (3.75 \text{ ft/s})\mathbf{i} + (v_B)_y \mathbf{j}$, determine $(v_B)_y$ and the velocity of $C$ immediately after impact.

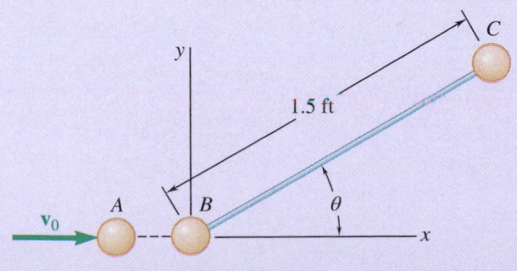

**Fig. P14.49**

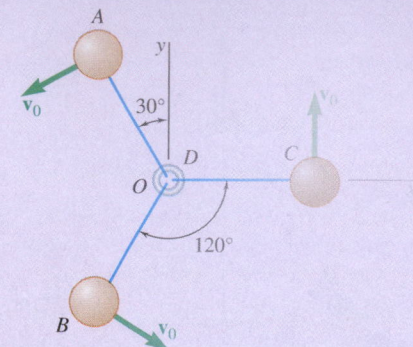

**Fig. P14.50**

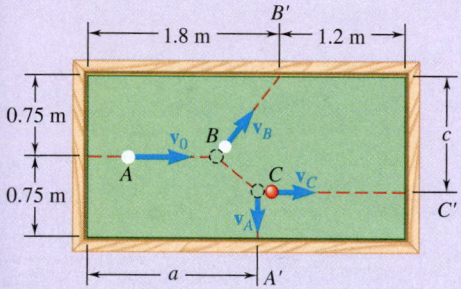

**Fig. P14.51**

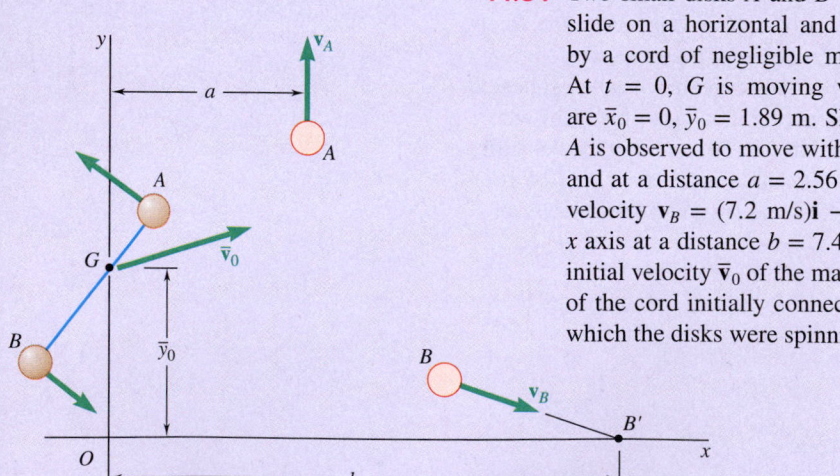

**Fig. P14.53 and P14.54**

**14.50** Three small spheres $A$, $B$, and $C$, each of mass $m$, are connected to a small ring $D$ of negligible mass by means of three inextensible, inelastic cords of length $l$. The spheres can slide freely on a frictionless horizontal surface and are rotating initially at a speed $v_0$ about ring $D$, which is at rest. Suddenly the cord $CD$ breaks. After the other two cords have again become taut, determine (a) the speed of ring $D$, (b) the relative speed at which spheres $A$ and $B$ rotate about $D$, (c) the fraction of the original energy of spheres $A$ and $B$ that is dissipated when cords $AD$ and $BD$ again became taut.

**14.51** In a game of billiards, ball $A$ is given an initial velocity $\mathbf{v}_0$ along the longitudinal axis of the table. It hits ball $B$ and then ball $C$, which are both at rest. Balls $A$ and $C$ are observed to hit the sides of the table squarely at $A'$ and $C'$, respectively, and ball $B$ is observed to hit the side obliquely at $B'$. Knowing that $v_0 = 4$ m/s, $v_A = 1.92$ m/s, and $a = 1.65$ m, determine (a) the velocities $\mathbf{v}_B$ and $\mathbf{v}_C$ of balls $B$ and $C$, (b) the point $C'$ where ball $C$ hits the side of the table. Assume frictionless surfaces and perfectly elastic impacts (i.e., conservation of energy).

**14.52** For the game of billiards of Prob. 14.51, it is now assumed that $v_0 = 5$ m/s, $v_C = 3.2$ m/s, and $c = 1.22$ m. Determine (a) the velocities $\mathbf{v}_A$ and $\mathbf{v}_B$ of balls $A$ and $B$, (b) the point $A'$ where ball $A$ hits the side of the table.

**14.53** Two small disks $A$ and $B$ of mass 3 kg and 1.5 kg, respectively, may slide on a horizontal, frictionless surface. They are connected by a cord, 600 mm long, and spin counterclockwise about their mass center $G$ at the rate of 10 rad/s. At $t = 0$, the coordinates of $G$ are $\bar{x}_0 = 0$, $\bar{y}_0 = 2$ m, and its velocity $\bar{\mathbf{v}}_0 = (1.2 \text{ m/s})\mathbf{i} + (0.96 \text{ m/s})\mathbf{j}$. Shortly thereafter the cord breaks; disk $A$ is then observed to move along a path parallel to the $y$ axis, and disk $B$ moves along a path that intersects the $x$ axis at a distance $b = 7.5$ m from $O$. Determine (a) the velocities of $A$ and $B$ after the cord breaks, (b) the distance $a$ from the $y$ axis to the path of $A$.

**14.54** Two small disks $A$ and $B$ of mass 2 kg and 1 kg, respectively, may slide on a horizontal and frictionless surface. They are connected by a cord of negligible mass and spin about their mass center $G$. At $t = 0$, $G$ is moving with the velocity $\bar{\mathbf{v}}_0$ and its coordinates are $\bar{x}_0 = 0$, $\bar{y}_0 = 1.89$ m. Shortly thereafter, the cord breaks and disk $A$ is observed to move with a velocity $\mathbf{v}_A = (5 \text{ m/s})\mathbf{j}$ in a straight line and at a distance $a = 2.56$ m from the $y$ axis, while $B$ moves with a velocity $\mathbf{v}_B = (7.2 \text{ m/s})\mathbf{i} - (4.6 \text{ m/s})\mathbf{j}$ along a path intersecting the $x$ axis at a distance $b = 7.48$ m from the origin $O$. Determine (a) the initial velocity $\bar{\mathbf{v}}_0$ of the mass center $G$ of the two disks, (b) the length of the cord initially connecting the two disks, (c) the rate in rad/s at which the disks were spinning about $G$.

**14.55** Three small identical spheres $A$, $B$, and $C$, which can slide on a horizontal, frictionless surface, are attached to three strings of length $l$, which are tied to a ring $G$. Initially the spheres rotate about the ring, which moves along the $x$ axis with a velocity $\mathbf{v}_0$. Suddenly the ring breaks, and the three spheres move freely in the $xy$ plane. Knowing that $\mathbf{v}_A = (8.66 \text{ ft/s})\mathbf{j}$, $\mathbf{v}_C = (15 \text{ ft/s})\mathbf{i}$, $a = 0.866$ ft, and $d = 0.5$ ft, determine ($a$) the initial velocity of the ring, ($b$) the length $l$ of the strings, ($c$) the rate in rad/s at which the spheres were rotating about $G$.

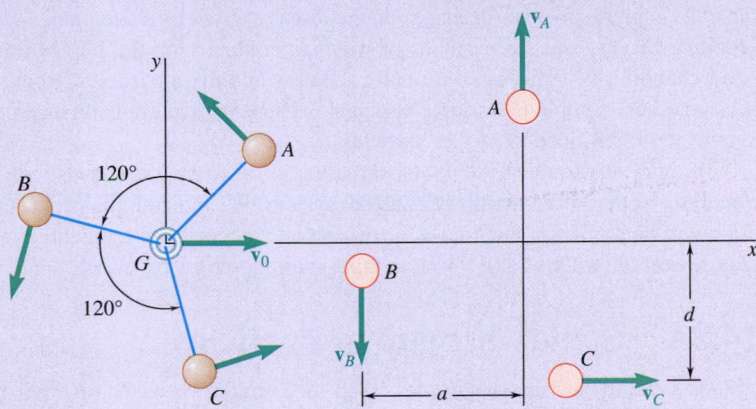

**Fig. P14.55 and P14.56**

**14.56** Three small identical spheres $A$, $B$, and $C$, which can slide on a horizontal, frictionless surface, are attached to three strings, 0.25 ft long, which are tied to a ring $G$. Initially the spheres rotate counterclockwise about the ring with a relative velocity of 2.5 ft/s and the ring moves along the $x$ axis with a velocity $\mathbf{v}_0 = (1.25 \text{ ft/s})\mathbf{i}$. Suddenly the ring breaks and the three spheres move freely in the $xy$ plane with $A$ and $B$ following paths parallel to the $y$ axis at a distance $a = 0.433$ ft from each other and $C$ following a path parallel to the $x$ axis. Determine ($a$) the velocity of each sphere, ($b$) the distance $d$.

# *14.3 VARIABLE SYSTEMS OF PARTICLES

All of the systems considered so far consisted of well-defined particles. These systems did not gain or lose any particles during their motion. In a large number of engineering applications, however, it is necessary to consider **variable systems of particles**, that is, systems that are continually gaining or losing particles, or doing both at the same time. Consider, for example, a hydraulic turbine. Its analysis involves determining the forces exerted by a stream of water on rotating blades, and the particles of water in contact with the blades form an ever-changing system that continually acquires and loses particles. Rockets furnish another example of variable systems, because their propulsion depends upon the continual ejection of fuel particles.

To analyze variable systems of particles, we must find a way to reduce the analysis to that of an auxiliary constant system. We indicate the procedure to follow in Secs. 14.3A and 14.3B for two broad categories of applications: a steady stream of particles and a system that is gaining or losing mass.

## *14.3A Steady Stream of Particles

Consider a steady stream of particles, such as a stream of water diverted by a fixed vane or a flow of air through a duct or through a blower. In order to determine the resultant of the forces exerted on the particles in contact with the vane, duct, or blower, we isolate these particles and define them to be a system $S$ (Fig. 14.9). Note that $S$ is a variable system of particles, because it continually gains particles flowing in and loses an equal number of particles flowing out. Therefore, the kinetics principles that we have established so far do not apply directly to $S$.

However, we can easily define an auxiliary system of particles that does remain constant for a short interval of time $\Delta t$. Consider at time $t$ the system $S$ *plus* the particles that will enter $S$ during the interval of time $\Delta t$ (Fig. 14.10$a$). Next, consider at time $t + \Delta t$ the system $S$ *plus* the particles that have left $S$ during the interval $\Delta t$ (Fig. 14.10$c$). Clearly, *the same particles are involved in both cases,* and we can apply the principle of impulse and momentum to those particles. Because the total mass $m$ of the system $S$ remains constant, the particles entering the system and those leaving the system in the time $\Delta t$ must have the

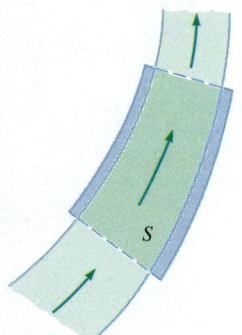

**Fig. 14.9** A system of particles in a steady stream.

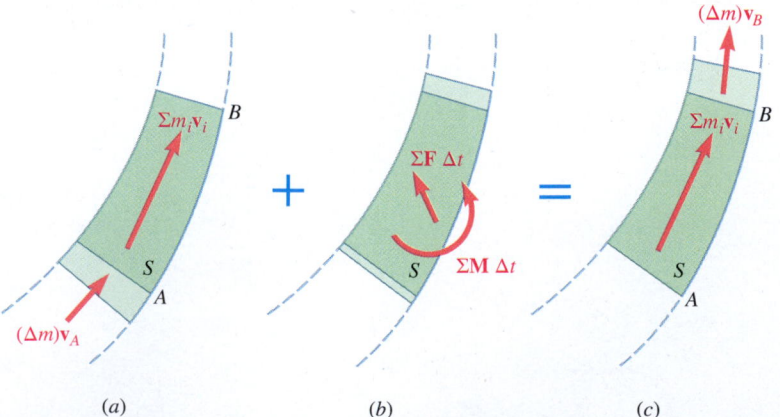

**Fig. 14.10** The impulse–momentum diagram for a stream of particles contains (*a*) momenta of particles entering and in the system $S$ plus (*b*) impulses during the time interval $\Delta t$ and (*c*) momenta of the particles in and leaving the system.

same mass $\Delta m$. Suppose we denote the velocities of the particles entering $S$ at $A$ and leaving $S$ at $B$ by $\mathbf{v}_A$ and $\mathbf{v}_B$, respectively. Then we can represent the momentum of the particles entering $S$ by $(\Delta m)\mathbf{v}_A$ (Fig. 14.10*a*) and the momentum of the particles leaving $S$ by $(\Delta m)\mathbf{v}_B$ (Fig. 14.10*c*). We also represent the momenta $m_i\mathbf{v}_i$ of the particles forming $S$ and the impulses of the forces exerted on $S$ by appropriate vectors. Then we indicate by blue plus and equal signs that the system of the momenta and impulses in parts *a* and *b* of Fig. 14.10 is equipollent to the system of the momenta in part *c* of the same figure.

The resultant $\Sigma m_i\mathbf{v}_i$ of the momenta of the particles of $S$ is found on both sides of the equal sign and thus can be omitted. We conclude:

**The system formed by the momentum $(\Delta m)\mathbf{v}_A$ of the particles entering $S$ in the time $\Delta t$ and the impulses of the forces exerted on $S$ during that time is equipollent to the momentum $(\Delta m)\mathbf{v}_B$ of the particles leaving $S$ in the same time $\Delta t$.**

Mathematically, we have

$$(\Delta m)\mathbf{v}_A + \Sigma \mathbf{F}\Delta t = (\Delta m)\mathbf{v}_B \qquad \textbf{(14.38)}$$

We can obtain a similar equation by taking the moments of the vectors involved (see Sample Prob. 14.7). Dividing all terms of Eq. (14.38) by $\Delta t$ and letting $\Delta t$ approach zero, we obtain at the limit

$$\Sigma \mathbf{F} = \frac{dm}{dt}(\mathbf{v}_B - \mathbf{v}_A) \qquad \textbf{(14.39)}$$

where $\mathbf{v}_B - \mathbf{v}_A$ represents the difference between the *vector* $\mathbf{v}_B$ and the *vector* $\mathbf{v}_A$.

If we use SI units, $dm/dt$ is expressed in kg/s and the velocities in m/s; we check that both sides of Eq. (14.39) are expressed in the same units (newtons). If we use U.S. customary units, $dm/dt$ must be expressed in slugs/s and the velocities in ft/s; we check again that both sides of the equation are expressed in the same units (pounds).[†]

We can use this principle to analyze a large number of engineering applications. Let's look at some of the more common of these applications.

**Fluid Stream Diverted by a Vane.**   If the vane is fixed, we can apply directly the method of analysis given here to find the force $\mathbf{F}$ exerted by the vane on the stream. Note that $\mathbf{F}$ is the only force we need to consider, because the pressure in the stream is constant (atmospheric pressure). The force exerted by the stream on the vane is equal and opposite to $\mathbf{F}$.

If the vane moves with a constant velocity, the stream is not steady. However, it will appear steady to an observer moving with the vane. We should therefore choose a system of axes moving with the vane. Because this system of axes is not accelerated, we can still use Eq. (14.38), but we must replace $\mathbf{v}_A$ and $\mathbf{v}_B$ by the *relative velocities* of the stream with respect to the vane (see Sample Prob. 14.8).

**Fluid Flowing Through a Pipe.**   We can determine the force exerted by the fluid on a pipe transition, such as a bend or a contraction, by considering the system of particles $S$ in contact with the transition. Because, in general, the pressure in the flow will vary, we should also consider the forces exerted on $S$ by the adjoining portions of the fluid.

---

[†]It is often convenient to express the mass rate of flow $dm/dt$ as the product $\rho Q$, where $\rho$ is the density of the stream (mass per unit volume) and $Q$ is its volume rate of flow (volume per unit time). If you use SI units, $\rho$ is in kg/m³ (for instance, $\rho = 1000$ kg/m³ for water) and $Q$ is in m³/s. However, if you use U.S. customary units, $\rho$ generally has to be computed from the corresponding specific weight $\gamma$ (weight per unit volume), $\rho = \gamma/g$. Because $\gamma$ is expressed in lb/ft³ (for instance, $\gamma = 62.4$ lb/ft³ for water), we obtain $\rho$ in slug/ft³. The volume rate of flow $Q$ is expressed in ft³/s.

**Jet Engine.**    In a jet engine, air enters the front of the engine with no velocity and leaves through the rear with a high velocity. The energy required to accelerate the air particles is obtained by burning fuel. The mass of the burned fuel in the exhaust gases is usually small enough compared with the mass of the air flowing through the engine that it can be neglected. Thus, the analysis of a jet engine reduces to that of an airstream. We can consider this stream as a steady stream if we measure all velocities with respect to the airplane. We assume, therefore, that the airstream enters the engine with a velocity **v** of magnitude equal to the speed of the airplane and leaves with a velocity **u** equal to the relative velocity of the exhaust gases (Fig. 14.11*a*). Because the intake and exhaust pressures are nearly atmospheric, the only external force we need to consider is the force exerted by the engine on the airstream. This force is equal and opposite to the thrust.[†]

**Fan.**    Consider the system of particles *S* shown in Fig. 14.11*b*. We assume the velocity $v_A$ of the particles entering the system is equal to zero, and the velocity $v_B$ of the particles leaving the system is the velocity of the *slipstream*. We can obtain the rate of flow by multiplying $v_B$ by the cross-sectional area of the slipstream. Because the pressure all around *S* is atmospheric, the only external force acting on *S* is the thrust of the fan.

**Helicopter.**    Determining the thrust created by the rotating blades of a hovering helicopter is similar to the determination of the thrust of a fan (Fig. 14.11*c*). We assume the velocity $v_A$ of the air particles as they approach the blades is zero, and we obtain the rate of flow by multiplying the magnitude of the velocity $v_B$ of the slipstream by its cross-sectional area.

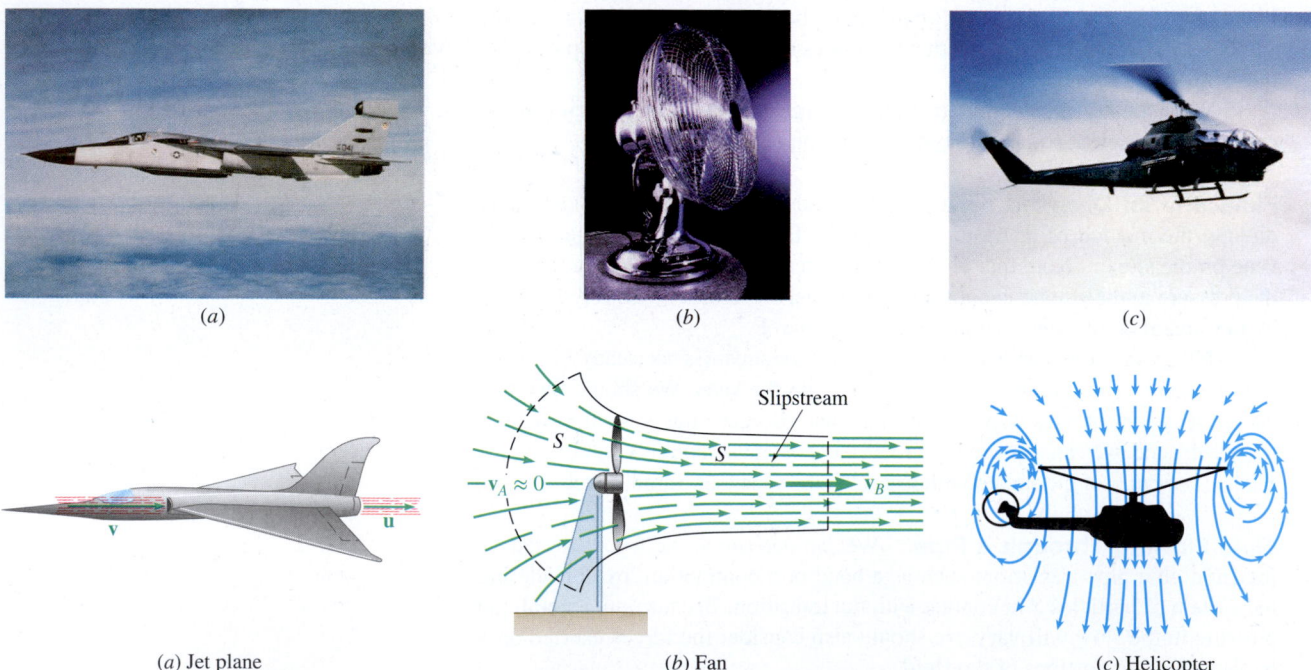

(*a*) Jet plane          (*b*) Fan          (*c*) Helicopter

**Fig. 14.11**    Applications of a steady stream of particles. (a) & (c) ©Purestock/SuperStock RF; (b) ©Design Pics/PunchStock RF

[†]Note that if the airplane is accelerating, we cannot use it as a newtonian frame of reference. However, we can obtain the same result for the thrust by using a reference frame at rest with respect to the atmosphere. In this frame, the air particles enter the engine with no velocity and leave it with a velocity of magnitude $u - v$.

# *14.3B  Systems Gaining or Losing Mass

Let us now analyze a different type of variable system of particles; namely, a system that gains mass by continually absorbing particles or loses mass by continually expelling particles. Consider the system $S$ shown in Fig. 14.12. Its mass, equal to $m$ at the instant $t$, increases by $\Delta m$ in the time interval $\Delta t$. In order to apply the principle of impulse and momentum to this system, we must consider at time $t$ the system $S$ *plus* the particles of mass $\Delta m$ that $S$ absorbs during the time interval $\Delta t$. The velocity of $S$ at time $t$ is denoted by $\mathbf{v}$, the velocity of $S$ at time $t + \Delta t$ is denoted by $\mathbf{v} + \Delta \mathbf{v}$, and the absolute velocity of the particles absorbed is denoted by $\mathbf{v}_a$. Applying the principle of impulse and momentum, we have

$$m\mathbf{v} + (\Delta m)\mathbf{v}_a + \Sigma\mathbf{F}\,\Delta t = (m + \Delta m)(\mathbf{v} + \Delta \mathbf{v}) \qquad \textbf{(14.40)}$$

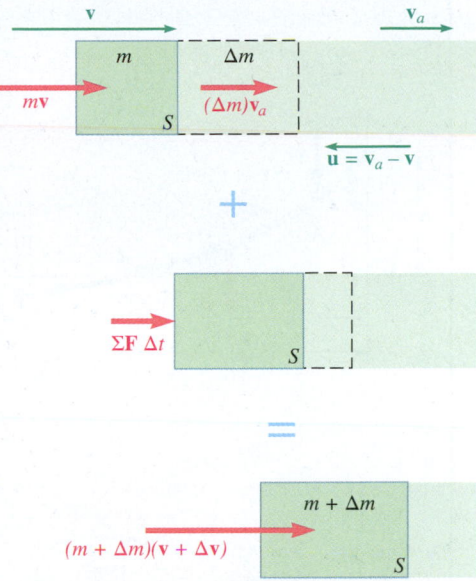

Solving for the sum $\Sigma\mathbf{F}\,\Delta t$ of the impulses of the external forces acting on $S$ (excluding the forces exerted by the particles being absorbed), we obtain

$$\Sigma\mathbf{F}\,\Delta t = m\Delta\mathbf{v} + \Delta m(\mathbf{v} - \mathbf{v}_a) + (\Delta m)(\Delta\mathbf{v}) \qquad \textbf{(14.41)}$$

Now we introduce the *relative velocity* $\mathbf{u}$ with respect to $S$ of the particles that are absorbed. We have $\mathbf{u} = \mathbf{v}_a - \mathbf{v}$ and note, because $v_a < v$, that the relative velocity $\mathbf{u}$ is directed to the left, as shown in Fig. 14.12. Neglecting the last term in Eq. (14.41), which is of the second order, we have

$$\Sigma\mathbf{F}\,\Delta t = m\,\Delta\mathbf{v} - (\Delta m)\mathbf{u}$$

Now we divide through by $\Delta t$ and let $\Delta t$ approach zero. In the limit we obtain[†]

$$\Sigma\mathbf{F} = m\frac{d\mathbf{v}}{dt} - \frac{dm}{dt}\mathbf{u} \qquad \textbf{(14.42)}$$

**Fig. 14.12**  Impulse–momentum diagram for a system that gains mass.

Rearranging terms and recalling that $d\mathbf{v}/dt = \mathbf{a}$, where $\mathbf{a}$ is the acceleration of the system $S$, we have

$$\Sigma\mathbf{F} + \frac{dm}{dt}\mathbf{u} = m\mathbf{a} \qquad \textbf{(14.43)}$$

This equation states that the action on $S$ of the particles being absorbed is equivalent to a thrust

$$\mathbf{P} = \frac{dm}{dt}\mathbf{u} \qquad \textbf{(14.44)}$$

that tends to slow down the motion of $S$, because the relative velocity $\mathbf{u}$ of the particles is directed to the left. If we use SI units, $dm/dt$ is expressed in kg/s, the relative velocity $u$ is in m/s, and the corresponding thrust is in newtons. If we use U.S. customary units, $dm/dt$ must be expressed in slug/s, $u$ in ft/s, and the corresponding thrust in pounds.

We can also use these equations to determine the motion of a system $S$ losing mass. In this case, the rate of change of mass is negative, and the action on $S$ of the particles being expelled is equivalent to a thrust in the direction of $-\mathbf{u}$; that is, in the direction opposite to that in which the particles are being expelled. A *rocket* represents a typical case of a system continually losing mass (see Sample Prob. 14.9).

**Photo 14.3**  As booster rockets are fired, the gas particles they eject provide the thrust required for liftoff. Source: NASA

[†]When the absolute velocity $\mathbf{v}_a$ of the particles absorbed is zero, $\mathbf{u} = -\mathbf{v}$ and formula (14.42) becomes

$$\Sigma\mathbf{F} = \frac{d}{dt}(m\mathbf{v})$$

Comparing this formula with Eq. (12.3) of Sec. 12.1B, we see that this is Newton's second law applied to a system gaining mass, *provided that the particles absorbed are initially at rest.* We can also apply it to a system losing mass, *provided that the velocity of the particles expelled is zero* with respect to the chosen frame of reference.

## Sample Problem 14.7

Grain falls from a hopper onto a chute $CB$ at the rate of 240 lb/s. It hits the chute at $A$ with a velocity of 20 ft/s and leaves at $B$ with a velocity of 15 ft/s, forming an angle of 10° with the horizontal. Knowing that the combined weight of the chute and of the grain it supports is a force $\mathbf{W}$ with a magnitude of 600 lb applied at $G$, determine the reaction at the roller support $B$ and the components of the reaction at the hinge $C$.

**STRATEGY:** Because we have a steady stream of particles, apply the principle of impulse and momentum for the time interval $\Delta t$.

**MODELING:** Choose a system that consists of the chute, the grain it supports, and the amount of grain that hits the chute in the interval $\Delta t$. The impulse–momentum diagram for this system is shown in Fig. 1. Because the chute does not move, it has no momentum. Note that the sum $\Sigma m_i \mathbf{v}_i$ of the momenta of the particles supported by the chute is the same at $t$ and $t + \Delta t$ and thus can be omitted.

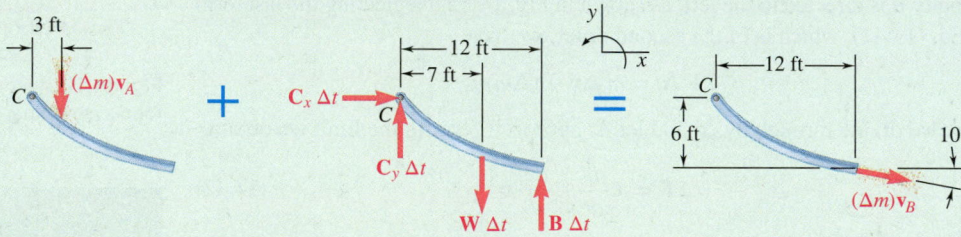

**Fig. 1** Impulse–momentum diagram for the system.

**ANALYSIS:** You can use the impulse-momentum diagram to obtain scalar equations for the $x$ and $y$ directions and for moments about point $C$.

$\xrightarrow{+}$ $x$ components:
$$C_x \Delta t = (\Delta m) v_B \cos 10° \tag{1}$$

$+\uparrow$ $y$ components:
$$-(\Delta m) v_A + C_y \Delta t - W \Delta t + B \Delta t = -(\Delta m) v_B \sin 10° \tag{2}$$

$+\circlearrowleft$ moments about $C$:
$$-3(\Delta m) v_A - 7(W \Delta t) + 12(B \Delta t)$$
$$= 6(\Delta m) v_B \cos 10° - 12(\Delta m) v_B \sin 10° \tag{3}$$

Using the given data, $W = 600$ lb, $v_A = 20$ ft/s, $v_B = 15$ ft/s, and $\Delta m / \Delta t = 240/32.2 = 7.45$ slug/s, and solving Eq. (3) for $B$ and Eq. (1) for $C_x$, you obtain

$$12B = 7(600) + 3(7.45)(20) + 6(7.45)(15)(\cos 10° - 2 \sin 10°)$$

$$12B = 5075 \qquad B = 423 \text{ lb} \qquad\qquad \mathbf{B} = 423 \text{ lb} \uparrow \ \blacktriangleleft$$

$$C_x = (7.45)(15) \cos 10° = 110.1 \text{ lb} \qquad \mathbf{C}_x = 110.1 \text{ lb} \rightarrow \ \blacktriangleleft$$

Substituting for $B$ and solving Eq. (2) for $C_y$, you end up with

$$C_y = 600 - 423 + (7.45)(20 - 15 \sin 10°) = 307 \text{ lb}$$

$$\mathbf{C}_y = 307 \text{ lb} \uparrow \ \blacktriangleleft$$

**REFLECT AND THINK:** This kind of situation is common in factory and storage settings. Being able to determine the reactions is essential for designing a proper chute that will support the stream safely. We can compare this situation to the case when there is no mass flow, which results in reactions of $B_y = 350$ lb, $C_y = 250$ lb, and $C_x = 0$ lb.

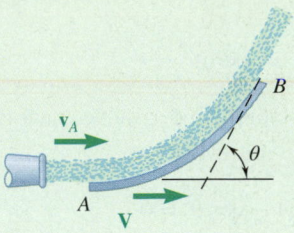

# Sample Problem 14.8

A nozzle discharges a stream of water of cross-sectional area $A$ with a velocity $\mathbf{v}_A$. The stream is deflected by a *single* blade that moves to the right with a constant velocity $\mathbf{V}$. Assuming that the water moves along the blade at constant speed, determine (*a*) the components of the force $\mathbf{F}$ exerted by the blade on the stream, (*b*) the velocity $\mathbf{V}$ for which maximum power is developed.

**STRATEGY:**  Because you have a steady stream of particles, apply the principle of impulse and momentum.

**MODELING:**  Choose the system to be the particles in contact with the blade and the particles striking the blade in the time $\Delta t$, and use a coordinate system that moves with the blade at a constant velocity $\mathbf{V}$. The particles of water strike the blade with a relative velocity $\mathbf{u}_A = \mathbf{v}_A - \mathbf{V}$ and leave the blade with a relative velocity $\mathbf{u}_B$, as shown in Fig. 1. Because the particles move along the blade at a constant speed, the relative velocities $\mathbf{u}_A$ and $\mathbf{u}_B$ have the same magnitude $u$. Denoting the density of water by $\rho$, the mass of the particles striking the blade during the time interval $\Delta t$ is $\Delta m = A\rho(v_A - V)\Delta t$; an equal mass of particles leaves the blade during $\Delta t$. The impulse–momentum diagram for this system is shown in Fig. 2.

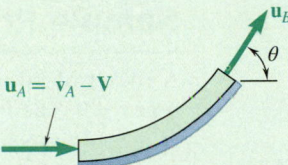

**Fig. 1**  Relative velocities of the water entering and leaving the blade.

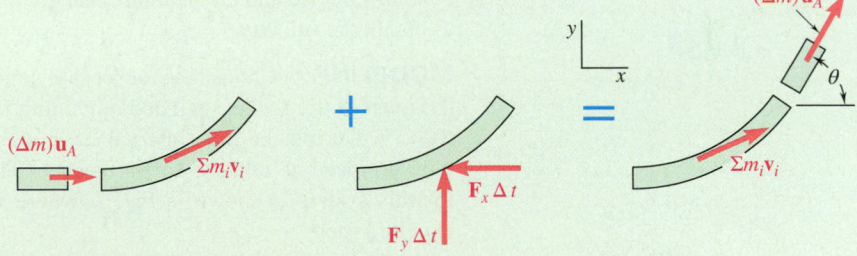

**Fig. 2**  Impulse–momentum diagram for the system.

## ANALYSIS:

### a. Components of Force Exerted on Stream.
Recalling that $\mathbf{u}_A$ and $\mathbf{u}_B$ have the same magnitude $u$ and omitting the momentum $\Sigma m_i \mathbf{v}_i$ that appears on both sides, applying the principle of impulse and momentum gives you

$\xrightarrow{+}$ $x$ components: $\qquad (\Delta m)u - F_x\Delta t = (\Delta m)u \cos\theta$

$+\uparrow$ $y$ components: $\qquad\qquad\quad +F_y\Delta t = (\Delta m)u \sin\theta$

Substituting $\Delta m = A\rho(v_A - V)\Delta t$ and $u = v_A - V$, you obtain

$$\mathbf{F}_x = A\rho(v_A - V)^2(1 - \cos\theta)\leftarrow \qquad \mathbf{F}_y = A\rho(v_A - V)^2 \sin\theta \uparrow$$

*(continued)*

**b. Velocity of Blade for Maximum Power.** You can obtain the power by multiplying the velocity $V$ of the blade by the component $F_x$ of the force exerted by the stream on the blade.

$$\text{Power} = F_x V = A\rho(v_A - V)^2(1 - \cos\theta)V$$

Differentiating the power with respect to $V$ and setting the derivative equal to zero, you have

$$\frac{d(\text{power})}{dv} = A\rho(v_A^2 - 4v_A V + 3V^2)(1 - \cos\theta) = 0$$

$$V = v_A \qquad V = \tfrac{1}{3}v_A \qquad \text{For maximum power } \mathbf{V} = \tfrac{1}{3}v_A \rightarrow \quad \blacktriangleleft$$

**REFLECT and THINK:** These results are valid only when a *single* blade deflects the stream. Different results appear when a series of blades deflects the stream, as in a Pelton-wheel turbine (see Prob. 14.81).

## Sample Problem 14.9

A rocket of initial mass $m_0$ (including shell and fuel) is fired vertically at time $t = 0$. The fuel is consumed at a constant rate $q = dm/dt$ and is expelled at a constant speed $u$ relative to the rocket. Derive an expression for the magnitude of the velocity of the rocket at time $t$, neglecting the resistance of the air.

**STRATEGY:** Because you have a system that is losing mass, apply the principle of impulse and momentum. This gives you an equation you can integrate to obtain the velocity.

**MODELING:** Choose the rocket shell and its fuel as your system. At time $t$, the mass of the rocket shell and remaining fuel is $m = m_0 - qt$, and the velocity is $\mathbf{v}$. During the time interval $\Delta t$, a mass of fuel $\Delta m = q\,\Delta t$ is expelled with a speed $u$ relative to the rocket. The impulse–momentum diagram for this system is shown in Fig. 1, where $\mathbf{v}_e$ is the absolute velocity of the expelled fuel.

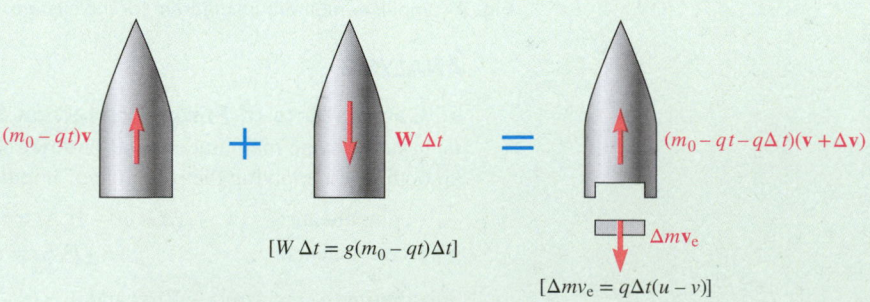

**Fig. 1** Impulse–momentum diagram for the system.

*(continued)*

**ANALYSIS:** Apply the principle of impulse and momentum between time $t$ and time $t + \Delta t$ to find

$$(m_0 - qt)v - g(m_0 - qt)\,\Delta t = (m_0 - qt - q\Delta t)(v + \Delta v) - q\Delta t(u - v)$$

Divide through by $\Delta t$ and let $\Delta t$ approach zero for

$$-g(m_0 - qt) = (m_0 - qt)\frac{dv}{dt} - qu$$

Separating variables and integrating from $t = 0$, $v = 0$ to $t = t$, $v = v$, you have

$$dv = \left(\frac{qu}{m_0 - qt} - g\right)dt$$

$$\int_0^v dv = \int_0^t \left(\frac{qu}{m_0 - qt} - g\right)dt$$

$$v = [-u \ln(m_0 - qt) - gt]_0^t \qquad\qquad v = u \ln\frac{m_0}{m_0 - qt} - gt \quad \blacktriangleleft$$

**REFLECT and THINK:** The mass remaining at time $t_f$, after all of the fuel has been expended, is equal to the mass of the rocket shell $m_s = m_0 - qt_f$, and the maximum velocity attained by the rocket is $v_m = u \ln(m_0/m_s) - gt_f$. Assuming that the fuel is expelled in a relatively short period of time, the term $gt_f$ is small, and we have $v_m \approx u \ln(m_0/m_s)$. In order to escape the gravitational field of the earth, a rocket must reach a velocity of 11.18 km/s. Assuming $u = 2200$ m/s and $v_m = 11.18$ km/s, we obtain $m_0/m_s = 161$. Thus, to project each kilogram of the rocket shell into space, it is necessary to consume more than 161 kg of fuel if we use a propellant yielding $u = 2200$ m/s.

# SOLVING PROBLEMS
# ON YOUR OWN

This section was devoted to the motion of **variable systems of particles**; that is, systems that are continually *gaining or losing particles* or doing both at the same time. The problems you will be asked to solve will involve (1) **steady streams of particles** and (2) **systems gaining or losing mass**.

**1. To solve problems involving a steady stream of particles** (Sample Probs. 14.7 and 14.8), consider a portion $S$ of the stream and express mathematically that the system formed by the momentum of the particles entering $S$ at $A$ in the time $\Delta t$ and that the impulses of the forces exerted on $S$ during that time is equipollent to the momentum of the particles leaving $S$ at $B$ in the same time $\Delta t$ (Fig. 14.10). Considering only the resultants of the vector systems involved, you can write the vector equation

$$(\Delta m)\mathbf{v}_A + \Sigma \mathbf{F} \ \Delta t = (\Delta m)\mathbf{v}_B \qquad \textbf{(14.38)}$$

You may also want to consider the angular momentum of the particle systems to obtain an additional equation (Sample Prob. 14.7). However, many problems can be solved using Eq. (14.38) or the equation obtained by dividing all terms by $\Delta t$ and letting $\Delta t$ approach zero,

$$\Sigma \mathbf{F} = \frac{dm}{dt}(\mathbf{v}_B - \mathbf{v}_A) \qquad \textbf{(14.39)}$$

Here $\mathbf{v}_B - \mathbf{v}_A$ represents a *vector subtraction,* and the mass rate of flow $dm/dt$ can be expressed as the product $\rho Q$ of the density $\rho$ of the stream (mass per unit volume) and the volume rate of flow $Q$ (volume per unit time). In U.S. customary units, $\rho$ is expressed as the ratio $\gamma/g$, where $\gamma$ is the specific weight of the stream and $g$ is the acceleration due to gravity.

Typical problems involving a steady stream of particles have been described in Sec. 14.3A. You may be asked to determine the following:

    **a. Thrust caused by a diverted flow.** Eq. (14.39) is applicable, but you will get a better understanding of the problem if you use a solution based on Eq. (14.38).

    **b. Reactions at supports of vanes or conveyor belts.** First draw a diagram showing on one side of the equal sign the momentum $(\Delta m)\mathbf{v}_A$ of the particles impacting the vane or belt in the time $\Delta t$, as well as the impulses of the loads and reactions at the supports during that time. On the other side, show the momentum $(\Delta m)\mathbf{v}_B$ of the particles leaving the vane or belt in the time $\Delta t$ (Sample Prob. 14.7). Equating the $x$ components, $y$ components, and moments of the quantities on both sides of the equal sign will yield three scalar equations that you can solve for three unknowns.

    **c. Thrust developed by a jet engine, a propeller, or a fan.** In most cases, a single unknown is involved, and you can obtain that unknown by solving the scalar equation derived from Eq. (14.38) or Eq. (14.39).

**2. To solve problems involving systems gaining mass,** consider the system $S$, which has a mass $m$ and is moving with a velocity $\mathbf{v}$ at time $t$, and the particles of mass $\Delta m$ with velocity $\mathbf{v}_a$ that $S$ absorbs in the time interval $\Delta t$ (Fig. 14.12). You will then express that the total momentum of $S$ and of the particles absorbed, *plus* the impulse of the external forces exerted on $S$, are equipollent to the momentum of $S$ at time $t + \Delta t$. Noting that the mass of $S$ and its velocity at that time are, respectively, $m + \Delta m$ and $\mathbf{v} + \Delta\mathbf{v}$, you will write the vector equation

$$m\mathbf{v} + (\Delta m)\mathbf{v}_a + \Sigma\mathbf{F}\,\Delta t = (m + \Delta m)(\mathbf{v} + \Delta\mathbf{v}) \qquad (14.40)$$

As we showed in Sec. 14.3B, if you introduce the relative velocity $\mathbf{u} = \mathbf{v}_a - \mathbf{v}$ of the particles being absorbed, you obtain the following expression for the resultant of the external forces applied to $S$:

$$\Sigma\mathbf{F} = m\frac{d\mathbf{v}}{dt} - \frac{dm}{dt}\mathbf{u} \qquad (14.42)$$

Furthermore, the action on $S$ of the particles being absorbed is equivalent to a thrust

$$\mathbf{P} = \frac{dm}{dt}\mathbf{u} \qquad (14.44)$$

exerted in the direction of the relative velocity of the particles being absorbed.

Examples of systems gaining mass are conveyor belts, moving railroad cars being loaded with gravel or sand, and chains being pulled out of a pile.

**3. To solve problems involving systems losing mass,** such as rockets and rocket engines, you can use Eqs. (14.40) through (14.44)—provided that you give negative values to the increment of mass $\Delta m$ and to the rate of change of mass $dm/dt$ (Sample Prob. 14.9). It follows that the thrust defined by Eq. (14.44) is exerted in a direction opposite to the direction of the relative velocity of the particles being ejected.

# Problems

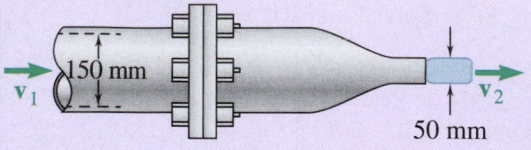

**Fig. P14.57**

**14.57** A stream of water with a density of $\rho = 1000$ kg/m$^3$ is discharged from a nozzle at the rate of 0.06 m$^3$/s. Using Bernoulli's equation, the gage pressure $P$ in the pipe just upstream from the nozzle is $P = 0.5\rho(v_2^2 - v_1^2)$. Knowing the nozzle is held to the pipe by six flange bolts, determine the tension in each bolt, neglecting the initial tension caused by the tightening of the nuts.

**14.58** A jet ski is placed in a channel and is tethered so that it is stationary. Water enters the jet ski with velocity $\mathbf{v}_1$ and exits with velocity $\mathbf{v}_2$. Knowing the inlet area is $A_1$ and the exit area is $A_2$, determine the tension in the tether.

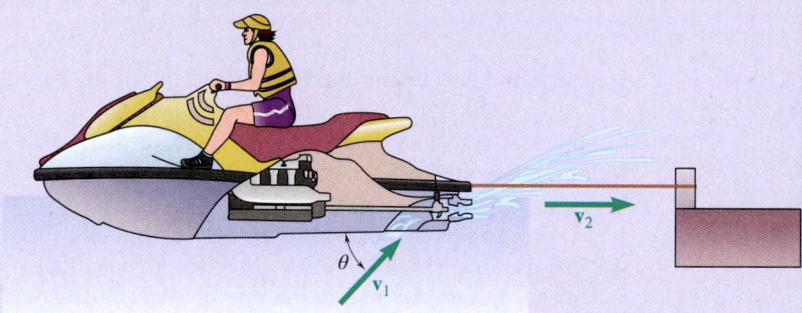

**Fig. P14.58**

**14.59** Tree limbs and branches are being fed at $A$ at the rate of 10 lb/s into a shredder which spews the resulting wood chips at $C$ with a velocity of 60 ft/s. Determine the horizontal component of the force exerted by the shredder on the truck hitch at $D$.

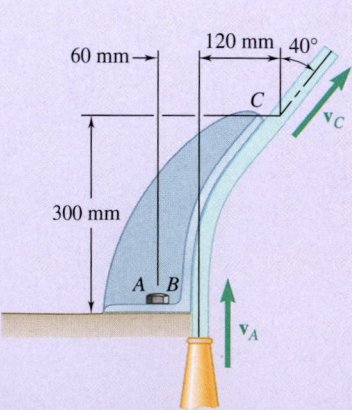

**Fig. P14.60**

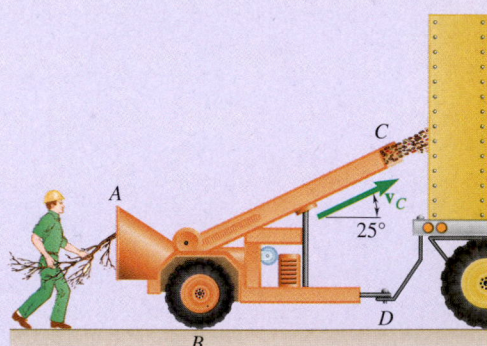

**Fig. P14.59**

**14.60** The nozzle shown discharges water at the rate of 800 L/min. Knowing that at both $B$ and $C$ the stream of water moves with a velocity of magnitude 30 m/s, and neglecting the weight of the vane, determine the force-couple system that must be applied at $A$ to hold the vane in place.

**14.61** A rotary power plow is used to remove snow from a level section of railroad track. The plow car is placed ahead of an engine that propels it at a constant speed of 20 km/h. The plow car clears 160 Mg of snow per minute, projecting it in the direction shown with a velocity of 12 m/s relative to the plow car. Neglecting friction, determine (a) the force exerted by the engine on the plow car, (b) the lateral force exerted by the track on the plow.

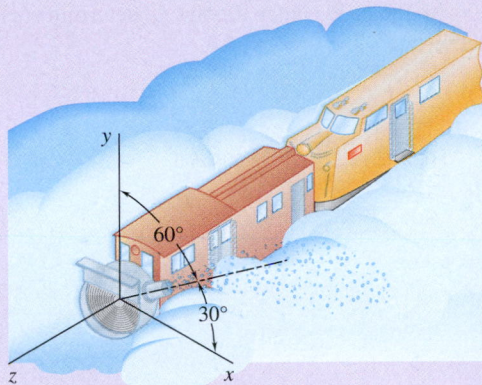

Fig. P14.61

**14.62** A hose discharges water at a rate of 8 m³/min with a velocity of 50 m/s from the bow of a fireboat. Determine the engine thrust necessary to keep the fireboat in a stationary position.

Fig. P14.62

**14.63** Sand falls from three hoppers onto a conveyor belt at a rate of 90 lb/s for each hopper. The sand hits the belt with a vertical velocity $v_1 = 10$ ft/s and is discharged at $A$ with a horizontal velocity $v_2 = 13$ ft/s. Knowing that the combined mass of the beam, belt system, and the sand it supports is 1300 lb with a mass center at $G$, determine the reaction at $E$.

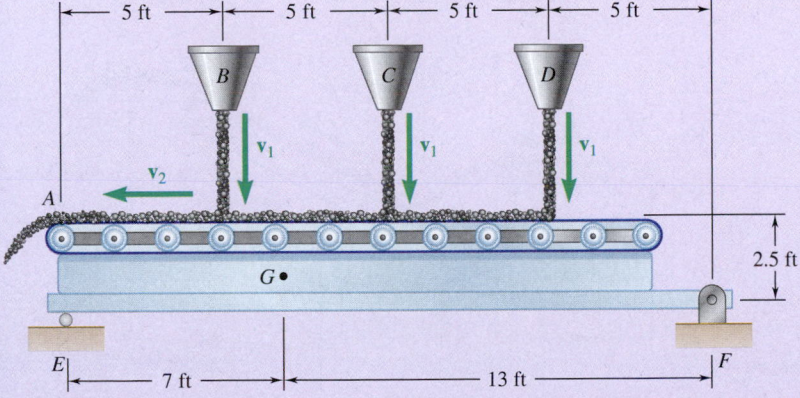

Fig. P14.63

**14.64** The stream of water shown flows at a rate of 550 L/min and moves with a velocity of magnitude 18 m/s at both $A$ and $B$. The vane is supported by a pin and bracket at $C$ and by a load cell at $D$ that can exert only a horizontal force. Neglecting the weight of the vane, determine the components of the reactions at $C$ and $D$.

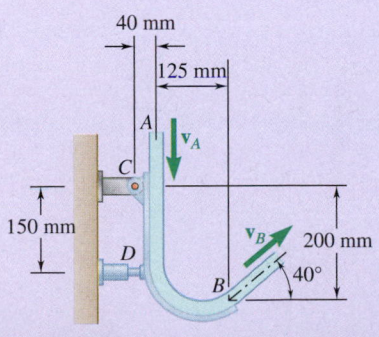

Fig. P14.64

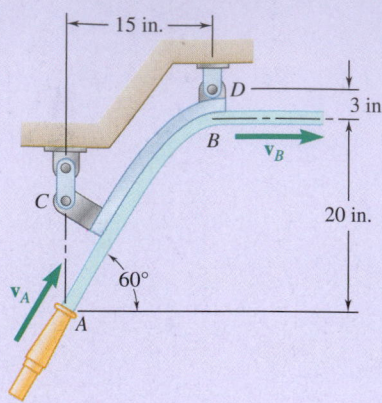

Fig. *P14.65*

**14.65** The nozzle shown discharges water at the rate of 40 ft³/min. Knowing that at both *A* and *B* the stream of water moves with a velocity of magnitude 75 ft/s and neglecting the weight of the vane, determine the components of the reactions at *C* and *D*.

**14.66** A stream of water flowing at a rate of 1.2 m³/min and moving with a speed of 30 m/s at both *A* and *B* is deflected by a vane welded to a hinged plate. Knowing that the combined mass of the vane and plate is 20 kg with the mass center at point *G*, determine (*a*) the angle θ, (*b*) the reaction at *C*.

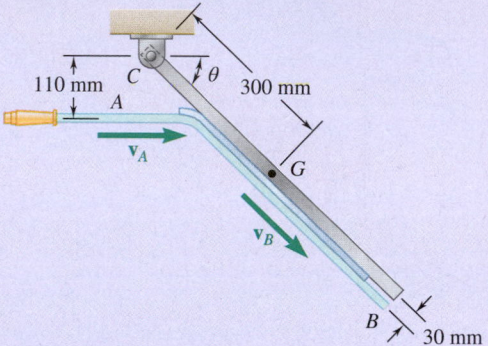

**Fig. P14.66 and P14.67**

**14.67** A stream of water flowing at a rate of 1.2 m³/min and moving with a speed of *v* at both *A* and *B* is deflected by a vane welded to a hinged plate. The combined mass of the vane and plate is 20 kg with the mass center at point *G*. Knowing that θ = 45°, determine (*a*) the speed *v* of the flow, (*b*) the reaction at *C*.

**14.68** Coal is being discharged from a first conveyor belt at the rate of 150 kg/s. It is received at *A* by a second belt that discharges it again at *B*. Knowing that $v_1$ = 3 m/s and $v_2$ = 4.25 m/s and that the second belt assembly and the coal it supports have a total mass of 500 kg, determine the components of the reactions at *C* and *D*.

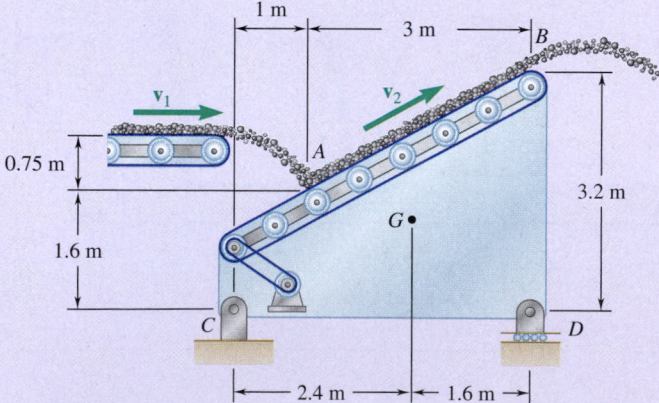

**Fig. P14.68**

**14.69** The total drag due to air friction on a jet airplane traveling at 900 km/h is 35 kN. Knowing that the exhaust velocity is 600 m/s relative to the airplane, determine the mass of air that must pass through the engine per second to maintain the speed of 900 km/h in level flight.

**14.70** While cruising in level flight at a speed of 540 mi/h, a jet plane takes in air at the rate of 150 lb/s and discharges it with a velocity of 2000 ft/s relative to the airplane. Determine the total drag due to air friction on the airplane.

**14.71** In order to shorten the distance required for landing, a jet airplane is equipped with movable vanes that partially reverse the direction of the air discharged by each of its engines. Each engine scoops in the air at a rate of 120 kg/s and discharges it with a velocity of 600 m/s relative to the engine. At an instant when the speed of the airplane is 270 km/h, determine the reverse thrust provided by each of the engines.

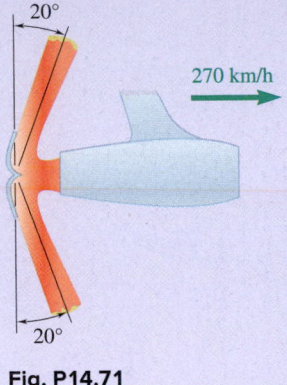

**Fig. P14.71**

**14.72** The helicopter shown can produce a maximum downward air speed of 80 ft/s in a 30-ft-diameter slipstream. Knowing that the weight of the helicopter and its crew is 3500 lb and assuming $\gamma = 0.076$ lb/ft$^3$ for air, determine the maximum load that the helicopter can lift while hovering in midair.

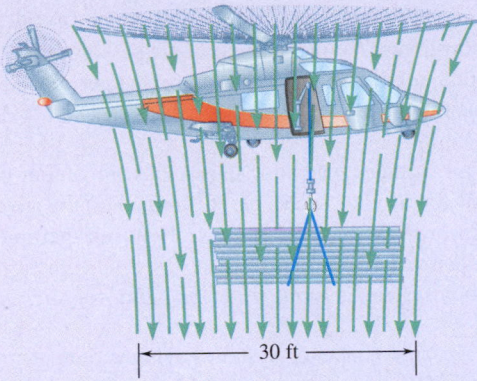

**Fig. P14.72**

**14.73** Prior to takeoff, the pilot of a 3000-kg twin-engine airplane tests the reversible-pitch propellers by increasing the reverse thrust with the brakes at point $B$ locked. Knowing that point $G$ is the center of gravity of the airplane, determine the velocity of the air in the two 2.2-m-diameter slipstreams when the nose wheel $A$ begins to lift off the ground. Assume $\rho = 1.21$ kg/m$^3$ and neglect the approach velocity of the air.

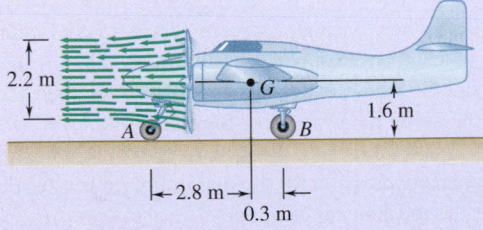

**Fig. P14.73**

**14.74** The jet engine shown scoops in air at $A$ at a rate of 200 lb/s and discharges it at $B$ with a velocity of 2000 ft/s relative to the airplane. Determine the magnitude and line of action of the propulsive thrust developed by the engine when the speed of the airplane is (*a*) 300 mi/h, (*b*) 600 mi/h.

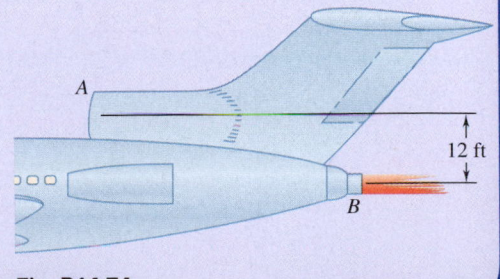

**Fig. P14.74**

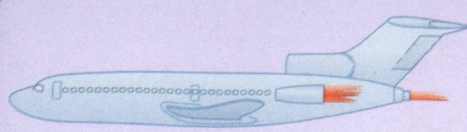

**Fig. P14.75**

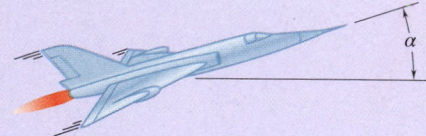

**Fig. P14.76**

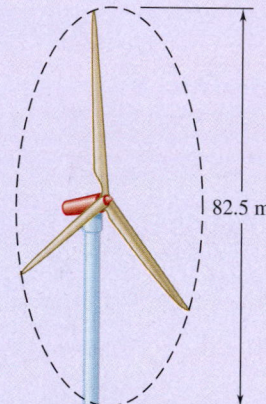

82.5 m

**Fig. P14.78 and P14.79**

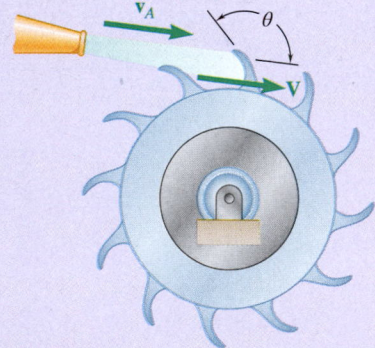

$v_A$

$\theta$

$V$

**Fig. P14.81**

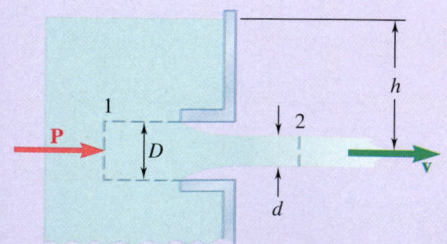

1

P

$D$

2

$h$

$d$

$v$

**Fig. P14.82**

**14.75** A jet airliner is cruising at a speed of 900 km/h with each of its three engines discharging air with a velocity of 800 m/s relative to the plane. Determine the speed of the airliner after it has lost the use of (a) one of its engines, (b) two of its engines. Assume that the drag due to air friction is proportional to the square of the speed and that the remaining engines keep operating at the same rate.

**14.76** A 16-Mg jet airplane maintains a constant speed of 774 km/h while climbing at an angle $\alpha = 18°$. The airplane scoops in air at a rate of 300 kg/s and discharges it with a velocity of 665 m/s relative to the airplane. If the pilot changes to a horizontal flight while maintaining the same engine setting, determine (a) the initial acceleration of the plane, (b) the maximum horizontal speed that will be attained. Assume that the drag due to air friction is proportional to the square of the speed.

**14.77** The propeller of a small airplane has a 2-m-diameter slipstream and produces a thrust of 3600 N when the airplane is at rest on the ground. Assuming $\rho = 1.225$ kg/m$^3$ for air, determine (a) the speed of the air in the slipstream, (b) the volume of air passing through the propeller per second, (c) the kinetic energy imparted per second to the air in the slipstream.

**14.78** The wind turbine generator shown has an output-power rating of 1.5 MW for a wind speed of 36 km/h. For the given wind speed, determine (a) the kinetic energy of the air particles entering the 82.5-m-diameter circle per second, (b) the efficiency of this energy conversion system. Assume $\rho = 1.21$ kg/m$^3$ for air.

**14.79** A wind turbine generator system having a diameter of 82.5 m produces 1.5 MW at a wind speed of 12 m/s. Determine the diameter of blade necessary to produce 10 MW of power assuming the efficiency is the same for both designs and $\rho = 1.21$ kg/m$^3$ for air.

**14.80** While cruising in level flight at a speed of 570 mi/h, a jet airplane scoops in air at a rate of 240 lb/s and discharges it with a velocity of 2200 ft/s relative to the airplane. Determine (a) the power actually used to propel the airplane, (b) the total power developed by the engine, (c) the mechanical efficiency of the airplane.

**14.81** In a Pelton-wheel turbine, a stream of water is deflected by a series of blades so that the rate at which water is deflected by the blades is equal to the rate at which water issues from the nozzle ($\Delta m/\Delta t = A\rho v_A$). Using the same notation as in Sample Prob. 14.8, (a) determine the velocity **V** of the blades for which maximum power is developed, (b) derive an expression for the maximum power, (c) derive an expression for the mechanical efficiency.

**14.82** A circular reentrant orifice (also called Borda's mouthpiece) of diameter $D$ is placed at a depth $h$ below the surface of a tank. Knowing that the speed of the issuing stream is $v = \sqrt{2gh}$ and assuming that the speed of approach $v_1$ is zero, show that the diameter of the stream is $d = D/\sqrt{2}$. (*Hint:* Consider the section of water indicated, and note that $P$ is equal to the pressure at a depth $h$ multiplied by the area of the orifice.)

**14.83** A railroad car with length $L$ and mass $m_0$ when empty is moving freely on a horizontal track while being loaded with sand from a stationary chute at a rate $dm/dt = q$. Knowing that the car was approaching the chute at a speed $v_0$, determine (a) the mass of the car and its load after the car has cleared the chute, (b) the speed of the car at that time.

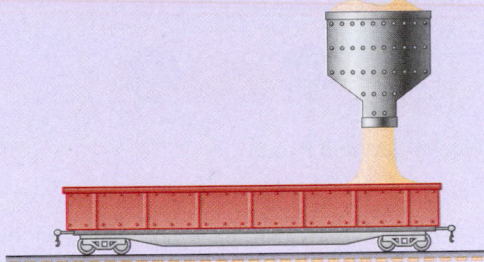

**Fig. P14.83**

**\*14.84** The depth of water flowing in a rectangular channel of width $b$ at a speed $v_1$ and a depth $d_1$ increases to a depth $d_2$ at a *hydraulic jump*. Express the rate of flow $Q$ in terms of $b$, $d_1$, and $d_2$.

**Fig. P14.84**

**\*14.85** Determine the rate of flow in the channel of Prob. 14.84, knowing that $b = 12$ ft, $d_1 = 4$ ft, and $d_2 = 5$ ft.

**14.86** A chain of length $l$ and mass $m$ lies in a pile on the floor. If its end $A$ is raised vertically at a constant speed $v$, express in terms of the length $y$ of chain that is off the floor at any given instant (a) the magnitude of the force **P** applied to $A$, (b) the reaction of the floor.

**14.87** Solve Prob. 14.86, assuming that the chain is being *lowered* to the floor at a constant speed $v$.

**14.88** The ends of a chain lie in piles at $A$ and $C$. When released from rest at time $t = 0$, the chain moves over the pulley at $B$, which has a negligible mass. Denoting by $L$ the length of chain connecting the two piles and neglecting friction, determine the speed $v$ of the chain at time $t$.

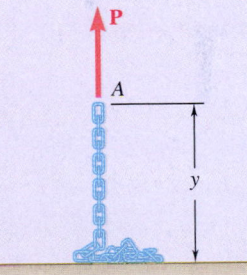

**Fig. P14.86**

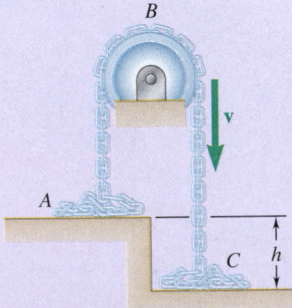

**Fig. P14.88**

**14.89** A toy car is propelled by water that squirts from an internal tank at a constant 6 ft/s relative to the car. The weight of the empty car is 0.4 lb and it holds 2 lb of water. Neglecting other tangential forces, determine the top speed of the car.

**Fig. P14.89 and P14.90**

**14.90** A toy car is propelled by water that squirts from an internal tank. The weight of the empty car is 0.4 lb and it holds 2 lb of water. Knowing the top speed of the car is 8 ft/s, determine the relative velocity of the water that is being ejected.

**14.91** The main propulsion system of a new space transport vehicle consists of three identical rocket engines that provide a total thrust of 6.3 MN in space and 2.4 MN at sea level. Knowing that the rate at which the hydrogen-oxygen propellant is burned by each of the three engines is 500 kg/s, determine the relative velocity of the ejected fuel in space and at sea level.

**14.92** The main propulsion system of a new space transport vehicle consists of three identical rocket engines, each of which produces a thrust of 470,000 lb in space. Determine the rate at which the hydrogen-oxygen propellant is burned by each of the three engines, knowing that it is ejected with a relative velocity of 14,000 ft/s.

**Fig. P14.91 and *P14.92***

**14.93** A rocket sled burns fuel at the constant rate of 120 lb/s. The initial weight of the sled is 1800 lb, including 360 lb of fuel. Assume that the track is lubricated and the sled is aerodynamically designed so that air resistance and friction are negligible. (*a*) Derive a formula for the acceleration *a* of the sled as a function of time *t* and the exhaust velocity $v_{ex}$ of the burned fuel relative to the sled. Plot the ratio $a/v_{ex}$ versus time *t* for the range $0 < t < 4$ s, and check the slope of the graph at $t = 0$ and $t = 4$ s using the formula for *a*. (*b*) Determine the ratio of the velocity $v_b$ of the sled at burnout to the exhaust velocity $v_{ex}$.

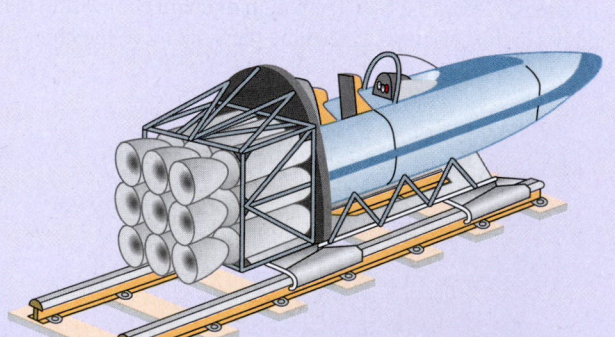

**Fig. *P14.93***

**14.94** A space vehicle describing a circular orbit about the earth at a speed of $24 \times 10^3$ km/h releases at its front end a capsule that has a gross mass of 600 kg, including 400 kg of fuel. If the fuel is consumed at the rate of 18 kg/s and ejected with a relative velocity of 3000 m/s, determine (a) the tangential acceleration of the capsule as its engine is fired, (b) the maximum speed attained by the capsule.

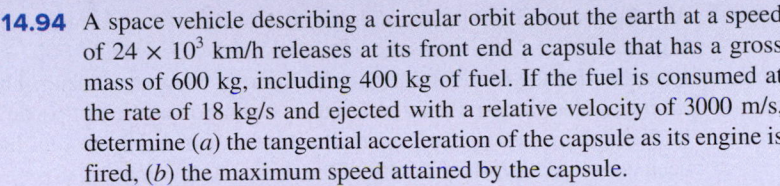

**Fig. P14.94**

**14.95** A 540-kg spacecraft is mounted on top of a rocket with a mass of 19 Mg, including 17.8 Mg of fuel. Knowing that the fuel is consumed at a rate of 225 kg/s and ejected with a relative velocity of 3600 m/s, determine the maximum speed imparted to the spacecraft if the rocket is fired vertically from the ground.

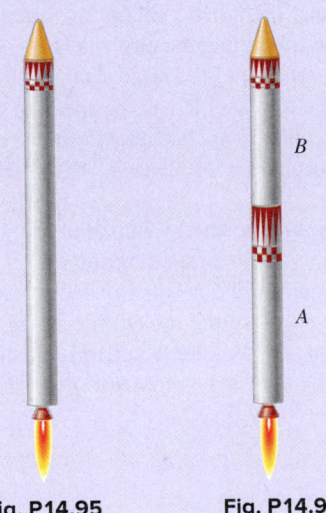

**Fig. P14.95**          **Fig. P14.96**

**14.96** The rocket used to launch the 540-kg spacecraft of Prob. 14.95 is redesigned to include two stages $A$ and $B$, each of mass 9.5 Mg, including 8.9 Mg of fuel. The fuel is again consumed at a rate of 225 kg/s and ejected with a relative velocity of 3600 m/s. Knowing that when stage $A$ expels its last particle of fuel, its casing is released and jettisoned, determine (a) the speed of the rocket at that instant, (b) the maximum speed imparted to the spacecraft.

**14.97** The weight of a spacecraft, including fuel, is 11,600 lb when the rocket engines are fired to increase its velocity by 360 ft/s. Knowing that 1000 lb of fuel is consumed, determine the relative velocity of the fuel ejected.

**14.98** The rocket engines of a spacecraft are fired to increase its velocity by 450 ft/s. Knowing that 1200 lb of fuel is ejected at a relative velocity of 5400 ft/s, determine the weight of the spacecraft after the firing.

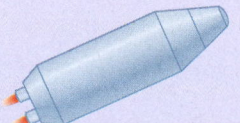
**Fig. P14.97 and P14.98**

**14.99** Determine the distance traveled by the spacecraft of Prob. 14.97 during the rocket engine firing, knowing that its initial speed was 7500 ft/s and the duration of the firing was 60 s.

**14.100** A rocket weighs 2600 lb, including 2200 lb of fuel, which is consumed at the rate of 25 lb/s and ejected with a relative velocity of 13,000 ft/s. Knowing that the rocket is fired vertically from the ground, determine (*a*) its acceleration as it is fired, (*b*) its acceleration as the last particle of fuel is being consumed, (*c*) the altitude at which all the fuel has been consumed, (*d*) the velocity of the rocket at that time.

**14.101** Determine the altitude reached by the spacecraft of Prob. 14.95 when all the fuel of its launching rocket has been consumed.

**14.102** For the spacecraft and the two-stage launching rocket of Prob. 14.96, determine the altitude at which (*a*) stage *A* of the rocket is released, (*b*) the fuel of both stages has been consumed.

**14.103** In a jet airplane, the kinetic energy imparted to the exhaust gases is wasted as far as propelling the airplane is concerned. The useful power is equal to the product of the force available to propel the airplane and the speed of the airplane. If $v$ is the speed of the airplane and $u$ is the relative speed of the expelled gases, show that the mechanical efficiency of the airplane is $\eta = 2v/(u + v)$. Explain why $\eta = 1$ when $u = v$.

**14.104** In a rocket, the kinetic energy imparted to the consumed and ejected fuel is wasted as far as propelling the rocket is concerned. The useful power is equal to the product of the force available to propel the rocket and the speed of the rocket. If $v$ is the speed of the rocket and $u$ is the relative speed of the expelled fuel, show that the mechanical efficiency of the rocket is $\eta = 2uv/(u^2 + v^2)$. Explain why $\eta = 1$ when $u = v$.

# Review and Summary

In this chapter, we analyzed the motion of **systems of particles**; that is, the motion of a large number of particles considered together. In the first part of the chapter, we considered systems consisting of well-defined particles, whereas in the second part, we analyzed systems that are continually gaining or losing particles or doing both at the same time.

## Newton's Second Law for a System of Particles

We showed that **the system of the external forces acting on the particles and the system of the $m_i\mathbf{a}_i$ terms of the particles are equipollent**; that is, both systems have the *same resultant* and the *same moment resultant* about $O$:

$$\sum_{i=1}^{n} \mathbf{F}_i = \sum_{i=1}^{n} m_i \mathbf{a}_i \tag{14.4}$$

$$\sum_{i=1}^{n} (\mathbf{r}_i \times \mathbf{F}_i) = \sum_{i=1}^{n} (\mathbf{r}_i \times m_i \mathbf{a}_i) \tag{14.5}$$

## Linear and Angular Momentum of a System of Particles

We defined the *linear momentum* $\mathbf{L}$ and the *angular momentum* $\mathbf{H}_O$ *about point* $O$ of the system of particles (Sec. 14.1B) as

$$\mathbf{L} = \sum_{i=1}^{n} m_i \mathbf{v}_i \qquad \mathbf{H}_O = \sum_{i=1}^{n} (\mathbf{r}_i \times m_i \mathbf{v}_i) \tag{14.6, 14.7}$$

Then, we showed that we can replace Eqs. (14.4) and (14.5) with the equations

$$\Sigma\mathbf{F} = \dot{\mathbf{L}} \qquad \Sigma\mathbf{M}_O = \dot{\mathbf{H}}_O \tag{14.10, 14.11}$$

Together, these equations state that **the sum of external forces is equal to the rate of change of the linear momentum, and the sum of the moments about $O$ is equal to the rate of change of the angular momentum about $O$.**

## Motion of the Mass Center of a System of Particles

In Sec. 14.1C, we defined the mass center of a system of particles as the point $G$ whose position vector $\bar{\mathbf{r}}$ satisfies the equation

$$m\bar{\mathbf{r}} = \sum_{i=1}^{n} m_i \mathbf{r}_i \tag{14.12}$$

where $m$ represents the total mass $m = \sum_{i=1}^{n} m_i$ of the particles. Differentiating both sides of Eq. (14.12) twice with respect to $t$, we obtained the relations

$$\mathbf{L} = m\bar{\mathbf{v}} \qquad \dot{\mathbf{L}} = m\bar{\mathbf{a}} \tag{14.14, 14.15}$$

where $\bar{\mathbf{v}}$ and $\bar{\mathbf{a}}$ represent, respectively, the velocity and the acceleration of the mass center $G$. Substituting for $\dot{\mathbf{L}}$ from Eq. (14.15) into Eq. (14.10), we obtained

$$\Sigma\mathbf{F} = m\bar{\mathbf{a}} \tag{14.16}$$

From this, we concluded that **the mass center of a system of particles moves as if the entire mass of the system and all of the external forces were concentrated at that point** (Sample Prob. 14.1).

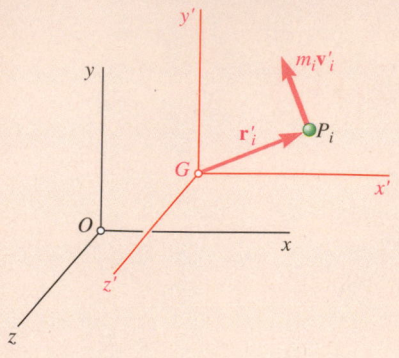
Fig. 14.13

## Angular Momentum of a System of Particles about Its Mass Center

In Sec. 14.1D, we considered the motion of the particles of a system with respect to a centroidal frame $Gx'y'z'$ attached to the mass center $G$ of the system and in translation with respect to the newtonian frame $Oxyz$ (Fig. 14.13). We defined the *angular momentum* of the system *about its mass center $G$* as the sum of the moments about $G$ of the momenta $m_i\mathbf{v}'_i$ of the particles relative to the frame $Gx'y'z'$. We also noted that we can obtain the same earlier result by considering the moments about $G$ of the momenta $m_i\mathbf{v}_i$ of the particles in their absolute motion. We therefore wrote

$$\mathbf{H}_G = \sum_{i=1}^{n} (\mathbf{r}'_i \times m_i \mathbf{v}_i) = \sum_{i=1}^{n} (\mathbf{r}'_i \times m_i \mathbf{v}'_i) \qquad (14.24)$$

and derived the relation

$$\Sigma \mathbf{M}_G = \dot{\mathbf{H}}_G \qquad (14.23)$$

This equation states that **the resultant moment about $G$ of the external forces is equal to the rate of change of the angular momentum about $G$ of the system of particles**. As you will see later, this relation is fundamental to the study of the motion of rigid bodies.

## Conservation of Momentum

When no external force acts on a system of particles (Sec. 14.1E), it follows from Eqs. (14.10) and (14.11) that the linear momentum $\mathbf{L}$ and the angular momentum $\mathbf{H}_O$ of the system are conserved (Sample Probs. 14.2 and 14.4). In problems involving central forces, the angular momentum of the system about the center of force $O$ is also conserved.

## Kinetic Energy of a System of Particles

The kinetic energy $T$ of a system of particles was defined as the sum of the kinetic energies of the particles (Sec. 14.2A):

$$T = \frac{1}{2} \sum_{i=1}^{n} m_i v_i^2 \qquad (14.28)$$

Using the centroidal frame of reference $Gx'y'z'$ of Fig. 14.13, we noted that we can also obtain the kinetic energy of the system by adding the kinetic energy $\frac{1}{2}m\bar{v}^2$ associated with the motion of the mass center $G$ and the kinetic energy of the system relative to the frame $Gx'y'z'$. Thus,

$$T = \frac{1}{2}m\bar{v}^2 + \frac{1}{2} \sum_{i=1}^{n} m_i v_i'^2 \qquad (14.29)$$

## Principle of Work and Energy

We applied the **principle of work and energy** to a system of particles, as well as to individual particles (Sec. 14.2B). We have

$$T_1 + U_{1\to2} = T_2 \qquad (14.30)$$

and noted that $U_{1\to2}$ represents the work of *all* of the forces acting on the particles of the system—internal as well as external.

## Conservation of Energy

If all of the forces acting on the particles of a system are *conservative*, we can determine the potential energy $V$ of the system and write

$$T_1 + V_1 = T_2 + V_2 \qquad (14.31)$$

which expresses the **principle of conservation of energy** for a system of particles.

## Principle of Impulse and Momentum

We saw in Sec. 14.2C that the **principle of impulse and momentum** for a system of particles can be expressed graphically, as shown in Fig. 14.14. The principle states that the momenta of the particles at time $t_1$ and the impulses of the external forces from $t_1$ to $t_2$ form a system of vectors equipollent to the system of the momenta of the particles at time $t_2$.

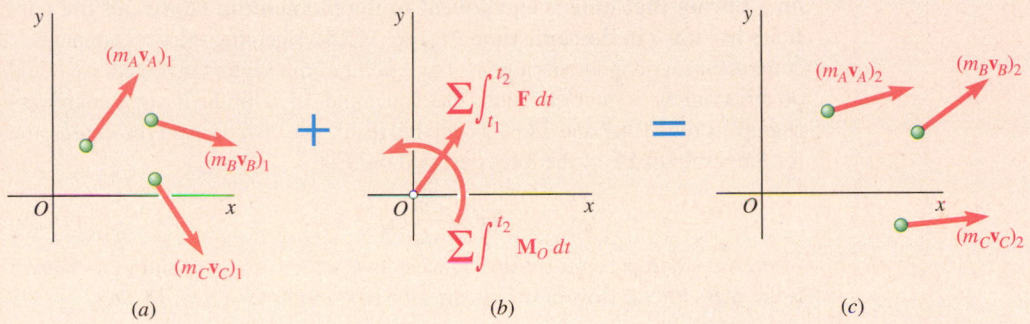

(a)  (b)  (c)

**Fig. 14.14**

If no external force acts on the particles of the system, the systems of momenta shown in parts *a* and *c* of Fig. 14.14 are equipollent, and we have

$$\mathbf{L}_1 = \mathbf{L}_2 \qquad (\mathbf{H}_O)_1 = (\mathbf{H}_O)_2 \qquad (14.36, 14.37)$$

## Use of Conservation Principles in the Solution of Problems Involving Systems of Particles

We can solve many problems involving the motion of systems of particles by applying simultaneously the principle of impulse and momentum and the principle of conservation of energy (Sample Prob. 14.5) or by expressing that the linear momentum, angular momentum, and energy of the system are conserved (Sample Prob. 14.6).

## Steady Stream of Particles

In the second part of the chapter, we considered **variable systems of particles.** First we considered a **steady stream of particles**, such as a stream of water diverted by a fixed vane or the flow of air through a jet engine (Sec. 14.3A). We applied the principle of impulse and momentum to a system $S$ of particles during a time interval $\Delta t$, including the particles that enter the system at $A$ during that time interval and those (of the same mass $\Delta m$) that leave the system at $B$. We concluded that **the system formed by the momentum $(\Delta m)\mathbf{v}_A$ of the particles entering $S$ in the time $\Delta t$ and the impulses of the forces exerted**

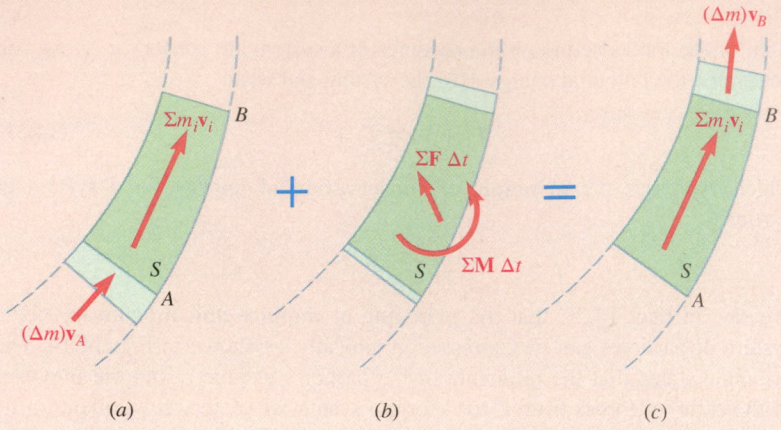

Fig. 14.15

on $S$ during that time is equipollent to the momentum $(\Delta m)\mathbf{v}_B$ of the particles leaving $S$ in the same time $\Delta t$ (Fig. 14.15). Equating the $x$ components, $y$ components, and moments about a fixed point of the vectors involved, we could obtain as many as three equations that you could solve for the desired unknowns (Sample Probs. 14.7 and 14.8). From this result, we also derived the expression for the resultant $\Sigma\mathbf{F}$ of the forces exerted on $S$ as

$$\Sigma\mathbf{F} = \frac{dm}{dt}(\mathbf{v}_B - \mathbf{v}_A) \tag{14.39}$$

where $\mathbf{v}_B - \mathbf{v}_A$ represents the difference between the *vectors* $\mathbf{v}_B$ and $\mathbf{v}_A$ and $dm/dt$ is the mass rate of flow of the stream (see first footnote in Sec. 14.3A).

## Systems Gaining or Losing Mass

We considered next a system of particles gaining mass by continually absorbing particles or losing mass by continually expelling particles (Sec. 14.3B), as in the case of a rocket. We applied the principle of impulse and momentum to the system during a time interval $\Delta t$, being careful to include the particles gained or lost during that time interval (Sample Prob. 14.9). We also noted that the action on a system $S$ of the particles being *absorbed* by $S$ was equivalent to a thrust

$$\mathbf{P} = \frac{dm}{dt}\mathbf{u} \tag{14.44}$$

where $dm/dt$ is the rate at which mass is being absorbed and $\mathbf{u}$ is the velocity of the particles *relative to S*. In the case of particles being *expelled* by $S$, the rate $dm/dt$ is negative, and the thrust $\mathbf{P}$ is exerted in a direction opposite to that in which the particles are being expelled.

# Review Problems

**14.105** Three identical cars are being unloaded from an automobile carrier. Cars $B$ and $C$ have just been unloaded and are at rest with their brakes off when car $A$ leaves the unloading ramp with a velocity of 5.76 ft/s and hits car $B$, which hits car $C$. Car $A$ then again hits car $B$. Knowing that the velocity of car $B$ is 5.04 ft/s after the first collision, 0.630 ft/s after the second collision, and 0.709 ft/s after the third collision, determine (a) the final velocities of cars $A$ and $C$, (b) the coefficient of restitution for each of the collisions.

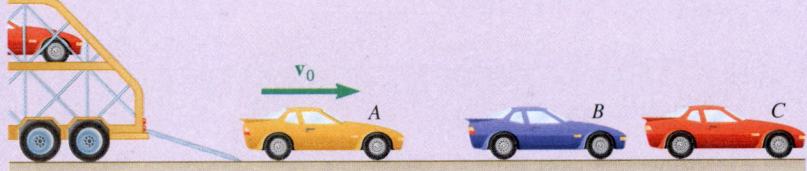

**Fig. P14.105**

**14.106** A 50-kg mother and her 26-kg son are sledding down a 20° incline when their 8-kg toboggan hits a small bump and the son falls off the sled (he was not holding onto the toboggan as instructed by his mother). After one second, the mother jumps off the toboggan to see if her son is unharmed. The son leaves the toboggan with zero relative velocity, and the mother has a relative velocity of 1 m/s up the hill when she jumps off the toboggan. Knowing that the toboggan is initially traveling at 5 m/s and that the coefficient of kinetic friction between the toboggan and the ground is 0.35, determine the speed of the toboggan immediately after the mother jumps off.

**Fig. P14.106**

**14.107** An 80-Mg railroad engine $A$ coasting at 6.5 km/h strikes a 20-Mg flatcar $C$ carrying a 30-Mg load $B$ that can slide along the floor of the car ($\mu_k = 0.25$). Knowing that the car was at rest with its brakes released and that it automatically coupled with the engine upon impact, determine the velocity of the car (a) immediately after impact, (b) after the load has slid to a stop relative to the car.

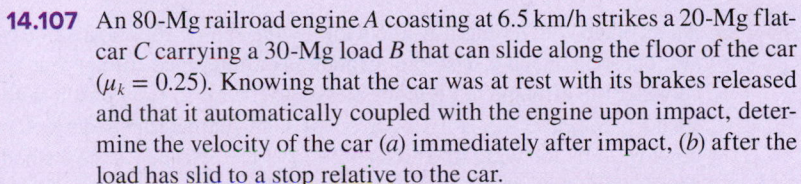

**Fig. P14.107**

**14.108** In a game of pool, ball $A$ is moving with a velocity $\mathbf{v}_0$ when it strikes balls $B$ and $C$, which are at rest and aligned as shown. Knowing that after the collision the three balls move in the directions indicated and that $v_0 = 12$ ft/s and $v_C = 6.29$ ft/s, determine the magnitude of the velocity of (a) ball $A$, (b) ball $B$.

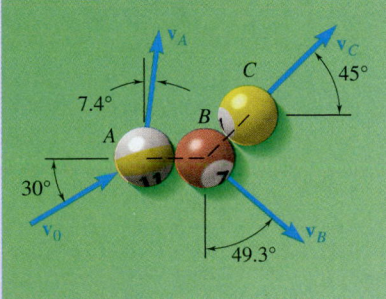

**Fig. P14.108**

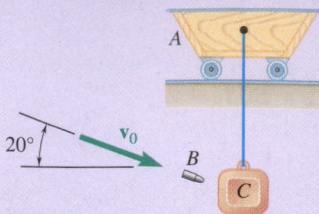

**Fig. P14.109**

**14.109** Mass $C$, which has a mass of 4 kg, is suspended from a cord attached to cart $A$, which has a mass of 5 kg and can roll freely on a frictionless horizontal track. A 60-g bullet is fired with a speed $v_0 = 500$ m/s and gets lodged in block $C$. Determine (*a*) the velocity of $C$ as it reaches its maximum elevation, (*b*) the maximum vertical distance $h$ through which $C$ will rise.

**14.110** A 15-lb block $B$ is at rest and a spring of constant $k = 72$ lb/in. is held compressed 3 in. by a cord. After 5-lb block $A$ is placed against the end of the spring, the cord is cut, causing $A$ and $B$ to move. Neglecting friction, determine the velocities of blocks $A$ and $B$ immediately after $A$ leaves $B$.

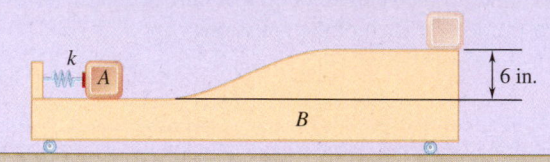

**Fig. P14.110**

**14.111** A 6000-kg dump truck has a 1500-kg stone block sitting in its bed when the operator accidently raises the bed to an angle of 30°. At this angle, the cables holding the block in place break, so the block slides down the bed and impacts the tailgate. Neglecting friction between the block and the bed and assuming the truck can roll freely, determine the speed of the truck and the block (*a*) immediately before the block hits the tailgate, (*b*) immediately after the block hits the tailgate. Assume a plastic impact.

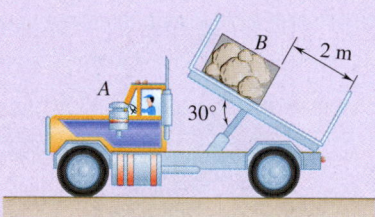

**Fig. P14.111**

**14.112** For the ceiling-mounted fan shown, determine the maximum allowable air velocity in the slipstream if the bending moment in the supporting rod $AB$ is not to exceed 80 ft·lb. Assume $\gamma = 0.076$ lb/ft$^3$ for air and neglect the approach velocity of the air.

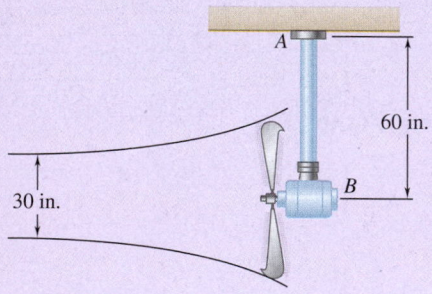

**Fig. P14.112**

**14.113** An airplane with a weight $W$ and a total wing span $b$ flies horizontally at a constant speed $v$. Use the airplane as a reference frame; that is, consider the airplane to be motionless and the air to flow past it with speed $v$. Suppose that a cylinder of air with diameter $b$ is deflected downward by the wing (the cross section of the cylinder is the dashed circle in the figure). Show that the angle through which the cylinder stream is deflected (called the *downwash angle*) is determined by the formula $\sin \theta = 4W/(\pi b^2 \rho v^2)$, where $\rho$ is the mass density of the air.

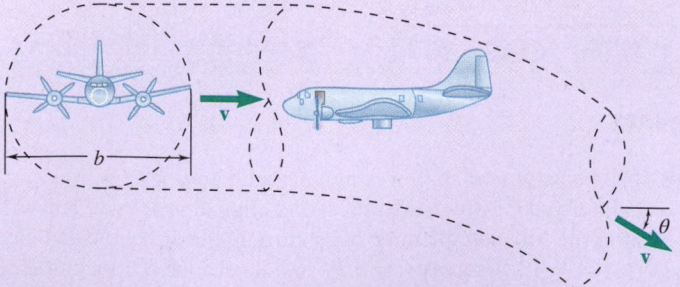

**Fig. P14.113**

**14.114** The final component of a conveyor system receives sand at a rate of 100 kg/s at $A$ and discharges it at $B$. The sand is moving horizontally at $A$ and $B$ with a velocity of magnitude $v_A = v_B = 4.5$ m/s. Knowing that the combined weight of the component and of the sand it supports is $W = 4$ kN, determine the reactions at $C$ and $D$.

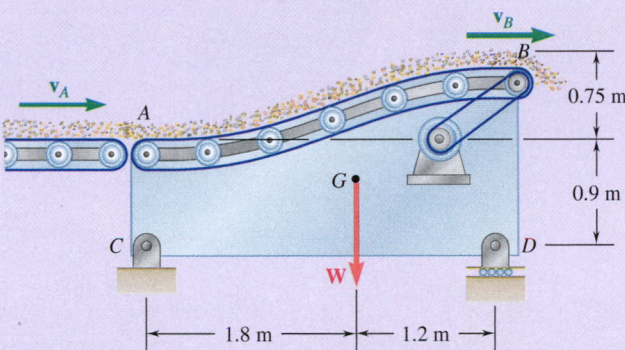

**Fig. P14.114**

**14.115** A garden sprinkler has four rotating arms, each of which consists of two horizontal straight sections of pipe forming an angle of 120° with each other. Each arm discharges water at a rate of 20 L/min with a velocity of 18 m/s relative to the arm. Knowing that the friction between the moving and stationary parts of the sprinkler is equivalent to a couple of magnitude $M = 0.375$ N·m, determine the constant rate at which the sprinkler rotates.

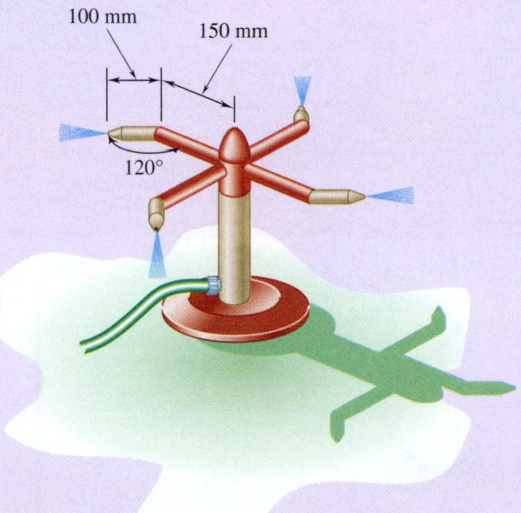

**Fig. P14.115**

**14.116** A chain of length $l$ and mass $m$ falls through a small hole in a plate. Initially, when $y$ is very small, the chain is at rest. In each case shown, determine (a) the acceleration of the first link $A$ as a function of $y$, (b) the velocity of the chain as the last link passes through the hole. In case 1, assume that the individual links are at rest until they fall through the hole. In case 2, assume that at any instant all links have the same speed. Ignore the effect of friction.

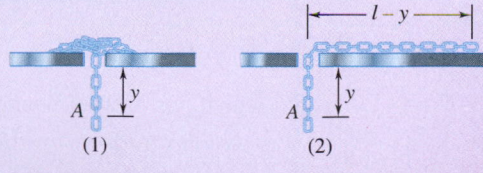

**Fig. P14.116**

©Ryan Pyle/Corbis

# 15

# Kinematics of
# Rigid Bodies

This huge crank belongs to a large diesel engine. In this chapter, you
will learn to perform the *kinematic* analysis of rigid bodies that undergo
*translation*, *fixed-axis rotation*, and *general plane motion*.

# Objectives

- **Describe** the five basic types of rigid body motion: translation, rotation about a fixed axis, general plane motion, motion about a fixed point, and general motion.

- **Use** angular kinematic relationships involving $\theta$, $\omega$, and $\alpha$ to determine the angular motion of a rigid body.

- **Identify** the directions of terms in the relative velocity and relative acceleration equations.

- **Calculate** the linear velocity and acceleration of any point on a rigid body undergoing translation, fixed-axis rotation, or general plane motion.

- **Solve** planar rigid-body kinematics problems using the relative velocity and relative acceleration equations.

- **Determine** the instantaneous center of rotation and use it to analyze the planar velocity kinematics of a rigid body.

- When appropriate, **define** a rotating coordinate frame and use it to solve planar and three-dimensional kinematics problems.

- **Determine** the angular velocity and angular acceleration of a body undergoing three-dimensional motion.

- **Calculate** the linear velocity and acceleration of any point on a rigid body undergoing three-dimensional motion.

# Introduction

In this chapter, we consider the kinematics of **rigid bodies**. We will investigate the relations between the time, the positions, the velocities, and the accelerations of the various particles forming a rigid body. As you will see, the various types of rigid-body motion can be conveniently grouped as follows:

1. **Translation.** A motion is said to be a translation if any straight line inside the body maintains the same orientation during the motion. In a translation, all of the particles forming the body move along parallel paths. If these paths are straight lines, the motion is called **rectilinear translation** (Fig. 15.1); if the paths are curved lines, the motion is called **curvilinear translation** (Fig. 15.2). ▶

2. **Rotation About a Fixed Axis.** In this motion, the particles forming the rigid body move in parallel planes along circles centered on the same fixed axis (Fig. 15.3). If this axis, called the **axis of rotation**, intersects the rigid body, the particles located on the axis have zero velocity and zero acceleration.

Be careful not to confuse rotation with certain types of curvilinear translation. For example, the plate shown in Fig. 15.4*a* is in curvilinear translation, with all of its particles moving along *parallel* circles, whereas the plate shown in Fig. 15.4*b* is in rotation, with all of its particles moving along *concentric* circles. In the first case, any given straight line drawn on the plate maintains the same direction, whereas in the second case, the orientation of the

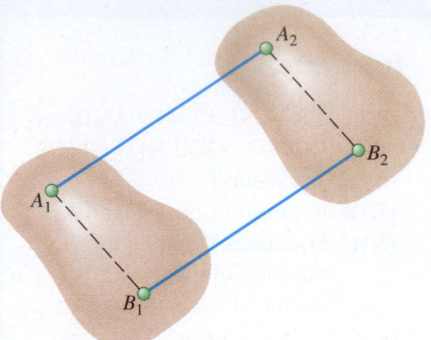

**Fig. 15.1** A rigid body in rectilinear translation.

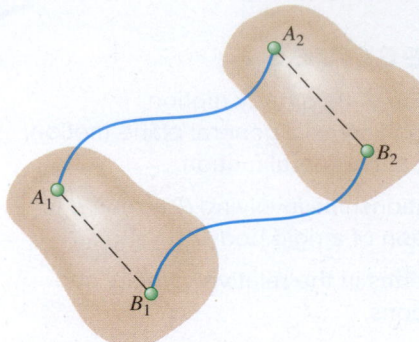

**Fig. 15.2** A rigid body in curvilinear translation.

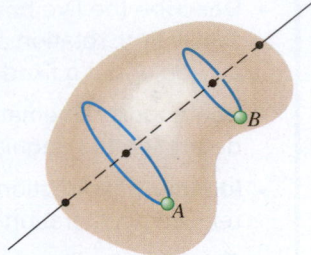

**Fig. 15.3** A rigid body rotating about a fixed axis.

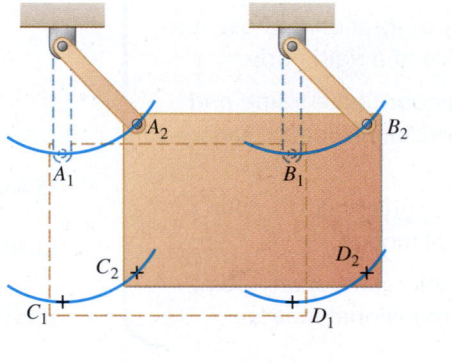

(*a*) Curvilinear translation

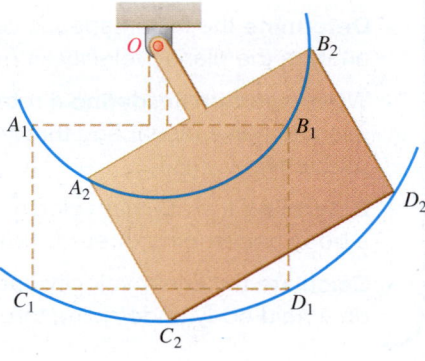

(*b*) Rotation

**Fig. 15.4** (*a*) In curvilinear motion, particles move along parallel circles, whereas (*b*) in fixed-axis rotation, particles move along concentric circles.

plate changes throughout the rotation. Because each particle moves in a given plane, the rotation of a body about a fixed axis is said to be a **plane motion**.

3. **General Plane Motion.** Many other types of plane motion can occur; that is, motions in which all the particles of the body move in a single plane. Any plane motion that is neither a rotation nor a translation is referred to as general plane motion. Fig. 15.5 shows two examples of general plane motion.

4. **Motion About a Fixed Point.** The three-dimensional motion of a rigid body attached at a fixed point *O,* such as the motion of a top on a rough floor (Fig. 15.6), is known as motion about a fixed point.

5. **General Motion.** Any motion of a rigid body that does not fall into any of these categories is referred to as a general motion.

After a brief discussion of the motion of translation, we consider the rotation of a rigid body about a fixed axis. We define the *angular velocity* and the *angular acceleration* of a rigid body rotating about a fixed axis, and you will see how to express the velocity and acceleration of a given point of the body in terms of its position vector and the angular velocity and angular acceleration of the body.

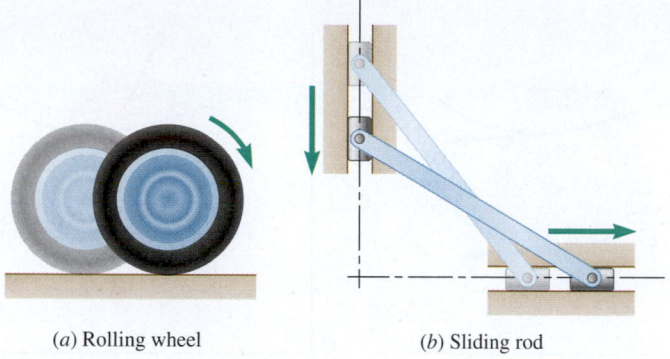

(*a*) Rolling wheel       (*b*) Sliding rod

**Fig. 15.5** (*a*) A rolling wheel and (*b*) a sliding rod are common examples of general plane motion.

Afterward, we study the general plane motion of a rigid body and apply the results to the analysis of mechanisms such as gears, connecting rods, and pin-connected linkages. If we resolve the plane motion of a rigid body into a translation and a rotation, we can then express the velocity of a point *B* of the body as the sum of the velocity of a reference point *A* and of the velocity of *B* relative to a frame of reference translating with *A* (i.e., moving with *A* but not rotating). We use the same approach later in Sec. 15.4 to express the acceleration of *B* in terms of the acceleration of *A* and of the acceleration of *B* relative to a frame translating with *A*. We also present an alternative method for analyzing velocities in plane motion based on the concept of the *instantaneous center of rotation,* and we discuss still another method of analysis based on the use of parametric expressions for the coordinates of a given point.

The motion of a particle relative to a rotating frame of reference and the concept of *Coriolis acceleration* are discussed in Sec. 15.5. We apply the results to the analysis of the plane motion of mechanisms containing parts that slide on each other.

In the remainder of this chapter, we analyze the three-dimensional motion of a rigid body: specifically, the motion of a rigid body with a fixed point and the general motion of a rigid body. We use a fixed frame of reference or a frame of reference in translation to carry out this analysis, and then we consider the motion of the body relative to a rotating frame or to a frame in general motion. Again, we use the concept of Coriolis acceleration.

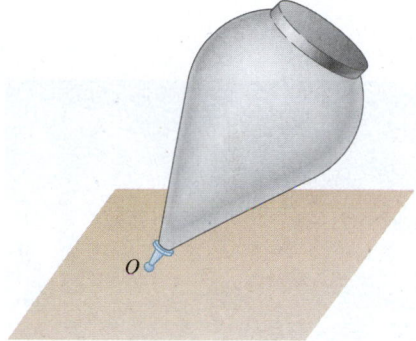

**Fig. 15.6** The motion of a spinning top on a rough surface is an example of three-dimensional motion about a fixed point.

# 15.1 TRANSLATION AND FIXED-AXIS ROTATION

We noted in the introduction that we can resolve a general plane motion into a translation and a rotation. Thus, our first step is to formulate the mathematical descriptions of simple translations and rotations.

## 15.1A Translation

Consider a rigid body in translation (either rectilinear or curvilinear translation), and let *A* and *B* be any two of its particles (Fig. 15.7*a*). Denoting the

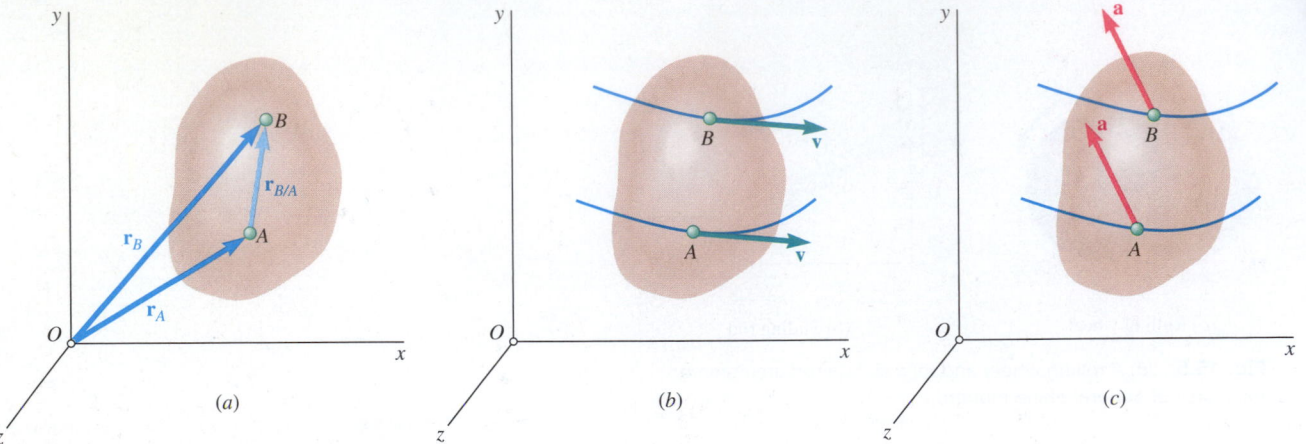

**Fig. 15.7** For a rigid body in translation, (*a*) the position vector between any two points is constant in magnitude and direction; (*b*) every point has the same velocity; (*c*) every point has the same acceleration.

**Photo 15.1** The horizontal linkage of a locomotive undergoes curvilinear translation. ©Alen Penton/Alamy RF

position vectors of *A* and *B* with respect to a fixed frame of reference by $\mathbf{r}_A$ and $\mathbf{r}_B$, respectively, and the vector from *A* to *B* by $\mathbf{r}_{B/A}$, we have

$$\mathbf{r}_B = \mathbf{r}_A + \mathbf{r}_{B/A} \tag{15.1}$$

To obtain the relationship between the velocities of *A* and *B*, we differentiate this expression with respect to *t*. Note that, from the very definition of a translation, the vector $\mathbf{r}_{B/A}$ must maintain a constant direction; its magnitude must also be constant, because *A* and *B* belong to the same rigid body. Thus, the derivative of $\mathbf{r}_{B/A}$ is zero, and we have

$$\mathbf{v}_B = \mathbf{v}_A \tag{15.2}$$

Differentiating once more, we obtain the relationship between the accelerations of *A* and *B* as

$$\mathbf{a}_B = \mathbf{a}_A \tag{15.3}$$

Thus, **when a rigid body is in translation, all the points of the body have the same velocity and the same acceleration at any given instant** (Fig. 15.7*b* and *c*). In the case of curvilinear translation, the velocity and acceleration change in direction, as well as in magnitude, at every instant. In the case of rectilinear translation, all particles of the body move along parallel straight lines, and their velocity and acceleration keep the same direction during the entire motion.

## 15.1B  Rotation About a Fixed Axis

Consider a rigid body that rotates about a fixed axis *AA'*. Let *P* be a point of the body and $\mathbf{r}$ be its position vector with respect to a fixed frame of reference. For convenience, let us assume that the frame is centered at point *O* on *AA'* and that the *z* axis coincides with *AA'* (Fig. 15.8). Let *B* be the projection of *P* on *AA'*. Because *P* must remain at a constant distance from *B*, it describes a circle with a center *B* and radius $r \sin \phi$, where $\phi$ denotes the angle formed by $\mathbf{r}$ and *AA'*.

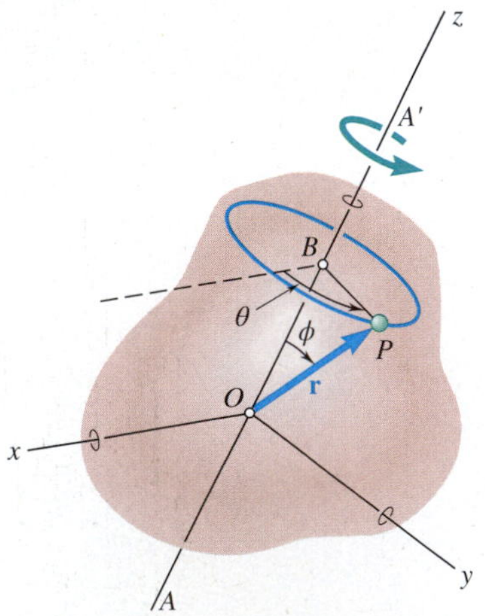

**Fig. 15.8** For a rigid body in rotation about a fixed axis, each point of the body moves in a circular path centered on the axis.

The position both of $P$ and of the entire body is completely defined by the angle $\theta$ that the line $BP$ forms with the $zx$ plane. The angle $\theta$ is known as the **angular coordinate** of the body and is defined as positive when viewed as counterclockwise from $A'$. The angular coordinate is expressed in radians (rad) or, occasionally, in degrees (°) or revolutions (rev). Recall that

$$1 \text{ rev} = 2\pi \text{ rad} = 360°$$

Recall from Sec. 11.4A that the velocity $\mathbf{v} = d\mathbf{r}/dt$ of a particle $P$ is a vector tangent to the path of $P$ and with a magnitude of $v = ds/dt$. The length $\Delta s$ of the arc described by $P$ when the body rotates through $\Delta\theta$ is

$$\Delta s = (BP)\, \Delta\theta = (r \sin \phi)\, \Delta\theta$$

Then dividing both members by $\Delta t$, we obtain in the limit, as $\Delta t$ approaches zero,

$$v = \frac{ds}{dt} = r\dot{\theta} \sin \phi \tag{15.4}$$

where $\dot{\theta}$ denotes the time derivative of $\theta$. (Note that the angle $\theta$ depends on the position of $P$ within the body, but the rate of change $\dot{\theta}$ is itself independent of $P$.) We conclude that the velocity $\mathbf{v}$ of $P$ is a vector perpendicular to the plane containing $AA'$ and $\mathbf{r}$, and of magnitude $v$ defined by Eq. (15.4). But this is precisely the result we would obtain if we drew a vector $\boldsymbol{\omega} = \dot{\theta}\mathbf{k}$ along $AA'$ and formed the vector product $\boldsymbol{\omega} \times \mathbf{r}$ (Fig. 15.9). We thus have

$$\mathbf{v} = \frac{d\mathbf{r}}{dt} = \boldsymbol{\omega} \times \mathbf{r} \tag{15.5}$$

The vector

$$\boldsymbol{\omega} = \omega \mathbf{k} = \dot{\theta}\mathbf{k} \tag{15.6}$$

is directed along the axis of rotation. It is called the **angular velocity** of the body and is equal in magnitude to the rate of change $\dot{\theta}$ of the angular coordinate. You can obtain the sense of the vector by using the right-hand rule (Sec. 3.1E); using your right hand, curl your fingers in the direction of the angular velocity, and your thumb will point in the direction of the vector.[†]

Now we can determine the acceleration $\mathbf{a}$ of particle $P$. Differentiating Eq. (15.5) and recalling the rule for the differentiation of a vector product (Sec. 11.4B), we have

$$\mathbf{a} = \frac{d\mathbf{v}}{dt} = \frac{d}{dt}(\boldsymbol{\omega} \times \mathbf{r})$$

$$= \frac{d\boldsymbol{\omega}}{dt} \times \mathbf{r} + \boldsymbol{\omega} \times \frac{d\mathbf{r}}{dt} \tag{15.7}$$

$$= \frac{d\boldsymbol{\omega}}{dt} \times \mathbf{r} + \boldsymbol{\omega} \times \mathbf{v}$$

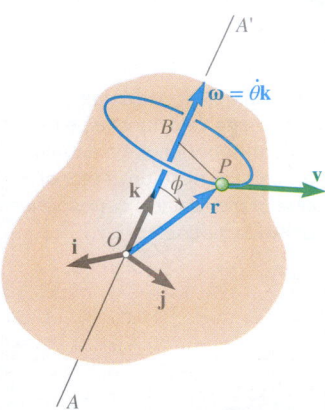

**Photo 15.2** For the central gear rotating about a fixed axis, the angular velocity and angular acceleration of that gear are vectors directed along the vertical axis of rotation.
©Luc Novovitch/Alamy RF

**Fig. 15.9** For a rigid body in rotation about a fixed axis, the velocity of a particle is the vector product of the angular velocity of the body and the position vector of the particle.

---

[†]We will show in Sec. 15.6 the more general case of a rigid body, rotating simultaneously about axes having different directions, where angular velocities obey the parallelogram law of addition and thus are actually vector quantities.

The vector $d\boldsymbol{\omega}/dt$ is denoted by $\boldsymbol{\alpha}$ and is called the **angular acceleration** of the body. Substituting for $\mathbf{v}$ from Eq. (15.5), we have

$$\mathbf{a} = \boldsymbol{\alpha} \times \mathbf{r} + \boldsymbol{\omega} \times (\boldsymbol{\omega} \times \mathbf{r}) \tag{15.8}$$

Differentiating Eq. (15.6) and recalling that $\mathbf{k}$ is constant in magnitude and direction, we have

$$\boldsymbol{\alpha} = \alpha\mathbf{k} = \dot{\omega}\mathbf{k} = \ddot{\theta}\mathbf{k} \tag{15.9}$$

Thus, the angular acceleration of a body rotating about a fixed axis is a vector directed along the axis of rotation and is equal in magnitude to the rate of change $\dot{\omega}$ of the angular velocity.

Returning to Eq. (15.8), we note that the acceleration of $P$ is the sum of two vectors. The first vector is equal to the vector product $\boldsymbol{\alpha} \times \mathbf{r}$; it is tangent to the circle described by $P$ and therefore represents the tangential component of the acceleration. The second vector is equal to the *vector triple product* $\boldsymbol{\omega} \times (\boldsymbol{\omega} \times \mathbf{r})$ obtained by forming the vector product of $\boldsymbol{\omega}$ and $\boldsymbol{\omega} \times \mathbf{r}$. Because $\boldsymbol{\omega} \times \mathbf{r}$ is tangent to the circle described by $P$, the vector triple product is directed toward the center $B$ of the circle and therefore represents the normal component of the acceleration.

### Rotation of a Representative Slab.

We can express the rotation of a rigid body about a fixed axis by examining the motion of a representative slab in a reference plane perpendicular to the axis of rotation. We choose the $xy$ plane as the reference plane and assume that it coincides with the plane of the figure with the $z$ axis pointing out of the page (Fig. 15.10). Recalling from Eq. (15.6) that $\boldsymbol{\omega} = \omega\mathbf{k}$, we note that a positive value of the scalar $\omega$ corresponds to a counterclockwise rotation of the representative slab, and a negative value corresponds to a clockwise rotation. Substituting $\omega\mathbf{k}$ for $\boldsymbol{\omega}$ in Eq. (15.5), we express the velocity of any given point $P$ of the slab as

$$\mathbf{v} = \omega\mathbf{k} \times \mathbf{r} \tag{15.10}$$

Because the vectors $\mathbf{k}$ and $\mathbf{r}$ are mutually perpendicular, the magnitude of the velocity $\mathbf{v}$ is

$$v = r\omega \tag{15.10'}$$

We can obtain its direction by rotating $\mathbf{r}$ through 90° in the sense of rotation of the slab.

If we substitute $\omega\mathbf{k}$ into Eq. (15.8), we obtain $\omega\mathbf{k} \times (\omega\mathbf{k} \times \mathbf{r})$, which simplifies to $-\omega^2\mathbf{r}$. This indicates that the direction of the normal acceleration is $-\mathbf{r}$, or toward the center of rotation, which is exactly what we expect. Using this expression and $\boldsymbol{\alpha} = \alpha\mathbf{k}$ in Eq. (15.8), we obtain

$$\mathbf{a} = \alpha\mathbf{k} \times \mathbf{r} - \omega^2\mathbf{r} \tag{15.11}$$

Resolving $\mathbf{a}$ into tangential and normal components (Fig. 15.11) gives

$$\begin{array}{ll} \mathbf{a}_t = \alpha\mathbf{k} \times \mathbf{r} & a_t = r\alpha \\ \mathbf{a}_n = -\omega^2\mathbf{r} & a_n = r\omega^2 \end{array} \tag{15.11'}$$

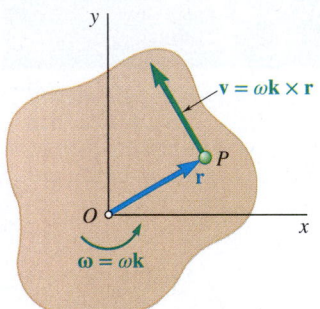

**Fig. 15.10**    For an object undergoing fixed-axis rotation, the velocity of a point $P$ equals the vector product of the angular velocity vector and the position vector of $P$. A positive value of the scalar $\omega$ corresponds to counterclockwise motion.

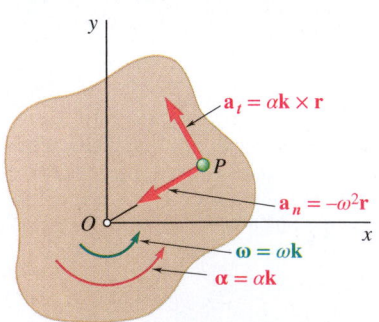

**Fig. 15.11**    For an object undergoing fixed-axis rotation, the acceleration of a point $P$ has a tangential component that depends on angular acceleration, as well as a normal component that depends on angular velocity.

The tangential component $\mathbf{a}_t$ points in the counterclockwise direction if the scalar $\alpha$ is positive and in the clockwise direction if $\alpha$ is negative. The normal component $\mathbf{a}_n$ always points in the direction opposite to that of $\mathbf{r}$; that is, toward $O$.

## 15.1C  Equations Defining the Rotation of a Rigid Body About a Fixed Axis

The motion of a rigid body rotating about a fixed axis $AA'$ is said to be *known* when we can express its angular coordinate $\theta$ as a known function of $t$. In practice, however, we can seldom describe the rotation of a rigid body by a relation between $\theta$ and $t$. More often, the conditions of motion are specified by the angular acceleration of the body. For example, $\alpha$ may be given as a function of $t$, as a function of $\theta$, or as a function of $\omega$. From the relations in Eqs. (15.6) and (15.9), we have

**Photo 15.3**  If the lower roll has a constant angular velocity, the speed of the paper being fed from it it decreases as the radius of the roll decreases. ©Ken Whitmore/Getty Images

$$\omega = \frac{d\theta}{dt} \qquad \textbf{(15.12)}$$

$$\alpha = \frac{d\omega}{dt} = \frac{d^2\theta}{dt^2} \qquad \textbf{(15.13)}$$

or solving Eq. (15.12) for $dt$ and substituting into Eq. (15.13), we have

$$\alpha = \omega \frac{d\omega}{d\theta} \qquad \textbf{(15.14)}$$

These equations are similar to those obtained in Chap. 11 for the rectilinear motion of a particle, so we can integrate them by following the procedures outlined in Sec. 11.1B.

Two particular cases of rotation occur frequently:

1. *Uniform Rotation.* This case is characterized by the fact that the angular acceleration is zero, so the angular velocity is constant and the angular position is given by

$$\theta = \theta_0 + \omega t \qquad \textbf{(15.15)}$$

2. *Uniformly Accelerated Rotation.* In this case, the angular acceleration is constant. We can derive the following formulas relating angular velocity, angular position, and time in a manner similar to that described in Sec. 11.2B. The similarity between the formulas derived here and those obtained for the rectilinear uniformly accelerated motion of a particle is apparent.

$$\omega = \omega_0 + \alpha t$$
$$\theta = \theta_0 + \omega_0 t + \tfrac{1}{2}\alpha t^2 \qquad \textbf{(15.16)}$$
$$\omega^2 = \omega_0^2 + 2\alpha(\theta - \theta_0)$$

We emphasize that you can use formula (15.15) only when $\alpha = 0$, and formulas (15.16) only when $\alpha = $ constant. In any other case, you need to use the general Eqs. (15.12), (15.13), and (15.14).

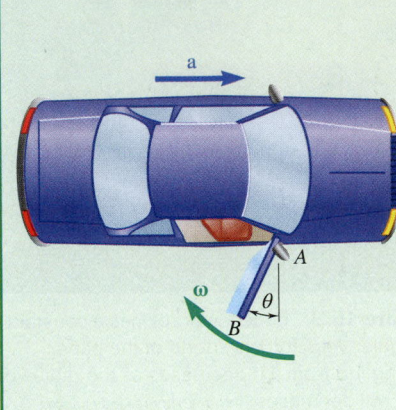

## Sample Problem 15.1

A driver starts his car with the door on the passenger's side wide open ($\theta = 0$). As the car moves forward with constant acceleration, the angular acceleration of the door is $\alpha = 2.5 \cos \theta$, where $\alpha$ is in rad/s². Determine the angular velocity of the door as it slams shut ($\theta = 90°$).

**STRATEGY:** You are given the angular acceleration as a function of $\theta$, so use the kinematic relationships between angular acceleration, angular velocity, angular position, and time.

**MODELING and ANALYSIS:** Model the door as a rigid body. Using the basic kinematic relationship gives

$$\alpha = \frac{d\omega}{dt} = \omega \frac{d\omega}{d\theta} = 2.5 \cos \theta$$

Separating variables gives

$$\omega \, d\omega = 2.5 \cos \theta \, d\theta$$

Integrating, using $\omega = 0$ when $\theta = 0$, you have

$$\int_0^{\omega} \omega \, d\omega = \int_0^{\theta} 2.5 \cos \theta \, d\theta$$

$$\frac{1}{2} \omega^2 = 2.5 \sin \theta \Big|_0^{\pi/2} = 2.5$$

$$\omega = 2.24 \text{ rad/s} \circlearrowleft \quad \blacktriangleleft$$

**REFLECT and THINK:** If the angular acceleration of the door had been a constant 2.5 rad/s², you would have found $\frac{1}{2}\omega^2 = 2.5|_0^{\pi/2}$ or $\omega = 2.80$ rad/s. Because $\alpha = 2.5 \cos \theta$ decreases as $\theta$ increases, it makes sense that the answer you found in this case is smaller than the case for constant angular acceleration.

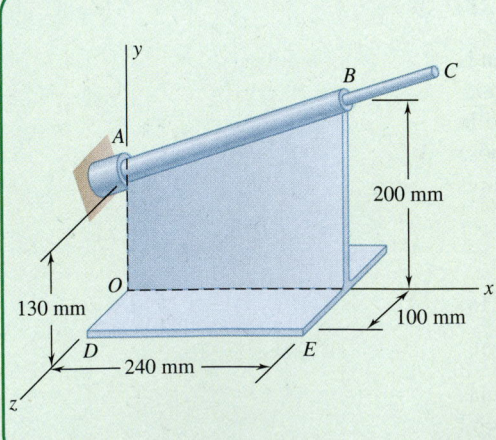

## Sample Problem 15.2

The assembly shown rotates about the rod $AC$. At the instant shown, the assembly has an angular velocity of 5 rad/s that is increasing with an angular acceleration of 25 rad/s². Knowing that the $y$ component of the velocity of corner $D$ is negative at this instant in time, determine the velocity and acceleration of corner $E$.

**STRATEGY:** You are interested in determining the velocity and acceleration of a point on a body undergoing fixed-axis rotation, so use rigid body kinematics.

*(continued)*

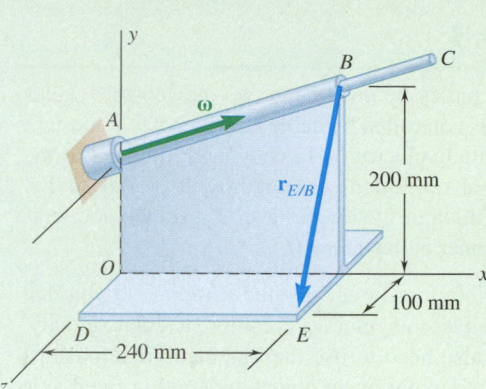

**Fig. 1**  Direction of the angular velocity and the position vector to point E.

**MODELING and ANALYSIS:**  Model the assembly as a rigid body. You can find the velocity and acceleration of E using

$$\mathbf{v}_E = \boldsymbol{\omega} \times \mathbf{r}_{E/B} \tag{1}$$

$$\mathbf{a}_E = \boldsymbol{\alpha} \times \mathbf{r}_{E/B} + \boldsymbol{\omega} \times (\boldsymbol{\omega} \times \mathbf{r}_{E/B}) = \boldsymbol{\alpha} \times \mathbf{r}_{E/B} + \boldsymbol{\omega} \times \mathbf{v}_E \tag{2}$$

To use these equations, you need the angular velocity vector, the angular acceleration vector, and the position vector. The direction of the angular velocity and acceleration vectors are along the axis of rotation. Because the corner D is moving downward and using the right-hand rule, you know $\boldsymbol{\omega}$ is in the direction shown in Fig. 1. Therefore, to write the angular velocity vector, you need a unit vector in this direction. You know that

$$\mathbf{AB} = (0.24 \text{ m})\mathbf{i} + (0.07 \text{ m})\mathbf{j}$$

so the unit vector from A to B is

$$\boldsymbol{\lambda}_{AB} = \frac{(0.24 \text{ m})\mathbf{i} + (0.07 \text{ m})\mathbf{j}}{\sqrt{(0.24 \text{ m})^2 + (0.07 \text{ m})^2}} = 0.960\mathbf{i} + 0.280\mathbf{j}$$

Thus, the angular velocity and angular acceleration are

$$\boldsymbol{\omega} = \omega \boldsymbol{\lambda}_{AB} = (5 \text{ rad/s})(0.960\mathbf{i} + 0.280\mathbf{j}) = (4.80 \text{ rad/s})\mathbf{i} + (1.40 \text{ rad/s})\mathbf{j}$$

$$\boldsymbol{\alpha} = \alpha \boldsymbol{\lambda}_{AB} = (25 \text{ rad/s})(0.960\mathbf{i} + 0.280\mathbf{j}) = (24.0 \text{ rad/s}^2)\mathbf{i} + (7.00 \text{ rad/s}^2)\mathbf{j}$$

The position vector of E with respect to B is

$$\mathbf{r}_{E/B} = (-0.20 \text{ m})\mathbf{j} + (0.10 \text{ m})\mathbf{k}$$

Substituting these expressions into Eqs. (1) and (2) gives

$$\mathbf{v}_E = \boldsymbol{\omega} \times \mathbf{r}_{E/B} = \begin{vmatrix} \mathbf{i} & \mathbf{j} & \mathbf{k} \\ 4.80 & 1.40 & 0 \\ 0 & -0.20 & 0.10 \end{vmatrix} = 0.140\mathbf{i} - 0.480\mathbf{j} - 0.960\mathbf{k}$$

$$\mathbf{v}_E = (0.14 \text{ m/s})\mathbf{i} - (0.480 \text{ m/s})\mathbf{j} - (0.960 \text{ m/s})\mathbf{k} \quad \blacktriangleleft$$

$$\mathbf{a}_E = \boldsymbol{\alpha} \times \mathbf{r}_{E/B} + \boldsymbol{\omega} \times \mathbf{v}_E = \begin{vmatrix} \mathbf{i} & \mathbf{j} & \mathbf{k} \\ 24.0 & 7.00 & 0 \\ 0 & -0.20 & 0.10 \end{vmatrix}$$

$$+ \begin{vmatrix} \mathbf{i} & \mathbf{j} & \mathbf{k} \\ 4.80 & 1.40 & 0 \\ 0.140 & -0.480 & -0.960 \end{vmatrix}$$

$$= 0.70\mathbf{i} - 2.40\mathbf{j} - 4.80\mathbf{k} - 1.344\mathbf{i} + 4.608\mathbf{j} + (-2.304 - 0.196)\mathbf{k}$$

$$\mathbf{a}_E = -(0.644 \text{ m/s}^2)\mathbf{i} + (2.21 \text{ m/s}^2)\mathbf{j} - (7.30 \text{ m/s}^2)\mathbf{k} \quad \blacktriangleleft$$

**REFLECT and THINK:**  The first term of Eq. (2) represents the tangential acceleration of point E. The second term of Eq. (2) represents the normal acceleration of point E and points toward the bar AB. Note that you could have chosen any point along the axis of rotation to define your position vector.

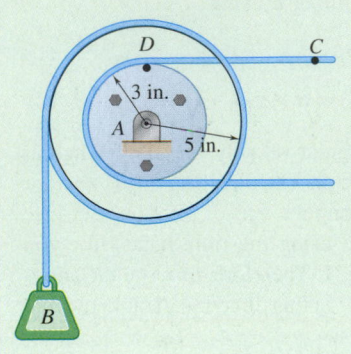

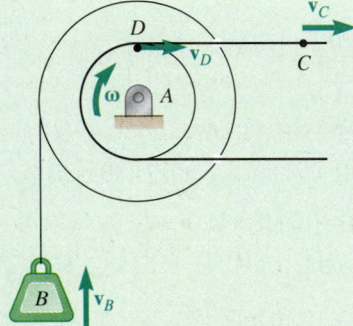

**Fig. 1** The velocity of two points on an inextensible cable are equal.

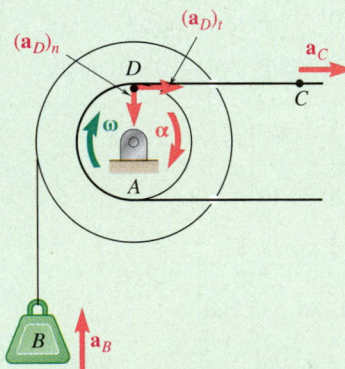

**Fig. 2** Acceleration of B, C, and D.

## Sample Problem 15.3

Load $B$ is connected to a double pulley by one of the two inextensible cables shown. The motion of the pulley is controlled by cable $C$, which has a constant acceleration of 9 in./s$^2$ and an initial velocity of 12 in./s, both directed to the right. Determine (a) the number of revolutions executed by the pulley in 2 s, (b) the velocity and change in position of the load $B$ after 2 s, (c) the acceleration of point $D$ on the rim of the inner pulley at $t = 0$.

**STRATEGY:** This is a case of uniformly accelerated rotation, so you can use the kinematic relationships between angular acceleration, angular velocity, angular position, and time. You also need to use the kinematic relationships for the velocity and acceleration of a point on an object undergoing fixed-axis rotation.

**MODELING and ANALYSIS:**

**a. Motion of Pulley.** You can model the pulley as a rigid body rotating about a fixed axis $A$. Because the cable is inextensible, the velocity of point $D$ is equal to the velocity of point $C$ (Fig. 1), and the tangential component of the acceleration of $D$ is equal to the acceleration of $C$ (Fig. 2).

$$(\mathbf{v}_D)_0 = (\mathbf{v}_C)_0 = 12 \text{ in./s} \rightarrow \qquad (\mathbf{a}_D)_t = \mathbf{a}_C = 9 \text{ in./s}^2 \rightarrow$$

The distance from $D$ to the center of the pulley is 3 in., so you have

$$(v_D)_0 = r\omega_0 \qquad 12 \text{ in./s} = (3 \text{ in.})\omega_0 \qquad \omega_0 = 4 \text{ rad/s } \circlearrowleft$$
$$(a_D)_t = r\alpha \qquad 9 \text{ in./s}^2 = (3 \text{ in.})\alpha \qquad \alpha = 3 \text{ rad/s}^2 \circlearrowleft$$

Using the equations of uniformly accelerated motion, for $t = 2$ s you obtain

$$\omega = \omega_0 + \alpha t = 4 \text{ rad/s} + (3 \text{ rad/s}^2)(2 \text{ s}) = 10 \text{ rad/s}$$
$$\boldsymbol{\omega} = 10 \text{ rad/s } \circlearrowleft$$
$$\theta = \omega_0 t + \tfrac{1}{2}\alpha t^2 = (4 \text{ rad/s})(2 \text{ s}) + \tfrac{1}{2}(3 \text{ rad/s}^2)(2 \text{ s})^2 = 14 \text{ rad}$$
$$\theta = 14 \text{ rad } \circlearrowleft$$

$$\text{Number of revolutions} = (14 \text{ rad})\left(\frac{1 \text{ rev}}{2\pi \text{ rad}}\right) = 2.23 \text{ rev} \quad \blacktriangleleft$$

**b. Motion of Load B.** The motion of load $B$ is the same as a point on the outer rim of the double pulley. Using $r = 5$ in., you have

$$v_B = r\omega = (5 \text{ in.})(10 \text{ rad/s}) = 50 \text{ in./s} \qquad \mathbf{v}_B = 50 \text{ in./s } \uparrow \quad \blacktriangleleft$$
$$\Delta y_B = r\theta = (5 \text{ in.})(14 \text{ rad}) = 70 \text{ in.} \qquad \Delta y_B = 70 \text{ in. upward} \quad \blacktriangleleft$$

**c. Acceleration of Point D at t = 0.** The acceleration of point $D$ has a tangential and a normal component (Fig. 2). The tangential component of the acceleration is

$$(\mathbf{a}_D)_t = \mathbf{a}_C = 9 \text{ in./s}^2 \rightarrow$$

Because, at $t = 0$, $\omega_0 = 4$ rad/s, the normal component of the acceleration is

$$(a_D)_n = r_D\omega_0^2 = (3 \text{ in.})(4 \text{ rad/s})^2 = 48 \text{ in./s}^2 \ (a_D)_n = 48 \text{ in./s}^2 \downarrow$$

*(continued)*

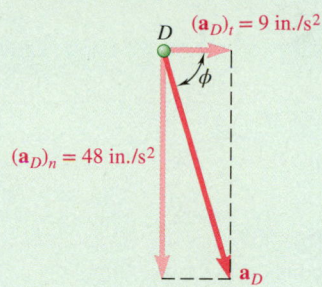

**Fig. 3** Vector triangle for resolving the acceleration vector into a magnitude and direction.

You can obtain the magnitude and direction of the total acceleration from Fig. 3.

$$\tan \phi = (48 \text{ in./s}^2)/(9 \text{ in./s}^2) \qquad \phi = 79.4°$$
$$a_D \sin 79.4° = 48 \text{ in./s}^2 \qquad a_D = 48.8 \text{ in./s}^2$$
$$\mathbf{a}_D = 48.8 \text{ in./s}^2 \; \text{\ensuremath{\angle}} \; 79.4° \quad \blacktriangleleft$$

**REFLECT and THINK:** A double pulley acts similarly to a system of gears; for every 3 inches that point $C$ moves to the right, point $B$ moves 5 inches upward. This is also similar to how your bicycle works; the size ratio of the front chainring to the rear sprocket controls the rotation of the rear tire.

## Sample Problem 15.4

Two friction wheels $A$ and $B$ are both rotating freely at 300 rpm clockwise when they are brought into contact. After 6 s of slippage, during which each wheel has a constant angular acceleration, wheel $A$ reaches a final angular velocity of 60 rpm clockwise. Determine the angular acceleration of each wheel during the period of slippage. ▶

**STRATEGY:** You are not given any masses or forces, so you can use kinematics to solve this problem.

**MODELING and ANALYSIS:** Model each wheel as a rigid body.

**Initial Data.** The initial angular velocities of the wheels are $(\omega_A)_0 = (\omega_B)_0 = 300$ rpm $= 31.42$ rad/s, both clockwise. After 6 s of slippage, the final angular velocity of $A$ is $\omega_A = 60$ rpm $= 6.28$ rad/s clockwise.

**Wheel A.** You are told the angular accelerations of the wheels are constant, so

$$\omega_A = (\omega_A)_0 + \alpha_A t: \quad 6.28 \text{ rad/s} = 31.42 \text{ rad/s} + \alpha_A(6 \text{ s})$$
$$\alpha_A = -4.19 \text{ rad/s}^2 \qquad \alpha_A = 4.19 \text{ rad/s}^2 \; \circlearrowright \quad \blacktriangleleft$$

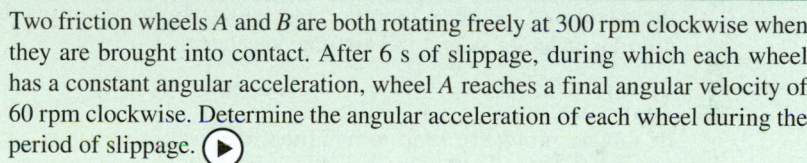

**Wheel B.** At $t = 6$ s, the wheels stop slipping and the two points in contact have the same velocity (Fig. 1). Thus,

$$r_A \omega_A = r_B \omega_B$$

so

$$\omega_B = \frac{r_A \omega_A}{r_B} = \frac{(125 \text{ mm})(6.28 \text{ rad/s})}{(75 \text{ mm})} = 10.47 \text{ rad/s} \; \circlearrowright$$

The angular acceleration of $B$ is constant, so

$$\omega_B = (\omega_B)_0 + \alpha_B t: \quad -10.47 \text{ rad/s} = 31.42 \text{ rad/s} + \alpha_B(6 \text{ s})$$
$$\alpha_B = -6.98 \text{ rad/s}^2$$

$$\alpha_B = 6.98 \text{ rad/s}^2 \; \circlearrowright \quad \blacktriangleleft$$

**REFLECT and THINK:** The initial angular velocity of $B$ is clockwise, and its final angular velocity is counterclockwise. There must be some time when this wheel has an angular velocity of zero and changes direction from rotating clockwise to rotating counterclockwise.

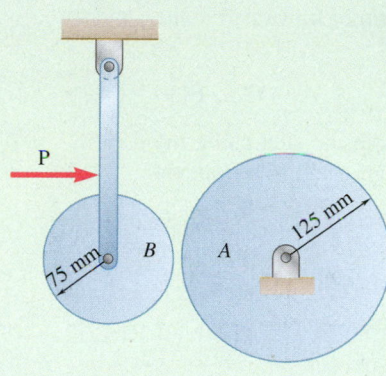

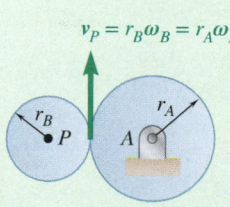

**Fig. 1** The wheels will stop slipping when the velocities of the points of contact are equal.

# SOLVING PROBLEMS
# ON YOUR OWN

In this section, we began the study of the motion of rigid bodies by considering two particular types of motion: **translation** and **rotation about a fixed axis**.

**1. Rigid body in translation.** At any given instant, all the points of a rigid body in translation have the *same velocity* and the *same acceleration* (Fig. 15.7).

**2. Rigid body rotating about a fixed axis.** The position of a rigid body rotating about a fixed axis is defined at any given instant by the **angular position** $\theta$, which is usually measured in radians. Selecting the unit vector $\mathbf{k}$ along the fixed axis in such a way that the rotation of the body appears counterclockwise as seen from the tip of $\mathbf{k}$, we define the **angular velocity** $\boldsymbol{\omega}$ and the **angular acceleration** $\boldsymbol{\alpha}$ of the body as

$$\boldsymbol{\omega} = \dot{\theta}\mathbf{k} \qquad \boldsymbol{\alpha} = \ddot{\theta}\mathbf{k} \qquad\qquad \textbf{(15.6, 15.9)}$$

In solving problems, keep in mind that the vectors $\boldsymbol{\omega}$ and $\boldsymbol{\alpha}$ are both directed along the fixed axis of rotation and that their sense can be obtained by the right-hand rule (Sample Prob. 15.2).

    **a. The velocity of a point $P$** of a body rotating about a fixed axis is

$$\mathbf{v} = \boldsymbol{\omega} \times \mathbf{r} \qquad\qquad \textbf{(15.5)}$$

where $\boldsymbol{\omega}$ is the angular velocity of the body and $\mathbf{r}$ is the position vector drawn from any point on the axis of rotation to point $P$ (Fig. 15.9).

    **b. The acceleration of point $P$** of a body rotating about a fixed axis is

$$\mathbf{a} = \boldsymbol{\alpha} \times \mathbf{r} + \boldsymbol{\omega} \times (\boldsymbol{\omega} \times \mathbf{r}) \qquad\qquad \textbf{(15.8)}$$

Because vector products are not commutative, *be sure to write the vectors in the order shown* when using either of the above two equations.

**3. Rotation of a representative slab.** In many problems, you will be able to reduce the analysis of the rotation of a three-dimensional body about a fixed axis to the case of the rotation of a representative slab in a plane perpendicular to the fixed axis. The $z$ axis should be directed along the axis of rotation and point out of the page. Thus, the representative slab rotates in the $xy$ plane about the origin $O$ of the coordinate system (Fig. 15.10).

To solve problems of this type, you should do the following steps.

    **a. Draw a diagram of the representative slab** showing its dimensions, its angular velocity and angular acceleration, and the vectors representing the velocities and accelerations of the points of the slab.

    **b. Relate the rotation of the slab and the motion of points of the slab** by writing

$$v = r\omega \qquad\qquad \textbf{(15.10}'\textbf{)}$$

$$a_t = r\alpha \qquad a_n = r\omega^2 \qquad\qquad \textbf{(15.11}'\textbf{)}$$

Remember that the velocity **v** and the component $\mathbf{a}_t$ of the acceleration of a point $P$ of the slab are tangent to the circular path described by $P$ (Sample Probs. 15.3 and 15.4). You can find the directions of **v** and $\mathbf{a}_t$ by rotating the position vector **r** through 90° in the sense indicated by **ω** and **α**, respectively. The normal component $\mathbf{a}_n$ of the acceleration of $P$ is always directed toward the axis of rotation.

**4. Equations defining the rotation of a rigid body.** Note the similarity between the equations defining the rotation of a rigid body about a fixed axis [Eqs. (15.12) through (15.16)] and those in Chap. 11 defining the rectilinear motion of a particle [Eqs. (11.1) through (11.8)]. All you have to do to obtain the new set of equations is to substitute $\theta$, $\omega$, and $\alpha$ for $x$, $v$, and $a$, respectively, in the equations of Chap. 11 (Sample Prob. 15.1).

# Problems

## CONCEPT QUESTIONS

**15.CQ1** A rectangular plate swings from arms of equal length as shown. What is the magnitude of the angular velocity of the plate?

**a.** 0 rad/s

**b.** 1 rad/s

**c.** 2 rad/s

**d.** 3 rad/s

**e.** Need to know the location of the center of gravity.

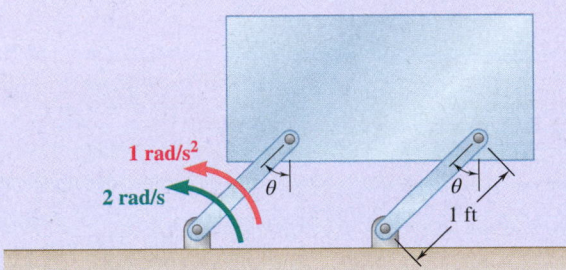

**Fig. P15.CQ1**

**15.CQ2** Knowing that wheel $A$ rotates with a constant angular velocity and that no slipping occurs between ring $C$ and wheel $A$ and wheel $B$, which of the following statements concerning the angular speeds of the three objects is true?

**a.** $\omega_a = \omega_b$

**b.** $\omega_a > \omega_b$

**c.** $\omega_a < \omega_b$

**d.** $\omega_a = \omega_c$

**e.** The contact points between $A$ and $C$ have the same acceleration.

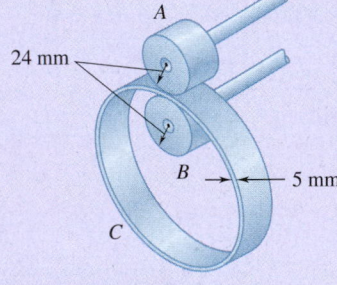

**Fig. P15.CQ2**

## END-OF-SECTION PROBLEMS

**15.1** The brake drum is attached to a larger flywheel that is not shown. The motion of the brake drum is defined by the relation $\theta = 100t - 5t^2$ where $\theta$ is expressed in radians and $t$ in seconds. Determine (*a*) the angular velocity at $t = 3$ s, (*b*) the number of revolutions executed by the brake drum before coming to rest.

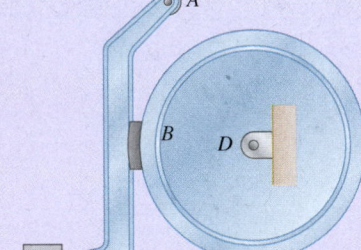

**Fig. P15.1**

**15.2** The motion of an oscillating flywheel is defined by the relation $\theta = \theta_0 e^{-3\pi t} \cos 4\pi t$, where $\theta$ is expressed in radians and $t$ in seconds. Knowing that $\theta_0 = 0.5$ rad, determine the angular coordinate, the angular velocity, and the angular acceleration of the flywheel when (*a*) $t = 0$, (*b*) $t = 0.125$ s.

**15.3** The motion of an oscillating flywheel is defined by the relation $\theta = \theta_0 e^{-7\pi t/6} \sin 4\pi t$, where $\theta$ is expressed in radians and $t$ in seconds. Knowing that $\theta_0 = 0.4$ rad, determine the angular coordinate, the angular velocity, and the angular acceleration of the flywheel when (*a*) $t = 0.125$ s, (*b*) $t = \infty$.

**Fig. P15.2 and P15.3**

**15.4** As steam is slowly injected into a turbine, the angular acceleration of the rotor is observed to increase linearly with the time $t$. Knowing that the rotor starts from rest at $t = 0$ and that after 10 s the rotor has completed 20 revolutions, write the equations of motion for the rotor and determine (a) the angular velocity at $t = 20$ s, (b) the time required for the rotor to complete its first 40 revolutions.

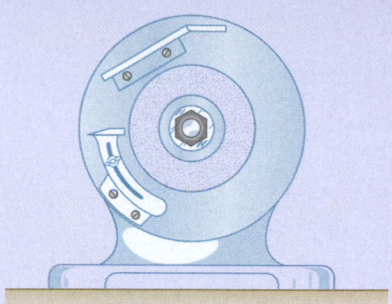

**Fig. P15.5**

**15.5** A small grinding wheel is attached to the shaft of an electric motor which has a rated speed of 3600 rpm. When the power is turned on, the unit reaches its rated speed in 5 s, and when the power is turned off, the unit coasts to rest in 70 s. Assuming uniformly accelerated motion, determine the number of revolutions that the motor executes (a) in reaching its rated speed, (b) in coasting to rest.

**15.6** A connecting rod is supported by a knife-edge at point $A$. For small oscillations the angular acceleration of the connecting rod is governed by the relation $\alpha = -6\theta$, where $\alpha$ is expressed in rad/s$^2$ and $\theta$ in radians. Knowing that the connecting rod is released from rest when $\theta = 20°$, determine (a) the maximum angular velocity, (b) the angular position when $t = 2$ s.

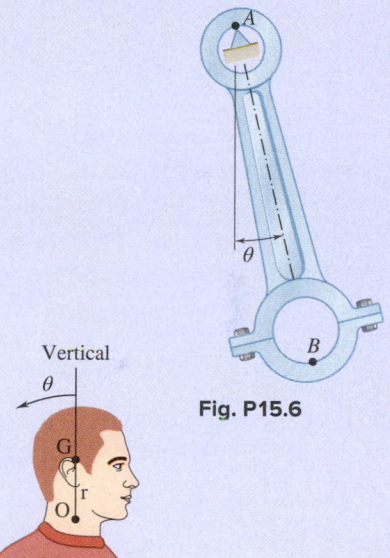

**Fig. P15.6**

**15.7** When studying whiplash resulting from rear-end collisions, the rotation of the head is of primary interest. An impact test was performed, and it was found that the angular acceleration of the head is defined by the relation $\alpha = 700 \cos \theta + 70 \sin \theta$, where $\alpha$ is expressed in rad/s$^2$ and $\theta$ in radians. Knowing that the head is initially at rest, determine the angular velocity of the head when $\theta = 30°$.

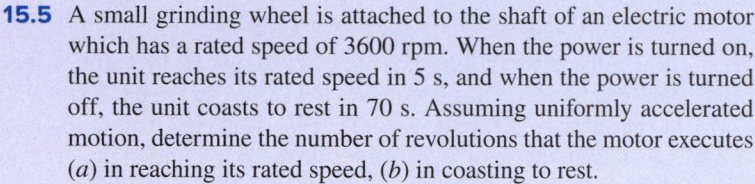
Vertical

**Fig. *P15.7***

**15.8** The angular acceleration of an oscillating disk is defined by the relation $\alpha = -k\theta$, where alpha is expressed in rad/s$^2$ and $\theta$ is expressed in radians. Determine (a) the value of $k$ for which $\omega = 12$ rad/s when $\theta = 0$ and $\theta = 6$ rad when $\omega = 0$, (b) the angular velocity of the disk when $\theta = 3$ rad.

**15.9** The angular acceleration of a shaft is defined by the relation $\alpha = -0.5\omega$, where $\alpha$ is expressed in rad/s$^2$ and $\omega$ in rad/s. Knowing that at $t = 0$ the angular velocity of the shaft is 30 rad/s, determine (a) the number of revolutions the shaft will execute before coming to rest, (b) the time required for the shaft to come to rest, (c) the time required for the angular velocity of the shaft to reduce to 2 percent of its initial value.

**15.10** The assembly shown consists of two rods and a rectangular plate $BCDE$ that are welded together. The assembly rotates about the axis $AB$ with a constant angular velocity of 10 rad/s. Knowing that the rotation is counterclockwise as viewed from $B$, determine the velocity and acceleration of corner $E$.

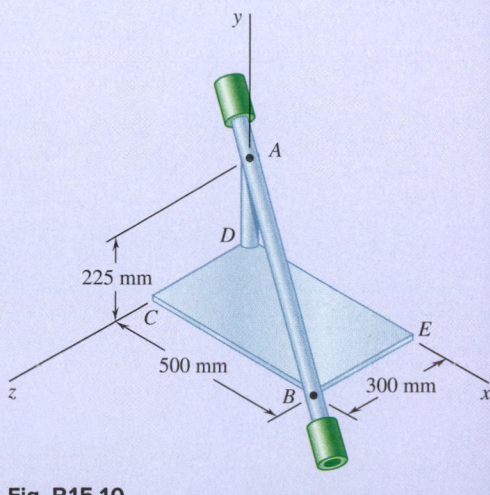

**15.11** In Prob. 15.10, determine the velocity and acceleration of corner $C$, assuming that the angular velocity is 10 rad/s and decreases at the rate of 20 rad/s$^2$.

**Fig. P15.10**

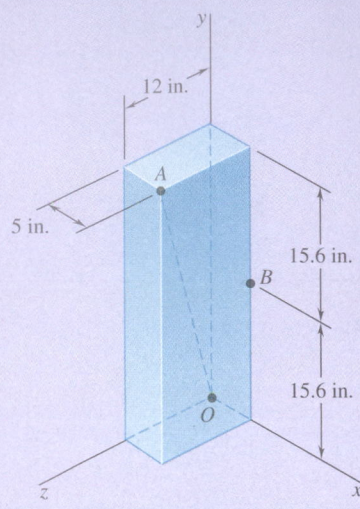

Fig. P15.12 and P15.13

**15.12** The rectangular block shown rotates about the diagonal *OA* with a constant angular velocity of 6.76 rad/s. Knowing that the rotation is counterclockwise as viewed from *A*, determine the velocity and acceleration of point *B* at the instant shown.

**15.13** The rectangular block shown rotates about the diagonal *OA* with an angular velocity of 3.38 rad/s that is decreasing at the rate of 5.07 rad/s². Knowing that the rotation is counterclockwise as viewed from *A*, determine the velocity and acceleration of point *B* at the instant shown.

**15.14** A circular plate of 120-mm radius is supported by two bearings *A* and *B* as shown. The plate rotates about the rod joining *A* and *B* with a constant angular velocity of 26 rad/s. Knowing that, at the instant considered, the velocity of point *C* is directed to the right, determine the velocity and acceleration of point *E*.

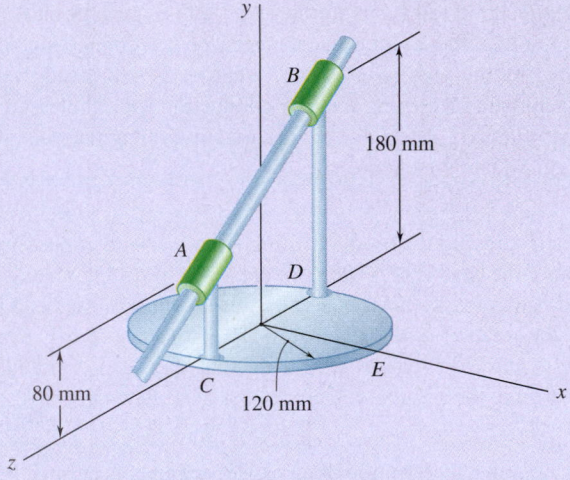

Fig. *P15.14*

**15.15** In Prob. 15.14, determine the velocity and acceleration of point *E*, assuming that the angular velocity is 26 rad/s and increases at the rate of 65 rad/s².

**15.16** The earth makes one complete revolution around the sun in 365.24 days. Assuming that the orbit of the earth is circular and has a radius of 93,000,000 mi, determine the velocity and acceleration of the earth.

**15.17** The earth makes one complete revolution on its axis in 23 h 56 min. Knowing that the mean radius of the earth is 3960 mi, determine the linear velocity and acceleration of a point on the surface of the earth (*a*) at the equator, (*b*) at Philadelphia, latitude 40° north, (*c*) at the North Pole.

**15.18** The sprocket wheel and chain shown are initially at rest. If the wheel has a uniform angular acceleration of 90 rad/s² counterclockwise, determine (*a*) the acceleration of point *A* of the chain, (*b*) the magnitude of the acceleration of point *B* of the wheel after 3 s.

**15.19** The sprocket wheel and chain shown are being operated at a speed of 600 rpm counterclockwise. When the power is turned off, it is observed that the wheel and chain coast to rest in 4 s. Assuming uniformly decelerated motion, determine the magnitude of the velocity and acceleration of point *B* of the wheel (*a*) immediately before the power is turned off, (*b*) 2.5 s later.

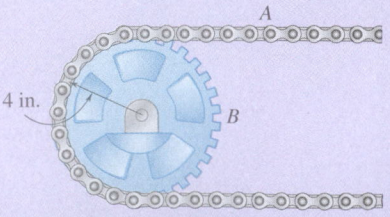

Fig. P15.18 and P15.19

**15.20** The belt sander shown is initially at rest. If the driving drum $B$ has a constant angular acceleration of 120 rad/s² counterclockwise, determine the magnitude of the acceleration of the belt at point $C$ when (a) $t = 0.5$ s, (b) $t = 2$ s.

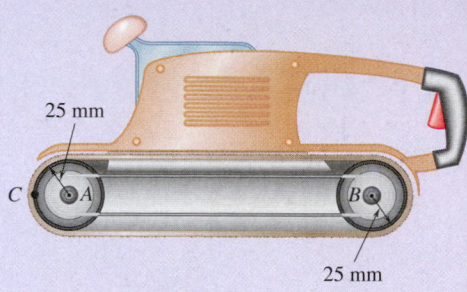

**15.21** The rated speed of drum $B$ of the belt sander shown is 2400 rpm. When the power is turned off, it is observed that the sander coasts from its rated speed to rest in 10 s. Assuming uniformly decelerated motion, determine the velocity and acceleration of point $C$ of the belt, (a) immediately before the power is turned off, (b) 9 s later.

**Fig. P15.20 and P15.21**

**15.22** The two pulleys shown may be operated with the V belt in any of three positions. If the angular acceleration of shaft $A$ is 6 rad/s² and if the system is initially at rest, determine the time required for shaft $B$ to reach a speed of 400 rpm with the belt in each of the three positions.

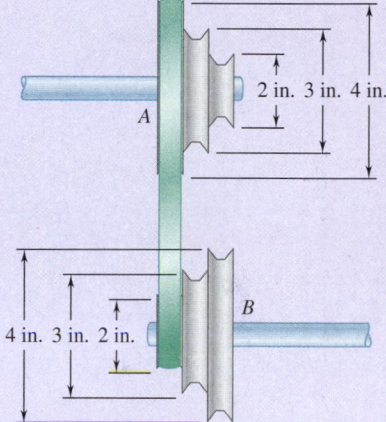

**15.23** A cyclist uses a stationary trainer during the winter to keep in shape. Knowing that she pushes down on her pedal with a velocity of 26 in./s that increases at a rate of 2 in./s², determine the velocity and acceleration of point $D$ on the bottom of the rear wheel.

**Fig. P15.22**

**Fig. P15.23**

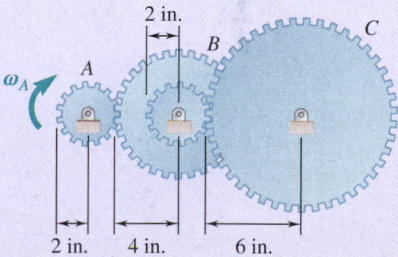

**Fig. P15.24**

**15.24** A gear reduction system consists of three gears $A$, $B$, and $C$. Knowing that gear $A$ rotates clockwise with a constant angular velocity $\omega_A = 600$ rpm, determine (a) the angular velocities of gears $B$ and $C$, (b) the accelerations of the points on gears $B$ and $C$ which are in contact.

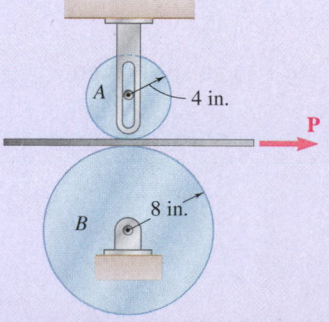

**15.25** A belt is pulled to the right between cylinders $A$ and $B$. Knowing that the speed of the belt is a constant 5 ft/s and no slippage occurs, determine (a) the angular velocities of $A$ and $B$, (b) the accelerations of the points that are in contact with the belt.

**Fig. P15.25**

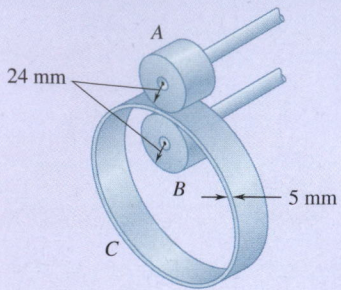

**Fig. P15.26**

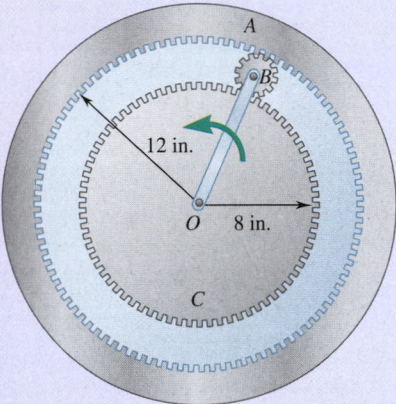

**Fig. P15.27**
Source: Boresi, P., and R. Schmidt,
*Advanced Mechanics of Materials,* 6e,
New York, Wiley: 2003.

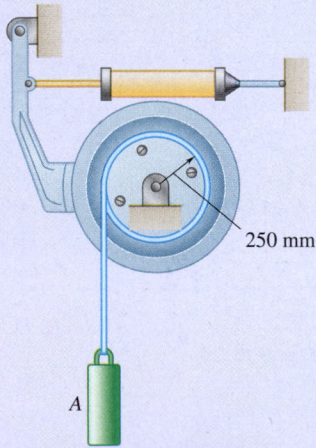

**Fig. P15.29 and P15.30**

**15.26** Ring $C$ has an inside radius of 55 mm and an outside radius of 60 mm and is positioned between two wheels $A$ and $B$, each of 24-mm outside radius. Knowing that wheel $A$ rotates with a constant angular velocity of 300 rpm and that no slipping occurs, determine (*a*) the angular velocity of ring $C$ and of wheel $B$, (*b*) the acceleration of the points on $A$ and $B$ that are in contact with $C$.

**15.27** At the instant shown, the angular velocity of crank $OB$ is 100 rpm in the counterclockwise direction. Knowing that gear $C$ is fixed, determine (*a*) the angular velocity of the planetary gear $B$, (*b*) the angular velocity of the ring gear $A$.

**15.28** A plastic film moves over two drums. During a 4-s interval, the speed of the tape is increased uniformly from $v_0 = 2$ ft/s to $v_1 = 4$ ft/s. Knowing that the tape does not slip on the drums, determine (*a*) the angular acceleration of drum $B$, (*b*) the number of revolutions executed by drum $B$ during the 4-s interval.

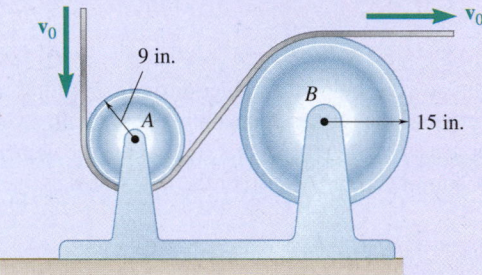

**Fig. P15.28**

**15.29** Cylinder $A$ is moving downward with a velocity of 3 m/s when the brake is suddenly applied to the drum. Knowing that the cylinder moves 6 m downward before coming to rest and assuming uniformly accelerated motion, determine (*a*) the angular acceleration of the drum, (*b*) the time required for the cylinder to come to rest.

**15.30** The system shown is held at rest by the brake-and-drum system shown. After the brake is partially released at $t = 0$, it is observed that the cylinder moves 5 m in 4.5 s. Assuming uniformly accelerated motion, determine (*a*) the angular acceleration of the drum, (*b*) the angular velocity of the drum at $t = 3.5$ s.

**15.31** A load is to be raised 20 ft by the hoisting system shown. Assuming gear $A$ is initially at rest, accelerates uniformly to a speed of 120 rpm in 5 s, and then maintains a constant speed of 120 rpm, determine (*a*) the number of revolutions executed by gear $A$ in raising the load, (*b*) the time required to raise the load.

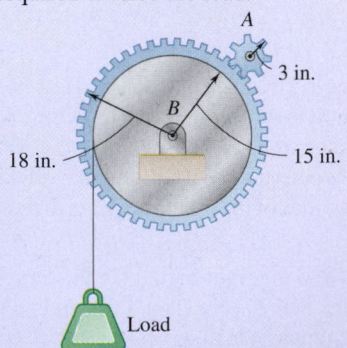

**Fig. P15.31**

**15.32** A simple friction drive consists of two disks $A$ and $B$. Initially, disk $B$ has a clockwise angular velocity of 500 rpm, and disk $A$ is at rest. It is known that disk $B$ will coast to rest in 60 s. However, rather than waiting until both disks are at rest to bring them together, disk $A$ is given a constant angular acceleration of 3 rad/s$^2$ counterclockwise. Determine (a) at what time the disks can be brought together if they are not to slip, (b) the angular velocity of each disk as contact is made.

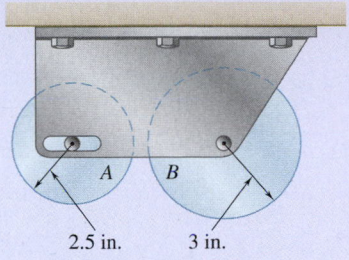

2.5 in.  3 in.

**Fig. P15.32 and P15.33**

**15.33** Two friction wheels $A$ and $B$ are both rotating freely at 300 rpm counterclockwise when they are brought into contact. After 12 s of slippage, during which time each wheel has a constant angular acceleration, wheel $B$ reaches a final angular velocity of 75 rpm counterclockwise. Determine (a) the angular acceleration of each wheel during the period of slippage, (b) the time at which the angular velocity of wheel $A$ is equal to zero.

**15.34** Two friction disks $A$ and $B$ are to be brought into contact without slipping when the angular velocity of disk $A$ is 240 rpm counterclockwise. Disk $A$ starts from rest at time $t = 0$ and is given a constant angular acceleration with a magnitude $\alpha$. Disk $B$ starts from rest at time $t = 2$ s and is given a constant clockwise angular acceleration, also with a magnitude $\alpha$. Determine (a) the required angular acceleration magnitude $\alpha$, (b) the time at which the contact occurs.

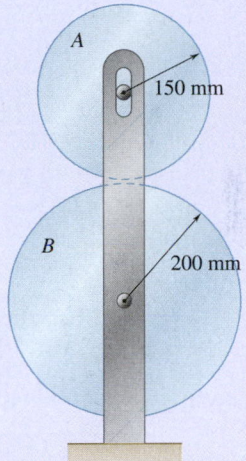

150 mm

200 mm

**15.35** Two friction disks $A$ and $B$ are brought into contact when the angular velocity of disk $A$ is 240 rpm counterclockwise and disk $B$ is at rest. A period of slipping follows and disk $B$ makes two revolutions before reaching its final angular velocity. Assuming that the angular acceleration of each disk is constant and inversely proportional to the cube of its radius, determine (a) the angular acceleration of each disk, (b) the time during which the disks slip.

**Fig. P15.34 and P15.35**

***15.36** Steel tape is being wound onto a spool that rotates with a constant angular velocity $\omega_0$. Denoting by $r$ the radius of the spool and tape at any given time and by $b$ the thickness of the tape, derive an expression for the acceleration of the tape as it approaches the spool.

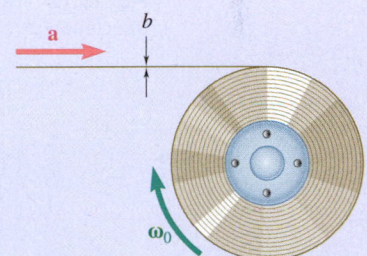

$a$   $b$

***15.37** In a continuous printing process, paper is drawn into the presses at a constant speed $v$. Denoting by $r$ the radius of the paper roll at any given time and by $b$ the thickness of the paper, derive an expression for the angular acceleration of the paper roll.

**Fig. P15.36**

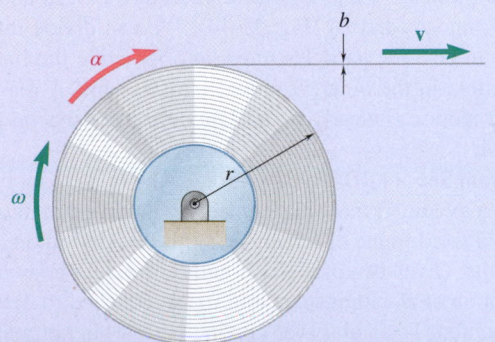

$b$   $v$

$\alpha$

$r$

$\omega$

**Fig. P15.37**

# 15.2 GENERAL PLANE MOTION: VELOCITY

As indicated in the chapter introduction, general plane motion describes a plane motion that is neither a pure translation nor a pure rotation. As you will presently see, however, **a general plane motion can always be considered as the sum of a translation and a rotation.**

## 15.2A Analyzing General Plane Motion

As an example of general plane motion, consider a wheel rolling on a straight track (Fig. 15.12). Over some interval of time, two given points $A$ and $B$ will have moved, respectively, from $A_1$ to $A_2$ and from $B_1$ to $B_2$. However, we could obtain the same result through a translation that would bring $A_1$ and $B_1$ into $A_2$ and $B_1'$ (the line $AB$ remaining vertical), followed by a rotation about $A$, bringing $B$ into $B_2$. The original rolling motion differs from the combination of translation and rotation when these motions are taken in succession, but we can duplicate the original motion exactly using a combination of simultaneous translation and rotation.

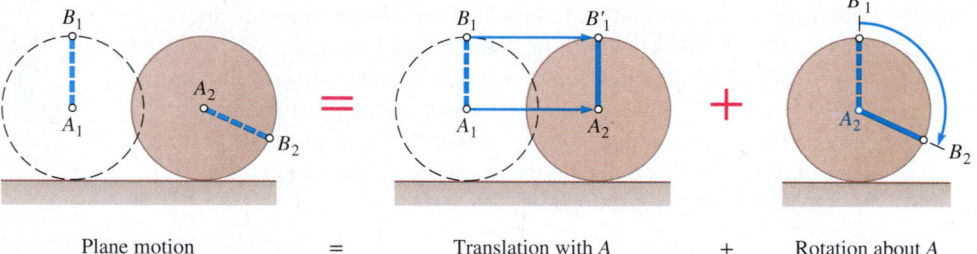

|  |  |  |  |  |
|---|---|---|---|---|
| Plane motion | = | Translation with $A$ | + | Rotation about $A$ |

**Fig. 15.12** The general plane motion of a rolling wheel can be analyzed as a combination of translation plus a fixed-axis rotation.

Another example of plane motion is shown in Fig. 15.13, which represents a rod whose ends slide along a horizontal and a vertical track. We can replace this motion using a horizontal translation and a rotation about $A$ (Fig. 15.13$a$) or using a vertical translation and a rotation about $B$ (Fig. 15.13$b$).

In the general case of plane motion, we consider a small displacement that brings two particles $A$ and $B$ of a representative rigid body, respectively, from $A_1$ and $B_1$ into $A_2$ and $B_2$ (Fig. 15.14). We can divide this displacement into two parts: in one, the particles move into $A_2$ while the line $AB$ maintains the same direction; in the other, $B$ moves into $B_2$ while $A$ remains fixed. The first part of the motion is clearly a translation, and the second part is clearly a rotation about $A$.

Recall from Sec. 11.4D the definition of the relative motion of a particle with respect to a moving frame of reference—as opposed to its absolute motion with respect to a fixed frame of reference. With that definition in mind, we can restate our results: Given two particles $A$ and $B$ of a rigid body in plane motion, the relative motion of $B$ with respect to a frame attached to $A$ and of fixed orientation is a rotation. To an observer moving with $A$ but not rotating, particle $B$ appears to describe an arc of a circle centered at $A$.

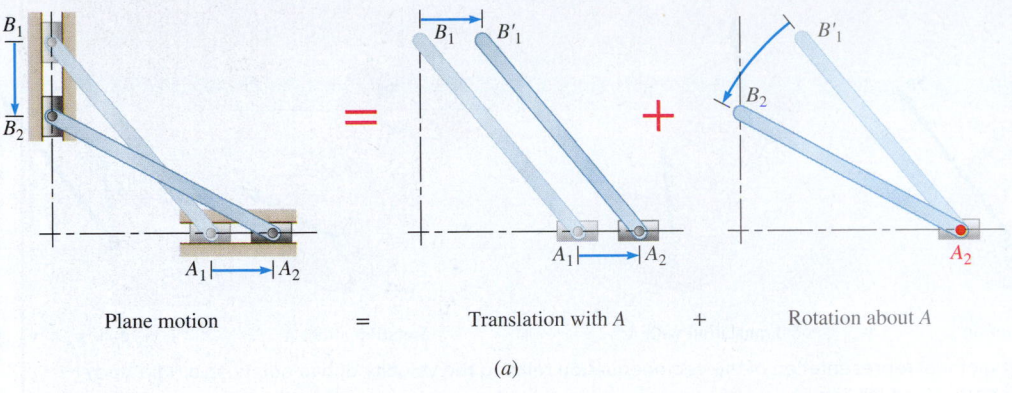

Plane motion   =   Translation with *A*   +   Rotation about *A*

(*a*)

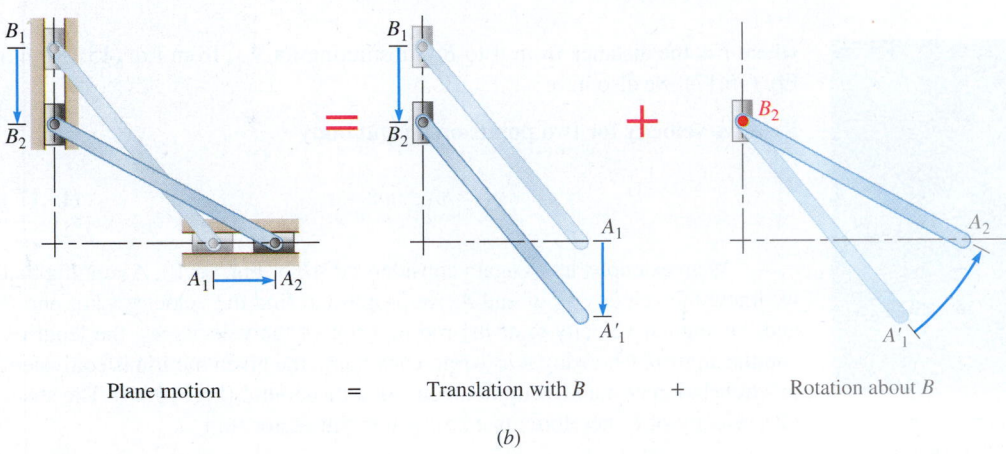

Plane motion   =   Translation with *B*   +   Rotation about *B*

(*b*)

**Fig. 15.13**   The general plane motion of this sliding rod can be analyzed as (*a*) a horizontal translation plus a fixed-axis rotation about *A* or (*b*) a vertical translation and a fixed-axis rotation about *B*. The results are the same either way. ▶

## 15.2B   Absolute and Relative Velocity in Plane Motion

We have just seen that any plane motion of a rigid body can be replaced by a translation of an arbitrary reference point *A* and a simultaneous rotation about *A*. We can obtain the absolute velocity $\mathbf{v}_B$ of a particle *B* of the rigid body from the relative velocity formula derived in Sec. 11.4D, as

$$\mathbf{v}_B = \mathbf{v}_A + \mathbf{v}_{B/A} \qquad (15.17)$$

where the right-hand side represents a vector sum. The velocity $\mathbf{v}_A$ corresponds to the translation of the rigid body with *A*, whereas the relative velocity $\mathbf{v}_{B/A}$ is associated with the rotation of the rigid body about *A* and is measured with respect to axes centered at *A* and of fixed orientation (Fig. 15.15). Denoting the position vector of *B* relative to *A* by $\mathbf{r}_{B/A}$ (which points from *A* to *B*) and the angular velocity of the rigid body with respect to axes of fixed orientation by $\omega\mathbf{k}$, we have from Eqs. (15.10) and (15.10′)

$$\mathbf{v}_{B/A} = \omega\mathbf{k} \times \mathbf{r}_{B/A} \qquad v_{B/A} = r\omega \qquad (15.18)$$

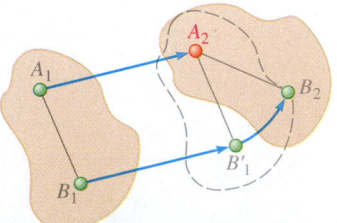

**Fig. 15.14**   General plane motion is a combination of a translation plus a fixed-axis rotation. To an observer moving with *A* but not rotating, particle *B* appears to travel in a circle centered at *A*.

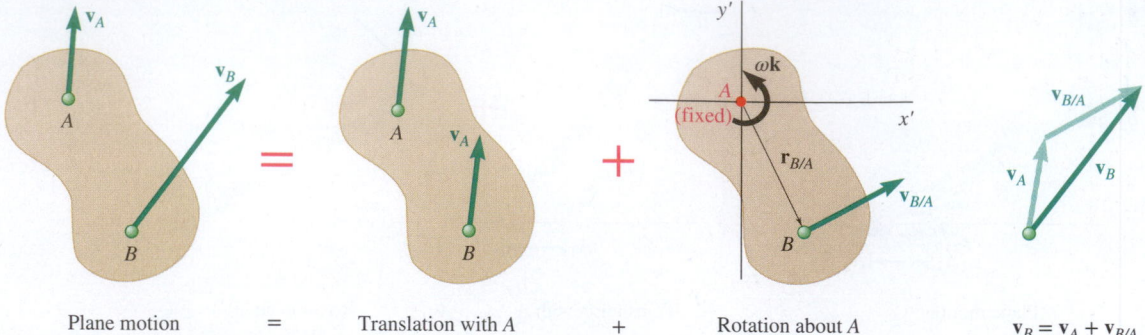

Plane motion   =   Translation with $A$   +   Rotation about $A$   $\mathbf{v}_B = \mathbf{v}_A + \mathbf{v}_{B/A}$

**Fig. 15.15**   A pictorial representation of the vector equation relating the velocity of two points on a rigid body undergoing general plane motion.

**Photo 15.4**   Planetary gear systems are used in applications requiring a large reduction ratio and a high torque-to-weight ratio. The small gears undergo general plane motion. ©Plus Pix/agefotostock

where $r$ is the distance from $A$ to $B$. Substituting for $\mathbf{v}_{B/A}$ from Eq. (15.18) into Eq. (15.17), we also have

**Relative velocity for two points on a rigid body**

$$\mathbf{v}_B = \mathbf{v}_A + \omega\mathbf{k} \times \mathbf{r}_{B/A} \qquad (15.17')$$

As an example, let us again consider rod $AB$ of Fig. 15.13. Assuming that we know the velocity $\mathbf{v}_A$ of end $A$, we propose to find the velocity $\mathbf{v}_B$ of end $B$ and the angular velocity $\boldsymbol{\omega}$ of the rod in terms of the velocity $\mathbf{v}_A$, the length $l$, and the angle $\theta$. Choosing $A$ as a reference point, the given motion is equivalent to a translation of $A$ and a simultaneous rotation about $A$ (Fig. 15.16). The absolute velocity of $B$ therefore must be equal to the vector sum

$$\mathbf{v}_B = \mathbf{v}_A + \mathbf{v}_{B/A} \qquad (15.17)$$

Note that although we know the direction of $\mathbf{v}_{B/A}$, its magnitude $l\omega$ is unknown. However, this is compensated for by the fact that the direction of $\mathbf{v}_B$ is known. We can therefore complete the vector diagram of Fig. 15.16. Solving for the magnitudes $v_B$ and $\omega$, we obtain

$$v_B = v_A \tan \theta \qquad \omega = \frac{v_{B/A}}{l} = \frac{v_A}{l \cos \theta} \qquad (15.19)$$

Alternatively, we can also solve this problem by using the vector relationship in Eq. (15.17'). Recognizing that point $A$ is constrained to move only in

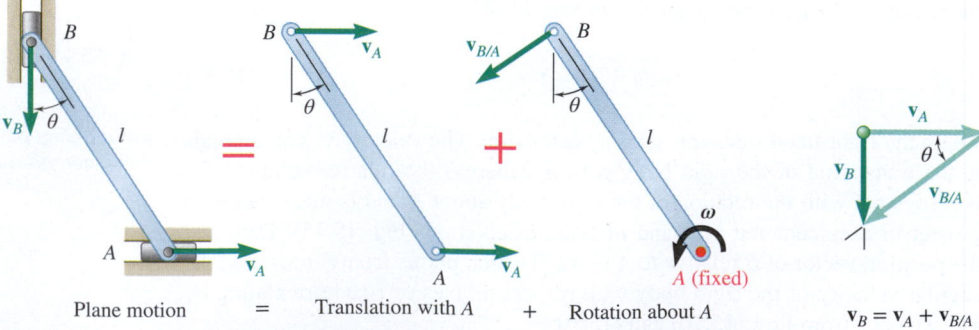

Plane motion   =   Translation with $A$   +   Rotation about $A$   $\mathbf{v}_B = \mathbf{v}_A + \mathbf{v}_{B/A}$

**Fig. 15.16**   Pictorial representation of Eq. (15.17) for a sliding rod. The relative velocity $\mathbf{v}_{B/A}$ is perpendicular to the line connecting $A$ and $B$.

the $x$ direction and $B$ moves only in the $y$ direction (assume it moves down), we can write

$$-v_B\mathbf{j} = v_A\mathbf{i} + \omega\mathbf{k} \times (-l \sin \theta\mathbf{i} + l \cos \theta\mathbf{j}) = (v_A - \omega l \cos \theta)\mathbf{i} - \omega l \sin \theta\mathbf{j}$$

Equating components in the $x$ direction, we obtain

$$v_A - \omega l \cos \theta = 0 \qquad \omega = \frac{v_A}{l \cos \theta}$$

Equating components in the $y$ direction, we obtain

$$v_B = \omega l \sin \theta = \left(\frac{v_A}{l \cos \theta}\right) l \sin \theta = v_A \tan \theta$$

These are the same results as we obtained in Eq. 15.19. We obtain the same result by using $B$ as a point of reference. Resolving the given motion into a translation of $B$ and a simultaneous rotation about $B$ (Fig. 15.17), we have the equation

$$\mathbf{v}_A = \mathbf{v}_B + \mathbf{v}_{A/B} = \mathbf{v}_B + \omega\mathbf{k} \times \mathbf{r}_{A/B} \qquad \textbf{(15.20)}$$

which is represented graphically in Fig. 15.17. Note that $\mathbf{v}_{A/B}$ and $\mathbf{v}_{B/A}$ have the same magnitude $l\omega$ but opposite sense. The sense of the relative velocity depends, therefore, upon the point of reference that we have selected and should be carefully ascertained from the appropriate diagram (Fig. 15.16 or 15.17).

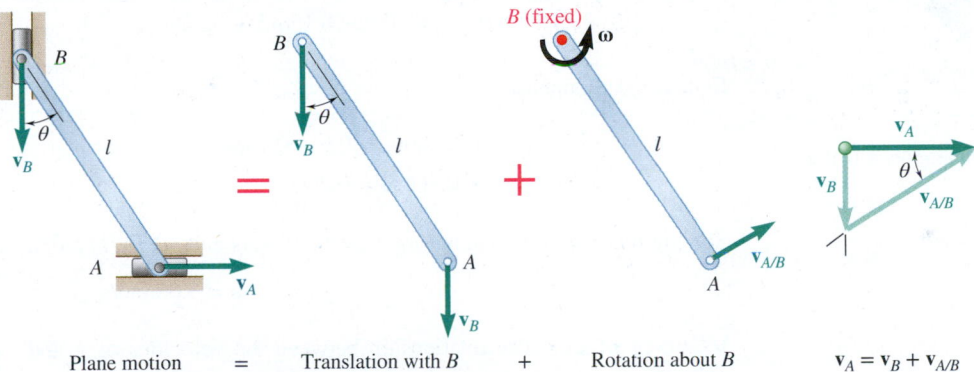

Plane motion   =   Translation with $B$   +   Rotation about $B$   $\mathbf{v}_A = \mathbf{v}_B + \mathbf{v}_{A/B}$

**Fig. 15.17**  Pictorial representation of Eq. (15.20) for a sliding rod. The relative velocity $\mathbf{v}_{A/B}$ is perpendicular to the line connecting $A$ and $B$.

Finally, observe that the angular velocity $\boldsymbol{\omega}$ of the rod in its rotation about $B$ is the same as in its rotation about $A$. It is measured in both cases by the rate of change of the angle $\theta$. This result is quite general; you should therefore bear in mind that

**The angular velocity $\boldsymbol{\omega}$ of a rigid body in plane motion is independent of the reference point.**

Most mechanisms consist not of one but of *several* moving parts. When the various parts of a mechanism are connected by pins, we can analyze the mechanism by considering each part as a rigid body, keeping in mind that the points where two parts are connected must have the same absolute velocity (see Sample Probs. 15.7 and 15.8). We can use a similar analysis when gears are involved, because the teeth in contact also must have the same absolute velocity. However, when a mechanism contains parts that slide on each other, the relative velocity of the parts in contact must be taken into account (see Sec. 15.5).

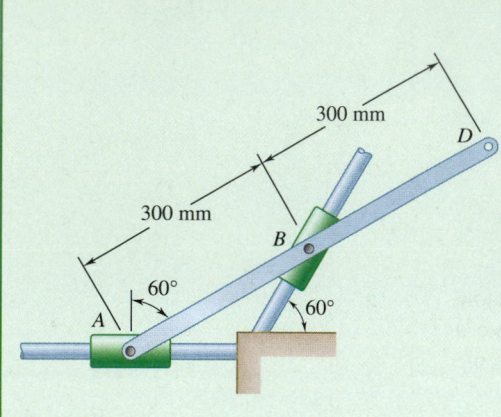

## Sample Problem 15.5

Collars *A* and *B* are pin-connected to bar *ABD* and can slide along fixed rods. Knowing that at the instant shown the velocity of *A* is 0.9 m/s to the right, determine (*a*) the angular velocity of *ABD*, (*b*) the velocity of point *D*.

**STRATEGY:** Use the kinematic equation that relates the velocity of two points on the same rigid body. Because you know the directions of the velocities of points *A* and *B*, choose these two points to relate.

**MODELING and ANALYSIS:** Model bar *ABD* as a rigid body. From kinematics you know

$$\mathbf{v}_B = \mathbf{v}_A + \mathbf{v}_{B/A} = \mathbf{v}_A + \boldsymbol{\omega} \times \mathbf{r}_{B/A}$$

Substituting in known values (Fig. 1) and assuming $\boldsymbol{\omega} = \omega\mathbf{k}$ gives you

$$v_B \cos 60°\mathbf{i} + v_B \sin 60°\mathbf{j} = (0.9)\mathbf{i} +$$
$$\omega\mathbf{k} \times [(0.3 \cos 30°)\mathbf{i} + (0.3 \sin 30°)\mathbf{j}]$$
$$0.500v_B\mathbf{i} + 0.866v_B\mathbf{j} = (0.9 - 0.15\omega)\mathbf{i} + 0.260\omega\mathbf{j}$$

Equating components,

$$\mathbf{i}: 0.500v_B = 0.9 - 0.15\omega$$
$$\mathbf{j}: 0.866v_B = 0.260\omega$$

Solving these equations gives you $v_B = 0.900$ m/s and $\omega = 3.00$ rad/s.

$$\boldsymbol{\omega} = 3.00 \text{ rad/s } \circlearrowleft \quad \blacktriangleleft$$

**Velocity of D.** The relationship between the velocities of *A* and *D* is

$$\mathbf{v}_D = \mathbf{v}_A + \mathbf{v}_{D/A} = \mathbf{v}_D + \boldsymbol{\omega} \times \mathbf{r}_{D/A}$$

Substituting in values from above gives

$$\mathbf{v}_D = 0.9\mathbf{i} + 3.00\mathbf{k} \times [(0.6 \cos 30°)\mathbf{i} + (0.6 \sin 30°)\mathbf{j}]$$
$$\mathbf{v}_D = (0.9 - 0.9)\mathbf{i} + 1.559\mathbf{j}$$

$$\mathbf{v}_D = 1.559 \text{ m/s } \uparrow \quad \blacktriangleleft$$

**REFLECT and THINK:** The velocity of point *D* is straight up at this instant in time, but as the bar continues to rotate counterclockwise, the direction of the velocity of *D* will continuously change.

**Fig. 1** Position vector and directions of the velocities of *A* and *B*.

## Sample Problem 15.6

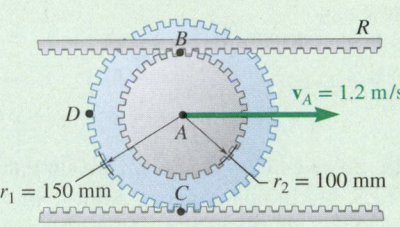

The double gear shown rolls on the stationary lower rack; the velocity of its center $A$ is 1.2 m/s directed to the right. Determine (a) the angular velocity of the gear, (b) the velocities of the upper rack $R$ and of point $D$ of the gear.

**STRATEGY:** The double gear is undergoing general motion, so use rigid body kinematics. Resolve the rolling motion into two component motions: a translation of point $A$ and a rotation about the center $A$ (Fig. 1). In the translation, all points of the gear move with the same velocity $\mathbf{v}_A$. In the rotation, each point $P$ of the gear moves about $A$ with a relative velocity $\mathbf{v}_{P/A} = \omega\mathbf{k} \times \mathbf{r}_{P/A}$, where $\mathbf{r}_{P/A}$ is the position vector of $P$ relative to $A$.

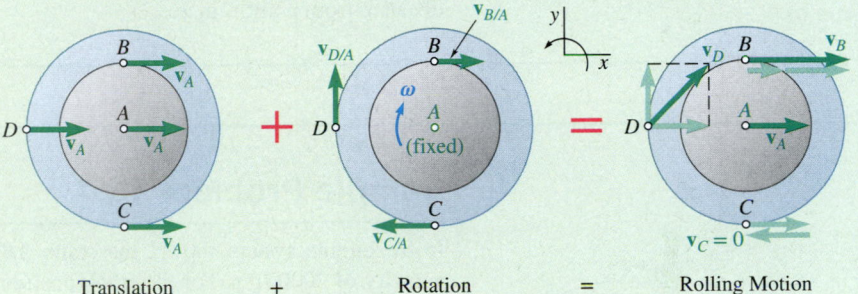

| Translation | + | Rotation | = | Rolling Motion |

**Fig. 1**   The gear motion can be modeled as a translation plus a rotation.

### MODELING and ANALYSIS:

**a. Angular Velocity of the Gear.** Because the gear rolls on the lower rack, its center $A$ moves through a distance equal to the outer circumference $2\pi r_1$ for each full revolution of the gear. Noting that 1 rev = $2\pi$ rad, and that when $A$ moves to the right ($x_A > 0$), the gear rotates clockwise ($\theta < 0$), you have

$$\frac{x_A}{2\pi r_1} = -\frac{\theta}{2\pi} \qquad x_A = -r_1\theta$$

Differentiating with respect to the time $t$ and substituting the known values $v_A = 1.2$ m/s and $r_1 = 150$ mm = 0.150 m, you obtain

$$v_A = -r_1\omega \qquad 1.2 \text{ m/s} = -(0.150 \text{ m})\omega \qquad \omega = -8 \text{ rad/s}$$

$$\boxed{\boldsymbol{\omega} = \omega\mathbf{k} = -(8 \text{ rad/s})\mathbf{k}} \qquad \blacktriangleleft$$

where $\mathbf{k}$ is a unit vector pointing out of the page.

**b. Velocity of Upper Rack.** The velocity of the upper rack is equal to the velocity of point $B$; you have

$$\begin{aligned}
\mathbf{v}_R = \mathbf{v}_B &= \mathbf{v}_A + \mathbf{v}_{B/A} = \mathbf{v}_A + \omega\mathbf{k} \times \mathbf{r}_{B/A} \\
&= (1.2 \text{ m/s})\mathbf{i} - (8 \text{ rad/s})\mathbf{k} \times (0.100 \text{ m})\mathbf{j} \\
&= (1.2 \text{ m/s})\mathbf{i} + (0.8 \text{ m/s})\mathbf{i} = (2 \text{ m/s})\mathbf{i}
\end{aligned}$$

$$\boxed{\mathbf{v}_R = 2 \text{ m/s} \rightarrow} \qquad \blacktriangleleft$$

*(continued)*

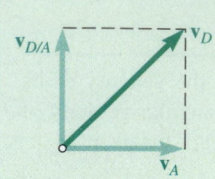

**Fig. 2** The two components of the velocity of D.

**Velocity of Point D.**  The velocity of point $D$ has two components (Fig. 2):

$$\mathbf{v}_D = \mathbf{v}_A + \mathbf{v}_{D/A} = \mathbf{v}_A + \omega\mathbf{k} \times \mathbf{r}_{D/A}$$
$$= (1.2 \text{ m/s})\mathbf{i} - (8 \text{ rad/s})\mathbf{k} \times (-0.150 \text{ m})\mathbf{i}$$
$$= (1.2 \text{ m/s})\mathbf{i} + (1.2 \text{ m/s})\mathbf{j}$$

$$\mathbf{v}_D = 1.697\text{m/s} \angle 45° \blacktriangleleft$$

**REFLECT and THINK:**  The principles involved in this problem are similar to those that you used in Sample Prob. 15.3, but in this problem, point $A$ was free to translate. Point $C$, because it is in contact with the fixed lower rack, has a velocity of zero. Every point along diameter $CAB$ has a velocity vector directed to the right (Fig. 1) and the magnitude of the velocity increases linearly as the distance from point $C$ increases.

## Sample Problem 15.7

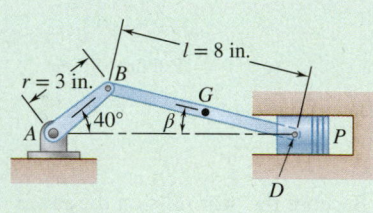

**Fig. 1**  Crank AB is undergoing fixed-axis rotation.

In the engine system shown, the crank $AB$ has a constant clockwise angular velocity of 2000 rpm. For the crank position shown, determine (*a*) the angular velocity of the connecting rod $BD$, (*b*) the velocity of the piston $P$. ▶

**STRATEGY:**  Connecting rod $BD$ is undergoing general motion, so use rigid-body kinematics. Crank $AB$ is undergoing fixed-axis rotation, and piston $P$ is translating. The motion of the piston is the same as the end $D$ of the connecting rod.

**MODELING and ANALYSIS:**

**Motion of Crank AB.**  The crank $AB$ rotates about point $A$. Expressing $\omega_{AB}$ in rad/s and writing $v_B = r\omega_{AB}$, you have (Fig. 1)

$$\omega_{AB} = \left(2000\,\frac{\text{rev}}{\text{min}}\right)\left(\frac{1 \text{ min}}{60 \text{ s}}\right)\left(\frac{2\pi \text{ rad}}{1 \text{ rev}}\right) = 209.4 \text{ rad/s}$$
$$v_B = (AB)\omega_{AB} = (3 \text{ in.})(209.4 \text{ rad/s}) = 628.3 \text{ in./s}$$
$$\mathbf{v}_B = 628.3 \text{ in./s} \angle 50°$$

**Motion of Connecting Rod BD.**  Consider this as a general plane motion. Using the law of sines, compute the angle $\beta$ between the connecting rod and the horizontal as

$$\frac{\sin 40°}{8 \text{ in.}} = \frac{\sin \beta}{3 \text{ in.}} \qquad \beta = 13.95°$$

The velocity $\mathbf{v}_D$ of point $D$ where the rod is attached to the piston must be horizontal, while the velocity of point $B$ is equal to the velocity $\mathbf{v}_B$ obtained previously. Expressing the relation between the velocities $\mathbf{v}_D$, $\mathbf{v}_B$, and $\mathbf{v}_{D/B}$, you have

$$\mathbf{v}_D = \mathbf{v}_B + \mathbf{v}_{D/B}$$

This equation is shown pictorially in Fig. 2 where the motion of $BD$ is resolved into a translation of $B$ and a rotation about $B$.

*(continued)*

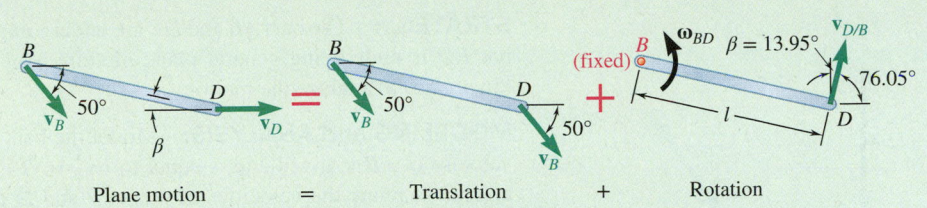

**Fig. 2** The general plane motion of the connecting rod can be modeled as a translation plus a rotation.

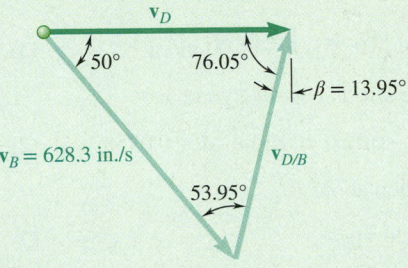

**Fig. 3** Vector triangle showing the relationship between the velocities of $B$ and $D$.

Draw the vector diagram corresponding to this equation (Fig. 3). Recalling that $\beta = 13.95°$, you can determine the angles of the triangle and write

$$\frac{v_D}{\sin 53.95°} = \frac{v_{D/B}}{\sin 50°} = \frac{628.3 \text{ in./s}}{\sin 76.05°}$$

$$v_{D/B} = 495.9 \text{ in./s} \qquad \mathbf{v}_{D/B} = 495.9 \text{ in./s} \measuredangle 76.05°$$
$$v_D = 523.4 \text{ in./s} = 43.6 \text{ ft/s} \qquad \mathbf{v}_D = 43.6 \text{ ft/s} \rightarrow$$
$$\mathbf{v}_P = \mathbf{v}_D = 43.6 \text{ft/s} \rightarrow \blacktriangleleft$$

Because $v_{D/B} = l\omega_{BD}$, you have

$$495.9 \text{ in./s} = (8 \text{ in.})\omega_{BD} \qquad \boldsymbol{\omega}_{BD} = 62.0 \text{ rad/s} \circlearrowleft \blacktriangleleft$$

**REFLECT and THINK:** Note that as the crank continues to move clockwise below the center line, the piston changes direction and starts to move to the left. Can you see what happens to the motion of the connecting rod at that point? You can also solve this problem using the vector relationship expressed in Eq. (15.17′); this type of approach is shown in Sample Prob. 15.8.

## Sample Problem 15.8

In the position shown, bar $AB$ has an angular velocity of 4 rad/s clockwise. Determine the angular velocity of bars $BD$ and $DE$.

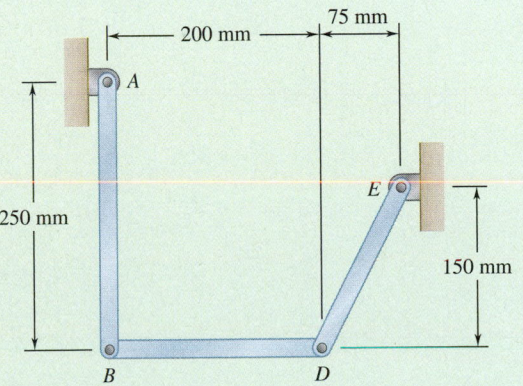

*(continued)*

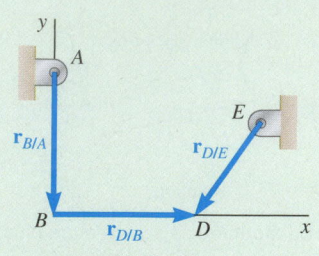

$\mathbf{r}_{B/A} = -(0.25\ m)\mathbf{j}$
$\mathbf{r}_{D/E} = -(0.075\ m)\mathbf{i} - (0.15\ m)\mathbf{j}$
$\mathbf{r}_{D/B} = (0.2\ m)\mathbf{i}$

**Fig. 1**  Relative position vectors for points B and D.

**STRATEGY:**   The bars AB and DE are undergoing fixed-axis rotation, whereas bar BD is undergoing general plane motion. You will need to use rigid-body kinematics to analyze the motion.

**MODELING and ANALYSIS:**   Model the bars as rigid bodies. The angular velocity of AB is given and is equal to $\omega_{AB} = -(4\ rad/s)\mathbf{k}$. You can use vector algebra to relate the velocities of points B and D on bar BD after you find the velocities of B and D from the connecting bars. Position vectors are defined in Fig. 1.

***Bar AB.***   (Rotation about A)

$$\mathbf{v}_B = \omega_{AB} \times \mathbf{r}_{B/A} = (-4\mathbf{k}) \times (-0.25\mathbf{j}) = -(1.00\ m/s)\mathbf{i} \qquad (1)$$

***Bar ED.***   (Rotation about E) Assuming $\omega_{DE}$ is positive, you have

$$\mathbf{v}_D = \omega_{DE}\mathbf{k} \times \mathbf{r}_{D/E} = \omega_{DE}\mathbf{k} \times (-0.075\mathbf{i} - 0.15\mathbf{j}) = 0.15\omega_{DE}\mathbf{i} - 0.075\omega_{DE}\mathbf{j} \qquad (2)$$

***Bar BD.***   (Translation with B and rotation about B.)

$$\mathbf{v}_D = \mathbf{v}_B + \mathbf{v}_{D/B} \qquad (3)$$

where you assume $\omega_{BD}$ is positive. The relative velocity is

$$\mathbf{v}_{D/B} = \omega_{BD}\mathbf{k} \times \mathbf{r}_{D/B} = \omega_{BD}\mathbf{k} \times 0.2\mathbf{i} = 0.2\omega_{BD}\mathbf{j} \qquad (4)$$

Substituting Eqs. (1), (2), and (4) into Eq. (3) gives

$$0.15\omega_{DE}\mathbf{i} - 0.075\omega_{DE}\mathbf{j} = -1.00\mathbf{i} + 0.2\omega_{BD}\mathbf{j}$$

Equating components allows you to solve for the unknown angular velocities:

**i:** $0.15\omega_{DE} = -1.00,$ $\omega_{DE} = -6.667\ rad/s$ $\qquad \boldsymbol{\omega}_{DE} = 6.67\ rad/s\ \circlearrowleft$ ◄

**j:** $-0.075\omega_{DE} = 0.2\omega_{BD},$ $\omega_{BD} = \dfrac{-(0.075)(-6.667)}{0.2}$

$$\boldsymbol{\omega}_{BD} = 2.50\ rad/s\ \circlearrowright\ ◄$$

**REFLECT and THINK:**   The vector algebra approach is very straightforward for problems like this. It makes sense that if AB is rotating clockwise, BD is rotating counterclockwise and DE is rotating clockwise.

# Case Study 15.1

The elliptical exercise machine shown in CS Photo 15.1 consists of two connected linkages, *ABDO* and *ABFH*, where each of the following are rigid bodies: *ABD*, *OD*, *EBF*, and *FHJ*, and pins are located at *B*, *D*, *F*, *H*, and *O*. Equipment designers often look to create different motion profiles by changing linkage lengths. Assume that you are asked to determine how changing the radius of the flywheel *OD* affects the velocity of the foot pedal, $v_E$.

**CS Photo 15.1**   Elliptical machine. ©Daniel Montoya

**STRATEGY:**   This is a complex problem that deals with detailed geometry. You can set up a system of relative velocity equations for the linkages, and then use a computer program to solve this system for different values of $\omega_{OD}$ and for different flywheel radii for all possible angles $\theta$ for the flywheel. Let $\theta$ measure the clockwise angle from the horizontal to the line *OD*, as shown in CS Fig. 15.1.

**MODELING:**   The physical system reduces to two linkages connected at point *B*, as shown in CS Fig. 15.1. Points *H* and *O* are both fixed, and point *A* is constrained to move only in the horizontal direction

**ANALYSIS:**   You can start by relating $\mathbf{v}_D$ to point *O*, then $\mathbf{v}_A$ to $\mathbf{v}_D$.
Link *OD:*

$$\mathbf{v}_D = \mathbf{v}_O + \omega_{OD}\mathbf{k} \times \mathbf{r}_{D/O} = \omega_{OD}\mathbf{k} \times \mathbf{r}_{D/O} \qquad (1)$$

Link *ABD:* Recognizing that $\mathbf{v}_A$ is constrained to the *x* direction and substituting from Eq. (1):

$$v_A\mathbf{i} = \mathbf{v}_D + \omega_{ABD}\mathbf{k} \times \mathbf{r}_{A/D} = \omega_{OD}\mathbf{k} \times \mathbf{r}_{D/O} + \omega_{ABD}\mathbf{k} \times \mathbf{r}_{A/D} \qquad (2)$$

Using $\omega_{OD}$ as a known input and equating components gives you two equations that allow you to solve for $\omega_{ABD}$ and $v_A$. For example, if you use

*(continued)*

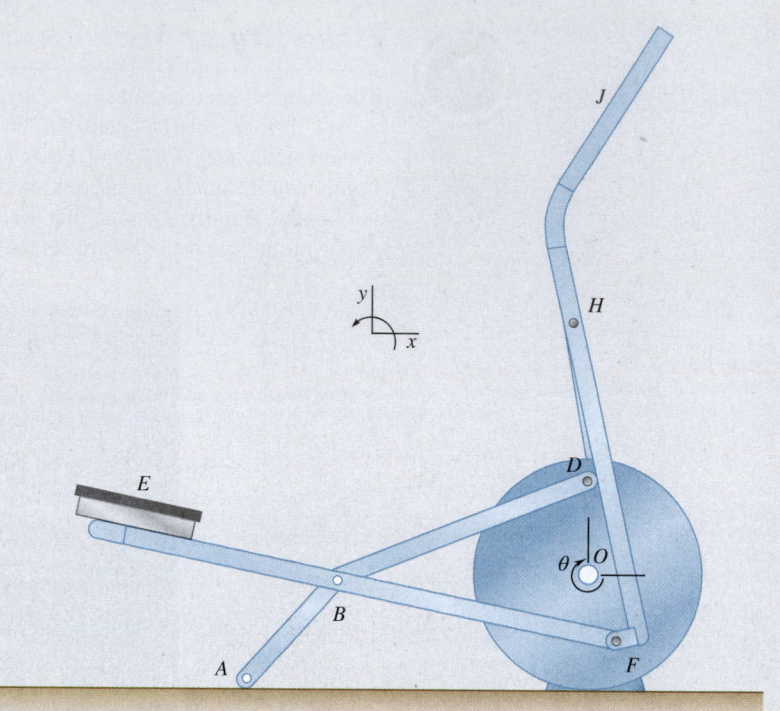

**CS Fig. 15.1** Linkage model of elliptical machine at $\theta = 270°$.

$\boldsymbol{\omega}_{OD} = -(5.5 \text{ rad/s})\mathbf{k}$ and calculate geometry when $\theta = 270°$ as shown in CS Fig. 15.1, then $\mathbf{r}_{D/O} = (10 \text{ in.})\mathbf{j}$ and $\mathbf{r}_{A/D} = -(29.354 \text{ in.})\mathbf{i} - (21.25 \text{ in.})\mathbf{j}$. Substituting into Eq. (2) gives you:

$$v_A\mathbf{i} = 55\mathbf{i} + 21.25\omega_{ABD}\mathbf{i} - 29.35\omega_{ABD}\mathbf{j}$$

Solving, we find that at this instant $v_A = 55$ in./s and $\omega_{ABD} = 0$.

Knowing $v_A$ and $\omega_{ABD}$, you can write an expression for $v_B$ in terms of $v_A$

$$\mathbf{v}_B = v_A\mathbf{i} + \omega_{ABD}\mathbf{k} \times \mathbf{r}_{B/A} \tag{3}$$

Now you can analyze the linkage *ABFH* by relating *B* to *F*,

$$\mathbf{v}_B = \mathbf{v}_F + \omega_{EBF}\mathbf{k} \times \mathbf{r}_{B/F} \tag{4}$$

and *F* to *H*,

$$\mathbf{v}_F = \mathbf{v}_H + \omega_{FH}\mathbf{k} \times \mathbf{r}_{F/H} = \omega_{FH}\mathbf{k} \times \mathbf{r}_{F/H} \tag{5}$$

Substituting Eq. (5) into Eq. (4), then equating this expression with Eq. (3) gives you

$$v_A\mathbf{i} + \omega_{ABD}\mathbf{k} \times \mathbf{r}_{B/A} = \omega_{FH}\mathbf{k} \times \mathbf{r}_{F/H} + \omega_{EBF}\mathbf{k} \times \mathbf{r}_{B/F} \tag{6}$$

Both $v_A$ and $\omega_{ABD}$ are known, so you can equate $x$ and $y$ components of Eq. (6) to solve for $\omega_{FH}$ and $\omega_{EBF}$.

At the position shown, $\mathbf{r}_{B/A} = (7.25 \text{ in.})\mathbf{i} + (9.57 \text{ in.})\mathbf{j}$, $\mathbf{r}_{F/H} = (6.70 \text{ in.})\mathbf{i} - (30.37 \text{ in.})\mathbf{j}$, and $\mathbf{r}_{B/F} = -(26.80 \text{ in.})\mathbf{i} + (5.69 \text{ in.})\mathbf{j}$.

*(continued)*

Substituting into Eq. (6) and solving gives you $\omega_{FDH} = 1.90$ rad/s and $\omega_{EBF} = 0.47$ rad/s. You can use $\omega_{EBF}$ to relate points $B$ and $E$ to determine the velocity of $E$.

$$\mathbf{v}_E = \mathbf{v}_B + \omega_{EBF}\mathbf{k} \times \mathbf{r}_{E/B}$$

Using $\mathbf{r}_{E/B} = -(14.58 \text{ in.})\mathbf{i} - (3.09 \text{ in.})\mathbf{j}$ and the value of $\omega_{EBF}$ found above, you obtain

$$\mathbf{v}_E = (53.6 \text{ in./s})\mathbf{i} - (6.85 \text{ in./s})\mathbf{j}, \text{ which has a magnitude of } v_E = 54.0 \text{ in./s.}$$

Finally, you can determine the magnitude of $v_E$ for any given $\omega_{OD}$ at every position. Using a computer program, you can also vary the radius of the flywheel to see how this affects the velocity profile of the foot pedal. Note that the different position vectors (e.g., $\mathbf{r}_{B/A}$, $\mathbf{r}_{F/H}$, and $\mathbf{r}_{B/F}$ will change as a function of the angle $\theta_{OD}$; this involves quite a bit of trigonometry).

A graph of the speed $|\mathbf{v}_E|$ as a function of flywheel radius is provided in CS Fig. 15.2, given a constant angular velocity $\boldsymbol{\omega}_{OD} = -(5.5 \text{ rad/s})\mathbf{k}$.

**REFLECT AND THINK:** There are many different designs to elliptical machines—for another linkage configuration, see Prob. 15.127. Note that the actual shape of the linkages of the machine don't really matter—the vector is the same whether the bars are bent or straight. Can you think of some reasons why the designers may have decided to use curved bars?

The end goal for this Case Study was to see how the flywheel radius affects the velocity at point $E$. We might also want to examine the accelerations at this point, to determine the forces at different pins, or to analyze the biomechanics of the athlete. We investigate this linkage further in Probs. 15.251 and 15.252.

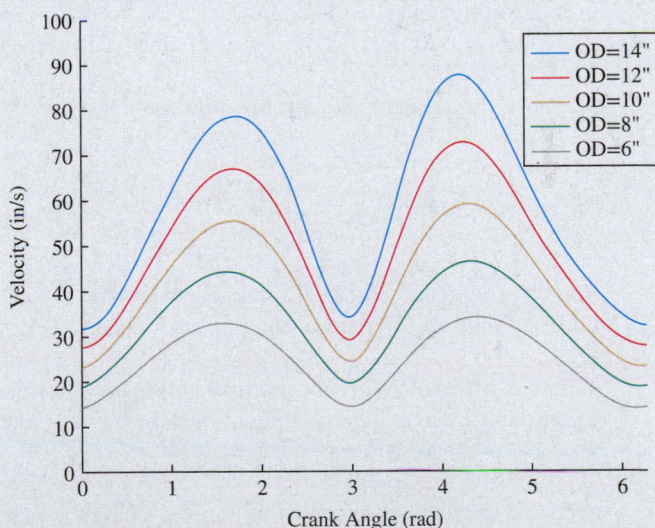

**CS Fig. 15.2** Magnitude of the velocity of point $E$ as a function of flywheel radius $OD$.

# SOLVING PROBLEMS ON YOUR OWN

In this section, you learned how to analyze the velocity of bodies in **general plane motion**. You found that you can always consider a general plane motion to be the sum of the two motions you studied in Sec. 15.1, namely, *a translation and a rotation.*

To solve a problem involving the velocity of a body in plane motion, you should take the following steps.

**1. Whenever possible, determine the velocity of the points of the body** where it is connected to another body whose motion is known (Sample Prob. 15.6). That other body may be an arm or crank rotating with a given angular velocity (Sample Probs. 15.7 and 15.8).

**2. Next, draw a diagram** to use in your solution (Figs. 15.15 and 15.16) if you are not using the vector algebra approach. This diagram consists of the following diagrams.
   **a. Plane motion diagram:** Draw a diagram of the body including all dimensions and showing those points for which you know or seek the velocity.
   **b. Translation diagram:** Select a reference point $A$ for which you know the direction and/or the magnitude of the velocity $\mathbf{v}_A$, and draw a second diagram showing the body in translation with all of its points having the same velocity $\mathbf{v}_A$.
   **c. Rotation diagram:** Consider point $A$ as a fixed point and draw a diagram showing the body in rotation about $A$. Show the angular velocity $\boldsymbol{\omega} = \omega\mathbf{k}$ of the body and the relative velocities with respect to $A$ of the other points, such as the velocity $\mathbf{v}_{B/A}$ of $B$ relative to $A$.

**3. Write the relative velocity formula as**

$$\mathbf{v}_B = \mathbf{v}_A + \mathbf{v}_{B/A} \tag{15.17}$$

or for plane motion as

$$\mathbf{v}_B = \mathbf{v}_A + \omega\mathbf{k} \times \mathbf{r}_{B/A} \tag{15.17'}$$

You can solve this vector equation analytically by writing the corresponding scalar equations, or you can solve it by using a vector triangle (Fig. 15.16).

**4. Use a different reference point to obtain an equivalent solution.** For example, if you select point $B$ as the reference point, the velocity of point $A$ is

$$\mathbf{v}_A = \mathbf{v}_B + \mathbf{v}_{A/B} = \mathbf{v}_B + \omega\mathbf{k} \times \mathbf{r}_{A/B} \tag{15.20}$$

Note that the relative velocities $\mathbf{v}_{B/A}$ and $\mathbf{v}_{A/B}$ have the same magnitude but opposite sense. Relative velocities, therefore, depend upon the reference point that you select. The angular velocity, however, is independent of the choice of reference point.

**5. Write additional relative velocity equations if you are analyzing a multi-body linkage.** For problems such as the crankshaft-piston in Sample Prob. 15.7, you may have to write multiple relative velocity equations. In that problem, you can express the velocity of $P$ with respect to $B$ and then the velocity of $B$ with respect to $A$. Generally, the ends of the linkages will have some type of constraint (e.g., the piston moving only in the $x$ direction).

# Problems

## CONCEPT QUESTIONS

**15.CQ3** The ball rolls without slipping on the fixed surface as shown. What is the direction of the velocity of point $A$?

    **a.** →   **b.** ↗   **c.** ↑   **d.** ↓   **e.** ↘

**15.CQ4** Three uniform rods—$ABC$, $DCE$, and $FGH$—are connected as shown. Which of the following statements concerning the angular speed of the three objects is true?

    **a.** $\omega_{ABC} = \omega_{DCE} = \omega_{FGH}$
    **b.** $\omega_{DCE} > \omega_{ABC} > \omega_{FGH}$
    **c.** $\omega_{DCE} < \omega_{ABC} < \omega_{FGH}$
    **d.** $\omega_{ABC} > \omega_{DCE} > \omega_{FGH}$
    **e.** $\omega_{FGH} = \omega_{DCE} < \omega_{ABC}$

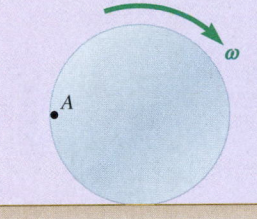

**Fig. P15.CQ3**

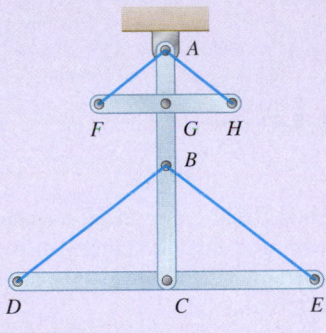

**Fig. P15.CQ4**

## END-OF-SECTION PROBLEMS

**15.38** An automobile travels to the right at a constant speed of 48 mi/h. If the diameter of a wheel is 22 in., determine the velocities of points $B$, $C$, $D$, and $E$ on the rim of the wheel.

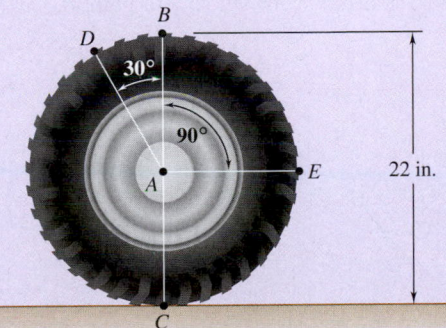

**Fig. P15.38**

**15.39** An overhead door is guided by wheels at $A$ and $B$ that roll in horizontal and vertical tracks. Knowing that when $\theta = 30°$ the velocity of wheel $B$ is 2 ft/s downward, determine (*a*) the angular velocity of the door, (*b*) the velocity of end $D$ of the door.

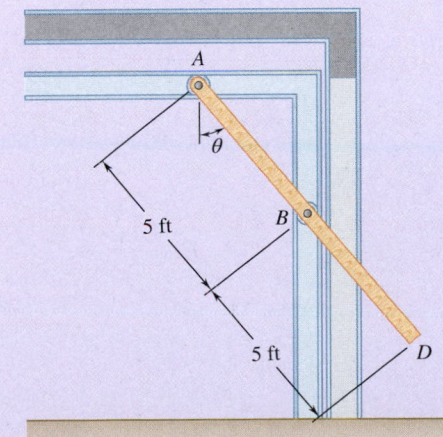

**Fig. P15.39**

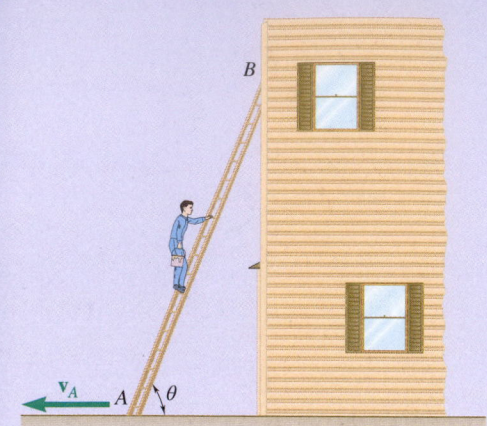

**Fig. P15.40**

**15.40** A painter is halfway up a 10-m ladder when the bottom starts sliding out from under him. Knowing that point $A$ has a velocity $\mathbf{v}_A = 2$ m/s directed to the left when $\theta = 60°$, determine (*a*) the angular velocity of the ladder, (*b*) the velocity of the painter.

**15.41** Rod $AB$ can slide freely along the floor and the inclined plane. At the instant shown, the velocity of end $A$ is 1.4 m/s to the left. Determine (*a*) the angular velocity of the rod, (*b*) the velocity of end $B$ of the rod.

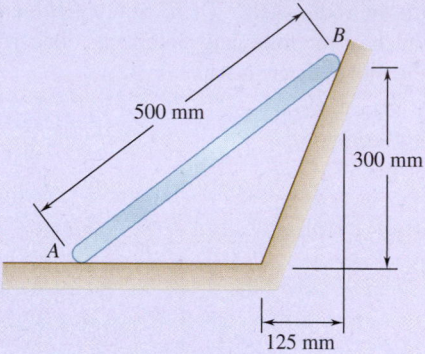

**Fig. P15.41 and *P15.42***

**15.42** Rod $AB$ can slide freely along the floor and the inclined plane. At the instant shown, the angular velocity of the rod is 4.2 rad/s counterclockwise. Determine (*a*) the velocity of end $A$ of the rod, (*b*) the velocity of end $B$ of the rod.

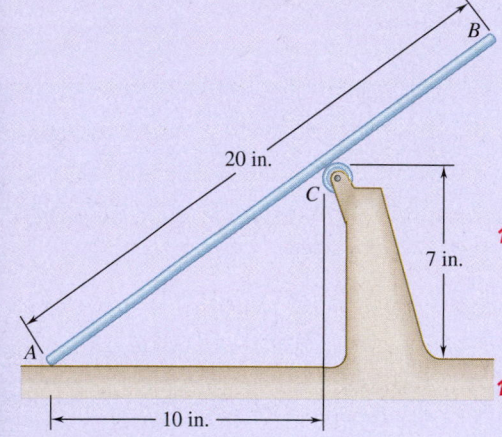

**Fig. *P15.43***

**15.43** Rod $AB$ moves over a small wheel at $C$ while end $A$ moves to the right with a constant velocity of 25 in./s. At the instant shown, determine (*a*) the angular velocity of the rod, (*b*) the velocity of end $B$ of the rod.

**15.44** The disk shown moves in the $xy$ plane. Knowing that $(v_A)_y = -7$ m/s, $(v_B)_x = -7.4$ m/s, and $(v_C)_x = -1.4$ m/s, determine (*a*) the angular velocity of the disk, (*b*) the velocity of point $B$.

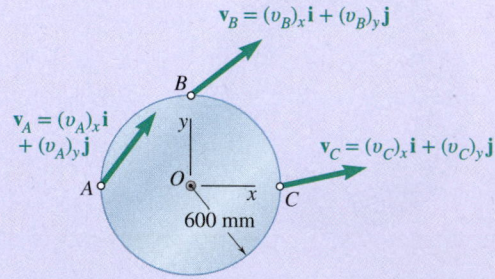

**Fig. P15.44 and P15.45**

**15.45** The disk shown moves in the $xy$ plane. Knowing that $(v_A)_y = -7$ m/s, $(v_B)_x = -7.4$ m/s, and $(v_C)_x = -1.4$ m/s, detemine (*a*) the velocity of point $O$, (*b*) the point of the disk with zero velocity.

**15.46** The plate shown moves in the $xy$ plane. Knowing that $(v_A)_x = 12$ in./s, $(v_B)_x = -4$ in./s, and $(v_C)_y = -24$ in./s, determine (*a*) the angular velocity of the plate, (*b*) the velocity of point $B$, (*c*) the point of the plate with zero velocity.

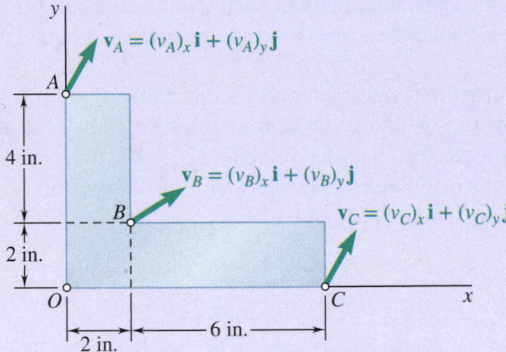

**Fig. P15.46**

**15.47** Velocity sensors are placed on a satellite that is moving only in the $xy$ plane. Knowing that at the instant shown the unidirectional sensors measure $(v_A)_x = 2$ ft/s, $(v_B)_x = -0.333$ ft/s, and $(v_C)_y = -2$ ft/s, determine (*a*) the angular velocity of the satellite, (*b*) the velocity of point $B$.

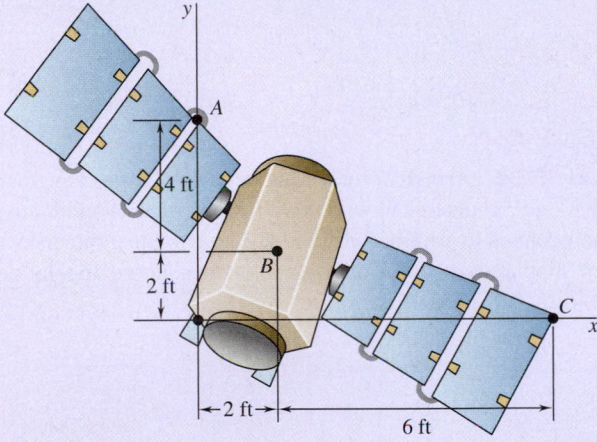

**Fig. P15.47**

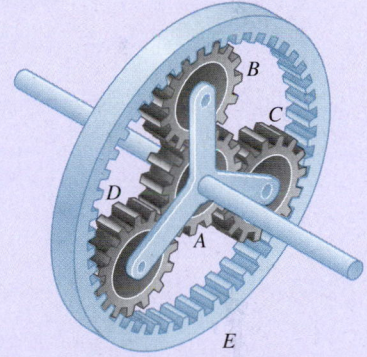

**Fig. P15.48 and P15.49**

**15.48** In the planetary gear system shown, the radius of gears $A$, $B$, $C$, and $D$ is $a$ and the radius of the outer gear $E$ is $3a$. Knowing that the angular velocity of gear $A$ is $\omega_A$ clockwise and that the outer gear $E$ is stationary, determine (*a*) the angular velocity of each planetary gear, (*b*) the angular velocity of the spider connecting the planetary gears.

**15.49** In the planetary gear system shown, the radius of gears $A$, $B$, $C$, and $D$ is 30 mm and the radius of the outer gear $E$ is 90 mm. Knowing that gear $E$ has an angular velocity of 180 rpm clockwise and that the central gear $A$ has an angular velocity of 240 rpm clockwise, determine (*a*) the angular velocity of each planetary gear, (*b*) the angular velocity of the spider connecting the planetary gears.

**15.50** The outer gear $C$ rotates with an angular velocity of 5 rad/s clockwise. Knowing that the inner gear $A$ is stationary, determine (*a*) the angular velocity of the intermediate gear $B$, (*b*) the angular velocity of the arm $AB$.

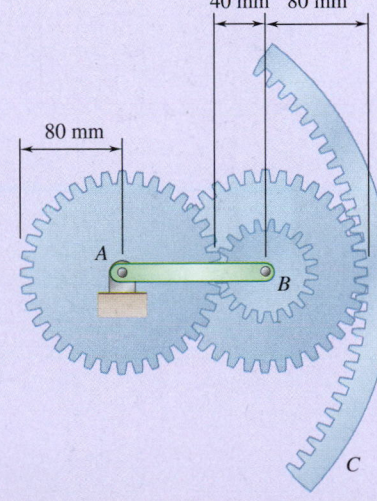

**Fig. P15.50**

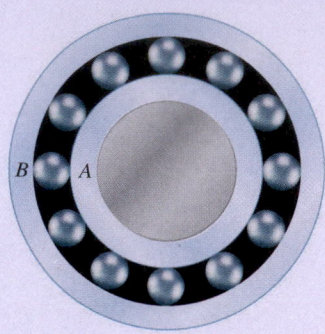

**Fig. P15.51**

**15.51** In the simplified sketch of a ball bearing shown, the diameter of the inner race $A$ is 60 mm and the diameter of each ball is 12 mm. The outer race $B$ is stationary, while the inner race has an angular velocity of 3600 rpm. Determine (*a*) the speed of the center of each ball, (*b*) the angular velocity of each ball, (*c*) the number of times per minute each ball describes a complete circle.

**15.52** A simplified gear system for a mechanical watch is shown. Knowing that gear $A$ has a constant angular velocity of 1 rev/h and gear $C$ has a constant angular velocity of 1 rpm, determine (*a*) the radius $r$, (*b*) the magnitudes of the accelerations of the points on gear $B$ that are in contact with gears $A$ and $C$.

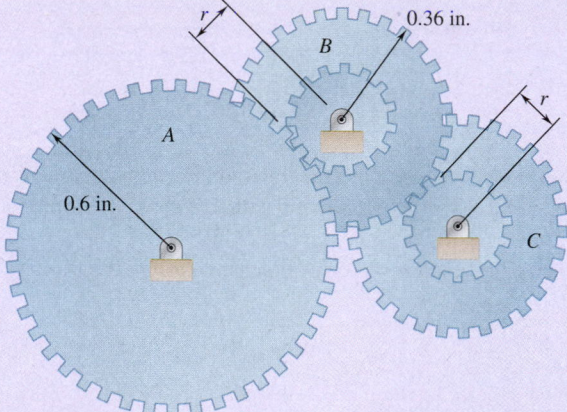

**Fig. *P15.52***

**15.53 and *15.54*** Arm $ACB$ rotates about point $C$ with an angular velocity of 40 rad/s counterclockwise. Two friction disks $A$ and $B$ are pinned at their centers to arm $ACB$ as shown. Knowing that the disks roll without slipping at surfaces of contact, determine the angular velocity of (*a*) disk $A$, (*b*) disk $B$.

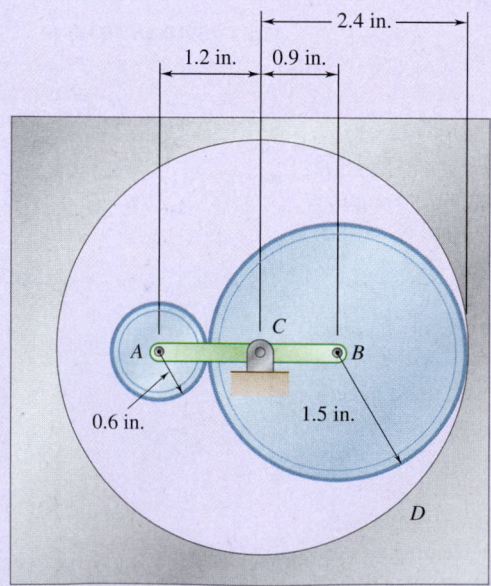

**Fig. P15.53**

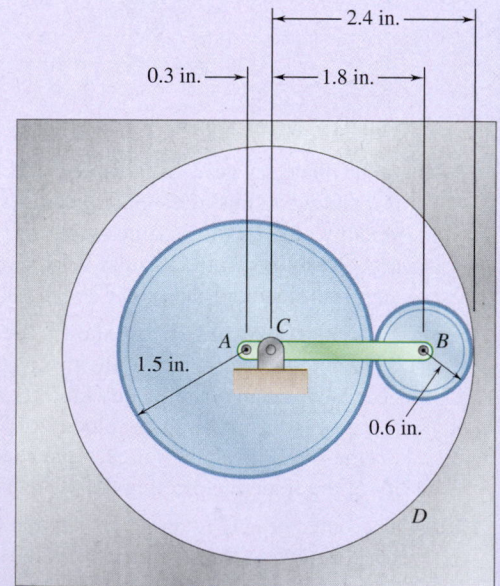

**Fig. *P15.54***

**15.55** Knowing that at the instant shown the angular velocity of rod *BE* is 4 rad/s counterclockwise, determine (*a*) the angular velocity of rod *AD*, (*b*) the velocity of collar *D*, (*c*) the velocity of point *A*.

**15.56** Knowing that at the instant shown the velocity of collar *D* is 1.6 m/s upward, determine (*a*) the angular velocity of rod *AD*, (*b*) the velocity of point *B*, (*c*) the velocity of point *A*.

**15.57** Knowing that the disk has a constant angular velocity of 15 rad/s clockwise, determine the angular velocity of bar *BD* and the velocity of collar *D* when (*a*) $\theta = 0$, (*b*) $\theta = 90°$, (*c*) $\theta = 180°$.

**15.58** The disk has a constant angular velocity of 20 rad/s clockwise. (*a*) Determine the two values of the angle $\theta$ for which the velocity of collar *D* is zero. (*b*) For each of these values of $\theta$, determine the corresponding value of the angular velocity of bar *BD*.

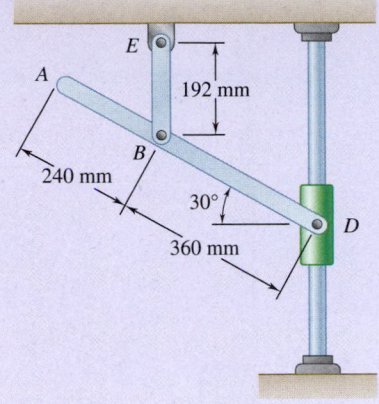

**Fig. P15.55 and P15.56**

**15.59** The test rig shown was developed to perform fatigue testing on fitness trampolines. A motor drives the 9-in.-radius flywheel *AB*, which is pinned at its center point *A*, in a counterclockwise direction. The flywheel is attached to slider *CD* by the 18-in. connecting rod *BC*. Knowing that the "feet" at *D* should hit the trampoline twice every second, at the instant when $\theta = 0°$, determine (*a*) the angular velocity of the connecting rod *BC*, (*b*) the velocity of *D*, (*c*) the velocity of midpoint *CB*.

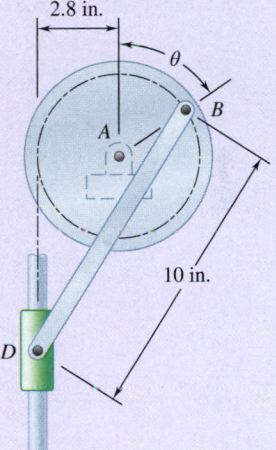

**Fig. P15.57 and P15.58**

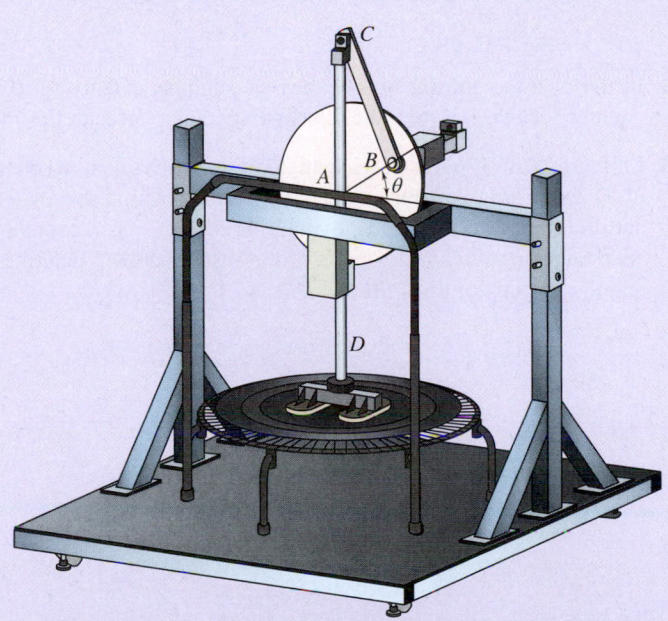

**Fig. P15.59**

**15.60** In the eccentric shown, a disk of 2-in. radius revolves about shaft *O* that is located 0.5 in. from the center *A* of the disk. The distance between the center *A* of the disk and the pin at *B* is 8 in. Knowing that the angular velocity of the disk is 900 rpm clockwise, determine the velocity of the block when $\theta = 30°$.

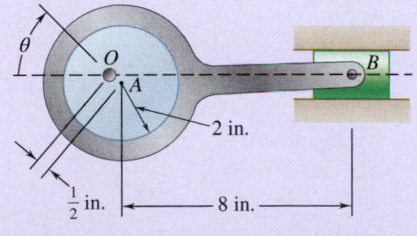

**Fig. P15.60**

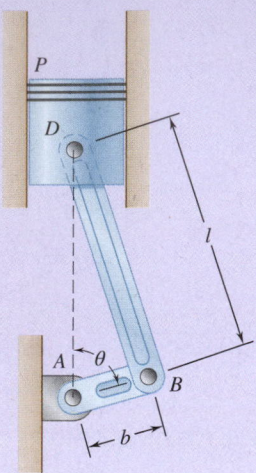

**Fig. P15.61 and P15.62**

**15.61** In the engine system shown, $l = 160$ mm and $b = 60$ mm. Knowing that the crank $AB$ rotates with a constant angular velocity of 1000 rpm clockwise, determine the velocity of the piston $P$ and the angular velocity of the connecting rod when (a) $\theta = 0$, (b) $\theta = 90°$.

**15.62** In the engine system shown, $l = 160$ mm and $b = 60$ mm. Knowing that crank $AB$ rotates with a constant angular velocity of 1000 rpm clockwise, determine the velocity of the piston $P$ and the angular velocity of the connecting rod when $\theta = 60°$.

**15.63** Knowing that the angular velocity of rod $DE$ is a constant 20 rad/s clockwise, determine in the position shown (a) the angular velocity of rod $BD$, (b) the velocity of the midpoint of rod $BD$.

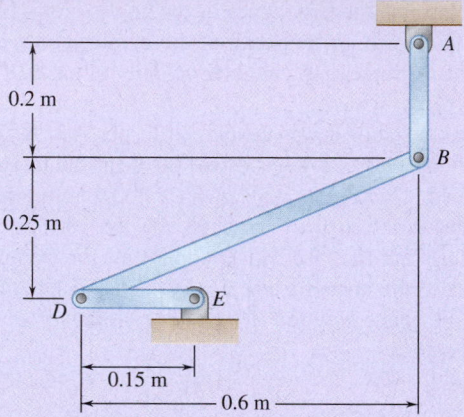

**Fig. P15.63**

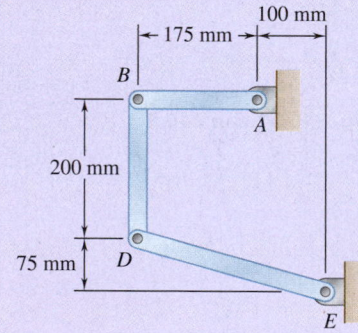

**Fig. P15.64**

**15.64** In the position shown, bar $AB$ has an angular velocity of 10 rad/s counterclockwise. Determine the angular velocity of bars $BD$ and $DE$.

**15.65** Linkage $DBEF$ is part of a windshield wiper mechanism, where points $O$, $F$, and $D$ are fixed pin connections. At the position shown, $\theta = 30°$ and link $EB$ is horizontal. Knowing that link $EF$ has a counterclockwise angular velocity of 4 rad/s at the instant shown, determine the angular velocity of links $EB$ and $DB$.

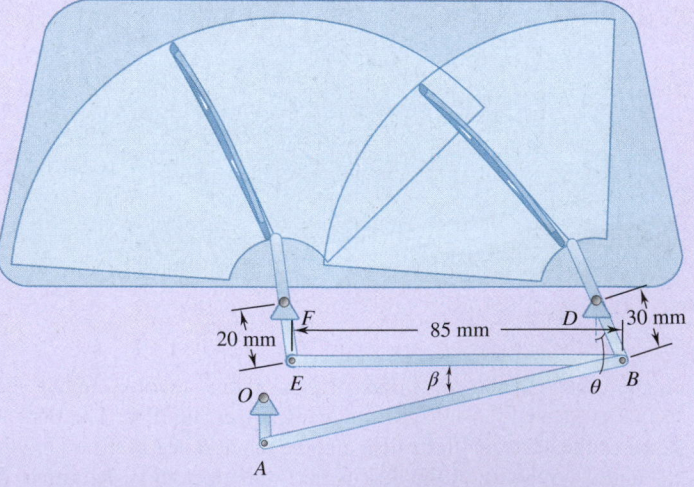

**Fig. P15.65**

**15.66** Roberts linkage is named after Richard Roberts (1789–1864) and can be used to draw a close approximation to a straight line by locating a pen at point $F$. The distance $AB$ is the same as $BF$, $DF$, and $DE$. Knowing that the angular velocity of bar $AB$ is 5 rad/s clockwise in the position shown, determine (*a*) the angular velocity of bar $DE$, (*b*) the velocity of point $F$.

**15.67** Roberts linkage is named after Richard Roberts (1789–1864) and can be used to draw a close approximation to a straight line by locating a pen at point $F$. The distance $AB$ is the same as $BF$, $DF$, and $DE$. Knowing that the angular velocity of plate $BDF$ is 2 rad/s counterclockwise when $\theta = 90°$, determine (*a*) the angular velocities of bars $AB$ and $DE$, (*b*) the velocity of point $F$. When $\theta = 90°$, point $F$ may be assumed to coincide with point $E$, with negligible error in the velocity analysis.

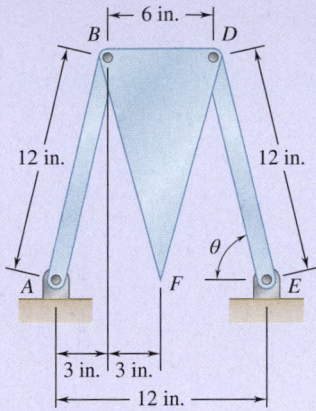

Fig. *P15.66* and *P15.67*

**15.68** For the oil pump rig shown, link $AB$ causes the beam $BCE$ to oscillate as the crank $OA$ revolves. Knowing that $OA$ has a radius of 0.6 m and a constant clockwise angular velocity of 10 rpm, determine (*a*) the angular velocity of arm $BCE$, (*b*) the velocity of point $D$ at the instant shown.

**15.69** For the oil pump rig shown, link $AB$ causes the beam $BCE$ to oscillate as the crank $OA$ revolves. Knowing that $OA$ has a radius of 0.6 m and at the instant shown the velocity of point $D$ is 1.5 m/s down, determine the angular velocity of arm $AB$ and $AO$.

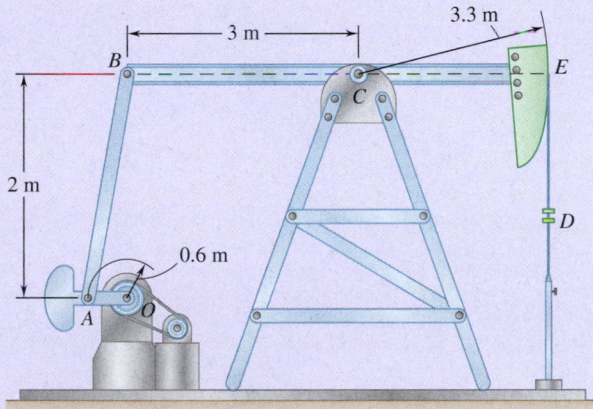

Fig. P15.68 and P15.69

**15.70** Both 6-in.-radius wheels roll without slipping on the horizontal surface. Knowing that the distance $AD$ is 5 in., the distance $BE$ is 4 in., and $D$ has a velocity of 6 in./s to the right, determine the velocity of point $E$.

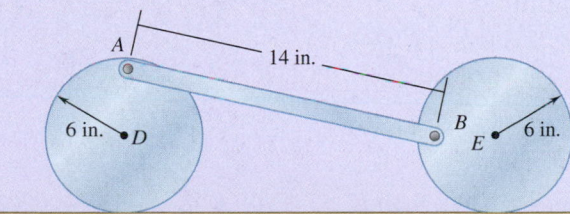

Fig. P15.70

**15.71** The 80-mm-radius wheel shown rolls to the left with a velocity of 900 mm/s. Knowing that the distance $AD$ is 50 mm, determine the velocity of the collar and the angular velocity of rod $AB$ when (a) $\beta = 0$, (b) $\beta = 90°$.

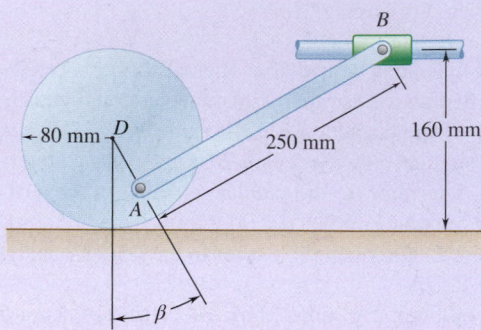

**Fig. P15.71**

**\*15.72** For the gearing shown, derive an expression for the angular velocity $\omega_C$ of gear $C$ and show that $\omega_C$ is independent of the radius of gear $B$. Assume that point $A$ is fixed and denote the angular velocities of rod $ABC$ and gear $A$ by $\omega_{ABC}$ and $\omega_A$, respectively.

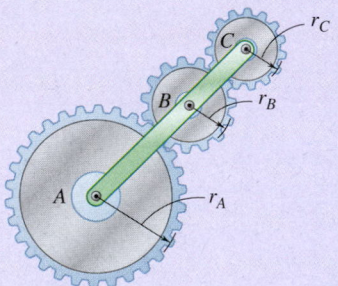

**Fig. P15.72**

# 15.3   INSTANTANEOUS CENTER OF ROTATION

Consider the general plane motion of a rigid body. We will show that, at any given instant, the velocities of the various particles of the rigid body are the same as if the body were rotating about an axis perpendicular to the plane of the body, called the **instantaneous axis of rotation**. This axis intersects the plane of the rigid body at a point $C$, called the **instantaneous center of rotation** of the body or the **instantaneous center of zero velocity**. This gives us an alternative method for solving problems involving the velocities of points on an object in plane motion, and it is sometimes simpler than using the equations in Sec. 15.2.

Recall that we can always replace the plane motion of a rigid body by a translation defined by the motion of an arbitrary reference point $A$ and by a rotation about $A$. As far as the velocities are concerned, the translation is characterized by the velocity $\mathbf{v}_A$ of the reference point $A$ and the rotation is characterized by the angular velocity $\boldsymbol{\omega}$ of the body (which is independent of the choice of $A$). Thus, the velocity $\mathbf{v}_A$ of point $A$ and the angular velocity $\boldsymbol{\omega}$ of the rigid body define completely the velocities of all the other particles of the body (Fig. 15.18$a$).

**Photo 15.5**   If the tires of this car are rolling without sliding, the instantaneous center of rotation of each tire is the point of contact between the road and the tire. ▶
©Glow Images RF

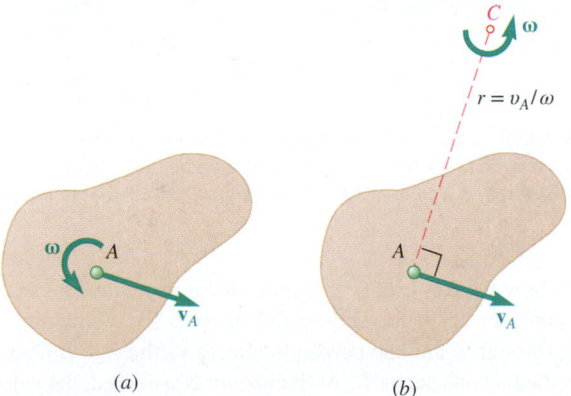

$(a)$ $\qquad\qquad$ $(b)$

**Fig. 15.18**   As far as velocities are concerned, at every instant in time the rigid body seems to rotate about a point called the instantaneous center $C$.

Now let us assume that $\mathbf{v}_A$ and $\boldsymbol{\omega}$ are known and that they are both different from zero. (If $\mathbf{v}_A = 0$, point $A$ is itself the instantaneous center of rotation, and if $\boldsymbol{\omega} = 0$, you have rigid body translation where all of the particles have the same velocity $\mathbf{v}_A$.) We could obtain these velocities by letting the rigid body rotate with the angular velocity $\boldsymbol{\omega}$ about a point $C$ located on the perpendicular to $\mathbf{v}_A$ at a distance $r = v_A/\omega$ from $A$, as shown in Fig. 15.18$b$. We check that the velocity of $A$ would be perpendicular to $AC$ and that its magnitude would be $r\omega = (v_A/\omega)\omega = v_A$. Thus, the velocities of all the other particles of the body are the same as originally defined. Therefore, *as far as the velocities are concerned, the rigid body seems to rotate about the instantaneous center $C$ at the instant considered.*

We can define the position of the instantaneous center in two other ways. If we know the directions of the velocities of two particles $A$ and $B$ of the rigid body and if they are different, we can obtain the instantaneous center $C$ by drawing the perpendicular to $\mathbf{v}_A$ through $A$ and the perpendicular to $\mathbf{v}_B$ through $B$. ▶ The point $C$ is where these two lines intersect (Fig. 15.19$a$). If the velocities $\mathbf{v}_A$ and $\mathbf{v}_B$ of two particles $A$ and $B$ are perpendicular to line $AB$ and we know their magnitudes, we can find the instantaneous center by intersecting line $AB$ with the line joining the ends of the vectors $\mathbf{v}_A$ and $\mathbf{v}_B$ (Fig. 15.19$b$). Note that if $\mathbf{v}_A$ and $\mathbf{v}_B$ were parallel in Fig. 15.19$a$ or if $\mathbf{v}_A$ and $\mathbf{v}_B$ had the same magnitude in Fig. 15.19$b$, the instantaneous center $C$ would be at an infinite distance and $\boldsymbol{\omega}$ would be zero; all points of the rigid body would have the same velocity.

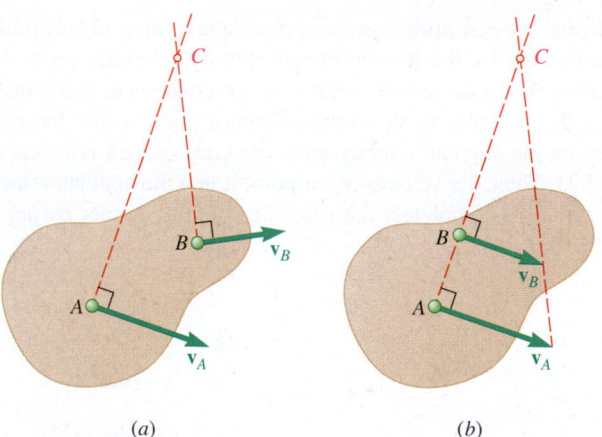

(a)                    (b)

**Fig. 15.19**   Locating the instantaneous center of rotation $C$
($a$) when you know the directions of the velocities of two points;
($b$) when the velocities of two points are perpendicular to line $AB$.

To see how we can use the concept of the instantaneous center of rotation, let us consider again the sliding rod of Sec. 15.2. Drawing the perpendicular to $\mathbf{v}_A$ through $A$ and the perpendicular to $\mathbf{v}_B$ through $B$ (Fig. 15.20), we obtain the instantaneous center $C$. At the instant considered, the velocities of all

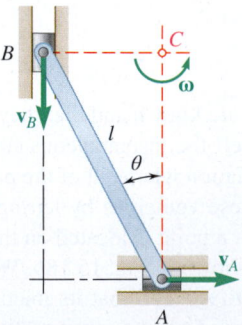

**Fig. 15.20**   Instantaneous center of rotation $C$
for the sliding rod $AB$.

the particles of the rod are thus the same as if the rod rotated about $C$. Now, if we know the magnitude $v_A$ of the velocity of $A$, we can obtain the magnitude $\omega$ of the angular velocity of the rod from

$$\omega = \frac{v_A}{AC} = \frac{v_A}{l \cos \theta}$$

Then we obtain the magnitude of the velocity of $B$ as

$$v_B = (BC)\omega = l \sin \theta \frac{v_A}{l \cos \theta} = v_A \tan \theta$$

Note that we used only *absolute* velocities in the computation.

The instantaneous center of a body in plane motion can be located either on the body or outside the body. If it is located on the rigid body, the particle $C$ coinciding with the instantaneous center at a given instant $t$ must have zero velocity at that instant. However, the instantaneous center of rotation is valid only at a given instant. Thus, particle $C$ of the rigid body that coincides with the instantaneous center at time $t$ generally does not coincide with the instantaneous center at time $t + \Delta t$. Its velocity is zero at time $t$, but it will probably be different from zero at time $t + \Delta t$. This means, in general, that particle $C$ *does not have zero acceleration* and, therefore, that the accelerations of the various particles of the rigid body cannot be determined as if the body were rotating about $C$.

As the motion of the rigid body proceeds, the instantaneous center moves in space. However, we just pointed out that the position of the instantaneous center on the body keeps changing. Thus, the instantaneous center describes one curve in space, called the *space centrode,* and another curve on the rigid body, called the *body centrode* (Fig. 15.21). It can be shown that at any instant, these two curves are tangent at $C$ and that as the rigid body moves, the body centrode appears to roll on the space centrode.

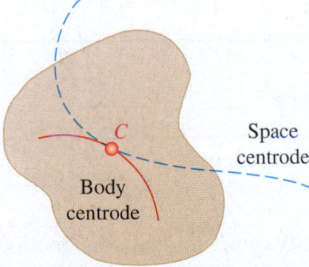

**Fig. 15.21** The space centrode and the body centrode are tangent to each other.

## Sample Problem 15.9

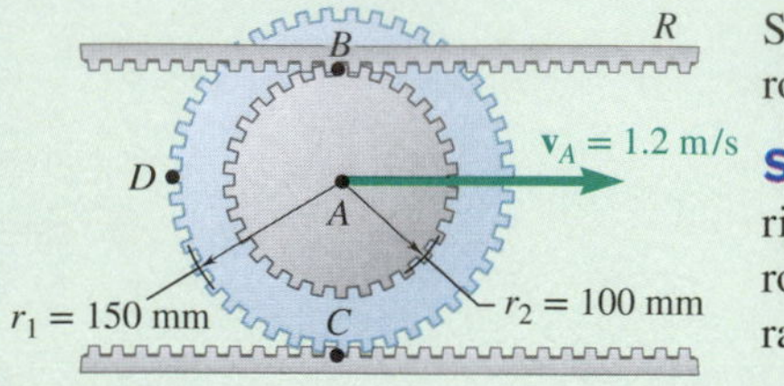

Solve Sample Prob. 15.6 using the method of the instantaneous center of rotation.

**STRATEGY:** You know the velocity direction of two points on the same rigid body, so you can find an instantaneous center of rotation. Because the gear rolls on the stationary lower rack, the point of contact $C$ of the gear with the rack has no velocity; point $C$ is therefore the instantaneous center of rotation.

**MODELING and ANALYSIS:**

**a. Angular Velocity of the Gear.** You can calculate the angular velocity directly from the data in Fig. 1.

$$v_A = r_A \omega \qquad\qquad 1.2 \text{ m/s} = (0.150 \text{ m})\omega$$

$$\omega = 8 \text{ rad/s} \circlearrowleft$$

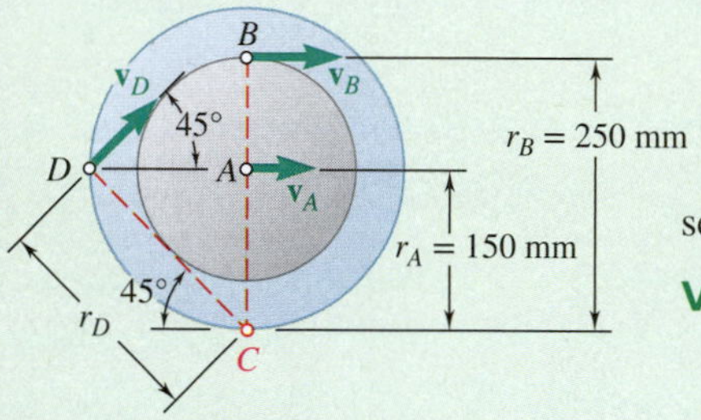

**b. Velocities.** As far as velocities are concerned, all points of the gear seem to rotate about the instantaneous center.

**Velocity of Upper Rack.** Recalling that $v_R = v_B$, you have

$$v_R = v_B = r_B \omega \qquad v_R = (0.250 \text{ m})(8 \text{ rad/s}) = 2 \text{ m/s}$$

$$\mathbf{v}_R = 2 \text{ m/s} \rightarrow$$

**Velocity of Point D.** Because $r_D = (0.150 \text{ m})\sqrt{2} = 0.2121 \text{ m}$, you obtain

$$v_D = r_D \omega \qquad v_D = (0.2121 \text{ m})(8 \text{ rad/s}) = 1.697 \text{ m/s}$$

$$\mathbf{v}_D = 1.697 \text{ m/s} \measuredangle 45°$$

**REFLECT and THINK:** The results are the same as in Sample Prob. 15.6, as you would expect, but it took much less computation to get them.

**Fig. 1** Distances from the instantaneous center of rotation to A, B, and D.

## Sample Problem 15.10

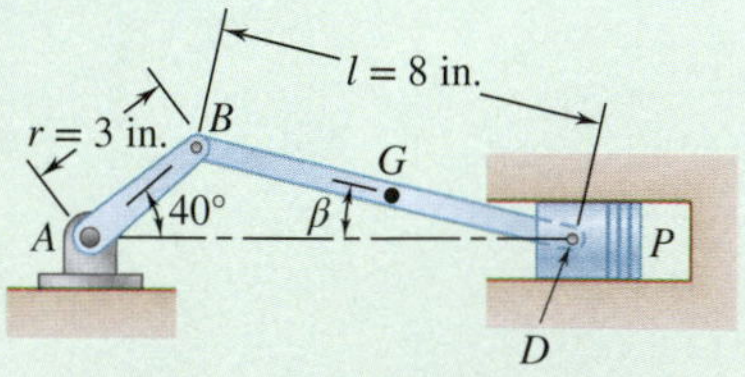

Solve Sample Prob. 15.7 using the method of the instantaneous center of rotation.

**STRATEGY:** You know the velocity of point $B$ from the motion of the crank (see Sample Prob. 15.7), and you know the direction of the velocity of point $D$. Therefore, you can find an instantaneous center of rotation.

**MODELING and ANALYSIS:**

**Motion of Crank AB.** Referring to Sample Prob. 15.7, you obtain the velocity of point $B$; $\mathbf{v}_B = 628.3$ in./s $\measuredangle 50°$.

*(continued)*

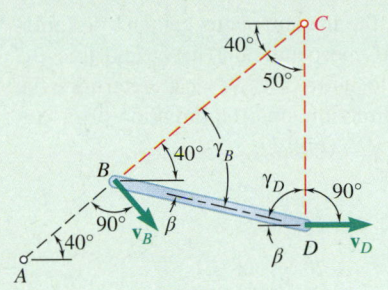

**Fig. 1** Instantaneous center of rotation for bar *BD*.

**Motion of the Connecting Rod BD.** First locate the instantaneous center *C* by drawing lines perpendicular to the absolute velocities $\mathbf{v}_B$ and $\mathbf{v}_D$ (Fig. 1). Recalling from Sample Prob. 15.7 that $\beta = 13.95°$ and that $BD = 8$ in., solve the triangle *BCD*.

$$\gamma_B = 40° + \beta = 53.95° \qquad \gamma_D = 90° - \beta = 76.05°$$

$$\frac{BC}{\sin 76.05°} = \frac{CD}{\sin 53.95°} = \frac{8 \text{ in.}}{\sin 50°}$$

$$BC = 10.14 \text{ in.} \qquad CD = 8.44 \text{ in.}$$

Because the connecting rod *BD* seems to rotate about point *C* at this instant, you have

$$v_B = (BC)\omega_{BD}$$

$$628.3 \text{ in./s} = (10.14 \text{ in.})\omega_{BD}$$

$$\boldsymbol{\omega}_{BD} = 62.0 \text{ rad/s} \circlearrowleft \quad \blacktriangleleft$$

$$v_D = (CD)\omega_{BD} = (8.44 \text{ in.})(62.0 \text{ rad/s})$$

$$= 523 \text{ in./s} = 43.6 \text{ ft/s}$$

$$\mathbf{v}_P = \mathbf{v}_D = 43.6 \text{ ft/s} \rightarrow \quad \blacktriangleleft$$

**REFLECT and THINK:** Often, the hardest part of solving a problem using the instantaneous center of rotation is the geometry. Remembering how to use the law of sines or the law of cosines is often helpful.

# Sample Problem 15.11

Two 20-in. rods *AB* and *DE* are connected as shown. Point *D* is the midpoint of rod *AB*, and at the instant shown, rod *DE* is horizontal. Knowing that the velocity of point *A* is 1 ft/s downward, determine (*a*) the angular velocity of rod *DE*, (*b*) the velocity of point *E*.

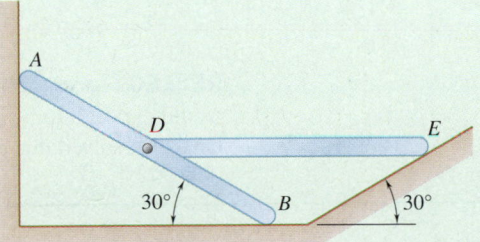

**STRATEGY:** You know the velocity directions of several points on these objects, so you can use instantaneous centers of rotation to solve this problem.

*(continued)*

**MODELING and ANALYSIS:** Locate the instantaneous center of rotation $C$ of bar $AB$ as the intersection of line $AC$ perpendicular to $\mathbf{v}_A$ and line $BC$ perpendicular to $\mathbf{v}_B$ (Fig. 1). Knowing the location of $C$, you can determine the direction of the velocity of $D$. From this direction, and the direction of $E$, you can find the instantaneous center, $I$, for bar $DE$ (Fig. 1).

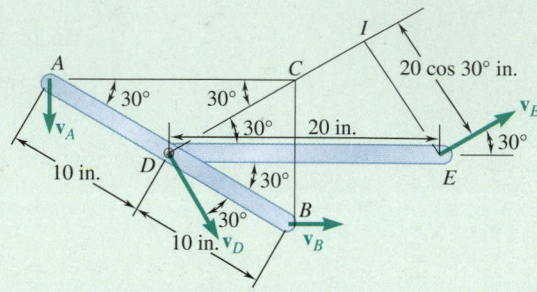

**Fig. 1** The instantaneous centers of rotation for bar $AB$ and $DE$ are $C$ and $I$, respectively.

**a. Angular velocity of DE.** From geometry, $r_{A/C} = (20 \cos 30°)$ in., so

$$\omega_{AB} = \frac{v_A}{r_{A/C}} = \frac{12 \text{ in./s}}{20 \cos 30° \text{ in.}} = 0.6928 \text{ rad/s} \; \circlearrowleft$$

Now you can find $v_D$ because $r_{D/C} = 10$ in.

$$v_D = \omega_{AB} r_{D/C} = (0.6928 \text{ rad/s})(10 \text{ in.}) = 6.928 \text{ in./s}$$

$$v_D = 6.928 \text{ in./s} \; \diagdown 30°$$

Now, because you know the directions of the velocities of $D$ and $E$, $\mathbf{v}_E = v_E \angle 30°$, you can find point $I$, which is the instantaneous center of bar $DE$. From geometry, $r_{D/I} = 20 \cos 30°$ in., and therefore

$$\omega_{DE} = \frac{v_D}{r_{D/I}} = \frac{6.298 \text{ in./s}}{20 \cos 30° \text{ in.}} = 0.400 \text{ rad/s} \qquad \boldsymbol{\omega_{DE} = 0.400 \text{ rad/s} \; \circlearrowleft} \; \blacktriangleleft$$

**b. Velocity of E.** Using this angular velocity, you can easily determine the velocity of $E$:

$$v_E = \omega_{DE} r_{E/I} = (0.400 \text{ rad/s})(20 \sin 30° \text{ in.}) = 4.00 \text{ in./s}$$

$$\boldsymbol{\mathbf{v}_E = 0.333 \text{ ft/s} \; \angle 30°} \; \blacktriangleleft$$

**REFLECT and THINK:** The direction of $\omega_{DE}$ makes intuitive sense; you would expect it to be rotating counterclockwise at the instant shown. You could have also solved this problem using vector equations.

# SOLVING PROBLEMS
# ON YOUR OWN

In this section, we introduced the **instantaneous center of rotation** in plane motion. This provides us with an alternative way of solving problems involving the *velocities* of the various points of a body in plane motion (Sample Probs. 15.9, 15.10, and 15.11). As its name suggests, the instantaneous center of rotation is the point about which you can assume a body is rotating at a given instant; you can use the instantaneous center to determine the velocity of any point on the body at that instant in time.

**A. To determine the instantaneous center of rotation** of a body in plane motion, you should use one of the following procedures.

**1. If you know both the velocity $v_A$ of a point $A$ and the angular velocity $\omega$ of the body** (Fig. 15.18):

   **a. Draw a sketch of the body,** showing point $A$, its velocity $v_A$, and the angular velocity $\omega$ of the body.

   **b. From A draw a line perpendicular to $v_A$** on the side of $v_A$ from which this velocity is viewed as having *the same sense as $\omega$.*

   **c. Locate the instantaneous center $C$** on this line at a distance $r = v_A/\omega$ from point $A$.

**2. If you know the directions of the velocities of two points $A$ and $B$ and they are different** (Fig. 15.19$a$):

   **a. Draw a sketch of the body** showing points $A$ and $B$ and their velocities $v_A$ and $v_B$.

   **b. From $A$ and $B$ draw lines perpendicular to $v_A$ and $v_B$, respectively.** The instantaneous center $C$ is located at the point where the two lines intersect.

   **c. If you know the velocity of one of the two points,** you can determine the angular velocity of the body at that instant in time. For example, if you know $v_A$, you can write $\omega = v_A/AC$, where $AC$ is the distance from point $A$ to the instantaneous center $C$.

**3. If you know the velocities of two points $A$ and $B$ and both are perpendicular to the line $AB$** (Fig. 15.19$b$):

   **a. Draw a sketch of the body,** showing points $A$ and $B$ with their velocities $v_A$ and $v_B$ *drawn to scale.*

   **b. Draw a line through points $A$ and $B$, and another line** through the tips of the vectors $v_A$ and $v_B$. The instantaneous center $C$ is located at the point where the two lines intersect.

   **c. Obtain the angular velocity of the body** by either dividing $v_A$ by $AC$ or $v_B$ by $BC$.

   **d. If the velocities $v_A$ and $v_B$ have the same magnitude,** the two lines drawn in part $b$ do not intersect; the instantaneous center $C$ is at an infinite distance. The angular velocity $\omega$ is zero and the body is in translation.

*(continued)*

**B. Once you have determined the instantaneous center and the angular velocity** of a body, you can determine the velocity $\mathbf{v}_P$ of any point $P$ of the body in the following way.

**1. Draw a sketch of the body,** showing point $P$, the instantaneous center of rotation $C$, and the angular velocity $\boldsymbol{\omega}$.

**2. Draw a line from $P$ to the instantaneous center $C$** and measure or calculate the distance from $P$ to $C$.

**3. The velocity $\mathbf{v}_P$ is a vector perpendicular to the line $PC$,** of the same sense as $\boldsymbol{\omega}$, and with a magnitude of $v_P = (PC)\omega$.

**Finally, keep in mind** that the instantaneous center of rotation can be used *only* to determine velocities at a specific instant in time. It cannot be used to determine accelerations.

# Problems

## CONCEPT QUESTIONS

**15.CQ5** The disk rolls without sliding on the fixed horizontal surface. At the instant shown, the instantaneous center of zero velocity for rod *AB* would be located in which region?

a. Region 1
b. Region 2
c. Region 3
d. Region 4
e. Region 5
f. Region 6

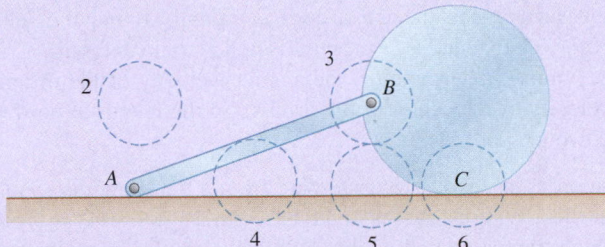

**Fig. P15.CQ5**

**15.CQ6** Bar *BDE* is pinned to two links, *AB* and *CD*. At the instant shown, the angular velocities of link *AB*, link *CD*, and bar *BDE* are $\omega_{AB}$, $\omega_{CD}$, and $\omega_{BDE}$, respectively. Which of the following statements concerning the angular speeds of the three objects is true at this instant?

a. $\omega_{AB} = \omega_{CD} = \omega_{BDE}$
b. $\omega_{BDE} > \omega_{AB} > \omega_{CD}$
c. $\omega_{AB} = \omega_{CD} > \omega_{BDE}$
d. $\omega_{AB} > \omega_{CD} > \omega_{BDE}$
e. $\omega_{CD} > \omega_{AB} > \omega_{BDE}$

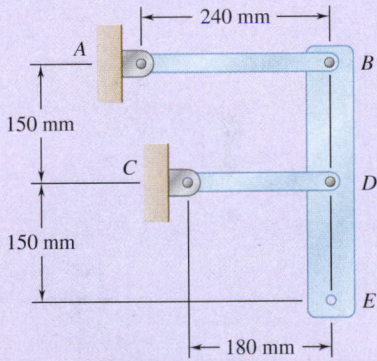

**Fig. P15.CQ6**

## END-OF-SECTION PROBLEMS

**15.73** A juggling club is thrown vertically into the air. The center of gravity *G* of the 20-in. club is located 12 in. from the knob. Knowing that at the instant shown, *G* has a velocity of 4 ft/s upward and the club has an angular velocity of 30 rad/s counterclockwise, determine (*a*) the speeds of points *A* and *B*, (*b*) the location of the instantaneous center of rotation.

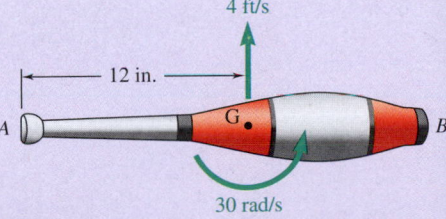

**Fig. P15.73**

**15.74** At the instant shown during deceleration, the velocity of an automobile is 40 ft/s to the right. Knowing that the velocity of the contact point *A* of the wheel with the ground is 5 ft/s to the right, determine (*a*) the instantaneous center of rotation of the wheel, (*b*) the velocity of point *B*, (*c*) the velocity of point *D*.

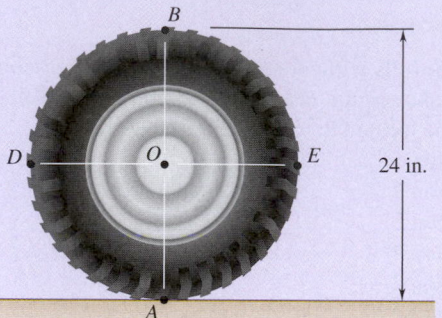

Fig. P15.74

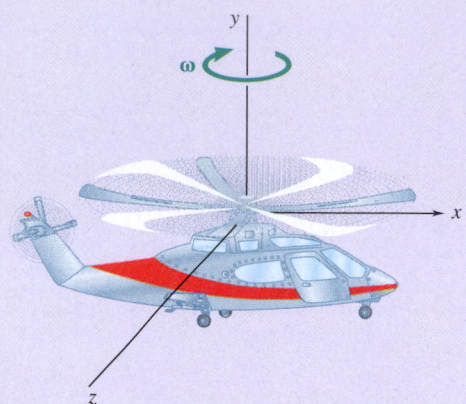

Fig. P15.75

**15.75** A helicopter moves horizontally in the *x* direction at a speed of 120 mi/h. Knowing that the main blades rotate clockwise when viewed from above with an angular velocity of 180 rpm, determine the instantaneous axis of rotation of the main blades.

**15.76 and 15.77** A 60-mm-radius drum is rigidly attached to a 100-mm-radius drum as shown. One of the drums rolls without sliding on the surface shown, and a cord is wound around the other drum. Knowing that end *E* of the cord is pulled to the left with a velocity of 120 mm/s, determine (*a*) the angular velocity of the drums, (*b*) the velocity of the center of the drums, (*c*) the length of cord wound or unwound per second.

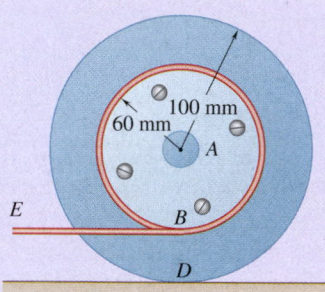

Fig. P15.76

**15.78** In order to uncoil electrical wire from a 0.6-m-radius spool fixed to a truck, a worker drives to the left with a speed of $v_A = 5$ m/s. At the same time, a second worker holds the cable as he walks to the right with a speed of $v_B = 3$ m/s. Knowing that at the instant shown the thickness of wire on the spool is 40 mm, determine (*a*) the instantaneous center of rotation of the spool, (*b*) the velocity of point *D* on the inside of the spool.

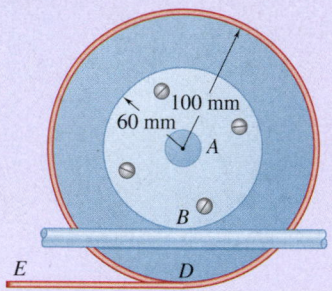

Fig. P15.77

**15.79** In order to uncoil electrical wire from a spool fixed to a truck, a worker drives to the left with a speed of $v_A = 4$ m/s. At the same time, a second worker holds the cable as he walks to the right. Knowing that at the instant shown the thickness of wire on the spool is 50 mm and that the 0.6-m-radius spool has an angular velocity of 8 rad/s, determine (*a*) the instantaneous center of rotation of the spool, (*b*) the velocity of point *D* on the spool, (*c*) the speed of the second worker *B*.

Fig. P15.78 and P15.79

**15.80** The arm *ABC* rotates with an angular velocity of 4 rad/s counter-clockwise. Knowing that the angular velocity of the intermediate gear *B* is 8 rad/s counterclockwise, determine (*a*) the instantaneous centers of rotation of gears *A* and *C*, (*b*) the angular velocities of gears *A* and *C*.

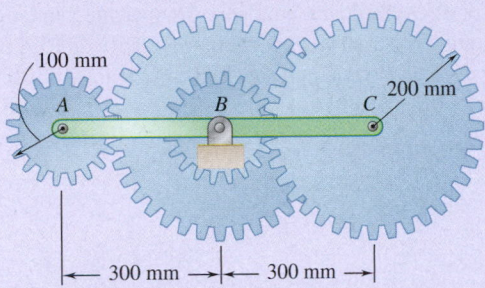

**Fig. P15.80**

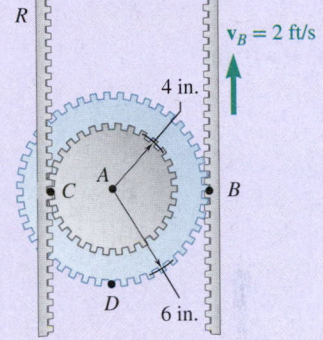

**Fig. P15.81**

**15.81** The double gear rolls on the stationary left rack *R*. Knowing that the rack on the right has a constant velocity of 2 ft/s, determine (*a*) the angular velocity of the gear, (*b*) the velocities of points *A* and *D*.

**15.82** An overhead door is guided by wheels at *A* and *B* that roll in horizontal and vertical tracks. Knowing that when $\theta = 40°$ the velocity of wheel *B* is 1.5 ft/s upward, determine (*a*) the angular velocity of the door, (*b*) the velocity of end *D* of the door.

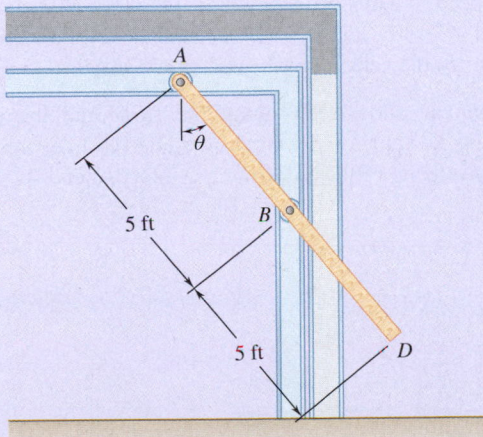

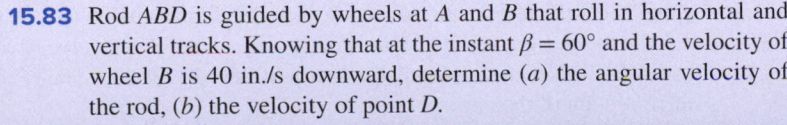

**Fig. P15.82**

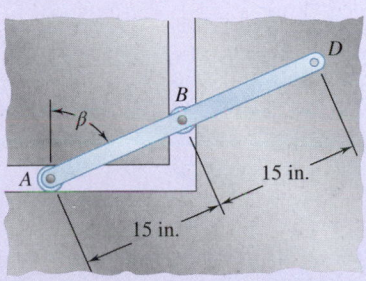

**Fig. P15.83**

**15.83** Rod *ABD* is guided by wheels at *A* and *B* that roll in horizontal and vertical tracks. Knowing that at the instant $\beta = 60°$ and the velocity of wheel *B* is 40 in./s downward, determine (*a*) the angular velocity of the rod, (*b*) the velocity of point *D*.

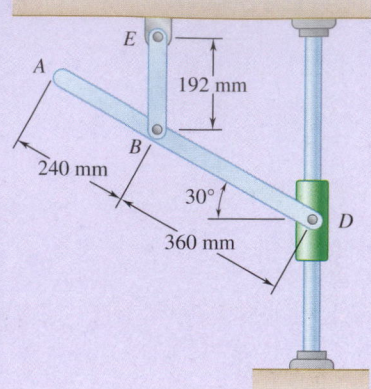

**Fig. P15.84 and P15.85**

**15.84** Knowing that at the instant shown the angular velocity of rod $BE$ is 4 rad/s counterclockwise, determine (*a*) the angular velocity of rod $AD$, (*b*) the velocity of collar $D$, (*c*) the velocity of point $A$.

**15.85** Knowing that at the instant shown the velocity of collar $D$ is 1.6 m/s upward, determine (*a*) the angular velocity of rod $AD$, (*b*) the velocity of point $B$, (*c*) the velocity of point $A$.

**15.86** A motor at $O$ drives the windshield wiper mechanism so that $OA$ has a constant counterclockwise angular velocity of 15 rpm. Knowing that at the instant shown linkage $OA$ is vertical, $\theta = 30°$, and $\beta = 15°$, determine (*a*) the angular velocity of bar $AB$, (*b*) the velocity of the center of bar $AB$.

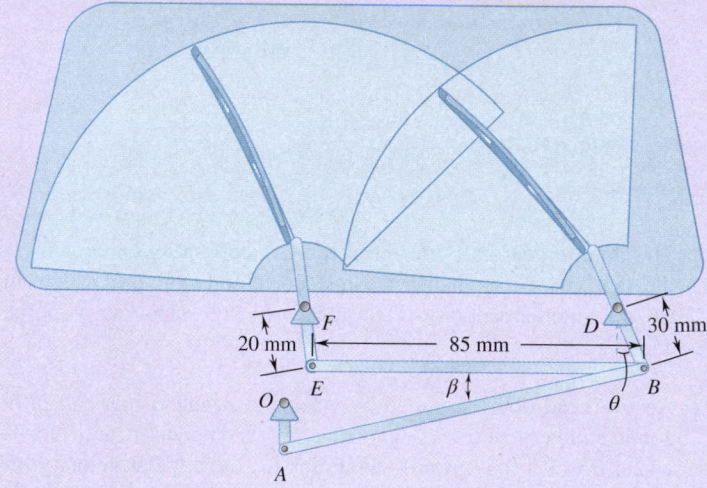

**Fig. P15.86 and P15.87**

**15.87** A motor at $O$ drives the windshield wiper mechanism so that point $B$ has a speed of 2 m/s. Knowing that at the instant shown linkage $OA$ is vertical, $\theta = 30°$, and $\beta = 15°$, determine (*a*) the angular velocity of bar $OA$, (*b*) the velocity of the center of bar $AB$.

**15.88** Rod $AB$ can slide freely along the floor and the inclined plane. Denoting the velocity of point $A$ by $\mathbf{v}_A$, derive an expression for (*a*) the angular velocity of the rod, (*b*) the velocity of end $B$.

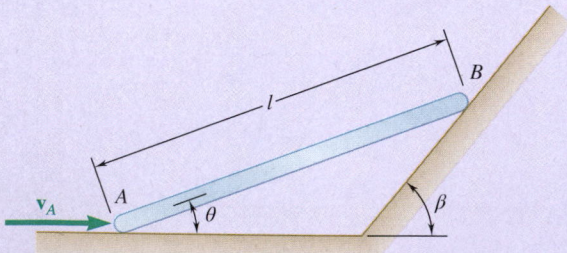

**Fig. P15.88**

**15.89** Small wheels have been attached to the ends of bar $AB$ and roll freely along the surfaces shown. Knowing that the velocity of wheel $B$ is 7.5 ft/s to the right at the instant shown, determine (*a*) the velocity of end $A$ of the bar, (*b*) the angular velocity of the bar, (*c*) the velocity of the midpoint of the bar.

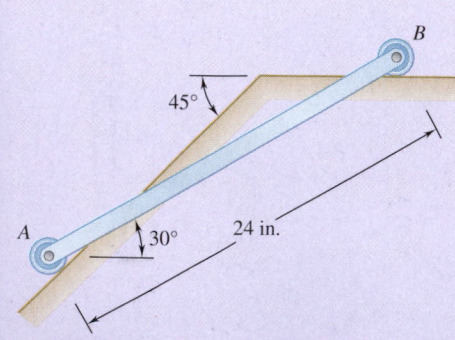

**Fig. P15.89**

**15.90** Two slots have been cut in plate *FG* and the plate has been placed so that the slots fit two fixed pins *A* and *B*. Knowing that at the instant shown the angular velocity of crank *DE* is 6 rad/s clockwise, determine (*a*) the velocity of point *F*, (*b*) the velocity of point *G*.

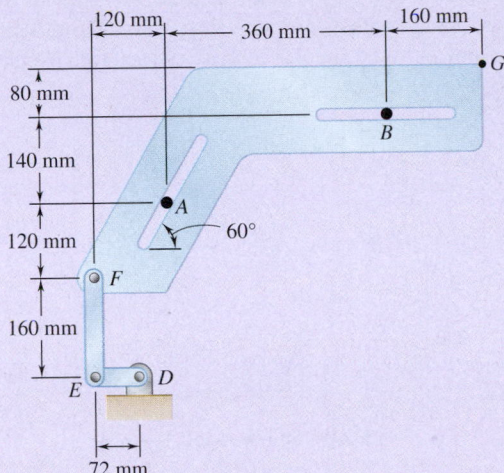

Fig. P15.90

**15.91** The disk is released from rest and rolls down the incline. Knowing that the speed of *A* is 1.2 m/s when $\theta = 0°$, determine at that instant (*a*) the angular velocity of the rod, (*b*) the velocity of *B*. (Only portions of the two tracks are shown.)

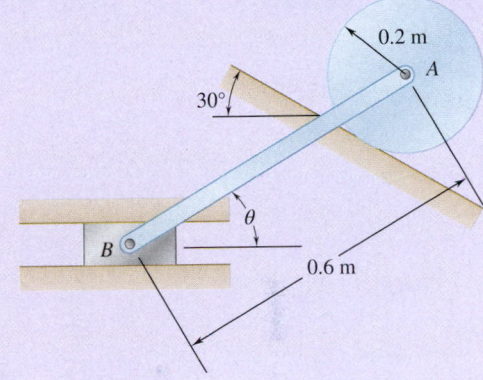

Fig. P15.91

**15.92** The pin at *B* is attached to member *ABD* and can slide freely along the slot cut in the fixed plate. Knowing that at the instant shown the angular velocity of arm *DE* is 3 rad/s clockwise, determine (*a*) the angular velocity of member *ABD*, (*b*) the velocity of point *A*.

**15.93** Two identical rods *ABF* and *DBE* are connected by a pin at *B*. Knowing that at the instant shown the velocity of point *D* is 200 mm/s upward, determine the velocity of (*a*) point *E*, (*b*) point *F*.

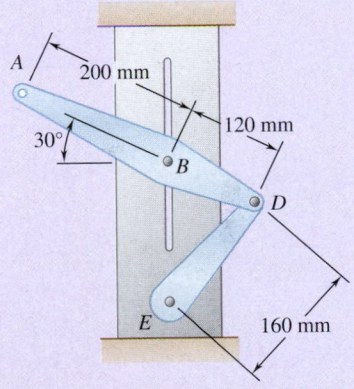

Fig. P15.92

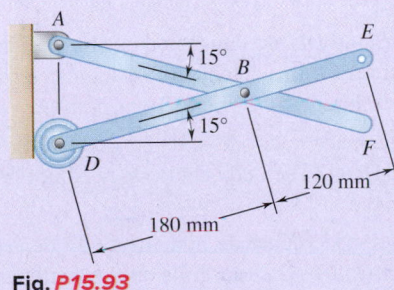

Fig. P15.93

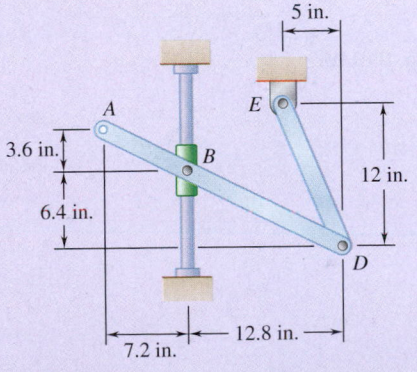

Fig. P15.94

**15.94** Arm *ABD* is connected by pins to a collar at *B* and to crank *DE*. Knowing that the velocity of collar *B* is 16 in./s upward, determine (*a*) the angular velocity of arm *ABD*, (*b*) the velocity of point *A*.

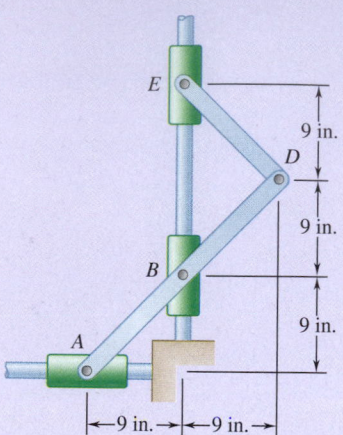

Fig. P15.95

**15.95** Two rods *ABD* and *DE* are connected to three collars as shown. Knowing that the angular velocity of *ABD* is 5 rad/s clockwise, determine at the instant shown (*a*) the angular velocity of *DE*, (*b*) the velocity of collar *E*.

**15.96** Two 500-mm rods are pin-connected at *D* as shown. Knowing that *B* moves to the left with a constant velocity of 360 mm/s, determine at the instant shown (*a*) the angular velocity of each rod, (*b*) the velocity of *E*.

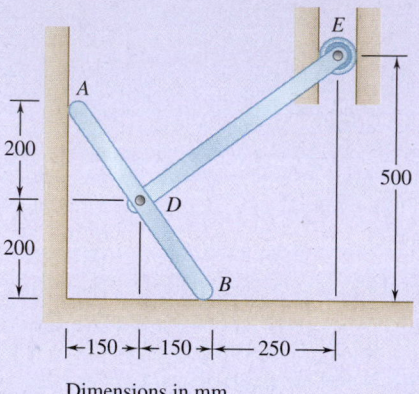

Dimensions in mm

Fig. P15.96

**15.97** At the instant shown, the velocity of collar *A* is 0.4 m/s to the right and the velocity of collar *B* is 1 m/s to the left. Determine (*a*) the angular velocity of bar *AD*, (*b*) the angular velocity of bar *BD*, (*c*) the velocity of point *D*.

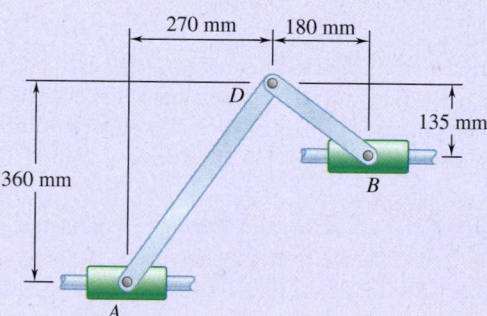

Fig. P15.97

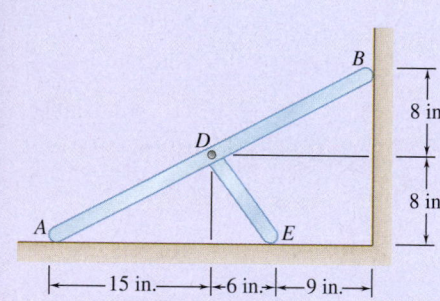

Fig. P15.98

**15.98** Two rods *AB* and *DE* are connected as shown. Knowing that point *B* moves downward with a velocity of 60 in./s, determine (*a*) the angular velocity of each rod, (*b*) the velocity of point *E*.

**15.99** Describe the space centrode and the body centrode of rod *ABD* of Prob. 15.83. (*Hint:* The body centrode need not lie on a physical portion of the rod.)

**15.100** Describe the space centrode and the body centrode of the gear of Sample Prob. 15.6 as the gear rolls on the stationary horizontal rack.

**15.101** Using the method of Sec. 15.3, solve Prob. 15.60.

**15.102** Using the method of Sec. 15.3, solve Prob. 15.64.

**15.103** Using the method of Sec. 15.3, solve Prob. 15.65.

**15.104** Using the method of Sec. 15.3, solve Prob. 15.38.

# 15.4   GENERAL PLANE MOTION: ACCELERATION

We saw in Sec. 15.2A that any plane motion can be replaced by a translation defined by the motion of an arbitrary reference point $A$ and a simultaneous rotation about $A$. We used this property in Sec. 15.2B to determine the velocity of the various points of a moving rigid body. We now use this same property to determine the acceleration of the points of the body.

## 15.4A   Absolute and Relative Acceleration in Plane Motion

We first recall that the absolute acceleration $\mathbf{a}_B$ of a particle of the rigid body can be obtained from the relative acceleration formula derived in Sec. 11.4D,

$$\mathbf{a}_B = \mathbf{a}_A + \mathbf{a}_{B/A} \tag{15.21}$$

where the right-hand side represents a vector sum. The acceleration $\mathbf{a}_A$ corresponds to the translation of the rigid body with $A$. The relative acceleration $\mathbf{a}_{B/A}$ is associated with the rotation of the body about $A$ and is measured with respect to axes centered at $A$ and with fixed orientation.

Recall from Sec. 15.1B that we can resolve the relative acceleration $\mathbf{a}_{B/A}$ into two components: a **tangential component** $(\mathbf{a}_{B/A})_t$ perpendicular to the line $AB$ and a **normal component** $(\mathbf{a}_{B/A})_n$ directed toward $A$ (Fig. 15.22). We denote the position vector of $B$ relative to $A$ by $\mathbf{r}_{B/A}$ and the angular velocity and angular acceleration of the rigid body with respect to axes of fixed orientation by $\omega\mathbf{k}$ and $\alpha\mathbf{k}$, respectively. Then we have

$$\begin{aligned}(\mathbf{a}_{B/A})_t &= \alpha\mathbf{k} \times \mathbf{r}_{B/A} & (a_{B/A})_t &= r\alpha \\ (\mathbf{a}_{B/A})_n &= -\omega^2\mathbf{r}_{B/A} & (a_{B/A})_n &= r\omega^2\end{aligned} \tag{15.22}$$

where $r$ is the distance from $A$ to $B$. Substituting the expressions obtained for the tangential and normal components of $\mathbf{a}_{B/A}$ into Eq. (15.21), we also have

**Relative acceleration for two points on a rigid body**

$$\mathbf{a}_B = \mathbf{a}_A + \alpha\mathbf{k} \times \mathbf{r}_{B/A} - \omega^2\mathbf{r}_{B/A} \tag{15.21'}$$

**Photo 15.6**   The central gear rotates about a fixed axis and is pin-connected to three bars in general plane motion. ©Lawrence Manning/Corbis RF

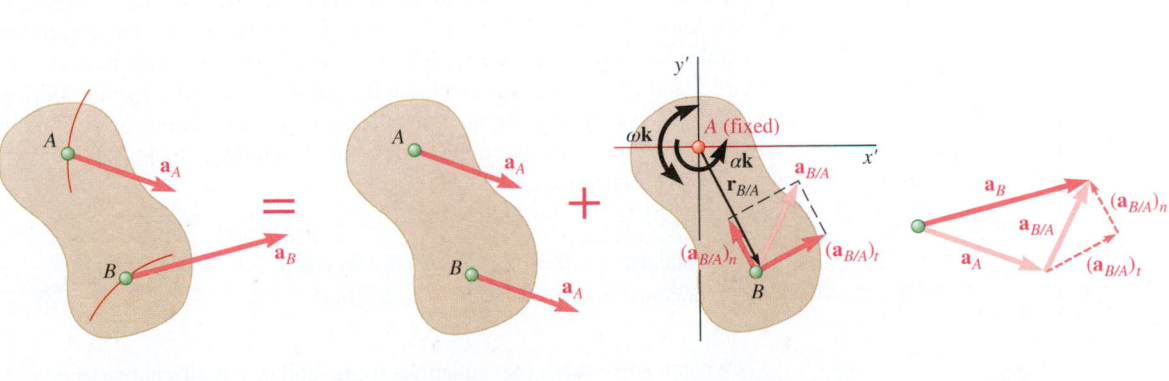

|Plane motion|=|Translation with $A$|+|Rotation about $A$|

**Fig. 15.22**   Pictorial representation of the vector equation relating the acceleration of two points on a rigid body undergoing general plane motion.

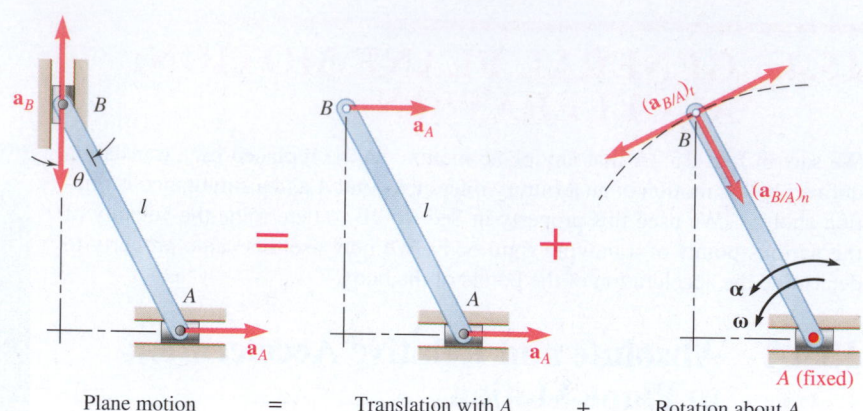

Plane motion    =    Translation with A    +    Rotation about A

**Fig. 15.23**  For a sliding rod in general plane motion, the acceleration of point B relative to point A may have a tangential component in either direction perpendicular to the rod. The normal acceleration of B relative to A will always point toward A.

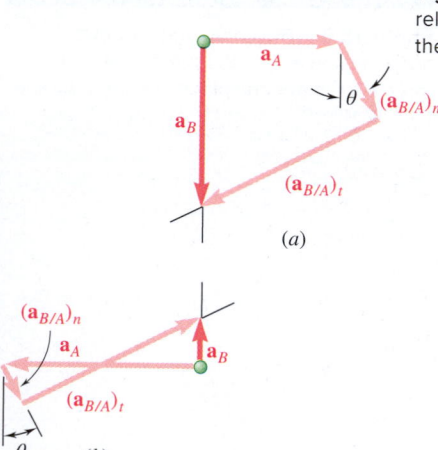

(a)

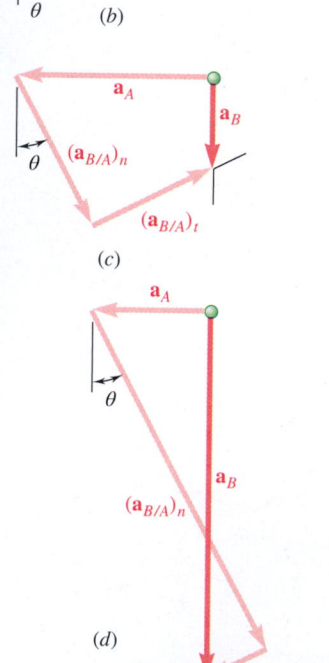

(b)

(c)

(d)

**Fig. 15.24**  Four possible vector polygons for the acceleration of the sliding rod.

As an example, let us again consider the rod $AB$ whose ends slide along a horizontal and a vertical track (Fig. 15.23). Assuming that we know the velocity $\mathbf{v}_A$ and the acceleration $\mathbf{a}_A$ of $A$, we propose to determine the acceleration $\mathbf{a}_B$ of $B$ and the angular acceleration $\boldsymbol{\alpha}$ of the rod. Choosing $A$ as a reference point, the given motion is equivalent to a translation with $A$ and a rotation about $A$. The absolute acceleration of $B$ must be equal to the sum

$$\mathbf{a}_B = \mathbf{a}_A + \mathbf{a}_{B/A}$$
$$= \mathbf{a}_A + (\mathbf{a}_{B/A})_n + (\mathbf{a}_{B/A})_t \qquad (15.23)$$

where $(\mathbf{a}_{B/A})_n$ has magnitude $l\omega^2$ and is *directed toward A*, while $(\mathbf{a}_{B/A})_t$ has the magnitude $l\alpha$ and is perpendicular to $AB$. Note that there is no way to tell whether the tangential component $(\mathbf{a}_{B/A})_t$ is directed to the left or to the right, and therefore, both possible directions for this component are indicated in Fig. 15.23. Similarly, both possible senses for $\mathbf{a}_B$ are indicated, because we do not know whether point $B$ is accelerated upward or downward.

We can illustrate Eq. (15.23) geometrically. Fig. 15.24 shows four different vector polygons, depending upon the sense of $\mathbf{a}_A$ and the relative magnitudes of $\mathbf{a}_A$ and $(\mathbf{a}_{B/A})_n$. To determine $a_B$ and $\alpha$ from one of these diagrams, we must know not only $a_A$ and $\theta$ but also $\omega$. Therefore, we need to determine the angular velocity of the rod separately, by one of the methods indicated in Secs. 15.2 and 15.3. Then we can obtain the values of $a_B$ and $\alpha$ by considering successively the $x$ and $y$ components of the vectors shown in Fig. 15.24. In the case of polygon $a$, we are assuming that $\boldsymbol{\alpha}$ is in the counter-clockwise direction and $\mathbf{a}_B$ is down. Therefore, we have

$\xrightarrow{+} x$ components:     $0 = a_A + l\omega^2 \sin\theta - l\alpha \cos\theta$

$+\uparrow y$ components:     $-a_B = -l\omega^2 \cos\theta - l\alpha \sin\theta$

We can solve these two equations for $a_B$ and $\alpha$. An alternative approach to drawing Fig. 15.24 is to use a vector algebra solution; that is, you substitute the vector quantities into Eq. (15.21'), take the cross product, and equate components to obtain the two scalar equations shown previously.

Clearly, the determination of accelerations is considerably more involved than the determination of velocities. Yet in the example considered here, the ends A and B of the rod were moving along straight tracks, and the diagrams drawn were relatively simple. If A and B had moved along curved tracks, it would have been necessary to resolve the accelerations $\mathbf{a}_A$ and $\mathbf{a}_B$ into normal and tangential components and the solution of the problem would have involved six different vectors.

When a mechanism consists of several moving parts that are pin-connected, we can analyze the mechanism by considering each part to be a rigid body, keeping in mind that the points at which two parts are connected must have the same absolute acceleration (see Sample Prob. 15.15). In the case of meshed gears (see Sample Prob. 15.13), the tangential components of the accelerations of the teeth in contact are equal, but their normal components are different.

## *15.4B   Analysis of Plane Motion in Terms of a Parameter

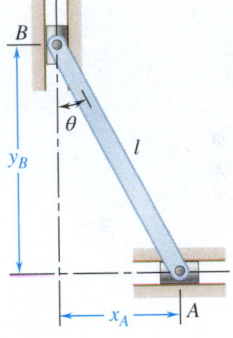

In analyzing some mechanisms, it is possible to express the coordinates x and y of all the significant points of the mechanism by means of simple analytic expressions containing a single parameter. It is sometimes advantageous in such a case to determine the absolute velocity and the absolute acceleration of the various points of the mechanism directly, because we can obtain the components of the velocity and of the acceleration of a given point by differentiating the coordinates x and y of that point.

Let us consider again the rod AB whose ends slide, respectively, in a horizontal and a vertical track (Fig. 15.25). We can express the coordinates $x_A$ and $y_B$ of the ends of the rod in terms of the angle $\theta$ that the rod forms with the vertical:

**Fig. 15.25**   The coordinates of the ends of the rod may be expressed in terms of the parameter $\theta$.

$$x_A = l \sin \theta \qquad y_B = l \cos \theta \qquad \text{(15.24)}$$

Differentiating Eqs. (15.24) twice with respect to $t$, we have

$$v_A = \dot{x}_A = l\dot{\theta} \cos \theta$$
$$a_A = \ddot{x}_A = -l\dot{\theta}^2 \sin \theta + l\ddot{\theta} \cos \theta$$

$$v_B = \dot{y}_B = -l\dot{\theta} \sin \theta$$
$$a_B = \ddot{y}_B = -l\dot{\theta}^2 \cos \theta - l\ddot{\theta} \sin \theta$$

Recalling that $\dot{\theta} = \omega$ and $\ddot{\theta} = \alpha$, we obtain

$$v_A = l\omega \cos \theta \qquad\qquad v_B = -l\omega \sin \theta \qquad \text{(15.25)}$$
$$a_A = -l\omega^2 \sin \theta + l\alpha \cos \theta \qquad a_B = -l\omega^2 \cos \theta - l\alpha \sin \theta \qquad \text{(15.26)}$$

Note that a positive sign for $v_A$ or $a_A$ indicates that the velocity $\mathbf{v}_A$ or the acceleration $\mathbf{a}_A$ is directed to the right; a positive sign for $v_B$ or $a_B$ indicates that $\mathbf{v}_B$ or $\mathbf{a}_B$ is directed upward. We can use Eqs. (15.25) to determine, for example, $v_B$ and $\omega$ when $v_A$ and $\theta$ are known. Substituting for $\omega$ in Eqs. (15.26), we can then determine $a_B$ and $\alpha$ if we know $a_A$.

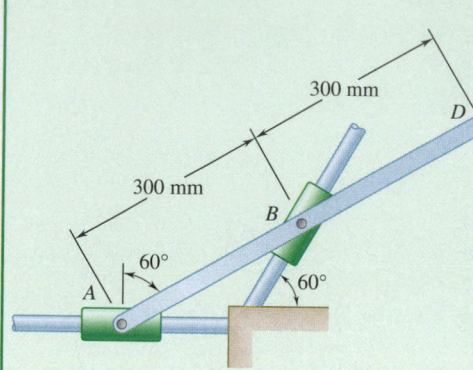

## Sample Problem 15.12

Collars $A$ and $B$ are pin-connected to bar $ABD$ and can slide along fixed rods. Knowing that, at the instant shown, the velocity of $A$ is a constant 0.9 m/s to the right, determine the angular acceleration of $AB$ and the acceleration of $B$.

**STRATEGY:** Use the kinematic equation that relates the acceleration of two points on the same rigid body. Because you know that the directions of the accelerations of $A$ and $B$ must be along the fixed rods, choose these two points to relate.

**MODELING and ANALYSIS:** Model bar $ABD$ as a rigid body. From Sample Prob. 15.5, you know $\boldsymbol{\omega} = 3.00$ rad/s $\circlearrowright$. The accelerations of $A$ and $B$ are related by

$$\mathbf{a}_B = \mathbf{a}_A + \mathbf{a}_{B/A} = \mathbf{a}_A + \boldsymbol{\alpha} \times \mathbf{r}_{B/A} - \omega^2 \mathbf{r}_{B/A}$$

Substituting in known values (Fig. 1) and assuming $\boldsymbol{\alpha} = \alpha \mathbf{k}$ gives

$$a_B \cos 60°\mathbf{i} + a_B \sin 60°\mathbf{j} = 0\mathbf{i} + \alpha \mathbf{k} \times [(0.3 \cos 30°)\mathbf{i} + (0.3 \sin 30°)\mathbf{j}]$$
$$- 3^2 [(0.3 \cos 30°)\mathbf{i} + (0.3 \sin 30°)\mathbf{j}]$$

$$0.500 a_B \mathbf{i} + 0.866 a_B \mathbf{j} = (0 - 0.15\alpha - 2.338)\mathbf{i} + (0.260\alpha - 1.350)\mathbf{j}$$

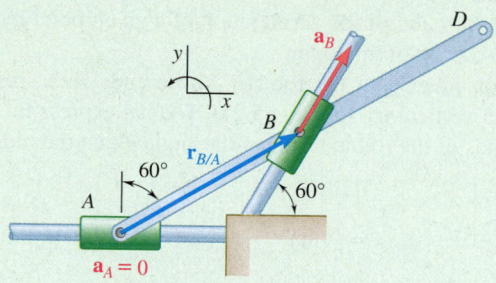

**Fig. 1** Position vector and the assumed direction of the acceleration of point $B$.

Equating components, you have

**i:** $\quad 0.500 a_B = -0.15\alpha - 2.338$

**j:** $\quad 0.866 a_B = 0.260\alpha - 1.350$

Solving these equations gives $a_B = -3.12$ m/s² and $\alpha = -5.20$ rad/s².

$$\boldsymbol{\alpha} = 5.20 \text{ rad/s}^2 \circlearrowright \quad \blacktriangleleft$$
$$\mathbf{a}_B = 3.12 \text{ m/s}^2 \nearrow 60° \quad \blacktriangleleft$$

**REFLECT and THINK:** Even though $A$ is traveling at a constant speed, bar $AB$ still has an angular acceleration, and $B$ has a linear acceleration. Just because one point on a body is moving at a constant speed doesn't mean the rest of the points on the body also have a constant speed.

## Sample Problem 15.13

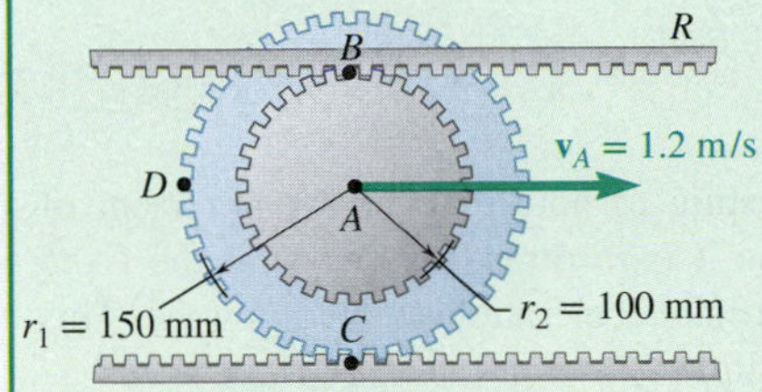

The center of the double gear of Sample Prob. 15.6 has a velocity of 1.2 m/s to the right and an acceleration of 3 m/s² to the right. Recalling that the lower rack is stationary, determine (a) the angular acceleration of the gear, (b) the acceleration of points B, C, and D of the gear.

**STRATEGY:**   The double gear is a rigid body undergoing general plane motion, so use acceleration kinematics. You can also differentiate the equation for the gear's velocity and use that to find the gear's acceleration.

**MODELING and ANALYSIS:**

**a. Angular Acceleration of the Gear.**   In Sample Prob. 15.6, you found that $x_A = -r_1\theta$ and $v_A = -r_1\omega$. Differentiating the second equation with respect to time, you obtain $a_A = -r_1\alpha$.

$$v_A = -r_1\omega \qquad 1.2\text{ m/s} = -(0.150\text{ m})\omega \qquad \omega = -8\text{ rad/s}$$
$$a_A = -r_1\alpha \qquad 3\text{ m/s}^2 = -(0.150\text{ m})\alpha \qquad \alpha = -20\text{ rad/s}^2$$

$$\boldsymbol{\alpha} = \alpha\mathbf{k} = -(20\text{ rad/s}^2)\mathbf{k} \quad \blacktriangleleft$$

**b. Accelerations.**   The relationship between the acceleration of any two points on a rigid body undergoing general plane motion is

$$\mathbf{a}_B = \mathbf{a}_A + \mathbf{a}_{B/A} = \mathbf{a}_A + (\mathbf{a}_{B/A})_t + (\mathbf{a}_{B/A})_n$$
$$= \mathbf{a}_A + \alpha\mathbf{k} \times \mathbf{r}_{B/A} - \omega^2\mathbf{r}_{B/A} \tag{1}$$

This equation indicates that the rolling motion of the gear can be thought of as a translation with A and a rotation about A (Fig. 1).

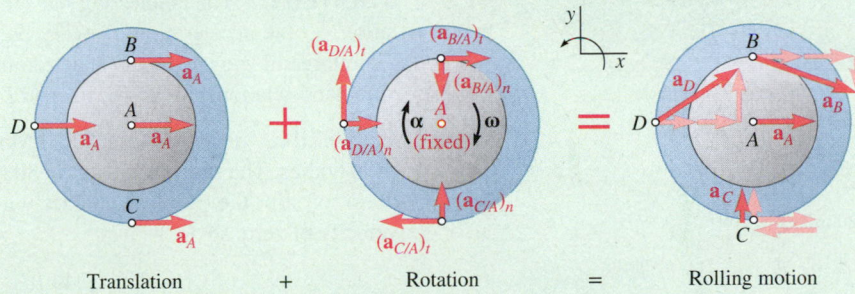

Translation        +        Rotation        =        Rolling motion

**Fig. 1**   A pictorial representation of Eq. 1.

**Acceleration of Point B.**   Substituting values into Eq. (1) gives

$$\mathbf{a}_B = \mathbf{a}_A + \mathbf{a}_{B/A} = \mathbf{a}_A + (\mathbf{a}_{B/A})_t + (\mathbf{a}_{B/A})_n$$
$$= \mathbf{a}_A + \alpha\mathbf{k} \times \mathbf{r}_{B/A} - \omega^2\mathbf{r}_{B/A}$$
$$= (3\text{ m/s}^2)\mathbf{i} - (20\text{ rad/s}^2)\mathbf{k} \times (0.100\text{ m})\mathbf{j} - (8\text{ rad/s})^2(0.100\text{ m})\mathbf{j}$$
$$= (3\text{ m/s}^2)\mathbf{i} + (2\text{ m/s}^2)\mathbf{i} - (6.40\text{ m/s}^2)\mathbf{j}$$

$$\mathbf{a}_B = 8.12\text{ m/s}^2 \; \measuredangle\; 52.0° \quad \blacktriangleleft$$

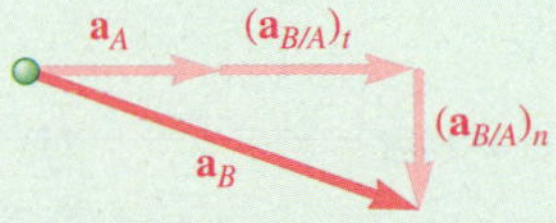

The vector triangle corresponding to this equation is shown in Fig. 2.

**Acceleration of Point C.**   Referring to Fig. 3,

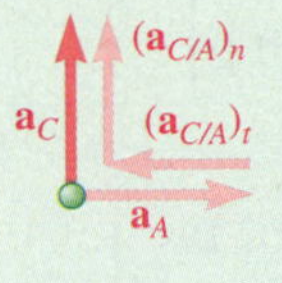

$$\mathbf{a}_C = \mathbf{a}_A + \mathbf{a}_{C/A} = \mathbf{a}_A + \alpha\mathbf{k} \times \mathbf{r}_{C/A} - \omega^2\mathbf{r}_{C/A}$$
$$= (3\text{ m/s}^2)\mathbf{i} - (20\text{ rad/s}^2)\mathbf{k} \times (-0.150\text{ m})\mathbf{j} - (8\text{ rad/s})^2(-0.150\text{ m})\mathbf{j}$$
$$= (3\text{ m/s}^2)\mathbf{i} - (3\text{ m/s}^2)\mathbf{i} + (9.60\text{ m/s}^2)\mathbf{j}$$

$$\mathbf{a}_C = 9.60\text{ m/s}^2 \uparrow \quad \blacktriangleleft$$

*(continued)*

---

$v_A = 1.2$ m/s

$r_1 = 150$ mm     $r_2 = 100$ mm

**Fig. 2**   Vector diagram relating the accelerations of A and B.

**Fig. 3**   Vector diagram of the equation relating the accelerations of A and C.

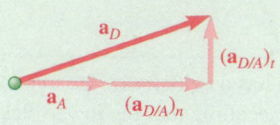

**Fig. 4** Vector diagram relating the accelerations of A and D.

### Acceleration of Point D.  (Fig. 4)

$$\mathbf{a}_D = \mathbf{a}_A + \mathbf{a}_{D/A} = \mathbf{a}_A + \alpha \mathbf{k} \times \mathbf{r}_{D/A} - \omega^2 \mathbf{r}_{D/A}$$
$$= (3 \text{ m/s}^2)\mathbf{i} - (20 \text{ rad/s}^2)\mathbf{k} \times (-0.150 \text{ m})\mathbf{i} - (8 \text{ rad/s})^2(-0.150 \text{ m})\mathbf{i}$$
$$= (3 \text{ m/s}^2)\mathbf{i} - (3 \text{ m/s}^2)\mathbf{j} + (9.60 \text{ m/s}^2)\mathbf{i}$$

$$\mathbf{a}_D = 12.95 \text{ m/s}^2 \; \measuredangle \; 13.4° \quad \blacktriangleleft$$

**REFLECT and THINK:**  It is interesting to note that the *x*-component of acceleration for point *C* is zero because it is in direct contact with the fixed lower rack. It does, however, have a normal acceleration pointed upward. This is also true for a wheel rolling without slip.

©Alen Penton/Alamy RF

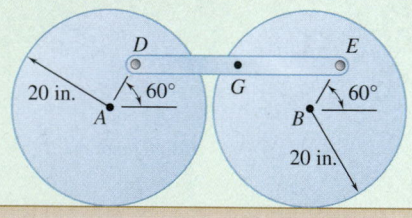

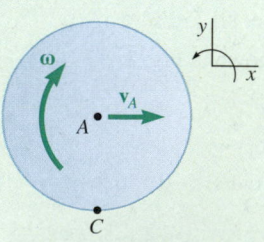

**Fig. 1** Velocity and angular velocity of the wheel.

## Sample Problem 15.14

Two adjacent identical wheels of a train can be modeled as rolling cylinders connected by a horizontal link. The distance between *A* and *D* is 10 in. Assume the wheels roll without sliding on the tracks. Knowing that the train is traveling at a constant 30 mph, determine the acceleration of the center of mass of *DE*. ▶

**STRATEGY:**  The connecting bar *DE* is undergoing curvilinear translation, so the acceleration of every point is identical; that is, $a_G = a_D$. Therefore, all you need to do is determine the acceleration of *D* using the kinematic relationship between *A* and *D*.

**MODELING and ANALYSIS:**  Model the wheels and bar *DE* as rigid bodies. The speed of *A* is $v_A = 30$ mph $= 44$ ft/s. Because the wheel does not slip, the point of contact with the ground, *C* (Fig. 1), has a velocity of zero, so

$$\omega = \frac{v_A}{r_{A/C}} = \frac{44 \text{ ft/s}}{(20/12) \text{ ft}} = 26.4 \text{ rad/s}$$

### Acceleration of D.   The acceleration of *D* is

$$\mathbf{a}_D = \mathbf{a}_A + \mathbf{a}_{D/A} = \mathbf{a}_A + \boldsymbol{\alpha} \times \mathbf{r}_{D/A} - \omega^2 \mathbf{r}_{D/A} \quad (1)$$

The train is traveling at a constant speed, so $a_A$ and $\alpha$ are both zero. Substituting known quantities into Eq. (1) gives

$$\mathbf{a}_D = 0 + 0 - (26.4 \text{ rad/s})^2 \left[ \left( \frac{10}{12}\cos 60° \text{ ft} \right)\mathbf{i} + \left( \frac{10}{12}\sin 60° \text{ ft} \right)\mathbf{j} \right]$$

$$= -(290.4 \text{ ft/s}^2)\mathbf{i} - (503.0 \text{ ft/s}^2)\mathbf{j}$$

$$\mathbf{a}_G = \mathbf{a}_D = -(290 \text{ ft/s}^2)\mathbf{i} - (503 \text{ ft/s}^2)\mathbf{j} \quad \blacktriangleleft$$

**REFLECT and THINK:**  Instead of using vector algebra, you could have recognized that the direction of $-\omega^2 \mathbf{r}_{D/A}$ is directed from *D* to *A*. So the final acceleration of *D* is simply $-\omega^2 \mathbf{r}_{D/A} \; \measuredangle \; 60°$.

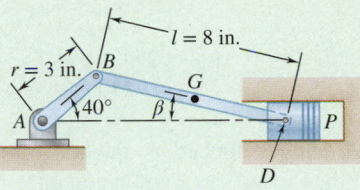

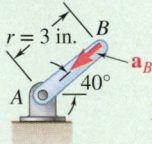

**Fig. 1** The acceleration of B is only in the normal direction.

## Sample Problem 15.15

Crank $AB$ of the engine system of Sample Prob. 15.7 has a constant clockwise angular velocity of 2000 rpm. For the crank position shown, determine the angular acceleration of the connecting rod $BD$ and the acceleration of point $D$. ▶

**STRATEGY:**   The linkage consists of two rigid bodies: crank $AB$ is rotating about a fixed axis and connecting rod $BD$ is undergoing general plane motion. Therefore, you need to use rigid-body kinematics.

### MODELING and ANALYSIS:

**Motion of Crank AB.**   Because the crank rotates about $A$ with constant $\omega_{AB} = 2000$ rpm $= 209.4$ rad/s, you have $\alpha_{AB} = 0$. The acceleration of $B$ is therefore directed toward $A$ (Fig. 1) and has the magnitude of

$$a_B = r\omega_{AB}^2 = \left(\frac{3}{12}\text{ ft}\right)(209.4 \text{ rad/s})^2 = 10{,}962 \text{ ft/s}^2$$

$$\mathbf{a}_B = 10{,}962 \text{ ft/s}^2 \,\diagdown\, 40°$$

**Motion of Connecting Rod BD.**   The angular velocity $\boldsymbol{\omega}_{BD}$ and the value of $\beta$ were obtained in Sample Prob. 15.7 using relative velocity equations:

$$\boldsymbol{\omega}_{BD} = 62.0 \text{ rad/s} \,\circlearrowleft \qquad \beta = 13.95°$$

Resolve the motion of $BD$ into a translation with $B$ and a rotation about $B$ (Fig. 2). Resolve the relative acceleration $\mathbf{a}_{D/B}$ into normal and tangential components:

$$(a_{D/B})_n = (BD)\omega_{BD}^2 = (\tfrac{8}{12}\text{ ft})(62.0 \text{ rad/s})^2 = 2563 \text{ ft/s}^2$$
$$(\mathbf{a}_{D/B})_n = 2563 \text{ ft/s}^2 \,\diagdown\, 13.95°$$
$$(a_{D/B})_t = (BD)\alpha_{BD} = (\tfrac{8}{12}\text{ ft})\alpha_{BD} = 0.6667\alpha_{BD}$$
$$(\mathbf{a}_{D/B})_t = 0.6667\alpha_{BD} \,\diagup\, 76.05°$$

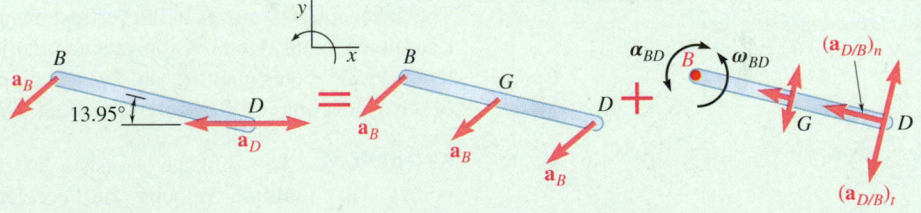

| Plane motion | = | Translation | + | Rotation |

**Fig. 2**   General plane motion is a translation plus a rotation.

Although $(\mathbf{a}_{D/B})_t$ must be perpendicular to $BD$, its sense is not known. Noting that the acceleration $\mathbf{a}_D$ must be horizontal, you have

$$\mathbf{a}_D = \mathbf{a}_B + \mathbf{a}_{D/B} = \mathbf{a}_B + (\mathbf{a}_{D/B})_n + (\mathbf{a}_{D/B})_t \tag{1}$$

$$[a_D \leftrightarrow] = [10{,}962 \,\diagdown\, 40°] + [2563 \,\diagdown\, 13.95°] + [0.6667\alpha_{BD} \,\diagup\, 76.05°]$$

Equating $x$ and $y$ components, you obtain the following scalar equations, as

$\xrightarrow{+}$ $x$ components:

$$-a_D = -10{,}962 \cos 40° - 2563 \cos 13.95° + 0.6667\alpha_{BD} \sin 13.95°$$

$+\uparrow$ $y$ components:

$$0 = -10{,}962 \sin 40° + 2563 \sin 13.95° + 0.6667\alpha_{BD} \cos 13.95°$$

*(continued)*

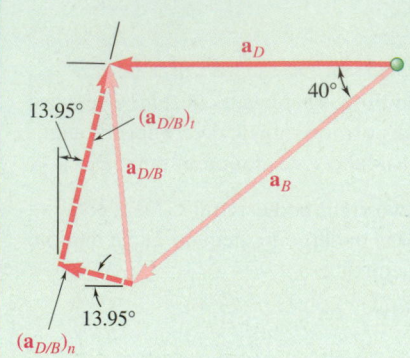

**Fig. 3** Vector polygon relating the accelerations of B and D.

Solving the equations simultaneously gives $\alpha_{BD} = +9940$ rad/s$^2$ and $a_D = +9290$ ft/s$^2$. The positive signs indicate that the senses shown on the vector polygon (Fig. 3) are correct.

$$\alpha_{BD} = 9940 \text{ rad/s}^2 \circlearrowleft \blacktriangleleft$$
$$a_D = 9290 \text{ ft/s}^2 \leftarrow \blacktriangleleft$$

**REFLECT and THINK:** In this solution, you looked at the magnitude and direction of each term in Eq. (1) and then found the $x$ and $y$ components. Alternatively, you could have assumed that $a_D$ was to the left, $\alpha_{BD}$ was positive, and then substituted in the vector quantities to get

$$\mathbf{a}_D = \mathbf{a}_B + \alpha\mathbf{k} \times \mathbf{r}_{D/B} - \omega^2\mathbf{r}_{D/B}$$

$$-a_D\mathbf{i} = -a_B \cos 40°\mathbf{i} - a_B \sin 40°\mathbf{j} + \alpha_{BD}\mathbf{k} \times (l \cos \beta\mathbf{i} - l \sin \beta\mathbf{j})$$
$$- \omega_{BD}^2(l \cos \beta\mathbf{i} - l \sin \beta\mathbf{j})$$
$$= -a_B \cos 40°\mathbf{i} - a_B \sin 40°\mathbf{j} + \alpha_{BD}\, l \cos \beta\mathbf{j} + \alpha_{BD}\, l \sin \beta\mathbf{i}$$
$$- \omega_{BD}^2\, l \cos \beta\mathbf{i} + \omega_{BD}^2\, l \sin \beta\mathbf{j}$$

Equating components gives

**i:** $-a_D = -a_B \cos 40° + \alpha_{BD}\, l \sin \beta - \omega_{BD}^2\, l \cos \beta$

**j:** $0 = -a_B \sin 40° + \alpha_{BD}\, l \cos \beta + \omega_{BD}^2\, l \sin \beta$

These are identical to the previous equations if you substitute in the numbers.

---

# Sample Problem 15.16

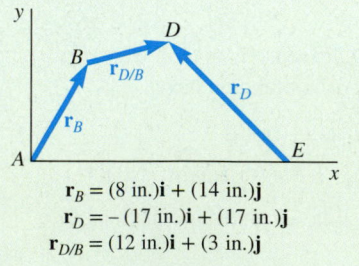

$\mathbf{r}_B = (8 \text{ in.})\mathbf{i} + (14 \text{ in.})\mathbf{j}$
$\mathbf{r}_D = -(17 \text{ in.})\mathbf{i} + (17 \text{ in.})\mathbf{j}$
$\mathbf{r}_{D/B} = (12 \text{ in.})\mathbf{i} + (3 \text{ in.})\mathbf{j}$

**Fig. 1** Position vectors for points B, D, and E.

The linkage *ABDE* moves in the vertical plane. Knowing that, in the position shown, crank *AB* has a constant angular velocity $\omega_1$ of 20 rad/s counterclockwise, determine the angular velocities and angular accelerations of the connecting rod *BD* and of the crank *DE*.

**STRATEGY:** The linkage consists of three interconnected rigid bodies. Use multiple velocity and acceleration kinematic equations to relate the motions of each body. You could solve this problem with the method used in Sample Prob. 15.15; however, we illustrate a vector approach, choosing position vectors $\mathbf{r}_B$, $\mathbf{r}_D$, and $\mathbf{r}_{D/B}$ as shown in Fig. 1.

**MODELING and ANALYSIS:**

**Velocities.** Assuming that the angular velocities of *BD* and *DE* are counterclockwise, you have

$$\boldsymbol{\omega}_{AB} = \omega_{AB}\mathbf{k} = (20 \text{ rad/s})\mathbf{k} \qquad \boldsymbol{\omega}_{BD} = \omega_{BD}\mathbf{k} \qquad \boldsymbol{\omega}_{DE} = \omega_{DE}\mathbf{k}$$

where $\mathbf{k}$ is a unit vector pointing out of the page. We can obtain the velocity of *D* by relating it to point *E*, as

$$\mathbf{v}_D = \mathbf{v}_E + \mathbf{v}_{D/E} = 0 + \omega_{DE}\mathbf{k} \times \mathbf{r}_D \qquad (1)$$

*(continued)*

We can obtain the velocity of $B$ by relating it to point $A$, as

$$\mathbf{v}_B = \mathbf{v}_A + \mathbf{v}_{B/A} = 0 + \omega_{AB}\mathbf{k} \times \mathbf{r}_B \qquad (2)$$

The relationship between the velocities of $D$ and $B$ is

$$\mathbf{v}_D = \mathbf{v}_B + \mathbf{v}_{D/B} \qquad (3)$$

Substituting Eqs. (1) and (2) into Eq. (3) and using $\mathbf{v}_{D/B} = \omega_{BD}\mathbf{k} \times \mathbf{r}_{D/B}$ gives

$$\omega_{DE}\mathbf{k} \times \mathbf{r}_D = \omega_{AB}\mathbf{k} \times \mathbf{r}_B + \omega_{BD}\mathbf{k} \times \mathbf{r}_{D/B}$$
$$\omega_{DE}\mathbf{k} \times (-17\mathbf{i} + 17\mathbf{j}) = 20\mathbf{k} \times (8\mathbf{i} + 14\mathbf{j}) + \omega_{BD}\mathbf{k} \times (12\mathbf{i} + 3\mathbf{j})$$
$$-17\omega_{DE}\mathbf{j} - 17\omega_{DE}\mathbf{i} = 160\mathbf{j} - 280\mathbf{i} + 12\omega_{BD}\mathbf{j} - 3\omega_{BD}\mathbf{i}$$

Equating the coefficients of the unit vectors $\mathbf{i}$ and $\mathbf{j}$, the following two scalar equations are

$$-17\omega_{DE} = -280 - 3\omega_{BD}$$
$$-17\omega_{DE} = +160 + 12\omega_{BD}$$

Solving these gives you $\quad \boldsymbol{\omega}_{BD} = -(29.33\,\text{rad/s})\mathbf{k} \qquad \boldsymbol{\omega}_{DE} = (11.29\,\text{rad/s})\mathbf{k}$ ◄

**Accelerations.** At the instant considered, crank $AB$ has a constant angular velocity, so you have

$$\boldsymbol{\alpha}_{AB} = 0 \qquad \boldsymbol{\alpha}_{BD} = \alpha_{BD}\mathbf{k} \qquad \boldsymbol{\alpha}_{DE} = \alpha_{DE}\mathbf{k} \qquad (4)$$
$$\mathbf{a}_D = \mathbf{a}_B + \mathbf{a}_{D/B}$$

Evaluate each term of Eq. (4) separately:

Bar $DE$: 
$$\mathbf{a}_D = \alpha_{DE}\mathbf{k} \times \mathbf{r}_D - \omega_{DE}^2\mathbf{r}_D$$
$$= \alpha_{DE}\mathbf{k} \times (-17\mathbf{i} + 17\mathbf{j}) - (11.29)^2(-17\mathbf{i} + 17\mathbf{j})$$
$$= -17\alpha_{DE}\mathbf{j} - 17\alpha_{DE}\mathbf{i} + 2170\mathbf{i} - 2170\mathbf{j}$$

Bar $AB$:
$$\mathbf{a}_B = \alpha_{AB}\mathbf{k} \times \mathbf{r}_B - \omega_{AB}^2\mathbf{r}_B = 0 - (20)^2(8\mathbf{i} + 14\mathbf{j})$$
$$= -3200\mathbf{i} - 5600\mathbf{j}$$

Bar $BD$:
$$\mathbf{a}_{D/B} = \alpha_{BD}\mathbf{k} \times \mathbf{r}_{D/B} - \omega_{BD}^2\mathbf{r}_{D/B}$$
$$= \alpha_{BD}\mathbf{k} \times (12\mathbf{i} + 3\mathbf{j}) - (29.33)^2(12\mathbf{i} + 3\mathbf{j})$$
$$= 12\alpha_{BD}\mathbf{j} - 3\alpha_{BD}\mathbf{i} - 10{,}320\mathbf{i} - 2580\mathbf{j}$$

Substituting into Eq. (4) and equating the coefficients of $\mathbf{i}$ and $\mathbf{j}$, you obtain

$$-17\alpha_{DE} + 3\alpha_{BD} = -15{,}690$$
$$-17\alpha_{DE} - 12\alpha_{BD} = -6010$$

Solving these gives you $\quad \boldsymbol{\alpha}_{BD} = -(645\,\text{rad/s}^2)\mathbf{k} \qquad \boldsymbol{\alpha}_{DE} = (809\,\text{rad/s}^2)\mathbf{k}$ ◄

**REFLECT and THINK:** The vector approach is preferred when there are more than two linkages. It is a very methodic approach and is easier to program when simulating mechanism movement over time.

# SOLVING PROBLEMS
# ON YOUR OWN

This section was devoted to determining the accelerations of the points of a rigid body in plane motion. As you did previously for velocities, you will again consider the plane motion of a rigid body as the sum of two motions, namely, a translation and a rotation.

To solve a problem involving accelerations in plane motion, use the following steps.

**1. Determine the angular velocity of the body.** To find $\omega$, you can either

**a.** Consider the motion of the body as the sum of a translation and a rotation, as you did in Sec. 15.2, or

**b.** Use the vector approach, as you did in Sec. 15.2, or the instantaneous center of rotation of the body, as you did in Sec. 15.3. However, keep in mind that you cannot use the instantaneous center to determine accelerations.

**2. A diagram** may be helpful to visualize the kinematics of the rigid bodies. The diagram will include the following diagrams (Fig. 15.22):

**a. Plane motion diagram.** Draw a sketch of the body, including all dimensions, as well as the angular velocity $\omega$. Show the angular acceleration $\alpha$ with its magnitude and sense if you know them. Also show those points for which you know or seek the accelerations, indicating all that you know about these accelerations.

**b. Translation diagram.** Select a reference point $A$ for which you know the direction, the magnitude, or a component of the acceleration $\mathbf{a}_A$. Draw a second diagram showing the body in translation with each point having the same acceleration as point $A$.

**c. Rotation diagram.** Considering point $A$ as a fixed reference point, draw a third diagram showing the body in rotation about $A$. Indicate the normal and tangential components of the relative accelerations of other points, such as the components $(\mathbf{a}_{B/A})_n$ and $(\mathbf{a}_{B/A})_t$ of the acceleration of point $B$ with respect to point $A$.

**3. Write the relative-acceleration formula relating two points of interest on the body being analyzed**

$$\mathbf{a}_B = \mathbf{a}_A + \mathbf{a}_{B/A} \quad \text{or} \quad \mathbf{a}_B = \mathbf{a}_A + (\mathbf{a}_{B/A})_n + (\mathbf{a}_{B/A})_t$$

**a. Graphical approach.** Select a point for which you know the direction, the magnitude, or a component of the acceleration and draw a vector diagram of the equation (Sample Prob. 15.15). Starting at the same point, draw all known acceleration components in tip-to-tail fashion for each member of the equation. Complete the diagram by drawing the two remaining vectors in appropriate directions and in such a way that the two sums of vectors end at a common point.

**b. Vector approach.** For a single rigid body, it is straightforward to apply

$$\mathbf{a}_B = \mathbf{a}_A + \boldsymbol{\alpha}_{AB} \times \mathbf{r}_{B/A} - \omega_{AB}^2 \mathbf{r}_{B/A}$$

For linkage type problems, you will need to write multiple relative acceleration equations relating the accelerations of points along the linkage (Sample Prob. 15.16).

**4. The analysis of plane motion in terms of a parameter** completed this section. This method should be used *only* if it is possible to express the coordinates $x$ and $y$ of all significant points of the body in terms of a single parameter (Sec. 15.4B). By differentiating the coordinates $x$ and $y$ of a given point twice with respect to $t$, you can determine the rectangular components of the absolute velocity and absolute acceleration of that point.

# Problems

**15.CQ7** A rear-wheel-drive car starts from rest and accelerates to the left so that the tires do not slip on the road. What is the direction of the acceleration of the point on the tire in contact with the road; that is, point $A$?

**a.** ← **b.** ↖ **c.** ↑ **d.** ↓ **e.** ↗

**Fig. P15.CQ7**

**END-OF-SECTION PROBLEMS**

**15.105** A 5-m steel beam is lowered by means of two cables unwinding at the same speed from overhead cranes. As the beam approaches the ground, the crane operators apply brakes to slow the unwinding motion. At the instant considered, the deceleration of the cable attached at $B$ is 2.5 m/s$^2$, while that of the cable attached at $D$ is 1.5 m/s$^2$. Determine (a) the angular acceleration of the beam, (b) the acceleration of points $A$ and $E$.

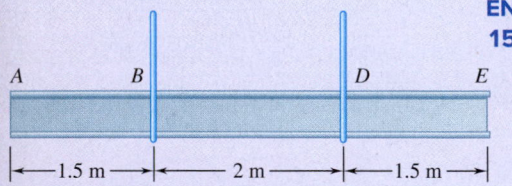

**Fig. P15.105 and P15.106**

**15.106** For a 5-m steel beam $AE$, the acceleration of point $A$ is 2 m/s$^2$ downward and the angular acceleration of the beam is 1.2 rad/s$^2$ counterclockwise. Knowing that at the instant considered the angular velocity of the beam is zero, determine the acceleration of (a) cable $B$, (b) cable $D$.

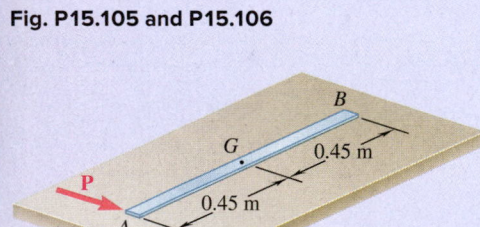

**Fig. P15.107 and P15.108**

**15.107** A 900-mm rod rests on a horizontal table. A force **P** applied as shown produces the following accelerations: $\alpha_A = 3.6$ m/s$^2$ to the right, $\alpha = 6$ rad/s$^2$ counterclockwise as viewed from above. Determine the acceleration of (a) point $G$, (b) point $B$.

**15.108** In Prob. 15.107, determine the point of the rod that (a) has no acceleration, (b) has an acceleration of 2.4 m/s$^2$ to the right.

**15.109** Knowing that point $A$ is moving to the right at a constant speed of 20 in./s, determine the acceleration of (a) point $B$, (b) point $D$.

**15.110** Knowing that at the instant shown crank $BC$ has a constant angular velocity of 45 rpm clockwise, determine the acceleration of (a) point $A$, (b) point $D$.

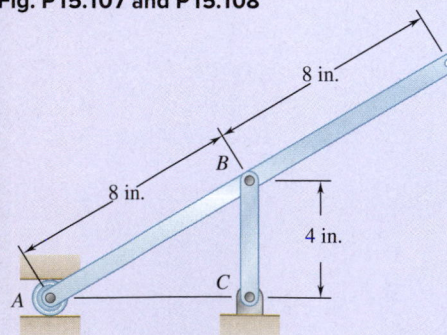

**Fig. P15.109 and P15.110**

**15.111** An automobile travels to the left at a constant speed of 90 km/h. Knowing that the diameter of the wheel is 650 mm, determine the acceleration of (*a*) point *B*, (*b*) point *C*, (*c*) point *D*.

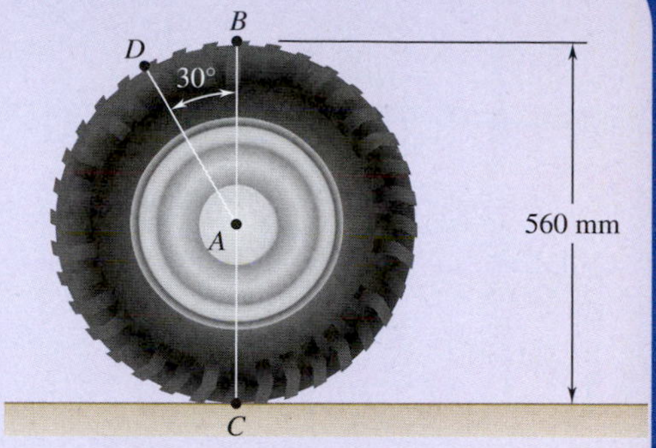

**15.112** The 18-in.-radius flywheel is rigidly attached to a 1.5-in.-radius shaft that can roll along parallel rails. Knowing that at the instant shown the center of the shaft has a velocity of 1.2 in./s and an acceleration of 0.5 in./s², both directed down to the left, determine the acceleration of (*a*) point *A*, (*b*) point *B*.

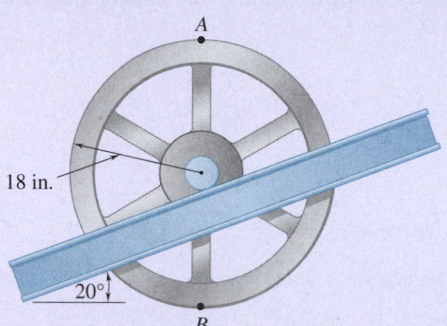

**Fig. P15.111**

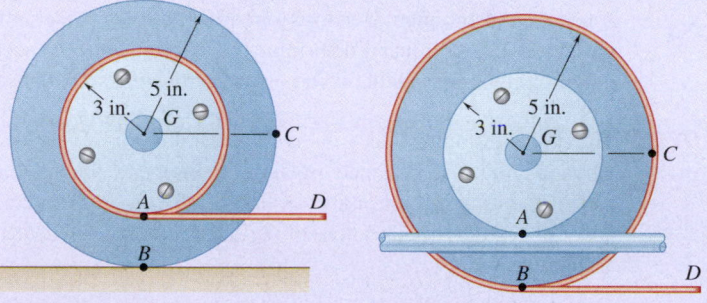

**Fig. P15.112**

**15.113 and 15.114** A 3-in.-radius drum is rigidly attached to a 5-in.-radius drum as shown. One of the drums rolls without sliding on the surface shown, and a cord is wound around the other drum. Knowing that at the instant shown end *D* of the cord has a velocity of 6 in./s and an acceleration of 20 in./s², both directed to the right, determine the accelerations of points *A*, *B*, and *C* of the drums.

**Fig. P15.113**          **Fig. P15.114**

**15.115** A heavy crate is being moved a short distance using three identical cylinders as rollers. Knowing that at the instant shown the crate has a velocity of 200 mm/s and an acceleration of 400 mm/s², both directed to the right, determine (*a*) the angular acceleration of the center cylinder, (*b*) the acceleration of point *A* on the center cylinder.

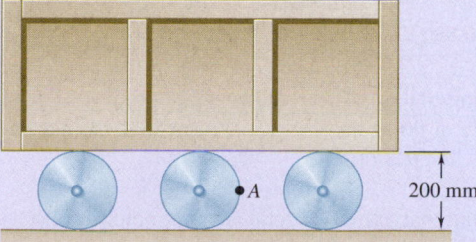

**Fig. P15.115**

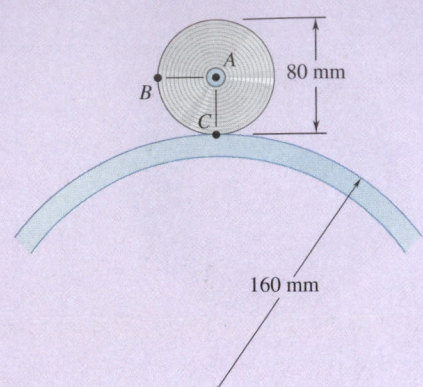

Fig. *P15.116*

**15.116** A wheel rolls without slipping on a fixed cylinder. Knowing that at the instant shown the angular velocity of the wheel is 10 rad/s clockwise and its angular acceleration is 30 rad/s² counterclockwise, determine the acceleration of (*a*) point *A*, (*b*) point *B*, (*c*) point *C*.

**15.117** The 100-mm-radius drum rolls without slipping on a portion of a belt that moves downward to the left with a constant velocity of 120 mm/s. Knowing that at a given instant the velocity and acceleration of the center *A* of the drum are as shown, determine the acceleration of point *D*.

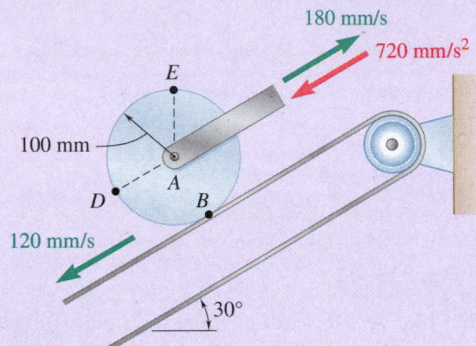

Fig. *P15.117*

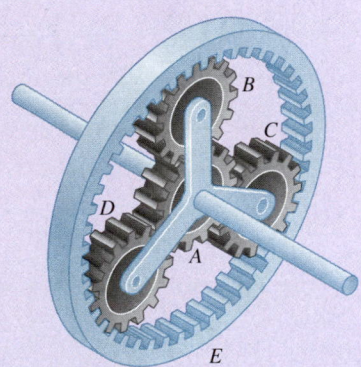

Fig. P15.118

**15.118** In the planetary gear system shown, the radius of gears *A*, *B*, *C*, and *D* is 3 in. and the radius of the outer gear *E* is 9 in. Knowing that gear *A* has a constant angular velocity of 150 rpm clockwise and that the outer gear *E* is stationary, determine the magnitude of the acceleration of the tooth of gear *D* that is in contact with (*a*) gear *A*, (*b*) gear *E*.

**15.119** The 200-mm-radius disk rolls without sliding on the surface shown. Knowing that the distance *BG* is 160 mm and that at the instant shown the disk has an angular velocity of 8 rad/s counterclockwise and an angular acceleration of 2 rad/s² clockwise, determine the acceleration of *A*.

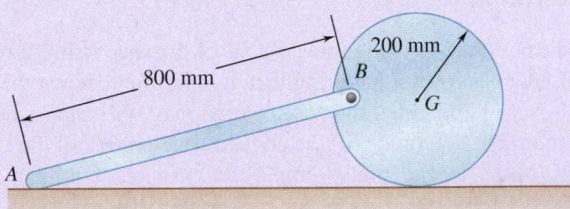

Fig. *P15.119*

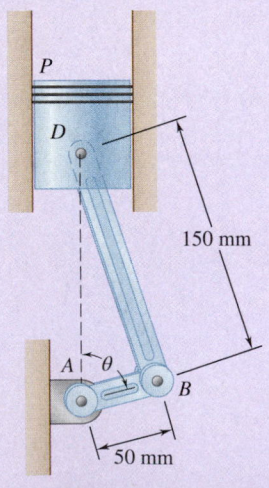

Fig. P15.120 and P15.121

**15.120** Knowing that crank *AB* rotates about point *A* with a constant angular velocity of 900 rpm clockwise, determine the acceleration of the piston *P* when *θ* = 60°.

**15.121** Knowing that crank *AB* rotates about point *A* with a constant angular velocity of 900 rpm clockwise, determine the acceleration of the piston *P* when *θ* = 120°.

**15.122** In the two-cylinder air compressor shown, the connecting rods $BD$ and $BE$ are each 190 mm long and crank $AB$ rotates about the fixed point $A$ with a constant angular velocity of 1500 rpm clockwise. Determine the acceleration of each piston when $\theta = 0$.

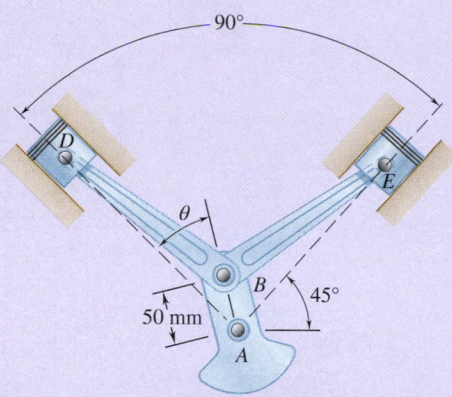

Fig. P15.122

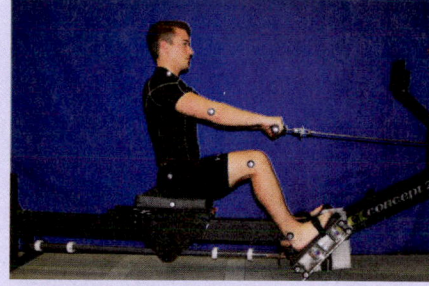

©Katherine Mavrommati

**15.123** The right leg of an athlete on a rowing machine can be modeled as a linkage as shown, where $A$ represents the ankle (which is stationary), $K$ the knee, and $H$ the hip. At the instant when $\theta = 75°$, the shank $AK$ has an angular velocity of 1 rad/s and an angular acceleration of 1.5 rad/s$^2$, both counterclockwise. Determine the velocity and acceleration of the hip $H$ at this instant in time.

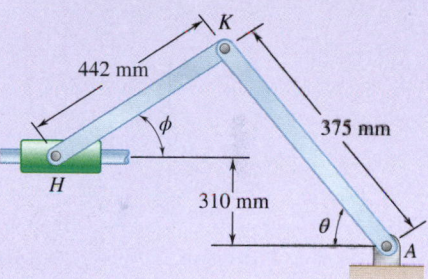

Fig. P15.123

**15.124** Arm $AB$ has a constant angular velocity of 16 rad/s counterclockwise. At the instant when $\theta = 90°$, determine the acceleration of (a) collar $D$, (b) the midpoint $G$ of bar $BD$.

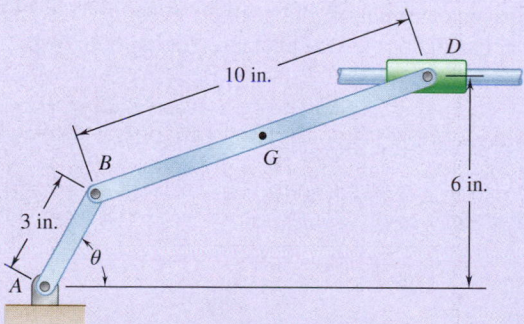

Fig. P15.124 and P15.125

**15.125** Arm $AB$ has a constant angular velocity of 16 rad/s counterclockwise. At the instant when $\theta = 60°$, determine the acceleration of collar $D$.

**15.126** A straight rack rests on a gear of radius $r = 3$ in. and is attached to a block $B$ as shown. Knowing that at the instant shown $\theta = 20°$, the angular velocity of gear $D$ is 3 rad/s clockwise, and it is speeding up at a rate of 2 rad/s$^2$, determine (a) the angular acceleration of $AB$, (b) the acceleration of block $B$.

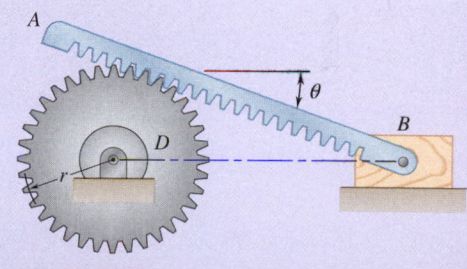

Fig. P15.126

**15.127** The elliptical exercise machine has fixed axes of rotation at points *A* and *E*. Knowing that at the instant shown the flywheel *AB* has a constant angular velocity of 6 rad/s clockwise, determine the acceleration of point *D*.

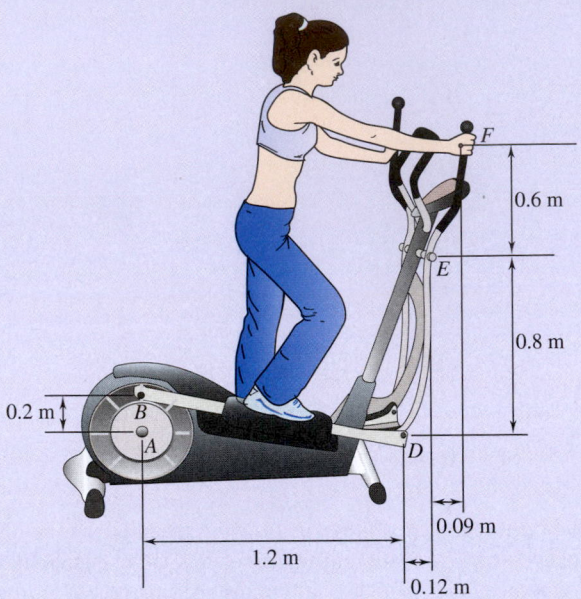

**Fig. P15.127 and P15.128**

**15.128** The elliptical exercise machine has fixed axes of rotation at points *A* and *E*. Knowing that at the instant shown the flywheel *AB* has a constant angular velocity of 6 rad/s clockwise, determine (*a*) the angular acceleration of bar *DEF*, (*b*) the acceleration of point *F*.

**15.129** Knowing that the angular velocity of rod *DE* is a constant 20 rad/s clockwise, determine in the position shown (*a*) the angular acceleration of bar *BD*, (*b*) the angular acceleration of bar *AB*.

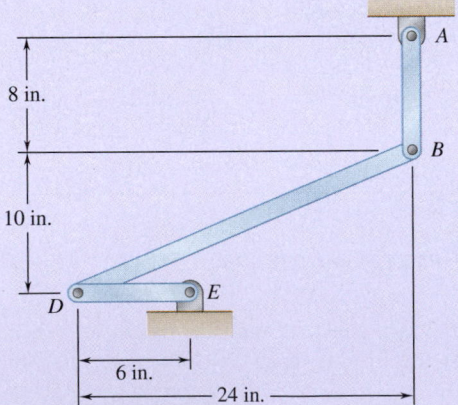

**Fig. P15.129 and P15.130**

**15.130** Knowing that at the instant shown bar *DE* has an angular velocity of 20 rad/s clockwise and an angular acceleration of 5 rad/s² counterclockwise, determine in the position shown (*a*) the angular acceleration of bar *BD*, (*b*) the angular acceleration of bar *AB*.

**15.131 and 15.132** Knowing that at the instant shown bar *AB* has a constant angular velocity of 10 rad/s clockwise, determine the angular acceleration of (*a*) bar *BD*, (*b*) bar *DE*.

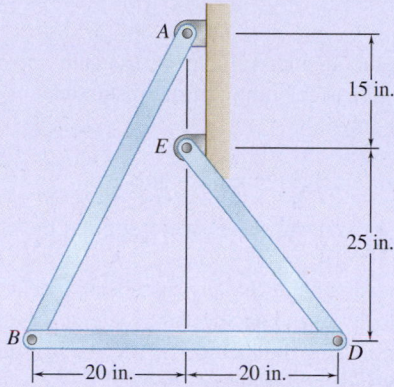

Fig. *P15.131* and P15.133

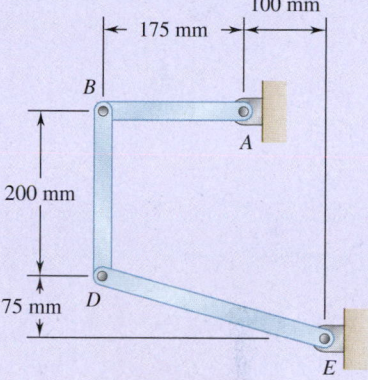

Fig. **P15.132 and P15.134**

**15.133 and 15.134** Knowing that at the instant shown bar *AB* has an angular velocity of 10 rad/s and an angular acceleration of 4 rad/s², both clockwise, determine the angular acceleration of (*a*) bar *BD*, (*b*) bar *DE* by using the vector approach as is done in Sample Prob. 15.16.

**15.135** Roberts linkage is named after Richard Roberts (1789–1864) and can be used to draw a close approximation to a straight line by locating a pen at point *F*. The distance *AB* is the same as *BF*, *DF*, and *DE*. Knowing that at the instant shown bar *AB* has a constant angular velocity of 4 rad/s clockwise, determine (*a*) the angular acceleration of bar *DE*, (*b*) the acceleration of point *F*.

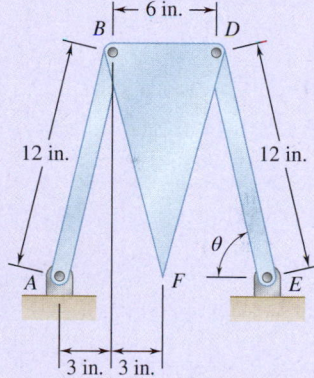

Fig. *P15.135*

**15.136** For the oil pump rig shown, link *AB* causes the beam *BCE* to oscillate as the crank *OA* revolves. Knowing that *OA* has a radius of 0.6 m and a constant clockwise angular velocity of 20 rpm, determine the velocity and acceleration of point *D* at the instant shown.

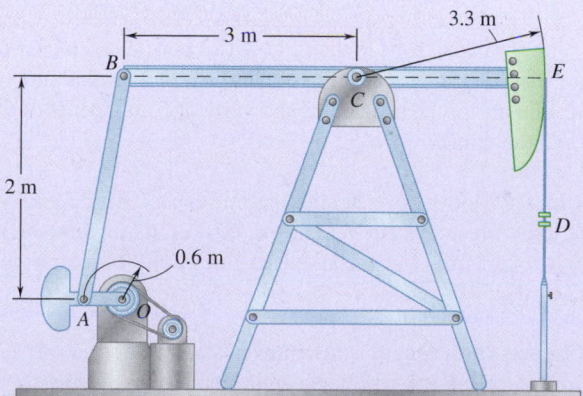

Fig. P15.136

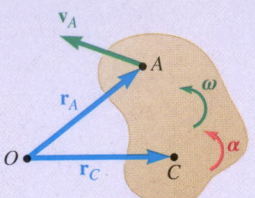

Fig. P15.137

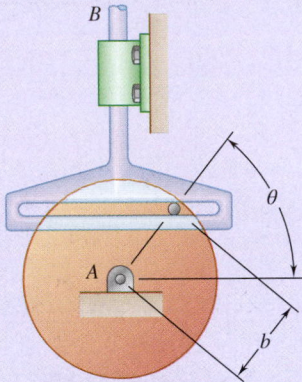

Fig. P15.138

**15.137** Denoting by $\mathbf{r}_A$ the position vector of a point $A$ of a rigid slab that is in plane motion, show that (a) the position vector $\mathbf{r}_C$ of the instantaneous center of rotation is

$$\mathbf{r}_C = \mathbf{r}_A + \frac{\boldsymbol{\omega} \times \mathbf{v}_A}{\omega^2}$$

where $\boldsymbol{\omega}$ is the angular velocity of the slab and $\mathbf{v}_A$ is the velocity of point $A$, (b) the acceleration of the instantaneous center of rotation is zero if, and only if,

$$\mathbf{a}_A = \frac{\alpha}{\omega} \mathbf{v}_A + \boldsymbol{\omega} \times \mathbf{v}_A$$

where $\boldsymbol{\alpha} = \alpha\mathbf{k}$ is the angular acceleration of the slab.

**\*15.138** The drive disk of the Scotch crosshead mechanism shown has an angular velocity $\boldsymbol{\omega}$ and an angular acceleration $\boldsymbol{\alpha}$, both directed counterclockwise. Using the method of Sec. 15.4B, derive expressions for the velocity and acceleration of point $B$.

**\*15.139** The wheels attached to the ends of rod $AB$ roll along the surfaces shown. Using the method of Sec. 15.4B, derive an expression for the angular velocity of the rod in terms of $v_B$, $\theta$, $l$, and $\beta$.

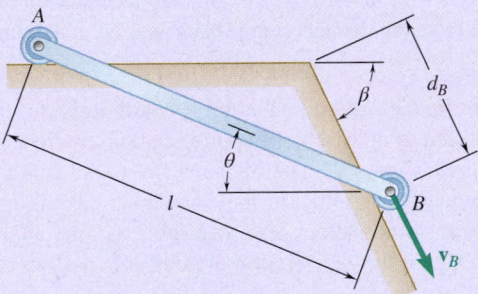

Fig. P15.139 and P15.140

**\*15.140** The wheels attached to the ends of rod $AB$ roll along the surfaces shown. Using the method of Sec. 15.4B and knowing that the acceleration of wheel $B$ is zero, derive an expression for the angular acceleration of the rod in terms of $v_B$, $\theta$, $l$, and $\beta$.

**\*15.141** A disk of radius $r$ rolls to the right with a constant velocity $\mathbf{v}$. Denoting by $P$ the point of the rim in contact with the ground at $t = 0$, derive expressions for the horizontal and vertical components of the velocity of $P$ at any time $t$.

**\*15.142** Ladder $AB$ moves over a smooth corner at $C$ while end $A$ moves to the left with a constant velocity $\mathbf{v}_A$. Using the method of Sec. 15.4B, derive expressions for the angular velocity and angular acceleration of the ladder.

**\*15.143** A ladder $AB$ of length $l$ moves over a smooth corner at $C$ while end $A$ moves to the left with a constant velocity $\mathbf{v}_A$. Using the method of Sec. 15.4B, derive expressions for the horizontal and vertical components of the velocity of point $B$.

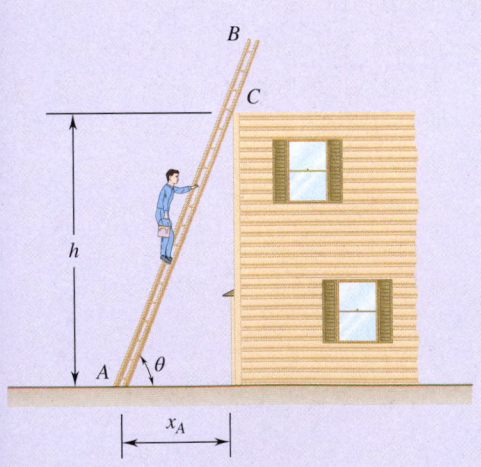

Fig. P15.142 and P15.143

**15.144** Crank $AB$ rotates with a constant clockwise angular velocity $\boldsymbol{\omega}$. Using the method of Sec. 15.4B, derive expressions for the angular velocity of rod $BD$ and the velocity of the point on the rod coinciding with point $E$ in terms of $\theta$, $\omega$, $b$, and $l$.

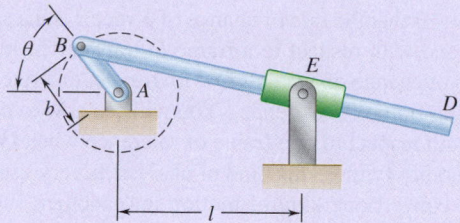

**Fig. P15.144 and P15.145**

**15.145** Crank $AB$ rotates with a constant clockwise angular velocity $\boldsymbol{\omega}$. Using the method of Sec. 15.4B, derive an expression for the angular acceleration of rod $BD$ in terms of $\theta$, $\omega$, $b$, and $l$.

**15.146** Solve the engine system from Sample Prob. 15.15 using the methods of Sec. 15.4B. (*Hint:* Define the angle between the horizontal and the crank $AB$ as $\theta$ and derive the motion in terms of this parameter.)

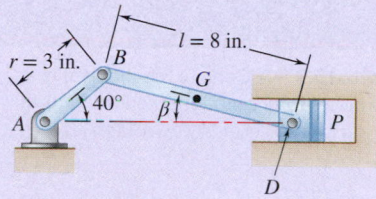

**Fig. P15.146**

**\*15.147** The position of rod $AB$ is controlled by a disk of radius $r$ that is attached to yoke $CD$. Knowing that the yoke moves vertically upward with a constant velocity $\mathbf{v}_0$, derive expressions for the angular velocity and angular acceleration of rod $AB$.

**\*15.148** A wheel of radius $r$ rolls without slipping along the inside of a fixed cylinder of radius $R$ with a constant angular velocity $\boldsymbol{\omega}$. Denoting by $P$ the point of the wheel in contact with the cylinder at $t = 0$, derive expressions for the horizontal and vertical components of the velocity of $P$ at any time $t$. (The curve described by point $P$ is a *hypocycloid*.)

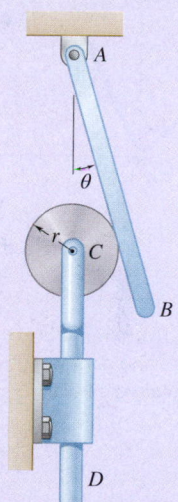

**Fig. P15.147**

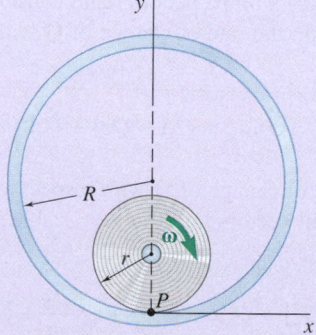

**Fig. P15.148**

**\*15.149** In Prob. 15.148, show that the path of $P$ is a vertical straight line when $r = R/2$. Derive expressions for the corresponding velocity and acceleration of $P$ at any time $t$.

# 15.5    ANALYZING MOTION WITH RESPECT TO A ROTATING FRAME

We saw in Sec. 11.4B that the rate of change of a vector is the same with respect to a fixed frame and with respect to a frame in translation. In this section, we consider the rates of change of a vector **Q** with respect to a fixed frame and with respect to a rotating frame of reference.[†] You will see how to determine the rate of change of **Q** with respect to one frame of reference when **Q** is defined by its components in another frame. This kind of analysis is very useful for designing mechanisms that convert one kind of motion into another, such as continuous rotation into intermittent rotation. It is also helpful when you have, say, an extending linear actuator that is also rotating.

**Photo 15.7**    A Geneva mechanism is used to convert rotary motion into intermittent motion. ©Purdue University/Physics/PRIME Lab

## 15.5A    Rate of Change of a Vector with Respect to a Rotating Frame

Consider two frames of reference centered at $O$: a fixed frame $OXYZ$ and a frame $Oxyz$ that rotates about the fixed axis $OA$. Let $\boldsymbol{\Omega}$ denote the angular velocity of the frame $Oxyz$ at a given instant (Fig. 15.26). Consider now a vector function $\mathbf{Q}(t)$ represented by the vector **Q** attached at $O$; as the time $t$ varies, both the direction and the magnitude of **Q** change. The variation of **Q** is viewed differently by an observer using $OXYZ$ as a frame of reference and by an observer using $Oxyz$, so we should expect the rate of change of **Q** to depend upon the frame of reference that has been selected. Therefore, we denote the rate of change of **Q** with respect to the fixed frame $OXYZ$ by $(\dot{\mathbf{Q}})_{OXYZ}$ and the rate of change of **Q** with respect to the rotating frame $Oxyz$ by $(\dot{\mathbf{Q}})_{Oxyz}$. We propose to determine the relation between these two rates of change.

Let us first resolve the vector **Q** into components along the $x$, $y$, and $z$ axes of the rotating frame. Denoting the corresponding unit vectors by **i**, **j**, and **k**, we have

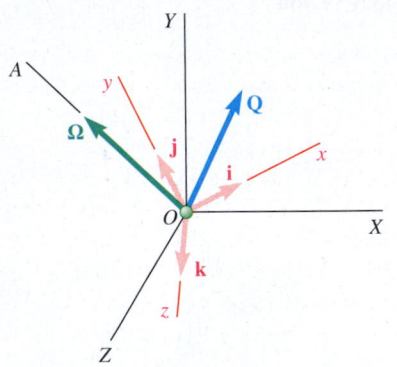

**Fig. 15.26**    A fixed frame of reference $OXYZ$ and a rotating frame $Oxyz$ with angular velocity $\boldsymbol{\Omega}$.

$$\mathbf{Q} = Q_x\mathbf{i} + Q_y\mathbf{j} + Q_z\mathbf{k} \qquad (15.27)$$

Differentiating Eq. (15.27) with respect to $t$ and considering the unit vectors **i**, **j**, **k** to be fixed, we obtain the rate of change of **Q** with respect to the rotating frame $Oxyz$, as

$$(\dot{\mathbf{Q}})_{Oxyz} = \dot{Q}_x\mathbf{i} + \dot{Q}_y\mathbf{j} + \dot{Q}_z\mathbf{k} \qquad (15.28)$$

To obtain the rate of change of **Q** with respect to the fixed frame $OXYZ$, we must consider the unit vectors **i**, **j**, **k** to be variable when differentiating Eq. (15.27). This gives

$$(\dot{\mathbf{Q}})_{OXYZ} = \dot{Q}_x\mathbf{i} + \dot{Q}_y\mathbf{j} + \dot{Q}_z\mathbf{k} + Q_x\frac{d\mathbf{i}}{dt} + Q_y\frac{d\mathbf{j}}{dt} + Q_z\frac{d\mathbf{k}}{dt} \qquad (15.29)$$

From Eq. (15.28), we observe that the sum of the first three terms in the right-hand side of Eq. (15.29) represents the rate of change $(\dot{\mathbf{Q}})_{Oxyz}$. We note,

[†] Recall that the selection of a fixed frame of reference is arbitrary. Any frame may be designated as "fixed"; all others are then considered as moving.

on the other hand, that the rate of change $(\dot{\mathbf{Q}})_{OXYZ}$ would reduce to the last three terms in Eq. (15.29) if vector $\mathbf{Q}$ were fixed within the frame $Oxyz$, because $(\dot{\mathbf{Q}})_{Oxyz}$ would then be zero. But in that case, $(\dot{\mathbf{Q}})_{OXYZ}$ would represent the velocity of a particle located at the tip of $\mathbf{Q}$ and belonging to a body rigidly attached to the frame $Oxyz$. Thus, the last three terms in Eq. (15.29) represent the velocity of that particle. Because the frame $Oxyz$ has an angular velocity $\boldsymbol{\Omega}$ with respect to $OXYZ$ at the instant considered, we have, by Eq. (15.5),

$$Q_x\frac{d\mathbf{i}}{dt} + Q_y\frac{d\mathbf{j}}{dt} + Q_z\frac{d\mathbf{k}}{dt} = \boldsymbol{\Omega} \times \mathbf{Q} \qquad (15.30)$$

Substituting from Eqs. (15.28) and (15.30) into Eq. (15.29), we obtain the fundamental relation

$$(\dot{\mathbf{Q}})_{OXYZ} = (\dot{\mathbf{Q}})_{Oxyz} + \boldsymbol{\Omega} \times \mathbf{Q} \qquad (15.31)$$

We conclude that the rate of change of vector $\mathbf{Q}$ with respect to the fixed frame $OXYZ$ consists of two parts: The first part represents the rate of change of $\mathbf{Q}$ with respect to the rotating frame $Oxyz$; the second part, $\boldsymbol{\Omega} \times \mathbf{Q}$, is induced by the rotation of the frame $Oxyz$.

The use of the relation in Eq. (15.31) simplifies the determination of the rate of change of a vector $\mathbf{Q}$ with respect to a fixed frame of reference $OXYZ$ when vector $\mathbf{Q}$ is defined by its components along the axes of a rotating frame $Oxyz$. In particular, this relation does not require separate computations of the derivatives of the unit vectors defining the orientation of the rotating frame.

## 15.5B   Plane Motion of a Particle Relative to a Rotating Frame

Consider two frames of reference with both centered at $O$ and both in the plane of the figure: a fixed frame $OXY$ and a rotating frame $Oxy$ (Fig. 15.27). Let $P$ be a particle moving in the plane of the figure. The position vector $\mathbf{r}$ of $P$ is the same in both frames, but its rate of change depends upon which frame of reference you select.

The absolute velocity $\mathbf{v}_P$ of the particle is defined as the velocity observed from the fixed frame $OXY$ and is equal to the rate of change $(\dot{\mathbf{r}})_{OXY}$ of $\mathbf{r}$ with respect to that frame. We can, however, express $\mathbf{v}_P$ in terms of the rate of change $(\dot{\mathbf{r}})_{Oxy}$ observed from the rotating frame if we make use of Eq. (15.31). Denoting the angular velocity of the frame $Oxy$ with respect to $OXY$ at the instant considered by $\boldsymbol{\Omega}$, we have

$$\mathbf{v}_P = (\dot{\mathbf{r}})_{OXY} = \boldsymbol{\Omega} \times \mathbf{r} + (\dot{\mathbf{r}})_{Oxy} \qquad (15.32)$$

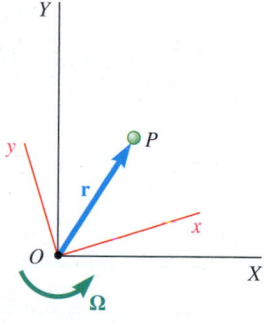

**Fig. 15.27**   We can express the motion of particle $P$ in either a fixed ($OXYZ$) or a rotating ($Oxyz$) frame of reference.

where $(\dot{\mathbf{r}})_{Oxy}$ defines the velocity of particle $P$ relative to the rotating frame $Oxy$ and is sometimes denoted as $\mathbf{v}_{\text{rel}}$. There also may be instances where point $O$ is not fixed and has a velocity denoted by $\mathbf{v}_O$. Therefore, an alternative way to express Eq. (15.32) is

$$\mathbf{v}_P = \mathbf{v}_O + \boldsymbol{\Omega} \times \mathbf{r} + \mathbf{v}_{\text{rel}} \qquad (15.32')$$

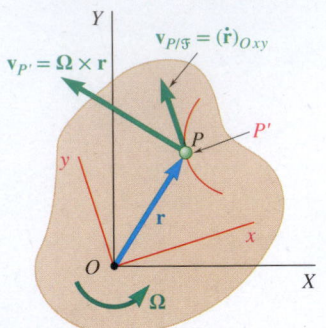

**Fig. 15.28** The velocity of a point $P$ is equal to the velocity of a point $P'$ coincident with $P$ but attached to the rotating frame plus the velocity of $P$ with respect to the rotating frame.

The relative velocity, $\mathbf{v}_{\text{rel}}$ or $(\dot{\mathbf{r}})_{Oxy}$, is the velocity of point $P$ with respect to the rotating frame. Denoting the rotating frame by $\mathcal{F}$, another way to represent the velocity $(\dot{\mathbf{r}})_{Oxy}$ of $P$ relative to the rotating frame is $\mathbf{v}_{P/\mathcal{F}}$. Let us imagine that a rigid body has been attached to the rotating frame. Then $\mathbf{v}_{P/\mathcal{F}}$ represents the velocity of $P$ along the path that it describes on that body (Fig. 15.28), and the term $\mathbf{\Omega} \times \mathbf{r}$ in Eq. (15.32) represents the velocity $\mathbf{v}_{P'}$ of the point $P'$ of the rigid body—or rotating frame—that coincides with $P$ at the instant considered. Thus, we have

$$\mathbf{v}_P = \mathbf{v}_{P'} + \mathbf{v}_{P/\mathcal{F}} \tag{15.33}$$

where

$\mathbf{v}_P$ = absolute velocity of particle $P$

$\mathbf{v}_{P'}$ = velocity of point $P'$ of moving frame $\mathcal{F}$ coinciding with $P$

$\mathbf{v}_{P/\mathcal{F}}$ = velocity of $P$ relative to moving frame $\mathcal{F}$

We define the absolute acceleration $\mathbf{a}_P$ of the particle as the rate of change of $\mathbf{v}_P$ with respect to the fixed frame $OXY$. Computing the rates of change with respect to $OXY$ of the terms in Eq. (15.32), we have

$$\mathbf{a}_P = \dot{\mathbf{v}}_P = \dot{\mathbf{\Omega}} \times \mathbf{r} + \mathbf{\Omega} \times \dot{\mathbf{r}} + \frac{d}{dt}[(\dot{\mathbf{r}})_{Oxy}] \tag{15.34}$$

where all derivatives are defined with respect to $OXY$, except where indicated otherwise. Referring to Eq. (15.31), we note that we can express the last term in Eq. (15.34) as

$$\frac{d}{dt}[(\dot{\mathbf{r}})_{Oxy}] = (\ddot{\mathbf{r}})_{Oxy} + \mathbf{\Omega} \times (\dot{\mathbf{r}})_{Oxy}$$

On the other hand, $\dot{\mathbf{r}}$ represents the velocity $\mathbf{v}_P$ and can be replaced by the right-hand side of Eq. (15.32). After completing these two substitutions into Eq. (15.34), we obtain

$$\mathbf{a}_P = \dot{\mathbf{\Omega}} \times \mathbf{r} + \mathbf{\Omega} \times (\mathbf{\Omega} \times \mathbf{r}) + 2\mathbf{\Omega} \times (\dot{\mathbf{r}})_{Oxy} + (\ddot{\mathbf{r}})_{Oxy} \tag{15.35}$$

As we had for the velocity expression, our reference point $O$ might also be accelerating. For plane motion,

$$\mathbf{a}_P = \mathbf{a}_O + \dot{\mathbf{\Omega}} \times \mathbf{r} - \Omega^2 \mathbf{r} + 2\mathbf{\Omega} \times \mathbf{v}_{\text{rel}} + \mathbf{a}_{\text{rel}} \tag{15.35'}$$

where

$\mathbf{a}_O$ = the linear acceleration of point $O$

$\dot{\mathbf{\Omega}}$ = angular acceleration of the rotating frame

$\mathbf{\Omega}$ = angular velocity of the rotating frame

$\mathbf{r}$ = position vector from the origin $O$ to point $P$

$\mathbf{v}_{\text{rel}}$ = relative velocity of point $P$ with respect to the rotating frame

$\mathbf{a}_{\text{rel}}$ = relative acceleration of point $P$ with respect to the rotating frame

From expression (15.8) obtained in Sec. 15.1B for the acceleration of a particle on a rigid body rotating about a fixed axis, we note that the sum of the first two terms in Eq. (15.35) represents the acceleration $\mathbf{a}_{P'}$ of the point $P'$ of the rotating frame that coincides with $P$ at the instant considered. The last term defines the acceleration $\mathbf{a}_{P/\mathcal{F}}$ of $P$ relative to the rotating frame. If it were not for the third term, which has not been accounted for, we could write a relation similar to Eq. (15.33) for the accelerations, and $\mathbf{a}_P$ could be expressed as the sum of $\mathbf{a}_{P'}$ and $\mathbf{a}_{P/\mathcal{F}}$. However, it is clear that *such a relation would be incorrect* and that we must include the additional term. This term, which we denote by $\mathbf{a}_C$, is called the **Coriolis acceleration**, after the French mathematician Gaspard de Coriolis (1792–1843). We have

$$\mathbf{a}_P = \mathbf{a}_{P'} + \mathbf{a}_{P/\mathcal{F}} + \mathbf{a}_C \qquad (15.36)$$

where

$\mathbf{a}_P$ = absolute acceleration of particle $P$

$\mathbf{a}_{P'}$ = acceleration of point $P'$ of moving frame $\mathcal{F}$ coinciding with $P$

$\mathbf{a}_{P/\mathcal{F}}$ = acceleration of $P$ relative to moving frame $\mathcal{F}$

$\mathbf{a}_C = 2\mathbf{\Omega} \times (\dot{\mathbf{r}})_{Oxy} = 2\mathbf{\Omega} \times \mathbf{v}_{P/\mathcal{F}}$

= Coriolis acceleration

Note the difference between Eq. (15.36) and Eq. (15.21). When we wrote

$$\mathbf{a}_B = \mathbf{a}_A + \mathbf{a}_{B/A} \qquad (15.21)$$

in Sec. 15.4A, we were expressing the absolute acceleration of point $B$ as the sum of the acceleration $\mathbf{a}_{B/A}$ relative to a frame in translation and the acceleration $\mathbf{a}_A$ of a point of that frame. We are now relating the absolute acceleration of point $P$ to its acceleration $\mathbf{a}_{P/\mathcal{F}}$ relative to a rotating frame $\mathcal{F}$ and to the acceleration $\mathbf{a}_{P'}$ of point $P'$ of that frame, which coincides with $P$. Eq. (15.36) shows that, because the frame is rotating, it is necessary to include an additional term to represent the Coriolis acceleration $\mathbf{a}_C$.

Note that because point $P'$ moves in a circle about the origin $O$, its acceleration $\mathbf{a}_{P'}$ has, in general, two components: $(\mathbf{a}_{P'})_t$ tangent to the circle and $(\mathbf{a}_{P'})_n$ directed toward $O$. Similarly, the acceleration $\mathbf{a}_{P/\mathcal{F}}$ generally has two components: $(\mathbf{a}_{P/\mathcal{F}})_t$ tangent to the path that $P$ describes on the rotating rigid body and $(\mathbf{a}_{P/\mathcal{F}})_n$ directed toward the center of curvature of that path. We further note that because the vector $\mathbf{\Omega}$ is perpendicular to the plane of motion, and thus to $\mathbf{v}_{P/\mathcal{F}}$, the magnitude of the Coriolis acceleration $\mathbf{a}_C = 2\mathbf{\Omega} \times \mathbf{v}_{P/\mathcal{F}}$ is equal to $2\Omega v_{P/\mathcal{F}}$, and its direction can be obtained by rotating the vector $\mathbf{v}_{P/\mathcal{F}}$ through 90° in the sense of rotation of the moving frame (Fig. 15.29). The Coriolis acceleration reduces to zero when either $\mathbf{\Omega}$ or $\mathbf{v}_{P/\mathcal{F}}$ is zero.

Consider a collar $P$ that is made to slide at a constant relative speed $u$ along a rod $OB$ rotating at a constant angular velocity $\boldsymbol{\omega}$ about $O$ (Fig. 15.30a). According to formula (15.36), we can obtain the absolute acceleration of $P$ by adding vectorially the acceleration $\mathbf{a}_A$ of the point $A$ of the rod coinciding with $P$, the relative acceleration $\mathbf{a}_{P/OB}$ of $P$ with respect to the rod, and the Coriolis acceleration $\mathbf{a}_C$.

Because the angular velocity $\boldsymbol{\omega}$ of the rod is constant, $\mathbf{a}_A$ reduces to its normal component $(\mathbf{a}_A)_n$ with a magnitude of $r\omega^2$; and because $u$ is constant, the relative acceleration $\mathbf{a}_{P/OB}$ is zero. According to the definition given previously, the Coriolis acceleration is a vector perpendicular to $OB$, has a magnitude of $2\omega u$, and is directed as shown in Fig. 15.30. The acceleration of the collar $P$ consists, therefore, of the two vectors shown in Fig. 15.30a. Note that you can check this result by applying the relation in Eq. (11.43).

To understand better the significance of the Coriolis acceleration, let us consider the absolute velocity of $P$ at time $t$ and at time $t + \Delta t$ (Fig. 15.30b).

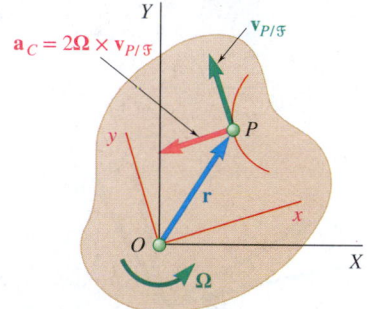

**Fig. 15.29**  The Coriolis acceleration is perpendicular to the relative velocity of $P$ with respect to the rotating frame.

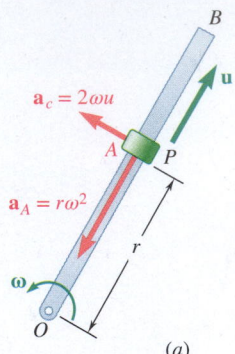

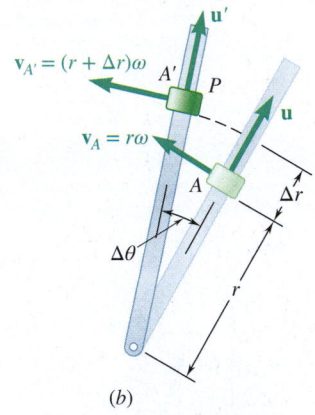

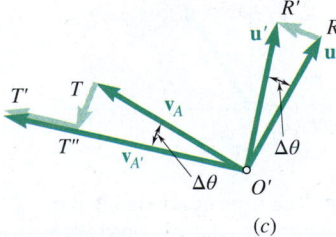

**Fig. 15.30** *(a)* A collar sliding at constant speed along a rotating rod; *(b)* velocities of the collar at two points in time; *(c)* the acceleration components equal the changes in velocity.

We can resolve the velocity at time $t$ into its components $\mathbf{u}$ and $\mathbf{v}_A$; we can resolve the velocity at time $t + \Delta t$ into its components $\mathbf{u}'$ and $\mathbf{v}_{A'}$. Drawing these components from the same origin (Fig. 15.30$c$), we note that the change in velocity during the time $\Delta t$ can be represented by the sum of three vectors: $\overrightarrow{RR'}$, $\overrightarrow{TT''}$, and $\overrightarrow{T''T'}$. The vector $\overrightarrow{TT''}$ measures the change in direction of the velocity $\mathbf{v}_A$, and the quotient $\overrightarrow{TT''}/\Delta t$ represents the acceleration $\mathbf{a}_A$ when $\Delta t$ approaches zero. We check that the direction of $\overrightarrow{TT''}$ is that of $\mathbf{a}_A$ when $\Delta t$ approaches zero and that

$$\lim_{\Delta t \to 0} \frac{TT''}{\Delta t} = \lim_{\Delta t \to 0} v_A \frac{\Delta \theta}{\Delta t} = r\omega\omega = r\omega^2 = a_A$$

The vector $\overrightarrow{RR'}$ measures the change in direction of $\mathbf{u}$ due to the rotation of the rod; the vector $\overrightarrow{T''T'}$ measures the change in magnitude of $\mathbf{v}_A$ due to the motion of $P$ on the rod. The vectors $\overrightarrow{RR'}$ and $\overrightarrow{T''T'}$ result from the *combined effect* of the relative motion of $P$ and of the rotation of the rod; they would vanish if *either* of these two motions stopped. It is easily verified that the sum of these two vectors defines the Coriolis acceleration. Their direction is that of $\mathbf{a}_C$ when $\Delta t$ approaches zero, and because $RR' = u\,\Delta\theta$ and $T''T' = v_{A'} - v_A = (r + \Delta r)\,\omega - r\omega = \omega\Delta r$, we check that $a_C$ is equal to

$$\lim_{\Delta t \to 0}\left(\frac{RR'}{\Delta t} + \frac{T''T'}{\Delta t}\right) = \lim_{\Delta t \to 0}\left(u\frac{\Delta\theta}{\Delta t} + \omega\frac{\Delta r}{\Delta t}\right) = u\omega + \omega u = 2\omega u$$

We can use formulas (15.33) and (15.36) to analyze the motion of mechanisms that contain parts sliding on each other. They make it possible, for example, to relate the absolute and relative motions of sliding pins and collars (see Sample Probs. 15.18, 15.19, and 15.20). The concept of Coriolis acceleration is also very useful in the study of long-range projectiles and of other objects whose motions are appreciably affected by the rotation of the earth. As we pointed out in Sec. 12.1A, a system of axes attached to the earth does not truly constitute a newtonian frame of reference; such a system of axes actually should be considered rotating. Thus, the formulas derived in this section facilitate the study of the motion of bodies with respect to axes attached to the earth.

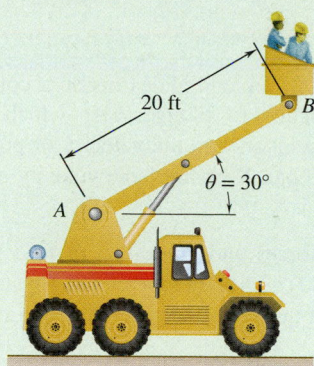

20 ft

$\theta = 30°$

$A$

$B$

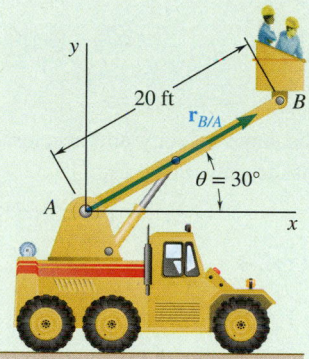

$y$

20 ft

$\mathbf{r}_{B/A}$

$B$

$\theta = 30°$

$A$

$x$

**Fig. 1** The rotating coordinate system is attached to the truck at $A$.

## Sample Problem 15.17

At the instant shown, the truck is moving forward with a speed of 2 ft/s and is slowing down at a rate of 0.25 ft/s². The length of the boom $AB$ is decreasing at a constant rate of 0.5 ft/s, the angular velocity of the boom is 0.1 rad/s, and the angular acceleration of the boom is 0.02 rad/s², both clockwise. Determine the velocity and acceleration of point $B$.

**STRATEGY:**  Because you are not given any forces and are asked to find the velocity and acceleration of a point, use rigid-body kinematics. The boom is moving with respect to the truck, so use a rotating reference frame.

**MODELING and ANALYSIS:**  Attach a rotating coordinate system to the boom housing with its origin at $A$ (Fig. 1).

**Velocity of B.**  From Eq. (15.32′), you know

$$\mathbf{v}_B = \mathbf{v}_A + \mathbf{\Omega} \times \mathbf{r}_{B/A} + \mathbf{v}_{\text{rel}} \qquad (1)$$

where $\mathbf{v}_A = (2 \text{ ft/s})\mathbf{i}$, $\mathbf{r}_{B/A} = (20 \cos 30° \text{ ft})\mathbf{i} + (20 \sin 30° \text{ ft})\mathbf{j}$, and $\mathbf{\Omega} = (-0.1 \text{ rad/s}^2)\mathbf{k}$. To find the relative velocity, ask yourself what the velocity of $B$ would be, assuming that the rotating coordinate system is not moving. In this case, $\mathbf{v}_{\text{rel}} = -(0.5 \cos 30° \text{ ft/s})\mathbf{i} - (0.5 \sin 30° \text{ ft/s})\mathbf{j}$. Substituting into Eq. (1) gives

$$\mathbf{v}_B = 2\mathbf{i} + (-0.1\mathbf{k}) \times (17.32\mathbf{i} + 10\mathbf{j}) - (0.433\mathbf{i} + 0.25\mathbf{j})$$

$$\mathbf{v}_B = (2.57 \text{ ft/s})\mathbf{i} - (1.982 \text{ ft/s})\mathbf{j} \quad \blacktriangleleft$$

**Acceleration of B.**  From Eq. (15.35′), you know

$$\mathbf{a}_B = \mathbf{a}_A + \dot{\mathbf{\Omega}} \times \mathbf{r}_{B/A} - \Omega^2 \mathbf{r}_{B/A} + 2\mathbf{\Omega} \times \mathbf{v}_{\text{rel}} + \mathbf{a}_{\text{rel}} \qquad (2)$$

where $\mathbf{a}_A = -(0.25 \text{ ft/s}^2)\mathbf{i}$, $\dot{\mathbf{\Omega}} = -(0.02 \text{ rad/s}^2)\mathbf{k}$, and $\mathbf{a}_{\text{rel}} = 0$. Substituting into Eq. (2) gives

$$\mathbf{a}_B = -0.25\mathbf{i} + (-0.02\mathbf{k}) \times (17.32\mathbf{i} + 10\mathbf{j}) - 0.1^2(17.32\mathbf{i} + 10\mathbf{j})$$
$$+ 2(-0.1\mathbf{k}) \times (-0.433\mathbf{i} - 0.25\mathbf{j}) + 0$$
$$= -0.25\mathbf{i} + (-0.3464\mathbf{j} + 0.2\mathbf{i}) - (0.1732\mathbf{i} + 0.10\mathbf{j}) + (0.0866\mathbf{j} - 0.05\mathbf{i}) + 0$$

$$\mathbf{a}_B = (-0.273 \text{ ft/s}^2)\mathbf{i} - (0.360 \text{ ft/s}^2)\mathbf{j} \quad \blacktriangleleft$$

**REFLECT and THINK:**  The biggest challenge with this problem is interpreting what you are given in the problem statement. After that, it is straightforward to substitute into the governing equations. The last four terms in Eq. (2) are analogous to the polar coordinate expressions we used in Chap. 11. The following terms represent the same physical quantities:

$\dot{\mathbf{\Omega}} \times \mathbf{r}_{B/A} \to r\ddot{\theta}$, $-\Omega^2 \mathbf{r}_{B/A} \to -r\dot{\theta}^2$, $2\mathbf{\Omega} \times \mathbf{v}_{\text{rel}} \to 2\dot{r}\dot{\theta}$, and $\mathbf{a}_{\text{rel}} \to \ddot{r}$.

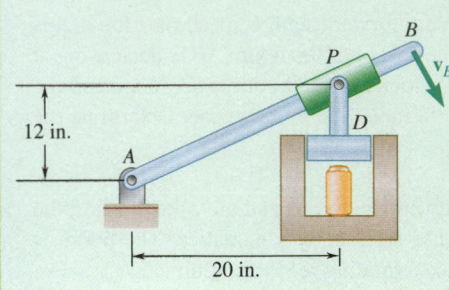

## Sample Problem 15.18

In a can crusher, bar $AB$ has a length of 30 in. and slides inside a collar located at point $P$. This collar is attached to plunger $DP$, which is constrained to move vertically. At the instant shown, the velocity of point $B$ is a constant 4 ft/s perpendicular to the bar. Determine the velocity and acceleration of the plunger $D$.

**STRATEGY:** You are not given any forces and are asked to find the velocity and acceleration of a point, so use rigid-body kinematics. Because the collar is moving with respect to the bar, use a rotating reference frame.

**MODELING and ANALYSIS:** Attach a rotating coordinate system to the bar with its origin at $A$ (Fig. 1).

**Angular Velocity of AB.** Rod $AB$ is undergoing fixed-axis rotation, so

$$\omega_{AB} = \frac{v_B}{r_{B/A}} = \frac{48 \text{ in./s}}{30 \text{ in.}} = 1.60 \text{ rad/s} \circlearrowleft$$

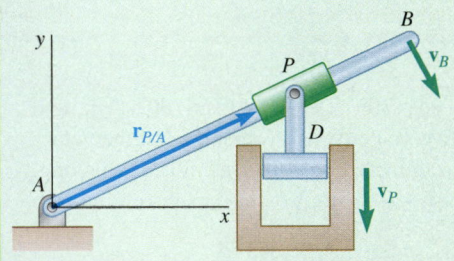

**Fig. 1** The rotating coordinate system is attached to arm $AB$.

**Velocity of P.** Points $D$ and $P$ have the same velocity and acceleration because the plunger is constrained to translate only. From Eq. (15.32′) you know

$$\mathbf{v}_P = \mathbf{v}_A + \mathbf{\Omega} \times \mathbf{r}_{P/A} + \mathbf{v}_{\text{rel}} \tag{1}$$

where $\mathbf{v}_A = 0$, $\mathbf{r}_{P/A} = (20 \text{ in.})\mathbf{i} + (12 \text{ in.})\mathbf{j}$, and $\mathbf{\Omega} = -(1.6 \text{ rad/s})\mathbf{k}$. To find the relative velocity, ask yourself what the velocity of $P$ would be, assuming that the rotating coordinate system is not moving. In this case, $\mathbf{v}_{\text{rel}} = v_{\text{rel}} \cos\theta\, \mathbf{i} + v_{\text{rel}} \sin\theta\, \mathbf{j}$ where $\theta = \tan^{-1}(12/20) = 30.96°$. Substituting into Eq. (1) gives

$$-v_P\mathbf{j} = 0 + (-1.6\mathbf{k}) \times (20\mathbf{i} + 12\mathbf{j}) + (v_{\text{rel}}\cos\theta\,\mathbf{i} + v_{\text{rel}}\sin\theta\,\mathbf{j})$$
$$= -32\mathbf{j} + 19.2\mathbf{i} + 0.8575v_{\text{rel}}\mathbf{i} + 0.5145v_{\text{rel}}\mathbf{j}$$

Equating components allows you to solve for the unknown velocities:

$$\begin{aligned} \mathbf{i}: \quad & 0 = 19.2 + 0.8575v_{\text{rel}} & \longrightarrow & \quad v_{\text{rel}} = -22.39 \text{ in./s} \\ \mathbf{j}: \quad & -v_P = -32 + 0.5145v_{\text{rel}} & \longrightarrow & \quad v_P = 43.53 \text{ in./s} \end{aligned}$$

$$\mathbf{v}_P = 43.53 \text{ in./s} \downarrow \quad \blacktriangleleft$$

**Acceleration of P.** From Eq. (15.35′), you know

$$\mathbf{a}_P = \mathbf{a}_A + \dot{\mathbf{\Omega}} \times \mathbf{r}_{P/A} - \Omega^2 \mathbf{r}_{P/A} + 2\mathbf{\Omega} \times \mathbf{v}_{\text{rel}} + \mathbf{a}_{\text{rel}} \tag{2}$$

where $\mathbf{a}_A = 0$, $\dot{\mathbf{\Omega}} = 0$, $\mathbf{a}_{\text{rel}} = a_{\text{rel}}\cos\theta\,\mathbf{i} + a_{\text{rel}}\sin\theta\,\mathbf{j}$. Substituting into Eq. (2) gives

$$-a_P\mathbf{j} = 0 + 0 - 1.6^2(20\mathbf{i} + 12\mathbf{j}) + 2(-1.6\mathbf{k}) \times$$
$$(-22.39\cos\theta\,\mathbf{i} - 22.39\sin\theta\,\mathbf{j}) + (a_{\text{rel}}\cos\theta\,\mathbf{i} + a_{\text{rel}}\sin\theta\,\mathbf{j})$$
$$= (-51.2\mathbf{i} - 30.72\mathbf{j}) + (61.44\mathbf{j} - 36.86\mathbf{i}) + (0.8575a_{\text{rel}}\mathbf{i} + 0.5145a_{\text{rel}}\mathbf{j})$$

Equating components allows you to solve for the unknown accelerations:

**i:** $\quad 0 = -51.2 - 36.86 + 0.8575 a_{\text{rel}} \qquad \longrightarrow \qquad a_{\text{rel}} = 102.7 \text{ in./s}^2$

**j:** $\quad -a_P = -30.72 + 61.44 + 0.5145 a_{\text{rel}} \qquad \longrightarrow \qquad a_P = -83.56 \text{ in./s}^2$

$$a_P = -83.6 \text{ in./s}^2 \downarrow \quad \blacktriangleleft$$

**REFLECT and THINK:** You used the same strategy for the telescoping boom in Sample Prob. 15.17 as you did for the sliding collar in this problem. For each case, the point of interest was moving with respect to a coordinate frame attached to a rigid body. The same strategy is used in problems where pins move within slotted bodies (such as the Geneva mechanism in Sample Prob. 15.19).

## Sample Problem 15.19

The Geneva mechanism shown is used in many counting instruments and in other applications where an intermittent rotary motion is required. Disk $D$ rotates with a constant counterclockwise angular velocity $\boldsymbol{\omega}_D$ of 10 rad/s. A pin $P$ is attached to disk $D$ and slides along one of several slots cut in disk $S$. It is desirable that the angular velocity of disk $S$ be zero as the pin enters and leaves each slot; in the case of four slots, this occurs if the distance between the centers of the disks is $l = \sqrt{2}R$.

At the instant when $\phi = 150°$, determine (a) the angular velocity of disk $S$, (b) the velocity of pin $P$ relative to disk $S$.

**STRATEGY:** You have two rigid bodies whose motions are related; therefore, use rigid-body kinematics. Because point $P$ is moving in a slot, use a rotating reference frame.

**MODELING and ANALYSIS:** Using geometry, you can solve triangle $OPB$, which corresponds to the position $\phi = 150°$ (Fig. 1). Using the law of cosines, you have

$$r^2 = R^2 + l^2 - 2Rl \cos 30° = 0.551R^2 \qquad r = 0.742R = 37.1 \text{ mm}$$

Then, from the law of sines, you have

$$\frac{\sin \beta}{R} = \frac{\sin 30°}{r} \qquad \sin \beta = \frac{\sin 30°}{0.742} \qquad \beta = 42.4°$$

Because pin $P$ is attached to disk $D$ and disk $D$ rotates about point $B$, the magnitude of the absolute velocity of $P$ is

$$v_P = R\omega_D = (50 \text{ mm})(10 \text{ rad/s}) = 500 \text{ mm/s}$$

$$\mathbf{v}_P = 500 \text{ mm/s} \; ⦨ \; 60°$$

*(continued)*

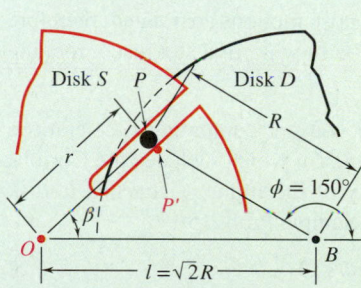

**Fig. 1** Distances and angles relating points $O$, $P$, and $B$.

Now consider the motion of pin $P$ along the slot in disk $S$. Denote the point of disk $S$ that coincides with $P$ by $P'$ at the instant considered and select a rotating frame $\mathcal{S}$ attached to disk $S$. Then from Eq. (15.33), you have

$$\mathbf{v}_P = \mathbf{v}_{P'} + \mathbf{v}_{P/\mathcal{S}} \qquad (1)$$

In Eq. (1), $\mathbf{v}_{P'}$ is perpendicular to the radius $OP$, and $\mathbf{v}_{P/\mathcal{S}}$ is directed along the slot. Draw the velocity triangle corresponding to Eq. (1) (see Fig. 2). From the triangle, you can compute

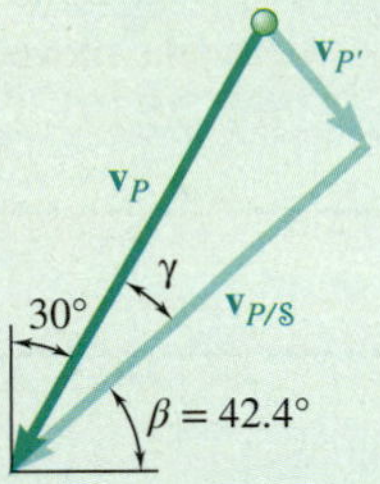

$$\gamma = 90° - 42.4 - 30° = 17.6°$$
$$v_{P'} = v_P \sin \gamma = (500 \text{ mm/s}) \sin 17.6°$$
$$\mathbf{v}_{P'} = 151.2 \text{ mm/s} \, \text{⦨} \, 42.4°$$
$$v_{P/\mathcal{S}} = v_P \cos \gamma = (500 \text{ mm/s}) \cos 17.6°$$

$$\mathbf{v}_{P/\mathcal{S}} = \mathbf{v}_{P/\mathcal{S}} = 477 \text{ mm/s} \, \text{⦨} \, 42.4° \quad \blacktriangleleft$$

Because $\mathbf{v}_{P'}$ is perpendicular to the radius $OP$, you have

$$v_{P'} = r\omega_\mathcal{S} \qquad 151.2 \text{ mm/s} = (37.1 \text{ mm})\omega_\mathcal{S}$$

$$\boldsymbol{\omega}_\mathcal{S} = \boldsymbol{\omega}_\mathcal{S} = 4.08 \text{ rad/s} \, \circlearrowleft \quad \blacktriangleleft$$

**Fig. 2** Vector diagram for the velocity of point $P$.

**REFLECT and THINK:** The result of the Geneva mechanism is that disk $S$ rotates ¼ turn each time pin $P$ engages, then it remains motionless while pin $P$ rotates around before entering the next slot. Disk $D$ rotates continuously, but disk $S$ rotates intermittently. An alternative approach to drawing the vector triangle is to use vector algebra, as was done in Sample Prob. 15.18.

## Sample Problem 15.20

In the Geneva mechanism of Sample Prob. 15.19, disk $D$ rotates with a constant counterclockwise angular velocity $\boldsymbol{\omega}_D$ of 10 rad/s. At the instant when $\phi = 150°$, determine the angular acceleration of disk $S$.

**STRATEGY:** You have two rigid bodies whose motions are related; therefore use rigid-body kinematics. Because point $P$ is moving in a slot, use a rotating reference frame.

**MODELING and ANALYSIS:** Because you are computing accelerations instead of velocities, you need to use Eq. (15.36), which includes the Coriolis acceleration. You found the angular velocity of the frame $\mathcal{S}$ attached to disk $S$ and the velocity of the pin relative to $\mathcal{S}$ in Sample Prob. 15.19:

$$\omega_\mathcal{S} = 4.08 \text{ rad/s} \, \circlearrowleft$$
$$\beta = 42.4° \qquad \mathbf{v}_{P/\mathcal{S}} = 477 \text{ mm/s} \, \text{⦨} \, 42.4°$$

Because pin $P$ moves with respect to the rotating frame $\mathcal{S}$, you have

$$\mathbf{a}_P = \mathbf{a}_{P'} + \mathbf{a}_{P/\mathcal{S}} + \mathbf{a}_c \qquad (1)$$

Investigate each term of this vector equation separately.

**Absolute Acceleration a$_P$.**   Because disk $D$ rotates with a constant angular velocity, the absolute acceleration $\mathbf{a}_P$ is directed toward $B$. This gives

$$a_P = R\omega_D^2 = (50 \text{ mm})(10 \text{ rad/s})^2 = 5000 \text{ mm/s}^2$$

$$\mathbf{a}_P = 5000 \text{ mm/s}^2 \; \measuredangle \; 30°$$

**Acceleration a$_{P'}$ of the Coinciding Point $P'$.**   Resolve into normal and tangential components the acceleration $\mathbf{a}_{P'}$ of the point $P'$ of the frame $\mathcal{S}$ that coincides with $P$ at the given instant. (Recall from Sample Prob. 15.19 that $r = 37.1$ mm.)

$$(a_{P'})_n = r\omega_{\mathcal{S}}^2 = (37.1 \text{ mm})(4.08 \text{ rad/s})^2 = 618 \text{ mm/s}^2$$

$$(\mathbf{a}_{P'})_n = 618 \text{ mm/s}^2 \; \measuredangle \; 42.4°$$

$$(a_{P'})_t = r\alpha_{\mathcal{S}} = 37.1\alpha_{\mathcal{S}} \quad (\mathbf{a}_{P'}) = 37.1\alpha_{\mathcal{S}} \; \measuredangle \; 42.4°$$

**Relative Acceleration a$_{P/\mathcal{S}}$.**   Because the pin $P$ moves in a straight slot cut in disk $S$, the relative acceleration $\mathbf{a}_{P/\mathcal{S}}$ must be parallel to the slot; that is, its direction must be $\measuredangle\,42.4°$.

**Coriolis Acceleration a$_C$.**   Rotating the relative velocity $\mathbf{v}_{P/\mathcal{S}}$ through 90° in the sense of $\boldsymbol{\omega}_{\mathcal{S}}$, you obtain the direction of the Coriolis component of the acceleration:

$$a_C = 2\omega_{\mathcal{S}} v_{P/\mathcal{S}} = 2(4.08 \text{ rad/s})(477 \text{ mm/s}) = 3890 \text{ mm/s}^2$$

$$\mathbf{a}_C = 3890 \text{ mm/s}^2 \; 42.4°$$

Rewrite Eq. (1) and substitute the accelerations found (Fig. 1):

$$\mathbf{a}_P = (\mathbf{a}_{P'})_n + (\mathbf{a}_{P'})_t + \mathbf{a}_{P/\mathcal{S}} + \mathbf{a}_C$$

$$[5000 \; \measuredangle \; 30°] = [618 \; \measuredangle \; 42.4°] + [37.1\alpha_{\mathcal{S}} \; \measuredangle \; 42.4°]$$

$$+ [a_{P/\mathcal{S}} \; \measuredangle \; 42.4°] + [3890 \; \measuredangle \; 42.4°]$$

Equating components in a direction perpendicular to the slot,

$$5000 \cos 17.6° = 37.1\alpha_{\mathcal{S}} - 3890$$

$$\boldsymbol{\alpha}_S = \alpha_{\mathcal{S}} = 233 \text{ rad/s}^2 \; \circlearrowleft \quad \blacktriangleleft$$

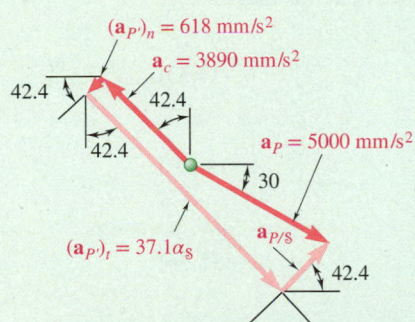

**Fig. 1**  Vector polygon for the acceleration of point $P$.

**REFLECT and THINK:**  It seems reasonable that, because disk $S$ starts and stops over the very short time intervals when pin $P$ is engaged in the slots, the disk must have a very large angular acceleration. An alternative approach would have been to use the vector algebra.

# SOLVING PROBLEMS
# ON YOUR OWN

In this section, you studied the rate of change of a vector with respect to a rotating frame and then applied that idea to the analysis of the plane motion of a particle relative to a rotating frame.

**1. Rate of change of a vector with respect to a fixed frame and with respect to a rotating frame.** Denoting the rate of change of a vector $\mathbf{Q}$ with respect to a fixed frame $OXYZ$ by $(\dot{\mathbf{Q}})_{OXYZ}$ and its rate of change with respect to a rotating frame $Oxyz$ by $(\dot{\mathbf{Q}})_{Oxyz}$, we obtained the fundamental relation

$$(\dot{\mathbf{Q}})_{OXYZ} = (\dot{\mathbf{Q}})_{Oxyz} + \mathbf{\Omega} \times \mathbf{Q} \tag{15.31}$$

where $\mathbf{\Omega}$ is the angular velocity of the rotating frame.

You can now apply this fundamental relation to the solution of two-dimensional problems.

**2. Plane motion of a particle relative to a rotating frame.** Using Eq. (15.31) and designating the rotating frame by $\mathscr{F}$, we obtained the following expressions for the velocity and the acceleration of a particle $P$:

$$\mathbf{v}_P = \mathbf{v}_{P'} + \mathbf{v}_{P/\mathscr{F}} \tag{15.33}$$

or

$$\mathbf{v}_P = \mathbf{v}_O + \mathbf{\Omega} \times \mathbf{r} + \mathbf{v}_{\text{rel}} \tag{15.32'}$$

and

$$\mathbf{a}_P = \mathbf{a}_{P'} + \mathbf{a}_{P/\mathscr{F}} + \mathbf{a}_C \tag{15.36}$$

or

$$\mathbf{a}_P = \mathbf{a}_O + \dot{\mathbf{\Omega}} \times \mathbf{r} - \Omega^2 \mathbf{r} + 2\mathbf{\Omega} \times \mathbf{v}_{\text{rel}} + \mathbf{a}_{\text{rel}} \tag{15.35'}$$

The notation in Eqs. (15.33) and (15.36) is as follows.

    **a.  The subscript $P$** refers to the absolute motion of the particle $P$; that is, to its motion with respect to a fixed or newtonian frame of reference $OXY$.

    **b.  The subscript $P'$** refers to the motion of the point $P'$ of the rotating frame $\mathscr{F}$ that coincides with $P$ at the instant considered.

    **c.  The subscript $P/\mathscr{F}$** refers to the motion of the particle $P$ relative to the rotating frame $\mathscr{F}$.

    **d.  The term $a_C$ represents the Coriolis acceleration of point $P$.** Its magnitude is $2\Omega v_{P/\mathscr{F}}$, and its direction is found by rotating $\mathbf{v}_{P/\mathscr{F}}$ through 90° in the sense of rotation of the frame $\mathscr{F}$.

You should keep in mind that you need to take the Coriolis acceleration into account whenever a point has a relative velocity in a rotating frame. The problems you will encounter in this section involve collars that slide on rotating rods, booms that extend from cranes rotating in a vertical plane, etc.

When solving a problem involving a rotating frame, you can either (a) draw vector diagrams representing Eqs. (15.33) and (15.36), respectively, and use these diagrams to obtain either an analytical or a graphical solution, or (b) use vector algebra.

# Problems

## CONCEPT QUESTIONS

**15.CQ8** A person walks radially inward on a platform that is rotating counterclockwise about its center. Knowing that the platform has a constant angular velocity $\omega$ and the person walks with a constant speed $\mathbf{u}$ relative to the platform, what is the direction of the acceleration of the person at the instant shown?
 a. Negative $x$
 b. Negative $y$
 c. Negative $x$ and positive $y$
 d. Positive $x$ and positive $y$
 e. Negative $x$ and negative $y$

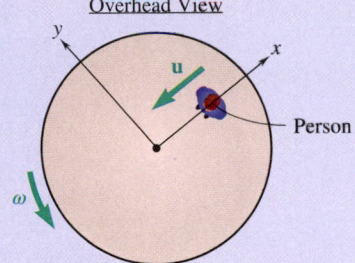

Fig. P15.CQ8

## END-OF-SECTION PROBLEMS

**15.150** The motion of pin $P$ is guided by slots cut in rods $AE$ and $BD$. Knowing that the rods rotate with the constant angular velocities $\omega_A = 4$ rad/s clockwise and $\omega_B = 5$ rad/s clockwise, determine the velocity of pin $P$ for the position shown.

**15.151** The motion of pin $P$ is guided by slots cut in rods $AE$ and $BD$. Knowing that the rods rotate with the constant angular velocities $\omega_A = 4$ rad/s clockwise and $\omega_B = 5$ rad/s counterclockwise, determine the velocity of pin $P$ for the position shown.

**15.152 and 15.153** Two rotating rods are connected by slider block $P$. The rod attached at $A$ rotates with a constant clockwise angular velocity $\omega_A$. For the given data, determine for the position shown (*a*) the angular velocity of the rod attached at $B$, (*b*) the relative velocity of slider block $P$ with respect to the rod on which it slides.
 **15.152**  $b = 8$ in., $\omega_A = 6$ rad/s.
 **15.153**  $b = 300$ mm, $\omega_A = 10$ rad/s.

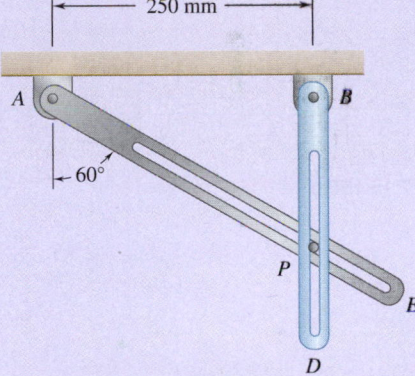

Fig. P15.150 and P15.151

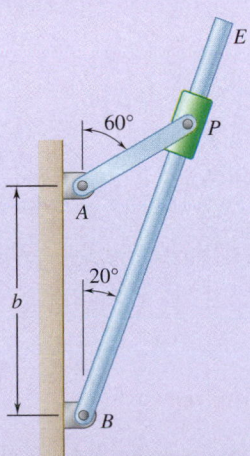

Fig. P15.152

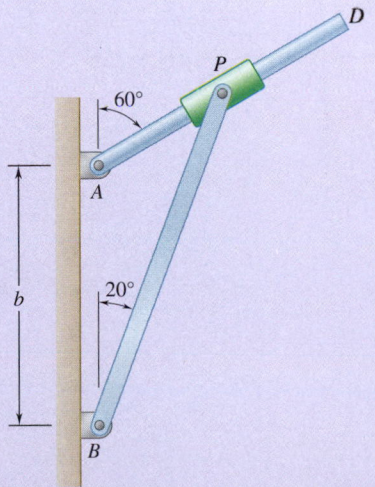

Fig. P15.153

**15.154** Pin $P$ is attached to the wheel shown and slides in a slot cut in bar $BD$. The wheel rolls to the right without slipping with a constant angular velocity of 20 rad/s. Knowing that $x = 480$ mm when $\theta = 0$, determine the angular velocity of the bar and the relative velocity of pin $P$ with respect to the rod when (a) $\theta = 0$, (b) $\theta = 90°$.

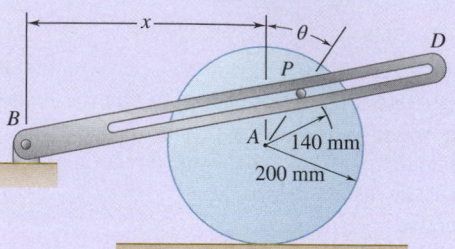

**Fig. P15.154**

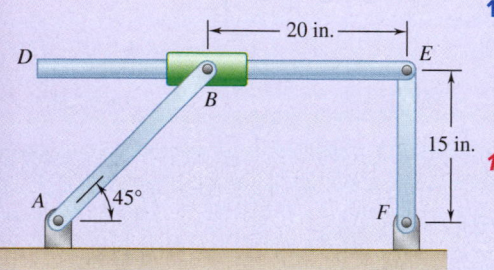

**Fig. P15.155 and *P15.156***

**15.155** Knowing that at the instant shown the angular velocity of bar $AB$ is 15 rad/s clockwise and the angular velocity of bar $EF$ is 10 rad/s clockwise, determine (a) the angular velocity of rod $DE$, (b) the relative velocity of collar $B$ with respect to rod $DE$.

**15.156** Knowing that at the instant shown the angular velocity of rod $DE$ is 10 rad/s clockwise and the angular velocity of bar $EF$ is 15 rad/s counterclockwise, determine (a) the angular velocity of bar $AB$, (b) the relative velocity of collar $B$ with respect to rod $DE$.

**15.157** The motion of pin $P$ is guided by slots cut in rods $AD$ and $BE$. Knowing that bar $AD$ has a constant angular velocity of 4 rad/s clockwise and bar $BE$ has an angular velocity of 5 rad/s counterclockwise and is slowing down at a rate of 2 rad/s², determine the velocity of $P$ for the position shown.

**15.158** Four pins slide in four separate slots cut in a circular plate as shown. When the plate is at rest, each pin has a velocity directed as shown and of the same constant magnitude $u$. If each pin maintains the same velocity relative to the plate when the plate rotates about $O$ with a constant counterclockwise angular velocity $\omega$, determine the acceleration of each pin.

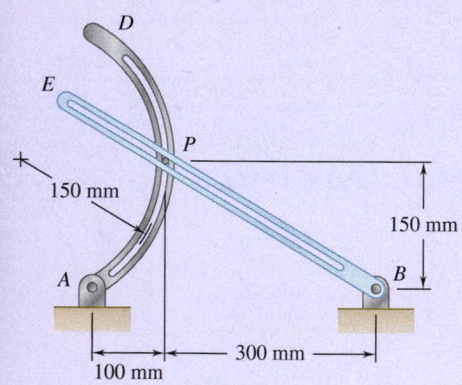

**Fig. *P15.157***

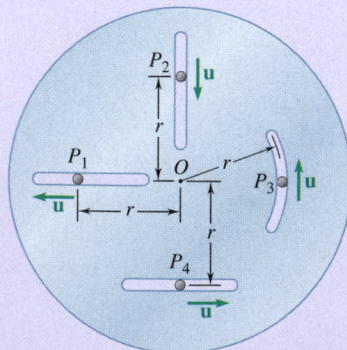

**Fig. *P15.158***

**15.159** Solve Prob. 15.158, assuming that the plate rotates about $O$ with a constant clockwise angular velocity $\omega$.

**15.160** The cage of a mine elevator moves downward at a constant speed of 12.2 m/s. Determine the magnitude and direction of the Coriolis acceleration of the cage if the elevator is located (*a*) at the equator, (*b*) at latitude 40° north, (*c*) at latitude 40° south.

**15.161** Pin *P* is attached to the collar shown; the motion of the pin is guided by a slot cut in bar *BD* and by the collar that slides on rod *AE*. Rod *AE* rotates with a constant angular velocity of 6 rad/s clockwise and the distance from *A* to *P* increases at a constant rate of 8 ft/s. Determine at the instant shown (*a*) the angular acceleration of bar *BD*, (*b*) the relative acceleration of pin *P* with respect to bar *BD*.

**15.162** A rocket sled is tested on a straight track that is built along a meridian. Knowing that the track is located at latitude 40° north, determine the Coriolis acceleration of the sled when it is moving north at a speed of 900 km/h.

**15.163** Solve the Geneva mechanism of Sample Prob. 15.20 using vector algebra.

**15.164** At the instant shown, the length of the boom *AB* is being *decreased* at the constant rate of 0.2 m/s and the boom is being lowered at the constant rate of 0.08 rad/s. Determine (*a*) the velocity of point *B*, (*b*) the acceleration of point *B*.

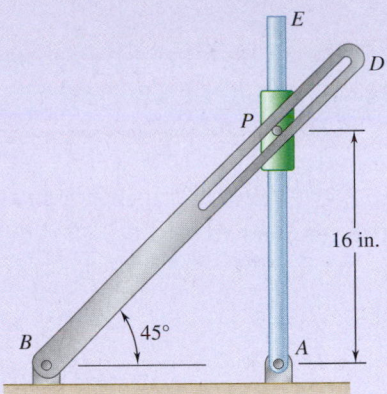

**Fig. P15.161**

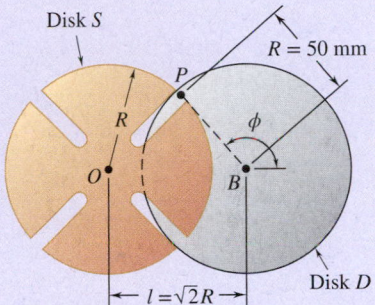

**Fig. P15.163**

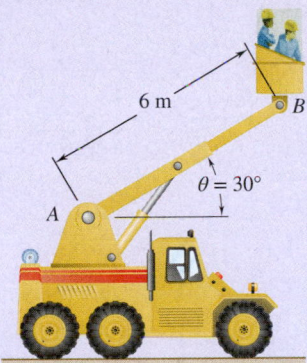

**Fig. P15.164 and P15.165**

**15.165** At the instant shown, the length of the boom *AB* is being *increased* at the constant rate of 0.2 m/s, the boom is being lowered at the constant rate of 0.08 rad/s, and the truck is moving forward with a speed of 0.3 m/s and is slowing down at 0.05 m/s². Determine (*a*) the velocity of point *B*, (*b*) the acceleration of point *B*.

**15.166** In the automated welding setup shown, the position of the two welding tips *G* and *H* is controlled by the hydraulic cylinder *D* and rod *BC*. The cylinder is bolted to the vertical plate that at the instant shown rotates counterclockwise about *A* with a constant angular velocity of 1.6 rad/s. Knowing that at the same instant the length *EF* of the welding assembly is increasing at the constant rate of 300 mm/s, determine (*a*) the velocity of tip *H*, (*b*) the acceleration of tip *H*.

**15.167** In the automated welding setup shown, the position of the two welding tips *G* and *H* is controlled by the hydraulic cylinder *D* and rod *BC*. The cylinder is bolted to the vertical plate that at the instant shown rotates counterclockwise about *A* with a constant angular velocity of 1.6 rad/s. Knowing that at the same instant the length *EF* of the welding assembly is increasing at the constant rate of 300 mm/s, determine (*a*) the velocity of tip *G*, (*b*) the acceleration of tip *G*.

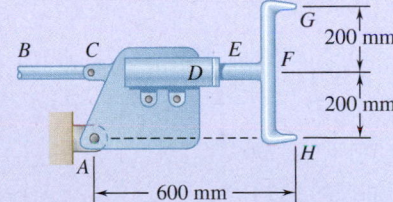

**Fig. P15.166 and P15.167**

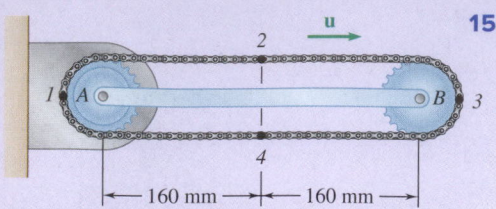

**Fig. P15.168 and P15.169**

**15.168 and 15.169** A chain is looped around two gears of radius 40 mm that can rotate freely with respect to the 320-mm arm *AB*. The chain moves about arm *AB* in a clockwise direction at the constant rate of 80 mm/s relative to the arm. Knowing that in the position shown arm *AB* rotates clockwise about *A* at the constant rate $\omega = 0.75$ rad/s, determine the acceleration of each of the chain links indicated.

      **15.168** Links 1 and 2
      **15.169** Links 3 and 4

**15.170** A basketball player shoots a free throw in such a way that his shoulder can be considered a pin joint at the moment of release as shown. Knowing that at the instant shown the upper arm *SE* has a constant angular velocity of 2 rad/s counterclockwise and the forearm *EW* has a constant clockwise angular velocity of 4 rad/s with respect to *SE*, determine the velocity and acceleration of the wrist *W*.

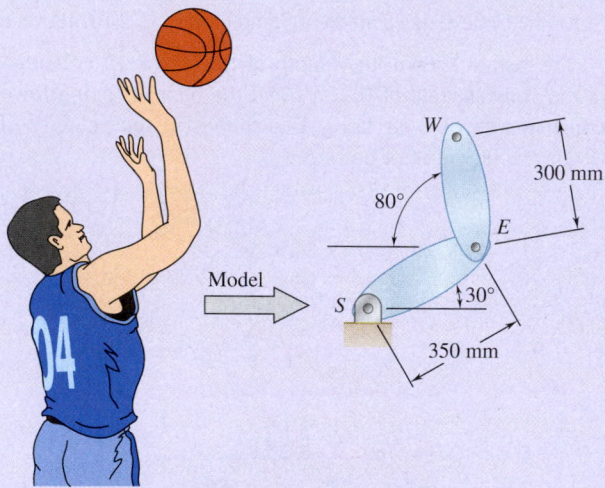

**Fig. P15.170**

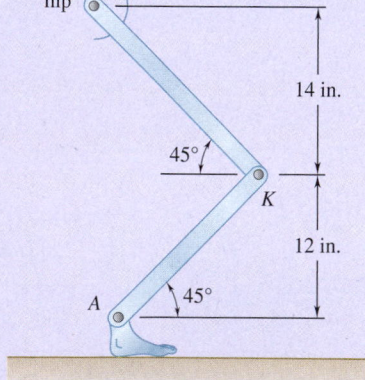

**Fig. P15.171**

**15.171** The human leg can be crudely approximated as two rigid bars (the femur and the tibia) connected with a pin joint. At the instant shown, the velocity of the ankle *A* is zero, the tibia *AK* has an angular velocity of 1.5 rad/s counterclockwise and an angular acceleration of 1 rad/s² counterclockwise. Determine the relative angular velocity and relative angular acceleration of the femur *KH* with respect to *AK* so that the velocity and acceleration of *H* are both straight up at this instant.

**15.172** The collar *P* slides outward at a constant relative speed *u* along rod *AB*, which rotates counterclockwise with a constant angular velocity of 20 rpm. Knowing that $r = 250$ mm when $\theta = 0$ and that the collar reaches *B* when $\theta = 90°$, determine the magnitude of the acceleration of the collar *P* just as it reaches *B*.

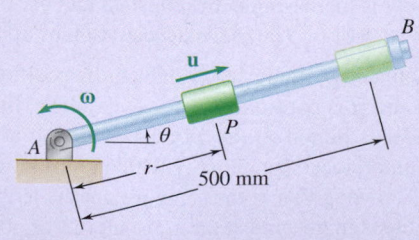

**Fig. P15.172**

**15.173** Pin $P$ slides in a circular slot cut in the plate shown at a constant relative speed $u = 90$ mm/s. Knowing that at the instant shown the plate rotates clockwise about $A$ at the constant rate $\omega = 3$ rad/s, determine the acceleration of the pin if it is located at (a) point $A$, (b) point $B$, (c) point $C$.

**15.174** Rod $AD$ is bent in the shape of an arc of a circle with a radius of $b = 150$ mm. The position of the rod is controlled by pin $B$ that slides in a horizontal slot and also slides along the rod. Knowing that at the instant shown pin $B$ moves to the right at a constant speed of 75 mm/s, determine (a) the angular velocity of the rod, (b) the angular acceleration of the rod.

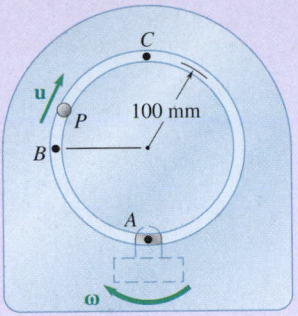

**Fig. P15.173**

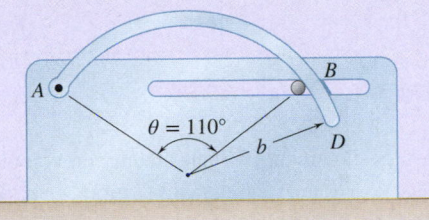

**Fig. P15.174**

**15.175** Solve Prob. 15.174 when $\theta = 90°$.

**15.176** Knowing that at the instant shown the rod attached at $A$ has an angular velocity of 5 rad/s counterclockwise and an angular acceleration of 2 rad/s² clockwise, determine the angular velocity and the angular acceleration of the rod attached at $B$.

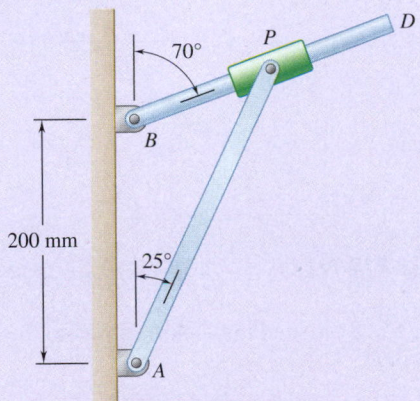

**Fig. P15.176**

**15.177** The Geneva mechanism shown is used to provide an intermittent rotary motion of disk $S$. Disk $D$ rotates with a constant counterclockwise angular velocity $\omega_D$ of 8 rad/s. A pin $P$ is attached to disk $D$ and can slide in one of the six equally spaced slots cut in disk $S$. It is desirable that the angular velocity of disk $S$ be zero as the pin enters and leaves each of the six slots; this will occur if the distance between the centers of the disks and the radii of the disks are related as shown. Determine the angular velocity and angular acceleration of disk $S$ at the instant when $\phi = 150°$.

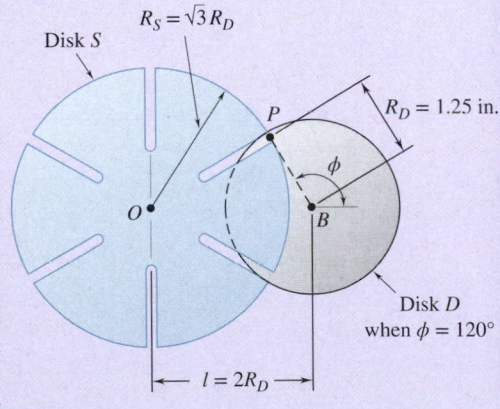

**Fig. P15.177**

**15.178** In Prob. 15.177, determine the angular velocity and angular acceleration of disk $S$ at the instant when $\phi = 135°$.

**15.179** At the instant shown, bar $BC$ has an angular velocity of 3 rad/s and an angular acceleration of 2 rad/s$^2$, both counterclockwise; determine the angular acceleration of the plate.

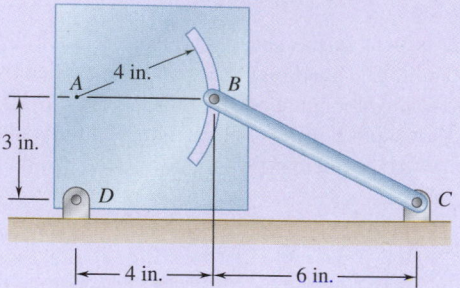

**Fig. P15.179 and P15.180**

**15.180** At the instant shown, bar $BC$ has an angular velocity of 3 rad/s and an angular acceleration of 2 rad/s$^2$, both clockwise; determine the angular acceleration of the plate.

**\*15.181** Rod $AB$ passes through a collar that is welded to link $DE$. Knowing that at the instant shown block $A$ moves to the right at a constant speed of 75 in./s, determine (a) the angular velocity of rod $AB$, (b) the velocity relative to the collar of the point of the rod in contact with the collar, (c) the acceleration of the point of the rod in contact with the collar. (*Hint:* Rod $AB$ and link $DE$ have the same $\omega$ and the same $\alpha$.)

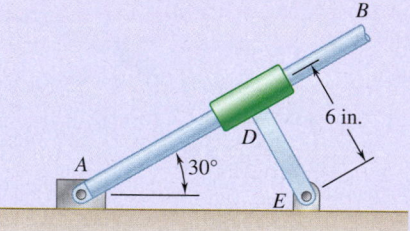

**Fig. P15.181**

**\*15.182** Solve Prob. 15.181 assuming block $A$ moves to the left at a constant speed of 75 in./s.

**\*15.183** In Prob. 15.157, determine the acceleration of pin $P$.

# *15.6 MOTION OF A RIGID BODY IN SPACE

Extending the study of motion in two dimensions to analyzing three-dimensional motion uses most of the same concepts as before, but with some added computational complexity. We introduce these ideas in this section and the next, and we will return to them when discussing kinetics of a rigid body in Chap. 18.

## 15.6A Motion About a Fixed Point

In Sec. 15.1B, we considered the motion of a rigid body constrained to rotate about a fixed axis. Here we examine the more general case of the three-dimensional motion of a rigid body that has a fixed point $O$. First, we prove:

> **The most general displacement of a rigid body with a fixed point $O$ is equivalent to a rotation of the body about an axis through $O$.**

This statement is known as Euler's theorem. We analyze the motion of a sphere with a center $O$; this analysis can be extended to a rigid body of any shape. Because three points define the position of a solid in space, we let the center $O$ and two points $A$ and $B$ on the surface of the sphere define the position of the sphere and thus the position of the body. Let $A_1$ and $B_1$ characterize the position of the sphere at one instant, and let $A_2$ and $B_2$ characterize its position at a later instant (Fig. 15.31a). Because the sphere is rigid, the lengths of the arcs of great circles $A_1B_1$ and $A_2B_2$ must be equal, but except for this requirement, the positions of $A_1$, $A_2$, $B_1$, and $B_2$ are arbitrary. We will show that the points $A$ and $B$ can be brought, respectively, from $A_1$ and $B_1$ into $A_2$ and $B_2$ by a single rotation of the sphere about an axis.

For convenience, and without loss of generality, we select point $B$ so that its initial position coincides with the final position of $A$; thus, $B_1 = A_2$ (Fig. 15.31b). We draw the arcs of great circles $A_1A_2$, $A_2B_2$, and the arcs bisecting, respectively, $A_1A_2$ and $A_2B_2$. Let $C$ be the point of intersection of these last two arcs. We complete the construction by drawing $A_1C$, $A_2C$, and $B_2C$. As pointed out above, because of the rigidity of the sphere, $A_1B_1 = A_2B_2$. Because $C$ is by construction equidistant from $A_1$, $A_2$, and $B_2$, we also have $A_1C = A_2C = B_2C$. As a result, the spherical triangles $A_1CA_2$ and $B_1CB_2$ are congruent, and the angles $A_1CA_2$ and $B_1CB_2$ are equal. Denoting the common value of these angles by $\theta$, we conclude that the sphere can be brought from its initial position into its final position by a single rotation through $\theta$ about the axis $OC$.

It follows that we can consider the motion during a time interval $\Delta t$ of a rigid body with a fixed point $O$ as a rotation through $\Delta\theta$ about a certain axis. Drawing a vector with a magnitude of $\Delta\theta/\Delta t$ along that axis and letting $\Delta t$ approach zero, we obtain in the limit the **instantaneous axis of rotation** and the angular velocity $\boldsymbol{\omega}$ of the body at the instant considered (Fig. 15.32). We can then obtain the velocity of a particle $P$ of the body, as in Sec. 15.1B, by forming the vector product of $\boldsymbol{\omega}$ and of the position vector $\mathbf{r}$ of the particle:

$$\mathbf{v} = \frac{d\mathbf{r}}{dt} = \boldsymbol{\omega} \times \mathbf{r} \tag{15.37}$$

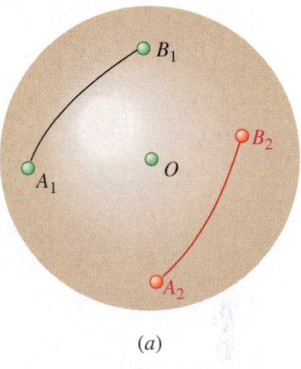

(a)

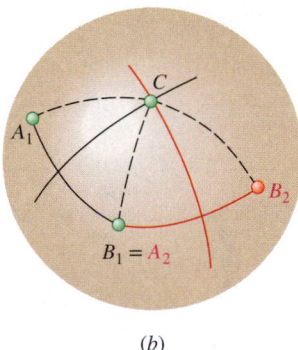

(b)

**Fig. 15.31** (a) Positions of two points on a rotating sphere; (b) the sphere can be brought into this new position by a single rotation.

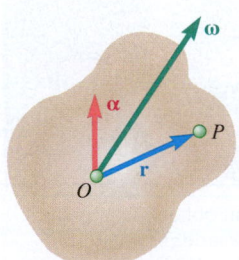

**Fig. 15.32** Angular velocity and angular acceleration of a rigid body moving about a fixed point $O$.

We obtain the acceleration of the particle by differentiating Eq. (15.37) with respect to *t*. As in Sec. 15.1B, we have

$$\mathbf{a} = \boldsymbol{\alpha} \times \mathbf{r} + \boldsymbol{\omega} \times (\boldsymbol{\omega} \times \mathbf{r}) \qquad (15.38)$$

Here we have defined the angular acceleration $\boldsymbol{\alpha}$ as the derivative

$$\boldsymbol{\alpha} = \frac{d\boldsymbol{\omega}}{dt} \qquad (15.39)$$

of the angular velocity $\boldsymbol{\omega}$.

In the case of the motion of a rigid body with a fixed point, the direction of $\boldsymbol{\omega}$ and of the instantaneous axis of rotation changes from one instant to the next. The angular acceleration $\boldsymbol{\alpha}$ therefore reflects the change in direction of $\boldsymbol{\omega}$, as well as its change in magnitude. Thus, in general, $\boldsymbol{\alpha}$ *is not directed along the instantaneous axis of rotation*. Although the particles of the body located on the instantaneous axis of rotation have zero velocity at the instant considered, they do not have zero acceleration. Also, the accelerations of the various particles of the body *cannot* be determined as if the body were rotating permanently about the instantaneous axis.

Recalling the definition of the velocity of a particle with position vector $\mathbf{r}$, we note that the angular acceleration $\boldsymbol{\alpha}$, as expressed in Eq. (15.39), represents the velocity of the tip of vector $\boldsymbol{\omega}$. This property may be useful in determining the angular acceleration of a rigid body. For example, it follows that vector $\boldsymbol{\alpha}$ is tangent to the curve described in space by the tip of vector $\boldsymbol{\omega}$.

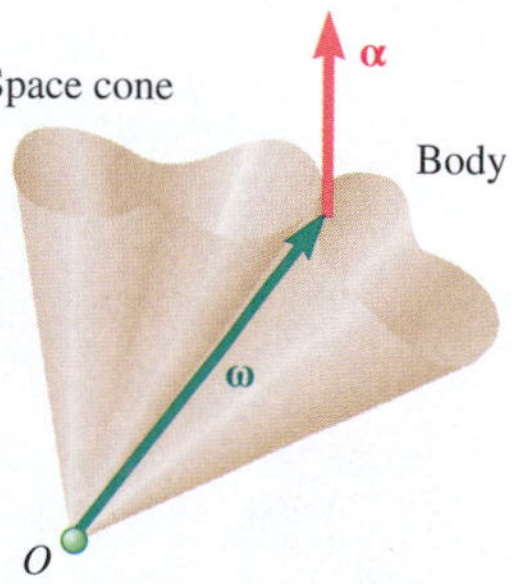

Note that vector $\boldsymbol{\omega}$ moves within the body, as well as in space. It thus generates two cones called, respectively, the **body cone** and the **space cone** (Fig. 15.33).[†] It can be shown that, at any given instant, the two cones are tangent along the instantaneous axis of rotation and that, as the body moves, the body cone appears to *roll* on the space cone.

Before concluding our analysis of the motion of a rigid body with a fixed point, we should prove that angular velocities are actually vectors. Some quantities, such as the *finite rotations* of a rigid body, have magnitude and direction but do not obey the parallelogram law of addition; these quantities cannot be considered to be vectors. In contrast, angular velocities (and also *infinitesimal rotations*), as we demonstrate presently, do obey the parallelogram law and thus are truly vector quantities.

Consider a rigid body with a fixed point *O* that rotates at a given instant simultaneously about the axes *OA* and *OB* with angular velocities $\boldsymbol{\omega}_1$ and $\boldsymbol{\omega}_2$ (Fig. 15.34*a*). We know that this motion must be equivalent at the instant considered to a single rotation of angular velocity $\boldsymbol{\omega}$. We propose to show that

$$\boldsymbol{\omega} = \boldsymbol{\omega}_1 + \boldsymbol{\omega}_2 \qquad (15.40)$$

that is, that we can obtain the resulting angular velocity by adding $\boldsymbol{\omega}_1$ and $\boldsymbol{\omega}_2$ using the parallelogram law (Fig. 15.34*b*).

Consider a particle *P* of the body, defined by the position vector $\mathbf{r}$. Denoting the velocity of *P* when the body rotates about *OA* only, about *OB* only, and about both axes simultaneously, by $\mathbf{v}_1$, $\mathbf{v}_2$, and $\mathbf{v}$, respectively, we have

$$\mathbf{v} = \boldsymbol{\omega} \times \mathbf{r} \qquad \mathbf{v}_1 = \boldsymbol{\omega}_1 \times \mathbf{r} \qquad \mathbf{v}_2 = \boldsymbol{\omega}_2 \times \mathbf{r} \qquad (15.41)$$

Space cone

α

Body cone

ω

O

**Fig. 15.33** The angular velocity vector generates a body cone and a space cone as it changes direction.

**Photo 15.8** You can obtain the angular velocity of a fire truck ladder rotating about its fixed base by adding the angular velocities that correspond to simultaneous rotations about two different axes. ©Syracuse Newspapers/M Greenlar/The Image Works

---

[†]Recall that a cone is, by definition, a surface generated by a straight line passing through a fixed point. In general, the cones considered here are not circular cones.

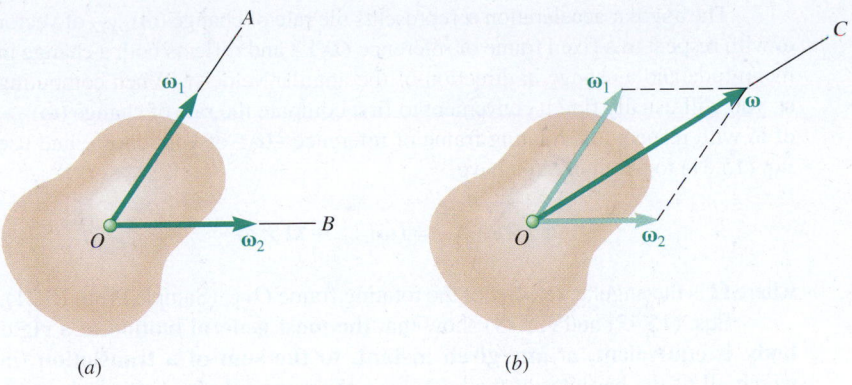

**Fig. 15.34** (a) A rigid body rotating about two axes simultaneously; (b) the motion is equivalent to a single rotation with angular velocity equal to the vector sum of the initial angular velocities.

But the vectorial character of *linear* velocities is well established (because they represent the derivatives of position vectors). We therefore have

$$\mathbf{v} = \mathbf{v}_1 + \mathbf{v}_2$$

where the plus sign indicates vector addition. Substituting from Eq. (15.41), we obtain

$$\boldsymbol{\omega} \times \mathbf{r} = \boldsymbol{\omega}_1 \times \mathbf{r} + \boldsymbol{\omega}_2 \times \mathbf{r}$$
$$\boldsymbol{\omega} \times \mathbf{r} = (\boldsymbol{\omega}_1 + \boldsymbol{\omega}_2) \times \mathbf{r}$$

where the plus sign still indicates vector addition. Because the relation obtained holds for an arbitrary **r**, we conclude that Eq. (15.40) must be true.

## *15.6B    General Motion

We now consider the most general motion of a rigid body in space. Let $A$ and $B$ be two particles of the body. Recall from Sec. 11.4D that we can express the velocity of $B$ with respect to the fixed frame of reference $OXYZ$ as

$$\mathbf{v}_B = \mathbf{v}_A + \mathbf{v}_{B/A} \tag{15.42}$$

where $\mathbf{v}_{B/A}$ is the velocity of $B$ relative to a frame $AX'Y'Z'$ attached to $A$ and of fixed orientation (Fig. 15.35). Because $A$ is fixed in this frame, the motion of the body relative to $AX'Y'Z'$ is the motion of a body with a fixed point. Therefore, we can obtain the relative velocity $\mathbf{v}_{B/A}$ from Eq. (15.37) after replacing **r** by the position vector $\mathbf{r}_{B/A}$ of $B$ relative to $A$. Substituting for $\mathbf{v}_{B/A}$ into Eq. (15.42), we have

$$\mathbf{v}_B = \mathbf{v}_A + \boldsymbol{\omega} \times \mathbf{r}_{B/A} \tag{15.43}$$

where $\boldsymbol{\omega}$ is the angular velocity of the body at the instant considered.

We can obtain the acceleration of $B$ by a similar reasoning. We first write

$$\mathbf{a}_B = \mathbf{a}_A + \mathbf{a}_{B/A}$$

and, from Eq. (15.38),

$$\mathbf{a}_B = \mathbf{a}_A + \boldsymbol{\alpha} \times \mathbf{r}_{B/A} + \boldsymbol{\omega} \times (\boldsymbol{\omega} \times \mathbf{r}_{B/A}) \tag{15.44}$$

where $\boldsymbol{\alpha}$ is the angular acceleration of the body at the instant considered.

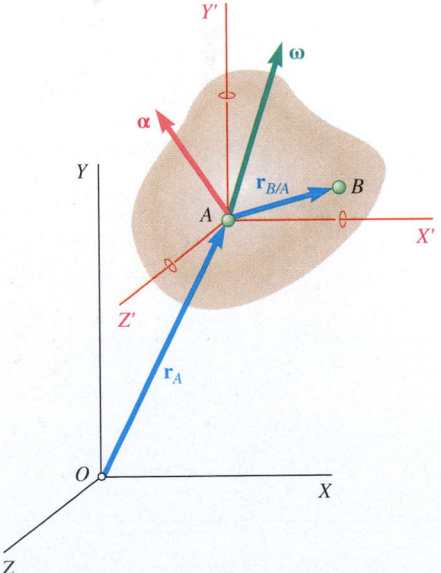

**Fig. 15.35** A rigid body moving relative to a fixed reference frame $OXYZ$ and a reference frame attached to the body but with fixed orientation, $OX'Y'Z'$.

The angular acceleration $\boldsymbol{\alpha}$ represents the rate of change $(\dot{\boldsymbol{\omega}})_{OXYZ}$ of vector $\boldsymbol{\omega}$ with respect to a fixed frame of reference $OXYZ$ and reflects both a change in magnitude and a change in direction of the angular velocity. When computing $\boldsymbol{\alpha}$, you will usually find it convenient to first compute the rate of change $(\dot{\boldsymbol{\omega}})_{Oxyz}$ of $\boldsymbol{\omega}$ with respect to a rotating frame of reference $Oxyz$ of your choice and use Eq. (15.31) to obtain $\boldsymbol{\alpha}$. You have

$$\boldsymbol{\alpha} = (\dot{\boldsymbol{\omega}})_{OXYZ} = (\dot{\boldsymbol{\omega}})_{Oxyz} + \boldsymbol{\Omega} \times \boldsymbol{\omega}$$

where $\boldsymbol{\Omega}$ is the angular velocity of the rotating frame $Oxyz$ (Sample Prob. 15.21).

Eqs. (15.43) and (15.44) show that **the most general motion of a rigid body is equivalent, at any given instant, to the sum of a translation** (in which all of the particles of the body have the same velocity and acceleration as a reference particle $A$) **and of a motion in which particle $A$ is assumed to be fixed.**[†]

By solving Eqs. (15.43) and (15.44) for $\mathbf{v}_A$ and $\mathbf{a}_A$, it can be shown that the motion of the body with respect to a frame attached to $B$ would be characterized by the same vectors $\boldsymbol{\omega}$ and $\boldsymbol{\alpha}$ as its motion relative to $AX'Y'Z'$. Thus, the angular velocity and angular acceleration of a rigid body at a given instant are independent of the choice of reference point. If $AX'Y'Z'$ is a non-rotating frame, you should keep in mind that whether the moving frame is attached to $A$ or to $B$, it should maintain a fixed orientation; that is, it should remain parallel to the fixed reference frame $OXYZ$ throughout the motion of the rigid body.

In many problems, it will be more convenient to use a moving frame that is allowed to rotate as well as to translate. We discuss the use of such moving frames in Sec. 15.7.

---

[†]Recall from Sec 15.6A that, in general, vectors $\boldsymbol{\omega}$ and $\boldsymbol{\alpha}$ are not collinear and that the accelerations of the particles of the body in their motion relative to the frame $AX'Y'Z'$ cannot be determined as if the body were rotating permanently about the instantaneous axis through $A$.

# Sample Problem 15.21

The crane shown rotates horizontally with a constant angular velocity $\omega_1$ of 0.30 rad/s. Simultaneously, the boom is being raised with a constant angular velocity $\omega_2$ of 0.50 rad/s relative to the cab. Knowing that the length of the boom $OP$ is $l = 12$ m, determine (*a*) the angular velocity $\omega$ of the boom, (*b*) the angular acceleration $\alpha$ of the boom, (*c*) the velocity $\mathbf{v}$ of the tip of the boom, (*d*) the acceleration $\mathbf{a}$ of the tip of the boom.

**STRATEGY:** There are multiple rotational axes, so you need to use the general motion velocity and acceleration kinematic equations. Add the given angular velocities vectorially to find the overall angular velocity of the boom, and differentiate that to find the angular acceleration.

## MODELING and ANALYSIS:

**a. Angular Velocity of Boom.** Add the angular velocity $\omega_1$ of the cab and the angular velocity $\omega_2$ of the boom relative to the cab to obtain the angular velocity $\omega$ of the boom at the instant considered:

$$\omega = \omega_1 + \omega_2 \qquad \omega = (0.30 \text{ rad/s})\mathbf{j} + (0.50 \text{ rad/s})\mathbf{k} \ \blacktriangleleft$$

**b. Angular Acceleration of Boom.** Obtain the angular acceleration $\alpha$ of the boom by differentiating $\omega$. Because the vector $\omega_1$ is constant in magnitude and direction, you have

$$\alpha = \dot{\omega} = \dot{\omega}_1 + \dot{\omega}_2 = 0 + \dot{\omega}_2$$

where the rate of change $\dot{\omega}_2$ is to be computed with respect to the fixed frame $OXYZ$. However, it is more convenient to use a frame $Oxyz$ attached to the cab and rotating with it, because the vector $\omega_2$ also rotates with the cab and therefore has zero rate of change with respect to that frame. Using Eq. (15.31) with $\mathbf{Q} = \omega_2$ and $\mathbf{\Omega} = \omega_1$, you have

$$(\dot{\mathbf{Q}})_{OXYZ} = (\dot{\mathbf{Q}})_{Oxyz} + \mathbf{\Omega} \times \mathbf{Q}$$
$$(\dot{\omega}_2)_{OXYZ} = (\dot{\omega}_2)_{Oxyz} + \omega_1 \times \omega_2$$
$$\alpha = (\dot{\omega}_2)_{OXYZ} = 0 + (0.30 \text{ rad/s})\mathbf{j} \times (0.50 \text{ rad/s})\mathbf{k}$$

$$\alpha = (0.15 \text{ rad/s}^2)\mathbf{i} \ \blacktriangleleft$$

**c. Velocity of Tip of Boom.** Noting that the position vector of point $P$ is $\mathbf{r} = (10.39 \text{ m})\mathbf{i} + (6 \text{ m})\mathbf{j}$ (Fig. 1) and using the expression found for $\omega$ in part (a), you get

$$\mathbf{v} = \omega \times \mathbf{r} = \begin{vmatrix} \mathbf{i} & \mathbf{j} & \mathbf{k} \\ 0 & 0.30 \text{ rad/s} & 0.50 \text{ rad/s} \\ 10.39 \text{ m} & 6 \text{ m} & 0 \end{vmatrix}$$

$$\mathbf{v} = -(3 \text{ m/s})\mathbf{i} + (5.20 \text{ m/s})\mathbf{j} - (3.12 \text{ m/s})\mathbf{k} \ \blacktriangleleft$$

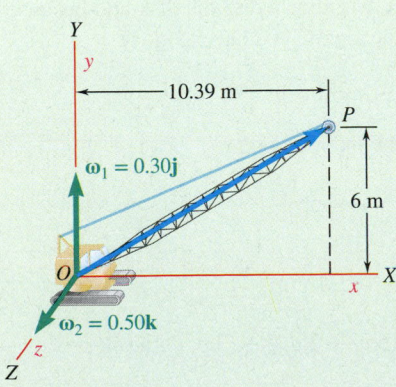

**Fig. 1** A rotating frame *xyz* is attached to the cab.

*(continued)*

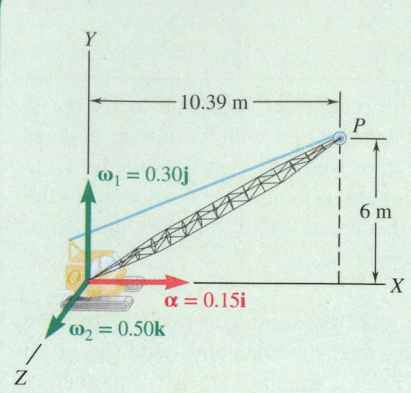

**Fig. 2** Angular velocities and accelerations of the boom.

**d. Acceleration of Tip of Boom.** Recall that $\mathbf{v} = \boldsymbol{\omega} \times \mathbf{r}$. Then, from Fig. 2,

$$\mathbf{a} = \boldsymbol{\alpha} \times \mathbf{r} + \boldsymbol{\omega} \times (\boldsymbol{\omega} \times \mathbf{r}) = \boldsymbol{\alpha} \times \mathbf{r} + \boldsymbol{\omega} \times \mathbf{v}$$

$$\mathbf{a} = \begin{vmatrix} \mathbf{i} & \mathbf{j} & \mathbf{k} \\ 0.15 & 0 & 0 \\ 10.39 & 6 & 0 \end{vmatrix} + \begin{vmatrix} \mathbf{i} & \mathbf{j} & \mathbf{k} \\ 0 & 0.30 & 0.50 \\ -3 & 5.20 & -3.12 \end{vmatrix}$$

$$= 0.90\mathbf{k} - 0.94\mathbf{i} - 2.60\mathbf{i} - 1.50\mathbf{j} + 0.90\mathbf{k}$$

$$\mathbf{a} = -(3.54 \text{ m/s}^2)\mathbf{i} - (1.50 \text{ m/s}^2)\mathbf{j} + (1.80 \text{ m/s}^2)\mathbf{k} \blacktriangleleft$$

**REFLECT and THINK:** The base of the cab acts as the fixed point of the motion. Even though both components of angular velocity are constant, there is an angular acceleration due to the change in direction of the angular velocity $\boldsymbol{\omega}_2$. The angular velocity vector $\boldsymbol{\omega}_2$ changes due to the rotation of the cab, $\boldsymbol{\omega}_1$.

## Sample Problem 15.22

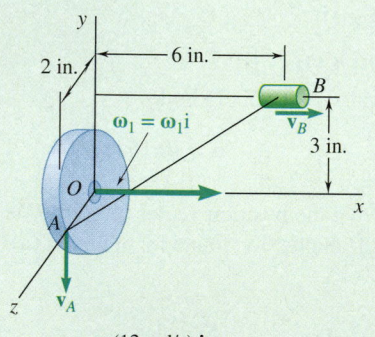

$\boldsymbol{\omega}_1 = (12 \text{ rad/s}) \mathbf{i}$
$\mathbf{r}_A = (2 \text{ in.})\mathbf{k}$
$\mathbf{r}_B = (6 \text{ in.})\mathbf{i} + (3 \text{ in.})\mathbf{j}$
$\mathbf{r}_{B/A} = (6 \text{ in.})\mathbf{i} + (3 \text{ in.})\mathbf{j} - (2 \text{ in.})\mathbf{k}$

**Fig. 1** Angular velocity of the disk and the direction of the velocities of $A$ and $B$.

The rod $AB$ has a length of 7 in. and is attached to the disk by a ball-and-socket connection and to the collar $B$ by a clevis. The disk rotates in the $yz$ plane at a constant rate of $\omega_1 = 12$ rad/s, while the collar is free to slide along the horizontal rod $CD$. For the position $\theta = 0$, determine (a) the velocity of the collar, (b) the angular velocity of the rod.

**STRATEGY:** Use the velocity and acceleration kinematic equations to relate the velocities of points $A$ and $B$.

**MODELING and ANALYSIS:**

**a. Velocity of Collar.** Because point $A$ is attached to the disk and because collar $B$ moves in a direction parallel to the $x$ axis, you have (Fig. 1)

$$\mathbf{v}_A = \boldsymbol{\omega}_1 \times \mathbf{r}_A = 12\mathbf{i} \times 2\mathbf{k} = -24\mathbf{j} \qquad \mathbf{v}_B = v_B\mathbf{i}$$

Denoting the angular velocity of the rod by $\boldsymbol{\omega}$, you obtain

$$\mathbf{v}_B = \mathbf{v}_A + \mathbf{v}_{B/A} = \mathbf{v}_A + \boldsymbol{\omega} \times \mathbf{r}_{B/A}$$

$$v_B\mathbf{i} = -24\mathbf{j} + \begin{vmatrix} \mathbf{i} & \mathbf{j} & \mathbf{k} \\ \omega_x & \omega_y & \omega_z \\ 6 & 3 & -2 \end{vmatrix}$$

$$v_B\mathbf{i} = -24\mathbf{j} + (-2\omega_y - 3\omega_z)\mathbf{i} + (6\omega_z + 2\omega_x)\mathbf{j} + (3\omega_x - 6\omega_y)\mathbf{k}$$

Equating the coefficients of the unit vectors, you get

$$v_B = \qquad -2\omega_y - 3\omega_z \tag{1}$$
$$24 = 2\omega_x \qquad +6\omega_z \tag{2}$$
$$0 = 3\omega_x - 6\omega_y \tag{3}$$

You have three equations and four unknowns in these equations. Fortunately, multiplying Eqs. (1), (2), (3), respectively, by 6, 3, −2 and adding gives you

$$6v_B + 72 = 0 \qquad v_B = -12 \qquad \mathbf{v}_B = -(12 \text{ in./s})\mathbf{i} \quad \blacktriangleleft$$

**b. Angular Velocity of Rod *AB*.**   Note that you cannot determine the angular velocity from Eqs. (1), (2), and (3) because the determinant formed by the coefficients of $\omega_x$, $\omega_y$, and $\omega_z$ is zero. You must therefore obtain an additional equation by considering the constraint imposed by the clevis at *B*.

The collar–clevis connection at *B* permits rotation of *AB* about rod *CD* and also about an axis perpendicular to the plane containing *AB* and *CD*. It prevents rotation of *AB* about the axis *EB*, which is perpendicular to *CD* and lies in the plane containing *AB* and *CD* (Fig. 2). Thus, the projection of **ω** on $\mathbf{r}_{E/B}$ must be zero, and you have

$$\boldsymbol{\omega} \cdot \mathbf{r}_{E/B} = 0$$

$$(\omega_x\mathbf{i} + \omega_y\mathbf{j} + \omega_z\mathbf{k}) \cdot (-3\mathbf{j} + 2\mathbf{k}) = 0$$

$$-3\omega_y + 2\omega_z = 0 \qquad\qquad (4)$$

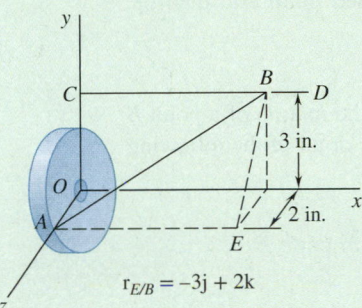

**Fig. 2**  The collar–clevis prevents rotation about *EB*.

$r_{E/B} = -3j + 2k$

Solving Eqs. (1), (2), (3), and (4) simultaneously, you obtain

$$v_B = -12 \qquad \omega_x = 3.69 \qquad \omega_y = 1.846 \qquad \omega_z = 2.77$$

$$\boldsymbol{\omega} = (3.69 \text{ rad/s})\mathbf{i} + (1.846 \text{ rad/s})\mathbf{j} + (2.77 \text{ rad/s})\mathbf{k} \quad \blacktriangleleft$$

**REFLECT and THINK:**   Note that the direction of *EB* is that of the vector triple product

$$\mathbf{r}_{B/C} \times (\mathbf{r}_{B/C} \times \mathbf{r}_{B/A})$$

so you could write

$$\boldsymbol{\omega} \cdot [\mathbf{r}_{B/C} \times (\mathbf{r}_{B/C} \times \mathbf{r}_{B/A})] = 0$$

This formulation would be particularly useful if the rod *CD* were not in a convenient direction.

# SOLVING PROBLEMS
# ON YOUR OWN

In this section, you started the study of the **kinematics of rigid bodies in three dimensions**. You first studied the **motion of a rigid body about a fixed point** and then the **general motion of a rigid body**.

**A. Motion of a rigid body about a fixed point.** To analyze the motion of a point $B$ of a body rotating about a fixed point $O$, you may have to take some or all of the following steps.

**1. Determine the position vector r** connecting the fixed point $O$ to point $B$.

**2. Determine the angular velocity ω of the body** with respect to a fixed frame of reference. You can often obtain the angular velocity $\boldsymbol{\omega}$ by adding two component angular velocities $\boldsymbol{\omega}_1$ and $\boldsymbol{\omega}_2$ (Sample Prob. 15.21).

**3. Compute the velocity of $B$** from the equation

$$\mathbf{v} = \boldsymbol{\omega} \times \mathbf{r} \tag{15.37}$$

Your computation is usually easier if you express the vector product as a determinant.

**4. Determine the angular acceleration α of the body.** The angular acceleration $\boldsymbol{\alpha}$ represents the rate of change $(\dot{\boldsymbol{\omega}})_{OXYZ}$ of vector $\boldsymbol{\omega}$ with respect to a fixed frame of reference $OXYZ$ and reflects both a change in magnitude and a change in direction of the angular velocity. When computing $\boldsymbol{\alpha}$, you will usually find it convenient to first compute the rate of change $(\dot{\boldsymbol{\omega}})_{Oxyz}$ of $\boldsymbol{\omega}$ with respect to a rotating frame of reference $Oxyz$ of your choice and use Eq. (15.31). You have

$$\boldsymbol{\alpha} = (\dot{\boldsymbol{\omega}})_{OXYZ} = (\dot{\boldsymbol{\omega}})_{Oxyz} + \boldsymbol{\Omega} \times \boldsymbol{\omega}$$

where $\boldsymbol{\Omega}$ is the angular velocity of the rotating frame $Oxyz$ (Sample Prob. 15.21).

**5. Compute the acceleration of $B$** by using the equation

$$\mathbf{a} = \boldsymbol{\alpha} \times \mathbf{r} + \boldsymbol{\omega} \times (\boldsymbol{\omega} \times \mathbf{r}) \tag{15.38}$$

Note that the vector product $(\boldsymbol{\omega} \times \mathbf{r})$ represents the velocity of point $B$ and was computed in Step 3. Also, the computation of the first vector product in Eq. (15.38) is often simpler if you express this product in determinant form. Remember that, as was the case with the plane motion of a rigid body, the instantaneous axis of rotation *cannot* be used to determine accelerations.

**B. General motion of a rigid body.** The general motion of a rigid body may be considered as *the sum of a translation and a rotation*. Keep the following in mind:

    **a. In the translation part of the motion,** all of the points of the body have the *same velocity $\mathbf{v}_A$ and the same acceleration $\mathbf{a}_A$* as point $A$ of the body that has been selected as the reference point.

    **b. In the rotation part of the motion,** the same reference point $A$ is treated as if it were a *fixed point*.

**1. To determine the velocity of a point $B$ of the rigid body** when you know the velocity $\mathbf{v}_A$ of the reference point $A$ and the angular velocity $\boldsymbol{\omega}$ of the body, you simply add $\mathbf{v}_A$ to the velocity $\mathbf{v}_{B/A} = \boldsymbol{\omega} \times \mathbf{r}_{B/A}$ of $B$ in its rotation about $A$:

$$\mathbf{v}_B = \mathbf{v}_A + \boldsymbol{\omega} \times \mathbf{r}_{B/A} \tag{15.43}$$

As indicated earlier, the computation of the vector product is usually simpler if you express this product in determinant form.

You can also use Eq. (15.43) to determine the magnitude of $\mathbf{v}_B$ when its direction is known, even if $\boldsymbol{\omega}$ is not known. Although the corresponding three scalar equations are linearly dependent and the components of $\boldsymbol{\omega}$ are indeterminate, you can eliminate these components and find $\mathbf{v}_A$ by using an appropriate linear combination of the three equations [Sample Prob. 15.22, part (a)]. Alternatively, you can assign an arbitrary value to one of the components of $\boldsymbol{\omega}$ and solve the equations for $\mathbf{v}_A$. However, you must seek an additional equation in order to determine the true values of the components of $\boldsymbol{\omega}$ [Sample Prob. 15.22, part (b)].

**2. To determine the acceleration of a point $B$ of the rigid body** when you know the acceleration $\mathbf{a}_A$ of the reference point $A$ and the angular acceleration $\boldsymbol{\alpha}$ of the body, you simply add $\mathbf{a}_A$ to the acceleration of $B$ in its rotation about $A$, as expressed by Eq. (15.38):

$$\mathbf{a}_B = \mathbf{a}_A + \boldsymbol{\alpha} \times \mathbf{r}_{B/A} + \boldsymbol{\omega} \times (\boldsymbol{\omega} \times \mathbf{r}_{B/A}) \tag{15.44}$$

Note that the vector product $(\boldsymbol{\omega} \times \mathbf{r}_{B/A})$ represents the velocity $\mathbf{v}_{B/A}$ of $B$ relative to $A$ and already may have been computed as part of your calculation of $\mathbf{v}_B$.

You can also use the three scalar equations associated with Eq. (15.44) to determine the magnitude of $\mathbf{a}_B$ when its direction is known, even if $\boldsymbol{\omega}$ and $\boldsymbol{\alpha}$ are not known. Although the components of $\boldsymbol{\omega}$ and $\boldsymbol{\alpha}$ are indeterminate, you can assign arbitrary values to one of the components of $\boldsymbol{\omega}$ and to one of the components of $\boldsymbol{\alpha}$ and solve the equations for $\mathbf{a}_B$.

# Problems

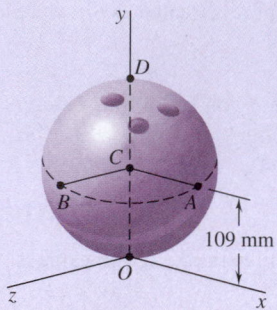

**Fig. P15.184 and P15.185**

**15.184** The bowling ball shown rolls without slipping on the horizontal $xz$ plane with an angular velocity $\boldsymbol{\omega} = \omega_x\mathbf{i} + \omega_y\mathbf{j} + \omega_z\mathbf{k}$. Knowing that $\mathbf{v}_A = (4.8 \text{ m/s})\mathbf{i} - (4.8 \text{ m/s})\mathbf{j} + (3.6 \text{ m/s})\mathbf{k}$ and $\mathbf{v}_D = (9.6 \text{ m/s})\mathbf{i} + (7.2 \text{ m/s})\mathbf{k}$, determine (a) the angular velocity of the bowling ball, (b) the velocity of its center $C$.

**15.185** The bowling ball shown rolls without slipping on the horizontal $xz$ plane with an angular velocity $\boldsymbol{\omega} = \omega_x\mathbf{i} + \omega_y\mathbf{j} + \omega_z\mathbf{k}$. Knowing that $\mathbf{v}_B = (3.6 \text{ m/s})\mathbf{i} - (4.8 \text{ m/s})\mathbf{j} + (4.8 \text{ m/s})\mathbf{k}$ and $\mathbf{v}_D = (7.2 \text{ m/s})\mathbf{i} + (9.6 \text{ m/s})\mathbf{k}$, determine (a) the angular velocity of the bowling ball, (b) the velocity of its center $C$.

**15.186** The blade assembly of an oscillating fan rotates with a constant angular velocity $\boldsymbol{\omega}_1 = -(450 \text{ rpm})\mathbf{i}$ with respect to the motor housing. Determine the angular acceleration of the blade assembly, knowing that at the instant shown the angular velocity and the angular acceleration of the motor housing are, respectively, $\boldsymbol{\omega}_2 = -(3 \text{ rpm})\mathbf{j}$ and $\boldsymbol{\alpha}_2 = 0$.

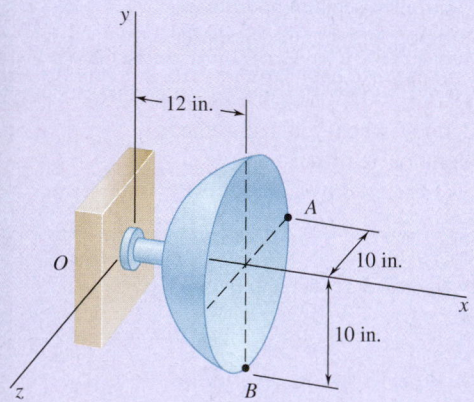

**Fig. P15.187**

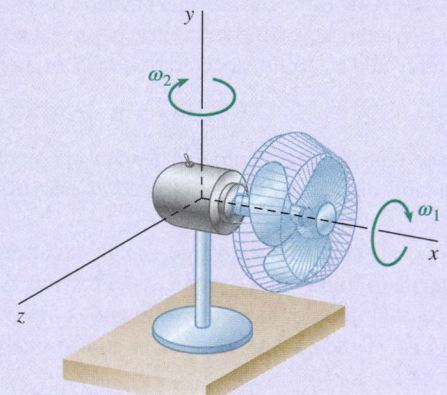

**Fig. P15.186**

**15.187** At the instant considered, the radar antenna shown rotates about the origin of coordinates with an angular velocity $\boldsymbol{\omega} = \omega_x\mathbf{i} + \omega_y\mathbf{j} + \omega_z\mathbf{k}$. Knowing that $(v_A)_y = 15$ in./s, $(v_B)_y = 9$ in./s, and $(v_B)_z = 18$ in./s, determine (a) the angular velocity of the antenna, (b) the velocity of point $A$.

**15.188** The rotor of an electric motor rotates at the constant rate $\omega_1 = 1800$ rpm. Determine the angular acceleration of the rotor as the motor is rotated about the $y$ axis with a constant angular velocity $\boldsymbol{\omega}_2$ of 6 rpm counterclockwise when viewed from the positive $y$ axis.

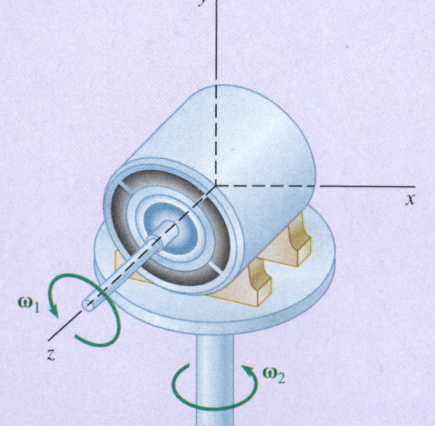

**Fig. P15.188**

**15.189** The disk of a portable sander rotates at the constant rate $\omega_1 = 4400$ rpm as shown. Determine the angular acceleration of the disk as a worker rotates the sander about the $z$ axis with an angular velocity of 0.5 rad/s and an angular acceleration of 2.5 rad/s², both clockwise when viewed from the positive $z$ axis.

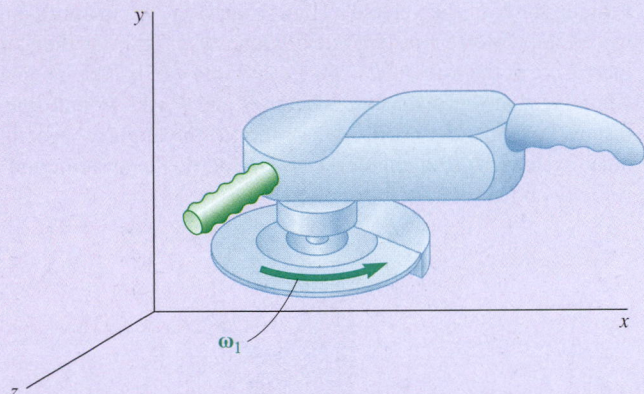

**Fig. P15.189**

**15.190** A flight simulator is used to train pilots on how to recognize spatial disorientation. It has four degrees of freedom, and can rotate around a planetary axis as well as in yaw, pitch, and roll. Knowing that the simulator is rotating around the planetary axis with a constant angular velocity of 20 rpm counterclockwise as seen from above, determine the angular acceleration of the cab if (a) the cab has a constant pitch angular velocity of $+3\mathbf{k}$ rad/s, (b) the cab has a constant roll angular velocity of $-4\mathbf{i}$ rad/s.

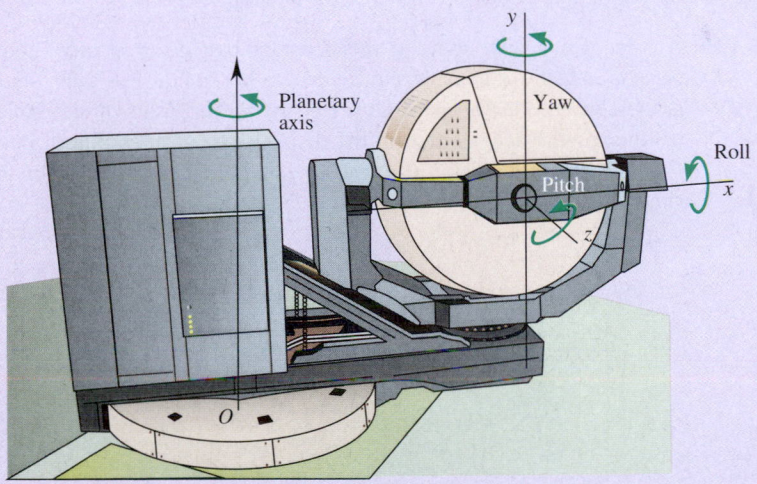

**Fig. P15.190**

**15.191** In the system shown, disk $A$ is free to rotate about the horizontal rod $OA$. Assuming that disk $B$ is stationary ($\omega_2 = 0$), and that shaft $OC$ rotates with a constant angular velocity $\boldsymbol{\omega}_1$, determine (a) the angular velocity of disk $A$, (b) the angular acceleration of disk $A$.

**15.192** In the system shown, disk $A$ is free to rotate about the horizontal rod $OA$. Assuming that shaft $OC$ and disk $B$ rotate with constant angular velocities $\boldsymbol{\omega}_1$ and $\boldsymbol{\omega}_2$, respectively, both counterclockwise, determine (a) the angular velocity of disk $A$, (b) the angular acceleration of disk $A$.

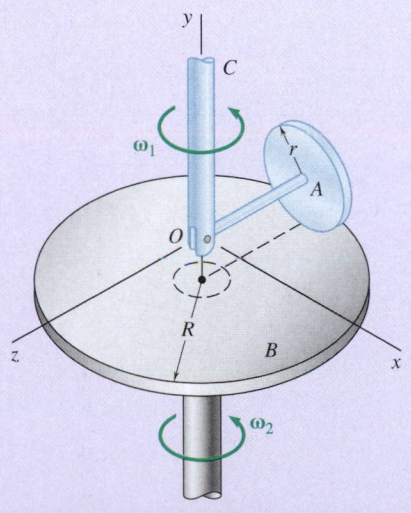

**Fig. P15.191 and P15.192**

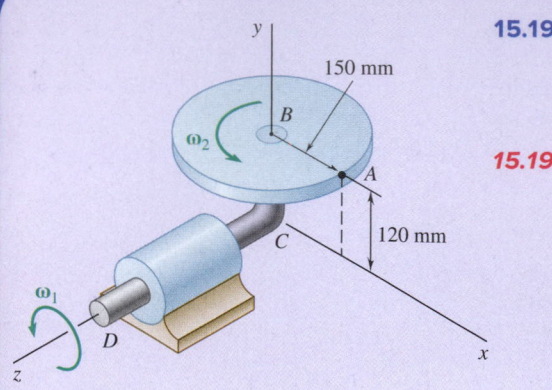

**Fig. P15.193**

**15.193** The L-shaped arm *BCD* rotates about the *z* axis with a constant angular velocity $\omega_1 = 5$ rad/s. Knowing that the 150-mm-radius disk rotates about *BC* with a constant angular velocity $\omega_2 = 4$ rad/s, determine (*a*) the velocity of point *A*, (*b*) the acceleration of point *A*.

**15.194** A radar system is used to track a new experimental space launch vehicle. Early in the vehicle's flight trajectory, the azimuth angle $\beta$ is increasing with the constant rate $d\beta/dt = 20°$/s. The elevation angle $\gamma$ is increasing at the rate $d\gamma/dt = 40°$/s, and this rate is increasing at $5°$/s². Knowing that the distance between *O* and *P* is 2 m and that at this instant $\beta = 0°$ and $\gamma = 30°$, determine (*a*) the angular velocity of the radar system, (*b*) the angular acceleration of the system, and (*c*) the velocity and acceleration of point *P*.

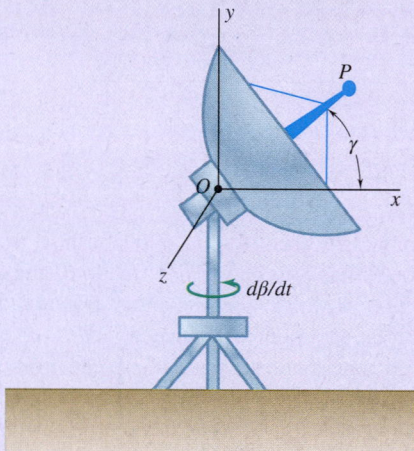

**Fig. *P15.194***

**15.195** A 3-in.-radius disk spins at the constant rate $\omega_2 = 4$ rad/s about an axis held by a housing attached to a horizontal rod that rotates at the constant rate $\omega_1 = 5$ rad/s. For the position shown, determine (*a*) the angular acceleration of the disk, (*b*) the acceleration of point *P* on the rim of the disk if $\theta = 0$, (*c*) the acceleration of point *P* on the rim of the disk if $\theta = 90°$.

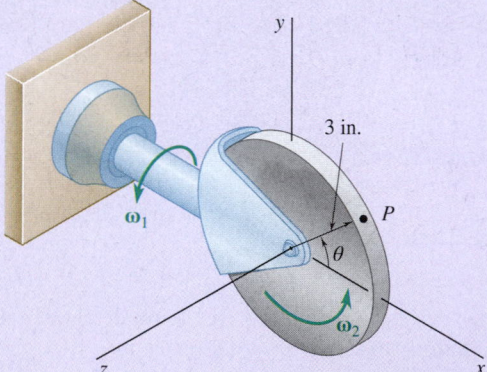

**Fig. P15.195 and P15.196**

**15.196** A 3-in.-radius disk spins at the constant rate $\omega_2 = 4$ rad/s about an axis held by a housing attached to a horizontal rod that rotates at the constant rate $\omega_1 = 5$ rad/s. Knowing that $\theta = 30°$, determine the acceleration of point *P* on the rim of the disk.

**15.197** The cone shown rolls on the $zx$ plane with its apex at the origin of coordinates. Denoting by $\boldsymbol{\omega}_1$ the constant angular velocity of the axis $OB$ of the cone about the $y$ axis, determine (a) the rate of spin of the cone about the axis $OB$, (b) the total angular velocity of the cone, (c) the angular acceleration of the cone.

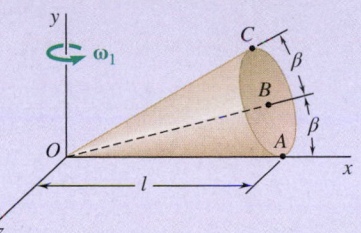

**Fig. P15.197**

**15.198** At the instant shown, the robotic arm $ABC$ is being rotated simultaneously at the constant rate $\omega_1 = 0.15$ rad/s about the $y$ axis, and at the constant rate $\omega_2 = 0.25$ rad/s about the $z$ axis. Knowing that the length of arm $ABC$ is 1 m, determine (a) the angular acceleration of the arm, (b) the velocity of point $C$, (c) the acceleration of point $C$.

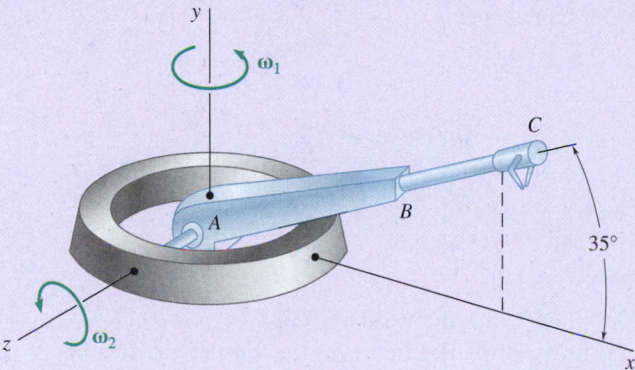

**Fig. P15.198**

**15.199** In the planetary gear system shown, gears $A$ and $B$ are rigidly connected to each other and rotate as a unit about shaft $FG$. Gears $C$ and $D$ rotate with constant angular velocities of 15 rad/s and 30 rad/s, respectively, both counterclockwise when viewed from the right. Choosing the $z$ axis pointing out of the plane of the figure, determine the common angular velocity of gears $A$ and $B$.

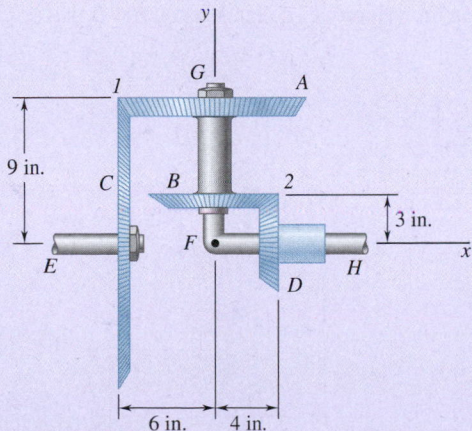

**Fig. P15.199**

**15.200** In Prob. 15.199, determine (a) the common angular acceleration of gears $A$ and $B$, (b) the acceleration of the tooth of gear $B$ that is in contact with gear $D$ at point 2.

**15.201** Several rods are brazed together to form the robotic guide arm shown that is attached to a ball-and-socket joint at $O$. Rod $OA$ slides in a straight inclined slot, while rod $OB$ slides in a slot parallel to the $z$ axis. Knowing that at the instant shown $\mathbf{v}_B = (9 \text{ in./s})\mathbf{k}$, determine ($a$) the angular velocity of the guide arm, ($b$) the velocity of point $A$, ($c$) the velocity of point $C$.

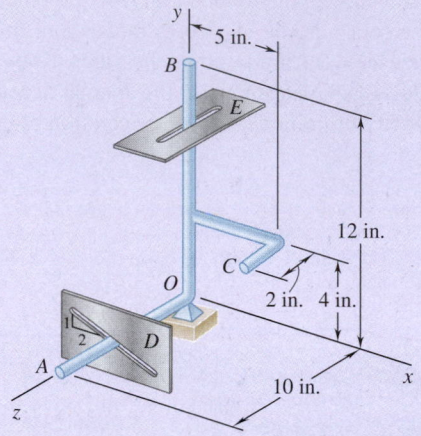

**Fig. P15.201**

**15.202** In Prob. 15.201, the speed of point $B$ is known to be constant. For the position shown, determine ($a$) the angular acceleration of the guide arm, ($b$) the acceleration of point $C$.

**15.203** Rod $AB$ of length 25 in. is connected by ball-and-socket joints to collars $A$ and $B$, which slide along the two rods shown. Knowing that collar $B$ moves toward point $E$ at a constant speed of 20 in./s, determine the velocity of collar $A$ as collar $B$ passes through point $D$.

**15.204** Rod $AB$ has a length of 13 in. and is connected by ball-and-socket joints to collars $A$ and $B$ that slide along the two rods shown. Knowing that collar $B$ moves toward point $D$ at a constant speed of 36 in./s, determine the velocity of collar $A$ when $b = 4$ in.

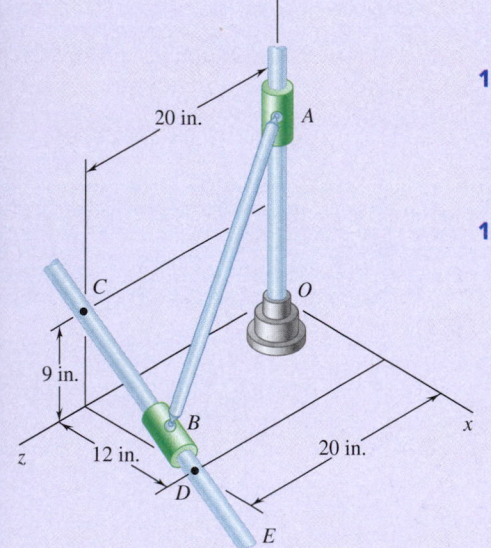

**Fig. P15.203**

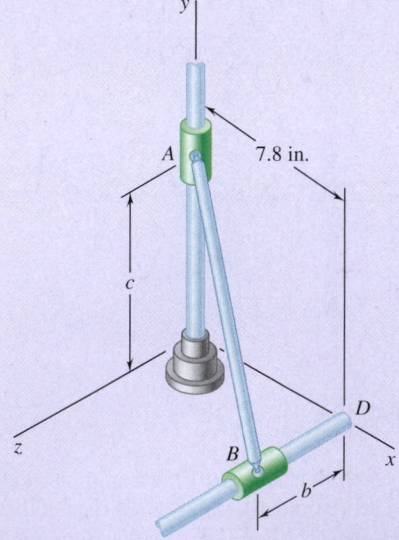

**Fig. P15.204**

**15.205** Rod *BC* and *BD* are each 840 mm long and are connected by ball-and-socket joints to collars that may slide on the fixed rods shown. Knowing that collar *B* moves toward *A* at a constant speed of 390 mm/s, determine the velocity of collar *C* for the position shown.

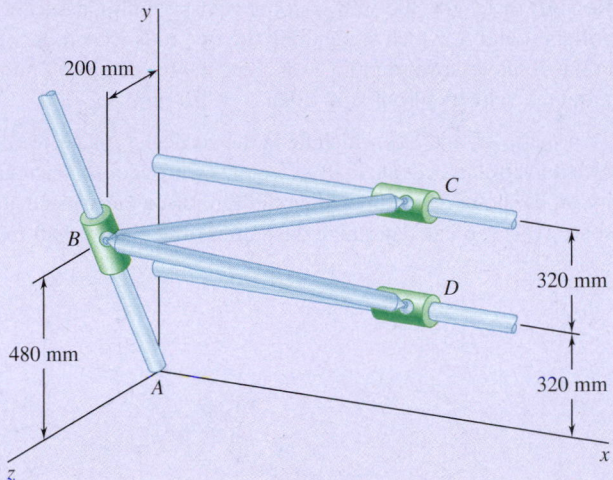

**Fig. P15.205**

**15.206** Rod *AB* is connected by ball-and-socket joints to collar *A* and to the 16-in.-diameter disk *C*. Knowing that disk *C* rotates counterclockwise at the constant rate $\omega_0 = 3$ rad/s in the *zx* plane, determine the velocity of collar *A* for the position shown.

**15.207** Rod *AB* of length 29 in. is connected by ball-and-socket joints to the rotating crank *BC* and to the collar *A*. Crank *BC* is of length 8 in. and rotates in the horizontal *xz* plane at the constant rate $\omega_0 = 10$ rad/s. At the instant shown, when crank *BC* is parallel to the *z* axis, determine the velocity of collar *A*.

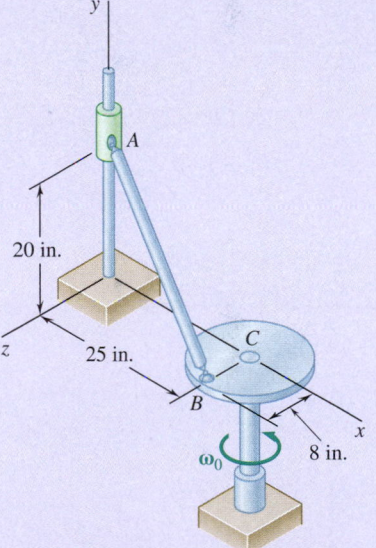

**Fig. P15.206**

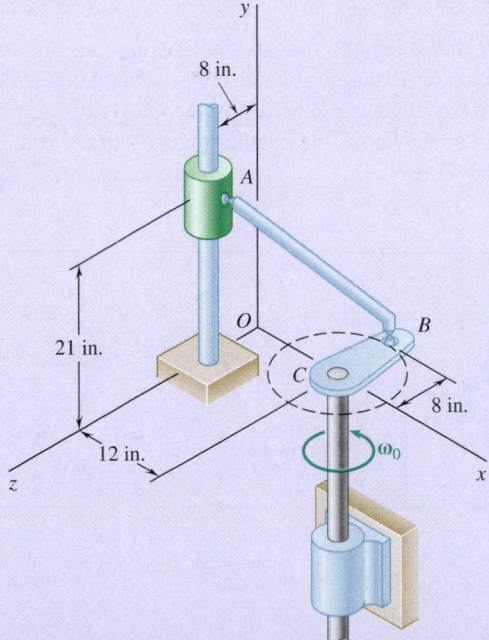

**Fig. P15.207**

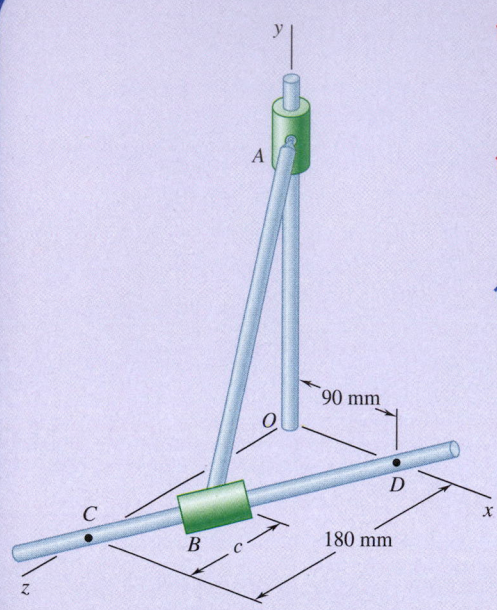

**Fig. P15.208 and P15.209**

**15.208** Rod *AB* of length 300 mm is connected by ball-and-socket joints to collars *A* and *B*, which slide along the two rods shown. Knowing that collar *B* moves toward point *D* at a constant speed of 50 mm/s, determine the velocity of collar *A* when $c = 80$ mm.

**15.209** Rod *AB* of length 300 mm is connected by ball-and-socket joints to collars *A* and *B*, which slide along the two rods shown. Knowing that collar *B* moves toward point *D* at a constant speed of 50 mm/s, determine the velocity of collar *A* when $c = 120$ mm.

**15.210** Two shafts *AC* and *EG*, which lie in the vertical *yz* plane, are connected by a universal joint at *D*. Shaft *AC* rotates with a constant angular velocity $\omega_1$ as shown. At a time when the arm of the crosspiece attached to shaft *AC* is vertical, determine the angular velocity of shaft *EG*.

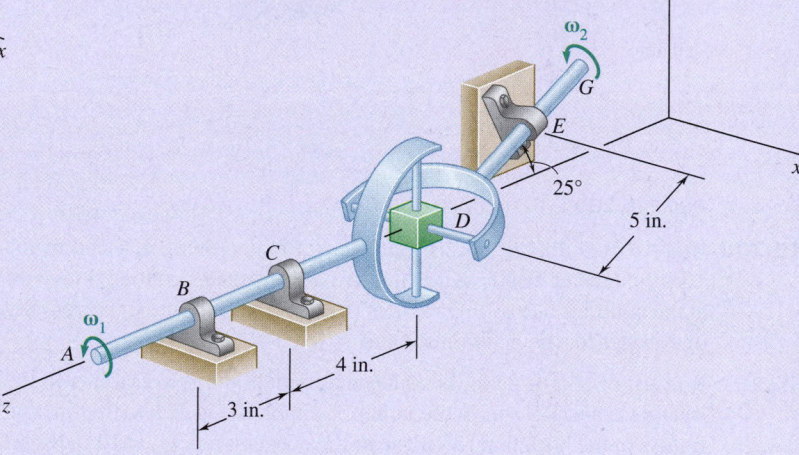

**Fig. P15.210**

**15.211** Solve Prob. 15.210, assuming that the arm of the crosspiece attached to shaft *AC* is horizontal.

**15.212** Rod *BC* has a length of 42 in. and is connected by a ball-and-socket joint to collar *B* and by a clevis connection to collar *C*. Knowing that collar *B* moves toward *A* at a constant speed of 19.5 in./s, determine at the instant shown (*a*) the angular velocity of the rod, (*b*) the velocity of collar *C*.

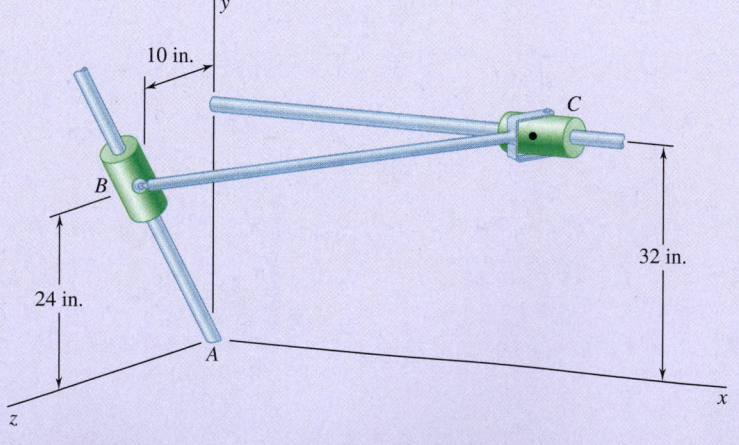

**Fig. P15.212**

**15.213** Rod $AB$ has a length of 275 mm and is connected by a ball-and-socket joint to collar $A$ and by a clevis connection to collar $B$. Knowing that collar $B$ moves down at a constant speed of 1.35 m/s, determine at the instant shown (*a*) the angular velocity of the rod, (*b*) the velocity of collar $A$.

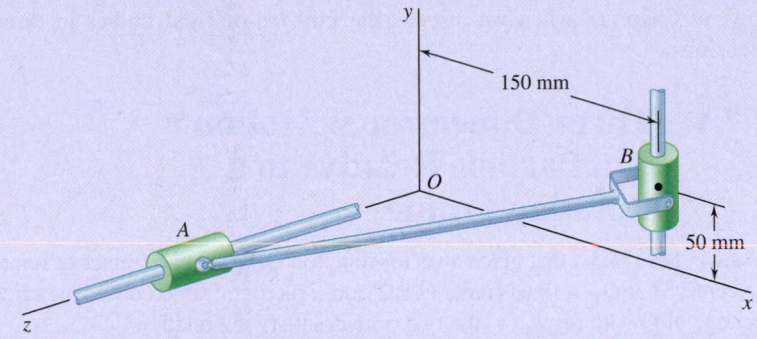

**Fig. P15.213**

**15.214** For the mechanism of Prob.15.204, determine the acceleration of collar $A$.

**15.215** In Prob. 15.205, determine the acceleration of collar $C$.

**15.216** In Prob. 15.206, determine the acceleration of collar $A$.

**15.217** In Prob. 15.207, determine the acceleration of collar $A$.

**15.218** In Prob. 15.208, determine the acceleration of collar $A$.

**15.219** In Prob. 15.209, determine the acceleration of collar $A$.

# *15.7 MOTION RELATIVE TO A MOVING REFERENCE FRAME

In this final section of the chapter, we describe motion relative to a moving reference frame—either rotating or in general motion. We will use these results in Chap. 18 when we discuss the kinetics of rigid bodies in three dimensions.

## 15.7A Three-Dimensional Motion of a Particle Relative to a Rotating Frame

We saw in Sec. 15.5A that given a vector function $\mathbf{Q}(t)$ and two frames of reference centered at $O$—a fixed frame $OXYZ$ and a rotating frame $Oxyz$—the rates of change of $\mathbf{Q}$ with respect to the two frames satisfy the relation

$$(\dot{\mathbf{Q}})_{OXYZ} = (\dot{\mathbf{Q}})_{Oxyz} + \mathbf{\Omega} \times \mathbf{Q} \tag{15.31}$$

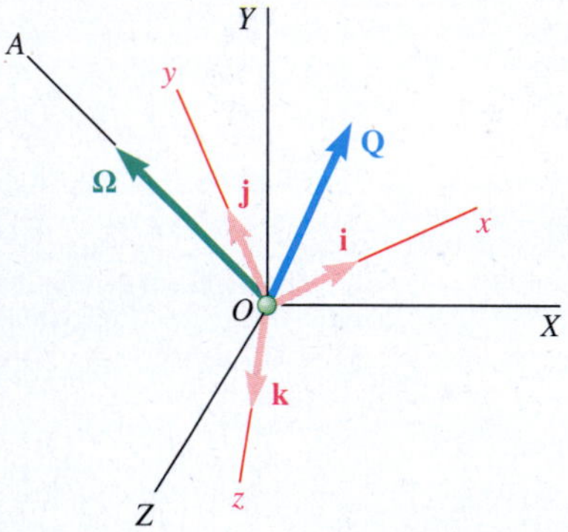

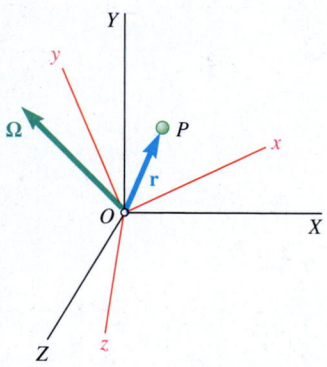

**Fig. 15.36** Reference frame $Oxyz$ rotating about an instantaneous axis in fixed frame $OXYZ$ with angular velocity $\mathbf{\Omega}$.

We had assumed at the time that the frame $Oxyz$ was constrained to rotate about a fixed axis $OA$. However, the derivation given in Sec. 15.5A remains valid when the frame $Oxyz$ is constrained to have only a fixed point $O$. Under this more general assumption, the axis $OA$ represents the *instantaneous* axis of rotation of the frame $Oxyz$ (Sec. 15.6A) and the vector $\mathbf{\Omega}$ represents its angular velocity at the instant considered (Fig. 15.36).

Let us now consider the three-dimensional motion of a particle $P$ relative to a rotating frame $Oxyz$ constrained to have a fixed origin $O$. Let $\mathbf{r}$ be the position vector of $P$ at a given instant, and let $\mathbf{\Omega}$ be the angular velocity of the frame $Oxyz$ with respect to the fixed frame $OXYZ$ at the same instant (Fig. 15.37). The derivations given in Sec. 15.5B for the two-dimensional motion of a particle can be readily extended to the three-dimensional case. Then we can express the absolute velocity $\mathbf{v}_P$ of $P$ (i.e., its velocity with respect to the fixed frame $OXYZ$) as

$$\mathbf{v}_P = \mathbf{\Omega} \times \mathbf{r} + (\dot{\mathbf{r}})_{Oxyz} \tag{15.45}$$

where $(\dot{\mathbf{r}})_{Oxyz}$ is the relative velocity of point $P$ with respect to the rotating frame. Sometimes this is also written as $\mathbf{v}_{\text{rel}}$. Denoting the rotating frame $Oxyz$ by $\mathscr{F}$, we can write this relation in the alternative form

$$\mathbf{v}_P = \mathbf{v}_{P'} + \mathbf{v}_{P/\mathscr{F}} \tag{15.46}$$

**Fig. 15.37** A particle $P$ moving relative to the rotating frame.

where $\mathbf{v}_P$ = absolute velocity of particle $P$
  $\mathbf{v}_{P'}$ = velocity of point $P'$ of moving frame $\mathscr{F}$ coinciding with $P$
  $\mathbf{v}_{P/\mathscr{F}}$ = velocity of $P$ relative to moving frame $\mathscr{F}$

The absolute acceleration $\mathbf{a}_P$ of $P$ can be expressed as

$$\mathbf{a}_P = \dot{\mathbf{\Omega}} \times \mathbf{r} + \mathbf{\Omega} \times (\mathbf{\Omega} \times \mathbf{r}) + 2\mathbf{\Omega} \times (\dot{\mathbf{r}})_{Oxyz} + (\ddot{\mathbf{r}})_{Oxyz} \tag{15.47}$$

An alternative form is

$$\mathbf{a}_P = \mathbf{a}_{P'} + \mathbf{a}_{P/\mathscr{F}} + \mathbf{a}_C \qquad (15.48)$$

where  $\mathbf{a}_P$ = absolute acceleration of particle $P$

$\quad \mathbf{a}_{P'}$ = acceleration of point $P'$ of moving frame $\mathscr{F}$ coinciding with $P$

$\quad \mathbf{a}_{P/\mathscr{F}}$ = acceleration of $P$ relative to moving frame $\mathscr{F}$

$\quad \mathbf{a}_C = 2\mathbf{\Omega} \times (\dot{\mathbf{r}})_{Oxyz} = 2\mathbf{\Omega} \times \mathbf{v}_{P/\mathscr{F}}$ = Coriolis acceleration

Note the difference between this equation and Eq. (15.21) of Sec. 15.4A, and recall the discussion following Eq. (15.36) of Sec. 15.5B.

Also note that the Coriolis acceleration is perpendicular to the vectors $\mathbf{\Omega}$ and $\mathbf{v}_{P/\mathscr{F}}$. However, because these vectors are usually not perpendicular to each other, the magnitude of $\mathbf{a}_C$ is in general *not* equal to $2\Omega v_{P/\mathscr{F}}$—as was the case for the plane motion of a particle. We further note that the Coriolis acceleration reduces to zero when the vectors $\mathbf{\Omega}$ and $\mathbf{v}_{P/\mathscr{F}}$ are parallel or when either of them is zero.

Rotating frames of reference are particularly useful in the study of the three-dimensional motion of rigid bodies. If a rigid body has a fixed point $O$—as was the case for the crane of Sample Prob. 15.21—we can use a frame $Oxyz$ that can rotate. Denoting the angular velocity of the frame $Oxyz$ by $\mathbf{\Omega}$, we then resolve the angular velocity $\boldsymbol{\omega}$ of the body into the components $\mathbf{\Omega}$ and $\boldsymbol{\omega}_{B/\mathscr{F}}$, where the second component represents the angular velocity of the body relative to the frame $Oxyz$ (see Sample Prob. 15.24). An appropriate choice of a rotating frame often leads to a simpler analysis of the motion of the rigid body than would be possible with axes of fixed orientation. This is especially true in the case of the general three-dimensional motion of a rigid body; that is, when the rigid body under consideration has no fixed point (see Sample Prob. 15.25).

## *15.7B   Frame of Reference in General Motion

Consider a fixed frame of reference $OXYZ$ and a frame $Axyz$ that moves in a known, but arbitrary, fashion with respect to $OXYZ$ (Fig. 15.38). Let $P$ be a particle moving in space. The position of $P$ is defined at any instant by the vector $\mathbf{r}_P$ in the fixed frame and by the vector $\mathbf{r}_{P/A}$ in the moving frame. Denoting the position vector of $A$ in the fixed frame by $\mathbf{r}_A$, we have

$$\mathbf{r}_P = \mathbf{r}_A + \mathbf{r}_{P/A} \qquad (15.49)$$

We obtain the absolute velocity $\mathbf{v}_P$ of the particle by differentiating, as

$$\mathbf{v}_P = \dot{\mathbf{r}}_P = \dot{\mathbf{r}}_A + \dot{\mathbf{r}}_{P/A} \qquad (15.50)$$

where the derivatives are defined with respect to the fixed frame $OXYZ$. The first term in the right-hand side of Eq. (15.50) thus represents the velocity $\mathbf{v}_A$ of the origin $A$ of the moving axes. Because the rate of change of a vector is the same with respect to both a fixed frame and a frame in translation (Sec. 11.4B),

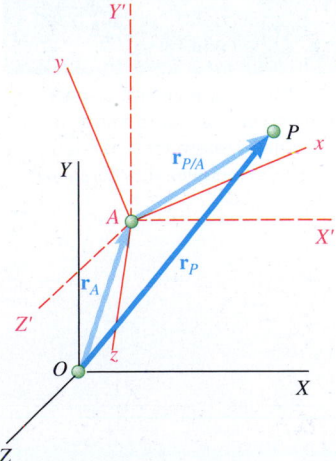

**Fig. 15.38**  Reference frame *Axyz* moves arbitrarily relative to fixed frame *OXYZ*.

we can regard the second term as the velocity $\mathbf{v}_{P/A}$ of $P$ relative to the frame $AX'Y'Z'$ with the same orientation as $OXYZ$ and the same origin as $Axyz$. We therefore have

$$\mathbf{v}_P = \mathbf{v}_A + \mathbf{v}_{P/A} \tag{15.51}$$

However, we can obtain the velocity $\mathbf{v}_{P/A}$ of $P$ relative to $AX'Y'Z'$ from Eq. (15.45) by substituting $\mathbf{r}_{P/A}$ for $\mathbf{r}$ in that equation. We get

$$\mathbf{v}_P = \mathbf{v}_A + \boldsymbol{\Omega} \times \mathbf{r}_{P/A} + (\dot{\mathbf{r}}_{P/A})_{Axyz} \tag{15.52}$$

where $\boldsymbol{\Omega}$ is the angular velocity of the frame $Axyz$ at the instant considered.

We obtain the absolute acceleration $\mathbf{a}_P$ of the particle by differentiating Eq. (15.51), as

$$\mathbf{a}_P = \dot{\mathbf{v}}_P = \dot{\mathbf{v}}_A + \dot{\mathbf{v}}_{P/A} \tag{15.53}$$

where the derivatives are defined with respect to either of the frames $OXYZ$ or $AX'Y'Z'$. Thus, the first term in the right-hand side of Eq. (15.53) represents the acceleration $\mathbf{a}_A$ of the origin $A$ of the moving axes, and the second term represents the acceleration $\mathbf{a}_{P/A}$ of $P$ relative to the frame $AX'Y'Z'$. We can obtain this acceleration from Eq. (15.47) by substituting $\mathbf{r}_{P/A}$ for $\mathbf{r}$. We therefore have

$$\mathbf{a}_P = \mathbf{a}_A + \dot{\boldsymbol{\Omega}} \times \mathbf{r}_{P/A} + \boldsymbol{\Omega} \times (\boldsymbol{\Omega} \times \mathbf{r}_{P/A})$$
$$+ 2\boldsymbol{\Omega} \times (\dot{\mathbf{r}}_{P/A})_{Axyz} + (\ddot{\mathbf{r}}_{P/A})_{Axyz} \tag{15.54}$$

Formulas (15.52) and (15.54) enable us to determine the velocity and acceleration of a given particle with respect to a fixed frame of reference when we know the motion of the particle with respect to a moving frame. These formulas become more significant, and considerably easier to remember, if we note that the sum of the first two terms in Eq. (15.52) represents the velocity of the point $P'$ of the moving frame that coincides with $P$ at the instant considered and that the sum of the first three terms in Eq. (15.54) represents the acceleration of the same point. Thus, relations in Eqs. (15.46) and (15.48) of the preceding section are still valid in the case of a reference frame in general motion, and we have

$$\mathbf{v}_P = \mathbf{v}_{P'} + \mathbf{v}_{P/\mathscr{F}} \tag{15.46}$$

$$\mathbf{a}_P = \mathbf{a}_{P'} + \mathbf{a}_{P/\mathscr{F}} + \mathbf{a}_C \tag{15.48}$$

where the various vectors involved were defined earlier.

Note that if the moving reference frame $\mathscr{F}$ (or $Axyz$) is in translation, the velocity and acceleration of the point $P'$ of the frame that coincides with $P$ become, respectively, equal to the velocity and acceleration of the origin $A$ of the frame. On the other hand, because the frame maintains a fixed orientation, $\mathbf{a}_c$ is zero, and the relations in Eqs. (15.46) and (15.48) reduce, respectively, to the relations in Eqs. (11.32) and (11.33) derived in Sec. 11.4D.

**Photo 15.9**    The motion of air particles in a hurricane can be considered as motion relative to a frame of reference attached to the Earth and rotating with it. ©StockTrek/Getty Images RF

## Sample Problem 15.23

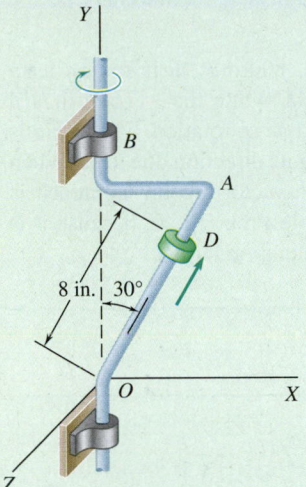

The bent rod *OAB* rotates about the vertical axis *OB*. At the instant considered, its angular velocity and angular acceleration are, respectively, 20 rad/s and 200 rad/s$^2$, which are both clockwise when viewed from the positive *Y* axis. The collar *D* moves along the rod, and at the instant considered, *OD* = 8 in. The velocity and acceleration of the collar relative to the rod are, respectively, 50 in./s and 600 in./s$^2$, where both are upward. Determine (*a*) the velocity of the collar, (*b*) the acceleration of the collar.

**STRATEGY:**   Use rigid-body kinematics with a rotating coordinate system because collar *D* is moving relative to the bent rod. Attach the rotating reference frame to the bent rod; then you can calculate its motion relative to the fixed frame and the collar's motion relative to the rotating frame.

**MODELING:**

**Frames of Reference.**   The angular velocity and angular acceleration of the bent rod (and rotating frame *Oxyz*) relative to the fixed frame *OXYZ* are $\mathbf{\Omega} = (-20 \text{ rad/s})\mathbf{j}$ and $\dot{\mathbf{\Omega}} = (-200 \text{ rad/s}^2)\mathbf{j}$, respectively (Fig. 1). The position vector of *D* is

$$\mathbf{r} = (8 \text{ in.})(\sin 30°\mathbf{i} + \cos 30°\mathbf{j}) = (4 \text{ in.})\mathbf{i} + (6.93 \text{ in.})\mathbf{j}$$

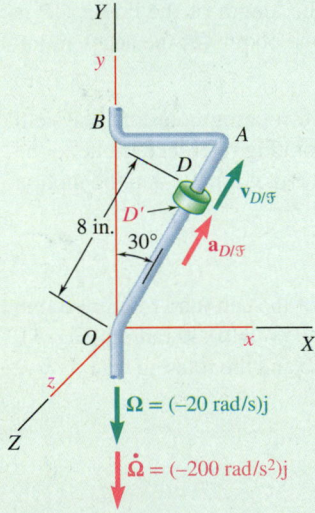

**Fig. 1**   The rotating coordinate system *xyz* is attached to rod *OAB*.

**ANALYSIS:**

**a. Velocity $\mathbf{V}_D$.**   Denote the point of the rod that coincides with *D* by *D′* and the rotating frame *Oxyz* by $\mathscr{F}$. Then from Eq. (15.46) you have

$$\mathbf{v}_D = \mathbf{v}_{D'} + \mathbf{v}_{D/\mathscr{F}} \qquad (1)$$

where

$$\mathbf{v}_{D'} = \mathbf{\Omega} \times \mathbf{r} = (-20 \text{ rad/s})\mathbf{j} \times [(4 \text{ in.})\mathbf{i} + (6.93 \text{ in.})\mathbf{j}] = (80 \text{ in./s})\mathbf{k}$$
$$\mathbf{v}_{D/\mathscr{F}} = (50 \text{ in./s})(\sin 30°\mathbf{i} + \cos 30°\mathbf{j}) = (25 \text{ in./s})\mathbf{i} + (43.3 \text{ in./s})\mathbf{j}$$

Substituting the values obtained for $\mathbf{v}_{D'}$ and $\mathbf{v}_{D/\mathscr{F}}$ into Eq. (1) gives

$$\mathbf{v}_D = (25 \text{ in./s})\mathbf{i} + (43.3 \text{ in./s})\mathbf{j} + (80 \text{ in./s})\mathbf{k} \quad \blacktriangleleft$$

**b. Acceleration $\mathbf{a}_D$.**   From Eq. (15.48), you have

$$\mathbf{a}_D = \mathbf{a}_{D'} + \mathbf{a}_{D/\mathscr{F}} + \mathbf{a}_C \qquad (2)$$

where

$$\mathbf{a}_{D'} = \dot{\mathbf{\Omega}} \times \mathbf{r} + \mathbf{\Omega} \times (\mathbf{\Omega} \times \mathbf{r})$$
$$= (-200 \text{ rad/s}^2)\mathbf{j} \times [(4 \text{ in.})\mathbf{i} + (6.93 \text{ in.})\mathbf{j}] - (20 \text{ rad/s})\mathbf{j} \times (80 \text{ in./s})\mathbf{k}$$
$$= +(800 \text{ in./s}^2)\mathbf{k} - (1600 \text{ in./s}^2)\mathbf{i}$$
$$\mathbf{a}_{D/\mathscr{F}} = (600 \text{ in./s}^2)(\sin 30°\mathbf{i} + \cos 30°\mathbf{j}) = (300 \text{ in./s}^2)\mathbf{i} + (520 \text{ in./s}^2)\mathbf{j}$$
$$\mathbf{a}_C = 2\mathbf{\Omega} \times \mathbf{v}_{D/\mathscr{F}}$$
$$= 2(-20 \text{ rad/s})\mathbf{j} \times [(25 \text{ in./s})\mathbf{i} + (43.3 \text{ in./s})\mathbf{j}] = (1000 \text{ in./s}^2)\mathbf{k}$$

*(continued)*

Substituting the values obtained for $\mathbf{a}_{D'}$, $\mathbf{a}_{D/\mathscr{F}}$, and $\mathbf{a}_C$ into Eq. (2), you obtain

$$\mathbf{a}_D = -(1300\ \text{in./s}^2)\mathbf{i} + (520\ \text{in./s}^2)\mathbf{j} + (1800\ \text{in./s}^2)\mathbf{k} \quad \blacktriangleleft$$

**REFLECT and THINK:** For this problem, the $(800\ \text{in./s}^2)\mathbf{k}$ in the $\mathbf{a}_{D'}$ term corresponds to a tangential acceleration due to $\dot{\Omega}$, while the $-(1600\ \text{in./s}^2)\mathbf{i}$ corresponds to a normal acceleration toward the axis of rotation. The Coriolis term reflects the fact that the $\mathbf{v}_{D/\mathscr{F}}$ term is changing its direction due to $\Omega$. When solving three-dimensional problems like this, the vector algebra approach is clearly superior to the method discussed in Sample Prob. 15.20, because it is very difficult to visualize the direction of the acceleration terms.

## Sample Problem 15.24

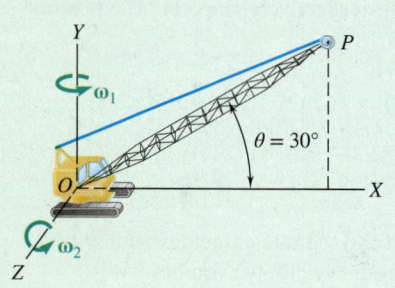

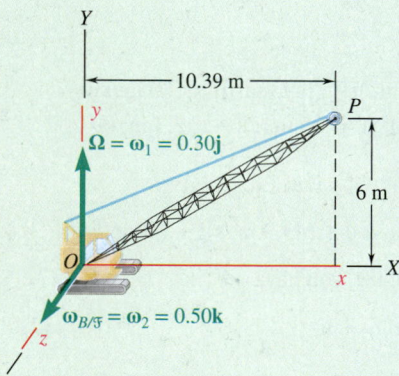

**Fig. 1** The rotating coordinate system *xyz* is attached to the cab.

The crane shown rotates with a constant angular velocity $\omega_1$ of 0.30 rad/s. Simultaneously, the boom is being raised with a constant angular velocity $\omega_2$ of 0.50 rad/s relative to the cab. Knowing that the length of the boom $OP$ is $l = 12$ m, determine (*a*) the velocity of the tip of the boom, (*b*) the acceleration of the tip of the boom.

**STRATEGY:** Use rigid body kinematics with a rotating coordinate system because $\omega_2$ is given relative to the cab. Attach a rotating reference frame to the cab; then you can calculate its motion relative to the fixed frame and the motion of the crane tip relative to the rotating frame.

**MODELING:**

**Frames of Reference.** The angular velocity of the cab (and rotating frame *Oxyz*) with respect to the fixed frame *OXYZ* is $\Omega = \omega_1 = (0.30\ \text{rad/s})\mathbf{j}$ (Fig. 1). The angular velocity of the boom relative to the cab and the rotating frame *Oxyz* (or $\mathscr{F}$ for short) is $\omega_{B/\mathscr{F}} = \omega_2 = (0.50\ \text{rad/s})\mathbf{k}$.

**ANALYSIS:**

**a. Velocity $\mathbf{v}_P$.** From Eq. (15.46), you have

$$\mathbf{v}_P = \mathbf{v}_{P'} + \mathbf{v}_{P/\mathscr{F}} \tag{1}$$

where $\mathbf{v}_{P'}$ is the velocity of the point $P'$ of the rotating frame that coincides with $P$ as

$$\mathbf{v}_{P'} = \Omega \times \mathbf{r} = (0.30\ \text{rad/s})\mathbf{j} \times [(10.39\ \text{m})\mathbf{i} + (6\ \text{m})\mathbf{j}] = -(3.12\ \text{m/s})\mathbf{k}$$

and where $\mathbf{v}_{P/\mathscr{F}}$ is the velocity of $P$ relative to the rotating frame *Oxyz*. However, you know that the angular velocity of the boom relative to *Oxyz* is $\omega_{B/\mathscr{F}} = (0.50\ \text{rad/s})\mathbf{k}$. The velocity of its tip $P$ relative to *Oxyz* is therefore

$$\mathbf{v}_{P/\mathscr{F}} = \omega_{B/\mathscr{F}} \times \mathbf{r} = (0.50\ \text{rad/s})\mathbf{k} \times [(10.39\ \text{m})\mathbf{i} + (6\ \text{m})\mathbf{j}]$$
$$= -(3\ \text{m/s})\mathbf{i} + (5.20\ \text{m/s})\mathbf{j}$$

*(continued)*

Substituting the values obtained for $\mathbf{v}_{P'}$ and $\mathbf{v}_{P/\mathcal{F}}$ into Eq. (1), you find

$$\mathbf{v}_P = -(3 \text{ m/s})\mathbf{i} + (5.20 \text{ m/s})\mathbf{j} - (3.12 \text{ m/s})\mathbf{k} \quad \blacktriangleleft$$

**b. Acceleration $\mathbf{a}_P$.** From Eq. (15.48), you have

$$\mathbf{a}_P = \mathbf{a}_{P'} + \mathbf{a}_{P/\mathcal{F}} + \mathbf{a}_C \qquad (2)$$

Because $\boldsymbol{\Omega}$ and $\boldsymbol{\omega}_{B/\mathcal{F}}$ are both constant, you obtain

$$\mathbf{a}_{P'} = \boldsymbol{\Omega} \times (\boldsymbol{\Omega} \times \mathbf{r}) = (0.30 \text{ rad/s})\mathbf{j} \times (-3.12 \text{ m/s})\mathbf{k} = -(0.94 \text{ m/s}^2)\mathbf{i}$$

$$\mathbf{a}_{P/\mathcal{F}} = \boldsymbol{\omega}_{B/\mathcal{F}} \times (\boldsymbol{\omega}_{B/\mathcal{F}} \times \mathbf{r})$$

$$= (0.50 \text{ rad/s})\mathbf{k} \times [-(3 \text{ m/s})\mathbf{i} + (5.20 \text{ m/s})\mathbf{j}]$$

$$= -(1.50 \text{ m/s}^2)\mathbf{j} - (2.60 \text{ m/s}^2)\mathbf{i}$$

$$\mathbf{a}_C = 2\boldsymbol{\Omega} \times \mathbf{v}_{P/\mathcal{F}}$$

$$= 2(0.30 \text{ rad/s})\mathbf{j} \times [-(3 \text{ m/s})\mathbf{i} + (5.20 \text{ m/s})\mathbf{j}] = (1.80 \text{ m/s}^2)\mathbf{k}$$

Substituting for $\mathbf{a}_{P'}$, $\mathbf{a}_{P/\mathcal{F}}$, and $\mathbf{a}_C$ into Eq. (2), you find

$$\mathbf{a}_P = -(3.54 \text{ m/s}^2)\mathbf{i} - (1.50 \text{ m/s}^2)\mathbf{j} + (1.80 \text{ m/s}^2)\mathbf{k} \quad \blacktriangleleft$$

**REFLECT and THINK:** You also could have attached your reference frame to rotate with the boom:

$$\boldsymbol{\Omega}_B = \boldsymbol{\omega}_{\mathcal{F}} + \boldsymbol{\omega}_{B/\mathcal{F}} = (0.30 \text{ rad/s})\mathbf{j} + (0.50 \text{ rad/s})\mathbf{k}$$

and used Eq. (15.52) for

$$\mathbf{v}_P = \boldsymbol{\Omega}_B \times \mathbf{r} = [(0.30 \text{ rad/s})\mathbf{j} + (0.5 \text{ rad/s})\mathbf{k}] \times [(10.39 \text{ m})\mathbf{i} + (6 \text{ m})\mathbf{j}]$$

$$= -(3.0 \text{ m/s})\mathbf{i} + (5.20 \text{ m/s})\mathbf{j} - (3.12 \text{ m/s})\mathbf{k}$$

which is the same answer you found previously. Similarly, you could use Eq. (15.54) to solve for the acceleration. If the crane were moving forward, you would just add its translational velocity and acceleration to those due to the rotations.

## Sample Problem 15.25

Disk $D$ has a radius $R$ and is pinned to end $A$ of the arm $OA$. $OA$ has a length $L$ and is located in the plane of the disk. The arm rotates about a vertical axis through $O$ at the constant rate $\omega_1$, and the disk rotates about $A$ at the constant rate $\omega_2$. Determine (a) the velocity of point $P$ located directly above $A$, (b) the acceleration of $P$, (c) the angular velocity and angular acceleration of the disk.

**STRATEGY:** Use rigid-body kinematics with a rotating coordinate system, because disk $D$ is moving relative to the arm $OA$.

**MODELING:**

**Frames of Reference.** Attach a moving frame $Axyz$ to arm $OA$. Its angular velocity with respect to the fixed frame $OXYZ$ is therefore $\boldsymbol{\Omega} = \omega_1\mathbf{j}$ (Fig. 1). The

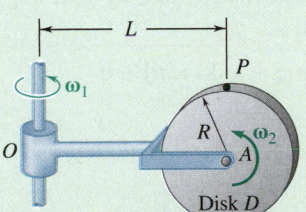

*(continued)*

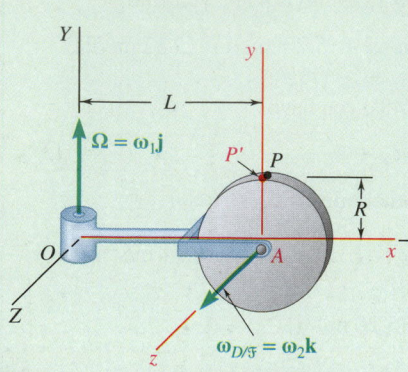

**Fig. 1** The rotating coordinate system *xyz* is attached to arm *OA* at point *A*.

angular velocity of disk $D$ relative to the moving frame $Axyz$ (or $\mathscr{F}$ for short) is $\boldsymbol{\omega}_{D/\mathscr{F}} = \omega_2\mathbf{k}$. The position vector of $P$ relative to $O$ is $\mathbf{r} = L\mathbf{i} + R\mathbf{j}$, and its position vector relative to $A$ is $\mathbf{r}_{P/A} = R\mathbf{j}$.

### ANALYSIS:

**a. Velocity $\mathbf{v}_P$.** Denote by $P'$ the point of the moving frame that coincides with $P$. Then from Eq. (15.46), you have

$$\mathbf{v}_P = \mathbf{v}_{P'} + \mathbf{v}_{P/\mathscr{F}} \qquad (1)$$

where $\mathbf{v}_{P'} = \boldsymbol{\Omega} \times \mathbf{r} = \omega_1\mathbf{j} \times (L\mathbf{i} + R\mathbf{j}) = -\omega_1 L\mathbf{k}$

$$\mathbf{v}_{P/\mathscr{F}} = \boldsymbol{\omega}_{D/\mathscr{F}} \times \mathbf{r}_{P/A} = \omega_2\mathbf{k} \times R\mathbf{j} = -\omega_2 R\mathbf{i}$$

Substituting the values obtained for $\mathbf{v}_{P'}$ and $\mathbf{v}_{D/\mathscr{F}}$ into Eq. (1), you obtain

$$\mathbf{v}_P = -\omega_2 R\mathbf{i} - \omega_1 L\mathbf{k} \quad \blacktriangleleft$$

**b. Acceleration $\mathbf{a}_P$.** From Eq. (15.48), you have

$$\mathbf{a}_P = \mathbf{a}_{P'} + \mathbf{a}_{P/\mathscr{F}} + \mathbf{a}_C \qquad (2)$$

Because $\boldsymbol{\Omega}$ and $\boldsymbol{\omega}_{D/\mathscr{F}}$ are both constant, you obtain

$$\mathbf{a}_{P'} = \boldsymbol{\Omega} \times (\boldsymbol{\Omega} \times \mathbf{r}) = \omega_1\mathbf{j} \times (-\omega_1 L\mathbf{k}) = -\omega_1^2 L\mathbf{i}$$
$$\mathbf{a}_{P/\mathscr{F}} = \boldsymbol{\omega}_{D/\mathscr{F}} \times (\boldsymbol{\omega}_{D/\mathscr{F}} \times \mathbf{r}_{P/A}) = \omega_2\mathbf{k} \times (-\omega_2 R\mathbf{i}) = -\omega_2^2 R\mathbf{j}$$
$$\mathbf{a}_C = 2\boldsymbol{\Omega} \times \mathbf{v}_{P/\mathscr{F}} = 2\omega_1\mathbf{j} \times (-\omega_2 R\mathbf{i}) = 2\omega_1\omega_2 R\mathbf{k}$$

Substituting these values into Eq. (2), you find

$$\mathbf{a}_P = -\omega_1^2 L\mathbf{i} - \omega_2^2 R\mathbf{j} + 2\omega_1\omega_2 R\mathbf{k} \quad \blacktriangleleft$$

**c. Angular Velocity and Angular Acceleration of Disk.**

$$\boldsymbol{\omega} = \boldsymbol{\Omega} + \boldsymbol{\omega}_{D/\mathscr{F}} \qquad \boldsymbol{\omega} = \omega_1\mathbf{j} + \omega_2\mathbf{k} \quad \blacktriangleleft$$

Using Eq. (15.31) with $\boldsymbol{\Omega} = \boldsymbol{\omega}$, you obtain

$$\boldsymbol{\alpha} = (\dot{\boldsymbol{\omega}})_{OXYZ} = (\dot{\boldsymbol{\omega}})_{Axyz} + \boldsymbol{\Omega} \times \boldsymbol{\omega}$$
$$= 0 + \omega_1\mathbf{j} \times (\omega_1\mathbf{j} + \omega_2\mathbf{k})$$

$$\boldsymbol{\alpha} = \omega_1\omega_2\mathbf{i} \quad \blacktriangleleft$$

**REFLECT and THINK:** Knowing the absolute angular velocity of the disk is equal to $\omega_1\mathbf{j} + \omega_2\mathbf{k}$, you could have determined the velocity of $P$ by attaching the rotating axes to the disk and using Eq. (15.52),

$$\mathbf{v}_P = \mathbf{v}_A + \boldsymbol{\Omega}_D \times \mathbf{r}_{P/A} + \mathbf{v}_{P/A} = \omega_1\mathbf{j} \times L\mathbf{i} + (\omega_1\mathbf{j} + \omega_2\mathbf{k}) \times R\mathbf{j} + 0$$
$$= -\omega_1 L\mathbf{k} - \omega_2 R\mathbf{i}$$

which is the same answer we found earlier. Similarly,

$$\mathbf{a}_P = \mathbf{a}_A + \dot{\boldsymbol{\Omega}}_D \times \mathbf{r}_{P/A} + \boldsymbol{\Omega}_D \times (\boldsymbol{\Omega}_D \times \mathbf{r}_{P/A}) + 2\boldsymbol{\Omega}_D \times \dot{\mathbf{r}}_{P/A} + \ddot{\mathbf{r}}_{P/A}$$
$$= -\omega_1^2 L\mathbf{i} + \omega_1\omega_2\mathbf{i} \times R\mathbf{j} + (\omega_1\mathbf{j} + \omega_2\mathbf{k}) \times [(\omega_1\mathbf{j} + \omega_2\mathbf{k}) \times R\mathbf{j}] + 0 + 0$$
$$= -\omega_1^2 L\mathbf{i} + \omega_1\omega_2 R\mathbf{k} + (\omega_1\mathbf{j} + \omega_2\mathbf{k}) \times (-\omega_2 R\mathbf{i})$$
$$= -\omega_1^2 L\mathbf{i} - \omega_2^2 R\mathbf{j} + 2\omega_1\omega_2 R\mathbf{k}$$

which, again, is the same answer shown previously.

# SOLVING PROBLEMS ON YOUR OWN

In this section, we concluded our presentation of the kinematics of rigid bodies by showing you how to use an auxiliary frame of reference $\mathcal{F}$ to analyze the three-dimensional motion of a rigid body. This auxiliary frame may be a *rotating frame* with a fixed origin $O$ or it may be a *frame in general motion*.

**A. Using a rotating frame of reference.** As you approach a problem involving the use of a rotating frame $\mathcal{F}$, you should take the following steps.

**1. Select the rotating frame** $\mathcal{F}$ that you wish to use and draw the corresponding coordinate axes $x$, $y$, and $z$ from the fixed point $O$.

**2. Determine the angular velocity** $\boldsymbol{\Omega}$ **of the frame** $\mathcal{F}$ with respect to a fixed frame $OXYZ$. In most cases, you will have selected a frame that is attached to some rotating element of the system; $\boldsymbol{\Omega}$ is then the angular velocity of that element.

**3. Designate as $P'$ the point of the rotating frame** $\mathcal{F}$ that coincides with the point $P$ of interest at the instant you are considering. Determine the velocity $\mathbf{v}_{P'}$ and the acceleration $\mathbf{a}_{P'}$ of point $P'$. Because $P'$ is part of $\mathcal{F}$ and has the same position vector $\mathbf{r}$ as $P$, you will find

$$\mathbf{v}_{P'} = \boldsymbol{\Omega} \times \mathbf{r} \qquad \text{and} \qquad \mathbf{a}_{P'} = \boldsymbol{\alpha} \times \mathbf{r} + \boldsymbol{\Omega} \times (\boldsymbol{\Omega} \times \mathbf{r})$$

where $\boldsymbol{\alpha}$ is the angular acceleration of $\mathcal{F}$.

**4. Determine the velocity and acceleration of point $P$** with respect to the frame $\mathcal{F}$. As you are trying to determine $\mathbf{v}_{P/\mathcal{F}}$ and $\mathbf{a}_{P/\mathcal{F}}$, you will find it useful to visualize the motion of $P$ on frame $\mathcal{F}$ when the frame is not rotating. If $P$ is a point of a rigid body $\mathcal{B}$ that has an angular velocity $\boldsymbol{\omega}_{\mathcal{B}}$ and an angular acceleration $\boldsymbol{\alpha}_{\mathcal{B}}$ relative to $\mathcal{F}$ (Sample Prob. 15.24), you will find that

$$\mathbf{v}_{P/\mathcal{F}} = \boldsymbol{\omega}_{\mathcal{B}} \times \mathbf{r} \qquad \text{and} \qquad \mathbf{a}_{P/\mathcal{F}} = \boldsymbol{\alpha}_{\mathcal{B}} \times \mathbf{r} + \boldsymbol{\omega}_{\mathcal{B}} \times (\boldsymbol{\omega}_{\mathcal{B}} \times \mathbf{r})$$

**5. Determine the Coriolis acceleration.** Considering the angular velocity $\boldsymbol{\Omega}$ of frame $\mathcal{F}$ and the velocity $\mathbf{v}_{P/\mathcal{F}}$ of point $P$ relative to that frame, which was computed in Step 4, you have

$$\mathbf{a}_C = 2\boldsymbol{\Omega} \times \mathbf{v}_{P/\mathcal{F}}$$

**6. The velocity and the acceleration of $P$** with respect to the fixed frame $OXYZ$ can now be obtained by adding the expressions you have determined:

$$\mathbf{v}_P = \mathbf{v}_{P'} + \mathbf{v}_{P/\mathcal{F}} \tag{15.46}$$

$$\mathbf{a}_P = \mathbf{a}_{P'} + \mathbf{a}_{P/\mathcal{F}} + \mathbf{a}_C \tag{15.48}$$

*(continued)*

**B. Using a frame of reference in general motion.** The steps that you will take differ only slightly from those listed under part A. They consist of the following:

**1. Select the frame $\mathscr{F}$ that you wish to use** and a reference point $A$ in that frame from which you will draw the coordinate axes, $x$, $y$, and $z$, defining that frame. Consider the motion of the frame as the sum of a **translation with $A$ and a rotation about $A$.**

**2. Determine the velocity $\mathbf{v}_A$ of point $A$ and the angular velocity $\Omega$ of the frame.** In most cases, you will have selected a frame that is attached to some element of the system; $\Omega$ is then the angular velocity of that element.

**3. Designate as $P'$ the point of frame** $\mathscr{F}$ that coincides with the point $P$ of interest at the instant you are considering, and determine the velocity $\mathbf{v}_{P'}$ and the acceleration $\mathbf{a}_{P'}$ of that point. In some cases, you can do this by visualizing the motion of $P$ if that point were prevented from moving with respect to $\mathscr{F}$ (Sample Prob. 15.25). A more general approach is to recall that the motion of $P'$ is the sum of a translation with the reference point $A$ and a rotation about $A$. You can obtain the velocity $\mathbf{v}_{P'}$ and the acceleration $\mathbf{a}_{P'}$ of $P'$; therefore, by adding $\mathbf{v}_A$ and $\mathbf{a}_A$, respectively, to the expressions found in part A, Step 3, and replacing the position vector $\mathbf{r}$ by the vector $\mathbf{r}_{P/A}$ drawn from $A$ to $P$:

$$\mathbf{v}_{P'} = \mathbf{v}_A + \Omega \times \mathbf{r}_{P/A} \qquad \mathbf{a}_{P'} = \mathbf{a}_A + \alpha \times \mathbf{r}_{P/A} + \Omega \times (\Omega \times \mathbf{r}_{P/A})$$

**4, 5, and 6 are the same as in part A of this summary,** except that the vector $\mathbf{r}$ should again be replaced by $\mathbf{r}_{P/A}$. Thus, Eqs. (15.46) and (15.48) can still be used to obtain the velocity and the acceleration of $P$ with respect to the fixed frame of reference $OXYZ$.

**C. Alternative approach using a frame of reference in general motion.** As shown in the sample problems, you can also use Eqs. (15.52) and (15.54) to determine the velocity and acceleration of point $P$, respectively.

$$\mathbf{v}_P = \mathbf{v}_A + \Omega \times \mathbf{r}_{P/A} + (\dot{\mathbf{r}}_{P/A})_{Axyz} \tag{15.52}$$

$$\mathbf{a}_P = \mathbf{a}_A + \dot{\Omega} \times \mathbf{r}_{P/A} + \Omega \times (\Omega \times \mathbf{r}_{P/A})$$
$$+ 2\Omega \times (\dot{\mathbf{r}}_{P/A})_{Axyz} + (\ddot{\mathbf{r}}_{P/A})_{Axyz} \tag{15.54}$$

You first need to determine a reference point $A$ and attach your rotating frame of reference at that point; generally this is attached to a specific part of the object under consideration (e.g., the cab or boom of a crane). Define the angular velocity of the frame as $\Omega$ and the angular acceleration of the frame as $\dot{\Omega}$. The terms $(\dot{\mathbf{r}}_{P/A})_{Axyz}$ and $(\ddot{\mathbf{r}}_{P/A})_{Axyz}$ represent the velocity and acceleration of point $P$ relative to the rotating frame of reference $A_{xyz}$.

# Problems

**15.220** A flight simulator is used to train pilots on how to recognize spatial disorientation. It has four degrees of freedom and can rotate around a planetary axis, as well as in yaw, pitch, and roll. The pilot is seated so that her head $B$ is located at $\mathbf{r} = (2 \text{ ft}) \mathbf{i} + (1 \text{ ft}) \mathbf{j}$ with respect to the center of the cab $A$. Knowing that the cab is rotating about the planetary axis with a constant angular velocity of 20 rpm counterclockwise as seen from above, and pitches with a constant angular velocity of $+3\mathbf{k}$ rad/s, determine (a) the velocity of the pilot's head, (b) the angular acceleration of the cab, (c) the acceleration of the pilot's head.

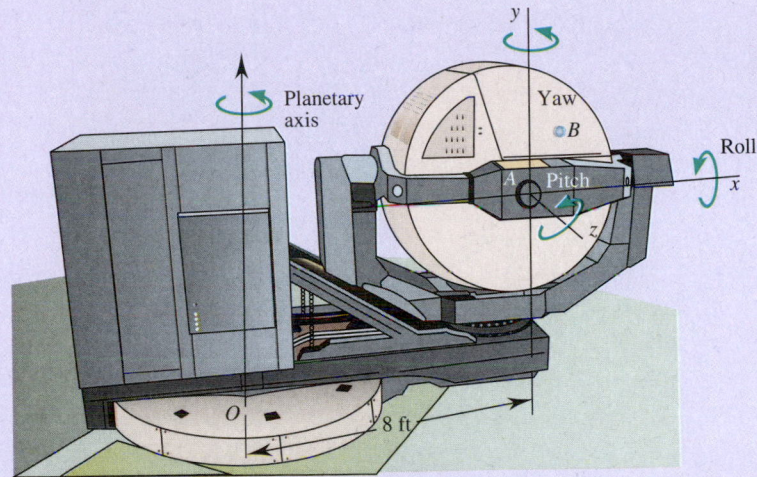

**Fig. P15.220 and P15.221**

**15.221** A flight simulator is used to train pilots on how to recognize spatial disorientation. It has four degrees of freedom and can rotate around a planetary axis, as well as in yaw, pitch, and roll. The pilot is seated so that her head $B$ is located at $r = (2 \text{ ft}) \mathbf{i} + (1 \text{ ft}) \mathbf{j}$ with respect to the center of the cab $A$. The cab is rotating about the planetary axis with an angular velocity of 20 rpm counterclockwise as seen from above, and this is increasing by 1 rad/s$^2$. Knowing that the cab rolls with a constant angular velocity of $(-4 \text{ rad/s})\mathbf{i}$, determine (a) the velocity of the pilot's head, (b) the angular acceleration of the cab, (c) the acceleration of the pilot's head.

**15.222** A square plate of side 360 mm is hinged at $A$ and $B$ to a clevis. The plate rotates at the constant rate $\omega_2 = 4$ rad/s with respect to the clevis, which itself rotates at the constant rate $\omega_1 = 3$ rad/s about the $Y$ axis. For the position shown, determine (a) the velocity of point $C$, (b) the acceleration of point $C$.

**15.223** A square plate of side 360 mm is hinged at $A$ and $B$ to a clevis. The plate rotates at the constant rate $\omega_2 = 4$ rad/s with respect to the clevis, which itself rotates at the constant rate $\omega_1 = 3$ rad/s about the $Y$ axis. For the position shown, determine (a) the velocity of corner $D$, (b) the acceleration of corner $D$.

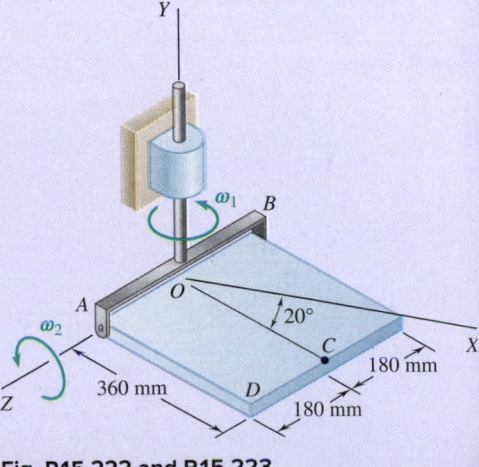

**Fig. P15.222 and P15.223**

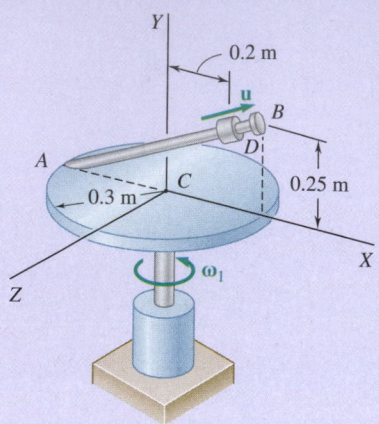

**Fig. P15.224**

**15.224** Rod $AB$ is welded to the 0.3-m-radius plate that rotates at the constant rate $\omega_1 = 6$ rad/s. Knowing that collar $D$ moves toward end $B$ of the rod at a constant speed $u = 1.3$ m/s, determine, for the position shown, (a) the velocity of $D$, (b) the acceleration of $D$.

**15.225** The bent rod shown rotates at the constant rate of $\omega_1 = 5$ rad/s and collar $C$ moves toward point $B$ at a constant relative speed of $u = 39$ in./s. Knowing that collar $C$ is halfway between points $B$ and $D$ at the instant shown, determine its velocity and acceleration.

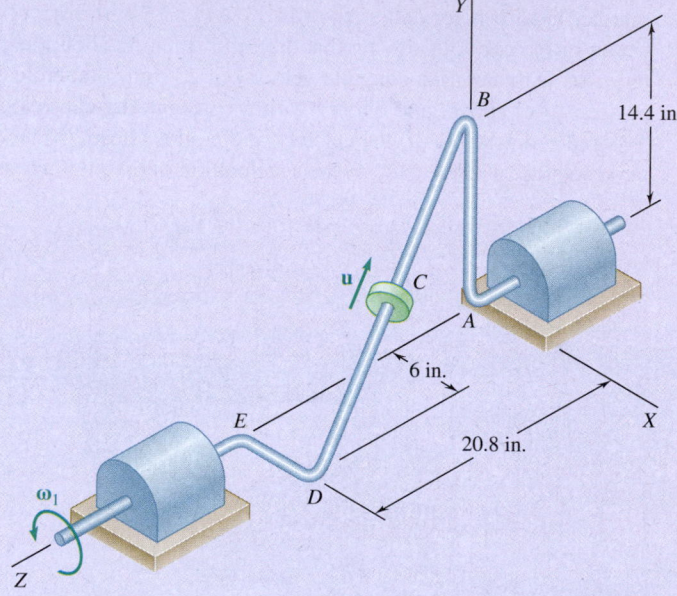

**Fig. P15.225**

**15.226** The bent pipe shown rotates at the constant rate $\omega_1 = 10$ rad/s. Knowing that a ball bearing $D$ moves in portion $BC$ of the pipe toward end $C$ at a constant relative speed $u = 2$ ft/s, determine at the instant shown (a) the velocity of $D$, (b) the acceleration of $D$.

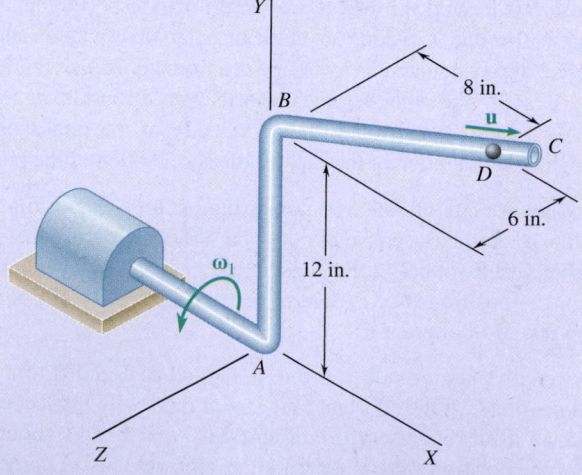

**Fig. P15.226**

**15.227** The circular plate shown rotates about its vertical diameter at the constant rate $\omega_1 = 10$ rad/s. Knowing that in the position shown the disk lies in the $XY$ plane and point $D$ of strap $CD$ moves upward at a constant relative speed $u = 1.5$ m/s, determine ($a$) the velocity of $D$, ($b$) the acceleration of $D$.

**15.228** Manufactured items are spray-painted as they pass through the automated work station shown. Knowing that the bent pipe $ACE$ rotates at the constant rate $\omega_1 = 0.4$ rad/s and that at point $D$ the paint moves through the pipe at a constant relative speed $u = 150$ mm/s, determine, for the position shown, ($a$) the velocity of the paint at $D$, ($b$) the acceleration of the paint at $D$.

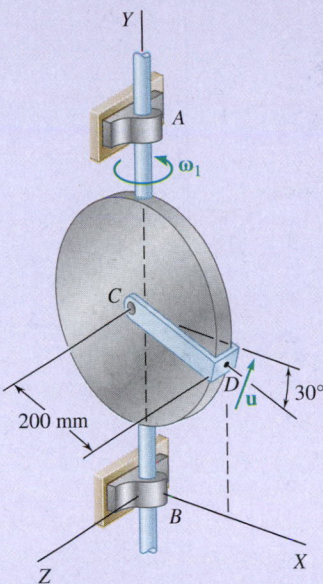

**Fig. P15.227**

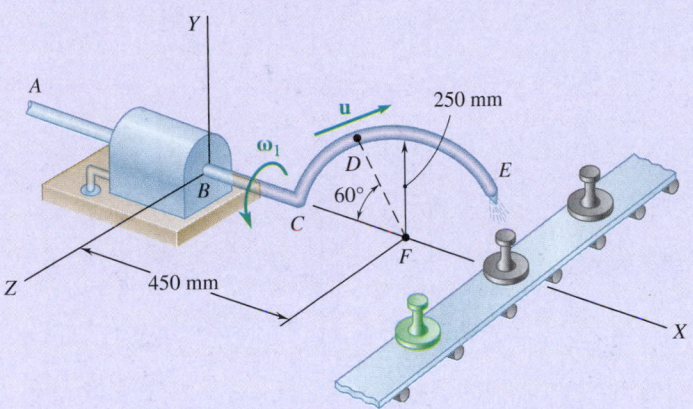

**Fig. P15.228**

**15.229** Solve Prob. 15.227, assuming that at the instant shown the angular velocity $\omega_1$ of the plate is 10 rad/s and is decreasing at the rate of 25 rad/s$^2$, while the relative speed $u$ of point $D$ of strap $CD$ is 1.5 m/s and is decreasing at the rate of 3 m/s$^2$.

**15.230** Solve Prob. 15.225, assuming that at the instant shown the angular velocity $\omega_1$ of the rod is 5 rad/s and is increasing at the rate of 10 rad/s$^2$, while the relative speed $u$ of the collar $C$ is 39 in./s and is decreasing at the rate of 260 in./s$^2$.

**15.231** Using the method of Sec. 15.7A, solve Prob. 15.192.

**15.232** Using the method of Sec. 15.7A, solve Prob. 15.196.

**15.233** Using the method of Sec. 15.7A, solve Prob. 15.198.

**15.234** The 400-mm bar $AB$ is made to rotate at the constant rate of $\omega_2 = d\theta/dt = 8$ rad/s with respect to the frame $CD$ that rotates at the constant rate of $\omega_1 = 12$ rad/s about the $Y$ axis. Knowing that $\theta = 60°$ at the instant shown, determine the velocity and acceleration of point $A$.

**15.235** The 400-mm bar $AB$ is made to rotate at the rate $\omega_2 = d\theta/dt$ with respect to the frame $CD$ that rotates at the rate $\omega_1$ about the $Y$ axis. At the instant shown, $\omega_1 = 12$ rad/s, $d\omega_1/dt = -16$ rad/s$^2$, $\omega_2 = 8$ rad/s, $d\omega_2/dt = 10$ rad/s$^2$, and $\theta = 60°$. Determine the velocity and acceleration of point $A$ at this instant.

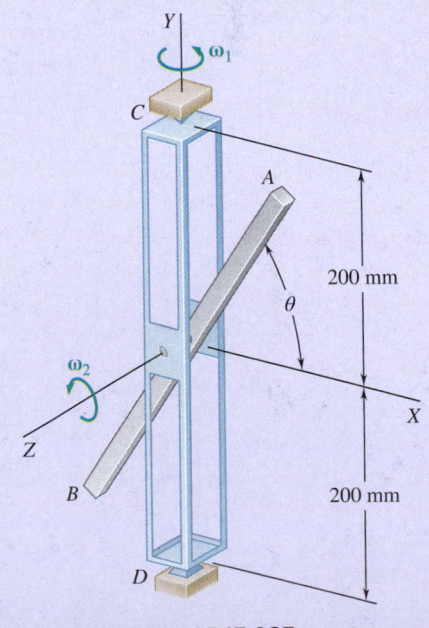

**Fig. P15.234 and P15.235**

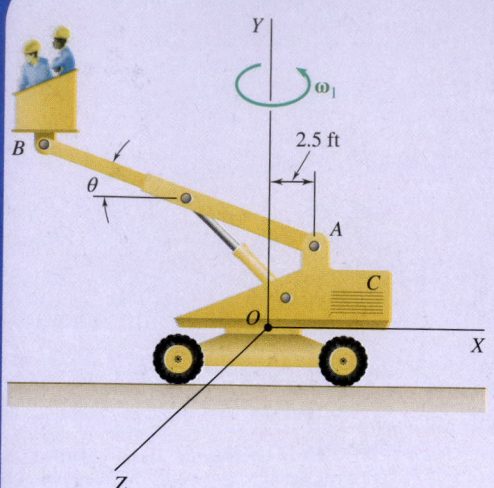

Fig. P15.236

**15.236** The arm $AB$ of length 16 ft is used to provide an elevated platform for construction workers. In the position shown, arm $AB$ is being raised at the constant rate $d\theta/dt = 0.25$ rad/s; simultaneously, the unit is being rotated about the $Y$ axis at the constant rate $\omega_1 = 0.15$ rad/s. Knowing that $\theta = 20°$, determine the velocity and acceleration of point $B$.

**15.237** The remote manipulator system (RMS) shown is used to deploy payloads from the cargo bay of space shuttles. At the instant shown, the whole RMS is rotating at the constant rate $\omega_1 = 0.03$ rad/s about the axis $AB$. At the same time, portion $BCD$ rotates as a rigid body at the constant rate $\omega_2 = d\beta/dt = 0.04$ rad/s about an axis through $B$ parallel to the $X$ axis. Knowing that $\beta = 30°$, determine (a) the angular acceleration of $BCD$, (b) the velocity of $D$, (c) the acceleration of $D$.

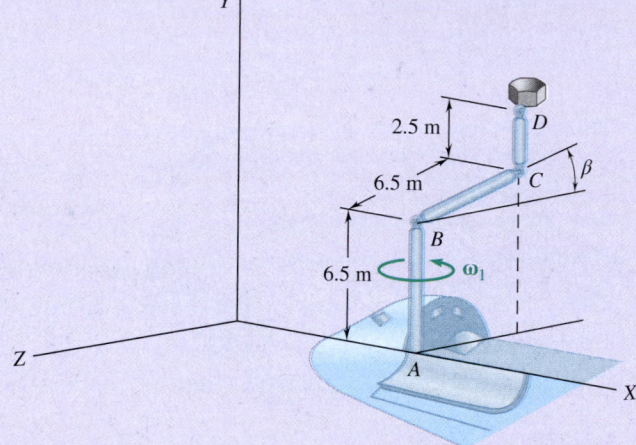

Fig. P15.237

**15.238** A disk with a radius of 120 mm rotates at the constant rate of $\omega_2 = 5$ rad/s with respect to the arm $AB$ that rotates at the constant rate of $\omega_1 = 3$ rad/s. For the position shown, determine the velocity and acceleration of point $C$.

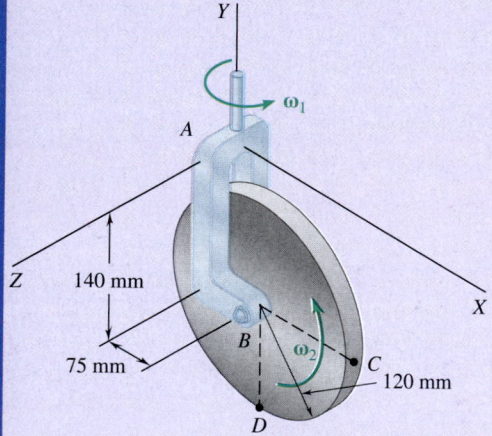

Fig. P15.238

**15.239** The crane shown rotates at the constant rate $\omega_1 = 0.25$ rad/s; simultaneously, the telescoping boom is being lowered at the constant rate $\omega_2 = 0.40$ rad/s. Knowing that at the instant shown the length of the boom is 20 ft and is increasing at the constant rate $u = 1.5$ ft/s, determine the velocity and acceleration of point $B$.

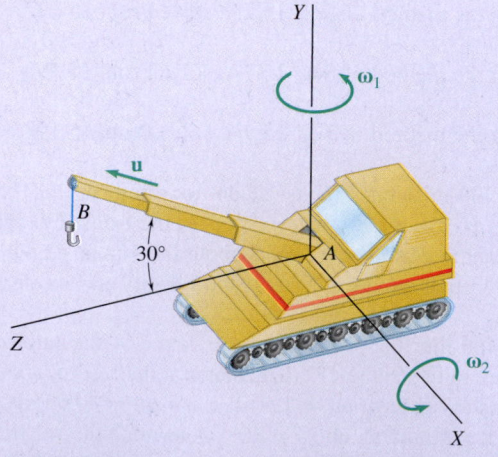

Fig. P15.239

**15.240** The vertical plate shown is welded to arm *EFG*, and the entire unit rotates at the constant rate $\omega_1 = 1.6$ rad/s about the *Y* axis. At the same time, a continuous link belt moves around the perimeter of the plate at a constant speed $u = 90$ mm/s. For the position shown, determine the acceleration of the link of the belt located (*a*) at point *A*, (*b*) at point *B*.

**15.241** The vertical plate shown is welded to arm *EFG*, and the entire unit rotates at the constant rate $\omega_1 = 1.6$ rad/s about the *Y* axis. At the same time, a continuous link belt moves around the perimeter of the plate at a constant speed $u = 90$ mm/s. For the position shown, determine the acceleration of the link of the belt located (*a*) at point *C*, (*b*) at point *D*.

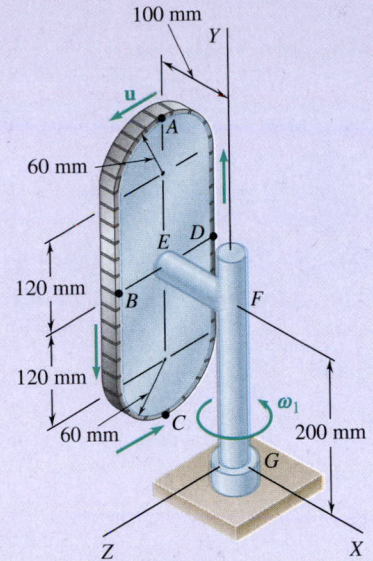

**Fig. P15.240 and P15.241**

**15.242** The cylinder shown rotates at the constant rate $\omega_2 = 8$ rad/s with respect to rod *CD*, which itself rotates at the constant rate $\omega_1 = 6$ rad/s about the *X* axis. For the position shown, determine the velocity and acceleration of point *A* on the edge of the cylinder.

**15.243** The cylinder shown rotates at the constant rate $\omega_2 = 8$ rad/s with respect to rod *CD*, which itself rotates at the constant rate $\omega_1 = 6$ rad/s about the *X* axis. For the position shown, determine the velocity and acceleration of point *B* on the edge of the cylinder.

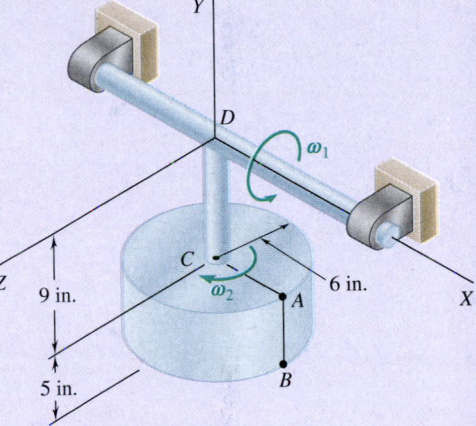

**Fig. P15.242 and P15.243**

**15.244** A square plate of side $2r$ is welded to a vertical shaft that rotates with a constant angular velocity $\omega_1$. At the same time, rod *AB* of length $r$ rotates about the center of the plate with a constant angular velocity $\omega_2$ with respect to the plate. For the position of the plate shown, determine the acceleration of end *B* of the rod if (*a*) $\theta = 0$, (*b*) $\theta = 90°$, (*c*) $\theta = 180°$.

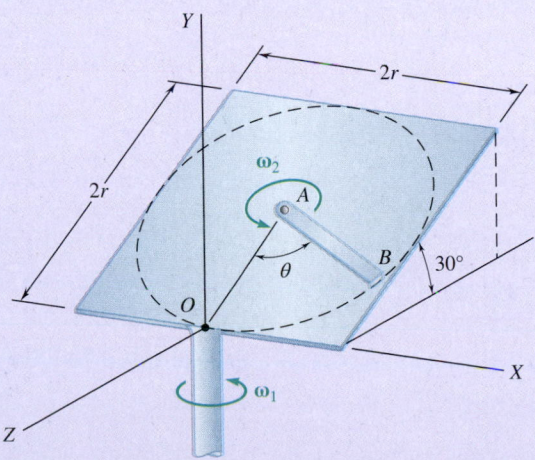

**Fig. P15.244**

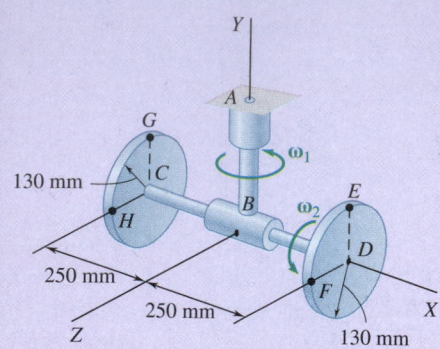

**Fig. P15.245**

**15.245** Two disks, each of 130-mm radius, are welded to the 500-mm rod *CD*. The rod-and-disks unit rotates at the constant rate $\omega_2 = 3$ rad/s with respect to arm *AB*. Knowing that at the instant shown $\omega_1 = 4$ rad/s, determine the velocity and acceleration of (*a*) point *E*, (*b*) point *F*.

**15.246** In Prob. 15.245, determine the velocity and acceleration of (*a*) point *G*, (*b*) point *H*.

**15.247** The position of the stylus tip *A* is controlled by the robot shown. In the position shown, the stylus moves at a constant speed $u = 180$ mm/s relative to the solenoid *BC*. At the same time, arm *CD* rotates at the constant rate $\omega_2 = 1.6$ rad/s with respect to component *DEG*. Knowing that the entire robot rotates about the *X* axis at the constant rate $\omega_1 = 1.2$ rad/s, determine (*a*) the velocity of *A*, (*b*) the acceleration of *A*.

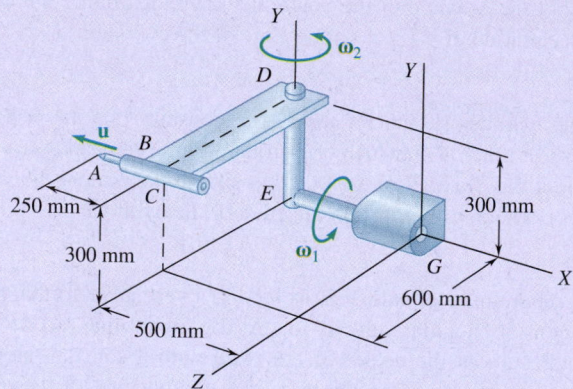

**Fig. P15.247**

# Review and Summary

This chapter was devoted to the study of the kinematics of rigid bodies.

## Rigid Body in Translation

We first considered the **translation** of a rigid body (Sec. 15.1A) and observed that in such a motion **all points of the body have the same velocity and the same acceleration at any given instant**.

## Rigid Body in Rotation About a Fixed Axis

We next considered the **rotation** of a rigid body about a fixed axis (Sec. 15.1B). The position of the body is defined by the angle $\theta$ that the line $BP$, drawn from the axis of rotation to a point $P$ of the body, forms with a fixed plane (Fig. 15.39). We found that the magnitude of the velocity of $P$ is

$$v = \frac{ds}{dt} = r\dot{\theta} \sin \phi \qquad (15.4)$$

where $\dot{\theta}$ is the time derivative of $\theta$. We then expressed the velocity of $P$ as

$$\mathbf{v} = \frac{d\mathbf{r}}{dt} = \boldsymbol{\omega} \times \mathbf{r} \qquad (15.5)$$

where the vector

$$\boldsymbol{\omega} = \omega\mathbf{k} = \dot{\theta}\mathbf{k} \qquad (15.6)$$

is directed along the fixed axis of rotation and represents the angular velocity of the body.

Denoting the derivative $d\boldsymbol{\omega}/dt$ of the angular velocity by $\boldsymbol{\alpha}$, we expressed the acceleration of $P$ as

$$\mathbf{a} = \boldsymbol{\alpha} \times \mathbf{r} + \boldsymbol{\omega} \times (\boldsymbol{\omega} \times \mathbf{r}) \qquad (15.8)$$

Differentiating Eq. (15.6) and recalling that $\mathbf{k}$ is constant in magnitude and direction, we found that

$$\boldsymbol{\alpha} = \alpha\mathbf{k} = \dot{\omega}\mathbf{k} = \ddot{\theta}\mathbf{k} \qquad (15.9)$$

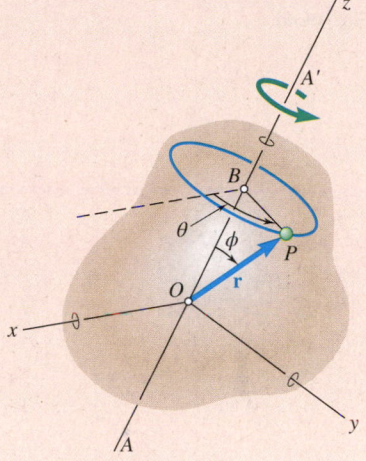

**Fig. 15.39**

The vector $\boldsymbol{\alpha}$ represents the angular acceleration of the body and is directed along the fixed axis of rotation (Sample Prob. 15.2).

## Rotation of a Representative Slab: Tangential and Normal Components

Next we considered the motion of a representative slab located in a plane perpendicular to the axis of rotation of the body (Fig. 15.40). Because the angular velocity is perpendicular to the slab, we expressed the velocity of a point $P$ of the slab as

$$\mathbf{v} = \omega\mathbf{k} \times \mathbf{r} \qquad (15.10)$$

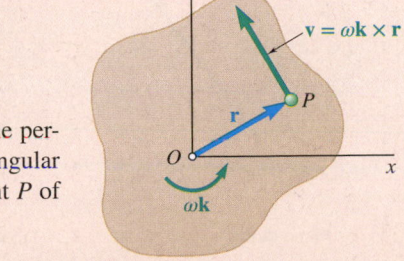

**Fig. 15.40**

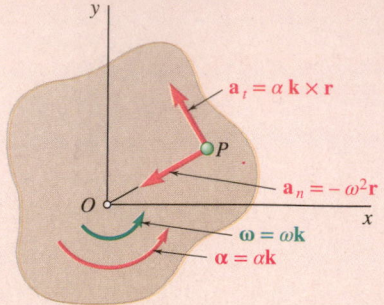

**Fig. 15.41**

where **v** is contained in the plane of the slab. Substituting $\boldsymbol{\omega} = \omega\mathbf{k}$ and $\boldsymbol{\alpha} = \alpha\mathbf{k}$ into Eq. (15.8), we found that we could resolve the acceleration of $P$ into tangential and normal components (Fig. 15.41) respectively equal to

$$\begin{aligned} \mathbf{a}_t &= \alpha\mathbf{k} \times \mathbf{r} & a_t &= r\alpha \\ \mathbf{a}_n &= -\omega^2\mathbf{r} & a_n &= r\omega^2 \end{aligned} \tag{15.11'}$$

### Angular Velocity and Angular Acceleration of a Rotating Rigid Body

Recalling Eqs. (15.6) and (15.9), we obtained the following expressions for the *angular velocity* and the *angular acceleration* of the rigid body (Sec. 15.1C):

$$\omega = \frac{d\theta}{dt} \tag{15.12}$$

$$\alpha = \frac{d\omega}{dt} = \frac{d^2\theta}{dt^2} \tag{15.13}$$

or

$$\alpha = \omega\frac{d\omega}{d\theta} \tag{15.14}$$

We noted that these expressions are similar to those obtained in Chap. 11 for the rectilinear motion of a particle.

Two particular cases of rotation are frequently encountered: *uniform rotation* and *uniformly accelerated rotation*. You can solve problems involving either of these motions by using equations similar to those used in Sec. 11.2 for the uniform rectilinear motion and the uniformly accelerated rectilinear motion of a particle, but where $x$, $v$, and $a$ are replaced by $\theta$, $\omega$, and $\alpha$, respectively (Sample Prob. 15.1).

### Velocities in Plane Motion

We can consider the **most general plane motion** of a rigid body as the **sum of a translation and a rotation** (Sec. 15.2A). For example, the body shown in Fig. 15.42 can be assumed to translate with point $A$, while simultaneously rotating about $A$. It follows (Sec. 15.2B) that the velocity of any point $B$ of the rigid body can be expressed as

$$\mathbf{v}_B = \mathbf{v}_A + \mathbf{v}_{B/A} \tag{15.17}$$

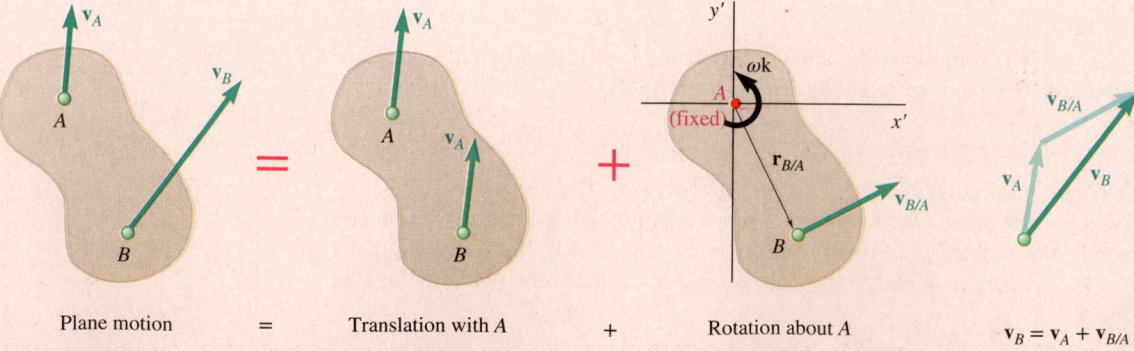

| Plane motion | = | Translation with $A$ | + | Rotation about $A$ | |
|---|---|---|---|---|---|

**Fig. 15.42**

where $\mathbf{v}_A$ is the velocity of $A$ and $\mathbf{v}_{B/A}$ is the relative velocity of $B$ with respect to $A$ or, more precisely, with respect to axes $x'y'$ translating with $A$. Denoting the position vector of $B$ relative to $A$ by $\mathbf{r}_{B/A}$, we found that

$$\mathbf{v}_{B/A} = \omega\mathbf{k} \times \mathbf{r}_{B/A} \qquad v_{B/A} = r\omega \qquad (15.18)$$

The fundamental equation (15.17) relating the absolute velocities of points $A$ and $B$ and the relative velocity of $B$ with respect to $A$ was expressed in the form of a vector diagram, which can be used to solve problems involving the motion of various types of mechanisms (Sample Probs. 15.6 and 15.7).

## Instantaneous Center of Rotation

We presented another approach to the solution of problems involving the velocities of the points of a rigid body in plane motion in Sec.15.3 and used it in Sample Probs. 15.9, 15.10, and 15.11. It is based on the determination of the **instantaneous center of rotation** $C$ of the rigid body (Fig. 15.43).

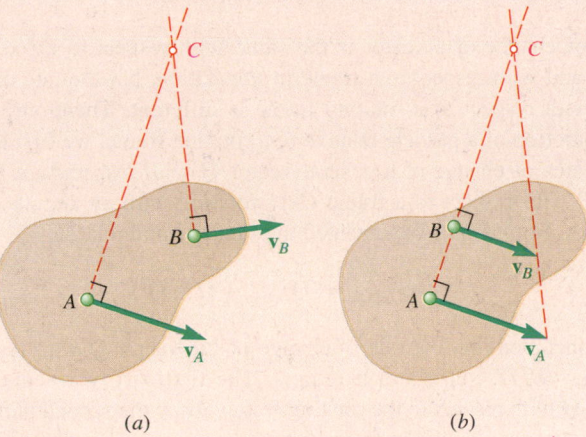

(a)           (b)

**Fig. 15.43**

## Accelerations in Plane Motion

In Sec. 15.4A, we used the fact that any plane motion of a rigid body can be considered the sum of a translation of the body with a reference point $A$ and a rotation about $A$. Knowing this, we can find the absolute acceleration of $A$ by adding the relative acceleration of $B$ with respect to $A$ to the absolute acceleration of $B$.

$$\mathbf{a}_B = \mathbf{a}_A + \mathbf{a}_{B/A} \qquad (15.21)$$

where $\mathbf{a}_{B/A}$ consisted of a *normal component* $(\mathbf{a}_{B/A})_n$ with a magnitude $r\omega^2$ directed toward $A$ and a *tangential component* $(\mathbf{a}_{B/A})_t$ with a magnitude $r\alpha$ perpendicular to the line $AB$ (Fig. 15.44). We expressed the fundamental relation

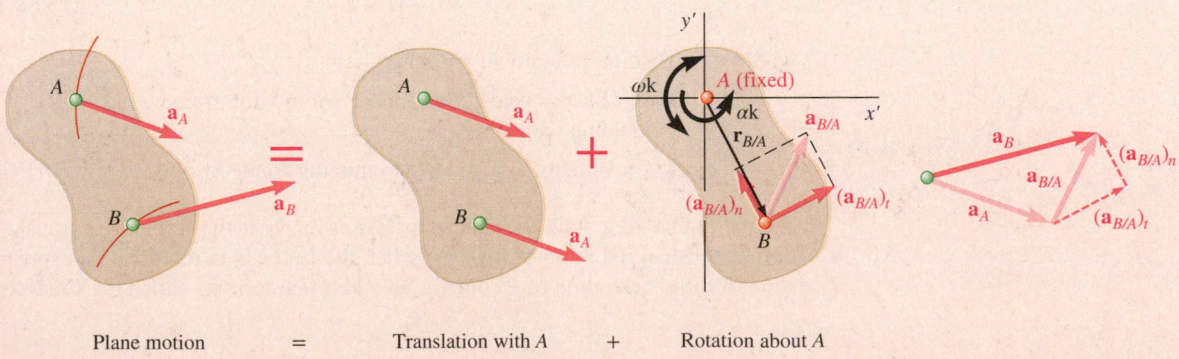

Plane motion    =    Translation with $A$    +    Rotation about $A$

**Fig. 15.44**

in Eq. (15.21) in terms of vector diagrams or vector equations and used them to determine the accelerations of given points of various mechanisms (Sample Probs. 15.12, 15.13, 15.14, 15.15, and 15.16). We noted that we cannot use the instantaneous center of rotation $C$ considered in Sec. 15.3 for the determination of accelerations, because point $C$, in general, does *not* have zero acceleration.

### Coordinates Expressed in Terms of a Parameter

In the case of certain mechanisms, it is possible to express the coordinates $x$ and $y$ of all significant points of the mechanism by means of simple analytic expressions containing a *single parameter*. We can obtain the components of the absolute velocity and acceleration of a given point by differentiating twice with respect to the time $t$ the coordinates $x$ and $y$ of that point (Sec. 15.4B).

### Rate of Change of a Vector with Respect to a Rotating Frame

The rate of change of a vector is the same with respect to a fixed frame of reference and with respect to a frame in translation, but the rate of change of a vector with respect to a rotating frame is different. Therefore, in order to study the motion of a particle relative to a rotating frame, we first had to compare the rates of change of a general vector $\mathbf{Q}$ with respect to a fixed frame $OXYZ$ and with respect to a frame $Oxyz$ rotating with an angular velocity $\mathbf{\Omega}$ (Sec. 15.5A, Fig. 15.45). We obtained the fundamental relation

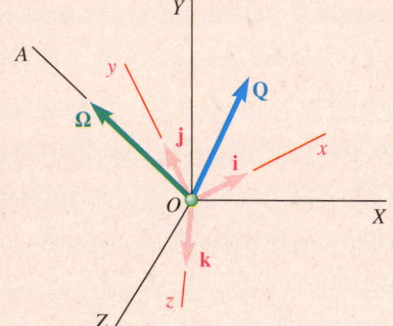

**Fig. 15.45**

$$(\dot{\mathbf{Q}})_{OXYZ} = (\dot{\mathbf{Q}})_{Oxyz} + \mathbf{\Omega} \times \mathbf{Q} \tag{15.31}$$

and we concluded that the rate of change of the vector $\mathbf{Q}$ with respect to the fixed frame $OXYZ$ consists of two parts: The first part represents the rate of change of $\mathbf{Q}$ with respect to the rotating frame $Oxyz$; the second part, $\mathbf{\Omega} \times \mathbf{Q}$, is induced by the rotation of the frame $Oxyz$.

### Plane Motion of a Particle Relative to a Rotating Frame

The next section (Sec. 15.5B) was devoted to the two-dimensional kinematic analysis of a particle $P$ moving with respect to a frame $\mathscr{F}$ rotating with an angular velocity $\mathbf{\Omega}$ about a fixed axis (Fig. 15.46). We found that the absolute velocity of $P$ could be expressed as

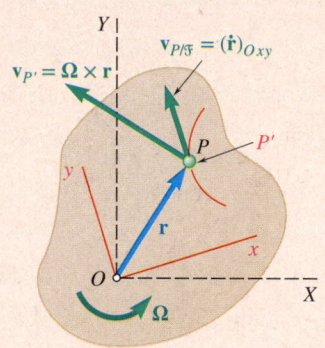

**Fig. 15.46**

$$\mathbf{v}_P = \mathbf{v}_{P'} + \mathbf{v}_{P/\mathscr{F}} \tag{15.33}$$

or

$$\mathbf{v}_P = \mathbf{v}_O + \mathbf{\Omega} \times \mathbf{r} + \mathbf{v}_{\text{rel}} \tag{15.32'}$$

where $\mathbf{v}_P$ = absolute velocity of particle $P$

$\mathbf{v}_{P'} = \mathbf{v}_O + \mathbf{\Omega} \times \mathbf{r}$ = velocity of point $P'$ of moving frame $\mathscr{F}$ coinciding with $P$

$\mathbf{v}_{P/\mathscr{F}} = \mathbf{v}_{\text{rel}}$ = velocity of $P$ relative to moving frame $\mathscr{F}$

We noted that we obtain the same expression for $\mathbf{v}_P$ if the frame is in translation rather than in rotation. However, when the frame is in rotation, the expression for the acceleration of $P$ contains an additional term $\mathbf{a}_C$ called the **Coriolis**

**acceleration**. We have

$$\mathbf{a}_P = \mathbf{a}_{P'} + \mathbf{a}_{P/\mathscr{F}} + \mathbf{a}_C \qquad (15.36)$$

or

$$\mathbf{a}_P = \mathbf{a}_O + \dot{\boldsymbol{\Omega}} \times \mathbf{r} - \Omega^2 \mathbf{r} + 2\boldsymbol{\Omega} \times \mathbf{v}_{\text{rel}} + \mathbf{a}_{\text{rel}}$$

where $\mathbf{a}_P =$ absolute acceleration of particle $P$
$\qquad \mathbf{a}_{P'} = \mathbf{a}_O + \dot{\boldsymbol{\Omega}} \times \mathbf{r} - \Omega^2 \mathbf{r} =$ acceleration of point $P'$ of moving frame $\mathscr{F}$ coinciding with $P$
$\qquad \mathbf{a}_{P/\mathscr{F}} = \mathbf{a}_{\text{rel}} =$ acceleration of $P$ relative to moving frame $\mathscr{F}$
$\qquad \mathbf{a}_C = 2\boldsymbol{\Omega} \times (\dot{\mathbf{r}})_{Oxyz} = 2\boldsymbol{\Omega} \times \mathbf{v}_{P/\mathscr{F}} = 2\boldsymbol{\Omega} \times \mathbf{v}_{\text{rel}} =$ Coriolis acceleration

Because $\boldsymbol{\Omega}$ and $\mathbf{v}_{P/\mathscr{F}}$ are perpendicular to each other in the case of plane motion, the Coriolis acceleration has a magnitude $a_C = 2\Omega v_{P/\mathscr{F}}$ and points in the direction obtained by rotating the vector $\mathbf{v}_{P/\mathscr{F}}$ through 90° in the sense of rotation of the moving frame. We can use formulas (15.33) and (15.36) to analyze the motion of mechanisms that contain parts sliding on each other (Sample Probs. 15.17 through 15.20).

## Motion of a Rigid Body with a Fixed Point

In the last part of this chapter, we studied the kinematics of rigid bodies in three dimensions. We first considered the motion of a rigid body with a fixed point (Sec. 15.6A). After proving that the most general displacement of a rigid body with a fixed point $O$ is equivalent to a rotation of the body about an axis through $O$, we were able to define the angular velocity $\boldsymbol{\omega}$ and the **instantaneous axis of rotation** of the body at a given instant. The velocity of a point $P$ of the body (Fig. 15.47) again could be expressed as

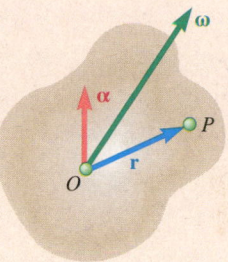

Fig. 15.47

$$\mathbf{v} = \frac{d\mathbf{r}}{dt} = \boldsymbol{\omega} \times \mathbf{r} \qquad (15.37)$$

Differentiating this expression gave

$$\mathbf{a} = \boldsymbol{\alpha} \times \mathbf{r} + \boldsymbol{\omega} \times (\boldsymbol{\omega} \times \mathbf{r}) \qquad (15.38)$$

However, because the direction of $\boldsymbol{\omega}$ changes from one instant to the next, the angular acceleration $\boldsymbol{\alpha}$ is, in general, not directed along the instantaneous axis of rotation (Sample Prob. 15.21).

## General Motion in Space

We showed in Sec. 15.6B that **the most general motion of a rigid body in space is equivalent, at any given instant, to the sum of a translation and a rotation**. Considering two particles $A$ and $B$ of the body, we found that

$$\mathbf{v}_B = \mathbf{v}_A + \mathbf{v}_{B/A} \qquad (15.42)$$

where $\mathbf{v}_{B/A}$ is the velocity of $B$ relative to a frame $AX'Y'Z'$ attached to $A$ and of fixed orientation (Fig. 15.48). Denoting by $\mathbf{r}_{B/A}$ the position vector of $B$ relative to $A$, we have

$$\mathbf{v}_B = \mathbf{v}_A + \boldsymbol{\omega} \times \mathbf{r}_{B/A} \qquad (15.43)$$

where $\boldsymbol{\omega}$ is the angular velocity of the body at the instant considered (Sample Prob. 15.22). We obtained the acceleration of $B$ using a similar reasoning. We first wrote

$$\mathbf{a}_B = \mathbf{a}_A + \mathbf{a}_{B/A}$$

and, recalling Eq. (15.38),

$$\mathbf{a}_B = \mathbf{a}_A + \boldsymbol{\alpha} \times \mathbf{r}_{B/A} + \boldsymbol{\omega} \times (\boldsymbol{\omega} \times \mathbf{r}_{B/A}) \qquad (15.44)$$

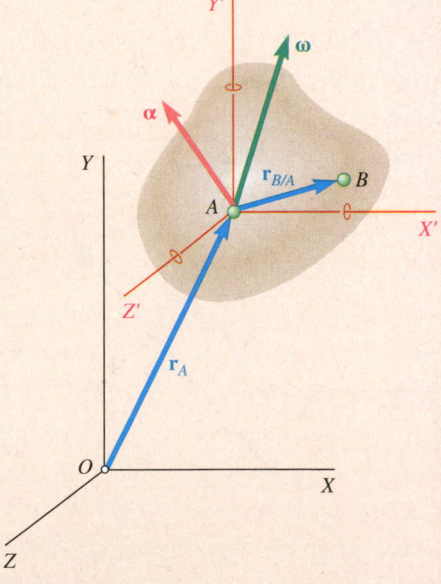

Fig. 15.48

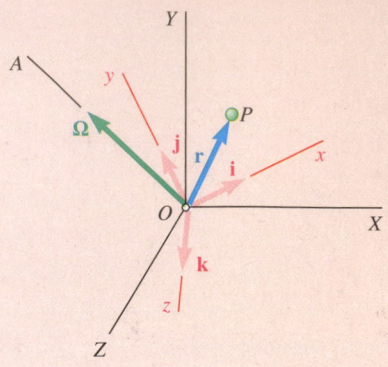

**Fig. 15.49**

## Three-Dimensional Motion of a Particle Relative to a Rotating Frame

In the final section of this chapter, we considered the three-dimensional motion of a particle $P$ relative to a frame $Oxyz$ rotating with an angular velocity $\mathbf{\Omega}$ with respect to a fixed frame $OXYZ$ (Fig. 15.49). In Sec. 15.7A, we expressed the absolute velocity $\mathbf{v}_P$ of $P$ as

$$\mathbf{v}_P = \mathbf{\Omega} \times \mathbf{r} + (\dot{\mathbf{r}})_{Oxyz} \qquad (15.45)$$

or alternatively as

$$\mathbf{v}_P = \mathbf{v}_{P'} + \mathbf{v}_{P/\mathscr{F}} \qquad (15.46)$$

where $\mathbf{v}_P$ = absolute velocity of particle $P$

$\mathbf{v}_{P'}$ = velocity of point $P'$ of moving frame $\mathscr{F}$ coinciding with $P$

$\mathbf{v}_{P/\mathscr{F}}$ = velocity of $P$ relative to moving frame $\mathscr{F}$

The absolute acceleration $\mathbf{a}_P$ of $P$ can be expressed as

$$\mathbf{a}_P = \dot{\mathbf{\Omega}} \times \mathbf{r} + \mathbf{\Omega} \times (\mathbf{\Omega} \times \mathbf{r}) + 2\mathbf{\Omega} \times (\dot{\mathbf{r}})_{Oxyz} + (\ddot{\mathbf{r}})_{Oxyz} \qquad (15.47)$$

or alternatively

$$\mathbf{a}_P = \mathbf{a}_{P'} + \mathbf{a}_{P/\mathscr{F}} + \mathbf{a}_C \qquad (15.48)$$

where $\mathbf{a}_P$ = absolute acceleration of particle $P$

$\mathbf{a}_{P'}$ = acceleration of point $P'$ of moving frame $\mathscr{F}$ coinciding with $P$

$\mathbf{a}_{P/\mathscr{F}}$ = acceleration of $P$ relative to moving frame $\mathscr{F}$

$\mathbf{a}_C = 2\mathbf{\Omega} \times (\dot{\mathbf{r}})_{Oxyz} = 2\mathbf{\Omega} \times \mathbf{v}_{P/\mathscr{F}}$ = Coriolis acceleration

We noted that the magnitude $a_C$ of the Coriolis acceleration is not equal to $2\Omega v_{P/\mathscr{F}}$ (Sample Prob. 15.23) except in the special case when $\mathbf{\Omega}$ and $\mathbf{v}_{P/\mathscr{F}}$ are perpendicular to each other. Additionally, we usually will have to use Eq. (15.31) to determine the angular acceleration $\dot{\mathbf{\Omega}}$ of the rotating frame.

## Frame of Reference in General Motion

We also observed (Sec. 15.7B) that Eqs. (15.46) and (15.48) remain valid when the frame $Axyz$ moves in a known—but arbitrary—fashion with respect to the fixed frame $OXYZ$ (Fig. 15.50), provided that the motion of $A$ is included in the terms $\mathbf{v}_{P'}$ and $\mathbf{a}_{P'}$ representing the absolute velocity and acceleration of the coinciding point $P'$. We obtained

$$\mathbf{v}_P = \mathbf{v}_A + \mathbf{\Omega} \times \mathbf{r}_{P/A} + (\dot{\mathbf{r}}_{P/A})_{Axyz} \qquad (15.52)$$

and

$$\mathbf{a}_P = \mathbf{a}_A + \dot{\mathbf{\Omega}} \times \mathbf{r}_{P/A} + \mathbf{\Omega} \times (\mathbf{\Omega} \times \mathbf{r}_{P/A}) + 2\mathbf{\Omega} \times (\dot{\mathbf{r}}_{P/A})_{Axyz} + (\ddot{\mathbf{r}}_{P/A})_{Axyz} \qquad (15.54)$$

Rotating frames of reference are particularly useful in the study of the three-dimensional motion of rigid bodies. Indeed, in many cases, an appropriate choice of the rotating frame leads to a simpler analysis of the motion of the rigid body than would be possible with axes of fixed orientation (Sample Probs. 15.24 and 15.25).

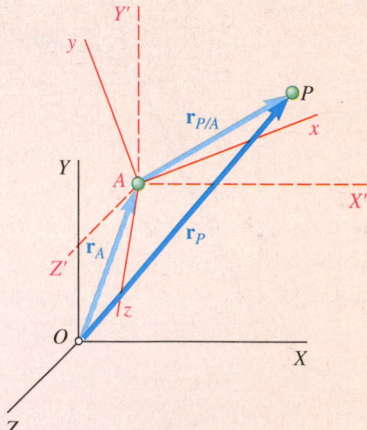

**Fig. 15.50**

# Review Problems

**15.248** A wheel moves in the $xy$ plane in such a way that the location of its center is given by the equations $x_O = 12t^3$ and $y_O = R = 2$, where $x_O$ and $y_O$ are measured in feet and $t$ is measured in seconds. The angular displacement of a radial line measured from a vertical reference line is $\theta = 8t^4$, where $\theta$ is measured in radians. Determine the velocity of point $P$ located on the horizontal diameter of the wheel at $t = 1$ s.

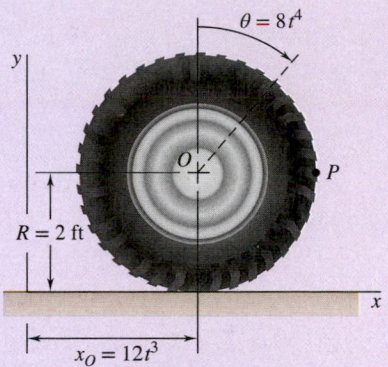

**Fig. P15.248**

**15.249** Two blocks and a pulley are connected by inextensible cords as shown. The relative velocity of block $A$ with respect to block $B$ is 2.5 ft/s to the left at time $t = 0$ and 1.25 ft/s to the left when $t = 0.25$ s. Knowing that the angular acceleration of the pulley is constant, find (a) the relative acceleration of block $A$ with respect to block $B$, (b) the distance block $A$ moves relative to block $B$ during the interval $0 \le t \le 0.25$ s.

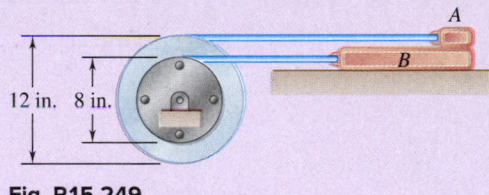

**Fig. P15.249**

**15.250** A baseball pitching machine is designed to deliver a baseball with a ball speed of 70 mph and a ball rotation of 300 rpm clockwise. Knowing that there is no slipping between the wheels and the baseball during the ball launch, determine the angular velocities of wheels $A$ and $B$.

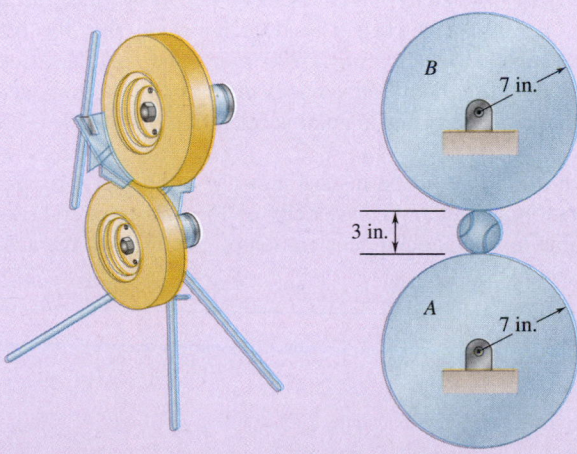

**Fig. P15.250**

**15.251** The flywheel *OD* on the elliptical machine analyzed in the case study for this chapter has a constant angular velocity of 5.5 rad/s clockwise. Note that bar *ABD* is one solid rigid body, connected to bar *EBF* at point *B*. The *x* and *y* locations of each point, expressed in inches, are provided in the table, where the reference origin is at point *O*. Note that both point *O* and *H* are fixed points, and *A* is a roller that can move in the horizontal plane. At the instant when $\theta = 180°$ as shown in the figure, determine (*a*) the velocity and acceleration of the roller *A*, (*b*) the velocity and acceleration of point *B*.

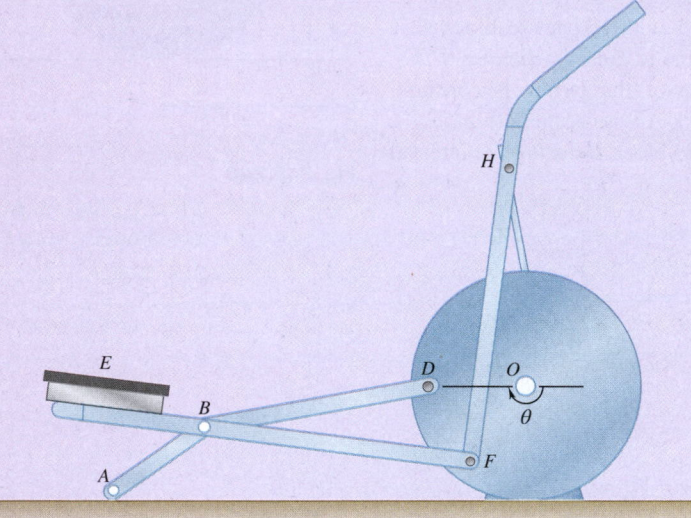

| | X (in.) | Y (in.) |
|---|---|---|
| A | −44.44 | −11.25 |
| B | −34.62 | −4.36 |
| D | −10.00 | 0 |
| E | −49.41 | −2.59 |
| F | −7.41 | −7.62 |
| H | −2.00 | 23.00 |

Fig. P15.251 and *P15.252*

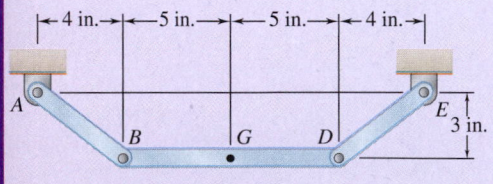

Fig. P15.253

**15.252** The roller at point *A* on the elliptical machine analyzed in the chapter case study has a velocity $v_A = 17.97$ in./s and acceleration $a_A = 400$ in./s$^2$ when $\theta = 180°$. Knowing that the bar *ABD* has an angular velocity of 1.5970 rad/s and angular acceleration of 0.833 rad/s$^2$, both counterclockwise, at this instant, determine the angular velocity and angular acceleration of bar *FH*.

**15.253** Knowing that at the instant shown rod *AB* has zero angular acceleration and an angular velocity of 15 rad/s counterclockwise, determine (*a*) the angular acceleration of arm *DE*, (*b*) the acceleration of point *D*.

**15.254** Rod $AB$ is attached to a collar at $A$ and is fitted with a wheel at $B$ that has a radius $r = 15$ mm. Knowing that when $\theta = 60°$ the collar has a velocity of 250 mm/s upward, and it is slowing down at a rate of 150 mm/s$^2$, determine (a) the angular acceleration of rod $AB$, (b) the angular acceleration of the wheel.

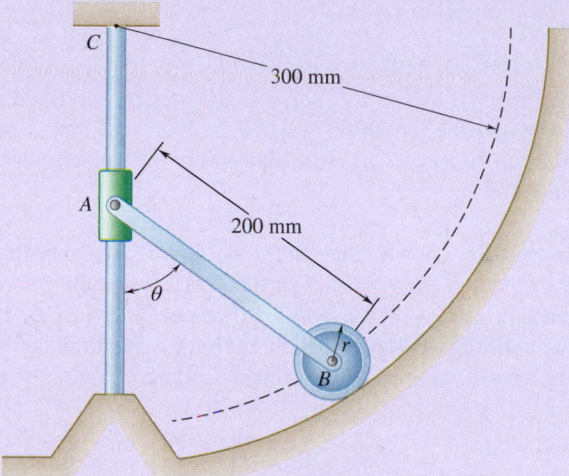

**Fig. P15.254**

**15.255** Water flows through a curved pipe $AB$ that rotates with a constant clockwise angular velocity of 90 rpm. If the velocity of the water relative to the pipe is 8 m/s, determine the total acceleration of a particle of water at point $P$.

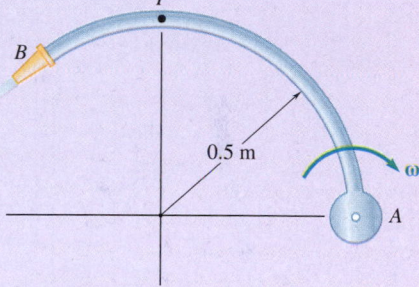

**Fig. P15.255**

**15.256** A disk of 0.15-m radius rotates at the constant rate $\omega_2$ with respect to plate $BC$, which itself rotates at the constant rate $\omega_1$ about the $y$ axis. Knowing that $\omega_1 = \omega_2 = 3$ rad/s, determine, for the position shown, the velocity and acceleration of (a) point $D$, (b) point $F$.

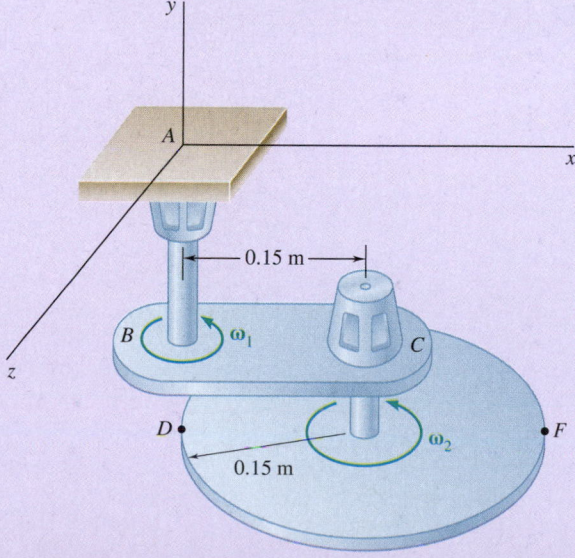

**Fig. P15.256**

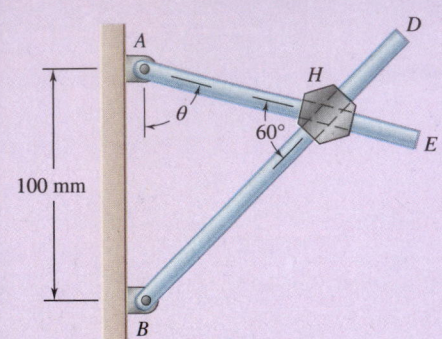

Fig. **P15.257**

**15.257** Two rods *AE* and *BD* pass through holes drilled into a hexagonal block. (The holes are drilled in different planes so that the rods will not touch each other.) Knowing that rod *AE* has an angular velocity of 20 rad/s clockwise and an angular acceleration of 4 rad/s² counterclockwise when $\theta = 90°$, determine (*a*) the relative velocity of the block with respect to each rod, (*b*) the relative acceleration of the block with respect to each rod.

**15.258** Rod *BC* of length 24 in. is connected by ball-and-socket joints to a rotating arm *AB* and to a collar *C* that slides on the fixed rod *DE*. Knowing that the length of arm *AB* is 4 in. and that it rotates at the constant rate $\omega_1 = 10$ rad/s, determine the velocity of collar *C* when $\theta = 0$.

**15.259** In the position shown, the thin rod moves at a constant speed $u = 3$ in./s out of the tube *BC*. At the same time, tube *BC* rotates at the constant rate $\omega_2 = 1.5$ rad/s with respect to arm *CD*. Knowing that the entire assembly rotates about the *X* axis at the constant rate $\omega_1 = 1.2$ rad/s, determine the velocity and acceleration of end *A* of the rod.

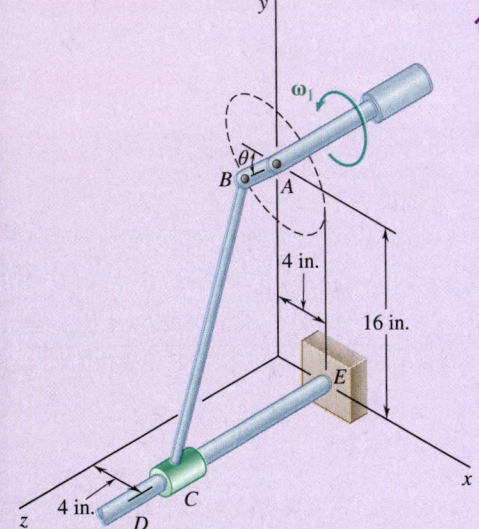

Fig. **P15.258**

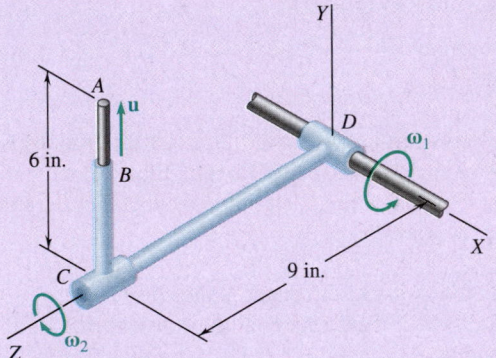

Fig. **P15.259**

# 16

# Plane Motion of Rigid Bodies: Forces and Accelerations

The blades of the wind turbines shown in this picture are subjected to large forces and moments during motion. In this chapter, you will learn to analyze the motion of a rigid body by considering the motion of its mass center, the motion relative to its mass center, and the external forces acting on it.

# Objectives

- **Discuss** how the mass and mass moment of inertia affect the linear and angular accelerations of a rigid body.

- **Model** physical systems involving rigid bodies by drawing correct free-body diagrams and kinetic diagrams.

- Using rigid-body kinetics principles, **determine** whether a body slips or tips and if a wheel rolls with or without slip.

- **Apply** appropriate kinetic equations and kinematics relationships to solve kinetics problems for a rigid body undergoing translation, centroidal rotation, or general plane motion.

- **Analyze** systems of connected rigid bodies using appropriate kinetic and kinematic equations.

- **Analyze** constrained motion of rigid bodies, including fixed-axis rotation and rolling disks and wheels.

# Introduction

In this chapter and in Chaps. 17 and 18, you will study the **kinetics of rigid bodies**; that is, the relations between the forces acting on a rigid body, the shape and mass of the body, and the motion produced. You studied similar relations in Chaps. 12 and 13, assuming then that you could consider the body as a particle, with its mass concentrated in one point and all forces acting at that point. Now you have to take into account the shape of the body, as well as the exact location of the points of application of the forces. You also will be concerned not only with the motion of the body as a whole but with the motion of the body about its mass center.

Our approach will be to consider rigid bodies as made up of large numbers of particles and to use the results obtained in Chap. 14 for the motion of systems of particles. Specifically, we use two equations from Chap. 14: Eq. (14.16), $\Sigma \mathbf{F} = m\bar{\mathbf{a}}$, which relates the resultant of the external forces and the acceleration of the mass center $G$ of the system of particles, and Eq. (14.23), $\Sigma \mathbf{M}_G = \dot{\mathbf{H}}_G$, which relates the resultant moment of the external forces and the angular momentum of the system of particles about $G$.

Except for Sec. 16.1A, which applies to the most general case of the motion of a rigid body, the results derived in this chapter are limited in two ways: (1) They are restricted to the *plane motion* of rigid bodies; that is, where all motion occurs in a single two-dimensional reference plane. (2) The rigid bodies considered consist only of plane rigid bodies and of bodies that are symmetrical with respect to a reference plane (or more generally, bodies that have a principal centroidal axis of inertia perpendicular to a reference plane). The study of the plane motion of nonsymmetrical three-dimensional bodies and, more generally, the motion of rigid bodies in three-dimensional space will be postponed until Chap. 18.

In Sec. 16.1B, we define the angular momentum of a rigid body in plane motion and show that the rate of change of the angular momentum $\dot{\mathbf{H}}_G$ about the mass center is equal to the product $\bar{I}\boldsymbol{\alpha}$ of the centroidal mass moment of inertia $\bar{I}$ and the angular acceleration $\boldsymbol{\alpha}$ of the body. We then prove that the external forces acting on a rigid body are equivalent to a vector $m\overline{\mathbf{a}}$ attached at the mass center and a couple of moment $\overline{I}\boldsymbol{\alpha}$.

We also derive the principle of transmissibility using only the parallelogram law and Newton's laws of motion, allowing us to remove this principle from the list of axioms (*Statics,* Sec. 1.2) required for the study of the statics and dynamics of rigid bodies. We then discuss the use of the free-body diagram and kinetic diagram in the solution of all problems involving the plane motion of rigid bodies.

We consider the plane motion of connected rigid bodies in Sec. 16.1F, which will prepare you to solve a variety of problems involving the translation, centroidal rotation, and unconstrained motion of rigid bodies. In the remaining part of this chapter, we present the solutions of problems involving noncentroidal rotation, rolling motion, and other partially constrained plane motions of rigid bodies.

# 16.1   KINETICS OF A RIGID BODY

As we saw in Chap. 15, we can generally consider the motion of a rigid body to be a combination of translation of the body and rotation about its mass center. We use this same idea to analyze the relationship between forces and moments acting on a rigid body and the body's linear and angular acceleration.

## 16.1A   Equations of Motion for a Rigid Body

Consider a rigid body acted upon by several external forces $\mathbf{F}_1$, $\mathbf{F}_2$, $\mathbf{F}_3$, . . . (Fig. 16.1). We can assume that the body is made of a large number $n$ of particles of mass $\Delta m_i$ ($i = 1, 2, \ldots, n$) and apply the results obtained in Chap. 14 for a system of particles (Fig. 16.2). Consider first the motion of the mass center $G$ of the body with respect to the newtonian frame of reference $Oxyz$. From Eq. (14.16), we have

**Translational equation
of motion**

$$\Sigma\mathbf{F} = m\overline{\mathbf{a}} \qquad\qquad \textbf{(16.1)}$$

where $m$ is the mass of the body and $\overline{\mathbf{a}}$ is the acceleration of the mass center $G$. Turning now to the motion of the body relative to the centroidal frame of reference $Gx'y'z'$, from Eq. (14.23), we have

**Rotational equation
of motion**

$$\Sigma\mathbf{M}_G = \dot{\mathbf{H}}_G \qquad\qquad \textbf{(16.2)}$$

where $\dot{\mathbf{H}}_G$ represents the rate of change of $\mathbf{H}_G$, which is the angular momentum about $G$ of the system of particles forming the rigid body. In the following discussion, we refer to $\mathbf{H}_G$ simply as the **angular momentum of the rigid body about its mass center $G$**. Together, Eqs. (16.1) and (16.2) express that

> **The system of the external forces and moments is equipollent to the system consisting of the vector $m\overline{\mathbf{a}}$ attached at $G$ and the couple of moment $\dot{\mathbf{H}}_G$ (Fig. 16.3).**

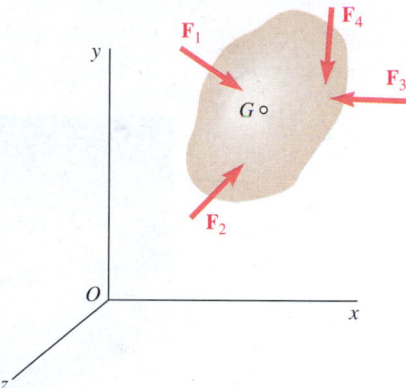

**Fig. 16.1**   A rigid body acted on by several external forces.

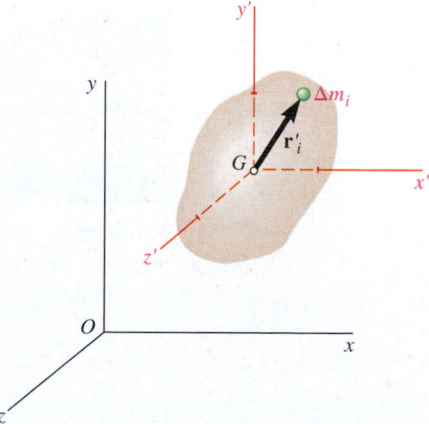

**Fig. 16.2**   A particle of a rigid body in relation to the mass center $G$.

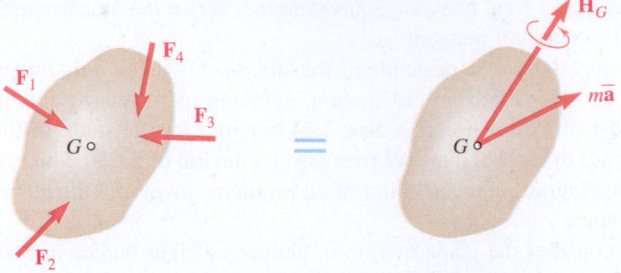

**Fig. 16.3** A system of external forces is equipollent to an inertial vector $m\bar{\mathbf{a}}$ and a couple of moment $\dot{\mathbf{H}}_G$ acting at the mass center.

As you will see in Chap. 18, Eqs. (16.1) and (16.2) apply to the general three-dimensional motion of a rigid body. In the rest of this chapter, however, we limit our analysis to the **plane motion** of rigid bodies; that is, to a motion in which each particle remains within a fixed reference plane. We also assume that the rigid bodies considered consist only of plane rigid bodies and of bodies that are symmetrical with respect to the plane of motion. Further study of the plane motion of nonsymmetrical three-dimensional bodies and of the motion of rigid bodies in three-dimensional space will be postponed until Chap. 18.

**Photo 16.1** The system of external forces acting on the man and wakeboard includes the weight, the tension in the tow rope, and the forces exerted by the water and the air. ©Chua Wee Boo/agefotostock

## 16.1B Angular Momentum of a Rigid Body in Plane Motion

Consider a rigid body in plane motion. Assume that the body is made of a large number $n$ of particles $P_i$ with a mass $\Delta m_i$. Then from Eq. (14.24) of Sec. 14.1D, we can compute the angular momentum $\mathbf{H}_G$ of the rigid body about its mass center $G$ by taking the moments about $G$ of the momenta of the particles of the body with respect to either of the frames $Oxy$ or $Gx'y'$ (Fig. 16.4). Choosing the second option gives

$$\mathbf{H}_G = \sum_{i=1}^{n} (\mathbf{r}_i' \times \mathbf{v}_i' \, \Delta m_i) \tag{16.3}$$

where $\mathbf{r}_i'$ and $\mathbf{v}_i'\Delta m_i$ denote, respectively, the position vector and the linear momentum of the particle $P_i$ relative to the centroidal frame of reference $Gx'y'$. However, because the particle is part of the rigid body, we have $\mathbf{v}_i' = \boldsymbol{\omega} \times \mathbf{r}_i'$, where $\boldsymbol{\omega}$ is the angular velocity of the body at the instant considered. We have

$$\mathbf{H}_G = \sum_{i=1}^{n} [\mathbf{r}_i' \times (\boldsymbol{\omega} \times \mathbf{r}_i') \, \Delta m_i] \tag{} $$

Referring to Fig. 16.4, we easily verify that this expression represents a vector of the same direction as $\boldsymbol{\omega}$ (i.e., perpendicular to the body) and with a magnitude of $\omega\Sigma r_i'^2 \, \Delta m_i$. Recalling that the sum $\Sigma r_i'^2 \, \Delta m_i$ represents the moment of inertia $\bar{I}$ of the rigid body about a centroidal axis perpendicular to the body, we conclude that the angular momentum $\mathbf{H}_G$ of the rigid body about its mass center is

**Angular momentum of a
rigid body about $G$**

$$\mathbf{H}_G = \bar{I}\boldsymbol{\omega} \tag{16.4}$$

Differentiating both sides of Eq. (16.4), we obtain

**Rate of change of angular
momentum about $G$**

$$\dot{\mathbf{H}}_G = \bar{I}\dot{\boldsymbol{\omega}} = \bar{I}\boldsymbol{\alpha} \tag{16.5}$$

Thus, the rate of change of the angular momentum of the rigid body is represented by a vector in the same direction as $\boldsymbol{\alpha}$ (i.e., perpendicular to the body) with a magnitude $\bar{I}\alpha$.

Keep in mind that the results obtained in this section have been derived for a rigid body in plane motion. As you will see in Chap. 18, they remain valid in the case of the plane motion of rigid bodies that are symmetrical with respect to a reference plane (or, more generally, bodies that have a principal centroidal axis of inertia perpendicular to a reference plane). However, they do not apply in the case of nonsymmetrical bodies or in the case of three-dimensional motion.

## 16.1C Plane Motion of a Rigid Body

Consider a rigid body with a mass $m$ moving under the action of several external forces $\mathbf{F}_1, \mathbf{F}_2, \mathbf{F}_3, \ldots$ contained in the plane of the body (Fig. 16.5). Substituting $\dot{\mathbf{H}}_G$ from Eq. (16.5) into Eq. (16.2) and writing the fundamental equations of motion from Eqs. (16.1) and (16.2) in scalar form, we have

$$\Sigma F_x = m\bar{a}_x \qquad \Sigma F_y = m\bar{a}_y \qquad \Sigma M_G = \bar{I}\alpha \tag{16.6}$$

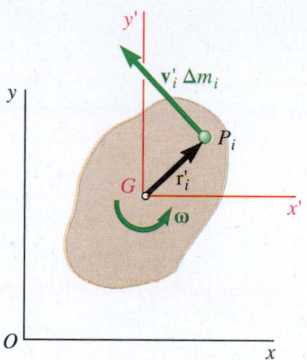

**Fig. 16.4** The angular momentum about $G$ of a particle of a rigid body is $\mathbf{r}_i' \times \mathbf{v}_i' \, \Delta m_i$.

**Photo 16.2** The hard disk and pick-up arm of a computer hard drive undergo fixed-axis rotation. Courtesy of Seagate Technology LLC

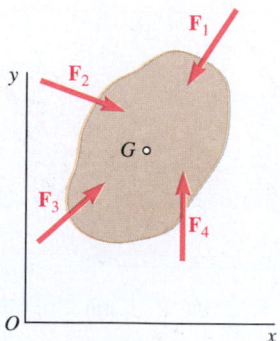

**Fig. 16.5** A rigid body acted upon by several external forces in the plane of the body.

Eqs. (16.6) show that we can obtain the acceleration of the mass center $G$ of the rigid body and its angular acceleration $\boldsymbol{\alpha}$ once we have determined the resultant of the external forces acting on the body and their moment resultant about $G$. Given appropriate initial conditions, we can then obtain the coordinates $\bar{x}$ and $\bar{y}$ of the mass center and the angular coordinate $\theta$ of the body by integration at any instant $t$. Thus,

**The motion of the rigid body is completely defined by the resultant force and resultant moment about $G$ acting on the body.**

Because the motion of a rigid body depends only upon the resultant and resultant moment of the external forces acting on it, it follows that **two systems of forces that are equipollent** (i.e., that have the same resultant and the same moment resultant) **are also equivalent**. That is, they have exactly the same effect on a given rigid body.

Consider in particular the system of external forces acting on a rigid body (Fig. 16.6*a*) and the system of inertial terms associated with the particles forming the rigid body (Fig. 16.6*b*). We showed in Sec. 14.1A that the two systems thus defined are equipollent. But because the particles we are considering now form a rigid body, it follows from the discussion above that the two systems are also equivalent. We can thus state that

**The external forces acting on a rigid body are equivalent to the inertial terms of the various particles forming the body.**

The fact that the system of external forces is equivalent to the system of inertial terms has been emphasized by the use of red equal signs in Fig. 16.6 and also in Fig. 16.7. Here, using results obtained earlier in this section, we replaced the inertial terms by a vector $m\bar{\mathbf{a}}$ attached at the mass center $G$ of the rigid body and the rotational inertial term $\bar{I}\boldsymbol{\alpha}$.

Let's look at three examples of rigid-body plane motion.

**Translation.** In the case of a body in translation, the angular acceleration of the body is equal to zero and its inertial terms reduce to the vector $m\bar{\mathbf{a}}$ attached at $G$ (Fig. 16.8). Thus, the resultant of the external forces acting on a rigid body in translation passes through the mass center of the body and is equal to $m\bar{\mathbf{a}}$.

**Centroidal Rotation.** When a rigid body, or more generally, a body symmetrical with respect to a reference plane, rotates about a fixed axis perpendicular to the reference plane and passing through its mass center $G$, we say that the body is in *centroidal rotation*. Because the acceleration $\bar{\mathbf{a}}$ is identically equal to zero, the inertial terms of the body reduce to the couple $\bar{I}\boldsymbol{\alpha}$ (Fig. 16.9). Thus, the external forces acting on a body in centroidal rotation are equivalent to the rotational inertia $\bar{I}\boldsymbol{\alpha}$.

**General Plane Motion.** Comparing Fig. 16.7 with Figs. 16.8 and 16.9, we observe that, from the point of view of *kinetics*, the most general plane motion of a rigid body symmetrical with respect to a reference plane can be replaced by the sum of a translation and a centroidal rotation. Note that this statement is more restrictive than the similar statement made earlier from the point of view of *kinematics* (Sec. 15.2A), because we now require that the mass center of the body be selected as the reference point.

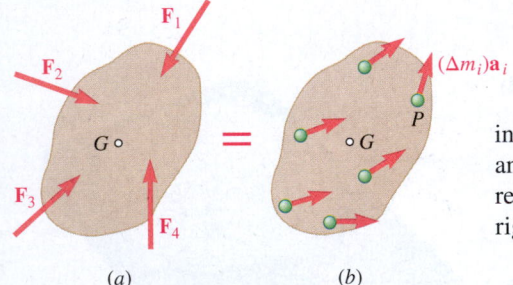

(a)                    (b)

**Fig. 16.6** The external forces acting on the rigid body are equivalent to the inertial terms of the particles of the body.

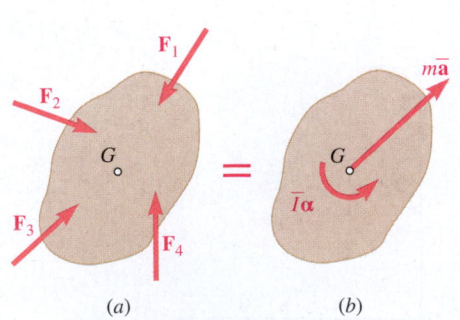

(a)                    (b)

**Fig. 16.7** The external forces acting on the rigid body are also equivalent to a vector $m\bar{\mathbf{a}}$ attached to the mass center $G$ and a rotational inertia $\bar{I}\boldsymbol{\alpha}$.

Referring to Eqs. (16.6), we observe that the first two equations are identical with the equations of motion of a particle of mass $m$ acted upon by the given forces $\mathbf{F}_1, \mathbf{F}_2, \mathbf{F}_3, \ldots$. We thus check that

> **The mass center $G$ of a rigid body in plane motion moves as if the entire mass of the body were concentrated at that point, and as if all the external forces act on it.**

Recall that we already obtained this result in Sec. 14.1C in the general case of a system of particles with the particles being not necessarily rigidly connected. We also note, as we did earlier, that the system of the external forces does not, in general, reduce to a single vector $m\overline{\mathbf{a}}$ attached at $G$. Therefore, in the general case of the plane motion of a rigid body, **the resultant of the external forces acting on the body does not pass through the mass center of the body**.

Finally, note that the last of Eqs. (16.6) would still be valid if the rigid body, while subjected to the same applied forces, were constrained to rotate about a fixed axis through $G$. Thus, **a rigid body in plane motion rotates about its mass center as if this point were fixed**.

## *16.1D    A Remark on the Axioms of the Mechanics of Rigid Bodies

The fact that two equipollent systems of external forces acting on a rigid body are also equivalent—that is, have the same effect on that rigid body—has already been established in *Statics,* Sec. 3.4B. However, there we derived it from the *principle of transmissibility,* which is one of the axioms used in our study of the statics of rigid bodies. We have not used this axiom in the present chapter because Newton's second and third laws of motion make its use unnecessary in the study of the dynamics of rigid bodies.

In fact, we can now derive the principle of transmissibility from the other axioms used in the study of mechanics. This principle stated, without proof (Sec. 3.1B), that the conditions of equilibrium or motion of a rigid body remain unchanged if a force $\mathbf{F}$ acting at a given point of the rigid body is replaced by a force $\mathbf{F}'$ of the same magnitude and same direction—but acting at a different point—provided that the two forces have the same line of action. But because $\mathbf{F}$ and $\mathbf{F}'$ have the same moment about any given point, it is clear that they form two equipollent systems of external forces. Thus, we may now *prove,* as a result of what we established in the preceding section, that $\mathbf{F}$ and $\mathbf{F}'$ have the same effect on the rigid body (see Fig. 3.2 repeated here).

We can therefore remove the principle of transmissibility from the list of axioms required for the study of the mechanics of rigid bodies. These axioms are reduced to the parallelogram law of addition of vectors and to Newton's laws of motion.

## 16.1E    Solution of Problems Involving the Motion of a Rigid Body

We saw in Sec. 16.1C that when a rigid body is in plane motion a fundamental relation exists between the forces $\mathbf{F}_1, \mathbf{F}_2, \mathbf{F}_3, \ldots$, acting on the body, the acceleration $\overline{\mathbf{a}}$ of its mass center, and the angular acceleration $\boldsymbol{\alpha}$ of the body. This

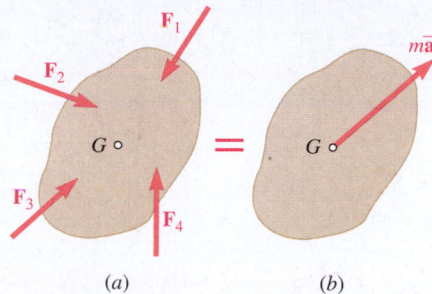

**Fig. 16.8**   A rigid body in translation has a vector $m\overline{\mathbf{a}}$ attached to the mass center $G$ but no rotational inertia $\overline{I}\boldsymbol{\alpha}$.

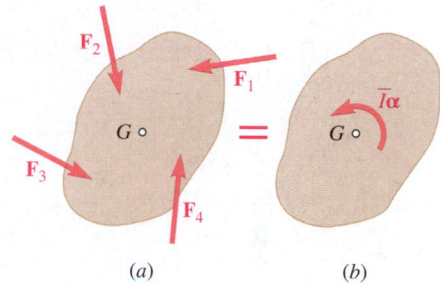

**Fig. 16.9**   A rigid body in centroidal rotation has a rotational inertia $\overline{I}\boldsymbol{\alpha}$ but no $m\overline{\mathbf{a}}$.

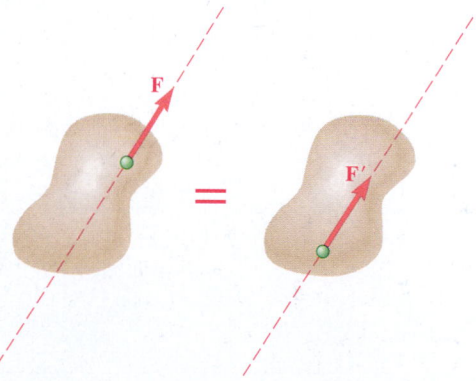

**Fig. 3.2**   (repeated)

relation is represented in Fig. 16.7 in the form of a **free-body diagram and a kinetic diagram**. We can use these diagrams to determine the acceleration $\bar{\mathbf{a}}$ and the angular acceleration $\boldsymbol{\alpha}$ produced by a given system of forces acting on a rigid body or, conversely, to determine the forces that produce a given motion of the rigid body.

We can use the three algebraic equations of Eq. (16.6) to solve problems of plane motion.[†] However, our experience in statics suggests that the solution of many problems involving rigid bodies can be simplified by an appropriate choice of the point about which we compute the moments of the forces. It is therefore preferable to remember the relation between the forces and the accelerations in the pictorial form shown in Fig. 16.7 and to derive from this fundamental relation the component or moment equations that best fit the solution of the problem under consideration.

Drawing a free-body diagram for rigid bodies follows the same basic steps as we discussed in Chap. 12. For rigid bodies, however, it is important to draw your forces at their location of action, because you will be summing moments about specific points. Labeling different dimensions on your free-body diagram is particularly helpful when summing these moments.

The kinetic diagram is also slightly modified from Chap. 12. The translational inertial term $m\bar{\mathbf{a}}$ is always located at the center of mass of the body. We are now concerned with the rotational inertia of the body, so we include an additional term on our kinetic diagram, $\bar{I}\boldsymbol{\alpha}$. This is also located at the center of mass of the body.

We can apply the steps from Chap. 12 to the pendulum shown in Fig. 16.10, where a moment $M$ is applied to the bar. These steps include:

1. Isolating the **body**
2. Defining the **axes**
3. Replacing constraints with **support forces**
4. Adding **applied forces and moments**, as well as **body forces** to the diagram
5. Labeling the free-body diagram with **dimensions**

For the kinetic diagram, we typically draw the translational inertial term in component form (e.g., $m\bar{a}_x$ and $m\bar{a}_y$) at the center of mass of the body and add the rotational inertial term $\bar{I}\alpha$. Using these steps gives you the free-body diagram and kinetic diagram shown in Fig. 16.11.

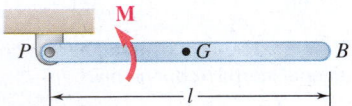

**Fig. 16.10** A pendulum with mass $m$, length $l$, and an applied moment **M**.

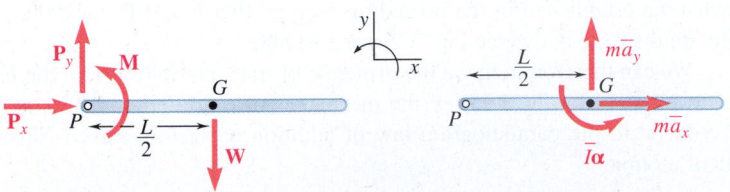

**Fig. 16.11** Free-body diagram and kinetic diagram for a pendulum with an external moment applied.

[†]Recall that the last of Eqs. 16.6 is valid only in the case of the plane motion of a rigid body symmetrical with respect to a reference plane. In all other cases, you need to use the methods of Chap. 18.

We use the pendulum shown in Fig. 16.10 to illustrate an alternative form of the moment equation. It is straightforward to apply Eqs. 16.6 to this problem, where the sum of moments about the center of mass results in

$$+\circlearrowleft \Sigma M_G = \bar{I}\alpha: \quad M - P_y\left(\frac{L}{2}\right) = \bar{I}\alpha$$

Alternatively, we could choose an arbitrary point $P$ about which to sum moments. If we choose $P$ to be at the left end of the rod, then we also have to sum the moments about $P$ due to the inertial terms. In this case, we obtain

$$+\circlearrowleft \Sigma M_P = \bar{I}\alpha + m\bar{a}d_\perp: \quad M - W\left(\frac{L}{2}\right) = \bar{I}\alpha + m\bar{a}_y\left(\frac{L}{2}\right) + m\bar{a}_x(0)$$

where $d_\perp$ is the perpendicular distance from point $P$ to the line of action of the resultant acceleration vector $\bar{\mathbf{a}}$. As in statics, you can also determine the moment about a point $P$ by using vector products, as

$$m\bar{a}d_\perp = \mathbf{r}_{G/P} \times m\bar{\mathbf{a}}$$

where $\mathbf{r}_{G/P}$ is the vector from point $P$ to the center of mass of the body. Therefore, we can also write Eqs. (16.6) as

$$\Sigma F_x = m\bar{a}_x \quad \Sigma F_y = m\bar{a}_y$$
$$\Sigma M_G = \bar{I}\alpha \quad \text{or} \quad \Sigma M_P = \bar{I}\alpha + m\bar{a}d_\perp \quad \text{or} \quad \Sigma M_P = \bar{I}\alpha + \mathbf{r}_{G/P} \times m\bar{\mathbf{a}} \tag{16.6'}$$

The use of a free-body diagram and a kinetic diagram, showing vectorially the relationship between the forces applied on the rigid body and the resulting linear and angular accelerations, presents considerable advantages over the blind application of formulas (16.6). We can summarize these advantages as follows.

1. The use of a pictorial representation provides a much clearer understanding of the effect of the forces on the motion of the body.

2. This approach makes it possible to divide the solution of a dynamics problem into two parts: In the first part, the analysis of the kinematic and kinetic characteristics of the problem leads to the free-body diagram and the kinetic diagram of Fig. 16.7; in the second part, you can use the diagrams to analyze the various forces and vectors involved.

3. A unified approach is provided for the analysis of the plane motion of a rigid body, regardless of the particular type of motion involved. Although the kinematics of the various motions considered may vary from one case to the other, the approach to the kinetics of the motion is consistently the same. In every case, you draw a diagram showing the external forces, the vector $m\bar{\mathbf{a}}$ associated with the motion of $G$, and the couple $\bar{I}\alpha$ associated with the rotation of the body about $G$.

4. The resolution of the plane motion of a rigid body into a translation and a centroidal rotation, which we use here, is a basic concept that can be applied effectively throughout the study of mechanics. We will use it again in Chap. 17 with both the method of work and energy and the method of impulse and momentum.

5. As you will see in Chap. 18, we can extend this approach to the study of the general three-dimensional motion of a rigid body. The motion of the body is again resolved into a translation and a rotation about the mass

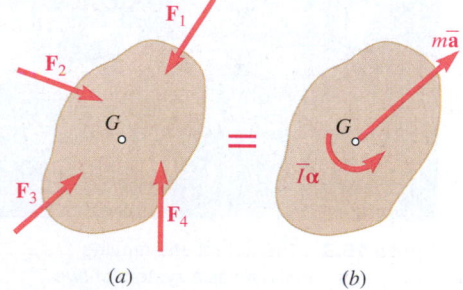

**Fig. 16.7** (repeated)

center, and we use free-body diagrams and kinetic diagrams to indicate the relationship between the external forces and the rates of change of the linear and angular momenta of the body.

## 16.1F Systems of Rigid Bodies

The method just described also can be used in problems involving the plane motion of several connected rigid bodies. For each part of the system, you draw a diagram similar to Fig. 16.7. You can obtain the equations of motion from these diagrams and solve them simultaneously.

In some cases, as in Sample Prob. 16.4, you can draw a single diagram for the entire system. This diagram should include all of the external forces, as well as the vectors $m\bar{\mathbf{a}}$ and the couples $\bar{I}\alpha$ associated with the various parts of the system. However, you can omit internal forces, such as the forces exerted by connecting cables, because they occur in pairs of equal and opposite forces and are thus equipollent to zero. The equations obtained by expressing that the system of external forces is equipollent to the system of inertial terms can be solved for the remaining unknowns (note that we cannot speak of equivalent systems because we are not dealing with a single rigid body). For systems involving multiple rigid bodies, the general equation of motion is written as

$$\Sigma\mathbf{F} = \Sigma m_i\bar{\mathbf{a}}_i \qquad \text{and} \qquad \Sigma\mathbf{M}_P = \dot{\mathbf{H}}_P$$

where

$$\dot{\mathbf{H}}_P = \Sigma\bar{I}_i\boldsymbol{\alpha}_i + \Sigma m_i\bar{\mathbf{a}}_i(d_\perp)_i = \Sigma\bar{I}_i\boldsymbol{\alpha}_i + \Sigma[(\mathbf{r}_{G/P})_i \times m_i\bar{\mathbf{a}}_i]$$

Historically, sometimes these equations have been written as

$$\Sigma\mathbf{F} = \Sigma\mathbf{F}_{\text{eff}} \qquad \text{and} \qquad \Sigma\mathbf{M}_P = \Sigma(\mathbf{M}_P)_{\text{eff}}$$

where the left-hand sides of these equations come from the free-body diagram and the right-hand sides come from the kinetic diagram. We have chosen not to use this notation because the terms on the right-hand side are due to the inertial terms and not due to external forces and moments.

It is not possible to include more than one rigid body in your system in problems involving more than three unknowns, because only three equations of motion are available when a single diagram is used. We will not elaborate upon this point, because the discussion involved would be completely similar to that given in Sec. 6.3B in the case of the equilibrium of a system of rigid bodies.

**Photo 16.3** The forklift and moving load can be analyzed as a system of two connected rigid bodies in plane motion.

©Tony Arruza/Corbis

# Sample Problem 16.1

When the forward speed of the van shown is 30 ft/s, the brakes are suddenly applied, causing all four wheels to stop rotating. The van skids to rest in 20 ft. Determine the magnitude of the normal reaction and of the friction force at each wheel as the van skids to rest.

**STRATEGY:**  You are given enough information to determine the acceleration and you want to find forces, so use Newton's second law. The motion described is pure translation, so the angular acceleration is zero.

**MODELING:**  Choose the van to be your system and model it as a rigid body. A free-body diagram and a kinetic diagram for this system are shown in Fig. 1. The external forces consist of the weight **W** of the truck and of the normal reactions and friction forces at the wheels. The vectors $N_A$ and $F_A$ represent the sum of the reactions at the rear wheels, while $N_B$ and $F_B$ represent the sum of the reactions at the front wheels. Because the truck is in translation, $\alpha = 0$ and the inertial terms reduce to the vector $m\bar{a}$ attached at $G$.

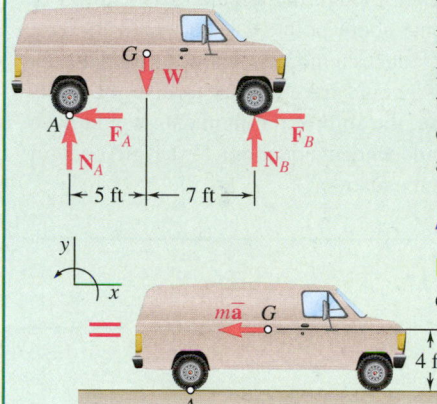

**Fig. 1**  Free-body diagram and kinetic diagram for the van.

**ANALYSIS:**

**Kinematics of Motion.**  Choose the positive sense to the right and use the equations of uniformly accelerated motion. You have

$$\bar{v}_0 = +30 \text{ ft/s} \qquad \bar{v}^2 = \bar{v}_0^2 + 2\bar{a}x \qquad 0 = (30)^2 + 2\bar{a}(20)$$
$$\bar{a} = -22.5 \text{ ft/s}^2 \qquad \mathbf{\bar{a}} = 22.5 \text{ ft/s}^2 \leftarrow$$

**Equations of Motion.**  You can obtain three equations of motion by expressing that the system of the external forces from your free-body diagram is equivalent to the inertial terms from your kinetic diagram. Applying Newton's second law in the $x$ and $y$ directions gives

$$+\uparrow \Sigma F_y = m\bar{a}_y: \qquad N_A + N_B - W = 0 \tag{1}$$

$$\overset{+}{\rightarrow} \Sigma F_x = m\bar{a}_x: \qquad -(F_A + F_B) = -m\bar{a} \tag{2}$$

Taking moments about any point gives you a third equation. For moments about point $A$, you find

$$+\circlearrowleft \Sigma M_A = \bar{I}\alpha + m\bar{a}d_\perp: \qquad -W(5 \text{ ft}) + N_B(12 \text{ ft}) = m\bar{a}(4 \text{ ft}) \tag{3}$$

In these three equations you have five unknowns: $N_A$, $N_B$, $F_A$, $F_B$, and $\bar{a}$. Because $F_A = \mu_k N_A$ and $F_B = \mu_k N_B$, where $\mu_k$ is the coefficient of kinetic friction, you have from Eq. (1)

$$F_A + F_B = \mu_k(N_A + N_B) = \mu_k W$$

Substituting into Eq. (2) and using $m = W/g$ gives

$$-\mu_k W = -\frac{W}{32.2 \text{ ft/s}^2}\bar{a} = -\frac{W}{32.2 \text{ ft/s}^2}(22.5 \text{ ft/s}^2)$$

*(continued)*

or $\mu_k = 0.699$. Solving Eq. (3) for $N_B$ gives you $N_B = 0.640W$. Substituting this into Eq. (1), you find $N_A = 0.350W$. The friction forces are easily determined once you know the normal forces $F_A = \mu_k N_A = (0.699)(0.350W) = 0.245W$ and $F_B = \mu_k N_B = (0.699)(0.650W) = 0.454W$.

**Reactions at Each Wheel.**   Recall that the values computed here represent the sum of the reactions at the two front wheels or the two rear wheels. You obtain the magnitude of the reactions at each wheel by writing

$$N_{\text{front}} = \tfrac{1}{2}N_B = 0.325W \qquad N_{\text{rear}} = \tfrac{1}{2}N_A = 0.175W \quad \blacktriangleleft$$
$$F_{\text{front}} = \tfrac{1}{2}F_B = 0.227W \qquad F_{\text{rear}} = \tfrac{1}{2}F_A = 0.122W \quad \blacktriangleleft$$

**REFLECT and THINK:**   Note that even though the angular acceleration of the van is zero, the sum of the moments about point $A$ is not equal to zero, because from the kinetic diagram, $m\bar{a}$ produces a moment about $A$. Rather than taking moments about point $A$, you also could have chosen to take moments about the center of mass, $G$. In this case, the sum of the moments would have been equal to zero. You only get three independent equations for a rigid body in plane motion: $\Sigma F_x$, $\Sigma F_y$, and one moment equation.

## Sample Problem 16.2

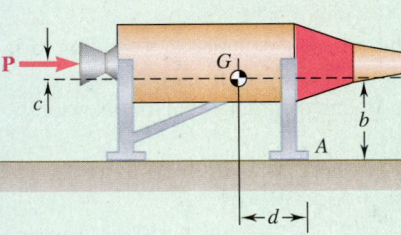

A sled is jet-propelled along a straight track by a force $P$ that increases linearly with time according to $P = kt$, where $k$ is a constant. The coefficient of sliding friction between the sled runners and the track is $\mu_k$, the coefficient of static friction is $\mu_s$, and the mass of the sled is $m$. Determine (a) the time at which the tip of the rocket begins to rotate downward, (b) the acceleration of the sled at this instant. Neglect loss of mass due to fuel consumption and assume that the sled will slide before it tips.

**STRATEGY:**   Because you are given a force, use Newton's second law to find the acceleration required for the rocket to begin rotating forward. You can then find the time using $P = kt$.

**MODELING:**   Choose the sled as your system and model it as a rigid body. The rocket force must overcome the static friction force before it begins moving. Define this time to be $t_0$. Fig. 1 shows a free-body diagram when the motion is impending. In this case, both of the friction forces are set equal to the maximum allowable friction force $\mu_s N$. Free-body and kinetic diagrams for when the sled is about to tip are shown in Fig. 2. Just as the sled starts to tip, the normal force on the rear of the sled goes to zero.

*(continued)*

**ANALYSIS:** Using Fig. 1 and applying Newton's second law in the $y$ and $x$ directions gives

$+\uparrow \Sigma F_y = m\bar{a}_y$:    $N_A + N_B - mg = 0$  or  $N_A + N_B = mg$

$\xrightarrow{+} \Sigma F_x = m\bar{a}_x$:  $kt_0 - (\mu_s N_A + \mu_s N_B) = 0$

or

$$kt_0 = \mu_s(N_A + N_B) = \mu_s mg \tag{1}$$

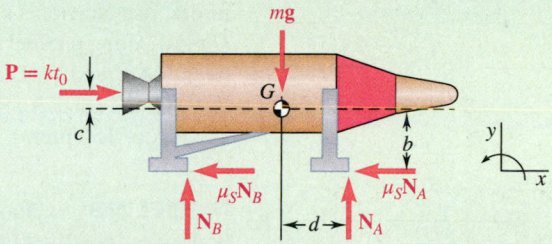

**Fig. 1**  Free-body diagram when motion is impending.

Now that you know when the sled begins to slide, you can determine the time it will start to tip using Fig. 2.

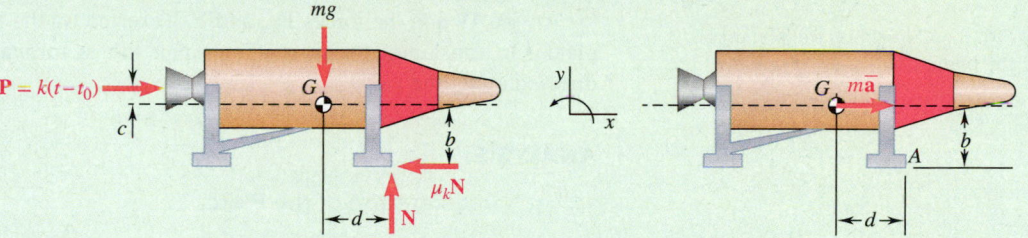

**Fig. 2**  Free-body diagram and kinetic diagram for the sled after it begins to move and just as it begins to move.

From this diagram, you can apply Newton's second law in the $x$ and $y$ directions and sum moments about any point. If you choose to take moments about $G$, you find

$\xrightarrow{+} \Sigma F_x = m\bar{a}_x$:  $k(t - t_0) - \mu_k N = m\bar{a}$  $\tag{2}$

$+\uparrow \Sigma F_y = m\bar{a}_y$:  $N - mg = 0$  $\tag{3}$

$+\circlearrowleft \Sigma M_G = \bar{I}\alpha$:  $Nd - \mu_k Nb - k(t - t_0)c = 0$  $\tag{4}$

Solving Eqs. (1), (2), (3), and (4) for $t_0$, $t$, $N$, and $\bar{a}$, you find $N = mg$, $t_0 = \mu_s mg/k$ and

$$t = \frac{mg(d + c\mu_s - b\mu_k)}{kc} \quad \blacktriangleleft$$

$$\bar{a} = \frac{g(d - c\mu_k - b\mu_k)}{c} \quad \blacktriangleleft$$

**REFLECT and THINK:** Rather than taking moments about $G$, you could have chosen any other point. For example, for moments about $A$, you have

$+\circlearrowleft \Sigma M_A = \bar{I}\alpha + m\bar{a}d$:    $mgd - k(t - t_0)(b + c) = -m\bar{a}b$

Using this equation rather than Eq. (4) will give you the same answer. To check the assumption that the sled slides before it tips, you would need to use Fig. 1 and show that both $N_A$ and $N_B$ are positive for the given value of $P = kt_0$.

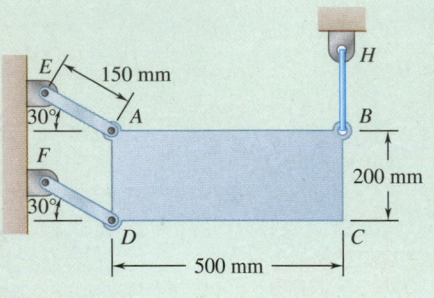

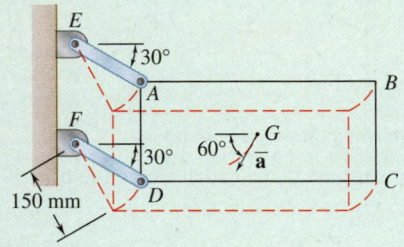

**Fig. 1** Curvilinear translation of the plate.

## Sample Problem 16.3

The thin plate $ABCD$ has a mass of 8 kg and is held in the position shown by the wire $BH$ and two links $AE$ and $DF$. Neglecting the mass of the links, determine immediately after wire $BH$ has been cut (a) the acceleration of the plate, (b) the force in each link. ▶

**STRATEGY:**   Because you are asked to determine the acceleration and forces, use Newton's second law. After wire $BH$ has been cut, corners $A$ and $D$ move along parallel circles, each with a radius of 150 mm centered, respectively, at $E$ and $F$. The motion of the plate is thus a curvilinear translation (Fig. 1); the particles forming the plate move along parallel circles, each with a radius of 150 mm.

**MODELING:**   Choose the plate to be your system and model it as a rigid body. To draw the kinetic diagram, you need to consider the kinematics of the motion. At the instant wire $BH$ is cut, the velocity of the plate is zero. Thus, the acceleration of the mass center $G$ of the plate is tangent to the circular path described by $G$ (Fig. 1). The free-body diagram and kinetic diagram for this system are shown in Fig. 2. The external forces consist of the weight $\mathbf{W}$ and the forces $\mathbf{F}_{AE}$ and $\mathbf{F}_{DF}$ exerted by the links. Because the plate is in translation, the kinetic diagram is the vector $m\bar{\mathbf{a}}$ attached at $G$ and directed along the $t$ axis.

## ANALYSIS:

### a.   Acceleration of the Plate.

$+\nearrow \Sigma F_t = m\bar{a}_t:$

$$W \cos 30° = m\bar{a}$$
$$mg \cos 30° = m\bar{a}$$
$$\bar{a} = g \cos 30° = (9.81 \text{ m/s}^2) \cos 30° \qquad (1)$$

$$\bar{\mathbf{a}} = 8.50 \text{ m/s}^2 \; \searrow \; 60° \;\; ◀$$

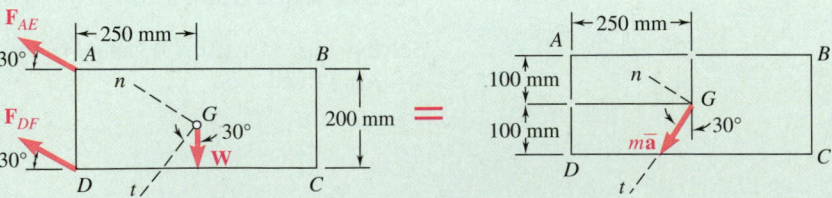

**Fig. 2**   Free-body diagram and kinetic diagram for the plate.

*(continued)*

### b.   Forces in Links *AE* and *DF*.

$$+\nwarrow \Sigma F_n = m\bar{a}_n: \qquad F_{AE} + F_{DF} - W\sin 30° = 0 \qquad \textbf{(2)}$$

$$+\circlearrowleft \Sigma M_G = \bar{I}\alpha:$$

$$(F_{AE}\sin 30°)(250\text{ mm}) - (F_{AE}\cos 30°)(100\text{ mm})$$

$$+ (F_{DF}\sin 30°)(250\text{ mm}) + (F_{DF}\cos 30°)(100\text{ mm}) = 0$$

$$38.4F_{AE} + 211.6F_{DF} = 0$$

$$F_{DF} = -0.1815F_{AE} \qquad \textbf{(3)}$$

Substituting $F_{DF}$ from Eq. (3) into Eq. (2), you have

$$F_{AE} - 0.1815F_{AE} - W\sin 30° = 0$$

$$F_{AE} = 0.6109W$$

$$F_{DF} = -0.1815(0.6109W) = -0.1109W$$

Noting that $W = mg = (8\text{ kg})(9.81\text{ m/s}^2) = 78.48\text{ N}$, you have

$$F_{AE} = 0.6109(78.48\text{ N}) \qquad F_{AE} = 47.9\text{ N } T \quad \blacktriangleleft$$

$$F_{DF} = -0.1109(78.48\text{ N}) \qquad F_{DF} = 8.70\text{ N } C \quad \blacktriangleleft$$

where bar *AE* is in tension and bar *DF* is in compression.

**REFLECT and THINK:**   If *AE* and *DF* had been cables rather than links, the answers you just determined indicate that *DF* would have gone slack (i.e., you can't push on a rope), because the analysis showed that it would be in compression. Therefore, the plate would not be undergoing curvilinear translation, but it would have been undergoing general plane motion. It is important to note that that there is always more than one way to solve problems like this, because you can choose to take moments about any point you wish. In this case, you took them about *G*, but you could have also chosen to take them about *A* or *D*.

## Sample Problem 16.4

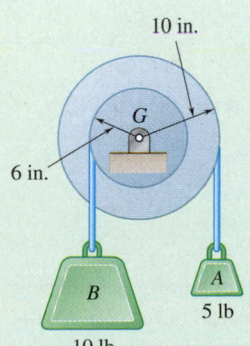

A pulley weighing 12 lb and having a radius of gyration of 8 in. is connected to two blocks as shown. Assuming no axle friction, determine the angular acceleration of the pulley and the acceleration of each block.

**STRATEGY:**   Because you want to determine accelerations and are given the weights, use Newton's second law.

**MODELING:**   Choose the pulley and the two blocks as a single system. The pulley moves in pure rotation and each block moves in pure translation.

*(continued)*

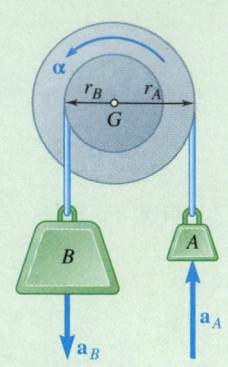

**Fig. 1** Acceleration directions assuming a CCW angular acceleration.

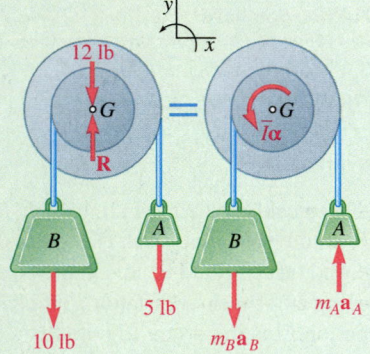

**Fig. 2** Free-body diagram and kinetic diagram for the system.

**Sense of Motion.** Although you can assume an arbitrary sense of motion as shown in Fig. 1 (because no friction forces are involved) and later check it by the sign of the answer, you may prefer to determine the actual sense of rotation of the pulley. First determine the weight of block B, $W'_B$, required to maintain the equilibrium of the pulley when it is acted upon by the 5-lb block A.

$$+\circlearrowleft \Sigma M_G = 0: \qquad W'_B(6 \text{ in.}) - (5 \text{ lb})(10 \text{ in.}) = 0 \qquad W'_B = 8.33 \text{ lb}$$

Because block B actually weighs 10 lb, the pulley rotates counterclockwise. The free-body and kinetic diagrams for this system are shown in Fig. 2. The forces external to the system consist of the weights of the pulley and the two blocks and of the reaction at G (Fig. 2). The forces exerted by the cables on the pulley and on the blocks are internal to the system and cancel out. Because the motion of the pulley is a centroidal rotation and the motion of each block is a translation, the inertial terms reduce to the couple $\bar{I}\alpha$ and the two vectors $m\mathbf{a}_A$ and $m\mathbf{a}_B$.

**ANALYSIS:**

**Kinematics of Motion.** Assuming $\alpha$ is counterclockwise and noting that $\alpha_A = r_A\alpha$ and $a_B = r_B\alpha$, you obtain

$$\mathbf{a}_A = (\tfrac{10}{12} \text{ ft})\boldsymbol{\alpha} \uparrow \qquad \mathbf{a}_B = (\tfrac{6}{12} \text{ ft})\boldsymbol{\alpha} \downarrow$$

**Equations of Motion.** The centroidal moment of inertia of the pulley is

$$\bar{I} = m\bar{k}^2 = \frac{W}{g}\bar{k}^2 = \frac{12 \text{ lb}}{32.2 \text{ ft/s}^2}\left(\frac{8}{12} \text{ ft}\right)^2 = 0.1656 \text{ lb·ft·s}^2$$

Because the system of external forces is equivalent to the system of inertial terms, you have

$$+\circlearrowleft \Sigma M_G = \dot{H}_G:$$

$$(10 \text{ lb})(\tfrac{6}{12} \text{ ft}) - (5 \text{ lb})(\tfrac{10}{12} \text{ ft}) = +\bar{I}\alpha + m_B a_B(\tfrac{6}{12} \text{ ft}) + m_A a_A(\tfrac{10}{12} \text{ ft})$$

$$(10)(\tfrac{6}{12}) - (5)(\tfrac{10}{12}) = 0.1656\alpha + \tfrac{10}{32.2}(\tfrac{6}{12}\alpha)(\tfrac{6}{12}) + \tfrac{5}{32.2}(\tfrac{10}{12}\alpha)(\tfrac{10}{12})$$

$$\alpha = +2.374 \text{ rad/s}^2 \qquad\qquad \boldsymbol{\alpha} = 2.37 \text{ rad/s}^2 \circlearrowleft \quad \blacktriangleleft$$

$$a_A = r_A\alpha = (\tfrac{10}{12} \text{ ft})(2.374 \text{ rad/s}^2) \qquad \mathbf{a}_A = 1.978 \text{ ft/s}^2\uparrow \quad \blacktriangleleft$$

$$a_B = r_B\alpha = (\tfrac{6}{12} \text{ ft})(2.374 \text{ rad/s}^2) \qquad \mathbf{a}_B = 1.187 \text{ ft/s}^2\downarrow \quad \blacktriangleleft$$

**REFLECT and THINK:** You could also solve this problem by considering the pulley and each block as separate systems, but you would have more resulting equations. You would have to use this approach if you wanted to know the forces in the cables.

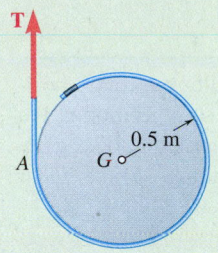

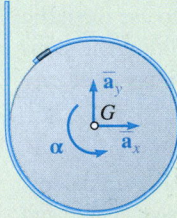

**Fig. 1**   Assumed directions for the angular acceleration and the acceleration of the center of mass.

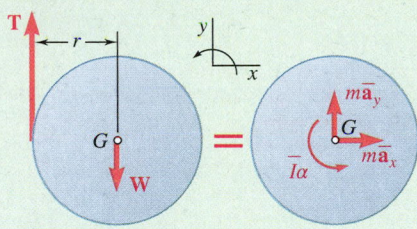

**Fig. 2**   Free-body diagram and kinetic diagram for the disk.

# Sample Problem 16.5

A cord is wrapped around a homogeneous disk with a radius of $r = 0.5$ m and a mass of $m = 15$ kg. If the cord is pulled upward with a force **T** of magnitude 180 N, determine (*a*) the acceleration of the center of the disk, (*b*) the angular acceleration of the disk, (*c*) the acceleration of the cord.

**STRATEGY:**   Because you have forces and are interested in determining accelerations, use Newton's second law.

**MODELING:**   Choose the disk and the cord as your system. Assume that the components $\bar{\mathbf{a}}_x$ and $\bar{\mathbf{a}}_y$ of the acceleration of the center are directed, respectively, to the right and upward and that the angular acceleration of the disk is counterclockwise (Fig. 1). A free-body diagram and kinetic diagram for this system are shown in Fig. 2. The external forces acting on the disk consist of its weight **W** and the force **T** exerted by the cord.

**ANALYSIS:**

**Equations of Motion.**   Applying Newton's second law in the *x* and *y* directions gives

$$\xrightarrow{+} \Sigma F_x = m\bar{a}_x: \qquad\qquad 0 = m\bar{a}_x \qquad\qquad\qquad \bar{\mathbf{a}}_x = 0 \ \blacktriangleleft$$

$$+\uparrow \Sigma F_y = m\bar{a}_y: \qquad\qquad T - W = m\bar{a}_y$$

$$\bar{a}_y = \frac{T - W}{m}$$

Using $T = 180$ N, $m = 15$ kg, and $W = (15 \text{ kg})(9.81 \text{ m/s}^2) = 147.1$ N, you have

$$\bar{a}_y = \frac{180 \text{ N} - 147.1 \text{ N}}{15 \text{ kg}} = +2.19 \text{ m/s}^2 \qquad \bar{\mathbf{a}}_y = 2.19 \text{ m/s}^2 \uparrow \ \blacktriangleleft$$

Now taking moments about the center of gravity, you get

$$+\circlearrowleft \Sigma M_G = \bar{I}\alpha: \qquad\qquad -Tr = \bar{I}\alpha$$

$$-Tr = (1/2 mr^2)\alpha$$

$$\alpha = -\frac{2T}{mr} = -\frac{2(180 \text{ N})}{(15 \text{ kg})(0.5 \text{ m})} = -48.0 \text{ rad/s}^2$$

$$\alpha = 48.0 \text{ rad/s}^2 \ \circlearrowright \ \blacktriangleleft$$

**Acceleration of Cord.**   The acceleration of the cord is equal to the tangential component of the acceleration of point *A* on the disk, so you have (Fig. 3)

$$\mathbf{a}_{\text{cord}} = (\mathbf{a}_A)_t = \bar{\mathbf{a}} + (\mathbf{a}_{A/G})_t$$

$$= [2.19 \text{ m/s}^2 \uparrow] + [(0.5 \text{ m})(48 \text{ rad/s}^2 \uparrow)]$$

$$\mathbf{a}_{\text{cord}} = 26.2 \text{ m/s}^2 \uparrow \ \blacktriangleleft$$

**REFLECT AND THINK:**   The angular acceleration is clockwise, as we would expect. A similar analysis would apply in many practical situations, such as pulling wire off a spool or paper off a roll. In such cases, you would need to be sure that the tension pulling on the disk is not larger than the tensile strength of the material.

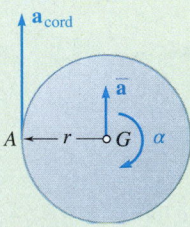

**Fig. 3**   Acceleration of points *A* and *G* on the disk.

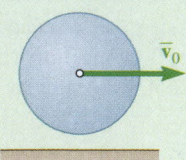

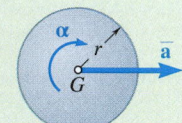

**Fig. 1** Assumed directions for the angular and linear acceleration of the sphere.

## Sample Problem 16.6

A uniform sphere with mass $m$ and radius $r$ is projected along a rough horizontal surface with a linear velocity $\overline{v}_0$ and no angular velocity. Denoting the coefficient of kinetic friction between the sphere and the floor by $\mu_k$, determine (*a*) the time $t_1$ at which the sphere starts rolling without sliding, (*b*) the linear velocity and angular velocity of the sphere at time $t_1$. ▶

**STRATEGY:** Because you have forces acting on the sphere, use Newton's second law. To relate the acceleration to the velocity, you need to use the basic kinematic relationships. The sphere starts out rotating and sliding; it stops sliding when the instantaneous point of contact with the ground has a velocity of zero.

**MODELING:** Choose the sphere as your system and model it as a rigid body. The assumed positive directions for the acceleration of the mass center and the angular acceleration are shown in Fig. 1. Free-body and kinetic diagrams for this system are shown in Fig. 2. Because the point of the sphere in contact with the surface is sliding to the right, the friction force **F** is directed to the left. While the sphere is sliding, the magnitude of the friction force is $F = \mu_k N$.

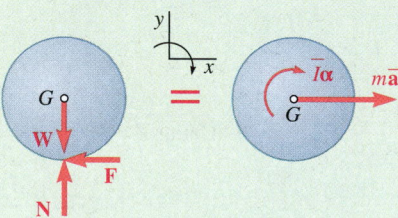

**Fig. 2** Free-body diagram and kinetic diagram for the sphere.

## ANALYSIS:

**Equations of Motion.** Applying Newton's second law in the $x$ and $y$ directions and recognizing that $F = \mu_k N$ during sliding gives

$$+\uparrow \Sigma F_y = m\overline{a}_y: \qquad\qquad N - W = 0$$

$$N = W = mg \qquad F = \mu_k N = \mu_k mg$$

$$\xrightarrow{+} \Sigma F_x = m\overline{a}_x: \qquad -F = m\overline{a} \qquad -\mu_k mg = m\overline{a} \qquad \overline{a} = -\mu_k g$$

Now taking moments about the center of gravity, you get

$$+\circlearrowleft \Sigma M_G = \overline{I}\alpha: \quad Fr = \overline{I}\alpha$$

Noting that $\overline{I} = \frac{2}{5}mr^2$ and substituting the given value for $F$, you have

$$(\mu_k mg)r = \tfrac{2}{5}mr^2 \alpha \qquad \alpha = \frac{5}{2}\frac{\mu_k g}{r}$$

*(continued)*

**Kinematics of Motion.**   As long as the sphere both rotates and slides, its linear and angular accelerations are constant. Therefore, you can use the constant-acceleration equations to relate these accelerations to the linear velocity and angular velocity.

$$t = 0, \bar{v} = \bar{v}_0 \qquad \bar{v} = \bar{v}_0 + \bar{a}t = \bar{v}_0 - \mu_k g t \qquad \textbf{(1)}$$

$$t = 0, \omega_0 = 0 \qquad \omega = \omega_0 + \alpha t = 0 + \left(\frac{5}{2}\frac{\mu_k g}{r}\right)t \qquad \textbf{(2)}$$

The sphere starts rolling without sliding when the velocity $\mathbf{v}_C$ of the point of contact $C$ is zero (Fig. 3). At that time, $t = t_1$, point $C$ becomes the instantaneous center of rotation, and you have

$$\bar{v}_1 = r\omega_1 \qquad \textbf{(3)}$$

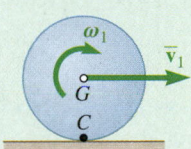

**Fig. 3**  The point of contact has zero velocity when the sphere starts rolling.

Substituting in Eq. (3) the values obtained for $\bar{v}_1$ and $\omega_1$ by making $t = t_1$ in Eqs. (1) and (2), respectively, you obtain

$$\bar{v}_0 - \mu_k g t_1 = r\left(\frac{5}{2}\frac{\mu_k g}{r}t_1\right) \qquad\qquad t_1 = \frac{2}{7}\frac{\bar{v}_0}{\mu_k g} \quad \blacktriangleleft$$

Substituting for $t_1$ into Eq. (2), you have

$$\omega_1 = \frac{5}{2}\frac{\mu_k g}{r}t_1 = \frac{5}{2}\frac{\mu_k g}{r}\left(\frac{2}{7}\frac{\bar{v}_0}{\mu_k g}\right) \qquad \omega_1 = \frac{5}{7}\frac{\bar{v}_0}{r} \qquad \boldsymbol{\omega}_1 = \frac{5}{7}\frac{\bar{v}_0}{r} \circlearrowleft \quad \blacktriangleleft$$

$$\bar{v}_1 = r\omega_1 = r\left(\frac{5}{7}\frac{\bar{v}_0}{r}\right) \qquad\qquad \bar{v}_1 = \frac{5}{7}\bar{v}_0 \qquad \mathbf{v}_1 = \frac{5}{7}\bar{v}_0 \rightarrow \quad \blacktriangleleft$$

**REFLECT and THINK:**   Notice we chose a different coordinate system than we usually do, with the positive rotation going clockwise. This means that you will not be able to use vector algebra solutions because it is not a right-handed coordinate system.

   You could use this type of analysis to determine how long it takes a bowling ball to begin to roll without slip or to see how the coefficient of friction affects this motion. Instead of taking moments about the center of gravity, you could have chosen to take moments about point $C$, in which case your third equation would have been $\Sigma M_C = \dot{H}_C \longrightarrow 0 = m\bar{a}r + \bar{I}\alpha$.

# SOLVING PROBLEMS
# ON YOUR OWN

This chapter deals with the *plane motion* of *rigid bodies*, and in this first section we considered rigid bodies that are free to move under the action of applied forces.

**1. Free-body diagram and kinetic diagram.** After choosing a system, your first step in the solution of a problem is to draw a *free-body diagram and a kinetic diagram*.

    **a. A free-body diagram** shows *the forces exerted on the body,* including the applied forces and moments, the reactions at the supports, and the weight of the body.

    **b. A kinetic diagram** shows the **inertial terms**: vector $m\bar{\mathbf{a}}$ and the couple $\bar{I}\boldsymbol{\alpha}$.

**2. Using your free-body diagram and kinetic diagram, generate the equations of motion for the system.** Drawing good free-body and kinetic diagrams will allow you to *sum components in any direction and to sum moments about any point.* For a single body, you can obtain a maximum of three independent equations (two translational and one moment) that can be used to help analyze the system. Noting that the external forces and moments are equivalent to the inertial terms, we wrote

$$\Sigma F_x = m\bar{a}_x \qquad \Sigma F_y = m\bar{a}_y$$
$$\Sigma M_G = \bar{I}\alpha \quad \text{or} \quad \Sigma M_P = \bar{I}\alpha + m\bar{a}d_\perp \quad \text{or} \quad \Sigma M_P = \bar{I}\alpha + \mathbf{r}_{G/P} \times m\bar{\mathbf{a}} \qquad \textbf{(16.6}')$$

where $G$ is the center of mass of the body, $P$ is any arbitrary point, and $d_\perp$ is the perpendicular distance between point $P$ and the line of action of the acceleration of the center of mass.

**3. Apply kinematic relationships.** Often, you will have more than three unknowns and will need to generate additional equations. You can usually do this by applying kinematic relationships, such as $a_n = r\omega^2$ and $a_t = r\alpha$ or for a rigid body undergoing fixed-axis rotation or the more general expression relating the acceleration of two points on a rigid body, as

$$\mathbf{a}_B = \mathbf{a}_A + \alpha\mathbf{k} \times \mathbf{r}_{B/A} - \omega^2\mathbf{r}_{B/A} \qquad \textbf{(15.21}')$$

**4. Plane motion of a rigid body.** The problems that you will be asked to solve will fall into one of the following categories.

    **a. Rigid body in translation.** For a body in translation, the angular acceleration is zero. The kinetic diagram, therefore, is simply the vector $m\bar{\mathbf{a}}$ applied at the mass center (Sample Probs. 16.1, 16.2, and 16.3).

    **b. Rigid body in centroidal rotation.** For a body in centroidal rotation, the linear acceleration of the mass center is zero. Therefore, the kinetic diagram is simply the couple $\bar{I}\boldsymbol{\alpha}$ (Sample Prob. 16.4).

    **c. Rigid body in general plane motion.** You can consider the general plane motion of a rigid body to be the sum of a translation and a centroidal rotation. The kinetic diagram contains the vector $m\bar{\mathbf{a}}$ and the couple $\bar{I}\boldsymbol{\alpha}$ (Sample Probs. 16.5 and 16.6).

**5. Plane motion of a system of rigid bodies.** You first should draw a free-body diagram and a kinetic diagram that includes all of the rigid bodies of the system. A vector $m\bar{a}$ and a couple $\bar{I}\alpha$ are attached to each body. However, the forces exerted on each other by the various bodies of the system can be omitted, because they occur in pairs of equal and opposite forces.

    **a. If no more than three unknowns are involved,** you can use the free-body and kinetic diagrams to sum components in any direction and sum moments about any point, obtaining equations that can be solved for the desired unknowns (Sample Prob. 16.4).

    **b. If more than three unknowns are involved,** you must choose a new system, use kinematics, or use additional information in the problem statement to find additional equations.

# Problems

**16.CQ1** Two pendulums, *A* and *B*, with the masses and lengths shown are released from rest. Which system has a larger mass moment of inertia about its pivot point?

**a.** *A*

**b.** *B*

**c.** They are the same.

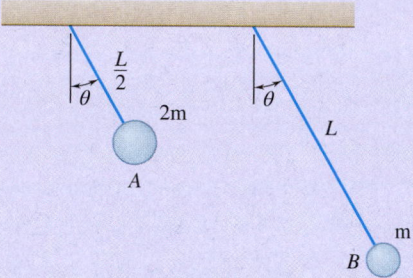

**Fig. P16.CQ1 and P16.CQ2**

**16.CQ2** Two pendulums, *A* and *B*, with the masses and lengths shown, are released from rest. Which system has a larger angular acceleration immediately after release?

**a.** *A*

**b.** *B*

**c.** They are the same.

**16.CQ3** Two solid cylinders, *A* and *B*, have the same mass *m* and the radii $2r$ and *r*, respectively. Each is accelerated from rest with a force applied as shown. In order to impart identical angular accelerations to both cylinders, what is the relationship between $F_1$ and $F_2$?

**a.** $F_1 = 0.5F_2$

**b.** $F_1 = F_2$

**c.** $F_1 = 2F_2$

**d.** $F_1 = 4F_2$

**e.** $F_1 = 8F_2$

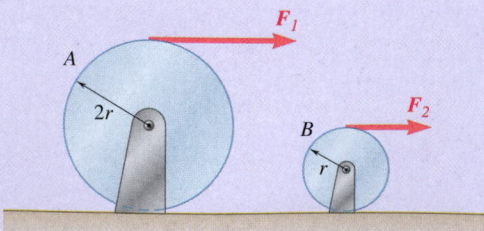

**Fig. P16.CQ3**

**16.F1** A 6-ft board is placed in a truck with one end resting against a block secured to the floor and the other leaning against a vertical partition. Draw the FBD and KD necessary to determine the maximum allowable acceleration of the truck if the board is to remain in the position shown.

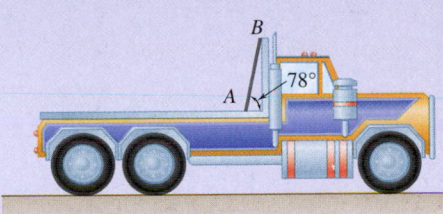

**Fig. P16.F1**

**16.F2** A uniform circular plate of mass 3 kg is attached to two links *AC* and *BD* of the same length. Knowing that the plate is released from rest in the position shown, in which lines joining *G* to *A* and *B* are, respectively, horizontal and vertical, draw the FBD and KD for the plate.

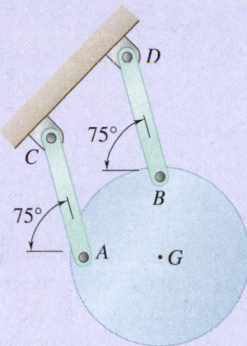

**Fig. P16.F2**

**16.F3** Two uniform disks and two cylinders are assembled as indicated. Disk *A* weighs 20 lb and disk *B* weighs 12 lb. Knowing that the system is released from rest, draw the FBD and KD for the whole system.

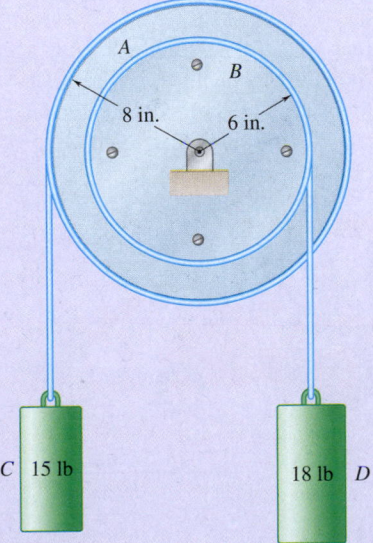

**Fig. P16.F3**

**16.F4** The 400-lb crate shown is lowered by means of two overhead cranes. Knowing the tension in each cable, draw the FBD and KD that can be used to determine the angular acceleration of the crate and the acceleration of the center of gravity.

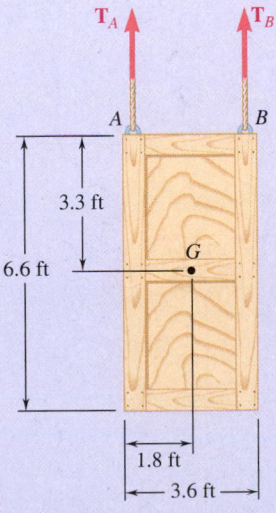

**Fig. P16.F4**

**16.1** A 60-lb uniform thin panel is placed in a truck with end *A* resting on a rough horizontal surface and end *B* supported by a smooth vertical surface. Knowing that the deceleration of the truck is 12 ft/s$^2$, determine (*a*) the reactions at ends *A* and *B*, (*b*) the minimum required coefficient of static friction at end *A*.

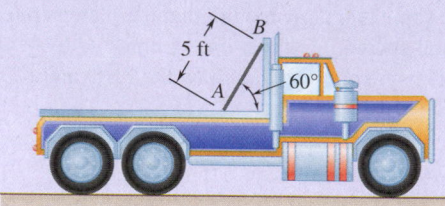

Fig. P16.1 and P16.2

**16.2** A 60-lb uniform thin panel is placed in a truck with end *A* resting on a rough horizontal surface and end *B* supported by a smooth vertical surface. Knowing that the panel remains in the position shown, determine (*a*) the maximum allowable acceleration of the truck, (*b*) the corresponding minimum required coefficient of static friction at end *A*.

**16.3** A loading car is at rest on a track forming an angle of 25° with the vertical. The gross weight of the car and its load is 5500 lb, and it acts at point *G*. Knowing the tension in the cable connected at *C* is 3000 lb, determine (*a*) the acceleration of the car, (*b*) the reaction at each pair of wheels.

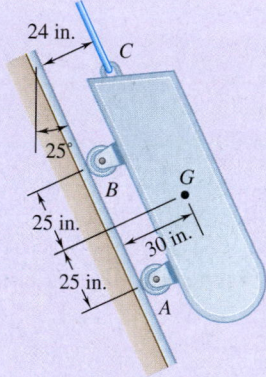

Fig. P16.3

**16.4** A 2100-lb rear-wheel-drive tractor carries a 900 lb load of gravel centered at point *L*. Knowing that the tractor starts from rest and accelerates forward at 2 ft/s$^2$, determine the reaction at each of the two (*a*) rear wheels *A*, (*b*) front wheels *B*.

Fig. P16.4

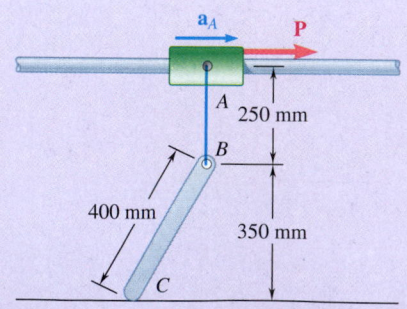

Fig. P16.5

**16.5** A uniform rod *BC* of mass 4 kg is connected to a collar *A* by a 250-mm cord *AB*. Neglecting the mass of the collar and cord, determine (*a*) the smallest constant acceleration **a**$_A$ for which the cord and the rod will lie in a straight line, (*b*) the corresponding tension in the cord.

**16.6** A 2000-kg truck is being used to lift a 400-kg boulder $B$ that is on a 50-kg pallet $A$. Knowing the acceleration of the rear-wheel-drive truck is 1 m/s$^2$, determine (*a*) the reaction at each of the front wheels, (*b*) the force between the boulder and the pallet.

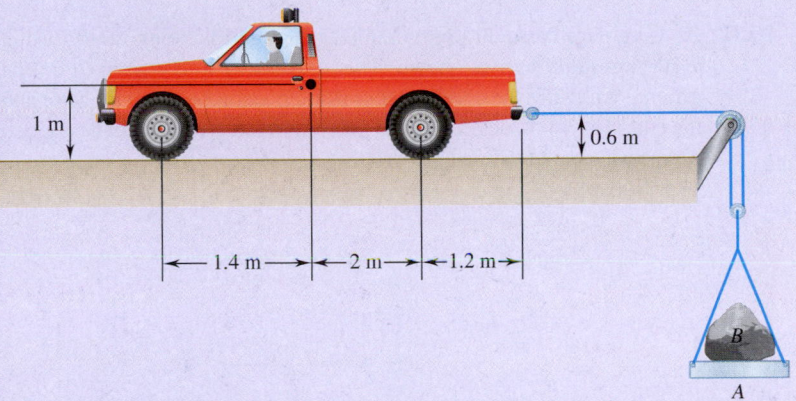

**Fig. P16.6**

**16.7** The support bracket shown is used to transport a cylindrical can from one elevation to another. Knowing that $\mu_s = 0.25$ between the can and the bracket, determine (*a*) the magnitude of the upward acceleration **a** for which the can will slide on the bracket, (*b*) the smallest ratio $h/d$ for which the can will tip before it slides.

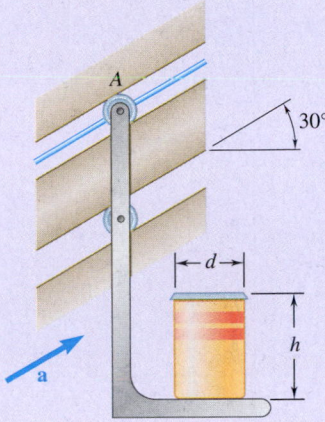

**Fig. *P16.7***

**16.8** A load of lumber weighing $W = 25$ kN is being raised by a crane. The weight of the boom $ABC$ is 3 kN and the combined weight of the truck and driver is 50 kN as shown. Determine the maximum vertical acceleration of the lumber so that the crane does not tip over.

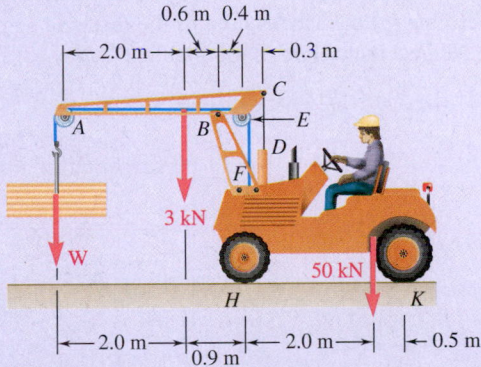

**Fig. P16.8**

**16.9** A 20-kg cabinet is mounted on casters that allow it to move freely ($\mu = 0$) on the floor. If a 100-N force is applied as shown, determine (*a*) the acceleration of the cabinet, (*b*) the range of values of $h$ for which the cabinet will not tip.

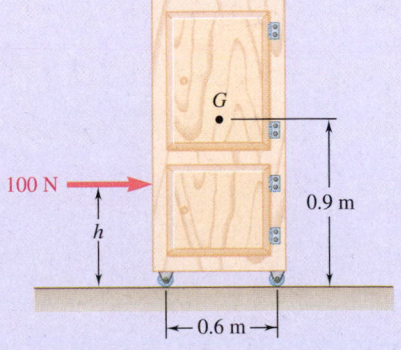

**Fig. P16.9**

**16.10** Solve Prob. 16.9, assuming that the casters are locked and slide on the rough floor ($\mu_k = 0.25$).

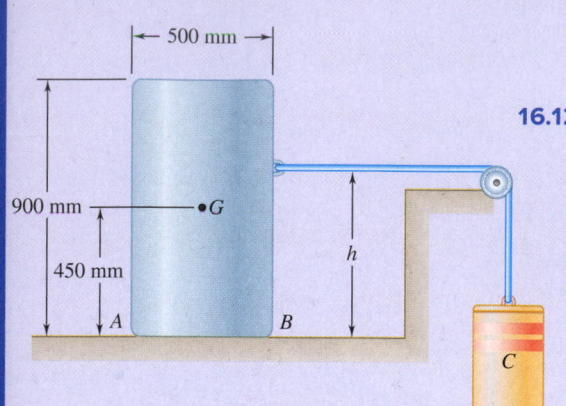

**16.11** A completely filled barrel and its contents have a combined mass of 90 kg and a center of mass at $G$. A cylinder $C$ with a mass of 200 kg is connected to the barrel as shown. Knowing $\mu_s = 0.40$ and $\mu_k = 0.35$, determine the maximum height $h$ so the barrel will not tip.

**16.12** A 40-kg vase has a 200-mm-diameter base and is being moved using a 100-kg utility cart as shown. The cart moves freely ($\mu = 0$) on the ground. Knowing the coefficient of static friction between the vase and the cart is $\mu_s = 0.4$, determine the maximum force $\mathbf{F}$ that can be applied if the vase is not to slide or tip.

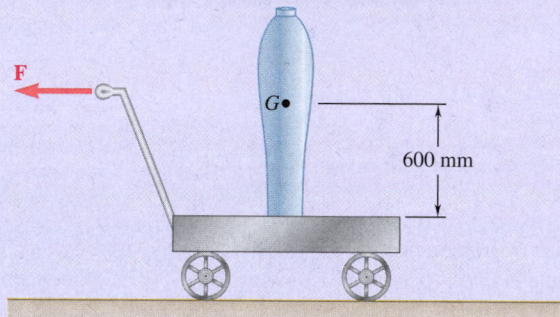

Fig. P16.12

**16.13** The retractable shelf shown is supported by two identical linkage-and-spring systems; only one of the systems is shown. A 20-kg machine is placed on the shelf so that half of its weight is supported by the system shown. If the springs are removed and the system is released from rest, determine (a) the acceleration of the machine, (b) the tension in link $AB$. Neglect the weight of the shelf and links.

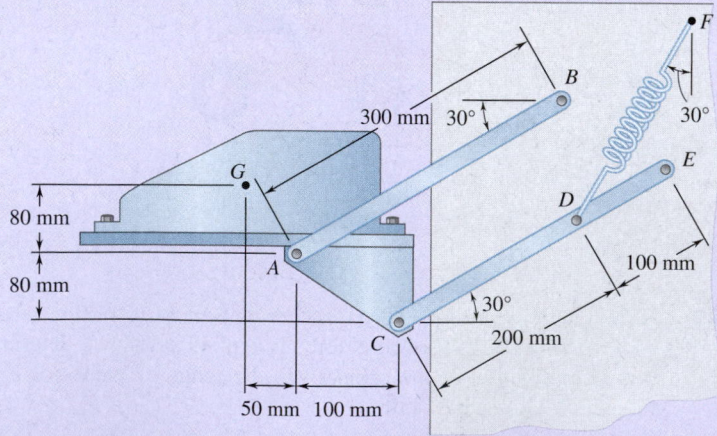

Fig. *P16.13*

**16.14** Bars *AB* and *BE,* each with a mass of 4 kg, are welded together and are pin-connected to two links *AC* and *BD.* Knowing that the assembly is released from rest in the position shown and neglecting the masses of the links, determine (*a*) the acceleration of the assembly, (*b*) the forces in the links.

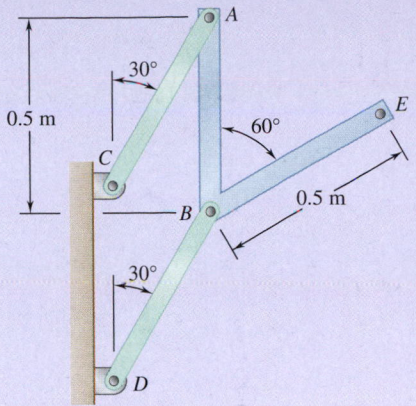

**Fig. P16.14**

**16.15** At the instant shown, the tensions in the vertical ropes *AB* and *DE* are 300 N and 200 N, respectively. Knowing that the mass of the uniform bar *BE* is 5 kg, determine, at this instant, (*a*) the force **P**, (*b*) the magnitude of the angular velocity of each rope, (*c*) the angular acceleration of each rope.

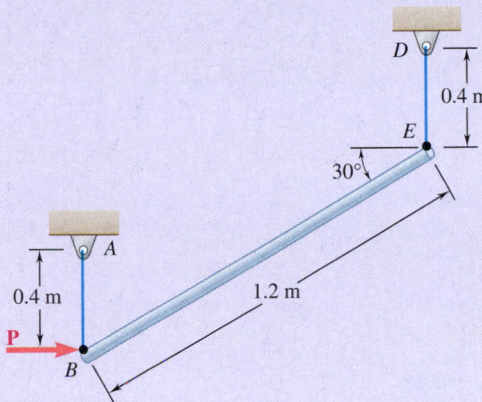

**Fig. P16.15**

**16.16** Three bars, each of mass 3 kg, are welded together and pin-connected to two links *BE* and *CF.* Neglecting the weight of the links, determine the force in each link immediately after the system is released from rest.

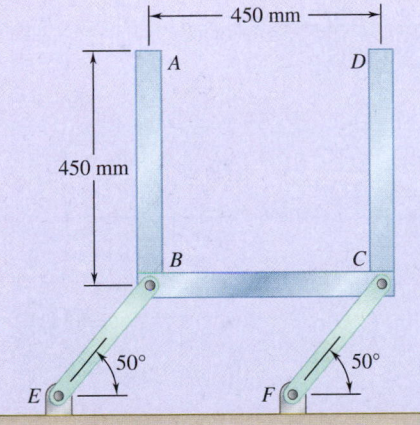

**Fig. P16.16**

**16.17** Members *ACE* and *DCB* are each 600 mm long and are connected by a pin at *C.* The mass center of the 10-kg member *AB* is located at *G.* Determine (*a*) the acceleration of *AB* immediately after the system has been released from rest in the position shown, (*b*) the corresponding force exerted by roller *A* on member *AB.* Neglect the weight of members *ACE* and *DCB.*

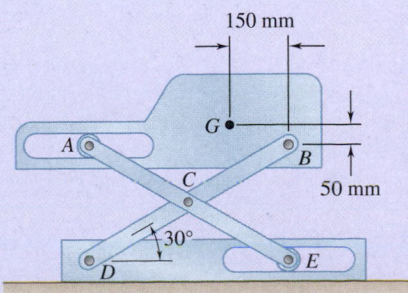

**Fig. P16.17**

**16.18** A prototype rotating bicycle rack is designed to save space at a train station. The combined weight of platform *BD* and the bicycle is 40 lb and is centered at 1 ft above the midpoint of the platform. The motor at *A* causes the support beam *AB* to have an angular velocity of 10 rpm and zero angular acceleration at $\theta = 60°$. At this instant, determine the vertical components of the forces exerted on platform *BD* by the pins at *B* and *D*.

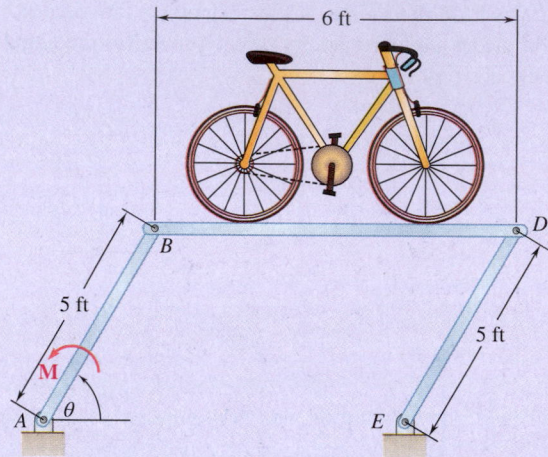

**Fig. P16.18**

**16.19** The control rod *AC* is guided by two pins that slide freely in parallel curved slots of radius 200 mm. The rod has a mass of l0 kg, and its mass center is located at point *G*. Knowing that for the position shown the *vertical* component of the velocity of *C* is 1.25 m/s upward and the *vertical* component of the acceleration of *C* is 5 m/s² upward, determine the magnitude of the force **P**.

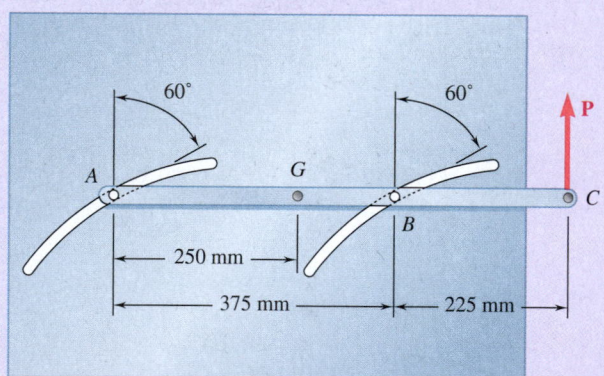

**Fig. P16.19**

**16.20** The coefficients of friction between the 30-lb block and the 5-lb platform *BD* are $\mu_s = 0.50$ and $\mu_k = 0.40$. Determine the accelerations of the block and of the platform immediately after wire *AB* has been cut.

**\*16.21** Draw the shear and bending-moment diagrams for the vertical rod *AB* of Prob. 16.16.

**\*16.22** Draw the shear and bending-moment diagrams for each of the bars *AB* and *BE* of Prob. 16.14.

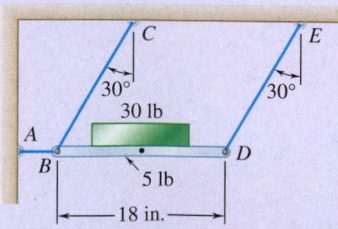

**Fig. P16.20**

**16.23** For a rigid body in translation, show that the system of the inertial terms consists of vectors $(\Delta m_i)\bar{\mathbf{a}}$ attached to the various particles of the body, where $\bar{\mathbf{a}}$ is the acceleration of the mass center $G$ of the body. Further show, by computing their sum and the sum of their moments about $G$, that the inertial terms reduce to a single vector $m\bar{\mathbf{a}}$ attached at $G$.

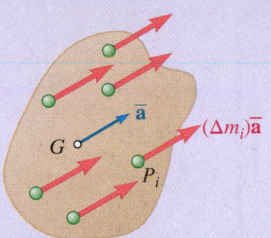

**Fig. P16.23**

**16.24** For a rigid body in centroidal rotation, show that the system of the inertial terms consists of vectors $-(\Delta m_i)\omega^2\mathbf{r}_i'$ and $(\Delta m_i)(\boldsymbol{\alpha} \times \mathbf{r}_i')$ attached to the various particles $P_i$ of the body, where $\boldsymbol{\omega}$ and $\boldsymbol{\alpha}$ are the angular velocity and angular acceleration of the body, and where $\mathbf{r}_i'$ denotes the position vector of the particle $P_i$ relative to the mass center $G$ of the body. Further show, by computing their sum and the sum of their moments about $G$, that the inertial terms reduce to a couple $\bar{I}\boldsymbol{\alpha}$.

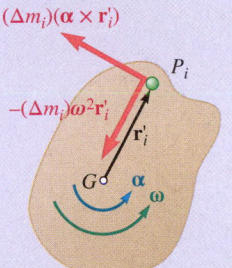

**Fig. P16.24**

**16.25** It takes 10 min for a 2.4-Mg flywheel to coast to rest from an angular velocity of 300 rpm. Knowing that the radius of gyration of the flywheel is 1 m, determine the average magnitude of the couple due to kinetic friction in the bearing.

**16.26** The rotor of an electric motor has an angular velocity of 3600 rpm when the load and power are cut off. The 120-lb rotor, which has a centroidal radius of gyration of 9 in., then coasts to rest. Knowing that kinetic friction results in a couple of magnitude 2.5 lb·ft exerted on the rotor, determine the number of revolutions that the rotor executes before coming to rest.

**16.27** The 10-in.-radius brake drum is attached to a larger flywheel that is not shown. The total mass moment of inertia of the drum and the flywheel about point $C$ is 15 lb·ft·s$^2$, and the coefficient of kinetic friction between the drum and the brake shoes is 0.35. When the hydraulic cylinder $F$ is actuated, it exerts a force of 30 lb directed to the right on point $B$ and to the left on point $E$. Knowing that the angular velocity of the flywheel is 360 rpm counterclockwise when $F$ is actuated, determine the number of revolutions executed by the flywheel before it comes to rest.

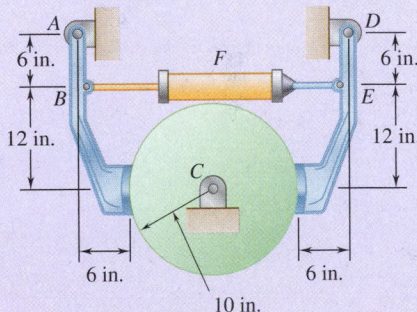

**16.28** The 10-in.-radius brake drum is attached to a larger flywheel that is not shown. The coefficient of kinetic friction between the drum and the brake shoe is 0.35 and the angular velocity of the flywheel is 360 rpm counterclockwise when the hydraulic cylinder shown exerts a force of 30 lb directed to the right on point $B$ and to the left on point $E$. Knowing that the drum comes to rest after 100 revolutions, determine the mass moment of inertia about point $C$ of the drum and the flywheel.

**Fig. P16.27 and P16.28**

**16.29** The 100-mm-radius brake drum is attached to a flywheel that is not shown. The drum and flywheel together have a mass of 300 kg and a radius of gyration of 600 mm. The coefficient of kinetic friction between the brake band and the drum is 0.30. Knowing that a force **P** of magnitude 50 N is applied at $A$ when the angular velocity is 180 rpm counterclockwise, determine the time required to stop the flywheel when $a = 200$ mm and $b = 160$ mm.

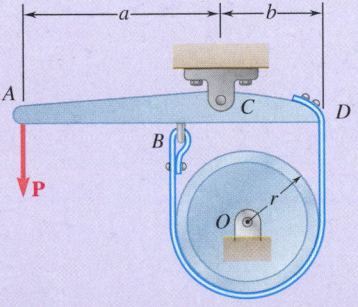

**Fig. P16.29**

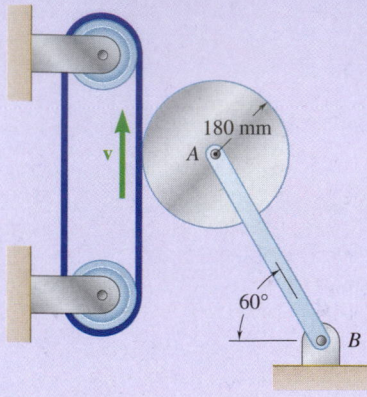

**Fig. P16.30**

**16.30** The 180-mm-radius disk is at rest when it is placed in contact with a belt moving at a constant speed. Neglecting the weight of the link *AB* and knowing that the coefficient of kinetic friction between the disk and the belt is 0.40, determine the angular acceleration of the disk while slipping occurs.

**16.31** Solve Prob. 16.30, assuming that the direction of motion of the belt is reversed.

**16.32** In order to determine the mass moment of inertia of a flywheel of radius 600 mm, a 12-kg block is attached to a wire that is wrapped around the flywheel. The block is released and is observed to fall 3 m in 4.6 s. To eliminate bearing friction from the computation, a second block of mass 24 kg is used and is observed to fall 3 m in 3.1 s. Assuming that the moment of the couple due to friction remains constant, determine the mass moment of inertia of the flywheel.

**Fig. *P16.32* and P16.33**

**16.33** The flywheel shown has a radius of 20 in., a weight of 250 lb, and a radius of gyration of 15 in. A 30-lb block *A* is attached to a wire that is wrapped around the flywheel, and the system is released from rest. Neglecting the effect of friction, determine (*a*) the acceleration of block *A*, (*b*) the speed of block *A* after it has moved 5 ft.

**16.34** Each of the double pulleys shown has a mass moment of inertia of 15 lb·ft·s$^2$ and is initially at rest. The outside radius is 18 in., and the inner radius is 9 in. Determine (*a*) the angular acceleration of each pulley, (*b*) the angular velocity of each pulley after point *A* on the cord has moved 10 ft.

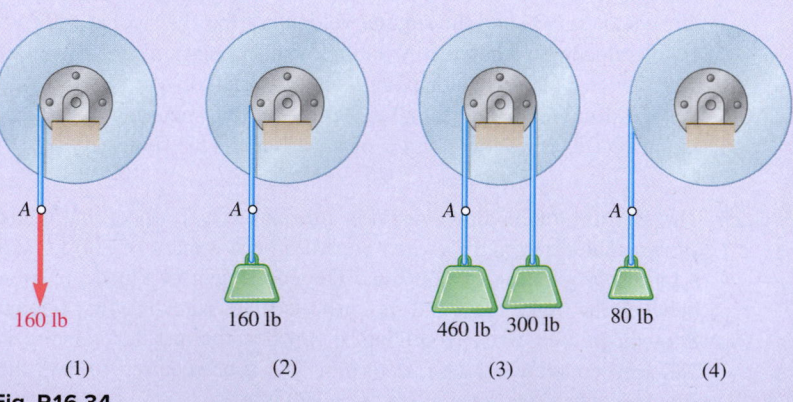

**Fig. P16.34**

**16.35** Two disks $A$ and $B$, of mass $m_A = 2$ kg and $m_B = 4$ kg, are connected by a belt as shown. Assuming no slipping between the belt and the disks, determine the angular acceleration of each disk if a 2.70-N·m counterclockwise couple **M** is applied to disk $A$.

**16.36** Two disks $A$ and $B$, of mass $m_A = 2$ kg and $m_B = 4$ kg, are connected by a belt as shown. Assuming no slipping between the belt and the disks, determine the angular acceleration of each disk if a 2.70-N·m counterclockwise couple **M** is applied to disk $B$.

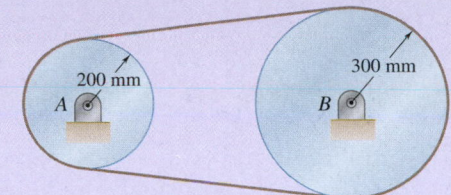

**Fig. P16.35 and P16.36**

**16.37** Gear $A$ weighs 1 lb and has a radius of gyration of 1.3 in.; gear $B$ weighs 6 lb and has a radius of gyration of 3 in.; gear $C$ weighs 9 lb and has a radius of gyration of 4.3 in. Knowing a couple **M** of constant magnitude of 40 lb·in. is applied to gear $A$, determine (a) the angular acceleration of gear $C$, (b) the tangential force that gear $B$ exerts on gear $C$.

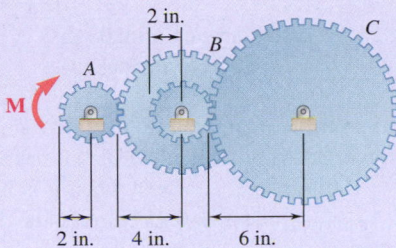

**Fig. P16.37**

**16.38** The 25-lb double pulley shown is at rest and in equilibrium when a constant 3.5-lb·ft couple **M** is applied. Neglecting the effect of friction and knowing that the radius of gyration of the double pulley is 6 in., determine (a) the angular acceleration of the double pulley, (b) the tension in each rope.

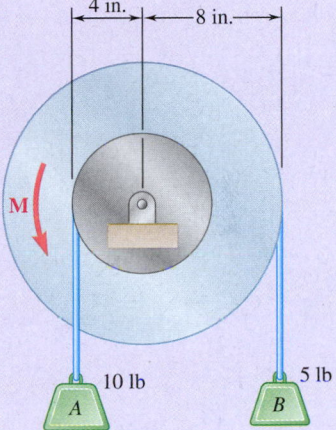

**Fig. P16.38**

**16.39** A belt of negligible mass passes between cylinders $A$ and $B$ and is pulled to the right with a force **P**. Cylinders $A$ and $B$ weigh, respectively, 5 and 20 lb. The shaft of cylinder $A$ is free to slide in a vertical slot and the coefficients of friction between the belt and each of the cylinders are $\mu_s = 0.50$ and $\mu_k = 0.40$. For $P = 3.6$ lb, determine (a) whether slipping occurs between the belt and either cylinder, (b) the angular acceleration of each cylinder.

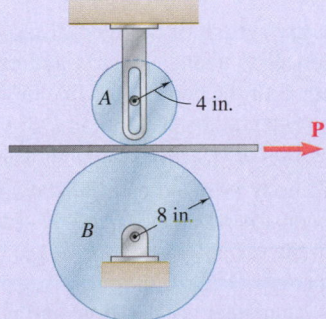

**Fig. P16.39**

**16.40** Solve Prob. 16.39 for $P = 2.00$ lb.

**16.41** Disk $A$ has a mass of 6 kg and an initial angular velocity of 360 rpm clockwise; disk $B$ has a mass of 3 kg and is initially at rest. The disks are brought together by applying a horizontal force of magnitude 20 N to the axle of disk $A$. Knowing that $\mu_k = 0.15$ between the disks and neglecting bearing friction, determine (a) the angular acceleration of each disk, (b) the final angular velocity of each disk.

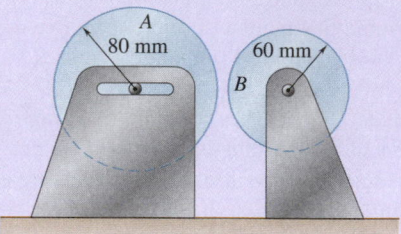

Fig. P16.41

**16.42** Solve Prob. 16.41, assuming that initially disk $A$ is at rest and disk $B$ has an angular velocity of 360 rpm clockwise.

**16.43** Disk $A$ has a mass $m_A = 4$ kg, a radius $r_A = 300$ mm, and an initial angular velocity $\omega_0 = 300$ rpm clockwise. Disk $B$ has a mass $m_B = 1.6$ kg, a radius $r_B = 180$ mm, and is at rest when it is brought into contact with disk $A$. Knowing that $\mu_k = 0.35$ between the disks and neglecting bearing friction, determine (a) the angular acceleration of each disk, (b) the reaction at the support $C$.

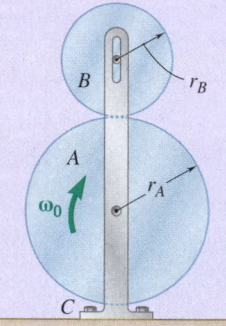

Fig. P16.43 and P16.44

**16.44** Disk $B$ is at rest when it is brought into contact with disk $A$, which has an initial angular velocity $\omega_0$. (a) Show that the final angular velocities of the disks are independent of the coefficient of friction $\mu_k$ between the disks as long as $\mu_k \neq 0$. (b) Express the final angular velocity of disk $A$ in terms of $\omega_0$ and the ratio of the masses of the two disks, $m_A/m_B$.

**16.45** Cylinder $A$ has an initial angular velocity of 720 rpm clockwise, and cylinders $B$ and $C$ are initially at rest. Disks $A$ and $B$ each weigh 5 lb and have radius $r = 4$ in. Disk $C$ weighs 20 lb and has a radius of 8 in. The disks are brought together when $C$ is placed gently onto $A$ and $B$. Knowing that $\mu_k = 0.25$ between $A$ and $C$ and no slipping occurs between $B$ and $C$, determine (a) the angular acceleration of each disk, (b) the final angular velocity of each disk.

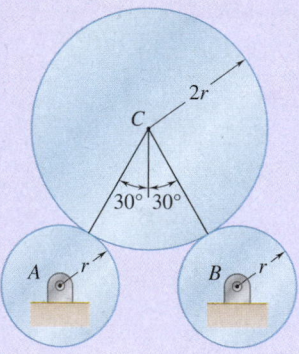

Fig. P16.45

**16.46** Show that the system of the inertial terms for a rigid body in plane motion reduces to a single vector, and express the distance from the mass center $G$ of the body to the line of action of this vector in terms of the centroidal radius of gyration $\overline{k}$ of the body, the magnitude $\overline{a}$ of the acceleration of $G$, and the angular acceleration $\alpha$.

**16.47** For a rigid body in plane motion, show that the system of the inertial terms consists of vectors $(\Delta m_i)\overline{\mathbf{a}}$, $-(\Delta m_i)\omega^2 \mathbf{r}'_i$, and $(\Delta m_i)(\boldsymbol{\alpha} \times \mathbf{r}'_i)$ attached to the various particles $P_i$ of the body, where $\overline{\mathbf{a}}$ is the acceleration of the mass center $G$ of the body, $\boldsymbol{\omega}$ is the angular velocity of the body, $\boldsymbol{\alpha}$ is its angular acceleration, and $\mathbf{r}'_i$ denotes the position vector of the particle $P_i$ relative to $G$. Further show, by computing their sum and the sum of their moments about $G$, that the inertial terms reduce to a vector $m\overline{\mathbf{a}}$ attached at $G$ and a couple $\overline{I}\alpha$.

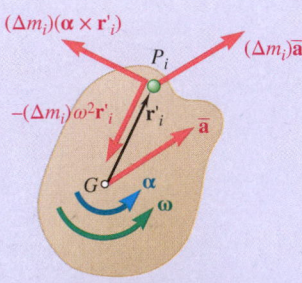

Fig. P16.47

**16.48** A uniform slender rod *AB* rests on a frictionless horizontal surface, and a force **P** of magnitude 0.75 lb is applied at *A* in a direction perpendicular to the rod. Knowing that the rod weighs 2 lb, determine (*a*) the acceleration of point *A*, (*b*) the acceleration of point *B*, (*c*) the location of the point on the bar that has zero acceleration.

**16.49** (*a*) In Prob. 16.48, determine the point of the rod *AB* at which the force **P** should be applied if the acceleration of point *B* is to be zero. (*b*) Knowing that *P* = 0.75 lb, determine the corresponding acceleration of point *A*.

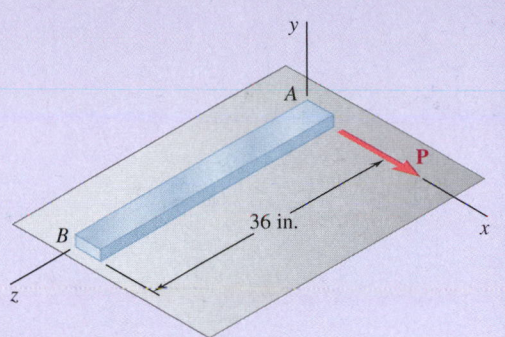

**Fig. P16.48**

**16.50 and 16.51** A force **P** with a magnitude of 3 N is applied to a tape wrapped around the body indicated. Knowing that the body rests on a frictionless horizontal surface, determine the acceleration of (*a*) point *A*, (*b*) point *B*.

   **16.50** A thin hoop of mass 2.4 kg.
   **16.51** A uniform disk of mass 2.4 kg.

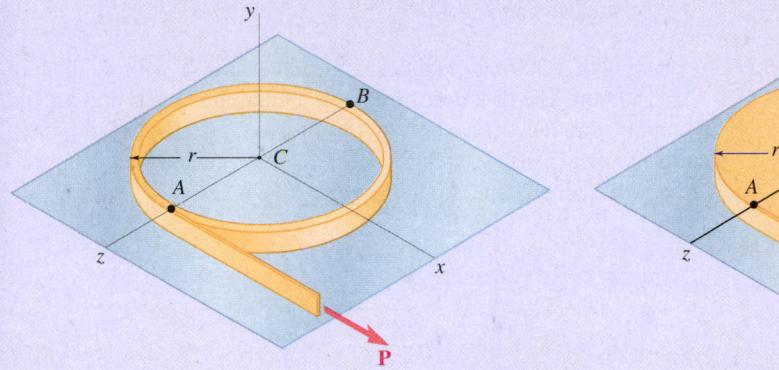

**Fig. P16.50**                                    **Fig. P16.51**

**16.52** A 250-lb satellite has a radius of gyration of 24 in. with respect to the *y* axis and is symmetrical with respect to the *zx* plane. Its orientation is changed by firing four small rockets—*A, B, C,* and *D*—each of which produces a 4-lb thrust **T** directed as shown. Determine the angular acceleration of the satellite and the acceleration of its mass center *G* (*a*) when all four rockets are fired, (*b*) when all rockets except *D* are fired.

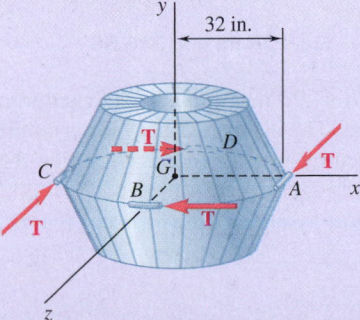

**Fig. P16.52**

**1147**

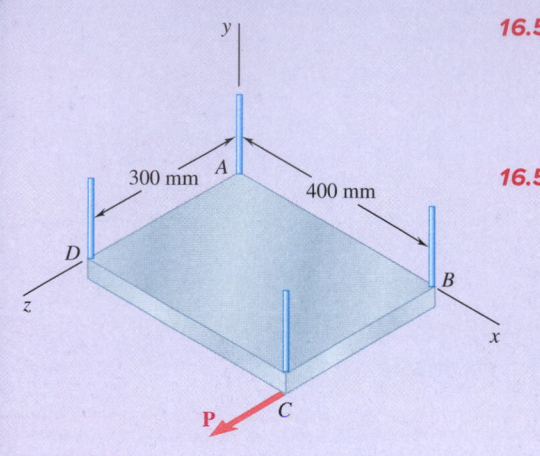

**Fig. P16.53**

**16.53** A rectangular plate of mass 5 kg is suspended from four vertical wires, and a force **P** of magnitude 6 N is applied to corner $C$ as shown. Immediately after **P** is applied, determine the acceleration of (a) the midpoint of edge $BC$, (b) corner $B$.

**16.54** A uniform semicircular plate with a mass of 6 kg is suspended from three vertical wires at points $A$, $B$, and $C$, and a force **P** with a magnitude of 5 N is applied to point $B$. Immediately after **P** is applied, determine the acceleration of (a) the mass center of the plate, (b) point $C$.

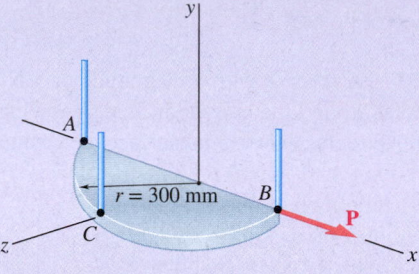

**Fig. P16.54**

**16.55** A drum with a 200-mm radius is attached to a disk with a radius of $r_A = 150$ mm. The disk and drum have a combined mass of 5 kg and a combined radius of gyration of 120 mm and are suspended by two cords. Knowing that $T_A = 35$ N and $T_B = 25$ N, determine the accelerations of points $A$ and $B$ on the cords.

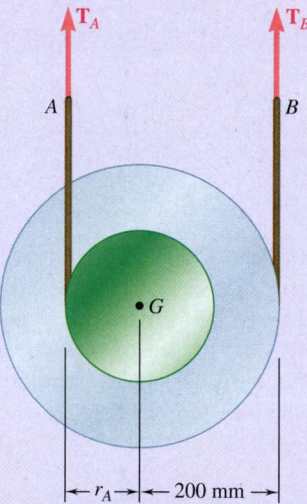

**Fig. P16.55 and P16.56**

**16.56** A drum with a 200-mm radius is attached to a disk with a radius of $r_A = 140$ mm. The disk and drum have a combined mass of 5 kg and are suspended by two cords. Knowing that the acceleration of point $B$ on the cord is zero, $T_A = 40$ N, and $T_B = 20$ N, determine the combined radius of gyration of the disk and drum.

**16.57** The 12-lb uniform disk shown has a radius of $r = 3.2$ in. and rotates counterclockwise. Its center $C$ is constrained to move in a slot cut in the vertical member $AB$, and an 11-lb horizontal force **P** is applied at $B$ to maintain contact at $D$ between the disk and the vertical wall. The disk moves downward under the influence of gravity and the friction at $D$. Knowing that the coefficient of kinetic friction between the disk and the wall is 0.12 and neglecting friction in the vertical slot, determine (a) the angular acceleration of the disk, (b) the acceleration of the center $C$ of the disk.

**16.58** The steel roll shown has a mass of 1200 kg, a centroidal radius of gyration of 150 mm, and is lifted by two cables looped around its shaft. Knowing that for each cable $T_A = 3100$ N and $T_B = 3300$ N, determine (a) the angular acceleration of the roll, (b) the acceleration of its mass center.

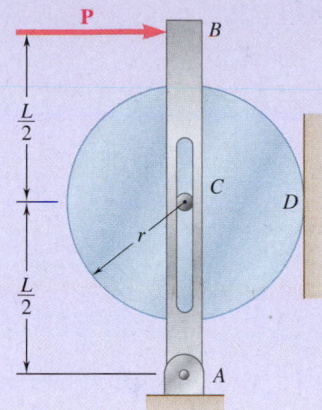

**Fig. P16.57**

**16.59** The steel roll shown has a mass of 1200 kg, has a centroidal radius of gyration of 150 mm, and is lifted by two cables looped around its shaft. Knowing that at the instant shown the acceleration of the roll is 150 mm/s² downward and that for each cable $T_A = 3000$ N, determine (a) the corresponding tension $T_B$, (b) the angular acceleration of the roll.

**16.60 and 16.61** The 400-lb crate shown is lowered by means of two overhead cranes. Knowing that at the instant shown the deceleration of cable $A$ is 3 ft/s² and that of cable $B$ is 1 ft/s², determine the tension in each cable.

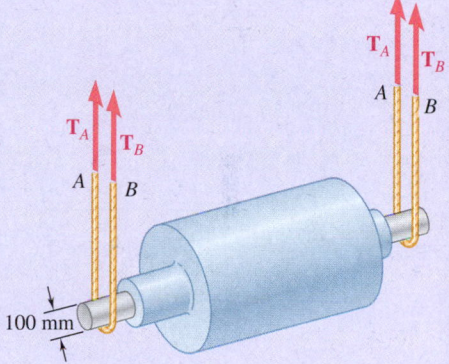

**Fig. P16.58 and P16.59**

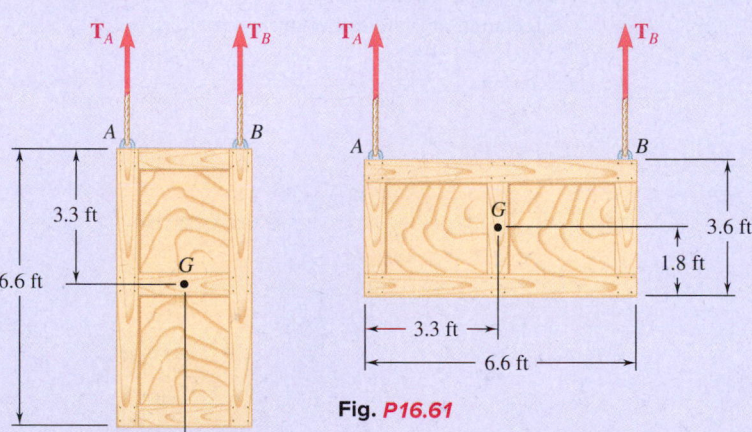

**Fig. P16.61**

**Fig. P16.60**

**16.62** Two uniform cylinders, each of weight $W = 14$ lb and radius $r = 5$ in., are connected by a belt as shown. If the system is released from rest, determine (a) the angular acceleration of each cylinder, (b) the tension in the portion of belt connecting the two cylinders, (c) the velocity of the center of the cylinder $A$ after it has moved through 3 ft.

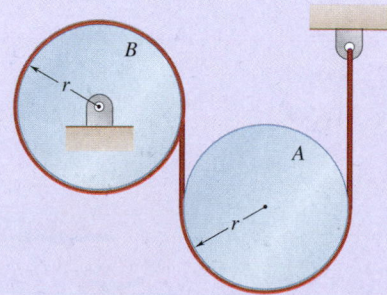

**Fig. P16.62**

**16.63 and 16.64** A beam $AB$ with a mass $m$ and of uniform cross-section is suspended from two springs as shown. If spring 2 breaks, determine at that instant (*a*) the angular acceleration of the beam, (*b*) the acceleration of point $A$, (*c*) the acceleration of point $B$.

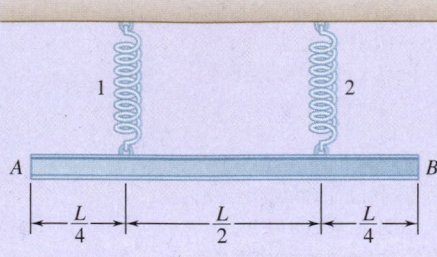

Fig. P16.63

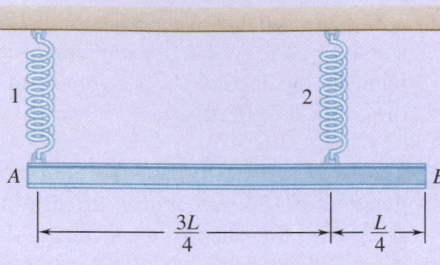

Fig. P16.64

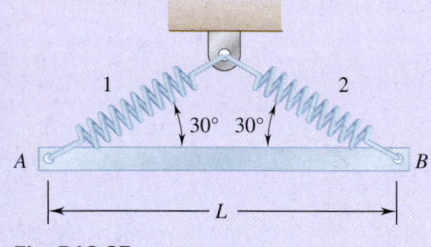

Fig. P16.65

**16.65** A uniform slender bar $AB$ with a mass $m$ is suspended from two springs as shown. If spring 2 breaks, determine at that instant (*a*) the angular acceleration of the bar, (*b*) the acceleration of point $A$, (*c*) the acceleration of point $B$.

**16.66 through 16.68** A thin plate of the shape indicated and of mass $m$ is suspended from two springs as shown. If spring 2 breaks, determine the acceleration at that instant of (*a*) point $A$, (*b*) point $B$.

**16.66** A square plate of side $b$
**16.67** A thin hoop of diameter $b$
**16.68** A rectangular plate of height $b$ and width $a$

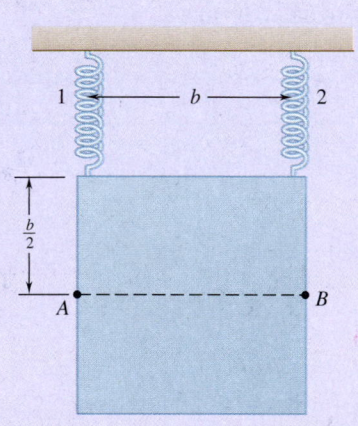

Fig. P16.66

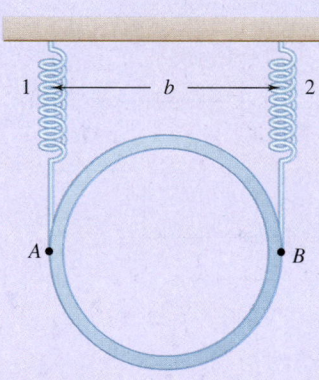

Fig. P16.67

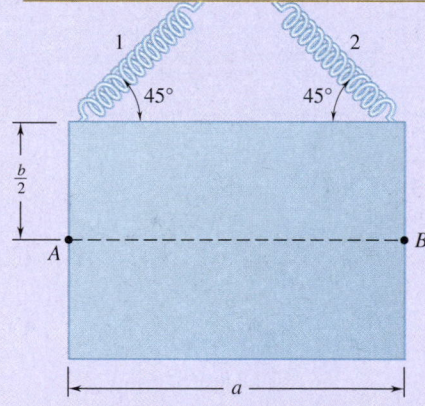

Fig. P16.68

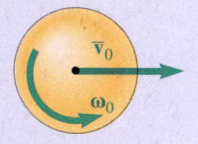

Fig. P16.69

**16.69** A sphere of radius $r$ and mass $m$ is projected along a rough horizontal surface with the initial velocities indicated. If the final velocity of the sphere is to be zero, express, in terms of $v_0$, $r$, and $\mu_k$, (*a*) the required magnitude of $\omega_0$, (*b*) the time $t_1$ required for the sphere to come to rest, (*c*) the distance the sphere will move before coming to rest.

**16.70** Solve Prob. 16.69, assuming that the sphere is replaced by a uniform thin hoop of radius $r$ and mass $m$.

**16.71** A bowler projects an 8-in.-diameter ball weighing 12 lb along an alley with a forward velocity $\mathbf{v}_0$ of 15 ft/s and a backspin $\boldsymbol{\omega}_0$ of 9 rad/s. Knowing that the coefficient of kinetic friction between the ball and the alley is 0.10, determine (a) the time $t_1$ at which the ball will start rolling without sliding, (b) the speed of the ball at time $t_1$, (c) the distance the ball will have traveled at time $t_1$.

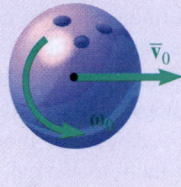

**Fig. P16.71**

**16.72** Solve Prob. 16.71, assuming that the bowler projects the ball with the same forward velocity but with a backspin of 18 rad/s.

**16.73** A uniform sphere of radius $r$ and mass $m$ is placed with no initial velocity on a belt that moves to the right with a constant velocity $\mathbf{v}_1$. Denoting by $\mu_k$ the coefficient of kinetic friction between the sphere and the belt, determine (a) the time $t_1$ at which the sphere will start rolling without sliding, (b) the linear and angular velocities of the sphere at time $t_1$.

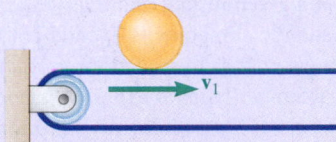

**Fig. P16.73**

**16.74** A sphere of radius $r$ and mass $m$ has a linear velocity $\mathbf{v}_0$ directed to the left and no angular velocity as it is placed on a belt moving to the right with a constant velocity $\mathbf{v}_1$. If after first sliding on the belt the sphere is to have no linear velocity relative to the ground as it starts rolling on the belt without sliding, determine in terms of $v_1$ and the coefficient of kinetic friction $\mu_k$ between the sphere and the belt (a) the required value of $v_0$, (b) the time $t_1$ at which the sphere will start rolling on the belt, (c) the distance the sphere will have moved relative to the ground at time $t_1$.

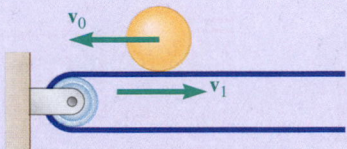

**Fig. P16.74**

# 16.2 CONSTRAINED PLANE MOTION

Most engineering applications deal with rigid bodies that are moving under given constraints. For example, cranks must rotate about a fixed axis, wheels must roll without sliding, and connecting rods must describe certain prescribed motions. In all such cases, definite relations exist between the components of the acceleration $\bar{a}$ of the mass center $G$ of the body considered and its angular acceleration $\alpha$. The corresponding motion is said to be a **constrained motion**.

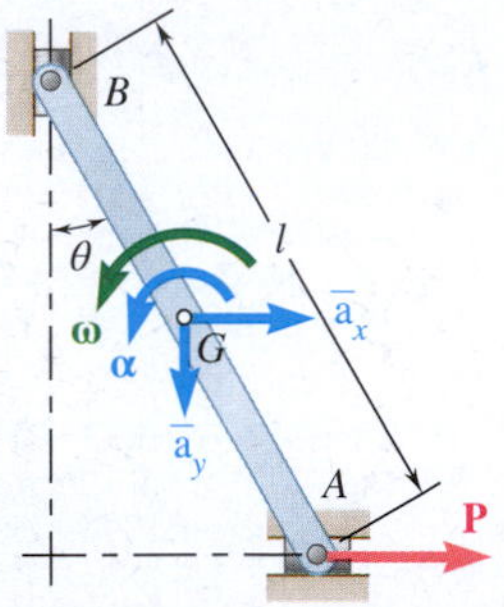

As discussed in the previous section, we draw our free-body and kinetic diagrams and then write the equations of motion. The solution of a problem involving a constrained plane motion also calls for a *kinematic analysis* of the problem. Consider, for example, a slender rod $AB$ with a length $l$ and a mass $m$, where the extremities are connected to blocks of negligible mass that slide along horizontal and vertical frictionless tracks. The rod is pulled by a force $\mathbf{P}$ applied at $A$ (Fig. 16.12). We know from Sec. 15.4A that we can determine the acceleration $\bar{a}$ of the mass center $G$ of the rod at any given instant from the position of the rod, its angular velocity, and its angular acceleration at that instant. Suppose, for example, that we know the values of $\theta$, $\omega$, and $\alpha$ at a given instant, and we wish to determine the corresponding value of the force $\mathbf{P}$ as well as the reactions at $A$ and $B$. We should first *determine the components $\bar{a}_x$ and $\bar{a}_y$ of the acceleration of the mass center $G$* using the method in Sec. 15.4A. We next solve our equations of motion using the expressions obtained for $\bar{a}_x$ and $\bar{a}_y$. We can then find the unknown forces $\mathbf{P}$, $\mathbf{N}_A$, and $\mathbf{N}_B$ by solving the appropriate equations.

Suppose now that we know the applied force $\mathbf{P}$, the angle $\theta$, and the angular velocity $\omega$ of the rod at a given instant and that we wish to find the angular acceleration $\alpha$ of the rod and the components $\bar{a}_x$ and $\bar{a}_y$ of the acceleration of its mass center at that instant, as well as the reactions at $A$ and $B$. The preliminary kinematic study of the problem will aim *to express the components $\bar{a}_x$ and $\bar{a}_y$ of the acceleration of G in terms of the angular acceleration $\alpha$ of the rod.* This is done by first expressing the acceleration of a suitable reference point such as $A$ in terms of the angular acceleration $\alpha$. We can then determine the components $\bar{a}_x$ and $\bar{a}_y$ of the acceleration of $G$ in terms of $\alpha$ and carry these expressions into Fig. 16.13. We can then derive three equations in terms of $\alpha$, $N_A$, and $N_B$ and solve for the three unknowns (see Sample Prob. 16.12).

When a mechanism consists of several moving parts, we can use the approach just described with each part of the mechanism. The procedure

**Fig. 16.12** Kinematic variables for a constrained rod pulled to the right.

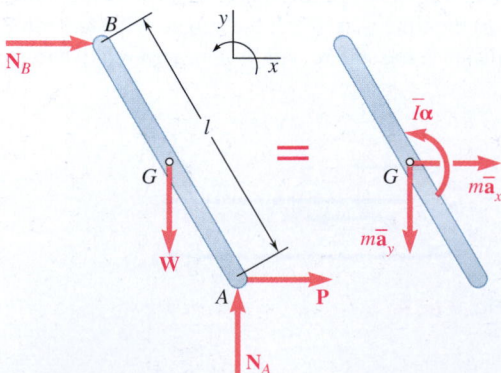

**Fig. 16.13** Free-body diagram and kinetic diagram for the rod in Fig. 16.12.

required to determine the various unknowns is then similar to the procedure followed in the case of the equilibrium of a system of connected rigid bodies (Sec. 6.3B).

Earlier, we analyzed two particular cases of constrained plane motion: translation of a rigid body, in which the angular acceleration of the body is constrained to be zero, and centroidal rotation, in which the acceleration $\bar{\mathbf{a}}$ of the mass center of the body is constrained to be zero. Two other particular cases of constrained plane motion are of special interest: *noncentroidal rotation* of a rigid body and *rolling motion* of a disk or wheel. We can analyze these two cases using one of the general methods described previously. However, in view of the range of their applications, they deserve a few special comments.

**Noncentroidal Rotation.**   The motion of a rigid body constrained to rotate about a fixed axis that does not pass through its mass center is called **noncentroidal rotation**. The mass center $G$ of the body moves along a circle with a radius $\bar{r}$ centered at point $O$, where the axis of rotation intersects the plane of reference (Fig. 16.14). Denoting the angular velocity and the angular acceleration of the line $OG$ by $\boldsymbol{\omega}$ and $\boldsymbol{\alpha}$, respectively, we obtain the following expressions for the tangential and normal components of the acceleration of $G$:

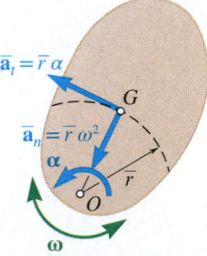

$$\bar{a}_t = \bar{r}\alpha \qquad \bar{a}_n = \bar{r}\omega^2 \qquad (16.7)$$

**Fig. 16.14**   For noncentroidal fixed-axis rotation, the center of mass has a tangential and a normal component of acceleration.

Because line $OG$ belongs to the body, its angular velocity $\boldsymbol{\omega}$ and its angular acceleration $\boldsymbol{\alpha}$ also represent the angular velocity and the angular acceleration of the body. Eqs. (16.7) define, therefore, the kinematic relation between the motion of the mass center $G$ and the motion of the body about $G$.

We obtain an interesting relation by equating the moments about the fixed point $O$ of the forces and vectors shown, respectively, in Fig. 16.15$a$ and $b$. We have

$$+\circlearrowleft \Sigma M_O = \bar{I}\alpha + (m\bar{r}\alpha)\bar{r} = (\bar{I} + m\bar{r}^2)\alpha$$

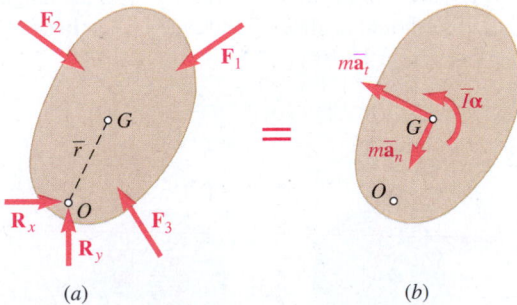

**Fig. 16.15**   Free-body diagram and kinetic diagram for the rigid body in Fig. 16.14.

But according to the parallel-axis theorem, we have $\bar{I} + m\bar{r}^2 = I_O$, where $I_O$ denotes the moment of inertia of the rigid body about the fixed axis. We therefore obtain

**Moments about a fixed axis**

$$\Sigma M_O = I_O\alpha \qquad (16.8)$$

Although formula (16.8) expresses an important relation between the sum of the moments of the external forces about the fixed point $O$ and the product $I_O\alpha$, we will still need to apply Eq. (16.1) to find the forces at $O$.

A particular case of noncentroidal rotation is of special interest—the case of *uniform rotation*, in which the angular velocity $\boldsymbol{\omega}$ is constant. Because $\alpha$ is zero, the inertia couple in Fig. 16.15 vanishes, and the inertia vector reduces to its normal component. This component (also called *centrifugal force* in layman's terms) represents the tendency of the rigid body to break away from the axis of rotation.

**Rolling Motion.** Another important case of plane motion is the motion of a disk or wheel rolling on a plane surface. If the disk is constrained to roll without sliding, the acceleration $\overline{\mathbf{a}}$ of its mass center $G$ and its angular acceleration $\alpha$ are not independent. Assuming that the disk is balanced so that its mass center and its geometric center coincide, the distance $\overline{x}$ traveled by $G$ during a rotation $\theta$ of the disk is $\overline{x} = r\theta$, where $r$ is the radius of the disk. Differentiating this relation twice, we have

$$\overline{a} = r\alpha \qquad\qquad (16.9)$$

Recall that the system of the inertial terms in plane motion reduces to a vector $m\overline{\mathbf{a}}$ and a couple $\overline{I}\boldsymbol{\alpha}$. We find that, in the particular case of the rolling motion of a balanced disk, these terms reduce to a vector of magnitude $mr\alpha$ attached at $G$ and to a couple with a magnitude of $\overline{I}\alpha$. We may thus say that the external forces are equivalent to the vector and couple shown in Fig. 16.16.

When a disk **rolls without sliding**, there is no relative motion between the point of the disk in contact with the ground and the ground itself. Thus, as far as the computation of the friction force $\mathbf{F}$ is concerned, a rolling disk can be compared with a block at rest on a surface. The magnitude $F$ of the friction force can have any value, as long as this value does not exceed the maximum value $F_m = \mu_s N$, where $\mu_s$ is the coefficient of static friction and $N$ is the magnitude of the normal force. In the case of a rolling disk, the magnitude $F$ of the friction force therefore should be determined independently of $N$ by solving the equation obtained from Fig. 16.16.

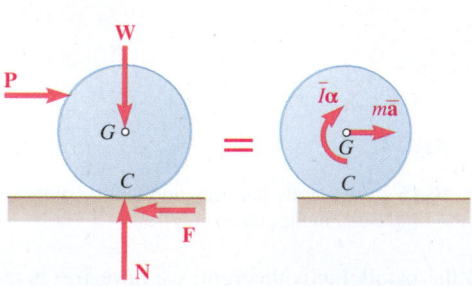

**Fig. 16.16** Free-body diagram and kinetic diagram for a disk rolling without slipping on a fixed surface.

**Photo 16.4** As the ball hits the bowling alley, it first spins and slides, then rolls without sliding. ©Doable/amanaimages/agefotostock RF

When *sliding is impending*, the friction force reaches its maximum value $F_m = \mu_s N$ and can be obtained after solving for $N$.

When the disk *rotates and slides* at the same time, a relative motion exists between the point of the disk in contact with the ground and the ground itself. The force of friction has the magnitude $F_k = \mu_k N$, where $\mu_k$ is the coefficient of kinetic friction. In this case, however, the motion of the mass center $G$ of the disk and the rotation of the disk about $G$ are independent, and $\bar{a}$ is not equal to $r\alpha$.

We can summarize these three different cases as

**Rolling, no sliding**            $F \leq \mu_s N$     $\bar{a} = r\alpha$

**Rolling, sliding impending**     $F = \mu_s N$     $\bar{a} = r\alpha$

**Rotating and sliding**           $F = \mu_k N$     $\bar{a}$ and $\alpha$ independent

If you do not know whether or not a disk slides, you should first assume that the disk rolls without sliding. You will then be able to solve your system of equations by assuming that $\bar{a} = r\alpha$. If $F$ is found to be smaller than or equal to $\mu_s N$, the assumption is proved correct. If $F$ is found to be larger than $\mu_s N$, the assumption is incorrect, and you should start the problem again, assuming rotating, sliding, and that $F = \mu_k N$.

When a disk is *unbalanced,* that is, when its mass center $G$ does not coincide with its geometric center $O$, the relation in Eq. (16.9) does not hold between $\bar{a}$ and $\alpha$. However, a similar relation holds between the magnitude $a_O$ of the acceleration of the geometric center and the angular acceleration $\alpha$ of an unbalanced disk that rolls without sliding. We have

$$a_o = r\alpha \qquad (16.10)$$

To determine $\bar{a}$ in terms of the angular acceleration $\alpha$ and the angular velocity $\omega$ of the disk, we can use the relative-acceleration formula, as

$$\begin{aligned}\bar{a} = \bar{a}_G &= a_O + a_{G/O} \\ &= a_O + (a_{G/O})_t + (a_{G/O})_n\end{aligned} \qquad (16.11)$$

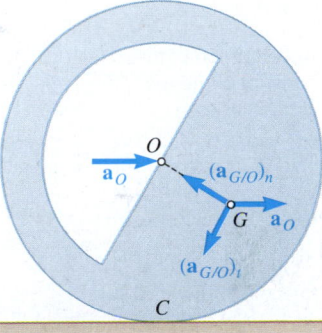

where the three component accelerations have the directions indicated in Fig. 16.17 and the magnitudes $a_O = r\alpha$, $(a_{G/O})_t = (OG)\,\alpha$, and $(a_{G/O})_n = (OG)\,\omega^2$. These terms also can be solved using the relationship between two points on a rigid body undergoing plane motion:

$$\bar{a} = a_O + \alpha \times r_{G/O} - \omega^2 r_{G/O} \qquad (16.12)$$

**Fig. 16.17** Accelerations of the geometric center $O$ and center of mass $G$ for a rolling unbalanced disk.

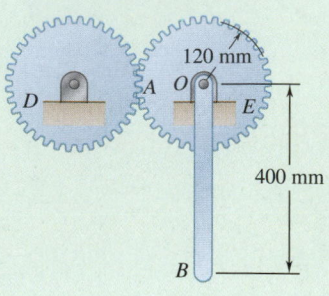

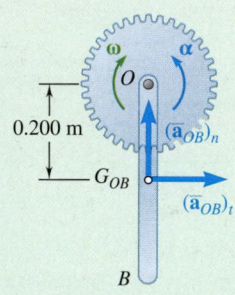

**Fig. 1** Acceleration of the center of gravity of the bar.

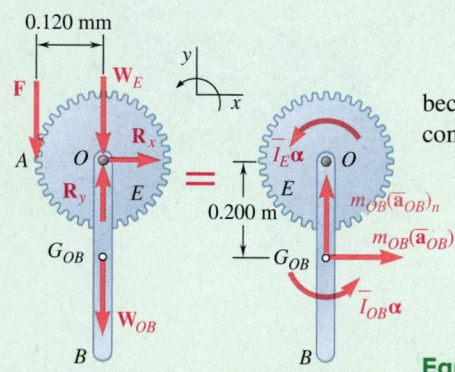

**Fig. 2** Free-body diagram and kinetic diagram for the system.

## Sample Problem 16.7

The portion $AOB$ of a mechanism consists of a 400-mm steel rod $OB$ welded to a gear $E$ with a radius of 120 mm that can rotate about a horizontal shaft $O$. It is actuated by a gear $D$ and, at the instant shown, has a clockwise angular velocity of 8 rad/s and a counterclockwise angular acceleration of 40 rad/s². Knowing that rod $OB$ has a mass of 3 kg and gear $E$ has a mass of 4 kg and a radius of gyration of 85 mm, determine (a) the tangential force exerted by gear $D$ on gear $E$, (b) the components of the reaction at point $O$ on the shaft.

**STRATEGY:** Because you are asked to determine forces, use Newton's second law.

**MODELING:** For your system, choose the single object that consists of the steel rod $OB$ and the gear $E$. Because these two objects are welded together, they have the same angular velocity and angular acceleration. Rather than finding the center of mass for this object, use the center of mass for gear $E$ and for rod $OB$ separately in your kinetic diagram. Therefore, first determine the components of the acceleration of the mass center $G_{OB}$ of the rod (Fig. 1) as

$$(\bar{a}_{OB})_t = \bar{r}\alpha = (0.200 \text{ m})(40 \text{ rad/s}^2) = 8 \text{ m/s}^2$$
$$(\bar{a}_{OB})_n = \bar{r}\omega^2 = (0.200 \text{ m})(8 \text{ rad/s}^2) = 12.8 \text{ m/s}^2$$

A free-body diagram and kinetic diagram for the system are shown in Fig. 2. The inertial terms on your kinetic diagram include a couple $\bar{I}_E\alpha$ (Because gear $E$ is in centroidal rotation), a couple $\bar{I}_{OB}\alpha$, and two vector components $m_{OB}(\bar{a}_{OB})_n$ and $m_{OB}(\bar{a}_{OB})_t$ at the mass center of $OB$.

**ANALYSIS:**

**Preliminary Calculations:** The magnitudes of the weights are

$$W_E = m_E g = (4 \text{ kg})(9.81 \text{ m/s}^2) = 39.2 \text{ N}$$
$$W_{OB} = m_{OB}g = (3 \text{ kg})(9.81 \text{ m/s}^2) = 29.4 \text{ N}$$

because you know the accelerations, you can compute the magnitudes of the components and couples on your kinetic diagram, as

$$\bar{I}_E\alpha = m_E\bar{k}_E^2\alpha = (4 \text{ kg})(0.085 \text{ m})^2(40 \text{ rad/s}^2) = 1.156 \text{ N·m}$$
$$m_{OB}(\bar{a}_{OB})_t = (3 \text{ kg})(8 \text{ m/s}^2) = 24.0 \text{ N}$$
$$m_{OB}(\bar{a}_{OB})_n = (3 \text{ kg})(12.8 \text{ m/s}^2) = 38.4 \text{ N}$$
$$\bar{I}_{OB}\alpha = (\tfrac{1}{12}m_{OB}L^2)\alpha = \tfrac{1}{12}(3 \text{ kg})(0.400 \text{ m})^2(40 \text{ rad/s}^2) = 1.600 \text{ N·m}$$

**Equations of Motion.** Setting the system of the external forces shown in your free-body diagram equal to the inertia terms in your kinetic diagram, you obtain the following equations, which you can solve as

$$+\circlearrowleft \Sigma M_O = \dot{H}_O:$$
$$F(0.120 \text{ m}) = \bar{I}_E\alpha + m_{OB}(\bar{a}_{OB})_t(0.200 \text{ m}) + \bar{I}_{OB}\alpha$$
$$F(0.120 \text{ m}) = 1.156 \text{ N·m} + (24.0 \text{ N})(0.200 \text{ m}) + 1.600 \text{ N·m}$$
$$F = 63.0 \text{ N} \qquad\qquad \mathbf{F = 63.0 \text{ N}} \downarrow \ \blacktriangleleft$$

*(continued)*

$$\xrightarrow{+}\ \Sigma F_x = \Sigma m\bar{a}_x: \qquad\qquad R_x = m_{OB}(\bar{a}_{OB})_t$$
$$R_x = 24.0\ \text{N} \qquad\qquad \mathbf{R}_x = 24.0\ \text{N} \rightarrow \ \blacktriangleleft$$

$$+\uparrow \Sigma F_y = \Sigma m\bar{a}_y: \qquad R_y - F - W_E - W_{OB} = m_{OB}(\bar{a}_{OB})_n$$
$$R_y - 63.0\ \text{N} - 39.2\ \text{N} - 29.4\ \text{N} = 38.4\ \text{N}$$
$$R_y = 170.0\ \text{N} \qquad\qquad \mathbf{R}_y = 170.0\ \text{N} \uparrow \ \blacktriangleleft$$

**REFLECT and THINK:**   When you drew your kinetic diagram, you put your inertia terms at the center of mass for the gear and for the rod. Alternatively, you could have found the center of mass for the system and put the vectors $\bar{I}_{AOB}\,\alpha$, $m_{AOB}\bar{a}_x$, and $m_{AOB}\bar{a}_y$ on the diagram. Finally, you could have found an overall $I_O$ for the combined gear and rod and used Eq. 16.8 to solve for force $F$.

## Sample Problem 16.8

A 6 × 8 in. rectangular plate weighing 60 lb is suspended from two pins $A$ and $B$. If pin $B$ is suddenly removed, determine (*a*) the angular acceleration of the plate, (*b*) the components of the reaction at pin $A$ immediately after pin $B$ has been removed. ▶

**STRATEGY:**   You are asked to determine forces and the angular acceleration of the plate, so use Newton's second law.

**MODELING:**   Choose the plate to be your system and model it as a rigid body. Observe that as the plate rotates about point $A$, its mass center $G$ describes a circle with a radius $\bar{r}$ and its center at $A$ (Fig. 1). The free-body diagram and kinetic diagram for this system are shown in Fig. 2. The plate is released from rest ($\omega = 0$), so the normal component of the acceleration of $G$ is zero. The magnitude of the acceleration $\bar{a}$ of the mass center $G$ is thus $\bar{a} = \bar{r}\alpha$.

**ANALYSIS:**

**a. Angular Acceleration.**   Using your free-body diagram and kinetic diagram, you can take moments about $A$ to find

$$+\circlearrowleft \Sigma M_A = \bar{I}\alpha + m\bar{a}d_\perp: \qquad W\bar{x} = \bar{I}\alpha + (m\bar{a})\bar{r}$$

Because $\bar{a} = \bar{r}\alpha$, you have

$$W\bar{x} = \bar{I}\alpha + (m\bar{r}\alpha)\bar{r} \qquad \alpha = \dfrac{W\bar{x}}{\dfrac{W}{g}\bar{r}^2 + \bar{I}} \qquad\qquad (1)$$

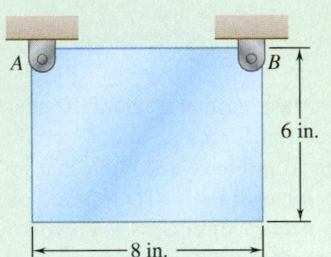

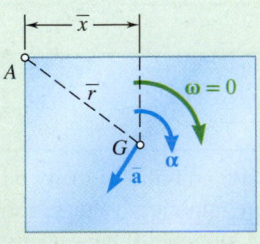

**Fig. 1**  The plate travels in a circle about *A*.

6 in.

8 in.

$\bar{x}$

$A$

$\bar{r}$

$\omega = 0$

$G$

$\alpha$

$\bar{a}$

*(continued)*

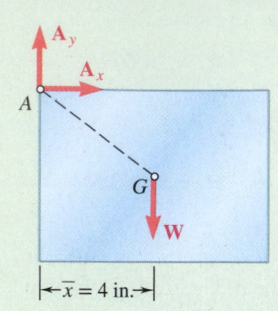

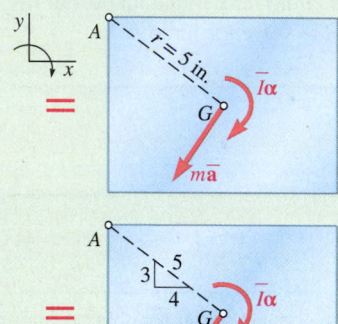

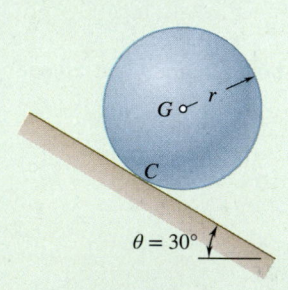

**Fig. 2** Free-body diagram and kinetic diagram for the plate.

The centroidal moment of inertia of the plate is

$$\bar{I} = \frac{m}{12}(a^2 + b^2) = \frac{60\ \text{lb}}{12(32.2\ \text{ft/s}^2)}[(\tfrac{8}{12}\ \text{ft})^2 + (\tfrac{6}{12}\ \text{ft})^2]$$

$$= 0.1078\ \text{lb·ft·s}^2$$

Substituting this value of $\bar{I}$ together with $W = 60$ lb, $\bar{r} = \frac{5}{12}$ ft, and $\bar{x} = \frac{4}{12}$ ft into Eq. (1), you obtain

$$\alpha = +46.4\ \text{rad/s}^2 \qquad\qquad \boldsymbol{\alpha = 46.4\ \text{rad/s}^2\ \circlearrowright} \quad \blacktriangleleft$$

**b. Reaction at A.** Using the computed value of $\alpha$, determine the magnitude of the vector $m\bar{a}$ attached at $G$ as

$$m\bar{a} = m\bar{r}\alpha = \frac{60\ \text{lb}}{32.2\ \text{ft/s}^2}\left(\frac{5}{12}\ \text{ft}\right)(46.4\ \text{rad/s}^2) = 36.0\ \text{lb}$$

Applying Newton's second law in the $x$ and $y$ directions gives

$$\xrightarrow{+}\ \Sigma F_x = m\bar{a}_x: \qquad A_x = -\tfrac{3}{5}(36\ \text{lb})$$
$$= -21.6\ \text{lb} \qquad\qquad \mathbf{A}_x = 21.6\ \text{lb} \leftarrow \quad \blacktriangleleft$$

$$+\uparrow\ \Sigma F_y = m\bar{a}_y: \qquad A_y - 60\text{lb} = -\tfrac{4}{5}(36\ \text{lb})$$
$$A_y = +31.2\ \text{lb} \qquad\qquad \mathbf{A}_y = 31.2\ \text{lb} \uparrow \quad \blacktriangleleft$$

**REFLECT and THINK:** If you had chosen to take moments about the center of gravity rather than point $A$, the two reaction forces $A_x$ and $A_y$ would have been in the resulting equation; that is, you would have had one equation and three unknowns, and you could not solve for $\alpha$ directly. Therefore, you would also need to use the equations from the $x$ and $y$ directions to solve for the three unknowns. Note that for convenience, we used a non-right-handed coordinate system.

## Sample Problem 16.9

A sphere with a radius $r$ and a weight $W$ is released with no initial velocity on an incline and rolls without slipping. Determine (a) the minimum value of the coefficient of static friction compatible with the rolling motion, (b) the velocity of the center $G$ of the sphere after the sphere has rolled 10 ft, (c) the velocity of $G$ if the sphere were to move 10 ft down a frictionless 30° incline. ▶

**STRATEGY:** Use Newton's second law to determine the acceleration of the center of gravity. Then determine the velocity from kinematics.

**MODELING:** Choose the sphere to be your system and model it as a rigid body. Recall that for rolling motion, the instantaneous point of contact has a velocity of zero, which leads to $\bar{a} = r\alpha$ (Fig. 1). A free-body diagram and kinetic diagram for this system are shown in Fig. 2. The external forces $\mathbf{W}$, $\mathbf{N}$, and $\mathbf{F}$ form a system equivalent to the inertial terms represented by the vector $m\bar{a}$ and the couple $\bar{I}\alpha$.

*(continued)*

**ANALYSIS:**

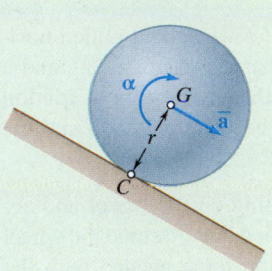

**Fig. 1** The acceleration of $G$ down the incline.

**a. Minimum $\mu_s$ for Rolling Motion.** Because the sphere rolls without sliding, you have $\bar{a} = r\alpha$ and can sum moments about $C$:

$$+\circlearrowleft \Sigma M_C = \bar{I}\alpha + m\bar{a}d_\perp: \qquad (W\sin\theta)r = \bar{I}\alpha + (m\bar{a})r$$

$$(W\sin\theta)r = \bar{I}\alpha + (mr\alpha)r$$

Noting that $m = W/g$ and $\bar{I} = \tfrac{2}{5}mr^2$, you have

$$(W\sin\theta)r = \frac{2}{5}\frac{W}{g}r^2\alpha + \left(\frac{W}{g}r\alpha\right)r \qquad \alpha = +\frac{5g\sin\theta}{7r}$$

$$\bar{a} = r\alpha = \frac{5g\sin\theta}{7} = \frac{5(32.2\text{ ft/s}^2)\sin 30°}{7} = 11.50\text{ ft/s}^2$$

Applying Newton's second law in the $x$ and $y$ directions gives

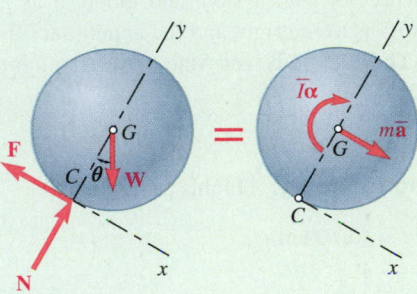

**Fig. 2** Free-body diagram and kinetic diagram for the sphere.

$$+\searrow \Sigma F_x = m\bar{a}_x: \qquad W\sin\theta - F = m\bar{a}$$

$$W\sin\theta - F = \frac{W}{g}\frac{5g\sin\theta}{7}$$

$$F = +\tfrac{2}{7}W\sin\theta = \tfrac{2}{7}W\sin 30° \qquad \mathbf{F} = 0.143W \nwarrow 30°$$

$$+\nearrow \Sigma F_y = m\bar{a}_y: \qquad N - W\cos\theta = 0$$

$$N = W\cos\theta = 0.866W \qquad \mathbf{N} = 0.866W \measuredangle 60°$$

$$\mu_s = \frac{F}{N} = \frac{0.143W}{0.866W} \qquad \mu_s = 0.165 \blacktriangleleft$$

**b. Velocity of Rolling Sphere.** This is a case of uniformly accelerated motion, so

$$\bar{v}_0 = 0 \qquad \bar{a} = 11.50\text{ ft/s}^2 \qquad \bar{x} = 10\text{ ft} \qquad \bar{x}_0 = 0$$

$$\bar{v}^2 = \bar{v}_0^2 + 2\bar{a}(\bar{x} - \bar{x}_0) \qquad \bar{v}^2 = 0 + 2(11.50\text{ ft/s}^2)(10\text{ ft})$$

$$\bar{v} = 15.17\text{ ft/s} \qquad \mathbf{\bar{v}} = 15.17\text{ ft/s} \searrow 30° \blacktriangleleft$$

**c. Velocity of Sliding Sphere.** Now assuming no friction, you have $F = 0$ and obtain

$$+\circlearrowleft \Sigma M_G = \bar{I}\alpha: \qquad 0 = \bar{I}\alpha \qquad \alpha = 0$$

$$+\searrow \Sigma F_X = m\bar{a}_x: \qquad W\sin 30° = m\bar{a} \qquad 0.50W = \frac{W}{g}\bar{a}$$

$$\bar{a} = +16.1\text{ ft/s}^2 \qquad \mathbf{\bar{a}} = 16.1\text{ ft/s}^2 \searrow 30°$$

Substituting $\bar{a} = 16.1\text{ ft/s}^2$ into the equations for uniformly accelerated motion, you obtain

$$\bar{v}^2 = \bar{v}_0^2 = 2\bar{a}(\bar{x} - \bar{x}_0) \qquad \bar{v}^2 = 0 + 2(16.1\text{ ft/s}^2)(10\text{ ft})$$

$$\bar{v} = 17.94\text{ ft/s} \qquad \mathbf{\bar{v}} = 17.94\text{ ft/s} \searrow 30° \blacktriangleleft$$

**REFLECT and THINK:** Note that the sphere moving down a frictionless surface has a higher velocity than the rolling sphere, as you would expect. It is also interesting to note that the expression you obtained for the acceleration of the center of mass—that is, $\bar{a} = 5g\sin\theta/7$—is independent of the radius of the sphere and the mass of the sphere. This means that any two solid spheres, as long they are rolling without sliding, have the same linear acceleration.

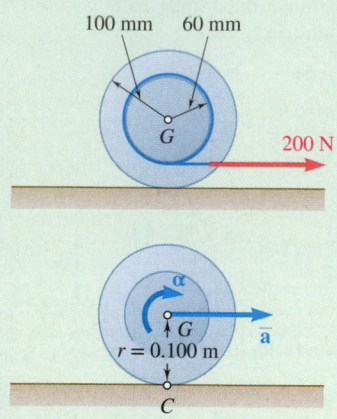

**Fig. 1**   Linear and angular acceleration of the wheel.

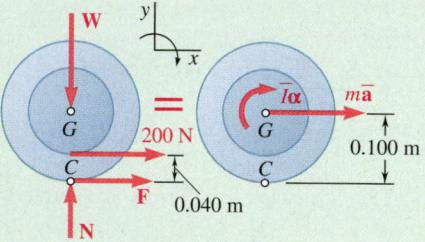

**Fig. 2**   Free-body diagram and kinetic diagram for the wheel assuming the friction force is to the right.

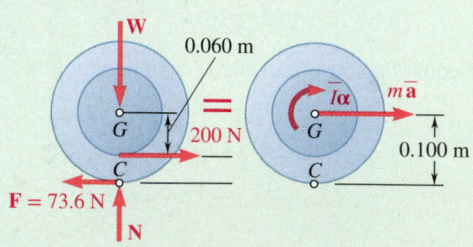

**Fig. 3**   Free-body diagram and kinetic diagram for the wheel when it is sliding and rotating.

# Sample Problem 16.10

A cord is wrapped around the inner drum of a wheel and pulled horizontally with a force of 200 N. The wheel has a mass of 50 kg and a radius of gyration of 70 mm. Knowing that the coefficients of friction are $\mu_s = 0.20$ and $\mu_k = 0.15$, determine the acceleration of $G$ and the angular acceleration of the wheel.

**STRATEGY:**   Because you have forces acting on the wheel and are interested in accelerations, use Newton's second law. Assume the wheel rolls without sliding and compare the friction force needed with the maximum possible friction force. If the force needed exceeds the force available, redo the problem assuming rotation and sliding.

**MODELING:**   Choose the wheel as your system and model it as a rigid body. The acceleration of $G$ is to the right and the angular acceleration is clockwise (Fig. 1). The free-body and kinetic diagrams for this system are shown in Fig. 2.

**ANALYSIS:**

**a. Assume Rolling without Sliding.**   In this case, you have

$$\bar{a} = r\alpha = (0.100 \text{ m})\alpha$$

The moment of inertia of the wheel is

$$\bar{I} = m\bar{k}^2 = (50 \text{ kg})(0.070 \text{ m})^2 = 0.245 \text{ kg·m}^2$$

**Equations of Motion.**   Setting the system of external forces in your free-body diagram equal to the system of inertial terms in your kinetic diagram, you obtain

$+\circlearrowright \Sigma M_C = \bar{I}\alpha + m\bar{a}d_\perp$:    $(200 \text{ N})(0.040 \text{ m}) = \bar{I}\alpha + (m\bar{a})(0.100 \text{ m})$

$8.00 \text{ N·m} = (0.245 \text{ kg·m}^2)\alpha + (50 \text{ kg})(0.100 \text{ m})\alpha(0.100 \text{ m})$

$$\alpha = +10.74 \text{ rad/s}^2$$

$$\bar{a} = r\alpha = (0.100 \text{ m})(10.74 \text{ rad/s}^2) = 1.074 \text{ m/s}^2$$

$\xrightarrow{+} \Sigma F_x = m\bar{a}_x$:    $F + 200 \text{ N} = m\bar{a}$

$F + 200 \text{ N} = (50 \text{ kg})(1.074 \text{ m/s}^2)$

$F = -146.3 \text{ N}$                **F = 146.3 N ←**

$+\uparrow \Sigma F_y = m\bar{a}_y$:

$N - W = 0$            $N - W = mg = (50 \text{ kg})(9.81 \text{ m/s}^2) = 490.5 \text{ N}$

**N = 490.5 N ↑**

**Maximum Possible Friction Force.**

$$F_{max} = \mu_s N = 0.20(490.5 \text{ N}) = 98.1 \text{ N}$$

Because $F > F_{max}$, the assumed motion is impossible.

**b. Rotating and Sliding.**   Because the wheel must rotate and slide at the same time, we draw new free-body and kinetic diagrams (Fig. 3), where $\bar{a}$ and $\alpha$ are independent and

$$F = F_k = \mu_k N = 0.15(490.5 \text{ N}) = 73.6 \text{ N}$$

From the computation of part (a), you found that **F** is directed to the left. You can obtain and solve the following equations of motion as

$\xrightarrow{+} \Sigma F_x = m\bar{a}_x:$     $200 \text{ N} - 73.6 \text{ N} = (50 \text{ kg})\bar{a}$

$\bar{a} = +2.53 \text{ m/s}^2$     $\mathbf{\bar{a} = 2.53 \text{ m/s}^2 \rightarrow}$  ◄

$+\circlearrowleft \Sigma M_G = \bar{I}\alpha:$

$(73.6 \text{ N})(0.100 \text{ m}) - (200 \text{ N})(0.060 \text{ m}) = (0.245 \text{ kg·m}^2)\alpha$

$\alpha = -18.94 \text{ rad/s}^2$     $\mathbf{\alpha = 18.94 \text{ rad/s}^2 \circlearrowleft}$  ◄

**REFLECT and THINK:**  The wheel has larger linear and angular accelerations under conditions of rotating while sliding than when rolling without sliding.

## Sample Problem 16.11

Overhead cranes are often used to move large containers in shipyards. A simplified model of a 60,000-lb container and crane is shown. The uniform container is at rest when the connection at *B* fails. Determine the tension in the cable connecting the pulley to the container at *A*. ▶

**STRATEGY:**  Because you are asked to find a tension, use Newton's second law.

**MODELING:**  Start by choosing the container to be your system. After the connection at *B* fails, the only external forces acting on the container are the tension in the cable at *A* and the weight. A free-body diagram and kinetic diagram for this system immediately after the connection at *B* fails are shown in Fig. 1. Because the container is undergoing general plane motion, in the kinetic diagram you can represent the acceleration of the center of mass as having a vertical and a horizontal component.

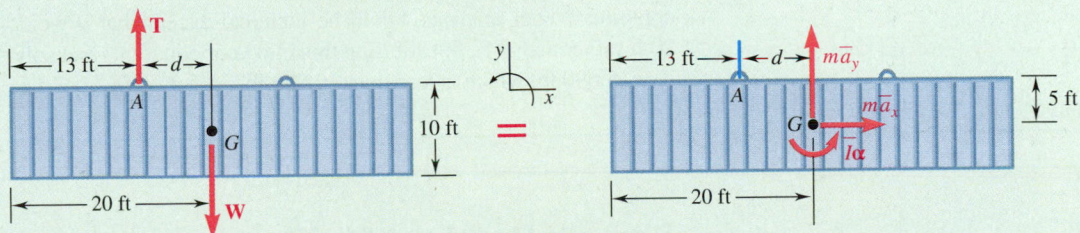

**Fig. 1**   Free-body diagram and kinetic diagram for the container.

**ANALYSIS:**  Using Fig. 1 and applying Newton's second law in the *x* direction and *y* direction and summing moments about point *G* gives you

$\xrightarrow{+} \Sigma F_x = m\bar{a}_x:$     $0 = m\bar{a}_x$  **(1)**

$+\uparrow \Sigma F_y = m\bar{a}_y:$     $T - W = m\bar{a}_y$  **(2)**

$+\circlearrowleft \Sigma M = \bar{I}\alpha:$     $-Td = \bar{I}\alpha$  **(3)**

*(continued)*

where

$$d = 7 \text{ ft}$$

$$m = \frac{W}{g} = \frac{60{,}000 \text{ lb}}{32.2 \text{ ft/s}^2} = 1863 \text{ lb} \cdot \text{s}^2/\text{ft}$$

$$\bar{I} = \tfrac{1}{12}m(b^2 + c^2) = \tfrac{1}{12}(1836 \text{ lb} \cdot \text{s}^2/\text{ft})[(40 \text{ ft})^2 + (10 \text{ ft})^2] = 26{,}400 \text{ lb} \cdot \text{ft} \cdot \text{s}^2$$

In Eqs. (1), (2), and (3), you have four unknowns: $T$, $\bar{a}_x$, $\bar{a}_y$, and $\alpha$. You can use kinematics to obtain additional equations. You want to relate the acceleration of the center of mass to that of another point on the container. At the instant the cable breaks, the angular velocity of the cable is zero, so point $A$ has no normal acceleration, but it has an acceleration perpendicular to the cable. The accelerations of $A$ and $G$ are related by

$$\mathbf{a}_G = \mathbf{a}_A + \mathbf{a}_{G/A} = \mathbf{a}_A + \boldsymbol{\alpha} \times \mathbf{r}_{G/A} - \omega^2 \mathbf{r}_{G/A}$$

Substituting in known values and letting $\boldsymbol{\omega} = 0$ and $\boldsymbol{\alpha} = \alpha\mathbf{k}$ gives you

$$\bar{a}_x\mathbf{i} + \bar{a}_y\mathbf{j} = a_A\mathbf{i} + \alpha\mathbf{k} \times (d\mathbf{i} - 5\mathbf{j}) - 0 = a_A\mathbf{i} + (d\alpha)\mathbf{j} + (5\alpha)\mathbf{i}$$

Equating components gives

**i:** $\quad \bar{a}_x = a_A + 5\alpha$ (4)

**j:** $\quad \bar{a}_y = d\alpha$ (5)

Solving Eqs. (1), (2), (3), (4), and (5) for $T$, $\bar{a}_y$, $\bar{a}_x$, $a_A$ and $\alpha$ gives you $T = 44{,}580$ lb, $\bar{a}_y = -8.275$ ft/s$^2$, $\bar{a}_x = 0$, $a_A = 5.911$ ft/s$^2$, and $\alpha = -1.182$ rad/s$^2$.

$$\mathbf{T} = 44{,}600 \text{ lb} \uparrow \quad \blacktriangleleft$$

**REFLECT and THINK:** You don't need all five equations to solve for the required unknowns; that is, you could have chosen to just use Eqs. (2), (3), and (5). The acceleration of the center of gravity is only in the vertical direction at the instant the cable breaks. When the container was at rest, the force in the cable at $A$ was 30,000 lb. The tension increased when the connection at $B$ failed. What would have happened if $A$ had been at the upper left edge of the container? Your analysis would be identical except that $d$ would be equal to 20 ft rather than 7 ft. Substituting this into your equations and solving gives you $T = 15{,}690$ lb, which is less than 30,000 lb.

## Sample Problem 16.12

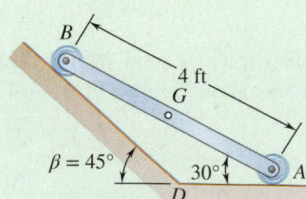

The ends of a 4-ft rod weighing 50 lb can move freely and with no friction along two straight tracks as shown. If the rod is released from rest at the position shown, determine (a) the angular acceleration of the rod, (b) the reactions at $A$ and $B$. ▶

**STRATEGY:** Because you are asked to determine forces and accelerations, use Newton's second law. The motion is constrained, so the acceleration of $G$ must be related to the angular acceleration $\boldsymbol{\alpha}$. To obtain this relation, first determine the magnitude of the acceleration of point $A$ in terms of $\alpha$.

*(continued)*

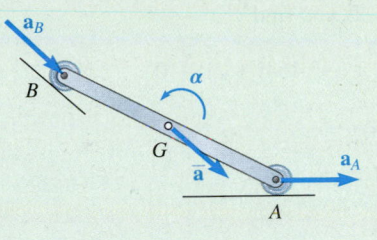

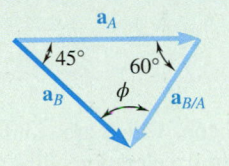

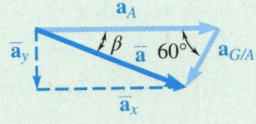

**Fig. 1** Vector diagrams for accelerations of points on the rod.

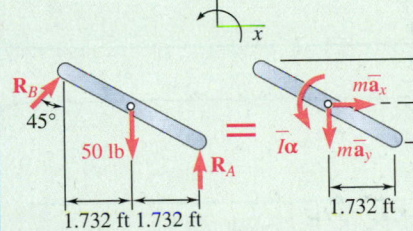

**Fig. 2** Free-body diagram and kinetic diagram for the rod assuming a downward acceleration.

**MODELING AND ANALYSIS:** Choose the rod to be your system and model it as a rigid body. Before drawing the kinetic diagram, you need to relate the acceleration of $G$ to the angular acceleration of the rod. You can do this using kinematics.

**Kinematics of Motion.** Assume that $\boldsymbol{\alpha}$ is directed counterclockwise. Noting that $a_{B/A} = 4\alpha$, you have (Fig. 1)

$$\mathbf{a}_B = \mathbf{a}_A + \mathbf{a}_{B/A}$$
$$[a_B \searrow 45°] = [a_A \rightarrow] + [4\alpha \nearrow 60°]$$

Noting that $\phi = 75°$ and using the law of sines, you obtain

$$a_A = 5.46\alpha \qquad a_B = 4.90\alpha$$

Now you can find the acceleration of $G$ from

$$\bar{\mathbf{a}} = \mathbf{a}_G = \mathbf{a}_A + \mathbf{a}_{G/A}$$
$$\bar{\mathbf{a}} = [5.46\alpha \rightarrow] + [2\alpha \nearrow 60°]$$

Resolving $\bar{\mathbf{a}}$ into $x$ and $y$ components, you obtain

$$\bar{a}_x = 5.46\alpha - 2\alpha \cos 60° = 4.46\alpha \qquad \bar{a}_x = 4.46\alpha \rightarrow$$
$$\bar{a}_y = -2\alpha \sin 60° = -1.732\alpha \qquad \bar{a}_y = 1.732\alpha \downarrow$$

**Kinetics of Motion.** Draw a free-body-diagram and kinetic diagram for your system (Fig. 2). Compute the following magnitudes.

$$\bar{I} = \tfrac{1}{12}ml^2 = \frac{1}{12}\frac{50 \text{ lb}}{32.2 \text{ ft/s}^2}(4 \text{ ft})^2 = 2.07 \text{ lb·ft·s}^2 \qquad \bar{I}\alpha = 2.07\alpha$$

$$m\bar{a}_x = \frac{50}{32.2}(4.46\alpha) = 6.93\alpha \qquad m\bar{a}_y = -\frac{50}{32.2}(1.732\alpha) = -2.69\alpha$$

**Equations of Motion.**

$+\circlearrowleft \Sigma M_B = \bar{I}\alpha + m\bar{a}d_\perp$:

$$R_A(4 \cos 30° \text{ ft}) - W(2\cos 30° \text{ft}) = \bar{I}\alpha + (m\bar{a}_x)(2\sin 30° \text{ ft})$$
$$- (m\bar{a}_y)(2\cos 30° \text{ ft})$$

$$R_A(3.464) - (50 \text{ lb})(1.732) = 2.07\alpha + (6.93\alpha)(1.000)$$
$$- (2.69\alpha)(1.732) \tag{1}$$

$\overset{+}{\rightarrow} \Sigma F_x = m\bar{a}_x$: $\qquad R_B \sin 45° = 69.3\alpha \tag{2}$

$+\uparrow \Sigma F_y = m\bar{a}_y$: $\qquad R_A + R_B \cos 45° - 50 = -2.69\alpha \tag{3}$

Solving these equations gives

$$\boldsymbol{\alpha} = 2.30 \text{ rad/s}^2 \circlearrowright \quad \blacktriangleleft$$
$$\mathbf{R}_B = 22.5 \text{ lb} \measuredangle 45° \quad \blacktriangleleft$$
$$\mathbf{R}_A = 27.9 \text{ lb} \uparrow \quad \blacktriangleleft$$

**REFLECT and THINK:** For the kinematics, you could have used the vector algebra approach rather than the method demonstrated in this example. Using the vector algebra approach, you can write

$$\mathbf{a}_B = \mathbf{a}_A + \alpha\mathbf{k} \times \mathbf{r}_{B/A} - \omega^2\mathbf{r}_{B/A}$$

*(continued)*

Substituting the directions assumed in Fig. 1, you find

$$\frac{a_B}{\sqrt{2}}\mathbf{i} - \frac{a_B}{\sqrt{2}}\mathbf{j} = a_A\mathbf{i} + \alpha\mathbf{k} \times (-3.464\mathbf{i} + 2\mathbf{j}) + 0$$

$$= a_A\mathbf{i} + (-3.464\alpha\mathbf{j} - 2\alpha\mathbf{i})$$

Equating components gives

i:  $\quad \dfrac{a_B}{\sqrt{2}} = a_A - 2\alpha$

j:  $\quad \dfrac{a_B}{\sqrt{2}} = -3.46\alpha$

Solving these, you find $a_B = 4.90\alpha$ and $a_A = 5.46\alpha$, which are similar to the approach shown previously. You can determine the acceleration of the center of gravity in terms of the angular acceleration using $\mathbf{a}_G = \mathbf{a}_A + \alpha\mathbf{k} \times r_{G/A} - \omega^2\mathbf{r}_{G/A}$. Substituting the directions assumed in Fig. 1, you find

$$\bar{a}_x\mathbf{i} + \bar{a}_y\mathbf{j} = a_A\mathbf{i} + \alpha\mathbf{k} \times (-1.732\mathbf{i} + 1\mathbf{j}) + 0 = a_A\mathbf{i} + (-1.732\alpha\mathbf{j} - 1\alpha\mathbf{i})$$

Equating components gives

i:  $\quad \bar{a}_x = a_A - 1\alpha = 4.46\alpha$
j:  $\quad \bar{a}_y = -1.732\alpha$

These are identical to the equations determined previously.

## Sample Problem 16.13

In the engine system from Sample Prob. 15.15, the crank $AB$ has a constant clockwise angular velocity of 2000 rpm. Knowing that the connecting rod $BD$ weighs 4 lb and the piston $P$ weighs 5 lb, determine the forces on the connecting rod at $B$ and $D$. Assume the center of mass of $BD$ is at its geometric center and it can be treated as a uniform, slender rod.

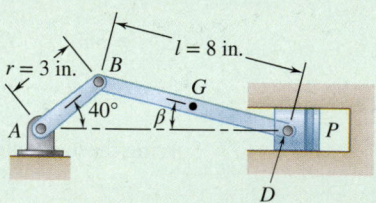

**STRATEGY:**  Because you are asked to find forces at the instant shown, use Newton's second law.

**MODELING:**  Because you want to determine the forces at $B$ and $D$, start by choosing the connecting rod $BD$ as your system. The pin forces at $B$ and $D$ are represented by horizontal and vertical components, and because the rod is undergoing general plane motion, you can represent the acceleration of the

*(continued)*

center of mass in the kinetic diagram as having a vertical and a horizontal component. The free-body and kinetic diagrams for this system are shown in Fig. 1, where $l = 8$ in. $= 0.6667$ ft and $\beta = 13.95°$.

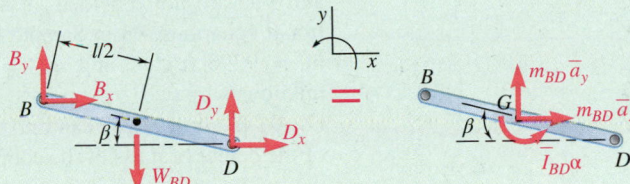

**Fig. 1** Free-body and kinetic diagrams for rod BD.

**ANALYSIS:** Using Fig. 1, applying Newton's second law in the $x$ direction and $y$ direction, and summing moments about point $G$ gives

$$\xrightarrow{+} \Sigma F_x = m\bar{a}_x: \quad B_x + D_x = m_{BD}\bar{a}_x \tag{1}$$

$$+\uparrow \Sigma F_y = m\bar{a}_y: \quad B_y + D_y - W_{BD} = m_{BD}\bar{a}_y \tag{2}$$

$$+\circlearrowleft \Sigma M_G = \bar{I}\alpha: \quad -B_y(l/2)\cos\beta - B_x(l/2)\sin\beta + D_y(l/2)\cos\beta$$
$$+ D_x(l/2)\sin\beta = \bar{I}_{BD}\alpha_{BD} \tag{3}$$

where

$$m_{BD} = \frac{W_{BD}}{g} = \frac{4 \text{ lb}}{32.2 \text{ ft/s}^2} = 0.1242 \text{ lb·s}^2/\text{ft}$$

$$\bar{I}_{BD} = \tfrac{1}{12}m_{BD}l^2 = \tfrac{1}{12}(0.1242 \text{ lb·s}^2/\text{ft})(0.6667 \text{ ft})^2 = 0.004601 \text{ lb·ft·s}^2$$

In Eqs. (1), (2), and (3), you have seven unknowns: $B_x$, $B_y$, $D_x$, $D_y$, $\bar{a}_x$, $\bar{a}_y$, and $\alpha_{BD}$. Therefore, you need more equations. You can get them from kinematics or by choosing another system. Choose the piston to be your system, model it as a particle, and draw its free-body and kinetic diagrams (Fig. 2).

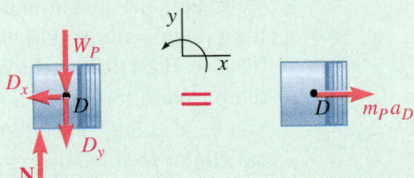

**Fig. 2** Free-body diagram and kinetic diagram for the piston.

Note that you must draw $D_x$ and $D_y$ in the opposite directions to what you drew for the connecting rod. Using Fig. 2 and applying Newton's second law in the $x$ direction and $y$ direction gives

$$\xrightarrow{+} \Sigma F_x = m\bar{a}_x: \quad -D_x = m_P a_D \tag{4}$$

$$+\uparrow \Sigma F_y = m\bar{a}_y: \quad -D_y + N - W_P = 0 \tag{5}$$

where

$$m_P = \frac{W_P}{g} = \frac{5 \text{ lb}}{32.2 \text{ ft/s}^2} = 0.1553 \text{ lb·s}^2/\text{ft}$$

*(continued)*

You now have five equations and nine unknowns: $N$, $B_x$, $B_y$, $D_x$, $D_y$, $\bar{a}_x$, $\bar{a}_y$, $\alpha_{BD}$, and $a_D$. You could choose crank $AB$ as another system, but because this will introduce three additional unknowns (the reactions at $A$ and the driving torque) and you are not provided its mass, you should turn to kinematics for additional equations. From Sample Prob. 15.15, you obtained $\omega_{BD} = 62.0$ rad/s $\circlearrowleft$, $a_D = 9290$ ft/s$^2\leftarrow$, and $\alpha_{BD} = 9940$ rad/s$^2$ $\circlearrowleft$. These reduce the number of unknowns by two, so you have five equations and seven unknowns: $N$, $B_x$, $B_y$, $D_x$, $D_y$, $\bar{a}_x$, and $\bar{a}_y$. You can find two more equations by relating the acceleration of the center of mass of the connecting rod to the acceleration of $D$,

$$\mathbf{a}_G = \mathbf{a}_D + \mathbf{a}_{G/D} = \mathbf{a}_D + \boldsymbol{\alpha} \times \mathbf{r}_{G/D} - \omega_{BD}^2 \mathbf{r}_{G/D}$$

Substituting in known and assumed values (Fig. 1) $\mathbf{a}_D = a_D \mathbf{i}$, where $a_D = -9290$ ft/s$^2$, and $\boldsymbol{\alpha}_{BD} = \alpha_{BD}\mathbf{k}$, where $\alpha_{BD} = 9940$ rad/s$^2$, gives

$$\bar{a}_x\mathbf{i} + \bar{a}_y\mathbf{j} = a_D\mathbf{i} + \alpha_{BD}\mathbf{k} \times [-\tfrac{l}{2}\cos\beta\mathbf{i} + \tfrac{l}{2}\sin\beta\mathbf{j}] - \omega_{BD}^2[-\tfrac{l}{2}\cos\beta\mathbf{i} + \tfrac{l}{2}\sin\beta\mathbf{j}]$$
$$= a_D\mathbf{i} - \alpha_{BD}\tfrac{l}{2}\cos\beta\mathbf{j} - \alpha_{BD}\tfrac{l}{2}\sin\beta\mathbf{i} + \omega_{BD}^2\tfrac{l}{2}\cos\beta\mathbf{i} - \omega_{BD}^2\tfrac{l}{2}\sin\beta\mathbf{j}$$

Equating components, you have

$$\mathbf{i}: \quad \bar{a}_x = a_D - \alpha_{BD}\tfrac{l}{2}\sin\beta + \omega_{BD}^2\tfrac{l}{2}\cos\beta \tag{6}$$

$$\mathbf{j}: \quad \bar{a}_y = -\alpha_{BD}\tfrac{l}{2}\cos\beta - \omega_{BD}^2\tfrac{l}{2}\sin\beta \tag{7}$$

You now have seven equations and seven unknowns. Substituting in numerical values and solving these equations using your calculator or software such as MathCad, Maple, Matlab, or Mathematica gives you $B_x = -2541$ lb, $B_y = 207.2$ lb, $D_x = 1442$ lb, $D_y = -641$ lb, $N = -636$ lb, $\bar{a}_x = -8845$ ft/s$^2$, and $\bar{a}_y = -3524$ ft/s$^2$.

| | |
|---|---|
| $\mathbf{B_x} = 2541$ lb $\leftarrow$ | $\mathbf{B_y} = 207$ lb $\uparrow$ $\blacktriangleleft$ |
| $\mathbf{D_x} = 1442$ lb $\rightarrow$ | $\mathbf{D_y} = 641$ lb $\downarrow$ $\blacktriangleleft$ |

**REFLECT and THINK:** The calculated forces are much larger than the weight of the piston and the connecting rod. This problem required multiple systems and rigid-body kinematics to solve, most of which was done in Sample Prob. 15.15. In problems like this, it is a good practice to focus on the problem formulation and to keep track of equations and unknowns. Once you have enough equations to solve for all the unknowns, using a computer or calculator to solve the resulting equations is often the easiest approach.

# Case Study 16.1

A biomechanics laboratory has been asked to investigate knee loads when athletes with prosthetic limbs use different exercise equipment, such as rowing machines (see CS Photo 16.1). A load cell measures forces applied to the right foot, and the velocity and acceleration of the seat are measured using a motion analysis system. You can use these measurements to determine the internal forces and moments at the knee during rowing.

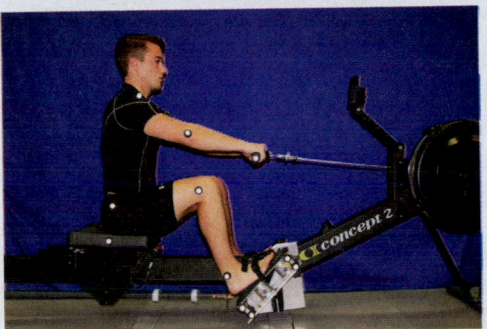

**CS Photo 16.1** Athlete with markers exercising on rowing machine. ©Katherine Mavrommati

**STRATEGY:** You will need to use rigid-body kinematics to determine the angular acceleration of the shank, as well as the linear acceleration of its center of mass. Then you will need to use Newton's second law to calculate the internal forces and moments acting on the knee joint.

**MODELING:** For the kinematic analysis, we can model the ankle as a hinge and the seat as a roller—so the analysis is basically just like a slider-crank mechanism as shown in CS Fig. 16.1. You can treat the thigh, *HK*, and shank, *KA*, as rigid bodies. To determine the knee loads, you should isolate the shank and draw its free-body and kinetic diagrams as in CS Fig. 16.2. The muscles that cross the joint can be modeled as a motor that applies an internal moment $\mathbf{M}_{int}$ to the knee, and the resultant joint forces from these muscles and from the contact within the knee joint can be broken into components $K_x$ and $K_y$.

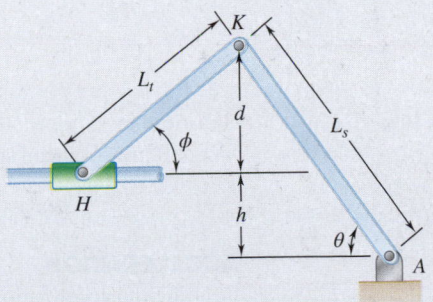

**CS Fig. 16.1** Schematic of rower's lower body.

*(continued)*

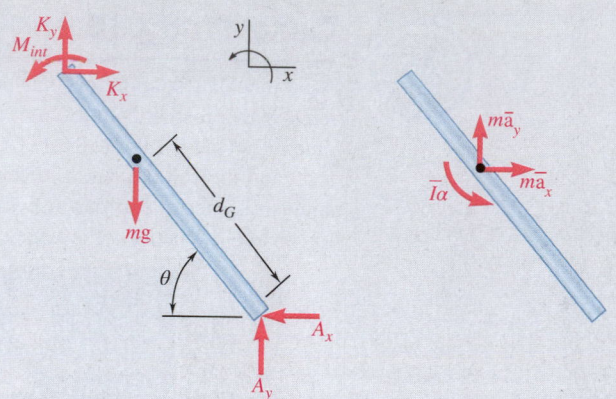

**CS Fig. 16.2** Free-body and kinetic diagrams of rower's shank.

**ANALYSIS:** Perform your analysis at the beginning of the push phase when the angle between the shank and the horizontal is 75°. At this instant, the load cells at the right foot read $A_x = 175$ N to the left and $A_y = 204$ N up and the seat is moving with a velocity $v_H = 506$ mm/s to the left with an acceleration of $a_H = 400$ m/s², also to the left. The length of the shaft and thigh are measured to be $L_s = 375$ mm and $L_t = 442$ mm, respectively, and the vertical distance between the ankle and hip is $h = 310$ mm. You will need to calculate the angular acceleration of the shank and the linear acceleration of its center of mass, so you should solve the kinematics of the linkage first.

**VELOCITY**

You can relate the velocity of the knee to the ankle:

$$\mathbf{v}_K = \mathbf{v}_A + \omega_{KA}\mathbf{k} \times \mathbf{r}_{K/A} = 0 + \omega_{KA}\mathbf{k} \times (-L_s \cos\theta\mathbf{i} + L_s \sin\theta\mathbf{j})$$
$$= -L_s \sin\theta\omega_{KA}\mathbf{i} - L_s \cos\theta\omega_{KA}\mathbf{j}$$

From the geometry you can find the angle $\phi = \arcsin\left(\dfrac{L_s \sin\theta - h}{L_t}\right)$. Moving onto link *HK,* you can relate $\mathbf{v}_K$ to $\mathbf{v}_H$:

$$\mathbf{v}_H = \mathbf{v}_K + \omega_{HK}\mathbf{k} \times \mathbf{r}_{H/K}$$
$$= -L_s \sin\theta\omega L_{KA}\mathbf{i} - L_s \cos\theta\omega L_{KA}\mathbf{j} + L_t \sin\phi\omega L_{HK}\mathbf{i} - L_t \cos\phi\omega L_{HK}\mathbf{j}$$

Realizing that $\mathbf{v}_H$ is only in the *x* direction, you can equate components to get

$$v_H = -L_s \sin\theta\omega_{KA} + L_t \sin\phi\omega_{HK}$$

$$0 = -L_s \cos\theta\omega_{KA} - L_t \cos\phi\omega_{HK}$$

Substituting in $v_H = -0.506$ m/s, $\theta = 75°$, $L_s = 375$ mm, $L_t = 442$ mm, and $h = 310$ mm, you can solve for $\omega_{KA}$ and $\omega_{HK}$.

$$\omega_{KA} = 1.354 \text{ rad/s} \qquad \omega_{HK} = -0.299 \text{ rad/s}$$

**ACCELERATION**

Relating the acceleration of the knee to the ankle, you get

$$\mathbf{a}_K = \mathbf{a}_A + \alpha_{KA}\mathbf{k} \times \mathbf{r}_{K/A} - \omega_{KA}^2\mathbf{r}_{K/A} = 0 + \alpha_{KA}\mathbf{k} \times (-L_s \cos\theta\mathbf{i} + L_s \sin\theta\mathbf{j})$$
$$- \omega_{KA}^2(-L_s \cos\theta\mathbf{i} + L_s \sin\theta\mathbf{j})$$
$$= -L_s \sin\theta\alpha L_{KA}\mathbf{i} - L_s \cos\theta\alpha L_{KA}\mathbf{j} + \omega_{KA}^2 L_s \cos\theta\mathbf{i} - \omega_{KA}^2 L_s \sin\theta\mathbf{j}$$

*(continued)*

Similarly, you can relate the hip to the knee:

$$\mathbf{a}_H = \mathbf{a}_K + \alpha_{HK}\mathbf{k} \times \mathbf{r}_{H/K} - \omega_{HK}^2 \mathbf{r}_{H/K}$$
$$= \mathbf{a}_K + \alpha_{HK}\mathbf{k} \times (-L_t \cos\phi\mathbf{i} - L_t \sin\phi\mathbf{j}) - \omega_{HK}^2(-L_t \cos\phi\mathbf{i} - L_t \sin\phi\mathbf{j})$$
$$\mathbf{a}_H\mathbf{i} = -L_s \sin\theta\alpha_{KA}\mathbf{i} - L_s \cos\theta\alpha_{KA}\mathbf{j} + \omega_{KA}^2 L_s \cos\theta\mathbf{i} - \omega_{KA}^2 L_s \sin\theta\mathbf{j}$$
$$+ L_t \sin\phi\alpha_{HK}\mathbf{i} - L_t \cos\phi\alpha_{HK}\mathbf{j} + \omega_{HK}^2 L_t \cos\phi\mathbf{i} + \omega_{HK}^2 L_t \sin\phi\mathbf{j}$$

Equating components like you did for velocity and substituting $a_H = -0.4$ m/s$^2$, you can solve for $\alpha_{AK}$ and $\alpha_{KH}$:

$$\alpha_{KA} = 1.441 \text{ rad/s}^2 \qquad \alpha_{HK} = -1.821 \text{ rad/s}^2$$

Note that you will also need to solve for the acceleration of the center of mass to use in Newton's second law. You can use some classic data collected by Dempster to determine the location of the center of mass of the shank, which is approximately $d_G = 212$ mm from the ankle. The acceleration of the center of mass of the shank can be related to the acceleration of the ankle:

$$\mathbf{a}_G = \mathbf{a}_A + \alpha_{KA}\mathbf{k} \times \mathbf{r}_{G/A} - \omega_{KA}^2 \mathbf{r}_{G/A}$$
$$= 0 + \alpha_{KA}\mathbf{k} \times (-d_G \cos\theta\mathbf{i} + d_G \sin\theta\mathbf{j}) - \omega_{KA}^2(-d_G \cos\theta\mathbf{i} + d_G \sin\theta\mathbf{j})$$
$$= -d_G \sin\theta\alpha_{KA}\mathbf{i} - d_G \cos\theta\alpha_{KA}\mathbf{j} + \omega_{KA}^2 d_G \cos\theta\mathbf{i} - \omega_{KA}^2 d_G \sin\theta\mathbf{j} \qquad \text{(1)}$$

Substituting in values gives you

$$\bar{\mathbf{a}} = -(0.194 \text{ m/s}^2)\,\mathbf{i} - (0.454 \text{ m/s}^2)\,\mathbf{j}$$

**KINETICS**

You can again use data developed by Dempster to estimate the mass and mass moments of inertia for different body segments. For the 61-kg, 170-cm-tall subject, the mass and mass moment of inertia about the ankle of the shank can be estimated as 2.84 kg and 0.165 kg·m$^2$, respectively.

Because you have the acceleration of the mass center of the shank, you can determine the joint forces at the knee.

$$\xrightarrow{+} \Sigma F_x = m\bar{a}_x: \qquad -A_x + K_x = m\bar{a}_x$$
$$K_x = m\bar{a}_x + A_x \qquad \text{(2)}$$

$$+\uparrow \Sigma F_y = m\bar{a}_y: \qquad A_y + K_y - mg = m\bar{a}_y$$
$$K_y = m\bar{a}_y + mg - A_y \qquad \text{(3)}$$

You can choose to sum moments about $G$ or about $A$. Because we are modeling the ankle as a fixed axis of rotation, you can use $\Sigma M_A = I_A \alpha_{KA}$.

$$\Sigma M_A = I_A\alpha_{KA}: \qquad mgd_G \cos\theta - K_x L_s \sin\theta - K_y L_s \cos\theta + M_{\text{int}} = I_A\alpha_{KA}$$
$$M_{\text{int}} = I_A\alpha_{KA} - mgd_G \cos\theta + K_x L_s \sin\theta + K_y L_s \cos\theta \qquad \text{(4)}$$

Substituting values from Eq. (1) into Eqs. (2) and (3) gives you the knee forces: $K_x = 174.4$ N, $K_y = -177.4$ N. Using these values in Eq. (4) gives you the internal moment: $M_{\text{int}} = -78.6$ N·m.

**REFLECT AND THINK:** You would want to plot the internal loads throughout the full exercise cycle to determine where dangerous loading patterns might occur, and you would probably want to obtain the knee forces in a coordinate system that is lined up with the long axis of the knee. Computer programs are used to analyze multiple subjects and automate the process. Analyses such as these are used to improve sports performance, to determine the efficacy of knee braces and orthoses, and to help plan surgical interventions in children with developmental disabilities. Although most analyses involve three-dimensional motion, many activities can be simplified to occur mainly in one plane.

# SOLVING PROBLEMS
# ON YOUR OWN

In this section, we considered the **plane motion of rigid bodies under constraints**. We found that the types of constraints involved in engineering problems vary widely. For example, a rigid body may be constrained to rotate about a fixed axis or to roll on a given surface, or it may be pin-connected to collars or to other bodies.

**1. Your solution of a problem involving the constrained motion of a rigid body** consists, in general, of three steps. First, you should model your system by drawing the free-body diagram and the kinetic diagram. Second, use these diagrams to write out your equations of motion. Finally, you will generally need to consider the *kinematics of the motion* to have enough equations to solve the problem. Sometimes it is helpful to examine the kinematics first to help you draw the kinetic diagram and choose an appropriate coordinate system.

**2. Free-body diagram and kinetic diagram.** Your first step in the solution of a problem is to draw a *free-body diagram* and a *kinetic diagram*.

    **a. A free-body diagram** shows *the forces exerted on the body,* including the applied forces, the reactions at the supports, and the weight of the body.

    **b. A kinetic diagram** shows the *inertial terms:* vector $m\bar{\mathbf{a}}$ and couple $\bar{I}\boldsymbol{\alpha}$.

**3. Using your free-body diagram and kinetic diagram, generate the equations of motion for the system.** Drawing good free-body and kinetic diagrams will allow you to *sum components in any direction and to sum moments about any point.* For a single body, you can obtain a maximum of three independent equations (two translational and one moment) that can be used to help analyze the system.

$$\Sigma F_x = m\bar{a}_x \qquad \Sigma F_y = m\bar{a}_y$$

$$\Sigma M_G = \bar{I}\alpha \quad \text{or} \quad \Sigma M_O = I_O\alpha \quad \text{or} \quad \Sigma M_P = \bar{I}\alpha + m\bar{a}d_\perp \quad \text{or} \quad \Sigma M_P = \bar{I}\alpha + \mathbf{r}_{G/P} \times m\bar{\mathbf{a}}$$

where $G$ is the center of mass of the body, $O$ is a fixed axis of rotation, $P$ is any arbitrary point, and $d_\perp$ is the perpendicular distance between point $P$ and the line of action of the acceleration of the center of mass.

**4. The kinematic analysis of the motion** uses the methods you learned in Chap. 15. Due to the constraints, linear and angular accelerations are related. You should establish relationships among the accelerations (angular as well as linear), and your goal should be to express all accelerations in terms of a single unknown acceleration.

    **a. For a body in noncentroidal rotation about a fixed axis,** the components of the acceleration of the mass center are $\bar{a}_t = \bar{r}\alpha$ and $\bar{a}_n = \bar{r}\omega^2$, where $\omega$ is generally known (Sample Probs. 16.7 and 16.8).

    **b. For a rolling disk or wheel,** the acceleration of the geometric center is $\bar{a} = r\alpha$ (Sample Prob. 16.9).

**c. For a body in general plane motion,** your best course of action if neither $\bar{a}$ nor $\alpha$ is known or readily obtainable is to express $\bar{a}$ in terms of $\alpha$ (Sample Probs. 16.10, 16.11, 16.12, and 16.13). This can be done by relating the acceleration of the center of mass to some reference point:

$$\bar{\mathbf{a}} = \mathbf{a}_A + \alpha\mathbf{k} \times \mathbf{r}_{G/A} - \omega^2\mathbf{r}_{G/A}$$

**5. When solving problems involving rolling disks or wheels,** keep in mind the following situations.

**a. If sliding is impending,** the friction force exerted on the rolling body has reached its maximum value, so $F_m = \mu_s N$, where $N$ is the normal force exerted on the body and $\mu_s$ is the coefficient of static friction between the surfaces of contact.

**b. If there is rolling without sliding,** the friction force $F$ can have any value smaller than $F_m$ and therefore should be considered an independent unknown. After you have determined $F$, be sure to check that it is smaller than $F_m$; if it is not, the body does not roll, but rotates and slides as described in the next paragraph.

**c. If the body rotates and slides at the same time,** then the body is not rolling, and the acceleration $\bar{a}$ of the mass center is *independent* of the angular acceleration $\alpha$ of the body: $\bar{a} \neq r\alpha$. On the other hand, the friction force has a well-defined value, $F = \mu_k N$, where $\mu_k$ is the coefficient of kinetic friction between the surfaces of contact.

**d. For an unbalanced rolling disk or wheel,** the relation $\bar{a} = r\alpha$ between the acceleration $\bar{a}$ of the mass center $G$ and the angular acceleration $\alpha$ of the disk or wheel does not hold any more. However, a similar relation holds between the acceleration $a_O$ of the geometric center $O$ and the angular acceleration $\alpha$ of the disk or wheel: $a_O = r\alpha$. This relation can be used to express $\bar{a}$ in terms of $\alpha$ and $\omega$ (Fig. 16.17).

**6. For a system of connected rigid bodies,** the goal of your kinematic analysis should be to determine all the accelerations from the given data or to express them all in terms of a single unknown. For systems with several degrees of freedom, you will need to use as many unknowns as there are degrees of freedom.

Your kinetic analysis will sometimes be carried out by drawing a free-body diagram and a kinetic diagram for the entire system. If you only have three unknowns, this is usually the best approach. In most cases, however, it will be necessary to analyze each rigid body separately in order to obtain enough equations to solve for all the unknown quantities in the problem.

# Problems

**16.CQ4** A cord is attached to a spool when a force **P** is applied to the cord as shown. Assuming the spool rolls without slipping, what direction does the spool move for each case?

Case 1: **a.** left   **b.** right   **c.** It would not move.
Case 2: **a.** left   **b.** right   **c.** It would not move.
Case 3: **a.** left   **b.** right   **c.** It would not move.

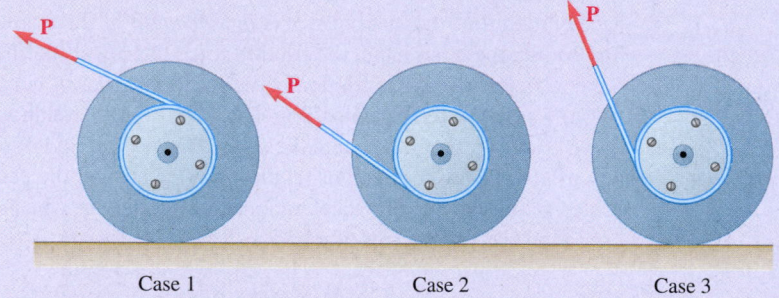

Case 1          Case 2          Case 3

**Fig. P16.CQ4 and P16.CQ5**

**16.CQ5** A cord is attached to a spool when a force **P** is applied to the cord as shown. Assuming the spool rolls without slipping, in what direction does the friction force act for each case?

Case 2: **a.** left   **b.** right   **c.** The friction force would be zero.
Case 3: **a.** left   **b.** right   **c.** The friction force would be zero.

**16.CQ6** A front-wheel-drive car starts from rest and accelerates to the right. Knowing that the tires do not slip on the road, what is the direction of the friction force the road applies to the front tires?

**a.** left
**b.** right
**c.** The friction force is zero.

**16.CQ7** A front-wheel-drive car starts from rest and accelerates to the right. Knowing that the tires do not slip on the road, what is the direction of the friction force the road applies to the rear tires?

**a.** left
**b.** right
**c.** The friction force is zero.

**16.F5**  A uniform 6 × 8-in. rectangular plate of mass *m* is pinned at *A*. Knowing the angular velocity of the plate at the instant shown is **ω**, draw the FBD and KD.

**16.F6**  Two identical 4-lb slender rods *AB* and *BC* are connected by a pin at *B* and by the cord *AC*. The assembly rotates in a vertical plane under the combined effect of gravity and a couple **M** applied to rod *AB*. Knowing that in the position shown the angular velocity of the assembly is **ω**, draw the FBD and KD that can be used to determine the angular acceleration of the assembly.

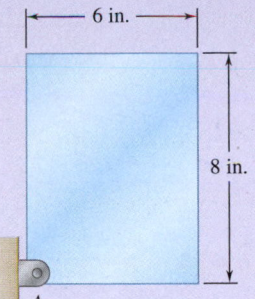

Fig. P16.F5

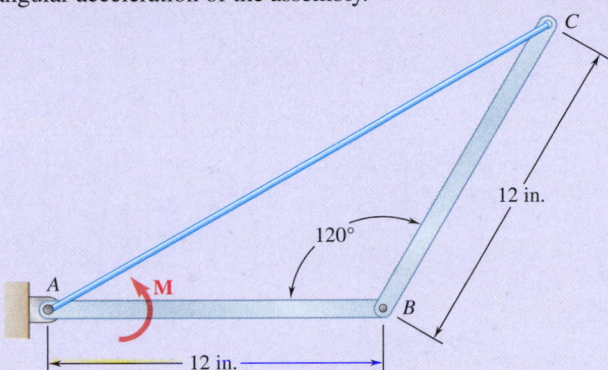

Fig. P16.F6

**16.F7**  The 4-lb uniform rod *AB* is attached to collars of negligible mass that slide without friction along the fixed rods shown. Rod *AB* is at rest in the position $\theta = 25°$ when a horizontal force **P** is applied to collar *A* causing it to start moving to the left. Draw the FBD and KD for the rod.

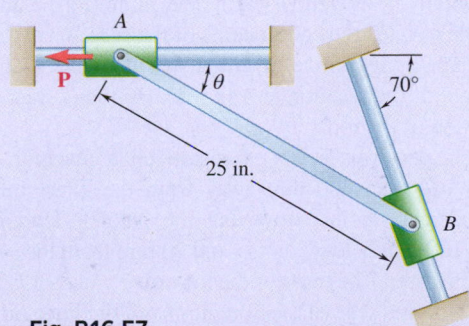

Fig. P16.F7

**16.F8**  A uniform disk of mass *m* = 4 kg and radius *r* = 150 mm is supported by a belt *ABCD* that is bolted to the disk at *B* and *C*. If the belt suddenly breaks at a point located between *A* and *B*, draw the FBD and KD for the disk immediately after the break.

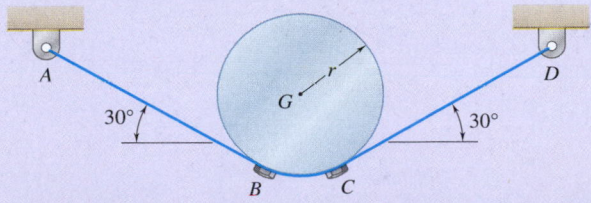

Fig. P16.F8

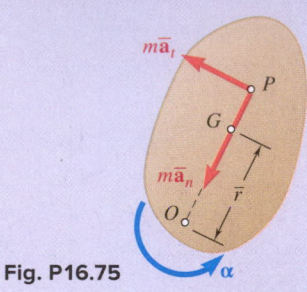

**Fig. P16.75**

**16.75** Show that the couple $\bar{I}\alpha$ of Fig. 16.15 can be eliminated by attaching the vectors $m\bar{a}_t$ and $m\bar{a}_n$ at a point $P$ called the *center of percussion*, located on line $OG$ at a distance $GP = \bar{k}^2/\bar{r}$ from the mass center of the body.

**16.76** A uniform slender rod of length $L = 900$ mm and mass $m = 4$ kg is suspended from a hinge at $C$. Knowing that a horizontal force **P** of magnitude 75 N is applied at end $B$, determine (a) the distance $\bar{r}$ for which the horizontal component of the reaction at $C$ is zero, (b) the corresponding angular acceleration of the rod.

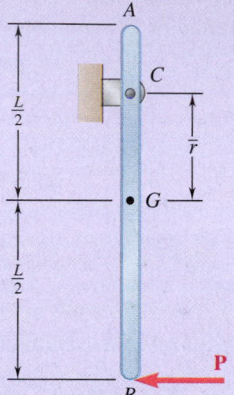

**Fig. P16.76**

**16.77** A crate of mass 80 kg is held in the position shown. Immediately after the man accidentally drops the right end of the crate, assuming that the box does not slip at point $E$, determine (a) the angular acceleration of the crate, (b) the reaction at point $E$.

**16.78** A uniform slender rod of length $L = 36$ in. and weight $W = 4$ lb hangs freely from a hinge at $A$. If a force **P** of magnitude 1.5 lb is applied at $B$ horizontally to the left ($h = L$), determine (a) the angular acceleration of the rod, (b) the components of the reaction at $A$.

**16.79** In Prob. 16.78, determine (a) the distance $h$ for which the horizontal component of the reaction at $A$ is zero, (b) the corresponding angular acceleration of the rod.

**16.80** An athlete performs a leg extension on a machine using a 20-kg mass at $A$ located 400 mm away from the knee joint at center $O$. Biomechanical studies show that the patellar tendon inserts at $B$, which is 100 mm below point $O$ and 20 mm from the center line of the tibia (see figure). The mass of the lower leg and foot is 5 kg, the center of gravity of this segment is 300 mm from the knee, and the radius of gyration about the knee is 350 mm. Knowing that the leg is moving at a constant angular velocity of 30 degrees per second when $\theta = 60°$, determine (a) the force **F** in the patellar tendon, (b) the magnitude of the joint force at the knee joint center $O$.

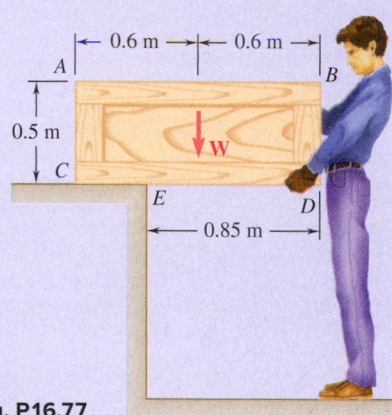

**Fig. P16.77**

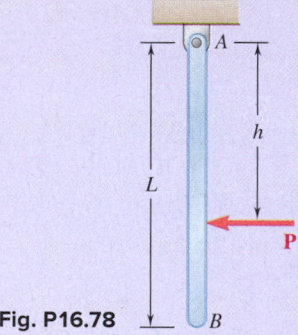

**Fig. P16.78**

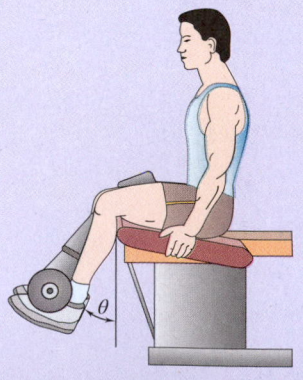

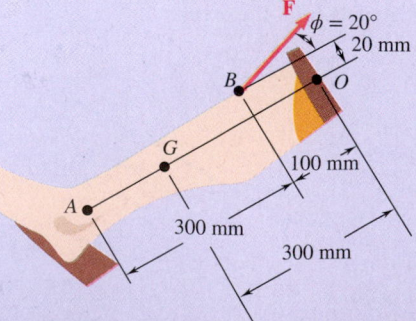

**Fig. P16.80**

**16.81** The shutter shown was formed by removing one quarter of a disk of 0.75-in. radius and is used to interrupt a beam of light emanating from a lens at C. Knowing that the shutter weighs 0.125 lb and rotates at the constant rate of 24 cycles per second, determine the magnitude of the force exerted by the shutter on the shaft at A.

**16.82** A turbine disk weighing 50 lb rotates at a constant rate of 9000 rpm. Knowing that the mass center of the disk coincides with the center of rotation O, determine the reaction at O immediately after a single blade at A, of weight 2 oz, becomes loose and is thrown off.

**16.83** The 80-lb tailgate of a car is supported by the hydraulic lift BC. If BC fails and the tailgate falls and hits the car with an angular velocity of 5 rad/s, determine the reactions at A immediately before the impact. The tailgate can be modeled as a 22-in.-long slender rod.

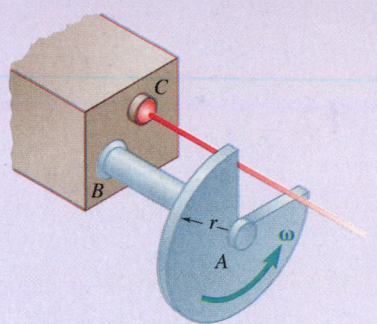

Fig. *P16.81*

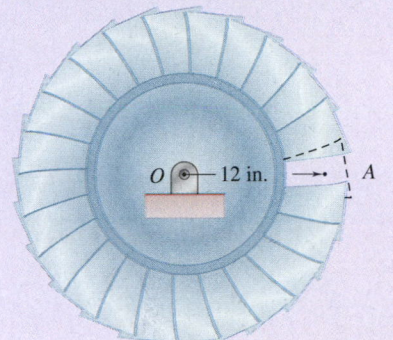

Fig. *P16.82*

Fig. P16.83

**16.84** A uniform rod of length L and mass m is supported as shown. If the cable attached at end B suddenly breaks, determine (a) the acceleration of end B, (b) the reaction at the pin support.

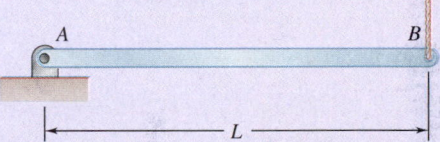

Fig. P16.84

**16.85** Three stage lights are mounted on a pipe fixture as shown. The lights at A and B each weigh 4.1 lb, while the one at C weighs 3.5 lb. The fixture weighs 6 lb and has a centroidal radius of gyration of 24 in. The centroid of the pipe can be assumed to be 42 in. from point D. If the connection at E fails so that the lights and fixture rotate about point D, determine the angular acceleration of the system immediately after the failure at E. Assume point D can be treated as a pinned connection and the lights can be modeled as particles.

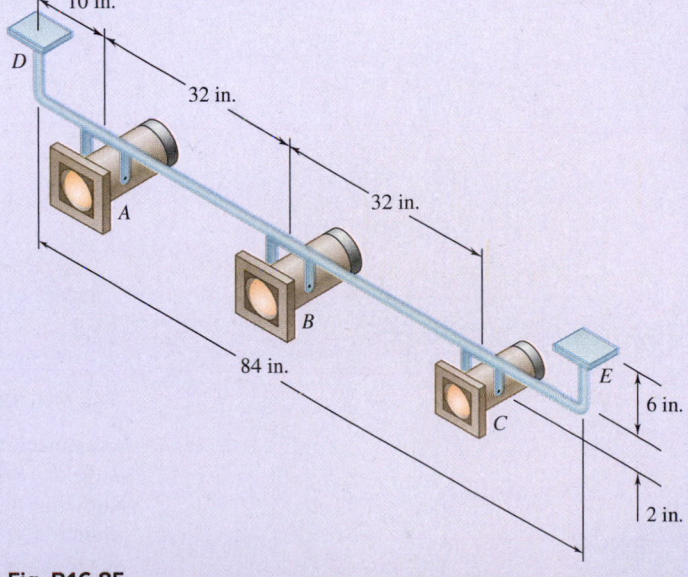

Fig. P16.85

**16.86** An adapted launcher uses a torsional spring about point $O$ to help people with mobility impairments throw a Frisbee®. Just after the Frisbee® leaves the arm, the angular velocity of the throwing arm is 200 rad/s and its acceleration is 10 rad/s²; both are counterclockwise. The rotation point $O$ is located 1 in. from the two sides. Assume that you can model the 2-lb throwing arm as a uniform rectangle. Just after the Frisbee® leaves the arm, determine (a) the moment about $O$ caused by the spring, (b) the forces on the pin at $O$.

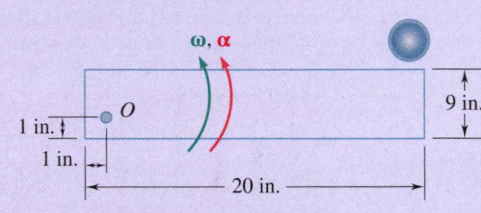

**Fig. P16.86**

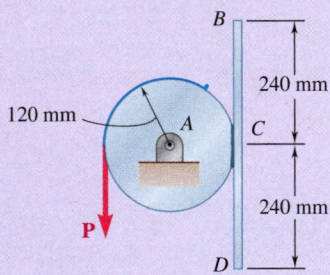

**Fig. P16.87**

**16.87** A 4-kg slender rod is welded to the edge of a 3-kg uniform disk as shown. The assembly rotates about $A$ in a vertical plane under the combined effect of gravity and of the vertical force **P**. Knowing that at the instant shown the assembly has an angular velocity of 12 rad/s and an angular acceleration of 36 rad/s², both counterclockwise, determine (a) the force **P**, (b) the components of the reaction at $A$.

**16.88** Two identical 4-lb slender rods $AB$ and $BC$ are connected by a pin at $B$ and by the cord $AC$. The assembly rotates in a vertical plane under the combined effect of gravity and a 6-lb·ft couple **M** applied to rod $AB$. Knowing that in the position shown the angular velocity of the assembly is zero, determine (a) the angular acceleration of the assembly, (b) the tension in cord $AC$.

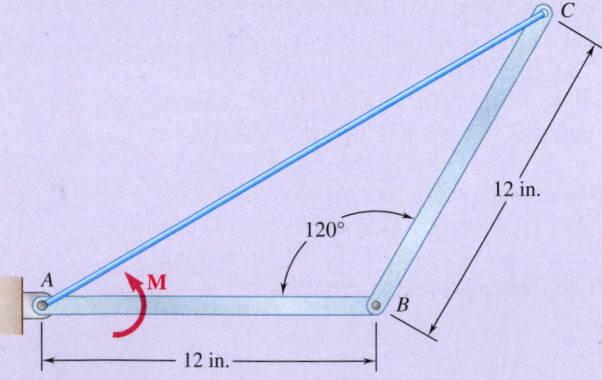

**Fig. P16.88**

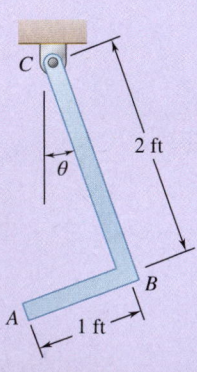

**Fig. P16.89**

**16.89** The object $ABC$ consists of two slender rods welded together at point $B$. Rod $AB$ has a weight of 2 lb and bar $BC$ has a weight of 4 lb. Knowing the magnitude of the angular velocity of $ABC$ is 10 rad/s when $\theta = 0°$, determine the components of the reaction at point $C$ at this location.

**16.90** A 3.5-kg slender rod $AB$ and a 2-kg slender rod $BC$ are connected by a pin at $B$ and by the cord $AC$. The assembly can rotate in a vertical plane under the combined effect of gravity and a couple $\mathbf{M}$ applied to rod $BC$. Knowing that in the position shown the angular velocity of the assembly is zero and the tension in cord $AC$ is equal to 25 N, determine (a) the angular acceleration of the assembly, (b) the magnitude of the couple $\mathbf{M}$.

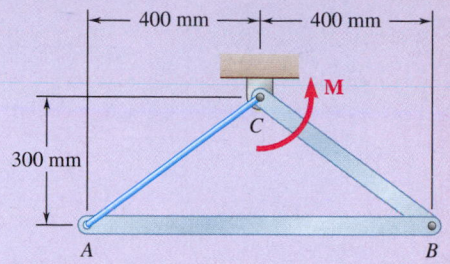

**Fig. P16.90**

**16.91** A 9-kg uniform disk is attached to the 5-kg slender rod $AB$ by means of frictionless pins at $B$ and $C$. The assembly rotates in a vertical plane under the combined effect of gravity and of a couple $\mathbf{M}$ that is applied to rod $AB$. Knowing that at the instant shown the assembly has an angular velocity of 6 rad/s and an angular acceleration of 25 rad/s$^2$, both counterclockwise, determine (a) the couple $\mathbf{M}$, (b) the force exerted by pin $C$ on member $AB$.

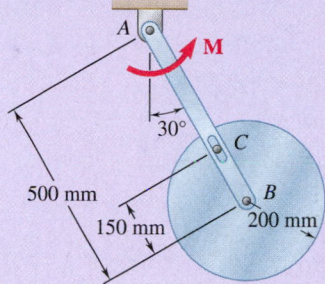

**Fig. P16.91**

**16.92** Derive the equation $\Sigma M_C = I_C \alpha$ for the rolling disk of Fig. 16.16, where $\Sigma M_C$ represents the sum of the moments of the external forces about the instantaneous center $C$, and $I_C$ is the moment of inertia of the disk about $C$.

**16.93** Show that in the case of an unbalanced disk, the equation derived in Prob. 16.92 is valid only when the mass center $G$, the geometric center $O$, and the instantaneous center $C$ happen to lie in a straight line.

**16.94** A wheel of radius $r$ and centroidal radius of gyration $\overline{k}$ is released from rest on the incline and rolls without sliding. Derive an expression for the acceleration of the center of the wheel in terms of $r$, $\overline{k}$, $\beta$, and $g$.

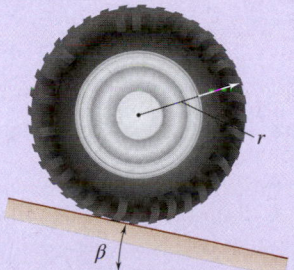

**Fig. P16.94**

**16.95** A homogeneous sphere $S$, a uniform cylinder $C$, and a thin pipe $P$ are in contact when they are released from rest on the incline shown. Knowing that all three objects roll without slipping, determine, after 4 s of motion, the distance between (a) the pipe and the cylinder, (b) the cylinder and the sphere.

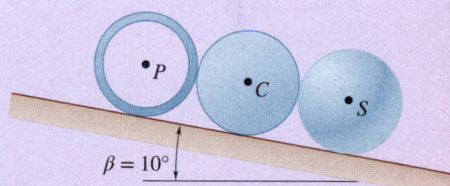

**Fig. P16.95**

**16.96** A 40-kg flywheel of radius $R = 0.5$ m is rigidly attached to a shaft of radius $r = 0.05$ m that can roll along parallel rails. A cord is attached as shown and pulled with a force $\mathbf{P}$ of magnitude 150 N. Knowing the centroidal radius of gyration is $\overline{k} = 0.4$ m, determine (a) the angular acceleration of the flywheel, (b) the velocity of the center of gravity after 5 s.

**16.97** A 40-kg flywheel of radius $R = 0.5$ m is rigidly attached to a shaft of radius $r = 0.05$ m that can roll along parallel rails. A cord is attached as shown and pulled with a force $\mathbf{P}$. Knowing the centroidal radius of gyration is $\overline{k} = 0.4$ m and the coefficient of static friction is $\mu_s = 0.4$, determine the largest magnitude of force $\mathbf{P}$ for which no slipping will occur.

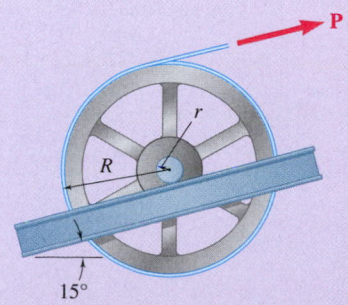

**Fig. P16.96 and P16.97**

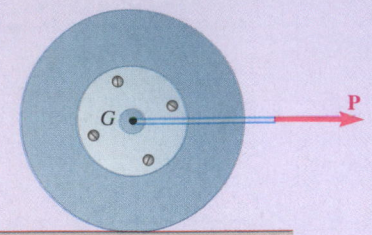

**Fig. P16.98 and P16.102**

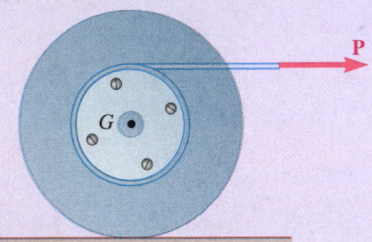

**Fig. P16.99 and P16.103**

**16.98 through *16.101*** A drum of 80-mm radius is attached to a disk of 160-mm radius. The disk and drum have a combined mass of 5 kg and a combined radius of gyration of 120 mm. A cord is attached as shown and pulled with a force **P** of magnitude 20 N. Knowing that the coefficients of static and kinetic friction are $\mu_s = 0.25$ and $\mu_k = 0.2$, respectively, determine (*a*) whether or not the disk slides, (*b*) the angular acceleration of the disk, (*c*) the acceleration of *G*.

**16.102 through 16.105** A drum of 4-in. radius is attached to a disk of 8-in. radius. The disk and drum have a total weight of 10 lb and a combined radius of gyration of 6 in. A cord is attached as shown and pulled with a force **P** of magnitude 5 lb. Knowing that the disk rolls without sliding, determine (*a*) the angular acceleration of the disk and the acceleration of *G*, (*b*) the minimum value of the coefficient of static friction compatible with this motion.

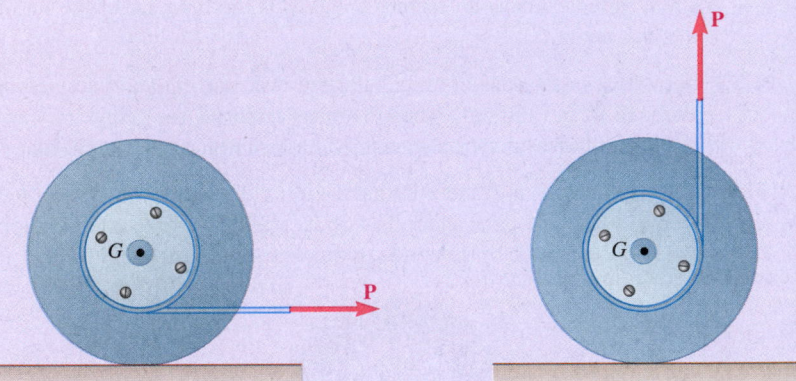

**Fig. *P16.100* and P16.104**          **Fig. *P16.101* and P16.105**

**16.106 and 16.107** A 12-in.-radius cylinder of weight 16 lb rests on a 6-lb carriage. The system is at rest when a force **P** of magnitude 4 lb is applied. Knowing that the cylinder rolls without sliding on the carriage and neglecting the mass of the wheels of the carriage, determine (*a*) the acceleration of the carriage, (*b*) the acceleration of point *A*, (*c*) the distance the cylinder has rolled with respect to the carriage after 0.5 s.

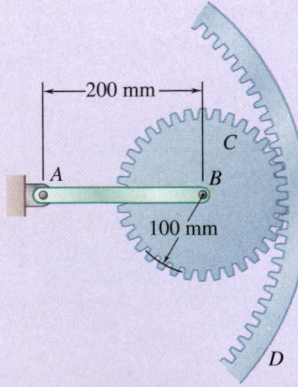

**Fig. P16.108**

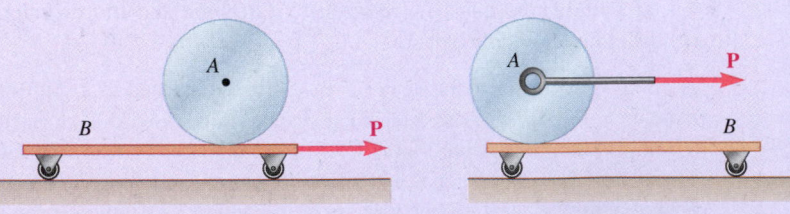

**Fig. *P16.106***          **Fig. P16.107**

**16.108** Gear *C* has a mass of 5 kg and a centroidal radius of gyration of 75 mm. The uniform bar *AB* has a mass of 3 kg and gear *D* is stationary. If the system is released from rest in the position shown, determine (*a*) the angular acceleration of gear *C*, (*b*) the acceleration of point *B*.

**16.109** Two uniform disks $A$ and $B$, each with a mass of 2 kg, are connected by a 2.5-kg rod $CD$ as shown. A counterclockwise couple $\mathbf{M}$ of moment 2.25 N·m is applied to disk $A$. Knowing that the disks roll without sliding, determine (a) the acceleration of the center of each disk, (b) the horizontal component of the force exerted on disk $B$ by pin $D$.

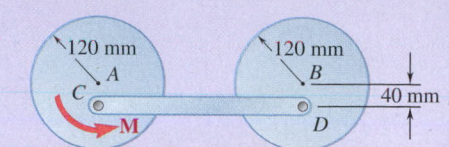

Fig. P16.109

**16.110** A single-axis personal transport device starts from rest with the rider leaning slightly forward. Together, the two wheels weigh 25 lbs, and each has a radius of 10 in. The mass moment of inertia of the wheels about the axle is 0.15 slug·ft². The combined weight of the rest of the device and the rider (excluding the wheels) is 200 lbs, and the center of gravity $G$ of this weight is located at $x = 4$ in. in front of axle $A$ and $y = 36$ in. above the ground. An initial clockwise torque $\mathbf{M}$ is applied by the motor to the wheels. Knowing that the coefficients of static and kinetic friction are 0.7 and 0.6, respectively, determine (a) the torque $\mathbf{M}$ that will keep the rider in the same angular position, (b) the corresponding linear acceleration of the rider.

**16.111** A hemisphere of weight $W$ and radius $r$ is released from rest in the position shown. Determine (a) the minimum value of $\mu_s$ for which the hemisphere starts to roll without sliding, (b) the corresponding acceleration of point $B$. (*Hint:* Note that $OG = \frac{3}{8}r$ and that, by the parallel-axis theorem, $\bar{I} = \frac{2}{5}mr^2 - m(OG)^2$.)

Fig. P16.110

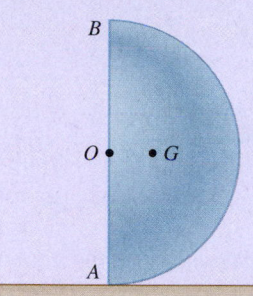

Fig. P16.111

**16.112** Solve Prob. 16.111, considering a half cylinder instead of a hemisphere. (*Hint:* Note that $OG = 4r/3\pi$ and that, by the parallel-axis theorem, $\bar{I} = \frac{1}{2}mr^2 - m(OG)^2$.)

**16.113** The center of gravity $G$ of a 1.5-kg unbalanced tracking wheel is located at a distance $r = 18$ mm from its geometric center $B$. The radius of the wheel is $R = 60$ mm and its centroidal radius of gyration is 44 mm. At the instant shown, the center $B$ of the wheel has a velocity of 0.35 m/s and an acceleration of 1.2 m/s², both directed to the left. Knowing that the wheel rolls without sliding and neglecting the mass of the driving yoke $AB$, determine the horizontal force $\mathbf{P}$ applied to the yoke.

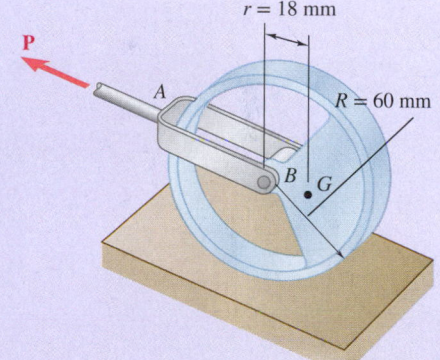

Fig. P16.113

**16.114** A small clamp of mass $m_B$ is attached at $B$ to a hoop of mass $m_h$. The system is released from rest when $\theta = 90°$ and rolls without sliding. Knowing that $m_h = 3m_B$, determine (a) the angular acceleration of the hoop, (b) the horizontal and vertical components of the acceleration of $B$.

**16.115** A small clamp of mass $m_B$ is attached at $B$ to a hoop of mass $m_h$. Knowing that the system is released from rest and rolls without sliding, derive an expression for the angular acceleration of the hoop in terms of $m_B$, $m_h$, $r$, and $\theta$.

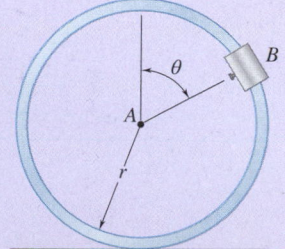

Fig. P16.114 and P16.115

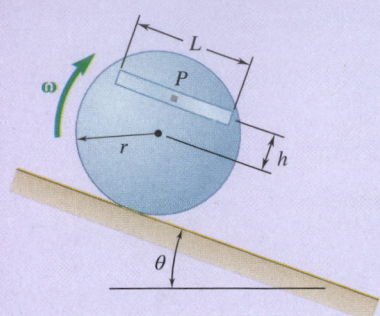

**Fig. P16.116**

**16.116** A 4-lb bar is attached to a 10-lb uniform cylinder by a square pin, P, as shown. Knowing that r = 16 in., h = 8 in., θ = 20°, L = 20 in., and ω = 2 rad/s at the instant shown, determine the reactions at P at this instant assuming that the cylinder rolls without sliding down the incline.

**16.117** The uniform rod AB with a mass m and a length of 2L is attached to collars of negligible mass that slide without friction along fixed rods. If the rod is released from rest in the position shown, derive an expression for (a) the angular acceleration of the rod, (b) the reaction at A.

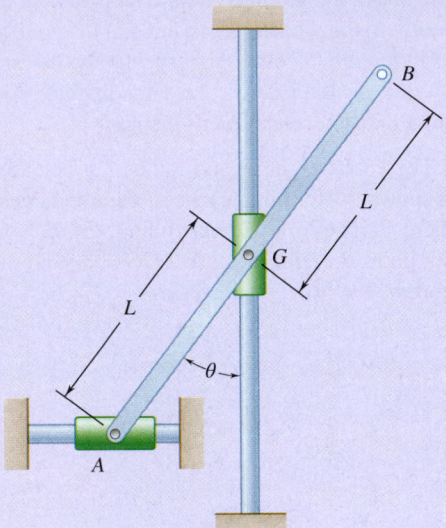

**Fig. P16.117 and P16.118**

**16.118** The 10-lb-uniform rod AB has a total length of 2L = 2 ft and is attached to collars of negligible mass that slide without friction along fixed rods. If rod AB is released from rest when θ = 30°, determine immediately after release (a) the angular acceleration of the rod, (b) the reaction at A.

**16.119** A 40-lb ladder rests against a wall when the bottom begins to slide out. The ladder is 30 ft long and the coefficient of kinetic friction between the ladder and all surfaces is 0.2. For θ = 40°, determine (a) the angular acceleration of the ladder, (b) the forces at A and B.

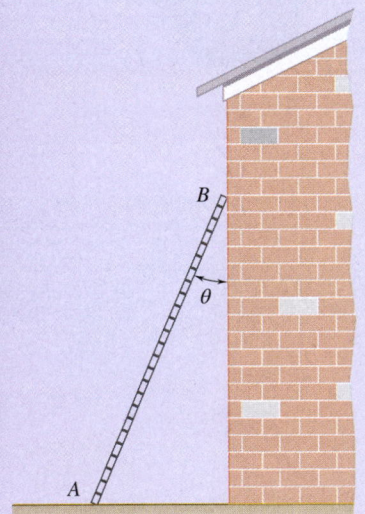

**Fig. P16.119**

**16.120** A beam AB of length L and mass m is supported by two cables as shown. If cable BD breaks, determine at that instant the tension in the remaining cable as a function of its initial angular orientation θ.

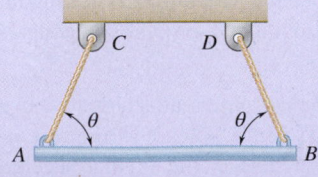

**Fig. P16.120**

**16.121** End *A* of the 6-kg uniform rod *AB* rests on the inclined surface, while end *B* is attached to a collar of negligible mass that can slide along the vertical rod shown. Knowing that the rod is released from rest when $\theta = 35°$ and neglecting the effect of friction, determine immediately after release (*a*) the angular acceleration of the rod, (*b*) the reaction at *B*.

**16.122** End *A* of the 6-kg uniform rod *AB* rests on the inclined surface, while end *B* is attached to a collar of negligible mass that can slide along the vertical rod shown. When the rod is at rest, a vertical force **P** is applied at *B*, causing end *B* of the rod to start moving upward with an acceleration of 4 m/s². Knowing that $\theta = 35°$, determine the force **P**.

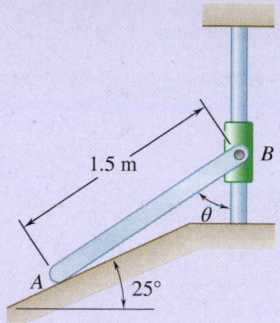

**Fig. P16.121 and *P16.122***

**16.123** End *A* of the 8-kg uniform rod *AB* is attached to a collar that can slide without friction on a vertical rod. End *B* of the rod is attached to a vertical cable *BC*. If the rod is released from rest in the position shown, determine immediately after release (*a*) the angular acceleration of the rod, (*b*) the reaction at *A*.

**16.124** The 4-kg uniform rod *ABD* is attached to the crank *BC* and is fitted with a small wheel that can roll without friction along a vertical slot. Knowing that at the instant shown crank *BC* rotates with an angular velocity of 6 rad/s clockwise and an angular acceleration of 15 rad/s² counterclockwise, determine the reaction at *A*.

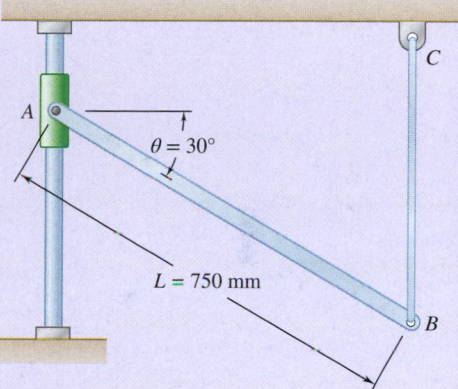

**Fig. *P16.123***

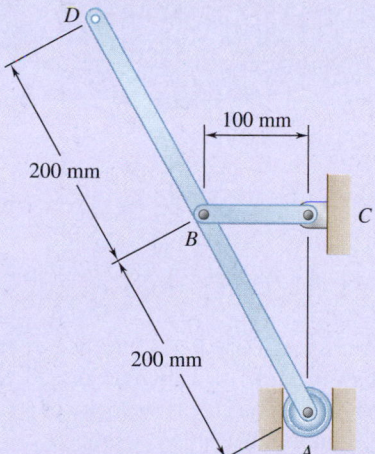

**Fig. P16.124**

**16.125** The 3-lb uniform rod *BD* is connected to crank *AB* and to a collar of negligible weight. A couple (not shown) is applied to crank *AB*, causing it to rotate with an angular velocity of 12 rad/s counterclockwise and an angular acceleration of 80 rad/s² clockwise at the instant shown. Neglecting the effect of friction, determine the reaction at *D*.

**16.126** The 3-lb uniform rod *BD* is connected to crank *AB* and to a collar of negligible weight. A couple (not shown) is applied to crank *AB* causing it to rotate. At the instant shown, crank *AB* has an angular velocity of 12 rad/s and an angular acceleration of 80 rad/s²; both are counterclockwise. Neglecting the effect of friction, determine the reaction at *D*.

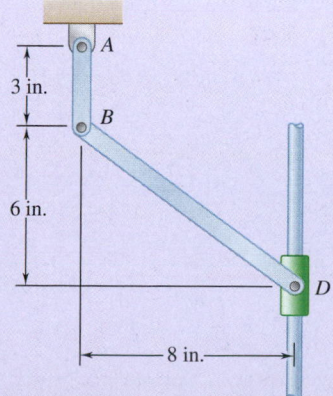

**Fig. P16.125 and P16.126**

**16.127** The test rig shown was developed to perform fatigue testing on fitness trampolines. A motor drives the 200-mm radius flywheel *AB*, which is pinned at its center point *A*, in a counterclockwise direction with a constant angular velocity of 120 rpm. The flywheel is attached to slider *CD* by the 400-mm connecting rod *BC*. The mass of the connecting rod *BC* is 5 kg, and the mass of the link *CD* and foot is 2 kg. At the instant when $\theta = 0°$ and the foot is just above the trampoline, determine the force exerted by pin *C* on rod *BC*.

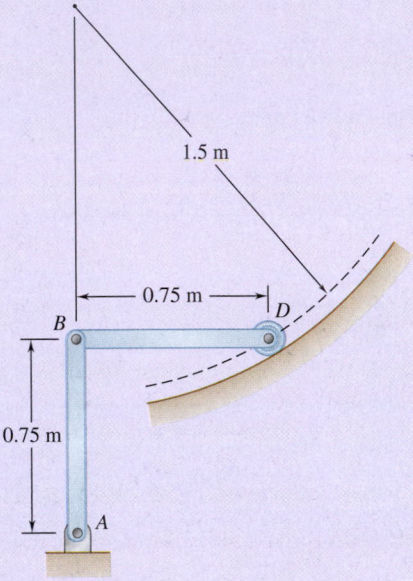

Fig. P16.129

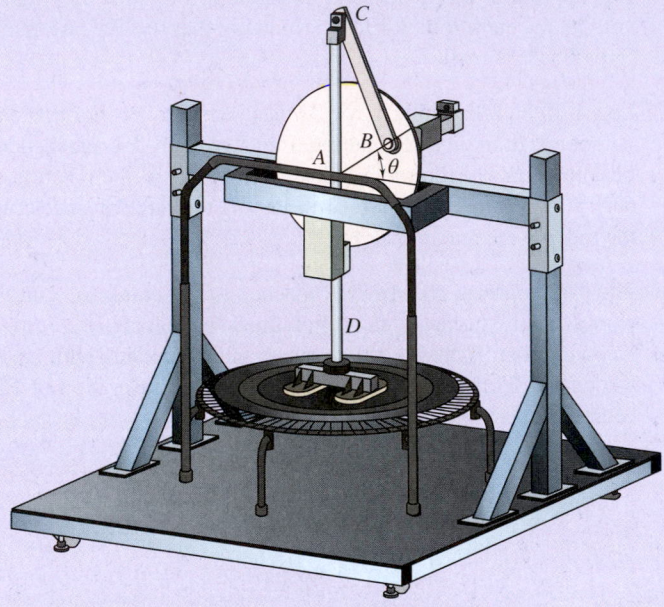

Fig. P16.127

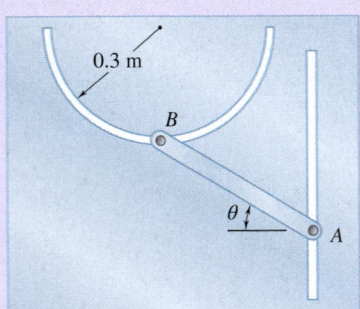

Fig. P16.130

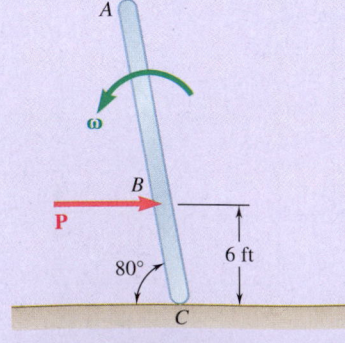

Fig. P16.131

**16.128** Solve Prob. 16.127 for $\theta = 90°$.

**16.129** The 4-kg uniform slender bar *BD* is attached to bar *AB* and a wheel of negligible mass that rolls on a circular surface. Knowing that at the instant shown bar *AB* has an angular velocity of 6 rad/s and no angular acceleration, determine the reaction at point *D*.

**16.130** The motion of the uniform slender rod of length $L = 0.5$ m and mass $m = 3$ kg is guided by pins at *A* and *B* that slide freely in frictionless slots, circular and horizontal, cut into a vertical plate as shown. Knowing that at the instant shown the rod has an angular velocity of 3 rad/s counterclockwise and $\theta = 30°$, determine the reactions at points *A* and *B*.

**16.131** At the instant shown, the 20-ft-long, uniform 100-lb pole *ABC* has an angular velocity of 1 rad/s counterclockwise and point *C* is sliding to the right. A 120-lb horizontal force **P** acts at *B*. Knowing the coefficient of kinetic friction between the pole and the ground is 0.3, determine at this instant (*a*) the acceleration of the center of gravity, (*b*) the normal force between the pole and the ground.

**16.132** A driver starts his car with the door on the passenger's side wide open ($\theta = 0$). The 100-lb door has a centroidal radius of gyration $\bar{k} = 12$ in. and its mass center is located at a distance $r = 20$ in. from its vertical axis of rotation. Knowing that the driver maintains a constant acceleration of 4 ft/s$^2$, determine the angular velocity of the door as it slams shut ($\theta = 90°$).

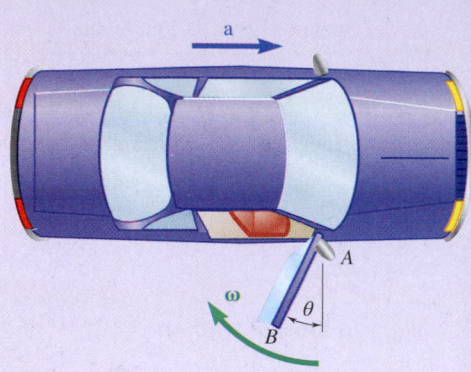

**Fig. P16.132**

**16.133** For the car of Prob. 16.132, determine the smallest constant acceleration that the driver can maintain if the door is to close and latch, knowing that as the door hits the frame, its angular velocity must be at least 1.5 rad/s for the latching mechanism to operate.

**16.134** The hatchback of a car is positioned as shown to help determine the appropriate size for a damping mechanism $AB$. The weight of the door is 40 lbs, and its mass moment of inertia about the center of gravity $G$ is 15 lb·ft·s$^2$. The linkage $DEFH$ controls the motion of the hatch and is shown in more detail in part ($b$) of the figure. Assume that the mass of the links $DE$, $EF$, and $FH$ are negligible, compared to the mass of the door. With $AB$ removed, determine ($a$) the initial angular acceleration of the 40-lb door as it is released from rest, ($b$) the force on link $FH$.

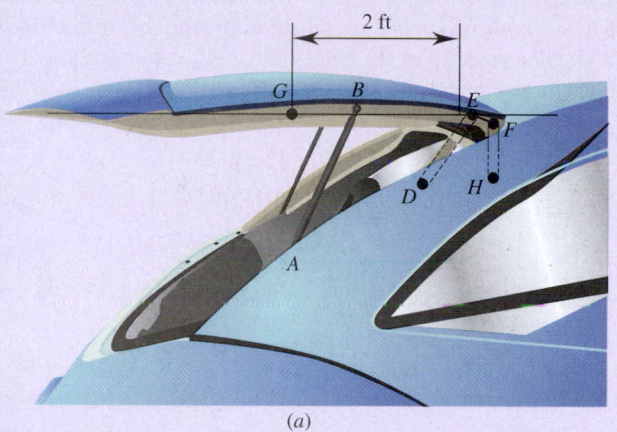

(a)

(b)

**Fig. P16.134**

**\*16.135** The 6-kg rod $BC$ connects a 10-kg disk centered at $A$ to a 5-kg rod $CD$. The motion of the system is controlled by the couple $\mathbf{M}$ applied to disk $A$. Knowing that at the instant shown disk $A$ has an angular velocity of 36 rad/s clockwise and no angular acceleration, determine ($a$) the couple $\mathbf{M}$, ($b$) the components of the force exerted at $C$ on rod $BC$.

**\*16.136** The 6-kg rod $BC$ connects a 10-kg disk centered at $A$ to a 5-kg rod $CD$. The motion of the system is controlled by the couple $\mathbf{M}$ applied to disk $A$. Knowing that at the instant shown disk $A$ has an angular velocity of 36 rad/s clockwise and an angular acceleration of 150 rad/s$^2$ counterclockwise, determine ($a$) the couple $\mathbf{M}$, ($b$) the components of the force exerted at $C$ on rod $BC$.

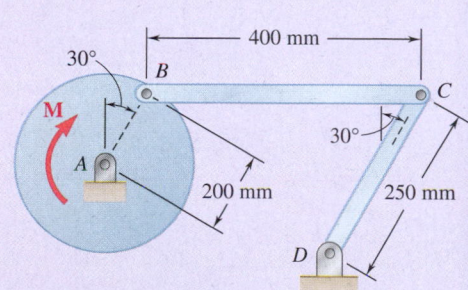

**Fig. P16.135 and P16.136**

**16.137** In the engine system shown, $l = 250$ mm and $b = 100$ mm. The connecting rod $BD$ is assumed to be a 1.2-kg uniform slender rod and is attached to the 1.8-kg piston $P$. During a test of the system, crank $AB$ is made to rotate with a constant angular velocity of 600 rpm clockwise with no force applied to the face of the piston. Determine the forces exerted on the connecting rod at $B$ and $D$ when $\theta = 180°$. (Neglect the effect of the weight of the rod.)

**16.138** Solve Prob. 16.137 when $\theta = 90°$.

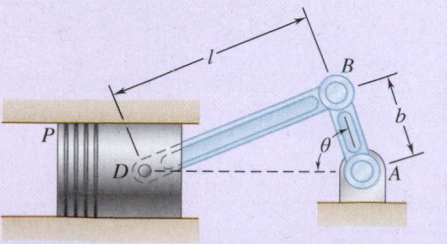

**Fig. P16.137**

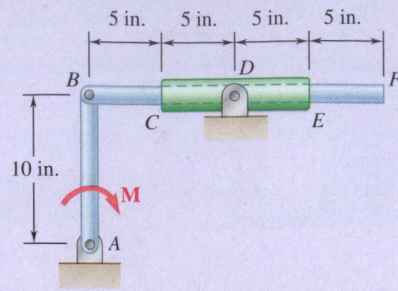

Fig. P16.139 and P16.140

**16.139** The 4-lb uniform slender rod *AB*, the 8-lb uniform slender rod *BF*, and the 4-lb uniform thin sleeve *CE* are connected as shown and move without friction in a vertical plane. The motion of the linkage is controlled by the couple **M** applied to rod *AB*. Knowing that at the instant shown the angular velocity of rod *AB* is 15 rad/s and the magnitude of the couple **M** is 5 ft·lb, determine (*a*) the angular acceleration of rod *AB*, (*b*) the reaction at point *D*.

**16.140** The 4-lb uniform slender rod *AB*, the 8-lb uniform slender rod *BF*, and the 4-lb uniform thin sleeve *CE* are connected as shown and move without friction in a vertical plane. The motion of the linkage is controlled by the couple **M** applied to rod *AB*. Knowing that at the instant shown the angular velocity of rod *AB* is 30 rad/s and the angular acceleration of rod *AB* is 96 rad/s² clockwise, determine (*a*) the magnitude of the couple **M**, (*b*) the reaction at point *D*.

**16.141** Two rotating rods in the vertical plane are connected by a slider block *P* of negligible mass. The rod attached at *A* has a weight of 1.6 lb and a length of 8 in. Rod *BP* weighs 2 lb and is 10 in. long and the friction between block *P* and *AE* is negligible. The motion of the system is controlled by a couple **M** applied to rod *BP*. Knowing that rod *BP* has a constant angular velocity of 20 rad/s clockwise, determine (*a*) the couple **M**, (*b*) the components of the force exerted on *AE* by block *P*.

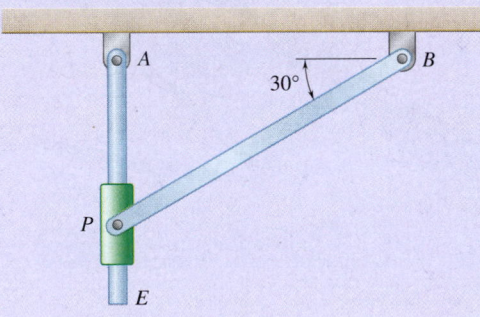

Fig. *P16.141* and *P16.142*

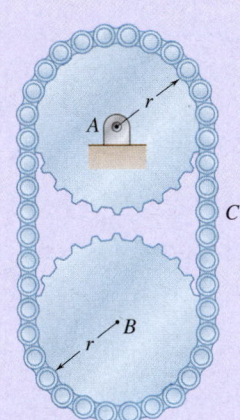

Fig. P16.143

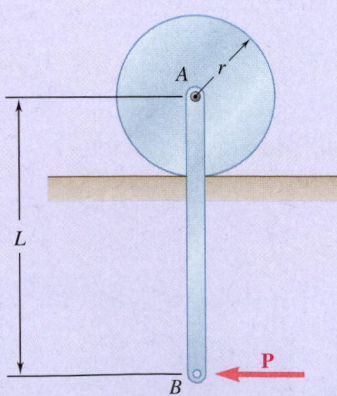

Fig. P16.144

**16.142** Two rotating rods in the vertical plane are connected by a slider block *P* of negligible mass. The rod attached at *A* has a mass of 0.8 kg and a length of 160 mm. Rod *BP* has a mass of 1 kg and is 200 mm long and the friction between block *P* and *AE* is negligible. The motion of the system is controlled by a couple **M** applied to bar *BP*. Knowing that at the instant shown rod *BP* has an angular velocity of 20 rad/s clockwise and an angular acceleration of 80 rad/s² clockwise, determine (*a*) the couple **M**, (*b*) the components of the force exerted on *AE* by block *P*.

**\*16.143** Two disks, each with a mass *m* and a radius *r*, are connected as shown by a continuous chain belt of negligible mass. If a pin at point *C* of the chain belt is suddenly removed, determine (*a*) the angular acceleration of each disk, (*b*) the tension in the left-hand portion of the belt, (*c*) the acceleration of the center of disk *B*.

**\*16.144** A uniform slender bar *AB* of mass *m* is suspended as shown from a uniform disk of the same mass *m*. Neglecting the effect of friction, determine the accelerations of points *A* and *B* immediately after a horizontal force **P** has been applied at *B*.

**16.145** A uniform rod *AB*, of mass 15 kg and length 1 m, is attached to the 20-kg cart *C*. Neglecting friction, determine immediately after the system has been released from rest, (*a*) the acceleration of the cart, (*b*) the angular acceleration of the rod.

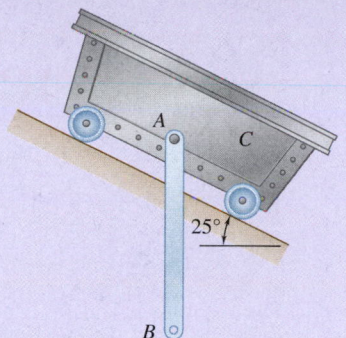
**Fig. P16.145**

*16.146** The uniform slender 2-kg bar *BD* is attached to the uniform 6-kg disk by a pin at *B* and released from rest in the position shown. Assuming that the disk rolls without slipping, determine (*a*) the initial reaction at the contact point *A*, (*b*) the corresponding smallest allowable value of the coefficient of static friction.

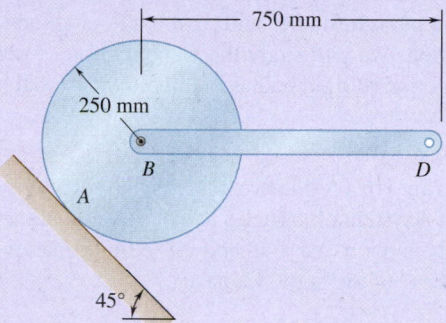

**Fig. P16.146**

*16.147 and *16.148** The 6-lb cylinder *B* and the 4-lb wedge *A* are held at rest in the position shown by cord *C*. Assuming that the cylinder rolls without sliding on the wedge and neglecting friction between the wedge and the ground, determine, immediately after cord *C* has been cut, (*a*) the acceleration of the wedge, (*b*) the angular acceleration of the cylinder.

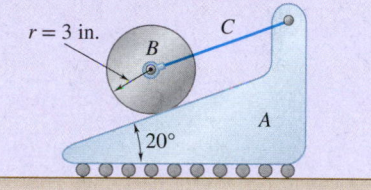
**Fig. P16.147**

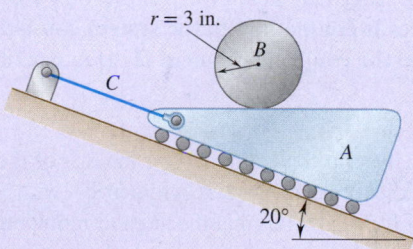

**Fig. P16.148**

*16.149** Each of the 3-kg bars *AB* and *BC* is of length *L* = 500 mm. A horizontal force **P** of magnitude 20 N is applied to bar *BC* as shown. Knowing that *b* = *L* (**P** is applied at *C*), determine the angular acceleration of each bar.

*16.150** Two identical uniform rods are connected by a pin at *B* and are held in a horizontal position by three wires as shown. If the wires attached at *A* and *B* are cut simultaneously, determine at that instant the acceleration of (*a*) point *A*, (*b*) point *B*.

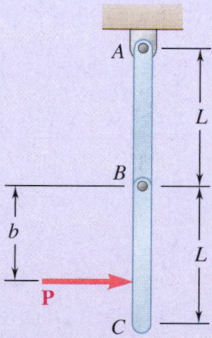
**Fig. P16.149**

*16.151** (*a*) Determine the magnitude and the location of the maximum bending moment in the rod of Prob. 16.78. (*b*) Show that the answer to part *a* is independent of the weight of the rod.

*16.152** Draw the shear and bending-moment diagrams for the rod of Prob. 16.84 immediately after the cable at *B* breaks.

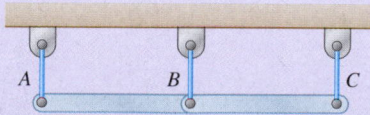
**Fig. P16.150**

# Review and Summary

In this chapter, we studied the **kinetics of rigid bodies**; that is, the relations between the forces acting on a rigid body, the shape and mass of the body, and the motion produced. Except for the first two sections, which apply to the most general case of the motion of a rigid body, our analysis was restricted to the **plane motion of rigid bodies** and rigid bodies symmetrical with respect to the plane of motion. We will study the plane motion of nonsymmetrical rigid bodies and the motion of rigid bodies in three-dimensional space in Chap. 18.

## Fundamental Equations of Motion for a Rigid Body

We first recalled (Sec. 16.1A) the two fundamental equations derived in Chap. 14 for the motion of a system of particles and observed that they apply in the most general case of the motion of a rigid body. The first equation defines the motion of the mass center $G$ of the body; we have

$$\Sigma \mathbf{F} = m\overline{\mathbf{a}} \tag{16.1}$$

where $m$ is the mass of the body and $\overline{\mathbf{a}}$ is the acceleration of $G$. The second equation is related to the motion of the body relative to a centroidal frame of reference; we have

$$\Sigma \mathbf{M}_G = \dot{\mathbf{H}}_G \tag{16.2}$$

where $\dot{\mathbf{H}}_G$ is the rate of change of the angular momentum $\mathbf{H}_G$ of the body about its mass center $G$. Together, Eqs. (16.1) and (16.2) state that **the system of the external forces is equipollent to the system consisting of the vector $m\overline{\mathbf{a}}$ attached at $G$ and the couple of moment $\dot{\mathbf{H}}_G$** (Fig. 16.18).

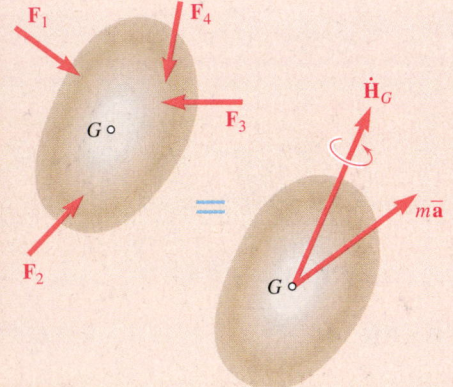

**Fig. 16.18**

## Angular Momentum in Plane Motion

Restricting our analysis at this point and for the rest of the chapter to the plane motion of rigid bodies and rigid bodies symmetrical with respect to the plane of motion, we showed (Sec. 16.1B) that the angular momentum of the body could be expressed as

$$\mathbf{H}_G = \overline{I}\boldsymbol{\omega} \tag{16.4}$$

where $\overline{I}$ is the moment of inertia of the body about a centroidal axis perpendicular to the reference plane and $\boldsymbol{\omega}$ is the angular velocity of the body. Differentiating both sides of Eq. (16.4), we obtained

$$\dot{\mathbf{H}}_G = \overline{I}\dot{\boldsymbol{\omega}} = \overline{I}\boldsymbol{\alpha} \tag{16.5}$$

which shows that, in the restricted case considered here, we can represent the rate of change of the angular momentum of the rigid body by a vector of the same direction as $\boldsymbol{\alpha}$ (i.e., perpendicular to the plane of reference) and of magnitude $\overline{I}\alpha$.

## Equations for the Plane Motion of a Rigid Body

It follows from (Sec. 16.1E) that the plane motion of a rigid body or of a rigid body symmetrical with respect to the reference plane is defined by the three scalar equations. You will have one equation for the $x$ direction, one for the $y$ direction, and one moment equation, as

$$\Sigma F_x = m\bar{a}_x \qquad \Sigma F_y = m\bar{a}_y$$
$$\Sigma M_G = \bar{I}\alpha \quad \text{or} \quad \Sigma M_O = I_O\alpha \quad \text{or} \quad \Sigma M_P = \bar{I}\alpha + m\bar{a}d_\perp \quad \text{or}$$
$$\Sigma M_P = \bar{I}\alpha + \mathbf{r}_{G/P} \times m\bar{\mathbf{a}}$$

where $G$ is the center of mass of the body, $O$ is a fixed axis of rotation, $P$ is any arbitrary point, and $d_\perp$ is the perpendicular distance between point $P$ and the line of action of the acceleration of the center of mass.

## Plane Motion of a Rigid Body

It further follows that *the external forces acting on the rigid body are actually equivalent to the inertial terms of the various particles forming the body.* This statement can be represented by a free-body diagram and kinetic diagram as shown in Fig. 16.19, where the inertial terms have been represented by a vector $m\bar{\mathbf{a}}$ attached at $G$ and a couple $\bar{I}\alpha$. In the particular case of a rigid body in *translation,* the inertial terms shown in part *b* of this figure reduce to the single vector $m\bar{\mathbf{a}}$, whereas in the particular case of a rigid body in *centroidal rotation,* they reduce to the single couple $\bar{I}\alpha$. In any other case of plane motion, both the vector $m\bar{\mathbf{a}}$ and the couple $\bar{I}\alpha$ should be included.

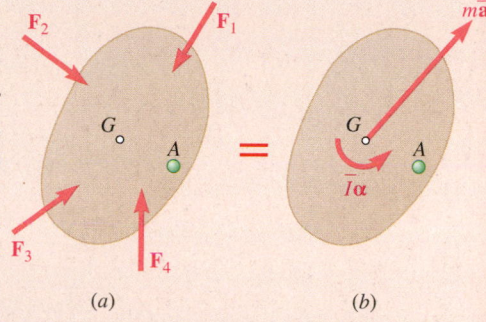

(a)     (b)

**Fig. 16.19**

## Free-Body Diagram and Kinetic Diagram

Any kinetics problem involving the plane motion of a rigid body may be solved by drawing a **free-body diagram and kinetic diagram** similar to that of Fig. 16.19 (Sec. 16.1E). You can then obtain three equations of motion (see previous equations) by equating the $x$ components, $y$ components, and moments about a chosen point (such as $G$ or some arbitrary point $P$) of the forces and vectors involved (Sample Probs. 16.1 through 16.5).

## Connected Rigid Bodies

We can also use the method described previously to solve problems involving the plane motion of several connected rigid bodies (Sec. 16.1F). You draw a free-body diagram and kinetic diagram for each system and solve the equations of motion simultaneously. In some cases, however, you can include multiple objects in your system and draw a single diagram for the entire system, including all of the external forces, as well as the vectors $m\bar{\mathbf{a}}$ and the couples $\bar{I}\alpha$ associated with the various parts of the system (Sample Prob. 16.4).

## Constrained Plane Motion

In the second section of this chapter, we were concerned with rigid bodies *moving under given constraints* (Sec. 16.2). Although the kinetic analysis of the constrained plane motion of a rigid body is the same as before, it must be supplemented by a *kinematic analysis* that aims to express the components $\bar{a}_x$ and $\bar{a}_y$ of the acceleration of the mass center $G$ of the body in terms of its angular acceleration $\alpha$. This often involves using analyses that we examined in Chap. 15, including the relationship between two points on a body undergoing general plane motion:

$$\bar{\mathbf{a}} = \mathbf{a}_A + \alpha\mathbf{k} \times \mathbf{r}_{G/A} - \omega^2\mathbf{r}_{G/A}$$

Problems solved in this way included the noncentroidal rotation of rods and plates (Sample Probs. 16.7 and 16.8), the rolling motion of spheres and wheels (Sample Probs. 16.9 and 16.10), the general plane motion of a body with no fixed point (Sample Probs. 16.11 and 16.12), and the plane motion of various types of linkages (Sample Prob. 16.13).

# Review Problems

**16.153** A cyclist is riding a bicycle at a speed of 20 mph on a horizontal road. The distance between the axles is 42 in., and the mass center of the cyclist and the bicycle is located 26 in. behind the front axle and 40 in. above the ground. If the cyclist applies the brakes only on the front wheel, determine the shortest distance in which he can stop without being thrown over the front wheel.

**16.154** The forklift truck shown weighs 3200 lb and is used to lift a 1700-lb crate. The forklift is moving to the right at a speed of 12 ft/s when the brakes are applied on the rear wheels. Knowing that the coefficient of static friction between the crate and the fork lift is 0.30, determine the smallest distance in which the truck can be brought to a stop without the crate sliding or the truck tipping forward.

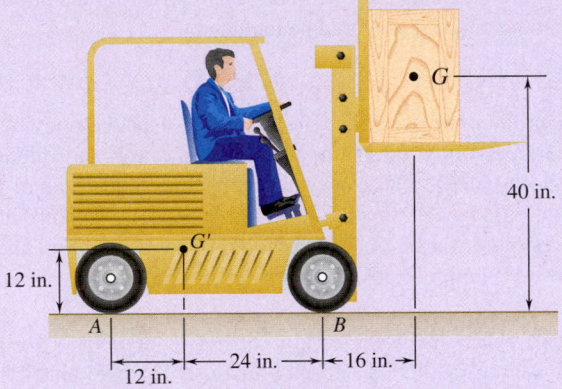

**Fig. P16.154**

**Fig. *P16.155***

**16.155** The total mass of the Baja car and driver, including the wheels, is 250 kg. Each pair of 58-cm radius wheels and the axle has a total mass of 20 kg and a mass moment of inertia of 2.9 kg·m². The center of gravity of the driver and Baja body (not including the wheels) is located $x = 0.70$ m from the rear axle $A$ and $y = 0.55$ m from the ground. The wheelbase is $L = 1.60$ m. If the engine exerts a torque of 500 N·m on the rear axle, what is the car's acceleration?

**16.156** Identical cylinders of mass $m$ and radius $r$ are pushed by a series of moving arms. Assuming the coefficient of friction between all surfaces to be $\mu < 1$ and denoting by $a$ the magnitude of the acceleration of the arms, derive an expression for ($a$) the maximum allowable value of $a$ if each cylinder is to roll without sliding, ($b$) the minimum allowable value of $a$ if each cylinder is to move to the right without rotating.

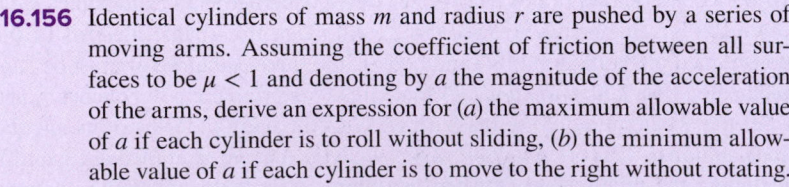

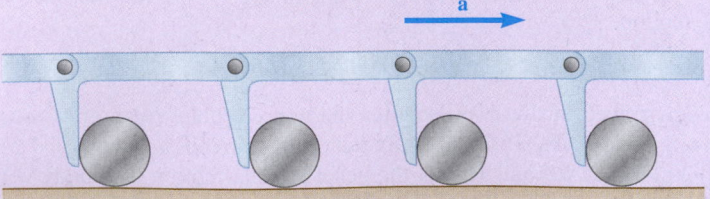

**Fig. P16.156**

**16.157** The uniform rod $AB$ of weight $W$ is released from rest when $\beta = 70°$. Assuming that the friction force between end $A$ and the surface is large enough to prevent sliding, determine immediately after release (a) the angular acceleration of the rod, (b) the normal reaction at $A$, (c) the friction force at $A$.

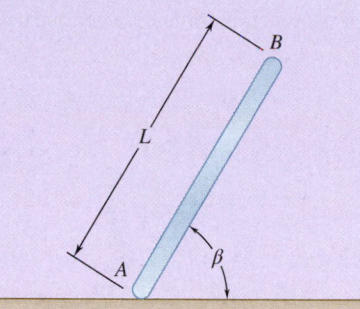

**Fig. P16.157 and P16.158**

**16.158** The uniform rod $AB$ of weight $W$ is released from rest when $\beta = 70°$. Assuming that the friction force is zero between end $A$ and the surface, determine immediately after release (a) the angular acceleration of the rod, (b) the acceleration of the mass center of the rod, (c) the reaction at $A$.

**16.159** A bar of mass $m = 5$ kg is held as shown between four disks, each of mass $m' = 2$ kg and radius $r = 75$ mm. Knowing that the normal forces on the disks are sufficient to prevent any slipping, for each of the cases shown determine the acceleration of the bar immediately after it has been released from rest.

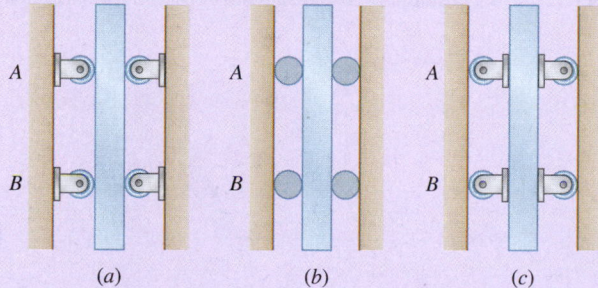

Fig. **P16.159**

**16.160** A uniform plate of mass $m$ is suspended in each of the ways shown. For each case, determine immediately after the connection $B$ has been released (a) the angular acceleration of the plate, (b) the acceleration of its mass center.

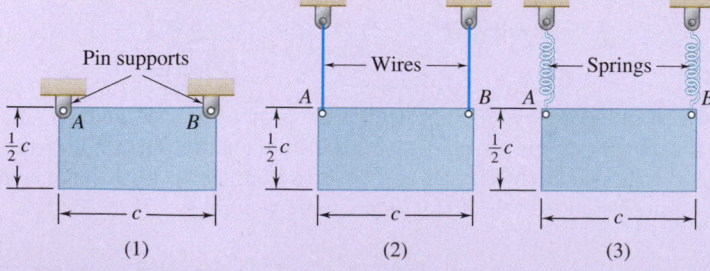

Fig. P16.160

1189

**16.161** A cylinder with a circular hole is rolling without slipping on a fixed curved surface as shown. The cylinder would have a weight of 16 lb without the hole, but with the hole it has a weight of 15 lb. Knowing that at the instant shown the disk has an angular velocity of 5 rad/s clockwise, determine (*a*) the angular acceleration of the disk, (*b*) the components of the reaction force between the cylinder and the ground at this instant.

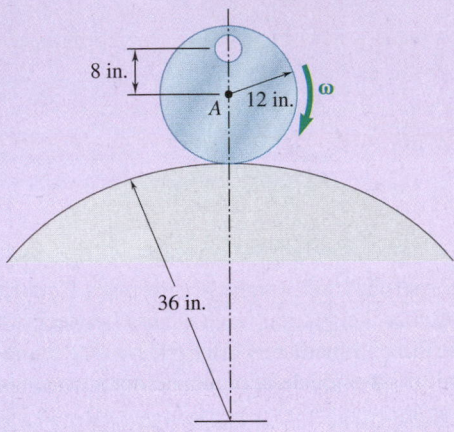

**Fig. *P16.161***

**16.162** Two 3-kg uniform bars are connected to form the linkage shown. Neglecting the effect of friction, determine the reaction at *D* immediately after the linkage is released from rest in the position shown.

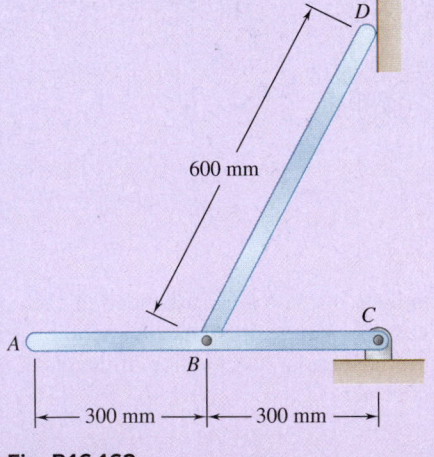

**Fig. P16.162**

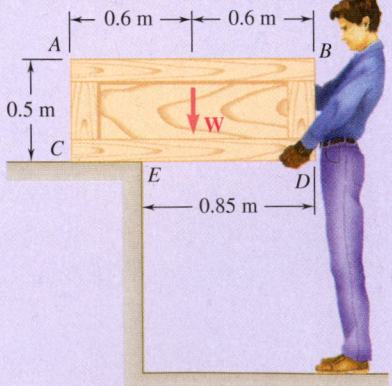

**Fig. P16.163**

**16.163** A crate of mass 80 kg is held in the position shown. If the man accidentally drops the right end of the crate and the surface between the crate and the horizontal surface is smooth, determine (*a*) the angular acceleration of the crate, (*b*) the force at point *E*.

**16.164** The Geneva mechanism shown is used to provide an intermittent rotary motion of disk $S$. Disk $D$ weighs 2 lb and has a radius of gyration of 0.9 in., and disk $S$ weighs 6 lb and has a radius of gyration of 1.5 in. The motion of the system is controlled by a couple $\mathbf{M}$ applied to disk $D$. A pin $P$ is attached to disk $D$ and can slide in one of the six equally spaced slots cut in disk $S$. It is desirable that the angular velocity of disk $S$ be zero as the pin enters and leaves each of the six slots; this will occur if the distance between the centers of the disks and the radii of the disks are related as shown. Knowing disk $D$ rotates with a constant counterclockwise angular velocity of 8 rad/s and the friction between the slot and pin $P$ is negligible, determine when $\phi = 150°$ (*a*) the couple $\mathbf{M}$, (*b*) the magnitude of the force pin $P$ applies to disk $S$.

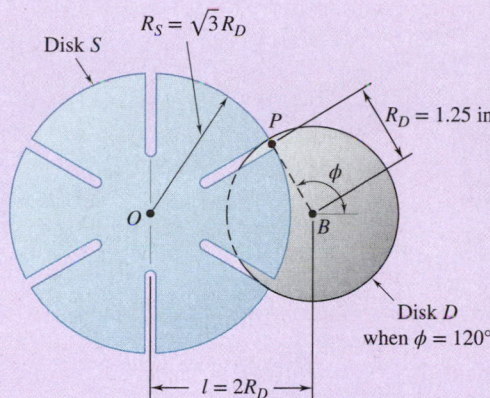

**Fig. P16.164**

# 17

# Plane Motion of Rigid Bodies: Energy and Momentum Methods

In this chapter the energy and momentum methods will be added to the tools available for your study of the motion of rigid bodies. We can analyze the transfer between potential and kinetic energy as the gymnast goes from a high position to a lower one, and we can use conservation of angular momentum to examine how changes in the gymnast's body position affect his angular velocity.

# Objectives

- **Calculate** the work done by a force or a moment on a rigid body.
- **Calculate** the kinetic energy of a rigid body in plane motion.
- **Solve** rigid body kinetics problems using the principle of work and energy.
- **Solve** rigid body kinetics problems using conservation of energy.
- **Calculate** the power of a mechanical system of rigid bodies.
- **Draw** complete and accurate impulse–momentum diagrams for problems involving rigid bodies.
- **Solve** rigid body kinetics problems using the principles of linear impulse and momentum and of angular impulse and momentum.
- **Solve** rigid body kinetics problems using conservation of angular momentum.
- **Solve** rigid body problems involving eccentric impact by using the principle of impulse and momentum and the coefficient of restitution.

# Introduction

In this chapter, we return to the method of work and energy and the method of impulse and momentum that were introduced in Chap. 13 in the context of particle kinetics. Here we use them to analyze the plane motion of rigid bodies and of systems of rigid bodies.

We consider the method of work and energy first. We define the work of a force and of a couple, and we obtain an expression for the kinetic energy of a rigid body in plane motion. Then we use the principle of work and energy to solve problems involving displacements and velocities. We also apply the principle of conservation of energy to solve a variety of engineering problems.

In the second section, we apply the principle of impulse and momentum to solve problems involving velocities and time. We also discuss the concept of conservation of angular momentum for rigid bodies in plane motion.

In the last section of this chapter, we consider problems involving the eccentric impact of rigid bodies. As we did in Chap. 13, where we analyzed the impact of particles, we use the coefficient of restitution between colliding bodies, together with the principle of impulse and momentum, to solve impact problems. We will show that the method used is applicable not only when the colliding bodies move freely after the impact, but also when the bodies are partially constrained in their motion.

# 17.1 ENERGY METHODS FOR A RIGID BODY

We now use the principle of work and energy to analyze the plane motion of rigid bodies. As we pointed out in Chap. 13, the method of work and energy is particularly well-adapted to solving problems involving velocities and displacements. Its main advantage is that the work of forces and the kinetic energy of objects are scalar quantities.

## 17.1A Principle of Work and Energy

To apply the principle of work and energy to the motion of a rigid body, we again assume that the rigid body is made up of a large number $n$ of particles of mass $\Delta m_i$. From Eq. (14.30) of Sec. 14.2B, we have

**Principle of work and energy, rigid body**

$$T_1 + U_{1\to2} = T_2 \tag{17.1}$$

where $T_1, T_2$ = the initial and final values of total kinetic energy of particles forming the rigid body

$U_{1\to2}$ = work of all forces acting on various particles of the body

Just as we did in Chap. 13, we can express the work done by nonconservative forces as $U_{1\to2}^{NC}$, and we can define potential energy terms for conservative forces. Then we can express Eq. (17.1) as

$$T_1 + V_{g_1} + V_{e_1} + U_{1\to2}^{NC} = T_2 + V_{g_2} + V_{e_2} \tag{17.1'}$$

where $V_{g_1}$ and $V_{g_2}$ are the initial and final gravitational potential energy of the center of mass of the rigid body with respect to a reference point or datum, and $V_{e_1}$ and $V_{e_2}$ are the initial and final values of the elastic energy associated with springs in the system.

We obtain the total kinetic energy

$$T = \frac{1}{2}\sum_{i=1}^{n}\Delta m_i v_i^2 \tag{17.2}$$

by adding positive scalar quantities, so it is itself a positive scalar quantity. You will see later how to determine $T$ for various types of motion of a rigid body.

The expression $U_{1\to2}$ in Eq. (17.1) represents the work of all the forces acting on the various particles of the body, whether these forces are internal or external. However, the total work of the internal forces holding together the particles of a rigid body is zero. To see this, consider two particles $A$ and $B$ of a rigid body and the two equal and opposite forces $\mathbf{F}$ and $-\mathbf{F}$ they exert on each other (Fig. 17.1). Although, in general, small displacements $d\mathbf{r}$ and $d\mathbf{r}'$ of the two particles are different, the components of these displacements along $AB$ must be equal; otherwise, the particles would not remain at the same distance from each other and the body would not be rigid. Therefore, the work of $\mathbf{F}$ is equal in magnitude and opposite in sign to the work of $-\mathbf{F}$,

**Photo 17.1**  The work done by friction reduces the kinetic energy of the wheel.
©Richard McDowell/Alamy RF

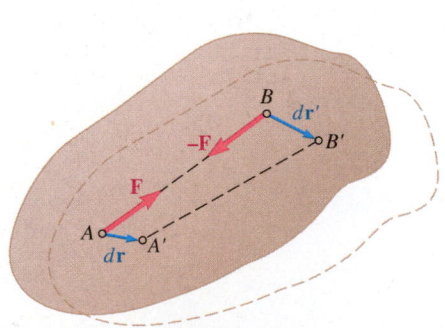

**Fig. 17.1**  The total work of the internal forces acting on the particles of a rigid body is zero.

and their sum is zero. Thus, the total work of the internal forces acting on the particles of a rigid body is zero, and the expression $U_{1\to2}$ in Eq. (17.1) reduces to the work of the external forces acting on the body during the displacement considered.

## 17.1B   Work of Forces Acting on a Rigid Body

We saw in Sec. 13.1A that the work of a force $\mathbf{F}$ during a displacement of its point of application from $A_1$ to $A_2$ is

**Work of a force**
$$U_{1\to2} = \int_{A_1}^{A_2} \mathbf{F}\cdot d\mathbf{r} \tag{17.3}$$

or

$$U_{1\to2} = \int_{s_1}^{s_2} (F\cos\alpha)\,ds \tag{17.3'}$$

where $F$ is the magnitude of the force, $\alpha$ is the angle it forms with the direction of motion of its point of application $A$, and $s$ is the variable of integration that measures the distance traveled by $A$ along its path.

In computing the work of the external forces acting on a rigid body, it is often convenient to determine the work of a couple without considering the work of each of the two forces forming the couple separately. Consider the two forces $\mathbf{F}$ and $-\mathbf{F}$ forming a couple of moment $\mathbf{M}$ and acting on a rigid body (Fig. 17.2). Any small displacement of the rigid body bringing $A$ and $B$, respectively, into $A'$ and $B''$ can be divided into two parts: in one part, points $A$ and $B$ undergo equal displacements $d\mathbf{r}_1$; in the other part, $A'$ remains fixed, while $B'$ moves into $B''$ through a displacement $d\mathbf{r}_2$ with a magnitude of $ds_2 = r\,d\theta$. In the first part of the motion, the work of $\mathbf{F}$ is equal in magnitude and opposite in sign to the work of $-\mathbf{F}$, and their sum is zero. In the second part of the motion, only force $\mathbf{F}$ works, and its work is $dU = F\,ds_2 = Fr\,d\theta$. But the product $Fr$ is equal to the magnitude $M$ of the moment of the couple. Thus, the work of a couple of moment $\mathbf{M}$ acting on a rigid body is

$$dU = M\,d\theta \tag{17.4}$$

where $d\theta$ is the small angle through which the body rotates and is expressed in radians. (We again note that work should be expressed in units obtained by multiplying units of force by units of length.) To obtain the work of the couple during a finite rotation of the rigid body, we integrate both members of Eq. (17.4) from the initial value $\theta_1$ of the angle $\theta$ to its final value $\theta_2$.

$$U_{1\to2} = \int_{\theta_1}^{\theta_2} M\,d\theta \tag{17.5}$$

*When the moment* $\mathbf{M}$ *of the couple is constant,* formula (17.5) reduces to

$$U_{1\to2} = M(\theta_2 - \theta_1) \tag{17.6}$$

We pointed out in Sec. 13.1A that some forces encountered in problems of kinetics *do no work.* These include forces applied to fixed points or acting

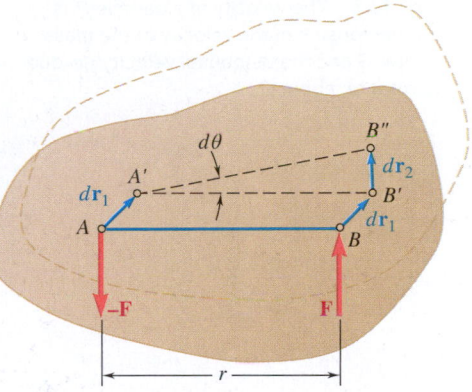

**Fig. 17.2**   The work of a couple acting on a rigid body equals the integral of the moment $\mathbf{M}$ of the couple with respect to the angular displacement of the body.

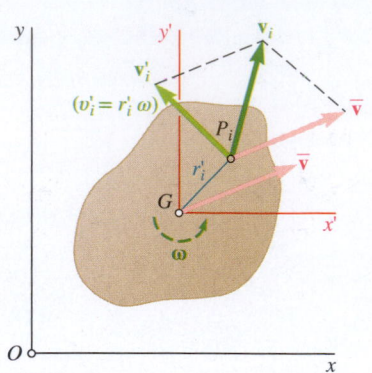

**Fig. 17.3** The velocity of a particle $P_i$ is the vector sum of the velocity of the mass center $G$ and the tangential velocity $r'_i\omega$ due to rotation about $G$.

in a direction perpendicular to the displacement of their point of application. Among these forces are the reaction at a frictionless pin when the body rotates about the pin; the reaction at a frictionless surface when the body in contact moves along the surface; and the weight of a body when its center of gravity moves horizontally. We can now add that

> **When a rigid body rolls without sliding on a fixed surface, the friction force $F$ at the point of contact $C$ does no work.**

The velocity $\mathbf{v}_C$ of the point of contact $C$ is zero, and the work of the friction force $\mathbf{F}$ during a small displacement of the rigid body is

$$dU = F\,ds_C = F(v_C\,dt) = 0$$

## 17.1C Kinetic Energy of a Rigid Body in Plane Motion

Consider a rigid body with a mass $m$ in plane motion. Recall from Sec. 14.2A that, if the absolute velocity $\mathbf{v}_i$ of each particle $P_i$ of the body is expressed as the sum of the velocity $\overline{\mathbf{v}}$ of the mass center $G$ of the body and of the velocity $\mathbf{v}'_i$ of the particle relative to a frame $Gx'y'$ attached to $G$ and of fixed orientation (Fig. 17.3), we can express the kinetic energy of the system of particles forming the rigid body in the form

$$T = \tfrac{1}{2}m\overline{v}^2 + \frac{1}{2}\sum_{i=1}^{n}\Delta m_i v_i'^{\,2} \tag{17.7}$$

As you can see in Fig. 17.3, $\mathbf{v}'_i$ of particle $P_i$ is equal to the product $r'_i\omega$, where $r'_i$ is the distance from $G$ to $P_i$ and $\omega$ is the angular velocity of the body at the instant considered. Substituting into Eq. (17.7), we have

$$T = \tfrac{1}{2}m\overline{v}^2 + \frac{1}{2}\left(\sum_{i=1}^{n} r_i'^{\,2}\,\Delta m_i\right)\omega^2 \tag{17.8}$$

The sum represents the moment of inertia $\overline{I}$ of the body about the axis through $G$, so we have

**Kinetic energy of a rigid body**

$$T = \tfrac{1}{2}m\overline{v}^2 + \tfrac{1}{2}\overline{I}\omega^2 \tag{17.9}$$

Note that, in the particular case of a body in translation ($\omega = 0$), this expression reduces to $\tfrac{1}{2}m\overline{v}^2$, whereas in the case of a centroidal rotation ($\overline{v} = 0$), it reduces to $\tfrac{1}{2}\overline{I}\omega^2$. We conclude that we can separate the kinetic energy of a rigid body in plane motion into two parts: (1) the kinetic energy $\tfrac{1}{2}m\overline{v}^2$ associated with the motion of the mass center $G$ of the body and (2) the kinetic energy $\tfrac{1}{2}\overline{I}\omega^2$ associated with the rotation of the body about $G$.

**Noncentroidal Rotation.** The relation in Eq. (17.9) is valid for any type of plane motion, so we can use it to express the kinetic energy of a rigid body rotating with an angular velocity $\boldsymbol{\omega}$ about a fixed axis through $O$ (Fig. 17.4). In that case, however, we can express the kinetic energy of the body more directly by noting that the speed $v_i$ of particle $P_i$ is equal to $r_i\omega$, where $r_i$ is the distance

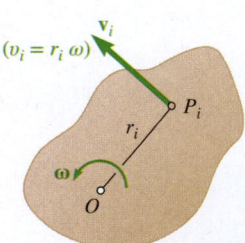

**Fig. 17.4** For noncentroidal rotation, the velocity of a particle $P_i$ is the tangential velocity $r_i\omega$ due to rotation about $O$.

from the fixed axis to $P_i$ and $\omega$ is the angular velocity of the body at the instant considered. Substituting into Eq. (17.2), we have

$$T = \frac{1}{2}\sum_{i=1}^{n}\Delta m_i(r_i\omega)^2 = \frac{1}{2}\left(\sum_{i=1}^{n}r_i^2\,\Delta m_i\right)\omega^2$$

The last sum represents the moment of inertia $I_O$ of the body about the fixed axis through $O$, so this equation reduces to

$$T = \tfrac{1}{2}I_O\omega^2 \qquad\qquad \textbf{(17.10)}$$

Note that these results are not limited to the motion of plane rigid bodies or to the motion of bodies that are symmetrical with respect to the reference plane—we can apply them to the study of the plane motion of any rigid body regardless of its shape. However, remember that Eq. (17.9) is applicable to any plane motion, whereas Eq. (17.10) is applicable only in cases involving rotating about a fixed axis.

## 17.1D   Systems of Rigid Bodies

When a problem involves several rigid bodies, we usually analyze all of the bodies together as a system instead of analyzing each individual rigid body separately. Adding the kinetic energies of all the rigid bodies and considering the work of all the forces involved, we can write the equation of work and energy for the entire system. We have

$$T_1 + U_{1\to2} = T_2 \qquad\qquad \textbf{(17.11)}$$

where $T$ represents the arithmetic sum of the kinetic energies of the rigid bodies forming the system (all terms are positive) and $U_{1\to2}$ represents the work of all the forces acting on the various bodies—whether these forces are *internal* or *external* from the point of view of the system as a whole.

The method of work and energy is particularly useful in solving problems involving pin-connected members, blocks and pulleys connected by inextensible cords, and meshed gears. In all of these cases, the internal forces occur in pairs of equal and opposite forces, and the points of application of the forces in each pair *move through equal distances* during a small displacement of the system. As a result, the work of the internal forces is zero, and $U_{1\to2}$ reduces to the work of the *forces external to the system.*

## 17.1E   Conservation of Energy

We saw in Sec. 13.2A that the work of conservative forces, such as the weight of a body or the force exerted by a spring, can be expressed as a change in potential energy. When a rigid body, or a system of rigid bodies, moves under the action of conservative forces, we can express the principle of work and energy in a modified form. Substituting for $U_{1\to2}$ from Eq. (13.19′) into Eq. (17.1), we have

**Conservation of energy, rigid body**

$$T_1 + V_1 = T_2 + V_2 \qquad\qquad \textbf{(17.12)}$$

In Chap. 13, we discussed two types of potential energy: gravitational potential energy, $V_g$, and elastic potential energy, $V_e$. Therefore, another way to write Eq. (17.12) is

$$T_1 + V_{g_1} + V_{e_1} = T_2 + V_{g_2} + V_{e_2} \qquad \textbf{(17.12')}$$

Formulas (17.12) and (17.12′) indicate that when a rigid body, or a system of rigid bodies, moves under the action of conservative forces, **the sum of the kinetic energy and of the potential energy of the system remains constant**. Note that, in the case of the plane motion of a rigid body, the kinetic energy of the body should include both the *translational* term $\frac{1}{2}m\bar{v}^2$ and the *rotational* term $\frac{1}{2}\bar{I}\omega^2$.

As an example of applying the principle of conservation of energy, let us consider a slender rod $AB$ with a length $l$ and a mass $m$, whose ends are connected to blocks of negligible mass sliding along horizontal and vertical tracks. We assume that the rod is released with no initial velocity from a horizontal position (Fig. 17.5a), and we wish to determine its angular velocity after it has rotated through an angle $\theta$ (Fig. 17.5b).

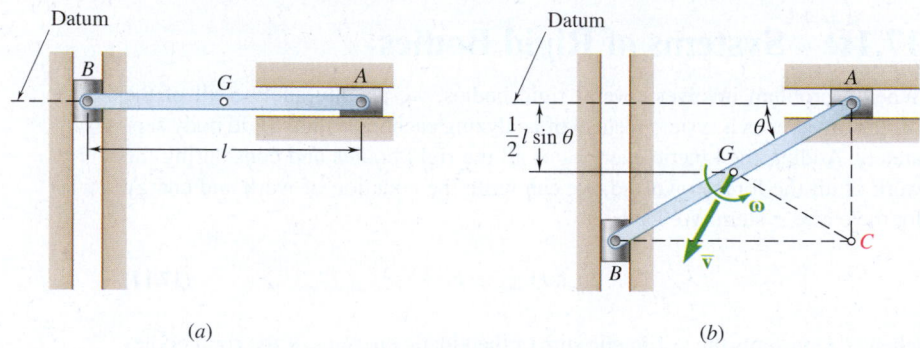

(a)                  (b)

**Fig. 17.5**   (a) Rod $AB$ in position 1 with the datum defined as shown. (b) Rod $AB$ in position 2 with an instantaneous center $C$.

Because the initial velocity is zero, we have $T_1 = 0$. Measuring the potential energy from the level of the horizontal track, we have $V_1 = 0$. After the rod has rotated through $\theta$, the center of gravity $G$ of the rod is at a distance $\frac{1}{2}l \sin \theta$ below the reference level, and we have

$$V_2 = -\tfrac{1}{2}Wl \sin \theta = -\tfrac{1}{2}mgl \sin \theta$$

In this position, the instantaneous center of the rod is located at $C$ and $CG = \frac{1}{2}l$, so $\bar{v}_2 = \frac{1}{2}l\omega$, and we obtain

$$T_2 = \tfrac{1}{2}m\bar{v}_2^2 + \tfrac{1}{2}\bar{I}\omega_2^2 = \tfrac{1}{2}m(\tfrac{1}{2}l\omega)^2 + \tfrac{1}{2}(\tfrac{1}{12}ml^2)\omega^2$$

$$= \frac{1}{2}\frac{ml^2}{3}\omega^2$$

Applying the principle of conservation of energy gives

$$T_1 + V_1 = T_2 + V_2$$

$$0 = \frac{1}{2}\frac{ml^2}{3}\omega^2 - \tfrac{1}{2}mgl \sin \theta$$

$$\omega = \left(\frac{3g}{l} \sin \theta\right)^{1/2}$$

The advantages of the method of work and energy, as well as its shortcomings, were indicated in Sec. 13.1C. Here we should add that we need to supplement the method of work and energy by the application of Newton's second law when we need to determine reactions at fixed axles, rollers, or sliding blocks. For example, in order to compute the reactions at the ends $A$ and $B$ of the rod of Fig. 17.5b, we need to draw a free-body diagram and a kinetic diagram to show that the system of the external forces applied to the rod is equivalent to both the vector $m\bar{\mathbf{a}}$ and the couple $\bar{I}\boldsymbol{\alpha}$. However, we first need to determine the angular velocity $\boldsymbol{\omega}$ of the rod using the method of work and energy before solving the equations of motion for the reactions. The complete analysis of the motion of the rod and of the forces exerted on the rod requires, therefore, the combined use of the method of work and energy and of the principle of equivalence of the external forces and moments and inertial terms.

## 17.1F   Power

We defined **power** in Sec. 13.1D as the time rate at which work is done. In the case of a body acted upon by a force $\mathbf{F}$ and moving with a velocity $\mathbf{v}$, we expressed the power as

$$\text{Power} = \frac{dU}{dt} = \mathbf{F}\cdot\mathbf{v} \tag{13.13}$$

In the case of a rigid body rotating with an angular velocity $\boldsymbol{\omega}$ and acted upon by a couple of moment $\mathbf{M}$ parallel to the axis of rotation, we have, by Eq. (17.4),

$$\text{Power} = \frac{dU}{dt} = \frac{M\,d\theta}{dt} = M\omega \tag{17.13}$$

The various units used to measure power, such as the watt and the horsepower, were defined in Sec. 13.1D.

## Sample Problem 17.1

A 240-lb block is suspended from an inextensible cable that is wrapped around a drum with a 1.25-ft radius that is rigidly attached to a flywheel. The drum and flywheel have a combined centroidal moment of inertia of $\bar{I} = 10.5$ lb·ft·s$^2$. At the instant shown, the velocity of the block is 6 ft/s directed downward. Knowing that the bearing at $A$ is poorly lubricated so that the bearing friction is equivalent to a couple **M** of magnitude 60 lb·ft, determine the velocity of the block after it has moved 4 ft downward.

**STRATEGY:** Because you have two positions and are interested in determining the velocity of the block, use the principle of work and energy.

**MODELING:** Consider the system formed by the flywheel and the block. Because the cable is inextensible, the work done by the internal forces exerted by the cable cancels out to zero. The initial and final positions of the system and the external forces acting on the system are shown in Fig. 1.

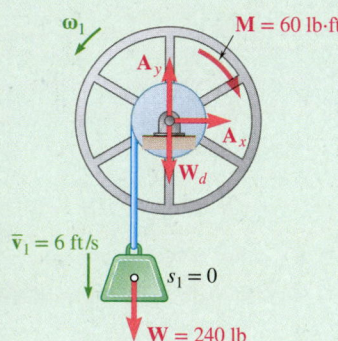

**ANALYSIS:** Apply the principle of work and energy

$$T_1 + U_{1\rightarrow2} = T_2 \tag{1}$$

**Kinetic Energy.** You need to calculate the initial and final kinetic energy and the work.

**Position 1.**

Block: $\quad\quad\quad\quad\quad\quad\quad \bar{v}_1 = 6$ ft/s

Flywheel: $\quad\quad\quad\quad\quad w_1 = \dfrac{\bar{v}_1}{r} = \dfrac{6 \text{ ft/s}}{1.25 \text{ ft}} = 4.80$ rad/s

$$
\begin{aligned}
T_1 &= \tfrac{1}{2}m\bar{v}_1^2 + \tfrac{1}{2}\bar{I}\omega_1^2 \\[4pt]
&= \frac{1}{2}\frac{240 \text{ lb}}{32.2 \text{ ft/s}^2}(6 \text{ ft/s}^2) + \tfrac{1}{2}(10.5 \text{ lb·ft·s}^2)(4.80 \text{ rad/s})^2 \\[4pt]
&= 255 \text{ ft·lb}
\end{aligned}
$$

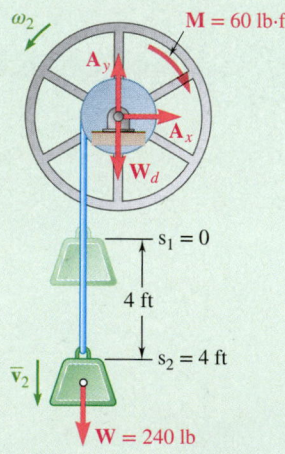

**Fig. 1** Free-body diagram of the system in positions 1 and 2.

**Position 2.** Noting that $\omega_2 = \bar{v}_2/1.25$, you have

$$
\begin{aligned}
T_2 &= \tfrac{1}{2}m\bar{v}_2^2 + \tfrac{1}{2}\bar{I}\omega_2^2 \\[4pt]
&= \frac{1}{2}\frac{240}{32.2}(\bar{v}_2)^2 + (\tfrac{1}{2})(10.5)\left(\frac{\bar{v}_2}{1.25}\right)^2 = 7.09\,\bar{v}_2^2
\end{aligned}
$$

*(continued)*

**Work.**  During the motion, only the weight **W** of the block and the friction couple **M** do work. Note that **W** does positive work, and the friction couple **M** does negative work. The total work done is

$$s_1 = 0 \qquad s_2 = 4 \text{ ft}$$

$$\theta_1 = 0 \qquad \theta_2 = \frac{s_2}{r} = \frac{4 \text{ ft}}{1.25 \text{ ft}} = 3.20 \text{ rad}$$

$$U_{1\to2} = W(s_2 - s_1) - M(\theta_2 - \theta_1)$$
$$= (240 \text{ lb})(4 \text{ ft}) - (60 \text{ lb·ft})(3.20 \text{ rad})$$
$$= 768 \text{ ft·lb}$$

Substituting these expressions into Eq. (1) gives

$$T_1 + U_{1\to2} = T_2$$
$$255 \text{ ft·lb} + 768 \text{ ft·lb} = 7.09\bar{v}_2^2$$
$$\bar{v}_2 = 12.01 \text{ ft/s} \qquad \blacktriangleleft \quad \bar{\mathbf{v}}_2 = 12.01 \text{ ft/s} \downarrow$$

**REFLECT and THINK:**  The speed of the block increases as it falls, but much more slowly than if it were in free fall. This seems like a reasonable result. Rather than calculating the work done by gravity, you could have also treated the effect of the weight using gravitational potential energy, $V_g$.

---

# Sample Problem 17.2

Gear *A* has a mass of 10 kg and a radius of gyration of 200 mm; gear *B* has a mass of 3 kg and a radius of gyration of 80 mm. The system is at rest when a couple **M** of magnitude 6 N·m is applied to gear *B*. Neglecting friction, determine (*a*) the number of revolutions executed by gear *B* before its angular velocity reaches 600 rpm, (*b*) the tangential force that gear *B* exerts on gear *A*.

**STRATEGY:**  You are given a couple and are asked to determine the position at a given angular velocity, so use the principle of work and energy.

**MODELING:**  For part (*a*), choose the system to be both gears and model each as a rigid body. In part (*b*), you are asked to determine an internal force, so you need to choose gear *A* as your system.

**ANALYSIS:**

**Kinematics.**  The velocity of the point of contact, *P*, is the same for both gears (Fig. 1), so you have

$$v_P = r_A \omega_A = r_B \omega_B \qquad \omega_A = \omega_B \frac{r_B}{r_A} = \omega_B \frac{100 \text{ mm}}{250 \text{ mm}} = 0.40\omega_B$$

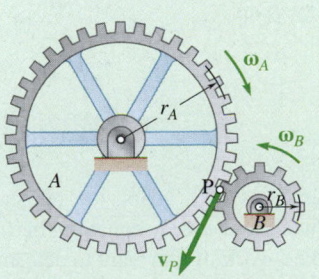

$r_A = 250$ mm

$r_B = 100$ mm

**Fig. 1**  The point of contact has the same velocity on each gear.

*(continued)*

**Calculations.**    For $\omega_B = 600$ rpm, you have

$$\omega_B = 62.8 \text{ rad/s} \qquad \omega_A = 0.40\omega_B = 25.1 \text{ rad/s}$$

$$\bar{I}_A = m_A \bar{k}_A^2 = (10 \text{ kg})(0.200 \text{ m})^2 = 0.400 \text{ kg·m}^2$$

$$\bar{I}_B = m_B \bar{k}_B^2 = (3 \text{ kg})(0.080 \text{ m})^2 = 0.0192 \text{ kg·m}^2$$

**Principle of Work and Energy:**    Apply the principle of work and energy

$$T_1 + U_{1\rightarrow 2} = T_2 \tag{1}$$

You need to calculate the initial and final kinetic energy and the work.

**Kinetic Energy.**    The system is initially at rest, so $T_1 = 0$. Adding the kinetic energies of the two gears when $\omega_B = 600$ rpm gives

$$T_2 = \tfrac{1}{2}\bar{I}_A \omega_A^2 + \tfrac{1}{2}\bar{I}_B \omega_B^2$$

$$= \tfrac{1}{2}(0.400 \text{ kg·m}^2)(25.1 \text{ rad/s})^2 + \tfrac{1}{2}(0.0192 \text{ kg·m}^2)(62.8 \text{ rad/s})^2$$

$$= 163.9 \text{ J}$$

**Work.**    Denote the angular displacement of gear $B$ by $\theta_B$. Then

$$U_{1\rightarrow 2} = M\theta_B = (6 \text{ N·m})(\theta_B \text{ rad}) = (6\theta_B) \text{ J}$$

Substituting these terms into Eq. (1) gives you

$$0 + (6\theta_B) \text{ J} = 163.9 \text{ J}$$

$$\theta_B = 27.32 \text{ rad} \qquad\qquad \theta_B = 4.35 \text{ rev} \;\blacktriangleleft$$

**Motion of Gear A.**

**Kinetic Energy.**    Initially, gear $A$ is at rest, so $T_1 = 0$. When $\omega_B = 600$ rpm, the kinetic energy of gear $A$ is

$$T_2 = \tfrac{1}{2}\bar{I}_A \omega_A^2 = \tfrac{1}{2}(0.400 \text{ kg·m}^2)(25.1 \text{ rad/s})^2 = 126.0 \text{ J}$$

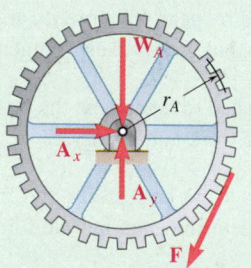

**Fig. 2**  Free-body diagram for gear A.

**Work.**    The forces acting on gear $A$ are shown in Fig. 2. The tangential force **F** does work equal to the product of its magnitude and of the length $\theta_A r_A$ of the arc described by the point of contact. Because $\theta_A r_A = \theta_B r_B$, you have

$$U_{1\rightarrow 2} = F(\theta_B r_B) = F(27.3 \text{ rad})(0.100 \text{ m}) = F(2.73 \text{ m})$$

Substituting these values into work and energy gives

$$T_1 + U_{1\rightarrow 2} = T_2$$

$$0 + F(2.73 \text{ m}) = 126.0 \text{ J}$$

$$F = +46.2 \text{ N} \qquad\qquad \mathbf{F} = 46.2 \text{ N} \;\nearrow \;\blacktriangleleft$$

**REFLECT and THINK:**  When the system was both gears, the tangential force between the gears did not appear in the work and energy equation, because it was internal to the system and therefore did no work. If you want to determine an internal force, you need to define a system where the force of interest is an external force. This problem, like most work and energy problems, also could have been solved using Newton's second law and kinematic relationships.

## Sample Problem 17.3

A sphere, a cylinder, and a hoop, each having the same mass and the same radius, are released from rest on an incline. Determine the velocity of each body after it has rolled through a distance corresponding to a change in elevation $h$. ▶

**STRATEGY:** You are given two positions, want to find the velocities, and the friction force **F** in rolling motion does no work, so use the conservation of energy. First solve the problem in general terms, and then find the results for each body. Denote the mass by $m$, the centroidal moment of inertia by $\bar{I}$ and the radius by $r$.

**MODELING:** Choose the rolling object as your system and model it as a rigid body. Because each body rolls, the instantaneous center of rotation is located at $C$ (Fig. 1). Free-body diagrams of the system at the two locations are shown in Fig. 2.

**ANALYSIS:**

### Conservation of Energy.

$$T_1 + V_{g_1} + V_{e_1} = T_2 + V_{g_2} + V_{e_2} \tag{1}$$

*Potential Energy.*
Because there is no spring in the system, $V_{e_1} = V_{e_2} = 0$. If you place your datum at the center of mass of the system when it is at position 2, you have $V_{g_2} = 0$ and $V_{g_1} = mgh$.

*Kinetic Energy.*

$$T_1 = 0$$
$$T_2 = \tfrac{1}{2}m\bar{v}^2 + \tfrac{1}{2}\bar{I}\omega^2$$

**Kinematics.** You need to relate $\bar{v}$ and $\omega$ using kinematics. Because each body rolls, the instantaneous center of rotation is located at $C$ (Fig. 1), which gives

$$\omega = \frac{\bar{v}}{r}$$

Substituting this into $T_2$ gives

$$T_2 = \tfrac{1}{2}m\bar{v}^2 + \tfrac{1}{2}\bar{I}\left(\frac{\bar{v}}{2}\right)^2 = \frac{1}{2}\left(m + \frac{\bar{I}}{r^2}\right)\bar{v}^2$$

Substituting these energy expressions into Eq. (1) gives

$$0 + mgh + 0 = \frac{1}{2}\left(m + \frac{\bar{I}}{r^2}\right)\bar{v}^2 + 0 + 0$$

Solving for the speed at position 2, you find

$$\bar{v}^2 = \frac{2gh}{1 + \bar{I}/mr^2}$$

*(continued)*

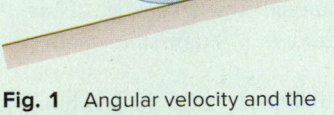

**Fig. 1** Angular velocity and the velocity of the center of mass of the rolling object.

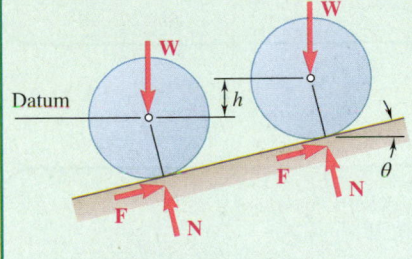

**Fig. 2** Free-body diagrams of the system in positions 1 and 2.

**Velocities of Sphere, Cylinder, and Hoop.**   Introducing the particular expressions for $\bar{I}$, you obtain

*Sphere:* $\qquad\qquad\qquad\qquad \bar{I} = \frac{2}{5}mr^2 \qquad\qquad \bar{v} = 0.845\sqrt{2gh}$ ◀

*Cylinder:* $\qquad\qquad\qquad\quad \bar{I} = \frac{1}{2}mr^2 \qquad\qquad \bar{v} = 0.816\sqrt{2gh}$ ◀

*Hoop:* $\qquad\qquad\qquad\qquad\; \bar{I} = mr^2 \qquad\qquad\; \bar{v} = 0.707\sqrt{2gh}$ ◀

**REFLECT and THINK:**   Comparing the results, we note that the velocity of the body is independent of both its mass and radius. However, the velocity does depend upon the quotient of $\bar{I}/mr^2 = \bar{k}^2/r^2$, which measures the ratio of the rotational kinetic energy to the translational kinetic energy. Thus, the hoop, which has the largest $\bar{k}$ for a given radius $r$, attains the smallest velocity.

Let us compare the results with the velocity attained by a frictionless block sliding through the same distance. The solution is identical to the previous solution except that $\omega = 0$; we find $\bar{v} = \sqrt{2gh}$. So, all the rolling objects are slower than one moving down a frictionless surface.

## Sample Problem 17.4

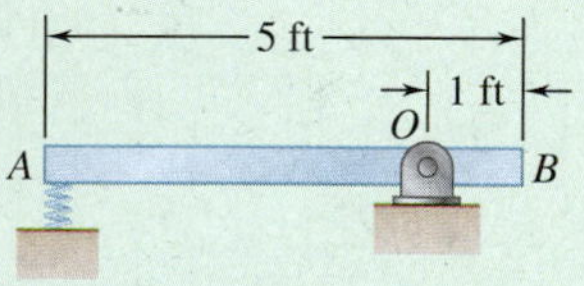

A 30-lb slender rod $AB$ is 5 ft long and is pivoted about a point $O$ that is 1 ft from end $B$. The other end is pressed against a vertical spring with a constant of $k = 1800$ lb/in. until the spring is compressed 1 in. The rod is then in a horizontal position. If the rod is released from this position, determine its angular velocity and the reaction at the pivot $O$ as the rod passes through a vertical position. ▶

**STRATEGY:**   Because you are given two positions, want to find the velocities, and no external forces do work, use the conservation of energy. To determine the reactions at position 2, use a free-body diagram and a kinetic diagram.

**MODELING:**   Choose the rod and the spring as your system and model the rod as a rigid body. Denote the initial position as position 1 and the vertical position as position 2 (Fig. 1). Choose your datum to be at position 1.

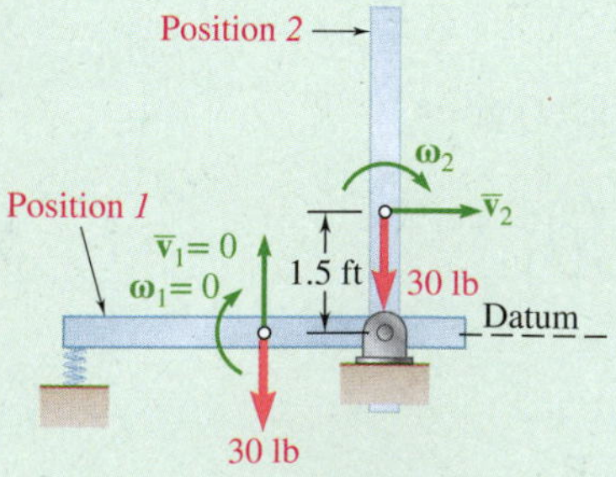

**ANALYSIS:**

**Conservation of Energy.**

$$T_1 + V_{g_1} + V_{e_1} = T_2 + V_{g_2} + V_{e_2} \qquad (1)$$

You need to calculate the energy at position 1 and position 2.

**Fig. 1**   The rod in positions 1 and 2.

*(continued)*

### Position 1.

***Potential Energy.*** The spring is compressed 1 in., so you have $x_1 = 1$ in. The elastic potential energy is

$$V_{e_1} = \tfrac{1}{2}kx_1^2 = \tfrac{1}{2}(1800 \text{ lb/in.})(1 \text{ in.})^2 = 900 \text{ in.·lb} = 75 \text{ ft·lb}$$

Because the datum is at position 1, you have $V_{g_1} = 0$.

***Kinetic Energy.*** The velocity in position 1 is zero, so you have $T_1 = 0$.

### Position 2.

***Potential Energy.*** The elongation of the spring is zero, so you have $V_{e_2} = 0$. Because the center of gravity of the rod is now 1.5 ft above the datum, you have

$$V_{g_2} = mgy = (30 \text{ lb})(1.5 \text{ ft}) = 45 \text{ ft·lb}$$

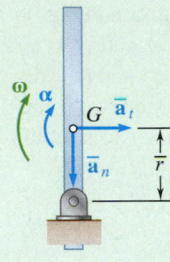

**Fig. 2** The acceleration of the center of mass and the angular velocity and acceleration of the rod.

***Kinetic Energy.*** Denote the angular velocity of the rod in position 2 by $\boldsymbol{\omega}_2$. The rod rotates about $O$, so you have $\bar{v}_2 = \bar{r}\omega_2 = 1.5\omega_2$ and

$$\bar{I} = \tfrac{1}{12}ml^2 = \frac{1}{12}\left(\frac{30 \text{ lb}}{32.2 \text{ ft/s}^2}\right)(5 \text{ ft})^2 = 1.941 \text{ lb·ft·s}^2$$

$$T_2 = \tfrac{1}{2}m\bar{v}_2^2 + \tfrac{1}{2}\bar{I}\omega_2^2 = \frac{1}{2}\frac{30}{32.2}(1.5\omega_2)^2 + \tfrac{1}{2}(1.941)\omega_2^2 = 2.019\omega_2^2$$

Substituting these expressions into Eq. (1) give

$$0 + 0 + 75 \text{ ft·lb} = 2.019\omega_2^2 + 45 \text{ ft·lb} + 0 \qquad \boldsymbol{\omega}_2 = 3.86 \text{ rad/s} \circlearrowleft \quad \blacktriangleleft$$

***Reaction.*** Because $\omega_2 = 3.86$ rad/s, the components of the acceleration of $G$ as the rod passes through position 2 are (Fig. 2)

$$\bar{a}_n = \bar{r}\omega_2^2 = (1.5 \text{ ft})(3.86 \text{ rad/s})^2 = 22.3 \text{ ft/s}^2 \qquad \bar{\mathbf{a}}_n = 22.3 \text{ ft/s}^2 \downarrow$$

$$\bar{a}_t = \bar{r}\alpha \qquad\qquad\qquad\qquad\qquad\qquad\qquad \bar{\mathbf{a}}_t = \bar{r}\alpha \rightarrow$$

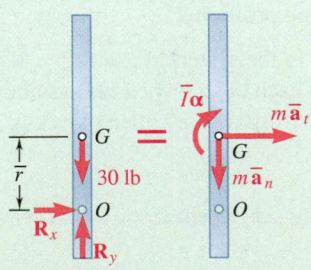

**Fig. 3** Free-body diagram and kinetic diagram for the rod.

Draw free-body and kinetic diagrams (Fig. 3) to express that the system of external forces is equivalent to the vector of components $m\bar{\mathbf{a}}_t$ and $m\bar{\mathbf{a}}_n$ attached at $G$ and the couple $\bar{I}\alpha$.

$$+\circlearrowleft \Sigma M_0 = \bar{I}\alpha + m\bar{a}d_\perp: \qquad 0 = \bar{I}\alpha + m(\bar{r}\alpha)\bar{r} \qquad\qquad \alpha = 0$$

$$\overset{+}{\rightarrow}\Sigma F_x = m\bar{a}_x: \qquad\qquad R_x = m(\bar{r}\alpha) \qquad\qquad\qquad R_x = 0$$

$$+\uparrow\Sigma F_y = m\bar{a}_y: \qquad\qquad R_y - 30 \text{ lb} = -m\bar{a}_n$$

$$R_y - 30 \text{ lb} = -\frac{30 \text{ lb}}{32.2 \text{ ft/s}^2}(22.3 \text{ ft/s}^2)$$

$$R_y = +9.22 \text{ lb} \qquad\qquad \mathbf{R} = 9.22 \text{ lb} \uparrow \quad \blacktriangleleft$$

**REFLECT and THINK:** This problem illustrates how you might need to supplement the conservation of energy with Newton's second law. What if the spring constant had been smaller, say 180 lb/in.? You would have found $V_{e_1} = 7.5$ ft·lb and then solved Eq. (1) to obtain $\omega_2^2 = -18.57$. This is clearly impossible and means that the rod would not make it to position 2 as assumed.

## Sample Problem 17.5

A large box with a mass $m$ and a flat bottom rests on two identical homogeneous cylindrical rollers, where each has radius $r$ and a mass half that of the crate. The system is released from rest on a plane that is inclined at angle $\phi$ to the horizontal. Determine the speed of the box at the instant when the rollers have turned through an angle $\theta$. Neglect rolling resistance and assume that the rollers do not slide. ▶

**STRATEGY:** You are interested in the velocity after the rollers have moved a specified distance, $r\theta$, and the friction force in rolling motion does no work, so use the conservation of energy.

**MODELING:** Choose the box and the two cylindrical rollers as your system and model them as rigid bodies. In order for you to draw the system in its initial and final positions, you need to know how far each mass travels. You can determine this by using the instantaneous center of rotation. The rollers do not slide, so the instantaneous center of rotation for each roller is located at the point of contact, $C$, between the roller and the ground (Fig. 1). Using this instantaneous center of velocity, you know $v_B = 2\omega r$ and $v_R = \omega r$. Therefore, the box moves down a distance $2h$ when the rollers move a distance $h$ (Fig. 2). Because you have three masses in the system (the two rollers and the box), you may define an individual datum for each mass to simplify the calculation of the gravitational potential energy.

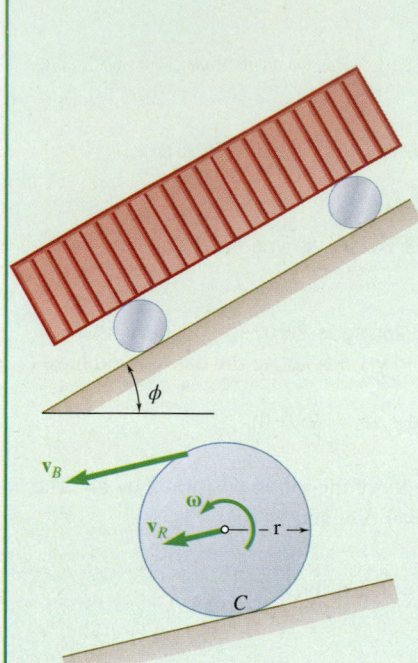

**Fig. 1** Velocity of various points on the roller.

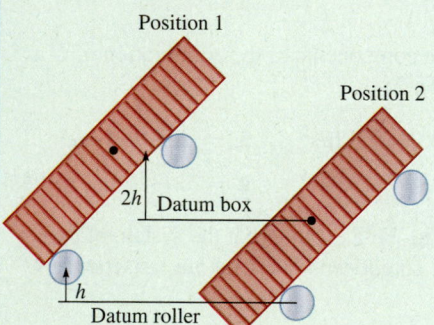

**Fig. 2** The system in positions 1 and 2.

### ANALYSIS:

**Conservation of Energy.**

$$T_1 + V_{g_1} + V_{e_1} = T_2 + V_{g_2} + V_{e_2} \tag{1}$$

You need to calculate the energy at position 1 and position 2.

*Potential Energy.* Because there is no spring in the system, $V_{e_1} = V_{e_2} = 0$. If you place your datum at the center of mass of each object when the system is at position 2, you have $V_{g_2} = 0$. The vertical distance a roller moves is $h = r\theta \sin\phi$, so

$$V_{g_1} = mg(2h) + 2(\tfrac{m}{2})g(h) = 3mgh = 3mgr\theta \sin\phi$$

*Kinetic Energy.* The velocity in position 1 is zero, so $T_1 = 0$.

At position 2,

$$T_2 = 2(\tfrac{1}{2}m_R v_R^2 + \tfrac{1}{2}\bar{I}\omega^2) + \tfrac{1}{2}mv_B^2$$

where $m_R$ is the mass of the roller and $\bar{I}$ is the mass moment of inertia of the roller about its center of gravity. Substituting $v_B = 2\omega r$, $v_R = \omega r$, and $\bar{I} = \tfrac{1}{2}m_R r^2 = \tfrac{1}{2}(\tfrac{m}{2})r^2 = \tfrac{1}{4}mr^2$ into $T_2$ gives

$$T_2 = \tfrac{11}{4}mr^2\omega^2$$

Substituting these expressions into Eq. (1), you find

$$0 + 3mgr\theta \sin \phi + 0 = \tfrac{11}{4} mr^2\omega^2 + 0 + 0$$

Solving for the angular velocity, $\omega = \sqrt{\dfrac{12g\theta \sin \phi}{11r}}$, so the velocity of the box at $v_B = 2\omega r$ is

$$v_B = 4 \sqrt{\frac{3}{11} gr\theta \sin \phi} \quad \triangleleft$$

**REFLECT and THINK:** If the rollers had been attached to the box by brackets, they would have traveled the same vertical distance as the box and the change in height of the centers of gravity of rollers and of the box would have been equal.

## Sample Problem 17.6

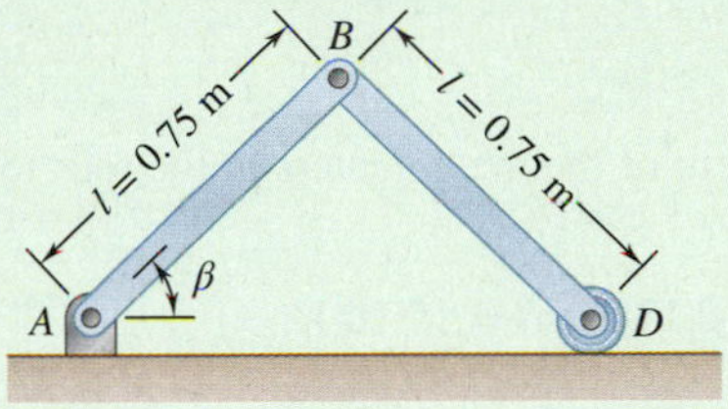

Each of the two slender rods shown is 0.75 m long and has a mass of 6 kg. If the system is released from rest with $\beta = 60°$, determine (a) the angular velocity of rod $AB$ when $\beta = 20°$, (b) the velocity of point $D$ at the same instant.

**STRATEGY:** You have two positions and are interested in velocities, so use the conservation of energy. You will also need to use kinematics to relate the velocity terms in the kinetic energy expression.

**MODELING:** Choose the system to be both bars and model them as rigid bodies.

**ANALYSIS:** To illustrate that the order in which you solve a problem doesn't matter, let's start with kinematics.

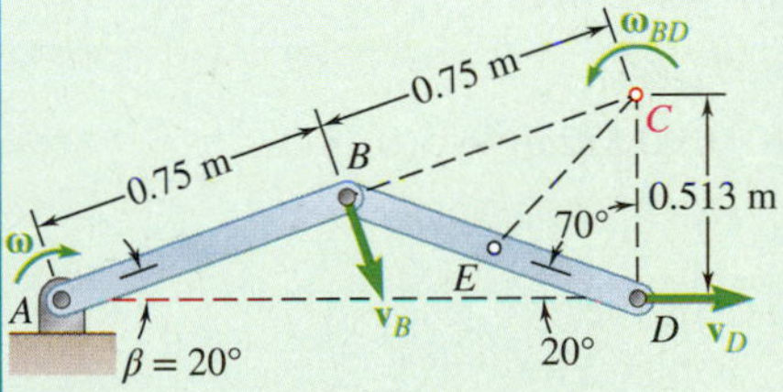

**Kinematics of Motion When $\beta = 20°$.** Because $\mathbf{v}_B$ is perpendicular to the rod $AB$ and $\mathbf{v}_D$ is horizontal, the instantaneous center of rotation of rod $BD$ is located at $C$ (Fig. 1). From the geometry of the figure, you obtain

$$BC = 0.75 \text{ m} \qquad CD = 2(0.75 \text{ m}) \sin 20° = 0.513 \text{ m}$$

Apply the law of cosines to triangle $CDE$, where $E$ is located at the mass center of rod $BD$. You find $EC = 0.522$ m. Denoting the angular velocity of rod $AB$ by $\omega$, you have (Fig. 2)

$$\overline{\mathbf{v}}_{AB} = (0.375 \text{ m})\omega \qquad \overline{\mathbf{v}}_{AB} = 0.375\omega \searrow$$
$$v_B = (0.75 \text{ m})\omega \qquad \mathbf{v}_B = 0.75\omega \searrow$$

Because rod $BD$ seems to rotate about point $C$, you have

$$v_B = (BC)\omega_{BD} \qquad (0.75 \text{ m})\omega = (0.75 \text{ m})\omega_{BD} \qquad \boldsymbol{\omega}_{BD} = \omega \circlearrowleft$$
$$\overline{v}_{BD} = (EC)\omega_{BD} = (0.522 \text{ m})\omega \qquad \overline{\mathbf{v}}_{BD} = 0.522\omega \searrow$$

*(continued)*

**Fig. 1** Instantaneous center of rotation $C$ for bar $BD$.

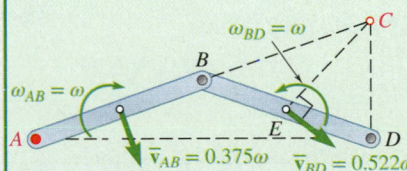

**Fig. 2** Velocities of the center of masses of $AB$ and $BD$ in terms of $\omega$.

**Conservation of Energy.**    Because there are no springs in the system

$$T_1 + V_{g_1} = T_2 + V_{g_2}$$

You first need to determine the energy at the two positions.

### Position 1.

*Potential Energy.*    Choose the datum as shown in Fig. 3, and observe that $W = (6 \text{ kg})(9.81 \text{ m/s}^2) = 58.86$ N. Then you have

$$V_{g_1} = 2W\overline{y}_1 = 2(58.86 \text{ N})(0.325 \text{ m}) = 38.26 \text{ J}$$

*Kinetic Energy.*    Initially, the system is at rest, so $T_1 = 0$.

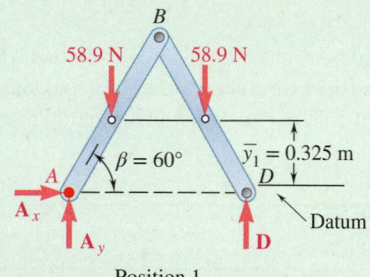

**Fig. 3**    Free-body diagram and distance from the datum in position 1.

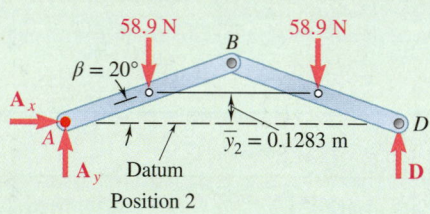

**Fig. 4**    Free-body diagram and distance from the datum in position 2.

### Position 2.

*Potential Energy.*    Compute the new height of the mass centers of the rods to be $\overline{y}_2 = 0.75\sin(20) = 0.1283$ m (Fig. 4).

$$V_{g_2} = 2W\overline{y}_2 = 2(58.86 \text{ N})(0.1283 \text{ m}) = 15.10 \text{ J}$$

*Kinetic Energy.*

$$\overline{I}_{AB} = \overline{I}_{BD} = \tfrac{1}{12}ml^2 = \tfrac{1}{12}(6 \text{ kg})(0.75 \text{ m})^2 = 0.281 \text{ kg·m}^2$$

$$
\begin{aligned}
T_2 &= \tfrac{1}{2}m\overline{v}_{AB}^2 + \tfrac{1}{2}\overline{I}_{AB}\omega_{AB}^2 + \tfrac{1}{2}m\overline{v}_{BD}^2 + \tfrac{1}{2}\overline{I}_{BD}\omega_{BD}^2 \\
&= \tfrac{1}{2}(6)(0.375\omega)^2 + \tfrac{1}{2}(0.281)\omega^2 + \tfrac{1}{2}(6)(0.522\omega)^2 + \tfrac{1}{2}(0.281)\omega^2 \\
&= 1.520\omega^2
\end{aligned}
$$

**Conservation of Energy.**    Now you can write

$$T_1 + V_{g_1} = T_2 + V_{g_2}$$
$$0 + 38.26 \text{ J} = 1.520\omega^2 + 15.10 \text{ J}$$
$$\omega = 3.90 \text{ rad/s} \qquad \boldsymbol{\omega}_{AB} = 3.90 \text{ rad/s} \; \circlearrowleft \quad \blacktriangleleft$$

**Velocity of Point D.**

$$v_D = (CD)\omega = (0.513 \text{ m})(3.90 \text{ rad/s}) = 2.00 \text{ m/s}$$
$$\mathbf{v}_D = 2.00 \text{ m/s} \rightarrow \quad \blacktriangleleft$$

**REFLECT and THINK:**    The only step in which you need to use forces is when calculating the gravitational potential energy in each position. However, it is good engineering practice to show the complete free-body diagram in each case to identify which, if any, forces do work. Rather than use the instantaneous center of rotation, you could have also used vector algebra to relate the velocities of the various objects.

# SOLVING PROBLEMS
# ON YOUR OWN

In this section, we introduced energy methods to determine the velocity of rigid bodies for various positions during their motion. As you saw previously in Chap. 13, energy methods are particularly useful for problems involving displacements and velocities.

**1. The method of work and energy,** when applied to all of the particles forming a rigid body, yields the equation

$$T_1 + U_{1\to2} = T_2 \tag{17.1}$$

where $T_1$ and $T_2$ are, respectively, the initial and final values of the total kinetic energy of the particles forming the body and $U_{1\to2}$ is the work done by the external forces exerted on the rigid body. If we express the work done by nonconservative forces as $U_{1\to2}^{NC}$ and define the potential energy terms for conservative forces, we can express Eq. (17.1) as

$$T_1 + V_{g_1} + V_{e_1} + U_{1\to2}^{NC} = T_2 + V_{g_2} + V_{e_2} \tag{17.1'}$$

where $V_{g_1}$ and $V_{g_2}$ are the initial and final gravitational potential energy of the center of mass of the rigid body and $V_{e_1}$ and $V_{e_2}$ are the initial and final values of the elastic energy associated with springs in the system. Recall that, for a linear spring, $V_e = \frac{1}{2}kx^2$, where $x$ is the deflection of the spring from its unstretched length. For a single rigid body, $V_g = mgy$, where $y$ is the elevation of the center of mass from a reference plane or datum.

   **a. Work of forces and couples.** To the expression for the work of a force (Chap. 13), we added the expression for the work of a couple and wrote

$$U_{1\to2} = \int_{A_1}^{A_2} \mathbf{F} \cdot d\mathbf{r} \qquad U_{1\to2} = \int_{\theta_1}^{\theta_2} M\,d\theta \tag{17.3, 17.5}$$

When the moment of a couple is constant, the work of the couple is

$$U_{1\to2} = M(\theta_2 - \theta_1) \tag{17.6}$$

where $\theta_1$ and $\theta_2$ are expressed in radians (Sample Probs. 17.1 and 17.2).

   **b. The kinetic energy of a rigid body in plane motion** was found by considering the motion of the body as the sum of a translation with its mass center and a rotation about the mass center. So

$$T = \tfrac{1}{2}m\bar{v}^2 + \tfrac{1}{2}\bar{I}\omega^2 \tag{17.9}$$

where $\bar{v}$ is the velocity of the mass center and $\omega$ is the angular velocity of the body (Sample Probs. 17.3 and 17.4). You will generally need to use kinematics to relate $\bar{v}$ and $\omega$.

*(continued)*

**2. For a system of rigid bodies** we again used the equation

$$T_1 + U_{1 \rightarrow 2} = T_2 \tag{17.1}$$

where $T$ is the sum of the kinetic energies of the bodies forming the system and $U$ is the work done by *all the forces acting on the bodies*—internal as well as external. Your computations will be simplified if you keep the following ideas in mind.

**a. The forces exerted on each other by pin-connected members or by meshed gears** are equal and opposite, and because they have the same point of application, they undergo equal small displacements. Therefore, *their total work is zero* and can be omitted from your calculations (Sample Prob. 17.2).

**b. The forces exerted by an inextensible cord** on the two bodies it connects have the same magnitude and their points of application move through equal distances, but the work of one force is positive and the work of the other is negative. Therefore, *their total work is zero* and again can be omitted from your calculations (Sample Prob. 17.1).

**c. The forces exerted by a spring** on the two bodies it connects also have the same magnitude, but their points of application generally move through different distances. Therefore, *their total work is usually not zero* and should be taken into account in your calculations. The easiest way to handle springs, therefore, is to use elastic potential energy.

**3. The principle of conservation of energy** can be expressed as

$$T_1 + V_1 = T_2 + V_2 \tag{17.12}$$

where $V$ represents the potential energy of the system. If you prefer to write this equation in terms of gravitational potential energy, $V_g$, and elastic potential energy, $V_e$, you get

$$T_1 + V_{g_1} + V_{e_1} = T_2 + V_{g_2} + V_{e_2} \tag{17.12'}$$

You can use this principle when a body or a system of bodies is acted upon by conservative forces, such as the force exerted by a spring or the force of gravity (Sample Probs. 17.4, 17.5, and 17.6).

**4. The last part of this section was devoted to power,** which is the time rate at which work is done. For a body acted upon by a couple of moment $\mathbf{M}$, the power can be expressed as

$$\text{Power} = M\omega \tag{17.13}$$

where $\omega$ is the angular velocity of the body, expressed in rad/s. As you did in Chap. 13, you should express power either in watts or in horsepower (1 hp = 550 ft·lb/s).

# Problems

CONCEPT QUESTIONS

**17.CQ1** A round object of mass $m$ and radius $r$ is released from rest at the top of a curved surface and rolls without slipping until it leaves the surface with a horizontal velocity as shown. Will a solid sphere, a solid cylinder, or a hoop travel the greatest distance $x$?

   **a.** Solid sphere
   **b.** Solid cylinder
   **c.** Hoop
   **d.** They will all travel the same distance.

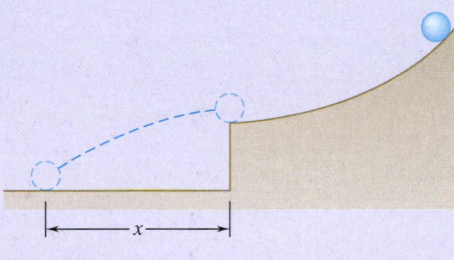

Fig. P17.CQ1

**17.CQ2** A solid steel sphere $A$ of radius $r$ and mass $m$ is released from rest and rolls without slipping down an incline as shown. After traveling a distance $d$, the sphere has a speed $v$. If a solid steel sphere of radius $2r$ is released from rest on the same incline, what will its speed be after rolling a distance $d$?

   **a.** $0.25v$
   **b.** $0.5v$
   **c.** $v$
   **d.** $2v$
   **e.** $4v$

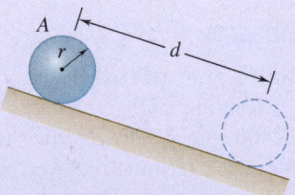

Fig. P17.CQ2

**17.CQ3** Slender bar $A$ is rigidly connected to a massless rod $BC$ in Case 1 and two massless cords in Case 2 as shown. The vertical thickness of bar $A$ is negligible compared to $L$. In both cases, $A$ is released from rest at an angle $\theta = \theta_0$. When $\theta = 0$, which system will have the larger kinetic energy?

**a.** Case 1
**b.** Case 2
**c.** The kinetic energy will be the same.

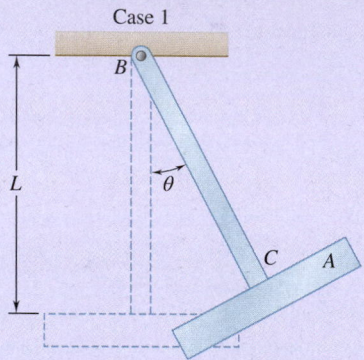
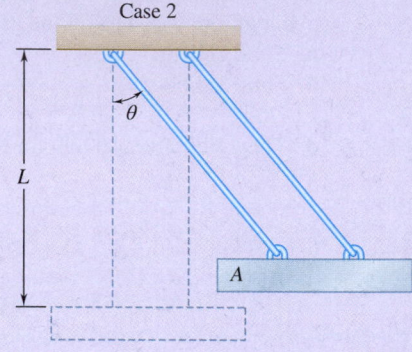

**Fig. P17.CQ3 and P17.CQ5**

**17.CQ4** In Prob. 17.CQ3, how will the speeds of the centers of gravity compare for the two cases when $\theta = 0$?

**a.** Case 1 will be larger.
**b.** Case 2 will be larger.
**c.** The speeds will be the same.

**17.CQ5** Slender bar $A$ is rigidly connected to a massless rod $BC$ in Case 1 and two massless cords in Case 2 as shown. The vertical thickness of bar $A$ is not negligible compared to $L$. In both cases, $A$ is released from rest at an angle $\theta = \theta_0$. When $\theta = 0$, which system will have the largest kinetic energy?

**a.** Case 1
**b.** Case 2
**c.** The kinetic energy will be the same.

**END-OF-SECTION PROBLEMS**

**17.1** A 200-kg flywheel is at rest when a constant 300 N·m couple is applied. After executing 560 revolutions, the flywheel reaches its rated speed of 2400 rpm. Knowing that the radius of gyration of the flywheel is 400 mm, determine the average magnitude of the couple due to kinetic friction in the bearing.

**17.2** The rotor of an electric motor has an angular velocity of 3600 rpm when the load and power are cut off. The 110-lb rotor, which has a centroidal radius of gyration of 9 in., then coasts to rest. Knowing that the kinetic friction of the rotor produces a couple with a magnitude of 2.5 lb·ft, determine the number of revolutions that the rotor executes before coming to rest.

**17.3** Two uniform disks of the same material are attached to a shaft as shown. Disk *A* has a weight of 10 lb and a radius of $r = 6$ in. Disk *B* is twice as thick as disk *A*. Knowing that a couple **M** with a magnitude of 22 lb·ft is applied to disk *A* when the system is at rest, determine the radius *nr* of disk *B* if the angular velocity of the system is to be 480 rpm after five revolutions.

**17.4** Two disks of the same material are attached to a shaft as shown. Disk *A* has a radius *r* and a thickness *b*, while disk *B* has a radius *nr* and a thickness 2*b*. A couple **M** with a constant magnitude is applied when the system is at rest and is removed after the system has executed two revolutions. Determine the value of *n* that results in the largest final speed for a point on the rim of disk *B*.

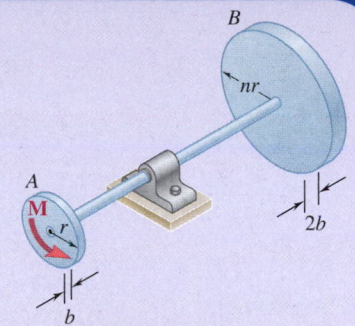

**Fig. P17.3 and P17.4**

**17.5** The flywheel of a punching machine has a weight of 650 lb and a radius of gyration of 30 in. Each punching operation requires 1800 ft·lb of work. (*a*) Knowing that the speed of the flywheel is 300 rpm just before a punching, determine the speed immediately after the punching. (*b*) If a constant 15 lb·ft couple is applied to the shaft of the flywheel, determine the number of revolutions executed before the speed is again 300 rpm.

**17.6** The flywheel of a small punching machine rotates at 360 rpm. Each punching operation requires 2034 J of work, and it is desired that the speed of the flywheel after each punching be no less than 95 percent of the original speed. (*a*) Determine the required moment of inertia of the flywheel. (*b*) Knowing that the initial velocity is to be 360 rpm at the start of each punching, determine the number of revolutions that must occur between two successive punchings if a constant 24.4 N·m couple is applied to the shaft of the flywheel.

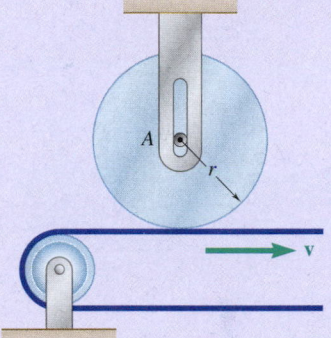

**Fig. P17.7**

**17.7** Disk *A* has a mass of 4 kg and a radius $r = 75$ mm; it is at rest when it is placed in contact with the belt, which moves at a constant speed $v = 18$ m/s. Knowing that $\mu_k = 0.25$ between the disk and the belt, determine the number of revolutions executed by the disk before it reaches a constant angular velocity.

**17.8** The uniform 4-kg cylinder *A* with a radius of $r = 150$ mm has an angular velocity of $\omega_0 = 50$ rad/s when it is brought into contact with an identical cylinder *B* that is at rest. The coefficient of kinetic friction at the contact point *D* is $\mu_k$. After a period of slipping, the cylinders attain constant angular velocities of equal magnitude and opposite direction at the same time. Knowing that cylinder *A* executes three revolutions before it attains a constant angular velocity and that cylinder *B* executes one revolution before it attains a constant angular velocity, determine (*a*) the final angular velocity of each cylinder, (*b*) the coefficient of kinetic friction $\mu_k$.

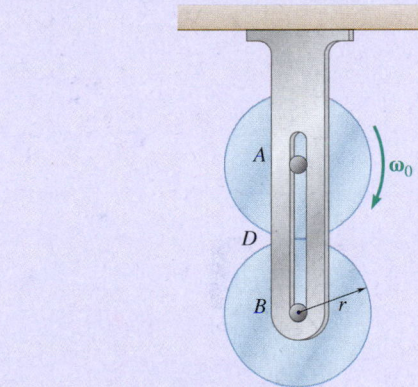

**Fig. P17.8**

**17.9** The 10-in.-radius brake drum is attached to a larger flywheel which is not shown. The total mass moment of inertia of the flywheel and drum is 16 lb·ft·s² and the coefficient of kinetic friction between the drum and the brake shoe is 0.40. Knowing that the initial angular velocity is 240 rpm clockwise, determine the force that must be exerted by the hydraulic cylinder if the system is to stop in 75 revolutions.

**17.10** Solve Prob. 17.9, assuming that the initial angular velocity of the flywheel is 240 rpm counterclockwise.

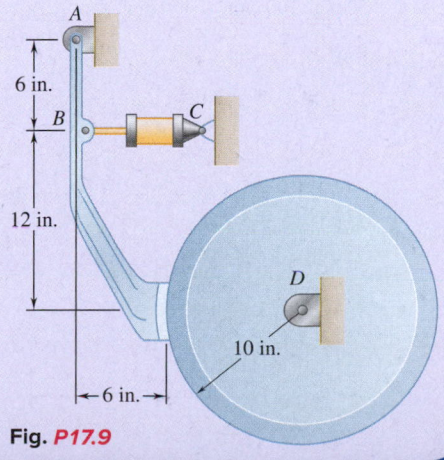

**Fig. P17.9**

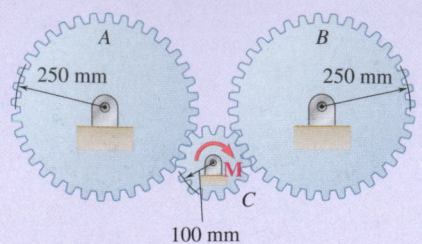

**Fig. P17.11**

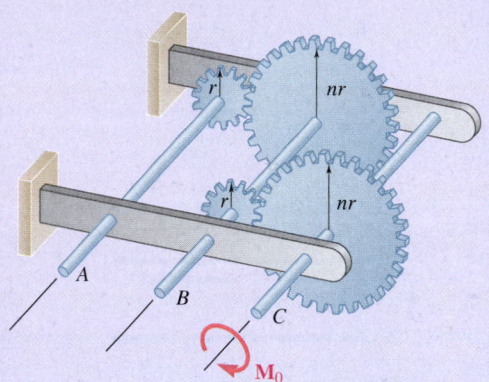

**Fig. P17.13**

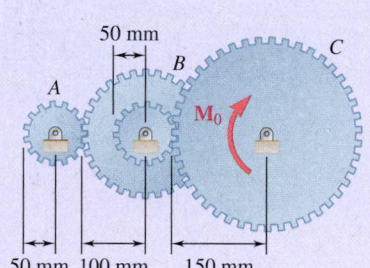

**Fig. P17.15**

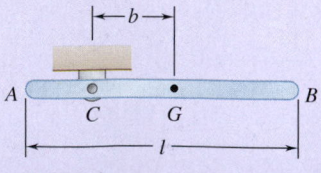

**Fig. P17.16**

**17.11** Each of the gears $A$ and $B$ has a mass of 10 kg and a radius of gyration of 190 mm, while gear $C$ has a mass of 2.5 kg and a radius of gyration of 80 mm. If a couple $\mathbf{M}$ of constant magnitude 6 N·m is applied to gear $C$, determine ($a$) the number of revolutions of gear $C$ required for its angular velocity to increase from 450 rpm to 1800 rpm, ($b$) the corresponding tangential force acting on gear $A$.

**17.12** Solve Prob. 17.11, assuming that the 6 N·m couple is applied to gear $B$.

**17.13** The gear train shown consists of four gears of the same thickness and of the same material; two gears are of radius $r$, and the other two are of radius $nr$. The system is at rest when the couple $\mathbf{M}_0$ is applied to shaft $C$. Denoting by $I_0$ the moment of inertia of a gear of radius $r$, determine the angular velocity of shaft $A$ if the couple $\mathbf{M}_0$ is applied for one revolution of shaft $C$.

**17.14** The double pulley shown has a mass of 15 kg and a centroidal radius of gyration of 160 mm. Cylinder $A$ and block $B$ are attached to cords that are wrapped on the pulleys as shown. The coefficient of kinetic friction between block $B$ and the surface is 0.2. Knowing that the system is at rest in the position shown when a constant force $\mathbf{P} = 200$ N is applied to cylinder $A$, determine ($a$) the velocity of cylinder $A$ as it strikes the ground, ($b$) the total distance that block $B$ moves before coming to rest.

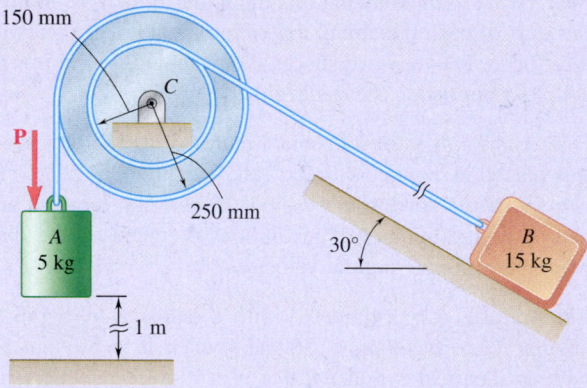

**Fig. *P17.14***

**17.15** Gear $A$ has a mass of 1 kg and a radius of gyration of 30 mm; gear $B$ has a mass of 4 kg and a radius of gyration of 75 mm; gear $C$ has a mass of 9 kg and a radius of gyration of 100 mm. The system is at rest when a couple $\mathbf{M}_0$ of constant magnitude 4 N·m is applied to gear $C$. Assuming that no slipping occurs between the gears, determine the number of revolutions required for disk $A$ to reach an angular velocity of 300 rpm.

**17.16** A slender rod of length $l$ and mass $m$ is pivoted about a point $C$ located at a distance $b$ from its center $G$. It is released from rest in a horizontal position and swings freely. Determine ($a$) the angular velocity of the rod as it passes through a vertical position if $b = l/2$, ($b$) the distance $b$ for which the angular velocity of the rod as it passes through a vertical position is maximum, ($c$) the corresponding values of its angular velocity and of the reaction at $C$ using the value of $b$ calculated.

**17.17** The 15-kg rear hatch of a vehicle opens as shown and can be modeled as a uniform 0.6-m long slender rod. Knowing that the tailgate is released from rest in the position shown, determine the angular velocity of the tailgate as it impacts the car body.

**17.18** A slender 9-lb rod can rotate in a vertical plane about a pivot at $B$. A spring of constant $k = 30$ lb/ft and of unstretched length 6 in. is attached to the rod as shown. Knowing that the rod is released from rest in the position shown, determine its angular velocity after it has rotated through 90°.

**17.19** An adapted golf device attaches to a wheelchair to help people with mobility impairments play putt-putt. The stationary frame $OD$ is attached to the wheelchair, and a club holder $OB$ is attached to the pin at $O$. Holder $OB$ is 6 in. long and weighs 8 oz, and the distance between $O$ and $D$ is $x = 1$ ft. The putter shaft has a length of $L = 36$ in. and weighs 10 oz, while the putter head at $A$ weighs 12 oz. Knowing that the 1-lb/in. spring between $D$ and $B$ is unstretched when $\theta = 90°$ and that the putter is released from rest at $\theta = 0$, determine the putter head speed when it hits the golf ball.

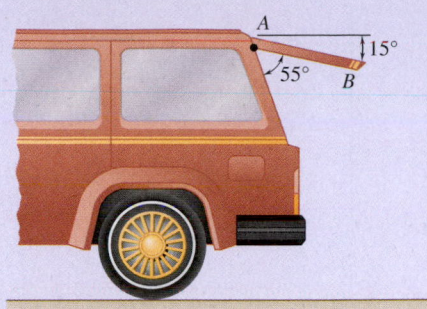

**Fig. P17.17**

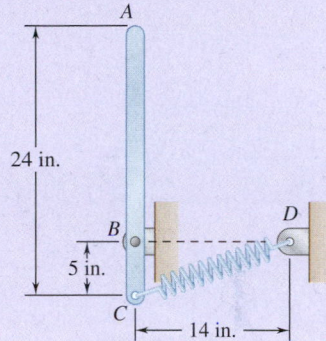

**Fig. P17.18**

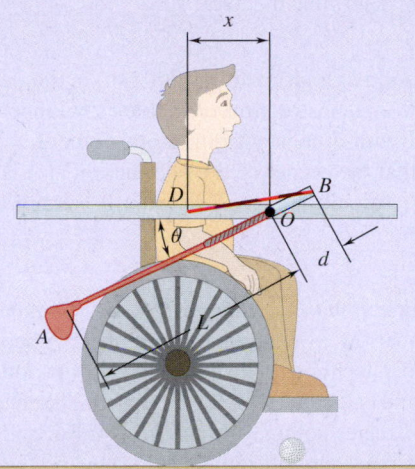

**Fig. P17.19**

**17.20** A 10-kg storm window measuring 900 × 1500 mm is held by hinges at $A$ and $B$. In the position shown, it is held away from the side of the house by a 600-mm stick $CD$. If the stick is suddenly removed, allowing the window to swing shut, determine immediately before impact (a) the angular velocity of the window, (b) the reactions at $A$ and $B$.

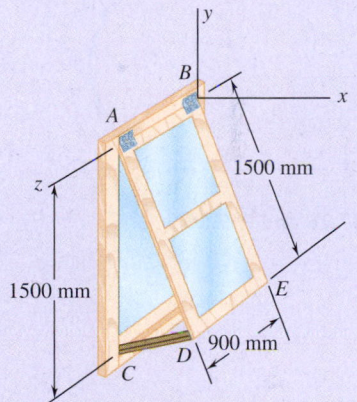

**Fig. P17.20**

**17.21** A collar with a mass of 1 kg is rigidly attached at a distance $d = 300$ mm from the end of a uniform slender rod $AB$. The rod has a mass of 3 kg and is of length $L = 600$ mm. Knowing that the rod is released from rest in the position shown, determine the angular velocity of the rod after it has rotated through 90°.

**17.22** A collar with a mass of 1 kg is rigidly attached to a slender rod $AB$ of mass 3 kg and length $L = 600$ mm. The rod is released from rest in the position shown. Determine the distance $d$ for which the angular velocity of the rod is maximum after it has rotated through 90°.

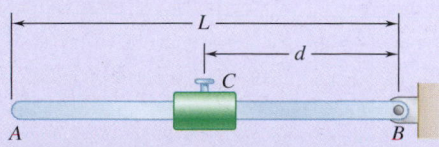

**Fig. *P17.21* and *P17.22***

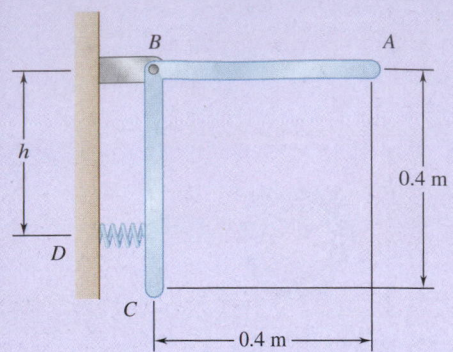

**Fig. P17.23**

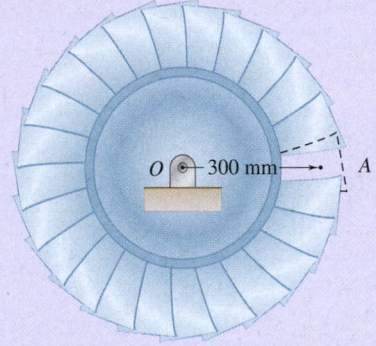

**Fig. P17.24**

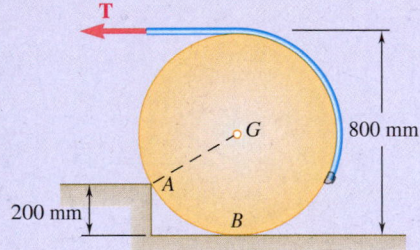

**Fig. P17.25 and P17.26**

**17.23** Two identical slender rods $AB$ and $BC$ are welded together to form an L-shaped assembly. The assembly is pressed against a spring at $D$ and released from the position shown. Knowing that the maximum angle of rotation of the assembly in its subsequent motion is 90° counterclockwise, determine the magnitude of the angular velocity of the assembly as it passes through the position where rod $AB$ forms an angle of 30° with the horizontal.

**17.24** The 30-kg turbine disk has a centroidal radius of gyration of 175 mm and is rotating clockwise at a constant rate of 60 rpm when a small blade of weight 0.5 N at point $A$ becomes loose and is thrown off. Neglecting friction, determine the change in the angular velocity of the turbine disk after it has rotated through (a) 90°, (b) 270°.

**17.25** A 100-kg solid cylindrical disk, 800 mm in diameter, is to be raised over a 200-mm obstruction. A cable is wrapped around the disk and pulled with a constant horizontal force $\mathbf{T}$ as shown. Knowing that the disk rotates about the corner of the obstruction and that the angular velocity of the disk is 1 rad/s when it has reached the top of the obstruction, determine the force $\mathbf{T}$.

**17.26** A 100-kg solid cylindrical disk, 800 mm in diameter, is to be raised over a 200-mm obstruction. A cable is wrapped around the disk and pulled with a constant horizontal force $\mathbf{T} = 260$ N as shown. Knowing that the corner of the obstruction at $A$ is rough, determine the angular velocity of the disk when it has reached the top of the obstruction.

**17.27** Greek engineers had the unenviable task of moving large columns from the quarries to the city. One engineer, Chersiphron, tried several different techniques to do this. One method was to cut pivot holes into the ends of the stone and then use oxen to pull the column. The 4-ft diameter column weighs 12,000 lbs, and the team of oxen generates a constant pull force of 1500 lbs on the center of the cylinder $G$. Knowing that the column starts from rest and rolls without slipping, determine (a) the velocity of its center $G$ after it has moved 5 ft, (b) the minimum static coefficient of friction that will keep it from slipping.

**Fig. P17.27**

**17.28** A small sphere of mass $m$ and radius $r$ is released from rest at $A$ and rolls without sliding on the curved surface to point $B$ where it leaves the surface with a horizontal velocity. Knowing that $a = 1.5$ m and $b = 1.2$ m, determine (a) the speed of the sphere as it strikes the ground at $C$, (b) the corresponding distance $c$.

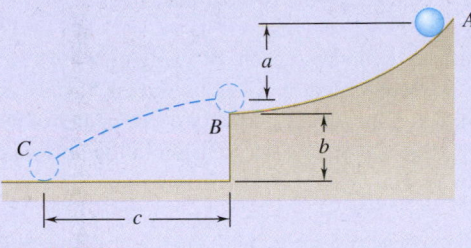

**Fig. P17.28**

**17.29** The mass center $G$ of a 3-kg wheel of radius $R = 180$ mm is located at a distance $r = 60$ mm from its geometric center $C$. The centroidal radius of gyration of the wheel is $\bar{k} = 90$ mm. As the wheel rolls without sliding, its angular velocity is observed to vary. Knowing that $\omega = 8$ rad/s in the position shown, determine (a) the angular velocity of the wheel when the mass center $G$ is directly above the geometric center $C$, (b) the reaction at the horizontal surface at the same instant.

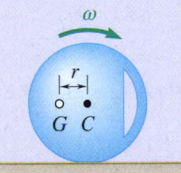

**Fig. P17.29**

**17.30** A half-cylinder with mass $m$ and radius $r$ is released from rest in the position shown. Knowing that the half-cylinder rolls without sliding, determine (a) its angular velocity after it has rolled through 90°, (b) the reaction at the horizontal surface at the same instant. (*Hint:* Note that $GO = 4r/3\pi$ and that, by the parallel-axis theorem, $\bar{I} = \frac{1}{2}mr^2 - m(GO)^2$.)

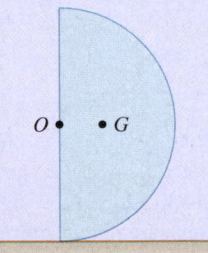

**Fig. P17.30**

**17.31** A sphere of mass $m$ and radius $r$ rolls without slipping inside a curved surface of radius $R$. Knowing that the sphere is released from rest in the position shown, derive an expression for (a) the linear velocity of the sphere as it passes through $B$, (b) the magnitude of the vertical reaction at that instant.

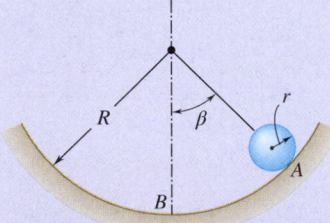

**Fig. P17.31**

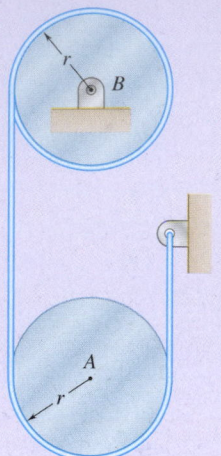

**Fig. P17.32 and P17.33**

**17.32** Two uniform cylinders, each of weight $W = 14$ lb and radius $r = 5$ in., are connected by a belt as shown. Knowing that at the instant shown the angular velocity of cylinder $B$ is 30 rad/s clockwise, determine (a) the distance through which cylinder $A$ will rise before the angular velocity of cylinder $B$ is reduced to 5 rad/s, (b) the tension in the portion of belt connecting the two cylinders.

**17.33** Two uniform cylinders, each of weight $W = 14$ lb and radius $r = 5$ in., are connected by a belt as shown. If the system is released from rest, determine (a) the velocity of the center of cylinder $A$ after it has moved through 3 ft, (b) the tension in the portion of belt connecting the two cylinders.

**17.34** A bar of mass $m = 5$ kg is held as shown between four disks, each of mass $m' = 2$ kg and radius $r = 75$ mm. Knowing that the forces exerted on the disks are sufficient to prevent slipping and that the bar is released from rest, for each of the cases shown, determine the velocity of the bar after it has moved through the distance $h$.

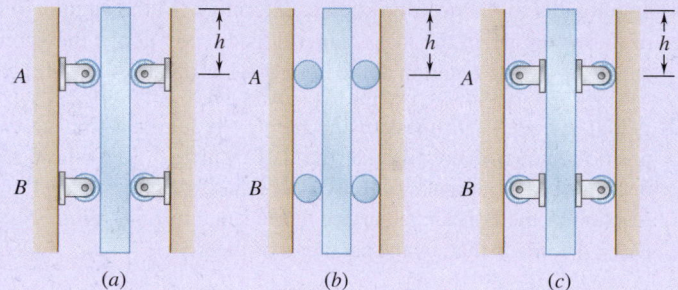

(a)          (b)          (c)

**Fig. *P17.34***

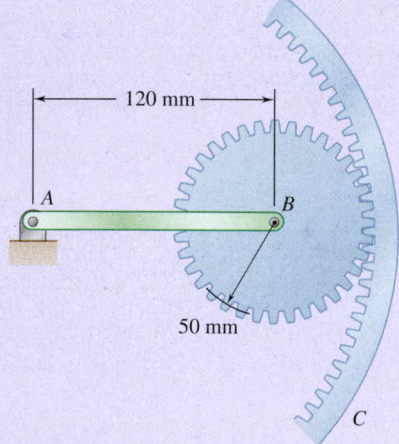

**Fig. P17.35**

**17.35** The 1.5-kg uniform slender bar $AB$ is connected to the 3-kg gear $B$ that meshes with the stationary outer gear $C$. The centroidal radius of gyration of gear $B$ is 30 mm. Knowing that the system is released from rest in the position shown, determine (a) the angular velocity of the bar as it passes through the vertical position, (b) the corresponding angular velocity of gear $B$.

**17.36** The motion of the uniform rod $AB$ is guided by small wheels of negligible mass that roll on the surface shown. If the rod is released from rest when $\theta = 0$, determine the velocities of $A$ and $B$ when $\theta = 30°$.

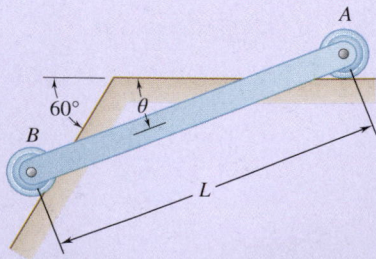

**Fig. P17.36**

**17.37** A 5-m-long ladder has a mass of 15 kg and is placed against a house at an angle $\theta = 20°$. Knowing that the ladder is released from rest, determine the angular velocity of the ladder and the velocity of end $A$ when $\theta = 45°$. Assume the ladder can slide freely on the horizontal ground and on the vertical wall.

**17.38** A long ladder of length $l$, mass $m$, and centroidal mass moment of inertia $\bar{I}$ is placed against a house at an angle $\theta = \theta_0$. Knowing that the ladder is released from rest, determine the angular velocity of the ladder when $\theta = \theta_2$. Assume the ladder can slide freely on the horizontal ground and on the vertical wall.

**17.39** The ends of a 9-lb rod $AB$ are constrained to move along slots cut in a vertical plate as shown. A spring of constant $k = 3$ lb/in. is attached to end $A$ in such a way that its tension is zero when $\theta = 0$. If the rod is released from rest when $\theta = 50°$, determine the angular velocity of the rod and the velocity of end $B$ when $\theta = 0$.

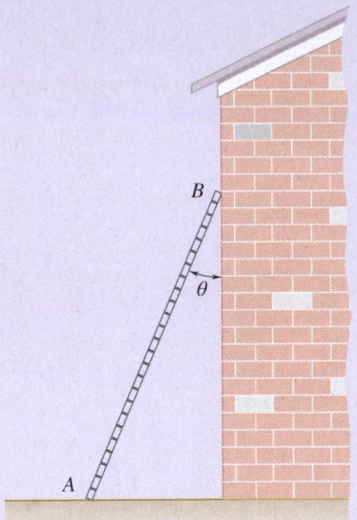

**Fig. P17.37 and P17.38**

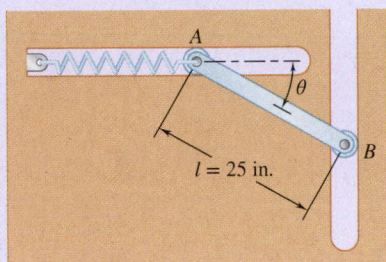

**Fig. P17.39**

**17.40** The mechanism shown is one of two identical mechanisms attached to the two sides of a 200-lb uniform rectangular door. Edge $ABC$ of the door is guided by wheels of negligible mass that roll in horizontal and vertical tracks. A spring with a constant of $k = 40$ lb/ft is attached to wheel $B$. Knowing that the door is released from rest in the position $\theta = 30°$ with the spring unstretched, determine the velocity of wheel $A$ just as the door reaches the vertical position.

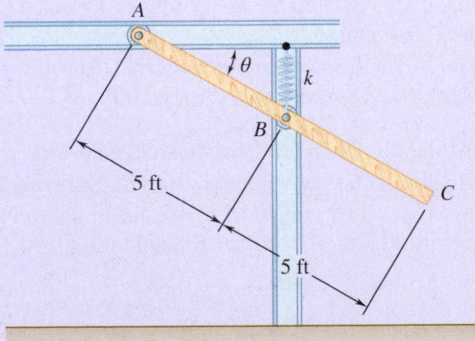

**Fig. P17.40 and P17.41**

**17.41** The mechanism shown is one of two identical mechanisms attached to the two sides of a 200-lb uniform rectangular door. Edge $ABC$ of the door is guided by wheels of negligible mass that roll in horizontal and vertical tracks. A spring with a constant $k$ is attached to wheel $B$ in such a way that its tension is zero when $\theta = 30°$. Knowing that the door is released from rest in the position $\theta = 45°$ and reaches the vertical position with an angular velocity of 0.6 rad/s, determine the spring constant $k$.

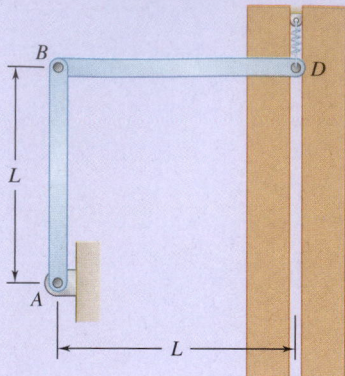

**Fig. P17.42**

**17.42** Each of the two rods shown is of length $L = 1$ m and has a mass of 5 kg. Point $D$ is connected to a spring of constant $k = 20$ N/m and is constrained to move along a vertical slot. Knowing that the system is released from rest when rod $BD$ is horizontal and the spring connected to point $D$ is initially unstretched, determine the velocity of point $D$ when it is directly to the right of point $A$.

**17.43** The 4-kg rod $AB$ is attached to a collar of negligible mass at $A$ and to a flywheel at $B$. The flywheel has a mass of 16 kg and a radius of gyration of 180 mm. Knowing that in the position shown the angular velocity of the flywheel is 60 rpm clockwise, determine the velocity of the flywheel when point $B$ is directly below $C$.

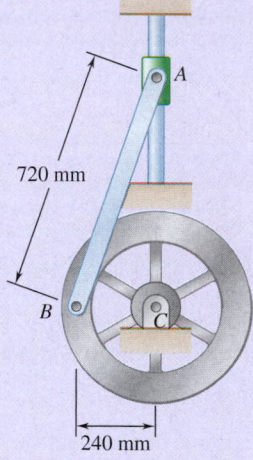

**Fig. P17.43 and P17.44**

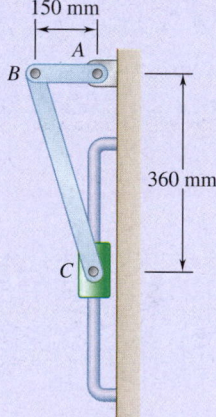

**Fig. P17.45**

**17.44** If in Prob. 17.43 the angular velocity of the flywheel is to be the same in the position shown and when point $B$ is directly above $C$, determine the required value of its angular velocity in the position shown.

**17.45** The uniform rods $AB$ and $BC$ are of mass 3 kg and 8 kg, respectively, and collar $C$ has a mass of 4 kg. Knowing that at the instant shown the velocity of collar $C$ is 0.9 m/s downward, determine the velocity of point $B$ after rod $AB$ has rotated through 90°.

**17.46** The uniform rods $AB$ and $BC$ weigh 2.4 kg and 4 kg, respectively, and the small wheel at $C$ is of negligible weight. Knowing that in the position shown the velocity of wheel $C$ is 2 m/s to the right, determine the velocity of pin $B$ after rod $AB$ has rotated through 90°.

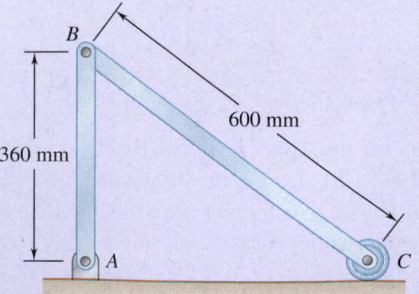

**Fig. P17.46**

**17.47** The 80-mm-radius gear shown has a mass of 5 kg and a centroidal radius of gyration of 60 mm. The 4-kg rod $AB$ is attached to the center of the gear and to a pin at $B$ that slides freely in a vertical slot. Knowing that the system is released from rest when $\theta = 60°$, determine the velocity of the center of the gear when $\theta = 20°$.

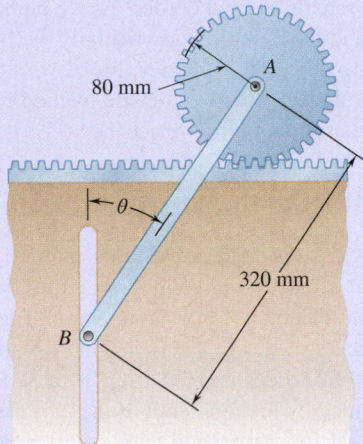

80 mm

$A$

$\theta$

320 mm

$B$

**Fig. P17.47**

**17.48** Knowing that the maximum allowable couple that can be applied to a shaft is 15.5 kip·in., determine the maximum horsepower that can be transmitted by the shaft at (a) 180 rpm, (b) 480 rpm.

**17.49** Three shafts and four gears are used to form a gear train that will transmit 7.5 kW from the motor at $A$ to a machine tool at $F$. (Bearings for the shafts are omitted from the sketch.) Knowing that the frequency of the motor is 30 Hz, determine the magnitude of the couple that is applied to shaft (a) $AB$, (b) $CD$, (c) $EF$.

**17.50** The experimental setup shown is used to measure the power output of a small turbine. When the turbine is operating at 200 rpm, the readings of the two spring scales are 10 and 22 lb, respectively. Determine the power being generated by the turbine.

**17.51** The drive belt on a vintage sander transmits ½ hp to a pulley that has a diameter of $d = 4$ in. Knowing that the pulley rotates at 1450 rpm, determine the tension difference $T_1 - T_2$ between the tight and slack sides of the belt.

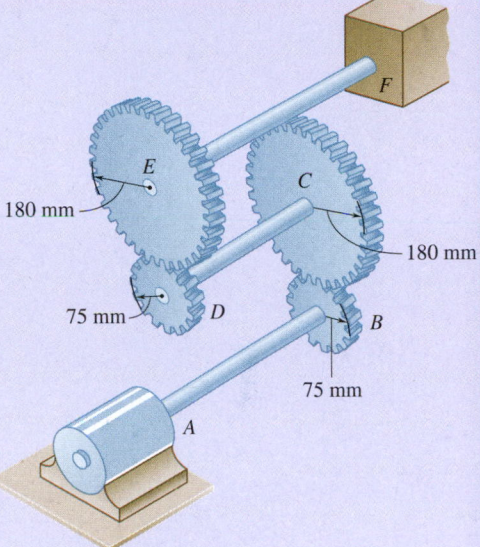

$F$

$E$

180 mm

$C$

180 mm

75 mm

$D$

$B$

75 mm

$A$

**Fig. P17.49**

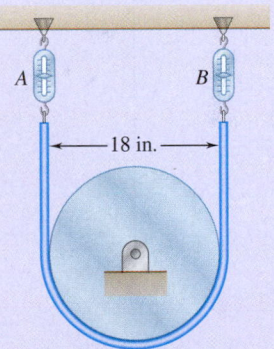

$A$

$B$

18 in.

**Fig. P17.50**

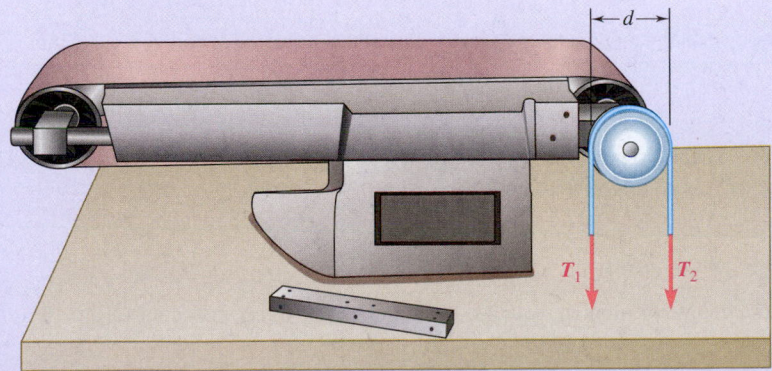

$d$

$T_1$

$T_2$

**Fig. P17.51**

# 17.2    MOMENTUM METHODS FOR A RIGID BODY

We now apply the principle of impulse and momentum to the plane motion of rigid bodies and of systems of rigid bodies. As we pointed out in Chap. 13, the method of impulse and momentum is particularly well-adapted to the solution of problems involving time and velocities. Moreover, the principle of impulse and momentum provides the only practicable method for the solution of problems involving impulsive motion or impact (Sec. 17.3).

## 17.2A    Principle of Impulse and Momentum

Consider again a rigid body made of a large number of particles $P_i$. Recall from Sec. 14.2C that impulse–momentum diagrams are a pictorial representation of the principle of impulse and momentum. They show (*a*) the system formed by the momenta of the particles at time $t_1$ and (*b*) the system of the impulses of the external forces applied from $t_1$ to $t_2$ are together equipollent to (*c*) the system formed by the momenta of the particles at time $t_2$ (Fig. 17.6). We can consider the vectors associated with a rigid body to be sliding vectors, so it follows (*Statics*, Sec. 3.4B) that the systems of vectors shown in Fig. 17.6 are not only equipollent, but they are truly *equivalent*. In other words, the vectors on the left-hand side of the equal sign can be transformed into the vectors on the right-hand side through the use of the fundamental operations listed in Sec. 3.3B. We therefore have

$$\textbf{Syst Momenta}_1 + \textbf{Syst Ext Imp}_{1\rightarrow2} = \textbf{Syst Momenta}_2 \qquad (17.14)$$

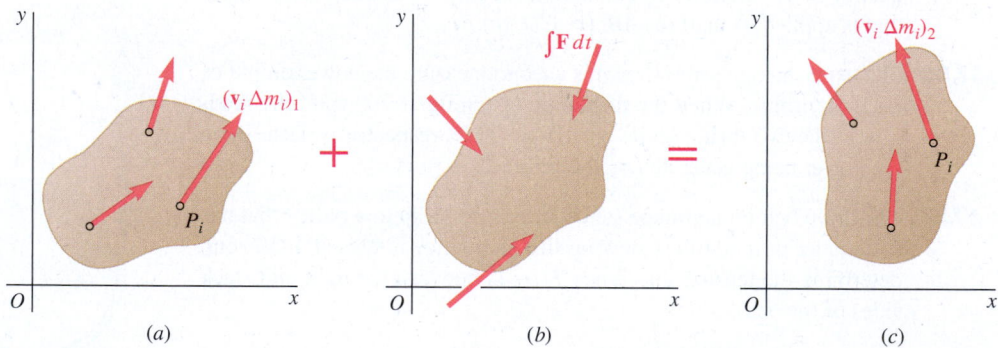

**Fig. 17.6**    For a rigid body in plane motion: (*a*) the system of particle momenta at time $t_1$ plus (*b*) the system of impulses of the external forces from time $t_1$ to $t_2$ is equivalent to (*c*) the system of particle momenta at time $t_2$.

The momenta $\mathbf{v}_i\,\Delta m_i$ of the particles can be reduced to a vector attached at $G$ that is equal to their sum

$$\mathbf{L} = \sum_{i=1}^{n} \mathbf{v}_i\,\Delta m_i$$

and a couple of moment equal to the sum of their moments about $G$, as

$$\mathbf{H}_G = \sum_{i=1}^{n} \mathbf{r}_i' \times \mathbf{v}_i\,\Delta m_i$$

Recall from Sec. 14.1B that $\mathbf{L}$ and $\mathbf{H}_G$ define, respectively, the linear momentum and the angular momentum about $G$ of the system of particles forming the rigid body. Also note from Eq. (14.14) that $\mathbf{L} = m\bar{\mathbf{v}}$. On the other hand, by restricting the present analysis to the plane motion of a rigid body or of a rigid body symmetrical with respect to the reference plane, we recall from Eq. (16.4) that $\mathbf{H}_G = \bar{I}\boldsymbol{\omega}$. We thus conclude that the system of the momenta $\mathbf{v}_i\,\Delta m_i$ is equivalent to the **linear momentum vector** $m\bar{\mathbf{v}}$ attached at $G$ and to the **angular momentum couple** $\bar{I}\boldsymbol{\omega}$ (Fig. 17.7).

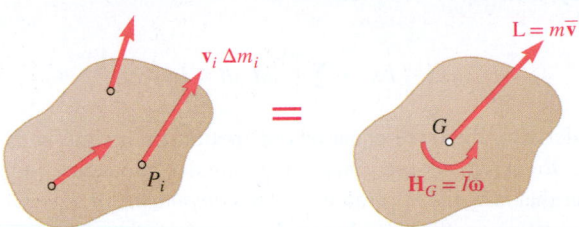

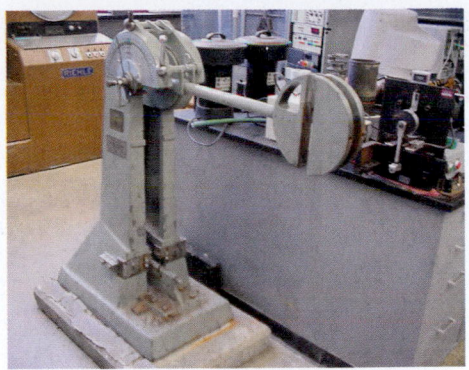

**Fig. 17.7** The system of momenta of a rigid body is equivalent to a linear momentum vector attached at $G$ and an angular momentum couple.

The system of momenta reduces to the vector $m\bar{\mathbf{v}}$ in the particular case of a translation ($\boldsymbol{\omega} = 0$) and to the couple $\bar{I}\boldsymbol{\omega}$ in the particular case of a centroidal rotation ($\bar{\mathbf{v}} = 0$). Thus, we verify once more that the plane motion of a rigid body that is symmetrical with respect to the reference plane can be resolved into a translation with the mass center $G$ and a rotation about $G$.

Replacing the system of momenta in Fig. 17.6a and c by the equivalent linear momentum vector and angular momentum couple, we obtain the three diagrams shown in Fig. 17.8. This impulse–momentum diagram is a visual representation of the fundamental relation in Eq. (17.14) in the case of the plane motion of a rigid body or of a rigid body symmetrical with respect to the reference plane.

We can derive three equations of motion from Fig. 17.8. Two equations come from summing and equating the *x* and *y components* of the momenta and impulses. The third equation is obtained by summing and equating the *moments* of these vectors *about any given point*. We can choose the coordinate axes to be fixed in space or allowed to move with the mass center of the body while maintaining a fixed direction. In either case, the point about which moments

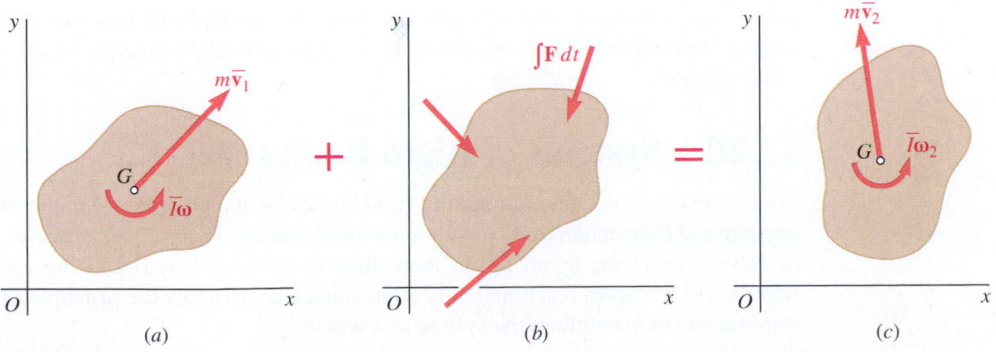

(a)      (b)      (c)

**Fig. 17.8** An impulse–momentum diagram is used for applying the principle of impulse and momentum.

are taken should keep the same position relative to the coordinate axes during the interval of time considered. If you choose to sum moments about a point *P*, Eq. (17.14) can be expressed as

$$\bar{I}\omega_1 + m\bar{v}_1 d_{\perp 1} + \sum \int_{t_1}^{t_2} M_P dt = \bar{I}\omega_2 + m\bar{v}_2 d_{\perp 2} \qquad \textbf{(17.14')}$$

where $d_\perp$ is the perpendicular distance from point *P* to the line of action of the linear velocity of *G*. If you choose to sum moments about the center of gravity of the body, then Eq. (17.14′) reduces to

$$\bar{I}\omega_1 + \sum \int_{t_1}^{t_2} M_G dt = \bar{I}\omega_2 \qquad \textbf{(17.14'')}$$

In deriving the three equations of motion for a rigid body, you should take care to avoid adding linear and angular momenta indiscriminately. Remember that $m\bar{v}_x$ and $m\bar{v}_y$ represent the *components of a vector,* namely, the linear momentum vector $m\bar{\mathbf{v}}$, whereas $\bar{I}\omega$ represents the *magnitude of a couple,* namely, the angular momentum couple $\bar{I}\boldsymbol{\omega}$. Thus, you should add the quantity $\bar{I}\omega$ only to the *moment* of the linear momentum $m\bar{\mathbf{v}}$—never to this vector itself nor to its components. All angular momentum quantities involved then will be expressed in the same units, namely N·m·s or lb·ft·s.

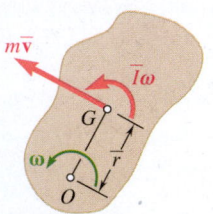

**Fig. 17.9**   The linear and angular momenta for a noncentroidal rotation.

**Noncentroidal Rotation.**   In this particular case of plane motion, the magnitude of the velocity of the mass center of the body is $\bar{v} = \bar{r}\omega$, where $\bar{r}$ represents the distance from the mass center to the fixed axis of rotation and $\boldsymbol{\omega}$ represents the angular velocity of the body at the instant considered. The magnitude of the momentum vector attached at *G* is thus $m\bar{v} = m\bar{r}\omega$. Summing the moments about *O* of the momentum vector and momentum couple (Fig. 17.9) and using the parallel-axis theorem for moments of inertia, we find that the angular momentum $\mathbf{H}_O$ of the body about fixed axis *O* has the magnitude[†]

$$\bar{I}\omega + (m\bar{r}\omega)\bar{r} = (\bar{I} + m\bar{r}^2)\omega = I_O\omega \qquad \textbf{(17.15)}$$

Equating the moments about *O* of the momenta and impulses in Eq. (17.14), we have

$$I_O\omega_1 + \sum \int_{t_1}^{t_2} M_O \, dt = I_O\omega_2 \qquad \textbf{(17.16)}$$

In the general case of plane motion of a rigid body symmetrical with respect to the reference plane, you can use Eq. (17.16) with respect to the instantaneous axis of rotation under certain conditions. We recommend, however, that all problems of plane motion be solved by the general method described earlier in this section.

## 17.2B   Systems of Rigid Bodies

We can analyze the motion of several rigid bodies by applying the principle of impulse and momentum to each body separately (Sample Prob. 17.7). However, in solving problems involving no more than three unknowns (including the impulses of unknown reactions), it is often convenient to apply the principle of impulse and momentum to the system as a whole.

---

[†]Note that the sum $\mathbf{H}_P$ of the moments about an arbitrary point *P* of the momenta of the particles of a rigid body is, in general, not equal to $I_P\boldsymbol{\omega}$ (see Prob. 17.67).

To do this, first draw impulse–momentum diagrams for the entire system of bodies. For each moving part of the system, the diagrams of momenta should include a linear momentum vector and a momentum couple. You can omit impulses of forces internal to the system from the diagram showing the impulses, because they occur in pairs of equal and opposite vectors. Summing and equating successively the $x$ components, $y$ components, and moments of all vectors involved, you obtain three relations expressing that the momenta at time $t_1$ and the impulses of the external forces form a system equipollent to the system of the momenta at time $t_2$. Again, you should take care not to add linear and angular momenta indiscriminately; check each equation to make sure that consistent units are used. This approach has been used in Sample Probs. 17.9 through 17.13.

## 17.2C   Conservation of Angular Momentum

When no external force acts on a rigid body or a system of rigid bodies, the impulses of the external forces are zero and the system of the momenta at time $t_1$ is equipollent to the system of the momenta at time $t_2$. Summing and equating successively the $x$ components, $y$ components, and moments of the momenta at times $t_1$ and $t_2$, we conclude that the total linear momentum of the system is conserved in any direction and that its total angular momentum is conserved about any point.

In many engineering applications, however, the linear momentum is not conserved, yet the angular momentum $\mathbf{H}_P$ of the system about a given point $P$ is conserved. That is,

$$(\mathbf{H}_P)_1 = (\mathbf{H}_P)_2 \qquad (17.17)$$

Such cases occur when the lines of action of all external forces pass through $P$ or, more generally, when the sum of the angular impulses of the external forces about $P$ is zero.

You can solve problems involving the **conservation of angular momentum** about a point $P$ using the general method of impulse and momentum; that is, by drawing impulse–momentum diagrams as described earlier. You then obtain Eq. (17.17) by summing and equating moments about $P$ (Sample Prob. 17.9). As you will see in Sample Prob. 17.11, you can obtain two additional equations by summing and equating the $x$ and $y$ components of the linear momentum; then you can use these equations to determine two unknown linear impulses, such as the impulses of the reaction components at a fixed point.

**Photo 17.3**   A figure skater at the beginning and at the end of a spin. By using the principle of conservation of angular momentum, you will find that her angular velocity is much higher at the end of the spin. ©Jill Braaten

## Sample Problem 17.7

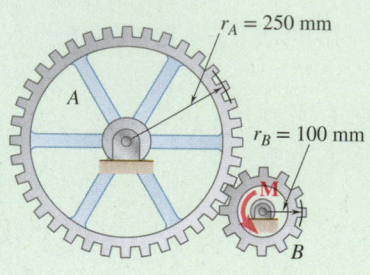

$r_A = 250$ mm

$r_B = 100$ mm

Gear $A$ has a mass of 10 kg and a radius of gyration of 200 mm, and gear $B$ has a mass of 3 kg and a radius of gyration of 80 mm. The system is at rest when a couple **M** with a magnitude of 6 N·m is applied to gear $B$. (These gears were considered in Sample Prob. 17.2.) Neglecting friction, determine (*a*) the time required for the angular velocity of gear $B$ to reach 600 rpm, (*b*) the tangential force that gear $B$ exerts on gear $A$.

**STRATEGY:**   Because you are given an angular velocity and are asked for time, use the principle of impulse and momentum.

**MODELING:**   You are asked to find the internal tangential force, so you need two systems for this problem; that is, gear $A$ and gear $B$. Model the gears as rigid bodies. Because all forces and couples are constant, you can obtain the impulses by multiplying the forces and moments by the unknown time $t$.

**ANALYSIS:**   Recall from Sample Prob. 17.2 that the centroidal moments of inertia and the final angular velocities are

$$\bar{I}_A = 0.400 \text{ kg·m}^2 \qquad \bar{I}_B = 0.0192 \text{ kg·m}^2$$

$$(\omega_A)_2 = 25.1 \text{ rad/s} \qquad (\omega_B)_2 = 62.8 \text{ rad/s}$$

**Principle of Impulse and Momentum for Gear A.**   The impulse–momentum diagram (Fig. 1) for gear $A$ shows the initial momenta, impulses, and final momenta.

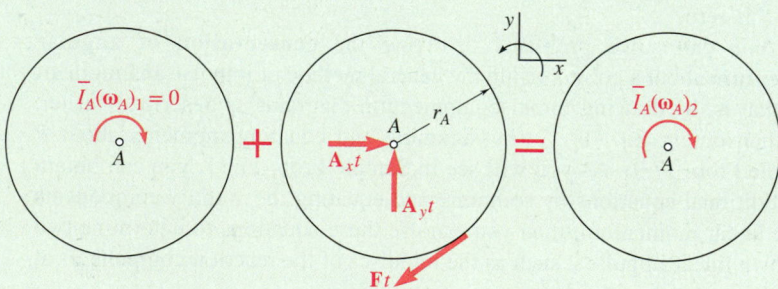

**Fig. 1**   Impulse–momentum diagram for gear $A$.

$$\textbf{Syst Momenta}_1 + \textbf{Syst Ext Imp}_{1\to2} = \textbf{Syst Momenta}_2$$

$+\circlearrowleft$ moments about $A$: 
$$0 - Ftr_A = -\bar{I}_A(\omega_A)_2$$

$$Ft(0.250 \text{ m}) = (0.400 \text{ kg·m}^2)(25.1 \text{ rad/s})$$

$$Ft = 40.2 \text{ N·s}$$

*(continued)*

**Principle of Impulse and Momentum for Gear *B*.**   Draw a separate impulse–momentum diagram for gear *B* (Fig. 2).

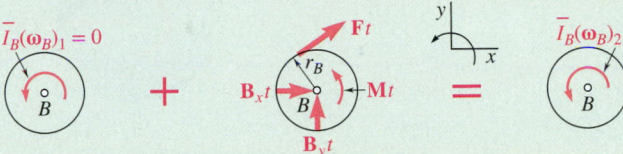

**Fig. 2**   Impulse–momentum diagram for gear *B*.

$$\text{Syst Momenta}_1 + \text{Syst Ext Imp}_{1\to2} = \text{Syst Momenta}_2$$

$+\circlearrowleft$ moments about *B*:   $0 + Mt - Ftr_B = \bar{I}_B(\omega_B)_2$

$+(6 \text{ N·m})t - (40.2 \text{ N·s})(0.100 \text{ m}) = (0.0192 \text{ kg·m}^2)(62.8 \text{ rad/s})$

$$t = 0.871 \text{ s} \quad \blacktriangleleft$$

Recall that $Ft = 40.2$ N·m, so you have

$$F(0.871 \text{ s}) = 40.2 \text{ N·s} \qquad F = +46.2 \text{ N}$$

Thus, the force exerted by gear *B* on gear *A* is

$$\mathbf{F} = 46.2 \text{ N} \nearrow \quad \blacktriangleleft$$

**REFLECT and THINK:**   This is the same answer obtained in Sample Prob. 17.2 by the method of work and energy, as you would expect. The difference is that in Sample Prob. 17.2, you were asked to find the number of revolutions, and in this problem, you were asked to find the time. What you are asked to find will often determine the best approach to use when solving a problem.

## Sample Problem 17.8

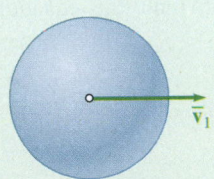

A uniform sphere with a mass $m$ and a radius $r$ is projected along a rough horizontal surface with a linear velocity $\bar{\mathbf{v}}_1$ and no angular velocity. Denote the coefficient of kinetic friction between the sphere and the surface by $\mu_k$. Determine (*a*) the time $t_2$ at which the sphere starts rolling without sliding, (*b*) the linear and angular velocities of the sphere at time $t_2$. ▶

**STRATEGY:**   You are asked to find the time, so use the principle of impulse and momentum. You can apply this principle to the sphere from the time $t_1 = 0$ when it is placed on the surface until the time $t_2 = t$ when it starts rolling without sliding.

**MODELING:**   Choose the sphere as your system and model it as a rigid body. While the sphere is sliding relative to the surface, it is acted upon by the normal force **N**, the friction force **F**, and its weight **W** with a magnitude of $W = mg$. An impulse–momentum diagram for this system is shown in Fig. 1.

*(continued)*

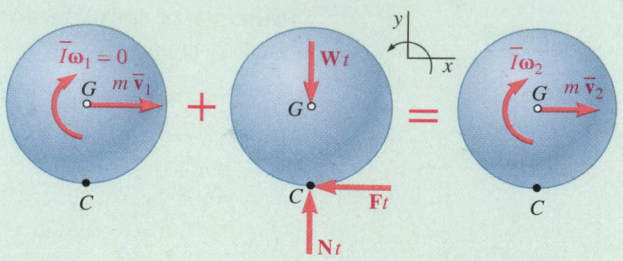

**Fig. 1**   Impulse–momentum diagram for the sphere.

## ANALYSIS:

### Principle of Impulse and Momentum.

Apply the principle of impulse and momentum for this system between time $t_1$ and $t_2$

$$\textbf{Syst Momenta}_1 + \textbf{Syst Ext Imp}_{1\rightarrow 2} = \textbf{Syst Momenta}_2$$

$+\uparrow y$ components:                $Nt - Wt = 0$                                                    (1)

$\overset{+}{\rightarrow} x$ components:                $m\bar{v}_1 - Ft = m\bar{v}_2$                                      (2)

$+\circlearrowleft$ moments about $G$:                $-Ftr = -\bar{I}\omega_2$                                      (3)

From Eq. (1) you obtain $N = W = mg$. During the entire time interval considered, sliding occurs at point $C$, and $F = \mu_k N = \mu_k mg$. Substituting this expression for $F$ into Eq. (2), you have

$$m\bar{v}_1 - \mu_k mgt = m\bar{v}_2 \qquad \bar{v}_2 = \bar{v}_1 - \mu_k gt \qquad (4)$$

Substituting $F = \mu_k mg$ and $\bar{I} = \frac{2}{5}mr^2$ into Eq. (3) gives

$$\mu_k mgtr = \frac{2}{5}mr^2\omega_2 \qquad \omega_2 = \frac{5}{2}\frac{\mu_k g}{r}t \qquad (5)$$

The sphere starts rolling without sliding when the velocity $\mathbf{v}_C$ of the point of contact is zero. At that time, point $C$ becomes the instantaneous center of rotation, and you have $\bar{v}_2 = r\omega_2$. Substituting Eqs. (4) and (5) into this equation, you obtain

$$\bar{v}_1 - \mu_k gt = r\left(\frac{5}{2}\frac{\mu_k g}{r}t\right) \qquad\qquad t = \frac{2}{7}\frac{\bar{v}_1}{\mu_k g} \blacktriangleleft$$

Substituting this expression for $t$ into Eq. (5), you have

$$\omega_2 = \frac{5}{2}\frac{\mu_k g}{r}\left(\frac{2}{7}\frac{\bar{v}_1}{\mu_k g}\right) \qquad \omega_2 = \frac{5}{7}\frac{\bar{v}_1}{r} \qquad \boldsymbol{\omega_2} = \frac{5}{7}\frac{\bar{v}_1}{r}\circlearrowright \blacktriangleleft$$

$$\bar{v}_2 = r\omega_2 \qquad \bar{v}_2 = r\left(\frac{5}{7}\frac{v_1}{r}\right) \qquad \bar{\mathbf{v}}_2 = \frac{5}{7}\bar{v}_1 \rightarrow \blacktriangleleft$$

**REFLECT and THINK:**   This is the same answer obtained in Sample Prob. 16.6 by first dealing directly with force and acceleration and then applying kinematic relationships.

## Sample Problem 17.9

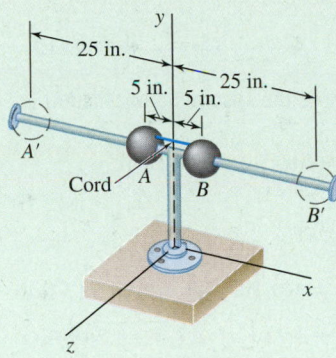

Two solid spheres with a radius 3 in. and weighing 2 lb each are mounted at $A$ and $B$ on the horizontal rod $A'B'$ that rotates freely about a vertical axis with a counterclockwise angular velocity of 6 rad/s. The spheres are held in position by a cord, which is suddenly cut. The centroidal moment of inertia of the rod and pivot is $\bar{I}_R = 0.25$ lb·ft·s$^2$. Determine (*a*) the angular velocity of the rod after the spheres have moved to positions $A'$ and $B'$, (*b*) the energy lost due to the plastic impact of the spheres and the stops at $A'$ and $B'$.

**STRATEGY:** You can first use the principle of impulse and momentum to find the angular velocity of the rod and then use the definition of kinetic energy to determine the change in energy.

**MODELING:** Choose the two solid spheres and the horizontal rod as your system and model these as rigid bodies. The impulse–momentum diagram for this system is shown in Fig. 1.

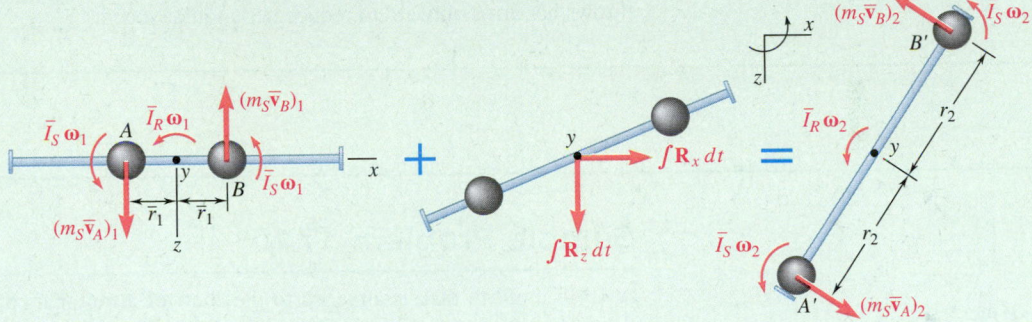

**Fig. 1** Impulse–momentum diagram for the system.

### ANALYSIS:

**a. Principle of Impulse and Momentum.** Apply the principle of impulse and momentum for this system between time $t_1$ (when the spheres are at $r_1$) and $t_2$ (when the spheres are at $r_2$)

$$\textbf{Syst Momenta}_1 + \textbf{Syst Ext Imp}_{1\rightarrow2} = \textbf{Syst Momenta}_2$$

The external forces consist of the weights and the reaction at the pivot, which have no moment about the $y$ axis. Noting that the rod is undergoing centroidal rotation and $\bar{v}_A = \bar{v}_B = \bar{r}\omega$, you can equate moments about the $y$ axis as

$$2(m_s\bar{r}_1\omega_1)\bar{r}_1 + 2\bar{I}_s\omega_1 + \bar{I}_R\omega_1 = 2(m_s\bar{r}_2\omega_2)\bar{r}_2 + 2\bar{I}_s\omega_2 + \bar{I}_R\omega_2$$
$$(2m_s\bar{r}_1^2 + 2\bar{I}_s + \bar{I}_R)\omega_1 = (2m_s\bar{r}_2^2 + 2\bar{I}_s + \bar{I}_R)\omega_2 \tag{1}$$

This states that the angular momentum of the system about the $y$ axis is conserved. You can now compute

$$\bar{I}_s = \tfrac{2}{5}m_sa^2 = \tfrac{2}{5}(2\text{ lb}/32.2\text{ ft/s}^2)(\tfrac{3}{12}\text{ ft})^2 = 0.00155\text{ lb·ft·s}^2$$
$$m_s\bar{r}_1^2 = (2/32.2)(\tfrac{5}{12})^2 = 0.0108 \qquad m_s\bar{r}_2^2 = (2/32.2)(\tfrac{25}{12})^2 = 0.2696$$

*(continued)*

Substituting these values, along with $\bar{I}_R = 0.25$ lb·ft·s$^2$ and $\omega_1 = 6$ rad/s, into Eq. (1) gives

$$0.275(6 \text{ rad/s}) = 0.792\omega_2 \qquad \boldsymbol{\omega_2 = 2.08 \text{ rad/s}} \circlearrowleft \ \blacktriangleleft$$

**b. Energy Lost.** The kinetic energy of the system at any instant is

$$T = 2(\tfrac{1}{2}m_s\bar{v}^2 + \tfrac{1}{2}\bar{I}_s\omega^2) + \tfrac{1}{2}\bar{I}_R\omega^2 = \tfrac{1}{2}(2m_s\bar{r}^2 + 2\bar{I}_s + \bar{I}_R)\omega^2$$

Using the numerical values found here, you have

$$T_1 = \tfrac{1}{2}(0.275)(6)^2 = 4.95 \text{ ft·lb} \qquad T_2 = \tfrac{1}{2}(0.792)(2.08)^2 = 1.713 \text{ ft·lb}$$

$$\Delta T = T_2 - T_1 = 1.71 - 4.95 \qquad \boldsymbol{\Delta T = -3.24 \text{ ft·lb}} \ \blacktriangleleft$$

**REFLECT and THINK:** As expected, when the spheres move outward, the angular velocity of the system decreases. This is similar to an ice skater who throws her arms outward to reduce her angular speed.

# Sample Problem 17.10

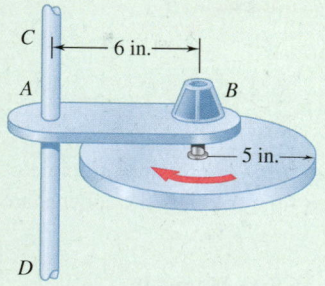

A 10-lb uniform disk is attached to the shaft of a motor mounted on arm *AB* that is free to rotate about the vertical axle *CD*. The arm-and-motor unit has a moment of inertia of 0.032 lb·ft·s$^2$ about axle *CD*. Knowing that the system is initially at rest, determine the angular velocities of the arm and of the disk when the motor reaches a speed of 360 rpm. ▶

**STRATEGY:** Because you have two times—when the system starts from rest and when the motor has reached a speed of 360 rpm—use the conservation of angular momentum. You cannot use the conservation of energy because the motor converts electrical energy into mechanical energy.

**MODELING:** Choose the arm *AB,* the motor, and the disk to be your system and model them as rigid bodies. The impulse–momentum diagram for this system is shown in Fig. 1.

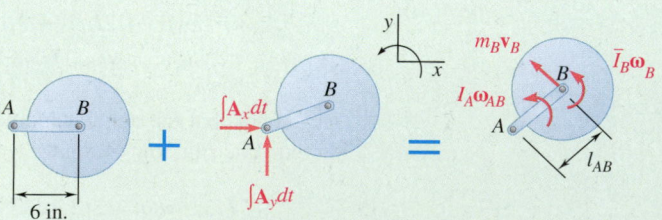

**Fig. 1** Impulse–momentum diagram for the system.

*(continued)*

**Moments of Inertia.** The mass moment of inertia of the arm and motor about the axle is $I_A = 0.032$ lb·ft·s$^2$, and the mass moment of inertia of disk $B$ about its center of mass is

$$\bar{I}_B = \frac{1}{2}\frac{W}{g}r^2 = \frac{1}{2}\left(\frac{10}{32.2}\right)\left(\frac{5}{12}\right)^2 = 0.02696 \text{ lb·ft·s}^2$$

**ANALYSIS:**

**Principle of Impulse and Momentum.** Apply the principle of impulse and momentum for this system between time $t_1$ (when the system is at rest) and $t_2$ (when the motor has an angular velocity of 360 rpm)

$$\textbf{Syst Momenta}_1 + \textbf{Syst Ext Imp}_{1\to2} = \textbf{Syst Momenta}_2$$

Taking moments about $A$ gives

$+\circlearrowleft$ moments about $A$: $\qquad 0 + 0 = (m_B v_B)l_{AB} + I_A\omega_{AB} + \bar{I}_B\omega_B \qquad$ **(1)**

**Kinematics.** You can relate the velocity of $B$ to the angular velocity of $AB$ using

$$v_B = l_{AB}\omega_{AB} = \tfrac{6}{12}\omega_{AB} \qquad \textbf{(2)}$$

The velocity of the motor is $\omega_M = 360$ rpm $= 12\pi$ rad/s, which is the angular velocity of the disk relative to the arm. Thus,

$$\omega_B = \omega_{AB} + \omega_M \qquad \textbf{(3)}$$

Substituting Eqs. (2) and (3) into Eq. (1) and solving for $\omega_{AB}$ gives

$$(m_B l_{AB}^2 + I_A)\omega_{AB} + \bar{I}_B(\omega_{AB} + \omega_M) = 0$$

$$\left[\left(\frac{10}{32.2}\right)\left(\frac{6}{12}\right)^2 + 0.032\right]\omega_{AB} + 0.02696(\omega_{AB} + 12\pi) = 0$$

$$\omega_{AB} = -7.44 \text{ rad/s}$$

$$\boldsymbol{\omega}_{AB} = 71.0 \text{ rpm } \circlearrowleft \quad \blacktriangleleft$$

The angular velocity of the disk is

$$\omega_B = -7.44 + 12\pi = 30.26 \text{ rad/s}$$

$$\boldsymbol{\omega}_B = 289 \text{ rpm } \circlearrowleft \quad \blacktriangleleft$$

**REFLECT and THINK:** When the motor spins the disk counterclockwise (as viewed from above), the arm $AB$ rotates in a clockwise direction. One key to solving this problem is recognizing that the angular velocity of the motor is the relative angular velocity of the disk with respect to the bar.

# SOLVING PROBLEMS
# ON YOUR OWN

In this section, we described how to use the method of impulse and momentum to solve problems involving the plane motion of rigid bodies. As you found out previously in Chap. 13, this method is most effective when used in the solution of problems involving velocities and time.

**1. The principle of impulse and momentum for the plane motion of a rigid body** is expressed by the vector equation:

$$\text{Syst Momenta}_1 + \text{Syst Ext Imp}_{1\to2} = \text{Syst Momenta}_2 \qquad (17.14)$$

where **Syst Momenta** represents the system of the momenta of the particles forming the rigid body and **Syst Ext Imp** represents the system of all the external impulses exerted during the motion.

   **a. The system of the momenta of a rigid body** is equivalent to a linear momentum vector $m\overline{\mathbf{v}}$ attached at the mass center of the body and an angular momentum couple about the center of mass $\overline{I}\boldsymbol{\omega}$ (Fig. 17.7).

   **b. You should draw an impulse–momentum diagram for the rigid body** to express the vector equation (17.14) graphically. Your diagram should consist of three sketches of the body representing, respectively, the initial momenta, the impulses of the external forces, and the final momenta. This shows that the system of the initial momenta and the system of the impulses of the external forces are together equivalent to the system of the final momenta (Fig. 17.8).

   **c. By using the impulse–momentum diagram,** you can sum components in any direction and sum moments about any point. For a single rigid body, if you choose to sum moments about an arbitrary point $P$, you can write Eq. (17.14) as

$$\overline{I}\omega_1 + m\overline{v}_1 d_{\perp 1} + \sum \int_{t_1}^{t_2} M_P dt = \overline{I}\omega_2 + m\overline{v}_2 d_{\perp 2} \qquad (17.14')$$

where $d_\perp$ is the perpendicular distance from point $P$ to the line of action of the linear velocity of $G$. If you choose to sum moments about the center of gravity of the body, Eq. (17.14′) reduces to

$$\overline{I}\omega_1 + \sum \int_{t_1}^{t_2} M_G dt = \overline{I}\omega_2 \qquad (17.14'')$$

If you choose to sum moments about a fixed point $O$, Eq. (17.14′) reduces to

$$I_O \omega_1 + \sum \int_{t_1}^{t_2} M_O dt = I_O \omega_2 \qquad (17.16)$$

where $I_O$ is the mass moment of inertia about point $O$. In most cases, you will be able to select and solve an equation that involves only one unknown.

**2. In problems involving a system of rigid bodies,** you can apply the principle of impulse and momentum to the system as a whole. Because internal forces occur in equal and opposite pairs, they should not be part of your solution (Sample Probs. 17.9 and 17.10).

**3. Conservation of angular momentum about a given axis** occurs when, for a system of rigid bodies, *the sum of the moments of the external impulses about that axis is zero.* You can indeed easily observe from the impulse–momentum diagram that the initial and final angular momenta of the system about that axis are equal and, thus, that the angular momentum of the system about the given axis is conserved. You can then sum the angular momenta of the various bodies of the system and the moments of their linear momenta about that axis to obtain an equation that you can solve for one unknown (Sample Probs. 17.9 and 17.10).

# Problems

**17.CQ6** Slender bar $A$ is rigidly connected to a massless rod $BC$ in Case 1 and two massless cords in Case 2 as shown. The vertical thickness of bar $A$ is negligible compared to $L$. If bullet $D$ strikes $A$ with a speed $v_0$ and becomes embedded in it, how will the speeds of the center of gravity of $A$ immediately after the impact compare for the two cases?

**a.** Case 1 will be larger.
**b.** Case 2 will be larger.
**c.** The speeds will be the same.

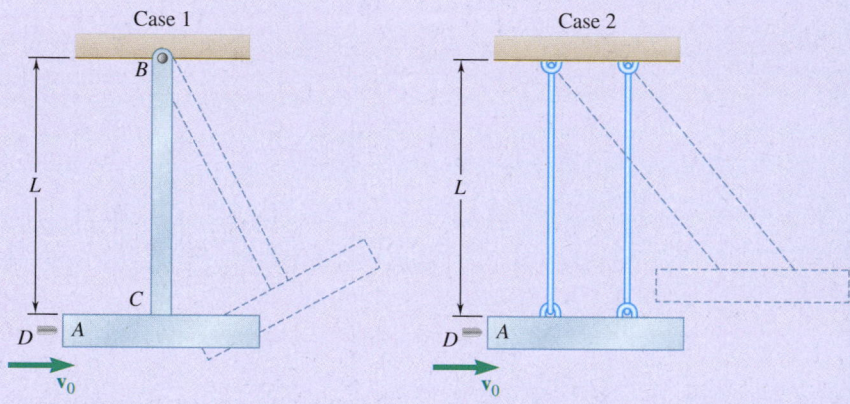

**Fig. P17.CQ6**

**17.CQ7** A 1-m-long uniform slender bar $AB$ has an angular velocity of 12 rad/s and its center of gravity has a velocity of 2 m/s as shown. About which point is the angular momentum of $A$ smallest at this instant?

**a.** $P_1$
**b.** $P_2$
**c.** $P_3$
**d.** $P_4$
**e.** It is the same about all the points.

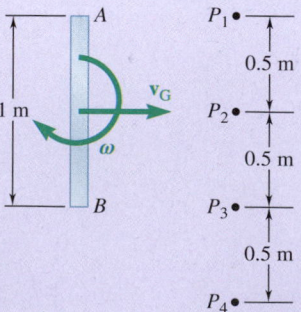

**Fig. P17.CQ7**

**17.F1** The 350-kg flywheel of a small hoisting engine has a radius of gyration of 600 mm. If the power is cut off when the angular velocity of the flywheel is 100 rpm clockwise, draw an impulse–momentum diagram that can be used to determine the time required for the system to come to rest.

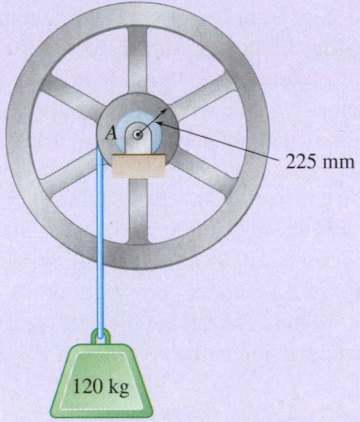

225 mm

120 kg

**Fig. P17.F1**

**17.F2** A sphere of radius $r$ and mass $m$ is placed on a horizontal floor with no linear velocity but with a clockwise angular velocity $\omega_0$. Denoting by $\mu_k$ the coefficient of kinetic friction between the sphere and the floor, draw the impulse–momentum diagram that can be used to determine the time $t_1$ at which the sphere will start rolling without sliding.

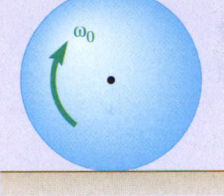

$\omega_0$

**Fig. P17.F2**

**17.F3** Two panels $A$ and $B$ are attached with hinges to a rectangular plate and held by a wire as shown. The plate and the panels are made of the same material and have the same thickness. The entire assembly is rotating with an angular velocity $\omega_0$ when the wire breaks. Draw the impulse–momentum diagram that is needed to determine the angular velocity of the assembly after the panels have come to rest against the plate.

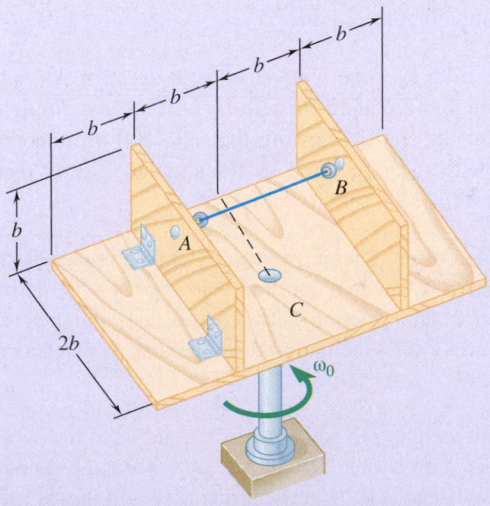

**Fig. P17.F3**

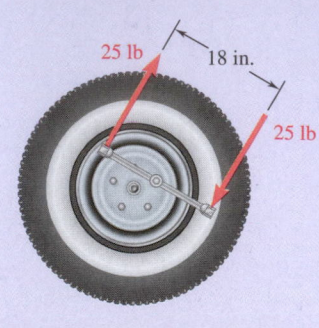

25 lb    18 in.

25 lb

**Fig. P17.53**

**17.52** The rotor of an electric motor has a mass of 25 kg, and it is observed that 4.2 min is required for the rotor to coast to rest from an angular velocity of 3600 rpm. Knowing that kinetic friction produces a couple of magnitude 1.2 N·m, determine the centroidal radius of gyration for the rotor.

**17.53** A bolt located 2 in. from the center of an automobile wheel is tightened by applying the couple shown for 0.10 s. Assuming that the wheel is free to rotate and is initially at rest, determine the resulting angular velocity of the wheel. The wheel weighs 42 lb and has a radius of gyration about its mass center of 10.8 in.

**17.54** A small grinding wheel is attached to the shaft of an electric motor which has a rated speed of 3600 rpm. When the power is turned off, the unit coasts to rest in 70 s. The grinding wheel and rotor have a combined mass of 3 kg and a combined radius of gyration about its mass center of 50 mm. Determine the average magnitude of the couple due to kinetic friction in the bearings of the motor.

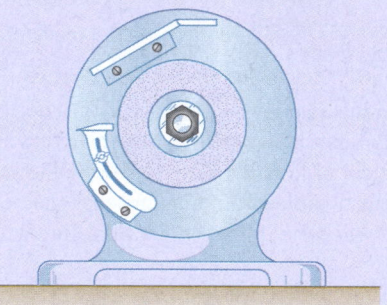

**Fig. P17.54**

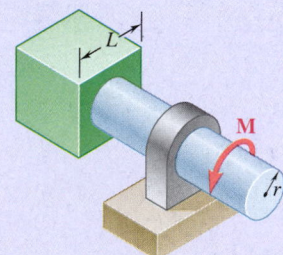

**Fig. P17.55 and *P17.56***

**17.55** A uniform 144-lb cube is attached to a uniform 136-lb circular shaft as shown, and a couple **M** with a constant magnitude is applied to the shaft when the system is at rest. Knowing that $r = 4$ in., $L = 12$ in., and the angular velocity of the system is 960 rpm after 4 s, determine the magnitude of the couple **M**.

**17.56** A uniform 75-kg cube is attached to a uniform 70-kg circular shaft as shown, and a couple **M** with a constant magnitude of 20 N·m is applied to the shaft. Knowing that $r = 100$ mm and $L = 300$ mm, determine the time required for the angular velocity of the system to increase from 1000 rpm to 2000 rpm.

**17.57** A disk of constant thickness, initially at rest, is placed in contact with a belt that moves with a constant velocity **v**. Denoting by $\mu_k$ the coefficient of kinetic friction between the disk and the belt, derive an expression for the time required for the disk to reach a constant angular velocity.

**17.58** Disk *A*, of weight 5 lb and radius $r = 3$ in., is at rest when it is placed in contact with a belt that moves at a constant speed $v = 50$ ft/s. Knowing that $\mu_k = 0.20$ between the disk and the belt, determine the time required for the disk to reach a constant angular velocity.

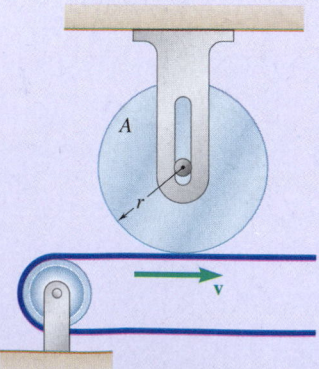

**Fig. *P17.57* and P17.58**

**17.59** A cylinder of radius $r$ and weight $W$ with an initial counterclockwise angular velocity $\boldsymbol{\omega}_0$ is placed in the corner formed by the floor and a vertical wall. Denoting by $\mu_k$ the coefficient of kinetic friction between the cylinder and the wall and the floor, derive an expression for the time required for the cylinder to come to rest.

**Fig. P17.59**

**17.60** Each of the double pulleys shown has a centroidal mass moment of inertia of 0.25 kg·m², an inner radius of 100 mm, and an outer radius of 150 mm. Neglecting bearing friction, determine (*a*) the velocity of the cylinder 3 s after the system is released from rest, (*b*) the tension in the cord connecting the pulleys.

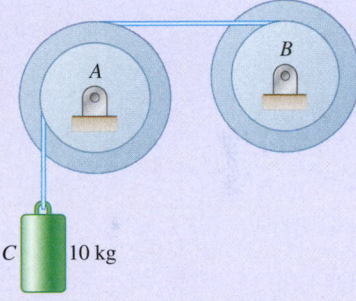

**Fig. P17.60**

**17.61** Each of the gears $A$ and $B$ has a mass of 675 g and a radius of gyration of 40 mm, while gear $C$ has a mass of 3.6 kg and a radius of gyration of 100 mm. Assume that kinetic friction in the bearings of gears $A$, $B$, and $C$ produces couples of constant magnitude 0.15 N·m, 0.15 N·m, and 0.3 N·m, respectively. Knowing that the initial angular velocity of gear $C$ is 2000 rpm, determine the time required for the system to come to rest.

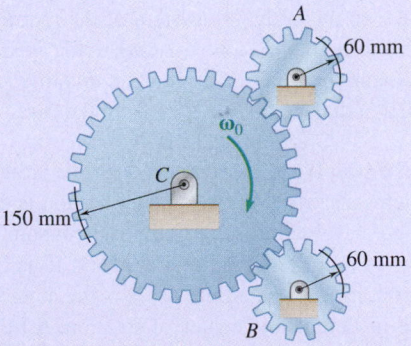

**Fig. P17.61**

**17.62** Two identical uniform cylinders of mass $m$ and radius $r$ are at rest at time $t = 0$ when a couple $\mathbf{M}$ of constant magnitude $M < mgr$ is applied to cylinder $A$. Knowing that the coefficient of kinetic friction between cylinder $B$ and the horizontal surface is $\mu_k < \dfrac{M}{2mgr}$ and that no slipping occurs between the two cylinders, derive an expression for the angular velocity of cylinder $B$ at time $t$.

**17.63** Two identical 16-lb uniform cylinders of radius $r = 4$ in. are at rest when a couple $\mathbf{M}$ of constant magnitude 4 lb·ft is applied to cylinder $A$. Slipping occurs between the two cylinders and between cylinder $B$ and the horizontal surface. Knowing that the coefficient of kinetic friction is 0.5 between the two cylinders and 0.2 between cylinder $B$ and the horizontal surface, determine the angular velocity of each cylinder after 5 s.

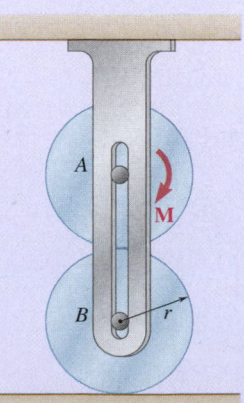

**Fig. P17.62 and P17.63**

**17.64** A tape moves over the two drums shown. Drum $A$ weighs 1.4 lb and has a radius of gyration of 0.75 in., while drum $B$ weighs 3.5 lb and has a radius of gyration of 1.25 in. In the lower portion of the tape, the tension is constant and equal to $T_A = 0.75$ lb. Knowing that the tape is initially at rest, determine (a) the required constant tension $T_B$ if the velocity of the tape is to be $v = 10$ ft/s after 0.24 s, (b) the corresponding tension in the portion of the tape between the drums.

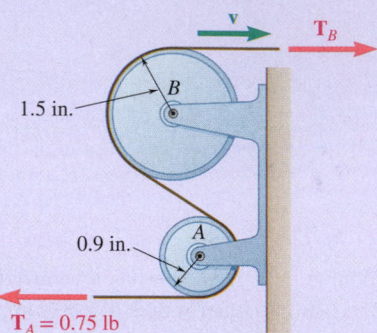

**Fig. P17.64**

**17.65** Show that the system of momenta for a rigid body in plane motion reduces to a single vector, and express the distance from the mass center $G$ to the line of action of this vector in terms of the centroidal radius of gyration $\bar{k}$ of the body, the magnitude $\bar{v}$ of the velocity of $G$, and the angular velocity $\boldsymbol{\omega}$.

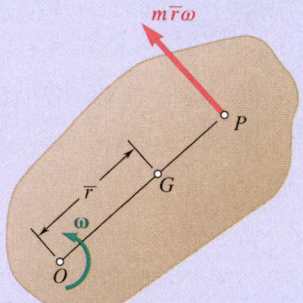

**Fig. P17.66**

**17.66** Show that, when a rigid body rotates about a fixed axis through $O$ perpendicular to the body, the system of the momenta of its particles is equivalent to a single vector of magnitude $m\bar{r}\omega$, perpendicular to the line $OG$, and applied to a point $P$ on this line, called the *center of percussion*, at a distance $GP = \bar{k}^2/\bar{r}$ from the mass center of the body.

**17.67** Show that the sum $\mathbf{H}_A$ of the moments about a point $A$ of the momenta of the particles of a rigid body in plane motion is equal to $I_A\omega$, where $\omega$ is the angular velocity of the body at the instant considered and $I_A$ the moment of inertia of the body about $A$, if and only if one of the following conditions is satisfied: (a) $A$ is the mass center of the body, (b) $A$ is the instantaneous center of rotation, (c) the velocity of $A$ is directed along a line joining point $A$ and the mass center $G$.

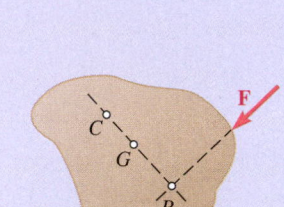

**Fig. P17.68**

**17.68** Consider a rigid body initially at rest and subjected to an impulsive force $\mathbf{F}$ contained in the plane of the body. We define the *center of percussion $P$* as the point of intersection of the line of action of $\mathbf{F}$ with the perpendicular drawn from $G$. (a) Show that the instantaneous center of rotation $C$ of the body is located on line $GP$ at a distance $GC = \bar{k}^2/GP$ on the opposite side of $G$. (b) Show that if the center of percussion were located at $C$, the instantaneous center of rotation would be located at $P$.

**17.69** A flywheel is rigidly attached to a 1.5-in.-radius shaft that rolls without sliding along parallel rails. Knowing that after being released from rest the system attains a speed of 6 in./s in 30 s, determine the centroidal radius of gyration of the system.

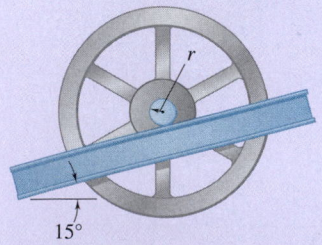

**Fig. P17.69**

**17.70** A wheel of radius $r$ and centroidal radius of gyration $\bar{k}$ is released from rest on the incline shown at time $t = 0$. Assuming that the wheel rolls without sliding, determine (*a*) the velocity of its center at time $t$, (*b*) the coefficient of static friction required to prevent slipping.

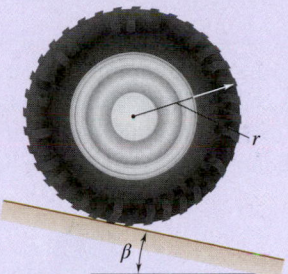

**Fig. P17.70**

**17.71** Cords are wrapped around two yo-yos with different proposed designs. Design 1 can be modeled as a thin-walled pipe and design 2 can be modeled as a solid cylinder as shown. Knowing that the two designs are each released from rest at time $t = 0$, determine at time $t$ the velocity of the center of (*a*) design 1, (*b*) design 2. Assume the yo-yos are going down.

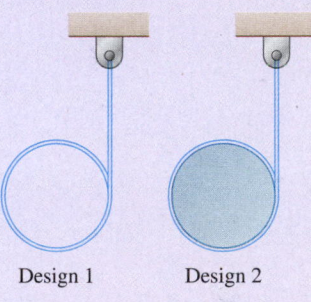

Design 1      Design 2

**Fig. P17.71**

**17.72 and 17.73** The 3-lb carriage $C$ is supported as shown by two uniform 2-lb disks, each having a radius of 3 in. Knowing that the carriage is initially at rest, determine the velocity of the carriage 0.5 s after the 0.2-lb force **P** has been applied. Assume that the disks roll without sliding.

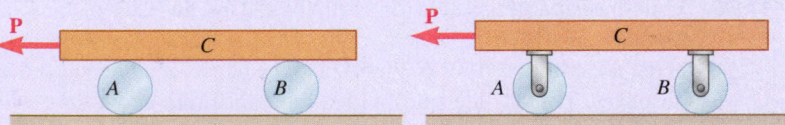

**Fig. P17.72**            **Fig. *P17.73***

**17.74** Two uniform cylinders, each of mass $m = 6$ kg and radius $r = 125$ mm, are connected by a belt as shown. If the system is released from rest when $t = 0$, determine (*a*) the velocity of the center of cylinder $B$ at $t = 3$ s, (*b*) the tension in the portion of belt connecting the two cylinders.

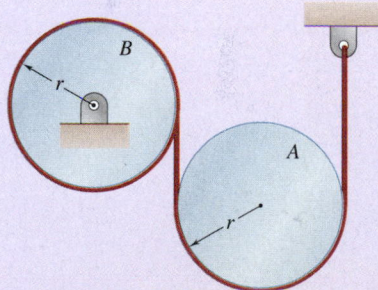

**Fig. P17.74 and P17.75**

**17.75** Two uniform cylinders, each of mass $m = 6$ kg and radius $r = 125$ mm, are connected by a belt as shown. Knowing that at the instant shown the angular velocity of cylinder $A$ is 30 rad/s counterclockwise, determine (*a*) the time required for the angular velocity of cylinder $A$ to be reduced to 5 rad/s, (*b*) the tension in the portion of belt connecting the two cylinders.

**17.76** In the gear arrangement shown, gears $A$ and $C$ are attached to rod $ABC$, which is free to rotate about $B$, while the inner gear $B$ is fixed. Knowing that the system is at rest, determine the magnitude of the couple **M** that must be applied to rod $ABC$, if 2.5 s later the angular velocity of the rod is to be 240 rpm clockwise. Gears $A$ and $C$ weigh 2.5 lb each and may be considered as disks of radius 2 in.; rod $ABC$ weighs 4 lb.

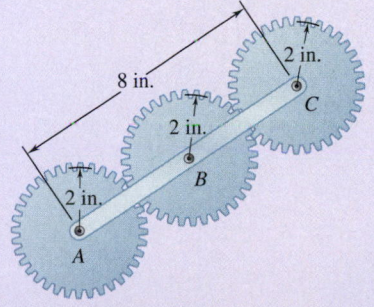

**Fig. *P17.76***

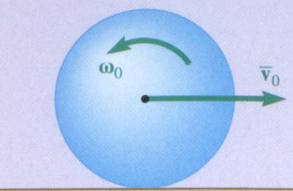

**Fig. P17.77**

**17.77** A sphere of radius $r$ and mass $m$ is projected along a rough horizontal surface with the initial velocities shown. If the final velocity of the sphere is to be zero, express (a) the required magnitude of $\omega_0$ in terms of $v_0$ and $r$, (b) the time required for the sphere to come to rest in terms of $v_0$ and the coefficient of kinetic friction $\mu_k$.

**17.78** A bowler projects an 8.5-in.-diameter ball weighing 16 lb along an alley with a forward velocity $\mathbf{v}_0$ of 25 ft/s and a backspin $\omega_0$ of 9 rad/s. Knowing that the coefficient of kinetic friction between the ball and the alley is 0.10, determine (a) the time $t_1$ at which the ball will start rolling without sliding, (b) the speed of the ball at time $t_1$.

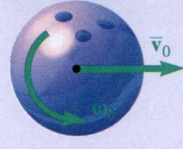

**Fig. P17.78**

**17.79** A semicircular panel with a radius $r$ is attached with hinges to a circular plate with a radius $r$ and initially is held in the vertical position as shown. The plate and the panel are made of the same material and have the same thickness. Knowing that the entire assembly is rotating freely with an initial angular velocity of $\omega_0$, determine the angular velocity of the assembly after the panel has been released and comes to rest against the plate.

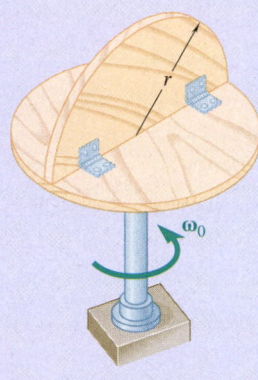

**Fig. P17.79**

**17.80** A satellite has a total weight (on Earth) of 250 lbs, and each of the solar panels weighs 15 lbs. The body of the satellite has a mass moment of inertia about the $z$ axis of 6 slug·ft$^2$, and the panels can be modeled as flat plates. The satellite spins with a rate of 10 rpm about the $z$ axis when the solar panels are positioned in the $xy$ plane. Determine the spin rate about $z$ after a motor on the satellite has rotated both panels to be positioned in the $yz$ plane (as shown in the figure).

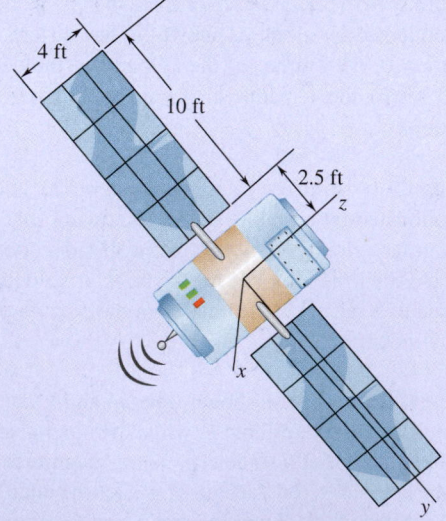

**Fig. P17.80**

**17.81** Two 10-lb disks and a small motor are mounted on a 15-lb rectangular platform that is free to rotate about a central vertical spindle. The normal operating speed of the motor is 180 rpm. If the motor is started when the system is at rest, determine the angular velocity of all elements of the system after the motor has attained its normal operating speed. Neglect the mass of the motor and of the belt.

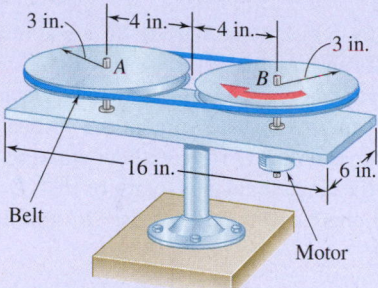

**Fig. P17.81**

**17.82** A 3-kg rod of length 800 mm can slide freely in the 240-mm cylinder *DE*, which in turn can rotate freely in a horizontal plane. In the position shown, the assembly is rotating with an angular velocity of magnitude $\omega = 40$ rad/s and end *B* of the rod is moving toward the cylinder at a speed of 75 mm/s relative to the cylinder. Knowing that the centroidal mass moment of inertia of the cylinder about a vertical axis is 0.025 kg·m² and neglecting the effect of friction, determine the angular velocity of the assembly as end *B* of the rod strikes end *E* of the cylinder.

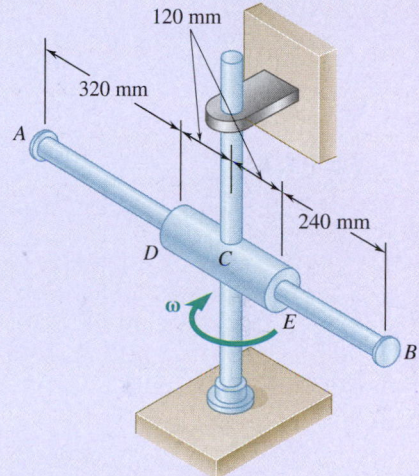

**Fig. P17.82**

**17.83** A 1.6-kg tube *AB* can slide freely on rod *DE*, which in turn can rotate freely in a horizontal plane. Initially, the assembly is rotating with an angular velocity of magnitude $\omega = 5$ rad/s and the tube is held in position by a cord. The moment of inertia of the rod and bracket about the vertical axis of rotation is 0.30 kg·m² and the centroidal moment of inertia of the tube about a vertical axis is 0.0025 kg·m². If the cord suddenly breaks, determine (*a*) the angular velocity of the assembly after the tube has moved to end *E*, (*b*) the energy lost during the plastic impact at *E*.

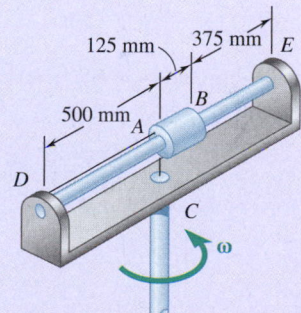

**Fig. P17.83**

**17.84** In the helicopter shown, a vertical tail propeller is used to prevent rotation of the cab as the speed of the main blades is changed. Assuming that the tail propeller is not operating, determine the final angular velocity of the cab after the speed of the main blades has been changed from 180 to 240 rpm. The speed of the main blades is measured relative to the cab, and the cab has a centroidal moment of inertia of 650 lb·ft·s². Each of the four main blades is assumed to be a slender 14-ft rod weighing 55 lb.

**17.85** Assuming that the tail propeller in Prob. 17.84 is operating and that the angular velocity of the cab remains zero, determine the final horizontal velocity of the cab when the speed of the main blades is changed from 180 to 240 rpm. The cab weighs 1250 lb and is initially at rest. Also, determine the force exerted by the tail propeller if the change in speed takes place uniformly in 12 s.

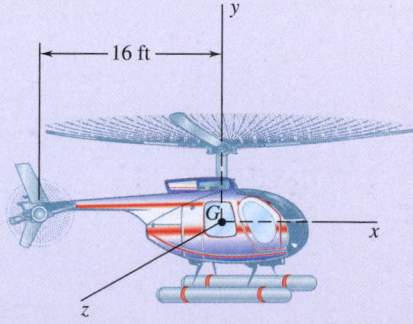

**Fig. P17.84**

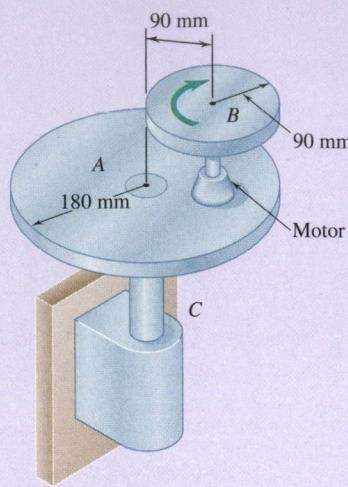

**Fig. P17.86**

**17.86** The 4-kg uniform disk $B$ is attached to the shaft of a motor mounted on plate $A$, which can rotate freely about the vertical shaft $C$. The motor–plate–shaft unit has a moment of inertia of 0.20 kg·m² with respect to the axis of the shaft. If the motor is started when the system is at rest, determine the magnitudes of the angular velocities of the disk and of the plate after the motor has attained its normal operating speed of 360 rpm.

**17.87** The 30-kg uniform disk $A$ and the bar $BC$ are at rest and the 5-kg uniform disk $D$ has an initial angular velocity of $\omega_1$ with a magnitude of 440 rpm when the compressed spring is released and disk $D$ contacts disk $A$. The system rotates freely about the vertical spindle $BE$. After a period of slippage, disk $D$ rolls without slipping. Knowing that the magnitude of the final angular velocity of disk $D$ is 176 rpm, determine the final angular velocities of bar $BC$ and disk $A$. Neglect the mass of bar $BC$.

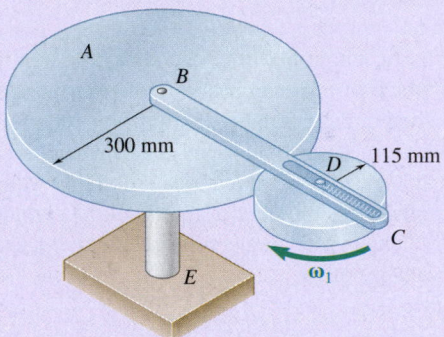

**Fig. P17.87**

**17.88** The 4-kg uniform rod $AB$ can slide freely inside the 6-kg tube. The rod was entirely within the tube ($x = 0$) and released with no initial velocity relative to the tube when the angular velocity of the assembly was 5 rad/s. Neglecting the effect of friction, determine the speed of the rod relative to the tube when $x = 400$ mm.

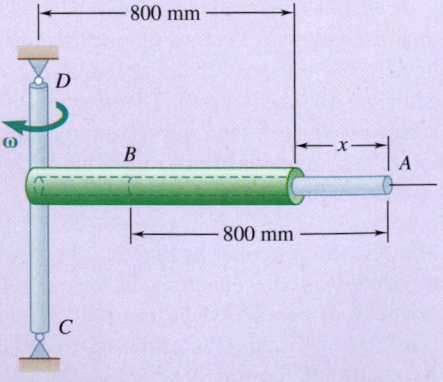

**Fig. P17.88**

**17.89** A 1.8-kg collar $A$ and a 0.7-kg collar $B$ can slide without friction on a frame, consisting of the horizontal rod $OE$ and the vertical rod $CD$, which is free to rotate about its vertical axis of symmetry. The two collars are connected by a cord running over a pulley that is attached to the frame at $O$. At the instant shown, the velocity $\mathbf{v}_A$ of collar $A$ has a magnitude of 2.1 m/s and a stop prevents collar $B$ from moving. The stop is suddenly removed and collar $A$ moves toward $E$. As it reaches a distance of 0.12 m from $O$, the magnitude of its velocity is observed to be 2.5 m/s. Determine at that instant the magnitude of the angular velocity of the frame and the moment of inertia of the frame and pulley system about $CD$.

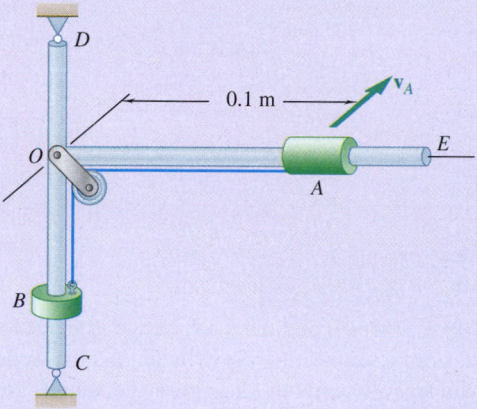

Fig. P17.89

**17.90** A 6-lb collar $C$ is attached to a spring and can slide on rod $AB$, which in turn can rotate in a horizontal plane. The mass moment of inertia of rod $AB$ with respect to end $A$ is 0.35 lb·ft·s². The spring has a constant $k = 15$ lb/in. and an undeformed length of 10 in. At the instant shown, the velocity of the collar relative to the rod is zero and the assembly is rotating with an angular velocity of 12 rad/s. Neglecting the effect of friction, determine (a) the angular velocity of the assembly as the collar passes through a point located 7.5 in. from end $A$ of the rod, (b) the corresponding velocity of the collar relative to the rod.

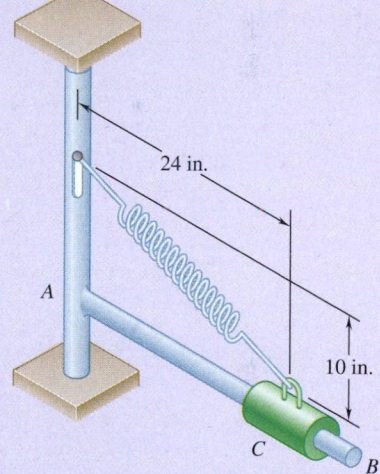

Fig. P17.90

**17.91** A small 4-lb collar $C$ can slide freely on a thin ring of weight 6 lb and radius 10 in. The ring is welded to a short vertical shaft, which can rotate freely in a fixed bearing. Initially, the ring has an angular velocity of 35 rad/s and the collar is at the top of the ring ($\theta = 0$) when it is given a slight nudge. Neglecting the effect of friction, determine (a) the angular velocity of the ring as the collar passes through the position $\theta = 90°$, (b) the corresponding velocity of the collar relative to the ring.

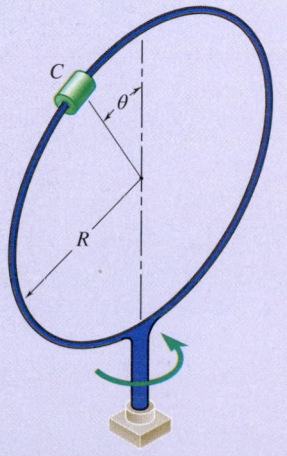

Fig. P17.91

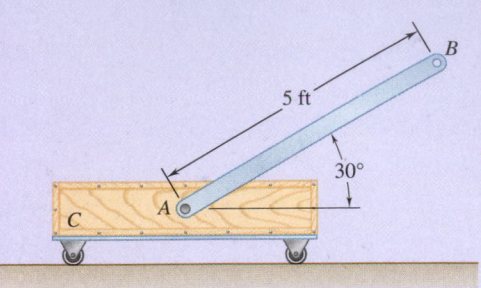

**17.92** Rod *AB* has a weight of 6 lb and is attached to a 10-lb cart *C*. Knowing that the system is released from rest in the position shown and neglecting friction, determine (*a*) the velocity of point *B* as rod *AB* passes through a vertical position, (*b*) the corresponding velocity of the cart *C*.

**17.93** A 3-kg uniform cylinder *A* can roll without sliding on a 5-kg cart *C* and is attached to a spring of constant $k = 100$ N/m as shown. The system is released from rest when the spring is stretched 20 mm. Neglecting the friction between the wheels and the floor, determine the velocity of the cart and the angular velocity of the cylinder when the spring first reaches its undeformed state.

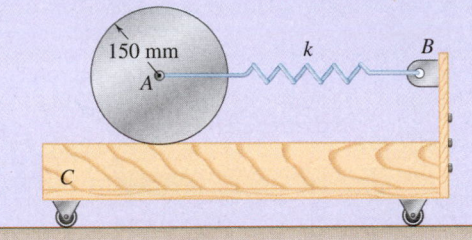

Fig. *P17.93*

**17.94** The 4-kg cylinder *B* and the 3-kg wedge *A* are at rest in the position shown. Cord *C* connecting the cylinder and the wedge is then cut and the cylinder rolls without sliding on the wedge. Neglecting friction between the wedge and the ground, determine (*a*) the angular velocity of the cylinder after it has rolled 180 mm down the wedge, (*b*) the corresponding velocity of the wedge.

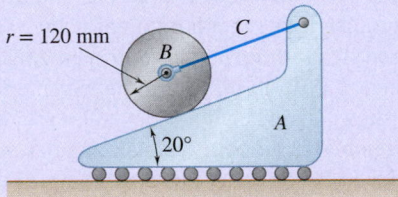

Fig. *P17.94*

**17.95** The 6-lb steel cylinder *A* of radius *r* and the 10-lb wooden cart *B* are at rest in the position shown when the cylinder is given a slight nudge, causing it to roll without sliding along the top surface of the cart. Neglecting friction between the cart and the ground, determine the velocity of the cart as the cylinder passes through the lowest point of the surface at *C*.

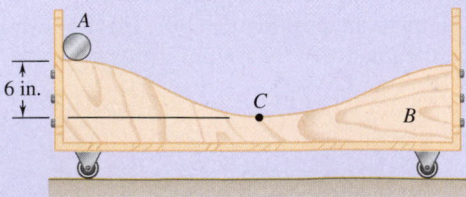

Fig. P17.95

# 17.3 ECCENTRIC IMPACT

You saw in Chap. 13 that the method of impulse and momentum is the only practicable method for solving problems involving the impulsive motion of a particle. Now you will see that problems involving the impulsive motion of a rigid body are particularly well-suited to a solution using the method of impulse and momentum. Because the time interval considered in the computation of linear impulses and angular impulses is very short, we can assume the bodies involved occupy the same position during that time interval, making the computation quite simple.

In Sec. 13.4, we described how to solve problems of **central impact**; that is, problems in which the mass centers of the two colliding bodies are located on the line of impact. We now analyze the **eccentric impact** of two rigid bodies.

Consider two colliding bodies and denote the velocities of the two points of contact $A$ and $B$ before impact by $\mathbf{v}_A$ and $\mathbf{v}_B$ (Fig. 17.10a). Under the impact, the two bodies *deform,* and at the end of the period of deformation, the velocities $\mathbf{u}_A$ and $\mathbf{u}_B$ of $A$ and $B$ have equal components along the line of

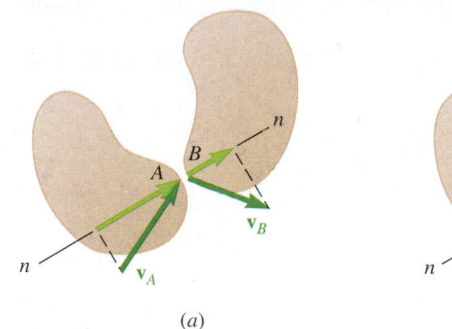
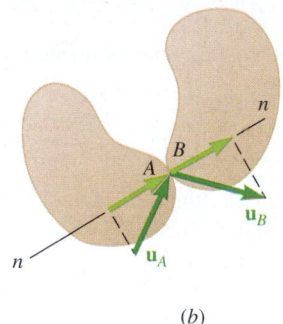
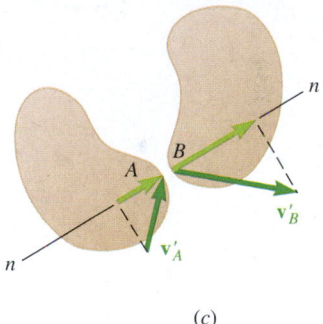

(a)          (b)          (c)

**Fig. 17.10** When two rigid bodies collide, (a) the velocities of the points of contact before impact (b) change during the period of deformation and (c) change again during the period of restitution.

impact $nn$ (Fig. 17.10b). A period of *restitution* then takes place, at the end of which points $A$ and $B$ have velocities of $\mathbf{v}'_A$ and $\mathbf{v}'_B$ (Fig. 17.10c). Assuming that the bodies are frictionless, we find that the forces they exert on each other are directed along the line of impact. We denote the magnitude of the impulse of one of these forces during the period of deformation by $\int P\,dt$ and the magnitude of its impulse during the period of restitution by $\int R\,dt$. Recall that we define the coefficient of restitution $e$ as the ratio of

$$e = \frac{\int R\,dt}{\int P\,dt} \qquad (17.18)$$

We propose to show that the relation established in Sec. 13.4 between the relative velocities of two particles before and after impact also holds between the components along the line of impact of the relative velocities of the two points of contact $A$ and $B$. That is, we want to show that

$$(v'_B)_n - (v'_A)_n = e[(v_A)_n - (v_B)_n] \qquad (17.19)$$

First, we assume that the motion of each of the two colliding bodies of Fig. 17.10 is unconstrained. Thus, the only impulsive forces exerted on the bodies during the impact are applied at $A$ and $B$, respectively. Consider the body to

**Photo 17.4** A swinging bat applies an impulsive force on contact with the ball. You can use the principle of impulse and momentum to determine the final velocities of the ball and bat. ©Tetra Images/Alamy RF

which point $A$ belongs and draw the impulse–momentum diagram corresponding to the period of deformation (Fig. 17.11). We denote the velocity of the mass center at the beginning and at the end of the period of deformation by $\overline{\mathbf{v}}$ and $\overline{\mathbf{u}}$, respectively, and we denote the angular velocity of the body at the

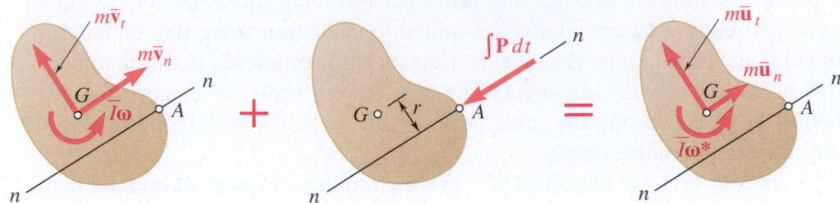

**Fig. 17.11** An impulse–momentum diagram for a body undergoing an eccentric impact during the period of deformation.

same instants by $\omega$ and $\omega^*$. Summing and equating the components of the momenta and impulses along the line of impact *nn,* we have

$$m\overline{v}_n - \int P\,dt = m\overline{u}_n \tag{17.20}$$

Summing and equating the moments about $G$ of the momenta and impulses, we also have

$$\overline{I}\omega - r\int P\,dt = \overline{I}\omega^* \tag{17.21}$$

where $r$ represents the perpendicular distance from $G$ to the line of impact. Considering now the period of restitution, we obtain in a similar way

$$m\overline{u}_n - \int R\,dt = m\overline{v}'_n \tag{17.22}$$

$$\overline{I}\omega^* - r\int R\,dt = \overline{I}\omega' \tag{17.23}$$

where $\overline{v}'_n$ and $\omega'$ represent, respectively, the velocity of the mass center in the *nn* direction and the angular velocity of the body after impact. First solving Eqs. (17.20) and (17.22) for the two impulses and substituting into Eq. (17.18) and then solving Eqs. (17.21) and (17.23) for the same two impulses and substituting again into Eq. (17.18), we obtain the two alternative expressions for the coefficient of restitution as

$$e = \frac{\overline{u}_n - \overline{v}'_n}{\overline{v}_n - \overline{u}_n} \qquad e = \frac{\omega^* - \omega'}{\omega - \omega^*} \tag{17.24}$$

Multiplying the numerator and denominator of the second expression for $e$ by $r$ and adding them, respectively, to the numerator and denominator of the first expression, we have

$$e = \frac{\overline{u}_n + r\omega^* - (\overline{v}'_n + r\omega')}{\overline{v}_n + r\omega - (\overline{u}_n + r\omega^*)} \tag{17.25}$$

Observe that $\overline{v}_n + r\omega$ represents the component $(v_A)_n$ along *nn* of the velocity of the point of contact $A$ and that, similarly, $\overline{u}_n + r\omega^*$ and $\overline{v}'_n + r\omega'$ represent, respectively, the components $(u_A)_n$ and $(v'_A)_n$. Thus, we have

$$e = \frac{(u_A)_n - (v'_A)_n}{(v_A)_n - (u_A)_n} \tag{17.26}$$

The analysis of the motion of the second body leads to a similar expression for $e$ in terms of the components along $nn$ of the successive velocities of point $B$. Recalling that $(u_A)_n = (u_B)_n$, and eliminating these two velocity components by a manipulation similar to the one used in Sec. 13.4, we obtain the relation in Eq. (17.19).

If one or both of the colliding bodies is constrained to rotate about a fixed point $O$—as in the case of a compound pendulum (Fig. 17.12a)—an impulsive reaction is exerted at $O$ (Fig. 17.12b). Let us verify that, although their

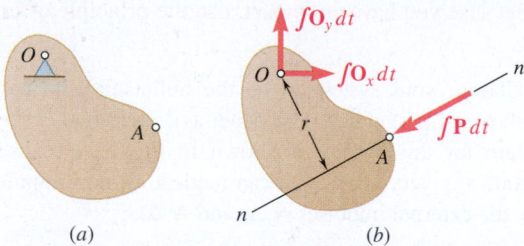

**Fig. 17.12**  (a) A rigid body constrained to rotate about a fixed point $O$; (b) impulsive reaction at $O$ resulting from an eccentric impact.

derivation must be modified, Eqs. (17.26) and (17.19) remain valid. Applying formula (17.16) to the period of deformation and to the period of restitution, we have

$$I_O \omega - r \int P \, dt = I_O \omega^* \tag{17.27}$$

$$I_O \omega^* - r \int R \, dt = I_O \omega' \tag{17.28}$$

where $r$ represents the perpendicular distance from the fixed point $O$ to the line of impact. We solve Eqs. (17.27) and (17.28) for the two impulses and substitute them into Eq. (17.18). Noting that $r\omega$, $r\omega^*$, and $r\omega'$ represent the components along $nn$ of the successive velocities of point $A$, we obtain

$$e = \frac{\omega^* - \omega'}{\omega - \omega^*} = \frac{r\omega^* - r\omega'}{r\omega - r\omega^*} = \frac{(u_A)_n - (v_A')_n}{(v_A)_n - (u_A)_n}$$

This verifies that Eq. (17.26) still holds. Thus, Eq. (17.19) remains valid when one or both of the colliding bodies is constrained to rotate about a fixed point $O$.

In order to determine the velocities of the two colliding bodies after impact, we need to use the relation in Eq. (17.19) in conjunction with one or several other equations obtained by applying the principle of impulse and momentum (Sample Probs. 17.11 and 17.13).

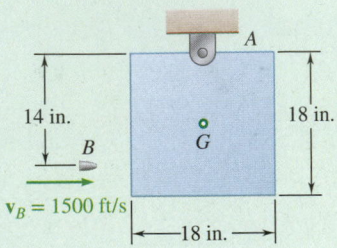

## Sample Problem 17.11

A 0.05-lb bullet $B$ is fired with a horizontal velocity of 1500 ft/s into the side of a 20-lb square panel suspended from a hinge at $A$. Knowing that the panel is initially at rest, determine (*a*) the angular velocity of the panel immediately after the bullet becomes embedded, (*b*) the impulsive reaction at $A$, assuming that the bullet becomes embedded in 0.0006 s.

**STRATEGY:**   Because you have an impact, use the principle of impulse and momentum.

**MODELING:**   Choose your system to be the bullet and the panel, where you model the bullet as a particle and the panel as a rigid body. The impulse–momentum diagram for this system is shown in Fig. 1. Because the time interval $\Delta t = 0.0006$ s is very short, you can neglect all non-impulsive forces and consider only the external impulses $\mathbf{A}_x\Delta t$ and $\mathbf{A}_y\Delta t$.

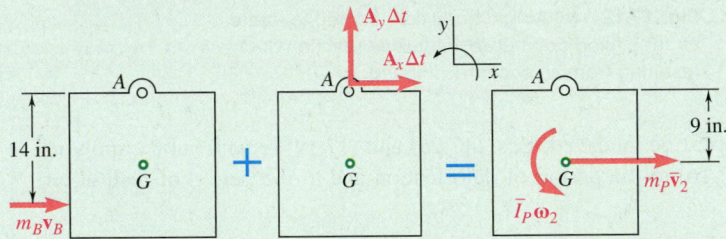

**Fig. 1**   Impulse–momentum diagram for the system. The bullet is neglected at time 2.

### ANALYSIS:

### Principle of Impulse and Momentum.

$$\textbf{Syst Momenta}_1 + \textbf{Syst Ext Imp}_{1\rightarrow 2} = \textbf{Syst Momenta}_2$$

$+\circlearrowleft$ moments about $A$: $\qquad m_B v_B(\tfrac{14}{12}\text{ ft}) + 0 = m_P \bar{v}_2(\tfrac{9}{12}\text{ ft}) + \bar{I}_P\omega_2$   **(1)**

$\overset{+}{\rightarrow} x$ components: $\qquad\qquad m_B v_B + A_x\,\Delta t = m_P\bar{v}_2$   **(2)**

$+\uparrow y$ components: $\qquad\qquad\quad 0 + A_y\,\Delta t = 0$   **(3)**

Note that the weight of the bullet is negligible compared to the weight of the panel, so we did not include it on the right-hand side of Eq. (1). The centroidal mass moment of inertia of the square panel is

$$\bar{I}_P = \tfrac{1}{6}m_P b^2 = \frac{1}{6}\left(\frac{20\text{ lb}}{32.2\text{ ft/s}^2}\right)\left(\frac{18}{12}\text{ ft}\right)^2 = 0.2329\text{ lb·ft·s}^2$$

Substituting this value as well as the given data into Eq. (1) and noting that from kinematics, you know

$$\bar{v}_2 = (\tfrac{9}{12}\text{ ft})\omega_2$$

*(continued)*

You now have

$$\left(\frac{0.05}{32.2}\right)(1500)(\tfrac{14}{12}) = 0.2329\omega_2 + \left(\frac{20}{32.2}\right)(\tfrac{9}{12}\omega_2)(\tfrac{9}{12})$$

$$\omega_2 = 4.67 \text{ rad/s} \qquad \boldsymbol{\omega}_2 = 4.67 \text{ rad/s} \circlearrowleft \quad \blacktriangleleft$$

$$\bar{v}_2 = (\tfrac{9}{12}\text{ ft})\omega_2 = (\tfrac{9}{12}\text{ ft})(4.67 \text{ rad/s}) = 3.50 \text{ ft/s}$$

Substituting $\bar{v}_2 = 3.50$ ft/s, $\Delta t = 0.0006$ s, and the given data into Eq. (2) gives you

$$\left(\frac{0.05}{32.2}\right)(1500) + A_x(0.0006) = \left(\frac{20}{32.2}\right)(3.50)$$

$$A_x = -259 \text{ lb} \qquad \mathbf{A}_x = 259 \text{ lb} \leftarrow \quad \blacktriangleleft$$

From Eq. (3), you find $A_y = 0$.

$$\mathbf{A}_y = 0 \quad \blacktriangleleft$$

**REFLECT and THINK:**   The speed of the bullet is in the range of a modern high-performance rifle. Notice that the reaction at $A$ is over 5000 times the weight of the bullet and over 10 times the weight of the plate.

## Sample Problem 17.12

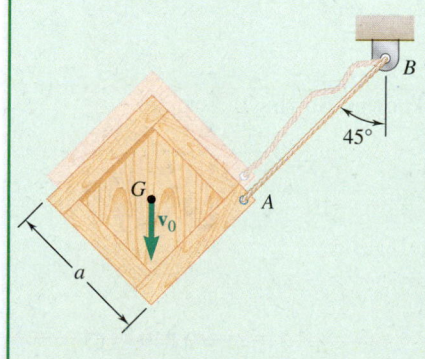

A uniformly loaded square crate is falling freely with a velocity $\mathbf{v}_0$ when cable $AB$ suddenly becomes taut. Assuming that the impact is perfectly plastic, determine the angular velocity of the crate and the velocity of its mass center immediately after the cable becomes taut.

**STRATEGY:**   Because impact occurs, use the principle of impulse and momentum.

**MODELING:**   Choose the crate as your system and model it as a rigid body. The impulse–momentum diagram for this system is shown in Fig. 1. The mass moment of inertia of the plate about $G$ is $\bar{I} = \tfrac{1}{6}ma^2$.

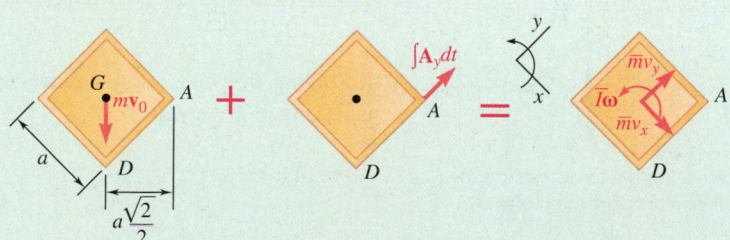

$$\textbf{Syst Momenta}_1 + \textbf{Syst Ext Imp}_{1\rightarrow 2} = \textbf{Syst Momenta}_2$$

**Fig. 1**   Impulse–momentum diagram for the crate.

*(continued)*

## ANALYSIS:

**Principle of Impulse and Momentum.** Applying the impulse–momentum principle in the $x$ direction and taking moments about $A$ gives

$+\circlearrowleft$ moments about $A$:
$$mv_0a\frac{\sqrt{2}}{2}+0=\bar{I}\omega+m\bar{v}_x\frac{a}{2}-m\bar{v}_y\frac{a}{2} \qquad (1)$$

$\overset{+}{\searrow}$ $x$ components:
$$mv_0\frac{\sqrt{2}}{2}+0=m\bar{v}_x \qquad (2)$$

There are three unknowns in these two equations: $\omega$, $\bar{v}_x$, and $\bar{v}_y$. For additional equations, you can use kinematics. Because you are told the impact is perfectly plastic, point $A$ has a velocity perpendicular to the rope (Fig. 2). Therefore, you can relate the acceleration of $A$ to that of $G$, as

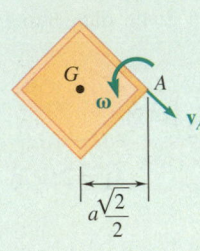

**Fig. 2**   Velocity of point $A$.

$$\bar{\mathbf{v}}=\mathbf{v}_G=\mathbf{v}_A+\mathbf{v}_{G/A}$$

$$=[v_A\searrow 45°]+\left[a\frac{\sqrt{2}}{2}\omega\downarrow\right]$$

Equating components in the $x$ and $y$ directions, you find

$\overset{+}{\searrow}$ $x$ components:
$$\bar{v}_x=v_A+a\frac{\sqrt{2}}{2}\omega\frac{\sqrt{2}}{2}=v_A+\frac{a\omega}{2} \qquad (3)$$

$\overset{+}{\nearrow}$ $y$ components:
$$\bar{v}_y=-a\frac{\sqrt{2}}{2}\omega\frac{\sqrt{2}}{2}=-\frac{a\omega}{2} \qquad (4)$$

You now have four equations and four unknowns. Solving these gives

$$\omega=\frac{3\sqrt{2}}{5}\frac{v_0}{a} \qquad \bar{v}_x=\frac{\sqrt{2}}{2}v_0 \qquad \bar{v}_y=-\frac{3\sqrt{2}}{10}v_0 \qquad v_A=\frac{\sqrt{2}}{5}v_0$$

So

$$\omega=0.849\frac{v_0}{d}\ \circlearrowleft \quad \blacktriangleleft$$

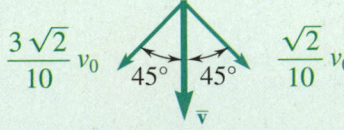

**Fig. 3**   Diagram to determine the magnitude and direction of $\bar{v}$

Resolving the velocity of the center of mass into a magnitude and direction using Fig. 3 gives you

$$\bar{\mathbf{v}}=0.825v_0\searrow 76.0° \quad \blacktriangleleft$$

**REFLECT and THINK:**   If the impact had not been plastic, point $A$ would have rebounded and the rope would have become slack. To solve the problem in this case, you would have needed to use the equation for the coefficient of restitution.

## Sample Problem 17.13

A 2-kg sphere moving horizontally to the right with an initial velocity of 5 m/s strikes the lower end of an 8-kg rigid rod $AB$. The rod is suspended from a hinge at $A$ and is initially at rest. Knowing that the coefficient of restitution between the rod and the sphere is 0.80, determine the angular velocity of the rod and the velocity of the sphere immediately after the impact.

**STRATEGY:**  Because you have an impact, use the principle of impulse and momentum.

**MODELING:**  Choose the sphere and the rod as your system; model the sphere as a particle and the rod as a rigid body. You also need to use the coefficient of restitution equation. The impulse–momentum diagram for this system is shown in Fig. 1. Note that the only impulsive force external to the system is the impulsive reaction at $A$.

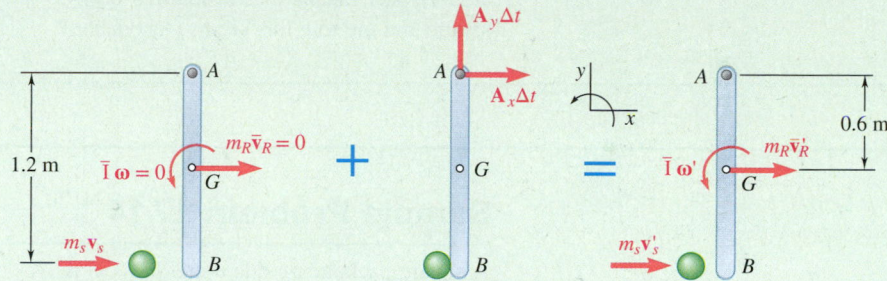

**Fig. 1**  Impulse–momentum diagram for the system.

## ANALYSIS:

### Principle of Impulse and Momentum.

$$\textbf{Syst Momenta}_1 + \textbf{Syst Ext Imp}_{1\rightarrow2} = \textbf{Syst Momenta}_2$$

$+\circlearrowleft$ moments about $A$:

$$m_s v_s (1.2 \text{ m}) = m_s v_s' (1.2 \text{ m}) + m_R \bar{v}_R'(0.6 \text{ m}) + \bar{I}\omega' \qquad (1)$$

In this case, the mass of the sphere is not negligible compared to the rod, so we must include it on the right-hand side of Eq. (1). Because the rod rotates about $A$, from kinematics, you know $\bar{v}_R' = \bar{r}\omega' = (0.6 \text{ m})\omega'$. Also,

$$\bar{I} = \tfrac{1}{12}mL^2 = \tfrac{1}{12}(8 \text{ kg})(1.2 \text{ m})^2 = 0.96 \text{ kg·m}^2$$

Substituting these values and the given data into Eq. (1), you obtain

$$(2 \text{ kg})(5 \text{ m/s})(1.2 \text{ m}) = (2 \text{ kg})v_s'(1.2 \text{ m}) + (8 \text{ kg})(0.6 \text{ m})\omega'(0.6 \text{ m})$$
$$+ (0.96 \text{ kg·m}^2)\omega'$$
$$12 = 2.4v_s' + 3.84\omega' \qquad (2)$$

*(continued)*

**Coefficient of Restitution.**   Choosing positive to the right, you have

$$v_B' - v_s' = e(v_s - v_B)$$

Substituting $v_s = 5$ m/s, $v_B = 0$, and $e = 0.80$ gives

$$v_B' - v_s' = 0.8(5 \text{ m/s} - 0) \qquad \text{(3)}$$

Again noting that the rod rotates about $A$, you have

$$v_B' = (1.2 \text{ m})\omega' \qquad \text{(4)}$$

Solving Eqs. (2), (3), and (4) simultaneously, you obtain

$$\omega' = 3.21 \text{ rad/s} \qquad \boldsymbol{\omega}' = 3.21 \text{ rad/s} \circlearrowleft \quad \blacktriangleleft$$
$$v_s' = -0.143 \text{ m/s} \qquad \mathbf{v}_s' = 0.143 \text{ m/s} \leftarrow \quad \blacktriangleleft$$

**REFLECT and THINK:**   The negative value for the velocity of the sphere after impact means that it bounces back to the left. Given the masses of the sphere and the rod, this seems reasonable.

## Sample Problem 17.14

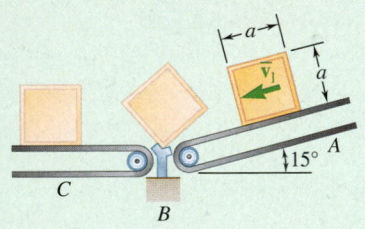

A square package of side $a$ and mass $m$ moves down a conveyor belt $A$ with a constant velocity $\overline{\mathbf{v}}_1$. At the end of the conveyor belt, the corner of the package strikes a rigid support at $B$. Assuming that the impact at $B$ is perfectly plastic, derive an expression for the smallest magnitude of the velocity $\overline{\mathbf{v}}_1$ for which the package will rotate about $B$ and reach conveyor belt $C$.

**STRATEGY:**   Because you have an impact, use the principle of impulse and momentum for when the package strikes the rigid support at $B$, and then apply the conservation of energy for the rotation of the package about the support $B$ after the impact.

**MODELING:**   Choose the package to be your system and model it as a rigid body. The impulse–momentum diagram for this system is shown in Fig. 1. Note that the only impulsive force external to the package is the impulsive reaction at $B$.

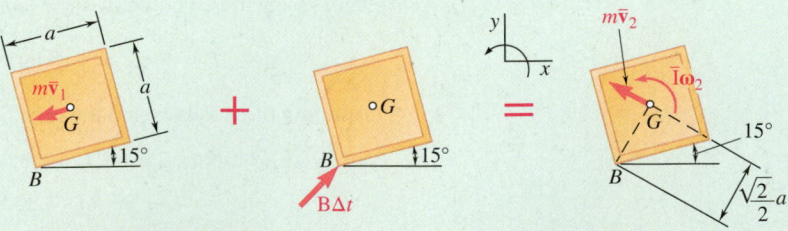

**Fig. 1**   Impulse–momentum diagram for the crate.

*(continued)*

Position 2

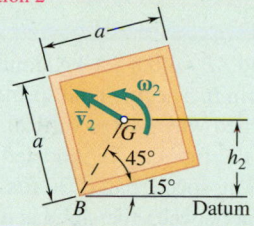

$GB = \frac{1}{2}2\overline{a} = 0.707a$

$h_2 = GB \sin (45° + 15°)$

$\quad\; = 0.612a$

Position 3

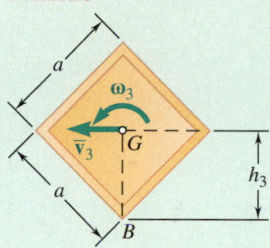

$h_3 = GB = 0.707a$

**Fig. 2** The crate in positions 2 and 3.

## ANALYSIS:

### Principle of Impulse and Momentum.

$$\text{Syst Momenta}_1 + \text{Syst Ext Imp}_{1\to2} = \text{Syst Momenta}_2$$

$+\circlearrowleft$ moments about $B$: $\qquad (m\overline{v}_1)(\frac{1}{2}a) + 0 = (m\overline{v}_1)(\frac{1}{2}\sqrt{2}a) + \overline{I}\omega_2$ **(1)**

Because the package rotates about $B$, from kinematics you have $\overline{v}_2 = (GB)\omega_2 = \frac{1}{2}\sqrt{2}a\omega_2$.

Substitute this expression, together with $\overline{I} = \frac{1}{6}ma^2$, into Eq. (1) for

$$(m\overline{v}_1)(\tfrac{1}{2}a) = m(\tfrac{1}{2}\sqrt{2}a\omega_2)(\tfrac{1}{2}\sqrt{2}a) + \tfrac{1}{6}ma^2\omega_2 \qquad \overline{v}_1 = \tfrac{4}{3}a\omega_2 \quad \textbf{(2)}$$

### Conservation of Energy.
Apply the principle of conservation of energy between position 2 and position 3 (Fig. 2) as

$$T_2 + V_2 = T_3 + V_3 \tag{3}$$

You need to determine the energy at these two positions.

*Position 2.* $V_2 = Wh_2$. Because $\overline{v}_2 = \frac{1}{2}\sqrt{2}\,a\,\omega_2$, you have

$$T_2 = \tfrac{1}{2}m\overline{v}_2^2 + \tfrac{1}{2}\overline{I}\omega_2^2 = \tfrac{1}{2}m(\tfrac{1}{2}\sqrt{2}a\omega_2)^2 + \tfrac{1}{2}(\tfrac{1}{6}ma^2)\omega_2^2 = \tfrac{1}{3}ma^2\omega_2^2$$

*Position 3.* The package must reach conveyor belt $C$, so it must pass through position 3 where $G$ is directly above $B$. Also, because you wish to determine the smallest velocity for which the package will reach this position, choose $\overline{v}_3 = \omega_3 = 0$. Therefore, $T_3 = 0$ and $V_3 = Wh_3$.

Substituting these into Eq. (3)

$$\tfrac{1}{3}ma^2\omega_2^2 + Wh_2 = 0 + Wh_3$$

$$\omega_2^2 = \frac{3W}{ma^2}(h_3 - h_2) = \frac{3g}{a^2}(h_3 - h_2) \tag{4}$$

Substituting the computed values of $h_2$ and $h_3$ into Eq. (4), you obtain

$$\omega_2^2 = \frac{3g}{a^2}(0.707a - 0.612a) = \frac{3g}{a^2}(0.095a) \qquad \omega_2 = \sqrt{0.285g/a}$$

$$\overline{v}_1 = \tfrac{4}{3}a\omega_2 = \tfrac{4}{3}a\sqrt{0.285g/a} \qquad\qquad\qquad \overline{v}_1 = 0.712\sqrt{ga} \;\blacktriangleleft$$

**REFLECT and THINK:** The combination of energy and momentum methods is typical of many design analyses. If you had been interested in determining the reaction at $B$ immediately after the impact or at some other point in the motion, you would have needed to draw a free-body diagram and a kinetic diagram and apply Newton's second law.

## Sample Problem 17.15

A soccer ball tester consists of a 15-kg slender rod $AB$ with a 1.1-kg simulated foot located at $A$ and a torsional spring located at pin $B$. The torsional spring has a spring constant of $k_t = 910$ N·m and is unstretched when $AB$ is vertical. The length of $AB$ is 0.9 m, and you can assume that the foot can be modeled as a point mass. Knowing that the velocity of the 0.45-kg soccer ball is 30 ft/s after impact, determine (*a*) the coefficient of restitution between the simulated foot and the ball, (*b*) the impulse at $B$ during the impact.

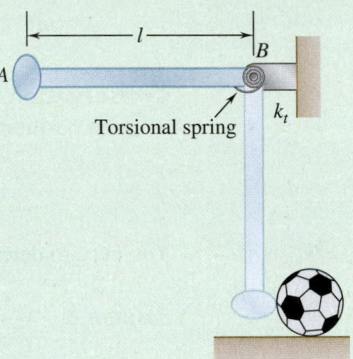

**STRATEGY:**   This problem can be broken into two distinct stages of motion. In stage 1, the arm moves downward under the influence of gravity and the torsional spring. You can use the conservation of energy for this stage. In stage 2, the foot hits the ball, and you need to use both the principle of impulse and momentum and the coefficient of restitution.

**MODELING:**   Each stage requires a different system. For stage 1, your system is rod $AB$, foot $A$, and the torsional spring. In stage 2, your system is rod $AB$, foot $A$, and the soccer ball. The appropriate diagrams are drawn in the analysis section. You can model $AB$ as a slender rod, so its mass moment of inertia is

$$\bar{I}_{AB} = \tfrac{1}{12}m_{AB}l^2 = \tfrac{1}{12}(15 \text{ kg})(0.9 \text{ m})^2 = 1.0125 \text{ kg·m}^2$$

**ANALYSIS:**

**Rod *AB* Moves Down.**   Apply the principle of conservation of energy

$$T_1 + V_{g_1} + V_{e_1} = T_2 + V_{g_2} + V_{e_2} \tag{1}$$

*(continued)*

***Position* 1.**   The system starts from rest, so $T_1 = 0$. Using the datum defined in Fig. 1, you know $V_{g_1} = 0$, and because the spring is unstretched at position 2, you find

$$V_{e_1} = \frac{1}{2}k_t\theta^2 = \frac{1}{2}(910\ \text{N·m})\left(\frac{\pi}{2}\right)^2 = 1123\ \text{J}$$

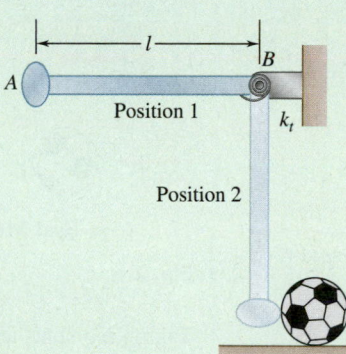

**Fig. 1**   The rod in positions 1 and 2.

***Position* 2.**   The elastic potential energy is $V_{e_2} = 0$, and the gravitational potential energy is

$$V_{g_2} = -m_{AB}g\frac{l}{2} - m_A g l = -(15\ \text{kg})(9.81\ \text{m/s}^2)(0.45\ \text{m}) - (1.1\ \text{kg})(9.81\ \text{m/s}^2)(0.9\ \text{m})$$
$$= -75.93\ \text{J}$$

The kinetic energy is

$$T_2 = \tfrac{1}{2}m_A v_A^2 + \tfrac{1}{2}m_{AB}v_G^2 + \tfrac{1}{2}\bar{I}_{AB}\omega^2$$

You can relate the velocity of the foot and the velocity of the center of gravity of the rod to the angular velocity of $AB$ by recognizing that $AB$ is undergoing fixed-axis rotation. Therefore, $v_G = \omega(\frac{l}{2})$ and $v_A = \omega l$. Substituting these into the expression for $T_2$ and putting in values gives

$$T_2 = \frac{1}{2}\left(m_A l^2 + m_{AB}\left(\frac{l}{2}\right)^2 + \bar{I}_{AB}\right)\omega^2 = 2.4705\omega^2$$

Substituting these energy terms into Eq. (1) gives

$$0 + 0 + 1123 = 2.4705\omega^2 - 75.93 + 0$$

Solving for the angular velocity, you find $\omega = 22.03$ rad/s. Knowing $\omega$, you can calculate the velocities $v_G = 9.912$ m/s and $v_A = 19.824$ m/s.

*(continued)*

**Foot *A* Impacts the Soccer Ball.** Impulse–momentum diagrams for the impact on the ball are shown in Fig. 2.

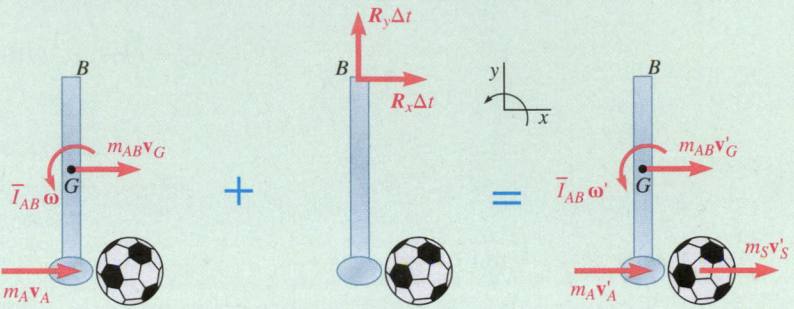

$$\text{Syst Momenta}_1 + \text{Syst Ext Imp}_{1\to2} = \text{Syst Momenta}_2$$

**Fig. 2**

Taking moments about *B* gives you

$+\circlearrowleft$ moments about *B*:

$$m_A v_A l + m_{AB} v_G \frac{l}{2} + \bar{I}_{AB}\omega + 0 = m_A v_A' l + m_{AB} v_G' \frac{l}{2} + \bar{I}_{AB}\omega' + m_S v_S' l \tag{2}$$

The equation for the coefficient of restitution is

$$v_S' - v_A' = e(v_A - 0) \tag{3}$$

where $v_S' = 30$ m/s. From kinematics, you know $v_A' = \omega'l$ and $v_G' = \omega'(l/2)$. Using these kinematic equations and Eqs. (2) and (3), you can solve for the unknown quantities

$$v_A' = 17.61 \text{ m/s} \qquad v_G' = 8.81 \text{ m/s} \qquad \omega' = 19.57 \text{ rad/s} \qquad e = 0.625$$

$$e = 0.625 \quad \blacktriangleleft$$

**Impulses During Impact.** Applying impulse–momentum in the *x* and *y* directions gives

$\xrightarrow{+} x$ components: $\qquad m_{AB} v_G + m_A v_A + R_x \Delta t = m_{AB} v_G' + m_A v_A' + m_S v_S' \tag{4}$

$+\uparrow y$ components: $\qquad\qquad\qquad 0 + R_x \Delta t = 0 \tag{5}$

Solving these equations, you find $R_x \Delta t = -5.53$ N and $R_y \Delta t = 0$.

$$\mathbf{R}\Delta t = 5.53 \text{ N} \leftarrow \quad \blacktriangleleft$$

**REFLECT and THINK:** This coefficient of restitution seems reasonable. As you decrease the pressure in the ball, you would expect the coefficient of restitution to decrease; therefore, the distance the ball travels will decrease. If you had been asked to determine the reactions at *B* after the impact, you would need to draw a free-body diagram and kinetic diagram for your system and apply Newton's second law.

## Case Study 17.1

As you learned in the Chap. 12 Case Study, people with disabilities can profit from the physical and social benefits of participating in sports. T-ball is a sport that can be great fun for athletes with mobility impairments, but it is often difficult for them to find equipment that functions well and provides them with some level of exercise. In the design shown in CS Photo 17.1, the athlete loads up a spring by cranking on a ratchet. This allows the athlete to apply a low level of force over a fairly small displacement. A helper can continue cranking the spring back if the athlete becomes too fatigued. The spring is attached to a cord that wraps around a pulley attached to the bat. Assuming you are on a team designing this device, determine the required spring to power the device for three different pulley radii—1, 2, and 3 in.

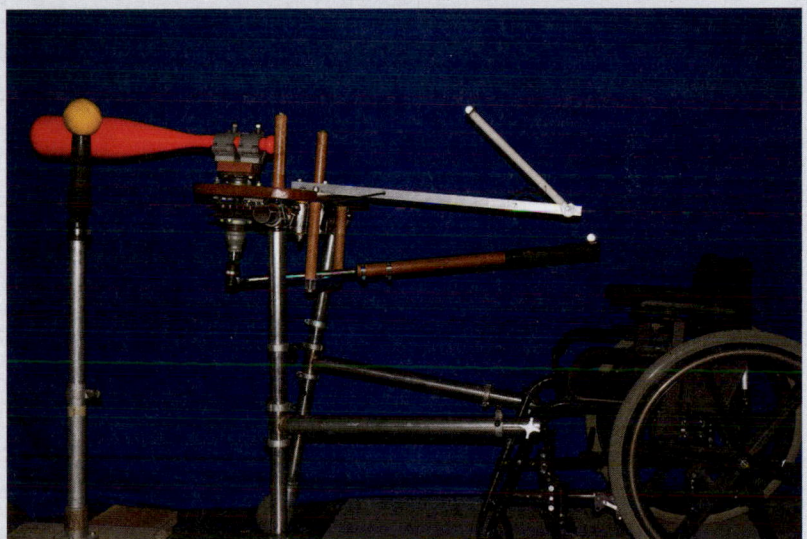

**CS Photo 17.1**   Adapted baseball device. ©Daniel Montoya

### STRATEGY

Use projectile motion to find the required velocity of the ball after the impact, use impulse and momentum for the impact, and use work and energy to analyze the motion of the bat before it hits the ball.

### MODELING AND ANALYSIS

You can treat the bat as a rigid body, model the ball as a particle, and neglect air drag during the flight of the ball. Assume that the bat hits the ball straight off the tee with no initial vertical velocity.

#### Projectile Motion

For the design of this device, we will assume the tee is placed a distance $h = 3$ ft above the ground, and the customer wants a ball that is hit straight on to go at

*(continued)*

least $d = 20$ ft in the air. Because there is constant acceleration and no initial vertical velocity, you can use the basic kinematic relationship

$$y = y_0 + (v_0)_y t - \frac{1}{2}gt^2: \qquad 0 = h + 0 - \frac{1}{2}(32.2 \text{ ft/s}^2)\, t^2$$

Solving for $t$ gives you $t = 0.43167$ s, which you can substitute into the equation governing the horizontal motion of the ball.

$$x = x_0 + (v_0)_x t: \qquad d = 0 + (v_0)_x (0.43167 \text{ s})$$

Solving for $(v_0)_x$ gives you $(v_0)_x = 46.332$ ft/s. This is the necessary velocity of ball $A$ for it to travel 20 ft in the air.

### Impulse and Momentum

For the impact, you can draw an impulse momentum diagram as shown in CS Fig. 17.1.

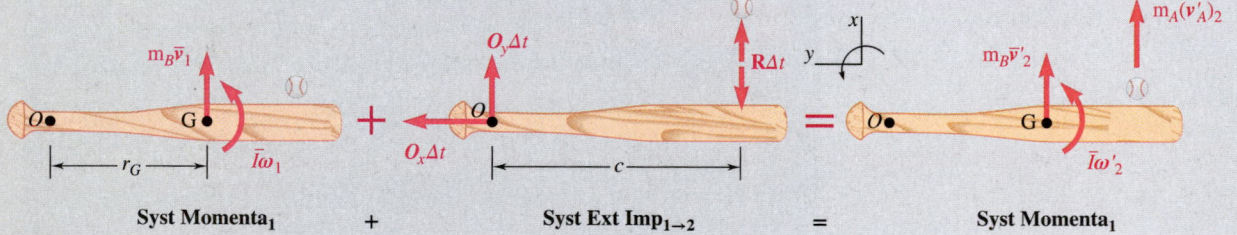

**Syst Momenta₁**          **+**          **Syst Ext Imp₁→₂**          **=**          **Syst Momenta₁**

**CS Fig. 17.1** Impulse–momentum diagram for the bat hitting the ball.

You will need to determine the necessary bat speed at the point of impact by using the coefficient of restitution equation along with the principle of impulse and momentum. Summing moments about $O$ from CS Fig. 17.1 gives you

$$\bar{I}\omega + m_B (\bar{v}_B) r_G + m_A (v_A)c = \bar{I}\omega' + m_B(\bar{v}_B')r_G + m_A(v_A')c \qquad (1)$$

Realizing that the bat is undergoing fixed axis rotation (and that experimental measurements of the mass moment of inertia are generally taken about the handle), that $(\bar{v}_A') = (v_0)_x$, and that the ball is initially at rest, you can simplify Eq. (1) to:

$$I_O\omega = I_O\omega' + m_A(v_0)_x c \qquad (2)$$

The coefficient of restitution equation relates the velocity of the impact point on the bat to that of the ball.

$$(v_B')_n - (v_A')_n = e[(v_A)_n - (v_B)_n] \qquad (3)$$

*(continued)*

Recognizing that the velocity of the impact point on the bat is $v_B = c\omega$, the velocity of the ball after the impact is $(v'_A)_n = (v_o)_x$, and that the ball is initially stationary, you can substitute terms into Eq. (3).

$$c\omega' - (v_o)_x = -c\omega e \quad \rightarrow \quad c\omega' + c\omega e = (v_o)_x \qquad \textbf{(4)}$$

There are two unknowns, $\omega$ and $\omega'$, in Eqs. (2) and (4). The weight of the baseball is 5 ounces and the moment of inertia about point $O$ of the bat and rotating parts is measured to be 0.2 slug·ft$^2$. The sweet spot of the bat is a distance $c = 2.0$ ft, and experimental measurements of the ball hitting the bat give you a value $e = 0.4$. Substituting in these values and solving gives you

$$\omega = 21.365 \text{ rad/s} \quad \text{and} \quad \omega' = 14.620 \text{ rad/s}$$

### Work and Energy

Now that you know the bat's angular velocity just before the impact, you can start to design the spring and pulley size. The spring is attached to the pulley as shown in CS Fig. 17.2, and this pulley rotates the mechanism that holds the bat. You can assume the maximum cocked position is at $\theta = 120°$ and that the spring is unstretched when the bat reaches position 2. You can now determine the appropriate spring constants for pulley radii of 1.0, 2.0, and 3.0 in.

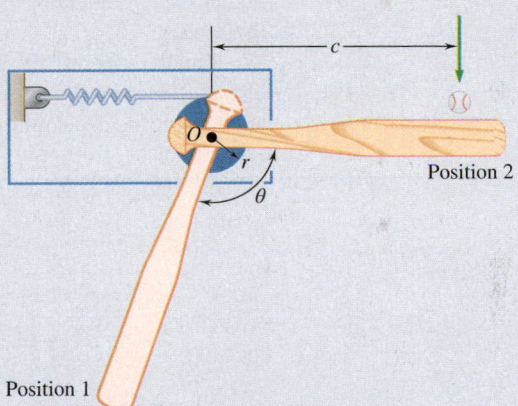

**CS Fig. 17.2** Top view of the baseball bat when cranked back (position 1) and just before it hits the ball (position 2).

Applying the principle of work and energy

$$T_1 + V_{g_1} + V_{e_1} + U_{1\to2}^{NC} = T_2 + V_{g_2} + V_{e_2}$$

Assuming that friction is negligible and recognizing that all terms are zero except $V_{e_1}$ and $T_2$, you get

$$\frac{1}{2}kx^2 = \frac{1}{2}I_O\omega^2 \rightarrow k = \frac{I_O\omega^2}{x^2} \qquad \textbf{(5)}$$

*(continued)*

Note that $x$ is the displacement of the spring when it is cranked back, and is equal to the arc length of the pulley, $x = r\theta$, where $r$ and $\theta$ are expressed in ft and radians, respectively. Using our result from earlier, $\omega = 21.365$ rad/s, and using different pulley radii, you get

| Pulley Radius $r$ (in.) | Spring Elongation $x$ (ft) | Spring Constant $k$ (lb/ft) |
|---|---|---|
| 1.0 | 0.175 | 3000 |
| 2.0 | 0.349 | 749 |
| 3.0 | 0.524 | 333 |

**REFLECT and THINK:**
There are many design trade-offs that must be considered before making a final decision on sizing the spring. A lower spring constant will mean lower forces on the device, contributing to longer life. A six-inch-diameter pulley, however, would add to the overall mass moment of inertia (we did not account for this in our analysis), might be large and more difficult to transport, and could disrupt the aesthetics of the device. You would also need to see what types of springs are commercially available, and should plan to purchase one that is somewhat stiffer than your design value to compensate for energy losses. You could plan ahead in case the athletes want to use baseball and softball bats that have a higher mass moment of inertia, and allow for the springs to be swapped out for those uses.

Finally, some of the athletes may want to hit a fly ball, in which case the bat would need to hit under the ball. This would require you to set a different $n$ and $t$ direction, depending on how high you would like to hit the ball. Batter up!

# SOLVING PROBLEMS ON YOUR OWN

This section was devoted to **impulsive motion** and to the **eccentric impact** of rigid bodies.

**1. Impulsive motion** occurs when a rigid body is subjected to a very large force **F** for a very short interval of time $\Delta t$; the resulting impulse $\mathbf{F}_{avg} \Delta t$ is both finite and different from zero. Such forces are referred to as **impulsive forces** and arise whenever an impact occurs between two rigid bodies. Forces for which the impulse is negligible are referred to as **non-impulsive forces**. As discussed in Chap. 13, you can assume the following forces to be non-impulsive: the weight of a body, the force exerted by a spring, and any other force that is known to be small by comparison with the impulsive forces. Unknown reactions, however, cannot be assumed to be non-impulsive.

**2. Eccentric impact of rigid bodies.** When two bodies collide, the velocity components along the line of impact of the points of contact $A$ and $B$ before and after impact satisfy

$$(v'_B)_n - (v'_A)_n = e[(v_A)_n - (v_B)_n] \qquad \textbf{(17.19)}$$

where the left-hand side is the *relative velocity after the impact* and the right-hand side is the product of the coefficient of restitution and the *relative velocity before the impact*.

This equation expresses the same relation between the velocity components of the points of contact before and after an impact that you used for particles in Chap. 13.

**3. To solve a problem involving an impact** you should use the *method of impulse and momentum* and take the following steps.

    **a. Draw an impulse–momentum diagram of the system** showing the momenta immediately before impact plus the impulses of the external forces acting during the impact; this sum is equivalent to the momenta immediately after impact.

    **b. Write the governing equations** for the angular momentum about some point. Depending on the problem type (especially when you want to find support impulsive reactions), you may also need to write the equations for linear momentum (Sample Prob. 17.11).

    **c. In the case of an impact in which e > 0,** the number of unknowns will be greater than the number of equations that you can write by summing components and moments. You should supplement the equations obtained from the impulse–momentum diagram with the coefficient of restitution from Eq. (17.19) that relates the relative velocities of the points of contact before and after impact (Sample Probs. 17.13 and 17.15).

    **d. During an impact, you must use the method of impulse and momentum.** However, *before and after the impact* you can, if necessary, use some of the other methods of solution that you have learned, such as the conservation of energy (Sample Probs. 17.14 and 17.15) or Newton's second law.

# Problems

**17.F4** A uniform slender rod $AB$ of mass $m$ is at rest on a frictionless horizontal surface when hook $C$ engages a small pin at $A$. Knowing that the hook is pulled upward with a constant velocity $\mathbf{v}_0$, draw the impulse–momentum diagram that is needed to determine the impulse exerted on the rod at $A$ and $B$. Assume that the velocity of the hook is unchanged and that the impact is perfectly plastic.

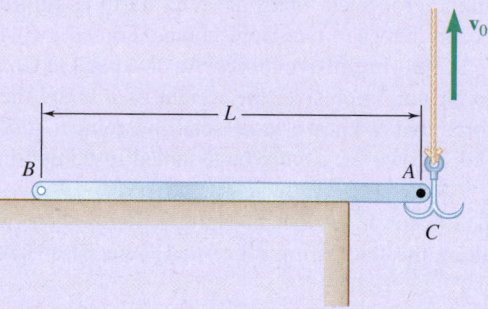

**Fig. P17.F4**

**17.F5** A uniform slender rod $AB$ of length $L$ is falling freely with a velocity $\mathbf{v}_0$ when cord $AC$ suddenly becomes taut. Assuming that the impact is perfectly plastic, draw the impulse–momentum diagram that is needed to determine the angular velocity of the rod and the velocity of its mass center immediately after the cord becomes taut.

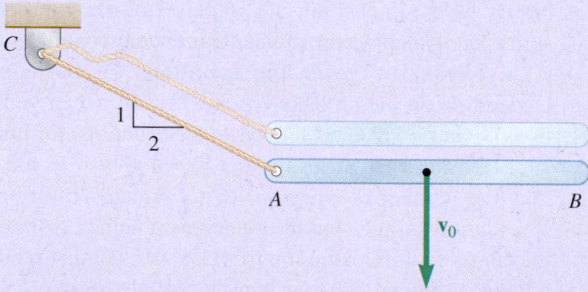

**Fig. P17.F5**

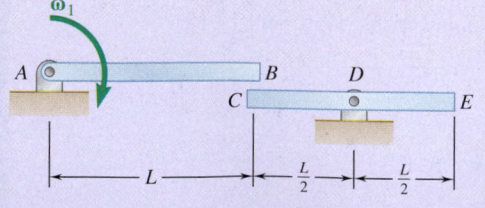

**Fig. P17.F6**

**17.F6** A slender rod $CDE$ of length $L$ and mass $m$ is attached to a pin support at its midpoint $D$. A second and identical rod $AB$ is rotating about a pin support at $A$ with an angular velocity $\boldsymbol{\omega}_1$ when its end $B$ strikes end $C$ of rod $CDE$. The coefficient of restitution between the rods is $e$. Draw the impulse–momentum diagrams that are needed to determine the angular velocity of each rod immediately after the impact.

**17.96** At what height $h$ above its center $G$ should a billiard ball of radius $r$ be struck horizontally by a cue if the ball is to start rolling without sliding?

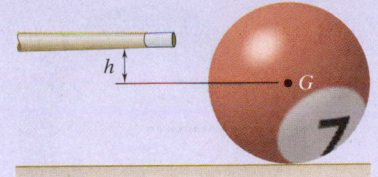

**Fig. P17.96**

**17.97** A bullet weighing 0.08 lb is fired with a horizontal velocity of 1800 ft/s into the lower end of a slender 15-lb bar of length $L = 30$ in. Knowing that $h = 12$ in. and that the bar is initially at rest, determine (*a*) the angular velocity of the bar immediately after the bullet becomes embedded, (*b*) the impulsive reaction at $C$, assuming that the bullet becomes embedded in 0.001 s.

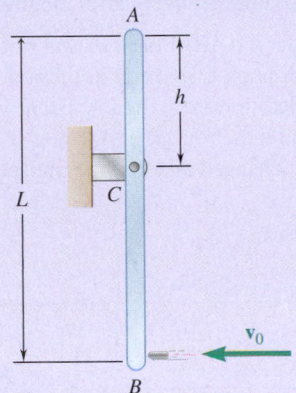

**Fig. P17.97**

**17.98** In Prob. 17.97, determine (*a*) the required distance $h$ if the impulsive reaction at $C$ is to be zero, (*b*) the corresponding angular velocity of the bar immediately after the bullet becomes embedded.

**17.99** A 16-lb wooden panel is suspended from a pin support at $A$ and is initially at rest. A 4-lb metal sphere is released from rest at $B$ and falls into a hemispherical cup $C$ attached to the panel at a point located on its top edge. Assuming that the impact is perfectly plastic, determine the velocity of the mass center $G$ of the panel immediately after the impact.

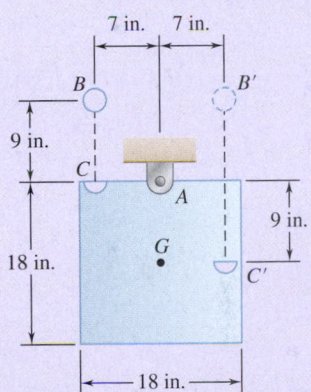

**Fig. *P17.99* and *P17.100***

**17.100** A 16-lb wooden panel is suspended from a pin support at $A$ and is initially at rest. A 4-lb metal sphere is released from rest at $B'$ and falls into a hemispherical cup $C'$ attached to the panel at the same level as the mass center $G$. Assuming that the impact is perfectly plastic, determine the velocity of the mass center $G$ of the panel immediately after the impact.

**17.101** A 45-g bullet is fired with a velocity of 400 m/s at $\theta = 30°$ into a 9-kg square panel of side $b = 200$ mm. Knowing that $h = 150$ mm and that the panel is initially at rest, determine (*a*) the velocity of the center of the panel immediately after the bullet becomes embedded, (*b*) the impulsive reaction at $A$, assuming that the bullet becomes embedded in 2 ms.

**17.102** A 45-g bullet is fired with a velocity of 400 m/s at $\theta = 5°$ into a 9-kg square panel of side $b = 200$ mm. Knowing that the panel is initially at rest, determine (*a*) the required distance $h$ if the horizontal component of the impulsive reaction at $A$ is to be zero, (*b*) the corresponding velocity of the center of the panel immediately after the bullet becomes embedded.

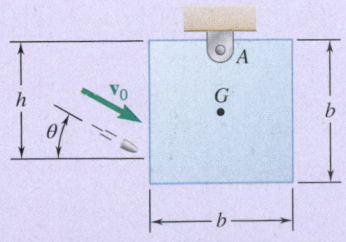

**Fig. P17.101 and P17.102**

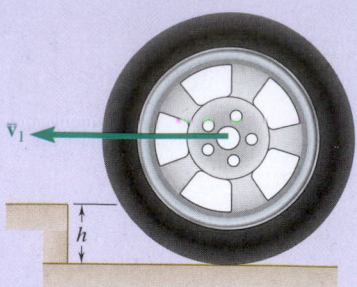

Fig. P17.103

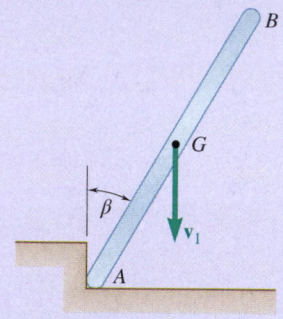

Fig. P17.104

**17.103** The tire shown has a radius $R = 300$ mm and a radius of gyration $\bar{k} = 200$ mm. The tire is rolling without sliding with a velocity $\bar{\mathbf{v}}_1$ of magnitude 3 m/s when it strikes a curb of height $h = 120$ mm. Because the tire is underinflated, no slipping occurs between the tire and the curb. Assuming a perfectly plastic impact, determine (a) the angular velocity of the tire immediately after the impact, (b) the angular velocity of the tire after it has rotated to the top of the step.

**17.104** The uniform slender rod $AB$ of weight 5 lb and length 30 in. forms an angle $\beta = 30°$ with the vertical as it strikes the smooth corner shown with a vertical velocity $\mathbf{v}_1$ of magnitude 8 ft/s and no angular velocity. Assuming that the impact is perfectly plastic, determine the angular velocity of the rod immediately after the impact.

**17.105** A uniform slender rod $AB$ of mass $m$ is at rest on a frictionless horizontal surface when hook $C$ embeds in the rod at point $A$. Knowing that the hook is pulled upward with a constant velocity $\mathbf{v}_0$, determine the impulse exerted on the rod (a) at $A$, (b) at $B$. Assume that the velocity of the hook is unchanged and that the impact is perfectly plastic.

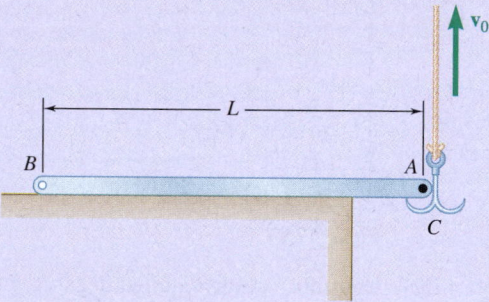

Fig. P17.105

**17.106** A prototype of an adapted bowling device is a simple ramp that attaches to a wheelchair. The bowling ball has a mass moment of inertia about its center of gravity of $cmr^2$, where $c$ is a unitless constant, $r$ is the radius, and $m$ is its mass. The athlete nudges the ball slightly from a height of $h$, and the ball rolls down the ramp without sliding. It hits the bowling lane, and after slipping for a short distance, it begins to roll again. Assuming that the ball does not bounce as it hits the lane, determine the angular velocity and the velocity of the mass center of the ball after it has resumed rolling.

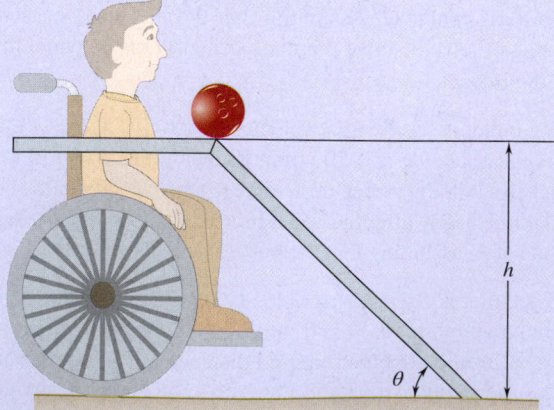

Fig. P17.106

**17.107** A uniform slender rod $AB$ is at rest on a frictionless horizontal table when end $A$ of the rod is struck by a hammer that delivers an impulse that is perpendicular to the rod. In the subsequent motion, determine the distance $b$ through which the rod will move each time it completes a full revolution.

**17.108** A bullet of mass $m$ is fired with a horizontal velocity $\mathbf{v}_0$ and at a height $h = \frac{1}{2}R$ into a wooden disk of much larger mass $M$ and radius $R$. The disk rests on a horizontal plane and the coefficient of friction between the disk and the plane is finite. (a) Determine the linear velocity $\overline{\mathbf{v}}_1$ and the angular velocity $\boldsymbol{\omega}_1$ of the disk immediately after the bullet has penetrated the disk. (b) Describe the ensuing motion of the disk and determine its linear velocity after the motion has become uniform.

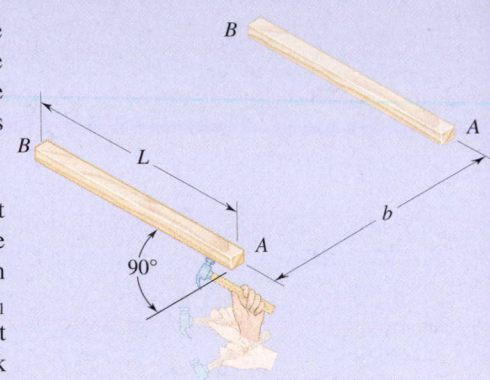

**Fig. P17.107**

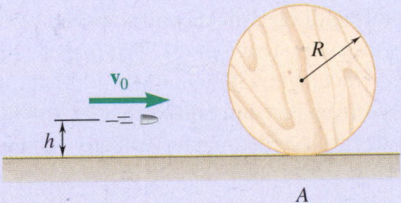

**Fig. P17.108**

**17.109** Determine the height $h$ at which the bullet of Prob. 17.108 should be fired (a) if the disk is to roll without sliding immediately after impact, (b) if the disk is to slide without rolling immediately after impact.

**17.110** A uniform slender bar of length $L = 200$ mm and mass $m = 0.5$ kg is supported by a frictionless horizontal table. Initially the bar is spinning about its mass center $G$ with a constant angular speed $\omega_1 = 6$ rad/s. Suddenly latch $D$ is moved to the right and is struck by end $A$ of the bar. Knowing that the coefficient of restitution between $A$ and $D$ is $e = 0.6$, determine the angular velocity of the bar and the velocity of its mass center immediately after the impact.

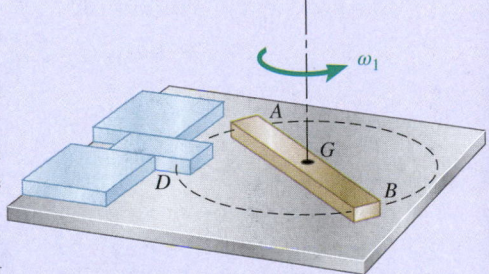

**Fig. P17.110**

**17.111** A uniform slender rod of length $L$ is dropped onto rigid supports at $A$ and $B$. Because support $B$ is slightly lower than support $A$, the rod strikes $A$ with a velocity $\overline{\mathbf{v}}_1$ before it strikes $B$. Assuming perfectly elastic impact at both $A$ and $B$, determine the angular velocity of the rod and the velocity of its mass center immediately after the rod (a) strikes support $A$, (b) strikes support $B$, (c) again strikes support $A$.

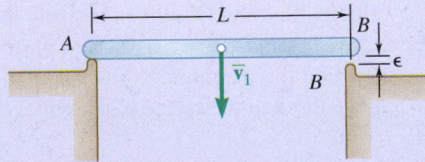

**Fig. P17.111**

**17.112** A uniform slender rod $AB$ has a mass $m$, a length $L$, and is falling freely with a velocity $\mathbf{v}_0$ when end $B$ strikes a smooth inclined surface as shown. Assuming that the impact is perfectly elastic, determine the angular velocity of the rod and the velocity of its mass center immediately after the impact.

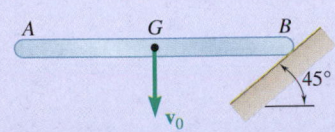

**Fig. P17.112**

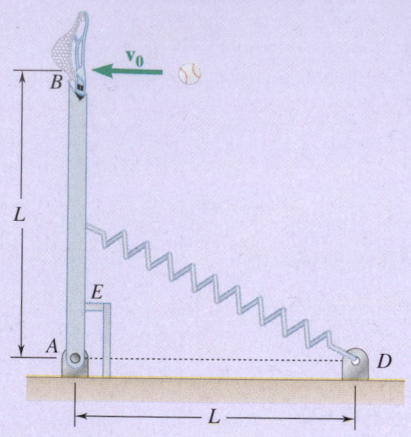

**Fig. P17.113**

**17.113** You have been hired to design a baseball "catcher" that consists of a 5-kg slender rod of length $L = 1.2$ m and a small net of negligible mass at point $B$ to catch the ball. A spring of unstretched length 0.3 m is attached to the midpoint of bar $AB$ at one end and to stationary point $D$ at the other. A stopper at point $E$ keeps the catcher in the vertical position before the pitch. Knowing the catcher just barely rotates through 90° when it catches a 40 m/s fastball of mass 0.145 kg, determine the required spring constant of the spring.

**17.114** The trapeze/lanyard air drop (t/LAD) launch is a proposed innovative method for airborne launch of a payload-carrying rocket. The release sequence involves several steps as shown in (1), where the payload rocket is shown at various instances during the launch. To investigate the first step of this process, where the rocket body drops freely from the carrier aircraft until the 2-m lanyard stops the vertical motion of $B$, a trial rocket is tested as shown in (2). The rocket can be considered a uniform $1 \times 7$-m rectangle with a mass of 4000 kg. Knowing that the rocket is released from rest and falls vertically 2 m before the lanyard becomes taut, determine the angular velocity of the rocket immediately after the lanyard is taut.

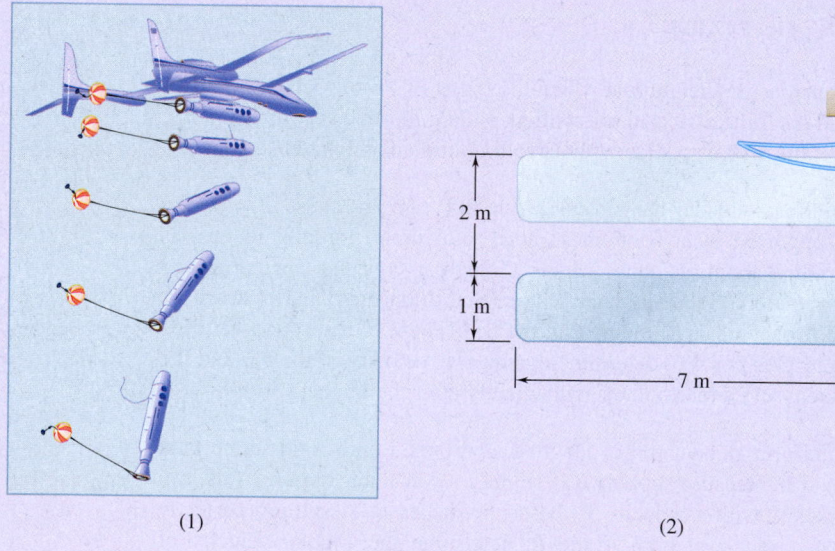

(1)                                (2)

**Fig. P17.114**

**17.115** The uniform rectangular block shown is moving along a frictionless surface with a velocity $\bar{\mathbf{v}}_1$ when it strikes a small obstruction at $B$. Assuming that the impact between corner $A$ and obstruction $B$ is perfectly plastic, determine the magnitude of the velocity $\bar{\mathbf{v}}_1$ for which the maximum angle $\theta$ through which the block will rotate will be 30°.

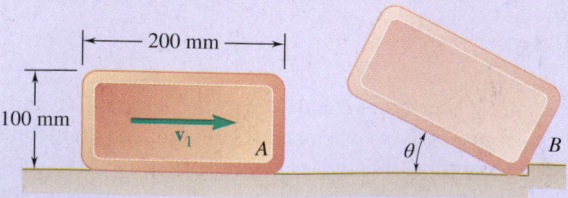

**Fig. P17.115**

**17.116** The 40-kg gymnast drops from her maximum height of $h = 0.5$ m straight down to the bar as shown. Her hands hit the bar and clasp it, and her body remains straight in the position shown. Her center of mass is 0.75 meters away from her hands, and her mass moment of inertia about her center of mass is 7.5 kg·m$^2$. Assuming that friction between the bar and her hands is negligible and that she remains in the same position throughout the swing, determine her angular velocity when she swings around to $\theta = 135°$.

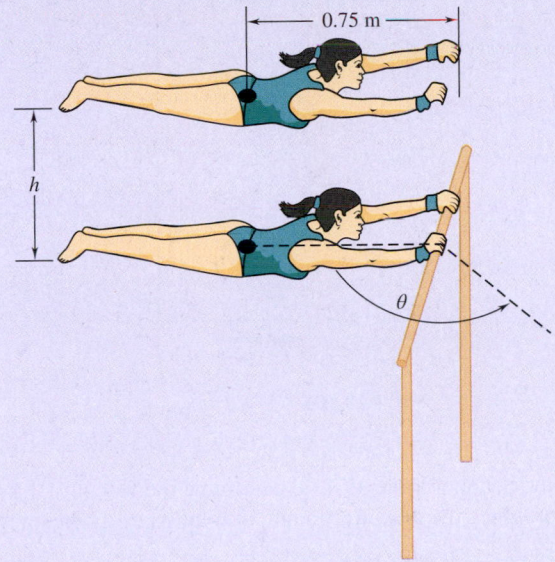

**Fig. P17.116**

**17.117** A uniform slender rod $AB$ of length $L = 600$ mm is placed with its center equidistant from two supports that are located at a distance $b = 100$ mm from each other. End $B$ of the rod is raised a distance $h_0 = 80$ mm and released; the rod then rocks on the supports as shown. Assuming that the impact at each support is perfectly plastic and that no slipping occurs between the rod and the supports, determine (a) the height $h_1$ reached by end $A$ after the first impact, (b) the height $h_2$ reached by end $B$ after the second impact.

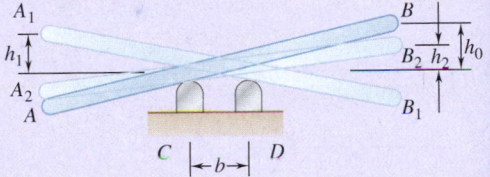

**Fig. P17.117**

**17.118** A uniformly loaded square crate is released from rest with its corner $D$ directly above $A$; it rotates about $A$ until its corner $B$ strikes the floor, and then rotates about $B$. The floor is sufficiently rough to prevent slipping and the impact at $B$ is perfectly plastic. Denoting by $\boldsymbol{\omega}_0$ the angular velocity of the crate immediately before $B$ strikes the floor, determine (a) the angular velocity of the crate immediately after $B$ strikes the floor, (b) the fraction of the kinetic energy of the crate lost during the impact, (c) the angle $\theta$ through which the crate will rotate after $B$ strikes the floor.

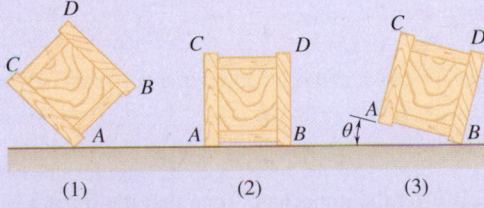

**Fig. P17.118**

**17.119** A 1-oz bullet is fired with a horizontal velocity of 750 mi/h into the 18-lb wooden beam *AB*. The beam is suspended from a collar of negligible mass that can slide along a horizontal rod. Neglecting friction between the collar and the rod, determine the maximum angle of rotation of the beam during its subsequent motion.

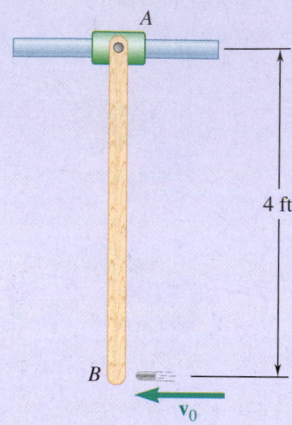

**Fig. P17.119**

**17.120** For the beam of Prob. 17.119, determine the velocity of the 1-oz bullet for which the maximum angle of rotation of the beam will be 90°.

**17.121** The plank *CDE* has a mass of 15 kg and rests on a small pivot at *D*. The 55-kg gymnast *A* is standing on the plank at *C* when the 70-kg gymnast *B* jumps from a height of 2.5 m and strikes the plank at *E*. Assuming perfectly plastic impact and that gymnast *A* is standing absolutely straight, determine the height to which gymnast *A* will rise.

**17.122** Solve Prob. 17.121, assuming that the gymnasts change places so that gymnast *A* jumps onto the plank while gymnast *B* stands at *C*.

**17.123** A slender rod *AB* is released from rest in the position shown. It swings down to a vertical position and strikes a second and identical rod *CD* that is resting on a frictionless surface. Assuming that the coefficient of restitution between the rods is 0.4, determine the velocity of rod *CD* immediately after the impact.

**17.124** A slender rod *AB* is released from rest in the position shown. It swings down to a vertical position and strikes a second and identical rod *CD* that is resting on a frictionless surface. Assuming that the impact between the rods is perfectly elastic, determine the velocity of rod *CD* immediately after the impact.

**17.125** Block *A* has a mass *m* and is attached to a cord that is wrapped around a uniform disk with a mass *M*. The block is released from rest and falls through a distance *h* before the cord becomes taut. Derive expressions for the velocity of the block and the angular velocity of the disk immediately after the impact. Assume that the impact is (*a*) perfectly plastic, (*b*) perfectly elastic.

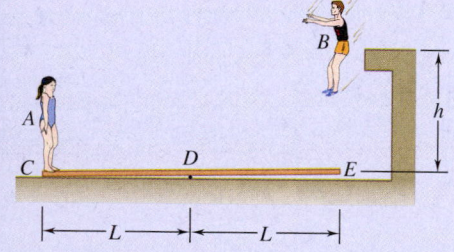

**Fig. P17.121**

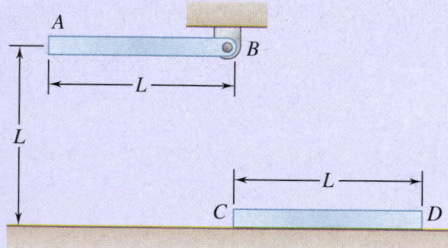

**Fig. P17.123 and P17.124**

**Fig. P17.125**

1268

**17.126** A 2-kg solid sphere of radius $r = 40$ mm is dropped from a height $h = 200$ mm and lands on a uniform slender plank $AB$ of mass 4 kg and length $L = 500$ mm that is held by two inextensible cords. Knowing that the impact is perfectly plastic and that the sphere remains attached to the plank at a distance $a = 40$ mm from the left end, determine the velocity of the sphere immediately after impact. Neglect the thickness of the plank.

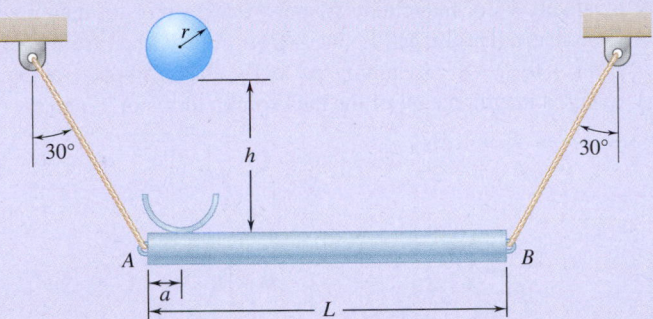

Fig. *P17.126*

**17.127 and 17.128** Member $ABC$ has a mass of 2.4 kg and is attached to a pin support at $B$. An 800-g sphere $D$ strikes the end of member $ABC$ with a vertical velocity $\mathbf{v}_1$ of 3 m/s. Knowing that each leg has a length $L = 750$ mm and that the coefficient of restitution between the sphere and member $ABC$ is 0.5, determine immediately after the impact (*a*) the angular velocity of member $ABC$, (*b*) the velocity of the sphere.

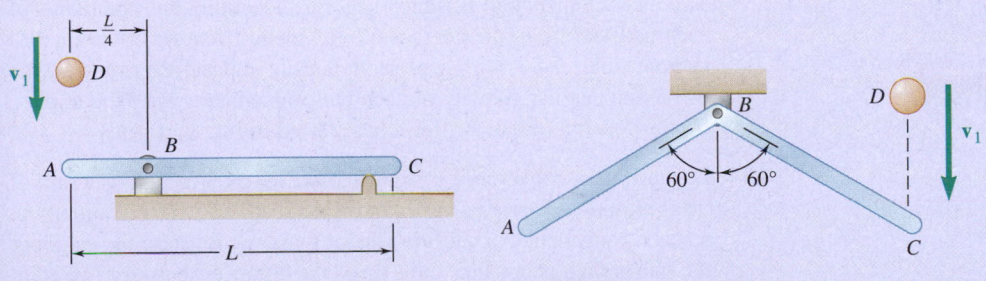

Fig. P17.127        Fig. P17.128

**17.129** Sphere $A$ of mass $m_A = 2$ kg and radius $r = 40$ mm rolls without slipping with a velocity $\overline{\mathbf{v}}_1 = 2$ m/s on a horizontal surface when it hits squarely a uniform slender bar $B$ of mass $m_B = 0.5$ kg and length $L = 100$ mm that is standing on end and is at rest. Denoting by $\mu_k$ the coefficient of kinetic friction between the sphere and the horizontal surface, neglecting friction between the sphere and the bar, and knowing the coefficient of restitution between $A$ and $B$ is 0.1, determine the angular velocities of the sphere and the bar immediately after the impact.

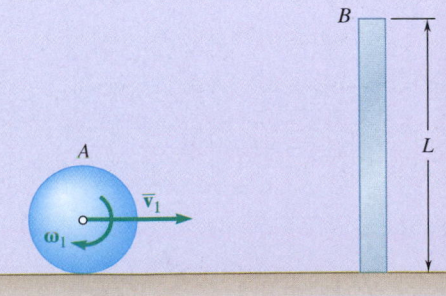

Fig. *P17.129*

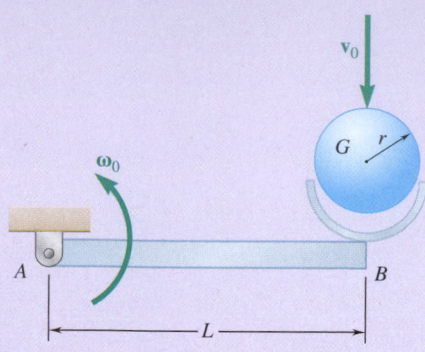

**Fig. P17.130**

**17.130** A large 3-lb sphere with a radius $r = 3$ in. is thrown into a light basket at the end of a thin, uniform rod weighing 2 lb and having length $L = 10$ in. as shown. Immediately before the impact, the angular velocity of the rod is 3 rad/s counterclockwise and the velocity of the sphere is 2 ft/s down. Assume the sphere sticks in the basket. Determine after the impact (*a*) the angular velocity of the bar and sphere, (*b*) the components of the reactions at *A*.

**17.131** A small rubber ball of radius *r* is thrown against a rough floor with a velocity $\bar{\mathbf{v}}_A$ of magnitude $\bar{v}_0$ and a backspin $\boldsymbol{\omega}_A$ of magnitude $\omega_0$. It is observed that the ball bounces from *A* to *B*, then from *B* to *A*, then from *A* to *B*, etc. Assuming perfectly elastic impact, determine the required magnitude $\omega_0$ of the backspin in terms of $\bar{v}_0$ and *r*.

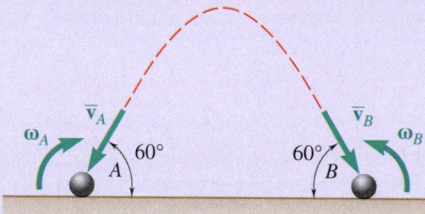

**Fig. P17.131**

**Fig. P17.132**

**17.132** Sphere *A* of mass *m* and radius *r* rolls without slipping with a velocity $\bar{\mathbf{v}}_1$ on a horizontal surface when it hits squarely an identical sphere *B* that is at rest. Denoting by $\mu_k$ the coefficient of kinetic friction between the spheres and the surface, neglecting friction between the spheres, and assuming perfectly elastic impact, determine (*a*) the linear and angular velocities of each sphere immediately after the impact, (*b*) the velocity of each sphere after it has started rolling uniformly.

**17.133** In a game of pool, ball *A* is rolling without slipping with a velocity $\bar{\mathbf{v}}_0$ as it hits obliquely ball *B*, which is at rest. Denoting by *r* the radius of each ball and by $\mu_k$ the coefficient of kinetic friction between a ball and the table, and assuming perfectly elastic impact, determine (*a*) the linear and angular velocity of each ball immediately after the impact, (*b*) the velocity of ball *B* after it has started rolling uniformly.

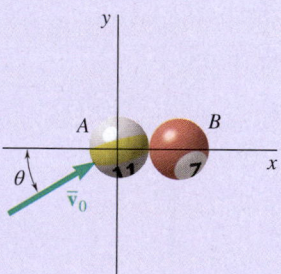

**Fig. P17.133**

**17.134** Luggage on a conveyance system is modeled as a rectangular slab of mass $m_S$ moving across a series of rollers, each of which is initially at rest and equivalent to a uniform disk of mass $m_R$. Because the length of the slab is slightly less than three times the distance *b* between two adjacent rollers, the slab leaves a roller just before it reaches another one. Each time a new roller enters into contact with the slab, slipping occurs between the roller and the slab for a short period of time (less than the time needed for the slab to move through the distance *b*). Knowing that the velocity of the slab in the position shown is $v_0$, determine the velocity of the slab after it has moved (*a*) a distance *b*, (*b*) a distance *nb*.

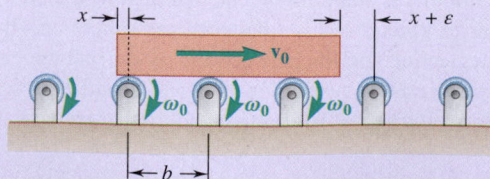

**Fig. P17.134**

# Review and Summary

In this chapter, we again considered the method of work and energy and the method of impulse and momentum. In the first section, we applied the method of work and energy to the analysis of the motion of rigid bodies and systems of rigid bodies.

The second section was devoted to the method of impulse and momentum and its application to the solution of various types of problems involving the plane motion of rigid bodies and rigid bodies symmetrical with respect to the reference plane.

## Principle of Work and Energy for a Rigid Body

In Sec. 17.1, we first expressed the principle of work and energy for a rigid body in the form

$$T_1 + U_{1 \to 2} = T_2 \tag{17.1}$$

where $T_1$ and $T_2$ represent the initial and final values of the kinetic energy of the rigid body and $U_{1 \to 2}$ represents the work of the external forces acting on it. If we express the work done by nonconservative forces as $U^{NC}_{1 \to 2}$ and define potential energy terms for conservative forces, we can express Eq. (17.1) as

$$T_1 + V_{g_1} + V_{e_1} + U^{NC}_{1 \to 2} = T_2 + V_{g_2} + V_{e_2} \tag{17.1'}$$

where $V_{g_1}$ and $V_{g_2}$ are the initial and final gravitational potential energy of the center of mass of the rigid body, and $V_{e_1}$ and $V_{e_2}$ are the initial and final values of the elastic energy associated with springs in the system, respectively.

## Work of a Force or a Couple

In Sec. 17.1B, we recalled the expression found in Chap. 13 for the work of a force $\mathbf{F}$ applied at a point $A$, namely

$$U_{1 \to 2} = \int_{A_1}^{A_2} \mathbf{F} \cdot d\mathbf{r} \tag{17.3}$$

or

$$U_{1 \to 2} = \int_{s_1}^{s_2} (F \cos \alpha) \, ds \tag{17.3'}$$

where $F$ is the magnitude of the force, $\alpha$ is the angle it forms with the direction of motion of $A$, and $s$ is the variable of integration measuring the distance traveled by $A$ along its path. We also derived the expression for the work of a couple of moment $\mathbf{M}$ applied to a rigid body during a rotation in $\theta$ of the rigid body as

$$U_{1 \to 2} = \int_{\theta_1}^{\theta_2} M \, d\theta \tag{17.5}$$

## Kinetic Energy in Plane Motion

We then derived an expression for the kinetic energy of a rigid body in plane motion (Sec. 17.1C):

$$T = \tfrac{1}{2} m \bar{v}^2 + \tfrac{1}{2} \bar{I} \omega^2 \tag{17.9}$$

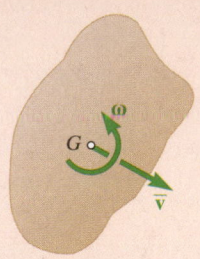

**Fig. 17.13**

where $\bar{v}$ is the speed of the mass center $G$ of the body, $\omega$ is the angular speed of the body, and $\bar{I}$ is its moment of inertia about an axis through $G$ perpendicular to the plane of reference (Fig. 17.13) (Sample Prob. 17.3). We noted that the kinetic energy of a rigid body in plane motion can be separated into two parts: (1) the kinetic energy $\frac{1}{2}m\bar{v}^2$ associated with the motion of the mass center $G$ of the body and (2) the kinetic energy $\frac{1}{2}\bar{I}\omega^2$ associated with the rotation of the body about $G$. You will generally need to use kinematics to relate $\bar{v}$ and $\omega$.

### Kinetic Energy in Rotation About a Fixed Axis

For a rigid body rotating about a fixed axis through $O$ with an angular velocity $\boldsymbol{\omega}$, we had

$$T = \tfrac{1}{2}I_O\omega^2 \tag{17.10}$$

where $I_O$ is the moment of inertia of the body about the fixed axis. We noted that this result is not limited to the rotation of plane rigid bodies or of bodies symmetrical with respect to the reference plane, but it also is valid regardless of the shape of the body or of the location of the axis of rotation.

### Systems of Rigid Bodies

Eq. (17.1) can be applied to the motion of systems of rigid bodies (Sec. 17.1D) as long as all the forces acting on the various bodies involved—internal as well as external to the system—are included in the computation of $U_{1\rightarrow2}$. However, in the case of systems consisting of pin-connected members or blocks and pulleys connected by inextensible cords or meshed gears, the points of application of the internal forces move through equal distances and the work of these forces cancels out (Sample Probs. 17.1, 17.2, and 17.6).

### Conservation of Energy

When a rigid body or a system of rigid bodies moves under the action of conservative forces, the principle of work and energy can be expressed in the form

$$T_1 + V_1 = T_2 + V_2 \tag{17.12}$$

or

$$T_1 + V_{g_1} + V_{e_1} = T_2 + V_{g_2} + V_{e_2} \tag{17.12'}$$

This is referred to as the *principle of conservation of energy* (Sec. 17.1E). We can use this principle to solve problems involving conservative forces such as the force of gravity or the force exerted by a spring (Sample Probs. 17.4, 17.5, and 17.6). However, if we need to determine a reaction, we must supplement the principle of conservation of energy by using Newton's second law (Sample Prob. 17.4).

### Power

In Sec. 17.1F, we extended the concept of power to a rotating body subjected to a couple as

$$\text{Power} = \frac{dU}{dt} = \frac{M\,d\theta}{dt} = M\omega \tag{17.13}$$

where $M$ is the magnitude of the couple and $\omega$ is the magnitude of the angular velocity of the body.

## Principle of Impulse and Momentum for a Rigid Body

In Sec. 17.2, we applied the principle of impulse and momentum as had been derived in Sec. 14.2C for a system of particles to the motion of a rigid body (Sec. 17.2A). We have

$$\text{Syst Momenta}_1 + \text{Syst Ext Imp}_{1\to2} = \text{Syst Momenta}_2 \quad (17.14)$$

Next we showed that, for a rigid body symmetrical with respect to the reference plane, the system of the momenta of the particles forming the body is equivalent to a vector $m\overline{\mathbf{v}}$ attached at the mass center $G$ of the body and a couple $\bar{I}\boldsymbol{\omega}$ (Fig. 17.14). The vector $m\overline{\mathbf{v}}$ is associated with the translation of the body with $G$ and represents the *linear momentum* of the body, whereas the couple $\bar{I}\boldsymbol{\omega}$ corresponds to the rotation of the body about $G$ and represents the *angular momentum* of the body about an axis through $G$.

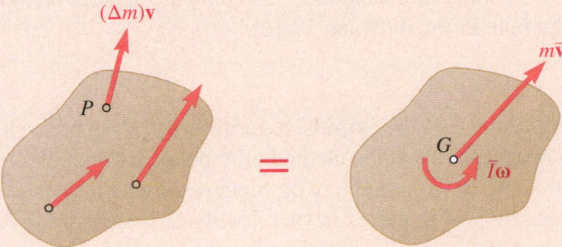

**Fig. 17.14**

We can express Eq. (17.14) graphically using an impulse–momentum diagram, as shown in Fig. 17.15. This diagram represents the system of the

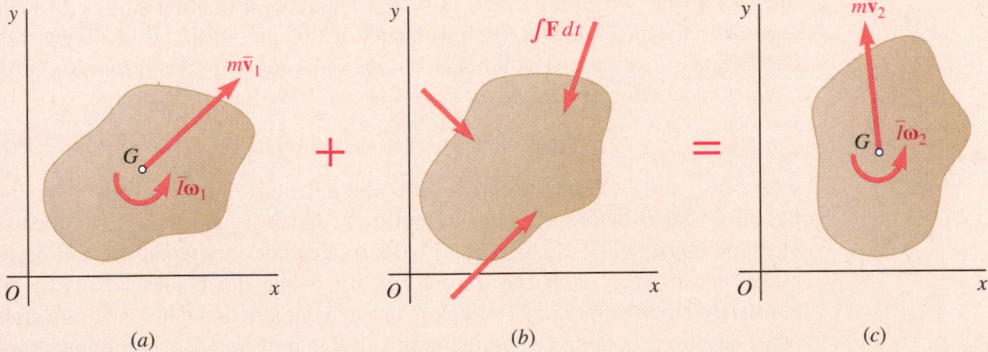

(a)        (b)        (c)

**Fig. 17.15**

initial momenta of the body, the impulses of the external forces acting on the body, and the system of the final momenta of the body, respectively. We can choose to sum moments about an arbitrary point $P$ using

$$\bar{I}\omega_1 + m\overline{v}_1 d_{\perp 1} + \sum\int_{t_1}^{t_2} M_P dt = \bar{I}\omega_2 + m\overline{v}_2 d_{\perp 2} \quad (17.14')$$

the center of mass $G$ using

$$\bar{I}\omega_1 + \sum\int_{t_1}^{t_2} M_G dt = \bar{I}\omega_2 \quad (17.14'')$$

or a fixed axis of rotation $O$ using

$$I_O\omega_1 + \sum\int_{t_1}^{t_2} M_O dt = I_O\omega_2 \quad (17.16)$$

Using one of these expressions and the *x and y components* of the linear impulse–momentum equation, we obtain three equations of motion that we can solve for the desired unknowns (Sample Probs. 17.7 and 17.8).

In problems dealing with several connected rigid bodies (Sec. 17.2B), we can consider each body separately (Sample Prob. 17.7), or if no more than three unknowns are involved, we can apply the principle of impulse and momentum to the entire system, considering the impulses of the external forces only (Sample Prob. 17.9).

### Conservation of Angular Momentum

When the lines of action of all the external forces acting on a system of rigid bodies pass through a given point *O*, the angular momentum of the system about *O* is conserved (Sec. 17.2C). We suggested that problems involving conservation of angular momentum be solved by the general method described previously (Sample Probs. 17.9 and 17.10).

### Impulsive Motion

Sec. 17.3 was devoted to the **impulsive motion** and the **eccentric impact** of rigid bodies. We recalled that the method of impulse and momentum is the only practicable method for the solution of problems involving impulsive motion and that the computation of impulses in such problems is straightforward (Sample Probs. 17.11 and 17.12).

### Eccentric Impact

We also recalled that the eccentric impact of two rigid bodies is defined as an impact in which the mass centers of the colliding bodies are *not* located on the line of impact. We showed that, in such a situation, a relation similar to that derived in Chap. 13 for the central impact of two particles and involving the coefficient of restitution *e* still holds, but *the velocities of points A and B where contact occurs during the impact should be used.* We have

$$(v'_B)_n - (v'_A)_n = e[(v_A)_n - (v_B)_n] \qquad \textbf{(17.19)}$$

where $(v_A)_n$ and $(v_B)_n$ are the components along the line of impact of the velocities of *A* and *B* before the impact, and $(v'_A)_n$ and $(v'_B)_n$ are their components after the impact (Fig. 17.16). Eq. (17.19) applies not only when the colliding bodies move freely after the impact but also when the bodies are partially constrained in their motion. You should use it in conjunction with one or several other equations obtained by applying the principle of impulse and momentum (Sample Prob. 17.13). We also considered problems where the method of impulse and momentum and the method of work and energy can be combined (Sample Probs. 17.14 and 17.15 ).

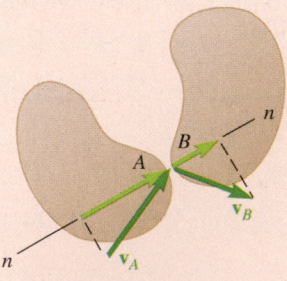

(*a*) Before impact          (*b*) After impact

**Fig. 17.16**

# Review Problems

**17.135** A uniform disk, initially at rest and of constant thickness, is placed in contact with the belt shown, which moves at a constant speed $v = 80$ ft/s. Knowing that the coefficient of kinetic friction between the disk and the belt is 0.15, determine (a) the number of revolutions executed by the disk before it reaches a constant angular velocity, (b) the time required for the disk to reach that constant angular velocity.

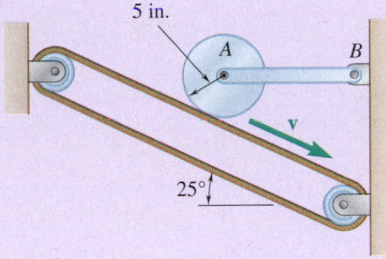

**Fig. P17.135**

**17.136** The 8-in.-radius brake drum is attached to a larger flywheel that is not shown. The total mass moment of inertia of the flywheel and drum is 14 lb·ft·s² and the coefficient of kinetic friction between the drum and the brake shoe is 0.35. Knowing that the initial angular velocity of the flywheel is 360 rpm counterclockwise, determine the vertical force **P** that must be applied to the pedal $C$ if the system is to stop in 100 revolutions.

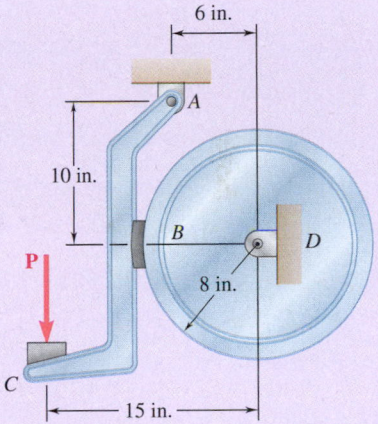

**Fig. P17.136**

**17.137** Charpy impact test pendulums are used to determine the amount of energy a test specimen absorbs during an impact (see ASTM Standard E23). The hammer weighs 71.2 lbs and has a mass moment of inertia about its center of gravity $G_H$ of 20.9 slug·in². The arm weighs 19.5 lbs and has a mass moment of inertia about its own center of gravity $G_A$ of 47.1 slug·in². The pendulum is released from rest from an initial position of $\theta = 39°$. Knowing that the friction at pin $O$ is negligible, determine (a) the impact speed when the hammer hits the test specimen, (b) the force on the pin $O$ just before the hammer hits the test specimen, (c) the amount of energy that the test specimen absorbs if the hammer swings up to a maximum of $\phi = 70°$ after the impact.

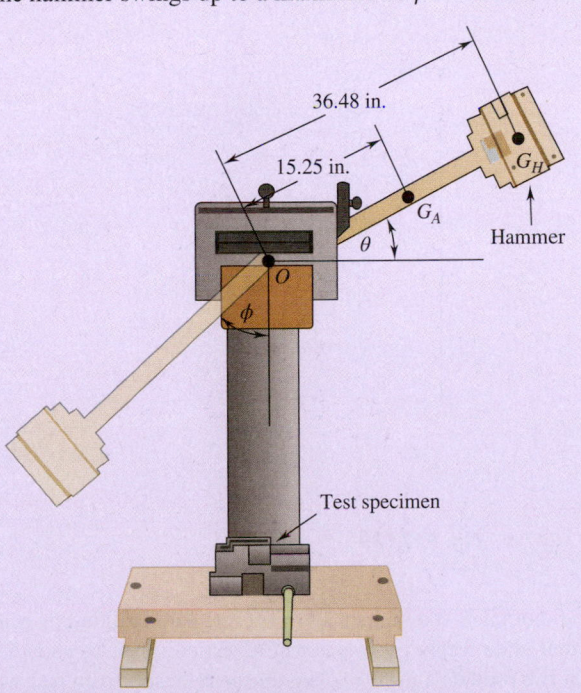

**Fig. P17.137**

**17.138** You are asked to analyze a catcher for a small drone. The catcher arm weighs 20 lb and is 8 ft long (you can model it as a slender rod); the net that catches the drone at $B$ has negligible mass. The 3-lb drone has a mass moment of inertia about its own center of mass of 0.01 slug·ft$^2$. Knowing that the arm swings up to an angle of 30° below the horizontal, determine (a) the initial velocity $v_0$ of the drone, (b) the forces at $A$ at that angle.

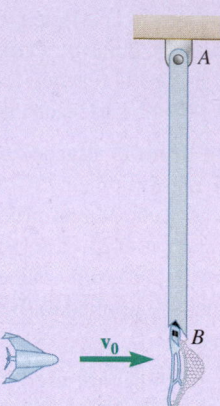

Fig. *P17.138*

**17.139** A uniform slender rod is placed at corner $B$ and is given a slight clockwise motion. Assuming that the corner is sharp and becomes slightly embedded in the end of the rod so that the coefficient of static friction at $B$ is very large, determine (a) the angle $\beta$ through which the rod will have rotated when it loses contact with the corner, (b) the corresponding velocity of end $A$.

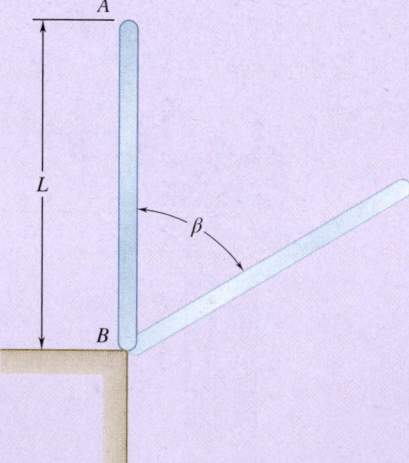

Fig. P17.139

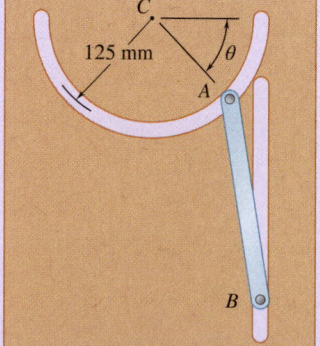

Fig. P17.140

**17.140** The motion of the slender 250-mm rod $AB$ is guided by pins at $A$ and $B$ that slide freely in slots cut in a vertical plate as shown. Knowing that the rod has a mass of 2 kg and is released from rest when $\theta = 0$, determine the reactions at $A$ and $B$ when $\theta = 90°$.

**17.141** A baseball attachment that helps people with mobility impairments play T-ball and baseball is powered by a spring that is unstretched at position 2. The spring is attached to a cord that is fastened to point $B$ on the 75-mm-radius pulley. The pulley is fixed at point $O$, rotates backward to the cocked position at $\theta = 120°$, and the rope wraps around the pulley and stretches the spring with a stiffness of $k = 2000$ N/m. The combined mass moment of inertia of all the rotating components about point $O$ is 0.40 kg·m². The swing is timed perfectly to strike a 145-gram baseball traveling with a speed of $v_0 = 10$ m/s at a distance of $h = 0.7$ m away from point $O$. Knowing that the coefficient of restitution between the bat and ball is 0.59, determine the velocity of the baseball immediately after the impact. Assume that the ball is traveling primarily in the horizontal plane and that its spin is negligible.

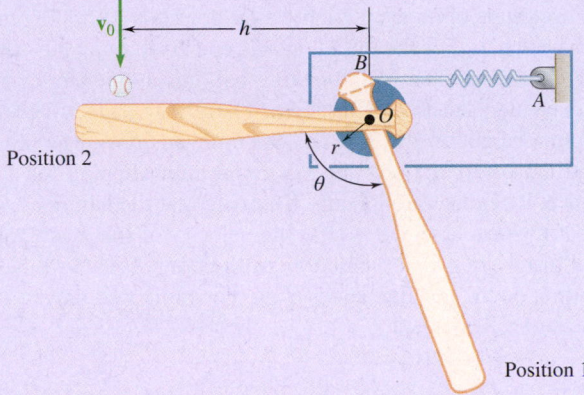

Position 2

Position 1

**Fig. P17.141**

**17.142** Two panels $A$ and $B$ are attached with hinges to a rectangular plate and held by a wire as shown. The plate and the panels are made of the same material and have the same thickness. The entire assembly is rotating with an angular velocity $\omega_0$ when the wire breaks. Determine the angular velocity of the assembly after the panels have come to rest against the plate.

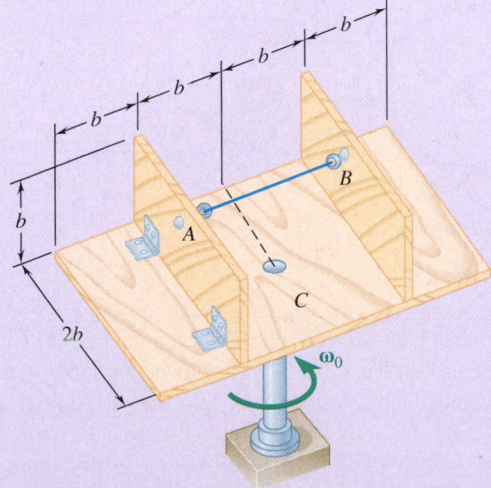

**Fig. P17.142**

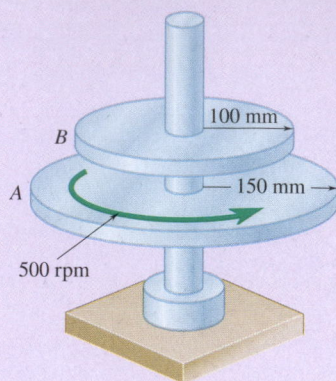

**Fig. P17.143**

**17.143** Disks $A$ and $B$ are made of the same material, are of the same thickness, and can rotate freely about the vertical shaft. Disk $B$ is at rest when it is dropped onto disk $A$, which is rotating with an angular velocity of 500 rpm. Knowing that disk $A$ has a mass of 8 kg, determine (a) the final angular velocity of the disks, (b) the change in kinetic energy of the system.

**17.144** A square block of mass $m$ is falling with a velocity $\bar{\mathbf{v}}_1$ when it strikes a small obstruction at $B$. Knowing that the coefficient of restitution for the impact between corner $A$ and the obstruction $B$ is $e = 0.5$, determine immediately after the impact (a) the angular velocity of the block, (b) the velocity of its mass center $G$.

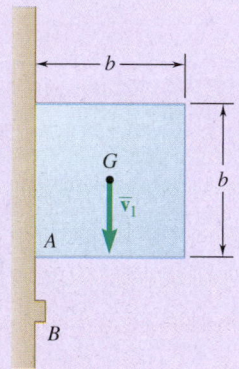

**Fig. P17.144**

**\* 17.145** A prototype of an adaptive bowling device is a simple ramp next to a wheelchair. For the initial design, the 20-lb ramp has small rollers on its bottom. The 15-lb bowling ball has a diameter of 8.5 in. and can be modeled as a solid sphere. The athlete nudges the ball slightly from a height of $h = 3$ ft, and the ball rolls down the ramp without sliding. It hits the bowling lane, and after slipping for a short distance, it begins to roll again. Knowing that the ball does not bounce as it hits the lane, determine the velocity of the mass center of the ball (a) when it is at the bottom of the ramp, (b) after it has resumed rolling on the bowling lane. (*Hint:* The ramp will move to the left as the ball rolls down it.)

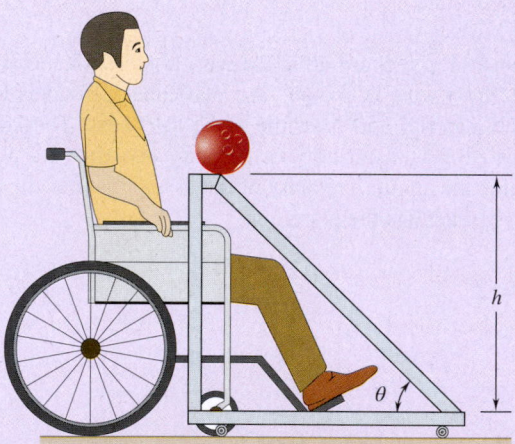

**Fig. P17.145**

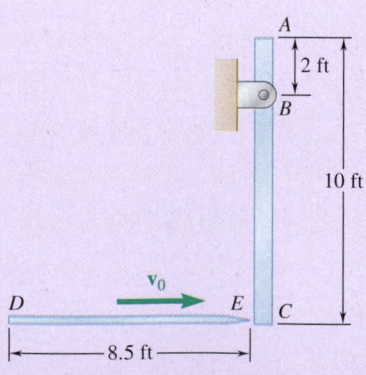

**Fig. P17.146**

**17.146** A 1.8-lb javelin $DE$ impacts a 10-lb slender rod $ABC$ with a horizontal velocity of $\mathbf{v}_0 = 30$ ft/s as shown. Knowing that the javelin becomes embedded into the end of the rod at point $C$ and does not penetrate very far into it, determine immediately after the impact (a) the angular velocity of the rod $ABC$, (b) the components of the reaction at $B$. Assume the javelin and the rod move as a single rigid body after the impact.

# 18

## Kinetics of Rigid Bodies in Three Dimensions

While the general principles that you learned in earlier chapters can be used again to solve problems involving the three-dimensional motion of rigid bodies, the solution of these problems requires a new approach and is considerably more involved than the solution of two-dimensional problems. One example is the determination of the forces acting on the robotic arm of the spacecraft.

# Objectives

- **Calculate** the angular momentum and kinetic energy of a rigid body undergoing general three-dimensional motion.
- **Define** the inertia tensor, products of inertia, and principal axes of inertia.
- **Apply** the principle of impulse and momentum to solve three-dimensional rigid body kinetics problems.
- **Solve** three-dimensional rigid body kinetics problems, including fixed point rotation, fixed axis rotation, and gyroscopic motion.
- **Describe** the relationship between applied moment, precession, and spin of a gyroscope undergoing steady precession.
- **Analyze** the motion of a rotating axisymmetric body under no external forces.

# Introduction

In Chaps. 16 and 17, we were concerned with the plane motion of rigid bodies and of systems of rigid bodies. In Chap. 16 and in the second half of Chap. 17 (impulse and momentum), our study was further restricted to the motion of plane rigid bodies and of bodies symmetrical with respect to the reference plane. However, many of the fundamental results obtained in these two chapters remain valid in the case of the motion of a rigid body in three dimensions. For example, the two fundamental equations

$$\Sigma\mathbf{F} = m\bar{\mathbf{a}} \tag{18.1}$$

$$\Sigma\mathbf{M}_G = \dot{\mathbf{H}}_G \tag{18.2}$$

on which we based the analysis of the plane motion of a rigid body remain valid in the most general case of motion of a rigid body. As indicated in Sec. 16.1, these equations express that the system of external forces is equipollent to the system consisting of the vector $m\bar{\mathbf{a}}$ attached at $G$ and the couple of moment $\dot{\mathbf{H}}_G$ (Fig. 18.1).

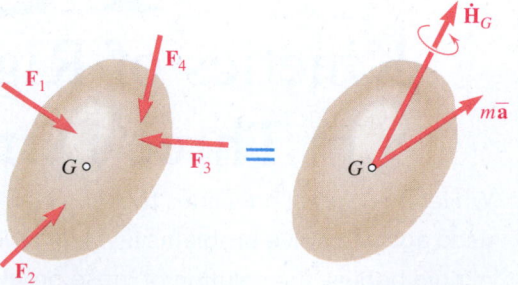

**Fig. 18.1**    The external forces acting on the rigid body are equipollent to a vector $m\bar{\mathbf{a}}$ attached to the mass center $G$ and a rotational inertia vector $\dot{\mathbf{H}}_G$.

The relation $\mathbf{H}_G = \bar{I}\boldsymbol{\omega}$ enabled us to determine the angular momentum of a rigid body and played an important part in the solution of problems involving the plane motion of rigid bodies and bodies symmetrical with respect to the reference plane. However, this equation ceases to be valid in the case of non-symmetrical bodies or three-dimensional motion. Thus, we need to develop a more general method for computing the angular momentum $\mathbf{H}_G$ of a rigid body in three dimensions.

Similarly, the main feature of the impulse–momentum method discussed in Sec. 17.2A is the reduction of the momenta of the particles of a rigid body to a linear momentum vector $m\bar{\mathbf{v}}$ attached at the mass center $G$ of the body and an angular momentum couple $\mathbf{H}_G$. This method remains valid in the more general case, but we must discard the relation $\mathbf{H}_G = \bar{I}\boldsymbol{\omega}$ and replace it with a more general relation before we can apply this method to the three-dimensional motion of a rigid body (Sec. 18.1B).

Also note that the work–energy principle and the principle of conservation of energy still apply in the case of the motion of a rigid body in three dimensions. However, we need to replace the expression obtained in Sec. 17.1C for the kinetic energy of a rigid body in plane motion with a new expression for a rigid body in three-dimensional motion.

In the second part of this chapter, you will learn to determine the rate of change $\dot{\mathbf{H}}_G$ of the angular momentum $\mathbf{H}_G$ of a three-dimensional rigid body using a rotating frame of reference where the moments and products of inertia of the body remain constant. Then you can express Eqs. (18.1) and (18.2) in the form of free-body and kinetic diagrams that you can use to solve various problems involving the three-dimensional motion of rigid bodies (Sec. 18.2).

The last part of this chapter (Sec. 18.3) is devoted to the study of the motion of gyroscopes or, more generally, of an axisymmetric body with a fixed point located on its axis of symmetry. We first consider the particular case of the steady precession of a gyroscope and then analyze the motion of an axisymmetric body subjected to no force except its own weight.

# 18.1   ENERGY AND MOMENTUM OF A RIGID BODY

All of the methods you studied in earlier chapters for analyzing the plane motion of a rigid body have corresponding versions for motion in three dimensions. However, some of the formulas for determining kinetic quantities such as energy and angular momentum need to be replaced by more general equations. In this section, we examine some of the basic quantities and equations needed for the study of motion in space.

## *18.1A   Angular Momentum of a Rigid Body in Three Dimensions

In this section, you will see how to determine the angular momentum $\mathbf{H}_G$ of a body about its mass center $G$ from the angular velocity $\boldsymbol{\omega}$ of the body in the case of three-dimensional motion.

According to Eq. (14.24), we can express the angular momentum of the body about $G$ as

$$\mathbf{H}_G = \sum_{i=1}^{n} (\mathbf{r}'_i \times \mathbf{v}'_i \, \Delta m_i) \tag{18.3}$$

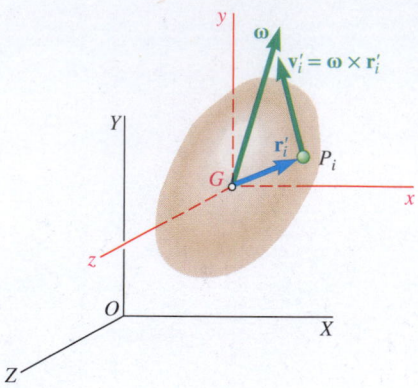

**Fig. 18.2** The velocity of particle $P_i$ is needed to derive the angular momentum of a rigid body in three dimensions.

where $\mathbf{r}_i'$ and $\mathbf{v}_i'$ denote, respectively, the position vector and the velocity of the particle $P_i$ with a mass $\Delta m_i$ that is relative to the centroidal frame $Gxyz$ (Fig. 18.2). However, $\mathbf{v}_i' = \boldsymbol{\omega} \times \mathbf{r}_i'$, where $\boldsymbol{\omega}$ is the angular velocity of the body at the instant considered. Substituting into Eq. (18.3), we have

$$\mathbf{H}_G = \sum_{i=1}^{n} [\mathbf{r}_i' \times (\boldsymbol{\omega} \times \mathbf{r}_i') \, \Delta m_i]$$

From the rule for determining the rectangular components of a vector product (*Statics*, Sec. 3.1D, or Appendix A), we obtain the following expression for the $x$ component of the angular momentum as

$$H_x = \sum_{i=1}^{n} [y_i(\boldsymbol{\omega} \times \mathbf{r}_i')_z - z_i(\boldsymbol{\omega} \times \mathbf{r}_i')_y] \, \Delta m_i$$

$$= \sum_{i=1}^{n} [y_i(\omega_x y_i - \omega_y x_i) - z_i(\omega_z x_i - \omega_x z_i)] \, \Delta m_i$$

$$= \omega_x \sum_i (y_i^2 + z_i^2) \, \Delta m_i - \omega_y \sum_i x_i y_i \, \Delta m_i - \omega_z \sum_i z_i x_i \, \Delta m_i$$

Replacing the sums by integrals in this expression and in the two similar expressions obtained for $H_y$ and $H_z$, we have

$$H_x = \omega_x \int (y^2 + z^2) \, dm - \omega_y \int xy \, dm - \omega_z \int zx \, dm$$
$$H_y = -\omega_x \int xy \, dm + \omega_y \int (z^2 + x^2) \, dm - \omega_z \int yz \, dm \qquad \textbf{(18.4)}$$
$$H_z = -\omega_x \int zx \, dm - \omega_y \int yz \, dm + \omega_z \int (x^2 + y^2) \, dm$$

Note that the integrals containing squares represent the *centroidal mass moments of inertia* of the body about the $x$, $y$, and $z$ axes, respectively (*Statics*, Sec. 9.5A, or Appendix B). That is,

$$\bar{I}_x = \int (y^2 + z^2) \, dm \qquad \bar{I}_y = \int (z^2 + x^2) \, dm$$
$$\bar{I}_z = \int (x^2 + y^2) \, dm \qquad\qquad\qquad \textbf{(18.5)}$$

Similarly, the integrals containing products of coordinates represent the *centroidal mass products of inertia* of the body (Sec. 9.6A); we have

$$\bar{I}_{xy} = \int xy \, dm \qquad \bar{I}_{yz} = \int yz \, dm \qquad \bar{I}_{zx} = \int zx \, dm \qquad \textbf{(18.6)}$$

Substituting from Eqs. (18.5) and (18.6) into Eq. (18.4), we obtain the components of the angular momentum $\mathbf{H}_G$ of the body about its mass center as

**Angular momentum about mass center**

$$H_x = +\bar{I}_x \omega_x - \bar{I}_{xy} \omega_y - \bar{I}_{xz} \omega_z$$
$$H_y = -\bar{I}_{yx} \omega_x + \bar{I}_y \omega_y - \bar{I}_{yz} \omega_z \qquad \textbf{(18.7)}$$
$$H_z = -\bar{I}_{zx} \omega_x - \bar{I}_{zy} \omega_y + \bar{I}_z \omega_z$$

The relations in Eqs. (18.7) show that the operation transforming the vector $\boldsymbol{\omega}$ into the vector $\mathbf{H}_G$ (Fig. 18.3) is characterized by the array of moments and products of inertia as

**Inertia tensor**

$$\begin{pmatrix} \bar{I}_x & -\bar{I}_{xy} & -\bar{I}_{xz} \\ -\bar{I}_{yx} & \bar{I}_y & -\bar{I}_{yz} \\ -\bar{I}_{zx} & -\bar{I}_{zy} & \bar{I}_z \end{pmatrix} \qquad \text{(18.8)}$$

The array in Eq. (18.8) defines the **inertia tensor** of the body at its mass center $G$.[†] We obtain a new array of moments and products of inertia if we use a different system of axes. The angular momentum $\mathbf{H}_G$ corresponding to a given angular velocity $\boldsymbol{\omega}$ is independent of the choice of the coordinate axes.

As we showed in *Statics*, Sec. 9.6, or in Appendix B, it is always possible to select a system of axes $Gx'y'z'$, called *principal axes of inertia*, with respect to which all the products of inertia of a given body are zero. The array of Eq. (18.8) then takes the diagonalized form as

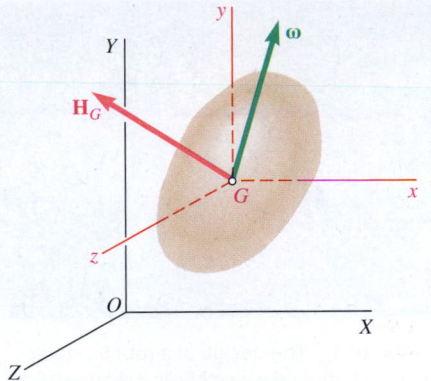

**Fig. 18.3**  In general, the angular momentum and the angular velocity are not in the same direction.

$$\begin{pmatrix} \bar{I}_{x'} & 0 & 0 \\ 0 & \bar{I}_{y'} & 0 \\ 0 & 0 & \bar{I}_{z'} \end{pmatrix} \qquad \text{(18.9)}$$

where $\bar{I}_{x'}, \bar{I}_{y'}, \bar{I}_{z'}$ represent the *principal centroidal moments of inertia* of the body, and the relations in Eq. (18.7) reduce to

$$H_{x'} = \bar{I}_{x'}\omega_{x'} \qquad H_{y'} = \bar{I}_{y'}\omega_{y'} \qquad H_{z'} = \bar{I}_{z'}\omega_{z'} \qquad \text{(18.10)}$$

Note that if the three principal centroidal moments of inertia $\bar{I}_{x'}, \bar{I}_{y'}, \bar{I}_{z'}$ are equal, the components $H_{x'}, H_{y'}, H_{z'}$ of the angular momentum about $G$ are proportional to the components $\omega_{x'}, \omega_{y'}, \omega_{z'}$ of the angular velocity, and the vectors $\mathbf{H}_G$ and $\boldsymbol{\omega}$ are collinear. In general, however, the principal moments of inertia are different, and the vectors $\mathbf{H}_G$ and $\boldsymbol{\omega}$ have different directions except when two of the three components of $\boldsymbol{\omega}$ happen to be zero; that is, when $\boldsymbol{\omega}$ is directed along one of the coordinate axes. Thus,

**The angular momentum $\mathbf{H}_G$ of a rigid body and its angular velocity $\boldsymbol{\omega}$ have the same direction if, and only if, $\boldsymbol{\omega}$ is directed along a principal axis of inertia.**[‡]

This condition is satisfied in the case of the plane motion of a rigid body that is symmetrical with respect to the reference plane, so in Secs. 16.1 and 17.2, we were able to represent the angular momentum $\mathbf{H}_G$ of such a body by the vector $\bar{I}\boldsymbol{\omega}$. We must realize, however, that this result cannot be extended to the case of the plane motion of a nonsymmetrical body or to the case of the three-dimensional motion of a rigid body. Except when $\boldsymbol{\omega}$ happens to be

[†]Setting $\bar{I}_x = I_{11}, \bar{I}_y = I_{22}, \bar{I}_z = I_{33}$, and $-\bar{I}_{xy} = I_{12}, -\bar{I}_{xz} = I_{13}$, etc., we can write the inertia tensor of Eq. (18.8) in the standard form

$$\begin{pmatrix} I_{11} & I_{12} & I_{13} \\ I_{21} & I_{22} & I_{23} \\ I_{31} & I_{32} & I_{33} \end{pmatrix}$$

[‡]In the particular case when $\bar{I}_{x'} = \bar{I}_{y'} = \bar{I}_{z'}$, any line through $G$ can be considered to be a principal axis of inertia, and the vectors $\mathbf{H}_G$ and $\boldsymbol{\omega}$ are always collinear.

directed along a principal axis of inertia, the angular momentum and angular velocity of a rigid body have different directions, and you must use the relation in Eq. (18.7) or (18.10) to determine $\mathbf{H}_G$ from $\boldsymbol{\omega}$.

### Reduction of the Momenta of the Particles of a Rigid Body to a Momentum Vector and a Couple at G.

We saw in Sec. 17.2A that we can reduce the system formed by the momenta of the various particles of a rigid body to a vector $\mathbf{L}$ that is attached at the mass center $G$ of the body, representing the linear momentum of the body, and to a couple $\mathbf{H}_G$, representing the angular momentum of the body about $G$ (Fig. 18.4). We are now in a position to determine the vector $\mathbf{L}$ and the couple $\mathbf{H}_G$ in the most general case of three-dimensional motion of a rigid body. As in the case of the two-dimensional motion considered earlier, the linear momentum $\mathbf{L}$ of the body is equal to the product $m\bar{\mathbf{v}}$ of its mass $m$ and velocity $\bar{\mathbf{v}}$ of its mass center $G$. However, we can no longer obtain the angular momentum $\mathbf{H}_G$ by simply multiplying the angular velocity $\boldsymbol{\omega}$ of the body by the scalar $\bar{I}$. Instead, we obtain it from the components of $\boldsymbol{\omega}$ and from the centroidal moments and products of inertia of the body through the use of Eq. (18.7) or (18.10).

We should also note that once we have determined the linear momentum $m\bar{\mathbf{v}}$ and the angular momentum $\mathbf{H}_G$ of a rigid body, we can obtain its angular momentum $\mathbf{H}_P$ about any given point $P$ by adding the moments about $P$ of vector $m\bar{\mathbf{v}}$ and of couple $\mathbf{H}_G$. We have

$$\mathbf{H}_P = \bar{\mathbf{r}} \times m\bar{\mathbf{v}} + \mathbf{H}_G \tag{18.11}$$

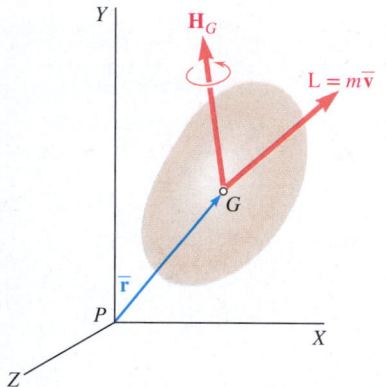

**Fig. 18.4** A momentum vector attached to the mass center of a rigid body and the angular momentum of the body about its mass center.

### Angular Momentum of a Rigid Body Constrained to Rotate about a Fixed Point.

In the particular case of a rigid body constrained to rotate in three-dimensional space about a fixed point $O$ (Fig. 18.5a), it is sometimes convenient to determine the angular momentum $\mathbf{H}_O$ of the body about $O$. Although we could obtain $\mathbf{H}_O$ by first computing $\mathbf{H}_G$ as indicated previously and then using Eq. (18.11), it is often advantageous to determine $\mathbf{H}_O$ directly from the angular velocity $\boldsymbol{\omega}$ of the body and its moments and products of inertia with respect to a frame $Oxyz$ centered at $O$. From Eq. (14.7), we have

$$\mathbf{H}_O = \sum_{i=1}^{n} (\mathbf{r}_i \times \mathbf{v}_i \, \Delta m_i) \tag{18.12}$$

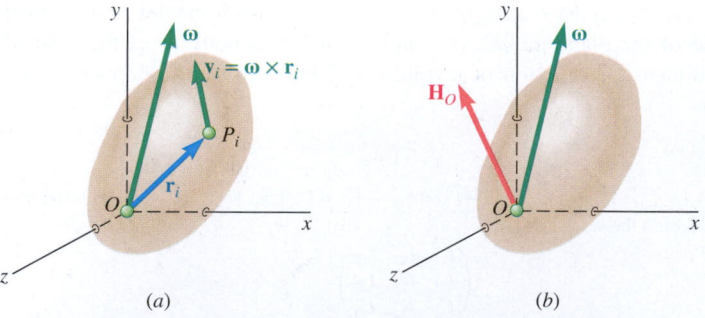

**Fig. 18.5** (a) The velocity of particle $P_i$ of a rigid body rotating with angular velocity $\boldsymbol{\omega}$; (b) angular velocity and angular momentum of a rigid body.

where $\mathbf{r}_i$ and $\mathbf{v}_i$ denote, respectively, the position vector and the velocity of particle $P_i$ with respect to the fixed frame $Oxyz$. Substituting $\mathbf{v}_i = \boldsymbol{\omega} \times \mathbf{r}_i$ and after making manipulations similar to those used in the earlier part of this section, we find that the components of the angular momentum $\mathbf{H}_O$ (Fig. 18.5$b$) are given by the relations

**Angular momentum about**
**a fixed point $O$**

$$H_x = +I_x\omega_x - I_{xy}\omega_y - I_{xz}\omega_z$$
$$H_y = -I_{yx}\omega_x + I_y\omega_y - I_{yz}\omega_z \qquad \textbf{(18.13)}$$
$$H_z = -I_{zx}\omega_x - I_{zy}\omega_y + I_z\omega_z$$

where we compute the moments of inertia $I_x$, $I_y$, $I_z$ and the products of inertia $I_{xy}$, $I_{yz}$, $I_{zx}$ with respect to the frame $Oxyz$ centered at the fixed point $O$.

## *18.1B   Applying the Principle of Impulse and Momentum to the Three-Dimensional Motion of a Rigid Body

Before we can apply the fundamental equation (18.2) to the solution of problems involving the three-dimensional motion of a rigid body, we must be able to compute the derivative of the vector $\mathbf{H}_G$. We show how to do this in Sec. 18.2A. However, we can use the results obtained already to solve problems using the impulse–momentum method.

Recall that the system formed by the momenta of the particles of a rigid body reduces to a linear momentum vector $m\overline{\mathbf{v}}$ attached at the mass center $G$ of the body and an angular momentum couple $\mathbf{H}_G$. We can represent the fundamental relation

**Syst Momenta**$_1$ + **Syst Ext Imp**$_{1\rightarrow2}$ = **Syst Momenta**$_2$    **(17.14)**

graphically by means of the impulse–momentum diagram shown in Fig. 18.6. To solve a given problem, we can use this diagram to write appropriate component and moment equations, keeping in mind that the components of the angular momentum $\mathbf{H}_G$ are related to the components of the angular velocity $\boldsymbol{\omega}$ by Eqs. (18.7).

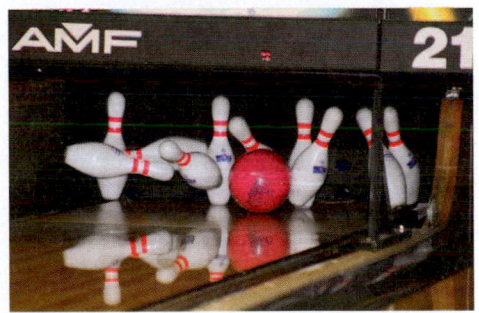

**Photo 18.2** As a result of the impulsive force applied by the bowling ball, a pin acquires both linear momentum and angular momentum. Source: Lance Cpl. Scott L. Tomaszycki

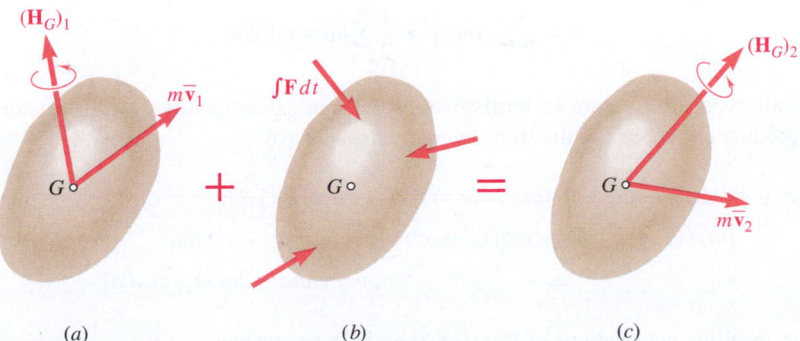

(a)          (b)          (c)

**Fig. 18.6** Impulse–momentum diagram for applying the principle of impulse and momentum to the motion of a rigid body in space.

In solving problems dealing with the motion of a body rotating about a fixed point $O$, it will be convenient to eliminate the impulse of the reaction at $O$ by writing an equation involving the moments of the momenta and impulses about $O$. Recall that you can obtain the angular momentum $\mathbf{H}_O$ of the body about the fixed point $O$ either directly from Eqs. (18.13) or by first computing its linear momentum $m\overline{\mathbf{v}}$ and its angular momentum $\mathbf{H}_G$ and then using Eq. (18.11).

## *18.1C  Kinetic Energy of a Rigid Body in Three Dimensions

Consider a rigid body with a mass $m$ in three-dimensional motion. Recall from Sec. 14.2A that, if we express the absolute velocity $\mathbf{v}_i$ of each particle $P_i$ of the body as the sum of velocity $\overline{\mathbf{v}}$ of the mass center $G$ of the body and velocity $\mathbf{v}_i'$ of the particle relative to a frame $Gxyz$ attached to $G$ and of fixed orientation (Fig. 18.7), we can write the kinetic energy of the system of particles forming the rigid body as

$$T = \tfrac{1}{2}m\overline{v}^2 + \tfrac{1}{2}\sum_{i=1}^{n}\Delta m_i v_i'^2 \qquad (18.14)$$

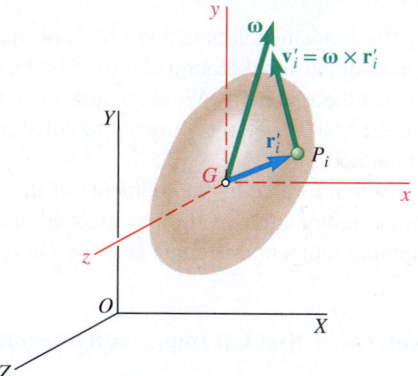

**Fig. 18.7** The relative velocity of a particle $P_i$ with respect to the mass center is $\boldsymbol{\omega} \times \mathbf{r}_i'$.

Here the last term represents the kinetic energy $T'$ of the body relative to the centroidal frame $Gxyz$. Because $v_i' = |\mathbf{v}_i'| = |\boldsymbol{\omega} \times \mathbf{r}_i'|$, we have

$$T' = \frac{1}{2}\sum_{i=1}^{n}\Delta m_i v_i'^2 = \frac{1}{2}\sum_{i=1}^{n}|\boldsymbol{\omega} \times \mathbf{r}_i'|^2 \Delta m_i$$

Expressing the square in terms of the rectangular components of the vector product and replacing the sums by integrals, we have

$$T' = \tfrac{1}{2}\int[(\omega_x y - \omega_y x)^2 + (\omega_y z - \omega_z y)^2 + (\omega_z x - \omega_x z)^2]\,dm$$
$$= \tfrac{1}{2}[\omega_x^2\int(y^2 + z^2)\,dm + \omega_y^2\int(z^2 + x^2)\,dm + \omega_z^2\int(x^2 + y^2)\,dm$$
$$- 2\omega_x\omega_y\int xy\,dm - 2\omega_y\omega_z\int yz\,dm - 2\omega_z\omega_x\int zx\,dm]$$

or recalling the relations of Eqs. (18.5) and (18.6), we have

$$T' = \tfrac{1}{2}(\overline{I}_x\omega_x^2 + \overline{I}_y\omega_y^2 + \overline{I}_z\omega_z^2 - 2\overline{I}_{xy}\omega_x\omega_y - 2\overline{I}_{yz}\omega_y\omega_z - 2\overline{I}_{zx}\omega_z\omega_x) \qquad (18.15)$$

Substituting Eq. (18.15) for the kinetic energy of the body relative to centroidal axes into Eq. (18.14), we obtain

**Kinetic energy of a rigid body**

$$T = \tfrac{1}{2}m\bar{v}^2 + \tfrac{1}{2}(\bar{I}_x\omega_x^2 + \bar{I}_y\omega_y^2 + \bar{I}_z\omega_z^2 - 2\bar{I}_{xy}\omega_x\omega_y \\ - 2\bar{I}_{yz}\omega_y\omega_z - 2\bar{I}_{zx}\omega_z\omega_x) \qquad \textbf{(18.16)}$$

If we choose the axes of coordinates so that they coincide with the principal axes $x'$, $y'$, $z'$ of the body at the instant considered, this relation reduces to

$$T = \tfrac{1}{2}m\bar{v}^2 + \tfrac{1}{2}(\bar{I}_{x'}\omega_{x'}^2 + \bar{I}_{y'}\omega_{y'}^2 + \bar{I}_{z'}\omega_{z'}^2) \qquad \textbf{(18.17)}$$

where $\bar{v}$ = magnitude of the velocity of the mass center
$\omega_{x'}, \omega_{y'}, \omega_{z'} = x'$, $y'$, and $z'$ components of the angular velocity
$m$ = mass of rigid body
$\bar{I}_{x'}, \bar{I}_{y'}, \bar{I}_{z'}$ = principal centroidal moments of inertia

These results enable us to apply the principles of work and energy (Sec. 17.1A) and the conservation of energy (Sec. 17.1E) to the three-dimensional motion of a rigid body.

### Kinetic Energy of a Rigid Body with a Fixed Point.

In the particular case of a rigid body rotating in three-dimensional space about a fixed point $O$, we can express the kinetic energy of the body in terms of its moments and products of inertia with respect to axes attached at $O$ (Fig. 18.8). Recalling the definition of the kinetic energy of a system of particles and substituting $v_i = |\mathbf{v}_i| = |\boldsymbol{\omega} \times \mathbf{r}_i|$, we have

$$T = \tfrac{1}{2}\sum_{i=1}^{n}\Delta m_i v_i^2 = \tfrac{1}{2}\sum_{i=1}^{n}|\boldsymbol{\omega} \times \mathbf{r}_i|^2 \Delta m_i \qquad \textbf{(18.18)}$$

Manipulations similar to those used to derive Eq. (18.15) yield

$$T = \tfrac{1}{2}(I_x\omega_x^2 + I_y\omega_y^2 + I_z\omega_z^2 - 2I_{xy}\omega_x\omega_y - 2I_{yz}\omega_y\omega_z - 2I_{zx}\omega_z\omega_x) \qquad \textbf{(18.19)}$$

or if we choose the principal axes $x'$, $y'$, $z'$ of the body at the origin $O$ as coordinate axes, we have

$$T = \tfrac{1}{2}(I_{x'}\omega_{x'}^2 + I_{y'}\omega_{y'}^2 + I_{z'}\omega_{z'}^2) \qquad \textbf{(18.20)}$$

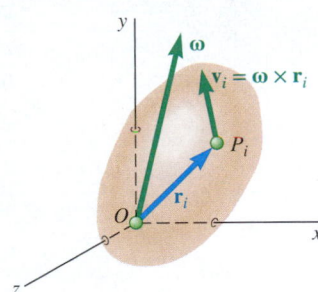

**Fig. 18.8** The velocity of every particle $P_i$ of a rigid body undergoing fixed-axis rotation is $\boldsymbol{\omega} \times \mathbf{r}_i$.

# Sample Problem 18.1

A rectangular plate with a mass $m$ is suspended from two wires at $A$ and $B$ and is hit at $D$ in a direction perpendicular to the plate. Denoting the impulse applied at $D$ by $\mathbf{F}\,\Delta t$, determine immediately after the impact (*a*) the velocity of the mass center $G$, (*b*) the angular velocity of the plate.

**STRATEGY:**   Because you have an impulse applied to the plate, use the principle of impulse and momentum.

**MODELING:**   Choose the plate to be your system and model it as a rigid body undergoing three-dimensional motion.

**ANALYSIS:**   Assume that the wires remain taut. Therefore, the components $\bar{v}_y$ of $\overline{\mathbf{v}}$ and $\omega_z$ of $\boldsymbol{\omega}$ are zero after the impact. Then you have

$$\overline{\mathbf{v}} = \bar{v}_x\mathbf{i} + \bar{v}_z\mathbf{k} \qquad \boldsymbol{\omega} = \omega_x\mathbf{i} + \omega_y\mathbf{j}$$

The $x$, $y$, $z$ axes are principal axes of inertia, so you have

$$\mathbf{H}_G = \bar{I}_x\omega_x\mathbf{i} + \bar{I}_y\omega_y\mathbf{j} \qquad \mathbf{H}_G = \tfrac{1}{12}mb^2\omega_x\mathbf{i} + \tfrac{1}{12}ma^2\omega_y\mathbf{j} \qquad \textbf{(1)}$$

**Principle of Impulse and Momentum.**   Because the initial momenta are zero, the system of the impulses must be equivalent to the system of the final momenta (Fig. 1).

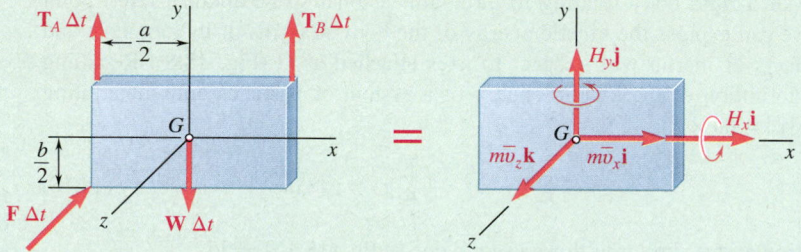

**Fig. 1**   Impulse–momentum diagram for the plate.

**a.  Velocity of Mass Center.**   Equate the components of the impulses and momenta in the $x$ and $z$ directions as

$x$ components:  $\qquad\qquad 0 = m\bar{v}_x \qquad \bar{v}_x = 0$

$z$ components:  $\qquad\qquad -F\,\Delta t = m\bar{v}_z \qquad \bar{v}_z = -F\,\Delta t/m$

$$\overline{\mathbf{v}} = \bar{v}_x\mathbf{i} + \bar{v}_z\mathbf{k} \qquad \overline{\mathbf{v}} = -(F\,\Delta t/m)\mathbf{k} \quad \blacktriangleleft$$

**b.  Angular Velocity.**   Equate the moments of the impulses and momenta about the $x$ and $y$ axes as

About $x$ axis:  $\qquad\qquad \tfrac{1}{2}bF\,\Delta t = H_x$

About $y$ axis:  $\qquad\qquad -\tfrac{1}{2}aF\,\Delta t = H_y$

$$\mathbf{H}_G = H_x\mathbf{i} + H_y\mathbf{j} \qquad \mathbf{H}_G = \tfrac{1}{2}bF\,\Delta t\,\mathbf{i} - \tfrac{1}{2}aF\,\Delta t\,\mathbf{j} \qquad \textbf{(2)}$$

*(continued)*

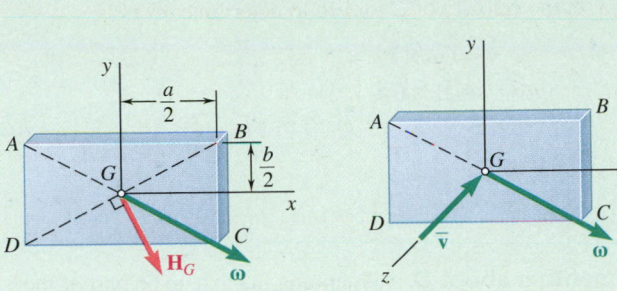

**Fig. 2** Directions of the angular velocity, angular momentum, and velocity of $G$ immediately after the impulse.

Comparing Eqs. (1) and (2), you can conclude that

$$\omega_x = 6F\,\Delta t/mb \qquad \omega_y = -6F\,\Delta t/ma$$

$$\boldsymbol{\omega} = \omega_x\mathbf{i} + \omega_y\mathbf{j} \qquad \boldsymbol{\omega} = (6F\,\Delta t/mab)(a\mathbf{i} - b\mathbf{j}) \;\blacktriangleleft$$

Note that $\boldsymbol{\omega}$ is directed along the diagonal $AC$ (Fig. 2).

**REFLECT and THINK:** Equating the $y$ components of the impulses and momenta and their moments about the $z$ axis, you can obtain two additional equations that yield $T_A = T_B = \frac{1}{2}W$. This verifies that the wires remain taut and that the initial assumption was correct. If the impulse was at $G$, this would reduce to a two-dimensional problem.

## Sample Problem 18.2

A homogeneous disk of radius $r$ and mass $m$ is mounted on an axle $OG$ of length $L$ and negligible mass. The axle is pivoted at the fixed point $O$, and the disk is constrained to roll on a horizontal floor. The disk rotates counterclockwise at the rate $\omega_1$ about the axle $OG$. Determine (a) the angular velocity of the disk, (b) its angular momentum about $O$, (c) its kinetic energy, (d) the linear momentum and angular momentum about $G$ of the disk.

**STRATEGY:** Recognizing that the wheel rolls without slip, you can use kinematics to calculate the angular velocity of the bar around $O$. Then you can determine the kinetic energy and momenta of the system.

**MODELING AND ANALYSIS:**

**a. Angular Velocity.** As the disk rotates about the axle $OG$, it also rotates with the axle about the $y$ axis at a rate of $\omega_2$ clockwise (Fig. 1). The total angular velocity of the disk is therefore

$$\boldsymbol{\omega} = \omega_1\mathbf{i} - \omega_2\mathbf{j} \tag{1}$$

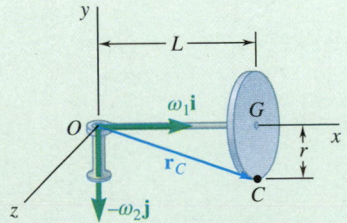

**Fig. 1** Angular velocity of the system.

*(continued)*

The disk is rolling, so set the velocity of $C$ to zero to determine $\omega_2$ as

$$\mathbf{v}_C = \boldsymbol{\omega} \times \mathbf{r}_C = 0$$
$$(\omega_1\mathbf{i} - \omega_2\mathbf{j}) \times (L\mathbf{i} - r\mathbf{j}) = 0$$
$$(L\omega_2 - r\omega_1)\mathbf{k} = 0 \qquad \omega_2 = r\omega_1/L$$

Substituting into Eq. (1) for $\omega_2$ gives

$$\boldsymbol{\omega} = \omega_1\mathbf{i} - (r\omega_1/L)\mathbf{j} \quad \blacktriangleleft$$

**b. Angular Momentum about O.**   Assuming the axle to be part of the disk, you can consider the disk to have a fixed point at $O$. Because the $x$, $y$, and $z$ axes are principal axes of inertia for the disk, you have

$$H_x = I_x\omega_x = (\tfrac{1}{2}mr^2)\omega_1$$
$$H_y = I_y\omega_y = (mL^2 + \tfrac{1}{4}mr^2)(-r\omega_1/L)$$
$$H_z = I_z\omega_z = (mL^2 + \tfrac{1}{4}mr^2)0 = 0$$
$$\mathbf{H}_O = \tfrac{1}{2}mr^2\omega_1\mathbf{i} - m(L^2 + \tfrac{1}{4}r^2)(r\omega_1/L)\mathbf{j} \quad \blacktriangleleft$$

**c. Kinetic Energy.**   Using the values obtained for the moments of inertia and the components of $\boldsymbol{\omega}$, you have

$$T = \tfrac{1}{2}(I_x\omega_x^2 + I_y\omega_y^2 + I_z\omega_z^2) = \tfrac{1}{2}[\tfrac{1}{2}mr^2\omega_1^2 + m(L^2 + \tfrac{1}{4}r^2)(-r\omega_1/L)^2]$$
$$T = \tfrac{1}{8}mr^2\left(6 + \frac{r^2}{L^2}\right)\omega_1^2 \quad \blacktriangleleft$$

**d. Linear Momentum and Angular Momentum about G.**   The linear momentum vector $m\overline{\mathbf{v}}$ and the angular momentum couple $\mathbf{H}_G$ are (Fig. 2)

$$m\overline{\mathbf{v}} = mr\omega_1\mathbf{k} \quad \blacktriangleleft$$

and

$$\mathbf{H}_G = \bar{I}_{x'}\omega_x\mathbf{i} + \bar{I}_{y'}\omega_y\mathbf{j} + \bar{I}_{z'}\omega_z\mathbf{k} = \tfrac{1}{2}mr^2\omega_1\mathbf{i} + \tfrac{1}{4}mr^2(-r\omega_1/L)\mathbf{j}$$
$$\mathbf{H}_G = \tfrac{1}{2}mr^2\omega_1\left(\mathbf{i} - \frac{r}{2L}\mathbf{j}\right) \quad \blacktriangleleft$$

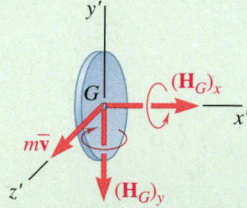

**Fig. 2**  Linear and angular momenta for the system.

**REFLECT and THINK:**   If the mass of the axle was not negligible and it was instead modeled as a slender rod with a mass $M_{\text{axle}}$, it would also contribute to the kinetic energy $T_{\text{axle}} = \tfrac{1}{2}(\tfrac{1}{3}M_{\text{axle}}L^2)\omega_2^2$ and to the angular momentum $\mathbf{H}_{\text{axle}} = -(\tfrac{1}{3}M_{\text{axle}}L^2)\omega_2\mathbf{j}$ of the system.

# SOLVING PROBLEMS ON YOUR OWN

In this section, you saw how to compute the **angular momentum of a rigid body in three dimensions** and to apply the principle of impulse and momentum to the three-dimensional motion of a rigid body. You also learned how to compute the **kinetic energy of a rigid body in three dimensions**. It is important for you to keep in mind that, except for very special situations, the angular momentum of a rigid body in three dimensions *cannot* be expressed as the product $\bar{I}\boldsymbol{\omega}$ and, therefore, does not have the same direction as the angular velocity $\boldsymbol{\omega}$ (Fig. 18.3).

**1. To compute the angular momentum $\mathbf{H}_G$ of a rigid body about its mass center $G$,** you must first determine the angular velocity $\boldsymbol{\omega}$ of the body with respect to a system of axes centered at $G$ and of fixed orientation. Because you will be asked in the problems to determine the angular momentum of the body *at a given instant only,* select the system of axes that will be most convenient for your computations.

  **a. If the principal axes of inertia of the body at $G$ are known,** use these axes as coordinate axes $x'$, $y'$, and $z'$, because the corresponding products of inertia of the body are equal to zero. Resolve $\boldsymbol{\omega}$ into components $\omega_{x'}$, $\omega_{y'}$, and $\omega_{z'}$ along these axes and compute the principal moments of inertia as $\bar{I}_{x'}, \bar{I}_{y'}, \bar{I}_{z'}$. The corresponding components of the angular momentum $\mathbf{H}_G$ are

$$H_{x'} = \bar{I}_{x'}\omega_{x'} \qquad H_{y'} = \bar{I}_{y'}\omega_{y'} \qquad H_{z'} = \bar{I}_{z'}\omega_{z'} \qquad\qquad (18.10)$$

  **b. If the principal axes of inertia of the body at $G$ are not known,** you must use Eqs. (18.7) to determine the components of the angular momentum $\mathbf{H}_G$. These equations require prior computation of the products of inertia of the body, as well as prior computation of its moments of inertia with respect to the selected axes.

  **c. The magnitude and direction cosines of $\mathbf{H}_G$** are obtained from formulas similar to those used in *Statics* (Sec. 2.4A). We have

$$H_G = \sqrt{H_x^2 + H_y^2 + H_z^2}$$

$$\cos\theta_x = \frac{H_x}{H_G} \qquad \cos\theta_y = \frac{H_y}{H_G} \qquad \cos\theta_z = \frac{H_z}{H_G}$$

  **d. Once you have determined $\mathbf{H}_G$,** you can obtain the angular momentum of the body about any given point $P$ by observing from Fig. (18.4) that

$$\mathbf{H}_P = \bar{\mathbf{r}} \times m\bar{\mathbf{v}} + \mathbf{H}_G \qquad\qquad (18.11)$$

where $\bar{\mathbf{r}}$ is the position vector of $G$ relative to $P$ and $m\bar{\mathbf{v}}$ is the linear momentum of the body.

**2. To compute the angular momentum $\mathbf{H}_O$ of a rigid body with a fixed point $O$,** follow the procedure described in paragraph 1, except that you should now use axes centered at the fixed point $O$. Alternatively, you can use Eq. 18.11.

  **a. If you know the principal axes of inertia of the body at $O$,** resolve $\boldsymbol{\omega}$ into components along these axes (Sample Prob. 18.2). Obtain the corresponding components of the angular momentum $\mathbf{H}_G$ from equations similar to Eqs. (18.10).

*(continued)*

**b. If you do not know the principal axes of inertia of the body at $O$,** you must compute the products as well as the moments of inertia of the body with respect to the axes that you have selected. Then use Eqs. (18.13) to determine the components of the angular momentum $\mathbf{H}_O$.

**3. To apply the principle of impulse and momentum** to the solution of a problem involving the three-dimensional motion of a rigid body, use the same vector equation that you used for plane motion in Chap. 17:

$$\textbf{Syst Momenta}_1 + \textbf{Syst Ext Imp}_{1\rightarrow2} = \textbf{Syst Momenta}_2 \qquad (17.14)$$

where the initial and final systems of momenta are each represented by a *linear momentum vector* $m\overline{\mathbf{v}}$ and an *angular momentum couple* $\mathbf{H}_G$. Now, however, these vector-and-couple systems should be represented in three dimensions, as shown in Fig. 18.6, and $\mathbf{H}_G$ should be determined as explained in paragraph 1.

**a. In problems involving the application of a known impulse to a rigid body,** draw the impulse–momentum diagram corresponding to Eq. (17.14). Equating the components of the vectors involved, you can determine the final linear momentum $m\overline{\mathbf{v}}$ of the body and, thus, the corresponding velocity $\overline{\mathbf{v}}$ of its mass center. Equating moments about $G$, you can determine the final angular momentum $\mathbf{H}_G$ of the body. Then substitute the values obtained for the components of $\mathbf{H}_G$ into Eq. (18.10) or (18.7) and solve for the corresponding values of the components of the angular velocity $\boldsymbol{\omega}$ of the body (Sample Prob. 18.1).

**b. In problems involving unknown impulses,** draw the impulse–momentum diagram corresponding to Eq. (17.14) and write equations that do not involve the unknown impulses. You can obtain such equations by equating moments about the point or line of impact.

**4. To compute the kinetic energy of a rigid body with a fixed point $O$,** resolve the angular velocity $\boldsymbol{\omega}$ into components along axes of your choice and compute the moments and products of inertia of the body with respect to these axes. As was the case for the computation of the angular momentum, use the principal axes of inertia $x'$, $y'$, and $z'$ if you can easily determine them. The products of inertia are then zero (Sample Prob. 18.2), and the expression for the kinetic energy reduces to

$$T = \tfrac{1}{2}(I_{x'}\omega_{x'}^2 + I_{y'}\omega_{y'}^2 + I_{z'}\omega_{z'}^2) \qquad (18.20)$$

If you must use axes other than the principal axes of inertia, express the kinetic energy of the body as shown in Eq. (18.19).

**5. To compute the kinetic energy of a rigid body in general motion,** consider the motion as the sum of a *translation with the mass center $G$ and a rotation about $G$.* The kinetic energy associated with the translation is $\tfrac{1}{2}m\overline{v}^2$. If you can use principal axes of inertia, express the kinetic energy associated with the rotation about $G$ in the form used in Eq. (18.20). The total kinetic energy of the rigid body is then

$$T = \tfrac{1}{2}m\overline{v}^2 + \tfrac{1}{2}(\overline{I}_{x'}\omega_{x'}^2 + \overline{I}_{y'}\omega_{y'}^2 + \overline{I}_{z'}\omega_{z'}^2) \qquad (18.17)$$

If you must use axes other than the principal axes of inertia to determine the kinetic energy associated with the rotation about $G$, express the total kinetic energy of the body as shown in Eq. (18.16).

# Problems

**18.1** A thin, homogeneous disk of mass $m$ and radius $r$ spins at the constant rate $\omega_1$ about an axle held by a fork-ended vertical rod that rotates at the constant rate $\omega_2$. Determine the angular momentum $\mathbf{H}_G$ of the disk about its mass center $G$.

**18.2** A thin homogeneous square plate of mass $m$ and side $a$ is welded to a vertical shaft $AB$ with which it forms an angle of 45°. Knowing that the shaft rotates with a constant angular velocity $\boldsymbol{\omega}$, determine the angular momentum $\mathbf{H}_A$ of the plate about point $A$.

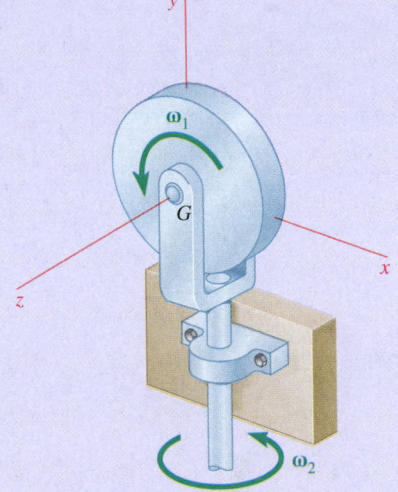

**Fig. P18.1**

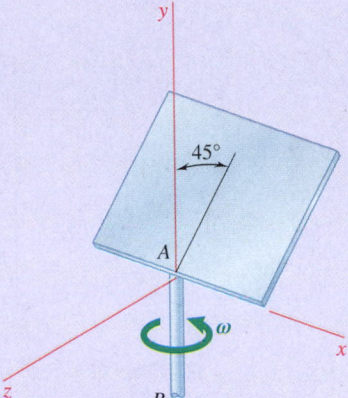

**Fig. P18.2**

**18.3** A uniform 3.6-lb rod $AB$ is welded at its midpoint $G$ to a vertical shaft $GD$. Knowing that the shaft rotates with an angular velocity of constant magnitude $\omega = 1200$ rpm, determine the angular momentum $\mathbf{H}_G$ of the rod about $G$.

**18.4** A homogeneous disk of weight $W = 6$ lb rotates at the constant rate $\omega_1 = 16$ rad/s with respect to arm $ABC$, which is welded to a shaft $DCE$ rotating at the constant rate $\omega_2 = 8$ rad/s. Determine the angular momentum $\mathbf{H}_A$ of the disk about its center $A$.

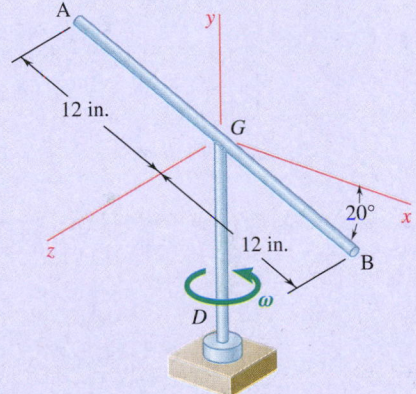

**Fig. P18.3**

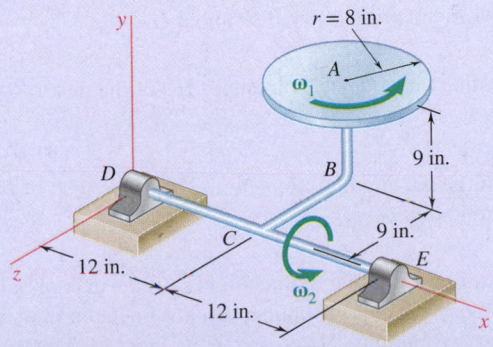

**Fig. P18.4**

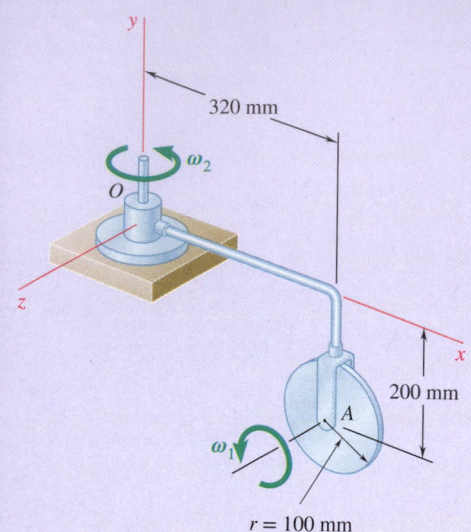

320 mm

200 mm

$r = 100$ mm

**Fig. P18.5**

**18.5** A homogeneous disk of mass $m = 8$ kg rotates at the constant rate $\omega_1 = 12$ rad/s with respect to arm $OA$, which itself rotates at the constant rate $\omega_2 = 4$ rad/s about the $y$ axis. Determine the angular momentum $\mathbf{H}_A$ of the disk about its center $A$.

**18.6** A solid rectangular parallelepiped of mass $m$ has a square base of side $a$ and a length $2a$. Knowing that it rotates at the constant rate $\omega$ about its diagonal $AC'$ and that its rotation is observed from $A$ as counterclockwise, determine ($a$) the magnitude of the angular momentum $\mathbf{H}_G$ of the parallelepiped about its mass center $G$, ($b$) the angle that $\mathbf{H}_G$ forms with the diagonal $AC'$.

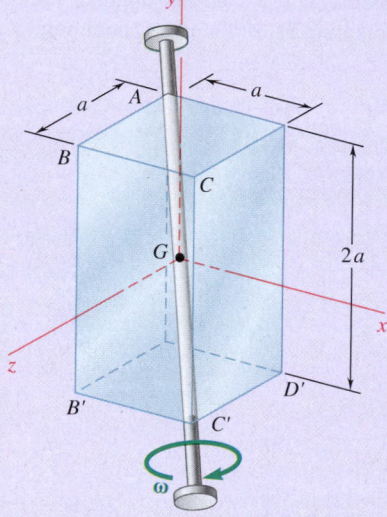

**Fig. P18.6**

**18.7** Solve Prob. 18.6, assuming that the solid rectangular parallelepiped has been replaced by a hollow one consisting of six thin metal plates welded together.

**18.8** A thin homogeneous disk with a mass $m$ and radius $r$ is mounted on the horizontal axle $AB$. The plane of the disk forms an angle of $\beta = 20°$ with the vertical. Knowing that the axle rotates with an angular velocity $\boldsymbol{\omega}$, determine the angle $\theta$ formed by the axle and the angular momentum of the disk about $G$.

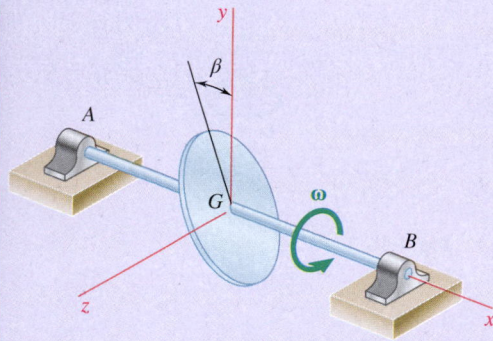

**Fig. P18.8**

**18.9** Determine the angular momentum $\mathbf{H}_D$ of the disk of Prob. 18.4 about point $D$.

**18.10** Determine the angular momentum $\mathbf{H}_O$ of the disk of Prob. 18.5 about the fixed point $O$.

**18.11** Determine the angular momentum $\mathbf{H}_O$ of the disk of Sample Prob. 18.2 from the expressions obtained for its linear momentum $m\overline{\mathbf{v}}$ and its angular momentum $\mathbf{H}_G$, using Eqs. (18.11). Verify that the result obtained is the same as that obtained by direct computation.

**18.12** The 100-kg projectile shown has a radius of gyration of 100 mm about its axis of symmetry $Gx$ and a radius of gyration of 250 mm about the transverse axis $Gy$. Its angular velocity $\boldsymbol{\omega}$ can be resolved into two components; one component, directed along $Gx$, measures the *rate of spin* of the projectile, while the other component, directed along $GD$, measures its *rate of precession*. Knowing that $\theta = 6°$ and that the angular momentum of the projectile about its mass center $G$ is $\mathbf{H}_G = (500 \text{ g·m}^2\text{/s})\mathbf{i} = (10 \text{ g·m}^2\text{/s})\mathbf{j}$, determine (*a*) the rate of spin, (*b*) the rate of precession.

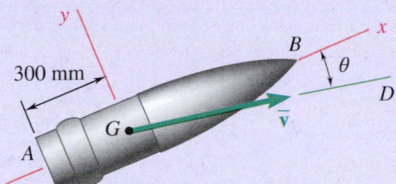

**Fig. P18.12**

**18.13** Determine the angular momentum $\mathbf{H}_A$ of the projectile of Prob. 18.12 about the center $A$ of its base, knowing that its mass center $G$ has a velocity $\overline{\mathbf{v}}$ of 750 m/s. Give your answer in terms of components respectively parallel to the $x$ and $y$ axes shown and to a third axis $z$ pointing toward you.

**18.14** (*a*) Show that the angular momentum $\mathbf{H}_B$ of a rigid body about point $B$ can be obtained by adding to the angular momentum $\mathbf{H}_A$ of that body about point $A$ the vector product of the vector $\mathbf{r}_{A/B}$ drawn from $B$ to $A$ and the linear momentum $m\overline{\mathbf{v}}$ of the body:

$$\mathbf{H}_B = \mathbf{H}_A + \mathbf{r}_{A/B} \times m\overline{\mathbf{v}}$$

(*b*) Further show that when a rigid body rotates about a fixed axis, its angular momentum is the same about any two points $A$ and $B$ located on the fixed axis ($\mathbf{H}_A = \mathbf{H}_B$) if, and only if, the mass center $G$ of the body is located on the fixed axis.

**18.15** Two L-shaped arms each have a mass of 5 kg and are welded at the one-third points of the 600-mm shaft $AB$ to form the assembly shown. Knowing that the assembly rotates at the constant rate of 360 rpm, determine (*a*) the angular momentum $\mathbf{H}_A$ of the assembly about point $A$, (*b*) the angle formed by $\mathbf{H}_A$ and $AB$.

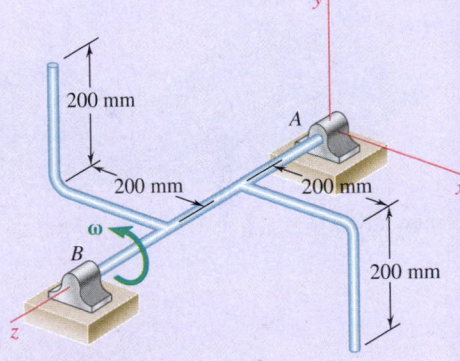

**Fig. P18.15**

**18.16** For the assembly of Prob. 18.15, determine (*a*) the angular momentum $\mathbf{H}_B$ of the assembly about point $B$, (*b*) the angle formed by $\mathbf{H}_B$ and $BA$.

**18.17** A 10-lb rod of uniform cross-section is used to form the shaft shown. Knowing that the shaft rotates with a constant angular velocity $\boldsymbol{\omega}$ of magnitude 12 rad/s, determine (*a*) the angular momentum $\mathbf{H}_G$ of the shaft about its mass center $G$, (*b*) the angle formed by $\mathbf{H}_G$ and the axis $AB$.

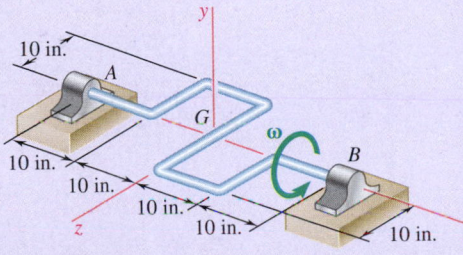

**18.18** Determine the angular momentum of the shaft of Prob. 18.17 about (*a*) point $A$, (*b*) point $B$.

**Fig. P18.17**

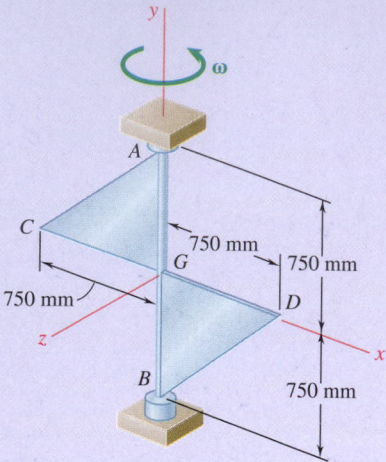

Fig. P18.19

**18.19** Two triangular plates each have a mass of 8 kg and are welded to a vertical shaft $AB$. Knowing that the system rotates at the constant rate of $\omega = 6$ rad/s, determine its angular momentum about $G$.

**18.20** The assembly shown consists of two pieces of sheet aluminum with a uniform thickness and total mass of 1.6 kg welded to a light axle supported by bearings $A$ and $B$. Knowing that the assembly rotates with an angular velocity of constant magnitude $\omega = 20$ rad/s, determine the angular momentum $\mathbf{H}_G$ of the assembly about point $G$.

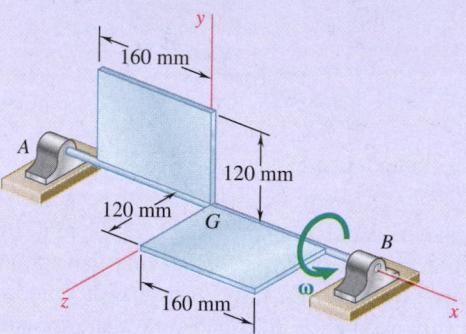

Fig. P18.20

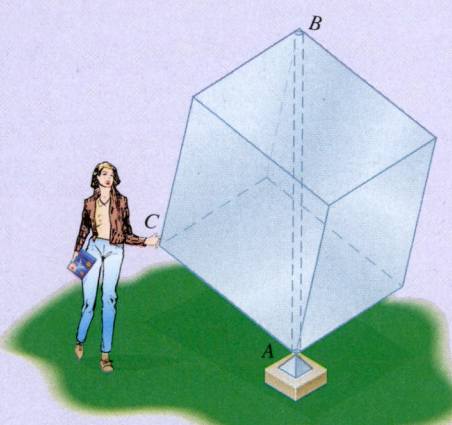

Fig. P18.21

**18.21** One of the sculptures displayed on a university campus consists of a hollow cube made of six aluminum sheets, each $1.5 \times 1.5$ m, welded together and reinforced with internal braces of negligible weight. The cube is mounted on a fixed base at $A$ and can rotate freely about its vertical diagonal $AB$. As she passes by this display on the way to a class in mechanics, an engineering student grabs corner $C$ of the cube and pushes it for 1.2 s in a direction perpendicular to the plane $ABC$ with an average force of 50 N. Having observed that it takes 5 s for the cube to complete one full revolution, she flips out her calculator and proceeds to determine the mass of the cube. What is the result of her calculation? (*Hint:* The perpendicular distance from the diagonal joining two vertices of a cube to any of its other six vertices can be obtained by multiplying the side of the cube by $\sqrt{2/3}$.)

**18.22** If the aluminum cube of Prob. 18.21 were replaced by a cube of the same size, made of six plywood sheets with mass 8 kg each, how long would it take for that cube to complete one full revolution if the student pushed its corner $C$ in the same way that she pushed the corner of the aluminum cube?

**18.23** A uniform rod of total mass $m$ is bent into the shape shown and is suspended by a wire attached at $B$. The bent rod is hit at $D$ in a direction perpendicular to the plane containing the rod (in the negative $z$ direction). Denoting the corresponding impulse by $\mathbf{F}\,\Delta t$, determine (*a*) the velocity of the mass center of the rod, (*b*) the angular velocity of the rod.

**18.24** Solve Prob. 18.23, assuming that the bent rod is hit at $C$.

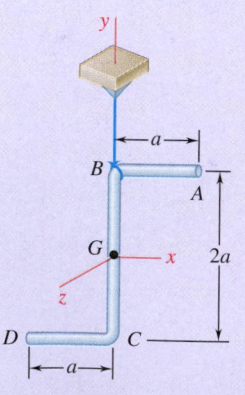

Fig. P18.23

**18.25** Three slender homogeneous rods, each of mass $m$ and length $d$, are welded together to form the assembly shown, which hangs from a wire attached at $G$. The assembly is hit at $A$ in a vertical downward direction. Denoting the corresponding impulse by $\mathbf{F}\,\Delta t$, determine immediately after the impact (a) the velocity of the mass center $G$, (b) the angular velocity of the assembly.

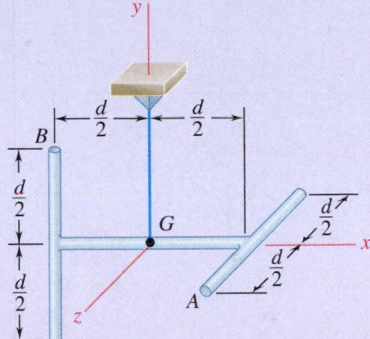

**Fig. P18.25**

**18.26** Solve Prob. 18.25, assuming that the assembly is hit at $B$ in a direction opposite to that of the $z$ axis.

**18.27** Two circular plates, each of mass 4 kg, are rigidly connected by a rod $AB$ of negligible mass and are suspended from point $A$ as shown. Knowing that an impulse $\mathbf{F}\,\Delta t = -(2.4\text{ N·s})\mathbf{k}$ is applied at point $D$, determine (a) the velocity of the mass center $G$ of the assembly, (b) the angular velocity of the assembly.

**18.28** Two circular plates, each of mass 4 kg, are rigidly connected by a rod $AB$ of negligible mass and are suspended from point $A$ as shown. Knowing that an impulse $\mathbf{F}\Delta t = (2.4\text{ N·s})\mathbf{j}$ is applied at point $D$, determine (a) the velocity of the mass center $G$ of the assembly, (b) the angular velocity of the assembly.

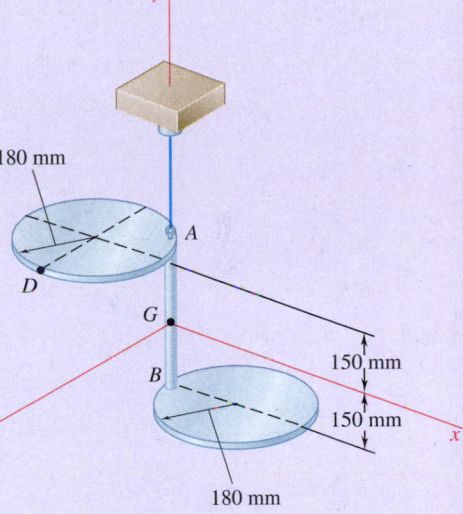

**Fig. P18.27 and P18.28**

**18.29** A circular plate of mass $m$ is falling with a velocity $\bar{\mathbf{v}}_0$ and no angular velocity when its edge $C$ strikes an obstruction. A line passing the origin and parallel to the line $CG$ makes a 45° angle with the $x$ axis. Assuming the impact to be perfectly plastic ($e = 0$), determine the angular velocity of the plate immediately after the impact.

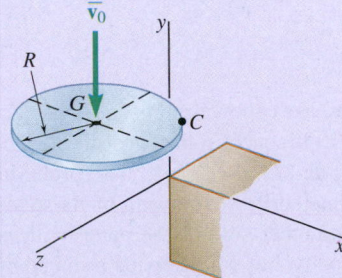

**Fig. P18.29**

**18.30** For the plate of Prob. 18.29, determine (a) the velocity of its mass center $G$ immediately after the impact, (b) the impulse exerted on the plate by the obstruction during the impact.

**18.31** A square plate of side $a$ and mass $m$ supported by a ball-and-socket joint at $A$ is rotating about the $y$ axis with a constant angular velocity $\boldsymbol{\omega} = \omega_0\mathbf{j}$ when an obstruction is suddenly introduced at $B$ in the $xy$ plane. Assuming the impact at $B$ to be perfectly plastic ($e = 0$), determine immediately after the impact (a) the angular velocity of the plate, (b) the velocity of its mass center $G$.

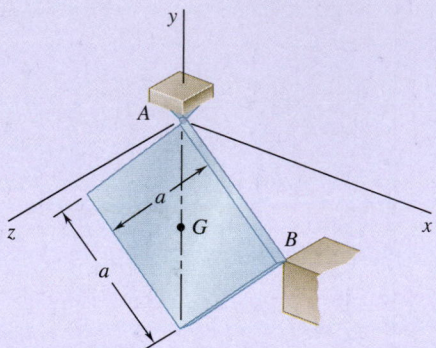

**Fig. P18.31**

**18.32** Determine the impulse exerted on the plate of Prob. 18.31 during the impact by (a) the obstruction at $B$, (b) the support at $A$.

**18.33** The coordinate axes shown represent the principal centroidal axes of inertia of a 3000-lb space probe whose radii of gyration are $k_x = 1.375$ ft, $k_y = 1.425$ ft, and $k_z = 1.250$ ft. The probe has no angular velocity when a 5-oz meteorite strikes one of its solar panels at point $A$ with a velocity $\mathbf{v}_0 = (2400 \text{ ft/s})\mathbf{i} - (3000 \text{ ft/s})\mathbf{j} + (3200 \text{ ft/s})\mathbf{k}$ relative to the probe. Knowing that the meteorite emerges on the other side of the panel with no change in the direction of its velocity, but with a speed reduced by 20 percent, determine the final angular velocity of the probe.

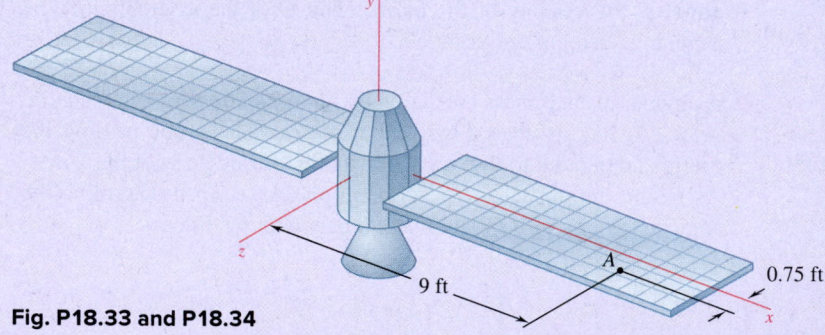

**Fig. P18.33 and P18.34**

**18.34** The coordinate axes shown represent the principal centroidal axes of inertia of a 3000-lb space probe whose radii of gyration are $k_x = 1.375$ ft, $k_y = 1.425$ ft, and $k_z = 1.250$ ft. The probe has no angular velocity when a 5-oz meteorite strikes one of its solar panels at point $A$ and emerges on the other side of the panel with no change in the direction of its velocity, but with a speed reduced by 25 percent. Knowing that the final angular velocity of the probe is $\boldsymbol{\omega} = (0.05 \text{ rad/s})\mathbf{i} - (0.12 \text{ rad/s})\mathbf{j} + \omega_z\mathbf{k}$ and that the $x$ component of the resulting change in the velocity of the mass center of the probe is $-0.675$ in./s, determine (a) the component $\omega_z$ of the final angular velocity of the probe, (b) the relative velocity $\mathbf{v}_0$ with which the meteorite strikes the panel.

**18.35** A 1200-kg satellite designed to study the sun has an angular velocity of $\boldsymbol{\omega}_0 = (0.050 \text{ rad/s})\mathbf{i} + (0.075 \text{ rad/s})\mathbf{k}$ when two small jets are activated at $A$ and $B$ in a direction parallel to the $y$ axis. Knowing that the coordinate axes are principal centroidal axes, that the radii of gyration of the satellite are $\bar{k}_x = 1.120$ m, $\bar{k}_y = 1.200$ m, and $\bar{k}_z = 0.900$ m, and that each jet produces a 50-N thrust, determine (*a*) the required operating time of each jet if the angular velocity of the satellite is to be reduced to zero, (*b*) the resulting change in the velocity of the mass center $G$.

**18.36** If jet $A$ in Prob. 18.35 is inoperative, determine (*a*) the required operating time of jet $B$ to reduce the $x$ component of the angular velocity of the satellite to zero, (*b*) the resulting final angular velocity, (*c*) the resulting change in the velocity of the mass center $G$.

**18.37** Denoting, respectively, by $\boldsymbol{\omega}$, $\mathbf{H}_O$, and $T$ the angular velocity, the angular momentum, and the kinetic energy of a rigid body with a fixed point $O$, (*a*) prove that $\mathbf{H}_O \cdot \boldsymbol{\omega} = 2T$; (*b*) show that the angle $\theta$ between $\boldsymbol{\omega}$ and $\mathbf{H}_O$ will always be acute.

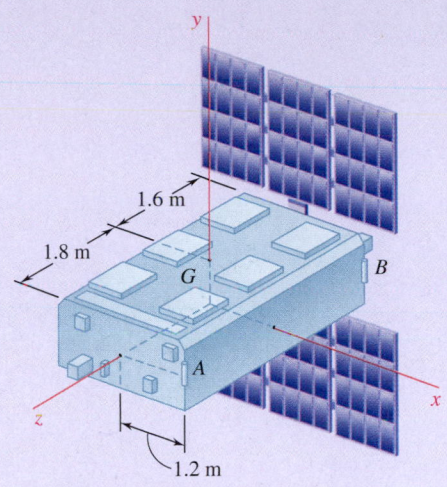

**Fig. P18.35**

**18.38** Show that the kinetic energy of a rigid body with a fixed point $O$ can be expressed as $T = \frac{1}{2}I_{OL}\omega^2$ where $\boldsymbol{\omega}$ is the instantaneous angular velocity of the body and $I_{OL}$ is its moment of inertia about the line of action $OL$ of $\boldsymbol{\omega}$. Derive this expression (*a*) from Eqs. (9.46) (or Eq. B.19 in the Appendix) and (18.19), (*b*) by considering $T$ as the sum of the kinetic energies of particles $P_i$ describing circles of radius $\rho_i$ about line $OL$.

**18.39** Determine the kinetic energy of the disk of Prob. 18.1.

**18.40** Determine the kinetic energy of the square plate of Prob. 18.2.

**18.41** Determine the kinetic energy of rod $AB$ of Prob. 18.3.

**18.42** Determine the kinetic energy of the disk of Prob. 18.4.

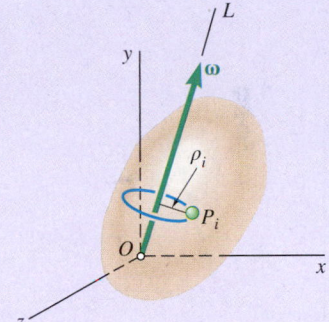

**Fig. P18.38**

**18.43** Determine the kinetic energy of the disk of Prob. 18.5.

**18.44** Determine the kinetic energy of the solid parallelepiped of Prob. 18.6.

**18.45** Determine the kinetic energy of the hollow parallelepiped of Prob. 18.7.

**18.46** Determine the kinetic energy of the disk of Prob. 18.8.

**18.47** Determine the kinetic energy of the assembly of Prob. 18.15.

**18.48** Determine the kinetic energy of the shaft of Prob. 18.17.

**18.49** Determine the kinetic energy of the assembly of Prob. 18.19.

**18.50** Determine the kinetic energy imparted to the cube of Prob. 18.21.

**18.51** Determine the kinetic energy lost when edge $C$ of the plate of Prob. 18.29 hits the obstruction.

**18.52** Determine the kinetic energy lost when the plate of Prob. 18.31 hits the obstruction at $B$.

**18.53** Determine the kinetic energy of the space probe of Prob. 18.33 in its motion about its mass center after its collision with the meteorite.

**18.54** Determine the kinetic energy of the space probe of Prob. 18.34 in its motion about its mass center after its collision with the meteorite.

## *18.2    MOTION OF A RIGID BODY IN THREE DIMENSIONS

As indicated in Sec. 18.1A, the fundamental equations

$$\Sigma \mathbf{F} = m\bar{\mathbf{a}} \tag{18.1}$$

$$\Sigma \mathbf{M}_G = \dot{\mathbf{H}}_G \tag{18.2}$$

remain valid in the most general case of the motion of a rigid body. Before we could apply Eq. (18.2) to the three-dimensional motion of a rigid body, however, it was necessary to derive Eqs. (18.7), which relate the components of the angular momentum $\mathbf{H}_G$ and those of the angular velocity $\boldsymbol{\omega}$. It still remains for us to find an effective and convenient way to compute the components of the derivative $\dot{\mathbf{H}}_G$ of the angular momentum. In this section, we do that first and then show how we can use the results to analyze motion of a rigid body in space.

### 18.2A    Rate of Change of Angular Momentum

The notation $\mathbf{H}_G$ represents the angular momentum of a rigid body in its motion relative to centroidal axes $GX'Y'Z'$ with a fixed orientation (Fig. 18.9). Because $\dot{\mathbf{H}}_G$ represents the rate of change of $\mathbf{H}_G$ with respect to the same axes, it would seem natural to use components of $\boldsymbol{\omega}$ and $\mathbf{H}_G$ along the axes $X'$, $Y'$, $Z'$ in writing the relations of Eq. (18.7). However, because the body rotates, its moments and products of inertia change continually, and it would be necessary to determine their values as functions of time. It is therefore more convenient to use axes $x$, $y$, $z$ attached to the body, ensuring that its moments and products of inertia maintain the same values during the motion. The angular velocity $\boldsymbol{\omega}$, however, still should be *defined* with respect to the frame $GX'Y'Z'$ with a fixed orientation. We can then *resolve* the vector $\boldsymbol{\omega}$ into components along the rotating $x$, $y$, and $z$ axes. Applying the relations of Eq. (18.7), we obtain the *components* of vector $\mathbf{H}_G$ along the rotating axes. Vector $\mathbf{H}_G$, however, represents the angular momentum about $G$ of the body *in its motion relative to the frame $GX'Y'Z'$*.

Differentiating the components of the angular momentum in Eq. (18.7) with respect to $t$, we define the rate of change of vector $\mathbf{H}_G$ with respect to the rotating frame $Gxyz$ as

$$(\dot{\mathbf{H}}_G)_{Gxyz} = \dot{H}_x \mathbf{i} + \dot{H}_y \mathbf{j} + \dot{H}_z \mathbf{k} \tag{18.21}$$

where $\mathbf{i}$, $\mathbf{j}$, and $\mathbf{k}$ are the unit vectors along the rotating axes. Recall from Sec. 15.5A that the rate of change $\dot{\mathbf{H}}_G$ of vector $\mathbf{H}_G$ with respect to the frame $GX'Y'Z'$ is found by adding $(\dot{\mathbf{H}}_G)_{Gxyz}$ to the vector product $\boldsymbol{\Omega} \times \mathbf{H}_G$, where $\boldsymbol{\Omega}$ denotes the angular velocity of the rotating frame. That is,

$$\dot{\mathbf{H}}_G = (\dot{\mathbf{H}}_G)_{Gxyz} + \boldsymbol{\Omega} \times \mathbf{H}_G \tag{18.22}$$

where $\mathbf{H}_G$ = angular momentum of the body with respect to frame $GX'Y'Z'$ with a fixed orientation

$(\dot{\mathbf{H}}_G)_{Gxyz}$ = rate of change of $\mathbf{H}_G$ with respect to rotating frame $Gxyz$ to be computed from the relations in Eqs. (18.7) and (18.21)

$\boldsymbol{\Omega}$ = angular velocity of rotating frame $Gxyz$

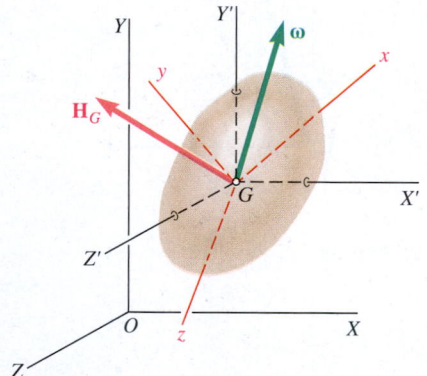

**Fig. 18.9** Angular velocity and angular momentum of a rigid body with centroidal axes $X'Y'Z'$ of fixed orientation and centroidal axes *xyz* attached to the body.

Substituting for $\dot{\mathbf{H}}_G$ from Eq. (18.22) into Eq. (18.2), we have

$$\Sigma\mathbf{M}_G = (\dot{\mathbf{H}}_G)_{Gxyz} + \boldsymbol{\Omega} \times \mathbf{H}_G \qquad (18.23)$$

If the rotating frame is attached to the body as we have assumed in this discussion, its angular velocity $\boldsymbol{\Omega}$ is identically equal to the angular velocity $\boldsymbol{\omega}$ of the body. In many applications, however, it is advantageous to use a frame of reference that is not actually attached to the body but rotates in an independent manner. For example, if the body considered is axisymmetric, as in Sample Prob. 18.5 or Sec. 18.3, it is possible to select a frame of reference where the moments and products of inertia of the body remain constant, but which rotate less than the body itself. As a result, it is possible to obtain simpler expressions for the angular velocity $\boldsymbol{\omega}$ and the angular momentum $\mathbf{H}_G$ of the body than we could have obtained if the frame of reference had actually been attached to the body. It is clear that in such cases the angular velocity $\boldsymbol{\Omega}$ of the rotating frame and the angular velocity $\boldsymbol{\omega}$ of the body are different.

## *18.2B   Euler's Equations of Motion

If we choose the $x$, $y$, and $z$ axes to coincide with the principal axes of inertia of the body, we can use the simplified relations in Eq. (18.10) to determine the components of the angular momentum $\mathbf{H}_G$. Omitting the primes from the subscripts, we have

$$\mathbf{H}_G = \bar{I}_x\omega_x\mathbf{i} + \bar{I}_y\omega_y\mathbf{j} + \bar{I}_z\omega_z\mathbf{k} \qquad (18.24)$$

where $\bar{I}_x$, $\bar{I}_y$, and $\bar{I}_z$ denote the principal centroidal moments of inertia of the body. Substituting for $\mathbf{H}_G$ from Eq. (18.24) into Eq. (18.23) and setting $\boldsymbol{\Omega} = \boldsymbol{\omega}$, we obtain the three scalar equations:

**Euler's equations
of motion**

$$
\begin{aligned}
\Sigma M_x &= \bar{I}_x\dot{\omega}_x - (\bar{I}_y - \bar{I}_z)\omega_y\omega_z \\
\Sigma M_y &= \bar{I}_y\dot{\omega}_y - (\bar{I}_z - \bar{I}_x)\omega_z\omega_x \\
\Sigma M_z &= \bar{I}_z\dot{\omega}_z - (\bar{I}_x - \bar{I}_y)\omega_x\omega_y
\end{aligned}
\qquad (18.25)
$$

We can use these equations, called **Euler's equations of motion** after the Swiss mathematician Leonhard Euler (1707–1783), to analyze the motion of a rigid body about its mass center. In the following sections, however, we will use Eq. (18.23) in preference to Eqs. (18.25), because Eq. (18.23) is more general, and the compact vectorial form in which it is expressed is easier to remember.

Writing Eq. (18.1) in scalar form, we obtain the three additional equations of

$$\Sigma F_x = m\bar{a}_x \qquad \Sigma F_y = m\bar{a}_y \qquad \Sigma F_z = m\bar{a}_z \qquad (18.26)$$

Together with Euler's equations, these form a system of six differential equations. Given appropriate initial conditions, these differential equations have a unique solution. Thus, the motion of a rigid body in three dimensions is completely defined by the resultant and the moment resultant of the external forces acting on it. This result is a generalization of a similar result obtained in Sec. 16.1C in the case of the plane motion of a rigid body. It follows that in three as well as in two dimensions, two systems of forces that are equipollent are also equivalent; that is, they have the same effect on a given rigid body.

Considering in particular the system of the external forces acting on a rigid body (Fig. 18.10*a*) and the system of the inertial terms associated with the particles forming the rigid body (Fig. 18.10*b*), we can state that the two systems—which were shown in Sec. 14.1A to be equipollent—are also equivalent. Replacing the inertia terms in Fig. 18.10*b* by $m\overline{\mathbf{a}}$ and $\dot{\mathbf{H}}_G$, we can verify that the system of the external forces acting on a rigid body in three-dimensional motion is equivalent to the system consisting of the vector $m\overline{\mathbf{a}}$ attached at the mass center $G$ of the body and the couple of moment $\dot{\mathbf{H}}_G$ (Fig. 18.11), where we obtain $\dot{\mathbf{H}}_G$ from the relations in Eqs. (18.7) and (18.22). Note that the equivalence of the systems of vectors shown in Figs. 18.10 and 18.11 has been indicated by *red* equal signs. You can solve problems involving the three-dimensional motion of a rigid body by considering the free-body diagram and kinetic diagram represented in Fig. 18.11 and by writing appropriate scalar equations relating the components or moments of the external forces and the inertial terms (see Sample Prob. 18.3).

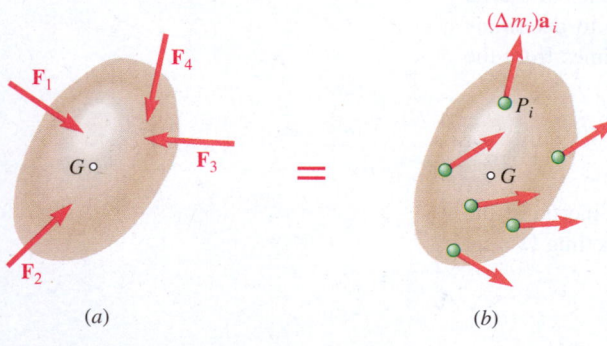

(*a*)

(*b*)

**Fig. 18.10** (*a*) The system of external forces acting on a rigid body is equivalent to (*b*) the system of inertia terms associated with the particles of the rigid body.

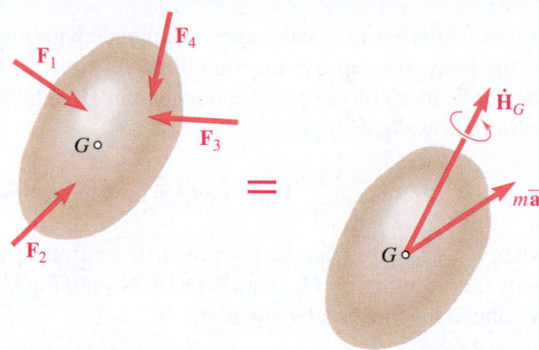

**Fig. 18.11** The free-body diagram and kinetic diagram show that the system of external forces is equivalent to the system consisting of the vectors $m\overline{\mathbf{a}}$ attached at the mass center $G$ and $\dot{\mathbf{H}}_G$

Similarly to what was developed in Eq. (16.6′), we can use Fig. 18.11 to sum moments about any point $P$ to obtain

$$\Sigma\mathbf{M}_P = \dot{\mathbf{H}}_G + r_{G/P} \times m\overline{\mathbf{a}}$$

where $\dot{\mathbf{H}}_G$ is obtained from Eq. 18.22.

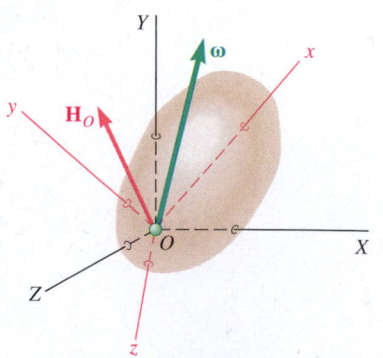

**Fig. 18.12** Angular velocity and angular momentum of a rigid body rotating about a fixed point.

## *18.2C Motion of a Rigid Body About a Fixed Point

If we want to analyze the motion of a rigid body constrained to rotate about a fixed point $O$, it is useful to write an equation involving the moments about $O$ of the external forces and of the inertial terms, because this equation contains the unknown reaction at $O$. Although we can obtain such an equation from Fig. 18.11, it may be more convenient to write it by considering the rate of change of the angular momentum $\mathbf{H}_O$ of the body about the fixed point $O$ (Fig. 18.12). Recalling Eq. (14.11), we have

$$\Sigma\mathbf{M}_O = \dot{\mathbf{H}}_O \qquad (18.27)$$

where $\dot{\mathbf{H}}_O$ denotes the rate of change of the vector $\mathbf{H}_O$ with respect to the fixed frame $OXYZ$. A derivation similar to that used in Sec. 18.2A enables us to relate $\dot{\mathbf{H}}_O$ to the rate of change $(\dot{\mathbf{H}}_O)_{Oxyz}$ of $\mathbf{H}_O$ with respect to the rotating frame $Oxyz$. Substitution into Eq. (18.27) leads to

$$\Sigma\mathbf{M}_O = (\dot{\mathbf{H}}_O)_{Oxyz} + \mathbf{\Omega} \times \mathbf{H}_O \qquad (18.28)$$

where $\Sigma\mathbf{M}_O$ = sum of moments about $O$ of forces applied to the rigid body

$\mathbf{H}_O$ = angular momentum of the body with respect to fixed frame $OXYZ$

$(\dot{\mathbf{H}}_O)_{Oxyz}$ = rate of change of $\mathbf{H}_O$ with respect to rotating frame $Oxyz$ to be computed from relations in Eq. (18.13)

$\mathbf{\Omega}$ = angular velocity of rotating frame $Oxyz$

If the rotating frame is attached to the body, its angular velocity $\mathbf{\Omega}$ is identically equal to the angular velocity $\mathbf{\omega}$ of the body. However, as indicated in the last paragraph of Sec. 18.2A, in many applications it is advantageous to use a frame of reference that is not actually attached to the body but rotates in an independent manner.

## *18.2D   Rotation of a Rigid Body About a Fixed Axis

We can use Eq. (18.28) to analyze the motion of a rigid body constrained to rotate about a fixed axis $AB$ (Fig. 18.13). First, we note that the angular velocity of the body with respect to the fixed frame $OXYZ$ is represented by the vector $\mathbf{\omega}$ directed along the axis of rotation. Attaching the moving frame of reference $Oxyz$ to the body, with the $z$ axis along $AB$, we have $\mathbf{\omega} = \omega\mathbf{k}$. Substituting $\omega_x = 0$, $\omega_y = 0$, and $\omega_z = \omega$ into the relations of Eq. (18.13), we obtain the components along the rotating axes of the angular momentum $\mathbf{H}_O$ of the body about $O$ as

$$H_x = -I_{xz}\omega \qquad H_y = -I_{yz}\omega \qquad H_z = I_z\omega$$

Because the frame $Oxyz$ is attached to the body, we have $\mathbf{\Omega} = \mathbf{\omega}$, and Eq. (18.28) yields

$$\begin{aligned}\Sigma\mathbf{M}_O &= (\dot{\mathbf{H}}_O)_{Oxyz} + \mathbf{\omega} \times \mathbf{H}_O \\ &= (-I_{xz}\mathbf{i} - I_{yz}\mathbf{j} + I_z\mathbf{k})\dot{\omega} + \omega\mathbf{k} \times (-I_{xz}\mathbf{i} - I_{yz}\mathbf{j} + I_z\mathbf{k})\omega \\ &= (-I_{xz}\mathbf{i} - I_{yz}\mathbf{j} + I_z\mathbf{k})\alpha + (-I_{xz}\mathbf{j} + I_{yz}\mathbf{i})\omega^2\end{aligned}$$

We can express this result by the three scalar equations

$$\begin{aligned}\Sigma M_x &= -I_{xz}\alpha + I_{yz}\omega^2 \\ \Sigma M_y &= -I_{yz}\alpha - I_{xz}\omega^2 \qquad (18.29) \\ \Sigma M_z &= I_z\alpha\end{aligned}$$

When the forces and moments applied to the body are known, you can obtain the angular acceleration $\alpha$ from the last of Eqs. (18.29). You can then determine the angular velocity $\omega$ by integration and substitute the values obtained for $\alpha$ and $\omega$ into the first two of Eqs. (18.29). You can then use these equations plus the three equations (18.26) that define the motion of the mass center of the body to determine the reactions at the bearings $A$ and $B$.

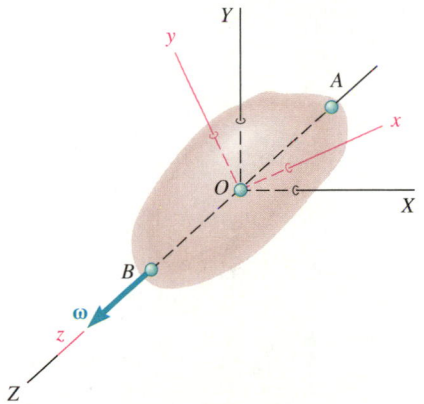

**Fig. 18.13**  Angular velocity of a rigid body rotating about a fixed axis $AB$.

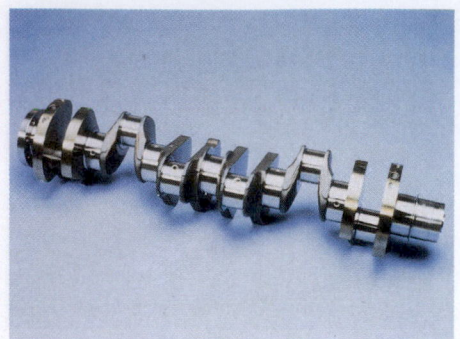

**Photo 18.4**  The rotating automobile crankshaft causes static and dynamic reactions on its bearings. The crankshaft can be designed to minimize dynamic imbalances and reduce these reaction forces. ©loraks/Getty Images RF

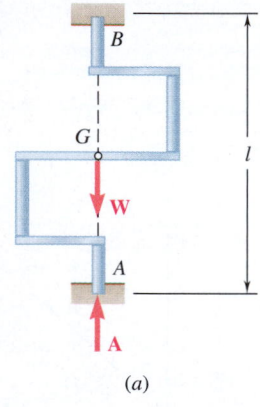

(a)

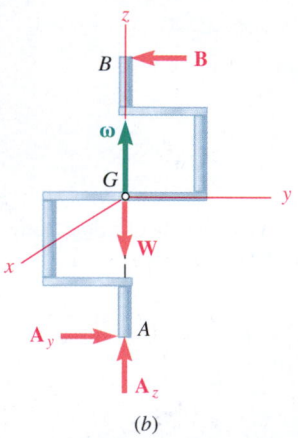

(b)

**Fig. 18.14**  (a) The crankshaft at rest is statically balanced; (b) the crankshaft rotating with constant angular velocity may or may not be dynamically balanced.

It is possible to select axes other than those shown in Fig. 18.13 to analyze the rotation of a rigid body about a fixed axis. In many cases, the principal axes of inertia of the body will be more advantageous. It is therefore a good idea to revert to Eq. (18.28) and select the system of axes that best fits the problem under consideration.

If the rotating body is symmetrical with respect to the $xy$ plane, the products of inertia $I_{xz}$ and $I_{yz}$ are equal to zero. Then Eqs. (18.29) reduce to

$$\Sigma M_x = 0 \qquad \Sigma M_y = 0 \qquad \Sigma M_z = I_z \alpha \qquad \text{(18.30)}$$

which is in agreement with the results obtained in Chap. 16. If, on the other hand, the products of inertia $I_{xz}$ and $I_{yz}$ are different from zero, the sum of the moments of the external forces about the $x$ and $y$ axes are also different from zero, even when the body rotates at a constant rate $\omega$. Indeed, in this case, Eqs. (18.29) yield

$$\Sigma M_x = I_{yz}\omega^2 \qquad \Sigma M_y = -I_{xz}\omega^2 \qquad \Sigma M_z = 0 \qquad \text{(18.31)}$$

This last observation leads us to discuss the **balancing of rotating shafts**. Consider, for instance, the crankshaft shown in Fig. 18.14*a* that is symmetrical about its mass center $G$. We first observe that, when the crankshaft is at rest, it exerts no lateral thrust on its supports, because its center of gravity $G$ is located directly above $A$. The shaft is said to be *statically balanced*. The reaction at $A$ is often referred to as a *static reaction* and is vertical, and its magnitude is equal to the weight $W$ of the shaft. Let us now assume that the shaft rotates with a constant angular velocity $\boldsymbol{\omega}$. Attaching our frame of reference to the shaft with its origin at $G$, the $z$ axis along $AB$, and the $y$ axis in the plane of symmetry of the shaft (Fig. 18.14*b*), we note that $I_{xz}$ is zero and that $I_{yz}$ is positive. According to Eqs. (18.31), there is an inertial term $I_{yz}\omega^2 \mathbf{i}$. Summing the moments about $G$ in the $x$ direction and applying Eq. (18.31), we have

$$\mathbf{A}_y = \frac{I_{yz}\omega^2}{l}\mathbf{j} \qquad \mathbf{B} = -\frac{I_{yz}\omega^2}{l}\mathbf{j} \qquad \text{(18.32)}$$

Because the bearing reactions are proportional to $\omega^2$, the shaft has a tendency to tear away from its bearings when rotating at high speeds. Moreover, because the bearing reactions $\mathbf{A}_y$ and $\mathbf{B}$, which are called *dynamic reactions,* are contained in the $yz$ plane, they rotate with the shaft and cause the structure supporting it to vibrate. These undesirable effects can be avoided if, by rearranging the distribution of mass around the shaft or by adding corrective masses, we let $I_{yz}$ become equal to zero. Then the dynamic reactions $\mathbf{A}_y$ and $\mathbf{B}$ vanish and the reactions at the bearings reduce to the static reaction $\mathbf{A}_z$—the direction of which is fixed. The shaft is then **dynamically as well as statically balanced**.

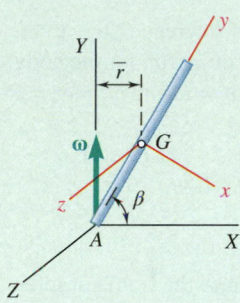

E

C                    B

$L = 8$ ft

$\beta = 60°$

A

D

**Fig. 1**   Angular velocity of the rod.

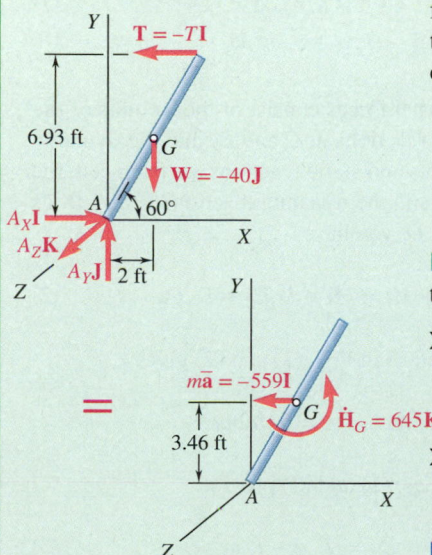

**Fig. 2**   Free-body diagram and kinetic diagram of the rod.

## Sample Problem 18.3

A slender rod $AB$ with a length of $L = 8$ ft and a weight of $W = 40$ lb is pinned at $A$ to a vertical axle $DE$ that rotates with a constant angular velocity $\boldsymbol{\omega}$ of 15 rad/s. The rod is maintained in position by means of a horizontal wire $BC$ attached to the axle and to end $B$ of the rod. Determine the tension in the wire and the reaction at $A$.

**STRATEGY:**   Because you have a rigid body that is not symmetrical with respect to the plane of motion, you need to use the three-dimensional form of Newton's second law.

**MODELING:**   Choose the rod $AB$ as your system. The angular velocity is shown in Fig. 1, and the free-body and kinetic diagrams consisting of the vector $m\bar{\mathbf{a}}$ attached at $G$ and the couple $\dot{\mathbf{H}}_G$ are shown in Fig. 2.

**ANALYSIS:**   Because $G$ describes a horizontal circle with a radius of $\bar{r} = \frac{1}{2}L\cos\beta$ and $BG$ rotates at the constant rate $\omega$ (Fig. 1), you have

$$\bar{\mathbf{a}} = \mathbf{a}_n = -\bar{r}\omega^2\mathbf{I} = -(\tfrac{1}{2}L\cos\beta)\omega^2\mathbf{I} = -(450\text{ ft/s}^2)\mathbf{I}$$

$$m\bar{\mathbf{a}} = \frac{40}{g}(-450\mathbf{I}) = -(559\text{ lb})\mathbf{I}$$

**Determination of $\dot{\mathbf{H}}_G$.**   First compute the angular momentum $\mathbf{H}_G$. Using the principal centroidal axes of inertia $x$, $y$, and $z$, you have

$$\bar{I}_x = \tfrac{1}{12}mL^2 \qquad \bar{I}_y = 0 \qquad \bar{I}_z = \tfrac{1}{12}mL^2$$
$$\omega_x = -\omega\cos\beta \qquad \omega_y = \omega\sin\beta \qquad \omega_z = 0$$
$$\mathbf{H}_G = \bar{I}_x\omega_x\mathbf{i} + \bar{I}_y\omega_y\mathbf{j} + \bar{I}_z\omega_z\mathbf{k}$$
$$\mathbf{H}_G = -\tfrac{1}{12}mL^2\omega\cos\beta\;\mathbf{i}$$

Obtain the rate of change $\dot{\mathbf{H}}_G$ of $\mathbf{H}_G$ with respect to axes of fixed orientation from Eq. (18.22). Observe that the rate of change $(\dot{\mathbf{H}}_G)_{Gxyz}$ of $\mathbf{H}_G$ with respect to the rotating frame $Gxyz$ is zero and that the angular velocity $\boldsymbol{\Omega}$ of that frame is equal to the angular velocity $\boldsymbol{\omega}$ of the rod. Thus, you have

$$\dot{\mathbf{H}}_G = (\dot{\mathbf{H}}_G)_{Gxyz} + \boldsymbol{\omega}\times\mathbf{H}_G$$
$$\dot{\mathbf{H}}_G = 0 + (-\omega\cos\beta\mathbf{i} + \omega\sin\beta\mathbf{j})\times(-\tfrac{1}{12}mL^2\omega\cos\beta\mathbf{i})$$
$$\dot{\mathbf{H}}_G = \tfrac{1}{12}mL^2\omega^2\sin\beta\cos\beta\mathbf{k} = (645\text{ lb·ft})\mathbf{k}$$

**Equations of Motion.**   The system of the external forces is equivalent to the inertia terms (Fig. 2). This gives

$$\Sigma\mathbf{M}_A = \dot{\mathbf{H}}_A = \mathbf{r}\times m\bar{\mathbf{a}} + \dot{\mathbf{H}}_G:$$
$$6.93\mathbf{J}\times(-T\mathbf{I}) + 2\mathbf{I}\times(-40\mathbf{J}) = 3.46\mathbf{J}\times(-559\mathbf{I}) + 645\mathbf{K}$$
$$(6.93T - 80)\mathbf{K} = (1934 + 645)\mathbf{K} \qquad\qquad T = 384\text{ lb} \;\blacktriangleleft$$
$$\Sigma\mathbf{F} = m\bar{\mathbf{a}}: \qquad A_X\mathbf{I} + A_Y\mathbf{J} + A_Z\mathbf{K} - 384\mathbf{I} - 40\mathbf{J} = -559\mathbf{I}$$
$$\mathbf{A} = -(175\text{ lb})\mathbf{I} + (40\text{ lb})\mathbf{J} \;\blacktriangleleft$$

**REFLECT and THINK:**   You could have obtained the value of $T$ from $\mathbf{H}_A$ and Eq. (18.28). Even though the rod rotates with a constant angular velocity, the asymmetry of the rod causes a moment about the $z$ axis. Note that we calculated the inertial term $\dot{\mathbf{H}}_A$ by adding $\mathbf{r}\times m\bar{\mathbf{a}}$ and the couple $\dot{\mathbf{H}}_G$.

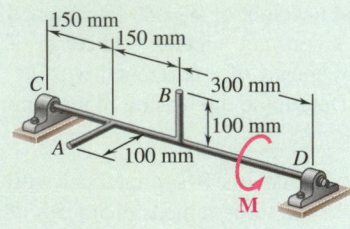

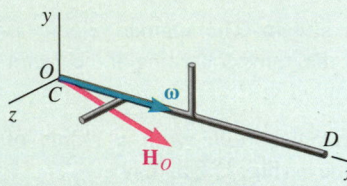

**Fig. 1**  Angular momentum and angular velocity of the system.

## Sample Problem 18.4

Two 100-mm rods $A$ and $B$ each have a mass of 300 g and are welded to shaft $CD$ that is supported by bearings at $C$ and $D$. If a couple $\mathbf{M}$ with a magnitude of 6 N·m is applied to the shaft, determine the components of the dynamic reactions at $C$ and $D$ at the instant when the shaft has reached an angular velocity of 1200 rpm. Neglect the moment of inertia of the shaft itself.

**STRATEGY:**  Use the three-dimensional form of Newton's second law in the form of Eq. (18.28) for the case of rotation about a fixed axis, where $\mathbf{\Omega} = \mathbf{\omega}$.

**MODELING:**  Choose the shaft and the two rods as your system. The angular momentum and angular velocity are shown in Fig. 1, and a free-body diagram is shown in Fig. 2.

**ANALYSIS:**

**Angular Momentum About O.**  Attach the frame of reference $Oxyz$ to the body and note that the axes chosen are not principal axes of inertia for the body. Because the body rotates about the $x$ axis, you know that $\omega_x = \omega$ and $\omega_y = \omega_z = 0$ (Fig. 1). Substituting into Eqs. (18.13), you have

$$H_x = I_x\omega \qquad H_y = -I_{xy}\omega \qquad H_z = -I_{xz}\omega$$
$$\mathbf{H}_O = (I_x\mathbf{i} - I_{xy}\mathbf{j} - I_{xz}\mathbf{k})\omega$$

**Moments of the External Forces About O.**  Because the frame of reference rotates with the angular velocity $\mathbf{\omega}$ and the only angular acceleration term is $\alpha_x = \alpha$, Eq. (18.28) gives

$$\begin{aligned}
\Sigma\mathbf{M}_O &= (\dot{\mathbf{H}}_O)_{Oxyz} + \mathbf{\omega} \times \mathbf{H}_O \\
&= (I_x\mathbf{i} - I_{xy}\mathbf{j} - I_{xz}\mathbf{k})\alpha + \omega\mathbf{i} \times (I_x\mathbf{i} - I_{xy}\mathbf{j} - I_{xz}\mathbf{k})\omega \qquad (1) \\
&= I_x\alpha\mathbf{i} - (I_{xy}\alpha - I_{xz}\omega^2)\mathbf{j} - (I_{xz}\alpha + I_{xy}\omega^2)\mathbf{k}
\end{aligned}$$

**Dynamic Reaction at D.**  The external forces consist of the weights of the shaft and rods, the couple $\mathbf{M}$, the static reactions at $C$ and $D$, and the dynamic reactions at $C$ and $D$. Because the weights and static reactions are balanced, the external forces reduce to the couple $\mathbf{M}$ and the dynamic reactions $\mathbf{C}$ and $\mathbf{D}$, as shown in Fig. 2. Taking moments about $O$, you have

$$\Sigma\mathbf{M}_O = L\mathbf{i} \times (D_y\mathbf{j} + D_z\mathbf{k}) + M\mathbf{i} = M\mathbf{i} - D_zL\mathbf{j} + D_yL\mathbf{k} \qquad (2)$$

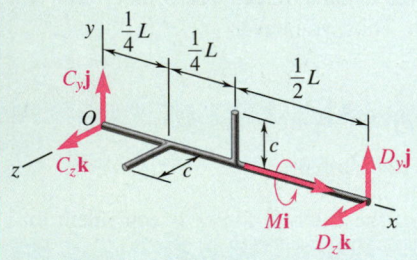

**Fig. 2**  Free-body diagram for the system.

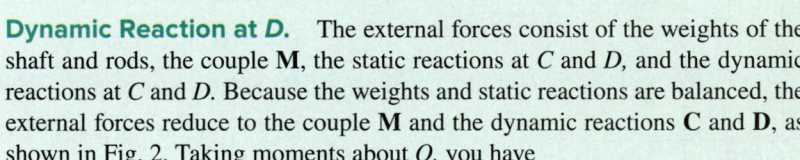

Equating the coefficients of the unit vector $\mathbf{i}$ in Eqs. (1) and (2) gives

$$M = I_x\alpha \qquad M = 2(\tfrac{1}{3}mc^2)\alpha \qquad \alpha = 3M/2mc^2$$

Equating the coefficients of $\mathbf{k}$ and $\mathbf{j}$ in Eqs. (1) and (2) provides

$$D_y = -(I_{xz}\alpha + I_{xy}\omega^2)/L \qquad D_z = (I_{xy}\alpha - I_{xz}\omega^2)/L \qquad (3)$$

*(continued)*

Using the parallel-axis theorem and noting that the product of inertia of each rod is zero with respect to their own centroidal axes, you obtain

$$I_{xy} = \Sigma m \overline{x}\,\overline{y} = m(\tfrac{1}{2}L)(\tfrac{1}{2}c) = \tfrac{1}{4}mLc$$
$$I_{xy} = \Sigma m \overline{x}\,\overline{z} = m(\tfrac{1}{4}L)(\tfrac{1}{2}c) = \tfrac{1}{8}mLc$$

Substituting into Eq. (3) the values found for $I_{xy}$, $I_{xz}$, and $\alpha$ gives

$$D_y = -\tfrac{3}{16}(M/c) - \tfrac{1}{4}mc\omega^2 \qquad D_z = \tfrac{3}{8}(M/c) - \tfrac{1}{8}mc\omega^2$$

Substituting $\omega = 1200$ rpm $= 125.7$ rad/s, $c = 0.100$ m, $M = 6$ N·m, and $m = 0.300$ kg, you have

$$D_y = -129.8 \text{ N} \qquad D_z = -36.8 \text{ N} \quad \blacktriangleleft$$

**Dynamic Reaction at C.** Using a frame of reference attached at $D$, you obtain equations similar to Eqs. (3) that yield

$$C_y = -152.2 \text{ N} \qquad C_z = -155.2 \text{ N} \quad \blacktriangleleft$$

**REFLECT and THINK:** The dynamic forces are larger at $C$ than at $D$. Rod $A$ is closer to this end of the bar, so you would expect it to affect this end more than the other end. Note that two small 300-g rods end up causing forces of over 150 N. You often have to account for these large forces when designing mechanical systems involving rotary equipment (e.g., automobiles, turbines, mills).

## Sample Problem 18.5

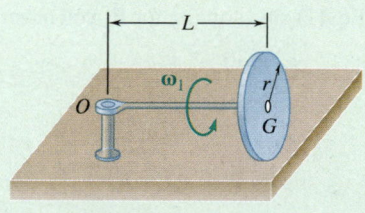

A homogeneous disk with radius $r$ and mass $m$ is mounted on an axle $OG$ with length $L$ and negligible mass. The axle is pivoted at the fixed point $O$, and the disk is constrained to roll on a horizontal surface. The disk rotates counterclockwise at the constant rate $\omega_1$ about the axle. Determine (a) the force (assumed vertical) exerted by the floor on the disk, (b) the reaction at the pivot $O$.

**STRATEGY:** Use the three-dimensional form of Newton's second law; that is, Eqs. (18.1) and (18.2).

**MODELING:** Choose the disk as your system and model it as a rigid body. The angular momentum and angular velocity are shown in Fig. 1, and free-body and kinetic diagrams consisting of the vector $m\overline{\mathbf{a}}$ attached at $G$ and the couple $\dot{\mathbf{H}}_G$ are shown in Fig. 2.

**ANALYSIS:** From Sample Prob. 18.2, the axle rotates about the $y$ axis at the rate $\omega_2 = r\omega_1/L$, so you have

$$m\overline{\mathbf{a}} = -mL\omega_2^2\mathbf{i} = -mL(r\omega_1/L)^2\mathbf{i} = -(mr^2\omega_1^2/L)\mathbf{i} \qquad (1)$$

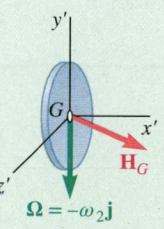

**Fig. 1** The angular momentum and angular velocity of the disk.

*(continued)*

**Determination of $\dot{\mathbf{H}}_G$.** Allow the $x$, $y$, $z$ axes to rotate with the bar $OG$ but not with the disk; the $x'$, $y'$, $z'$ axes rotate with both the bar and the disk. Recall from Sample Prob. 18.2 that the angular momentum of the disk about $G$ is

$$\mathbf{H}_G = \tfrac{1}{2}mr^2\omega_1\left(\mathbf{i} - \frac{r}{2L}\mathbf{j}\right)$$

where $\mathbf{H}_G$ is resolved into components along the rotating axes $x'$, $y'$, $z'$; $x'$ is along $OG$; and $y'$ is vertical at the instant shown (Fig. 1). Obtain the rate of change $\dot{\mathbf{H}}_G$ of $\mathbf{H}_G$ with respect to axes of fixed orientation from Eq. (18.22). Note that the rate of change $(\dot{\mathbf{H}}_G)_{Gx'y'z'}$ of $\mathbf{H}_G$ with respect to the rotating frame is zero and that the angular velocity $\boldsymbol{\Omega}$ of that frame is

$$\boldsymbol{\Omega} = -\omega_2\mathbf{j} = -\frac{r\omega_1}{L}\mathbf{j}$$

Then you have

$$\dot{\mathbf{H}}_G = (\dot{\mathbf{H}}_G)_{Gx'y'z'} + \boldsymbol{\Omega} \times \mathbf{H}_G$$

$$= 0 - \frac{r\omega_1}{L}\mathbf{j} \times \tfrac{1}{2}mr^2\omega_1\left(\mathbf{i} - \frac{r}{2L}\mathbf{j}\right)$$

$$= \tfrac{1}{2}mr^2(r/L)\omega_1^2\,\mathbf{k} \qquad (2)$$

**Equations of Motion.** The system of the external forces is equivalent to the system of the inertial terms (Fig. 2), so you have

$$\Sigma\mathbf{M}_O = \dot{\mathbf{H}}_G: \qquad L\mathbf{i} \times (N\mathbf{j} - W\mathbf{j}) = \dot{\mathbf{H}}_G$$

$$(N - W)L\mathbf{k} = \tfrac{1}{2}mr^2(r/L)\omega_1^2\mathbf{k}$$

$$N = W + \tfrac{1}{2}mr(r/L)^2\omega_1^2 \qquad \mathbf{N} = \left[W + \tfrac{1}{2}mr(r/L)^2\omega_1^2\right]\mathbf{j} \quad (3) \blacktriangleleft$$

$$\Sigma\mathbf{F} = m\bar{\mathbf{a}}: \qquad \mathbf{R} + N\mathbf{j} - W\mathbf{j} = m\bar{\mathbf{a}}$$

Substituting for $N$ from Eq. (3), for $m\bar{\mathbf{a}}$ from Eq. (1), and solving for $\mathbf{R}$, you have

$$\mathbf{R} = -(mr^2\omega_1^2/L)\mathbf{i} - \tfrac{1}{2}mr(r/L)^2\omega_1^2\mathbf{j}$$

$$\mathbf{R} = -\frac{mr^2\omega_1^2}{L}\left(\mathbf{i} + \frac{r}{2L}\mathbf{j}\right) \blacktriangleleft$$

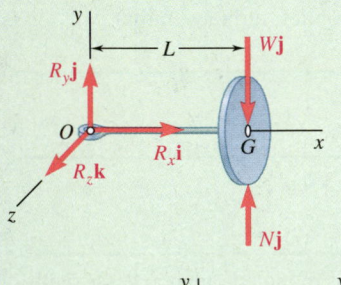

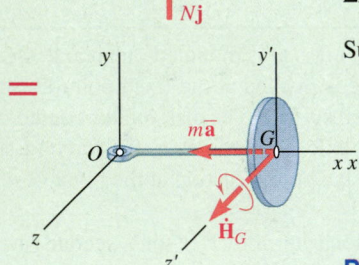

**Fig. 2** Free-body diagram and kinetic diagram for the system.

**REFLECT and THINK:** This is a case where the coordinate system attached to the rotating object has its own angular velocity. The change in direction of the angular momentum of the disk ends up increasing the normal force.

## CASE STUDY 18.1

Motion-based flight simulators can help train pilots to recognize situations in which they may become disoriented. The simulator shown in CS Fig. 18.1 has four degrees of freedom and can rotate a full 360 degrees about the planetary, the roll ($x$), the yaw ($y$), and the pitch ($z$) axes. The principal axes are aligned with the $xyz$ on the figure and $\bar{I}_x = 400$ slug·ft$^2$, $\bar{I}_y = 450$ slug·ft$^2$, and $\bar{I}_z = 900$ slug·ft$^2$. At the instant shown, the gondola is rotating at a constant 30 rpm about the roll axis, and 20 rpm about the pitch axis. The angular velocity about the pitch axis is increasing at a rate of 90 deg/s$^2$. Determine the dynamic loads at the roll gimbals at the instant shown.

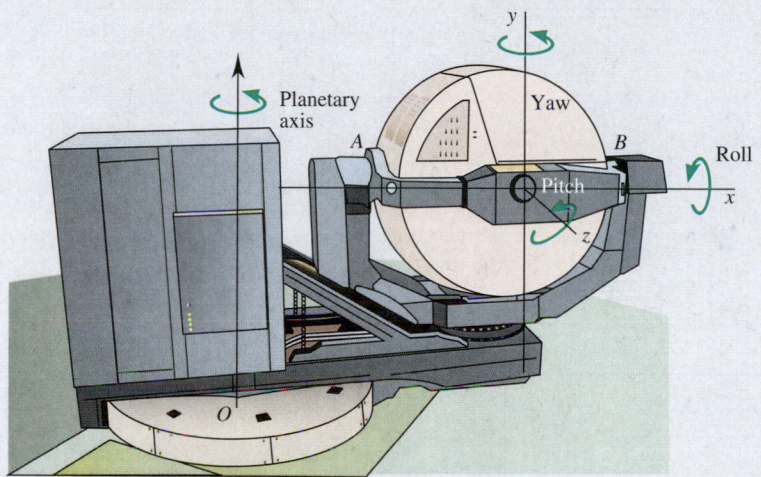

**CS Fig. 18.1**    Rotational capabilities of the flight simulator.

**STRATEGY:**    Forces will develop on the bearings $A$ and $B$ due to the motion of the simulator because of the changing angular momentum. You will need to determine the time derivative of the angular momentum of the gondola about its center of mass $G$ and equate this to the sum of moments about $G$.

**MODELING:**    Assuming that the manufacturers placed the center of mass at the intersection of the roll and pitch axes, we can draw our free-body and kinetic diagrams as shown in CS Fig. 18.2. You can place a moment $M_x$ at $A$ to represent the torque applied by the roll motor.

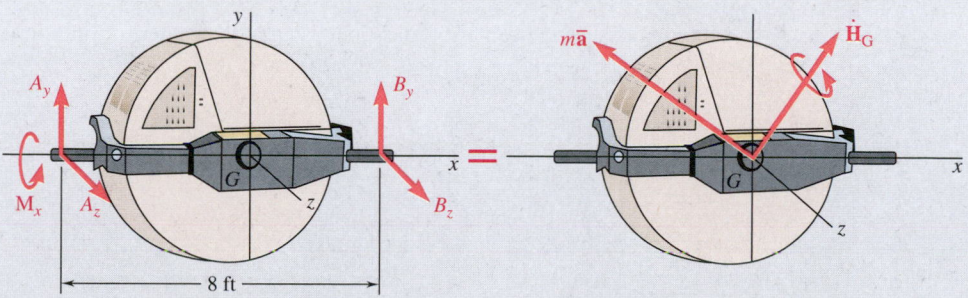

**CS Fig. 18.2**    Free-body and kinetic diagrams of the simulator.

*(continued)*

**ANALYSIS:** You can choose which point to sum moments about; the center of gravity $G$, point $A$, or point $B$ would all work well for this problem. Choosing $G$, you have

$$\Sigma \mathbf{M}_G = (\dot{\mathbf{H}}_G)_{Gxyz} + \mathbf{\Omega} \times \mathbf{H}_G \qquad (1)$$

Substituting in $\omega_{roll} = 30$ rpm $= 3.1416$ rad/s and $\omega_{pitch} = 20$ rpm $= 2.0944$ rad/s, the angular velocity of the gondola is

$$\mathbf{\omega} = \omega_{roll}\mathbf{i} + \omega_{pitch}\mathbf{k} = (3.1416 \text{ rad/s})\mathbf{i} + (2.0944 \text{ rad/s})\mathbf{k} \qquad (2)$$

Recognizing that $90$ deg/s$^2 = 1.5708$ rad/s$^2$, you can obtain the angular acceleration of the gondola:

$$\dot{\mathbf{\omega}} = (\dot{\mathbf{\omega}})_{xyz} + \mathbf{\omega}_{roll} \times \mathbf{\omega}_{pitch} = (1.5708 \text{ rad/s})\mathbf{k} + (3.1416 \text{ rad/s})\mathbf{i} \times (2.0944 \text{ rad/s})\mathbf{k}$$
$$\dot{\mathbf{\omega}} = (-6.2507 \text{ rad/s})\mathbf{j} + (1.5708 \text{ rad/s})\mathbf{k} \qquad (3)$$

To find the angular momentum of the gondola about point $G$, you will need to use the inertial properties of the device. Use these to obtain

$$\mathbf{H}_G = I_x\omega_{roll}\mathbf{i} + I_z\omega_{pitch}\mathbf{k} = (400 \text{ slug·ft}^2)(3.1416 \text{ rad/s})\mathbf{i} + (900 \text{ slug·ft}^2)(2.0944 \text{ rad/s})\mathbf{k}$$
$$\mathbf{H}_G = (1256.6 \text{ lb·ft·s})\mathbf{i} + (1885.0 \text{ lb·ft·s})\mathbf{k} \qquad (4)$$

Define a set of rotating axes aligned with the gondola so that the mass moments of inertia will be invariant with respect to your frame. Therefore, $\mathbf{\Omega} = \mathbf{\omega} = \omega_{roll}\mathbf{i} + \omega_{pitch}\mathbf{k}$. Now you can substitute Eqs. (3) and (4) into Eq. (1)

$$\Sigma \mathbf{M}_G = I_x\dot{\omega}_x\mathbf{i} + I_y\dot{\omega}_y\mathbf{j} + I_z\dot{\omega}_z\mathbf{k} + (\omega_{roll}\mathbf{i} + \omega_{pitch}\mathbf{k}) \times \mathbf{H}_G$$

Substituting in values,

$$\Sigma \mathbf{M}_G = (450 \text{ slug·ft}^2)(-6.2507 \text{ rad/s})\mathbf{j} + (900 \text{ slug·ft}^2)(1.5708 \text{ rad/s})\mathbf{k}$$
$$+ [(3.1416 \text{ rad/s})\mathbf{i} + (2.0944 \text{ rad/s})\mathbf{k}] \times [(1256.6 \text{ lb·ft·s})\mathbf{i} + (1885.0 \text{ lb·ft·s})\mathbf{k}]$$
$$\Sigma \mathbf{M}_G = (-6250.7 \text{ ft·lb})\mathbf{j} + (1413.7 \text{ ft·lb})\mathbf{k} \qquad (5)$$

Because the center of mass is located at the center of rotation, it isn't accelerating and you have $m\bar{\mathbf{a}} = 0$ and

$$\Sigma \mathbf{F} = 0: \quad A_y = -B_y \quad \text{and} \quad A_z = -B_z \qquad (6)$$

*(continued)*

From the free-body diagram, you can now determine the left-hand side of Eq. (5).

$$\Sigma \mathbf{M}_G = \mathbf{r}_{A/G} \times \mathbf{A} + \mathbf{r}_{B/G} \times \mathbf{B} + M_x = (-4 \text{ ft})\mathbf{i} \times (A_y\mathbf{j} + A_z\mathbf{k}) + (4 \text{ ft})\mathbf{i} \times (B_y\mathbf{j} + B_z\mathbf{k}) + M_x\mathbf{i}$$
$$= 4A_z\mathbf{j} - 4A_y\mathbf{k} - 4B_z\mathbf{j} + 4B_y\mathbf{k} + M_x\mathbf{i} \tag{7}$$

Substituting Eq. (6) into Eq. (7) and setting the result equal to Eq. (5) to get the dynamic forces at $B$ gives you

$$-8B_z\mathbf{j} + 8B_y\mathbf{k} + M_x\mathbf{i} = (-6250.7 \text{ ft·lb})\mathbf{j} + (1413.7 \text{ ft·lb})\mathbf{k}$$

Because there are no inertial components in the $x$ direction, $M_x = 0$. Solving for $B_y$ and $B_z$, you obtain $B_y = 176.71$ lb and $B_z = 781.34$ lb.
Using Eq. (6), $A_y = -176.71$ lb and $A_z = -781.34$ lb.

**REFLECT and THINK:**  Dynamic loads on bearings have to be considered when designing equipment with multiple axes of rotation. Products of inertia might introduce additional loads, and could even require additional torque to be input to the motors to obtain desired motion profiles. Also, if you add rotation about the planetary axis, then the linear acceleration of the center of mass will no longer be zero and the angular velocity and angular acceleration will need to be calculated and included in the force equations. When designing this system, placing the center of mass at the center of rotation and minimizing the products of inertia reduces the reaction forces on the bearings; this is part of the machine balancing process that you typically do in rotating machinery.

# SOLVING PROBLEMS ON YOUR OWN

In this section, you were asked to solve problems involving the three-dimensional motion of rigid bodies. The method you used is basically the same one you used in Chap. 16 in your study of the plane motion of rigid bodies. You made free-body and kinetic diagrams showing that the system of the external forces is equivalent to the system of the inertia terms. You equated sums of components and sums of moments on both sides of this equation. Now, however, the system of the inertia terms is represented by the vector $m\bar{\mathbf{a}}$ and a couple vector $\dot{\mathbf{H}}_G$, which are explained next in paragraphs 1 and 2.

To solve a problem involving the three-dimensional motion of a rigid body, you should take the following steps.

**1. Determine the angular momentum $\mathbf{H}_G$ of the body about its mass center $G$** from its angular velocity $\boldsymbol{\omega}$ with respect to a frame of reference $GX'Y'Z'$ of fixed orientation. This is an operation you learned to perform in Sec. 18.1. However, because the configuration of the body is changing with time, it is now necessary for you to use an auxiliary system of axes $Gx'y'z'$ (Fig. 18.9) to compute the components of $\boldsymbol{\omega}$ and the moments and products of inertia of the body. These axes may be rigidly attached to the body, in which case their angular velocity is equal to $\boldsymbol{\omega}$ (Sample Probs. 18.3 and 18.4), or they may have an angular velocity $\boldsymbol{\Omega}$ of their own (Sample Prob. 18.5).

Recall the following ideas from the preceding section.

**a. If you know the principal axes of inertia of the body at $G$,** use these axes as coordinate axes $x'$, $y'$, and $z'$, because the corresponding products of inertia of the body are equal to zero. (Note that if the body is axisymmetric, these axes do not need to be rigidly attached to the body.) Resolve $\boldsymbol{\omega}$ into components $\omega_{x'}$, $\omega_{y'}$, and $\omega_{z'}$ along these axes and compute the principal moments of inertia $\bar{I}_{x'}, \bar{I}_{y'}$, and $\bar{I}_{z'}$. The corresponding components of the angular momentum $\mathbf{H}_G$ are

$$H_{x'} = \bar{I}_{x'}\omega_{x'} \qquad H_{y'} = \bar{I}_{y'}\omega_{y'} \qquad H_{z'} = \bar{I}_{z'}\omega_{z'} \qquad \textbf{(18.10)}$$

**b. If you do not know the principal axes of inertia of the body at $G$,** you must use Eqs. (18.7) to determine the components of the angular momentum $\mathbf{H}_G$. These equations require your prior computation of the products of inertia of the body—as well as of its moments of inertia—with respect to the selected axes.

**2. Compute the rate of change $\dot{\mathbf{H}}_G$ of the angular momentum $\mathbf{H}_G$ with respect to the frame $GX'Y'Z'$.** Note that this frame has a *fixed orientation*, whereas the frame $Gx'y'z'$ you used when calculating the components of the vector $\boldsymbol{\omega}$ was a *rotating frame*. (Review the discussion in Sec. 15.5A of the rate of change of a vector with respect to a rotating frame.) Recalling Eq. (15.31), you can express the rate of change $\dot{\mathbf{H}}_G$ as

$$\dot{\mathbf{H}}_G = (\dot{\mathbf{H}}_G)_{Gx'y'z'} + \boldsymbol{\Omega} \times \mathbf{H}_G \tag{18.22}$$

The first term in the right-hand side of Eq. (18.22) represents the rate of change of $\mathbf{H}_G$ with respect to the rotating frame $Gx'y'z'$. This term drops out if $\boldsymbol{\omega}$—and thus $\mathbf{H}_G$—remains constant in both magnitude and direction when viewed from that frame. On the other hand, if any of the time derivatives $\dot{\omega}_{x'}, \dot{\omega}_{y'}$, or $\dot{\omega}_{z'}$ is different from zero, $(\dot{\mathbf{H}}_G)_{Gx'y'z'}$ is also different from zero, and you should determine its components by differentiating Eqs. (18.10) with respect to $t$. Finally, we remind you that if the rotating frame is rigidly attached to the body, its angular velocity is the same as that of the body, and $\boldsymbol{\Omega}$ can be replaced by $\boldsymbol{\omega}$.

**3. Draw the free-body and kinetic diagrams for the rigid body** showing that the system of the external forces exerted on the body is equivalent to the vector $m\bar{\mathbf{a}}$ applied at $G$ and the couple vector $\dot{\mathbf{H}}_G$ (Fig. 18.11). By equating components in any direction and moments about any point, you can write as many as six independent scalar equations of motion (Sample Probs. 18.3 and 18.5). You can sum moments about any point $P$ by applying the equation

$$\Sigma\mathbf{M}_P = \dot{\mathbf{H}}_G + \mathbf{r}_{G/P} \times m\bar{\mathbf{a}}$$

**4. When solving problems involving the motion of a rigid body about a fixed point $O$,** you may find it convenient to use the following equation that was derived in Sec. 18.2C and eliminates the components of the reaction at the support $O$. So

$$\Sigma\mathbf{M}_O = (\dot{\mathbf{H}}_O)_{Oxyz} + \boldsymbol{\Omega} \times \mathbf{H}_O \tag{18.28}$$

Here the first term on the right-hand side represents the rate of change of $\mathbf{H}_O$ with respect to the rotating frame $Oxyz$, and $\boldsymbol{\Omega}$ is the angular velocity of that frame.

**5. When determining the reactions at the bearings of a rotating shaft,** use Eq. (18.28) and take the following steps.

    **a. Place the fixed point $O$ at one of the two bearings supporting the shaft** and attach the rotating frame $Oxyz$ to the shaft with one of the axes directed along it. Assuming, for instance, that the $x$ axis has been aligned with the shaft, you will have $\boldsymbol{\Omega} = \boldsymbol{\omega} = \omega\mathbf{i}$ (Sample Prob. 18.4).

    **b. Because the selected axes are usually not the principal axes of inertia at $O$,** you must compute the products of inertia of the shaft—as well as its moments of inertia—with respect to these axes and use Eqs. (18.13) to determine $\mathbf{H}_O$. Assuming again that the $x$ axis has been aligned with the shaft, Eqs. (18.13) reduce to

$$H_x = I_x\omega \qquad H_y = -I_{yx}\omega \qquad H_z = -I_{zx}\omega \tag{18.13'}$$

These equations show that $\mathbf{H}_O$ is not directed along the shaft.

*(continued)*

**c. To obtain $\dot{\mathbf{H}}_O$, substitute these expressions into Eq. (18.28),** and let $\Omega = \omega = \omega\mathbf{i}$. If the angular velocity of the shaft is constant, the first term in the right-hand side of the equation drops out. However, if the shaft has an angular acceleration $\boldsymbol{\alpha} = \alpha\mathbf{i}$, the first term is not zero and must be determined by differentiating the expressions in Eq. (18.13′) with respect to $t$. The result will be equations similar to Eqs. (18.13′) with $\omega$ replaced by $\alpha$. The result also can be expressed by the three scalar equations of Eq. (18.29).

**d. Because point $O$ coincides with one of the bearings,** you can solve the three scalar equations corresponding to Eq. (18.28) for the components of the dynamic reaction at the other bearing. If the mass center $G$ of the shaft is located on the line joining the two bearings, the inertial term $m\bar{\mathbf{a}}$ is zero. Drawing the free-body diagram and kinetic diagram of the shaft, you then observe that the components of the dynamic reaction at the first bearing must be equal and opposite to those you have just determined. If $G$ is not located on the line joining the two bearings, you can determine the reaction at the first bearing by placing the fixed point $O$ at the second bearing and repeating the earlier procedure (Sample Prob. 18.4); or you can obtain additional equations of motion from the free-body and kinetic diagrams of the shaft, making sure to first determine and include the inertial term $m\bar{\mathbf{a}}$ applied at $G$.

**e. Most problems call for the determination of the "dynamic reactions"** at the bearings; that is, for the additional forces exerted by the bearings on the shaft when the shaft is rotating. When determining dynamic reactions, ignore the effect of static loads, such as the weight of the shaft.

# Problems

**18.55** Determine the rate of change $\dot{\mathbf{H}}_G$ of the angular momentum $\mathbf{H}_G$ of the disk of Prob. 18.1.

**18.56** Determine the rate of change $\dot{\mathbf{H}}_A$ of the angular momentum $\mathbf{H}_A$ of the square plate of Prob. 18.2.

**18.57** Determine the rate of change $\dot{\mathbf{H}}_G$ of the angular momentum $\mathbf{H}_G$ of rod *AB* of Prob. 18.3.

**18.58** Determine the rate of change $\dot{\mathbf{H}}_A$ of the angular momentum $\mathbf{H}_A$ of the disk of Prob. 18.4.

**18.59** Determine the rate of change $\dot{\mathbf{H}}_A$ of the angular momentum $\mathbf{H}_A$ of the disk of Prob. 18.5.

**18.60** Determine the rate of change $\dot{\mathbf{H}}_G$ of the angular momentum $\mathbf{H}_G$ of the disk of Prob. 18.8 for an arbitrary value of $\beta$, knowing that its angular velocity $\boldsymbol{\omega}$ remains constant.

**18.61** Determine the rate of change $\dot{\mathbf{H}}_D$ of the angular momentum $\mathbf{H}_D$ of the 3-lb rod *CDE*, assuming that at the instant considered the assembly has an angular velocity $\boldsymbol{\omega} = (12 \text{ rad/s})\mathbf{i}$ and an angular acceleration $\boldsymbol{\alpha} = -(96 \text{ rad/s}^2)\mathbf{i}$.

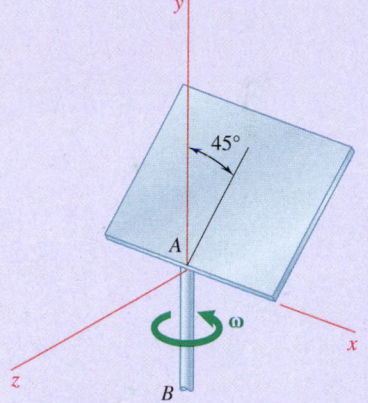

**Fig. P18.63**

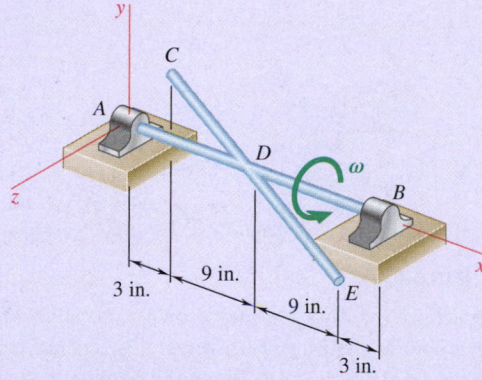

**Fig. P18.61 and P18.62**

**18.62** Determine the rate of change $\dot{\mathbf{H}}_D$ of the angular momentum $\mathbf{H}_D$ of the 3-lb rod *CDE*, assuming that at the instant considered the assembly has an angular velocity $\boldsymbol{\omega} = (12 \text{ rad/s})\mathbf{i}$ and an angular acceleration $\boldsymbol{\alpha} = (96 \text{ rad/s}^2)\mathbf{i}$.

**18.63** A thin, homogeneous square of mass $m$ and side $a$ is welded to a vertical shaft *AB* with which it forms an angle of 45°. Knowing that the shaft rotates with an angular velocity $\boldsymbol{\omega} = \omega\mathbf{j}$ and an angular acceleration $\boldsymbol{\alpha} = \alpha\mathbf{j}$, determine the rate of change $\dot{\mathbf{H}}_A$ of the angular momentum $\mathbf{H}_A$ of the plate assembly.

**18.64** Determine the rate of change $\dot{\mathbf{H}}_G$ of the angular momentum $\mathbf{H}_G$ of the disk of Prob. 18.8 for an arbitrary value of $\beta$, knowing that the disk has an angular velocity $\boldsymbol{\omega} = \omega\mathbf{i}$ and an angular acceleration $\boldsymbol{\alpha} = \alpha\mathbf{i}$.

**18.65** A slender, uniform rod *AB* of mass $m$ and a vertical shaft *CD*, each of length $2b$, are welded together at their midpoints $G$. Knowing that the shaft rotates at the constant rate $\omega$, determine the dynamic reactions at $C$ and $D$.

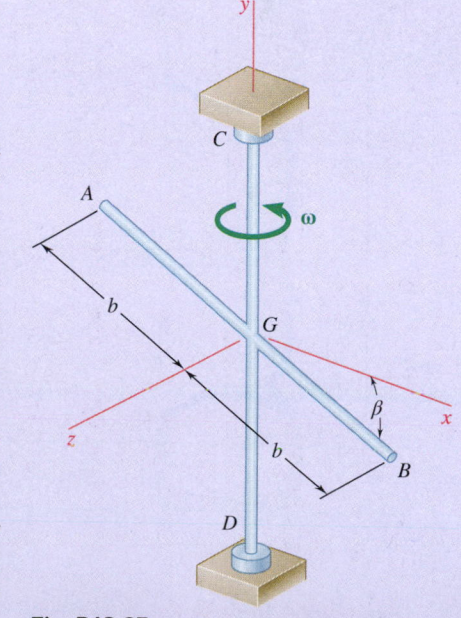

**Fig. P18.65**

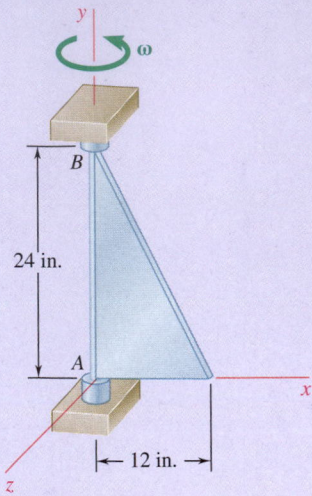

**Fig. P18.66**

**18.66** A thin, homogeneous triangular plate of weight 10 lb is welded to a light, vertical axle supported by bearings at $A$ and $B$. Knowing that the plate rotates at the constant rate $\omega = 8$ rad/s, determine the dynamic reactions at $A$ and $B$.

**18.67** The assembly shown consists of pieces of sheet aluminum of uniform thickness and of total weight 2.7 lb welded to a light axle supported by bearings at $A$ and $B$. Knowing that the assembly rotates at the constant rate $\omega = 240$ rpm, determine the dynamic reactions at $A$ and $B$.

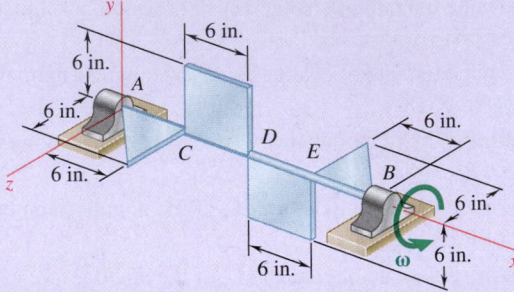

**Fig. P18.67**

**18.68** The 8-kg shaft shown has a uniform cross-section. Knowing that the shaft rotates at the constant rate $\omega = 12$ rad/s, determine the dynamic reactions at $A$ and $B$.

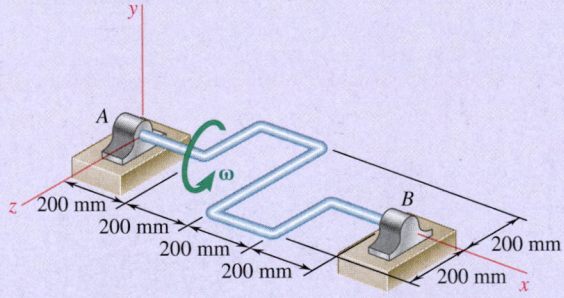

**Fig. P18.68**

**18.69** After attaching the 18-kg wheel shown to a balancing machine and making it spin at the rate of 15 rev/s, a mechanic has found that to balance the wheel both statically and dynamically, he should use two corrective masses, a 170-g mass placed at $B$ and a 56-g mass placed at $D$. Using a right-handed frame of reference rotating with the wheel (with the $z$ axis perpendicular to the plane of the figure), determine before the corrective masses have been attached (*a*) the distance from the axis of rotation to the mass center of the wheel and the products of inertia $I_{xy}$ and $I_{zx}$, (*b*) the force-couple system at $C$ equivalent to the forces exerted by the wheel on the machine.

**18.70** When the 18-kg wheel shown is attached to a balancing machine and made to spin at a rate of 12.5 rev/s, it is found that the forces exerted by the wheel on the machine are equivalent to a force-couple system consisting of a force $\mathbf{F} = (160 \text{ N})\mathbf{j}$ applied at $C$ and a couple $\mathbf{M}_C = (14.7 \text{ N·m})\mathbf{k}$, where the unit vectors form a triad that rotates with the wheel. (*a*) Determine the distance from the axis of rotation to the mass center of the wheel and the products of inertia $I_{xy}$ and $I_{zx}$. (*b*) If only two corrective masses are to be used to balance the wheel statically and dynamically, what should these masses be and at which of the points $A$, $B$, $D$, or $E$ should they be placed?

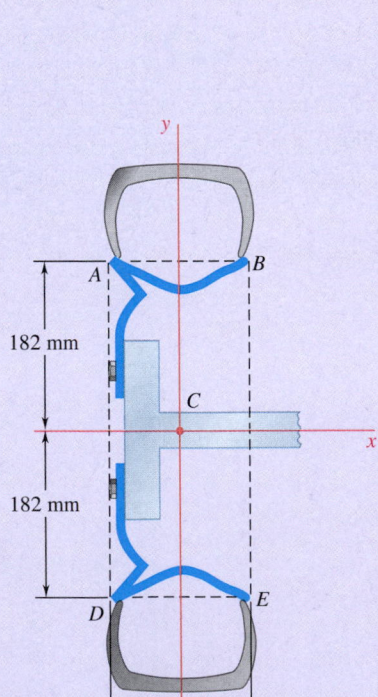

**Fig. P18.69 and P18.70**

**18.71** Knowing that the assembly of Prob. 18.65 is initially at rest ($\omega = 0$) when a couple of moment $\mathbf{M}_0 = M_0\mathbf{j}$ is applied to shaft $CD$, determine (*a*) the resulting angular acceleration of the assembly, (*b*) the dynamic reactions at $C$ and $D$ immediately after the couple is applied.

**18.72** Knowing that the plate of Prob. 18.66 is initially at rest ($\omega = 0$) when a couple of moment $\mathbf{M}_0 = (0.75 \text{ ft·lb})\mathbf{j}$ is applied to it, determine (*a*) the resulting angular acceleration of the plate, (*b*) the dynamic reactions $A$ and $B$ immediately after the couple has been applied.

**18.73** The assembly of Prob. 18.67 is initially at rest ($\omega = 0$) when a couple $\mathbf{M}_0$ is applied to axle $AB$. Knowing that the resulting angular acceleration of the assembly is $\boldsymbol{\alpha} = (150 \text{ rad/s}^2)\mathbf{i}$, determine (*a*) the couple $\mathbf{M}_0$, (*b*) the dynamic reactions at $A$ and $B$ immediately after the couple is applied.

**18.74** The shaft of Prob. 18.68 is initially at rest ($\omega = 0$) when a couple $\mathbf{M}_0$ is applied to it. Knowing that the resulting angular acceleration of the shaft is $\boldsymbol{\alpha} = (20 \text{ rad/s}^2)\mathbf{i}$, determine (*a*) the couple $\mathbf{M}_0$, (*b*) the dynamic reactions at $A$ and $B$ immediately after the couple is applied.

**18.75** The assembly shown weighs 12 lb and consists of 4 thin 16-in.-diameter semicircular aluminum plates welded to a light 40-in.-long shaft $AB$. The assembly is at rest ($\omega = 0$) at time $t = 0$ when a couple $\mathbf{M}_0$ is applied to it as shown, causing the assembly to complete one full revolution in 2 s. Determine (*a*) the couple $\mathbf{M}_0$, (*b*) the dynamic reactions at $A$ and $B$ at $t = 0$.

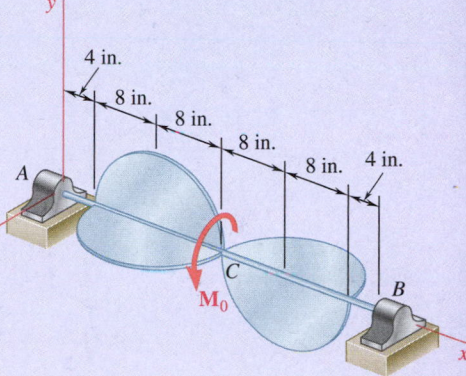

**Fig. P18.75**

**18.76** For the assembly of Prob. 18.75, determine the dynamic reactions at $A$ and $B$ at $t = 2$ s.

**18.77** The sheet-metal component shown is of uniform thickness and has a mass of 600 g. It is attached to a light axle supported by bearings at $A$ and $B$ located 150 mm apart. The component is at rest when it is subjected to a couple $\mathbf{M}_0$ as shown. If the resulting angular acceleration is $\boldsymbol{\alpha} = (12 \text{ rad/s}^2)\mathbf{k}$, determine (*a*) the couple $\mathbf{M}_0$, (*b*) the dynamic reactions $A$ and $B$ immediately after the couple has been applied.

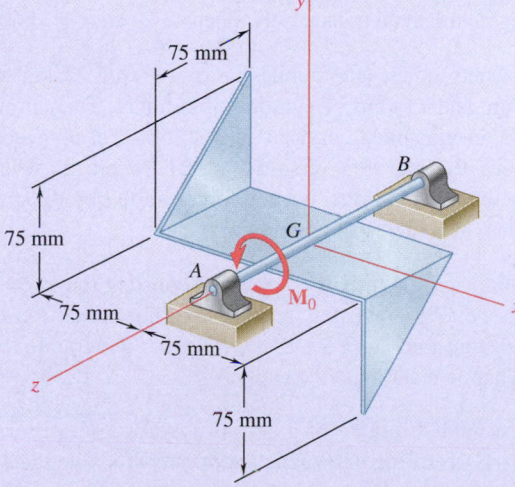

**Fig. *P18.77***

**18.78** For the sheet-metal component of Prob. 18.77, determine (*a*) the angular velocity of the component 0.6 s after the couple $\mathbf{M}_0$ has been applied to it, (*b*) the magnitude of the dynamic reactions at $A$ and $B$ at that time.

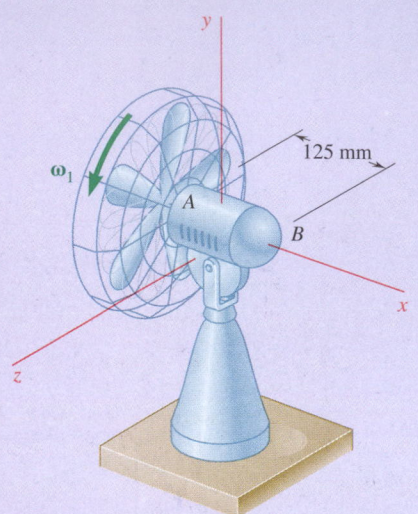

Fig. P18.79

**18.79** The blade of an oscillating fan and the rotor of its motor have a total mass of 300 g and a combined radius of gyration of 75 mm. They are supported by bearings at $A$ and $B$, 125 mm apart, and rotate at the rate $\omega_1 = 1800$ rpm. Determine the dynamic reactions at $A$ and $B$ when the motor casing has an angular velocity $\omega_2 = (0.6 \text{ rad/s})\mathbf{j}$.

**18.80** The blade of a portable saw and the rotor of its motor have a total weight of 2.5 lb and a combined radius of gyration of 1.5 in. Knowing that the blade rotates as shown at the rate $\omega_1 = 1500$ rpm, determine the magnitude and direction of the couple $\mathbf{M}$ that a worker must exert on the handle of the saw to rotate it with a constant angular velocity $\omega_2 = -(2.4 \text{ rad/s})\mathbf{j}$.

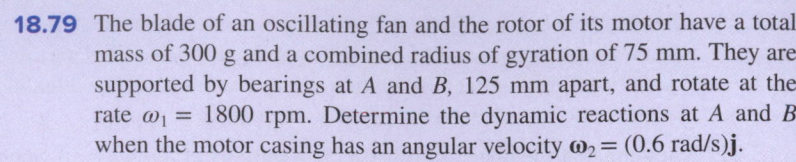

Fig. *P18.80*

**18.81** The flywheel of an automobile engine, which is rigidly attached to the crankshaft, is equivalent to a 400-mm-diameter, 15-mm-thick steel plate. Determine the magnitude of the couple exerted by the flywheel on the horizontal crankshaft as the automobile travels around an unbanked curve of 200-m radius at a speed of 90 km/h, with the flywheel rotating at 2700 rpm. Assume the automobile to have (*a*) a rear-wheel drive with the engine mounted longitudinally, (*b*) a front-wheel drive with the engine mounted transversely. (Density of steel = 7860 kg/m³)

**18.82** Each wheel of an automobile has a mass of 22 kg, a diameter of 575 mm, and a radius of gyration of 225 mm. The automobile travels around an unbanked curve of radius 150 m at a speed of 95 km/h. Knowing that the transverse distance between the wheels is 1.5 m, determine the additional normal force exerted by the ground on each outside wheel due to the motion of the car.

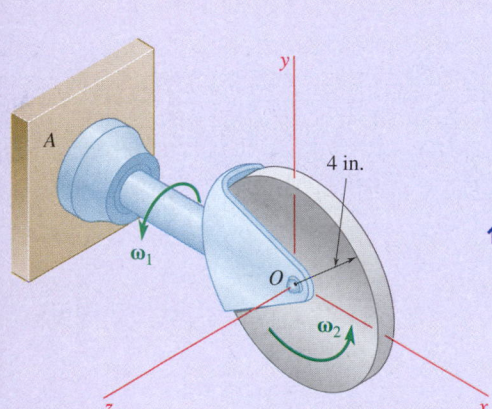

Fig. *P18.83*

**18.83** The uniform, thin 5-lb disk spins at a constant rate $\omega_2 = 6$ rad/s about an axis held by a housing attached to a horizontal rod that rotates at the constant rate $\omega_1 = 3$ rad/s. Determine the couple that represents the dynamic reaction at the support $A$.

**18.84** The essential structure of a certain type of aircraft turn indicator is shown. Each spring has a constant of 500 N/m, and the 200-g uniform disk of 40-mm radius spins at the rate of 10 000 rpm. The springs are stretched and exert equal vertical forces on yoke $AB$ when the airplane is traveling in a straight path. Determine the angle through which the yoke will rotate when the pilot executes a horizontal turn of 750-m radius to the right at a speed of 800 km/h. Indicate whether point $A$ will move up or down.

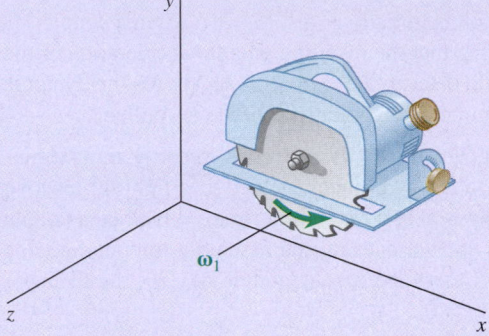

Fig. P18.84

**18.85** A model of a type of crusher is shown. A disk of weight $W$ is mounted on a shaft $AB$ about which it can rotate freely. Shaft $AB$ is attached by means of a clevis to a vertical shaft, which is made to rotate at a constant angular velocity $\omega_1$. The disk rolls on the inside of a vertical cylinder (only one half of the cylinder is shown). Determine the minimum angular velocity $\omega_1$ for which contact is maintained between the disk and the cylinder.

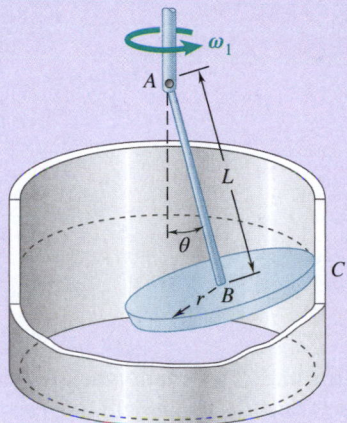

**Fig. P18.85**

**18.86** A uniform square plate with side $a = 225$ mm is hinged at points $A$ and $B$ to a clevis that rotates with a constant angular velocity $\omega$ about a vertical axis. Determine (*a*) the constant angle $\beta$ that the plate forms with the horizontal $x$ axis when $\omega = 12$ rad/s, (*b*) the largest value of $\omega$ for which the plate remains vertical ($\beta = 90°$).

**18.87** A uniform square plate with side $a = 300$ mm is hinged at points $A$ and $B$ to a clevis that rotates with a constant angular velocity $\omega$ about a vertical axis. Determine (*a*) the value of $\omega$ for which the plate forms a constant angle $\beta = 60°$ with the horizontal $x$ axis, (*b*) the largest value of $\omega$ for which the plate remains vertical ($\beta = 90°$).

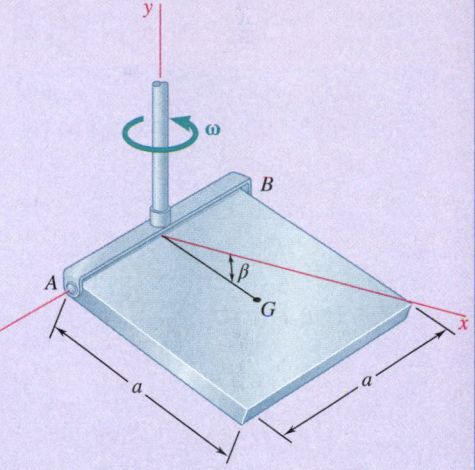

**Fig. P18.86 and P18.87**

**18.88** The 2-lb gear $A$ is constrained to roll on the fixed gear $B$ but is free to rotate about axle $AD$. Axle $AD$ has a length of 20 in., a negligible weight, and is connected by a clevis to the vertical shaft $DE$ that rotates as shown with a constant angular velocity $\omega_1$. Assuming that gear $A$ can be approximated by a thin disk with a radius of 4 in., determine the largest allowable value of $\omega_1$ if gear $A$ is not to lose contact with gear $B$.

**18.89** Determine the force $\mathbf{F}$ exerted by gear $B$ on gear $A$ of Prob. 18.88 when shaft $DE$ rotates with the constant angular speed of $\omega_1 = 4$ rad/s. (*Hint:* The force $\mathbf{F}$ must be perpendicular to the line drawn from $D$ to $C$.)

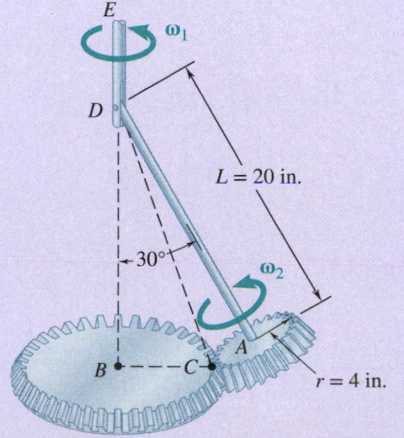

**Fig. P18.88**

**18.90 and 18.91** The slender rod $AB$ is attached by a clevis to arm $BCD$ that rotates with a constant angular velocity $\omega$ about the centerline of its vertical portion $CD$. Determine the magnitude of the angular velocity $\omega$.

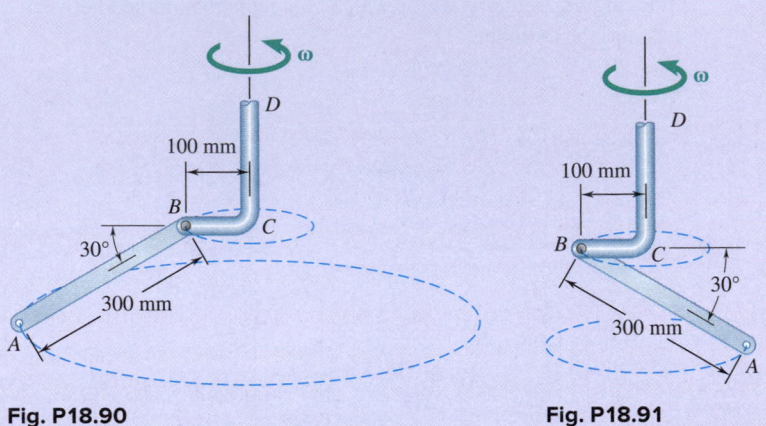

Fig. P18.90          Fig. P18.91

**18.92** The essential structure of a certain type of aircraft turn indicator is shown. Springs $AC$ and $BD$ are initially stretched and exert equal vertical forces at $A$ and $B$ when the airplane is traveling in a straight path. Each spring has a constant of 600 N/m and the uniform disk has a mass of 250 g and spins at the rate of 12 000 rpm. Determine the angle through which the yoke will rotate when the pilot executes a horizontal turn of 800-m radius to the right at a speed of 720 km/h. Indicate whether point $A$ will move up or down.

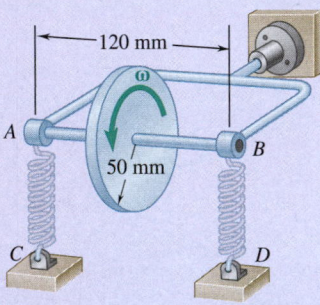

Fig. *P18.92*

**18.93** The 10-oz disk shown spins at the rate $\omega_1 = 750$ rpm, while axle $AB$ rotates as shown with an angular velocity $\omega_2$ of magnitude 6 rad/s. Determine the dynamic reactions at $A$ and $B$.

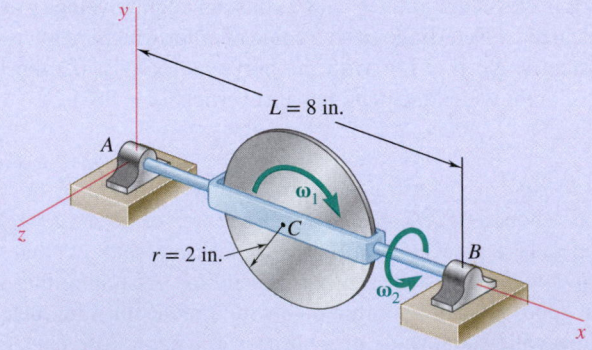

Fig. P18.93 and P18.94

**18.94** The 10-oz disk shown spins at the rate $\omega_1 = 750$ rpm, while axle $AB$ rotates as shown with an angular velocity $\omega_2$. Determine the maximum allowable magnitude of $\omega_2$ if the dynamic reactions at $A$ and $B$ are not to exceed 0.25 lb each.

**18.95** Two disks each have a mass of 5 kg and a radius 300 mm. They spin as shown at the rate of $\omega_1 = 1200$ rpm about a rod $AB$ of negligible mass that rotates about the horizontal $z$ axis at the rate of $\omega_2 = 60$ rpm. (a) Determine the dynamic reactions at points $C$ and $D$. (b) Solve part (a) assuming that the direction of spin of disk $A$ is reversed.

**18.96** Two disks each have a mass of 5 kg and a radius of 300 mm. They spin as shown at the rate of $\omega_1 = 1200$ rpm about a rod $AB$ of negligible mass that rotates about the horizontal $z$ axis at the rate $\omega_2$. Determine the maximum allowable value of $\omega_2$ if the magnitudes of the dynamic reactions at points $C$ and $D$ are not to exceed 350 N each.

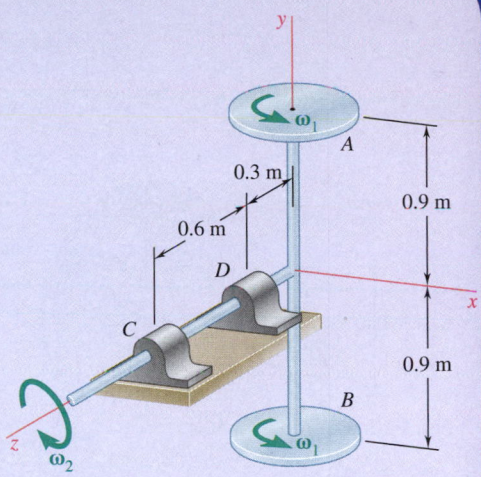

**Fig. P18.95 and P18.96**

**18.97** A stationary horizontal plate is attached to the ceiling by means of a fixed vertical tube. A wheel of radius $a$ and mass $m$ is mounted on a light axle $AC$ that is attached by means of a clevis at $A$ to a rod $AB$ fitted inside the vertical tube. The rod $AB$ is made to rotate with a constant angular velocity $\Omega$, causing the wheel to roll on the lower face of the stationary plate. Determine the minimum angular velocity $\Omega$ for which contact is maintained between the wheel and the plate. Consider the particular cases (a) when the mass of the wheel is concentrated in the rim, (b) when the wheel is equivalent to a thin disk of radius $a$.

**18.98** Assuming that the wheel of Prob. 18.97 weighs 8 lb, has a radius $a = 4$ in., and a radius of gyration of 3 in., and that $R = 20$ in., determine the force exerted by the plate on the wheel when $\Omega = 25$ rad/s.

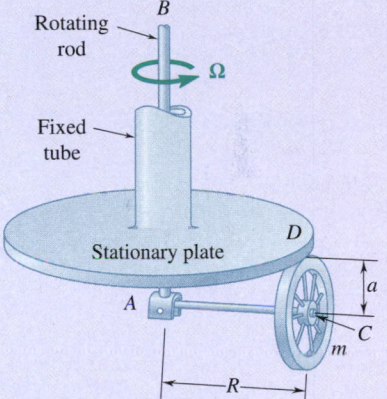

**Fig. P18.97**

**18.99** A thin disk of mass $m = 4$ kg rotates with an angular velocity $\omega_2$ with respect to arm $ABC$, which itself rotates with an angular velocity $\omega_1$ about the $y$ axis. Knowing that $\omega_1 = 5$ rad/s and $\omega_2 = 15$ rad/s and that both are constant, determine the force-couple system representing the dynamic reaction at the support at $A$.

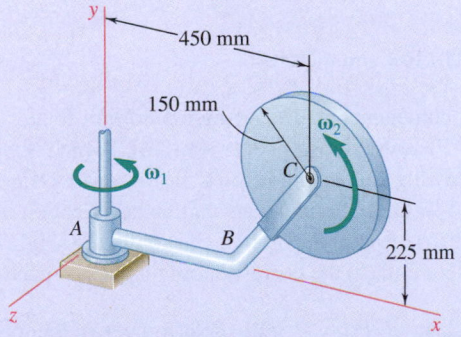

**Fig. P18.99**

**18.100** An experimental Fresnel-lens solar-energy concentrator can rotate about the horizontal axis $AB$ that passes through its mass center $G$. It is supported at $A$ and $B$ by a steel framework that can rotate about the vertical $y$ axis. The concentrator has a mass of 30 Mg, a radius of gyration of 12 m about its axis of symmetry $CD$, and a radius of gyration of 10 m about any transverse axis through $G$. Knowing that the angular velocities $\omega_1$ and $\omega_2$ have constant magnitudes equal to 0.20 rad/s and 0.25 rad/s, respectively, determine for the position $\theta = 60°$ (a) the forces exerted on the concentrator at $A$ and $B$, (b) the couple $M_2\mathbf{k}$ applied to the concentrator at that instant.

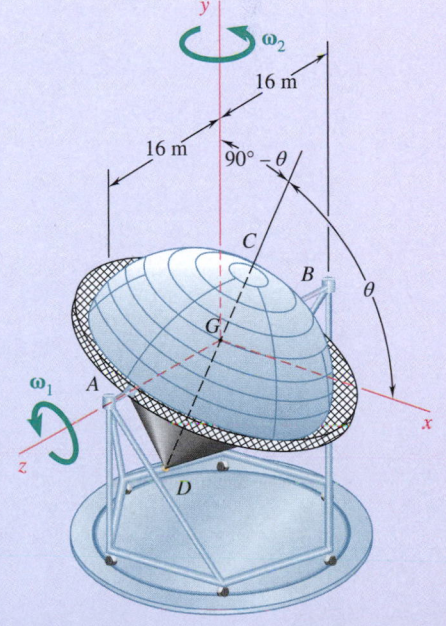

**Fig. P18.100**

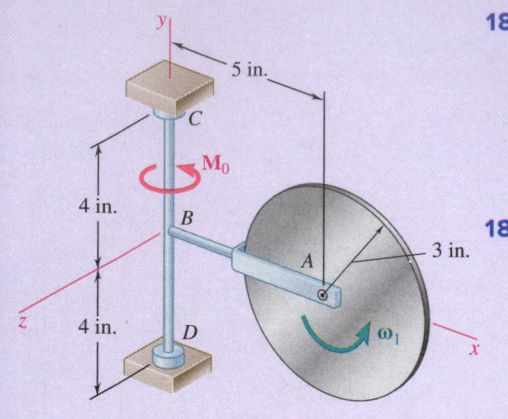

Fig. P18.101 and P18.102

**18.101** A 6-lb homogeneous disk of radius 3 in. spins as shown at the constant rate $\omega_1 = 60$ rad/s. The disk is supported by the fork-ended rod $AB$, which is welded to the vertical shaft $CBD$. The system is at rest when a couple $\mathbf{M}_0 = (0.25 \text{ ft·lb})\mathbf{j}$ is applied to the shaft for 2 s and then removed. Determine the dynamic reactions at $C$ and $D$ after the couple has been removed.

**18.102** A 6-lb homogeneous disk of radius 3 in. spins as shown at the constant rate $\omega_1 = 60$ rad/s. The disk is supported by the fork-ended rod $AB$, which is welded to the vertical shaft $CBD$. The system is at rest when a couple $\mathbf{M}_0$ is applied as shown to the shaft for 3 s and then removed. Knowing that the maximum angular velocity reached by the shaft is 18 rad/s, determine (a) the couple $\mathbf{M}_0$, (b) the dynamic reactions at $C$ and $D$ after the couple has been removed.

**18.103** A 2.5-kg homogeneous disk of radius 80 mm rotates with an angular velocity $\boldsymbol{\omega}_1$ with respect to arm $ABC$, which is welded to a shaft $DCE$ rotating as shown at the constant rate $\omega_2 = 12$ rad/s. Friction in the bearing at $A$ causes $\omega_1$ to decrease at the rate of 15 rad/s$^2$. Determine the dynamic reactions at $D$ and $E$ at a time when $\omega_1$ has decreased to 50 rad/s.

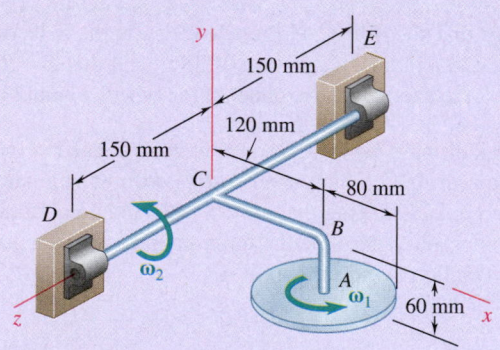

Fig. P18.103 and P18.104

**18.104** A 2.5-kg homogeneous disk of radius 80 mm rotates at the constant rate $\omega_1 = 50$ rad/s with respect to arm $ABC$, which is welded to a shaft $DCE$. Knowing that at the instant shown, shaft $DCE$ has an angular velocity $\boldsymbol{\omega}_2 = (12 \text{ rad/s})\mathbf{k}$ and an angular acceleration $\boldsymbol{\alpha}_2 = (8 \text{ rad/s}^2)\mathbf{k}$, determine (a) the couple that must be applied to shaft $DCE$ to produce that acceleration, (b) the corresponding dynamic reactions at $D$ and $E$.

**18.105** For the disk of Prob. 18.99, determine (a) the couple $M_1\mathbf{j}$ that should be applied to arm $ABC$ to give it an angular acceleration $\boldsymbol{\alpha}_1 = -(7.5 \text{ rad/s}^2)\mathbf{j}$ when $\omega_1 = 5$ rad/s, knowing that the disk rotates at the constant rate $\omega_2 = 15$ rad/s, (b) the force-couple system representing the dynamic reaction at $A$ at that instant. Assume that $ABC$ has a negligible mass.

**\*18.106** A slender homogeneous rod $AB$ of mass $m$ and length $L$ is made to rotate at a constant rate $\omega_2$ about the horizontal $z$ axis, while frame $CD$ is made to rotate at the constant rate $\omega_1$ about the $y$ axis. Express as a function of the angle $\theta$ (a) the couple $\mathbf{M}_1$ required to maintain the rotation of the frame, (b) the couple $\mathbf{M}_2$ required to maintain the rotation of the rod, (c) the dynamic reactions at the supports $C$ and $D$.

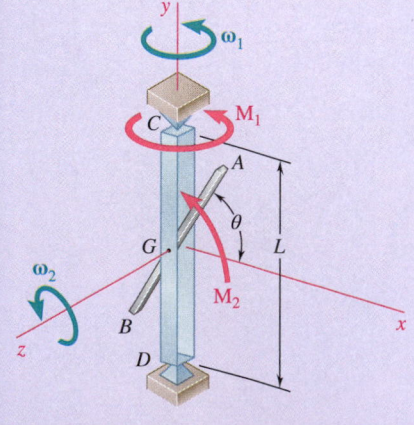

Fig. *P18.106*

# *18.3 MOTION OF A GYROSCOPE

A **gyroscope** consists essentially of a rotor that can spin freely about its geometric axis. When mounted in a Cardan's suspension (Fig. 18.15), a gyroscope can assume any orientation, but its mass center must remain fixed in space. Because a gyroscope can measure its orientation in space and maintain that orientation, it has become an indispensable part of modern navigational equipment. In this section, we examine the motion of a gyroscope as a practical example of analyzing the motion of a rigid body in three dimensions.

## 18.3A Eulerian Angles

In order to define the position of a gyroscope at a given instant, let us select a fixed frame of reference $OXYZ$ with the origin $O$ located at the mass center of the gyroscope and the $Z$ axis directed along the line defined by the bearings $A$ and $A'$ of the outer gimbal. We consider a reference position of the gyroscope in which the two gimbals and a given diameter $DD'$ of the rotor are located in the fixed $YZ$ plane (Fig. 18.15a). The gyroscope can be brought from this reference position into any arbitrary position (Fig. 18.15b) by means of the following steps.

1. A rotation of the outer gimbal through an angle $\phi$ about the axis $AA'$.
2. A rotation of the inner gimbal through $\theta$ about $BB'$.
3. A rotation of the rotor through $\psi$ about $CC'$.

The angles $\phi$, $\theta$, and $\psi$ are called the **Eulerian angles**; they completely characterize the position of the gyroscope at any given instant. Their derivatives $\dot\phi$, $\dot\theta$, and $\dot\psi$ define, respectively, the rate of **precession**, the rate of **nutation**, and the rate of **spin** of the gyroscope at the instant considered. Precession is the revolution of the axis $BB'$ about the $Z$ axis, and nutation is the back-and-forth motion of $CC'$ as the object precesses.

In order to compute the components of the angular velocity and of the angular momentum of the gyroscope, we will use a rotating system of axes $Oxyz$ *attached to the inner gimbal*, with the $y$ axis along $BB'$ and the $z$ axis along $CC'$ (Fig. 18.16). These axes are principal axes of inertia for the gyroscope. Although they follow it in its precession and nutation, they do not spin with $\dot\psi$; for that reason, they are more convenient to use than axes actually attached to the gyroscope. The angular velocity $\boldsymbol{\omega}$ of the gyroscope with respect to the fixed frame of reference $OXYZ$ now can be expressed as the sum of three partial angular velocities that correspond to the precession, the nutation, and the spin of the gyroscope, respectively. Denoting the unit vectors along the rotating axes by $\mathbf{i}$, $\mathbf{j}$, and $\mathbf{k}$ and the unit vector along the fixed $Z$ axis by $\mathbf{K}$, we have

$$\boldsymbol{\omega} = \dot\phi\mathbf{K} + \dot\theta\mathbf{j} + \dot\psi\mathbf{k} \qquad (18.33)$$

Because the vector components obtained for $\boldsymbol{\omega}$ in Eq. (18.33) are not orthogonal (Fig. 18.16), we resolve the unit vector $\mathbf{K}$ into components along the $x$ and $z$ axes; we obtain

$$\mathbf{K} = -\sin\theta\mathbf{i} + \cos\theta\mathbf{k} \qquad (18.34)$$

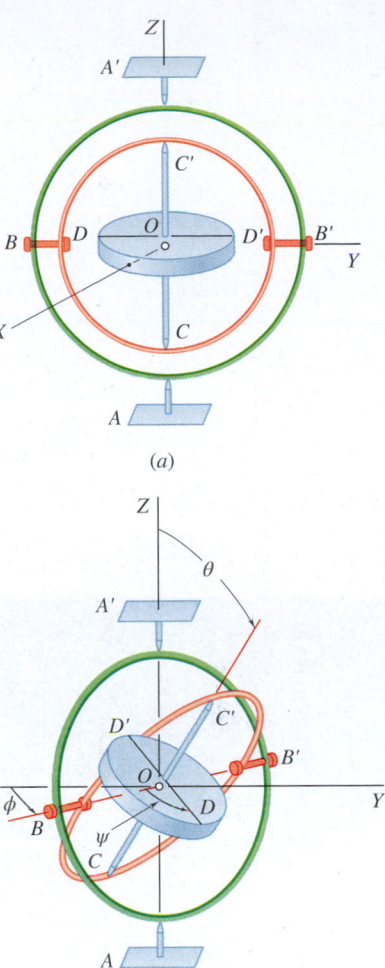

(a)

(b)

**Fig. 18.15** (a) Reference position of a gyroscope; (b) arbitrary position of the gyroscope by rotation through the three Eulerian angles.

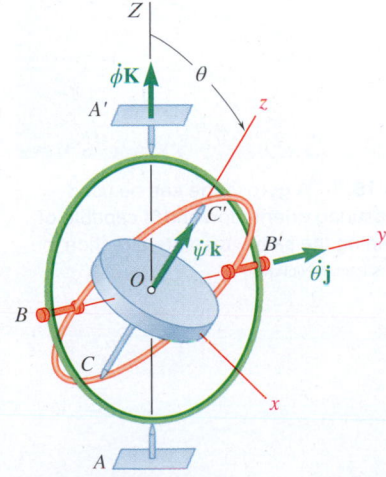

**Fig. 18.16** Precession $\dot\phi$, nutation $\dot\theta$, and spin $\dot\psi$ of a gyroscope.

Then, substituting for **K** into Eq. (18.33), we have

$$\boldsymbol{\omega} = -\dot{\phi}\sin\theta\,\mathbf{i} + \dot{\theta}\,\mathbf{j} + (\dot{\psi} + \dot{\phi}\cos\theta)\mathbf{k} \qquad (18.35)$$

The coordinate axes are principal axes of inertia, so we can obtain the components of the angular momentum $\mathbf{H}_O$ by multiplying the components of $\boldsymbol{\omega}$ by the moments of inertia of the rotor about the $x$, $y$, and $z$ axes, respectively. Denoting the moment of inertia of the rotor about its spin axis by $I$, its moment of inertia about a transverse axis through $O$ by $I'$, and neglecting the mass of the gimbals, we have

$$\mathbf{H}_O = -I'\dot{\phi}\sin\theta\,\mathbf{i} + I'\dot{\theta}\,\mathbf{j} + I(\dot{\psi} + \dot{\phi}\cos\theta)\mathbf{k} \qquad (18.36)$$

Recalling that the rotating axes are attached to the inner gimbal and thus do not spin with $\dot{\psi}$, we express their angular velocity as the sum

$$\boldsymbol{\Omega} = \dot{\phi}\mathbf{K} + \dot{\theta}\mathbf{j} \qquad (18.37)$$

or substituting for **K** from Eq. (18.34), we have

$$\boldsymbol{\Omega} = -\dot{\phi}\sin\theta\,\mathbf{i} + \dot{\theta}\,\mathbf{j} + \dot{\phi}\cos\theta\,\mathbf{k} \qquad (18.38)$$

Substituting for $\mathbf{H}_O$ and $\boldsymbol{\Omega}$ from Eqs. (18.36) and (18.38) into the equation gives

$$\Sigma\mathbf{M}_O = (\dot{\mathbf{H}}_O)_{Oxyz} + \boldsymbol{\Omega} \times \mathbf{H}_O \qquad (18.28)$$

We now obtain the three differential equations

$$\Sigma M_x = -I'(\ddot{\phi}\sin\theta + 2\dot{\theta}\dot{\phi}\cos\theta) + I\dot{\theta}(\dot{\psi} + \dot{\phi}\cos\theta)$$
$$\Sigma M_y = I'(\ddot{\theta} - \dot{\phi}^2\sin\theta\cos\theta) + I\dot{\phi}\sin\theta(\dot{\psi} + \dot{\phi}\cos\theta) \qquad (18.39)$$
$$\Sigma M_z = I\frac{d}{dt}(\dot{\psi} + \dot{\phi}\cos\theta)$$

Eqs. (18.39) define the motion of a gyroscope subjected to a given system of forces when the mass of its gimbals is neglected. We can also use them to define the motion of an **axisymmetric body** (or body of revolution) attached at a point on its axis of symmetry and to define the motion of an axisymmetric body about its mass center. The gimbals of the gyroscope helped us visualize the Eulerian angles, but it is clear that we can use these angles to define the position of any rigid body with respect to axes centered at a point of the body—regardless of the way in which the body is actually supported.

Because Eqs. (18.39) are nonlinear, it is not possible to express the Eulerian angles $\phi$, $\theta$, and $\psi$ as analytical functions of time $t$ in general, and you may need to use numerical methods of solution. However, as you will see in the rest of this section, several particular cases of interest can be analyzed easily.

## *18.3B   Steady Precession of a Gyroscope

Let us now investigate the particular case of gyroscopic motion in which the angle $\theta$, the rate of precession $\dot{\phi}$, and the rate of spin $\dot{\psi}$ remain constant. We propose to determine the forces that must be applied to the gyroscope to maintain this motion, which is known as the **steady precession** of a gyroscope.

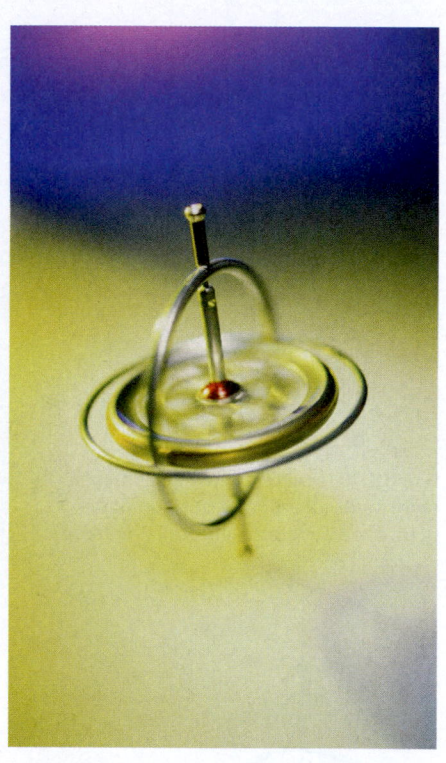

**Photo 18.5** A gyroscope can be used for measuring orientation and is capable of maintaining the same absolute direction in space. ©Ingram Publishing RF

Instead of applying the general equations (18.39), we determine the sum of the moments of the required forces by computing the rate of change of the angular momentum of the gyroscope in the particular case considered. We first note that the angular velocity $\boldsymbol{\omega}$ of the gyroscope, its angular momentum $\mathbf{H}_O$, and the angular velocity $\boldsymbol{\Omega}$ of the rotating frame of reference (Fig. 18.17) reduce, respectively, to

$$\boldsymbol{\omega} = -\dot{\phi}\sin\theta\,\mathbf{i} + \omega_z\mathbf{k} \qquad (18.40)$$

$$\mathbf{H}_O = -I'\dot{\phi}\sin\theta\,\mathbf{i} + I\omega_z\mathbf{k} \qquad (18.41)$$

$$\boldsymbol{\Omega} = -\dot{\phi}\sin\theta\,\mathbf{i} + \dot{\phi}\cos\theta\,\mathbf{k} \qquad (18.42)$$

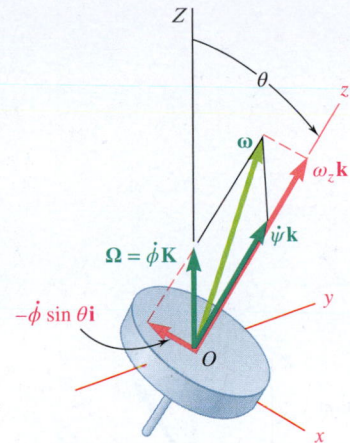

**Fig. 18.17**   Kinematic quantities used to determine the steady rate of precession of a gyroscope.

where $\omega_z = \dot{\psi} + \dot{\phi}\cos\theta$ is the rectangular component along the spin axis of the total angular velocity of the gyroscope.

Because $\theta$, $\phi$, and $\dot{\psi}$ are constant, the vector $\mathbf{H}_O$ is constant in magnitude and direction with respect to the rotating frame of reference. Therefore its rate of change $(\dot{\mathbf{H}}_O)_{Oxyz}$ with respect to that frame is zero. Thus, Eq. (18.28) reduces to

$$\Sigma\mathbf{M}_O = \boldsymbol{\Omega} \times \mathbf{H}_O \qquad (18.43)$$

which yields, after substitutions from Eqs. (18.41) and (18.42),

$$\Sigma\mathbf{M}_O = (I\omega_z - I'\dot{\phi}\cos\theta)\dot{\phi}\sin\theta\,\mathbf{j} \qquad (18.44)$$

The mass center of the gyroscope is fixed in space, so using Eq. (18.1), we have $\Sigma\mathbf{F} = 0$. Thus, the forces that must be applied to the gyroscope to maintain its steady precession reduce to a couple of moment equal to the right-hand side of Eq. (18.44). Note that *this couple should be applied about an axis perpendicular to the precession axis and to the spin axis of the gyroscope* (Fig. 18.18).

In the particular case when the precession axis and the spin axis are at a right angle to each other, we have $\theta = 90°$, and Eq. (18.44) reduces to

$$\Sigma\mathbf{M}_O = I\dot{\psi}\dot{\phi}\,\mathbf{j} \qquad (18.45)$$

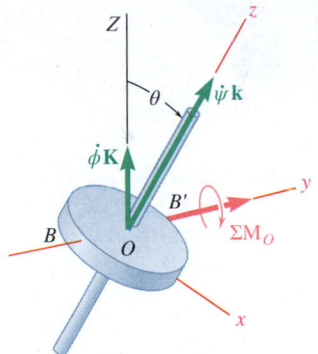

**Fig. 18.18**   To maintain a gyroscope in steady precession, a couple must be applied about an axis perpendicular to the precession and spin axis.

Thus, if we apply a couple $\mathbf{M}_O$ to the gyroscope about an axis perpendicular to its axis of spin, the gyroscope precesses about an axis perpendicular to both the spin axis and the couple axis. The sense of the precession is such that the vectors representing the spin, the couple, and the precession, respectively, form a right-handed triad (Fig. 18.19). The relationship of this triad also can be represented by writing Eq. (18.45) as the vector equation of

$$\Sigma\mathbf{M}_O = \dot{\boldsymbol{\phi}} \times I\dot{\boldsymbol{\psi}} \qquad (18.45')$$

Because of the relatively large couples required to change the orientation of their axles, gyroscopes are used as stabilizers in torpedoes and ships. Spinning bullets and shells remain tangent to their trajectory because of gyroscopic action. Also, a bicycle is easier to keep balanced at high speeds because of the stabilizing effect of its spinning wheels. However, gyroscopic action is not always welcome; it must be taken into account in the design of bearings

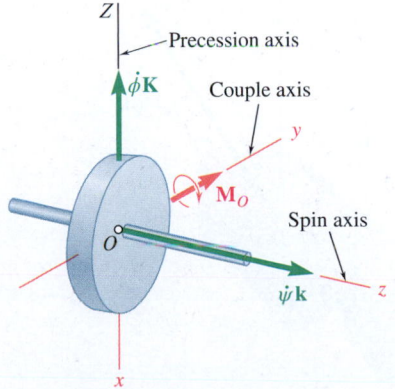

**Fig. 18.19**   A right-handed triad of the spin, couple, and precession axes.

supporting rotating shafts subjected to forced precession. The reactions exerted on an airplane by its propellers, which changes the direction of flight, also must be taken into consideration and compensated for whenever possible.

## *18.3C    Motion of an Axisymmetric Body Under No Force

We can now analyze the motion of an axisymmetric body about its mass center under no force except its own weight. Examples of such motion are furnished by projectiles (if air resistance is neglected) and by satellites and space vehicles after the burnout of their launching rockets.

The sum of the moments of the external forces about the mass center $G$ of the body is zero, so Eq. (18.2) yields $\dot{\mathbf{H}}_G = 0$ . It follows that the angular momentum $\mathbf{H}_G$ of the body about $G$ is constant. Thus, the direction of $\mathbf{H}_G$ is fixed in space and can be used to define the $Z$ axis, or axis of precession (Fig. 18.20). Let us select a rotating system of axes $Gxyz$ with the $z$ axis along the axis of symmetry of the body, the $x$ axis in the plane defined by the $Z$ and $z$ axes, and the $y$ axis pointing away from you (Fig. 18.21). This gives us

$$H_x = -H_G \sin \theta \qquad H_y = 0 \qquad H_z = H_G \cos \theta \qquad \textbf{(18.46)}$$

where $\theta$ represents the angle formed by the $Z$ and $z$ axes, and $H_G$ denotes the constant magnitude of the angular momentum of the body about $G$. Because the $x$, $y$, and $z$ axes are principal axes of inertia for the body considered, we have

$$H_x = I' \omega_x \qquad H_y = I' \omega_y \qquad H_z = I \omega_z \qquad \textbf{(18.47)}$$

where $I$ denotes the moment of inertia of the body about its axis of symmetry and $I'$ denotes its moment of inertia about a transverse axis through $G$. It follows from Eqs. (18.46) and (18.47) that

$$\omega_x = -\frac{H_G \sin \theta}{I'} \qquad \omega_y = 0 \qquad \omega_z = \frac{H_G \cos \theta}{I} \qquad \textbf{(18.48)}$$

The second of these relations shows that the angular velocity $\boldsymbol{\omega}$ has no component along the $y$ axis; that is, along an axis perpendicular to the $Z$-$z$ plane. Thus, the angle $\theta$ formed by the $Z$ and $z$ axes remains constant and *the body is in steady precession about the $Z$ axis.*

Dividing the first of the relations in Eqs. (18.48) by the third, and observing from Fig. 18.21 that $-\omega_x/\omega_z = \tan \gamma$, we obtain the following relation between the angles $\gamma$ and $\theta$ that the vectors $\boldsymbol{\omega}$ and $\mathbf{H}_G$, respectively, form with the axis of symmetry of the body.

$$\tan \gamma = \frac{I}{I'} \tan \theta \qquad \textbf{(18.49)}$$

Two particular cases of motion of an axisymmetric body under no force involve no precession.

1. If the body is set to spin about its axis of symmetry, we have $\omega_x = 0$ and, by Eq. (18.47), $H_x = 0$. Thus, the vectors $\boldsymbol{\omega}$ and $\mathbf{H}_G$ have the same orientation, and the body keeps spinning about its axis of symmetry (Fig. 18.22a).

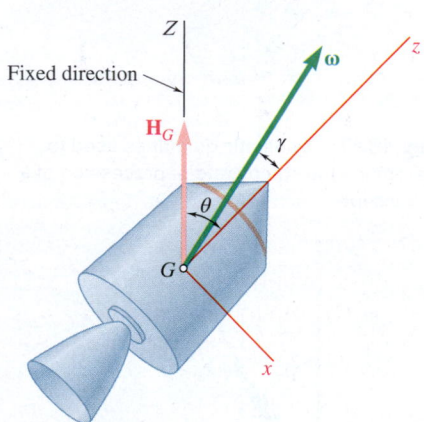

**Fig. 18.20** For an axisymmetric body under no force other than its own weight, the angular momentum has a constant direction.

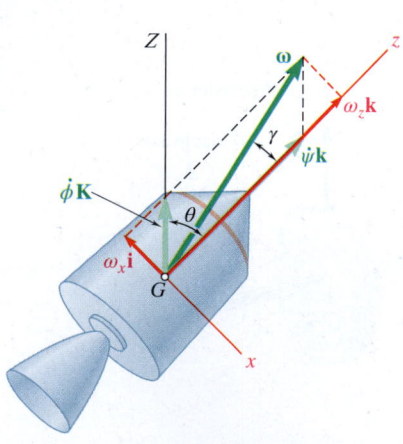

**Fig. 18.21** The angular velocity of an axisymmetric body expressed in terms of body-fixed coordinates *xyz*.

**2.** If the body is set to spin about a transverse axis, we have $\omega_z = 0$ and, by Eq. (18.47), $H_z = 0$. Again $\boldsymbol{\omega}$ and $\mathbf{H}_G$ have the same orientation, and the body keeps spinning about the given transverse axis (Fig. 18.22*b*).

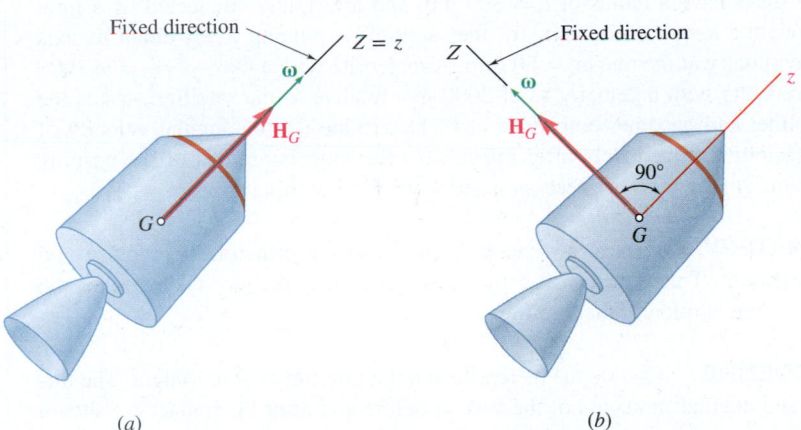

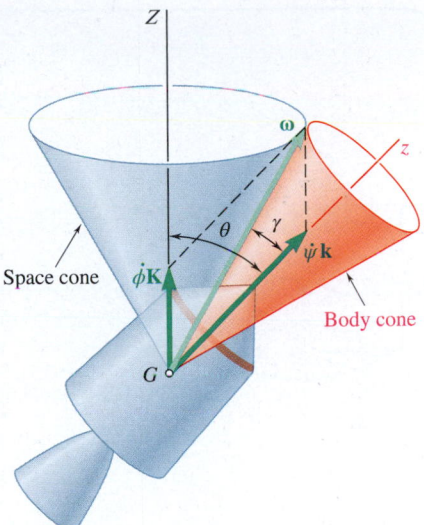

**Fig. 18.23**  Space cone and body cone for an elongated body ($l < l'$) in direct precession.

(*a*)

(*b*)

**Fig. 18.22**  (*a*) A body spinning about its axis of symmetry; (*b*) a body spinning about a transverse axis.

Considering now the general case represented in Fig. 18.21, recall from Sec. 15.6A that we can represent the motion of a body about a fixed point—or about its mass center—by the motion of a body cone rolling on a space cone. In the case of steady precession, the two cones are circular, because the angles $\gamma$ and $\theta - \gamma$ that the angular velocity $\boldsymbol{\omega}$ forms, respectively, with the axis of symmetry of the body and with the precession axis are constant. Two cases should be distinguished.

**1.** $l < l'$. This is the case of an elongated body, such as the space vehicle of Fig. 18.23. From Eq. (18.49), we have $\gamma < \theta$. The vector $\boldsymbol{\omega}$ lies inside the angle $ZGz$; the space cone and the body cone are tangent externally; and the spin and the precession are both observed as counterclockwise from the positive $z$ axis. The precession is said to be *direct*.

**2.** $l > l'$. This is the case of a flattened body, such as the satellite of Fig. 18.24. From Eq. (18.49), we have $\gamma > \theta$. Because the vector $\boldsymbol{\omega}$ must lie outside the angle $ZGz$, the vector $\dot{\psi}\mathbf{k}$ has a sense opposite to that of the $z$ axis; the space cone is inside the body cone; and the precession and the spin have opposite senses. The precession is said to be *retrograde*.

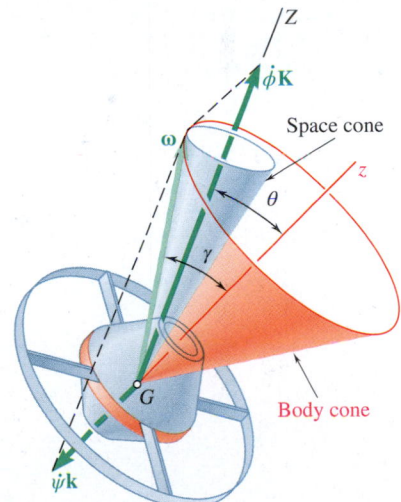

**Fig. 18.24**  Space cone and body cone for a flattened body ($l > l'$) in retrograde precession.

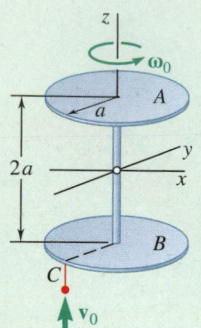

# Sample Problem 18.6

A space satellite with mass $m$ can be modeled as two thin disks of equal mass. The disks have a radius of $a = 800$ mm and are rigidly connected by a light rod with a length of $2a$. Initially, the satellite is spinning freely about its axis of symmetry at the rate $\omega_0 = 60$ rpm. A meteorite with a mass of $m_0 = m/1000$ is traveling with a velocity $\mathbf{v}_0$ of 2000 m/s relative to the satellite, strikes the satellite, and becomes embedded at $C$. Determine (a) the angular velocity of the satellite immediately after impact, (b) the precession axis of the ensuing motion, (c) the rates of precession and spin of the ensuing motion.

**STRATEGY:** Because an impact occurs, use the principle of impulse and momentum. Then you can use the relations in this section to determine the gyroscopic motion of the satellite.

**MODELING:** Choose the meteorite and the satellite as your system. The linear and angular momenta of the system before and after the impact are shown in Fig. 1.

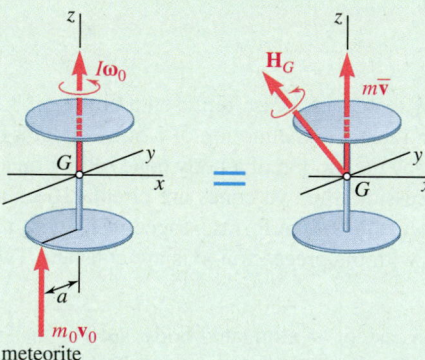

**Fig. 1** Momenta before and after the impact.

## ANALYSIS:

**Moments of Inertia.** Note that the axes shown are principal axes of inertia for the satellite. Thus, you have

$$I = I_z = \tfrac{1}{2}ma^2 \qquad I' = I_x = I_y = 2[\tfrac{1}{4}(\tfrac{1}{2}m)a^2 + (\tfrac{1}{2}m)a^2] = \tfrac{5}{4}ma^2$$

**Principle of Impulse and Momentum.** Because no external force acts on the system, the momenta before and after impact are equal (Fig. 1). Taking moments about $G$, you have

$$-a\mathbf{j} \times m_0v_0\mathbf{k} + I\omega_0\mathbf{k} = \mathbf{H}_G$$
$$\mathbf{H}_G = -m_0v_0a\mathbf{i} + I\omega_0\mathbf{k} \qquad (1)$$

*(continued)*

**Angular Velocity After Impact.** Substitute the values obtained for the components of $\mathbf{H}_G$ in Eq. (1) and for the moments of inertia into

$$H_x = I_x \omega_x \qquad H_y = I_y \omega_y \qquad H_z = I_z \omega_z$$

The result is

$$-m_0 v_0 a = I' \omega_x = \tfrac{5}{4} m a^2 \omega_x \qquad 0 = I' \omega_y \qquad I \omega_0 = I \omega_z$$

$$\omega_x = -\frac{4}{5}\frac{m_0 v_0}{ma} \qquad \omega_y = 0 \qquad \omega_z = \omega_0 \qquad \text{(2)}$$

For the satellite considered, you have $\omega_0 = 60$ rpm $= 6.283$ rad/s, $m_0/m = 1/1000$, $a = 0.800$ m, and $v_0 = 2000$ m/s. You obtain

$$\omega_x = -2 \text{ rad/s} \qquad \omega_y = 0 \qquad \omega_z = 6.283 \text{ rad/s}$$

$$\omega = \sqrt{\omega_x^2 + \omega_z^2} = 6.594 \text{ rad/s} \qquad \tan \gamma = \frac{-\omega_x}{\omega_z} = +0.3183$$

$$\omega = 63.0 \text{ rpm} \qquad \gamma = 17.7° \quad \blacktriangleleft$$

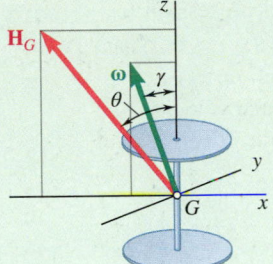

**Fig. 2** Angles between the z axis and the angular velocity and the angular momentum.

**Precession Axis.** In free motion, the direction of the angular momentum $\mathbf{H}_G$ is fixed in space, so the satellite precesses about this direction. The angle $\theta$ formed by the precession axis and the z axis is (Fig. 2)

$$\tan \theta = \frac{-H_x}{H_z} = \frac{m_0 v_0 a}{I \omega_0} = \frac{2 m_0 v_0}{m a \omega_0} = 0.796 \qquad \theta = 38.5° \quad \blacktriangleleft$$

**Rates of Precession and Spin.** Sketch the space and body cones for the free motion of the satellite (Fig. 3). Using the law of sines, compute the rates of precession and spin.

$$\frac{\omega}{\sin \theta} = \frac{\dot{\phi}}{\sin \gamma} = \frac{\dot{\psi}}{\sin (\theta - \gamma)}$$

$$\dot{\phi} = 30.8 \text{ rpm} \qquad \dot{\psi} = 35.9 \text{ rpm} \quad \blacktriangleleft$$

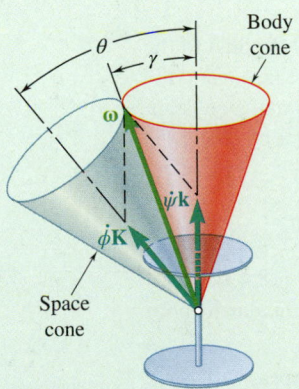

**Fig. 3** Space and body cones for the satellite.

**REFLECT and THINK:** If you applied the principle of impulse and momentum in the z direction, you would find that $P\Delta t = m\overline{v}$ where $P\Delta t$ is the impulse the meteorite applies to the satellite. In this problem, we were interested in the three-dimensional rotation of the satellite and modeled it as a rigid body. In Chap. 12, we were concerned with the orbits of satellites over the earth and modeled the satellite as a particle. As engineers, how we model a system depends on what type of problem we are trying to solve.

# SOLVING PROBLEMS
# ON YOUR OWN

In this section, we analyzed the motion of **gyroscopes** and of other **axisymmetric bodies** with a fixed point $O$. In order to define the position of these bodies at any given instant, we introduced the three **Eulerian angles** $\phi$, $\theta$, and $\psi$ (Fig. 18.15) and noted that their time derivatives define, respectively, the rate of **precession**, the rate of **nutation**, and the rate of **spin** (Fig. 18.16). The problems you encountered in this section fall into one of the following categories.

**1. Steady precession.** This is the motion of a gyroscope or other axisymmetric body with a fixed point located on its axis of symmetry in which the angle $\theta$, the rate of precession $\dot{\phi}$, and the rate of spin $\dot{\psi}$ all remain constant.

   **a.  Using the rotating frame of reference** $Oxyz$ shown in Fig. 18.17, which *precesses* with the body, *but does not spin* with it, we obtained the expressions for the angular velocity $\boldsymbol{\omega}$ of the body, its angular momentum $\mathbf{H}_O$, and the angular velocity $\boldsymbol{\Omega}$ of the frame $Oxyz$ as

$$\boldsymbol{\omega} = -\dot{\phi}\,\sin\theta\,\mathbf{i} + \omega_z\mathbf{k} \tag{18.40}$$

$$\mathbf{H}_O = -I'\dot{\phi}\,\sin\theta\,\mathbf{i} + I\,\omega_z\mathbf{k} \tag{18.41}$$

$$\boldsymbol{\Omega} = -\dot{\phi}\,\sin\theta\,\mathbf{i} + \dot{\phi}\,\cos\theta\,\mathbf{k} \tag{18.42}$$

where $I$ = moment of inertia of body about its axis of symmetry
   $I'$ = moment of inertia of body about a transverse axis through $O$
   $\omega_z$ = rectangular component of $\boldsymbol{\omega}$ along $z$ axis = $\dot{\psi} + \dot{\phi}\cos\theta$

   **b.  The sum of the moments about** $O$ **of the forces applied to the body** is equal to the rate of change of its angular momentum, as expressed by Eq. (18.28). But, because $\theta$ and the rates of change $\dot{\phi}$ and $\dot{\psi}$ are constant, it follows from Eq. (18.41) that $\mathbf{H}_O$ remains constant in magnitude and direction when viewed from the frame $Oxyz$. Thus, its rate of change is zero with respect to that frame, and you have

$$\Sigma\mathbf{M}_O = \boldsymbol{\Omega} \times \mathbf{H}_O \tag{18.43}$$

where $\boldsymbol{\Omega}$ and $\mathbf{H}_O$ are defined by Eqs. (18.42) and (18.41), respectively. Eq. (18.43) shows that the moment resultant at $O$ of the forces applied to the body is perpendicular to both the axis of precession and the axis of spin (Fig. 18.18).

   **c.  Keep in mind that the method described applies** not only to gyroscopes, where the fixed point $O$ coincides with the mass center $G$, but also to any axisymmetric body with a fixed point $O$ located on its axis of symmetry. This method, therefore, can be used to analyze the *steady precession of a top* on a rough floor.

   **d.  When an axisymmetric body has no fixed point but is in steady precession about its mass center** $G$, you should draw a free-body diagram and a kinetic diagram showing that the system of the external forces exerted on the body (including the body's weight) is equivalent to the vector $m\bar{\mathbf{a}}$ applied at $G$ and the couple vector $\dot{\mathbf{H}}_G$.

You can use Eqs. (18.40), (18.41), and (18.42), replacing $\mathbf{H}_O$ with $\mathbf{H}_G$, and express the moment of the couple as

$$\dot{\mathbf{H}}_G = \boldsymbol{\Omega} \times \mathbf{H}_G$$

You can then use the free-body and kinetic diagrams to write as many as six independent scalar equations.

**2. Motion of an axisymmetric body under no force, except its own weight.** We have $\Sigma\mathbf{M}_G = 0$ and thus $\dot{\mathbf{H}}_G = 0$; it follows that the angular momentum $\mathbf{H}_G$ is constant in magnitude and direction (Sec. 18.3C). The body is in **steady precession** with the precession axis $GZ$ directed along $\mathbf{H}_G$ (Fig. 18.20). Using the rotating frame $Gxyz$ and denoting by $\gamma$ the angle that $\boldsymbol{\omega}$ forms with the spin axis $Gz$ (Fig. 18.21), we obtained the relation between $\gamma$ and the angle $\theta$ formed by the precession and spin axes as

$$\tan \gamma = \frac{I}{I'} \tan \theta \tag{18.49}$$

The precession is said to be *direct* if $I < I'$ (Fig. 18.23) and *retrograde* if $I > I'$ (Fig. 18.24).

   **a. In many problems** dealing with the motion of an axisymmetric body under no force, you will be asked to determine the precession axis and the rates of precession and spin of the body when given the magnitude of its angular velocity $\boldsymbol{\omega}$ and the angle $\gamma$ that it forms with the axis of symmetry $Gz$ (Fig. 18.21). From Eq. (18.49), determine the angle $\theta$ that the precession axis $GZ$ forms with $Gz$ and resolve $\boldsymbol{\omega}$ into its two oblique components $\dot{\phi}\mathbf{K}$ and $\dot{\psi}\mathbf{k}$. Using the law of sines, you then can determine the rate of precession $\dot{\phi}$ and the rate of spin $\dot{\psi}$.

   **b. In other problems,** the body is subjected to a given impulse and you will first determine the resulting angular momentum $\mathbf{H}_G$. Using Eqs. (18.10), you can calculate the rectangular components of the angular velocity $\boldsymbol{\omega}$, its magnitude $\omega$, and the angle $\gamma$ that it forms with the axis of symmetry. You then determine the precession axis and the rates of precession and spin as described previously (Sample Prob. 18.6).

**3. General motion of an axisymmetric body with a fixed point $O$ located on its axis of symmetry, and subjected only to its own weight.** This is a motion in which the angle $\theta$ is allowed to vary. At any given instant you should take into account the rate of precession $\dot{\phi}$, the rate of spin $\dot{\psi}$, and the rate of nutation $\dot{\theta}$—none of which will remain constant. An example of such a motion is the motion of a top, which is discussed in Probs. 18.137 and 18.138. The rotating frame of reference $Oxyz$ that you will use is still the one shown in Fig. 18.18, but this frame now rotates about the $y$ axis at the rate $\dot{\theta}$. Eqs. (18.40), (18.41), and (18.42), therefore, should be replaced by

$$\boldsymbol{\omega} = -\dot{\phi} \sin \theta \, \mathbf{i} + \dot{\theta} \, \mathbf{j} + (\dot{\psi} + \dot{\phi} \cos \theta)\mathbf{k} \tag{18.40$'$}$$

$$\mathbf{H}_O = -I'\dot{\phi} \sin \theta \, \mathbf{i} + I'\dot{\theta} \, \mathbf{j} + I(\dot{\psi} + \dot{\phi} \cos \theta)\mathbf{k} \tag{18.41$'$}$$

$$\boldsymbol{\Omega} = -\dot{\phi} \sin \theta \, \mathbf{i} + \dot{\theta} \, \mathbf{j} + \dot{\phi} \cos \theta \, \mathbf{k} \tag{18.42$'$}$$

*(continued)*

Because substituting these expressions into Eq. (18.44) would lead to nonlinear differential equations, it is preferable, whenever feasible, to apply the following conservation principles.

**a. Conservation of energy.** Denoting the distance between the fixed point $O$ and the mass center $G$ of the body by $c$ and the total energy by $E$, you have

$$T + V = E: \qquad \tfrac{1}{2}(I'\omega_x^2 + I'\omega_y^2 + I\omega_z^2) + mgc \cos \theta = E$$

Then substitute the expressions obtained in Eq. (18.40′) for the components of $\boldsymbol{\omega}$. Note that $c$ is positive or negative depending upon the position of $G$ relative to $O$. Also, $c = 0$ if $G$ coincides with $O$; the kinetic energy is then conserved.

**b. Conservation of the angular momentum about the axis of precession.** Because the support at $O$ is located on the $Z$ axis and the weight of the body and the $Z$ axis are both vertical and thus are parallel to each other, it follows that $\Sigma M_Z = 0$. Thus, $H_Z$ remains constant. We can express this by writing that the scalar product $\mathbf{K} \cdot \mathbf{H}_O$ is constant, where $\mathbf{K}$ is the unit vector along the $Z$ axis.

**c. Conservation of the angular momentum about the axis of spin.** Because the support at $O$ and the center of gravity $G$ are both located on the $z$ axis, it follows that $\Sigma M_z = 0$ and, thus, that $H_z$ remains constant. Thus, the coefficient of the unit vector $\mathbf{k}$ in Eq. (18.41′) is constant. Note that this last conservation principle cannot be applied when the body is restrained from spinning about its axis of symmetry, but in that case, the only variables are $\theta$ and $\phi$.

# Problems

**18.107** A uniform thin disk with a 6-in. diameter is attached to the end of a rod $AB$ of negligible mass that is supported by a ball-and-socket joint at point $A$. Knowing that the disk is observed to precess about the vertical axis $AC$ at the constant rate of 36 rpm in the sense indicated and that its axis of symmetry $AB$ forms an angle $\beta = 60°$ with $AC$, determine the rate at which the disk spins about rod $AB$.

**18.108** A uniform thin disk with a 6-in. diameter is attached to the end of a rod $AB$ of negligible mass that is supported by a ball-and-socket joint at point $A$. Knowing that the disk is spinning about its axis of symmetry $AB$ at the rate of 2100 rpm in the sense indicated and that $AB$ forms an angle $\beta = 45°$ with the vertical axis $AC$, determine the two possible rates of steady precession of the disk about the axis $AC$.

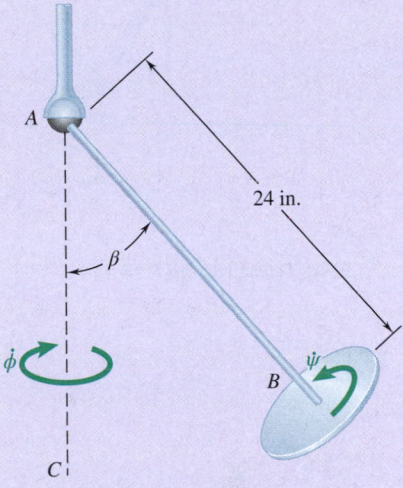

**Fig. P18.107 and *P18.108***

**18.109** The 85-g top shown is supported at the fixed point $O$. The radii of gyration of the top with respect to its axis of symmetry and with respect to a transverse axis through $O$ are 21 mm and 45 mm, respectively. Knowing that $c = 37.5$ mm and that the rate of spin of the top about its axis of symmetry is 1800 rpm, determine the two possible rates of steady precession corresponding to $\theta = 30°$.

**18.110** The top shown is supported at the fixed point $O$ and its moments of inertia about its axis of symmetry and about a transverse axis through $O$ are denoted, respectively, by $I$ and $I'$. (*a*) Show that the condition for steady precession of the top is

$$(I\omega_z - I'\dot{\phi}\cos\theta)\,\dot{\phi} = Wc$$

where $\dot{\phi}$ is the rate of precession and $\omega_z$ is the rectangular component of the angular velocity along the axis of symmetry of the top. (*b*) Show that if the rate of spin $\dot{\psi}$ of the top is very large compared with its rate of precession $\dot{\phi}$, the condition for steady precession is $I\dot{\psi}\dot{\phi} \approx Wc$. (*c*) Determine the percentage error introduced when this last relation is used to approximate the slower of the two rates of precession obtained for the top of Prob. 18.109.

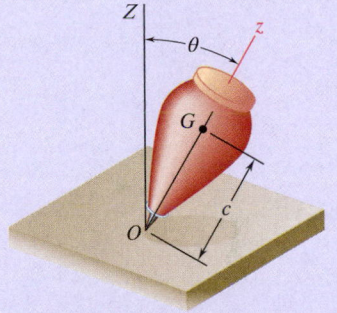

**Fig. P18.109 and *P18.110***

**18.111** A solid cone of height 9 in. with a circular base of radius 3 in. is supported by a ball-and-socket joint at $A$. Knowing that the cone is observed to precess about the vertical axis $AC$ at the constant rate of 40 rpm in the sense indicated and that its axis of symmetry $AB$ forms an angle $\beta = 40°$ with $AC$, determine the rate at which the cone spins about the axis $AB$.

**18.112** A solid cone of height 9 in. with a circular base of radius 3 in. is supported by a ball-and-socket joint at $A$. Knowing that the cone is spinning about its axis of symmetry $AB$ at the rate of 3000 rpm and that $AB$ forms an angle $\beta = 60°$ with the vertical axis $AC$, determine the two possible rates of steady precession of the cone about the axis $AC$.

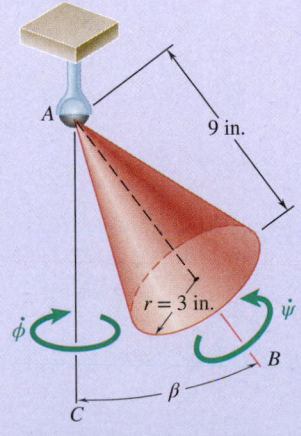

**Fig. P18.111 and P18.112**

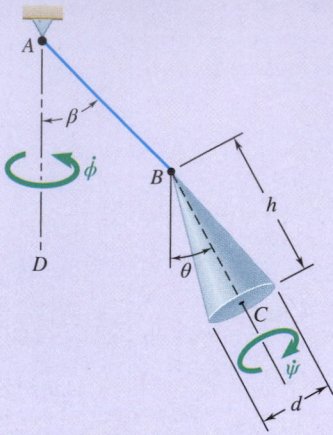

**Fig. P18.113 and P18.114**

**18.113** A homogeneous cone with a height $h$ and a base with a diameter $d < h$ is attached as shown to a cord $AB$. The cone spins about its axis $BC$ at the constant rate $\dot{\psi}$ and precesses about the vertical through $A$ at the constant rate $\dot{\phi}$. Determine the angle $\beta$ for which the axis $BC$ of the cone is aligned with cord $AB$ $(\theta = \beta)$.

**18.114** A homogeneous cone with a height of $h = 12$ in. and a base with a diameter of $d = 6$ in. is attached as shown to a cord $AB$. Knowing that the angles that cord $AB$ and the axis $BC$ of the cone form with the vertical are, respectively, $\beta = 45°$ and $\theta = 30°$ and that the cone precesses at the constant rate $\dot{\phi} = 8$ rad/s in the sense indicated, determine (a) the rate of spin $\dot{\psi}$ of the cone about its axis $BC$, (b) the length of cord $AB$.

**18.115** A homogeneous sphere of radius $c$ is attached as shown to a cord $AB$. The cord forms an angle $\beta$ with the vertical and precesses at the constant rate $\dot{\phi}$, while the sphere spins at the constant rate $\dot{\psi}$ about its diameter $BC$. Determine the angle $\theta$ that $BC$ forms with the vertical.

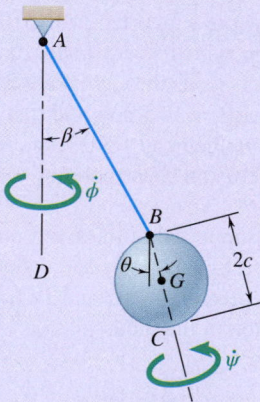

**Fig. P18.115 and P18.116**

**18.116** A homogeneous sphere of radius $c = 40$ mm is attached as shown to a cord $AB$. The cord forms an angle $\beta = 30°$ with the vertical and is observed to precess at the constant rate $\dot{\phi} = 5$ rad/s about the vertical through $A$. Determine the angle $\theta$ that the diameter $BC$ forms with the vertical, knowing that the sphere (a) has no spin, (b) spins about its diameter $BC$ at the rate $\dot{\psi} = 30$ rad/s, (c) spins about $BC$ at the rate $\dot{\psi} = -30$ rad/s.

**18.117** A high-speed photographic record shows that a certain projectile was fired with a horizontal velocity $\overline{\mathbf{v}}$ of 2000 ft/s and with its axis of symmetry forming an angle $\beta = 3°$ with the horizontal. The rate of spin $\dot{\psi}$ of the projectile was 6000 rpm, and the atmospheric drag was equivalent to a force $\mathbf{D}$ of 25 lb acting at the center of pressure $C_P$ located at a distance $c = 6$ in. from $G$. (a) Knowing that the projectile has a weight of 45 lb and a radius of gyration of 2 in. with respect to its axis of symmetry, determine its approximate rate of steady precession. (b) If it is further known that the radius of gyration of the projectile with respect to a transverse axis through $G$ is 8 in., determine the exact values of the two possible rates of precession.

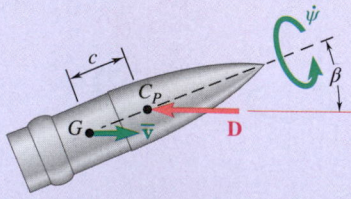

**Fig. P18.117**

**18.118** The propeller of an air boat rotates at 1800 rpm. The moment of inertia about its spin axis is 4 slug·ft$^2$. Knowing the boat makes a circular turn of radius 40 ft at a speed of 40 mi/h, determine the moment exerted on the boat due to the gyroscopic effect of the propeller.

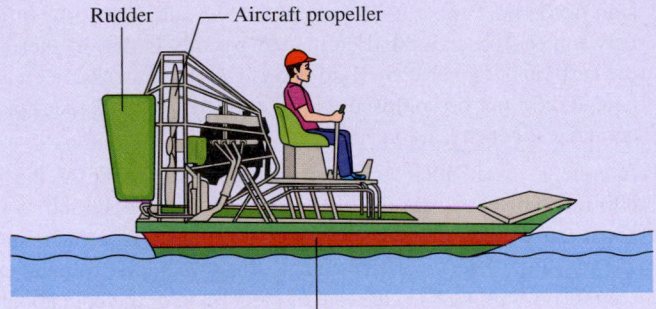

Rudder — Aircraft propeller

Flat hull for skimming over water

**Fig. P18.118**

**18.119** Show that for an axisymmetric body under no force, the rates of precession and spin can be expressed, respectively, as

$$\dot{\phi} = \frac{H_G}{I'}$$

and

$$\dot{\psi} = \frac{H_G \cos \theta (I' - I)}{I I'}$$

where $H_G$ is the constant value of the angular momentum of the body.

**18.120** (a) Show that for an axisymmetric body under no force, the rate of precession can be expressed as

$$\dot{\phi} = \frac{I \omega_2}{I' \cos \theta}$$

where $\omega_2$ is the rectangular component of $\boldsymbol{\omega}$ along the axis of symmetry of the body. (b) Use this result to check that the condition (18.44) for steady precession is satisfied by an axisymmetric body under no force.

**18.121** Show that the angular velocity vector $\boldsymbol{\omega}$ of an axisymmetric body under no force is observed from the body itself to rotate about the axis of symmetry at the constant rate

$$n = \frac{I' - I}{I'} \omega_2$$

where $\omega_2$ is the rectangular component of $\boldsymbol{\omega}$ along the axis of symmetry of the body.

**18.122** For an axisymmetric body under no force, prove (a) that the rate of retrograde precession can never be less than twice the rate of spin of the body about its axis of symmetry, (b) that in Fig. 18.24 the axis of symmetry of the body can never lie within the space cone.

**18.123** Using the relation given in Prob. 18.121, determine the period of precession of the north pole of the earth about the axis of symmetry of the earth. The earth may be approximated by an oblate spheroid of axial moment of inertia $I$ and of transverse moment of inertia $I' = 0.9967I$. (*Note:* Actual observations show a period of precession of the north pole of about 432.5 mean solar days; the difference between the observed and computed periods is due to the fact that the earth is not a perfectly rigid body. The free precession considered here should not be confused with the much slower precession of the equinoxes, which is a forced precession.)

**18.124** A coin is tossed into the air. It is observed to spin at the rate of 600 rpm about an axis $GC$ perpendicular to the coin and to precess about the vertical direction $GD$. Knowing that $GC$ forms an angle of $15°$ with $GD$, determine (*a*) the angle that the angular velocity $\boldsymbol{\omega}$ of the coin forms with $GD$, (*b*) the rate of precession of the coin about $GD$.

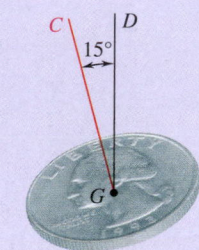

Fig. P18.124

**18.125** The angular velocity vector of a football that has just been kicked is horizontal, and its axis of symmetry $OC$ is oriented as shown. Knowing that the magnitude of the angular velocity is 200 rpm and that the ratio of the axis and transverse moments of inertia is $I/I' = \frac{1}{3}$, determine (*a*) the orientation of the axis of precession $OA$, (*b*) the rates of precession and spin.

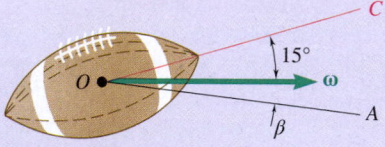

Fig. P18.125

**18.126** A space station consists of two sections $A$ and $B$ of equal masses that are rigidly connected. Each section is dynamically equivalent to a homogeneous cylinder with a length of 15 m and a radius of 3 m. Knowing that the station is precessing about the fixed direction $GD$ at the constant rate of 2 rev/h, determine the rate of spin of the station about its axis of symmetry $CC'$.

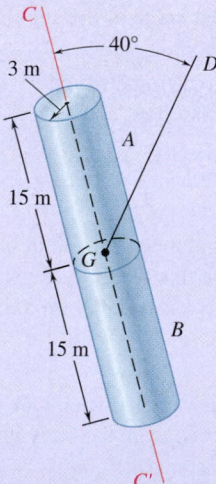

Fig. P18.126 and P18.127

**18.127** If the connection between sections $A$ and $B$ of the space station of Prob. 18.126 is severed when the station is oriented as shown and if the two sections are gently pushed apart along their common axis of symmetry, determine (*a*) the angle between the spin axis and the new precession axis of section $A$, (*b*) the rate of precession of section $A$, (*c*) its rate of spin.

**18.128** Solve Sample Prob. 18.6, assuming that the meteorite strikes the satellite at $C$ with a velocity $\mathbf{v}_0 = (2000 \text{ m/s})\mathbf{i}$.

**18.129** An 800-lb geostationary satellite is spinning with an angular velocity $\boldsymbol{\omega}_0 = (1.5 \text{ rad/s})\mathbf{j}$ when it is hit at $B$ by a 6-oz meteorite traveling with a velocity $\mathbf{v}_0 = -(1600 \text{ ft/s})\mathbf{i} + (1300 \text{ ft/s})\mathbf{j} + (4000 \text{ ft/s})\mathbf{k}$ relative to the satellite. Knowing that $b = 20$ in. and that the radii of gyration of the satellite are $\bar{k}_x = \bar{k}_z = 28.8$ in. and $\bar{k}_y = 32.4$ in., determine the precession axis and the rates of precession and spin of the satellite after the impact.

**18.130** Solve Prob. 18.129, assuming that the meteorite hits the satellite at $A$ instead of $B$.

**18.131** A homogeneous disk of mass $m$ is connected at $A$ and $B$ to a fork-ended shaft of negligible mass that is supported by a bearing at $C$. The disk is free to rotate about its horizontal diameter $AB$, and the shaft is free to rotate about a vertical axis through $C$. Initially, the disk lies in a vertical plane ($\theta_0 = 90°$) and the shaft has an angular velocity $\dot{\phi}_0 = 16$ rad/s. If the disk is slightly disturbed, determine for the ensuing motion (a) the minimum value of $\dot{\phi}$, (b) the maximum value of $\dot{\theta}$.

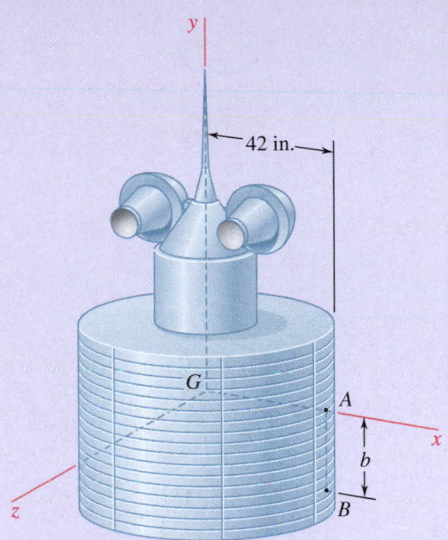

**Fig. P18.129**

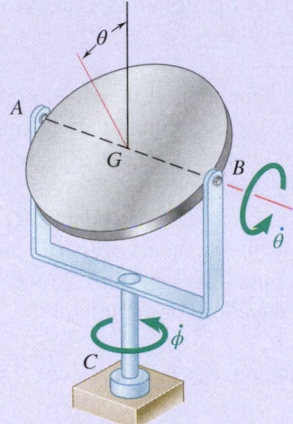

**Fig. P18.131 and P18.132**

**18.132** A homogeneous disk of mass $m$ is connected at $A$ and $B$ to a fork-ended shaft of negligible mass that is supported by a bearing at $C$. The disk is free to rotate about its horizontal diameter $AB$, and the shaft is free to rotate about a vertical axis through $C$. Knowing that initially $\theta_0 = 40°$, $\dot{\theta}_0 = 0$, and $\dot{\phi}_0 = 16$ rad/s, determine for the ensuing motion (a) the range of values of $\theta$, (b) the minimum value of $\dot{\phi}$, (c) the maximum value of $\dot{\theta}$.

**18.133** A homogeneous square plate with a mass $m$ and side $c$ is held at points $A$ and $B$ by a frame of negligible mass that is supported by bearings at points $C$ and $D$. The plate is free to rotate about $AB$, and the frame is free to rotate about the vertical $CD$. Knowing that, initially, $\theta_0 = 45°$, $\dot{\theta}_0 = 0$, and $\dot{\phi}_0 = 8$ rad/s, determine for the ensuing motion (a) the range of values of $\theta$, (b) the minimum value of $\dot{\phi}$, (c) the maximum value of $\dot{\theta}$.

**18.134** A homogeneous square plate with a mass $m$ and side $c$ is held at points $A$ and $B$ by a frame of negligible mass that is supported by bearings at points $C$ and $D$. The plate is free to rotate about $AB$, and the frame is free to rotate about the vertical $CD$. Initially, the plate lies in the plane of the frame ($\theta_0 = 90°$), and the frame has an angular velocity of $\dot{\phi} = 8$ rad/s. If the plate is slightly disturbed, determine for the ensuing motion (a) the minimum value of $\dot{\phi}$, (b) the maximum value of $\dot{\theta}$.

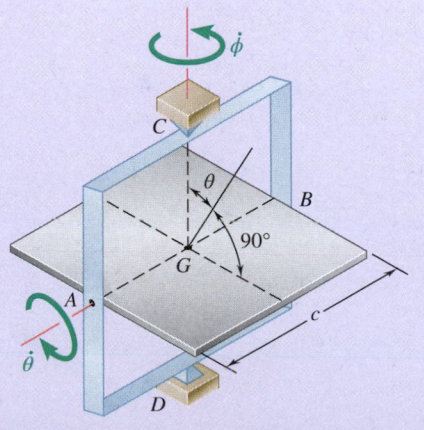

**Fig. P18.133 and P18.134**

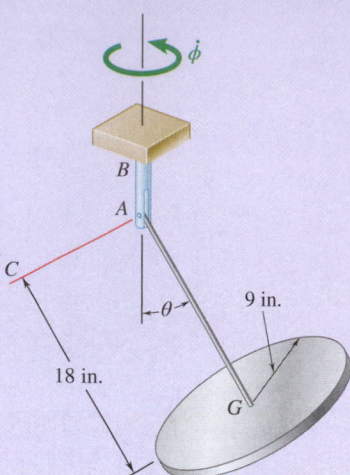

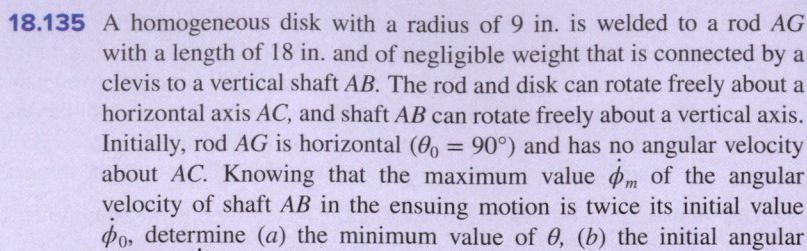

**Fig. P18.135 and P18.136**

**18.135** A homogeneous disk with a radius of 9 in. is welded to a rod $AG$ with a length of 18 in. and of negligible weight that is connected by a clevis to a vertical shaft $AB$. The rod and disk can rotate freely about a horizontal axis $AC$, and shaft $AB$ can rotate freely about a vertical axis. Initially, rod $AG$ is horizontal ($\theta_0 = 90°$) and has no angular velocity about $AC$. Knowing that the maximum value $\dot{\phi}_m$ of the angular velocity of shaft $AB$ in the ensuing motion is twice its initial value $\dot{\phi}_0$, determine (a) the minimum value of $\theta$, (b) the initial angular velocity $\dot{\phi}_0$ of shaft $AB$.

**18.136** A homogeneous disk with a radius of 9 in. is welded to a rod $AG$ with a length of 18 in. and of negligible weight that is connected by a clevis to a vertical shaft $AB$. The rod and disk can rotate freely about a horizontal axis $AC$, and shaft $AB$ can rotate freely about a vertical axis. Initially, rod $AG$ is horizontal ($\theta_0 = 90°$) and has no angular velocity about $AC$. Knowing that the smallest value of $\theta$ in the ensuing motion is 30°, determine (a) the initial angular velocity of shaft $AB$, (b) its maximum angular velocity.

**\*18.137** The top shown is supported at the fixed point $O$. Denoting by $\phi$, $\theta$, and $\psi$ the Eulerian angles defining the position of the top with respect to a fixed frame of reference, consider the general motion of the top in which all Eulerian angles vary.

(a) Observing that $\Sigma M_Z = 0$ and $\Sigma M_z = 0$, and denoting by $I$ and $I'$, respectively, the moments of inertia of the top about its axis of symmetry and about a transverse axis through $O$, derive the two first-order differential equations of motion

$$I'\dot{\phi}\sin^2\theta + I(\dot{\psi} + \dot{\phi}\cos\theta)\cos\theta = \alpha \qquad (1)$$

$$I(\dot{\psi} + \dot{\phi}\cos\theta) = \beta \qquad (2)$$

where $\alpha$ and $\beta$ are constants depending upon the initial conditions. These equations express that the angular momentum of the top is conserved about both the $Z$ and $z$ axes; that is, that the rectangular component of $\mathbf{H}_O$ along each of these axes is constant.

(b) Use Eqs. (1) and (2) to show that the rectangular component $\omega_z$ of the angular velocity of the top is constant and that the rate of precession $\dot{\phi}$ depends upon the value of the angle of nutation $\theta$.

**\*18.138** (a) Applying the principle of conservation of energy, derive a third differential equation for the general motion of the top of Prob. 18.137.

(b) Eliminating the derivatives $\dot{\phi}$ and $\dot{\psi}$ from the equation obtained and from the two equations of Prob. 18.137, show that the rate of nutation $\dot{\theta}$ is defined by the differential equation $\dot{\theta}^2 = f(\theta)$, where

$$f(\theta) = \frac{1}{I'}\left(2E - \frac{\beta^2}{I} - 2mgc\cos\theta\right) - \left(\frac{\alpha - \beta\cos\theta}{I'\sin\theta}\right)^2 \qquad (1)$$

(c) Further show, by introducing the auxiliary variable $x = \cos\theta$, that the maximum and minimum values of $\theta$ can be obtained by solving the cubic equation for $x$

$$\left(2E - \frac{\beta^2}{I} - 2mgcx\right)(1 - x^2) - \frac{1}{I'}(\alpha - \beta x)^2 = 0 \qquad (2)$$

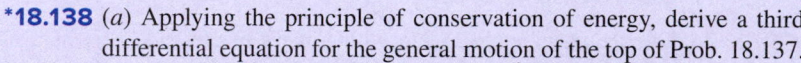

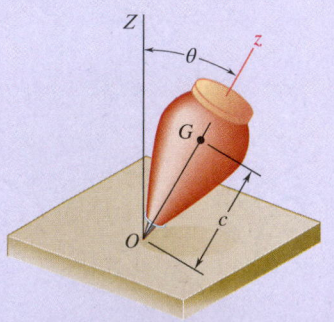

**Fig. P18.137 and P18.138**

**\*18.139** A solid cone of height 180 mm with a circular base of radius 60 mm is supported by a ball and socket at $A$. The cone is released from the position $\theta_0 = 30°$ with a rate of spin $\dot{\psi}_0 = 300$ rad/s, a rate of precession $\dot{\phi}_0 = 20$ rad/s, and a zero rate of nutation. Determine (a) the maximum value of $\theta$ in the ensuing motion, (b) the corresponding values of the rates of spin and precession. (*Hint:* Use Eq. (2) of Prob. 18.138; you can either solve this equation numerically or reduce it to a quadratic equation, given that one of its roots is known.)

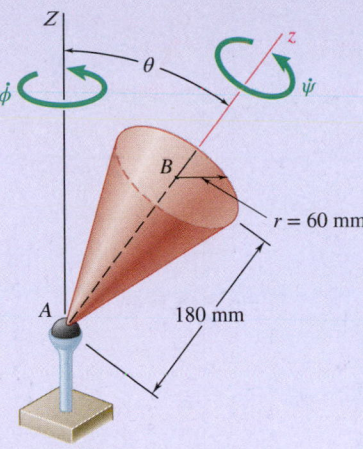

**\*18.140** A solid cone of height 180 mm with a circular base of radius 60 mm is supported by a ball and socket at $A$. The cone is released from the position $\theta_0 = 30°$ with a rate of spin $\dot{\psi}_0 = 300$ rad/s, a rate of precession $\dot{\phi}_0 = -4$ rad/s, and a zero rate of nutation. Determine (a) the maximum value of $\theta$ in the ensuing motion, (b) the corresponding values of the rates of spin and precession, (c) the value of $\theta$ for which the sense of the precession is reversed. (See hint in Prob. 18.139.)

**Fig. P18.139 and P18.140**

**\*18.141** A homogeneous sphere of mass $m$ and radius $a$ is welded to a rod $AB$ of negligible mass, which is held by a ball-and-socket support at $A$. The sphere is released in the position $\beta = 0$ with a rate of precession $\dot{\phi} = \sqrt{17g/11a}$ with no spin or nutation. Determine the largest value of $\beta$ in the ensuing motion.

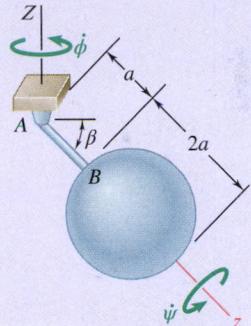

**\*18.142** A homogeneous sphere of mass $m$ and radius $a$ is welded to a rod $AB$ of negligible mass, which is held by a ball-and-socket support at $A$. The sphere is released in the position $\beta = 0$ with a rate of precession $\dot{\phi} = \dot{\phi}_0$ with no spin or nutation. Knowing that the largest value of $\beta$ in the ensuing motion is $30°$, determine (a) the rate of precession $\dot{\phi}_0$ of the sphere in its initial position, (b) the rates of precession and spin when $\beta = 30°$.

**Fig. P18.141 and P18.142**

**\*18.143** Consider a rigid body of arbitrary shape that is attached at its mass center $O$ and subjected to no force other than its weight and the reaction of the support at $O$.

    (a) Prove that the angular momentum $\mathbf{H}_O$ of the body about the fixed point $O$ is constant in magnitude and direction, that the kinetic energy $T$ of the body is constant, and that the projection along $\mathbf{H}_O$ of the angular velocity $\boldsymbol{\omega}$ of the body is constant.

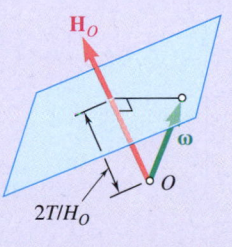

    (b) Show that the tip of the vector $\boldsymbol{\omega}$ describes a curve on a fixed plane in space (called the *invariable plane*), which is perpendicular to $\mathbf{H}_O$ and at a distance $2T/H_O$ from $O$.

    (c) Show that with respect to a frame of reference attached to the body and coinciding with its principal axes of inertia, the tip of the vector $\boldsymbol{\omega}$ appears to describe a curve on an ellipsoid of equation

$$I_x\omega_x^2 + I_y\omega_y^2 + I_z\omega_z^2 = 2T = \text{constant}$$

The ellipsoid (called the *Poinsot ellipsoid*) is rigidly attached to the body and is of the same shape as the ellipsoid of inertia, but of a different size.

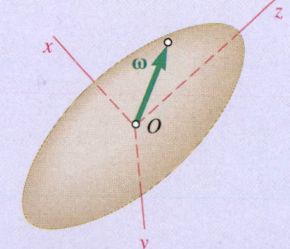

**Fig. P18.143**

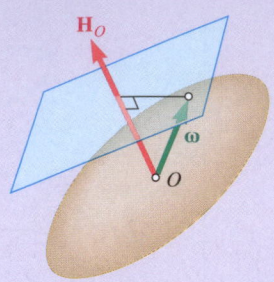

**Fig. P18.144**

*18.144 Referring to Prob. 18.143, (*a*) prove that the Poinsot ellipsoid is tangent to the invariable plane, (*b*) show that the motion of the rigid body must be such that the Poinsot ellipsoid appears to roll on the invariable plane. (*Hint:* In part *a*, show that the normal to the Poinsot ellipsoid at the tip of **ω** is parallel to **H**$_O$. It is recalled that the direction of the normal to a surface of equation $F(x, y, z) = $ constant at a point $P$ is the same as that of **grad** $F$ at point $P$.)

*18.145 Using the results obtained in Probs. 18.143 and 18.144, show that for an axisymmetric body attached at its mass center $O$ and under no force other than its weight and the reaction at $O$, the Poinsot ellipsoid is an ellipsoid of revolution and the space and body cones are both circular and are tangent to each other. Further show that (*a*) the two cones are tangent externally, and the precession is direct, when $I < I'$, where $I$ and $I'$ denote, respectively, the axial and transverse moment of inertia of the body, (*b*) the space cone is inside the body cone, and the precession is retrograde, when $I > I'$.

*18.146 Refer to Probs. 18.143 and 18.144.

(*a*) Show that the curve (called *polhode*) described by the tip of the vector **ω** with respect to a frame of reference coinciding with the principal axes of inertia of the rigid body is defined by the equations

$$I_x\omega_x^2 + I_y\omega_y^2 + I_z\omega_z^2 = 2T = \text{constant} \tag{1}$$

$$I_x^2\omega_x^2 + I_y^2\omega_y^2 + I_z^2\omega_z^2 = H_O^2 = \text{constant} \tag{2}$$

and that this curve can, therefore, be obtained by intersecting the Poinsot ellipsoid with the ellipsoid defined by Eq. (2).

(*b*) Further show, assuming $I_x > I_y > I_z$, that the polhodes obtained for various values of $H_O$ have the shapes indicated in the figure.

(*c*) Using the result obtained in part *b*, show that a rigid body under no force can rotate about a fixed centroidal axis if, and only if, that axis coincides with one of the principal axes of inertia of the body, and that the motion will be stable if the axis of rotation coincides with the major or minor axis of the Poinsot ellipsoid ($z$ or $x$ axis in the figure) and unstable if it coincides with the intermediate axis ($y$ axis).

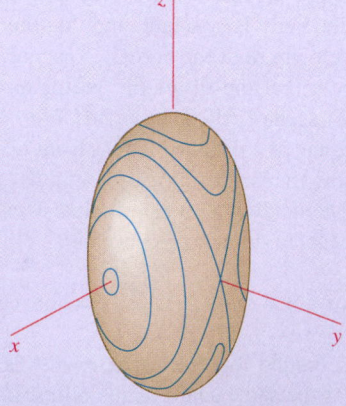

**Fig. P18.146**

# Review and Summary

This chapter was devoted to the kinetic analysis of the motion of rigid bodies in three dimensions.

## Fundamental Equations of Motion for a Rigid Body

We first noted that the two fundamental equations derived in Chap. 14 for the motion of a system of particles,

$$\Sigma \mathbf{F} = m\overline{\mathbf{a}} \qquad (18.1)$$

$$\Sigma \mathbf{M}_G = \dot{\mathbf{H}}_G \qquad (18.2)$$

provide the foundation of our analysis, just as they did in Chap. 16 in the case of the plane motion of rigid bodies. The computation of the angular momentum $\mathbf{H}_G$ of the body and of its derivative $\dot{\mathbf{H}}_G$, however, are now considerably more involved.

## Angular Momentum of a Rigid Body in Three Dimensions

In Sec. 18.1A, we saw that we can express the rectangular components of the angular momentum $\mathbf{H}_G$ of a rigid body in terms of the components of its angular velocity $\boldsymbol{\omega}$ and of its centroidal moments and products of inertia:

$$\begin{aligned}
H_x &= +\bar{I}_x\omega_x - \bar{I}_{xy}\omega_y - \bar{I}_{xz}\omega_z \\
H_y &= -\bar{I}_{yx}\omega_x + \bar{I}_y\omega_y - \bar{I}_{yz}\omega_z \\
H_z &= -\bar{I}_{zx}\omega_x - \bar{I}_{zy}\omega_y + \bar{I}_z\omega_z
\end{aligned} \qquad (18.7)$$

If we use **principal axes of inertia** $Gx'y'z'$, these relations reduce to

$$H_{x'} = \bar{I}_{x'}\omega_{x'} \qquad H_{y'} = \bar{I}_{y'}\omega_{y'} \qquad H_{z'} = \bar{I}_{z'}\omega_{z'} \qquad (18.10)$$

We observed that, in general, *the angular momentum $\mathbf{H}_G$ and the angular velocity $\boldsymbol{\omega}$ do not have the same direction* (Fig. 18.25). They do, however, have the same direction if $\boldsymbol{\omega}$ is directed along one of the principal axes of inertia of the body.

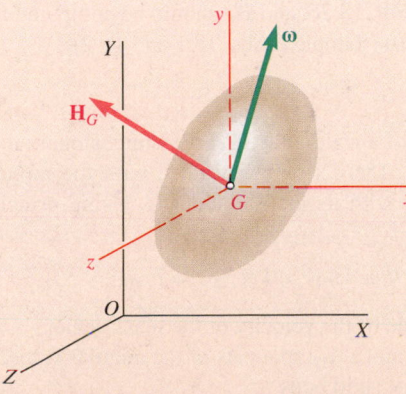

**Fig. 18.25**

## Angular Momentum About a Given Point

Recalling that the system of the momenta of the particles forming a rigid body can be reduced to the vector $m\bar{\mathbf{v}}$ attached at $G$ and the couple $\mathbf{H}_G$ (Fig. 18.26), we noted that, once we have determined the linear momentum $m\bar{\mathbf{v}}$ and the angular momentum $\mathbf{H}_G$ of a rigid body, we can obtain the angular momentum $\mathbf{H}_P$ of the body about any given point $P$ from

$$\mathbf{H}_P = \bar{\mathbf{r}} \times m\bar{\mathbf{v}} + \mathbf{H}_G \tag{18.11}$$

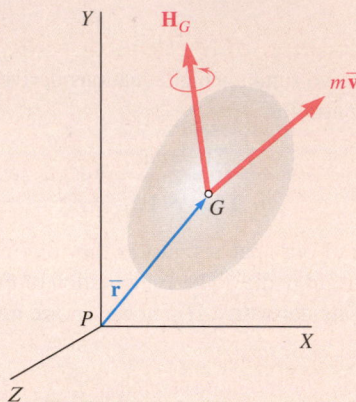

**Fig. 18.26**

## Rigid Body with a Fixed Point

In the particular case of a rigid body *constrained to rotate about a fixed point O*, we can obtain the components of the angular momentum $\mathbf{H}_O$ of the body about $O$ directly from the components of its angular velocity and from its moments and products of inertia with respect to axes through $O$.

$$\begin{aligned}
H_x &= +I_x\omega_x - I_{xy}\omega_y - I_{xz}\omega_z \\
H_y &= -I_{yx}\omega_x + I_y\omega_y - I_{yz}\omega_z \\
H_z &= -I_{zx}\omega_x - I_{zy}\omega_y + I_z\omega_z
\end{aligned} \tag{18.13}$$

## Principle of Impulse and Momentum

The *principle of impulse and momentum* for a rigid body in three-dimensional motion (Sec. 18.1B) is expressed by the same fundamental formula that was used in Chap. 17 for a rigid body in plane motion as

$$\textbf{Syst Momenta}_1 + \textbf{Syst Ext Imp}_{1 \to 2} = \textbf{Syst Momenta}_2 \tag{17.14}$$

However, the systems of the initial and final momenta should now be represented as shown in Fig. 18.26, and $\mathbf{H}_G$ should be computed from the relations in Eqs. (18.7) or (18.10) (Sample Probs. 18.1 and 18.2).

## Kinetic Energy of a Rigid Body in Three Dimensions

The kinetic energy of a rigid body in three-dimensional motion can be divided into two parts (Sec. 18.1C): one associated with the motion of its mass center $G$ and the other with its motion about $G$. Using principal centroidal axes $x'$, $y'$, $z'$, we wrote

$$T = \tfrac{1}{2}m\bar{v}^2 + \tfrac{1}{2}(\bar{I}_{x'}\omega_{x'}^2 + \bar{I}_{y'}\omega_{y'}^2 + \bar{I}_{z'}\omega_{z'}^2) \tag{18.17}$$

where $\bar{v}$ = magnitude of the velocity of the mass center
$\omega_{x'}, \omega_{y'}, \omega_{z'}$ = $x'$, $y'$, and $z'$ components of the angular velocity
$\quad\quad m$ = mass of rigid body
$\bar{I}_{x'}, \bar{I}_{y'}, \bar{I}_{z'}$ = principal centroidal moments of inertia

We also noted that, in the case of a rigid body *constrained to rotate about a fixed point O,* we can express the kinetic energy of the body as

$$T = \tfrac{1}{2}(I_{x'}\omega_{x'}^2 + I_{y'}\omega_{y'}^2 + I_{z'}\omega_{z'}^2) \qquad \textbf{(18.20)}$$

where the $x'$, $y'$, and $z'$ axes are the principal axes of inertia of the body about $O$. These results make it possible to extend the application of the principle of work and energy and the principle of conservation of energy to the three-dimensional motion of a rigid body.

## Using a Rotating Frame to Write the Equations of Motion of a Rigid Body in Space

Sec. 18.2 was devoted to applying the fundamental equations

$$\Sigma\mathbf{F} = m\overline{\mathbf{a}} \qquad \textbf{(18.1)}$$

$$\Sigma\mathbf{M}_G = \dot{\mathbf{H}}_G \qquad \textbf{(18.2)}$$

to the motion of a rigid body in three dimensions. We first recalled (Sec. 18.2A) that $\mathbf{H}_G$ represents the angular momentum of the body relative to a centroidal frame $GX'Y'Z'$ of fixed orientation (Fig. 18.27) and that $\dot{\mathbf{H}}_G$ in Eq. (18.2)

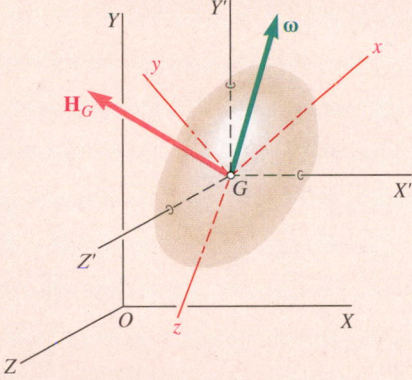

**Fig. 18.27**

represents the rate of change of $\mathbf{H}_G$ with respect to that frame. We noted that, as the body rotates, its moments and products of inertia with respect to the frame $GX'Y'Z'$ change continually. Therefore, it is more convenient to use a rotating frame $Gxyz$ when resolving $\boldsymbol{\omega}$ into components and computing the moments and products of inertia that are used to determine $\mathbf{H}_G$ from Eqs. (18.7) or (18.10). However, because $\dot{\mathbf{H}}_G$ in Eq. (18.2) represents the rate of change of $\mathbf{H}_G$ with respect to the frame $GX'Y'Z'$ of fixed orientation, we must use the method of Sec. 15.5A to determine its value. Recalling Eq. (15.31), we wrote

$$\dot{\mathbf{H}}_G = (\dot{\mathbf{H}}_G)_{Gxyz} + \boldsymbol{\Omega} \times \mathbf{H}_G \qquad \textbf{(18.22)}$$

where $\mathbf{H}_G$ = angular momentum of body with respect to frame $GX'Y'Z'$ of fixed orientation

$(\dot{\mathbf{H}}_G)_{Gxyz}$ = rate of change of $\mathbf{H}_G$ with respect to rotating frame $Gxyz$ to be computed from relations in Eq. (18.7)

$\boldsymbol{\Omega}$ = angular velocity of the rotating frame $Gxyz$

Substituting for $\dot{\mathbf{H}}_G$ from Eq. (18.22) into Eq. (18.2), we obtained

$$\Sigma \mathbf{M}_G = (\dot{\mathbf{H}}_G)_{Gxyz} + \Omega \times \mathbf{H}_G \qquad (18.23)$$

If the rotating frame is actually attached to the body, its angular velocity $\Omega$ is identically equal to the angular velocity $\omega$ of the body. In many applications, however, it is advantageous to use a frame of reference that is not attached to the body but rotates in an independent manner (Sample Prob. 18.5).

### Euler's Equations of Motion

Setting $\Omega = \omega$ in Eq. (18.23), using principal axes, and writing this equation in scalar form, we obtained **Euler's equations of motion** (Sec.18.2B). Then we extended Newton's second law to the three-dimensional motion of a rigid body, showing that the system of the external forces acting on the rigid body is not only equipollent but actually *equivalent* to the inertial terms of the body represented by the vector $m\overline{\mathbf{a}}$ and the couple $\dot{\mathbf{H}}_G$ (Fig. 18.28). You can solve problems involving the three-dimensional motion of a rigid body by considering the free-body and kinetic diagrams represented in Fig. 18.28 and writing appropriate scalar equations relating the components or moments of the external forces and inertial terms (Sample Probs. 18.3 and 18.5). You can use Fig. 18.28 to help you sum moments about any point $P$ using the expression

$$\Sigma \mathbf{M}_P = \dot{\mathbf{H}}_G + \mathbf{r}_{G/P} \times m\overline{\mathbf{a}}$$

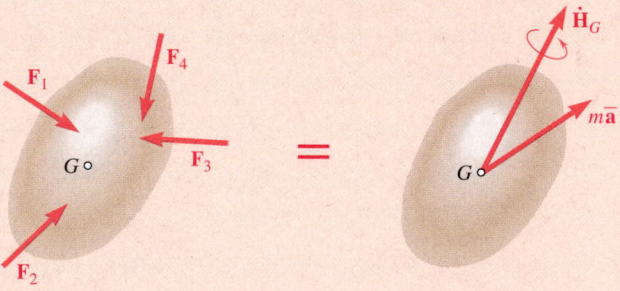

Fig. 18.28

### Rigid Body with a Fixed Point

In the case of a rigid body *constrained to rotate about a fixed point O*, we can use an alternative method of solution involving the moments of the forces and the rate of change of the angular momentum about point $O$. We wrote (Sec. 18.2C)

$$\Sigma \mathbf{M}_O = (\dot{\mathbf{H}}_O)_{Oxyz} + \Omega \times \mathbf{H}_O \qquad (18.28)$$

You can use this approach to solve some types of problems involving the rotation of a rigid body about a fixed axis (Sec. 18.2D), such as an unbalanced rotating shaft (Sample Prob. 18.4).

### Motion of a Gyroscope

In Sec. 18.3, we considered the motion of **gyroscopes** and other *axisymmetric bodies*. We introduced the **Eulerian angles** $\phi$, $\theta$, and $\psi$ to define the position of a gyroscope (Fig. 18.29), and we observed that their derivatives $\dot{\phi}$, $\dot{\theta}$, and $\dot{\psi}$ represent, respectively, the rates of **precession, nutation**, and **spin** of the

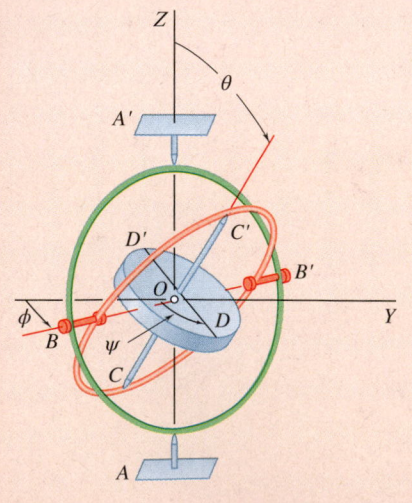

Fig. 18.29

gyroscope (Sec. 18.3A). Expressing the angular velocity **ω** in terms of these derivatives, we wrote

$$\boldsymbol{\omega} = -\dot{\phi} \sin \theta \mathbf{i} + \dot{\theta} \mathbf{j} + (\dot{\psi} + \dot{\phi} \cos \theta) \mathbf{k} \qquad (18.35)$$

where the unit vectors are associated with a frame $Oxyz$ attached to the inner gimbal of the gyroscope (Fig. 18.30). These vectors rotate, therefore, with the angular velocity

$$\boldsymbol{\Omega} = -\dot{\phi} \sin \theta \mathbf{i} + \dot{\theta} \mathbf{j} + \dot{\phi} \cos \theta \mathbf{k} \qquad (18.38)$$

Denoting the moment of inertia of the gyroscope with respect to its spin axis $z$ by $I$ and its moment of inertia with respect to a transverse axis through $O$ by $I'$, we wrote

$$\mathbf{H}_O = -I' \dot{\phi} \sin \theta \mathbf{i} + I' \dot{\theta} \mathbf{j} + I(\dot{\psi} + \dot{\phi} \cos \theta) \mathbf{k} \qquad (18.36)$$

Substituting for $\mathbf{H}_O$ and $\boldsymbol{\Omega}$ into Eq. (18.28) led us to the differential equations defining the motion of a gyroscope.

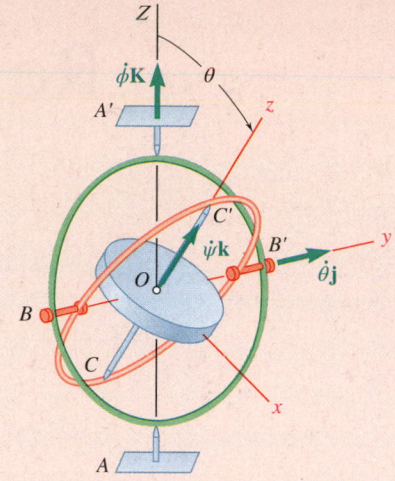

**Fig. 18.30**

## Steady Precession

In the particular case of the **steady precession** of a gyroscope (Sec. 18.3B), the angle $\theta$, the rate of precession $\dot{\phi}$, and the rate of spin $\dot{\psi}$ remain constant. We saw that such a motion is possible only if the moments of the external forces about $O$ satisfy the relation

$$\Sigma \mathbf{M}_O = (I\omega_z - I' \dot{\phi} \cos \theta) \dot{\phi} \sin \theta \mathbf{j} \qquad (18.44)$$

that is, if the external forces reduce to a couple of moment equal to the right-hand side of Eq. (18.44) and applied about an axis perpendicular to the precession axis and to the spin axis (Fig. 18.31). This chapter ended with a discussion of the motion of an axisymmetric body spinning and precessing under no force (Sec. 18.3C; Sample Prob. 18.6).

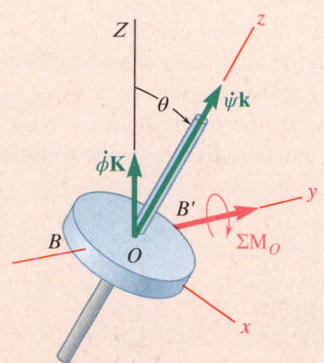

**Fig. 18.31**

# Review Problems

**18.147** Three 25-lb rotor disks are attached to a shaft that rotates at 720 rpm. Disk $A$ is attached eccentrically so that its mass center is $\frac{1}{4}$ in. from the axis of rotation, while disks $B$ and $C$ are attached so that their mass centers coincide with the axis of rotation. Where should 2-lb weights be bolted to disks $B$ and $C$ to balance the system dynamically?

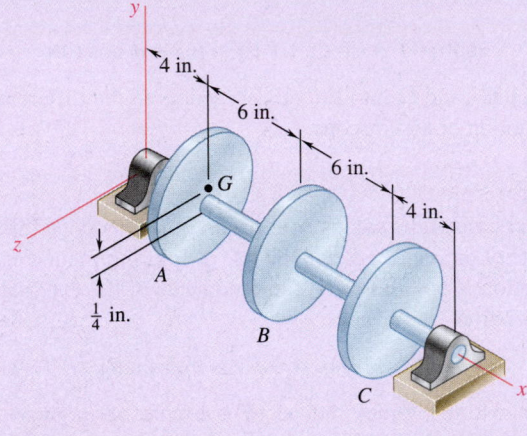

**Fig. P18.147**

**18.148** A homogeneous disk of mass $m = 5$ kg rotates at the constant rate $\omega_1 = 8$ rad/s with respect to the bent axle $ABC$, which itself rotates at the constant rate $\omega_2 = 3$ rad/s about the $y$ axis. Determine the angular momentum $\mathbf{H}_C$ of the disk about its center $C$.

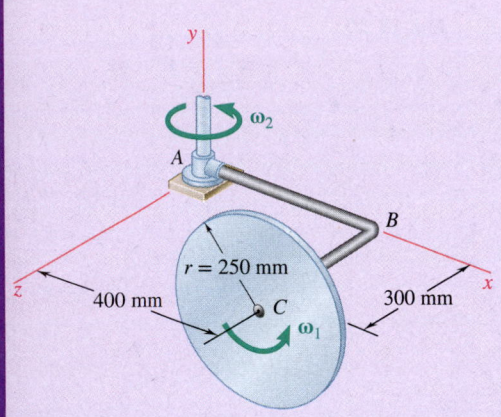

**Fig. P18.148**

**18.149** A rod of uniform cross-section is used to form the shaft shown. Denoting by $m$ the total mass of the shaft and knowing that the shaft rotates with a constant angular velocity $\boldsymbol{\omega}$, determine (a) the angular momentum $\mathbf{H}_G$ of the shaft about its mass center $G$, (b) the angle formed by $\mathbf{H}_G$ and the axis $AB$, (c) the angular momentum of the shaft about point $A$.

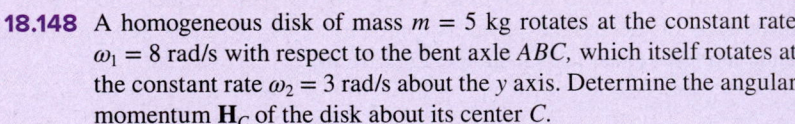
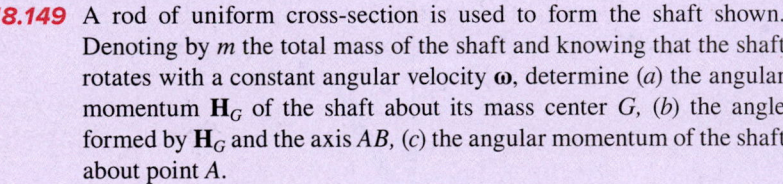

**Fig. P18.149**

**18.150** A uniform rod of mass $m$ and length $5a$ is bent into the shape shown and is suspended from a wire attached at point $B$. Knowing that the rod is hit at point $A$ in the negative $y$ direction and denoting the corresponding impulse by $-(F\,\Delta t)\mathbf{j}$, determine immediately after the impact (a) the velocity of the mass center $G$, (b) the angular velocity of the rod.

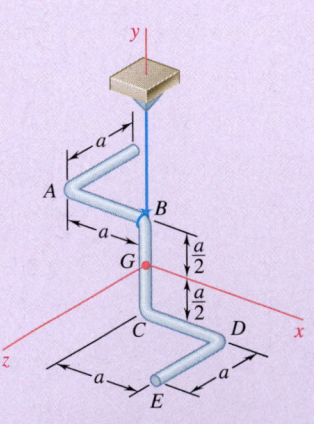

**Fig. P18.150**

**18.151** A four-bladed airplane propeller has a mass of 160 kg and a radius of gyration of 800 mm. Knowing that the propeller rotates at 1600 rpm as the airplane is traveling in a circular path of 600-m radius at 540 km/h, determine the magnitude of the couple exerted by the propeller on its shaft due to the rotation of the airplane.

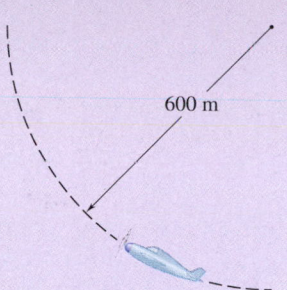

**Fig. P18.151**

**18.152** A 2.4-kg piece of sheet steel with dimensions $160 \times 640$ mm was bent to form the component shown. The component is at rest ($\omega = 0$) when a couple $\mathbf{M}_0 = (0.8 \text{ N·m})\mathbf{k}$ is applied to it. Determine (a) the angular acceleration of the component, (b) the dynamic reactions at $A$ and $B$ immediately after the couple is applied.

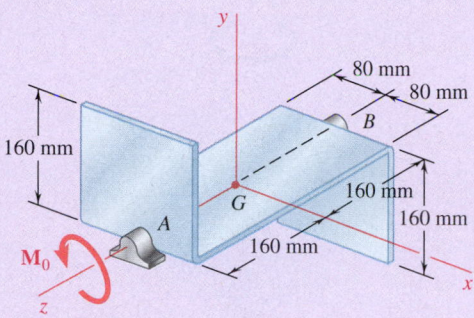

**Fig. *P18.152***

**18.153** A homogeneous disk of weight $W = 6$ lb rotates at the constant rate $\omega_1 = 16$ rad/s with respect to arm $ABC$, which is welded to a shaft $DCE$ rotating at the constant rate $\omega_2 = 8$ rad/s. Determine the dynamic reactions at $D$ and $E$.

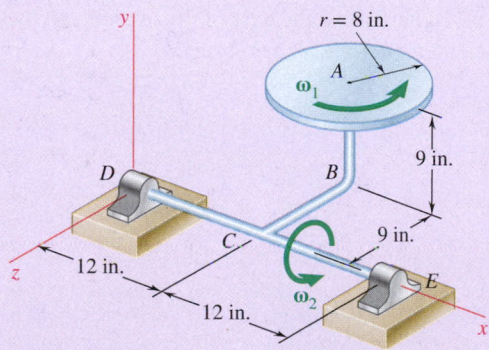

**Fig. P18.153**

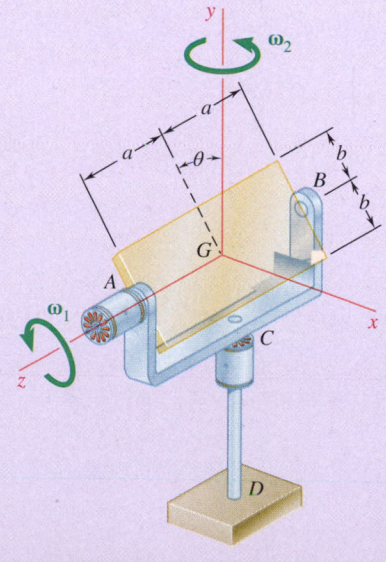

**Fig. P18.154**

**18.154** A 48-kg advertising panel of length $2a = 2.4$ m and width $2b = 1.6$ m is kept rotating at a constant rate $\omega_1$ about its horizontal axis by a small electric motor attached at $A$ to frame $ACB$. This frame itself is kept rotating at a constant rate $\omega_2$ about a vertical axis by a second motor attached at $C$ to the column $CD$. Knowing that the panel and the frame complete a full revolution in 6 s and 12 s, respectively, express, as a function of the angle $\theta$, the dynamic reaction exerted on column $CD$ by its support at $D$.

**18.155** A 2500-kg satellite is 2.4 m high and has octagonal bases of sides 1.2 m. The coordinate axes shown are the principal centroidal axes of inertia of the satellite, and its radii of gyration are $k_x = k_z = 0.90$ m and $k_y = 0.98$ m. The satellite is equipped with a main 500-N thruster $E$ and four 20-N thrusters $A$, $B$, $C$, and $D$ that can expel fuel in the positive $y$ direction. The satellite is spinning at the rate of 36 rev/h about its axis of symmetry $Gy$, which maintains a fixed direction in space, when thrusters $A$ and $B$ are activated for 2 s. Determine (a) the precession axis of the satellite, (b) its rate of precession, (c) its rate of spin.

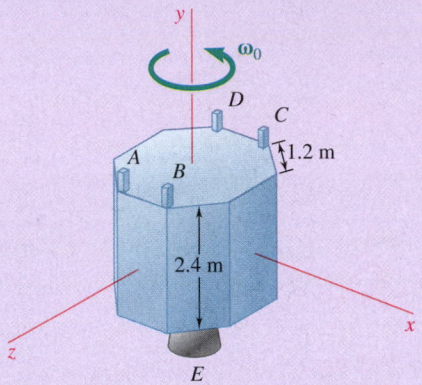

**Fig. P18.155**

**18.156** The space capsule has no angular velocity when the jet at $A$ is activated for 1 s in a direction parallel to the $x$ axis. Knowing that the capsule has a mass of 1000 kg, that its radii of gyration are $\bar{k}_z = \bar{k}_y = 1.00$ m and $\bar{k}_z = 1.25$ m, and that the jet at $A$ produces a thrust of 50 N, determine the axis of precession and the rates of precession and spin after the jet has stopped.

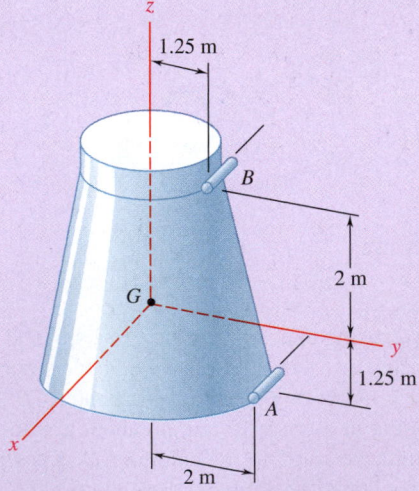

**Fig. P18.156**

**18.157** A homogeneous rectangular plate of mass $m$ and sides $c$ and $2c$ is held at $A$ and $B$ by a fork-ended shaft of negligible mass, which is supported by a bearing at $C$. The plate is free to rotate about $AB$, and the shaft is free to rotate about a horizontal axis through $C$. Knowing that initially $\theta_0 = 40°$, $\dot{\theta}_0 = 0$, and $\dot{\phi}_0 = 10$ rad/s, determine for the ensuing motion ($a$) the range of values of $\theta$, ($b$) the minimum value of $\dot{\phi}$, ($c$) the maximum value of $\dot{\theta}$.

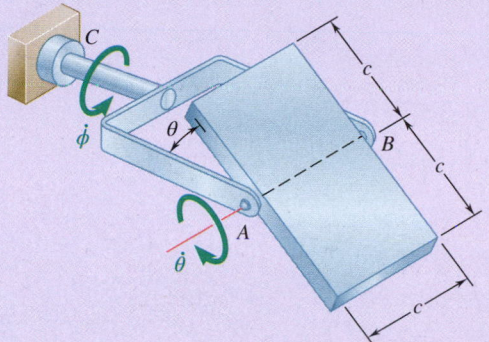

**Fig. P18.157**

**18.158** The essential features of the gyrocompass are shown. The rotor spins at the rate $\dot{\psi}$ about an axis mounted in a single gimbal, which may rotate freely about the vertical axis $AB$. The angle formed by the axis of the rotor and the plane of the meridian is denoted by $\theta$, and the latitude of the position on the earth is denoted by $\lambda$. We note that line $OC$ is parallel to the axis of the earth, and we denote by $\boldsymbol{\omega}_e$ the angular velocity of the earth about its axis.

($a$) Show that the equations of motion of the gyrocompass are

$$I'\ddot{\theta} + I\omega_z\omega_e \cos \lambda \sin \theta - I'\omega_e^2 \cos^2 \lambda \sin \theta \cos \theta = 0$$
$$I\dot{\omega}_z = 0$$

where $\omega_z$ is the rectangular component of the total angular velocity $\boldsymbol{\omega}$ along the axis of the rotor, and $I$ and $I'$ are the moments of inertia of the rotor with respect to its axis of symmetry and a transverse axis through $O$, respectively.

($b$) Neglecting the term containing $\omega_e^2$, show that for small values of $\theta$, we have

$$\ddot{\theta} + \frac{I\omega_z\omega_e \cos \lambda}{I'}\theta = 0$$

and that the axis of the gyrocompass oscillates about the north–south direction.

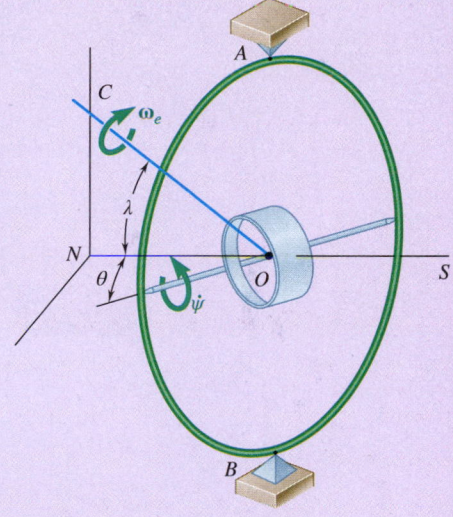

**Fig. P18.158**

©Peter Tsai Photography

# 19

## Mechanical Vibrations

The Wind Damper inside of a building helps protect against typhoons and earthquakes by reducing the effects of wind and vibrations on the building. Mechanical systems may undergo *free vibrations* or they may be subject to *forced vibrations*. The vibrations are *damped* when there is energy dissipation and *undamped* otherwise. This chapter is an introduction to many fundamental concepts in vibration analysis.

# Objectives

- **Define, compare, and contrast** simple harmonic motion, undamped free and forced vibrations, and damped free and forced vibrations.

- Using Newton's second law, **determine** the differential equation of motion of a particle or a rigid body undergoing vibratory motion.

- Using the conservation of energy, **determine** the differential equation of motion of a particle or a rigid body undergoing vibratory motion.

- **Calculate** the natural circular frequency, period, and natural frequency for a system undergoing simple harmonic motion.

- **Calculate** the maximum amplitude and the magnification factor for a body undergoing forced vibrations.

- **Compare and contrast** the vibration responses of under-damped, critically damped, and overdamped systems.

# Introduction

A **mechanical vibration** is the motion of a particle or body that oscillates about a position of equilibrium. Most vibrations in machines and structures are undesirable because of the increased stresses and energy losses that accompany them. Appropriate design therefore aims to eliminate or reduce vibrations as much as possible. The analysis of vibrations has become increasingly important in recent years owing to the current trend toward higher-speed machines and lighter structures. There is every reason to expect that this trend will continue and that an even greater need for vibration analysis will develop in the future.

The analysis of vibrations is a very extensive subject to which entire texts have been devoted. Our present study is limited to the simplest types of vibrations—namely, the vibrations of a body or a system of bodies with one degree of freedom.

A mechanical vibration generally results when a system is displaced from a position of stable equilibrium. The system tends to return to this position under the action of restoring forces (either elastic forces, as in the case of a mass attached to a spring, or gravitational forces, as in the case of a pendulum). But the system generally reaches its original position with an acquired velocity that carries it beyond that position. Because the process can be repeated indefinitely, the system keeps moving back and forth across its position of equilibrium. The time interval required for the system to complete a full cycle of motion is called the **period** of the vibration. The number of cycles per unit time defines the **frequency**, and the maximum displacement of the system from its position of equilibrium is called the **amplitude** of the vibration.

When the motion is maintained by the restoring forces only, the vibration is said to be a **free vibration**. When a periodic force is applied to the system, the resulting motion is described as a **forced vibration**. If we can neglect the effects of friction, the vibrations are said to be **undamped**. However, all vibrations are actually **damped** to some degree. If a free vibration is only slightly damped, its amplitude slowly decreases until, after a certain time, the motion comes to a stop. But if damping is large enough to prevent any true vibration, the system then slowly regains its original position. A damped forced vibration is maintained as long as the periodic force that produces the vibration is applied. The amplitude of the vibration, however, is affected by the magnitude of the damping forces.

In this chapter, we first examine vibrations without damping, studying vibrations of particles, rigid bodies, and forced vibrations. Then we will look at damped vibrations, including both free and forced vibrations.

# 19.1 VIBRATIONS WITHOUT DAMPING

The first step in analyzing vibrations is to formulate an equation of motion for the simple case of a particle in free vibration. We will modify this equation as we consider more complicated situations, such as damped and forced vibrations.

## 19.1A Simple Harmonic Motion and Free Vibrations of Particles

Consider a body with a mass $m$ attached to a spring with a constant $k$ (Fig. 19.1$a$). At the moment, we are concerned only with the motion of its mass center, so we will refer to this body as a particle. When the particle is in static equilibrium, the forces acting on it are its weight **W** and the force **T** exerted by the spring, which has a magnitude $T = k\delta_{st}$, where $\delta_{st}$ denotes the static elongation of the spring from its unstretched length. We therefore have

$$W = k\delta_{st}$$

Suppose now that the particle is displaced through a distance $x_m$ from its equilibrium position and released with no initial velocity. If we have chosen $x_m$ to be smaller than $\delta_{st}$, the particle moves back and forth through its equilibrium position; a vibration with an amplitude $x_m$ is generated. Note that we can also produce a vibration by imparting an initial velocity to the particle when it is in its equilibrium position $x = 0$ or, more generally, by starting the particle from any given position $x = x_0$ with a given initial velocity $\mathbf{v}_0$.

To analyze the vibration, let us consider the particle in a position $P$ at some arbitrary time $t$ (Fig. 19.1$b$). Denoting the displacement $OP$ measured from the equilibrium position $O$ (positive downward) by $x$, we note that the forces acting on the particle are its weight **W** and the force **T** exerted by the spring. In this position, the spring force has a magnitude $T = k(\delta_{st} + x)$. Recalling that $W = k\delta_{st}$, we find that the magnitude of the resultant **F** of the two forces (positive downward) is

$$F = W - k(\delta_{st} + x) = -kx \tag{19.1}$$

Thus, the resultant of the forces exerted on the particle is proportional to the displacement $OP$ **measured from the equilibrium position**. Recalling the sign convention, we note that **F** is always directed *toward* the equilibrium

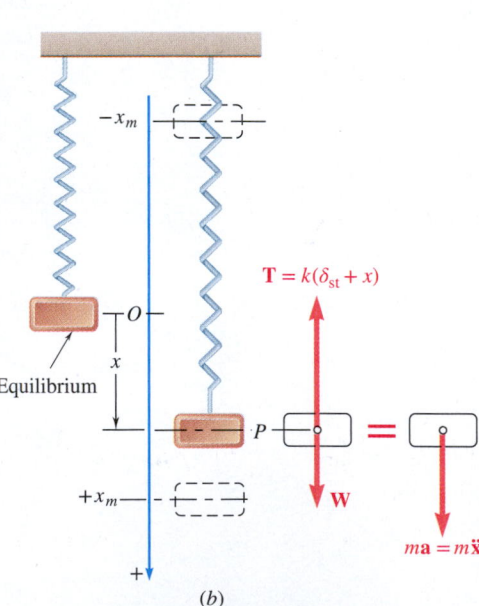

**Fig. 19.1** ($a$) At the equilibrium position, the spring force is equal to the weight; ($b$) the block at position $P$ with its free-body diagram and kinetic diagram.

position $O$. Substituting for $F$ into the fundamental equation $F = ma$ and recalling that $a$ is the second derivative $\ddot{x}$ of $x$ with respect to $t$, we have

**Equation of motion for
simple harmonic motion**

$$m\ddot{x} + kx = 0 \tag{19.2}$$

Note that we use the same sign convention for the acceleration $\ddot{x}$ and for the displacement $x$, namely, positive downward. By measuring the displacement from the static equilibrium point, we get a homogeneous differential equation; that is, the right-hand side is equal to zero.

The motion defined by Eq. (19.2) is called **simple harmonic motion**. It is characterized by the fact that **the acceleration is proportional to the displacement and in the opposite direction**. We can verify that each of the functions

$$x_1 = \sin\left(\sqrt{k/m}\; t\right) \quad \text{and} \quad x_2 = \cos\left(\sqrt{k/m}\; t\right)$$

satisfies Eq. (19.2). These functions, therefore, constitute two *particular solutions* of the differential equation (19.2). We can obtain the *general solution* of Eq. (19.2) by multiplying each of the particular solutions by an arbitrary constant and adding. Thus, the general solution is

$$x = C_1 x_1 + C_2 x_2 = C_1 \sin\left(\sqrt{\frac{k}{m}}\; t\right) + C_2 \cos\left(\sqrt{\frac{k}{m}}\; t\right) \tag{19.3}$$

Note that $x$ is a **periodic function** of the time $t$ and therefore represents a vibration of the particle $P$. The coefficient of $t$ in the expression we have obtained is referred to as the **natural circular frequency** of the vibration and is denoted by $\omega_n$. We have

$$\text{Natural circular frequency} = \omega_n = \sqrt{\frac{k}{m}} \tag{19.4}$$

Substituting for $\sqrt{k/m}$ into Eq. (19.3) gives

$$x = C_1 \sin \omega_n t + C_2 \cos \omega_n t \tag{19.5}$$

This is the general solution of the differential equation

$$\ddot{x} + \omega_n^2 x = 0 \tag{19.6}$$

that we can obtain from Eq. (19.2) by dividing both terms by $m$ and observing that $k/m = \omega_n^2$. Differentiating both sides of Eq. (19.5) twice with respect to $t$, we obtain the expressions for the velocity and the acceleration at time $t$ as

$$v = \dot{x} = C_1 \omega_n \cos \omega_n t - C_2 \omega_n \sin \omega_n t \tag{19.7}$$

$$a = \ddot{x} = -C_1 \omega_n^2 \sin \omega_n t - C_2 \omega_n^2 \cos \omega_n t \tag{19.8}$$

The values of the constants $C_1$ and $C_2$ depend upon the *initial conditions* of the motion. For example, we have $C_1 = 0$ if the particle is displaced from its equilibrium position and released at $t = 0$ with no initial velocity. Also, we have $C_2 = 0$ if the particle starts from $O$ at $t = 0$ with a given initial velocity. In general, substituting $t = 0$ and the initial values $x_0$ and $v_0$ of the displacement and the velocity into Eqs. (19.5) and (19.7), we find that $C_1 = v_0/\omega_n$ and $C_2 = x_0$.

We can write these expressions for the displacement, velocity, and acceleration of a particle in a more compact form if we note that Eq. (19.5) says that the displacement $x = OP$ is the sum of the $x$ components of two vectors $\mathbf{C}_1$ and $\mathbf{C}_2$, respectively, with magnitudes of $C_1$ and $C_2$ that are directed as shown in Fig. 19.2a. As $t$ varies, both vectors rotate clockwise; we also note that the magnitude of their resultant $\overrightarrow{OQ}$ is equal to the maximum displacement $x_m$. Thus, we can obtain the simple harmonic motion of $P$ along the $x$ axis by projecting on this axis the motion of a point $Q$ describing an *auxiliary circle* of radius $x_m$ with a constant angular velocity $\omega_n$. This explains the name natural *circular frequency* given to $\omega_n$. Denoting the angle formed by the vectors $\overrightarrow{OQ}$ and $\mathbf{C}_1$ by $\phi$, we have

$$OP = OQ \sin{(\omega_n t + \phi)} \tag{19.9}$$

This leads to new expressions for the displacement, velocity, and acceleration of $P$:

$$x = x_m \sin{(\omega_n t + \phi)} \tag{19.10}$$

$$v = \dot{x} = x_m \omega_n \cos{(\omega_n t + \phi)} \tag{19.11}$$

$$a = \ddot{x} = -x_m \omega_n^2 \sin{(\omega_n t + \phi)} \tag{19.12}$$

The displacement–time curve is represented by a sine curve (Fig. 19.2b); the maximum value $x_m$ of the displacement is called the **amplitude** of the vibration, and the angle $\phi$ that defines the initial position of $Q$ on the circle is called the **phase angle**. As we can see from Fig. 19.2, a full cycle occurs every $2\pi$ rad. The corresponding value of $t$ is denoted by $\tau_n$. This is called the **period** of the free vibration and is measured in seconds. We have

$$\text{Period} = \tau_n = \frac{2\pi}{\omega_n} \tag{19.13}$$

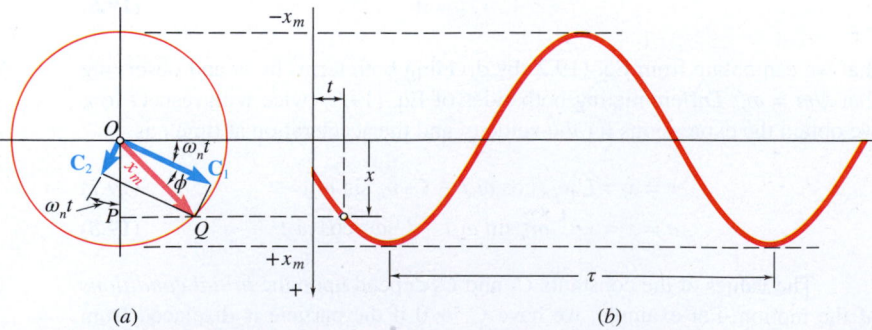

(a)                  (b)

**Fig. 19.2** (*a*) Auxiliary circle of simple harmonic motion: the resultant *OQ* rotates at constant angular velocity $\omega_n$; (*b*) the graph of displacement versus time is a sine curve.

The number of cycles described per unit of time is denoted by $f_n$ and is known as the **natural frequency** of the vibration. We have

$$\text{Natural frequency} = f_n = \frac{1}{\tau_n} = \frac{\omega_n}{2\pi} \qquad \textbf{(19.14)}$$

The unit of frequency is called a *hertz* (Hz). It also follows from Eq. (19.14) that a frequency of 1 s$^{-1}$ or 1 Hz corresponds to a circular frequency of $2\pi$ rad/s. In problems involving angular velocities expressed in revolutions per minute (rpm), we have 1 rpm $= \frac{1}{60}$ s$^{-1} = \frac{1}{60}$ Hz, or 1 rpm $= (2\pi/60)$ rad/s.

Recall that we defined $\omega_n$ in Eq. (19.4) in terms of the constant $k$ of the spring and the mass $m$ of the particle. Thus, the period and the frequency are independent both of the initial conditions and of the amplitude of the vibration. Also, $\tau_n$ and $f_n$ depend on the *mass* rather than on the *weight* of the particle and thus are independent of the value of $g$.

We can represent the velocity–time and acceleration–time curves using sine curves of the same period as the displacement–time curve—but with different amplitudes and different phase angles. From Eqs. (19.11) and (19.12), the maximum values of the magnitudes of the velocity and acceleration are

$$v_m = x_m \omega_n \qquad a_m = x_m \omega_n^2 \qquad \textbf{(19.15)}$$

The point $Q$ describes the auxiliary circle with a radius $x_m$ at the constant angular velocity $\omega_n$, so its velocity and acceleration are equal, respectively, to the expressions of Eq. (19.15). Recalling Eqs. (19.11) and (19.12), we can find the velocity and acceleration of $P$ at any instant by projecting vectors of magnitudes $v_m = x_m \omega_n$ and $a_m = x_m \omega_n^2$ on the $x$ axis. These two vectors represent the velocity and acceleration of $Q$, respectively, at the same instant (Fig. 19.3).

These results are not limited to the solution of the problem of a mass attached to a spring. We can use them to analyze the rectilinear motion of a particle whenever the resultant $\mathbf{F}$ of the forces acting on the particle is proportional to the displacement $x$ and directed toward $O$. In such a case, we can write the fundamental equation of motion $F = ma$ in the form of Eq. (19.6), which is characteristic of a simple harmonic motion. Observing that the coefficient of $x$ must be equal to $\omega_n^2$, we can easily determine the natural circular frequency $\omega_n$ of the motion. Substituting the value obtained for $\omega_n$ into Eqs. (19.13) and (19.14), we then obtain the period $\tau_n$ and the natural frequency $f_n$ of the motion.

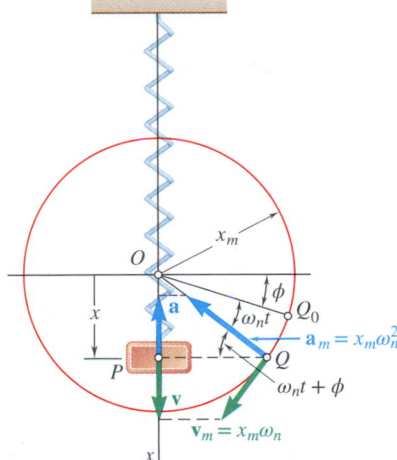

**Fig. 19.3** Auxiliary circle of simple harmonic motion showing the maximum values of velocity and acceleration.

## 19.1B Simple Pendulum (Approximate Solution)

Many of the vibrations encountered in engineering applications can be represented using simple harmonic motion. Many others can be *approximated* by a simple harmonic motion—provided that their amplitude remains small. Consider, for example, a **simple pendulum** consisting of a bob with a mass $m$ attached to a cord of length $l$ that can oscillate in a vertical plane (Fig. 19.4a). At a given time $t$, the cord forms an angle $\theta$ with the vertical. The forces acting

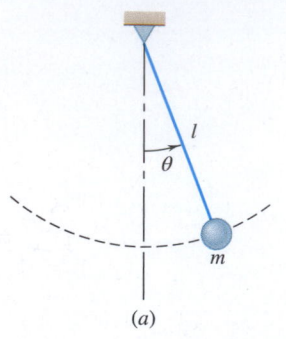

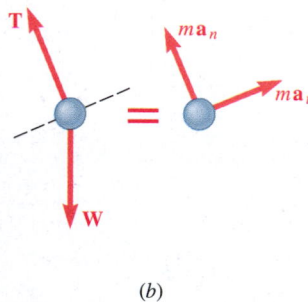

**Fig. 19.4** (a) A simple pendulum consists of a bob of mass $m$ at the end of a cord of length $l$; (b) free-body diagram and kinetic diagram of the simple pendulum.

on the bob are its weight $\mathbf{W}$ and the force $\mathbf{T}$ exerted by the cord (Fig. 19.4b). Resolving the vector $m\mathbf{a}$ into tangential and normal components, with $m\mathbf{a}_t$ directed to the right (i.e., in the direction corresponding to increasing values of $\theta$), and observing that $a_t = l\alpha = l\ddot{\theta}$, we have

$$\Sigma F_t = ma_t: \qquad\qquad -W \sin\theta = ml\ddot{\theta}$$

Noting that $W = mg$ and dividing through by $ml$, we obtain

$$\ddot{\theta} + \frac{g}{l}\sin\theta = 0 \qquad\qquad \textbf{(19.16)}$$

For oscillations of small amplitude, we can replace $\sin\theta$ by $\theta$, which is expressed in radians, obtaining

$$\ddot{\theta} + \frac{g}{l}\theta = 0 \qquad\qquad \textbf{(19.17)}$$

Comparison with Eq. (19.6) shows that the differential equation (19.17) is that of a simple harmonic motion with a natural circular frequency $\omega_n$ equal to $(g/l)^{1/2}$. Thus, we can express the general solution of Eq. (19.17) as

$$\theta = \theta_m \sin\left(\omega_n t + \phi\right)$$

where $\theta_m$ is the amplitude of the oscillations and $\phi$ is the phase angle. Substituting the value obtained for $\omega_n$ into Eq. (19.13), we get the expression for the period of the small oscillations of a pendulum of length $l$ as

$$\tau_n = \frac{2\pi}{\omega_n} = 2\pi\sqrt{\frac{l}{g}} \qquad\qquad \textbf{(19.18)}$$

## *19.1C   Simple Pendulum (Exact Solution)

Eq. (19.18) is only approximate. To obtain an exact expression for the period of the oscillations of a simple pendulum, we must return to Eq. (19.16). Multiplying both terms by $2\dot{\theta}$ and integrating from an initial position corresponding to the maximum deflection (that is, $\theta = \theta_m$ and $\dot{\theta} = 0$), we have

$$\left(\frac{d\theta}{dt}\right)^2 = \frac{2g}{l}(\cos\theta - \cos\theta_m)$$

We replace $\cos\theta$ by $1 - 2\sin^2(\theta/2)$ and $\cos\theta_m$ by a similar expression, solve for $dt$, and integrate over a quarter period from $t = 0$, $\theta = 0$ to $t = \tau_n/4$, $\theta = \theta_m$. This gives

$$\tau_n = 2\sqrt{\frac{l}{g}}\int_0^{\theta_m}\frac{d\theta}{\sqrt{\sin^2(\theta_m/2) - \sin^2(\theta/2)}}$$

The integral on the right-hand side is known as an *elliptic integral;* it cannot be expressed in terms of the usual algebraic or trigonometric functions. However, setting

$$\sin(\theta/2) = \sin(\theta_m/2)\sin\phi$$

we can write

$$\tau_n = 4\sqrt{\frac{l}{g}}\int_0^{\pi/2}\frac{d\phi}{\sqrt{1-\sin^2(\theta_m/2)\sin^2\phi}} \tag{19.19}$$

We can calculate this integral, commonly denoted by $K$, by using a numerical method of integration. It also can be found using computer programs such as Maple, Mathematica, or Matlab or in *tables of elliptic integrals* for various values of $\theta_m/2$.[†]

In order to compare this result with that of the preceding section, we write Eq. (19.19) in the form

$$\tau_n = \frac{2K}{\pi}\left(2\pi\sqrt{\frac{l}{g}}\right) \tag{19.20}$$

Eq. (19.20) shows that we can obtain the actual value of the period of a simple pendulum by multiplying the approximate value given in Eq. (19.18) by the correction factor $2K/\pi$. Values of the correction factor are given in Table 19.1 for various values of the amplitude $\theta_m$. Note that for ordinary engineering computations, the correction factor can be omitted as long as the amplitude does not exceed 10°.

**Table 19.1   Correction Factor for the Period of a Simple Pendulum**

| $\theta_m$ | 0° | 10° | 20° | 30° | 60° | 90° | 120° | 150° | 180° |
|---|---|---|---|---|---|---|---|---|---|
| $K$ | 1.571 | 1.574 | 1.583 | 1.598 | 1.686 | 1.854 | 2.157 | 2.768 | $\infty$ |
| $2K/\pi$ | 1.000 | 1.002 | 1.008 | 1.017 | 1.073 | 1.180 | 1.373 | 1.762 | $\infty$ |

[†]See, for example, *Standard Mathematical Tables and Formulae*, CRC Press, Cleveland, Ohio.

## Sample Problem 19.1

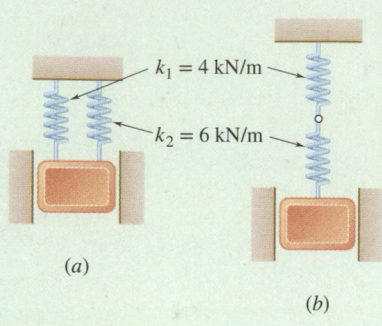

$k_1 = 4$ kN/m

$k_2 = 6$ kN/m

(a)

(b)

A 50-kg block moves between vertical guides as shown. The block is pulled 40 mm down from its equilibrium position and released. For each spring arrangement, determine the period of the vibration, the maximum velocity of the block, and the maximum acceleration of the block.

**STRATEGY:**    You first need to calculate the equivalent spring constant for each arrangement of the springs. Then you can use the information in this section to determine the motion.

**MODELING and ANALYSIS:**

**a. Springs Attached in Parallel.**    First determine the constant $k$ of a single spring equivalent to the two springs by finding the magnitude of the force **P** required to cause a given deflection $\delta$ (Fig. 1). Because for a deflection $\delta$ the magnitudes of the forces exerted by the springs are, respectively, $k_1\delta$ and $k_2\delta$, you have

$$P = k_1\delta + k_2\delta = (k_1 + k_2)\delta$$

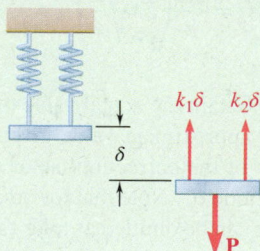

$k_1\delta$    $k_2\delta$

$\delta$

**P**

**Fig. 1**   Springs in parallel elongated a distance $\delta$.

Thus, the constant $k$ of the single equivalent spring is

$$k = \frac{P}{\delta} = k_1 + k_2 = 4 \text{ kN/m} + 6 \text{ kN/m} = 10 \text{ kN/m} = 10^4 \text{ N/m}$$

**Period of Vibration:**   Because $m = 50$ kg, Eq. (19.4) yields

$$\omega_n^2 = \frac{k}{m} = \frac{10^4 \text{ N/m}}{50 \text{ kg}} \qquad \omega_n = 14.14 \text{ rad/s}$$

$$\tau_n = 2\pi/\omega_n \qquad\qquad\qquad \tau_n = 0.444 \text{ s} \quad \blacktriangleleft$$

**Maximum Velocity:**

$$v_m = x_m\omega_n = (0.040 \text{ m})(14.14 \text{ rad/s})$$

$$v_m = 0.566 \text{ m/s} \qquad \mathbf{v}_m = 0.566 \text{ m/s} \updownarrow \quad \blacktriangleleft$$

**Maximum Acceleration:**

$$a_m = x_m\omega_n^2 = (0.040 \text{ m})(14.14 \text{ rad/s})^2$$

$$a_m = 8.00 \text{ m/s}^2 \qquad \mathbf{a}_m = 8.00 \text{ m/s}^2 \updownarrow \quad \blacktriangleleft$$

*(continued)*

**b. Springs Attached in Series.** In this case, determine the constant $k$ of a single spring equivalent to the two springs by finding the total elongation $\delta$ of the springs under a given static load **P** (Fig. 2).

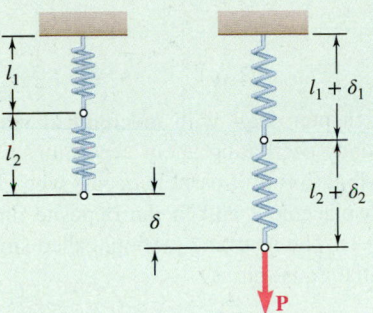

**Fig. 2** Springs in series elongated a distance $\delta$.

To facilitate the computation, you can use an arbitrary static load with a magnitude of $P = 12$ kN (this number is chosen because it has four and six as divisors). You obtain

$$\delta = \delta_1 + \delta_2 = \frac{P}{k_1} + \frac{P}{k_2} = \frac{12\text{ kN}}{4\text{ kN/m}} + \frac{12\text{ kN}}{6\text{ kN/m}} = 5\text{ m}$$

$$k = \frac{P}{\delta} = \frac{12\text{ kN}}{5\text{ m}} = 2.4\text{ kN/m} = 2400\text{ N/m}$$

**Period of Vibration:**

$$\omega_n^2 = \frac{k}{m} = \frac{2400\text{ N/m}}{50\text{ kg}} \qquad\qquad \omega_n = 6.93\text{ rad/s}$$

$$\tau_n = \frac{2\pi}{\omega_n} \qquad\qquad \tau_n = 0.907\text{ s} \quad\blacktriangleleft$$

**Maximum Velocity:**

$$v_m = x_m\omega_n = (0.040\text{ m})(6.93\text{ rad/s})$$

$$v_m = 0.277\text{ m/s} \qquad \mathbf{v}_m = 0.277\text{ m/s}\;\updownarrow \quad\blacktriangleleft$$

**Maximum Acceleration:**

$$a_m = x_m\omega_n^2 = (0.040\text{ m})(6.93\text{ rad/s})^2$$

$$a_m = 1.920\text{ m/s}^2 \qquad \mathbf{a}_m = 1.920\text{ m/s}^2\;\updownarrow \quad\blacktriangleleft$$

**REFLECT and THINK:** The problem did not ask you to determine the expression for combining springs in series, but from this analysis, it is clear that $\delta = \frac{P}{k_1} + \frac{P}{k_2} = \frac{P}{k}$ or $\frac{1}{k} = \frac{1}{k_1} + \frac{1}{k_2}$. Thus, for springs in series, $\frac{1}{k} = \frac{1}{k_1} + \frac{1}{k_2}$, and for springs in parallel, $k = k_1 + k_2$.

# SOLVING PROBLEMS
# ON YOUR OWN

This chapter deals with **mechanical vibrations**; that is, with the motion of a particle or body oscillating about a position of equilibrium. In this first section, we saw that a **free vibration** of a particle occurs when the particle is subjected to a force proportional to its displacement and in the opposite direction, such as the force exerted by a spring (Fig. 19.1). The resulting motion, called **simple harmonic motion**, is characterized by the differential equation as

$$m\ddot{x} + kx = 0 \tag{19.2}$$

where $x$ is the displacement of the particle from the equilibrium point, $\ddot{x}$ is its acceleration, $m$ is its mass, and $k$ is the constant of the spring. We found the solution of this differential equation to be

$$x = x_m \sin(\omega_n t + \phi) \tag{19.10}$$

where $x_m$ = amplitude of the vibration
 $\omega_n = \sqrt{k/m}$ = natural circular frequency (rad/s)
 $\phi$ = phase angle (rad)

We defined the **period** of the vibration as the time $\tau_n = 2\pi/\omega_n$ needed for the particle to complete one cycle. The **natural frequency** is the number of cycles per second, $f_n = 1/\tau_n = \omega_n/2\pi$, expressed in Hz or $s^{-1}$. Differentiating Eq. (19.10) twice yields the velocity and the acceleration of the particle at any time. We found the maximum values of the velocity and acceleration to be

$$v_m = x_m \omega_n \qquad a_m = x_m \omega_n^2 \tag{19.15}$$

To determine the parameters in Eq. (19.10), you can follow these steps:

**1. Draw a free-body diagram showing the forces exerted on the particle** when the particle is at a distance $x$ from its position of equilibrium. The resultant of these forces is proportional to $x$, and its direction is opposite to the positive direction of $x$ [Eq. (19.1)].

**2. Write the differential equation of motion** by equating $m\ddot{x}$ to the resultant of the forces found in Step 1. Note that once you have chosen a positive direction for $x$, you should use the same sign convention for the acceleration $\ddot{x}$. After transposition, you will obtain an equation of the form of Eq. (19.2).

**3. Determine the natural circular frequency** $\omega_n$ by dividing the coefficient of $x$ by the coefficient of $\ddot{x}$ in this equation and taking the square root of the result. Make sure that $\omega_n$ is expressed in rad/s.

**4. Determine the amplitude $x_m$ and the phase angle $\phi$** by substituting the value obtained for $\omega_n$ and the initial values of $x$ and $\ddot{x}$ into Eq. (19.10) and the equation obtained by differentiating Eq. (19.10) with respect to $t$.

You can now use Eq. (19.10) and the two equations obtained by differentiating Eq. (19.10) twice with respect to $t$ to find the displacement, velocity, and acceleration of the particle at any time. Eqs. (19.15) yield the maximum velocity $v_m$ and the maximum acceleration $a_m$.

**5. For the small oscillations of a simple pendulum,** the angle $\theta$ that the cord of the pendulum forms with the vertical satisfies the differential equation

$$\ddot{\theta} + \frac{g}{l}\theta = 0 \tag{19.17}$$

where $l$ is the length of the cord and $\theta$ is expressed in radians (Sec. 19.1B). This equation defines again a simple harmonic motion, and its solution is of the same form as Eq. (19.10) as

$$\theta = \theta_m \sin(\omega_n t + \phi)$$

where the natural circular frequency $\omega_n = \sqrt{g/l}$ is expressed in rad/s. The determination of the various constants in this expression is carried out in a manner similar to that described previously. Remember that the velocity of the bob is tangent to the path and that its magnitude is $v = l\dot{\theta}$, whereas the acceleration of the bob has a tangential component $\mathbf{a}_t$ with a magnitude of $a_t = l\ddot{\theta}$ and a normal component $\mathbf{a}_n$ directed toward the center of the path and with a magnitude of $a_n = l\dot{\theta}^2$.

# Problems

**19.1** A particle moves in simple harmonic motion. Knowing that the maximum velocity is 200 mm/s and the maximum acceleration is 4 m/s², determine the amplitude and frequency of the motion.

**19.2** A particle moves in simple harmonic motion. Knowing that the amplitude is 0.21 in. and the maximum acceleration is 175 ft/s², determine the maximum velocity of the particle and the frequency of its motion.

**19.3** Determine the amplitude and maximum acceleration of a particle that moves in simple harmonic motion with a maximum velocity of 4 ft/s and a frequency of 6 Hz.

**19.4** A 32-kg block is attached to a spring and can move without friction in a slot as shown. The block is in its equilibrium position when it is struck by a hammer that imparts to the block an initial velocity of 250 mm/s. Determine (*a*) the period and frequency of the resulting motion, (*b*) the amplitude of the motion and the maximum acceleration of the block.

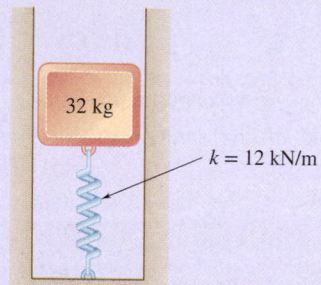

Fig. P19.4

**19.5** An instrument package *A* is bolted to a shaker table as shown. The table moves vertically in simple harmonic motion at the same frequency as the variable-speed motor that drives it. The package is to be tested at a peak acceleration of 20 m/s². Knowing that the amplitude of the shaker table displacement is 10 mm, determine (*a*) the required speed of the motor in rpm, (*b*) the maximum velocity of the table.

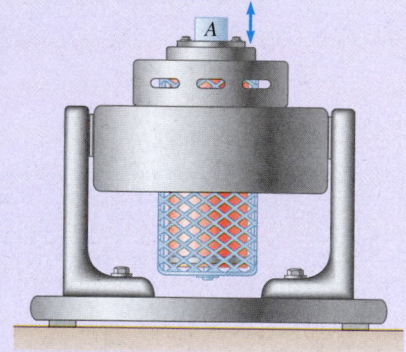

Fig. P19.5

**19.6** A 20-lb block is initially held so that the vertical spring attached as shown is undeformed. Knowing that the block is suddenly released from rest, determine (*a*) the amplitude and frequency of the resulting motion, (*b*) the maximum velocity and maximum acceleration of the block.

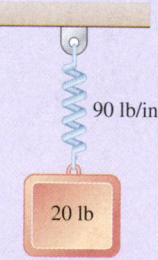

90 lb/in.

20 lb

Fig. P19.6

**19.7** A simple pendulum consisting of a bob attached to a cord oscillates in a vertical plane with a period of 1.5 s. Assuming simple harmonic motion and knowing that the maximum velocity of the bob is 0.5 m/s, determine (*a*) the amplitude of the motion in degrees, (*b*) the maximum tangential acceleration of the bob.

**19.8** A simple pendulum consisting of a bob attached to a cord of length *l* = 500 mm oscillates in a vertical plane. Assuming simple harmonic motion and knowing that the bob is released from rest when $\theta = 7°$, determine (*a*) the frequency of oscillation, (*b*) the maximum velocity of the bob.

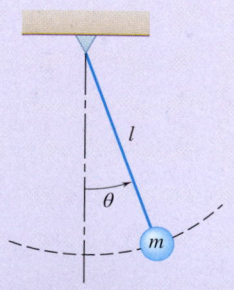

Fig. P19.7 and P19.8

**19.9** A 10-lb block $A$ rests on a 40-lb plate $B$ that is attached to an unstretched spring with a constant of $k = 60$ lb/ft. Plate $B$ is slowly moved 2.4 in. to the left and released from rest. Assuming that block $A$ does not slip on the plate, determine (a) the amplitude and frequency of the resulting motion, (b) the corresponding smallest allowable value of the coefficient of static friction between $A$ and $B$.

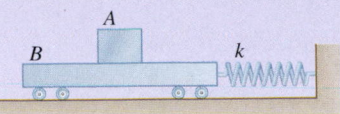

Fig. P19.9

**19.10** A 5-kg fragile glass vase is surrounded by packing material in a cardboard box of negligible weight. The packing material has negligible damping and a force-deflection relationship as shown. Knowing that the box is dropped from a height of 1 m and the impact with the ground is perfectly plastic, determine (a) the amplitude of vibration for the vase, (b) the maximum acceleration the vase experiences in g's.

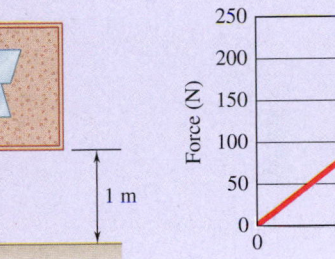

Fig. P19.10

**19.11** A 3-lb block is supported as shown by a spring of constant $k = 2$ lb/in. that can act in tension or compression. The block is in its equilibrium position when it is struck from below by a hammer that imparts to the block an upward velocity of 90 in./s. Determine (a) the time required for the block to move 3 in. upward, (b) the corresponding velocity and acceleration of the block.

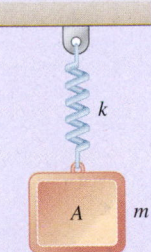

Fig. P19.11

**19.12** In Prob. 19.11, determine the position, velocity, and acceleration of the block 0.90 s after it has been struck by the hammer.

**19.13** The bob of a simple pendulum of length $l = 40$ in. is released from rest when $\theta = 5°$. Assuming simple harmonic motion, determine 1.6 s after release (a) the angle $\theta$, (b) the magnitudes of the velocity and acceleration of the bob.

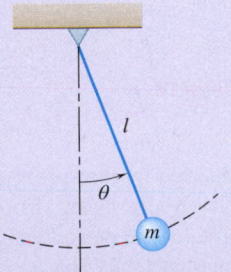

Fig. P19.13

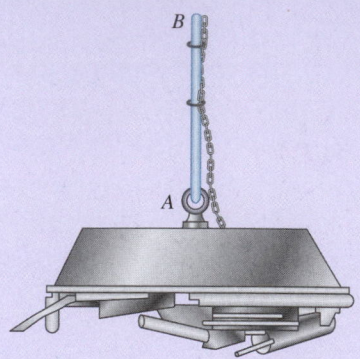

**Fig. P19.14**

**19.14** A 150-kg electromagnet is at rest and is holding 100 kg of scrap steel when the current is turned off and the steel is dropped. Knowing that the cable and the supporting crane have a total stiffness equivalent to a spring of constant 200 kN/m, determine (*a*) the frequency, the amplitude, and the maximum velocity of the resulting motion, (*b*) the minimum tension that will occur in the cable during the motion, (*c*) the velocity of the magnet 0.03 s after the current is turned off.

**19.15** A 5-kg collar *C* is released from rest in the position shown and slides without friction on a vertical rod until it hits a spring with a constant of $k = 720$ N/m that it compresses. The velocity of the collar is reduced to zero, and the collar reverses the direction of its motion and returns to its initial position. The cycle is then repeated. Determine (*a*) the period of the motion of the collar, (*b*) the velocity of the collar 0.4 s after it was released. (*Note:* This is a periodic motion, but it is not simple harmonic motion.)

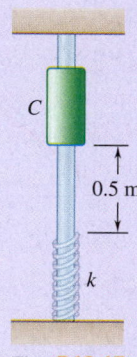

**Fig. P19.15**

**19.16** A small bob is attached to a cord of length 1.2 m and is released from rest when $\theta_A = 5°$. Knowing that $d = 0.6$ m, determine (*a*) the time required for the bob to return to point *A*, (*b*) the amplitude $\theta_C$.

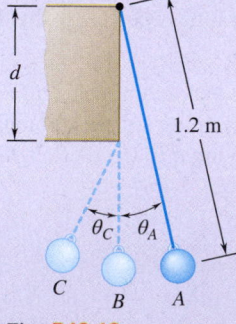

**Fig. P19.16**

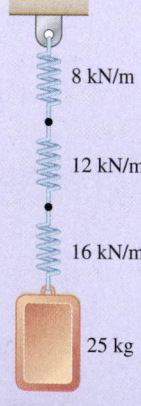

**Fig. P19.17**

**19.17** A 25-kg block is supported by the spring arrangement shown. If the block is moved vertically downward from its equilibrium position and released, determine (*a*) the period and frequency of the resulting motion, (*b*) the maximum velocity and acceleration of the block if the amplitude of the motion is 30 mm.

**19.18** An 11-lb block is attached to the lower end of a spring whose upper end is fixed and vibrates with a period of 7.2 s. Knowing that the constant $k$ of a spring is inversely proportional to its length (e.g., if you cut a 10 lb/in. spring in half, the remaining two springs each have a spring constant of 20 lb/in.), determine the period of a 7-lb block that is attached to the center of the same spring if the upper and lower ends of the spring are fixed.

**19.19** Block $A$ has a mass $m$ and is supported by the spring arrangement as shown. Knowing that the mass of the pulley is negligible and that the block is moved vertically downward from its equilibrium position and released, determine the frequency of the motion.

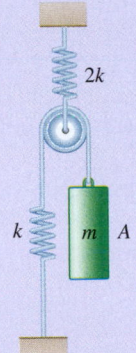

**Fig. P19.19**

**19.20** A 13.6-kg block is supported by the spring arrangement shown. If the block is moved from its equilibrium position 44 mm vertically downward and released, determine (*a*) the period and frequency of the resulting motion, (*b*) the maximum velocity and acceleration of the block.

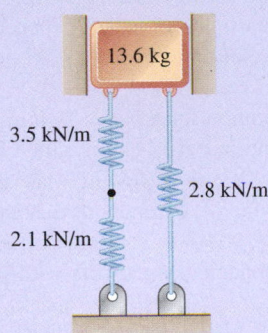

**Fig. P19.20**

**19.21 and 19.22** A 50-kg block is supported by the spring arrangement shown. The block is moved vertically downward from its equilibrium position and released. Knowing that the amplitude of the resulting motion is 60 mm, determine (*a*) the period and frequency of the motion, (*b*) the maximum velocity and maximum acceleration of the block.

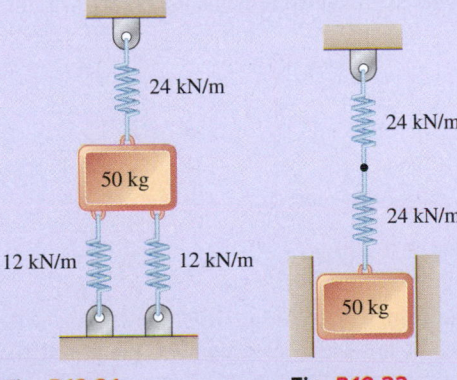

**Fig. P19.21**     **Fig. P19.22**

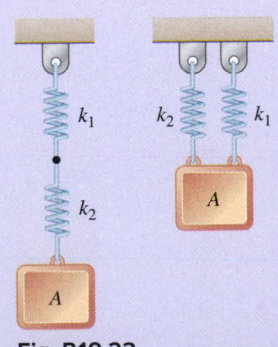

**19.23** Two springs with constants $k_1$ and $k_2$ are connected in series to a block $A$ that vibrates in simple harmonic motion with a period of 5 s. When the same two springs are connected in parallel to the same block, the block vibrates with a period of 2 s. Determine the ratio $k_1/k_2$ of the two spring constants.

**Fig. P19.23**

**19.24** The period of vibration of the system shown is observed to be 0.8 s. If block $A$ is removed, the period is observed to be 0.7 s. Determine (a) the mass of block $C$, (b) the period of vibration when both blocks $A$ and $B$ have been removed.

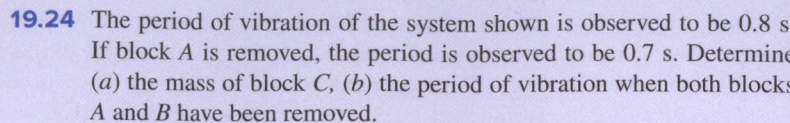
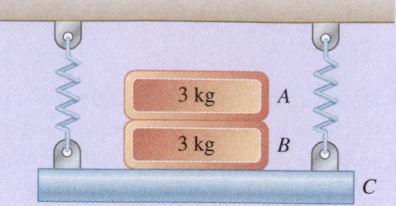

**Fig. P19.24**

**19.25** The 100-lb platform $A$ is attached to springs $B$ and $D$, each of which has a constant $k = 120$ lb/ft. Knowing that the frequency of vibration of the platform is to remain unchanged when an 80-lb block is placed on it and a third spring $C$ is added between springs $B$ and $D$, determine the required constant of spring $C$.

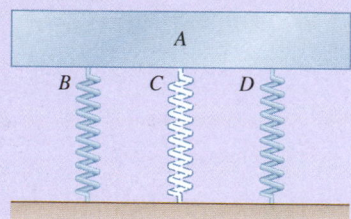

**Fig. P19.25**

**19.26** The period of vibration for a barrel floating in salt water is found to be 0.58 s when the barrel is empty and 1.8 s when it is filled with 55 gallons of crude oil. Knowing that the density of the oil is 900 kg/m$^3$, determine (a) the mass of the empty barrel, (b) the density of the salt water, $\rho_{sw}$. (*Hint:* The force of the water on the bottom of the barrel can be modeled as a spring with constant $k = \rho_{sw}gA$.)

**Fig. P19.26**

**19.27** From mechanics of materials, it is known that for a simply supported beam of uniform cross-section, a static load $P$ applied at the center will cause a deflection of $\delta_A = PL^3/48EI$, where $L$ is the length of the beam, $E$ is the modulus of elasticity, and $I$ is the moment of inertia of the cross-sectional area of the beam. Knowing that $L = 15$ ft, $E = 30 \times 10^6$ psi, and $I = 2 \times 10^{-3}$ ft$^4$, determine (a) the equivalent spring constant of the beam, (b) the frequency of vibration of a 1500-lb block attached to the center of the beam. Neglect the mass of the beam and assume that the load remains in contact with the beam.

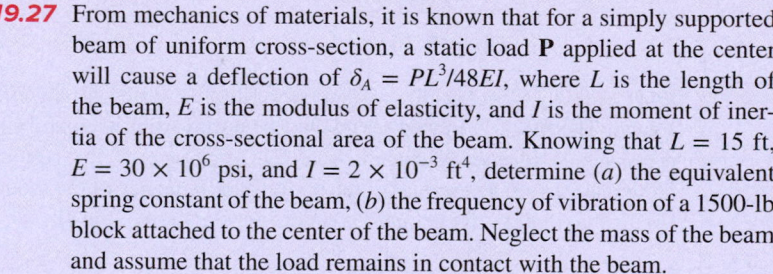

**Fig. *P19.27***

**19.28** From mechanics of materials it is known that when a static load $P$ is applied at the end $B$ of a uniform metal rod fixed at end $A$, the length of the rod will increase by an amount $\delta = PL/AE$, where $L$ is the length of the undeformed rod, $A$ is its cross-sectional area, and $E$ is the modulus of elasticity of the metal. Knowing that $L = 450$ mm and $E = 200$ GPa and that the diameter of the rod is 8 mm, and neglecting the mass of the rod, determine (a) the equivalent spring constant of the rod, (b) the frequency of the vertical vibrations of a block of mass $m = 8$ kg attached to end $B$ of the same rod.

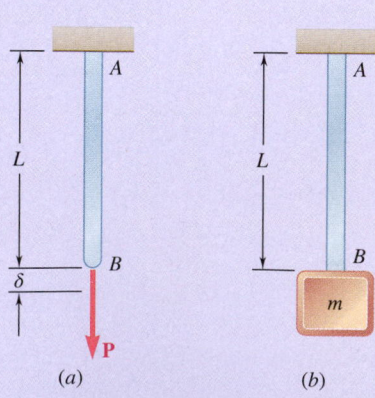

**Fig. P19.28**

**19.29** Denoting by $\delta_{st}$ the static deflection of a beam under a given load, show that the frequency of vibration of the load is

$$f = \frac{1}{2\pi}\sqrt{\frac{g}{\delta_{st}}}$$

Neglect the mass of the beam, and assume that the load remains in contact with the beam.

**19.30** A 40-mm deflection of the second floor of a building is measured directly under a newly installed 3500-kg piece of rotating machinery that has a slightly unbalanced rotor. Assuming that the deflection of the floor is proportional to the load it supports, determine (a) the equivalent spring constant of the floor system, (b) the speed in rpm of the rotating machinery that should be avoided if it is not to coincide with the natural frequency of the floor-machinery system.

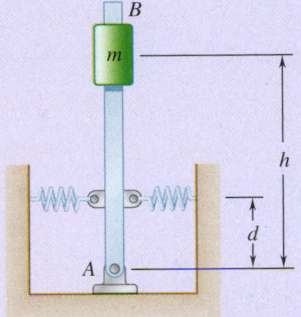

**19.31** If $h = 700$ mm and $d = 500$ mm and each spring has a constant $k = 600$ N/m, determine the mass $m$ for which the period of small oscillations is (a) 0.50 s, (b) infinite. Neglect the mass of the rod and assume that each spring can act in both tension and compression.

**Fig. P19.31**

*19.32* The force-deflection equation for a nonlinear spring fixed at one end is $F = 2.5x^{1/2}$, where $F$ is the force, expressed in pounds, and $x$ is the deflection, expressed in feet. (a) Determine the static deflection $x_0$ if a 6-oz block is suspended from the spring. (b) Assuming that the slope of the force-deflection curve at the point corresponding to this loading can be used as an equivalent spring constant, determine the frequency of vibration of the block if it is given a very small downward displacement from its equilibrium position and released.

*19.33** Expanding the integrand in Eq. (19.19) of Sec. 19.1C into a series of even powers of $\sin \phi$ and integrating, show that the period of a simple pendulum of length $l$ may be approximated by the formula

$$\tau = 2\pi\sqrt{\frac{l}{g}}\left(1 + \tfrac{1}{4}\sin^2\frac{\theta_m}{2}\right)$$

where $\theta_m$ is the amplitude of the oscillations.

*19.34** Using the formula given in Prob. 19.33, determine the amplitude $\theta_m$ for which the period of a simple pendulum is $\frac{1}{2}$ percent longer than the period of the same pendulum for small oscillations.

*19.35** Using the data of Table 19.1, determine the period of a simple pendulum of length $l = 750$ mm (a) for small oscillations, (b) for oscillations of amplitude $\theta_m = 60°$, (c) for oscillations of amplitude $\theta_m = 90°$.

*19.36** Using the data of Table 19.1, determine the length in inches of a simple pendulum that oscillates with a period of 2 s and an amplitude of 90°.

## 19.2 FREE VIBRATIONS OF RIGID BODIES

The analysis of the vibrations of a rigid body (or of a system of rigid bodies) possessing a single degree of freedom is similar to the analysis of the vibrations of a particle. We choose an appropriate variable, such as a distance $x$ or an angle $\theta$, to define the position of the body or system of bodies and write an equation relating this variable and its second derivative with respect to $t$. If the equation is of the same form as Eq. (19.6)—that is, if we have

$$\ddot{x} + \omega_n^2 x = 0 \qquad \text{or} \qquad \ddot{\theta} + \omega_n^2 \theta = 0 \qquad \textbf{(19.21)}$$

then the vibration is a simple harmonic motion. We can obtain the period and natural frequency of the vibration by identifying $\omega_n$ and substituting its value into Eqs. (19.13) and (19.14).

In general, a simple way to obtain one of Eqs. (19.21) is to use Newton's second law. To do this, first draw free-body and kinetic diagrams for the system displaced in the positive direction. The acceleration in your kinetic diagram needs to be in the same positive direction you defined for the displacement. From your drawn diagrams, it is straightforward to write the appropriate equation of motion. Recall that the goal should be the determination of the coefficient of the variable $x$ or $\theta$—*not* the determination of the variable itself or of the derivative $\ddot{x}$ or $\ddot{\theta}$. Setting this coefficient equal to $\omega_n^2$, we obtain the natural circular frequency $\omega_n$ from which we can determine $\tau_n$ and $f_n$.

This method can be used to analyze vibrations that are truly represented by a simple harmonic motion or by vibrations of small amplitude that can be *approximated* by a simple harmonic motion. As an example, let us determine the period of the small oscillations of a square plate with a side $2b$ that is suspended from the midpoint $O$ of one of its sides (Fig. 19.5a). We consider the plate in an arbitrary position defined by the angle $\theta$ that the line $OG$ forms with the vertical. Then we draw free-body and kinetic diagrams to express that the weight $\mathbf{W}$ of the plate and the components $\mathbf{R}_x$ and $\mathbf{R}_y$ of the reaction at $O$ are equivalent to the vectors $m\mathbf{a}_t$ and $m\mathbf{a}_n$ and to the couple $\bar{I}\alpha$ (Fig. 19.5b). Because the angular velocity and angular acceleration of the plate are equal to $\dot{\theta}$ and $\ddot{\theta}$, respectively, the magnitudes of the two vectors $m\mathbf{a}_t$ and $m\mathbf{a}_n$ are $mb\ddot{\theta}$ and $mb\dot{\theta}^2$, respectively, and the moment of the couple is $\bar{I}\ddot{\theta}$. In previous applications of this method (Chap. 16), we tried whenever possible to assume the correct sense for the acceleration. Here, however, we must assume the same positive sense for $\theta$ and $\ddot{\theta}$ in order to obtain an equation of the form in Eq. (19.21). Consequently, we assume the angular acceleration $\ddot{\theta}$ is positive counterclockwise—even though this assumption is obviously unrealistic. Equating moments about $O$, we have

$$-W(b \sin \theta) = (mb\ddot{\theta})b + \bar{I}\ddot{\theta}$$

Noting that

$$\bar{I} = \tfrac{1}{12} m[(2b)^2 + (2b)^2] = \tfrac{2}{3} mb^2 \text{ and } W = mg$$

we obtain

$$\ddot{\theta} + \frac{3}{5} \frac{g}{b} \sin \theta = 0 \qquad \textbf{(19.22)}$$

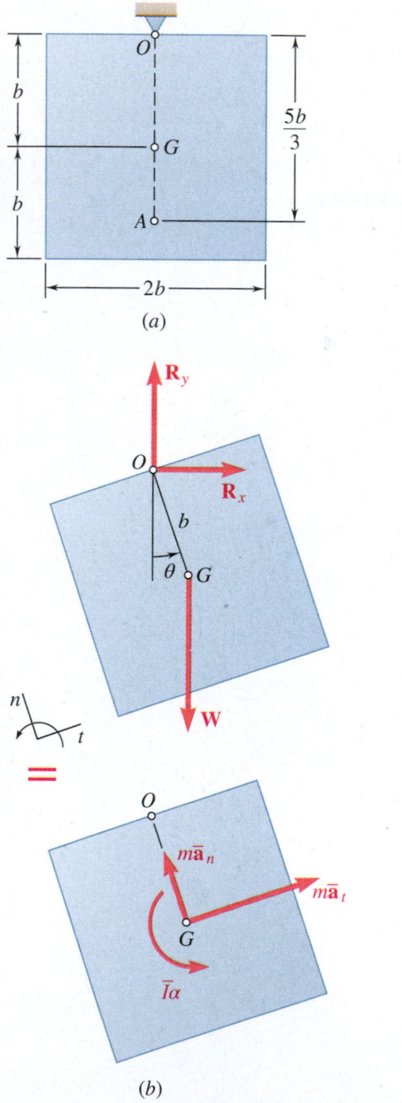

**Fig. 19.5** (a) A square plate of side $2b$ suspended from the midpoint of one of its sides; (b) free-body diagram and kinetic diagram for the plate.

For oscillations of small amplitude, we can replace $\sin \theta$ by $\theta$, expressed in radians, which gives

$$\ddot{\theta} + \frac{3}{5}\frac{g}{b}\theta = 0 \qquad (19.23)$$

Comparison with Eq. (19.21) shows that this equation is that of a simple harmonic motion and that the natural circular frequency $\omega_n$ of the oscillations is equal to $(3g/5b)^{1/2}$. Substituting into Eq. (19.13), we find that the period of the oscillations is

$$\tau_n = \frac{2\pi}{\omega_n} = 2\pi\sqrt{\frac{5b}{3g}} \qquad (19.24)$$

This result is valid only for oscillations of small amplitude. A more accurate description of the motion of the plate is obtained by comparing Eqs. (19.16) and (19.22). Note that the two equations are identical if we choose $l$ equal to $5b/3$. This means that the plate oscillates as a simple pendulum with a length of $l = 5b/3$, and we can use the results of Sec. 19.1C to correct the value of the period given in Eq. (19.24). Point $A$ of the plate located on line $OG$ at a distance $l = 5b/3$ from $O$ is defined as the **center of oscillation** corresponding to $O$ (Fig. 19.5a).

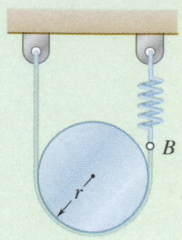

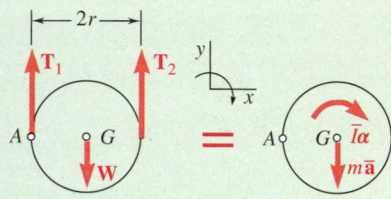

**Fig. 1**   Free-body diagram and kinetic diagram for the cylinder.

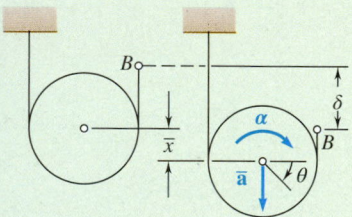

**Fig. 2**   Linear and angular displacements and linear and angular accelerations of the cylinder.

## Sample Problem 19.2

A cylinder with weight $W$ and radius $r$ is suspended from a looped cord as shown. One end of the cord is attached directly to a rigid support, and the other end is attached to a spring with a constant $k$. Determine the period and natural frequency of the vibrations of the cylinder.

**STRATEGY:**   First choose a coordinate to describe the motion, and then use Newton's second law to determine the equations of motion.

**MODELING:**   Choose the cylinder to be your system, and model it as a rigid body. The system of external forces acting on the cylinder consists of the weight $W$ and of the forces $T_1$ and $T_2$ exerted by the cord. Draw free-body and kinetic diagrams (Fig. 1) to express that this system is equivalent to the system represented by the vector $m\bar{\mathbf{a}}$ attached at $G$ and the couple $\bar{I}\alpha$.

**ANALYSIS:**

**Kinematics of Motion.**   Express the linear displacement and the acceleration of the cylinder in terms of the angular displacement $\theta$. Choosing the positive sense clockwise and measuring the displacements from the equilibrium position (Fig. 2), you have

$$\bar{x} = r\theta \qquad \delta = 2\bar{x} = 2r\theta$$
$$\alpha = \ddot{\theta} \circlearrowright \qquad \bar{a} = r\alpha = r\ddot{\theta} \qquad \bar{a} = r\ddot{\theta} \downarrow \tag{1}$$

**Equations of Motion.**   Newton's second law gives you (Fig. 1)

$$+\circlearrowright \Sigma M_A = m\bar{a}d_\perp + \bar{I}\alpha: \qquad Wr - T_2(2r) = m\bar{a}r + \bar{I}\alpha \tag{2}$$

When the cylinder is in its position of equilibrium, the tension in the cord is $T_0 = \frac{1}{2}W$. Note that for an angular displacement $\theta$, the magnitude of $T_2$ is

$$T_2 = T_0 + k\delta = \tfrac{1}{2}W + k\delta = \tfrac{1}{2}W + k(2r\theta) \tag{3}$$

Substituting from Eqs. (1) and (3) into Eq. (2) and recalling that $\bar{I} = \frac{1}{2}mr^2$, you have

$$Wr - (\tfrac{1}{2}W + 2kr\theta)(2r) = m(r\ddot{\theta})r + \tfrac{1}{2}mr^2\ddot{\theta}$$

$$\ddot{\theta} + \frac{8}{3}\frac{k}{m}\theta = 0$$

The motion is simple harmonic, and you have

$$\omega_n^2 = \frac{8}{3}\frac{k}{m} \qquad \omega_n = \sqrt{\frac{8}{3}\frac{k}{m}}$$

$$\tau_n = \frac{2\pi}{\omega_n} \qquad\qquad \tau_n = 2\pi\sqrt{\frac{3}{8}\frac{m}{k}} \quad \blacktriangleleft$$

$$f_n = \frac{\omega_n}{2\pi} \qquad\qquad f_n = \frac{1}{2\pi}\sqrt{\frac{8}{3}\frac{k}{m}} \quad \blacktriangleleft$$

**REFLECT and THINK:**   If the cylinder had been smooth, it would not have rotated when displaced downward. Also note that the answers you obtained are independent of $r$.

## Sample Problem 19.3

A circular disk weighs 20 lb, has a radius of 8 in., and is suspended from a wire as shown. The disk is rotated (thus twisting the wire) and then released; the period of the torsional vibration is observed to be 1.13 s. A gear is then suspended from the same wire, and the period of torsional vibration for the gear is observed to be 1.93 s. Assuming that the moment of the couple exerted by the wire is proportional to the angle of twist, determine (a) the torsional spring constant of the wire, (b) the centroidal moment of inertia of the gear, (c) the maximum angular velocity reached by the gear if it is rotated through 90° and released.

**STRATEGY:**   Use Newton's second law to obtain the equation of motion. From this, you can find the circular natural frequency in terms of the torsional spring constant and the centroidal moment of inertia. You can determine the torsional spring constant for the wire from the analysis of the disk. Then you can use that to describe the motion of the gear.

**MODELING:**   Choose the disk (or gear) as your system, and model it as a rigid body. The kinematic variables are shown in Fig. 1, and the free-body and kinetic diagrams are shown in Fig. 2.

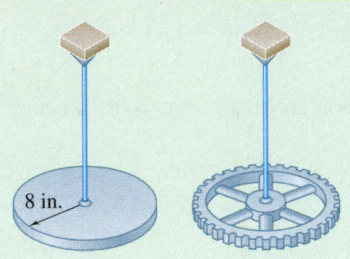

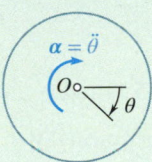

**Fig. 1** Angular displacement and acceleration for the disk (or gear).

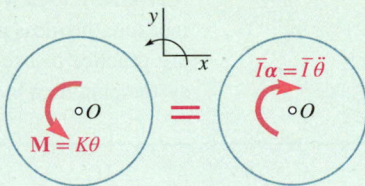

**Fig. 2** Free-body diagram and kinetic diagram for the disk (or gear).

**ANALYSIS:**

**a. Vibration of Disk.**   Denoting the angular displacement of the disk by $\theta$ (Fig. 1), you can express that the magnitude of the couple exerted by the wire is $M = K\theta$, where $K$ is the torsional spring constant of the wire. Applying Newton's second law, you have

$$+\circlearrowleft \Sigma M_O = \bar{I}\alpha: \qquad\qquad +K\theta = -\bar{I}\ddot{\theta}$$

$$\ddot{\theta} + \frac{K}{\bar{I}}\theta = 0$$

The motion is simple harmonic, so you have

$$\omega_n^2 = \frac{K}{\bar{I}} \qquad \tau_n = \frac{2\pi}{\omega_n} \qquad \tau_n = 2\pi\sqrt{\frac{\bar{I}}{K}} \qquad\qquad (1)$$

For the disk,

$$\tau_n = 1.13\text{ s} \qquad \bar{I} = \tfrac{1}{2}mr^2 = \frac{1}{2}\left(\frac{20\text{ lb}}{32.2\text{ ft/s}^2}\right)\left(\frac{8}{12}\text{ ft}\right)^2 = 0.138\text{ lb·ft·s}^2$$

*(continued)*

Substituting into Eq. (1), you obtain

$$1.13 = 2\pi \sqrt{\frac{0.138}{K}} \qquad\qquad K = 4.27 \text{ lb·ft/rad} \quad \blacktriangleleft$$

**b. Vibration of Gear.** The period of vibration of the gear is 1.93 s and $K = 4.27$ lb·ft/rad, so Eq. (1) yields

$$1.93 = 2\pi \sqrt{\frac{\bar{I}}{4.27}} \qquad\qquad \bar{I}_{\text{gear}} = 0.403 \text{ lb·ft·s}^2 \quad \blacktriangleleft$$

**c. Maximum Angular Velocity of Gear.** Because it is simple harmonic motion, you have

$$\theta = \theta_m \sin \omega_n t \qquad \omega = \theta_m \omega_n \cos \omega_n t \qquad \omega_m = \theta_m \omega_n$$

Recalling that $\theta_m = 90° = 1.571$ rad and $\tau = 1.93$ s, you have

$$\omega_m = \theta_m \omega_n = \theta_m \left(\frac{2\pi}{\tau}\right) = (1.571 \text{ rad})\left(\frac{2\pi}{1.93 \text{ s}}\right)$$

$$\omega_m = 5.11 \text{ rad/s} \quad \blacktriangleleft$$

**REFLECT and THINK:** A torsional spring is often used experimentally to measure the mass moment of inertia of different objects. It is common engineering practice to use one situation to determine the dynamic characteristics of a system and then to use those parameters to analyze a slightly different situation.

# SOLVING PROBLEMS
# ON YOUR OWN

In this section, you saw that a rigid body, or a system of rigid bodies, whose position can be defined by a single coordinate $x$ or $\theta$, executes a simple harmonic motion if the differential equation obtained by applying Newton's second law is of the form

$$\ddot{x} + \omega_n^2 x = 0 \qquad \text{or} \qquad \ddot{\theta} + \omega_n^2 \theta = 0 \tag{19.21}$$

Your goal should be to determine $\omega_n$, from which you can obtain the period $\tau_n$ and the natural frequency $f_n$. Taking into account the initial conditions, you can then write an equation of the form

$$x = x_m \sin(\omega_n t + \phi) \tag{19.10}$$

where you should replace $x$ by $\theta$ if a rotation is involved. To solve the problems in this section, you should follow these steps:

**1. Choose a coordinate that measures the displacement of the body** from its equilibrium position. You will find that many of the problems in this section involve the rotation of a body about a fixed axis and that the angle measuring the rotation of the body from its equilibrium position is the most convenient coordinate to use. In problems involving the general plane motion of a body, where a coordinate $x$ (and possibly a coordinate $y$) is used to define the position of the mass center $G$ of the body and a coordinate $\theta$ is used to measure its rotation about $G$, kinematic relations will allow you to express $x$ (and $y$) in terms of $\theta$ (Sample Prob. 19.2).

**2. Draw a free-body diagram and a kinetic diagram** to express that the system of the external forces is equivalent to the vector $m\bar{\mathbf{a}}$ and the couple $\bar{I}\boldsymbol{\alpha}$ where $\bar{a} = \ddot{x}$ and $\alpha = \ddot{\theta}$. Be sure that each applied force or couple is drawn in a direction consistent with the assumed displacement and that the senses of $\bar{\mathbf{a}}$ and $\boldsymbol{\alpha}$ are those in which the coordinates $x$ and $\theta$ are increasing.

**3. Write the differential equations of motion** by equating the sums of the components of the external forces and the inertial terms in the $x$ and $y$ directions and the sums of their moments about a given point. If necessary, use the kinematic relations developed in Step 1 to obtain equations involving only the coordinate $\theta$. If $\theta$ is a small angle, replace $\sin \theta$ by $\theta$ and $\cos \theta$ by 1 if these functions appear in your equations. Eliminating any unknown reactions, you will obtain an equation of the type of Eqs. (19.21). Note that, in problems involving a body rotating about a fixed axis, you can immediately obtain such an equation by equating the moments of the external forces and inertial terms about the fixed axis.

*(continued)*

**4. Comparing the equation you have obtained with one of Eqs. (19.21),** you can identify $\omega_n^2$ and thus determine the natural circular frequency $\omega_n$. Remember that the object of your analysis is *not to solve* the differential equation you have obtained *but to identify* $\omega_n^2$.

**5. Determine the amplitude and the phase angle $\phi$** by substituting the value obtained for $\omega_n$ and the initial values of the coordinate and its first derivative into Eq. (19.10) and the equation obtained by differentiating Eq. (19.10) with respect to $t$. From Eq. (19.10) and the two equations obtained by differentiating Eq. (19.10) twice with respect to $t$ and using the kinematic relations developed in Step 1, you will be able to determine the position, velocity, and acceleration of any point of the body at any given time.

**6. In problems involving torsional vibrations,** the torsional spring constant $K$ is expressed in N·m/rad or lb·ft/rad. The product of $K$ and the angle of twist $\theta$, where $\theta$ is expressed in radians, yields the moment of the restoring couple, which should be equated to the inertial terms in the system (Sample Prob. 19.3).

# Problems

**19.37** The 9-kg uniform rod $AB$ is attached to springs at $A$ and $B$, each of constant 850 N/m, which can act in both tension and compression. If end $A$ of the rod is depressed slightly and released, determine (a) the frequency of vibration, (b) the amplitude of the angular motion of the rod, knowing that the maximum velocity of point $A$ is 1.1 mm/s.

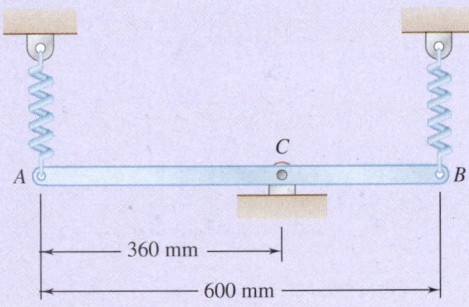

**Fig. P19.37**

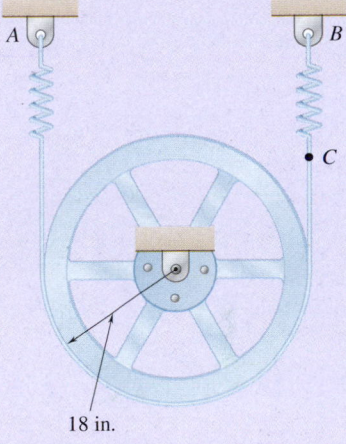

18 in.

**Fig. P19.38**

**19.38** A belt is placed around the rim of a 500-lb flywheel and attached as shown to two springs, each of constant $k = 85$ lb/in. If end $C$ of the belt is pulled 1.5 in. down and released, the period of vibration of the flywheel is observed to be 0.5 s. Knowing that the initial tension in the belt is sufficient to prevent slipping, determine (a) the maximum angular velocity of the flywheel, (b) the centroidal radius of gyration of the flywheel.

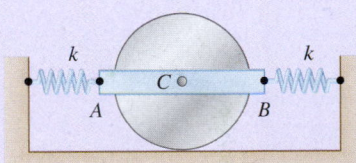

**Fig. P19.39 and *P19.40***

**19.39** A 6-kg uniform cylinder can roll without sliding on a horizontal surface and is attached by a pin at point $C$ to the 4-kg horizontal bar $AB$. The bar is attached to two springs, each having a constant of $k = 5$ kN/m, as shown. Knowing that the bar is moved 12 mm to the right of the equilibrium position and released, determine (a) the period of vibration of the system, (b) the magnitude of the maximum velocity of bar $AB$.

**19.40** A 6-kg uniform cylinder is assumed to roll without sliding on a horizontal surface and is attached by a pin at point $C$ to the 4-kg horizontal bar $AB$. The bar is attached to two springs, each having a constant of $k = 3.5$ kN/m, as shown. Knowing that the coefficient of static friction between the cylinder and the surface is 0.5, determine the maximum amplitude of the motion of point $C$ that is compatible with the assumption of rolling.

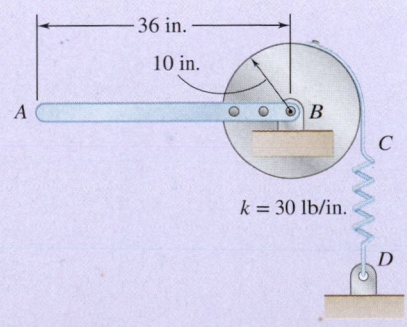

**19.41** A 15-lb slender rod $AB$ is riveted to a 12-lb uniform disk as shown. A belt is attached to the rim of the disk and to a spring that holds the rod at rest in the position shown. If end $A$ of the rod is moved 0.75 in. down and released, determine (a) the period of vibration, (b) the maximum velocity of end $A$.

**Fig. P19.41**

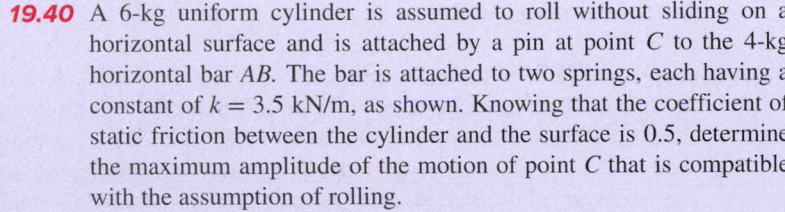

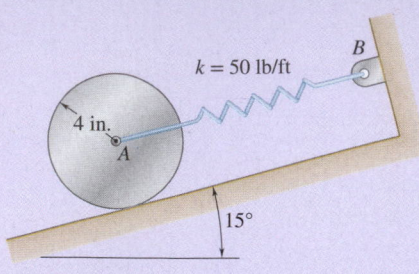

**Fig. P19.42**

**19.42** A 20-lb uniform cylinder can roll without sliding on an incline and is attached to spring $AB$ as shown. If the center of the cylinder is moved 0.5 in. down the incline from the equilibrium position and released from rest, determine (a) the period of vibration, (b) the maximum velocity of the center of the cylinder.

**19.43** A square plate of mass $m$ is held by eight springs, each of constant $k$. Knowing that each spring can act in either tension or compression, determine the frequency of the resulting vibration if (a) the plate is given a small vertical displacement and released, (b) the plate is rotated through a small angle about $G$ and released.

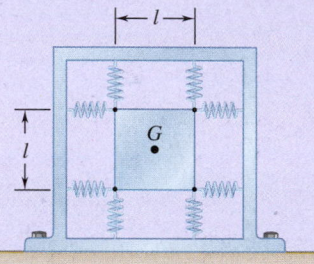

**Fig. P19.43**

**19.44** Two small weights $w$ are attached at $A$ and $B$ to the rim of a uniform disk of radius $r$ and weight $W$. Denoting by $\tau_0$ the period of small oscillations when $\beta = 0$, determine the angle $\beta$ for which the period of small oscillations is $2\tau_0$.

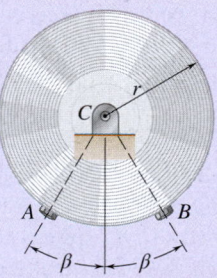

**Fig. P19.44 and P19.45**

**19.45** Two 40-g weights are attached at $A$ and $B$ to the rim of a 1.5-kg uniform disk of radius $r = 100$ mm. Determine the frequency of small oscillations when $\beta = 60°$.

**19.46** A three-blade wind turbine used for research is supported on a shaft so that it is free to rotate about $O$. One technique to determine the centroidal mass moment of inertia of an object is to place a known weight at a known distance from the axis of rotation and to measure the frequency of oscillations after releasing it from rest with a small initial angle. In this case, a weight of $W_{add} = 50$ lb is attached to one of the blades at a distance $R = 20$ ft from the axis of rotation. Knowing that when the blade with the added weight is displaced slightly from the vertical axis, and the system is found to have a period of 7.6 s, determine the centroidal mass moment of inertia of the three-blade rotor.

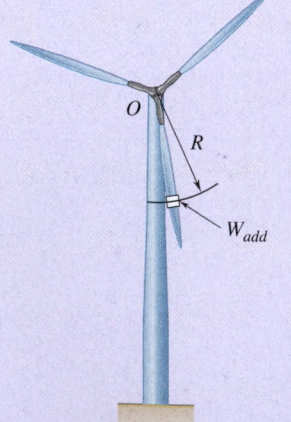

**Fig. P19.46**

**19.47** A connecting rod is supported by a knife-edge at point A; the period of its small oscillations is observed to be 0.87 s. The rod is then inverted and supported by a knife edge at point B and the period of its small oscillations is observed to be 0.78 s. Knowing that $r_a + r_b = 10$ in., determine (a) the location of the mass center G, (b) the centroidal radius of gyration $\bar{k}$.

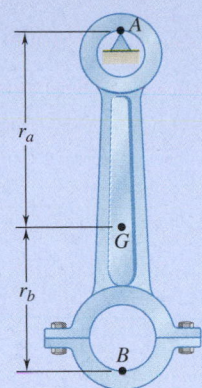

**Fig. P19.47**

**19.48** A semicircular hole is cut in a uniform square plate that is attached to a frictionless pin at its geometric center O. Determine (a) the period of small oscillations of the plate, (b) the length of a simple pendulum that has the same period.

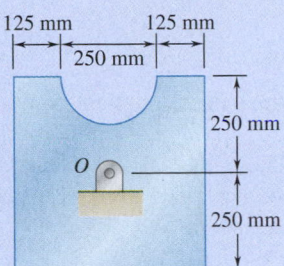

125 mm    125 mm
250 mm
250 mm
O
250 mm

**Fig. P19.48**

**19.49** A uniform disk of radius $r = 250$ mm is attached at A to a 650-mm rod AB of negligible mass that can rotate freely in a vertical plane about B. Determine the period of small oscillations (a) if the disk is free to rotate in a bearing at A, (b) if the rod is riveted to the disk at A.

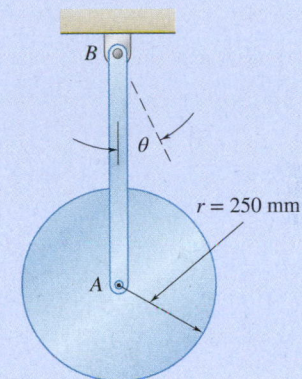

B
θ
r = 250 mm
A

**Fig. P19.49**

**19.50** A small collar of mass 1 kg is rigidly attached to a 3-kg uniform rod of length $L = 750$ mm. Determine (a) the distance d to maximize the frequency of oscillation when the rod is given a small initial displacement, (b) the corresponding period of oscillation.

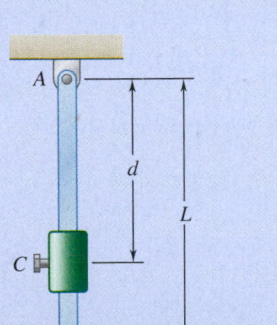

A
d
L
C
B

**Fig. P19.50**

**19.51** A thin homogeneous wire is bent into the shape of an isosceles triangle of sides b, b, and 1.6b. Determine the period of small oscillations if the wire (a) is suspended from point A as shown, (b) is suspended from point B.

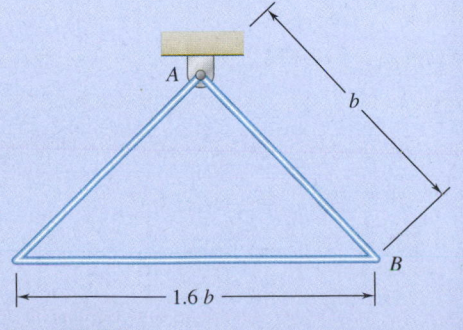

A
b
B
1.6 b

**Fig. P19.51**

**19.52** A *compound pendulum* is defined as a rigid body that oscillates about a fixed point $O$, called the center of suspension. Show that the period of oscillation of a compound pendulum is equal to the period of a simple pendulum of length $OA$, where the distance from $A$ to the mass center $G$ is $GA = \bar{k}^2/\bar{r}$. Point $A$ is defined as the center of oscillation and coincides with the center of percussion defined in Prob. 17.66.

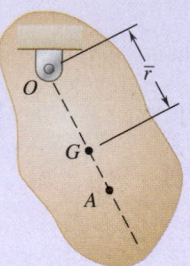

Fig. P19.52 and P19.53

**19.53** A rigid slab oscillates about a fixed point $O$. Show that the smallest period of oscillation occurs when the distance $\bar{r}$ from point $O$ to the mass center $G$ is equal to $\bar{k}$.

**19.54** Show that if the compound pendulum of Prob. 19.52 is suspended from $A$ instead of $O$, the period of oscillation is the same as before and the new center of oscillation is located at $O$.

**19.55** The 8-kg uniform bar $AB$ is hinged at $C$ and is attached at $A$ to a spring of constant $k = 500$ N/m. If end $A$ is given a small displacement and released, determine ($a$) the frequency of small oscillations, ($b$) the smallest value of the spring constant $k$ for which oscillations will occur.

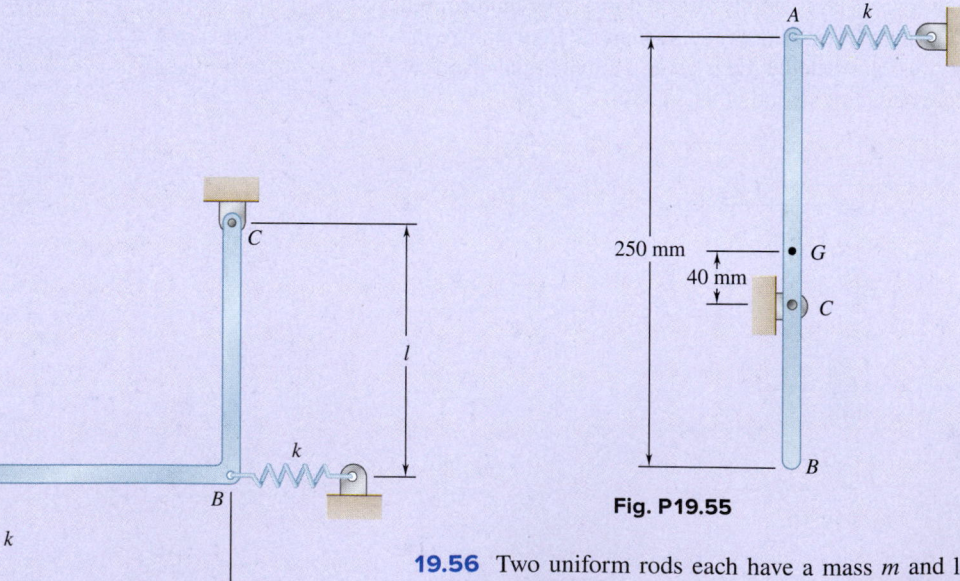

Fig. P19.56

Fig. P19.55

**19.56** Two uniform rods each have a mass $m$ and length $l$ and are welded together to form an L-shaped assembly. The assembly is constrained by two springs, each with a constant $k$, and is in equilibrium in a vertical plane in the position shown. Determine the frequency of small oscillations of the system.

**19.57** A uniform disk with radius $r$ and mass $m$ can roll without slipping on a cylindrical surface and is attached to bar $ABC$ with a length $L$ and negligible mass. The bar is attached at point $A$ to a spring with a constant $k$ and can rotate freely about point $B$ in the vertical plane. Knowing that end $A$ is given a small displacement and released, determine the frequency of the resulting vibration in terms of $m$, $L$, $k$, and $g$.

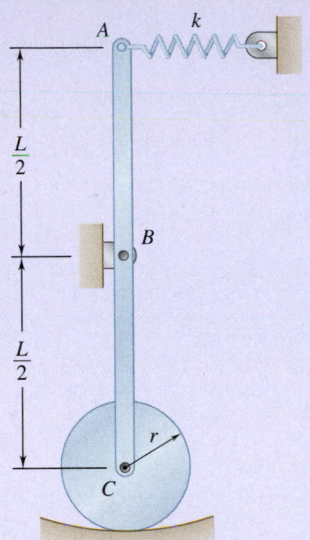

**Fig. P19.57**

**19.58** A 1300-kg sports car has a center of gravity $G$ located a distance $h$ above a line connecting the front and rear axles. The car is suspended from cables that are attached to the front and rear axles as shown. Knowing that the periods of oscillation are 4.04 s when $L = 4$ m and 3.54 s when $L = 3$ m, determine $h$ and the centroidal radius of gyration.

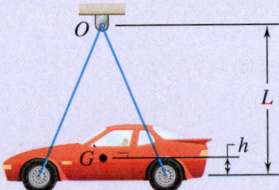

**Fig. P19.58**

**19.59** A 6-lb slender rod is suspended from a steel wire that is known to have a torsional spring constant $K = 1.5$ ft·lb/rad. If the rod is rotated through 180° about the vertical and released, determine (a) the period of oscillation, (b) the maximum velocity of end $A$ of the rod.

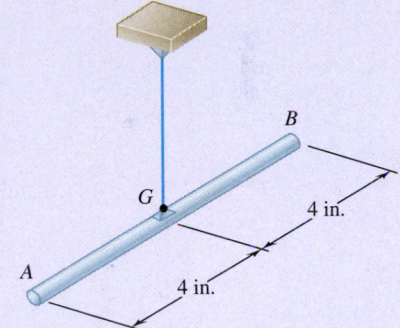

**Fig. P19.59**

**19.60** A uniform disk of radius $r = 250$ mm is attached at $A$ to a 650-mm rod $AB$ of negligible mass that can rotate freely in a vertical plane about $B$. If the rod is displaced 2° from the position shown and released, determine the magnitude of the maximum velocity of point $A$, assuming that the disk is (a) free to rotate in a bearing at $A$, (b) riveted to the rod at $A$.

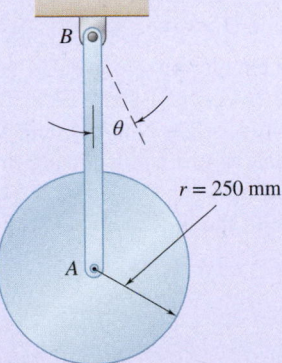

**Fig. P19.60**

**19.61** Two uniform rods, each of weight $W = 24$ lb and length $L = 40$ in., are welded together to form the assembly shown. Knowing that the constant of each spring is $k = 50$ lb/ft and that end $A$ is given a small displacement and released, determine the frequency of the resulting motion.

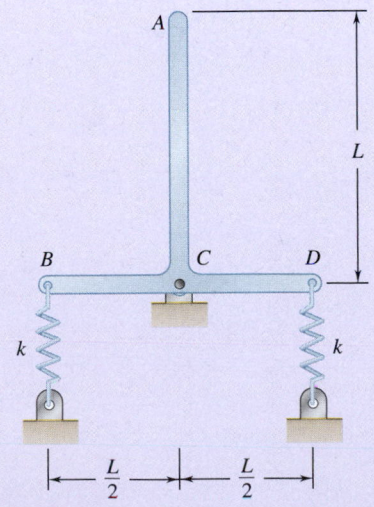

**Fig. P19.61**

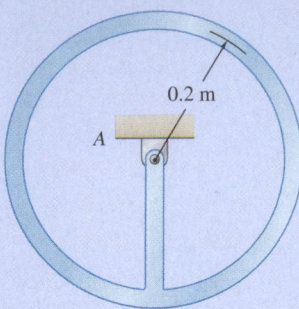

Fig. P19.62

**19.62** A homogeneous rod of mass per unit length equal to 0.4 kg/m is used to form the assembly shown, which rotates freely about pivot $A$ in a vertical plane. Knowing that the assembly is displaced 2° clockwise from its equilibrium position and released, determine its angular velocity and angular acceleration 5 s later.

**19.63** A horizontal platform $P$ is held by several rigid bars that are connected to a vertical wire. The period of oscillation of the platform is found to be 2.2 s when the platform is empty and 3.8 s when an object $A$ of unknown moment of inertia is placed on the platform with its mass center directly above the center of the plate. Knowing that the wire has a torsional constant $K = 27$ N·m/rad, determine the centroidal moment of inertia of object $A$.

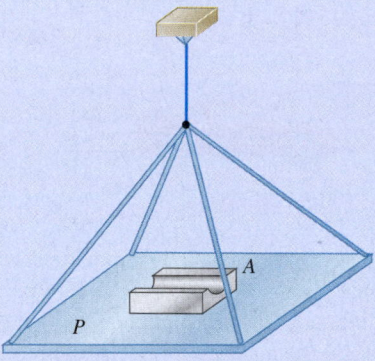

Fig. P19.63

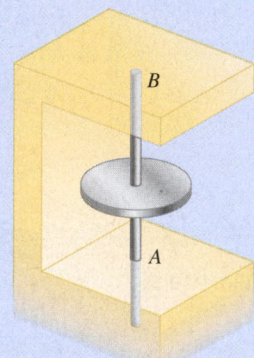

Fig. P19.64

**19.64** A uniform disk of radius $r = 120$ mm is welded at its center to two elastic rods of equal length with fixed ends at $A$ and $B$. Knowing that the disk rotates through an 8° angle when a 500-mN·m couple is applied to the disk and that it oscillates with a period of 1.3 s when the couple is removed, determine (a) the mass of the disk, (b) the period of vibration if one of the rods is removed.

**19.65** A 60-kg uniform circular plate is welded to two elastic rods, which have fixed ends at supports $A$ and $B$ as shown. The torsional spring constant of each rod is 200 N·m/rad, and the system is in equilibrium when the plate is vertical. Knowing that the plate is rotated 2° about axis $AB$ and released, determine (a) the period of oscillation, (b) the magnitude of the maximum velocity of the mass center $G$ of the plate.

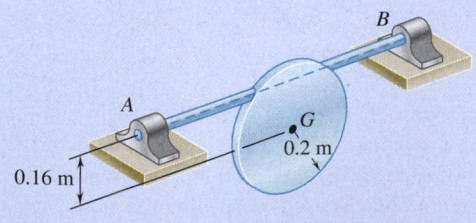

Fig. P19.65

**19.66** A uniform equilateral triangular plate with a side *b* is suspended from three vertical wires of the same length *l*. Determine the period of small oscillations of the plate when (*a*) it is rotated through a small angle about a vertical axis through its mass center *G*, (*b*) it is given a small horizontal displacement in a direction perpendicular to *AB*.

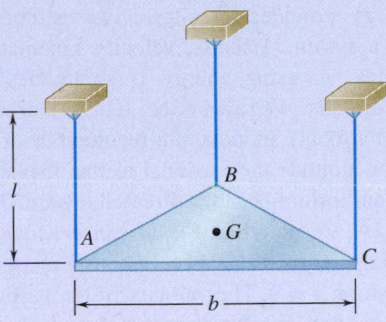

Fig. *P19.66*

**19.67** A period of 6.00 s is observed for the angular oscillations of a 4-oz gyroscope rotor suspended from a wire as shown. Knowing that a period of 3.80 s is obtained when a 1.25-in.-diameter steel sphere is suspended in the same fashion, determine the centroidal radius of gyration of the rotor. (Specific weight of steel = 490 lb/ft³)

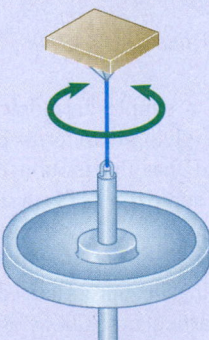

Fig. P19.67

**19.68** The centroidal radius of gyration $\overline{k}_y$ of an airplane is determined by suspending the airplane by two 12-ft-long cables as shown. The airplane is rotated through a small angle about the vertical through *G* and then released. Knowing that the observed period of oscillation is 3.3 s, determine the centroidal radius of gyration $\overline{k}_y$.

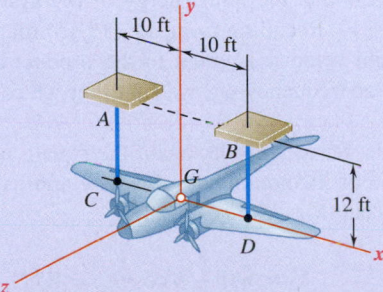

Fig. P19.68

# 19.3   APPLYING THE PRINCIPLE OF CONSERVATION OF ENERGY

Conservation of energy provides an alternative method to determine the natural frequency of a system. Usually, velocity kinematics are easier than acceleration kinematics, so using energy is sometimes easier than using Newton's second law directly. We saw in Sec. 19.1A that, when a particle with mass $m$ is in simple harmonic motion, the resultant **F** of the forces exerted on the particle has a magnitude proportional to the displacement $x$ measured from the position of equilibrium $O$ and is directed toward $O$; we have $F = -kx$. Referring to Sec. 13.2A, we note that **F** is a *conservative force* and that the corresponding potential energy is $V = \frac{1}{2}kx^2$, where $V$ is assumed equal to zero in the equilibrium position $x = 0$. The velocity of the particle is equal to $\dot{x}$, so its kinetic energy is $T = \frac{1}{2}m\dot{x}^2$. We can state that the total energy of the particle is conserved by writing

$$T + V = \text{constant} \qquad \tfrac{1}{2}m\dot{x}^2 + \tfrac{1}{2}kx^2 = \text{constant}$$

Dividing through by $m/2$ and recalling from Sec. 19.1A that $k/m = \omega_n^2$, where $\omega_n$ is the natural circular frequency of the vibration, we have

$$\dot{x}^2 + \omega_n^2 x^2 = \text{constant} \tag{19.25}$$

Eq. (19.25) is characteristic of a simple harmonic motion, because we can obtain it from Eq. (19.6) by multiplying both terms by $2\dot{x}$ and integrating.

Once we have established that the motion of the system is a simple harmonic motion or that it can be approximated by a simple harmonic motion, the principle of conservation of energy provides a convenient way for determining the period of vibration of a rigid body or of a system of rigid bodies possessing a single degree of freedom. Choosing an appropriate variable, such as a distance $x$ or an angle $\theta$, we consider two particular positions of the system:

1. **The displacement of the system is maximum.** We have $T_1 = 0$, and we can express $V_1$ in terms of the amplitude $x_m$ or $\theta_m$ (choosing $V = 0$ in the equilibrium position).
2. **The system passes through its equilibrium position.** We have $V_2 = 0$, and we can express $T_2$ in terms of the maximum velocity $\dot{x}_m$ or the maximum angular velocity $\dot{\theta}_m$.

We then express that the total energy of the system is conserved and write $T_1 + V_1 = T_2 + V_2$. Recalling from Eq. (19.15) that for simple harmonic motion the maximum velocity is equal to the product of the amplitude and of the natural circular frequency $\omega_n$, we find that we can solve this equation for $\omega_n$.

As an example, let us consider again the square plate of Sec. 19.2 and determine the period of its motion with this new approach. In the position of maximum displacement (Fig. 19.6$a$), we have

$$T_1 = 0 \qquad V_1 = W(b - b\cos\theta_m) = Wb(1 - \cos\theta_m)$$

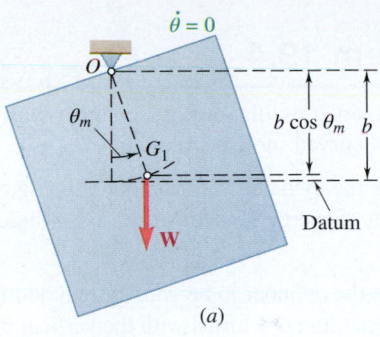

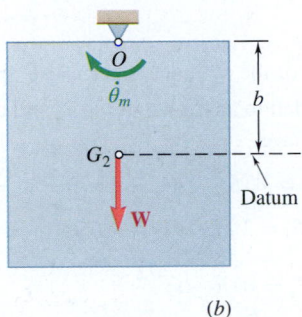

**Fig. 19.6** A square plate: (*a*) in the position of maximum displacement; (*b*) as it passes through its equilibrium position.

or because $1 - \cos \theta_m = 2 \sin^2(\theta_m/2) \approx 2(\theta_m/2)^2 = \theta_m^2/2$ for oscillations of small amplitude,

$$T_1 = 0 \qquad V_1 = \tfrac{1}{2}Wb\theta_m^2 \qquad \textbf{(19.26)}$$

As the plate passes through its position of equilibrium (Fig. 19.6*b*), its velocity is maximum, and we have

$$T_2 = \tfrac{1}{2}m\overline{v}_m^2 + \tfrac{1}{2}\overline{I}\omega_m^2 = \tfrac{1}{2}mb^2\dot{\theta}_m^2 + \tfrac{1}{2}\overline{I}\dot{\theta}_m^2 \qquad V_2 = 0$$

or recalling from Sec. 19.2 that $\overline{I} = \dfrac{2}{3}mb^2$,

$$T_2 = \tfrac{1}{2}(\tfrac{5}{3}mb^2)\dot{\theta}_m^2 \qquad V_2 = 0 \qquad \textbf{(19.27)}$$

Substituting from Eqs. (19.26) and (19.27) into $T_1 + V_1 = T_2 + V_2$ and noting that the maximum velocity $\dot{\theta}_m$ is equal to the product $\theta_m\omega_n$, we have

$$\tfrac{1}{2}Wb\theta_m^2 = \tfrac{1}{2}(\tfrac{5}{3}mb^2)\theta_m^2\omega_n^2 \qquad \textbf{(19.28)}$$

This gives us $\omega_n^2 = 3g/5b$ and

$$\tau_n = \frac{2\pi}{\omega_n} = 2\pi\sqrt{\frac{5b}{3g}} \qquad \textbf{(19.29)}$$

as obtained earlier in Sec. 19.2.

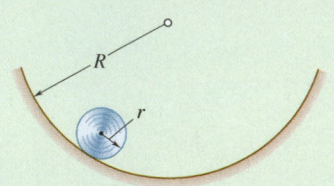

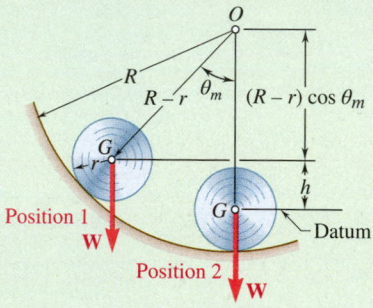

**Fig. 1** The cylinder in positions 1 and 2.

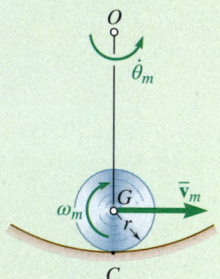

**Fig. 2** Kinematic quantities to describe the motion of the disk.

## Sample Problem 19.4

Determine the period of small oscillations of a cylinder with radius $r$ that rolls without slipping inside a curved surface with radius $R$. ▶

**STRATEGY:**   Because the cylinder rolls without slipping, you can apply the principle of conservation of energy between position 1, where $\theta = \theta_m$, and position 2, where $\theta = 0$.

**MODELING:**   Choose the cylinder to be your system and model it as a rigid body. Denote the angle that line $OG$ forms with the vertical by $\theta$ (Fig. 1).

**ANALYSIS:**

### Position 1.

**Kinetic Energy.**    The velocity of the cylinder is zero, so $T_1 = 0$.

**Potential Energy.**    Choose a datum as shown in Fig. 1 and denote the weight of the cylinder by $W$. Then you have

$$V_1 = Wh = W(R - r)(1 - \cos \theta)$$

For small oscillations, $(1 - \cos \theta) = 2\sin^2(\theta/2) \approx \theta^2/2$, so you have

$$V_1 = W(R - r)\frac{\theta_m^2}{2}$$

**Position 2.**   Denote the angular velocity of line $OG$ as the cylinder passes through position 2 by $\dot{\theta}_m$, and observe that point $C$ is the instantaneous center of rotation of the cylinder (Fig. 2). Then

$$\bar{v}_m = (R - r)\dot{\theta}_m \qquad \omega_m = \frac{\bar{v}_m}{r} = \frac{R - r}{r}\dot{\theta}_m$$

**Kinetic Energy.**

$$\begin{aligned} T_2 &= \tfrac{1}{2}m\bar{v}_m^2 + \tfrac{1}{2}\bar{I}\omega_m^2 \\ &= \tfrac{1}{2}m(R - r)^2\dot{\theta}_m^2 + \tfrac{1}{2}(\tfrac{1}{2}mr^2)\left(\frac{R - r}{r}\right)^2\dot{\theta}_m^2 \\ &= \tfrac{3}{4}m(R - r)^2\dot{\theta}_m^2 \end{aligned}$$

**Potential Energy.**

$$V_2 = 0$$

**Conservation of Energy.**

$$T_1 + V_1 = T_2 + V_2$$

$$0 + W(R - r)\frac{\theta_m^2}{2} = \tfrac{3}{4}m(R - r)^2\dot{\theta}_m^2 + 0$$

Because $\dot{\theta}_m = \omega_n\theta_m$ and $W = mg$, you have

$$mg(R - r)\frac{\theta_m^2}{2} = \tfrac{3}{4}m(R - r)^2(\omega_n\theta_m)^2 \qquad \omega_n^2 = \frac{2}{3}\frac{g}{R - r}$$

$$\tau_n = \frac{2\pi}{\omega_n} \qquad\qquad \tau_n = 2\pi\sqrt{\frac{3}{2}\frac{R - r}{g}} \blacktriangleleft$$

**REFLECT and THINK:**   This answer makes sense, because as the radius $R$ increases, the period also increases. In the limit as $R$ goes to infinity, the period also goes to infinity; in other words, the system would not oscillate. This is the case of a cylinder on a horizontal surface. The small angle approximation, $(1 - \cos \theta) = 2\sin^2(\theta/2) \approx \theta^2/2$, is often used in problems like this one.

# SOLVING PROBLEMS
# ON YOUR OWN

$\mathbf{I}$n the problems that follow, you will be asked to use the *principle of conservation of energy* to determine the period or natural frequency of the simple harmonic motion of a particle or rigid body. Assuming that you choose an angle $\theta$ to define the position of the system (with $\theta = 0$ in the equilibrium position), as you will in most of the problems in this section, you will express that the total energy of the system is conserved using $T_1 + V_1 = T_2 + V_2$ between position 1 of maximum displacement ($\theta_1 = \theta_m$, $\dot{\theta}_1 = 0$) and position 2 of maximum velocity ($\dot{\theta}_2 = \dot{\theta}_m$, $\theta_2 = 0$). It follows that $T_1$ and $V_2$ are both zero, and the energy equation reduces to $V_1 = T_2$, where $V_1$ and $T_2$ are homogeneous quadratic expressions in $\theta_m$ and $\dot{\theta}_m$, respectively. Recalling that for a simple harmonic motion, $\dot{\theta}_m = \theta_m \omega_n$, and substituting this product into the energy equation, after reduction you obtain an equation that you can solve for $\omega_n^2$. Once you have determined the natural circular frequency $\omega_n$, you can obtain the period $\tau_n$ and the natural frequency $f_n$ of the vibration.

The steps that you should take are as follows:

**1. Calculate the potential energy $V_1$ of the system in its position of maximum displacement.** Draw a sketch of the system in its position of maximum displacement and express the potential energy of all the forces involved in terms of the maximum displacement $x_m$ or $\theta_m$.

   **a. The potential energy associated with the weight $W$ of a body is $V_g = Wy$,** where $y$ is the elevation of the center of gravity $G$ of the body above its equilibrium position. If the problem you are solving involves the oscillation of a rigid body about a horizontal axis through a point $O$ located at a distance $b$ from $G$ (Fig. 19.6), express $y$ in terms of the angle $\theta$ that the line $OG$ forms with the vertical: $y = b(1 - \cos \theta)$. For small values of $\theta$, you can replace this expression with $y = \frac{1}{2} b\theta^2$ (Sample Prob. 19.4). Therefore, when $\theta$ reaches its maximum value $\theta_m$ and for oscillations of small amplitude, you can express $V_g$ as

$$V_g = \tfrac{1}{2} Wb\theta_m^2$$

Note that *if $G$ is located above $O$* in its equilibrium position (instead of below $O$, as we have assumed), the vertical displacement $y$ is negative and should be approximated as $y = -\frac{1}{2} b\theta^2$, which results in a negative value for $V_g$. In the absence of other forces, the equilibrium position is unstable, and the system does not oscillate. (See, for instance, Prob. 19.89.)

   **b. The potential energy associated with the elastic force exerted by a spring is $V_e = \frac{1}{2} kx^2$,** where $k$ is the constant of the spring and $x$ is its deflection. In problems involving the rotation of a body about an axis, you generally have $x = a\theta$, where $a$ is the distance from the axis of rotation to the point of the body where the spring is attached, and $\theta$ is the angle of rotation. Therefore, when $x$ reaches its maximum value $x_m$ and $\theta$ reaches its maximum value $\theta_m$, you can express $V_e$ as

$$V_e = \tfrac{1}{2} kx_m^2 = \tfrac{1}{2} ka^2\theta_m^2$$

**c. The potential energy $V_1$ of the system in its position of maximum displacement** is obtained by adding the various potential energies that you have computed. It is equal to the product of a constant and $\theta_m^2$.

**2. Calculate the kinetic energy $T_2$ of the system in its position of maximum velocity.** Note that this position is also the equilibrium position of the system.

   **a. If the system consists of a single rigid body,** the kinetic energy $T_2$ of the system is the sum of the kinetic energy associated with the motion of the mass center $G$ of the body and the kinetic energy associated with the rotation of the body about $G$. Therefore, you can write

$$T_2 = \tfrac{1}{2}m\bar{v}_m^2 + \tfrac{1}{2}\bar{I}\omega_m^2$$

Assuming that the position of the body has been defined by an angle $\theta$, express $\bar{v}_m$ and $\omega_m$ in terms of the rate of change $\dot{\theta}_m$ of $\theta$ as the body passes through its equilibrium position. The kinetic energy of the body is thus expressed as the product of a constant and $\dot{\theta}_m^2$. Note that if $\theta$ measures the rotation of the body about its mass center, as was the case for the plate of Fig. 19.6, then $\omega_m = \dot{\theta}_m$. In other cases, however, the kinematics of the motion should be used to derive a relation between $\omega_m$ and $\dot{\theta}_m$ (Sample Prob. 19.4).

   **b. If the system consists of several rigid bodies,** repeat the previous computation for each of the bodies using the same coordinate $\theta$ and add the results.

**3. Equate the potential energy $V_1$ of the system to its kinetic energy $T_2$,**

$$V_1 = T_2$$

and recalling the first of Eqs. (19.15), replace $\dot{\theta}_m$ in the right-hand term with the product of the amplitude $\theta_m$ and the circular frequency $\omega_n$. Because both terms now contain the factor $\theta_m^2$, you can cancel this factor and solve the resulting equation for the circular frequency $\omega_n$.

# Problems

**19.69** Two blocks each have a mass 1.5 kg and are attached to links that are pin-connected to bar *BC* as shown. The masses of the links and bar are negligible, and the blocks can slide without friction. Block *D* is attached to a spring of constant $k = 720$ N/m. Knowing that block *A* is at rest when it is struck horizontally with a mallet and given an initial velocity of 250 mm/s, determine the magnitude of the maximum displacement of block *D* during the resulting motion.

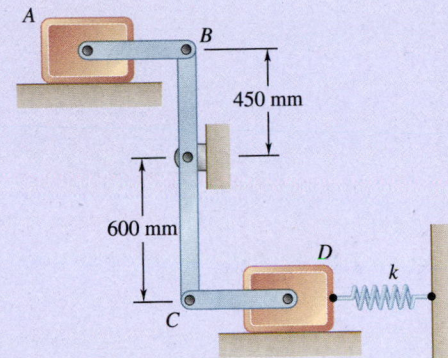

**Fig. P19.69**

**19.70** Two small spheres, *A* and *C*, each have a mass *m* and are attached to rod *AB* that is supported by a pin and bracket at *B* and by a spring *CD* with constant *k*. Knowing that the mass of the rod is negligible and that the system is in equilibrium when the rod is horizontal, determine the frequency of the small oscillations of the system.

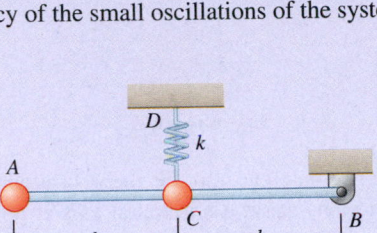

**Fig. P19.70**

**19.71** A 20-lb block is attached to spring *A* and connected to spring *B* by a cord and pulley. The block is held in the position shown with both springs unstretched when the support is removed and the block is released with no initial velocity. Neglecting friction and the masses of the pulley and the springs, determine (*a*) the period of the resulting vibration, (*b*) the maximum velocity achieved by the block.

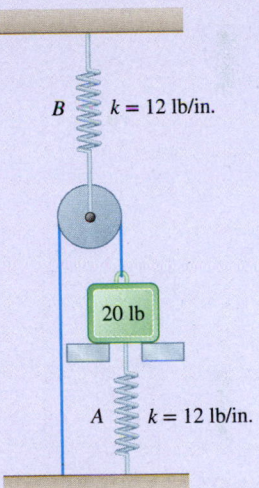

**Fig. P19.71**

**19.72** Determine the period of small oscillations of a small particle that moves without friction inside a cylindrical surface of radius *R*.

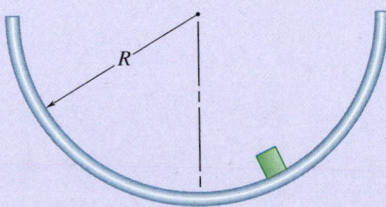

**Fig. P19.72**

**19.73** The inner rim of an 85-lb flywheel is placed on a knife edge, and the period of its small oscillations is found to be 1.26 s. Determine the centroidal moment of inertia of the flywheel.

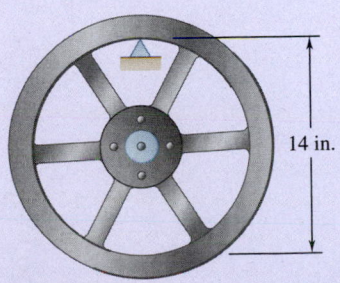

**Fig. P19.73**

1387

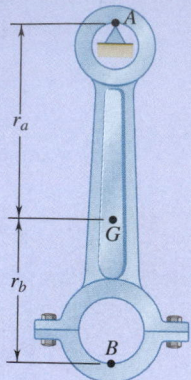

Fig. *P19.74*

**19.74** A connecting rod is supported by a knife edge at point *A;* the period of its small oscillations is observed to be 1.03 s. Knowing that the distance $r_a$ is 6 in., determine the centroidal radius of gyration of the connecting rod.

**19.75** A uniform rod *AB* can rotate in a vertical plane about a horizontal axis at *C* located at a distance *c* above the mass center *G* of the rod. For small oscillations, determine the value of *c* for which the frequency of the motion will be maximum.

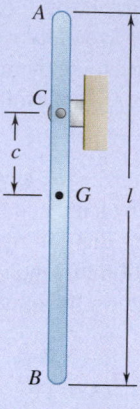

Fig. P19.75

**19.76** A thin uniform plate cut into the shape of a quarter circle can rotate in a vertical plane about a horizontal axis at point *O*. Determine the period of the small oscillations of the plate.

**19.77** A uniform disk of radius *r* and mass *m* can roll without slipping on a cylindrical surface and is attached to bar *ABC* of length *L* and negligible mass. The bar is attached to a spring of constant *k* and can rotate freely in the vertical plane about point *B*. Knowing that end *A* is given a small displacement and released, determine the frequency of the resulting oscillations in terms of *m, L, k,* and *g*.

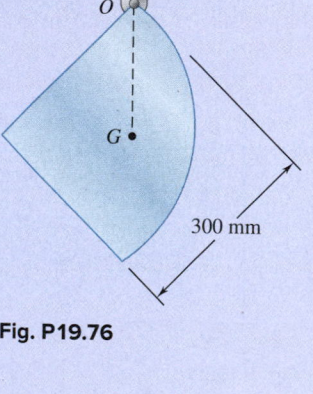

Fig. P19.76

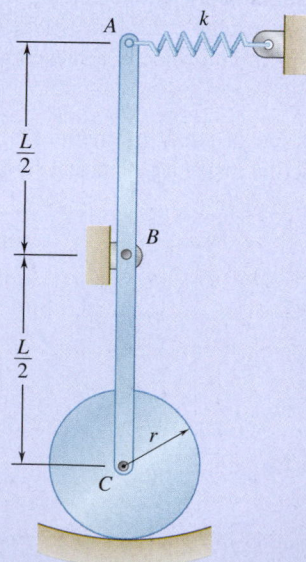

Fig. P19.77

**19.78** Blade *AB* of the experimental wind-turbine generator shown is to be temporarily removed. Motion of the turbine generator about the *y* axis is prevented, but the remaining three blades may oscillate as a unit about the *x* axis. Assuming that each blade can be modeled as a 100-ft slender rod, determine the period of small oscillations of the blades.

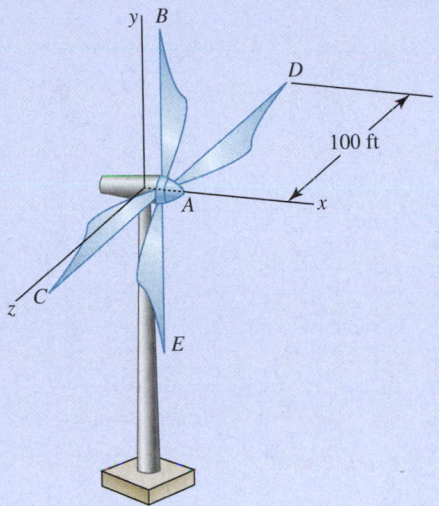

Fig. P19.78

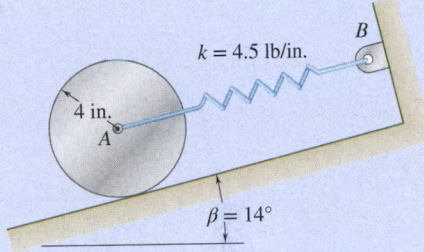

Fig. P19.79

**19.79** A 15-lb uniform cylinder can roll without sliding on an incline and is attached to a spring *AB* as shown. If the center of the cylinder is moved 0.4 in. down the incline and released, determine (*a*) the period of vibration, (*b*) the maximum velocity of the center of the cylinder.

**19.80** A 3-kg slender rod *AB* is bolted to a 5-kg uniform disk. A spring of constant 280 N/m is attached to the disk and is unstretched in the position shown. If end *B* of the rod is given a small displacement and released, determine the period of vibration of the system.

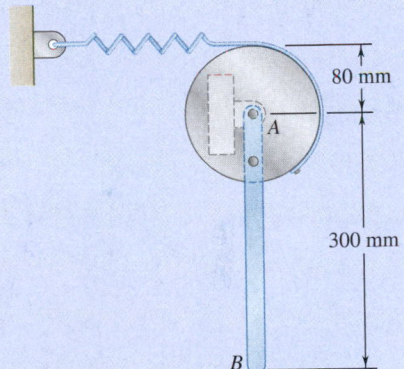

Fig. P19.80

**19.81** A slender 10-kg bar *AB* with a length of *l* = 0.6 m is connected to two collars of negligible weight. Collar *A* is attached to a spring with a constant of *k* = 1.5 kN/m and can slide on a horizontal rod, while collar *B* can slide freely on a vertical rod. Knowing that the system is in equilibrium when bar *AB* is vertical and that collar *A* is given a small displacement and released, determine the period of the resulting vibrations.

**19.82** A slender 5-kg bar *AB* with a length of *l* = 0.6 m is connected to two collars, each of mass 2.5 kg. Collar *A* is attached to a spring with a constant of *k* = 1.5 kN/m and can slide on a horizontal rod, while collar *B* can slide freely on a vertical rod. Knowing that the system is in equilibrium when bar *AB* is vertical and that collar *A* is given a small displacement and released, determine the period of the resulting vibrations.

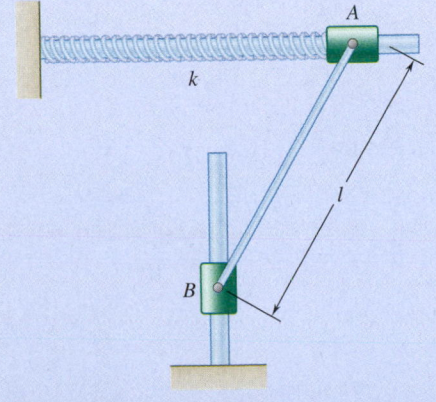

Fig. *P19.81* and *P19.82*

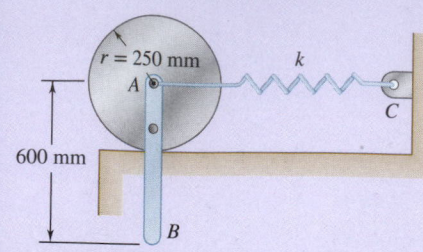

**Fig. P19.83**

**19.83** An 800-g rod $AB$ is bolted to a 1.2-kg disk. A spring of constant $k = 12$ N/m is attached to the center of the disk at $A$ and to the wall at $C$. Knowing that the disk rolls without sliding, determine the period of small oscillations of the system.

**19.84** Three identical 3.6-kg uniform slender bars are connected by pins as shown and can move in a vertical plane. Knowing that bar $BC$ is given a small displacement and released, determine the period of vibration of the system.

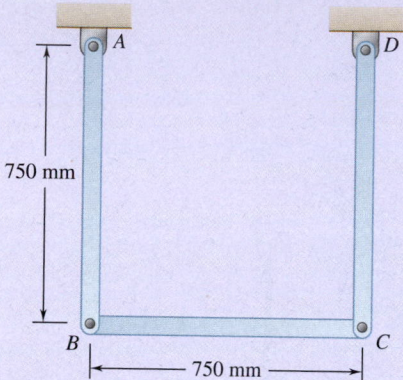

**Fig. *P19.84***

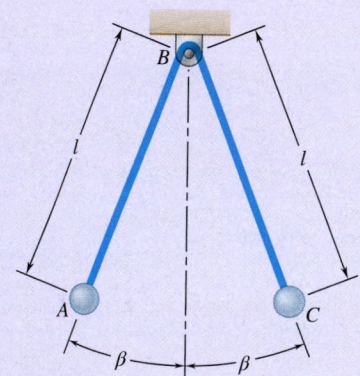

**Fig. P19.85**

**19.85** A homogeneous rod of weight $W$ and length $2l$ is bent as shown, and two spheres $A$ and $C$, each of weight $W$, are attached to its ends. The system is allowed to oscillate about a frictionless pin at $B$. Knowing that $\beta = 40°$ and $l = 25$ in., determine the frequency of small oscillations.

**19.86** A 10-lb uniform rod $CD$ is welded at $C$ to a shaft of negligible mass that is welded to the centers of two 20-lb uniform disks $A$ and $B$. Knowing that the disks roll without sliding, determine the period of small oscillations of the system.

**19.87 and *19.88*** Two uniform rods $AB$ and $CD$, each of length $l$ and mass $m$, are attached to gears as shown. Knowing that the mass of gear $C$ is $m$ and that the mass of gear $A$ is $4m$, determine the period of small oscillations of the system.

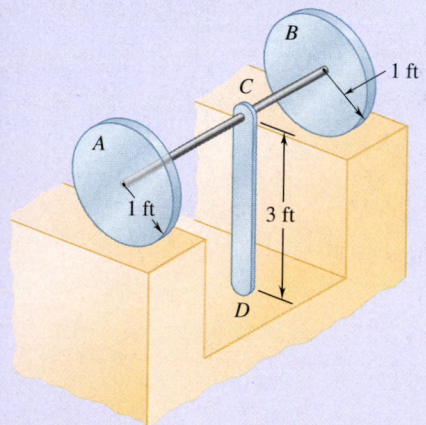

**Fig. P19.86**

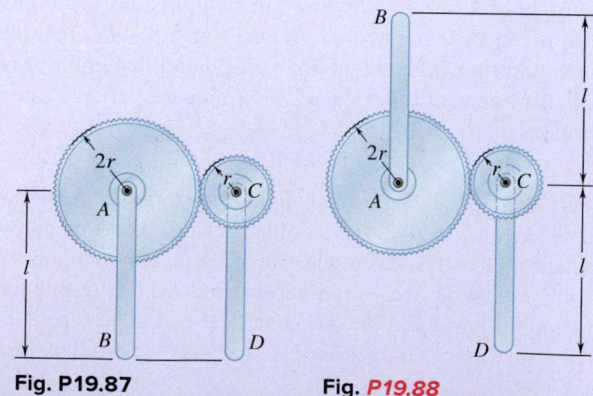

**Fig. P19.87**     **Fig. *P19.88***

**19.89** An inverted pendulum consisting of a 2-kg sphere and a 1.5-kg bar *ABC* of length $l = 0.7$ m is supported by a pin and bracket at *C*. A spring of constant $k = 200$ N/m is attached to the bar at *B* and is undeformed when the bar is in the vertical position shown. Determine (*a*) the frequency of small oscillations for $a = 0.5$ m, (*b*) the smallest value of *a* for which oscillations will occur.

**19.90** Two 12-lb uniform disks are attached to the 20-lb rod *AB* as shown. Knowing that the constant of the spring is 30 lb/in. and that the disks roll without sliding, determine the frequency of vibration of the system.

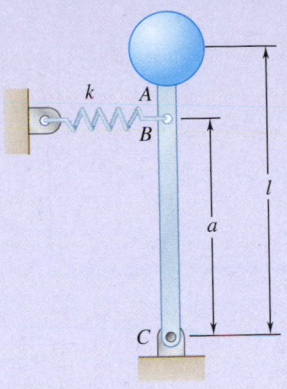

**Fig. P19.89**

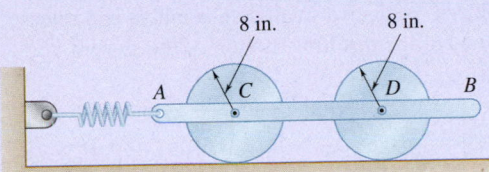

**Fig. P19.90**

**19.91** Two 6-lb uniform semicircular plates are attached to the 4-lb rod *AB* as shown. Knowing that the plates roll without sliding, determine the period of small oscillations of the system.

**19.92** A half section of a uniform cylinder of radius *r* and mass *m* rests on two casters *A* and *B*, each of which is a uniform cylinder of radius *r*/4 and mass *m*/8. Knowing that the half cylinder is rotated through a small angle and released and that no slipping occurs, determine the frequency of small oscillations.

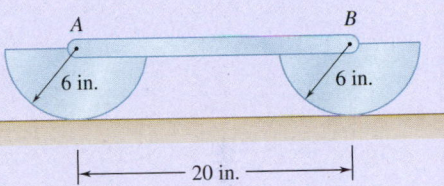

**Fig. P19.91**

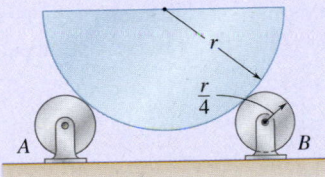

**Fig. P19.92**

**19.93** The motion of the uniform rod *AB* is guided by the cord *BC* and by the small roller at *A*. Determine the frequency of oscillation when the end *B* of the rod is given a small horizontal displacement and released.

**19.94** A uniform rod of length *L* is supported by a ball-and-socket joint at *A* and by a vertical wire *CD*. Derive an expression for the period of oscillation of the rod if end *B* is given a small horizontal displacement and then released.

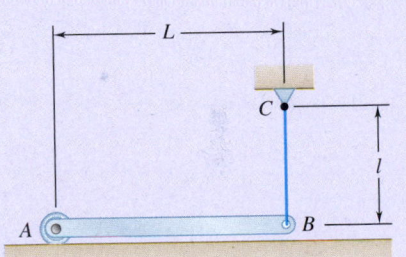

**Fig. P19.93**

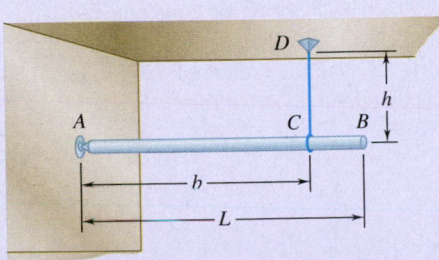

**Fig. P19.94**

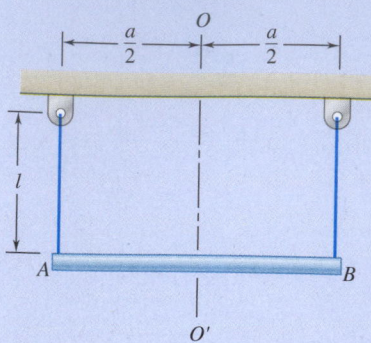

**Fig. P19.95**

**19.95** A section of uniform pipe is suspended from two vertical cables attached at $A$ and $B$. Determine the frequency of oscillation when the pipe is given a small rotation about the centroidal axis $OO'$ and released.

**19.96** Three collars each have a mass $m$ and are connected by pins to bars $AC$ and $BC$, each having length $l$ and negligible mass. Collars $A$ and $B$ can slide without friction on a horizontal rod and are connected by a spring of constant $k$. Collar $C$ can slide without friction on a vertical rod and the system is in equilibrium in the position shown. Knowing that collar $C$ is given a small displacement and released, determine the frequency of the resulting motion of the system.

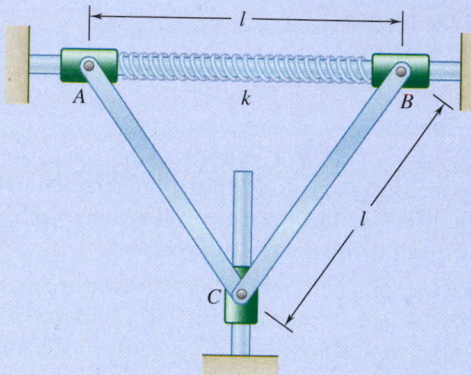

**Fig. P19.96**

**\*19.97** A thin plate of length $l$ rests on a half cylinder of radius $r$. Derive an expression for the period of small oscillations of the plate.

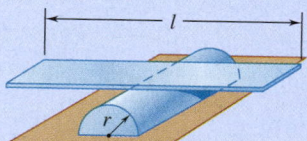

**Fig. P19.97**

**Fig. P19.98**

**\*19.98** As a submerged body moves through a fluid, the particles of the fluid flow around the body and thus acquire kinetic energy. In the case of a sphere moving in an ideal fluid, the total kinetic energy acquired by the fluid is $\frac{1}{4}\rho Vv^2$, where $\rho$ is the mass density of the fluid, $V$ is the volume of the sphere, and $v$ is the velocity of the sphere. Consider a 500-g hollow spherical shell of radius 80 mm that is held submerged in a tank of water by a spring of constant 500 N/m. (*a*) Neglecting fluid friction, determine the period of vibration of the shell when it is displaced vertically and then released. (*b*) Solve part *a*, assuming that the tank is accelerated upward at the constant rate of 8 m/s$^2$.

# 19.4  FORCED VIBRATIONS

From the point of view of engineering applications, the most important vibrations are the **forced vibrations** of a system. These vibrations occur when a system is subjected to a periodic force or when it is elastically connected to a support that has an alternating motion.

Consider first the case of a body of mass $m$ suspended from a spring and subjected to a periodic force **P** with a magnitude of $P = P_m \sin \omega_f t$, where $\omega_f$ is the circular frequency of **P** and is referred to as the **forced circular frequency** of the motion (Fig. 19.7). This force may be an actual external force applied to the body, or it may be a result of the rotation of some unbalanced part of the body (see Sample Prob. 19.5). Denoting the displacement of the body measured from its equilibrium position by $x$, the equation of motion is obtained from the free-body diagram and kinetic diagram in Fig. 19.7 as

$$+\downarrow \Sigma F = ma: \qquad P_m \sin \omega_f t + W - k(\delta_{st} + x) = m\ddot{x}$$

Recalling that $W = k\delta_{st}$, we have

$$m\ddot{x} + kx = P_m \sin \omega_f t \qquad (19.30)$$

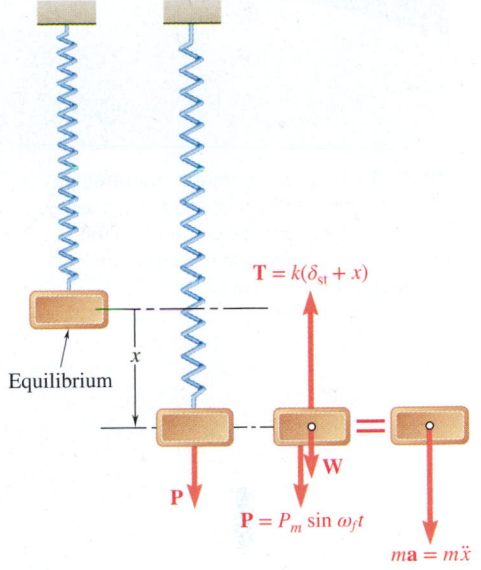

**Fig. 19.7** Free-body diagram and kinetic diagram of a block suspended from a spring and subjected to a periodic force.

Next we consider the case of a body with a mass $m$ suspended from a spring attached to a moving support whose displacement $\delta$ is equal to $\delta_m \sin \omega_f t$ (Fig. 19.8). Measuring the displacement $x$ of the body from the position of static equilibrium corresponding to $\omega_f t = 0$, we find that the total elongation of the spring at time $t$ is $\delta_{st} + x - \delta_m \sin \omega_f t$. The equation of motion is thus

$$+\downarrow \Sigma F = ma: \qquad W - k(\delta_{st} + x - \delta_m \sin \omega_f t) = m\ddot{x}$$

Again recalling that $W = k\delta_{st}$, we have

$$m\ddot{x} + kx = k\delta_m \sin \omega_f t \qquad (19.31)$$

Note that Eqs. (19.30) and (19.31) are of the same form and that a solution of the first equation will satisfy the second if we set $P_m = k\delta_m$.

A differential equation such as Eq. (19.30) or (19.31), possessing a right-hand side different from zero, is said to be *nonhomogeneous*. We can obtain its general solution by adding a particular solution of the given equation to the general solution of the corresponding *homogeneous* equation (with the right-hand side equal to zero). We can obtain a *particular solution* of Eq. (19.30) or (19.31) by trying a solution of the form

$$x_{part} = x_m \sin \omega_f t \qquad (19.32)$$

Substituting $x_{part}$ for $x$ into Eq. (19.30), we find

$$-m\omega_f^2 x_m \sin \omega_f t + kx_m \sin \omega_f t = P_m \sin \omega_f t$$

We can solve this equation for the amplitude as

$$x_m = \frac{P_m}{k - m\omega_f^2}$$

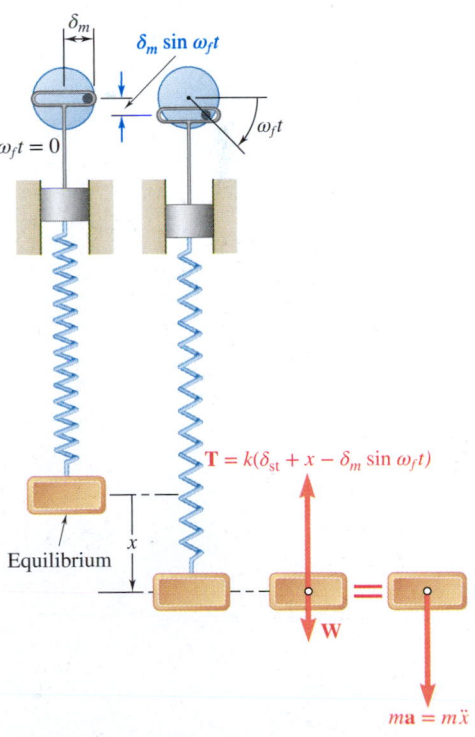

**Fig. 19.8** Free-body diagram and kinetic diagram of a block suspended from a spring attached to a harmonically moving support.

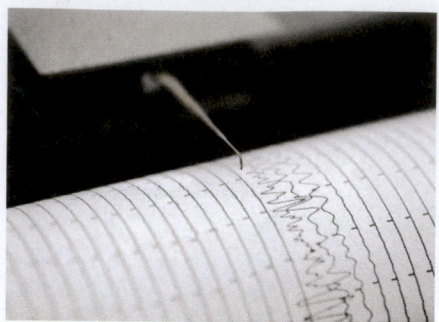

**Photo 19.1** A seismometer operates by measuring the amount of electrical energy needed to keep a mass centered in the housing in the presence of strong ground vibration. ©kickers/Getty Images RF

Recall from Eq. (19.4) that $k/m = \omega_n^2$, where $\omega_n$ is the natural circular frequency of the system. Then we have

$$x_m = \frac{P_m/k}{1 - (\omega_f/\omega_n)^2} \qquad (19.33)$$

If we define the frequency ratio, $r$, as $r = \omega_f/\omega_n$, we can write this equation as

$$x_m = \frac{P_m/k}{1 - r^2}$$

In a similar way, substituting from Eq. (19.32) into Eq. (19.31), we obtain

$$x_m = \frac{\delta_m}{1 - (\omega_f/\omega_n)^2} \qquad (19.33')$$

or

$$x_m = \frac{\delta_m}{1 - r^2}$$

The homogeneous equation corresponding to Eq. (19.30) or (19.31) is Eq. (19.2), which defines the free vibration of the body. We found its general solution, called the *complementary function*, in Sec. 19.1A:

$$x_{\text{comp}} = C_1 \sin \omega_n t + C_2 \cos \omega_n t \qquad (19.34)$$

Adding the particular solution of Eq. (19.32) to the complementary function of Eq. (19.34), we obtain the **general solution** of Eqs. (19.30) and (19.31) as

$$x = C_1 \sin \omega_n t + C_2 \cos \omega_n t + x_m \sin \omega_f t \qquad (19.35)$$

Note that this vibration consists of two superposed vibrations. The first two terms in Eq. (19.35) represent a free vibration of the system. The frequency of this vibration is the *natural frequency* of the system, which depends only upon the constant $k$ of the spring and the mass $m$ of the body, and the constants $C_1$ and $C_2$ can be determined from the initial conditions. This free vibration is also called a **transient** vibration, because in actual practice, it is soon damped out by friction forces (Sec. 19.5B).

The last term in Eq. (19.35) represents the **steady-state** vibration produced and maintained by the impressed force or impressed support movement. Its frequency is the **forced frequency** imposed by this force or movement, and its amplitude $x_m$, defined by Eq. (19.33) or (19.33′), depends upon the **frequency ratio** $r = \omega_f/\omega_n$. Dividing the amplitude $x_m$ of the steady-state vibration by $P_m/k$ in the case of a periodic force, or by $\delta_m$ in the case of an oscillating support, we obtain the **magnification factor**. From Eqs. (19.33) and (19.33′), we obtain

$$\text{Magnification factor} = \frac{x_m}{P_m/k} = \frac{x_m}{\delta_m} = \frac{1}{1 - (\omega_f/\omega_n)^2} \qquad (19.36)$$

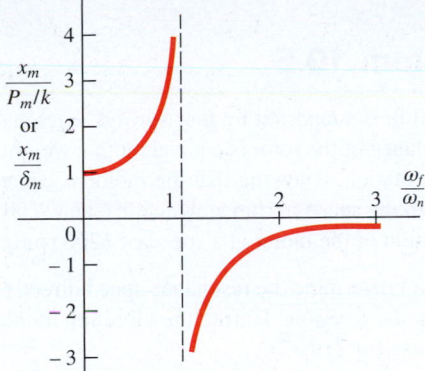

**Fig. 19.9** For an undamped system, the magnification factor becomes infinite at a forcing frequency equal to the natural frequency.

In Fig. 19.9, we have plotted the magnification factor against the frequency ratio $\omega_f/\omega_n$. Note that when $\omega_f = \omega_n$, the amplitude of the forced vibration becomes infinite. The impressed force or impressed support movement is said to be in **resonance** with the given system. Actually, the amplitude of the vibration remains finite because of damping forces (Sec. 19.5B); nevertheless, such a situation should be avoided, and the forced frequency should not be chosen too close to the natural frequency of the system. Also note that for $\omega_f < \omega_n$, the coefficient of $\sin \omega_f t$ in Eq. (19.35) is positive, whereas for $\omega_f > \omega_n$, this coefficient is negative. In the first case, the forced vibration is *in phase* with the impressed force or impressed support movement, while in the second case, it is 180° *out of phase*.

Finally, observe that we can obtain the velocity and acceleration of the steady-state vibration by differentiating the last term of Eq. (19.35) twice with respect to $t$. The maximum values are given by expressions similar to those of Eqs. (19.15) of Sec. 19.1A, except that these expressions now involve the amplitude and the circular frequency of the forced vibration:

$$v_m = x_m\omega_f \qquad a_m = x_m\omega_f^2 \qquad\qquad\qquad \textbf{(19.37)}$$

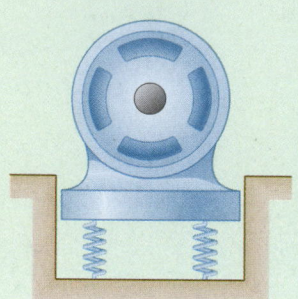

## Sample Problem 19.5

A motor weighing 350 lb is supported by four springs, each having a constant of 750 lb/in. The unbalance of the rotor is equivalent to a weight of 1 oz located 6 in. from the axis of rotation. Knowing that the motor is constrained to move vertically, determine (*a*) the speed in rpm at which resonance will occur, (*b*) the amplitude of the vibration of the motor at a speed of 1200 rpm.

**STRATEGY:**  You can determine the resonance speed directly from the given data because you know $\omega_n = \sqrt{k/m}$. To find the vibration amplitude at a speed of 1200 rpm, you can use Eq. (19.33).

**MODELING:**  Choose the motor to be your system, and model it as a single-degree-of-freedom particle undergoing forced oscillation.

**ANALYSIS:**

**a. Resonance Speed.**  The resonance speed is equal to the natural circular frequency $\omega_n$ (in rpm) of the free vibration of the motor. The mass of the motor, $M$, and the equivalent constant of the supporting springs are

$$M = \frac{350 \text{ lb}}{32.2 \text{ ft/s}^2} = 10.87 \text{ lb·s}^2/\text{ft}$$

$$k = 4(750 \text{ lb/in.}) = 3000 \text{ lb/in.} = 36{,}000 \text{ lb/ft}$$

$$\omega_n = \sqrt{\frac{k}{M}} = \sqrt{\frac{36{,}000}{10.87}} = 57.5 \text{ rad/s} = 549 \text{ rpm}$$

**Resonance speed = 549 rpm** ◄

**b. Amplitude of Vibration at 1200 rpm.**  The angular velocity of the motor and the mass $m$ of the equivalent 1-oz weight are

$$\omega = 1200 \text{ rpm} = 125.7 \text{ rad/s}$$

$$m = (1 \text{ oz})\frac{1 \text{ lb}}{16 \text{ oz}}\frac{1}{32.2 \text{ ft/s}^2} = 0.001941 \text{ lb·s}^2/\text{ft}$$

To find the equivalent of an applied force, you can draw a free-body diagram and kinetic diagram (Fig. 1).

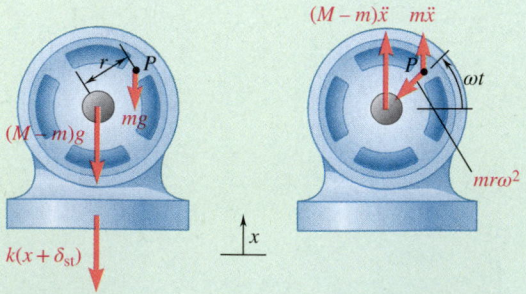

**Fig. 1**  Free-body diagram and kinetic diagram for the system.

*(continued)*

Applying Newton's second law in the vertical direction gives

$$-(M - m)g - mg - k(x + \delta_{st}) = (M - m)\ddot{x} + m\ddot{x} - mr\omega^2 \sin \omega t$$

Recognizing that $Mg = k\delta_{st}$, this equation simplifies to

$$M\ddot{x} + kx = mr\omega^2 \sin \omega t$$

Thus, the rotating unbalanced mass is equivalent to an applied force

$$P_m = mr\omega^2 = (0.001941 \text{ lb·s}^2/\text{ft})(\tfrac{6}{12} \text{ ft})(125.7 \text{ rad/s})^2 = 15.33 \text{ lb}$$

The static deflection that would be caused by a constant load $P_m$ is

$$\frac{P_m}{k} = \frac{15.33 \text{ lb}}{3000 \text{ lb/in.}} = 0.00511 \text{ in.}$$

The forced circular frequency $\omega_f$ of the motion is the angular velocity of the motor,

$$\omega_f = \omega = 125.7 \text{ rad/s}$$

Substituting the values of $P_m/k$, $\omega_f$, and $\omega_n$ into Eq. (19.33), we obtain

$$x_m = \frac{P_m/k}{1 - (\omega_f/\omega_n)^2} = \frac{0.00511 \text{ in.}}{1 - (125.7/57.5)^2} = -0.001352 \text{ in.}$$

$$x_m = 0.001352 \text{ in. (out of phase)} \quad \blacktriangleleft$$

**REFLECT and THINK:** In problems involving an unbalanced mass, the result of the imbalance is equivalent to an applied force of $P_m = mr\omega^2$. In this problem, because $\omega_f > \omega_n$, the vibration is 180° out of phase with the force due to the unbalance of the rotor. For example, when the unbalanced mass is directly below the axis of rotation, the position of the motor is $x_m = 0.001352$ in. above the position of equilibrium.

# SOLVING PROBLEMS
# ON YOUR OWN

In this section, we analyzed the **forced vibrations** of a mechanical system. These vibra-tions occur either when the system is subjected to a periodic force **P** (Fig. 19.7) or when it is elastically connected to a support that has an alternating motion (Fig. 19.8). In the first case, the motion of the system is defined by the differential equation

$$m\ddot{x} + kx = P_m \sin \omega_f t \qquad (19.30)$$

where the right-hand side represents the magnitude of the force **P** at a given instant. In the second case, the motion is defined by the differential equation

$$m\ddot{x} + kx = k\delta_m \sin \omega_f t \qquad (19.31)$$

where the right-hand side is the product of the spring constant $k$ and the displacement of the support at a given instant.

You will be concerned only with the **steady-state** motion of the system, which is defined by a *particular solution* of Eqs. (19.30) and (19.31), of the form

$$x_{\text{part}} = x_m \sin \omega_f t \qquad (19.32)$$

**1. If the forced vibration is caused by a periodic force P** with an amplitude $P_m$ and circular frequency $\omega_f$, the amplitude of the vibration is

$$x_m = \frac{P_m/k}{1 - (\omega_f/\omega_n)^2} \qquad (19.33)$$

where $\omega_n$ is the *natural circular frequency* of the system, $\omega_n = \sqrt{k/m}$, and $k$ is the spring constant. Note that the circular frequency of the vibration is $\omega_f$ and that the amplitude $x_m$ does not depend upon the initial conditions. For $\omega_f = \omega_n$, the denominator in Eq. (19.33) is zero, and $x_m$ is infinite (Fig. 19.9); the impressed force **P** is said to be in **resonance** with the system. Also, for $\omega_f < \omega_n$, $x_m$ is positive and the vibration is *in phase* with **P**, whereas for $\omega_f > \omega_n$, $x_m$ is negative and the vibration is *out of phase*.

  **a. In the problems that follow,** you may be asked to determine one of the parameters in Eq. (19.33) when the others are known. We suggest that you keep Fig. 19.9 in front of you when solving these problems. For example, if you are asked to find the frequency at which the amplitude of a forced vibration has a given value, but you do not know whether the vibration is in or out of phase with respect to the impressed force, you should note from Fig. 19.9 that there can be two frequencies satisfying this requirement. One frequency cor-responds to a positive value of $x_m$ and to a vibration in phase with the impressed force, and the other corresponds to a negative value of $x_m$ and to a vibration out of phase with the impressed force.

**b. Once you have obtained the amplitude** $x_m$ of the motion of a component of the system from Eq. (19.33), you can use Eqs. (19.37) to determine the maximum values of the velocity and acceleration of that component:

$$v_m = x_m \omega_f \qquad a_m = x_m \omega_f^2 \tag{19.37}$$

**c. When the impressed force P is due to the unbalance of the rotor of a motor,** its maximum value is $P_m = mr\omega_f^2$ where $m$ is the mass of the rotor, $r$ is the distance between its mass center and the axis of rotation, and $\omega_f$ is equal to the angular velocity $\omega$ of the rotor expressed in rad/s (Sample Prob. 19.5).

**2. If the forced vibration is caused by the simple harmonic motion of a support** with an amplitude $\delta_m$ and a circular frequency $\omega_f$, the amplitude of the vibration is

$$x_m = \frac{\delta_m}{1 - (\omega_f/\omega_n)^2} \tag{19.33'}$$

where $\omega_n$ is the *natural circular frequency* of the system and $\omega_n = \sqrt{k/m}$. Again, note that the circular frequency of the vibration is $\omega_f$ and that the amplitude $x_m$ does not depend upon the initial conditions.

**a. Be sure to read our comments** in paragraphs 1, 1a, and 1b, because they apply equally well to a vibration caused by the motion of a support.

**b. If the maximum acceleration** $a_m$ **of the support is specified,** rather than its maximum displacement $\delta_m$, remember that, because the motion of the support is a simple harmonic motion, you can use the relation $a_m = \delta_m \omega_f^2$ to determine $\delta_m$; then substitute this value into Eq. (19.33').

# Problems

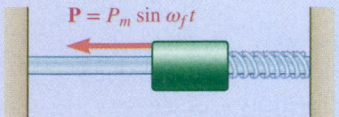

**Fig. P19.99, P19.100, and P19.101**

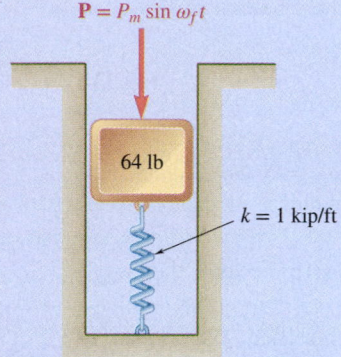

**Fig. P19.102**

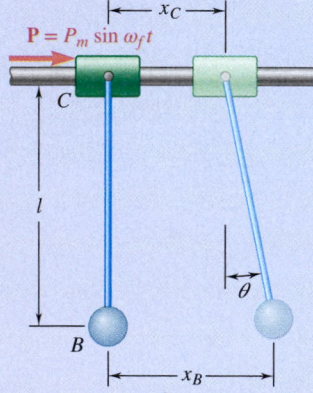

**Fig. P19.103**

**19.99** A 4-kg collar can slide on a frictionless horizontal rod and is attached to a spring with a constant of 450 N/m. It is acted upon by a periodic force with a magnitude of $P = P_m \sin \omega_f t$, where $P_m = 13$ N. Determine the amplitude of the motion of the collar if (a) $\omega_f = 5$ rad/s, (b) $\omega_f = 10$ rad/s.

**19.100** A 4-kg collar can slide on a frictionless horizontal rod and is attached to a spring with constant $k$. It is acted upon by a periodic force of magnitude $P = P_m \sin \omega_f t$, where $P_m = 9$ N and $\omega_f = 5$ rad/s. Determine the value of the spring constant $k$ knowing that the motion of the collar has an amplitude of 150 mm and is (a) in phase with the applied force, (b) out of phase with the applied force.

**19.101** A collar with mass $m$ that slides on a frictionless horizontal rod is attached to a spring with constant $k$ and is acted upon by a periodic force with a magnitude of $P = P_m \sin \omega_f t$. Determine the range of values of $\omega_f$ for which the amplitude of the vibration exceeds three times the static deflection caused by a constant force with a magnitude of $P_m$.

**19.102** A 64-lb block is attached to a spring with a constant of $k = 1$ kip/ft and can move without friction in a vertical slot as shown. It is acted upon by a periodic force with a magnitude of $P = P_m \sin \omega_f t$, where $\omega_f = 10$ rad/s. Knowing that the amplitude of the motion is 0.75 in., determine $P_m$.

**19.103** The 1.2-kg bob of a simple pendulum of length $l = 600$ mm is suspended from a 1.4-kg collar $C$. Knowing that the collar is acted upon by a periodic force $\mathbf{P} = P_m \sin \omega_f t$, where $P_m = 0.5$ N and $\omega_f = 3$ rad/s, determine the amplitude and phase of the motion of the bob. Assume small angles.

**19.104** An 8-kg uniform disk of radius 200 mm is welded to a vertical shaft with a fixed end at $B$. The disk rotates through an angle of 3° when a static couple of magnitude 50 N·m is applied to it. If the disk is acted upon by a periodic torsional couple of magnitude $T = T_m \sin \omega_f t$, where $T_m = 60$ N·m, determine the range of values of $\omega_f$ for which the amplitude of the vibration is less than the angle of rotation caused by a static couple of magnitude $T_m$.

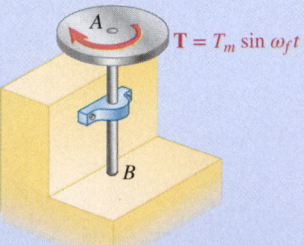

**Fig. P19.104**

**19.105** A precision experiment sits on an optical table that is isolated from ground motion resulting from nearby equipment vibrations. The system can be modeled with a single mass and spring as shown. The weight of the table and experiment, modeled by block $A$, is 1200 lb, and the displacement of the ground $B$ is 0.03 in. at a frequency of 100 Hz. Knowing that the resulting amplitude of the table is to be restricted to less than $x_{max} = 4 \times 10^{-5}$ in., determine ($a$) the required spring stiffness, ($b$) the range of forcing frequencies, using this stiffness, for which the table will have an amplitude less than $x_{max}$.

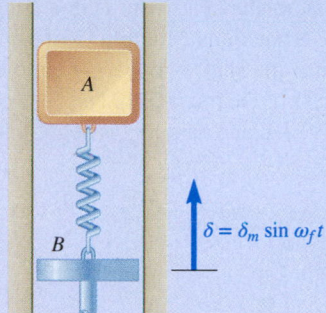

**Fig. P19.105**

**19.106** A beam $ABC$ is supported by a pin connection at $A$ and by rollers at $B$. A 120-kg block placed on the end of the beam causes a static deflection of 15 mm at $C$. Assuming that the support at $A$ undergoes a vertical periodic displacement $\delta = \delta_m \sin \omega_f t$, where $\delta_m = 10$ mm and $\omega_f = 18$ rad/s, and the support at $B$ does not move, determine the maximum acceleration of the block at $C$. Neglect the weight of the beam and assume that the block does not leave the beam.

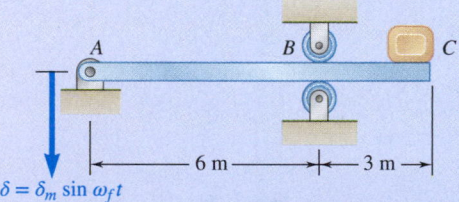

**Fig. P19.106**

**19.107** A small 2-kg sphere $B$ is attached to the bar $AB$ of negligible mass that is supported at $A$ by a pin and bracket and connected at $C$ to a moving support $D$ by means of a spring of constant $k = 3.6$ kN/m. Knowing that support $D$ undergoes a vertical displacement $\delta = \delta_m \sin \omega_f t$, where $\delta_m = 3$ mm and $\omega_f = 15$ rad/s, determine ($a$) the magnitude of the maximum angular velocity of bar $AB$, ($b$) the magnitude of the maximum acceleration of sphere $B$.

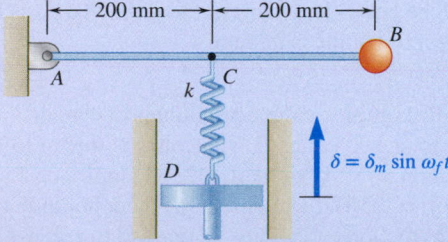

**Fig. P19.107**

**19.108** The crude-oil pumping rig shown is driven at 20 rpm. The inside diameter of the well pipe is 2 in., and the diameter of the pump rod is 0.75 in. The length of the pump rod and the length of the column of oil lifted during the stroke are essentially the same, and equal to 6000 ft. During the downward stroke, a valve at the lower end of the pump rod opens to let a quantity of oil into the well pipe, and the column of oil is then lifted to obtain a discharge into the connecting pipeline. Thus, the amount of oil pumped in a given time depends upon the stroke of the lower end of the pump rod. Knowing that the upper end of the rod at $D$ is essentially sinusoidal with a stroke of 45 in. and the specific weight of crude oil is 56.2 lb/ft$^3$, determine (a) the output of the well in ft$^3$/min if the shaft is rigid, (b) the output of the well in ft$^3$/min if the stiffness of the rod is 2210 N/m, the equivalent mass of the oil and shaft is 290 kg, and damping is negligible.

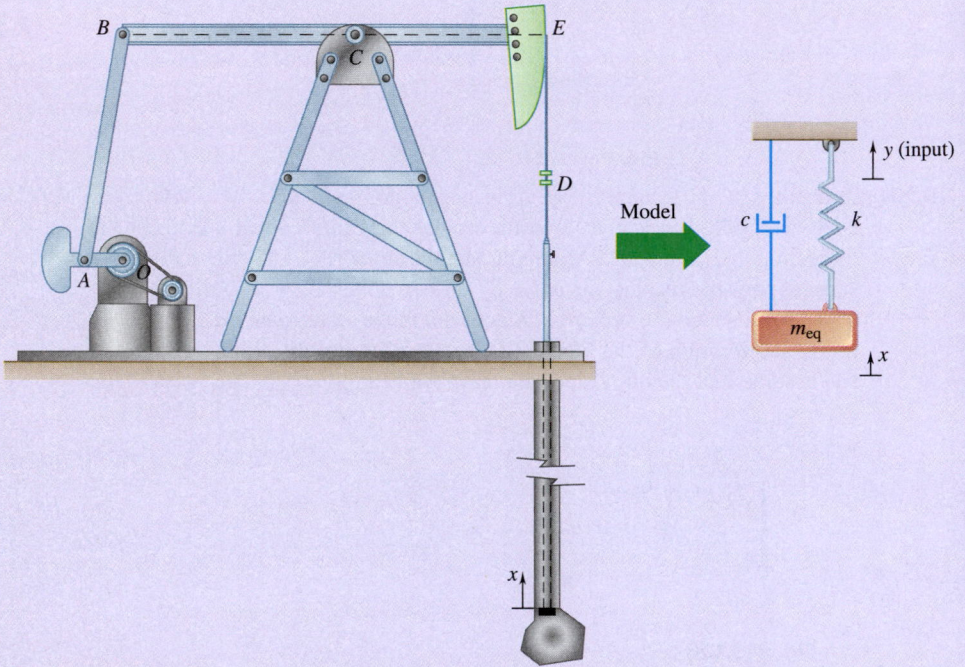

**Fig. P19.108**

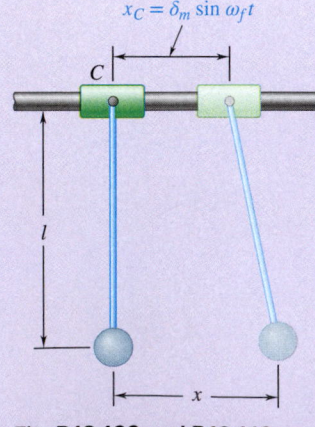

**Fig. P19.109 and P19.110**

**19.109** A simple pendulum of length $l$ is suspended from collar $C$ that is forced to move horizontally according to the relation $x_C = \delta_m \sin \omega_f t$. Determine the range of values of $\omega_f$ for which the amplitude of the motion of the bob is less than $\delta_m$. (Assume that $\delta_m$ is small compared with the length $l$ of the pendulum.)

**19.110** The 2.75-lb bob of a simple pendulum of length $l = 24$ in. is suspended from a 3-lb collar $C$. The collar is forced to move according to the relation $x_C = \delta_m \sin \omega_f t$, with an amplitude $\delta_m = 0.4$ in. and a frequency $f_f = 0.5$ Hz. Determine (a) the amplitude of the motion of the bob, (b) the force that must be applied to collar $C$ to maintain the motion.

**19.111** An 18-lb block $A$ slides in a vertical frictionless slot and is connected to a moving support $B$ by means of a spring $AB$ of constant $k = 8$ lb/ft. Knowing that the acceleration of the support is $a = a_m \sin \omega_f t$, where $a_m = 5$ ft/s$^2$ and $\omega_f = 6$ rad/s, determine (a) the maximum displacement of block $A$, (b) the amplitude of the fluctuating force exerted by the spring on the block.

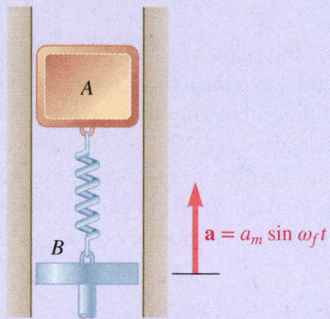

$\mathbf{a} = a_m \sin \omega_f t$

**Fig. P19.111**

**19.112** Rod $AB$ is rigidly attached to the frame of a motor running at a constant speed. When a collar of mass $m$ is placed on the spring, it is observed to vibrate with an amplitude of 15 mm. When two collars, each of mass $m$, are placed on the spring, the amplitude is observed to be 18 mm. What amplitude of vibration should be expected when three collars, each of mass $m$, are placed on the spring? (Obtain two answers.)

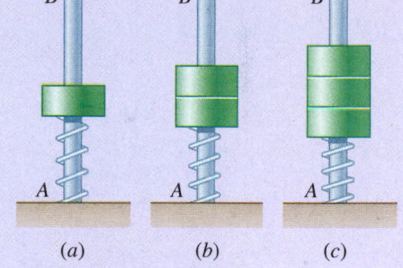

(a)  (b)  (c)

**Fig. P19.112**

**19.113** A motor of mass $M$ is supported by springs with an equivalent spring constant $k$. The unbalance of its rotor is equivalent to a mass $m$ located at a distance $r$ from the axis of rotation. Show that when the angular velocity of the motor is $\omega_f$, the amplitude $x_m$ of the motion of the motor is

$$ x_m = \frac{r(m/M)(\omega_f/\omega_n)^2}{1 - (\omega_f/\omega_n)^2} $$

where $\omega_n = \sqrt{k/M}$.

**19.114** As the rotational speed of a spring-supported 100-kg motor is increased, the amplitude of the vibration due to the unbalance of its 15-kg rotor first increases and then decreases. It is observed that as very high speeds are reached, the amplitude of the vibration approaches 3.3 mm. Determine the distance between the mass center of the rotor and its axis of rotation. (*Hint:* Use the formula derived in Prob. 19.113.)

**19.115** A motor of weight 100 lb is supported by four springs, each of constant 250 lb/in. The motor is constrained to move vertically, and the amplitude of its motion is observed to be 0.04 in. at a speed of 1200 rpm. Knowing that the weight of the rotor is 5 lb, determine the distance between the mass center of the rotor and the axis of the shaft.

**19.116** A motor weighing 400 lb is supported by springs having a total constant of 1200 lb/in. The unbalance of the rotor is equivalent to a 1-oz weight located 8 in. from the axis of rotation. Determine the range of allowable values of the motor speed if the amplitude of the vibration is not to exceed 0.06 in.

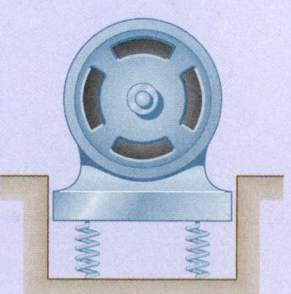

**Fig. P19.115 and P19.116**

**Fig. P19.117**

**19.117** A 180-kg motor is bolted to a light horizontal beam. The unbalance of its rotor is equivalent to a 28-g mass located 150 mm from the axis of rotation, and the static deflection of the beam due to the weight of the motor is 12 mm. The amplitude of the vibration due to the unbalance can be decreased by adding a plate to the base of the motor. If the amplitude of vibration is to be less than 60 $\mu$m for motor speeds above 300 rpm, determine the required mass of the plate.

**19.118** The unbalance of the rotor of a 400-lb motor is equivalent to a 3-oz weight located 6 in. from the axis of rotation. In order to limit to 0.2 lb the amplitude of the fluctuating force exerted on the foundation when the motor is run at speeds of 100 rpm and above, a pad is to be placed between the motor and the foundation. Determine (*a*) the maximum allowable spring constant $k$ of the pad, (*b*) the corresponding amplitude of the fluctuating force exerted on the foundation when the motor is run at 200 rpm.

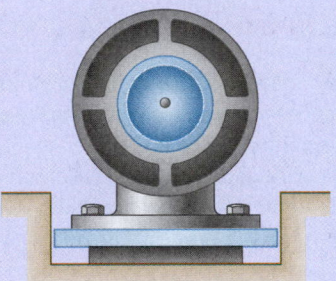

**Fig. *P19.118***

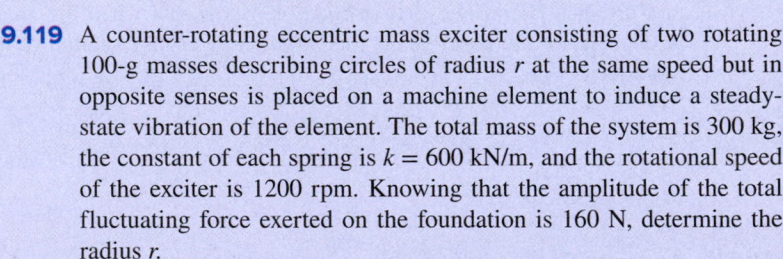

**Fig. P19.119**

**19.119** A counter-rotating eccentric mass exciter consisting of two rotating 100-g masses describing circles of radius *r* at the same speed but in opposite senses is placed on a machine element to induce a steady-state vibration of the element. The total mass of the system is 300 kg, the constant of each spring is $k = 600$ kN/m, and the rotational speed of the exciter is 1200 rpm. Knowing that the amplitude of the total fluctuating force exerted on the foundation is 160 N, determine the radius *r*.

**19.120** One of the tail rotor blades of a helicopter has an unbalanced weight of 1 lb at a distance of $e = 6$ in. from the axis of rotation. The fuselage can be considered to be fixed, and the tail rotor and tail can be modeled as an equivalent stiffness of 8800 lb/ft and an equivalent weight of 165 lb. Knowing that the blades rotate at 500 rpm, determine the forced response of the tail section.

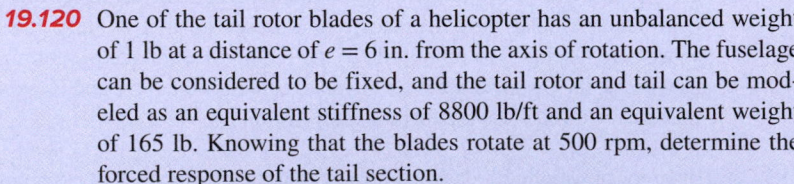

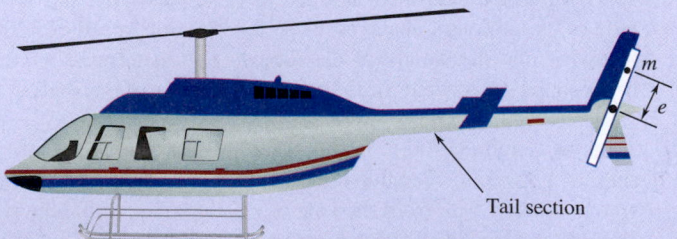

Tail section

**Fig. *P19.120***

**19.121** Figs. (1) and (2) show how springs can be used to support a block in two different situations. In Fig. (1), they help decrease the amplitude of the fluctuating force transmitted by the block to the foundation. In Fig. (2), they help decrease the amplitude of the fluctuating displacement transmitted by the foundation to the block. The ratio of the transmitted force to the impressed force or the ratio of the transmitted displacement to the impressed displacement is called the *transmissibility*. Derive an equation for the transmissibility for each situation. Give your answer in terms of the ratio $\omega_f/\omega_n$ of the frequency $\omega_f$ of the impressed force or impressed displacement to the natural frequency $\omega_n$ of the spring–mass system. Show that in order to cause any reduction in transmissibility, the ratio $\omega_f/\omega_n$ must be greater than $\sqrt{2}$.

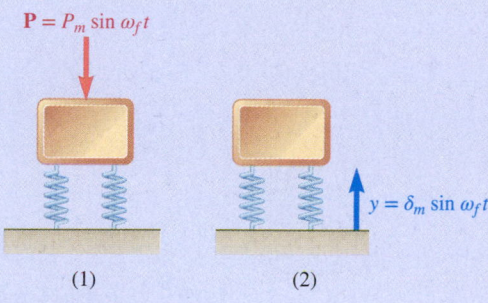

$\mathbf{P} = P_m \sin \omega_f t$

$y = \delta_m \sin \omega_f t$

(1)     (2)

**Fig. P19.121**

**19.122** A vibrometer used to measure the amplitude of vibrations consists essentially of a box containing a mass-spring system with a known natural frequency of 120 Hz. The box is rigidly attached to a surface that is moving according to the equation $y = \delta_m \sin \omega_f t$. If the amplitude $z_m$ of the motion of the mass relative to the box is used as a measure of the amplitude $\delta_m$ of the vibration of the surface, determine (a) the percent error when the frequency of the vibration is 600 Hz, (b) the frequency at which the error is zero.

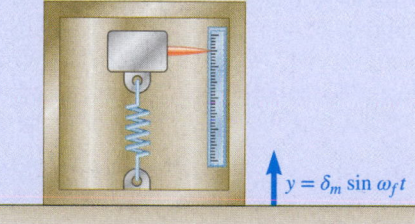

$y = \delta_m \sin \omega_f t$

**Fig. P19.122 and P19.123**

**19.123** A certain accelerometer consists essentially of a box containing a mass–spring system with a known natural frequency of 2200 Hz. The box is rigidly attached to a surface that is moving according to the equation $y = \delta_m \sin \omega_f t$. If the amplitude $z_m$ of the motion of the mass relative to the box times a scale factor $\omega_n^2$ is used as a measure of the maximum acceleration $a_m = \delta_m \omega_f^2$ of the vibrating surface, determine the percent error when the frequency of the vibration is 600 Hz.

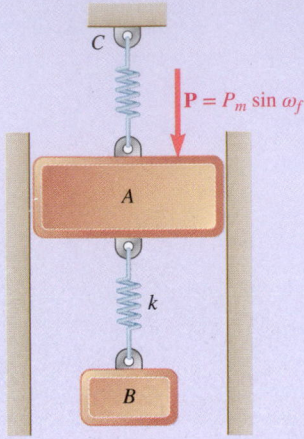

**Fig. P19.124**

**19.124** Block A can move without friction in the slot as shown and is acted upon by a vertical periodic force of magnitude $P = P_m \sin \omega_f t$, where $\omega_f = 2$ rad/s and $P_m = 20$ N. A spring of constant $k$ is attached to the bottom of block A and to a 22-kg block B. Determine (a) the value of the constant $k$ that will prevent a steady-state vibration of block A, (b) the corresponding amplitude of the vibration of block B.

**19.125** A 60-lb disk is attached with an eccentricity $e = 0.006$ in. to the midpoint of a vertical shaft AB that revolves at a constant angular velocity $\omega_f$. Knowing that the spring constant $k$ for horizontal movement of the disk is 40,000 lb/ft, determine (a) the angular velocity $\omega_f$ at which resonance will occur, (b) the deflection $r$ of the shaft when $\omega_f = 1200$ rpm.

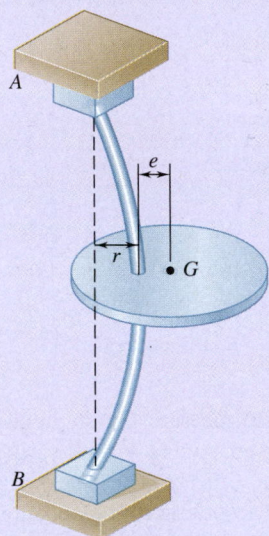

**Fig. P19.125**

**19.126** A small trailer and its load have a total mass of 250 kg. The trailer is supported by two springs, each of constant 10 kN/m, and is pulled over a road, the surface of which can be approximated by a sine curve with an amplitude of 40 mm and a wavelength of 5 m (i.e., the distance between successive crests is 5 m and the vertical distance from crest to trough is 80 mm). Determine (a) the speed at which resonance will occur, (b) the amplitude of the vibration of the trailer at a speed of 50 km/h.

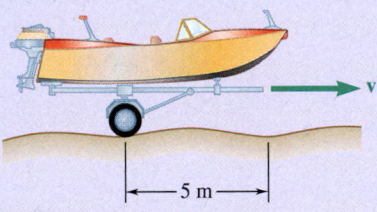

**Fig. P19.126**

# 19.5   DAMPED VIBRATIONS

The vibrating systems considered in the first part of this chapter were assumed free of damping. Actually, all vibrations are damped to some degree by friction forces. These forces can be caused by *dry friction*, or *Coulomb friction*, between rigid bodies; by *fluid friction* when a rigid body moves in a fluid; or by *internal friction* between the molecules of a seemingly elastic body. A type of damping of special interest is the *viscous damping* caused by fluid friction at low and moderate speeds. We will first consider free vibrations with viscous damping and then examine the effect of viscous damping on forced vibrations.

## *19.5A   Damped Free Vibrations

Viscous damping is characterized by the fact that the friction force is *directly proportional and opposite in direction to the velocity* of the moving body. As an example, let us again consider a body with mass $m$ suspended from a spring of constant $k$, assuming that the body is attached to the plunger of a dashpot (Fig. 19.10). The magnitude of the friction force exerted on the plunger by the surrounding fluid is equal to $c\dot{x}$, where the constant $c$, expressed in N·s/m or lb·s/ft and known as the *coefficient of viscous damping*, depends upon the physical properties of the fluid and the construction of the dashpot. Examining the free-body and kinetic diagrams, the equation of motion is

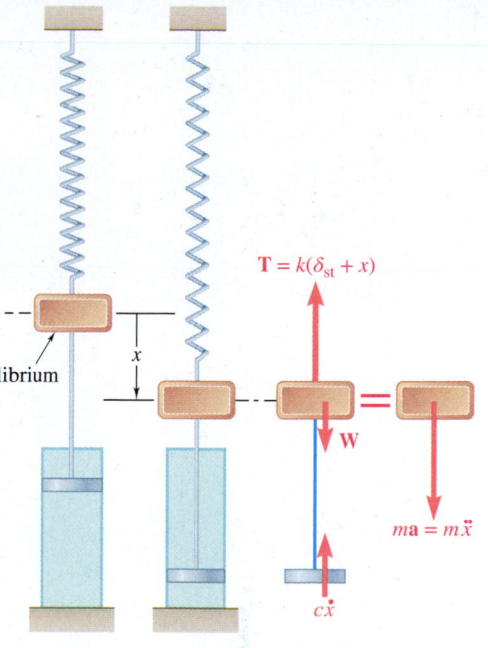

**Fig. 19.10**   Free-body diagram and kinetic diagram of a spring–mass–damper system.

$$+\downarrow \Sigma F = ma: \qquad W - k(\delta_{st} + x) - c\dot{x} = m\ddot{x}$$

Recalling that $W = k\delta_{st}$, we have

$$m\ddot{x} + c\dot{x} + kx = 0 \qquad (19.38)$$

If we substitute $x = e^{\lambda t}$ into Eq. (19.38) and divide through by $e^{\lambda t}$, we obtain

**Characteristic equation**

$$m\lambda^2 + c\lambda + k = 0 \qquad (19.39)$$

and obtain the roots

$$\lambda = -\frac{c}{2m} \pm \sqrt{\left(\frac{c}{2m}\right)^2 - \frac{k}{m}} \qquad (19.40)$$

Defining the *critical damping coefficient* $c_c$ as the value of $c$ that makes the radical in Eq. (19.40) equal to zero, we have

$$\left(\frac{c_c}{2m}\right)^2 - \frac{k}{m} = 0 \qquad c_c = 2m\sqrt{\frac{k}{m}} = 2m\omega_n \qquad (19.41)$$

where $\omega_n$ is the natural circular frequency of the system in the absence of damping. We can distinguish three different cases of damping, depending upon the value of the coefficient $c$.

1. **Overdamped:** $c > c_c$. The roots $\lambda_1$ and $\lambda_2$ of the characteristic equation (19.39) are real and distinct, and the general solution of the differential equation (19.38) is

$$x = C_1 e^{\lambda_1 t} + C_2 e^{\lambda_2 t} \qquad (19.42)$$

This solution corresponds to a nonvibratory motion. Because $\lambda_1$ and $\lambda_2$ are both negative, $x$ approaches zero as $t$ increases indefinitely. However, the system actually regains its equilibrium position after a finite time.

2. **Critically damped:** $c = c_c$. The characteristic equation has a double root $\lambda = -c_c/2m = -\omega_n$, and the general solution of Eq. (19.38) is

$$x = (C_1 + C_2 t)e^{-\omega_n t} \qquad (19.43)$$

This motion is again nonvibratory. Critically damped systems are of special interest in engineering applications because they regain their equilibrium position in the shortest possible time without oscillation.

3. **Underdamped:** $c < c_c$. The roots of Eq. (19.39) are complex and conjugate, and the general solution of Eq. (19.38) is of the form

$$x = e^{-(c/2m)t} (C_1 \sin \omega_d t + C_2 \cos \omega_d t) \qquad (19.44)$$

where $\omega_d$ is defined by the relation

$$\omega_d^2 = \frac{k}{m} - \left( \frac{c}{2m} \right)^2$$

Substituting $k/m = \omega_n^2$ and recalling Eq. (19.41), we have

$$\omega_d = \omega_n \sqrt{1 - \left( \frac{c}{c_c} \right)^2} \qquad (19.45)$$

where the constant $c/c_c$ is known as the **damping factor** or the **damping ratio**. This quantity is often denoted by $\zeta$. Even though the motion does not actually repeat itself, the constant $\omega_d$ is commonly referred to as the *damped circular frequency.* In terms of the damping ratio, the damped circular frequency is

$$\omega_d = \omega_n \sqrt{1 - \zeta^2} \qquad (19.45')$$

A substitution similar to the one used in Sec. 19.1A enables us to write the general solution of Eq. (19.38) in the form

$$x = x_0 e^{-(c/2m)t} \sin (\omega_d t + \phi) \qquad (19.46)$$

or

$$x = x_0 e^{-\zeta \omega_n t} \sin (\omega_d t + \phi) \qquad (19.46')$$

The motion defined by Eq. (19.46) is vibratory with diminishing amplitude (Fig. 19.11). The time interval $\tau_d = 2\pi/\omega_d$ separating two successive points where the curve defined by Eq. (19.46) touches one of the limiting curves shown in Fig. 19.11 is commonly referred to as the *period of the damped vibration.* Recalling Eq. (19.45), we observe that $\omega_d < \omega_n$ and, thus, that $\tau_d$ is larger than the period of vibration $\tau_n$ of the corresponding undamped system.

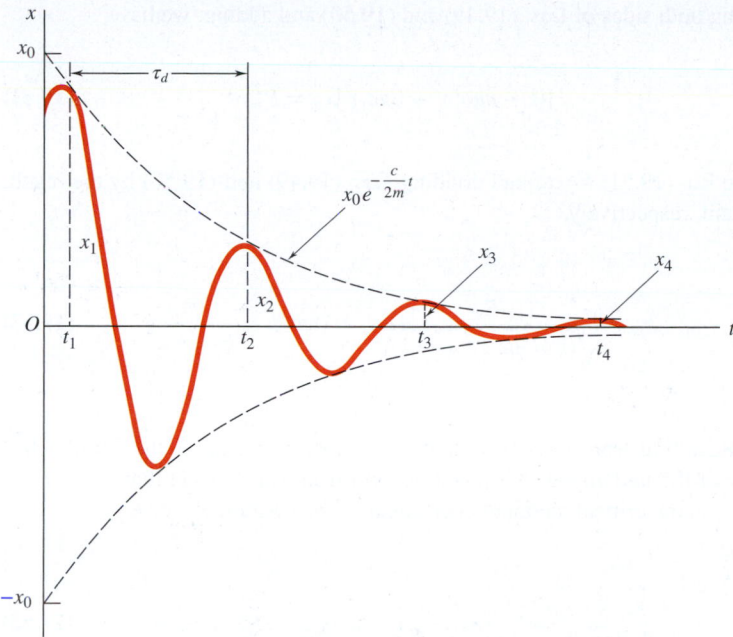

**Fig. 19.11** The free response of a viscously damped system decays exponentially and oscillates with a frequency $\omega_d$.

## *19.5B   Damped Forced Vibrations

If the system considered in the preceding section is subjected to a periodic force **P** of magnitude $P = P_m \sin \omega_f t$, the equation of motion becomes

$$m\ddot{x} + c\dot{x} + kx = P_m \sin \omega_f t \qquad \textbf{(19.47)}$$

We can obtain the general solution of Eq. (19.47) by adding a particular solution of Eq. (19.47) to the complementary function or general solution of the homogeneous equation (19.38). The complementary function is given by Eq. (19.42), (19.43), or (19.44), depending upon the type of damping considered. It represents a *transient* motion that is eventually damped out.

Our interest in this section is centered on the steady-state vibration represented by a particular solution of Eq. (19.47) of the form

$$x_{\text{part}} = x_m \sin (\omega_f t - \phi) \qquad \textbf{(19.48)}$$

Substituting $x_{\text{part}}$ for $x$ into Eq. (19.47), we obtain

$$-m\omega_f^2 x_m \sin (\omega_f t - \phi) + c\omega_f x_m \cos (\omega_f t - \phi) + kx_m \sin (\omega_f t - \phi)$$
$$= P_m \sin \omega_f t$$

Making $\omega_f t - \phi$ successively equal to 0 and to $\pi/2$ gives

$$c\omega_f x_m = P_m \sin \phi \qquad \textbf{(19.49)}$$
$$(k - m\omega_f^2) x_m = P_m \cos \phi \qquad \textbf{(19.50)}$$

**Photo 19.2** The automobile suspension shown consists essentially of a spring and a shock absorber, which will cause the body of the car to undergo *damped forced vibrations* when the car is driven over an uneven road.
©McGraw-Hill Education/Sabina Dowell

**Photo 19.3** This truck is experiencing damped forced vibration in the vehicle dynamics test. Image courtesy of MTS Systems Corporation

Squaring both sides of Eqs. (19.49) and (19.50) and adding, we have

$$[(k - m\omega_f^2)^2 + (c\omega_f)^2]x_m^2 = P_m^2 \qquad (19.51)$$

Solving Eq. (19.51) for $x_m$ and dividing Eqs. (19.49) and (19.50) by the result, we obtain, respectively,

$$x_m = \frac{P_m}{\sqrt{(k - m\omega_f^2)^2 + (c\omega_f)^2}} \qquad \tan\phi = \frac{c\omega_f}{k - m\omega_f^2} \qquad (19.52)$$

Recalling from Eq. (19.4) that $k/m = \omega_n^2$, where $\omega_n$ is the circular frequency of the undamped free vibration, and from Eq. (19.41) that $2m\omega_n = c_c$, where $c_c$ is the critical damping coefficient of the system, we have

$$\frac{x_m}{P_m/k} = \frac{x_m}{\delta_m} = \frac{1}{\sqrt{[1 - (\omega_f/\omega_n)^2]^2 + [2(c/c_c)(\omega_f/\omega_n)]^2}} \qquad (19.53)$$

$$\tan\phi = \frac{2(c/c_c)(\omega_f/\omega_n)}{1 - (\omega_f/\omega_n)^2} \qquad (19.54)$$

Defining the frequency ratio $r = \omega_f/\omega_n$, we can write the steady-state response of a viscously damped system in terms of the frequency ratio and the damping ratio as

$$\frac{x_m}{P_m/k} = \frac{x_m}{\delta_{st}} = \frac{1}{\sqrt{(1 - r^2)^2 + (2\zeta r)^2}} \qquad (19.53')$$

$$\tan\phi = \frac{2\zeta r}{1 - r^2} \qquad (19.54')$$

We can use these equations to determine the amplitude of the steady-state vibration produced by an impressed force of magnitude $P = P_m \sin\omega_f t$ or by an impressed support movement $\delta = \delta_m \sin\omega_f t$. Using these same parameters, Eq. (19.54) defines the *phase difference* $\phi$ between the impressed force or impressed support movement and the resulting steady-state vibration of the damped system. The magnification factor has been plotted against the frequency ratio in Fig. 19.12 for various values of the damping ratio. ⏵ Note that we can keep the amplitude of a forced vibration small by choosing a large coefficient of viscous damping $c$ or by keeping the natural and forced frequencies far apart.

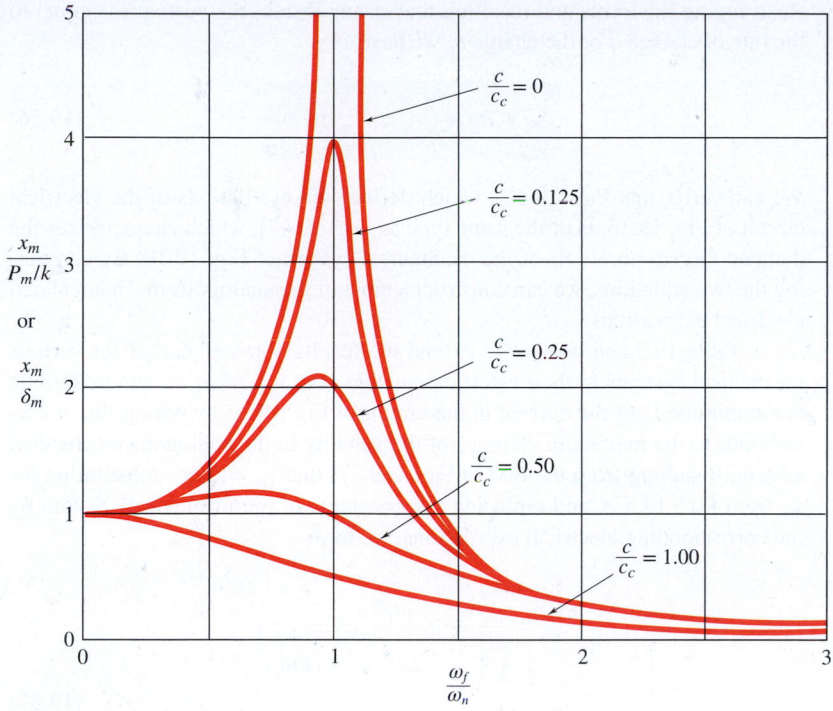

**Fig. 19.12** Graph of magnification factor as a function of frequency ratio for several values of the damping ratio. ▶

## *19.5C   Electrical Analogs

Oscillating electrical circuits are characterized by differential equations of the same type as those just discussed. Their analysis is therefore similar to that of a mechanical system, and the results obtained for a given vibrating system can be readily extended to the equivalent circuit. Conversely, any result obtained for an electrical circuit also applies to the corresponding mechanical system.

Consider an electrical circuit consisting of an inductor of inductance $L$, a resistor of resistance $R$, and a capacitor of capacitance $C$, connected in series with a source of alternating voltage $E = E_m \sin \omega_f t$ (Fig. 19.13). Elementary circuit theory[†] says that if $i$ denotes the current in the circuit and $q$ denotes the electric charge on the capacitor, the drop in potential is $L\,(di/dt)$ across the inductor, $Ri$ across the resistor, and $q/C$ across the capacitor. The algebraic sum of the applied voltage and of the drops in potential around the circuit loop must be zero, so we have

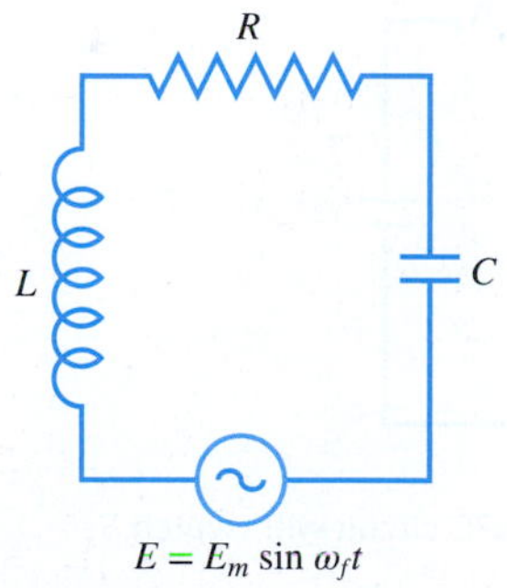

$$E_m \sin \omega_f t - L\frac{di}{dt} - Ri - \frac{q}{C} = 0 \qquad \textbf{(19.55)}$$

**Fig. 19.13**   An electrical circuit with inductance $L$, resistance $R$, capacitance $C$, and a source of alternating voltage $E$.

(circuit diagram label: $E = E_m \sin \omega_f t$)

[†]See C. R. Paul, S. A. Nasar, and L. E. Unnewehr, *Introduction to Electrical Engineering,* 2nd ed., McGraw-Hill, New York, 1992.

Rearranging the terms and recalling that at any instant the current $i$ is equal to the rate of change $\dot{q}$ of the charge $q$, we have

$$L\ddot{q} + R\dot{q} + \frac{1}{C}q = E_m \sin \omega_f t \qquad (19.56)$$

We can verify that Eq. (19.56), which defines the oscillations of the electrical circuit of Fig. 19.13, is of the same type as Eq. (19.47), which characterizes the damped forced vibrations of the mechanical system of Fig. 19.10. By comparing the two equations, we can construct a table of the analogous mechanical and electrical expressions.

Table 19.2 can be used to extend the results obtained earlier for various mechanical systems to their electrical analogs. For instance, we can determine the amplitude $i_m$ of the current in the circuit of Fig. 19.13 by noting that it corresponds to the maximum value $v_m$ of the velocity in the analogous mechanical system. Recalling from the first of Eqs. (19.37) that $v_m = x_m \omega_f$, substituting for $x_m$ from Eq. (19.52), and replacing the constants of the mechanical system by the corresponding electrical expressions, we have

$$i_m = \frac{\omega_f E_m}{\sqrt{\left(\dfrac{1}{C} - L\omega_f^2\right)^2 + (R\omega_f)^2}}$$

$$i_m = \frac{E_m}{\sqrt{R^2 + \left(L\omega_f - \dfrac{1}{C\omega_f}\right)^2}} \qquad (19.57)$$

The radical term in this expression is known as the *impedance* of the electrical circuit.

The analogy between mechanical systems and electrical circuits holds for transient as well as steady-state oscillations. The oscillations of the circuit shown in Fig. 19.14, for instance, are analogous to the damped free vibrations of the system of Fig. 19.10. As far as the initial conditions are concerned, we should note that closing the switch $S$ when the charge on the capacitor is $q = q_0$ is equivalent to releasing the mass of the mechanical system with no initial velocity from the position $x = x_0$. Also note that, if a battery of constant voltage $E$ is introduced in the electrical circuit of Fig. 19.14, closing the switch $S$ is equivalent to suddenly applying a force of constant magnitude $P$ to the mass of the mechanical system of Fig. 19.10.

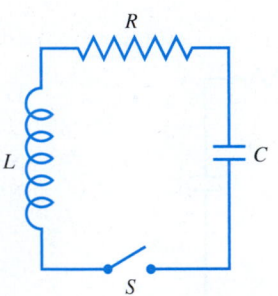

**Fig. 19.14** An *LRC* circuit with switch *S*.

**Table 19.2    Characteristics of a Mechanical System and of Its Electrical Analog**

| Mechanical System | | Electrical Circuit | |
|---|---|---|---|
| $m$ | Mass | $L$ | Inductance |
| $c$ | Coefficient of viscous damping | $R$ | Resistance |
| $k$ | Spring constant | $1/C$ | Reciprocal of capacitance |
| $x$ | Displacement | $q$ | Charge |
| $v$ | Velocity | $I$ | Current |
| $P$ | Applied force | $E$ | Applied voltage |

This discussion would be of questionable value if its only result were to make it possible for mechanics students to analyze electrical circuits without learning the elements of circuit theory. We hope that this discussion will instead encourage students to apply to the solution of problems in mechanical vibrations the mathematical techniques they may learn in later courses in circuit theory. The chief value of the concept of electrical analogs, however, resides in its application to *experimental methods* for determining the characteristics of a given mechanical system. Indeed, an electrical circuit is much more easily constructed than is a mechanical model, and the fact that we can modify its characteristics by varying the inductance, resistance, or capacitance of its various components makes the use of the electrical analog particularly convenient.

To determine the electrical analog of a given mechanical system, we focus our attention on each moving mass in the system and observe which springs, dashpots, or external forces are applied directly to it. We can then construct an equivalent electrical loop to match each of these mechanical units; the various loops obtained in this way will together form the desired circuit. Consider, for instance, the mechanical system of Fig. 19.15. The mass $m_1$ is acted upon by two springs with constants $k_1$ and $k_2$ and by two dashpots characterized by the coefficients of viscous damping $c_1$ and $c_2$. The electrical circuit should therefore include a loop consisting of an inductor of inductance $L_1$ proportional to $m_1$; of two capacitors of capacitance $C_1$ and $C_2$ inversely proportional to $k_1$ and $k_2$, respectively; and of two resistors of resistance $R_1$ and $R_2$, proportional to $c_1$ and $c_2$, respectively. Because the mass $m_2$ is acted upon by the spring $k_2$ and the dashpot $c_2$, as well as by the force $P = P_m \sin \omega_f t$, the circuit should also include a loop containing the capacitor $C_2$, the resistor $R_2$, the new inductor $L_2$, and the voltage source $E = E_m \sin \omega_f t$ (Fig. 19.16).

To check that the mechanical system of Fig. 19.15 and the electrical circuit of Fig. 19.16 actually satisfy the same differential equations, we first derive the equations of motion for $m_1$ and $m_2$. Denoting the displacements of $m_1$ and $m_2$ from their equilibrium positions by $x_1$ and $x_2$, respectively, we observe that the elongation of the spring $k_1$ (measured from the equilibrium position) is equal to $x_1$, while the elongation of the spring $k_2$ is equal to the relative displacement $x_2 - x_1$ of $m_2$ with respect to $m_1$. The equations of motion for $m_1$ and $m_2$ are therefore

$$m_1\ddot{x}_1 + c_1\dot{x}_1 + c_2(\dot{x}_1 - \dot{x}_2) + k_1x_1 + k_2(x_1 - x_2) = 0 \quad \textbf{(19.58)}$$

$$m_2\ddot{x}_2 + c_2(\dot{x}_2 - \dot{x}_1) + k_2(x_2 - x_1) = P_m \sin \omega_f t \quad \textbf{(19.59)}$$

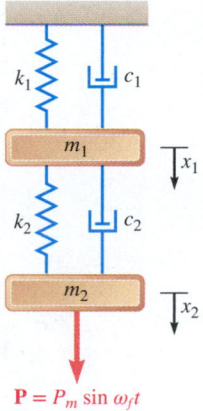

**Fig. 19.15**   Model of a two-degree-of-freedom harmonically excited system.

Now consider the electrical circuit of Fig. 19.16; we denote the current in the first and second loops by $i_1$ and $i_2$, respectively, and by $q_1$ and $q_2$ the integrals $\int i_1 \, dt$ and $\int i_2 \, dt$. Noting that the charge on the capacitor $C_1$ is $q_1$ and the charge on $C_2$ is $q_1 - q_2$, we can state that the sum of the potential differences in each loop is zero and obtain

$$L_1\ddot{q}_1 + R_1\dot{q}_1 + R_2(\dot{q}_1 - \dot{q}_2) + \frac{q_1}{C_1} + \frac{q_1 - q_2}{C_2} = 0 \quad \textbf{(19.60)}$$

$$L_2\ddot{q}_2 + R_2(\dot{q}_2 - \dot{q}_1) + \frac{q_2 - q_1}{C_2} = E_m \sin \omega_f t \quad \textbf{(19.61)}$$

We easily check that Eqs. (19.60) and (19.61) reduce to Eqs. (19.58) and (19.59), respectively, after performing the substitutions indicated in Table 19.2.

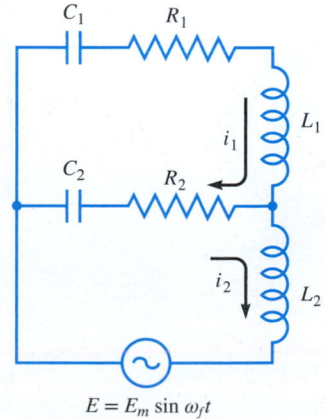

**Fig. 19.16**   An electrical circuit analogous to the mechanical system in Fig. 19.15.

# Case Study 19.1
# Vibration Isolation

Industrial equipment can have large vibration levels, which can cause significant vibration disturbance to the ground, thereby affecting nearby equipment and people. These vibrations are often caused by rotating machines where the center of mass is not coincident with the center of rotation. CS Fig. 19.1a shows that when the machine is bolted directly to the ground, the force transmitted is equal to the force generated by the machine (often equal to $me\omega^2$ where $m$ is the unbalanced mass and $e$ is the distance between the location of $m$ and the axis of rotation). Some ways of reducing the force transmitted are by putting the machine on vibration isolators (basically springs with damping) as shown in CS Fig. 19.1b, or by adding an inertial mass in addition to the vibration isolators, as shown in CS Fig. 19.1c. There are many types of isolators, including neoprene pads, coiled metal springs, and molded and bonded rubber mounts.

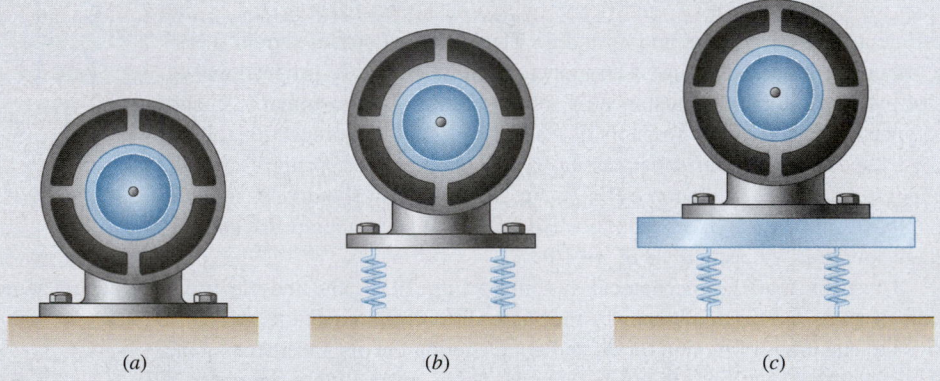

(a)　　　　(b)　　　　(c)

**CS Fig. 19.1**   Options for isolating a piece of vibrating equipment.

One of your first assignments as a new engineer is to design a vibration isolator for a metal tumbling drum that is driven at 1080 rpm. The drum, motor, and support base weighs 400 lb and the system has four mounting points for the isolators. The required isolation is 80%.

**STRATEGY:**   Newton's second law can be used to determine the equation of motion that can be used to determine an equation for the force transmitted.

**MODELING:**   You can model the system as a single degree of freedom and the isolator as a spring and a damper. The free-body diagram for this system is shown in CS Fig. 19.2.

**ANALYSIS:**   The equation of motion for this system is shown in Eq. 19.53

$$x(t) = \frac{\frac{F_0}{k}}{\sqrt{(1-r^2)^2 + (2\zeta r)^2}} \sin(\omega_f t - \phi)$$

where $r = \omega_f/\omega_n$, $\omega_n = \sqrt{k/m}$, $\phi = \tan^{-1}[2\zeta r/(1-r^2)]$, and $\zeta$ is the damping ratio.

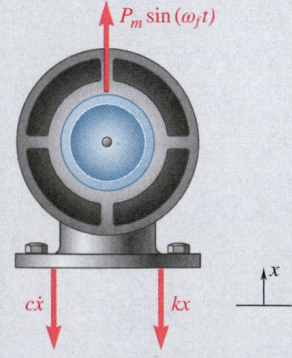

$P_m \sin(\omega_f t)$

$c\dot{x}$　　$kx$

**CS Fig. 19.2**   Free-body diagram.

*(continued)*

The magnitude of the force transmitted to the ground is

$$F_T = |kx + c\dot{x}| = \frac{F_0\sqrt{1 + (2\zeta r)^2}}{\sqrt{(1 - r^2)^2 + (2\zeta r)^2}}$$

The transmissibility, $T$, is defined to be the magnitude of the force transmitted divided by the input force, giving you

$$T = \frac{\sqrt{1 + (2\zeta r)^2}}{\sqrt{(1 - r^2)^2 + (2\zeta r)^2}} \qquad (1)$$

CS Fig. 19.3 shows a plot of the transmissibility as a function of the frequency ratio for different damping ratios. From CS Fig. 19.3, it is clear that in order for the force transmitted to be smaller than the input force, we need $r > \sqrt{2}$.

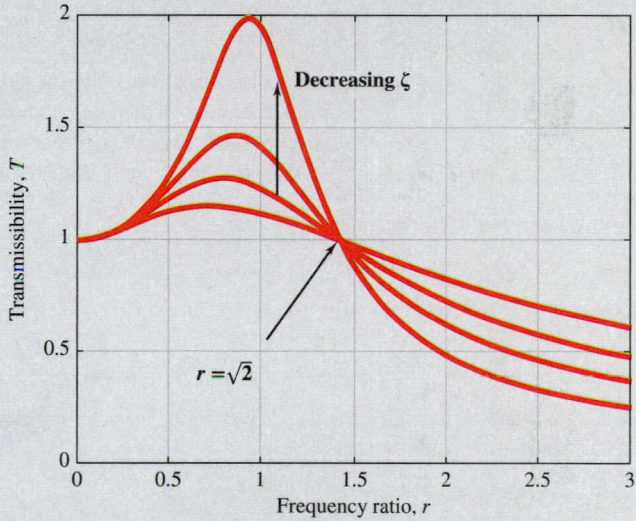

**CS Fig. 19.3**   Transmissibility as a function of frequency ratio.

Now let's design an isolator for this system. The transmissibility corresponding to an 80% reduction in the force is $T = 0.2$. Using information provided by vibration isolator manufacturers, we know neoprene or rubber isolators have a damping ratio of approximately 0.05. Substituting these values for $T$ and $\zeta$ into Eq. 1 and solving for the required frequency ratio, we find $r = 2.479$. Using a forcing frequency of 1080 rpm (113.097 rad/s), you can find the required natural frequency:

$$\omega_n = \omega_f/r = \frac{\left(113.097\frac{\text{rad}}{\text{s}}\right)}{2.479} = 45.62\frac{\text{rad}}{\text{s}} = 7.26\text{Hz}$$

*(continued)*

Knowing the natural frequency, we can now find the required stiffness using

$$k = m\omega_n^2 = \left(\frac{400}{32.2}\,\frac{\text{lb·s}^2}{\text{ft}}\right)(45.62\ \text{rad/s})^2 = 25{,}856\ \text{lb/ft}$$

and because we are using four isolators $k_{\text{reqd}} = 6464$ lb/ft. This will be the minimum spring constant required for our isolator. Rather than providing spring constant information, however, isolator manufacturers will often provide information on the static deflection, $\delta_{\text{ST}}$ which can be found from the weight.

$$mg = k\delta_{\text{ST}}$$

or

$$\delta_{\text{ST}} = \frac{mg}{k} = \frac{g}{\omega_n^2} = \frac{386.4\ \text{in./s}^2}{(45.62\ \text{rad/s})^2} = 0.186\ \text{in.}$$

An isolator that best fits these calculated values can be found using product information provided by manufacturers. In this case, Barry Part 633A-100, shown in CS Fig. 19.4, meets these requirements. The information for this case study came from an "Isolator Selection Guide" by Barry Controls.

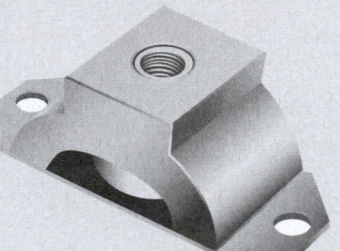

**CS Fig. 19.4**   Selected isolator.

# SOLVING PROBLEMS ON YOUR OWN

In this section, we developed a more realistic model of a vibrating system by including the effect of the **viscous damping** caused by fluid friction. We represented viscous damping in Fig. 19.10 by the force exerted on the moving body by a plunger moving in a dashpot. This force is equal in magnitude to $c\dot{x}$, where the constant $c$, expressed in N·s/m or lb·s/ft, is known as the *coefficient of viscous damping*. Keep in mind that the same sign convention should be used for $x$, $\dot{x}$, and $\ddot{x}$

**1. Damped free vibrations.** The differential equation defining this motion was found to be

$$m\ddot{x} + c\dot{x} + kx = 0 \qquad (19.38)$$

To obtain the solution of this equation, calculate the *critical damping coefficient* $c_c$, using the formula

$$c_c = 2m\sqrt{k/m} = 2m\omega_n \qquad (19.41)$$

where $\omega_n$ is the natural circular frequency of the undamped system.

**a. If $c > c_c$ (overdamped),** the solution of Eq. (19.38) is

$$x = C_1 e^{\lambda_1 t} + C_2 e^{\lambda_2 t} \qquad (19.42)$$

where

$$\lambda_{1,2} = -\frac{c}{2m} \pm \sqrt{\left(\frac{c}{2m}\right)^2 - \frac{k}{m}} \qquad (19.40)$$

and where the constants $C_1$ and $C_2$ can be determined from the initial conditions $x(0)$ and $\dot{x}(0)$. This solution corresponds to a nonvibratory motion.

**b. If $c = c_c$ (critically damped),** the solution of Eq. (19.38) is

$$x = (C_1 + C_2 t)e^{-\omega_n t} \qquad (19.43)$$

which also corresponds to a nonvibratory motion. Critically damped systems are of special interest in engineering applications because they regain their equilibrium position in the shortest possible time without oscillation.

**c. If $c < c_c$ (underdamped),** the solution of Eq. (19.38) is

$$x = x_0 e^{-(c/2m)t} \sin(\omega_d t + \phi) \qquad (19.46)$$

or in terms of the damping ratio $\zeta = c/c_{cr}$,

$$x = x_0 e^{-\zeta\omega_n t} \sin(\omega_d t + \phi) \qquad (19.46')$$

where

$$\omega_d = \omega_n\sqrt{1 - \left(\frac{c}{c_c}\right)^2} \qquad (19.45)$$

or

$$\omega_d = \omega_n \sqrt{1 - \zeta^2} \qquad \textbf{(19.45')}$$

and where $x_0$ and $\phi$ can be determined from the initial conditions $x(0)$ and $\dot{x}(0)$. This solution corresponds to oscillations of decreasing amplitude and of period $\tau_d = 2\pi/\omega_d$ (Fig. 19.11).

**2. Damped forced vibrations.** These vibrations occur when a system with viscous damping is subjected to a periodic force **P** with a magnitude of $P = P_m \sin \omega_f t$ or when it is elastically connected to a support with an alternating motion of $\delta = \delta_m \sin \omega_f t$. In the first case, the motion is defined by the differential equation

$$m\ddot{x} + c\dot{x} + kx = P_m \sin \omega_f t \qquad \textbf{(19.47)}$$

and in the second case, by a similar equation obtained by replacing $P_m$ with $k\delta_m$. You will be concerned only with the *steady-state* motion of the system, which is defined by a *particular solution* of these equations of the form

$$x_{\text{part}} = x_m \sin (\omega_f t - \phi) \qquad \textbf{(19.48)}$$

where

$$\frac{x_m}{P_m/k} = \frac{x_m}{\delta_m} = \frac{1}{\sqrt{[1 - (\omega_f/\omega_n)^2]^2 + [2(c/c_c)(\omega_f/\omega_n)]^2}} \qquad \textbf{(19.53)}$$

and

$$\tan \phi = \frac{2(c/c_c)(\omega_f/\omega_n)}{1 - (\omega_f/\omega_n)^2} \qquad \textbf{(19.54)}$$

The expression given in Eq. (19.53) is referred to as the *magnification factor* and has been plotted against the frequency ratio $\omega_f/\omega_n$ in Fig. 19.12 for various values of the damping ratio $c/c_c$. Eqs. (19.53) and (19.54) can be written in terms of the damping ratio $\zeta$ and frequency ratio $r$ as shown in Eqs. (19.53') and (19.54'). In the problems that follow, you may be asked to determine one of the parameters in Eqs. (19.53) and (19.54) when the others are known.

# Problems

**19.127** Show that in the case of heavy damping ($c > c_c$), a body never passes through its position of equilibrium $O$ if it is (*a*) released with no initial velocity from an arbitrary position, (*b*) started from $O$ with an arbitrary initial velocity.

**19.128** Show that in the case of heavy damping ($c > c_c$), a body released from an arbitrary position with an arbitrary initial velocity cannot pass more than once through its equilibrium position.

**19.129** In the case of light damping ($c < c_c$), the displacements $x_1$, $x_2$, and $x_3$ shown in Fig. 19.11 may be assumed equal to the maximum displacements. Show that the ratio of any two successive maximum displacements $x_n$ and $x_{n+1}$ is a constant and that the natural logarithm of this ratio, called the *logarithmic decrement*, is

$$\ln \frac{x_n}{x_{n+1}} = \frac{2\pi(c/c_c)}{\sqrt{1 - (c/c_c)^2}}$$

**19.130** In practice, it is often difficult to determine the logarithmic decrement of a system with light damping defined in Prob. 19.129 by measuring two successive maximum displacements. Show that the logarithmic decrement can also be expressed as $(1/k) \ln(x_n/x_{n+k})$, where $k$ is the number of cycles between readings of the maximum displacement.

**19.131** In an underdamped system ($c < c_c$), the period of vibration is commonly defined as the time interval $\tau_d = 2\pi/\omega_d$ corresponding to two successive points, where the displacement–time curve touches one of the limiting curves shown in Fig. 19.11. Show that the interval of time (*a*) between a maximum positive displacement and the following maximum negative displacement is $\frac{1}{2}\tau_d$, (*b*) between two successive zero displacements is $\frac{1}{2}\tau_d$, (*c*) between a maximum positive displacement and the following zero displacement is greater than $\frac{1}{4}\tau_d$.

**19.132** A loaded railroad car weighing 30,000 lb is rolling at a constant velocity $\mathbf{v}_0$ when it couples with a spring and dashpot bumper system (Fig. 1). The recorded displacement–time curve of the loaded railroad car after coupling is as shown (Fig. 2). Determine (*a*) the damping constant, (*b*) the spring constant. (*Hint:* Use the definition of logarithmic decrement given in Prob. 19.129.)

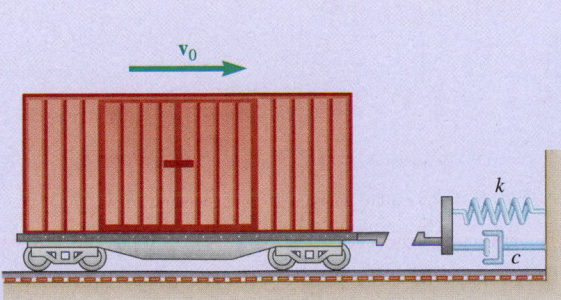

(1)

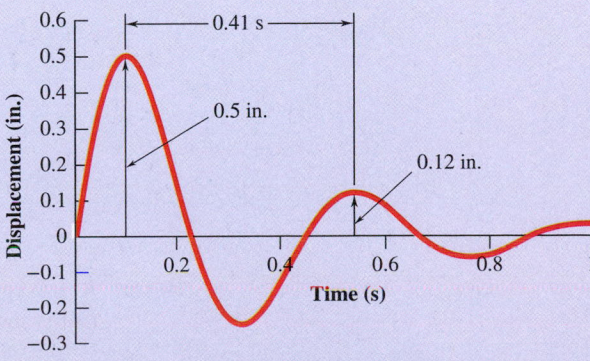

(2)

Fig. P19.132

1419

**19.133** A torsional pendulum has a centroidal mass moment of inertia of 0.3 kg·m² and when given an initial twist and released is found to have a frequency of oscillation of 200 rpm. Knowing that when this pendulum is immersed in oil and when given the same initial condition it is found to have a frequency of oscillation of 180 rpm, determine the damping constant for the oil.

**19.134** The barrel of a field gun weighs 1800 lb and is returned into firing position after recoil by a recuperator of constant $c = 1320$ lb·s/ft. Determine (*a*) the constant $k$ which should be used for the recuperator to return the barrel into firing position in the shortest possible time without any oscillation, (*b*) the time needed for the barrel to move back two-thirds of the way from its maximum-recoil position to its firing position. (*Hint:* Use the logarithmic decrement discussed in Probs. 19.129 and 19.130.)

**19.135** A 2-kg block is supported by a spring with a constant of $k = 128$ N/m and a dashpot with a coefficient of viscous damping of $c = 0.6$ N·s/m. The block is in equilibrium when it is struck from below by a hammer that imparts to the block an upward velocity of 0.4 m/s. Determine (*a*) the logarithmic decrement, (*b*) the maximum upward displacement of the block from equilibrium after two cycles.

**19.136** A 150-kg electromagnet is at rest and is holding 100 kg of scrap steel when the current is turned off, thus dropping the steel. Knowing that the cable and the supporting crane have a total stiffness equivalent to a spring of constant $k = 60$ kN/m and a damper $c = 300$ N·s/m, determine (*a*) the frequency of oscillation, (*b*) the maximum upward displacement.

**19.137** A 0.9-kg block *B* is connected by a cord to a 2.4-kg block *A* that is suspended as shown from two springs, each with a constant of $k = 180$ N/m, and a dashpot with a damping coefficient of $c = 7.5$ N·s/m. Knowing that the system is at rest when the cord connecting *A* and *B* is cut, determine the minimum tension that will occur in each spring during the resulting motion.

**19.138** A 0.9-kg block *B* is connected by a cord to a 2.4-kg block *A* that is suspended as shown from two springs, each with a constant of $k = 180$ N/m, and a dashpot with a damping coefficient of $c = 60$ N·s/m. Knowing that the system is at rest when the cord connecting *A* and *B* is cut, determine the velocity of block *A* after 0.1 s.

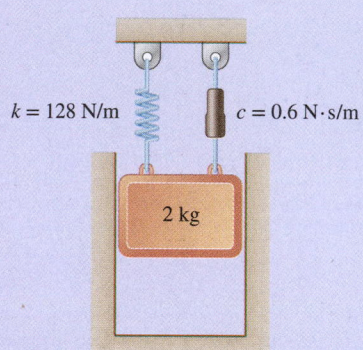

$k = 128$ N/m  $c = 0.6$ N·s/m

2 kg

**Fig. P19.135**

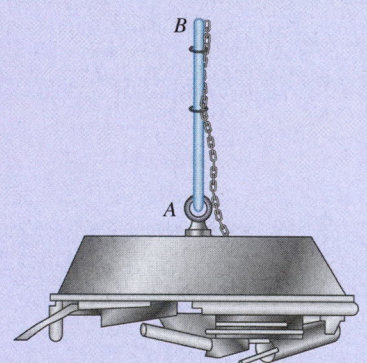

*B*

*A*

**Fig. P19.136**

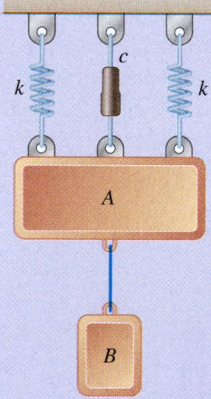

*c*

*k*    *k*

*A*

*B*

**Fig. P19.137 and P19.138**

**19.139** A machine element weighing 500 lb is supported by two springs, each having a constant of 250 lb/in. A periodic force of maximum value 20 lb is applied to the element with a frequency of 2 cycles per second. Knowing that the coefficient of damping is 10 lb·s/in., determine the amplitude of the steady-state vibration of the element.

**19.140** In Prob. 19.139, determine the required value of the coefficient of damping if the amplitude of the steady-state vibration of the element is to be 0.05 in.

**19.141** In the case of the forced vibration of a system, determine the range of values of the damping factor $c/c_c$ for which the magnification factor will always decrease as the frequency ratio $\omega_f/\omega_n$ increases.

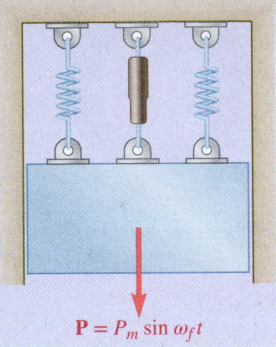

$$P = P_m \sin \omega_f t$$

**Fig. P19.139**

**19.142** Show that for a small value of the damping factor $c/c_c$, the maximum amplitude of a forced vibration occurs when $\omega_f \approx \omega_n$ and that the corresponding value of the magnification factor is $\frac{1}{2}(c/c_c)$.

**19.143** A counter-rotating eccentric mass exciter consisting of two rotating 14-oz weights describing circles of 6-in. radius at the same speed but in opposite senses is placed on a machine element to induce a steady-state vibration of the element and to determine some of the dynamic characteristics of the element. At a speed of 1200 rpm, a stroboscope shows the eccentric masses to be exactly under their respective axes of rotation and the element to be passing through its position of static equilibrium. Knowing that the amplitude of the motion of the element at that speed is 0.6 in. and that the total weight of the system is 300 lb, determine (*a*) the combined spring constant $k$, (*b*) the damping factor $c/c_c$.

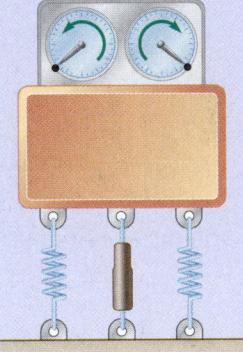

**Fig. P19.143**

**19.144** A 36-lb motor is bolted to a light horizontal beam that has a static deflection of 0.075 in. due to the weight of the motor. Knowing that the unbalance of the rotor is equivalent to a weight of 0.64 oz located 6.25 in. from the axis of rotation, determine the amplitude of the vibration of the motor at a speed of 900 rpm, assuming (*a*) that no damping is present, (*b*) that the damping factor $c/c_c$ is equal to 0.055.

**Fig. P19.144**

**19.145** One of the tail rotor blades of a helicopter has an unbalanced weight of 1 lb at a distance of $e = 6$ in. from the axis of rotation. The fuselage can be considered to be fixed and the tail rotor and tail can be modeled as an equivalent stiffness of 8800 lb/ft, weight of 165 lb, and damping ratio of 0.15. Knowing that the blades rotate at 500 rpm, determine the amplitude of vibration of the tail section.

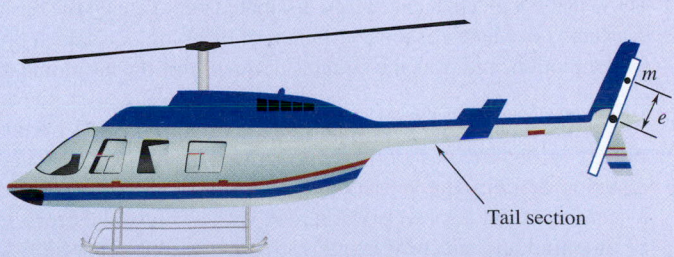

Tail section

**Fig. P19.145**

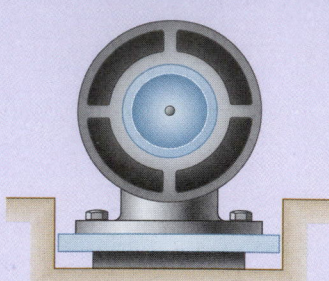

Fig. P19.146

**19.146** The unbalance of the rotor of a 180-kg motor is equivalent to a mass of 85 g located 150 mm from the axis of rotation. The pad that is placed between the motor and the foundation is equivalent to a spring with a constant of $k = 7.5$ kN/m in parallel with a dashpot with constant $c$. Knowing that the magnitude of the maximum acceleration of the motor is 9 mm/s$^2$ at a speed of 100 rpm, determine the damping factor $c/c_c$.

**19.147** A machine element is supported by springs and is connected to a dashpot as shown. Show that if a periodic force of magnitude $P = P_m \sin \omega_f t$ is applied to the element, the amplitude of the fluctuating force transmitted to the foundation is

$$F_m = P_m \sqrt{\frac{1 + [2(c/c_c)(\omega_f/\omega_n)]^2}{[1 - (\omega_f/\omega_n)^2]^2 + [2(c/c_c)(\omega_f/\omega_n)]^2}}$$

$P = P_m \sin \omega_f t$

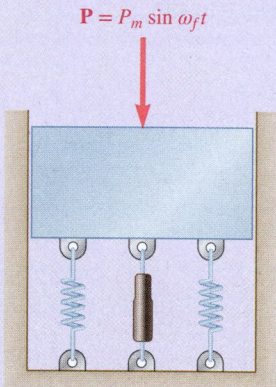

Fig. P19.147 and P19.148

**19.148** A 91-kg machine element supported by four springs, each of constant $k = 175$ N/m, is subjected to a periodic force of frequency 0.8 Hz and amplitude 89 N. Determine the amplitude of the fluctuating force transmitted to the foundation if (a) a dashpot with a coefficient of damping $c = 365$ N·s/m is connected to the machine element and to the ground, (b) the dashpot is removed.

**19.149** A simplified model of a washing machine is shown. A bundle of wet clothes forms a weight $w_b$ of 20 lb in the machine and causes a rotating unbalance. The rotating weight is 40 lb (including $w_b$) and the radius of the washer basket $e$ is 9 in. Knowing the washer has an equivalent spring constant $k = 70$ lb/ft and damping ratio $\zeta = c/c_c = 0.05$ and during the spin cycle the drum rotates at 250 rpm, determine the amplitude of the motion and the magnitude of the force transmitted to the sides of the washing machine.

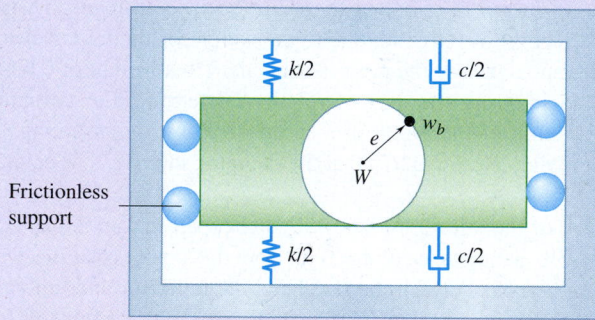

Frictionless support

Fig. P19.149

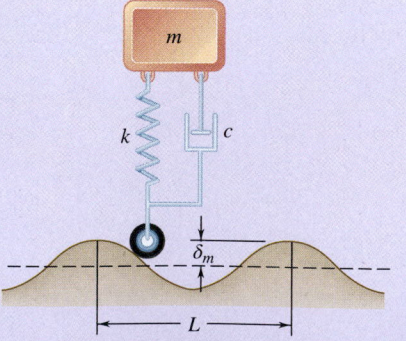

Fig. P19.151

**\*19.150** For a steady-state vibration with damping under a harmonic force, show that the mechanical energy dissipated per cycle by the dashpot is $E = \pi c x_m^2 \omega_f$, where $c$ is the coefficient of damping, $x_m$ is the amplitude of the motion, and $\omega_f$ is the circular frequency of the harmonic force.

**\*19.151** The suspension of an automobile can be approximated by the simplified spring-and-dashpot system shown. (a) Write the differential equation defining the vertical displacement of the mass $m$ when the system moves at a speed $v$ over a road with a sinusoidal cross-section of amplitude $\delta_m$ and wavelength $L$. (b) Derive an expression for the amplitude of the vertical displacement of the mass $m$.

*19.152 Two blocks $A$ and $B$, each of mass $m$, are supported as shown by three springs of the same constant $k$. Blocks $A$ and $B$ are connected by a dashpot, and block $B$ is connected to the ground by two dashpots, each dashpot having the same coefficient of damping $c$. Block $A$ is subjected to a force of magnitude $P = P_m \sin \omega_f t$. Write the differential equations defining the displacements $x_A$ and $x_B$ of the two blocks from their equilibrium positions.

19.153 Express in terms of $L$, $C$, and $E$ the range of values of the resistance $R$ for which oscillations will take place in the circuit shown when switch $S$ is closed.

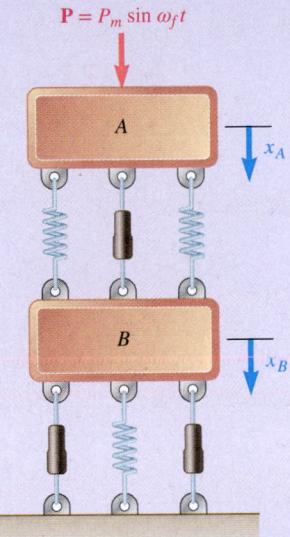

Fig. **P19.152**

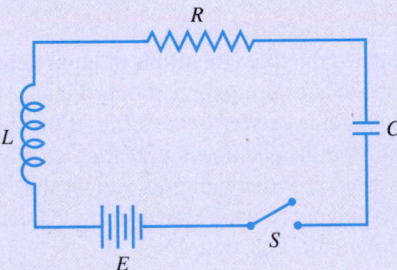

Fig. P19.153

19.154 Consider the circuit of Prob. 19.153 when the capacitor $C$ is removed. If switch $S$ is closed at time $t = 0$, determine (a) the final value of the current in the circuit, (b) the time $t$ at which the current will have reached $(1 - 1/e)$ times its final value. (The desired value of $t$ is known as the *time constant* of the circuit.)

19.155 and 19.156 Draw the electrical analog of the mechanical system shown. (*Hint:* Draw the loops corresponding to the free bodies $m$ and $A$.)

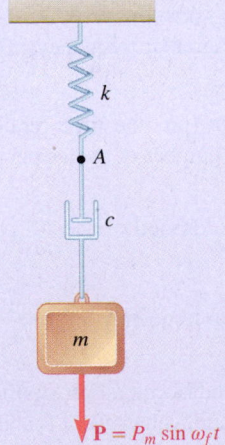

Fig. P19.155 and P19.157

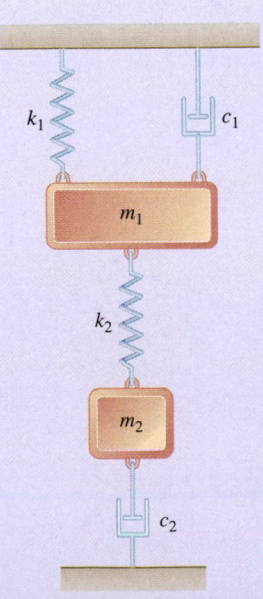

Fig. P19.156 and *P19.158*

19.157 and *19.158* Write the differential equations defining (a) the displacements of the mass $m$ and of the point $A$, (b) the charges on the capacitors of the electrical analog.

1423

# Review and Summary

This chapter was devoted to the study of **mechanical vibrations**; that is, to the analysis of the motion of particles and rigid bodies oscillating about a position of equilibrium. In the first part of the chapter (Secs. 19.1, 19.2, 19.3, and 19.4), we considered *vibrations without damping,* while the second part was devoted to *damped vibrations* (Sec. 19.5).

## Free Vibrations of a Particle

In Sec. 19.1, we considered the **free vibrations of a particle**; that is, the motion of a particle $P$ subjected to a restoring force proportional to the displacement of the particle—such as the force exerted by a spring. If the displacement $x$ of the particle $P$ is measured from its equilibrium position $O$ (Fig. 19.17), the resultant $\mathbf{F}$ of the forces acting on $P$ (including its weight) has a magnitude $kx$ and is directed toward $O$. Applying Newton's second law $F = ma$ and recalling that $a = \ddot{x}$, we wrote the differential equation

$$m\ddot{x} + kx = 0 \tag{19.2}$$

or, setting $\omega_n^2 = k/m$,

$$\ddot{x} + \omega_n^2 x = 0 \tag{19.6}$$

The motion defined by this equation is called **simple harmonic motion**.

The solution of Eq. (19.6), which represents the displacement of the particle $P$, was expressed as

$$x = x_m \sin(\omega_n t - \phi) \tag{19.10}$$

where $x_m$ = amplitude of the vibration
$\omega_n = \sqrt{k/m}$ = natural circular frequency
$\phi$ = phase angle

The **period of the vibration** (i.e., the time required for a full cycle) and its **natural frequency** (i.e., the number of cycles per second) were expressed as

$$\text{Period} = \tau_n = \frac{2\pi}{\omega_n} \tag{19.13}$$

$$\text{Natural frequency} = f_n = \frac{1}{\tau_n} = \frac{\omega_n}{2\pi} \tag{19.14}$$

We obtained the velocity and acceleration of the particle by differentiating Eq. (19.10), and their maximum values were found to be

$$v_m = x_m \omega_n \qquad a_m = x_m \omega_n^2 \tag{19.15}$$

Because all of the above parameters depend directly upon the natural circular frequency $\omega_n$, and thus upon the ratio $k/m$, it is essential in any given problem to calculate the value of the constant $k$. This can be done by determining the

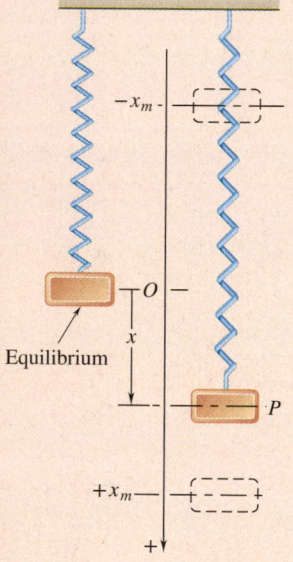

Equilibrium

**Fig. 19.17**

relation between the restoring force and the corresponding displacement of the particle (Sample Prob. 19.1).

It was also shown that we can represent the oscillatory motion of particle $P$ by the projection on the $x$ axis of the motion of a point $Q$ describing an auxiliary circle of radius $x_m$ with the constant angular velocity $\omega_n$ (Fig. 19.18). Then we can obtain the instantaneous values of the velocity and acceleration of $P$ by projecting on the $x$ axis the vectors $\mathbf{v}_m$ and $\mathbf{a}_m$ representing, respectively, the velocity and acceleration of $Q$.

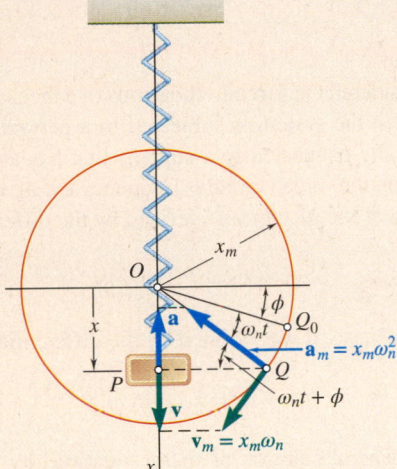

**Fig. 19.18**

## Simple Pendulum

Although the motion of a **simple pendulum** is not truly a simple harmonic motion, we can use the formulas given previously with $\omega_n^2 = g/l$ to calculate the period and natural frequency of the *small oscillations* of a simple pendulum (Sec. 19.1B). Large-amplitude oscillations of a simple pendulum were discussed in Sec. 19.1C.

## Free Vibrations of a Rigid Body

We can analyze the **free vibrations of a rigid body** by choosing an appropriate variable, such as a distance $x$ or an angle $\theta$, to define the position of the body. We then draw a free-body diagram and kinetic diagram to express the equivalence of the external forces and inertial terms and write an equation relating the selected variable and its second derivative (Sec. 19.2). If the equation obtained is of the form

$$\ddot{x} + \omega_n^2 x = 0 \qquad \text{or} \qquad \ddot{\theta} + \omega_n^2 \theta = 0 \qquad \textbf{(19.21)}$$

the vibration considered is a simple harmonic motion, and its period and natural frequency can be obtained *by identifying* $\omega_n$ and substituting its value into Eqs. (19.13) and (19.14) (Sample Probs. 19.2 and 19.3).

## Using the Principle of Conservation of Energy

We can use the *principle of conservation of energy* as an alternative method for determining the period and natural frequency of the simple harmonic motion of a particle or rigid body (Sec. 19.3). Choosing again an appropriate variable, such as $\theta$, to define the position of the system, we express that the total energy of the system is conserved, using $T_1 + V_1 = T_2 + V_2$, between the position of maximum displacement ($\theta_1 = \theta_m$) and the position of maximum velocity ($\dot{\theta}_2 = \dot{\theta}_m$). If the motion considered is simple harmonic, the two sides of the equation obtained consist of homogeneous quadratic expressions in $\theta_m$ and $\dot{\theta}_m$, respectively. Substituting $\dot{\theta}_m = \theta_m \omega_n$ in this equation, we can factor out $\theta_m^2$ and solve for the circular frequency $\omega_n$ (Sample Prob. 19.4). It is important to note that if the motion can be approximated only by a simple harmonic motion, such as for the small oscillations of a body under gravity, we must approximate the potential energy by a quadratic expression in $\theta_m$ (Sample Prob. 19.4).

## Forced Vibrations

In Sec. 19.4, we considered the **forced vibrations** of a mechanical system. These vibrations occur when the system is subjected to a periodic force (Fig. 19.19) or when it is elastically connected to a support that has an alternating motion (Fig. 19.20). Denoting the forced circular frequency by $\omega_f$, we found that, in the first case, the motion of the system was defined by the differential equation

$$m\ddot{x} + kx = P_m \sin \omega_f t \tag{19.30}$$

and that in the second case, it was defined by the differential equation

$$m\ddot{x} + kx = k\delta_m \sin \omega_f t \tag{19.31}$$

We can obtain the general solution of these equations by adding a particular solution of the form

$$x_{\text{part}} = x_m \sin \omega_f t \tag{19.32}$$

to the general solution of the corresponding homogeneous equation. The particular solution of Eq. (19.32) represents a **steady-state vibration** of the system,

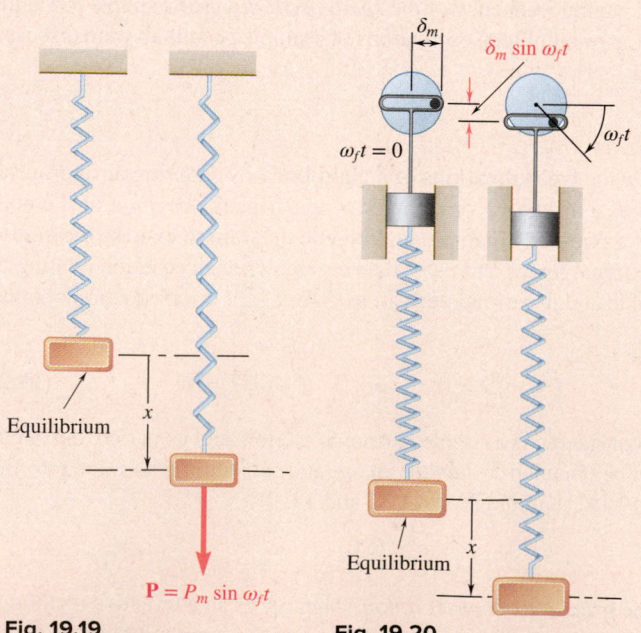

**P** = $P_m \sin \omega_f t$

Equilibrium

**Fig. 19.19**

$\delta_m$

$\delta_m \sin \omega_f t$

$\omega_f t = 0$

$\omega_f t$

Equilibrium

**Fig. 19.20**

whereas the solution of the homogeneous equation represents a **transient free vibration** that can generally be neglected.

Dividing the amplitude $x_m$ of the steady-state vibration by $P_m/k$ in the case of a periodic force or by $\delta_m$ in the case of an oscillating support, we defined the **magnification factor** of the vibration and found that

$$\text{Magnification factor} = \frac{x_m}{P_m/k} = \frac{x_m}{\delta_m} = \frac{1}{1 - (\omega_f/\omega_n)^2} \quad \textbf{(19.36)}$$

According to Eq. (19.36), the amplitude $x_m$ of the forced vibration becomes infinite when $\omega_f = \omega_n$; that is, when the forced frequency is equal to the natural frequency of the system. The impressed force or impressed support movement is then said to be in **resonance** with the system (Sample Prob. 19.5). (Actually, the amplitude of the vibration remains finite, due to damping forces.)

### Damped Free Vibrations

In Sec. 19.5, we considered the **damped vibrations** of a mechanical system. First, we analyzed the damped free vibrations of a system with **viscous damping** (Sec. 19.5A). We found that the motion of such a system was defined by the differential equation

$$m\ddot{x} + c\dot{x} + kx = 0 \quad \textbf{(19.38)}$$

where $c$ is a constant called the *coefficient of viscous damping*. Defining the *critical damping coefficient* $c_c$ as

$$c_c = 2m\sqrt{\frac{k}{m}} = 2m\omega_n \quad \textbf{(19.41)}$$

where $\omega_n$ is the natural circular frequency of the system in the absence of damping, we distinguished three different cases of damping, namely, (1) *overdamped,* when $c > c_c$; (2) *critically damped,* when $c = c_c$; and (3) *underdamped,* when $c < c_c$. In the first two cases, the system when disturbed tends to regain its equilibrium position without any oscillation. In the third case, the motion is vibratory with diminishing amplitude. For an underdamped system, the transient response is

$$x = x_0 e^{-(c/2m)t} \sin(\omega_d t + \phi) \quad \textbf{(19.46)}$$

where

$$\omega_d = \omega_n \sqrt{1 - \left(\frac{c}{c_c}\right)^2} \quad \textbf{(19.45)}$$

### Damped Forced Vibrations

In Sec. 19.5B, we considered the **damped forced vibrations** of a mechanical system. These vibrations occur when a system with viscous damping is subjected to a periodic force **P** of magnitude $P = P_m \sin \omega_f t$ or when it is elastically connected to a support with an alternating motion of $\delta = \delta_m \sin \omega_f t$. In the first case, the motion of the system was defined by the differential equation

$$m\ddot{x} + c\dot{x} + kx = P_m \sin \omega_f t \quad \textbf{(19.47)}$$

and in the second case, by a similar equation obtained by replacing $P_m$ by $k\delta_m$ in Eq. (19.47).

The *steady-state vibration* of the system is represented by a particular solution of Eq. (19.47) of the form

$$x_{\text{part}} = x_m \sin(\omega_f t - \phi) \tag{19.48}$$

Dividing the amplitude $x_m$ of the steady-state vibration by $P_m/k$ in the case of a periodic force or by $\delta_m$ in the case of an oscillating support, we obtained the following expression for the magnification factor as

$$\frac{x_m}{P_m/k} = \frac{x_m}{\delta_m} = \frac{1}{\sqrt{[1 - (\omega_f/\omega_n)^2]^2 + [2(c/c_c)(\omega_f/\omega_n)]^2}} \tag{19.53}$$

or

$$\frac{x_m}{P_m/k} = \frac{x_m}{\delta_{st}} = \frac{1}{\sqrt{(1 - r^2)^2 + (2\zeta r)^2}}$$

where $\omega_n = \sqrt{k/m}$ = natural circular frequency of undamped system
$c_c = 2m\omega_n$ = critical damping coefficient
$c/c_c = \zeta$ = damping ratio
$r = \omega_f/\omega_n$ = frequency ratio

We also found that the *phase difference $\phi$* between the impressed force or support movement and the resulting steady-state vibration of the damped system was defined by the relation

$$\tan \phi = \frac{2(c/c_c)(\omega_f/\omega_n)}{1 - (\omega_f/\omega_n)^2} \tag{19.54}$$

or

$$\tan \phi = \frac{2\zeta r}{1 - r^2} \tag{19.54'}$$

## Electrical Analogs

This chapter ended with a discussion of *electrical analogs* (Sec. 19.5C) in which we showed that the vibrations of mechanical systems and the oscillations of electrical circuits are defined by the same differential equations. Electrical analogs of mechanical systems therefore can be used to study or predict the behavior of these systems.

# Review Problems

**19.159** An automobile wheel-and-tire assembly of total weight 47 lb is attached to a mounting plate of negligible weight that is suspended from a steel wire. The torsional spring constant of the wire is known to be $K = 0.40$ lb·in/rad. The wheel is rotated through 90° about the vertical and then released. Knowing that the period of oscillation is observed to be 30 s, determine the centroidal mass moment of inertia and the centroidal radius of gyration of the wheel-and-tire assembly.

**Fig. P19.159**

**19.160** A firefighting helicopter carries a bucket of water weighing 7000 lb. The distance $L$ between the helicopter and the bucket is 30 ft. Knowing that the total stiffness of the cables in the vertical direction is 14,000 lb/ft, determine (*a*) the frequency of oscillation in the horizontal direction, (*b*) the frequency of oscillation in the vertical direction.

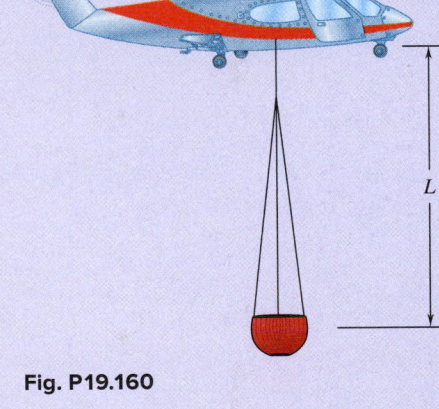

**Fig. P19.160**

**19.161** Disks $A$ and $B$ weigh 30 lb and 12 lb, respectively, and a small 5-lb block $C$ is attached to the rim of disk $B$. Assuming that no slipping occurs between the disks, determine the period of small oscillations of the system.

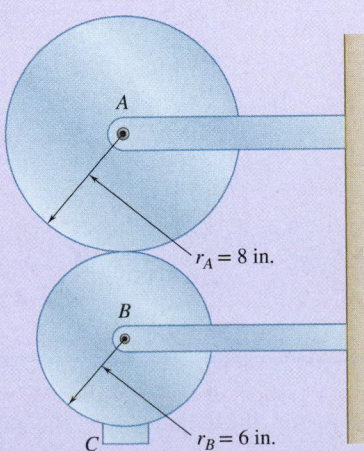

**Fig. P19.161**

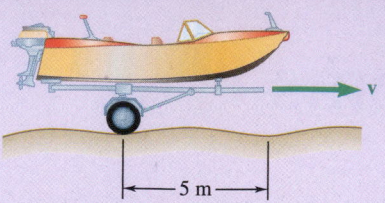

**Fig. P19.162**

*19.162 A small trailer and its load have a total mass of 250 kg. The trailer is supported by two springs, and when the boat is given an initial displacement $x_0$ downward and released from rest, the amplitude of oscillation is reduced to $\frac{1}{4} x_0$ after one half cycle and the period of the oscillation is 0.8 s. The trailer is then pulled over a road, the surface of which can be approximated by a sine curve with an amplitude of 35 mm and a wavelength of 5 m (i.e., the distance between successive crests is 5 m and the vertical distance from crest to trough is 70 mm). Determine (a) the damping ratio, (b) the spring stiffness, (c) the amplitude of the vibration of the trailer at a speed of 50 km/h. (*Hint:* Use the logarithmic decrement discussed in Probs. 19.129 and 19.130.)

**19.163** A 0.8-lb ball is connected to a paddle by means of an elastic cord $AB$ of constant $k = 5$ lb/ft. Knowing that the paddle is moved vertically according to the relation $\delta = \delta_m \sin \omega_f t$, where $\delta_m = 8$ in., determine the maximum allowable circular frequency $\omega_f$ if the cord is not to become slack.

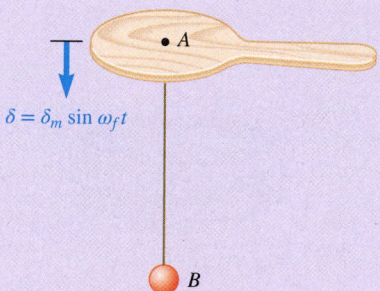

$\delta = \delta_m \sin \omega_f t$

**Fig. P19.163**

**19.164** A 3-kg slender rod $AB$ is bolted to a 5-kg uniform disk. A dashpot with a damping coefficient of $c = 9$ N·s/m is attached to the disk as shown. Determine (a) the differential equation of motion for small oscillations, (b) the damping factor $c/c_c$.

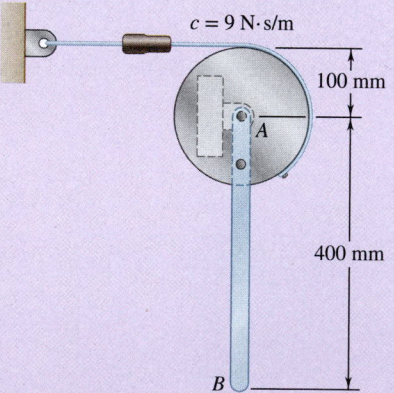

$c = 9$ N·s/m

100 mm

$A$

400 mm

$B$

**Fig. P19.164**

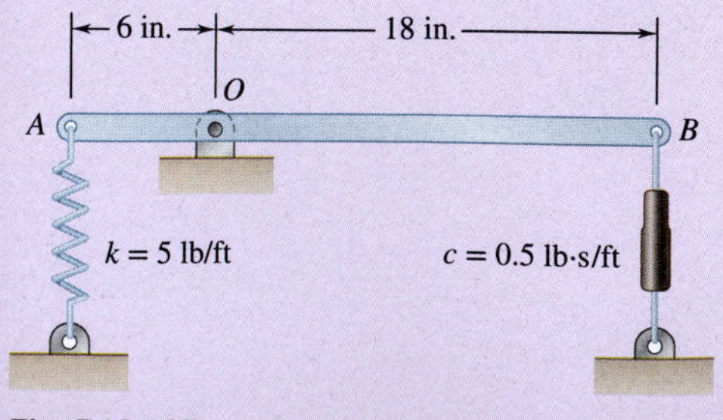

**19.165** A 4-lb uniform rod is supported by a pin at $O$ and a spring at $A$ and is connected to a dashpot at $B$. Determine (a) the differential equation of motion for small oscillations, (b) the angle that the rod will form with the horizontal 5 s after end $B$ has been pushed 0.9 in. down and released.

— 6 in. — — 18 in. —

$O$

$A$                        $B$

$k = 5$ lb/ft            $c = 0.5$ lb·s/ft

**Fig. P19.165**

**19.166** A 400-kg motor supported by four springs, each of constant 150 kN/m, and a dashpot of constant $c = 6500$ N·s/m is constrained to move vertically. Knowing that the unbalance of the rotor is equivalent to a 23-g mass located at a distance of 100 mm from the axis of rotation, determine for a speed of 800 rpm (a) the amplitude of the fluctuating force transmitted to the foundation, (b) the amplitude of the vertical motion of the motor.

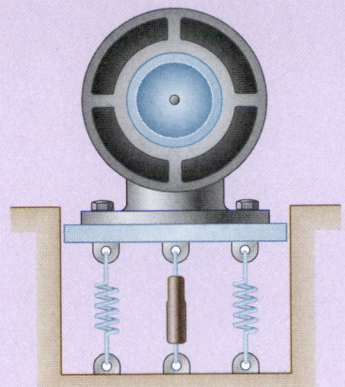

**19.167** The compressor shown has a mass of 250 kg and operates at 2000 rpm. At this operating condition, the force transmitted to the ground is excessively high and is found to be $mr\omega_f^2$, where $mr$ is the unbalance and $\omega_f$ is the forcing frequency. To fix this problem, it is proposed to isolate the compressor by mounting it on a square concrete block separated from the rest of the floor as shown. The density of concrete is 2400 kg/m$^3$ and the spring constant for the soil is found to be $80 \times 10^6$ N/m. The geometry of the compressor leads to choosing a block that is 1.5 m by 1.5 m. Determine the depth $h$ that will reduce the force transmitted to the ground by 75 percent.

**Fig. P19.166**

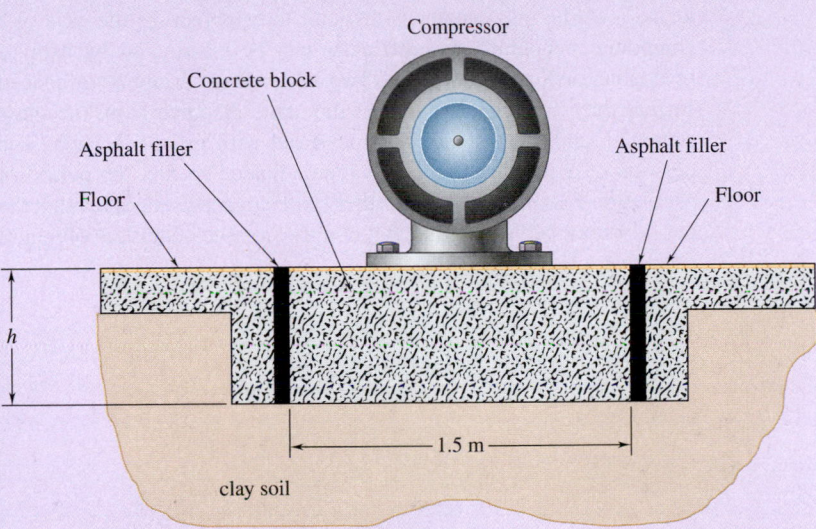

**Fig. P19.167**

**19.168** A small ball of mass $m$ attached at the midpoint of a tightly stretched elastic cord of length $l$ can slide on a horizontal plane. The ball is given a small displacement in a direction perpendicular to the cord and released. Assuming the tension $T$ in the cord to remain constant, (a) write the differential equation of motion of the ball, (b) determine the period of vibration.

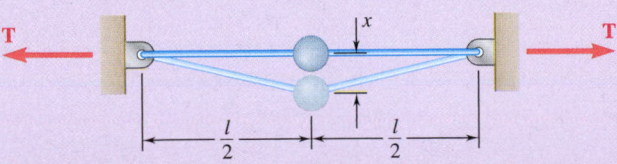

**Fig. P19.168**

**19.169** A certain vibrometer used to measure vibration amplitudes consists essentially of a box containing a slender rod to which a mass $m$ is attached; the natural frequency of the mass–rod system is known to be 5 Hz. When the box is rigidly attached to the casing of a motor rotating at 600 rpm, the mass is observed to vibrate with an amplitude of 0.06 in. relative to the box. Determine the amplitude of the vertical motion of the motor.

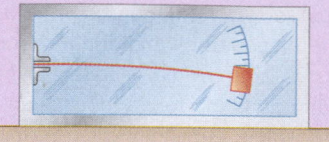

**Fig. P19.169**

**19.170** If either a simple or a compound pendulum is used to determine experimentally the acceleration of gravity $g$, difficulties are encountered. In the case of the simple pendulum, the string is not truly weightless, while in the case of the compound pendulum, the exact location of the mass center is difficult to establish. In the case of a compound pendulum, the difficulty can be eliminated by using a reversible, or Kater, pendulum. Two knife edges $A$ and $B$ are placed so that they are obviously not at the same distance from the mass center $G$, and the distance $l$ is measured with great precision. The position of a counterweight $D$ is then adjusted so that the period of oscillation $\tau$ is the same when either knife edge is used. Show that the period $\tau$ obtained is equal to that of a true simple pendulum of length $l$ and that $g = (4\pi^2 l)/\tau^2$.

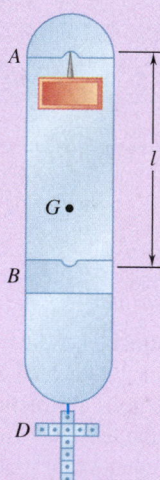

**Fig. P19.170**

# Fundamentals of Engineering Examination

Engineers are required to be licensed when their work directly affects the public health, safety, and welfare. The intent is to ensure that engineers have met minimum qualifications involving competence, ability, experience, and character. The licensing process involves an initial exam, called the *Fundamentals of Engineering Examination;* professional experience; and a second exam, called the *Principles and Practice of Engineering.* Those who successfully complete these requirements are licensed as a *Professional Engineer.* The exams are developed under the auspices of the *National Council of Examiners for Engineering and Surveying.*

The first exam, the *Fundamentals of Engineering Examination,* can be taken just before or after graduation from a four-year accredited engineering program. The exam stresses subject material in a typical undergraduate engineering program, including statics. The topics included in the exam cover much of the material in this book. The following is a list of the main topic areas, with references to the appropriate sections in this book. Also included are problems that can be solved to review this material.

**Concurrent Force Systems (2.1–2.2; 2.4)**
Problems: 2.31, 2.35, 2.36, 2.37, 2.77, 2.83, 2.92, 2.94, 2.97

**Vector Forces (3.1–3.2)**
Problems: 3.17, 3.18, 3.26, 3.33, 3.37, 3.39

**Equilibrium in Two Dimensions (2.3; 4.1–4.2)**
Problems: 4.1, 4.13, 4.14, 4.17, 4.31, 4.33, 4.67, 4.77

**Equilibrium in Three Dimensions (2.5; 4.3)**
Problems: 4.99, 4.101, 4.103, 4.108, 4.115, 4.117, 4.127, 4.129, 4.135

**Centroids of Areas and Volumes (5.1–5.2; 5.4)**
Problems: 5.9, 5.16, 5.30, 5.35, 5.41, 5.55, 5.62, 5.96, 5.102, 5.103, 5.125

**Analysis of Trusses (6.1–6.2)**
Problems: 6.3, 6.4, 6.32, 6.43, 6.44, 6.53

**Equilibrium of Two-Dimensional Frames (6.3)**
Problems: 6.75, 6.81, 6.85, 6.93, 6.94

**Shear and Bending Moment (7.1–7.3)**
Problems: 7.22, 7.30, 7.36, 7.41, 7.45, 7.49, 7.71, 7.79

**Friction (8.1–8.2; 8.4)**
Problems: 8.11, 8.18, 8.19, 8.30, 8.49, 8.52, 8.103, 8.104, 8.105

**Moments of Inertia (9.1–9.4)**
Problems: 9.6, 9.31, 9.32, 9.33, 9.72, 9.74, 9.80, 9.83, 9.98, 9.103

**Kinematics (11.1–11.2; 11.4–11.5; 15.1–15.4)**
Problems: 11.3, 11.10, 11.34, 11.35, 11.97, 11.102, 15.4, 15.6, 15.28, 15.39, 15.61, 15.63, 15.82, 15.111, 15.112

**Force, Mass, and Acceleration (12.1; 16.1–16.2)**

Problems: 12.5, 12.6, 12.11, 12.23, 12.36, 12.44, 12.45, 12.50, 16.1, 16.3, 16.9, 16.26, 16.27, 16.50, 16.60, 16.63, 16.78, 16.84

**Work and Energy (13.1–13.2; 13.8; 17.1)**

Problems: 13.3, 13.6, 13.13, 13.17, 13.40, 13.42, 13.47, 13.64, 13.66, 13.68, 17.1, 17.2, 17.16, 17.20

**Impulse and Momentum (13.3–13.4; 17.2–17.3)**

Problems: 13.119, 13.120, 13.129, 13.134, 13.146, 13.155, 13.163, 13.169, 17.53, 17.58, 17.70, 17.72, 17.96, 17.97, 17.104

**Vibration (19.1; 19.2–19.4)**

Problems: 19.1, 19.2, 19.11, 19.18, 19.23, 19.28, 19.50, 19.55, 19.64, 19.79, 19.99, 19.101, 19.105, 19.116

**Friction** (Problems involving friction occur in many of the preceding subjects.)

# Answers to Problems

## CHAPTER 2

**2.1** 1391 N ∡ 47.8°.
**2.2** 906 lb ∡ 26.6°.
**2.4** 77.1 lb ↗ 85.4°.
**2.5** (*a*) 101.4 N. (*b*) 196.6 N.
**2.6** (*a*) 853 lb. (*b*) 567 lb.
**2.8** (*a*) $T_{AC}$ = 2.60 kN. (*b*) *R* = 4.26 kN.
**2.9** (*a*) $\mathbf{T}_{AC}$ = 2.66 kN ↘ 34.3°.
**2.10** (*a*) 37.1°. (*b*) 73.2 N.
**2.11** (*a*) 392 lb. (*b*) 346 lb.
**2.13** (*a*) 368 lb →. (*b*) 213 lb.
**2.14** (*a*) 21.1 N ↓. (*b*) 45.3 N.
**2.15** 5380 lb ∡ 8.94°.
**2.16** 1391 N ∡ 47.8°.
**2.17** 8.03 kips ↗ 3.8°.
**2.19** 21.8 kN ↘ 86.6°.
**2.21** (29 lb) 21.0 lb, 20.0 lb; (50 lb) −14.00 lb, 48.0 lb; (51 lb) 24.0 lb, −45.0 lb.
**2.23** (350 N) 317 N, 147.9 N; (800 N) 274 N, 752 N; (600 N) −300 N, 520 N.
**2.24** (80 lb) 69.3 lb, −40.0 lb; (120 lb) 31.1 lb, −115.9 lb; (150 lb) −114.9 lb, −96.4 lb.
**2.26** (*a*) 523 lb. (*b*) 428 lb.
**2.27** (*a*) 621 N. (*b*) 160.8 N.
**2.28** (*a*) 610 lb. (*b*) 500 lb.
**2.29** (*a*) 2190 N. (*b*) 2060 N.
**2.31** 38.6 lb ∡ 36.6°.
**2.32** 1449 N ∡ 78.4°.
**2.34** 654 N ↘ 21.5°.
**2.35** 309 N ↗ 86.6°.
**2.36** 226 N ↗ 62.3°.
**2.37** 203 lb ∡ 8.46°.
**2.39** (*a*) 28.1°. (*b*) 155.0 N.
**2.40** (*a*) 580 N. (*b*) 300 N.
**2.42** (*a*) 56.3°. (*b*) 204 lb.
**2.43** (*a*) 352 lb. (*b*) 261 lb.
**2.44** (*a*) 5.22 kN. (*b*) 3.45 kN.
**2.46** (*a*) 305 N. (*b*) 514 N.
**2.48** (*a*) 1244 lb. (*b*) 115.4 lb.
**2.49** $T_{CA}$ = 134.6 N; $T_{CB}$ = 110.4 N.
**2.50** 179.3 N < *P* < 669 N.
**2.51** $T_A$ = 231 lb; $T_B$ = 577 lb.
**2.53** $F_C$ = 6.40 kN; $F_D$ = 4.80 kN.
**2.54** $F_B$ = 15.00 kN; $F_C$ = 8.00 kN.
**2.55** (*a*) $T_{ACB}$ = 269 lb. (*b*) $T_{CD}$ = 37.0 lb.
**2.57** (*a*) $\alpha$ = 35.0°; $T_{AC}$ = 4.91 kN; $T_{BC}$ = 3.44 kN, (*b*) $\alpha$ = 55.0°; $T_{AC}$ = $T_{BC}$ = 3.66 kN.
**2.58** (*a*) 784 N. (*b*) $\alpha$ = 71.0°.
**2.59** (*a*) $\alpha$ = 5.00°. (*b*) 104.6 lb.
**2.61** 1.250 m.
**2.62** 75.6 mm.
**2.63** (*a*) 10.98 lb. (*b*) 30.0 lb.

**2.65** (*a*) 2450 N. (*b*) 2220 N.
**2.67** (*a*) 300 lb. (*b*) 300 lb. (*c*) 200 lb. (*d*) 200 lb. (*e*) 150.0 lb.
**2.68** (*a*) 200 lb. (*b*) 150.0 lb.
**2.69** (*a*) 1293 N. (*b*) 2220 N.
**2.71** (*a*) +278 N, +383 N, +160.7 N. (*b*) 56.2°, 40.0°, 71.3°.
**2.72** (*a*) −115.6 N, +752 N, +248 N. (*b*) 98.3°, 20.0°, 71.9°.
**2.73** (*a*) −175.8 N, −257 N, 251 N. (*b*) 116.1°, 130.0°, 51.1°.
**2.74** (*a*) 350 N, −169.0 N, 93.8 N. (*b*) 28.9°, 115.0°, 76.4°.
**2.75** (*a*) −78.6 lb, 282 lb, −66.0 lb. (*b*) 105.2°, 20.0°, 102.7°.
**2.77** (*a*) −1861 lb, 3360 lb, 677 lb. (*b*) 118.5°, 30.5°, 80.0°.
**2.79** *F* = 900 N; $\theta_x$ = 73.2°, $\theta_y$ = 110.8°, $\theta_z$ = 27.3°.
**2.81** (*a*) $F_x$ = 105.7 lb, $F_y$ = 191.5 lb; $F_z$ = 121.2 lb. (*b*) 61.0°.
**2.82** (*a*) 118.2°. (*b*) $F_x$ = 36.0 lb, $F_y$ = −90.0 lb; *F* = 110.0 lb.
**2.84** (*a*) 34.3°. (*b*) 48.4 N.
**2.85** −240 N, +444 N, −192.0 N.
**2.87** −1.260 kips; 1.213 kips; 0.970 kips.
**2.88** −0.820 kips; 0.978 kips; −0.789 kips.
**2.89** 192.0 N; 288 N; −216 N.
**2.91** 515 N; $\theta_x$ = 70.2°; $\theta_y$ = 27.6°; $\theta_z$ = 71.5°.
**2.92** 515 N; $\theta_x$ = 79.8°; $\theta_y$ = 33.4°; $\theta_z$ = 58.6°.
**2.94** 913 lb; $\theta_x$ = 50.6°; $\theta_y$ = 117.6°; $\theta_z$ = 51.8°.
**2.95** 1171 N; $\theta_x$ = 89.5°, $\theta_y$ = 36.2°, $\theta_z$ = 126.2°.
**2.96** $T_{AB}$ = 490 N; $T_{AD}$ = 515 N.
**2.97** 130.0 lb.
**2.99** 13.98 kN.
**2.101** 926 N ↑.
**2.103** 2030 lb.
**2.104** 3380 lb.
**2.106** $T_{AB}$ = $T_{AC}$ = 3.66 lb; $T_{AD}$ = 5.18 lb.
**2.107** 960 N.
**2.108** 0 ≤ *Q* < 300 N.
**2.109** 845 N.
**2.110** 768 N.
**2.112** 2000 lb.
**2.113** $T_{AB}$ = 30.8 lb; $T_{AC}$ = 62.5 lb.
**2.115** $T_{AB}$ = 510 N; $T_{AC}$ = 56.2 N; $T_{AD}$ = 536 N.
**2.116** $T_{AB}$ = 1340 N; $T_{AC}$ = 1025 N; $T_{AD}$ = 915 N.
**2.117** $T_{AB}$ = 1431 N; $T_{AC}$ = 1560 N; $T_{AD}$ = 183.0 N.
**2.118** $T_{DA}$ = 8.50 kN; $T_{DB}$ = 19.50 kN; $T_{DC}$ = 14.00 kN.
**2.119** $T_{AB}$ = 974 lb; $T_{AC}$ = 531 lb; $T_{AD}$ = 533 lb.
**2.121** 378 N.
**2.123** *P* = 36.0 lb; *Q* = 54.0 lb.
**2.124** *W* = 180.0 lb; *P* = 24.0 lb.
**2.125** (*a*) 1155 N. (*b*) 1012 N.
**2.127** 104.4 N ↘ 86.7°.
**2.128** (80 N) 61.3 N, 51.4 N; (120 N) 41.0 N, 112.8 N; (150 N) −112.9 N, 86.0 N.
**2.130** (*a*) 172.7 lb. (*b*) 231 lb.
**2.131** (*a*) 312 N. (*b*) 144.0 N.
**2.133** (*a*) 56.4 lb; −103.9 lb; −20.5 lb. (*b*) 62.0°, 150.0°, 99.8°.
**2.135** 940 N; 65.7°, 28.2°, 76.4°.
**2.136** $T_{BAC}$ = 76.7 lb; $T_{AD}$ = 26.9 lb; $T_{AE}$ = 49.2 lb.
**2.137** (*a*) 125.0 lb. (*b*) 45.0 lb.

# CHAPTER 3

**3.1**  1.277 N·m ↺.
**3.2**  1.277 N·m ↺.
**3.4**  (a) 41.7 N·m ↺. (b) 147.4 N ∡ 45.0°.
**3.5**  (a) 41.7 N·m ↺. (b) 334 N. (c) 176.8 N ∡ 58.0°.
**3.6**  186.6 lb·in. ↺.
**3.7**  8.97 lb ⦨ 19.98°.
**3.9**  (a) 292 N·m ↺. (b) 292 N·m ↺.
**3.11**  116.2 lb·ft ↺.
**3.12**  128.2 lb·ft ↺.
**3.13**  140.0 N·m ↺.
**3.17**  (a) $-0.428\mathbf{i} + 0.1515\mathbf{j} - 0.891\mathbf{k}$. (b) $(-3\mathbf{i} - 3\mathbf{j} + \mathbf{k})/\sqrt{19}$.
**3.18**  1.886 m.
**3.20**  (a) $-58\mathbf{i} + 4\mathbf{j} + 32\mathbf{k}$. (b) $6\mathbf{i} - 4\mathbf{k}$. (c) $-30\mathbf{i} + 12\mathbf{j}$.
**3.22**  $(2400\ \text{lb·ft})\mathbf{j} + (1440\ \text{lb·ft})\mathbf{k}$.
**3.23**  $(7.50\ \text{N·m})\mathbf{i} - (6.00\ \text{N·m})\mathbf{j} - (10.39\ \text{N·m})\mathbf{k}$.
**3.25**  $(-25.4\ \text{lb·ft})\mathbf{i} - (12.60\ \text{lb·ft})\mathbf{j} - (12.60\ \text{lb·ft})\mathbf{k}$.
**3.26**  $(1200\ \text{N·m})\mathbf{i} - (1500\ \text{N·m})\mathbf{j} - (900\ \text{N·m})\mathbf{k}$.
**3.27**  7.37 ft.
**3.28**  70.8 mm.
**3.29**  269 mm.
**3.30**  5.17 ft.
**3.32**  2.36 m.
**3.33**  1.491 m.
**3.35**  $\mathbf{P} \cdot \mathbf{Q} = 0$; $\mathbf{P} \cdot \mathbf{S} = -11$; $\mathbf{Q} \cdot \mathbf{S} = 2$.
**3.37**  38.7°.
**3.39**  (a) 59.0°. (b) 144.0 lb.
**3.40**  (a) 70.5°. (b) 60.0 lb.
**3.41**  (a) 52.9°. (b) 326 N.
**3.43**  26.8°.
**3.44**  33.3°.
**3.45**  (a) 150.0. (b) 138.0.
**3.46**  4.
**3.47**  $M_x = 78.9$ kN·m, $M_y = 13.15$ kN·m, $M_z = -9.86$ kN·m.
**3.48**  3.04 kN.
**3.49**  $\phi = 24.6°$; $d = 34.6$ in.
**3.51**  $M_x = -31.2$ N·m, $M_y = 13.20$ N·m, $M_z = -2.42$ N·m.
**3.52**  $M_x = -25.6$ N·m, $M_y = 10.80$ N·m, $M_z = 40.6$ N·m.
**3.53**  283 lb.
**3.55**  3090 lb·in.
**3.57**  −90.0 N·m.
**3.58**  −111.0 N·m.
**3.59**  +2.28 N·m.
**3.60**  −9.50 N·m.
**3.61**  $aP/\sqrt{2}$.
**3.64**  9.50 in.
**3.65**  4.97 in.
**3.67**  0.249 m.
**3.68**  0.1198 m.
**3.70**  (a) 7.33 N·m ↺. (b) 91.6 mm.
**3.71**  16.39 N·m ↺.
**3.73**  1.125 in.
**3.74**  (a) 26.7 N. (b) 50.0 N. (c) 23.5 N.
**3.76**  $M = 604$ lb·in.; $\theta_x = 72.8°$, $\theta_y = 27.3°$, $\theta_z = 110.5°$.
**3.77**  $M = 1170$ lb·in.; $\theta_x = 81.2°$, $\theta_y = 13.70°$, $\theta_z = 100.4°$.
**3.78**  (a) $M = 13.63$ N·m; $\theta_x = 27.8°$, $\theta_y = 62.2°$, $\theta_z = 90.0°$.
(b) 18.17 N ⦨ 62.2° at B and 18.17 N ⦩ 62.2° at C.
**3.79**  $M = 8.78$ N·m; $\theta_x = 84.8°$, $\theta_y = 43.6°$; $\theta_z = 133.1°$.
**3.80**  $M = 2150$ lb·ft; $\theta_x = 113.0°$, $\theta_y = 92.7°$, $\theta_z = 23.2°$.
**3.82**  $\mathbf{F} = 16.00$ kips ↓; $\mathbf{M} = 192.0$ kip·in. ↺.
**3.83**  $\mathbf{F}_A = 389$ N ⦨ 60.0°; $\mathbf{F}_C = 651$ N ⦨ 60.0°.
**3.84**  (a) $\mathbf{F} = 30.0$ lb ↓; $\mathbf{M} = 150.0$ lb·in. ↺. (b) $\mathbf{B} = 50.0$ lb ←; $\mathbf{C} = 50.0$ lb →.
**3.86**  $\mathbf{F}_A = 168.0$ N ⦩ 50.0°; $\mathbf{F}_C = 192.0$ N ⦩ 50.0°.
**3.87**  $\mathbf{F} = 900$ N ↓; $x = 50.0$ mm.
**3.89**  (a) $\mathbf{F} = 48.0$ lb ∡ 65.0°; $\mathbf{M} = 490$ lb·in. ↺.
(b) $\mathbf{F} = 48.0$ lb ∡ 65.0° applied 17.78 in. to the left of B.
**3.90**  (a) 48.0 N intersecting line AB 144.0 mm to the right of A.
(b) 77.7° or −15.72°.
**3.91**  $(0.227\ \text{lb})\mathbf{i} + (0.1057\ \text{lb})\mathbf{k}$; 63.6 in. to the right of B.
**3.93**  $\mathbf{F} = -(250\ \text{kN})\mathbf{j}$; $\mathbf{M} = (15.00\ \text{kN·m})\mathbf{i} + (7.50\ \text{kN·m})\mathbf{k}$.
**3.95**  $\mathbf{F} = -(122.9\ \text{N})\mathbf{j} - (86.0\ \text{N})\mathbf{k}$; $\mathbf{M} = (22.6\ \text{N·m})\mathbf{i} + (15.49\ \text{N·m})\mathbf{j} - (22.1\ \text{N·m})\mathbf{k}$.
**3.96**  $\mathbf{F} = (5.00\ \text{N})\mathbf{i} + (150.0\ \text{N})\mathbf{j} - (90.0\ \text{N})\mathbf{k}$; $\mathbf{M} = (77.4\ \text{N·m})\mathbf{i} + (61.5\ \text{N·m})\mathbf{j} + (106.8\ \text{N·m})\mathbf{k}$.
**3.97**  $\mathbf{F} = (36.0\ \text{lb})\mathbf{i} - (28.0\ \text{lb})\mathbf{j} - (6.00\ \text{lb})\mathbf{k}$; $\mathbf{M} = -(157.0\ \text{lb·ft})\mathbf{i} + (22.5\ \text{lb·ft})\mathbf{j} - (240\ \text{lb·ft})\mathbf{k}$.
**3.98**  $\mathbf{F} = -(28.5\ \text{N})\mathbf{i} + (106.3\ \text{N})\mathbf{k}$; $\mathbf{M} = (12.35\ \text{N·m})\mathbf{i} - (19.16\ \text{N·m})\mathbf{j} - (5.13\ \text{N·m})\mathbf{k}$.
**3.99**  $\mathbf{F} = -(450\ \text{lb})\mathbf{i} + (180.0\ \text{lb})\mathbf{j} - (300\ \text{lb})\mathbf{k}$; $\mathbf{M} = (3600\ \text{lb·ft})\mathbf{j} + (2160\ \text{lb·ft})\mathbf{k}$.
**3.101**  (a) *Loading a:* $\mathbf{R} = 600$ N ↓; $\mathbf{M} = 1000$ N·m ↺.
*Loading b:* $\mathbf{R} = 600$ N ↓; $\mathbf{M} = 900$ N·m ↺.
*Loading c:* $\mathbf{R} = 600$ N ↓; $\mathbf{M} = 900$ N·m ↺.
*Loading d:* $\mathbf{R} = 400$ N ↑; $\mathbf{M} = 900$ N·m ↺.
*Loading e:* $\mathbf{R} = 600$ N ↓; $\mathbf{M} = 200$ N·m ↺.
*Loading f:* $\mathbf{R} = 600$ N ↓; $\mathbf{M} = 800$ N·m ↺.
*Loading g:* $\mathbf{R} = 1000$ N ↓; $\mathbf{M} = 1000$ N·m ↺.
*Loading h:* $\mathbf{R} = 600$ N ↓; $\mathbf{M} = 900$ N·m ↺.
(b) *Loadings c and h.*
**3.102**  Loading f.
**3.104**  Equivalent force-couple system at D.
**3.105**  (a) 2.00 ft to the right of C. (b) 2.31 ft to the right of C.
**3.106**  (a) 39.6 in. to the right of D. (b) 33.1 in.
**3.108**  44.7 lb ⦨ 26.6°; 10.61 in. to the left of C and 5.30 in. below C.
**3.110**  (a) 224 N ⦩ 63.4°. (b) 130.0 mm to the left of B and 260 mm below B.
**3.111**  329 kN ⦩ 61.7°; 6.82 m to the right of A.
**3.113**  (a) $\mathbf{R} = 758$ lb →; $\mathbf{M} = 15{,}190$ lb·ft ↺. (b) 758 lb →, 20.0 ft below DE.
**3.114**  (a) 29.9 lb ⦨ 23.0°. (b) AB: 10.30 in. to the left of B; BC: 4.36 in. below B.
**3.115**  (a) 60.2 lb·in. ↺. (b) 200 lb·in. ↺. (c) 20.0 lb·in. ↺.
**3.116**  (a) 0.365 m above G. (b) 0.227 m to the right of G.
**3.117**  (a) 0.299 m above G. (b) 0.259 m to the right of G.
**3.118**  (a) $R = F$ ⦩ $\tan^{-1}(a^2/2bx)$; $M = 2Fb^2(x - x^3/a^2)/\sqrt{a^4 + 4b^2x^2}$ ↺. (b) 0.369 m.
**3.119**  $\mathbf{R} = -(300\ \text{N})\mathbf{i} - (240\ \text{N})\mathbf{j} + (25.0\ \text{N})\mathbf{k}$; $\mathbf{M} = -(3.00\ \text{N·m})\mathbf{i} + (13.50\ \text{N·m})\mathbf{j} + (9.00\ \text{N·m})\mathbf{k}$.
**3.120**  $\mathbf{R} = (420\ \text{N})\mathbf{j} - (339\ \text{N})\mathbf{k}$; $\mathbf{M} = (1.125\ \text{N·m})\mathbf{i} + (163.9\ \text{N·m})\mathbf{j} - (109.9\ \text{N·m})\mathbf{k}$.
**3.122**  (a) 60.0° (b) $(20.0\ \text{lb})\mathbf{i} - (34.6\ \text{lb})\mathbf{j}$; $(520\ \text{lb·in.})\mathbf{i}$.
**3.124**  $\mathbf{R} = -(420\ \text{N})\mathbf{i} - (50.0\ \text{N})\mathbf{j} - (250\ \text{N})\mathbf{k}$; $\mathbf{M} = (30.8\ \text{N·m})\mathbf{j} - (22.0\ \text{N·m})\mathbf{k}$.
**3.125**  (a) $\mathbf{B} = -(75.0\ \text{N})\mathbf{k}$, $\mathbf{C} = -(25.0\ \text{N})\mathbf{i} + (37.5\ \text{N})\mathbf{k}$.
(b) $R_y = 0$, $R_z = -37.5$ N. (c) When the slot is vertical.
**3.126**  $\mathbf{A} = (1.600\ \text{lb})\mathbf{i} - (36.0\ \text{lb})\mathbf{j} + (2.00\ \text{lb})\mathbf{k}$, $\mathbf{B} = -(9.60\ \text{lb})\mathbf{i} + (36.0\ \text{lb})\mathbf{j} + (2.00\ \text{lb})\mathbf{k}$.
**3.127**  200 N at $y = 63.4$ mm, $z = 200$ mm.
**3.128**  72.2 N.
**3.129**  405 lb; 12.60 ft to the right of AB and 2.94 ft below BC.
**3.130**  $a = 0.722$ ft; $b = 20.6$ ft.

**3.133** (*a*) $P\sqrt{3}$; $\theta_x = \theta_y = \theta_z = 54.7°$. (*b*) $-a$. (*c*) Axis of the wrench is diagonal *OA*.

**3.134** (*a*) *P*; $\theta_x = 90.0°$, $\theta_y = 90.0°$, $\theta_z = 0°$. (*b*) $5a/2$. (*c*) Axis of the wrench is parallel to the *z* axis at $x = a$, $y = -a$.

**3.136** (*a*) $-(21.0\ \text{lb})\mathbf{j}$. (*b*) 0.571 in. (*c*) At $x = 0$, $z = 1.667$ in; and is parallel to the *y* axis.

**3.137** (*a*) $-(84.0\ \text{N})\mathbf{j} - (80.0\ \text{N})\mathbf{k}$. (*b*) 0.477 m. (*c*) $x = 0.526$ m, $y = 0$, $z = -0.1857$ m.

**3.140** (*a*) $3P(2\mathbf{i} - 20\mathbf{j} - \mathbf{k})/25$. (*b*) $-0.0988a$. (*c*) $x = 2.00a$, $y = 0$, $z = -1.990a$.

**3.141** $\mathbf{R} = (20.0\ \text{N})\mathbf{i} + (30.0\ \text{N})\mathbf{j} - (10.00\ \text{N})\mathbf{k}$; $y = -0.540$ m, $z = -0.420$ m.

**3.143** $\mathbf{F}_A = (M/b)\mathbf{i} + R[1 + (a/b)]\mathbf{k}$; $\mathbf{F}_B = -(M/b)\mathbf{i} - (aR/b)\mathbf{k}$.

**3.147** (*a*) 20.5 N·m $\circlearrowleft$. (*b*) 68.4 mm.

**3.148** 760 N·m $\circlearrowleft$.

**3.150** 43.6°.

**3.151** 23.0 N·m.

**3.153** $\mathbf{M} = 4.50$ N·m; $\theta_x = 90.0°$, $\theta_y = 177.1°$, $\theta_z = 87.1°$.

**3.154** $\mathbf{F} = 260$ lb $\not\!\searrow 67.4°$; $M_c = 200$ lb·in. $\circlearrowleft$.

**3.156** (*a*) 135.0 mm. (*b*) $\mathbf{F}_2 = (42.0\ \text{N})\mathbf{i} + (42.0\ \text{N})\mathbf{j} - (49.0\ \text{N})\mathbf{k}$; $\mathbf{M}_2 = -(25.9\ \text{N·m})\mathbf{i} + (21.2\ \text{N·m})\mathbf{j}$.

**3.158** (*a*) $\mathbf{B} = (2.50\ \text{lb})\mathbf{i}$, $\mathbf{C} = (0.1000\ \text{lb})\mathbf{i} - (2.47\ \text{lb})\mathbf{j} - (0.700\ \text{lb})\mathbf{k}$. (*b*) $\mathbf{R}_y = -2.47$ lb; $\mathbf{M}_x = 1.360$ lb·ft.

# CHAPTER 4

**4.1** 42.0 N ↑.
**4.2** 0.264 m.
**4.4** (*a*) 245 lb ↑. (*b*) 140.0 lb.
**4.5** (*a*) 34.0 kN ↑. (*b*) 4.96 kN ↑.
**4.6** (*a*) 81.1 kN. (*b*) 134.1 kN ↑.
**4.7** (*a*) 23.4 lb ↑. (*b*) 78.3 lb ↑.
**4.9** 1.250 kN ≤ Q ≤ 27.5 kN.
**4.12** 6.00 kips ≤ P ≤ 27.0 kips.
**4.13** 150.0 mm ≤ d ≤ 400 mm.
**4.14** 2.00 in. ≤ a ≤ 10.00 in.
**4.15** (*a*) 500 N ↓. (*b*) 1300 N $\not\!\measuredangle 22.6°$.
**4.17** (*a*) 125.0 lb. (*b*) 261 lb $\not\!\measuredangle 69.8°$.
**4.18** 230 lb.
**4.19** (*a*) 2.00 kN. (*b*) 2.32 kN $\not\!\measuredangle 46.4°$.
**4.22** (*a*) 400 N. (*b*) 458 N $\not\!\measuredangle 49.1°$.
**4.23** (*a*) $\mathbf{A} = 44.7$ lb $\measuredangle 26.6°$; $\mathbf{B} = 30.0$ lb ↑. (*b*) $\mathbf{A} = 30.2$ lb $\measuredangle 41.4°$; $\mathbf{B} = 34.6$ lb $\measuredangle 60.0°$.
**4.24** (*a*) $\mathbf{A} = 20.0$ lb ↑; $\mathbf{B} = 50.0$ lb $\measuredangle 36.9°$. (*b*) $\mathbf{A} = 23.1$ lb $\measuredangle 60.0°$; $\mathbf{B} = 59.6$ lb $\measuredangle 30.2°$.
**4.25** (*a*) 190.9 N. (*b*) 142.3 N $\measuredangle 18.43°$.
**4.26** (*a*) 324 N. (*b*) 270 N →.
**4.28** (*a*) $\mathbf{A} = 225$ N ↑; $\mathbf{C} = 641$ N $\not\!\searrow 20.6°$. (*b*) $\mathbf{A} = 365$ N $\measuredangle 60.0°$; $\mathbf{C} = 844$ N $\not\!\searrow 22.0°$.
**4.31** $T = 2P/3$; $\mathbf{C} = 0.577P$ →.
**4.32** $T = 0.586P$; $\mathbf{C} = 0.414P$ →.
**4.33** (*a*) 117.0 lb. (*b*) 129.8 lb $\not\!\searrow 56.3°$.
**4.34** (*a*) 195.0 lb. (*b*) 225 lb $\not\!\searrow 45.0°$.
**4.35** (*a*) 1432 N. (*b*) 1100 N ↑. (*c*) 1400 N ←.
**4.36** $T_{BE} = 196.2$ N; $\mathbf{A} = 73.6$ N →; $\mathbf{D} = 73.6$ N ←.
**4.39** $\mathbf{A} = \mathbf{D} = 0$; $\mathbf{B} = 964$ N ←; $\mathbf{C} = 140.2$ N →.
**4.40** $\mathbf{A} = \mathbf{C} = 0$; $\mathbf{B} = 469$ N ←; $\mathbf{D} = 50.2$ N ←.
**4.41** $\mathbf{B} = P/2 \measuredangle 45°$; $\mathbf{C} = 3P/2 \measuredangle 45°$; $\mathbf{D} = P/\sqrt{2}$ ↓.
**4.42** 26.6° ≤ α ≤ 153.4°.
**4.43** (*a*) $\mathbf{E} = 8.80$ kips ↑; $\mathbf{M}_E = 36.0$ kip·ft $\circlearrowleft$. (*b*) $\mathbf{E} = 4.80$ kips ↑; $\mathbf{M}_E = 51.0$ kip·ft $\circlearrowleft$.
**4.45** $T_{max} = 2240$ N; $T_{min} = 1522$ N.

**4.46** $\mathbf{C} = 1951$ N $\measuredangle 88.5°$; $\mathbf{M}_C = 75.0$ N·m $\circlearrowright$.
**4.47** 1.232 kN ≤ T ≤ 1.774 kN.
**4.48** (*a*) $\mathbf{D} = 20.0$ lb ↓; $\mathbf{M}_D = 20.0$ lb·ft $\circlearrowright$. (*b*) $\mathbf{D} = 10.00$ lb ↓; $\mathbf{M}_D = 30.0$ lb·ft $\circlearrowright$.
**4.50** (*a*) $\mathbf{A} = 5540$ N $\measuredangle 87.3°$; $\mathbf{C} = 683$ N $\not\!\searrow 67.4°$. (*b*) $\mathbf{A} = 4900$ N ↑; $\mathbf{M}_A = 1890$ N·m $\circlearrowright$. (*c*) $\mathbf{A} = 6740$ N $\measuredangle$; 87.3°; $\mathbf{M}_A = 3510$ N·m $\circlearrowright$; $\mathbf{C} = 1950$ N $\not\!\searrow 67.4°$.
**4.51** $\theta = \sin^{-1}(2M \cot \alpha/Wl)$.
**4.52** $\theta = \tan^{-1}(Q/3P)$.
**4.53** (*a*) $T = (W/2)/(1 - \tan \theta)$. (*b*) 39.8°.
**4.54** (*a*) $\theta = 2\cos^{-1}\left[\frac{1}{4}(\frac{W}{P} \pm \sqrt{\frac{W^2}{P^2} + 8})\right]$. (*b*) 65.1°.
**4.55** (*a*) $\theta = 2\sin^{-1}(W/2P)$. (*b*) 29.0°.
**4.57** 141.1°.
**4.59** (1) completely constrained; determinate; $\mathbf{A} = \mathbf{C} = 196.2$ N ↑. (2) completely constrained; determinate; $\mathbf{B} = 0$, $\mathbf{C} = \mathbf{D} = 196.2$ N ↑. (3) completely constrained; indeterminate; $\mathbf{A}_x = 294$ N →; $\mathbf{D}_x = 294$ N ←. (4) improperly constrained; indeterminate; no equilibrium. (5) partially constrained; determinate; equilibrium; $\mathbf{C} = \mathbf{D} = 196.2$ N ↑. (6) completely constrained; determinate; $\mathbf{B} = 294$ N →, $\mathbf{D} = 491$ N $\measuredangle 53.1°$. (7) partially constrained; no equilibrium. (8) completely constrained; indeterminate; $\mathbf{B} = 196.2$ N ↑, $\mathbf{D}_y = 196.2$ N ↑.
**4.61** $T = 289$ lb; $\mathbf{A} = 577$ lb $\measuredangle 60.0°$.
**4.62** $\mathbf{A} = 534$ N $\measuredangle 69.4°$; $\mathbf{E} = 187.5$ N ←.
**4.63** (*a*) 36.9°. (*b*) $\mathbf{A} = 400$ N ↑; $\mathbf{E} = 300$ N ←.
**4.65** $\mathbf{B} = 501$ N $\measuredangle 56.3°$; $\mathbf{C} = 324$ N $\not\!\searrow 31.0°$.
**4.66** $\mathbf{A} = 269$ lb $\measuredangle 21.8°$; $\mathbf{E} = 269$ lb $\measuredangle 21.8°$.
**4.67** $\mathbf{B} = 888$ N $\not\!\searrow 41.3°$; $\mathbf{D} = 943$ N $\measuredangle 45.0°$.
**4.69** (*a*) 499 N. (*b*) 457 N $\measuredangle 26.6°$.
**4.71** (*a*) 5.63 kips. (*b*) 4.52 kips $\not\!\searrow 4.76°$.
**4.72** $\mathbf{A} = 50.0$ lb $\measuredangle 30.0°$; $\mathbf{T} = 50.0$ lb.
**4.73** $\mathbf{A} = 244$ N →; $\mathbf{D} = 344$ N $\measuredangle 22.2°$.
**4.75** $\mathbf{A} = 170.0$ N $\measuredangle 33.9°$; $\mathbf{C} = 160.0$ N $\measuredangle 28.1°$.
**4.77** (*a*) 100.0 lb. (*b*) 125.0 lb $\not\!\searrow 36.9°$.
**4.78** (*a*) 400 N. (*b*) 458 N $\measuredangle 49.1°$.
**4.79** (*a*) $2P \measuredangle 60.0°$. (*b*) $1.239P \not\!\searrow 36.2°$.
**4.80** (*a*) $1.55\ P \measuredangle 30.0°$. (*b*) $1.086\ P \measuredangle 22.9°$.
**4.81** $\mathbf{A} = 163.1$ N $\not\!\searrow 74.1°$; $\mathbf{B} = 258$ N $\measuredangle 65.0°$.
**4.83** 60.0 mm.
**4.84** $\tan \theta = 2\tan \beta$.
**4.85** (*a*) 49.1°. (*b*) $\mathbf{A} = 45.3$ N ←; $\mathbf{B} = 90.6$ N $\measuredangle 60.0°$.
**4.86** $\alpha = 73.9°$; $T_A = 4160$ lb; $T_B = 2310$ lb.
**4.88** (*a*) 225 mm. (*b*) 23.1 N. (*c*) 12.21 N →.
**4.90** (*a*) 59.4°. (*b*) $\mathbf{A} = 8.45$ lb →; $\mathbf{B} = 13.09$ lb $\measuredangle 49.8°$.
**4.91** $\mathbf{A} = (120.0\ \text{N})\mathbf{j} + (133.3\ \text{N})\mathbf{k}$; $\mathbf{D} = (60.0\ \text{N})\mathbf{j} + (166.7\ \text{N})\mathbf{k}$.
**4.93** (*a*) 96.0 lb. (*b*) $\mathbf{A} = (2.40\ \text{lb})\mathbf{j}$; $\mathbf{B} = (214\ \text{lb})\mathbf{j}$.
**4.94** (*a*) 37.5 lb. (*b*) $\mathbf{B} = (33.8\ \text{lb})\mathbf{j} - (70.0\ \text{lb})\mathbf{k}$; $\mathbf{D} = (33.8\ \text{lb})\mathbf{j} + (28.0\ \text{lb})\mathbf{k}$.
**4.95** (*a*) 78.5 N. (*b*) $\mathbf{A} = -(27.5\ \text{N})\mathbf{i} + (58.9\ \text{N})\mathbf{j}$; $\mathbf{B} = (106.0\ \text{N})\mathbf{i} + (58.9\ \text{N})\mathbf{j}$.
**4.97** $T_A = 22.5$ lb; $T_B = 7.50$ lb; $T_C = 30.0$ lb.
**4.99** (*a*) 121.9 N. (*b*) $-46.2$ N. (*c*) 100.9 N.
**4.100** (*a*) 95.6 N. (*b*) $-7.36$ N. (*c*) 88.3 N.
**4.101** $T_A = 23.5$ N; $T_C = 11.77$ N; $T_D = 105.9$ N.
**4.102** (*a*) 0.480 in. (*b*) $T_A = 23.5$ N; $T_C = 0$; $T_D = 117.7$ N.
**4.103** (*a*) $T_A = 6.00$ lb; $T_B = T_C = 9.00$ lb. (*b*) 15.00 in.

**4.105** $T_{BD} = 1100$ lb; $T_{BE} = 1100$ lb; $\mathbf{A} = (1200 \text{ lb})\mathbf{i} - (560 \text{ lb})\mathbf{j}$.

**4.106** $T_{BD} = 780$ N; $T_{BE} = 390$ N; $\mathbf{A} = -(195.0 \text{ N})\mathbf{i} + (1170 \text{ N})\mathbf{j} + (130.0 \text{ N})\mathbf{k}$.

**4.107** $T_{BD} = 525$ N; $T_{BE} = 105.0$ N; $\mathbf{A} = -(105.0 \text{ N})\mathbf{i} + (840 \text{ N})\mathbf{j} + (140.0 \text{ N})\mathbf{k}$.

**4.108** (a) $T_{CD} = T_{CE} = 3.96$ kN. (b) $\mathbf{A} = (6.67 \text{ kN})\mathbf{i} + (1.667 \text{ kN})\mathbf{j}$.

**4.109** (a) $T_{CD} = 0.954$ kN; $T_{CE} = 5.90$ kN. (b) $\mathbf{A} = (5.77 \text{ kN})\mathbf{i} + (1.443 \text{ KN})\mathbf{j} - (0.833 \text{ kN})\mathbf{k}$.

**4.110** $T_{EBF} = 990$ lb; $T_{CD} = 765$ lb.

**4.113** $F_{CD} = 19.62$ N; $\mathbf{A} = -(19.22 \text{ N})\mathbf{i} + (45.1 \text{ N})\mathbf{j}$; $\mathbf{B} = (49.1 \text{ N})\mathbf{j}$.

**4.115** $\mathbf{A} = -(56.3 \text{ lb})\mathbf{i}$; $\mathbf{B} = -(56.2 \text{ lb})\mathbf{i} + (150.0 \text{ lb})\mathbf{j} - (75.0 \text{ lb})\mathbf{k}$; $F_{CE} = 202$ lb.

**4.116** (a) 116.6 lb. (b) $\mathbf{A} = -(72.7 \text{ lb})\mathbf{j} - (38.1 \text{ lb})\mathbf{k}$; $\mathbf{B} = (37.5 \text{ lb})\mathbf{j}$.

**4.117** (a) 345 N. (b) $\mathbf{A} = (114.4 \text{ N})\mathbf{i} + (377 \text{ N})\mathbf{j} + (141.5 \text{ N})\mathbf{k}$; $\mathbf{B} = (113.2 \text{ N})\mathbf{j} + (185.5 \text{ N})\mathbf{k}$.

**4.119** $F_{CD} = 19.62$ N; $\mathbf{B} = -(19.22 \text{ N})\mathbf{i} + (94.2 \text{ N})\mathbf{j}$; $\mathbf{M}_B = -(40.6 \text{ N·m})\mathbf{i} - (17.30 \text{ N·m})\mathbf{j}$.

**4.120** $\mathbf{A} = -(112.5 \text{ lb})\mathbf{i} + (150.0 \text{ lb})\mathbf{j} - (75.0 \text{ lb})\mathbf{k}$; $\mathbf{M}_A = (600 \text{ lb·ft})\mathbf{i} + (225 \text{ lb·ft})\mathbf{j}$; $F_{CE} = 202$ lb.

**4.121** (a) 5.00 lb. (b) $\mathbf{C} = -(5.00 \text{ lb})\mathbf{i} + (6.00 \text{ lb})\mathbf{j} - (5.00 \text{ lb})\mathbf{k}$; $\mathbf{M}_C = (8.00 \text{ lb·in.})\mathbf{j} - (12.00 \text{ lb·in})\mathbf{k}$.

**4.122** $T_{CF} = 200$ N; $T_{DE} = 450$ N; $\mathbf{A} = (160.0 \text{ N})\mathbf{i} + (270 \text{ N})\mathbf{k}$; $\mathbf{M}_A = -(16.20 \text{ N·m})\mathbf{i}$.

**4.123** $T_{BD} = 7.80$ kN; $T_{BE} = 6.50$ kN; $T_{CF} = 6.50$ kN; $\mathbf{A} = (19.20 \text{ kN})\mathbf{i} - (3.00 \text{ kN})\mathbf{k}$.

**4.124** $T_{BE} = 6.50$ kN; $T_{CE} = 7.16$ kN; $T_{CF} = 0$; $\mathbf{A} = (12.00 \text{ kN})\mathbf{i} - (3.00 \text{ kN})\mathbf{k}$.

**4.127** $\mathbf{A} = (120.0 \text{ lb})\mathbf{j} - (150.0 \text{ lb})\mathbf{k}$; $\mathbf{B} = (180.0 \text{ lb})\mathbf{i} + (150.0 \text{ lb})\mathbf{k}$; $\mathbf{C} = -(180.0 \text{ lb})\mathbf{i} + (120.0 \text{ lb})\mathbf{j}$.

**4.128** $\mathbf{A} = (20.0 \text{ lb})\mathbf{j} + (25.0 \text{ lb})\mathbf{k}$; $\mathbf{B} = (30.0 \text{ lb})\mathbf{i} - (25.0 \text{ lb})\mathbf{k}$; $\mathbf{C} = -(30.0 \text{ lb})\mathbf{i} - (20.0 \text{ lb})\mathbf{j}$.

**4.129** $T_{BE} = 975$ N; $T_{CF} = 600$ N; $T_{DG} = 625$ N; $\mathbf{A} = (2100 \text{ N})\mathbf{i} + (175.0 \text{ N})\mathbf{j} - (375 \text{ N})\mathbf{k}$.

**4.131** $T_B = -0.366P$; $T_C = 1.219P$; $T_D = -0.853P$; $F = -0.345P\mathbf{i} + P\mathbf{j} - 0.862P\mathbf{k}$.

**4.133** 360 N.

**4.135** 420 lb.

**4.136** 180.0 lb.

**4.137** $(45.0 \text{ lb})\mathbf{j}$.

**4.138** 343 N.

**4.140** (a) $x = 4.00$ ft, $y = 8.00$ ft. (b) 10.73 lb.

**4.141** (a) $x = 0$, $y = 16.00$ ft. (b) 11.31 lb.

**4.142** (a) 1761 lb ↑. (b) 689 lb ↑.

**4.143** (a) 150.0 lb. (b) 225 lb ⦨ 32.3°.

**4.145** (a) 130.0 N. (b) 224 ⦨ 2.05°.

**4.146** $T = 80.0$ N; $\mathbf{A} = 160.0$ N ⦨ 30.0°; $\mathbf{C} = 160.0$ N ⦨ 30.0°.

**4.148** $\mathbf{A} = 680$ N ⦨ 28.1°; $\mathbf{B} = 600$ N ←.

**4.149** $\mathbf{A} = 63.6$ lb ⦨ 45.0°; $\mathbf{C} = 87.5$ lb ⦨ 59.0°.

**4.151** $T_A = 5.63$ lb; $T_B = 16.88$ lb; $T_C = 22.5$ lb.

**4.153** (a) $\mathbf{A} = 0.745P$ ⦨ 63.4°; $\mathbf{C} = 0.471P$ ⦨ 45.0°. (b) $\mathbf{A} = 0.812P$ ⦨ 60.0°; $\mathbf{C} = 0.503P$ ⦨ 36.2°. (c) $\mathbf{A} = 0.448P$ ⦨ 60.0°; $\mathbf{C} = 0.652P$ ⦨ 69.9°. (d) Improperly constrained; no equilibrium.

# CHAPTER 5

**5.1** $\overline{X} = 175.6$ mm, $\overline{Y} = 94.4$ mm.

**5.2** $\overline{X} = 19.28$ in., $\overline{Y} = 6.94$ in.

**5.3** $\overline{X} = 55.4$ mm, $\overline{Y} = 93.8$ mm.

**5.4** $\overline{X} = 52.0$ mm, $\overline{Y} = 65.0$ mm.

**5.5** $\overline{X} = 3.18$ in., $\overline{Y} = 6.00$ in.

**5.6** $\overline{X} = -10.00$ mm, $\overline{Y} = 87.5$ mm.

**5.9** $\overline{X} = \overline{Y} = 16.75$ mm.

**5.10** $\overline{X} = 10.11$ in., $\overline{Y} = 3.88$ in.

**5.11** $\overline{X} = 120.0$ mm, $\overline{Y} = 60.0$ mm.

**5.13** $\overline{X} = \overline{Y} = 9.00$ in.

**5.14** $\overline{X} = 11.91$ mm, $\overline{Y} = 28.8$ mm.

**5.16** $\overline{Y} = \left(\dfrac{2}{3}\right)\left(\dfrac{r_2^3 - r_1^3}{r_2^2 - r_1^2}\right)\left(\dfrac{2 \cos \alpha}{\pi - 2\alpha}\right)$.

**5.17** $\overline{Y} = (r_1 + r_2)(\cos \alpha)/(\pi - 2\alpha)$.

**5.19** 0.520.

**5.20** 648 N.

**5.21** $(Q_x)_1 = 25.0$ in$^3$; $(Q_x)_2 = -25.0$ in$^3$.

**5.23** (a) $b(c^2 - y^2)/2$. (b) $y = 0$; $Q_x = bc^2/2$.

**5.24** $\overline{X} = 172.5$ mm, $\overline{Y} = 97.5$ mm.

**5.26** $\overline{X} = 3.19$ in., $\overline{Y} = 6.00$ in.

**5.29** (a) 125.3 N. (b) 137.0 N ⦨ 56.7°.

**5.30** 120.0 mm.

**5.31** 99.5 mm.

**5.32** (a) $0.513a$. (b) $0.691a$.

**5.34** $\bar{x} = 2a/3$, $\bar{y} = h/3$.

**5.35** $\bar{x} = 2a/5$, $\bar{y} = b/2$.

**5.37** $\bar{x} = a(3 - 4 \sin \alpha)/6(1 - \alpha)$, $\bar{y} = 0$.

**5.39** $\bar{x} = 2a/3(4 - \pi)$, $\bar{y} = 2b/3(4 - \pi)$.

**5.40** $\bar{x} = a/4$, $\bar{y} = 3b/10$.

**5.41** $\bar{x} = 3a/5$, $\bar{y} = 12b/35$.

**5.43** $\bar{x} = 17a/130$, $\bar{y} = 11b/26$.

**5.44** $\bar{x} = a$, $\bar{y} = 17b/35$.

**5.45** $2a/5$.

**5.46** $-2\sqrt{2}r/3\pi$.

**5.48** $\bar{x} = -9.27a$, $\bar{y} = 3.09a$.

**5.49** $\bar{x} = 0.236L$, $\bar{y} = 0.454a$.

**5.51** $\bar{x} = \bar{y} = 1.027$ in.

**5.52** (a) $V = 6.19 \times 10^6$ mm$^3$; $A = 458 \times 10^3$ mm$^2$. (b) $V = 16.88 \times 10^6$ mm$^3$; $A = 1.171 \times 10^6$ mm$^2$.

**5.53** (a) $V = 169.0 \times 10^3$ in$^3$; $A = 28.4 \times 10^3$ in$^2$. (b) $V = 88.9 \times 10^3$ in$^3$; $A = 15.48 \times 10^3$ in$^2$.

**5.54** (a) $V = 2.26 \times 10^6$ mm$^3$; $A = 116.3 \times 10^3$ mm$^2$. (b) $V = 1.471 \times 10^6$ mm$^3$; $A = 116.3 \times 10^3$ mm$^2$.

**5.55** $V = \pi^2 Rr^2$; $A = 2\pi^2 Rr$.

**5.58** 308 in$^2$.

**5.60** 31.9 liters.

**5.62** $V = 3.96$ in$^3$, $W = 1.211$ lb.

**5.63** 14.52 in$^2$.

**5.64** $300 \times 10^3$ mm$^3$.

**5.66** (a) $\mathbf{R} = 6000$ N ↓, $\bar{x} = 3.60$ m. (b) $A = 6000$ N ↑, $M_A = 21.6$ kN·m ↺.

**5.67** (a) $\mathbf{R} = 32.0$ kN ↓, 3.00 m to the right of $A$. (b) $\mathbf{A} = 20.0$ kN ↑, $\mathbf{B} = 12.00$ kN ↑.

**5.69** $\mathbf{A} = 10,800$ lb ↑; $\mathbf{B} = 3600$ lb ↑.

**5.70** $\mathbf{A} = 90.0$ lb ↑; $\mathbf{B} = 240$ lb ↓.

**5.71** $\mathbf{B} = 1200$ N ↑; $\mathbf{M}_B = 800$ N·m ↺.

**5.73** $\mathbf{A} = 3.00$ kN ↑; $\mathbf{M}_A = 12.60$ kN·m. ↺.

**5.74** (a) 0.536 m. (b) $\mathbf{A} = \mathbf{B} = 761$ N ↑.

**5.76** $\mathbf{B} = 150.0$ lb ↑; $\mathbf{C} = 5250$ lb ↑.

**5.77** (a) 100.0 lb/ft. (b) 4950 lb ↑.

**5.78** $w_A = 10.00$ kN/m; $w_B = 50$ kN/m.

**5.80** (a) $\mathbf{H} = 10.11$ kips →, $\mathbf{V} = 37.8$ kips ↑. (b) 10.48 ft to the right of $A$. (c) $\mathbf{R} = 10.66$ kips ⦨ 18.43°.

**5.81** (a) $\mathbf{H} = 44.1$ kN →, $\mathbf{V} = 228$ kN ↑. (b) 1.159 m to the right of $A$. (c) $\mathbf{R} = 59.1$ kN ⦨ 41.6°.

**5.82** 6.98%.

**5.84** $\mathbf{T} = 3.70$ kips ↑.

**5.85** 5.88 ft.

**5.87** $T = 6.72$ kN ←; $A = 141.2$ kN ←.

**5.88** $A = 1197$ N ⦨ 53.1°; $B = 1511$ N ⦨ 53.1°.

**5.89** 3570 N.

**5.90** 6.00 ft.

**5.92** 0.683 m.

**5.93** 0.0711 m.

**5.94** 208 lb.

**5.96** (a) 0.548L. (b) $2\sqrt{3}$.

**5.97** $21h/16$ above the vertex of the cone.

**5.98** $\bar{x} = 0$, $\bar{y} = -0.608h$, $\bar{z} = 0$.

**5.99** 27.8 mm above base of cone.

**5.100** $\overline{X} = 46.8$ mm.

**5.102** 0.610 in.

**5.103** −2.34 in.

**5.104** −19.02 mm.

**5.106** $\overline{X} = 125.0$ mm, $\overline{Y} = 167.0$ mm, $\overline{Z} = 33.5$ mm.

**5.107** $\overline{X} = 0.295$ m, $\overline{Y} = 0.423$ m, $\overline{Z} = 1.703$ m.

**5.109** $\overline{X} = \overline{Z} = 4.21$ in., $\overline{Y} = 7.03$ in.

**5.110** $\overline{X} = 180.2$ mm, $\overline{Y} = 38.0$ mm, $\overline{Z} = 193.5$ mm.

**5.111** $\overline{X} = 17.00$ in., $\overline{Y} = 15.68$ in., $\overline{Z} = 14.16$ in.

**5.113** $\overline{X} = 93.9$ mm, $\overline{Y} = 3.40$ mm, $\overline{Z} = 0$.

**5.114** $\overline{X} = 0.909$ m, $\overline{Y} = 0.1842$ m, $\overline{Z} = 0.884$ m.

**5.116** $\overline{X} = 0.410$ m, $\overline{Y} = 0.510$ m, $\overline{Z} = 0.1500$ m.

**5.117** $\overline{X} = 0$, $\overline{Y} = 10.05$ in., $\overline{Z} = 5.15$ in.

**5.118** $\overline{X} = 61.1$ mm from the end of the handle.

**5.119** $\overline{Y} = 0.526$ in. above the base.

**5.121** $\overline{Y} = 421$ mm above the floor.

**5.122** $(\bar{x}_1) = 21a/88$; $(\bar{x}_2) = 27a/40$.

**5.123** $(\bar{x}_1) = 21h/88$; $(\bar{x}_2) = 27h/40$.

**5.124** $(\bar{x}_1) = 2h/9$; $(\bar{x}_2) = 2h/3$.

**5.125** $\bar{x} = 2.34$ m; $\bar{y} = \bar{z} = 0$.

**5.128** $\bar{x} = 1.297a$; $\bar{y} = \bar{z} = 0$.

**5.129** $\bar{x} = \bar{z} = 0$; $\bar{y} = 0.374b$.

**5.132** (a) $\bar{x} = \bar{z} = 0$, $\bar{y} = -121.9$ mm. (b) $\bar{x} = \bar{z} = 0$, $\bar{y} = -90.2$ mm.

**5.134** $\bar{x} = 0$, $\bar{y} = 5h/16$, $\bar{z} = -b/4$.

**5.135** $\bar{x} = a/2$, $\bar{y} = 8h/25$, $\bar{z} = b/2$.

**5.136** $V = 688$ ft³; $\bar{x} = 15.91$ ft.

**5.137** $\overline{X} = 5.67$ in., $\overline{Y} = 5.17$ in.

**5.138** $\overline{X} = 92.0$ mm, $\overline{Y} = 23.3$ mm.

**5.139** (a) 5.09 lb. (b) 9.48 lb ⦨ 57.5°.

**5.141** $\bar{x} = 2L/5$, $\bar{y} = 12h/25$.

**5.143** $A = 2860$ lb ↑; $B = 740$ lb ↑.

**5.144** $w_{BC} = 2810$ N/m; $w_{DE} = 3150$ N/m.

**5.146** $-(2h^2 - 3b^2)/2(4h - 3b)$.

**5.148** $\overline{X} = \overline{Z} = 0$, $\overline{Y} = 83.3$ mm above the base.

## CHAPTER 6

**6.1** $F_{AB} = 1600$ lb C; $F_{AC} = 2000$ lb T; $F_{BC} = 1709$ lb T.

**6.2** $F_{AB} = 3900$ N T; $F_{AC} = 4500$ N C; $F_{BC} = 3600$ N C.

**6.3** $F_{AB} = 375$ lb C; $F_{AC} = 780$ lb C; $F_{BC} = 300$ lb T.

**6.4** $F_{AD} = 125.0$ kN T; $F_{CD} = 120.0$ kN C; $F_{AB} = 175.0$ kN T; $F_{AC} = 84.0$ kN C; $F_{BC} = 120.0$ kN C.

**6.6** $F_{AC} = 80.0$ kN T; $F_{CE} = 45.0$ kN T; $F_{DE} = 51.0$ kN C; $F_{BD} = 51.0$ kN C; $F_{CD} = 48.0$ kN T; $F_{BC} = 19.00$ kN C.

**6.8** $F_{AB} = 4.80$ kN T; $F_{AC} = 5.20$ kN T; $F_{AD} = 4.00$ kN C; $F_{BD} = 5.20$ kN C; $F_{CD} = 4.80$ kN C.

**6.9** $F_{AB} = 15.00$ kN T; $F_{AD} = 17.00$ kN C; $F_{BC} = 15.00$ kN T; $F_{CE} = 8.00$ kN T; $F_{EF} = 8.00$ kN T; $F_{DF} = 17.00$ kN C; $F_{BE} = F_{BD} = F_{DE} = 0$.

**6.11** $F_{AB} = F_{FH} = 1500$ lb C; $F_{AC} = F_{CE} = F_{EG} = F_{GH} = 1200$ lb T; $F_{BC} = F_{FG} = 0$; $F_{BD} = F_{DF} = 1200$ lb C; $F_{BE} = F_{EF} = 60.0$ lb C; $F_{DE} = 72.0$ lb T.

**6.12** $F_{AB} = F_{FH} = 1500$ lb C; $F_{AC} = F_{CE} = F_{EG} = F_{GH} = 1200$ lb T; $F_{BC} = F_{FG} = 0$; $F_{BD} = F_{DF} = 1000$ lb C; $F_{BE} = F_{EF} = 500$ lb C; $F_{DE} = 600$ lb T.

**6.13** $F_{AB} = 6.24$ kN C; $F_{AC} = 2.76$ kN T; $F_{BC} = 2.50$ kN C; $F_{BD} = 4.16$ kN C; $F_{CD} = 1.867$ kN T; $F_{CE} = 2.88$ kN T; $F_{DE} = 3.75$ kN C; $F_{DF} = 0$; $F_{EF} = 1.200$ kN C.

**6.15** $F_{AB} = F_{FG} = 7.50$ kips C; $F_{AC} = F_{EG} = 4.50$ kips T; $F_{BC} = F_{EF} = 7.50$ kips T; $F_{BD} = F_{DF} = 9.00$ kips C; $F_{CD} = F_{DE} = 0$; $F_{CE} = 9.00$ kips T.

**6.17** $F_{AB} = 47.2$ kN C; $F_{AC} = 44.6$ kN T; $F_{BC} = 10.50$ kN C; $F_{BD} = 47.2$ kN C; $F_{CD} = 17.50$ kN T; $F_{CE} = 30.6$ kN T; $F_{DE} = 0$.

**6.18** $F_{AB} = 2250$ N C; $F_{AC} = 1200$ N T; $F_{BC} = 750$ N T; $F_{BD} = 1700$ N C; $F_{BE} = 400$ N C; $F_{CE} = 850$ N C; $F_{CF} = 1600$ N T; $F_{DE} = 1500$ N T; $F_{EF} = 2250$ N T.

**6.19** $F_{AB} = F_{FH} = 7.50$ kips C; $F_{AC} = F_{GH} = 4.50$ kips T; $F_{BC} = F_{FG} = 4.00$ kips T; $F_{BD} = F_{DF} = 6.00$ kips C; $F_{BE} = F_{EF} = 2.50$ kips T; $F_{CE} = F_{EG} = 4.50$ kips T; $F_{DE} = 0$.

**6.21** $F_{AB} = 9.90$ kN C; $F_{AC} = 7.83$ kN T; $F_{BC} = 0$; $F_{BD} = 7.07$ kN C; $F_{BE} = 2.00$ kN C; $F_{CE} = 7.83$ kN T; $F_{DE} = 1.000$ kN T; $F_{DF} = 5.03$ kN C; $F_{DG} = 0.559$ kN T; $F_{EG} = 5.59$ kN T.

**6.22** $F_{AB} = 3610$ lb C; $F_{AC} = 4110$ lb T; $F_{BC} = 768$ lb C; $F_{BD} = 3840$ lb C; $F_{CD} = 1371$ lb T; $F_{CE} = 2740$ lb T; $F_{DE} = 1536$ lb C.

**6.23** $F_{DF} = 4060$ lb C; $F_{DG} = 1371$ lb T; $F_{EG} = 2740$ lb T; $F_{FG} = 768$ lb T; $F_{FH} = 4290$ lb C; $F_{GH} = 4110$ lb T.

**6.24** $F_{AB} = F_{DF} = 2.29$ kN T; $F_{AC} = F_{EF} = 2.29$ kN C; $F_{BC} = F_{DE} = 0.600$ kN C; $F_{BD} = 2.21$ kN T; $F_{BE} = F_{EH} = 0$; $F_{CE} = 2.21$ kN C; $F_{CH} = F_{EJ} = 1.200$ kN C.

**6.27** $F_{AB} = F_{BC} = F_{CD} = 36.0$ kips T; $F_{AE} = 57.6$ kips T; $F_{AF} = 45.0$ kips C; $F_{BF} = F_{BG} = F_{CG} = F_{CH} = 0$; $F_{DH} = F_{FG} = F_{GH} = 39.0$ kips C; $F_{EF} = 36.0$ kips C.

**6.28** $F_{AB} = 128.0$ kN T; $F_{AC} = 136.7$ kN C; $F_{BD} = F_{DF} = F_{FH} = 128.0$ kN T; $F_{CE} = F_{EG} = 136.7$ kN C; $F_{GH} = 192.7$ kN C; $F_{BC} = F_{BE} = F_{DE} = F_{DG} = F_{FG} = 0$.

**6.29** Truss of Prob. 6.33a is the only simple truss.

**6.30** All three trusses are simple trusses.

**6.32** (a) AI, BJ, CK, DI, EI, FK, GK. (b) FK, IO.

**6.34** (a) BC, HI, IJ, JK. (b) BF, BG, CG, CH.

**6.35** $F_{AB} = F_{AD} = 244$ lb C; $F_{AC} = 1040$ lb T; $F_{BC} = F_{CD} = 500$ lb C; $F_{BD} = 280$ lb T.

**6.36** $F_{AB} = F_{AD} = 861$ N C; $F_{AC} = 676$ N C; $F_{BC} = F_{CD} = 162.5$ N T; $F_{BD} = 244$ N T.

**6.37** $F_{AB} = F_{AD} = 2810$ N T; $F_{AC} = 5510$ N C; $F_{BC} = F_{CD} = 1325$ N T; $F_{BD} = 1908$ N C.

**6.38** $F_{AB} = F_{AC} = 1061$ lb C; $F_{AD} = 2500$ lb T; $F_{BC} = 2100$ lb T; $F_{BD} = F_{CD} = 1250$ lb C; $F_{BE} = F_{CE} = 1250$ lb C; $F_{DE} = 1500$ lb T.

**6.39** $F_{AB} = 840$ N C; $F_{AC} = 110.6$ N C; $F_{AD} = 394$ N C; $F_{AE} = 0$; $F_{BC} = 160.0$ N T; $F_{BE} = 200$ N T; $F_{CD} = 225$ N T; $F_{CE} = 233$ N C; $F_{DE} = 120.0$ N T.

**6.40** $F_{AB} = F_{AE} = F_{BC} = 0$; $F_{AC} = 995$ N T; $F_{AD} = 1181$ N C; $F_{BE} = 600$ N T; $F_{CD} = 375$ N T; $F_{CE} = 700$ N C; $F_{DE} = 360$ N T.

**6.43** $F_{BD} = 216$ kN T; $F_{DE} = 270$ kN T.

**6.44** $F_{DG} = 459$ kN T; $F_{EG} = 216$ kN C.

**6.45** $F_{BD} = 36.0$ kips C; $F_{CD} = 45.0$ kips C.

**6.46** $F_{DF} = 60.0$ kips C; $F_{DG} = 15.00$ kips C.

**6.49** $F_{CD} = 20.0$ kN $C$; $F_{DF} = 52.0$ kN $C$.

**6.50** $F_{CE} = 36.0$ kN $T$; $F_{EF} = 15.00$ kN $C$.

**6.51** $F_{DE} = 25.0$ kips $T$; $F_{DF} = 13.00$ kips $C$.

**6.52** $F_{EG} = 16.00$ kips $T$; $F_{EF} = 6.40$ kips $C$.

**6.53** $F_{DF} = 91.4$ kN $T$; $F_{DE} = 38.6$ kN $C$.

**6.54** $F_{CD} = 64.2$ kN $T$; $F_{CE} = 92.1$ kN $C$.

**6.55** $F_{CE} = 8.00$ kN $T$; $F_{DE} = 4.50$ kN $C$; $F_{DF} = 10.00$ kN $C$.

**6.56** $F_{FH} = 10.00$ kN $C$; $F_{FI} = 4.92$ kN $T$; $F_{GI} = 6.00$ kN $T$.

**6.59** $F_{AD} = 3.38$ kips $C$; $F_{CD} = 0$; $F_{CE} = 14.03$ kips $T$.

**6.60** $F_{DG} = 18.75$ kips $C$; $F_{FG} = 14.03$ kips $T$; $F_{FH} = 17.43$ kips $T$.

**6.61** 22.5 kN $C$.

**6.62** $F_{DG} = 75.0$ kN $C$; $F_{FH} = 75.0$ kN $T$.

**6.65** (a) $CJ$. (b) 1.026 kN $T$.

**6.66** (a) $IO$. (b) 2.05 kN $T$.

**6.67** $F_{BE} = 10.00$ kips $T$; $F_{DE} = 0$; $F_{EF} = 5.00$ kips $T$.

**6.68** $F_{BE} = 2.50$ kips $T$; $F_{DE} = 1.500$ kips $C$; $F_{DG} = 2.50$ kips $T$.

**6.69** (a) Improperly constrained. (b) Completely constrained, determinate. (c) Completely constrained, indeterminate.

**6.70** (a) Completely constrained, determinate. (b) Partially constrained. (c) Improperly constrained.

**6.71** (a) Completely constrained, determinate. (b) Completely constrained, indeterminate. (c) Improperly constrained.

**6.72** (a) Partially constrained. (b) Completely constrained, determinate. (c) Completely constrained, indeterminate.

**6.75** $F_{BD} = 1750$ N $C$; $C_x = 1400$ N ←; $C_y = 700$ N ↓.

**6.76** $F_{BD} = 300$ lb $T$; $C_x = 150.0$ lb ←; $C_y = 180.0$ lb ↑.

**6.77** (a) 125.0 N ↖ 36.9°. (b) 125.0 N ↘ 36.9°.

**6.79** (a) 200 N ↙ 60.0°. (b) 574 N ↖ 89.0°.

**6.80** (a) 828 N $T$. (b) 1197 N ↗ 86.2°.

**6.81** $A = 150.0$ lb →; $B_x = 150.0$ lb ←, $B_y = 60.0$ lb ↑; $C = 20.0$ lb ↑; $D = 80.0$ lb ↓.

**6.83** (a) $A_x = 2700$ N →, $A_y = 200$ N ↑; $E_x = 2700$ N ←, $E_y = 600$ N ↑. (b) $A_x = 300$ N →, $A_y = 200$ N ↑; $E_x = 300$ N ←, $E_y = 600$ N ↑.

**6.85** (a) $A_x = 300$ N ←, $A_y = 660$ N ↑; $E_x = 300$ N →, $E_y = 90.0$ N ↑. (b) $A_x = 300$ N ←, $A_y = 150.0$ N ↑; $E_x = 300$ N →, $E_y = 600$ N ↑.

**6.87** (a) $A = 48.0$ lb ↓; $B = 108.0$ lb ↑. (b) $A_x = 80.0$ lb →; $A_y = 48.0$ lb ↓; $B_x = 80.0$ lb ←; $B_y = 108.0$ lb ↑.

**6.88** (a) and (c) $B_x = 32.0$ lb →, $B_y = 10.00$ lb ↑; $F_x = 32.0$ lb ←, $F_y = 38.0$ lb ↑. (b) $B_x = 32.0$ lb ←, $B_y = 34.0$ lb ↑; $F_x = 32.0$ lb →, $F_y = 14.00$ lb ↑.

**6.89** (a) and (c) $B_x = 24.0$ lb ←, $B_y = 7.50$ lb ↓; $F_x = 24.0$ lb →, $F_y = 7.50$ lb ↑. (b) $B_x = 24.0$ lb ←, $B_y = 10.50$ lb ↑; $F_x = 24.0$ lb →, $F_y = 10.50$ lb ↓.

**6.91** $D_x = 13.60$ kN →, $D_y = 7.50$ kN ↑; $E_x = 13.60$ kN ←, $E_y = 2.70$ kN ↓.

**6.92** $A_x = 45.0$ N ←, $A_y = 30.0$ N ↓; $B_x = 45.0$ N →, $B_y = 270$ N ↑.

**6.93** $A_x = 176.3$ lb ←; $A_y = 60.0$ lb ↓; $G_x = 56.3$ lb →; $G_y = 510$ lb ↑.

**6.94** $A_x = 56.3$ lb ←; $A_y = 157.5$ lb ↓; $G_x = 56.3$ lb →; $G_y = 383$ lb ↑.

**6.95** (a) $A = 982$ lb ↑; $B = 935$ lb ↑; $C = 733$ lb ↑. (b) $\Delta B = +291$ lb; $\Delta C = -72.7$ lb.

**6.96** (a) 572 lb. (b) $A = 1070$ lb ↑; $B = 709$ lb ↑; $C = 870$ lb ↑.

**6.99** $B = 152.0$ lb ↓; $C_x = 60.0$ lb ←, $C_y = 200$ lb ↑; $D_x = 60.0$ lb →, $D_y = 42.0$ lb ↑.

**6.100** $B = 108.0$ lb ↓; $C_x = 90.0$ lb ←, $C_y = 150.0$ lb ↑; $D_x = 90.0$ lb →, $D_y = 18.00$ lb ↑.

**6.101** $A_x = 13.00$ kN ←, $A_y = 4.00$ kN ↓; $B_x = 36.0$ kN →, $B_y = 6.00$ kN ↑; $E_x = 23.0$ kN ←, $E_y = 2.00$ kN ↓.

**6.102** $A_x = 2025$ N ←, $A_y = 1800$ kN ↓; $B_x = 4050$ N →, $B_y = 1200$ N ↑; $E_x = 2025$ N ←, $E_y = 600$ N ↑.

**6.103** (a) $C_x = 100.0$ lb ←; $C_y = 100.0$ lb ↑; $D_x = 100.0$ lb →; $D_y = 20.0$ lb ↓. (b) $E_x = 100.0$ lb ←; $E_y = 180.0$ lb ↑.

**6.104** (a) $C_x = 100.0$ lb ←; $C_y = 60.0$ lb ↑; $D_x = 100.0$ lb →; $D_y = 20.0$ lb ↑. (b) $E_x = 100.0$ lb ←; $E_y = 140.0$ lb ↑.

**6.107** (a) $A_x = 200$ kN →, $A_y = 122.0$ kN ↑. (b) $B_x = 200$ kN ←, $B_y = 10.00$ kN ↓.

**6.108** (a) $A_x = 205$ kN →, $A_y = 134.5$ kN ↑. (b) $B_x = 205$ kN ←, $B_y = 5.50$ kN ↑.

**6.109** $B = 98.5$ lb ↗ 24.0°; $C = 90.6$ lb ↖ 6.34°.

**6.110** $B = 25.0$ lb ↑; $C = 79.1$ lb ↖ 18.43°.

**6.112** $F_{AF} = P/4$ $C$; $F_{BG} = F_{DG} = P/\sqrt{2}$ $C$; $F_{EH} = P/4$ $T$.

**6.113** $F_{AG} = \sqrt{2}P/6$ $C$; $F_{BF} = 2\sqrt{2}P/3$ $C$; $F_{DI} = \sqrt{2}P/3$ $C$; $F_{EH} = \sqrt{2}P/6$ $T$.

**6.115** $F_{AF} = M_0/4a$ $C$; $F_{BG} = F_{DG} = M_0/\sqrt{2}a$ $T$; $F_{EH} = 3M_0/4a$ $C$.

**6.116** $F_{AF} = \sqrt{2}M_0/3a$ $C$; $F_{BG} = M_0/a$ $T$; $F_{DG} = M_0/a$ $C$; $F_{EH} = 2\sqrt{2}M_0/3a$ $T$.

**6.117** $A = P/15$ ↑; $D = 2P/15$ ↑; $E = 8P/15$ ↑; $H = 4P/15$ ↑.

**6.118** $E = P/5$ ↓; $F = 8P/5$ ↑; $G = 4P/5$ ↓; $H = 2P/5$ ↑.

**6.120** (a) $A = 2.06P$ ↗ 14.04°; $B = 2.06P$ ↖ 14.04°; frame is rigid. (b) Frame is not rigid. (c) $A = 1.25P$ ↗ 36.9°. $B = 1.031P$ ↗ 14.04°; frame is rigid.

**6.122** (a) 2860 N ↓. (b) 2700 N ↗ 68.5°.

**6.123** 564 lb →.

**6.124** 275 lb →.

**6.125** 764 N ←.

**6.127** (a) 746 N ↓. (b) 565 N ↙ 61.3°.

**6.129** 832 lb·in. ↻.

**6.130** 360 lb·in. ↻.

**6.131** 195.0 kN·m ↻.

**6.132** 40.5 kN·m ↻.

**6.133** (a) 160.8 N·m ↻. (b) 155.9 N·m ↻.

**6.134** (a) 117.8 N·m ↻. (b) 47.9 N·m ↻.

**6.137** 18.43 N·m ↻.

**6.138** 208 N·m ↻.

**6.139** $F_{AE} = 800$ N $T$; $F_{DG} = 100.0$ N $C$.

**6.140** $P = 120.0$ N ↓; $Q = 110.0$ N ←.

**6.141** $C = 4.65$ kips →; $E = 6.14$ kips ↗ 40.7°.

**6.143** $D = 9360$ N ←; $F = 6770$ N ↙ 15.12°.

**6.144** $A_x = 210$ N ←; $A_y = 2400$ N ↓; $B = 2720$ N ↗ 61.9°; $C = 1070$ N ←.

**6.145** (a) 475 lb. (b) 528 lb ↖ 63.3°.

**6.147** 44.8 kN.

**6.148** 1200 N.

**6.149** 140.0 N.

**6.151** 315 lb.

**6.152** (a) 312 lb. (b) 135.0 lb·in. ↻.

**6.153** (a) 3000 lb $T$. (b) $H_x = 2400$ lb ←; $H_y = 4800$ lb ↓.

**6.154** $F_{AB} = 18.97$ kips $C$; $F_{CD} = 4.27$ kips $T$; $F_{EF} = 9.61$ kips $C$.

**6.155** (a) 9.29 kN ↖ 44.4°. (b) 8.04 kN ↙ 34.4°.

**6.159** (a) $(12.5$ N·m$)\mathbf{i}$. (b) $G = 0$, $\mathbf{M}_G = -(45.5$ N·m$)\mathbf{i}$; $H = 0$, $\mathbf{M}_H = (13.00$ N·m$)\mathbf{i}$.

**6.160** (a) 27.0 mm. (b) 40.0 N·m ↻.

**6.163** $E_x = 100.0$ kN →, $E_y = 154.9$ kN ↑; $F_x = 26.5$ kN →, $F_y = 118.1$ kN ↓; $H_x = 126.5$ kN ←, $H_y = 36.8$ kN ↓.

**6.164** $F_{AB} = 4.00$ kN $T$; $F_{AD} = 15.00$ kN $T$; $F_{BD} = 9.00$ kN $C$; $F_{BE} = 5.00$ kN $T$; $F_{CD} = 16.00$ kN $C$; $F_{DE} = 4.00$ kN $C$.

**6.165** $F_{AB} = 7.83$ kN $C$; $F_{AC} = 7.00$ kN $T$; $F_{BC} = 1.886$ kN $C$; $F_{BD} = 6.34$ kN $C$; $F_{CD} = 1.491$ kN $T$; $F_{CE} = 5.00$ kN $T$; $F_{DE} = 2.83$ kN $C$; $F_{DF} = 3.35$ kN $C$; $F_{EF} = 2.75$ kN $T$;

$F_{EG} = 1.061$ kN $C$; $F_{EH} = 3.75$ kN $T$; $F_{FG} = 4.24$ kN $C$;
$F_{GH} = 5.30$ kN $C$.

**6.166** $F_{AB} = 8.20$ kips $T$; $F_{AG} = 4.50$ kips $T$; $F_{FG} = 11.60$ kips $C$.

**6.168** $\mathbf{A}_x = 900$ lb ←, $\mathbf{A}_y = 75.0$ lb ↑; $\mathbf{B} = 825$ lb ↓;
$\mathbf{D}_x = 900$ lb →, $\mathbf{D}_y = 750$ lb ↑.

**6.170** $\mathbf{B}_x = 700$ N ←, $\mathbf{B}_y = 200$ N ↓; $\mathbf{E}_x = 700$ N →, $\mathbf{E}_y = 500$ N ↑.

**6.171** $\mathbf{C}_x = 78.0$ lb →, $\mathbf{C}_y = 28.0$ lb ↑; $\mathbf{F}_x = 78.0$ lb ←,
$\mathbf{F}_y = 12.00$ lb ↑.

**6.172** $\mathbf{A} = 327$ lb →; $\mathbf{B} = 827$ lb ←; $\mathbf{D} = 621$ lb ↑; $\mathbf{E} = 246$ lb ↑.

**6.174** (*a*) 21.0 kN ←. (*b*) = 52.5 kN ←.

## CHAPTER 7

**7.1** (On *JE*) $\mathbf{F} = 120.0$ lb ←; $\mathbf{V} = 30.0$ lb ↓; $\mathbf{M} = 120.0$ lb·in. ↺.

**7.2** (On *JD*) $\mathbf{F} = 0$; $\mathbf{V} = 80.0$ lb ↑; $\mathbf{M} = 480.0$ lb·in. ↺.

**7.3** (On *JCD*) $\mathbf{F} = 4.80$ kN ←; $\mathbf{V} = 1.400$ kN ↓;
$\mathbf{M} = 1.380$ kN·m ↻.

**7.4** (On *JCD*) $\mathbf{F} = 3.00$ kN ←; $\mathbf{V} = 0$; $\mathbf{M} = 0.600$ kN·m ↻.

**7.7** $\mathbf{F} = 23.6$ lb ⦨ 76.0°; $\mathbf{V} = 29.1$ lb ⦨ 14.04°; $\mathbf{M} = 540$ lb·in. ↺.

**7.8** (*a*) 30.0 lb at *C*. (*b*) 33.5 lb at *B* and *D*. (*c*) 960 lb·in. at *C*.

**7.9** $\mathbf{F} = 103.9$ N ⦨ 60.0°; $\mathbf{V} = 60.0$ N ⦨ 30.0°;
$\mathbf{M} = 18.71$ N·m ↺ (On *AJ*).

**7.10** $\mathbf{F} = 60.0$ N ⦨ 30.0°; $\mathbf{V} = 103.9$ N ⦨ 60.0°;
$\mathbf{M} = 10.80$ N·m ↺ (On *BK*).

**7.11** $\mathbf{F} = 194.6$ N ⦨ 60.0°; $\mathbf{V} = 257$ N ⦨ 30.0°; $\mathbf{M} = 24.7$ N·m ↺
(On *AJ*).

**7.12** 45.2 N·m for $\theta = 82.9°$.

**7.15** (*a*) (On *CA*) $\mathbf{F} = 480$ N ←; $\mathbf{V} = 576$ N ↑; $\mathbf{M} = 0$.
(*b*) (On *JA*) $\mathbf{F} = 480$ N ←; $\mathbf{V} = 576$ N ↑; $\mathbf{M} = 57.6$ N·m ↺.

**7.16** (*a*) (On *CA*) $\mathbf{F} = 520$ N ←; $\mathbf{V} = 540$ N ↑; $\mathbf{M} = 0$.
(*b*) (On *JA*) $\mathbf{F} = 520$ N ←; $\mathbf{V} = 540$ N ↑; $\mathbf{M} = 54.0$ N·m ↺.

**7.17** 150.0 lb·in. at *D*.

**7.18** 105.0 lb·in. at *E*.

**7.19** $\mathbf{F} = 200$ N ⦨ 36.9°; $\mathbf{V} = 120.0$ N ⦨ 53.1°;
$\mathbf{M} = 120.0$ N·m ↺ (On *BJ*).

**7.20** $\mathbf{F} = 520$ N ←; $\mathbf{V} = 120.0$ N ↓; $\mathbf{M} = 96.0$ N·m↺ (On *AK*).

**7.23** $0.0557Wr$ (On *AJ*).

**7.24** $0.1009Wr$ for $\theta = 57.3°$.

**7.25** $0.289Wr$ (On *BJ*).

**7.26** $0.417Wr$ (On *BJ*).

**7.29** (*b*) $Pab/L$.

**7.30** (*b*) $wa^2/2$.

**7.31** (*b*) $Pa$.

**7.32** (*b*) $3wL^2/8$.

**7.35** (*b*) $|V|_{max} = 40.0$ kN; $|M|_{max} = 55.0$ kN·m.

**7.36** (*b*) $|V|_{max} = 50.5$ kN; $|M|_{max} = 39.8$ kN·m.

**7.39** (*b*) $|V|_{max} = 64.0$ kN; $|M|_{max} = 92.0$ kN·m.

**7.40** (*b*) $|V|_{max} = 60.0$ kN; $|M|_{max} = 72.0$ kN·m.

**7.41** (*b*) $|V|_{max} = 18.00$ kips; $|M|_{max} = 48.5$ kip·ft.

**7.42** (*b*) $|V|_{max} = 600$ lb; $|M|_{max} = 1200$ lb·ft.

**7.45** (*b*) $|V|_{max} = 6.00$ kips; $|M|_{max} = 12.00$ kip·ft.

**7.46** (*b*) $|V|_{max} = 4.00$ kips; $|M|_{max} = 6.00$ kip·ft.

**7.47** (*b*) $|V|_{max} = 10.00$ kN; $|M|_{max} = 20.0$ kN·m.

**7.48** (*b*) $|V|_{max} = 10.00$ kN; $|M|_{max} = 20.0$ kN·m.

**7.49** (*a*) 400 N; 160.0 N·m. (*b*) −200 N; 40.0 N·m.

**7.50** $|V|_{max} = 240$ N; $|M|_{max} = 48.0$ N·m.

**7.51** $|V|_{max} = 7.50$ kips; $|M|_{max} = 7.20$ kip·ft.

**7.52** $|V|_{max} = 112.5$ lb; $|M|_{max} = 1020$ lb·in.

**7.55** (*a*) 54.5°. (*b*) 675 N·m.

**7.56** (*a*) 1.236. (*b*) $0.1180wa^2$.

**7.57** (*a*) 40.0 mm. (*b*) 1.600 N·m.

**7.58** (*a*) 0.840 m. (*b*) 1.680 N·m.

**7.59** $0.207L$.

**7.62** (*a*) $0.414wL$; $0.0858wL^2$. (*b*) $0.250wL$; $0.250wL^2$.

**7.69** (*a*) $|V|_{max} = 12.00$ kN; $|M|_{max} = 4.64$ kN·m.

**7.70** (*b*) $|V|_{max} = 41.4$ kN; $|M|_{max} = 35.3$ kN·m.

**7.77** (*b*) 44.7 kN·m, 3.28 m from *A*.

**7.78** (*b*) 90.0 kN·m, at *B*.

**7.79** (*b*) 9.00 kN·m, 1.700 m from *A*.

**7.80** (*b*) 5.76 kN·m, 2.40 m from *A*.

**7.81** (*b*) 3130 lb·ft, 5.00 ft from *A*.

**7.82** (*b*) 16.20 kip·ft, 13.50 ft from *A*.

**7.86** (*a*) $V = (w_0/6L)(3x^2 - 6Lx + 2L^2)$; $M = (w_0/6L)(x^3 - 3Lx^2 + 2L^2x)$. (*b*) $0.0642w_0L^2$, at $x = 0.423L$.

**7.87** (*a*) $V = (w_0L/4)[3(x/L)^2 - 4(x/L) + 1]$; $M = (w_0L^2/4)\,[(x/L)^3 - 2(x/L)^2 + (x/L)]$. (*b*) $w_0L^2/27$, at $x = L/3$.

**7.89** (*a*) $\mathbf{P} = 300$ N ↓; $\mathbf{Q} = 300$ N ↓. (*b*) $M_D = -75.0$ N·m.

**7.90** (*a*) $\mathbf{P} = 175.0$ N ↓; $\mathbf{Q} = 175.0$ N ↓. (*b*) $M_D = -75.0$ N·m.

**7.91** (*a*) $\mathbf{P} = 23.0$ kips ↓; $\mathbf{Q} = 79.0$ kips ↓. (*b*) $V_{max} = 63.0$ kips
at *B*; $M_{max} = 252$ kip·ft at *F*.

**7.92** (*a*) $\mathbf{P} = 20.0$ kips ↓; $\mathbf{Q} = 75.0$ kips ↓. (*b*) $V_{max} = 67.0$ kips
at *B*; $M_{max} = 252$ kip·ft at *F*.

**7.93** (*a*) $\mathbf{E}_x = 10.00$ kN →, $\mathbf{E}_y = 7.00$ kN↑. (*b*) 12.21 kN.

**7.94** 1.667 m.

**7.95** $\mathbf{E}_x = 1000$ lb →; $\mathbf{E}_y = 600$ lb ↑.

**7.96** (*a*) $\mathbf{E}_x = 2000$ lb →, $\mathbf{E}_y = 700$ lb ↑. (*b*) 2120 lb.

**7.97** (*a*) $d_B = 4.40$ m; $d_D = 3.90$ m. (*b*) 21.9 kN.

**7.98** (*a*) 3.33 m. (*b*) $\mathbf{E}_x = 24.0$ kN →; $\mathbf{E}_y = 20.0$ kN ↑.

**7.101** 196.2 N.

**7.102** 157.0 N.

**7.103** (*a*) 240 lb. (*b*) 9.00 ft.

**7.104** $a = 7.50$ ft; $b = 17.50$ ft

**7.107** (*a*) 1775 N. (*b*) 60.1 m.

**7.109** (*a*) 50,200 kips. (*b*) 3580 ft.

**7.110** 3.75 ft.

**7.111** (*a*) 56,400 kips. (*b*) 4284 ft.

**7.112** (*a*) 6.75 m. (*b*) $T_{AB} = 615$ N; $T_{BC} = 600$ N.

**7.114** (*a*) $\sqrt{3L\Delta/8}$. (*b*) 12.25 ft.

**7.115** $h = 27.6$ mm; $\theta_A = 25.5°$; $\theta_C = 27.6°$.

**7.116** (*a*) 4.05 m. (*b*) 16.41 m. (*c*) $A_x = 5890$ N ←, $A_y = 5300$ N ↑.

**7.117** (*a*) 58,900 kips, (*b*) 29.2°.

**7.118** (*a*) 16.00 ft to the left of *B*. (*b*) 2000 lb.

**7.125** $Y = h[1 - \cos(\pi x/L)]$; $T_{min} = w_0L^2/h\pi^2$; $T_{max} = (w_0L/\pi)\sqrt{(L^2/h^2\pi^2)} + 1$

**7.127** (*a*) 12.36 ft. (*b*) 15.38 lb.

**7.128** (*a*) 412 ft. (*b*) 875 lb.

**7.129** (*a*) 35.6 m. (*b*) 49.2 kg.

**7.130** 49.86 ft.

**7.133** (*a*) 5.89 m. (*b*) 10.89 N →.

**7.134** 10.05 ft.

**7.135** (*a*) 56.3 ft. (*b*) 2.36 lb/ft.

**7.136** (*a*) 30.2 m. (*b*) 56.6 kg.

**7.139** 31.8 N.

**7.140** 29.8 N.

**7.143** (*a*) $a = 79.0$ ft; $b = 60.0$ ft. (*b*) 103.9 ft.

**7.144** (*a*) $a = 65.8$ ft; $b = 50.0$ ft. (*b*) 86.6 ft.

**7.145** 119.1 N →.

**7.146** 177.6 N →.

**7.147** 3.50 ft.

**7.148** 5.71 ft.

**7.151** 0.394 m and 10.97 m.

**7.152** 0.1408.

**7.153** (*a*) 0.338. (*b*) 56.5°; $0.755wL$.

**7.154** (On *AJ*) $\mathbf{F} = 750$ N ↑; $\mathbf{V} = 400$ N ←; $\mathbf{M} = 130.0$ N·m ↺.

**7.156** (On *BJ*) $\mathbf{F} = 12.50$ lb ∡ $30.0°$; $\mathbf{V} = 21.7$ lb ⤡ $60.0°$; $\mathbf{M} = 75.0$ lb·in. ↺.

**7.157** (*a*) (On *AJ*) $\mathbf{F} = 500$ N ←; $\mathbf{V} = 500$ N↑; $\mathbf{M} = 300$ N·m ↺. (*b*) (On *AK*) $\mathbf{F} = 970$ N ↑; $\mathbf{V} = 171.0$ N ←; $\mathbf{M} = 446$ N·m ↺.

**7.158** (*a*) 40.0 kips. (*b*) 40.0 kip·ft.

**7.161** (*a*) 18.00 kip·ft, 3.00 ft from *A*. (*b*) 34.1 kip·ft, 2.25 ft from *A*.

**7.163** (*a*) 2.28 m. (*b*) $\mathbf{D}_x = 13.67$ kN →; $\mathbf{D}_y = 7.80$ kN ↑. (*c*) 15.94 kN.

**7.164** (*a*) 138.1 m. (*b*) 602 N.

**7.165** (*a*) 4.22 ft. (*b*) 80.3°.

## CHAPTER 8

**8.1** Equilibrium; $\mathbf{F} = 48.3$ N ↘.

**8.2** Block moves; $\mathbf{F} = 103.5$ N ↘.

**8.3** Block moves; $\mathbf{F} = 48.3$ lb ↗.

**8.4** Equilibrium; $\mathbf{F} = 61.5$ lb ↗.

**8.5** $222$ N $\leq P \leq 479$ N.

**8.7** (*a*) 29.7 N ←. (*b*) 20.9 N →.

**8.9** (*a*) 462 N. (*b*) 447 N. (*c*) 276 N.

**8.10** (*a*) 453 N ∡ $49.0°$. (*b*) 215 N ∡ $21.0°$.

**8.11** 31.0°.

**8.12** 46.4°.

**8.13** Package *C* does not move; $\mathbf{F}_C = 10.16$ N ↗. Package *A* and *B* move; $\mathbf{F}_A = 7.58$ N ↗; $\mathbf{F}_B = 3.03$ N ↗.

**8.14** All packages move; $\mathbf{F}_A = \mathbf{F}_C = 7.58$ N ↗; $\mathbf{F}_B = 3.03$ N ↗.

**8.17** (*a*) 75.0 lb. (*b*) Pipe will slide.

**8.18** 35.7 lb →.

**8.19** $P = 8.34$ lb.

**8.20** $P = 7.50$ lb.

**8.21** (*a*) $0.300Wr$. (*b*) $0.349Wr$.

**8.22** $M = Wr\mu_s(1 + \mu_s)/(1 + \mu_s^2)$.

**8.23** 0.955 lb.

**8.25** 0.208.

**8.27** 664 N ↓.

**8.29** (*a*) Plate in equilibrium. (*b*) Plate moves downward.

**8.30** $10.00$ lb $< P < 36.7$ lb.

**8.32** 0.860.

**8.34** (*a*) 3.08 in. (*b*) 1.170 in.

**8.35** (*a*) 3.08 in. (*b*) 3.08 in.

**8.36** $0.818WL \leq M_0 \leq 1.048WL$.

**8.37** (*a*) 2.94 N. (*b*) 4.41 N.

**8.39** (*a*) $\theta = 73.3°$; $\mathbf{P} = 102.4$ N. (*b*) $\theta = 59.0°$; $\mathbf{P} = 76.3$ N.

**8.40** (*a*) 77.5 N. (*b*) Tube slides.

**8.41** 135.0 lb.

**8.43** (*a*) System slides; $P = 62.8$ N. (*b*) System rotates about *B*; $P = 73.2$ N.

**8.44** 35.8°.

**8.45** 20.5°.

**8.46** $1.225W$.

**8.47** $46.4° \leq \theta \leq 52.4°$ and $67.6° \leq \theta \leq 79.4°$.

**8.48** (*a*) 283 N ←. (*b*) $\mathbf{B}_x = 413$ N ←; $\mathbf{B}_y = 480$ N ↓.

**8.49** (*a*) 107.0 N ←. (*b*) $\mathbf{B}_x = 611$ N ←; $\mathbf{B}_y = 480$ N ↓.

**8.52** (*a*) 15.26 kips. (*b*) 5.40 kips.

**8.53** (*a*) 6.88 kips. (*b*) 5.40 kips.

**8.54** 9.86 kN ←.

**8.55** 9.13 N ←.

**8.56** (*a*) 28.1°. (*b*) 728 N ∡ $14.04°$.

**8.57** 28.1°; 47.1 lb.

**8.59** 143.4 N.

**8.60** 1.400 lb.

**8.62** (*a*) 90.0 lb. (*b*) Base moves.

**8.63** (*a*) 89.4 lb. (*b*) Base does not move.

**8.64** (*b*) 283 N ←.

**8.65** 0.442.

**8.66** 3.03 kN.

**8.67** (*a*) The wedge will slide up and out from the slot. (*b*) The wedge binds in the slot.

**8.71** 693 lb·ft.

**8.72** 35.8 N·m.

**8.73** 9.02 N·m.

**8.74** (*a*) Screw *A*. (*b*) 14.06 lb·in.

**8.77** 0.226.

**8.78** 0.0980.

**8.79** 450 N.

**8.80** 412 N.

**8.81** 334 N.

**8.82** 376 N.

**8.84** $T_{AB} = 77.5$ lb; $T_{CD} = 72.5$ lb. $T_{EF} = 67.8$ lb.

**8.86** (*a*) 4.80 kN. (*b*) 1.375°.

**8.88** 22.0 lb ←.

**8.89** 1.948 lb ↓.

**8.90** 18.01 lb ←.

**8.92** 0.277.

**8.93** 3.75 lb.

**8.98** 10.87 lb.

**8.99** 0.0600 in.

**8.100** 154.4 N.

**8.101** 300 mm.

**8.102** (*a*) 1.288 kN. (*b*) 1.058 kN.

**8.103** 2.34 ft.

**8.104** (*a*) 0.329. (*b*) 2.67 turns.

**8.105** $14.23$ kg $\leq m \leq 175.7$ kg.

**8.106** (*a*) 0.292. (*b*) 310 N.

**8.109** 31.8 N·m ↺.

**8.110** (*a*) $T_A = 8.40$ lb; $T_B = 19.60$ lb. (*b*) 0.270.

**8.111** (*a*) $T_A = 11.13$ lb; $T_B = 20.9$ lb. (*b*) 91.3 lb·in. ↺.

**8.112** 35.1 N·m.

**8.113** (*a*) 27.0 N·m. (*b*) 675 N.

**8.114** (*a*) 4.97 N·m ↺. (*b*) 42.3 N.

**8.117** 4.49 in.

**8.118** (*a*) 11.66 kg. (*b*) 38.6 kg. (*c*) 34.4 kg.

**8.119** (*a*) 9.46 kg. (*b*) 167.2 kg. (*c*) 121.0 kg.

**8.120** (*a*) 10.39 lb. (*b*) 58.5 lb.

**8.121** (*a*) 28.9 lb. (*b*) 28.9 lb.

**8.124** 5.97 N.

**8.125** 9.56 N.

**8.126** 0.350.

**8.128** (*a*) 30.3 lb·in. ↺. (*b*) 3.78 lb ↓.

**8.129** (*a*) 17.23 lb·in. ↺. (*b*) 2.15 lb ↑.

**8.133** (*a*) 51.0 N·m. (*b*) 875 N.

**8.134** (*a*) 353 N ←. (*b*) 196.2 N ←.

**8.136** (*a*) 136.0 lb →. (*b*) 30.0 lb →. (*c*) 12.86 lb →.

**8.137** $6.35 \leq L/a \leq 10.81$.

**8.138** 151.5 N·m.

**8.140** 0.225.

**8.141** 313 lb →.

**8.143** 6.44 N·m.

**8.144** (*a*) 0.238. (*b*) 218 N ↓.

## CHAPTER 9

**9.1** $a^3b/30$.

**9.2** $3a^3b/10$.

**9.3** $2a^3b/15$.

**9.4** $a^3b/6$.

**9.6** $ab^3/6$.

**9.8** $ab^3/30$.

**9.9** $ab^3/15$.

**9.10** $ab^3/20$.

**9.11** $0.1056ab^3$.

**9.12** $3.43a^3b$.

**9.15** $ab^3/9$; $b\sqrt{5/27}$.

**9.16** $\pi ab^3/8$; $b/2$.

**9.17** $3a^3b/11$; $a\sqrt{5/11}$.

**9.18** $\pi a^3b/8$; $a/2$.

**9.21** (a) $4a^4/3$; $a\sqrt{2/3}$. (b) $17a^4/6$; $a\sqrt{17/12}$.

**9.22** $35a^4/48$; $a\sqrt{35/72}$.

**9.23** $64a^4/15$; $1.265a$.

**9.25** (a) $(\pi/2)(R_2^4 - R_1^4)$. (b) $(\pi/4)(R_2^4 - R_1^4)$.

**9.26** (b) for $t/R_m = 1$, $-10.56\%$; for $t/R_m = 1/2$, $-2.99\%$; for $t/R_m = 1/10$, $-0.1250\%$.

**9.28** $bh(12h^2 + b^2)/48$; $\sqrt{(12h^2 + b^2)/24}$.

**9.31** $614 \times 10^3$ mm$^4$; $19.01$ mm.

**9.32** $28.0$ in$^4$; $2.26$ in.

**9.33** $1.894 \times 10^6$ mm$^4$; $33.4$ mm.

**9.34** $6.99$ in$^4$; $1.127$ in.

**9.37** $25.0$ in$^2$; $400$ in$^4$.

**9.39** $\bar{I}_x = 1.500 \times 10^6$ mm$^4$; $\bar{I}_y = 3.00 \times 10^6$ mm$^4$.

**9.40** $A = 6000$ mm$^2$; $\bar{J}_c = 7.20 \times 10^6$ mm$^4$; $d = 80.0$ mm.

**9.41** $\bar{I}_x = 46.8 \times 10^6$ mm$^4$; $\bar{I}_y = 13.89 \times 10^6$ mm$^4$.

**9.42** $\bar{I}_x = 479 \times 10^3$ mm$^4$; $\bar{I}_y = 149.7 \times 10^3$ mm$^4$.

**9.43** $\bar{I}_x = 191.3$ in$^4$; $\bar{I}_y = 75.2$ in$^4$.

**9.44** $\bar{I}_x = 204$ in$^4$; $\bar{I}_y = 135.0$ in$^4$.

**9.47** (a) $159.5 \times 10^6$ mm$^4$. (b) $31.9 \times 10^6$ mm$^4$.

**9.48** (a) $12.16 \times 10^6$ mm$^4$. (b) $9.73 \times 10^6$ mm$^4$.

**9.49** $\bar{I}_x = 255 \times 10^6$ mm$^4$; $\bar{k}_x = 134.1$ mm; $\bar{I}_y = 100.0 \times 10^6$ mm$^4$; $\bar{k}_y = 83.9$ mm.

**9.50** $1.077$ in.

**9.51** $\bar{I}_x = 250$ in$^4$; $\bar{k}_x = 4.10$ in.; $\bar{I}_y = 141.9$ in$^4$; $\bar{k}_y = 3.09$ in.

**9.52** $\bar{I}_x = 260 \times 10^6$ mm$^4$; $\bar{k}_x = 144.6$ mm; $\bar{I}_y = 17.53$ mm$^4$; $\bar{k}_y = 37.6$ mm.

**9.54** $\bar{I}_x = 745 \times 10^6$ mm$^4$; $\bar{I}_y = 91.3 \times 10^6$ mm$^4$.

**9.55** $\bar{I}_x = 3.55 \times 10^6$ mm$^4$; $\bar{I}_y = 49.8 \times 10^6$ mm$^4$.

**9.57** $h/2$.

**9.58** $15h/14$.

**9.59** $3\pi r/16$.

**9.60** $3\pi b/16$.

**9.63** $5a/8$.

**9.64** $80.0$ mm.

**9.67** $a^4/2$.

**9.68** $a^2(h_1^2 + 2h_1h_2 + 3h_2^2)/24$.

**9.69** $a^2b^2/12$.

**9.71** $-1.760 \times 10^6$ mm$^4$.

**9.72** $2.40 \times 10^6$ mm$^4$.

**9.74** $-0.380$ in$^4$.

**9.75** $471 \times 10^3$ mm$^4$.

**9.76** $-9010$ in$^4$.

**9.78** $2.54 \times 10^6$ mm$^4$.

**9.79** (a) $\bar{I}_{x'} = 0.482a^4$; $\bar{I}_{y'} = 1.482a^4$; $\bar{I}_{x'y'} = -0.589a^4$. (b) $\bar{I}_{x'} = 1.120a^4$; $\bar{I}_{y'} = 0.843a^4$; $\bar{I}_{x'y'} = 0.760a^4$.

**9.80** $\bar{I}_{x'} = 2.12 \times 10^6$ mm$^4$; $\bar{I}_{y'} = 8.28 \times 10^6$ mm$^4$; $\bar{I}_{x'y'} = -0.532 \times 10^6$ mm$^4$.

**9.81** $\bar{I}_{x'} = 1033$ in$^4$; $\bar{I}_{y'} = 2020$ in$^4$; $\bar{I}_{x'y'} = -873$ in$^4$.

**9.83** $\bar{I}_{x'} = 0.236$ in$^4$; $\bar{I}_{y'} = 1.244$ in$^4$; $\bar{I}_{x'y'} = 0.1132$ in$^4$.

**9.85** $20.2°$ and $110.2°$; $1.754a^4$; $0.209a^4$.

**9.86** $25.1°$ and $115.1°$; $\bar{I}_{max} = 8.32 \times 10^6$ mm$^4$; $\bar{I}_{min} = 2.08 \times 10^6$ mm$^4$.

**9.87** $29.7°$ and $119.7°$; $2530$ in$^4$; $524$ in$^4$.

**9.89** $-23.7°$ and $66.3°$; $1.257$ in$^4$; $0.224$ in$^4$.

**9.91** (a) $\bar{I}_{x'} = 0.482a^4$; $\bar{I}_{y'} = 1.482a^4$; $\bar{I}_{x'y'} = -0.589a^4$. (b) $\bar{I}_{x'} = 1.120a^4$; $\bar{I}_{y'} = 0.843a^4$; $\bar{I}_{x'y'} = 0.760a^4$.

**9.92** $\bar{I}_{x'} = 2.12 \times 10^6$ mm$^4$; $\bar{I}_{y'} = 8.28 \times 10^6$ mm$^4$; $\bar{I}_{x'y'} = -0.532 \times 10^6$ mm$^4$.

**9.93** $\bar{I}_{x'} = 1033$ in$^4$; $\bar{I}_{y'} = 2020$ in$^4$; $\bar{I}_{x'y'} = -873$ in$^4$.

**9.95** $\bar{I}_{x'} = 0.236$ in$^4$; $\bar{I}_{y'} = 1.244$ in$^4$; $\bar{I}_{x'y'} = 0.1132$ in$^4$.

**9.97** $20.2°$; $1.754a^4$; $0.209a^4$.

**9.98** $23.9°$; $8.33 \times 10^6$ mm$^4$; $1.465 \times 10^6$ mm$^4$.

**9.99** $33.4°$; $22.1 \times 10^3$ in$^4$; $2490$ in$^4$.

**9.100** $29.7°$; $2530$ in$^4$; $524$ in$^4$.

**9.103** (a) $-1.146$ in$^4$. (b) $29.1°$ clockwise. (c) $3.39$ in$^4$.

**9.104** $23.8°$ clockwise; $0.524 \times 10^6$ mm$^4$; $0.0917 \times 10^6$ mm$^4$.

**9.105** $19.54°$ counterclockwise; $4.34 \times 10^6$ mm$^4$; $0.647 \times 10^6$ mm$^4$.

**9.106** (a) $25.3°$. (b) $1459$ in$^4$; $40.5$ in$^4$.

**9.107** (a) $88.0 \times 10^6$ mm$^4$. (b) $96.3 \times 10^6$ mm$^4$; $39.7 \times 10^6$ mm$^4$.

**9.111** (a) $\bar{I}_{AA'} = \bar{I}_{BB'} = ma^2/24$. (b) $ma^2/12$.

**9.112** (a) $m(r_1^2 + r_2^2)/4$. (b) $m(r_1^2 + r_2^2)/2$.

**9.113** (a) $ma^2/4$, $mb^2/4$. (b) $m(a^2 + b^2)/4$.

**9.114** (a) $mb^2/7$. (b) $m(7a^2 + 10b^2)/70$.

**9.117** (a) $I_{AA'} = mb^2/24$; $I_{BB'} = mh^2/18$. (b) $m(3b^2 + 4h^2)/72$.

**9.118** $I_{DD'} = m(b^2 + 24d^2)/24$; $I_{EE'} = m(h^2 + 18d^2)/18$.

**9.119** $m(3a^2 + L^2)/12$.

**9.120** $1.329mh^2$.

**9.121** $ma^2(2n + 1)/2(4n + 1)$.

**9.122** $m(2b^2 + h^2)/10$.

**9.124** $ma^2/3$; $a/\sqrt{3}$.

**9.126** $I_x = I_y = ma^2/4$; $I_z = ma^2/2$.

**9.127** $1.160 \times 10^{-6}$ lb·ft·s$^2$; $0.341$ in.

**9.128** $837 \times 10^{-9}$ kg·m$^2$; $6.92$ mm.

**9.130** $m(3a^2 + 2h^2)/6$.

**9.131** (a) $\rho h(a^4 + 6a^2r^2 - 9\pi r^4)/6$. (b) $a\sqrt{1/3\pi}$. (c) $\rho a^4 h(\pi + 1)/6\pi$; $a\sqrt{(\pi + 1)/4\pi}$.

**9.132** (a) $\pi pl^2[6a^2t(5a^2/3l^2 + 2a/l + 1) + d^2l/4]$. (b) $0.1851$.

**9.133** (a) $27.5$ mm to the right of $A$. (b) $32.0$ mm.

**9.135** $I_x = 7.11 \times 10^{-3}$ kg·m$^2$; $I_y = 16.96 \times 10^{-3}$ kg·m$^2$; $I_z = 15.27 \times 10^{-3}$ kg·m$^2$.

**9.136** $I_x = 175.5 \times 10^{-3}$ kg·m$^2$; $I_y = 309 \times 10^{-3}$ kg·m$^2$; $I_z = 154.4 \times 10^{-3}$ kg·m$^2$.

**9.138** $I_x = 334 \times 10^{-6}$ lb·ft·s$^2$; $I_y = I_z = 1.356 \times 10^{-3}$ lb·ft·s$^2$.

**9.139** $I_x = I_z = 36.1 \times 10^{-3}$ lb·ft·s$^2$; $I_y = 30.0 \times 10^{-3}$ lb·ft·s$^2$.

**9.141** (a) $13.99 \times 10^{-3}$ kg·m$^2$. (b) $20.6 \times 10^{-3}$ kg·m$^2$. (c) $14.30 \times 10^{-3}$ kg·m$^2$.

**9.142** $I_x = 28.3 \times 10^{-3}$ kg·m$^2$; $I_y = 183.8 \times 10^{-3}$ kg·m$^2$; $k_x = 42.9$ mm; $k_y = 109.3$ mm.

**9.143** $30.5 \times 10^{-3}$ lb·ft·s$^2$.

**9.145** (a) $26.4 \times 10^{-3}$ kg·m$^2$. (b) $31.2 \times 10^{-3}$ kg·m$^2$. (c) $8.58 \times 10^{-3}$ kg·m$^2$.

**9.147** $I_x = 0.0392$ lb·ft·s$^2$; $I_y = 0.0363$ lb·ft·s$^2$; $I_z = 0.0304$ lb·ft·s$^2$.

**9.148** $I_x = 0.323$ kg·m$^2$; $I_y = I_z = 0.419$ kg·m$^2$.

**9.149** $I_{xy} = 2.50 \times 10^{-3}$ kg·m$^2$; $I_{yz} = 4.06 \times 10^{-3}$ kg·m$^2$; $I_{zx} = 8.81 \times 10^{-3}$ kg·m$^2$.

**9.150** $I_{xy} = 286 \times 10^{-6}$ kg·m$^2$; $I_{yz} = I_{zx} = 0$.

**9.151** $I_{xy} = I_{zx} = 0$; $I_{yz} = -3.53 \times 10^{-3}$ lb·ft·s$^2$.

**9.152** $I_{xy} = -538 \times 10^{-6}$ lb·ft·s$^2$; $I_{yz} = -171.4 \times 10^{-6}$ lb·ft·s$^2$; $I_{zx} = 1120 \times 10^{-6}$ lb·ft·s$^2$.

**9.155** $I_{xy} = -8.04 \times 10^{-3}$ kg·m$^2$; $I_{yz} = 12.90 \times 10^{-3}$ kg·m$^2$; $I_{zx} = 94.0 \times 10^{-3}$ kg·m$^2$.

**9.156** $I_{xy} = 0$; $I_{yz} = 48.3 \times 10^{-6}$ kg·m$^2$; $I_{zx} = -4.43 \times 10^{-3}$ kg·m$^2$.

**9.157** $I_{xy} = 47.9 \times 10^{-6}$ kg·m$^2$; $I_{yz} = 102.1 \times 10^{-6}$ kg·m$^2$; $I_{zx} = 64.1 \times 10^{-6}$ kg·m$^2$.

**9.158** $I_{xy} = -m'R_1^3/2$; $I_{yz} = m'R_1^3/2$; $I_{zx} = -m'R_2^3/2$.

**9.159** $I_{xy} = wa^3(1 - 5\pi)g$; $I_{yz} = -11\pi wa^3/g$; $I_{zx} = 4wa^3(1 + 2\pi)/g$.

**9.160** $I_{xy} = -11wa^3/g$; $I_{yz} = wa^3(\pi + 6)/2g$; $I_{zx} = -wa^3/4g$.

**9.162** (a) $mac/20$. (b) $I_{xy} = mab/20$, $I_{yz} = mbc/20$.

**9.165** $18.17 \times 10^{-3}$ kg·m$^2$.

**9.166** $11.81 \times 10^{-3}$ kg·m$^2$.

**9.167** $5Wa^2/18g$.

**9.168** $4.41\gamma ta^4/g$.

**9.169** $281 \times 10^{-3}$ kg·m$^2$.

**9.170** $0.354$ kg·m$^2$.

**9.173** (a) $1/\sqrt{3}$. (b) $\sqrt{7/12}$.

**9.174** (a) $b/a = 2$; $c/a = 2$. (b) $b/a = 1$; $c/a = 0.5$.

**9.175** (a) 2. (b) $\sqrt{2/3}$.

**9.179** (a) $K_1 = 0.363ma^2$; $K_2 = 1.583ma^2$; $K_3 = 1.720ma^2$.
(b) $(\theta_x)_1 = (\theta_z)_1 = 49.7°$, $(\theta_y)_1 = 113.7°$; $(\theta_x)_2 = 45.0°$, $(\theta_y)_2 = 90.0°$, $(\theta_z)_2 = 135.0°$; $(\theta_x)_3 = (\theta_z)_3 = 73.5°$, $(\theta_y)_3 = 23.7°$.

**9.180** (a) $K_1 = 14.30 \times 10^{-3}$ kg·m$^2$; $K_2 = 13.96 \times 10^{-3}$ kg·m$^2$; $K_3 = 20.6 \times 10^{-3}$ kg·m$^2$. (b) $(\theta_x)_1 = (\theta_y)_1 = 90.0°$, $(\theta_z)_1 = 0°$; $(\theta_x)_2 = 3.42°$, $(\theta_y)_2 = 86.6°$, $(\theta_z)_2 = 90.0°$; $(\theta_x)_3 = 93.4°$, $(\theta_y)_3 = 3.43°$, $(\theta_z)_3 = 90.0°$

**9.182** (a) $K_1 = 0.1639Wa^2/g$; $K_2 = 1.054Wa^2/g$; $K_3 = 1.115Wa^2/g$.
(b) $(\theta_x)_1 = 36.7°$, $(\theta_y)_1 = 71.6°$, $(\theta_z)_1 = 59.5°$; $(\theta_x)_2 = 74.9°$, $(\theta_y)_2 = 54.5°$, $(\theta_z)_2 = 140.5°$; $(\theta_x)_3 = 57.5°$, $(\theta_y)_3 = 138.8°$, $(\theta_z)_3 = 112.4°$.

**9.183** (a) $K_1 = 2.26\gamma ta^4/g$; $K_2 = 17.27\gamma ta^4/g$; $K_3 = 19.08\gamma ta^4/g$.
(b) $(\theta_x)_1 = 85.0°$, $(\theta_y)_1 = 36.8°$, $(\theta_z)_1 = 53.7°$; $(\theta_x)_2 = 81.7°$, $(\theta_y)_2 = 54.7°$, $(\theta_z)_2 = 143.4°$; $(\theta_x)_3 = 9.70°$, $(\theta_y)_3 = 99.0°$, $(\theta_z)_3 = 86.3°$.

**9.185** $I_x = 16ah^3/105$; $I_y = ha^3/5$.

**9.186** $\pi a^3 b/8$; $a/2$.

**9.188** $\bar{I}_x = 1.874 \times 10^6$ mm$^4$; $\bar{I}_y = 5.82 \times 10^6$ mm$^4$.

**9.189** (a) $3.13 \times 10^6$ mm$^4$. (b) $2.41 \times 10^6$ mm$^4$.

**9.191** $-2.81$ in$^4$.

**9.193** (a) $ma^2/3$. (b) $3ma^2/2$.

**9.195** $I_x = 0.877$ kg·m$^2$; $I_y = 1.982$ kg·m$^2$; $I_z = 1.652$ kg·m$^2$.

**9.196** $0.0442$ lb·ft·s$^2$.

# CHAPTER 10

**10.1** 65.0 N ↓.

**10.2** 132.0 lb →.

**10.3** 39.0 N·m. ↺.

**10.4** 1320 lb·in. ↺.

**10.5** (a) 60.0 N $C$, 8.00 mm ↓. (b) 300 N $C$, 40.0 mm ↓.

**10.6** (a) 120.0 N $C$, 16.00 mm ↓. (b) 300 N $C$, 40.0 mm ↓.

**10.9** $T = Wa \cot \theta/(a + b)$.

**10.10** $Q = P[(l/a) \sin^3 \theta - 1]$.

**10.12** $Q = 2P \sin \theta/\cos (\theta/2)$.

**10.14** $Q = 3P \tan \theta$.

**10.15** $M = 2Pl \cot \theta$.

**10.16** $M = Pl(\sin \theta + \cos \theta)$.

**10.17** $M = \frac{1}{2}Wl \tan \alpha \sin \theta$.

**10.18** (a) $M = Pl \sin 2\theta$. (b) $M = 3Pl \cos \theta$. (c) $M = Pl \sin \theta$.

**10.19** 85.2 lb·ft ↺.

**10.20** 22.8 lb ⬋ 70.0°.

**10.23** 38.7°.

**10.24** 68.0°.

**10.27** 36.4°.

**10.28** 57.5°.

**10.30** 25.0°.

**10.31** 39.7° and 69.0°.

**10.32** 390 mm.

**10.33** 330 mm.

**10.35** 38.7°.

**10.36** 52.4°.

**10.37** 22.6°.

**10.38** 51.1°.

**10.39** 59.0°.

**10.40** 78.7°, 324°, 379°.

**10.43** 12.03 kN ↘.

**10.44** 20.4°.

**10.45** 2370 lb ↖.

**10.46** 2550 lb ↖.

**10.48** 300 N·m, 81.8 N·m.

**10.49** $\eta = 1/(1 + \mu \cot \alpha)$.

**10.50** $\eta = \tan \theta/\tan (\theta + \phi_s)$.

**10.52** 37.6 N, 31.6 N.

**10.53** $\mathbf{A} = 250$ N ↑; $\mathbf{M}_A = 450$ N·m ↺.

**10.54** 1050 N ↑.

**10.57** 0.625 in. ↓.

**10.58** 0.469 in. →.

**10.69** $\theta = -45.0°$, unstable; $\theta = 135.0°$, stable.

**10.70** $\theta = -63.4°$, unstable; $\theta = 116.6°$, stable.

**10.71** $\theta = 90.0°$ and $\theta = 270°$, unstable; $\theta = 22.0°$ and $\theta = 158.0°$, stable.

**10.72** $\theta = 0$ and $\theta = 180.0°$, unstable; $\theta = 75.5°$ and $\theta = 284°$, stable.

**10.73** 59.0°, stable.

**10.74** 78.7°, stable; 324°, unstable; 379°, stable.

**10.77** (a) $(1 - \cos \theta) \tan \theta = 2mg/kl$. (b) 52.0°, stable.

**10.78** (a) $(0.866 - \cos \theta) \tan \theta = 2mg/kl$. (b) 56.9°, stable.

**10.80** 9.39° and 90.0°, stable; 34.2°, unstable.

**10.81** 17.11°, stable; 72.9°, unstable.

**10.83** 49.1°.

**10.86** 16.88 m.

**10.87** 54.8°.

**10.88** 37.4°.

**10.89** $P < kl/2$.

**10.91** $k > 6.94$ lb/in.

**10.92** 15.00 in.

**10.93** $P < 2kL/9$.

**10.94** $P < kL/18$.

**10.96** $P < 160.0$ N.

**10.98** $P < 573$ lb.

**10.100** $P < 4kl/5$.

**10.101** 60.0 lb ↓.

**10.102** 600 lb·in. ↺.

**10.103** 500 N ↑.

**10.105** $M = 7Pa \cos \theta$

**10.107** 19.40°.

**10.108** 7.13 in.

**10.110** $\theta = 0$, unstable; $\theta = 137.8°$, stable.

**10.112** (a) 22.0°. (b) 30.6°.

## CHAPTER 11

**11.1** 97.5 ft, 49.5 ft/s, 17 ft/s$^2$.

**11.2** $t = 2$ s or 6 s
for $t = 2$ s, $x = 62$ ft, $a = -12$ ft/s$^2$
for $t = 6$ s, $x = 30$ ft, $a = 12$ ft/s$^2$

**11.3** (a) $x = -0.578$ mm, $v = 8.22$ mm/s, $a = 57.8$mm/s$^2$;
(b) $v_{max} = 10.05$ mm/s, $a_{max} = 100.5$ mm/s$^2$.

**11.4** (a) 0 mm, 960 mm/s →, 9220 mm/s$^2$ or 9.22 m/s$^2$ ←.
(b) 14.16 mm ←, 87.9 mm/s →, 3110 mm/s$^2$
or 3.11 m/s$^2$ →.

**11.5** (a) 11,980ft, (b) 4660 ft, (c) 4460 ft.

**11.7** (a) 0.586 s and 3.414 s. (b) 0 m. (c) 3.656 m.

**11.9** (a) 77.5 ft/s. (b) 7.75 s.

**11.10** 1.427 ft/s, 0.363 ft.

**11.11** (a) 3 s; (b) 13 ft, −28 ft/s; (c) 32.5 ft.

**11.12** (a) $v(t) = 20 - 2.5t^{3/2}$, $x(t) = 20t - t^{5/2}$;
(b) 48 m.

**11.15** 1067 m/s$^2$ ↑.

**11.16** (a) $-4.00 \times 10^6$ ft/s$^2$, (b) $4.48{\times}10^{-4}$ s.

**11.17** $x = 0.271$ m, $v = 0.455$ m/s.

**11.18** 167.1 mm/s$^2$ ↑, 15.19 m/s$^2$ ↑.

**11.21** (a) 2.52 m$^2$/s$^2$, 4.70 m/s.

**11.22** $x = 22.5$ ft, $v = 38.4$ ft/s.

**11.23** (a) 42.0 ft. (b) 12.86 ft/s.

**11.24** (a) 29.3 m/s. (b) 0.947 s.

**11.25** (a) 4.76 mm/s. (b) 0.171 s.

**11.26** 1.995 m/s$^2$.

**11.27** (a) $-0.0525$ m/s$^2$. (b) 6.17 s.

**11.28** (a) $x = 7.15$ km, (b) $a = -52.1 \times 10^{-6}$ m/s$^2$,
(c) $t = 49.9$ min.

**11.31** (a) 2.36 $v_0T$, $\pi v_0/T$. (b) 0.363 $v_0$.

**11.32** $r + \dfrac{d_{max}}{2} \cos\theta,\ -v_{max}\sin\theta,\ -\dfrac{d_{max}}{2}\ddot\theta\sin\theta - \dfrac{2v_{max}^2}{d_{max}}\cos\theta$

**11.33** (a) 2.5 m/s$^2$, (b) 1125 m.

**11.34** (a) 3.05 m/s$^2$, (b) 8.77 m/s$^2$.

**11.35** (a) 6.0 s. (b) 180.0 ft.

**11.36** (a) 252 ft/s. (b) 1076 ft.

**11.39** 11.60 s, 50.4 m.

**11.40** (a) 1.563 m/s$^2$. (b) 3.13 m/s$^2$.

**11.41** (a) $-3.20$ ft/s$^2$ and 3.72 ft/s$^2$. (b) 3.41 s before $A$ reaches the exchange zone.

**11.42** (a) 15.05 s, 734 ft from the initial point of $A$.
(b) $A$: 42.5 mi/h; $B$: 23.7 mi/h.

**11.43** (a) $\mathbf{a}_A = 0.767$ ft/s$^2$ ←, $\mathbf{a}_B = 0.834$ ft/s$^2$ →. (b) 20.7 s.
(c) 51.8 mi/h.

**11.44** (a) 1.330 s. (b) 4.68 m below the man.

**11.47** (a) 8.00 m/s ↑. (b) 4.00 m/s ↑. (c) 12.00 m/s ↑.
(d) 8.00 m/s ↑.

**11.48** (a) $\mathbf{a}_E = 2.40$ ft/s$^2$ ↑, $\mathbf{a}_C = 4.80$ ft/s$^2$ ↓.
(b) 12.00 ft/s ↑.

**11.49** (a) 0.125 m/s ↑. (b) 0.5154 m/s ⦩ 14°.

**11.50** (a) 18 ft/s$^2$ ←, 6 ft/s$^2$ ↑. (b) 9 ft/s ←, 2.25 ft ←.

**11.51** (a) 600 mm/s →, (b) 1200 mm/s ←,
(c) 900 mm/s ←.

**11.52** (a) $a_A = 50.8$ mm/s$^2$ →, $a_B = 25.4$ mm/s$^2$ ←.
(b) $v_B = 152.2$ mm/s ←, $\Delta x_B = 458$ mm ←.

**11.55** (a) 2.5 s. (b) 7.5 in. ↓.

**11.56** (a) $(4.02 - 1.5t^2)$ mm/s$^2$ ↓, (b) 7.96 mm.

**11.57** (a) $\mathbf{a}_A = 345$ mm/s$^2$ ↓, $\mathbf{a}_B = 240$ mm/s$^2$ ↑.
(b) $(v_A)_0 = 43.3$ mm/s ↑, $(v_C)_0 = 130.0$ mm/s →.
(c) 728 mm →.

**11.58** (a) 10.00 mm/s →. (b) $\mathbf{a}_A = 2.00$ mm/s$^2$ ↑,
$\mathbf{a}_C = 6.00$ mm/s$^2$ →. (c) 175.0 mm ↑.

**11.61** 88 ft.

**11.62** (b) 5.83 s.

**11.63** (a) 10 s to 26 s, $a = -5.00$ m/s$^2$; 41 s to 46 s, $a = 3.00$ m/s$^2$;
otherwise $a = 0$. (b) 1383 m. (c) 9.00 s, 49.5 s.

**11.64** (a) Same as Prob. 11.63. (b) 420 m.
(c) 10.69 s, 40.0 s.

**11.65** (a) 162 ft. (b) 18 s and 30 s.

**11.66** (a) 44.8 s. (b) 103.3 m/s$^2$.

**11.69** (a) 0.600 s. (b) 0.200 m/s, 2.84 m.

**11.70** (a) 60.0 m/s, 1194 m. (b) 59.3 m/s.

**11.71** (a) $A$: 52.2 s, $B$: 52.0 s. (b) 1.879 m.

**11.72** 9.39 s.

**11.73** 8.54 s, 58.3 mi/h.

**11.74** 77.5 ft.

**11.75** 5.67 s.

**11.78** (a) 18.00 s. (b) 178.8 m. (c) 34.7 km/h.

**11.79** (a) 5.01 min. (b) 19.18 mi/h.

**11.80** (a) 2.00 s. (b) 1.200 ft/s, 0.600 ft/s.

**11.83** (a) 2.96 s. (b) 224 ft.

**11.84** (a) 163.0 in/s$^2$. (b) 114.3 in/s$^2$.

**11.85** (a) 15.49 s. (b) 4.65 m/s. (c) 2.90 m/s, 8.50 m.

**11.89** (a) 6.28 m/s ⦩ 37.2°. (b) 7.49 m.

**11.90** (a) 67.1 mm/s ⦨ 63.4°, 256 mm/s$^2$ ⦩ 69.4°.
(b) 8.29 mm/s ⦨ 36.2°, 336 mm/s$^2$ ⦩ 86.6°.

**11.91** (a) 9.42 ft/s ↑, (b) 7.26 ft/s ←, (c) 3.14 ft/s ↓.

**11.92** (a) max: 15.00 ft/s, min: 5.00 ft/s. (b) min: $t = 2\pi N$ s,
$x = 20\pi N$ ft, $y = 5$ ft, $v_x = 5$ ft/s, $v_y = 0$, $\theta = 0$; max:
$t = (2N + 1)\,\pi$ s, $x = 20\pi(N + 1)$ ft, $y = 15$ ft, $v_x = 15$ ft/s,
$v_y = 0$, $\theta = 0$.

**11.95** $\sqrt{R^2\left(1 + w_n^2t^2\right) + c^2}$, $Rw_n\sqrt{4 + w_n^2t^2}$.

**11.97** 723 ft

**11.98** (a) 2.94 s. (b) 84.9 m. (c) 10.62 m.

**11.99** (a) 115.3 km/h $\leq v_0 \leq$ 148.0 km/h. (b) $h = 0.788$ m,
$\alpha = 6.66°$; $h = 1.068$ m, $\alpha = 4.05°$.

**11.100** 15.38 ft/s $< v_0 <$ 35.0 ft/s.

**11.102** (a) Meets max. height requirement.
(b) 0.937 m.

**11.103** (a) Ball clears the net.
(b) 7.01 m from the net.

**11.105** 22.9 ft/s.

**11.106** 16.20 m/s $< v_0 <$ 21.0 m/s.

**11.107** (a) 29.8 ft/s. (b) 29.6 ft/s.

**11.108** 37.7 m/s $< v_0 <$ 44.3 m/s.

**11.111** (a) 4.72°, (b) 4.07°

**11.112** (a) 4.17°. (b) 285 m. (c) 15.89 s.

**11.113** (a) 14.89°, (b) 0.1254 s

**11.114** (a) 4.98 m. (b) 23.8°.

**11.117** 17.80 ft/s ⦩ 50.9°.

**11.118** $\mathbf{v}_A = 125$ mm/s ↑, $\mathbf{v}_B = 75$ mm/s ↓, $\mathbf{v}_C = 175$ mm/s ↓.

**11.119** (a) 91.0 ft/s ⦩ 47.0°. (b) 364 ft ⦩ 47.0°.
(c) 293 ft.

**11.120** 3.20 km/h ⦩ 17.8°.

**11.123** (a) 4 ft/s ↑. (b) 6 ft/s$^2$ ↓.

**11.124** (a) 8.53 in/s ⦨ 54.1°. (b) 6.40 in/s ⦨ 54.1°.

**11.125** (*a*) 0.979 m. (*b*) 12.55 m/s ⦨ 86.5°.
**11.126** (*a*) 0.835 mm/s² ⦨ 75°. (*b*) 8.35 mm/s ⦨ 75°.
**11.127** (*a*) 3.26 ft/s, (*b*) 0.300 ft/s.
**11.128** 10.54 ft/s ⦨ 81.3°.
**11.129** 5.96 m/s ⦨ 82.8°.
**11.131** 22.3 km/h ⦨ 9.04°
**11.133** 2.89 m/s²
**11.134** 97.6 km/h.
**11.135** 15.66 m/s
**11.136** (*a*) 0.407 ft/s². (*b*) 0.0333 ft/s². (*c*) 0.00593 ft/s².
**11.137** 8.56 s.
**11.138** (*a*) 10.20 mm/s². (*b*) 25.2 s.
**11.139** (*a*) 178.9 m. (*b*) 1.118 m/s².
**11.141** (*a*) 189.5 km/h ⦨ 54.0°. (*b*) 21.8 m/s² ⦨ 5.3°.
**11.143** (*a*) $1.047\mathbf{i} - 33.726\mathbf{j}$ m/s². (*b*) $-47.55\mathbf{i} - 8.64\mathbf{j}$ m/s.
**11.144** (*a*) 0.800 in./s², (*b*) 1.084 in./s².
**11.145** (*a*) 281 m. (*b*) 209 m.
**11.146** (*a*) 21.8 ft, (*b*) 12.0 ft.
**11.147** (*a*) 0.634 m. (*b*) 9.07 m.
**11.149** (*a*) 14.48 m/s. (*b*) 21.3 m.
**11.151** $(R^2 + c^2)/2w_nR$.
**11.152** 2.50 ft.
**11.153** 149.8 Gm.
**11.154** 1425 Gm.
**11.155** 16,200 mi/h.
**11.156** 7740 mi/h.
**11.159** 1.606 h.
**11.161** $\mathbf{v} = -(3.93 \text{ ft/s})\mathbf{e}_\theta$, $\mathbf{a} = (13.57 \text{ ft/s}^2)\mathbf{e}_r - (3.93 \text{ ft/s}^2)\mathbf{e}_\theta$.
**11.162** (*a*) $-(0.398 \text{ ft/s})\mathbf{e}_r$, (*b*) $(0.797 \text{ ft/s}^2)\mathbf{e}_r + (13.19 \text{ ft/s}^2)\mathbf{e}_\theta$
**11.163** 13.280 m/s ⦨ 27.08°, 0.2437 m/s² ⦨ 30.00°.
**11.164** (*b*) 1.787 m/s².
**11.165** (*a*) $\mathbf{v} = bk\mathbf{e}_\theta$, $\mathbf{a} = -(bk^2/2)\mathbf{e}_r$. (*b*) $\mathbf{v} = 2bk\mathbf{e}_r + 2bk\mathbf{e}_\theta$, $\mathbf{a} = 2bk^2\mathbf{e}_r + 4bk^2\mathbf{e}_\theta$.
**11.166** (*a*) $a = 4b\dot{\theta}^2$. (*b*) directed toward point *A*.
**11.169** $\dot{r} = 370$ ft/s, $\ddot{r} = 57.9$ ft/s², $\dot{\theta} = -0.0924$ rad/s, $\ddot{\theta} = 0.0315$ rad/s².
**11.170** $v = 409$ mi/h, $\alpha = 3.80°$
**11.171** 232 km/h.
**11.172** 61.8 mi/h, 49.7°.
**11.175** $be^{\frac{1}{2}\theta^2}\theta(\theta^2 + 4)^{\frac{1}{2}}\omega^2$.
**11.176** $\dfrac{b}{\theta^4}(36 + 4\theta^2 + \theta^4)^{\frac{1}{2}}\omega^2$.
**11.177** $v = \frac{1}{2}A\sqrt{16\pi^2 + t^2}$, $a = \frac{1}{2}A\sqrt{64\pi^4 + 1}$
**11.179** $t = 0$: $v = \sqrt{A^2 + A^2B^2 + C^2}$, $a = \sqrt{4A^2 + A^2B^4 + 4C^2}$
$t = \infty$: $v = 0$, $a = 0$
**11.180** $\tan^{-1}[R(2 + w_n^2t^2)/c\sqrt{4 + w_n^2t^2}]$.
**11.181** (*a*) $\theta_x = 90°$, $\theta_y = 123.7°$, $\theta_z = 33.7°$. (*b*) $\theta_x = 103.4°$, $\theta_y = 134.3°$, $\theta_z = 47.4°$.
**11.182** (*a*) 80.0 s; (*b*) 14.67 m/s; (*c*) $h_{max} = 600$ m, $d = 678$ m.
**11.183** (*a*) 9.6 s. (*b*) 543.0 m.
**11.185** (*a*) 111.4 km/h ⦨ 10.50°. (*b*) 2.96 km.
**11.187** (*a*) 2.15 s, (*b*) $-(7.88 \text{ m/s}^2)\mathbf{i} - (11.84 \text{ m/s}^2)\mathbf{j}$.
**11.188** (*a*) 38.1 m/s, 20.4 m. (*b*) 41.1 m/s, 29.6 m.
**11.189** (*a*) 3.21 ft/s² ⦨ 22.4°. (*b*) 6.43 ft/s² ⦨ 22.4°.
**11.190** $1.097\,\mathbf{e}_t + 19.71\mathbf{e}_n$ m/s².
**11.191** (*a*) 23.4 ft/s. (*b*) 103.2 ft.

## CHAPTER 12

**12.1** 37.5 N.
**12.2** (*a*) 0.1554 slugs; (*b*) 4.987 lb, 4.992 lb, 5.002 lb.

**12.3** (*a*) $m = 59.0$ lb·s²/ft, (*b*) $W = 1901$ lb.
**12.5** (*a*) 0.0896 ft/s², (*b*) 17.92 ft, (*c*) 1.108 s.
**12.6** (*a*) 1160 ft, (*b*) 8.64 s.
**12.7** (*a*) $v = 110.5$ km/h, (*b*) $v = 85.6$ km/h, (*c*) $v = 69.9$ km/h.
**12.8** (*a*) 18.84 s. (*b*) 36.1 m.
**12.11** (*a*) 1.598 km. (*b*) 45.8 s.
**12.12** (*a*) 234 m. (*b*) 3.33 kN (tension).
**12.15** (*a*) (1): 10.73 ft/s² ↓, (2): 16.10 ft/s² ↓, (3): 0.749 ft/s² ↓. (*b*) (1): 14.65 ft/s ↓, (2): 17.94 ft/s ↓, (3): 3.87 ft/s ↓. (*c*) (1): 1.864 s, (2): 1.242 s, (3): 26.7 s.
**12.16** $\mathbf{a}_A = 0.997$ ft/s² ⦨ 15°, $\mathbf{a}_B = 1.619$ ft/s² ⦨ 15°.
**12.17** (*a*) 765 lb. (*b*) 1016 lb.
**12.18** (*a*) 0.986 m/s² ⦨ 25°. (*b*) 51.7 N.
**12.19** (*a*) 1.794 m/s² ⦨ 25°. (*b*) 58.2 N.
**12.20** (*a*) 16.19 kN. (*b*) 2.45 m/s².
**12.23** $\mathbf{a}_1 = 19.53$ m/s² ⦨ 65°, $\mathbf{a}_2 = 4.24$ m/s² ⦨ 65°.
**12.24** (*a*) 1.598 km, (*b*) 45.8 s.
**12.25** $v = 191.8$ km/h.
**12.27** $\sqrt{k/m}\,(\sqrt{l^2 + x_0^2} - l)$.
**12.28** (*a*) 10.00 N. (*b*) 103.1 N.
**12.29** (*a*) 8.94 ft/s² ←, 18.06 lb. (*b*) 12.38 ft/s² ←, 15.38 lb. (*c*) Same as (*b*).
**12.30** 20.26 kg.
**12.31** (*a*) 2.43 lb. (*b*) $\mathbf{a}_A = 3.14$ ft/s² →, $\mathbf{a}_B = 0.881$ m/s² →, $\mathbf{a}_C = 5.41$ m/s² ↓.
**12.34** (*a*) 8.63 ft/s² ←, (*b*) 32.2 ft/s² ⦨ 25°.
**12.35** (*a*) 5.94 m/s² ⦨ 75.6°. (*b*) 3.74 m/s ⦨ 20°.
**12.36** (*a*) 10.54 rpm, (*b*) 389 lb.
**12.37** (*a*) 80.4 N. (*b*) 1.103 m.
**12.38** (*a*) 22.55 s. (*b*) 6.379°.
**12.39** 3.47 m/s.
**12.40** 3.01 m/s ≤ $v$ ≤ 3.85 m/s.
**12.42** 0.732 m/s ≤ $v$ ≤ 4.34 m/s.
**12.43** 2.72 m/s ≤ $v$ ≤ 3.92 m/s.
**12.44** (*a*) 122.2 lb. (*b*) 145.6 lb.
**12.45** (*a*) 668 ft. (*b*) 120.0 lb ↑.
**12.46** 434 N.
**12.47** (*a*) 4.63 m/s². (*b*) 1.962 m/s². (*c*) 0.1842 m/s².
**12.48** 77.23 rpm.
**12.49** (*a*) 2.91 N. (*b*) 13.09°.
**12.50** (*a*) 1.534 m/s². (*b*) 924 N ⦨ 29.5°.
**12.51** (*a*) 12.19 m/s. (*b*) 2290 N.
**12.53** (*a*) 0.1858 W. (*b*) 10.28°.
**12.55** 7.67 m/s.
**12.56** (*a*) 12.00 m/s. (*b*) $2.05 \times 10^{-3}$ N.
**12.57** 0.236.
**12.58** 3.71 m.
**12.60** 6.18 ft/s ≤ $v$ ≤ 13.05 ft/s.
**12.61** 0.400.
**12.62** (*a*) 0.1834. (*b*) left: 10.39°, right 169.6°.
**12.63** (*a*) 2.98 ft/s. (*b*) left: 19.29°, right 160.7°.
**12.64** 0°, 180°, and 69.6°.
**12.65** (*a*) no sliding, 0.611 N ⦨ 75°. (*b*) sliding, 0.957 N ⦨ 40°.
**12.66** (*a*) 289.1 lb.
**12.67** $-2.17$ lb and 64.9 lb.
**12.68** 2.00 s.
**12.69** (*a*) 7.47 N ⦨ 45°. (*b*) 6.94 m/s² ⦨ 45°.
**12.71** 10,350 N ⦨ 33.2°.
**12.72** (*a*) 475 N, (*b*) 61.4 N →.
**12.74** $v_r = v_0 \sin 2\theta/\sqrt{\cos 2\theta}$, $v_\theta = v_0\sqrt{\cos 2\theta}$.
**12.77** (*a*) 0. (*b*) $8m\,v_0^2/r_0$.
**12.78** $413 \times 10^{21}$ lb·s²/ft.

**12.79** $383 \times 10^3$ km, $238 \times 10^3$ mi.

**12.80** (*a*) 35 800 km, 22,200 mi. (*b*) 3.07 km/s, $10.09 \times 10^3$ ft/s.

**12.81** (*b*) 24.8 m/s$^2$.

**12.82** (*a*) $1.998 \times 10^{30}$ kg. (*b*) 276 m/s$^2$.

**12.85** (*a*) 1684 N. (*b*) 2510 km. (*c*) 1.620 m/s$^2$.

**12.86** (*a*) 1551 m/s. (*b*) −15.8 m/s.

**12.87** 2.64 km/s.

**12.88** (*a*) 5280 ft/s. (*b*) 8000 ft/s.

**12.89** (*a*) $5.12 \times 10^3$ ft/s. (*b*) 97.0 ft/s.

**12.90** (*a*) $(a_A)_r = (a_A)_\theta = 0$. (*b*) 38.4 m/s$^2$. (*c*) 0.800 m/s.

**12.91** (*a*) $(a_B)_r = (a_B)_\theta = 0$. (*b*) 61.4 ft/s$^2$. (*c*) 2.98 ft/s.

**12.100** (*a*) 10.13 km/s. (*b*) 2.97 km/s.

**12.101** 1.147.

**12.103** $\sqrt{2/(2 + \alpha)}$.

**12.104** (*a*) $1.637 \times 10^3$ m/s. (*b*) 725 m/s. (*c*) 0.333.

**12.107** (*a*) $52.4 \times 10^3$ ft/s. (*b*) *A*: 1318 ft/s, *B'*: 3900 ft/s.

**12.108** $5.31 \times 10^9$ km.

**12.109** $91.8 \times 10^3$ yr.

**12.112** 4.95 h.

**12.113** 50 min 55 s.

**12.114** $\cos^{-1}[(1 - n\beta^2)/(1 - \beta^2)]$.

**12.115** (*a*) 4.00 km/s. (*b*) 0.684.

**12.124** (*a*) 21.0 ft/s$^2$ ⦦ 37.9°, (*b*) 14.61 lb.

**12.125** (*a*) 1.088 ft/s$^2$ ←. (*b*) 233 lb.

**12.126** (*a*) 11.17 m/s. (*b*) $T_{\text{before}} = 1873$ N, $T_{\text{after}} = 4370$ N, so the cable breaks. (*c*) He clears the pit by 0.38 m, so he saves Jane.

**12.127** 18.4 kN ⦨ 31.97°.

**12.128** 0.950 N ⦨ 87.7°.

**12.132** 54.0°.

**12.133** (*a*) 0.500 m, 0. (*b*) 0.270 m, −84.1 N.

# CHAPTER 13

**13.1** 6.17 GJ.

**13.2** (*a*) $T_2 = 70.1$ ft·lb, $h = 140.1$ ft. (*b*) $T_2 = 70.1$ ft·lb, $h = 850$ ft.

**13.5** 8.11 ft/s.

**13.6** 7.66 ft.

**13.7** (*a*) 110.5 km/h. (*b*) 90.2 km/h. (*c*) 142.6 km/h.

**13.8** (*a*) 17.54 m/s. (*b*) 0.893.

**13.9** 4.01 ft/s.

**13.11** 6.71 m.

**13.12** (*a*) 4.65 m/s. (*b*) 4.49 m.

**13.15** (*a*) 57.8 m. (*b*) 154 N →.

**13.16** (*a*) 7.41 kN. (*b*) 5.56 kN (tension).

**13.17** (*a*) 124.1 ft. (*b*) *A* to *B*: 19.38 kips (tension); *B* to *C*: 8.62 kips (tension).

**13.18** (*a*) 279 ft. (*b*) *A* to *B*: 19.38 kips (compression); *B* to *C*: 8.62 kips (compression).

**13.19** 1.350 ft/s

**13.20** (*a*) 7.43 ft/s. (*b*) 0.800 ft.

**13.22** (*a*) 4.63 m/s ↑. (*b*) 4.63 m/s ↓.

**13.23** 1014 lb.

**13.24** 1.190 m/s.

**13.25** $v_1 = 3.45$ m/s, $v_2 = 4.72$ m/s.

**13.26** (*a*) 3.29 m/s. (*b*) 1.533 m.

**13.27** (*a*) 3.29 m/s. (*b*) 1.472 m.

**13.28** (*a*) 8.83 lb/in. (*b*) 5.13 in.

**13.29** (*a*) 0.159. (*b*) 5.92 ft/s.

**13.32** (*a*) 0.521 ft/s. (*b*) 48.3 ft/s$^2$.

**13.33** (*a*) 13.43 ft. (*b*) 386 ft/s$^2$.

**13.34** *A*: 5.37 in.; *B*: 7.21 in.

**13.36** (*a*) 10.39 km/s. (*b*) 11.14 km/s. (*c*) 11.18 km/s.

**13.37** (*a*) 0.0316%. (*b*) 25.4%.

**13.38** 364 m.

**13.39** 14.00°.

**13.40** (*a*) $\sqrt{3gl}$. (*b*) $\sqrt{2gl}$.

**13.41** 41.8°.

**13.44** 2.30 m/s.

**13.45** (*a*) 27.4°. (*b*) 3.81 ft.

**13.46** (*a*) 57.2 kW. (*b*) 269 kW.

**13.47** (*a*) 2.75 kW. (*b*) 3.35 kW.

**13.48** 14.80 kN.

**13.51** (*a*) 14.95 kW. (*b*) 45.4 kW.

**13.52** (*a*) 375 kW. (*b*) 5.79 km/h.

**13.54** (*a*) 8.00 hp. (*b*) 7.91 hp.

**13.55** (*a*) $k_1 k_2/(k_1 + k_2)$. (*b*) $k_1 + k_2$.

**13.57** (*a*) 5.12 m/s. (*b*) 4.20 m/s.

**13.58** 4.27 ft/s.

**13.59** 3.99 ft/s.

**13.62** (*a*) 533 lb/ft. (*b*) 37.0 ft.

**13.64** (*a*) 2.71 m/s. (*b*) 3.60 m/s.

**13.65** (*a*) 2.92 m/s. (*b*) (−33.9 N)**i** + (33.3 N)**j**.

**13.66** (*a*) 43.5°. (*b*) 8.02 ft/s ↓.

**13.68** 0.269 m.

**13.69** 0.201 m.

**13.70** $N_B = 3760$ N, $N_D = 818$ N, $N_E = -58.9$ N (which means the rider needs a constraint to stay in the seat).

**13.71** $15\ \text{m} \le h \le 18.5\ \text{m}$.

**13.72** 14.34 ft/s, 13.77 lb ↑.

**13.74** Loop 1: (*a*) 25.1 ft/s. (*b*) 1.500 lb ←. Loop 2: (*a*) 24.1 ft/s. (*b*) 1.000 lb.

**13.76** Loop 1: (*a*) $\sqrt{5gr}$. (*b*) 3 W →. Loop 2: (*a*) $\sqrt{4gr}$. (*b*) 2 W →.

**13.77** 0.488 m.

**13.78** $\sqrt{gl}$

**13.80** $V = -\ln xyz$.

**13.81** (*a*) Not conservative, $\dfrac{\pi k a^2}{4}$. (*b*) Conservative, 0.

**13.82** (*a*) $P_x = x/R$, $P_y = y/R$, $P_z = z/R$, where $R = (x^2 + y^2 + z^2)^{1/2}$. (*b*) $U_{OABD} = -\Delta V_{OD} = a\sqrt{3}$.

**13.85** (*a*) 62.5 MJ/kg. (*b*) 11.18 km/s.

**13.86** (*a*) 9.56 km/s. (*b*) 2.39 km/s.

**13.87** (*a*) $50.1 \times 10^9$ ft·lb. (*b*) $115.9 \times 10^9$ ft·lb.

**13.88** (*a*) $1.918 \times 10^6$ ft·lb/lb. (*b*) $10.51 \times 10^6$ ft·lb/lb.

**13.89** 25.1 Mm/h.

**13.90** 6.48 km/s.

**13.93** $v_r = \pm 3.87$ m/s, $v_\theta = 1.000$ m/s.

**13.94** (*a*) 0.720 m. (*b*) 0.834 m/s.

**13.95** 3.77 in., $(28.04\ \text{ft/s})\mathbf{e}_r + (7.96\ \text{ft/s})\mathbf{e}_r$.

**13.96** (*a*) 14.36 ft/s. (*b*) 1.225 ft.

**13.97** (*a*) 4.14 ft/s. (*b*) 16.58 ft/s.

**13.100** $27.6 \times 10^3$ km/h.

**13.101** (*a*) 7960 ft/s. (*b*) 4820 ft/s.

**13.102** (*a*) 16,800 ft/s. (*b*) 32,700 ft/s.

**13.103** 14.20 km/s.

**13.106** (*a*) 7.35 km/s. (*b*) 45.0°.

**13.107** 68.9°.

**13.108** $r_{\max} = r_0(1 + \sin \alpha)$, $r_{\min} = (1 - \sin \alpha)r_0$.

**13.109** 3450 m/s.

**13.110** (*a*) $11.32 \times 10^3$ ft/s. (*b*) $13.68 \times 10^3$ ft/s.

**13.111** $30.9 \times 10^3$ ft/s, 58.9°.

**13.115** (*b*) $v_{\text{esc}}\sqrt{\alpha/(1 + \alpha)} < v_0 < v_{\text{esc}}\sqrt{(1 + \alpha)/(2 + \alpha)}$.

**13.119** 4 min 19 s.

**13.120** (a) 3.64 s. (b) 27.3 s.

**13.121** 17.86 lb.

**13.123** 6.26 s.

**13.124** (a) 2280 lb. (b) 3.00 s.

**13.125** 0.278.

**13.126** (a) 18.16 s. (b) 1.94 km.

**13.129** (a) 5.64 s. (b) $F_{AB}$ = 19,380 lb (T), $F_{BC}$ = 8620 lb (T).

**13.130** (a) 12.69 s. (b) $F_{AB}$ = 19,380 lb (C), $F_{BC}$ = 8620 lb (C).

**13.131** (a) 5.28 s. (b) 17.05 kN (compression).

**13.132** 6.52 s.

**13.134** (a) 3730 lb. (b) 7450 lb.

**13.136** (a) 53.7 ft/s. (b) 8.34 s.

**13.138** 15.36 mi/h.

**13.139** 76.9 lb.

**13.140** 1.449 kips.

**13.141** 6.21 W.

**13.142** 2.68 kN.

**13.145** (a) 5.73 ft/s. (b) 2030 lb·s.

**13.146** (a) Car A. (b) 115.2 km/h.

**13.147** 65.0 kN.

**13.148** (a) 9.32 ft·lb, 0.932 lb·s. (b) 7.99 ft·lb, 0.799 lb·s.

**13.149** 497 ft/s.

**13.150** (a) 2.80 ft/s ←. (b) 0.229 ft/s ←.

**13.151** (a) 1.694 m/s ↓. (b) 0.1619 J.

**13.152** (a) 778.9 m/s. (b) 4.65 J. (c) 19.74 N.

**13.155** (a) $v'_A$ = 2.3 m/s ←, $v'_B$ = 2.2 m/s →. (b) $\Delta E$ = 2.84 J.

**13.156** 0.875.

**13.157** $0.728 \le e \le 0.762$.

**13.158** $\frac{1}{4}mv^2(1 - e^2)$.

**13.161** (a) $v'_A = \frac{v_0(1 - e)}{2}$, $v'_B = \frac{v_0(1 + e)}{2}$.

(b) $v''_B = \frac{v_0(1 - e)^2}{4}$, $v'_C = \frac{v_0(1 + e)^2}{4}$.

(c) $v'_n = \frac{v_0(1 + e)^{(n-1)}}{2^{(n-1)}}$. (d) $0.732v_0$

**13.163** 0.294 m/s ←.

**13.164** (a) $v'_A$ = 0.711 $v_0$ ⦨ 39.3°, $v'_B$ = 0.636 $v_0$ ⦩ 45°.

**13.165** $v'_A$ = 0.474$v_0$ ⦨ 43.4°, $v'_B$ = 0.757$v_0$ ⦩ 63.0°.

**13.166** $v'_A$ = 6.37 m/s ⦨ 77.2°, $v'_B$ = 1.802 m/s ⦨ 40°.

**13.167** $v'_A$ = 1.322 m/s ⦨ 70.9°, $v'_B$ = 3.85 m/s ⦩ 27.0°.

**13.168** (a) $\cos^{-1}\frac{2r}{d}$.

(b) $\angle ABC > \cos^{-1}\frac{2r}{d}$ and $\angle ACB > \sin^{-1}\frac{2r}{d}$.

**13.169** 0.837.

**13.172** 13.09 m/s ⦨ 26.6°.

**13.174** (a) 20.6 mi/h. (b) 0.203.

**13.175** (a) 0.294 m. (b) 54.4 mm.

**13.176** (a) 0.324. (b) 14.30 ft/s.

**13.177** (a) 2.90 m/s. (b) 100.5 J.

**13.179** (a) 8.89 mm. (b) 3758 N.

**13.180** (a) 0.588. (b) 148.7 kN/m.

**13.182** (a) $v_A'$ = 0, $v'_B$ = 0.

(b) $v'_A$ = 1.201 m/s ←, $v'_B$ = 0.400 m/s →.

**13.183** 0.226.

**13.184** (a) 26.65 ft/s ⦨ 30°. (b) 31.93 ft/s ⦨ 39.0°.

**13.185** 3.47 in.

**13.186** (a) 0.923. (b) 1.278 m.

**13.188** (a) $v'_A$ = 2.36 ft/s ⦩ 83.88°, $v'_B$ = 3.23 ft/s →. (b) 1.97 in.

**13.190** 111.6 kips.

**13.191** $213 \times 10^6$ ft·lb.

**13.193** Minimum (just above B) 731 N; maximum (at D) 5520 N.

**13.194** (a) 8.43 in. (b) 19.38 ft/s.

**13.195** (a) 13.31 N →. (b) 4.49 N ↓. (c) 13.31 N ←.

**13.197** 7.35 m.

**13.198** (a) $v'_A = v'_B = v'_C$ = 1.368 m/s. (b) 0.668 m. (c) 1.049 m.

**13.200** 0.107 m.

## CHAPTER 14

**14.1** (a) 4.46 m/s ←. (b) 0.409 m/s ←.

**14.2** 10.67 km/h ←, 4.27 km/h ←, and 4.27 km/h ←.

**14.3** (a) 20.0 lb. (b) 3.60 ft/s →.

**14.4** 0.792 ft/s →.

**14.7** (a) 3.79 km/h →, 2.77 km/h →. (b) 5.54 km/h →, 2.77 km/h →. (c) 5.54 km/h →, 3.60 km/h →.

**14.8** $v_A$ = 1.013 m/s ←, $v_B$ = 0.338 m/s ←, $v_C$ = 0.150 m/s ←.

**14.9** $-(600$ kg·m²/s$)\mathbf{i} - (1070.0$ kg·m²/s$)\mathbf{j} + (370.0$ kg·m²/s$)\mathbf{k}$.

**14.10** (a) $(22.78$ m$)\mathbf{i} + (15.00$ m$)\mathbf{j} + (11.67$ m$)\mathbf{k}$. (b) $(38.0$ kg·m/s$)\mathbf{i} + (32.0$ kg·m/s$)\mathbf{j} + (40.0$ kg·m/s$)\mathbf{k}$. (c) $-(826.67$ kg·m²/s$)\mathbf{i} - (602.22$ kg·m²/s$)\mathbf{j} + (211.11$ kg·m²/s$)\mathbf{k}$.

**14.11** (a) $\mathbf{v}_A$ = (4.00 ft/s)$\mathbf{j}$, $\mathbf{v}_B$ = (1.000 ft/s)$\mathbf{i}$, $\mathbf{v}_C$ = (3.00 ft/s)$\mathbf{k}$. (b) (1.20 ft·lb·s)$\mathbf{i}$ + (0.60 ft·lb·s)$\mathbf{j}$ − (2.40 ft·lb·s)$\mathbf{k}$.

**14.12** (a) $\mathbf{v}_A$ = (10.00 ft/s)$\mathbf{j}$, $\mathbf{v}_B$ = (5.00 ft/s)$\mathbf{i}$, $\mathbf{v}_C$ = (10.00 ft/s)$\mathbf{k}$. (b) (6.00 ft·lb·s)$\mathbf{i}$ + (3.00 ft·lb·s)$\mathbf{j}$ − (6.00 ft·lb·s)$\mathbf{k}$.

**14.15** $(114.4$ m$)\mathbf{i} − (76.1$ m$)\mathbf{j} + (8.75$ m$)\mathbf{k}$.

**14.16** $(1180$ m$)\mathbf{i} + (140$ m$)\mathbf{j} + (155$ m$)\mathbf{k}$.

**14.19** (a) 1.300 s. (b) 75 mi/h.

**14.20** $x_D$ = 47.5 ft, $y_D$ = 50.6 ft.

**14.21** $(81.5$ ft$)\mathbf{i} + (351$ ft$)\mathbf{k}$.

**14.22** (a) 8.00 ft/s →. (b) 36.6°, $v_C$ = 10.39 ft/s, $v_D$ = 8.72 ft/s.

**14.24** $v_A$ = 431 m/s, $v_B$ = 395 m/s, $v_C$ = 528 m/s.

**14.25** $v_A$ = 646 m/s, $v_B$ = 789 m/s, $v_C$ = 176 m/s.

**14.26** $v_A$ = 919 m/s, $v_B$ = 717 m/s, $v_C$ = 619 m/s.

**14.31** friction: 2.97 J, first impact: 3007 J, second impact: 24.3 J.

**14.32** (a) 15.09 ft·lb. (b) 5.03 ft·lb.

**14.33** (woman) 382 ft·lb, (man) 447 ft·lb.

**14.35** (b) $E_A$ = 180.0 kJ, $E_B$ = 320 kJ.

**14.37** (a) $\mathbf{v}_B = \frac{m_A v_0}{m_A + m_B}$ →.

(b) $h = \frac{m_A}{m_A + m_B}\frac{v_0^2}{2g}$.

**14.38** (a) $v_A = v_B = \frac{1}{2}v_0$. (b) $v_A = v_0$, $v_B = 0$ or $v_A = 0$, $v_B = v_0$. (c) $v_A = v_B = \frac{1}{2}v_0$.

**14.39** (a) $\mathbf{v}_{B/A}$ = 11.59 ft/s ⦨ 30°. (b) $\mathbf{v}_A$ = 3.76 ft/s →.

**14.40** $\mathbf{v}_A$ = 3.11 ft/s ←, $\mathbf{v}_B$ = 4.66 ft/s →.

**14.41** $v_A$ = 7.50 ft/s, $v_B$ = 6.50 ft/s, $v_C$ = 11.25 ft/s.

**14.42** $v_A$ = 10.61 ft/s, $v_B$ = 9.19 ft/s, $v_C$ = 5.30 ft/s.

**14.45** $v_A$ = 0.218 m/s ⦨ 53.1 and $v_B$ = 1.813 m/s ⦩ 43.8.

**14.46** $(200$ ft/s$)\mathbf{i} + (172$ ft/s$)\mathbf{j} + (1560$ ft/s$)\mathbf{k}$.

**14.47** (a) $v_C$ = 11.00 ft/s, $v_D$ = 5.50 ft/s. (b) 0.786.

**14.48** $x$ = 181.7 mm, $y$ = 0, $z$ = 139.4 mm.

**14.51** (a) $\mathbf{v}_B$ = 2.40 m/s ⦨ 53.1°, $\mathbf{v}_C$ = 2.56 m/s →. (b) $c$ = 1.059 m.

**14.52** (a) $\mathbf{v}_A$ = 2.40 m/s ↓, $\mathbf{v}_B$ = 3.00 m/s ⦨ 53.1°. (b) $a$ = 1.864 m.

**14.55** (a) 5.00 ft/s →. (b) 0.500 ft. (c) 20.0 rad/s ↻.

**14.56** (a) $v_A$ = 2.17 ft/s ↑, $v_B$ = 2.17 ft/s ↓, $v_C$ = 3.75 ft/s →. (b) 0.250 ft.

**14.57** 1086.5 N.

**14.58** $\rho A_2 v_2^2 - \rho A_1 v_1^2 \cos\theta$.

**14.59** 16.89 lb ←.

**14.60** $\mathbf{M}_A$ = 46.0 N·m ↻, $\mathbf{A}$ = 274 N ⦩ 20.0°.

**14.61** (a) 14.8 kN. (b) 27.7 kN.

**14.62** 5.46 kN.

**14.64** $D_x = 329$ N, $D_y = 0$, $C_x = -203$ N, $C_y = 271$ N.

**14.66** (a) $\theta = 35.4°$. (b) 187.3N ↘ 53.8°.

**14.67** (a) 26.0 m/s. (b) 230 N ↘ 48.4°.

**14.68** D = 3120 N ↑, $C_x = 112.5$ N, $C_y = 2660$ N.

**14.69** 100 kg/s.

**14.70** 5.63 kips.

**14.71** 33.6 kN ←.

**14.72** 7180 lb.

**14.74** (a) 9690 lb, 3.38 ft. (b) 6960 lb, 9.43 ft.

**14.76** (a) 3.03 m/s² ∡ 18°. (b) 922 km/h.

**14.77** (a) 30.6 m/s. (b) 96.1 m³/s. (c) 55 100 N·m/s.

**14.78** (a) 3.23 MW. (b) 0.464.

**14.79** 213 m.

**14.80** (a) 15 450 hp. (b) 28 060 hp. (c) 0.551.

**14.83** (a) $m_0 e^{qL/m_0 v_0}$. (b) $v_0 e^{-qL/m_0 v_0}$.

**14.86** (a) $m(v^2 + gy)/l$. (b) $\mathbf{R} = mg(1 - y/l)$↑.

**14.87** (a) $mgy/l$. (b) $m[g(l - y) + v^2]/l$↑.

**14.88** $\sqrt{gh}$ tan h($\sqrt{gh}$ $t/L$).

**14.89** 10.10 ft/s.

**14.90** 4.75 ft/s.

**14.91** Space: 4200 m/s; sea level: 1600 m/s.

**14.94** (a) 90.0 m/s². (b) $35.9 \times 10^3$ km/h.

**14.95** 7930 m/s.

**14.96** (a) 1800 m/s. (b) 9240 m/s.

**14.99** 87.2 mi.

**14.100** (a) 92.8 ft/s² ↑. (b) 780 ft/s² ↑. (c) 119.3 mi. (d) 14660 mi/h.

**14.101** 186.8 km/h.

**14.102** (a) 31.2 km. (b) 197.5 km.

**14.106** 5.99 m/s.

**14.107** (a) 5.20 km/h →. (b) 4.00 km/h →.

**14.108** (a) 6.05 ft/s. (b) 6.81 ft/s.

**14.110** $\mathbf{v}_A = 15.38$ ft/s →, $\mathbf{v}_B = 5.13$ ft/s ←.
(a) $v_B = 4.10$ m/s, $v_A = 0.832$ m/s. (b) $v_A = v_B = 0$.

**14.111** (a) $v_B = 4.10$ m/s, $v_A = 0.832$ m/s. (b) $v_A = v_B = 0$.

**14.112** 37.2 ft/s.

**14.114** D = 2.29 kN ↑, C = 1.712 kN ↑.

**14.115** 414 rpm.

**14.116** Case 1: (a) 0.333 g ↓. (b) $0.817\sqrt{gl}$.
Case 2: (a) $gy/l$ ↓. (b) $\sqrt{gl}$.

## CHAPTER 15

**15.1** (a) 70.0 rad/s. (b) 79.6 rev.

**15.2** (a) 0.50 rad, −4.71 rad/s, −34.50 rad/s². (b) 0, −1.934 rad/s, 36.46 rad/s².

**15.3** (a) 0.253 rad, −0.927 rad/s, −36.55 rad/s². (b) 0, 0, 0.

**15.4** (a) 1440 rpm. (b) 12.60 s.

**15.5** (a) 150 rev. (b) 2100 rev.

**15.6** (a) 0.855 rad/s. (b) 3.71°.

**15.9** (a) 9.55 rev. (b) ∞. (c) 7.82 s.

**15.10** $\mathbf{v}_E = (1.080$ m/s$)\mathbf{i} + (2.40$ m/s$)\mathbf{j}◄$, $\mathbf{a}_E = -(11.52$ m/s²$)\mathbf{i} + (5.18$ m/s²$)\mathbf{j} + (23.1$ m/s²$)\mathbf{k}◄$.

**15.11** $\mathbf{v}_C = -(2.40$ m/s$)\mathbf{j} - (1.800$ m/s$)\mathbf{k}◄$, $\mathbf{a}_C = (18.00$ m/s²$)\mathbf{i} + (19.20$ m/s²$)\mathbf{j} - (15.60$ m/s²$)\mathbf{k}◄$.

**15.12** $-(37.4$ in/s$)\mathbf{i} + (12.00$ in/s$)\mathbf{j} - (15.60$ in/s$)\mathbf{k}$, $-(126.1$ in/s²$)\mathbf{i} - (74.3$ in/s²$)\mathbf{j} + (246$ in/s²$)\mathbf{k}$.

**15.13** $-(18.72$ in/s$)\mathbf{i} + (6.00$ in/s$)\mathbf{j} - (7.80$ in/s$)\mathbf{k}$, $-(3.46$ in/s²$)\mathbf{i} - (27.6$ in/s²$)\mathbf{j} + (73.1$ in/s²$)\mathbf{k}$.

**15.16** 66,700 mi/h, $19.47 \times 10^{-3}$ ft/s².

**15.17** (a) 1525 ft/s, 0.1112 ft/s². (b) 1168 ft/s, 0.0852 ft/s². (c) 0, 0.

**15.18** (a) 30 ft/s² ←. (b) 24,300 ft/s².

**15.19** (a) 20.9 ft/s, 1316 ft/s². (b) 7.85 ft/s, 185.1 ft/s².

**15.22** left: 3.49 s; middle: 6.98 s; right: 13.96 s.

**15.23** $v_D = 112$ in./s, $a_{D)tan} = 8.161$ in./s², $a_{D)norm} = 896$ in./s².

**15.24** (a) 300 rpm ↺, 100 rpm ↻. (b) $\mathbf{a}_B = 1974$ in/s² ←, $\mathbf{a}_C = 658$ in/s² →.

**15.25** (a) A: 15.00 rad/s ↻; B: 7.50 rad/s ↺. (b) A: 75.0 ft/s² ↑; B: 37.5 ft/s² ↺.

**15.26** (a) C: 120 rpm; B: 275 rpm. (b) A: 23.7 m/s² ↑; B: 19.90 m/s² ↓.

**15.27** $\omega_B = 52.4$ rad/s ↺. $\omega_A = 17.5$ rad/s ↺.

**15.28** (a) 0.400 rad/s² ↻. (b) 1.528 rev.

**15.29** (a) 3.00 rad/s² ↻. (b) 4.00 s.

**15.30** (a) 1.975 rad/s² ↻. (b) 6.91 rad/s ↻.

**15.31** (a) 15.28 rev. (b) 10.14 s.

**15.32** (a) 15.52 s. (b) $\omega_A = 445$ rpm ↺, $\omega_B = 371$ rpm ↻.

**15.33** (a) $\alpha_A = 3.40$ rad/s² ↻, $\alpha_B = 1.963$ rad/s² ↻. (b) 9.23 s.

**15.36** $bw_0^2/2\pi$ →.

**15.37** $bv^2/2\pi r^3$ ↻.

**15.38** $\mathbf{v}_B = 140.8$ ft/s →, $\mathbf{v}_C = 0$, $\mathbf{v}_0 = 136.0$ ft/s ∡ 158, $\mathbf{v}_E = 99.6$ ft/s ↘ 45°.

**15.39** (a) 0.800 rad/s ↻. (b) $(-3.46$ ft/s$)\mathbf{i} - (4.00$ ft/s$)$ $\mathbf{j}$.

**15.40** (a) 0.231 rad/s ↻. (b) $-(1.00$ m/s$)\mathbf{i} - (0.577$ m/s$)\mathbf{j}$.

**15.41** (a) 3.00 rad/s ↻. (b) 1.30 m/s ↗ 67.4°.

**15.44** (a) 10.00 rad/s ↻. (b) $-(7.40$ m/s$)\mathbf{i} - (1.00$ m/s$)\mathbf{j}$.

**15.45** (a) $-(1.40$ m/s$)\mathbf{i} - (1.00$ m/s$)\mathbf{j}$. (b) $x = 100.0$ mm, $y = -140.0$ mm.

**15.47** (a) 0.583 rad/s ↻. (b) 1.537 ft/s ↖ 77.48°.

**15.48** (a) $\omega_B = \omega_C = \omega_D = \frac{1}{2}\omega_A$ ↺. (b) $\omega_S = 0.25$ $\omega_A$ ↺.

**15.49** (a) $\omega_B = \omega_C = \omega_D = 150$ rpm ↻. (b) $\omega_S = 195$ rpm ↻.

**15.50** (a) 8.33 rad/s ↻. 2.78 rad/s ↻.

**15.51** (a) 5.65 m/s ↑. (b) 9000 rpm, (c) 1500.

**15.53** (a) 200 rad/s ↻. (b) 24.0 rad/s ↻.

**15.55** (a) $\omega_{AD} = 4.27$ rad/s ↻. (b) $\mathbf{v}_D = 1.330$ m/s ↓. (c) $\mathbf{v}_A = 1.557$ m/s ∡ 34.7°.

**15.56** (a) $\omega_{AD} = 5.13$ rad/s ↻. (b) $\mathbf{v}_B = 0.924$ m/s ←. (c) $\mathbf{v}_A = 1.873$ m/s ↘ 34.7°.

**15.57** (a) 4.38 rad/s ↻, 12.25 in./s↑. (b) 0, 42.0 in./s↓. (c) 4.38 rad/s ↻, 12.25 in./s↓.

**15.58** (a) 22.9° and 192.6°. (b) 5.60 rad/s ↻ and 5.60 rad/s ↺.

**15.61** (a) $\mathbf{v}_P = 0$, $\omega_{BD} = 39.3$ rad/s ↻. (b) $\mathbf{v}_P = 6.28$ m/s ↓, $\omega_{BD} = 0$.

**15.62** $\mathbf{v}_P = 6.52$ m/s↓, $\omega_{BD} = 20.8$ rad/s ↻.

**15.63** (a) 5.00 rad/s ↻. (b) $(0.625$ m/s$)\mathbf{i} + (1.5$ m/s$)\mathbf{j}$.

**15.64** $\omega_{DE} = 6.35$ rad/s ↺, $\omega_{BD} = 2.39$ rad/s ↻

**15.65** $\omega_{BD} = 4.00$ rad/s ↻, $\omega_{EB} = 0.600$ rad/s ↻.

**15.68** (a) 0.209 rad/s ↻. (b) 0.691 m/s ↓.

**15.69** $\omega_{AB} = 0$, $\omega_{A0} = 2.27\dfrac{\text{rad}}{\text{s}}$ ↻.

**15.70** 14.76 in/s →.

**15.71** (a) 338 mm/s ←, 0. (b) 710 mm/s ←, 2.37 rad/s ↻.

**15.72** $(1 - r_A/r_C)\omega_{ABC}$.

**15.74** (a) 1.714 in. below A. (b) 75.0 ft/s →. (c) 53.2 ft/s ∡ 41.2°.

**15.75** $x = 0$, $z = 9.34$ ft.

**15.76** (a) 3.00 rad/s ↻. (b) 300 mm/s ←. (c) 180.0 mm/s (wound).

**15.77** (a) 3.00 rad/s ↻. (b) 180 mm/s →. (c) 300 mm/s (unwound).

**15.78** (a) 0.4 m above the center of the spool. (b) $v_D = 12.5$ m/s ←.

**15.79** (a) 0.5 m above the center of the spool. (b) $v_D = 8.8$ m/s ←. (c) $v_B = 1.2$ m/s →.

**15.80** (a) A: 300 mm to the left of A. C: 600 mm to the left of C. (b) $\omega_A = 4.00$ rad/s ↻, $\omega_C = 2.00$ rad/s ↺.

**15.82** (a) 0.467 rad/s ↺. (b) 3.49 ft/s ∡ 59.2°.

**15.83** (a) 3.08 rad/s ↺. (b) 83.3 m/s ⦨ 73.9°.

**15.84** (a) $\omega_{AD}$ = 4.27 rad/s ↻. (b) $\mathbf{v}_D$ = 1.330 m/s ↓. (c) $\mathbf{v}_A$ = 1.557 m/s ⦨ 34.7°.

**15.85** (a) $\omega_{AD}$ = 5.13 rad/s ↺. (b) $\mathbf{v}_D$ = 0.924 m/s ←. (c) $\mathbf{v}_A$ = 1.870 m/s ⦨ 34.7°.

**15.86** (a) 0.122 rad/s ↺. (b) 22.76 mm/s ⦨ 15°.

**15.87** (a) 0.133 rad/s ↺. (b) 18.22 mm/s ⦨ 15°.

**15.88** (a) $(v_A/l)\sin\beta/\cos(\beta-\theta)$. (b) $v_A\cos\theta/\cos(\beta-\theta)$.

**15.89** (a) 6.72 ft/s ⦨ 45°. (b) 2.75 rad/s ↺. (c) 6.57 ft/s ⦨ 21.2°.

**15.90** (a) 0.900 rad/s ↺. (b) 411 mm/s ⦨ 20.5°.

**15.91** (a) 1.00 rad/s ↺. (b) 1.04 m/s →.

**15.94** (a) 1.58 rad/s ↺. (b) 28.0 in./s ⦨ 78.3°.

**15.95** (a) 5 rad/s ↺. (b) 135 in./s ↓.

**15.96** (a) 0.60 rad/s ↺. (b) 105 mm/s ↓.

**15.97** (a) 2.49 rad/s ↺. (b) 3.73 rad/s ↺. (c) 0.835 m/s ⦨ 53.6°.

**15.98** (a) $\omega_{AB}$ = 2 rad/s ↺, $\omega_{DE}$ = 5 rad/s ↺. (b) $\mathbf{v}_E$ = 24 in./s →.

**15.99** Space centrode: quarter circle, $r$ = 15 in., centered at $O$. Body centrode: semicircle, $r$ = 7.5 in., centered midway between.

**15.100** Space centrode: lower rack. Body centrode: circumference of gear.

**15.102** $\omega_{BD}$ = 2.39 rad/s ↺, $\omega_{DE}$ = 6.35 rad/s ↺.

**15.103** $\omega_{BD}$ = 4.000 rad/s ↺, $\omega_{EB}$ = 0.600 rad/s ↺.

**15.105** (a) 0.50 rad/s² ↺. (b) $\mathbf{a}_A$ = 3.25 m/s² ↑, $\mathbf{a}_E$ = 0.75 m/s² ↑.

**15.106** (a) 0.20 m/s² ↓. (b) 2.20 m/s² ↑.

**15.107** (a) 0.900 m/s² →. (b) 1.800 m/s² ←.

**15.108** (a) 0.600 m from $A$. (b) 0.200 m from $A$.

**15.109** (a) $-(57.5 \text{ in./s}^2)\mathbf{i} - (100 \text{ in./s}^2)\mathbf{j}$. (b) $-(115.4 \text{ in./s}^2)\mathbf{i} - (200 \text{ in./s}^2)\mathbf{j}$.

**15.110** (a) $\mathbf{a}_A$ = 51.3 in./s² ←. (b) $\mathbf{a}_D$ = 184.9 in./s² ⦨ 73.9°.

**15.111** (a) 1923 m/s² ↓. (b) 1923 m/s² ↑. (c) 1923 m/s² ⦨ 60°.

**15.112** (a) 13.35 in/s² ⦨ 61.0°. (b) 12.62 in/s² ⦨ 64.0°.

**15.113** $\mathbf{a}_A$ = 33.6 in/s² ⦨ 53.5°. $\mathbf{a}_B$ = 45 in./s² ↑. $\mathbf{a}_C$ = 50.2 in./s² ⦨ 84.3°.

**15.114** $\mathbf{a}_A$ = 27 in./s² ↑. $\mathbf{a}_B$ = 49.2 in./s² ⦨ 66.0°. $\mathbf{a}_C$ = 87.5 in./s² ⦨ 33.7°.

**15.115** (a) 2.00 rad/s² ↺. (b) 0.224 m/s² ⦨ 63.4°.

**15.118** (a) 92.5 in./s². (b) 278 in./s².

**15.120** 148.3 m/s² ↓.

**15.121** 296 m/s² ↑.

**15.122** $\mathbf{a}_D$ = 1558 m/s² ⦨ 45°, $\mathbf{a}_E$ = 337 m/s² ⦨ 45°.

**15.123** $\mathbf{v}_A$ = 0.374 m/s ←, $\mathbf{a}_A$ = 0.485 m/s² ←.

**15.124** (a) 242 in/s² ←. (b) 403 in/s² ⦨ 72.5°.

**15.125** 694 in/s² ←.

**15.127** 2.10 m/s² ⦨ 47.1°.

**15.128** (a) 1.47 rad/s² ↺. (b) 1.575 m/s² ⦨ 47.1°.

**15.129** (a) 23.4 rad/s² ↺, (b) 195.7 rad/s² ↺.

**15.130** (a) 24.7 rad/s² ↺. (b) 194.1 rad/s² ↺.

**15.132** (a) 26.1 rad/s² ↺. (b) 15.19 rad/s² ↺.

**15.133** (a) 62.0 rad/s² ↺. (b) 8.00 rad/s² ↺.

**15.134** (a) 25.2 rad/s² ↺. (b) 17.73 rad/s² ↺.

**15.136** $\mathbf{v}_D$ = 1.382 m/s ↓. $\mathbf{a}_D$ = 0.695 m/s² ↓.

**15.138** $v_B = b\omega\cos\theta$, $a_B = b\alpha\cos\theta - b\omega^2\sin\theta$.

**15.139** $v_B\sin\beta/l\cos\theta$.

**15.140** $(v_B\sin\beta/l)^2(\sin\theta/\cos^3\theta)$.

**15.141** $v_x = v[1 - \cos(vt/r)]$, $v_y = v\sin(vt/r)$.

**15.142** $\omega = \dfrac{hv_A}{h^2 + x_A^2}$ ↺, $\alpha = \dfrac{2hx_Av_A^2}{(h^2 + x_A^2)^2}$ ↺.

**15.143** $(v_B)_x = \dfrac{lh^2v_A}{(h^2 + x_A^2)^{3/2}} - v_A$ →, $(v_B)_y = \dfrac{lhx_Av_A}{(h^2 + x_A^2)^{3/2}}$ ↓.

**15.144** $\omega_{BD} = b\omega(b + l\cos\theta)/(l^2 + b^2 + 2bl\cos\theta)$ ↺, $v_E = bl\omega\sin\theta/(l^2 + b^2 + 2bl\cos\theta)$ ⦨ $\tan^{-1}[(b\sin\theta/(l + b\cos\theta)]$

**15.145** $bl\omega^2(l^2 - b^2)\sin\theta/(l^2 + b^2 + 2bl\cos\theta)$ ↺.

**15.147** $\omega = v_0\sin^2\theta/r\cos\theta$ ↺, $\alpha = (v_0/r)^2(1 + \cos^2\theta)\tan^3\theta$ ↺.

**15.148** $(v_\rho)_x = r\omega\left[\cos\dfrac{r\omega t}{R - r} - \cos\omega t\right]$, $(v_\rho)_y = r\omega\left[\sin\dfrac{r\omega t}{R - r} + \sin\omega t\right]$,

**15.149** Path is the $y$ axis. $\mathbf{v} = (R\omega\sin\omega t)\mathbf{j}$, $\mathbf{a} = (R\omega^2\cos\omega t)\mathbf{j}$.

**15.150** $\mathbf{v}_P$ = 1167 mm/s ⦨ 51.8°.

**15.151** $\mathbf{v}_P$ = 1893 mm/s² ⦨ 67.6°.

**15.152** (a) $\omega_{BE}$ = 1.815 rad/s ↺. (b) $\mathbf{v}_{P/BE}$ = 16.42 in./s ⦨ 70°.

**15.153** (a) $\omega_{BD}$ = 5.16 rad/s ↺. (b) $\mathbf{v}_{P/AD}$ = 1.339 m/s ⦨ 30°.

**15.154** (a) 3.81 rad/s ↺, 6.53 m/s ⦨ 16.26°. (b) 3.00 rad/s ↺, 4.00 m/s →.

**15.155** (a) 11.25 rad/s ↺. (b) 75.0 in./s →.

**15.160** (a) $1.78 \times 10^{-3}$ m/s² west. (b) $1.36 \times 10^{-3}$ m/s² west. (c) $1.36 \times 10^{-3}$ m/s² west.

**15.161** (a) 54 rad/s² ↺. (b) 33.9 ft/s² ⦨ 45°.

**15.162** 0.0234 m/s² west.

**15.164** (a) 0.520 m/s ⦨ 82.6°. (b) 50.0 mm/s² ⦨ 9.8°.

**15.165** (a) 0.603 m/s ⦨ 31.6°. (b) 82.0 mm/s² ⦨ 34.9°.

**15.166** (a) 1006 mm/s ⦨ 72.6°. (b) 1811 mm/s² ⦨ 32.0°.

**15.167** (a) 1018 mm/s ⦨ 70.5°. (b) 1537 mm/s² ⦨ 2.4°.

**15.168** (1) 303 mm/s² →; (2) 168.5 mm/s² ⦨ 57.7°.

**15.169** (3) 483 mm/s² ←; (4) 168.5 mm/s² ⦨ 57.7°.

**15.170** 0.750 m/s ⦨ 71.3°, 2.13 m/s² ⦨ 61.9°.

**15.171** 2.79 rad/s ↺, 2.13 rad/s² ↺.

**15.174** (a) 0.436 rad/s ↺. (b) 0.271 rad/s² ↺.

**15.175** (a) 0.354 rad/s ↺. (b) 0.125 rad/s² ↺.

**15.176** 7.86 rad/s ↺, 81.1 rad/s² ↺.

**15.177** 3.81 rad/s ↺, 81.4 rad/s² ↺.

**15.178** 1.526 rad/s ↺, 57.6 rad/s² ↺.

**15.181** (a) 3.61 rad/s ↺. (b) 86.6 in./s ⦨ 30°. (c) 563 in./s² ⦨ 46.1°.

**15.182** (a) 3.61 rad/s ↺. (b) 86.6 in./s ⦨ 30°. (c) 563 in./s² ⦨ 46.1°.

**15.183** 51.5 m/s² ⦨ 44.4°.

**15.184** (a) $(33.0 \text{ rad/s})\mathbf{i} - (44.0 \text{ rad/s})\mathbf{k}$. (b) $(4.80 \text{ m/s})\mathbf{i} + (3.60 \text{ m/s})\mathbf{k}$.

**15.185** (a) $(44.0 \text{ rad/s})\mathbf{i} - (33.0 \text{ rad/s})\mathbf{k}$. (b) $(3.60 \text{ m/s})\mathbf{i} + (4.80 \text{ m/s})\mathbf{k}$.

**15.186** $\boldsymbol{\alpha} = -(14.80 \text{ rad/s}^2)\mathbf{k}$.

**15.187** (a) $(0.60 \text{ rad/s})\mathbf{i} - (2.00 \text{ rad/s})\mathbf{j} + (0.75 \text{ rad/s})\mathbf{k}$. (b) $(20.0 \text{ in./s})\mathbf{i} + (15.0 \text{ in./s})\mathbf{j} + (24.0 \text{ in./s})\mathbf{k}$.

**15.188** $(118.4 \text{ rad/s}^2)\mathbf{i}$.

**15.189** $(230 \text{ rad/s}^2)\mathbf{i} - (2.5 \text{ rad/s}^2)\mathbf{k}$.

**15.190** (a) $(6.28 \text{ rad/s}^2)\mathbf{i}$. (b) $(8.38 \text{ rad/s}^2)\mathbf{k}$.

**15.193** (a) $-(0.600 \text{ m/s})\mathbf{i} + (0.750 \text{ m/s})\mathbf{j} - (0.600 \text{ m/s})\mathbf{k}$. (b) $-(6.15 \text{ m/s}^2)\mathbf{i} - (3.00 \text{ m/s}^2)\mathbf{j}$.

**15.195** (a) $-(20.0 \text{ rad/s}^2)\mathbf{j}$. (b) $-(4.00 \text{ ft/s}^2)\mathbf{i} + (10.00 \text{ ft/s}^2)\mathbf{k}$. (c) $-(10.25 \text{ ft/s}^2)\mathbf{j}$.

**15.196** $-(3.46 \text{ ft/s}^2)\mathbf{i} - (5.13 \text{ ft/s}^2)\mathbf{j} + (8.66 \text{ ft/s}^2)\mathbf{k}$.

**15.197** (a) $\omega_1/\sin\beta$. (b) $\omega_1/\tan\beta\mathbf{i}$. (c) $\omega_1^2/\tan\beta\mathbf{k}$.

**15.198** (a) $(0.0375 \text{ rad/s}^2)\mathbf{i}$. (b) $-(0.1434 \text{ m/s})\mathbf{i} + (0.204 \text{ m/s})\mathbf{j} - (0.1228 \text{ m/s})\mathbf{k}$. (c) $-(0.696 \text{ m/s}^2)\mathbf{i} - (0.0358 \text{ m/s}^2)\mathbf{j} + (0.0430 \text{ m/s}^2)\mathbf{k}$.

**15.199** (a) $\boldsymbol{\omega} = (20.0 \text{ rad/s})\mathbf{i} - (7.50 \text{ rad/s})\mathbf{j}$.

**15.200** (a) $\alpha = -(150.0 \text{ rad/s}^2)\mathbf{k}$. (b) $\mathbf{a}_2 = -(18.75 \text{ ft/s}^2)\mathbf{i} -$ (200 ft/s$^2$)$\mathbf{j}$.

**15.203** $-(33.3 \text{ in./s})\mathbf{j}$.

**15.204** $(15.0 \text{ in./s})\mathbf{j}$.

**15.205** $-(34.5 \text{ mm/s})\mathbf{i}$.

**15.206** $-(30.0 \text{ in./s})\mathbf{j}$.

**15.207** $(45.7 \text{ in./s})\mathbf{j}$.

**15.210** $(\omega_2/\cos 25°) (-\sin 25°\mathbf{i} + \cos 25° \mathbf{k})$.

**15.211** $(\omega_1 \cos 25°) (-\sin 25°\mathbf{i} + \cos 25°\mathbf{k})$.

**15.212** (a) $(1.463 \text{ rad/s})\mathbf{i} + (0.1052 \text{ rad/s})\mathbf{j} + (0.0841 \text{ rad/s})\mathbf{k}$. (b) $-(1.725 \text{ in./s})\mathbf{i}$.

**15.213** (a) $-(4.15 \text{ rad/s})\mathbf{i} + (0.615 \text{ rad/s})\mathbf{j} - (2.77 \text{ rad/s})\mathbf{k}$. (b) $(0.30 \text{ m/s})\mathbf{k}$.

**15.216** $-(45.0 \text{ in./s}^2)\mathbf{j}$.

**15.217** $(205 \text{ in./s}^2)\mathbf{j}$.

**15.218** $-(9.51 \text{ mm/s}^2)\mathbf{j}$.

**15.219** $-(8.76 \text{ mm/s}^2)\mathbf{j}$.

**15.220** (a) $(-3.00 \text{ ft/s})\mathbf{i} + (6.00 \text{ ft/s})\mathbf{j} - (20.94 \text{ ft/s})\mathbf{k}$. (b) $(6.28 \text{ rad/s}^2)\mathbf{i}$. (c) $(-62.87 \text{ ft/s}^2)\mathbf{i} - (9.00 \text{ ft/s}^2)\mathbf{j} +$ (12.57 ft/s$^2$)$\mathbf{k}$.

**15.221** (a) $-(24.94 \text{ ft/s})\mathbf{k}$. (b) $(1.00 \text{ rad/s}^2)\mathbf{j} + (8.38 \text{ rad/s}^2)\mathbf{k}$. (c) $-(60.62 \text{ ft/s}^2)\mathbf{i} - (16.00 \text{ ft/s}^2)\mathbf{j} - (10.00 \text{ ft/s}^2)\mathbf{k}$.

**15.222** (a) $\mathbf{v}_C = (0.493 \text{ m/s})\mathbf{i} + (1.353 \text{ m/s})\mathbf{j} - (1.015 \text{ m/s})\mathbf{k}$. (b) $\mathbf{a}_C = -(8.46 \text{ m/s}^2)\mathbf{i} + (1.970 \text{ m/s}^2)\mathbf{j} - (2.96 \text{ m/s}^2)\mathbf{k}$.

**15.223** (a) $\mathbf{v}_D = (1.033 \text{ m/s})\mathbf{i} + (1.353 \text{ m/s})\mathbf{j} - (1.015 \text{ m/s})\mathbf{k}$. (b) $\mathbf{a}_D = -(8.46 \text{ m/s}^2)\mathbf{i} + (1.970 \text{ m/s}^2)\mathbf{j} - (4.58 \text{ m/s}^2)\mathbf{k}$.

**15.224** (a) $(1.200 \text{ m/s})\mathbf{i} + (0.500 \text{ m/s})\mathbf{j} - (1.200 \text{ m/s})\mathbf{k}$. (b) $-(7.20 \text{ m/s}^2)\mathbf{i} - (14.40 \text{ m/s}^2)\mathbf{k}$.

**15.227** (a) $(0.750 \text{ m/s})\mathbf{i} + (1.299 \text{ m/s})\mathbf{j} - (1.732 \text{ m/s})\mathbf{k}$. (b) $(27.1 \text{ m/s}^2)\mathbf{i} + (5.63 \text{ m/s}^2)\mathbf{j} - (15.00 \text{ m/s}^2)\mathbf{k}$.

**15.228** (a) $(129.9 \text{ mm/s})\mathbf{i} + (75.0 \text{ mm/s})\mathbf{j} + (86.6 \text{ mm/s})\mathbf{k}$. (b) $(45.0 \text{ mm/s}^2)\mathbf{i} - (112.6 \text{ mm/s}^2)\mathbf{j} + (60.0 \text{ mm/s}^2)\mathbf{k}$.

**15.230** $\mathbf{v}_C = -(45.0 \text{ in./s})\mathbf{i} + (36.6 \text{ in./s})\mathbf{j} - (31.2 \text{ in./s})\mathbf{k}$, $\mathbf{a}_C = -(303 \text{ in./s}^2)\mathbf{i} - (384 \text{ in./s}^2)\mathbf{j} + (208 \text{ in./s}^2)\mathbf{k}$.

**15.231** (a) $\omega_1 + (R/r) (\omega_1 - \omega_2)\mathbf{k}$. (b) $\omega_1(\omega_1 - \omega_2) (R/r)\mathbf{j}$.

**15.232** $-(41.6 \text{ in/s}^2)\mathbf{i} - (61.5 \text{ in/s}^2)\mathbf{j} + (103.9 \text{ in/s}^2)\mathbf{k}$.

**15.233** (a) $(0.0375 \text{ rad/s}^2)\mathbf{i}$. (b) $-(0.143 \text{ m/s})\mathbf{i} + (0.205 \text{ m/s})\mathbf{j} - (0.123 \text{ m/s})\mathbf{k}$. (c) $-(0.0696 \text{ m/s}^2)\mathbf{i} - (0.0358 \text{ m/s}^2)\mathbf{j} + (0.0430 \text{ m/s}^2)\mathbf{k}$.

**15.234** $\mathbf{v}_A = -(1.39 \text{ m/s})\mathbf{i} + (0.80 \text{ m/s})\mathbf{j} - (1.20 \text{ m/s})\mathbf{k}$, $\mathbf{a}_A = -(20.8 \text{ m/s}^2)\mathbf{i} - (11.09 \text{ m/s}^2)\mathbf{j} + (33.3 \text{ m/s}^2)\mathbf{k}$.

**15.235** $\mathbf{v}_A = -(1.39 \text{ m/s})\mathbf{i} + (0.80 \text{ m/s})\mathbf{j} - (1.20 \text{ m/s})\mathbf{k}$, $\mathbf{a}_A = -(22.5 \text{ m/s}^2)\mathbf{i} - (10.09 \text{ m/s}^2)\mathbf{j} + (34.9 \text{ m/s}^2)\mathbf{k}$.

**15.236** (a) $-(1.37 \text{ ft/s})\mathbf{i} + (3.76 \text{ ft/s})\mathbf{j} + (1.88 \text{ ft/s})\mathbf{k}$. (b) $(1.22 \text{ ft/s}^2)\mathbf{i} - (0.342 \text{ ft/s}^2)\mathbf{j} - (0.410 \text{ ft/s}^2)\mathbf{k}$.

**15.239** (a) $(4.33 \text{ ft/s})\mathbf{i} - (6.18 \text{ ft/s})\mathbf{j} + (5.30 \text{ ft/s})\mathbf{k}$. (b) $(2.65 \text{ ft/s}^2)\mathbf{i} - (2.64 \text{ ft/s}^2)\mathbf{j} - (3.25 \text{ ft/s}^2)\mathbf{k}$.

**15.240** (a) $\mathbf{a}_A = (544 \text{ mm/s}^2)\mathbf{i} - (135.0 \text{ mm/s}^2)\mathbf{j}$. (b) $\mathbf{a}_B = (256 \text{ mm/s}^2)\mathbf{i} - (153.6 \text{ mm/s}^2)\mathbf{k}$.

**15.241** (a) $\mathbf{a}_C = -(32.0 \text{ mm/s}^2)\mathbf{i} + (135.0 \text{ mm/s}^2)\mathbf{j}$. (b) $\mathbf{a}_D = (256 \text{ mm/s}^2)\mathbf{i} + (153.6 \text{ mm/s}^2)\mathbf{k}$.

**15.242** $\mathbf{v}_A = -(0.500 \text{ ft/s})\mathbf{k}$. $\mathbf{a}_A = -(32.0 \text{ ft/s}^2)\mathbf{i} - (21.0 \text{ ft/s}^2)\mathbf{j}$.

**15.243** $\mathbf{v}_B = -(3.00 \text{ ft/s})\mathbf{k}$, $\mathbf{a}_B = -(32.0 \text{ ft/s}^2)\mathbf{i} - (6.00 \text{ ft/s}^2)\mathbf{j}$.

**15.244** (a) $r\omega_2^2 \sin 30° j - (r\omega_2^2 \cos 30° + 2r\omega_1\omega_2)k$. (b) $-r(\omega_1^2 + \omega_2^2 + 2\omega_1\omega_2 \cos 30°)i + r\omega_1^2 \cos 30°k$. (c) $-r\omega_2^2 \sin 30° j + r(2\omega_1^2 \cos 30° + \omega_2^2 \cos 30° + 2\omega_1\omega_2)k$.

**15.245** (a) $(0.610 \text{ m/s})\mathbf{k}, -(0.880 \text{ m/s}^2)\mathbf{i} + (1.170 \text{ m/s}^2)\mathbf{j}$. (b) $(5.20 \text{ m/s})\mathbf{i} - (0.390 \text{ m/s})\mathbf{j} - (1.000 \text{ m/s})\mathbf{k}$, $-(4.00 \text{ m/s}^2)\mathbf{i} - (3.25 \text{ m/s}^2)\mathbf{k}$.

**15.248** $(36.0 \text{ ft/s})\mathbf{i} - (64.0 \text{ ft/s})\mathbf{j}$.

**15.249** (a) $5.00 \text{ ft/s}^2 \rightarrow$. (b) 5.63 in. $\leftarrow$.

**15.251** (a) $\mathbf{v}_A = 17.97 \text{ in./s} \rightarrow$, $\mathbf{a}_A = 400 \text{ in./s}^2 \rightarrow$. (b) $\mathbf{v}_B = (6.96 \text{ in./s})\mathbf{i} + (15.68 \text{ in./s})\mathbf{j}$, $\mathbf{a}_B = (369 \text{ in./s}^2)\mathbf{i} - (9.39 \text{ in./s}^2)\mathbf{j} \rightarrow$.

**15.253** (a) $1080 \text{ rad/s}^2 \circlearrowleft$. (b) $460 \text{ ft/s}^2 \measuredangle 64.9°$.

**15.255** $49.4 \text{ m/s}^2 \measuredangle 26.0°$.

**15.256** (a) $(0.450 \text{ m/s})\mathbf{k}, (4.05 \text{ m/s}^2)\mathbf{i}$. (b) $-(1.350 \text{ m/s})\mathbf{k}$, $-(6.75 \text{ m/s}^2)\mathbf{i}$.

**15.258** $(40.0 \text{ in./s})\mathbf{k}$.

**15.259** $(9.00 \text{ in./s})\mathbf{i} - (7.80 \text{ in./s})\mathbf{j} + (7.20 \text{ in./s})\mathbf{k}, (9.00 \text{ in./s}^2)\mathbf{i} - (22.1 \text{ in./s}^2)\mathbf{j} - (5.76 \text{ in./s}^2)\mathbf{k}$.

## CHAPTER 16

**16.1** (a) $R_A = 60.31 \text{ lb} \measuredangle 84.2°$ and $N_B = 28.5 \text{ lb} \leftarrow$. (b) $\mu = 0.1023$.

**16.2** (a) $18.59 \text{ ft/s}^2 \rightarrow$. (b) 0.577.

**16.3** (a) $11.62 \text{ ft/s}^2 \measuredangle 65°$. (b) $N_A = 1522 \text{ lb}, N_B = 802 \text{ lb}$.

**16.4** (a) $377 \text{ lb} \measuredangle 75.7°$. (b) 1135 lb↑.

**16.5** (a) $4.09 \text{ m/s}^2$. (b) 42.5 N.

**16.6** (a) 5270 N ↑. (b) 4120 N.

**16.9** (a) $5.00 \text{ m/s}^2 \rightarrow$. (b) $0.311 \text{ m} \leq h \leq 1.489 \text{ m}$.

**16.10** (a) $2.55 \text{ m/s}^2 \rightarrow$. (b) $h \leq 1.047 \text{ m}$.

**16.11** 349 mm.

**16.12** 229 N.

**16.14** (a) $4.91 \text{ m/s}^2 \measuredangle 30°$. (b) $F_A = 0, F_B = 68.0 \text{ N compression}$.

**16.15** (a) $173.2 \text{ N} \rightarrow$. (b) 15.02 rad/s. (c) $86.6 \text{ rad/s}^2 \circlearrowleft$.

**16.18** $B_y = 16.48 \text{ lb}, D_y = 17.62 \text{ lb}$.

**16.19** 1381 N ↑.

**16.20** Block: $17.01 \text{ ft/s}^2 \measuredangle 58.5°$; platform: $31.3 \text{ ft/s}^2 \measuredangle 30°$.

**16.25** 125.7 N·m.

**16.26** 9480 rev.

**16.27** 193.9 rev.

**16.28** 7.74 lb·ft·s$^2$.

**16.29** 74.5 s.

**16.30** $20.4 \text{ rad/s}^2 \circlearrowleft$.

**16.31** $32.7 \text{ rad/s}^2 \circlearrowleft$.

**16.33** (a) $5.66 \text{ ft/s}^2 \downarrow$. (b) 7.52 ft/s ↓.

**16.34** (1): (a) $8.00 \text{ rad/s}^2 \circlearrowleft$. (b) $14.61 \text{ rad/s} \circlearrowleft$. (2): (a) $6.74 \text{ rad/s}^2 \circlearrowleft$. (b) $13.41 \text{ rad/s} \circlearrowleft$. (3): (a) $4.24 \text{ rad/s}^2 \circlearrowleft$. (b) $10.64 \text{ rad/s} \circlearrowleft$. (4): (a) $5.83 \text{ rad/s}^2 \circlearrowleft$. (b) $8.82 \text{ rad/s} \circlearrowleft$.

**16.36** $\alpha_A = 15.00 \text{ rad/s}^2 \circlearrowleft, \alpha_B = 10.00 \text{ rad/s}^2 \circlearrowleft$.

**16.39** (a) No slipping on $A$; slipping on $B$. (b) $\alpha_A = 61.8 \text{ rad/s}^2 \circlearrowleft$; $\alpha_B = 9.66 \text{ rad/s}^2 \circlearrowleft$.

**16.40** (a) No slipping at either cylinder. (b) $\alpha_A = 15.46 \text{ rad/s}^2 \circlearrowleft$, $\alpha_B = 7.73 \text{ rad/s}^2 \circlearrowleft$.

**16.41** (a) $\alpha_A = 12.50 \text{ rad/s}^2 \circlearrowleft, \alpha_B = 33.3 \text{ rad/s}^2 \circlearrowleft$. (b) $\omega_A = 240 \text{ rpm} \circlearrowleft, \omega_B = 320 \text{ rpm} \circlearrowleft$.

**16.42** (a) $\alpha_A = 12.50 \text{ rad/s}^2 \circlearrowleft, \alpha_B = 33.3 \text{ rpm} \circlearrowleft$. (b) $\omega_A = 90.0 \text{ rpm} \circlearrowleft, \omega_B = 120.0 \text{ rpm} \circlearrowleft$.

**16.43** (a) $\alpha_A = 9.16 \text{ rad/s}^2 \circlearrowleft, \alpha_B = 38.2 \text{ rad/s}^2 \circlearrowleft$. (b) $\mathbf{C} = 54.9 \text{ N} \uparrow, \mathbf{M}_C = 2.64 \text{ N·m} \circlearrowleft$.

**16.44** (b) $\omega_0/(1 + m_B/m_A) \circlearrowleft$.

**16.48** (a) $48.3 \text{ ft/in}^2/\text{s}^2 \rightarrow$. (b) $24.2 \text{ ft/in}^2/\text{s}^2 \leftarrow$. (c) 24 in. from end $A$.

**16.49** (a) 12 in. from end $A$. (b) $24.2 \text{ ft/in}^2/\text{s}^2 \rightarrow$.

**16.50** (a) $2.50 \text{ m/s}^2 \rightarrow$. (b) 0.

**16.51** (a) $3.75 \text{ m/s}^2 \rightarrow$. (b) $1.25 \text{ m/s}^2 \leftarrow$.

**16.52** (a) $0, -1.374 \text{ rad/s}^2 \mathbf{j}$. (b) $-(0.515 \text{ ft/s}^2)\mathbf{i}, -1.030 \text{ rad/s}^2 \mathbf{j}$.

**16.55** $\mathbf{a}_A = 2.71 \text{ m/s}^2 \uparrow, \mathbf{a}_B = 1.496 \text{ m/s}^2 \uparrow$.

**16.56** 170.9 mm.

**16.57** (a) $53.1 \text{ rad/s}^2 \circlearrowleft$. (b) $\mathbf{a} = 39.3 \text{ ft/s}^2 \downarrow$.

**16.58** (a) 0.741 rad/s² ↻. (b) 0.857 m/s².
**16.59** (a) 2800 N. (b) 15.11 rad/s² ↻.
**16.60** $T_A$ = 221 lb, $T_B$ = 203 lb.
**16.63** (a) $\dfrac{3g}{2L}$ ↻. (b) $\dfrac{g}{4}$ ↑. (c) $\dfrac{5g}{4}$ ↓.
**16.64** (a) $\dfrac{2g}{L}$ ↻. (b) $\dfrac{g}{3}$ ↑. (c) $\dfrac{5g}{3}$ ↓.
**16.65** (a) $\dfrac{3g}{L}$ ↻. (b) 1.323g ∡ 49.1°. (c) 2.18g ⦨ 66.6°.
**16.66** (a) 0.25g ↑. (b) 5g/4 ↓.
**16.67** (a) 0. (b) g ↓.
**16.69** (a) $5v_0/2r$ ↻. (b) $v_0/\mu_k g$. (c) $v_0^2/2\mu_k g$.
**16.70** (a) $v_0/r$ ↻. (b) $v_0/\mu_k g$. (c) $v_0^2/2\mu_k g$.
**16.71** (a) 1.597 s. (b) 9.86 ft/s. (c) 19.85 ft.
**16.72** (a) 1.863 s. (b) 9.00 ft/s. (c) 22.4 ft.
**16.76** (a) 150 mm. (b) 125 rad/s² ↻.
**16.77** (a) 5.41 rad/s² ↻. (b) 637 N ∡ 74.4°.
**16.78** (a) 12.08 rad/s² ↻. (b) 0.750 lb ←, 4.00 lb ↑.
**16.79** (a) 8.05 rad/s² ↻. (b) 24.0 in.
**16.80** (a) 1522.9 N. (b) 1341.8 N.
**16.83** $A_x$ = 38.8 lb ←, $A_y$ = 126.5 lb ↑.
**16.84** (a) 1.5 g ↓. (b) 0.25 mg ↑.
**16.85** 0.248 rad/s² ↻.
**16.86** (a) 0.6727 ft·lb. (b) 1999.2 lb.
**16.87** (a) 86.0 N ↓. (b) $A_x$ = 69.1 N ←, $A_y$ = 172.0 N ↑.
**16.88** (a) 3.72 rad/s² ↻. (b) 1.462 lb.
**16.94** $r^2 g \sin \beta/(r^2 + \bar{k}^2)$.
**16.95** (a) 2.27 m (7.46 ft). (b) 0.649 m (2.13 ft).
**16.98** (a) No sliding. (b) 16.0 rad/s² ↻. (c) 2.56 m/s² →.
**16.99** (a) No sliding. (b) 24.0 rad/s² ↻. (c) 3.84 m/s² →.
**16.102** (a) 15.46 rad/s² ↻, 10.30 ft/s² →. (b) 0.180.
**16.103** (a) 23.2 rad/s² ↻, 15.46 ft/s² →. (b) 0.0200.
**16.104** (a) 7.73 rad/s² ↻, 5.15 ft/s² →. (b) 0.340.
**16.105** (a) 7.73 rad/s² ↻, 5.15 ft/s² ←. (b) 0.320.
**16.107** (a) 6.63 ft/s² →. (b) 3.79 ft/s² →. (c) 0.355 ft →.
**16.108** (a) 72.4 rad/s² ↻. (b) 7.24 m/s² ↓.
**16.109** (a) 2.64 m/s² ←. (b) 11.87 N ←.
**16.111** (a) 0.298. (b) 0.536 g →.
**16.112** (a) 0.322. (b) 0.566 g →.
**16.113** 8.26 N ←.
**16.114** (a) 0.125 g/r ↻. (b) 0.125 g →, 0.125 g ↓.
**16.115** $m_B g \sin \theta/[2r \{m_h + m_B (1 + \cos \theta)\}]$.
**16.116** 3.43 lb ∡ 70.5°, 0.1550 ft·lb ↻.
**16.117** (a) $\dfrac{g}{L}\left[\dfrac{\sin \theta}{\frac{1}{3} + \sin^2 \theta}\right]$ ↻. (b) $\dfrac{mg}{1 + 3 \sin^2 \theta}$ ↑.
**16.118** (a) 27.6 rad/s² ↻. (b) 5.714 lb ↑.
**16.119** (a) 0.510 rad/s² ↻. (b) $F_A$ = 31.80 lb ∡ 78.7°, $F_B$ = 13.79 lb ∡ 11.3°.
**16.120** $mg \sin \theta/(1 + 3 \sin \theta)$.
**16.121** (a) 6.26 rad/s² ↻. (b) 13.22 N ←.
**16.124** 6.40 N ←.
**16.125** 7.10 lb →.
**16.126** 5.51 lb →.
**16.127** 67.62 N ⦨ 56.0°.
**16.128** 75.13 N ↑.
**16.129** 25.9 N ∡ 60°.
**16.131** (a) 37.8 ft/s² ⦨ 26.1°. (b) 48.4 lb ↑.
**16.132** 1.879 rad/s.
**16.134** (a) 4.36 rad/s² ↻. (b) 31.36 lb ↑.
**16.135** (a) 36.3 N·m ↻. (b) 231 N ←, 524 N ↑.
**16.136** (a) 82.3 N·m ↻. (b) 147.2 N ←, 479 N ↑.
**16.137** B = 805 N ←, D = 426 N →.

**16.138** B = 525 N ⦨ 38.1°, D = 322 N ⦧ 15.7°.
**16.139** (a) 24.8 rad/s² ↻. (b) 29.5 lb ↑.
**16.140** (a) 19.3 ft·lb ↻. (b) 81.9 lb ↑.
**16.143** (a) $\boldsymbol{\alpha}_A = \dfrac{2g}{5r}$ ↺ and $\boldsymbol{\alpha}_B = \dfrac{2g}{5r}$ ↻. (b) $\dfrac{1}{5} mg$. (c) $\dfrac{4}{5} g$ ↓.
**16.144** (a) $a_A = \dfrac{2P}{7m}$ →. (b) $a_B = \dfrac{22P}{7m}$ ←.
**16.145** (a) $\mathbf{a}_C$ = 5.63 m/s² ⦨ 25°. (b) $\boldsymbol{\alpha}$ = 7.66 rad/s² ↻.
**16.146** (a) 50.2 N ∡ 60.3°. (b) 0.273.
**16.148** (a) 17.03 ft/s² ⦨ 20°. (b) 42.7 rad/s² ↻.
**16.151** (a) $M_{max}$ = 10.39 lb·in. located 20.8 in. below $A$.
**16.153** 20.6 ft.
**16.154** 7.45 ft.
**16.156** (a) $2\mu g/(1 + 3\mu)$. (b) 1.000 g.
**16.157** (a) 0.513 g/L ↻. (b) 0.912 mg ↑. (c) 0.241 mg →.
**16.158** (a) 1.519 g/L ↻. (b) 0.260 g ↓. (c) 0.740 mg ↑.
**16.160** (1): (a) 1.200 g/c ↻. (b) 0.671 g ⦨ 63.4°.
 (2): (a) 24 g/17c ↻, (b) 12 g/17 ↓.
 (3): (a) 2.40 g/c ↻, (b) 0.500 g ↓.
**16.162** 1.214 N ←.
**16.163** (a) 12.06 rad/s² ↻. (b) 544 N ↑.

## CHAPTER 17

**17.1** 12.77 N·m.
**17.2** 8690 rev.
**17.3** 9.60 in.
**17.4** 0.841.
**17.5** (a) 296 rpm. (b) 19.10 rev.
**17.6** (a) 29.4 kg·m². (b) 13.27 rev.
**17.7** 70.1 rev.
**17.10** 109.4 lb →.
**17.11** (a) 58.1 rev. (b) 26.4 N.
**17.12** (a) 145.3 rev. (b) 10.54 N.
**17.13** $\omega_A = \dfrac{2n}{n^2 + 1}\sqrt{\dfrac{\pi M_0}{\bar{I}_0}}$.
**17.16** (a) $\omega_2 = \sqrt{\dfrac{3g}{l}}$ ↻. (b) $b = \dfrac{l}{\sqrt{12}}$. (c) $\boldsymbol{\omega}_2 = 1.861\sqrt{\dfrac{g}{l}}$ ↻. C = 2mg ↑.
**17.17** 5.78 rad/s ↻.
**17.18** 11.52 rad/s ↺.
**17.19** 16.23 ft/s.
**17.20** (a) 1.253 rad/s ↺. (b) 109.9 N ↑.
**17.23** 7.09 rad/s.
**17.24** (a) −0.250 rpm. (b) 0.249 rpm.
**17.25** 272 N.
**17.26** 0.481 rad/s ↺.
**17.27** (a) 5.18 ft/s. (b) 0.042.
**17.29** (a) 5.00 rad/s. (b) 24.9 N ↑.
**17.30** (a) $1.142\sqrt{\dfrac{g}{r}}$ ↻. (b) 1.553 mg ↑.
**17.31** (a) $[10g (R - r) (1 - \cos \beta)/7]^{1/2}$. (b) $mg(17 - 10 \cos \beta)/7$.
**17.32** (a) 2.06 ft. (b) 4.00 lb.
**17.33** (a) 7.43 ft/s ↓. (b) 4.00 lb.
**17.35** (a) 11.57 rad/s ↻. (b) 27.8 rad/s ↻.
**17.36** $v_A = 0.775\sqrt{gl}$ ←, $v_B = 0.775\sqrt{gl}$ ⦨ 60°.
**17.37** 1.170 rad/s ↻, 5.07 m/s ←.
**17.38** $[3g (\cos \theta_0 - \cos \theta_2)/L]^{1/2}$ ↻.
**17.39** 3.71 rad/s ↺, 7.74 ft/s ↑.

**17.40** 15.54 ft/s →.
**17.42** 2.69 m/s ↓.
**17.43** 84.7 rpm ↺.
**17.44** 110.8 rpm ↺.
**17.45** 3.87 m/s →.
**17.46** 3.87 m/s →.
**17.47** 0.770 m/s ←.
**17.48** (a) 44.3 hp. (b) 118.1 hp.
**17.49** (a) 39.8 N·m. (b) 95.5 N·m. (c) 229 N·m.
**17.50** 0.343 hp.
**17.51** 10.87 lb.
**17.52** 179.1 mm.
**17.53** 3.55 rad/s ↺.
**17.54** 0.0404 N·m.
**17.55** 24.6 ft·lb.
**17.58** 3.88 s.
**17.59** $(1 + \mu_k^2)\, r\omega_0/[2\mu_k(1 + \mu_k)g]$.
**17.62** $\left(\dfrac{M}{mr^2} - \dfrac{2\mu_k g}{r}\right)t$.
**17.63** $(\omega_A)_2 = 242$ rad/s ↺, $(\omega_B)_2 = 96.6$ rad/s ↺.
**17.64** (a) 5.15 lb. (b) 2.01 lb.
**17.65** $X = mv,\ d = \overline{k}^2\omega/\overline{v}$.
**17.69** 2.79 ft.
**17.70** (a) $r^2gt \sin \beta/(r^2 + \overline{k}^2)$ ⊻ $\beta$. (b) $u_s \geq \overline{k}^2 \tan \beta/(r^2 + \overline{k}^2)$.
**17.71** (a) $\dfrac{1}{2}gt\downarrow$. (b) $\dfrac{2}{3}gt\downarrow$.
**17.72** 0.716 ft/s.
**17.74** (a) 8.41 m/s ↓. (b) 16.82 N.
**17.75** (a) 0.557 s. (b) 16.82 N.
**17.77** (a) $2.50\,\overline{v}_0/r$. (b) $\overline{v}_0/\mu_k g$.
**17.78** (a) 2.50 s. (b) 16.95 ft/s.
**17.79** $\dfrac{5}{6}\omega_0$.
**17.80** 10.19 rpm.
**17.81** A and B: 159.1 rpm ↺; platform 20.9 rpm ↺.
**17.82** 18.07 rad/s.
**17.83** (a) 2.54 rad/s. (b) 1.902 J.
**17.86** $\omega_B = 337$ rpm, $\omega_A = 32.5$ rpm.
**17.87** $\boldsymbol{\omega}_{BC} = 36.6$ rpm ↺ and $\boldsymbol{\omega}_A = 16.87$ rpm ↺.
**17.88** 2.51 m/s.
**17.89** 18.83 rad/s, 0.0508 kg·m².
**17.90** (a) 31.1 rad/s. (b) 18.13 ft/s.
**17.91** (a) 15.00 rad/s. (b) 20.5 ft/s.
**17.94** 1.542 m/s.
**17.95** 2.01 ft/s ←.
**17.96** 0.400 r.
**17.97** (a) 24.4 rad/s ↺. (b) 1545 lb →.
**17.98** (a) 10.00 in. (b) 22.6 rad/s ↺.
**17.101** (a) 2.16 m/s →. (b) 4.87 kN ⊿ 66.9°.
**17.102** (a) 158.0 mm. (b) 1.992 m/s →.
**17.103** (a) 7.23 rad/s ↺. (b) 5.85 rad/s ↺.
**17.104** 2.40 rad/s ↺.
**17.105** (a) $mv_0/3$ ↑. (b) $mv_0/6$ ↑.
**17.106** $\omega = \dfrac{\sqrt{2gh}(c + \cos \theta)}{r(1 + c)^{\frac{3}{2}}}$ ↺ and $v = \dfrac{\sqrt{2gh}(c + \cos \theta)}{(1 + c)^{\frac{3}{2}}}$ →.
**17.107** $\dfrac{\pi}{3}L$.
**17.108** (a) $mv_0/M$ →. (b) $mv_0/MR$ ↺.
**17.109** (a) 1.500 R. (b) 1.000 R.
**17.112** $\omega = 2.4\dfrac{v_0}{L}$ ↺ and $\overline{v} = 0.721\,v_0$ ⊽ 56.3°.
**17.115** 2.38 m/s.

**17.116** 4.867 rad/s ↺.
**17.117** (a) 57.3 mm. (b) 41.0 mm.
**17.118** (a) 0.250 $\omega_0$ ↺. (b) 0.9375. (c) 1.50°.
**17.119** 48.7°.
**17.120** 1887 ft/s.
**17.121** 725 mm.
**17.122** 447 mm.
**17.123** $0.606\sqrt{gL}$ →.
**17.124** $0.866\sqrt{gL}$ →.
**17.127** (a) 3.00 rad/s ↺. (b) 0.938 m/s ↑.
**17.128** (a) 2.60 rad/s ↺. (b) 1.635 m/s ⊻ 53.4°.
**17.131** $1.250\,v_0/r$.
**17.132** (a) $\mathbf{v}_A = 0,\ \omega_A = v_1/r$ ↺, $\mathbf{v}_B = v_1$ →, $\omega_B = 0$.
(b) $\mathbf{v}'_A = 0.286\,v_1$ →, $v'_B = 0.514\,v_1$ →.
**17.133** (a) $\mathbf{v}_A = (v_0 \sin \theta)\mathbf{j}$, $\mathbf{v}_B = (v_0 \cos \theta)\mathbf{i}$, $\boldsymbol{\omega}_A = (v_0/r)$ $(-\sin \theta \mathbf{i} + \cos \theta \mathbf{j})$, $\omega_B = 0$. (b) $0.714\,v_0 \cos \theta \mathbf{i}$.
**17.134** (a) $\left[\dfrac{m_s + m_R}{m_s + \frac{3}{2}m_R}\right]v_0$. (b) $\left[\dfrac{m_s + m_R}{m_s + \frac{3}{2}m_R}\right]^n v_0$.
**17.135** (a) 106.7 rev. (b) 6.98 s.
**17.136** 70.1 lb ↓.
**17.137** (a) 18.22 ft/s. (b) 359.7 lb ↑. (c) 234.2 ft·lbs.
**17.139** (a) 53.1°. (b) (b) $1.095\sqrt{gL}$ ⊻ 53.1°.
**17.140** $\mathbf{A} = 100.1$ N ↑, $\mathbf{B} = 43.9$ N →.
**17.142** $0.778\,\omega_0$.
**17.143** (a) 418 rpm. (b) −20.4 J.
**17.145** (a) 10.27 ft/s ⊻ 64.4°. (b) 6.99 ft/s.

# CHAPTER 18

**18.1** $0.250\, mr^2\, \omega_2\mathbf{j} + 0.500\, mr^2\, \omega_1\mathbf{k}$.
**18.2** $\mathbf{H}_A = \dfrac{ma^2\omega}{12}(3\mathbf{j} + 2\mathbf{k}◄)$.
**18.3** $\mathbf{H}_G = (1.505\text{ lb·s·ft})\mathbf{i} + (4.14\text{ lb·s·ft})\mathbf{j}$.
**18.5** $\mathbf{H}_A = (0.0800\text{ kg·m}^2/\text{s})\mathbf{j} + (0.480\text{ kg·m}^2/\text{s})\mathbf{k}◄$.
**18.7** $0.432\, ma^2\omega$, 20.2°.
**18.8** 9.7°.
**18.9** $(1.843\text{ lb·ft·s})\mathbf{i} - (0.455\text{ lb·ft·s})\mathbf{j} + (1.118\text{ lb·ft·s})\mathbf{k}$.
**18.10** $\mathbf{H}_O = (2.05\text{ kg·m}^2/\text{s})\mathbf{i} + (3.36\text{ kg·m}^2/\text{s})\mathbf{j} + (0.480\text{ kg·m}^2/\text{s})\mathbf{k}◄$.
**18.11** $0.500\, mr^2\omega_1\mathbf{i} - m(L^2 + 0.250\, r^2)\,(r\omega_1/L)\mathbf{j}$.
**18.12** (a) 0.485 rad/s. (b) 0.01531 rad/s.
**18.15** (a) $(5.65\text{ kg·m}^2/\text{s})\mathbf{i} - (1.885\text{ kg·m}^2/\text{s})\mathbf{j} + (12.57\text{ kg·m}^2/\text{s})\mathbf{k}$. (b) 25.4°.
**18.16** (a) $(5.65\text{ kg·m}^2/\text{s})\mathbf{i} - (1.885\text{ kg·m}^2/\text{s})\mathbf{j} + (12.57\text{ kg·m}^2/\text{s})\mathbf{k}$. (b) 154.6°.
**18.17** (a) $(1.078\text{ lb·s·ft})\mathbf{i} - (0.647\text{ lb·s·ft})\mathbf{k}$. (b) 31.0°.
**18.18** (a) $(1.078\text{ lb·s·ft})\mathbf{i} - (0.647\text{ lb·s·ft})\mathbf{k}$. (b) $(1.078\text{ lb·s·ft})\mathbf{i} - (0.647\text{ lb·s·ft})\mathbf{k}$.
**18.21** 93.6 kg.
**18.22** 2.57 s.
**18.25** (a) $\overline{\mathbf{v}} = 0$. (b) $\boldsymbol{\omega} = \left(\dfrac{3F\Delta t}{md}\right)\mathbf{i} - \left(\dfrac{3F\Delta t}{4md}\right)\mathbf{k}$.
**18.26** (a) $\overline{\mathbf{v}} = -\left(\dfrac{F\Delta t}{3m}\right)\mathbf{k}$. (b) $\boldsymbol{\omega} = -\left(\dfrac{3F\Delta t}{md}\right)\mathbf{i} - \left(\dfrac{3F\Delta t}{4md}\right)\mathbf{j}$.
**18.27** (a) −(0.300 m/s)i. (b) −(0.962 rad/s)i − (0.577 rad/s)j.
**18.28** (a) (0.300 m/s)j. (b) −(3.46 rad/s)i + (1.923 rad/s)j − (0.857 rad/s)k.
**18.31** (a) $0.1250\,\omega_0\,(-\mathbf{i} + \mathbf{j})$. (b) $0.0884\,a\omega_0\mathbf{k}$.
**18.32** (a) $0.1031\,ma\omega_0\mathbf{k}$. (b) $-0.01473\,ma\omega_0\mathbf{k}$.
**18.33** (0.0248 rad/s)i − (0.277 rad/s)j −(0.360 rad/s)k.
**18.34** (a) −0.726 rad/s. (b) −(2160 ft/s)i − (4860 ft/s)j + (860 ft/s)k.

**18.35** (a) $t_A = 0.129$ s, $t_B = 1.086$ s. (b) $-(50.6$ mm/s$)\mathbf{j}$.
**18.36** (a) 0.941 s. (b) $(0.0169$ rad/s$)\mathbf{j}$. (c) $-(39.2$ mm/s$)\mathbf{j}$.
**18.39** $0.1250\,mr^2\,(\omega_2^2 + 2\omega_1^2)$.
**18.40** $T = \frac{1}{8}ma^2\omega^2$.
**18.41** $T = 260$ ft·lb.
**18.42** 12.67 ft·lb.
**18.43** $T = 9.59$ J.
**18.44** $0.1250\,ma^2\omega^2$.
**18.45** $0.203\,ma^2\omega^2$.
**18.47** 237 **J**.
**18.49** 27.0 **J**.
**18.50** 46.2 **J**.
**18.51** $0.1000\,m\bar{v}_0^2$.
**18.53** 16.75 ft·lb.
**18.54** 39.9 ft·lb.
**18.55** $0.500\,mr^2\omega_1\omega_2\mathbf{i}$.
**18.56** $\dot{\mathbf{H}}_A = \frac{1}{6}ma^2\omega^2\mathbf{i}$.
**18.57** $\dot{\mathbf{H}}_G = -(189.1$ lb·ft$)k$.
**18.58** $(5.30$ lb·ft$)\mathbf{k}$.
**18.59** $\mathbf{H}_A = (1.920$ N·m$)\mathbf{i}$.
**18.60** $14mr^2\omega^2 \sin\gamma \cos\gamma\mathbf{k}$.
**18.61** $\dot{\mathbf{H}}_D = -(1.304$ lb·ft$)\mathbf{i} - (1.479$ lb·ft$)\mathbf{j} + (2.22$ lb·ft$)\mathbf{k}$.
**18.62** $\dot{\mathbf{H}}_D = (1.304$ lb·ft$)\mathbf{i} - (1.479$ lb·ft$)\mathbf{j} + (2.22$ lb·ft$)\mathbf{k}$.
**18.64** $\frac{1}{4}mr^2\alpha \sin\beta \cos\beta\mathbf{j} + \frac{1}{4}mr^2\omega^2 \sin\beta \cos\beta\mathbf{k}$.
**18.65** $\mathbf{C} = 0.1667\,mb\omega^2 \sin\beta \cos\beta\mathbf{i}$.
$\mathbf{D} = -0.1667\,mb\omega^2 \sin\beta \cos\beta\mathbf{i}$.
**18.66** $\mathbf{A} = -(4.97$ lb$)\mathbf{i}$, $\mathbf{B} = -(1.656$ lb$)\mathbf{i}$.
**18.67** $\mathbf{A} = -(1.103$ lb$)\mathbf{j} - (0.920$ lb$)\mathbf{k}$. $\mathbf{B} = (1.103$ lb$)\mathbf{j} + (0.920$ lb$)\mathbf{k}$.
**18.68** $\mathbf{A} = (14.4$ N$)\mathbf{k}$, $\mathbf{B} = -(14.4$ N$)\mathbf{k}$.
**18.71** (a) $3M_0/mb^2 \cos^2\beta$. (b) $\mathbf{C} = -\mathbf{D} = (M_0 \tan\beta/2b)\mathbf{k}$.
**18.72** (a) $(14.49$ rad/s$^2)\mathbf{j}$. (b) $\mathbf{A} = -(1.125$ lb$)\mathbf{k}$, $\mathbf{B} = -(0.375$ lb$)\mathbf{k}$.
**18.73** (a) $(0.873$ lb·ft$)\mathbf{i}$. (b) $\mathbf{A} = -\mathbf{B} = -(0.218$ lb$)\mathbf{j} + (0.262$ lb$)\mathbf{k}$.
**18.74** (a) $(2.67$ N·m$)\mathbf{i}$. (b) $\mathbf{A} = -\mathbf{B} = (2.00$ N$)\mathbf{j}$.
**18.75** (a) $(0.1301$ lb·ft$)\mathbf{i}$. (b) $\mathbf{A} = -\mathbf{B} = -(0.0331$ lb$)\mathbf{i} + (0.0331$ lb$)\mathbf{j}$.
**18.76** $\mathbf{A} = -\mathbf{B} = -(0.449$ lb$)\mathbf{j} - (0.383$ lb$)\mathbf{k}$.
**18.79** $\mathbf{A} = -\mathbf{B} = (1.527$ N$)\mathbf{j}$.
**18.81** (a) 10.47 N·m. (b) 10.47 N·m.
**18.82** 24.0 N ↑.
**18.84** 1.138°; up.
**18.85** $\omega_1 = \sqrt{\dfrac{g/L}{\left[1 + \left(\dfrac{r}{2L}\right)^2\right]\cos\theta + \left(\dfrac{r}{2L}\right)\sin\theta}}$
**18.86** (a) 27.0°. (b) 8.09 rad/s.
**18.87** (a) 7.53 rad/s. (b) 7.00 rad/s.
**18.88** 4.84 rad/s.
**18.90** 7.89 rad/s.
**18.91** 15.24 rad/s.
**18.93** $\mathbf{A} = -\mathbf{B} = (0.1906$ lb$)\mathbf{k}$.
**18.94** 7.87 rad/s.
**18.95** (a) $\mathbf{C} = -(592$ N$)\mathbf{j}$ and $\mathbf{D} = (592$ N$)\mathbf{j}$. (b) $\mathbf{C} = \mathbf{D} = 0$.
**18.96** 35.5 rpm.
**18.99** $-(45.0$ N$)\mathbf{i}$, $(3.38$ N·m$)\mathbf{i} + (10.13$ N·m$)\mathbf{k}$.
**18.100** (a) $\mathbf{A} = (1.786$ kN$)\mathbf{i} + (143.5$ kN$)\mathbf{j}$, $\mathbf{B} = -(1.786$ kN$)\mathbf{i} + (150.8$ kN$)\mathbf{j}$. (b) $-(35.7$ kN·m$)\mathbf{k}$.
**18.101** $\mathbf{C} = -(7.81$ lb$)\mathbf{i} + (7.43$ lb$)\mathbf{k}$, $\mathbf{D} = -(7.81$ lb$)\mathbf{i} - (7.43$ lb$)\mathbf{k}$.
**18.102** $\mathbf{C} = -(12.58$ lb$)\mathbf{i} + (9.43$ lb$)\mathbf{k}$, $\mathbf{D} = -(12.58$ lb$)\mathbf{i} - (9.43$ lb$)\mathbf{k}$.
**18.103** $\mathbf{D} = -(22.0$ N$)\mathbf{i} + (26.8$ N$)\mathbf{j}$, $\mathbf{E} = -(21.2$ N$)\mathbf{i} - (5.20$ N$)\mathbf{j}$.
**18.104** (a) $(0.392$ N·m$)\mathbf{k}$. (b) $\mathbf{D} = -(21.0$ N$)\mathbf{i} + (28.0$ N$)\mathbf{j}$, $\mathbf{E} = -(21.0$ N$)\mathbf{i} - (4.00$ N$)\mathbf{j}$.

**18.107** 2930 rpm.
**18.109** 45.9 rpm, 533 rpm.
**18.111** 1666 rpm.
**18.112** $\dot{\phi} = 27.2$ rpm, $-370$ rpm◄.
**18.113** $\cos^{-1}\left[\dfrac{2d^2\dot{\psi}}{(d^2 + h^2)\dot{\phi}}\right]$.
**18.114** (a) 128.3 rad/s. (b) 2.17 in.
**18.115** $\theta = \tan^{-1}\left[\dfrac{5g \tan\beta}{5g + 2c\dot{\psi}\dot{\phi}}\right]$.
**18.116** (a) $\theta = 30.0°$. (b) $\theta = 24.9°$. (c) $\theta = 37.4°$.
**18.117** (a) 4.89 rpm. (b) 4.96 rpm, 396 rpm.
**18.116** 1105.8 lb·ft.
**18.118** M = 1105.8 lb·ft.
**18.124** (a) 13.19°. (b) 1242 rpm (retrograde).
**18.126** 24.8 rev/h.
**18.127** (a) 12.85°. (b) 5.78 rev/h. (c) 20.7 rev/h.
**18.128** (a) 109.4 rpm, $\gamma_x = 90°$, $\gamma_y = 100.05°$, $\gamma_z = 10.05°$. (b) $\theta_x = 90°$, $\theta_y = 113.9°$, $\theta_z = 23.9°$. (c) precession: 47.1 rpm; spin: 64.6 rpm.
**18.130** (a) $\theta_x = 90.0°$, $\theta_y = 26.0°$, $\theta_z = 64.0°$. (b) precession, 0.847 rad/s (retrograde); spin: 0.1593 rad/s.
**18.131** (a) $\dot{\phi}_{\min} = 8.00$ rad/s. (b) $\dot{\theta}_{\max} = 11.31$ rad/s.
**18.132** (a) $-40° < \theta < 40°$. (b) $\dot{\phi}_{\min} = 12.69$ rad/s. (c) $\dot{\theta}_{\max} = 9.16$ rad/s.
**18.135** (a) 41.2°. (b) 5.52 rad/s.
**18.136** (a) 4.23 rad/s. (b) 12.50 rad/s.
**18.139** (a) 47.0°. (b) precession: 15.25 rad/s; spin: 307 rad/s.
**18.140** (a) 76.3°. (b) precession: 9.62 rad/s; spin: 294 rad/s. (c) 36.5°.
**18.148** $(0.234$ kg·m$^2$/s$)\mathbf{j} + (1.250$ kg·m$^2$/s$)\mathbf{k}$.
**18.150** (a) 0. (b) $(F\Delta t/ma)$ $(2.50\mathbf{i} - 1.454\mathbf{j} + 2.19\mathbf{k})$.
**18.151** 4.29 kN·m.
**18.153** $\mathbf{D} = -(7.12$ lb$)\mathbf{j} + (4.47$ lb$)\mathbf{k}$, $\mathbf{E} = -(1.822$ lb$)\mathbf{j} + (4.47$ lb$)\mathbf{k}$.
**18.154** $\mathbf{D} = 0$; $\mathbf{M}_D = (11.23$ N·m$)\cos^2\theta\mathbf{i} + (11.23$ N·m$)\sin\theta\cos\theta\mathbf{j} - (2.81$ N·m$)\sin\theta\cos\theta\mathbf{k}$.
**18.155** (a) $\theta_x = 52.5°$, $\theta_y = 37.5°$, $\theta_z = 90°$. (b) 53.8 rev/h. (c) 6.68 rev/h.
**18.156** axis: 32.0°, precession: 1.126 rpm, and spin: 0.344 rpm.
**18.157** (a) $-40° < \theta < 140°$. (b) $\dot{\phi}_{\min} = 5.31$ rad/s. (c) $\dot{\theta}_{\max} = 5.58$ rad/s.

## CHAPTER 19

**19.1** 10 mm, 3.18 Hz.
**19.2** $v_m = 1.750$ ft/s, $f_n = 15.92$ Hz.
**19.3** 1.273 in., 150.8 ft/s$^2$.
**19.4** (a) 0.324 s, 3.08 Hz. (b) 12.91 mm, 4.84 m/s$^2$.
**19.5** (a) 427 rpm. (b) 0.447 m/s.
**19.6** (a) $x_m = 0.222$ in., $f = 6.64$ Hz. (b) $v_m = 9.27$ in./s, $a_m = 32.2$ ft/s$^2$.
**19.7** (a) 12.23°. (b) 2.09 m/s$^2$.
**19.8** (a) 0.705 Hz. (b) 271 mm/s.
**19.11** (a) 0.0352 s. (b) 6.34 ft/s ↑, 64.4 ft/s$^2$ ↓.
**19.12** 0.445 ft ↑, 2.27 ft/s ↓, 114.7 ft/s$^2$ ↓.
**19.13** (a) 1.288°. (b) 0.874 ft/s, 0.760 ft/s$^2$.
**19.14** (a) 4.91 mm, 5.81 Hz, 0.1791 m/s. (b) 491 N. (c) 0.1592 m/s ↑.
**19.17** (a) 0.517 s, 1.934 Hz. (b) 0.365 m/s, 4.43 m/s$^2$.
**19.18** 2.87 s.
**19.19** $\sqrt{\dfrac{k}{2\,m}}$.

**19.20** (*a*) 0.361 s, 2.77 Hz. (*b*) 0.765 m/s, 13.30 m/s$^2$.
**19.23** 4.
**19.24** (*a*) 6.80 kg. (*b*) 0.583 s.
**19.25** 192 lb/ft.
**19.26** (*a*) 21.7 kg. (*b*) 1011 kg/m$^3$.
**19.28** (*a*) 22.3 MN/m. (*b*) 266 Hz.
**19.30** (*a*) 858 N/mm. (*b*) 149.5 rpm.
**19.31** (*a*) 3.56 kg. (*b*) 43.7 kg.
**19.34** 16.26°.
**19.35** (*a*) 1.737 s. (*b*) 1.864 s. (*c*) 2.05 s.
**19.36** 28.1 in.
**19.37** (*a*) 3.65 Hz. (*b*) 0.0076°.
**19.38** (*a*) 1.047 rad/s. (*b*) 16.42 in.
**19.39** (*a*) 0.227 s. (*b*) 333 mm/s.
**19.41** (*a*) 0.491 s. (*b*) 9.60 in/s.
**19.42** (*a*) 0.858 s. (*b*) 3.66 in./s.
**19.44** 75.5°.
**19.45** 0.346 Hz.
**19.48** (*a*) 2.79 s. (*b*) 1.933 m.
**19.49** (*a*) 1.617 s. (*b*) 1.676 s.
**19.50** (*a*) 227 mm. (*b*) 1.352 s.
**19.51** (*a*) $6.33\sqrt{\dfrac{b}{g}}$. (*b*) $6.67\sqrt{\dfrac{b}{g}}$.
**19.55** (*a*) 2.21 Hz. (*b*) 115.3 N/m.
**19.56** $\dfrac{1}{2\pi}\sqrt{\dfrac{6k}{5m}+\dfrac{9g}{10l}}$ Hz.
**19.57** $\dfrac{1}{2\pi}\sqrt{\dfrac{2k}{3m}+\dfrac{4g}{3L}}$ Hz.
**19.59** (*a*) 0.426 s. (*b*) 15.44 ft/s.
**19.60** (*a*) 88.1 mm/s. (*b*) 85.1 mm/s.
**19.62** $\omega = 0.01345$ rad/s ↺, $\alpha = 0.1268$ rad/s$^2$ ↺.
**19.63** 6.57 kg·m$^2$.
**19.64** (*a*) 21.3 kg. (*b*) 1.836 s.
**19.67** 0.672 in.
**19.68** 8.60 ft.
**19.69** 19.02 mm.
**19.70** $\dfrac{1}{2\pi}\sqrt{\dfrac{k}{5m}}$ Hz.
**19.71** (*a*) 0.369 s. (*b*) 1.892 ft/s.
**19.72** $6.28\sqrt{R/g}$.
**19.75** $l/\sqrt{12}$.
**19.76** 1.048 s.
**19.77** $0.159\sqrt{(2k/3m)+(4g/3L)}$.
**19.78** 15.66 s.
**19.79** (*a*) 0.715 s. (*b*) 0.293 ft/s.
**19.80** 0.821 s.
**19.83** 1.327 s.
**19.85** 0.567 Hz.
**19.86** 2.39 s.
**19.87** $2\pi\sqrt{(12r^2 + 2l^2)/3gl}$.
**19.89** (*a*) 0.802 Hz. (*b*) 0.307 m.
**19.90** 2.29 Hz.
**19.91** 1.192 s.
**19.92** $0.1312\sqrt{g/r}$.
**19.95** $0.276\sqrt{g/l}$.
**19.96** $\dfrac{1}{2\pi}\sqrt{\dfrac{12k}{7m}+\dfrac{8g}{7\sqrt{3}l}}$ Hz.
**19.97** $1.814l\sqrt{gr}$.

**19.98** 0.352 s.
**19.99** (*a*) 37.1 mm. (*b*) 260 mm.
**19.100** (*a*) 160.0 N/m. (*b*) 40.0 N/m.
**19.101** $\sqrt{\dfrac{2k}{3m}} < \omega_f < \sqrt{\dfrac{4k}{3m}}$.
**19.102** 50.1 lb.
**19.103** 30.4 mm, out of phase (180°).
**19.105** (*a*) 19,590 lb/ft. (*b*) $f > 100$ Hz.
**19.106** 3.21 m/s$^2$.
**19.107** (*a*) 0.450 rad/s. (*b*) 2.70 m/s$^2$.
**19.109** $\omega_f > \sqrt{2g/l}$.
**19.110** (*a*) 1.034 in. (*b*) $-0.1033 \sin \pi t$ (lb).
**19.112** 22.5 mm for $x > 0$ and $-5.63$ mm for $x < 0$.
**19.114** 22.0 mm.
**19.115** 0.609 in.
**19.116** $\omega_f < 322$ rpm.
**19.117** 39.1 kg.
**19.119** 149.3 mm.
**19.121** Force transmissibility: $1/(1 - \omega_f^2/\omega_n^2)$; displacement transmissibility: $1/(1 - \omega_f^2/\omega_n^2)$.
**19.122** (*a*) 4.17%. (*b*) 84.9 Hz.
**19.123** 8.04%.
**19.125** (*a*) 1399 rpm. (*b*) 0.01670 in.
**19.132** (*a*) 6.49 kip·s/ft. (*b*) 230 kips/ft.
**19.133** 5.48 N·m·s.
**19.134** (*a*) 7790 lb/ft. (*b*) 0.1939 s.
**19.135** (*a*) 0.118. (*b*) 38.4 mm.
**19.136** (*a*) 19.97 rad/s. (*b*) 7.73 mm up from the starting point.
**19.137** 8.82 N.
**19.138** 106.5 mm/s ↑.
**19.139** 0.0623 in.
**19.140** 21.4 lb·s/in.
**19.141** $\geq 0.707$.
**19.143** (*a*) 147 kip/ft. (*b*) 0.0292.
**19.144** 0.0162 in.
**19.145** 0.0821 in.
**19.146** 0.487.
**19.148** (*a*) 71.8 N. (*b*) 39.0 N.
**19.149** (*a*) 4.90 in. (*b*) 30.3 lb.
**19.151** (*a*) $m\ddot{x} + c\dot{x} + kx = (k \sin \omega_f t + c\omega_f \cos \omega_f t)\delta_m$.
(*b*) $x = x_m \sin (\omega_f t - \varphi + \psi)$, where
$x_m = \delta_m\sqrt{k^2 + (c\omega_f)^2}/\sqrt{(k - m\omega_f^2)^2 + (c\omega_f)^2}$,
$\tan \varphi = c\omega_f/(k - m\omega_f^2)$, $\tan \psi = c\omega_f/k$.
**19.153** $R < 2\sqrt{L/C}$.
**19.154** (*a*) $E/R$. (*b*) $L/R$.
**19.157** (*a*) $c(\dot{x}_A - \dot{x}_m) + kx_A = 0$
$m\ddot{x}_m + c(\dot{x}_m - \dot{x}_A) = P_m \sin \omega_f t$
(*b*) $R(\dot{q}_A - \dot{q}_m) + (1/C)q_A = 0$
$L\ddot{q}_m + R(\dot{q}_m - \ddot{q}_A) = E_m \sin \omega_f t$
**19.159** 0.760 lb·s$^2$· ft, 8.66 in.
**19.160** (*a*) 0.1649 Hz. (*b*) 1.277 Hz.
**19.161** 1.785 s.
**19.162** (*a*) 0.404. (*b*) 18,420 N/m. (*c*) 19.02 mm.
**19.165** (*a*) $0.07246\ddot{\theta} + 0.3375\dot{\theta} + 1.25\theta = 0$.
(*b*) $-19.05 \times 10^{-6}$ degrees.
**19.168** (*a*) $m\ddot{x} + 2T(2x/l) = 0$. (*b*) $\pi\sqrt{ml/T}$.
**19.169** 0.045 in.

# Index

# W

# Z

## Centroids of Common Shapes of Areas and Lines

| Shape | | $\bar{x}$ | $\bar{y}$ | Area |
|---|---|---|---|---|
| Triangular area | | | $\dfrac{h}{3}$ | $\dfrac{bh}{2}$ |
| Quarter-circular area | | $\dfrac{4r}{3\pi}$ | $\dfrac{4r}{3\pi}$ | $\dfrac{\pi r^2}{4}$ |
| Semicircular area | | 0 | $\dfrac{4r}{3\pi}$ | $\dfrac{\pi r^2}{2}$ |
| Semiparabolic area | | $\dfrac{3a}{8}$ | $\dfrac{3h}{5}$ | $\dfrac{2ah}{3}$ |
| Parabolic area | | 0 | $\dfrac{3h}{5}$ | $\dfrac{4ah}{3}$ |
| Parabolic spandrel | | $\dfrac{3a}{4}$ | $\dfrac{3h}{10}$ | $\dfrac{ah}{3}$ |
| Circular sector | | $\dfrac{2r\sin\alpha}{3\alpha}$ | 0 | $\alpha r^2$ |
| Quarter-circular arc | | $\dfrac{2r}{\pi}$ | $\dfrac{2r}{\pi}$ | $\dfrac{\pi r}{2}$ |
| Semicircular arc | | 0 | $\dfrac{2r}{\pi}$ | $\pi r$ |
| Arc of circle | | $\dfrac{r\sin\alpha}{\alpha}$ | 0 | $2\alpha r$ |

## Moments of Inertia of Common Geometric Shapes

### Rectangle

$$\bar{I}_{x'} = \tfrac{1}{12}bh^3$$
$$\bar{I}_{y'} = \tfrac{1}{12}b^3h$$
$$I_x = \tfrac{1}{3}bh^3$$
$$I_y = \tfrac{1}{3}b^3h$$
$$J_C = \tfrac{1}{12}bh(b^2 + h^2)$$

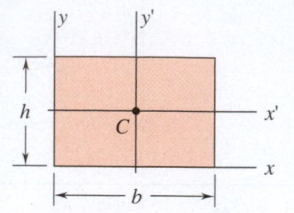

### Triangle

$$\bar{I}_{x'} = \tfrac{1}{36}bh^3$$
$$I_x = \tfrac{1}{12}bh^3$$

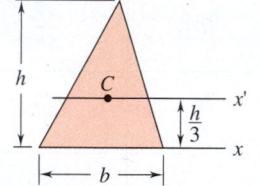

### Circle

$$\bar{I}_x = \tfrac{1}{4}\pi r^4$$
$$J_O = \tfrac{1}{2}\pi r^4$$

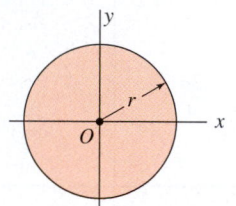

### Semicircle

$$I_x = \tfrac{1}{8}\pi r^4$$
$$J_O = \tfrac{1}{4}\pi r^4$$

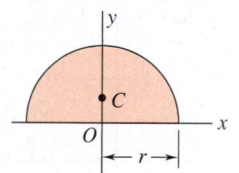

### Quarter circle

$$I_x = I_y = \tfrac{1}{16}\pi r^4$$
$$J_O = \tfrac{1}{8}\pi r^4$$

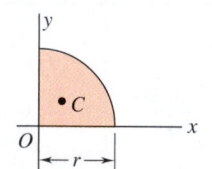

### Ellipse

$$\bar{I}_x = \tfrac{1}{4}\pi ab^3$$
$$\bar{I}_y = \tfrac{1}{4}\pi a^3b$$
$$J_O = \tfrac{1}{4}\pi ab(a^2 + b^2)$$

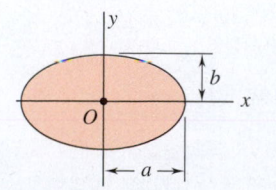

## Mass Moments of Inertia of Common Geometric Shapes

### Slender rod

$$I_y = I_z = \tfrac{1}{12}mL^2$$
$$I_{y'} = I_{z'} = \tfrac{1}{3}mL^2$$

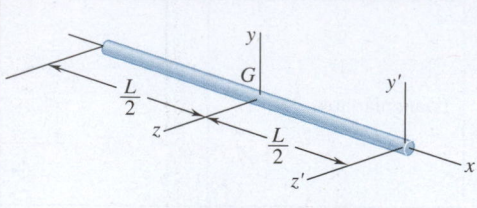

### Thin rectangular plate

$$I_x = \tfrac{1}{12}m(b^2 + c^2)$$
$$I_y = \tfrac{1}{12}mc^2$$
$$I_z = \tfrac{1}{12}mb^2$$

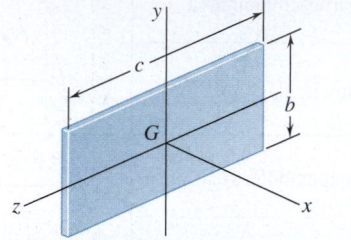

### Rectangular prism

$$I_x = \tfrac{1}{12}m(b^2 + c^2)$$
$$I_y = \tfrac{1}{12}m(c^2 + a^2)$$
$$I_z = \tfrac{1}{12}m(a^2 + b^2)$$

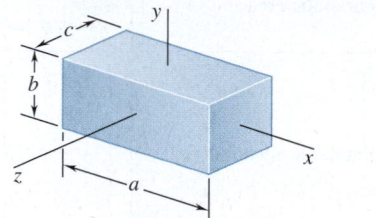

### Thin disk

$$I_x = \tfrac{1}{2}mr^2$$
$$I_y = I_z = \tfrac{1}{4}mr^2$$

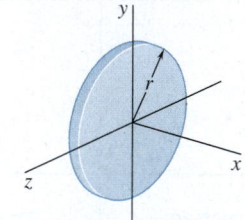

### Circular cylinder

$$I_x = \tfrac{1}{2}ma^2$$
$$I_y = I_z = \tfrac{1}{12}m(3a^2 + L^2)$$
$$I_{y'} = I_{z'} = \tfrac{1}{4}ma^2 + \tfrac{1}{3}mL^2$$

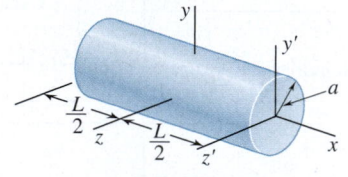

### Circular cone

$$I_x = \tfrac{3}{10}ma^2$$
$$I_y = I_z = \tfrac{3}{5}m\left(\tfrac{1}{4}a^2 + h^2\right)$$

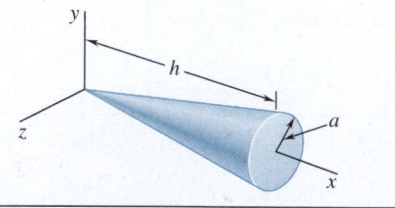

### Sphere

$$I_x = I_y = I_z = \tfrac{2}{5}ma^2$$

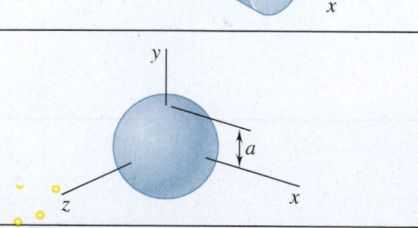